GENERA PALMARUM

The Evolution and Classification of Palms

GENERA PALMARUM

The Evolution and Classification of Palms

John Dransfield, Natalie W. Uhl, Conny B. Asmussen, William J. Baker, Madeline M. Harley and Carl E. Lewis

Kew Publishing
Royal Botanic Gardens, Kew

First published in 2008 by
Royal Botanic Gardens, Kew,
Richmond, Surrey, TW9 3AB, UK
www.kew.org

Published in association with the International Palm Society and
The L.H. Bailey Hortorium, Cornell University

ISBN 978-1-84246-182-2

British Library Cataloguing in Publication Data
A catalogue record for this book is available from the British Library

Production editor: Sharon Whitehead
Typesetting and page layout: Christine Beard
Cover design: John Stone
Publishing & Media Resources, Royal Botanic Gardens, Kew

Front cover: *Hedyscepe canterburyana* growing on Mount Gower, Lord Howe Island. (Photo: I. Hutton)

Frontispiece: *Pritchardia viscosa*, one of the world's most endangered palms, Powerline Trail, Kauai, Hawaii. (Photo: J. Dransfield)

Printed in Italy by Printer Trento

For information or to purchase all Kew titles please visit
www.kewbooks.com or email publishing@kew.org

Kew's mission is to inspire and deliver science-based plant conservation worldwide, enhancing the quality of life.

CONTENTS

FOREWORD

The appearance of a new edition of a standard reference inevitably invites comparisons with the older edition. In this case, this is the successful first edition of *Genera Palmarum* (1987), the standard reference and a milestone in palm systematics. The improvements over the first edition — readers will find them on every page of the second edition — reflect the major advances in biology of the past two decades. Botanical exploration has advanced, especially in poorly known but palm-rich areas of South America, Madagascar, Asia and Malesia. Automated sequencing, geographic information systems and digital imaging have become standard tools for the exploration of biodiversity. Most importantly, new data in the form of permanently preserved, pressed specimens and DNA sequences have made their way into museums, laboratories and herbaria around the world. Since the publication of *Genera Palmarum*, palm research has flourished on nearly every continent. New exploration, new data and techniques, and twenty years worth of new research are all synthesised in this second edition.

Many of the refinements in this second edition of *Genera Palmarum* fill in the gaps in our knowledge of palms identified by the first edition. Even after the release of the first *Genera Palmarum*, some palm genera were incompletely known and a few, like *Medemia* and *Carpoxylon*, assumed almost mythical status. These genera have since been rediscovered, fully described and placed within the new classification. Gaps in the descriptions of several genera from Madagascar led to focused botanical exploration on that island, which resulted in the inclusion of several genera within *Dypsis* and the discovery of three new genera, *Lemurophoenix*, *Satranala* and *Voanioala*. An astounding fourth new genus, *Tahina*, was discovered just as the manuscript for this new volume was being prepared for publication!

I cannot overemphasise the significance of recent molecular analyses in palm systematics and the ripple effect they have produced in studies of morphology and anatomy, palynology, biogeography and evolution. Molecular analyses have revealed some startling generic relationships, such as the inclusion of *Chamaedorea* and its kin in the Arecoideae, and the placement of *Phytelephas* and related genera within the Ceroxyloideae. At the same time, molecular studies have reaffirmed many relationships that were recognised in the first *Genera Palmarum* on the basis of morphology and anatomy: groups like the Ptychospermatinae, Caryoteae, Borasseae and the Bactridinae are confirmed as natural lineages.

The coming decades will undoubtedly see refinements and modifications to the classification of palms, but that work will have as a starting point the work presented here. Future palm systematists will be tremendously indebted to the authors of the second edition of *Genera Palmarum*, and I have no doubt that this book will be every bit the milestone and stimulus to research that the first edition was.

SCOTT ZONA
Florida International University
Miami, Florida, USA

INTRODUCTION

Since the publication of the first edition of *Genera Palmarum* in 1987, there has been an explosion of interest in this quintessentially tropical flowering plant family. Palms tend to attract attention, perhaps because of their recognisable rather simple growth form, their ecological and economic importance, or indeed simply because of their association with the exotic. The numerous biological intrigues within the family explain the abundance of palm research across the world, as demonstrated by the vast body of palm literature published since 1987 that is listed in the bibliography of this book. *Genera Palmarum* Edition 1 provided the basic taxonomic framework on which such palm studies could be based. Over the same period, worldwide interest in the cultivation of palms has also increased dramatically, and the range of species now available to the amateur grower is truly astonishing.

In the past twenty years, there have been major advances in our understanding of the palm family. New exploration, particularly in Madagascar and the West Pacific, has uncovered remarkable new genera, even at the very moment that the manuscript of this new edition was completed in April 2007. Moreover, several genera that were known only from fragmentary type material have been rediscovered. Exploration throughout the tropics has provided new observations that have allowed detailed reassessment of genera.

At the same time a vast new range of phylogenetically informative characters has become available, principally through analysis of DNA. Despite some challenging properties inherent within palm genomes, a substantial body of literature has been amassed since the publication of the first molecular phylogenetic study of palms in 1995. More than any other source, these molecular characters have allowed us to gain a thorough understanding of palm phylogenetics and to obtain sometimes dramatic new insights into both relationships and morphology. The first edition of *Genera Palmarum*, which was based on painstaking comparison across the family, provided a hypothesis of relationships that has now been comprehensively tested using the most up-to-date tools and methods. The new classification that we present here provides confirmation of some of the groupings recognised in the first edition, but has also revealed substantial differences and at times quite surprising discrepancies when compared with previous classifications.

This new edition of *Genera Palmarum* is more than just a revision of the first edition. In almost every respect, it is an entirely new book. The taxonomic core of the book, namely the generic descriptions have been extensively rewritten, restructured and augmented. All other content, the notes associated with each genus, the introductory chapters and so on, have been written *de novo*. We have amassed the most extensive and complete selection of palm images ever published, thanks to the generosity of many contributors; only a few of these featured in the first edition. Even the illustrations have been revised, with some errors corrected and new genera illustrated by Lucy T. Smith in the characteristic style of Marion Ruff Sheehan.

The team for this new *Genera Palmarum* came together for the first time at a meeting held at the Montgomery Botanical Center on 17 January 2003, the details of the phylogenetic classification were thrashed out at a week long summit held in the L.H. Bailey Hortorium in early June 2004. Responsibilities for each part of the book were shared out among the authors as follows. JD took on the revision of the taxonomy, descriptions and keys. He also contributed many of the notes and introductory chapters (Natural History and Conservation and the introduction to the Classification of Palms) and supplementary materials, such as the geographical listings and the literature cited. NWU supported JD closely in this commitment, and took the lead on the rewriting of Chapter 1, The Structure of Palms, with collaboration from JD and WJB. MMH contributed all chapters, notes and descriptive information on pollen and fossils. WJB wrote the introductory chapters on Chromosomes and Cytogenetics, Phylogeny and Evolution (with CBA-L) and Biogeography. CBA-L, WJB and CEL together prepared all the phylogenetic notes for the taxonomic accounts. CEL provided the Chemistry chapter and new illustrations for the glossary, which he revised with the other co-authors. JD took on the role of collating the manuscript and co-ordinating with Kew Publishing and co-authors.

This new *Genera Palmarum* will, we hope, provide a robust framework for the studies of the palm family for many years to come. There are still areas of uncertainty, but we believe that the higher-level classification of subfamilies, tribes and subtribes is largely robust and unlikely to require substantial change. Uncertainties in inter-generic relationships and generic circumscription remain, and the field of species-level phylogenetics remains wide open. We hope that our work has the same invigorating affect that the first edition of *Genera Palmarum* appeared to have and look forward to reflecting, in another two decades, on the triumphs of a new generation of palm biologists.

JOHN DRANSFIELD, NATALIE UHL, CONNY ASMUSSEN-LANGE,
BILL BAKER, MADELINE HARLEY AND CARL LEWIS

ABOUT THE AUTHORS

John Dransfield PhD studied for his doctorate at the University of Cambridge, working on two Malaysian palm genera, *Johannesteijsmannia* and *Eugeissona*. In 1970 he took up a job for the British Government's Overseas Development Administration, stationed in Bogor, Indonesia, where he developed a deep interest in the climbing palms, the rattans. In 1975, he joined the staff of the Royal Botanic Gardens, Kew, first as a research fellow and then as head of palm research. His main interests focus on the global classification of palms and the palm floras of Malesia, Africa and Madagascar. Having retired in 2005, he continues his research as an honorary research fellow at Kew. He is author of over 250 scientific papers and several books on palms, including (with Natalie Uhl) the first edition of *Genera Palmarum*. He currently co-edits *Palms* (formerly *Principes*), journal of the International Palm Society. He has been awarded the Founder's Medal by Fairchild Tropical Botanic Garden, the David Fairchild Medal for Plant Exploration by the National Tropical Botanic Garden, the Engler Medal by the International Association of Plant Taxonomists and the Linnean Medal for Botany by the Linnean Society of London.

Natalie W. Uhl PhD received her MSc and doctorate from Cornell University and is presently Professor Emeritus at Cornell University. Her early research was on xylem and phloem in monocotyledons and floral anatomy of Najadales (Helobiae). She began her research on palms in 1963 and spent 17 years working in collaboration with Professor Harold E. Moore, Jr. towards a *Genera Palmarum* as envisioned by Liberty Hyde Bailey. She continued this work with John Dransfield, after Harold Moore's death in 1980, *Genera Palmarum* finally being published in 1987. She co-edited *Principes* (now *Palms*) for 20 years with John Dransfield. She continues to work on the systematics of palms with a focus on inflorescence and floral anatomy and development but has also published on leaves and stems, and has worked in collaboration with John Dransfield and others in phylogenetic investigations. Her palm work has earned her many awards including the Founder's Medal from Fairchild Tropical Botanic Garden, the Dent Smith Award from The International Palm Society, the Asa Gray Award from the American Society of Plant Taxonomists, the Allerton Award by the National Tropical Botanic Garden, the Engler Medal from the International Association of Plant Taxonomists and a Merit Award from The Botanical Society of America.

Conny B. Asmussen-Lange PhD completed her doctoral research on legumes at the University of Aarhus. She developed her interests in palm phylogenetics during postdoctoral positions at Cornell University and the Royal Botanic Gardens, Kew, focusing first on the geonomoid palms and subsequently on the higher level relationships of palms. She has been Associate Professor at the Faculty of Life Sciences, University of Copenhagen since 2000. Her current research centres on the molecular phylogenetics of palms and legumes. In addition, she teaches plant systematics and is the curator of plants at the herbarium of the Faculty of Life Sciences, University of Copenhagen.

William Baker PhD is head of palm research at the Royal Botanic Gardens, Kew and is an authority on the systematics, evolution and biogeography of the palm family. He completed his PhD on the phylogenetics of subfamily Calamoideae at the University of Reading in 1997 and joined the staff of Kew in 1998. Since that time, he has conducted a wide range of collaborative research focused on the application of DNA data to systematic and evolutionary questions in palms. He has studied palms in the field in Malesia, the Pacific, Africa and Madagascar, and is an authority on the taxonomy of the palms of Papuasia. In addition, he has served on committees of the Systematics Association, Royal Horticultural Society and Royal Geographical Society. His work on palms was recognised by the Linnean Society of London in the award of the Society's Bicentenary Medal in May 2008.

Madeline Harley PhD has enjoyed a long career at the Royal Botanic Gardens, Kew, during which she specialised in the pollen morphology of flowering plants. She completed her doctorate on palm pollen and the fossil record at the University of East London in 1996. Her work uses pollen data to improve our understanding of the relationships and evolution of flowering plants. She has authored or co-authored numerous scientific papers on pollen, and has presented her work at many international conferences. Her deep interest in communicating the fascination of pollen to a wider audience led to the publication of a widely acclaimed book, *Pollen: the Hidden Sexuality of Flowers*, first published in 2004. Now retired but with an honorary research fellowship at Kew, she continues to take a keen interest in many aspects of pollen morphology.

Carl Lewis PhD obtained his doctorate at Cornell University. His research focused on the systematics of palm subtribes Oncospermatinae and Verschaffeltiinae and resulted in important technical developments for palm molecular phylogenetics. He is currently employed at Fairchild Tropical Botanic Garden as a researcher in palm systematics and conservation. His interests lie in the molecular systematics, evolution, and biogeography of palms, particularly those of the Caribbean. He is the author of Fairchild Tropical Botanic Garden's innovative on-line *Guide to Palms*.

ACKNOWLEDGEMENTS

This book has such a long and involved history that almost everybody in the palm world (and very many beyond it) has had some kind of contact with its evolution. We highlight many of these people below, any omissions being entirely accidental.

Throughout the project, we have received the strongest of support from our respective institutions: the Royal Botanic Gardens, Kew, the L.H. Bailey Hortorium, the Fairchild Tropical Botanic Garden and the Faculty of Life Sciences, University of Copenhagen.

We acknowledge the constant interest of the Board of Directors of the International Palm Society in the co-publication of this book; in particular, we thank Paul Craft, Libby Besse and Randy Moore.

Throughout our work on palms, we have been given generous access to many of the best living collections in botanic gardens throughout the world, allowing us to make direct observations of living palms and to collect material for micromorphological and molecular studies. We make special mention of the unparalleled facilities provided by the Montgomery Botanical Center (MBC), Miami. From 1997, JD and NWU visited Miami annually for seven years in a row, making ourselves at home in the MBC guest house and making extensive use of the fantastic palm collections at both the MBC and the neighbouring Fairchild Tropical Botanic Garden. We often received visits from the remaining co-authors and we are left with many happy memories of those times, of the camaraderie engendered by our meetings, and of the concentrated peace and quiet in which we were able to discuss and develop ideas on palm morphology.

Individuals and institutions far too numerous to mention have aided our efforts in studying and collecting palms in their natural environments, providing authorisations and assistance in navigating local politics, collaborating in the fieldwork itself, guiding us to wonderful palm localities, and helping to process and distribute our materials. Without these people, we would be immeasurably less well informed on the subject of palms!

Colleagues Henrik Balslev, Anders Barfod, Finn Borchsenius, John Dowe, Jack Fisher, Rafaël Govaerts, Bee Gunn, Lynda Hanson, Andrew Henderson, Ilia Leitch, Larry Noblick, Jean-Christophe Pintaud, Mijoro Rakotoarinivo, Martin Röser, Paula Rudall, Vincent Savolainen, Philipp Trénel, Mary C. Uhl, Johan van Valkenburg and Scott Zona have supported us with stimulating discussions, comments on manuscripts and specialist input of various kinds. Xavier Metz provided material of and assisted with access to the most recently described genus, *Tahina*.

CBA-L, WJB and CEL are particularly grateful to all those colleagues who have supported our molecular phylogenetic work on palms, collaborating and sharing both data and material. In particular, we thank Anders Barfod, Sasha Barrow, Ross Bayton, Finn Borchsenius, Mark Chase, Thomas Couvreur, Argelia Cuenca Navarro, Vinnie Deichmann, John Dowe, Felix Forest, Bee Gunn, Bill Hahn, Charlie Heatubun, Jette Teilmann Knudsen, Adrian Loo, Andres Moreira, Jean-Christope Pintaud, Maria Norup, Julissa Roncal, Fred Stauffer, Terry Sunderland, Murphy Thomas, Philipp Trénel, Johan van Valkenberg, Tomas Wilmot and Mark Wilkinson.

In building up the huge body of data related to palm pollen, MMH acknowledges several sandwich course students (Caroline Batchelor, Andrew Blakey, Peter Cade, James Clarkson, Ian Flawn, Darroch Hall, Alison Havard, Stephen Hoad, Fiona Page, Helen Sanderson and Phyllis Whittington-Vaughan) who worked successively over many years under her supervision on palm pollen, as well as two post-doctoral visitors, Neliya Mendis and Unsook Song. Tom Balm and Chrissie Prychid allowed MMH to photograph their fossil *Nypa* fruit from Sheppey. Robert (Bob) Morley provided valuable discussion on the palm fossil record in southeast Asia. Eckart Shrank provided useful dialogue on the dating of the 'Aptian' Egyptian *Hyphaene* fossil fruits.

At Kew, Head of Publishing, Gina Fullerlove, took an immediate interest in the publication of the book. Lloyd Kirton and John Harris developed the mode and schedule of publication. Sharon Whitehead meticulously copy-edited our manuscript and John Stone designed the cover and prepared the maps. Christine Beard patiently designed and set the pages, always cheerful and willing to discuss changes; we are immensely grateful to her. Paul Little carefully digitised vast numbers of transparencies on our behalf. In the Palm Room of the Kew Herbarium, Helen Sanderson, Melinda Trudgen, Soraya Villalba and Jovita Yesilyurt, provided willing help at various times. Craig Brough (Kew Library) and Sarah Sworder (Palaeontological Library, Natural History Museum London) provided unstinting help in responding to requests for articles from obscure sources relating to palm fossils. In the Palm House and Temperate House at Kew, David Cooke, Emma Fox, Wesley Shaw and their staff provided valuable palm material.

At the L.H. Bailey Hortorium, Cornell University, Robert Dirig, Peter Fraissinet, Kent Loffler and Sherry Vance provided invaluable assistance in many ways.

We acknowledge the great dedication shown by the late Marion Ruff Sheehan in producing the superb analytical plates of palm genera and thank Lucy T. Smith for accepting the challenge of preparing additional plates in the same style.

The following kindly allowed us to reproduce their photographs of palms: Susyn Andrews, Aino Askgaard, Henrik Balslev, Anders Barfod, Sasha Barrow, Ross Bayton, John Beaman, Henk Beentje, Rodrigo Bernal, Finn Borchsenius, Felipe Castaño, David Cooke, Wally Donovan, John Dowe, Tad Dyer, Michael Ferrero, Maria Gandolfo, Martin Gibbons, Anne Giddey, Lorena Guevara, Lynda Hanson, Andrew Henderson, Don Hodel, Sheila Hooper, Jay Horn, Ian Hutton, Margaret Johnson, Anders Kjaer, Xavier Metz, Jean-Christophe Pintaud, Mijoro Rakotoarinivo, Julissa Roncal, Paula Rudall, Ruth Ryder, Franco Simonetti, Toby Spanner, Fred Stauffer, Terry Sunderland, Johan van Valkenburg, Paul Wilkin, B.J. Wood and the National Museums, Liverpool.

Finally, we thank our families, friends and colleagues for their tolerance of the demands made on the authors during the preparation of the manuscript.

HOW TO USE THIS BOOK

The main body of this book consists of descriptions of genera arranged within a phylogenetic classification of subfamilies, tribes and subtribes. As a background to the classification, we provide introductory chapters that explain palm morphology and anatomy, cytology, pollen, chemistry, the fossil record, phylogeny, biogeography, natural history and the history of palm classification. This introductory material provides a general introduction to the biology of palms.

Each genus is introduced with a short diagnosis that highlights distinctive features. This is followed by a full nomenclatural citation that includes all synonyms and the current disposition of the type species of each generic name. Author abbreviations follow Brummitt and Powell (1992). Abbreviations of journals follow Botanico Periodicum Huntianum (Bridson 1991) and those of books follow Taxonomic Literature II (Stafleu & Cowan 1976–1998). The etymology of the accepted generic name is then provided. The generic description follows a sequence from habit to stem, to leaves, inflorescence, flowers, pollen, fruit, seed, seedling and finally, where known, cytology. Numbers in square brackets after the pollen description within the generic description refer to the number of species examined out of a total number for the genus; thus [12/18] means that the pollen grains of 12 species out of a total of 18 species have been examined. After the generic description, the distribution and number of species currently accepted in the World Checklist of Palms (Govaerts & Dransfield 2005, updated at http://www.kew.org/wcsp/monocots/) and brief ecological notes are provided and the anatomy where studied is briefly discussed. Authorities for species names not cited in the nomenclatural citations are not cited; the reader is referred to the World Checklist for full citations. Relationships as recovered in the light of recent phylogenetic research are discussed in detail. The notes on common names and uses include a selection of common names rather than an exhaustive listing, and only the most important uses are highlighted. References to the most recent monographs for each genus are given, followed by information on what is known of the fossil record. Finally under Notes, brief items of particular interest are discussed. For each genus, there is an annotated wash drawing in which details of flowers and fruits are clearly illustrated. These drawing were developed by the late Marion Ruff Sheehan. Lucy T. Smith adapted her technique, matching the original wash drawings as closely as possible, to prepare plates of new genera or newly accepted genera and to make additions in some instances to the original plates. In addition to the wash drawings, we provide photographs of the living palms and of the pollen of each genus. Additional plates are scattered throughout the book in the introductory chapters, as frontispieces to each section and in the glossary. Where appropriate, these are numbered and cross-referenced under each genus, the compound number referring to the figure number within the numbered chapter — viz. Fig. 1.3 refers to Figure 3 in Chapter 1, and Glossary fig. 3 refers to Figure 3 in the Glossary.

The glossary provides an illustrated explanation of the specialist terms we have found necessary to use in the description and explanation of palms.

The keys in this book are based on the technical differences between genera. These differences are very often characters of flowers and fruit that are not always present, and without them the genera would sometimes be impossible to key out. The keys are not intended to be used for field identification. In the field in any one geographic area, one encounters a small subset of palm genera that are usually easy to key out using obvious vegetative characters. Our keys will not work in this way. For the most part, they require complete material of flowers and fruit, nevertheless, they should at least guide the user to the correct genus. In the future, we hope to make available an interactive multi-access key.

At the end of the book are listings of geographical areas with their recorded genera and a complete bibliography of all items of literature cited in the book.

Chapter 1

THE STRUCTURE OF PALMS

Palms are the icons of the tropics. They are seldom mistaken for members of other families, yet they are more diverse in form than other monocotyledon families, perhaps even more than any other family of seed plants. P.B. Tomlinson (1990) wrote, "Palms are not then merely emblematic of the tropics, they are emblematic of the way in which the structural biology of plants must be understood before evolutionary scenarios can be reconstructed." Our goals in this first chapter are to describe and clarify the exceptional structural variation of palms as an introduction to a broadly based new hierarchy and to updated descriptions of the genera. To discuss the root, stem, leaves, flowers and other organs in detail would require a complete chapter for each. This overview aims to assist readers who might be unfamiliar with terms used in the descriptions, and to provide advanced readers with introductions to the latest interpretations of structure and to the areas where more research is needed. The illustrated glossary should be used with this chapter.

ARCHITECTURE

The form or structural design of a palm depends upon three parameters: the branching pattern, the morphology of the shoot, and the reproductive strategy.

Branching patterns

The quintessential palm is a solitary stem, a shoot, bearing a crown of leaves. This shoot has a single apical meristem that produces leaves, initially in the seedling in a two-ranked or distichous arrangement. Distichy persists to the adult stage in a few palms, such as *Oenocarpus distichus*, *Orania disticha* and *Wallicha disticha*, but in most, adult leaf arrangement becomes helical or spiral (polystichous) and phyllotaxis follows the primary Fibonacci series. In some palms, orthostichies can be observed in the crown, as in the triangle palm, *Dypsis decaryi*, which exhibits a $^{1}/_{3}$ phyllotaxy and three orthostichies. Most small palms have a $^{2}/_{5}$ phyllotaxy but larger palms may have higher numbers that can be determined by contra-rotating contact parastichies in the crown. Phyllotaxy is occasionally of taxonomic importance as noted above. (For a further discussion of phyllotaxy see Tomlinson [1990].)

In the palm shoot, the axil of each leaf potentially subtends a bud, and elongation of the stem between leaf insertions produces an internode. The resulting quadripartite structure — comprising a node, the associated leaf, an axillary bud and an internode — has been considered a unit of growth and termed a module (Tomlinson 1990) or a metamer (Henderson 2002b). The concept of a modular growth pattern helps to interpret the construction of a shoot and can be useful for certain investigations. In palms, however, the metamer is a unit of construction but not of growth. Meristematically, growth is not modular as the meristems are uninterrupted (Fisher & French 1976). Furthermore, in palms, positional gradients in inflorescences or within individual flowers can lead to quantitative or qualitative variation in form or function (Diggle 2003), in which cases, metamers would not be homologous. In some other monocotyledons (Poaceae, Cyperaceae, Commelinaceae and others) where there are nodal or intercalary meristems (Fisher & French 1976), the metamer concept is applicable to growth.

Different palm habits result from the development of the axillary bud. The bud may become a new shoot, form an inflorescence, or abort. The common clustered habit results when new shoots are formed from buds at basal nodes on the stem. In clustered species, axillary buds at the base of the plant are usually vegetative whereas distal buds develop into inflorescences. Many genera have both solitary and clustered species, and there are even species that display both solitary and clustered individuals. The formation of successive new shoots, each essentially equivalent to the axis from which it is derived, was considered by Holttum (1955) to represent a sympodial habit in monocotyledons as a whole. This habit has been further related to the unique vascular organization of monocotyledons (Tomlinson 1995).

The position of the stem can influence the ultimate habit. If the stem is prostrate or plagiotropic (Henderson 2002b), it may root on the lower side and turn distally to support a crown of leaves bearing axillary inflorescences (e.g., *Nypa fruticans*, *Johannesteijsmannia altifrons*, *Brahea decumbens*, *Ammandra*, *Phytelephas* spp. [Bernal 1998], *Elaeis oleifera, Calyptrogyne ghiesbreghtiana* and many others). The axillary branch may remain prostrate for a short distance and develop scale-like leaves before turning erect; several branches result in a colonial habit in which branches are dimorphic.

Axillary buds usually develop into vegetative shoots at the base of the stem and inflorescences distally. Exceptions occur and can be characteristic of a few genera. In *Serenoa repens* (Fig. 1.1e), either inflorescences or vegetative shoots are produced in leaf axils along a mostly prostrate stem (Fisher & Tomlinson

Opposite: The coconut, ***Cocos nucifera***, Queensland, Australia. (Photo: J. Dransfield)

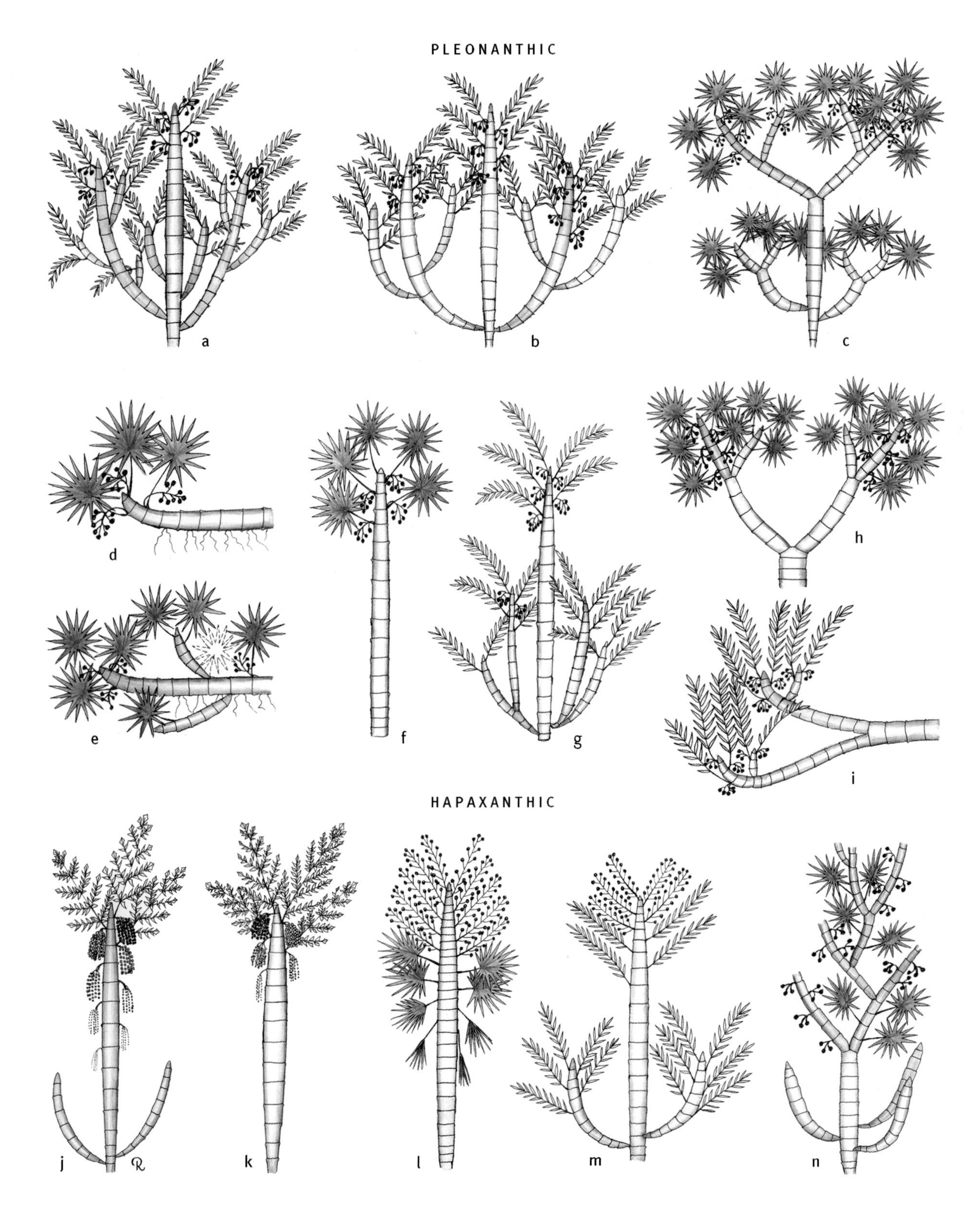

Fig. 1.1 Diagrammatic drawings of palm architectures. **a**, non-axillary branching, e.g., *Dypsis lutescens*; **b**, colonial (clustering) by rhizome or stolon, e.g., *Bactris coloniata*, *Chamaedorea stolonifera*; **c**, dichotomous branching and basal axillary branching, e.g., *Hyphaene coriacea*; **d**, prostrate stem, e.g., *Brahea decumbens*; **e**, prostrate stem, some axillary buds producing inflorescences, others producing vegetative shoots, *Serenoa repens*; **f**, stem solitary, e.g., *Trachycarpus fortunei* and many others; **g**, stem erect or climbing, caespitose (clustering), e.g., *Chamaedorea seifrizii* and many others; **h**, erect, dichotomously branching stem, e.g., *Hyphaene compressa*; **i**, dichotomous branching in prostrate stem, e.g., *Nypa fruticans*; **j**, basipetal hapaxanthic flowering, clustering stems, e.g., *Caryota mitis*; **k**, basipetal hapaxanthic flowering, solitary stem, e.g., *Caryota urens*; **l**, acropetal hapaxanthic flowering, solitary erect stem, e.g., all species of *Corypha*; **m**, acropetal hapaxanthic flowering and basal branching, e.g., *Metroxylon sagu* (can also be acaulescent, e.g., *Eleiodoxa conferta*); **n**, acropetal hapaxanthic flowering, aerial dichotomous branching, and basal axillary branching, shrub or vine, e.g., *Nannorrhops ritchiana* and *Korthalsia* spp. (Drawn by Marion Ruff Sheehan)

1973). In the hapaxanthic *Plectocomia*, bulbils are sometimes formed at the nodes and, on the death of the shoot, may develop into new plants (Dransfield 1978b, Isnard 2006). Inflorescences in some calamoid palms root distally to produce shoots (e.g., *Salacca flabellata*, *S. wallichiana*, *Calamus pygmaeus* and *Daemonorops ingens*) (Dransfield 1992b). In *Pinanga rivularis*, the axillary buds form inflorescences and internodal buds develop as vegetative shoots (Dransfield 1992a).

Axillary branching

Branching patterns are further determined by the position of the bud. Most frequently, it is centred in the axil. In some genera, however, the position of the bud varies giving rise to different branching patterns (Fisher & Maidman 1999). The bud may be captured by the developing subtending leaf and then emerge through a slit on the abaxial side of the leaf sheath, as in all species of *Salacca* (Fisher & Mogea 1980) and in the monotypic *Kerriodoxa*. Alternatively, the bud may be displaced longitudinally, as in the Calaminae in which the ensuing inflorescence is adnate to the internode and leaf sheath above the node of origin (Fisher & Dransfield 1977). Multiple inflorescences form when several buds develop on an original crescent-shaped primordium. The prophylls of multiple buds may be separate (as in *Wettinia* and *Chamaedorea alternans*) or fused (as in *Chamaedorea pauciflora*, *Calyptrocalyx* and *Howea forsteriana*). Vegetative buds are leaf-opposed in *Daemonorops* and borne on the abaxial leaf base in *Oncosperma* and *Dypsis*. In *Myrialepis* the vegetative bud is borne at an angle of 130° from the leaf axil. A complete circle of adventitious buds occurs in leaf axils in some species of *Plectocomia*. Internodal buds, unrelated to leaf axils, occur only rarely, e.g., in *Pinanga rivularis* mentioned above.

Apical branching

Some palms are characterised by a special form of branching resulting from actual division of the apex (Fisher & Zona 2006, Table 1.1). The division may be into equal halves, a true dichotomy, or may be unequal. Equal division occurs in *Nypa*, certain species of *Hyphaene*, and *Chamaedorea cataractarum*; unequal division occurs in other species of *Hyphaene*, *Nannorrhops*, *Oncosperma* and *Dypsis*. Fisher and Maidman (1999) considered these developmental variations to be apomorphies, although perhaps some act as synapomorphies within species groups of certain genera (e.g., *Hyphaene*).

TABLE 1.1
Dichotomous and unusual lateral branching, both vegetative and reproductive

DICHOTOMOUS BRANCHING	UNUSUAL LATERAL BRANCHING
I CALAMOIDEAE *Eugeissona* *Korthalsia* *Raphia* II NYPOIDEAE *Nypa* III CORYPHOIDEAE CHUNIOPHOENICEAE *Nannorrhops* BORASSEAE *Hyphaene* V ARECOIDEAE CHAMAEDOREEAE *Chamaedorea cataractarum* COCOSEAE *Allagoptera* spp. MANICARIEAE *Manicaria saccifera* ARECEAE *Dypsis* *Oncosperma*	**By multiple axillary buds (inflorescences only)** III CORYPHOIDEAE CARYOTEAE – *Arenga* spp. IV CEROXYLOIDEAE – *Ravenea* 7 spp. V ARECOIDEAE IRIARTEEAE – *Wettinia* spp. CHAMAEDOREEAE – *Chamaedorea* 15 spp. COCOSEAE – *Aiphanes* spp. ARECEAE – *Calyptrocalyx* spp., *Howea forsteriana*, *Masoala madagascariensis* **By displaced axillary buds** I CALAMOIDEAE CALAMEAE **Radial displacement (inflorescences)** Salaccinae – *Salacca* all spp. **Longitudinal displacement (inflorescences)** Calaminae – 5 genera of rattans Plectocomiinae – all genera **Leaf opposed buds (vegetative)** Korthalsiinae – *Korthalsia rigida* (in part) Plectocomiinae – *Myrialepis* Calaminae – *Daemonorops* spp. **Adventitious buds (vegetative)** Plectocomiinae – *Plectocomia* spp. III CORYPHOIDEAE CHUNIOPHOENICEAE **Radial displacement (inflorescences)** *Kerriodoxa*

Shoot morphology

Size and form

Ultimate form may depend on size. Tree palms, shrub palms, acaulescent palms, and climbing palms are terms that describe habit (Dransfield 1978b). In this book, we have used small, dwarf, moderate, tall, large, or massive to describe the size of palms.

Many organs of palms exhibit gigantism and have set records for plants (Tomlinson 2006a). Palm species have the widest stems achieved by primary growth, the largest leaves, the longest unbranched aerial stems and the largest inflorescences and seeds among angiosperms. Palms also express axial conformity, i.e., thicker branches bear thicker appendages (Tomlinson 2006a). Higher orders of branching decrease in size. Within *Geonoma*, Chazdon (1991) found that leaf size and complexity of form increase with stem diameter and crown height. Solitary species have larger leaves and larger stem diameters than clustered species.

Climbing

A climbing habit appears in two subfamilies. It is most frequent among Calamoideae, and it characterises *Korthalsia* (Korthalsiinae), the Plectocomiinae, the three genera of the Ancistrophyllinae, and the six genera of the Calaminae, notwithstanding apparent reversals to non-climbing habits in the last. Elsewhere, climbing occurs in three unrelated tribes of Arecoideae: in *Chamaedorea elatior* (Chamaedoreeae), in *Desmoncus* (Cocoseae) and in *Dypsis scandens* (Areceae). Clearly these represent convergent origins of the habit. It has been estimated that the climbing habit has evolved on between two and four separate occasions in the Calamoideae alone (Baker *et al.* 2000a). No palmate-leaved climbing palms are known, perhaps because these leaves, unlike pinnate leaves, are not predisposed to adaptation for climbing.

Structures that are associated with climbing are often taxonomically useful. Climbing is facilitated by the elongation of stems, the presence of spines or reflexed leaflets, or a combination of all of these features. Spines on leaf rachises and climbing organs may be stout and recurved. They can be scattered, but are sometimes organised in whorls and thus behave like grapnels. Weak climbers in Calamoideae (e.g., *Retispatha*) may have only specialised leaf rachises, but two unique structures, the cirrus and the flagellum, are found in strong climbers.

The cirrus is a whip-like extension of the leaf rachis. Two forms of cirri are recognised (Baker *et al.* 1999b). In the African rattan subtribe Ancistrophyllinae, the leaf rachis is greatly elongated between terminal leaflets, these being modified as reflexed, thorn-like acanthophylls, sometimes occurring in association with recurved rachis spines. A near-identical cirrus type occurs in *Desmoncus*. The non-spiny *Chamaedorea elatior* and *Dypsis scandens* are weak climbers; their leaf rachises are elongated in a similar manner to those of Ancistrophyllinae, but to a lesser extent, and their terminal leaflets are only partially reduced. Some species of *Calamus* approach the condition observed in Ancistrophyllinae, but to a much lesser extent, resulting in a structure termed a subcirrus. An entirely different form of the cirrus develops when the rachis extends beyond rather than between the terminal leaflets; this form occurs in most of the rattan genera of tribe Calameae (*Korthalsia*, *Daemonorops*, *Calamus*, *Ceratolobus* and the Plectocomiinae).

A second climbing organ, the flagellum, which is functionally similar to the cirrus but morphologically different, is found in the genus *Calamus*. The flagellum has been identified as a sterile inflorescence and may bear grapnel spines. Like the inflorescences of all climbing *Calamus* species, the flagellum is adnate to the internode and leaf sheath above its node of origin. No rattan species produces both a flagellum and a cirrus, although some cirrate species (e.g., *C. pogonacanthus*) produce a vestigial flagellum.

Palms are passive climbers. Neither their stems nor their climbing organs are capable of active twining. They become anchored to supporting vegetation by chance and their reflexed spines and acanthophylls perform a ratchet-like function, bringing their elongated stems closer to their support as a result of wind movement or other perturbations. Putz (1990) has described the dynamics of climbing in Australian *Calamus* species.

Architectural models

Hallé (2004) designated four of 23 architectural models of trees for palms. Holttum's model refers to unbranched hapaxanthic palms, such as *Metroxylon salomonense* or *Corypha* spp. (Fig.1.1l) Unbranched pleonanthic palms, exemplified by species of *Sabal*, *Phoenix* and many common genera, are characteristic of Corner's model (Fig.1.1f). Clustered pleonanthic palms are represented by Tomlinson's model (Fig. 1.1g), whereas single stems with dichotomous branching are described by Schoute's model (Fig. 1.1h). Corner's or Tomlinson's models may typify different individuals of the same species. Some palms exhibit combinations of the models (Tomlinson 1990, Henderson 2002b). Thus, clustered hapaxanthic palms, such as *Metroxylon sagu* (Fig. 1.1m), combine Holttum's and Tomlinson's models. Hallé (2004) states that he considers *Serenoa* (Fig. 1.1e) to represent a fifth model in palms, which he has not named. Size is not considered in distinguishing the different models although, as noted above, it contributes significantly to the overall appearance of palms.

Life-history strategies

Two reproductive strategies further influence architecture in palms. In some genera, the stems undergo a long period of vegetative growth followed by a distinct and relatively short reproductive phase, after which the shoot dies. These palms are termed hapaxanthic (Table 1.2) and are in a functional sense 'determinate': their shoot apical meristems appear to abort. The majority of palms are pleonanthic, continuing to grow vegetatively and to produce inflorescences simultaneously throughout their life spans, and are thus 'indeterminate.'

Two other sets of terms have been used to describe palm life history (Henderson 2002b). The common distinction between monocarpy (flowering once) and polycarpy (flowering several times) does not work well for palms. For example, the solitary *Corypha* is hapaxanthic and monocarpic, but clustered hapaxanthic rattans are polycarpic because dead axes are replaced by younger shoots. Some authors prefer semelparity (birth once) and iteroparity (repeated birth) because these terms are more widely used in the life-history literature. Nevertheless, here we retain hapaxanthy and pleonanthy, which are well-established terms in palm literature. We accept that a range of variation exists within pleonanthic palms, but regard hapaxanthy to be a particularly useful term to highlight the extreme life-histories of genera such as *Corypha*, *Metroxylon* and *Caryota*. (For a further discussion of terminology, see Henderson 2002b, Sunderland 2002.)

Hapaxanthy is found in 16 genera representing two subfamilies. These genera include *Corypha*, *Nannorrhops*, *Tahina* and all members of Caryoteae (Coryphoideae) except for a few species of *Arenga* (Moore 1960, Moore & Meijer 1965, Dransfield & Mogea 1984). Hapaxanthy is most frequent in the Calamoideae, in which eight genera are completely hapaxanthic and two exhibit both conditions (Table 1.2). In *Metroxylon*, only one of the seven species, *M. amicarum*, is pleonanthic, whereas in *Daemonorops* only a few species, including *D. calicarpa* and its relatives, are hapaxanthic. Hapaxanthy may be an adaptation to colonizing temporary habitats by rapid production of large amounts of fruit (Dransfield 1978b, Baker *et al.* 2000a). However, other ecological factors, such as fruit size and dispersal agents, appear to be correlated with hapaxanthy in Africa (Sunderland 2002). In most hapaxanthic palms, the sequence of inflorescence maturation is acropetal, that is in a direction away from the base of the stem (Fig. 1.1l,m). In hapaxanthic members of Caryoteae inflorescence maturation is basipetal (Fig. 1.1j,k). After prolonged vegetative growth, stems begin to flower, the first inflorescences appearing in a downwards sequence, node after node. This sequence of flowering suggests that inflorescences were initiated in the leaf axils as the stem developed, but are somehow suppressed until the stem reaches flowering size, after which there is a progressive release of suppression down the stem. The control of flowering in Caryoteae has never been investigated.

TABLE 1.2
The hapaxanthic habit

I CALAMOIDEAE
- EUGEISSONEAE
  - *Eugeissona*
- LEPIDOCARYEAE
  - Ancistrophyllinae
    - *Laccosperma*
  - Raphiinae
    - *Raphia*
- CALAMEAE
  - Korthalsiinae
    - *Korthalsia*
  - Salaccinae
    - *Eleiodoxa*
  - Metroxylinae
    - *Metroxylon* (except *M. amicarum*)
  - Plectocomiinae
    - *Myrialepis*
    - *Plectocomia*
    - *Plectocomiopsis*
  - Calaminae
    - *Daemonorops calicarpa* and related species

III CORYPHOIDEAE
- CHUNIOPHOENICEAE
  - *Nannorrhops*
  - *Tahina*
- CORYPHEAE
  - *Corypha*
- CARYOTEAE (all genera)
  - *Arenga* (not all species)
  - *Caryota*
  - *Wallichia*

CELLULAR INCLUSIONS

Silica bodies

The systematics and biology of silica bodies in monocotyledons has been reported by Prychid *et al.* (2004). Silica (hydrated silicon dioxide) is abundant in many palm organs and adds to the hardness of tissues. It is soluble only in hydrofluoric acid and causes great difficulty in obtaining thin sections of palms and in using palm wood. Silica often occurs in special cells, known as stegmata, which are borne in files adjacent to vascular or non-vascular fibres. Silica bodies develop in the vacuoles of the stegmata (Schmitt *et al.* 1995). Subsequently, the wall of the stegmata adjacent to the fibre becomes thickened and the silica body partially embedded in the thickened wall. Files of stegmata are not always continuous. The parenchyma cell side does not develop a thickened wall and the cytoplasm degenerates after the wall thickening. Stegmata do not normally form in epidermal cells although these cells frequently contain silica.

Two forms of silica bodies or silica cells occur in palms. In Nypoideae, in Caryoteae (Coryphoideae) and in Chamaedoreeae, Iriarteeae and Bactridinae (Arecoideae), conical, hat-shaped bodies with a smooth base, a rough or spiny surface, and a basal cell with a wall that is slightly thickened and not evidently pitted occur. The second form of irregular, more or less spherical bodies with rough or spiny surfaces is found in Calamoideae, in Sabaleae, Phoeniceae and Borasseae (Coryphoideae), in Phytelepheae (Ceroxyloideae) and in other Arecoideae. Only in *Lodoicea* have both types been seen together, large hat-shaped bodies being present in the lamina and spherical bodies elsewhere (Tomlinson 1961, Prychid *et al.* 2004).

In palm fruits, silica bodies can be found in the locular epidermis, in brachysclereids, and in some fibres (Essig 1999). Stegmata are diagnostic for palms, including palm fossils, but also occur in other monocotyledons (Bromeliaceae, Cyperaceae, Poaceae, Orchidaceae, Restionaceae and Strelitziaceae). Prasad *et al.* (2005) record the presence of palm and grass silica bodies (phytoliths) in dinosaur coprolites (fossil dung) in the Late Cretaceous, indicating that dinosaurs browsed on palms.

Raphides

Raphides are bundles of elongate needle-shaped crystals with pointed ends composed of calcium oxalate monohydrate. They are borne in specialised cells throughout the monocotyledons. Raphides have long been thought to serve as protection against herbivores and as a store for calcium oxylate or hydrogen peroxide (D'Arcy *et al.* 1996). In palms, they are frequent in roots, stems, leaves, flowers, fruits, seeds and occasionally in anthers and epidermal layers. They often appear to shield ovules as they occur around the locule (e.g., in *Chamaedorea*) (Uhl & Moore 1973, Uhl & Moore 1977a). However, raphides are found mixed with pollen in Bromeliaceae, Araceae, Lemnaceae, Liliaceae and several dicotyledonous families, in which they are considered to be a reward for pollinators rather than a means of mechanical protection (D'Arcy *et al.* 1996). Two reports suggest such a function in palms. Pohl (1941) reported crystal needles in anthers of the oil palm. In *Aphandra*, raphides resemble pollen grains in shape and occur in epidermal and subepidermal layers of the pseudopedicel, and mixed with the pollen in anthers of the staminate flower (Barfod & Uhl 2001). Research is needed to check other palms and to determine whether an alternate function, as a reward, may be appropriate for some. In general, raphides are too ubiquitous to be of taxonomic value, but Zona (2004) found their occurrence in palm embryos noteworthy. Raphides are found in embryos of Coryphoideae, being especially numerous in the Caryoteae. They are rare in Ceroxyloideae but numerous in the Areceae (Arecoideae). They were not observed in embryos of Calamoideae or Phytelepheae (Ceroxyloideae). Pseudoraphides or styloids, i.e., prismatic crystals borne singly in cells, may occur in some palm flowers (Morrow 1965) and in fruits (Essig pers. comm.).

Sieve element plastids

Cuneate protein crystals, common to all monocots, are present in sieve element plastids in all palms examined. Twenty-four genera described by Behnke (2000) all have type P2sc plastids, which contain crystals and starch. Only *Aiphanes* differed, its plastids also contain filaments and therefore represent a different type (P2 cfs).

THE STEM

Size and shape

"....one of the most striking aspects of palms is their great range in stem size, noticeable even among closely related taxa" (Henderson 2002b). The height of different palm species varies from 12–25 cm (species of *Chamaedorea* and *Iguanura*) to over 50–60 m (*Ceroxylon quindiuense*); and diameters from 3 mm (*Dypsis tenuissima*) to 1.0 m or more (*Borassus* and *Jubaea chilensis*). Henderson (2002b) found a 'slenderness ratio', the comparison of diameter to height (d/h), useful for contrasting size and shape. The height and diameter of stems can vary extensively both within genera and between closely related genera. Compare, for example, the pencil-stemmed palmlet *Dypsis delicatula* with the massive canopy emergent *D. bejofo*, the acaulescent *Acrocomia hassleri* with its tall, sometimes ventricose-stemmed congeners, or the dwarf saw palmetto *Serenoa repens* with its robust sister genus *Acoelorrhaphe wrightii*. These traits thus appear to be very labile phylogenetically.

The longest palm stems are found in the rattans; a length of nearly 200 m has been reported for *Calamus manan*, the longest, unbranched aerial plant stem on record (Burkill 1966). The exceptionally long internodes of climbing palms can reach a meter or more. A quantitative study of shoot development in *Daemonorops angustifolia*, *Calamus manan* and *C. insignis* showed that all internodes develop from an uninterrupted (distal) region of the internode; maturation is acropetal from the base (Fisher 1978). It is all the more striking then that *Calamus* contains many slender, short-stemmed and even acaulescent species.

When the crown of leaves appears to be at ground level, the palm is described as acaulescent. Taxa that have this morphology are not stemless as the term implies but have a short erect or prostrate stem. Thus, forms of acaulescence are not always homologous. The condition has been identified as derived in phylogenies of *Sabal* (Zona 1990), *Phoenix* (Barrow 1998) and in the Calamoideae (Baker *et al.* 2000b).

Most stems are cylindrical, although usually tapering distally with age. Bottle palms have a large swelling, which may be either at the base of the stem (*Hyophorbe* and *Pseudophoenix*) or in the middle (e.g., *Acrocomia crispa* and *Colpothrinax wrightii*). Fluctuations in stem diameter may be present or absent in a single species (*Iriartea deltoidea* and *Acrocomia aculeata*). In *I. deltoidea*, swollen stems are more frequent among palms growing at lower elevations (Henderson *et al.* 1995). Research has shown expanded parenchyma in these stems, perhaps functioning to store water (Fisher *et al.* 1996).

Growth and development

Palms and most other monocotyledons lack a cambium and do not develop large stems or plant bodies by secondary growth as do large dicotyledonous plants. In palms, the formation of a mature shoot from a seedling can be broken down into phases (Tomlinson 1990).

Phasic growth

Most palms are plants of the tropics and subtropics and display no dormancy. Growth is continuous and although the rate may fluctuate from, for example, wet season to dry season, growth processes are largely uninterrupted. Nevertheless, it is useful to distinguish stages within the life cycle of an individual palm. Tomlinson (1990) distinguished five phases — the embryo, the seedling, establishment, adult vegetative and adult reproductive (Fig. 1.2). The change from embryo to seedling involves the breaking of any dormancy. Thereafter, major changes between one phase and the next involve changes in size of the stem and leaves, changes in the form of the leaf (increase in numbers of leaflets or segments), and the eventual production of inflorescences, flowers and fruit.

During establishment growth, the third phase, each successive internode, beginning with the plumule, gradually increases in diameter until the diameter of the mature palm is attained. Internodes are short at first and gradually become longer. This process forms an inverted cone at the base of the shoot and eventually establishes the mature size of the individual. Typically, a conspicuous aerial stem does not develop during this phase. Subsequent elongation of internodes creates height. Usually, the palm stem is of more or less uniform diameter, the diameter not increasing with age (but see below). Atypical elongation of internodes during establishment phase occurs in stilt palms (e.g., *Iriartea deltoidea*) in which stilt or prop roots develop to support the tapering, obconical base of the stem. As the stem increases in diameter, the leaves increase in size, and the leaflets and segments increase in number. Once the aerial stem begins to elongate, internodes become longer and leaves may actually decrease in size.

Diffuse secondary growth

Known also as sustained primary growth, diffuse secondary growth occurs in some taxa and leads to an increase in stem diameter, largely as a result of an increase in fibrousness. The diameters of fibres may increase, as may the wall thickness of individual cells. Likewise, both the length of fibres and the number of fibres in bundle sheaths can become greater. There may also be some cell division in parenchyma cells. Large air spaces, lacunae, are formed in some stems.

Saxophone stems

In some genera, the plumule is positively geotropic, growing obliquely downward into the soil. The seedling leaf bases are reoriented and the leaf base develops a bulbous protuberance,

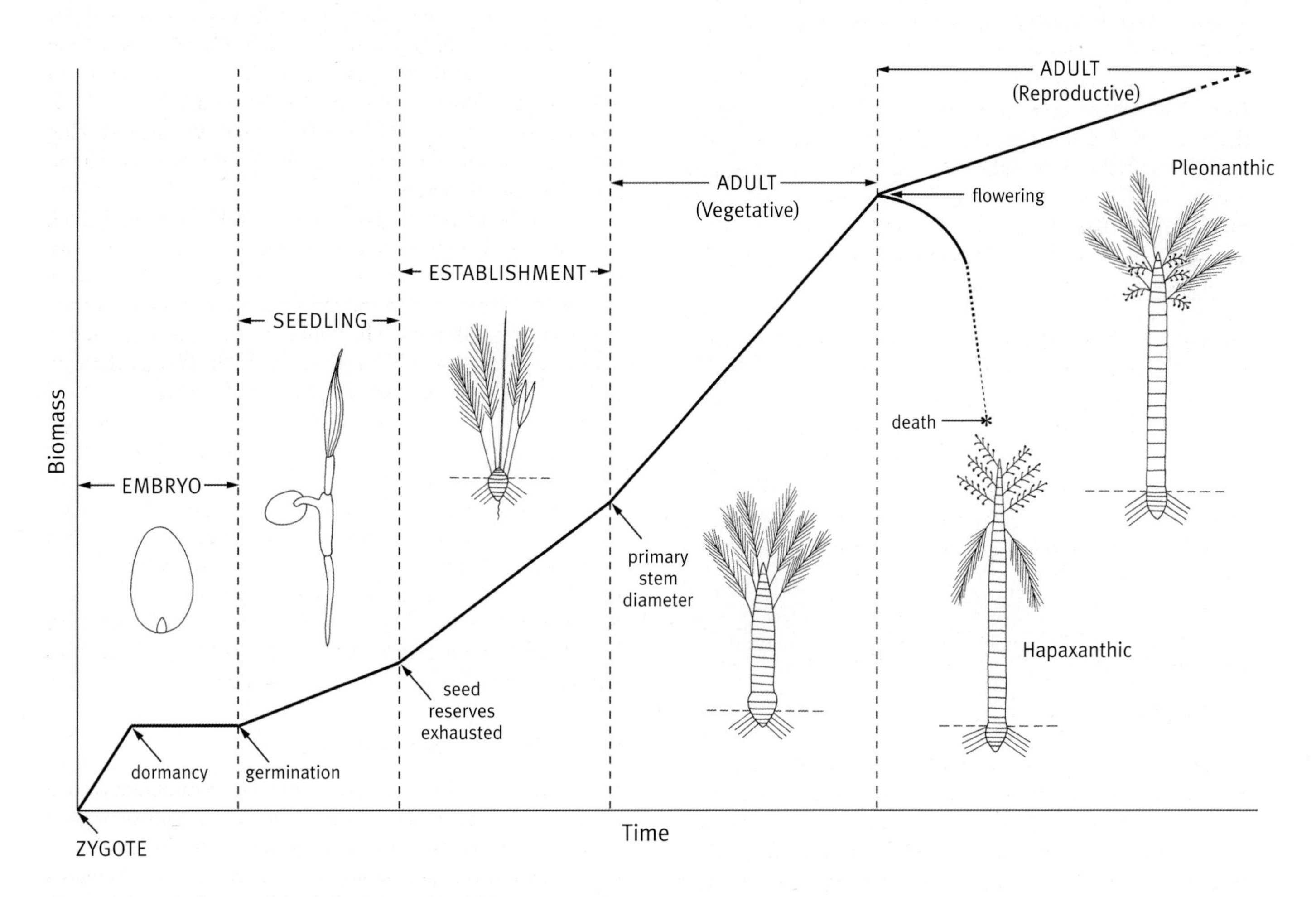

Fig. 1.2 Schematic diagram of phasic development in palms. (Redrawn from Tomlinson [1990] by Lucy T. Smith)

which apparently aids in forcing the axis into the soil. The apex descends 20–40 cm before turning to an obliquely erect position. Such growth produces a sharply curved bulbous protuberance and is known as a saxophone stem. Saxophone stems occur in species of *Sabal*, *Attalea*, *Syagrus*, *Acrocomia*, and perhaps other genera. This phenomenon may serve to facilitate establishment growth while impeding the development of an unstable aerial stem or protecting the apex in fire-affected habitats.

Stem surface

Stems may be covered in the remains of leaf sheaths and petioles, which sometimes provide useful diagnostic patterns (see Glossary fig. 5). Split petiole bases characterise *Sabal* and other genera, whorls of sheath spines *Zombia* and species of *Guihaia* and *Maxburretia*, and a network of fibres *Coccothrinax* and many others. Internodes of trunks may bear spines (Bactridinae and Oncospermatinae) and root spines characterise *Cryosophila*, the stilt roots of some *Iriarteeae* and several other palms. Leaf scars also provide characteristic patterns; they vary in size, shape, and distance apart. Some are larger and much more obvious than others; most are wider dorsally, reflecting the dorsal bulge of the leaf base (see Glossary fig. 4). Henderson (2002b) proposes that palms with distinctive internodes that often remain shiny and green are typical of moist habitats. By contrast, palms such as many coryphoids and cocosoids, which have less conspicuous internodes resulting from the erosion of surface layers of stems, cork development, and sclerotic and lignified surfaces, are characteristic of drier, more seasonal habitats. There are many exceptions to these patterns.

Anatomy

A cross-section of an internode reveals that the palm stem can be divided into three areas: an epidermis, a cortex, and a central cylinder of vascular bundles. The single-layered epidermis is usually thick walled, variously sclerotic, often heavily cutinised and lignified, and may include hairs or stomata. A survey of 13 genera, including 218 species of rattan palms (Weiner & Liese 1993), found an epidermis of a single layer of unlignified cells including stomata covered by a silica layer. The shapes of epidermal cells varied in surface view and lumina shapes differed in different genera. In older palm stems, the epidermis is sometimes eroded to expose the ends of fibrous bundles, resulting in what is known as 'porcupine wood'. Arcs of cork cambium develop on the surface of many palms, forming a rough stem that is frequently longitudinally fissured, as in *Sabal palmetto* (Glossary fig. 4g, i). If secondary tissues do not form, the stem becomes roughened and discoloured by sclerosis and suberisation of underlying tissues.

The cortex is a minor part of the palm stem, which has no delimiting characters, is often very narrow and may contain fibrous strands, sclerotic parenchyma, various forms of sclerenchyma, tannins, silica, and raphides. The large central cylinder has sparsely or densely arranged vascular bundles of various types distributed variously throughout ground tissue.

Vascular bundles

Individual vascular bundles contain xylem and phloem surrounded by a fibrous sheath. The sheath is composed of thick-walled, long narrow cells and may enclose only the phloem. Alternatively, a (usually larger) sheath may surround the entire bundle. Contiguous bundle sheaths are often merged, especially at the stem periphery where vascular bundles may be concentrated, and contribute to stem hardness and mechanical strength.

Phloem

Phloem is composed of sieve tubes and companion cells. Vascular bundles have either a single phloem strand or, more frequently, two strands often with a central lacuna where the protophloem occurred.

A striking correlation has been found between the numbers of phloem strands in the central vascular bundles of the petiole (Parthasarathy 1968) and taxonomic groups. There is a single phloem strand in all tribes of Coryphoideae except Trachycarpeae, and in Nypoideae, Ceroxyleae, and Iriarteeae. Two phloem strands occur in Calamoideae and tribes Trachycarpeae, Cyclospatheae, Phytelepheae, Cocoseae, Geonomateae and Areceae.

The structure of the sieve tube in monocotyledons (Cheadle & Whitford 1941, Cheadle 1948) forms a series from very oblique end walls, having only compound sieve plates, to transverse or slightly oblique end walls and simple sieve plates. A study of 374 species in 163 genera of palms (Parthasarathy 1968) found sieve elements in roots to have compound sieve plates on very oblique to oblique end walls, whereas those in stems and petioles have end walls varying from very oblique to transverse with compound to simple sieve plates. In the species studied, compound sieve plates predominate except among the Calamoideae and Nypoideae, which exhibit a high percentage of simple sieve plates on transverse to oblique end walls in stems and petioles. Otherwise in palms, the inclination of the end wall of sieve-tube elements shows a progressive change from oblique in the roots to transverse in the leaf lamina. Thus, in general, studies of palms support the conclusions about sieve tubes in other monocotyledons.

There is also some correlation between major categories of palms and the inclination of the end wall in sieve elements. Phytelepheae, Cocoseae and Areceae exhibit a high percentage of transverse to slightly oblique compound sieve plates in the sieve elements of petioles, whereas the Corypheae (*sensu* Uhl & Dransfield 1987) and to a lesser extent the Borasseae have sieve elements with a high percentage of oblique to very oblique end walls.

It is also noteworthy that sieve elements are long-lived, presumably functioning for the life of the palm, perhaps decades or a century or more. During this time, there is little change in structure except for an apparent decrease in the amount of endoplasmic reticulum and increased degeneration of mitochondria and plastids in the older elements (Parthasarathy 1974, Parthasarathy & Tomlinson 1967).

Indeed, palms have been described as histologically astounding because their sieve cells and companion cells remain indefinitely functional (Tomlinson 2006a). The sieve elements of palms differ structurally from those of most dicotyledons in the absence of obvious quantities of P-protein and in their failure to produce definitive callose except after leaf fall (Parthasarathy & Klotz 1976).

Xylem

Xylem is composed of two parts: the metaxylem having one, two or several large vessels; and the protoxylem, which has several to many narrow elements, mostly tracheids with distinctive annular or spiral thickenings (Klotz 1978a). A bundle may contain protoxylem alone, metaxylem alone, or metaxylem and differing numbers of protoxylem elements. Elegant studies designed to determine the course of vascular bundles and transport in palms (Tomlinson 1990) have explained why a variety of vascular bundles are seen in a cross-section of a palm stem (see below).

Relative lengths and certain features of the end walls of xylem vessel elements may be useful in determining relationships between palm taxa. The variously sculptured end walls of vessels have been divided into three categories: those bearing long scalariform (ladder-like) perforation plates, those with shorter plates with fewer bars, and porous plates. A survey by Klotz (1978b) of large vessels in the roots, stems, and petioles of 209 species in 169 genera of palms has shown that vessels are mostly porous in roots, intermediate in stems, and have longest scalariform plates in leaves. This agrees with trends established for monocotyledons as a whole (Cheadle 1943, 1944).

Klotz (1978a) further found that differences between groups of palms could be stated as differences in averages rather than in absolute degrees, perhaps indicating that vessels have specialised independently within some major subdivisions of palms. Vessel specialisation appears to be prominent in Calamoideae and successively less prevalent in the Coryphoideae, Areceae, and Nypoideae. *Nypa* is among the palms with the least-specialised vessels and the only subfamily lacking simple perforation plates. Scalariform perforation plates in roots, stems, and petioles are also present in most Chamaedoreeae and some Iriarteeae, but some members of these groups have more advanced vessels. Relatively recent observations on xylem structure, such as those made using scanning electron microscopy (SEM), and to some extent transmission electron microscopy (TEM), have provided information on the presence and extent of pit membranes, the extent of conductive areas on end versus lateral walls of vessels, and the morphology of perforation plates (Carlquist & Schneider 2002).

Some palm groups of similar habit are alike in vessel structure. The small slender palms of tribe Chamaedoreeae have uniformly scalariform perforation plates and the calamoid rattans are all highly specialised. Not all vines within the family follow this pattern, however, as the climbing *Chamaedorea elatior* and species of *Desmoncus* both have multiple perforation plates in their stems. Vessels in stems vary in width according to habits; those of lianas are widest, those of rhizomatous species narrowest, and those of erect stems intermediate.

The number of wide vessels in the major bundles of the petiole may be of taxonomic significance (Klotz 1978b). Most taxonomic categories have one, two, or a combination of one and two wide vessels per bundle. A few species are distinguished by more than two wide vessels per bundle. Coryphoid palms are most variable in this regard.

Longitudinal continuity of vascular bundles

Determination of the longitudinal continuity of vascular bundles in palm stems was undertaken from the 1960s onwards by P. B. Tomlinson and the late M. H. Zimmermann of Harvard University (Tomlinson 1990). These investigators had to develop techniques to accommodate a very large number of bundles over long distances. Some 20,000 bundles are present in the cross-section of a coconut stem, and hence Tomlinson and Zimmermann chose a smaller palm, *Rhapis excelsa*, with only about 1000 bundles in the stem. They adapted cinematography to photograph serial sections or sequentially cut surfaces on movie film, and hence to determine the longitudinal continuity of the vascular bundles. Cinematography proved excellent for analysing complex three-dimensional structures, and was subsequently applied to the rhizome, juvenile phase, apex, inflorescence, root, branch, and flower (Uhl *et al.* 1969) of *Rhapis*, resulting in an exemplary series of papers (e.g., Tomlinson & Zimmermann 2003). Other palm genera (e.g., *Sabal*, *Acoelorrhaphe*, *Washingtonia*, *Phoenix*, *Chamaedorea* and *Geonoma*) were also analysed (Zimmermann & Tomlinson 1974).

When a single bundle is followed through a palm stem, a relatively simple system can be discerned. As followed distally, a major bundle in the periphery of the stem curves toward the centre. From the stem centre, the bundle recurves towards its periphery and, before entering a leaf, branches in an upward direction, producing a continuing axial bundle. During its course, the original bundle may also produce short branches to neighbouring bundles and to other branches that become traces to axillary buds and inflorescences. Within the stem, the axial bundle follows a helical course. This pattern, determined by extensive analysis of *Rhapis excelsa*, has become known as the '*Rhapis* principle' (Tomlinson 1990, Tomlinson & Zimmermann 2003). The '*Rhapis* principle' is fundamental in interpreting vascular patterns in other palms and monocotyledons. A somewhat different system has been discovered in the stems of *Calamus*, a genus of climbing palms, which have extensive internodal elongation, a small amount of thickening growth and a discontinuous axial vascular system. In these palms, stem bundles end blindly below and are connected to each other only by narrow transverse commissures (Tomlinson *et al.* 2001, Tomlinson & Spangler 2002, Tomlinson 2006a, 2006b). Analysis of stem vasculature architecture (Tomlinson & Zimmermann 2003, Tomlinson

2006b) in the unrelated climber *Desmoncus* (Cocoseae) revealed a pattern resembling that of *Rhapis*, and an intermediate pattern was found in the related *Daemonorops*. The authors suggest that the unusual vasculature of *Calamus* may be a factor in the success of this most-species-rich genus of palms.

Zimmermann and Tomlinson (Tomlinson 1990) discovered that the diversity of bundle types in a palm stem can be accounted for by the anatomical changes that occur throughout the course of a single bundle. An axial bundle at the periphery of the stem has a fibrous sheath around the phloem, and possibly also around the xylem, but lacks protoxylem. As the bundle progresses toward the centre of the stem, the sheath becomes smaller and one or two elements of protoxylem appear. At the stem centre, the sheath is very small, there are several large vessels in parallel, and protoxylem is well developed. As the bundle recurves toward the periphery of the stem, an axial branch containing only metaxylem and phloem is formed, as are several similar short connections, containing only metaxylem, to other axial bundles and bundles that will be inflorescence traces. When the bundle becomes a leaf trace, it contains only protoxylem with phloem and a reduced fibrous sheath.

The majority of palms remain to be studied histologically. The number of large vessels and the shape of the fibrous sheath may aid in the identification of a sample as a palm (Tomlinson 1961, Klotz 1978b). It is, therefore, relatively easy to recognise a fossil palm stem by the large bundles of characteristic form, but it is usually impossible to relate the palm to a modern genus. Fossil roots, if present, may be more useful (see below).

Wood

The wood of dicotyledons consists of secondary tissues formed by a meristematic layer, the cambium, which usually produces yearly increments throughout the life of a tree. By contrast, the 'wood' of palms is primary in origin. A typical palm stem attains a maximum diameter, and subsequently shows relatively little increase in girth. The mechanical properties of palm stems, however, change with age. Rich (1987a, 1987b) found that the stiffness of palm stems increases with height. The trunks become stronger as they age, possibly due to increasing sclerenchyma. The characteristics of palm wood reflect the structure of the cortex and central vascular cylinder. The vascular bundles may be crowded in a peripheral ring, their fibrous sheaths united, so that the stem is markedly heterogeneous with a hard peripheral and a soft central layer. In other palms, such as the coconut, bundles are more evenly scattered and the stem relatively homogeneous. The structure of the central cylinder determines the usability of the wood (see Tomlinson 1990). Uniformly solid small stems are used as poles (*Geonoma* spp.); larger solid stems, such as those of coconut, are sawn into planking. Stems with softer centres are used for gutters or split into slightly concave planks. Rattan stems are long, light, pliable, easily split and can be bent under steam pressure, making them suitable for use in furniture and many other products.

Age

The age of a temperate dicotyledonous tree can easily be determined by counting the annual rings in a cross-section. Palms do not have growth rings, and so no simple direct reliable method exists for determining the age of a palm (Tomlinson 1990). Estimates can be made, however, by measuring the time it takes for a leaf to emerge and be replaced by the next leaf, the so-called plastochron. Age is estimated by multiplying the plastochron by the number of nodal scars plus the number of green leaves in the crown. An allowance has to be made for establishment growth, in which no internodes are visible. The crude estimates of age obtained in this way do not take account of varying growth rates during the life of the palm. Furthermore, size is of little value in age estimates: a vigorous young palm may be larger than an older palm that is environmentally stressed. Internode length can be a measure of stem vigour rather than age (Waterhouse & Quinn 1978).

THE ROOT

Structure

In dicotyledons, the tap root of the seedling is increased by secondary growth, whereas in palms, the tap or primary root is very small and functions for a short time only in the seedling. The primary root of palms is subsequently replaced by lateral adventitious roots (Glossary fig. 3d), which develop on the internodes of the stem and increase in size as the internodes enlarge. Roots may branch to four orders, but two is most common. First-order roots emerge directly from the stem, second-order roots emerge as branches of the first and so on. Large masses of relatively short roots are formed at ground level or slightly below. In sandy soil, roots of the first order may extend horizontally for up to 40 m, as reported for *Euterpe oleracea* and *Mauritia flexuosa* (De Granville 1974, Seubert 1997), or in some genera, may descend for several metres into the ground.

Stilt roots

Stilt roots are first-order, aerial, adventitious prop roots that develop on internodes (Glossary fig. 3e). Larger stilt roots on an extended obconical axis are characteristic of the Iriarteeae. Such roots develop in a circle around the trunk and slant to the ground providing the palm with a cone-like base. A second higher cone of roots is sometimes formed (Avalos 2004). Stilt roots in *Iriartea deltoidea* are tightly clustered at the base of the stem, have lenticels and lack thorns, whereas those of *Socratea* are separated, have thorns and can reach 4 m in length. Shorter stilt roots sometimes develop at the base of the trunk (e.g., *Gaussia*, *Podococcus*, *Euterpe*, *Drymophloeus*, *Cyphophoenix*, *Physokentia*, *Verschaffeltia* and *Roscheria*). The ecological significance of stilt roots has been debated. They have been related to growth in swampy habitats, to light environment and, for *Cyphophoenix fulcita* and *Verschaffeltia*, to stabilisation in a rocky habitat. It should be noted that reports of stilt palms being able to 'walk' (e.g., Leopold 2000) are entirely spurious and are not based on empirical evidence. The term 'stilt-root walking', as coined by Bodley and Benson (1980), refers to the ability of stilt palms to stabilise themselves after being disturbed by a tree fall or other stochastic event. Stilt roots may form along a leaning stem, providing support while the apex turns upwards and resumes vertical growth.

Modifications

Modifications of roots afford protection and/or gaseous exchange. Root tubercles are short, modified lateral roots. They are of two forms: either more rounded or elongate and somewhat bottle-shaped. They are very conspicuous in *Pritchardia*, *Acoelorrhaphe* and *Serenoa*. Root spines are also modified short, second-order roots, which may be blunt or pointed. In root spines, cortical tissues are highly lignified, cell walls are thickened, and aerenchyma is lacking. The vascular cylinder is fibrous and/or lignified and extends as a sharp point. Spine-like roots are diagnostic for *Cryosophila* (Glossary fig. 4k) but are found in many genera of palms (Seubert 1996a, 1996b, 1997, 1998b). Second-order roots may extend distally for 10–15 cm and can develop as pneumatophores (breathing roots), which are especially common in swamp palms (*Mauritia*, *Metroxylon*, *Phoenix* and *Raphia*). Other modifications include pneumathodes, white mealy patches that are modified parts of the outer cortex, perhaps comparable to lenticels. Spine roots that have pneumathodes are termed pneumatorhiza. In coryphoid palms, pneumatophores occur only in *Phoenix*, but pneumathodes are frequent both on pneumatophores and on normal first- and second-order roots (Seubert 1997).

Anatomy

The anatomy of palm roots is remarkably variable and, until recently, has been sparsely studied. Seubert (1997), who undertook a masterful survey of the family, described "... an unforeseen obstacle: that is the chaotic use of histological concepts by writers hitherto engaged with the study of palm roots."

In cross-section, roots have an outer layer, the rhizodermis, which may be uniseriate or consist of several layers of thin-walled cells. The rhizodermis is also referred to as a 'velamen' (a term more commonly used in orchid root anatomy), especially where it is multiple layered. From the rhizodermis inward, the following tissues are found: an outer cortex, an inner cortex, an endodermis, a pericycle, and a vascular cylinder. Aerenchyma frequently occurs in the inner cortex; its development can be either lysigenous or schizogenous. Each one of these tissues can be ontogenetically and phylogenetically modified in various ways (e.g., Seubert 1997). Root hairs occur in some palms (Coryphoideae, Ceroxyleae, Chamaedoreeae and perhaps others). They are most easily seen in young roots.

Cork, a layer consisting of thick-walled, suberised, lignified cells, usually containing tannin, is found in only a few monocots. This tissue is derived from the cells of the inner cortex. Monocot cork appears to differ in development from dicot cork in that it is of the etagen-type, i.e., lacking a discrete row of initials (Tomlinson 1990, Seubert 1997). Cork develops in many genera of palms; in Chamaedoreeae, it is formed in older roots of relatively large diameter.

Evolution

In her wide ranging studies of the palm family, Seubert found that some groups and/or genera can be distinguished by characters of root anatomy. Most importantly, she found that although genera vary in root structure, the species of individual genera vary only slightly.

Although we have not included root characters in the morphology/anatomy dataset, they do support the new classification in numerous ways. The Calamoideae (Asmussen *et al.* 2006, Baker *et al.* in review) is described by Seubert (1996a) as a homogeneous group that does not show any close relationship to coryphoid palms, thus supporting an isolated position for this subfamily.

Seubert (1996b) found that *Nypa* is isolated from other palms in root structure, an observation that is perhaps in part related to the mangrove habitat of this subfamily. Nevertheless, this finding concurs with *Nypa*'s position in the classification.

The 36 coryphoid genera that Seubert (1997) studied can each be distinguished by root anatomy. The Cryosophileae form a distinct group that has affinities with Sabaleae. The Rhapidinae and other Trachycarpeae show similarities in root anatomy. Members of the Borasseae tribe are alike in root structure, and exhibit characters considered by Seubert to be advanced. It seems especially noteworthy that the root anatomy of *Corypha* is unlike that of other members of the Chuniophoeniceae but resembles that of the Borasseae, supporting recent findings that *Corypha* is closely related to the Borasseae, and not to Chuniophoeniceae genera as previously indicated (Uhl & Dransfield 1987).

Root anatomy does not support the placement of the Caryoteae in Coryphoideae, although the tribe also shows a striking similarity to the Iriarteeae and certain Bactridinae: *Acrocomia* and *Aiphanes*. *Pseudophoenix* shows similarities to the Ceroxyleae, which on the basis of root anatomy characters is a natural group. However, placement of the Phytelepheae in Ceroxyloideae is not supported. *Aphandra* was not studied but *Ammandra* and *Phytelephas* approach the Geonomateae in root structure (Seubert 1998b).

In Arecoideae, root hairs are absent except in the Chamaedoreeae. In most arecoids, mucilage cells containing tannins or raphides are present in the inner cortex of the roots where stone cells also often occur. Arecoids have two types of fibres in their roots: the first type, borne singly or in bundles, has fibres with wide lumina containing tannin and moderately thick lignified walls. The second type, usually borne singly, has thick birefringent walls. More regular fibre bundles may be surrounded by stegmata. The Chamaedoreeae is a unified group, structurally different from the Ceroxyleae to which it was previously thought to be related (Uhl & Dransfield 1987).

In the Attaleinae, root anatomy confirms the relationship of *Butia* and *Jubaea*. *Parajubaea* is well defined, as are most genera of the subtribe, but *Butia*, *Jubaea* and *Parajubaea* are closer to each other than to the other genera of the group. The root anatomy of *Allagoptera* is described as hardly distinct (Seubert 1998b), supporting the combination of these genera in *Allagoptera*. Several genera now reduced to synonymy with *Attalea* were found to be extremely similar in root anatomy. The root anatomy of *Elaeis* resembles that of other Cocoseae, especially *Attalea*, *Butia*, *Jubaea*, *Parajubaea* and *Allagoptera*. The genera of Bactridinae, however, share many characters that make the subtribe distinct from other Cocoseae. Nevertheless, these genera are also very well separated.

Linospadicinae is very homogeneous in root anatomy. A combination of root characters unites the genera of Ptychospermatinae, suggests affinities with Linospadicinae and Dypsidinae, and sets Ptychospermatinae genera apart from Euterpeae. For more on relationships indicated by root anatomy, see listed papers by Seubert (1996a, 1996b, 1997, 1998a, 1998b).

ARMATURE

Many palms are armed, often fiercely, with spines that come in a variety of forms and represent a diversity of plant parts (Uhl & Moore 1973, Glossary figs 6, 7; Table 1.3). In *Phoenix*, the presence of stout spine-like, induplicate leaflets (acanthophylls) on the basal portion of the blade separates the genus from all other palms that have pinnate leaves. Long needle-like spines formed from the fibres of the disintegrated leaf sheath characterise several other coryphoids (*Rhapidophyllum*, *Zombia* [Glossary fig. 5e], *Trithrinax acanthocoma*, *Guihaia grossifibrosa* and *Maxburretia furtadoana*). As noted above, *Cryosophila*, some members of Iriarteeae and species of the calamoids *Eugeissona* and *Mauritiella* are characterised in part by the presence of usually numerous, frequently branched spines that are derived from roots.

Spines that do not result from any organ but that develop from epidermal and underlying tissues are known as emergent spines. In many members of Borasseae, *Corypha* (Corypheae) and in many species in Trachycarpeae, large teeth develop along the margins of the petiole and sometimes the blade (Glossary fig. 7). The Calamoideae exhibits an immense diversity of emergent spines on leaves, leaf sheaths and inflorescences (Glossary fig. 6). In *Calamus* and *Daemonorops*, spines on the leaf sheaths of some species form elaborate ant galleries. The Bactridinae (Arecoideae: Cocoseae), the Oncospermatinae and the Verschaffeltiinae (Arecoideae: Areceae) are distinguished by slender to very stout, emergent spines on the trunk, leaves, inflorescences, and sometimes (in Bactridinae) even on flowers and fruit. Certain genera are armed with spines of multiple origins; for example, *Elaeis* has spine-like rachilla tips, marginal teeth on the petiole (Glossary fig. 7g) and spine-like bases on the midribs of the leaflets.

TABLE 1.3
Distribution and derivation of armature

	Pinnae	Sheaths	Roots	Teeth	Emergences
I CALAMOIDEAE	+		+		+
III CORYPHOIDEAE					
CRYOSOPHILEAE		+	+		
PHOENICEAE	+				
TRACHYCARPEAE		+		+	
CORYPHEAE				+	
BORASSEAE				+	
V ARECOIDEAE					
IRIARTEEAE			+		
COCOSEAE	+	+		+	+
ARECEAE					+

LEAVES

The leaf is the distinguishing character of the palms. As in most monocotyledons, it has three components: a sheathing base, a petiole, and a blade. However, the leaves of palms are far from simple in structure. Indeed, as Tomlinson (1990) states, the leaf of palms "represents the most complex determinate organ built by plants." The initial complexity results from two factors: all palm leaves are plicate or corrugated and a majority are dissected in one way or another. Within these parameters, palm leaves exhibit many forms. It has long been known that the compound leaves of palms arise by a unique two-step process (Tomlinson 1990, Arunika *et al.* 2006), which involves, first, the origin of folds or plications, and second, division of the blade.

Size

Palm leaves differ markedly in the size and shape of all stages from the eophylls (the first leaves with blades in seedlings) throughout juvenile leaves to the mature blade (Tomlinson 1960). Among the smallest are the bifid or undivided leaves of *Chamaedorea tenella* and *C. tuerckheimii*, with blades sometimes less than 15 cm long. By contrast, the pinnate blades of *Raphia regalis*, reach a length of over 25 m (Hallé 1977), the largest leaves in the plant kingdom. Simple leaves reach lengths of 5 m in *Manicaria* and *Marojejya darianii* and 4 m in *Verschaffeltia*. The largest palmate leaves, those of *Corypha umbraculifera*, can be 8 m in diameter and have a petiole that is 5 m long. Large size characterises all forms of palm leaves, making the leaf a primary exemplar of the 'gigantism' expressed in all organs of palms (Tomlinson 2006a).

Early development

In both dicotyledons and monocotyledons, the leaf originates as a dorsiventral primordium with a discernible upper and lower leaf zone. The blade and petiole are initiated in dicotyledons from the distal leaf zone and the leaf base from the proximal zone (Arunika *et al.* 2006). In most monocotyledons, the reverse is true: the blade arises from the lower zone. In a few monocotyledon families, including the palms (Kaplan *et al.* 1982a), however, leaf blades are derived from an upper zone. An alternative interpretation has been brought forward by Rudall and Burzgo (2002) who found that blade development takes place in a meristematic transition zone between the leaf sheath and its precursor tip. Few palms have been examined and further investigation is needed.

Leaf sheath

The sheath or leaf base of palms is extremely varied in form. It is an important factor in support, transport, and palm architecture. The sheath completely encircles the stem from early in development, when the leaf primordium is about 1 mm in diameter. It may be fibrous, leathery, or woody with different parts varying in texture. Tomlinson (1990) identified four stresses that the leaf sheath must accommodate during its development and that contribute to its ultimate form. These are: the expansion of younger enclosed leaves, the enlargement of the stem, the weight of the blade and the weight of axillary inflorescences.

In many palms, the sheath splits and becomes open at maturity, so that its originally closed nature is no longer evident. Mature sheaths are tubular or non-tubular, and tubular sheaths are also of two types. In some relatively small palms (e.g., species of *Calamus*, *Pinanga* and *Chamaedorea*), the stems are small and the sheath may remain attached to the stem as it ages. In larger palms (e.g., *Roystonea* and *Veitchia*), the tubular sheath often forms a crownshaft that has two abscission zones, a vertical one along the adaxial line (i.e., opposite the petiole base) and a second horizontal one around the stem at the base of the sheath. In these palms, splitting of the tubular sheath releases the whole leaf, which falls to the ground. The presence of a crownshaft is a putative synapomorphy for the Areceae, although there are numerous reversals within the clade and independent origins elsewhere (Norup *et al.* 2006).

Non-tubular sheaths are even more diverse. Tomlinson (1990) has identified six types, distinguished by how the sheath disintegrates. The *Hyphaene* type, in which the leaf base splits widely on the dorsal side, is found in a number of large fan-leaved palms (e.g., *Sabal*, *Washingtonia*, *Borassus*, *Hyphaene*, and *Latania*) and is an important diagnostic character. In the *Phoenix* type, the ventral tissue of the leaf base erodes without becoming conspicuously fibrous. By contrast, the ventral tissues of the *Cocos* type separate as a fibrous network from the dorsal part of the sheath. In the *Trachycarpus* type, the ventral leaf sheath persists as a fibrous mat. In the *Zombia* type, the leaf base disintegrates into spines. Finally, in the *Caryota* type, found in *C. mitis* and smaller species of *Arenga*, the sheath splits ventrally with the surfaces connected by persistent fibres. Intermediates of these various forms may occur.

The anatomy of the leaf base has been studied in very few species and varies extensively among the genera that have been studied. A detailed investigation of the coconut by Tomlinson (1964) demonstrates the complexity of the leaf sheath, which displays a warp and weft system of vascular and fibrous bundles.

Leaf sheaths obviously function to protect the delicate tissues of the stem apex and axillary buds of various types. The leaf sheaths of two climbers, *Calamus tetradactylus* and *Desmoncus orthacanthos*, contribute in different ways to flexural stiffness and are important in biomechanical adaptation for climbing (Rowe *et al.* 2004). Many other functions of leaf sheaths doubtless await discovery.

Leaf blade

Shape

Palm leaf blades are fundamentally of two types, pinnate and palmate. In pinnate blades, the axis of the leaf is continuous from the petiole into the blade as an elongated rachis that bears pinnae or leaflets, the whole leaf thus appears feather-like in outline. An elaboration of the pinnate leaf type is represented by the bipinnate leaf, known only in and diagnostic for *Caryota*. Here the rachis of the blade bears not discrete leaflets but secondary rachises, these in turn bearing the leaflets (Periasamy 1966a). In palmate leaves, the blade is fan-shaped in outline, the ribs radiating from the base of the blade. Truly palmate leaves are in fact rather unusual (e.g., *Thrinax*). A more widespread type of fan-shaped leaf is the so-called costapalmate leaf, in which the axis of the leaf is extended into the blade to varying degrees as a major rib or costa. The ribs of the blade still radiate from the base although a few are inserted on the costa.

Palmate (including costapalmate) blades are characteristic of Coryphoideae, whose members all have palmate leaves apart from Caryoteae and Phoeniceae. Palmate leaves also characterise Mauritiinae (Calamoideae). All other palms, including Caryoteae and Phoeniceae, have pinnate leaves.

In most palmate and pinnate leaf types, the blade is divided into leaflets or segments; in some species, however, the blade remains undivided or is divided just along the line of the rachis near the leaf tip, so the blade is bifid. Among the fan-leaved palms, some species of *Licuala* (e.g., *L. grandis*, *L. cordata* and *L. orbicularis*) and all species of *Johannesteijsmannia* have blades that are essentially simple (entire), being only very shallowly notched along the margins. Among pinnate-leaved palms, there are many examples of leaf blades that are simple, although usually bifid at the tip (e.g., *Manicaria*, *Pelagodoxa*, species of *Cyphosperma*, *Marojejya darianii* and *Pinanga tomentella*). The leaves of *Iguanura sanderiana* and of some forms of *I. minor* and *Areca dayung*, which have a scarcely discernible apical notch, are virtually entire.

Dissected leaves are the form that occurs most frequently throughout the palm family. In mature palms, they may be regularly or irregularly divided into one-ribbed or several-ribbed units (Fig. 1.3). These units are called segments in palmate and costapalmate leaves, and leaflets or pinnae in pinnate leaves. Segment or leaflet tips are acute to acuminate or briefly and sometimes unequally bifid at the tip. Expanded, toothed (praemorse) apices occur in many groups (see Glossary fig. 15; Table 1.4).

Division

It is generally agreed that the compound nature of the palm leaf originates from the dissection of a simple blade (Arunika *et al.* 2006). Splitting of the blade occurs in two places: along folds and around the outer edges of the leaf, where the shedding of a marginal strip releases the tips of the segments. The segments or leaflets may each have only a single fold or, in both palmate and pinnate leaves, the blade may be divided into unequal segments or leaflets, some or all with several folds (see Glossary figs 10, 12). Fan leaves are most often divided regularly into uniform, single-fold leaflets, but variation in the depth and position of splits does occur and some palms bear leaves with both shallow and deep splits (Glossary fig. 10). The arrangement of leaflets is particularly varied in pinnate-leaved palms, in which leaflets may be inserted singly or in groups: they may be distributed regularly or irregularly and they may lie in one plane or multiple planes (Glossary fig. 13).

Multilayered leaves and dense radial packing of leaflets along the rachis allow palms to adjust foliar displays in response to light reception. Such arrangements also permit development of massive crowns without branching (Rich *et al.* 1995). If the distribution of leaf forms is examined, it is clear that palms that have adaxially split, induplicate, palmate, and costapalmate blades show a marked association with less tropical or more seasonal climates or with more open habitats (as does also *Phoenix*). In general, the reduplicately folded or pinnate leaf is associated with palms of wet forests and less-seasonal tropical climates. There are important exceptions to these trends, however, such as the Attaleinae, in which genera such as *Jubaea* and *Butia* and many species of *Syagrus*, are found in seasonal subtropical and drier climates.

Plication

Palm leaf blades are always plicate. Whether the folds or plications originate by splitting or by differential growth has been debated for over a century (Kaplan *et al.* 1982a, Padmanaban 1998, Arunika *et al.* 2006). Answering this question has required extensive research, involving careful attention to the planes of sectioning in three-dimensional primordia, the analysis of serial sections, and the use of scanning electron microscopy to determine that the folds originate by intercalary growth in submarginal positions in the pinnate leaf of *Dypsis* (as *Chrysalidocarpus*; Dengler *et al.* 1982) and in the palmate leaf of *Rhapis* (Kaplan *et al.* 1982b). This research confirmed previous reports that plications originate

TABLE 1.4
Distribution of praemorse leaflets or blade margins

Subfamily	Tribe	Subtribe	Genus/species
I CALAMOIDEAE			
	CALAMEAE		
		Korthalsiinae	
			Korthalsia
		Calaminae	
			Calamus caryotoides
			Ceratolobus spp.
III CORYPHOIDEAE			
	CARYOTEAE		
			Arenga
			Caryota
			Wallichia
V ARECOIDEAE			
	IRIARTEEAE		
			Dictyocaryum
			Iriartea
			Iriartella
			Socratea
			Wettinia
	CHAMAEDOREEAE		
			Chamaedorea tenerrima
	PODOCOCCEAE		
			Podococcus
	ORANIEAE		
			Orania
	REINHARDTIEAE		
			Reinhardtia
	COCOSEAE		
		Bactridinae	
			Aiphanes
			Bactris caryotifolia
	MANICARIEAE		
			Manicaria
	ARECEAE		
		Basseliniinae	
			Physokentia
		Dypsidinae	
			Dypsis thiryana
			Dypsis trapezoidea
		Linospadicinae	
			Linospadix spp.
		Ptychospermatinae	
			Adonidia
			Balaka
			Carpentaria
			Drymophloeus
			Brassiophoenix
			Normanbya
			Ponapea
			Ptychococcus
			Ptychosperma
			Solfia
			Veitchia (not all)
			Wodyetia
		Oncospermatinae	
			Verschaffeltia
			Roscheria
		Unplaced	
			Iguanura
			Hydriastele
			Loxococcus

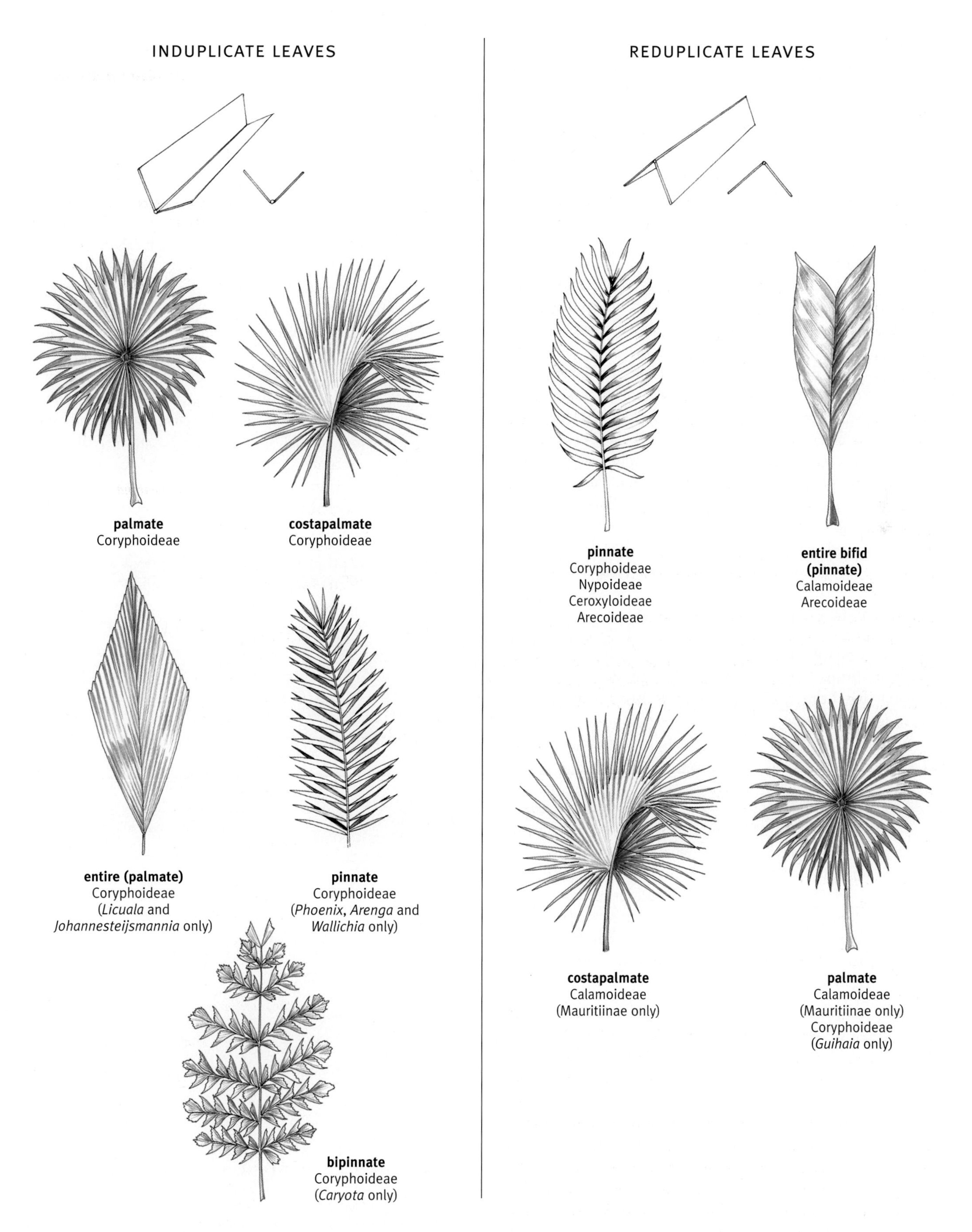

Fig. 1.3 The distribution of major leaf forms and splitting types across the subfamilies of palms. (Drawn by Marion Ruff Sheehan)

by differential growth in four other palms, *Cocos nucifera*, *Phoenix sylvestris*, *Borassus flabellifer* and *Caryota mitis* (Periasamy 1962, 1965, 1966a, 1966b, 1967). Although only six genera have been studied in detail, they represent phylogenetically diverse groups within the family (Asmussen *et al.* 2006, Baker *et al.* in review). The origin of plication by differential growth is widely accepted.

Induplicate or reduplicate form

As a rule, the adaxial (upper) ribs of palm leaves are stouter than the abaxial (lower) and contain more vascular bundles. When primary splitting occurs along the adaxial ('upper') ribs, segments or leaflets that are V-shaped in section (induplicate) are formed (Fig. 1.3, Glossary fig. 9). When splitting occurs along the abaxial ('lower') ribs, segments or leaflets that are Λ-shaped in section (reduplicate) result (Tomlinson 1960; Fig. 1.3; see Glossary fig. 9). Characteristically, the rachis continues through the central segment of induplicately folded leaves, whereas the blade of reduplicately folded leaves is bifid at the apex. Induplicate leaves characterise subfamily Coryphoideae, with one exception, the genus *Guihaia*, which has reduplicate leaves. All other groups of palms have reduplicate leaves (Fig. 1.3). Although the position of splitting seems to define induplicate and reduplicate taxa, the difference may be more profound, as seen in *Rhapidophyllum* and *Rhapis* where the splitting is between the folds but the leaves are clearly induplicate (Glossary fig. 9).

Processes of splitting

Plication formation has been studied in detail, but little research has been done to determine precise methods of splitting along folds. Two processes that separate the plications into individual leaflets have been noted (Arunika *et al.* 2006). In the first, a mucilaginous disintegration results in separation. In the second process, schizogeny, splitting does not extend across the blade but a narrow isthmus of tissue remains to hold the leaflets together (Dengler *et al.* 1982). The zone of schizogeny, where observed in abaxial splitting of folds and in abscission of the marginal strip (Arunika *et al.* 2006), affects ground and dermal tissue only. When induplicate folds are formed, the abscission zone extends along both sides of the vascular bundles, which often persist as interfold filaments and are apparent and thread-like in many coryphoid palms (e.g., *Sabal*, *Pritchardia*, *Brahea* and *Hyphaene*) (Glossary fig. 15).

The shedding of the marginal strip is considered an abscission process but its cell biology has not been determined. Programmed cell death or simple schizogeny have been suggested to explain it (Arunika *et al.* 2006). In palmate leaves, the marginal strip may be relatively ephemeral and might disintegrate without being freed as a separate entity. In many pinnate leaves, the strips are fibrous, even occasionally vascularised. They may be shed as long strips sometimes seen hanging from the leaves and referred to as 'reins' (Eames 1953).

The induplicate leaf

As noted above, induplicate leaves are restricted to subfamily Coryphoideae and are mostly palmate, with the exception of leaves of Caryoteae and Phoeniceae. Different patterns of shape and division appear in group after group (Table 1.5). The palmate blade varies from flabellate to cuneate in outline, and from nearly lacking a rachis to possessing such a long one that it approaches a pinnately ribbed form (Glossary figs 8, 10). Leaves are usually regularly divided into single-fold segments by splits along the adaxial ridges. In some palms, such as *Trachycarpus fortunei*, the adaxial splits are irregular, extending to different depths from the leaf margin. The central unsplit area of a fan-leaf is called a palman. In some genera (*Licuala* and *Pholidocarpus*), the blade is divided by a few deep splits into compound (several-fold) segments, which are again regularly divided distally by much shorter splits into single-fold induplicate segments. The strongly costapalmate leaves of *Livistona lorophylla* and *L. decipiens* have very deep adaxial splits. In these species, the distal portion of the blade is divided almost to the rachis into single-fold segments, making the apical parts of the leaf appear remarkably similar to the pinnate leaf of *Phoenix*.

Abaxial splitting of induplicate blades

Guihaia is exceptional among the Coryphoideae in having clearly reduplicate leaves with rather shallow segments of one or two folds; by contrast, its close relatives in the Rhapidinae have induplicate leaves. In several coryphoid genera, another type of abaxial splitting occurs. In

TABLE 1.5
Distribution of leaf types

	Induplicate		Reduplicate	
	Palmate[1]	Pinnate[2]	Palmate[1]	Pinnate
I CALAMOIDEAE				
EUGEISSONEAE				+
CALAMEAE				+
LEPIDOCARYEAE				
Ancistrophyllinae				+
Raphiinae				+
Mauritiinae			+	
II NYPOIDEAE				+
III CORYPHOIDEAE				
SABALEAE	+			
CRYOSOPHILEAE	+			
PHOENICEAE		+		
TRACHYCARPEAE	+		+	
CHUNIOPHOENICEAE	+			
CARYOTEAE		+		
CORYPHEAE	+			
BORASSEAE	+			
IV CEROXYLOIDEAE				+
V ARECOIDEAE				+

[1] including costapalmate
[2] including bipinnate

Chelyocarpus, *Itaya*, species of *Cryosophila* (Glossary fig. 10e) and *Trithrinax*, and in juvenile leaves of some species of *Sabal*, the otherwise induplicate palmate blade is divided along the midline or just to one side of it by a deep abaxial split, effectively separating the leaf into two halves (see Glossary fig. 10). In *S. palmetto* leaf development, this deep central split occurs before the adaxial splits (Uhl, pers. obs.).

Abaxial splitting appears to be an apomorphy in *Licuala*, in which most species have basically induplicate leaf blades, as can be seen by the lobing of the segment tips. However, superimposed on this pattern are abaxial splits, which divide the blade into compound or, very rarely (sometimes in *L. bidentata*), single-fold reduplicate segments. As in *Sabal*, these abaxial splits occur very early in the development of the leaf (Dransfield 1970, Dransfield *et al.* 1990); the adaxial splits that are typical in Coryphoideae occur much later. The abaxial splits of *Sabal* and *Licuala* are unusual in being the only ones in the subfamily to reach the insertion of the blade, except for the central divisions noted above. Adaxial splits may be very deep but there is always, even in *Phoenix*, a thin flange of lamina tissue between the split and the costa. Further developmental work is needed to understand these apparently anomalous forms.

Splitting between the folds

Another unusual type of splitting occurs in *Rhapis* and *Rhapidophyllum*, where, although the distal margin of the blade may be shallowly and regularly induplicately lobed, the major divisions of the leaf into segments occur between the folds rather than along the adaxial or abaxial ribs (see Glossary fig. 9). Once again, the unusual splits occur relatively early in the leaf's development (Dransfield 1970, Dransfield *et al.* 1990).

The leaf of* Phoenix *and the haut

The leaf of *Phoenix*, which is clearly pinnate but induplicate, is yet another example of the remarkable diversity of leaf form in the Coryphoideae. The origin of the leaflets in *Phoenix* is unique: a thin membranous layer, known as the haut, develops by proliferation of the adaxial ridges, forming a sheet of tissue that eventually contains vascular bundles and is shed along with the unplicate marginal strip to free the leaflets. The significance of the haut remains in question (Periasamy 1967, Padmanaban 1998). In the differential elongation of the costa, the leaf of *Phoenix* parallels reduplicate pinnate leaves.

The reduplicate leaf

The reduplicate leaf occurs in many more taxa than the induplicate leaf. It also exhibits many different forms (Fig. 1.3, Glossary figs 9, 12, 13). The majority of reduplicate leaves are pinnate, but three genera (*Mauritia*, *Mauritiella* and *Lepidocaryum* in Calamoideae) have palmate leaves. In large groups of palms, such as tribes Calameae, Cocoseae and Areceae, the pinnate leaf has nearly the full range of dissection from undivided to regularly pinnate. Leaflets may be in groups in one plane or in several planes (see Glossary fig. 13), sometimes with both arrangements occurring within the same genus (e.g., *Salacca* [Calamoideae], *Chamaedorea* and *Bactris* [Arecoideae]). In some climbing species, distal leaflets may be modified into hooks (acanthophylls) that vary from straight to recurved as noted above (Glossary fig. 6).

Hastula

Almost all members of palmate-leaved coryphoid palms have leaves that display a hastula, an adaxial flap or crest that appears to arise as an extension of the petiole at its junction with the blade (see Glossary fig. 11). An abaxial hastula also occurs sometimes. The origin and adaptive nature of the hastula is as yet not clear. A small flange occurs on the rachis of some pinnate leaves in *Oraniopsis*, *Ceroxylon* and *Cocos*, and may be homologous. Palmate-leaved Calamoideae (Mauritiinae) do not possess a hastula, instead, a low crest is present on the adaxial surface of the junction of the blade and petiole.

Ligule

A structure in the position of a ligule at the top of the sheath in front of the petiole, or sometimes lateral or opposite to the petiole, is found in the leaves of many palms (*Livistona*, *Arenga*, and species of Arecoideae, including some Ptychospermatinae; see Glossary fig. 6). In Calamoideae it is termed an ocrea and is almost always in front of the petiole; it occurs in some species of *Calamus* and *Daemonorops*, in all species of *Plectocomiopsis*, *Eremospatha*, *Oncocalamus*, *Laccosperma*, and conspicuously in *Korthalsia* (Glossary fig. 6), where the ocrea is a diagnostic feature at the specific level and often harbours ants (Beccari 1918, Dransfield 1981a). Ants are also present in the ocreas of *Laccosperma* in Africa and in a few species of *Calamus* in the Philippines and New Guinea. *Pogonotium*, unlike any other calamoid genus, has two ear-like processes, one on either side of the petiole at its base (Dransfield 1980a).

Veins

Most palm leaflets or segments have veins that are fixed in number at maturity. The veins extend more or less parallel to the midrib from the insertion to the acute, acuminate, bifid or praemorse tips. In leaflets that have praemorse apices, veins also diverge from the base. Secondary splitting between thickened veins gives rise to the mature leaf of *Socratea exorrhiza* or of *Normanbya normanbyi*. Although the midrib of the leaflet is usually central and most prominent, in a few genera (*Iriartea*, *Normanbya*, *Podococcus* and *Socratea*), the midrib is eccentric and distinguished from other veins or ribs only by its uppermost position at the attachment on the rachis.

The three genera of Caryoteae are exceptional in that all the veins do not run parallel or diverge from the base of the leaflet: some diverge from the midrib of the leaflet. All tend to terminate along the margin, and there may be an increase in the number of veins toward the apex. In the Iriarteeae and in *Normanbya* and *Wodyetia* (Areceae), as in Caryoteae, the leaflets of juvenile plants are simple, with veins that diverge toward the margin from the base and midrib. Secondary splitting between veins of this sort of leaflet produces leaflets along an extended axis. Although this is presumably the manner in which leaflets of *Caryota* were derived (see Periasamy 1967), further developmental studies are needed for confirmation.

Anatomy

Tomlinson (1961) described the anatomy of the segments or leaflets of many genera and outlined general trends within the family. Subsequent studies of lamina anatomy for several groups have proven diagnostically useful. Leaf anatomy was valuable in monographing *Euterpe*, *Prestoea* and *Neonicholsonia* (Henderson & Galeano 1996). Anatomical characters of leaves were found to distinguish species of *Thrinax* (Read 1975). The species of *Chelyocarpus* (Uhl 1972c) and of *Hyophorbe* (Uhl 1978c) can be identified from transections of the lamina. Leaf anatomy of the New Caledonian palms helped to identify species from both herbarium and fresh material, and revealed some remarkable similarities in indigenous members of the Archontophoenicinae. An update of earlier work on the vegetative anatomy of palms (Tomlinson 1961) by Drs P. B. Tomlinson and J. Fisher at Fairchild Tropical Garden is in preparation, and is certain to provide valuable systematic information and structural understanding for many taxa.

Eophylls

The simplest eophyll is linear to elliptic and entire. This form occurs in most members of the Coryphoideae (including the recently added Caryoteae), in some Ceroxyloideae, in Podococceae and in some members of three other tribes: Iriarteeae, Cocoseae and Areceae of the Arecoideae. Undivided eophylls of both induplicate (e.g., *Phoenix*) and reduplicate genera (e.g., *Clinostigma haerostigma* and *Roystonea*) are basically similar (see Glossary fig. 25). Bifid eophylls are most common in palms that have reduplicate leaves, and usually have an obvious rachis. The eophylls are divided to the base in some coryphoids (*Chelyocarpus*, *Arenga hastata* and *Caryota*). Compound eophylls occur in *Latania* (Borasseae: Coryphoideae), *Tahina* (Chuniophoeniceae: Coryphoideae), several species of *Chamaedorea*, e.g., *C. elegans* (Chamaedoreeae: Arecoideae) and *Rhopaloblaste* (Areceae: Arecoideae). In *Nypa*, the eophyll may sometimes appear imparipinnate.

The eophylls are followed by a progression of juvenile leaves. The size and nature of the mature blade develops through an increase in the number of lateral ribs and through the division of the blade into segments or leaflets. The bifid eophyll may represent the first step in the division of the leaf. The leaf of *Syagrus smithii*, for example, may not divide until it has reached a length of about 3 m (Moore 1963a). The mature leaf of *S. smithii* has numerous leaflets arranged in groups and dispersed in several planes, but during the juvenile interval, the blade is bifid and resembles that of geonomoid palms such as *Asterogyne martiana*.

Evolution

Dissected leaves are found in four orders of monocotyledons: Alismatales, Dioscoreales, Pandanales and Arecales. These four orders lack a common ancestor and their mechanisms of structural division differ markedly, indicating that each order has a separate origin of dissected leaves (Arunika *et al.* 2006). In dicotyledons, dissected leaves appear to have arisen some 29 times, with multiple reversions to simple leaves (Barathan *et al.* 2002).

Angiosperm phylogenies suggest that ancestral forms had simple leaves (Taylor & Hickey 1996). The development of palm leaves shows that the dissected leaf originates by splitting of a simple, plicate lamina. It can be conjectured that differential expansion, elongation, and dissection of a simple leaf, which resembles the undivided eophyll of most coryphoid palms and of other genera such as *Pseudophoenix*, *Roystonea* and several genera of Cocoseae, could result in the different forms of the palm leaf. The pinnately nerved or divided leaf could develop through elongation of the central axis together with the intercalation of additional ribs and laminar tissue. Inhibition or partial development of the central axis would yield the costapalmate and palmate leaf. Some evidence points toward this interpretation (Uhl & Dransfield 1987) but, for the most part, the processes to verify such developmental patterns have not been investigated. Furthermore, palm phylogenies (Asmussen *et al.* 2006, Baker *et al.* in review) demonstrate that simple-leaved forms arise independently within unrelated groups and suggest that the condition is not plesiomorphic.

Phylogenies indicate that the various forms of the palm leaf have arisen more than once (Table 1.5). The palmate leaf seems to have developed separately in the coryphoid and calamoid subfamilies, especially as it is reduplicate in the latter. The pinnate form is most likely plesiomorphic within the family, with two reversals from palmate to pinnate occurring within the Coryphoideae (in *Phoenix*) and the Caryoteae, both with distinct and different splitting mechanisms. Secondary division of the pinnae has arisen in the Caryoteae (*Caryota*) and in Iriarteeae and Areceae within Arecoideae.

Phylogenies (Baker *et al.* in review, Asmussen *et al.* 2006) have confirmed the importance of splitting as a taxonomic character. Induplicate plications support the monophyly of the Coryphoideae, including the previously proposed placement of the pinnate-leaved *Phoenix* within this group (Uhl & Dransfield 1987) and, more surprising, the removal of the anomalously induplicate-leaved Caryoteae from Arecoideae to Coryphoideae. Although the evolution of leaf form involves somewhat greater convergence, phylogenies indicate that this character suite too is of fundamental importance in palm systematics. When understood in greater depth, palm leaf development may shed new light on the phylogenetic structure of the family.

Research

Palm leaves have not been studied in sufficient detail. In particular, more developmental studies would be helpful. The compound leaf in palms, which results from the splitting of an entire lamina, is very different from compound dicot leaves, in which leaflets arise from marginal meristems as independent primordia. So much so, that it is questionable whether palm pinnae or segments are properly referred to as leaflets. All aspects of palm leaves require further investigation. The development of plication has been studied in only a few genera. The haut in *Phoenix* has not been subjected to modern analyses (Padmanaban 1998). The extent and identity of the 'rein' is unclear, and the types of meristems found in leaves are also in question. Neither are the mechanisms that cause splitting into leaflets understood.

INFLORESCENCE

Form

The palm inflorescence is often large and appears exceedingly complex, but when carefully analysed is straightforward in structure and composed of only a few variable organs (Tomlinson & Moore 1968). In the physiological sense of a reproductive event or flowering, inflorescences are discussed above under pleonanthy and hapaxanthy (Fig.1.4). In a strictly morphological sense, the inflorescence is a monopodial branching system that develops in the axil of a leaf. As noted above, every leaf potentially bears an axillary bud, although these buds may abort at any stage. Usually, the buds in scale leaves at the base of a shoot develop into new shoots, whereas buds in the axils of vegetative leaves develop into reproductive branches or inflorescence units. An inflorescence (Fig. 1.5) has a main axis that is usually unbranched proximally and branched distally, the branches bearing the flowers. The unbranched, stalk-like part of the main axis is termed the peduncle, while the branch-bearing portion is known as the rachis. Some inflorescences are completely unbranched (spicate).

Position

The most common position of the inflorescence, found in some 77% of palms (Henderson 2002b), is among the leaves in the crown (interfoliar) (Fig. 1.4). A second position is found in many Arecoideae, in which the inflorescence is exposed in bud by the fall of a deciduous subtending leaf and its long tubular sheath. The inflorescence then becomes positioned below the crown and is described as infrafoliar. In Iriarteeae, Areceae and less frequently in some larger species of Chamaedoreeae and Geonomateae, infrafoliar inflorescences are frequently found in species with larger stems and sustained primary growth (Henderson 2002b). Third, in a few genera, such as *Corypha* and *Metroxylon*, inflorescences are produced in the axils of reduced leaves (bracts) above the crown. This results in a compound inflorescence that is described as suprafoliar.

Bracts

Series from leaves to reduced leaves to bracts have been observed in many palms. These are described in *Nannorrhops*, *Metroxylon* and *Corypha* (Tomlinson & Moore 1968, Tomlinson 1971, Tomlinson & Soderholm 1975), and the homology of inflorescence bracts to leaves is well established. The apex of the main axis of the inflorescence first develops a prophyll. It then initiates none to several empty bracts, termed peduncular bracts, and finally a number of bracts that subtend branches, known as rachis bracts (Figs 1.5, 1.6).

Prophyll

The first bract formed by the inflorescence apex is always an adaxial, two-keeled prophyll, which is a synapomorphy for the palms. Although the prophyll varies in form, it is always bicarinate. It may be shorter or longer than subsequent bracts on the peduncle, have a rounded, u-shaped or beaked tip, and split apically, marginally, adaxially or abaxially. In some palms, such as *Hyophorbe* (Arecoideae) and the genera of the *Cocoseae* (Arecoideae), the prophyll matures when the inflorescence is in early stages of development and opens to allow sequential emergence of the peduncular bracts (see Fig. 1.5), each of which in turn encloses the inflorescence as it enlarges. The prophyll remains as a permanent shield around the flowering inflorescence in *Ceratolobus* (Calamoideae) and in two species of *Pinanga* (Arecoideae)

TABLE 1.6
Inflorescences with incomplete prophyll

IV CEROXYLOIDEAE			
	CEROXYLEAE		
			Ceroxylon
			Ravenea
			Oraniopsis
V ARECOIDEAE			
	ARECEAE		
		Basseliniinae	*Basselinia*
			Burretiokentia
			Cyphophoenix
			Cyphosperma
			Lepidorrhachis
			Physokentia

(Dransfield 1980b). In other palms, e.g., many Areceae, the prophyll encloses the inflorescence entirely until the inflorescence opens, at which point the prophyll abscises neatly at the base and falls, usually together with a similar peduncular bract. The prophyll may enlarge much more rapidly than other bracts and branches internal to it, as in *Phytelephas* and *Welfia*. It may be leathery or fibrous and is net-like in some species, e.g., *Cryosophila albida*, *Elaeis oleifera*, *Manicaria saccifera*, *Pholidostachys dactyloides* and *Sclerosperma* spp. In a few genera, when young stages do not completely close or when an abaxial opening forms at an early stage, the prophyll is incomplete, only partly encircling the peduncle (Moore 1973a, Table 1.6). Incomplete prophylls are a synapomorphy for the genera of Basseliniinae.

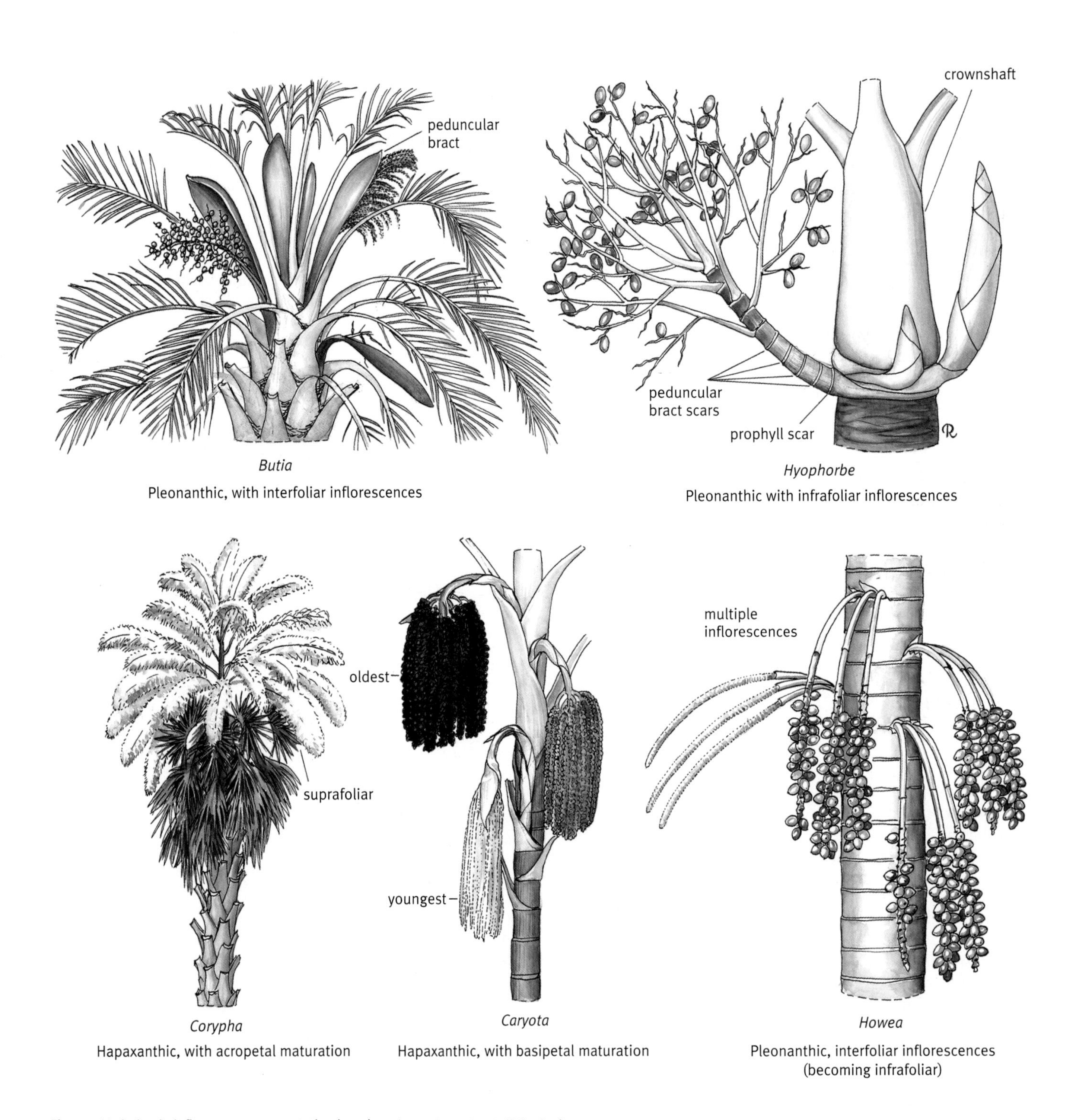

Fig. 1.4 Variation in inflorescence presentation in palms. (Drawn by Marion Ruff Sheehan)

Peduncular bracts

Peduncular bracts are the empty bracts present on the peduncle above the prophyll. They exhibit many forms. In most coryphoid palms, peduncular bracts are small, numerous, tubular and sheathing, opening early at the apex. In tribe Phytelepheae (Ceroxyloideae) and in Cocoseae, Geonomateae, Oranieae and Areceae (Arecoideae), there is usually only a much enlarged single peduncular bract (Fig. 1.6). This bract is particularly striking in Attaleinae (Cocoseae), in which the peduncular bract is woody and often cowl-like, and may be deeply plicate in bud and long-persistent above the inflorescence. Other large peduncular bracts occur in many arecoid genera, where they are often hard and beaked or thin, fibrous and sometimes deciduous. Peduncular bracts are not always borne close to the prophyll. In *Beccariophoenix* (Arecoideae: Cocoseae), the peduncular bract is thick, woody, circumscissile and borne at the tip of the peduncle.

Peduncular bracts are absent in nine genera representing calamoid, coryphoid and arecoid subfamilies, in which the prophyll is then the largest and only bract (Table 1.7). Elsewhere, the number of peduncular bracts is diagnostic for genera and ranges from zero to one, two, three, and four to seven (or possibly more). Three to seven peduncular bracts characterise some calamoid genera (*Mauritia*), most genera in Coryphoideae, and Iriarteeae, Podococceae and Chamaedoreeae (Arecoideae). They may persist on the inflorescence, as in Caryoteae, most Chamaedoreeae, *Socratea* (Iriarteeae) and Podococceae, or they may be caducous, as in other Chamaedoreeae and Iriarteeae. Even the prophyll may sometimes subtend a first-order branch (Tomlinson & Moore 1968, Tomlinson & Soderholm 1975), and so it is probable that the branch primordia subtended by peduncular bracts are aborted or suppressed.

TABLE 1.7
Inflorescences bearing an enlarged prophyll but no peduncular or rachis bracts

Subfamily	Tribe	Subtribe	Genus
I CALAMOIDEAE			
	CALAMEAE		
		Calaminae	
			Ceratolobus
			Pogonotium
III CORYPHOIDEAE			
	PHOENICEAE		
			Phoenix
	TRACHYCARPEAE		
		Rhapidinae	
			Chamaerops
		Livistoninae	
			Licuala spp.
	BORASSEAE		
		Lataniinae	
			Borassus
V ARECOIDEAE			
	ARECEAE		
		Arecinae	
			Areca
			Nenga
			Pinanga

Rachis bracts

Peduncular bracts are followed by bracts designated as rachis bracts (Fig. 1.6), which subtend branches of the first and subsequent orders. They too vary in form and may be completely sheathing, as in *Cryosophila* and *Nypa*, or elongate (*Washingtonia*). Frequently, they are relatively small and scale-like.

Branching

Except for the prophyll and peduncular bracts (by definition empty) and occasional proximal bracts on branches, there is a 1:1 relationship between bracts and branches including, with a few exceptions (see flowering units), those ultimately bearing flowers (Fig. 1.5). A simple system for referring to the orders of branches and of bracts was devised for *Nannorrhops* (Tomlinson & Moore 1968). First-order branches, which sometimes also bear prophylls, may again branch one to several times, each successive branch subtended by a bract, with the ultimate axes or rachillae bearing solitary flowers or flower clusters. Inflorescences may bear up to six orders of branching (*Sabal mauritiiformis*). A tendency toward an apparent suppression of branching is seen, however, in the spicate inflorescences that are found within certain major divisions and genera (Table 1.8). Inflorescence branching may differ within a species, as in several calamoid genera: *Salacca*, *Calamus*, *Daemonorops*, *Ceratolobus*, *Pogonotium* and *Retispatha*. In these genera, pistillate inflorescences are usually branched to two orders and staminate inflorescences to three. In *Chamaedorea ernesti-augusti* and *C. metallica*, also, the pistillate inflorescences are normally spicate, but the staminate are branched with numerous rachillae of the second order.

Rachillae

Branches of all orders end as a flower-bearing part, a rachilla (Fig. 1.6, Glossary fig. 16). Rachillae are structurally very distinctive parts of palm inflorescences. Developmental studies (Uhl 1976a and unpublished) show that when the inflorescence primordium is very small, rachillae are initiated in their entirety — each with a complete series of the bracts that will subtend flowers or flower clusters. The branches that terminate in the rachillae originate later. Mature rachillae may be pendulous or otherwise positioned and shorter or longer than the rachis. Their bracts are tubular or open.

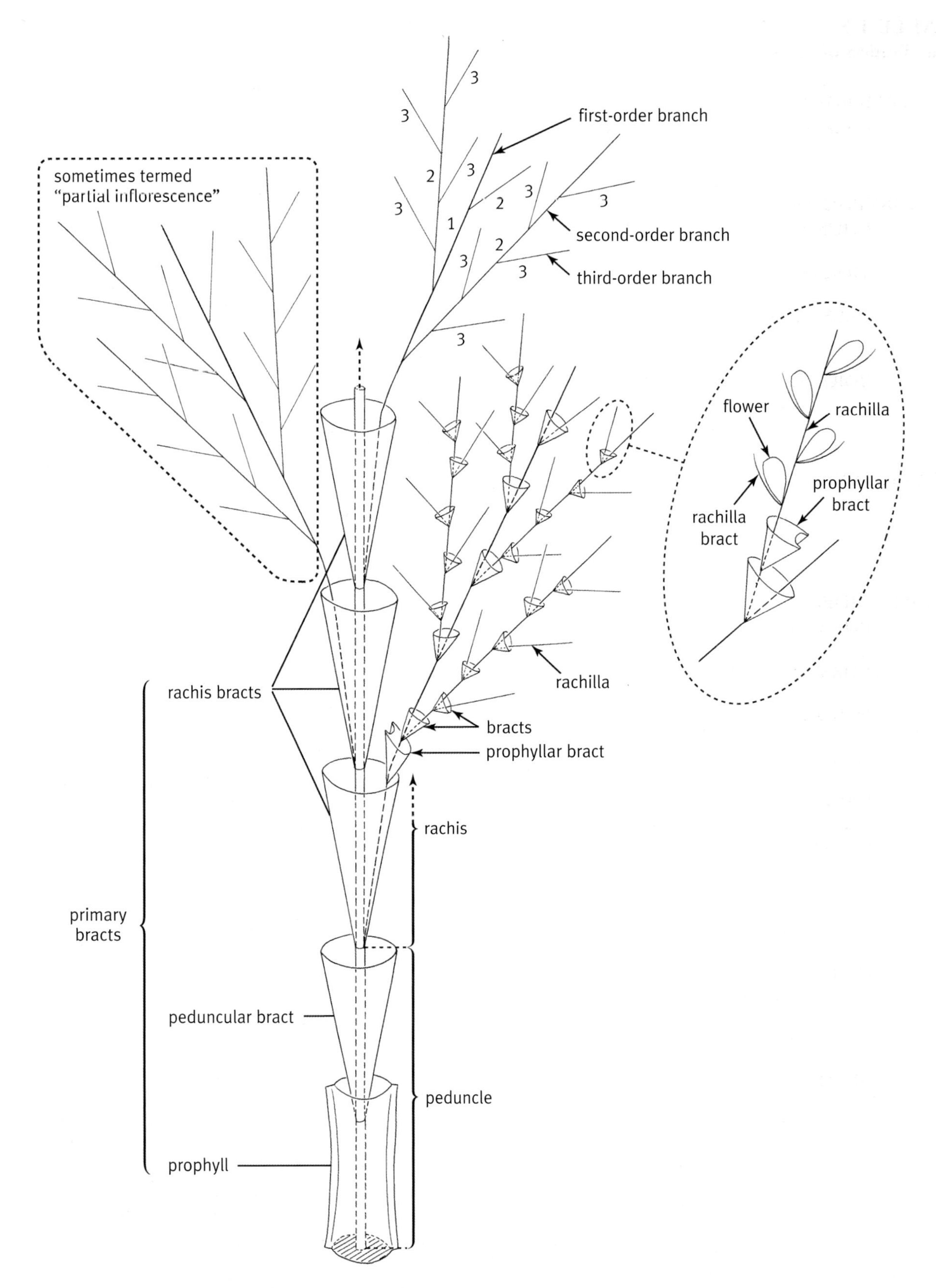

Fig. 1.5 Schematic diagram of a generalised palm inflorescence. Note prophyllar bracts may be borne at or near the base of all inflorescence axes. However, the one that is most often referred to as 'the prophyll' in palm descriptions is the prophyllar bract at the very base of the inflorescence, which is often systematically informative. Only one peduncular bract is shown here, whereas there may be more than one or none present. (Drawn by Lucy T. Smith)

TABLE 1.8
Distribution of spicate inflorescence units in palms

I CALAMOIDEAE
- CALAMEAE
  - *Korthalsia tenuissima*
  - *Salacca* (rarely)

III CORYPHOIDEAE
- CHUNIOPHOENICEAE
  - *Chuniophoenix humilis* (not always)
- TRACHYCARPEAE
  - *Licuala* spp.
- CARYOTEAE
  - *Arenga* spp.
  - *Caryota monostachya*
- BORASSEAE
  - *Borassodendron* (♀)
  - *Borassus* (♀, sometimes ♂)
  - *Lodoicea* (not always)

IV CEROXYLOIDEAE
- PHYTELEPHEAE
  - *Ammandra* (♀)
  - *Aphandra* (♀)
  - *Phytelephas* (♀ and ♂)

V ARECOIDEAE
- IRIARTEEAE
  - *Wettinia* spp.
- CHAMAEDOREEAE
  - *Chamaedorea* spp.
- PODOCOCCEAE
  - *Podococcus*
- SCLEROSPERMEAE
  - *Sclerosperma*
- REINHARDTIEAE
  - *Reinhardtia koschnyana*
- COCOSEAE
  - *Aiphanes* spp.
  - *Allagoptera*
  - *Bactris* spp.
  - *Cocos nucifera* (rarely)
  - *Syagrus* spp.
- GEONOMATEAE
  - *Asterogyne martiana*
  - *Calyptrogyne*
  - *Geonoma* spp.
  - *Pholidostachys pulchra*
- ARECEAE
  - *Areca chaiana*
  - *Calyptrocalyx*
  - *Dypsis* spp.
  - *Hydriastele* spp.
  - *Howea*
  - *Iguanura* spp.
  - *Laccospadix*
  - *Linospadix*
  - *Neonicholsonia*
  - *Pinanga* spp.
  - *Prestoea* spp.

Structural variation

Inflorescence structure is illustrated in Figs 1.5, 1.6 and Glossary fig. 16. The palm inflorescence may be described as polytelic as the branches are not terminated by flowers (Weberling 1998). The question of whether some inflorescences are determinate has not been completely resolved. In hapaxanthic inflorescences, such as those of *Corypha*, an inflorescence unit may appear to terminate the palm stem, but in most instances, careful examination reveals evidence of the apical bud of the shoot aborting or becoming suppressed. In *Arenga*, Tucker (1991) found that the terminal bud aborts and the terminal-appearing inflorescence is actually lateral.

Multiple inflorescences

Several inflorescences rather than the usual one at each node are produced in three subfamilies (Table 1.9): in some members of tribes Caryoteae (Coryphoideae), in Ceroxyleae (Ceroxyloideae), and in Iriarteeae, Chamaedoreeae and Areceae (Arecoideae). In *Arenga*, staminate inflorescences are often multiple, several units occurring at a node, but pistillate inflorescences occur singly. In *Wettinia*, three or more similar units, each with its own prophyll and peduncular bracts, develop in a leaf axil. The central one is often pistillate, the lateral ones staminate (Moore 1973a, Uhl & Moore 1973). A central apex on an axillary meristem gives rise to the single inflorescence in most palms. By studying development, Fisher and Moore (1977) showed that multiple inflorescence buds develop from an axillary meristem. In *Calyptrocalyx* (Arecoideae), two or three separate apices arise and develop in parallel. Elsewhere, multiple inflorescences develop on an expanded hypodium as a centrifugal series of new apices on either side of the central apex. The new lateral buds are initiated externally to the prophyll of the central bud, and each lateral bud apex develops its own prophyll.

The prophylls remain distinct in species of *Arenga*, *Chamaedorea* and *Wettinia*. In *Calyptrocalyx* and one species of *Chamaedorea*, the separate prophylls are united laterally during later growth to form a chambered structure that encloses all the bud apices. In these species, the collateral inflorescences are interpreted as separate inflorescences that are superimposed on the single original inflorescence meristem during early development. *Howea forsteriana* and a species of *Aiphanes* probably follow the same pattern, although they have not yet been studied developmentally. In the staminate inflorescence of *Ravenea madagascariensis* and *R. sambiranensis*, the prophyll of the inflorescence is incompletely sheathing, remaining open abaxially, and undulate and lobed distally, but the individual axes within are discrete (Dransfield & Beentje 1995b). Developmental study of this genus is needed.

Several species of *Livistona*, e.g., *L. woodfordii* and *L. rotundifolia*, and species of *Pritchardia* produce inflorescences that appear to be multiple, but the two to three axes are enclosed in a single prophyll, showing that the lateral branches

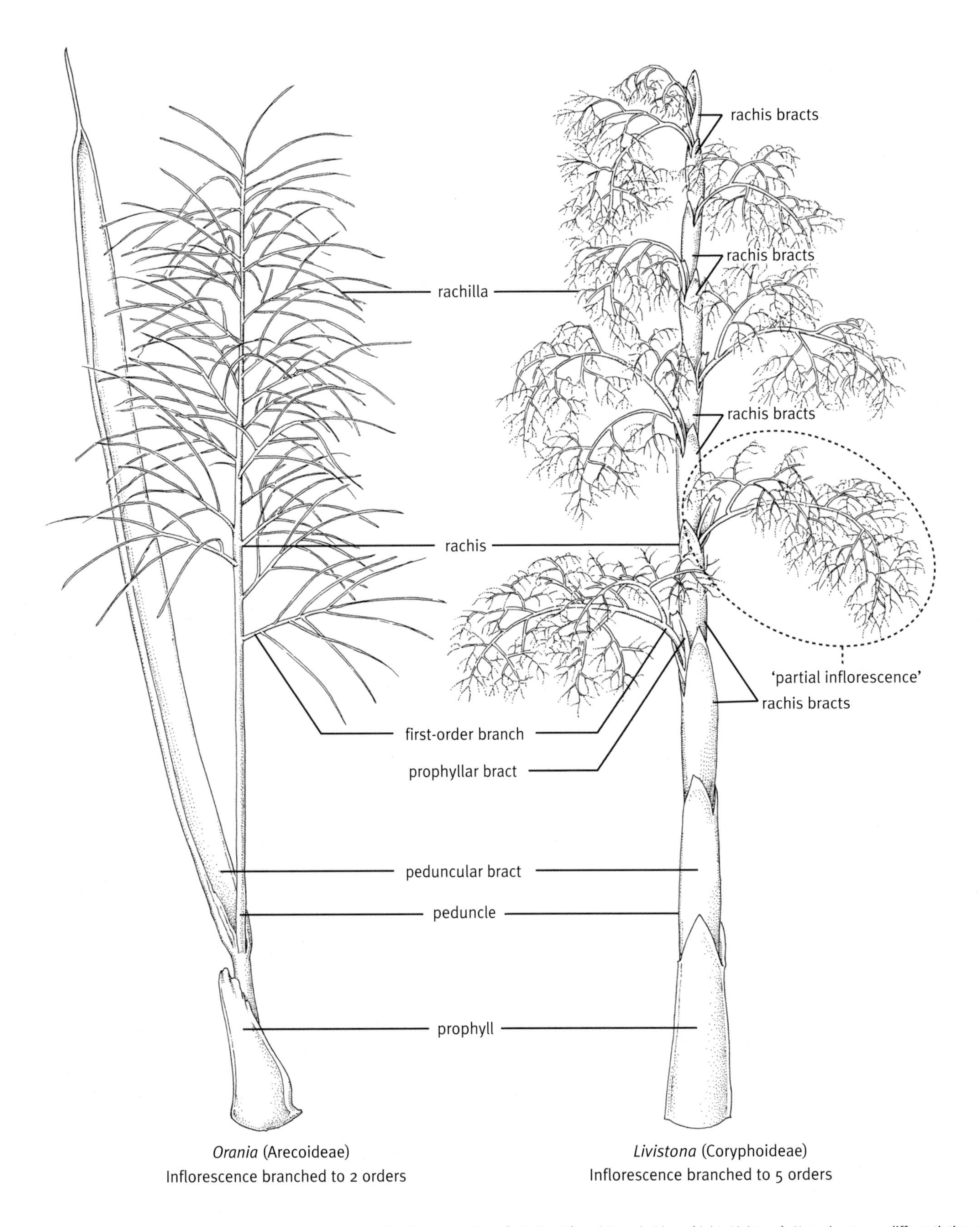

Fig. 1.6 Contrasting inflorescence morphologies in two subfamilies, Arecoideae (left, *Orania*) and Coryphoideae (right, *Livistona*). Note the strong differentiation of primary bracts (prophyll and peduncular bract) in *Orania* that is not evident in *Livistona*. The rachis bracts of *Orania*, while present early in development, are highly reduced or invisible at maturity. (Drawn by Lucy T. Smith)

TABLE 1.9
Palms with multiple inflorescences

III CORYPHOIDEAE
CARYOTEAE
Arenga spp.

IV CEROXYLOIDEAE
CEROXYLEAE
Ravenea – 7 out of 17 spp. (Dransfield & Beentje 1995b)

V ARECOIDEAE
IRIARTEEAE
Wettinia spp.
CHAMAEDOREEAE
Chamaedorea – ca. 15 of 92 spp. (Hodel 1992)
COCOSEAE
Aiphanes spp.
ARECEAE
Masoala – seen once (Dransfield & Beentje 1995b)
Calyptrocalyx spp.
Howea forsteriana

are equal in size to the main axis and are not separate inflorescence branches.

Multiple inflorescences increase the number of staminate flowers and the amount of pollen produced. They also extend the flowering period and offer increased protection of reproductive structures against herbivory (Fisher & Moore 1977).

Specialization

Specialization of the inflorescence is frequent in Calamoideae, which has the greatest range in inflorescence length. Within the palms, Coryphoideae includes the whole spectrum of variation in peduncular bracts from none in *Chamaerops* (Trachycarpeae) and *Phoenix* (Phoeniceae) to several in Sabaleae. This range reveals a structural potential that is not always evident in other groups. Most Arecoideae have only one peduncular bract. Adnation of inflorescences to branches or to branches and leaf sheaths is important in some groups. In Calamoideae, adnation between the inflorescence and internode above occurs in *Korthalsia*, *Myrialepis*, *Plectocomia* and *Plectocomiopsis*, and adnation between the inflorescence and both the internode and leaf sheath above occurs in *Calamus*, *Ceratolobus* and *Daemonorops* (Fisher & Dransfield 1977), and in *Retispatha* and *Pogonotium* (Dransfield 1979c, 1980a). No such adnation is found in the inflorescences of *Laccosperma*, *Eremospatha* and *Oncocalamus*. Substantial basal adnation of axes to those of the internode above also occurs in the Coryphoideae and Nypoideae (Tomlinson & Moore 1968, Tomlinson 1971, Uhl 1972a), but this form is not found in the inflorescences of Ceroxyloideae or Arecoideae.

Function

However the prophyll, peduncular bracts and rachis bracts are modified, they appear to function during pollination (Henderson 2002b, Uhl & Moore 1977a). The timing of bract opening can be correlated with both floral structure and pollination events (Uhl & Moore 1977a, Henderson 1986a). Henderson (1984) found that the large inflated rachis bracts open distally to provide access for pollinators in *Cryosophila*. In *Socratea*, erect peduncular bracts provide a funnel that guides pollinators into the inflorescence when pistillate flowers are receptive (Henderson 1985).

Henderson (2002b) divides palm inflorescences into condensed and elongate forms, the latter having much longer internodes and in general more open branching. A whole suite of morphological characters is associated with each type of inflorescence. Condensed inflorescences have shorter peduncles, rachises and rachillae and are spicate or branched to one order. In addition, in these inflorescences, flowers and flower clusters are more congested and staminate flowers are asymmetrical. Elongate inflorescences are more branched with longer internodes, more orders of branching, and symmetrical staminate flowers. These two different forms of inflorescences tend to attract different pollinators. The condensed inflorescences are pollinated largely by beetles, and the elongate inflorescences by bees. *Masoala* provides an example of an elongate inflorescence and related *Marojejya* of a condensed inflorescence. These two inflorescence forms and associated characters can occur in related species, as in *Pholidostachys synanthera* (elongate) and *P. dactyloides* (condensed).

Evolution

Tubular inflorescence bracts are found in Calamoideae, *Nypa* and most Coryphoideae, suggesting that this is a plesiomorphic condition in palms. In some Calamoideae, in *Nypa* and in Coryphoideae, prophylls are borne on both lateral branches and the main axis. Their occurrence on lateral branches in these lineages suggests the possibility that the development of a prophyll on each branch may also be plesiomorphic; however, expanded prophylls on branches of the second and higher orders may, in some cases, be associated with elaboration of the inflorescence.

FLORAL COMPLEXES (Glossary fig. 18)

The basic flowering unit in palms is a solitary flower, subtended by a bract and with a pedicel, which bears a bracteole or two-keeled prophyll. The flower may be sessile and the subtending bract or bracteole is often reduced or missing. In Borasseae (Coryphoideae), Calamoideae and perhaps elsewhere, some flowers have two bracteoles. These may represent a prophyll and a floral bracteole. Solitary flowers are usually borne in spirals but are often nearly decussate basally on rachillae. In the staminate spikes and the pistillate head of *Nypa fruticans*, the spicate inflorescences of *Sclerosperma* and elsewhere, the flowers or flower clusters are closely appressed on a thick axis and may be in parastichies.

Flower clusters

Sympodial clusters

If the floral bracteole bears a second flower, the bracteole of that flower a third flower, and so on, a cincinnus or sympodial cluster results. The form of the cincinnus depends on the relative positions of the bracteole and its subtended flower, and is characteristic of different groups (Fig. 1.7, Table 1.10). Short irregular cincinni and longer, regularly two-ranked forms are found in the Coryphoideae. Small irregular cincinni of 2–5 flowers, termed glomerules, occur in several coryphoid genera (*Chamaerops*, *Trachycarpus* and *Livistona*; Fig. 1.7). Tubular bracteoles are rare but present in the cincinnus of *Nannorrhops* (Uhl 1969a) and in Calamoideae; in the cincinnus of *Corypha,* flowers are sessile and bracteoles are adnate to the rachilla. The largest and most regular cincinni in Coryphoideae are the recurved branches borne in pits formed by thick bracts in the staminate inflorescences of *Borassus* and *Lodoicea*.

Dyad

A remarkable series of variations in a pair of flowers, a dyad, representing a two-flowered sympodial unit (Uhl & Dransfield 1984) occurs in Calamoideae (Dransfield 1970, Baker *et al.* 1999b, Fig. 1.7a–j). A dyad may be composed of two perfect, two staminate, one staminate and one perfect or a pistillate and a neuter (sterile staminate) flower. Bracts and bracteoles are usually tubular throughout the subfamily. The number and position of bracteoles show that solitary pistillate flowers in some genera, e.g., *Korthalsia*, are reduced from dyads (Fig. 1.7d). In a few species of *Calamus*, flowers in the pistillate inflorescence may be arranged in threes, composed of two pistillate flowers and a sterile staminate, or in fours, composed of two pistillate flowers and two sterile staminate flowers (Dransfield & Baker 2003). The most unusual flowering unit in the subfamily is that of *Oncocalamus* (Fig. 1.7b), in which the flower cluster has 0–2 central pistillate flowers and two lateral cincinni, each with 2–4 staminate flowers.

Triad

A triad consisting of a cluster of a central pistillate (Fig. 1.7ag, ah) and two usually lateral staminate flowers is found in the Caryoteae (Coryphoideae) and all genera of Arecoideae. The triad, therefore, characterises the majority of palms. Detailed studies in Caryoteae, Iriarteeae, Ptychospermatinae and Geonomateae (Uhl 1966, 1976a, and unpublished) have confirmed the sympodial nature of this unit. The bracteoles of the three flowers vary in form and may be of diagnostic value (see the generic drawings). The form of the triad varies in minor ways, depending on the positions of two staminate flowers relative to the pistillate flower. In Geonomateae, the pistillate flower is adaxial to the staminate flowers, whereas in Podococceae, it is abaxial to the staminate flowers. In other genera, staminate flowers are variously positioned lateral or rarely terminal to the pistillate. The pistillate and/or one staminate flower is frequently missing in Arecoideae, producing paired or solitary staminate flowers distally on rachillae. Unisexual inflorescences result when staminate or pistillate flowers are suppressed (*Wallichia*, some species of *Arenga*, *Wettinia*, *Marojejya*, *Lepidorrhachis* [Baker & Hutton 2006], *Leopoldinia*, *Elaeis* and *Attalea*).

TABLE 1.10
Distribution of flower clusters

	Solitary	Sympodial	Monopodial
I CALAMOIDEAE	+	+	
II NYPOIDEAE	+		
III CORYPHOIDEAE			
SABALEAE	+		
CRYOSOPHILEAE	+	+	
PHOENICEAE	+		
TRACHYCARPEAE	+	+	
CHUNIOPHOENICEAE		+	
CARYOTEAE		+	
CORYPHEAE		+	
BORASSEAE	+	+	
IV CEROXYLOIDEAE			
CYCLOSPATHEAE	+		
CEROXYLEAE	+		
PHYTELEPHEAE	+(Pi)		+(Stam)
V ARECOIDEAE			
CHAMAEDOREEAE	+	+	
All other tribes		+	

(Pi, pistillate; Stam, staminate)

Acervulus

The monoecious genera of tribe Chamaedoreeae (Arecoideae) are distinguished by distinct lines of closely two-ranked unisexual flowers, called acervuli (Fig. 1.7aj, Fig. 1.8). The arrangement represents a reversed cincinnus, with flowers adnate to the rachillae, the proximal flower usually pistillate and the distal staminate, and the bracteoles mostly reduced or not evident at anthesis (Uhl & Moore 1978). The acervulus of *Gaussia maya* consists of only three flowers (Fig. 1.7ai), and so resembles the triad superficially, but the position and anatomy of the flowers differ from those of the triad and indicate a basic similarity to the acervulus. In the dioecious *Chamaedorea*, the pistillate flowers are borne singly; in some species, staminate flowers are also borne singly (Fig. 1.7ak,al), whereas in others (e.g., *C. microspadix*),

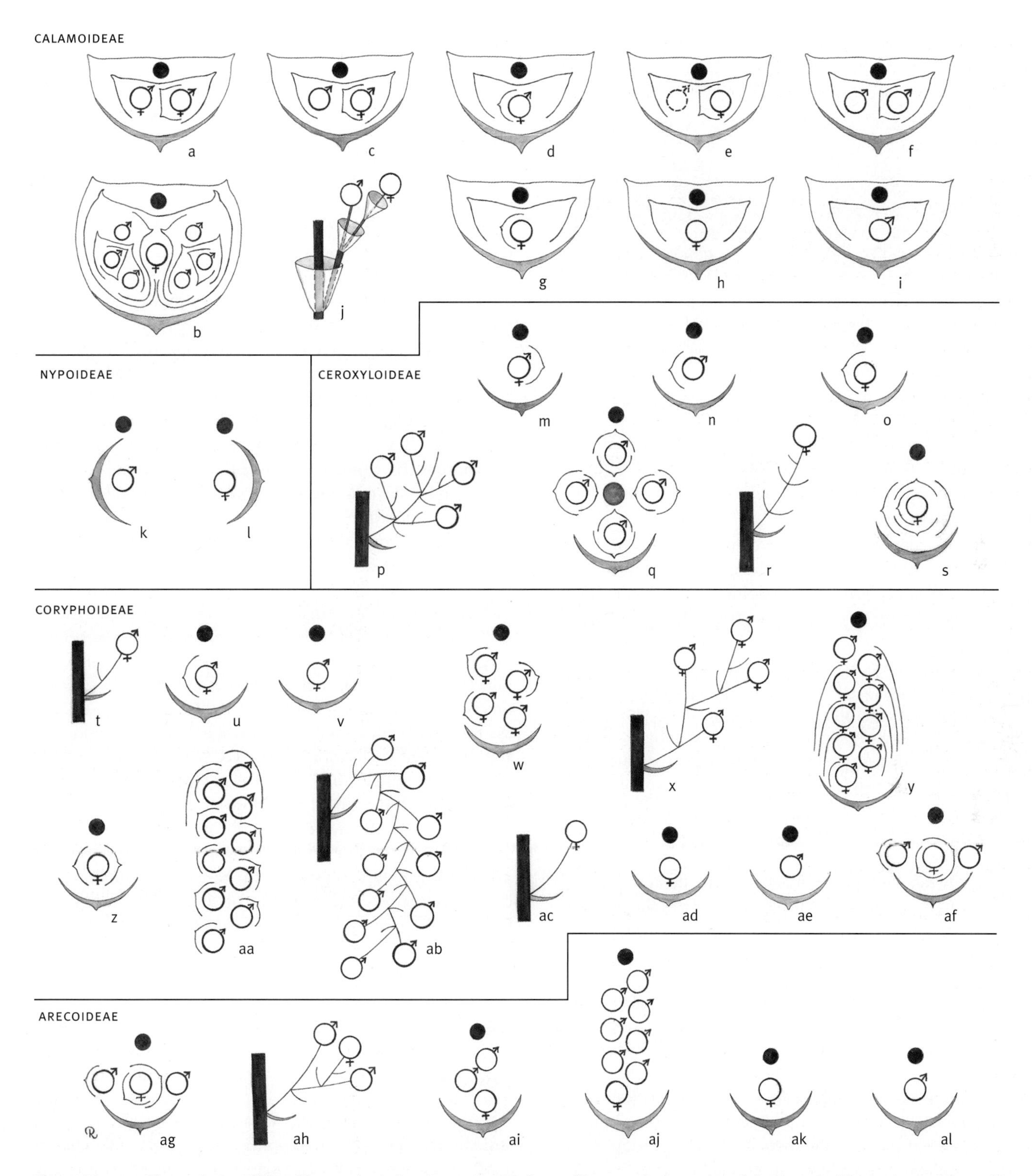

Fig. 1.7 Diagram of flower clusters. **Calamoideae**: **a**, dyad of two hermaphroditic flowers (*Eremospatha*, *Laccosperma*); **b**, the unusual flower cluster of *Oncocalamus*; **c**, dyad of staminate and hermaphroditic flowers (*Eugeissona*, *Metroxylon*); **d**, solitary hermaphroditic flower (*Korthalsia*); **e**, dyad of sterile staminate and fertile pistillate flowers (e.g., in pistillate inflorescences of most Salaccinae and Calaminae); **f**, dyad of paired staminate flowers (e.g., in staminate inflorescences of Salaccinae, Pigafettinae and *Plectocomia*); **g**, single pistillate flower (e.g., *Raphia*); **h**, single pistillate flower (e.g., Pigafettinae, Plectocomiinae); **i**, single staminate flower (*Raphia*); **j**, dyad showing tubular bracts. **Nypoideae**: **k**, **l**, staminate and pistillate flowers of *Nypa*. **Ceroxyloideae**: **m**, hermaphroditic flower of *Pseudophoenix*; **n**, **o**, solitary staminate and pistillate flowers of e.g., *Ravenea*; **p**, **q**, monopodial staminate flower cluster (*Phytelephas*); **r**, **s**, solitary pistillate flower, bracts and bracteoles (*Phytelephas*). **Coryphoideae**: **t**, **u**, solitary flower with bracteole; **v**, solitary flower without bracteole; **w**, **x**, cincinnus as in species of *Livistona*; **y**, adnate cincinnus (e.g., *Corypha*); **z**, bibracteolate pistillate flower of Borasseae; **aa**, **ab**, cincinnus of staminate flowers in, e.g., *Borassus* and *Lodoicea*; **ac**, **ad**, **ae**, solitary staminate and pistillate flowers in *Phoenix*; **af**, triad in Caryoteae. **Arecoideae**: **ag**, **ah**, the triad; **ai**, triad lacking obvious bracteoles (*Gaussia*), **aj**, the acervulus (e.g., *Hyophorbe*); **ak**, **al**, solitary staminate and pistillate flowers lacking bracteoles (e.g., *Chamaedorea*). (Drawn by Marion Ruff Sheehan)

Fig. 1.8 The acervulus of *Synechanthus warscewiczianus*. The acervulus is a remarkable floral cluster characteristic of some genera in tribe Chamaedoreeae (see text). Scale bar = 1 mm. (Photo: R. Ryder & P. Rudall)

staminate flowers are in few-flowered, curved adnate cincinni. Reflexed vascular bundles suggest that the solitary flower in this genus has been derived from an acervulate unit.

Monopodial clusters

Bosch (1947) postulated that the condensed floral units in palms represent monopodial branching systems. Only in one tribe, the Phytelepheae (Ceroxyloideae), have the floral clusters been found to be monopodial (Uhl & Dransfield 1984, Barfod & Uhl 2001; Fig. 1.7). Developmental investigations (Uhl & Dransfield 1984, Barfod 1991, Barfod & Uhl 2001) have established that phytelephantoid palms differ markedly from other palms in inflorescence and flower structure. In *Phytelephas aequatorialis*, staminate flower clusters consist of a short axis bearing four flowers in two close sub-opposite pairs (Fig. 1.7p,q). At very young stages, each flower is subtended by a bract and bears a bracteole. The apex of the axis bearing the flowers becomes obscured as the pedicels of the flowers elongate and become partially united, and as subtending bracts of the flowers become united in an irregular shallow ring at the base of the pedicels. These four-flowered monopodial units are arranged in parastichies on the flattened main axis of the inflorescence. The other two phytelephantoid genera have structurally different stalk-like bases of the flower clusters. In *Ammandra*, the staminate inflorescences resemble those of *P. aequatorialis* but the branch bearing the four flowers, rather than the united floral pedicels, forms the stalk of the cluster. In *Aphandra*, flower clusters have four (1–6) staminate flowers. A pseudopedicel develops in each flower by elongation of the floral receptacle, and the partially united pseudopedicels of the four flowers provide yet another different stalk for the flower cluster.

The pistillate inflorescences in *Phytelephas aequatorialis* and *Aphandra* are similar in structure but have only one flower in each unit. The flower is terminal and surrounded by bracts, which suggests that it represents one flower of a four-flowered cluster, similar to that in the staminate inflorescences (Fig. 1.7r,s).

Pollination

Although a review of pollination is beyond the scope of this book, some brief remarks are helpful in understanding reproductive structures. The inflorescence is the functional unit in pollination. The development and maturation of bracts, branches, and floral organs are coordinated to provide rewards for pollinators and protection of various organs as needed. A multitude of insects are associated with palms. Insects use palm reproductive structures in many ways, including ovipositing and use of fleshy parts and pollen as food for larvae and adults. As pollination syndromes are understood for a growing number of species, two principal modes of insect pollination can be distinguished (Silberbauer-Gottsberger 1990, Henderson 2002b). In palms pollinated by bees, flies, and wasps, inflorescences are usually elongate (as in *Sabal* or *Licuala*), with rachillae loosely spaced and few bracts at anthesis; flowers are often hermaphrodite and generally protandrous. By contrast, palms that are pollinated by beetles have condensed inflorescences, exemplified by *Chamaerops*, *Cryosophila* or *Rhapidophyllum*, with closely spaced, short rachillae, with bracts more or less covering flowers at anthesis, and with flowers often unisexual and protogynous. Henderson (2002b) concludes that the story of palm pollination is one of constant shifts from one of these systems to the other. It should be noted that although bees or beetles may be designated as principal pollinators, a large number of insects visit all palm inflorescences and perform various functions. Silberbauer-Gottsberger (1990) hypothesised that the basic pollination mode in palms was general entomophily and that beetle pollination is a derived form. Considering the diversity we find in inflorescence and floral structure, it is perhaps not surprising that a host of different pollination modes are being discovered. The impact of pollination mode and pollinators on morphology and anatomy is a subject that invites much further study. In relation to pollination, characters such as the presence or absence of bracts, the location of nectaries, the form of filaments and anthers, and gynoecial structure provide important clues to the relationships of pollinators and floral structure.

THE FLOWER

Form

Although palms are specialised in many ways, they retain a striking number of plesiomorphic characters (Tomlinson 1990). Nowhere is this more evident than in the flower. The simple palm flower exemplifies the basic form in monocotyledons (Endress 1994). It has three slightly imbricate sepals, three slightly imbricate petals, six stamens in two whorls and three distinct uniovulate carpels (see Glossary fig. 17). As noted above, the flower may be sessile or have a short stalk, the stalk bearing one (rarely two) bracteoles. An abscission zone occurs between the sepals and the bracteole in the flowers of all palms except those of the genera of Phytelepheae. Trimerous unspecialised flowers are seen in *Chelyocarpus chuco*, which has similar sepals and petals. *Trithrinax* is similar but has the sepals connate basally. This basic form is modified in many ways. Every floral organ, including the receptacle, displays remarkable variation (Fig. 1.9). The diversity is increased by connation and/or adnation, reduction or loss of parts, and by differential growth. Trimery is the principal pattern, but the number of parts in single floral whorls may vary. Only in the three species of *Chelyocarpus*, however, is there variation in all whorls of the flower: dimery in *C. ulei* (Fig. 1.9f,g), trimery in *C. chuco* and tetramery in *C. dianeurus* (Moore 1972). Elsewhere, variation in single whorls is rare. Tetramery occurs in the perianth of staminate flowers of Phytelepheae, although the pattern is largely obscured by the time of anthesis (Uhl & Dransfield 1984), and five to ten carpels are present in the pistillate flowers. The perianth is trimerous in *Attalea* but a four- to seven-parted gynoecium occurs in some species.

Sexuality

In the Calamoideae, hermaphrodite flowers occur in *Eugeissona*, *Laccosperma*, *Eremospatha*, *Metroxylon* and *Korthalsia*. Otherwise, hermaphrodite flowers characterise 20 genera of coryphoid palms (Fig. 1.9d,e,j,k). The genera of Rhapidinae vary in the sexuality of their flowers: *Chamaerops*, *Guihaia, Maxburretia* and *Rhapis* are dioecious; *Trachycarpus* is polygamous; and *Rhapidophyllum* is dioecious, polygamodioecious or rarely monoecious. In other coryphoid palms, dioecy occurs rarely in species of *Livistona* and *Licuala*, and almost always in *Chuniophoenix*, *Kerriodoxa*, *Phoenix*, *Maxburretia*, *Rhapis* and all eight genera of the Borasseae. The monogeneric first tribe of Ceroxyloideae, Cyclospatheae, has hermaphroditic flowers, with occasional staminate flowers borne distally on rachillae. The other two tribes of Ceroxyloideae, Ceroxyleae and Phytelepheae, are dioecious. The remainder of the palms bear staminate and pistillate flowers in triads, or in acervuli in Chamaedoreeae, and are monoecious or very rarely dioecious (*Chamaedorea*). Thus, in palms as a whole, 30 genera have hermaphroditic flowers; 37 genera are completely dioecious and five genera occasionally so, and all other genera (some 122) are monoecious.

Evidence abounds for the modification of hermaphrodite flowers to unisexual forms. Developmental studies of *Phoenix* confirm both the loss of the gynoecium from staminate flowers and of the androecium from pistillate flowers (De Mason *et al.* 1982). In some Coryphoideae, Ceroxyleae (Ceroxyloideae) and most Calamoideae, staminate and pistillate flowers are very similar, differing only late in ontogeny. Throughout the family, however, staminate and pistillate flowers are often dimorphic and distinctly so in *Nypa*, in the Phytelepheae, and in most Arecoideae (Fig. 1.9b,c,n,o).

In the Eugeissoneae and some Coryphoideae (e.g., *Schippia* and *Rhapidophyllum*), the unisexual flowers are very similar to the hermaphroditic flowers except where organs are aborted or appropriate organs are lacking. *Nypa*, on the other hand, is exceptional in several respects. It is monoecious and its flowers are bizarre (Fig. 1.9b,c). The staminate and pistillate flowers are dimorphic in structure. The pistillate is apocarpous with a carpel that is more cupular than those of other palms. The staminate flower, in which the filaments of three stamens are united, has some similarities to staminate flowers in certain Triuridaceae, which are thought to be early monocots (Gandolfo *et al.* 2000).

Floral receptacle

The floral receptacle varies in expression in different groups, exhibiting remarkable plasticity. The tip of the floral receptacle is frequently present above the origin of the carpels and can be observed in young stages (*Phoenix*, Calamoideae and others) (Morrow 1965, Uhl & Moore 1971 and unpubl.). It may be elongate between any two whorls of organs, and is frequently so at the base of the flower. The receptacle elongates in different positions to exsert flowers that are borne in pits: between the calyx and corolla in *Welfia* but between corolla and stamens in *Lodoicea*. In the pistillate flowers of Geonomateae, however, it is the style rather than the receptacle that elongates to exsert the stigmas, and the fruit develops within the pit, ultimately forcing it open. In *Chuniophoenix*, flowers are exserted from tubular bracts by elongation of a receptacular stalk. In *Aphandra*, elongation of the floral receptacle forms a solid funnel-shaped pseudopedicel that elevates the androecium (Barfod & Uhl 2001) (Fig. 1.9n). Read (1968) found that the sepals are adnate to an elongate receptacle in the floral stalk of *Pseudophoenix*. Various modes of receptacular expansion provide for different forms of polyandry (Uhl & Moore 1980).

Perianth (Glossary figs 19, 20)

In dicotyledons, there are commonly two series of perianth parts. The outer series, the sepals, is mainly protective, whereas the inner, the petals, functions to attract pollinators (Endress 1994). This is true for some palms, but in most, the petals are also protective. The perianth of palms comprises two alternating, structurally different, trimerous whorls recognisable as sepals and petals with two exceptions. First, in a few genera, e.g., *Nypa*, *Chelyocarpus* and *Elaeis* (especially in the pistillate flower), the members of the two whorls are similar and therefore are considered tepals (Fig. 1.9b). The promising developmental genetic research of Adam *et al.* (2007) looking at MADS-box gene expression has, however, revealed distinct identities for sepals and petals in the oil palm. A second exception occurs in certain Cryosophileae: *Zombia*, *Coccothrinax*, *Hemithrinax*, *Leucothrinax* and *Thrinax*, and in the staminate flowers of *Phytelephas*, where the entire perianth is reduced to an irregularly lobed whorl with minute segments (Fig. 1.9d). In staminate flowers of the Phytelepheae, these tiny segments contrast sharply in size with the petals of the pistillate flowers, which may reach a length of 10 cm or more.

The size and shape of sepals varies from group to group; connation occurs frequently, especially in staminate flowers. Petals also vary, common trends being toward the valvate state at maturity in the hermaphrodite and staminate flower, and toward extreme imbrication in the pistillate flower. Gamopetalous, often valvately lobed, corollas are present in some genera in all subfamilies except Nypoideae, frequently with connation and/or adnation of stamen filaments or staminodes in the Coryphoideae and Geonomateae. Such adnation occurs more frequently in the pistillate than in the staminate flower. The petals of the pistillate flowers in many palms have short valvate apices above extremely imbricate lower parts (e.g., some *Hydriastele* species), and in Geonomateae and *Pritchardia* (Trachycarpeae), the whole free portion is valvate and sometimes circumscissile (see plates of *Calyptrogyne* and *Calyptronoma* [Geonomateae] and *Pritchardia* [Trachycarpeae]).

In dicotyledons, sepals usually have a broad base and three to five traces, whereas petals often have a narrow base, a broad tip, and are vascularised by a single trace. In palms, the vasculature of each perianth part varies from none to many bundles, and their histology from parenchymatous to extremely tanniniferous and fibrous, with fibres present in bundle sheaths or as separate strands (Morrow 1965, Uhl 1969b, 1972b, 1978b, Uhl *et al.* 1969). Raphides can be abundant, often occurring in specific locations. Variation in floral vasculature and histology may be correlated with environmental factors, activities of vectors during pollination, growth sequences, and other factors (Uhl & Moore 1977a).

Androecium

As is true for the stamens of many angiosperms (Endress 1996), the structural diversity of the palm stamen is largely unexplored. Although all palms exhibit the basic ground plan of the angiosperm stamen (D'Arcy 1996), there are many structural variations in both filaments and anthers (Glossary fig. 19). In Calamoideae, the stamen of *Eugeissona* has a short erect filament with a narrow and elongate, basifixed, latrorse to introrse anther. In Calamoideae as in the rest of the family, the majority of anthers are latrorse, but less frequently they may also be introrse (Plectocomiinae) or extrorse (*Korthalsia*). In the arecoid *Areca furcata*, they are even poricidal (Glossary fig. 19). Three calamoid genera (*Raphia*, *Korthalsia* and *Calamus*) have introrse or latrorse species. Other features of the stamen are more variable throughout the palms. Filaments are commonly dorsifixed but may be medifixed (*Eremospatha*, *Laccosperma* and *Metroxylon*). Some are versatile (*Itaya*, *Washingtonia* and *Nannorrhops*) or inflexed (*Bactris*). Sometimes they are fused into a tube (*Raphia*) or a thick-walled cup (*Chamaerops*, *Eremospatha* and *Plectocomiopsis*). They may be adnate to petals, singly or united, as in many Coryphoideae, some Areceae and Geonomateae. *Guihaia* is unique in the family in that anthers that lack filaments completely are adnate to the petals (Glossary fig. 19). *Hyospathe* is distinguished by two different whorls of stamens: the antesepalous three with shorter filaments and the antepetalous whorl adnate to the pistillode. In some species of *Dypsis*, filaments are also adnate to the pistillode, a complex synorganisation (Rudall *et al.* 2003). In some genera, the connective tissue is prolonged into a tip, often tanniniferous (*Howea*, *Welfia* and *Hydriastele*). Anthers likewise vary in form; they may be short and rounded (*Nephrosperma*), didymous (species of *Dypsis*), or sagittate basally (*Washingtonia*). Rarely, separate thecae are borne on a bifid connective (*Asterogyne* and *Geonoma*; see Glossary fig. 19, Fig. 1.9s). The relation of these characters of filaments and anthers to function in pollination modes and elsewhere is largely unstudied, but would be valuable to systematics (Endress 1996).

Staminodes (see Glossary fig. 20) may have abortive anthers, may be reduced to filaments or minute teeth, or may be united in a cupule, sometimes large and fibrous (*Attalea*, Cocoseae) and surrounding the base of the ovary. The number and arrangement of staminodes is diagnostic for some species. In some Geonomateae (*Calyptrogyne* and *Calyptronoma*, see illustrations of genera), staminodes form a unique, distally expanded tube around the stigmas, which in *Calyptrogyne ghiesbreghtiana* serves as food for bat pollinators (Cunningham 1995). In some genera (e.g., *Neonicholsonia*), staminodes fail to develop.

Stamens and staminodes have one to several vascular bundles and may be fibrous or otherwise histologically specialised. The expansion of filaments and an increase in the number of traces per stamen appear to be characteristic of many groups.

Both reduction and increase from the usual number of stamens occur. Reduction to three stamens occurring opposite the sepals occurs in *Nypa*, in which the filaments are fused to form a column (Uhl 1972a). Three stamens also characterise one species of *Synechanthus* (Ceroxyloideae) (Fig. 1.9p) and some species of *Orania*, *Dypsis*, *Areca*, *Astrocaryum* and *Geonoma* (Arecoideae). In *Dypsis*, stamen morphology and position indicate that reduction to three stamens has occurred several times (Dransfield & Beentje 1995b, Rudall *et al.* 2003). Polyandry is much more common. More than six stamens occur in some 64 genera, including members of all subfamilies except Nypoideae. Tribes in which all taxa have only six stamens are very few: the monogeneric Phoeniceae, Cyclospatheae and Podococceae (Table 1.11). Elsewhere the number of stamens ranges from seven or eight or nine to many. The largest numbers of stamens within the palms are found in the Phytelepheae, in which samples of 849–954 have been counted in *Phytelephas*, 419–521 in *Ammandra*, and 400 to over 650 in *Aphandra* (Fig. 1.9n).

Increase in stamen number appears to have occurred in different ways and perhaps in response to different factors in different groups of palms (Uhl & Moore 1980; Table 1.13). Androecial development has been studied in all but two of the groups in which many stamens occur (Uhl & Moore 1980). After petal inception, the floral apex expands in diameter and/or height in a different way in each group to

TABLE 1.11
Reduction and elaboration in the androecium

	Stamens 3	Stamens more than 6
I CALAMOIDEAE		
EUGEISSONEAE		
Eugeissona		+
LEPIDOCARYEAE		
Raphia		+
II NYPOIDEAE		
Nypa	+	
III CORYPHOIDEAE		
CRYOSOPHILEAE		
Chelyocarpus		+
Itaya		+
Thrinax		+
Coccothrinax		+
Zombia		+
CARYOTEAE		
Arenga		+
Caryota		+
Wallichia	+	
BORASSEAE		
Lataniinae		
Latania		+
Lodoicea		+
Borassodendron		+
IV CEROXYLOIDEAE		
CEROXYLEAE		
Ceroxylon		+
PHYTELEPHEAE		
Ammandra		+
Phytelephas		+
V ARECOIDEAE		
IRIARTEEAE		
Iriartea		+
Socratea		+
Wettinia		+
CHAMAEDOREEAE		
Synechanthus	+	
ORANIEAE		
Orania	+	+
ROYSTONEEAE		
Roystonea		+
REINHARDTIEAE		
Reinhardtia		+
MANICARIEAE		
Manicaria		+
EUTERPEAE		
Oenocarpus		+
COCOSEAE		
Attaleinae		
Beccariophoenix		+
Jubaea		+
Jubaeopsis		+
Parajubaea		+
Allagoptera		+
Attalea		+
Bactridinae		
Astrocaryum	+	
Bactris		+
GEONOMATEAE		
Asterogyne		+
Geonoma	+	
Welfia		+
ARECEAE		
Archontophoenicinae		
Actinorhytis		+
Actinokentia		+
Archontophoenix		+
Chambeyronia		+
Kentiopsis		+
Arecinae		
Areca	+	+
Pinanga		+
Clinospermatinae		
Cyphokentia		+
Dypsidinae		
Dypsis	+	+
Linospadicinae		
Calyptrocalyx		+
Howea		+
Laccospadix		+
Linospadix		+
Oncospermatinae		
Acanthophoenix		+
Deckenia		+
Ptychospermatinae		
Adonidia		+
Balaka		+
Brassiophoenix		+
Carpentaria		+
Drymophloeus		+
Normanbya		+
Ponapea		+
Ptychococcus		+
Ptychosperma		+
Solfia		+
Veitchia		+
Rhopalostylidinae		
Hedyscepe		+
Verschaffeltiinae		
Nephrosperma		+
Oncosperma		+
Phoenicophorium		+
Unplaced		
Cyrtostachys		+
Heterospathe		+
Hydriastele		+
Loxococcus		+

accommodate the larger number of stamens. In Phytelepheae, stamens arise centrifugally on a circular apex (Uhl & Moore 1977b). Elsewhere, stamens develop in different arrangements but in antesepalous and antepetalous positions. The larger number of stamens in *Thrinax*, *Coccothrinax* and *Zombia* (Cryosophileae) seems to be associated with wind pollination, but the numerous stamens in *Chelyocarpus* and *Itaya* of the same subtribe may be related to pollination by beetles (Silberbauer-Gottsberger 1990). Structurally, the androecia of these two palms are similar, with the stamens arranged in an irregular ring that has a larger number of stamens in wider antepetalous positions. In Borasseae, Caryoteae, and Areceae, antesepalous whorls always consist of only three stamens, one opposite each sepal, but more stamens develop in wider antepetalous positions. *Eugeissona* (Calameae) has a similar pattern except that the last-formed stamens develop centrifugally (Uhl & Dransfield 1984). Floral apices may also expand in height, with stamens in alternating whorls that have one stamen above each sepal and several above each petal. The only polyandric ceroxyloid genus, *Ceroxylon*, has large primary primordia on which two to three stamens develop opposite each petal; in species of *Ceroxylon* with more than 12 stamens, two to three stamens also develop opposite each sepal. Iriarteeae are characterised by distinctive truncate primordial, bearing up to three or more stamens opposite each sepal and petal. Stamens arise in antesepalous and antepetalous arcs of about seven stamens each on a flat trilobed apex in *Welfia* (Geonomateae). In *Allagoptera*, multiple stamens are formed on the sides of a rather tall irregularly chunky apex (Uhl 1988).

All multi-staminate taxa except phytelephantoid genera exhibit an underlying trimery. The different patterns of apical expansion and stamen arrangement indicate that polyandry has arisen separately in several tribes, including the Iriarteeae, Geonomateae, Cocoseae and Areceae of Arecoideae (Uhl & Dransfield 1984). In the development of multistaminate androecia, the modes of apical expansion and the resulting stamen patterns appear to conform to pressures exerted by bracts and perianth segments. Thus, it seems reasonable to suggest that specialisation of the inflorescence bracts and perianth occurred before the advent of multistaminy in the palms as a whole. The presence of a trimerous perianth and gynoecium in the multistaminate taxa (except in Phytelepheae) further supports this conclusion.

Gynoecium

The gynoecium reflects the diversity of the palm family and provides excellent diagnostic and systematic data. Trimery in the gynoecium is basic, as it is elsewhere in the flower, but there are exceptions (Table 1.12). Among the 18 genera of apocarpous palms (Table 1.13), six genera, each specialised in some way, develop only a single carpel (Fig. 1.9d). The gynoecium of *Thrinax*, *Coccothrinax* and *Zombia* can be identified by vasculature as a single carpel (Morrow 1965, Uhl & Moore 1971). There is some minor evidence of trimery in some of the unicarpellate taxa and this needs confirmation. Morrow (1965) found a vascular configuration below the gynoecium of *Coccothrinax* to be similar to that in the tricarpellate *Johannesteijsmannia*, and he also shows that the carpel of *Hemithrinax* has three plicate stigmatic branches. Among the Arecoideae, pseudomonomery is exhibited in 68 genera in five of the 14 tribes: Sclerospermeae, Roystoneeae, some Geonomateae, Euterpeae, Pelagodoxeae and Areceae (including 10 unplaced genera) (Fig. 1.9u,w, Glossary fig. 20). In these taxa, one carpel develops completely and bears a fertile ovule, but only parts of the two other carpels are present. Pseudomonomery is anticipated to occur in apocarpous trimerous genera because of the development of only one carpel into a fruit, as is also the case in many syncarpous genera (e.g., *Sabal*, *Socratea*, *Podococcus* and Cocoseae, where two ovules frequently abort). When techniques become available, evolutionary developmental geneticists will find fertile ground in palms. More than three carpels occur in only five genera: *Chelyocarpus* (Crysophileae: Coryphoideae), *Attalea* (Cocoseae: Arecoideae) and all three genera of Phytelepheae (Ceroxyloideae) (Glossary fig. 20).

Opposite: **Fig. 1.9** Floral diversity in palms. **Calamoideae**: **a**, *Mauritiella armata* (Lepidocaryeae), pistillate flower, gynoecium, note reflexed scales. **Nypoideae**: **b–c**, *Nypa fruticans*, **b**, staminate flower, perianth partially removed to reveal three connate stamens; **c**, one carpel isolated from pistillate flower. **Coryphoideae**: **d**, *Leucothrinax morrisii* (Cryosophileae), hermaphroditic flower, stamens and perianth partially removed to show unicarpellate gynoecium; **e**, *Cryosophila warscewiczii* (Cryosophileae), hermaphroditic flower, perianth removed to show staminal tube and exserted stigmas; **f–g**, *Chelyocarpus ulei* (Cryosophileae), hermaphroditic flower, **f**, androecium and **g**, apocarpous gynoecium; **h**, *Rhapis subtilis* (Trachycarpeae), staminate flower, androecium; **i**, *Rhapis excelsa* (Trachycarpeae), pistillate flower, apocarpous gynoecium surrounded by conspicuous staminodes; **j**, *Licuala peltata* (Trachycarpeae), hermaphroditic flower, corolla removed to show stamens and exserted style; **k–l**, *Copernicia* × *burretiana* (Trachycarpeae), hermaphroditic flower, **k**, perianth and androecium partially removed to reveal gynoecium comprising three carpels free at the base, united distally, **l**, gynoecium with one carpel dislodged to reveal darker distal area of carpel connation and united style; **m**, *Hyphaene coriacea* (Borasseae), pistillate flower, perianth removed, syncarpous gynoecium with conspicuous staminodes. **Ceroxyloideae**. **n**, *Aphandra natalia* (Phytelepheae), staminate flower, note reduced, irregular perianth and multistaminate androecium containing 400–650 stamens; **o**, *Phytelephas macrocarpa* (Phytelepheae), pistillate flower, perianth partially removed, note the large size of this flower, the numerous staminodes surrounding the ovary, the syncarpous pentamerous gynoecium (4–10 carpels in *Phytelephas*) comprising rounded ovary, and elongate style and stigmas. **Arecoideae**. **p**, *Synechanthus warscewiczianus* (Chamaedoreeae), staminate flower sepal removed, androecium comprising three stamens only; **q**, *Chamaedorea hooperiana* (Chamaedoreeae), staminate flower, one petal removed; **r**, *Chamaedorea brachypoda* (Chamaedoreeae), staminate flower, one petal removed, note variation in stamen and pistillode morphology in **q** and **r**; **s**, *Asterogyne martiana* (Geonomateae), staminate flower; **t**, *Calyptrogyne ghiesbreghtiana* (Geonomateae), pistillate flower, trimerous gynoecium (see **u**); **u**, *Geonoma simplicifrons* (Geonomateae), pistillate flower, pseudomonomerous gynoecium (see **t**, **w**); **v–w**, *Pelagodoxa henryana* (Pelagodoxeae), **v**, staminate flower, perianth removed to show androecium and pistillode, **w**, pistillate flower, perianth and androecium removed to show pseudomonomerous gynoecium (see **u**) and staminodes. Scale bars: a–c, j, q–s = 1 mm; d–i, k–m, p, t–w = 500 µm; n, o = 1 cm. (Photos: a, L. Guevara; b–c, j, A. S. Barfod & F. W. Stauffer; d, k–m, p, R. Ryder & P. Rudall; e–g, F. Castaño & F. W. Stauffer; h–i, A. Giddey & F. W. Stauffer; n–o, A. S. Barfod; q–r, A. Askgaard; s–w, F. W. Stauffer).

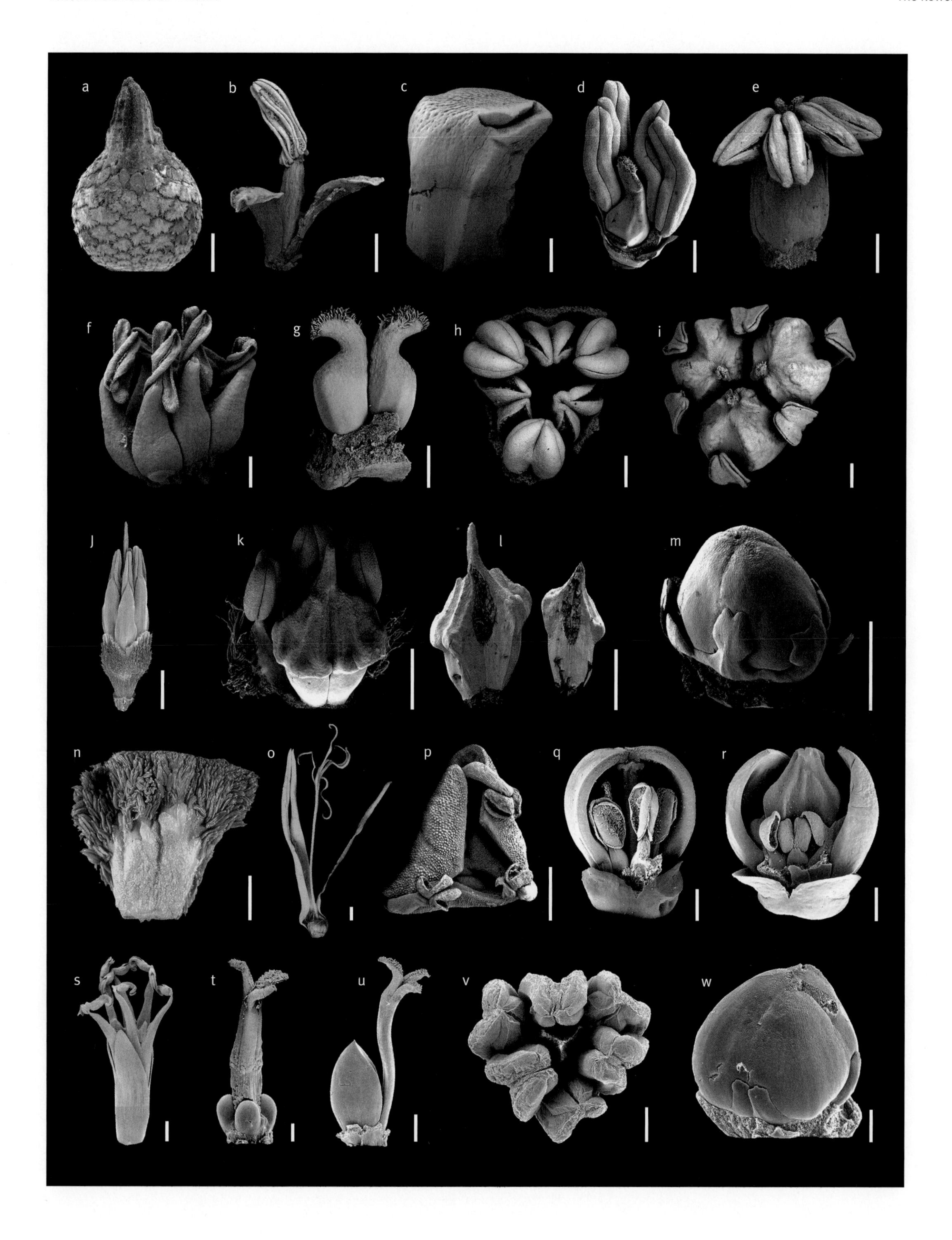
a
b
c
d
e
f
g
h
i
j
k
l
m
n
o
p
q
r
s
t
u
v
w

Form

The numerous forms of palm carpels have not been appreciated. Palm carpels, except for those of Calamoideae, have at least a shallow closed cupular base and can be described as ascidiate or peltate (Endress 1994). Some are weakly ascidiate (*Trachycarpus*), others (*Thrinax*, *Coccothrinax* and *Zombia*) are strongly so, and many are intermediate. During development, carpels are crescentic in shape, and adaxial meristematic activity forms an adaxial lip and thus a shallow ascidiate base. The ovule usually arises ventrally on one side of the shallow cup-like carpel primordium (Uhl & Moore 1971).

The carpel varies in shape, often with dorsal bulges, as seen in the apocarpous Coryphoideae: Cryosophileae, Rhapidinae and *Phoenix*. Carpel shape characterises some groups. In Rhapidinae, carpels have a prominent dorsal bulge and short, more or less reflexed stigmas. In Trachycarpeae, carpels have more angular dorsal bulges and the united styles are terete and erect, some very long (*Washingtonia* and *Serenoa*). The remaining apocarpous genus, *Nypa* (Uhl 1972a) has a more cupular form and the distal stylar parts of carpels are always symplicate or conduplicate (Endress 1994) (Glossary fig. 20 and generic plates). In Calamoideae, although the carpels are joined peripherally, the ventral slits or sutures of the carpels are not closed in the centre of the gynoecium, which can be said to be incomplete. Developmental observations show that the ovules are initiated on the central floral apex in positions axillary to crescentic carpel primordia (*Salacca* [Van Heel 1977, 1988], *Plectocomiopsis* [Uhl & Moore 1971] and *Eugeissona* [Uhl & Dransfield 1984]). This is evident in the earliest developmental stages. Carpel primordia of *Eugeissona minor*, *E. tristis* and *Salacca zalacca* (Uhl & Dransfield 1984, Van Heel 1988) are crescentic in shape whereas the carpel primordia of *Phoenix* (De Mason *et al.* 1982) are small mounds, each of which develops a more or less central depression and later becomes plicate. The calamoid carpels appear to be plicate or conduplicate throughout, but are united peripherally by postgenital fusion. Development of the ascidiate base apparently takes place later, possibly by congenital fusion but this has not been ascertained. In all genera studied, carpels are developmentally separate in early stages even if they become syncarpous later. This has been reported to characterise monocotyledons and to allow for the development of septal nectaries (Van Heel 1988, Endress 1994). The ascidiate or peltate carpel is comparable to that considered basic in angiosperms (Endress 1994). The highly vascularised carpel of *Nypa*, with its cupular shape, two-crested distal opening with open but more-or-less plicate margins, diffuse internal stigmatic surfaces and laminar to submarginal placentation, may, however, represent a different type of carpel (Uhl 1972a, Igersheim *et al.* 2001).

Anatomy of the carpel

The vasculature of the carpel usually consists of a dorsal bundle, two ventral bundles (Eames 1961), and one to several pairs of lateral bundles. Although not currently used for describing carpel vasculature (Endress 1994, Igersheim *et al.* 2001), 'ventral bundles' are useful for describing palms in that they are

TABLE 1.12
Exceptions to trimery in the gynoecium

	Carpel number		
	1	2	4–10
III CORYPHOIDEAE			
CRYOSOPHILEAE			
Schippia	+		
Zombia	+		
Coccothrinax	+		
Hemithrinax	+		
Leucothrinax	+		
Thrinax	+		
Chelyocarpus		+	+
Itaya	+		
IV CEROXYLOIDEAE			
PHYTELEPHEAE			
Ammandra			+
Aphandra			+
Phytelephas			+
V ARECOIDEAE			
COCOSEAE			
Attalea			+

TABLE 1.13
Distribution of apocarpy
(number of carpels in parentheses)

II NYPOIDEAE
Nypa (3)

III CORYPHOIDEAE
CRYOSOPHILEAE
Trithrinax (3)
Chelyocarpus (2, 3, 4)
Cryosophila (3)
Itaya (1)
Schippia (1)
Thrinax (1)
Coccothrinax (1)
Hemithrinax (1)
Leucothrinax (1)
Zombia (1)
TRACHYCARPEAE
Trachycarpus (3)
Rhapidophyllum (3)
Chamaerops (3)
Maxburretia (3)
Guihaia (3)
Rhapis (3)
PHOENICEAE
Phoenix (3)

distinguished by length and position from other lateral bundles (Uhl & Moore 1971). In thick-walled carpels, accessory bundle systems may also be present. Thickness of the ovarian wall is correlated with different morphological and anatomical features in different groups; for example, scales in Calamoideae, raphides but few fibres in Ceroxyloideae, and sclerenchyma, tannins, and other idioblasts in various patterns in Coryphoideae and Arecoideae (see Uhl & Moore 1971, 1973).

Throughout the palms, the carpel varies in thickness of the walls, in position of the locule, and in the size and shape of the style and stigma (Uhl & Moore 1971). The locule varies in position from median (Cryosophileae and Caryoteae), to basal (Borasseae and Cocoseae), and distal (Phytelepheae).

Style

In all palm carpels, a canal extends from the locule to the stigma. This appears to be open in tubular styles, such as those of *Thrinax*, *Coccothrinax* and *Zombia*, but for the most part is slit-like with stigmatoid or otherwise definite epidermal layers (Uhl & Moore 1971). In the Coryphoideae (Glossary fig. 20), styles are short and tapering (*Trachycarpus*), tubular (*Thrinax*), abaxially expanded (*Rhapis*), and elongate (*Sabal*). Upper gynoecial regions are extremely wide in Borasseae (Fig. 1.9m) and Caryoteae; styles are moderate in size in Ceroxyloideae, and either long and wide (*Socratea*) or more slender and elongate (*Wettinia*: Iriarteeae). The most expanded upper gynoecial areas are found in *Nypa* (Fig. 1.9c) and in Cocoseae (species of *Attalea*, where the basal ovarian portion represents only about $^{1}/_{16}$ the carpel length at anthesis). The most elongate styles characterise Geonomateae (Stauffer & Endress 2003) and Phytelepheae (see genus plates) and represent synapomorphies for these groups (Fig. 1.9o,t,u)

Stigma

Stigmas of angiosperms are classified as wet or dry, and further distinguished by the relief of the surface and the structure of stigmatic papillae (Endress 1994). Palms have moist to wet stigmas. Their shape varies (Glossary fig. 20) and can be tapering and recurved (*Trithrinax*), bilabiate and funnel-shaped (*Thrinax* and *Nypa*) (Fig 1.9c), sessile (*Sabal*), relatively undeveloped (Borasseae (Fig 1.9m) and Caryoteae), large and fibrous (some Cocoseae), or elongate (Phytelepheae) (Fig. 1.9o). Stigmas in Calamoideae are trifid or pyramidal. All Geonomateae have long styles and reflexed symplicate stigmas (Fig. 1.9e,u), those of *Pholidostachys* being unusually long. Stigmas are unicellular papillate in many monocots (Igersheim *et al.* 2001), but in palms they are variable. In all Geonomatae, unicellular papillae cover the stigmatic surfaces, except in *Welfia* where they are multicellular (Stauffer & Endress 2003). Elsewhere, a variety of stigmatoid tissues occur: cuboidal cells (*Trachycarpus*), elongate strands (*Sabal*), two-celled trichomes with basal bulbous cells (*Chamaedorea*), unicellular trichomes with bulbous bases (*Socratea*), uniseriate multicellular trichomes (*Calyptrocalyx* as *Paralinospadix*), unicellular trichomes (*Butia*), and unicellular or bicellular trichomes (*Elaeis* [Uhl & Moore 1971]).

Syncarpy

Syncarpy is structurally and developmentally different in various groups of palms. Whether or not the gynoecium is syncarpous, carpels arise as separate primordia in all of the palms studied developmentally (Stauffer & Endress 2003, Uhl 1976b, Uhl & Dransfield 1984, Uhl & Moore 1971, Van Heel 1988). Gynoecia become syncarpous by postgenital and/or congenital development. In *Nannorrhops*, carpels are separate initially but are later joined ventrally by interlocking epidermal layers (Uhl 1969a). Fusion does not occur in their stipitate bases, which are separated by a septal nectary. Carpels of Livistoninae and the unplaced genera of Trachycarpeae are joined by their styles with ventral fusion of epidermal layers in the lower regions of some (Fig. 1.9k,l). Carpels of *Corypha* are united basally and distally nearly to the end of the long styles. Locular canals are separate and stigmas are scarcely differentiated. In *Sabal*, the long single stylar canal represents the united plicate styles of the three carpels. In several groups of palms (Caryoteae, Borasseae and Cocoseae), gynoecia are completely syncarpous except for relatively short or small stigmatic branches. Different patterns of syncarpy characterise the pseudomonomerous genera (Uhl & Moore 1971, Stauffer & Endress 2003). Syncarpy is thus structurally different in different groups and appears to have arisen several times in the family, although detailed character analysis is needed for confirmation.

Compitum

The most obvious advantage of syncarpy is considered to be a compitum, which is a common pollen tube transmitting tract (PTTT) for all carpels of a flower (Carr & Carr 1961, Endress 1994). The compitum allows both an even distribution of pollen tubes among carpels and distribution in a single transmitting tract. Little is known as yet about PTTTs in palms. They were described by Rao (1959a, 1959b) in a number of genera and more recently in species of *Dypsis* (Rudall *et al.* 2003), Geonomateae (Stauffer & Endress 2003), and *Euterpe* (Küchmeister *et al.* 1997). A compitum is apparently absent in some palms. In *Nannorrhops*, the locular canals of the three carpels are never joined, each opening separately in the appropriate stylar lobe (Uhl 1969a). In several other coryphoids (*Brahea*, *Livistona*, *Acoelorrhaphe*, *Serenoa*, *Pritchardia*, *Colpothrinax* and *Corypha*), stylar canals are also separate (Morrow 1965), whereas in related genera (*Licuala*, *Copernicia*, *Johannesteisjmannia*, *Washingtonia* and *Sabal*), locular canals are joined (Morrow 1965). This seeming enigma requires further research. In *Sabal*, a canal extends from the locule of each carpel toward the centre of the gynoecium, forming a compitum at about the top of the ovary in the base of the long central stylar canal (Morrow 1965, Uhl unpublished). In *Corypha*, canals from each locule join in upper regions of the rather long style (Uhl & Moore 1971).

Septal nectaries

In the angiosperms, septal or gynopleural nectaries, regions of nectar-producing cells on the sides or at the bases of carpels, occur in monocotyledons but are not known in dicotyledons (Van Heel 1988, Endress 1994). Nectaries develop between lateral or basal carpel walls. Their occurrence is facilitated by free or partly free carpels. Monocotyledons are apocarpous in early stages, even when they are syncarpous later as noted above for palms. In a number of palms, a triradiate cavity lined by a secretory epithelium is present in the base of the gynoecium below the locules or at higher levels between the carpels (Daumann 1970, Schmid 1985, Uhl & Moore 1971, Uhl & Moore 1977a). Openings or canals from this cavity occur at various levels. In *Sabal*, the nectary is triradiate in the base of the gynoecium and openings occur to the outside at about the upper level of the locules. A similar nectary in *Corypha* has openings just below the locules (Uhl & Moore 1971). In *Latania* and *Arenga*, canals from a triradiate basal nectariferous cavity open by pores on the surface of the gynoecium. Openings are between the carpels at the bases of the stigmas in *Butia* (Uhl & Moore 1971). Nectaries have been reported in *Eugeissona* and *Salacca* (Calamoideae); the latter case is unusual as pollination is by beetles, but the beetles are known to lick the nectar (Silberbauer-Gottsberger 1990). In *Euterpe precatoria* (Küchmeister *et al.* 1997), septal nectaries are present in both staminate and pistillate flowers. In the pistillate flower, there is a triradiate cavity below the ovary from which canals extend to the upper style, where they join the compitum formed by the locular canals. The nectary in the staminate flower is similar but smaller. Staminate flowers of *Dypsis bejofo* have unusual distally bulbous prominent supralocular nectaries (Rudall *et al.* 2003). Septal nectaries are reported for: *Arenga*, *Caryota*, *Socratea*, *Euterpe*, *Prestoea*, *Archontophoenix*, *Ptychosperma*, *Butia*, *Syagrus*, *Cocos*, *Allagoptera*, *Pseudophoenix*, *Hyophorbe*, *Chamaedorea* (Küchmeister *et al.* 1997) and all Geonomateae (Stauffer & Endress 2003, Stauffer *et al.* 2003). Labyrinthine nectaries, interpreted as complex septal nectaries created by undulation and convolution, have been found in *Licuala* (Stauffer & Barfod, pers. comm.). Nectary production is reported to be external in *Chamaerops*, *Livistona*, *Licuala*, *Pritchardia* and *Corypha*, where septal nectaries do not occur. Other palms apparently lack nectaries completely: *Chelyocarpus*, *Cryosophila*, *Rhapidophyllum*, *Thrinax*, *Nypa*, *Attalea*, *Elaeis*, *Acrocomia*, *Astrocaryum*, *Bactris* and *Phytelephas* (Silberbauer-Gottsberger 1990). Lack of nectaries has been suggested to be plesiomorphic in palms (see Smets *et al.* 2000 and references therein), but much remains to be learned about nectaries in the family.

Ovule

Anatropous, hemianatropous, campylotropous and orthotropous ovules occur in palms (see Glossary fig. 21). The ovule also varies in position; it is variously oriented within the locule and may be partly adnate to the locular wall (Glossary fig. 20). Many palm ovules have what has been called an aril (Morrow 1965, Uhl & Moore 1971) or obturator (Van Heel 1988, Rao 1959a, 1959b) on the base of the funiculus. This can be shaped like a small bract or perianth part (as in *Chelyocarpus* — Glossary Fig. 20), more irregular in shape (*Trachycarpus* and *Areca*), or just a small rim. The aril is thought to direct the pollen tube to the micropyle, but its morphology and function are in question. The inner integument is usually narrow, frequently only 2–3 cells wide, and relatively short, so that unless sections are cut directly through the micropyle, it may not be apparent. By contrast, the outer integument is wide (more than four cells). One or both integuments may form the micropyle. The outer integument extends beyond the inner to form the micropyle in *Thrinax*, *Bactris*, *Butia*, *Syagrus* and *Jubaeopsis*, but in *Hyophorbe* the inner integument protrudes. The outer integument may be enlarged around the micropyle (e.g., *Rhapidophyllum*, *Trachycarpus* and *Schippia*). Some ovules are highly vascularised with numerous bundles entering and often branching in the large outer integument. Nine bundles supply the ovule in *Zombia*, a tenth entering the aril or obturator (Morrow 1965). Numerous integumentary bundles are thought to supply ovules that develop into large seeds (Endress 1994). The form of the ovule and its position in the carpel is diagnostic for some genera or groups of genera. In the Calamoideae, for example, all evidence indicates that ovules are anatropus and the funicles are twisted so that the micropyles face the centre of the gynoecium. The ovule of *Nypa* is also anatropous. Distinctive hemianatropous ovules characterise the genera of the Chamaedoreeae and are different from those of Ceroxyleae. By contrast, all forms of ovules are found in the Coryphoideae. Orthotropy occurs in certain Cryosophileae and in Borasseae and Cocoseae; hemianatropy and campylotropy occur within the Ceroxyloideae and various Arecoideae. The presence or form of the aril or obturator, the size and degree of fusion of the two integuments, nucellar structure and vascular supply also vary. These characters are not known for many genera but would be useful phylogenetically (Endress 2003).

Floral anatomy

Both developmental and anatomical studies of flowers have been exceptionally useful in interpreting structure and function, and in indicating evolutionary patterns. Certain histological and vascular patterns distinguish palm genera or groups of related genera. Arrangements of sclerenchyma and tannins may be diagnostic. The presence of similar histology and vasculature in *Chelyocarpus*, *Cryosophila* and *Itaya* (Uhl 1972b), of different patterns in the species of *Maxburretia* (Uhl 1978a), and of still other patterns in the ceroxyloid genera *Juania*, *Ravenea* and *Ceroxylon* (Uhl 1969b) helped to relate and circumscribe these genera. All species of *Hyophorbe* have been studied anatomically and show rather remarkable variation in the vasculature and histology of their perianth parts but consistency in gynoecial and ovule structure (Uhl 1978b). Similar patterns in *Pelagodoxa* and *Sommieria* support their placement in the Pelagodoxeae and their potential relationship to the Indo-Pacific pseudomonomerous clade (Stauffer *et al.* 2004). Floral anatomy in *Dypsis* has revealed multiple origins of three stamens (Rudall *et al.* 2003). Thus the value of comparative studies of floral structure has been repeatedly demonstrated. Flowers of many genera have yet to be studied anatomically, and such studies seem certain to have diagnostic, biological, and evolutionary significance.

Evolution

An evolutionary evaluation of gynoecial characters can now be based on the phylogenetic framework provided by molecular systematics. Although the precise placement of palms among other commelinid monocots is still to be confirmed, the broadest analyses confirm that they diverge from other groups near to or at the basal node of the commelinid clade (Chase *et al*. 2006). Although the evolutionary origin of the angiosperm carpel remains in doubt, gynoecial structure in basal angiosperms has been assessed (Endress & Igersheim 2000, Igersheim *et al*. 2001). When their position in the commelinids is considered, it is perhaps not surprising that palm carpels exhibit several features of basal angiosperms. Carpel closure, the hallmark of angiospermy, is divided into four types. In palms, this distinction cannot be definitely made without more observations; however, palms appear to exhibit carpels of the first two types. Angiospermy type 1, by secretion only, could apply to *Nypa*, as the wide open-mouth-like stigmatic suture of the carpel is filled with gelatinous material in our specimens. Carpels of the unicarpellate coryphoids (*Thrinax*, *Coccothrinax* and *Zombia*) are possibly angiospermy type 2, with congenitally and partly postgenetically fused peripheries but with an unfused canal in the centre. The carpel of *Trachycarpus* also has a distal ventral slit-like opening, closed only by trichomes. As in basal angiosperms, the carpels of many palms are shortly stipitate (*Nannorrhops* [Uhl 1969a]; *Arenga* and *Corypha* [Uhl & Moore 1971]); *Thrinax*, *Coccothrinax* and *Zombia* [Morrow 1965]). The three carpels of *Trithrinax*, *Rhapidophyllum* and *Trachycarpus* have common gynoecial stalk-like bases (Morrow 1965). Ovule structure in basal monocots is reportedly diverse, but anatropous ovules are most common as they are in palms. Some unspecialised palms (*Thrinax*, *Coccothrinax* and *Zombia*) have orthotropous ovules, which are also found in basal monocots (*Acorus* and some Alismatales [Igersheim *et al*. 2001]). Ovules of basal monocots are crassinucellate as are all palm ovules. In short, palm carpels exhibit an astonishing amount of variation including several features attributed to basal angiosperms.

THE FRUIT

Structural variation in palm fruits occurs in shape (Glossary fig. 22), size, surface texture, composition of the mesocarp, extent of the endocarp, number of seeds, features of the endosperm and position of the embryo. If the fruit has a fleshy or fibrous mesocarp and one seed (rarely it may have more, from two to ten) and a thin endocarp, it could be considered a berry. If the endocarp is sclerotic and thick, the fruit is by definition a drupe. The shape is most frequently obovoid to ellipsoidal rarely obpyriform, oblate or top-shaped. The size of fruits varies enormously from about 4.5 mm in diameter (species of *Geonoma*) to some 50 cm long. In general, larger fruits occur in Borasseae, Cocoseae and Phytelepheae. The colour of fruits varies widely and can be correlated with pollinators and dispersal agents (Henderson 2002b). Although most fruits are smooth, the texture of surfaces varies. Scales cover all fruits of Calamoideae (Glossary figs 22d, 23). Hairs appear more rarely (*Rhapidophyllum* and *Wettinia*), as do prickles (some species of *Astrocaryum* (Glossary fig. 22g) and *Bactris*). Warty processes that arise by cracking of the pericarp (Dransfield 1970) characterise fruits of one species of *Chelyocarpus*, *Itaya*, *Johannesteijsmannia*, possibly one species of *Livistona*, a few species of *Licuala* and all but one species of *Pholidocarpus*. In the Arecoideae, warty fruits are a synapomorphy for the two genera that comprise the Pelagodoxeae (*Pelagodoxa* [Glossary fig. 22i] and *Sommieria*), confirming DNA evidence of the unexpected relationship of these palms, and also occur in *Lemurophoenix* and *Manicaria*. In Ceroxyloideae, warty fruits distinguish the three genera of Phytelepheae.

Fruits of a few genera are exceptions to the general pattern: those of the Hyphaeninae are irregular, often shouldered (Glossary fig. 22f), and have a rather dry but sweet mesocarp. The very large fruit of *Lodoicea* has a heavily fibrous mesocarp. Fruits of the Phytelepheae resemble war clubs, having a fibrous mesocarp that eventually disintegrates to free the seeds. In *Lytocaryum*, fruits have a thin mesocarp that splits into three valves to expose the endocarp; and in some species of *Astrocaryum* and *Socratea*, the mesocarp splits irregularly at the apex and recurves to expose coloured or white flesh. Fruits of the coconut (*Cocos nucifera*, Cocoseae: Arecoideae), a strand plant, and *Nypa* (Nypoideae), the mangrove palm, have heavily fibrous walls that are adapted for floating and distribution by water. Fruits of some palms persist on the tree for a long time; Moore (pers. comm.) reported the presence of 23 fruits on one tree of *Phytelephas aequatorialis*, these ranged from young stages, with gelatinous endosperm in the seeds, to old fruits that had disintegrated so as to release the seeds.

Endocarp

The endocarp of some palms is very thin and crustaceous (e.g., many species of *Licuala*, such as *L. spinosa*, *L. glabra* and *L. grandis*). In other palms, it is very thick and hard, e.g., in *Eugeissona* (Calamoideae), *Nypa* (Nypoideae), Borasseae, a few species of *Licuala* (e.g., *L. beccariana*) (Coryphoideae), *Ptychococcus*, the Cocoseae (Arecoideae), and the Phytelepheae (Ceroxyloideae). Three types of endocarp have been described (Murray 1973). The coryphoid type differentiates from the inner part of the fruit wall; the chamaedoreoid type forms solely from the locular epidermis; and the cocosoid-arecoid type develops from the locular epidermis, bundle sheaths, and intervening parenchyma. Different developmental patterns, described as continuous, discontinuous, and basipetal, are correlated with the endocarp types.

Several different forms of endocarp occur in the family. The brittle, sometimes highly sculptured endocarps of the Verschaffeltiinae and Basseliniinae are distinctive (Glossary fig. 24). Cage-like endocarps of *Leopoldinia*, consisting of inner and outer layers of massive branched fibres (Kubitski 1991), wash up on river banks in South America. *Ravenea musicalis* fruits make musical sounds as they hit the water, and have spongy floating endocarps containing germinated seeds that are ready to develop (Dransfield & Beentje 1995b). The exceptional endocarp of *Satranala* splits into two valves as the seed germinates. *Eugeissona* is unusual in that it develops two 'endocarp' layers. The 'true' endocarp develops when the inner layer of the pericarp becomes meristematic and eventually cuts off cells in regular files (Dransfield 1970). The 'false' endocarp is a hard fibrous layer, presumably representing the vascularised central part of the carpel. Oddly, the 'false' endocarp is functional and the 'true' endocarp disintegrates as the seed enlarges. In a few palms, hard seed-like bodies, pyrenes, are formed when an endocarp develops separately in the wall of each carpel in a syncarpous ovary. The pyrene forms a hard, sometimes sculptured outer layer around each seed; the sculpturing of pyrenes is diagnostic in *Latania*. Thinner areas in the wall of the endocarp (pores) (Glossary fig. 24) are characteristic features of the hard endocarps of Cocoseae. Usually there are three of these, but in multicarpellate taxa (species of *Attalea*) there are more, with the embryo or embryos opposite them. When one embryo develops, its pore usually becomes larger than those of other embryos.

Anatomy

Essig and colleagues (see references under Essig) have carried out extensive histological studies of the pericarp in palms. They have found that anatomical variation occurs in a great many features: in the outer epidermis, the locular epidermis, the extent of sclerified parenchyma, patterns of brachysclereids, arrangements of fibres in bundle sheaths or separate bundles, distribution of tanniniferous cells, and locations of raphides and silica (Essig 1999). The extensive variation in these characters appears to be related to protection and dispersal. Genera and species are usually distinct in terms of pericarp anatomy, different combinations of characters occur in different groups but parallelism interferes with assessing relationships of higher groups. The data may help to determine relationships within tribes and subtribes (Essig 2002, Essig *et al.* 2001), and pericarp anatomy provides substantial support for some groups in the classification. The division of the former Oncospermatinae (*sensu* Uhl & Dransfield 1987) into the Oncospermatinae and Verschaffeltiinae is well supported by pericarp data (Essig *et al.* 2001), as is the separation of Rhopalostylidinae from the Archontophoenicinae (Essig & Hernandez 2002). Subtribes Basseliniinae and Clinospermatinae (Essig *et al.* 1999) are also well supported. Essig's research suggests strongly that incorporation of anatomical pericarp data into formal phylogenetic analyses will be very productive (Essig & Litten 2004).

THE SEED

Palm seeds, like palm fruits, come in a wide range of sizes. They are usually ovoid, ellipsoidal or globose in shape (Glossary fig. 24. They are provided with abundant nuclear endosperm that is often homogeneous but may be dissected by ingrowths of the seed coat and thereby ruminate (see Glossary fig. 24). The endosperm is the main contributor to seed weight. Fifty-three palm genera have ruminate endosperm; some are heterogenous with both conditions. *Nypa fruticans*, *Ptychococcus paradoxus* and *Synechanthus fibrosus* have individuals with both types (Zona 2004). In a few genera (e.g., *Cocos*), the endosperm is hollow. The seed coat is heavily vascularised in Iriarteeae, in some Areceae, and in the Cocoseae. A sarcotesta, a fleshy outer seed coat (see Glossary fig. 23), is well developed in many Calamoideae. The endosperm is penetrated by the seed coat in many coryphoids, or it may be dissected (*Coccothrinax* and *Zombia*) or perforate (*Thrinax*). The seed may reach 30 cm in length and weigh 5–18 kg in *Lodoicea* (Spanner, pers. comm.) where weights as high as 25 kg have also been recorded (Tomlinson 2006a). The shape of the seed distinguishes some taxa. Those of some genera in the Ptychospermatinae, some Borasseae, and the calamoid genus *Eugeissona* are angled or ridged and grooved (see generic drawings of these groups). Seeds of some genera of the Basseliniinae are elaborately sculptured in conformity with the endocarp (Glossary fig. 24).

Embryo

With a few exceptions, palm embryos are very small, cylindrical to conical in shape and late-maturing (see Glossary figs 23, 24), but *Nypa* is unusual (Murray 1971). Zona (2004) has found that raphides are present but rare in embryos of Coryphoideae, but common in Caryoteae and Ceroxyloideae and frequent in Areceae (Arecoideae). Raphides were not found in embryos of Calamoideae neither in the Phytelepheae (Arecoideae).

Germination (Glossary fig. 25)

Chavez (2003) conducted a detailed study of palm germination and of seedlings. During germination, either the entire cotyledon or its distal tip remains inside the seed. All three types of germination described by Boyd (1932) occur in palms. In our descriptions, three types of germination are recognised on the basis of different combinations of two characters: 1) presence or absence of a cotyledonary petiole, which pushes the seed away from the developing seedling; and 2) presence or absence of a cotyledonary ligule. Germination is described as remote-tubular, remote ligular and adjacent ligular (Table 1.14; see Chavez 2003 for discussion).

TABLE 1.14
Germination

	Remote-tubular	Remote-ligular	Adjacent-ligular
I CALAMOIDEAE			
EUGEISSONEAE		+	
LEPIDOCARYEAE			+
CALAMEAE			+
II NYPOIDEAE			+
III CORYPHOIDEAE			
SABALEAE		+	
CRYOSOPHILEAE*	+	+	
PHOENICEAE	+		
TRACHYCARPEAE*	+		
CHUNIOPHOENICEAE	+	+	
CARYOTEAE	+		
CORYPHEAE	+		
BORASSEAE	+		
IV CEROXYLOIDEAE			
CYCLOSPATHEAE	+		
CEROXYLEAE		+	+
PHYTELEPHEAE		+	
V ARECOIDEAE			
IRIARTEEAE			+
CHAMAEDOREEAE			+
PODOCOCCEAE			+
ORANIEAE	+		
SCLEROSPERMEAE	+		
ROYSTONEEAE			+
REINHARDTIEAE			+
COCOSEAE	+		+
MANICARIEAE			
EUTERPEAE			+
GEONOMATEAE			+
LEOPOLDINIEAE			+
PELAGODOXEAE			+
ARECEAE			+

* In Cryosophileae, germination is not recorded for *Zombia* or *Hemithrinax*. In Trachycarpeae, *Rhapidophyllum*, *Acoelorrhaphe*, *Brahea*, and *Washingtonia* are remote-ligular and germination for *Maxburretia* is unknown; other genera are remote-tubular as noted.

SEEDLING

The seedling has three parts: the plumule, the cotyledon, and the primary root. The plumule or epicotyl is the shoot apical meristem and its leaf primordia, which terminates the axis of the embryonic plant. The cotyledon consists of three parts: a haustorium, a middle part or petiole, which may be very short, and a sheath. The haustorium and the petiole together make up the hyperphyll. The cotyledonary sheath may have a tubular extension, a coleoptile or ligule. The primary root may be straight, oblique, or perpendicular to the plumular axis. Orientation of the primary root is diagnostic for certain groups. The coryphoids as circumscribed here exhibit a straight orientation, the calamoids and ceroxyloids have oblique roots, and the arecoids have primary roots oriented at right angles. As the seedling develops, its apex may form one or more scale leaves or cataphylls before the origin of the eophyll, the first leaf with a blade.

Evolution

Seedling morphology and anatomy have been studied (Chavez 2003, Henderson & Stevenson 2006) and 49 characters useful for cladistic analyses found. Cladograms from seedling data, including anatomical parameters, resolve the Coryphoideae as circumscribed here as monophyletic. A non-pinnate venation pattern characterises eophylls of all coryphoids. Tillich (2000) reported that the seedlings of palms exhibit the complete lower evolutionary level within monocotyledons. Along with Poaceae, Cyperaceae and Hyacinthaceae, palms have radiated extensively, increased in genus number and occupied different ecological locations, but have not developed derived seedling types.

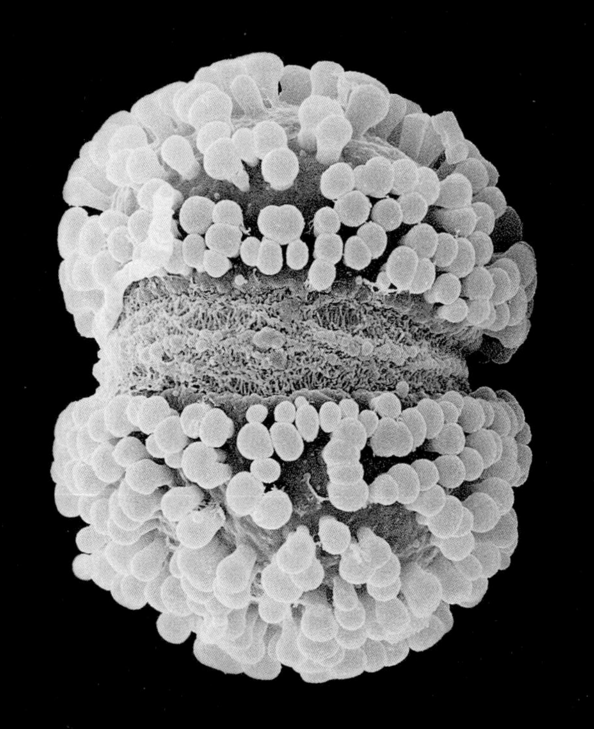
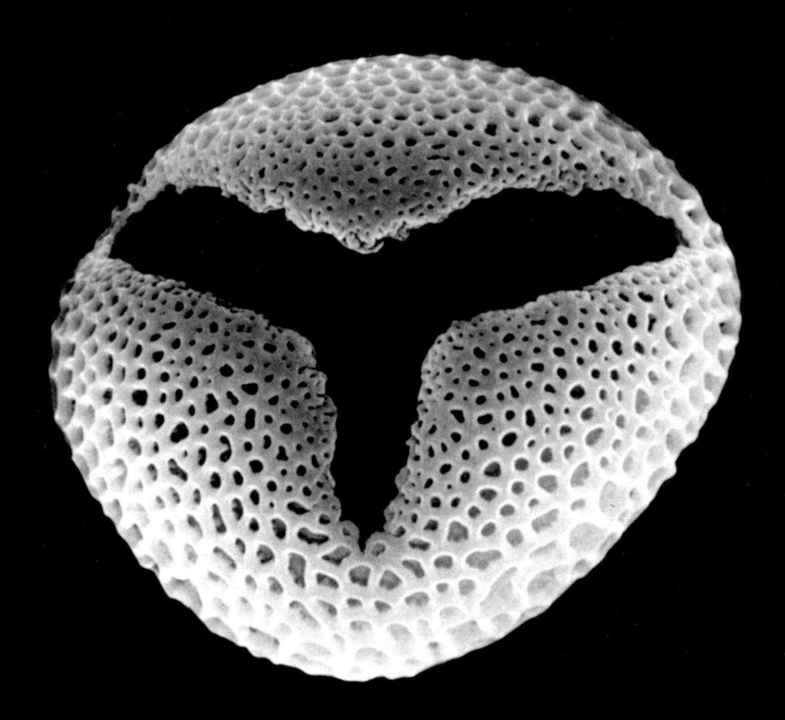
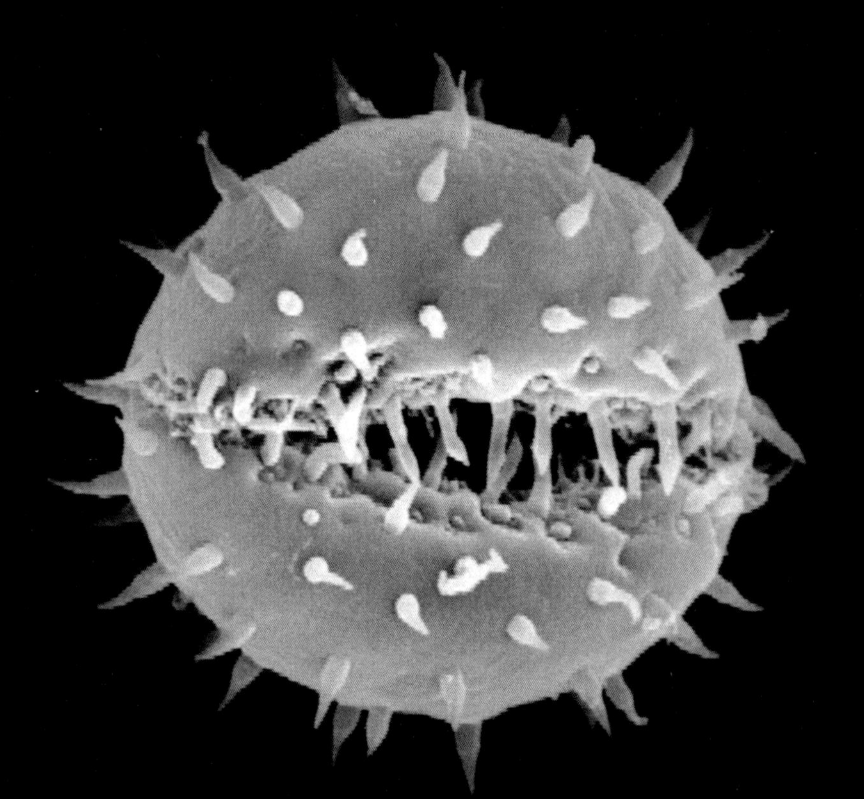
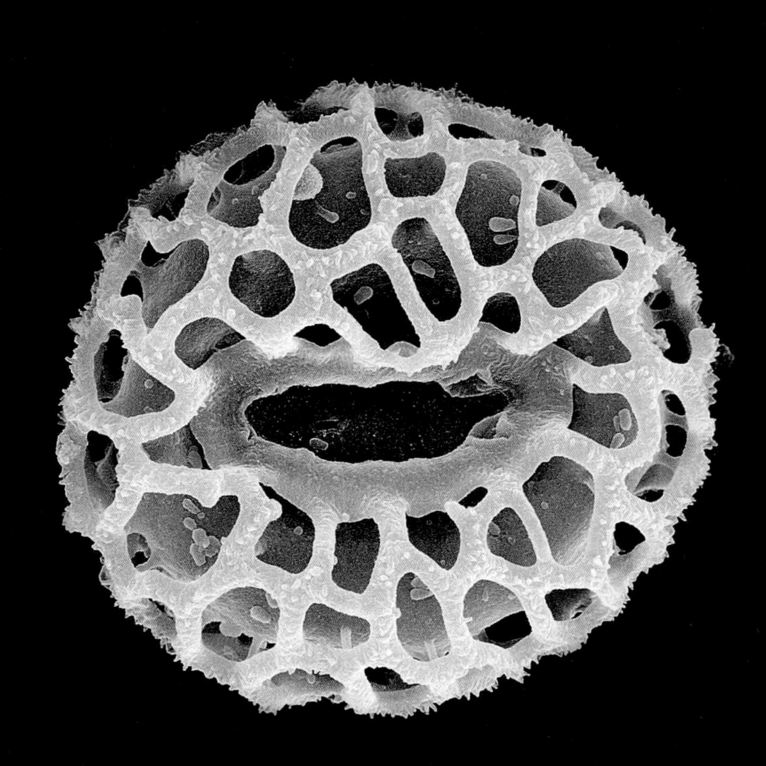

Chapter 2
PALM POLLEN

INTRODUCTION

Palm pollen has generally conservative characteristics but shows more variation than the pollen of any other monocotyledonous family. Since the publication of the first edition of *Genera Palmarum* in 1987, substantial progress has been made in the studies of palm pollen morphology instigated at the Royal Botanic Gardens, Kew in the early 1980s (Anon. 1981). By 1987, representative species of just over half the currently recognised genera had been studied at Kew using light microscopy (LM), scanning electron microscopy (SEM) and transmission electron microscopy (TEM); pollen data for the remaining genera were either non-existent or relied on previously published descriptions from light microscopy only, especially those of Thanikaimoni (1970). Now, there are data from the Kew pollen project for all genera. This comprehensive resource has enabled a much better understanding of many of the aperture variants that have been described, although the significance of the wide range of elaborate modifications to the outer pollen wall (ectexine) remains largely obscure. The descriptions of pollen morphology in this new edition, with rare exceptions such those provided by Sannier *et al*. (2006, 2007), draw almost exclusively on the data accumulated at Kew.

The pollen of the palm family has been the subject of numerous taxonomic and systematic studies during the past forty years. Pollen surveys of the family have been published by Thanikaimoni (1966, 1970), Sowunmi (1968, 1972) and Kedves (1980); of these, the study by Thanikaimoni (1970) was the most extensive in its coverage of genera and species. Before 1975, studies of pollen were based predominantly on light microscopy, whereas more recent studies have added many new data, especially from SEM and TEM (e.g., Ferguson 1986, Ferguson & Harley 1993, Harley & Baker 2001). Electron microscopes enable pollen to be viewed at far higher magnifications (ca. 50,000) than light microscopes (max. 1500), but the value of LM data must not be underestimated. Data from electron microscopy complement those from light microscopy rather than negating them. Scanning electron microscopy is particularly valuable for resolving details of the pollen surface and ornamentation, whereas TEM resolves the fine structure of the pollen wall. Nevertheless, light microscopy remains the only accurate way of measuring pollen, and it is also very important for interpreting shape and aperture configuration.

Important systematic studies of palm pollen include those of subfamily Coryphoideae by Dransfield *et al*. (1990), of tribe Borasseae (Coryphoideae) by Ferguson *et al*. (1987) and, in subfamily Arecoideae, of the cocosoid and the geonomatoid palms (Punt & Wessels Boer 1966a, 1966b), of subtribes Oncospermatinae and Verschaffeltiinae (Mendis *et al*. 1987) and of tribe Phytelepheae (Barfod 1988). Descriptions of pollen at generic level include those of *Licuala* (Coryphoideae) (Ambwani & Kumar 1993), *Ravenea* (Ceroxyloideae) (Ferguson *et al*. 1983, 1988), *Voanioala* (Arecoideae) (Harley 1989), and *Bentinckia* (Harley 1998).

Nearly all palm pollen studies are based on acetolysed pollen. Acetolysis is an aggressive technique that removes both the internal cellular fraction and the external lipid-based coatings within and on the surface of the ectexine. The technique was developed (Erdtman 1934, 1960) before the advent of electron microscopy. Its purpose is to clear the pollen grain exine and aperture area(s) so that the morphology is more easily resolved by light microscopy. Acetolysis also gives a quality to the pollen that is comparable to that of fossilised pollen, although extant pollen grains usually remain more 3-dimensional than their fossil counterparts. Indiscriminate use of acetolysis in pollen studies that are related to plant systematics and evolution is avoided nowadays, but, for the reasons outlined, its use remains valid for studies in which extant pollen is compared with fossil pollen. The application of acetolysis to studies of monocotyledonous pollen should be carefully monitored, however, because the chemical composition of the pollen wall in many monocotyledonous groups is usually less acetolysis-resistant than that in dicotyledons. In some groups of monocotyledons, the pollen wall may even be destroyed by acetolysis (Hesse *et al*. 2000), and alternative techniques are used to preserve structural integrity. Weak resistance to the process of burial and fossilisation is reflected in the poor fossil pollen record for many monocotyledonous families (Muller 1979, 1981, Collinson *et al*. 1993, Herendeen & Crane 1995). Among monocotyledons, the pollen of palms has an unusually tough ectexine, which can not only withstand acetolysis but also the process of fossilisation; consequently, there is a substantial fossil pollen record for the palm family (Chapter 5).

Opposite: Some of the pollen morphological diversity found in palms.
Top left: ***Korthalsia furtadoana***: zonasulcate gemmate pollen grain, *Matusop* 7427, SEM × 3,650.
Top right: ***Chamaedorea* sp.**: trichotomosulcate reticulate pollen grain, distal face, cultivated RBG Kew s.n. (AH44), SEM × 5,000.
Bottom left: ***Wallichia oblongifolia***: monosulcate pollen grain, long sharp spines with basal swellings, distal face, cultivated RBG Kew s.n., SEM × 5,500.
Bottom right: ***Nenga gajah***: monosulcate pollen grain, coarsely reticulate with spinose muri, distal face, *Gundersen et al.* LVG29, SEM × 2,600.

APERTURES

Throughout the palm family, pollen apertures are basically sulcate or porate; inaperturate pollen grains occur very rarely. Nevertheless, there are 16 variations on the configuration of aperture number, position and symmetry within the family; and the systematic distribution (Table 2.1) of these variants is phylogenetically significant (Harley & Baker 2001). Pollen grains that have sulcate apertures are far more common than porate pollen grains, and inaperturate pollen grains are unique to *Pigafetta* (Calamoideae: Calameae: Pigafettinae).

Bi-symmetric ellipsoid pollen grains with a single bi-symmetric and distally positioned sulcus are termed monosulcate (Fig. 2.1a) and are common to all subfamilies except Nypoideae. Slightly to highly asymmetric (basically ellipsoid) pollen grains with a symmetric to highly asymmetric distal sulcus (Fig. 2.1b) are also monosulcate and are common in subfamilies Coryphoideae, Ceroxyloideae and Arecoideae but never occur in subfamily Calamoideae. A single distal sulcus, where the apices extend meridionally beyond the equator towards the proximal face (e.g., Figs b,d, p.145; a,b, p.152), occurs sporadically in Calamoideae (*Eugeissona* and *Eremospatha*), in Arecoideae (*Areca*, *Pinanga* and *Hydriastele*) and rarely in Coryphoideae (*Licuala*). Pollen in which the sulcus is very short (brevi) (e.g., Figs a, p.157; a,d, p.328; a,d, p.513) also has a small and sporadic distribution: Calamoideae (*Eugeissona*, *Raphia* and *Mauritia*), Coryphoideae (*Borassodendron*), Ceroxyloideae (*Ammandra*) and Arecoideae (*Nenga*).

Relative to the phylogeny of the palms, a distal trichotomosulcate aperture (Fig. 2.2) is one of the two most significant aperture types. It is almost always found in association with asymmetric monosulcate pollen grains and, furthermore, both trichotomosulcate and monosulcate forms may occur within the same anther, although the trichotomosulcate pollen is almost always the lesser proportion (<25%).

TABLE 2.1
Systematic distribution of pollen aperture types

The four most frequent aperture types are highlighted in colour of different intensity: from the most frequent highlighted in the darkest colour to the fourth most frequent in the lightest shade. (N.B. More than one aperture type might occur within a genus.)

Aperture type	CALAMOIDEAE (21 genera)	NYPOIDEAE (1 genus)	CORYPHOIDEAE (46 genera)	CEROXYLOIDEAE (8 genera)	ARECOIDEAE (107 genera)
1a. Monosulcate, bi-symmetric	5	0	42	7	109
1b. Monosulcate, asymmetric	0	0	31	5	96
1c. Monosulcate, brevi	3	0	1	1	2
1d. Monosulcate, extended	2	0	1	0	3
2a. Trichotomosulcate, symmetric	0	0	0	0	4
2b. Trichotomosulcate, asymmetric	0	0	5	1	26
3. Zonasulcate, incomplete equatorial	0	0	0	0	1
4. Zonasulcate, meridional	2	1	0	0	1
5. Disulcate, distal	0	0	1	0	1
6. Disulcate, equatorial	8	0	0	0	0
7. Monoporate, distal	1	0	1	1	1
8. Monoporate, asymmetric equatorial	1	0	0	0	0
9. Diporate, equatorial	1	0	0	0	0
10. Diporate, subequatorial	2	0	0	0	0
11. Triporate, equatorial	0	0	0	0	1
12. Triporate-operculate, subequatorial	0	0	0	0	1
13. Inaperturate	1	0	0	0	0

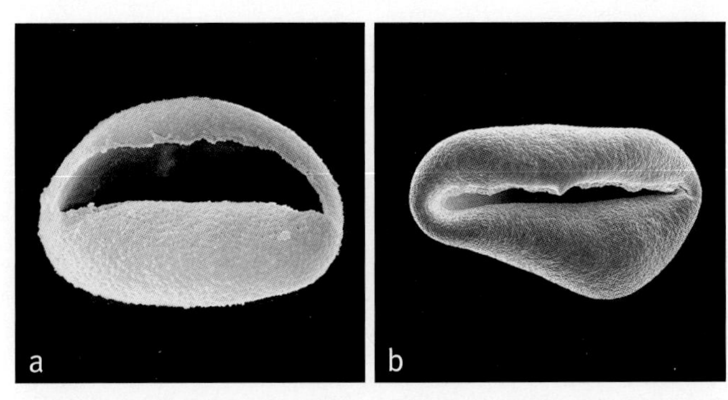

Fig. 2.1 **a,** ***Sabal minor*** symmetric monosulcate pollen grain SEM × 1500. *Drummond* 357; **b,** ***Ponapea ledermanniana*** asymmetric pollen grain SEM × 1000. *Kanehira* 1361.

Trichotomosulcate apertures are associated with Coryphoideae, Ceroxyloideae and Arecoideae, but never with Calamoideae or the monotypic Nypoideae. In Coryphoideae, the combination of mono- and trichotomosulcate pollen occurs in five genera: *Thrinax*, *Chelyocarpus* and *Cryosophila* in tribe Cryosophileae; and *Chamaerops* (in association with more or less symmetric distal disulcate apertures) and *Pritchardia* in tribe Trachycarpeae. In Ceroxyloideae, the mono- and trichotomosulcate combination occurs in *Ceroxylon*. In Arecoideae, this combination is widespread and is recorded in at least some species of *Hyophorbe* and *Chamaedorea* (Chamaedoreeae), *Roystonea* (Roystoneeae), *Reinhardtia* (Reinhardtieae), *Voanioala* and *Attalea* (Cocoseae: Attaleinae), *Astrocaryum* and *Bactris* (Cocoseae: Bactridinae), *Manicaria* (Manicarieae), *Oenocarpus* (Euterpeae), *Calyptrogyne* (Geonomateae), *Sommieria* (Pelagodoxeae), *Kentiopsis* (Areceae: Archontophoenicinae), *Areca* and *Pinanga* (Areceae: Arecinae), *Cyphophoenix* and *Physokentia* (Areceae: Basseliniinae), *Carpoxylon* (Areceae: Carpoxylinae), *Clinosperma* (Areceae: Clinospermatinae), *Dypsis*, *Marojejya* and *Masoala* (Areceae: Dypsidinae), *Drymophloeus* (Areceae: Ptychospermatinae) and *Rhopalostylis* (Areceae: Rhopalostylidinae), as well as in the currently unplaced genera, *Clinostigma*, *Cyrtostachys*, *Heterospathe*, *Hydriastele* and *Iguanura*. The frequency of the co-occurrence of asymmetric mono- and trichotomosulcate pollen suggests that it is a normal result of microsporogenesis in many palms and probably does not affect pollen viability, although this remains to be tested scientifically. Rarely, symmetric trichotomosulcate pollen occur (Fig. 2.2); the known examples are all from tribe Cocoseae. In *Voanioala* (subtribe Attaleinae), this pollen type occurs in combination with symmetric monosulcate pollen (Harley 1989). In subtribe Bactridinae, the pollen grains are almost exclusively trichotomosulcate, although monosulcate pollen are frequent in a few species of the genus *Bactris* (e.g., *B. hirta*). In *Astrocaryum*, some asymmetric tetrachotomosulcate apertures have also been noted, although these do appear to be a developmental abnormality. In *Elaeis guineensis*, where the trichotomosulcate apertures are usually symmetric, asymmetry is occasionally observed; the pollen of *Elaeis oleifera*, however, is usually asymmetric monosulcate.

Incomplete equatorially positioned zonasulcate apertures (Figs h,n, p.510) are found in the pollen of a few species of *Areca* (Arecoideae): *A. abdulrahmanii*, *A. andersonii*, *A. chaiana* and *A. klingkangensis* (Harley & Dransfield 2003). By contrast, meridional zonasulcate apertures (e.g., Figs a, p.175; a,d, p.213) are known for two genera of Calamoideae (i.e., most species of *Salacca* [Calameae: Salaccinae] and some species of *Korthalsia* [Calameae: Korthalsiinae]), for *Nypa fruticans* (Nypoideae) and a rare arecoid occurrence in *Pinanga celebica* (Arecoideae: Areceae: Arecinae).

Distal disulcate ('pontoperculate') apertures (Figs a, p.249; a, p.359) comprise paired sulci separated by a band of ectexine, which is more or less unmodified and is continuous with the main exine. They are known only in two genera, *Chamaerops* (Coryphoideae: Trachycarpeae: Rhapidinae) and *Iriartella* (Arecoideae: Iriarteeae). Ultra-thin sections of the pollen of *Chamaerops humilis* have demonstrated that the underlying distribution of intine is continuous under the two sulci (Furness & Rudall 2003), suggesting that this is a rare modification of the basic monosulcate aperture.

Equatorial disulcate apertures — two sulci positioned on the two opposing short equatorial axes of the pollen grain (Fig. 2.3) — occur only in subfamily Calamoideae. They are typical of a number of genera: *Metroxylon* (Calameae: Metroxylinae), *Plectocomia*, *Myrialepis* and *Plectocomiopsis* (Calameae: Plectocomiinae), *Calamus*, *Retispatha*, *Daemonorops* and *Ceratolobus* (Calameae: Calaminae). Notably, if the systematic distributions of trichotomosulcate and equatorial disulcate pollen are compared (Table 2.1), they are undoubtedly the most phylogenetically significant aperture types within the family.

A single distal porate aperture (e.g., Figs a,d, p.166) occurs in the spheroidal pollen of *Mauritiella* (Calamoideae: Lepidocaryeae: Mauritiinae) and *Borassodendron* (Coryphoideae: Borasseae: Lataniinae); in both these genera, the apertures range from porate to brevi-ellipsoid. All species of *Ravenea* (Ceroxyloideae: Ceroxyleae) have distal porate pollen, and the oblate-spheroidal pollen of *Areca caliso* (Arecoideae: Arecinae) also has a distal pore (Figs g,i, p.511). The single subequatorially positioned pore in one of the two

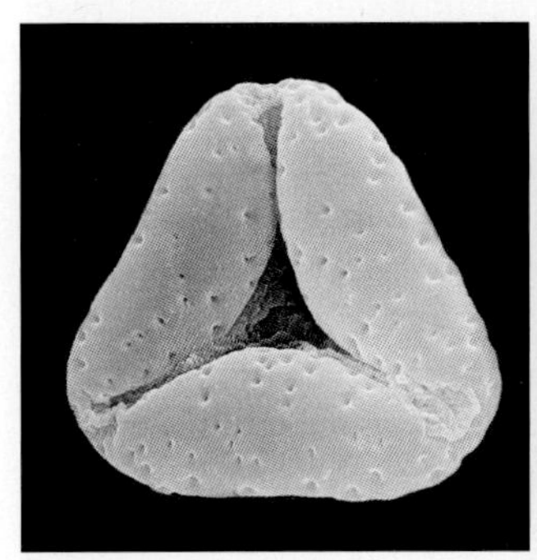

Fig. 2.2 ***Bactris concinna*** trichotomosulcate pollen grain SEM × 1000. *Krukoff* 6497.

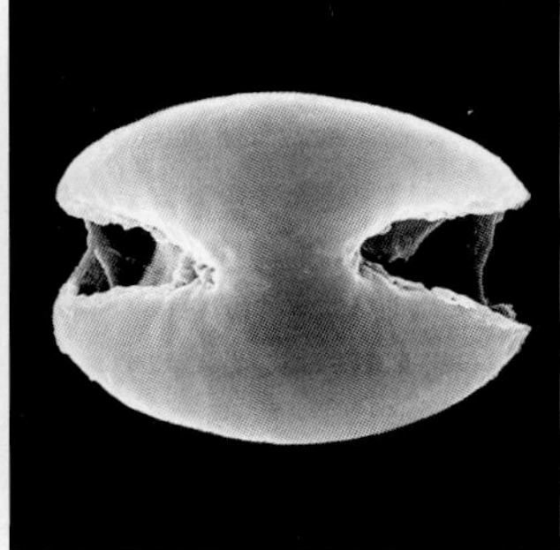

Fig. 2.3 ***Metroxylon sagu*** disulcate pollen grain, distal face SEM × 1000. *Floyd* HHRBK 5407.

short equatorial axes of the ellipsoid pollen of *Pogonotium* (Calamoideae: Calameae: Calaminae) is not only unique within the family (Figs a,b, p.206) but also within the monocotyledons, with the exception of a rare occurence in Pandanaceae: *Pandanus eydouxia* Balf.f. (Huynh 1991).

The spheroidal or spheroidal-ellipsoid pollen of *Daemonorops oblata* (Fig. 2.8j) and of several species of *Korthalsia* (Calamoideae) have two equatorial, symmetrically positioned pores (Fig. a, p.168). By contrast, the ellipsoid pollen of four other calamoid species: *Daemonorops oxycarpa*, *D. sparsiflora*, *D. verticillaris* and *Eleiodoxa conferta*, have a subequatorially positioned pore on the two opposing short axes of the grains (Figs c–e, p.203).

Oblate triangular pollen with three regularly spaced pores occurs in just two arecoid genera, *Areca* and *Sclerosperma*. This condition has been observed in only one collection of *Areca klingkangensis* (Areceae: Arecinae): here the pores are regularly positioned around the perimeter of the rounded trianguloid pollen (Figs f,m, p.56). Pollen of another collection of this species has incomplete zonasulcate apertures (Harley & Dransfield 2003). In *Sclerosperma* (Sclerospermeae), all species have three operculate pores subequatorially positioned near the apices of the acutely trianguloid pollen (Figs a–d, p.391).

Within the palms, there are two notable examples of parallel pollen evolution in systematically separate genera. Not only does the pollen in each example have similar apertures and ectexine, it is also similar in size and shape. In *Borassodendron* (Coryphoideae) and *Ammandra* (Ceroxyloideae), the pollen grains are rounded ellipsoid to oblate spheroidal with brevi sulcate or monoporate apertures, a foveolate or reticulate ectexine and a size range of 60–85 µm (c.f. Figs a–d, p.328; a,b, p.348). In *Chamaerops* (Coryphoideae) and *Iriartella* (Arecoideae), the ellipsoid pollen grains have two parallel distal sulci, a perforate or finely reticulate ectexine and a size range of 22–31 µm (c.f. Figs a, p.249; a, p.359).

The three most frequently occurring aperture configurations are monosulcate, trichotomosulcate and equatorial disulcate, including zonasulcate and extended monosulcate, which appear to have a developmental or evolutionary link to equatorial disulcate pollen types (Harley & Baker 2001). Apart from these widely occurring configurations, there remains an intriguing range of rare or very rare aperture types. Two of these, brevi monosulcate and distal monoporate, are found in all subfamilies except Nypoideae and seem to be closely associated, either with each other (in *Ammandra*, *Mauritia* and *Mauritiella* [Calamoideae: Lepidocaryeae] and in *Borassodendron*) or with monosulcate apertures (in *Raphia* [Calamoideae] and *Nenga* [Arecoideae]). The asymmetric monoporate aperture in the ellipsoid pollen of *Pogonotium* (Calamoideae) seems to be a rare variant of the unusual subequatorial diporate apertures in the ellipsoid pollen of *Daemonorops sparsiflora*, *D. verticillaris* and *Eleiodoxa conferta* (Calamoideae: Calaminae). The hypothesis that diporate pollen may be an evolutionary development from equatorial disulcate is supported by its restricted occurrence within tribe Calameae (Calamoideae), where all known examples of equatorial disulcy occur. Furthermore, it would not be unreasonable to extrapolate that the rare examples of triporate pollen in subfamily Arecoideae could be evolutionary developments from distal trichotomosulcate (Erdtman & Singh 1957).

MICROSPOROGENESIS

Studying microsporogenesis requires very different techniques from those employed for pollen morphology (see for example Harley 1999a, Sannier *et al.* 2006, 2007). However, the data from studies of microsporogenesis are proving to be very important in evolutionary and systematic studies of palms. In summary, although the pollen grains of all palms are free individuals at maturity, while still in the callose envelope (Fig. 2.4) of the original pollen mother cell (PMC) the four daughter cells (young pollen grains) which develop from the PMC during meiosis remain briefly held together in one of three or four tetrad configurations: usually tetragonal (Figs 2.5a,b) or tetrahedral (Figs 2.6a,b) or, infrequently to rarely, rhomboidal or decussate. Even at maturity (free microspore stage) a Y-shaped or linear impression mark (Harley 1990, 1996) often remains on the proximal face of the pollen grain and these impression marks are indicative of the previous post-meiotic tetrad form: (a linear mark indicating a tetragonal, rhomboidal or decussate configuration; a Y-shaped mark indicating a tetrahedral configuration; Figs 2.5c; 2.6c). Tetragonal tetrads are usually associated with successive cytokinesis in which the cell plates extend centrifugally, whereas tetrahedral tetrads are

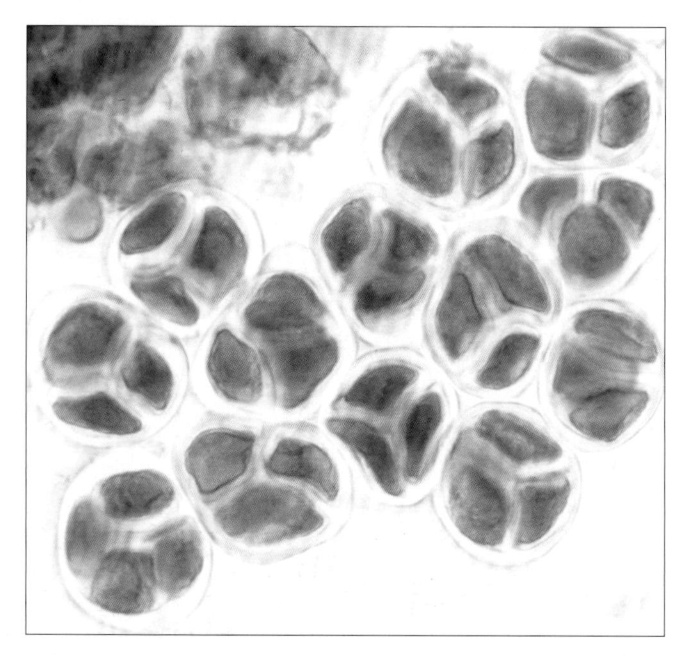

Fig. 2.4 ***Chamaedorea seifrizii*** tetrahedral tetrads with the 4 daughter cells still enclosed within the callose envelope of the pollen mother cell LM × 500 [fourth daughter cell is concealed from view]. Cultivated RBG Kew 362-66.36221.

usually associated with simultaneous cytokinesis, in which the cell plates advance centripetally. Rhomboidal and decussate tetrads tend to be associated with tetragonal tetrads although, unusually, decussate and/or tetragonal tetrads have also been noted in predominantly tetrahedral samples, for example, of *Washingtonia* (Harley 1999a), *Chamaedorea* (Rao 1959a, González-Cervantes *et al.* 1997, Harley 1999a), *Hyphaene* (Rao 1959a, Harley 1999a), *Areca catechu* (Rao 1959b) and *Dypsis* (Rao 1959b, Harley 1999a). Recent research (Sannier *et al.* 2006, 2007) indicates that microsporogenesis in palms is complex, and that the key processes during microsporogenesis are not well understood. It is already known (Harley 1996, Harley & Baker 2001) that within palms, distal monosulcate pollen may result from either successive or simultaneous microsporogenesis. In more recent work led by Sannier (Sannier *et al.* 2006, 2007), three key characteristics of microsporogenesis have been closely observed in relation to each other: cytokinesis type, callose deposition between the microspores and the resultant tetrad type. Results from studies of microsporogenesis in selected species of *Sabal*, *Trachycarpus*, *Serenoa*, *Copernicia*, *Caryota*, *Bismarckia* and *Hyphaene* (Coryphoideae) and *Hyophorbe*, *Synechanthus*, *Chamaedorea*, *Gaussia*, *Allagoptera*, *Butia*, *Areca*, *Dypsis*, *Ptychosperma* and *Veitchia* (Arecoideae), in which tetrads are predominantly tetrahedral, shows that during cytokinesis, the intersporal wall formation and resulting tetrad form are frequently very variable within species. Although predominantly tetrahedral (ca. 70–97%), a smaller percentage of tetragonal (1–18%) and a very small percentage (0.3–7.2%) of rhomboidal tetrads are also present. This highly irregular type of microsporogenesis conforms to the 'modified simultaneous model' with 'ephemeral cell plates' recognised by Blackmore *et al.* (1987). At present, the data of Sannier *et al.* (2006, 2007) are

TABLE 2.2

Systematic distribution of pollen tetrad types

'Predominant tetrad type' is either tetragonal or tetrahedral; however, if ca. >10% but <50% of post-meiotic tetrads in a generic sample were of the non-predominant type, then that genus is also listed both under the predominant type and in parentheses under the non-predominant type. (Bold indicates genera examined by Sannier *et al.* 2006, 2007.)

Predominant tetrad type	CALAMOIDEAE (21 genera)	NYPOIDEAE (1 genus)	CORYPHOIDEAE (46 genera)	CEROXYLOIDEAE (8 genera, no data)	ARECOIDEAE (107 genera)
Tetragonal	*Eugeissona* *Metroxylon* *Salacca* *Calamus*	*Nypa*	**(*Sabal*)** **(*Trachycarpus*)** ***Licuala*** ***Serenoa*** **(*Caryota*)** **(*Hyphaene*)**		**(*Chamaedorea*)** **(*Butia*)** **(*Ptychosperma*)**
Tetrahedral			***Sabal*** *Chamaerops* ***Trachycarpus*** *Rhapis* **(*Serenoa*)** ***Copernicia*** *Washingtonia* *Chuniophoenix* ***Caryota*** *Arenga* ***Bismarckia*** ***Hyphaene***		*Iriartella* ***Hyophorbe*** *Wendlandiella* ***Synechanthus*** ***Chamaedorea*** ***Gaussia*** *Sclerosperma* *Voanioala* ***Allagoptera**** ***Butia*** *Syagrus* *Calyptronoma* ***Areca*** *Pinanga* ***Dypsis*** ***Ptychosperma*** ***Veitchia*** *Carpentaria* *Cyrtostachys* *Hydriastele* *Iguanura*

restricted to Coryphoideae and Arecoideae, and the species examined all have more or less bisymmetrical or highly asymmetric monosulcate pollen. Nevertheless, the phylogenetic implications of these data are of great interest in relation to the influence of cytokinesis on pollen aperture type, number and arrangement (Tables 2.2, 2.3).

Data from observations of microsporogenesis in the palms by Rao (1959a,b), Thanikaimoni (1970), González-Cervantes *et al.* (1997), Harley (1999a), Sannier *et al.* (2006, 2007), as well as previously unpublished observations (M.H.), represent 39 genera. These data, plus observations of impression marks on mature pollen grains, show striking systematic distribution patterns for the two main tetrad types. At the pollen mother cell stage, *Nypa* pollen and all calamoid pollen examined to date (*Eugeissona*, *Salacca*, *Metroxylon* and *Calamus*) have tetragonal tetrads; coryphoid and arecoid pollen are predominantly from tetrahedral tetrads, but often with lower proportions of tetragonal and/or rhomboidal tetrads. No data are yet available from the ceroxyloid palms (Table 2.2). The data also show a remarkable correlation between aperture type and tetrad type (Table 2.3), particularly with regard to aperture number and position (distal or equatorial). It is notable in this context that in the arecoid species *Pinanga subruminata*, which has extended sulcate pollen, the tetrads are, predictably, tetragonal. In *Pinanga*, however, there are a number of species in which trichotomosulcate pollen occurs and, although not yet demonstrated, it is highly probable that the tetrads would be predominantly tetrahedral.

Aperture types that are associated with tetragonal, rhomboidal or decussate tetrads and/or linear impression marks are: symmetric monosulcate, extended monosulcate, equatorial disulcate, asymmetric monoporate, and meridional zonasulcate. It is also assumed, but not proven, that equatorial diporate and subequatorial diporate aperture types are

TABLE 2.3

Association between tetrad type and aperture type

Note: the strong association of a single distal aperture with tetrahedral tetrads (bright blue), two equatorial apertures with tetragonal tetrads (mid-blue) and three equatorial apertures with tetrahedral tetrads (light blue). [(✓) = tetrad type not proven.]

Aperture type	Aperture number	Aperture position	Tetragonal	Rhomboidal or decussate	Tetrahedral
1a. Monosulcate, bi-symmetric	1	Distal	✓	3	3
1b. Monosulcate, asymmetric	1	Distal			3
1c. Monosulcate, brevi	1	Distal	(✓)		(3)
1d. Monosulcate, extended	1	Distal>equatorial	✓	✓	✓
2a. Trichotomosulcate, symmetric	1	Distal			✓
2b. Trichotomosulcate, asymmetric	1	Distal			✓
3. Zonasulcate, incomplete equatorial	1	Equatorial			✓
4. Zonasulcate, meridional	1	Meridional	✓		
5. Disulcate, distal	2	Distal			✓
6. Equatorial, disulcate	2	Equatorial	✓		
7. Monoporate, distal	1	Distal	(✓)		(✓)
8. Equatorial, monoporate, asymmetric	1	Equatorial	✓		
9. Equatorial, diporate	2	Equatorial	✓		
10. Sub-equatorial, diporate	2	Subequatorial	✓		
11. Triporate	3	Equatorial			✓
12. Triporate-operculate, subequatorial	3	Subequatorial			✓
13. Inaperturate	0	n/a	(✓)		

associated with successive cytokinesis and tetragonal tetrads. Aperture types that are associated with tetrahedral tetrads and/or Y-shaped impression marks are: symmetric or asymmetric monosulcate, extended monosulcate, distal disulcate, symmetric and asymmetric trichotomosulcate, apical and subapical triporate, and incomplete equatorial zonasulcate. Aperture types for which there are no tetrad data are brevi-monosulcate and symmetric monoporate. Systematically, the sporadic occurrence of these last two aperture types in all subfamilies except Nypoideae, combined with differences in pollen characteristics between the species that are associated with each aperture type, suggests that brevi-sulcate and distal monoporate pollen could result from either successive, simultaneous or modified simultaneous cytokinesis. The type of cytokinesis for inaperturate pollen is unknown, but the isolated occurrence of this pollen type in Calamoideae, where only successive cytokinesis is known to date, suggests an association between inaperturate pollen and successive cytokinesis.

POLLEN SHAPE

Pollen shape ranges from bi-symmetrical or asymmetrical ellipsoid, bi-symmetrical or asymmetrical trianguloid, through oblate spheroidal to spheroidal. To some extent, shape is related to aperture type. Ellipsoid grains usually have a single distal sulcus or paired sulci on the equatorial short axes. Oblate trianguloid pollen grains usually have a distal trichotomosulcus or, rarely (*Areca klingkangensis* and *Sclerosperma* spp.), three equatorial or subequatorial pores on or near the three apices of the triangle. Spheroidal grains are, most frequently, monoporate or zonasulcate; whereas asymmetric, basically ellipsoid pollen grains, also throw up a number of random shapes, the most frequent being lozenge-shaped, asymmetric elongate, trianguloid or pyriform.

ECTEXINE

The unmodified surface of the pollen ectexine in palms is usually psilate (smooth) or slightly scabrate (roughened). With the exception of the smooth, featureless pollen of *Metroxylon sagu* (Fig. a, p.178), however, the ectexine is variously modified, from simple perforations to elaborate structures.

Thirteen types of tectate or intectate ectexine topology are recognised. 1. Perforate (e.g., Figs b, p.223; a, p.244;): this includes a continuum of variation involving perforation size and density, and absence or presence at varying densities of micro-channels or micro-fossulae. 2. Foveolate (e.g., Figs c, p.264; d, p.279): basically large rounded holes (lumina) in the tectum. 3. Reticulate (e.g., Figs c, p.339; b, p.513; g,h,i, p.516): with a net-like appearance. 4. Ring murate (Fig. f, p. 516): superficially similar to reticulate but each lumen in the tectum (sometimes two or three holes together) is surrounded by a discrete ring or loop of tectum pressed against neighbouring rings (rare). 5. Rugulate (e.g., Figs e, p.380; g, p.144; b, p.451): elongated irregular crumpled/wrinkled elements, sometimes with perforations or micro-fossulae/micro-channels between. The distinction of strictly rugulate from strictly perforate or reticulate is often not discrete, thus an intermediate surface topology may be described, for example, as 'perforate-rugulate' or 'rugulate-reticulate'. 6. Striate (Fig. b, p.174): narrow, elongate and closely set, generally parallel elements. 7. Verrucate (e.g., Figs j, p.195; d, p.480; b, p.267): wart-like elements where breadth exceeds height. 8. Granular (e.g., Figs b, p.361; h, p.380; g, p.549): small rounded elements <1 µm. 9. Gemmate: discrete more-or-less spheroidal elements that may be attached to the tectum (supratectate) (Figs b,d, p.317) or to the foot layer (intectate) (Figs a,b, p.195). 10. Clavate: a club-shaped element, diameter less than height and thicker at apex than at base, they may be supratectate (e.g., Figs c,d, p.569) or intectate (e.g., Figs a,b, p.363). 11. Spinose: elements tapering from the base with a length of 2 µm or more, usually pointed, less frequently blunt; they may be supratectate (e,g, Figs b,c, p.213) or intectate (e.g., Figs a–d, p.306 00). 12. Spinulose (tectate): with small spines, <2 µm in length (e.g., Figs a,b, p.513). 13. Urceolate (Fig. d, p.516): with intectate urn-shaped elements (Ferguson *et al.* 1983).

The tectate-perforate type occurs widely throughout all subfamilies with the exception of Nypoideae. It is difficult to qualify discrete variations of this widespread type, but an understanding of the variation, particularly when combined with ultrastructural characters, is of taxonomic and systematic value (Harley 1990, 1996, 1999b; Harley & Morley 1995).

The semitectate foveolate type occurs sporadically in Calamoideae: *Eugeissona tristis* (Eugeissoneae), *Calamus arborescens*, *Daemonorops formicaria*, *D. leptopus*, *D. robusta* and *Pogonotium ursinum* (Calameae: Calaminae). It also occurs in Coryphoideae (*Chelyocarpus chuco*, *Cryosophila stauracantha* and *Cryosophila warscewiczii* [Cryosophileae]), in some species of *Phoenix* (Phoeniceae), in *Trachycarpus fortunei* (Trachycarpeae: Rhapidinae) and in subtribe Livistoninae (*Licuala peltata*, *Colpothrinax wrightii* and *Washingtonia robusta*); in tribe Borasseae, it is found in subtribe Lataniinae, (*Borassodendron borneense* and *B. machadonis*). The semitectate foveolate type also occurs in species from four genera of Ceroxyloideae: *Pseudophoenix vinifera* (Cyclospatheae), *Oraniopsis appendiculata* and *Ravenea albicans* (Ceroxyleae) and *Phytelephas* (Phytelepheae). In Arecoideae, it is known in five genera: *Neonicholsonia watsonii* (Euterpeae), *Pinanga dumetosa* (Areceae: Arecinae), and sometimes in *Dypsis moorei* (Areceae: Dypsidinae), *Acanthophoenix rubra* (Areceae: Oncospermatinae) and *Hydriastele rostrata* (Areceae: unplaced).

The semitectate reticulate type is systematically widespread and has many variants (Harley 1996). The lumina of the reticulum range from small to very large. In different species, the muri may be rounded or angular and, although usually psilate, they may be granular as in *Dypsis fibrosa* (Areceae: Dypsidinae), spinulose as in *Nenga gajah* (Areceae: Arecinae),

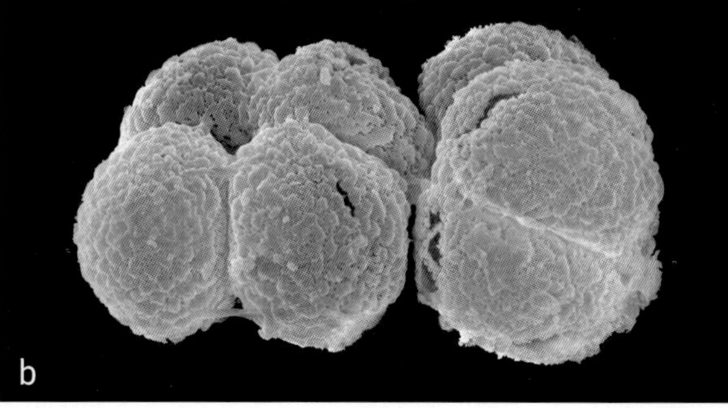

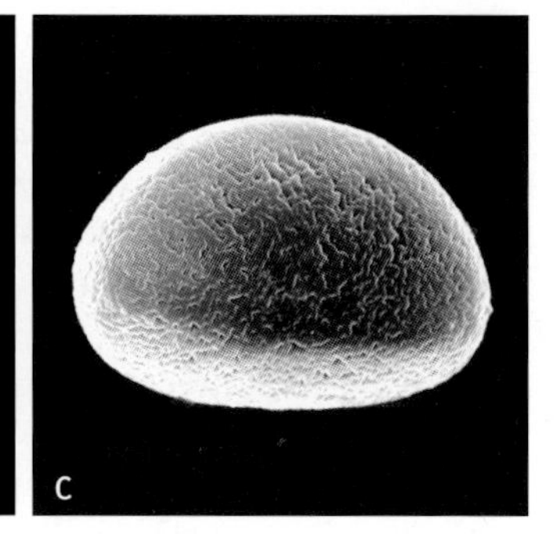

Fig. 2.5 **a**, ***Pinanga subruminata*** tetragonal tetrad SEM × 800. *Meng Kuang* s.n.; **b**, ***Calamus densiflorus*** tetragonal tetrad SEM × 1000. *Dransfield* 5373; **c**, ***Raphia hookeri*** proximal face of pollen grain to show the linear impression mark remaining from tetrad stage SEM × 1000. *Letouzey* 14594.

or even perforate (Fig. g, p.516) as in some species of *Pinanga* (Areceae: Arecinae) and *Hydriastele* (Areceae: unplaced). The semitectate reticulate ectexine type is found in all subfamilies, including the basically spiny *Nypa* (Nypoideae) where the tectum ranges from coarsely perforate to finely reticulate. In Calamoideae, this ectexine type is present in some species of *Laccosperma* (Lepidocaryeae: Ancistrophyllinae), in *Metroxylon salomonense* (Calameae: Metroxylinae), in *Pigafetta* (Calameae: Pigafettinae), in *Plectocomiopsis triquetra* (Calameae: Plectocomiinae), in a few species of *Calamus* and *Daemonorops*, and in *Ceratolobus* (Calameae: Calaminae). In Coryphoideae, it occurs in six genera: most species of *Phoenix* (Phoeniceae), the monotypic *Chamaerops* and also *Trachycarpus* (Trachycarpeae: Rhapidinae), *Kerriodoxa* and *Nannorrhops* (Chuniophoeniceae), and *Corypha* (Corypheae); in Ceroxyloideae, it is found in some species of *Pseudophoenix* (Cyclospatheae) and *Ceroxylon* (Ceroxyleae), and also in the monotypic *Ammandra* (Phytelepheae). The reticulate ectexine is, however, most strongly represented in the Arecoideae although, interestingly, in a restricted number of genera: in numerous species of *Chamaedorea* (Chamaedoreeae) where, in a few species (e.g., *Chamaedorea fragrans* and *C. ibarrae*), the proximal face of the pollen grain is tectate perforate psilate (Fig. a, p.380), and sometimes in *Sclerosperma mannii* (Sclerospermeae), *Syagrus harleyi* (Cocoseae), *Neonicholsonia watsonii* (Euterpeae), *Geonoma longivaginata* (Geonomateae), numerous species of *Areca* and *Pinanga* (Areceae: Arecinae), three species of *Dypsis* (Areceae: Dypsidinae), *Linospadix palmeriana* (Linospadicinae), and numerous species of *Hydriastele* (Areceae: unplaced).

The semitectate ring murate type is known only in *Pinanga gracilis* (Arecoideae: Arecinae). It has been postulated that this tectum formation is an extreme modification of the perforated muri that are typical in the reticulate pollen of a number of species in *Pinanga* (Harley 1999b).

The rugulate type is the second most widespread ectexine type in the palms (Table 2.4), it occurs in all subfamilies except Nypoideae. In Ceroxyloideae, it is represented in the same number of genera as the perforate type, but elsewhere in the family, it is represented in notably fewer genera. There are, however, varying degrees of intergradation between rugulate and perforate types, which sometimes make it difficult to make a clear distinction between the two; the topology is then better described as perforate-rugulate.

The finely striate pollen of *Wendlandiella gracilis* (Arecoideae: Chamaedoreeae) is unique within the palm family.

The tectate verrucate type is uncommon except in *Calamus* (Calamoideae: Calameae) where it is noted in about 17 species. In Coryphoideae, it is known in *Hyphaene* (Borasseae: Hyphaeninae), and in Arecoideae, there are four genera in which some species have a verrucate or rugulate-verrucate pollen surface: *Chamaedorea carchensis* (Chamaedoreeae), *Aiphanes* (Cocoseae: Bactridinae), *Prestoea* (Euterpeae) and *Cyrtostachys* (Areceae: unplaced). The verrucate type appears to be transitional between the rugulate and tectate gemmate types: three of the six genera in which tectate verrucate pollen occur, *Calamus*, *Hyphaene* and *Cyrtostachys*, also have species that have tectate gemmate pollen.

A distinctly granular ectexine that is not dominated by other more distinctive features, such as gemmae or spines, is almost exclusive to subfamily Arecoideae, with the exception of *Salacca zalacca* (Calamoideae: Calameae: Salaccinae). Within Arecoideae, it is found in *Dictyocaryum* (Iriarteeae), a few species of *Chamaedorea* (Chamaedoreeae) and many species of *Dypsis* (Areceae: Dypsidinae). In *Laccospadix australasicus* (Linospadicinae), the ectexine is finely granular-rugulate, and similar examples occur in some species of *Chamaedorea*, such as *C. cataractarum* and *C. geonomiformis*.

The gemmate type is one of three ectexine types that have supratectate and intectate forms. Supratectate gemmae occur on the ectexine of five species of *Calamus* (Calamoideae: Calameae) and in two genera of Coryphoideae (Borasseae), *Hyphaene* (Hyphaeninae) and *Borassus* (Lataniinae). In Arecoideae, the occurrence of supratectate gemmae is rare and sporadic: in one species each in *Attalea* (Cocoseae: Attaleinae), *Bactris* (Cocoseae: Bactridinae) and *Cyrtostachys* (Arecaceae: unplaced). With the exception of *Calamus lambirensis* and *C. lobbianus*, the systematic distribution of pollen that has intectate gemmae is more or less disjunct from that of supratectate gemmate pollen. Intectate gemmae are found in one species of *Ceroxylon* (Ceroxyloideae) and, in Arecoideae, in *Dictyocaryum* and *Iriartea* (Iriarteeae) and *Voanioala* (Cocoseae: Attaleinae).

The clavate type also has both tectate and intectate forms. Supratectate clavae may be striate, angular and block-like (Fig. a, p.569), or psilate-scabrate and somewhat elongate with rounded apices, or elongate, irregularly-sized and interspersed with smaller, shorter clavae (Fig. l, p.516). Intectate clavae may be psilate, scabrate or striate; broad and mushroom-like with flat rounded or rounded triangular heads (Fig. a, p.363); coarse and angular or spinulose; discrete or anastomosed; small to large in proportion to overall grain size; regularly or irregularly sized; arranged in a reticulate pattern (Fig. c, p.571); and widely to densely packed. Supratectate clavate pollen is found in Calamoideae (Calameae) (*Salacca* [Salaccinae] and *Daemonorops* [Calaminae]) and in Arecoideae (Areceae) (some species of *Pinanga* [Arecinae] and *Oncosperma* [Oncospermatinae]). Intectate clavate pollen is also found in Calamoideae (Calameae): in *Korthalsia* (Korthalsiinae) and *Plectocomiopsis* (Plectocomiinae). In Coryphoideae, it is known in some species of *Caryota* and *Arenga hastata* (Caryoteae); whereas in subfamily Arecoideae, it occurs in *Iriartea* spp. (Iriarteeae), some species of *Pinanga* (Areceae: Arecinae) and *Deckenia* (Areceae: Oncospermatinae).

The spinose type is the third pollen ectexine type that has both supratectate and intectate forms. The range of spine types, and their systematic distribution within the family, is extraordinary; the different spine types are often genus- or even species-specific. Spines in palm pollen are defined as being more than 2 µm in height; the ratio of length to basal width of spines varies between genera and species. In pollen with supratectate spines, the surrounding tectum is usually fine to coarse perforate or reticulate (e.g., *Nypa fruticans*) (Figs a–d, p.213). The spines are usually erect and psilate, average to widely spaced, conical and gradually tapering, with either

TABLE 2.4
Systematic distribution of pollen ectexine types

Numbers within the table represent the type of ectexine for each subfamily and for all genera ('total genera') in which the ectexine type is known to occur. The four most frequent ectexine types are highlighted in colour of different intensity: from the most frequent highlighted in the darkest colour to the fourth most frequent in the lightest shade.

Ectexine type	Total genera	CALAMOIDEAE (21 genera)	NYPOIDEAE (1 genus)	CORYPHOIDEAE (46 genera)	CEROXYLOIDEAE (8 genera)	ARECOIDEAE (107 genera)
1. Tectate perforate	129	13	0	30	2	84
2. Semitectate foveolate	22	4	0	9	4	5
3. Semitectate reticulate	27	7	0	7	4	9
4. Semitectate ring murate	1	0	0	0	0	1
5. Tectate rugulate	68	5	0	13	2	48
6. Tectate striate	1	0	0	0	0	1
7. Tectate verrucate	6	1	0	1	0	4
8. Tectate/semitectate granular	7	2	0	0	0	5
9a. Supratectate gemmate	6	1	0	2	0	3
9b. Intectate gemmate	7	1	0	0	1	5
10a. Supratectate clavate	4	2	0	0	0	2
10b. Intectate clavate	7	2	0	2	0	3
11a. Supratectate spinose	8	1	1	0	1	5
11b. Intectate spinose	11	5	0	3	0	3
12. Tectate spinulose	5	3	0	0	0	2
13. Intectate urceolate	1	0	0	0	0	1

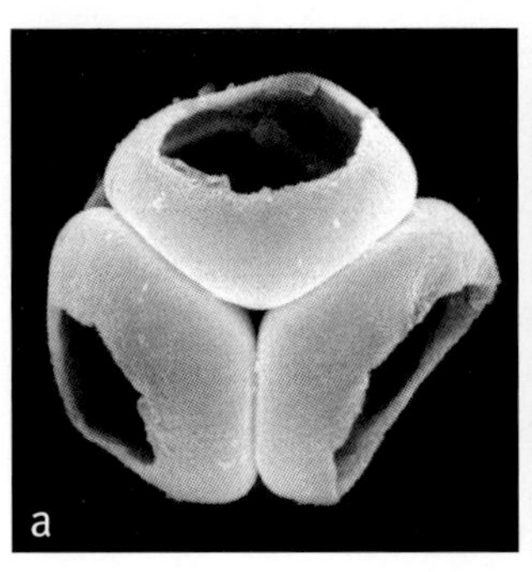

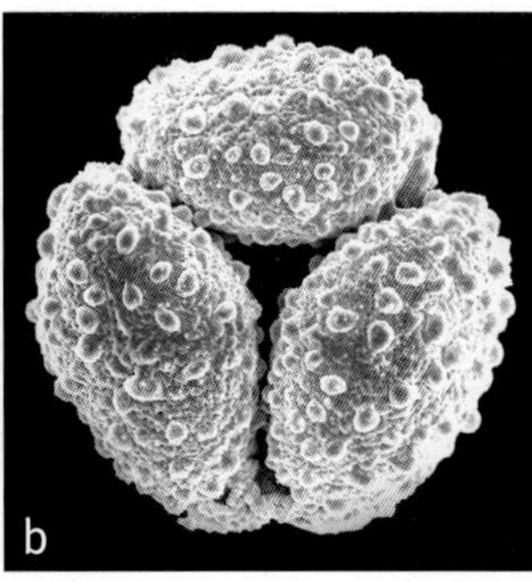

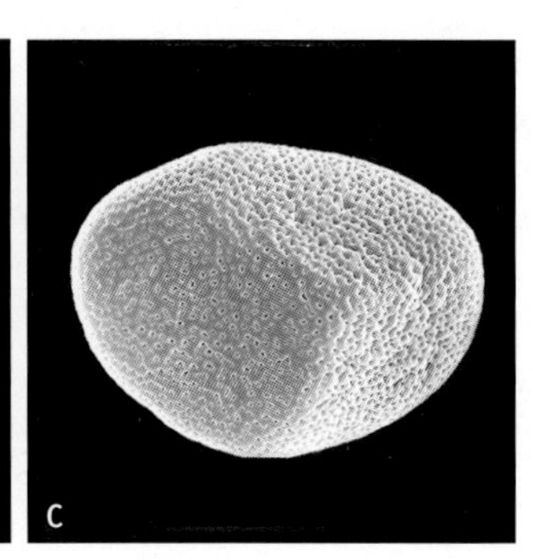

Fig. 2.6 **a,** ***Chamaedorea pumila*** tetrahedral tetrad SEM × 1000 [fourth daughter cell is concealed from view]. *N.E. Brown* s.n.; **b,** ***Cyrtostachys loriae*** tetrahedral tetrad SEM × 800 [fourth daughter cell is concealed from view]. *Darbyshire* 867; **c,** ***Geonoma macrostachys*** proximal face of pollen grain to show Y-shaped impression mark remaining from tetrad stage SEM × 1500. *Prance et al.* 2801.

pointed (Figs a,b, p.175) or blunt apices (Fig. e, p.144), or they are vertically ridged with pointed apices (Fig. a, p.198). The spines may have swollen bases (Fig. b, p.213), or be irregularly shaped (Fig. b, p.475). In *Bactris longiseta*, the spines are unusually long and often prostrate or semiprostrate (Figs d,l, p.444). Pollen that has supratectate spines is found in the subfamily Calamoideae (Calameae) in most species of *Salacca* (Salaccinae) and *Retispatha dumetosa* (Calaminae); in *Nypa fruticans* (Nypoideae); in most species of *Ravenea* (Ceroxyloideae: Ceroxyleae); and in several Arecoideae species including one species of *Aiphanes*, *Bactris acanthocarpoides* and *B. longiseta* (Cocoseae: Bactridinae), *Pholidostachys dactyloides* (Geonomateae), *Pinanga cleistantha* (Areceae: Arecinae) and *Phoenicophorium borsigianum* (Areceae: Verschaffeltiinae).

In pollen with intectate spines, the upper surface of the surrounding foot layer may be psilate to scabrate or densely covered by fine granulae or very small clavae. The spines may be psilate and simply conical with an irregular basal cavity; conical and elongate (Fig. b, p.300); conical with a basal 'brim' loosely attached to small exposed columellae or granulae (Figs a,b, p.366); or vertically ridged, directly attached to foot layer, and frequently bifurcate. The spines may be either bulbous (bottle-shaped) or elongate, deeply set into (and loosely connected to) cavities in a very thick foot layer that is swollen beneath the spines, and with pronounced lamellae underlying the inner margin of the foot layer (Figs b, p.259; a,b, p.163; b, p.166). Alternatively, spines may be wide-based conical and slightly inset into a thick foot layer that is distinctly swollen beneath the spines or more-or-less pointed or blunt (truncheon-like), with a distinct swelling slightly above the base (Figs a–d, p.306). Pollen with intectate spines is found in Calamoideae — *Lepidocaryum*, *Mauritia* and *Mauritiella* (Lepidocaryeae: Mauritiinae), some species of *Korthalsia* (Calameae: Korthalsiinae) and *Daemonorops oblata* (Calameae: Calaminae); in Coryphoideae — *Caryota maxima* and *C. ochlandra*, *Arenga* and *Wallichia* (Caryoteae); in Arecoideae — *Socratea* and *Wettinia* (Iriarteeae), and some species of *Pinanga* (Areceae: Arecinae): *P. cleistantha, P. scortechinii* and *P. polymorpha.*

Tectate spinulose cannot be strictly defined as an ectexine 'type' in palms because it is very uncommon and because each example differs noticeably from the other examples. To date, it has been recorded in three calamoid genera: *Oncocalamus mannii* (Lepidocaryeae: Ancistrophyllinae), *Raphia regalis* (Lepidocaryeae: Raphiinae) and *Salacca dubia* (Calameae: Salaccinae). In Arecoideae, it is only known in *Nenga gajah* (Areceae: Arecinae), in which the muri of the ectexine are spinulose, and in some collections of *Dictyosperma album* (Areceae: unplaced), in which the appearance of the spinules (Fig. c, p.629) suggests that they may be a form of elongate orbicule (which might explain why these spinulae are not present in all collections).

Intectate urceolate structures are the most extraordinary of all ectexine elements, not only within the family but for angiosperms generally, where they are unknown outside the palms. They occur only in *Pinanga aristata* and *P. pilosa* (Arecoideae: Areceae: Arecinae) (Ferguson *et al.* 1983). These structures do not indicate an obvious modification of any other ectexine structure known within the family.

The most frequently occurring (Table 2.4) and apparently least-specialised ectexine types are perforate (129 genera) and the closely linked rugulate type (68 genera). Reticulate (27 genera) and foveolate (22 genera) are the next most commonly encountered types, while spinose pollen, tectate or intectate, is known in 19 genera. The remaining ectexine types — ring murate, striate, verrucate, granular, gemmate, clavate, spinulose and urceolate — occur infrequently or in single species. The more elaborate types are found most frequently in the Calamoideae (13 genera) and the Arecoideae (20 genera). With the exceptions of *Salacca* (granular, clavate, spinose and spinulose) and *Pinanga* (ring murate, gemmate, clavate, spinose and urceolate), however, genera that have more than two of these unusual types are unknown. Most of the genera in which the less frequent types occur also have species that have perforate, foveolate, reticulate or rugulate pollen, or some combination of two or more of these. The rare exceptions to this are *Korthalsia* (Calamoideae), *Caryota* and *Arenga* (Coryphoideae), and *Dictyocaryum* and *Iriartea* (Arecoideae). Interestingly, if tribe Caryoteae is excluded, there are no other genera in which intectate (or spinose) pollen occur in subfamily Coryphoideae.

Without doubt, the genus *Pinanga* has the widest range of differing ectexine types, including two of the three rarest: ring murate and urceolate. Notable occurrences of unusual ectexine features that are shared by different genera include the 'interrupted' reticulate exine (Figs a, p.181; b, p.292) shared by *Pigafetta* (Calamoideae) and *Kerriodoxa* (Coryphoideae), the tectate gemmate pollen of the coryphoid genera *Hyphaene* (Borasseae: Hyphaeninae) and *Borassus* (Borasseae: Lataniinae), and intectate spiny pollen shared by some species of *Caryota*, *Arenga* and *Wallichia*. An intectate clavate ectexine also occurs in other species of *Caryota* and in *Arenga hastata* (Coryphoideae: Caryoteae). Another intectate spiny type occurs in *Socratea* and *Wettinia* (Arecoideae: Iriarteeae), whereas the curious tectate block-like striate clavae (Figs j, p.516; c, p.569) of *Pinanga variegata* pollen (Areceae: Arecinae) also occur in the pollen of *Oncosperma horridum* (Areceae: Oncospermatinae). A similarity is noted between the reticulate ectexine of the pollen of some species of *Areca*, *Nenga* and *Pinanga* (Arecinae) and the pollen ectexine of some species in the re-circumscribed *Hydriastele* (Arecinae: unplaced), most strikingly between *Nenga* (Fig. a, p.513) and some of the species previously included in *Gronophyllum* (Fig. a, p.637).

EXINE ULTRASTRUCTURE

In the majority of palm species, the pollen wall comprises a foot layer, a columellate infratectum and a tectum (e.g., Fig. c, p.145; Harley 1990, 1999b). The ectexine (Fig. 2.7) may have a tectum, which is either tectate or semitectate, and comprises an inner foot layer (infratectum), a middle layer (columellar layer) and an outer layer (tectum) on which there may be supratectate structures (gemmae, clavae, spines or spinulae). Alternatively, the exine may have no tectum (intectate) and simply comprise a foot layer on which the intectate structures (modified infratectal elements) are situated; these include gemmae, clavae and spines but not, in palm pollen, spinulae.

In common with most monocotyledons, no endexine *sensu stricto* has been observed in acetolysed palm pollen. There are a few species of palms, particularly in the Calamoideae (e.g., *Daemonorops sparsiflora*), in which lamellae underlie and are loosely attached to the foot layer; these lamellae are more pronounced in the aperture region. In tribe Lepidocaryeae subtribe Mauritiinae, the lamellae are very pronounced (Figs b, p.159; b, p.166). Distinguishing specific or even generic differences between ultrastructural characteristics in the less-specialised exine types is problematic. Nevertheless, differences in the underlying ultrastructure of superficially similar perforate or rugulate ectexines are useful in distinguishing possible relationships within the palms (Harley 1990, 1999b). For example, significant distinctions have been made between the ultrastructure of *in situ* fossil (Harley 1997) and dispersed fossil palm pollen (Harley & Morley 1995). The distributions of six simple perforate to perforate-rugulate ectexine types were defined by Harley (1996, 1999b) and their systematic distributions plotted through the family. Results showed that although most types were sporadically distributed, one type is almost exclusive to the coryphoid palms while another is more or less restricted to the arecoid palms (Harley 1999b). The most unusual pollen ultrastructure is probably that of *Raphia* (Calamoideae), in which the wall has a very weakly defined infratectum, so dense in some species that the wall is a more or less solid layer (Harley 1999b). The ultrastructure of the diporate pollen of *Daemonorops verticillaris* is very unusual; ultrathin wall sections show a thick tectum surrounding a thinner foot layer, these layers are attached only at the aperture margins, between the layers there are virtually no infratectal structures (Fig. d, p.203).

POLLEN SIZE

Based on dimensions of the pollen grain long axis, the smallest palm pollen (15–19 µm) is that of *Maxburretia* (Coryphoideae) and the overall largest (70–90 µm) is produced by the three genera comprising tribe Phytelepheae (Ceroxyloideae). Within genera or species, size ranges are difficult to define because of overlap. Nevertheless, an overview of pollen size within the palm family, based on maximum size for each genus, shows a clear trend towards larger pollen in the Ceroxyloideae and Arecoideae (Table 2.5).

The difference in the long axis dimension between smallest and largest pollen within a genus ranges from 1 µm in the monotypic *Pritchardiopsis* (Coryphoideae) to 50 µm in the largest palm genus, *Calamus* (Calamoideae). Of the ca. 13 genera with more than 40 species, eight have size differences between species ranging from 33–50 µm: *Calamus* and *Daemonorops* (Calamoideae), *Attalea*, *Astrocaryum*, *Areca*, *Pinanga*, *Dypsis* and *Hydriastele* (Arecoideae). In monotypic genera, the size difference within the species is usually 10 µm or less; exceptionally, the pollen of *Nypa fruticans* has a size range of 37–80 µm.

In Calamoideae, there are six genera, *Raphia*, *Plectocomiopsis*, *Retispatha*, *Calamus*, *Daemonorops* and *Pogonotium*, in which the pollen may be less than 20 µm but is usually not more than 35 µm, although a small proportion of species in *Daemonorops* (up to 55 µm) and *Calamus* (up to 67 µm) have much larger pollen. Other calamoid genera that have pollen grains larger than 50 µm include *Eugeissona* (Eugeissoneae), *Eremospatha*, *Laccosperma* (Lepidocaryeae: Ancistrophyllinae), *Mauritia* and *Mauritiella* (Lepidocaryeae: Mauritiinae), *Korthalsia* (Calameae: Korthalsiinae) and *Metroxylon* (Calameae: Metroxylinae).

Fig. 2.7 ***Salacca wallichiana*** typical tectate wall stratification, non-acetolysed pollen, showing, from inside to out, a pale inner band of finely lamellated intine, a well-defined foot layer, a columellate infratectum, and a tectum. In addition this pollen grain has supratectal spines TEM × 8000. *Furtado* 33024.

TABLE 2.5
Systematic distribution of maximum pollen size ranges.
Measurements are of the greatest dimension for pollen of each genus. The number of genera in each size range is shown for each subfamily. Colour depth shows the frequency of each size range for each subfamily: the darkest colour indicating the most frequent and the lightest colour the most infrequent. There is an overall trend toward larger pollen grains in the Arecoideae.

Size range for largest grains of each species in a genus	CALAMOIDEAE (21 genera)	NYPOIDEAE (1 genus)	CORYPHOIDEAE (46 genera)	CEROXYLOIDEAE (8 genera)	ARECOIDEAE (107 genera)
<20 µm			1		
20–39 µm	10		22	1	19
40–59 µm	4		19	4	60
60–74 µm	6		2		28
75 µm or more	1	1	1	3	5

In Coryphoideae, most pollen falls within a size range of 25–55 µm; exceptions below this range include the pollen of *Maxburretia*, *Guihaia* and *Pritchardiopsis* (15–24 µm), whereas *Lodoicea* (60–66 µm), *Borassodendron* (60–85 µm), and some collections of *Colpothrinax wrightii* (up to 66 µm) have pollen sizes above this range.

In Ceroxyloideae, there is a notable pollen size distinction between tribes Cyclospatheae and Ceroxyleae (24–55 µm) and tribe Phytelepheae (70–90 µm).

In Arecoideae, genera that have small pollen (<36 µm) include *Iriartella*, *Dictyocaryum* and *Iriartea* (Iriarteeae), *Wendlandiella*, *Synechanthus*, *Chamaedorea* and *Gaussia* (Chamaedoreeae), *Podococcus* (Podococceae), *Aiphanes* (Cocoseae: Bactridineae), *Leopoldinia* (Leopoldinieae), *Pelagodoxa* and *Sommieria* (Pelagodoxeae), *Deckenia* (Oncospermatinae), *Phoenicophorium* and *Roscheria* (Verschaffeltiinae). Genera with an overall size range of between 50–85 µm include *Roystonea* (Roystoneeae), *Attalea*, *Jubaeopsis* and *Cocos* (Cocoseae: Attaleinae), *Neonicholsonia* (Euterpeae), *Cyphophoenix* (Areceae: Basseliniinae), *Lemurophoenix* (Areceae: Dypsidinae), *Solfia*, *Wodyetia*, *Drymophloeus* and *Normanbya* (Areceae: Ptychospermatinae), *Rhopalostylis* and *Hedyscepe* (Areceae: Rhopalostylidinae); of these genera the largest pollen (>80 µm) occurs in *Jubaeopis*, *Attalea* and *Normanbya*. In all other arecoid genera, although there may be some species with notably small pollen or large pollen, a greater number of species tend to have pollen within the average 35–50 µm size range.

Fig. 2.8 (*opposite*) Some of the variation in pollen exine surface morphology found in palms. **a**, ***Pinanga rupestris***: semitectate with elongate psilate clavae, interspersed with smaller clavae, SEM × 12,000. *Dransfield* JD5917; **b**, ***Dypsis ambilaensis***: tectate and finely reticulate-crotonoid, SEM × 12,000. *Dransfield* et al. JD6444; **c**, ***Licuala peltata***: semitectate foveolate, SEM × 12,000. *Kerr* 11726; **d**, ***Caryota ophiopellis***: semitectate gemmate-reticulate, SEM × 12,000. *Dowe* 130; **e**, ***Nenga gajah***: semitectate, coarsely reticulate with spinulose muri, SEM × 8,625. *Gundersen et al.* LVG29; **f**, ***Deckenia nobilis***: semitectate with angular striate clavae, separate or coalesced, SEM × 12,000. *Dransfield* JD6236; **g**, ***Rhopaloblaste augusta***: tectate, psilate-perforate, SEM × 12,000. Furtado 30946; **h**, ***Calyptronoma occidentalis***: tectate, coarsely rugulate-verrucate, finely rugulate in depressions between, SEM × 12,000. *Harris* 9842; **i**, ***Pinanga andamanensis***: semitectate reticulate with finely reticulate muri, SEM × 12,000. *Nair* s.n.; **j**, ***Daemonorops oblata***: intectate spinose, SEM × 12,000. *Dransfield et al.* JD5712; **k**, ***Roscheria melanochaetes:*** tectate, rugulate-reticulate with finely ridged muri, SEM × 15,000. *Jeffrey* 728; **l**, ***Pinanga disticha***: intectate gemmate, small and large gemmae interspersed, SEM × 9,000. cultivated RBG Kew 085-85 01527.

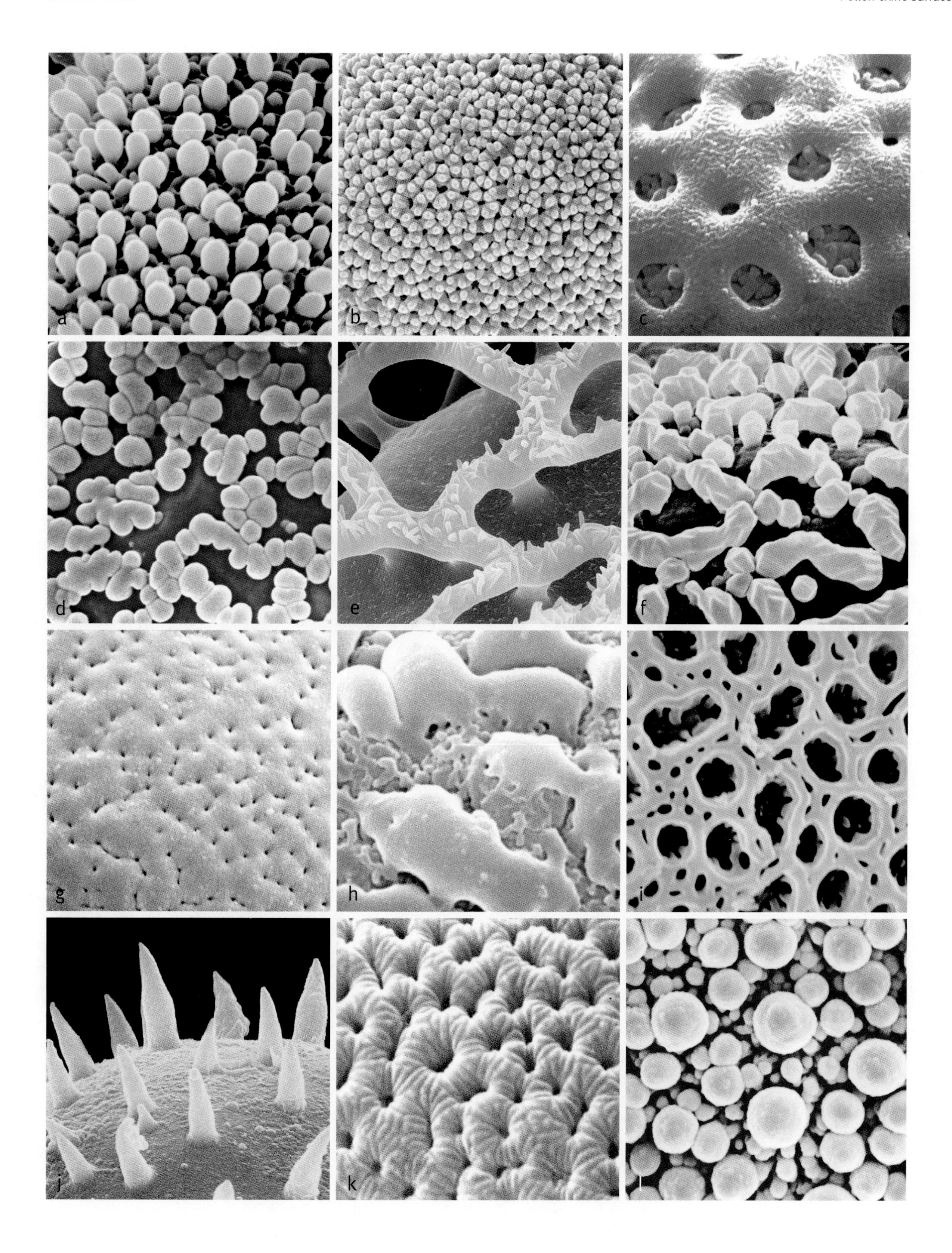
a
b
c
d
e
f
g
h
i
j
k
l

Chapter 3

CHROMOSOMES AND CYTOGENETICS

A substantial number of authors have examined the chromosomes of palms (see references in Table 3.1). Although current knowledge of the comparative cytogenetics of the Arecaceae remains limited, no other large family of tropical woody angiosperms has been studied in greater detail. To date, chromosome numbers have been published for approximately 330 species in 126 genera (Table 3.1). Chromosome morphology and genome size has been studied in only a fraction of these species.

Palms are not attractive subjects to cytogeneticists. Plant material is often scarce and the availability of growing roots from which chromosome preparations can be made is usually limited. Technical problems are caused by tough tissues in the root and by the poor stainability of chromosomes caused by tannins that are present in the vacuoles of meristematic tissues (Röser 2000). Moreover, original material may be incorrectly identified with no means of subsequent verification. Older counts must be treated with caution, principally because early methods of chromosome preparation that were based on microtome-sectioning of dividing cells sometimes yielded inaccurate counts. The modern method of squashing entire cells is less likely to disrupt the integrity of the chromosome complement. Fortunately, the careful work of more recent researchers has yielded a sizeable body of reliable data on the chromosomes of palms (see papers by Essig, Read, Johnson and Röser).

All reliable chromosome counts show clearly that diploid chromosome numbers in palms range from 2n = 26–36. Counts outside this range have been published, but these have not been confirmed by recent studies. Supernumerary chromosomes have also been recorded in *Chelyocarpus*, *Chamaerops*, *Trachycarpus*, *Pritchardia* and *Desmoncus* (Read 1967, Röser 1993, 1994). Chromosome number is usually uniform within genera and sometimes within larger groups. However, there are well-documented cases of chromosome number variation within *Phoenix*, *Chamaedorea*, *Ravenea* and *Dypsis*.

The sequence of basic chromosome numbers that occurs in palms (n = 13–18) is known as a dysploid series. It is thought to arise through chromosome mutations that result in the translocation of all vital parts of one original chromosome to other chromosomes, with the non-essential remainder being subsequently lost. Consequently, the basic gametic number is reduced by one.

CALAMOIDEAE

The chromosomes of the Calamoideae are relatively poorly known, with only half of the genera being studied to date. In the Calameae, 2n = 26 appears to be restricted to the Calaminae (*Calamus*, *Daemonorops* and *Ceratolobus*) and *Metroxylon*, with 2n = 28 in the two remaining members of the tribe in which chromosomes have been counted (*Pigafetta* and *Salacca*). We regard early records of 2n = 28 in *Calamus* as unreliable. Counts of 2n = 32 have been published for subtribe Korthalsiinae, although more recent records are not available. Of the Lepidocaryeae, only *Raphia* (2n = 28) and the genera of the Mauritiinae (2n = 30) have been investigated. No information is available from the Eugeissoneae, Ancistrophyllinae or the Plectocomiinae.

NYPOIDEAE

Numerous authors have recorded 2n = 34 from the Nypoideae.

CORYPHOIDEAE

Extensive sampling within the Coryphoideae indicates that 2n = 36 is by far the most common diploid chromosome count. Lower numbers are found only in some borassoids (*Lodoicea* 2n = 34, *Latania* 2n = 28), in all of the caryotoids (2n = 32, 34), in certain species of *Phoenix* (e.g., *P. pusilla*, 2n = 32) and in both *Johannesteijsmannia* (2n = 34) and *Licuala* (2n = 28).

CEROXYLOIDEAE

Chromosome numbers in the Ceroxyloideae range between 2n = 30 and 2n = 36. Different counts are found in each tribe, with 2n = 36 in all Phytelepheae, 2n = 34 in *Pseudophoenix*, and 2n = 30, 32 and 36 in Ceroxyleae.

ARECOIDEAE

In the Arecoideae, counts span the entire range of palm diploid chromosome numbers, although 2n = 32 appears to be the most common number, occurring in more than half of the species for which published counts are available. Tribe Areceae is almost exclusively recorded as 2n = 32, with a very rare exception in one species of *Dypsis* (2n = 34). In the

Opposite: ***Pinanga javana***, infructescences, cultivated, Indonesia. (Photo: J. Dransfield)

TABLE 3.1 Published chromosome counts and nuclear DNA amounts for palms
Asterisks indicate dubious counts that require confirmation. References given in square brackets.

SUBFAMILY Tribe Subtribe	Genus	Species	Chromosome number (2n)	Nuclear 4C DNA amount (pg)
CALAMOIDEAE				
Lepidocaryeae				
Raphiinae	5. ***Raphia***	*R. africana*	**28** [1]	
		R. farinifera	32* [2]	
		R. hookeri	**28** [1]	
		R. longiflora	**28** [3]	
		R. mambillensis	**28** [1]	
		R. regalis	**28** [1]	
		R. sudanica	**28** [1]	
		R. taedigera	**28** [4]	
		R. vinifera	**28** [1,3]	
Mauritiinae	6. ***Lepidocaryum***	*L. tenue* var. *tenue*	**30** [5,6]	**16.37** [6]
	7. ***Mauritia***	*M. flexuosa*	**30** [6]; 36* [3]	**18.88** [6]
	8. ***Mauritiella***	*M. aculeata*	**30** [6]; 36* [3]	**29.05** [6]
Calameae				
Korthalsiinae	9. ***Korthalsia***	*K. laciniosa*	**32** [3]	
		K. rostrata	**32** [3]	
Salaccinae	11. ***Salacca***	*S. affinis*	**28** [3,7]	
		S. dransfieldiana	**28** [6]	**6.71** [6]
		S. glabrescens	**28** [6]	**8.01** [6]
		S. zalacca	**28** [3,8]	**5.2** [9]
Metroxylinae	12. ***Metroxylon***	*M. sagu*	**26** [10]; **32** [3]	
Pigafettinae	13. ***Pigafetta***	*P. filaris*	**28** [11]	
Calaminae	17. ***Calamus***	*C. arborescens*	**26** [12,13]; 28* [14]	
		C. caesius	**26** [5]	**5.8** [9]
		C. caryotoides	**26** [15]; 28* [14,16]	
		C. ciliaris	**26** [6,12,17]	**10.43** [6]
		C. erectus	**26** [12,17]	
		C. guruba	**26** [12,17]	
		C. khasianus	28* [14]	
		C. leptospadix	**26** [12,17]; 28* [14]	
		C. longisetus	**26** [12]	
		C. muelleri	**26** [15]	
		C. ornatus	**26** [5]	
		C. rotang	**26** [12,13,17]; 28* [14]	
		C. scipionum	28* [18]	
		C. subinermis		**8.6** [9]
		C. viminalis	**26** [6,12,19]	**9.27** [6]
		Calamus sp.	28*	

SUBFAMILY **Tribe** Subtribe	**Genus**	**Species**	**Chromosome number** (2n)	**Nuclear 4C DNA amount** (pg)
Calaminae	19. ***Daemonorops***	*D. angustifolia*		**7.32** [9]
		D. calicarpa	28* [3]	
		D. cristata	**26** [11]	
		D. formicaria	**26** [11]	
		D. grandis	28* [20]	
		D. longipes	28* [3]	
		D. verticillaris	**26** [6]	**11.1** [6]
	20. ***Ceratolobus***	*C. concolor*	**26** [21]	
		C. glaucescens	**26** [21]	
		C. pseudoconcolor	**26** [21]	
NYPOIDEAE	22. ***Nypa***	*N. fruticans*	16* [3]; **34** [4–6, 22]	**4.74** [6]
CORYPHOIDEAE				
Sabaleae	23. ***Sabal***	*S. bermudana*	**36** [2]	
		S. causiarum	**36** [23]	
		S. gretherae	**36** [24]	
		S. maritima	**36** [4]	
		S. mauritiiformis	**36** [6,20,25]	**9.02** [6]
		S. mexicana	**36** [14,24,25]	
		S. minor	**36** [2,14,23,24,26,27]	
		S. palmetto	**36** [2,16,23,28]	
		S. pumos	**36** [3]	
		S. uresana	**36** [20]	
		S. yapa	**36** [25]	
Cryosophileae	24. ***Schippia***	*S. concolor*	**36** [4]	
	25. ***Trithrinax***	*T. brasiliensis*	**36** [2,5,22,29]	
		T. campestris	**36** [5,30]	
	26. ***Zombia***	*Z. antillarum*	**36** [4,5,27]	
	27. ***Coccothrinax***	*C. argentata*	**36** [6,22,27]	**29.75** [6]
		C. argentea	**36** [2,18,28]	
		C. barbadensis	**36** [20]	
		C. crinita subsp. *brevicrinis*	**36** [4,26]	
		C. fragrans		**25.92** [9]
		C. inaguensis	**36** [4]	
		C. littoralis	**36** [5,26]	
		C. miraguama	**36** [4]	
	29. ***Leucothrinax***	*L. morrisii*	**36** [16,22]	
	30. ***Thrinax***	*T. excelsa*	**36** [28,31]	
		T. parviflora	**36** [22]	
		T. parviflora subsp. *parviflora*	**36** [31]	
		T. parviflora subsp. *puberula*	**36** [31]	
		T. radiata	**36** [4,29,31]	

SUBFAMILY **Tribe** Subtribe	**Genus**	**Species**	**Chromosome number** (2n)	**Nuclear 4C DNA amount** (pg)
Cryosophileae	31. ***Chelyocarpus***	*Chelyocarpus* sp.	**36 + satellite** [32]	
	32. ***Cryosophila***	*C. stauracantha*	**c. 36** [5]	
		Cryosophila sp.	**36** [27]	
	33. ***Itaya***	*I. amicorum*	**36** [5]	
Phoeniceae	34. ***Phoenix***	*P. canariensis*	**36** [2,14,26,33]	
		P. dactylifera	**36** [14,26,33–36]	**3.8** [37]
		P. dactylifera cultivars	28*, 34*, **36**, 40* [34]	
		P. loureiroi	**36** [14,38–40]	
		P. paludosa	**36** [14,33,41]	
		P. pusilla	**32** [5]	
		P. reclinata	**36** [4,19,20,33,41]	
		P. roebelenii	28* [18]; **36** [22]	**6.1** [6]
		P. rupicola	**36** [14,42]	**6.01**[6]
		P. sylvestris	**36** [14,33,42]	
		P. theophrasti	**36** [5,6]	**5.27** [6]
		Phoenix sp.	**36** [3]	
Trachycarpeae				
Rhapidinae	35. ***Chamaerops***	*C. humilis*	**36** [2,3,20,26,27]; **36 + 1** [26]; **36 + 2** [26]	
	36. ***Guihaia***	*G. argyrata*	**36** [6,26]	**23.77** [6]
	37. ***Trachycarpus***	*T. fortunei*	**36** [20,26,38,39,43,44]	
		T. martianus	**36** [3,20]	
		T. nanus	**36** [6]	**22.18** [6]
	38. ***Rhapidophyllum***	*R. hystrix*	**36** [22]	
	40. ***Rhapis***	*R. excelsa*	32* [7]; **36** [2,4,14,28]	**19.2** [45]
		R. humilis	**36** [2,14]; **72** [19]	
		R. subtilis	**36** [5]	
Livistoninae	41. ***Livistona***	*L. australis*	**36** [14,18,20]	
		L. chinensis	**36** [2,14,18,26-28,46]	
		L. rotundifolia	**36** [5,14,28]	
		Livistona sp.	**36** [6]	**6.6** [6]
	42. ***Licuala***	*L. grandis*	16* [14,28]	
		L. paludosa	16* [14]	
		L. peltata	**28** [14,28]	
		L. spinosa	**28** [14,46]	
		Licuala sp.	**28** [5,6]	**6.03** [6]
	43. ***Johannesteijsmannia***	*J. altifrons*	32* [3]; **34** [5]	
		Johannesteijsmannia sp.		**6.5** [6]
Unplaced	46. ***Acoelorrhaphe***	*A. wrightii*	**36** [13,22]	
	47. ***Serenoa***	*S. repens*	**36** [27]	

SUBFAMILY Tribe Subtribe	Genus	Species	Chromosome number (2n)	Nuclear 4C DNA amount (pg)
Unplaced	48. ***Brahea***	*B. aculeata*	**36** [22]	
		B. armata	**36** [3]	
		B. calcarea	**36** [26]	
		B. dulcis		**4.22** [9]
		B. edulis	**36** [26]	
	50. ***Copernicia***	*C. alba*	**36** [20,47]	
		C. baileyana	**36** [48]	
		C. ×*burretiana*	**36** [48]	
		C. cowellii	**36** [48]	
		C. glabrescens	**36** [48]	
		C. hospita	**36** [48]	
		C. macroglossa	**36** [26,48]	
		C. prunifera	**36** [3,16,47]	
		C. rigida	**36** [48]	
		C. ×*vespertilionum*	**36** [48]	
		C. yarey	**36** [27,48]	
	51. ***Pritchardia***	*P. pacifica*	**36** [14,28]	
		P. thurstonii	**36** [22,26,27]; **36 + 2** [26]	
	52. ***Washingtonia***	*W. filifera*	24* [3]; **36** [4,6,14,20,28]	**6.21** [6]
		W. robusta	**36** [26,27]	
Chuniophoeniceae	53. ***Chuniophoenix***	*C. hainanensis*	10* [38,39]	
		C. nana	**36** [6]	**6.15** [6]
	55. ***Nannorrhops***	*N. ritchiana*	**36** [3,27]	
Caryoteae	57. ***Caryota***	*C. mitis*	32* [46]	
		C. ochlandra	30* [38,49]	
		C. rumphiana	**34** [5]	
		C. urens	32* [2,14]; **34** [5]	**26.44** [6]
		Caryota sp.	**34** [4]	
	58. ***Arenga***	*A. caudata*	**64** [4]	
		A. engleri	**32** [2,50,51]	
		A. obtusifolia	**32** [14,41,42,50]	
		A. pinnata	26* [16]; **32** [2,4,14,28,50]	
		A. aff. *porphyrocarpa*	**32** [4]	
		A. tremula	**32** [4]	
		A. undulatifolia	**32** [19,50]	
		A. westerhoutii	**32** [5]	
		A. wightii	**32** [4]	
	59. ***Wallichia***	*W. densiflora*	**32** [15,19,46]	
		W. disticha	**32** [3,52]	
Corypheae	60. ***Corypha***	*C. umbraculifera*	**36** [14,53]	
		C. utan	**36** [14,20]	

SUBFAMILY **Tribe** Subtribe	**Genus**	**Species**	**Chromosome number** (2n)	**Nuclear 4C DNA amount** (pg)
Borasseae				
Hyphaeninae	61. ***Bismarckia***	*B. nobilis*	**36** [22,26]	**8.11** [6]
	63. ***Hyphaene***	*H. coriacea*	**36** [22]	
		H. dichotoma	**36** [46,54]	
		H. guineensis	**36** [20]	
		H. petersiana		**13.6** [9]
		H. thebaica	**36** [3]	
		Hyphaene sp.	**36** [2]	
	64. ***Medemia***	*M. argun*		**14.52** [9]
Lataniinae	65. ***Latania***	*L. loddigesii*	**28** [4]; 36* [3]	
		L. lontaroides	**28** [4,6]; 32* [14,28]	**14** [6]
		L. verschaffeltii	**28** [4]; 32* [7]	
	66. ***Lodoicea***	*L. maldivica*	**34** [4]; **36** [3]	
	67. ***Borassus***	*B. flabellifer*	**36** [4,6,7,14,28,42,55,56]	**34.39** [6]
CEROXYLOIDEAE				
Cyclospatheae	69. ***Pseudophoenix***	*P. sargentii*	**34** [4]	**11.32** [6]
		P. vinifera	**34** [4,5,57]	
Ceroxyleae	70. ***Ceroxylon***	*C. alpinum*	**36** [3]	
		C. parvifrons	**36** [3]	
		Ceroxylon sp.		**15.42** [6]
	73. ***Ravenea***	*R. glauca*	**32** [6]	**9.22** [6]
		R. hildebrandtii	**30** [6]	**8.69** [6]
		R. madagascariensis	**30** [6]	**9.43** [6]
		R. musicalis	**32** [6]	**12.12** [6]
Phytelepheae	75. ***Aphandra***	*A. natalia*	**36** [5,58]	
	76. ***Phytelephas***	*P. aequatorialis*	**36** [6,58]	**3.89** [6]
		P. macrocarpa	24* [3]; **36** [6,20]	**3.99** [6]; **4.22** [6]
		P. seemannii	**36** [58]	
		Phytelephas sp.	32* [4]	
ARECOIDEAE				
Iriarteeae	79. ***Iriartea***	*I. deltoidea*	**32** [3]	**49.12** [6]
	80. ***Socratea***	*S. exorrhiza*	**36** [6]	**18.22** [6]
Chamaedoreeae	82. ***Hyophorbe***	*H. amaricaulis*	36* [3]	
		H. indica	36* [3]	
		H. lagenicaulis	**32** [4]	
		H. verschaffeltii	**32** [2,4]; 36* [3]	
	83. ***Wendlandiella***	*W. gracilis*	**28** [5,6]	**11.44** [6]

SUBFAMILY Tribe Subtribe	Genus	Species	Chromosome number (2n)	Nuclear 4C DNA amount (pg)
Chamaedoreeae	84. ***Synechanthus***	*S. fibrosus*	**32** [4]	
	85. ***Chamaedorea***	*C. alternans*	**32** [4]	
		C. aff. *arenbergiana*	**32** [4]	
		C. brachypoda	**26** [59]	
		C. cataractarum	**26** [20]; **32** [4]	
		C. elatior	**26** [60]	
		C. elegans	**26** [4,5,16,29]	
		C. ernesti-augusti	24* [20]; **26** [4]	
		C. glaucifolia	**26** [5,60]; 32* [3]	
		C. klotzschiana	**26** [6]	**16.34** [6]
		C. liebmannii	**32** [3]	
		C. microspadix	**26** [4]	
		C. oblongata	**26** [4,20,61]	**16.81** [6]
		C. oreophila	**26** [4]	
		C. parvisecta	**26** [59]	
		C. pinnatifrons	**26** [6]	**17.81** [6]
		C. pochutlensis	**26** [14,59]	
		C. pumila	**26** [18]	
		C. radicalis	**26** [4]	
		C. sartorii	**26** [4]	
		C. schiedeana	**26** [4,5]	
		C. seifrizii	**26** [4]	
		C. tenella	**26** [6,59]	**18.94** [6]
		C. tepejilote	**32** [20]	
		C. aff. *tepejilote*	**32** [4]	
		C. cf. *woodsoniana*		**16.25** [6]
		Chamaedorea sp.	**26** [4,6]	**20.22** [6]
	86. ***Gaussia***	*G. attenuata*	**28** [4]	
		G. maya	**28** [4]	
Oranieae	88. ***Orania***	*O. palindan*	**32** [20]	
		O. sylvicola	**32** [20]	
Roystoneeae	90. ***Roystonea***	*R. altissima*	**36** [4]	
		R. oleracea	**36** [14,62]	
		R. princeps	**36** [4]	
		R. regia	28* [39]; 30* [38,39]; **36** [4,5,14,20,28,63]; 38* [2,38]	
		Roystonea sp.	**36** [4,6]	**19.17** [6]
Cocoseae				
Attaleinae	92. ***Beccariophoenix***	*B. madagascariensis*	**36** [6]	**7.21** [6]
	93. ***Jubaeopsis***	*J. caffra*	**160–200** [64]	
	94. ***Voanioala***	*V. gerardii*	**>550** [65]; **>596** [66]; **606 ± 3** [5]	**156.4** [6]

SUBFAMILY **Tribe** Subtribe	**Genus**	**Species**	**Chromosome number** (2n)	**Nuclear 4C DNA amount** (pg)
Attaleinae	95. ***Allagoptera***	*A. arenaria*	**32** [4]	
		A. caudescens	**32** [4,5]	
	96. ***Attalea***	*A. allenii*	**32** [32]	
		A. cohune	**32** [2,14]	
		A. aff. *cohune*	**32** [4]	
		A. eichleri	**32** [67]	
		A. speciosa	**32** [67,68]	
		A. spectabilis	**32** [68]	
		Attalea sp.	**32** [4]	
	97. ***Butia***	*B. capitata*	**32** [2,4,16]	
		B. paraguayensis	16* [69]	
		B. stolonifera	**32** [3]	
	98. ***Cocos***	*C. nucifera*	**32** [2,4,6,14,16,68,70]	**14.19** [6]
	99. ***Jubaea***	*J. chilensis*	**32** [2–6]	**10.2** [6]
	100. ***Lytocaryum***	*L. weddellianum*	**32** [18]	
	101. ***Syagrus***	*S. amara*	**32** [4]	
		S. comosa	**32** [4,8]	
		S. coronata	**32** [4]	
		S. glaucescens	**32** [6]	**13.76** [6]
		S. orinocensis	**32** [5]	
		S. romanzoffiana	**32** [2–6,8,14,16]	**12.2** [6]
		S. schizophylla	30* [4]; **32** [14,28]	
		Syagrus sp.	30* [4]	
Bactridinae	103. ***Acrocomia***	*A. aculeata*	**30** [4,6]	**13.54** [6]
		A. crispa	**30** [4]	
	104. ***Astrocaryum***	*A. mexicanum*	**30** [4]	
	105. ***Aiphanes***	*A. aculeata*	32* [2,14,28]	
		A. minima	**30** [4]; 32* [14]; 36* [68]	
	106. ***Bactris***	*B. gasipaes*	28* [3]; **30** [4,5]	
		B. hondurensis	**30** [6]	**16.27** [6]
		B. major	28* [3]	
		Bactris sp.	**30** [4]	
		Bactris sp.		**13.22** [6]
	107. ***Desmoncus***	*D. polyacanthos*	**30** [6]	**11.99** [6]
		Desmoncus sp.	**30** [5]	
		Desmoncus sp.	**30 + 1B** [5,6]	**11.71** [6]
Elaeidinae	109. ***Elaeis***	*E. guineensis*	**32** [2,14,16,28,71]	**4** [72]
		E. oleifera	**32** [4,5]	

SUBFAMILY **Tribe** Subtribe	**Genus**	**Species**	**Chromosome number** (2n)	**Nuclear 4C DNA amount** (pg)
Euterpeae	112. ***Euterpe***	*E. oleracea*	**36** [32]	
		E. precatoria	**36** [6]	**21.24** [6]
		E. precatoria var. *longivaginata*	**36** [5]	
	113. ***Prestoea***	*P. decurrens*	**36** [32]	
		P. longepetiolata	**36** [32]	
	114. ***Neonicholsonia***	*N. watsonii*	**36** [4]	
	115. ***Oenocarpus***	*O. bataua*	**36** [6]	**15.7** [6]
Geonomateae	118. ***Calyptrogyne***	*C. ghiesbreghtiana*	**28** [6]	**13.67** [6]
	119. ***Calyptronoma***	*C. occidentalis*	**28** [4]	
		C. plumeriana	**28** [5]	
		C. rivalis	**28** [11]	
	121. ***Geonoma***	*G. camana*	**28** [4]	
		G. cuneata	32* [3]	
		G. interrupta	**28** [6]; 32* [3]	**14.5** [6]
		Geonoma sp.	**28** [6]	**14.66** [6]
		Geonoma sp.		**19.94** [6]
		Geonoma sp.		**14.6** [6]
Pelagodoxeae	123. ***Pelagodoxa***	*P. henryana*	**32** [32]	
	124. ***Sommieria***	*S. leucophylla*	**34** [6,11]	**23.17** [6]
Areceae				
Archontophoenicinae	126. ***Archontophoenix***	*A. alexandrae*	28* [20]; **32** [3,4,15]; 36* [3]	
		A. cunninghamiana	28* [20]; **32** [5,6]	**18.97** [6]
	128. ***Chambeyronia***	*C. macrocarpa*	**32** [11]	
	129. ***Kentiopsis***	*K. oliviformis*	**32** [5]	
Arecinae	130. ***Areca***	*A. catechu*	**32** [2,4,14,73]	
		A. concinna	**32** [6]	**24.04** [6]
		A. macrocalyx	**32** [74]	
		A. minuta	**32** [6]	**24.35** [6]
		A. triandra	**32** [5,14,16,50,73,75]	
	132. ***Pinanga***	*P. celebica*	**32** [6]	**26.81** [6]
		P. coronata	28* [18,20]; **32** [3,4,6]	**35.42** [6]
		P. patula	**32** [3]	
		P. subintegra	**32** [6]	**55.62** [6]
Basseliniinae	136. ***Cyphosperma***	*C. trichospadix*	**32** [15]	
	138. ***Physokentia***	*P. dennisii*	**c. 32** [5]	

SUBFAMILY **Tribe** Subtribe	**Genus**	**Species**	**Chromosome number** (2n)	**Nuclear 4C DNA amount** (pg)
Dypsidinae	144. ***Dypsis***	*D. baronii*	**ca. 34** [5]	
		D. beentjei	**32** [6]	**11.73** [6]
		D. cabadae	**32** [6]	**10.03** [6]
		D. catatiana	**32** [6]	**10.86** [6]
		D. decaryi	**32** [4]	
		D. louvelii	**32** [4–6,14,28]	**11.57** [6]
		D. lutescens	28* [20, 68]	**6.12** [6]
		D. madagascariensis	**32** [4,41]; 38* [14]	
		D. mananjarensis		**9.89** [6]
		D. mcdonaldiana	**32** [6]	**10.56** [6]
		D. pilulifera	**32** [6]	**10.15** [6]
		D. pinnatifrons	**32** [6]	**10.94** [6]
		D. scottiana	**32** [6]	**9.88** [6]; **10.29** [6]
		D. utilis	**32** [5]	
		Dypsis sp.	**34** [5]	
		Dypsis sp.		**10.19** [6]
		Dypsis sp.		**10.78** [6]
	146. ***Marojejya***	*M. darianii*	**32** [5]	
	147. ***Masoala***	*M. madagascariensis*	**32** [6]	**10.71** [6]
Linospadicinae	148. ***Calyptrocalyx***	*C. forbesii*	**32** [4]	
		C. spicatus	**32** [3]	
	149. ***Linospadix***	*L. minor*	**32** [5]	
	150. ***Howea***	*H. belmoreana*	**32** [76]	
		H. forsteriana	**32** [14,28,76]	**13.7** [77]
	151. ***Laccospadix***	*L. australasicus*	**32** [5,15]	
Oncospermatinae	152. ***Oncosperma***	*O. tigillarium*	**32** [14,28]; **32** (+ 2–4) [5]	
	154. ***Acanthophoenix***	*A. rubra*	36* [3]	
		Acanthophoenix sp.	**32** [16,78]	
Ptychospermatinae	156. ***Ptychosperma***	*P. elegans*	**32** [15,20]	
		P. macarthurii	**32** [2,4,14]; 36* [3]	
		P. sanderianum	**32** [14,20,28]	
	158. ***Adonidia***	*A. merrillii*	**32** [4]	
	159. ***Solfia***	*S. samoensis*	**32** [78]	
	161. ***Veitchia***	*V. arecina*	**32** [4]	
		V. filifera	**32** [15]	
		V. vitiensis	**32** [15]	
	162. ***Carpentaria***	*C. acuminata*	**32** [15]	
	163. ***Wodyetia***	*W. bifurcata*	**32** [5,11]	
	164. ***Drymophloeus***	*D. litigiosus*	**32** [4,75]	

SUBFAMILY Tribe Subtribe	Genus	Species	Chromosome number (2n)	Nuclear 4C DNA amount (pg)
Ptychospermatinae	165. ***Normanbya***	*N. normanbyi*		**19.42** [6]
	166. ***Brassiophoenix***	*B. schumannii*	**32** [11]	
		Brassiophoenix sp.	**32** [4]	
	167. ***Ptychococcus***	*P. lepidotus*	**32** [15]	
Rhopalostylidinae	168. ***Rhopalostylis***	*R. baueri*	**32** [20]	
Verschaffeltiinae	170. ***Nephrosperma***	*N. van-houtteanum*	**32** [4,14,28]	
	171. ***Phoenicophorium***	*P. borsigianum*	**32** [3,4,75]	
	172. ***Roscheria***	*R. melanochaetes*	**32** [3]	
	173. ***Verschaffeltia***	*V. splendida*	36* [3]	
Unplaced	174. ***Bentinckia***	*B. condapanna*	**32** [3]	**11.11** [6]
		B. nicobarica	**32** [4]	
	175. ***Clinostigma***	*C. exorrhizum*	**32** [5]	
		C. savoryanum	18* [79]; **32** [2]	
	176. ***Cyrtostachys***	*C. renda*	**32** [3,75]	
	177. ***Dictyosperma***	*D. album*	**32** [4,14,28]	
		D. album var. *aureum*	**32** [4]	
	179. ***Heterospathe***	*H. elata*	**32** [4,14,28]	
		H. humilis	**32** [4,15]	
		H. woodfordiana	**32** [5]	
	180. ***Hydriastele***	*H. beguinii*	**32** [4]	
		H. costata	**32** [15]	
		H. hombronii	**32** [15]	
		H. macrospadix	**32** [5]	
		Hydriastele sp.	**32** [4]	
		Hydriastele sp.	**ca. 32** [5]	
	181. ***Iguanura***	*I. wallichiana*	**32** [11]	
		Iguanura sp.	**32** [5]	
	182. ***Loxococcus***	*L. rupicola*	**32** [3]	**14.3** [6]
	183. ***Rhopaloblaste***	*R. ceramica*	**32** [3,15]	

References: [1]Okolo 1988, [2]Satô 1946, [3]Sarkar 1970, [4]Read 1966, [5]Röser 1994, [6]Röser *et al.* 1997, [7]Bosch 1947, [8]Sarkar 1958, [9]Bennett & Leitch 2005, [10]Rauwerdink 1985, [11]Johnson 1985, [12]Sarkar & Datta 1985, [13]Sarkar *et al.* 1976, [14]Sharma & Sarkar 1956 [15]Read 1965b, [16]Darlington & Janaki Ammal 1945, [17]Sarkar 1986, [18]Eichorn 1953, [19]Sarkar *et al.* 1978a, [20]Eichorn 1957, [21]Johnson 1979, [22]Read 1964, [23]Bowden 1945, [24]Palomino & Quero 1992, [25]Romo *et al.* 1988, [26]Röser 1993, [27]Read 1965a, [28]Venkatasubban 1945, [29]Read 1963, [30]Di Fulvio 1966, [31]Read 1975, [32]Read 1967, [33]Beal 1937, [34]Al-Mayah 1986, [35]Murín & Chaudhri 1970, [36]Soliman & Al-Mayah 1978, [37]Olszewska & Osiecka 1982, [38]Huang *et al.* 1989, [39]Huang *et al.* 1985, [40]Hsu 1967, [41]Sarkar *et al.* 1977, [42]Sarkar 1957, [43]Huang *et al.* 1986b, [44]Ge & Li 1989, [45]Bharathan *et al.* 1994, [46]Sarkar *et al.* 1978b, [47]Dahlgren & Glassman 1961, [48]Dahlgren & Glassman 1963, [49]Huang *et al.* 1986a, [50]Sarkar 1987, [51]Hsu 1972, [52]Sawant 1975, [53]Darlington & Wylie 1955, [54]Mahabalé & Chennaveeraiah 1958, [55]Stephen 1974, [56]Rangasamy & Devasahayam 1972, [57]Read 1968, [58]Barfod 1991, [59]Essig 1970, [60]Suessenguth 1921, [61]Söderberg 1919, [62]Simmonds 1954, [63]Shibata 1962, [64]Robertson 1976a, [65]Johnson 1989, [66]Johnson *et al.* 1989, [67]De Carvalho & Bandel 1986, [68]Gassner 1941, [69]Cocucci 1964, [70]Ninan & Raveendranath 1975, [71]Guerra 1986, [72]Jones *et al.* 1982, [73]Bavappa & Raman 1965, [74]Nair & Ratnambal 1978, [75]Al-Rawi 1945, [76]Savolainen *et al.* 2006, [77]Zonneveld *et al.* 2005, [78]Essig 1971b, [79]Ono & Masuda 1981.

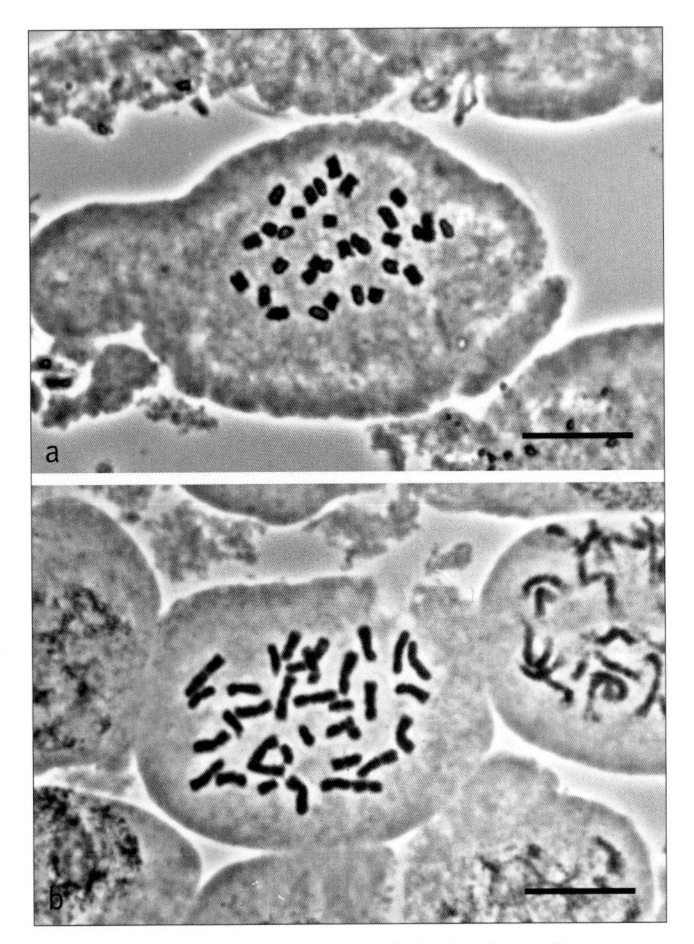

Fig. 3.1 Examples of palm chromosomes, mitotic metaphase cells. a, ***Ravenea robustior*** (2n = 30). **b, *Sommieria leucophylla*** (2n = 34). Scale bar = 10 µm. (Photos: L. Hanson)

Pelagodoxeae, *Pelagodoxa* shares 2n = 32 with the Areceae, whereas 2n = 34 has been confirmed in *Sommieria* (Fig. 3.1), a noteworthy finding in the context of the striking phylogenetic relationship of the Pelagodoxeae and the Areceae (Lewis & Doyle 2002). Some other major groups in the Arecoideae are as consistent as the Areceae; for example, Euterpeae (2n = 36) and Geonomateae (2n = 28). Other groups display considerable diversity in chromosome counts. For example, within the Cocoseae, we find 2n = 30 in the Bactridinae and 2n = 32 in the majority of Attaleinae and Elaeidinae. Notable exceptions exist in the non-Neotropical genera of Attaleinae, namely *Beccariophoenix* (2n = 36) and the two striking polyploids, *Jubaeopsis* and *Voanioala* (Fig. 3.2). In the Chamaedoreeae, 2n = 28 is found in *Gaussia* and *Wendlandiella*, and in *Hyophorbe* and *Synechanthus* 2n = 32, whereas *Chamaedorea* displays both 2n = 26 and 2n = 32. Only a few counts are available in other arecoid tribes: 2n = 32 and 36 in Iriarteeae, the former representing an early and potentially dubious record; an early record of 2n = 32 for Oranieae; and numerous publications of 2n = 36 for Roystoneeae. Counts for tribes Podococceae, Sclerospermeae, Reinhardtieae, Manicarieae and Leopoldinieae have not yet been published.

KARYOTYPE EVOLUTION

A comparison of the karyotypes of the closely related genera *Johannesteijsmannia*, *Licuala* and *Livistona* has shed some light on mechanisms of change in the chromosome complement in palms (Röser 1995). Although *Licuala* carries 22% fewer chromosomes than *Livistona*, its genome size is only 8% smaller. Similarly, *Johannesteijsmannia* possesses 5% fewer chromosomes, but only 2% less DNA than *Livistona*. This suggests that aneuploid change, the gain or loss of one or more individual chromosomes in addition to the basic set, has not played a role and that dysploidy may indeed be the mechanism that underlies karyotype evolution in this case. Röser (1995) hypothesised that repeated reciprocal translocations of unequal chromosome segments in the ancestors of *Johannesteijsmannia* and *Licuala* would have resulted in the unusual chromosome numbers in the Livistoninae, and would also explain contrasts between the structure of the chromosomes in these genera and those in *Livistona*.

In the light of general concepts of mechanisms of dysploidy and the detailed information now available in palm cytogenetics, it has been hypothesised that 2n = 36 is the primitive condition in palms and that descending dysploid series have occurred in parallel in each of the subfamilies with no evidence of reversal (Moore & Uhl 1982, Röser 1994, 1995, 1999). This hypothesis has been influenced by the traditional notion that the coryphoids, in which 2n = 36 predominates, carry more primitive character states than other palms (Uhl & Dransfield 1987). Although this hypothesis is certainly plausible, it has not yet been explored within the phylogenetic framework that is now available. Indeed, the position of the Coryphoideae relative to the Calamoideae and Nypoideae (in which 2n = 36 has not yet been recorded) suggests that the highest diploid number may not necessarily represent the plesiomorphic condition. Careful character analyses and further chromosome counts, especially in the Calamoideae, are required to explore the evolution of the palm chromosome complement in greater detail.

POLYPLOIDY

Polyploidy is rare in palms; only four reliable cases have been published. Two of these, *Arenga caudata* (Read 1966) and *Rhapis humilis* (Sarkar *et al.* 1978a), are tetraploids in otherwise diploid genera. The remaining two are spectacular polyploids in monotypic genera in tribe Cocoseae, namely *Jubaeopsis* from South Africa (2n = 160–200) and *Voanioala* from Madagascar (2n = ca. 600), the latter yielding the highest chromosome number in the monocotyledons (Robertson 1976a, Johnson 1989, Johnson *et al.* 1989, Röser 1994). Though not a polyploid, *Beccariophoenix*, another monotypic genus from Madagascar, is also karyologically distinct within the Cocoseae, having a diploid count of 2n = 36 as opposed to 2n = 30 or 32. It is intriguing that unusual chromosome numbers in the Cocoseae are concentrated in

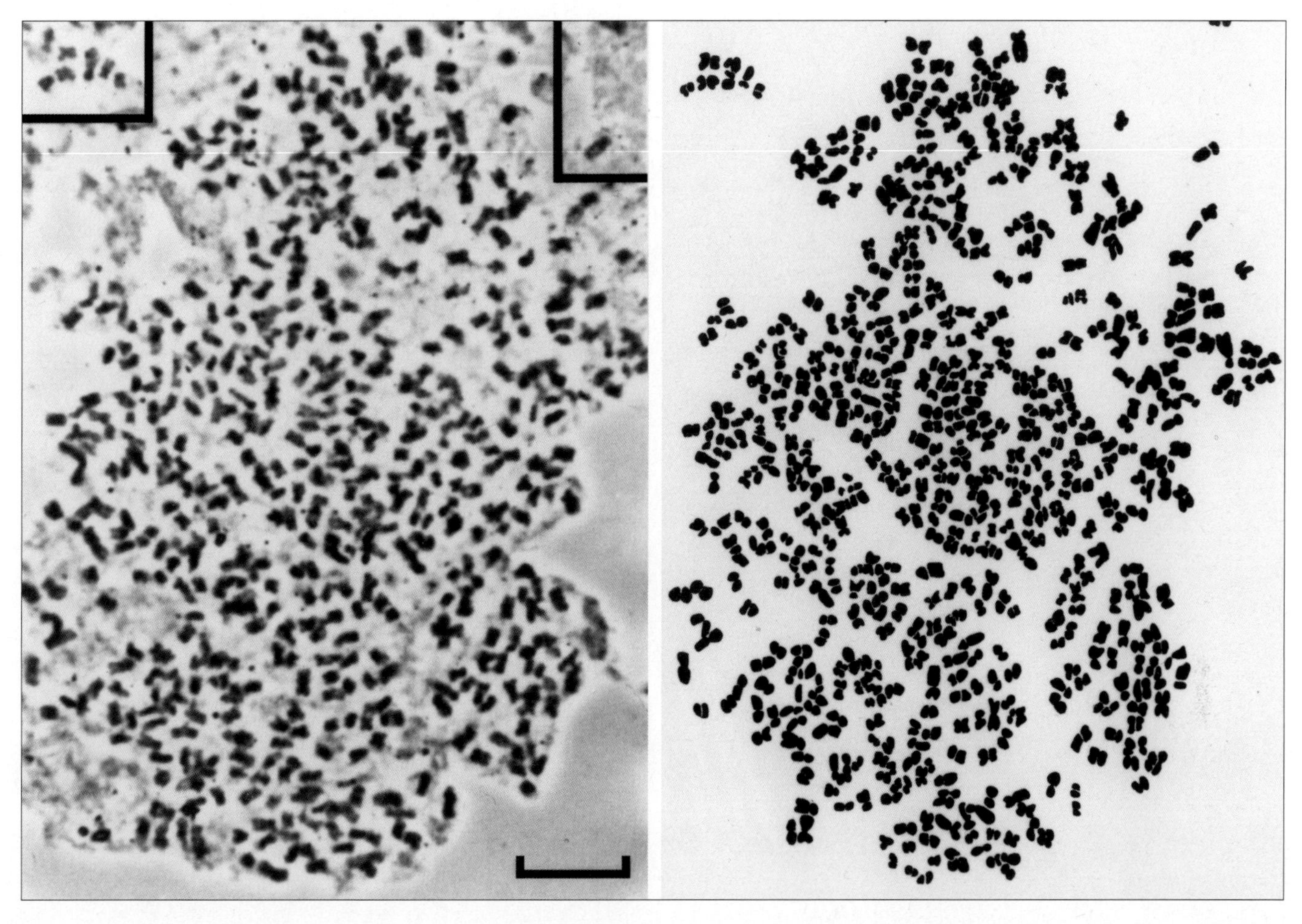

Fig. 3.2 Photograph and drawing of a mitotic metaphase cell of *Voanioala gerardii* showing 2n = 596 chromosomes, the highest known chromosome number in the monocotyledons (Johnson *et al.* 1989). Scale bar = 10 µm. (Reproduced with permission).

the small number of genera that occur outside the tribe's principal distribution in the Neotropics.

DNA C-VALUES

Nuclear DNA amounts have been quantified in almost 100 palm species (Table 3.1; Bennett & Leitch 2005). The amount of DNA in an unreplicated gametic nucleus of an organism is referred to as its C-value, irrespective of the ploidy level of the taxon. C-value research has revealed that, in contrast to the relatively modest variation in chromosome number, there is remarkable diversity in DNA C-values. Up to 14.6-fold variation has been found in diploid species ranging from 1C = 0.95 pg in *Phoenix dactylifera* to 1C = 13.90 pg in *Pinanga subintegra*. Striking variation can even be found within genera. For example, the genomes of two species of *Pinanga*, *P. subintegra* and *P. coronata*, are substantially different having DNA C-values of 1C = 6.70 pg and 13.90 pg, respectively. The former contains almost 60% more DNA than the latter, and yet both have 2n = 32 chromosomes and very similar chromosome morphologies. The marked difference may be due to a largely proportionate amplification of dispersed repetitive DNA sequences in each chromosome and chromosome arm (Castilho *et al.* 2000), although it is not certain if this is the case in *Pinanga*. The phenomenon demands closer attention across the family (Röser 1999).

CHROMOSOME BANDING

The structure of chromosomes themselves has been explored by chromosome-banding methods (Röser 1993, 1994, 1995, 1999, 2000). In summary, the structure of interphase nuclei and the longitudinal differentiation of prophase and metaphase chromosomes is highly differentiated in the palm family; a more detailed review is beyond the scope of this book. Röser has repeatedly drawn links between putatively derived karyological traits, in terms of chromosome structure, number or genome size, and large, species-rich genera such as *Bactris*, *Chamaedorea*, *Geonoma*, *Licuala* and *Pinanga*. He also notes that these genera typically represent radiations in the forest understorey. These correlations have not been specifically tested, but it is reasonable to suspect that large radiations and potentially increased rates of lineage diversification might be accompanied or even caused by genomic innovation.

Chapter 4
CHEMISTRY

Although palms possess a wide range of structural defences against herbivory, they are not known to be rich in chemical defences. Several economically important chemical products are derived from palms, but these are generally restricted to single species or small groups of species. Emerging data from the few existing family-wide surveys of palm chemistry suggest that chemical data may be useful for refining our phylogenetic hypotheses at low taxonomic levels. More detailed studies reveal the diversity of chemical constituents that can occur within single species or small groups of taxa.

FAMILY-WIDE SURVEYS

Li and Willaman (1968) detected alkaloids in six of the 69 palm species they tested, suggesting that alkaloids are present but rare in the family. The utility of alkaloids as taxonomic characters in palms remains to be tested.

Williams *et al.* (1973) found flavonoids to be common in palm leaves, present in 114 of the 125 species tested. The distribution of 18 different flavonoid compounds among palm taxa provided insight into relationships within the family. For example, similarities in flavonoid composition suggested a relationship between tribes Caryoteae, Borasseae and Corypheae (subfamily Coryphoideae). The distinctness of subfamilies Nypoideae and Arecoideae and tribe Phytelephanteae was confirmed by the lack of similarity among the flavonoid profiles of these groups. *Hyophorbe* (as *Mascarena*) and *Gaussia* (as *Opsiandra*), two genera of the arecoid tribe Chamaedoreeae, were found to share a similar form of the flavonoid compound luteolin. Williams *et al.* (1973) detected a wide range of variation in flavonoid composition within subfamilies Coryphoideae and Arecoideae, suggesting that a more intensive study of these groups may yield additional taxonomic information.

A more recent synthesis of flavonoid data (Harborne & Williams 1991) revealed striking incongruence with recent classifications of the palm family (Uhl & Dransfield 1987; this volume). This included a lack of similarity between *Chamaedorea* and the other genera of the Chamaedoreeae, a lack of support for placing *Phoenix* in subfamily Coryphoideae, and a separation between the Cocoseae and the other tribes of subfamily Arecoideae (Harborne & Williams 1991).

Lewis and Zona (2000) surveyed the occurrence of cyanide-producing compounds in 155 palm species, representing all subfamilies and tribes. They found that cyanogenic glycosides are very rare in the family and probably have limited use for systematics. However, two closely related species of *Drymophloeus* (*D. subdistichus* and *D. pachycladus*) were found to have high levels of cyanogenic glycosides in their leaves.

In a family-wide chromatographic survey, coumarins were detected in 27 of 133 palm species sampled (Zona *et al.* unpublished). Coumarins were found in all five subfamilies, but were most common in tribes Borasseae, Caryoteae, Chuniophoeniceae and Corypheae of subfamily Coryphoideae. Further characterization by high pressure liquid chromatography (HPLC) is necessary in order to determine the identity and distribution of coumarins in palms.

A survey of 14 palm species found that the chemical constituents of floral scent were correlated with pollination mode (Knudsen *et al.* 2001).

STUDIES OF SINGLE SPECIES OR SMALL GROUPS OF TAXA

Calamoideae: Several species of *Daemonorops* exude 'dragon's blood', a flavonoid-rich resin, from developing fruit. A diverse assemblage of its constituent flavonoids has been characterised (Arnone *et al.* 1997). Ohtsuki *et al.* (2006) isolated and characterised the biological activity of steroid saponins from *Calamus insignis*.

Nypoideae: The chemistry of *Nypa fruticans* has been examined as part of larger surveys of taxonomically diverse mangrove plants (Bandaranayake 2002, Azuma *et al.* 2002). Among eight common mangrove taxa, *Nypa* was found to have the greatest complexity in floral scent chemistry, with 25 individual compounds detected (Azuma *et al.* 2002).

Coryphoideae: A study of flavonoid and phenolic acid composition in the genus *Coccothrinax* suggests that their patterns of variation may be useful for delimiting species and assessing relationships (Kowalska *et al.* 1991). Several flavonoids and steroid saponins have been isolated from six species of *Phoenix* (Asami *et al.* 1991, Idaka *et al.* 1991, Hong *et al.* 2006), and others have been characterised in *Chamaerops* (Hirai *et al.* 1986), *Licuala* (Asami *et al.* 1991), *Rhapis* (Hirai *et al.* 1984b), *Sabal* (Idaka *et al.* 1988) and *Trachycarpus* (Hirai *et al.* 1984a, 1986).

Ceroxyloideae: Ervik *et al.* (1999) characterised the patterns of floral scent chemistry in the Phytelepheae and found that scent chemistry was useful for differentiating sympatric species. It was correlated with pollinator specificity, and appeared to be a mechanism of reproductive isolation among closely related taxa. Further studies of floral scent chemistry in palms may yield new information on species delimitation and pollination ecology. Mannans are insoluble polysaccharide polymers that are often found in the endosperm of palms. Their structures have been characterised in *Phytelephas macrocarpa*, the ivory nut palm (Aspinall *et al.* 1953, Timell 1957, Aspinall *et al.* 1958).

Arecoideae: Knudsen (1999a, 1999b) documented variation in floral scent chemistry within tribe Geonomateae, including that in infraspecific taxa of *Geonoma* that are difficult to differentiate on the basis of morphology. Major and minor alkaloid constituents of *Areca catechu* and their properties as stimulants have been characterised by Holdsworth *et al.* (1998). Results of flavonoid surveys in the arecoid tribe Cocoseae (Glassman *et al.* 1981, Williams *et al.* 1983) may be useful for refining the emerging phylogenies of the tribe (e.g., Gunn 2004).

Opposite: ***Chamaerops humilis***, habit, Mallorca. (Photo: J. Dransfield)

Chapter 5
THE FOSSIL RECORD OF PALMS

INTRODUCTION

The palm fossil literature

Published records of fossil palms make up a remarkably large body of literature, starting in the early 19th Century with mainly European records and increasing as geological surveys got underway in the late 19th Century, most notably in the USA. Descriptions of palm fossils and their suggested affinities within the family are found in a wide range of publications, including not just geological survey papers but also journals for earth sciences, palaeoclimate and environment, palaeobotany and botany. Some of the older, particularly 19th Century records, although referred to by later authors, are difficult to access for re-examination; in this circumstance, the article referred to is noted as 'not seen' in 'Literature cited'.

Previous review articles specific to palm fossils include those by Stenzel (1904; palm stem wood), Tralau (1964; *Nypa*), Read and Hickey (1972; leaves), Rao and Achuthan (1973; general review of fossil palm remains in India), Prakash (1974; Indian Palaeogene palm stems, roots and petioles), Daghlian (1978; the coryphoid palms), Muller (1979; pollen of the family), Gregor (1980; a critical review of fossil palms from the younger Tertiary sediments of southern Germany), Gee (1990; *Nypa*), Ediger *et al.* (1990; *Calamus*-like disulcate pollen) and Harley and Baker (2001; pollen of the family). Exceptionally, the whole (macro plus micro) fossil record has been reviewed by Uhl and Dransfield (1987) and, more recently, by Harley (2006). Noé (1936) and Lamotte (1952) list fossil palms for North America, and Muller (1981) includes fossil pollen of Arecaceae. Collinson *et al.* (1993) provide a short review of key macro and pollen records for the Arecaceae in their overview of fossil records for the Angiospermae. The African fossil record is summarised by Pan *et al.* (2006). In Morley (2000), there are numerous references to palm fossils in relation to the origin and evolution of tropical rain forests. The Australian fossil record for Cretaceous and Tertiary monocotyledons, including palms, is summarised by Greenwood and Conran (2000).

Types of fossil

Palm fossils can be separated into two groups: macro-fossils are visible with the naked eye and include anything from whole stems, leaves and fruits to fragments that are observable with a hand lens; micro-fossils are those requiring examination with a microscope. Examination of the microscopic detail of larger fossils, particularly stem, leaf and fruit anatomy, is also of crucial importance. (See, for example, Koch [1972], Daghlian [1978], Van der Burgh [1984], Schaarschmidt and Wilde [1986], Suzuki [1989], Erwin & Stockey [1991]. For pollen, see Harley & Baker [2001] and Harley [2006] for overviews.)

The majority of fossils that have been identified as palms have not been equated with taxa below family or subfamily level, although their overall appearance usually leaves little doubt regarding their affinity with the family. For example, Reid and Chandler (1933) and Chandler (1961a, 1961b) describe numerous palm seeds, seed casts and fruits from the Lower Tertiary London Clay flora. Some are assigned to modern genera (see generic accounts) but the majority can only be assigned to the fossil genus *Palmospermum* (Reid & Chandler 1933). Nevertheless, some of these fossils are clearly distinguishable from each other, and are further described as species: *P. bracknellense*, *P. cooperi*, *P. davisi*, *P. elegans*, *P. excavatum*, *P. jenkinsi*, *P. minimum*, *P. minutum*, *P. ornatum*, *P. ovale*, *P. parvum*, *P. pulchrum*, *P. pusillum* and *P. subglobulare*. There are hundreds of reports of fossilised palm stem wood from many regions of the world, and many species have been described (see, for example, Stenzel [1904] and Prakash [1974]) but it is extremely difficult to assign the fossils below family level (Mahabalé 1958, Kaul 1960) because developmental features peculiar to the palm stem "... tend to transcend systematic differences" (Tomlinson 1990). In regional fossil pollen surveys, palm-like, simple tectate-monosulcate, slightly asymmetric pollen is frequently recorded, but this pollen type is so common in the family that only ultrastructural study can help to refine possible relationships (see, for example, Harley & Morley [1995]). However, there are also many very distinctive pollen forms in the palms (Chapter 2), some of which are genus-specific and allow good resolution of affinity (Harley & Baker 2001, Harley 2006).

ASSESSING POSSIBLE RELATIONSHIPS WITH MODERN GENERA

Equating any palm fossil to a modern genus is at best tenuous. Furthermore, it has not, of course, been possible to re-evaluate every fossil reported in the present account. Generic affiliations proposed in the literature should be treated with great scepticism. Comprehensive or complete reference collections of recent palms have rarely been at the disposal of the geologists and/or palaeobotanists making the generic

Opposite: ***Sabalites***, leaf fossil. (Photo: National Museums, Liverpool)

assumptions, and there are many problems with parallelisms in the morphology of modern palms, especially when, as is usual in fossil studies, the comparison is based only on one part of the whole plant.

In the generic accounts in this volume, we have cited work in which the authors have compared fossil records with living genera. The further removed a record is from recent time, however, the more difficult it is to make a like-for-like comparison; over time, some degree of morphological modification will most probably have occurred. An elegant example is the comparison of *Spinizonocolpites* (Fig. 5.1) and recent *Nypa* pollen (p. 213); in this particular example, the fossils provide evidence that *Nypa* was probably not always a monotypic genus. In *Spinizonocolpites*, there is a range of spine variation (i.e., in length and distribution) that is not apparent in the pollen of modern *Nypa* (see, for example, Muller 1968, Harley *et al.* 1991) but is reflected in numerous species of *Spinizonocolpites*. Furthermore, there is no obvious relationship between this variation and palaeogeographic distribution. Changes to pollen morphology over geologically long periods of time are probably more conservative than changes to the macromorphology of the plant, because the developing pollen is enclosed within the flower, protected from external conditions until the moment of release from the anther, and presumably does not have to adapt in response to the outside environment to the same extent as flowers, leaves, stems and roots. The generally slow rate of morphological change in vascular plants through time poses difficulties in stratigraphic alignment and a need for constant re-assessment of dating. Nevertheless, it is quite remarkable that we are able to compare the leaves of living *Sabal* with costapalmate leaves from the Upper Cretaceous, a period when highly evolved plant-eating dinosaurs still roamed the earth and perhaps even browsed on palms (Prasad *et al.* 2005, Ambwani & Dutta 2005).

The success of the angiosperms, from their beginnings in the Lower Cretaceous through to the present, is well-documented; as are the evolutionary developments of the herbivorous dinosaurs during the Cretaceous, even though the end of this period witnessed their unexplained extinction. Although never conclusively resolved, the most frequently discussed theories to explain the demise of dinosaurs at the Cretaceous–Tertiary boundary, in relation to the rapid evolution of angiosperms during the same period, have been ably summarised by a number of authors including Desmond (1975), Wing and Tiffney (1987) and Coe *et al.* (1987).

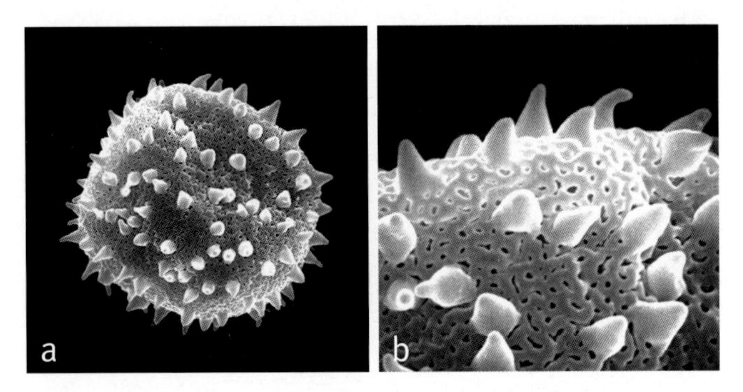

Fig. 5.1 ***Spinizonocolpites echinatus*** Muller, London Clay, Isle of Wight, UK. *Harley, Hill & Ferguson* s.n. (Royal Botanic Gardens, Kew). **a,** Middle Eocene zonasulcate pollen grain SEM × 1000; **b,** Middle Eocene close up of spiny surface '1a' SEM × 4000.

Fossils are now frequently used to calibrate internal nodes in molecular phylogenies, and thus to estimate divergence times of extant angiosperm lineages (e.g., Sanderson *et al.* [2004]); palms have been included in three such studies (see Chapter 7). For reasons outlined above, the selection of published fossil data for inclusion in such analyses should be from sources where stratigraphic dating, comparison with recent taxa and provenance can be checked and verified and, if possible, the original fossils re-examined. With regard to the 'first appearance' of a taxon in the fossil record, it is obvious that the oldest fossil record of a taxon only provides a minimum estimate of its age; nevertheless, this has to be the age accepted for phylogenetic analyses and there can be no open licence to extend ranges back tens of millions of years earlier (Pole 1994).

How far we can we go in assessing relationships of palm fossils with modern genera varies considerably and depends on a number of factors. These include the degree of detail that has been preserved in the fossil and the extent of variation in that detail between similar organs in living taxa: the greater the variation, the higher the potential for resolution of fossil affinity. It is important to be aware that there are no complete palm fossils, only isolated parts, and even these are often incomplete. Most fossil leaves (or large fragments of) and fossil stem wood referred to palms are highly recognisable at family level. However, it is comparatively rare for either leaves (Read & Hickey 1972) or stem wood (Stenzel 1904, Mahabalé 1958, Kaul 1960) to be convincingly associated with modern genera or, in the case of stem wood, even subfamilies. Two studies in the 1960s focussed on cuticular material: a palmate leaf from the Polish Miocene, with details of stomata in the epidermis, was referred to *Chamaerops* by Szafer (1961), while Litke (1966) described *Chamaerops* cuticle from the Upper Tertiary brown coals of Germany.

Read and Hickey (1972) surveyed fossil palm leaves and established five form genera. These authors pointed out that among all palms only the genus *Phoenix*, which has an induplicate pinnate leaf with the basal pinnae modified into spines, can be definitely identified to genus from leaf remains. They suggested that some or all of the following characters can be used to identify a fossil leaf as a palm: pinnate or palmate form or venation, leaflets with a strong mid-vein bounded on each side by at least two orders of parallel veins, a hastula at least adaxially, and a well-organised primary costa. These distinctions do not apply to all palm leaves. The leaves of some palms are not strongly plicate. In several genera, such as *Podococcus*, *Iriartea* and *Iriartella*, the leaflets lack midribs; in others, such as *Rhapidophyllum* and *Rhapis*, the ribs are not central in the segments; and still others, *Chuniophoenix*, *Nannorrhops*, *Lodoicea* and *Medemia*, lack hastulae. Nevertheless, Read and Hickey's work provides a way to determine whether fossil leaves are from palms. In a somewhat later study, Daghlian (1978) used epidermal characters to assess costapalmate and palmate fossil leaves from the Eocene of the southeastern USA. Daghlian (1978) described five different taxa, but only one of them (*Sabal dortchii*) has been referred to a modern genus. Daghlian's work highlighted the potential of studying cuticular material, and other microscopic details, to achieve better resolution of fossil leaf affinity. Nevertheless, only a few studies of fossil palms since the 1960s and 1970s, such as those by Van der Burgh (1984) and Schaarschmidt & Wilde (1986), appear to have employed such techniques.

Depending on their state of preservation, fossilised fruits or seeds can sometimes be related to modern genera; for example, fossil seeds comparable with those of *Phoenix* and *Nipadites* endocarps closely resembling those of *Nypa* (Fig. 5.5); and endocarps resembling tiny coconuts can be equated with tribe Butiinae, although not with the large endocarps of *Cocos*. Inflorescences, rachillae, peduncles and flowers are less common than most other palm organs in the fossil record, and they are also usually difficult to assign to genera, even where the flowers have the added advantage of smaller components such as calyces, androecia and/or gynoecia and perhaps even pollen (e.g., Harley [1997]). Nevertheless, among the palm flowers described in the literature, some of those preserved in amber show exquisite three-dimensional detail (Poinar 2002a, 2002b).

The wide range of differences in the pollen of living palms is well-reflected in the pollen fossil record. This has allowed a considerable number of comparisons at generic level, particularly in the calamoid and arecoid palms (see summaries in Harley & Baker 2001 and Harley 2006).

Taxonomy of palm fossils

The taxonomy of palm fossils has been recently summarised (Harley 2006). For most palm fossils, a lack of distinctive morphological variation in organs makes it difficult to assign affinity to taxonomic units below the family level; for example, almost all described fossil palm stems are included in the single genus *Palmoxylon* (see, for example, Fig. 5.2), or *Rhizopalmoxylon* for rhizomes. Exceptionally, at the other end of the scale, over 50 fossil pollen genera have been described as having a clear or suspected affinity to palms. This is partly because the range of variation in palm-like fossil pollen reflects quite a lot of the variation seen in the pollen of living palms (Harley & Baker 2001, Ramanujam *et al.* 2001, Ramanujam 2004).

There are many fossil records for costapalmate leaves, which are frequently referred to the fossil genus *Sabalites*. This does not necessarily infer an affinity with the extant genus *Sabal*; there are a number of other modern, predominantly coryphoid genera, that have costapalmate leaves. The re-circumscription of *Sabalites* by Read and Hickey (1972) embraces any palmate leaf "with a definite costa or extension of the petiole into the blade", whereas *Palmacites* is recommended for palm-like leaves that are "pure palmate, lacking a costa or extension of the petiole into the blade".

Features of the pinnate leaves of the modern genus *Phoenix* — induplicate plication, basal pinnae modified as spines (acanthophylls) — are particularly distinctive characteristics. However, the fossil genus *Phoenicites* has been applied somewhat indiscriminately to fossil pinnate leaves. Therefore, Read and Hickey (1972) redefined the fossil genus *Phoenicites* to define pinnate leaves with reduplicate plication and lowermost pinnae not spine-like. Their taxonomic recommendation that *Phoenicites* should, henceforth, be reserved or applied to non-*Phoenix*-like pinnate leaves is potentially confusing.

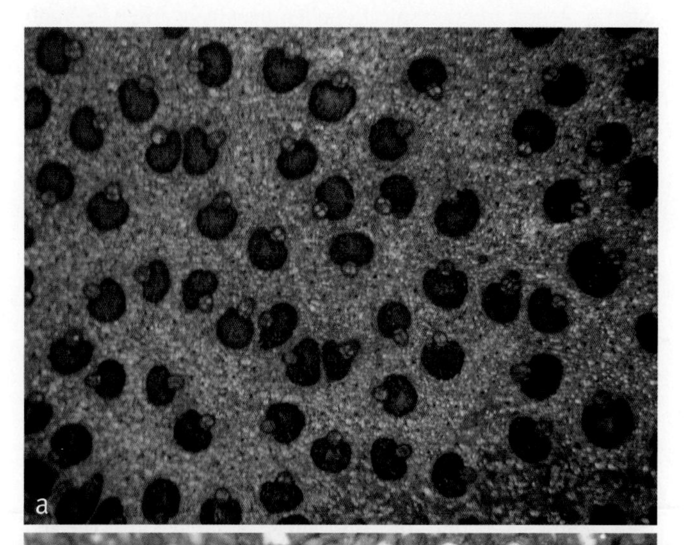

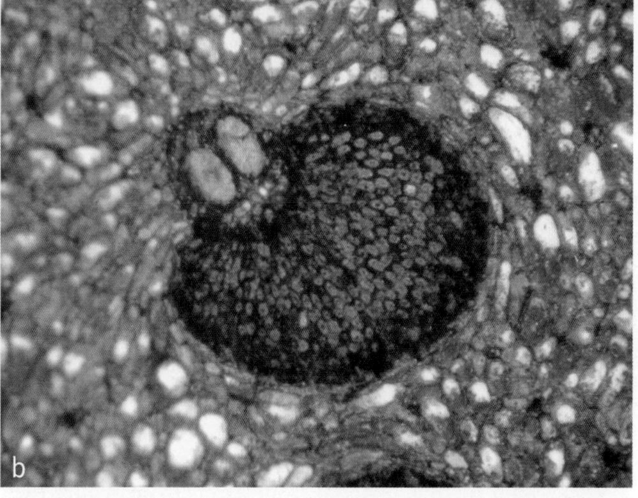

Fig. 5.2 ***Palmoxylon cheyennense*** Wieland, Pierre farm, Valley of Cheyenne, east of Black Hills. Part of type specimen. CUPC# 1656 (Cornell University, USA). **a,** Late Cretaceous palm stem: central region of central cylinder LM × 25; **b,** Late Cretaceous palm stem: close up of a vascular bundle, surrounded by ground tissue, central region of central cylinder LM × 66.

Fossils that are associated with modern *Nypa*, especially fruits, are frequently referred to as *Nipa* Thunb. or *Nipadites* Bowerbank. In fossil terminology, seed or endocarp-like structures are often loosely referred to as 'nuts', and the term 'spathes' is used to describe inflorescences in either the flowering or the fruiting stage; neither of these terms is, however, currently accepted.

Longapertites van Hoeken-Klinkenberg (1964) was originally used to describe extended sulcate pollen from the Maastrichtian of Nigeria. There are isolated examples of species with extended sulcate pollen among recent palms, for example, Calamoideae (*Eugeissona* and *Eremospatha*), Coryphoideae (*Licuala* all spp.) and Arecoideae (*Areca*, *Pinanga* and *Hydriastele*). In the fossil literature, records of *Longapertites* are numerous, and it is likely that a proportion of these are from palms. However, some *Longapertites* probably represent equatorial disulcate (disulculate) pollen with a broken distal 'bridge', a situation that frequently occurs in extant disulculate palm pollen. Extended sulcate pollen may have arisen through the loss of the distal bridge (Harley & Baker 2001) or, alternatively, via the extended sulcate model by development of a distal bridge.

Two final notes of caution relating to the fossil palm literature need stressing. The generic names *Phoenicites* and *Sabalites* are used for either leaves or fruits, and the fossil pollen genus *Arecipites* is widely used for small more-or-less symmetric monosulcate Arecaceae-like dispersed fossil pollen; its use for *Areca*-like fossil pollen (e.g., Salard-Cheboldaeff 1979) is exceptional.

THE EARLIEST OCCURRENCES

In the fossil record, palms are one of the earliest recognisable modern angiosperm families and, without doubt, they have the richest and one of the oldest fossil records of any monocotyledonous family (Herendeen & Crane 1995, Harley 2006; Tables 5.1 and 5.2). The range of palm fossil organs is remarkably diverse, including leaves, leaf rachides, petioles, stems, spiny bark, loose spines, rhizomes, roots, fruits, endocarps, seeds, pollen and, more rarely, peduncles, rachillae, inflorescences or individual flowers (Tables 5.2–5.5).

It must be noted that all reports of pre-Cretaceous occurrences of palms have proven incorrect. Among fossil records of palm leaves, those of Berry (1914b) do not seem to have yet been superseded by any of earlier age. When Brown (1956a) discovered fossil palm-like leaves from the Triassic of Colorado (*Sanmiguelia*, type species *S. lewisii*), it seemed for a while that the earliest fossil record for the Palmae might stretch back as far as the Late Triassic. Subsequent to Brown's discovery, however, Tidwell *et al.* (1977) found further material from Texas from which they were able to reconstruct the habit of a semi-aquatic plant. Later still, Cornet (1986, 1989a, 1989b) described reproductive organs from the same beds as the leaf fossils, and he considered these organs to belong to *Sanmiguelia*. The pollen recovered from the anthers is described as "small, psilate, and monosulcate with a tectate-granular exine" (Cornet 1989b). Summarising the anatomical evidence, however, Cornet (1989b) suggests that *Sanmiguelia lewisii* "appears to be a very primitive pre-magnoliid dicotyledon possessing many monocotyledon-like characters". He emphasised the need for more material to be described to substantiate his theories. Other palaeobotanists believe that *Sanmiguelia* can be more accurately described as a gnetophyte, or cycadophyte (see for example, Stewart and Rothwell [1993]). *Propalmophyllum* O. Lignier, described from the Lower Jurassic of France, is not palm-like and is too fragmentary to be assigned to any group of plants (Read & Hickey 1972). *Palmoxylon simperi* and *P. pristina*, originally described as mid-Jurassic (Tidwell *et al.* 1970) have been shown to be Tertiary (Scott *et al.* 1972).

There are some equivocal palm fossil records that have been dated to the late Lower–early Upper Cretaceous (Aptian–Turonian). The fossil palm leaf *Eolirion primigenium* Schenk, from the Cretaceous (Urgonian) of Austria (Schenk 1869), was reported to be the earliest palm fossil record (Collinson *et al.* 1993). However, Read and Hickey (1972) expressed doubts that this leaf was from a palm. Furthermore, the Urgonian, which is not a chronostratigraphic time stage but set in the Barremian to earliest Aptian of southeastern France, is apparently a coral limestone facies (http://www.cretaceousfossils.com/). Fruits, *Hyphaenocarpon aegypticum* (Vaudois-Miéja & Lejal-Nicol 1987), from the Lower Cretaceous (Aptian) of Egypt appear to be the oldest generically associated palm fossils. However, the geological formation in which the fossils were found "north of Lake Aswan (i.e., Lake Nasser), near Abu Simbel" is not clearly defined. The Lake Nasser Formation, which is found in the surroundings of Abu Simbel is regarded as Aptian (Klitzsch *et al.* 1987, Hermina *et al.* 1989) and as a near-shore equivalent of the Abu Ballas Formation. However, the fossil assemblage described by Vaudois-Miéja and Lejal-Nicol (1987) is not a typical Abu Ballas assemblage; it is probably younger (Late Cretaceous) and further research on this matter is needed (Schrank 1992, and pers. comm.). A notably early record of a *Nypa* fruit, *Nipadites burtini* also from the Aptian, was published in a brief note by Jahnichen (1990) but with insufficient detail to be a reliable record. The fossil endocarps *Cocoopsis* and *Astrocaryopsis*, from the Lower and Upper Cenomanien of France (Fliche 1894, 1896), bear some resemblance to cocosoid endocarps but the geology of these records needs re-visiting.

Fossil palm stems, *Palmoxylon andevagense* and *P. ligerinum* (Crié 1892), were recovered from the lower Upper Cretaceous (Turonian) of France. The anatomy and stratigraphy of these stems has been revised (Kvaček & Hermann 2004) and "... they appear as the oldest palm (stem) remains". A later fossil stem is recorded from the Upper Cretaceous (Coniacian–Santonian) of New Jersey, *Palmoxylon cliffwoodensis*

TABLE 5.1

Geological timescale with earliest occurrences of the major fossil categories. 'Leaves' includes leaf axes, petioles, cuticle etc. (In this column, the chronological extent of palmate leaves is further — possibly by ca.10 myr [lighter green] — than that of pinnate leaves [very dark green].) 'Stems' includes rhizomes, bark and roots. 'Fruits' includes endocarps and seeds. 'Flowers' includes rachillae, peduncles and inflorescences.

ERA	PERIOD	SUB-PERIOD	EPOCH		STAGE	AGE (myr)	LEAVES	STEMS	FRUITS	FLOWERS	POLLEN
↑ CAINOZOIC ↓	QUATERNARY		Holocene		Versilian	0.01					
			Pleistocene	Upper	Tyrrhenian	0.1					
					Milazzian	0.4					
					Sicilian	0.7					
				Lower	Emilian	1.0					
					Calabrian	1.8					
	TERTIARY	Neogene	Pliocene	Upper	Piacenzian	3.5					
				Lower	Zanclian	5					
			Miocene	Upper	Messinian	7					
					Tortonian	11					
				Middle	Seravallian	12					
					Langhian	15					
				Lower	Burdigalian	20					
					Aquitanian	22.5					
		Palaeogene	Oligocene	Upper	Chattian	33					
				Lower	Rupelian	37					
			Eocene	Upper	Bartonoan	40					
					Priabonian	43					
				Middle	Lutetian	50					
				Lower	Ypriesian	55					
			Palaeocene	Upper	Thanetian	57					
				Lower	Montian	62					
					Danian	65					
MESOZOIC ↓	CRETACEOUS		Upper	Senonian	Maastrichtian	70					
					Campanian	76					?
					Santonian	82					?
					Coniacian	86					?
					Turonian	95					?
					Cenomanian	100					?
			Lower		Albian	106					?
					Aptian	112					?
					Barremian	120					?
				Neocomian	Hauterivian	125					
					Valanginian	130					
					Berriasian	141					
	JURASSIC ↓		Upper	Malm	Tithonian	145					
					Kimmeridgian	150					
					Oxfordian	160					
			Middle ↓	↓	↓	↓					
					↓	↓					
					↓	↓					

TABLE 5.2
Summary of Cretaceous palm fossils associated with modern genera

PROVENANCE	EPOCH	STAGE	PALAEOTAXON	FOSSIL TYPE	AUTHOR(S)
USA	Senonian	Upper Cretaceous (stage not given)	*Sabalites eocenica*	Palmate leaf	Dorf 1942
USA			*Sabalites montana*	Palmate leaf	Dorf 1942
Japan			*Sabalites ooaraiensis*	Palmate leaf	Ôyama & Matsuo 1964
South America (Argentina)			*Palmoxylon valchetense*	Stem	Ancibor 1995
Globally widespread		Maastrichtian	*Spinizonocolpites*	Pollen	Tralau 1964, Gee 1990
Africa (Nigeria, Somalia)			*Mauritiidites*	Pollen	Hoeken-Klinkenberg 1964, Schrank 1994
Africa (Somalia)			*Dicolpopollis malesianus*	Pollen	Schrank 1994
Canada			cf. *Livistona*	Pollen	Jarzen 1978
Africa (Cameroon)			*Retimonocolpites pluribaculatus*	Pollen	Salard-Cheboldaeff 1978
Africa (Cameroon)			*Arecipites lusaticus*	Pollen	Salard-Cheboldaeff 1979, 1981
Africa (Cameroon)			*Arecipites convexus*	Pollen	
Europe (Austria)		Lower Campanian	*Sabalites longirhachis*	Palmate leaf	Kvaček & Herman 2004
USA			cf. *Phoenix*	Pinnate leaf	Crabtree 1987
USA		Santonian	*Sabalites magothiensis*	Palmate leaf	Berry 1905, 1911
USA		Coniacian–Santonian	*Palmoxylon cliffwoodensis*	Stem wood	Berry 1916a
USA			*Sabalites carolinensis*	Palmate leaf	Berry 1914b
Europe (France)	early Upper Cretaceous	Turonian	*Palmoxylon andevagense*	Stem wood	Crié 1892
Europe (France)			*Palmoxylon ligerinum*		
EARLIER FOSSILS — DATING NEEDS TO BE CHECKED					
India	Upper Cretaceous	(Deccan Intertrappean)	*Nypa*	Fruit	Singh 1999
Central America (Mexico)		Lower Maastrichtian	cf. *Manicaria*	Inflorescence	Weber 1978
Europe (France)	late Lower Cretaceous–early Upper Cretaceous	Albian–Cenomanian (?)	*Cocos cocopsis*	Endocarp	Fliche 1896
Europe (France)		Albian–Cenomanian (?)	*Astrocaryum astrocaryopsis*	Endocarp	Fliche 1894, 1896
Egypt	Lower Cretaceous	Aptian (?)	*Hyphaenocarpon aegypticum*	Fruit	Vaudois-Miéja & Lejal-Nicol 1987
Europe (Germany)			*Nypa burtini*	Fruit	Jahnichen 1990

(Berry 1916a). None of these early fossil stems, in common with most fossil palm stems, are assigned to a modern palm genus or even subfamily.

The earliest unequivocal records of systematically assignable palms are of costapalmate leaves that are clearly assignable to subfamily Coryphoideae, although their association with *Sabal* has not been proven. *Sabalites carolinensis* is from the late Coniacian–early Santonian of South Carolina (USA) (Berry 1914b); *S. magothiensis* Berry is slightly later and from the Santonian Magothy Formation, New Jersey and Maryland (USA) (Berry 1905, 1911). *Sabalites longirhachis* (Unger) J. Kvaček & Hermann (Kvaček & Hermann 2004) is reported from the Lower Campanian of Austria. The earliest record of pinnate leaves appears to be from the Lower Campanian of northern Montana (USA) (Crabtree 1987). It is not assigned to a genus but is highlighted because Crabtree (1987) records the pinnate palm fronds as, "dominat(ing) the channel margin facies throughout the outcrop (of the 'Two Medicine' Formation)... The fronds are up to two meters long and are incompletely divided into induplicate segments". Crabtree (1987) is clearly confused by Read and Hickey's (1972) redefinition of *Phoenicites* for re-duplicate non-*Phoenix*-like pinnate leaves, believing it to be for the use of *Phoenix*-like fossil leaves, as he continues, "The fossil palm could be assigned to the form genus *Phoenicites* (*sensu* Read & Hickey 1972) were it not for the absence of spines at the base of the frond."

There are no records of fruits, flowers, rachillae, peduncles or inflorescences before the Maastrichtian. However, Weber (1978) published a note relating to an inflorescence from the lower Maastrichtian of northeastern Mexico that was compared with *Manicaria*. Unfortunately, a published description has not yet been traced. All other records are likely to be at least Palaeocene.

The earliest undisputed records for pollen are those of *Spinizonocolpites*, widespread from the Maastrichtian (see Gee 1990, Morley 2000); *Mauritiidites* (van Hoeken-Klinkenberg 1964, Schrank 1994) from the Maastrichtian of Nigeria and northern Somalia; and *Dicolpopollis malesianus* (Fig. 5.3), also from the Maastrichtian of northern Somalia (Schrank 1994). Nevertheless, the palm pollen fossil record is probably at least as old as the earliest macrofossil records. The problem is that the frequently recorded simple tectate, monosulcate pollen grains with a columellate infratectum and no endexine are quite common among monocotyledons, especially in the Asparagales and Liliales. To add to the problem, this type of pollen has a very wide size range among the palms: 16–90 µm. Nevertheless, among the Lower Cretaceous dispersed fossil pollen described in the literature there are two genera, *Liliacidites* and *Retimonocolpites* (Walker & Walker 1984, 1986), that have small to average-sized pollen (long axis minimum 15 µm, average 20–30 µm, maximum 42 µm) and form species very similar to some modern palm pollen. For example, *Retimonocolpites peroreticulatus* (Brenner) Doyle (Walker & Walker 1984), from the Middle Albian of the eastern North America Potomac group, is notably like *Nenga* pollen (p. 513), as are the 'Retisulc-dubdent'/'Retisulc-dentat' groups of Hughes (1994), from the English Barremian. The characteristics of the Lower Cretaceous *Liliacidites* series from the Potomac group were redefined by Walker and Walker (1984, 1986). These grains are basically reticulate, endexine-less and monosulcate, but a number of exine modifications that are rare or even unknown in modern monocotyledons occur in the pollen of modern palms (see Chapter 2). For example, trichotomosulcate pollen is frequent in subfamily Arecoideae; small lumina in the muri between the main lumina are a common feature in the pollen of a number of species of *Pinanga* (Fig. g, p.516); a reticulum, finely modified around the sulcus, is a very common feature of palm pollen; a reticulum modified on the non-apertural face is typical of the pollen of a number of species of *Chamaedorea* (Fig. d, p.380); and finally, polygonal lumina (as opposed to rounded lumina) are known to occur in the pollen of *Hydriastele* (Fig. a, p.637), *Nenga* (Figs a–d, p.513) and *Pinanga* (Figs g–h, p.516). (See Glossary for explanation of pollen terminology.)

The generically associated Upper Cretaceous fossil palm record is summarised in Table 5.2.

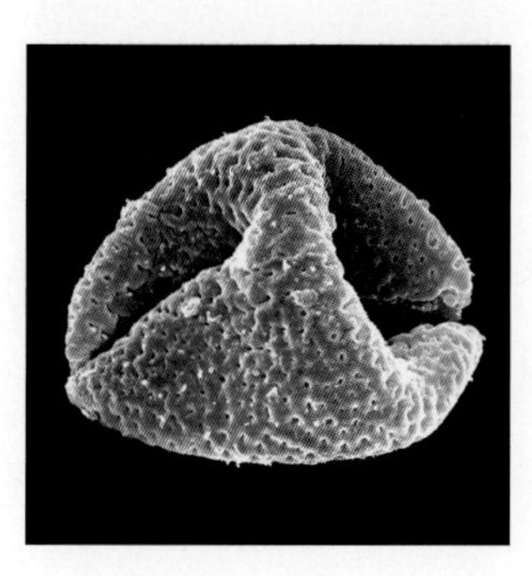

Fig. 5.3 ***Dicolpopollis malesianus*** Muller, Eocene disulcate pollen grain, equatorial view, long axis × 1000. Nanggulan Formation, Central Java. *Morley* XIV-H.

TERTIARY PALMS AND THEIR PALAEO-GEOGRAPHIC RANGE

The number of records for palm fossils increases by the early Tertiary (Palaeocene) and, by the Middle Eocene, reaches a peak of geographic distribution, notably in regions of the world where palms do not occur naturally today, such as Greenland, British Columbia, northern Europe, Russia and Japan. The regional summaries that follow include many key references but they are not exhaustive. For succinctness, the summaries concentrate on fossils that have been attributed to modern genera.

North America

Many localities have yielded palm remains (see Noé [1936], Lamotte [1952]). The most northerly record is probably a palmate leaf, *Flabellaria alaskana*, from the early Tertiary (Hollick 1936), which was later included in the re-circumscription of *Palmacites* by Read and Hickey (1972).

TABLE 5.3
Summary of Tertiary palm fossils from southern England associated with modern genera

EPOCH	GEOLOGICAL FORMATION	PALAEOTAXON	FOSSIL TYPE	AUTHOR(S)
Oligocene	Bembridge Beds	*Palaeothrinax mantelli* (= *Trachycarpus*)	Leaves	Reid & Chandler 1926
		cf. *Caryota rumphiana*	Pollen	Pallot 1961
	Bovey Tracey Lake Basin, Devon	cf. *Calamus*, *Calamus daemonorops*	Spines, seeds, fruits, fruiting axes, male and female flowers, *in situ* disulcate pollen	Chandler 1957
Middle Eocene	Bracklesham Beds	*Spinizonocolpites*	Pollen	Harley *et al.* 1991
	Bournemouth Beds	*Calamus daemonorops*, cf. *Calamus*, cf. *Daemonorops*	Spines or thorns, spine bases, fragments, spine bundles, thorny bark, epidermis	Chandler 1963
		cf. *Phoenix*	Female inflorescence	Gardner 1882, Chandler 1963
Lower Eocene	London Clay Flora	*Nypa burtini*, *Nipadites*	Fruits	Bowerbank 1840, Chandler 1961b
		Oncosperma anglicum	Fruits	Reid & Chandler 1933, Chandler 1964
		Sabalites, *Sabal* sp.	Palmate leaves, leaf fragments	Reid & Chandler 1933, Chandler 1961b, 1961c, 1962, 1963
		Sabal sp.	Pollen	Khin Sein 1961
		Thrinax tranquillus	Pollen	
		cf. *Phoenix*	Pollen	
		cf. *Livistona*	Pollen	
		Livistona(?) *minima*	Seeds	Reid & Chandler 1933
		Caryotispermum	Seeds	
		Sabal grandisperma	Fruit	
		Serenoa eocenica	Fruit	
		Corypha wilkinsonii	Seed	
		cf. *Trachycarpus*	Fruit	Chandler 1978
	Lower Bagshot Beds — Dorset Pipe Clay Series	*Sabal* sp.	Palmate leaf fragment	
		Trachycarpus raphifolia	Palmate leaves	Chandler 1962
		Calamus daemonorops	Spines or prickles	
	Woolwich, London Clay & Lower Bagshot Beds	*Dicolpopollis*	Pollen	Khin Sein 1961, Gruas-Cavagnetto 1976

Stem and leaf fragments from the Middle Eocene Princeton Chert have been compared with *Rhapidophyllum* and *Brahea* (Erwin & Stockey 1991) and, from the Lower Palaeocene of Greenland, numerous seeds have been associated with *Corypha* (Koch 1972).

Tschudy (1973) records *Dicolpopollis* from the Eocene of the Mississippi Basin. There are a number of Lower Eocene records for fossil *Nypa* fruits from southeastern North America (Berry 1914c, 1916b) and from the Middle Eocene Claibornian of Texas (Arnold 1952). From the Middle Eocene to earliest Oligocene, *Spinizonocolpites* is recovered from some of the Gulf Coast states (these records are reviewed by Gee 1990).

From the Palaeocene onwards, records of *Sabalites* (leaves and some fruits) become more frequent and widespread in North America (see Fig. 5.4). Records for the Rocky Mountains and Great Plains have been reviewed by Brown (1962), whereas those for the Middle and Upper Eocene floras of southeastern North America were reviewed by Berry (1924) and also by Daghlian (1978). Details of leaf epidermis (stomata) allowed Daghlian (1978) to make a direct comparison between the epidermis of the Eocene species *Sabal dortchii* Daghlian and the leaf epidermis of living *Sabal*. Pollen, *Sabalpollenites* sp., is described from the Eocene of Tennessee (Potter 1976). Other coryphoid records include: costapalmate leaves, *Thrinax eocenica*, from the Mississippi Claiborne Flora (Berry 1914b, 1924); stem wood (cf. *Phoenix*) from the Oligocene-Miocene of Louisiana (Schmidt 1994); *Phoenix* seeds from the Tertiary of Texas (Berry 1914a); and petioles and seeds (cf. *Serenoa*) from the Pleistocene of Florida (Berry 1917). Pollen from the Upper Eocene of Texas, *Liliacidites tritus* (Frederiksen 1980, 1981), is compared with the genus *Pseudophoenix* (Ceroxyloideae); and in the arecoid palms, Eocene pinnate leaves, *Chamaedorea danai* (Berry 1916b) and *Bactrites pandanifoliolus* (Berry 1924), have been compared with pinnate leaves from the modern genera *Chamaedorea* and *Bactris*, respectively, but these affinities are uncertain. An affinity with *Attalea* is suggested by Berry (1929) for an endocarp (*Attalea gunteri*) from the Upper Eocene of Florida.

Central America

It is noticeable that comparisons of fossil palms with modern genera in the Central American region are most frequently with subfamily Arecoideae, whereas in North America, a higher percentage of palm fossils show affinity with coryphoid genera. Another difference between the palm fossils recovered from these two major regions is that fewer leaf fossils but more fruits, endocarps and seeds have been recorded from Central America.

A sandstone fruit cast from the Upper Miocene–Lower Pliocene of Ecuador has been assigned to *Phytelephas olssonii* (Ceroxyloideae). From the Middle Oligocene of Puerto Rico, Hollick (1928) described a number of palm endocarps and seeds and suggested affinities with *Arenga*, *Iriartea*, *Cocos*, *Acrocomia*, *Astrocaryum*, *Bactris* and *Elaeis*. Fruits and endocarps have been recorded from the Miocene of Trinidad (Berry 1937, Menendez 1969) but these fossils are not associated with modern genera. A coryphoid affinity has been proposed for two records of flowers in amber from the Middle Eocene–Middle Miocene La Toca mine in the Dominican Republic: *Palaeoraphe dominicana* (cf. *Brahea*) and *Trithrinax dominicana* (Poinar 2002a, 2002b). Other flowers in amber are reported from the La Toca Mine (cf. *Roystonea*; Poinar 2002b) and from the Upper Oligocene to Lower Miocene amber mines of Chiapas State, Mexico: *Socratea brownii* and cf. *Neonicholsonia* (Poinar 2002a). There are coryphoid pollen records from the Paraje Solo Flora, Mexico (cf. *Brahea*; Graham 1976) and from the Gatun Lake Formation, Panama (cf. *Cryosophila*; Graham 1991). Other arecoid pollen grains reported from the Upper Miocene Paraje Solo Flora include cf. *Chamaedorea*, cf. *Attalea* and cf. *Astrocaryum* (Graham 1976). The Upper Miocene sediments of the Gatun Lake Formation, Panama have yielded a seed (cf. *Iriartea*; Berry 1921a), and monosulcate pollen grain which has been compared to pollen of *Reinhardtia* (Graham 1991). A large specimen of stem wood, *Palmoxylon iriarteum*, from the West Indies (Antigua), now in the palaeontology collections of the Naturhistoriska Riksmuseet, Stockholm, was described in anatomical detail by Stenzel (1897) who considered it ancestral to *Iriartea*; no age was given.

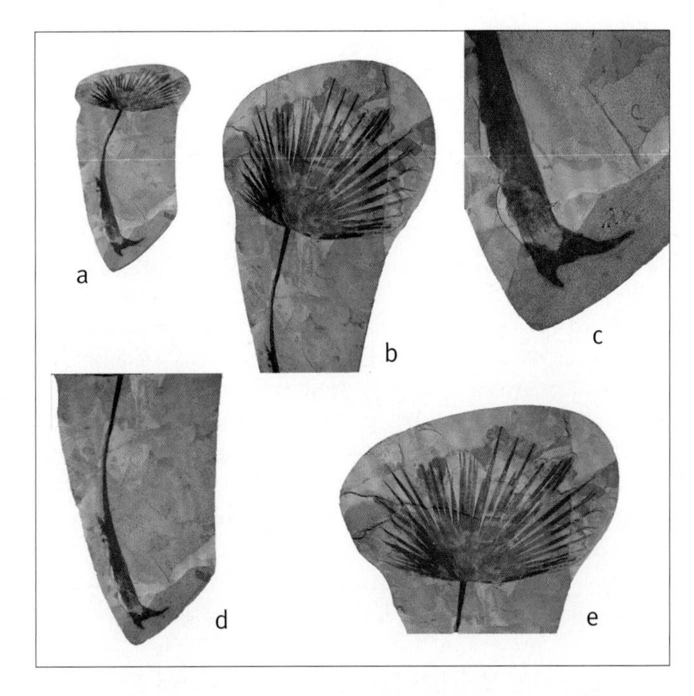

Fig. 5.4 ***Sabalites*** Saporta, Fossil Butte Member, Green River Formation, Wyoming, USA. LIV 2006.56 (Liverpool Museum, UK). **a,** palmate leaf and petiole × ca. $^1/_{50}$; **b,** palmate leaf and part of petiole × ca. $^1/_{40}$; **c,** close-up of petiole base × ca. $^1/_8$; **d,** petiole × ca. $^1/_{20}$; **e,** palmate leaf × ca. $^1/_{40}$.

South America

From Colombia, there are Upper Palaeocene and Middle Eocene records for *Spinizonocolpites* (Jaramillo & Dilcher 2001) and Palaeocene–Middle Eocene records for *Mauritiidites* (Sole de Porta 1961, van der Hammen & Garcia de Mutis 1966, González-Guzmán 1967, Schuler & Doubinger 1970, Jaramillo & Dilcher 2001). In Brazil, *Nypa* fruits are recorded from the Palaeocene of Pernambuco (Dolianiti 1955) and *Mauritia* pollen is known from a Pleistocene swamp in central Brazil (Ferraz-Vicentini & Salgado-Labouriau 1996). A leaflet from the Tertiary of Venezuela, assigned by Berry (1921a) to *Sabalites*, somewhat resembles a leaflet of *Iriartea*. *Mauritia* pollen is known from the Upper Tertiary of Venezuela (Lorente 1986). From the Middle Eocene of Peru, fossil fruit resembling *Astrocaryum* (*A. olsoni*) are recorded (Berry 1926a); and, from the Miocene of northwestern Peru (Berry 1919b), leaf fragments are compared with *Iriartea* (*Iriartites tumbezensis*).

Europe: the British Isles, Belgium and France

The European record of Tertiary palms that have been assigned to modern genera is extensive. In the British Isles, the records (see Table 5.3) are from sedimentary beds in southern England. Many are reported from the London Clay Flora, one of the world's most varied fruit and seed floras and the only diverse flora from Lower Eocene strata in Europe (Collinson 1983).

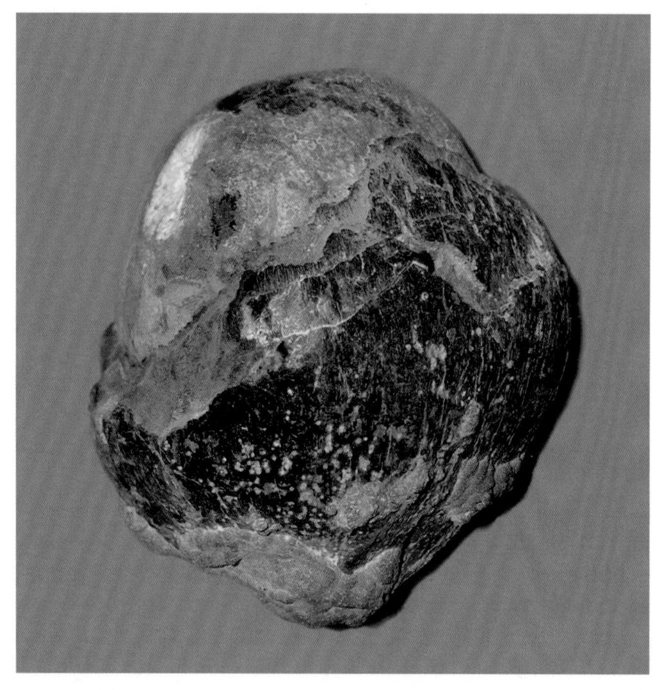

Fig. 5.5 ***Nipadites*** Bowerbank, Late Eocene fruit showing remains of calyx × 1.1. London Clay, Sheppey, UK. *Balm & Prychid* s.n. (Royal Botanic Gardens, Kew, UK)

Nypa fruits (*Nipadites* [Fig. 5.5]) and pollen (*Spinizonocolpites*) and various *Calamus* fossils, including pollen (*Dicolpopollis*), are well known from the Lower Eocene Basins of southern England, Belgium and France (Tralau 1964, Gruas-Cavagnetto 1976, Roche 1982, Ollivier-Pierre *et al.* 1987, Schuler *et al.* 1992), the three Northwest European basins influenced by the English Channel (La Manche) and the North Sea. Ollivier-Pierre *et al.* (1987) discuss the principal geological basins in North West Europe and note correlations between the Lower Eocene fossil records for the Paris and Belgium Basins and those of southern England during the Ypresian and Lutetian. Ollivier-Pierre *et al.* (1987) suggest that these correlations indicate a rise in temperature in the Northwest European seaboard (the English Channel and the North sea), leading to the development of a continental flora, including the development of mangroves, in a particularly warm and humid climate.

Saporta (1865) describes both costapalmate leaves (*Sabalites*) and a leaf frond and male inflorescence (*Phoenix aymardii*; Saporta 1878, 1879, 1889) from the French Eocene. From the Neogene "probably Miocene" lignites of the Landes District of France, Huard (1967) describes spines and spiny leaf sheaths (cf. *Calamus* or *Daemonorops*).

Europe: Central Europe

Of the many palm fossils described from Germany, the Czech Republic and Poland, most are found in the Miocene lignite strata (the brown coals) (Table 5.4). Brown coal deposits in Germany are geographically widespread from the Lower Rhenish Plain in the west to the Halle (Geiseltal)/ Leipzig/Bitterfeld region and the Niederlausitz and Oberlausitz regions of the southeast near the Polish border. The Silesian Zittau Basin overlaps the borders of all three countries (Teodoridis 2003). The Turow mines in Poland and Czechoslovakia, which have yielded a number of calamoid and coryphoid palm fossils (Czeczott & Juchniewicz 1975, 1980), are part of the Zittau Basin. The North Bohemian coal basin (Konzalová 1971, Knobloch *et al.* 1996) has also yielded calamoid and coryphoid pollen types. (See also Gregor [1980] for a review of palm remains from the Miocene–Pliocene of southern Germany.)

From the Middle Eocene to Lower Miocene of Germany, the Czech Republic, Hungary, Poland and Switzerland, there are records of *Sabal*-like fruit (Mai 1976), *Sabal*-like pollen (Macko 1957), costapalmate leaves (*Sabalites*: Mai & Walther 1978, Van der Burgh 1984, Büchler 1990, Kvaček 1998) and a *Sabal*-like inflorescence with pollen (*Tuzsonia hungarica*; Andréanszky 1949). However, despite the important record of *Sabalites longirhachis* from the Lower Campanian of Austria (Kvaček & Hermann 2004), there are no Tertiary palm fossils from Austria associated with *Sabal* or *Sabalites*.

Fossils compared with *Phoenix* have been recovered from the Middle Eocene of Germany: fruits and seeds from the brown coals of Geiseltal (Mai 1976); a flower (affinity

TABLE 5.4
Tertiary of Central Europe: summary of palm fossils attributed to genera

PROVENANCE	EPOCH	GEOLOGICAL FORMATION	PALAEOTAXON	FOSSIL TYPE	AUTHOR(S)
Germany	Middle Miocene–Pliocene		*Calamus daemonorops*	Thorn-like fragments	Gregor 1982
	Middle Miocene	Oberlausitz	*Calamus daemonorops*	Thorn-like fragments	Mai 1964
	Miocene	Lower Rhenish Plain	*Sabalites*	Costapalmate leaves, vascular bundles, cuticle, scalariform pits	Van der Burgh 1984
		Niederlausitz brown coals	cf. *Chamaerops*	Cuticle	Litke 1966
	Lower–middle Oligocene	Haselbach Flora	*Calamus daemonorops*	Thorn-like fragments	Mai & Walther 1978
			Sabalites	Costapalmate leaf	Mai & Walther 1978
	Middle Eocene	Geiseltal	*Phoenix hercynica*	Fruits and seeds	Mai 1976
			Livistona atlantica	Seed	
			Serenoa carbonaria	Fruit	
			Sabal bracknellense		
		Messel oil shales	cf. *Calamoid* palms	Pinnate leaf and spine Bundles	Schaarschmidt & Wilde 1986
			Palmaemargosulcites fossperforatus cf. *Phoenix* etc.	*in situ* Pollen	Harley 1997
			Palmaemargosulcites insulatus cf. *Dictyocaryum*	*in situ* Pollen	Harley 1997
		Bernstein Flora — Baltic amber	*Phoenix eichleri*	Flower	Conwentz 1886
	Lower–Middle Eocene	Lower Saxony	*Spinizonocolpites*	Pollen	see Gee 1990
Switzerland	Upper Oligocene	Ebnat-Kappel section	cf. *Sabal minor*	Costapalmate leaves	Büchler 1990
Austria	Palaeocene–Eocene	Krappfeld sedimentary succession	*Punctilongisulcites microechinatus* cf. *Salacca*	Pollen	Hofmann & Zetter 2001, Zetter & Hofmann 2001
			cf. *Daemonorops*		
			Dicolpopollis cf. *Calamus*		
			Spinizonocolpites cf. *Nypa*	Pollen	Zetter & Hofmann 2001
Hungary	Upper Oligocene	Tarnoc	*Calamus noszkyi* cf. *Calamus*	Scaly fruit, pinnate leaf fragments	Jablonszky 1914
	Oligocene		*Attaleinites*	Fruit	Tuzson 1913
		Eger	*Tuzsonia hungarica* cf. *Sabal*	Inflorescence, male flowers and pollen	Andréanszky 1949
	Upper Eocene	Dudarról	*Nipadites burtini*	Fruit	Rásky 1949
		Mt Martinovics, Budapest	cf. *Actinorhytis*	Fruit	Rásky 1956
	Palaeocene–Middle Miocene		*Spinizonocolpites*	Pollen	Rákosi 1976

(contd.)

TABLE 5.4 (continued)

PROVENANCE	EPOCH	GEOLOGICAL FORMATION	PALAEOTAXON	FOSSIL TYPE	AUTHOR(S)
Czech Republic	Miocene	Zittau Basin (Czech part)	cf. *Calamus* or *Daemonorops*	Spines	Teodoris 2003
		North Bohemian Brown Coal Basin	*Dicolpopollis*, *Sabalpollenites*	Pollen cf. *Calamus*, cf. coryphoid pollen	Konzalová 1971, Knobloch *et al.* 1996
		Turów Basin (Czech part — Hrádek on Nisa)	*Spinophyllum daemonorops* cf. *Calamus* or *Daemonorops*	Fragmentary spathe with setae	Czeczott & Juchniewicz 1975
			Trachycarpus raphifolia	Costapalmate leaf	
			cf. *Livistona*	Palmate leaf and seed cast	
	Lower Miocene	Tuchořice	*Phoenicites bohemica* cf. *Phoenix*	Date seeds	Bůžek 1977
		Bílina	*Sabalites*	Costapalmate leaves	Kvaček 1998
Poland	Miocene	Turów Basin (Polish part — Bogatinia)	*Spinophyllum daemonorops*, cf. *Astrocaryum aculeatum* and cf. *Iriartea* (aerial roots)	Spiny stem fragments, thorn bundles, epidermis, small roots and ?aerial root	Czeczott & Juchniewicz 1980
	Lower Miocene	Gliwice, Upper Silesia	cf. *Sabal palmetto*, *Copernicia*, *Corypha*, *Trachycarpus*	Pollen	Macko 1957
			cf. *Chamaerops*	Leaf fragment, damaged seed	Szafer 1961

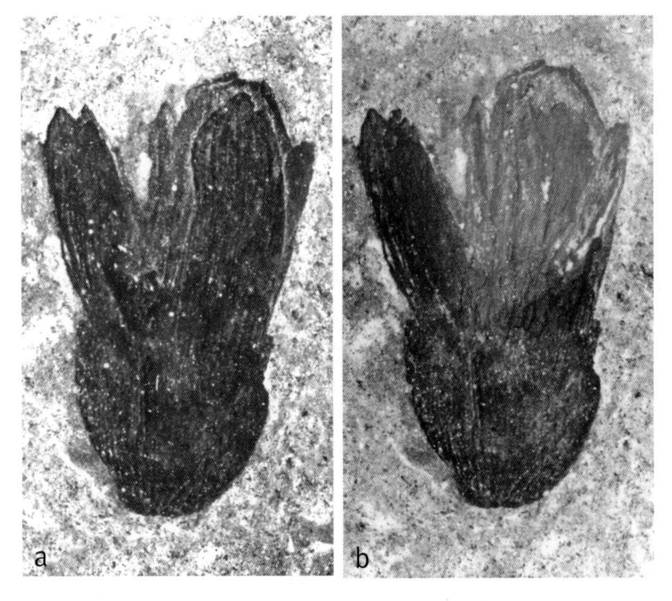

Fig. 5.6 cf. ***Phoenix***, Messel oil shales, Germany. *SM.B Me* 10281 (Senckenberg Museum, Frankfurt, Germany). **a,** Middle Eocene flower × 8; **b,** Middle Eocene flower, as (a) but with facing petal removed to show anthers × 8.

tenuous) from the Baltic amber of Bernstein (Conwentz 1886; Fig. 5.6); and a pinnate leaf and spine bundles (Schaarschmidt & Wilde 1986) as well as flowers (Fig. 5.6) and *in situ* pollen (Harley 1997) from the oil shales of Messel. Date seeds are also described from the Lower Miocene Tuchořice fresh water limestones in the Czech Republic (Bůžek 1977).

Southern Europe and the Balkans

Spinizonocolpites is recorded from the Eocene to Middle Miocene in Northern Spain (Pyrenees) (Haseldonckx 1972). *Nypa* fruits, *Nypa burtini* (see Tralau 1964), are known mainly from the Upper Eocene of Italy. There are also many species of pinnate leaf fossils described from the Italian Tertiary: *Phoenicites*, *Hemiphoenicites* and *Geonomites* (de Visiani 1864), *Manicarites*, *Kentites* and *Pritchardites* (Bureau 1896), all of which were placed in synonomy with *Phoenicites sensu* Read and Hickey (1972). From the Lower Miocene of Sardinia, stems and roots resembling those of *Chamaerops* have been reported by Biondi and Filigheddu (1990). From the Pleistocene of Greece (Santorini), leaves are compared with *Phoenix* and leaves attached to a 'trunk' are compared with *Chamaerops* (Lacroix 1896).

Eastern Europe and Russia

From Eastern Europe, leaf remnants from the Miocene of Moldavia are compared with *Chamaerops humilis* (Stephyrtza 1972). *Nipadites burtini* was recorded from the Upper Eocene of Russia (Odessa) (Kryshtofovich 1927) and from some other localities (see Tralau 1964, Gee 1990). *Sabalites* occurs from the Eocene (e.g., Budantsev 1979) through the Oligocene (e.g., Akhmetiev 1989) to the Miocene (e.g., Takhtajan 1958). Flowers in Baltic amber from the Kalingrad mines have been compared with *Phoenix* (Poinar 2002b) and *Trachycarpus* is found in the Oligocene and Miocene (Takhtajan 1958, Akhmetiev 1989).

Middle East

Pollen is recorded from the Lower Eocene Shumaysi Formation of Saudi Arabia (cf. *Mauritia*; Srivastava & Binda 1991).

African Continent

Spinizonocolpites occurs in the Lower to Middle Eocene of Morocco (see Gee 1990). *Nipadites burtini* is known from the Palaeocene to Eocene of Egypt (Bonnet 1904, Kräusel 1939, Chandler 1954, Gregor & Hagn 1982). *Medemia argun* subfossil fruits are recorded from Egyptian tombs (Kunth 1826, Newton 2001). From the Oligocene of Ethiopia (Guang River Flora), there is an incomplete leaflet of *Eremospatha*, and spiny petioles and incomplete petioles of *Hyphaene* (Pan *et al.* 2006). From Senegal, Fritel (1921) reports Eocene fruits of *Nypa*; Médus (1975) records an uncertain record of *Mauritiidites franciscoi*, *Elaeis*-like pollen from the Middle Eocene to Lower Miocene, and pollen grains that strongly resemble *Sclerosperma* from the Miocene; Ergo (1997) describes seeds of *Elaeis* also from the Upper Miocene. In the Eocene of Nigeria, *Mauritiidites* and *Spinizonocolpites* are recorded by Jan du Chêne *et al.* (1978). From the Lower Eocene of Cameroon, Salard-Cheboldaeff (1978, 1981) compares *Echimonocolpites rarispinosus* with *Lepidocaryum* pollen but this association is somewhat uncertain. The palm fossil record for Africa is well-documented in Pan *et al.* (2006) (see also Morley [2000]).

Indian Subcontinent

Spinizonocolpites and *Longapertites* from the Palaeocene of Pakistan are reported by Frederiksen (1994). *Dicolpopollis* has been described from Eocene lignites in Burma (Potonié 1960).

The palm fossil record for the Indian subcontinent during the Tertiary is very rich, possibly representing at least 17 modern genera. *Nypa* fruits, leaf fragments and cuticle, together with *Nypa*-like pollen (*Spinizonocolpites*), are reported from the Upper Cretaceous to the Miocene in a number of localities (see Singh 1999). The volcanic Deccan Intertrappean beds, which occur in Madhya Pradesh (north central India) and Maharashtra State (west India), are a rich source of fossil plants (frequently petrified), including palms. The age of the Deccan Intertrappeans is controversial. Some geologists consider their age span to be of short duration whereas others, including Morley (2000), suggest that they could extend from the Upper Cretaceous to the mid-Oligocene. We accept Morley's opinion here. Other key strata for palm fossils include many of India's large Tertiary (Lower Eocene–Lower Miocene) lignite deposits, including the Makum and Dilli Jeypore coalfields of Assam in the extreme north east, the Rajpardi and Panandho lignites of Gujarat in the extreme west, the Ratnagiri lignites of Maharashtra, and the southeastern Neyveli lignites (Saxena 1992) of the Cauvery Basin, Tamil Nadu (Neyveli, South Arcot and Tiruchirapalli districts). Many Miocene palm fossils have also been provided by cores from bore wells for oil, natural gas or water drilling in a number of regions; for example, the Cauvery Basin, the Godavari-Krishna Basin (Andhra Pradesh) and the Warkalli and Quilon Beds of the Kerala Basin. (See also Rao & Achuthan [1973], Prakash [1974], Morley [2000]; Table 5.5.)

Malay Islands

From the Middle Eocene Nanggulan Formation of Central Java, *Dicolpopollis* is recorded (Takahashi 1982, Harley & Morley 1995) and different collections of *Palmaepollenites* (see for example Fig. 5.7) have been compared with *Pritchardia*, *Actinorhytis*, *Basselinia*, *Burretiokentia*, *Cyphophoenix*, *Cyphosperma* and *Cyphokentia* (Harley &

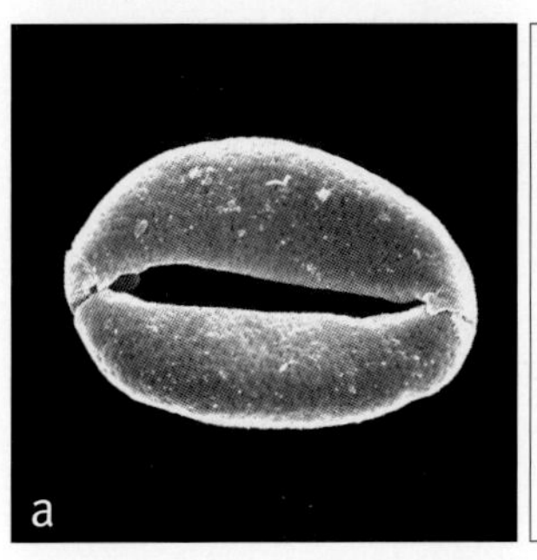
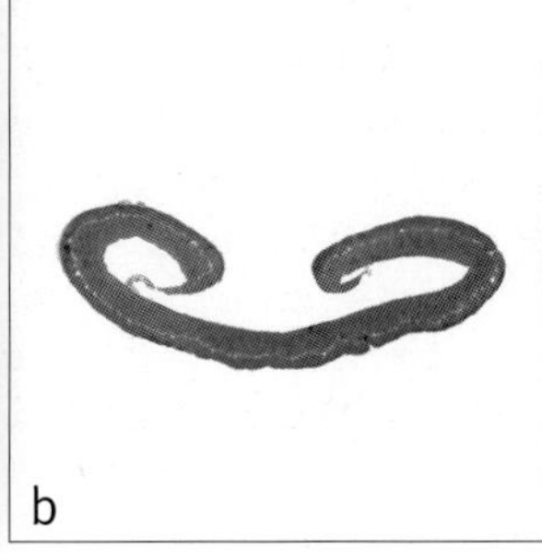
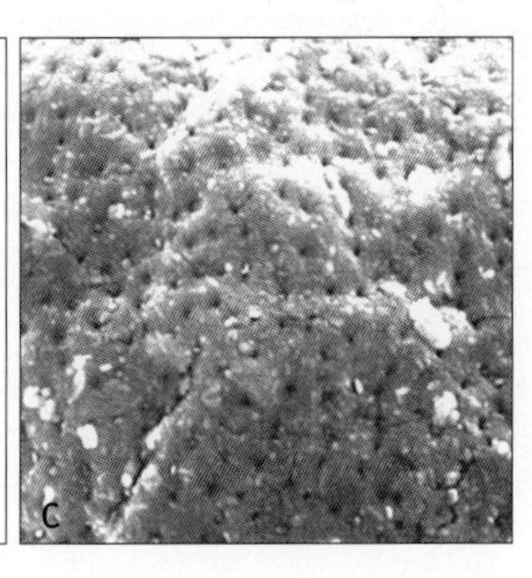

Fig. 5.7 ***Palmaepollenites kutchensis*** Venkatachala & Kar., Nanggulan Formation, Central Java. *Morley* XIV-H. **a**, Eocene monosulcate pollen grain SEM × 800; **b**, Eocene, ultrathin section through monosulcate pollen grain shown in (a) TEM × 1000; **c**, Eocene, surface of monosulcate pollen grain shown in (a) SEM × 5000.

TABLE 5.5
Tertiary of India: summary of palm fossils attributed to genera

REGION OF INDIA	EPOCH	GEOLOGICAL FORMATION	PALAEOTAXON	FOSSIL TYPE	AUTHOR(S)
NE: Assam	Oligocene–Lower Miocene	Makum & Dilli Jeypore coalfield	cf. *Nypa*	Fruits and leaf fragments	Mehrotra *et al.* 2003
NE: West Bengal	Miocene	Equivalent to Tipam Series	*Palmoxylon coronatum* cf. *Borassus*	Stem wood	Roy & Ghosh 1980
NE: Meghalaya	Miocene	Shillong Plateau	*Dicolpopollis*	Pollen	Salujha *et al.* 1973a, 1973b
NW: (extreme) Jammu and Kashmir	Miocene	Upper Indus valley; Liyan Formation	*Trachycarpus ladakhensis*	Palmate leaves	Lakhanpal *et al.* 1984
NW: Rajasthan	Eocene	Fuller's Earth mines	*Cocos sahnii*	Endocarp	Kaul 1951, Guleria *et al.* 1996
N Central: Madhya Pradesh	Cretaceous–mid Oligocene	Deccan Intertrappean	*Amesoneuron borassoides*	Fragment of large palmate/costapalmate leaf	Bonde 1986a
			cf. *Sabal*	Leaf axis	Trivedi & Verma 1981
			Phoenicicaulon mahabalei	Sheathing leaf base	Bonde *et al.* 2000
			Palmostroboxylon arengoideum	Peduncle	Ambwani 1984
			Hyphaenocarpon indicum	Fruit	Bande *et al.* 1982
			cf. *Cocos*	Small cocoid-like petrified fruit	Tripathi *et al.* 1999
			Cocos intertrappeansis	Small cocoid-like fruit	Patil & Uphadhye 1984
			Palmoxylon sundaram cf. *Cocos*	Very large, almost complete stem	Sahni 1946
			Palmoxylon ghuguensis, cf. *Dypsis*	Petrified stem	Ambwani & Prakash 1983
			Palmoxylon toroides cf. *Corypha*	Stem wood	Ambwani & Mehrotra 1989
			Palmoxylon shahpuraensis cf. *Licuala*	Stem wood	Ambwani 1983
			Palmoxylon coronatum cf. *Borassus*	Stem wood	Sahni 1964
			cf. *Nypa*	Stems and rhizomes	Verma 1974
W: Gujarat	Lower Eocene	Naredi Formation: Panandhro lignite	*Arengapollenites echinatus*	Pollen	Kar 1985, Kar & Bhattacharya 1992
		Kutch Basin: Rajpardi lignite	*Arengapollenites echinatus*, *Arengapollenites ovatus*	Pollen	Kar & Bhattacharya 1992
W: Maharashtra State	Lower Miocene	Ratnagiri lignite	cf. *Nypa*	Cuticle	Kulkarni & Phadtare 1980
			Eugeissonocarpon	Fruit	Shinde & Kulkarni 1989
			Quilonipollenites cf. *Eugeissona*	Pollen	Phadtare & Kulkarni 1980, 1984

TABLE 5.5 (continued)

REGION OF INDIA	EPOCH	GEOLOGICAL	PALAEOTAXON	FOSSIL TYPE	AUTHOR(S)
W: Maharashtra State	Cretaceous–mid Oligocene	Deccan Intertrappean	*Sabalophyllum livistonoides*	Palmate leaf (transverse section)	Bonde 1986b
			Palmocaulon hyphaenoides	Petrified petiole	Shete & Kulkarni 1980
			cf. *Licuala*	Silicified fruits	Shete & Kulkarni 1985
			cf. *Phoenix*	Seeds	Ambwani & Dutta 2005
			Palmoxylon kamalan cf. *Roystonea*	Stem 'trunk'	Kulkarni & Mahabalé 1971
			Palmoxylon livistonoides	Stem wood	Prakash & Ambwani 1980
			cf. *Borassus flabellifer*	Root	Ambwani 1981
			Dicolpopollis	Pollen	Saxena & Misra 1990
E: Andhra Pradesh	Miocene	Krishna-Godavari Basin	*Dicolpopollis*	Pollen	Ramanujam *et al.* 1986, 2001
SE: Tamil Nadu	Lower-Middle Miocene	Cauvery Basin: Neyveli, South Arcot and Tiruchirapalli districts	*Dicolpopollis*	Pollen	Sarma *et al.* 1984, Ramanujam *et al.* 1986
			Gemmamonocolpites hyphaenoides	Pollen	Ramanujam *et al.* 2001
			Jacobipollenites cf. *Borassodendron*	Pollen	Ramanujam 1966, Ramanujam *et al.* 1998
			Quilonipollenites cf. *Eugeissona*	Pollen	Ramanujam & Reddy 1984, Sarma *et al.* 1984
SW: Kerala State	Lower Miocene	Kerala Basin: Warkalli & Quilon Beds	*Dicolpopollis*	Pollen	Ramanujam 1987, Ramanujam & Rao 1977, Ramanujam *et al.* 1986, 1991a, 1991b, 1992, 2001, Rao & Ramanujam 1975, 1978, Singh & Rao 1990, Srisailam & Ramanujam 1982, Varma *et al.* 1986
			Paravuripollis cf. *Korthalsia*	Pollen	Ramanujam 1987, Ramanujam *et al.* 1992
			Quilonipollenites cf. *Eugeissona*	Pollen	Rao & Ramanujam 1978, Ramanujam *et al.* 1991a, b, 1992
			Spinizonocolpites	Pollen	Ramanujam 1987, Rao & Ramanujam 1975, 1978, Srisailam & Ramanujam 1982

Morley 1995). In Borneo, *Nypa* fruits are known from the Eocene (Kräusel 1923), and *Dicolpopollis* is recorded from the Palaeocene (Muller 1968) and Miocene (Muller 1979). Other pollen records from Borneo include cf. *Oncosperma* from the Oligocene (Muller 1964, 1972, 1979), cf. *Arenga* (Muller 1972, 1979) and cf. *Nenga* (Muller 1964, 1972, 1979) from the Lower Miocene, cf. *Cyrtostachys* from the Upper Miocene (Muller 1972) and cf. *Borassodendron machadonis* from the Pliocene to Lower Quaternary (Caratini & Tissot 1985). Pollen has also been reported from the Middle Miocene of Brunei (cf. *Cyrtostachys*; Morley 2000) and from the Middle Miocene Malawa Formation in Southern Sulawesi (cf. *Eugeissona*; Morley 1998). (See Morley [2000] for more detailed commentary.)

Eastern Asia

Four species of *Sabalites* leaves are reported from the Tertiary of China (Peking Institute of Botany and Nanjing Institute of Geology and Palaeontology 1978). Pollen records from the Palaeocene–Middle Miocene of China include *Spinizonocolpites*, *Dicolpopollis* and *Sabalpollenites areolatus* (Song *et al.* 1999).

Australia and Papua New Guinea

There are records of *Spinizonocolpites* from the Palaeocene of the Perth Basin (Churchill 1973) and from Western Tasmania (Cookson & Eisenack 1967), *Nypa* pollen from the Lower to Middle Eocene of the Gippsland Basin in southeast Victoria (Stover & Evans 1973, Stover & Partridge 1973), and *Nypa* pollen, fruits and leaves from the Eocene of Tasmania (Pole & Macphail 1996). There is also a record of a silicified cocosoid fruit from the Upper Pliocene of Queensland (Rigby 1995). *Dicolpopollis bungonensis* from the Eocene of New South Wales is favourably compared with the pollen of *Calamus moti* F.M. Bailey (Truswell & Owen 1988). (See also Greenwood & Conran [2000] for a summary of macro fossil records.)

There is an Upper Miocene record of *Dicolpopollis metroxylonoides* (Khan 1976) from Papua New Guinea.

New Zealand

In New Zealand, North Island, there are records of small cocosoid-like endocarps, *Cocos zeylandica*, from the Miocene and Pliocene (Berry 1926b, Ballance *et al.* 1981) and of Pliocene pollen (cf. *Cocos* and *Rhopalostylis*; Couper 1952, 1953). On South Island, there are records of cocosoid-like endocarps from the Middle Eocene to Lower Oligocene (Campbell *et al.* 2000), and a rare record of *Dicolpopollis metroxylonoides* from the Miocene of Central Otago (Mildenhall & Pocknall 1989).

South East Polynesia

Subfossil cocosoid endocarps, *Paschalococos disperta*, have been described from a cave floor on Easter Island (Dransfield *et al.* 1984).

CONCLUSIONS

The palms are very well-represented in the fossil record (Table 5.6). Nevertheless, it is reasonable to suppose that, for any one geological epoch, the fossil record represents no more than a tantalizingly small proportion of the palm species that existed during that time. The earliest unequivocal fossil palm, although not assignable below family level, is apparently fossilised stem wood from the Turonian of France. This is followed by costapalmate leaves, which are clearly associated with subfamily Coryphoideae, from the Coniacian and Santonian of eastern USA (Carolina and Maryland) and then by pinnate leaves from the Campanian of north western USA (Montana). By the latest Cretaceous–earliest Tertiary, leaves, stems, fruits, inflorescences and pollen, undoubtedly representing the Arecaceae, are present. By this time, there is also a range of variation within each organ category, suggesting that the palms are already a well-established lineage rather than an emergent group.

The presence of unmistakable coryphoid, costapalmate leaves around 86 million years ago, and of highly distinctive palm pollen (*Spinizonocolpites*, *Mauritiidites* and *Longapertites*) ca. 70 million years ago, provokes speculation regarding the true age of the palms. Monosulcate forms resembling palm pollen, although almost impossible to identify to genera, are known from at least the latest Lower Cretaceous (Barremian–Cenomanian). The much later Maastrichtian forms, such as *Spinizonocolpites* and *Mauritiidites* (both spiny but with strikingly different ultrastructure) appear, by comparison, to be more specialised (i.e., developmentally evolved) structures.

Fossil pollen records for *Nypa* fruits and pollen allow us to trace the expansion and contraction of the geographic distribution of the *Nypa* lineage from the Maastrichtian until the Upper Miocene. The pollen record for *Nypa* also suggests that it may not always have been a monotypic genus. The demise of palms from latitudes exceeding 44° from the equator (see Chapter 7) is largely explained by cooling episodes that occurred from the Miocene onwards, but this does not account for the demise of palms in some regions of the tropics and subtropics, notably in Africa (Dransfield 1988b; Pan *et al.* 2006) and India, where fossil evidence suggests that the palm floras have been historically much richer.

One of the problems in providing a comprehensive fossil record is the almost serendipitous nature of fossil recovery, and the limits this imposes on the range and quantity of recovered fossils. For economic reasons, Cretaceous and Tertiary palaeobotany relies mainly on bedding planes or cores made available either by natural events or by drilling, mining and quarrying. Work on fossilised pollen is less affected by these limitations than are studies of macrofossil organs because, for example, coring for pollen samples is a less expensive undertaking than quarrying to expose possibly plant-rich bedding planes. Nevertheless, the records of both pollen and macro fossils are richer in the Northern Hemisphere than in the Southern, largely reflecting the more extensive development and commercial exploitation of land in the Northern Hemisphere. The African continent, for example, has few fossil records south of the Equator (see Pan *et al.* 2006). The present richness of the Madagascar palm flora, when compared with that of mainland Africa, is not reflected in its almost non-existent palm fossil record. This discrepancy could reflect a curious reversal of fortunes between the palm floras of Madagascar and mainland Africa or, perhaps more plausibly, Madagascar's lack of prevailing conditions in the past to effect rapid burial of plant material into suitably protective

TABLE 5.6
Systematic distribution of genera with confirmed or suggested fossil counterparts

SUBFAMILY	TRIBE	SUBTRIBE	GENUS
CALAMOIDEAE	**Eugeissoneae**		1. ***Eugeissona***
	Lepidocaryeae	Ancistrophyllinae	3. ***Eremospatha***
		Mauritiinae	6. ***Lepidocaryum***
			7. ***Mauritia***
	Calameae	Korthalsiinae	9. ***Korthalsia***
		Salaccinae	11. ***Salacca***
		Metroxylinae	12. ***Metroxylon***
		Calaminae	17. ***Calamus***
			19. ***Daemonorops***
NYPOIDEAE			22. ***Nypa***
CORYPHOIDEAE	**Sabaleae**		23. ***Sabal***
	Cryosophileae		25. ***Trithrinax***
			30. ***Thrinax***
			32. ***Cryosophila***
	Phoeniceae		34. ***Phoenix***
	Trachycarpeae	Rhapidinae	35. ***Chamaerops***
			37. ***Trachycarpus***
			38. ***Rhapidophyllum***
		Livistoninae	41. ***Livistona***
			42. ***Licuala***
		Unplaced	47. ***Serenoa***
			48. ***Brahea***
			51. ***Pritchardia***
	Chuniophoeniceae		55. ***Nannorrhops***
	Caryoteae		57. ***Caryota***
			58. ***Arenga***
	Corypheae		60. ***Corypha***
	Borasseae	Hyphaeninae	63. ***Hyphaene***
			64. ***Medemia***
		Lataniinae	67. ***Borassodendron***
			68. ***Borassus***
CEROXYLOIDEAE	**Cyclospatheae**		69. ***Pseudophoenix***
	Ceroxyleae		70. ***Ceroxylon***
	Phytelepheae		76. ***Phytelephas***
ARECOIDEAE	**Iriarteeae**		78. ***Dictyocaryum***
			79. ***Iriartea***
			80. ***Socratea***

(contd.)

TABLE 5.6 (continued)

SUBFAMILY	TRIBE	SUBTRIBE	GENUS
ARECOIDEAE	**Chamaedoreeae**		85. ***Chamaedorea***
	Sclerospermeae		89. ***Sclerosperma***
	Roystoneeae		90. ***Roystonea***
	Reinhardtieae		91. ***Reinhardtia***
	Cocoseae	Attaleinae	96. ***Attalea***
			97. ***Butia***
			98. ***Cocos***
			99. ***Jubaea***
			101. ***Syagrus***
		Bactridinae	103. ***Acrocomia***
			104. ***Astrocaryum***
			105. ***Aiphanes***
			106. ***Bactris***
		Elaeidinae	109. ***Elaeis***
	Manicarieae		110. ***Manicaria***
	Euterpeae		114. ***Neonicholsonia***
	Geonomateae		121. ***Geonoma***
	Areceae	Archontophoenicinae	125. ***Actinorhytis***
		Arecinae	130. ***Areca***
			131. ***Nenga***
		Basseliniinae	133. ***Basselinia***
			134. ***Burretiokentia***
			135. ***Cyphophoenix***
			136. ***Cyphosperma***
		Clinospermatinae	142. ***Cyphokentia***
		Dypsidinae	144. ***Dypsis***
		Oncospermatinae	152. ***Oncosperma***
		Rhopalostylidinae	168. ***Rhopalostylis***
		Verschaffeltiinae	172. ***Roscheria***
		Unplaced	176. ***Cyrtostachys***
			180. ***Hydriastele***

anaerobic conditions followed by the slow process of fossilisation. A recent expedition to search for fossil pollen evidence of palms and other plants in northwest Madagascar (Zavada & Harley unpublished) was wholly unsuccessful because all the Tertiary exposures visited proved to be coarse sands, which are notoriously poor for pollen preservation (Horowitz 1992).

An impressive body of palm fossil data from plant-bearing geological strata in many regions of the world has been summarised here. It represents the accumulation of numerous observations and detailed studies made during the past two hundred years. However, many mysteries about the history of palms remain to be solved; to mention but a few, the origin of *Cocos*; the apparent origin of *Mauritia* in Africa, and its subsequent dispersal to South America followed by extinction in Africa; the probable species extinctions indicated by the fossil record of *Nypa* pollen; and the palaeogeography of the calamoid palms and of *Phoenix* in Europe and elsewhere. Further fossil evidence might provide greater insight into some of these questions as well as providing new data for other taxa. Clearly there is scope, not only for carefully focused re-exploration of some of the known palm fossil localities, as advocated for North America by Uhl and Dransfield (1987), but also for exploring the potential of other areas, including the less commercially exploited Southern Hemisphere.

A positive input into phylogenetic analyses from the fossil record for palms (or any other plant group) relies on highly recognisable fossils and well-dated strata. Therefore, opportunities to re-consider unresolved questions regarding the morphology and relationships of some of the important collections of palm macro fossils and pollen already described must not be overlooked. Notably, many of the existing collections of fossil fruits, seeds and endocarps are in need of careful attention from palm experts, as are the induplicate pinnate leaves from the Campanian of Montana. Then there are all the pinnate leaves that cannot be attributed to *Phoenix*, now grouped together as *Phoenicites* — forever? The published data for living palms and the many excellent reference collections of living and herbarium material provide the tools and the challenge to palm specialists to apply their expertise to some of the outstanding issues regarding the affinities of fossils with living members of the Arecaceae. It is time for palm specialists, palaeobotanists and palynostratigraphers to work together to develop a more synthetic approach to palm fossils and the origins of the Arecaceae. Perhaps then, it will only be a matter of time before unequivocal palm fossils predating the Turonian are discovered.

Chapter 6
PHYLOGENY AND EVOLUTION

AN INTRODUCTION TO PHYLOGENETIC CLASSIFICATION

Over recent decades, systematic biology has been transformed by a phylogenetic revolution. Phylogenetics is the part of the broader field of systematics that focuses on the evolutionary relationships between organisms. Modern phylogenetic studies have superseded earlier studies of evolutionary relationships because they are based on explicit analyses of clearly formulated datasets of characters rather than on subjective interpretations. No other major group of organisms has benefited more from phylogenetic research than the angiosperms (flowering plants). Nowadays, most systematic studies include some kind of phylogenetic component, and phylogenies have far-reaching implications beyond systematics, underpinning comparative and evolutionary biology as well as much applied research, for example, in human health and agriculture.

Phylogenetic analyses yield phylogenetic trees (also termed phylogenies), which are dichotomously branching diagrams that are sometimes likened to a family tree or a 'tree of life'. The point of divergence between two branches is termed a node and is taken to represent a hypothesis of shared ancestry of the resultant lineages. Phylogeneticists use various methods to evaluate their confidence in these hypotheses of relationship; the most common of these are bootstrap and jackknife analyses. High bootstrap or jackknife values justify greater confidence in a particular relationship.

Phylogenetic trees can be used to test existing classifications and to build new ones. Different types of grouping can be recognised within the structure of a phylogenetic tree. Monophyletic groups consist of all descendants of a common ancestor. They are informative in classification because they can be readily defined, and they are deemed meaningful and 'natural' in an evolutionary context because shared ancestry implies that the members of the group share common properties and history. A group that contains some, but not all, descendants of a common ancestor is termed paraphyletic. Paraphyletic groups are not useful in classification because they do not reflect history and are defined arbitrarily by the characters and taxa that a worker chooses to exclude subjectively. Though paraphyletic groups are not desirable or meaningful in a phylogenetic context, they can possess a clear-cut morphology, defined for the convenience of the human eye. Reptiles, for example, are easily recognised but are paraphyletic because birds are embedded within them. Taxonomic changes to such groups can be both controversial and unpopular. Polyphyletic groups are more readily discounted, comprising organisms that do not share a common ancestor, but have acquired superficially similar morphology through evolutionary convergence that has been misinterpreted as evidence of relationship. In a phylogenetic classification, only monophyletic groups are recognised.

The classification of palms used in this book is phylogenetic. Many groups in the classification are unchanged, at least in circumscription, from the previous classification (Uhl & Dransfield 1987) because their monophyly has been confirmed. However, phylogenetic research has identified that some previously recognised groups of palms were not monophyletic, and has thus resulted in some significant amendments to the classification.

HISTORY OF PALM PHYLOGENETICS

Studies of palm evolution pre-date the first use of formal phylogenetic methods in palms (e.g., Moore 1973a, Moore & Uhl 1982, Dransfield *et al.* 1990). Among these, the first two studies have been particularly influential. Harold E. Moore's seminal paper (1973a), "The major groups of palms and their distribution", represented the first explicit attempt to build a classification of the family that reflected notions of evolutionary relationship among the genera. The groups were proposed as hypotheses of relationship that might stimulate further discussion and analysis. Thus, Moore avoided formal nomenclature in this paper, giving each group informal names only. The arrangement of genera was built in part around concepts of transformation from primitive to derived states in morphological and anatomical characters, presented as "criteria used in evaluating specialization in palms". These so-called "major trends of evolution" were described and justified in greater detail subsequently (Moore & Uhl 1982). Although the "major trends" lack a formal cladistic foundation, they were based on extensive comparative studies within the family and also within monocots as a whole. Modern phylogenetic approaches call many of the major trends into question but they remain as important syntheses of information and as valuable stimuli for hypothesis testing. These papers were also significant for their contribution to the classification and rationale of the first edition of *Genera Palmarum* (Uhl & Dransfield 1987).

Opposite: ***Pritchardia viscosa***, Kauai, Hawaii. (Photo: J. Dransfield)

TABLE 6.1 An overview of published phylogenetic studies of palms
The studies are grouped according to the taxonomic rank that they span. Within each taxonomic rank, the papers are listed chronologically.

Taxonomic level	Reference	Group	Plastid DNA	Nuclear DNA	Morphology
Family	Uhl *et al.* 1995	Arecaceae	RFLP		✓
	Baker *et al.* 1999a	Arecaceae	*trn*L-*trn*F		
	Asmussen *et al.* 2000	Arecaceae	*rps*16, *trn*L-*trn*F		
	Asmussen & Chase 2001	Arecaceae	*rbc*L, *rps*16, *trn*L-*trn*F		
	Lewis & Doyle 2001	Arecaceae		*ms*	
	Hahn 2002a	Arecaceae	*atp*B, *rbc*L	18S	✓
	Asmussen *et al.* 2006	Arecaceae	*mat*K, *rbc*L, *rps*16, *trn*L-*trn*F		
	Henderson & Stevenson 2006	Arecaceae			✓
Subfamily	Baker *et al.* 1999b	Calamoideae			✓
	Baker *et al.* 2000a	Calamoideae	*rps*16	ITS	✓
	Baker *et al.* 2000b	Calamoideae	*rps*16	ITS	
	Hahn 2002b	Ceroxyloideae Arecoideae	*atp*B, *rbc*L, *ndh*F, *trn*D-*trn*T, *trn*Q-*rps*16		
	Savolainen *et al.* 2006	Ceroxyloideae Arecoideae		*prk*, *rpb*2	
	Trénel *et al.* 2007	Ceroxyloideae	*mat*K, *ndh*F, *trn*D-*trn*T	*prk*, *rpb*2	
Tribe	Henderson 1990	Iriarteeae			✓
	Barfod 1991	Phytelepheae			✓
	Asmussen 1999a	Geonomateae	*rpl*16		
	Asmussen 1999b	Geonomateae	*rps*16		
	Barfod *et al.* 1999	Phytelepheae			✓
	Henderson 1999a	Euterpeae			✓
	Lewis & Doyle 2002	Areceae		*ms*, *prk*	
	Gunn 2004	Cocoseae		*prk*	
	Roncal *et al.* 2005	Geonomateae		*prk*, *rpb*2	
	Norup *et al.* 2006	Areceae		*prk*, *rpb*2	
	Cuenca & Asmussen-Lange 2007	Chamaedoreeae	*mat*K, *ndh*F, *rps*16, *trn*D-*trn*T		
	Roncal et al. 2008	Cryosophileae		*prk*, *rpb*2	
Subtribe	Pintaud 1999a	Archontophoenicinae			✓
	Zona 1999a	Ptychospermatinae			✓
	Baker *et al.* 2000c	Calaminae		5S Spacer	
	Lewis 2002	Oncospermatinae Verschaffeltiinae			✓
	Loo *et al.* 2006	Arecinae		*prk*, *rpb*2	

(Contd.)

TABLE 6.1 (continued)

Taxonomic level	Reference	Group	Plastid DNA	Nuclear DNA	Morphology
Genus	Zona 1990	*Sabal*			✓
	Evans 1995	*Cryosophila*			✓
	Salzman & Judd 1995	*Bactris*			✓
	Zona 1996	*Roystonea*			✓
	Barrow 1998	*Phoenix*		5S Spacer	✓
	Barrow 1999	*Phoenix*		5S Spacer	✓
	Ferreira 1999	*Bactris*			✓
	Fuller 1999	*Physokentia*			✓
	Hahn & Systma 1999	*Caryota*	RFLP		
	McClatchey 1999	*Metroxylon*			✓
	Zona 1999b	*Drymophloeus*			✓
	Henderson 2002a	*Reinhardtia*			✓
	Stauffer *et al.* 2003	*Asterogyne*			✓
	Loo *et al.* 2006	*Hydriastele*		*prk*, *rpb2*	
	Norup *et al.* 2006	*Heterospathe* *Rhopaloblaste*		*prk*, *rpb2*	
	Thomas *et al.* 2006	*Chamaedorea*		*prk*, *rpb2*	
	Couvreur *et al.* 2007	*Bactris*	*trn*D-*trn*T, *trn*Q-*rps*16, *psb*C-*trn*S, *trn*S-*trnf*M	Microsatellites	
Total number	**46 studies**		**15 studies, 11 DNA regions, 2 RFLP studies**	**17 studies, 6 DNA regions, 1 microsatellite study**	**24 studies**

Palm researchers have now been working with phylogenies for almost two decades and have published phylogenies of many palm groups, both at lower and higher taxonomic levels. This relatively early start on palm phylogeny reconstruction, followed by a focused collaborative effort, has led directly to the phylogenetic basis for the classification in this book, an earlier version of which was published by Dransfield *et al.* (2005).

Following the publication of the first phylogenies of a group of palms (Henderson 1990, Zona 1990), the publication of palm phylogenies was slow for most of the 1990s (Barfod 1991, Evans 1995, Salzman & Judd 1995, Uhl *et al.* 1995). In 1999, the number of palm phylogenies increased dramatically with the appearance of 14 new papers, largely thanks to the publication of the proceedings of a conference held at the New York Botanical Garden in 1997 entitled "Evolution, Variation, and Classification of Palms" (Henderson & Borchsenius 1999). Since then, a continuous flow of palm phylogenies has emerged (Table 6.1).

Most of the earlier palm phylogenies were based on morphological characters alone. Molecular approaches to palm phylogenetic research were first exploited by Uhl *et al.* (1995) with their comprehensive family-wide study based on plastid DNA restriction fragment length polymorphisms (RFLP). The first DNA sequence-based phylogeny of the palm family was published a few years later (Baker *et al.* 1999a), initiating a series of studies that have been influential in higher level palm classification (Asmussen *et al.* 2000, Asmussen & Chase 2001, Asmussen *et al.* 2006). More than thirty molecular phylogenetic studies of palms have now been completed, each contributing in different ways to advancing the systematics of the family.

MORPHOLOGICAL DATA

The inexorable rise of phylogenetic systematics can be attributed to conceptual, theoretical and methodological advances, coupled with the exponential increase both in available computing power and the ease with which large datasets, principally of DNA sequences, may be generated. Phylogenetic analyses may be based on many different types of data, morphology (including micromorphology) and DNA sequences being the most common. As the cost and effort required to generate DNA sequence data has fallen, the morphological studies that characterised the early days of phylogenetic research (studies of palms being no exception) have been overtaken by molecular phylogenetic approaches. It is often claimed that the dominance of large DNA sequence datasets over morphological datasets has had a detrimental impact on the study of plant morphology in general. It is true that DNA sequences have certain advantages, for example, large amounts of data can be generated quickly and inexpensively, and DNA characters appear, at least superficially, to be less prone to problems of interpretation. The construction of a morphological dataset presents specific conceptual challenges, particularly in homology assessment and the unambiguous definition of characters and character states (Scotland *et al.* 2003). Incongruence between morphological and molecular datasets (e.g., Baker *et al.* 2000a) also promotes the belief that morphological data constrain accurate phylogenetic inference. In practical terms, however, DNA datasets are now providing ever larger and increasingly robust phylogenetic frameworks within which morphological data may be incorporated and explored in various ways, and have thus opened dramatic new avenues for studies of the evolution of plant form and function, among other aspects of evolutionary biology.

To date, morphological data have been utilised in 24 palm phylogenetic studies (Table 6.1). These studies tend to date from the earlier days of palm phylogenetic research and are also biased towards lower taxonomic levels, especially the intrageneric level. Only a small number of studies feature both morphological and molecular data (Uhl *et al.* 1995, Barrow 1998, 1999, Baker *et al.* 2000a, Hahn 2002a) and thus little can be said regarding the relative performance of the two data types. It is clear, however, that groups that are well-defined by unambiguous morphological synapomorphies can be recovered as easily with morphological evidence as with DNA data. More cryptic and unexpected relationships, especially those at deeper nodes, are less readily recovered by morphology alone.

DNA DATA

Plant cells contain three genomes, the largest within the nucleus and much smaller genomes resident in each of two organelles, the plastid and the mitochondrion. These three genomes have different qualities that affect both the practical ease with which they can be explored and the amount of information that they can contribute to phylogenetic research. The rate of evolution of a DNA region is of particular interest because it determines the degree of sequence divergence between taxa and thus the taxonomic level at which the region may prove informative. In general, the much larger nuclear genome evolves faster than the plastid and the mitochondrial genomes; and the plastid genome evolves faster than the mitochondrial genome. This generalisation may also be true for palms, but sufficiently broad comparative analyses based on numerous taxa and DNA regions are lacking. In those few studies in which direct comparisons have been made between plastid and nuclear DNA regions (Baker *et al.* 2000b, Hahn 2002a, Cuenca & Asmussen-Lange 2007, Trenel *et al.* 2007), however, nuclear regions consistently yield more variable sites per length of DNA sequenced than plastid regions, and thus appear to evolve more rapidly.

Molecular phylogenetic studies of palms have progressed tremendously since the first review of the molecular phylogeny of the Arecaceae (Hahn 1999). At that time, only six studies had been published and just a few others were under way. On the basis of this information, Hahn predicted that the future would bring palm phylogenies based on a combination of plastid, nuclear, and mitochondrial DNA regions. He also suggested that the plastid genes *mat*K and *ndh*F, the nuclear gene *adh*, and the mitochondrial gene *atp*A would be included in these future studies. Of these regions, only *mat*K and *ndh*F have actually been utilised to date; *adh* has not been used in palm phylogenies, probably because of the presence of multiple paralogues, and mitochondrial genes have not been used at all for any palm phylogeny. Nevertheless, many studies based on combinations of multiple datasets have now been completed and published.

Plastid DNA

Plastid DNA regions are attractive as molecular markers in plant phylogenetics for three main reasons. First, the plastid genome has been sequenced in its full length for a number of plant taxa, providing invaluable information for the development of new DNA regions for phylogeny reconstruction. Second, each plant cell contains numerous plastids. Thus, every cell contains numerous identical copies of the plastid genome whereas there is only one copy of the nuclear genome per cell. The abundance of identical copies of the plastid genome makes the isolation of plastid DNA regions by polymerase chain reaction (PCR) amplification more straightforward. Third, many comparative studies of plastid DNA regions are available that facilitate informed choices of appropriate regions for a given study.

Initially, RFLPs from the plastid genome rather than plastid DNA sequences were used for phylogenetic purposes (Uhl *et al.* 1995, Hahn & Sytsma 1999). This approach explores the genetic diversity of the entire plastid genome rather than the diversity of selected regions only. However, DNA sequences have inevitably superseded RFLP data because they represent direct observations of DNA variation, they raise fewer problems of homology assessment and they pose fewer technical challenges, reducing the quality of the DNA that is required to secure good results.

Eleven different plastid DNA regions have been used to reconstruct palm phylogenies in addition to plastid DNA RFLPs (Table 6.1). The DNA regions used to date are the coding regions *rbc*L, *atp*B, *mat*K and *ndh*F; the introns of *rps*16 and *rpl*16; and the intergenic spacers *trn*L-*trn*F, *trn*D-*trn*T, *trn*Q-*rps*16, *psb*C-*trn*S and *trn*S-*trnf*M, some of which include parts of short transfer RNA genes that may also include introns (e.g., the *trn*L intron).

A number of comparative studies have concluded that the plastid genome of palms has a slow rate of nucleotide substitution when compared with those of other monocots (Wilson *et al.* 1990, Gaut *et al.* 1992, 1996), and this conclusion has been cited by numerous subsequent palm researchers (e.g., Asmussen 1999a, Baker *et al.* 1999a). Initial estimates of substitution rate based on plastid RFLP data indicated that nucleotide substitution rates in palms are 5–13 times lower than those of annual monocot and dicot species (Wilson *et al.* 1990). Subsequently, relative rate tests based on *rbc*L data from a range of monocot families suggested that substitution rates in palms are lower than those of many other monocot groups and are more than 5 times lower than those of grasses (Gaut *et al.* 1992). Though widely accepted, these conclusions require further verification for two reasons (Cuenca & Asmussen-Lange 2007). First, the DNA sequence-based comparisons were focused on *rbc*L alone and thus did not take into account variation between DNA regions. Second, more recent studies have focused increasingly on rate differences between palms and grasses (Gaut *et al.* 1992, 1996, Morton *et al.* 1996). Life-history characteristics, evolutionary history and selection may account for heterogeneous substitution rates among lineages (Gaut *et al.* 1992, 1996, Felsenstein 2004). Differences in these factors between palms and grasses, such as their contrasting life histories, may bias the comparison. Moreover, grasses appear to evolve rapidly relative to other groups that have similar life histories and their molecular evolution is unusually complicated. Clearly, a broader survey of plastid DNA regions across a range of different families is required.

The low rate of plastid DNA evolution in palms has been suggested as the primary limitation of plastid DNA as a tool to resolve relationships between closely related species (Asmussen 1999a, 1999b, Hahn 2002b, Lewis & Doyle 2002, Gunn 2004, Roncal *et al.* 2005, Thomas *et al.* 2006). Slow rates of molecular evolution in plastid DNA remain to be demonstrated convincingly for palms. Nevertheless, practical experience has shown that some plastid DNA regions that have been used at the species level in other angiosperm families, such as *trn*L-*trn*F and the *rps*16 intron (e.g., Compton *et al.* 1998), are informative only at higher taxonomic levels in palms (Baker *et al.* 1999a, Asmussen *et al.* 2000, Baker *et al.* 2000b). However, the unexpectedly high resolution at the species level achieved in recent studies of Chamaedoreeae and Borasseae that were based on multiple plastid DNA regions (Bayton 2005, Cuenca & Asmussen-Lange 2007) questions this general view. These studies suggest a more complex system of DNA evolution where rates may vary among groups within the family and between plastid DNA regions.

Nuclear DNA

Six different nuclear DNA regions have been used in palm phylogenetics (Table 6.1). These fall into two categories: multicopy nuclear ribosomal DNA and low-copy nuclear DNA. From the former category, all of which encode ribosomal subunits, the 18S gene (Hahn 2002a), the internal transcribed spacers (ITS) between the 18S and 26S genes (including the 5.8S gene) (Baker *et al.* 2000a, 2000b) and the 5S gene (Barrow 1998, 1999, Baker *et al.* 2000c) have been utilised. The low-copy nuclear regions studied in palms comprise part of the gene that encodes malate synthase (*ms*, exons 2, 3 and 4, introns 2 and 3; Lewis & Doyle 2001, 2002), the second and third intron of the phosphoribokinase gene (*prk*) (Lewis & Doyle 2002; Table 6.1) and intron 23 of *rpb*2, the gene encoding the second largest subunit of RNA polymerase II (Roncal *et al.* 2005; Table 6.1). The low-copy nuclear gene *adh* has been explored with respect to its rate of evolution in palms compared to that in grasses, but this gene has never been used seriously in a palm phylogeny (Gaut *et al.* 1996, Morton *et al.* 1996).

In general, nuclear DNA presents more technical and molecular evolutionary challenges than plastid DNA. Nuclear ribosomal DNA is present in numerous copies in tandem arrays in the nuclear genome, which typically facilitates easy isolation of ribosomal genes by PCR amplification. Sometimes, however, concerted evolution, the set of genomic processes that homogenise the tandem repeats of nuclear ribosomal DNA sequences, breaks down, resulting in different copy types within individual nuclear genomes. This has been observed in the ITS and 5S spacer of calamoid palms (Baker *et al.* 2000b, 2000c) and corroborated by pilot studies in some arecoids (Lewis *et al.* unpublished). By contrast, low-copy nuclear DNA ideally occurs in the nuclear genome as single copies, although in reality more than one copy is often present. Compared to the high-copy number of repetitive ribosomal DNA, the small copy number of nuclear DNA regions reduces the chances of successful PCR amplification. Moreover, where duplicate copies (termed paralogues) occur, they can diverge and evolve independently of each other. The presence of multiple, divergent copy types of individual genes within the nuclear genome can be problematic for phylogeny reconstruction, often increasing homoplasy and demanding thorough sampling of the genome to secure a representative sample of copy types. In this way, the likelihood of the so-called gene tree accurately reflecting species history is compromised. Such inaccuracies may be obvious, for example, when different gene copies from single individuals do not resolve together in phylogenetic trees, or when relationships that are highly supported by other data are anomalously contradicted. Paralogues have now been well documented in low-copy nuclear DNA sequences of palms (Lewis & Doyle 2002, Loo *et al.* 2006, Norup *et al.* 2006, Thomas *et al.* 2006). The apparently spurious placement of *Barcella* in a phylogeny of Cocoseae based on *prk* (Gunn 2004) may be attributable to paralogy. Nevertheless, although nuclear DNA regions should be used with some caution, many evolve rapidly and are thus potentially more informative than plastid regions, especially at lower taxonomic levels. The large number of published palm phylogeny studies that are based on low-copy nuclear DNA is indicative of the current importance of this data type to the field. Nuclear ribosomal DNA has perhaps been abandoned prematurely, following reports of multiple copy types (e.g., ITS, 5S: Baker *et al.* 2000b, 2000c) and low rates of molecular evolution (18S: Hahn 2002a). Recent studies (Borchsenius & Trénel pers. comm.) suggest that some of these regions may yet prove valuable in reconstructing palm phylogenies.

Like the substitution rate of plastid DNA, the rate of nuclear DNA sequence divergence in palms has been identified as slow relative to that of other monocots (Gaut *et al.* 1996, Morton *et al.* 1996). However, recent palm studies have produced highly resolved phylogenies based on a single or two short nuclear DNA regions (Lewis & Doyle 2002, Baker & Loo 2004, Gunn 2004, Roncal *et al.* 2005, Baker *et al.* 2006, Loo *et al.* 2006, Norup *et al.* 2006, Savolainen *et al.* 2006), indicating the need for further exploration of the rate of nuclear DNA evolution in palms.

Mitochondrial DNA

Mitochondrial DNA is less well studied even though numerous mitochondria, like plastids, are present in each cell, each with its own identical genome. This is mainly because of the substantial changes in genome structure, size and configuration that are reported from the mitochondrial genome, which result in significant challenges for accurate phylogeny reconstruction (Petersen *et al.* 2006). Like plastid and nuclear DNA, mitochondrial DNA has been said to evolve slowly in palms (Eyre-Walker & Gaut 1997), but to date, no phylogeny of palms based on mitochondrial DNA has been produced and further conclusions will have to await proper evaluation across a broad range of palm species.

COMBINING DATA

The merits of combining datasets in simultaneous analysis are widely accepted. In recent years, all published studies of palm phylogenetics have been based on more than one dataset. However, remarkably few published molecular studies include combinations of data from more than one genome (Baker *et al.* 2000b, Hahn 2002a, Trénel *et al.* 2007). This is perhaps due to the fact that plastid and nuclear regions have tended to be used for different kinds of studies, the former being generally limited to higher-level studies, while the latter have potential at lower levels as well. It is also striking that rather few palm phylogenetic studies have combined molecular and morphological data directly (Uhl *et al.* 1995, Barrow 1998, 1999, Baker *et al.* 2000a, Hahn 2002a). Nevertheless, trends appear to be changing, with several of the most recent molecular phylogenetic projects drawing on multiple genomes and, in some cases, combining these data with morphological evidence (Bayton 2005, Cuenca-Navarro 2007, Trénel *et al.* 2007). Valuable insights are also being gained through so-called supermatrix (de Queiroz & Gatesy 2007) and supertree (Bininda-Emonds *et al.* 2002) analyses of the family (Baker *et al.* in review). These analyses build on the substantial legacy of existing phylogenetic data by incorporating all published datasets regardless of discrepancies in sampling. The resultant phylogenies have not only been influential in building the phylogenetic classification of palms presented in this book, but also promise to be valuable for future comparative studies of palms.

PALMS AND THEIR RELATIONSHIPS TO OTHER MONOCOTS

Phylogenetic research has radically changed long-standing opinions regarding the relationships of palms to other monocots. Traditionally, palms have been linked to three other monocot families in numerous classifications of flowering plants, the Araceae, Cyclanthaceae and Pandanaceae (e.g., Cronquist 1981; see also summary of classifications in Dahlgren & Clifford 1982). This putative relationship was justified on the grounds of superficial resemblances in habit, leaf and reproductive morphology, and became widely accepted (e.g., Heywood 1978). Latterly, yet prior to the dawn of molecular phylogenetics, several authors suspected that these four families were not as closely related as previously believed and recognised this in their classifications (Thorne 1983, Dahlgren *et al.* 1985).

From the earliest higher-level molecular systematic studies of angiosperms, it has been clear that palms are not closely related to Araceae, Cyclanthaceae or Pandanaceae (Chase *et al.* 1993, 1995a, 1995b, 2000, 2006, Chase 2004, Davis *et al.* 2004, 2006, Graham *et al.* 2006). The palms consistently resolve within a major clade of families now termed the commelinid monocots. The current phylogenetic classification of angiosperms (Soltis *et al.* 2005) includes four orders within the commelinids — Arecales (Arecaceae only), Commelinales (e.g., Commelinaceae, Hanguanaceae and Pontederiaceae), Poales (e.g., Bromeliaceae, Cyperaceae, Eriocaulaceae, Juncaceae, Poaceae and Restionaceae) and Zingiberales (e.g., Costaceae, Heliconiaceae, Musaceae and Zingiberaceae) — as well as a single unplaced family, Dasypogonaceae. The relationship among the commelinid monocots is reflected in a suite of shared chemical and micromorphological characters, specifically silica bodies, starchy endosperm, *Strelitzia*-type epicuticular waxes and UV-fluorescent ferulic acids within cell walls (Dahlgren *et al.* 1985, Chase 2004, Soltis *et al.* 2005). By contrast, the Araceae is well-supported as a member of the Alismatales, an order dominated by aquatic families, whereas the Cyclanthaceae and Pandanales are placed within a remarkable clade, the Pandanales, which also contains Stemonaceae, Triuridaceae and Velloziaceae (Soltis *et al.* 2005).

Relationships among the diverse orders of the commelinid monocots remain uncertain. Only the sister group relationship between the Commelinales and Zingiberales is widely supported (Chase *et al.* 1995a, 1995b, 2000, 2006, Davis *et al.* 2004, 2006, Graham *et al.* 2006). Palms have resolved as sister to all other commelinids (Chase *et al.* 1995a, 2000, 2006), sister to the Commelinales/Zingiberales clade (Chase *et al.* 2006, Davis *et al.* 2006), sister to a group comprising the Commelinales/Zingiberales clade plus Poales (Davis *et al.* 2004), sister to the Poales (Graham *et al.* 2006) and sister to the Dasypogonaceae (Chase *et al.* 1995b, 2000). None of these relationships is strongly supported by bootstrap analysis or equivalent support measures. However, the studies that contain both the greatest taxon and data sampling (Chase *et al.* 2006, Davis *et al.* 2006) place palms as sister to all other commelinids or to the Commelinales/Zingiberales clade, suggesting that the correct relationship is more likely to be found among these two resolutions than among alternatives. Large amounts of molecular data have already been utilised in these studies, and yet it seems unlikely that resolution will be improved without obtaining sequences from still more DNA regions and larger numbers of taxa.

PHYLOGENETIC RELATIONSHIPS AMONG PALMS

All palm genera have now been included in at least one, if not several phylogenetic analyses. The body of phylogenetic research that has built up since the early 1990s means that many of the most fundamental relationships among palms are now well established, and that we are able to say something about the phylogenetic context of all palms (Fig. 6.1). All available phylogenetic information for every subfamily, tribe, subtribe and genus has been summarised in this book and is presented alongside the treatment for each of these groups. However, a synthetic, integrated overview is required and is provided in the remainder of this chapter.

Palm subfamilies: monophyly, relationships and the root node of the family

A complete understanding of the earliest branching events at the base of the palm family tree has at times been frustratingly elusive. Confidence in the monophyly of the palm subfamilies that we now recognise and the relationships among them has been achieved only recently (Fig. 6.1). Current knowledge of these basal nodes results from a succession of studies aimed specifically at exploring the highest-level relationships among palms. The first of these palm family-wide studies was based on a combination of morphological and plastid RFLP datasets (Uhl *et al.* 1995). This was followed by a series of papers based on plastid DNA sequence data (Baker *et al.* 1999a, Asmussen *et al.* 2000, Asmussen & Chase 2001, Asmussen *et al.* 2006). Each paper in the sequence included another plastid DNA region and additional taxon sampling, starting with a phylogeny of *trn*L-*trn*F sequences for 65 palm species (Baker *et al.* 1999a), followed by the addition of *rps*16 intron data for the same number of species (Asmussen *et al.* 2000), then incorporating *rbc*L sequences and increasing sampling to 94 species (Asmussen & Chase 2001) and, most recently, including a fourth plastid DNA region, *mat*K, and substantially denser sampling of 178 palm species covering 86% of genera accepted at the time (Asmussen *et al.* 2006). Each time another plastid DNA region and more taxa has been included, the resulting phylogenies have become more resolved and better supported. Although palm phylogeneticists have not returned to gather more RFLP data, it is striking that the plastid DNA sequence-based phylogenies are largely congruent with the RFLP-based phylogeny of Uhl *et al.* (1995). In fact, several of the conspicuous alterations to the classification in the first edition of *Genera Palmarum* first emerged from the study of Uhl *et al.* (1995) and have been corroborated by subsequent molecular studies. These differences include, among other things, the relocation of tribe Caryoteae from subfamily Arecoideae to subfamily Coryphoideae and the removal of tribe Chamaedoreeae from subfamily Ceroxyloideae to subfamily Arecoideae.

Several further palm family phylogenies provide a valuable contrast to the studies based on plastid DNA described above because they are based in part or entirely on nuclear DNA. Lewis and Doyle (2001) analysed sequences of *ms* from 45 palm species to test the utility of this gene for phylogeny reconstruction. These data resolve some clades but, like most plastid data, are insufficiently informative to provide a well-supported resolution throughout the palms. Lewis and Doyle (2001) recommend using *ms* in combination with plastid DNA sequences, additional nuclear DNA regions and morphology. Hahn (2002a) gathered data from two plastid DNA regions (*atp*B and *rbc*L) and from the nuclear ribosomal gene 18S. He explored various combinations of these sequences and incorporated published data from *rps*16, *trn*L-*trn*F, plastid RFLPs and morphology (Uhl *et al.* 1995, Baker *et al.* 1999a, Asmussen *et al.* 2000, Asmussen & Chase 2001) in reduced analyses determined by taxon sampling compatibility between data partitions. Many of the tribes and subtribes of the new phylogenetic classification of palms are present in one or several of the many phylogenies presented by Hahn (2002a). The wide range of alternative data combinations and analytical approaches presented by Hahn

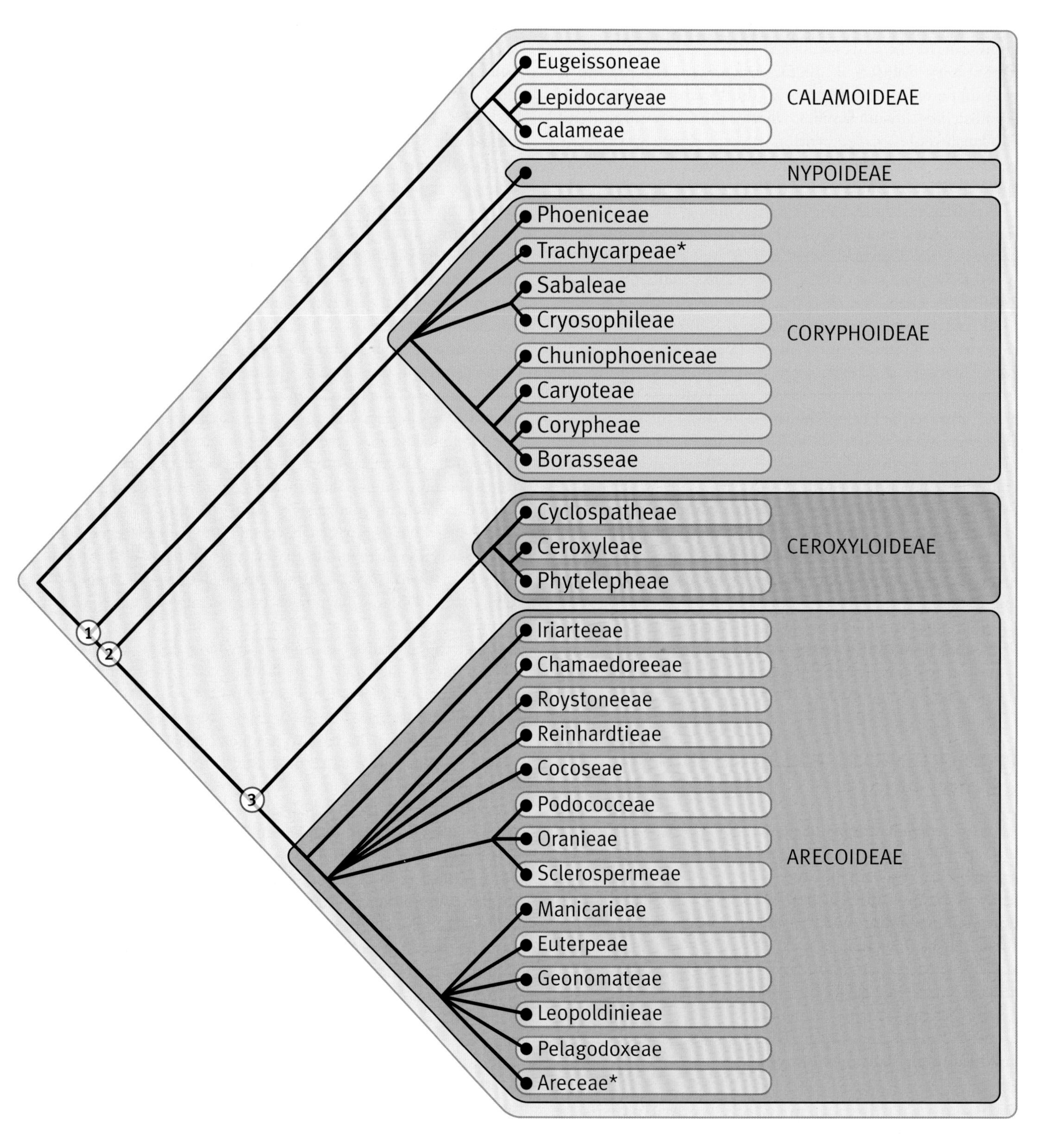

Fig. 6.1 Summary tree showing relationships among subfamilies and tribes of the Arecaceae that have been resolved with moderate (50%–89%) to high (≥ 90%) bootstrap or jackknife support in multiple phylogenetic analyses. Labelled nodes are resolved in the following studies — **node 1:** Asmussen & Chase 2001, Asmussen *et al.* 2006, Baker *et al.* in review; **node 2:** Uhl *et al.* 1995, Asmussen & Chase 2001, Hahn 2002a, Asmussen *et al.* 2006, Baker *et al.* in review; **node 3:** Asmussen & Chase 2001, Asmussen *et al.* 2006, Savolainen *et al.* 2006, Baker *et al.* in review, in prep. Asterisks indicate tribes containing genera that cannot yet be placed to subtribe.

confuses the extent to which general conclusions may be drawn. Nevertheless, those topologies based on the broadest data sampling are strongly congruent with the classification presented here.

The supertree and supermatrix analyses of Baker *et al.* (in review) have provided a useful mechanism for arbitrating between the alternative phylogenetic hypotheses presented in the published literature. On the basis of evidence from 16 different data partitions, covering all of those listed above and including plastid data, nuclear data and morphology, Baker *et al.* (in review) built phylogenies containing all accepted palm genera. Their methodological approaches have limitations; in the case of the supermatrix approach, large amounts of missing data are admitted to analyses, whereas supertree approaches are controversial because the data are derived from tree topologies rather than from raw data themselves. Nevertheless, these synthetic approaches are complimentary and provide a formalised way of unifying the information content of otherwise disparate studies and lend considerable support to the classification presented here. Some other, slightly narrower, studies that cover one or more subfamilies have also contributed in the clarification of subfamily relationships (Hahn 2002b, Savolainen *et al.* 2006, Trénel *et al.* 2007, Baker *et al.* in prep.).

The monophyly of two of the five palm subfamilies has never been in doubt. In the case of the Calamoideae, strong support for this highly distinctive group has been obtained consistently (Uhl *et al.* 1995, Baker *et al.* 1999a, 2000a, 2000b, Asmussen *et al.* 2000, 2006, Asmussen & Chase 2001, Lewis & Doyle 2001, Hahn 2002a). In the case of the Nypoideae, monophyly has never been relevant because the group comprises a single monotypic genus. However, the remaining three subfamilies have resolved less readily. In particular, the Coryphoideae, though monophyletic in the study of Uhl *et al.* (1995), resolved equivocally in most DNA-sequence-based studies. Hahn (2002a) recovered a well-supported monophyletic Coryphoideae, but only in the analysis with the broadest data sampling and narrowest taxon sampling. In the series of four studies of plastid DNA sequence, only the last (Asmussen *et al.* 2006) recovers a monophyletic Coryphoideae, though with very high bootstrap support. Baker *et al.* (in review) concur that the Coryphoideae is indeed a robust monophyletic group, as had been suspected all along despite the lack of resolution.

Evidence of the monophyly of the Ceroxyloideae was first found by Asmussen and Chase (2001), but without bootstrap support. Meanwhile, other studies indicated that the group was not monophyletic, mainly because of varied resolutions of *Pseudophoenix* (Uhl *et al.* 1995, Lewis & Doyle 2001, Hahn 2002b). The monophyletic circumscription of the subfamily remained in doubt until a subsequent study (Asmussen *et al.* 2006) recovered moderate support for it. The most recent and heavily sampled studies provide high support for the monophyly of the Ceroxyloideae (Trénel *et al.* 2007, Baker *et al.* in review, in prep.). The Arecoideae has been less problematic, with groups corresponding to the currently accepted subfamily being recovered by many authors (Asmussen *et al.* 2000, 2006, Asmussen & Chase 2001, Lewis & Doyle 2002, Loo *et al.* 2006, Savolainen *et al.* 2006, Baker *et al.* in review, in prep). As for the Ceroxyloideae, plastid data alone provide only moderate bootstrap support for the Arecoideae (Asmussen *et al.* 2000, 2006, Asmussen & Chase 2001), but more recent data corroborate these findings with high support (Baker *et al.* in review, in prep).

The relationships among the five palm subfamilies are now well-established in the most densely sampled phylogenies of palms (Asmussen *et al.* 2006, Baker *et al.* in review) (Fig. 6.1). These studies indicate that the Arecoideae and Ceroxyloideae are sister to each other, and together form the sister group of the Coryphoideae. The Nypoideae is sister to this group of three subfamilies, with the Calamoideae being sister to all other palms. This relationship can be summarised in bracket notation as follows: (Calamoideae(Nypoideae(Coryphoideae (Ceroxyloideae, Arecoideae)))). The palm phylogeny is thus rooted on the branch that connects the Calamoideae with the rest of the family (Fig. 6.2). Evidence in support of these relationships has been garnered incrementally over the years. For example, before concrete evidence of the individual monophyly of Arecoideae and Ceroxyloideae had been obtained, many studies supported the monophyly of the broader group of two subfamilies (Baker *et al.* 1999a, Asmussen *et al.* 2000, Asmussen & Chase 2001, Hahn 2002a, 2002b). Indeed, the group had been recognised without the assistance of formal analysis many years previously in Harold Moore's informal classification of the family, in which it was termed "the arecoid line" (Moore 1973a). Similarly, although the monophyly of the Coryphoideae was not recovered until recently, many studies hinted at the shared ancestry of the coryphoid groups, the Arecoideae and the Ceroxyloideae (Uhl *et al.* 1995, Baker *et al.* 1999a, Asmussen *et al.* 2000, Asmussen & Chase 2001, Hahn 2002a, 2002b).

Confidence in the relationships among subfamilies has been compromised in the past by uncertainty regarding the rooting of palm phylogenies. Early studies shed little light on this important issue. Although the study of Uhl *et al.* (1995) was ahead of its time in many respects, with the benefit of hindsight, the choice of *Dioscorea* (Dioscoreaceae) as an outgroup was unfortunate as this genus is only distantly related to palms. Moreover, *Dioscorea* was only included in the morphological dataset; RFLP data were not obtained from the genus. The conclusion that *Nypa* is sister to all other palms (Uhl *et al.* 1995) coloured some subsequent studies (Baker *et al.* 1999a, Asmussen *et al.* 2000) in which non-palm outgroups were not included and *Nypa* was used to root topologies. Some additional evidence of *Nypa* as sister to all other palms was found by Lewis and Doyle (2001), but their study contained only a sparse sample of other palm genera and provided other resolutions that have not been supported elsewhere. Subsequently, more broadly sampled studies began

Fig. 6.2 Subfamily Calamoideae, which is characterised by its scaly fruit as shown here in ***Ceratolobus concolor***, is now accepted as the sister group of all remaining palms. (Photo: W.J. Baker)

to find evidence that the Calamoideae is sister to all remaining palms (Asmussen & Chase 2001, Hahn 2002a), though bootstrap support for these findings remained moderate to low until recently (Asmussen *et al.* 2006, Baker *et al.* in review). The placement of the Calamoideae is now widely accepted and represents an excellent working hypothesis. Nevertheless, the rooting of the palm family would benefit from further analyses based on additional data, especially from non-plastid DNA sources.

Subfamily Calamoideae

Subfamily Calamoideae (Fig. 6.2) has been the focus of thorough phylogenetic studies based on both morphological data and sequence data from plastid and nuclear DNA (Baker *et al.* 1999b, 2000a, 2000b, in prep.). This research facilitated a revised phylogenetic classification (Baker *et al.* 2000a) that subsequent published analyses have corroborated (Asmussen & Chase 2001, Asmussen *et al.* 2006). Subfamily Calamoideae now includes three tribes — Eugeissoneae, Lepidocaryeae and Calameae — of which the latter two are further subdivided into three and six subtribes, respectively. This arrangement differs substantially from that described in the first edition of *Genera Palmarum* (Uhl & Dransfield 1987), in which only two tribes were recognised, one of which corresponded to the current subtribe Mauritiinae (Lepidocaryeae *sensu* Uhl & Dransfield 1987), while Calameae *sensu* Uhl and Dransfield (1987) included all remaining taxa.

Most evidence indicates that the Eugeissoneae is sister to a clade comprising tribes Lepidocaryeae and Calameae (Fig. 6.1; page 141; Baker *et al.* 2000a, 2000b, in prep., Asmussen *et al.* 2006). These relationships have not yet received high support and remain to be confirmed. However, this arrangement does reflect the apparent isolation of *Eugeissona*, with its unusual inflorescence, flower and fruit morphology, among Calamoideae. Within the Lepidocaryeae, the Raphiinae appears to be sister to the Mauritiinae, though again the relationship is not highly supported (Baker *et al.* 2000a, 2000b, Asmussen *et al.* 2006). Within the Calameae, relationships among subtribes are uncertain; only the sister relationship between Calaminae and Plectocomiinae is well supported (Baker *et al.* 2000a, 2000b). The remaining four subtribes, Korthalsiinae, Metroxylinae, Pigafettinae and Salaccinae, share distinctive inflorescence morphology, in particular, catkin-like rachillae. However, evidence of the relationships among these groups is equivocal, and further data are required for clarification.

Subtribe Calaminae has received additional phylogenetic attention, primarily because it includes the genus *Calamus*, the largest of all palm genera (Baker *et al.* 2000c). Data from the 5S spacer indicate that *Calamus* is paraphyletic and that all remaining genera of Calaminae are nested within it. Apart from reducing the monotypic *Calospatha* into synonymy with *Calamus* (Baker & Dransfield 2008), taxonomic action has not yet been taken to resolve this issue because, to date, evidence from only one DNA region has been obtained for a relatively limited taxon sample.

Subfamily Nypoideae

Containing just one species, the Nypoideae has only been included in phylogenetic studies addressing broader questions across the family (discussed above) or as an outgroup (Fig. 6.1). The taxonomic status of the Nypoideae remains unchanged from that described in the previous edition of *Genera Palmarum* (Uhl & Dransfield 1987).

Subfamily Coryphoideae

Subfamily Coryphoideae has never been studied in detail in its own right outside the context of broader phylogenetic evaluations of the entire palm family, with the exception of the preliminary study of Barrow (1996). This is perhaps an inevitable consequence of the fact that the monophyly of the subfamily was only recently demonstrated. Fortunately, several higher-level studies include substantial samplings of coryphoid genera (e.g., Asmussen *et al.* 2006, Baker *et al.* in review) and thus higher-level relationships are understood up to a point. The Coryphoideae now includes eight tribes, two of which are further subdivided into subtribes. Of the eight tribes, only two — the Borasseae and the Phoeniceae — correspond directly to tribes accepted in subfamily Coryphoideae by Uhl and Dransfield (1987). A third tribe, the Caryoteae, remains unchanged in its delimitation but its placement in Coryphoideae represents one of the most striking discoveries of modern palm phylogenetic research; the group had formerly been placed in Arecoideae by Uhl and Dransfield (1987) (Fig. 6.3). Tribe Sabaleae represents only a change of rank from subtribe Sabalinae (*sensu* Uhl & Dransfield 1987), but the remaining tribes are delimited anew.

Four distinct lineages can be recognised within the Coryphoideae (Fig. 6.1; page 215), representing relationships that are consistently observed across phylogenetic studies (e.g., Uhl *et al.* 1995, Asmussen *et al.* 2006, Baker *et al.* in review). First, the New World thatch palm clade comprises the sister taxa, tribes Sabaleae and Cryosophileae. Second, the syncarpous clade is made up of four tribes of coryphoids that have completely syncarpous gynoecia (excluding the syncarpous Sabaleae): Chuniophoeniceae, Caryoteae, Corypheae and Borasseae. The remaining two lineages each comprise single tribes, the Phoeniceae and the Trachycarpeae. Unfortunately, the relationships among these four lineages vary from study to study, and as yet little can be said with confidence regarding which topology is most likely to be accurate.

Within the New World thatch palm clade, the phylogenetics of tribe Cryosophileae have been studied by Roncal *et al.* (2008), resulting in the resurrection of *Hemithrinax* and the recognition of a new genus *Leucothrinax*. The syncarpous clade has been studied in detail by Bayton (2005). His results, which are congruent with those of Asmussen *et al.* (2006), indicate that Borasseae and Corypheae are sister taxa, forming a group that is sister to the Caryoteae. The Chuniophoeniceae is sister to the three other tribes of the syncarpous clade. Both studies also support the monophyly and sister group status of the two subtribes of Borasseae.

The monophyly of tribe Trachycarpeae is well established (Uhl *et al.* 1995, Asmussen & Chase 2001, Hahn 2002a, Asmussen *et al.* 2006, Baker *et al.* in review). The group contains all members of Uhl and Dransfield's (1987) subtribe Livistoninae and half of their subtribe Thrinacinae (the Old World genera and *Rhapidophyllum*). These former members of the Thrinacinae constitute subtribe Rhapidinae. However, despite their unique floral morphology, the members of the Livistoninae *sensu* Uhl and Dransfield (1987) are poorly resolved by current evidence and do not form a monophyletic group. Here, we recognise a more narrowly defined subtribe Livistoninae that is supported as monophyletic (Asmussen & Chase 2001, Hahn 2002a, Asmussen *et al.* 2006, Baker *et al.* in review) containing the non-American genera *Livistona*, *Licuala*, *Johannesteijsmannia*, *Pholidocarpus* and *Pritchardiopsis*. This leaves the poorly resolved remainder, the American genera and the Pacific genus *Pritchardia*, unplaced within the Trachycarpeae. We do not exclude the possibility that these unplaced Trachycarpeae may ultimately be included within the Livistoninae in the future. Nevertheless, further data are required to provide the necessary evidence that this should be the case, or that some other group or groups should be erected to accommodate them within the tribe.

Fig. 6.3 The position of the pinnate- and bipinnate-leaved Caryoteae within the fan palm subfamily Coryphoideae would never have been identified without the aid of molecular phylogenetic methods. This astonishing relationship, which places the Caryoteae in the syncarpous clade that also includes the Borasseae, Chuniophoeniceae and Corypheae, is here illustrated by ***Caryota obtusa*** (left, Caryoteae) and ***Latania verschaffeltii*** (right, Borasseae). (Photo left: W.J. Baker; photo right: J. Dransfield)

Subfamily Ceroxyloideae

The Ceroxyloideae has benefited from much phylogenetic attention, within both focused, intensive studies (Savolainen *et al.* 2006, Trénel *et al.* 2007, Baker *et al.* in prep.) and broader family-wide projects (Uhl *et al.* 1995, Lewis & Doyle 2001, Asmussen & Chase 2001, Hahn 2002a, 2002b, Asmussen *et al.* 2006, Baker *et al.* in review). Despite some conflict between the findings of earlier studies, the limits of the subfamily and the three tribes that it contains are now robust. Since the first edition of *Genera Palmarum* (Uhl & Dransfield 1987), the subfamily has lost one tribe, the Chamaedoreeae (= Hyophorbeae *sensu* Uhl & Dransfield 1987), which is widely supported as a member of the Arecoideae, and gained another, the Phytelepheae, which equates to subfamily Phytelephantoideae of Uhl and Dransfield (1987). The Ceroxyloideae is perhaps now the most morphologically heterogeneous of palm subfamilies, but is widely supported by current phylogenetic evidence. It represents a particularly exciting example of morphological and biogeographic differentiation within the family and is worthy of extensive comparative study in the future.

There is wide agreement that the Phytelepheae and Ceroxyleae are sister to each other, forming a group that is, in turn, sister to the Cyclospatheae (Fig. 6.1; page 333; Hahn 2002b, Asmussen *et al.* 2006, Trénel *et al.* 2007, Baker *et al.* in review, in prep). The broad sampling of all three tribes presented by Trénel *et al.* (2007) provides robust evidence that all genera of Ceroxyloideae are monophyletic.

Subfamily Arecoideae

Among the palm subfamilies, Arecoideae presents the greatest phylogenetic challenges because it is the largest and because, despite broad sampling in many phylogenetic studies (e.g., Asmussen *et al.* 2006, Savolainen *et al.* 2006, Baker *et al.* in review, in prep.), many uncertainties remain regarding the more fundamental relationships towards the base of the subfamily. In the current classification, 14 tribes and 14 subtribes have been recognised. Three of the subtribes fall within the Cocoseae, the remainder within the Areceae. Half of the tribes consist of a single genus only (Roystoneeae, Reinhardtieae, Podococceae, Oranieae, Sclerospermeae, Manicarieae and Leopoldinieae), representing isolated and highly distinct lineages within the subfamily. Eleven of the tribes find equivalents within the classification of Arecoideae of Uhl and Dransfield (1987), although several have been raised from subtribal status. All but four subtribes, however, are newly delimited groups. Of all the tribes, the Areceae has changed most substantially in circumscription, now relating only to the major radiation of pseudomonomerous Arecoideae that occurs in the Indo-Pacific region. The Chamaedoreeae, previously placed in the Ceroxyloideae *sensu* Uhl and Dransfield (1987) as tribe Hyophorbeae, is a new addition to the Arecoideae.

Much conflict exists between alternative hypotheses of relationships among arecoid tribes. Only two major intertribal relationships have been established with confidence (Fig. 6.1; page 355). First, a group of six tribes, comprising Areceae, Euterpeae, Geonomateae, Leopoldinieae, Manicarieae and Pelagodoxeae, has been resolved with high support in several studies (Hahn 2002a, 2002b, Lewis & Doyle 2002, Norup *et al.* 2006, Baker *et al.* in review, in prep.); here, we term this group the core arecoid clade. Second, a group comprising Oranieae, Podococceae and Sclerospermeae has been reported frequently (Uhl *et al.* 1995, Hahn 2002a, Lewis & Doyle 2002, Loo *et al.* 2006, Baker *et al.* in review, in prep.). In addition to these two major relationships, some intertribal relationships are better supported than others, while remaining in doubt. The most broadly sampled palm family studies (Asmussen *et al.* 2006, Baker *et al.* in review) moderately support the placement of the Iriarteeae as sister to all other Arecoideae. However, this is questioned by other studies that are based on narrower sampling and the low-copy nuclear genes *prk* and *rpb*2 (Lewis & Doyle 2002, Loo *et al.* 2006, Baker *et al.* in prep.). These studies suggest an alternative placement for the tribe, albeit with less support, as sister to the core arecoid clade. Several studies indicate a close relationship between the Cocoseae and tribes Reinhardtieae and Roystoneeae (Hahn 2002b, Lewis & Doyle 2002, Loo *et al.* 2006, Baker *et al.* in review, in prep.), but they differ on whether Reinhardtieae (Hahn 2002b, Baker *et al.* in review) or a clade of Reinhardtieae and Roystoneeae (Lewis & Doyle 2002, Loo *et al.* 2006) is the sister group of the Cocoseae. Hahn (2002a) also suggests that Oranieae might be the sister of the Cocoseae. Within the Cocoseae, it is well established that subtribe Bactridinae is sister to subtribe Elaeidinae (Gunn 2004 [in part], Asmussen *et al.* 2006, Baker *et al.* in review, in prep.).

Of all developments in arecoid phylogenetics, the identification of the Indo-Pacific pseudomonomerous clade, now recognised as tribe Areceae, is perhaps the most significant (Hahn 2002b, Lewis & Doyle 2002, Loo *et al.* 2006, Asmussen *et al.* 2006, Norup *et al.* 2006, Baker *et al.* in review, in prep.). Although this clade was not explicitly acknowledged until 2002 (Hahn 2002b, Lewis & Doyle 2002), it had in fact been recovered in earlier studies (Uhl *et al.* 1995, Asmussen & Chase 2001, Lewis & Doyle 2001). This group, containing some 660 species and 59 genera, is the largest of palm tribes. Its excision from the wide array of diversity previously included in the complex and non-monophyletic tribe Areceae *sensu* Uhl and Dransfield (1987) represented an important step towards understanding morphological diversity in the subfamily and how to classify it. Through focused research (e.g., Lewis & Doyle 2002, Norup *et al.* 2006), many new subtribes have been delimited within the Areceae and some higher relationships have been established, such as the presence of a clade comprising almost all western Pacific genera of Areceae. Nevertheless, our knowledge of the phylogenetics of Areceae remains limited to the extent that several genera remain unplaced to subtribe and the monophyly of some subtribes, especially Basseliniinae, Dypsidinae and Linospadicinae, remains poorly established.

LOWER-LEVEL PHYLOGENETIC STUDIES OF PALMS

Much of this chapter has focused on the relationship between genera, subtribes, tribes and subfamilies that are crucial to the construction of a robust infrafamilial classification, the primary focus of this book. However, many studies contain substantial sampling of species within genera, such that the monophyly of genera may be tested and the relationships among their species diagnosed. The study of Trénel *et al.* (2007), for example, is broadly focused on subfamily Ceroxyloideae and yet species sampling is high within genera, providing useful estimates of species relationships. Palm groups for which published phylogenetic studies include substantial and informative sampling at the species level are as follows: Ceroxyloideae (Trénel *et al.* 2007), Chamaedoreeae (Cuenca & Asmussen-Lange 2007), Euterpeae (Henderson 1999a), Geonomateae (Roncal *et al.* 2005), Iriarteeae (Henderson 1990), Phytelepheae (Barfod 1991, Barfod *et al.* 1999), Archontophoenicinae

(Pintaud 1999a), Arecinae (Loo *et al.* 2006), Calaminae (Baker *et al.* 2000c), Oncospermatinae (Lewis 2002), Verschaffeltiinae (Lewis 2002), *Asterogyne* (Stauffer *et al.* 2003), *Bactris* (Salzman & Judd 1995, Ferreira 1999, Couvreur *et al.* 2007), *Caryota* (Hahn & Systma 1999), *Chamaedorea* (Thomas *et al.* 2006), *Cryosophila* (Evans 1995), *Drymophloeus* (Zona 1999b), *Heterospathe* (Norup *et al.* 2006), *Hydriastele* (Loo *et al.* 2006), *Metroxylon* (McClatchey 1999), *Phoenix* (Barrow 1998, 1999), *Physokentia* (Fuller 1999), *Reinhardtia* (Henderson 2002a), *Rhopaloblaste* (Norup *et al.* 2006), *Roystonea* (Zona 1996), and *Sabal* (Zona 1990).

Thus, the species relationships of approximately one quarter of palm genera have been investigated at least to some extent. Compared to higher-level studies, a high proportion of these studies are based on morphology alone. Molecular data, where utilised, are most often limited to one or sometimes two nuclear DNA regions. Clearly, species-level phylogenetic studies have not developed to the same extent as higher-level studies wherein the exploitation of numerous alternative DNA sources is becoming routine. The recent works of Bayton (2005), Cuenca-Navarro (2007) and Trénel *et al.* (2007) are particularly promising, however, because they have successfully explored lower-level relationships among palms by constructing large datasets containing numerous DNA regions from both plastid and nuclear genomes and, in some instances, from morphological data. Their work demonstrates that perceived problems in reconstructing lower-level palm phylogenies are not insurmountable and that a bright future for species-level studies lies ahead.

PROSPECTS FOR PALM PHYLOGENETICS

The phylogenetic foundations of the classification of palms presented in this book promise to provide a level of stability in palm systematics that has not been achievable previously. Refinements to the classification are inevitable; unplaced taxa will be allocated to groups, and the proposed monophyly of some higher groups may even be overturned. Nevertheless, we expect the broad structure of this classification to persist. Although we have been able to define potentially stable units of classification, as discussed above, many key questions about the phylogenetic relationships among palms have not yet been answered. The root of the palm family requires further confirmation. The higher-level relationships within subfamilies, especially the Calamoideae, Coryphoideae and Arecoideae, will remain poorly understood without the analysis of new data. Relationships within many larger tribes, such as the Trachycarpeae and Areceae, are still poorly resolved, and there is wide scope for the study of species-level relationships, especially in response to focused questions and hypotheses.

Plastid DNA, which has revealed so much so far, is still providing new insights, as demonstrated by the impact of the addition *matK* sequences to a long-standing dataset of *trnL-trnF*, *rps16* and *rbcL* sequences (Asmussen *et al.* 2006). The plastid genome may yet become widely exploited for lower-level studies, as suggested by Cuenca and Asmussen-Lange (2007), and should not be abandoned. However, it is essential that the use of DNA regions from the nuclear genome is broadened. Currently, nuclear ribosomal DNA is avoided and a limited number of low-copy nuclear genes (*ms*, *prk* and *rpb2*), representing a minute fraction of the nuclear genome, are being exploited to the exclusion of all others. New low-copy nuclear regions (e.g. Bacon *et al.*, 2008) must be developed for palm phylogenetic studies, not only so that untapped potential is realised but also to ensure that molecular evolutionary biases are not introduced by the widespread use of a small number of DNA regions. Further investigations are also required into nuclear ribosomal DNA, especially the ITS regions and the 5S spacer, which have already been informative in palms (Barrow 1998, 1999, Baker *et al.* 2000b, 2000c). The role of mitochondrial DNA in palm phylogenetics remains uncertain, but merits assessment.

The majority of new phylogenetic findings will inevitably come from molecular data, but the role of morphology has never been more important. Many new insights into palm evolution will be gained from analysis of morphology with molecular data, either in direct combined analysis or by exploring focused morphological questions in the context of a well-supported molecular phylogeny (e.g., Norup *et al.* 2006). It is only now becoming possible for long-standing issues in palm biology, such as the evolution of the palm gynoecium and the origins of apocarpy, to be resolved, and palm researchers are encouraged to capitalise on this opportunity.

Finally, phylogenetic methodology is diversifying rapidly. Analytical approaches, optimality criteria, molecular evolutionary models and dating methods abound. In the past, palm phylogeneticists have tended to restrict their methodology almost entirely to simple parsimony analysis. More recently, a broader range of approaches have featured, for example, those incorporating Bayesian methods and maximum likelihood. This is indicative of a healthy and inquisitive field of research, but should not detract from the fact that data quality is paramount. No amount of methodological experimentation can substitute for well-verified, vouchered and legally sourced DNA material, carefully authenticated DNA sequences, cautious assessment of DNA and morphological homologies, and a thorough understanding of the biology of our study organisms. By maintaining these perspectives and an eye on the future, the next twenty years of palm phylogenetics should be every bit as exciting and productive as the first.

Chapter 7
BIOGEOGRAPHY

GLOBAL EXTENT AND THE LIMITS OF PALM DISTRIBUTION

Palms are distributed throughout the tropical and subtropical regions of the world. Tropical rain forest habitats sustain the most prolific palm floras, but some seasonal and semi-arid habitats, such as the cerrado of central Brazil, are also relatively palm rich. Palms are also characteristic of some desert floras, but such habitats are never rich in palm species, palms being found only where ground water exists (Fig. 7.1). A strong association of palm distribution with water availability has been demonstrated in detailed analyses of the American palm flora (Bjorholm *et al.* 2005, Kreft *et al.* 2006), with energy and productivity also found in the latter study to be important variables in determining the suitability of a habitat for palms. Although these studies are yet to be extended to a global scale, they support the generalisations presented here.

Despite their equatorial bias, palms extend to relatively high latitudes. They reach their northern extreme at around 44° N in Mediterranean southern France (*Chamaerops humilis*) and their southerly limit at more than 44° S in the Chatham Islands near to New Zealand (*Rhopalostylis sapida*) (Endt 1998) (Fig. 7.2). These two species significantly exceed latitudinal outliers from other parts of the world. For example, *Sabal minor* reaches almost 36° N in North Carolina in eastern North America, whereas on the western side, *Washingtonia filifera* attains 37° N in California (Tripp & Dexter 2006). In South America, *Jubaea chilensis*, at 35° S in Chile (Henderson *et al.* 1995), represents the southern-most extreme for the family in the Americas (Fig. 7.2), although *Trithrinax campestris*, *Syagrus romanzoffiana* and some species of *Butia* occur almost as far south. In mainland Asia, *Nannorrhops ritchiana* attains the highest latitude at approximately 34° N in Afghanistan and Pakistan. In Australia, *Livistona australis* occurs beyond 37° S in eastern Victoria, whereas in Africa, *Jubaeopsis caffra* is found around 31° S on the eastern coast of South Africa. It has been suggested that palms are generally unable to exceed these limits because they cannot undergo dormancy due to the cells of the stem remaining permanently active, and damage to these cells from freezing would be irreversible (Tomlinson 2006). Horticultural evidence, however, indicates that many species can survive freezing temperatures, with species of *Brahea*, *Butia*, *Jubaea*, *Nannorrhops*, *Sabal*, *Trachycarpus*, *Trithrinax* and *Washingtonia* reported to withstand temperatures of -10°C or lower (Meerow 2005). *Trachycarpus fortunei* is reported to have naturalised in southern Switzerland (Walther 2003), more than two degrees further north than the most northerly natural populations of *C. humilis*. This event could be indicative of global climate change, which may have an increasing influence on the distributions of native and introduced palms throughout the world.

The palms that occur at the family's latitudinal extremes are largely restricted to subfamily Coryphoideae and tribe Cocoseae (Arecoideae), *Rhopalostylis* (Areceae: Arecoideae) in New Zealand being the only exception. Species that are tolerant of high elevations, however, occur in a broader range of taxonomic groups. The highest elevation record for any palm is for *Ceroxylon parvifrons*, which reaches almost 3600 m in the Ecuadorian Andes (Borchsenius & Skov 1997). Several species of *Geonoma*, such as *G. densa*, *G. jussieuana* and *G. orbignyana*, grow at elevations exceeding 3000 m in South America (Henderson *et al.* 1995, Borchsenius & Skov 1997). *Phoenix reclinata* occurs up to 3000 m (Dransfield 1986) in Africa, while in SE Asia, *Calamus gibbsianus* has been recorded at 3000 m on Mt Kinabalu (Dransfield 1984b). Throughout mountainous parts of the Asian and American tropics, several other genera, such as *Aiphanes* and *Heterospathe*, are biased towards montane habitats, albeit at slightly lower elevations than the extremes described here. A case study of New Guinea palms demonstrated that mid-elevation habitats can sustain some of the richest palm floras (Bachman *et al.* 2004). In summary, it appears that the latitudinal boundaries of the family are determined by a predisposition to hardiness in a small number of palm clades, notably Coryphoideae and Cocoseae, whereas the elevational extremes of palms result from adaptations within lineages native at a given location.

REGIONAL DISTRIBUTION OF PALM DIVERSITY

The diversity of palm genera and species varies greatly across the world (Table 7.1). Comparisons between areas that share similar climates and vegetation types reveal some striking contrasts.

Malesia is home to the largest palm flora with 992 species and 50 genera. Expanding this region to cover all of tropical Asia (from the Indian Subcontinent to the Solomon Islands) brings the total to over 1200 species in 57 genera. Many palm groups are present in the region, but the predominant tribes include Calameae (Calamoideae), Trachycarpeae and

Opposite: Dense forest of ***Howea forsteriana***, Lord Howe Island. (Photo: W.J. Baker)

TABLE 7.1

Distribution of genus and species diversity of palms throughout the world

Based on Govaerts & Dransfield (2005) with modifications.

	GENERA	SPECIES
The Americas	**65**	**730**
South America	50	437
Central America[i]	39	251
Caribbean	26	132
Cuba	14	76
Hispaniola	15	30
North America[ii]	9	14
Africa	**16**	**65**
Western Indian Ocean[iii]	**25**	**193**
Madagascar	15	172
Western Indian Ocean Islands[iv]	15	23
Europe[v]	**2**	**2**
Mainland Asia[vi]	**43**	**354**
Western Asia[vii] and Arabian Peninsula	4	8
Indian Subcontinent[viii]	20	88
Sri Lanka	9	18
Indochina[ix]	37	229
China	18	92
Malesia[x]	**50**	**992**
Malay Peninsula	32	233
Sumatra	25	163
Java	17	46
Borneo	27	274
Philippines	19	139
Sulawesi	15	58
Papuasia[xi]	30	287
New Guinea[xii]	29	251
Bismarck Archipelago	17	21
Solomon Islands[xiii]	18	34
Pacific[xiv]	**30**	**128**
New Caledonia	10	38
Fiji	11	22
Vanuatu	13	18
Hawaii	1	23
Australasia[xv]	**21**	**58**
Australia[xvi]	21	57
New Zealand	1	2

[i] Including Mexico
[ii] Excluding Mexico
[iii] Mascarenes, Seychelles, Comoro Islands and Madagascar
[iv] Excluding Madagascar
[v] Excluding Canary Islands
[vi] Excluding Malesia, including Sri Lanka
[vii] Sinai Peninsula, Turkey to Afghanistan
[viii] Including Pakistan and Sri Lanka
[ix] Myanmar, Laos, Vietnam, Thailand, Cambodia, excluding Andaman and Nicobar Islands
[x] *Sensu* Flora Malesiana: Sumatra, Malay Peninsula, Philippines to New Guinea and Solomon Islands
[xi] New Guinea, Bismarck Archipelago and Solomon Islands
[xii] Excluding Bismarck Archipelago
[xiii] Including Bougainville
[xiv] Excluding Solomon Islands and New Zealand
[xv] Australia to New Zealand, including Lord Howe Island
[xvi] Including Lord Howe Island and Norfolk Island

Caryoteae (Coryphoideae), and Areceae (Arecoideae), with four genera alone, *Calamus*, *Daemonorops*, *Licuala* and *Pinanga*, accounting for more than half of the species totals for both Malesia and tropical Asia as a whole. The high level of diversity can be explained by prolific speciation across complex island archipelagos and the admixture of distinct northern and southern hemisphere floras during the juxtaposition of the eastern and western ends of Malesia in the latter half of the Tertiary (Dransfield 1981b, 1987, Baker *et al.* 1998, Hall 1998, Hahn & Sytsma 1999, Loo *et al.* 2006). Some groups have successfully spanned Wallace's Line (e.g., *Calamus*), whereas others remain endemic to one side or the other (e.g., *Hydriastele*). In addition, some parts of the region experienced prolonged periods of climatic stability in the Tertiary, allowing uninterrupted diversification of tropical flora (Morley 2000).

Despite having a considerably larger land surface area, the Americas sustain fewer species than Malesia (730), but are richer in genera (65). The contrast in species numbers may, in part, be an artefact of differences in species concepts between workers (Dransfield 1999, Henderson 1999b). Recent monographs suggest that the number of species that can be recognised in the Americas may yet rise significantly (Henderson 2005). Numerous tribes of subfamily Arecoideae dominate the American palm flora, especially the Iriarteeae, Chamaedoreeae, Cocoseae, Euterpeae and Geonomateae. Three genera of Arecoideae, *Chamaedorea*, *Geonoma* and *Bactris*, account for one third of the American palm species (Henderson *et al.* 1995). The Ceroxyloideae is also important in the Americas, as are tribes Cryosophileae and Trachycarpeae (Coryphoideae) in more northerly parts of the region. The Calamoideae is also significant because of the abundance of individuals of some species (e.g., *Mauritia flexuosa*), but is represented by just seven species from tribe Lepidocaryeae. The ecological biogeography of the American palm flora has been studied at a range of spatial scales (Skov & Borchsenius 1997, Svenning 1999, 2000, Vormisto *et al.* 2000, Svenning 2001a, 2001b, Vormisto 2002, Vormisto *et al.* 2004a, Vormisto *et al.* 2004b, Bjorholm *et al.* 2005, Kreft *et al.* 2006, Montufar & Pintaud 2006). At the broadest scale, Bjorholm *et al.* (2006) demonstrated that the species richness patterns of the American palm flora reflect the diverse histories of its constituent lineages; for example, the Coryphoideae shows a strong spatial bias reflecting its putative boreotropical invasion route, whereas the Arecoideae is more influenced by environmental variables as a result of its apparent long residency in the region.

The African palm flora is anomalous among other regional palm floras. With only 65 species in 16 genera, the palm diversity of Africa is depauperate given the size of the continent and the great extent of potential palm habitats. In addition, African palms do not match the elevational range that Asian and American palms achieve, being almost entirely restricted to areas below 1000 m (Moore 1973b), with the exception of *Phoenix reclinata*. Of the comparable continental palm floras, only Australia has fewer palm species (58), though it has more genera (22) and a much smaller area of habitat suitable for palms. Despite its small size, however, the African palm flora contains substantial diversity and endemism at higher taxonomic levels. It includes representatives of tribes Lepidocaryeae and Calameae (Calamoideae), tribes Phoeniceae, Trachycarpeae and Borasseae (Coryphoideae), and tribes Podococceae, Sclerospermeae, Cocoseae and Areceae (Arecoideae). More than half of the African palm species are calamoid, with notable examples including the largest African genus, *Raphia*, and the three genera of subtribe Ancistrophyllinae. Borassoid palms, especially those of the genus *Hyphaene*, are also important in Africa. Several authors concur that the low species numbers combined with the wide systematic spectrum of palms in Africa is a consequence of extinction in the Neogene, specifically, the result of aridification and retraction of rain forest in the Pleistocene (Moore 1973b, Uhl & Dransfield 1987, Dransfield 1988b, Morley 2000). On the basis of a survey of the African palm fossil record, however, Pan *et al.* (2006) infer that Africa and South America shared similar palm diversity until a decline at the Cretaceous–Tertiary boundary that was followed by a more significant turnover and decline at the Eocene–Oligocene boundary (also observed by Morley [2000]). Their data do not support a Neogene decline, and they suggest that extinctions in the Tertiary were compounded by slow diversification rates at that time.

The richness of the palm floras in the western Indian Ocean islands further emphasises the incongruity of the family's distribution in Africa. Although the majority of the diversity in the western Indian Ocean (192 species, 25 genera) is located in Madagascar (172 species, 15 genera), the small island groups of the Seychelles, Mascarenes and, to a lesser extent, the Comoros, are rich in endemics. All genera of palms in the Seychelles and Mascarenes are placed in tribes Areceae or Borasseae, with Chamaedoreeae (Arecoideae) also in the Mascarenes (see below), and all are endemic to their respective archipelagos. The palm flora of Madagascar is dominated by the genus *Dypsis* (Areceae), which accounts for 80% of the species and includes outlying species in the Comoro Islands and Pemba. However, Ceroxyleae (Ceroxyloideae), Cocoseae, Oranieae (Arecoideae) and Borasseae are also significant. The presence of borassoid and cocosoid palms in the western Indian Ocean reflects a relationship with the palm flora of Africa, whereas the numerous genera of Areceae represent the western extreme of this important Indo-Pacific clade. It has been suggested that the palm flora of Madagascar is a remnant of an assemblage that was once present in Africa (Dransfield & Beentje 1995b), but the fossil record does not support this (Pan *et al.* 2006).

The palm flora of Madagascar is remarkable, especially for the dramatic radiation in the Dypsidinae. However, this diversity is in proportion with that of the two larger tropical islands, Borneo and New Guinea. Both contain more than 250 palm species, although surprisingly New Guinea, the world's largest tropical island, has a slightly smaller total species richness.

The palm flora of Australasia, discussed above, has affinities with those of both the Pacific and SE Asia. It includes a wide array of palm groups, including Calameae, Nypoideae, Trachycarpeae, Corypheae and Areceae, as well as an outpost of tribe Ceroxyleae. The Pacific palm flora (127 species, 36 genera) is characterised by small genera and high levels of generic endemism. Only three of the 36 genera include more than 10 species, and approaching 50% are endemic to specific islands or island groups. The majority of Pacific palms belong to tribe Areceae, but small components of tribes Pelagodoxeae (Arecoideae), Trachycarpeae, Calameae, Cocoseae and Nypoideae are also present. The most significant centres of palm diversity in the Pacific are found in New Caledonia, Fiji and Vanuatu, although the palm floras of much smaller islands can be equally important on a smaller scale. For example, Lord Howe Island, which has a surface area of less than 12 km^2 is home to four palm species in three endemic genera, each belonging to a different subtribe of Areceae. The three genera undoubtedly represent three independent invasions of this oceanic island with the two species of *Howea* resulting from sympatric speciation *in situ* (Savolainen *et al.* 2006). The Hawaiian Islands are also significant because 23 of the 26 species of *Pritchardia* occur there, a classic example of an island radiation, comparable to other well-known Hawaiian groups such as the silversword alliance (Asteraceae). No other palm genus, however, is native to Hawaii.

HISTORICAL BIOGEOGRAPHY OF THE PALMS

Several authors have explored the biogeographic evolution of the palms on the basis of earlier concepts of palm relationships and data from the fossil record (Corner 1966, Moore 1973b, Dransfield 1981b, 1987, 1988b, Uhl & Dransfield 1987, Henderson 1990). These inferences are compromised, however, by the lack of an explicit and objective phylogenetic framework, which can lead to erroneous biogeographic speculation based on spurious hypotheses of relationship. However, increasingly robust phylogenetic hypotheses, developments in the use of molecular clocks to establish absolute timescales for divergence events, and recent reviews of the palm fossil record provide a stronger foundation for biogeographic inference. Here, we present an interpretation of the history of the family in time and space that draws on all of these emerging sources of evidence. Readers are encouraged to refer to the maps given for each subfamily, tribe and genus in conjunction with the text of this chapter.

Molecular clocks and the age of the palm lineage

Palms have been included in three studies in which molecular phylogenies have been calibrated to estimate the ages of divergence of extant angiosperm lineages (Bremer 2000, Wikstrom *et al.* 2001, Janssen & Bremer 2004). The estimates of Wikstrom *et al.*, who used a three-gene dataset, suggest that the palm lineage is younger than indicated by the other studies, which were based on single gene datasets. Wikstrom *et al.* suggest that palms diverged from their closest relatives around 91–99 mya, with the radiation of modern palm lineages commencing 73–63 mya. These results are, however, partly inconsistent with the earliest palm fossil records, which pre-date at least the latter estimate. Bremer's study suggests that the palms are more than 100 myr old (Bremer 2000), but even earlier estimates are provided by Janssen and Bremer (2004), who suggested that palms diverged from their relatives 120 mya with the modern lineages diverging 110 mya. Although these ages vary quite widely, all studies agree that the palms diverged from other commelinid monocots in the Middle Cretaceous. All estimates for the date of palm divergence comfortably post-date the earliest known monocot fossil, which is of Upper Barremian or Lower Aptian age (110–120 mya) (Friis *et al.* 2004).

The key outcome of these estimates from molecular clocks and the fossil data is that palms appear to have originated after the initial breakup of Gondwana, which was well under way by 130 mya. Sea gaps probably existed between most of the constituent land masses by the time the palms began to diversify (Smith *et al.* 1994). Therefore, oceanic dispersal has undoubtedly played an important role in establishing palms as a component of tropical and subtropical ecosystems worldwide.

Calamoideae

The historical biogeography of the Calamoideae is perhaps interpreted more readily than those of the other palm subfamilies because this subfamily has an abundant fossil record, which is often assignable to tribe, subtribe or genus, and well-resolved phylogenies are available (Baker *et al.* 2000a, 2000b). A biogeographic narrative based on these data has been published (Baker & Dransfield 2000), but much of the scenario described was placed within the context of vicariance and the break-up of Gondwana. Though sampling is limited, the analyses of Wikstrom *et al.* (2001) suggest that calamoids diverged from other palms at the Cretaceous–Tertiary boundary (73–63 mya). This coincides with the appearance of the first convincing calamoid fossils in the Maastrichtian, but mediates against Gondwanan vicariance as a likely explanation of biogeographic patterns.

Eugeissoneae

Eugeissona, the sole genus of the Eugeissoneae, is endemic to the palm-rich Sunda shelf. The genus occurs throughout southern Thailand, Peninsular Malaysia and Borneo, but is absent from Sumatra. This apparent anomaly is particularly surprising in light of the numerous palm species shared by parts of northern Sumatra and Peninsular Malaysia, which may reflect land connections that once existed between the two areas (Baker & Dransfield 2000).

Fig. 7.1 Palms are most commonly associated with tropical and subtropical humid forests, but can occur in much more extreme environments. Despite the extreme heat and aridity, ground water at this oasis in the Nubian desert of Egypt sustains three species of palm, ***Hyphaene thebaica*** (foreground), ***Phoenix dactylifera*** (far distance) and *Medemia argun* (not shown). Beyond the oasis, the land is almost devoid of vegetation. (Photo: W.J. Baker)

However, palaeopalynological evidence suggests that *Eugeissona* once occurred outside its current range, for example in India (Muller 1981, Phadtare & Kulkarni 1984, Morley 1998). The phylogenetic position of the tribe as sister to all of the remaining Calamoideae is also biogeographically significant. Despite occurring in the centre of calamoid diversity, *Eugeissona* does not fall within the principally Asian tribe Calameae, which contains all remaining Sundaic calamoid taxa. Its placement as sister to a clade comprising tribes Calameae and Lepidocaryeae suggests that its divergence occurred at a place and time that permitted migration between Asia, Africa and South America (Baker & Dransfield 2000). As calamoid fossils are present from the Upper Cretaceous of Africa and the Lower Tertiary of South America (Rull 1998, Harley 2006), a Gondwanan vicariance scenario has been proposed to account for this evidence (Baker & Dransfield 2000). This hypothesis invokes an Early Cretaceous divergence between *Eugeissona* and the remaining Calamoideae in south-west Gondwana, followed by the tribe rafting on India and dispersing into the Sunda region during the Tertiary, becoming extinct later in India. This scenario is, however, at odds with phylogenetic age estimates; furthermore, it does not account for the fact that concrete fossil evidence of the origin of the palm family and subfamily Calamoideae is not found until the Upper Cretaceous (Morley 2000, Harley 2006). Moreover, it ignores indications that Early Cretaceous climates in India would not have been favourable for tropical palms (Morley 2003). The geographical structure of the scenario could be correct, but the temporal frame and the role of dispersal may require reconsideration. Alternatively, *Eugeissona* may have reached India from SE Asia during the Tertiary, which would demand an entirely different scenario.

Lepidocaryeae

This morphologically diverse tribe displays an amphi-Atlantic distribution pattern. Subtribe Mauritiinae is found in northern South America whereas subtribe Ancistrophyllinae occurs only in the humid rain forests of Africa. The sole genus of subtribe Raphiinae, *Raphia*, is largely restricted to humid Africa. One species, *R. farinifera*, is common in Madagascar but is not thought to be native there (Dransfield & Beentje 1995b). *Raphia taedigera* is disjunctly distributed between Africa, Central America and the Amazon estuary. Until recently, *Raphia* was considered to have been introduced to the Americas by humans, as was the case in

Madagascar (Otedoh 1977, Tuley 1995), but recent evidence suggests that it was present in the Americas prior to transatlantic trade and may represent a recent long-distance dispersal from Africa (Urquhart 1997, Renner 2004).

Fossil pollen with a remarkably close resemblance to the unusual spiny pollen of the Mauritiinae (*Mauritiidites*) is well documented in the Tertiary fossil records of both Africa and South America (Rull 1998, Morley 2000, Harley & Baker 2001, Harley 2006, Pan *et al.* 2006). The earliest records come from the Upper Cretaceous of Africa (van Hoeken-Klinkenberg 1964, Schrank 1994), but Mauritiinae-like pollen is also well known from the Palaeocene onwards of South America (Rull 1998). This widespread fossil record has been interpreted as a consequence of the Gondwanan land connection between Africa and South America; the divergence between the sister groups Mauritiinae and Raphiinae then being linked to the opening of the Atlantic Ocean in the mid-Cretaceous (Baker & Dransfield 2000). More recently, it has been suggested that Late Cretaceous dispersal along putative transatlantic island chains would result in a similar pattern of dispersal for the Mauritiinae (Pennington & Dick 2004), a more plausible scenario given the fossil record and phylogenetic age estimates. Pollen evidence indicates that transatlantic plant dispersals were in fact frequent in the Late Cretaceous and Early Tertiary, despite the separation of Africa and South America (Morley 2003).

The three genera of Ancistrophyllinae, though very strongly supported as a monophyletic group, are morphologically divergent from each other in reproductive characters to an extent not observed in other calamoid subtribes. This has been interpreted as support for the hypothesis that the African palm flora has experienced significant extinctions, resulting in striking morphological diversity (Moore 1973b, Uhl & Dransfield 1987, Dransfield 1988b). The fossil record of the group is limited to a recently described fossil leaflet from the Upper Oligocene of Ethiopia, *Eremospatha chilgaensis* (Pan *et al.* 2006). The somewhat ambiguous pollen genus *Longapertites*, an extended sulcate grain that is recorded from the Upper Cretaceous and Lower Tertiary of Africa, may be attributed to *Eremospatha* but matches several other palm groups as well as groups within other plant families (Harley 2006). Despite the limited fossil evidence, the position of the Ancistrophyllinae as sister to the remaining Lepidocaryeae suggests that the ancestors of the group must be at least as old as the putative ancestors of Mauritiinae described above, but there is no evidence that the lineage has ever occurred outside Africa.

Calameae

This widespread tribe occurs throughout humid tropical mainland Asia from India eastwards, extending through the Malesian archipelago to Australia and into the Pacific as far as Fiji. It also includes a single disjunct species in Africa, *Calamus deerratus*. Despite its wide distribution, the tribe clearly shows a bicentric pattern of diversity across Wallace's line; the Sunda shelf represents the primary centre of species and generic richness but there is a secondary centre in Papuasia (Dransfield 1981b, 1987, Baker *et al.* 1998).

The unusual equatorial disulcate pollen produced by most genera within Calameae renders the tribe relatively conspicuous in the fossil record from the Upper Cretaceous onwards (Schrank 1994). From the Lower Tertiary, equatorial disulcate fossil pollen is recorded within the modern distributional range of the Calameae (Baker & Dransfield 2000, Harley 2006). In addition, records from the Palaeocene and Eocene of Europe and North America, respectively (Ediger *et al.* 1990), accompanied in some cases by macrofossils, suggest that the tribe has a long history in Laurasian areas. Records of such pollen from the Upper Cretaceous of Africa and South America also exist (Frederiksen *et al.* 1985, Ediger *et al.* 1990, Schrank 1994). Nevertheless, the abundance of the fossil record in the Northern hemisphere and the richness of modern Calameae in mainland tropical Asia and the Sunda region suggest that the diversification of the tribe occurred principally in Laurasia.

The presence of one species of *Calamus* in Africa could be explained by a recent long-distance dispersal, but its phylogenetic position as sister to a large clade of *Calamus* species suggests that the event may not be recent. Vicariance has been suggested as a distribution mechanism (Baker *et al.* 2000a), but dispersal from Laurasia into Africa during the Tertiary is more plausible (Baker & Dransfield 2000).

Owing to a lack of phylogenetic resolution among the subtribes of Calameae (Baker *et al.* 2000b, 2000c) and a lack of fossils that are clearly identifiable to subtribe, it is not yet possible to develop a more detailed biogeographic narrative for the Calameae group in Malesia and the Pacific. Nevertheless, the distribution patterns are certainly strongly linked to dispersal opportunities across Wallace's Line during or after the Neogene evolution of the Malesian archipelago (Hall 1998).

Nypoideae

The distinctive pollen of modern *Nypa*, with its spines and meridional zonasulcus, is widely recorded in palaeotropical palynofloras of the Upper Cretaceous and Tertiary. Nypoid pollen first appears almost simultaneously in the Maastrichtian of South America, Africa, India and Malaysia (Gee 1990), a distribution that is biased towards (but not exclusively to) Gondwanan regions. By the Eocene, *Nypa* was widespread in North and South America, Eurasia, SE Asia, Africa and Australia (Baker *et al.* 1998), perhaps as a result of long-distance dispersal by water as is characteristic of modern *Nypa*. Macrofossils also form a significant part of the record (Harley 2006). Together with the early fossil record for Calamoideae and Coryphoideae, the Nypoideae fossil record clearly demonstrates how important palms must have been in the vegetation of the Late Cretaceous onwards.

Climatic cooling at the end of the Eocene coincided with the decline of *Nypa* in many parts of its palaeodistribution

(Gee 1990, Baker *et al.* 1998, Morley 2000, Harley 2006, Pan *et al.* 2006). A detailed explanation of the retraction of the Nypoideae to its modern range from India to Australia and the Solomon Islands has not been given, but climate and geological change almost certainly played a major role. Our interpretation is, however, potentially hampered by the assumption that fossil *Nypa* occupied the same distinctive ecological niche as it does today. The strong representation of the lineage in the fossil record may be attributed to the favourable conditions for fossilisation provided by mangrove swamps, but it is also possible that *Nypa* once displayed a broader range of habits and ecologies, which, if known, would inform our understanding of nypoid biogeography. There is also evidence from variation in the morphology of fossil *Nypa* pollen that several species may have existed previously.

Coryphoideae

The coryphoid palms have achieved a very wide distribution that corresponds closely to the distribution of the palm family as a whole, and they have pioneered a wider range of habitats and climatic extremes than any other subfamily of palms. Despite the wide distribution of the Coryphoideae, there is strong evidence from their long fossil record and from phylogenetic hypotheses that the subfamily originated in the Northern hemisphere and probably diversified in boreotropical vegetations through the Late Cretaceous and Tertiary.

Sabaleae

The distribution of *Sabal*, the only genus of tribe Sabaleae, is restricted to southern USA, the Caribbean, Bermuda and Central America, although one species, *S. mauritiiformis*, is found in Central America and northern-most South America. An overwhelming body of palaeobotanical evidence from North America, Europe, Russia, India and Japan, summarised by Zona (1990) and elsewhere in this volume, indicates that the ancestors of *Sabal* originated in the Northern hemisphere, probably as a component of the well-documented boreotropical floras of the Cretaceous and Tertiary (Morley 2000). Although the identity of the numerous coryphoid leaf fossils that are assigned to the genera *Sabal* and *Sabalites* cannot be confirmed unequivocally, it is highly likely that at least some are genuine representatives of the lineage (Read & Hickey 1972, Daghlian 1978, 1981, Harley 2006). The presence of a single species in northern South America is probably a consequence of the formation of the Panama Isthmus in the Pliocene, although dispersals between North and South America at earlier times are well documented (Morley 2003).

Cryosophileae

Many of the generalisations that are applicable to tribe Sabaleae are equally relevant to the Cryosophileae. The two tribes are sister taxa and share similar distributions, and it is possible that the coryphoid leaf fossil record includes ancestors of both tribes. The distribution of the Cryosophileae overlaps almost entirely with that of Sabaleae, although its northern limit in the Bahamas and South Florida is further south than that of *Sabal*. The Cryosophileae also extends much further south in South America than Sabaleae, reaching Argentina and Uruguay.

Only three genera of Cryosophileae, *Chelyocarpus*, *Itaya* and *Trithrinax*, are endemic to mainland South America. The first two have patchy distributions largely around the western Amazon region, the last occurs in more seasonal parts of central South America. The remaining genera are restricted to Central America and the Caribbean, except for *Cryosophila kalbreyeri*, which extends into Colombia. The Cryosophileae is especially rich in the Caribbean islands, where *Coccothrinax* has radiated dramatically, especially in Cuba and Hispaniola. It has been suggested that the presence of tricarpellate, apocarpous genera of Cryosophileae in South America reflects an origin in western Gondwana for the family, as these taxa have been perceived as primitive groups according to intuitive evolutionary trends (Moore 1973b, Moore & Uhl 1982). Phylogenetic evidence does not, however, support the notion that Cryosophileae represent an early lineage, neither is it clear that the tricarpellate, apocarpous condition is plesiomorphic within Cryosophileae or palms as a whole (Asmussen *et al.* 2006, Roncal *et al.* 2008). One of the endemic South American genera of Cryosophileae (*Itaya*) is in fact unicarpellate, like the majority of the Caribbean and Central American genera, whereas *Cryosophila* is tricarpellate but is principally Central American. Moreover, flowers in mid-Tertiary amber from the Dominican Republic have been plausibly identified as *Trithrinax*, suggesting that the genus was once distributed outside South America (Poinar 2002a).

The fact that Cryosophileae and Sabaleae are so similar in modern geographic distribution and are also sister taxa suggests that both genera share a common biogeographic history and probable Laurasian origins. It appears that the diversification of these two groups took place in parallel in similar locations; indeed, representatives of each tribe can be found growing sympatrically today in shared parts of their modern ranges. It is intriguing that the tribe Sabaleae is relatively uniform and modest in size (16 species) whereas Cryosophileae has diversified extensively in terms of morphology and species (approximately 70 species), despite the fact that both tribes appear to have experienced similar adaptive opportunities in the island archipelagos of the Caribbean. Further studies of the temporal scale of these two contrasting radiations may shed some light on this enigma.

Phoeniceae

Tribe Phoeniceae, which includes the genus *Phoenix* alone, is widespread in Africa, Madagascar, the Arabian Peninsula, India and SE Asia, reaching its eastern limits in Sumatra, the Malay Peninsula, Taiwan and the Philippines. The western limits of the genus have been somewhat confused by the debate surrounding the origins of the cultivated date palm, *P. dactylifera*, the distribution of which has been strongly

influenced by human activity (Barrow 1998). Nevertheless, *P. reclinata* reaches Senegal, *P. canariensis* is endemic to the Canary Islands and *P. atlantica* is endemic to the Cape Verde Islands (Henderson *et al.* 2006), the western-most extreme of the modern distribution.

The fossil record of the Phoeniceae should arguably be convincing, given the distinctiveness of the tribe's induplicately pinnate leaves with leaflets reduced to acanthophylls at the base. However, ambiguity has been introduced by the use of the fossil leaf genus *Phoenicites*, which has been applied generally to fossil pinnate leaves (Read & Hickey 1972, Harley 2006). A careful review of all fossil records applied to both *Phoenicites* and *Phoenix* is required to clarify the palaeodistribution of the Phoeniceae. Although modern *Phoenix* has a wide distribution, the fossil record, summarised elsewhere in this volume, suggests that the group has a long history in the northern hemisphere.

Trachycarpeae

The complex, modern distribution of tribe Trachycarpeae is strongly suggestive of a group that has diversified in the Northern hemisphere and dispersed on several occasions over long distances into new island and continental locations. Only three of the 18 genera (*Pritchardiopsis*, *Pritchardia* and *Copernicia*) are not found on Laurasian land masses, and 10 of the remaining genera are strictly Laurasian. As for Cryosophileae and Sabaleae, the northern hemisphere bias in the fossil record for coryphoids provides additional support for this assertion. However, fossils that can be linked unequivocally to the tribe and its genera are scarce.

Subtribe Rhapidinae is largely restricted to mainland Eurasia but includes *Rhapidophyllum* in North America, a possible example of the disjunction between eastern Asia and eastern North America that has been well documented in numerous plant genera (Zona 1997). The relationship between *Rhapidophyllum* and Asian Rhapidinae (Asmussen *et al.* 2006) favours eastward migration of the subtribe into North America from Asia. Subtribe Livistoninae again shows a Laurasian bias in SE Asia but includes three genera that span Wallace's line, with *Licuala* and *Livistona* displaying striking bimodal patterns of species richness across this biogeographic divide (Dransfield 1981b). *Livistona* is also represented by *L. carinensis* in the horn of Africa and the Arabian Peninsula. The New Caledonian endemic genus *Pritchardiopsis* is exceptional within subtribe Livistoninae in being entirely extra-Laurasian. Thus, the distribution of the subtribe is most likely explained by multiple dispersals from Laurasia across Wallace's Line during the Miocene evolution of the Malesian Archipelago, with subsequent radiations in *Livistona* and *Licuala* in both Papuasia and Australia. The presence of *Livistona* in Africa may be attributed to long-distance dispersal.

A lack of confidence in the phylogenetic relationships within the Trachycarpeae has resulted in several genera remaining unplaced in the subtribal classification. Of these taxa, only one, *Pritchardia*, is distributed outside the Americas and the Caribbean. *Pritchardia* has radiated extensively in Hawaii, but is also found in the SW Pacific. Available phylogenetic evidence, though poorly supported, suggests that the genus is most closely related to American members of the Trachycarpeae (Uhl *et al.* 1995, Asmussen *et al.* 2006, Baker *et al.* in review), suggesting that the tribe may have dispersed into the Pacific from both the east and the west.

Chuniophoeniceae

Tribe Chuniophoeniceae is remarkable because it is very strongly supported by phylogenetic analyses and by shared floral morphology, and yet its four genera are highly divergent in many other aspects of morphology and ecology. The genera are disjunctly distributed between Madagascar, the Middle East, Peninsular Thailand and a limited part of Indochina. Climatic differentiation and major geological changes, such as the uplift of the Himalayas, may explain this apparently vicariant pattern. Long-distance dispersal almost certainly explains the presence of *Tahina* in Madagascar.

Caryoteae

The distribution of tribe Caryoteae is very similar to that of the Calameae and Livistoninae. Patterns of genus and species richness in the tribe are strongly biased towards the Laurasian land masses. Fossil seeds (*Caryotispermum*) in the Eocene of Europe have been compared with *Caryota* (Reid & Chandler 1933), and pollen records from India (summarised elsewhere in this volume) resemble *Arenga*. Like the Calameae and Livistoninae, the Caryoteae span Wallace's Line, with a small number of species reaching into Australasia and the SW Pacific. Unlike Calameae and Livistoninae, however, the Caryoteae has not speciated dramatically east of Wallace's line, although some divergent taxa have arisen there, such as *Caryota zebrina* in New Guinea and *C. ophiopellis* in Vanuatu. Once again, this pattern is most plausibly attributed to the migration of a Laurasian group across Malesia in the Miocene. Biogeographic interpretations within the genus *Caryota* are also concordant with this scenario (Hahn & Sytsma 1999).

Opposite: **Fig. 7.2** ***Chamaerops humilis*** (top left), achieves the highest latitude of all palms in the northern hemisphere (44° N, southern France), whereas ***Rhopalostylis sapida*** (top right) reaches the latitudinal limit of the family in the southern hemisphere (44° S, Chatham Islands, New Zealand). ***Jubaea chilensis*** (bottom) is the most southerly occurring palm in the Americas (35° S, Chile). All palms reaching latitudinal extremes belong to either subfamily Coryphoideae or tribe Cocoseae (Arecoideae), *Rhopalostylis* of tribe Areceae (Arecoideae) being the sole exception. (Photos: top left, J. Dransfield; top right, S. Andrews; bottom, F. Simonetti)

Corypheae

Tribe Corypheae combines the same broad pattern of species richness bias in Laurasia and presence to the east of Wallace's Line that we have described for Caryoteae, Calameae and Livistoninae. As for these other groups, Tertiary fossils from Europe have also been attributed to *Corypha*, including fossil fruits and seeds from the Palaeocene of Greenland (Koch 1972). In the case of *Corypha*, however, the species that is present to the east of Wallace's Line, *C. utan*, is widespread, ranging from India to northern Australia. This is not surprising given that *Corypha* is a pioneer in successional habitats and can reproduce on a massive scale. The lack of endemism in the south-eastern extreme of the tribe's distribution may indicate that it has invaded these areas only recently.

Borasseae

Unlike other coryphoid tribes, the Borasseae is most strongly represented in Africa and Madagascar, as well as on adjacent Indian Ocean islands and the Indian subcontinent itself. Subtribe Hyphaeninae is almost entirely restricted to Africa and Madagascar, with just two species of *Hyphaene* extending beyond these areas. The biogeography of subtribe Lataniinae is more complex. It includes one of the most widespread of all palm genera, *Borassus*, which is present throughout tropical sub-Saharan Africa and tropical Asia to New Guinea, *Borassodendron* in the Malay Peninsula and Borneo, and two genera, *Latania* and *Lodoicea*, that are endemic to the Mascarenes and the Seychelles, respectively.

Despite the bias towards the Gondwanan land masses, tribe Borasseae is nested within the clade of syncarpous coryphoid tribes that otherwise display a strong Laurasian bias. The ancestors of the Borasseae thus most likely originated in Northern hemisphere tropical environments. The modern distribution and phylogenetic evidence (Bayton 2005) for Borasseae agree that the tribe diversified primarily in Gondwana, or fragments thereof, and that its presence in the Arabian Peninsula, mainland Asia, and Malesia as far east as New Guinea is a result of migration from Gondwana. Previous authors have directly linked the distribution of Borasseae to the breakup of Gondwana (Uhl & Dransfield 1987, Whitmore 1998), but preliminary attempts to date divergences imply that the events coincide with a warm period in the Eocene rather than with major continental movements (Bayton 2005). It is therefore likely that the distribution of the Borasseae owes much to oceanic dispersal.

Ceroxyloideae

Containing a small Caribbean lineage (Cyclospatheae), a trans-Andean radiation (Phytelepheae), and a Gondwanan disjunction (Ceroxyleae), subfamily Ceroxyloideae is as diverse in its biogeography as in structural biology. Each tribal distribution is interesting in its own right, but even more striking in combination. The Cyclospatheae is disjunct from the other two tribes and does not reach South America. By contrast, the Phytelepheae is a near endemic of north-western South America and overlaps with the Ceroxyleae, which is distributed across widely separated Gondwanan land masses. These patterns and the placement of the Cyclospatheae as the sister of remaining Ceroxyloideae indicate that a major disjunction occurred early in the history of subfamily, perhaps as a result of the breaking of a land bridge between North and South America after the end of the Palaeocene (Morley 2000, 2003). Dates of divergence estimated from molecular clocks correlate well with this event (Savolainen *et al.* 2006). This scenario implies that the subfamily originated in the Americas, and that it has subsequently dispersed from there. However, long branches in molecular phylogenies, major morphological disjunctions between the tribes and the great estimated age of the subfamily, which may have diverged from its sister group as early as the Late Cretaceous, present significant challenges for the inference of biogeographic history from the extant diversity of the group.

Cyclospatheae

Pseudophoenix is a small, morphologically isolated genus within the family and the sole member of tribe Cyclospatheae. *Pseudophoenix sargentii* is distributed from Florida and the Bahamas, through the Greater Antilles to Dominica, Mexico and Belize. The remaining three species in the genus are endemic to small regions of Hispaniola. Although there are few extant species in the genus, long branches in emerging phylogenetic hypotheses (e.g., Asmussen & Chase 2001) suggest that the lineage is relatively old. Molecular clock estimates indicate that *Pseudophoenix* may have diverged from its sister group in the Palaeocene (Savolainen *et al.* 2006) and yet the modern radiation of the genus dates from the Pliocene, corresponding well with the origin of the modern Caribbean (Trénel *et al.* 2007). It is possible that *Pseudophoenix* represents a lineage that was more diverse in the past, but any groups that may have diverged between the Palaeocene and Pliocene now appear to be extinct.

Ceroxyleae

This small tribe of four genera presents an apparently classic example of a disjunct Gondwanan distribution. With *Ceroxylon* in the Andes, *Juania* in the Juan Fernandez Islands, *Ravenea* in Madagascar and the Comoro Islands, and *Oraniopsis* in northern Australia, it is tempting to follow previous accounts in attributing this pattern to the break-up of Gondwana (Dransfield *et al.* 1985a, Uhl & Dransfield 1987). A route to Australia via the Antarctic route (Uhl & Dransfield 1987) has also been proposed as an adequate explanation for components of the distribution, whereas Moore (1973b) invoked stepping-stone dispersal to the Indian Ocean via Africa and subsequent extinction (see also Chamaedoreeae below). Although the role of Africa cannot be excluded, it has not been substantiated by robust fossil evidence (Pan *et al.* 2006).

Molecular clock estimates suggest that the divergence between the Ceroxyleae and their sister group Phytelepheae occurred in the Early Tertiary (Savolainen *et al.* 2006, Trénel *et al.* 2007), thus ruling out Gondwanan vicariance. The divergence between *Ravenea* and other Ceroxyleae recovered by Trénel *et al.* (2007) is dated as Middle Eocene to Late Oligocene, corresponding well with the hypothesised origins of the rain forests of Madagascar, to which most extant *Ravenea* species are native, from the Eocene onwards. Thus, Trénel *et al.* conclude that a trans-Atlantic/trans-African dispersal to Madagascar must have occurred to account for the timing of these divergences. Their estimated dates for the divergence of *Oraniopsis* from a clade of *Ceroxylon* and *Juania* range from Middle Eocene to Early Oligocene, and coincide with a period of global cooling after the Eocene climatic maximum. Nevertheless, Trénel *et al.* regard a Late Eocene dispersal route from South America to Australia via Antarctica as plausible because the Ceroxyleae (including modern *Oraniopsis*) is a relatively cool hardy lineage of palms and would very likely have tolerated putative austral climatic conditions at that time. Notably, estimated dates for the divergence of *Juania* from *Ceroxylon* substantially pre-date the geological origins of the volcanic Juan Fernandez Islands, and thus the genus appears to have dispersed to the archipelago having diverged elsewhere (Trénel *et al.* 2007).

Phytelepheae

The three genera of tribe Phytelepheae are restricted to north-western South America, with one species extending into Panama, most likely a northward invasion following the closure of the Isthmus. The relatively restricted distribution of such a morphologically remarkable tribe is curious, especially given the considerable estimated age of its divergence from its sister group (see Ceroxyleae), but perhaps more interesting is the disjunction of the tribe across the Andes. No species of *Phytelephas* is shared by both eastern and western sides of the Andes, *Aphandra* is monotypic and restricted to the east of the Andes. (The two disjunct species of *Ammandra* have recently been reduced to one species [Bernal *et al.* 2001]). Although this pattern has been linked to Pleistocene refugia (Moore 1973b), it is more widely accepted that these taxa speciated allopatrically as a result of the Andean uplift during the Late Neogene (Prance 1982, Uhl & Dransfield 1987, Barfod 1991, Morley 2000). This conclusion is also supported by molecular clock estimates (Trénel *et al.* 2007).

Arecoideae

The Arecoideae, the largest subfamily of palms, has a worldwide distribution in both the tropics and subtropics, and is most species-rich in rain forest habitats. It is estimated that the subfamily diverged from its sister group, the Ceroxyloideae, in the Late Cretaceous (Savolainen *et al.* 2006), but despite the relatively great age and modern diversity of the subfamily, its fossil record is limited and inconclusive, except for the Cocoseae.

The modern distribution of the subfamily is heavily biased towards the Americas, though there is much differentiation within this region, with Caribbean, Central American and South American distributions being displayed at tribal and generic level. It is likely that the group originated in the Americas. However, the phylogenetic placement of groups that are principally Central American or Caribbean in distribution calls into question the traditional view that the Arecoideae originated in South America (Moore 1973b, Uhl & Dransfield 1987). Most Old World Arecoideae are nested within broader groups of American taxa, suggesting that the subfamily dispersed repetitively from the Americas, perhaps seven times or more.

Iriarteeae

This distinctive group of five genera is restricted to northern South America, although a few species reach southern Central America, presumably due to the closure of the Isthmus of Panama. Much of the species diversity occurs on the slopes of the Andes and adjacent lowlands, especially in *Wettinia* and to some extent *Socratea*. Like the Phytelepheae, isolation as a result of the Neogene Andean uplift and subsequent speciation has probably been a major driver of diversification in the group (Henderson 1990). It should be noted, however, that fossil flowers resembling *Socratea* have been discovered in Mexican Amber of the Upper Oligocene–Lower Eocene age (Poinar 2002a).

Chamaedoreeae

Tribe Chamaedoreeae displays one of the most extraordinary disjunctions in the palm family (Fig. 7.3). The majority of the tribe's species diversity and four of the five genera occur in Central America, and to a lesser extent in northern South America and the Caribbean. The fifth genus, *Hyophorbe*, comprises five species that are endemic to the Mascarene Islands, a volcanic archipelago in the Indian Ocean that arose in the Miocene. As for Ceroxyleae, Moore (1973b) suggested that the tribe once occurred in Africa but is now extinct there. Nevertheless, molecular clock estimates and the relatively recent origin of the Mascarene Islands invalidate Gondwanan vicariance as an explanation for the current distribution (Cuenca *et al.* 2008). Long-distance dispersal must certainly have played a key role.

Within the American genera, contrasting patterns of diversity exist: three genera comprise fewer than five species whereas *Chamaedorea*, the largest neotropical palm genus, contains over 100 species. *Chamaedorea* is particularly species-rich in Central America, where two centres of diversity occur, the first in Mexico and Guatemala and the second in Costa Rica and Panama (Henderson *et al.* 1995).

Fig. 7.3 Tribe Chamaedoreeae displays a remarkable disjunct distribution. The genus *Hyophorbe* (***H. lagenicaulis***, left) occurs in the Mascarene Islands of the Indian Ocean, while the remaining four genera, represented here by ***Gaussia attenuata*** (right), are variously distributed in tropical America (including the Caribbean). (Photos: left, W.J. Baker; right, C.E. Lewis)

Podococceae, Oranieae and Sclerospermeae

A clade comprising the three monogeneric tribes Podococceae, Oranieae and Sclerospermeae has been resolved in several phylogenetic analyses (Uhl *et al.* 1995, Hahn 2002b, Lewis & Doyle 2002, Loo *et al.* 2006, Baker *et al.* in review, in prep.). *Podococcus* and *Sclerosperma* are restricted to humid tropical Africa, whereas *Orania* displays a disjunct distribution with three species endemic to Madagascar and more than 20 distributed from Thailand, through Malesia to New Guinea, where the genus is particularly species-rich. A dispersal route incorporating India and perhaps Africa during the middle Eocene may have resulted in this pattern (Morley 2003), a hypothesis that is supported by the African affinities of the closest relatives of *Orania*.

Roystoneeae and Reinhardtieae

Resolved as relatives in some phylogenetic studies (Lewis & Doyle 2002, Loo *et al.* 2006), tribes Roystoneeae and Reinhardtieae display similar Caribbean distributions. Tribe Roystoneeae comprises the single genus *Roystonea* with 10 species in the Caribbean Islands and adjacent parts of Florida, Central America and northern South America. The genus is particularly rich in Cuba, with four of the species being strict endemics in the eastern-most tip of the island. Like the Roystoneeae, tribe Reinhardtieae contains a single genus, *Reinhardtia*, with five species in Central America and a sixth in Hispaniola.

Cocoseae

Tribe Cocoseae has long been of biogeographic interest, first because of uncertainty surrounding the origin of the coconut and, second, on account of the numerous cocosoid fossils that have been found within and beyond the tribe's modern natural range. Fossil evidence indicates that the tribe was present in South America, India and New Zealand during the Tertiary (Gunn 2004). Today, the Cocoseae is the most widespread palm tribe in the Americas; the majority of cocosoid genera and species occur there and their diverse ecological adaptations allow them to occupy both seasonal and ever-wet habitats. However, the tribe is also represented in the Old World by two genera in Africa and two in Madagascar. The coconut, *Cocos nucifera*, has a pantropical distribution, which is due to movement by man and water dispersal. Prevailing opinion suggests that it originated in the western Pacific (Harries 1978), but many other locations have been suggested (Gunn 2004). In addition, subfossil remains indicate that a genus of cocosoids, *Paschalococos*, a member of the Attaleinae closely allied to *Jubaea*, was present on Easter Island approximately 800 years ago (Dransfield *et al.* 1984).

The Cocoseae is divided into three subtribes, the Bactridinae and Elaeidinae, which are sister taxa, and the Attaleinae. Molecular clock estimates suggest that the divergence between the Bactridinae–Elaeidinae clade and the Attaleinae took place in the Palaeocene or Early Eocene (Gunn 2004, Savolainen *et al.* 2006).

The Bactridinae is widespread in Central and South America and the Caribbean. The Elaeidinae contains two small genera, the monotypic Brazilian endemic *Barcella* and *Elaeis*, which has one species in Central and South America and a second, the widely cultivated oil palm of commerce, *E. guineensis*, in Africa. It is most likely that *Elaeis* reached Africa by dispersal from America (Renner 2004) before the end of the Miocene, at which point *Elaeis* pollen appears in the African fossil flora (Pan *et al.* 2006).

The Attaleinae contains mostly American taxa, but also includes the Madagascan genera, *Beccariophoenix* and *Voanioala*, and the South African endemic *Jubaeopsis*. Well-sampled phylogenies concur that the American Attaleinae and *Cocos* form a clade that is embedded within the African and Madagascan genera (Hahn 2002a, Baker *et al.* in review); the study of Gunn (2004) is broadly congruent with these relationships. These relationships imply that the early radiation of the Attaleinae occurred outside the Americas, and that the group diverged in Africa, Madagascar and perhaps elsewhere, which may account for fossils beyond the modern range of Cocoseae. These events are estimated to have occurred in the Eocene or Early Oligocene (Gunn 2004, Savolainen *et al.* 2006), which coincides well with the earliest cocosoid fossil records. The presence of a rich diversity of Attaleinae in the Americas would appear to be the result of a single colonisation after those early divergences in the subtribe.

The Caribbean distribution of the putative immediate relatives of the Cocoseae, tribes Reinhardtieae and Roystoneeae, the distributions of the cocosoid subtribes, and the phylogenetic relationships between their genera call into question the assertion that the tribe originated in west Gondwana or South America (Moore 1973b, Uhl & Dransfield 1987, Gunn 2004). It is reasonable to suggest that the group is most likely to have originated in the Americas in the broader sense and that at least three dispersals have occurred: 1) to Africa in Elaeidinae, 2) to Madagascar, Africa and possibly beyond early in the history of the Attaleinae, and 3) to Easter Island. This pattern cannot, as has been previously suggested (Uhl & Dransfield 1987, Hahn 2002a), be linked to Gondwanan break-up.

The coconut almost certainly has a complex history of natural dispersal, now clouded by human intervention. We may never know with certainty where the coconut evolved and precisely how it achieved its modern extent. However, the deep embedding of *Cocos* within the neotropical Attaleinae (Hahn 2002a, Gunn 2004, Baker *et al.* in review) indicates that its immediate ancestors were almost certainly South American.

Manicarieae, Euterpeae, Geonomateae and Leopoldinieae

Within the subfamily Arecoideae, a strongly supported group of tribes, known as the core arecoid clade, is resolved, consisting of tribes Manicarieae, Euterpeae, Geonomateae, Leopoldinieae, Pelagodoxeae and Areceae (Hahn 2002a, 2002b, Lewis & Doyle 2002, Norup *et al.* 2006, Baker *et al.* in review, in prep.). The first four of these tribes are tropical American lineages that are largely distributed in South America. Although phylogenetic studies do not agree on the relationships among these tribes, they all indicate a probable American origin of the core arecoid clade, from which the two remaining tribes, Pelagodoxeae and Areceae, dispersed independently.

The monotypic genus *Manicaria*, the only member of tribe Manicarieae, has a scattered distribution throughout central America and northern South America (Henderson 1995, Henderson *et al.* 1995). The genus may be disjunctly distributed between these two areas, perhaps partly because of the presence of intervening upland areas. Tribes Euterpeae and Geonomateae are widespread in South and Central America as well as in the Caribbean. The Geonomateae shows some regional endemism at the generic level, with *Calyptronoma* restricted to the Caribbean and *Calyptrogyne* almost exclusively occurring in Central America. These two genera are extremely closely related (Roncal *et al.* 2005) and together represent a lineage that has radiated around the Caribbean, perhaps originating as a dispersal out of South America before the closure of the Isthmus of Panama. Tribe Leopoldinieae is much more restricted, comprising a single genus of three species that grow in the central Amazon region only, most often on the margins of blackwater rivers in seasonally inundated areas.

Pelagodoxeae

Subtribe Pelagodoxeae contains two monotypic genera, *Sommieria*, a New Guinea endemic, and *Pelagodoxa*, a very rare genus from the Marquesas. Individuals of *Pelagodoxa* have also been recorded in Fiji, the Solomon Islands and Vanuatu (Chapin *et al.* 2001), but always near to villages. In addition, the population in the Marquesas is located in an archaeological site. It is impossible to say to which of these locations *Pelagodoxa* is truly native (Stauffer *et al.* 2004), though it is reasonable to assume that it is a native of the Pacific region.

The relationships of the Pelagodoxeae are surprising as there is no obvious morphological explanation for its exclusion from the Areceae, with which it has much in common biogeographically. Nevertheless, no phylogenetic study has placed the Pelagodoxeae either as sister to or within the Areceae (Lewis & Doyle 2002, Loo *et al.* 2006, Norup *et al.* 2006, Baker *et al.* in review, in prep.). We conclude that this small western Pacific lineage has originated as a dispersal from the Americas independently from tribe Areceae.

Areceae

The Areceae is one of the largest and most diverse groups of palms, including more than 660 species in 11 subtribes and 59 genera that are distributed throughout the Indo-Pacific region on islands and continental areas. The palm floras of many Pacific and some Indian Ocean islands outside the Malesian region are almost entirely composed of Areceae. All but one of the native palm species of New Caledonia are placed in the Areceae, and even the rich palm flora of Madagascar is largely accounted for by four genera from the tribe.

The Areceae has been the focus of several recent phylogenetic studies (Hahn 2002a, Lewis & Doyle 2002, Norup *et al.* 2006, Baker *et al.* in prep.) that have provided many insights into the relationships within the tribe, although several genera remain unplaced to subtribe. Some studies have resolved a moderately supported western Pacific clade within the Areceae comprising subtribes Archontophoenicinae, Basseliniinae, Carpoxylinae, Clinospermatinae, Linospadicinae, Ptychospermatinae and Rhopalostylidinae (Lewis & Doyle 2002, Norup *et al.* 2006). Norup *et al.* (2006) also resolved a principally Indian Ocean clade including, among other taxa, Arecinae, Dypsidinae, Oncospermatinae and Verschaffeltiinae, although this group is not well supported. Most interestingly, however, this Indian Ocean clade included several taxa such as *Areca*, *Rhopaloblaste* and *Cyrtostachys* that occur in both the Pacific and Indian Oceans, as well as the strictly Pacific genus *Clinostigma*. By contrast, the western Pacific clade scarcely crosses to the west of Wallace's line, with only *Heterospathe* and *Adonidia* reaching the Philippines and, in the case of the latter, northern Borneo. Regardless of the monophyly of the Indian Ocean clade, it appears that tribe Areceae has penetrated the Pacific region on several independent occasions, whereas the same may not be true for Areceae in the Indian Ocean.

The dated phylogeny of Savolainen *et al.* (2006) suggests that the Areceae began to radiate in the Oligocene, which may not be consistent with the hypothesis that the tribe has radiated very rapidly (Hahn 2002a). Further dating analyses are required to test this hypothesis thoroughly. The very wide distribution that the tribe has now achieved must have resulted from numerous long-distance dispersal events between the islands and continents of the Indo-Pacific region. For example, the three endemic genera of palms on Lord Howe Island represent three separate subtribes, all of which have dispersed there separately in the 6.9 million years since the island's formation. Despite this ability to disperse, some groups show high levels of endemism at the sub-tribal level; examples include the New Caledonian Clinospermatinae, Verschaffeltiinae in Seychelles, and the Dypsidinae, which are found only in Madagascar and the neighbouring Comoro Islands and Pemba.

SYNTHESIS

Previous analyses of palm biogeography have aimed to identify a single region as a place of origin of the family. For example, Moore (1973b) proposed an origin in western Gondwana with subsequent northward dispersal into Laurasia via Africa, as well as some potential dispersal along austral routes. He reached this conclusion on the basis of the distribution of the major groups of palms that he recognised (Moore 1973a) and, in particular, because of the presence of certain apocarpous coryphoid genera, which he believed to be the most primitive living palms, in South America. Subsequently, this view was contested in light of the Northern hemisphere bias in the modern and fossil distributions of those coryphoid groups, which displayed the putative primitive character state of apocarpy. This led to the hypothesis that the palm family has a Laurasian origin prior to or at the separation of Laurasia and Gondwana (Dransfield 1987, Uhl & Dransfield 1987). However, neither of these hypotheses can be upheld by current evidence from the fossil record, molecular clock estimates, tectonic reconstructions nor by our current understanding of the evolution of reproductive characters.

It is not possible to determine precisely a date for the origin of the palm family. Nevertheless, some key conclusions can be drawn. First, while some fossils assigned to Arecaceae occur earlier, the first unequivocal palms appear in the Upper Cretaceous of North America and Europe. Other groups appear by the Upper Cretaceous and Lower Tertiary in both northern and Southern hemisphere localities. Molecular clock studies are, for the most part, consistent with the fossil record and lead to the conclusion that palms diverged from their closest relatives in the Middle Cretaceous, and that the initial radiation of lineages that we see today occurred through the Late Cretaceous. Second, by these dates, Laurasia and Gondwana had largely fragmented into constituent continents and yet, despite significant sea gaps, palms rapidly dispersed from an unknown point of origin throughout these areas to achieve the widespread distribution observed in the fossil record by the beginning of the Tertiary. Third, distinct regional biases have arisen, giving rise to a marked north–south bicentricity observed by Uhl and Dransfield (1987). For example, the distribution patterns observed for most tribes of Coryphoideae are skewed towards the Northern hemisphere, whereas various tribes of Calamoideae, Ceroxyloideae and Arecoideae display apparently Gondwanan distributions. These patterns may have been promoted by increased isolation due to widening sea gaps, regionally favourable environmental conditions or intercontinental dispersals, for example via Antarctica for austral patterns or along the Walvis Ridge for South American–African distributions (Morley 2003). These patterns have become less clear where regional floras intermix, as in Malesia, the Caribbean and Central America.

Fourth, the temporal and spatial framework indicates that long-distance dispersal events must have played a substantial role (de Queiroz 2005). It appears that the large seeds of many palms may not be an obstacle to stochastic dispersal events over such long time frames, and therefore the family may not necessarily be a good indicator of past geological events, as has been suggested previously (Dransfield 1981b, Uhl & Dransfield 1987). Finally, the distribution of palms is strongly constrained by climatic factors that are likely to have been as influential in determining palaeodistributions as they are in explaining palm diversity today.

Chapter 8
NATURAL HISTORY AND CONSERVATION

INTRODUCTION

The way palms interact with their environment is a huge subject that has attracted much interest. In this chapter, we draw attention to some of the more unusual aspects of palm natural history, and go on to mention briefly how man has interacted with these supremely useful plants. Within the confines of limited space, this survey cannot be exhaustive; instead, by focussing on aspects that we have found particularly interesting, we hope to whet the reader's appetite to delve more deeply.

Since the publication of the first edition of *Genera Palmarum*, there has been an explosion of research on the natural history of palms. To review all of these studies is beyond the scope of this book. Besides, there are additional excellent sources of information. Tomlinson's *Structural Biology of Palms* (1990) touches on many aspects of the natural history of palms. Henderson (2002b) published his review of the ecology of palms, a book in which he attempts to correlate aspects of palm morphology, in particular size and growth form, with ecology, and draws evolutionary inferences. Henderson's book is particularly useful for its extensive bibliography. In this chapter, the range of studies carried out to investigate the ecology of palms will be suggested, but the reader is referred to Tomlinson (1990) and Henderson (2002b) for initial access to the growing literature on the subject.

REGIONAL VARIATION IN GROWTH FORMS IN PALM COMMUNITIES

The distribution of palms is discussed in Chapter 7. There are substantial and remarkable differences between the major tropical rain forest blocks not only in the phylogenetic composition of the palm flora but also in habit composition. Tree palms that reach the canopy or subcanopy are relatively abundant in South America, Madagascar and New Guinea. In Madagascar, for example, ten or more large tree species of *Dypsis* can occur in the same locality. In western Malesia, on the other hand, such tall canopy palm species are few in number; however, a single tree species, such as *Livistona endauensis*, can occur locally in huge stands comprising thousands of individuals. Climbing palms are particularly abundant in Malesia, but much less so or completely absent elsewhere. The variation in palm floras in different regions of the world contributes to the distinctive appearance of their forests.

LOCAL PATTERNS OF DIVERSITY AND ABUNDANCE

The local patterns of palm species composition have been studied in some detail in South America. The differences in composition of the palm flora on wet and dry sites are particularly striking in Amazonia (see for example Balslev *et al.* 1987, Kahn & Mejia 1991, Vormisto *et al.* 2004a, 2004b, Montufar & Pintaud 2006). No such studies have been conducted in the rain forests of the Old World tropics. However, some detailed studies have been made of the local distribution of the economically important rattans, with a view to assessing the economic value of areas of rain forest and the potential for sustainable harvesting (Bøgh 1996a, Nur Supardi *et al.* 1998). The effects of logging on the abundance and regeneration of palm communities have also been studied in lowland West Malaysia (Nur Supardi *et al.* 1998).

PALM-DOMINATED COMMUNITIES

Several palms occur in such vast natural stands as to dominate the vegetation completely. In the eastern tropics, *Nypa fruticans* often occurs in dense stands on estuarine mud with dicotyledonous mangrove species almost excluded by the palm; in other situations, it may occur mixed with dicotyledonous mangrove trees. Pure stands of *Nypa* may extend for several hundred hectares in extent, especially in parts of Borneo, eastern Sumatra. Such *Nypa* forests are of immense importance, not only for the wealth of products *Nypa* can provide but also in terms of gas exchange and mud stabilization. On the landward fringe of the mangrove in West Malesia occurs another vegetation type that is almost completely dominated by palms, such as *Oncosperma tigillarium*, *Calamus erinaceus* and, in Borneo, *Daemonorops longispatha*. There is no direct equivalent of the *Nypa*-dominated, brackish-water vegetation types in the New World, though Richards (1952) considers *Manicaria* to be analogous to *Nypa*. *Raphia taedigera* grows on the landward fringe of the mangrove in the Amazon estuary (Bouillenne 1930), and species of *Raphia* occur in similar habitats in West Africa (Ainslie 1926). *Bactris major* also occurs in mangrove forests in central America, but does not dominate the vegetation. *Metroxylon sagu* dominates fresh-water swamplands in parts of the lowlands of New Guinea, its occurrence often influenced by man. *Mauritia flexuosa* occurs in vast numbers, dominating

Opposite: ***Medemia argun***, habit, Egypt. (Photo: W.J. Baker)

the landscape in parts of the Amazon basin. In Paraguay, and in adjoining parts of Brazil, Argentina and Bolivia, *Copernicia alba* forms huge stands on areas subject to periodic flooding and periodic drought; the largest populations have been estimated to contain half a billion individuals (Markley 1955). *Copernicia prunifera* occurs in almost equally great concentrations in northeastern Brazil. In Africa, *Hyphaene compressa* grows in great abundance in alluvial flats and coastal plains; its abundance may be partly due to the activity of man and, when left undisturbed, *Hyphaene* woodland may revert to closed forest in places (Dransfield 1978b). In upland areas of Borneo, *Eugeissona utilis* may grow in locally dense stands to the exclusion of other trees, and in the montane forests of the Caribbean, *Prestoea montana* may produce a palm-dominated vegetation type. In the high Andes of South America, *Ceroxylon* spp. once occurred in vast natural stands but these have increasingly been decimated by land clearance and disease. Dense forests of *Howea forsteriana* occur patchily on calcareous soils on Lord Howe Island, the vegetation is totally dominated by the palms and appears almost plantation-like (Savolainen *et al.* 2006). These are a few examples of vegetation types in which palms are the most conspicuous and apparently dominant plant. When the particular palms are useful, these natural stands become immensely significant in either local or international economies; indeed such stands may, as in the case of *Hyphaene*, owe their survival or part of their development to man's activities.

PALM COMMUNITIES ON UNUSUAL SOIL TYPES

Lowland tropical rain forest developed on nutrient-poor white sands is often physiognomically very distinctive. In Borneo, such heath forest is known locally as *kerangas* and carries a varied, peculiar and important palm flora. *Areca insignis* var. *moorei*, *A. brachypoda*, *Calamus corrugatus*, *Licuala bidentata*, *L. orbicularis* and *Pogonotium divaricatum* and many other palms are exclusive to this vegetation type. *Johannesteijsmannia altifrons* is confined to *kerangas* in Sarawak, yet the same species occurs in Dipterocarp forest in Malaya. In New Guinea, a similar type of lowland heath forest occurs in foothills of the high mountains on river terraces, such as at the southern foot of Mt Jaya. Here, the palm flora includes species of *Livistona*, *Hydriastele* and *Calamus* that are apparently restricted to the habitat (Dransfield *et al.* 2000). In Madagascar, white sand forest occurs patchily and includes palms such as *Dypsis saintelucei* and *Beccariophoenix madagascariensis*. In South America, *Mauritia carana* and *Bactris ptariana* are restricted to forest on white sands (Henderson *et al.* 1995).

Ultramafic rocks, with their usually toxic soils that are rich in heavy metals such as manganese, chromium, iron and copper, usually support a restricted but peculiar palm flora. Distinctive palm floras on ultramafics occur in Cuba, New Caledonia, Sabah, Palawan, Madagascar and elsewhere. Pintaud *et al.* (1999) contrast the palm flora on ultramafics with that on schistose rocks in New Caledonia. The ecology of these palms has scarcely been investigated beyond the observation that they seem to be confined to soils that are rich in heavy metals.

In more seasonal areas of the New World, there is a great diversity of palms on limestone, including species of, for example, *Brahea*, *Gaussia*, *Pseudophoenix*, *Coccothrinax* and *Thrinax* spp. Within the perhumid tropics, limestone appears to have a more restricted palm flora, but those palms that do occur are often of considerable interest or peculiarity (e.g., *Maxburretia* in southern Thailand and Peninsular Malaysia, and *Chamaedorea fragrans* in Peru).

THE EFFECTS OF PALMS ON SOIL AND OTHER PLANTS

Furley (1975) demonstrated the importance of *Attalea cohune* in the development of the soil profile. The stems of this palm are initially geotropic and establishment growth of the stem occurs at depths of up to 1 m. When the palm eventually dies, there is thus a void in the soil that becomes infilled with soil and detritus. Over a large period of time, a population of this palm may be responsible for considerable turn-over of the soil. In Peninsular Malaya, *Eugeissona tristis* is locally abundant in hill Dipterocarp forest, where its massive leaf litter and deep shade tend to prevent regeneration of dicotyledonous trees. Disturbance of such forest tends to encourage *E. tristis* with the effect that extensive, almost pure stands of the palm develop to the exclusion of commercially important timber trees. *Oncosperma horridum* also produces a dense leaf litter that prevents regeneration (House 1984). Peters *et al.* (2004) demonstrated the importance of falling palm fronds of *Iriartea deltoidea* in determining the composition and structure of South American rain forest.

MYCORRHIZAE

Mycorrhizal associations have been demonstrated in several palms (see, for example, Janos 1977, Broschat & Meerow 2000, Carillo *et al.* 2002). Their occurrence is likely to be very widespread. The presence of vesicular-arbuscular mycorrhizae is known to increase the uptake of nutrients, especially of elements such as phosphorus in nutrient-poor soils (Pfleger & Linderman 1994). Janos (1977) and Jaizme-Vega and Díaz-Pérez (1999) demonstrated enhanced growth of palm seedlings in cultivation after inoculation with mycorrhizae.

The close association of other fungi with palms has been studied by Hyde and associated researchers, who have demonstrated a huge diversity of fungi living on palms (Frohlich & Hyde 1999, Frohlich *et al.* 2000).

GROWTH FORMS

The growth forms of palms are discussed in Chapter 1, and elsewhere in some detail by Dransfield (1978b) and Hallé *et al.* (1978). We may conveniently distinguish tree palms, shrub palms, and acaulescent and climbing habits. Superimposed on these growth forms are branching and flowering behaviour, as discussed in Chapter 1. Some major growth forms have not evolved in the family. There are no parasitic palms and no deciduous species, although a few fire-tolerant species, such as *Calamus acanthophyllus* (Evans & Sengdala 2001), are defoliated by burning on an annual basis. There are no true epiphytes, although seedlings or even adults of some palms (e.g., oil palm) are sometimes found as chance-sown individuals in pockets of soil in the crown of trees.

Climbing palms

The morphology of climbing palms and their distribution within the family are discussed in Chapter 1. How the adaptations to the climbing habit interact with the environment has been the subject of studies by Putz (1990). He studied species of *Calamus* in Queensland and showed how rattans grow into the forest canopy, where they maintain themselves as their stems slip downwards under their own weight and their climbing organs are constantly replaced. The maturing cane accumulates on the forest floor. Tree falls result in whole rattan crowns being carried back to ground level, whence they begin to climb to the canopy once more. Recent biomechanical research (Isnard *et al.* 2005, Isnard 2006) has demonstrated the varying importance of the leaf sheaths in providing mechanical support for the very flexible stems. The physics of water movement through the immensely long rattan stems has been studied by Cobb (2006) and by Tomlinson (2006b).

Fig. 8.1 ***Calamus scipionum***, a high climbing rattan with flagella to 10 m long, Indonesia. (Photo: J. Dransfield)

Aquatics and rheophytes

There are several aquatic palms, such as *Nypa fruticans* whose seedlings can be submerged twice daily, and palms such as *Leopoldinia pulchra*, which grows in swamp forest beside black water rivers in South America, can be partially submerged for substantial periods during annual river floods (Kubitski 1991). *Calamus trachycoleus*, similarly, can survive partially submerged during floods lasting two to three months in southwest Borneo (Dransfield 1977). Seedlings of the remarkable *Ravenea musicalis* in southeast Madagascar grow as completely submerged aquatics with flaccid leaves for several years, a growth form not known to occur in the family when the first edition of *Genera Palmarum* was published (Beentje 1993).

Rheophytes are plants that are adapted to growing in the flood zone of fast-flowing rocky rivers (van Steenis 1981). Many palms, particularly in Amazonia, can withstand prolonged seasonal flooding, but these are not regarded as true rheophytes. True rheophytes usually possess narrow flexible leaves or leaflets that allow them to withstand rapidly flowing flood water. Rheophytes have evolved in many flowering plant families, including the palms, but there are only a few rheophytic palms (Dransfield 1992a). In southern China and Laos, *Phoenix roebelenii* is restricted to the rocky flood-zone of the Mekong and its tributaries, where it is abundant locally (Barrow 1994). In the Malesian region, *Hydriastele rheophytica* (Dowe & Ferrero 2000), *Calamus reticulatus* and *Heterospathe macgregori* (Baker 1997) occur in New Guinea and *Pinanga tenella* var. *tenella*, *P. tenella* var. *calamifrons*, *P. rivularis* and *Areca rheophytica* in Borneo (Dransfield 1992a). In the New World, *Geonoma linearis* occurs in Colombia (Galeano & Skov 1989) and one species of *Chamaedorea* (*C. cataractarum*) in Mexico (Hodel 1992), though there must surely be more palms adapted to this habitat in such a palm-rich area. In Madagascar, *Ravenea musicalis*, *Dypsis aquatilis* and *D. crinita* are all rheophytic. In *Pinanga rivularis* and *Geonoma linearis*, fibres in the fruit become hooked on the disintegration of the mesocarp and it may well be that the hooks play some role in the dispersal of the seed (Dransfield 1992a). Similar hooked fibres have also been reported in *Euterpe oleracea* (Moegenburg 2003),

Fig. 8.2 ***Ravenea musicalis***, above, adult palms; below, submerged seedlings with flaccid leaves, Madagascar. (Photos: J. Dransfield)

a river-bank species of Amazonia, which may also be considered to be a rheophyte. In the case of *Ravenea musicalis* (Fig. 8.2), the seed germinates while still borne on the inflorescence, the fruit falling into the stream below, rupturing and releasing the seeds; the first scale leaves of the seedling are hook-like and it is postulated that they help anchor the seedling to the streambed (Beentje 1993).

Litter-trappers

Most palms are constructed in such a way that leaf litter and debris falling into the crown is shed by the palm. The debris falling on pinnate leaves falls off between the leaflets. In the case of fan leaves, little accumulation takes place as debris falls to the ground between the elongate petioles. In some palms, however, absence of petioles or the presence of long spines along the petiole margins results in the accumulation of litter within the crown. Here, it composts, presumably making some of the nutrients released available to the palm. While the palm is still in the rosette stage, roots of other plants, particularly dicotyledonous trees, often snake their way into the composting litter in the palm crown, following the outlines of the leaf sheaths. As the palm continues to grow, these zig-zag roots become exposed when the leaf sheaths rot (Dransfield & Beentje 1995b). Good examples of litter trappers are species of *Marojejya* (Fig. 8.3) and several rattans. In *Marojejya darianii*, the apex of the leaf sheath forms a rounded auricle that partially encloses the leaves within, preventing the shedding of rain water and debris. Water accumulates in the crown along with debris and a diverse assemblage of invertebrates. In a way, this species can be regarded as a tank plant, somewhat analogous to the bromeliads of the New World. Adventitious roots grow from the internodes into the accumulating compost. The natural history of litter trapping has scarcely been studied (but see the studies of Rickson and Rickson, mentioned below).

PALM DEMOGRAPHY

Tomlinson (1979) pointed out that palms are ideal subjects for demographic research because they are easily recognized as being palms (though difficulty may be experienced in providing specific names). This usually makes recognising and charting the individuals of all ages of a given palm in one area much easier than for many dicotyledonous trees. It is also apparently much easier to estimate the growth rates of palms than those of other plants. The leaf scars are usually easily counted, and observations over a relatively short period can give an estimation of the size of the plastochron, the time interval between the production of successive leaves. This, in turn, can be extrapolated to give an estimate of the age of an individual that is crude but probably not too inaccurate, given an estimate of the number of leaves developing in the crown and the time required for establishment growth of the seedling (Corner 1966). Only very rarely (in well-documented cultivated palms) do we know absolute ages of individual palms. Estimates of the ages of palms made in the method just described have given some very surprising results. Kiew (1972) estimated that 2 m tall individuals of the Malayan undergrowth palm *Iguanura wallichiana* were 100 years old. Bullock (1980a) estimated the longevity of the undergrowth palm *Podococcus barteri* in Cameroon as 63–74 years. Sarukhán (1978) has estimated that *Astrocaryum mexicanum* can live to 70 years old. The last example, involving a moderately large palm, is less unexpected than the first two. In a study of the population structure of *Livistona eastonii*, Hnatiuk (1977) estimated that exceptionally tall individuals of this Australian species might be about 720 years old. Studies made by Savage and Ashton (1983) in the Seychelles indicated that *Lodoicea* may reach a maximum age of about 350 years, perhaps less than had been expected for this slow-growing massive species. As a final example of longevity, van Valen (1975) estimated *Prestoea montana* (as *Euterpe globosa*) to reach an age of 150 years. All the studies cited have attempted, to varying degrees, to analyze population structure and to investigate life expectancy and vulnerability of different age classes. This wide range of studies indicates the attraction of palms as subjects for demographic research; despite this, few major generalizations can yet be made.

Demographic studies have also been applied to understanding the potential harvest of economically important palms, such as the rattans and *Hyphaene*, as a basis for applying sustainable management. During studies of the West Malesian rattans, astonishing growth rates of over 7 m a year were observed for *Calamus caesius* and *C. trachycoleus* (Tan & Woon 1992).

PALM POLLINATION

Pollination in palms has been reviewed by Henderson (1986a) and evolutionary trends in pollination examined by Silberbauer-Gottsberger (1990).

For a long time, it was assumed that palms are wind-pollinated (Delpino 1870), but a growing number of detailed studies show that wind pollination is an exception and that the overwhelming majority of palms studied are insect pollinated. There is also a single well-documented instance of bat pollination in *Calyptrogyne ghiesbreghtiana* (Beach 1986, Cunningham 1995, Tschapka 2003), and a remarkable example of pollination by primates in *Eugeissona*, as yet not fully documented (Wiens, pers. comm.). It seems reasonable to predict that most palms will be shown to be insect pollinated. Large quantities of pollen, a feature of the wind-pollination syndrome, seem, at least in some palms, to be an adaptation to predation by insects. True wind-pollination does indeed occur, for example, in *Thrinax* (Read 1975) and *Howea* (Savolainen *et al.* 2006), but even the date palm, *Phoenix dactylifera*, may in fact be pollinated only secondarily by wind.

Bees and wasps appear to be rather non-specific generalist pollinators of many palms (e.g., in species of *Licuala* [Barfod *et al.* 2003] and *Calamus* [Bøgh 1996b]). A wide range of bees and beetles have been observed visiting inflorescences of *Euterpe precatoria*, many of which, together with wind, are predicted to effect pollination (Küchmeister *et al.* 1997). *Nypa*, at least in New Guinea, is pollinated by drosophilid flies (Essig 1973). Syrphid flies are responsible for pollinating *Asterogyne martiana* (Schmid 1970), whereas bees have been

Fig. 8.3 ***Marojejya insignis***, a squat litter-trapping palm, Madagascar. (Photo: J. Dransfield)

shown to pollinate *Sabal palmetto* (Brown 1976), *Ptychosperma macarthurii* (Essig 1973) and *Iriartea deltoidea* (Henderson 1985).

Beetle pollination has been observed in *Rhapidophyllum* (Shuey & Wunderlin 1977), *Hydriastele microspadix* (Essig 1973), *Pinanga coronata* (Dransfield & Saeruddin unpublished), *Salacca zalacca* (Mogea 1978), *Bactris gasipaes* and *B. porschiana* (Beach 1984), *B. guineensis* and *B. major* (Essig 1971a), *Cryosophila albida* (Henderson 1984), *Socratea exorrhiza* (Henderson 1985), *Aphandra* and *Phytelephas* (where the role of scent has been studied in detail by Ervik *et al.* 1999) and *Attalea* (Küchmeister *et al.* 1993). This method of pollination is also suggested by casual observations for many species of *Salacca*, *Daemonorops*, *Pinanga*, *Ceratolobus*, *Manicaria*, *Johannesteijsmannia* and *Pogonotium* (Dransfield unpublished). Reports by Hartley (1977) that *Elaeis guineensis* is anemophilous (i.e., wind pollinated) have been disproved by Syed (1979) who has shown the oil palm to be pollinated by at least twelve different insects in Cameroon, West Africa; among these pollinators is a weevil, *Elaeiodobius kamerunicus*, which has been successfully introduced into western Malaysia to assist in pollination in oil palm plantations (Ooi 1982). Thrips (*Thrips hawaiiensis*) had previously been the major pollinator in Malaysia, but hand pollination had been necessary to obtain high yields before the West African weevil increased fruit set. The similarity in the overall form of the rachillae in *Elaeis* and the quite unrelated *Salacca* is noteworthy, as is the similar position of the inflorescences, which are borne down among the protected leaf bases in both genera, especially as weevils are implicated as pollinators of both. Several studies have implicated similar derelomid beetles in the pollination of palms such as *Phoenix canariensis* (Meekijjaroenroj & Anstett 2003). Perhaps the most remarkable example is that of *Chamaerops humilis*. A derelomid beetle species is attracted in large numbers to the inflorescences of *C. humilis*, not by a floral scent but by a scent produced by the leaves that subtend flowers at anthesis. The minute beetles travel down the petioles to the inflorescences where they mate, the females laying eggs in inflorescence axes. The only larvae that develop further are those in the staminate inflorescences, where they eventually pupate and emerge as adults the following spring (Anstett 1998, Dufay *et al.* 2003).

Microlepidoptera have been implicated in the pollination of several species of *Calamus* in Sabah, Malaysia (Lee 1995). Flower visitation occurs at night when the flowers emit a sickly smell. Nectar is produced by the staminate flowers and by the sterile staminate flowers in the pistillate inflorescence; pollen is carried passively from the staminate to the pistillate plant. It is interesting to note that in Thailand, in different species of *Calamus*, Bøgh observed diurnal production of nectar and postulated pollination by bees (Bøgh 1996b).

Some correlations between pollen structure and pollination syndromes can be made. Beetle pollination usually seems to involve protogyny, anthesis in the whole inflorescence occurs over a relatively short period and the inflorescence is usually rather condensed or bears crowded flowers. Bee pollination seems to involve protandry, anthesis may be over an extended period, and the inflorescence is often laxly branched or the flowers are rather distant. Diverse sculpturing of the exine has been recorded in genera (such as *Pinanga* and *Daemonorops*) that we suggest are beetle pollinated, whereas closely related genera that seem to be bee pollinated have pollen with more uniform, low relief exine sculpturing. *Iriartea*, shown by Henderson (1984) to be bee pollinated, has finely reticulate exine (Ferguson 1987) whereas beetle-pollinated *Socratea* has spiny pollen grains. The significance of the extraordinary variation in exine ornamentation in *Pinanga* (Ferguson *et al.* 1983) is not known, but casual observations suggest that beetle pollination is widespread in the genus. In *Hydriastele*, an extraordinary and complex situation occurs; within the one genus, protandry and protogyny occur with differing floral morphology. Phylogenetic studies (Loo *et al.* 2006) suggest that the change from protandry to protogyny may have arisen from one to five times within *Hydriastele*, and with as many as four separate reversals to protandry. Unfortunately pollination studies have been carried out on just one species, *H. microspadix* (Essig 1973); it is protogynous and pollinated by beetles. Pollen exine patterning is known to be quite diverse within *Hydriastele*, and clearly it would be exciting to study pollination more widely in this genus.

PALM DISPERSAL

Aspects of dispersal in palms are reviewed by Zona and Henderson (1989) and in Henderson (2002b). A few species have fruits dispersed by water. The best-known examples are the coconut (*Cocos nucifera*) and *Nypa fruticans*, the very thick, fibrous mesocarps of both are seen as adaptations to floating in sea water. There are probably many other palms adapted to dispersal by freshwater, such as some of the rheophytes mentioned earlier in this chapter. However, the greatest number of palms seem to be adapted to dispersal by animals. For instance, all members of Calamoideae, except for *Eugeissona*, have either a fleshy mesocarp or a well-developed sarcotesta surrounding the seed; and the fruits of an overwhelming number of members of the Arecoideae, Coryphoideae and Ceroxyloideae have fleshy mesocarps and brightly coloured epicarps that are attractive to vertebrates. In Java, civet cats are known to feed on and defecate viable seeds of *Caryota maxima*, *Arenga pinnata* and *Pinanga coronata* (Dransfield unpublished). Gibbons feed on *Arenga obtusifolia* (Whitten 1980). Hornbills in the Malay Peninsula are known to ingest whole fruits of *Korthalsia* (Rubeli in Dransfield 1981a); they have also been observed in Borneo feeding on the fruit of *Caryota no* (Dransfield 1974). Coyotes eat fruits of *Washingtonia* (Bullock 1980b) and passage through the gut of a coyote enhances the germination of these fruits (Cornett 1985). Elephant dung in East Africa is frequently full of pyrenes of *Hyphaene* and *Borassus*, and although it is not known whether germination is enhanced, clusters of seedlings suggest, at least, that it is not impaired. Agoutis feed on fruit of *Socratea* (Yeaten 1979). Parrots feed on both mesocarp and endosperm of many cocosoid and probably other arecoid palms (see below). Palm fruits are eaten by several species of fish in Amazonia and are thereby dispersed (Gottsberger 1978). These few examples indicate the range of palms whose fruit is eaten by animals. There is, however, a great need for detailed studies of the fate of fruit to indicate whether dispersal is effected by animals or not. The study of dispersal of *Welfia regia* by Vandermeer *et al.* (1979) discusses the interplay of dispersal and predation and its demographic consequences. Seed predation has been studied in detail in *Attalea butyracea* by Janzen (1971). There often seems to be a fine balance between predation and dispersal (see discussion of predation and dispersal in cocosoid and other arecoid palms in Henderson 2002b).

Corky-warted fruits are found in a few isolated and mostly quite unrelated genera; they are known in *Chelyocarpus*, *Lemurophoenix*, *Pelagodoxa*, *Sommieria*, *Manicaria*, *Ammandra*, *Aphandra*, *Phytelephas*, *Johannesteijsmannia*, most species of *Pholidocarpus* and some species of *Licuala*. The adaptive significance of the warts is not understood. The fruits appear dull brown and inconspicuous, and usually have no indication of fruit ripeness. *A priori* it might be thought that such palms are unattractive to animals and perhaps not animal dispersed. However, fruits of *Johannesteijsmannia* and cultivated *Pelagodoxa* at least are predated upon by squirrels and rats long before they are ripe; the warts do not seem to provide protection against predation; in the case of *Johannesteijsmannia*, the warts of the ripe fruit often seem to be chewed. The more substantial, rigid, fibrous warts of the fruits of Phytelephanteae probably afford much greater protection.

OTHER PALM AND ANIMAL INTERACTIONS

Anyone who has collected herbarium specimens of palms in the tropics will be aware of the abundance of arthropods that inhabit different parts of the palm. Lepesme (1947) described in detail the various habitats on the palm that may be occupied by insects. An excellent modern overview has been recently published (Howard *et al.* 2001). In some instances, the arthropod feeds directly on the palm, and such organisms are frequently host specific. They can be of enormous economic importance to cultivated palms. Other arthropods, and indeed many other animals, inhabit palms without actually feeding on them. In a few instances, there are direct precise relationships between the palm and animal that involves morphological adaptations in the palm. Ants and aphids are involved in a very close association with some rattan species, and the morphological adaptations involved with this association have been discussed by Dransfield (1979a, 1981a). The most striking adaptations are found in some species of *Korthalsia* and in a few New Guinean and Philippine species of *Calamus*. *Laccosperma* in Africa possibly behaves similarly. In these species, the ocrea is either swollen and inflated around the sheathed stem or diverges from the sheathed stem at an acute angle, the margins inrolling and forming a chamber. Ants occupy the chambers, using them for nesting and rearing aphids on the young palm tissue toward the apparent tip of the stem (the youngest expanded leaves). The aphids are milked for honey-dew and are apparently moved on to younger tissue by the ants as the ocreas mature. When the rattans *Korthalsia hispida* and *K. robusta* are disturbed, the ants make an extraordinary sound by banging their mandibles against the dry ocreas, the ants of one ocrea followed by those of the next ocrea, the resulting pulses of hissing noises being one of the remarkable sounds of the West Malesian rain forest (Dransfield 1979a). A functionally similar adaptation associated with an ant–rattan relationship is found in several species of *Daemonorops* (e.g., *D. formicaria*, *D. verticillaris*, *D. macrophylla*, *D. crinita* and others) and one species of *Calamus* (*C. polystachys*). In these species, the leaf sheath bears collars that are tipped with fine horse-hair-like spines; some collars are reflexed, whereas other upward-pointing adjacent collars and spines sometimes interlock to form galleries in which ants make their nests. Galleries of successive sheaths are often interconnected by tunnels constructed of 'carton', a sort of papier maché made from indumentum or spicules. It has always been assumed that the close ant–rattan–aphid arrangement provides protection for the rattan against herbivores; indeed, there is some circumstantial evidence to suggest that the presence of ants deters herbivores. A novel interpretation of the relationship has recently been made by Rickson & Rickson (1986) who demonstrated the uptake of nutrients from litter accumulating in the crown of *D. verticillaris* and from the detritus and carton of the ants' nests in *D. verticillaris* and *D. macrophylla*. In these two species at any rate, the rattans obtain enhanced nutrient accumulation from the presence of ants and protection against herbivores remains a possibility. Other associations between ants and rattans are discussed in Dransfield (1979a).

PALMS AND MAN

Palms are of immense significance to man. Their uses range from material for the construction of buildings, both temporary and permanent, material for weaving, clothing, fuel, sources of food and medicine to ornamentals. They are regarded with the grasses and legumes as one of the three most economically important families of flowering plants. Whole human communities, particularly in the tropics, depend on palms for their survival. Some palms, like the coconut, date and African oil palm, have been fully domesticated and are cultivated intensively on a commercial scale. Others, such as the palmyra (*Borassus flabellifer*) which occurs in vast stands but receives only rudimentary cultivation, are nevertheless of crucial economic importance, although this importance may not be reflected in government statistics. Rural communities living near to rain forest may depend very heavily on the ability to harvest a wide range of products from the varied palm flora of the nearby forest. It is this multiplicity of uses that underlies the great significance of palms to the survival of some of the poorest people of the tropics. Throughout the tropics and subtropics, palms are under threat of over-exploitation.

The effect of human activity on palms

As tropical rain forest is cleared for agriculture, larger palms are frequently spared the axe, either because they are useful species or because their stems are too hard for easy felling. Such palms often survive the burning of the felled trees. Unlike dicotyledonous trees, palm stems do not posses a cambium and thus, lacking this superficial, easily damaged, vital zone, can survive even if the stem is partly burned. Residual palms may survive for many years giving a false appearance of vigour; they may flower and fruit freely, but they are usually unable to regenerate in the surrounding cleared land. Some palms seem genuinely adapted to surviving fire. *Cerrado*, a major vegetation type of Brazil, consisting of open woodland, scrub or savannah is subject to burning, and although these communities are edaphic rather than strictly fire climax, *cerrado* certainly seems to be adapted to fire (Eiten 1972). The palms of the *cerrado* are remarkable in their ability to survive fire. Most of the species are acaulescent with quite deep subterranean stems that grow, at least initially, positively geotropically rather than upwards. *Acrocomia hassleri, Allagoptera campestris* and *Syagrus loefgrenii* (de Medeiros-Costa & Panizza 1983), as well as *Butia* species such as *B. leptospatha* and *B. exospatha* (Noblick 2006), provide good examples. In the Indochinese region, *Calamus acanthophyllus* behaves similarly: it grows in fire-prone grasslands and scrub, it has a subterranean stem, and only its leaves and inflorescences are exposed above ground level. It can survive having its leaves completely burnt and can then resprout on the onset of rains (Evans & Sengdala 2001).

In many parts of the tropics, palms are selectively harvested from rain forest, even from supposedly fully protected areas. The maintenance of forest cover gives a spurious impression that the flora is still intact. In Madagascar, for example, palms are cut down in abundance to provide palm hearts for consumption, and in areas near to habitation, the larger palms are often completely decimated. In Laos, people utilise the forest intensively; not only are the rattans harvested for furniture and handicrafts, but the young shoots of rattans along with many other species of palms are indiscriminately harvested for sale as food, with little thought for sustainability. Where forest is selectively logged, there may be a dramatic effect on the remaining palm flora; the opening of major light gaps can stimulate the growth of some species but scorch and kill others. The effects of logging on the abundance and regeneration of palm communities have been studied in detail in lowland West Malaysia (Nur Supardi *et al.* 1998).

The effects of fragmentation of forest on palms and their dispersal agents have been studied by Galetti *et al.* (2006). They showed that the dispersal of *Astrocaryum aculeatissimum*, which is normally dispersed by scatter-hoarding rodents, became seriously affected when forest fragments became too small to sustain viable populations of the dispersers. Fruit accumulated beneath parent trees then became much more prone to predation by seed-boring insects. Similarly, forest fragmentation has had a disastrous effect on the intricate parrot dispersal systems of South American palms. Yamashita and co-workers have studied the close interdependence of parrots and palms in South America, distinguishing between palm-fruit specialists and palm-nut specialists. For example, Blue-throated macaws (*Ara glaucogularis*) are palm-fruit specialists that feed on the mesocarp of *Attalea phalerata* fruit (Yamashita & de Barros 1997). By contrast, palm-nut specialists, such as *Anodorhynchus* macaws (Yamashita & Valle 1993) that feed on *Butia* fruits, often tackle a very narrow range of palm-fruit sizes and specialise in endosperm extraction. When the palm populations are fragmented by land clearance, the food supply for the macaws falls below a critical level and the parrots become extinct within the area affected. Similarly, the Yellow-eared parrot (*Ognorhynchus icterotis*) is heavily dependent for survival on *Ceroxylon quindiuense* in Colombia (Krabbe 2000).

Conservation problems

Land-use pressure and overexploitation are currently the major threats to the survival of many species of palms, whereas in the future, we may expect climate change to play an increasingly serious role in the survival of palms. The prospects for some species are extremely poor and it seems likely that some palms have already become extinct or are dangerously near becoming so. Yet the arguments for saving palms from extinction are compelling. The immense economic importance of the family is very clear and wild palm populations have great importance as the source of genetic material for the agriculture of the future. This has been underlined by recent interest throughout the tropics in the development of sylvicultural or agroforestry systems including previously wild palms, such as *Oenocarpus bataua*, *Calamus manan* and *Attalea* species. Besides, whole human communities living within rain forest or on the forest boundary depend for much of their livelihood on wild palms. Although rarely quantified and entering government statistics,

Fig. 8.4 ***Lepidorrhachis mooreana***, confined to a narrow band of vegetation at the very summit of Mt. Gower, Lord Howe Island. (Photo: W.J. Baker)

production from these palms is the basis of human survival in such communities. The prospect of palm extinction should thus be of great concern. Cultivation of rare palms may well save them from total extinction, and the International Palm Society has certainly played a very important role in the popularising and wide dissemination of rare and unusual species. However, the only way to save viable populations of endangered palms from extinction is by the protection of their habitats. This is a great challenge for all palm enthusiasts in the next decade. The arguments for palm conservation can be easily made, but governments and international agencies require education and persuasion.

Currently, only eight palms are listed under the Convention on Trade in Endangered Species (CITES). One (*Dypsis decipiens*) is listed under Appendix I; trade in this species is permitted only under exceptional circumstances. The other seven are listed under Appendix II, under which seeds, the usually traded form of the listed palms, are exempt. It is difficult to see how the listing of these species, all from Madagascar, will have any positive effect on their conservation. Indeed, perhaps the only effect of the listing will be to make it more difficult to collect and exchange scientific specimens of these taxa, activities that are of fundamental importance in the collection of baseline data aimed at their conservation.

The impatience of palm growers, particularly those involved in the development of instant landscape projects using mature or semi-mature palms, coupled with lax quarantine, has resulted in the transfer of serious pests and diseases of palms in some parts of the world. In particular, the weevil *Rhynchophorus ferrugineus* and the moth *Paysandisia archon* are beginning to have a disastrous effect on populations of cultivated palms in Europe. The former was introduced to southern Europe with mature *Phoenix dactylifera* plants (Ferry & Gómez 2002), whereas the latter was probably introduced to Europe from Argentina with mature *Trithrinax campestris* (Drescher & Dufay 2002). There is also evidence that native *Chamaerops humilis* in the wild may be under threat from these pests.

When the first edition of *Genera Palmarum* went to press in 1986, several palms had not been seen in the wild for several decades. For example, *Medemia argun* had not been seen alive in the wild since 1963 (Boulos 1968) and *Beccariophoenix madagascariensis* not since 1947, both were considered possibly to be extinct. *Medemia argun* has since been rediscovered in Sudan by Gibbons and Spanner (1996), its persistence in small populations in Egypt being confirmed subsequently (Ibrahim & Baker pers. comm.). *Beccariophoenix madagascariensis* was rediscovered in Madagascar in 1986 (Dransfield 1988a). Since its rediscovery, *Beccariophoenix* has been found in at least six localities. However, the populations of these palms in all localities are fragile and severely threatened by exploitation for palm heart, despite the fact that one population occurs within a national park. In Madagascar, palms known in 1995 only from fragmentary herbarium specimens and assumed to be on the verge of extinction if not already extinct have subsequently been rediscovered in viable populations. There is, however, no doubt that palms throughout the tropics are threatened by a wide range of human activities. In many instances, conservation status can only be inferred from the knowledge of a few fieldworkers, and in some instances, the population of the palm may be so restricted and the locality and habitat so well-known that we can say with certainty that the palm is on the verge of extinction. *Pritchardia viscosa* in Kauaii, Hawaii, a beautiful species, is now restricted to a mere four mature individuals and some seedlings, while *Hyophorbe amaricaulis* is now restricted to a single individual in cultivation in Curepipe Botanic Garden, Mauritius, where all attempts to propagate it have failed, even those using micropropagation techniques.

Palm seed can be recalcitrant (dies when dried) or orthodox (survives drying) in their germination behaviour, but generally do not lend themselves well to conservation in seed banks due to the typically large size of the seeds and the fact that a significant number do not withstand long-term storage (Pritchard *et al*. 2004). The creation of protected areas that are properly controlled is undoubtedly the most effective way of conserving palms. However, at present, we have little idea of the resilience of palms to climate change and it may well be that the family will suffer massive extinctions in the near future as the world's climate changes. For example, *Lepidorrhachis mooreana* (Fig. 8.4), confined to a very narrow band of vegetation at the summit of two peaks on Lord Howe Island, is extremely vulnerable to any vegetation change that may occur as a result of the shifting of the cloud base during climate change (Baker & Hutton 2006). *Ex situ* cultivation may prove the last resort for the conservation of some species, but there are serious problems with this conservation method, such as maintenance of genetic diversity within the cultivated population of a species and prevention of hybridisation with a related species, which are not easily addressed in living collections. Furthermore, palms in living collections in botanic gardens and private collections are often vulnerable to pest attack (e.g., the decimation caused by lethal yellowing), freak weather (such as hurricanes or cold snaps), lack of continuity in funding and, in the case of private collections, the death of the owner. Nevertheless, an astonishing number of palm species are already maintained in botanic gardens collections (Maunder *et al*. 2001).

The genetic diversity of highly reduced populations of endangered palms is an essential consideration in the drawing up of management plans for conservation. A few studies have been carried out, for example the work by Shapcott on *Pinanga* species, *Ptychosperma bleeseri* (= *P. macarthurii*) and *Carpentaria acuminata* (Shapcott 1998a, 1998b, 1999, 2000). Shapcott has made studies on populations of *Beccariophoenix* in Madagascar, and such information can give indications of genetic erosion and population viability (Shapcott *et al*. 2007).

In 1996, the Palm Specialist group of the Species Survival Commission of IUCN produced an action plan for the conservation of the world's palms (Johnson 1996). This plan has done much to highlight and publicise those palms that are most seriously threatened. What is now needed is the implementation of conservation recommendations.

Chapter 9

CLASSIFICATION OF PALMS

HISTORY AND DEVELOPMENT OF PALM TAXONOMY

The pre-eminence and beauty of palms in tropical landscapes ensured that early European naturalists visiting the tropics would marvel at them as plants and be aware of their many uses. Pre-Linnean studies of natural history, such as Rumphius' *Herbarium Amboinense* (1741–1755), van Rheede tot Drakenstein's *Hortus Indicus Malabaricus* (1678–1693) and Kaempfer's *Amoenitatum exoticarum* (1712) introduced European scholars to the details of palms. Some of the early ethnobotanical descriptions of the coconut by Rheede and of the sago palm by Rumphius read with a freshness and accuracy that has scarcely been improved upon. Furthermore, Rumphius was aware of a considerable diversity of palms. He described or referred to over fifty different taxa, a number that must include most of the palms to be found in the relatively rich area of the central Moluccas.

Linnaeus knew only *Chamaerops* from living material and had seen herbarium material of *Calamus*. Other palms he knew of from the writings of these great early naturalists, and he based the few names of palms that he published on their accounts, representing nine ethnobotanically important genera (Linnaeus 1753). In recognizing these palms as being related to each other but isolated from other flowering plants, Linnaeus laid the basis for future classification. Linnean palm names have now been typified because it has been possible to decide the elements on which he probably based his names (Moore & Dransfield 1979). Names published just after Linnaeus' time by botanists such as Giseke (1792) are far more problematical; impressed by the order and convenience of the Linnean binomial system, these botanists published numerous names by putting into binomials names encountered in pre-Linnean works. No types exist for most of these names, and interpretation of their origins is difficult; yet they are validly published and should be respected under the International Code of Botanical Nomenclature.

In the late eighteenth and early nineteenth centuries, exploration by European naturalists such as Alexander von Humboldt increased. Through accounts of their travels and through their specimens, the palms gradually became better known and described in detail. Botanists began to assemble a framework within the family to reflect differences and similarities. In 1824, Martius made the first significant attempt to order the palms in *Palmarum Familia*, arranging the genera in six series: Series I *Sabalinae*, Series II *Coryphinae*, Series III *Lepidocarya*, Series IV *Borasseae*, Series V *Arecinae* and Series VI *Cocoinae*. Although these 'series' might be interpreted as subfamilies, they are unfortunately not validly published at that rank. Later, when Martius published the monumental *Historia Naturalis Palmarum* (1823–1850), the six 'series' of palms were elevated to the rank of family, the order *Principes* thus comprising six families. Martius' series were based on the number of bracts and their completeness, and the number of carpels or locules. The series were further divided using leaf characters, but they were not named. The resulting series seem very heterogeneous to our modern eye. Martius' work was followed by the efforts of others: Lindley (1830), Beilschmied (1833) and Burnett (1835) divided the palms into many groups referred to as sections or tribes. It could be argued that these divisions are implicitly subfamilies, but with the confusion of different names and different ranks it seems preferable to use only those names explicitly published as subfamilies. Such names appear not to have been published until William Griffith (1844, 1845) wrote his account of the Palms of British India in the *Calcutta Journal of Natural History*. He regarded the palms as a single family and clearly divided the Asiatic palms with which he was familiar into subfamilies. His names are among the first published explicitly as subfamilies (although his endings '-inae' do not accord with modern usage). It is to him that we owe the first use of Calamoideae, Nypoideae and Coryphoideae, explicitly as subfamilies (as Calaminae, Coryphinae and Nipinae).

While Griffith was working on Indian and Malayan palms, Carl Ludwig Blume worked in Java, describing (among many other flowering plants) the palms of Java, Sumatra, and other islands of the then Dutch East Indies. Although he ordered the palms, he published no acceptable names of higher categories. At the generic and species level, however, his contribution to palm botany is very significant. The nomenclatural problems caused by the almost simultaneous publication of new names by Martius, Griffith and Blume have been clarified by Dransfield and Moore (1982).

In the New World, Alfred Russel Wallace and Richard Spruce carried out major explorations of the South American tropics in the 1850s and early 1860s. Both naturalists were much impressed by the palms and their articles on them are of lasting significance. Wallace (1853) published a popular book on palms of the Amazon and their uses, whereas Spruce (1871) presented careful detailed descriptions of many South American palms.

The accumulating body of information about palms formed

Opposite: ***Orania trispatha***, Masoala Peninsula, Madagascar. (Photo: J. Dransfield)

the basis of the most significant system of classification after William Griffith's, that of J.D. Hooker (1883), which was published in Bentham and Hooker's *Genera Plantarum*. Hooker arranged all palm genera known to him in six tribes, further divided into subtribes. He saw no reason to recognize a third rank and his tribes are equivalent to Griffith's subfamilies. To Hooker, we owe many of the subtribes recognized in the present system. Hooker's system is notable for its thoroughness and conciseness and for the accuracy of the descriptions. The alignment of the 132 genera known to him has been followed by many subsequent authors and even the composition of some subtribes (e.g., Linospadicinae) has not changed at all in the present work. Hooker's system of classification synthesizes the work of his predecessors. The importance of his account is not so much the originality of his concepts as the clear, concise order he introduced into the family.

The first phylogenetic classfication of the family was that of Drude (1887) in Engler and Prantl's *Die Natürlichen Pflanzenfamilien*. Drude recognized the unspecialized nature of the Coryphoideae, Calamoideae and Nypoideae.

Towards the end of the nineteenth century, Barbosa Rodrigues made a detailed floristic study of the palms of Brazil. His work culminated in two magnificent folio works closely modelled on Martius' *Historia Naturalis Palmarum*, which were published as *Sertum Palmarum Brasiliensium* in 1903.

The greatest palm taxonomist of the latter half of the 19th and the early 20th century, was undoubtedly the Italian botanist Beccari. Beccari's considerable field experience in the Malesian region, in particular in Borneo, is very evident in his writing. He was certainly aware of variation in palm species and when some of his taxa have proven to be conspecific with previously described species, his descriptions are usually so careful and complete that critical realignment is easily accomplished. Beccari concentrated on the palms of Asia, and to a lesser extent on those of Africa and Madagascar. His results were published in a wide range of journals but culminated in a magnificent series of volumes of the *Annals of the Royal Botanic Garden, Calcutta*, illustrated, for the most part, by natural-size photographic plates of types or representative specimens. These monographs of the rattan palms and of the Corypheae remain the most important works on those groups.

Although Beccari published a prodigious number of exceptional works, he left many projects unfinished when he died in 1920. One of these was a monograph of the genera of Old World arecoid palms. Beccari began his work on Old World arecoid palms by continuing the studies initiated by Scheffer (1873, 1876) of arecoid genera cultivated in the Botanic Gardens at Bogor. Beccari gradually accumulated an unparalleled knowledge of this most complex subfamily. His account was finally edited and published by Pichi-Sermolli (Beccari & Pichi-Sermolli 1955). This extremely important work contains a synopsis of Beccari's classification of the whole family and a very detailed classification of subfamily Arecoideae (*sensu* Beccari).

Burret, working in Berlin and with relatively very little experience in the field, replaced Beccari in the 1920s and 1930s as the most prolific palm student. He concentrated initially on palms from the New World but soon monographed those of Arabia, China, and the Pacific. His specific and to some extent generic concepts are generally regarded today as being narrow, and many Burret names have been reduced to synonymy. However, his published works on palms are very extensive. Burret too was working towards a *Genera Palmarum*. A virtually completed manuscript of the family for *Die Natürlichen Pflanzenfamilien* was destroyed during the second World War. It appears that Burret never recovered from the demoralizing effects of the loss of the manuscript, but an outline of his system was published by Eva Potztal (1964). The classification is certainly of considerable significance.

Saakov (1954) published a system of classification based primarily on the form of the leaf — whether palmate or pinnate — and secondarily on the method of germination. This system results in the wide separation of genera that, based on characters other than germination type, are manifestly closely related (e.g., *Butia* and *Jubaea*). Although Saakov's system seems artificial and less usable, we owe to him the publication of the subtribal name Livistoninae.

Satake (1962) also published a classification of the family based largely on Burret's system, but with rather greater emphasis on leaf vernation. Leaf vernation has proved to be less reliable as a major taxonomic character than was originally thought, and Satake's system, in relying on this character, also separates some apparently closely related taxa. We owe the names Ceroxyleae and Trachycarpeae to Satake.

Corner (1966) did much to vitalise interest in the palm family with his inspiring book on the natural history of palms, but did not make any direct contribution to the formal taxonomy of palms.

The next most significant event in the development of palm classification is the publication of Moore's paper, 'The major groups of palms and their distribution' (Moore 1973a). Moore arranged the genera, as well as his major groups in order of assumed specialization, based on his immense experience and the best techniques then available. He used an included list of characters to justify the arrangement. Moore's groupings were widely followed; for more than a decade they formed a basis for serious palm students, as they did for the first edition of *Genera Palmarum* (Uhl & Dransfield 1987) and for two other contributions to palm classification (Dahlgren *et al.* 1985; Imchanitzkaja 1985). The major classifications of the family are compared in Table 9.1.

Moore recognized five major lines of evolution: the first including the coryphoid, phoenicoid and borassoid major groups, the second comprising just the lepidocaryoid major group, the third consisting of the nypoid major group, the fourth comprising the caryotoid major group, and the remaining major groups forming the fifth line. Within this fifth line, Moore recognized nine major groups that he arranged in three further groups. Thus, within Moore's classification, there was a considerable amount of structure. None of Moore's major groups had any formal nomenclatural status. In not indicating ranks for the entities, he left the way open for a decision on whether to divide the whole family into several families or whether to leave it as one and recognize all fifteen major groups as subfamilies or tribes.

Basing their work on Moore's paper, Dransfield and Uhl (1986) and Uhl and Dransfield (1987) reassessed all the groupings in Moore's informal classification, modified generic placements and the hierarchy of groupings, and produced a new formal classification of the whole family. The 1986 paper formalized the nomenclature of the classification, which was then published in full detail in the first edition of *Genera Palmarum*. Dransfield and Uhl maintained Moore's first three lines as subfamilies Coryphoideae, Calamoideae, and Nypoideae. His fourth line, the Caryotoid major group, they included in the large subfamily Arecoideae. Finally, they split Moore's fifth line, the large arecoid group, into three subfamilies: Ceroxyloideae, Arecoideae and Phytelephantoideae. The major difference between Moore's classification and that of Uhl and Dransfield (1987) is the position of the Caryotoid palms. Although considered by Moore (1973a) as the fourth separate line of evolution, Uhl and Dransfield felt that the group had so many similarities with the Arecoideae that they included it therein as a tribe. Subsequent research, described in detail elsewhere in this volume, has shown that the caryotoid palms are misplaced within the Arecoideae. Uhl and Dransfield (1987) accepted six subfamilies in all: Coryphoideae, Calamoideae, Nypoideae, Ceroxyloideae, Arecoideae and Phytelephantoideae.

A summary of the classification published in the first edition of *Genera Palmarum*, with some additional genera and new synonymy, was published by Dransfield and Uhl in volume 4 of the *Families and Genera of Vascular Plants* (Dransfield & Uhl 1998).

The wealth of phylogenetic research described in Chapter 6 has brought us to an important point in palm systematics. The available phylogenetic evidence is now robust enough to form the basis of the classification that we present in this book. The new phylogenetic classification is significant for two reasons. First, it is more informative than previous classifications because it reflects evolutionary relationships determined through rigorous analysis. Although previous classifications were based on meticulous comparative studies of morphology and anatomy, the groups that they contain arose from subjective interpretations of the data. In many instances, these groupings have been corroborated by phylogenetic studies, but they are vulnerable to error primarily due to misleading parallelisms and convergences. Phylogenetic classifications are by no means infallible, but they are more explicit and repeatable. Second, the new classification is potentially more stable than previous classifications because it is built around a robust phylogenetic framework upon which numerous independent datasets have converged.

How was this classification devised? The authors of this book met for one week in early June 2004 at the L.H. Bailey Hortorium, Cornell University. During this week, we collectively evaluated all published phylogenetic research on palms, as well as the results of some ongoing, as-yet-unpublished research. We balanced the phylogenetic evidence against our own knowledge of palm systematics, morphology and anatomy. The outcome of this meeting was published the following year as a new phylogenetic classification of palms (Dransfield *et al.* 2005), and we have continued to refine the classification, resulting in some modifications to orthography, nomenclature and circumscription that are summarised below.

Our classification consists of three parts: a set of accepted genera, a linear sequence of these genera and a hierarchy of nested, higher level groups into which the genera are placed. Branching diagrams do not readily submit to the production of a linear classification sequence and compromises are inevitably made. We found that the following method ensured that related genera were generally placed close to each other in a logical manner. We worked through topologies node by node, from the base of a tree towards its tip. At each node, we dealt with the smaller of the two alternative sister groups first. For example, at the basal node of the family, we were faced with the divergence between subfamily Calamoideae and the remainder of the family, thus we tackled subfamily Calamoideae first. Within the Calamoideae, the most robust phylogenies indicated that *Eugeissona* is sister to all other Calamoideae and was thus picked off first, in this manner becoming genus number one in the classification. A similar approach has been taken to produce a linear, phylogenetic sequence of angiosperm families (Haston *et al.* 2007). In cases where poor resolution hampered this logical progression through topologies, genera and, in some cases, higher groups were ordered alphabetically. The sequence, therefore, merely reflects our attempt to devise a linear sequence, with all its compromises, and has nothing to do with perceptions of how 'primitive' or 'derived' any given genus is.

In determining which groups to recognise at the subfamily, tribe or subtribe level, we employed a number of criteria. First and foremost, we demanded evidence of a group's monophyly, preferably with high support from a confidence measure, such as bootstrap support. In two cases (Linospadicinae and Basseliniinae), we have recognised groups of genera that have not been resolved as monophyletic in any phylogenetic analysis. Such decisions are justified by morphological evidence and by potential failings in the relevant phylogenetic analyses. Second, wherever possible, we endeavoured to maintain stability by recognising groups that are familiar to users from the previous classification, even though a change of rank may have been required. Third, we have tried to recognise only groups that can be diagnosed morphologically. This has been challenging in some instances (e.g., Ceroxyloideae, Carpoxylinae and Dypsidinae), but we have been broadly successful in this respect. In two instances, in tribes Areceae and Trachycarpeae, we were able to place

TABLE 9.1
Comparison of palm classifications

Present classification (Subfamilies and tribes)	Uhl & Dransfield 1987 (based on Dransfield & Uhl 1986, see also Dransfield & Uhl 1998)	Moore 1973 (without rank)	Potztal 1964
I. Calamoideae	II. Calamoideae	IV. Lepidocaryoid palms	4. Lepidocaryoideae
Eugeissoneae	Calameae (in part)		
Lepidocaryeae	Lepidocaryeae and Calameae (in part)		
Calameae	Calameae		
II. Nypoideae	III. Nypoideae	V. Nypoid palms	2. Nypoideae
III. Coryphoideae	I. Coryphoideae	I. Coryphoid palms	5. Coryphoideae
Sabaleae	Corypheae (in part)		
Cryosophileae	Corypheae (in part)		
Phoeniceae	Phoeniceae	II. Phoenicoid palms	6. Phoenicoideae
Trachycarpeae	Corypheae (in part)		
Chuniophoeniceae	Corypheae (in part)		
Caryoteae	Caryoteae (V. Arecoideae)	VI. Caryotoid palms	8. Caryotoideae
Corypheae	Corypheae (in part)		
Borasseae	Borasseae	III. Borassoid palms	3. Borassoideae
IV. Ceroxyloideae	IV. Ceroxyloideae	Arecoid Line	
Cyclospatheae	Cyclospatheae	VII. Pseudophoenicoid palms	Ceroxyleae (Arecoideae)
Ceroxyleae	Ceroxyleae	VIII. Ceroxyloid palms	Ceroxyleae (Arecoideae)
Phytelepheae	VI. Phytelephantoideae	XV. Phytelephantoid palms	9. Phytelephantoideae
V. Arecoideae	V. Arecoideae	Arecoid Line (continued)	7. Arecoideae
Iriarteeae	Iriarteeae	X. Iriarteoid palms	Iriarteeae (Arecoideae)
Chamaedoreeae	Hyophorbeae (VI. Ceroxyloideae)	IX. Chamaedoreoid palms	Ceroxyleae and Chamaedoreeae (Arecoideae)
Podococceae	Podococceae	XI. Podococcoid palms	Dypsidae (Arecoideae)
Oranieae	Areceae (in part)		Areceae (Arecoideae)
Sclerospermeae	Areceae (in part)		Areceae (Arecoideae)
Roystoneeae	Areceae (in part)		Areceae (Arecoideae)
Reinhardtieae	Areceae (in part)		Areceae (Arecoideae)
Cocoseae	Cocoeae	XIII. Cocosoid palms	1. Cocosoideae
Manicarieae	Areceae (in part)		Areceae (Arecoideae)
Euterpeae	Areceae (in part)		Areceae (Arecoideae)
Geonomateae	Geonomeae	XIV. Geonomoid palms	Geonomeae (Arecoideae)
Leopoldinieae	Areceae (in part)		Areceae (Arecoideae)
Pelagodoxeae	Areceae (in part)		Areceae (Arecoideae)
Areceae	Areceae (in part)		Areceae (Arecoideae)

Satake 1962 (as subfamilies)	Drude 1887 (as subfamilies)	Hooker 1883 (as tribes)	Martius 1849–1853 (as families)
	III. Lepidocaryinae	IV. Lepidocaryeae	5. Lepidocaryinae
IV. Calamoideae			
III. Lepidocaryoideae	Mauritieae		
IV. Calamoideae	Calameae		
X. Nypoideae	V. Phytelephantinae	I. Areceae dubiae affinitatis	6. Palmae Heteroclitae
II. Coryphoideae	I. Coryphinae	III. Corypheae	3. Coryphinae
Corypheae	Coryphinae tribe Sabaleae		
Corypheae			
V. Phoenicoideae	Coryphinae tribe Phoeniceae	II. Phoeniceae	Phoenicinae
Trachycarpeae			
Corypheae			
VII. Caryotoideae	Ceroxylinae-Arecinae subtribe Caryoteae	I. Areceae subtribe Caryotideae	Caryotinae (Arecinae)
Corypheae			
I. Borassoideae	II. Borassinae	V. Borasseae	2. Borassinae
Ceroxyleae (Arecoideae)	Ceroxylinae-Arecinae-Morenieae		
Ceroxyleae (Arecoideae)	Ceroxylinae-Arecinae-Iriarteeae	I. Areceae subtribe Ceroxyleae	1. Arecinae
VIII. Phytelephantoideae	V. Phytelephantinae	I. Areceae dubiae affinitatis	6. Palmae Heteroclitae
VI. Arecoideae			1. Arecinae
Iriarteeae	Ceroxylinae-Arecinae-Iriarteeae	I. Areceae subtribes Iriarteeae, Wettinieae	
Ceroxyleae and Chamaedoreeae	Ceroxylinae-Arecinae-Morenieae	I. Areceae subtribe Chamaedoreeae	
Dypsideae	Ceroxylinae-Arecinae-Geonomeae	I. Areceae subtribe Geonomeae	
Oranieae	Ceroxylinae-Arecinae-Caryoteae	I. Areceae subtribe Caryotideae	
Oranieae	Ceroxylinae-Arecinae-Caryoteae	I. Areceae subtribe Caryotideae	
Dypsideae	Ceroxylinae-Arecinae-Areceae	I. Areceae subtribe Oncospermeae	
Ptychospermeae	Ceroxylinae-Arecinae-Morenieae	I. Areceae subtribe Malortieae	
IX. Cocosoideae	IV. Ceroxylinae tribe Cocoinae	VI. Cocoineae	4. Cocoinae
Phytelephanteae (Phytelephantoideae)	Ceroxylinae-Arecinae-Geonomeae	I. Areceae dubiae affinitatis	
Ptychospermeae	Ceroxylinae-Arecinae-Areceae		
Geonomeae	Ceroxylinae-Arecinae-Geonomeae	I. Areceae subtribe Geonomeae	Arecinae Alveolares
Oranieae	Ceroxylinae-Arecinae-Geonomeae	I. Areceae dubiae affinitatis	
Phytelephanteae (Phytelephantoideae)			
Arecoideae tribe Areceae	IV. Ceroxylinae tribe Arecinae	I. Areceae	

genera to tribe, but lacked sufficient evidence to place them all to subtribe. We have left these poorly understood genera unplaced within their respective tribes in the hope that future research will clarify their relationships. In one case, this has already happened; since the publication of the new phylogenetic classification (Dransfield *et al.* 2005), in which *Lepidorrhachis* was included among the unplaced Areceae, sufficiently convincing morphological evidence has been obtained to justify the inclusion of the genus within subtribe Basseliniinae (Pintaud & Baker 2008).

We acknowledge that we recognise many genera whose monophyly has not yet been tested. We also accept genera, such as *Calamus*, that we know not to be monophyletic; in this case, we lack sufficient phylogenetic evidence to suggest an alternative classification of this genus and its close relatives. These decisions are pragmatic and were made because phylogenetic knowledge at the species level is inadequate for the majority of palm genera.

In addition to the implicit phylogenetic information content of the classification, we have included a summary of the phylogenetic results available for each subfamily, tribe, subtribe and genus. In preparing this overview, we have drawn on all available studies on the phylogeny of palms, more than 40 sources in total. Where choices existed, we have worked mostly from maximum parsimony trees with bootstrap support. When numerous trees were presented (e.g., Hahn 2002a), we have mostly deferred to the topologies based on all available data, the so-called total evidence trees. When ambiguity exists, we favour the phylogenies based on the largest datasets or the most complete sampling of taxa. In describing the level of support as low, moderate or high, we are referring explicitly to formal evaluations of confidence through bootstrap or jackknife analysis. By 'low support', we mean that a given node has less than 50% bootstrap or jackknife support, or that a support value is not available. 'Moderate support' relates to nodes with 50–89% bootstrap or jackknife support, and 'high support' refers to those nodes with values of 90% or more.

FAMILY DESCRIPTION

Order: **Arecales** Bromhead, Mag. Nat. Hist., n.s. 4: 333 (1840). (*Principes* Endl., Gen. pl. 244 (1837), alternative name).

Family: **ARECACEAE** Schultz Sch., Nat. Syst. Pflanzenr. 317 (1832) (conserved name).
Palmae Juss., Gen. pl. 37 (1789).

Small, medium-sized or large, solitary or clustered, armed or unarmed, hapaxanthic or pleonanthic, hermaphroditic, polygamous, monoecious or dioecious plants. *Stems* 'woody', slender to massive, very short to very tall, creeping, subterranean, climbing or erect, usually unbranched aerially, rarely branching dichotomously, lacking cambium but sometimes increasing in diameter by diffuse growth, sometimes ventricose, internodes very short to elongate, leaf scars conspicuous or not, stilt roots present or absent, roots adventitious, sometimes modified into spines. *Leaves* alternate, spirally arranged, rarely distichous, later frequently splitting, unarmed or armed with spines or prickles, glabrous or variously scaly or hairy, sometimes with a ligule-like appendage on either side or in front of the petiole, sheaths sometimes forming a crownshaft; petiole usually present, terete or variously channelled or ridged, unarmed or bearing spines or teeth, glabrous or variously scaly or hairy; hastulae present or absent; blade palmate, costapalmate, pinnate, bipinnate or bifid, or entire and pinnately veined, plicate in bud, splitting along the adaxial folds (induplicate) or abaxial folds (reduplicate), rarely splitting between the folds, or not splitting; segments or leaflets lanceolate or linear to rhomboid, or wedge-shaped, V-shaped in cross-section (induplicate) or Λ- shaped (reduplicate), single-fold or many-fold, a midrib and numerous parallel secondary veins usually present, segments very rarely splitting further between the secondary veins, tips acute, acuminate, truncate, oblique or bifid, praemorse or irregularly toothed or lobed, sometimes armed with spines or bristles along the margins and/or main veins, variously scaly and hairy, transverse veinlets conspicuous or obscure; proximal leaflets sometimes modified as spines (acanthophylls), rachis prolonged distally into a climbing whip (cirrus) in many climbing palms, sometimes also bearing acanthophylls. *Inflorescence* axillary, solitary or multiple, infrafoliar, interfoliar or aggregated into a suprafoliar compound inflorescence, spicate or branched to up to 6 orders, usually maturing acropetally, rarely basipetally, in some species of *Calamus* inflorescence modified as a climbing whip (flagellum); peduncle short to long; prophyll usually 2-keeled, very varied in shape and size, rarely subtending a first-order branch; peduncular bracts 0–many, very varied in shape and size; rachis shorter or longer than the peduncle; rachis bracts similar to peduncular bracts, or dissimilar, or much reduced; rachillae (flower-bearing branches) short to long, slender to massive, rachilla bracts conspicuous or minute or apparently lacking, sometimes connate laterally and adnate to the rachilla to form pits containing the flowers. *Flowers* hermaphroditic or unisexual, then similar or dimorphic, sessile or stalked, borne singly or in cincinni of various forms including dyads, triads or acervuli, or rarely in short monopodial clusters; perianth rarely of similar parts, usually differentiated into sepals and petals, rarely uniseriate with a variable number of lobes; sepals (2) 3 (rarely more), distinct or variously connate, usually imbricate or basally connate, rarely valvate or widely separated; petals (2) 3 (rarely more), distinct or variously connate, valvate, imbricate or imbricate with briefly valvate tips; stamens (1–3) 6 (or many, up to 950 or more), filaments erect or inflexed in bud, free, or variously connate, or adnate to the petals, or both connate and adnate, anthers basifixed or dorsifixed, rarely didymous, or with widely separated anther sacs, straight or rarely twisted, introrse, latrorse, extrorse, or very rarely opening by pores; pollen circular or elliptic in polar view, inaperturate, monoporate, diporate, triporate, monosulcate, disulcate, tri- or tetrachotomosulcate, exine intectate or tectate, very varied in ornamentation; staminodes ranging from tooth-like to well developed, distinct or connate, sometimes adnate to the petals or gynoecium, rarely absent; gynoecium apocarpous with (1–2) 3 (4) carpels, or variously syncarpous with 3 or rarely more (to 10) locules, or pseudomonomerous with 1 fertile locule, carpels follicular or rarely ascidiform, glabrous, variously hairy, or covered with imbricate scales, or spiny, styles distinct or connate or not clearly differentiated, stigmas erect or recurved, rarely indistinct; ovule solitary in each locule, anatropous, hemianatropous, campylotropous or orthotropous, basally, laterally or apically attached, crassinucellate, integuments 2, the outer wide, the inner narrow, the outer or the inner or both integuments forming the micropyle; pistillode present or absent in the staminate flower, ranging from minute and often trifid to large, bottle-shaped, exceeding the stamens. *Fruit* usually 1-seeded, sometimes 2–3–10-seeded, ranging from small to very large, stigmatic remains apical, lateral or basal; epicarp smooth or hairy, prickly, corky-warted, or covered with imbricate scales, mesocarp fleshy, fibrous or dry, endocarp not differentiated, or thin, sometimes with an operculum over the embryo, or thick and then often with 1–3 or more pores at, below or above the middle. *Seed* adhering to the pericarp or free, with thin or sometimes fleshy testa (sarcotesta), endosperm homogeneous or ruminate, sometime penetrated by the testa; embryo apical, lateral or basal. *Germination* adjacent-ligular, remote-ligular or remote-tubular; eophyll simple and entire, bifid, palmate, or pinnate. Type: **Areca** L.

The family is divided into five subfamilies:
Calamoideae, **Nypoideae**, **Coryphoideae**, **Ceroxyloideae** and **Arecoideae**.

Key to subfamilies

1. Ovary and fruit covered in imbricate scales; flowers hermaphroditic or unisexual, but only rarely dimorphic, arranged singly or in dyads or rarely in cincinni . . . . **Calamoideae**
1. Ovary and fruit glabrous, or with peltate or basifixed scales, hairs, corky warts, or spines, but not with imbricate scales; flowers hermaphroditic or unisexual, often dimorphic, borne singly or in triads or in pairs derived from triads, or in cincinni 2
2. Pistillate flowers borne in a terminal head, each flower with 3(–4) free, large, asymmetrical carpels and 6 minute perianth segments; staminate flowers crowded on spikes at the tips of the inflorescence branches, below the pistillate head, each flower with 6 linear distinct perianth segments and 3 anthers borne on a solid stalk . . . . . . **Nypoideae**
2. Pistillate flowers not borne in a terminal head, or if so, then plants dioecious and flowers multi-parted; staminate flowers with stamens filaments free or variously connate, very rarely forming a solid stalk 3
3. Leaves splitting along adaxial folds to give induplicate segments or leaflets, the leaf palmate or pinnate, or bipinnate, rarely splitting between folds or along abaxial folds (and then the leaf always palmate), or entire but with apical lobing representing very short induplicate segments . . . . **Coryphoideae**
3. Leaves splitting along abaxial folds to give reduplicate leaflets, the leaf always pinnate, or entire bifid, pinnately ribbed . . 4
4. Flowers usually unisexual and the plants dioecious, rarely hermaphroditic, borne singly or in monopodial clusters . . . . **Ceroxyloideae**
4. Flowers always unisexual, the plants usually monoecious, rarely dioecious, the flowers borne in groups of 3 (triads), each with a pistillate and 2 staminate flowers, or in groups derived from triads, very rarely in longitudinal lines (acervuli), even more rarely by reduction solitary . . . . **Arecoideae**

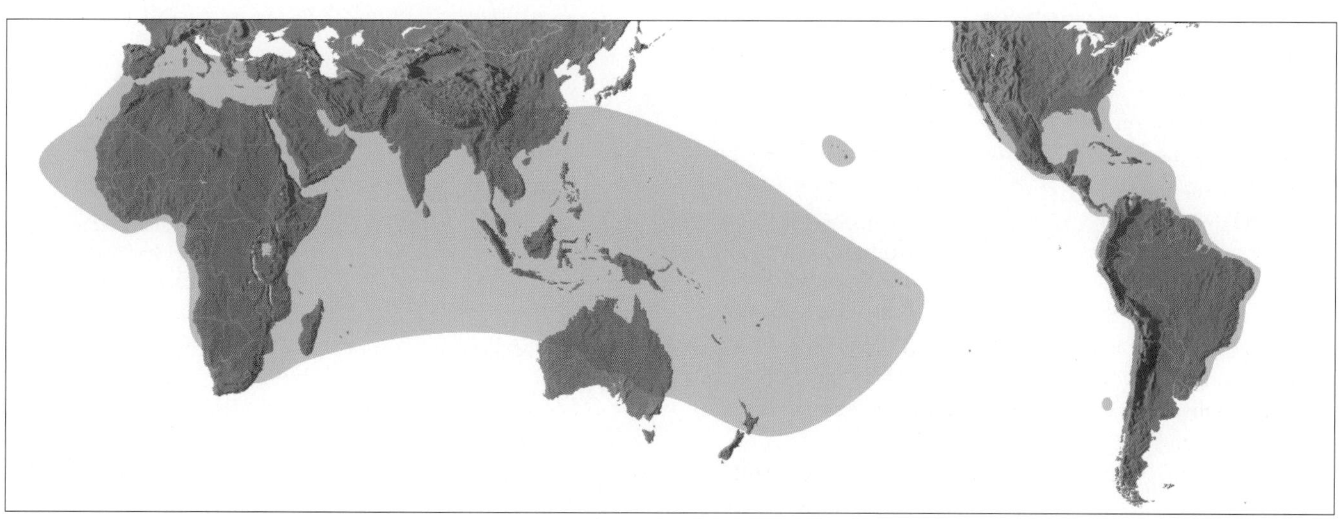

The distribution of living palms

SYNOPSIS OF THE CLASSIFICATION

I CALAMOIDEAE

Eugeissoneae
1. Eugeissona

Lepidocaryeae

Ancistrophyllinae
2. Oncocalamus
3. Eremospatha
4. Laccosperma

Raphiinae
5. Raphia

Mauritiinae
6. Lepidocaryum
7. Mauritia
8. Mauritiella

Calameae

Korthalsiinae
9. Korthalsia

Salaccinae
10. Eleiodoxa
11. Salacca

Metroxylinae
12. Metroxylon

Pigafettinae
13. Pigafetta

Plectocomiinae
14. Plectocomia
15. Myrialepis
16. Plectocomiopsis

Calaminae
17. Calamus
18. Retispatha
19. Daemonorops
20. Ceratolobus
21. Pogonotium

II NYPOIDEAE
22. Nypa

III CORYPHOIDEAE

Sabaleae
23. Sabal

Cryosophileae
24. Schippia
25. Trithrinax
26. Zombia
27. Coccothrinax
28. Hemithrinax
29. Leucothrinax
30. Thrinax
31. Chelyocarpus
32. Cryosophila
33. Itaya

Phoeniceae
34. Phoenix

Trachycarpeae

Rhapidinae
35. Chamaerops
36. Guihaia
37. Trachycarpus
38. Rhapidophyllum
39. Maxburretia
40. Rhapis

Livistoninae
41. Livistona
42. Licuala
43. Johannesteijsmannia
44. Pholidocarpus
45. Pritchardiopsis

Unplaced members of Trachycarpeae
46. Acoelorrhaphe
47. Serenoa
48. Brahea
49. Colpothrinax
50. Copernicia
51. Pritchardia
52. Washingtonia

Chuniophoeniceae
53. Chuniophoenix
54. Kerriodoxa
55. Nannorrhops
56. Tahina

Caryoteae
57. Caryota
58. Arenga
59. Wallichia

Corypheae
60. Corypha

Borasseae

Hyphaeninae
61. Bismarckia
62. Satranala
63. Hyphaene
64. Medemia

Lataniinae
65. Latania
66. Lodoicea
67. Borassodendron
68. Borassus

IV CEROXYLOIDEAE

Cyclospatheae
69. Pseudophoenix

Ceroxyleae
70. Ceroxylon
71. Juania
72. Oraniopsis
73. Ravenea

Phytelepheae
74. Ammandra
75. Aphandra
76. Phytelephas

V ARECOIDEAE

Iriarteeae
77. Iriartella
78. Dictyocaryum
79. Iriartea
80. Socratea
81. Wettinia

Chamaedoreeae
82. Hyophorbe
83. Wendlandiella
84. Synechanthus
85. Chamaedorea
86. Gaussia

Podococceae
87. Podococcus

Oranieae
88. Orania

Sclerospermeae
89. Sclerosperma

Roystoneeae
90. Roystonea

Reinhardtieae
91. Reinhardtia

Cocoseae	
Attaleinae	92. Beccariophoenix 93. Jubaeopsis 94. Voanioala 95. Allagoptera 96. Attalea 97. Butia 98. Cocos 99. Jubaea 100. Lytocaryum 101. Syagrus 102. Parajubaea
Bactridinae	103. Acrocomia 104. Astrocaryum 105. Aiphanes 106. Bactris 107. Desmoncus
Elaeidinae	108. Barcella 109. Elaeis
Manicarieae	110. Manicaria
Euterpeae	111. Hyospathe 112. Euterpe 113. Prestoea 114. Neonicholsonia 115. Oenocarpus
Geonomateae	116. Welfia 117. Pholidostachys 118. Calyptrogyne 119. Calyptronoma 120. Asterogyne 121. Geonoma
Leopoldinieae	122. Leopoldinia
Pelagodoxeae	123. Pelagodoxa 124. Sommieria
Areceae	
Archontophoenicinae	125. Actinorhytis 126. Archontophoenix 127. Actinokentia 128. Chambeyronia 129. Kentiopsis
Arecinae	130. Areca 131. Nenga 132. Pinanga
Basseliniinae	133. Basselinia 134. Burretiokentia 135. Cyphophoenix 136. Cyphosperma 137. Lepidorrhachis 138. Physokentia
Carpoxylinae	139. Carpoxylon 140. Satakentia 141. Neoveitchia
Clinospermatinae	142. Cyphokentia 143. Clinosperma
Dypsidinae	144. Dypsis 145. Lemurophoenix 146. Marojejya 147. Masoala
Linospadicinae	148. Calyptrocalyx 149. Linospadix 150. Howea 151. Laccospadix
Oncospermatinae	152. Oncosperma 153. Deckenia 154. Acanthophoenix 155. Tectiphiala
Ptychospermatinae	156. Ptychosperma 157. Ponapea 158. Adonidia 159. Solfia 160. Balaka 161. Veitchia 162. Carpentaria 163. Wodyetia 164. Drymophloeus 165. Normanbya 166. Brassiophoenix 167. Ptychococcus
Rhopalostylidinae	168. Rhopalostylis 169. Hedyscepe
Verschaffeltiinae	170. Nephrosperma 171. Phoenicophorium 172. Roscheria 173. Verschaffeltia
Unplaced members of Areceae	174. Bentinckia 175. Clinostigma 176. Cyrtostachys 177. Dictyosperma 178. Dransfieldia 179. Heterospathe 180. Hydriastele 181. Iguanura 182. Loxococcus 183. Rhopaloblaste

I. CALAMOIDEAE

CALAMOIDEAE Griff., Calcutta J. Nat. Hist. 5: 4 (1844).
Lepidocaryoideae Mart. ex Horan., Char. Ess. Fam.: 42 (1847).

Hermaphroditic, monoecious, dioecious or polygamous; hapaxanthic or pleonanthic; often fiercely armed; leaves pinnate, pinnately ribbed or rarely palmate, reduplicate; inflorescence often highly branched, frequently displaying adnation, the bracts of all orders usually tubular; flowers almost always borne in dyads or dyad derivatives, unisexual flowers only slightly dimorphic; ovary and fruit covered in reflexed scales, usually arranged in neat vertical rows; ovary incompletely trilocular, ovules anatropous, with the micropyles facing the centre of the gynoecium; pericarp usually thin at maturity, endocarp not differentiated (except in *Eugeissona*); seeds 1–3, usually with a thick sarcotesta.

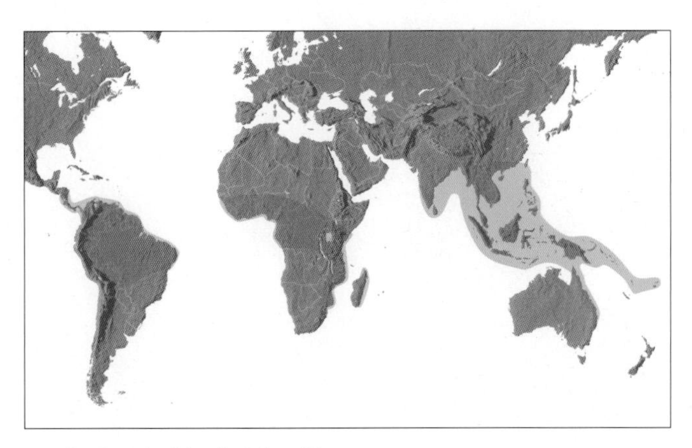
Distribution of subfamily **Calamoideae**

This subfamily includes almost a quarter of all known palms and contains the largest genus, *Calamus*. Four genera, *Mauritia*, *Mauritiella*, *Lepidocaryum* and *Raphia* are present in the New World Tropics (the last represented by a single species in the New World but otherwise diverse in Africa). All other genera belong in the Old World, where they are particularly diverse in western Malesia and neighbouring parts of Southeast Asia.

Members of the subfamily are immediately distinguished by the overlapping scales that cover ovaries and fruit.

Subfamily Calamoideae is universally highly supported as monophyletic (Uhl *et al.* 1995, Baker *et al.* 1999a, 2000a, 2000b, Asmussen *et al.* 2000, 2006, Asmussen & Chase 2001, Lewis & Doyle 2001, Hahn 2002a). The Calamoideae is highly supported as sister to the rest of the palms in the study of Asmussen *et al.* (2006) and with some support found by Baker *et al.* (in review). Earlier studies also resolve Calamoideae as sister to the rest of the palms (e.g., Asmussen & Chase 2001). However, Uhl *et al.* (1995) and Lewis & Doyle (2001) resolved the Calamoideae as sister to all palms except the Nypoideae. The analyses of Hahn (2002a) support numerous alternative resolutions, but the combined analysis of all available sequence data resolves the Calamoideae as sister to the rest of the palms.

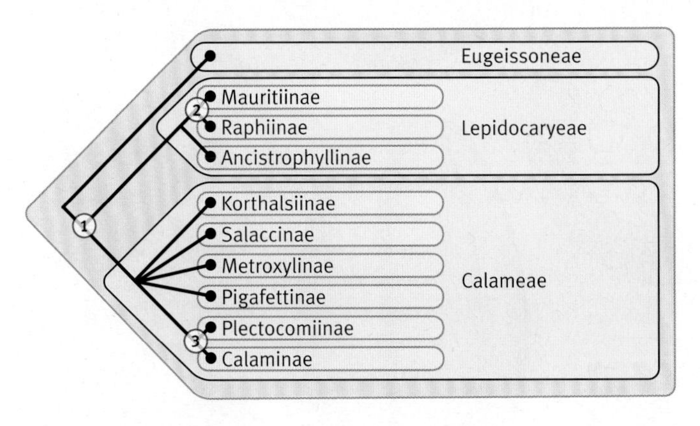

Summary tree showing relationships among tribes and subtribes of the Calamoideae that have been resolved with moderate to high support in multiple phylogenetic analyses. Labelled nodes are resolved in the following studies: **node 1** — Baker *et al.* 2000a, 2000b, in prep., Asmussen *et al.* 2006; **node 2** — Baker *et al.* 2000a, 2000b, Asmussen *et al.* 2006; **node 3** — Baker *et al.* 2000a, 2000b, in review. For further explanation, see discussions under tribe and subtribe treatments.

Key to Tribes of Calamoideae

1. Flowers very large, borne in a tight cupule of 7–11 overlapping bracts; petals ± woody, sharply pointed; stamens 21–70; fruit ovoid, beaked, covered with minute scales; mesocarp fibrous; endocarp thick, woody, with 6 or 12 internal longitudinal flanges penetrating the seed . . . . **Eugeissoneae**
1. Flowers not borne in a cupule of bracts; petals not sharply pointed; stamens not more than 12; fruit lacking an endocarp 2
2. Arborescent, acaulescent or climbing, dioecious, monoecious or hermaphroditic, hapaxanthic or pleonanthic palms; leaves pinnate or palmate, spiny or unarmed, if spiny spines generally not organised into groups; stigmas pyramidal. New World and African palms . . . . **Lepidocaryeae**
2. Arborescent, acaulescent or climbing, mostly dioecious, hapaxanthic or pleonanthic palms; leaves pinnate, spiny, with spines regularly or subregularly organised into whorls or partial whorls; stigmas pyramidal or trifid. Palms of Asia, Malesia and Pacific with one outlier in Africa . . . . **Calameae**

Opposite: ***Calamus zeylanicus*** and ***C. ovoideus***, Sri Lanka. (Photo: J. Dransfield)

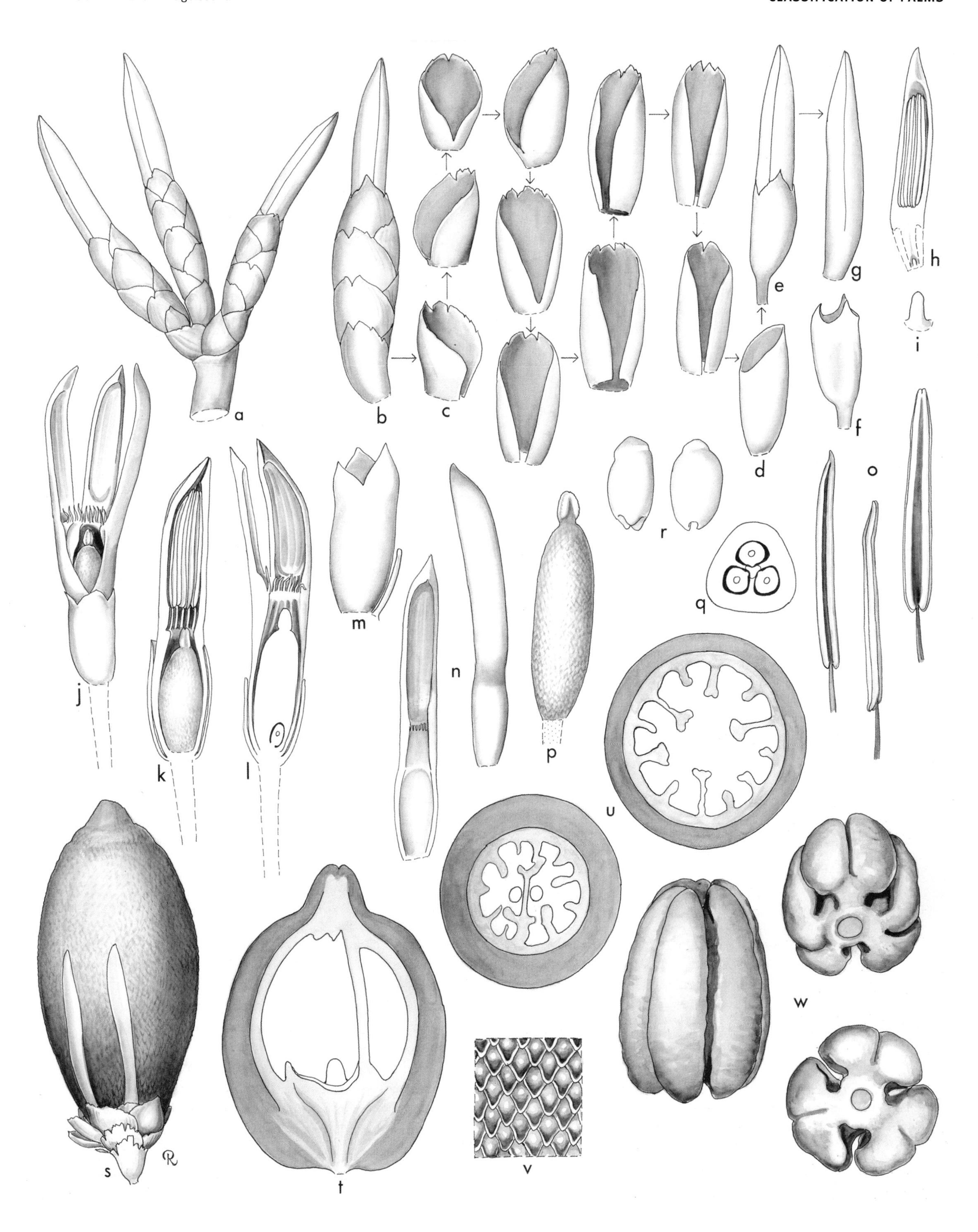
a
b
c
d
e
f
g
h
i
j
k
l
m
n
o
p
q
r
s
t
u
v
w

Tribe Eugeissoneae W.J. Baker & J. Dransf., Syst. Bot. 25: 318 (2000). Type: **Eugeissona**.

Acaulescent or arborescent, polygamous, hapaxanthic palms; leaves pinnate, spiny; inflorescences with rachillae bearing a single dyad in the axil of the distal-most bract, other rachilla bracts empty and forming a cupule surrounding the dyad; flowers multistaminate with more than 20 stamens; fruit covered in numerous minute scales; endocarp developing from a middle layer of the pericarp and bearing 6 or 12 internal flanges penetrating the homogeneous endosperm; germination remote.

The tribe consists of the single genus *Eugeissona*, confined to the Malay Peninsula and Borneo. The dyads, each comprising a staminate and a hermaphroditic flower surrounded by a cupule of empty bracts, are most unusual. Furthermore, this is the only member of the subfamily to display a well-developed endocarp. Unlike endocarps in other subfamilies, this hard layer develops from within the mesocarp rather than from the innermost part of the pericarp; at fruit maturity, tissue internal to the hard layer dries out and become compressed by the developing seed.

Tribe Eugeissoneae resolves as sister to the rest of the Calamoideae with low-to-moderate support (Baker *et al.* 2000a, 2000b, in prep., Asmussen *et al.* 2006). Alternative positions have been recovered, but they lack bootstrap support (e.g., Baker *et al.* in review).

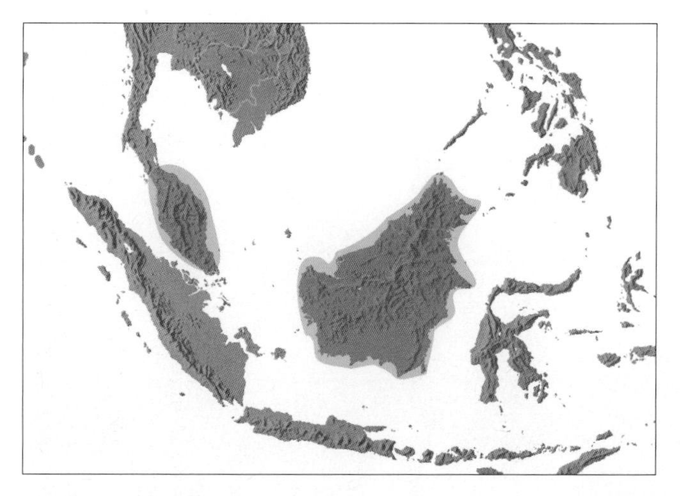

Distribution of ***Eugeissona***

1. EUGEISSONA

Stemless or short-stemmed often viciously spiny hapaxanthic pinnate-leaved palms of the Malay Peninsula and Borneo; flowers very large, with spine-tipped woody petals, borne in pairs of a staminate and a hermaphroditic flower in a cupule of overlapping leathery bracts. The fruit is most unusual in having myriads of minute scales and a thick endocarp.

Eugeissona Griff., Calcutta J. Nat. Hist. 5: 101 (1844). Type: **E. tristis** Griff.

Eu — *good,* geisson — *cornice of a roof, referring to the use of the leaves of* E. tristis *as thatch.*

Moderate to robust, clustering, spiny, hapaxanthic, polygamous, acaulescent or tree palms. *Stem* subterranean, erect, or borne on robust stilt roots at heights up to 3 m above the ground, branching sympodially by basal (?axillary) suckers, internodes very short to moderately elongate, usually covered by rotting leaf sheaths, becoming exposed in tree species and sometimes bearing short spine-like adventitious roots; cortex very hard, pith soft with abundant starch deposition before flowering. *Leaves* pinnate, spirally arranged or markedly 3-ranked; sheath usually splitting opposite the petiole, unarmed at the very base, bearing black, flattened spines distally, sometimes also bearing scales and branched hairs, sheath margin ligule-like distally; petiole well developed, adaxially deeply channelled in proximal portion, distally ± rounded in section, the abaxial surface sparsely to densely armed with black, flattened spines, scattered, paired or in longitudinal rows, scales and hairs usually abundant between spines, sometimes absent; rachis armed as the petiole but more sparsely; leaflets single-fold, numerous, linear to lanceolate, entire, regularly arranged or grouped and fanned within the group to give the leaf a plumose appearance, frequently bearing bristles along the main veins or the margins, and irregular bands of caducous indumentum, midribs prominent, transverse veinlets moderately conspicuous. *Inflorescence* erect, composed of branches equivalent to the axillary inflorescences of pleonanthic palms, each branched to the fourth-order, and subtended by leaves with much reduced blades or by tubular, apiculate, spiny or unarmed, dull brown, longitudinally imbricate bracts; branches of all orders bearing a tubular, 2-keeled prophyll, and terminating in a cupule of dull brown, longitudinally striate, spirally arranged, or more often subdistichous, tightly sheathing bracts enclosing a flower pair; cupule bracts 11–13, rarely 1 or 2 more, the 1–3 most proximal and 1–3 most distal each with an abortive axillary bud, the rest empty, the most proximal up to 5 tubular, the rest open; flower pair consisting of a large staminate and lateral to it a large hermaphroditic flower, the staminate appearing first, then pushed out of the cupule by the developing hermaphroditic. *Staminate flowers* borne on a short flattened pedicel, the flower base compressed on one side by the developing hermaphroditic bud; calyx tubular, coriaceous, striate, dull brown, with 3 short, pointed lobes; corolla tubular in the basal $^1/_4$ to $^1/_3$, distally with 3 narrow, elongate, woody, valvate lobes terminating in hard, sharp, spine-like tips; stamens 20–70 borne just above the mouth of the corolla tube, filaments short, erect, anthers narrow, elongate, basifixed, dull yellow to purple, latrorse to introrse, deciduous after anthesis; pistillode minute. *Pollen* ellipsoidal, bi-symmetric; aperture distal, brevi- or extended sulcate; ectexine tectate, coarsely perforate, or foveolate, aperture margin similar; infratectum columellate; longest axis 41–73 µm; post-meiotic tetrads tetragonal [2/6]. *Hermaphroditic flowers* protandrous, sessile, bearing a 2-keeled, coriaceous, tubular prophyll, similar to the cupule bracts, the whole flower very similar in size and shape to the staminate except for the apex, flattened on one side by pressure of the staminate flower in bud; calyx, corolla and androecium like those of the

Opposite: ***Eugeissona.*** **a**, branch bearing three rachillae × $^3/_4$; **b**, single rachilla with bracts and staminate flower × 1; **c–d**, bracts of single rachilla, expanded × 1; **e**, staminate flower × 1; **f**, staminate calyx × 1; **g**, staminate flower, calyx removed × 1; **h**, staminate flower, calyx removed, in vertical section × 1; **i**, pistillode × 3; **j**, hermaphrodite flower, stamens fallen × 1; **k**, hermaphrodite flower before anthesis in vertical section × 1; **l**, hermaphrodite flower, stamens fallen, in vertical section × 1; **m**, hermaphrodite calyx and remains of stalk of staminate flower × 2; **n**, hermaphrodite petal in 2 views × 1; **o**, stamens in 3 views × 2; **p**, gynoecium × $1^1/_2$; **q**, ovary in cross-section × 3; **r**, ovule in 2 views × 5; **s**, fruit × $^3/_4$; **t**, fruit in vertical section × $^3/_4$; **u**, fruit in cross-section at 2 levels × $^3/_4$; **v**, portion of epicarp showing imbricate scales, much enlarged; **w**, seed in 3 views × $^3/_4$. *Eugeissona tristis*: **a–r**, *Moore & Pennington* 9059; *E. utilis*: **s–w**, *Moore & Meijer* 9219; *E. tristis*: **s–w**, *Moore et al.* 9105. (Drawn by Marion Ruff Sheehan)

staminate; gynoecium tricarpellate, triovulate, ovary columnar, faintly 3-angled, covered in vertical rows of minute reflexed scales, stigma conical to pyramidal with 3 glandular angles, ovules basally attached, anatropous. *Fruit* ovoid, beaked, stigmatic remains apical, cupule bracts, calyx, and usually the corolla persisting; epicarp covered in irregular vertical rows of very small reflexed, fringed scales, mesocarp somewhat corky at maturity, traversed by longitudinal fibre bundles, endocarp developing from a layer external to the locule wall, dark brown to blackish, very hard and thick, sometimes linked to the fibres of the mesocarp, with 3 + 3, or 3 + 3 + 6 flanges penetrating into the fruit cavity, forming symmetrical, incomplete partitions. *Seed* basally attached, single, filling the fruit cavity and closely adhering to the endocarp and thus indented by the incomplete partitions, seed coat thin, dry, endosperm homogeneous; embryo basal. *Germination* remote-ligular; eophyll pinnate. *Cytology* not studied.

Distribution and ecology: Six species, two confined to the Malay Peninsula, and four in Borneo, one of which (*E. ambigua*) is still known only from its type. In Borneo, *E. insignis*, *E. utilis* and *E. minor* usually seem to be associated with poor soils that have abundant humus. They are particularly conspicuous on scarp faces or sharp ridgetops. *Eugeissona minor* and *E. insignis* are also found in low-lying 'kerangas' (heath) forest. *Eugeissona ambigua*, known only from the type, was collected on a slightly raised heathy area in the Kapuas lake district of West Kalimantan. In the Malay Peninsula, *E. brachystachys* is associated with richer soils on hillslopes, particularly where some flushing takes place, and *E. tristis* is found in a wide range of forest types, from swamp margins to hilltops but grows in greatest abundance on ridgetops up to 1000 m altitude. All but *E. ambigua*, about which very little is known, grow in large colonies.

Anatomy: Leaf (Tomlinson 1961) distinguished from those of other calamoid genera by short foliar sclereids in some species, perhaps supporting the rather isolated position of the genus. Floral development indicates the apparently terminal flower pair to be lateral and to consist of a typical dyad. The gynoecium consists of three carpels with ventral sutures open, and ovules initiated directly on the large apex of the floral axis (Uhl & Dransfield 1984). The flowers are among the largest in the palms, and also unusual is the development of first the staminate and later the hermaphroditic flowers, so that at any one time the plant seems to bear large solitary flowers.

Relationships: *Eugeissona* is strongly supported as monophyletic (Baker *et al.* 2000a, 2000b). For relationships, see Eugeissoneae. The dyad of a hermaphroditic and a staminate flower is found elsewhere only in *Metroxylon*.

Common names and uses: *Bertam* (*Eugeissona tristis*), wild Bornean sago palm (*E. utilis*). All species have a wide range of local uses. Sago from the stems of *E. utilis* forms the staple of the nomadic Penan people of Borneo and *E. insignis* can be used similarly. Leaves of all species may provide thatch.

Top: ***Eugeissona insignis***, habit, Sarawak. (Photo: J. Dransfield)

Bottom left: ***Eugeissona brachystachys***, flower, Peninsular Malaysia. (Photo: J. Dransfield)

Bottom right: ***Eugeissona tristis***, fruit, Peninsular Malaysia. (Photo: J. Dransfield)

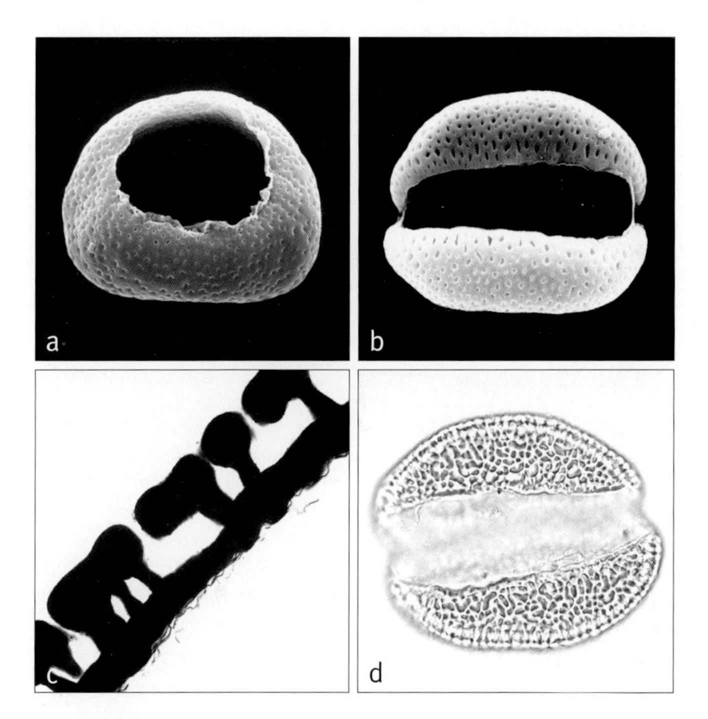

Above: ***Eugeissona***
a, brevisulcate pollen grain, distal face SEM × 1000;
b, extended sulcate pollen grain, distal face SEM × 650;
c, ultrathin section through pollen wall, proximal face TEM × 5000;
d, extended sulcate pollen grain LM × 1000.
Eugeissona brachystachys: **a** & **c**, *Ridley* 16294;
E. tristis: **b** & **d**, *Holttum* 9772.

Right: ***Eugeissona minor***, stilt roots, Brunei Darussalam. (Photo: J. Dransfield)

Petioles are used in the manufacture of blinds, blowpipe darts and toys, and the pith of the petioles for the occlusions on blowpipe darts. The young endosperm is edible and even the pollen has been eaten. Stilt roots of *E. minor* make excellent walking sticks. *Eugeissona tristis* has become a serious pest of Hill Dipterocarp forest in Malaya, where it dominates the undergrowth after logging, thereby preventing regeneration of commercially important timber trees.

Taxonomic account: Dransfield (1970).

Fossil record: The large, thick-walled, extended monosulcate pollen of some species of *Eugeissona* is distinctive. Similar pollen, with uniformly coarse reticulate pollen grains, referred to *E. minor* (Muller 1972, 1979, 1981), is known from the 'oldest' Middle Miocene of Java, the Nanggulan Formation (Morley 1998), and from the Upper Oligocene onwards in Sarawak (Muller 1972). In addition, similar extended sulcate pollen grains, reticulate but with a more coarsely reticulate sulcus margin, described as closely similar to *E. insignis*, *E. tristis* and *E. utilis*, are known from the Middle Miocene of Sarawak (Muller 1972, 1979, 1981). Numerous records of *Eugeissona*-type fossil pollen grains, similar to those described from Sarawak, have been recovered from India. They are generally described under the fossil genus *Quilonipollenites* (Rao & Ramanujam 1978), which is named for the Lower Miocene limestone Quilon Beds, of Kerala State, southwestern India. *Quilonipollenites* occurs throughout the Lower to Middle Miocene deposits of the Kerala Basin, it has also been recovered further north in the Lower Miocene Ratnagiri lignite beds of Maharashtra State (Phadtare & Kulkarni 1980, 1984), where fossil fruits attributed to *Eugeissona* have also been described (*Eugeissonocarpon indicum* [Shinde & Kulkarni 1989]). In southwestern India, *Quilonipollenites* is also known from the Lower to Middle Eocene Neyveli lignites of South Arcot District, Tamil Nadu (Sarma *et al.* 1984; Ramanujam & Reddy 1984). Morley (1998) speculates on the origins of *Eugeissona* on the basis of the extended sulcus characteristic that is shared with *Longapertites*, "Although today *Eugeissona* is endemic to Borneo and Malaysia, its pollen shows some similarities to members of the fossil genus *Longapertites* (see Frederiksen 1994), which is recorded widely in the uppermost Cretaceous and Lower Tertiary of South America, West Africa and India, raising the possibility that *Eugeissona* may be derived from the parent taxon of this group, which is of very ancient origin, with a former pantropical distribution." Indeed Frederiksen (1994) thinks that,

"...*Quilonipollenites* should be considered a synonym of *Longapertites*." However, Frederiksen further comments that some palynologists, "... prefer that *Quilonipollenites* should be a separate genus based on its coarse ornamentation." Extended monosulcate pollen, however, is systematically widespread but sporadic in the palms, occurring in the coryphoids (*Licuala*) and in the arecoids (*Areca*, *Pinanga* and *Hydriastele*), as well as in *Eugeissona* and *Eremospatha* (Calamoideae).

Notes: In Peninsular Malaysia, slow lorises (*Nyctecebus coucang*) have been recorded visiting the flowers of *Eugeissona tristis* at night time, sipping the fermented nectar and possibly distributing pollen from one inflorescence to another (Wiens, pers. comm.)

Additional figures: Fig. 1.7.

Tribe Lepidocaryeae Mart. ex Dumort., Anal. Fam. Pl.: 55 (1829). Type: **Lepidocaryum**.

Arborescent, acaulescent or climbing, dioecious, monoecious or hermaphroditic, hapaxanthic or pleonanthic palms; leaves pinnate or palmate, spiny or unarmed; inflorescences with usually distichous bracts; stigmas pyramidal.

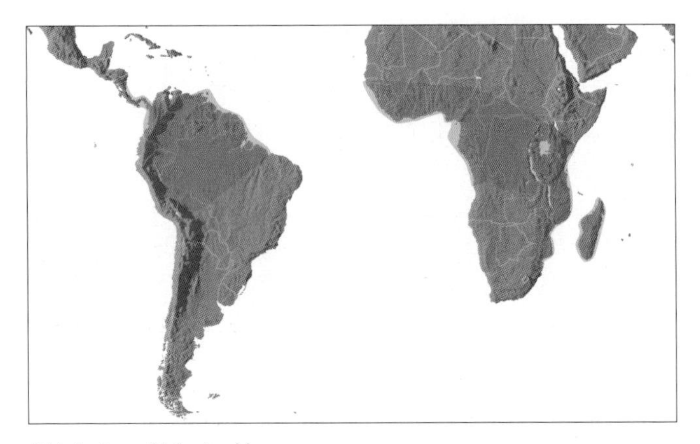

Distribution of tribe **Lepidocaryeae**

Three subtribes are recognised. Tribe Lepidocaryeae is moderately to highly supported as monophyletic in the majority of studies (Baker *et al.* 2000a, 2000b, Asmussen & Chase 2001, Asmussen *et al.* 2006). In one study, *Eugeissona* is nested within the tribe, but this topology is poorly supported (Baker *et al.* in review). The tribe is moderately supported as sister to the Calameae (Baker *et al.* 2000a, 2000b, in prep., Asmussen *et al.* 2006).

Key to Subtribes of Lepidocaryeae

1. Leaves palmate with usually numerous reduplicate segments . . . . **Mauritiinae**
1. Leaves pinnate or, rarely, simply forked but with pinnate venation 2
2. Climbing palms, the leaves terminating in a cirrus armed with acanthophylls; flowers hermaphroditic and borne in pairs, or unisexual and borne in a complex dichasial cincinnus . . . . **Ancistrophyllinae**
2. Massive acaulescent or tree palms, leaves not terminating in a cirrus; rachillae bearing pistillate flowers basally and staminate flowers distally . . . . **Raphiinae**

Subtribe Ancistrophyllinae Becc., Ann. Roy. Bot. Gard. (Calcutta) 12(2): 209 (1918). Type: *Ancistrophyllum* (= **Laccosperma**).

Oncocalaminae J. Dransf. & N.W. Uhl, Principes 30: 5 (1986). Type: *Oncocalamus*.

Climbing, hermaphroditic or monoecious, hapaxanthic or pleonanthic palms; leaves pinnate, spiny, with cirri bearing reflexed acanthophylls; inflorescences with the terminal flower of the floral dyad hermaphroditic or pistillate.

The subtribe consists of three genera of tropical African climbing palms. Vegetatively, they are all remarkably similar but the inflorescences display considerable diversity in structure and sexuality. In the first edition of *Genera Palmarum*, *Oncocalamus* was included in its own subtribe, Oncocalaminae, reflecting the unique structure of its flower clusters. However, the work of Baker *et al.* (2000a, 2000c) and subsequent workers provides compelling phylogenetic evidence that these African rattans are closely related to each other.

This well-defined group is highly supported as monophyletic in most studies (Baker *et al.* 2000a, 2000b, in review, Asmussen & Chase 2001, Asmussen *et al.* 2006). The Ancistrophyllinae has most often been resolved as sister to a clade of Mauritiinae and Raphiinae with moderate to high support (Baker *et al.* 2000a, 2000b, Asmussen *et al.* 2006), although one phylogeny provides weak evidence for its placement as sister to all other Calamoideae (Baker *et al.* in review).

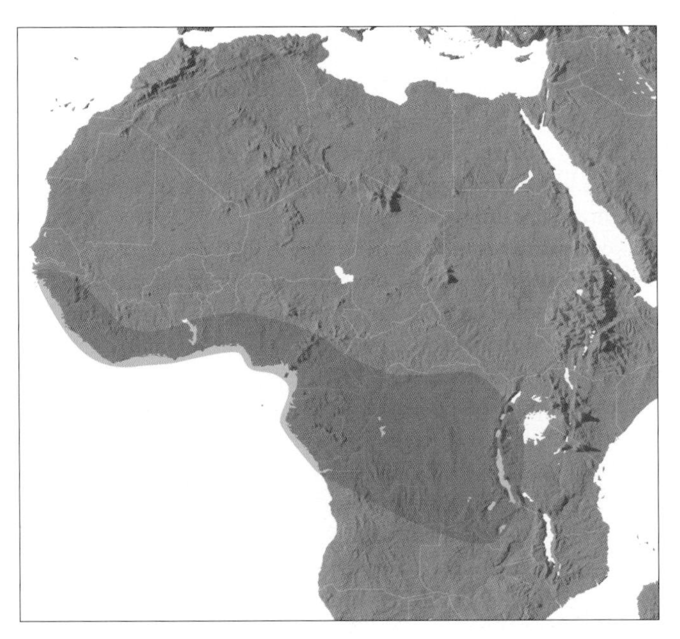

Distribution of subtribe **Ancistrophyllinae**

Key to Genera of Ancistrophyllinae

1. Flowers borne in clusters of ca. 5–11 in the axils of overlapping distichous bracts, the central 1 or 3 pistillate, the rest staminate in 2 lateral cincinni. Africa . . . . **2. Oncocalamus**
1. Flowers hermaphroditic, borne in pairs (very rarely in threes) 2
2. Pleonanthic rattans; rachilla bracts minute, incomplete; floral bracteoles absent; corolla tubular, very thick, with 3 short lobes; filaments united in a fleshy epipetalous ring. Africa . . . . **3. Eremospatha**
2. Hapaxanthic rattans; rachilla bracts conspicuous, complete; floral bracteoles present; corolla coriaceous, tubular at the base, with 3 boat-shaped lobes; filaments not united in a fleshy ring. Africa . . . . **4. Laccosperma**

2. ONCOCALAMUS

Clustering high-climbing pinnate-leaved rattan palms of Equatorial West Africa; sheaths armed with detachable spines; pleonanthic and monoecious, the flowers are borne in paired cincinni within conspicuous bracts, the basal 1–few pistillate, the others staminate, an arrangement unique in the family.

Oncocalamus (G. Mann & H. Wendl.) G. Mann & H. Wendl. ex Hook.f. in Benth. & Hook.f., Gen. Pl. 3: 881, 936 (1883). Type: **O. mannii** (H. Wendl.) H. Wendl. ex Drude (*Calamus mannii* H. Wendl.).

Calamus subgenus *Oncocalamus* G. Mann & H. Wendl., Trans. Linn. Soc. London 24: 436 (1864).

Onkos — *hook*, calamus — *reed, but no explanation was given for the derivation*.

Clustered, spiny, high-climbing, pleonanthic, monoecious, rattan palms. *Stem* eventually becoming bare, circular in cross-section, with long internodes. *Leaves* pinnate, bifid in juveniles, with a terminal cirrus; sheath strictly tubular, bearing scattered, black, bulbous-based, triangular, brittle spines and scattered, thin, caducous indumentum; ocrea conspicuous, tightly sheathing, neatly truncate, armed as the sheath; knee absent; petiole present but usually very short, absent in mature flowering stems; rachis armed with scattered spines as the leaf sheath; cirrus bearing neat pairs of reflexed acanthophylls; leaflets few to numerous, usually single-fold, sometimes with 2 or more folds, entire, acute, linear, lanceolate or somewhat sigmoid, regularly arranged, usually armed along the thickened margins with robust spines, midribs evident, other large veins rather distant, transverse veinlets conspicuous; proximal few leaflets sometimes smaller than the rest, heavily armed and reflexed across the sheathed stem. *Inflorescences* branched to 1 order; peduncle enclosed within the leaf sheath and emerging from its mouth, ± hemispherical in cross-section; prophyll tubular, tightly sheathing, 2-keeled, 2-lobed at its tip, much shorter than the sheath; peduncular bracts ca. 4, ± distichous, tightly sheathing at first, later splitting longitudinally, each with a short triangular lobe; rachis longer than the peduncle; rachis bracts like the peduncular, rather close; first-order branches pendulous or spreading with a basal 2-keeled tubular prophyll and numerous distichous, short, tubular, somewhat inflated, striate bracts, each enclosing a flower cluster, after anthesis eventually irregularly splitting and tattering; flower cluster partially covered by a tubular 2-keeled prophyll and consisting of up to 11 flowers arranged in a group with a central 1 or 3 pistillate flowers and 2 lateral cincinni of 2–4 staminate flowers, each flower, apart from the central pistillate bearing an open, spathulate, 2-keeled, prophyllar bracteole (the precise arrangement of the flowers not yet understood). *Staminate flowers* symmetrical; calyx membranous, striate basally, stalked, tubular, with 3 short triangular, apiculate lobes; corolla apparently only slightly exceeding the calyx, divided almost to the base into 3 elongate, striate, valvate petals; stamens 6, filaments united to form a thick, fleshy androecial tube, free from the corolla, tipped with 6 shallow lobes, bearing pendulous, rounded, latrorse anthers on the inside; pistillode very narrow, conical, slightly exceeding the androecial tube. *Pollen* ellipsoidal, bi-symmetric; aperture a distal sulcus; ectexine tectate, very finely perforate, interspersed with very small spinulae, aperture margin similar; infratectum columellate; longest axis 23–29 µm [1/4]. *Pistillate flowers* superficially very similar to the staminate except slightly broader; the calyx and corolla similar; staminodal tube bearing minute empty anthers; gynoecium tricarpellate, triovulate, ± ellipsoidal, covered in reflexed scales, style long, narrow, 3-angled, ovule form unknown. *Fruit* ± spherical, stigmatic remains minute, conical, apical; epicarp covered in vertical rows of rather thin reflexed scales, mesocarp very thin, almost

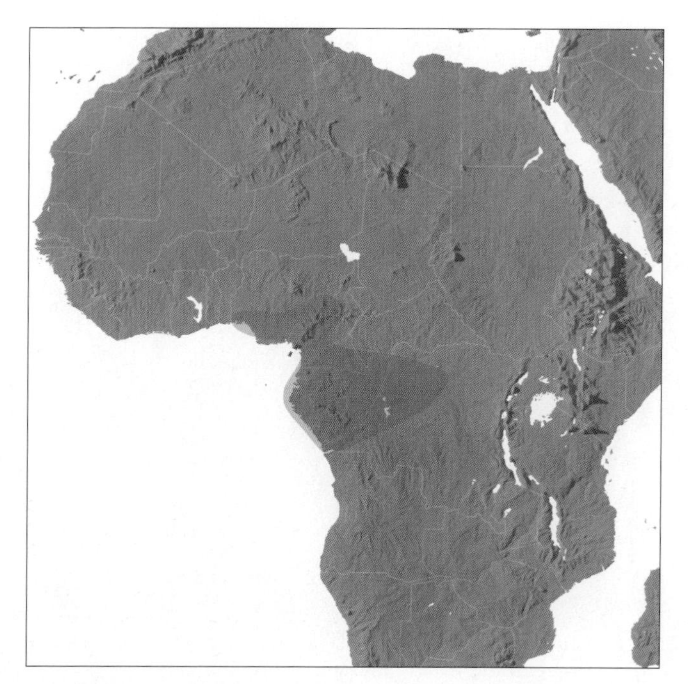

Distribution of ***Oncocalamus***

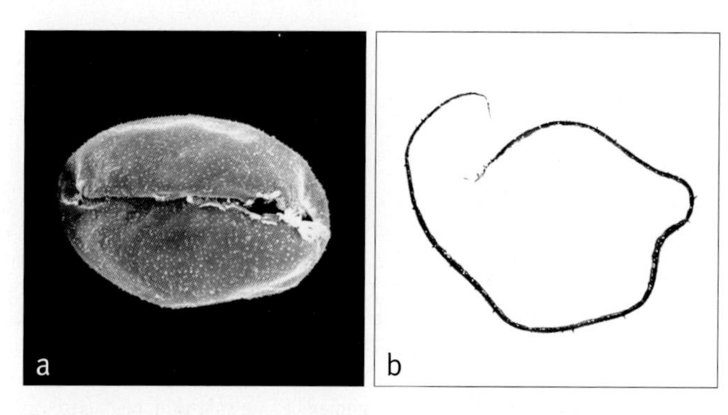

Above: ***Oncocalamus***
a, monosulcate pollen grain, distal face SEM × 1500;
b, ultrathin section, whole pollen grain, polar plane TEM × 2500.
Oncocalamus mannii: **a** & **b**, *Mann* 1044.

Left: ***Oncocalamus tuleyi***, young rachillae, Camaeroon. (Photo: J. Dransfield)
Middle: ***Oncocalamus tuleyi***, leaf sheaths, Cameroon. (Photo: J. Dransfield)
Right: ***Oncocalamus tuleyi***, young fruit, Cameroon. (Photo: J. Dransfield)

obsolescent at maturity, endocarp not differentiated. *Seed* single, basally attached with an oval hilum, covered with a ?thick sarcotesta, endosperm homogeneous, laterally deeply penetrated by a smooth-margined mass of inner seed coat; embryo lateral opposite the intrusion. *Germination* and eophyll unknown. *Cytology* not studied.

Distribution and ecology: Five species described from equatorial West Africa and the Congo Basin, confined to low-lying tropical rain forest.

Anatomy: Not studied.

Relationships: The monophyly of *Oncocalamus* has not been tested. The genus resolves as sister to a clade of *Eremospatha* and *Laccosperma* with moderate support (Baker *et al.* 2000a, 2000b, Asmussen *et al.* 2006).

Common names and uses: Common names numerous (Sunderland 2001). No local uses have been specifically recorded but the stems are probably used as a source of cane.

Taxonomic account: Sunderland (2001, 2007).

Fossil record: No generic records found.

Notes: The strange flower cluster of *Oncocalamus* is unique, not only in the subfamily but within the whole family. Vegetatively, *Oncocalamus* is very similar to *Eremospatha* and *Laccosperma*.

Opposite: ***Oncocalamus.*** **a**, portion of rachilla with one entire flower cluster × 6; **b**, flower cluster, schematic to show bract and flower arrangement × 6; **c**, flower cluster in bud, outer bract removed × 6; **d**, flower cluster in adaxial view, pistillate flower central, staminate flowers lateral, outer bract removed × 6; **e**, flower cluster in abaxial view, pistillate flower central, staminate flowers lateral, outer bract removed × 6; **f**, flower cluster in lateral view, pistillate flower right, staminate flowers left, outer bract removed × 6; **g**, flower cluster in tangential view, pistillate flower center back, staminate flowers front, outer bract removed × 6; **h**, flower cluster in median view, pistillate flower left (note lack of bracts), staminate flowers right, outer bract removed × 6; **i**, staminate bud × 9; **j**, staminate bud, outer portion of calyx removed × 9; **k**, staminate bud in vertical section × 9; **l**, staminate sepal in 2 views × 9; **m**, staminate bud, calyx removed × 9; **n**, staminate petal × 9; **o**, androecium × 9; **p**, androecium, expanded × 15; **q**, stamen in 3 views × 18; **r**, pistillode × 15; **s**, pistillate bud × 9; **t**, pistillate bud in vertical section × 9; **u**, pistillate sepal × 9; **v**, pistillate petal in 2 views × 9; **w**, staminodial ring × 9; **x**, staminodial ring, expanded × 9; **y**, gynoecium × 15; **z**, gynoecium in vertical section × 15; **aa**, ovary in cross-section × 15; **bb**, fruit × 2 1/4; **cc**, stigmatic remains × 6; **dd**, scales of fruit × 3; **ee**, fruit in vertical section × 2 1/4; **ff**, seed in 2 views × 2 1/4. *Oncocalamus macrospathus*: all from *Letouzey* 11889. (Drawn by Marion Ruff Sheehan)

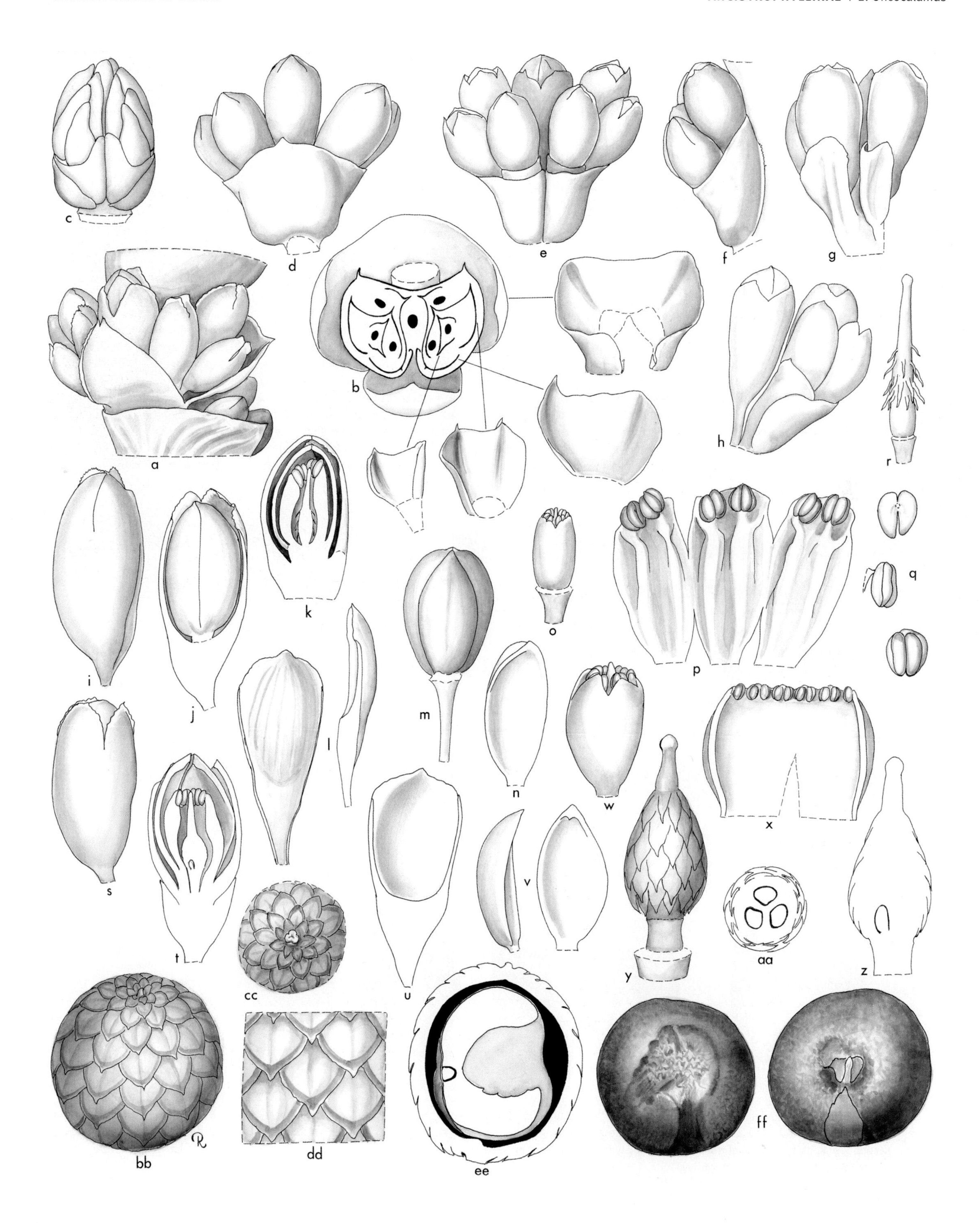
c
d
e
f
g
b
a
h
r
k
o
p
q
i
j
m
l
n
w
x
s
t
v
y
aa
z
u
cc
bb
dd
ee
ff

3. EREMOSPATHA

Clustering, high-climbing, pinnate-leaved rattan palms of humid Equatorial West and Central Africa; sheaths are always unarmed; pleonanthic and hermaphroditic, the flowers are borne in pairs and are distinctive in the almost inflated fleshy to leathery tubular calyx and corolla.

Eremospatha (G. Mann & H. Wendl.) H. Wendl. in Kerch., Palmiers 244 (1878). Lectotype: **E. hookeri** (G. Mann & H. Wendl.) H. Wendl. (*Calamus hookeri* G. Mann & H. Wendl.) (see Moore 1963c).

Calamus subgenus *Eremospatha* G. Mann & H. Wendl., Trans. Linn. Soc. London 24: 433 (1864).

Eremos — *destitute of*, spatha — *bract, referring to the lack of conspicuous bracts in the inflorescence*.

Clustered (?always), spiny, high-climbing, pleonanthic, hermaphroditic rattan palms. *Stem* eventually becoming bare, with long internodes, usually circular in cross-section, sometimes obscurely 3-angled, juvenile stem apparently much more slender than the adult, sucker shoots apparently axillary. *Leaves* pinnate, bifid in juveniles, with a terminal cirrus; sheath strictly tubular, unarmed, longitudinally striate, sometimes with a thin caducous cover of indumentum; ocrea conspicuous, tightly sheathing, neatly truncate (?always); knee present in mature climbing stems, but rather inconspicuous; petiole present in juvenile stems, absent in mature climbing stems; rachis usually armed with reflexed spines, and sometimes bearing caducous indumentum; cirrus bearing neat pairs (rarely not paired) of reflexed acanthophylls, sometimes also with scattered reflexed spines; leaflets few to numerous, single-fold except, rarely, in juvenile leaves where lamina undivided, praemorse or abruptly narrowed to a pointed tip, or entire, linear to rhomboid, usually somewhat plicate, regularly arranged, variously indumentose, sometimes white tomentose beneath, usually armed along the thickened margins with conspicuous robust, distally pointing or reflexed spines, transverse veinlets moderately conspicuous; proximal few leaflets on each side of the rachis frequently very much smaller than the rest, strap-like, heavily armed along margins, and reflexed across the sheathed stem. *Inflorescence* arching outward, branched to 1 order, branches horizontal, peduncle enclosed within the leaf sheath and emerging from its mouth, flattened, not adnate to the internode, the surface usually minutely papillose; bracts throughout the inflorescence very inconspicuous; prophyll absent?; peduncular bracts absent; rachis much longer than the peduncle; rachis bracts low, triangular, striate, ± opposite or alternate, often united to form an incomplete sheathing collar; rachillae adnate to the inflorescence axis a short distance above the bract, either opposite (in which case subtended by a double bract) or alternate (in which case subtended by a single triangular bract), distal rachillae always alternate, distichous, rachillae minutely papillose, bearing ± distichous, minute, triangular,

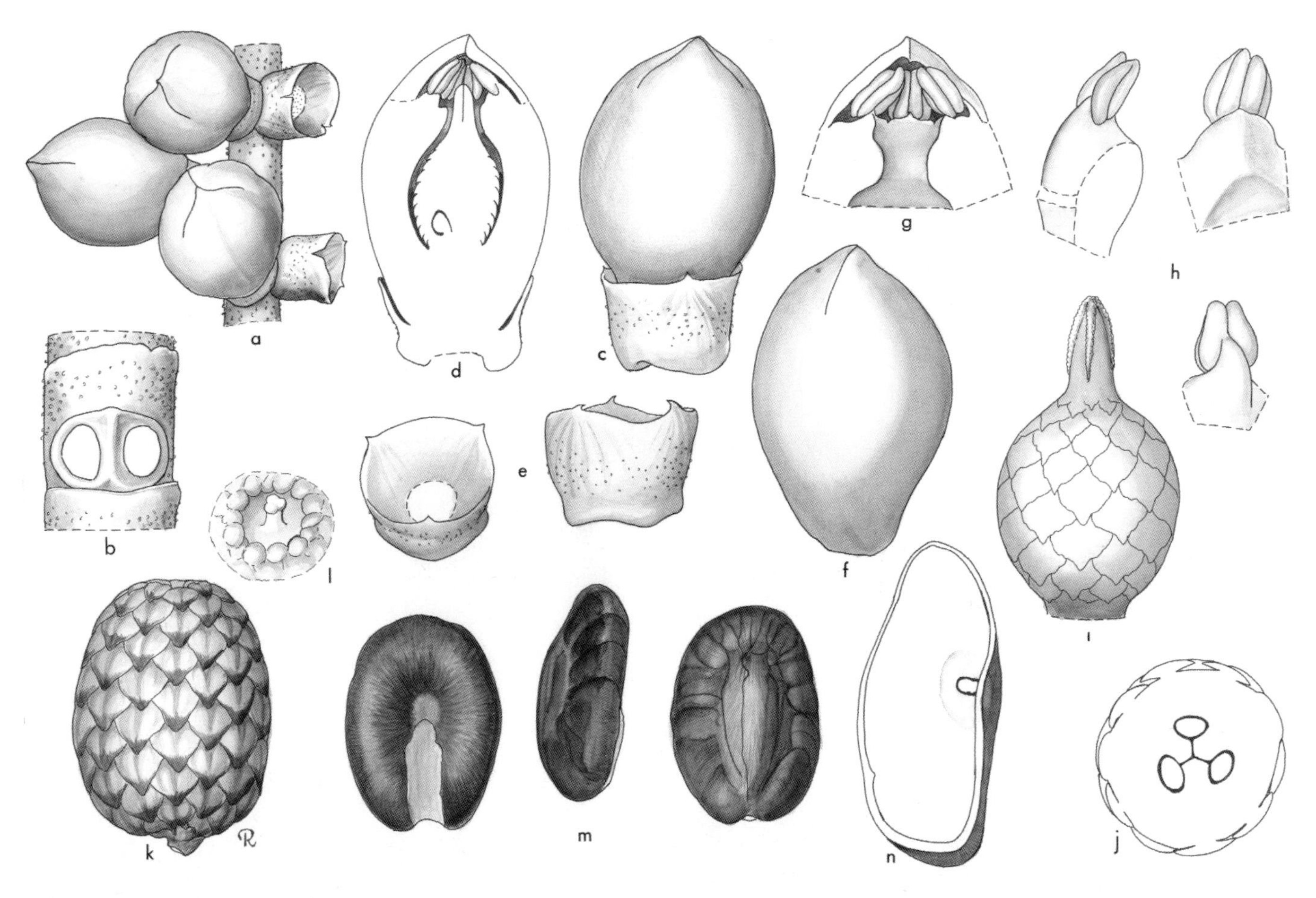

***Eremospatha*. a**, portion of rachilla × 3; **b**, portion of rachilla, pair of flowers removed to show scars × 4$^1/_2$; **c**, flower in late bud × 4 $^1/_2$; **d**, flower in vertical section × 4 $^1/_2$; **e**, calyx in 2 views × 4 $^1/_2$; **f**, flower, calyx removed × 4 $^1/_2$; **g**, tip of petal with stamens × 7$^1/_2$; **h**, stamen in 3 views × 9; **i**, gynoecium × 9; **j**, ovary in cross-section × 9; **k**, fruit × 1; **l**, tip of fruit with stigmatic remains, enlarged; **m**, seed in 3 views × 1; **n**, seed in vertical section × 1 $^1/_2$. *Eremospatha macrocarpa*: all from *Moore* 9891. (Drawn by Marion Ruff Sheehan)

incomplete bracts, each subtending a pair of equal flowers without bracteoles. *Flowers* pale in colour, very fragrant; calyx thick, coriaceous, very shallowly 3-lobed distally, obscurely veined, minutely papillose; corolla very thick, coriaceous, divided at the apex to $^1/_4$ to $^1/_3$ its length into 3 short, triangular, valvate lobes, remaining approximate even at anthesis, the lobes then separating slightly; stamens 6, united into a massive, fleshy, epipetalous ring, ± occluding the mouth of the flower, clasping the gynoecium, free filaments angled, very short, anthers enclosed within the flower, ± medifixed, very short, somewhat sagittate, latrorse; gynoecium tricarpellate, triovulate, rounded, covered in reflexed scales, tipped by a columnar or tapered, ± 3-angled style, apically with 3 stigmatic angles, ovule basally attached, anatropous. *Pollen* ellipsoidal, bi-symmetric; aperture an extended distal sulcus; ectexine tectate, coarsely perforate, or rugulate-reticulate, aperture margin usually much finer; infratectum columellate; longest axis 32–63 µm [4/10]. *Fruit* 1–3 seeded, stigmatic remains minute, apical, perianth whorls persistent; epicarp covered in vertical rows of reddish-brown reflexed scales with fringed margins, mesocarp apparently fleshy at maturity, endocarp not differentiated. *Seed* subbasally attached, from the shape of $^1/_3$ of a sphere to hemispherical or ellipsoidal depending on the number of seeds developing, sometimes slightly lobed or grooved, with a conspicuous abaxial ridge opposite the embryo, seed coat thin, scarcely fleshy, endosperm homogeneous; embryo lateral. *Germination* adjacent-ligular; eophyll bifid. *Cytology* not studied.

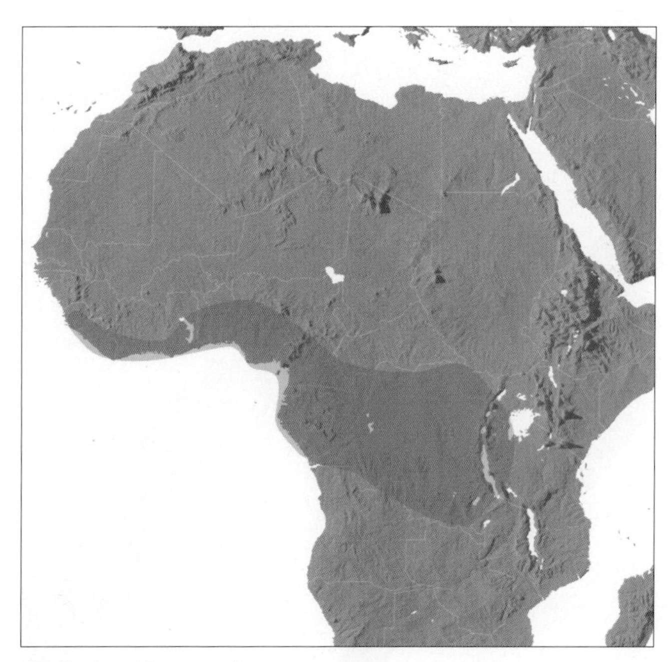

Distribution of ***Eremospatha***

Distribution and ecology: Ten species confined to humid rain forest of West Africa, the Congo Basin, and eastward to Tanzania. Apparently most abundant in rain forest on swampy soils.

Anatomy: Leaf, stem (Tomlinson 1961), root (Seubert 1996a), floral (Uhl & Moore 1973).

Relationships: The monophyly of *Eremospatha* has not been tested. *Eremospatha* is resolved as sister to *Laccosperma* with moderate support (Baker *et al.* 2000a, 2000b, Asmussen *et al.* 2006).

Common names and uses: Common names numerous (Sunderland 2001, 2007). Stems are used as a source of cane.

Taxonomic account: Sunderland (2001, 2007).

Below: ***Eremospatha macrocarpa***, fruit, Cameroon. (Photo: J. Dransfield)

Right: ***Eremospatha macrocarpa***, leaf sheaths and basal leaflets, Cameroon. (Photo: J. Dransfield)

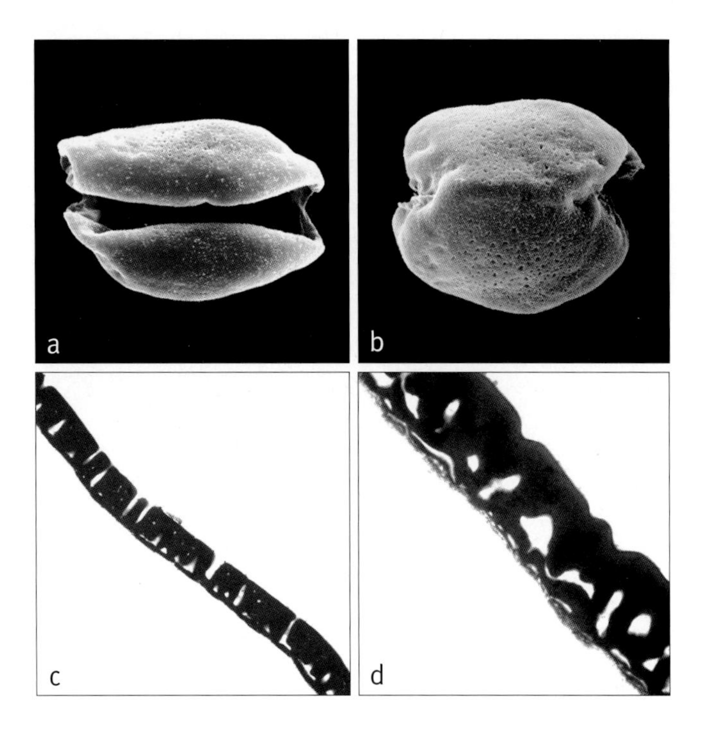

Left: ***Eremospatha***
a, extended sulcate pollen grain, distal face SEM × 1750;
b, extended sulcate pollen grain, proximal face SEM × 1750;
c, ultrathin section through pollen wall, proximal face TEM × 500;
d, ultrathin section through pollen wall, proximal face TEM × 500.
Eremospatha macrocarpa: **a**, **b** & **d**, *Mann* 2338; *E. cuspidata*: **c**, *Mann* 1043.

Fossil record: A single incomplete leaflet with a single spine near the acute, robust base (*Eremospatha chilgaensis* [Pan *et al.* 2006]) is described from the Oligocene Guang River flora of Chilga Woreda, Ethiopia. After comparison with morphologically similar leaf bases in subtribes Ancistrophyllinae and Raphiinae, the authors conclude that this fossil bears notable similarity to *Eremospatha*.

Notes: The flowers of *Eremospatha* are remarkable for the thickness of the corolla tube and the small gaps between their short lobes; the androecial tube is also thick, fibrous, and extensively vascularised (Uhl & Moore 1973). The flowers are very like those of *Plectocomiopsis*, but this similarity is a parallel development. Although the dyad of two hermaphroditic flowers in *Eremospatha* is similar to that of *Laccosperma*, the two genera are otherwise rather different. *Eremospatha* is pleonanthic and has only minute rachilla bracts; *Laccosperma* is hapaxanthic and rachilla bracts are conspicuous.

Additional figures: Fig. 1.7.

4. LACCOSPERMA

Clustering high-climbing pinnate-leaved rattan palms of Equatorial West Africa; sheaths densely armed; hapaxanthic and hermaphroditic, the flowers are borne in pairs, with leathery perianths but not inflated.

Laccosperma (G. Mann & H. Wendl.) Drude, Bot. Zeit. 35: 632, 635 (1877). Type: **L. opacum** (G. Mann & H. Wendl.) Drude (*Calamus opacus* G. Mann & H. Wendl.).

Calamus subgenus *Laccosperma* G. Mann & H. Wendl., Trans. Linn. Soc. London 24: 430 (1864).

Ancistrophyllum (G. Mann & H. Wendl.) H. Wendl. (non Göpp. 1841) in Kerch., Les Palmiers: 230 (1878). Type: *A. secundiflorum* (P. Beauv.) H. Wendl. (*Calamus secundiflorus* P. Beauv.) (= *Laccosperma secundiflorum* [P. Beauv.] Kuntze).

Calamus subgenus *Ancistrophyllum* G. Mann & H. Wendl., Trans. Linn. Soc. London 24: 432 (1864).

Ancistrophyllum subgenus *Laccosperma* (G. Mann & H. Wendl.) Hook.f. in Benth. & Hook.f., Gen. Pl. 3: 937 (1883).

Ancistrophyllum subgenus *Ancistrophyllum* Hook.f. in Benth. & Hook.f., Gen. Pl. 3: 937 (1883).

Neoancistrophyllum Rauschert, Taxon 31: 557 (1982). Type: *N. secundiflorum* (P. Beauv.) Rauschert (superfluous substitute name).

Laccos — *hole or pit*, sperma — *seed, referring to the deeply pitted seeds of some of the species.*

Clustered, spiny, high-climbing, hapaxanthic, hermaphroditic, rattan palms. *Stem* sometimes branching aerially, eventually becoming bare, with long internodes, circular in cross-section, sucker shoots apparently axillary. *Leaf* pinnate with a cirrus; sheath strictly tubular, variously armed with scattered spines and abundant caducous indumentum; ocrea conspicuous, split opposite the leaf, scarcely sheathing, sometimes slightly inflated with inrolled edges and ant-infested, unarmed or armed like the sheath; knee absent; petiole present, usually armed with scattered or grouped spines abaxially and along margins, and frequently indumentose, rarely unarmed; rachis armed like the petiole; cirrus armed with reflexed spines and bearing neat pairs of reflexed acanthophylls; leaflets few to very numerous, 1–4-fold, entire, linear to sigmoid, regularly or irregularly arranged, often fiercely armed with short spines along the margins and the main ribs, midribs prominent adaxially, transverse veinlets conspicuous or inconspicuous. *Inflorescences* produced simultaneously in the axils of the

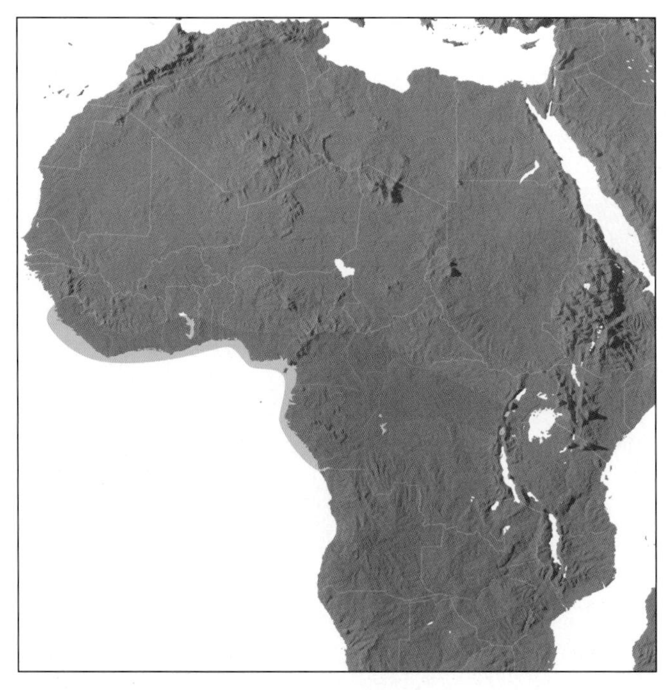

Distribution of ***Laccosperma***

most distal few frequently reduced leaves, branched to 1 order; peduncle enclosed within the leaf sheath and emerging from its mouth or bursting through the sheath, not adnate to the internode, ± hemispherical in cross-section; prophyll strictly tubular, 2-keeled, enclosed within the leaf sheath; peduncular bracts 1–3; rachis longer than the peduncle; rachis bracts distichous, strictly tubular with a triangular limb, without spines, sparsely indumentose, becoming tattered, each subtending a pendulous or spreading rachilla; rachilla prophyll tubular, 2-keeled, included within the subtending bract, rachilla bracts distichous, tubular with apiculate, triangular limb, striate, sparsely indumentose, the margin sometimes ciliate, each, except sometimes for the basal 1–2, subtending a flower cluster. *Flowers* very rarely borne in triads, usually in dyads, sometimes solitary towards the tips of the rachillae, the flower cluster bearing a tubular 2-keeled prophyll and 0, 1, or 2, 2-keeled bracteoles (depending on the number of flowers); calyx slightly to strongly stalk-like at the base, often bent at right angles, incompletely divided distally into 3 triangular striate lobes; corolla tubular at the very base, divided above into 3 oblong, narrow, triangular, valvate lobes; stamens 6, borne at the very base of the corolla, filaments distinct, much swollen, angular, scarcely narrowed at the connective, anthers medifixed, oblong, latrorse; gynoecium tricarpellate, triovulate, ovary covered with scales, those at the base of the style minute, spine-like, style elongate, 3-angled, stigma minute, pyramidal, ovules basally attached, anatropous. *Pollen* ellipsoidal, bi-symmetric; aperture a distal sulcus; ectexine tectate, finely to coarsely perforate, or rugulate-reticulate, aperture margin finer; infratectum columellate; longest axis 37–75 µm [4/5]. *Fruit* 1-seeded (?always), tipped with the base of the style, the rest of the style usually breaking off early in fruit development, the perianth whorls persistent; epicarp covered in vertical rows of reflexed scales with fringed margins, mesocarp fleshy and sweet at maturity, endocarp not differentiated. *Seed* attached subbasally at one side, ovoid and laterally flattened, or rounded and deeply scalloped, with a very shallow to very deep lateral pit, seed coat apparently sometimes fleshy, endosperm homogeneous; embryo lateral, opposite the pit. *Germination* adjacent-ligular; eophyll bifid. *Cytology* not studied.

Distribution and ecology: Five species confined to humid rain forest of West Africa and the Congo Basin. Apparently most abundant in rain forest on swampy soils.

Anatomy: Leaf, stem, root (Tomlinson 1961).

Relationships: *Laccosperma* is strongly supported as monophyletic (Baker *et al.* 2000a, 2000c). It is resolved as sister to *Eremospatha* with moderate support (Baker *et al.* 2000a, 2000b, Asmussen *et al.* 2006).

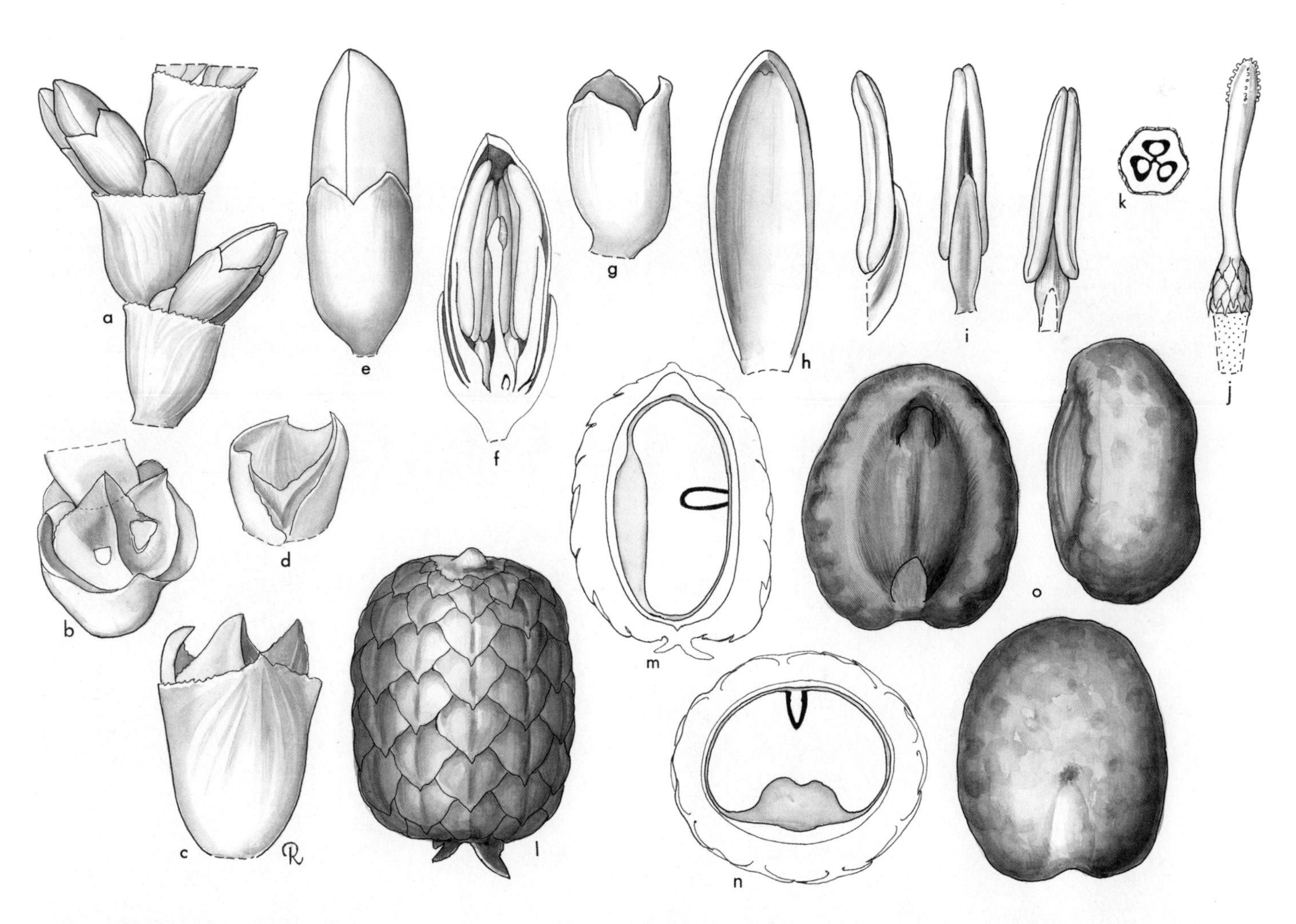

Laccosperma. **a**, portion of rachilla × 3; **b**, portion of rachilla, paired flowers removed to show scars and bracts × 6; **c**, bract subtending flower pair × 6; **d**, floral bracteole × 6; **e**, bud × 4 1/2; **f**, bud in vertical section × 4 1/2; **g**, calyx × 4 1/2; **h**, petal × 6; **i**, stamen in 3 views × 6; **j**, gynoecium × 6; **k**, ovary in cross-section × 12; **l**, fruit in cross-section × 2 1/4; **m**, fruit in vertical section × 2 1/4; **n**, fruit in cross-section × 2 1/4; **o**, seed in 3 views × 3. *Laccosperma acutiflorum*: **j** & **k**, Soyaux 155; *L. secundiflorum*: **a–i** and **l–o**, *Enti s.n.* and *Moore & Enti* 9886, respectively. (Drawn by Marion Ruff Sheehan)

Common names and uses: Common names numerous and varied throughout the area of occurrence (Sunderland 2001). Stems are used as a source of cane.
Taxonomic account: Sunderland (2001, 2007).
Fossil record: No generic records found.
Notes: Very rarely in *Laccosperma*, triads of hermaphroditic flowers are present and bract arrangement indicates a sympodial nature for the triad and the dyad.
Additional figures: Fig. 1.7, Glossary fig. 19.

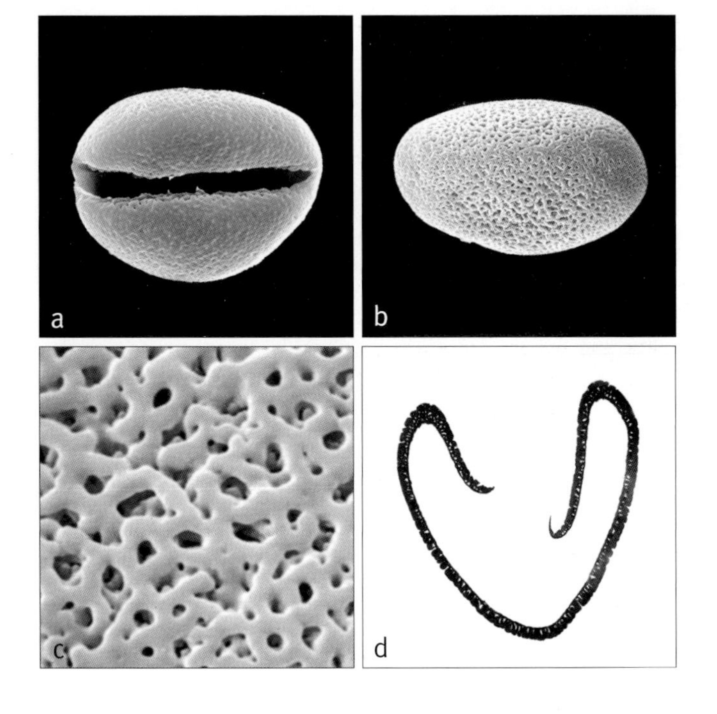

Right: ***Laccosperma***
a, monosulcate pollen grain, distal face SEM × 1000;
b, monosulcate pollen grain, proximal face SEM × 1000;
c, close-up, pollen surface SEM × 5000;
d, ultrathin section, whole pollen grain, polar plane TEM × 1500.
Laccosperma opacum: **a,** *Irvine* 2075; *L. laeve*: **b** & **c**, *Mann* 1045; *L. secundiflorum*: **d**, *Tuley* 851.

Below left: ***Laccosperma secundiflorum***, flower buds, Cameroon. (Photo: J. Dransfield)
Below right: ***Laccosperma secundiflorum***, fruit, Cameroon. (Photo: J. Dransfield)

Subtribe Raphiinae H. Wendl., J. Bot. 3: 383 (1865). Type: **Raphia**.

Acaulescent or arborescent, monoecious, hapaxanthic palms; leaves pinnate, rachis and petiole lacking spines; inflorescences with rachillae distally staminate and proximally pistillate.

The subtribe comprises the single genus *Raphia*, usually massive hapaxanthic pinnate-leaved palms from Africa with one species in the New World; spines are restricted to the leaflets. The fruit are generally large, with conspicuous large scales, and the endosperm has rather few large ruminations.

Subtribe Raphiinae is most often resolved as sister to the Mauritiinae with moderate support (Baker *et al.* 2000a, 2000b, Asmussen *et al.* 2006), but one study places it as sister to the Eugeissoneae with low support (Baker *et al.* in review).

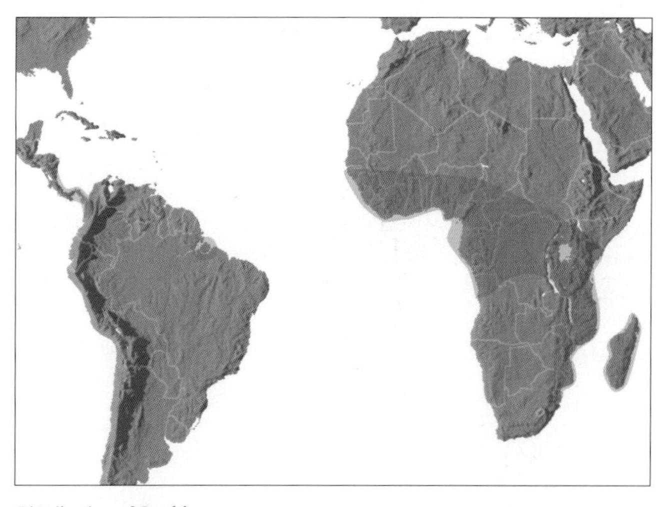

Distribution of ***Raphia***

5. RAPHIA

Generally massive acaulescent or tree palms of Equatorial Africa, Madagascar and South and Central America, with huge pinnate leaves and often fibrous leaf sheaths; hapaxanthic and monoecious, the rachilllae bear solitary pistillate flowers near the base and solitary staminate flowers distally. The fruit is usually very large.

Raphia P. Beauv., Fl. Oware 1: 75 t. 44–46 (1806). Lectotype: **R. vinifera** Palisot de Beauvois.

Sagus Gaertn., Fruct. Sem. Pl. 1: 27 (1788). Type: *S. palma-pinus* Gaertn. (= *R. palma-pinus* [Gaertn.] Hutch.) (non *Sagus* Steck 1757 = *Metroxylon* Rottb.).

Derived from the Malagasy vernacular name, rofia.

Massive, solitary or clustered, armed, hapaxanthic, monoecious, acaulescent or tree palms. *Stem* subterranean to erect, usually partly obscured by the marcescent leaf bases, the internodes sometimes bearing short, negatively geotropic, ± spine-like roots, cortex hard, pith soft. *Leaves* massive, pinnate, marcescent; sheath unarmed, splitting opposite the petiole, with or without a conspicuous ligule, disintegrating into thin sheets or sometimes partly into black fibre bundles ('piassava'); petiole short to very long, unarmed, usually deeply channelled adaxially only at the base, rounded distally; rachis unarmed, angled adaxially, rounded abaxially; leaflets single-fold, linear, numerous, regularly arranged or grouped and fanned within the groups to give the leaf a plumose appearance, often whitish beneath, armed with short spines along the margins and the midrib, the margins frequently greatly thickened, midribs very large, transverse veinlets conspicuous or inconspicuous. *Inflorescences* branched to 2 orders, produced simultaneously in the axils of the most distal few leaves, either interfoliar and pendulous or aggregated into a massive, erect, suprafoliar, compound inflorescence; peduncle short; prophyll tubular, 2-keeled, closely sheathing to inflated, sometimes splitting opposite the keels, with 1 or 2 short triangular lobes; peduncular bracts several (ca. 6) and inflated basally with triangular limbs; rachis much longer than the peduncle; rachis bracts distichous or in 4 ranks, tubular, closely sheathing to somewhat inflated, usually each subtending a first-order branch, rarely empty; first-order branches reflexed or variously spreading, sometimes scarcely exserted from the bract, bearing a basal, 2-keeled, tubular prophyll, and distichous or 4 ranked, tubular bracts with triangular limbs, each, except for the prophyll and 1–few basal-most, subtending a rachilla; rachillae very crowded to distant and sometimes not exserted; rachilla prophyll tightly sheathing, 2-keeled; subsequent rachilla bracts tending to be distichous or in 4 ranks, tubular, tightly sheathing, with short, striate triangular limbs; distal 1–3 bracts empty, of the remaining, the proximal bracts from $^1/_4$–$^2/_3$ the rachilla length each subtending a pistillate flower and 2 prophyllar bracteoles, the distal, each subtending a staminate flower with a single prophyllar bracteole, very rarely at the junction between staminate and pistillate parts of the rachilla, the bract subtending a dyad of 1 staminate and 1

Right: ***Raphia farinifera***, habit, Madagascar. (Photo: W.J. Baker)

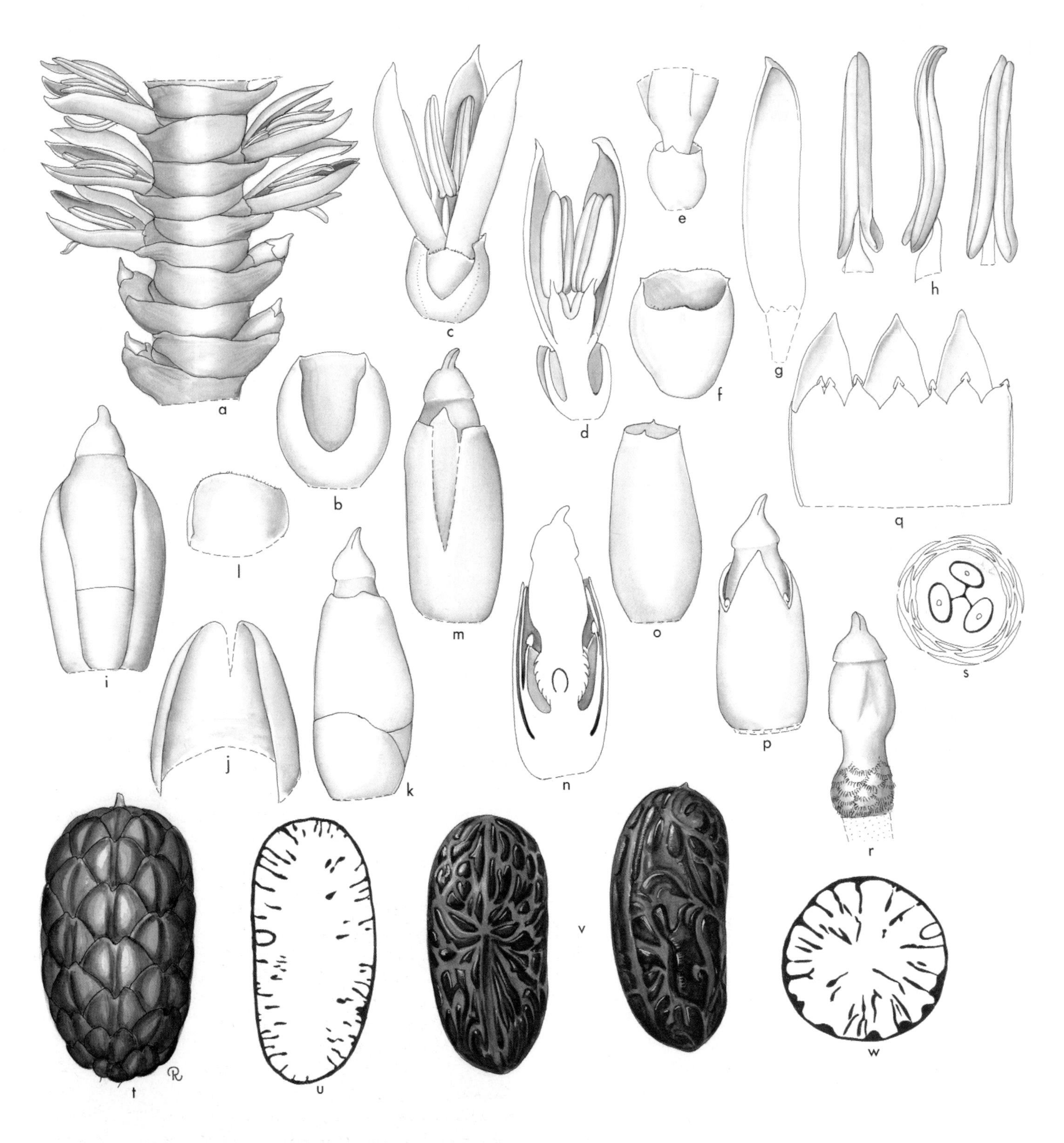

Raphia. **a**, portion of rachilla, staminate flowers above, pistillate flowers below × 3; **b**, bracteole of staminate flower × 6; **c**, staminate flower with bracteole × $4^1/_2$; **d**, staminate flower in vertical section × $4^1/_2$; **e**, staminate calyx and portion of corolla × $4^1/_2$; **f**, staminate calyx × 6; **g**, staminate petal × 6; **h**, stamen × 9; **i**, pistillate flower with two bracteoles × 6; **j**, outer bracteole of pistillate flower × 6; **k**, pistillate flower with inner bract × 6; **l**, inner bracteole of pistillate flower × 6; **m**, pistillate flower, calyx split to show corolla × 6; **n**, pistillate flower in vertical section × 6; **o**, pistillate calyx × 6; **p**, pistillate flower, calyx removed × 6; **q**, pistillate corolla and staminodial ring, expanded, interior view × 6; **r**, gynoecium × 6; **s**, ovary in cross-section × 12; **t**, fruit × $^3/_4$; **u**, seed in vertical section × $^3/_4$; **v**, seed in 2 views × $^3/_4$; **w**, seed in cross-section × $^3/_4$. *Raphia farinifera*: ***a–s***, from Fairchild Tropical Garden; *R. palma-pinus*: **t–w**, *Moore* 9889. (Drawn by Marion Ruff Sheehan)

pistillate flower; rarely in apical portions of the inflorescence, rachillae bearing staminate flowers only. *Staminate flowers* conspicuously exserted; calyx tubular, shallowly 3-lobed; corolla greatly exceeding the calyx, sometimes glossy, tubular at the base, with 3 elongate, triangular, sometimes spine-like, valvate lobes; stamens 6–30, filaments narrowly spindle-shaped, joined to the corolla near the base, variously distinct or connate into a fleshy tube, abruptly contracted at the connective, anthers elongate, sagittate basally, uneven distally, introrse or latrorse; pistillode absent or minute. *Pollen* ellipsoidal, bi-symmetric; aperture a distal sulcus, often notably shorter than long axis; ectexine tectate, scabrate, perforate, perforate-rugulate, rugulate, or granular-rugulate or, rarely, ectexine intectate and sparsely spinulose, aperture margin similar; infratectum usually very dense and narrow, barely columellate, although in some species interrupted by wide cavities; longest axis 17–35 µm [13/20]. *Pistillate flowers* sometimes only partly exserted from the rachilla bracts; calyx tubular, ± truncate or shallowly 3-lobed, later splitting; corolla exceeding or scarcely longer than the calyx, tubular in proximal ca. $^1/_2$, distally with 3 valvate, triangular lobes; staminodes united into an epipetalous ring, with 6–16 irregular teeth of varying lengths bearing the flattened, sagittate, short, empty anthers; gynoecium tricarpellate, triovulate, ovoid to somewhat conical, with a short style and conical 3-lobed stigma, locule partitions incomplete, ovule basally attached, anatropous. *Fruit* usually large, elliptical, 1-

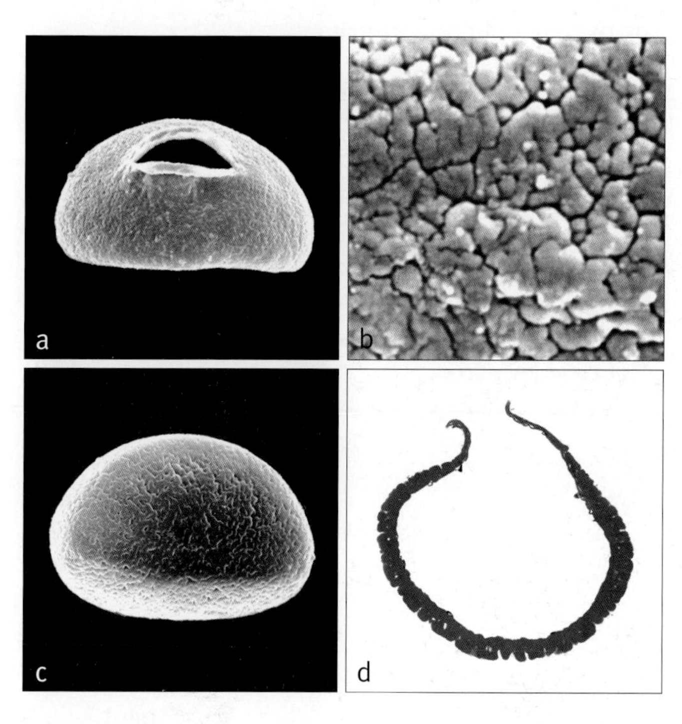

Above: ***Raphia***
a, brevi monosulcate pollen grain, distal face SEM × 1000;
b, close-up, pollen surface SEM × 5000;
c, brevi monosulcate pollen grain, proximal face SEM × 1000;
d, ultrathin section, whole pollen grain, polar plane TEM × 1800.
Raphia australis: **a**, *Strey* 7493; *R. rostrata*: **b**, *Luali* 9091;
R. hookeri: **c**, *Letouzey* 14594; *R. matombe*: **d**, *Cango* 8683.

Top right: ***Raphia humilis***, plant in flower, cultivated, Fairchild Tropical Botanic Garden, Florida. (Photo: J. Dransfield)

Middle right: ***Raphia humilis***, inflorescence, cultivated, Fairchild Tropical Botanic Garden, Florida. (Photo: J. Dransfield)

Bottom right: ***Raphia humilis***, fruit, cultivated, Fairchild Tropical Botanic Garden, Florida. (Photo: J. Dransfield)

seeded, with apical stigmatic remains; epicarp covered in neat vertical rows of large reflexed scales, mesocarp thick, mealy, oil-rich, endocarp not differentiated. *Seed* subbasally attached with dry seed coat, variously coarsely furrowed, endosperm with a few large ruminations; embryo lateral. *Germination* adjacent-ligular; eophyll usually pinnate, more rarely bifid. *Cytology*: 2n = 28.

Distribution and ecology: Twenty species, throughout the more humid areas of Africa; one species in Madagascar possibly introduced; one species, *Raphia taedigera*, in tropical America but stated by Otedoh (1977) also to occur in West Africa and to be introduced in tropical America (this, however, seems unlikely). Most species of *Raphia* seem to be plants of swamplands, but *R. regalis* occurs on hillslopes in humid tropical rain forest.
Anatomy: Leaf, petiole, stem, root (Tomlinson 1961), root (Seubert 1996a).
Relationships: The monophyly of *Raphia* has not been tested. For relationships, see Raphiinae.
Common names and uses: Raphia palms. Species of *Raphia* are of extreme economic importance. Raphia fibre is obtained by stripping off the cuticle and hypodermis from the emerging leaflets; locally it is used for a wide range of purposes, such as basketware and twine, and it is exported for garden twine and weaving. Piassava is obtainable from the leaf sheaths of *R. hookeri*. Petioles of many species are used as a substitute for bamboo in house and furniture construction and leaflets are used for thatch. Palm wine can be obtained by tapping the stem apex. The mesocarp of some species provides a source of cooking oil. The kernels and stem apex are sometimes eaten. The fruit of some species is used as a fish poison.
Taxonomic accounts: Beccari (1910), Russell (1965) and Otedoh (1982).
Fossil record: No generic records found.
Notes: The leaves reaching 25 m in *Raphia regalis* (Hallé 1977) are considered to be the longest in the plant kingdom.
Additional figures: Figs 1.7, 2.5c, Glossary fig. 22.

Subtribe Mauritiinae Meisn., Plant. Vasc. Gen. 265 (1842). Type **Mauritia**.

Arborescent, dioecious, pleaonanthic palms; leaves palmate, lacking spines; pollen spheroidal with intectate processes; seed with small, knob-like appendage at apex.

The three genera of the Mauritiinae are outstanding in their possession of palmate rather than pinnate reduplicate leaves. Apart from one species of *Raphia* (*R. taedigera*) they are the only members of the subfamily occurring naturally in the New World.

Highly supported as a monophyletic group (Baker *et al.* 1999b, 2000a, 2000b, in review), subtribe Mauritiinae is resolved as sister to the Raphiinae with moderate support (Baker *et al.* 2000a, 2000b, Asmussen *et al.* 2006). One study places the subtribe as sister to a clade of Eugeissoneae and Raphiinae, but this position is poorly supported (Baker *et al.* in review).

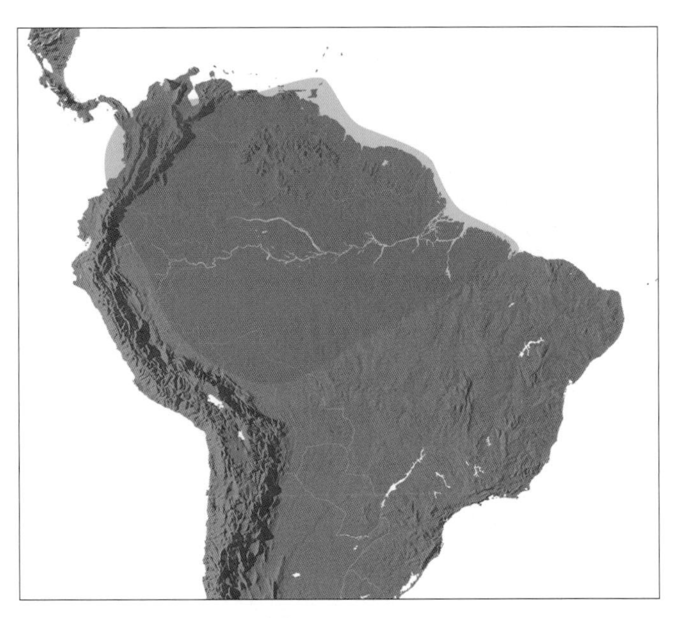

Distribution of subtribe **Mauritiinae**

Key to Genera of Mauritiinae

1. Slender, unarmed, clustered, undergrowth palms; leaf usually irregularly divided into 1–several-fold segments; rachillae short and compressed; staminate flowers in dyads borne distichously on the rachillae. South America . . . . **6. Lepidocaryum**
1. Moderate to robust, solitary or clustered, armed or unarmed palms; leaf regularly divided into single-fold segments; the staminate rachillae and sometimes pistillate rachillae catkin-like, bearing spirally arranged flowers, singly or in pairs 2
2. Moderate, clustered palms with spiny stems; pistillate flowers borne on short catkin-like branches; staminate flowers solitary in the axil of each rachilla bract. South America . . . . **8. Mauritiella**
3. Massive, solitary palms with unarmed stems; pistillate flowers very few or single, borne on scarcely catkin-like branches; staminate flowers borne in dyads. South America . . . . **7. Mauritia**

6. LEPIDOCARYUM

Clustering unarmed palms of South America with slender erect stems and palmate leaves with segments of varying width; inflorescences small with staminate and pistillate rachillae bearing solitary flowers at each bract.

Lepidocaryum Mart., Hist. Nat. Palm. 2: 49 (1824). Lectotype. **L. gracile** Mart. (see H.E. Moore 1963c).

Lepidos — *scale,* caryon — *nut, referring to the scaly fruits.*

Slender, clustered, unarmed, pleonanthic, dioecious, undergrowth palms. *Stem* erect, colonial by slender rhizomes, partly obscured by marcescent leaf sheaths above, becoming bare basally, with rather short internodes and inconspicuous nodal scars. *Leaves* small, reduplicately palmate; sheath splitting opposite the petiole, frequently covered with very dense, caducous tomentum; petiole conspicuous, ± rounded in cross-section except at the base where channelled adaxially; hastulae mostly absent, a low crest sometimes present adaxially; blade flabellate or ± orbicular with a very short costa, divided along a few abaxial folds to the insertion into few broad or narrow single-fold or compound, spathulate, acuminate segments, tips sometimes bristly, blade surfaces similar in colour, midribs more prominent abaxially, transverse veinlets conspicuous, rather distant and somewhat sinuous; young leaves sometimes reddish-tinged. *Inflorescences* solitary, interfoliar, the staminate and pistillate superficially similar, branched to 2 orders; prophyll tubular, closely sheathing, 2-keeled, with 2 short triangular lobes; peduncle elongate, bearing several (ca. 6) closely sheathing tubular bracts with short triangular limbs; rachis longer than the peduncle; rachis bracts like the peduncular, but tending to split at the tip, each subtending a first-order branch; first-order branches few, rather short, bearing a basal, tubular, 2-keeled prophyll and sometimes 1 empty tubular bract, and distichously arranged, tubular, triangular-tipped bracts, each subtending a rachilla; staminate rachilla short, becoming recurved, with a basal, membranous, striate, 2-keeled prophyll and few (up to 12) distichous, membranous, apiculate, cup-shaped bracts, each, including sometimes the prophyll, subtending a solitary staminate flower bearing a 2-keeled bracteole, or a pair of staminate flowers enclosed within a 2-keeled, explanate bracteole, one of the pair bearing a second bracteole; pistillate rachilla usually very short, sometimes scarcely exserted from the subtending bract, bearing a basal, membranous, 2-keeled, striate prophyll and few (up to ca. 8) distichous, membranous, cup-like, apiculate bracts, each, including sometimes the prophyll, subtending a solitary pistillate flower, bearing a 2-keeled bracteole and a minute, ovate, flattened second bracteole. *Staminate flowers* symmetrical; calyx tubular, briefly 3-lobed, ± striate; petals much exceeding the calyx, basally connate, the tips valvate; stamens 6, borne at the very base of the petals, filaments thick, fleshy, ± angled, the anthers small, basifixed, latrorse; pistillode minute or short, columnar. *Pollen* ellipsoidal, bi-symmetric; aperture a distal sulcus; ectexine intectate, surface very finely granular, interspersed with bottle-shaped spines set in, and loosely connected to, cavities in a wide foot layer, bulging noticeably inward beneath each spine, the inner face of the foot layer clearly lamellate, aperture margin similar; longest axis 28–41 µm [1/1]. *Pistillate flowers* larger than the staminate; calyx tubular, 3-lobed, splitting somewhat irregularly after fertilization; corolla much exceeding the calyx, tubular in basal ca. 1/3, with 3 elongate, valvate lobes; staminodes 6, adnate to the base of the corolla lobes, the filaments somewhat angled, the empty anthers minute; gynoecium incompletely trilocular, triovulate, ± rounded, covered in vertical rows of reflexed scales, style conical, briefly 3-lobed, ovule anatropous, ?basally attached. *Fruit* rounded or oblong, usually 1-seeded, with apical stigmatic remains; epicarp covered in vertical rows of reflexed reddish-brown scales, mesocarp thin, endocarp not differentiated. *Seed* attached near the base at one side, with a shallow furrow along the raphe, testa ?fleshy, endosperm homogeneous; embryo lateral. *Germination* adjacent-ligular; eophyll bifid. *Cytology*: 2n = 30.

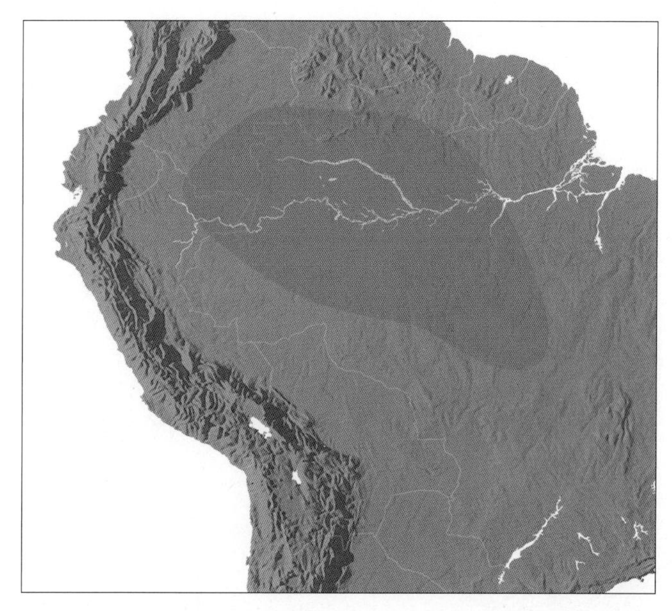

Distribution of ***Lepidocaryum***

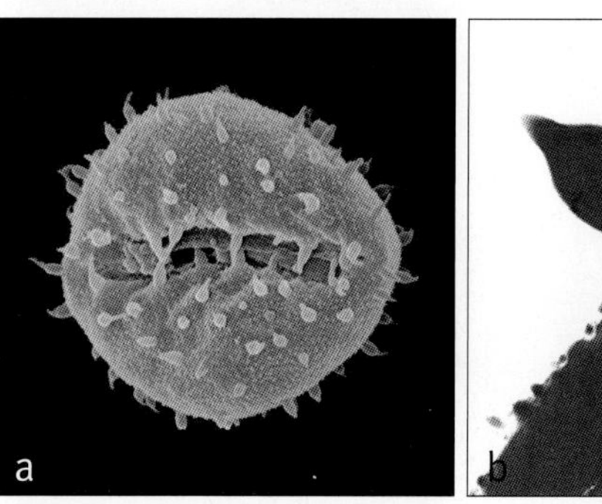

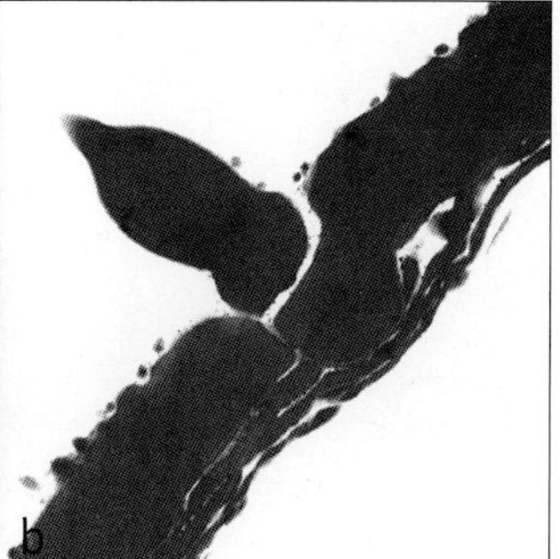

Above: ***Lepidocaryum***
a, monosulcate pollen grain, distal face SEM × 1000;
b, ultrathin section through pollen wall TEM × 10,000.
Lepidocaryum tenue: **a** & **b**, *Schomburgk* s.n.

Left: ***Lepidocaryum tenue***, habit, Peru. (Photo: A.J. Henderson)

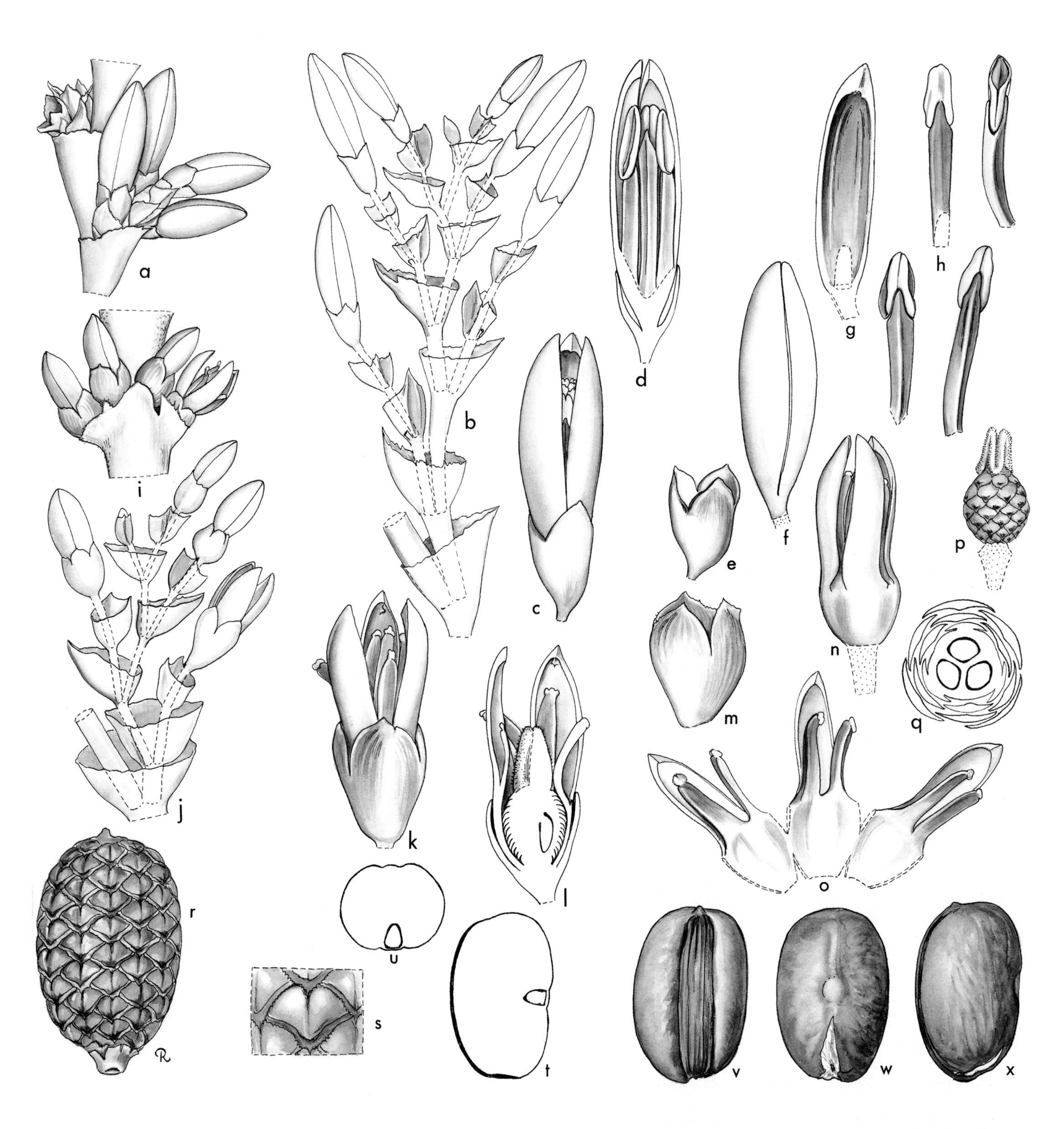

Lepidocaryum. **a**, portion of staminate inflorescence × 3; **b**, portion of staminate inflorescence, lateral axis expanded to show arrangement of bracts with flowers × 3; **c**, staminate bud × 6; **d**, staminate bud in vertical section × 6; **e**, staminate calyx × 6; **f**, staminate bud, calyx removed × 6; **g**, staminate petal, interior view × 6; **h**, stamen in 4 views × 6; **i**, portion of pistillate inflorescence × 3; **j**, portion of pistillate inflorescence, lateral axis expanded to show arrangement of bracts and flowers × 3; **k**, pistillate flower × 6; **l**, pistillate flower in vertical section × 6; **m**, pistillate calyx × 6; **n**, pistillate flower, calyx removed × 6; **o**, pistillate corolla with staminodes, interior view, expanded × 6; **p**, gynoecium × 6; **q**, ovary in cross-section × 12; **r**, fruit × $1^1/_2$; **s**, scales × 3; **t**, seed in vertical section × $1^1/_2$; **u**, seed in cross-section × $1^1/_2$; **v**, **w**, **x**, seed in 3 views × $1^1/_2$. *Lepidocaryum tenue* var. *gracile*: **a–h**, *Schultes & Cabrera* 15198; **r–x**, *Schultes & Cabrera* 15756; *L. tenue* var. *tenue*: **i–g**, *Schultes & Cordeiro* 6511. (Drawn by Marion Ruff Sheehan)

Distribution and ecology: One species with three varieties, distributed in the wetter parts of Colombia, Peru, Venezuela, Guyana and Brazil, growing in the undergrowth of lowland tropical rain forest.

Anatomy: Leaf (Tomlinson 1961), root (Seubert 1996a).

Relationships: *Lepidocaryum* is strongly supported as sister to a robust clade comprising *Mauritia* and *Mauritiella* (Baker *et al.* 2000a, 2000b, in review).

Common names and uses: *Poktamiu*. Leaves are used as thatch.

Taxonomic account: Henderson (1995), Henderson *et al.* (1995).

Fossil record: Monosulcate spiny pollen grains resembling those of *Lepidocaryum gracile* ('or *Nuphar luteum*') are described from the Upper Cretaceous (Senonian) of Gabon (Boltenhagen 1967). The size of the grains (28–41 μm) is more comparable to *Lepidocaryum* pollen than to pollen of *N. luteum* (>45 μm). Furthermore, unlike *Lepidocaryum*, *Nuphar* is associated with cold or cool climates; Salard-Cheboldaeff (1978, 1981) associated *Echimonocolpites rarispinosus* (44–45 μm long axis) from the Lower Eocene of Cameroon with *Lepidocaryum*, although this is an uncertain comparison. The pollen grains of *Mauritia*, although similar morphologically, are much larger (54–65 μm) than those of *Lepidocaryum*. Neither *Lepidocaryum* nor *Mauritia* are present in the African continent today, although *Mauritia* has a notable pollen fossil record in Africa.

Notes: *Lepidocaryum* is the smallest member of the subtribe in terms of habit and inflorescence, but the three genera, *Mauritia*, *Mauritiella*, and *Lepidocaryum*, are clearly very closely related.

7. MAURITIA

Massive stately solitary palms of South America, with robust erect stems and huge palmate leaves with segments of equal width; inflorescences are robust and the staminate rachillae catkin-like, each rachilla bract subtending a pair of staminate flowers.

Mauritia L.f., Suppl. Pl. 70, 454 (1782). Type: **M. flexuosa** L.f.

Orophoma Spruce, J. Linn. Soc., Bot. 11: 93 (1869 ['1871']). Type: *Mauritia carana* Wallace.

Commemorating Count Johan Mauritz van Nassau-Siegen (1604–1679), once governor of the Netherlands West India Company in Brazil.

Massive, solitary, unarmed, pleonanthic, dioecious, tree palms. *Stem* erect, partly obscured by marcescent leaf sheaths above, becoming bare basally, cortex hard, pith soft. *Leaves* large, reduplicate, briefly costapalmate; sheath tubular at first, splitting opposite the petiole, the margins sometimes bearing coarse fibres; petiole conspicuous, adaxially channelled near the base, otherwise circular in cross-section, smooth, unarmed; blade bearing a low crest adaxially at the base, abaxially with a low ridge; blade orbicular, divided along abaxial folds almost to the insertion into numerous crowded single-fold segments, very shortly bifid at their tips, midribs prominent, transverse veinlets not conspicuous. *Inflorescences* solitary, interfoliar, the staminate and pistillate superficially similar; prophyll short, tubular, 2-keeled, with 2 short triangular lobes, striate; peduncle shorter than the rachis, elliptical in cross-section, bearing numerous overlapping distichous, tubular, striate, peduncular bracts, each with a short, triangular, dorsal limb and a shallow point on opposite side; rachis bracts numerous, completely sheathing the branches, distichous, as the peduncular, each subtending a ± pendulous or spreading first-order branch; the first-order branch bearing a short, 2-keeled, striate, tubular prophyll, and 1–few empty distichous bracts, subsequent bracts tubular, flaring, short, each subtending a very short or moderate, straight or recurved rachilla; staminate rachilla catkin-like, bearing a basal, tubular, 2-keeled prophyll and crowded, spirally inserted bracts, each subtending a pair of staminate flowers, each flower bearing a basal 2-keeled bracteole; pistillate rachilla very short, not catkin-like, bearing a basal, tubular, 2-keeled prophyll, and subdistichous, ± explanate bracts, each subtending a solitary pistillate flower with a flattened, 2-keeled bracteole, and often also bearing a minute spathulate, second bracteole, with 2 minute flanges on its abaxial surface. *Staminate flowers* with calyx tubular, shortly 3-lobed, often densely scaly; petals 3, elongate, much exceeding the calyx, valvate, coriaceous, joined briefly at the base; stamens 6, the filaments ± free, thick, ± angled, elongate, anthers elongate, basifixed, latrorse; pistillode minute. *Pollen* spheroidal, symmetric; aperture either a large distal pore or a short sulcus; ectexine intectate, very finely clavate, interspersed with bottle-shaped spines set in, and loosely connected to, cavities in a wide foot layer bulging strongly inwards beneath each spine, the inner face of the foot layer finely lamellate, aperture margins similar; longest axis 54–65 μm [1/2]. *Pistillate flowers* larger than the staminate; calyx tubular, striate, shortly 3-lobed, often densely scaly; corolla tubular in the basal $^1/_3 - ^1/_2$, with 3 valvate, elongate lobes distally; staminodes 6, connate laterally by their flattened broad filaments and adnate to the corolla at the mouth of the tube; gynoecium trilocular, triovulate, ± rounded, covered in vertical rows of reflexed scales, style short, conical, stigmas 3, ovules anatropous, basally attached. *Fruit* ± rounded, very large, usually 1-seeded, with apical stigmatic remains; epicarp covered in many neat vertical rows of reddish-brown, reflexed scales, mesocarp rather thick, fleshy, endocarp not differentiated. *Seed* rounded, attached near the base, apically with a blunt beak, testa thin, endosperm homogeneous; embryo basal. *Germination* adjacent-ligular; eophyll with a pair of divergent leaflets (?always). *Cytology*: 2n = 30.

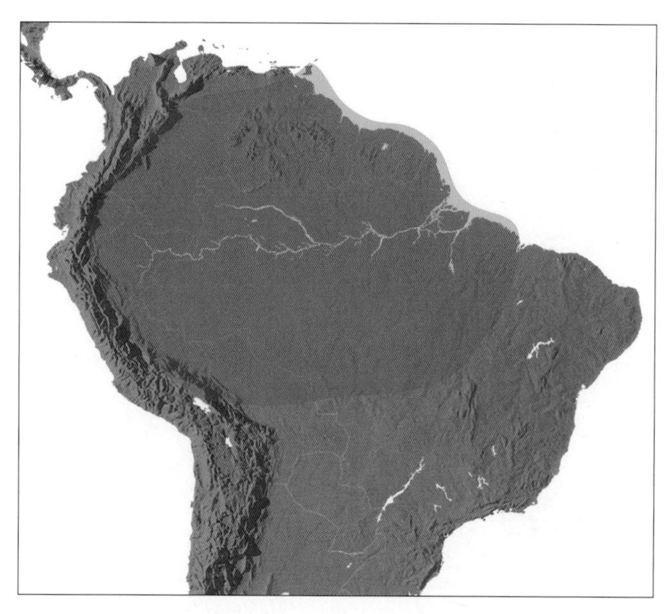

Distribution of ***Mauritia***

Distribution and ecology: Two species distributed in wetter parts of Trinidad, Colombia, Ecuador, Peru, Venezuela, Guyana, Surinam, French Guiana and Brazil. Occurring often in vast natural stands in periodically inundated areas in the lowlands.

Anatomy: Leaf, petiole, stem (Tomlinson 1961), root (Seubert 1996a), stegmata (Killmann & Hong 1989).

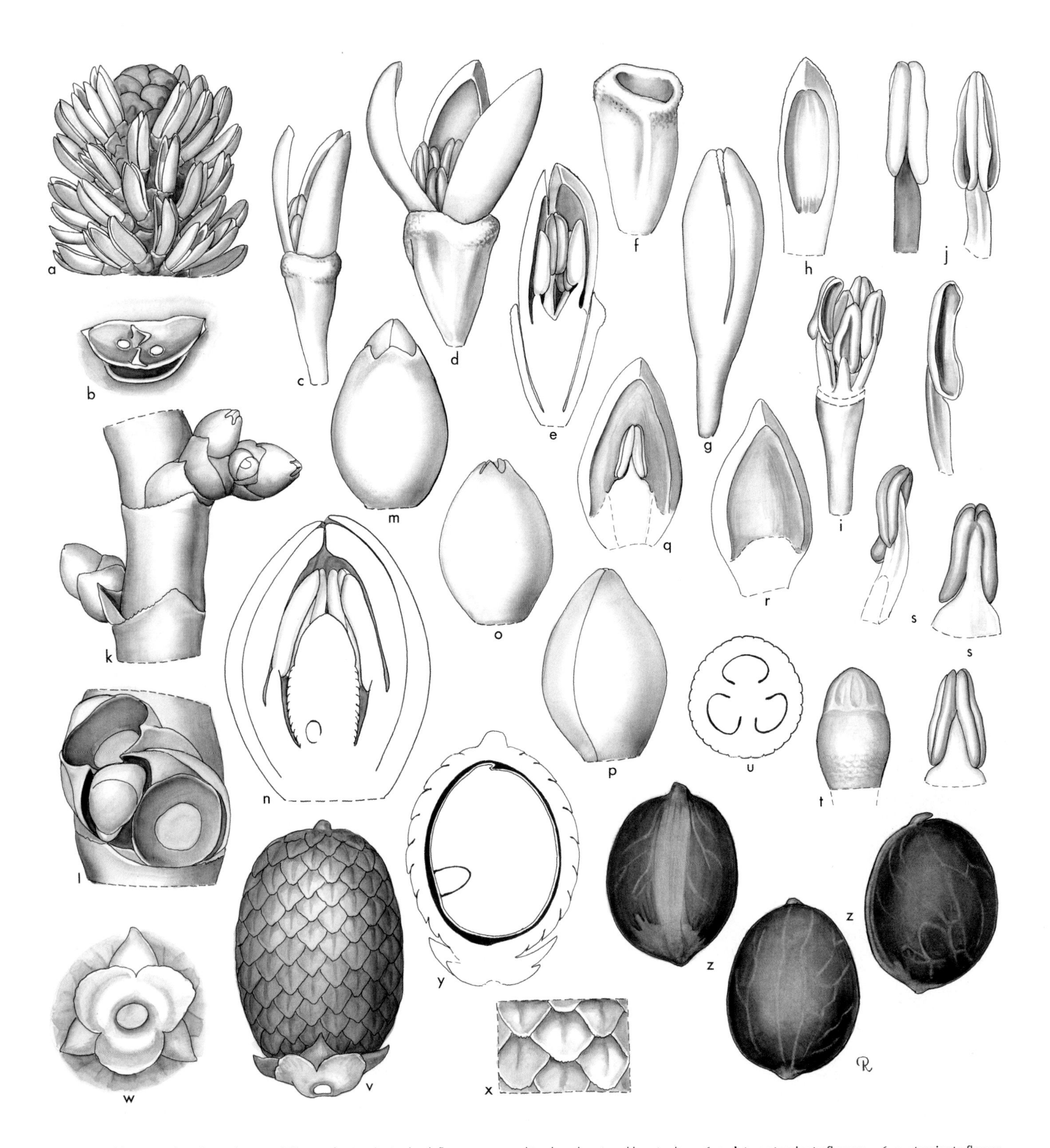

Mauritia. **a**, portion of staminate rachilla × 2; **b**, staminate dyad, flowers removed to show bract and bracteoles × 6; **c**, **d**, two staminate flowers × 6; **e**, staminate flower in vertical section × 6; **f**, staminate calyx × 6; **g**, staminate flower, calyx removed × 6; **h**, staminate petal, interior view × 6; **i**, androecium × 6; **j**, stamen in 3 views × 9; **k**, portion of pistillate rachilla × $1^{1}/_{2}$; **l**, portion of pistillate rachilla, two buds removed to show bracts and bracteoles × 6; **m**, pistillate bud × $4^{1}/_{2}$; **n**, pistillate bud in vertical section × 6; **o**, pistillate calyx × $4^{1}/_{2}$; **p**, pistillate bud, calyx removed × $4^{1}/_{2}$; **q**, pistillate petal and staminode, interior view × $4^{1}/_{2}$; **r**, pistillate petal, interior view × $4^{1}/_{2}$; **s**, staminode in 3 views × 6; **t**, gynoecium × 6; **u**, ovary in cross-section × 9; **v**, fruit × 1; **w**, base of fruit with perianth × 1; **x**, scales of fruit × $2^{1}/_{4}$; **y**, fruit in vertical section × 1; **z**, seed in 3 views × 1. *Mauritia carana*: **a–j**, *Schultes & Cabrera* 18315; *M. flexuosa*: **k–z**, *G.P. Lewis* s.n. (Drawn by Marion Ruff Sheehan)

Relationships: The monophyly of *Mauritia* has not been tested. For relationships, see *Lepidocaryum*.

Common names and uses: Mauritia palms, moriche palms, *buriti*. The buriti palms are immensely useful. They may be a source of oil and starch, wine, timber, cork, fibre for weaving and tying, and palm hearts.

Taxonomic account: Henderson (1995), Henderson *et al.* (1995).

Fossil record: Large (49–68 µm long axis) monosulcate finely clavate pollen grains with distinctive inset spines, subtended by a swelling of the foot layer — *Mauritiidites* (nov. gen. van Hoeken-Klinkenberg 1964, syn. *Monocolpites franciscoi* van der Hammen 1956) — are recorded in the Upper Cretaceous to Lower Tertiary of West Africa: Nigeria (Hoeken-Klinkenberg 1964; Jan Du Chêne 1978). Also in Babajide Salami (1985) where, although the pollen grain is smaller, the spines are not clearly inset, possibly *Lepidocaryum* but aperture type not described. *Mauritiidites* is also described from the Upper Cretaceous of Somalia (Schrank 1994). However, the record from the Zinguinchor borehole (Middle Eocene to Lower Miocene) in Senegal (Médus 1975) shows a narrow columellate infratectum that is not present in *Mauritia* (intectate) and, furthermore, the spines are not characteristic for *Mauritia*. In Saudi Arabia, Srivastava and Binda (1991) describe pollen that closely resembles *Mauritia* from the Lower Eocene Shumaysi Formation. In a survey of subsurface Miocene sediments from the east coast of southern India, Ramanujam *et al.* (1986) include spiny pollen bearing some resemblance to *Mauritiidites* but it is not conclusive. In northwestern South America, there are a number of Palaeocene and Eocene records for *Mauritiidites*, especially from Colombia (Sole de Porta 1961; van der Hammen & Garcia de Mutis 1966 — in this paper, the larger grain(s) are rather more convincing than the smaller ones; González-Guzmán 1967 — pollen not described but illustration shows an asymmetric, thick-walled grain very reminiscent of *Attalea* in all respects excepting the presence of sparse spinulae; Schuler & Doubinger 1970; Jaramillo & Dilcher 2001 — an excellent record, the sunken spines with underlying bulges in the foot layer clearly visible). There is also a convincing record from Venezuela (Lorente 1986). A Pleistocene core from

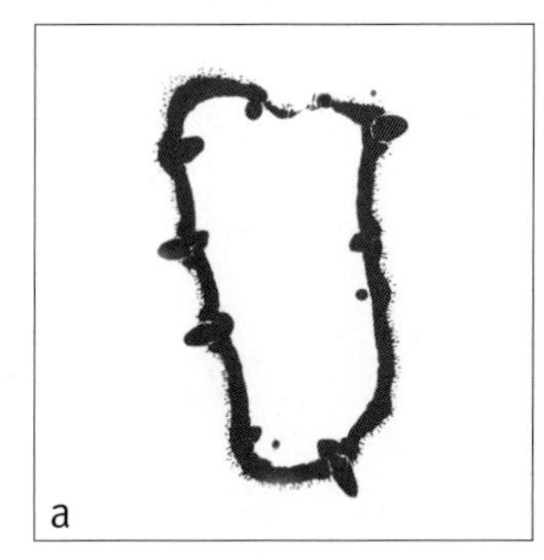

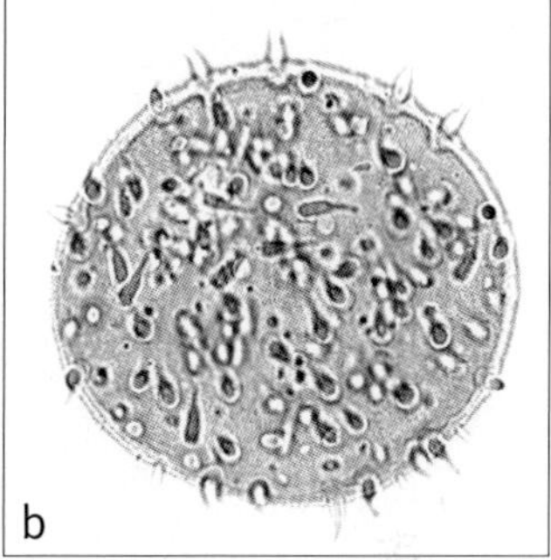

Above: ***Mauritia***
a, ultrathin section through whole pollen grain, polar plane TEM × 800;
b, pollen grain, mid-focus LM × 1000.
Mauritia flexuosa: **a** & **b**, *Heringer et al.* 3153.

Below left: ***Mauritia flexuosa***, habit, Ecuador. (Photo: F. Borchsenius)
Middle: ***Mauritia flexuosa***, staminate rachillae, Brazil. (Photo: A.J. Henderson)
Right: ***Mauritia flexuosa***, fruit, Colombia (Photo: A.J. Henderson)

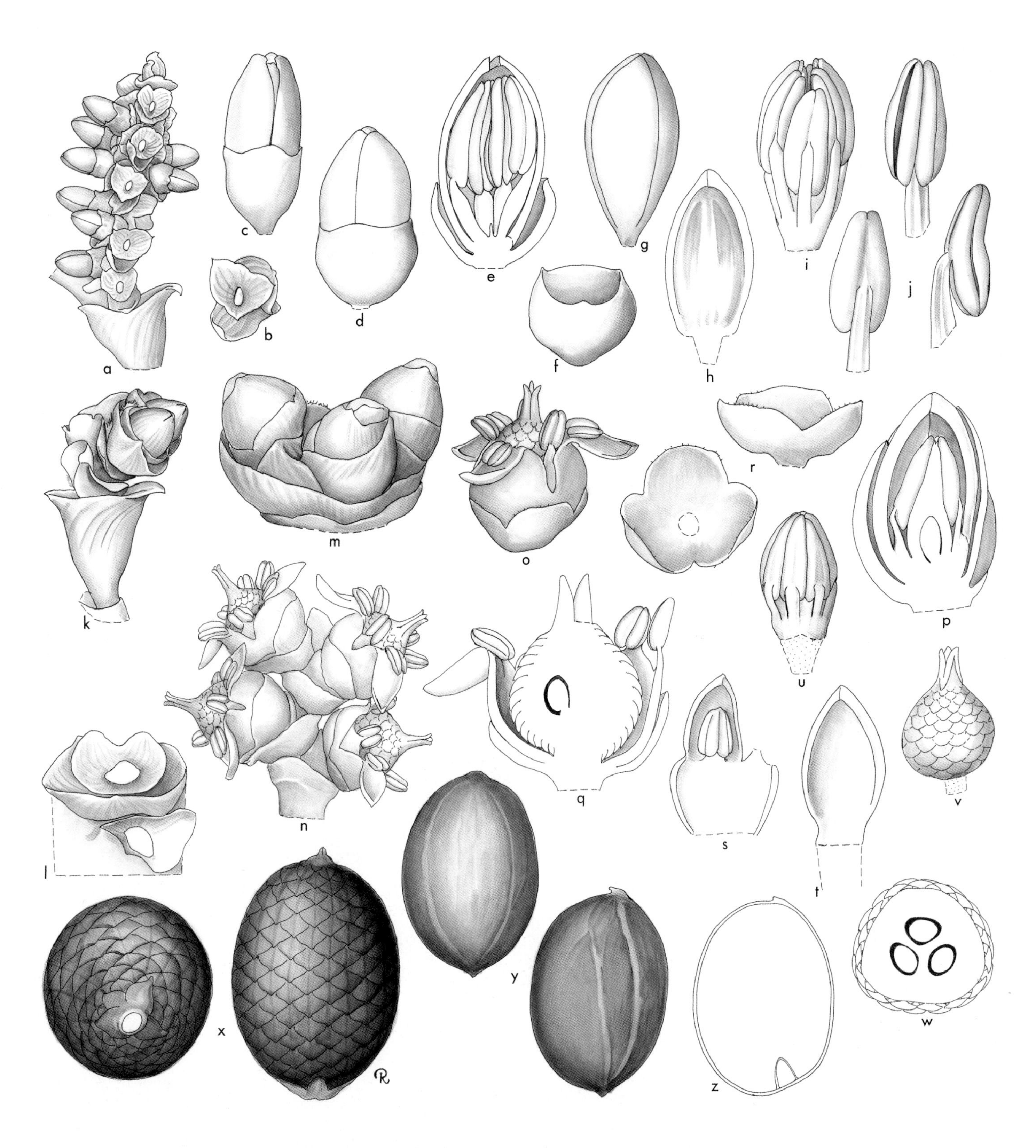

Mauritiella. **a**, portion of staminate rachilla × 3; **b**, bract and bracteole of staminate flower × 6; **c**, **d**, two staminate flowers × 7 $^1/_2$; **e**, staminate bud in vertical section × 9; **f**, staminate calyx × 9; **g**, staminate bud, calyx removed × 9; **h**, staminate petal, interior view × 9; **i**, androecium × 9; **j**, stamen in 3 views × 9; **k**, rachilla of pistillate inflorescence with prophyll × 3; **l**, portion of pistillate rachilla, flowers removed to show bracts and bracteoles × 3; **m**, portion of pistillate rachilla in bud × 3; **n**, portion of pistillate rachilla in flower × 2 $^1/_4$; **o**, pistillate flower × 3; **p**, pistillate bud in vertical section × 4 $^1/_2$; **q**, pistillate flower in vertical section × 4 $^1/_2$; **r**, pistillate calyx in 2 views × 3; **s**, pistillate petal with staminode, interior view × 4 $^1/_2$; **t**, pistillate petal, interior view × 7 $^1/_2$; **u**, staminodes × 4 $^1/_2$; **v**, gynoecium × 3; **w**, ovary in cross-section × 4 $^1/_2$; **x**, fruit in 2 views × 1 $^1/_2$; **y**, seed in 2 views × 1 $^1/_2$; **z**, seed in vertical section × 1 $^1/_2$. *Mauritiella armata*: ***a–w***, *Balick et al.* 913; **x–z**, *Moore et al.* 9537.
(Drawn by Marion Ruff Sheehan)

Central Brazil, representing 28,000 years, traces the demise and re-appearance of a *Mauritia* palm swamp (Ferraz-Vicentini & Salgado-Labouriau 1996). Rull (1998) provides an overview of biogeographical and evolutionary considerations for *Mauritia* that is based on the pollen evidence.

Notes: *Mauritia* and *Mauritiella* have been combined (Balick 1981) but there are consistent differences in flower clusters and habits. In *Mauritiella*, pistillate flowers are borne on short branches and staminate flowers are solitary, whereas in *Mauritia*, pistillate flowers are borne singly or on very short branches and staminate flowers are in pairs. Species of *Mauritia* are solitary with unarmed stems, and species of *Mauritiella* clustering with spiny stems.

Additional figures: Glossary fig. 10.

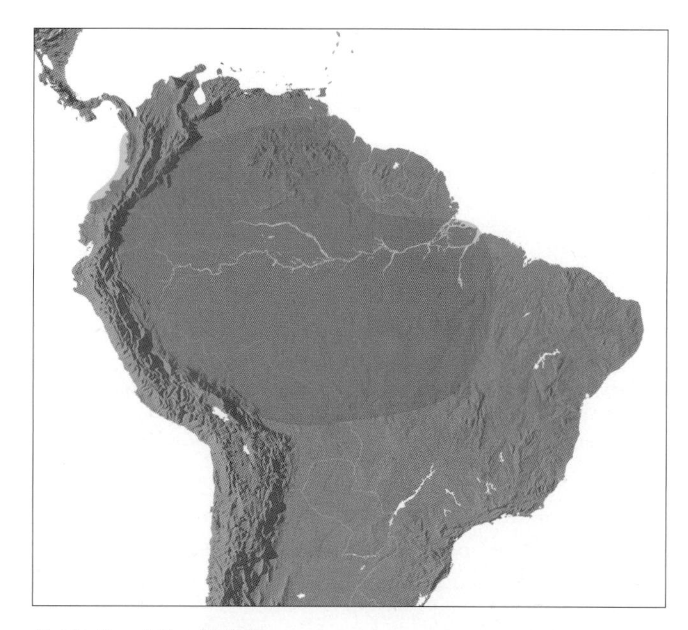

Distribution of ***Mauritiella***

8. MAURITIELLA

Moderate clustered palms of South America with erect stems armed with root spines; leaves palmate, with segments of equal width; inflorescences are robust, the staminate rachillae catkin-like, each rachilla bract subtending a single staminate flower.

Mauritiella Burret, Notizbl. Bot. Gart. Berlin-Dahlem 12: 609 (1935). Type: **M. aculeata** (Kunth) Burret (*Mauritia aculeata* Kunth).

Lepidococcus H. Wendl. & Drude ex A.D. Hawkes, Arq. Bot. Estado São Paulo ser. 2, 2: 173 (1952) (non *Lepidococca* Turcz. 1848). Type: *L. aculeatus* (Kunth) H. Wendl. & Drude ex A.D. Hawkes (*Mauritia aculeata* Kunth) (superfluous name).

Combining the generic name Mauritia *with the diminutive ending* — ella.

Moderate clustered, armed, pleonanthic, dioecious, tree palms. *Stem* erect, partly obscured by marcescent leaf sheaths above, becoming bare at the base, the internodes frequently bearing spine-like adventitious roots. *Leaves* moderate, reduplicate, briefly costapalmate; sheath splitting opposite the petiole; petiole conspicuous, adaxially channelled near the base, otherwise circular in cross-section, smooth, unarmed, frequently waxy; a hastula-like crest present adaxially at the base of the blade; blade ± orbicular in outline, divided along abaxial folds almost to the insertion into numerous crowded single-fold segments, very briefly bifid at their tips, adaxial surface glabrous, abaxial surface usually covered with white wax and short bifid scales, midribs prominent, transverse veinlets inconspicuous. *Inflorescences* solitary, interfoliar, the staminate and pistillate superficially similar; peduncle short, ± elliptical in cross-section; prophyll short, tubular, 2-keeled, with 2 short, striate, triangular lobes; peduncular bracts numerous, overlapping, distichous, striate, each with a triangular limb; rachis much longer than the peduncle; rachis bracts similar to the peduncular, each subtending a ± pendulous or spreading first-order branch; the first-order branch bearing a short 2-keeled, striate, tubular prophyll and 1–few empty distichous bracts; subsequent bracts tubular, short, ± flaring, each subtending a very short, straight or recurved rachilla; staminate rachilla catkin-like, bearing a basal, tubular, 2-keeled prophyll and crowded, ± rounded, spirally arranged rachilla bracts, connate shortly at the base, each subtending a single staminate flower bearing a tubular, 2-keeled bracteole; pistillate rachilla very short, ± catkin-like, bearing a basal, 2-keeled prophyll and spiral to subdistichous rachilla bracts, each subtending a solitary pistillate flower bearing a flattened 2-keeled bracteole. *Staminate* flower symmetrical; calyx tubular, briefly 3-lobed, often scaly; corolla tubular at the very base with 3 elongate, valvate, leathery lobes much exceeding the calyx; stamens 6, the filaments distinct, thick, ± angled, elongate, anthers elongate, basifixed, latrorse; pistillode minute. *Pollen* spheroidal; aperture monoporate; ectexine

Above: ***Mauritiella armata***, habit, Brazil. (Photo: A.J. Henderson)

intectate, surface very finely granular, interspersed with long, thin, slightly bottle-shaped spines set in, and loosely connected to cavities in a wide foot layer, distinctly separated into an upper typically solid layer and a slightly wider strongly lamellate inner layer bulging slightly beneath each spine, aperture margin similar; longest axis 40–55 µm [2/3]. *Pistillate* flowers larger than the staminate; calyx tubular, striate, briefly 3-lobed, often scaly; corolla tubular in the basal $^1/_3$–$^1/_2$ with 3 elongate, valvate lobes; staminodes 6, connate laterally by their flattened broad filaments and adnate to the corolla at the mouth of the tube; gynoecium trilocular, triovulate, ± rounded, covered in vertical rows of reflexed scales, style short, conical, stigmas 3, ovules anatropous, basally attached. *Fruit* ± rounded, usually 1-seeded, with apical stigmatic remains, perianth persistent; epicarp covered in many neat vertical rows of reddish-brown reflexed scales, mesocarp rather thick, fleshy, endocarp scarcely differentiated. *Seed* ± rounded to ellipsoidal, attached basally, apically with an elongate knob, and thin testa, endosperm homogeneous; embryo basal. *Germination* adjacent ligular; eophyll with a pair of divergent leaflets (?always). *Cytology*: 2n = 30.

Distribution and ecology: Three species in northern South America. Species of *Mauritiella* are predominantly lowland palms, often characteristic of the banks of black-water rivers.
Anatomy: Root (Seubert 1996a).
Relationships: The monophyly of *Mauritiella* has not been tested. For relationships, see *Lepidocaryum*.
Common names and uses: *Buriti*. Leaves are used for thatching and the fruit eaten.
Taxonomic account: Henderson 1995, Henderson *et al*. 1995.
Fossil record: No generic records found.
Notes: See under *Mauritia*. Fossil record not differentiated from *Mauritia*.
Additional figures: Fig 1.9, Glossary fig. 4.

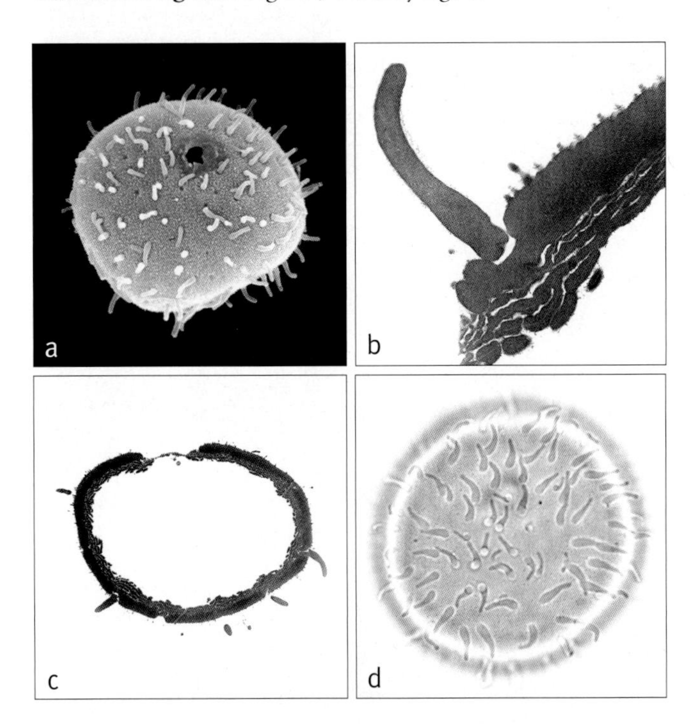

Above: ***Mauritiella***
a, monoporate pollen grain, distal face SEM × 750;
b, ultrathin section through pollen wall adjacent to pore TEM × 10,000;
c, ultrathin section, whole pollen grain, polar plane TEM × 750;
d, pollen grain, low-focus LM × 1000.
Mauritiella macroclada: **a** & **d**, *Dransfield et al.* 4866.

Tribe Calameae Kunth ex Lecoq. & Juillet, Dict. Rais. Term. Bot. 98 (1831). Type: **Calamus**.

Arborescent, acaulescent or climbing, mostly dioecious, hapaxanthic or pleonanthic palms; leaves pinnate, spiny, with spines regularly or subregularly organised into whorls or partial whorls; seed usually with sarcotesta; pollen usually diaperturate.

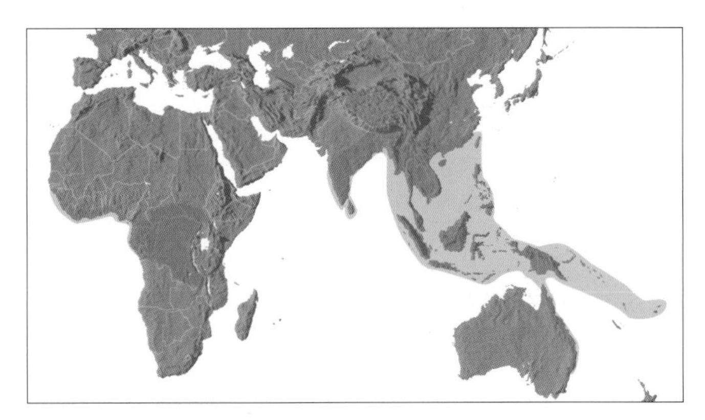

Distribution of tribe **Calameae**

The tribe is restricted to the Old World tropics and subtropics and is divided into six subtribes.

Tribe Calameae is moderately to highly supported as monophyletic (Baker *et al*. 2000a, 2000b, in review, in prep., Asmussen *et al*. 2006). The Calameae is resolved as sister to tribe Lepidocaryeae with moderate support in several studies (Baker *et al*. 2000a, 2000b, in prep., Asmussen *et al*. 2006).

Key to Subtribes of Calameae

1. Hermaphroditic or polygamous palms; stigmas pyramidal 2
1. Dioecious palms; stigmas trifid. 3
2. Climbing palms; flowers solitary; leaflets lanceolate to broadly diamond-shaped, praemorse . . . **Korthalsiinae**
2. Massive tree palms; leaflets linear to lanceolate, acute, entire . . . . **Metroxylinae**
3. Acaulescent palms, the leaves lacking cirri, flowers of both sexes borne in catkin-like rachillae . . . . **Salaccinae**
3. Acaulescent, erect or climbing palms, rachillae not usually catkin-like 4
4. Very tall, pleonanthic, tree palms . . . . **Pigafettinae**
4. Climbing palms, usually with climbing whips, more rarely erect or acaulescent 5
5. Hapaxanthic rattans; leaf sheaths lacking a knee-shaped swelling; pistillate rachillae bearing solitary flowers; staminate flowers solitary or paired . . . **Plectocomiinae**
5. Almost always pleonanthic rattans; leaf sheaths usually bearing a knee-shaped swelling below the insertion of the petiole; pistillate rachillae bearing dyads of one pistillate flower together with a sterile staminate flower, soon falling (very rarely 2 pistillate and 1 sterile staminate or more), or if sterile staminate lacking, then pleonanthic; staminate rachilla bracts subtending solitary fertile staminate flowers . . . . **Calaminae**

Subtribe Korthalsiinae Becc., Ann. Roy. Bot. Gard. (Calcutta) 12(2): 209 (1918). Type: **Korthalsia**.

Climbing, hermaphroditic, hapaxanthic palms; stems often aerially branching; leaves with cirri lacking acanthophylls; leaflet apices praemorse; inflorescence adnate to internode; seed lacking sarcotesta.

The subtribe consists of the single genus *Korthalsia*, a distinctive, isolated and easily recognised genus of rattans.

The relationships of four subtribes in tribe Calameae, Korthalsiinae, Salaccinae, Metroxylinae and Pigafettinae, remain highly ambiguous. Given the striking morphological similarity of the catkin-like rachillae borne by all species of these subtribes, it is perhaps surprising that they do not form a strongly supported monophyletic group. Two studies involving morphological characters place all four subtribes in a single clade (Baker *et al.* 1999b, 2000a), but with low support. No phylogenies based on DNA data alone recover this group. More often the subtribes are resolved as a paraphyletic group within the Calameae, with differing relationships among them. Of the various arrangements, it appears that the placement of Korthalsiinae as sister to all other Calameae, with Salaccinae as sister to all Calameae excluding Korthalsiinae, may be the relationships best supported by available evidence (Baker *et al.* 2000b, Asmussen *et al.* 2006). However, the numerous alternatives arising from different combinations of data and taxon sampling erode any confidence in these conclusions (Baker *et al.* 2000a, 2000b, in review, in prep.).

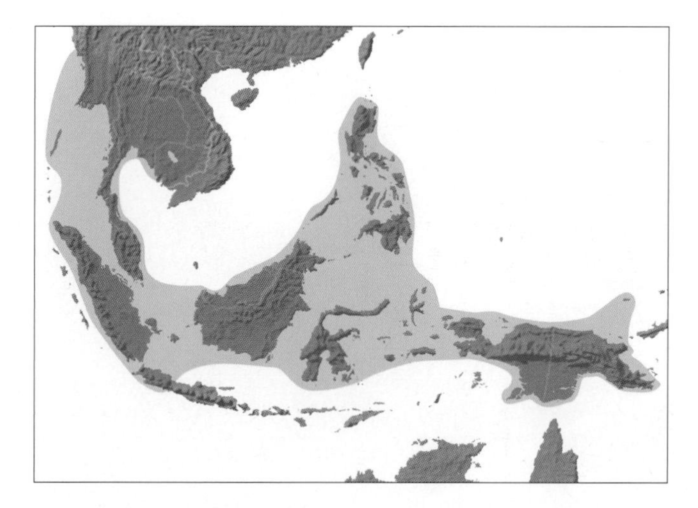

Distribution of ***Korthalsia***

Below: ***Korthalsia zippelii***, habit, Papua New Guinea. (Photo: W.J. Baker)

9. KORTHALSIA

Clustering high-climbing and aerially branching pinnate-leaved hapaxanthic rattan palms of Southeast Asia and Malesia; the sheaths end in an elaborated ocrea and leaflets are usually rhomboid and praemorse at their tips; flowers are hermaphroditic and borne singly in catkin-like rachillae.

Korthalsia Blume, Rumphia 2: 166 ([‘1836’] 1843). Lectotype: **K. rigida** Blume (see H.E. Moore 1963c).

Calamosagus Griff., Calcutta J. Nat. Hist. 5: 22 (1844). Lectotype: *C. laciniosus* Griff. (= *Korthalsia laciniosa* [Griff.] Mart.) (see H.E. Moore 1963c).

Commemorates Pieter Willem Korthals (1807–1892), Dutch botanist and explorer in Indonesia.

Slender to moderate, clustered, spiny, high-climbing and aerially branching, hapaxanthic, hermaphroditic rattan palms. *Stem* eventually becoming partly bare, the inner epidermis of the leaf sheaths tending to adhere to the stem surface, internodes elongate, nodal scars often very uneven, shallowly hollowed, aerial branching possibly due to equal forking (dichotomy), basal suckering leaf opposed at 130° from the petiole in at least one species. *Leaves* pinnate, with a cirrus; sheath tubular, sometimes splitting longitudinally opposite the petiole, unarmed, or variously armed with spines, usually with abundant scales and floccose indumentum; knee absent; ocrea always well developed, unarmed or variously spiny, tightly sheathing, or expanded into a loose funnel-shaped net of fibres, or sheathing but distally grossly swollen to form an ant nest-chamber, or diverging from the stem with inrolled margins, also forming an ant chamber; petiole present or absent; rachis and cirrus armed with scattered and grouped, reflexed grapnel spines; leaflets relatively few, single-fold, linear, lanceolate to rhomboid, praemorse, often

densely white indumentose beneath, regularly arranged, distant, very rarely a single pair only, frequently borne on short pseudopetiolules ('ansae'), midrib inconspicuous, the main veins radiating from the leaflet base, transverse veinlets conspicuous or obscure. *Inflorescences* produced simultaneously in the axils of the most distal few, frequently reduced leaves; sometimes bursting through the leaf sheaths, rarely unbranched, usually branching to 1–2 orders; peduncle adnate to the internode above the subtending leaf; prophyll 2-keeled, tightly sheathing, usually included within the leaf sheath, sometimes subtending a branch; rachis much longer than the peduncle; rachis bracts tubular, tightly sheathing, sparsely armed or unarmed, frequently densely covered with indumentum; bracts on first-order branches similar to rachis bracts; rachillae usually distant, rarely aggregated into a head, cylindrical, and catkin-like, bearing a few empty basal bracts and a tight spiral of imbricate bracts, connate laterally to each other, or, more rarely, distinct, the rachillae then with a looser appearance, each rachilla bract forming a pit, usually densely filled with multicellular hairs, and including a 2-keeled membranous bracteole, densely hairy on the abaxial surface, a minute triangular bracteole, and a single flower. *Flowers* apparently protandrous; calyx tubular at the base, with 3 valvate lobes distally; corolla tubular basally with 3 valvate lobes apically, circumscissile in fruit at the level of the ovary equator, carried up on the top of the developing fruit, disintegrating or persisting to mature fruiting; stamens 6–9, borne at the mouth of the corolla tube, filaments fleshy, elongate, anthers short to elongate, introrse or latrorse; gynoecium tricarpellate, triovulate, rounded, scaly, style conical or narrow pyramidal with 3 stigmatic lines, ovule anatropous, basally attached. *Pollen* spheroidal; apertures either equatorial di-porate or presumed meridional zonasulcate; ectexine intectate, psilate clavae or gemmae, striate and/or spinulose gemmae, or granular and interspersed with long pointed or apically branched spines, occasionally vertically ridged, aperture margins similar to surrounding ectexine; longest axis 25–60 µm [12/26]. *Fruit* globose to ovoid, 1-seeded, stigmatic remains apical; epicarp covered with vertical rows of reflexed, imbricate scales, mesocarp thinly fleshy, sweet, endocarp not differentiated. *Seed* attached basally, seed coat thin, not fleshy, endosperm homogeneous or ruminate, with a conspicuous pit; embryo lateral. *Germination* adjacent-ligular; eophyll undivided or bifid, margins praemorse. *Cytology*: 2n = 32.

Top: ***Korthalsia echinometra***, swollen ocreas with ants, Brunei Darussalam. (Photo: J. Dransfield)

Middle: ***Korthalsia*** **laciniosa**, flower buds, Peninsular Malaysia. (Photo: J. Dransfield)

Bottom left: ***Korthalsia*** **laciniosa**, flowers at anthesis, Peninsular Malaysia. (Photo: J. Dransfield)

Bottom right: ***Korthalsia*** **echinometra**, fruit, Brunei Darussalam. (Photo: J. Dransfield)

Below: ***Korthalsia***
a, one face of diporate pollen grain SEM × 1000;
b, zonasulcate pollen grain SEM × 1500.
Korthalsia merrillii: **a**, *Fernando* 638; *K. furtadoana*: **b**, *Matusop* 7427.

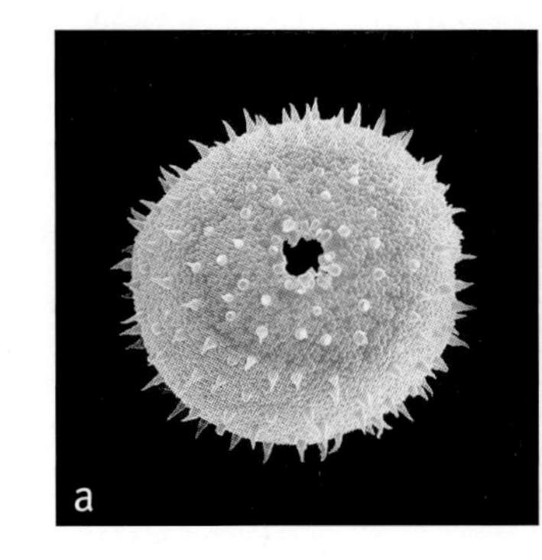

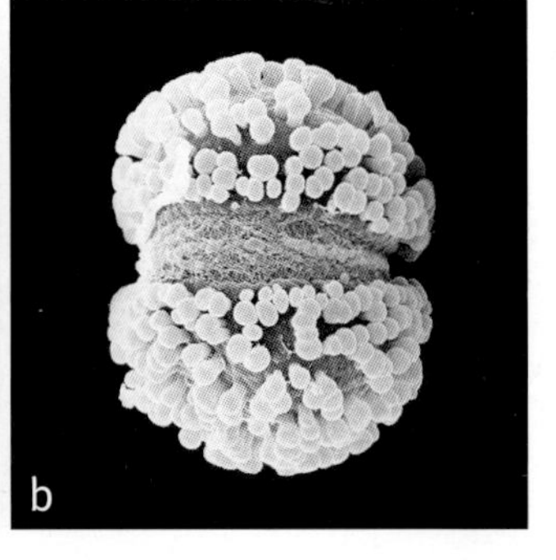

Distribution and ecology: About 26 species, centred on the perhumid areas of the Sunda Shelf with outliers north as far as Indochina, Burma, and the Andaman Inlands, and southeastward to Celebes and New Guinea. All species are confined to lowland and hill tropical rain forest and are absent in montane forest. Most species have, however, a very wide ecological range, and although abundant in primary forest, also seem to be peculiarly well adapted to withstanding forest disturbance: they are a conspicuous feature of old secondary forest or regenerated logged forest. It has been suggested that the hapaxanthic habit may be an adaptation to colonizing secondary habitats (Dransfield 1978b). A few species have very narrow ecological limits, e.g., *Korthalsia concolor*, which seems to be confined to forest on ultrabasic rock in Sabah, Borneo. Several species in which the ocrea is inflated or divergent from the stem have very close associations with ants. The ants husband aphids on young tissue within the ocreas of the distal portion of the stem, using the older dry ocreas as brood chambers; in some species (e.g., *K. robusta* and *K. hispida*), the ants produce alarm signals by banging their mandibles on the dry ocreas. The significance of the ant–rattan relationship has not been fully investigated, but there is much to suggest that the relationship provides protection of the rattan against herbivores (Dransfield 1981a). Bees have been observed visiting the flowers of *K. laciniosa*, and Southern pied hornbills (*Anthracoceros convexus*) feed on the ripe fruit of the same species (Rubeli in Dransfield 1981a).

Anatomy: Leaf (Tomlinson 1961), root (Seubert 1996a).

Relationships: The monophyly of *Korthalsia* has not been tested. For relationships, see *Korthalsiinae*.

Common names and uses: Ant rattans; for Malay names see Dransfield (1979a). Species of *Korthalsia* produce very hard durable canes much used in local basketware and for binding in house construction. The cane, however, is disfigured by large irregular nodal scars and the inner epidermis of the sheaths closely adheres to the cane surface; these two cane features are responsible for the limited importance of *Korthalsia* in the rattan trade.

Taxonomic account: Dransfield (1981a).

Fossil record: Dicolpate clavate pollen apparently referable to present-day *Korthalsia*, *K. rigida* or *K. laciniosa* (Muller 1979), has been recorded from the Upper Miocene of northwestern

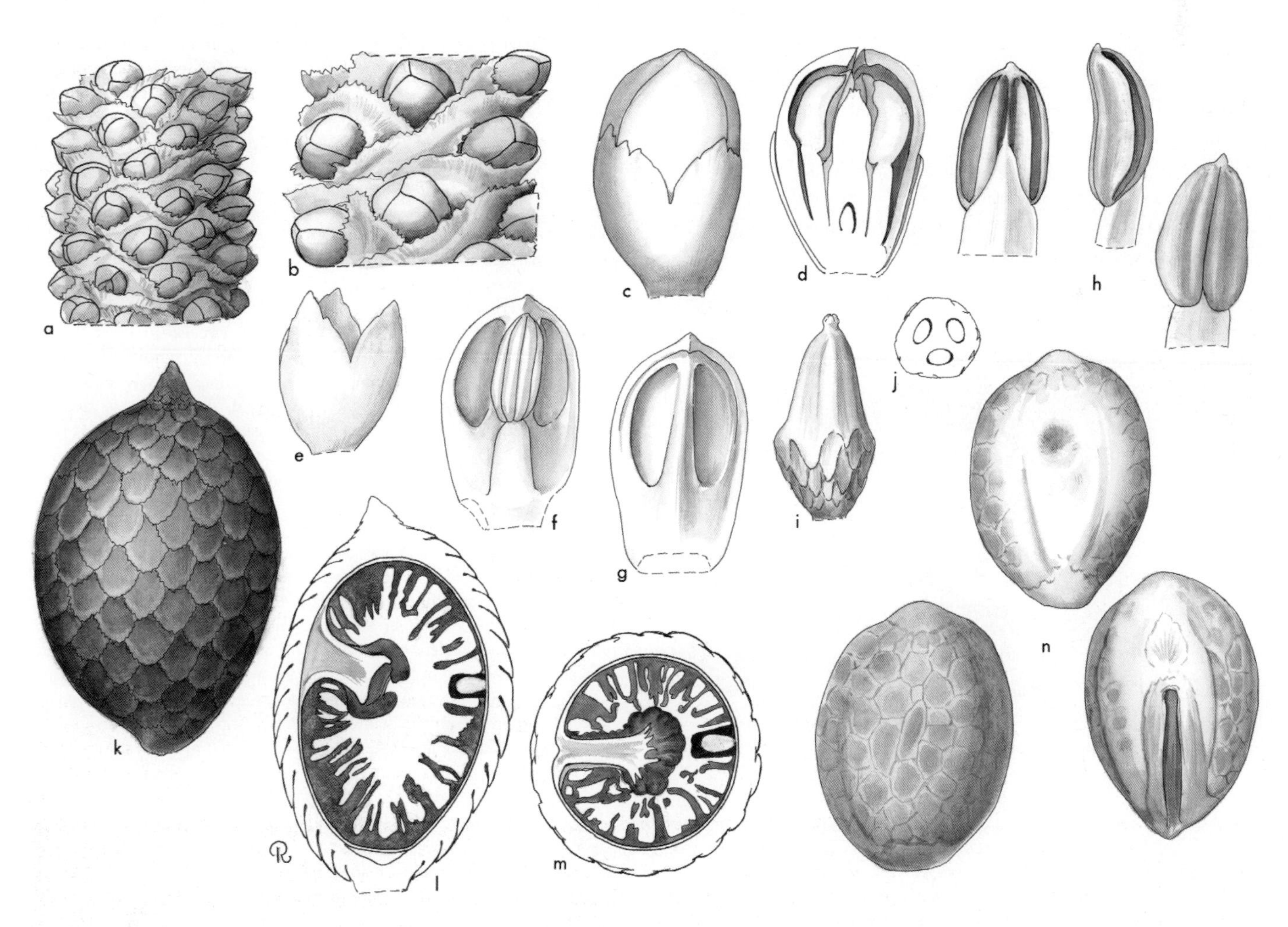

***Korthalsia*.** **a**, portion of rachilla × 3; **b**, portion of rachilla × 6; **c**, bud × 9; **d**, bud in vertical section × 9; **e**, calyx × 9; **f**, petal with stamen × 9; **g**, petal × 9; **h**, stamen in 3 views × 15; **i**, gynoecium × 9; **j**, ovary in cross-section × 9; **k**, fruit × 2 1/4; **l**, fruit in vertical section × 2 1/4; **m**, fruit in cross-section × 2 1/4; **n**, seed in 3 views × 2 1/4. *Korthalsia zippelii*: **a–j**, *Zeick* 36204; *K. echinometra*: **k–n**, *Corner* 30467. (Drawn by Marion Ruff Sheehan)

Borneo. However, Thanikaimoni (1970) described the pollen of both of these species as diporate and more or less spheroidal (see also Harley & Baker 2001). Nevertheless, it is highly probable that the fossil pollen does represent a species of *Korthalsia*, but this cannot be confirmed without access to the fossil grain or images. The small clavate zonasulcate *Paravuripollis* first described by Rao and Ramanujam (1978) from the Miocene Quilon beds of Kerala State, India, closely resembles the pollen of some of species of *Korthalsia. Paravuripollis* is well known from a number of Tertiary sites in Kerala State (Ramanujam 1987, Ramanujam *et al.* 1991b, 1992, Ramanujam & Rao 1977, Rao & Ramanujam 1978).

Notes: *Korthalsia* is distinguished by the solitary hermaphroditic flowers borne in catkin-like rachillae and by hapaxanthy.

Additional figures: Figs 1.1, 1.7, Ch. 2 frontispiece, Glossary figs 2, 6, 19.

Subtribe Salaccinae Becc., Ann. Roy. Bot. Gard. (Calcutta) 12(2): 207 (1918). Type: **Salacca**.

Acaulescent, dioecious, hapaxanthic or pleonanthic palms; inflorescence with prophyll splitting abaxially or adaxially.

This subtribe consists of two Southeast Asian and West Malesian genera, *Eleiodoxa* and *Salacca*. In the first edition of *Genera Palmarum*, these two genera were included in Calaminae, but recent phylogenetic analyses (Baker *et al.* 2000a) provide consistent support for the separation of Salaccinae.

Subtribe Salaccinae is monophyletic with moderate to high support (Baker *et al.* 1999b, 2000a, 2000b, in review). For relationships, see Korthalsiinae.

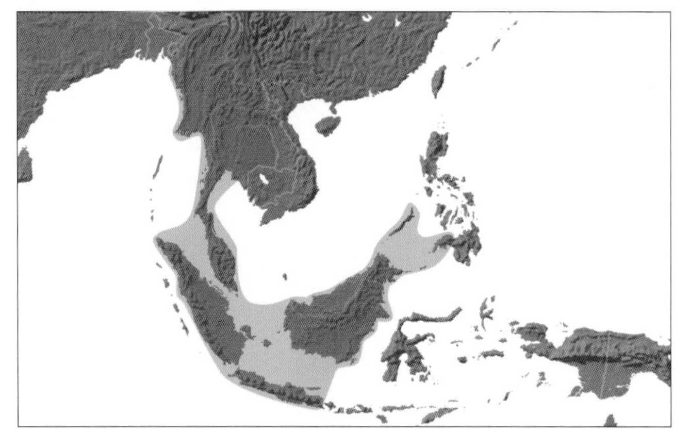

Distribution of subtribe **Salaccinae**

Key to Genera of Salaccinae

1. Pleonanthic palms (but see *Salacca secunda* and *Salacca sp. nov.* Henderson); inflorescences short, in bud enclosed within the subtending leaf sheath, at anthesis emerging from a vertical groove on the abaxial surface of the sheath; terminal leaflets compound, or the whole leaf flabellate; sarcotesta easily separable from rest of seed. Southeast Asia to West Malesia . . . . **11. Salacca**
1. Hapaxanthic palm; inflorescence not enclosed in a vertical groove in the surface of the sheath; terminal leaflets not compound; sarcotesta adhering to rest of seed. Thailand, Malaya, Sumatra and Borneo . . . . **10. Eleiodoxa**

10. ELEIODOXA

Acaulescent, clustering palm forming dense thickets, in swamps in Southeast Asia and West Malesia; the sheaths and petiole are densely armed with long spines; flowering is hapaxanthic; the fruit has a sarcotesta that is difficult to separate from the rest of the seed.

Eleiodoxa (Becc.) Burret, Notizbl. Bot. Gart. Berlin-Dahlem 15: 733 (1942). Lectotype: **E. conferta** (Griff.) Burret (*Salacca conferta* Griff.) (see H.E. Moore 1963c).

Salacca section *Eleiodoxa* Becc., Ann. Roy. Bot. Gard. (Calcutta) 12(2): 71 (1918).

Eleio — *swamp,* doxa — *glory, referring to the habitat.*

Moderate, acaulescent, clustering, armed, hapaxanthic, dioecious palm. *Stem* subterranean with short internodes, bearing strictly axillary sucker shoots. *Leaves* robust, pinnate, marcescent; sheath splitting opposite the petiole, unarmed at the extreme base, otherwise armed with neat partial whorls of robust spines and abundant caducous scales, the sheath mouth bearing a tattering ligule-like structure; petiole well developed, channelled adaxially in proximal part, rounded abaxially, circular in cross-section distally, armed with neat, somewhat oblique, partial whorls of slender, rigid spines; rachis armed as the petiole, but more sparsely so; leaflets single-fold, linear-lanceolate, regularly arranged, the apical pair slender, very rarely partly united to the penultimate pair, the margins armed with short spines, surfaces similar in colour, transverse veinlets distinct. *Inflorescences* aggregated into a terminal compound inflorescence, held erect at ground level between the leaf bases, the staminate and pistillate superficially similar; first-order branches of the compound inflorescence (i.e., the axillary inflorescences) each subtended by a highly reduced leaf or tattering bract, and bearing an empty, short, tubular, 2-keeled prophyll, quickly tattering, and short, tubular, tattering bracts with triangular limbs, each subtending a robust, erect, cylindrical, catkin-like rachilla; rachilla bearing a basal, 2-keeled, tubular prophyll and a few empty bracts at the base and at the very tip, otherwise bearing a tight spiral of imbricate, laterally adnate, low triangular-tipped bracts, each enclosing a dyad of flowers, comprising in the staminate inflorescence, 2 staminate flowers, and in the pistillate, 1 sterile staminate and 1 fertile pistillate flower, each

Opposite: ***Eleiodoxa.*** **a**, portion of staminate rachilla × 3; **b**, dyad of staminate flowers × 4 1/2; **c**, dyad of staminate flowers, one flower removed to show bracteole and hairs × 4 1/2; **d**, staminate bud × 7 1/2; **e**, staminate bud in vertical section × 7 1/2; **f**, staminate calyx × 7 1/2; **g**, staminate bud, calyx removed × 7 1/2; **h**, staminate petal × 7 1/2; **i**, androecium × 7 1/2; **j**, stamen in 3 views × 7 1/2; **k**, portion of pistillate rachilla × 2 1/4; **l**, dyad of a pistillate and sterile staminate flower × 4 1/2; **m**, dyad of a pistillate and sterile staminate flower, flowers removed to show bracteoles and hairs × 4 1/2; **n**, sterile staminate flower × 6; **o**, sterile staminate flower in vertical section × 6; **p**, sterile staminate petal and staminodes × 6; **q**, pistillate flower × 6; **r**, pistillate flower in vertical section × 6; **s**, pistillate calyx × 6; **t**, pistillate flower, calyx removed × 6; **u**, pistillate petal and staminodes, interior view × 6; **v**, gynoecium × 6; **w**, ovary in cross-section × 7 1/2; **x**, fruit in 3 views × 1; **y**, fruit in vertical section × 1; **z**, sarcotesta × 1; **aa**, seed in 2 views × 1; **bb**, seed in 2 views × 1 1/2; **cc**, seed in vertical section × 1. *Eleiodoxa conferta*: **a–j**, *Dransfield* 757; **k–w**, *Dransfield s.n.*; **x–cc**, *Dransfield* 724. (Drawn by Marion Ruff Sheehan)

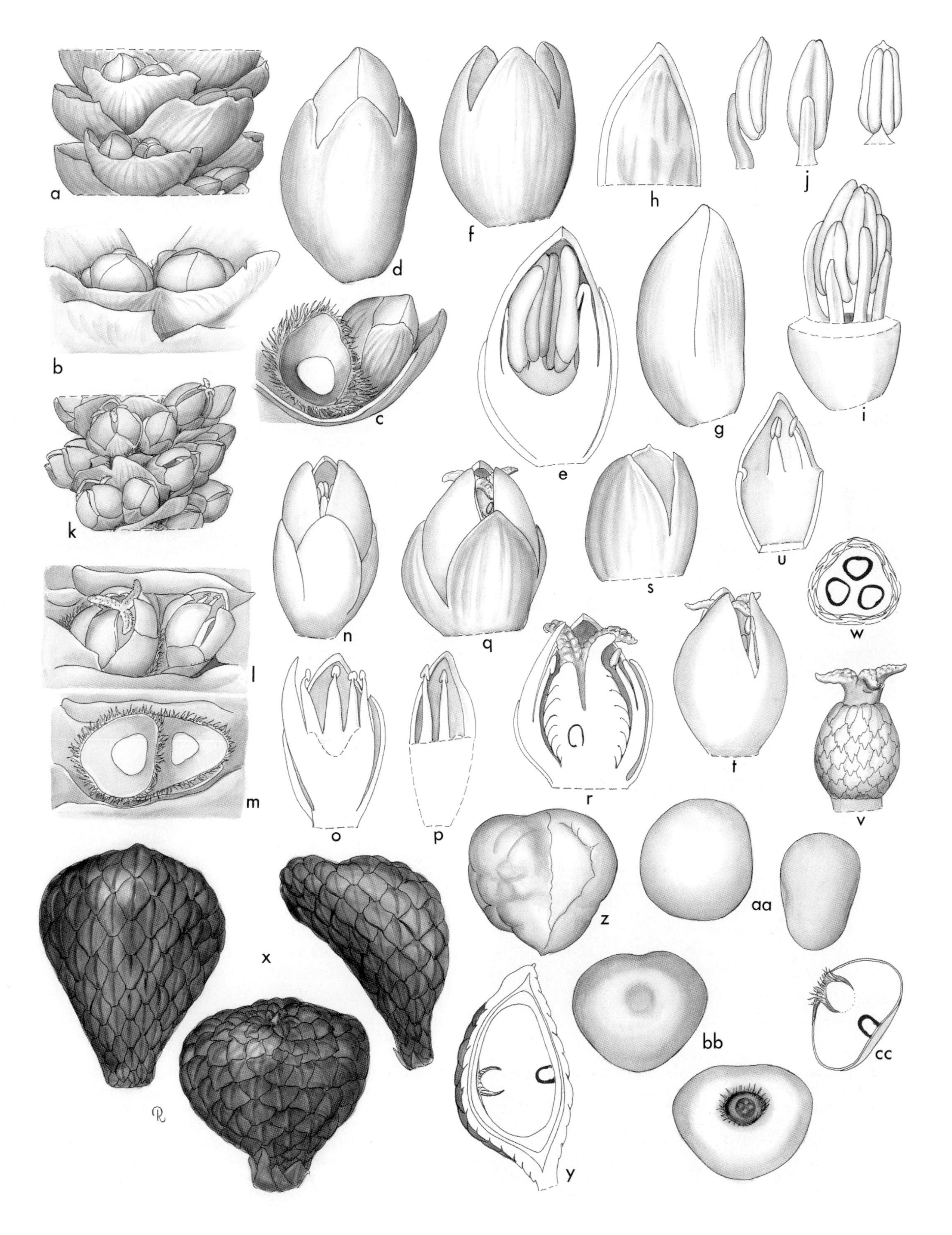
a
b
c
d
e
f
g
h
i
j
k
l
m
n
o
p
q
r
s
t
u
v
w
x
y
z
aa
bb
cc

flower bearing a prophyllar bracteole and surrounded by a dense pile of hairs. *Staminate flowers* pinkish-tinged at anthesis; calyx cupular, striate, with 3 triangular lobes; corolla tubular at the base, split to about $^4/_5$ its length into 3 triangular, valvate petals; stamens 6, borne at the mouth of the corolla tube, filaments fleshy, elongate, abruptly contracted and inflexed at the tip, anthers elongate, introrse; pistillode absent. *Pollen* ellipsoidal, bi-symmetric; apertures subequatorial diporate; ectexine tectate, coarsely perforate, aperture margins similar; infratectum columellate; longest axis 24–27 µm [1/1]. *Sterile staminate flowers* like the fertile but with fleshier filaments, not abruptly contracted, and with empty anthers. *Pistillate flowers* superficially similar to the staminate but larger, the corolla tubular in the basal ca. $^1/_3$; staminodes 6, borne at the mouth of the corolla tube, filaments closely appressed to the corolla, empty anthers somewhat sagittate; gynoecium tricarpellate, triovulate, globose, covered in reflexed scales, stigmas 3, reflexed, sinuous, in bud compressed into a pyramid, locules incomplete, ovules basally attached, anatropous. *Fruit* almost always 1-seeded, stylar remains apical; epicarp covered in neat vertical rows of reflexed scales, mesocarp somewhat spongy, endocarp not differentiated. *Seed* ± rounded, sarcotesta thick, sour, closely adhering to the inner integument and difficult to separate from it due to the presence of short radiating fibres, endosperm homogeneous, ± disc-like or very broadly oblate, with large, wide pit at the apex; embryo basal or lateral due to distortion of the fruit. *Germination* adjacent-ligular; eophyll bifid with narrow, entire lobes. *Cytology* not studied.

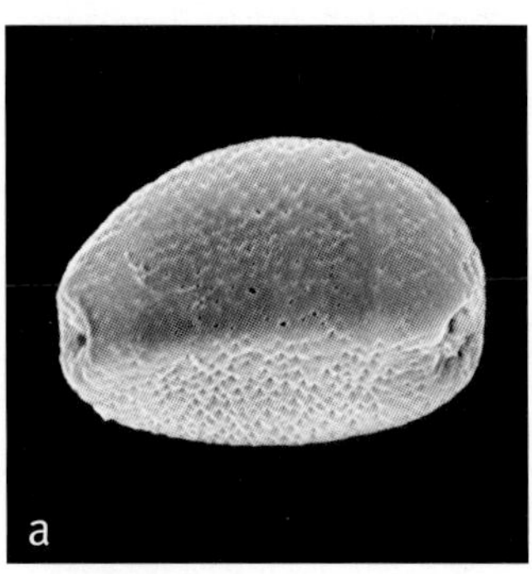

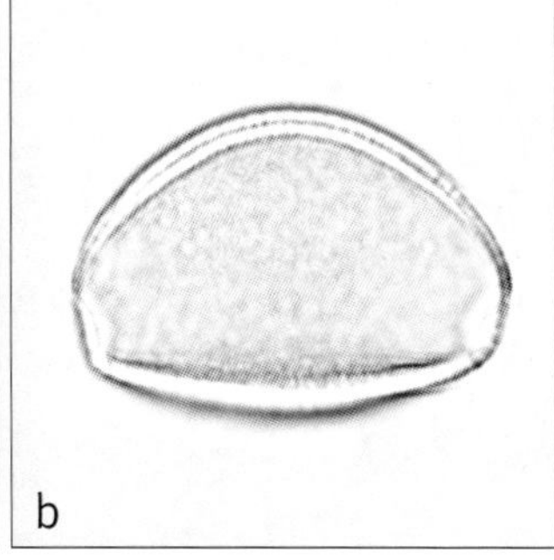

Above: ***Eleiodoxa***
a, diporate pollen grain, proximal face SEM × 2000;
b, diporate pollen grain, equatorial view, mid-focus LM × 2000.
Eleiodoxa conferta: **a** & **b** *Whitmore et al.* TCW 3104.

Below: ***Eleiodoxa conferta***, infructescence, Brunei Darussalam.
(Photo: J. Dransfield)

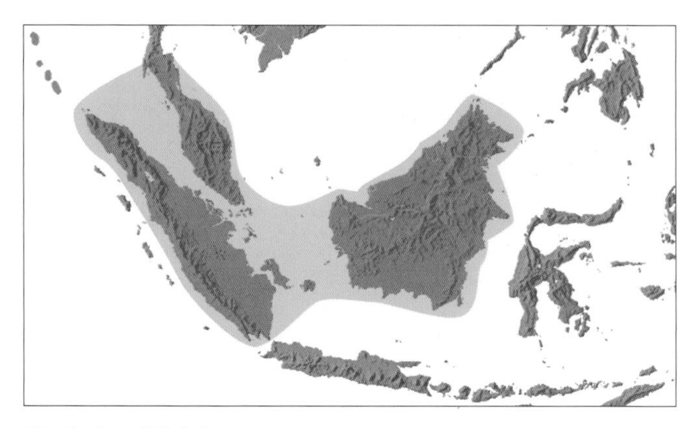

Distribution of ***Eleiodoxa***

Distribution and ecology: Although five names have been published, there appears to be only one widespread species, known from South Thailand, Sumatra, Malay Peninsula, and Borneo. *Eleiodoxa conferta* is a highly characteristic, gregarious, undergrowth palm of lowland fresh water swamps, being particularly abundant in facies of peat swamp forest where a certain amount of water movement occurs.
Anatomy: Not studied.
Relationships: *Eleiodoxa* is strongly supported as sister to *Salacca* (Baker *et al.* 2000a, 2000b).
Common names and uses: *Kelubi* or *asam paya*. Leaves are occasionally used for temporary thatching. The extremely sour sarcotesta is used throughout the range of the palm as a substitute for tamarind in cooking and with sugar is sometimes made into a sweetmeat.
Taxonomic accounts: Beccari (1918), Burret (1942).
Fossil record: No generic records found.
Notes: See *Salacca*.
Additional figures: Fig. 1.1.

11. SALACCA

Acaulescent, clustering palms, often forming dense thickets, in Southeast Asia and West Malesia; sheaths and petiole are densely armed with long spines; flowering is pleonanthic and the inflorescence emerges from a slit in the back of the leaf that subtends it; the sarcotesta is easily separated from the rest of the seed.

Salacca Reinw., Syll. Ratisb. 2: 3 (1826 ['1828']). Type: *S. edulis* Reinw. = **Salacca zalacca** (Gaertn.) Voss ex Vilmorin (*Calamus zalacca* Gaertner) '*Zalacca*'.

Lophospatha Burret, Notizbl. Bot. Gart. Berlin-Dahlem 15: 752 (1942). Type: *L. borneensis* Burret (= *Salacca lophospatha* J. Dransf. & Mogea).

Latinisation of the Malay vernacular name, salak.

Minute to rather robust, usually acaulescent, clustered, spiny, pleonanthic, dioecious palms. *Stem* subterranean, decumbent or very short and erect, usually obscured by the leaf bases, the internodes short, often with abundant adventitious roots, rarely with short stilt roots, sucker buds leaf opposed at angles of 90–180° from the subtending leaf, very rarely developing as whip-like shoots rooting at their tips. *Leaves* very small to robust, pinnate, or entire, bifid, with pinnate venation, marcescent; sheath splitting opposite the petiole, unarmed at the extreme base, otherwise very sparsely to very densely armed with robust, scattered or whorled spines and usually with abundant caducous or persistent scales, sheath mouth frequently bearing a tattered ligule-like structure; petiole channelled adaxially near the base, rounded distally and abaxially, variously armed with spines and indumentum, often fiercely so; rachis armed as the petiole, but more sparsely; leaflets single-fold, where pinnate, except for the terminal pair, linear or sigmoid, acuminate or very rarely deeply lobed at the tip, regularly arranged or grouped and fanned within the groups, variously armed with short bristles along the main veins and margins, terminal pair compound, joined along the midline, deeply lobed at the tip, where leaf entire, bifid, the apical margins deeply lobed or almost entire, abaxial blade surface often with a dense covering of powdery indumentum, midribs prominent adaxially, transverse veinlets usually conspicuous. *Inflorescences* axillary but enclosed within the sheath of the subtending leaf, and emerging through a slit along the midline of the abaxial surface of the sheath, inflorescences usually short, sometimes spicate, more often with 1 or 2 orders of crowded or spreading branches, occasionally hidden by detritus, sometimes arching out of the crown, very rarely whip-like with the tip metamorphosing into a vegetative axis, rooting and becoming established as an independent plant, staminate inflorescences usually branched to at least 1 more order than the pistillate; peduncle usually short; prophyll usually rather inconspicuous, partly enclosed within the leaf sheath slit, tubular, 2-keeled, irregularly tattering; peduncular bracts several, tubular at the base with irregularly tattering, frequently densely scaly limbs; rachis usually longer than the peduncle; rachis bracts like the peduncular bracts; rachillae cylindrical, catkin-like, exposed or hidden by the bracts, bearing a tight spiral of imbricate, triangular or low, rounded bracts, sometimes connate laterally to form a continuous spiral, sometimes very small and scarcely imbricate, each, except for the proximal and most distal few, subtending flowers and sometimes also a dense pile of hairs. *Staminate flowers* borne in dyads with 2 small prophyllar bracteoles, these sometimes split and variously connate to each other, the flowers exserted from the pit at anthesis; calyx tubular, variously split to give 3 lobes, sometimes distinct almost to the base, chaffy, striate; corolla with a short stalk-like base, and a long proximal tube, bearing 3 triangular, ± hooded, valvate lobes; stamens 6, borne at the mouth of the corolla tube, filaments short, wide basally, anthers rounded to elongate, introrse; pistillode minute or absent. *Pollen* spheroidal or oblate-spheroidal; aperture a presumed meridional zonasulcus, occasionally incompletely so, rarely equatorial disulcate; ectexine tectate and spinose, rarely spinulose, or psilate and sparsely perforate or, semitectate and spinulose, or gemmate, aperture margins similar to surrounding ectexine; infratectum columellate; longest axis 22–34 μm; post-meiotic tetrads tetragonal [16/20]. *Pistillate flowers* either solitary (Section *Leiosalacca*) or borne in a dyad with a sterile staminate flower (Section *Salacca*) similar to the fertile but with empty anthers; calyx of pistillate flower tubular at the base, distally with 3, triangular, striate lobes; corolla similar with 3, triangular, valvate lobes; staminodes 6, borne at the mouth of the corolla tube, the filaments usually elongate, anthers ± sagittate, empty; gynoecium tricarpellate, triovulate, covered in flattened, smooth, or erect spine-tipped scales, stigmas 3, fleshy, reflexed at anthesis, locules incomplete, ovules basifixed, anatropous. *Fruit* (1–)(2–)3-seeded, globose to pear-shaped or ellipsoidal, with apical stigmatic remains; epicarp covered in somewhat irregular vertical rows of reflexed scales, the scale tips smooth (section *Leiosalacca*) or spine-like and upward pointing (Section *Salacca*), mesocarp very thin at maturity, endocarp not differentiated. *Seeds* basally attached, conforming to $^1/_3$ or $^1/_2$ of a sphere (depending on number reaching maturity), sarcotesta very thick, sour or sweet, inner seed coat very thin, endosperm homogeneous, the apex with a pit; embryo basal. *Germination* adjacent-ligular; eophyll entire bifid. *Cytology*: 2n = 28.

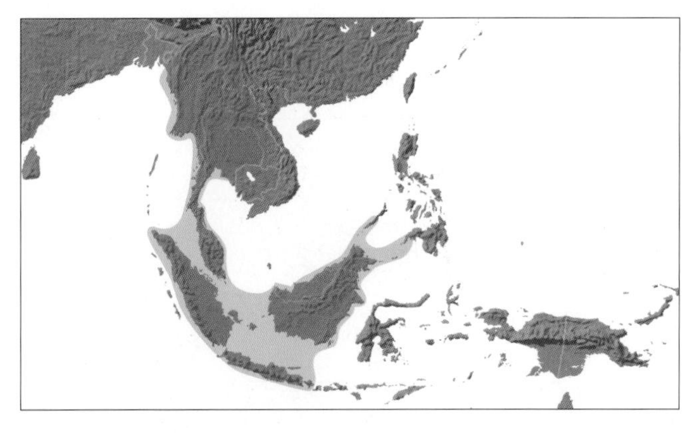

Distribution of ***Salacca***

Distribution and ecology: About 20 recognised species, but several more remain to be described; distributed from Burma and Indochina, south and eastwards to Borneo, Java, and the Philippines. *Salacca zalacca*, wild in Java and Sumatra, has been introduced into the Malay Peninsula, Borneo, Celebes, the Moluccas, and Bali for its excellent fruit. The greatest number of species and morphological diversity is found in the Malay Peninsula and Borneo. *Salacca* species are plants of the undergrowth of primary tropical rain forest. However, they may be left after the surrounding forest has been destroyed to persist in the open, because of their usefulness as a source of fruit and building materials. Many species favour swampy valley bottoms where they form rather dense spiny thickets. Other species may be found on hillslopes or ridgetops. *S. rupicola* grows in the crevices of limestone cliffs and in the dwarf forest on limestone hilltops.

Right: ***Salacca ramosiana***, habit, Sabah. (Photo: J. Dransfield)

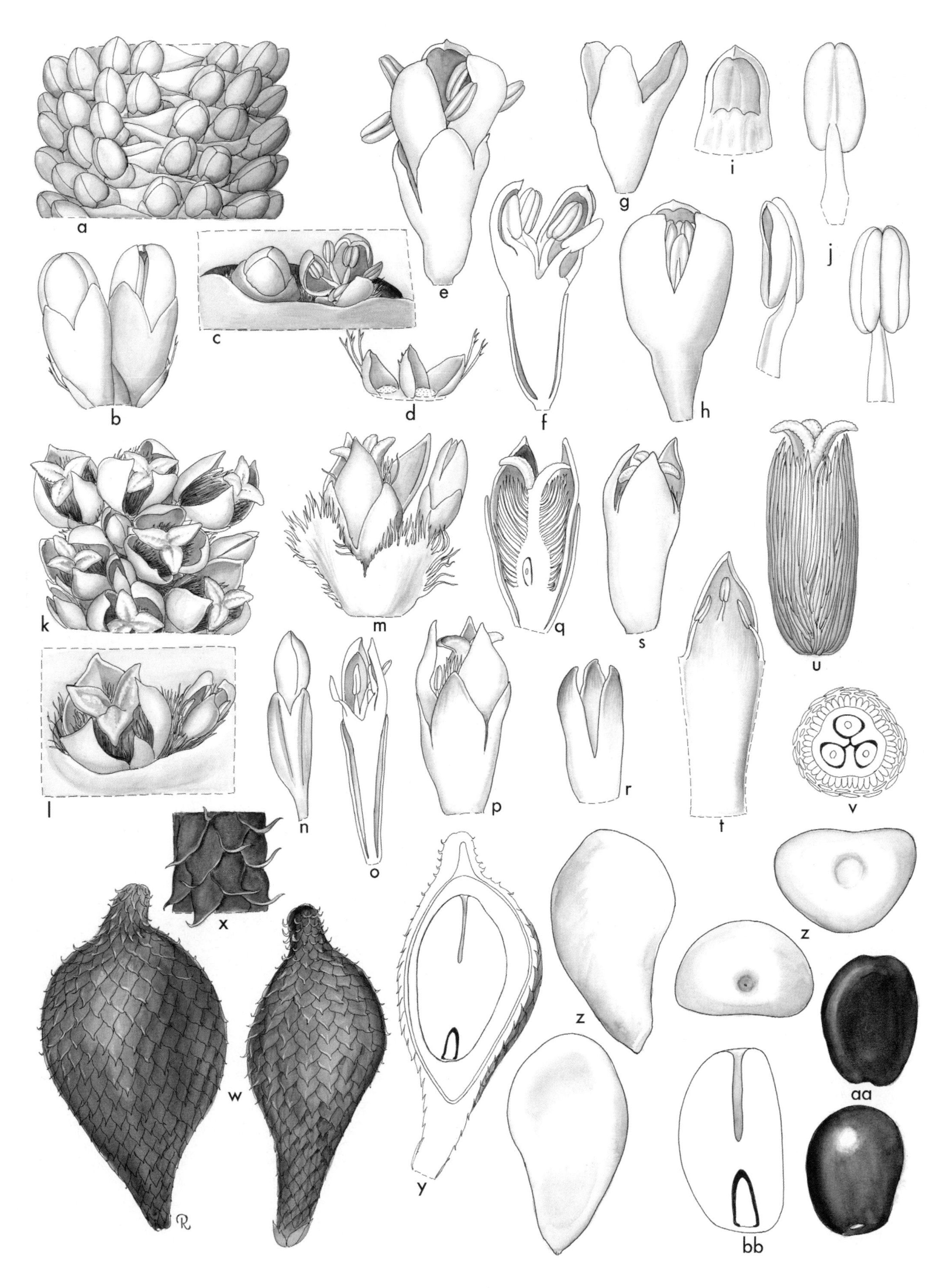
a
b
c
d
e
f
g
h
i
j
k
l
m
n
o
p
q
r
s
t
u
v
w
x
y
z
z
aa
bb

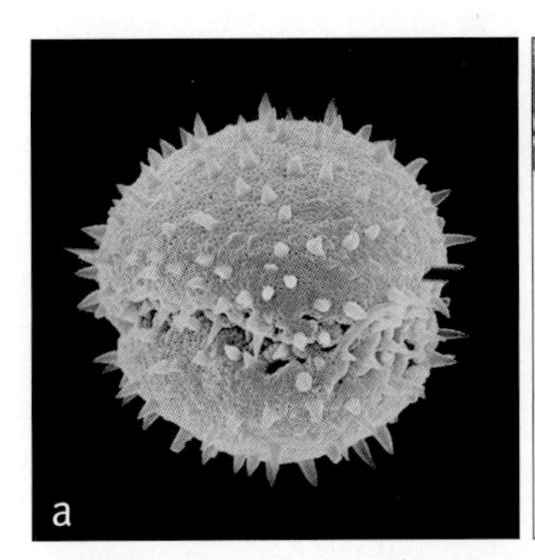

Left: ***Salacca***
a, zonasulcate pollen grain SEM × 1500;
b. ultrathin section through pollen wall TEM × 8000.
Salacca clemensiana: **a**, *Clemens* 26380; *S. wallichiana*: **b**, *Furtado* 33024.

Below left: ***Salacca clemensiana***, pistillate rachillae at anthesis, Sabah. (Photo: J. Dransfield)

Below right: ***Salacca glabrescens***, young staminate rachillae, Peninsular Malaysia. (Photo: J. Dransfield)

Bottom: ***Salacca* sp.**, fruit, Peninsular Malaysia. (Photo: J. Dransfield)

Anatomy: Leaf, petiole, stem, root (Tomlinson 1961), root (Seubert 1996a), differing from other Calamoideae in having very large epidermal cells with sinuous walls in the leaf.
Relationships: Published phylogenies resolve *Salacca* as a strongly supported monophyletic group (Baker *et al.* 2000a, 2000b). However, the problematic species *S. secunda*, which displays features of both *Salacca* and *Eleiodoxa* has not yet been sampled. *Salacca* is strongly supported as sister to *Eleiodoxa* (Baker *et al.* 2000a, 2000b).
Common names and uses: *Salak*. *Salacca zalacca* is cultivated for its excellent fruit, and fruits of other species are eaten although they are sometimes very sour. Petioles of *S. wallichiana* are sometimes used in house construction in Thailand (Dransfield 1981c). The leaves of several species provide temporary thatch. The spiny petioles of *S. zalacca* are occasionally employed in Java to prevent thieves from climbing fruit trees and bats from roosting on roof beams.
Taxonomic accounts: Beccari (1918), Furtado (1949), Mogea (1980).
Fossil record: Fossil pollen, *Punctilongisulcites microechinatus* Theile Pfeiffer (1988), from the Middle Eocene of Germany, is favourably compared with *Salacca affinis* by Zetter and Hofmann (2001). *Paravuripollis* has been compared with *Salacca* pollen (Ramanujam 1987), but its affinities are more likely to be with *Korthalsia*.
Notes: *Salacca* is neatly divisable into two sections: section *Salacca* with spine-tipped scales and pistillate rachillae bearing dyads of a fertile pistillate and a sterile staminate flower and section *Leiosalacca* with smooth scales and pistillate rachillae bearing solitary pistillate flowers. The little-known *S. secunda* and *Salacca* sp. nov. Henderson are hapaxanthic and are palynologically different from other *Salacca* species. These species call into question the currently accepted generic limits in Salaccinae, but, in the absence of a detailed phylogenetic study of the subtribe, we retain the traditional circumscriptions here.
Additional figures: Fig. 2.7, Glossary fig. 7.

Opposite: ***Salacca*. a**, portion of staminate rachilla × 4 1/2; **b**, **c**, dyad of staminate flowers in 2 views × 6; **d**, dyad of staminate flowers, flowers removed to show bracteoles × 6; **e**, staminate flower × 7 1/2; **f**, staminate flower in vertical section × 7 1/2; **g**, staminate calyx × 7 1/2; **h**, staminate flower, calyx removed × 7 1/2; **i**, staminate petal, interior view × 7 1/2; **j**, stamen in 3 views × 15; **k**, portion of pistillate rachilla × 3; **l**, dyad of a pistillate and sterile staminate flower × 4 1/2; **m**, dyad of a pistillate and sterile staminate flower with bracteoles × 3; **n**, sterile staminate flower × 3; **o**, sterile staminate flower in vertical section × 4 1/2; **p**, pistillate flower × 3; **q**, pistillate flower in vertical section × 3; **r**, pistillate calyx × 3; **s**, pistillate flower, calyx removed × 3; **t**, pistillate petal with staminodes, interior view × 4 1/2; **u**, gynoecium × 4 1/2; **v**, ovary in cross-section × 7 1/2; **w**, fruit in 2 views × 1; **x**, scales of fruit × 2 1/4; **y**, fruit in vertical section × 1; **z**, sarcotesta in 4 views × 1; **aa**, seed in 2 views × 1; **bb**, seed in vertical section × 1 1/2. *Salacca glabrescens*: **a–v**, *Dransfield* 682; *S. zalacca*: **w–bb**, *Dransfield* 921. (Drawn by Marion Ruff Sheehan)

Subtribe Metroxylinae Blume, Rumphia 2: 157 (1843). Type: **Metroxylon** Rottb.

Arborescent polygamous, hapaxanthic (rarely pleonanthic) palms; inflorescence adnate to internode.

The subtribe comprises one genus restricted to Melanesia and the western Pacific, but now widespread in cultivation throughout the Asian region.

For relationships, see Korthalsiinae.

12. METROXYLON

The sago palms. Massive solitary or clustered tree palms of the Moluccas, New Guinea and the western Pacific islands, although widely cultivated elsewhere; sheaths, rachis and leaflet margins are armed with spines, rarely sheaths unarmed; hapaxanthic or pleonanthic, the inflorescences are interfoliar or more usually suprafoliar; rachillae are catkin-like, bearing spirally arranged bracts, filled with hairs, each subtending a pair of flowers, one staminate the other hermaphroditic.

Metroxylon Rottb., Nye Saml. Kongel. Dansk Vidensk. Selsk. Skr. 2: 527 (1783) (conserved name). Type: **M. sagu** Rottb.

Sagus Steck, Sagu 21 (1757). Lectotype: *S. genuina* Giseke (see H.E. Moore 1962b) (non *Sagus* Gaertn. = *Raphia* P. Beauv.).

Coelococcus H. Wendl., Bonplandia 10: 199 (1862). Type: *C. vitiensis* H. Wendl. (*Metroxylon vitiense* [H. Wendl.] Benth. & Hook.f.).

Metra — *pith,* xylon — *wood, referring to the well-developed pith, filled with sago.*

Robust to massive, solitary or clustered, armed or unarmed, hapaxanthic or pleonanthic, polygamous tree palms. *Stem* erect, usually partly obscured by the marcescent leaf bases, the internodes sometimes bearing adventitious roots, these usually spine-like; cortex hard, pith soft, rich in starch. *Leaves* large, pinnate, marcescent or sometimes neatly abscising; sheath splitting opposite the petiole, unarmed, or armed with partial whorls of rather slender spines united by their bases to form low collars, and covered with caducous indumentum; petiole well developed, unarmed or armed as the sheath, channelled adaxially in proximal part, becoming rounded distally, rounded abaxially throughout; rachis like the petiole, but angled adaxially; leaflets numerous, single-fold, linear, regularly arranged or grouped and fanned within the groups to give the leaf a plumose appearance, rarely bearing white wax on abaxial surface, usually armed with inconspicuous short spines along the margins and main vein, midribs prominent adaxially, transverse veinlets usually conspicuous. *Inflorescences* branched to 2 orders, either interfoliar in pleonanthic *Metroxylon amicarum* or aggregated into a suprafoliar, compound inflorescence, with branches equivalent to axillary inflorescences, each subtended by a reduced leaf or bract and sometimes emerging through a split in its mid-line; peduncle very short; prophyll tubular, tightly sheathing, 2-keeled, 2-lobed; peduncular bract 1–several, tubular; rachis much longer than peduncle; rachis bracts ± distichous, tubular, tightly sheathing, with a triangular limb, unarmed or rarely with a few scattered spines; first-order branches horizontal or pendulous, each with a basal tubular, 2-keeled, 2-lobed empty prophyll and ± distichous, tightly sheathing, tubular bracts, unarmed or armed with few scattered spines, all but the proximal 1–ca. 3 subtending a catkin-like rachilla (second-order branch of inflorescence); rachillae robust, cylindrical, with a short proximal, bare, stalk-like portion, and a dense spiral of imbricate wide, rounded or apiculate, striate bracts, the proximal and distal few empty, the rest each enclosing a dyad of a small staminate and a similar hermaphroditic flower, in bud partly obscured by a dense pile of hairs, except in *M. amicarum* where hairs sparse, dyad prophyll completely tubular, with 2 keels and 2 triangular lobes, usually bearing dense hairs on the abaxial surface, inner bracteole with 2 keels and dense hairs. *Staminate flowers* opening before the hermaphroditic; calyx tubular, ± striate, with 3 triangular lobes; corolla usually ± twice the length of the calyx, divided to ± $^{2}/_{3}$ its length into 3 oblong, valvate, smooth petals with triangular tips; stamens 6, borne on the base of the corolla, filaments fleshy, abruptly contracted and reflexed, anthers medifixed, oblong, latrorse; pistillode conical. *Pollen* ellipsoidal, bi-symmetric; apertures equatorially disulcate; ectexine tectate, completely psilate, finely to coarsely perforate, or coarsely reticulate, aperture margins similar to surrounding ectexine; infratectum columellate; longest axis 44–64 µm, post-meiotic tetrads tetragonal [5/7]. *Hermaphroditic flowers* superficially similar to the staminate but somewhat fatter; calyx and corolla like the staminate; stamens like the staminate, but with filaments united proximally to form an androecial tube surrounding the ovary; gynoecium tricarpellate, triovulate, rounded, covered in vertical rows of minute scales, and bearing a conical style with 3 stigmatic angles, ovule basally attached, anatropous. *Fruit* rounded, usually large, 1-seeded, with apical stigmatic remains; epicarp covered in neat vertical rows of straw- to chestnut-coloured reflexed scales, mesocarp rather thick, corky or spongy, endocarp not differentiated. *Seed* globose, basally attached, deeply invaginated apically, enveloped in a thin to thick sarcotesta, endosperm homogeneous; embryo basal. *Germination* adjacent-ligular; eophyll bifid or pinnate. *Cytology*: 2n = 26.

Above: ***Metroxylon sagu***, habit, Papua. (Photo: W.J. Baker)

Distribution and ecology: About seven species native to east Malesia, the Solomon Islands, New Hebrides, Samoa, Fiji, and the Carolines; one species *Metroxylon sagu*, thought to be native to New Guinea and the Moluccas, is now widespread and naturalised throughout the Southeast Asian region as a source of sago and thatching material. Most species are plants of lowland swamps, where they may grow gregariously in great numbers. *M. amicarum* also grows in deep valleys and high in the mountains in Micronesia (Moore & Fosberg 1956).

Anatomy: Leaf (Tomlinson 1961), root (Seubert 1996a).

Relationships: *Metroxylon* is strongly supported as monophyletic (Baker *et al.* 2000a, 2000b). Relationships between the species have been examined by McClatchy (1999). For higher-level relationships, see Metroxylinae.

Common names and uses: True sago palm, ivory nut palms, *sagu*, *rumbia*. The two major uses of *Metroxylon* species are as sources of sago and as materials for house construction, leaves are used for thatching and woven walling is made from split petioles. Sago from *Metroxylon* is of great importance as a

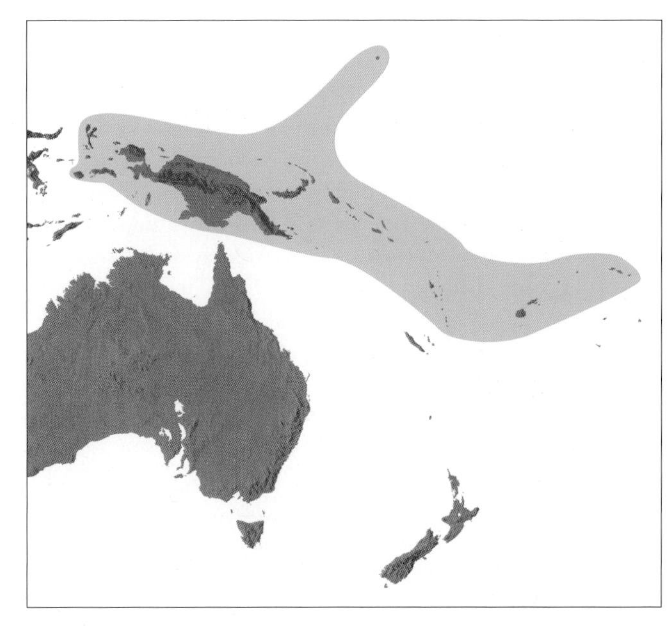

Distribution of ***Metroxylon***

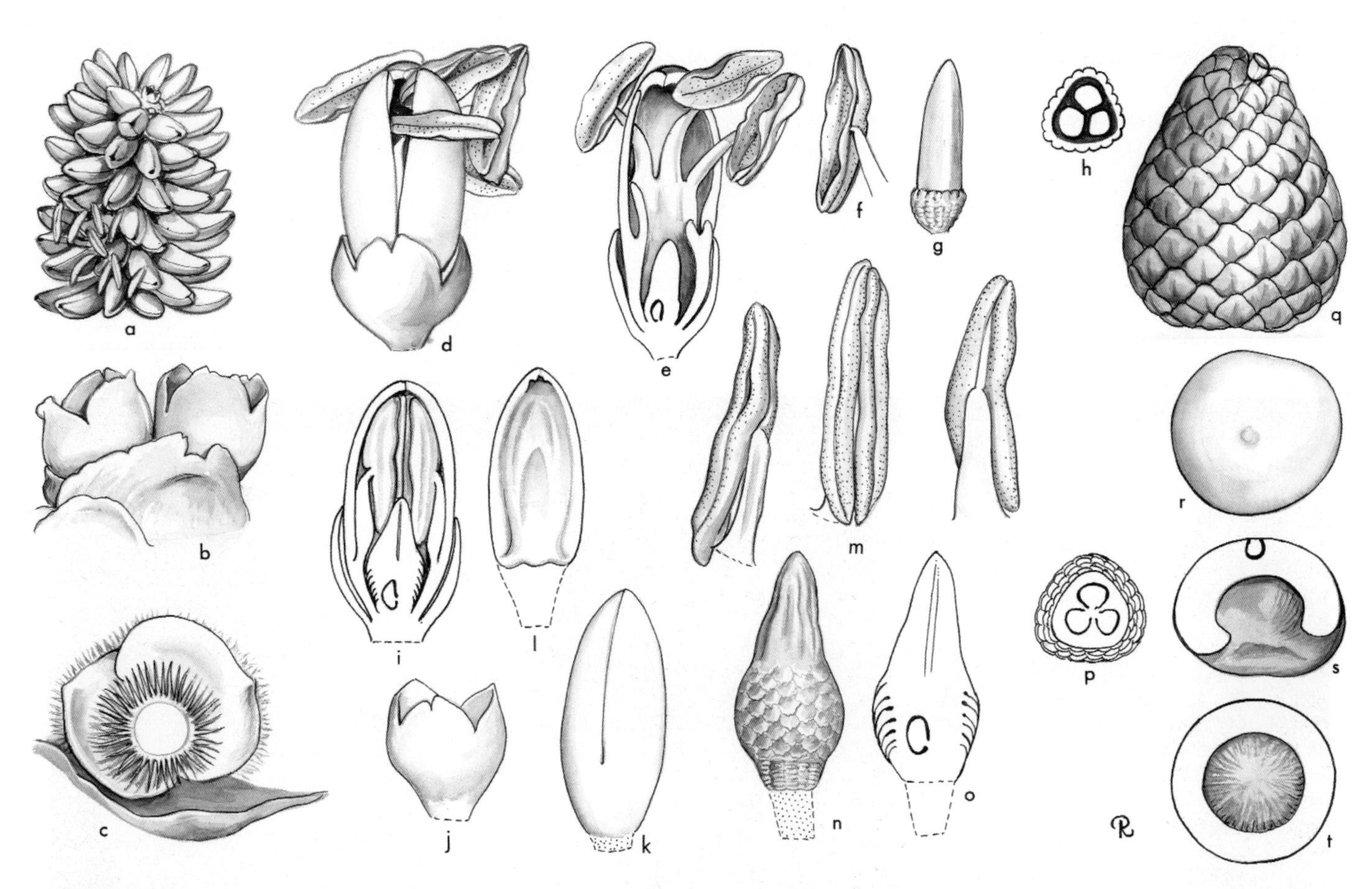

Metroxylon. **a,** portion of rachilla at anthesis × 1; **b,** portion of rachilla to show paired arrangement of flowers with corollas removed × 3; **c,** bracteoles subtending flower × 6; **d,** staminate flower × 3; **e,** staminate flower in vertical section × 3; **f,** stamen × 3; **g,** pistillode × 3; **h,** pistillode in cross-section × 12; **i,** hermaphrodite flower in vertical section × 3; **j,** hermaphrodite calyx × 3; **k,** hermaphrodite flower, calyx removed × 3; **l,** hermaphrodite petal × 3; **m,** stamen from hermaphrodite flower in 3 views × 6; **n,** gynoecium × 6; **o,** gynoecium in vertical section × 6; **p,** ovary in cross-section × 6; **q,** fruit × $^1/_2$; **r,** seed × $^1/_2$; **s,** seed in vertical section × $^1/_2$; **t,** seed in cross-section × $^1/_2$. *Metroxylon warburgii*: all from *Moore* 9319. (Drawn by Marion Ruff Sheehan)

staple in parts of the Moluccas, New Guinea, and the western Pacific. The apparatus used in the production of sago is, itself, often made from *Metroxylon*. Seeds of some species furnish a form of vegetable ivory used in the past but of little value now. Felled trunks are used for rearing sago grubs, which are curculionid beetle larvae, a highly esteemed food.

Taxonomic accounts: Beccari (1918), Rauwerdink (1985), McClatchey (1999).

Fossil record: *Dicolpopollis metroxylonoides* in the Tertiary sediments of Papua New Guinea (Khan 1976) is compared with pollen of *Metroxylon americanum* [*sic.*] (= *M. amicarum*). However, the long axis in *M. amicarum* pollen is 50–65 µm, whereas that of the fossil is 30–40 µm. There is also a rare record of *D. metroxylonoides* from the Miocene of South Island, New Zealand (Mildenhall & Pocknall 1989). Ramanujam *et al.* (1997) describe 'dicolpate' palm pollen from the Neogene deposits of Godavari (Krishna Basin), they consider that there may be an affinity between *Disulcipollis psilatus* and *Metroxylon*, based on the psilate exine. However, the *D. psilatus* pollen is smaller than that of recent *Metroxylon* spp. (22–32 µm cf. 44–64 µm), and its wall thickness is notably thinner (1.5 µm cf. 2.5–5.0 µm). The size range for *Daemonorops* pollen (16–55 µm) and the occurrence in the genus of pollen with psilate exines suggest that the affinity of the fossil grain is more probably with *Daemonorops*. Rao & Ramanujam (1978) consider an affinity between *Dicolpopollis elegans* (Muller 1968) from the Neogene Quilon beds of Kerala State, south India, with the pollen of *Metroxylon*. However, the size range of *Metroxylon* pollen is notably larger than that of their fossil (17.5 × 18.5 µm), again suggesting a more probable affinity with *Daemonorops* or *Calamus*.

Notes: A report on flowering and inflorescence structure is given by Tomlinson (1971) for *Metroxylon vitiense* and *M. sagu*. The form of the rachillae is very characteristic.

Additional figures: Figs 1.1, 1.7, 2.2.

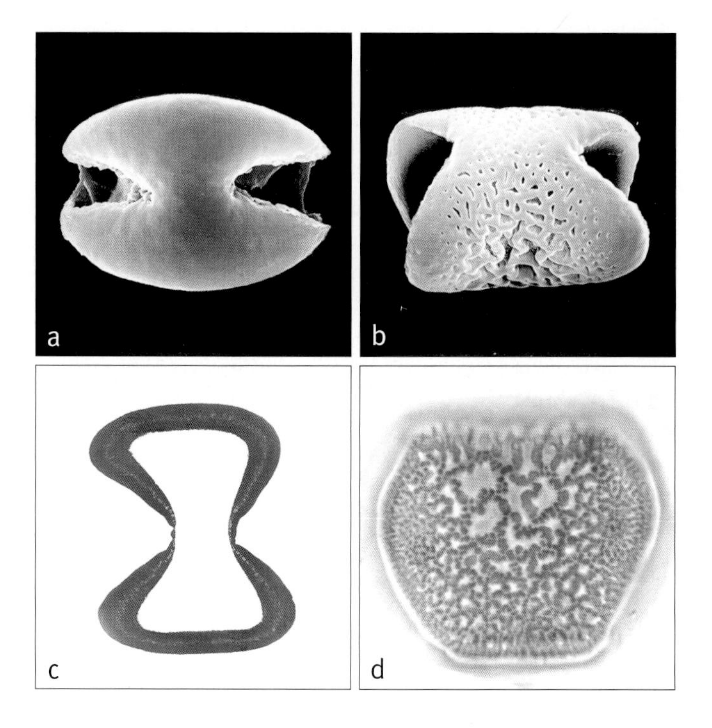

Above: ***Metroxylon***
a, equatorial disulcate pollen grain, distal face SEM × 1000;
b, equatorial disulcate pollen grain, tilted distal/equatorial face SEM × 1000;
c, ultrathin section through whole pollen grain, equatorial plane TEM × 750;
d, equatorial disulcate pollen grain, equatorial face LM × 1000.
Metroxylon sagu: **a** & **c**, *Floyd* HHRBK 5407; *M. salomonense*: **b** & **d**, *Whitmore* BSIP 1818.

Below left: ***Metroxylon sagu***, rachilla at staminate anthesis, Papua. (Photo: W.J. Baker)

Below right: ***Metroxylon sagu***, fruit, Papua. (Photo: W.J. Baker)

Subtribe Pigafettinae J. Dransf. & N.W. Uhl, Principes 30: 5 (1986). Type: **Pigafetta**.

Pleonanthic, dioecious, tree palms; rachillae slender bearing highly reduced bracts; staminate flowers paired; pistillate flowers solitary; pollen inaperturate; sarcotesta well developed.

This subtribe consists of the single genus *Pigafetta*, confined to Sulawesi, Moluccas and New Guinea. Although there is a superficial resemblance in habit and leaf to *Metroxylon*, the structure of the rachillae, the dioecism, the flower structure and the much smaller fruit are very different.

For relationships, see Korthalsiinae.

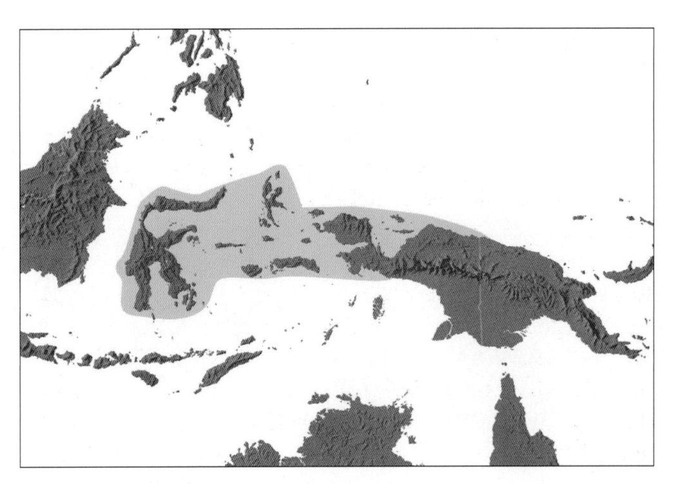

Distribution of ***Pigafetta***

13. PIGAFETTA

Massive, elegant, solitary tree palms of the Sulawesi, Moluccas and New Guinea; sheaths, rachis and leaflet margins armed with soft spines; the palms are pleonanthic and dioecious, with inflorescences interfoliar, sometimes becoming infrafoliar in fruit; fruits are remarkably small in comparison with those of other Calamoideae.

Pigafetta (Blume) Becc., Malesia 1: 89 (1877) (conserved name). ('*Pigafettia*'). Lectotype: **P. filaris** (Giseke) Becc. (*Sagus filaris* Giseke).

Sagus section *Pigafetta* Blume, Rumphia 2: 154 (1843).

Metroxylon section *Pigafetta* (Blume) Mart., Hist. Nat. Palm. 3: 213 (2nd Edn) (1845).

Commemorates Antonio Pigafetta (ca.1491–ca.1534), Italian mariner, who circumnavigated the world with Magellan.

Below: ***Pigafetta filaris***, habit, Papua. (Photo: J. Dransfield)

Massive, solitary, armed, pleonanthic, dioecious, tree palms. *Stem* erect, becoming very tall, with conspicuous nodal scars, internodes glossy, green in distal areas, becoming brown with age, usually bearing abundant, somewhat spine-like, adventitious roots near the base, the cortex very hard, pith soft. *Leaves* pinnate, strongly curved, abscising neatly in trunked individuals; sheath splitting opposite the petiole, bearing a tattered ligule around the mouth or two conspicuous auricles, the sheath unarmed at the very base, distally armed abaxially with low collars bearing abundant soft flexible spines, both sheath surface and spines with a dense caducous felt of tomentum; petiole (true petiole) absent (*P. filaris*) or massive (*P. elata*), channelled adaxially, rounded and armed like the sheath abaxially; rachis angled adaxially, rounded and armed with sparse collars and spines abaxially; leaflets single-fold, numerous, regularly arranged, curving, slender, elongate, acuminate, armed with short marginal bristles and long bristles on the main veins, midribs large, one other pair of large veins evident, transverse veinlets sinuous, moderately conspicuous. *Inflorescences* axillary, interfoliar at anthesis, sometimes becoming infrafoliar after abscission of the subtending leaf, ± horizontal, tending to diverge obliquely from the plane of the leaf, branched to 2 orders; prophyll tightly sheathing, with 2 triangular limbs and dense caducous indumentum; peduncular bracts (ca. 8), tubular, in the staminate inflorescence somewhat inflated, with short triangular limbs and dense caducous indumentum; rachis much longer than the peduncle; rachis bracts subtending pendulous first-order branches, partly adnate to the primary axis above the bract; first-order branches with a tightly sheathing prophyll and numerous tubular bracts each subtending a ± pendulous rachilla; flower-bearing part of the rachilla exserted from the bracts on a bare flattened basal stalk; rachilla prophyll minute, 2-keeled, empty, borne at the distal end of the rachilla stalk; rachilla bracts minute, membranous, striate, triangular, in the staminate inflorescence subtending a dyad of fertile staminate flowers, and in the pistillate subtending a single fertile pistillate flower; bracteoles

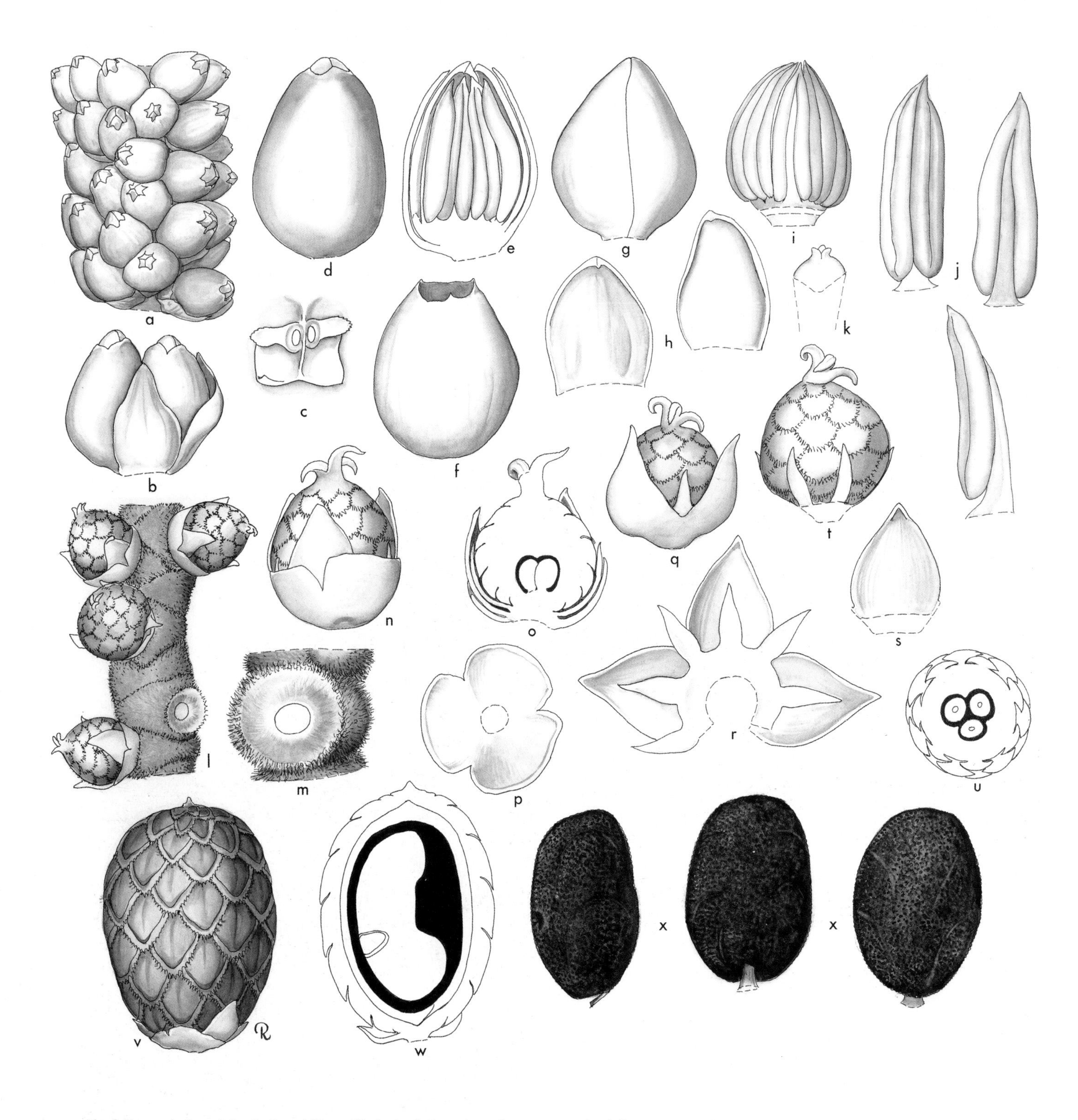

Pigafetta. **a**, portion of staminate rachilla × $4^{1}/_{2}$; **b**, dyad of staminate flowers × 9; **c**, dyad, flowers removed to show bracts and bracteoles × 6; **d**, staminate bud × 12; **e**, staminate bud in vertical section × 12; **f**, staminate calyx × 12; **g**, staminate bud, calyx removed × 12; **h**, staminate petal in 2 views × 12; **i**, androecium × 12; **j**, stamen in 3 views × 18; **k**, pistillode × 30; **l**, portion of pistillate rachilla × 4; **m**, scar of pistillate flower × 8; **n**, pistillate flower × 8; **o**, pistillate flower in vertical section × 8; **p**, pistillate calyx × 8; **q**, pistillate flower, calyx removed × 8; **r**, pistillate corolla and staminodial ring, expanded, interior view × 8; **s**, pistillate petal × 8; **t**, gynoecium and staminodes × 8; **u**, ovary in cross-section × 8; **v**, fruit × 4; **w**, fruit in vertical section × 4; **x**, seed covered with sarcotesta in 3 views × 4. *Pigafetta elata*: **a–k**, *Dransfield* 3867; **l–u**, *Mogea* 5353; *P. filaris*: **v–x**, *Essig* 55078. (Drawn by Marion Ruff Sheehan)

Left: ***Pigafetta elata***, young staminate inflorescences, Sulawesi. (Photo: J. Dransfield)
Right: ***Pigafetta filaris***, infructescence, Papua. (Photo: J. Dransfield)

in both sexes minute, obscured by a deep pile of crowded hairs. *Staminate flowers* symmetrical; calyx striate, cupular, with 3 very shallow lobes; corolla exceeding the calyx, tubular at the base, divided distally into 3 striate, narrow, triangular valvate lobes; stamens 6, borne at the mouth of the corolla tube, the filaments adnate laterally in a low ring, the anthers elongate, somewhat sagittate, apparently latrorse; pistillode minute. *Pollen* grains spheroidal, symmetric; inaperturate; ectexine tectate to semitectate, reticulate with frequently interrupted, angular or spinulose muri; infratectum columellate; longest axis 29–36 µm [2/2]. *Pistillate flowers* ± globose; calyx cupular with 3 very short lobes, later splitting; corolla divided ± to the base into 3 broad triangular lobes; staminodes 6, united by their filaments into a short ring with 6 narrow lobes bearing flattened empty sagittate anthers; gynoecium incompletely trilocular, triovulate, globose, covered in reflexed scales, stigmas 3, short, reflexed, ovules anatropous, basally attached. *Fruit* relatively very small, ovoid, single-seeded; epicarp covered in neat vertical rows of reflexed scales, mesocarp thin, endocarp thin, not differentiated. *Seed* basally attached, laterally somewhat flattened, covered in thick sweet sarcotesta, endosperm homogeneous with very shallow depressions and laterally with a shallow pit; embryo lateral, opposite the pit. *Germination* adjacent-ligular; eophyll bifid, with very narrow leaflets and softly spiny petiole. *Cytology*: 2n = 28.

Distribution and ecology: Two species, *Pigafetta filaris*, confined to Moluccas and New Guinea and *P. elata* endemic to Sulawesi. *Pigafetta elata* seems to behave as a pioneer palm of disturbed habitats in the mountains, where it is most abundant between 300 and 1500 m. It grows on old landslips, old lava flows and river banks, and also seems to colonise cultivated land reverting to secondary forest. Seedlings appear to require high light intensities. *Pigafetta filaris* also seems to be a pioneer species, but rather less is known of its behaviour in the wild. Both species have very small seeds for the size of the palm; that of *P. elata* shows staggered germination. These are the tallest recorded palm species in the Asian tropics, individuals sometimes reaching 50 m in height; both are also very fast growing (see Dransfield 1976b).

Anatomy: Not studied.

Relationships: *Pigafetta* is strongly supported as monophyletic (Baker *et al.* 2000a, 2000b). For relationship, see Pigafettinae.

Common names and uses: Pigafetta palm, *wanga*. The trunks of *P. elata* are used for the piles of houses, floor boards and for simple furniture in parts of Sulawesi (Rotinsulu 2001). These beautiful palms have been much sought after by collectors.

Taxonomic accounts: Beccari (1918), Dransfield (1998).

Fossil record: No generic records found.

Notes: There is a remarkable similarity between the rachillae of *Pigafetta* and those of *Plectocomia*. Otherwise, the two genera are very different, *Pigafetta* being tall solitary pleonanthic trees and *Plectocomia* hapaxanthic rattans. *Metroxylon* and *Pigafetta* show similarity in habit and leaves but the structure of the inflorescences, flowers and fruit is very different. Until recently, there was thought to be but a single species, but two taxa can be clearly differentiated (Dransfield 1998).

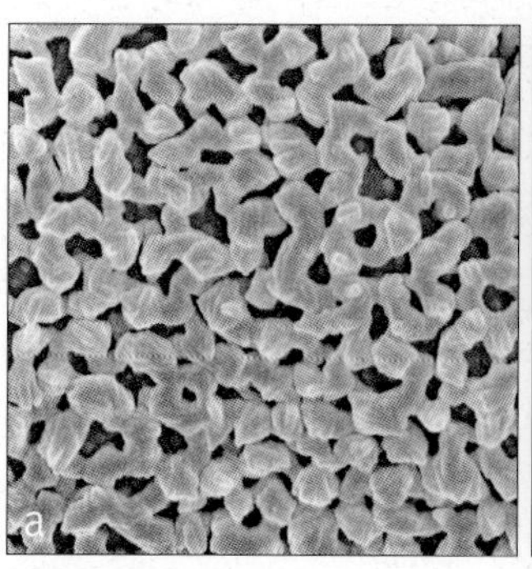

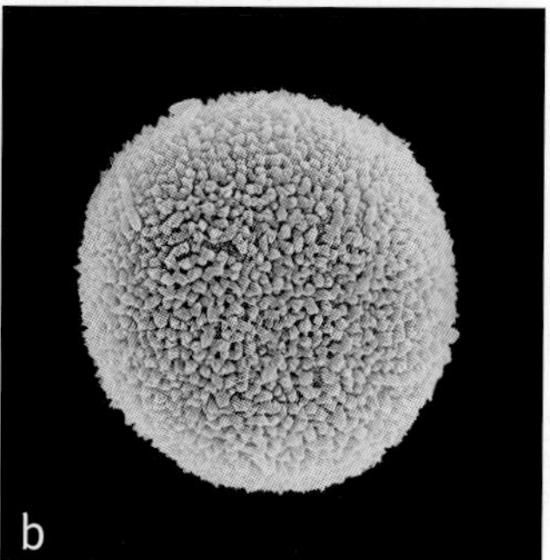

Above: ***Pigafetta***
a, inaperturate pollen grain SEM × 1500;
b, close-up, pollen surface SEM × 7500.
Pigafetta elata: **a** & **b**, *Migliaccio* s.n.

Subtribe Plectocomiinae J. Dransf. & N.W. Uhl, Principes 30: 5 (1986). Type: **Plectocomia**.

Climbing, dioecious, hapaxanthic palms; leaves with cirri lacking acanthophylls; inflorescence adnate to internode.

The subtribe contains three closely related rattan genera restricted to the Asian tropics and subtropics to the north and west of Wallace's Line. They are vegetatively distinct from members of Calaminae in lacking a knee on the leaf sheath, a character almost always present in climbing members of the Calaminae.

Subtribe Plectocomiinae is consistently resolved as monophyletic with moderate to high support (Baker *et al.* 2000a, 2000b, in review). Most studies place the subtribe as sister to Calaminae with moderate to high support (Baker *et al.* 2000a, 2000b, in review), although weak evidence in another study suggests a sister relationship to the Metroxylinae (Asmussen *et al.* 2006).

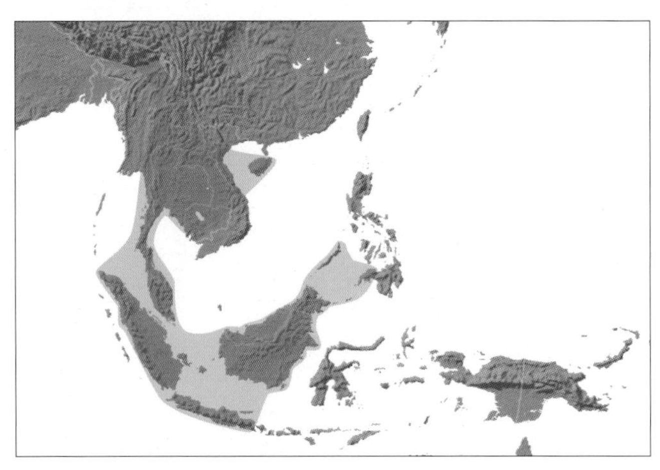

Distribution of subtribe **Plectocomiinae**

Key to Genera of Plectocomiinae

1. First-order inflorescence branches pendulous, bearing strictly distichous, conspicuous, overlapping or not, boat-shaped bracts, often completely enclosing the rachillae; rachilla bracts obscure (except in monstrous forms); staminate flowers borne in pairs; tips of fruit scales often reflexed. Himalayas to Bali . . . . **14. Plectocomia**
1. First-order inflorescence branches pendulous or not, bearing bracts scarcely differentiated from other inflorescence bracts, not obscuring rachillae; rachilla bracts conspicuous; staminate flowers borne singly; tips of fruit scales not reflexed 2
2. Leaf sheath without ocrea; inflorescences of both sexes diffusely branched, usually to 3 orders; flowers with membranous perianth whorls; anthers borne on long, fine, inflexed filaments, pendulous from the lobes of the androecial ring; fruit covered with innumerable, minute, irregularly arranged scales. Indo China to Malaya Peninsula and Sumatra . . . . **15. Myrialepis**
2. Leaf sheath bearing an ocrea, sometimes disintegrating; inflorescences of both sexes branching to 2 orders only, very rarely diffusely branched to 3 orders; flowers with thick coriaceous perianth whorls; anthers ± sessile, pendulous from the tips of the lobes of the androecial ring; fruit covered in large, regularly arranged scales. Thailand, Burma, Malay Peninsula, Sumatra, Borneo . . . . **16. Plectocomiopsis**

14. PLECTOCOMIA

Clustering high-climbing pinnate-leaved rattan palms of Southeast Asia and West Malesia; sheaths lack knees and the ocrea is very short or absent; hapaxanthic and dioecious, the first-order branches of the inflorescence are pendulous and bear conspicuous dictichous bracts, partially enclosing the very short rachillae.

Plectocomia Mart. ex Blume in J.J. Roemer & J.A. Schultes, Syst. Veg. 7: 1333 (1830). Type: **P. elongata** Mart. ex Blume.

Plectos — *plaited*, come — *hair, referring to the appearance of the inflorescence branches*.

Robust (very rarely slender), solitary or clustered, high-climbing, spiny, hapaxanthic, dioecious, rattan palms. *Stem* eventually becoming bare, with long internodes and conspicuous nodal scars, sometimes bearing multiple, bulbil-like shoots in the proximal part of the lower internodes, clear gum frequently exuding from cut surfaces. *Leaves* of mature climbing stems usually massive, cirrate, pinnate; sheath tubular, unarmed or sparsely to very densely armed with spines, usually borne in partial whorls, and also bearing abundant, caducous, floccose hairs on and between spines; knee absent; ocrea absent; flagellum absent; petiole present or absent, it and the proximal portion of the rachis deeply channelled and sparsely to densely armed; cirrus and distal part of rachis armed abaxially with regular groups of massive reflexed grapnel spines; leaflets numerous, single-fold, entire, usually lanceolate, regularly arranged or grouped and fanned within the groups, adaxial surface often with scattered bands of caducous indumentum, the abaxial often bearing white tomentum, midribs and submarginal ribs slightly larger than other veins, transverse veinlets not evident. *Inflorescences* produced simultaneously in the axils of the most distal 2–20, frequently reduced, leaves, axis of inflorescence adnate to the proximal part of the internode above the subtending leaf, emerging from the leaf-sheath mouth, branching to (1) 2 (very rarely 3) orders; peduncle short; prophyll empty, tightly sheathing, 2-keeled, usually included within the leaf sheaths; peduncular bracts few, tubular, tightly sheathing; rachis much longer than the peduncle; rachis bracts similar to the peduncular, ± distichous, each subtending a pendulous first-order branch; first-order branches 3–20 or more, each with a basal tubular 2-keeled prophyll and 1–several empty tubular bracts, distal bracts very conspicuous, distichous, tubular at first, before anthesis splitting longitudinally almost to the base opposite the insertion, spreading or usually remaining imbricate and partially or completely enclosing the rachillae, or very rarely subtending second-order branches bearing similar bracts subtending the rachillae (?teratological), bract texture thin to very thick, coriaceous or subwoody, glabrous or with scattered hairs or scales; rachillae unbranched, slender, frequently densely covered in trichomes, in the pistillate inflorescence bearing from (1) 2–10 flowers, in the staminate from 2–ca. 100 flowers; staminate flowers sometimes borne in distinct dyads, otherwise basic dyad arrangement obscured by overcrowding and elongation of pedicels, rachilla bracts minute; floral bracteoles minute or obscure. *Staminate flower* sessile or with a stalk-like base; calyx tubular, much shorter than the corolla, with 3 short lobes; corolla tubular at the very base with 3 triangular to lanceolate lobes; stamens usually 6, rarely to 12, filaments sometimes connate basally, borne near the mouth of the corolla tube, anthers usually elongate, latrorse or introrse; pistillode minute or absent.

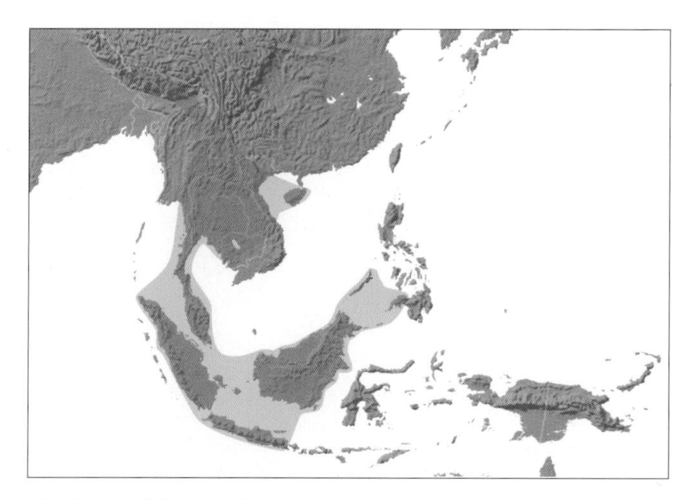

Distribution of ***Plectocomia***

Pollen ellipsoidal, bi-symmetric; apertures equatorially disulcate; ectexine tectate, finely or coarsely perforate (perforations sometimes sparsely distributed), perforate-rugulate, or granular-rugulate, aperture margins similar; infratectum columellate; longest axis 26–49 µm [6/16]. *Pistillate flowers* solitary, often pedicellate with the rachilla bract carried by adnation and subsequent growth on to the pedicel, each flower also bearing a 2-keeled bracteole; calyx tubular at the base, with 3, very short to very long, triangular lobes, usually splitting further after anthesis; corolla tubular at the base with 3, narrow to broad, triangular lobes, slightly to greatly exceeding the calyx, becoming flattened out in fruit; staminodes 6 with flattened filaments and empty anthers; gynoecium rounded, scaly, stigmas 3, usually very long, flexuous, sometimes with a well-developed style, locules 3, incomplete, ovules 3, basally attached, anatropous. *Fruit* 1 (rarely 2–3)-seeded, with apical stigmatic remains; epicarp covered in numerous vertical rows of reflexed scales, the scale tips frequently fringed and upward pointing, mesocarp thin, fibrous, endocarp not differentiated. *Seed* attached near the base, sarcotesta thick but not juicy, endosperm homogeneous; embryo basal. *Germination* adjacent-ligular; eophyll where known, simple, lanceolate, plicate, not split. *Cytology* not studied.

Distribution and ecology: About 16 species distributed from the Himalayas, south China and Hainan, south through Burma and Indochina to the Sunda Shelf and the Philippines; not known east of Wallace's Line. *Plectocomia elongata* and *P. mulleri* are the two most widespread species and also appear to have the widest altitudinal range, being found from sea level up to ca. 2000 m in the mountains; the former is characteristic of disturbed sites on poor soils and the latter is found in similar habitats but also as a conspicuous component of some facies of peat-swamp forest and heath forest ('kerangas'). Other species are less well known and seem to be more restricted in distribution; however, many do appear to be characteristic of seral forest on poor soils. In the Himalayas, *P. himalayana* has been recorded at altitudes of about 2000 m. There are indications that the hapaxanthic habit may be an adaptation to the colonization of temporary habitats such as landslips. There are suggestions that, in some localities, *P. elongata* may flower gregariously (see Dransfield 1979a). Trigonid bees and nitidulid and staphylinid beetles have been observed visiting the intensely fragrant staminate flowers of *P. dransfieldiana*. Nothing is known of fruit dispersal. Neither is anything known of the time taken to reach flowering size.

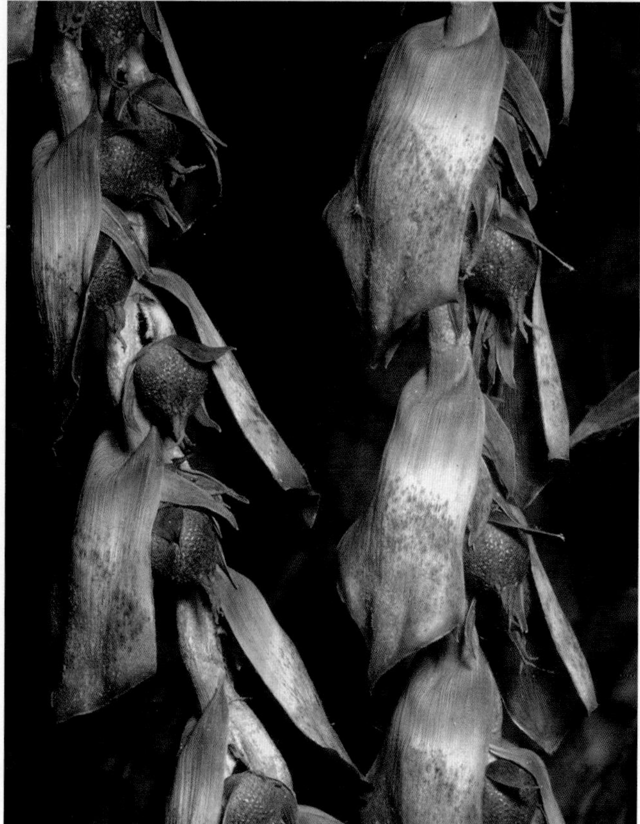

Top: ***Plectocomia kerriana***, habit, Thailand. (Photo: J. Dransfield)

Bottom: ***Plectocomia kerriana***, close-up of fruiting branches, Thailand. (Photo: J. Dransfield)

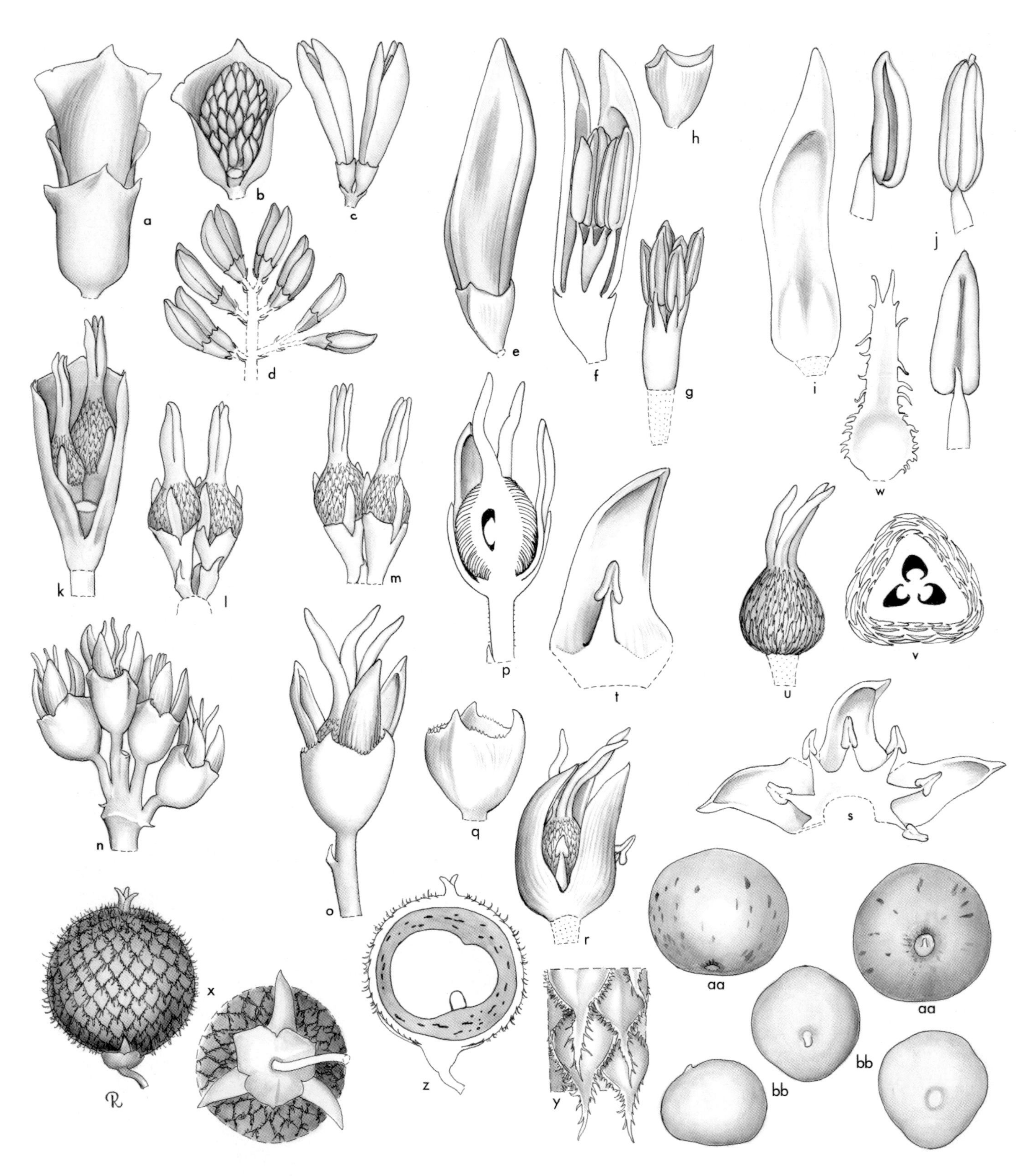

Plectocomia. **a**, portion of staminate spike × 1 1/2; **b**, staminate axis and bract × 1 1/2; **c**, paired staminate flowers × 6; **d**, portion of staminate axis, extended to show arrangement of paired units × 3; **e**, staminate bud × 12; **f**, staminate bud in vertical section × 12; **g**, androecium × 12; **h**, staminate calyx × 12; **i**, staminate petal, interior view × 12; **j**, stamen in 3 views × 12; **k**, paired pistillate flowers in bract × 1 1/2; **l, m**, paired pistillate flowers, bract removed, in 2 views × 1 1/2; **n**, pistillate flowers, bract removed × 1 1/2; **o**, pistillate flower × 3; **p**, pistillate flower in vertical section × 3; **q**, pistillate calyx × 3; **r**, pistillate flower, calyx removed × 3; **s**, pistillate corolla and staminodes, expanded, interior view × 3; **t**, pistillate petal and staminode, interior view × 6; **u**, gynoecium × 3; **v**, ovary in cross-section × 6; **w**, scale × 12; **x**, fruit in 2 views × 1 1/2; **y**, scales of fruit, much enlarged; **z**, fruit in vertical section × 1 1/2; **aa**, sarcotesta in 2 views × 1 1/2; **bb**, seed in 3 views × 1 1/2. *Plectocomia mulleri*: **a–m** and **x–bb**, *Moore et al.* 9158; *P. elongata*: **n–w**, *Rahmat Si Boea* 7417. (Drawn by Marion Ruff Sheehan)

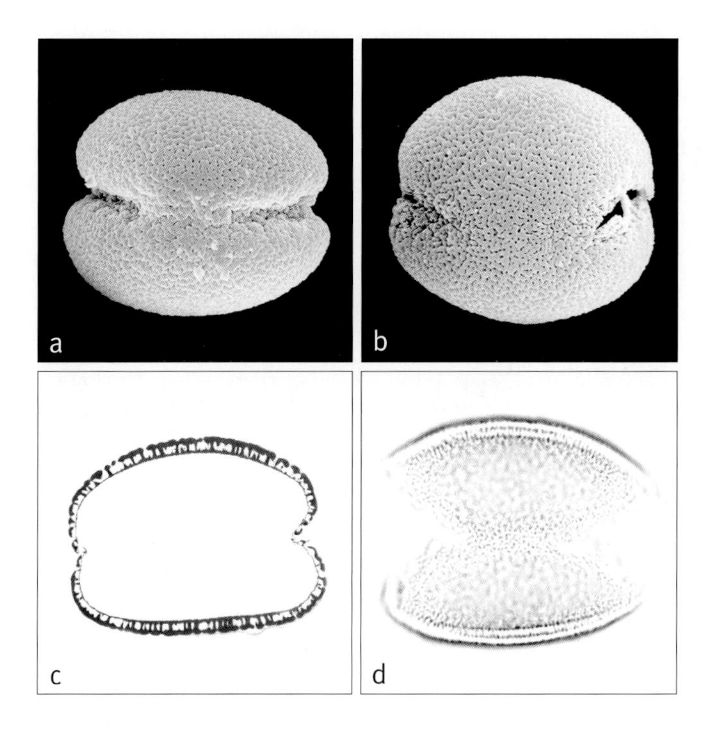

Above: ***Plectocomia***
a, equatorial disulcate pollen grain, distal face SEM × 1000;
b, equatorial disulcate pollen grain, proximal face SEM × 1000;
c, ultrathin section through whole pollen grain, equatorial plane TEM × 750;
d, equatorial disulcate pollen grain, distal face, high-focus LM × 1000.
Plectocomia elongata: **a** & **b**, *Curtis* 2436, **d**, *Dransfield & Saerudin* JD1873; **c**, *P. kerriana*: *Kerr* 1871.

Anatomy: Leaf, petiole, stem, root (Tomlinson 1961), root (Seubert 1996a); similar to *Myrialepis* in having large epidermal cells. The biomechanics of the stems of *Plectocomia* have been studied by Rowe, Isnard and co-workers (Rowe *et al.* 2004, Isnard *et al.* 2005, Isnard 2006).

Relationships: *Plectocomia* is strongly supported as monophyletic (Baker *et al.* 2000a, 2000b). The relationships between the three genera, *Plectocomia*, *Myrialepis* and *Plectocomiopsis*, are somewhat ambiguous in published studies. *Plectocomiopsis* has been resolved as sister to *Plectocomia* (Baker *et al.* 2000b) or to *Myrialepis* (Baker *et al.* 2000a). However, additional DNA data (Baker *et al.* in prep.) strongly support the latter resolution.

Common names and uses: Rattan, *rotan*. The stems of *Plectocomia* species have a soft pith that renders them unsuitable for fine furniture construction. They are used only for coarse or temporary basketry. The dried inflorescences are sometimes used as ornaments.

Taxonomic account: Madulid (1981).

Fossil record: No generic records found.

Notes: The inflorescence of *Plectocomia* is striking and beautiful; the long pendulous first-order branches with large distichous boat-shaped bracts are unlike those of any other palm. Of particular note are the much reduced rachillae with scarcely evident bracts. The distinctive first-order branches and paired staminate flowers separate *Plectocomia* from the other hapaxanthic, high-climbing rattans, *Myrialepis* and *Plectocomiopsis*.

Additional figures: Fig. 1.7, Glossary fig. 16.

Below left: ***Plectocomia dransfieldiana***, staminate flowers, Peninsular Malaysia. (Photo: J. Dransfield)
Right: ***Plectocomia dransfieldiana***, staminate inflorescence, Peninsular Malaysia. (Photo: J. Dransfield)

15. MYRIALEPIS

Clustering high-climbing pinnate-leaved rattan palms of Southeast Asia and West Malesia; sheaths are densely armed with whorls of spines, ocrea absent; hapaxanthic and dioecious, the rachillae bear solitary staminate flowers or solitary pistillate flowers; perianths are membranous and the fruit relatively large, covered with minute irregularly arranged scales.

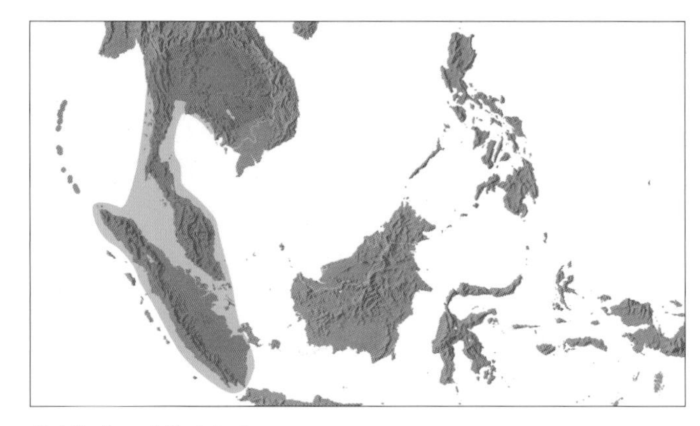

Distribution of ***Myrialepis***

Myrialepis Becc. in Hook.f., Fl. Brit. India 6: 480 (1893). Type: *M. scortechinii* Becc. (= **M. paradoxa** [H. Kurz] J. Dransf. [*Calamus paradoxus* H. Kurz]).

Bejaudia Gagnep., Notul. Syst. (Paris) 6(3): 149 (1937). Type: *B. cambodiensis* Gagnep. (= *Myrialepis paradoxa* [H. Kurz] J. Dransf.).

Myrioi — *very many,* lepis — *scale, referring to the countless minute scales on the fruit.*

Robust, clustered, high-climbing, spiny, hapaxanthic, dioecious, rattan palm. *Stem* eventually becoming bare with long internodes and conspicuous nodal scars, basal vegetative branches borne opposite the leaves. *Leaves* on mature climbing stems usually massive, pinnate, cirrate; sheath tubular, sparsely armed with neat whorls of large spines in juveniles, with scattered spines in adults, indumentum abundant on sheath surfaces; knee absent; ocrea absent; flagellum absent; petiole very short to well developed, deeply channelled, sparsely armed; rachis sparsely armed proximally, distally armed as the cirrus with regular groups of grapnel spines on the abaxial surface; leaflets numerous, lanceolate, entire, regularly arranged or rather indistinctly grouped, sometimes armed with short marginal spines, adaxial surface bearing bands of caducous indumentum, abaxial surface with scattered minute peltate scales, midribs slightly larger than other veins, transverse veinlets not evident. *Inflorescences* produced simultaneously in the axils of the most distal few, often reduced leaves, inflorescence axis adnate to the proximal part of the internode above the subtending leaf, emerging from the leaf-sheath mouth, branched to 3 orders; peduncle short; prophyll tubular, 2-keeled, with 2 triangular lobes, included within the leaf sheaths; peduncular bracts absent (?always); rachis much longer than the peduncle; rachis bracts tubular, subdistichous, each subtending a first-order branch; each order of branching with a tubular 2-keeled prophyll and subdistichous tubular bracts, each subtending a branch; rachillae very short, formed from branches of several orders, often somewhat curved, staminate rachillae bearing groups of up to 8 distichous flowers, each subtended by a cup-like rachilla bract and bearing a minute 2-keeled bracteole, pistillate rachillae bearing groups of 2–7 flowers, each subtended by a tubular rachilla bract and bearing a 2-keeled cup-like bracteole. *Staminate flowers* symmetrical; calyx membranous, tubular, with 3 apical lobes; corolla membranous, divided almost to the base into 3 triangular lobes; stamens 6, borne at the base

Below left: ***Myrialepis paradoxa***, habit, Thailand. (Photo: J. Dransfield)
Middle: ***Myrialepis paradoxa***, leaf sheaths, Peninsular Malaysia. (Photo: J. Dransfield)
Right: ***Myrialepis paradoxa***, fruit, Peninsular Malaysia. (Photo: J. Dransfield)

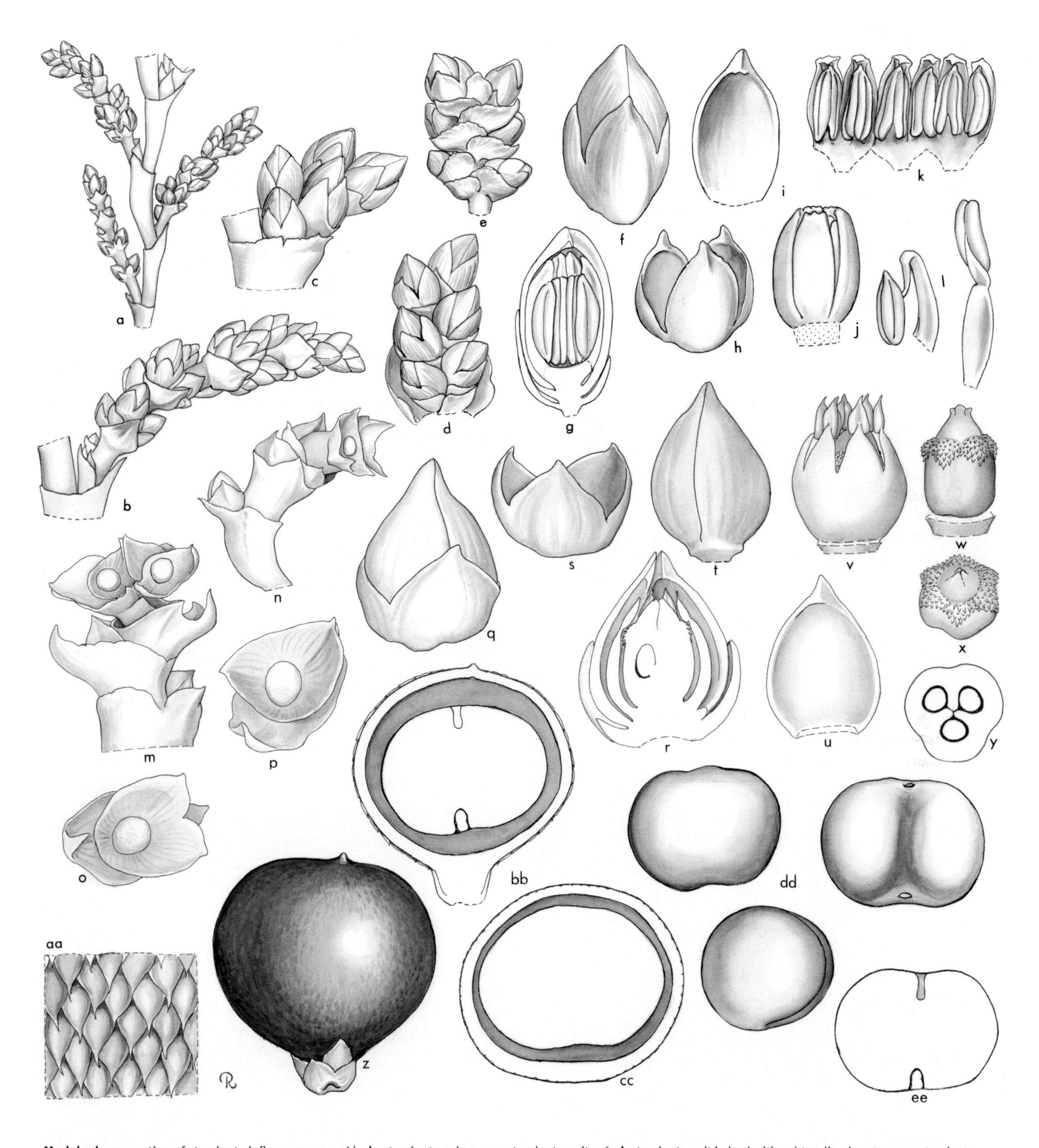

Myrialepis. **a**, portion of staminate inflorescence × 1$^1/_2$; **b**, staminate axis × 4; **c**, staminate unit × 6; **d**, staminate unit in bud with subtending bract × 4; **e**, staminate unit, abaxial view × 4; **f**, staminate bud × 12; **g**, staminate bud in vertical section × 12; **h**, staminate calyx × 12; **i**, staminate petal × 12; **j**, androecium in bud × 12; **k**, androecium in bud, expanded, interior view × 12; **l**, stamen in 2 views × 12; **m, n**, portion of pistillate inflorescence in 2 views × 5; **o, p**, scar, bract and bracteole of pistillate flower × 9; **q**, pistillate bud × 9; **r**, pistillate bud in vertical section × 9; **s**, pistillate calyx × 9; **t**, pistillate bud, calyx removed × 9; **u**, pistillate petal × 9; **v**, staminodial ring and gynoecium × 9; **w, x**, gynoecium in 2 views × 9; **y**, ovary in cross-section × 12; **z**, fruit × 1$^1/_2$; **aa**, scales of fruit, much enlarged; **bb**, fruit in vertical section × 1$^1/_2$; **cc**, fruit in cross-section × 1$^1/_2$; **dd**, seed in 3 views × 1$^1/_2$; **ee**, seed in cross-section × 1$^1/_2$. *Myrialepis paradoxa*: **a–l**, *Moore* 9075; **m–y**, *Ridley* 12500; **z–ee**, *Rahmat Si Boea* 8009. (Drawn by Marion Ruff Sheehan)

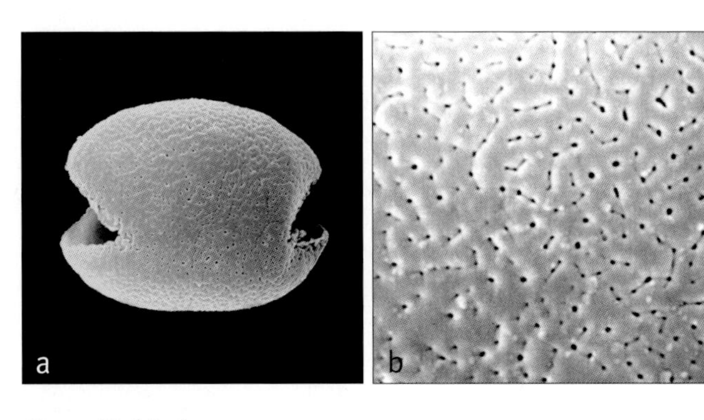

Above: ***Myrialepis***
a, equatorial disulcate pollen grain, proximal face SEM × 1500;
b, close-up, pollen surface SEM × 8000.
Myrialepis paradoxa: **a** & **b**, *Moore* 9075.

of the corolla lobes, filaments adnate laterally near the base, free portions gradually narrowing, inflexed at the tip bearing pendulous, medifixed, oblong, somewhat sagittate, introrse anthers; pistillode minute. *Pollen* ellipsoidal, bi-symmetric; apertures equatorially disulcate; ectexine tectate, perforate-rugulate, aperture margins similar; infratectum columellate; longest axis 31–38 µm [1/1]. *Pistillate flowers* larger than the staminate; calyx membranous, tubular, divided into 3 triangular lobes; corolla membranous, divided almost to the base into 3 triangular lobes; staminodal ring borne at the base of the corolla, bearing 6 triangular lobes, each tipped with a short slender filament bearing empty anthers; gynoecium incompletely trilocular, triovulate, ± spherical, covered with minute hair-like scales, stigmas 3 short, apical, ovule basally attached, anatropous. *Fruit* 1-seeded, perianth whorls persistent, stigmatic remains very small, apical; epicarp covered in minute, ± irregularly arranged scales, mesocarp thin, fibrous, endocarp not differentiated. *Seed* attached near the base, sarcotesta thick but not juicy, endosperm homogeneous; embryo basal. *Germination* adjacent ligular, eophyll bifid. *Cytology* not studied.

Distribution and ecology: A single species widespread in Indochina, Burma, Thailand, Malay Peninsula and Sumatra. *Myrialepis paradoxa* is found at altitudes from near sea level to about 1000 m in the mountains; it tends to form extensive thickets and, like the other Asiatic hapaxanthic rattan genera, appears to show a preference for disturbed sites in primary forest, such as large light gaps or old landslips. Nothing is known of pollination or dispersal.
Anatomy: Leaf, stem (Tomlinson 1961).
Relationships: For relationships, see *Plectocomia*.
Common names and uses: Rattan, *rotan kertong*. The stems are of poor quality and used only in coarse basketry.
Taxonomic account: Dransfield (1982b).
Fossil record: No generic records found.

Notes: *Myrialepis* is very similar to *Plectocomiopsis* but lacks a sheathing ocrea. Other differences are the diffuse branching of the inflorescence, the much thinner androecium and the minute, irregularly arranged scales on the fruit in contrast to the large, regularly arranged scales of *Plectocomiopsis*.

16. PLECTOCOMIOPSIS

Clustering high-climbing pinnate-leaved rattan palms of Southeast Asia; sheaths are densely to very sparsely armed and an ocrea is present; hapaxanthic and dioecious, the staminate flowers are borne in dense clusters representing condensed rachillae, each flower solitary, the pistillate flowers also in clusters, but with fewer flowers, each individual flower solitary; fruit relatively large, covered with conspicuous scales.

Plectocomiopsis Becc. in Hook.f., Fl. Brit. India 6: 479 (1893). Lectotype: **P. geminiflora** (Griff.) Becc. (*Calamus geminiflorus* Griff.) (see H.E. Moore 1963c).

Combining the generic name Plectocomia *with the ending* -opsis — *similar to*.

Moderate to robust, clustered, high-climbing, spiny, hapaxanthic, dioecious, rattan palms. *Stem* eventually becoming bare, often 3-angled, with long internodes and conspicuous nodal scars, basal branches borne in an axillary position. *Leaves* of mature climbing stems pinnate, cirrate; leaf sheath tubular, rather sparsely armed or unarmed, usually bearing caducous indumentum; knee absent; ocrea present, sometimes tattering, but usually remaining entire and conspicuous; flagellum absent; petiole present or absent, it and the proximal part of the rachis deeply channelled and sparsely spiny; cirrus and distal part of rachis abaxially armed with regularly arranged groups of reflexed grapnel spines; leaflets few to numerous, single-fold, lanceolate, entire, regularly arranged, sometimes armed along margins and/or midrib with conspicuous bristles and bands of caducous scales, midribs evident, transverse veinlets conspicuous. *Inflorescences* produced simultaneously from the axils of the most distal, often reduced leaves, branched to 2 (3) orders, inflorescence axis adnate to the proximal part of the internode above the subtending node, emerging from the leaf sheath mouth; peduncle short; prophyll tubular, 2-keeled, included within the leaf sheath; peduncular bracts absent (?always); rachis much longer than the peduncle; rachis bracts tubular, ± distichous, each subtending a horizontal or ± pendulous, first-order branch; first-order branches with basal, 2-keeled prophyll and close, distichous, tubular bracts with triangular limbs, each subtending a flower cluster (except in *Plectocomiopsis corneri* where inflorescences diffusely branched to 3 orders, and flowers borne on axes of all 3 orders); flower cluster monopodial, representing a condensed rachilla. *Staminate flowers* borne ± distichously in clusters of up to 32 flowers, each flower in the axil of a cup-like rachilla bract and bearing a cup-like, 2-keeled bracteole; calyx thick, very leathery, tubular, with 3 short lobes, the abaxial surface often covered in scale-like trichomes; corolla thick, leathery, tubular through most of its length, split distally to form 3 approximate triangular lobes, covered in scale-like trichomes; stamens 6, epipetalous, united laterally to form a tube tipped with 6 short, reflexed, free filaments bearing short, rounded to oblong, medifixed, introrse

Opposite: ***Plectocomiopsis***. **a**, portion of staminate inflorescence × 1 1/2; **b**, staminate rachilla, some flowers removed to show scars and bracteoles × 3; **c**, staminate flower × 3; **d**, staminate flower in vertical section × 6; **e**, staminate calyx × 3; **f**, staminate flower, calyx removed × 3; **g**, androecium × 3; **h**, stamen in 3 views × 12; **i**, pistillate inflorescence × 1 1/2; **j**, pair of pistillate flowers × 1 1/2; **k**, pair of pistillate flowers, one flower removed to show bract and bracteole × 1 1/2; **l**, bract and bracteole of pistillate flower × 3; **m**, pair of pistillate flowers, flowers removed to show bracteoles × 1 1/2; **n**, pistillate flower × 3; **o**, pistillate flower in vertical section × 3; **p**, pistillate calyx × 3; **q**, pistillate flower, calyx removed × 3; **r**, pistillate petal × 3; **s**, pistillate flower, perianth removed to show staminodial ring × 3; **t**, pistillate flower, perianth removed and staminodial ring cut open to show gynoecium × 3; **u**, ovary in cross-section × 6; **v**, fruit × 1 1/2; **w**, base of fruit × 3; **x**, scales of fruit × 6; **y**, fruit in vertical section × 1 1/2; **z**, fruit in cross-section × 1 1/2; **aa**, seed in 2 views × 1 1/2. *Plectocomiopsis geminiflora*: **a–h**, *Moore* 9602; **i–u**, *Moore* 9203; **v–aa**, *Moore* 9186. (Drawn by Marion Ruff Sheehan)

anthers; pistillode minute. *Pollen* ellipsoidal, bi-symmetric; apertures equatorially disulcate; ectexine tectate or semitectate, finely perforate-rugulate, or reticulate, alternatively exine intectate with angular clavae interspersed with granulae, aperture margins similar to or slightly finer than surrounding ectexine; infratectum columellate; longest axis 19–34 µm [5/5]. *Pistillate flowers* rarely solitary, or borne in groups of 2–4 (rarely more), each in the axil of a cup-like rachilla bract and bearing a 2-keeled, cup-like bracteole and rarely a minute second bracteole; calyx tubular, thick, leathery, divided into 3 low lobes, frequently bearing scale-like trichomes, persisting into fruiting stage, enlarging, splitting and cracking irregularly; corolla thick, leathery, densely scaly, divided into 3 short lobes, later splitting irregularly; staminodial ring epipetalous, bearing 6 very short lobes and pendulous empty anthers; gynoecium ovoid to cylindrical, at anthesis scaly only near the base of the 3 apical stigmas, locules 3, incomplete, each with 1 anatropous, basally attached ovule. *Fruit* 1 (very rarely 2)-seeded, perianth whorls persistent and enlarging, stigmatic remains apical; epicarp covered in somewhat irregular vertical rows of reflexed scales, mesocarp thin, endocarp not differentiated. *Seed* basally attached, usually depressed, globose, sarcotesta thick, but not juicy, endosperm homogeneous; embryo basal. *Germination* adjacent-ligular; eophyll bifid, lamina composed of several folds. *Cytology* not studied.

Distribution and ecology: Five species distributed in south Thailand (1 species), the Malay Peninsula (4), Sumatra (2) and Borneo (3). All species, like *Myrialepis*, seem to be adapted to colonizing disturbed habitats within primary forest.

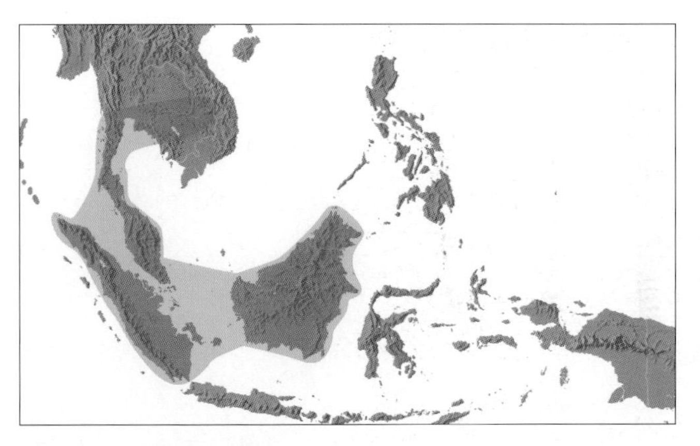

Distribution of ***Plectocomiopsis***

Plectocomiopsis triquetra and *P. wrayi* are confined to lowland peat-swamp forest in Borneo and the Malay Peninsula, respectively; *P. corneri* and *P. mira* are local plants of lowland dipterocarp forest up to about 700 m altitude, whereas *P. geminiflora* is found in a wide range of forest types up to about 1200 m altitude.
Anatomy: Leaf (Tomlinson 1961).
Relationships: The monophyly of *Plectocomiopsis* has not been tested. For relationships, see *Plectocomia*.

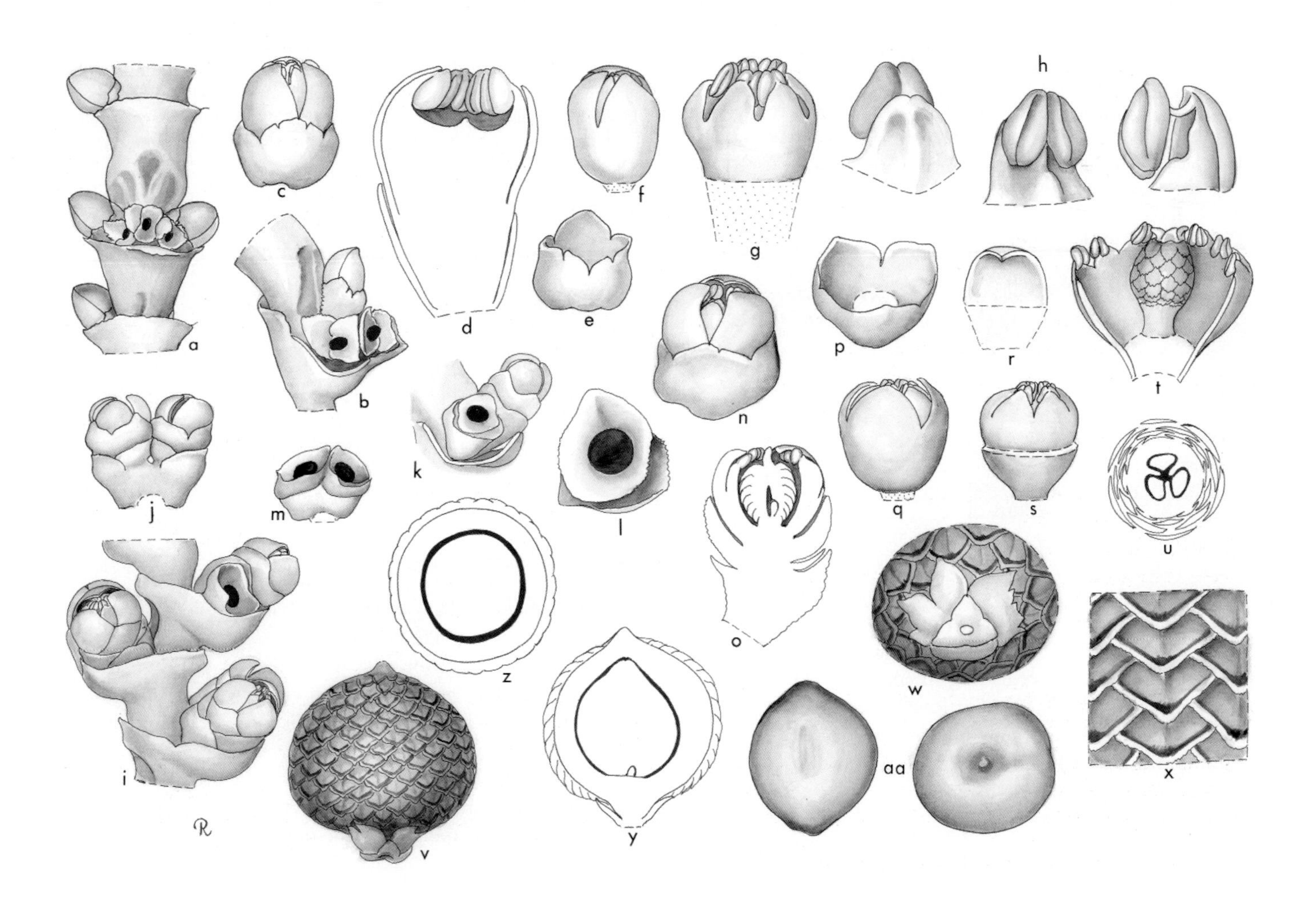

Above left: ***Plectocomiopsis geminiflora***, habit, Peninsular Malaysia. (Photo: J. Dransfield)

Right: ***Plectocomiopsis mira***, fruit, Sabah. (Photo: J. Dransfield)

Below: ***Plectocomiopsis***
a, equatorial disulcate pollen grain, distal face SEM × 2000;
b, ultrathin section through whole pollen grain, equatorial plane TEM × 2000.
Plectocomiopsis triquetra: **a**, *Brunig* S5700; *P. wrayi*: **b**, *Dransfield* JD5365.

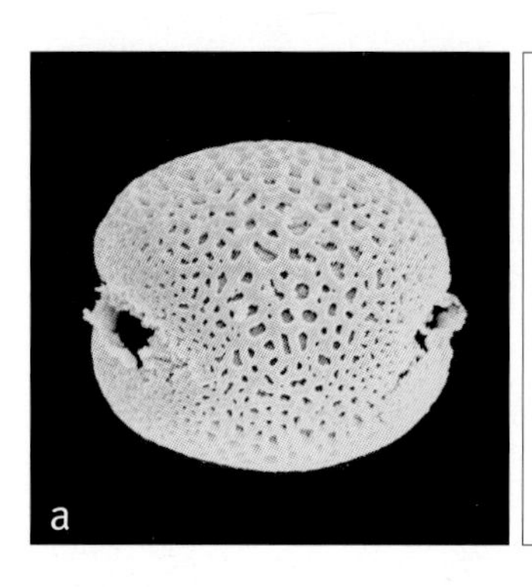

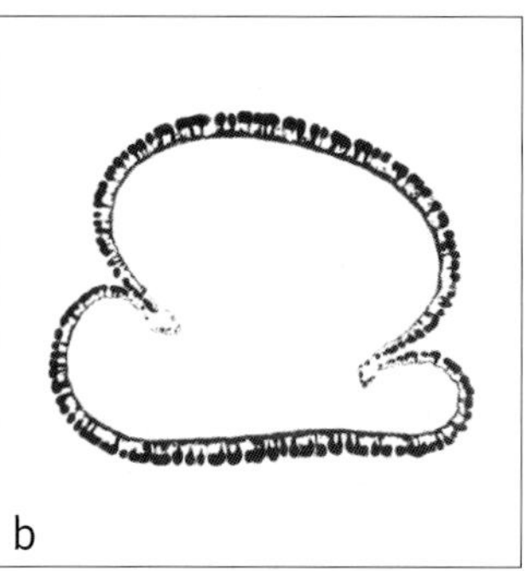

Common names and uses: Rattan, *rotan*. The apex of *Plectocomiopsis geminiflora*, though bitter, is highly esteemed in Borneo as a vegetable; opinions vary concerning other species, with some villagers regarding them as good, others as poisonous. The stems are of little value being prone to split when bent and are used only in coarse basketry.
Taxonomic account: Dransfield (1982b).
Fossil record: No generic records found.
Notes: See *Myrialepis*.

Subtribe Calaminae Meisn., Plant. Vasc. Gen.: Tab. Diagn. 354, Comm. 265 (1842). Type: **Calamus**.

Climbing or acaulescent, dioecious, pleonanthic (rarely hapaxanthic) palms; leaves with cirri, where present, lacking acanthophylls; inflorescence adnate to internode and leaf sheath; leaf sheath usually with knee-like swelling below insertion of petiole; anthers dorsifixed; stigmas divergent.

This subtribe includes most of the climbing palms encountered in Southeast Asia and Malesia. The economically most important species of rattan are furthermore members of genera in this subtribe. With an enormous number of species, the subtribe represents an astonishing radiation.

Subtribe Calaminae is consistently supported as monophyletic with moderate to high support (Baker *et al.* 1999b, 2000a, 2000b, in review). The Calaminae is sister to subtribe Plectocomiinae with moderate to high support (Baker *et al.* 2000a, 2000b, in review), although one study provides poor evidence for a sister relationship to subtribe Pigafettinae (Asmussen *et al.* 2006).

Key to Genera of Calaminae

1. Pistillate rachillae bearing solitary flowers; rachillae enclosed by net-like bracts. Borneo . . . . **18. Retispatha**
1. Pistillate rachillae bearing dyads of one pistillate flower together with a sterile staminate flower (very rarely 2 pistillate and 1 sterile staminate or more); rachillae hardly enclosed in net-like bracts 2
2. Entire inflorescence enclosed within the prophyll 3
2. Inflorescence not enclosed within the prophyll, or prophyll sheathing only the base of the inflorescence 5

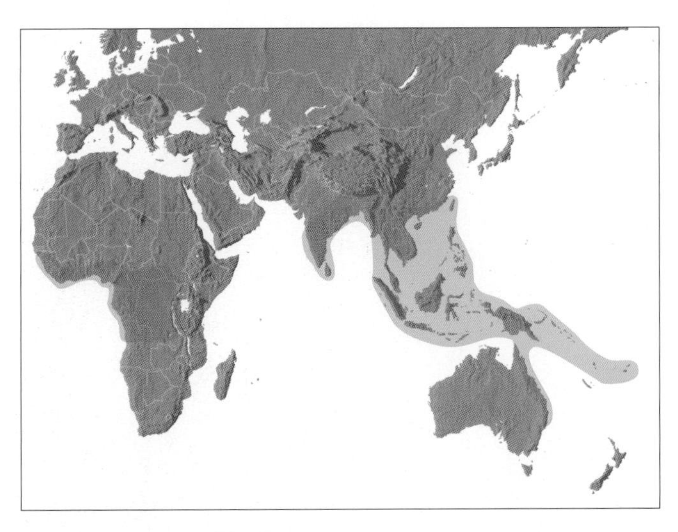

Distribution of subtribe **Calamineae**

3. Prophyll opening by a pair of pores near the apex, splitting along its length only after anthesis or accidentally. Malay Peninsula, Sumatra, Java, Borneo .... **20. Ceratolobus**
3. Prophyll splitting along its length at anthesis 4
4. Prophyll much larger than any other inflorescence bract; two erect ear-like appendages (auricles) present at the mouth of the leaf sheath; climbing whips absent; endosperm homogeneous. Malay Peninsula, Borneo .... **21. Pogonotium**
4. Prophyll enclosing a series of similar rachis bracts, but with their tips enfolded in the beak of the prophyll; auricles absent; cirrus almost always present; endosperm ruminate. India–New Guinea .... **19. Daemonorops** .... (in part, subgenus **Daemonorops**)
5. Prophyll and rachis bracts splitting along their lengths to the very base, usually neatly falling at anthesis except for the prophyll, which may persist long after anthesis; floral bracteoles usually low, often rather inconspicuous; endosperm always ruminate. India–New Guinea .... **19. Daemonorops** (in part, subgenus **Piptospatha**)
5. Prophyll and rachis bracts strictly tubular, or, if splitting, then persistent and with a tubular base; floral bracteoles inconspicuous to conspicuous; endosperm ruminate or homogeneous .... **17. Calamus**

17. CALAMUS

Immensely variable genus of mostly climbing palms, some acaulescent or erect, found in equatorial Africa, India, Himalayan foothills to south China, throughout Southeast Asia to the western Pacific Islands and Australia; sheaths, petioles and rachis usually densely armed, leaf often terminating in a cirrus armed with spines, or cirrus absent; flagellum sometimes present (sterile inflorescence modified as a climbing organ); pleonanthic and dioecious, the inflorescence is very varied but bracts are usually tubular, sometimes splitting, but if so, never to the base and never caducous.

Right: ***Calamus manan***, habit, in plantation, Sarawak. (Photo: J. Dransfield)

Calamus L., Sp. Pl. 325 (1753). Type: **C. rotang** L.

Rotanga Boehm. in Ludw., Defin. Gen. Pl. (3rd Edn): 395 (1760).

Rotang Adans., Fam. Pl. 2: 24, 599 (1763).

Palmijuncus Kuntze, Revis. Gen. Pl. 2: 731 (1891).

Calospatha Becc., Ann. Roy. Bot. Gard. (Calcutta) 12(1): 232 (1911). Type: *C. scortechinii* Becc. (= *Calamus calospathus* (Ridl.) W.J. Baker & J. Dransf.).

Zalaccella Becc., Ann. Roy. Bot. Gard. (Calcutta) 11(1): 496 (1908). Type: *Z. harmandii* (Pierre ex Becc.) Becc. (*Calamus harmandii* Pierre ex Becc.).

Schizospatha Furtado, Gard. Bull. Singapore 14: 525 (1955). Type: *S. setigera* (Burret) Furtado (*Calamus setigerus* Burret = *Calamus anomalus* Burret).

Cornera Furtado, Gard. Bull. Singapore 14: 518 (1955). Type: *C. pycnocarpa* Furtado (= *Calamus pycnocarpus* [Furtado] J. Dransf.).

From the Latin, calamus *— a reed.*

Solitary or clustered, spiny, acaulescent, erect, or high-climbing, pleonanthic, dioecious, rattan palms. *Stem* eventually becoming bare, with short to long internodes, sucker shoots strictly axillary. *Leaves* pinnate, rarely bifid, sometimes with a terminal cirrus; sheath splitting in acaulescent species, in the exposed area usually densely armed with scattered or whorled spines, in one species (*Calamus polystachys*) the spines

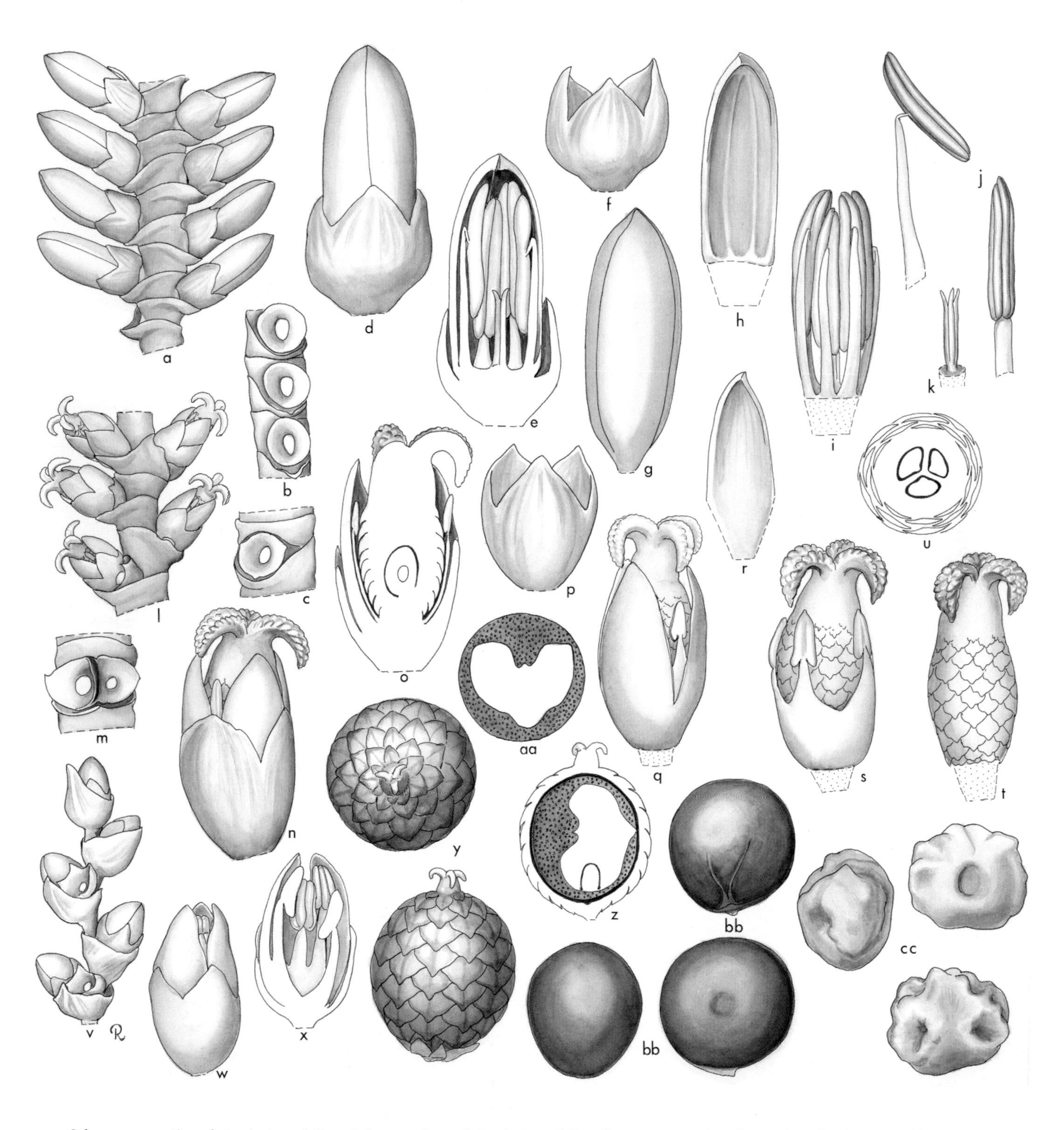

Calamus. **a**, portion of staminate rachilla × 6; **b**, **c**, portions of staminate rachillae, flowers removed to show subtending bracts and bracteoles × 6; **d**, staminate bud × 12; **e**, staminate bud in vertical section × 12; **f**, staminate calyx × 12; **g**, staminate bud, calyx removed × 12; **h**, staminate petal, interior view × 12; **i**, androecium × 12; **j**, stamen in 2 views × 12; **k**, pistillode × 12; **l**, portion of pistillate rachilla with dyads of pistillate and sterile staminate flowers, sterile staminate flowers removed × 6; **m**, dyad, flowers removed to show subtending bract and bracteoles × 9; **n**, pistillate flower × 15; **o**, pistillate flower in vertical section × 15; **p**, pistillate calyx × 15; **q**, pistillate flower, calyx removed × 15; **r**, pistillate petal × 15; **s**, gynoecium and staminodial ring × 15; **t**, gynoecium × 15; **u**, ovary × 15; **v**, portion of pistillate rachilla with dyads of pistillate and sterile staminate flowers, pistillate flowers removed × 6; **w**, sterile staminate flower × 12; **x**, sterile staminate flower in vertical section × 12; **y**, fruit in 2 views × $4^{1}/_{2}$; **z**, fruit in vertical section × 12; **aa**, endocarp in cross-section × $4^{1}/_{2}$; **bb**, endocarp in 3 views × $4^{1}/_{2}$; **cc**, seed in 3 views × $7^{1}/_{2}$. *Calamus javensis*: **a–k** *Dransfield* 4519; *C. acuminatus*: **l–u** and **y–cc**, *Moore* 9208; **v–x**, SAN 23490. (Drawn by Marion Ruff Sheehan)

interlocking to form galleries occupied by ants, indumentum often abundant on sheath surface; ocrea often present, sometimes greatly elaborated, papery and disintegrating, or coriaceous, rarely greatly swollen or diverging with inrolled margins and occupied by ants; knee present in most climbing species; flagellum (climbing whip derived from a sterile inflorescence) often present in species lacking cirri, very rarely a small vestigial flagellum present in cirrate species (e.g., *C. pogonacanthus*); petiole absent or well developed, flattened adaxially, rounded abaxially, variously armed; rachis often armed with distant groups of reflexed grapnel spines; cirrus when present armed with scattered (rarely) or grouped reflexed spines; leaflets few to very numerous, single-fold, entire or in 1 species praemorse (*C. caryotoides*), linear to lanceolate or rhomboid, sometimes the terminal pair partially joined along their inner margins forming a terminal compound leaflet or flabellum, regularly arranged or irregular, grouped, sometimes fanned within the groups, concolourous or discolourous, variously bearing hairs, bristles, spines, and scales, midribs conspicuous or not, transverse veinlets conspicuous or obscure. *Inflorescences* axillary but adnate to the internode and leaf sheath of the following leaf, staminate and pistillate superficially similar, but the staminate usually branching to 3 orders and the pistillate to 2 orders, the inflorescence frequently flagelliform, very rarely rooting at its tip and producing a new vegetative shoot; peduncle absent or present, sometimes very long, erect or pendulous, variously armed; prophyll usually inconspicuous, 2-keeled, tubular, tightly sheathing, variously armed or unarmed, rarely inflated, papery or coriaceous, splitting down one side, usually empty; rachis bracts persistent, like the prophyll, close or sometimes very distant, variously armed, usually strictly tubular, even where splitting remaining tubular at the base, rarely irregularly tattering in the distal part, each subtending a first-order branch or 'partial inflorescence', this frequently adnate to the rachis above the bract axil, very rarely bursting through the bract; first-order branch bearing a 2-keeled, tubular prophyll and ± subdistichous, tubular bracts, unarmed or variously armed, each subtending a second-order branch, usually adnate to the first-order branch above the bract node; rachillae very varied within the genus, spreading to very short and crowded, bearing a basal, 2-keeled prophyll and conspicuous, usually distichous, tubular bracts with triangular tips, variously armed or unarmed, very rarely the bracts

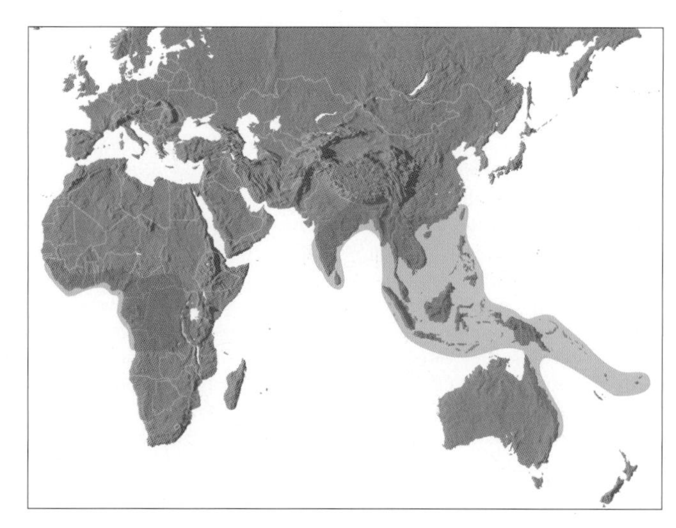

Distribution of ***Calamus***

highly condensed and spiral, in staminate rachilla, each bract subtending a solitary staminate flower bearing a prophyllar bracteole, in pistillate rachilla each bract subtending a dyad of a sterile staminate and a fertile pistillate flower and 2, usually quite conspicuous, prophyllar bracteoles, very rarely each bract subtending a triad of 2 lateral pistillate flowers and a central sterile staminate flower. *Staminate flowers* symmetrical; calyx tubular at the base, 3-lobed distally; corolla usually exceeding the calyx, divided into 3, valvate lobes except at the tubular base; stamens 6 (12 in *Calamus ornatus*), borne at the mouth of the corolla tube, filaments often fleshy, elongate, sometimes abruptly narrowed, anthers medifixed, short to elongate, latrorse or introrse; pistillode minute to quite conspicuous. *Pollen* ellipsoidal, bi-symmetric; apertures equatorially disulcate; ectexine tectate or semi-tectate, psilate, perforate, coarsely perforate, foveolate, finely to coarsely reticulate, reticulate-rugulate, verrucate, gemmate or,

Below left: ***Calamus manan***, leaf sheath, Peninsular Malaysia. (Photo: J. Dransfield)
Middle: ***Calamus muricatus***, leaf sheath, Brunei Darussalam. (Photo: J. Dransfield)
Right: ***Calamus blumei***, leaflets, Sabah. (Photo: J. Beaman)

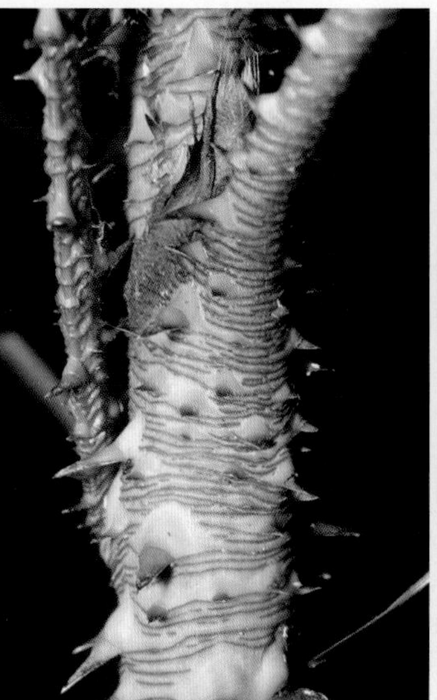

rarely, ectexine intectate with large, loosely attached, psilate gemmae, aperture margins usually similar to surrounding ectexine; infratectum columellate; longest axis 17–67 µm; post-meiotic tetrads tetragonal [104/363]. *Sterile staminate flowers* like the fertile but with empty anthers. *Pistillate flowers* usually larger than the staminate; calyx tubular, shallowly 3-lobed; corolla rarely exceeding the calyx, divided more deeply than the calyx into 3 valvate lobes; staminodes 6, epipetalous, the filaments distinct or united into a short ring, anthers empty; gynoecium tricarpellate, triovulate, spherical to ellipsoidal, covered in reflexed scales, stigmas 3, apical, fleshy, reflexed, sometimes borne on a beak, locules incomplete, ovules basal, anatropous. *Fruit* usually 1-seeded, rarely consistently 2- or 3-seeded, stigmatic remains apical; epicarp covered in neat vertical rows of reflexed scales, mesocarp usually very thin at maturity, endocarp not differentiated. *Seed* with thick sweet, sour, or astringent sarcotesta, inner part of the seed rounded, grooved, angled, or sharply winged, endosperm homogeneous or ruminate; embryo basal or lateral. *Germination* adjacent-ligular; eophyll bifid or pinnate. *Cytology*: 2n = 26.

Distribution and ecology: With about 374 species, *Calamus* is the largest palm genus. It has a very wide distribution, occurring in the humid tropics of Africa (one variable species), India, Burma, and south China through the Malay Archipelago to Queensland and Fiji, reaching greatest diversity and number of species in the Sunda Shelf area (especially Borneo), with a second centre of diversity in New Guinea. The ecology is very varied as might be expected in such a large genus, but, although some species are adapted to seasonally dry habitats such as monsoon forest, there are no species in semi-arid habitats. There are species adapted to sub-mangrove conditions (*C. erinaceus*). Other species have narrow ecological requirements, such as limestone or ultrabasic soils. In altitude, the genus ranges from sea-level to over 3000 m (*C. gibbsianus* on Mt Kinabalu).

Anatomy: Leaf, petiole, stem, root (Tomlinson 1961), root (Seubert 1996a). Similar in anatomy to *Ceratalobus*, *Myrialepis*, and *Plectocomiopsis*.

Relationships: All pertinent studies provide strong evidence that *Calamus* is not monophyletic and that the other four genera of Calaminae are nested within it (Baker *et al.* 2000a, 2000b, 2000c). The most densely sampled study (Baker *et al.* 2000c) suggests that *Calamus* breaks into two major clades, with *Retispatha* embedded in one and a clade of *Daemonorops*, *Ceratolobus* and *Pogonotium* sister to the other. The disjunct African species *C. deerratus* may constitute a third distinct lineage in the genus. The inter-generic relationships resolved within the clade of *Daemonorops*, *Ceratolobus* and *Pogonotium* are not well supported by the data, but all studies suggest that *Daemonorops* is also non-monophyletic (Baker *et al.* 2000a, 2000b, 2000c). In the absence of more widely sampled phylogenies based on multiple data sets, however, an alternative classification of these genera cannot be proposed at this time.

In practice, there is usually little difficulty in assigning fertile material to one of the five genera recognised here. For practical reasons, we therefore maintain the genera recognised in the first edition of *Genera Palmarum*, except for monotypic *Calospatha*, which is clearly nested within a small group of *Calamus* (including species formerly included in *Cornera*). When the results of further analyses are available, it will be necessary to review the group and further changes to generic delimitation can be expected.

Top: ***Calamus castaneus***, staminate flower visited by a wasp, a fly and a mosquito, Peninsular Malaysia. (Photo: J. Dransfield)

Middle: ***Calamus spectatissimus***, pistillate rachilla at anthesis, Sumatra. (Photo: J. Dransfield)

Bottom: ***Calamus nanodendron***, fruit, Sarawak. (Photo: J. Dransfield)

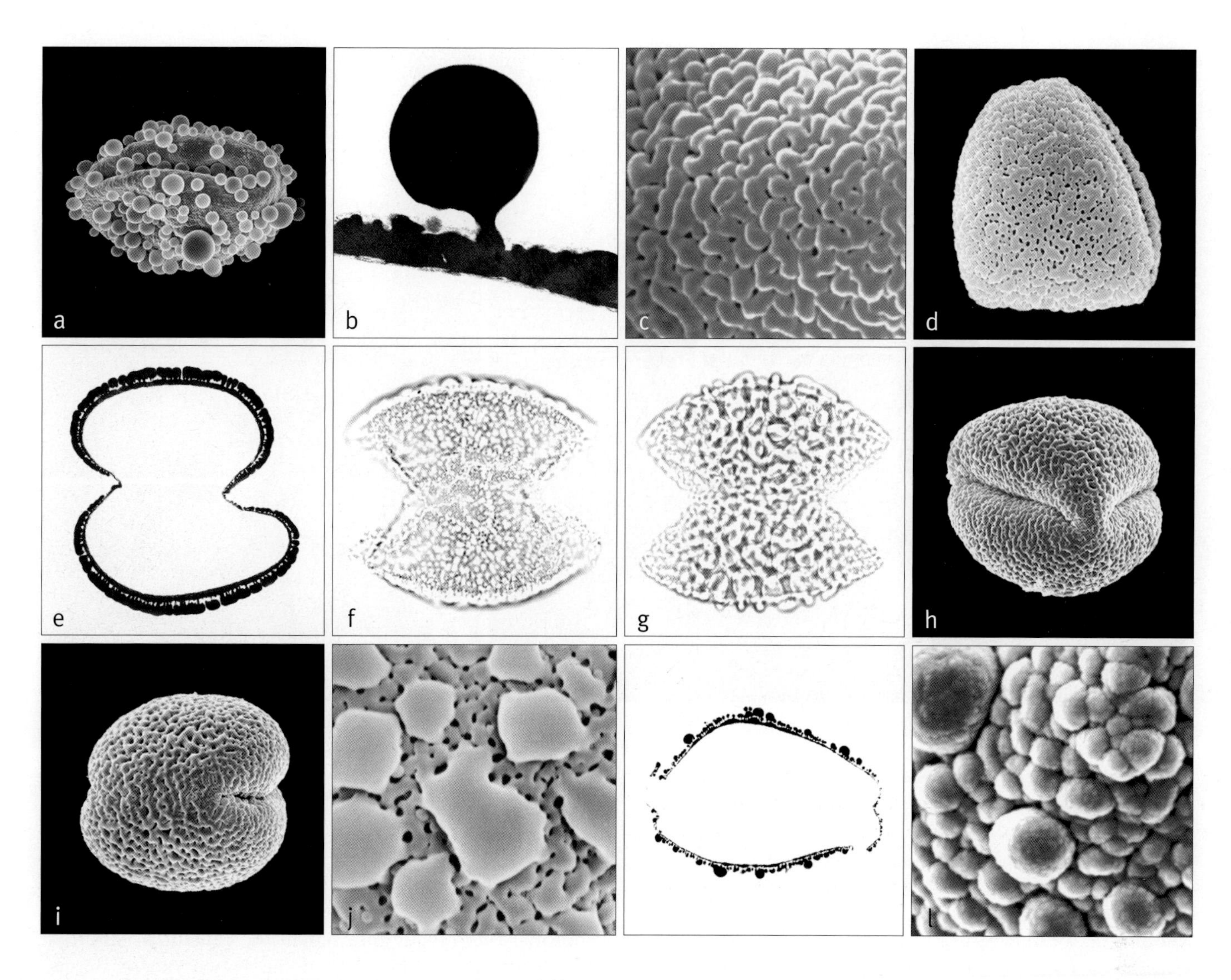

Above: ***Calamus***
a, equatorial disulcate pollen grain, collapsed, intectate gemmate surface SEM × 1000;
b, ultrathin section through a gemma TEM × 10,000;
c, close-up, finely rugulate surface SEM × 8000;
d, equatorial disulcate pollen grain, equatorial face, sulcus margins infolded SEM × 1500;
e, ultrathin section through whole pollen grain, equatorial plane TEM × 1500;
f, equatorial disulcate pollen grain, proximal face, mid-focus LM × 800;
g, equatorial disulcate pollen grain, distal face, high-focus LM × 800;
h, equatorial disulcate pollen grain, distal face SEM × 2000;
i, equatorial disulcate pollen grain, proximal face SEM × 2000;
j, close-up, reticulate/coarsely verrucate surface SEM × 8000;
k, ultrathin section through whole pollen grain, equatorial plane TEM × 2000;
l, close-up, densely verrucate/gemmate surface SEM × 8000.
Calamus lobbianus: **a** & **b**, *Dransfield* JD5875; *C. amplijugus*: **c**, *Chai* SAN31671; *C. scipionum*: **d**, *Furtado* S.F. 30904; *C. mesilauensis*: **e**, *Dransfield* JD5698; *C. ornatus*: **f**, *King* 3951; *C. bousigonii* subsp. *smitinandii*: **g**, *Charoenphol et al.* 3721; *C. deerratus*: **h** & **i**, *Tuley* 846; *C. viminalis*: **j**, *Lobb s.n.*; *C. exilis*: **k**, *Dransfield* JD4510; *C. psilocladus*: **l**, *Dransfield* JD6135.

Common names and uses: Rattan, *rotan*; for local names see Dransfield 1979a. The finest kinds of rattan are all species of *Calamus*. *C. manan*, *C. caesius*, and *C. trachycoleus*, in particular, dominate world trade in rattans. Other species are almost as important. For further details of rattans and their exploitation see Dransfield 1979a. Species of *Calamus* have a wide range of uses apart from entering the rattan trade. Leaves are used for thatch, spines in various ways, cirri have been used for constructing fish traps, fruits are eaten and may even be sold in local markets, and some species may be medicinally valuable.

Fossil record: Fossils attributed to *Calamus* include pinnate leaf fragments, spines and spine bundles, spiny bark, fruits, male and female flowers and pollen, but the leaf, spines and fruit could equally be attributed to other members of Calaminae or indeed to spiny members of Arecoideae. The oldest records of macro 'organs' are probably those from the Lower Eocene of southern England (Chandler 1957, 1962); under the cautionary name *C. daemonorops*, Chandler (1957) assigns spines, fruiting axes, young and older fruits, male flowers, one with disulcate pollen, and female flowers from the Oligocene flora of the Bovey Tracey Lake Basin, UK, and a pair of spines/prickles from the Lower Eocene Bagshot Beds

Above left: ***Calamus calospathus***, pistillate inflorescence, Peninsular Malaysia. (Photo: J. Dransfield)
Middle: ***Calamus* sp.**, staminate rachillae at anthesis, Papua. (Photo: W.J. Baker)
Right: ***Calamus wailong***, mature fruit, Thailand. (Photo: J. Dransfield)

(Chandler 1962). From the Landes District of France, Huard (1967) describes spines from Neogene lignite ("probably Miocene"), concluding that the closest comparison is with *Calamus* or *Daemonorops*. Other macro records include reduplicate pinnate leaf fragments in association with a small "tessellated" fruit, *C. noszkyi* (Jablonszky 1914) from the Lower Miocene of Tarnóc, Hungary. From the upper brown coal seam of the Miocene Turów Basin, Poland, Czeczott and Juchniewicz (1980) describe fragments of spiny bark, epidermis and loose spine bundles, *Spinophyllum daemonorops*, which they conclude are probably from *Calamus* or *Daemonorops*. From the Miocene of the Czech Republic (Zittau Basin), Teodoridis (2003) also reports *Calamus/Daemonorops*-type spines. From the German Middle Miocene, Mai (1964), Mai and Walther (1978) and Gregor (1982) describe palm-like spine fragments, which they include in *C. daemonorops*. At least some of these fossil spine fragments could represent spiny members of tribe Cocoseae. Fossil records of palm-like equatorial disulcate (dicolpate) pollen are numerous; most are included in the fossil genus *Dicolpopollis* (Pflanzl 1956), but less frequently the names *Disulcites* or *Dicolpites* are used. The earliest record is probably that of Schrank (1994) who records *Dicolpopollis* from the Upper Cretaceous Yesomma Formation, northern Somalia. Van der Hammen (1954) records two species of *Dicolpites* from the Maastrichtian of Colombia, but the drawings of these grains do not readily compare with calamoid disulcate pollen and no suggestions regarding their possible affinity were made by the author. Two types of reticulate equatorial disulcate *Dicolpopollis* are described from the Palaeocene of Borneo (Muller 1968), and a verrucate pollen type is recorded from the Miocene of Borneo (Muller 1979). *Dicolpopollis* is widespread in Europe, for example: UK — Lower Tertiary, Woolwich, London Clay and Bagshot Beds (Khin Sein 1961, Gruas-Cavagnetto 1976); Austria — Lower Eocene, Krappfeld area (Hofmann & Zetter 2001, Zetter & Hofmann 2001); Czech Republic — Miocene, North Bohemian brown coal basin (Konzalová 1971); France — Lower Eocene, Paris Basin (Ollivier Pierre *et al.* 1987); Belgium — Lower Eocene, Argile de Merelbeke (Roche 1982). The Miocene sediments of the west coast of southern India are rich in *Dicolpopollis*, notably in Kerala State (Ramanujam 1987, Ramanujam & Rao 1977, Ramanujam *et al.* 1986, 1991b, 1992, 2001, Rao & Ramanujam 1975, 1978, Varma *et al.* 1986, Srisailam & Ramanujam 1982, Singh & Rao 1990) but there are also records from Maharashtra State, for example, Saxena and Misra (1990). *Dicolpopollis* also occurs in southeast India — Tamil Nadu (Sarma *et al.* 1984, Ramanujam *et al.* 1986, 2001); eastern India — Andhra Pradesh (Ramanujam *et al.* 1986, 2001); and in north eastern India — Shillong Plateau, Meghalaya (Salujha *et al.* 1973a, 1973b); it is also known from the Eocene of Burma (Potonié 1960). *Dicolpopollis* is abundant

in the Eocene Nanggulan Formation of Central Java (Takahashi 1982, Harley & Morley 1995 — cf. *Calamus paspalanthus*), whereas *D. metroxylonoides* in the Tertiary sediments of Papua New Guinea (Khan 1976) is considered to have an affinity with *Metroxylon*. *Dicolpopollis bungonensis* from the Eocene of New South Wales, Australia, is favourably compared with the pollen of *C. moti* F.M. Bailey (Truswell & Owen 1988). *Dicolpopollis* is well known in some Tertiary deposits of China (Song *et al.* 1999). The first American record of *Dicolpopollis* (Tschudy 1973) occurs in the Eocene Wilcox Group, Mississippi Basin, southern USA where it is present in appreciable numbers (ca. 10% of sample) and has an "... apparently very short stratigraphic range making it a potentially very useful fossil." The images in Tschudy (1973) are typical *Dicolpopollis* and can confidently be assumed to have a close affinity with *Calamus*. A review of the palaeopalynology and palaeoecology of *Dicolpopollis* is provided in Ediger *et al.* (1990).
(N.B. The fossil genus *Calamuspollenites* should not be confused with *Calamus*; it is monosulcate and was placed in synonomy with *Arecipites* [= Arecaceae-like monosulcate pollen] by Nichols *et al.* [1973].)

Taxonomic accounts: Beccari (1908, 1913b), Dransfield (1979a, 1984a, 1992c), Evans *et al.* (2002).

Notes: *Calamus*, the largest genus in the family, has yet to be treated satisfactorily at the subgeneric level. Beccari's informal groupings remain the most useful subgeneric categories.

Draco Crantz, De Duabus Draconis Arboribus Botanicorum 13 (1768), is occasionally cited as a synonym of *Calamus*. We have been unable to locate a satisfactory lectotypification of the generic name; of the six specific names published by Crantz, two refer to *Dracaena draco*, two to *Pterocarpus* spp., one (*Draco thaa* Crantz) is without equivalent in Index Kewensis, and one refers to *C. rotang* in synonymy. We suggest it more appropriate to select *D. clusii* Crantz as the type of the genus, *Draco* thus becoming a synonym of *Dracaena* (Liliaceae). *Calamus rotang* does not produce dragon's blood, nor does any other species of the genus.

Additional figures: Figs 2.5b, 8.1, Glossary figs 6, 18.

18. RETISPATHA

Erect or shortly climbing palm of Borneo; sheaths, petioles and rachis are densely armed, the sheaths lacking a knee and a cirrus absent; inflorescences are interfoliar and pendulous, all bracts net-like.

Retispatha J. Dransf., Kew Bull. 34(3): 529 (1979c). Type: **R. dumetosa** J. Dransf.

Rete — *network*, spatha — *bract, in reference to the net-like inflorescence bracts*.

Moderate, clustered, erect or briefly climbing, spiny, pleonanthic, dioecious, rattan palm. *Stem* eventually becoming bare, relatively robust, with conspicuous nodal scars and relatively short internodes, short bulbil-like shoots sometimes present at the nodes in the lower part of the stem, adventitious roots also abundant at lower nodes. *Leaves* without cirrus, pinnate; sheath tubular, densely armed with slender spines in whorls and partial whorls, and dense indumentum; ocrea absent; knee absent; flagellum absent; petiole well developed, channelled

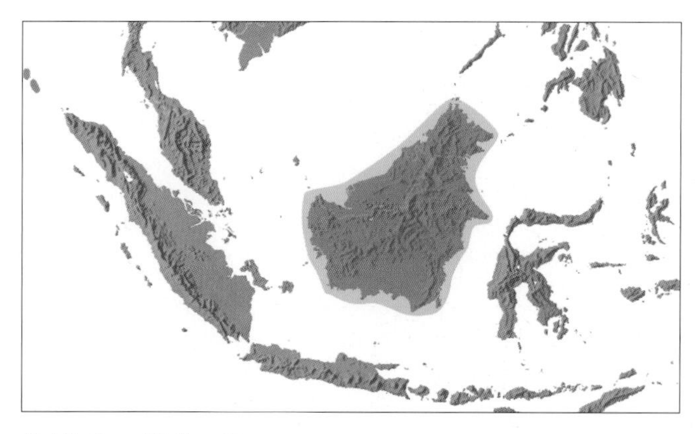

Distribution of ***Retispatha***

adaxially, armed with abundant lateral and abaxial groups of spines; rachis armed with reflexed grapnel spines in groups of up to 5, and bearing abundant indumentum; leaflets numerous, regularly arranged, single-fold, linear, armed along margins and main veins with bristles, midribs prominent, transverse veinlets conspicuous. *Inflorescences* axillary, but adnate to the internode and leaf sheath of the following leaf, erect at first, becoming pendulous, staminate and pistillate superficially similar, branched to 3 orders in staminate, to 1 (rarely 2) orders in pistillate; prophyll large, tubular in proximal $^{1}/_{2}$, splitting and tattering distally, densely covered with black spines in partial whorls; peduncular bracts absent; rachis bracts similar to the prophyll; first-order branches

Below: ***Retispatha dumetosa***, leaf sheaths and inflorescence, Brunei Darussalam. (Photo: J. Dransfield)

becoming pendulous at anthesis, bearing distichous, imbricate, unarmed bracts, tubular in proximal 2/3, with a triangular limb, composed of close criss-cross fibres producing a fine network, each net-like bract subtending and partially or wholly enclosing in the staminate inflorescence a catkin-like condensed branching system; each second-order branch subtended by a triangular, membranous, ciliate-margined tubular bract; third-order branchlets (rachillae) bearing membranous ciliate-margined tubular bracts, each subtending a 2-keeled ciliate-margined, tubular bracteole and a single staminate flower. *Staminate flowers* very small, ± symmetrical; calyx tubular with 3, triangular lobes tipped with hairs; corolla about twice as long as the calyx in bud, tubular only at the very base, lobes 3, striate, valvate, at anthesis the receptacle elongating, the corolla then appearing tubular in the basal 1/3; stamens 6, free from the corolla in bud, becoming briefly epipetalous, filaments briefly connate laterally, anthers oblong to ovate, dorsifixed near the base, latrorse; pistillode trifid, very small. *Pollen* ellipsoidal, bi-symmetric; apertures equatorially disulcate; ectexine tectate, perforate-rugulate with supratectal, often vertically ridged spines, aperture margins similar; infratectum columellate; longest axis 19–24 µm [1/1]. *Pistillate inflorescences* sometimes with 1 major branch near the base, the rachillae borne on the axis and on this major branch, or more frequently the rachillae borne on the main axis only; rachillae subtended by distichous, imbricate, net-like bracts as in the staminate inflorescence, rachillae usually concealed by the bracts, bearing up to ca. 20 distichous, tubular, ciliate-margined bracts, each enclosing a 2-keeled prophyllar bracteole, a tubular, second bracteole and 1 pistillate flower; sterile staminate flowers lacking (see notes). *Pistillate flowers* much larger than the staminate; calyx tubular, with 3 short triangular, valvate lobes, splitting after fertilization; corolla tubular, slightly shorter than the calyx, with 3 short valvate lobes, also splitting further; staminodes 6, briefly epipetalous, filaments connate laterally to form a short tube, empty anthers flattened; gynoecium incompletely trilocular, triovulate, ovoid, scaly, stigmas 3, conspicuous, reflexed, fleshy, borne on a non-scaly style, ovule basally attached, anatropous. *Fruit* 1-seeded, partially concealed by the net-like bracts, ovoid to slightly obpyriform, beaked, stigmatic remains apical; epicarp with neat vertical rows of reflexed scales, mesocarp thin, endocarp not differentiated. *Seed* basally attached with thin sweet sarcotesta, endosperm obscurely angled, homogeneous; embryo basal. *Germination* adjacent-ligular; eophyll pinnate with ca. 4 ciliate-hairy leaflets on each side of the rachis. *Cytology* not studied.

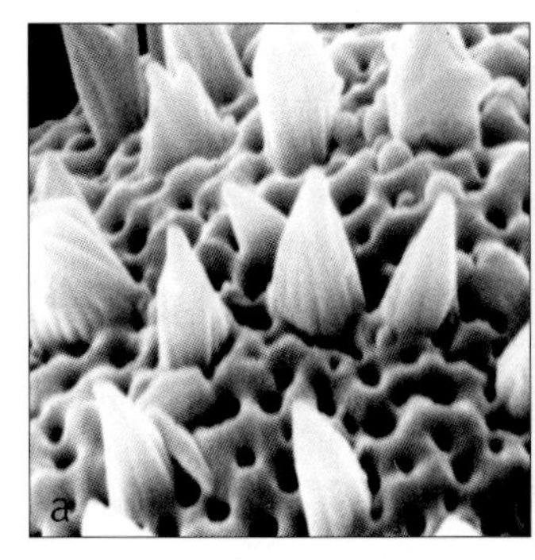

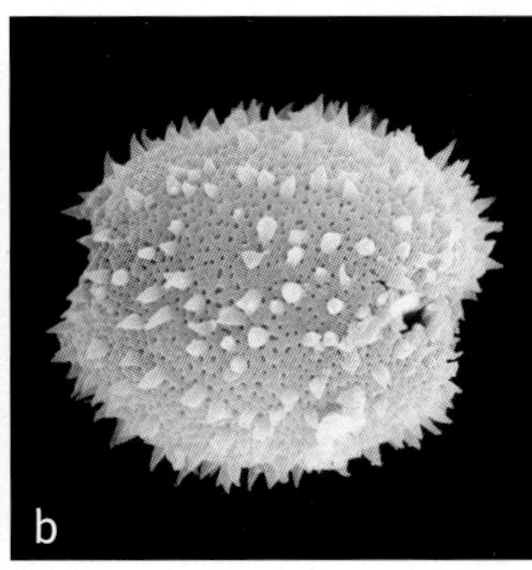

Above: ***Retispatha***
a, close-up, coarsely peforate-rugulate surface with vertically ridged spines SEM × 8000;
b, equatorial disulcate pollen grain, proximal face SEM × 2000.
Retispatha dumetosa: **a** & **b**, *Dransfield* JD5985.

Distribution and ecology: A single species endemic to Borneo, where it is has been recorded from scattered localities throughout the island. *Retispatha dumetosa* forms thickets on hillslopes and valley bottoms in hill Dipterocarp forest. It is absent from montane and heath forest and although its habitat is widespread, it is a rare palm.

Anatomy: Not studied.

Relationships: For relationships, see *Calamus*.

Common names and uses: *Wi tebu bruang* (Malay, the bear's sugar-cane). No local uses have been recorded.

Taxonomic account: Dransfield (1979c).

Fossil record: No generic records found.

Notes: *Retispatha* is distinguished by the extraordinary net-like bracts that subtend the flowering branches and by the lack of both cirrus and flagellum. Although the pistillate rachillae lack the sterile staminate flowers usual in the Calaminae, the affinities of *Retispatha* seem to be with *Calamus* rather than with the Plectocomiinae. Marion Sheehan, in preparing the plate, discovered a single sterile staminate flower borne with an apparently abortive pistillate flower at the very tip of one pistillate rachilla. We have been unable to find any trace of another sterile staminate flower. This single sterile staminate flower can be regarded as an unusual occurrence that confirms the affinity with the Calaminae. The cane of *Retispatha* is remarkably heavy and stiff. As a juvenile, it can stand as a small erect tree.

Additional figures: Glossary fig. 18.

Left: ***Retispatha dumetosa***, staminate rachillae and bracts, Sabah.
(Photo: J. Dransfield)

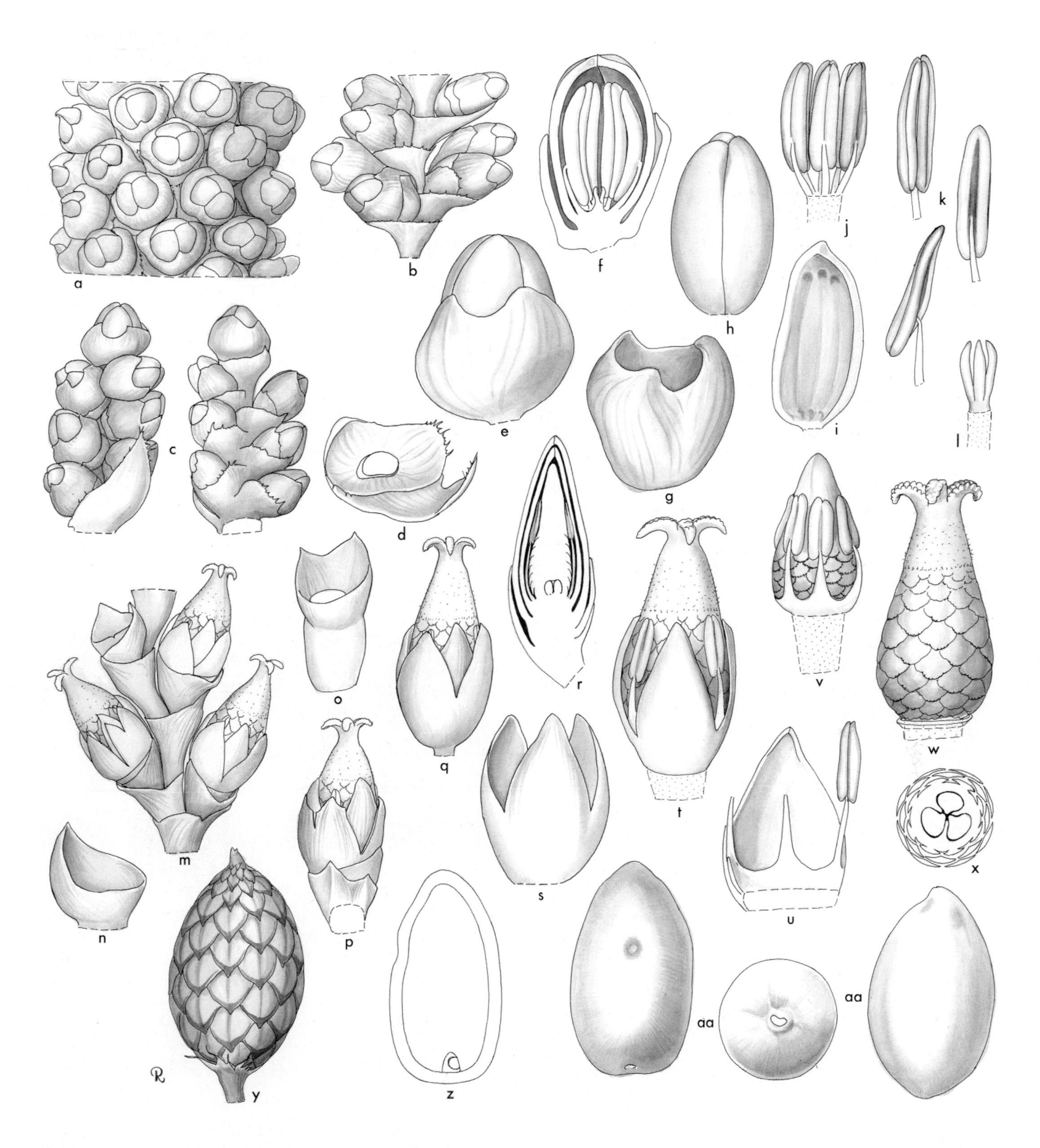

Retispatha. **a**, **b**, portion of staminate inflorescence, second-order branch, in 2 views × 4 1/2; **c**, portion of staminate inflorescence, third-order branch, in 2 views × 4 1/2; **d**, scar and bracts of staminate flower × 12; **e**, staminate bud × 12; **f**, staminate bud in vertical section × 12; **g**, staminate calyx × 12; **h**, staminate bud, calyx removed × 12; **i**, staminate petal, interior view × 12; **j**, androecium × 12; **k**, stamen in 3 views × 15; **l**, pistillode × 24; **m**, portion of pistillate inflorescence × 2 1/4; **n**, outer bracteole of pistillate flower × 3; **o**, inner bracteole of pistillate flower × 3; **p**, pistillate flower with bracteoles × 3; **q**, pistillate flower × 3; **r**, pistillate bud in vertical section × 6; **s**, pistillate calyx × 4 1/2; **t**, pistillate flower, calyx removed × 4 1/2; **u**, pistillate petal and portion of staminodial ring, interior view × 6; **v**, young gynoecium and staminodial ring × 7 1/2; **w**, gynoecium × 4 1/2; **x**, ovary in cross-section × 12; **y**, fruit × 1 1/2; **z**, seed in vertical section × 2 1/4; **aa**, seed in 3 views × 2 1/4. *Retispatha dumetosa*: **a–l**, *Dransfield* 5595; **m–aa**, *Dransfield* 2840. (Drawn by Marion Ruff Sheehan)

19. DAEMONOROPS

Variable genus of mostly climbing palms, some acaulescent and erect, found in Himalayan foothills to south China, throughout Southeast Asia to New Guinea; sheaths, petioles and rachis are usually densely armed and the leaf usually terminates in a cirrus; pleonanthic (rarely hapaxanthic) and dioecious, the inflorescences are either short, with all bracts splitting but remaining enclosed in the prophyll, or longer, with bracts splitting to their bases and mostly caducous.

Daemonorops Blume in J.J. Roemer & J.A. Schultes, Syst. Veg. 7: 1333 (1830). Type: **D. melanochaetes** Blume.

Daemon — *malignant spirit*, rhops — *bush, in reference to the often vicious armature*.

Solitary or clustered, spiny, acaulescent, erect, or high-climbing, hapaxanthic or pleonanthic, dioecious, rattan palms. *Stem* eventually becoming bare, with short to long internodes, branching at the base from axillary or leaf-opposed buds. *Leaves* pinnate, very rarely bifid, usually with a terminal cirrus except in a few acaulescent species and in juvenile individuals; sheath splitting in acaulescent species, in the exposed area densely armed with spines, these frequently organised into whorls and in a few species forming interlocking galleries occupied by ants, scaly or floccose indumentum often abundant between the spines and along their margins; ocrea rarely present; knee present in climbing species; flagellum absent; petiole usually well developed, grooved to rounded adaxially, rounded abaxially, variously armed; rachis and cirrus, except in acaulescent species, armed with grouped reflexed grapnel spines and scattered caducous tomentum; leaflets single-fold, entire, linear to broadly lanceolate, regularly arranged or grouped, rarely fanned within the groups, variously armed with bristles along the longitudinal veins and margins, midribs prominent, 1 pair of lateral veins sometimes large, transverse veinlets short, often conspicuous. *Inflorescences* axillary but adnate to the internode and leaf sheath of the following leaf, very rarely several inflorescences produced simultaneously from the axils of the most distal leaves, the stem then hapaxanthic, branching to 2–3 orders, staminate and pistillate inflorescences superficially similar, but the staminate usually branching to 1 order more than the pistillate; peduncle absent or present, sometimes very long, erect or pendulous, variously armed; prophyll conspicuous, 2-keeled, woody, coriaceous, membranous or papery, variously armed, tubular at first, later splitting along ± its entire length; peduncular bracts usually absent; rachis bracts ± distichous, similar to the prophyll, also splitting along their entire length, sometimes with the tips remaining enclosed within the tip of the prophyll to form a beak, the flowers at anthesis thus enclosed, or with the tips free, and all bracts but the prophyll normally falling at anthesis, the flowers then exposed or very rarely the bracts persisting; prophyll often empty, sometimes subtending a first-order branch as the other rachis bracts; first-order branches usually covered with abundant floccose indumentum, and bearing very small, truncate, ± distichous bracts, more rarely bracts larger and tattering, each bract subtending a second-order branch adnate to the first-order branch above the node; second-order branches in pistillate inflorescence bearing dyads, in staminate inflorescence branched a further time to give 3 orders of branching, each branch subtended by a bract, staminate inflorescence with flowers sometimes strictly distichous and crowded, or ± distant and subdistichous, sometimes arranged distantly along one side of the rachilla, each flower subtended by a small triangular scale-like bract, more rarely by a short tubular bract, the bracts then ± imbricate. *Staminate flowers* bearing a short, tubular, 2-keeled prophyll (the involucre of Beccari) sometimes ± stalk-like, frequently very inconspicuous; calyx cupular, striate, shallowly 3-lobed; corolla exceeding the calyx, usually at least twice as long, divided almost to the base into 3 narrow triangular petals; stamens 6, borne at the mouth of the tubular corolla base, usually ± equal, rarely of 2 sizes, filaments slender to rather broad, fleshy, terminating in slender to broad connectives, anthers narrow elongate to broad and somewhat sinuous, introrse; pistillode short, trifid to elongate, slender and unlobed, or absent. *Pollen* ellipsoidal, bi-symmetric; apertures equatorially disulcate or, rarely, equatorially or subequatorially di-porate; ectexine tectate or semi-tectate, psilate, finely to coarsely perforate, foveolate, rugulate, finely to coarsely reticulate, or densely spinulose or clavate or, rarely, ectexine intectate with long spines on a thick foot layer, aperture margins usually similar to surrounding ectexine; infratectum columellate, longest axis 16–55 µm [56/101]. *Pistillate inflorescences* like the staminate but with more robust rachillae, pistillate flowers borne in a dyad with a sterile staminate flower; dyad prophyll (involucrophore) usually conspicuously angular, stalk-like; prophyll of pistillate flower (involucre) inconspicuous or cup-like, forming a cushion bearing the flower. Sterile staminate flower quickly shed, as the fertile but with empty anthers. *Pistillate flowers* only slightly larger than the staminate; calyx cupular, striate, shallowly 3-lobed; corolla ± twice as long as the calyx, divided to ± $^1/_2$ into 3 triangular valvate petals; staminodes 6, borne at the mouth of the corolla tube, with empty anthers; gynoecium incompletely trilocular, triovulate, ovary variable in shape, scaly, stigmas 3, recurved, fleshy, ovules basally attached, anatropous. *Fruit* variously rounded, obpyriform, turbinate, cylindrical or oblate with apical stigmatic remains; epicarp covered in neat vertical rows of reflexed, sometimes resinous scales, mesocarp thin, endocarp not differentiated. *Seed*, usually only 1 reaching maturity, angular or rounded, covered with thick, sweet or sour and bitter sarcotesta, endosperm deeply ruminate; embryo basal. *Germination* adjacent-ligular; eophyll usually pinnate, sometimes with congested leaflets and appearing almost palmate. *Cytology*: 2n = 26.

Right: ***Daemonorops calicarpa***, hapaxanthic habit, Peninsular Malaysia (Photo: J. Dransfield)

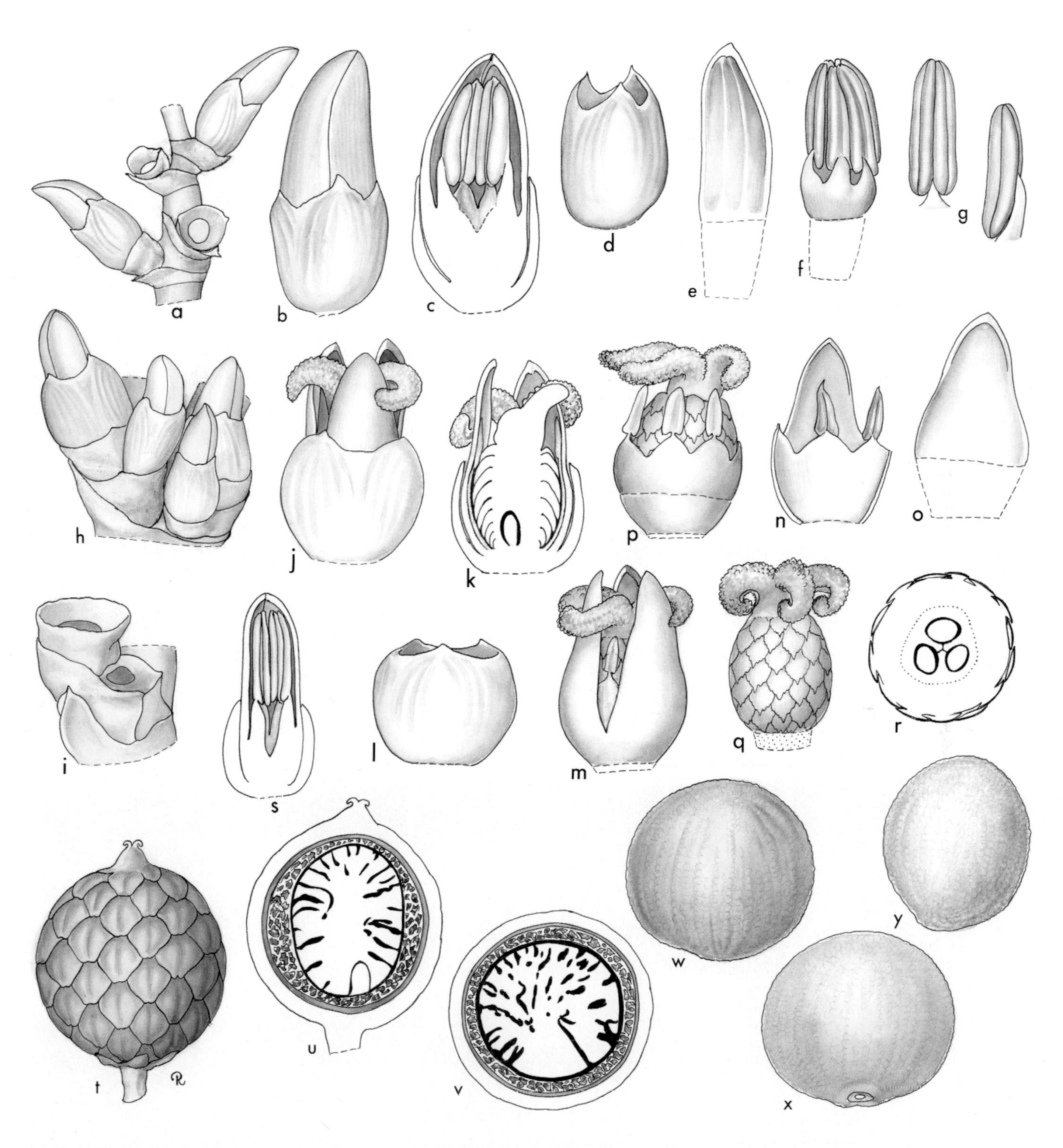

Daemonorops. **a**, portion of staminate rachilla, some flowers removed to show subtending bract and bracteole × $4^1/_2$; **b**, staminate bud × 9; **c**, staminate bud in vertical section × 9; **d**, staminate calyx × 9; **e**, staminate petal, interior view × 9; **f**, androecium × 9; **g**, stamen in 2 views × 12; **h**, portion of pistillate rachilla with dyads of pistillate and sterile staminate flowers × $4^1/_2$; **i**, dyad, flowers removed to show bracteoles × 6; **j**, pistillate flower × 6; **k**, pistillate flower in vertical section × 6; **l**, pistillate calyx × 6; **m**, pistillate flower, calyx removed × 6; **n**, pistillate petal and staminodes, interior view × 6; **o**, pistillate petal, interior view × 6; **p**, gynoecium and staminodial ring × 6; **q**, gynoecium × 6; **r**, ovary in cross-section × 6; **s**, sterile staminate flower in vertical section × 6; **t**, fruit × $2^1/_4$; **u**, fruit in vertical section × $2^1/_4$; **v**, fruit in cross-section × $2^1/_4$; **w, x, y**, seed in 3 views × $2^1/_4$. *Daemonorops grandis*: **a–g**, *Moore* 9044; *D. angustifolia*: **h–y**, *Moore* 9067. (Drawn by Marion Ruff Sheehan)

Distribution and ecology: 101 species; distributed from India and south China through the Malay Archipelago to New Guinea where represented by one species; the greatest morphological diversity and number of species is in the Malay Peninsula, Sumatra, and Borneo. Species are mostly confined to primary tropical rain forest on a great variety of soils, some species with narrow ecological requirements. A few are of a rather more weedy nature, abundant in forest habitats with high-light intensities such as riverbanks; one species in Borneo, *Daemonorops longispatha*, grows on the landward margin of mangrove forest. Some species are strictly montane, occurring at altitudes up to ca. 2500 m above sea level. Several species in Borneo are confined to heath forest, or to limestone or serpentine rock. With so many species, it is difficult to give more precise ecological data.

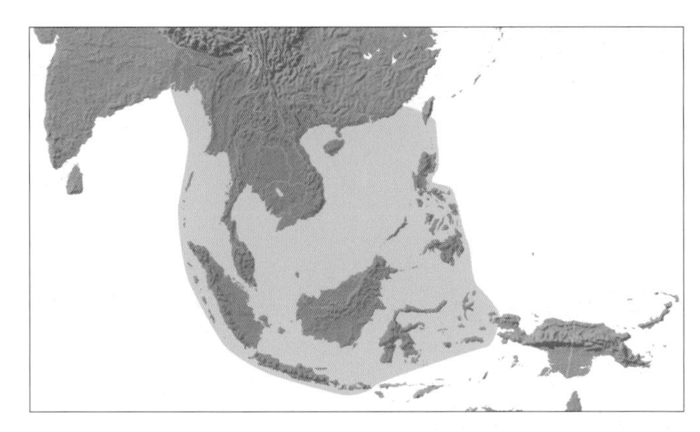

Distribution of ***Daemonorops***

Anatomy: Leaf, petiole, stem, and root (Tomlinson 1961), root (Seubert 1996a). Leaves often distinguished from *Calamus* by structure of guard cells, by more frequent and longer fibres around the transverse veinlets and by presence of large tannin cells.

Relationships: For relationships, see *Calamus*.

Common names and uses: Rattan, *rotan*; for local names see Dransfield (1979a). The canes of many species are of good quality and enter the rattan trade as well as being used locally. The apices of several species are sought after for food. In the past, red resin from the fruits of species related to and including *Daemonorops draco* and *D. rubra* was collected as 'dragon's blood' for medicinal or dyeing purposes. At one time, 'dragon's blood' was an item of trade between Borneo, Sumatra and Malay Peninsula, and China.

Taxonomic accounts: Beccari (1911), Furtado (1953) and Dransfield (1979a).

Fossil record: From the Neogene of the Landes (France), a number of spiny leaf sheaths associated with flattened stems were described by Huard (1967) under the new genus and species *Spinophyllum lepidocaryoides*. From anatomical comparison, he suggested that their affinity is calamoid, possibly *Calamus* or more probably *Daemonorops*. From the Miocene of Czech Republic, Czeczott and Juchniewicz (1975) described spiny fragmentary bracts which, in their opinion, compare most favourably with modern *D. geniculata*.

Below left: ***Daemonorops fissa***, inflorescence, Sarawak. (Photo: J. Dransfield)
Middle: ***Daemonorops verticillaris***, leaf sheath with ant gallery, Peninsular Malaysia. (Photo: J. Dransfield)
Right: ***Daemonorops didymophylla***, pistillate rachilla, Sarawak. (Photo: J. Dransfield)

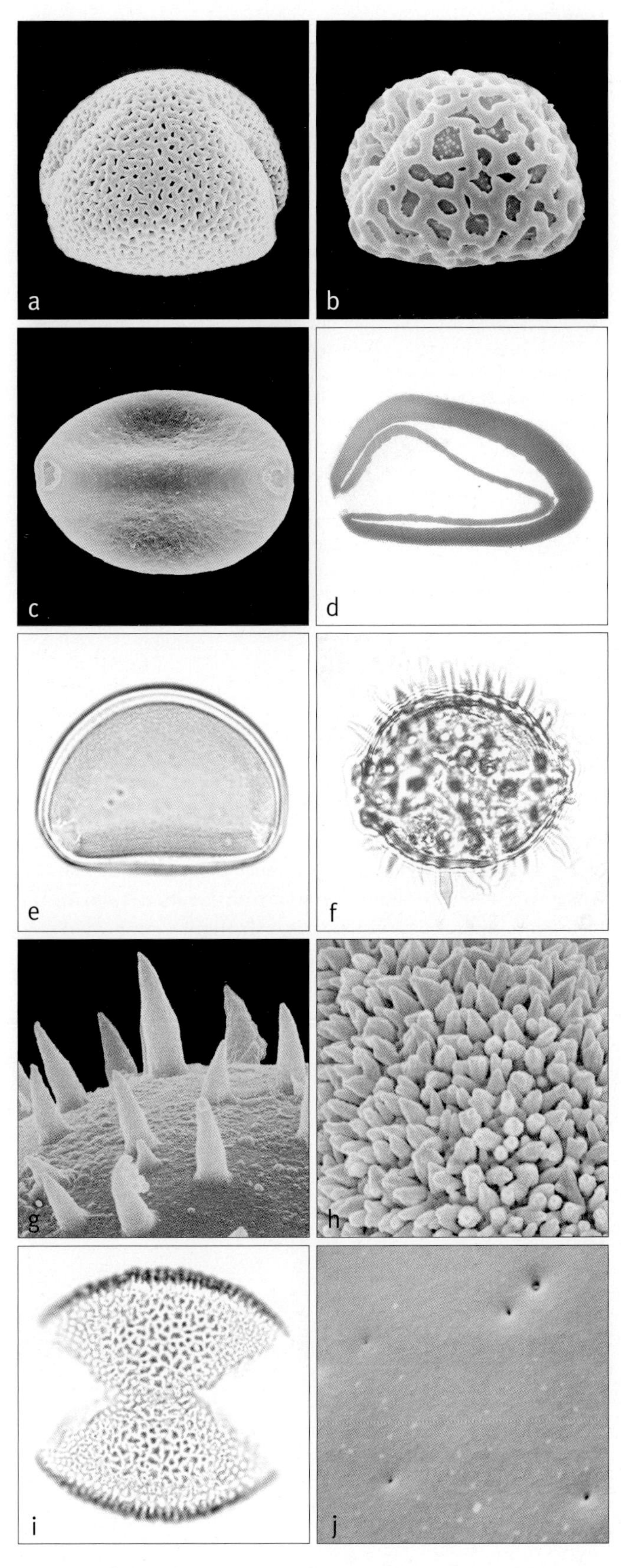

Zetter and Hoffmann (2001) revive the controversy over whether *Diporoconia iszkaszentgyoergyi* Kedves has an affinity with the diporate pollen of *D. sparsiflora* and *D. verticillaris*, a topic previously discussed by Frederiksen *et al.* (1985). They present new finds of *D. iszkaszentgyoergyi* from the Lower Eocene of Krappfeld, Austria, which they compare with modern diporate *Daemonorops*. However, based on pore size, ectexine detail around the pore perimeter, and wall thickness, the data are not entirely convincing; ultra-thin sections would be of great value in resolving the proposed affinity. See also under *Metroxylon* and *Calamus*.

Notes: The genus is divisable into two sections based on the arrangement and persistence of the rachis bracts of the inflorescence. Palynologically, so-called section '*Cymbospatha*' (correctly section *Daemonorops*) is very uniform, whereas section *Piptospatha* shows a truly remarkable diversity in pollen form. Fossil record, see under *Calamus*.

Additional figures: Figs 2.8j, Glossary figs 7, 20, 23.

Left: ***Daemonorops***

a, equatorial disulcate pollen grain, finely reticulate-rugulate surface, equatorial face SEM × 2000;

b, equatorial disulcate pollen grain, coarsely reticulate surface, equatorial face SEM × 1500;

c, subequatorial diporate pollen grain, proximal face SEM × 1500;

d, oblique ultrathin section cutting through one pore of a subequatorial diporate pollen grain TEM × 1500;

e, subequatorial diporate pollen grain, proximal face LM × 1500;

f, spinose, equatorial diporate pollen grain LM × 1000;

g, close-up, intectate spiny pollen surface (cf. **'f'**) SEM × 8000;

h, close-up, densely clavate pollen surface SEM × 8000;

i, equatorial disulcate pollen grain, distal face, low-focus LM × 800;

j, close-up, psilate, sparsely perforate pollen surface SEM × 8000.

Daemonorops unijuga: **a**, *Dransfield* JD5895; *D. micracantha*: **b**, *Beccari* 3644; *D. sparsiflora*: **c**, *Tiggi* S. 3319; *D. verticillaris*: **d**, *Kunstler* 576, **e**, *Ridley s.n.*; *D. oblata*: **f** & **g**, *Dransfield et al.* JD5712; *D. lewisiana*: **h**, *Kloss et al.* 6957; *D. formicaria*: **i**, *Chai* S. 39600; *D. crinita*: **j**, *Dransfield* JD6126.

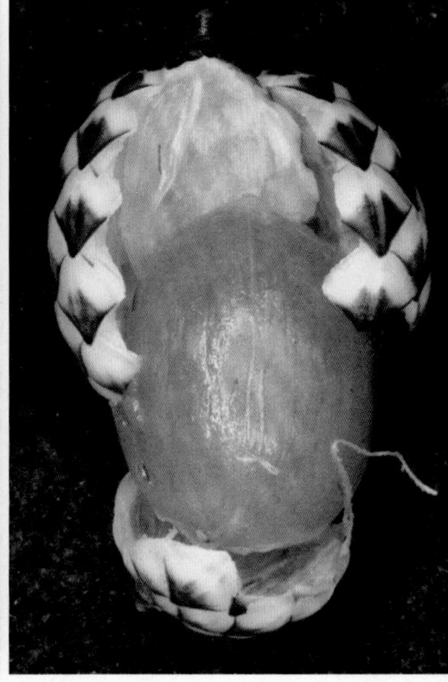

Right: ***Daemonorops micracantha***, fruit with dragon's blood, Peninsular Malaysia. (Photo: J. Dransfield)

Far right: ***Daemonorops scapigera***, ripe fruit, note sarcotesta, Sarawak. (Photo: W.J. Baker)

20. CERATOLOBUS

Slender clustering climbing palms, found in South Thailand, Malay Peninsula, Borneo, Sumatra and Java, immediately distinguished by the single large bract (the prophyll) that covers the entire inflorescence and opens by two apical slits.

Ceratolobus Blume in J.J. Roemer & J.A. Schultes, Syst. Veg. 7: 1334 (1830). Type: **C. glaucescens** Blume.

Keras — *horn*, lobos — *pod, referring to the shape of the inflorescence bract (prophyll)*.

Slender to moderate, clustered, spiny, climbing, pleonanthic, dioecious, rattan palms. *Stem* eventually becoming bare, with long internodes and conspicuous nodal scars, white mucilage sometimes exuding from cut surfaces. *Leaves* of mature climbing stems cirrate, pinnate; leaf sheath tubular, armed with spines and/or spicules, frequently organised into whorls, and often with abundant indumentum; knee present, sometimes rather weakly developed; ocrea inconspicuous; flagellum absent; petiole present or absent, if present, flat adaxially, rounded abaxially, armed with spines and sometimes with spicules; cirrus and distal part of rachis armed with regular groups of grapnel spines on the abaxial surface; leaflets relatively few, linear to lanceolate and entire, or rhomboid and praemorse, concolourous or discolourous, regularly arranged or grouped; emerging leaf pink-tinged. *Inflorescences* axillary but adnate to the internode and sheath of the following leaf, sessile and erect or pendulous on a long slender, unarmed or spiny peduncle, the whole inflorescence much shorter than the leaves; staminate inflorescence branching to 3 orders, pistillate to 2 orders, prophyll persistent, membranous to subwoody, flattened-tubular with lateral wings and a terminal beak, entirely enclosing the inflorescence, opening at anthesis by 2 narrow, lateral slits in the beak, this remaining the only access to the flowers during anthesis and young fruiting stage, prophyll unarmed or rarely armed with scattered spines, frequently bearing caducous indumentum, prophyll often splitting longitudinally in fruit, very rarely falling completely, staminate and pistillate inflorescences indistinguishable without splitting the prophyll; peduncular bracts absent; rachis bracts inconspicuous, tubular, with triangular limbs, each (and the prophyll) subtending a first-order branch, usually adnate to the axis for a short distance above the bract node. *Staminate flowers* borne singly, subdistichously, rather distant from each other, each subtended by a membranous triangular bract and a 2-keeled, prophyllar bracteole, the latter ± forming a cushion beneath the flower; calyx tubular, with 3 short triangular lobes; petals 3, boat-shaped, valvate, briefly joined basally; stamens 6, borne at the base of the corolla, filaments short, fleshy, anthers linear, latrorse; pistillode trifid, minute. *Pollen* ellipsoidal, bi-symmetric; apertures equatorially disulcate; ectexine tectate or semi-tectate, finely or coarsely perforate, or foveolate-reticulate, aperture margins similar or finer; infratectum columellate; longest axis 22–30 µm [3/6]. *Pistillate flowers* borne with a sterile staminate flower and 2 similar 2-keeled prophyllar bracteoles, in a cup formed by the subtending bract. *Sterile staminate flower* like the fertile but usually rather distorted by close packing and with empty anthers and borne on a short to long stalk. *Pistillate flowers* larger than the staminate; calyx tubular, with 3 short, triangular lobes; corolla partially divided into 3, valvate, triangular lobes; staminodes 6, epipetalous, flattened; gynoecium incompletely trilocular, triovulate, globose, or ellipsoidal, covered in scales, stigmas 3, fleshy, recurved, borne on a short style, ovule basally attached, anatropous. *Fruit* l-seeded, globose to ellipsoidal, stigmatic remains apical, epicarp covered in vertical rows of reflexed scales, mesocarp becoming thin and papery as fruit ripens, endocarp not differentiated. *Seed* attached basally, globose to ellipsoidal, with thin or thick, sour or sweet sarcotesta and homogeneous or ruminate endosperm; embryo basal. *Germination* adjacent-ligular; eophyll with 4–6 praemorse or entire, crowded leaflets displayed in a fan. *Cytology*: 2n = 26.

Distribution and ecology: Six species confined to the perhumid areas of the Sunda Shelf (Malay Peninsula [2], Sumatra [4], Borneo [3], Java [2]). All species are found in lowland and hill tropical rain forest and do not occur above 1000 m altitude; all are usually confined to dipterocarp forest but *Ceratolobus subangulatus* also occurs in heath forest in Sarawak, where it is the most conspicuous rattan in some facies. The pollination ecology of the extraordinary closed inflorescence deserves further study.

Anatomy: Leaf (Tomlinson 1961), root (Seubert 1996a).

Relationships: *Ceratolobus* is strongly supported as monophyletic (Baker *et al.* 2000c). For relationships, see *Calamus*.

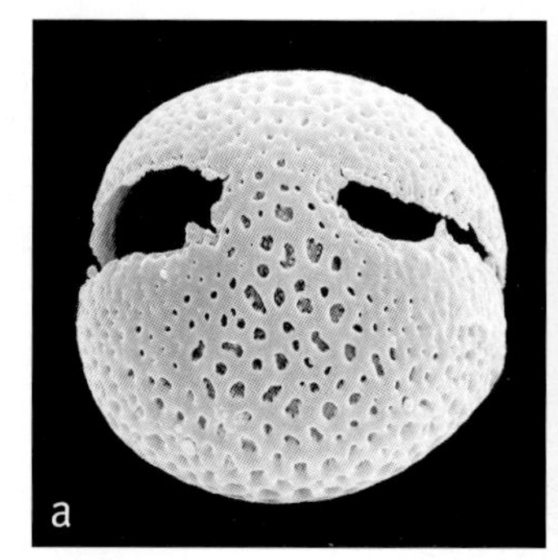

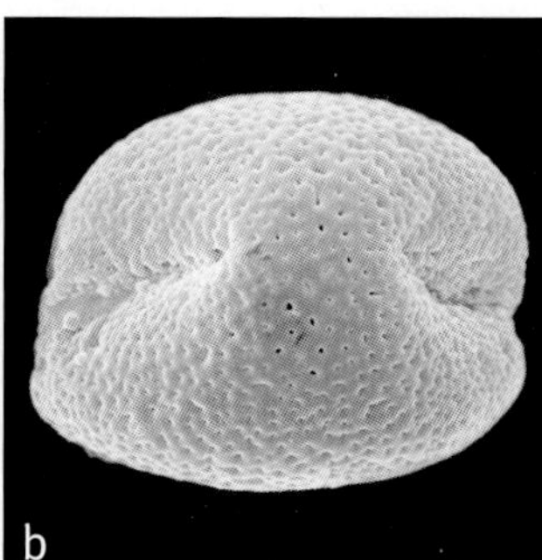

Above: ***Ceratolobus***
a, equatorial disulcate pollen grain, distal face, reticulate surface SEM × 2000;
b, equatorial disulcate pollen grain, distal face, perforate-rugulate surface SEM × 2000.
Ceratolobus discolor: **a**, *Dransfield* JD6077; *C. subangulatus*: **b**, *Dransfield* JD 5963.

Far left: ***Ceratolobus subangulatus***, young fruit, Peninsular Malaysia. (Photo: J. Dransfield)

Left: ***Ceratolobus concolor***, ripe fruit, Sarawak. (Photo: W.J. Baker)

Common names and uses: Rattan, *rotan*. Species of *Ceratolobus* appear to have weak, not durable canes and for this reason are scarcely ever utilised.
Taxonomic account: Dransfield (1979b).
Fossil record: No generic records found.
Notes: *Ceratolobus* has the most reduced inflorescence of all the Calaminae, having only one major bract, the prophyll, which remains closed around the inflorescence, but splitting and sometimes falling in fruit. The presence of knees on the leaf sheaths allows those species with rhomboid leaflets to be distinguished in the sterile state from *Korthalsia*, but without inflorescences, the remaining species (*C. subangulatus*) cannot be separated easily from *Calamus* and *Daemonorops*.
Additional figures: Fig. 6.2.

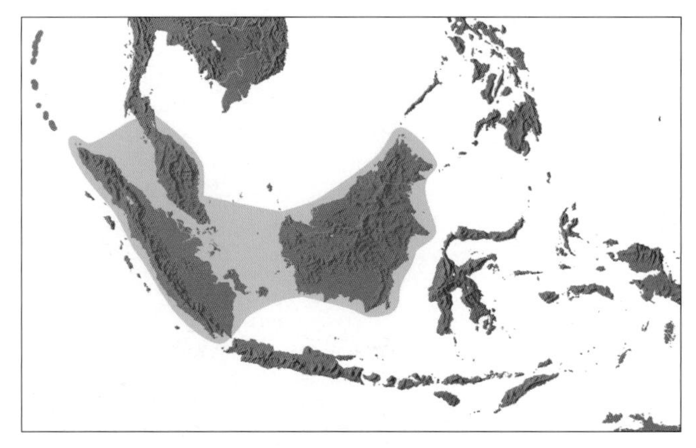

Distribution of ***Ceratolobus***

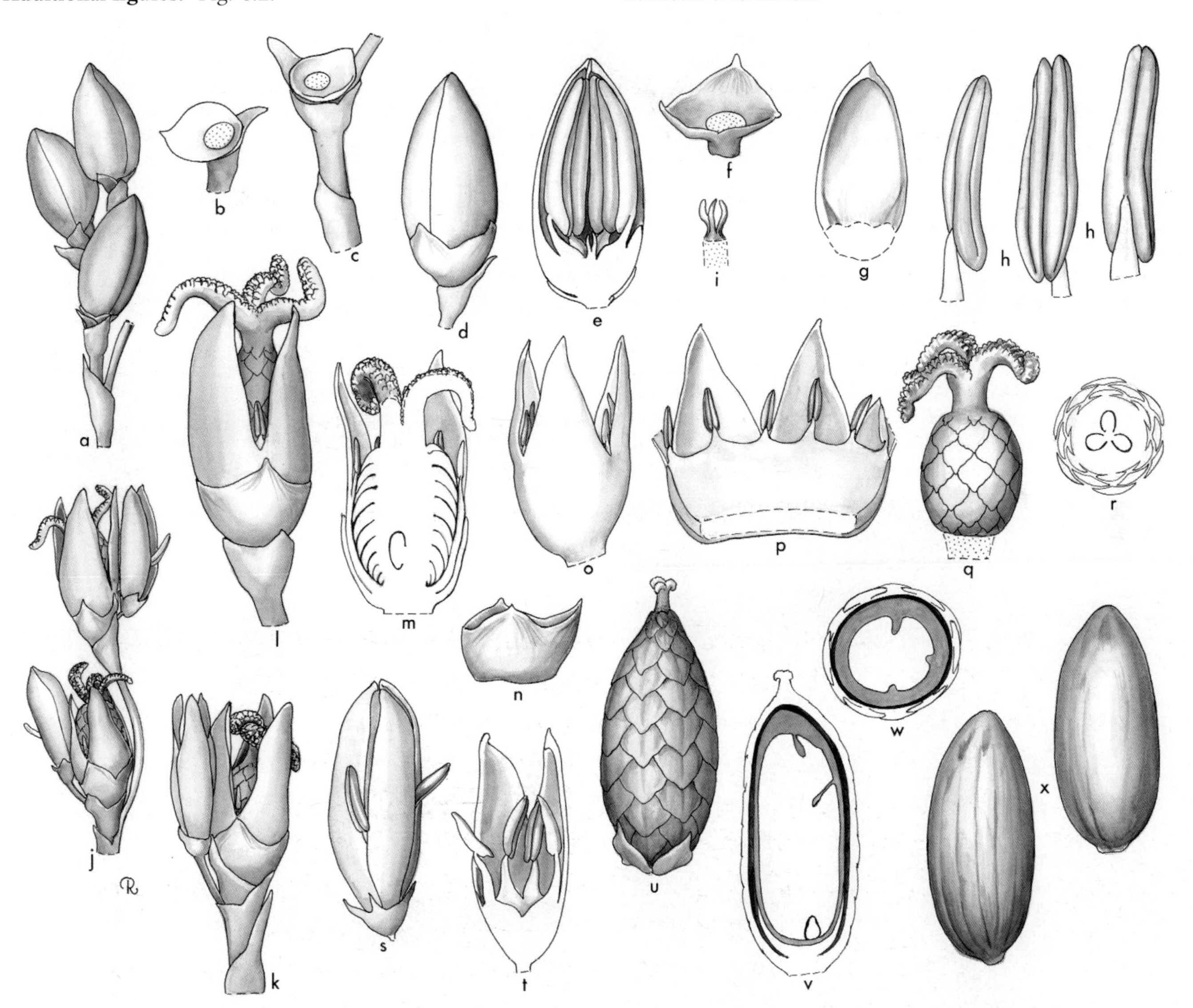

***Ceratolobus*. a**, portion of staminate inflorescence × $4\frac{1}{2}$; **b**, bract and bracteole subtending staminate flower × 9; **c**, prophyll, bract, bracteole, and pedicel of staminate flower × 9; **d**, staminate bud × $7\frac{1}{2}$; **e**, staminate bud in vertical section × 9; **f**, staminate calyx × 9; **g**, staminate petal × $7\frac{1}{2}$; **h**, stamen in 3 views × 12; **i**, pistillode × 15; **j**, pistillate inflorescence with dyads of pistillate and sterile staminate flowers × $4\frac{1}{2}$; **k**, portion of pistillate inflorescence with pistillate flower, sterile staminate flower, and subtending bracteoles × 6; **l**, pistillate flower × $7\frac{1}{2}$; **m**, pistillate flower in vertical section × $7\frac{1}{2}$; **n**, pistillate calyx × $7\frac{1}{2}$; **o**, pistillate flower, calyx removed × $7\frac{1}{2}$; **p**, pistillate corolla and staminodial ring, expanded, interior view × $7\frac{1}{2}$; **q**, gynoecium × $7\frac{1}{2}$; **r**, ovary in cross-section × $7\frac{1}{2}$; **s**, sterile staminate flower × $7\frac{1}{2}$; **t**, sterile staminate flower in vertical section × $7\frac{1}{2}$; **u**, fruit × 3; **v**, fruit in vertical section × 3; **w**, fruit in cross-section × 3; **x**, seed in 2 views × 3. *Ceratolobus glaucescens*: all from *Moore* 9950. (Drawn by Marion Ruff Sheehan)

21. POGONOTIUM

Slender solitary or clustering erect or short climbing palms, found in Malay Peninsula and Borneo; immediately distinguished by the two strange erect spiny slender ear-like lobes (auricles) at the base of the petiole.

Pogonotium J. Dransf., Kew Bull. 34(4): 763 (1980a). Type: **P. ursinum** (Becc.) J. Dransf. (*Daemonorops ursina* Becc.).

Pogon — *beard*, otion — *a little ear, referring to the finely spiny auricles*.

Solitary or clustered, spiny, erect or short-climbing, pleonanthic, dioecious, rattan palms. *Stem* with short internodes. *Leaves* pinnate, without cirrus; sheath tubular, densely armed with whorled and scattered spines and caducous tomentum, terminating in 2 erect, narrow auricles, 1 on each side of the petiole, the auricles variously armed like the sheath; knee absent or poorly developed; flagellum absent; petiole well developed, flat adaxially, rounded abaxially, armed with reflexed grapnel spines and various papillae and hairs; rachis armed as the petiole; leaflets few to very numerous, linear, single-fold, regularly arranged, very crowded to very distant, the surface covered with a variety of bristles and scales, midribs prominent adaxially, transverse veinlets short, conspicuous. *Inflorescences* axillary but adnate to the internode and leaf sheath of the following leaf, held erect between the 2 auricles of the subtending leaf, ± sessile, the pistillate branching to 2 orders, the staminate to 3 orders; prophyll enclosing the inflorescence, boat-shaped, with a flattened beak, armed or unarmed, splitting longitudinally along the mid adaxial or abaxial line, thus exposing the flowers; rachis bracts very much smaller than the prophyll, with free tips, each and the prophyll subtending a variously hairy branch; bracts on first-order branches tubular at the base, with triangular limbs. *Staminate flowers* solitary, borne subdistichously on branches of the second or third-order, each subtended by a minute, short, tubular, triangular bract and bearing a 2-keeled bracteole; calyx tubular in proximal part, striate, with 3 triangular lobes; corolla split almost to the base into 3 triangular valvate lobes; stamens 6, borne at the base of the corolla lobes, filaments fleshy, elongate, inflexed at the tips, anthers oblong, medifixed, sagittate basally, latrorse; pistillode minute. *Pollen* spheroidal, bi-symmetric; monoporate with pore on one of two short axes of grain; ectexine tectate or semi-tectate, perforate-rugulate or foveolate-reticulate, aperture margin usually similar; infratectum columellate; longest axis 18–32 µm [2/3]. *Pistillate flowers* in dyads, each subtended by a small, triangular bract, and consisting of a pistillate and a sterile staminate flower and two 2-keeled cup-like bracteoles. Sterile staminate flower like the fertile but tending to be contorted by close-packing and with empty flattened anthers and a variable pistillode. Pistillate flowers larger than the staminate; calyx cupular, striate, lobes triangular, valvate; corolla split almost to the base into 3 triangular, valvate lobes; staminodes 6, epipetalous; gynoecium incompletely trilocular, triovulate, ovoid, scaly, stigmas 3, fleshy, rugose, divergent, ovule basally attached, anatropous. *Fruit* 1-seeded, globose or ovoid, beaked, stigmatic remains apical; epicarp covered in neat vertical rows of reflexed magenta to chestnut-coloured scales, mesocarp becoming thin and dry at maturity, endocarp not differentiated. *Seed* basally attached, sarcotesta thick, sweet, endosperm homogeneous; embryo basal. *Germination* adjacent-ligular; eophyll pinnate or with a single pair of divergent leaflets. *Cytology* not studied.

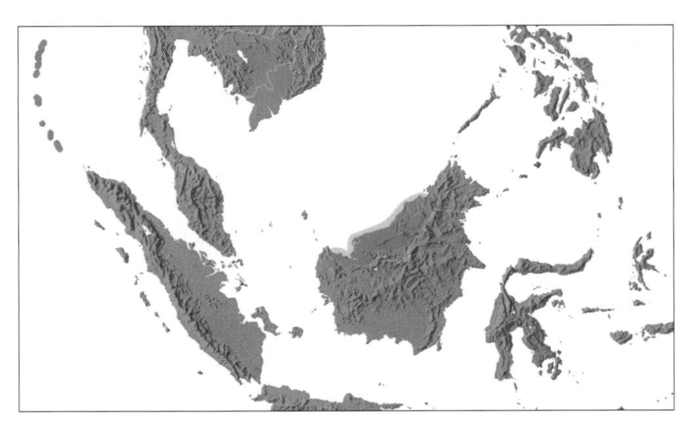

Distribution of ***Pogonotium***

Far left: ***Pogonotium divaricatum***, habit, Brunei Darussalam. (Photo: J. Dransfield)

Left: ***Pogonotium divaricatum***, fruit, Brunei Darussalam. (Photo: J. Dransfield)

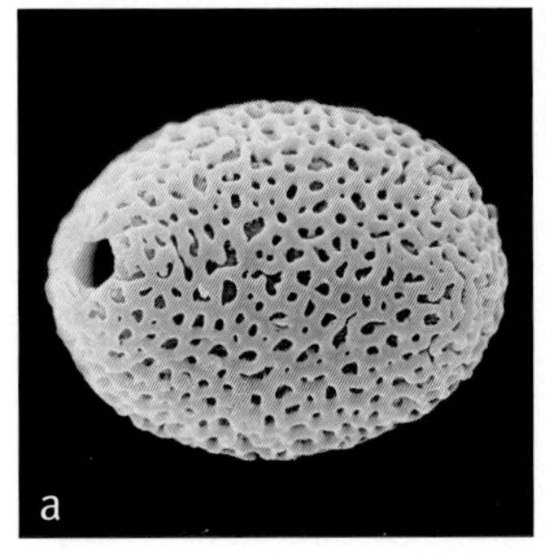

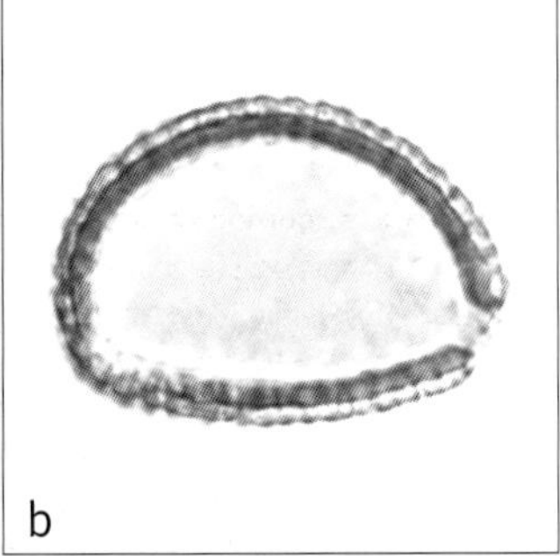

Below: ***Pogonotium***

a, monoporate pollen grain, a pore on only one of the short equatorial axes SEM × 2000;

b, monoporate pollen grain, mid-focus LM × 1800.

Pogonotium ursinum: **a** & **b**, *Dransfield* JD5877.

Distribution and ecology: Three species, one in Malay Peninsula and Sarawak (Borneo), the other two confined to Sarawak. In Borneo, all three species occur as small populations confined to podsolised soils on ridgetops at about 700–1000 m altitude (at the transition between lowland and montane forest) or to some facies of 'kerangas' forest (heath forest). In the Malay Peninsula, *Pogonotium ursinum* has been found in lowland dipterocarp forest.

Anatomy: Not studied.

Relationships: The monophyly of *Pogonotium* has not been tested. For relationships, see *Calamus*.

Common names and uses: Common names not recorded. *Pogonotium ursinum* is very decorative, especially when young, but has not been introduced into general cultivation.

Taxonomic accounts: Dransfield (1980a, 1982c).

Fossil record: No generic records found.

Notes: The highly reduced inflorescence nestling between the two auricles makes this a distinctive and unusual rattan genus.

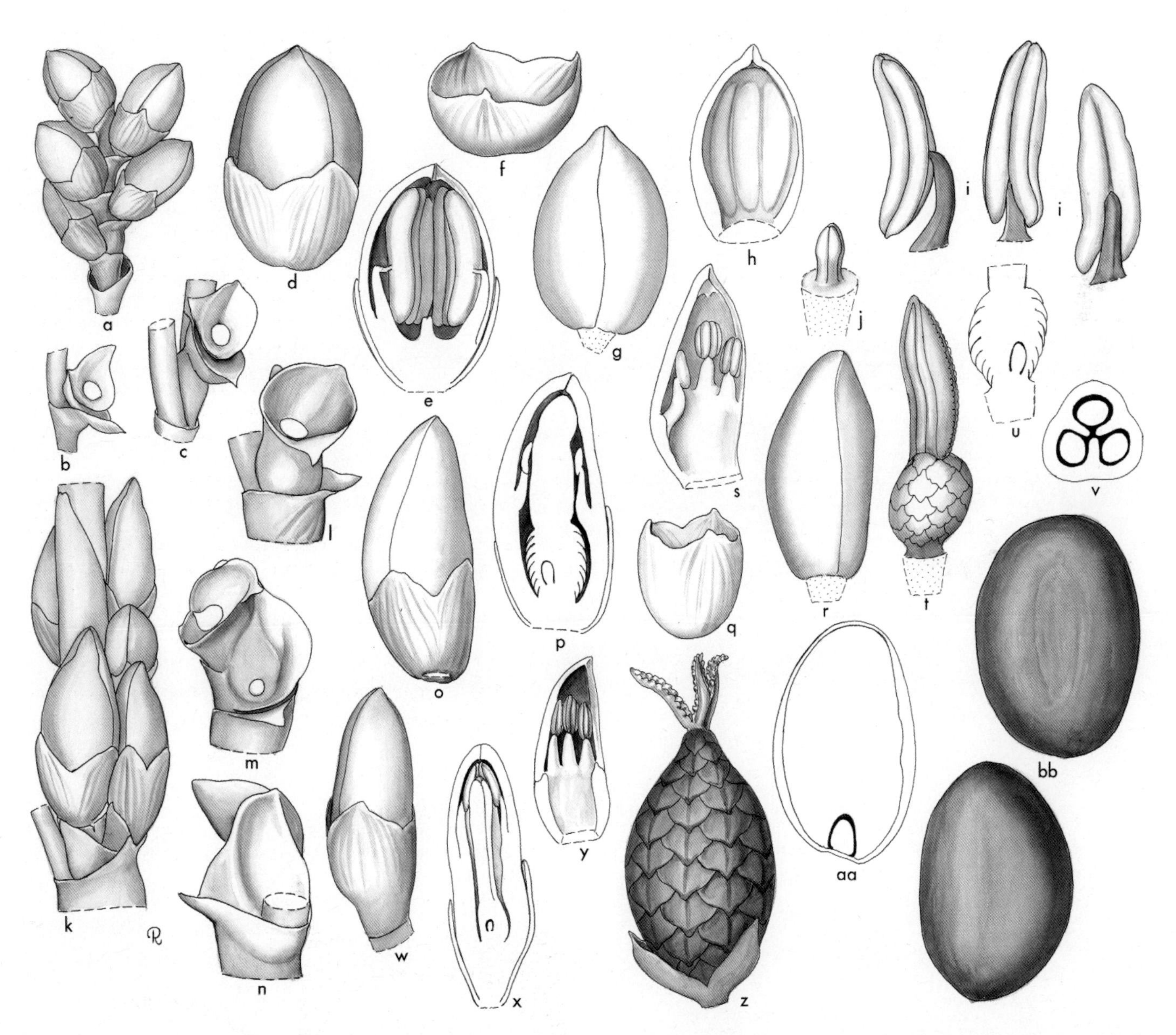

***Pogonotium*.** **a**, portion of staminate rachilla × $4^{1}/_{2}$; **b**, **c**, portions of staminate rachilla, flowers removed to show subtending bracts and bracteoles × $7^{1}/_{2}$; **d**, staminate bud × 9; **e**, staminate bud in vertical section × 9; **f**, staminate calyx × 9; **g**, staminate bud, calyx removed × 9; **h**, staminate petal, interior view × 9; **i**, stamen in 3 views × 15; **j**, pistillode × 18; **k**, portion of pistillate inflorescence with dyads of pistillate and sterile staminate flowers × $4^{1}/_{2}$; **l**, portion of pistillate inflorescence, pistillate flower removed to show subtending bracteole × $7^{1}/_{2}$; **m**, portion of pistillate inflorescence, pistillate (top) and sterile staminate (bottom) flowers removed to show subtending bracteoles × $7^{1}/_{2}$; **n**, back view of **m** showing portion of axis × $7^{1}/_{2}$; **o**, pistillate bud × 6; **p**, pistillate bud in vertical section × 6; **q**, pistillate calyx × 6; **r**, pistillate bud, calyx removed × 6; **s**, pistillate petal and staminodes, interior view × 6; **t**, gynoecium × 6; **u**, ovary in vertical section × 6; **v**, ovary in cross-section × 9; **w**, sterile staminate flower × 6; **x**, sterile staminate flower in vertical section × 6; **y**, sterile staminate petal with staminodes × 6; **z**, fruit × 3; **aa**, seed in vertical section × 3; **bb**, seed in 2 views × 3. *Pogonotium ursinum*: **a–j**, *Dransfield* 5877; *P. moorei*: **k–bb**, *Dransfield* 6102. (Drawn by Marion Ruff Sheehan)

II. NYPOIDEAE

NYPOIDEAE Griff., Palms Br. India 7 (1850) ('Nipinae'). Type: **Nypa**.

Monoecious, pleonanthic; stem prostrate, branching dichotomously; unarmed; leaves pinnate, reduplicate; inflorescence terminated by a pistillate head with lateral branches ending in short spikes of staminate flowers; flowers strongly dimorphic; perianth parts 6, distinct, linear, similar in both sexes; stamens 3, filaments united in a column, pistillode absent; carpels 3, distinct, irregularly polyhedric, curved and angled, with a funnel-shaped stylar opening; fruiting head large, globose, composed of irregularly compressed, enlarged carpels, with or without seeds.

The subfamily contains the single monotypic genus *Nypa*, a mangrove palm of Asia and the west Pacific.

Most recent studies resolve subfamily Nypoideae as sister to all palms except subfamily Calamoideae (Asmussen & Chase 2001, Asmussen *et al.* 2006, Baker *et al.* in review). There is high support for this position in Asmussen *et al.* (2006) and moderate support in Baker *et al.* (in review). Hahn (2002a) presents numerous alternative resolutions, but those trees based on the broadest data sampling are congruent with Nypoideae being sister to all palms except the Calamoideae. In some studies, however, the Nypoideae is resolved as sister to all other palms (including the Calamoideae) with low or moderate support (Uhl *et al.* 1995, Lewis & Doyle 2001).

22. NYPA

Pinnate-leaved mangrove palm with horizontally creeping dichotomously branching stem, usually concealed by mud, distinctive in the erect inflorescence with orange-yellow bracts and the massive round head of shiny grooved fruits.

Nypa Steck, Sagu 15 (1757). Type: **N. fruticans** Wurmb

Nipa Thunb., Kong. Vetensk. Acad. Nya Handl. 3: 234 (1782). Type: *N. fruticans* Thunb. (= *Nypa fruticans* Wurmb).

Derived from nipah, *the Malay vernacular name.*

Large, creeping, unarmed, pleonanthic, monoecious palm. *Stem* stout, prostrate or subterranean, branching dichotomously, curved leaf scars evident above, roots borne along the lower side. *Leaves* few, very large, erect, reduplicately pinnate; sheath soon splitting, glabrous; petiole stout, elongate, wide basally, channelled adaxially, terete distally, the base often persistent as a conical stub after the blade has disintegrated; rachis terete basally, becoming angled distally; leaflets numerous, single-fold, regularly arranged, acute, coriaceous, midrib prominent bearing distinctive, shining, chestnut-coloured, membranous ramenta abaxially, transverse veinlets not evident. *Inflorescences* solitary, interfoliar, erect, branching to 5(–6) orders, protogynous; peduncle terete; prophyll 2-keeled, tubular; peduncular bract tubular, somewhat inflated, pointed, rubbery, splitting longitudinally; rachis usually shorter than the peduncle, terete, terminating in a head of pistillate flowers and below

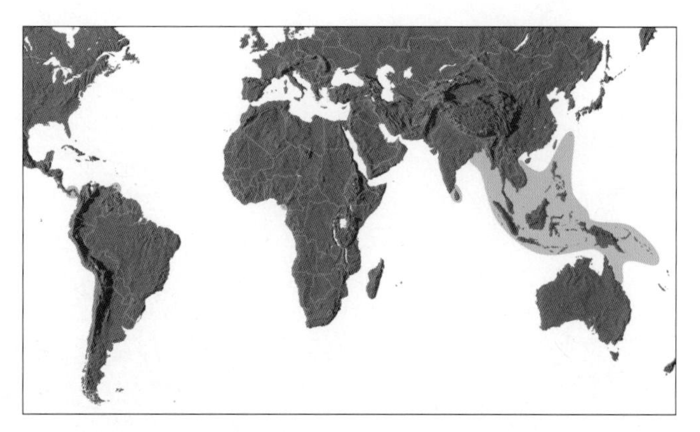

Distribution of ***Nypa***, including naturalised records in Africa and the Americas.

this bearing 7–9 spirally arranged, closed, ± inflated, tubular bracts each subtending a first-order branch; first-order branches adnate ca. $^{1}/_{2}$ their length above the subtending bracts, each bearing and enclosed by a tubular prophyll in bud; subsequent branches all bearing a complete, tubular, closed prophyll and ending in a short catkin-like rachilla, bearing densely crowded, spirally arranged, solitary staminate flowers, each subtended by a small bract. *Staminate flowers* sessile; sepals 3, distinct, narrow, oblanceolate; petals 3, distinct, slightly imbricate, similar to the sepals but slightly larger, both loosely closed over the stamens in bud; stamens 3, filaments and connectives connate in a solid stalk, anthers elongate, extrorse; pistillode lacking. *Pollen* spheroidal, bi-laterally symmetric; aperture a meridional zonasulcus; ectexine semi-tectate, finely reticulate with wide-based supratectal spines; infratectum columellate; diameter 37–80 µm; post-meiotic tetrads tetragonal [1/1]. *Pistillate flowers* very different from the staminate; sepals 3, distinct, irregularly oblanceolate, petals 3, similar to those of the staminate flower; staminodes lacking; carpels 3(–4), distinct, much longer than and obscuring the perianth at maturity, ± obovoid, asymmetrical, angled by mutual pressure, ± acute distally, and with a ± lateral, funnel-shaped stigmatic opening, ovule anatropous, attached dorsally or submarginally near the base of the locule. *Fruit* borne in ± globose head, fertile and partially developed fruits intermixed, 1–3 carpels per flower maturing a seed; fruit developing from 1 carpel, compressed and irregularly angled, stigmatic remains terminal, pyramidal; epicarp smooth, mesocarp fibrous, endocarp thick, composed of interwoven fibrous strands, with an adaxial internal longitudinal ridge intruded into the seed. *Seed* broadly ovoid, grooved adaxially, hilum basal, raphe branches ascending from the base, endosperm homogeneous or rarely ruminate, with a central hollow; embryo basal. *Germination* on the fruiting head with the plumule exserted and pushing the fruit away; eophyll bifid or with several leaflets. *Cytology*: 2n = 34.

Distribution and ecology: A single species, *Nypa fruticans*, occurring from Sri Lanka and the Ganges Delta to Australia, the Solomon Islands and the Ryukyu Islands. Introduced in the late 19th Century to the Niger Delta in West Africa, *Nypa* has now spread thence to western Cameroon. It has also been reported recently as naturalised in Panama (Duke 1991) and Trinidad (Bacon 2001), possibly having arrived from West

Opposite: ***Nypa fruticans***, in mangrove forest, *Oncosperma tigillarium* behind, Sumatra. (Photo: J. Dransfield)

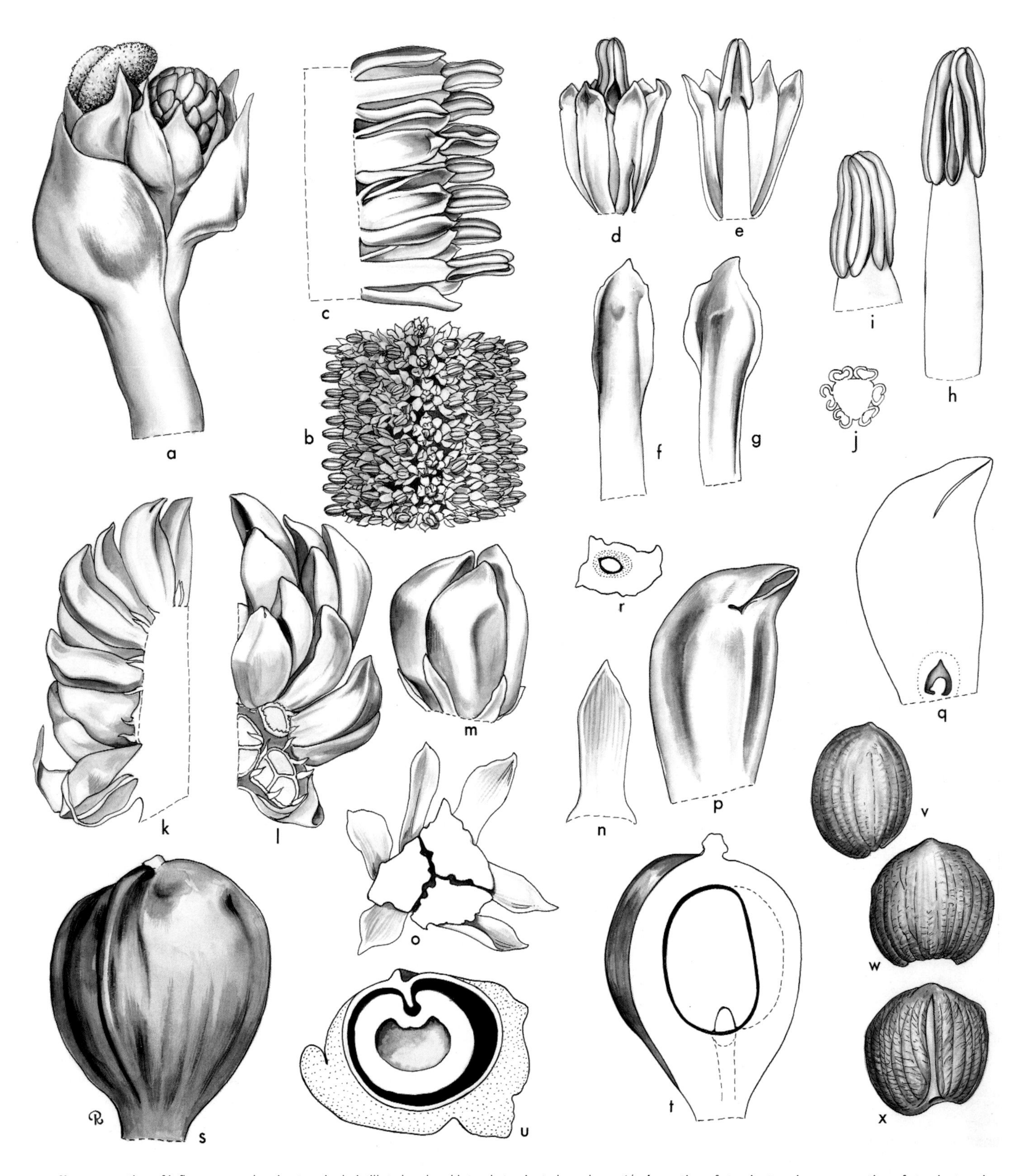

Nypa. **a**, portion of inflorescence showing terminal pistillate head and lateral staminate branches × $^1/_2$; **b**, portion of staminate axis × 3; **c**, portion of staminate axis in vertical section × 6; **d**, staminate flower × 6; **e**, staminate flower in vertical section × 6; **f**, staminate sepal × 12; **g**, staminate petal × 12; **h**, androecium × 12; **i**, anthers × 12; **j**, anthers in cross-section × 6; **k**, portion of pistillate head in vertical section × 1 $^1/_2$; **l**, portion of pistillate head, some flowers removed × 1 $^1/_2$; **m**, pistillate flower × 3; **n**, pistillate perianth segment × 6; **o**, pistillate flower, carpels removed × 6; **p**, carpel × 6; **q**, carpel in vertical section × 6; **r**, ovary in cross-section × 3; **s**, fruit × $^1/_2$; **t**, fruit in vertical section × $^1/_2$; **u**, fruit in cross-section × $^1/_2$; **v**, **w**, **x**, seed in 3 views × $^1/_2$. *Nypa fruticans*: **a–r**, *Moore* 5846; **s–x**, *L.H. Bailey* 520.
(Drawn by Marion Ruff Sheehan)

Africa by ocean currents. *Nypa* is strictly a mangrove palm, occurring in a variety of estuarine situations; it usually grows in soft mud, often in vast natural stands. Pollination appears to be by drosophilid flies in New Guinea (Essig 1973), but Hoppe (2005) suggests a combination of pollination by various different insects and possibly also wind; correlations of pollination with floral anatomy and development have been noted (Uhl & Moore 1977a).

Relationships: For relationships, see subfamily Nypoideae.

Anatomy: Leaf (Tomlinson 1961); root (Seubert 1996b); silica bodies hat-shaped; central vascular bundles of petioles with a single phloem strand; floral (Uhl 1972a, Uhl & Moore 1977a).

Common names and uses: *Nipah*, mangrove palm. *Nypa fruticans* is ethnobotanically very important. The leaves are one of the most important sources for the production of palm shingles ('atap') for thatching, and also have minor uses such as for cigarette papers and fishing floats. The inflorescences are tapped for sap for sugar and alcohol production. The large natural stands of *Nypa* remain a greatly underexploited resource for fuel alcohol. The young endosperm is eaten, usually boiled in syrup, as a sweetmeat. The great but passive potential of *Nypa* as a stabiliser of estuarine mud in preventing coastal erosion should not be underestimated. For details of the utilisation of *Nypa*, see Burkill (1966), Brown and Merrill (1919) and Fong (1987, 1989).

Taxonomic account: Tralau (1964).

Fossil record: The fossil record for *Nypa* is outstanding among palms (see for example, Gee 1990). Most of the records are of fruits (e.g., Bowerbank 1840, Rendle 1894, Tralau 1964) or pollen (e.g., Muller 1968, 1979, 1981, Morley 2000), although occasionally leaf (Chandler 1961a, Pole & McPhail 1996, Mehrotra *et al.* 2003), flowering parts (e.g., Chandler 1961c), leaf epidermis (e.g., Kulkarni & Phadtare 1980) or root (e.g., Verma 1974) material is reported. Furthermore, the fossils are globally distributed through tropic and temperate zones until the global climatic deterioration at the end of the Middle Miocene. Records of possible affinity with *Nypa* for pinnate palm leaves are rare. From the Middle Eocene Bournemouth Freshwater Beds, Gardner (1882) recovered an abundance of pinnate palm leaves, which he considered to resemble *Iriartea* more than any other genus. However, later in a Guide to Fossil Plants in the British Museum (Natural History) it was written of one large specimen that, "it might possibly belong to *Nypa* whose fruits have been found at Bournemouth." (see Chandler 1963 p. 7). Occasionally pinnate leaves have been found in close association with other fossil *Nypa* organs, suggesting that they could be *Nypa* leaves (Pole & McPhail 1996, Mehrotra *et al.* 2003). Leaf cuticle considered to be from *Nypa* has been

Top left: ***Nypa fruticans***, inflorescence, Peninsular Malaysia. (Photo: J. Dransfield)

Bottom left: ***Nypa fruticans***, staminate flowers visited by drosophilid flies, Peninsular Malaysia. (Photo: J. Dransfield)

Above: ***Nypa fruticans***, infructescence, cultivated, Montgomery Botanic Center, Florida (Photo: J. Dransfield)

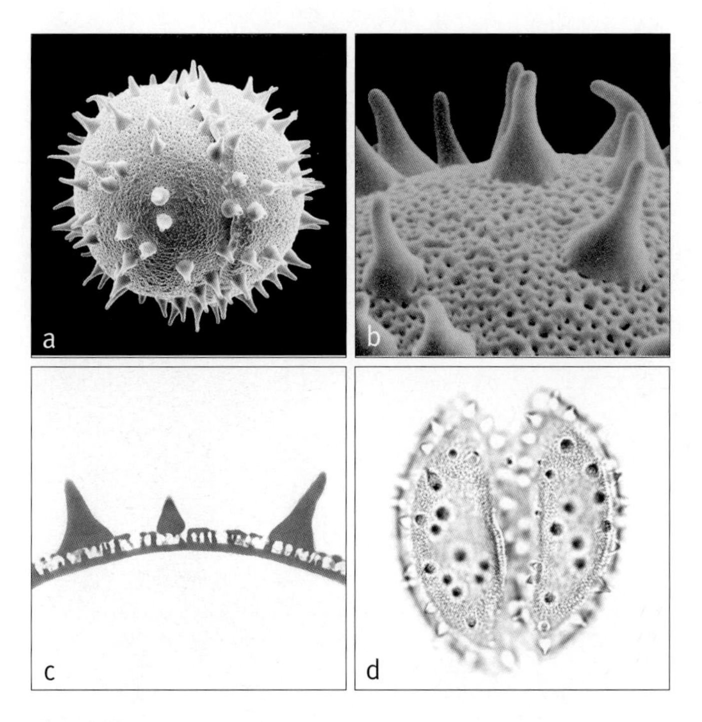

Above: ***Nypa***
a, meridional zonasulcate pollen grain, supratectate spines SEM × 1000;
b, close-up, typical spines with swollen bases SEM × 8000;
c, ultrathin section through pollen wall TEM × 5000;
d, meridional zonasulcate pollen grain, supratectate spines LM × 1000.
Nypa fruticans: **a** & **b**, *Whitmore* 596, **c**, *Letouzey* 14944, **d**, *Bryan Jr.* 1277.

found in close association with *Spinizonocolpites* in the Ratnagiri lignite Beds in India (Kulkarni & Phadtare 1980, 1981). The earliest records for fossil fruits and seeds (*Nypa*, *Nipadites*, *Nipa*) seem to be those of Gregor and Hagn (1982) recovered from the earliest Palaeocene (Danian) of Egypt (Bir Abu Munqar Formation) and of Dolianiti (1955) for the Palaeocene of Brazil or, possibly, the remarkably early Lower Cretaceous (Aptian) record of Jahnichen (1990; see comments in Chapter 5: The Fossil Record of Palms). Fossil fruits are also recorded from the Lower Eocene of southeastern North America (e.g., Berry 1914c, 1916b), and Middle Eocene (Claibornian) of Texas (Arnold 1952); from the Eocene of Europe — Belgium, France, The Netherlands, Italy, Hungary (see Tralau 1964), and the Lower Eocene London Clay Flora (e.g., Bowerbank 1840, Chandler 1961b). They are also present in the Upper Eocene of Russia (Kryshtofovich 1927). On the African continent, fruits are known from the Eocene of Egypt (Bonnet 1904, Kräusel 1939) and of Senegal (Fritel 1921). From the Eocene of Borneo, Kräusel (1923) described a fossil fruit, *Nipadites borneensis*. *Nypa* fruits are also recorded from the Upper Cretaceous in India (Deccan Intertrappean — although the age span of these volcanic deposits is controversial, see Chapter 5: The Fossil Record of Palms) and these records are summarised by Singh (1999). The earliest records of *Spinizonocolpites* (see Muller 1981 and Gee 1990 for summaries and authorship) are from the Maastrichtian, and occur almost simultaneously in South America (Colombia, Brazil and Venezuela), Africa (Ivory Coast, Senegal and Cameroon), India and Malesia (Borneo; Muller 1968, although there is some doubt about the age of this record, Morley 1998). Other Upper Cretaceous records not reviewed by Muller (1981) or Gee (1990) include those from Somalia (Schrank 1994) and India (Singh 1999). From the Palaeocene until the Middle Miocene, *Spinizonocolpites* is also recovered from deposits in current temperate zones, as well as from those of the tropics and subtropics (see Gee 1990, 2001) — USA (Gulf Coast), Europe (southern England, Belgium, France, Spain, Germany and Hungary), Russia, Africa (Senegal, Morocco and Egypt), India, Australia and New Zealand. The Indian record is extensive (see Singh 1999) although a proportion of the spinose pollen described and illustrated needs further re-appraisal. Further records include those from Colombia (Jaramillo & Dilcher 2001), southern England (Harley *et al.* 1991), France and Belgium (Paris and Belgian Basins; Schuler *et al.* 1992), Austria (Hofmann & Zetter 2001, Zetter & Hofmann 2001), Hungary (Ràkosi 1976), Nigeria (Jan du Chêne *et al.* 1978, Babajide Salami 1985), Pakistan (Frederiksen 1994), China (Song *et al.* 1999) and Tasmania (Pole & McPhail 1996). Rarely pollen, fruits or other fossil *Nypa* 'organs' are found in close association (e.g., Kulkarni & Phadtare [1980, 1981], Pole & McPhail [1996], Mehrotra *et al.* [2003]).

Notes: In both vegetative and reproductive characters, *Nypa* differs markedly from other palms. The erect inflorescence (Uhl 1972a) bearing a terminal head of pistillate flowers and lateral spikes of staminate flowers is unique in the family. Similar sepals and petals in both staminate and pistillate flowers, lack of staminodes and pistillodes, an androecium of only three stamens with united filaments, and three separate carpels of unusual form are exceptional floral characters. The adaptability of the fruit for floating is also noteworthy.

The pistillate head was once thought similar to those of phytelephantoid palms; early scholars (Drude 1887, Martius 1823–1850) classified *Nypa* with the phytelephantoid palms, but both the arrangement and the structure of the flowers are markedly different.

The combination of solitary flowers, distinct and similar sepals and petals, free carpels of cupular form, a chromosome number of 17, the reduced stamen number and fusion of the filaments, the lack of staminodes and pistillodes and the adaptation of the fruit for floating occurs nowhere else in the family.

The pistillate flowers appear to be structurally and developmentally unique within the family. Although the perianth encloses the carpels in early stages, by anthesis the carpels much exceed the perianth, members of which are displayed and obscured. The carpel is basally tubular with a mouth-like stigmatic opening and appears to represent a different, unspecialised form.

Additional figures: Figs 1.1, 1.7, 1.9, Glossary figs 7, 14.

III. CORYPHOIDEAE

CORYPHOIDEAE Griff., Calcutta J. Nat. Hist. 5: 311 (1844), ('*Coryphinae*').

Hermaphroditic, monoecious, dioecious or polygamous; hapaxanthic or pleonanthic; sometimes armed but spines never emergences; leaves palmate, costapalmate, pinnate, pinnately ribbed or rarely bipinnate, with very few exceptions induplicate; inflorescence often highly branched, the bracts sometimes tubular; flowers solitary or borne in cincinni, rarely in triads of a central pistillate and 2 lateral staminate, hermaphroditic or unisexual; ovary apocarpous or syncarpous; ovules very varied in form; pericarp smooth or broken into corky warts, endocarp usually thin at maturity, more rarely thick; seeds 1–3.

This large subfamily includes most of the palmate-leaved palms. It also contains four genera of pinnate-leaved palms, which are remarkable for their induplicate leaflets. In fact, all induplicate-leaved palms belong to this group; a few members, however, display reduplicate or anomalous leaf-splitting.

Many thorough analyses have produced equivocal resolutions with respect to the monophyly of subfamily Coryphoideae (e.g., Asmussen & Chase 2001). A monophyletic subfamily Coryphoideae is, however, resolved by Uhl *et al.* (1995), Asmussen *et al.* (2006) and Baker *et al.* (in review). The monophyly is highly supported in the latter two studies (Asmussen *et al.* 2006, Baker *et al.* in review). Hahn (2002a) supports numerous alternative resolutions, but his total evidence dataset resolves the Coryphoideae as monophyletic. When monophyletic, the Coryphoideae is almost always resolved as sister to a clade comprising the Arecoideae and Ceroxyloideae (Hahn 2002a, Asmussen *et al.* 2006, Baker *et al.* in review). Subfamily Coryphoideae is divided into four major clades: 1) the New World thatch palm clade, consisting of tribes Sabaleae and Cryosophileae; 2) the syncarpous clade, consisting of tribes Chuniophoeniceae, Caryoteae, Corypheae and Borasseae; 3) tribe Phoeniceae; and 4) tribe Trachycarpeae (Asmussen *et al.* 2006, Baker *et al.* in review). The relative positions of these four clades vary among phylogenies (e.g., compare Uhl *et al.* 1995, Asmussen *et al.* 2006 and Baker *et al.* in review).

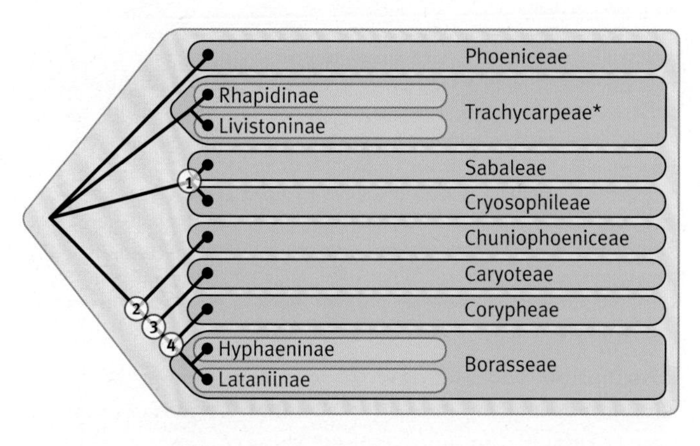

Summary tree showing relationships among tribes and subtribes of the Coryphoideae that have been resolved with moderate to high support in multiple phylogenetic analyses. Labelled nodes are resolved in the following studies: **node 1** (the New World thatch palm clade) — Uhl *et al.* 1995, Asmussen & Chase 2001, Asmussen *et al.* 2006, Baker *et al.* in review; **node 2** (the syncarpous clade) — Bayton 2005, Asmussen *et al.* 2006; **node 3** — Bayton 2005, Asmussen *et al.* 2006; **node 4** — Bayton 2005, Asmussen *et al.* 2006, Baker *et al.* in review. For further explanation, see discussions under tribe and subtribe treatments. An asterisk indicates a tribe containing genera that cannot yet be placed into a subtribe.

Key to Tribes of Coryphoideae

1. Leaves pinnate, bipinnate or entire 2
1. Leaves palmate, costapalmate or entire, leaf and segment tips not conspicuously praemorse, rarely shallowly toothed . . 3
2. Leaf or leaflet margins strongly and irregularly praemorse; basal-most leaflets not modified as spines; monoecious or very rarely dioecious; inflorescences with an inconspicuous prophyll and numerous large peduncular bracts; flowers unisexual, arranged in triads of a central pistillate and two lateral staminate flowers, or reduced by abortion to paired staminate and solitary pistillate flowers . . . . . . **Caryoteae**
2. Leaflets acute, not praemorse; the basal-most leaflets modified as spines; dioecious; inflorescences bearing a single large primary bract (prophyll); flowers solitary . . . . **Phoeniceae**
3. Dioecious palms; flowers strongly dimorphic; staminate and sometimes pistillate flowers borne in deep pits formed by connation and adnation of rachilla bracts; endocarp very thick and hard, sometimes heavily ornamented, with an apical pore or splitting into 2 valves . . . . . **Borasseae**
3. Hermaphroditic, polygamodioecious or rarely strictly dioecious; flowers not borne in pits; endocarp not usually thick, lacking a pore and never splitting into 2 valves . . 4
4. Gynoecium composed of 3 carpels, united at the base, with free styles, or with a common style 5
4. Gynoecium composed of 1–4 carpels, free throughout their length, or 3 carpels basally free but with united styles . . 7
5. Rachillae bearing complete tubular bracts subtending solitary or grouped flowers; seed with basal or subbasal embryo . . . . **Chuniophoeniceae**
5. Rachilla bracts minute, not tubular 6
6. Hapaxanthic palms with huge suprafoliar aggregation of inflorescences; petioles armed with regular teeth; flowers grouped in cincinni; style with 3 separate canals; seed with apical embryo. . . . . **Corypheae**

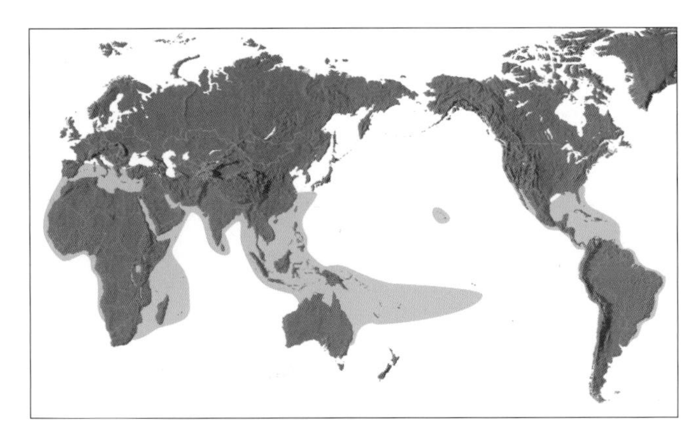

Distribution of subfamily **Coryphoideae**

6. Pleonanthic palms with interfoliar inflorescences; petiole unarmed; flowers solitary; style with a single common canal; seed with lateral or subapical embryo . . **Sabaleae**
7. Leaf sometimes with a central abaxial split, never with any other abaxial splits; gynoecium apocarpous, with 3–4 carpels free throughout their length, sometimes only a single carpel present; strictly New World palms . . . . . **Cryosophileae**
7. Leaf usually with only adaxial splits, sometimes with regular abaxial splits or with splits between the folds; gynoecium apocarpous, with 3 carpels free throughout their length, or free basally and with connate styles; palms of the Old and New World tropics and subtropics . . . . **Trachycarpeae**

Tribe Sabaleae Mart. ex Dumort., Anal. Fam. Pl.: 55 (1829).

Pleonanthic, hermaphroditic palms; leaves induplicate, palmate or costapalmate; leaf base with a central split; flowers solitary; gynoecium syncarpous with 3 carpels united throughout their length and with a long common stylar canal.

The tribe contains the single genus *Sabal*, found in the Caribbean region and northwards into the southern USA.

Tribe Sabaleae is strongly supported as monophyletic (Asmussen *et al.* 2006), and resolved as sister to tribe Cryosophileae with moderate to high support (Uhl *et al.* 1995, Asmussen *et al.* 2006, Baker *et al.* in review), forming a group termed here the New World thatch palm clade. Some resolutions of Hahn (2002a) are in agreement, though other resolutions that show a relationship to *Corypha* could be an artefact of taxon sampling.

23. SABAL

Single-stemmed fan palms of the Caribbean and neighbouring mainland of Central and North America, usually with conspicuously costapalmate leaves with unarmed petioles that have a triangular cleft at the base. The highly branched inflorescences bear solitary hermaphroditic flowers and usually rather small blackish fruit.

Sabal Adans., Fam. pl. 2: 495, 599 (1763). Lectotype: *S. adansonii* Guers. (illegitimate name), (*Corypha minor* Jacq. = **S. minor** [Jacq.] Pers.).

Inodes O.F. Cook, Bull. Torrey Bot. Club 28: 529 (1901). Type: *I. causiarum* O.F. Cook (= *Sabal causiarum* [O.F. Cook] Becc.).

Derivation unknown, perhaps based on a vernacular name.

Dwarf, moderate or tall, usually robust, solitary, acaulescent or erect, unarmed, pleonanthic, hermaphroditic palms. *Stem* often descending shortly and recurved, covered with leaf bases, rough, striate, and obscurely ringed, or becoming ± smooth, grey, and bare with age. *Leaves* induplicate, marcescent, shortly to prominently costapalmate; sheath later with a conspicuous cleft below the petiole, margins fibrous; petiole often very long, channelled adaxially, rounded abaxially, sometimes bearing caducous indumentum; adaxial hastula short and truncate, or usually elongate and acute or acuminate, margins sharp, abaxial hastula sometimes distinguishable as a low ridge; blade flat to mostly arched, divided along the central abaxial fold to the middle or nearly to the costa, further divided along adaxial folds into drooping linear, ± even, rarely uneven, single-fold segments, briefly to rather deeply bifid, sometimes filiferous, segments with midribs prominent abaxially, interfold filaments sometimes present, glaucous or not, sometimes paler beneath, often with caducous indumentum along the major ribs, midribs prominent, transverse veinlets obscure or conspicuous. *Inflorescence* shorter, as long as or longer than the leaves, interfoliar, branching to 4 orders; prophyll short, 2-keeled, 2-lobed; peduncular bracts several, tubular below with a conspicuous, short to long and narrow tip, variously caducously tomentose; rachis equalling or longer than the peduncle; rachis bracts like peduncular bracts, decreasing in size distally; bracts of the second and third order well developed, tubular, decreasing in size distally; prophylls present on most branches; rachillae slender, with spirally arranged bracts, each subtending a low spur branch bearing a solitary flower. *Flowers* symmetrical; calyx somewhat thickened at the base, tubular, shallowly 3-lobed, often prominently nerved when dried; corolla tubular below, lobes elliptic, slightly imbricate in bud, spreading to suberect with incurved membranous margins at anthesis, becoming strongly inrolled when dry; stamens 6, the filaments rather fleshy, flattened, united in a tube about as high as the calyx, adnate up to the mouth of the corolla tube, then distinct and awl-shaped, not inflexed at the apex, anthers erect in bud, dorsifixed, ± versatile or erect, narrowly elliptic, latrorse; carpels 3, completely connate, ovarian part trilobed and only slightly broader than the elongate 3-grooved style, stigma capitate, trilobed, papillose, ovule basal, anatropous. *Pollen* ellipsoidal, slightly asymmetric; aperture a distal sulcus; ectexine tectate, finely to coarsely perforate, or perforate and micro-channelled, aperture margin similar or slightly finer; infratectum columellate; longest axis 33–50 µm; post-meiotic tetrads usually tetrahedral, sometimes tetragonal or, rarely, rhomboidal [8/16]. *Fruit* usually developing from 1 carpel, sometimes from 2 or 3, globose to pyriform, stigmatic scar and abortive carpels basal; epicarp smooth, mesocarp fleshy without fibres, endocarp thin, membranous. *Seed* free from endocarp, shining brown, depressed-globose, usually concave below when dry, raphe and hilum basal, endosperm homogeneous with a shallow intrusion of seed coat; embryo lateral or subdorsal. *Germination* remote-ligular; eophyll entire, elongate. *Cytology*: 2n = 36.

Distribution and ecology: 16 species. One of the larger coryphoid genera confined to the central Western hemisphere from Colombia to northeastern Mexico, the southeastern USA and the Caribbean basin. Some species (*Sabal minor*) grow in swampy areas, others in sandy coastal regions and dry open lands.

Anatomy: Leaf (Tomlinson 1961, Zona 1990), root (Seubert 1997), floral (Morrow 1965), gynoecium (Uhl & Moore 1971).

Relationships: *Sabal* is monophyletic with strong support (Zona 1990, Asmussen *et al.* 2006). For relationships, see under tribe Sabaleae.

Common names and uses: Palmetto, variously designated as bush (*Sabal minor*), cabbage (*S. palmetto*) and so on. Formerly used for making brooms and locally as a source of thatch. Many species are important ornamentals.

Taxonomic accounts: Zona (1990) and Quero (1991).

Fossil record: There are many fossil records for costapalmate leaves; they are the earliest type of fossil palm leaves recovered to date. The leaves are usually described under the fossil genus *Sabalites*, even though there are a number of other modern, predominantly coryphoid, genera, that have costapalmate leaves. The re-circumscription of *Sabalites* by Read and Hickey (1972) embraces any palmate leaf, 'with a definite costa or extension of the petiole into the blade', whereas *Palmacites* is recommended for palm-like leaves that are, 'pure palmate, lacking a costa or extension of the petiole into the blade'. New fossil genera for costapalmate leaves, *Costapalma*, and for palmate leaves, *Palustrapalma*, were published by Daghlian (1978), but these genera have not been widely adopted. Other *Sabal*-like fossils include small monosulate pollen grains and, rarely, leaf cuticle and fruits. The fossil genus *Sabalites* is not only used frequently as a generic name for costapalmate leaves but also applied to fruits. The oldest record of *Sabalites* appears to be *S. carolinensis* Berry (Berry 1914b) from the Upper Cretaceous (Coniacian-Lower Santonian Black Creek Formation [Middendorf arkose member] of South Carolina [USA]); this is also the oldest palm fossil assignable below family level. *Sabalites magothiensis* (Berry) Berry (Berry 1905, 1911) from the Santonian of Maryland and New Jersey (USA) and *S. longirhachis* (Unger) J. Kvaček & Hermann (Hermann & Kvaček 2002; Kvaček & Hermann 2004) from the Lower

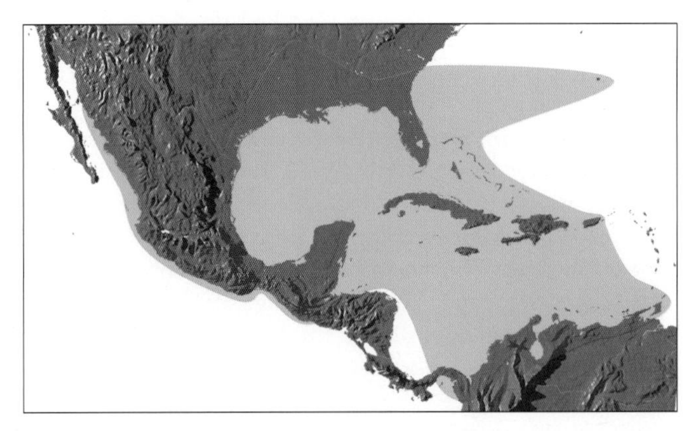

Distribution of ***Sabal***

Campanian of Austria are somewhat younger. Other Upper Cretaceous records include, from North America (Wyoming), *S. eocenica* (originally described from an Eocene location) and *S. montana* (Dorf 1942); and fragments of a large fan leaf, *Sabalites* sp. (Berry 1919a), from Tennessee (Ripley Formation). An incompletely preserved costapalmate leaf, *S. ooaraiensis*, (Ôyama & Matsuo 1964) is described from the fluvial deposits of the Upper Cretaceous Ôarai flora on the coast of Naka-gawa. From the Palaeocene onwards, records of costapalmate leaves become more frequent and widespread. In North America, records for the Rocky Mountains and Great Plains have been reviewed by Brown (1962), whereas records for the Middle and Upper Eocene floras of southeastern North America were reviewed by (Berry 1924) and also by Daghlian (1978). Details of leaf venation, adaxial and abaxial epidermal cells and stomatal cell

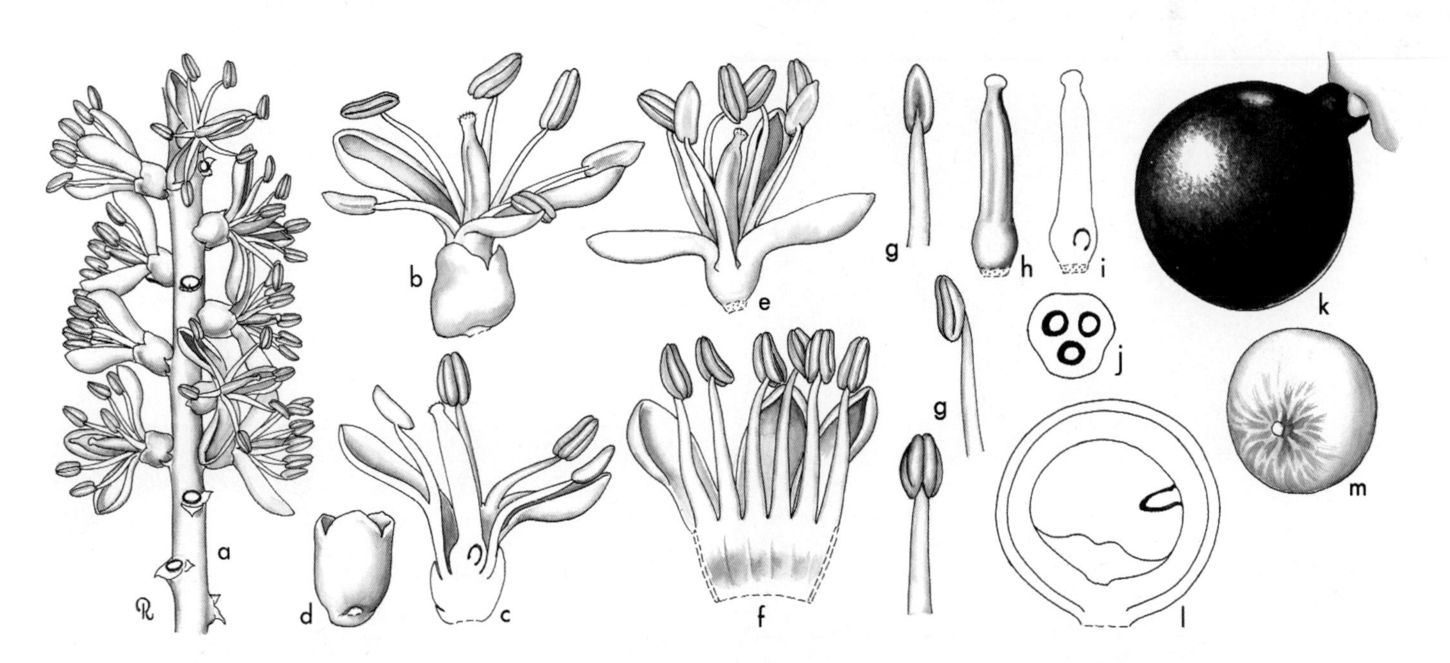

Sabal. **a**, portion of rachilla × 3; **b**, flower × 6; **c**, flower in vertical section × 6; **d**, calyx × 6; **e**, flower, calyx removed × 6; **f**, corolla and androecium (**e** expanded), interior view × 6; **g**, stamen in 3 views × 6; **h**, gynoecium × 6; **i**, gynoecium in vertical section × 6; **j**, ovary in cross-section × 12; **k**, fruit × 3; **l**, fruit in vertical section × 3; **m**, seed × 3. *Sabal palmetto*: all from *Read* s.n. (Drawn by Marion Ruff Sheehan)

arrangement allowed Daghlian (1978) to make direct comparison between the Eocene *Sabal dortchii* Daghlian and the modern genus. Other Tertiary records come from Europe: southern England (Palaeocene) (Reid & Chandler 1933, Chandler 1961b, 1961c, 1962, 1963); France, Tertiary (Saporta 1865); Germany, Upper Eocene–Miocene (Mai & Walther 1978), Miocene (Van der Burgh 1984); Switzerland, Upper Oligocene (Büchler 1990); Czech Republic, Miocene (Kvaček 1998); and Hungary, Oligocene (Andreánszky 1949). A relationship with the fossil *Sabal major* Unger was suggested for the inflorescence and rachilla of *Tuzsonia hungarica* (Andreánszky 1949), based on associated palm leaf fragments. However, the pollen described from individual flowers is probably too small, and the exine pattern unlike that of modern *Sabal* pollen. A number of Tertiary records have also been found: in Russia, Caucasus, Middle Miocene (Takhtajan 1958); Kamchatka Peninsula, Eocene (Budantsev 1979); and Transcaucasia, Oligocene (Akhmetiev 1989; leaf and associated monosulcate pollen); and in India, a leaf axis (Trivedi & Verma 1981) from the Deccan Intertrappean of Madhya Pradesh (although the age span of these volcanic deposits is controversial; see Chapter 5). Four species of *Sabalites* leaves are reported from the Tertiary of China (Peking Institute of Botany, and Nanjing Institute of Geology and Palaeontology 1978) and, from the Eocene of Japan, leaf fragments have been recorded from lignite mines in Hokkaido: *Sabalites nipponica* (Kryshtofovich 1918). Subfossil fragments of clasping petiole bases and seeds ('stones') are reported by Berry (1917) from the Pleistocene of Vero (Florida). Fossil fruits are reported from southern England, Lower Eocene as *Sabal grandisperma* (Reid & Chandler 1933); and Germany (Geiseltal), Eocene as *S. bracknellense* (Mai 1976). Few dispersed monosulcate fossil palm-like pollen can be confidently assigned to *Sabal*. This is because the pollen of many genera in tribes Sabaleae, Cryosophileae, Trachycarpeae and Chuniophoeniceae share similar size ranges and exine characteristics. Records include *Sabalpollenites* sp. from the Eocene of North America (Tennessee) (Potter 1976), *Sabal* sp. from the Lower Eocene of southern England (Khin Sein 1961), and three new Miocene species of finely reticulate *Sabalpollenites* from the Czech Republic (Konzalova 1971) (this association is questionable because the exine of *Sabal* pollen tends to be finely to coarsely perforate, or perforate and micro-channelled). There are also records from the Lower Miocene of Poland (Macko 1957) and from the Tertiary of China (*Sabalpollenites areolatus*; Song *et al.* 1999).

Notes: The occurrence of prophylls on inflorescence branches is useful in delimiting species. *Sabal* is distinguished from other genera that have costapalmate leaves by the lack of spines on the petiole and by the split leaf bases.

Additional figures: Fig. 2.1a, Glossary figs 2, 3, 4, 8, 11, 15, 25.

Left: ***Sabal domingensis***, habit, Dominican Republic. (Photo: J. Dransfield)

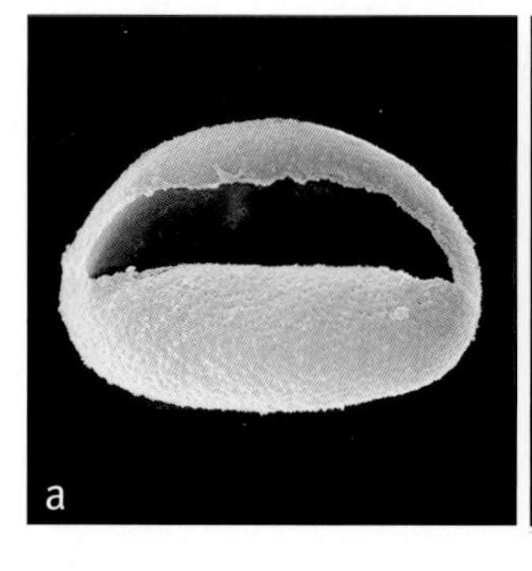

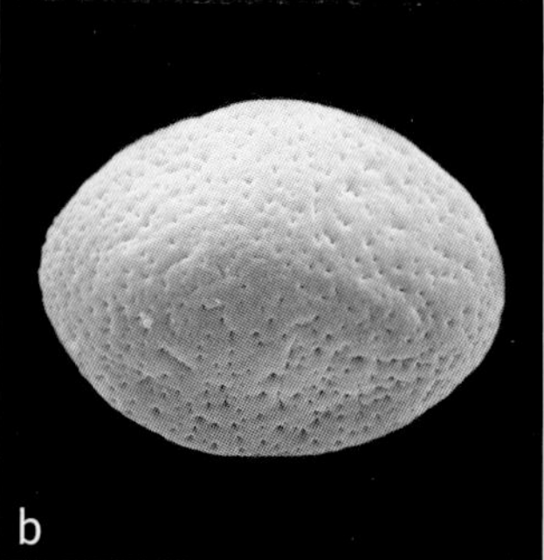

Below: ***Sabal***
a, monosulcate pollen grain, distal face SEM × 1500;
b, monosulcate pollen grain, proximal face × 1500.
Sabal minor: **a**, *Drummond* 357; *S. yapa*: **b**, *Gentle* 1156.

Tribe Cryosophileae J. Dransf., N.W. Uhl, C. Asmussen, W.J. Baker, M.M. Harley & C. Lewis., Kew Bull. 60: 561 (2005). Type: **Cryosophila**.

Pleonanthic, hermaphroditic (rarely polygamous) palms; leaves induplicate, palmate; leaf base with or without a central split; flowers solitary; gynoecium apocarpous with 1–4 carpels; large vascular bundes of petiole usually with 1 phloem strand.

The tribe includes ten genera restricted to the New World tropics and subtropics. These genera were formerly included in subtribe Thrinacinae, together with a group of Old and New World genera, now separated into subtribe Rhapidinae of tribe Trachycarpeae. The tribe displays considerable variation in floral structure. Large vascular bundles of the petiole usually contain a single phloem strand.

Tribe Cryosophileae is highly supported as monophyletic (Asmussen & Chase 2001, Hahn 2002a, Asmussen *et al.* 2006, Baker *et al.* in review, Roncal *et al.* 2008). The tribe is moderately to highly supported as sister to tribe Sabaleae (Uhl *et al.* 1995, Asmussen & Chase 2001, Asmussen *et al.* 2006, Baker *et al.* in review), forming a group termed the New World thatch palm clade. Some resolutions in Hahn (2002a) show a relationship with Sabaleae and others show relationships with tribes Phoeniceae and Trachycarpeae.

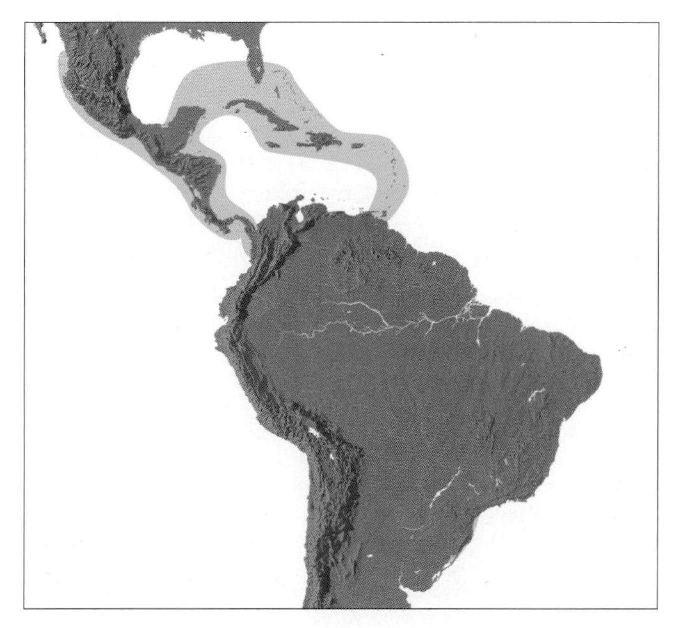

Distribution of tribe **Cryosophileae**

Key to Genera of Cryosophileae

1. Perianth biseriate; carpels 1–3; stamens 5–25 2
1. Perianth uniseriate, usually 6-toothed; carpel 1; stamens 5–15 6
2. Carpel 1 3
2. Carpels 2, 3, or 4 4
3. Flowers with a long stalk-like base; proximal flowers on each rachilla hermaphroditic, dital flowers staminate; stamens 6 . . . . **24. Schippia**
3. Flowers lacking a stalk-like base, all hermaphroditic; stamens 18–24 . . . . **33. Itaya**
4. Stem lacking distinctive root spines; stamen filaments free 5
4. Stem bearing distinctive root spines; stamen filaments connate in a ring . . . . **32. Cryosophila**
5. Sepals united to approximately half their length; stamens with distinct long filaments, long exserted, more than twice as long as the corolla; seed with lobed intrusion of the seed coat . . . . **25. Trithrinax**
5. Sepals imbricate, free nearly to base; stamens with fleshy wide filaments, only slightly exerted; seed lacking a conspicuous intrusion of seed coat . . . **31. Chelyocarpus**
6. Base of leaf sheath split in petiolar region 7
6 Base of leaf sheath not split in petiolar region 8
7. Leaf sheaths disintegrating into a regular fibrous network, the fibre tips spine-like and reflexed; flowers subdistichous on the rachillae; fruit white at maturity; seed deeply bilobed, the lobes themselves irregularly bilobed . . . . **26. Zombia**
7. Leaf sheaths very rarely reflexed and spine-like; fruit brown, pink, red or black, very rarely white; seed deeply furrowed, not bilobed . . . . **27. Coccothrinax**
8. Flowers sessile or borne on low protuberances
8. Flowers borne on conspicuous stalks . . . . . **30. Thrinax**
9. Stamens inflexed in bud; filaments basally connate in a low ring, connective very broad, anthers sessile, oblong-elliptic in outline, extrorse . . . . **28. Hemithrinax**
9. Stamens erect in bud; filaments more elongate, basally broadly connate in a ring equaling the perianth, free portion very slender, anthers elongate, latrorse . . . . **29. Leucothrinax**

24. SCHIPPIA

Monotypic fan palm found only in Belize and Guatemala, distinct in the long-pedicellate flowers and large pale green fruit.

Schippia Burret, Notizbl. Bot. Gart. Berlin-Dahlem 11: 867 (1933b). Type: **S. concolor** Burret.

Commemorating William A. Schipp (1891–1967), who collected many plants in Belize, including Schippia.

Moderate, solitary, unarmed, pleonanthic, hermaphroditic or polygamo-monoecious palm. *Stem* slender, longitudinally striate, rough, with raised, close, oblique leaf scars. *Leaves* induplicate, palmate to very shortly costapalmate; sheath split basally, densely tomentose, disintegrating to form a thick fibrous network; petiole very long, narrow, adaxially channelled and abaxially keeled basally, biconvex distally, sparsely to densely tomentose, margins acute; adaxial hastula triangular or rounded, rather large, abaxial hastula a low rounded ridge; blade divided to below the middle into narrow, tapering, single-fold segments, with pointed, unequal, very shortly bifid apices, lighter coloured beneath, appearing glabrous on both surfaces, midrib conspicuous abaxially, transverse veinlets inconspicuous, very short. *Inflorescences* solitary, interfoliar, much shorter than the leaves, branched to 2 (rarely 3) orders; peduncle moderate, dorsiventrally flattened; prophyll tubular, elongate, adaxially flattened, abaxially rounded, 2-keeled laterally, pointed, splitting unevenly apically, densely tomentose; peduncular bracts 3, ± like the prophyll but ± keeled dorsally; rachis about as long as the peduncle, glabrous; rachis bracts tubular, pointed, splitting adaxially, densely tomentose; other bracts membranous, pointed, small and inconspicuous; first-order branches adnate above the subtending bracts; rachillae short, narrow, spirally arranged, spreading, becoming smaller distally, bearing small triangular bracts each subtending a solitary flower. *Flowers* of two kinds, one hermaphoditic, the other staminate; calyx and receptacle forming a long pseudopedicel, calyx tubular basally with 3 triangular-lanceolate lobes; petals 3, distinct, much larger than the calyx, slightly imbricate; stamens 6, filaments distinct, elongate, anthers linear, sagittate basally, dorsifixed, versatile, latrorse; gynoecium unicarpellate, ovoid, style elongate, tubular, open distally with stigmatic area around the rim, ovule basal, probably hemianatropous. *Pollen* ellipsoidal, with slight to obvious asymmetry; aperture a distal sulcus; ectexine tectate, finely rugulate and micro-channelled, aperture margin psilate or scabrate; infratectum columellate; longest axis 30–35 µm [1/1]. *Fruit* globose with apical stigmatic scar; epicarp smooth, mesocarp thin, fleshy, endocarp smooth, membranous with anastomosing bundles. *Seed* globose with indistinct basal raphe, endosperm homogeneous; embryo nearly apical. *Germination* remote; eophyll simple. *Cytology*: 2n = 36.

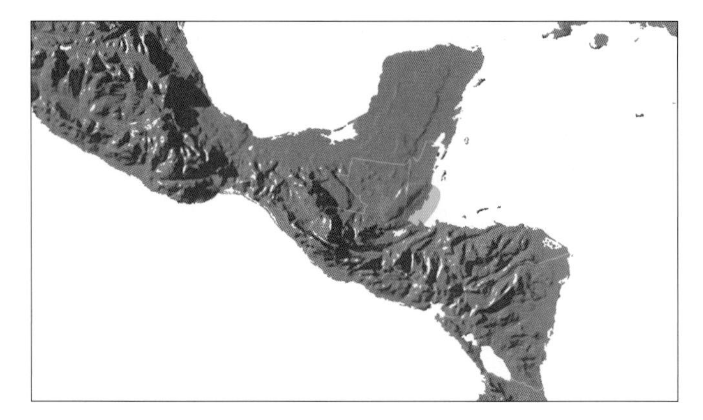

Distribution of ***Schippia***

Distribution and ecology: One species in Belize and Guatemala, occurring in the undergrowth of tropical rain forest.

Anatomy: Leaf anatomy not studied, roots (Seubert 1997), floral (Morrow 1965).

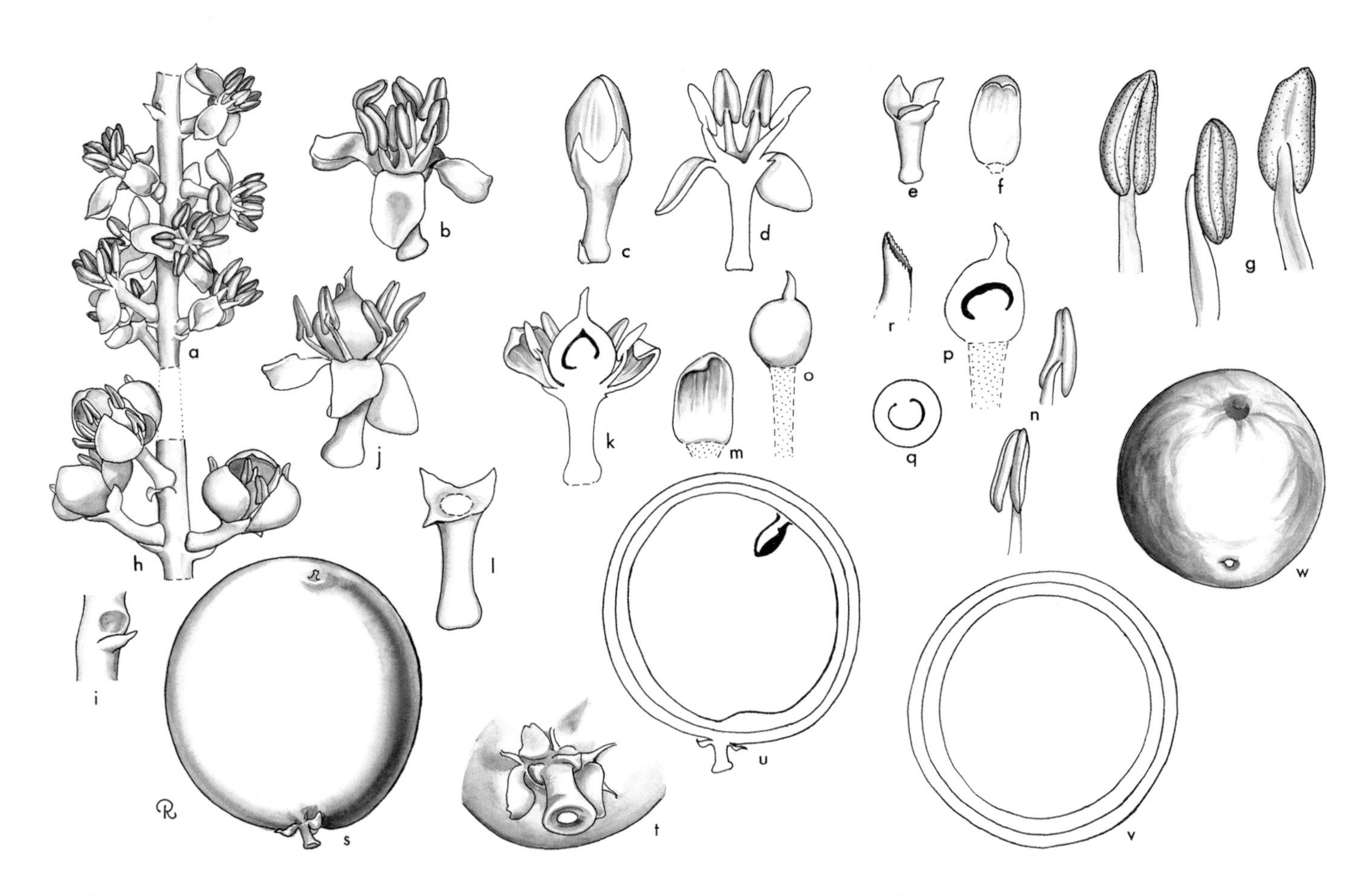

Schippia. **a**, apical portion of rachilla with staminate flowers × 3; **b**, staminate flower, expanded × 6; **c**, staminate bud × 6; **d**, staminate flower in vertical section × 6; **e**, staminate calyx × 6; **f**, staminate petal, interior view × 6; **g**, stamen in 3 views × 12; **h**, lower portion of rachilla with pistillate flowers × 3; **i**, bracteole subtending pistillate flower × 6; **j**, pistillate flower × 6; **k**, pistillate flower in vertical section × 6; **l**, pistillate calyx × 6; **m**, pistillate petal, interior view × 6; **n**, staminodes in 2 views × 12; **o**, gynoecium × 6; **p**, gynoecium in vertical section × 6; **q**, ovary in cross-section × 6; **r**, stigma × 12; **s**, fruit × 1 1/2; **t**, base of fruit × 3; **u**, fruit in vertical section × 1 1/2; **v**, fruit in cross-section × 1 1/2; **w**, seed × 1 1/2. *Schippia concolor*: all from *A.C. & M. Langlois* s.n. (Drawn by Marion Ruff Sheehan)

Relationships: There are two different hypotheses on the phylogenetic position of *Schippia* in the Cryosophileae. *Schippia* is weakly supported as sister to *Cryosophila* (Roncal *et al.* 2008) or as sister to a clade of *Zombia*, *Coccothrinax*, *Hemithrinax*, *Leucothrinax* and *Thrinax* with low support (Baker *et al.* in review).

Common names and uses: Pimento palm. Occasionally cultivated.

Taxonomic accounts: Burret (1933b); see also Balick & Johnson (1994).

Fossil record: No generic records found.

Notes: Pinheiro (2001) describes the remote germination of *Schippia*. During germination, the cotyledonary axis elongates rapidly, followed by translocation of food reserves from the endosperm into the cotyledonary axis, which then becomes swollen; the first foliar organs then emerge through the cotyledonary sheath.

Additional figures: Glossary figs 4, 10, 16.

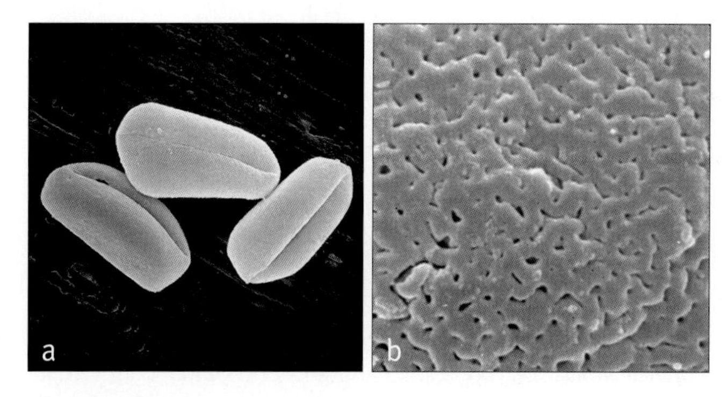

Above: ***Schippia***
a, group of monosulcate pollen grains, distal faces SEM × 750;
b, close-up, finely perforate rugulate pollen surface SEM × 8000.
Schippia concolor: **a** & **b**, *Whiteford* 2277.

Below left: ***Schippia concolor***, in flower, cultivated, Fairchild Tropical Botanic Garden, Florida. (Photo: C.E. Lewis)

Below right: ***Schippia concolor***, in fruit, cultivated, Fairchild Tropical Botanic Garden, Florida. (Photo: C.E. Lewis)

25. TRITHRINAX

Solitary or clustering hermaphroditic fan palms of warm temperate parts of eastern South America; the leaf sheaths end in fibre spines and the unspecialised trimerous flowers have stamens greatly exceeding the petals in length.

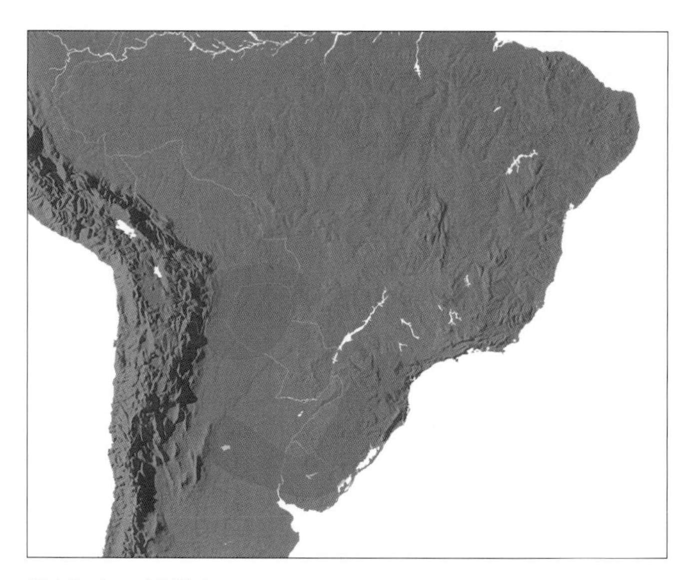

Distribution of ***Trithrinax***

Trithrinax Mart., Hist. nat. palm. 2: 149 (1837). Type: **T. brasiliensis** Mart.

Diodosperma H. Wendl., Bot. Zeit. 36: 118 (1878). Type: *D. burity* H. Wendl. (= *Trithrinax schizophylla* Drude).

Chamaethrinax H. Wendl. ex R. Pfister., Beitr. Verg. Anat. Sabaleenblatter: 19 (1891) (invalid name). Type: *C. hookeriana* H. Wendl. ex R. Pfister (= *Trithrinax campestris* [Burmeist.] Drude & Griseb.)

Combines tri — *three, with the generic name* Thrinax, *but why is not explained.*

Moderate, solitary or sometimes clustering, armed, pleonanthic, hermaphroditic palms. *Stem* erect, clothed with persistent, fibrous, sometimes spiny leaf sheaths, eventually becoming bare, rough, and longitudinally striate. *Leaves* induplicate, palmate, marcescent; sheath tubular, drying into a fibrous, often ± woody network, the upper fibres becoming stout rigid spines; petiole adaxially shallowly channelled or rounded, abaxially rounded, the margins entire, sharp; adaxial hastula triangular or deltoid usually with a definite point, abaxial hastula similar, often smaller; blade fan-shaped to nearly circular, not or only slightly costapalmate, nearly regularly divided beyond the middle (*Trithrinax biflabellata* divided centrally, almost to the base) into numerous single-fold, stiff segments with shallowly to deeply bifid, apiculate to sharp tips, adaxially glabrous, abaxially lightly waxy and tomentose, midribs more prominent abaxially, other veins numerous, small, transverse veinlets not evident. *Inflorescences* solitary, interfoliar, rather short to moderate, robust, curved, creamy-white when young, branched to 3 orders; peduncle short; prophyll and 2(–3) peduncular bracts similar, inflated, tubular at base, expanded and split along one side, slightly keeled dorsally toward the apex, with short solid tips, glabrous or densely but irregularly tomentose; rachis longer than the peduncle; rachis bracts like peduncular bracts but becoming smaller, absent distally, each subtending a first-order branch; first-order branches adnate to the rachis and often to the tubular base of the next higher bract, stout, curved, bearing chartaceous, small, triangular bracts subtending rachillae; rachillae spirally arranged, ± equal in length, much shorter than first-order branches, bearing small elongate triangular bracts each subtending a flower. *Flowers* spirally arranged, solitary on short stalks, slightly asymmetrical; sepals 3, very shortly united basally, ovate; petals 3, ± twice as long as the sepals, ovate, imbricate, fleshy, acute; stamens 6, exserted, filaments distinct, twice as long as the petals, slender, tapering, anthers linear oblong, versatile, latrorse; carpels 3, distinct, ovarian part obovoid, attenuate to a tubular, short to long, erect or recurved style with apical stigma, ovule basal, hemianatropous, with aril. *Pollen* grains ellipsoidal, with slight to obvious asymmetry;

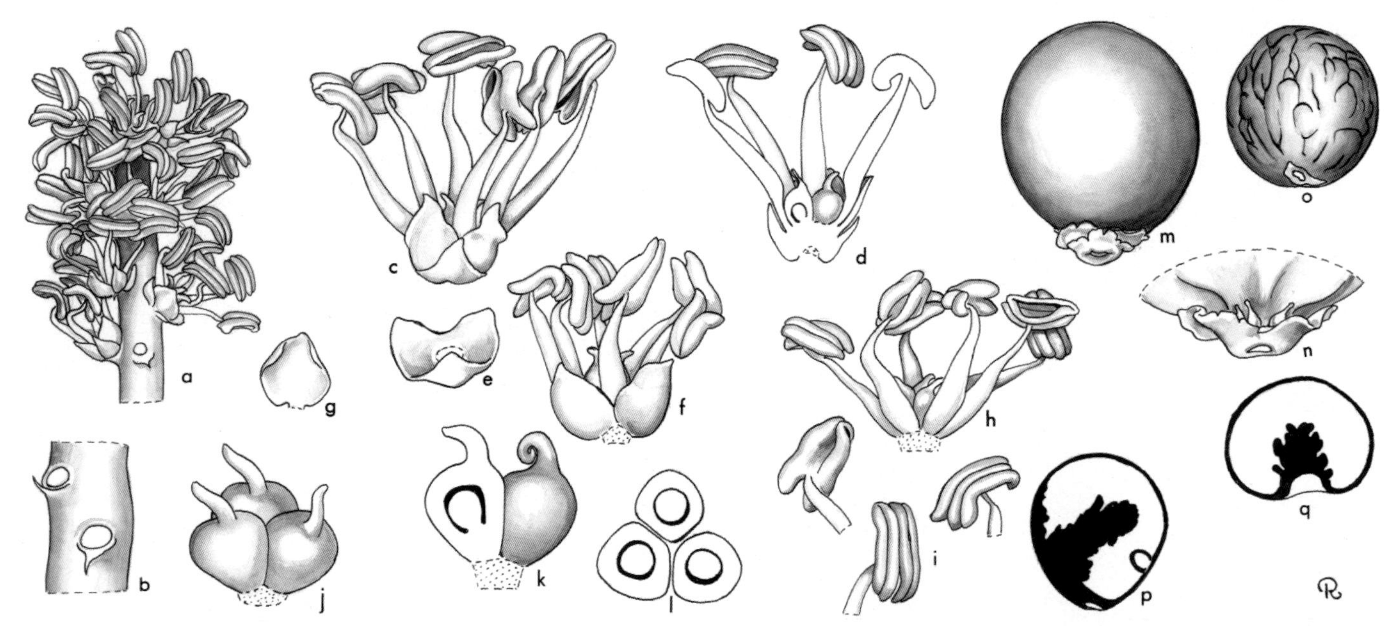

Trithrinax. **a**, portion of rachilla × 3; **b**, portion of rachilla showing floral scars and bracts × 6; **c**, flower × 6; **d**, flower in vertical section × 6; **e**, calyx × 6; **f**, flower, calyx removed × 6; **g**, petal, interior view × 6; **h**, androecium and gynoecium × 6; **i**, stamen in 3 views × 6; **j**, gynoecium × 12; **k**, gynoecium in vertical section × 12; **l**, carpels in cross-section × 12; **m**, fruit × 1 1/2; **n**, base of fruit and abortive carpels × 3; **o**, seed × 1 1/2; **p**, seed in vertical section × 1 1/2; **q**, seed in cross-section × 1 1/2. *Trithrinax brasiliensis*: **a–l**, *Read* 723, **m–q**, *Hertrich* s.n. (Drawn by Marion Ruff Sheehan)

aperture a distal sulcus; ectexine tectate, finely perforate, perforate and micro-channelled, or perforate-rugulate, aperture margin slightly finer or psilate; infratectum columellate; longest axis 25–45 µm [2/3]. *Fruit* 1-seeded, white, globose, stigmatic scar apical, abortive carpels basal; epicarp smooth, mesocarp fleshy, endocarp thin, papery. *Seed* becoming free, globose, hilum circular, basal with ascending branches, endosperm homogeneous with deeply intruded seed coat below the raphe; embryo lateral, opposite the raphe. *Germination* remote-tubular (Chavez 2003); eophyll simple. *Cytology*: 2n = 36.

Distribution and ecology: Three species in Bolivia, western tropical and southern Brazil, Paraguay, Uruguay and Argentina. *Trithrinax schizophylla* is reported from sandy marshes and along river banks. The other species occur in dry areas.

Anatomy: Leaf (Tomlinson 1961), roots (Seubert 1997), floral (de Magnano 1973, Morrow 1965).

Relationships: The monophyly of *Trithrinax* has not been tested. Baker *et al.* (in review) find moderate support for a sister relationship between *Trithrinax* and *Chelyocarpus*.

Common names and uses: *Caranday*, for other local names see Glassman (1972). Stems are used in construction and leaves as thatch. The leaf sheaths have been used as filters. The fruit are eaten fresh or fermented and the seed can be a source of oil. *Trithrinax campestris* is a much sought-after ornamental (Gibbons 2001).

Taxonomic accounts: A new treatment is needed. See Beccari (1931) and Henderson *et al.* (1995).

Fossil record: From amber deposits in the northern Dominican Republic, 13 hermaphroditic palm flowers have been described as *Trithrinax dominicana* (Poinar 2002a); the age of the amber is estimated to be somewhere between mid Eocene and mid Miocene.

Notes: Species in cultivation are notably resistant to cold and drought.

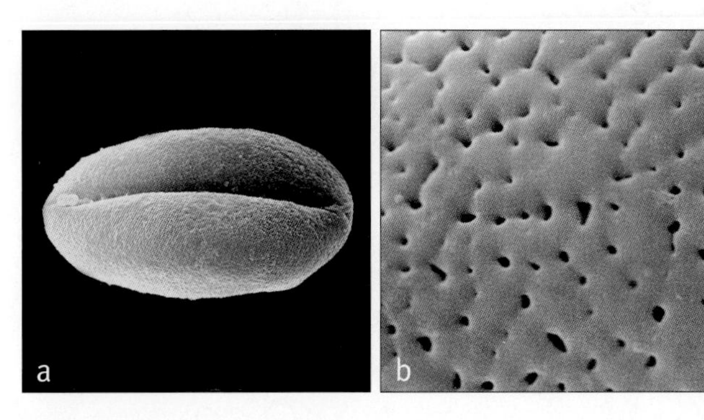

Above: ***Trithrinax***
a, monosulcate pollen grain, distal face SEM × 1500;
b, close-up, coarsely perforate pollen surface SEM × 8000.
Trithrinax brasiliensis: **a**, *Glaziou* 9014; *T. campestris*: **b**, *Harrison Wright* 3/10/35.

Top left: ***Trithrinax campestris***, habit, Argentina. (Photo: M. Gibbons & T. Spanner)

Bottom left: ***Trithrinax campestris***, detail of fibre spines, Argentina. (Photo: M. Gibbons & T. Spanner)

26. ZOMBIA

Distinctive clustering hermaphroditic fan palm of Hispaniola, with close erect stems almost completely obscured by spines formed from the fibres of the expanded leaf sheaths; petiole base not split; fruit white, the seed deeply bilobed, each lobe further lobed.

Zombia L.H. Bailey, Gentes Herb. 4: 240 (1939). Type: **Z. antillarum** (Descourt. ex B.D. Jacks.) L.H. Bailey (*Chamaerops antillarum* Descourt. ex B.D. Jacks.).

Coccothrinax subgenus *Oothrinax* Becc., Feddes Repert Spec. Nov. Regni Veg. 6: 95 (1908) (invalid name). Type: *C. anomala* Becc. (= *Zombia antillarum*).

Oothrinax (Becc.) O.F. Cook, Natl. Hort. Mag. 20: 21 (1941) (illegitimate name).

From the Haitian local name, latanier zombie*, itself alluding to zombies.*

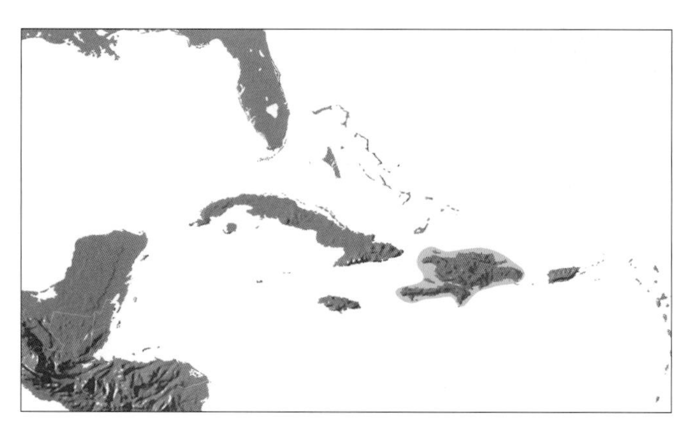

Distribution of ***Zombia***

Moderate, clustering, spiny, pleonanthic, hermaphroditic palm. *Stem* erect, slender, covered with persistent, overlapping spiny sheaths, few erect spine-like pneumatophores emerging from the ground near the stem base. *Leaves* induplicate, palmate, marcescent; sheath expanding into a regular network of fibres, distal fibres reflexed, forming a partial whorl of spines; petiole very slender, elongate, unarmed, semicircular in cross-section; adaxial hastula 3-lobed, center lobe pointed, lateral lobes rounded becoming irregularly tattered, abaxial hastula also pointed with very shallow lateral ridges; blade irregularly divided along the adaxial folds $^1/_2-^2/_3$ to the base into single-fold, lanceolate, rather thin, shortly bifid segments, shiny dark green above, whitish beneath, caducous indumentum along ribs on both surfaces, midrib prominent abaxially, interfold ribs prominent adaxially, transverse veinlets scarcely discernible. *Inflorescences* interfoliar, shorter than the leaves, branched to 2 orders; peduncle short; prophyll partly included in the subtending the leaf sheath, tubular, 2-keeled, opening and ± 2-lobed distally, longitudinally striate; peduncular bracts lacking; rachis longer than the peduncle, densely tomentose; rachis bracts tubular, longitudinally striate, sparsely tomentose, each with a short pointed lobe, sometimes with short apical splits; first-order branches distant, each bearing a basal, tubular, 2-keeled, 2-lobed, longitudinally striate, sparsely tomentose prophyll; rachillae (second-order branches) sparse, short, spreading, proximal 2 ± subopposite, subtended by narrow bracts, distal rachillae subopposite or alternate, subtending bracts minute, inconspicuous, rachillae slender, glabrous, minutely papillose, bearing rather distant, solitary, subdistichous flowers each subtended by a minute triangular bract. *Flowers* cream-coloured; perianth shallow, cup-shaped with 6 short, membranous points; stamens 9–12, the filaments short, variable, slender, anthers basifixed, erect, elongate, latrorse; gynoecium obpyriform, unicarpellate, tapered to a large, laterally compressed, cup-like stigma, ovule basal, orthotropous. *Pollen* ellipsoidal, slight to obvious asymmetry; aperture a distal sulcus; ectexine tectate, perforate, aperture margin ± psilate; infratectum columellate; longest axis 30–38

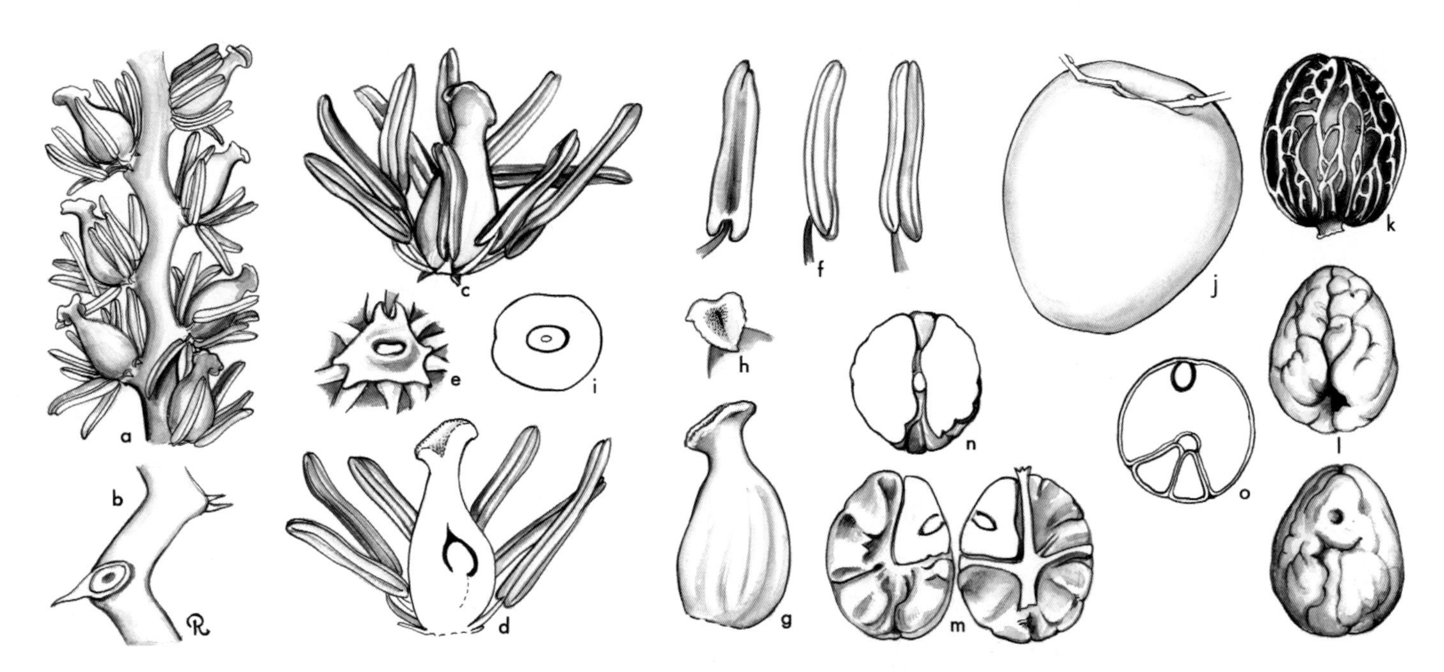

Zombia. **a**, portion of rachilla × 3; **b**, floral scar and bracts × 6; **c**, flower × 6; **d**, flower in vertical section × 6; **e**, perianth × 6; **f**, stamen in 3 views × 6; **g**, gynoecium × 6; **h**, stigma × 6; **i**, ovary in cross-section × 6; **j**, fruit × $1^1/_2$; **k**, seed with adhering endocarp fibers × $1^1/_2$; **l**, seed in 2 views × $1^1/_2$; **m**, seed in vertical section × $1^1/_2$; **n, o**, seed in cross-section through middle of embryo × $1^1/_2$. *Zombia antillarum*: all from *Read* 721. (Drawn by Marion Ruff Sheehan)

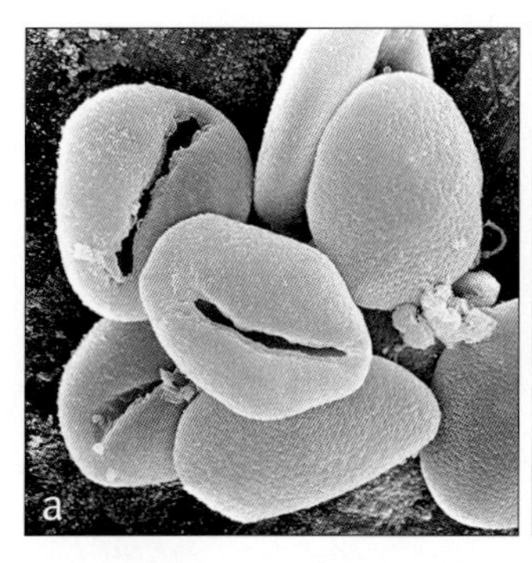

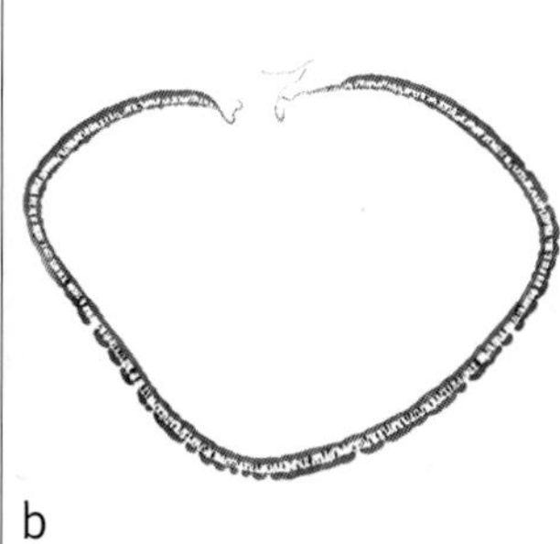

Above: ***Zombia***

a, group of monosulcate pollen grains: distal left and centre, proximal right and bottom SEM × 750;

b, ultrathin section, whole pollen grain, polar plane TEM × 2000.

Zombia antillarum: **a** & **b**, *Ekman* 4570.

Below left: ***Zombia antillarum***, habit, cultivated, Montgomery Botanical Centre, Florida. (Photo: J. Dransfield)

Middle: ***Zombia antillarum***, close-up of leaf sheaths, cultivated, Montgomery Botanical Centre, Florida. (Photo: J. Dransfield)

Right: ***Zombia antillarum***, fruit, cultivated, Fairchild Tropical Botanic Garden, Florida. (Photo: C.E. Lewis)

µm [1/1]. *Fruit* falling with the perianth attached and stamen bases often obvious, oblong-globose, white, fleshy, large, with apical stigmatic remains; epicarp smooth except for the stigmatic remains, mesocarp fleshy without obvious fibres, endocarp crustaceous. *Seed* basally attached, very deeply bilobed, the lobes again divided irregularly in a bilobed fashion, the 2 halves connected by a portion in which the embryo lies centrally, endosperm homogeneous within the lobes. *Germination* and eophyll not recorded. *Cytology*: 2n = 36.

Distribution and ecology: One species in Hispaniola. Open and bushy slopes of very dry hills.

Anatomy: Leaf (Tomlinson 1961), roots (Seubert 1997), floral (Morrow 1965, Moore & Uhl 1973).

Relationships: Baker *et al*. (in review) find high support for a sister relationship between *Zombia* and a clade of *Thrinax*, *Coccothrinax*, *Hemithrinax* and *Leucothrinax*. Roncal *et al*. (2008) place *Zombia* as sister to a clade of *Coccothrinax*, *Hemithrinax* and *Leucothrinax* with weak support.

Common names and uses: Zombi palm. A striking ornamental. Fruit are said to be fed to hogs.

Taxonomic account: Bailey (1939b).

Fossil record: No generic records found.

Notes: An unusual palm easily distinguished by the bizarre, persistent, spiny leaf sheaths. Differing from *Coccothrinax* by having white fruits and from *Thrinax*, *Hemithrinax* and *Leucothrinax* by having unsplit petioles.

Additional figures: Glossary figs 5, 11.

27. COCCOTHRINAX

Small to moderate, solitary or clustering hermaphroditic fan palms occurring widely in the Caribbean, particularly diverse on Cuba; leaf sheaths very varied, fibrous, sometimes spectacularly so, or even spiny, petiole bases not split; fruit usually purplish black at maturity, rarely pink or white, the seed deeply grooved.

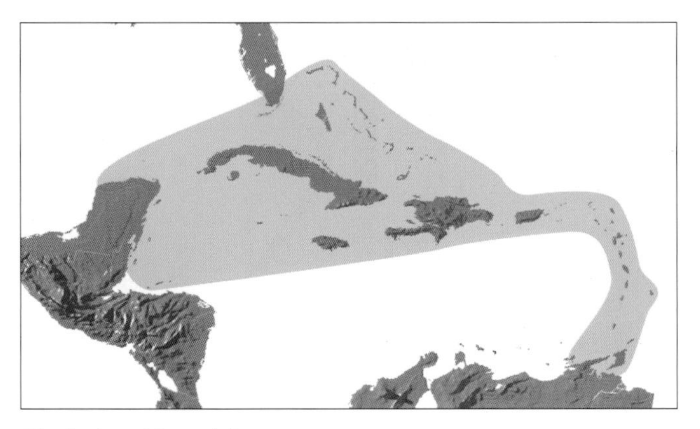

Distribution of ***Coccothrinax***

Coccothrinax Sarg., Bot. Gaz. 27: 87 (1899). Type: **C. jucunda** Sarg.

Haitiella L.H. Bailey, Contrib. Gray Herb. 165: 7 (1947). Type: *H. ekmanii* (Burret) L.H. Bailey (= *Coccothrinax ekmanii* Burret).

Thrincoma O.F. Cook, Bull. Torrey Bot. Club 28: 539 (1901). Type: *T. alta* O.F. Cook (= *Coccothrinax alta* [O.F. Cook] Becc.).

Thringis O.F. Cook, Bull. Torrey Bot. Club 28: 544 (1901). Lectotype: *T. latifrons* O.F. Cook (see H.E. Moore 1963c) (= *Coccothrinax alta* [O.F. Cook] Becc.).

Derivation not explained but presumably from coccus *— a berry, and the palm generic name* Thrinax.

Small to moderate, solitary or clustered, unarmed or partly armed, pleonanthic, hermaphroditic palms. *Stem* slender, at first covered with fibrous leaf sheaths, then with a regular fibrous network or masses of long slender fibres or stout spines, eventually becoming bare, and closely ringed with narrow leaf scars. *Leaves* induplicate, palmate, ascending to spreading, marcescent; sheath sometimes long persistent and disintegrating into a regular fibrous network or masses of long slender fibres, or becoming ± spiny, covered with dense, deciduous tomentum (?always); petiole long, slender, flat to ridged adaxially, rounded abaxially, densely tomentose or glabrous; hastula prominent adaxially, triangular to ± rounded, absent or a very narrow ridge abaxially; blade fan-shaped, irregularly folded when large, divided to about the middle into long, rather narrow, pointed segments, tips usually bifid, glabrous adaxially, silvery, punctate, with hairs or glabrous abaxially; midribs prominent, lateral ribs sometimes conspicuous, transverse veinlets evident or inconspicuous on one or both surfaces. *Inflorescences* shorter than the leaves, slender, branched to 2 orders; peduncle rather short, slender; prophyll tubular, 2-keeled, pointed, opening apically; peduncular bracts several, like the prophyll but lacking keels, closely sheathing and overlapping; rachis longer than the peduncle, slender, bearing spirally arranged, tubular, overlapping, pointed bracts subtending rachillae; rachillae rather short, slender, bearing very small, spirally arranged, thin, pointed bracts (?bracts appear to be borne on the floral stalk where seen), each subtending a flower. *Flowers* solitary, sessile(?) or usually pedicellate; perianth broadly and shallowly cup-shaped with several (5–9) short points; stamens 9 (6–13), filaments slender, flat, shortly connate basally, not inflexed at the apex, anthers oblong or sometimes sagittate, dorsifixed near the base, latrorse, apically acute to briefly bifid; gynoecium of 1 carpel, unilocular, uniovulate, globose basally, attenuate in a long style terminating in a cup-like, ± laterally compressed stigma, ovule basal,

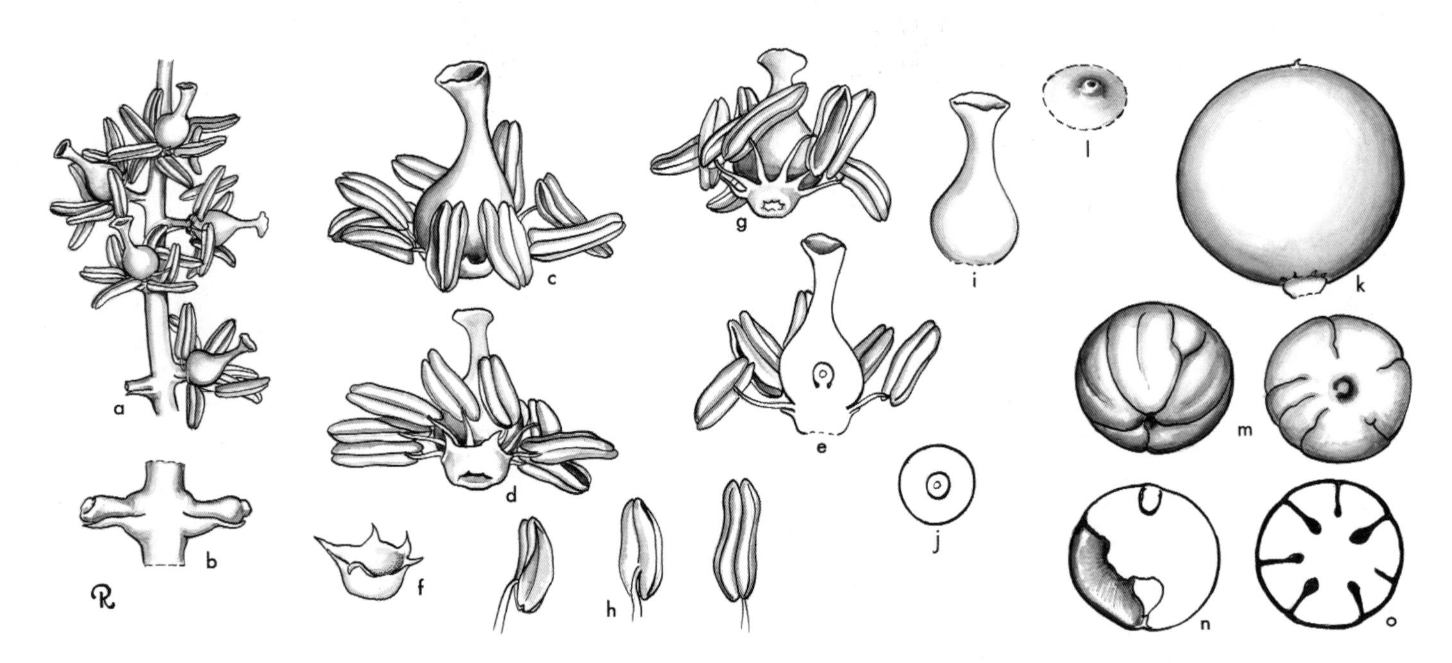

Coccothrinax. **a**, portion of rachilla × 3; **b**, portion of rachilla with floral scars and bracts × 6; **c**, flower × 6; **d**, flower from below × 6; **e**, flower in vertical section × 6; **f**, perianth × 6; **g**, flower, perianth removed × 6; **h**, stamen in 3 views × 6; **i**, gynoecium × 6; **j**, ovary in cross-section × 6; **k**, fruit × 3; **l**, apex of fruit × 6; **m**, seed in 2 views × 3; **n**, seed in vertical section × 3; **o**, seed in cross-section × 3. *Coccothrinax argentata*: **a–j**, *Read* 612; **k–o**, *Dahlberg* s.n. (Drawn by Marion Ruff Sheehan)

erect, nearly orthotropous. *Pollen* ellipsoidal, with slight to obvious asymmetry, aperture a distal sulcus; ectexine tectate, perforate, or perforate-rugulate, aperture margin similar; infratectum columellate; longest axis 31–44 µm [3/50]. *Fruit* globose, purplish-black at maturity, rarely pink or white, stigmatic remains apical; epicarp smooth or rough, mesocarp thin or somewhat fleshy with flat, slender, anastomosing fibres next to the membranous endocarp. *Seed* globose, attached basally, deeply grooved, hilum rounded, basal, endosperm homogeneous except for grooves; embryo apical or subapical. *Germination* remote-tubular; eophyll entire, very narrow. *Cytology*: 2n = 36.

Distribution and ecology: About 50 species occurring from Florida south to Colombia, mostly on islands of the West Indies, with the greatest diversity (about 34 species) in Cuba. Restricted to limestone or serpentine rocks, usually in dry and often exposed highlands, sometimes in valleys and on coasts.

Anatomy: Leaf (Tomlinson 1961) for *Coccothrinax barbadensis* and *C. argentea* but identification questioned, roots (Seubert 1997), floral (Uhl & Moore 1977a), fruit (Murray 1973).

Top left: ***Coccothrinax boschiana***, habit, Dominican Republic. (Photo: J. Dransfield)

Top right: ***Coccothrinax fissa***, habit, Dominican Republic. (Photo: J. Dransfield)

Bottom left: ***Coccothrinax ekmanii***, crown, Dominican Republic. (Photo: J. Dransfield)

Bottom right: ***Coccothrinax miraguama***, inflorescence, cultivated, Montgomery Botanical Centre, Florida. (Photo: J. Dransfield)

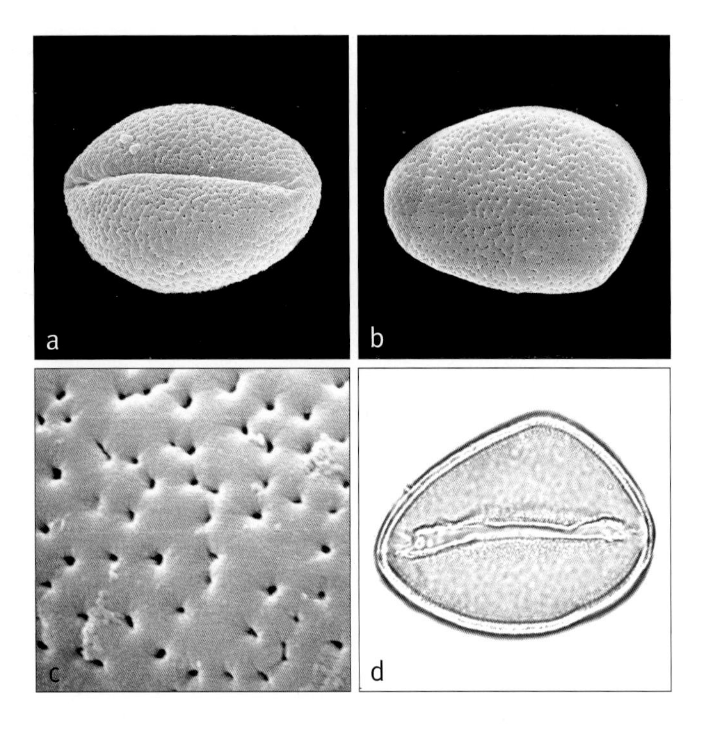

Left: ***Coccothrinax***
a, monosulcate pollen grain, distal face SEM × 1000;
b, monosulcate pollen grain, proximal face SEM × 1500;
c, close-up, perforate pollen surface SEM × 8000;
d, monosulcate pollen grain, high–mid-focus LM × 1000.
Coccothrinax argentata: **a**, *Small & Carter* s.n.; *C. fragrans*: **b** & **c**, *Wright* 3218; *C. rigida*: **d**, *Wright* 3220.

Below, top: ***Hemithrinax ekmaniana***, habit, Cuba. (Photo: C.E. Lewis)
Bottom: ***Hemithrinax ekmaniana***, infructescence, Cuba. (Photo: C.E. Lewis)

Relationships: *Coccothrinax* is resolved as monophyletic with high support within a clade that also includes *Hemithrinax* and *Leucothrinax* (Roncal *et al*. 2008).
Common names and uses: Broom, silver, and thatch palms; used for thatch and for making brooms, also grown as ornamentals.
Taxonomic accounts: A new treatment is much needed. See Bailey & Moore (1949), Leon (1939, 1946), Muñiz & Borhidi (1982), Quero (1980) and Read (1980).
Fossil record: No generic records found.
Notes: O.F. Cook published the generic names *Antia*, *Beata* and *Pithodes*, based on species of *Coccothrinax*, in the National Horticultural Magazine Volume 20. The names appear without Latin descriptions and postdate 1935, and are thus invalid and lacking any botanical standing.

28. HEMITHRINAX

Small to moderate, solitary hermaphroditic fan palms endemic to Cuba; leaf sheaths fibrous, petiole bases deeply split at base; fruit very small, white, the seed smooth or shallowly grooved.

Hemithrinax Hook.f. in Benth. & Hook.f., Gen. pl. 3: 930 (1883). Type: *Trithrinax compacta* Griseb. & H. Wendl. = **Hemithrinax compacta** (Griseb. & H. Wendl.) Hook.f. ex Salomon.

Combining hemi — *half, with the palm generic name* Thrinax, *presumably referring to the fact that the palms are not quite true* Thrinax.

Small to moderate, solitary, unarmed, pleonanthic, hermaphroditic palms. *Stem* erect, sometimes very short, columnar, smooth or fibrous, sometimes obscured by a skirt of marcescent leaves, obscurely ringed with leaf scars, usually with a basal mass of fibrous roots. *Leaves* very crowded and congested or more laxly arranged, induplicate, palmate, often irregular; sheath becoming split both opposite the petiole and abaxially, sometimes scarcely fibrous, or disintegrating into irregular curled robust fibres, or fibres free proximally and united distally in a short or greatly elongated ligule-like structure, covered in thick, deciduous tomentum, margins fibrous; petiole very short to moderately elongate, slender, rounded to shallowly ridged both adaxially and abaxially, margins rather sharp; adaxial hastula prominent, triangular or rounded, abaxially hastula ridge-like, moderately conspicuous, rounded or triangular; blade fan-shaped, split to about half to two-thirds its length into single-fold segments, segments lanceolate, pointed and usually bifid apically, glabrous adaxially, sometimes densely covered in wax, abaxially variously scaly, sometimes white, midrib and marginal ribs conspicuous, abaxial surface of lamina with minute dot-like scales, transverse veinlets obscure. *Inflorescences* interfoliar, lax or congested, scarcely longer than the sheaths and protruding through the split petiole base, or erect, longer than the leaves, branched to 1–3 orders; peduncle rather slender, laterally compressed or rounded in cross-section; prophyll coriaceous, 2-keeled, with triangular tip, splitting apically and abaxially along its entire length (?always); peduncular bracts to 5 or apparently absent in *Hemithrinax compacta*, elongate, membranous or coriaceous, variously indumentose, with bifid tips, closely sheathing the peduncle and splitting only at the tip; rachis bracts 2 or 3 (*H. rivularis*, *H. ekmaniana*) or numerous to ca. 10 (*H. compacta*), spirally arranged, overlapping and very closely sheathing the rachis, each subtending a first-order branch, variously indumentose; first-order branches tapering, bearing spirally arranged bracts similar to but smaller than the rachis bracts, distally bearing minute triangular bracts subtending second-order branches or rachillae; rachillae slender, very short, much shorter or slightly longer than the rachis, glabrous or variously indumentose, stiff, bearing spirally arranged, small narrow triangular bracts subtending very small solitary flowers, bracteoles apparently lacking. *Flowers* sessile or borne on very short protuberances; perianth a single cupule with 6 triangular lobes; stamens 6–8, filaments short, basally connate in a ring, connective very broad, anthers oblong-elliptic in outline, dorsifixed and facing outwards at anthesis, forming a low ring around the gynoecium, extrorse; gynoecium consisting of 1 carpel, unilocular, uniovulate, conical, tapering into a short style, approximately twice as long as the androecium, stigma conduplicate, ovule basally attached, erect, campylotropous with a basal aril. *Pollen* ellipsoidal asymmetric, sometimes pyriform; aperture a distal sulcus; ectexine tectate, coarsely perforate and micro-channelled, wide psilate and sparsely perforate aperture margin; infratectum columellate; longest axis 24–26 µm. [1/3]. *Fruit* very small to moderate, white at maturity, stigmatic remains apical, perianth persistent; epicarp smooth when fresh, drying pebbled, mesocarp thin, mealy, endocarp very thin, papery. *Seed* depressed-globose, smooth or shallowly and sparsely grooved, hilum round, impressed, forming a basal intrusion, raphe branches sparse, shallowly impressed, endosperm homogeneous; embryo subapical. *Germination* not recorded; eophyll unknown. *Cytology* not studied.

Above: ***Hemithranax ekmaniana***, crown with inflorescences, Cuba. (Photo: C.E. Lewis)

Hemithrinax. **a**, portion of rachilla × 2; **b**, flower from below × 8; **c**, flower from above × 8; **d**, flower in vertical section × 8; **e**, corolla, interior view, stamens and gynoecium removed × 8; **f**, androecium × 8; **g**, androecium expanded, interior view × 8; **h**, stamens in 3 views × 8; **i**, gynoecium × 8; **j**, ovary in cross-section × 8; **k**, fruit × 1; **l**, apex of fruit × 4; **m**, seed × 2; **n**, seed in cross-section × 2; **o**, seed in vertical section × 2. *Hemithrinax ekmaniana*: **a–j** *Victorin* 17637; *H. savannarum*: **k–o** *Leon* 20101. (Drawn by Marion Ruff Sheehan)

Distribution of ***Hemithrinax***

Distribution and ecology: Three species, all very local endemics in Cuba (Muñiz & Borhidi 1982, Borhidi & Muñiz 1985). Found only on limestone outcrops (*Hemithrinax compacta* and *H. ekmaniana*) up to about 400 m elevation and on ultramafic soils near streams at low elevation (*H. rivularis*).
Anatomy: Not studied.
Relationships: *Hemithrinax* is monophyletic with high support (Roncal *et al.* 2008) and sister to *Leucothrinax* with moderate support (Asmussen *et al.* 2006, Baker *et al.* in review).
Common names and uses: Not recorded.
Taxonomic accounts: Leon (1941) and Borhidi & Muñiz (1985).
Fossil record: No generic records found.
Notes: These remain rather poorly known palms, not often cultivated.

29. LEUCOTHRINAX

Small to moderate, solitary hermaphroditic fan palms endemic to the northern Caribbean; leaf sheaths fibrous, petiole bases deeply split at base; flowers and fruit sessile; fruit very small, white, the seed smooth.

Leucothrinax C.E. Lewis & Zona, Palms 52: 87 (2008). Type: **L. morrisii** (H. Wendl.) C.E. Lewis & Zona (*Thrinax morrisii* H. Wendl.).

Combining leucon — *white, with the palm generic name* Thrinax, *in reference to the whitish colour of the leaves.*

Small to moderate, solitary, unarmed, pleonanthic, hermaphroditic palm. *Stem* erect, columnar, smooth, grey, obscurely ringed with leaf scars, usually with a basal mass of fibrous roots. *Leaves* induplicate, palmate; sheath becoming split both opposite the petiole and abaxially to emit the inflorescence, disintegrating into a network of irregular fibres, covered in thick, deciduous tomentum, margins fibrous; petiole long, slender, rounded to shallowly ridged both adaxially and abaxially, densely covered with caduceus white indumentum, margins rather sharp; adaxial hastula prominent, rounded to triangular, densely covered in caducous hairs, abaxial hastula very short, apically membranous and disintegrating; blade fan-shaped, with irregularly folded segments, not held in the same plane, split apically to ca. $^1/_2$ their length or more into lanceolate, pointed and usually bifid segments, glabrous adaxially, abaxially covered in white wax and bearing minute punctiform scales, midrib and marginal ribs conspicuous, transverse veinlets evident. *Inflorescences* interfoliar, slender, erect to arching, equalling or exceeding the leaves, branched to 2 orders, primary branches pendulous; peduncle moderate, rather slender, round in

Below left: ***Leucothrinax morrisii***, crown, cultivated, Fairchild Tropical Botanic Garden, Florida. (Photo: C.E. Lewis)

Below right: ***Leucothrinax morrisii***, infructescence, cultivated, Fairchild Tropical Botanic Garden, Florida. (Photo: J. Dransfield)

cross-section; prophyll short, tubular, 2-keeled, pointed, opening distally, tomentose; peduncular bracts several (ca. 3), like the prophyll but lacking keels, overlapping and very closely sheathing the peduncle; rachis longer than the peduncle, slender, tapering, bearing spirally arranged, long, tubular, pointed distally and obliquely open primary bracts subtending first-order branches; first-order branches each with a short to long basal bare portion, bearing a 2-keeled, bifid prophyll and spirally arranged, narrow, triangular bracts subtending rachillae; rachillae slender, rather short, stiff, bearing spirally arranged, small triangular bracts subtending solitary flowers, bracteoles apparently lacking. *Flowers* ± sessile; perianth a single cupule with 6 lobes or teeth; stamens mostly 6, filaments basally broadly connate in a ring equalling the perianth, free portion very slender, anthers elongate, dorsifixed near the base, emarginate apically, latrorse; gynoecium consisting of 1 carpel, unilocular, uniovulate, ovule basally attached, erect, campylotropous but tilted so that the micropyle faces the upper dorsal wall of the locule, and with a basal aril. *Pollen* ellipsoidal with slight to obvious asymmetry; aperture a distal sulcus; ectexine tectate,

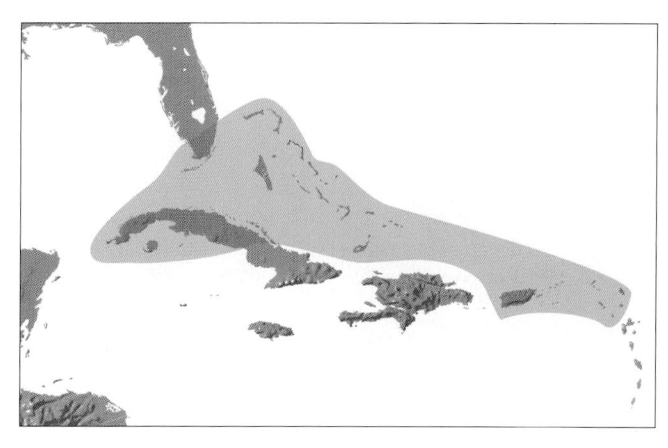

Distribution of ***Leucothrinax***

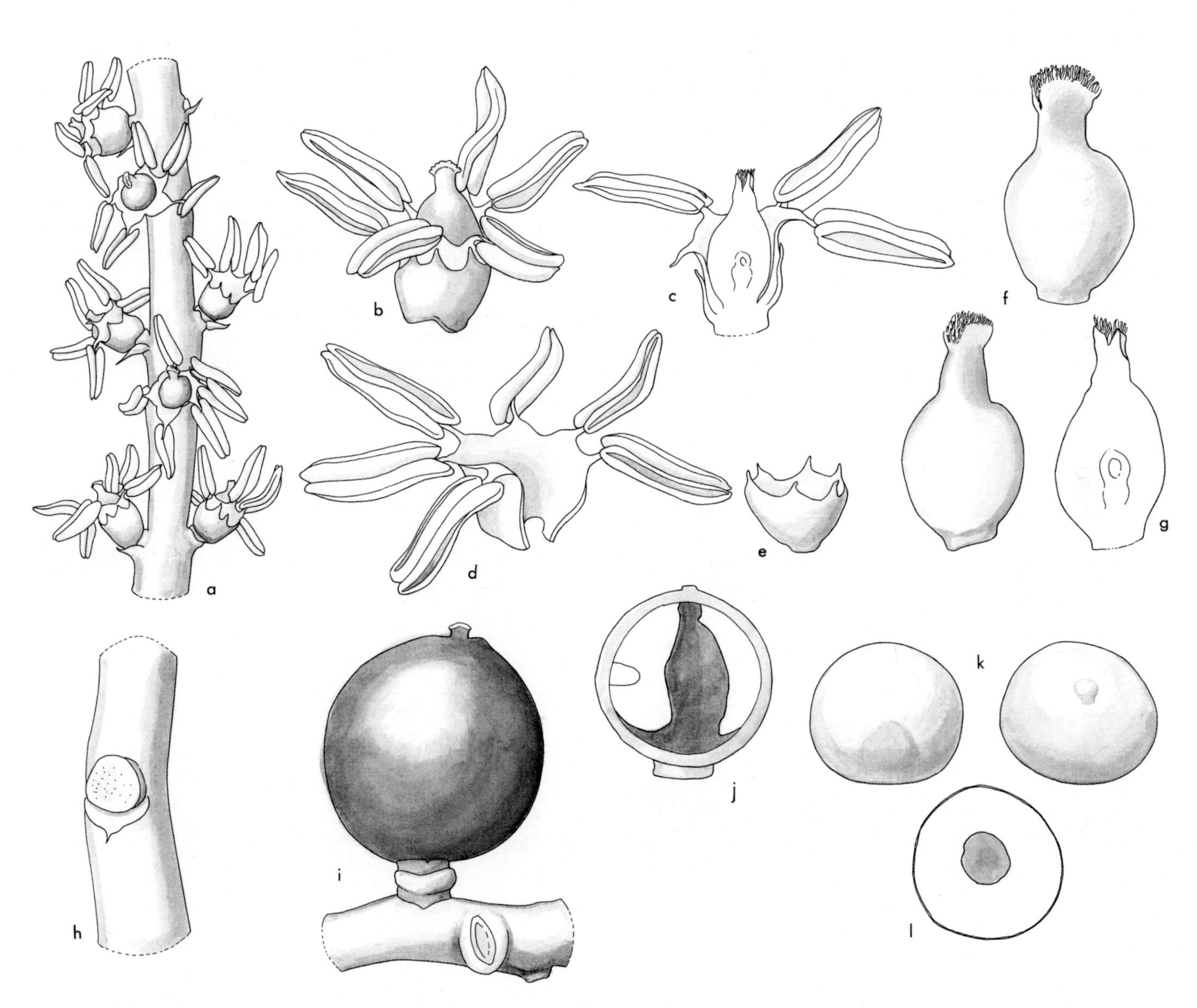

***Leucothrinax*.** **a**, portion of rachilla × 6; **b**, flower × 10; **c**, flower in vertical section × 10; **d**, staminal ring × 10; **e**, perianth × 10; **f**, gynoecium, 2 views × 16; **g**, gynoecium in vertical section × 16; **h**, rachilla and bract × 18; **i**, fruit × 5; **j**, fruit in vertical section × 4; **k**, seed, 2 views × 4; **l**, seed in transverse section × 4. *Leucothrinax morrisii*: all from *Roncal* 37. (Drawn by Lucy T. Smith)

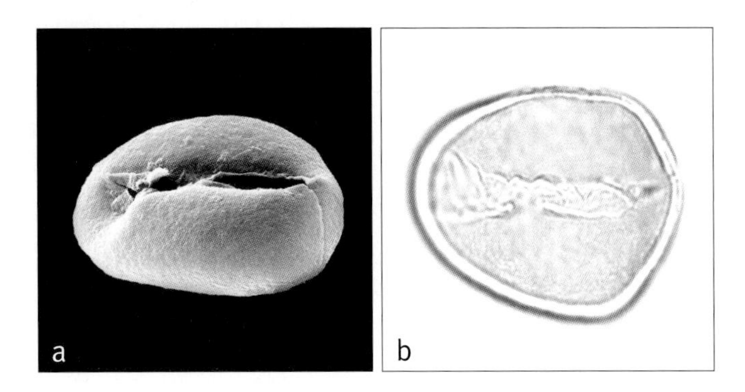

Above: ***Leucothrinax***
a, monosulcate pollen grain, distal face SEM × 2000;
b, monosulcate pollen grain, low-focus LM × 1000.
Leucothrinax morrisii: **a** & **b**, *Wright* 3865.

finely perforate, aperture margin similar; infratectum columellate; longest axis 21–31 µm [1/1]. *Fruit* very small, white at maturity, stigmatic remains apical, perianth persistent; epicarp smooth when fresh, mesocarp thin, endocarp very thin, papery. *Seed* depressed-globose, smooth, hilum round, impressed, forming a basal intrusion, raphe branches scarcely impressed, endosperm homogeneous; embryo subapical. *Germination* remote-tubular; eophyll narrow, lanceolate. *Cytology*: 2n = 36.

Distribution and ecology: A single species occurring in the northern part of the Caribbean as far east as the Virgin Islands and Anguilla, occurring on coralline sands and limestone near the sea.

Anatomy: Leaf anatomy (Read 1975); floral anatomy (Morrow 1965, Uhl & Moore 1971); correlations of floral anatomy and wind pollination suggested by Uhl and Moore (1977a).

Relationships: *Leucothrinax* is resolved as sister to *Hemithrinax* with moderate support (Asmussen *et al.* 2006, Baker *et al.* in review).

Common names and uses: Brittle thatch, Keys thatch palm; leaves are used for making brooms. Widely cultivated in Florida.

Taxonomic account: Lewis & Zona (2008).

Fossil record: No generic records found.

Notes: O.F. Cook published the name *Simpsonia* O.F. Cook, Science n.s. 85: 332–333 (1937b). This name, based on *Thrinax microcarpa*, is invalid as it was published without Latin description.

Additional figures: Glossary figs 5, 8, 25.

30. THRINAX

Small to moderate, solitary hermaphroditic fan palms found in Caribbean islands and neighbouring coastal mainland; leaf sheaths fibrous, petiole bases deeply split at base; flowers and fruit stalked; fruit very small, white, the seed smooth.

Thrinax L.f. ex Sw., Prodr. 4: 57 (1788). Type: **T. parviflora** Sw.

Porothrinax H. Wendl. ex Griseb., Cat. Pl. Cub.: 221 (1866). Type: *P. pumilio* H. Wendl. ex Griseb. (= *Thrinax radiata* Lodd. ex Schult. & Schult.f.).

Thrinax — *a trident, thought to be a reference to sharply pointed divided tips of the leaf segments.*

Small to moderate, solitary, unarmed, pleonanthic, hermaphroditic palms. *Stem* erect, columnar, smooth or fibrous, tan or grey, obscurely ringed with leaf scars, usually with a basal mass of fibrous roots. *Leaves* induplicate, palmate, often irregular; sheath becoming split both

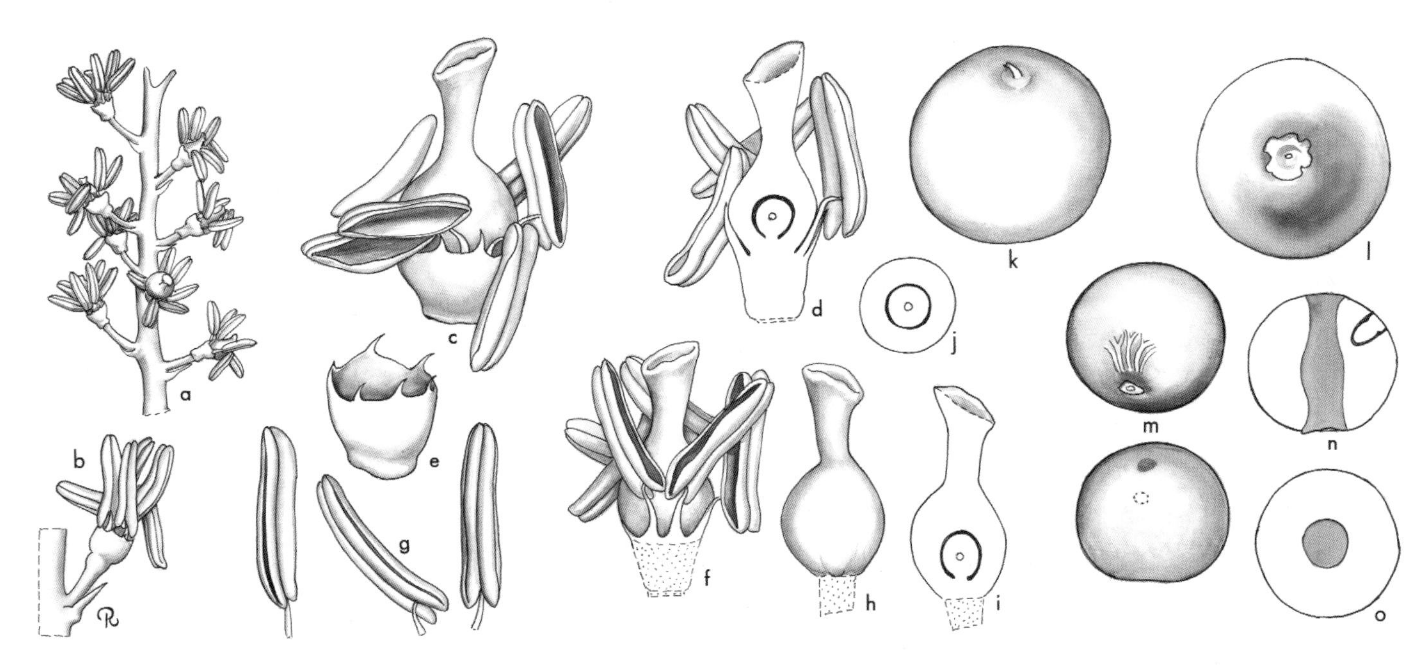

Thrinax. **a**, portion of rachilla × 3; **b**, flower and pedicel with bract × 6; **c**, flower × 12; **d**, flower in vertical section × 12; **e**, perianth × 12; **f**, flower, perianth removed × 12; **g**, anther in 3 views × 12; **h**, carpel × 12; **i**, carpel in vertical section × 12; **j**, carpel in cross-section × 12; **k**, fruit × 3 1/2; **l**, base of fruit × 3 1/2; **m**, seed in 2 views × 3 1/2; **n**, seed in vertical section × 3 1/2; **o**, seed in cross-section × 3 1/2. *Thrinax parviflora*: all from *Read* 714. (Drawn by Marion Ruff Sheehan)

opposite the petiole and abaxially to emit the inflorescence, disintegrating into irregular fibres, covered in thick, deciduous tomentum, margins fibrous; petiole long, slender, rounded to shallowly ridged both adaxially and abaxially, margins rather sharp; adaxial hastula prominent, long pointed, frequently inrolled, short and blunt at high elevations, abaxial hastula less conspicuous, rounded or triangular, lacking or very small at high elevations; blade fan-shaped, often irregularly folded segments united basally $^1/_2$ their length or less, lanceolate, pointed and usually bifid apically, glabrous adaxially, abaxially variously scaly, sometimes white, midrib and marginal ribs conspicuous, transverse veinlets evident. *Inflorescences* interfoliar, slender, erect to arching, branched to 2 orders, primary branches pendulous; peduncle moderate, rather slender, round in cross-section; prophyll short, tubular, 2-keeled, pointed, opening distally, tomentose; peduncular bracts several (ca. 4), like the prophyll but lacking keels, overlapping and very closely sheathing the peduncle; rachis longer than the peduncle, slender, tapering, bearing spirally arranged, long, tubular, pointed distally and obliquely open primary bracts subtending first-order branches; first-order branches each with a short basal bare portion, bearing a 2-keeled, bifid prophyll and spirally arranged, narrow, triangular bracts subtending rachillae; rachillae slender, rather short, stiff, bearing spirally arranged, small triangular bracts subtending solitary flowers, bracteoles apparently lacking. *Flowers* borne on conspicuous stalks; perianth a single cupule with 6 lobes or teeth; stamens mostly 6–12 (5–15), filaments very slender, sometimes partly united basally, anthers elongate, dorsifixed near the base, emarginate apically, latrorse; gynoecium consisting of 1 carpel, unilocular, uniovulate, ovule basally attached, erect, campylotropous but tilted so that the micropyle faces the upper dorsal wall of the locule, and with a basal aril. *Pollen* ellipsoidal, less frequently oblate triangular, with slight to obvious asymmetry; aperture a distal sulcus, less frequently a trichotomosulcus; ectexine tectate, perforate, or perforate-rugulate, aperture margin similar or slightly finer; infratectum columellate; longest axis 24–46 µm [2/3]. *Fruit* very small, white at maturity, stigmatic remains apical, perianth often persistent; epicarp smooth when fresh, sometimes drying pebbled, mesocarp thin, mealy, endocarp very thin, papery. *Seed* depressed-globose, smooth, hilum round, impressed, forming a basal intrusion, raphe branches deeply impressed forming peripheral ruminations, otherwise endosperm homogeneous; embryo lateral to subapical. *Germination* remote-tubular; eophyll narrow, lanceolate. *Cytology*: 2n = 36.

Below left: ***Thrinax radiata***, crown, cultivated, Fairchild Tropical Botanic Garden, Florida. (Photo: C.E. Lewis)

Below right: ***Thrinax excelsa***, infructescence, cultivated, Fairchild Tropical Botanic Garden, Florida. (Photo: C.E. Lewis)

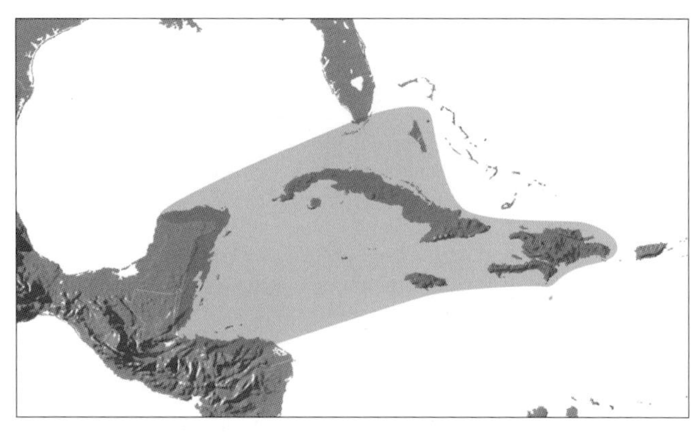

Distribution of ***Thrinax***

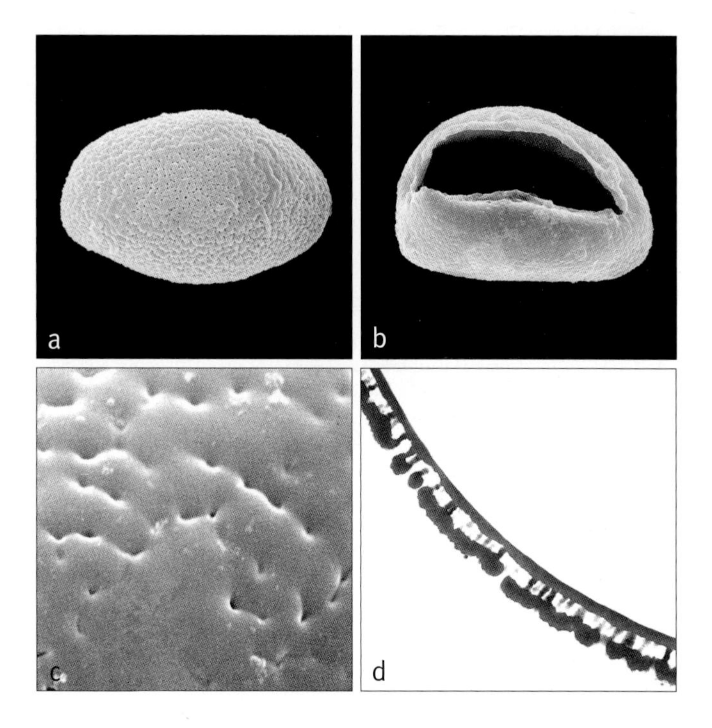

Right: ***Thrinax***
a, monosulcate pollen grain, proximal face SEM × 1500;
b, monosulcate pollen grain, distal face SEM × 1500;
c, close-up, perforate pollen surface × 8000;
d, ultrathin section through pollen wall TEM × 8000.
Thrinax parviflora: **a** & **d**, *March* 1730; *T. radiata*: **b** & **c**, *Gaumer* 2176.

Distribution and ecology: Three species; of which two, *Thrinax parviflora* and *T. excelsa*, are endemic to Jamaica. *Thrinax radiata* also occurs in Jamaica but is widely distributed in littoral habitats of Atlantic Honduras, Belize, Mexico and the northern Caribbean. Each of the three species in Jamaica is confined to one habitat: *T. parviflora* to dry evergreen woodland or thicket, *T. excelsa* to lower montane rain forest, and *T. radiata,* which thrives under exposure to salt-laden winds, to littoral woodland or thicket.

Anatomy: Leaf (Read 1975), roots (Seubert 1997), floral (Morrow 1965, Uhl & Moore 1971), fruit (Murray 1973, Reddy & Kulkarni 1982, Essig 1999); correlations of floral anatomy and wind pollination suggested by Uhl & Moore (1977a).

Relationships: *Thrinax* is monophyletic with high support and resolved as sister to a clade of *Schippia* and *Cryosophila* with low support (Roncal *et al.* 2008).

Common names and uses: Thatch palms, Key palms. Leaves are used for thatch, fibre for basketry, and other purposes. The heart of some species is eaten.

Taxonomic account: Read (1975).

Fossil record: Fossil leaves that have been compared with those of *Thrinax* include *T. eocenica* from the Middle Eocene of southeastern North America (Claiborne flora) (Berry 1914b, 1924); and *Palaeothrinax mantelli* from the Lower-Middle Oligocene of the Isle of Wight (UK) (Bembridge flora) (Reid & Chandler 1926). From the Lower Eocene London Clay flora, Khin Sein (1961) described dispersed, irregularly rounded monosulate pollen as *T. tranquillus*. It is not possible to comment further than to say that the pollen grain is a typical small asymmetric monosulcate palm, a type frequent in Coryphoideae, including *Thrinax*, and also in Arecoideae.

Notes: Distinctive in the split petiole bases and stalked flowers and fruit.

Additonal figures: Figs G3, G19, G20.

31. CHELYOCARPUS

Solitary or clustered short-stemmed or creeping hermaphroditic fan palms of humid tropical rain forests of Amazonia and western Colombia; leaves often discolorous, often divided by deep and shallow splits; petiole base not split; fruit sometimes corky-warted.

Chelyocarpus Dammer, Notizbl. Bot. Gart. Berlin-Dahlem 7: 395 (1920). Type: **C. ulei** Dammer.

Tessmanniophoenix Burret, Notizbl. Bot. Gart. Berlin-Dahlem 10: 397 (1928). Lectotype: *T. longibracteata* Burret (see Burret 1941) (= *Chelyocarpus ulei* Dammer).

Tessmanniodoxa Burret, Notizbl. Bot. Gart. Berlin-Dahlem 15: 336 (1941). Lectotype: *T. chuco* (Mart.) Burret (*Thrinax chuco* Mart.) (see H.E. Moore 1963c) (= *Chelyocarpus chuco* [Mart.] H.E. Moore).

Chelys — *tortoise,* carpos — *fruit, in reference to the cracked surface of the fruit that resembles a tortoise carapace.*

Moderate, solitary or clustered, unarmed, pleonanthic, hermaphroditic palms. *Stem* erect or procumbent (*Chelyocarpus repens*), slender, naked except for fibrous remains of leaf sheaths below the crown, closely ringed with narrow leaf scars. *Leaves* spreading, induplicate, palmate, or shortly costapalmate (*C. chuco*), sheath fibrous, not splitting opposite the petiole, densely velvety-hairy, golden-brown when young, with a prominent fibrous ligule on each side of the petiole at its apex, this disintegrating into loose fibres in age; petiole elongate, unarmed, not splitting basally, adaxially channelled basally, becoming angled distally, abaxially rounded, margins thin; adaxial hastula often large, erect, deltoid, abaxial hastula narrow, ridge-like; blade flat, thin, divided along the central abaxial fold to well beyond the middle or nearly to the base, each half further divided adaxially into paired or irregularly grouped, rather wide, single-fold segments (*C. chuco*), or divided to the base into elongate, wedge-shaped, many-fold segments, these again divided into several acute or very briefly bifid, single-fold, 1-ribbed segments, midribs raised abaxially, blade strongly discolorous or concolorous, transverse veinlets evident. *Inflorescences* interfoliar, pendulous, branching to 1 or 2 orders; peduncle flattened, short; prophyll flattened, tubular, with long fibrous beak, shortly 2-keeled laterally, surfaces and margins covered in dense soft tomentum; peduncular bracts 3(–4), like the prophyll but lacking keels; rachis flattened; first-order branches several, recurved, flattened, basally adnate to the rachis, each subtended by a prominent rachis bract similar to those on the peduncle but progressively smaller distally, at least the lower first-order branches bearing a membranous prophyll (*C. chuco*), or the lower rachillae sometimes fasciculate or subfasciculate on short branches but the bracts subtending first-order branches, small and not like those on the peduncle; rachillae usually adnate for some distance above an acute, sometimes elongate subtending bract and bearing spirally arranged, small to prominent, acute bracts each subtending a sessile or shortly pedicellate flower. *Flowers* (at least in *C. ulei*) strongly scented; sepals 2 or 3, distinct or briefly connate basally, or 4, distinct, and slightly imbricate; petals like the sepals; stamens 5, 6, 7, 8, or 9, one opposite each sepal, remainder opposite the petals, filaments erect, distinct, fleshy, thick and broad below, ± abruptly narrowed at the tip, anthers exserted at anthesis, dorsifixed at the middle, bifid at apex and base, latrorse; carpels 3 or 2,

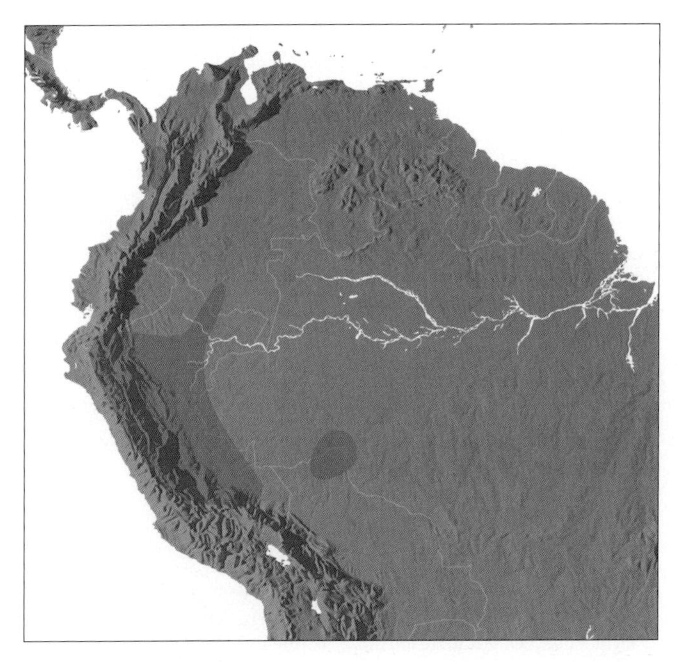

Distribution of ***Chelyocarpus***

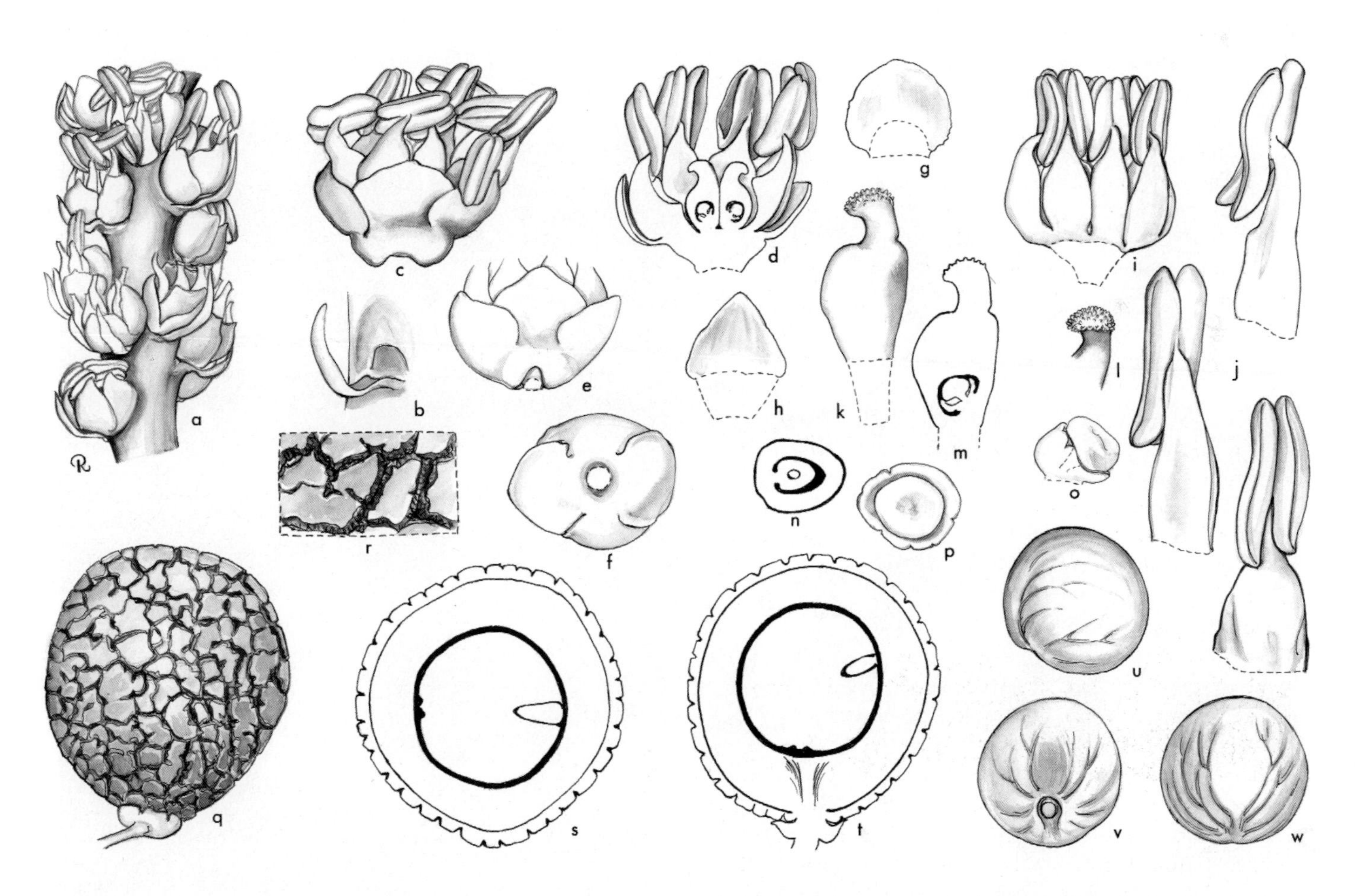

Chelyocarpus. **a**, portion of rachilla × 3; **b**, bracteole and floral scar × 6; **c**, flower × 6; **d**, flower in vertical section × 6; **e**, **f**, perianth in lateral and bottom views × 6; **g**, sepal × 6; **h**, petal × 6; **i**, flower, perianth removed × 6; **j**, stamen in 3 views × 12; **k**, carpel × 12; **l**, stigma × 12; **m**, carpel in vertical section × 12; **n**, ovary in cross-section × 12; **o**, **p**, ovule in lateral and top views × 24; **q**, fruit × 1 1/2; **r**, surface of fruit × 3; **s**, fruit in cross-section × 1 1/2; **t**, fruit in vertical section × 1 1/2; **u**, **v**, **w**, seed in 3 views × 1 1/2. *Chelyocarpus ulei*: **a– p**, *Moore & Salazar* 9494; **q–w**, *Moore* 9548. (Drawn by Marion Ruff Sheehan)

rarely 1 or 4, follicular, style short, somewhat recurved, stigma papillose, ovule hemianatropous, attached adaxially at the base, an aril present and basally fused to the locular wall. *Pollen* ellipsoidal, ± bisymmetric, or with slight to obvious asymmetry, less frequently, oblate triangular; aperture a distal sulcus or, infrequently, a trichotomosulcus; ectexine tectate, coarsely perforate or foveolate-reticulate, aperture margin finely perforate; longest axis 22–36 µm [3/4]. *Fruit* usually developing from only 1 carpel, globose with eccentrically apical stigmatic remains; epicarp smooth or coarsely corky-warted, mesocarp thick, dry, endocarp membranous. *Seed* basally attached, globose, hilum basal, circular, raphe slightly impressed along the length of the seed and with ascending branches, endosperm homogeneous; embryo below or above the middle opposite the raphe. *Germination* remote-tubular; eophyll bifid. *Cytology*: 2n = 36 ± satellite.

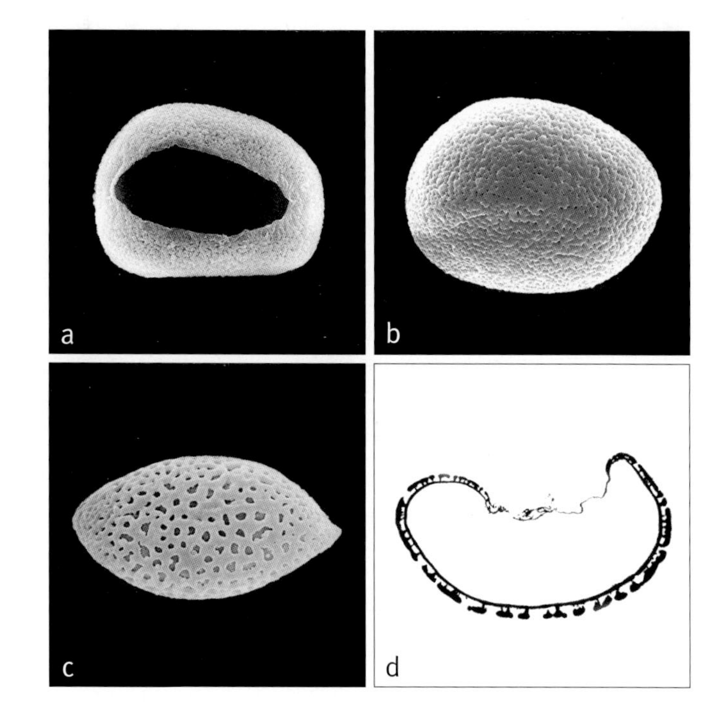

Right: ***Chelyocarpus***
a, monosulcate pollen grain, distal face SEM × 2000;
b, monosulcate pollen grain, proximal face SEM × 2000;
c, monosulcate pollen grain, proximal face SEM × 1000;
d, ultrathin section, whole pollen grain, polar plane TEM × 1500.
Chelyocarpus ulei: **a** & **b**, *Henderson et al.* 1669; *C. chuco*: **c** & **d**, *Prance et al.* 8717.

Below left: ***Chelyocarpus dianeurus***, habit, Colombia. (Photo: A.J. Henderson)
Below right: ***Chelyocarpus ulei***, inflorescences, Brazil. (Photo: A.J. Henderson)

Distribution and ecology: Four species in Amazonian Bolivia, Brazil, Peru, Ecuador and western Colombia. The species occur at low elevations in areas of high rainfall. It is particularly interesting to note that the distribution of these species corresponds with three of nine refugia — Madeira-Tapajos, East Peruvian, Choco — postulated by Haffer (1969) as regions where rain forest persisted during drier epochs of the Pleistocene.

Anatomy: Leaf (Uhl 1972c), roots (Seubert 1997), floral (Uhl 1972b).

Relationships: *Chelyocarpus* is moderately supported as sister to all remaining Cryosophileae (Roncal *et al.* 2008).

Common names and uses: Not recorded.

Taxonomic accounts: Moore (1972) and Kahn & Mejia (1988).

Fossil record: No generic records found.

Notes: *Chelyocarpus* has very characteristic distinct carpels and deeply divided palmate leaves. The genus was initially set apart because of the corky-warted surface of the fruit of the type species, which so resembled a turtle's carapace as to suggest the generic name. Dammer commented on the dimerous or two-parted perianth of *C. ulei*, but without flowers to examine thought his material perhaps atypical. More complete collections, however, have shown that the perianth is normally dimerous in the species. Within *Chelyocarpus*, there is apparent simplification of the inflorescence. There are also modifications of the flower that suggest directions of evolution toward connation and reduction in the perianth, elaboration of the androecium, and reduction of the gynoecium: characters that are found in the related genus *Itaya* and in the more specialised group of genera that includes *Thrinax* and related taxa.

Additional figures: Fig. 1.9, Glossary figs 18, 20.

32. CRYOSOPHILA

Solitary moderate hermaphroditic fan palms of humid and monsoon tropical rain forests of Central America and northern South America; stems are often covered with root spines and the leaves are discolorous, with the blade divided by a central deep split and the petiole base not split.

Cryosophila Blume, Rumphia 2: 53 (1838 ['1836']). Type: **C. nana** (Kunth) Blume ex Salomon (*Corypha nana* Kunth).

Acanthorrhiza H. Wendl., Gartenflora 18: 241 (1869). Type: *A. aculeata* (Liebm.) H. Wendl. (*Trithrinax aculeata* Liebmann) (= *Cryosophila nana* [Kunth] Blume ex Salomon).

Derivation obscure; crios — *goat,* phila — *loving, but this meaning seems absurd.*

Moderate, solitary, armed, pleonanthic, hermaphroditic palms. *Stem* slender to rather stout, bearing often-branched root spines, prickly stilt roots sometimes developed. *Leaves* induplicate, palmate; sheath fibrous, densely floccose-tomentose, becoming split basally and splitting opposite the petiole, margins fibrous; petiole elongate, unarmed, channelled adaxially, rounded abaxially, margins sharp throughout their length; adaxial hastula deltoid, elevated, sometimes grooved, abaxial hastula narrow or lacking; blade divided centrally to or nearly to the base (?except in *C. williamsii*), each half further deeply divided into elongate, wedge-shaped, many-fold segments, these again divided into 2 or more, single-fold, acute or briefly bifid segments, white-tomentose abaxially, interfold ribs prominent adaxially, midribs prominent abaxially, transverse veinlets evident. *Inflorescences* interfoliar, curved or pendulous, branching to 2

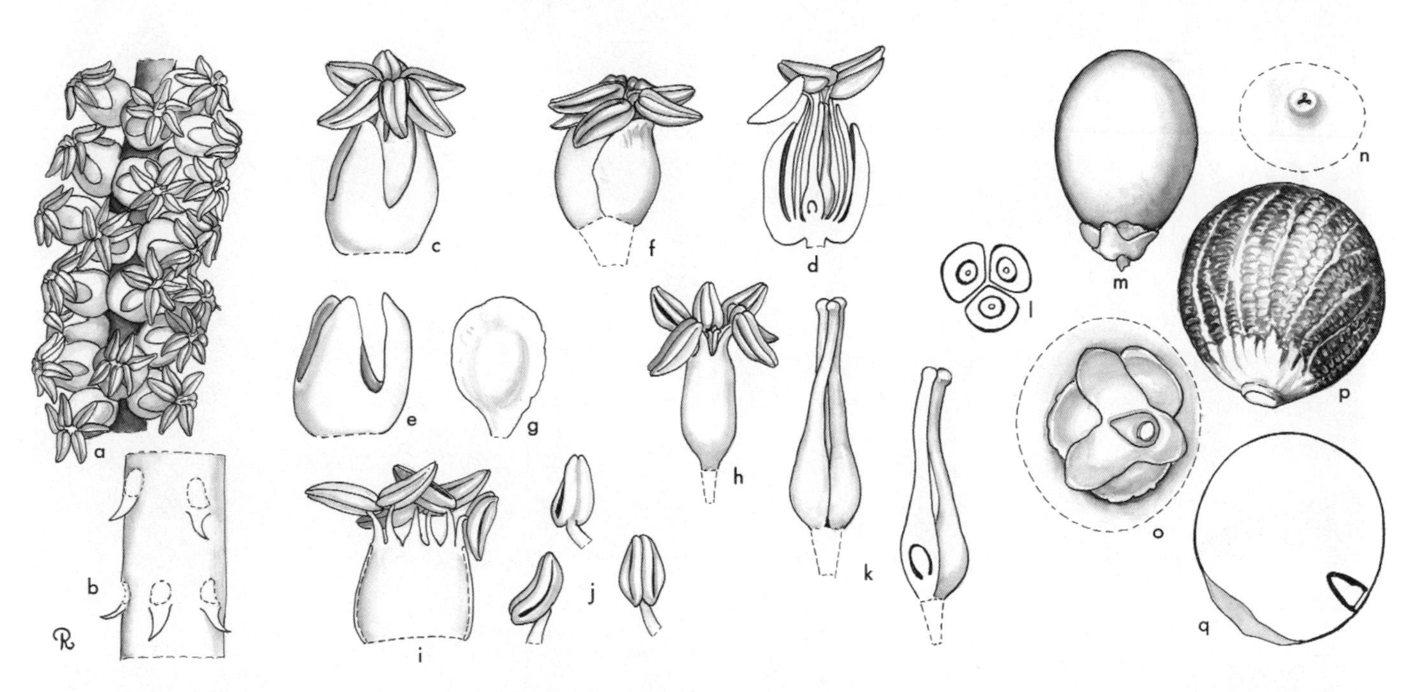

Cryosophila. **a**, portion of rachilla × 3; **b**, portion of rachilla showing floral scars and bracts × 6; **c**, flower × 6; **d**, flower in vertical section × 6; **e**, calyx × 6; **f**, flower, calyx removed × 6; **g**, petal × 6; **h**, androecium × 6; **i**, androecium expanded, interior view × 6; **j**, anther in 3 views × 6; **k**, gynoecium in exterior view and vertical section × 12; **l**, carpels in cross-section × 12; **m**, fruit × $1^1/_2$; **n**, apex of fruit × 12; **o**, base of fruit × 6; **p**, seed × 3; **q**, seed in vertical section × 3. *Cryosophila stauracantha*: **a–l**, *Read* 689; *C. nana*: **m–q**, *McVaugh* 900. (Drawn by Marion Ruff Sheehan)

orders; peduncle short, terete, tomentose; prophyll short, tubular, pointed, 2-keeled, tomentose; peduncular bracts several (to 6 or more), much larger than the prophyll, tubular at base, inflated distally, splitting laterally, densely tomentose; rachis about as long as the peduncle, somewhat angled, bearing several to many, recurved, first-order branches, only the proximal or all (in *C. guagara*) subtended by prominent bracts like those of the peduncle but progressively smaller, or the upper branches with reduced bracts only; first-order branches usually with obvious, not, or only slightly adnate bases, but rarely (*C. cookii*) the peduncle and rachis very short and rachillae fastigiately grouped along the rachis; rachillae little or not adnate above an acute subtending bract, spirally arranged, flowers borne on brief pedicels each subtended by a small, acute bract. *Flowers* small; sepals 3, narrowly ovate to deltoid, briefly connate basally; petals 3, distinct, scarcely longer than sepals, imbricate, rounded at apex; stamens 6, filaments flat, connate basally in a tube $^{1}/_{2}$ their length or more, apically distinct and strap-shaped or terete, anthers exserted and spreading at an angle of 90°, dorsifixed near the base, briefly bifid at base and apex, latrorse; carpels 3, distinct, narrowly follicular, styles elongate, exserted, stigma scarcely expanded, ovule campylotropous, inserted adaxially at the base, with a small aril on the funicle. *Pollen* ellipsoidal, less frequently oblate triangular, with slight to obvious asymmetry; aperture a distal

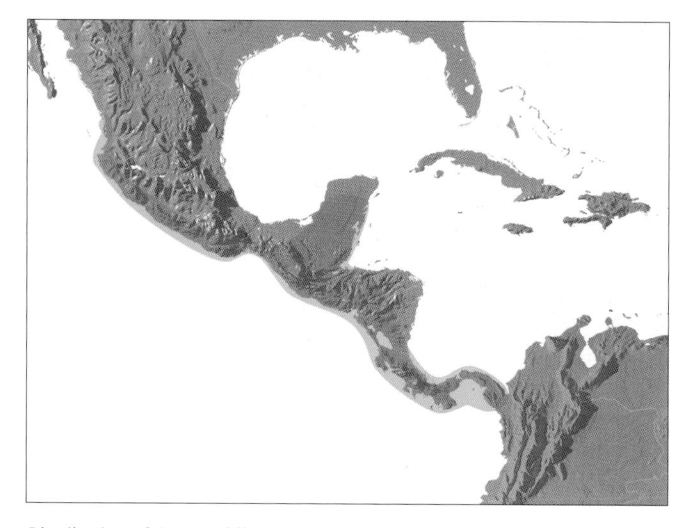

Distribution of ***Cryosophila***

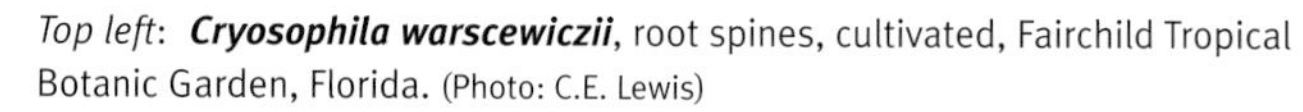

Top left: ***Cryosophila warscewiczii***, root spines, cultivated, Fairchild Tropical Botanic Garden, Florida. (Photo: C.E. Lewis)

Bottom left: ***Cryosophila warscewiczii***, leaf, cultivated, Fairchild Tropical Botanic Garden, Florida. (Photo: C.E. Lewis)

Below: ***Cryosophila stauracantha***, inflorescence, cultivated, Montgomery Botanical Center, Florida. (Photo: J. Dransfield)

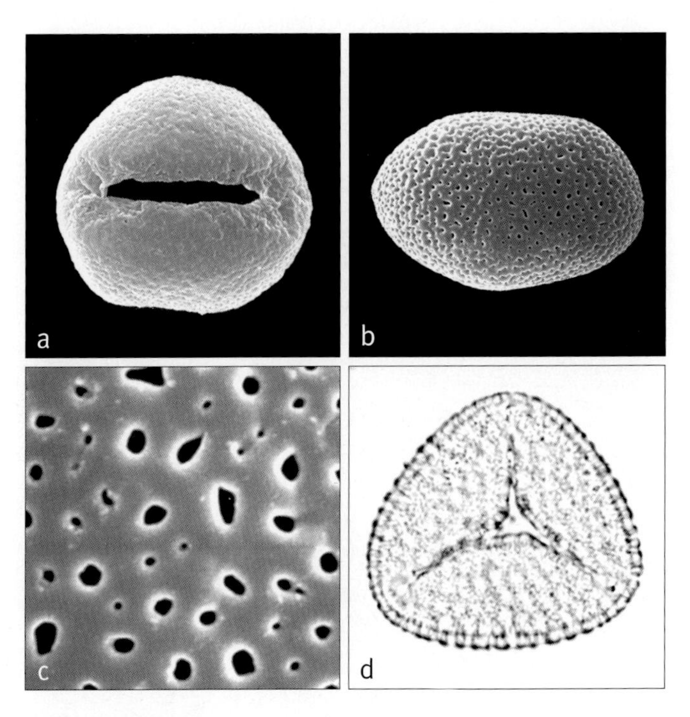

Above: ***Cryosophila***
a, monosulcate pollen grain, distal face SEM × 1000;
b, monosulcate pollen grain, proximal face SEM × 1000;
c, close-up, foveolate-reticulate pollen surface SEM × 8000;
d, trichotomosulcate pollen grain, mid-focus LM x 1200.
Cryosophila warscewiczii: **a–c**, cultivated Herrenhausen Botanic Garden;
C. stauracantha: **d**, *Gentle* 1392.

sulcus, or a trichotomosulcus; ectexine tectate or semitectate, foveolate or reticulate on proximal face, perforate, or coarsely perforate on distal face, including aperture margins; infratectum columellate; longest axis 22–36 µm [3/9]. *Fruit* developing from 1 carpel, white at maturity, stigmatic remains apical; epicarp smooth, mesocarp somewhat fleshy, endocarp membranous. *Seed* globose, not adherent to the endocarp, with round basal hilum, raphe branches impressed, ascending and anastamosing from the base, endosperm homogeneous with slight intrusion of seed coat adaxially at the base; embryo lateral, at or below the middle. *Germination* adjacent-ligular or remote-tubular (Chavez 2003); eophyll simple, entire, white abaxially. *Cytology*: 2n = 36.

Distribution and ecology: Ten species from western Mexico to northern Colombia. Species occur at moderate to low elevations in dry woods (*Cryosophila nana*) to rain forest.

Anatomy: Leaf, root (Tomlinson 1961), leaf (Uhl 1972c), roots (Seubert 1997), flower (Uhl 1972b).

Relationships: *Cryosophila* is weakly supported as sister to *Schippia* (Roncal *et al*. 2008).

Common names and uses: *Palma de escola*, rootspine palms. Widely cultivated as ornamentals.

Taxonomic account: Evans (1995).

Fossil record: Two monosulcate palm pollen types are recovered from the Pliocene, Gatun Lake Formation, Panama (Graham 1991). The second of these types, bisymmetrical and finely reticulate, is compared with *Colpothrinax*, *Cryosophila* and *Neonicholsonia*. Of these, *Cryosophila* is the most probable.

Notes: The branched root spines on the trunk are distinctive.

Additonal figures: Figs 1.9, G4, G10, G11.

33. ITAYA

Solitary moderate hermaphroditic fan palms of Amazonian rain forest, stems becoming bare; leaves are discolorous, with the blade divided by a central deep split and petiole base split; stamens are numerous and the fruit minutely roughened.

Itaya H.E. Moore, Principes 16: 85 (1972). Type: **I. amicorum** H.E. Moore.

Named after the River Itaya, a tributary of the Amazon near which the genus was discovered.

Moderate, solitary, unarmed, pleonanthic, hermaphroditic palm. *Stem* smooth (drying roughened), bare except for fibrous remains of sheaths and a lattice of long persistent, split petioles below the crown. *Leaves* spreading, induplicate, palmate; sheaths short, fibrous, split opposite the petiole, persisting as fibrous margins on the bases of the petioles; petiole elongate, unarmed, the base prominently split, channelled adaxially, rounded abaxially near the base, becoming biconvex and rhomboid in section distally, margins obtuse; adaxial hastula deltoid, often large, basally grooved, abaxial hastula narrow; blade held in one plane, thin, orbicular, divided to ca. $^{3}/_{4}$ the radius at the middle, each half again deeply divided into several (4–7) elongate, wedge-shaped, 4–7-fold segments, these very shallowly divided apically into briefly bifid, 1-fold segments, abaxially lighter, midribs very prominent abaxially, transverse veinlets of 2 sizes very prominent. *Inflorescences* interfoliar, elongate, curved, branched to 3 orders basally, to 1 order distally; peduncle terete; prophyll short, 2-keeled, abaxially split, peduncular bracts ca. 5, bases tubular, apices inflated, acute, coriaceous, persistent, larger than the prophyll, marcescent, lightly tomentose, split on one side; rachis about as long as the peduncle, tapering, ± angled, tomentose; first-order branches each subtended by a persistent, marcescent bract similar to the peduncular bracts but progressively smaller and the upper-most scarcely tubular at the base, the branches ± flattened, adnate to the rachis often

Below: ***Itaya amicorum***, habit, cultivated, Fairchild Tropical Botanic Garden, Florida. (Photo: C.E. Lewis)

nearly to the succeeding bract; rachillae short, rather distant, slightly sinuous, each subtended by a linear acute bract, rachillae bearing spirally arranged, solitary flowers, each on a very short pedicel subtended by a small acute bract. *Flowers* creamy-white; sepals 3, connate in an acutely 3-lobed cup; petals 3, connate ca. $^1/_2$ their length, the 3 lobes rounded and erect at anthesis, probably valvate in bud; stamens 18–24, one or two opposite each sepal, the remainder opposite the petals, filaments connate basally in a fleshy tube less than $^1/_2$ their length, slightly adnate to corolla basally, fleshy and ± awl-shaped distally, anthers exserted at anthesis, oblong, dorsifixed at the middle, versatile, bifid at apex and base, latrorse; gynoecium of 1 carpel, eccentrically ovoid, narrowed to a slender curved style and oblique papillose stigma, ovule hemianatropous, attached adaxially at the base, the short funicle bearing a large oblique aril. *Pollen* ellipsoidal, with slight to obvious asymmetry; aperture a distal sulcus; ectexine tectate, coarsely perforate, aperture margin finely perforate; infratectum columellate; longest axis 36–41 µm [1/1]. *Fruit* oblong-ovoid or subglobose with eccentrically apical stigmatic remains; epicarp minutely granular-roughened and with minute perforations, mesocarp thick, white, dry, with anastamosing fibres and a peripheral layer of sclerosomes, endocarp not differentiated. *Seed* oblong-ovoid, hilum elliptic, subbasal, raphe branches ascending-spreading, endosperm homogeneous; embryo eccentrically basal. *Germination* remote-tubular (Chavez 2003); eophyll undivided, elliptic. *Cytology*: 2n = 36.

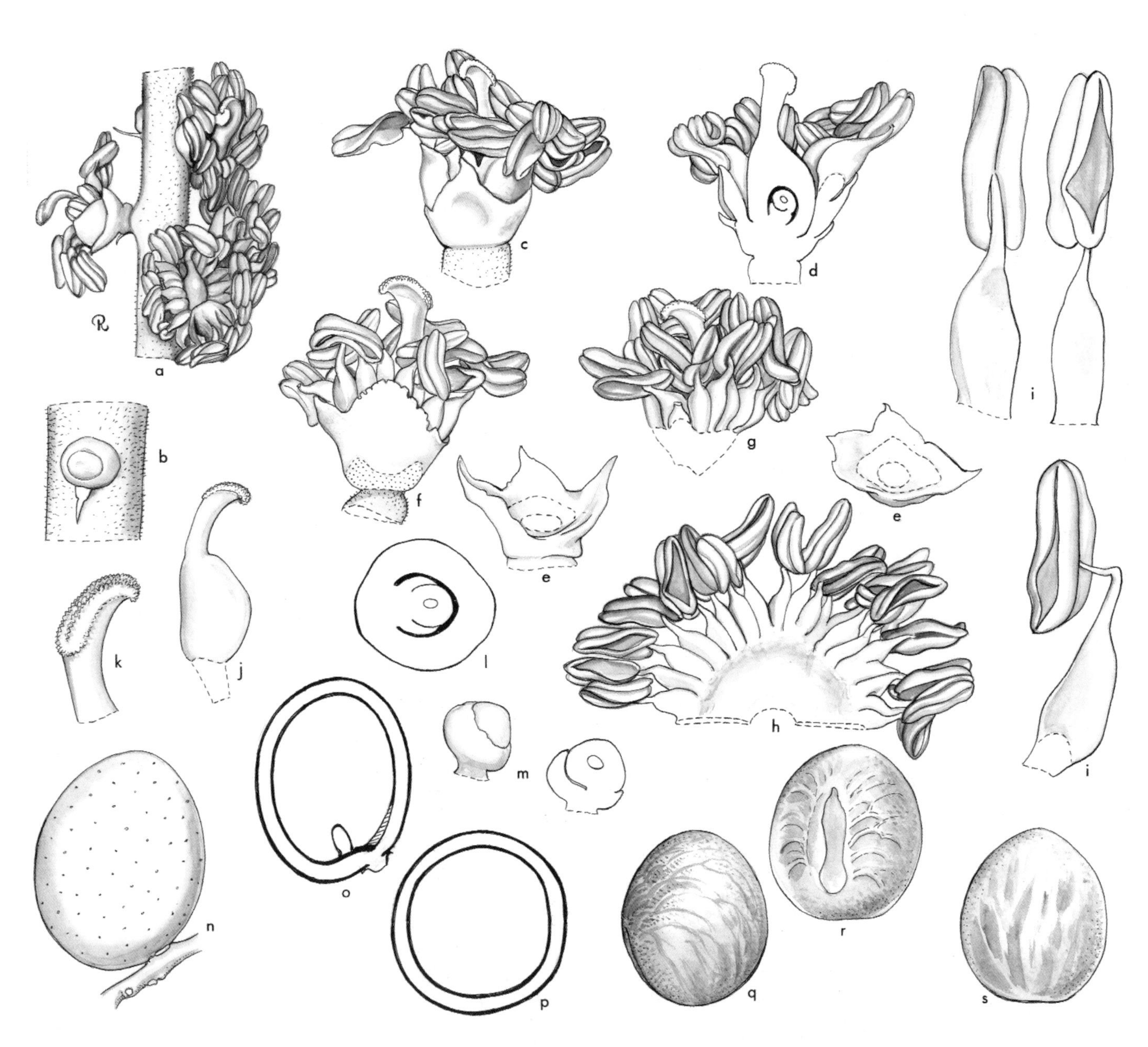

Itaya. **a**, portion of rachilla × 3; **b**, pedicel and bract × 6; **c**, flower × 6; **d**, flower in vertical section × 6; **e**, calyx in 2 views × 6; **f**, flower, calyx removed × 6; **g**, flower, perianth removed × 6; **h**, androecium expanded, interior view × 6; **i**, stamen in 3 views × 12; **j**, gynoecium × 6; **k**, stigma × 12; **l**, ovary in cross-section × 12; **m**, ovule in lateral view and vertical section × 12; **n**, fruit × $1^1/_2$; **o**, fruit in vertical section × $1^1/_2$; **p**, fruit in cross-section × $1^1/_2$; **q**, **r**, **s**, seed in 3 views × $1^1/_2$. *Itaya amicorum*: **a–m**, *Moore et al.* 9509; **n–s**, *Gutierrez* 1940. (Drawn by Marion Ruff Sheehan)

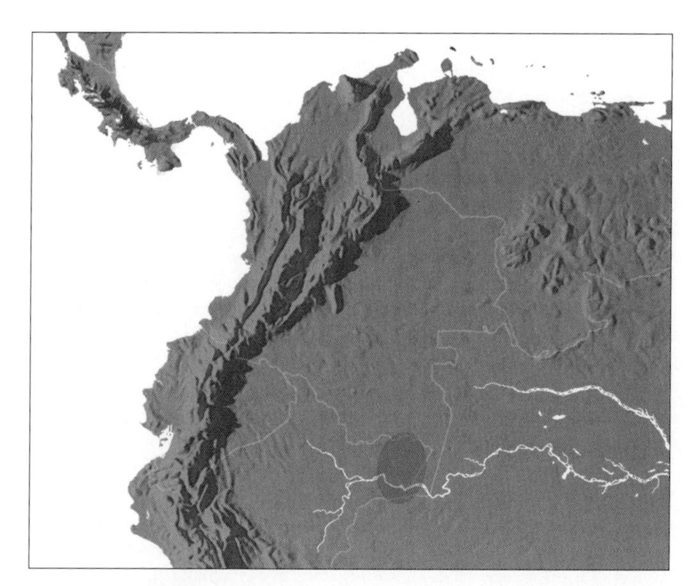

Distribution of ***Itaya***

Distribution and ecology: One species, known from a few localities in Amazonian Colombia, Ecuador, Peru and Brazil. *Itaya amicorum* occurs in rain forest at low elevations.
Anatomy: Leaf (Uhl 1972c), roots (Seubert 1997), floral (Uhl 1972b).
Relationships: Roncal *et al.* (2008) resolve *Itaya* as sister to all Cryosophileae except *Chelyocarpus* with low support.
Common names and uses: Not recorded. *Itaya* is a promising horticultural subject because of its large and handsome leaves much resembling those of *Licuala* species, its moderate stature, and creamy-white inflorescences and flowers.
Fossil record: No generic records found.
Taxonomic account: Moore (1972).
Notes: *Itaya* appears to be most closely related to *Chelyocarpus* and *Cryosophila*. It differs, however, in the connation and adnation of sepals and petals, in its numerous stamens, in its unicarpellate gynoecium, and in the presence of two phloem strands in central vascular bundles of the petiole. The split in the petiole base has been commented on for *Thrinax* by Read (1975). It is an immediately recognizable field character but is not useful with herbarium material, which usually lacks leaf bases.
Additional figures: Glossary fig. 20.

Above, top: ***Itaya amicorum***, inflorescence, cultivated, Fairchild Tropical Botanic Garden, Florida. (Photo: J. Dransfield)

Bottom: ***Itaya amicorum***, infructescence, cultivated, Fairchild Tropical Botanic Garden, Florida. (Photo: J. Dransfield)

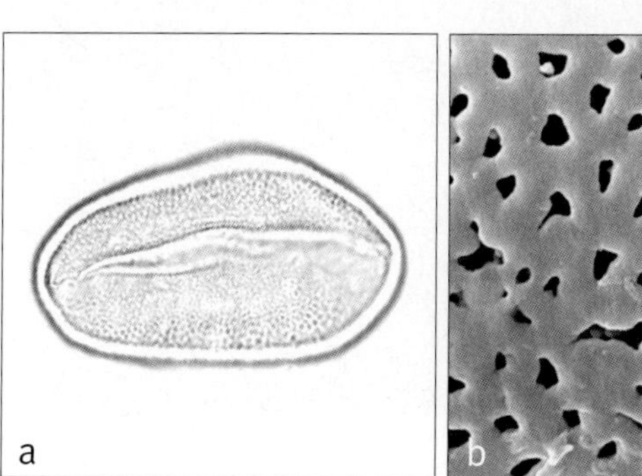

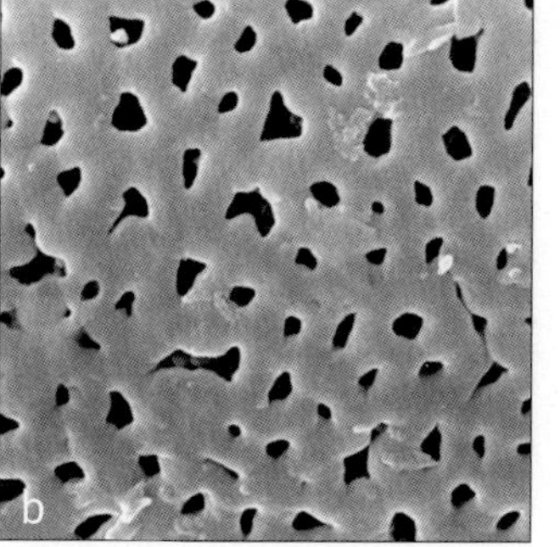

Left: ***Itaya***
a, monosulcate pollen grain, mid–low-focus LM × 1000;
b, close-up, coarsely-perforate/finely reticulate surface SEM × 8000.
Itaya amicorum: **a** & **b**, *Moore et al.* 9509.

Tribe Phoeniceae J. Presl, Wšobecný Rostl. 2: 1609 (1846). Type **Phoenix.**

Dioecious; leaves pinnate, induplicate, leaflets with acute tips, proximal leaflets developed as spines; inflorescence bearing a prophyll only and branching to 1 order; flowers solitary, dimorphic; endocarp membranous; seed with a deep longitudinal furrow.

The tribe consists of the single, wide ranging but strictly Old World genus, *Phoenix*. Not only is the pinnate induplicate leaf of *Phoenix* unique in gross morphology but also its development is unlike that of any other palm. The function and homology of a covering layer, the 'haut', in the developing leaf are as yet not understood. Staminate and pistillate flowers are alike until late in development (DeMason *et al.* 1982). The free carpels are follicular and ovules are anatropous.

Tribe Phoeniceae is strongly supported as monophyletic (Asmussen *et al.* 2006). The tribe is resolved in various sister relationships; for example, as sister to tribe Trachycarpeae with moderate support (Asmussen *et al.* 2006), sister to tribes Sabaleae and Cryosophileae (Uhl *et al.* 1995), sister to a clade consisting of tribes Sabaleae, Cryosophileae and Trachycarpeae with low support (Baker *et al.* in review), or in a clade with tribes Cryosophileae and Trachycarpeae with moderate support (Hahn 2002a).

34. PHOENIX

The Date Palms. Solitary or clustering dioecious pinnate-leaved palms of the Old World, usually in arid or semi-arid areas, sometimes in mangrove or monsoon forest, instantly recognisable by the induplicate leaflets with spine-like tips, and the acanthophylls at the leaf base; inflorescence with a single large bract.

Phoenix L., Sp. pl. 1188 (1753). Type: **P. dactylifera** L.

Elate L., Sp. pl. 1189 (1753). Type: *E. sylvestris* L. (= *Phoenix sylvestris* [L.] Roxb.).

Palma Mill., Gard. Dict. abr. ed. 4 (1754). Lectotype: *P. dactylifera* (L.) Mill. (*Phoenix dactylifera* L.) (see Moore 1963b).

Phoniphora Neck., Elem. Bot. 3: 302 (1790). Type not designated.

Dachel Adans., Fam. pl. 2: 25, 548 (1763).

Fulchironia Lesch. in Desf., Tabl. École Bot., ed. 3: 29 (1829). Type not designated.

Zelonops Raf., Fl. Tellur. 2: 102 (1837). Type: *Z. pusilla* (Gaertn.) Raf. (*Phoenix pusilla* Gaertn.).

Latin transcription of phoinix, *date palm or palm; the name is often used in combination with other epithets in palm generic names.*

Dwarf or creeping to large, solitary or clustered, armed, pleonanthic, dioecious palms. *Stem*, when developed, often clothed with spirally arranged leaf bases. *Leaves* induplicate, pinnate, usually marcescent; sheath forming a fibrous network; petiole very short to well developed, adaxially channelled to flattened or ridged, abaxially rounded; rachis elongate, tapering, adaxially rounded or flat to angled, abaxially rounded to flat, usually terminating in a leaflet; leaflets single-fold, acute, regularly arranged or variously grouped, the proximal few modified as spines (acanthophylls), parallel-veined, midrib usually evident abaxially, often bearing scales, emergent leaves frequently with brown floccose indumentum and/or wax, transverse veinlets obscure. *Inflorescences* interfoliar, branching to 1 order, the staminate and pistillate superficially similar; peduncle flattened, short to elongate, in the pistillate frequently elongating after fertilization, bearing an often caducous, sometimes bivalved, 2-keeled, glabrous or floccose-hairy prophyll; other bracts inconspicuous; rachis flattened, usually shorter than the peduncle; rachillae unbranched, numerous, often in groups in a spiral along the rachis, somewhat adnate above small triangular bracts, the rachillae bearing spirally arranged, low triangular bracts, each subtending a solitary flower. *Staminate flowers* with 3 sepals connate in a low cupule; petals 3, ± valvate, acute or rounded, much exceeding the calyx; stamens usually 6 (rarely 3 or 9), filaments short, erect, the anthers linear, latrorse; pistillode absent, or of 3 abortive carpels, or a minute trifid vestige. *Pollen* ellipsoidal, bi-symmetric or very slightly asymmetric; aperture a distal sulcus; ectexine tectate, coarsely perforate, finely reticulate, foveolate, or perforate-rugulate; aperture margin slightly finer, psilate or scabrate; infratectum columellate; longest axis 17–30 µm [11/13]. *Pistillate flowers* globose; sepals connate in a 3-lobed cupule; petals imbricate, strongly-nerved, about twice as long as the calyx or more; staminodes usually 6, scale-like or connate in a low cupule; carpels 3, distinct, follicular, ± ovoid, narrowed into a short, recurved, exserted stigma, ovule attached adaxially at the base, anatropous. *Fruit* usually developing from 1 carpel, ovoid to oblong with apical stigmatic remains; epicarp smooth, mesocarp fleshy, endocarp membranous. *Seed* elongate, terete or plano-convex, and deeply grooved with intruded seed coat below the elongate raphe, hilum basal, rounded, endosperm homogeneous or rarely ruminate (*Phoenix anadamanensis*); embryo lateral or subbasal. *Germination* remote-tubular; eophyll undivided, narrowly lanceolate. *Cytology*: 2n = 32, 36.

Above: ***Phoenix dactylifera***, date grove, Oman. (Photo: J. Dransfield)

Distribution and ecology: 14 species ranging from the Atlantic islands through Africa, Crete, the Middle East and India to Hong Kong, Taiwan, Philippines, Sumatra and Malaya. Widely cultivated as ornamentals, one species, *Phoenix dactylifera*, the date palm, is a major economic plant, now widespread in semi-arid areas as a fruit tree. For a summary of uses, see Johnson (1985). Most species are plants of semi-arid regions but grow near water courses, oases, or underground water sources; a few species are found in tropical monsoonal areas. *Phoenix paludosa* occurs in the Asian perhumid regions, where it is confined to the landward fringe of mangrove forest. *Phoenix roebelenii* grows as a rheophyte on the banks of the Mekong and some of its tributaries.

Anatomy: Central vascular bundles of the petioles with a single phloem strand (Parthasarathy 1968). Leaf (Tomlinson 1961), roots (Seubert 1997), floral (Uhl & Moore 1971, 1977a, DeMason *et al.* 1982), axillary bud, inflorescence, offshoot (Hilgeman 1954), seed (Werker 1997).

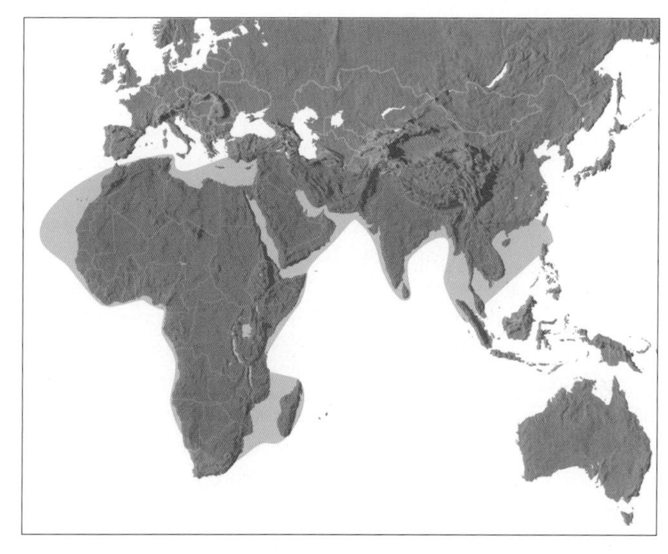

Distribution of ***Phoenix***

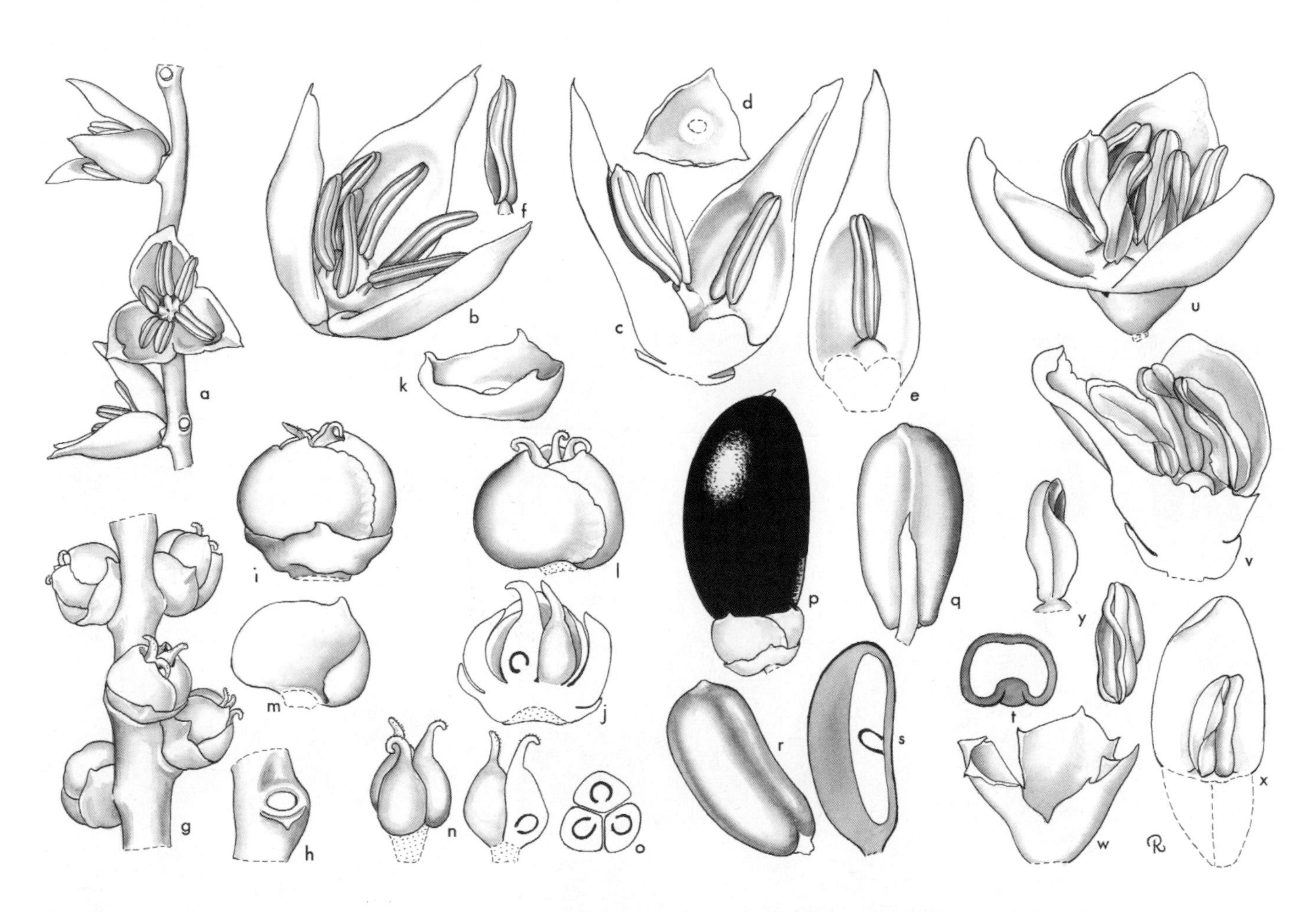

Phoenix. **a**, portion of rachilla with staminate flowers × 3; **b**, staminate flower × 6; **c**, staminate flower in vertical section × 6; **d**, staminate calyx × 6; **e**, staminate petal with adnate stamen, interior view × 6; **f**, stamen × 6; **g**, portion of rachilla with pistillate flowers × 3; **h**, bract and scar of pistillate flower × 6; **i**, pistillate flower × 6; **j**, pistillate flower in vertical section × 6; **k**, pistillate calyx × 6; **l**, pistillate flower, calyx removed × 6; **m**, pistillate petal, interior view × 6; **n**, gynoecium, entire and in vertical section × 6; **o**, carpels in cross-section × 6; **p**, fruit × 3; **q**, **r**, seed in 2 views × 3; **s**, seed in vertical section × 3; **t**, seed in cross-section × 3; **u**, staminate flower × 6; **v**, staminate flower in vertical section × 6; **w**, staminate calyx × 6; **x**, staminate petal with adnate stamen, interior view × 6; **y**, stamen in 2 views × 6. *Phoenix roebelenii*: **a–t**, *Read* 748 and 749; *P. canariensis*: **u–y**, *Read* 777. (Drawn by Marion Ruff Sheehan)

Relationships: Asmussen *et al.* (2006) found strong support for the monophyly of *Phoenix*. Interspecific relationships in the genus are discussed by Barrow (1998, 1999). For relationships, see tribe Phoeniceae.

Common names and uses: Variously designated as date palms, as wild date (*Phoenix sylvestris*), roebelin or miniature date (*P. roebelenii*). The genus is immensely important from an economic point of view. It includes not only the date palm, the major crop of several Middle Eastern countries and of lesser importance elsewhere, but also other species that are widely used as sources of fibre for weaving, starch, sugar, and a multiplicity of purposes such as thatch and fuel. Many species are widely grown as ornamentals. *Phoenix roebelenii* is commercially important as a pot plant. Species are known to hybridise freely. For references on uses, see Johnson (1983a, 1984).

Taxonomic account: Barrow (1998).

Fossil record: The oldest records of *Phoenix*-like fossils are probably from the Deccan Intertrappean of India (although the age span of these volcanic deposits is controversial, see Chapter 5). These records include a sheathing leaf base of

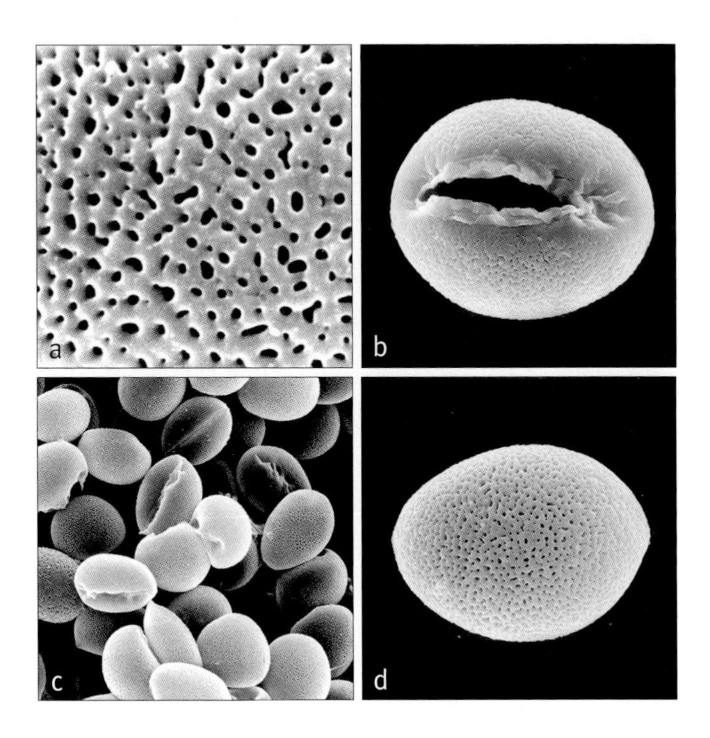

Above: ***Phoenix***
a, close-up, coarsely perforate pollen surface SEM × 8000;
b, monosulcate pollen grain, distal face SEM × 2000;
c, pollen mass SEM × 400;
d, monosulcate pollen grain, proximal face SEM × 2000.
Phoenix canariensis: **a–d**, *Smith* 9/1895.

Below, left: ***Phoenix canariensis***, leaf bases, cultivated, Montgomery Botanical Center. Florida. (Photo: J. Dransfield)

Middle: ***Phoenix theophrasti***, pistillate inflorescences, Crete. (Photo: J. Dransfield)

Right: ***Phoenix reclinata***, staminate flowers, Tanzania. (Photo: J. Dransfield)

Phoenicicaulon mahabalei (Bonde *et al.* 2000) and seeds (Ambwani & Dutta 2005). Ancibor (1995) describes fruit and stem wood, *Palmoxylon valchetense*, from the Upper Cretaceous of Argentina, although identifying stem wood to generic level is rarely conclusive. Fossils attributed to *Phoenix* occur throughout the Tertiary; for example, a leaf and staminate inflorescence from the Eocene of France recorded as *Phoenix aymardii* (Saporta 1878, 1879, 1889). It is noted that the fossil leaf *Hemiphoenicites* from the Tertiary of Italy (Visiani 1864) was placed by Read and Hickey (1972) in the synonomy of *Phoenicites*, a genus redefined by these authors as not having induplicate leaves (see below). Fossil palm wood from the Oligocene–Miocene of Louisiana (USA) has been compared with *Phoenix* (Schmidt 1994), although palm wood is notoriously difficult to define below family level. Fruits and seeds are described from the Middle to Upper Eocene and Lower Oligocene of southeastern and eastern USA, *Phoenicites occidentalis* (Berry 1914a, 1924); from the Middle Eocene of Germany, *Phoenix hercynica* (Mai 1976); and from the Lower Miocene of Central Europe, *Phoenicites bohemica* (Bůžek 1977). A pistillate inflorescence from the Eocene, Bournemouth Beds, UK was regarded as date palm by Gardener (1882, cited in Chandler 1963); and from the Bernstein flora of Germany, a palm-like flower embedded in Eocene Baltic amber is questionably assigned to *Phoenix*: *P. eichleri* (Conwentz 1886). Notably, Poinar (2002b) also describes a palm flower from the Baltic amber, which, "resembles the one described by Conwentz (1886)", although he does not comment on affinity. Small monosulcate grains, *Palmaemargosulcites fossperforatus*, from palm flower compression fossils, recovered from the Middle Eocene oil shales of Messel, Germany, have been compared with pollen of a number of coryphoid genera, but most favourably with *Phoenix* (Harley 1997); and small dispersed *Phoenix*-like pollen occurs sporadically in the UK Eocene London Clay Basin (Khin Sein 1961). Pleistocene subfossils of *Phoenix* leaves (Lacroix 1896) were discovered in ancient volcanic ash deposits on the island of Phira (Santorini).

(N.B. Read & Hickey [1972] redefined the fossil genus *Phoenicites* as pinnate with reduplicate plication and lowermost pinnae not spine-like. Their taxonomic recommendation that *Phoenicites* should, henceforth, be reserved or applied to non-*Phoenix*-like pinnate leaves is, perhaps, more than a little confusing to the unwary.)

Notes: Easily distinguished from all other palms by the induplicate pinnate leaf with the lower leaflets modified as spines. A modern developmental study of the unique leaf is badly needed.

Additional figures: Figs 1.3, 5.6, 7.1, Glossary figs 4, 7, 12, 15.

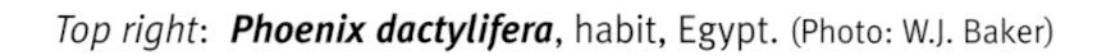

Top right: ***Phoenix dactylifera***, habit, Egypt. (Photo: W.J. Baker)

Tribe Trachycarpeae Satake, Hikobia 3: 121 (1962). Type: **Trachycarpus**

Livistoneae J. Dransf., N.W. Uhl, C. Asmussen, W.J. Baker, M.M. Harley & C. Lewis., Kew Bull. 60: 561 (2005). Type: *Livistona* (invalid name).

Pleonanthic, hermaphroditic or dioecious palms; leaves induplicate or rarely reduplicate, or splitting anomalously, a central abaxial split never present, palmate or costapalmate or unsplit; leaf base also lacking a central split; flowers solitary or in simple cincinni; gynoecium tricarpellate, apocarpous or syncarpous.

This large pantropical and subtropical tribe is divisable into two contrasting subtribes, Rhapidinae and Livistoninae.

Tribe Trachycarpeae is monophyletic with moderate to high support (Uhl *et al.* 1995, Asmussen & Chase 2001, Hahn 2002a, Asmussen *et al.* 2006, Baker *et al.* in review). Four different hypotheses for the sister group relationships of Trachycarpeae have been obtained. The Trachycarpeae has been resolved as 1) sister to tribe Phoeniceae with moderate support (Asmussen *et al.* 2006), 2) sister to tribes Sabaleae and Cryosophileae with low support (Baker *et al.* in review), 3) moderately supported in a clade with tribes Phoeniceae and Cryosophileae (Hahn 2002a), and 4) sister to the rest of the Coryphoideae excluding tribe Caryoteae (Uhl *et al.* 1995). Several genera of tribe Trachycarpeae, principally from the Neotropics, cannot yet be placed in subtribes because of a lack of resolution. Various relationships between subtribes Rhapidinae, Livistoninae and unplaced members of tribe Trachycarpeae remain poorly supported by the available data (Asmussen *et al.* 2006, Baker *et al.* in review).

Unfortunately, in publishing tribe Livistoneae (Dransfield *et al.* 2005), we overlooked the earlier name Trachycarpeae.

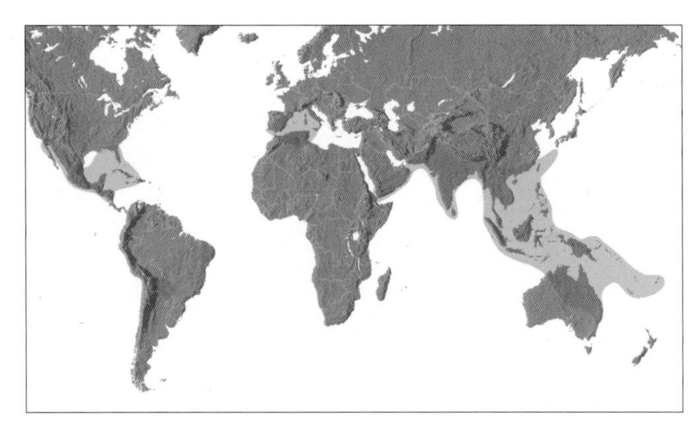

Distribution of tribe **Trachycarpeae**

Key to Genera of Trachycarpeae

1. Gynoecium composed of 3 carpels free throughout their length (**Subtribe Rhapidinae**) 2
1. Gynoecium composed of 3 carpels free at their bases but united apically in a common style (**Livistoninae** and **unplaced Trachycarpeae**) 6
2. Leaf split along abaxial folds for part of their length to produce reduplicate segments . . . . **36. Guihaia**
2. Leaf splitting along adaxial folds or between the folds . 3
3. Stamen filaments free 4
3. Stamen filaments connate in a ring or epipetalous 5
4. Leaf sheath fibres not spine-like; leaf blade splitting along adaxial folds to varying lengths; inflorescences exserted from sheaths; seed with deep intrusion of seed coat . . . . **37. Trachycarpus**
4. Leaf sheath fibres spine-like; leaf blade splitting between the folds; inflorescences borne deep among the leaf sheaths; seed without deep intrusion of seed coat . . . . **38. Rhapidophyllum**
5. Stems slender, cane-like; leaves divided between the folds into several-many fold segments, usually with truncate tips; perianth whorls fleshy to thick coriaceous; sepals connate; stamen filaments free, epipetalous . . . . **40. Rhapis**
5. Stems stouter, scarcely cane-like; leaves divided along adaxial folds into single-fold acute or bifid segments; perianth whorls membranous to coriaceous; sepals distinct, imbricate; stamen filaments connate in an epipetalous ring or separate and epipetalous . . . . **39. Maxburretia**
6. Leaf blade undivided, diamond shaped, pinnately nerved, the costa prominent; petiole and basal margins of the blade armed with hooked teeth; inflorescences each with several inflated peduncular bracts; fruit corky-warted; endosperm deeply and irregularly penetrated by seed coat . . . . **43. Johannesteijsmannia**
6. Leaf blade divided, or if undivided, scarcely diamond-shaped and basal margins unarmed; fruit smooth, if corky-warted, then leaf blade divided 7
7. Leaf blades divided along abaxial folds to the insertion into single-fold (rarely) or multi-fold, usually wedge-shaped segments or undivided . . . . **42. Licuala**
7. Leaf blades divided along adaxial folds into single-fold or several-fold but not truncate segments 8
8. Inflorescence with generally long peduncle, branched only towards the tip, or branched at the very base to produce 2–4 ± equal long axes, also branched only towards the tip, giving the appearance of more than 1 inflorescence per axil; corolla lobes caducous at anthesis exposing the prominent androecial tube; seed coat not intruded . . . . **51. Pritchardia**
8. Inflorescence branching not confined to the tip; corolla lobes not usually caducous at anthesis, very rarely tardily deciduous 9
9. Rachis bracts split on one side and pendulous, sword-shaped; petals relatively large, flat, chaffy, without grooves on the inner surfaces, becoming strongly reflexed at anthesis; style elongate, about 3 times as long as the ovarian part of the gynoecium; flowers and fruit pedicellate; fruit with persistent chaffy calyx . . . . **52. Washingtonia**
9. Inflorescence bracts tubular, not splitting into sword-shaped blades at anthesis; petals neither flat nor chaffy and with grooves matching anthers on inner surfaces; styles as long as the ovarian part or rarely 2–2$^1/_2$ times as long (*Serenoa*) 10
10. Fruit large and smooth (ca. 5 cm diameter at maturity); mesocarp spongy with short fibres near and closely adhering to the endocarp, endocarp with a triangular basal projection, rounded on 1 side, keeled on the other 3 sides; embryo eccentrically apical, endosperm with a conspicuous lobed intrusion of the seed coat . . . **45. Pritchardiopsis**
10. Fruit various, if large then usually corky-warted; endocarp smooth, not keeled, without a basal projection; embryo lateral or basal; seed coat intrusion present or absent. 11
11. Petioles and often margins and/or veins of leaves with stout teeth; abortive carpels apical in fruit; seeds with ruminate endosperm . . . . **50. Copernicia**
11. Petioles armed or not, leaves lacking teeth; abortive carpels usually basal in fruit; seeds with homogeneous endosperm, sometimes penetrated by a lobed intrusion 12
12. Endosperm intruded by seed coat 13
12. Endosperm not intruded by seed coat 15
13. Trunk solitary, sometimes ventricose at or near middle; petiole unarmed; androecial tube projecting beyond the corolla lobes . . . . **49. Colpothrinax**
13. Trunks clustered, erect or creeping and branching; petiole spiny, armed with fine or coarse teeth; androecial tube not projecting beyond the corolla lobes 14
14. Stems erect, densely clustered; flowers in cincinni, but paired or solitary towards the apex of the rachillae; sepals distinct or basally connate, imbricate; stamen filaments united in a basal ring, free portions short, abruptly narrowed; fruit globose; raphe short, scarcely more than half as long as the seed . . . . **46. Acoelorrhaphe**
14. Stems prostrate, rarely erect; flowers solitary or occasionally paired; calyx tubular, shortly 3-lobed; stamen

filaments shortly united at the base, free portions gradually narrowed; fruit ellipsoidal; raphe branches extending the length of the seed .... **47. Serenoa**

15. Sepals distinct, imbricate .... **48. Brahea**
15. Sepals united basally 16
16. Blade divided by deep splits into compound segments, further divided by shallow splits into single-fold segments; androecial ring conspicuous, thick, almost free at the base, filaments where separate, slender not broadly rounded; gynoecium top-shaped; fruit very large (at least 6 cm in diameter) usually corky-warted ... **44. Pholidocarpus**
16. Blade regularly divided into single-fold segments, very rarely into compound segments (*L. saribus*, *L. exigua*); androecial ring broadly scalloped apically, epipetalous; gynoecium widest above the locules, abruptly narrowed to the style; fruit small (rarely exceeding 4 cm in diameter at maturity), smooth .... **41. Livistona**

Subtribe Rhapidinae J. Dransf., N.W. Uhl, C. Asmussen, W.J. Baker, M.M. Harley & C. Lewis., Kew Bull. 60: 561 (2005). Type: **Rhapis**.

Mostly dioecious or polygamodioecious, rarely hermaphroditic palms; leaves palmate, induplicate or reduplicate, or the lamina splitting between the folds; flowers solitary or borne in simple cincinni; gynoecium apocarpous, with 3 carpels.

The six genera of this subtribe are largely subtropical palms, all but *Rhapidophyllum* occurring in the Old World. Some of the smallest of all palm flowers are found in this subtribe. There is remarkable variation in the way the leaf blade splits; *Chamaerops*, *Trachycarpus* and *Maxburretia* have induplicately splitting leaves of a form commonly found among most members of Coryphoideae; in *Rhapis* and *Rhapidophyllum*, the blade is dissected by splits that occur between the folds, whereas in *Guihaia*, the splitting occurs abaxially to give reduplicate segments.

Subtribe Rhapidinae is resolved as monophyletic in all phylogenies, with moderate support in some studies (Hahn 2002a, Asmussen & Chase 2001, Asmussen *et al.* 2006) and with high support in Baker *et al.* (in review). For relationships, see tribe Trachycarpeae. The relationships between genera remain poorly understood.

35. CHAMAEROPS

Dwarf dioecious fan palm of the Mediterranean region, with stems covered with leaf bases; leaves generally rather stiff; inflorescence with a single large bract. Fruit reddish brown smelling of vomit.

Chamaerops L., Sp. pl. 1187 (1753). Type: **C. humilis** L.
Chamaeriphe Steck, Sagu 20 (1757).
Chamaeriphes Ponted. ex Gaertn., Fruct. sem. pl. 1: 25 (1788). Type: *C. major* Gaertn. (illegitimate name) (= *Chamaerops humilis* L.).

Chamai — *low*, rhops — *a bush, referring to the dwarf habit.*

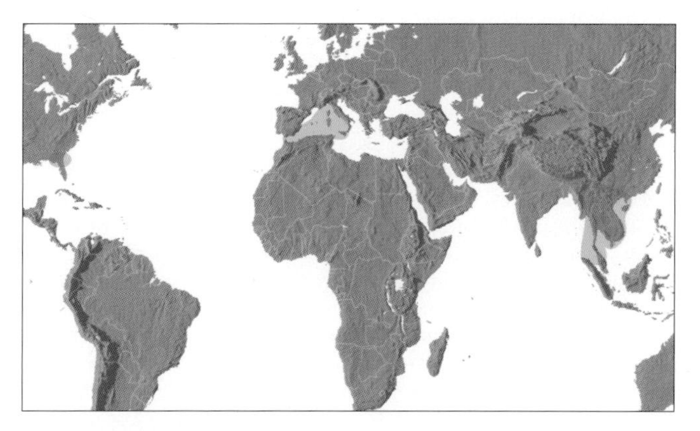

Distribution of subtribe **Rhapidinae**

Dwarf, rarely moderate, clustering, acaulescent or shrubby, armed, pleonanthic, polygamous or dioecious palm. *Stem* ± erect, clothed with very close, persistent petiole bases and fibrous sheaths, eventually becoming ± bare. *Leaves* induplicate, palmate, marcescent; sheath disintegrating into a mass of fine fibres; petiole elongate, slender, adaxially flattened or slightly rounded, abaxially rounded or angled, densely covered with caducous white tomentum, armed along the margins with robust, bulbous-based spines pointing toward the leaf tip; adaxial hastula well developed, ± acute, abaxial hastula poorly developed; blade divided to $^2/_3$–$^3/_4$ the radius into single-fold segments, the segments further divided to ca. $^1/_2$–$^2/_3$ along the

Above: ***Chamaerops humilis***, habit, Mallorca. (Photo: J. Dransfield)

abaxial folds, folds longitudinally striate, tips ± rounded or pointed, abaxially sparsely to densely covered with caducous tomentum, midribs prominent abaxially, transverse veinlets obscure. *Inflorescences* solitary, interfoliar, very short, branching to 2 orders, staminate, pistillate, and hermaphroditic inflorescences similar; peduncle very short, oval in cross-section; prophyll conspicuous, tubular, somewhat inflated, laterally 2-keeled, splitting apically into 2 triangular lobes, covered with dense tomentum especially along keels; peduncular bracts absent; rachis short, but longer than the peduncle; rachis bracts inconspicuous, the proximal enclosed within the prophyll; first-order branches bearing minute slender bracts; rachillae short, very crowded, glabrous. *Flowers* solitary, spirally arranged, borne on short tubercles subtended by minute bracts; abnormalities in all floral parts frequent; sepals 3, triangular, low, glabrous, united at the base; petals 3, similar in staminate, pistillate, and hermaphroditic flowers, imbricate, distally triangular, united at the very base; stamens 6, united by their broad triangular filaments to form a conspicuous staminal ring folded in young stages, much expanded at anthesis, anthers yellow, oblong, medifixed, latrorse; staminodes similar to stamens but less-well-developed filaments and empty anthers; carpels 3, distinct, follicular, glabrous, with conspicuous recurved apical stigmas, ovule hemianatropous, attached basally; pistillodes 1–3 minute carpels, or absent. *Pollen* ellipsoidal, bi-symmetric or slightly asymmetric, or infrequently, oblate triangular; aperture comprising 2 parallel distal sulci, narrowly separated by an ectexinous bridge, less frequently a trichotomosulcus; ectexine tectate densely perforate or finely and densely reticulate, outer aperture margins similar, tectum between sulci sometimes similar or psilate-perforate; infratectum columellate; longest axis 27–31 µm; post-meiotic tetrads tetrahedral [1/1]. *Fruit* developing from 1–3 carpels, product of each carpel globose to oblong, ellipsoidal, rich brown, pale-dotted, stigmatic remains apical; epicarp smooth, mesocarp thin, ± fleshy, rich in butyric acid, endocarp scarcely developed. *Seed* globose to ellipsoidal, basally attached; endosperm ruminate, also with a conspicuous lateral intrusion of seed coat; embryo lateral. *Germination* remote-tubular; eophyll entire, narrow, plicate. *Cytology*: 2n = 36, or 36 ± 1, or 36 ± 2.

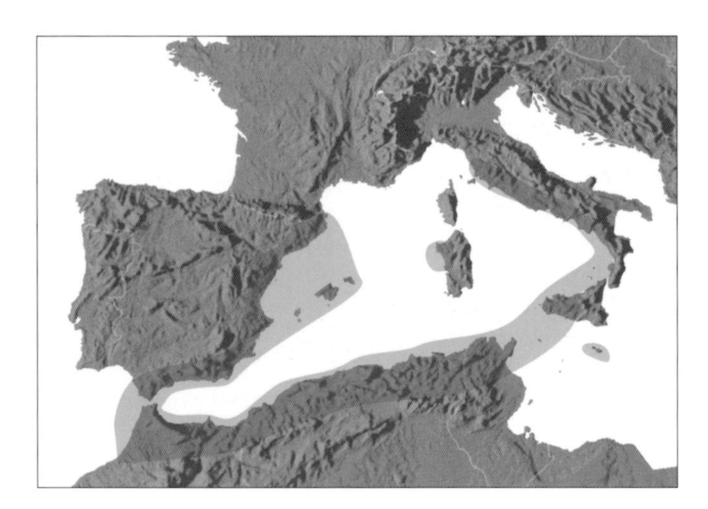

Distribution of ***Chamaerops***

Distribution and ecology: One species, *Chamaerops humilis*, native to coastal areas of the western Mediterranean, both in Europe and North Africa, becoming rarer eastwards to Malta. Widely cultivated and very variable, particularly in leaf form and fruit shape. It is found on sandy or rocky ground, usually near the sea but up to 600 m altitude or more on coastal hills, usually acaulescent in the wild, but in the absence of burning, producing a well-developed trunk as in cultivated specimens.

Anatomy: Stem (Schweingruber 1990), leaf (Tomlinson 1961), roots (Seubert 1997), floral (Morrow 1965).

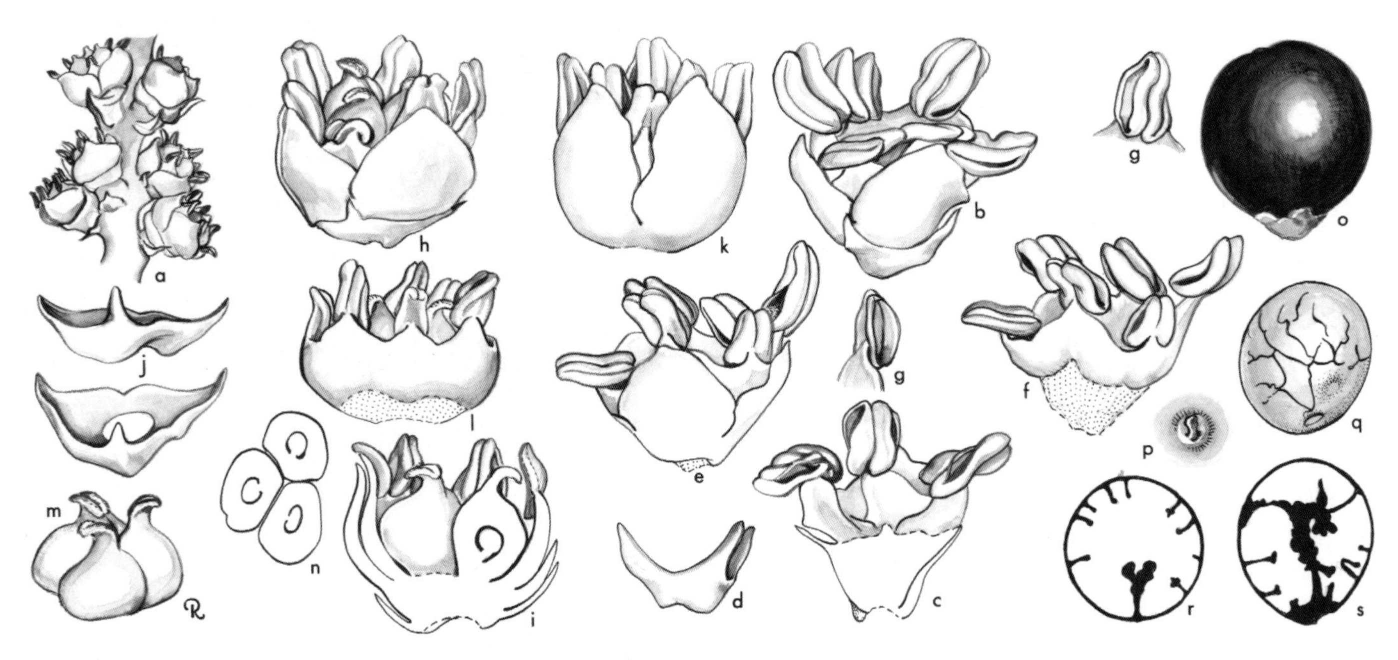

Chamaerops. **a**, portion of pistillate rachilla × 3; **b**, staminate flower × 6; **c**, staminate flower in vertical section × 6; **d**, staminate calyx × 6; **e**, staminate flower, calyx removed × 6; **f**, androecium × 6; **g**, anther in 2 views × 6; **h**, pistillate flower × 6; **i**, pistillate flower in vertical section × 6; **j**, pistillate calyx in 2 views × 6; **k**, pistillate flower, calyx removed × 6; **l**, pistillate flower, perianth removed × 6; **m**, gynoecium × 6; **n**, carpels in cross-section × 6; **o**, fruit × 1 1/2; **p**, apex of fruit × 6; **q**, seed × 1 1/2; **r**, seed in cross-section × 1 1/2; **s**, seed in vertical section × 1 1/2. *Chamaerops humilis*: **a** and **h–s**, *Read* 700; **b–g**, *Read* 724. (Drawn by Marion Ruff Sheehan)

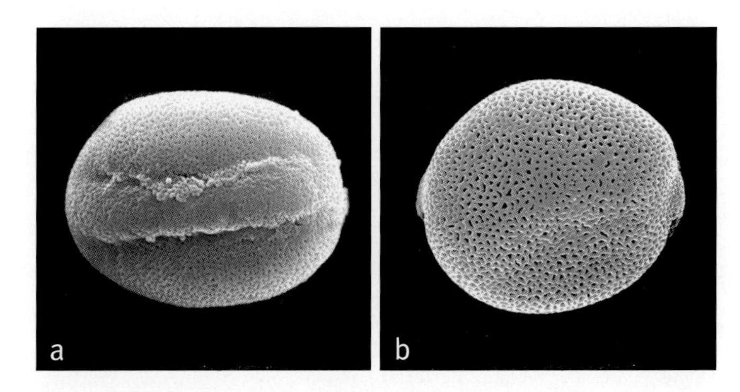

Above: ***Chamaerops***
a, distal disulcate pollen grain, distal face SEM × 2000;
b, distal disulcate pollen grain, proximal face × 2000.
Chamaerops humilis: **a**, *Moris* s.n.; **b**, *Brummitt & Ernst* 5936.

Relationships: *Chamaerops* is resolved as sister to the rest of the Rhapidinae with low or moderate support (Asmussen & Chase 2001, Asmussen *et al.* 2006, Baker *et al.* in review).
Common names and uses: European fan palm, windmill palm. Fibre has been used for cordage and woven articles. Forms with abnormal leaves have been selected in cultivation; one such in the garden of R.O. Douglas, California, bears leaves with the blade divided between as well as along the folds.
Taxonomic account: Beccari (1931).
Fossil record: Cuticle from the brown coals of Niederlausitz, Upper Tertiary of Germany, is attributed to *Chamaerops* (Litke 1966). From the Miocene of Poland (Gliwice, Upper Silesia), Szafer (1961) compares a large fragment of a palmate leaf, including details of stomata in the epidermis, and also a somewhat damaged seed, with *Chamaerops* aff. *humilis*. After studying the morphology and anatomy of fossil palm leaf remnants from Miocene of Moldavia, Stephyrtza (1972) considered that the fossil material represented *C. humilis* L. var. *fossilis* Kolak. Biondi and Filigheddu (1990) describe the results of anatomical studies on a silicified palm fossil from the Lower Miocene of Sardinia, *Palmoxylon homeochamaerops*. The fossil material comprises part of a stem and surrounding roots, which the authors consider closely related to *C. humilis* L. Pleistocene leaf subfossils of *Chamaerops* strongly attached to a short trunk (Lacroix 1896) were discovered in ancient volcanic ash deposits on the island of Phira (Santorini).
Notes: The pollination of this familiar palm has only recently been elucidated (Anstett 1998). Anstett and her co-workers demonstrated that derelomid weevils are responsible for transfer of pollen between the staminate and pistillate plants. They are attracted by floral-type odours produced by the leaves subtending the inflorescences rather than by the flowers themselves (see Chapter 8). Floral abnormalities are of frequent occurrence; for example, the following variation may be found within the same inflorescence: petals 2 instead of 3; stamens fewer than 6, and both fertile and infertile anthers within a single flower; carpels 3, 2, or 1 fertile, with or without fertile stamens, and pistillodes present or absent.
Additional figures: Figs Ch. 4 frontispiece, 7.2, Glossary figs 7, 20.

36. GUIHAIA

Dwarf clustering ± stemless dioecious fan palms of karst limestone in southern China and northern Vietnam, instantly recognizable by the leaf blades partially split into reduplicate segments.

Guihaia J. Dransf., S.K. Lee & F.N. Wei, Principes 29: 7 (1985). Type: **G. argyrata** (S.K. Lee & F.N. Wei) S.K. Lee, F.N. Wei & J. Dransf. (*Trachycarpus argyratus* S.K. Lee & F.N. Wei).

Named after the area referred to as Gui Hai *in old Chinese literature and including the karst limestone areas of Guangxi.*

Dwarf, clustering, acaulescent, unarmed or armed, pleonanthic, dioecious palms. *Stem* decumbent or erect, very short, clothed with persistent petiole bases and sheaths. *Leaves* reduplicate, palmate, marcescent; sheath disintegrating into an interwoven mass of coarse, erect, black, spine-like fibres or into a tongue-shaped lattice of coarse flat fibres; petiole moderate, unarmed, abaxially rounded, adaxially flattened or slightly rounded, the margins quite sharp, bearing caducous woolly hairs; adaxial hastula rounded, glabrous or bearded with woolly hairs; lamina orbicular or cuneate, rather small, divided to $^3/_4$ to $^4/_5$ the radius, or nearly to the insertion along the abaxial ribs, into several (ca. 20), ± linear, single or rarely 2-fold reduplicate segments, minutely bifid at the tips, the outermost segments consisting of 1–2 folds only, margins of the segments minutely toothed or smooth, lamina adaxially dark green, glabrous except for scales along

Above: ***Guihaia argyrata***, habit, cultivated, Fairchild Tropical Botanic Garden, Florida. (Photo: J. Dransfield)

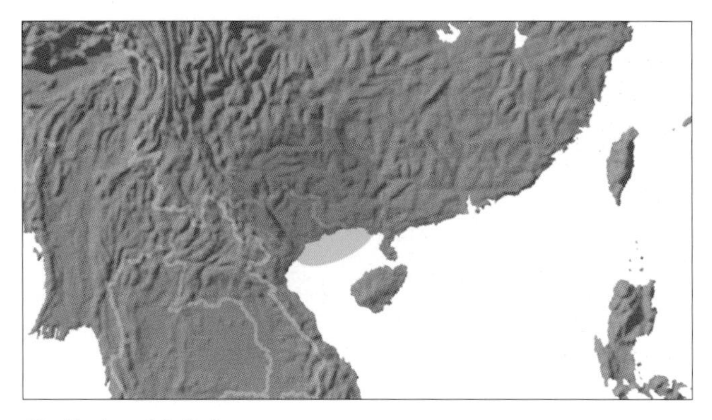

Distribution of ***Guihaia***

the ribs, abaxially covered with a dense felt of silvery woolly hairs or glabrous except for scattered dot-like scales, transverse veinlets obscure or evident. *Inflorescences* solitary, axillary, interfoliar, branching to 4 orders, staminate and pistillate superficially similar; prophyll elongate, tubular, 2-keeled, thin, somewhat coriaceous, apically splitting along 2 sides, glabrous or bearing caducous hairs; peduncle elongate, ± flattened, caducously scaly; peduncular bracts absent; rachis longer or shorter than the peduncle; rachis bracts ca. 2–5, similar to the prophyll but not 2-keeled, with tattering limb; first-order branches 4–5, adnate to the rachis to just below the insertion of the following bract; subsequent bracts minute, scarcely evident; rachillae spreading, few to numerous, very slender, ± straight, glabrous, or bearing scattered caducous scales and spirally arranged solitary flowers borne on very low swellings. *Staminate flowers* extremely small, symmetrical; sepals 3, distinct except at the very base, basally imbricate, ± rounded to ovate, abaxially bearing hairs and fringed with wool-like hairs; petals longer than the sepals, basally connate ca. $^1/_3$–$^1/_2$ their length, with rounded lobes, glabrous; stamens 6, the filaments not forming a staminal tube, but completely adnate to the corolla, anthers ± rounded, didymous, apparently inserted directly on the corolla, latrorse; pistillode absent. *Pollen* ellipsoidal, bi-symmetric or slightly asymmetric; aperture a distal sulcus; ectexine tectate, perforate, or foveolate-reticulate, aperture margin slightly finer; infratectum columellate; longest axis 17–24 µm [2/2]. *Pistillate flowers* similar to the staminate but perhaps more rounded; sepals as in the staminate; petals only slightly longer than to more than twice as long as the sepals, joined in the basal ca. $^1/_3$; staminodes 6, borne directly on the petals; carpels 3, distinct, glabrous, ± abruptly narrowed to a short style, ovule basally attached. *Fruit* developing from only 1 carpel, rounded to ellipsoidal, blue-black and bearing thin white wax, the stigmatic remains apical, the abortive carpels basal; epicarp glabrous, mesocarp very thin, fleshy, endocarp papery. *Seed* ± flattened on one side with lateral hilum and a well-defined, rounded intrusion of integument, endosperm homogeneous; embryo lateral. *Germination* remote-tubular; eophyll entire, plicate, very narrow. *Cytology*: 2n = 36.

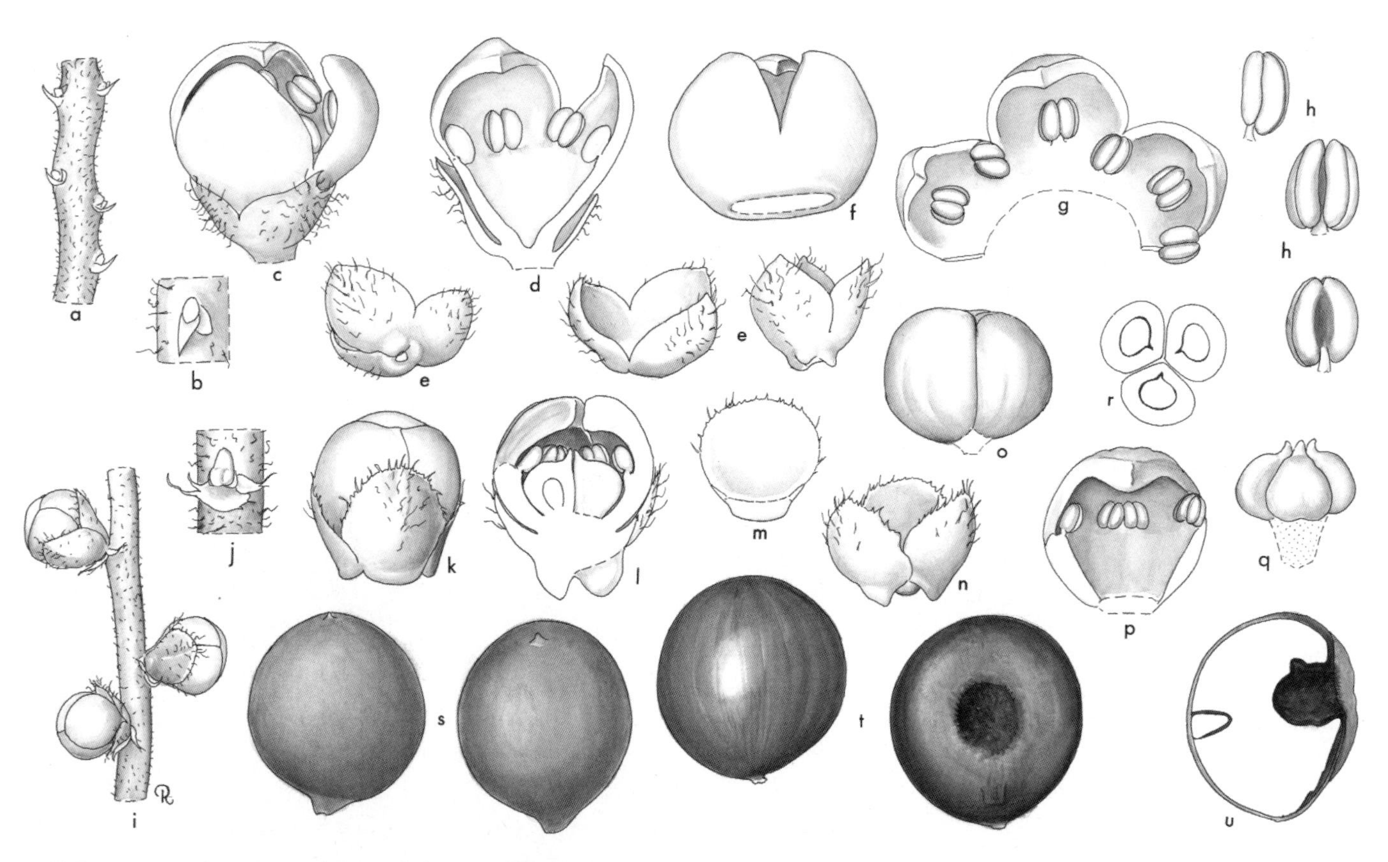

Guihaia. **a**, portion of staminate rachilla × 7$^1/_2$; **b**, scar and bracts of staminate flower × 9; **c**, staminate flower × 15; **d**, staminate flower in vertical section × 15; **e**, staminate calyx in 3 views × 15; **f**, staminate corolla × 15; **g**, staminate corolla with epipetalous stamens, interior view, expanded × 15; **h**, stamen in 3 views × 30; **i**, portion of pistillate rachilla × 7$^1/_2$; **j**, scar, bract, and bracteole of pistillate flower × 9; **k**, pistillate bud × 15; **l**, pistillate bud in vertical section × 18; **m**, pistillate sepal, interior view × 15; **n**, pistillate calyx × 15; **o**, pistillate flower, sepals removed × 15; **p**, pistillate petal with staminodes, interior view × 18; **q**, gynoecium × 18; **r**, carpels in cross-section × 18; **s**, fruit in 2 views × 4$^1/_2$; **t**, seed in 2 views × 6; **u**, seed in vertical section × 6. *Guihaia argyrata*: **a–h**, *Wei* 937; **i–r**, *Wei* 1524; **s–u**, *Wei* 1513. (Drawn by Marion Ruff Sheehan)

Distribution and ecology: Two species, *Guihaia argyrata* endemic to south China (Guangxi and Guangdong), *G. grossifibrosa* in northern Vietnam and southwestern Guangxi. Confined to steep karst limestone hill slopes and crevices in warm temperate to subtropical climates, at low elevations (ca. 200 m above sea level), occurring to about 26° N.
Anatomy: Roots (Seubert 1997).
Relationships: The monophyly of *Guihaia* has not been tested. There are two different hypotheses on the placement of the genus within subtribe Rhapidinae. One places the genus as sister to *Rhapis* with high support (Uhl *et al.* 1995, Baker *et al.* in review) and the other resolves *Guihaia* as sister to *Trachycarpus* with moderate support (Asmussen *et al.* 2006).
Common names and uses: Common names not recorded. These elegant palms make fine ornamentals, but remain rare in cultivation; as far as is known, they have no local uses.
Fossil record: No generic records found.
Taxonomic account: Dransfield *et al.* (1985b).
Notes: The extraordinary reduplicate leaf immediately sets the genus apart from all other coryphoid palms (but see *Licuala*).
Additional figures: Figs 1.3, Glossary figs 11, 18, 19.

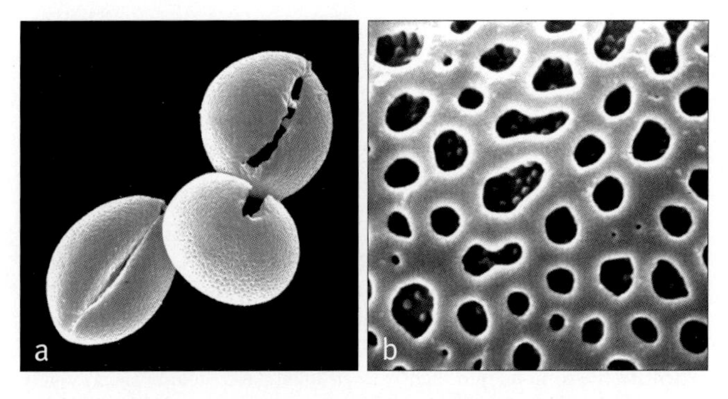

Above: ***Guihaia***
a, three monosulcate pollen grains: left & right distal faces, centre short equatorial axis SEM × 1500;
b, close-up, foveolate-reticulate surface SEM × 5000.
Guihaia argyrata: **a**, *Wei* 937; *G. grossifibrosa*: **b**, *Wei* 937.

Left: ***Guihaia argyrata***, infructescence and fibre spines, cultivated, Fairchild Tropical Botanic Garden, Florida. (Photo: J. Dransfield)

37. TRACHYCARPUS

Moderate or dwarf, very rarely clustering dioecious fan palms of warm temperate parts of northeastern Indian Subcontinent, Burma, Thailand, Vietnam and China; leaves induplicate; fruit often kidney-shaped.

Trachycarpus H. Wendl. in J. Gay, Bull. Soc. Bot. France 8: 429 (1863(?) ['1861']). Lectotype: **T. fortunei** (Hook.) H. Wendl. (*Chamaerops fortunei* Hook.) (see H.E. Moore 1963c).

Trachus – *rough,* karpos – *fruit, referring to the irregularly shaped fruit.*

Dwarf or moderate, solitary or clustering, acaulescent or erect, unarmed or lightly armed, pleonanthic, dioecious or polygamous palms. *Stem* decumbent or erect, becoming bare and marked with conspicuous, rather close, oblique leaf scars, or clothed with persistent petiole bases and fibrous sheaths, or obscured by a skirt of dead leaves. *Leaves* induplicate, palmate, marcescent; sheath disintegrating into a mass of fine and coarse fibres, the upper margin ribbon-like, becoming twisted; petiole elongate, narrow, adaxially flattened or slightly rounded, abaxially rounded or angled, bearing scattered, deciduous indumentum or glabrous, armed along the margins with very fine teeth or unarmed; adaxial hastula well developed, rounded or triangular, abaxial hastula absent; blade fan-shaped to almost circular, equally or unequally divided along adaxial ribs into single-fold segments, shallowly bifid at the tips, longitudinally striate or not, the abaxial surface sometimes glaucous, sometimes dotted with minute brown scales, midrib conspicuous abaxially, transverse veinlets conspicuous or not. *Inflorescences* solitary, interfoliar, arching or ± erect, copiously branching to 4 orders; peduncle oval in cross-section, bearing sparse indumentum; prophyll complete, conspicuous, with a tubular base, and inflated distally, 2-keeled laterally, splitting apically and along one side, covered with deciduous indumentum; peduncular bracts 1–3, as the prophyll but single-keeled, rachis shorter or longer than the peduncle, bearing spirally arranged bracts similar to the peduncular bracts, but each subtending a first-order branch; bracts of subsequent orders inconspicuous, triangular, not sheathing; rachillae slender, stiff, short, very crowded, bright yellow to greenish, glabrous or sparsely hairy, bearing spirally arranged flowers, which are solitary or in clusters of 2–3, sessile or borne on low tubercles, each flower bearing a minute, apiculate, membranous bracteole. *Flowers* similar in both sexes; sepals 3, united at the base, triangular, short or long, glabrous; petals usually considerably exceeding the sepals, 3, distinct, imbricate, ovate, triangular-tipped or rounded, glabrous; stamens 6, filaments distinct, fleshy, ± parallel-sided, anthers short, oblong, sometimes slightly pointed, latrorse; staminodes when present, similar to fertile stamens but with flattened filaments and empty anthers, sometimes with filaments connate at the very base; carpels 3, distinct, follicular, hairy, ventral sutures partially open, stylar projections short, ovule basally attached, hemianatropous, surrounded dorsally and ventrally by a fleshy aril; pistillodes when present similar to, but much smaller than, the fertile carpels. *Pollen* ellipsoidal, with slight to obvious asymmetry; aperture a distal sulcus; ectexine tectate, finely rugulate-perforate, foveolate or reticulate, aperture margin slightly finer; infratectum columellate; longest axis 22–32 µm; post-meiotic tetrads usually tetrahedral, sometimes tetragonal or, rarely, rhomboidal [2/9]. *Fruit* usually developing from 1 carpel, purplish-black with a pale bloom, kidney-shaped to oblong, slightly grooved on the adaxial side with lateral or subapical stigmatic remains; epicarp thin, hairy in immature fruit, becoming glabrous in mature fruit, mesocarp thin with scattered layer of tannin cells, endocarp crustaceous. *Seed* kidney-shaped to oblong, endosperm homogeneous with a shallow to deep lateral intrusion of seed coat, sometimes also with very shallow ruminations; embryo lateral. *Germination* remote-tubular; eophyll simple, narrow, plicate. *Cytology*: 2n = 36.

Distribution and ecology: Nine species recorded (two of these perhaps only cultivars) ranging from the Himalayas in northern India to northern Thailand, Vietnam and China. *Trachycarpus nanus* and *T. oreophilus* have both been recorded from limestone hills; the latter has been found at altitudes up to 2400 m and appears not to be exclusive to limestone.

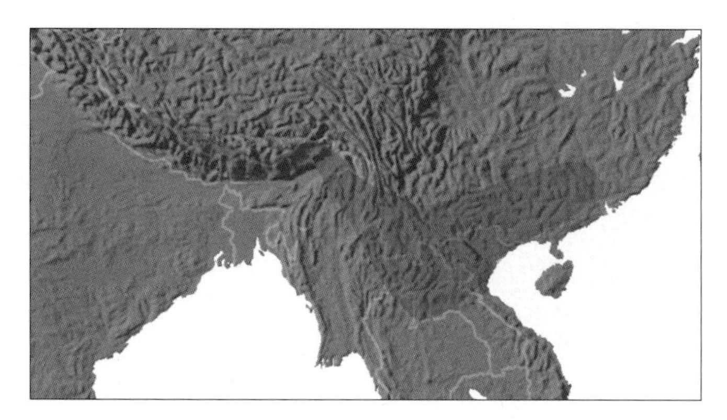

Distribution of ***Trachycarpus***

Trachycarpus takil has been reported from damp oak forests at 2400 m altitude, where the ground is under snow from November to March. *Trachycarpus fortunei*, one of the most cold tolerant of all cultivated palms, is hardy in the British Isles.

Anatomy: Leaf (Tomlinson 1961), roots (Seubert 1997), floral (Morrow 1965, Uhl & Moore 1971).

Relationships: Preliminary analyses indicate that *Trachycarpus* is monophyletic (Stührk 2006). The genus is sister to all other Rhapidinae except *Chamaerops* with low support (Baker *et al.* in review). Alternative placements include sister to *Guihaia* with moderate support (Asmussen *et al.* 2006) and sister to a clade of *Rhapidophyllum*, *Guihaia* and *Rhapis* (Uhl *et al.* 1995).

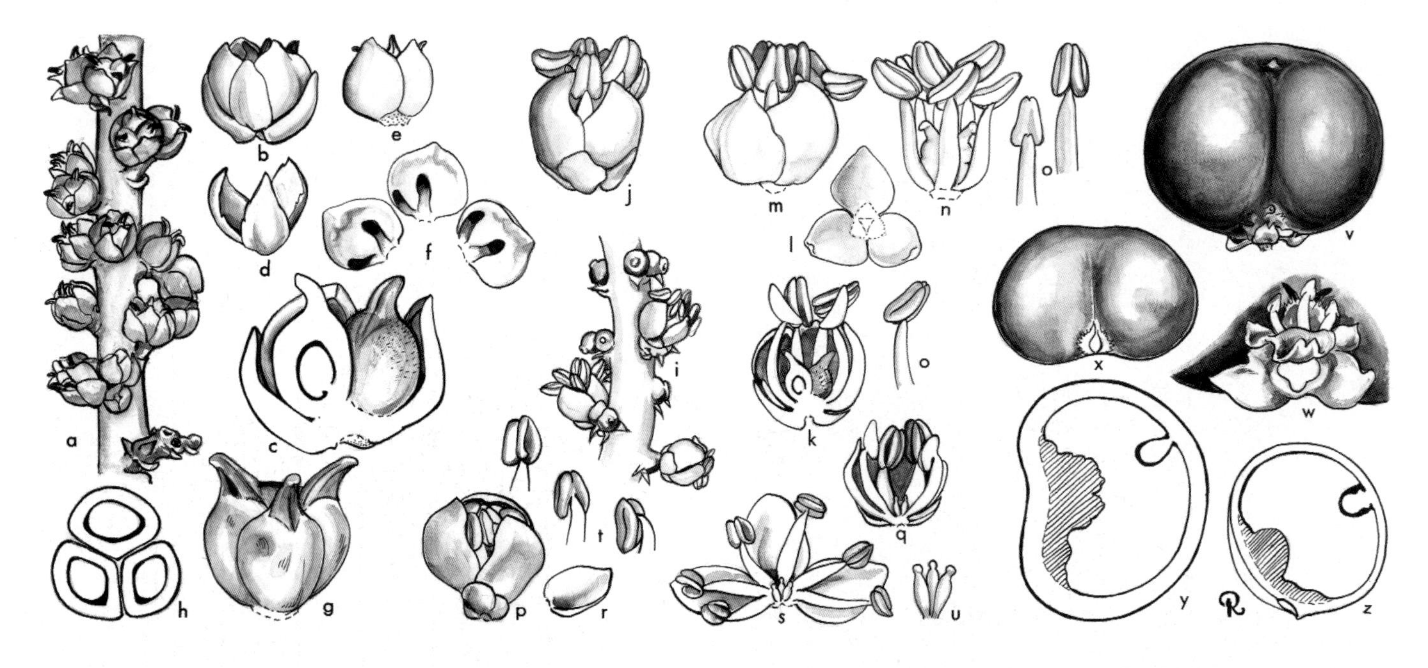

***Trachycarpus*.** **a**, portion of pistillate rachilla × 3; **b**, pistillate flower × 6; **c**, pistillate flower in vertical section × 12; **d**, pistillate calyx × 6; **e**, pistillate flower, calyx removed × 6; **f**, pistillate petals and staminodes, interior view × 6; **g**, gynoecium × 12; **h**, carpels in cross-section × 12; **i**, portion of rachilla with morphologically hermaphrodite flowers × 3; **j**, hermaphrodite flower × 6; **k**, hermaphrodite flower in vertical section × 6; **l**, calyx of hermaphrodite flower, interior view × 6; **m**, hermaphrodite flower, calyx removed × 6; **n**, androecium and gynoecium × 6; **o**, stamen or staminode in 3 views × 6; **p**, staminate flower × 6; **q**, staminate flower in vertical section × 6; **r**, staminate sepal × 6; **s**, staminate flower, expanded, interior view × 6; **t**, stamen in 3 views × 6; **u**, pistillode × 12; **v**, fruit × 3; **w**, base of fruit with abortive carpels × 6; **x**, seed × 3; **y**, fruit in cross-section × 3; **z**, fruit in vertical section × 3. *Trachycarpus fortunei*: **a–h**, *Read* 756; **i–o**, *Bailey & Bailey* s.n.; **p–u** and **w–y**, *Bailey & Bailey* s.n.; **v, z**, *Read* s.n. (Drawn by Marion Ruff Sheehan)

Common names and uses: Chinese windmill palm, Chusan palm (*Trachycarpus fortunei*). Stems are used as posts in China, and fibres of leaf sheath and stem are used for brushes, plaiting and raincoats; seeds are used medicinally and are believed to have anticancer properties. Grown as ornamentals in cooler climates.

Taxonomic accounts: Beccari (1931), see also Kimnach (1977). Several new species have been described recently by Gibbons *et al.* (1995, 2003), Spanner *et al.* (1997) and Gibbons & Spanner (1997, 1998).

Fossil record: Palmate leaves referred to *Trachycarpus* include: southern England, Lower Eocene (Bagshot Beds), *T. raphifolia* (Sternberg) Takhtajan (Chandler 1962); Russia, Caucasus, Middle Miocene (Takhtajan 1958; who considers that most specimens of fan palms from Russia should be referred to *T. raphifolia*), and Transcaucasia, Oligocene (Akhmetiev 1989); Czech Republic, Lower Miocene, *T. rhapifolia* (Czeczott & Juchniewicz 1975); northern India, upper Indus Valley, Miocene, *T. ladakhensis* (Lakhanpal *et al.* 1984). Chandler (1962) casts doubt on Takhtajan's inclusion of *Palaeothrinax mantelli* (Reid & Chandler 1926) in *Trachycarpus* because, "the margins of the pinnules are thickened and there is no marked midrib." A fruit containing a seed, from the Lower Eocene (London Clay) of southern England, is compared with *Trachycarpus* fruit (Chandler 1978). Some *Monocolpopollenites* pollen from the Tertiary of Hungary (Kedves & Bohony 1966) and monocolpate pollen from the Lower Miocene of Poland (Macko 1957) have been compared to *Trachycarpus* pollen but the pollen is of too general a coryphoid type to be conclusive.

Notes: The follicular carpels, with open ventral sutures, are among the least specialised in the family.

Additional figures: Fig. 1.1.

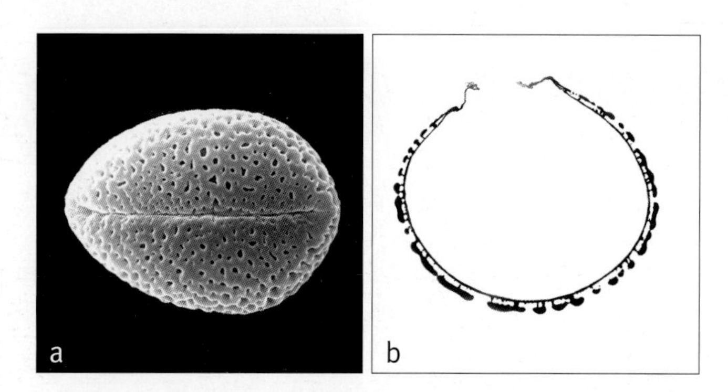

Above: ***Trachycarpus***
a, monosulcate pollen grain, distal face SEM × 2000;
b, ultrathin section through whole pollen grain, polar plane TEM × 2250.
Trachycarpus fortunei: **a**, *Gamble* 22733, **b**, *Forrest* 21168.

Top left: ***Trachycarpus martianus***, habit, Nepal.
(Photo: M. Gibbons & T. Spanner)

Bottom left: ***Trachycarpus martianus***, infructescences, Nepal.
(Photo: M. Gibbons & T. Spanner)

38. RHAPIDOPHYLLUM

Clustering ± stemless dioecious or polygamous fan palm of southeastern USA, immediately distinguished by the sharp leaf sheath spines and the leaf blade divided into segments between the folds; inflorescences short, hidden among the sheath spines.

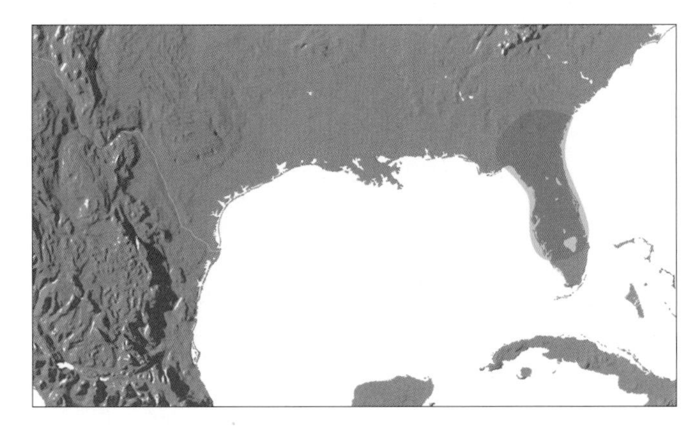

Distribution of ***Rhapidophyllum***

Rhapidophyllum H. Wendl. & Drude in Drude, Bot. Zeit. 34: 803 (1876). Type: **R. hystrix** (Pursh) H. Wendl. & Drude (*Chamaerops hystrix* Pursh).

Combines the palm generic name Rhapis *with* phyllon — *leaf, perhaps in allusion to the similarity in leaf splitting in the two genera.*

Small to moderate, clustering, armed, pleonanthic, dioecious, polygamodioecious or rarely monoecious palm. *Stem* very short, decumbent or erect, upper $^3/_4$ or more covered with needle-like spines and fibres, basally becoming bare, ringed with leaf scars. *Leaves* induplicate, shortly costapalmate, marcescent; sheath persistent, soft-fibrous with protruding elongate, stout, needle-like spines; petiole adaxially flat, abaxially rounded, the margins entire, sharp, rough; adaxial hastula very short, triangular to truncate, rounded, membranous, abaxial hastula lacking; blade very shortly costapalmate, narrow basally, divided between the folds nearly to the base into linear, mostly 2–4-veined, ± stiff segments, each composed of part of 1 adaxial and 1 abaxial fold and having 2 large ribs, 1 near a margin, tips ± rounded, shortly bifid, adaxial surface with small deciduous scales and wax, abaxial surface with white-waxy small brown scales, sometimes both surfaces glaucous, transverse veinlets inconspicuous. *Inflorescences* interfoliar, staminate and pistillate similar but pistillate stouter (polygamous not seen), very short, scarcely or not exserted from the leaf sheaths, usually once-branched, rarely unbranched; peduncle short; prophyll laterally 2-keeled, inflated, chartaceous, opening distally; peduncular bracts 4–6, short, slightly inflated, tubular at the base, splitting apically, rachis about equalling the peduncle at maturity in pistillate inflorescences, longer in staminate inflorescences, bearing spirally arranged, very small, narrow, elongate bracts subtending rachillae; rachillae short, bearing similar but smaller bracts, each subtending 2–4 small flowers in only slightly elevated cincinni. *Staminate flowers* globose, predominantly in clusters of 3; sepals 3, distinct to slightly connate at the base, deltoid; petals 3, distinct, fleshy, imbricate, ovate, acute; stamens 6, filaments distinct, slender, slightly exceeding the petals, anthers linear-oblong, dorsifixed near the base, latrorse; pistillodes 3, minute, resembling the carpels. *Pollen* ellipsoidal, with slight to obvious asymmetry; aperture a distal sulcus; ectexine tectate, coarsely perforate, aperture margin finely

Below left: ***Rhapidophyllum hystrix***, habit, cultivated, Fairchild Tropical Botanic Garden, Florida. (Photo: J. Dransfield)

Right: ***Rhapidophyllum hystrix***, inflorescence, cultivated, Fairchild Tropical Botanic Garden, Florida. (Photo: C.E. Lewis)

perforate; infratectum columellate; longest axis 23–29 µm [1/1]. *Pistillate flowers* similar to the staminate, predominantly paired; perianth as in the staminate; sterile stamens 6, shorter than the fertile; carpels 3, distinct, follicular, basally tomentose, styles short, recurved, terminating in punctiform stigmas, ovule erect, basal, hemianatropous. *Fruit* globose to globose-ovoid, covered with deciduous hairs, brownish, slightly flattened on ventral surface with ventrally eccentric, apical stigmatic scar; epicarp drying in a scale-like pattern, mesocarp thin, fleshy, sweet, endocarp cartilaginous. *Seed* narrowly ovoid to elliptical, laterally attached, becoming free from endocarp, hilum basal, raphe ventral, prominent, unbranched, endosperm homogeneous, seed coat somewhat thickened and intruded below the raphe; embryo lateral opposite the raphe. *Germination* remote-ligular (Chavez 2003); eophyll simple, lanceolate, tip truncate. *Cytology*: 2n = 36.

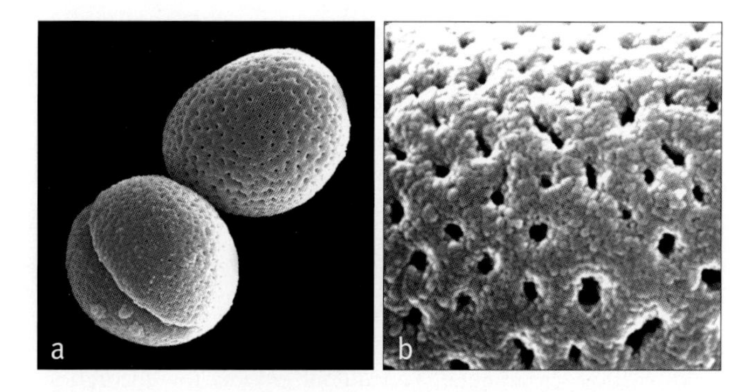

Above: ***Rhapidophyllum***
a, two pollen grains, distal face left, proximal face right SEM × 1000;
b, close-up, coarsely perforate pollen surface SEM × 6000.
Rhapidophyllum hystrix: **a** & **b**, *Curtiss* 2680.

Distribution and ecology: One species in southeastern USA. *Rhapidophyllum hystrix* is found in low, moist to wet areas with rich humus, calcareous clay, or sandy soils in woods and swamps. It may occur in limestone sinks and shaded pinelands. It will thrive in well-drained sites in the sun if sufficient moisture is provided. *Rhapidophyllum* grows very slowly in the wild, presumably because of low light conditions. Growth is accelerated when adequate light is provided but the long period (four to six years) taken to grow a mature plant from seed has discouraged its cultivation. Reproduction also occurs by suckering. The lower end of the stem decays as the palm grows; the rotting often separates suckers from the main plant but also tends to eliminate anchoring roots, causing the trunk to lean. During development, flowers and seeds are well protected by the sheath spines. It is unclear how seeds are dispersed and they often germinate close to their parent, where seedlings have little chance of survival. The spines are so conspicuous that *Rhapidophyllum* has been called the 'vegetable porcupine' (Small 1923). Pollination appears to be primarily by an undescribed species of *Notolomus* (Curculionidae). Beetles are attracted to both staminate and pistillate flowers by a musky odor and feed on pollen and flower parts. The species seems self-compatible, although pollinator visits have not been completely deciphered (Shuey & Wunderlin 1977).

Anatomy: Leaf (Tomlinson 1961), roots (Seubert 1997), floral (Morrow 1965), fruit (Murray 1973).

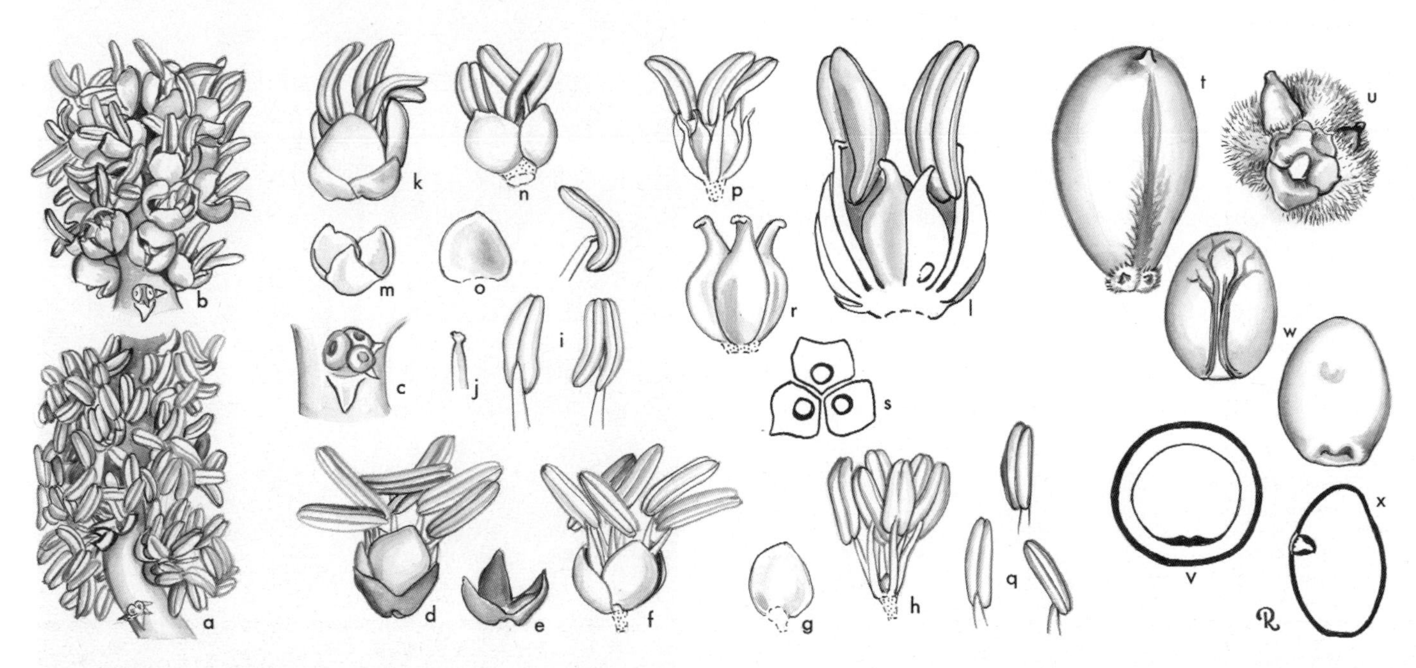

***Rhapidophyllum*.** **a**, portion of rachilla with staminate flowers × 3; **b**, portion of rachilla with hermaphrodite or pistillate flowers × 3; **c**, floral scars and bracts × 6; **d**, staminate flower × 6; **e**, staminate calyx × 6; **f**, staminate flower, calyx removed × 6; **g**, staminate petal × 6; **h**, androecium × 6; **i**, stamen in 3 views × 6; **j**, pistillode × 6; **k**, pistillate flower × 6; **l**, pistillate flower in vertical section × 12; **m**, pistillate calyx × 6; **n**, pistillate flower, calyx removed × 6; **o**, pistillate petal × 6; **p**, pistillate flower, perianth removed × 6; **q**, staminodes in 3 views × 6; **r**, gynoecium × 6; **s**, carpels in cross-section × 6; **t**, fruit × 1$^1/_2$; **u**, base of fruit and abortive carpels × 3; **v**, fruit in cross-section × 1$^1/_2$; **w**, seed in 2 views × 1$^1/_2$; **x**, seed in vertical section × 1. *Rhapidophyllum hystrix*: **a–s**, *Read* 773; **t–x**, *Moore* 7822. (Drawn by Marion Ruff Sheehan)

Relationships: The genus appears to be related to *Guihaia*, *Maxburretia* and *Rhapis*, but its exact position with respect to these genera is unclear. Baker *et al.* (in review) place *Rhapidophyllum* as sister to a clade of *Guihaia*, *Maxburretia* and *Rhapis* with low support. Asmussen *et al.* (2006) place the genus as sister to a clade of *Maxburretia* and *Rhapis* with low support. Uhl *et al.* (1995) resolve it as sister to a clade of *Guihaia* and *Rhapis*. There is also moderate support for a sister relationship between *Rhapidophyllum* and *Rhapis* (Asmussen & Chase 2001).
Common names and uses: Needle palm. Cultivated as a cold-tolerant ornamental. Cutting of crowns for decoration during the late 19th and early 20th century decimated many populations and some exploitation by nurseries continues.
Taxonomic accounts: Shuey & Wunderlin (1977), Small (1923) and Zona (1997).
Fossil record: A study of the Middle Eocene Princeton chert of British Columbia, Canada (Erwin & Stockey 1991) shows palm vegetative organs to be the most common elements: five stems up to 9cm wide with attached petiole bases and roots, plus numerous additional isolated petioles, midribs and laminae. A comparison with extant palms suggests that fossil stem and leaf anatomy is most similar to two coryphoid genera, *Rhapidophyllum* and *Brahea*.
Notes: *Rhapidophyllum hystrix* is considered a relict genus (Shuey & Wunderlin 1977).
Additional figures: Glossary fig. 9.

39. MAXBURRETIA

Dwarf clustering ± stemless or erect dioecious fan palms of karst limestone in southern Thailand and Peninsular Malaysia, with fibrous sheaths, the fibres sometimes spine-like; leaves induplicate; inflorescences slender with very small flowers and fruit.

Maxburretia Furtado, Gardens' Bull. Straits Settlem. 11: 240 (1941). Type: **M. rupicola** (Ridl.) Furtado (*Livistona rupicola* Ridl.).

Livistona subgenus *Livistonella* Becc., Webbia 5: 16 (1921). Type: *Livistona rupicola* Ridl.

Symphyogyne Burret, Notizbl. Bot. Gart. Berlin-Dahlem 15: 316 (1941). Type: *S. gracilis* Burret (= *Maxburretia gracilis* [Burret] J. Dransf.) (non *Symphyogyna* Nees & Mont. [1836]).

Liberbaileya Furtado, Gard. Bull. Straits Settlem. 11: 238 (1941). Type: *L. lankawiensis* Furtado (= *Maxburretia gracilis* [Burret] J. Dransf.).

Commemorates German palm botanist Karl Ewald Maximilian Burret (1883–1964).

Small, clustering, acaulescent or shrubby, unarmed, pleonanthic, hermaphroditic or dioecious palms. *Stem* moderate if present, with very close leaf scars, usually completely obscured by persistent leaf sheaths. *Leaves* induplicate, palmate, marcescent; sheath expanding into a mass of discrete fibres, irregular or neatly joined at the tips opposite the petiole,

Top right: ***Maxburretia furtadoana***, habit, Thailand. (Photo: J. Dransfield)

Bottom right: ***Maxburretia furtadoana***, fibre spines and staminate inflorescences, Thailand. (Photo: J. Dransfield)

or developed as rigid spines; petiole well developed, unarmed, ± semicircular in cross-section; adaxial hastula ± triangular or rounded, sometimes hairy, abaxial hastula obscure; blade neatly divided to ca. $^2/_3$ of its radius into slender, single-fold, usually glaucous segments, tips shallowly split along the folds, surfaces often slightly dissimilar, scattered scales sometimes present on abaxial surface, midribs prominent abaxially, transverse veinlets obscure. *Inflorescences* solitary, interfoliar, arching out of the crown, branching to 1–3 orders; prophyll tubular, 2-keeled, narrow, elongate, usually obscured by the leaf sheaths; peduncular bracts 1–3 or more, similar to the prophyll; rachis bracts closely tubular with triangular limbs, each subtending a first-order branch; subsequent orders of bracts minute, inconspicuous; rachillae slender, bearing distant, spirally arranged, minute, triangular bracts subtending solitary or, rarely, groups of 2–3 flowers. *Flowers* very small; where plants dioecious, staminate and pistillate flowers superficially similar; sepals 3, distinct, imbricate, ovate or triangular, glabrous; petals 3, joined for $^1/_3$ to $^1/_2$ their length at the base, somewhat imbricate in midportion, valvate near the tips, elongate, usually with somewhat thickened tips; stamens in staminate and hermaphroditic flowers 6, adnate to the petals, the filaments forming a thin or thick staminal cupule, or distinct, anthers rather short, latrorse; staminodes in pistillate flower similar to the stamens but with thinner cupule and smaller, empty anthers; carpels 3, distinct, follicular, united for a very short distance at the base, with triangular style, the carpel surface hairy distally in 2 species, glabrous in the third, ovules basally attached, anatropous or intermediate between anatropous and hemianatropous with basal funicular arils; pistillode of staminate flower minute, 3-lobed. *Pollen* ellipsoidal, usually bi-symmetric; aperture a distal sulcus; ectexine tectate, perforate, or perforate-rugulate, aperture margin slightly finer; infratectum columellate; longest axis 15–19 µm [3/3]. *Fruit* usually developing from only 1 carpel, ellipsoidal (?always), with apical stigmatic remains, perianth whorls persistent; epicarp silky hairy when young, the hairs falling off at maturity (?always), mesocarp thin, fleshy, endocarp scarcely developed. *Seed* basally attached, endosperm homogeneous, with a thin lateral intrusion of seed coat; embryo lateral opposite the intrusion. *Germination* not known; eophyll simple, entire, plicate. *Cytology* not studied.

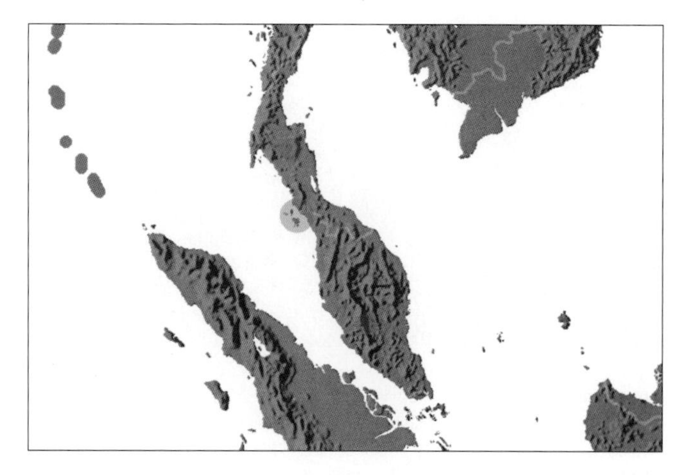

Distribution of ***Maxburretia***

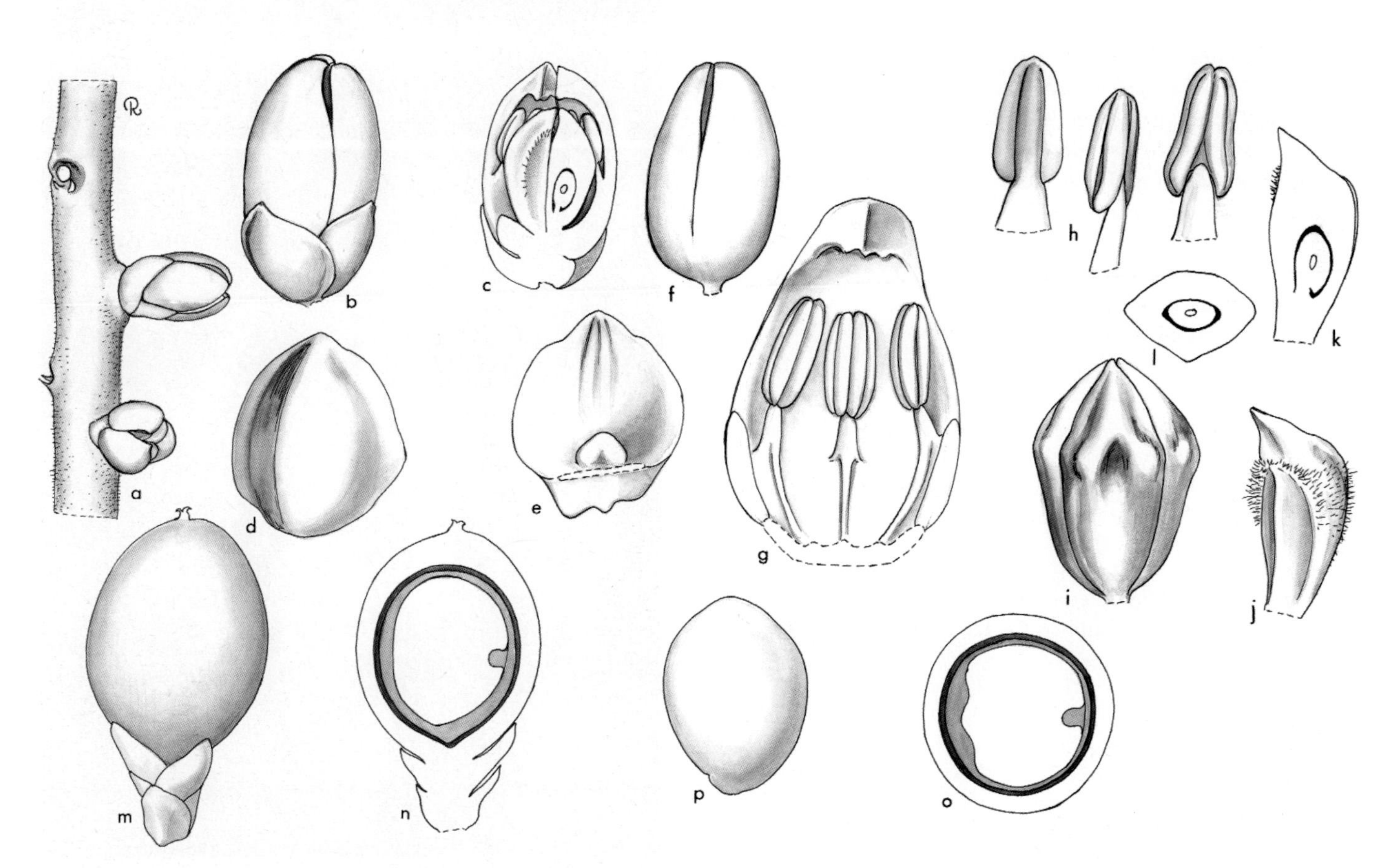

Maxburretia. **a**, portion of rachilla × 6; **b**, flower × 12; **c**, flower in vertical section × 12; **d**, sepal, exterior view × 24; **e**, sepal, interior view × 24; **f**, flower, corolla removed × 12; **g**, petal and stamens, interior view × 24; **h**, stamen in 3 views × 24; **i**, gynoecium × 16; **j**, carpel, view × 24; **k**, carpel in vertical section × 24; **l**, carpel in cross-section × 24; **m**, fruit × 6; **n**, fruit in vertical section × 6; **o**, fruit in cross-section × 6; **p**, seed × 6. *Maxburretia rupicola*: **a–l**, *Whitmore* FRI 699; **m–p**, *Dransfield* 910. (Drawn by Marion Ruff Sheehan)

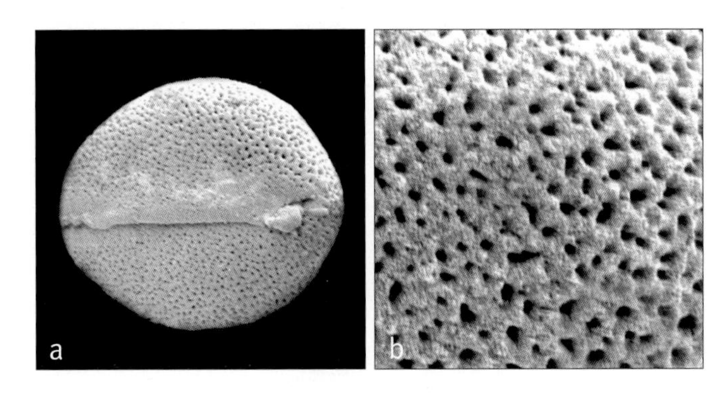

Above: ***Maxburretia***
a, monosulcate pollen grain, distal face SEM × 2000;
b, close-up, coarsely perforate pollen surface SEM × 8000.
Maxburretia furtadoana: **a** & **b**, *Dransfield* JD2349.

Distribution and ecology: Three species in West Malaysia and Peninsular Thailand: *Maxburretia rupicola* at Batu Caves, Bukit Takun and Bukit Anak Takun in Selangor, *M. gracilis* on Pulau Dayang Bunting in the Langkawi Islands and one locality in southern Thailand, and *M. furtadoana* at a few localities near Surat Thani. All three species are palms of the low forest on exposed sides and summits of limestone hills.
Anatomy: Leaf anatomy not studied; all species very similar in floral anatomy (Uhl 1978a).
Relationships: The monophyly of *Maxburretia* has not been tested. *Maxburretia* is resolved as sister to *Rhapis* with low support (Asmussen *et al.* 2006) or as sister to a clade of *Guihaia* and *Rhapis* with low support (Baker *et al.* in review).
Common names and uses: Serdang-batu. Uses not recorded.
Taxonomic account: Dransfield (1978a).
Fossil record: No generic records found.
Notes: The three species of *Maxburretia* survive as relics on limestone hills in Southeast Asia.

40. RHAPIS

Slender clustering dioecious or polygamous fan palms with reed-like stems, from Southern China to Thailand and North Sumatra, often on limestone; instantly recognizable from the leaves divided between the folds into segments.

Rhapis L.f. ex Aiton, Hort. Kew. 3: 473 (1789). Lectotype: *R. flabelliformis* L'Heritier ex W. Aiton (illegitimate name) (*Chamaerops excelsa* Thunberg = **R. excelsa** (Thunberg) A. Henry ex Rehder).

Rhapis — *rod, presumably alluding to the rod-like slender stems.*

Small, clustering, unarmed, pleonanthic, dioecious or polygamous palms. *Stems* slender, reed-like, erect, covered with persistent leaf sheaths, eventually becoming bare, conspicuously ringed with leaf scars. *Leaves* induplicate, palmate, marcescent, divided to the base or to ca. $^3/_4$ the radius between the folds into several-ribbed segments, apices divided along and between the folds to form shallow teeth; sheath composed of numerous, interwoven, black or grey-brown fibres, when young bearing sparse, caducous brown indumentum; petiole elongate, slender, ± elliptic in cross-section, margins smooth; adaxial hastula small, ± triangular,

Top: ***Rhapis subtilis***, cultivated, Fairchild Tropical Botanic Garden, Florida. (Photo: J. Dransfield)
Bottom: ***Rhapis laosensis***, staminate inflorescence, Laos. (Photo: J. Dransfield)

sometimes tomentose, abaxial hastula absent; blade palmate to deeply bifid, segments usually variable in number of ribs, position of splits precise, usually at a position about 2/3 the width of the interfold nearer the abaxial fold, the segment margins minutely toothed, blade glabrous, transverse veinlets conspicuous. *Inflorescences* interfoliar, usually very short, branching to 1–2 orders in pistillate, up to 3 orders in staminate; peduncle short, frequently entirely enclosed by the leaf sheaths; prophyll tubular, 2-keeled, splitting along the abaxial midline; peduncular bracts absent; rachis longer than the peduncle, bearing 1–2 large, tubular, single-keeled bracts, distal rachis bracts much smaller; rachis bracts each subtending a first-order branch adnate to the axis above the bract node and bearing very inconspicuous, narrow triangular bracts subtending second-order branches; second-order branches adnate to the first-order branches; rachillae glabrous or hairy, lax, spreading in pistillate and polygamous inflorescences, more crowded in staminate, rachillae bearing spirally arranged, solitary or rarely paired flowers in the axils of minute apiculate bracts. *Staminate flowers* symmetrical; calyx cup-shaped, thick, shallowly 3-lobed distally, the lobes somewhat irregular, triangular, glabrous or hairy; corolla fleshy, tubular, inserted above calyx and appearing ± stalked basally, the 3 lobes ± triangular, valvate, usually very short, sometimes ciliate at the margins; stamens 6, filaments elongate, but adnate along ± the entire length of the corolla tube, free at their very tip, anthers short, rounded, latrorse; pistillode minute, 3-lobed. *Pistillate and rare hermaphroditic flowers* superficially similar; sepals as in staminate but more fleshy; corolla with a short or long stalk-like base, a short tube, and 3 fleshy triangular, basally imbricate, apically valvate, fleshy lobes; staminodes as the stamens but smaller and with empty anthers; carpels 3,

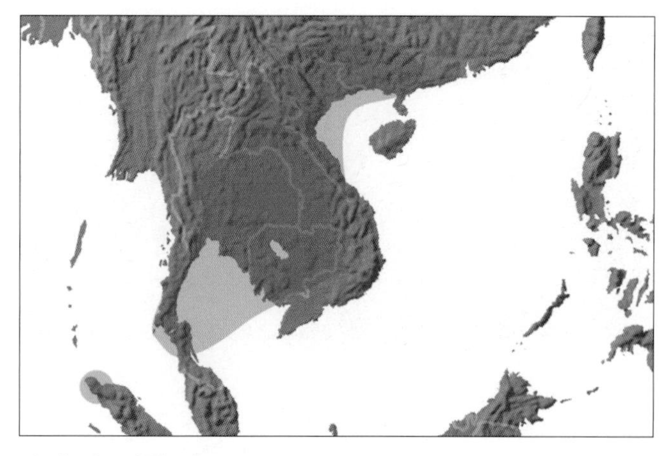

Distribution of ***Rhapis***

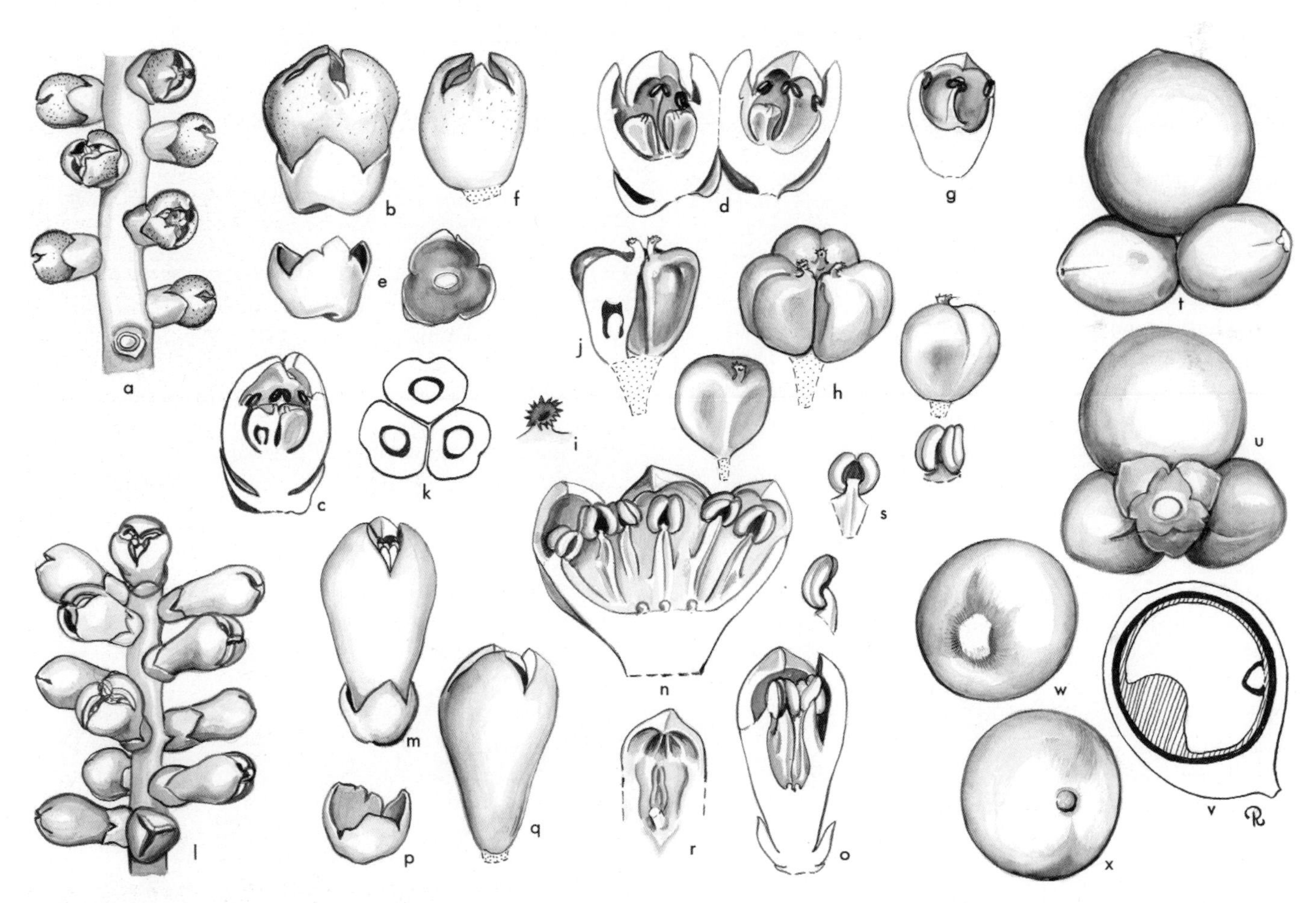

Rhapis. **a**, portion of pistillate rachilla × 3; **b**, pistillate flower × 6; **c**, pistillate flower in vertical section × 6; **d**, pistillate flower, expanded × 6; **e**, pistillate calyx, exterior and interior views × 6; **f**, pistillate flower, calyx removed × 6; **g**, pistillate petal and staminodes, interior view × 6; **h**, single carpel in 2 views and whole gynoecium × 12; **i**, stigma × 24; **j**, gynoecium in vertical section × 12; **k**, carpels in cross-section × 12; **l**, portion of staminate rachilla × 3; **m**, staminate flower × 6; **n**, staminate flower, expanded × 6; **o**, staminate flower in vertical section × 6; **p**, staminate calyx × 6; **q**, staminate flower, calyx removed × 6; **r**, apical portion of staminate petal with base of stamen, interior view × 6; **s**, stamen in 3 views × 6; **t**, **u**, fruit with 1 mature and 2 partially developed carpels from above and below × 3; **v**, fruit in vertical section × 3; **w**, **x**, seed in 2 views × 3. *Rhapis excelsa*: all from *Read* 774. (Drawn by Marion Ruff Sheehan)

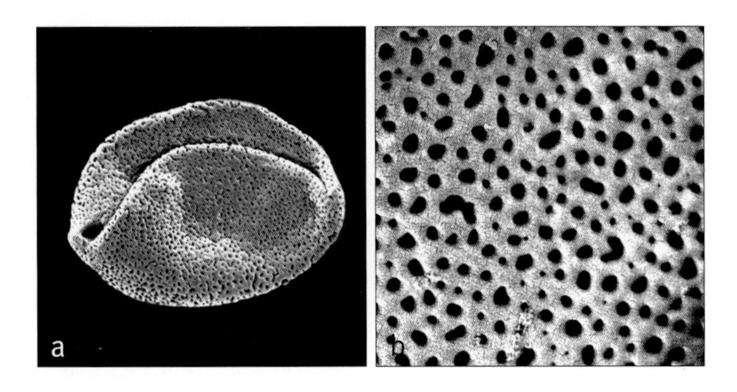

Above: ***Rhapis***
a, monosulcate pollen grain, equatorial face SEM × 1800;
b, close-up, coarsely perforate pollen surface SEM × 8000.
Rhapis humilis: **a** & **b**, *Fitt* s.n.

distinct, wedge-shaped, each with a short apical style, distally expanded into a conduplicate, fimbriate, tube-shaped stigma, ovules basally attached, 1 in each carpel, hemianatropous, with a basal fleshy aril. *Pollen* ellipsoidal, usually with slight to obvious asymmetry; aperture a distal sulcus; ectexine tectate, coarsely perforate, aperture margin slightly finer; infratectum columellate; longest axis 21–40 µm; post-meiotic tetrads tetrahedral [2/8]. *Fruit* usually developing from 1 carpel with apical stigmatic remains, more rarely 2 or 3 carpels developing, sometimes the stalk-like corolla base persisting and becoming a sub-woody fruit stalk; epicarp becoming purplish-brown or white, mesocarp fleshy, somewhat fibrous, endocarp thin, brittle. *Seed* with short lateral raphe, endosperm homogeneous, laterally penetrated by the seed coat; embryo subbasal or lateral. *Germination* remote-tubular; eophyll entire, slender, strap-shaped, plicate. *Cytology*: 2n = 36, 4n = 72.

Distribution and ecology: About eight species in southern China, southwards through Indochina to peninsular Thailand, one species in northernmost Sumatra. Undergrowth palms of dry evergreen forest; *Rhapis subtilis* and some other species seem to be confined to forest on limestone hills.

Anatomy: Leaf (Tomlinson 1961); plication development in the leaf (Kaplan *et al.* 1982b); stem vasculature (Zimmermann & Tomlinson 1965), roots (Seubert 1997), floral (Uhl *et al.* 1969).

Relationships: The monophyly of *Rhapis* has not been tested. There is high support for a sister relationship with *Guihaia* (Uhl *et al.* 1995, Baker *et al.* in review), moderate support for a sister relationship with *Rhapidophyllum* (Asmussen & Chase 2001) and low support for a sister relationship with *Maxburretia*.

Common names and uses: Lady palms. Widely grown as ornamentals; many dwarf varieties have been developed in Japan. See also McKamey (1983). Stems are used as sticks and canes.

Taxonomic account: Hastings (2003).

Fossil record: No generic records found.

Notes: *Rhapis* is distinguished by the leaf with few to manyfold truncate segments and divisions between the folds, and by the fleshy flowers with sepals and petals united basally, and stamens borne on the corolla.

Subtribe Livistoninae Saakov, Palms and their culture in USSR 193 (1954). Type: **Livistona**.

Leaves induplicate, palmate or costapalmate, or entire, or with complete secondary abaxial splits; flowers solitary or borne in simple cincinni; gynoecium syncarpous with three carpels free at their bases but united by their styles, with separate stylar canals, or stylar canals distally briefly united; large vascular bundes of petiole usually with 2 phloem strands.

This subtribe is widely distributed throughout the tropics and subtropics. The differences between the genera are mostly small. Although the leaves are usually palmate or costapalmate, there are two very distinctive leaf forms unknown elsewhere in the Coryphoideae — the extraordinary diamond-shaped leaf of *Johannesteijsmannia*, and the wedge-shaped segments with 'reduplicate' margins present in almost all species of *Licuala*. A gynoecium of three carpels connate by their styles is present throughout; stylar canals may be separate or shortly united distally.

Subtribe Livistoninae is monophyletic with low support (Asmussen *et al.* 2006) or with moderate support (Asmussen & Chase 2001, Baker *et al.* in review). For relationships, see Trachycarpeae. The relationships between genera remain poorly understood.

41. LIVISTONA

Usually tall, single-stemmed fan-palms of the Horn of Africa and Arabia, and Himalayas to Australia; there are a few dwarf species; most are hermaphroditic but a few dioecious species are known.

Livistona R. Br., Prodr. 267 (1810). Lectotype: **L. humilis** R. Br. (see H.E. Moore 1963c).

Saribus Blume, Rumphia 2: 48 (1838 ['1836']). Lectotype: *S. rotundifolius* (Lam.) Blume (*Corypha rotundifolia* Lam.) = *Livistona rotundifolia* (Lam.) Mart. (see H.E. Moore 1963c).

Wissmannia Burret, Bot. Jahrb. Syst. 73: 184 (1943). Type: *W. carinensis* (Chiov.) Burret (*Hyphaene carinensis* Chiov.) (= *Livistona carinensis* [Chiov.] J. Dransf. & N. Uhl).

Honours Patrick Murray, Baron Livingstone, who laid out a garden on his estate at Livingstone, west of Edinburgh, Scotland, in the latter part of the seventeenth century.

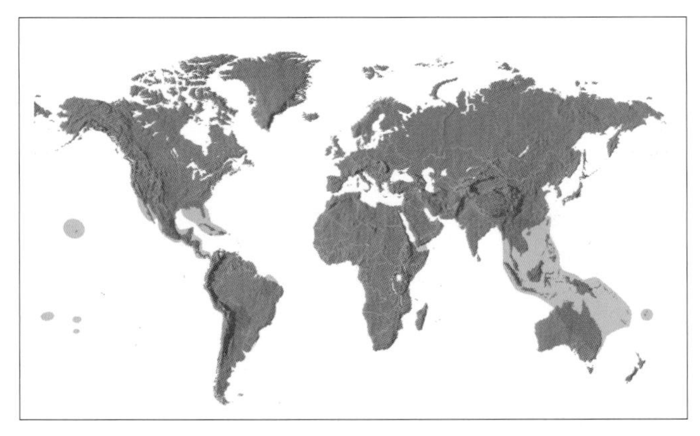

Distribution of subtribe ***Livistoninae***

Slender (rarely) to robust, solitary, armed or unarmed, pleonanthic, hermaphroditic (rarely dioecious), shrub or tree palms. *Stem* erect, obscured at first by persistent sheaths, later becoming bare or covered with persistent petiole bases, conspicuously or obscurely ringed with leaf scars. *Leaves* induplicate, palmate or costapalmate, marcescent or deciduous under their own weight, a skirt of dead leaves sometimes developing; sheath disintegrating into a conspicuous interwoven, often cloth-like, reddish brown mass of broad and fine fibres; petiole well developed, grooved or flattened adaxially, rounded or angled abaxially, sparsely covered with indumentum or not, expanded and sometimes bulbous at the occasionally persistent base, the margins unarmed or armed with inconspicuous to robust horizontal spines or teeth; adaxial hastula well developed, abaxial hastula poorly developed or absent; blade divided along adaxial ribs to varying depths to form single or, very rarely, multiple-fold segments, these further divided for a short to long distance along abaxial folds near the tip, rarely the adaxial splits almost reaching the hastula and the costa, the segments then all single-fold and very fine; segments stiff or pendulous, interfold filaments sometimes present, scattered caducous indumentum present along ribs, wax sometimes present on the abaxial surface, more rarely waxy on both surfaces, midribs conspicuous, transverse veinlets obscure or conspicuous. *Inflorescences* interfoliar, solitary, branched to 5 orders, sometimes immediately trifurcating to give 3 equal 'inflorescences' enclosed within a common prophyll, each branch with its own prophyll (e.g. *L. rotundifolia*); peduncle elongate; prophyll 2-keeled, tubular, closely sheathing, variously covered with indumentum or not, frequently tattering at the tip; peduncular bracts 1–few, tubular, like the prophyll; rachis usually longer than the peduncle; rachis bracts variously covered with indumentum, each subtending a first-order branch; bracts of subsequent orders generally inconspicuous; rachillae erect, pendulous or divaricate, glabrous or hairy, usually numerous, bearing spirally arranged flowers, singly or in cincinni of up to 5, sessile or on low tubercles or slender stalks, each group subtended by a minute rachilla bract and each flower bearing a minute bracteole. *Flowers* small to very small, usually cream-coloured; calyx with receptacle often producing a short, broad stalk, tubular above, tipped with 3 triangular lobes, these sometimes imbricate at the very base, glabrous or hairy; corolla shallow, tubular at the base, apically with 3 triangular, valvate lobes; stamens 6, epipetalous, the filaments connate to form a fleshy ring, tipped with short, slender distinct filaments, anthers medifixed, rounded or oblong, latrorse; gynoecium tricarpellate, the carpels wedge-shaped, distinct in the ovarian region, connate distally to form a common, slender style, with an apical, dot-like or minutely 3-lobed stigma, ovule basally attached, anatropous; where dioecious, anthers or ovules not developing but otherwise as in the hermaphroditic. *Pollen* ellipsoidal, bi-symmetric, occasionally slightly asymmetric; aperture a distal sulcus; ectexine tectate, psilate and sparsely perforate, finely perforate, perforate, perforate-rugulate, foveolate or finely reticulate, aperture margin sometimes slightly finer; infratectum columellate; longest axis 19–37 µm [10/33]. *Fruit* usually developing from 1 carpel, globose to ovoid, pyriform, or ellipsoidal, small to medium-sized, variously coloured, green,

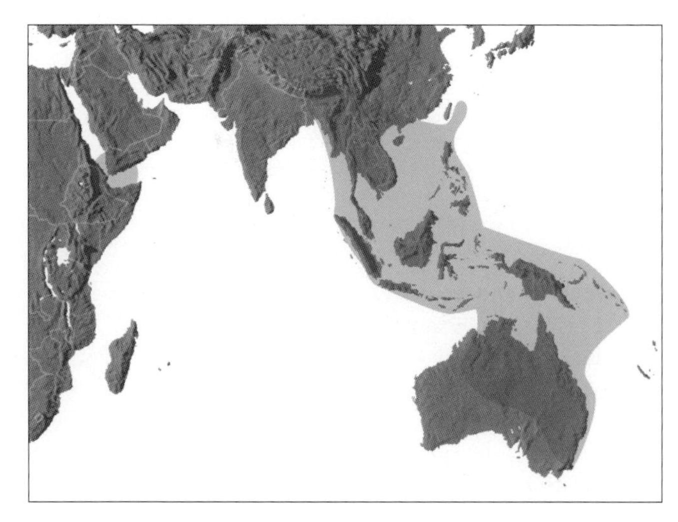

Distribution of ***Livistona***

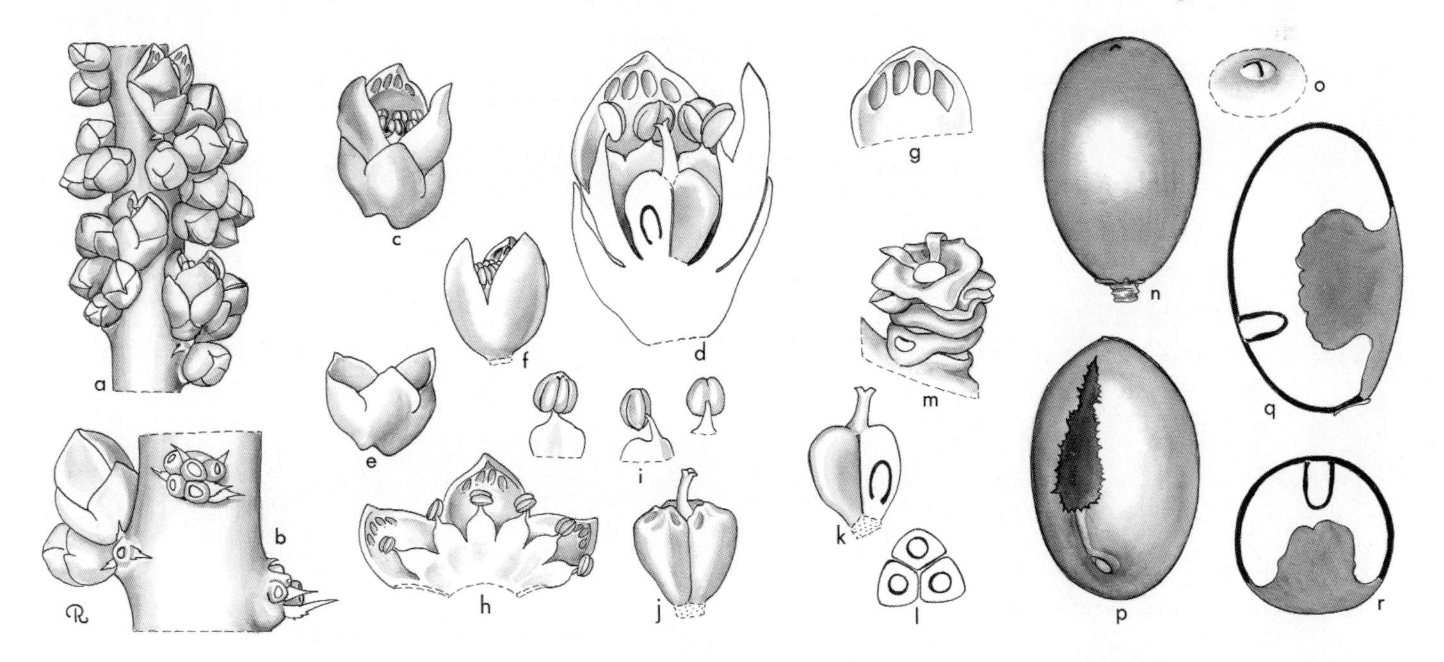

Livistona. **a**, portion of rachilla × 3; **b**, detail of rachilla × 6; **c**, flower × 6; **d**, flower in vertical section × 12; **e**, calyx × 6; **f**, flower, calyx removed × 6; **g**, corolla lobe × 12; **h**, corolla and androecium, expanded × 6; **i**, stamen in 3 views × 12; **j**, gynoecium × 12; **k**, gynoecium in vertical section × 12; **l**, gynoecium in cross-section × 12; **m**, fruiting perianth and scars of flowers × 6; **n**, fruit × $1^{1}/_{2}$; **o**, stigmatic remains, enlarged; **p**, seed × 3; **q**, seed in vertical section × 3; **r**, seed in cross-section × 3. *Livistona chinensis*: all from *Read* 712. (Drawn by Marion Ruff Sheehan)

scarlet, blue-green, blue-black, black or dark brown, stigmatic remains apical, sterile carpel remains basal; epicarp smooth, dull or shining, often with a wax bloom, mesocarp thin or thick, fleshy or dry, somewhat fibrous, usually easily separated from the bony or woody endocarp. *Seed* ellipsoidal or globose, basally attached, hilum circular or ± elongate, raphe branches few or lacking, endosperm homogeneous, penetrated laterally by a variable, frequently convoluted intrusion of seed coat; embryo lateral. *Germination* remote-tubular; eophyll lanceolate, plicate, minutely toothed apically. *Cytology*: 2n = 36.

Distribution and ecology: About 35 species, ranging from the Horn of Africa and Arabia (*Livistonia carinensis*), to the Himalayas and Ryukyu Islands, south through Indochina and Malesia to New Guinea, the Solomon Islands and Australia, where there is a great diversity of species. The ecology is very varied. There are species adapted to fresh water and peat swamp forest (*L. saribus*), montane forest (*L. tahanensis* and *L. jenkinsiana*), undergrowth of tropical rain forest (*L. exigua*), dry savannah woodland (*L. humilis* and *L. lorophylla*), canyon bottoms with more or less permanent water in desert areas (*L. mariae* and *L. carinensis*), and subtropical woodland (*L. chinensis* and *L. australis*). Species are frequently gregarious, the tallest species often occurring in spectacularly beautiful groves (e.g., *L. rotundifolia* in Celebes and elsewhere).

Anatomy: Leaf (*L. australis* and *L. chinensis*; Tomlinson 1961), roots (Seubert 1997), floral (Morrow 1965); stegmata (Killmann & Hong 1989).

Relationships: The monophyly of *Livistona* has not been tested. *Livistona* is resolved as sister to the rest of the Livistoninae with low support (Asmussen *et al.* 2006, Baker *et al.* in review).

Common names and uses: Cabbage palm (*Livistona australis*), Chinese fan palm (*L. chinensis*), *serdang* (West Malesian species). Many are planted as ornamentals. Leaves of several species are used for thatch, their segments for umbrellas, and fibres for rope and cloth. Trunks have been used for wood. The 'cabbage' of *L. australis* is edible.

Left: ***Livistona lanuginosa***, habit, Australia. (Photo: J. Dransfield)

Top right: ***Livistona decora***, crown, cultivated, Montgomery Botanical Center, Florida. (Photo: J. Dransfield)

Bottom right: ***Livistona chinensis***, infructescences, cultivated, Montgomery Botanical Center, Florida. (Photo: J. Dransfield)

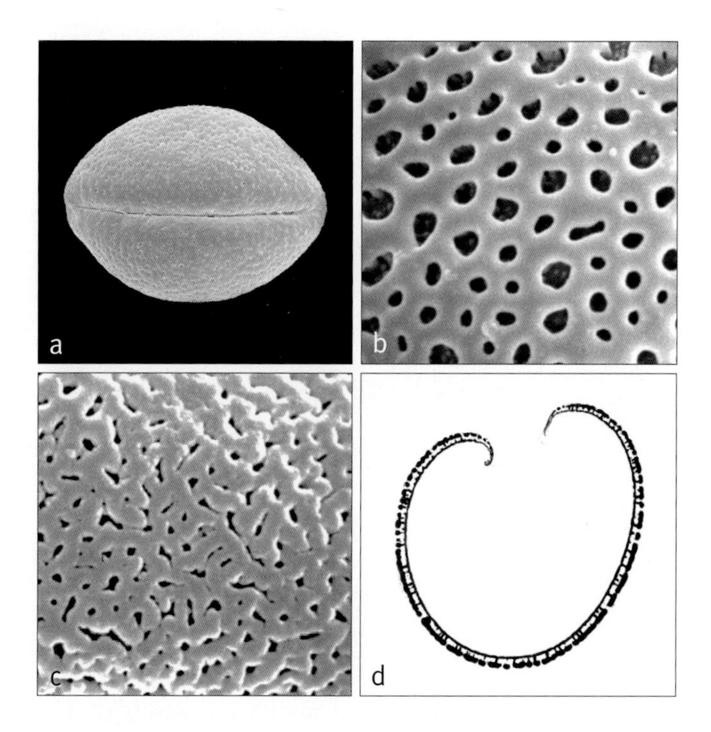

Above: ***Livistona***
a, monosulcate pollen grain, equatorial face SEM × 2000;
b, close-up, foveolate-reticulate pollen surface SEM × 8000;
c, close-up, finely perforate-rugulate pollen surface SEM × 8000;
d, ultrathin section through whole pollen grain, polar plane TEM × 2250.
Livistona humilis: **a**, *Latz* 3729; *L. jenkinsiana*: **b**, *Kerr* 5600; *L. chinensis*: **c**, *Melgrito* 10029; *L. cf. jenkinsiana*: **d**, *Kerr* 3430.

Taxonomic accounts: Beccari (1931). See also Dransfield & Uhl (1983b). The genus has recently been revised by John Dowe (in prep.).

Fossil record: Leaves from India, Maharashtra State, were described as *Sabalophyllum livistonoides* (Bonde 1986b) and compared with *Livistona* (although the age span of the volcanic deposits in which they were found is controversial, see Chapter 5). Reid & Chandler (1933) report seeds, *Livistona* (?)*minima* [sic], from the Eocene (London Clay); and a seed, *L. atlantica*, is described from Germany, Middle Eocene (Geiseltal) (Mai 1976: the diagnosis for *L. minima* [Reid & Chandler 1933] is emended). From the Czech Republic, Turów, Lower Miocene, a seed, *L. australis*, is reported (Czeczott & Juchniewicz 1975). Stem wood, *Palmoxylon*, is difficult to identify to generic level; however, the Deccan Intertrappean beds of India (Maharashtra State) have yielded *P. livistonoides* (Prakash & Ambwani 1980) and *P. arcotense* (Ramanujam 1953), both of which are compared with *Livistona*. It needs to be stated that the generic attribution of all of these records is doubtful. Pollen (Jarzen 1978) from the Maastrichtian of Canada (Saskatchewan) is comparable to that of *Livistona*, and at least some of the small perforate or finely reticulate monosulcate pollen grains described by Khin Sein (1961) from southern England, Lower Eocene (London Clay) are probably assigned correctly to *Livistona*. Small monosulcate grains from palm flower compression fossils, *Palmaemargosulcites fossperforatus*, recovered from the Middle Eocene oil shales of Messel, Germany, are compared with pollen of a number of coryphoid genera, including *Livistona* (Harley 1997).

Notes: A large and variable genus distinguished by flower structure, in particular by the gynoecium of three carpels connate only by their styles, by united sepals, by petals with internal grooves, by the usually small fruits with apical stigmatic remains and basal carpel remains, by seed with homogeneous endosperm, and by a large intrusion of seedcoat.

42. LICUALA

Very small to moderate, solitary or clustered, hermaphroditic or dioecious fan palms of Southeast Asia to the western Pacific and Australia, usually immediately recognisable by the leaf being divided along the abaxial folds all the way to the petiole into wedge-shaped segments; there are a few species with undivided leaves.

Licuala Thunb., Kong. Vetensk. Acad. Nya Handl. 3: 286 (1782). Type: **L. spinosa** Thunb.

Pericycla Blume, Rumphia 2: 47 (1838 ['1836']). Type: *P. penduliflora* Blume (= *Licuala penduliflora* [Blume] Miq.).

Dammera Lauterb. & K. Schum. in K. Schum. & Lauterb., Fl. Schutzgeb. Südsee 201 (1900 ['1901']). Lectotype: *D. ramosa* Lauterb. & K. Schum. (see H.E. Moore 1963c) (= *Licuala ramosa* (Lauterb. & K. Schum.) Becc. [non Blume 1850] = *L. beccariana* Furtado).

Latinization of the vernacular name, leko wala, *supposedly used for* Licuala spinosa *in Makassar, Sulawesi.*

Below: ***Licuala paludosa***, habit, Thailand. (Photo: P. Wilkin)

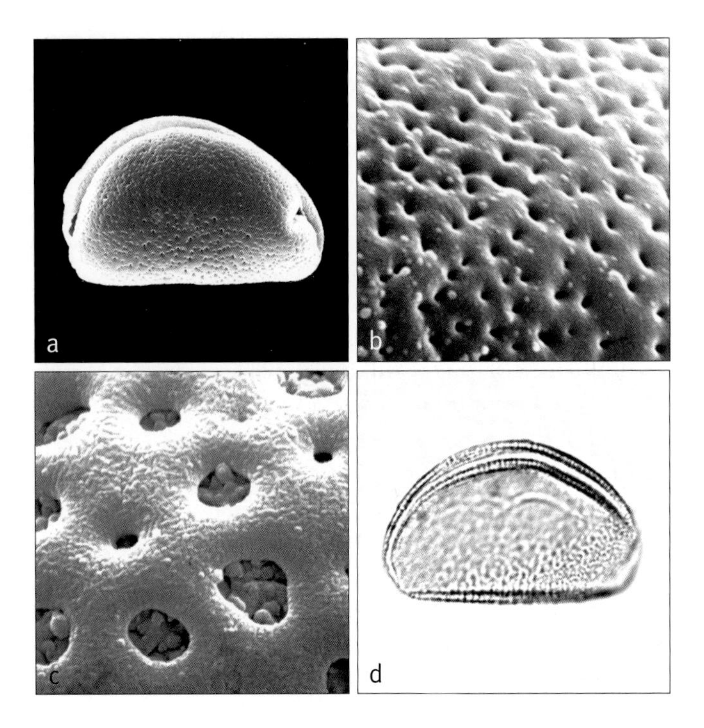

Left: ***Licuala***
a, extended sulcate pollen grain, equatorial face SEM × 2000;
b, close-up, perforate pollen surface SEM × 8000;
c, close-up, foveolate pollen surface SEM × 8000;
d, extended sulcate pollen grain, equatorial face, mid–low-focus LM × 2000.
Licuala kunstleri: **a**, *Kerr* 7572; *L. mattanensis* var. *mattanensis*: **b**, *Dransfield* JD6022; *L. peltata*: **c**, *Kerr* 11726; *L. fordiana*: **d**, *McClure* 20050.

Very small to moderate, solitary or clustered, acaulescent to shrubby, rarely tree-like, armed or unarmed, pleonanthic, hermaphroditic (very rarely dioecious) palms. *Stem* very short and subterranean, creeping or erect, ringed with close leaf scars, partly obscured by remains of leaf sheaths, sometimes bearing short bulbil-like shoots at the nodes. *Leaves* palmate, marcescent; leaf sheath disintegrating into a weft of fibres, the margin sometimes remaining as a broad, ligule-like ribbon or tongue; petiole adaxially channelled near the base, rounded or channelled distally, abaxially rounded or angled, armed along margins with close sharp teeth or triangular spines, or unarmed, caducous indumentum often abundant; adaxial hastula well developed, usually triangular, abaxial hastula absent; blade entire or split variously along the abaxial ribs to the very base to produce single to multiple-fold, wedge-shaped reduplicate segments, these in turn with very short splits along the abaxial folds and slightly longer splits along adaxial folds, the central segment usually entire, sometimes bifid, sometimes borne on a stalk-like extension, the ribs often with caducous indumentum, transverse veinlets usually conspicuous. *Inflorescences* interfoliar, much shorter to much longer than the leaves, very varied in aspect and degree of branching, from spicate to branched to 3 orders; peduncle short to very long, bearing a basal, 2-keeled tubular prophyll, and 0–5 or more, similar, tubular, closely sheathing or inflated, glabrous or tomentose, peduncular bracts; rachis bracts subtending usually distant, first-order branches adnate to the inflorescence axis above the bract mouth; subsequent orders of bracts minute; first-order branches spicate or branched further; rachillae few to ca. 30 or more, crowded or spreading, glabrous to variously scaly or hairy, bearing spirally arranged, distant or very crowded flowers. *Flowers* solitary or in groups of 2–3, sessile or borne on short to long spurs, each subtended by a minute triangular bract; calyx sometimes stalk-like at the base, tubular, truncate, irregularly splitting, or with 3 neat triangular lobes, glabrous or variously hairy; corolla usually considerably exceeding the calyx, tubular at the base, divided into 3 rather thick, triangular, valvate lobes, glabrous to variously hairy, usually marked near the tip on the adaxial face with the impressions of the anthers; stamens 6, epipetalous, the filaments distinct, somewhat flattened, or united into a conspicuous tube tipped with 6 equal, short to moderate teeth bearing erect or pendulous anthers, or androecial ring 3-lobed, 3 anthers borne on short distinct filaments, 3 borne at the sinuses between the lobes, anthers rounded or oblong, very small to moderate, latrorse; gynoecium tricarpellate, glabrous or variously

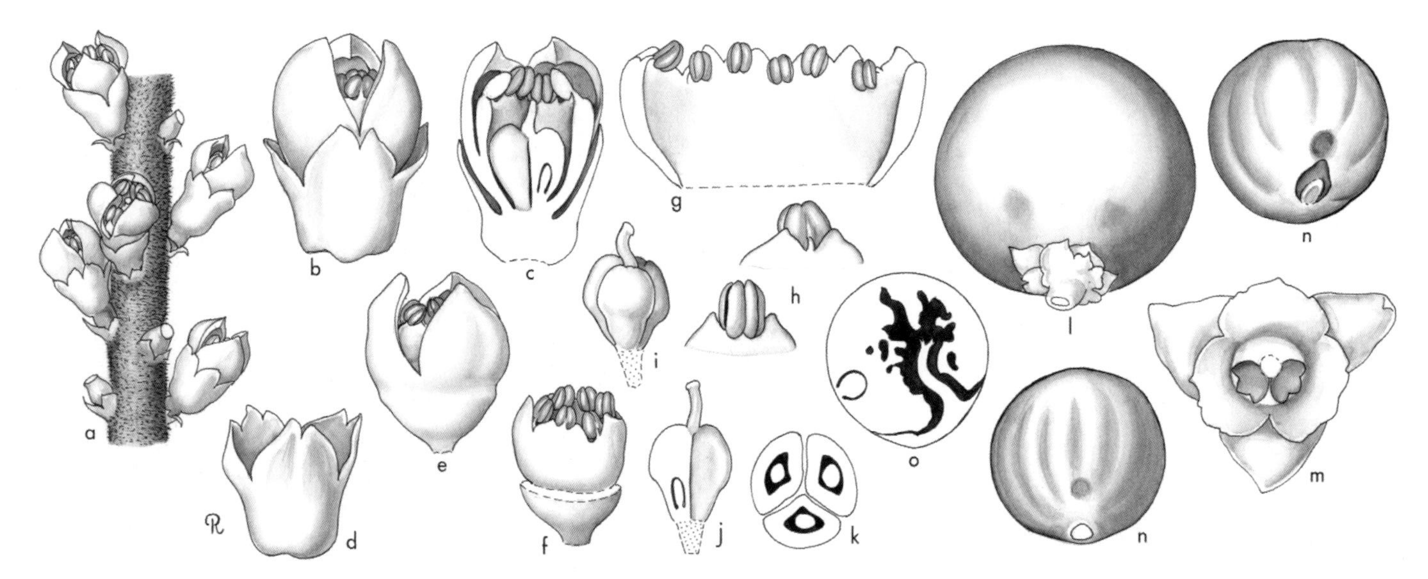

Licuala. **a**, portion of rachilla × 3; **b**, flower × 6; **c**, flower in vertical section × 6; **d**, calyx × 6; **e**, flower, calyx removed × 6; **f**, flower, perianth removed × 6; **g**, androecium expanded × 6; **h**, anther in 2 views × 12; **i**, gynoecium × 6; **j**, gynoecium in vertical section × 6; **k**, carpels in cross-section × 12; **l**, fruit × 3; **m**, fruiting perianth with abortive carpels, interior view × 6; **n**, seed in 2 views × 3; **o**, seed in vertical section × 3. *Licuala grandis*: **a–k**, *Bailey* 1640; **l–o**, *Furtado* 36537. (Drawn by Marion Ruff Sheehan)

hairy, carpels wedge-shaped, distinct in the ovarian region, united distally in a long, slender columnar style tipped with a minute dot-like stigma, ovules basally attached, anatropous. *Pollen* ellipsoidal, usually bisymmetric; aperture an extended distal sulcus; ectexine tectate, psilate, scabrate, perforate, perforate-rugulate, foveolate or finely reticulate, aperture margin slightly finer or similar; infratectum columellate; longest axis 28–50 µm; post-meiotic tetrads tetragonal or decussate [32/134]. *Fruit* globose, ovoid, narrow, straight, spindle-shaped or curved, perianth whorls usually persistent, 1–3 discrete carpels developing, abortive carpels frequently carried with the stigmatic remains at the tip of the fertile carpel, otherwise remaining at the base; epicarp frequently brightly coloured, dull or shining, rarely corky-warted, mesocarp fleshy, somewhat fibrous, thin to thick, endocarp thin, crustaceous. *Seed* basally attached, endosperm homogeneous or rarely ruminate, penetrated by a smooth or greatly lobed intrusion of seed coat, in species with spindle-shaped fruit the intrusion running ± the length of the seed in the middle; embryo lateral. *Germination* remote-tubular; eophyll strap-shaped, plicate, ± truncate and minutely lobed at the apex. *Cytology*: 2n = 28.

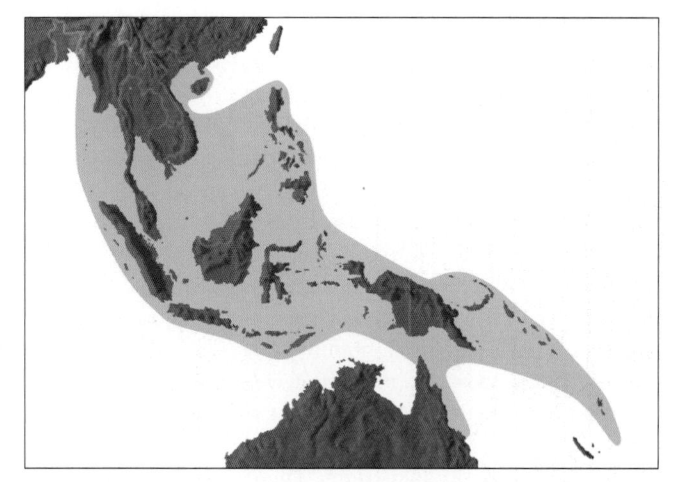
Distribution of ***Licuala***

Distribution and ecology: About 134 species, ranging from India and southern China through Southeast Asia to Malesia, Queensland, the Solomon Islands and New Hebrides, the greatest diversity being in Malay Peninsula, Borneo, and New Guinea. The species are mostly plants of the forest undergrowth; some are gregarious and lend a distinctive appearance to certain forest types, others are very local and occur as scattered individuals. A few species, e.g., *L. calciphila*, are strict calcicoles; *L. spinosa*, the most widespread species, occurs in forest on the landward fringe of mangrove and *L. paludosa* is common in peat swamp forest. A remarkable feature of some Bornean forest types is the abundance of *Licuala* spp. that grow sympatrically.

Anatomy: Leaf (Tomlinson 1961), roots (Seubert 1997), floral (Morrow 1965).

Relationships: Preliminary analyses show that *Licuala* is monophyletic (Look 2007). Uhl *et al.* (1995) place the genus as sister to *Johannesteijsmannia*. Baker *et al.* (in review) and Asmussen *et al.* (2006) found low support for a sister relationship between *Licuala* and a clade comprising *Johannesteijsmannia*, *Pholidocarpus* and *Pritchardiopsis*.

Common names and uses: Licuala palms, *palas*. Leaves of some species are used for thatching and for making sleeping mats. The sword leaf of some may be used for wrapping food before or after cooking. Smaller stems are used for walking sticks and larger ones as palisades in building. Many species are highly decorative but appear generally to be slow growing. Pith and stem apices are edible.

Taxonomic accounts: Beccari (1931) and Furtado (1940). The latter provides the most recent attempt at a subgeneric and sectional delimitation. See also the thorough revision for Peninsular Malaysia by Saw (1997).

Top left: ***Licuala cordata***, habit, cultivated, Sarawak. (Photo: J. Dransfield)

Bottom left: ***Licuala valida***, leaf and infructescence, Brunei Darussalam. (Photo: A. McRobb)

Fossil record: From the Indian Deccan Intertrappean (although the age span of these volcanic deposits is controversial, see Chapter 5), Ambwani (1983) reports palm wood, *Palmoxylon shahpuraensis*, from Madhya Pradesh. He considers it comparable to a number of species of *Licuala*, although wood identifications to genus should always be viewed with caution. Silicified fruits, *Palmocarpon coryphoidium*, which, "come very close to some species of *Pritchardia* and *Licuala* specially [sic] the latter" (Shete & Kulkarni 1985) are reported from the Deccan Intertrappean of Maharashtra. Again, the generic affinity of these fossil fruits remains doubtful.

Notes: Easily recognizable by the wedge-shaped marginally reduplicate segments of the leaves of most species. Those with undivided leaves (*Licuala grandis* and *L. orbicularis*) are unlikely to be confused with other palms. The species examined differ from other coryphoid palms in having large transverse fibre-sclereids in the mesophyll of the leaf; in some epidermal characters, they resemble *Pholidocarpus*, *Livistona* and *Johannesteijsmannia*.

Left: ***Licuala kunstleri***, flowers, Peninsular Malaysia. (Photo: J. Dransfield)

Top right: ***Licuala ferruginea***, fruit, Peninsular Malaysia. (Photo: J. Dransfield)

Bottom right: ***Licuala*** sp. aff. ***L. bintulensis***, fruit, Brunei Darussalam. (Photo: J. Dransfield)

43. JOHANNESTEIJSMANNIA

Spectacular stemless or short-stemmed hermaphroditic palms of the forest undergrowth in South Thailand, Peninsular Malaysia, Sumatra and Borneo, instantly recognisable by the large diamond-shaped undivided leaf; fruits are corky warted.

Johannesteijsmannia H.E. Moore, Principes 5: 116 (1961). Type: **J. altifrons** (Rchb.f. & Zoll.) H.E. Moore (*Teysmannia altifrons* Rchb.f. & Zoll.).

Teysmannia Rchb.f. & Zoll. in Zoll., Linnaea 28: 657 (1858 ['1856']) (non *Teysmannia* Miq. [1857]).

Named for Johannes Elias Teijsmann (1808–1882), Dutch gardener and botanist at the Buitenzorg Botanical Garden, Java (now Kebun Raya Indonesia, Bogor).

Moderate, solitary, armed, acaulescent or short-trunked, pleonanthic, hermaphroditic palms. *Stem* very short, decumbent, or erect, ringed with close leaf scars. *Leaves* large, entire, diamond-shaped, subpinnately ribbed, marcescent; sheath tubular at first, later drying and disintegrating into an interwoven mass of fibres; petiole well developed, ± triangular in cross-section, adaxially flattened, armed along the margins with small, sharp teeth, caducous tomentum present on very young petioles; adaxial hastula present on developing leaves, oblique, disappearing before leaf expansion; costa extending almost to the leaf apex, more prominent abaxially than adaxially; blade subpinnately ribbed, glabrous or the abaxial surface densely covered with white indumentum, lower margins thickened, armed with teeth like the petiole, the upper margins alternately notched, notches short along abaxial ribs, long along adaxial ribs, giving margins an irregularly stepped appearance, and perhaps representing highly reduced induplicate leaflets, midrib raised abaxially, transverse veinlets conspicuous. *Inflorescences* interfoliar, short, usually partly obscured by leaf litter, branching to 1–5 orders; peduncle well developed, usually curved, tomentose; prophyll tubular, ± inflated, 2-keeled, usually densely tomentose; peduncular bracts conspicuous, up to 7 in number, cream at first, later cinnamon-brown, tubular, ± inflated, distichous in origin but all and the prophyll splitting along the side nearest the ground, allowing the inflorescence to curve; rachis shorter than the peduncle, the branches subtended by minute triangular bracts, the branches tending to form a condensed mass; rachillae 3–6, very thick, or very numerous and slender, glabrous or tomentose, bearing spirally arranged, minute, apiculate bracts subtending flowers, the flowers solitary or in groups of 2–4 arranged in a cincinnus with minute bracteoles, on a short tubercle, or ± sessile. *Flowers* cream-coloured, strongly scented; calyx cup-shaped with 3 low, glabrous, triangular lobes; corolla divided to $^2/_3$ or almost to the base into 3 thin or very thick, fleshy, triangular, glabrous, sometimes densely papillose, valvate lobes; stamens 6, epipetalous, filaments very broad, fleshy, angled, connate basally to form an androecial ring, abruptedly narrowed to short, very slender, distinct tips, anthers minute, rounded, introrse; gynoecium tricarpellate, the carpels distinct at the base, united by their tips in a common slender elongate style, stigma dot-like, ovule basally attached, anatropous. *Pollen* ellipsoidal, bi-symmetric or slightly asymmetric; aperture a distal sulcus; ectexine tectate, scabrate, finely perforate, or perforate; aperture margin similar; infratectum columellate; longest axis 20–32 µm [4/4]. *Fruit* rounded, usually developing from 1 carpel, but sometimes 2 or 3 carpels developing, the fruit then 2 or 3-lobed; epidermis of fruit dying early in development, the mesocarp then cracking to produce thick, corky, pyramidal warts at maturity, chestnut brown in colour, endocarp moderately thick, crustaceous. *Seed* basally attached, endosperm homogeneous, but penetrated by a convoluted mass of seed coat at the base; embryo lateral. *Germination* remote-tubular; eophyll simple, plicate, minutely dentate at the tip. *Cytology*: 2n = 34.

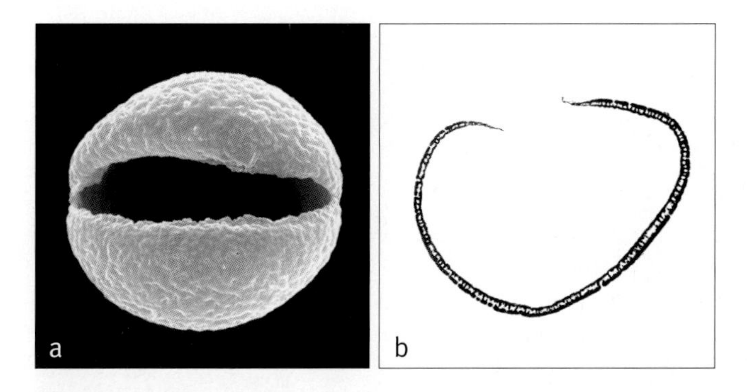

Above: ***Johannesteijsmannia***
a, monosulcate pollen grain, distal face SEM × 2000;
b, ultrathin section through whole pollen grain, polar plane TEM × 2000.
Johannesteijsmannia perakensis: **a**, *Dransfield* JD871; *J. altifrons*: **b**, *Dransfield* JD919.

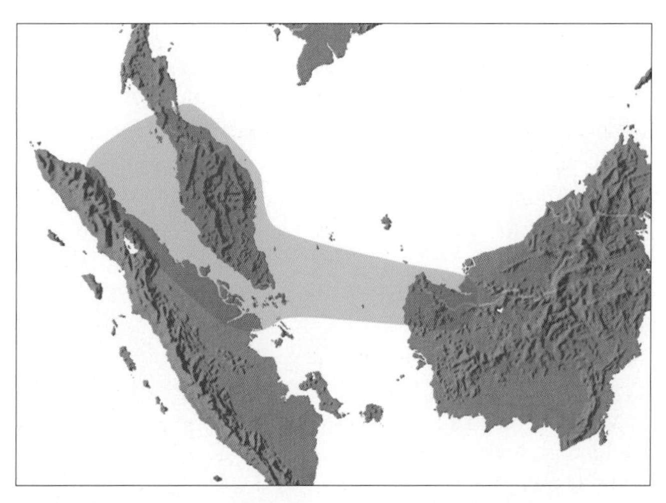

Distribution of ***Johannesteijsmannia***

Distribution and ecology: Four species, one widespread but very local in south Thailand, West Malaysia, Sumatra, and the western part of Borneo, the other three endemic to West Malaysia, where they are also very local. These magnificent palms are plants of the undergrowth of primary rain forest, and seem very intolerant of disturbance; in Sarawak *Johannesteijsmannia altifrons* appears confined to some facies of 'kerangas' (heath forest); elsewhere, it seems less restricted but avoids wet valley bottom soils. *Johannesteijsmannia magnifica* and *J. lanceolata* are plants of hillslopes and *J. perakensis* is a plant of hillslopes and ridgetops. The distribution is remarkably disjunct, species being absent from apparently suitable forest.

Anatomy: Leaf (Tomlinson 1961), roots (Seubert 1997), floral (Morrow 1965).

Relationships: Preliminary analyses show that *Johannesteijsmannia* is monophyletic (Look 2007). The genus is placed as sister to *Licuala* (Uhl *et al.* 1995) or as sister to *Pholidocarpus* with low support (Asmussen *et al.* 2006, Baker *et al.* in review).

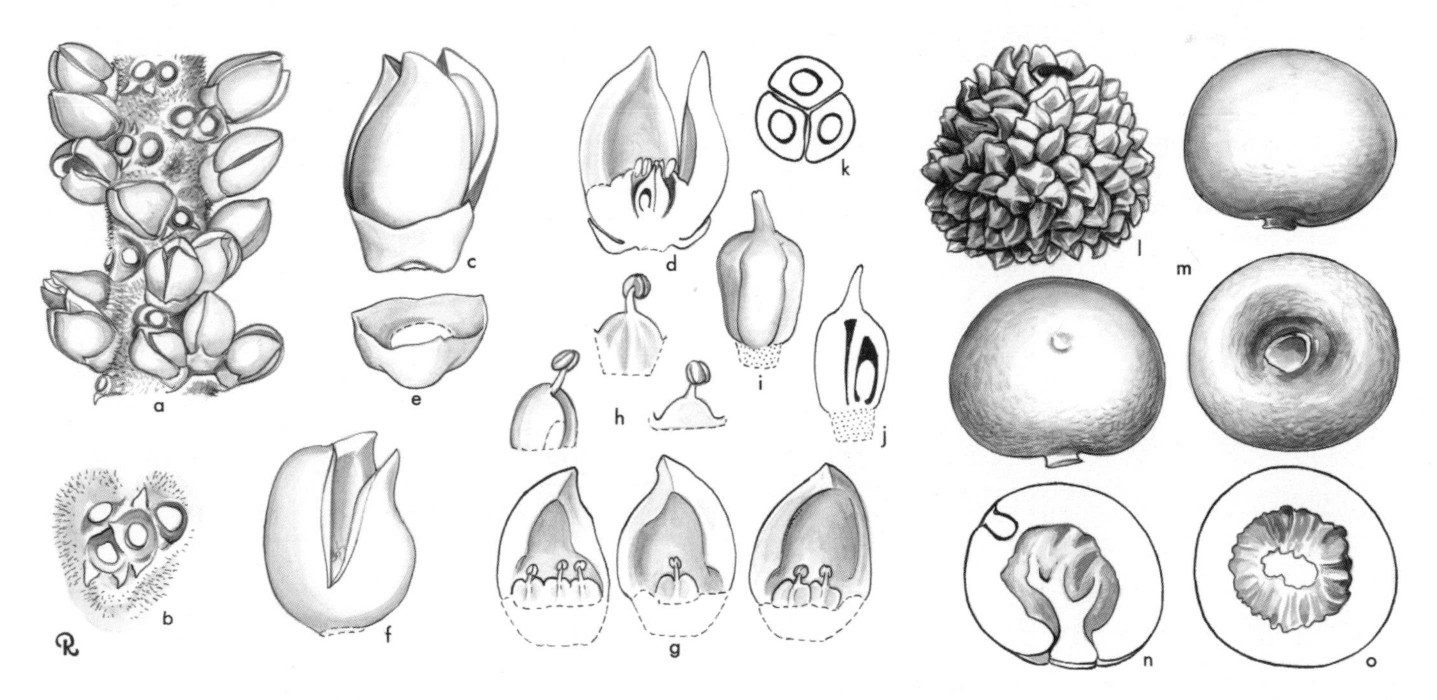

Johannesteijsmannia. **a**, portion of rachilla × 3; **b**, flower cluster with flowers removed × 6; **c**, flower × 6; **d**, flower in vertical section × 6; **e**, calyx × 6; **f**, flower, calyx removed × 6; **g**, corolla and androecium expanded × 6; **h**, stamen in 3 views × 12; **i**, gynoecium × 12; **j**, gynoecium in vertical section × 12; **k**, carpels in cross-section × 12; **l**, fruit × $1^1/_2$; **m**, seed in 3 views × $1^1/_2$; **n**, seed in vertical section × $1^1/_2$; **o**, seed in cross-section × $1^1/_2$. *Johannesteijsmannia altifrons*: **a–k**, *Moore & Pennington* 9051; **l–o** *Moore* 9110. (Drawn by Marion Ruff Sheehan)

Left: ***Johannesteijsmannia magnifica***, habit, Peninsular Malaysia. (Photo: J. Dransfield)
Top right: ***Johannesteijsmannia altifrons***, inflorescence, Peninsular Malaysia. (Photo: J. Dransfield)
Bottom right: ***Johannesteijsmannia altifrons***, fruit, Peninsular Malaysia. (Photo: J. Dransfield)

Common names and uses: *Daun payung*. Leaves are used for thatch and shelters.
Taxonomic account: Dransfield (1972b).
Fossil record: No generic records found.
Notes: Distinguished by the striking diamond-shaped leaves and the structure of the inflorescence. Similar to *Licuala* but differing in inflorescence and leaf shape, and in hooked teeth on the lower margins of the leaf blade.

44. PHOLIDOCARPUS

Tall, single-stemmed hermaphroditic fan-palms of South Thailand and Malesia eastwards to the Moluccas; leaves have fiercely toothed petioles and blades that are divided by deep and shallow splits to form deeply lobed segments; most species have large corky-warted fruit.

Pholidocarpus Blume in J.J. Roemer & J.A. Schultes, Syst. veg. 7: 1308 (1830). Type: *P. rumphii* Meisn. (illegitimate name) (*Borassus ihur* Giseke = **P. ihur** [Giseke] Blume).

Pholidos — *scale*, carpos — *fruit, referring to the corky-warted fruit of most species*.

Robust, solitary, armed, pleonanthic, hermaphroditic, tree palms. *Stem* erect, ringed with inconspicuous, close leaf scars. *Leaves* induplicate, costapalmate, marcescent in immature individuals, abscising under their own weight in trunked individuals; sheath disintegrating into a conspicuous interwoven mass of reddish-brown fibres; petiole long, robust, particularly in juveniles, bearing thin caducous indumentum, slightly channelled adaxially, abaxially rounded or angular, frequently with 2 lateral yellowish lines, the margins armed with very robust, bulbous-based, horizontal spines; adaxial hastula well developed, triangular, ring-like, abaxial hastula inconspicuous or lacking; blade divided by splits along adaxial folds almost to the hastula, producing 3–4-fold segments, these further divided along adaxial folds to $^2/_3$ or $^1/_2$ the radius into single-fold segments, also split very shallowly along abaxial folds, lowermost segments overlapping, interfold filaments not persisting, scattered caducous indumentum present along ribs in young leaves, segments longitudinally striate, midrib prominent, transverse veinlets conspicuous. *Inflorescences* interfoliar, emerging from the leaf sheath mouths and arching out of the crown, branching to 4 orders, several axils (up to ca. 5) producing inflorescences simultaneously; peduncle robust; prophyll tubular, 2-keeled, somewhat inflated; peduncular bracts 1–ca. 5, conspicuous, robust, tubular, tending to split rather irregularly with age; rachis longer than the peduncle; rachis bracts rather distant, each subtending a first-order branch, adnate to the inflorescence axis; subsequent bracts disintegrating or very inconspicuous; rachillae glabrous or hairy, ± spreading, bearing spirally arranged flowers, solitary or in clusters of 2–3 on low tubercles, subtended by minute triangular bracts and each bearing a minute bracteole. *Flowers* sessile, golden-yellow; calyx cup-shaped, shallowly 3-lobed, glabrous or sparsely hairy; corolla divided almost to the base into 3 triangular, valvate, glabrous or sparsely hairy petals; stamens 6, filaments united to form a conspicuous tube, free from the corolla, shallowly 6-lobed, tipped with short, slender, distinct filaments, bearing ± rounded or oblong, dorsifixed, introrse anthers; gynoecium tricarpellate, distinctly conical, hairy, the carpels distinct from each other basally, united apically in a long slender style, tipped with a dot-like stigma, ovule basally attached, anatropous. *Pollen* ellipsoidal, symmetric or slightly asymmetric; aperture a distal sulcus; ectexine tectate, coarsely perforate, perforate or finely reticulate, aperture margin occasionally finer; infratectum columellate; longest axis 34–37 µm [1/6]. *Fruit* developing from 1 carpel, very large, globose, stigmatic remains scarcely visible but apical; pericarp massive, the epicarp smooth (*P. kingiana*), or cracked into numerous low corky brown warts, mesocarp thick, ± fleshy, frequently traversed by radiating fibres, endocarp crustaceous. *Seed* attached laterally or near the base, endosperm massive, homogeneous, but penetrated on one side by a large convoluted intrusion of seed coat; embryo subbasal or lateral. *Germination* remote-tubular; eophyll entire, lanceolate, plicate. *Cytology* not studied.

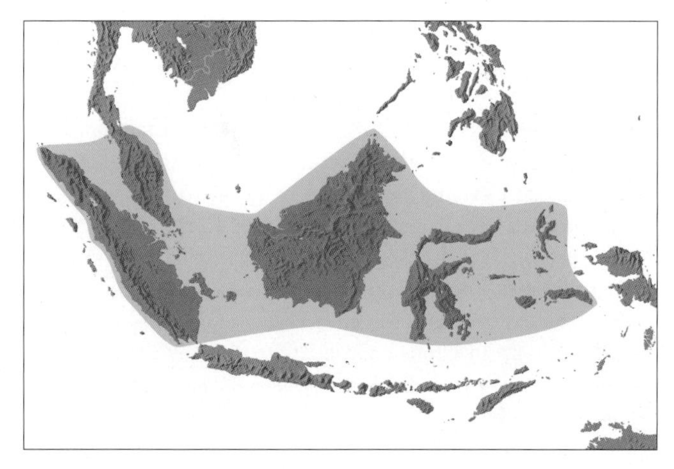

Distribution of ***Pholidocarpus***

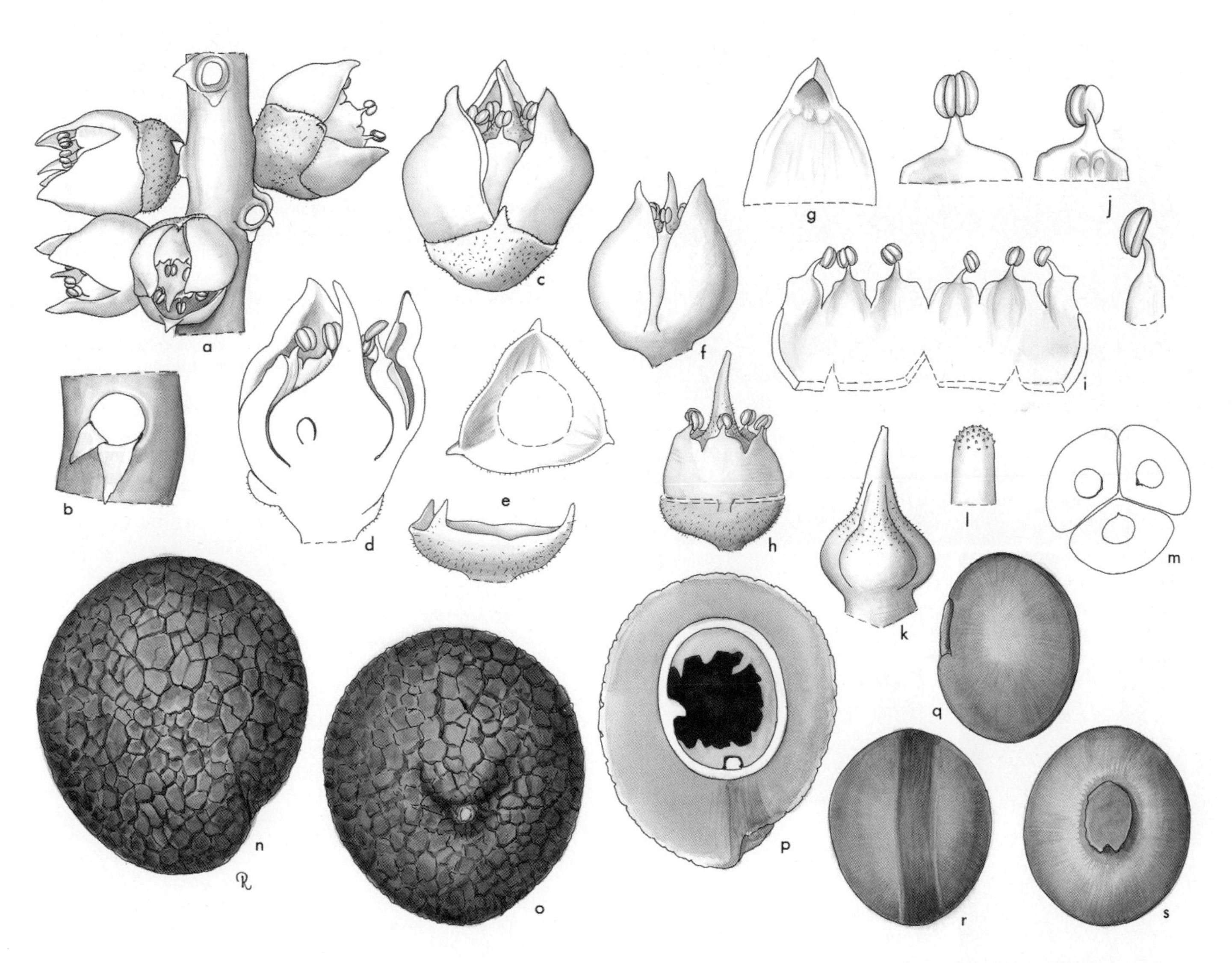

***Pholidocarpus*. a**, portion of rachilla × 4 1/2; **b**, portion of rachilla with flower removed to show scar, subtending bract, and floral bracteole × 6; **c**, flower × 6; **d**, flower in vertical section × 7 1/2; **e**, calyx in 2 views × 6; **f**, flower, calyx removed × 6; **g**, petal × 6; **h**, flower, upper portion of perianth removed to show staminal tube and gynoecium × 6; **i**, staminal tube, expanded × 6; **j**, stamen in 3 views × 12; **k**, gynoecium × 6; **l**, stigma, much enlarged; **m**, carpels in cross-section × 7 1/2; **n, o**, fruit in 2 views × 1/2; **p**, fruit in vertical section × 1/2; **q, r, s**, seed in 3 views × 1/2. *Pholidocarpus sumatranus*: **a–m**, *Dransfield* 2432; *P. macrocarpus*: **n–s**, *Corner* Al. (Drawn by Marion Ruff Sheehan)

Distribution and ecology: Six recognised species, but possibly fewer, from south Thailand, Malay Peninsula, Sumatra, Borneo, Sulawesi, and the Moluccas. In Malay Peninsula, Sumatra, and Borneo, species of *Pholidocarpus* are conspicuous palms of the lowlands, usually found in fresh water and peat swamp forest, rarely away from waterlogged soils. They may reach great heights (e.g., 45 m).

Anatomy: Leaf (Tomlinson 1961), floral anatomy not studied.

Relationships: The monophyly of *Pholidocarpus* has not been tested. *Pholidocarpus* is resolved as sister to *Johannesteijsmannia* with low support (Asmussen *et al*. 2006, Baker *et al*. in review).

Common names and uses: *Serdang*, *kepau* (*Pholidocarpus macrocarpus*). The leaves may be used for thatch.

Taxonomic accounts: Beccari (1931). See also Dransfield & Uhl (1983a).

Fossil record: No generic records found.

Notes: Distinguished by the deeply divided leaves with compound segments and by very large smooth or corky-warted fruits. Leaf anatomy shows some affinities with *Licuala*.

Top: ***Pholidocarpus macrocarpus***, crown, cultivated, Lae Botanic Garden, Papua New Guinea. (Photo: J. Dransfield)

Bottom left: ***Pholidocarpus sumatranus***, petioles, Sumatra. (Photo: J. Dransfield)

Bottom right: ***Pholidocarpus sumatranus***, fruit, Sumatra. (Photo: J. Dransfield)

45. PRITCHARDIOPSIS

Moderate hermaphroditic fan palm endemic to New Caledonia, remarkable for its large fruit, the endocarp with a keel and a basal extension.

Pritchardiopsis Becc., Webbia 3: 131 (1910). Type: **P. jeanneneyi** Becc. (see Moore & Uhl 1984).

Derived from Pritchardia *and* opsis — *similar to*.

Moderate, solitary, armed, pleonanthic, hermaphroditic, tree palm. *Stem* erect, smooth, ringed with conspicuous leaf scars. *Leaves* induplicate, briefly costapalmate, neatly abscising; sheath disintegrating into a network of fine rusty-brown fibres; petiole elongate, adaxially flat to ridged, abaxially rounded, margins distally smooth, basally with short recurved spines in juvenile plants, smooth in adults, adaxial hastula short, rounded, abaxial hastula lacking; blade stiff, regularly divided to or beyond the middle into single-fold, briefly bifid, lanceolate, spreading segments, glabrous on both surfaces, midribs and intercostal ribs prominent, transverse veinlets conspicuous. *Inflorescences* interfoliar, branched to 4 orders, branches angled; peduncle very short, flattened; prophyll not seen, inserted above the base; peduncular bracts lacking; rachis very short, deeply divided into 3 elongate first-order branches, each branch bearing 1 peduncular bract below 2 unilateral second-order branches subtended by tubular, chartaceous bracts, bifid and flaring at the apex; proximal two second-order branches flattened, adnate to the first-order branch, subsequent second-order branches not subtended by tubular bracts, not adnate; bracts not evident at bases of rachillae; rachillae short, ± clustered at the ends of second-order branches, bearing rather distant cincinni of 3 flowers basally and solitary flowers distally. *Flowers* sessile, ebracteolate; calyx tubular, adnate basally to the receptacle, with 3 short free lobes; petals 3, briefly connate basally, valvate and adaxially hollowed distally, persistent; stamens 6, inserted at throat of the corolla, filaments erect, briefly connate and adnate to the petals basally, anthers erect, subglobose, sagittate, introrse; gynoecium globose-trilobate, trilocular, triovulate, carpels distinct in ovarian region, connate through styles, style slender, short, awl-shaped, stigma dot-like, ovule inserted basally, anatropous. *Pollen* ellipsoidal, occasionally slightly asymmetric; aperture a distal sulcus; ectexine tectate, aperture margin similar; infratectum columellate; longest axis 23–24 µm [1/1]. *Fruit* large, 1-seeded, globose, with apical stigmatic remains; epicarp smooth, purplish, mesocarp fleshy to fibrous near endocarp, endocarp woody, rounded on one side, keeled on the opposite side, laterally elongate and attenuate basally. *Seed* globose, erect, hilum basal, raphe orbicular, adjacent to hilum, endosperm homogeneous, deeply hollowed out basally with a large, erect intrusion of seed coat; embryo eccentrically apical. *Germination* remote-tubular; eophyll entire, oblanceolate, toothed distally. *Cytology* not studied.

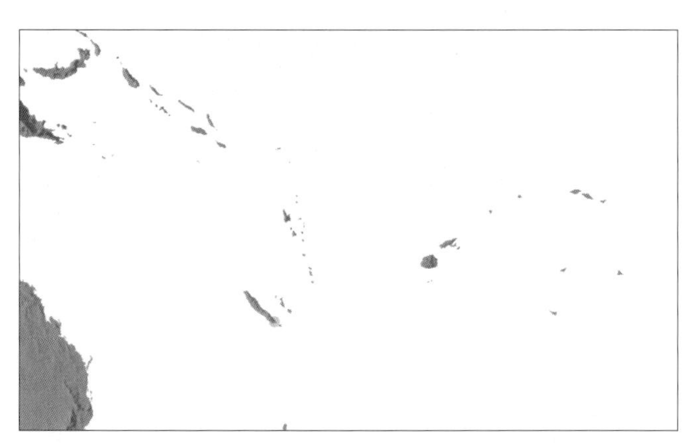

Distribution of ***Pritchardiopsis***

Distribution and ecology: A single species in New Caledonia. Restricted to south-eastern New Caledonia on steep slopes with serpentine soils at ca. 200 m in the vicinity of the Bay of Prony, once thought to be extinct but recently relocated in three very small populations.

Common names and uses: Common names not recorded. The apex is edible and destructive exploitation has resulted in near extinction.

Anatomy: Leaf with coryphoid midrib, irregular 2-layered adaxial hypodermis, and indistinct or lacking palisade layers (Uhl & Martens 1980); floral (Uhl pers. obs.).

Relationships: There is low bootstrap support for a sister relationship between *Pritchardiopsis* and a clade of *Johannesteijsmannia* and *Pholidocarpus* (Asmussen *et al.* 2006, Baker *et al.* in review).

Taxonomic accounts: Moore & Uhl (1984) and Hodel & Pintaud (1998).

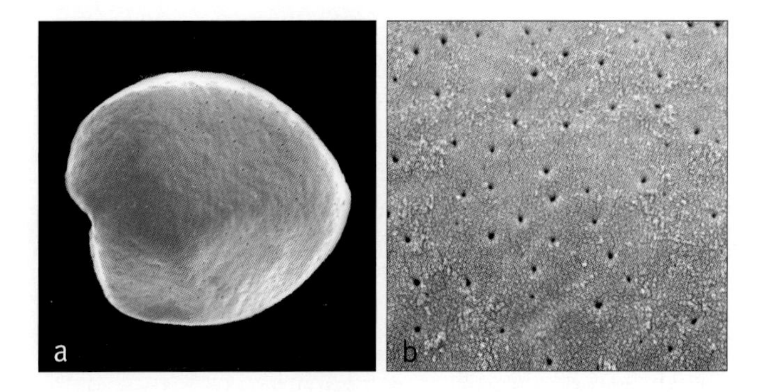

Above: ***Pritchardiopsis***
a, monosulcate pollen grain, proximal face SEM × 2000;
b, close-up, finely perforate pollen surface SEM × 8000.
Pritchardiopsis jeanneneyi: **a** & **b**, *Jeanneneyi* s.n.

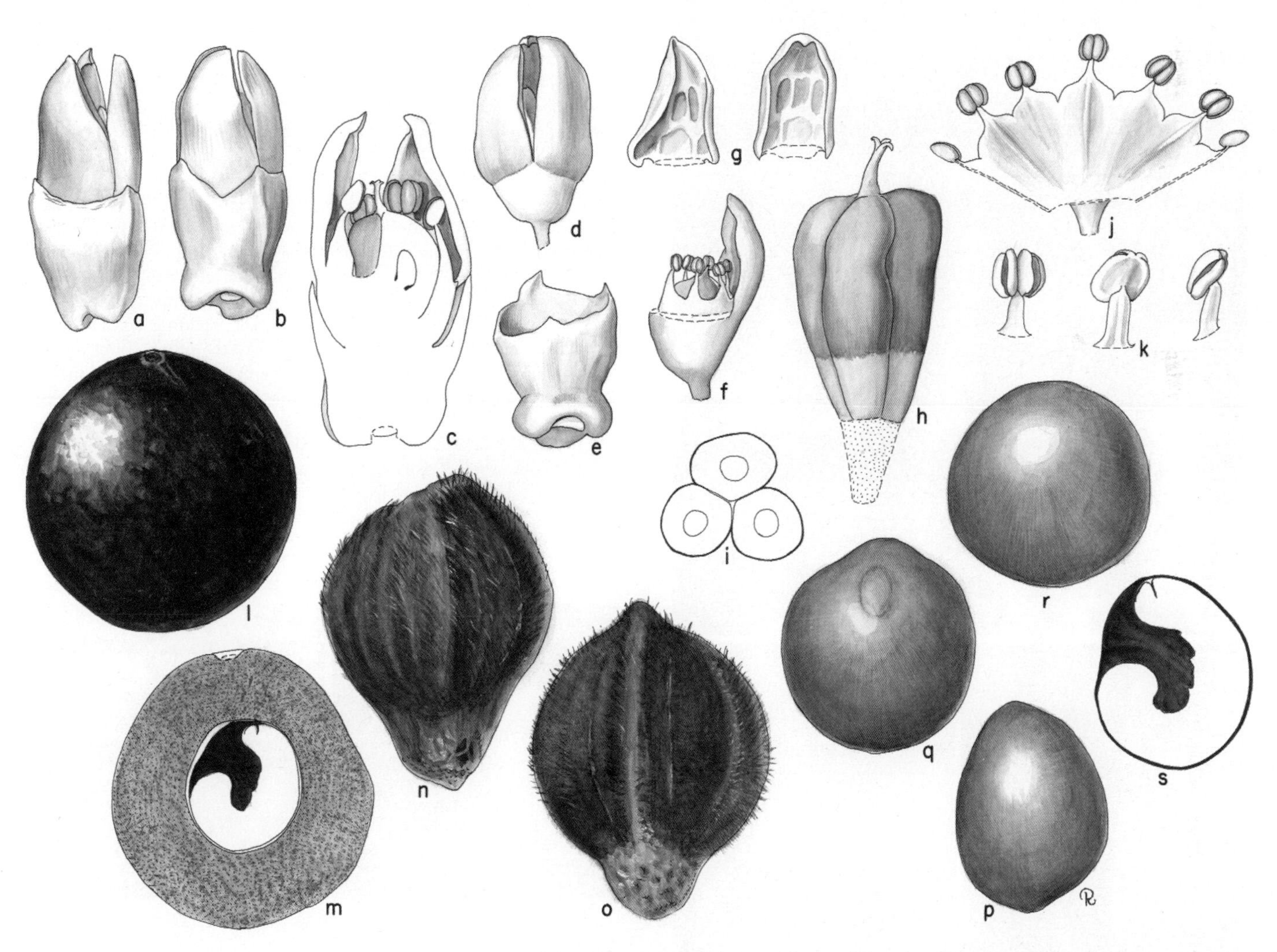

Pritchardiopsis. **a**, **b**, flowers × 9; **c**, flower in vertical section × 12; **d**, flower, sepals removed × 9; **e**, calyx × 9; **f**, flower, sepals and 2 petals removed × 9; **g**, 2 petals, interior views × 9; **h**, gynoecium × 22 $^1/_2$; **i**, gynoecium in cross-section × 22 $^1/_2$; **j**, androecium × 12; **k**, stamen in 3 views × 22 $^1/_2$; **l**, fruit × 1; **m**, fruit in cross-section × 1; **n**, **o**, endocarp in 2 views × 1 $^1/_2$; **p**, **q**, **r**, seed in 3 views × 1 $^1/_2$; **s**, seed in vertical section × 1 $^1/_2$. *Pritchardiopsis jeanneneyi*: **a–k**, *Jeanneney* s.n.; **l–s**, *MacKee* 38038.
(Drawn by Marion Ruff Sheehan)

Fossil record: No generic records found.

Notes: *Pritchardiopsis* differs from most other Old World members of the Trachycarpeae in lacking any kind of armature on the petioles. The endocarp is unlike that of any other member of the group. In habit, it resembles *Livistona*, and in the branching of the inflorescence, it is much like *L. woodfordii*.

Below: ***Pritchardiopsis jeanneneyi***, habit, New Caledonia. (Photo: J.-C. Pintaud)

Bottom left: ***Pritchardiopsis jeanneneyi***, petiole margin, New Caledonia. (Photo: J.-C. Pintaud)

Bottom right: ***Pritchardiopsis jeanneneyi***, detail of leaf blade, New Caledonia. (Photo: J.-C. Pintaud)

Unplaced members of Trachycarpeae

46. ACOELORRHAPHE

Clustering shrubby hermaphroditic fan palm occurring in swampy coastal areas of the northern Caribbean, distinctive in its long erect or pendulous inflorescences bearing numerous very small blue-black fruit.

Acoelorrhaphe H. Wendl., Bot. Zeit. 37: 148 (1879). Type: **A. wrightii** (Griseb. & H. Wendl.) H. Wendl. ex Becc. (*Copernicia wrightii* Griseb. & H. Wendl.) ('*Acoelorhaphe*', '*Acoelorraphe*').

Paurotis O.F. Cook, in Northr., Mem. Torrey Bot. Club 12: 21 (1902). Type: *P. androsana* O.F. Cook.

Acanthosabal Prosch, Gard. Chron., series 3. 77: 91 (1925). Type: *A. caespitosa* Prosch.

A — *without*, coelos — *hollow*, rhaphe — *seam, alluding to the fact that the seed lacks the impressed raphe common in many coryphoid genera.*

Moderate, armed, clustered, pleonanthic, hermaphroditic palm. *Stem* slender, erect, clothed with persistent leaf sheaths and petiole bases, older stems naked below, internodes very short. *Leaves* rather small, induplicate, very briefly costapalmate; sheath disintegrating into an interwoven mass of coarse, rich brown fibres; petiole moderate, slightly channelled or flattened adaxially, rounded abaxially, fiercely armed with robust, triangular, reflexed or inflexed spines, adaxial hastula conspicuous, irregularly lobed, abaxial hastula a low ridge; blade nearly orbicular, relatively flat regularly divided to below the middle into narrow, single-fold, deeply bifid, stiff segments, usually silvery abaxially due to small scales, midribs prominent abaxially, transverse veinlets conspicuous. *Inflorescences* slender, solitary, interfoliar, exceeding the leaves, branched to 4 orders; peduncle slender, elongate, elliptical in cross-section, usually erect; prophyll short, partly to completely enclosed by the leaf sheaths, tubular, 2-keeled laterally, splitting apically into short, irregular lobes; peduncular bracts 2, like the prophyll but much longer and more shallowly keeled; rachis about as long as the peduncle, ± glabrous, completely sheathed by tubular bracts, other branches densely tomentose; rachis bracts like the peduncular bract but smaller and decreasing in size distally; first-order branches bearing a 2-keeled membranous prophyll ± included within the primary bract; subsequent bracts very inconspicuous, triangular, membranous; rachillae slender, bearing spirally arranged, minute bracts each subtending a low spur

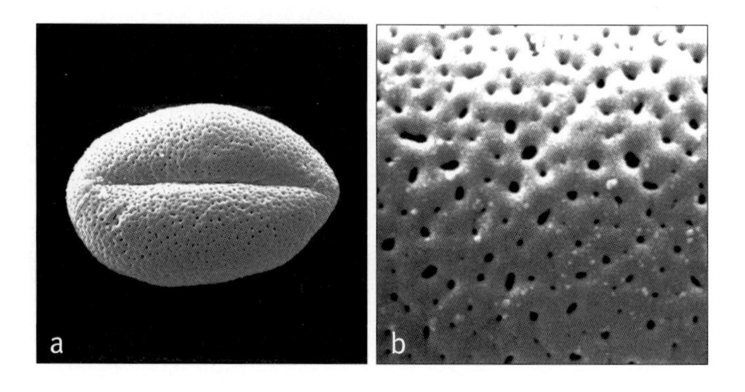

Above: ***Acoelorrhaphe***
a, monosulcate pollen grain, distal face SEM × 2000;
b, close-up, perforate pollen surface × 8000.
Acoelorrhaphe wrightii: **a** & **b**, *Curtiss* 449.

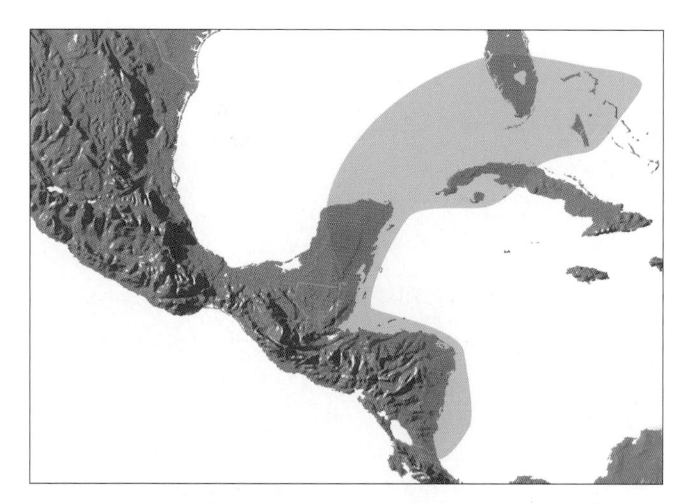

Distribution of ***Acoelorrhaphe***

Left: ***Acoelorrhaphe wrightii***, habit, cultivated, Montgomery Botanical Center, Florida. (Photo: J. Dransfield)

bearing a cluster (cincinnus) of (1)–2–3 flowers, each flower subtended by a small thin bract. *Flowers* cream-coloured; sepals 3, fleshy and slightly connate basally; petals 3, united in a basal tube for $^{1}/_{4}$ their length; stamens 6, borne at the mouth of corolla tube, filaments connate in a shallow cup at the base, free portions abruptly narrowed to a filiform apex, not inflexed in bud; anthers dorsifixed, short, rounded, versatile at anthesis, latrorse; gynoecium of 3 glabrous, follicular carpels, connate in stylar regions, ovule basal, erect, anatropous. *Pollen* ellipsoidal, with slight to obvious asymmetry; aperture a distal sulcus; ectexine tectate, finely

Acoelorrhaphe. **a**, portion of rachilla × 3; **b**, portion of rachilla showing floral scar and bract × 6; **c**, flower × 12; **d**, flower in vertical section × 12; **e**, sepals, interior and exterior views × 12; **f**, flower, sepals removed × 12; **g**, free portion of petal, interior view × 12; **h**, corolla and androecium expanded, interior view × 12; **i**, flower, sepals and free portions of petals removed × 12; **j**, stamen in 3 views × 12; **k**, gynoecium, exterior view and vertical section × 12; **l**, carpels in cross-section × 12; **m**, fruit × 3; **n**, base of fruit and abortive carpels × 6; **o**, seed × 3; **p**, seed in vertical section × 3; **q**, seed in cross-section × 3. *Acoelorrhaphe wrightii*: all from *Read* 745. (Drawn by Marion Ruff Sheehan)

perforate, perforate, micro-channelled, and slightly rugulate, aperture margin slightly finer; infratectum columellate, longest axis 29–33 µm [1/1]. *Fruit* small, rounded, developing from 1 carpel, black, stigmatic scar apical, abortive carpels basal; epicarp smooth, mesocarp thinly fleshy with prominent longitudinal fibres, endocarp thin, crustaceous. *Seed* with basal hilum, endosperm homogeneous penetrated by a thin intrusion of the seed coat at one side; embryo lateral near base on the antirapheal side. *Germination* remote-ligular (Chavez 2003); eophyll simple, narrow, lanceolate. *Cytology*: 2n = 36.

Distribution and ecology: One species in southern Florida, the West Indies, and parts of the Caribbean coast of Central America, forming clumps in brackish swamps.
Anatomy: Leaf (Tomlinson 1961), roots (Seubert 1997), floral (Morrow 1965).
Relationships: *Acoelorrhaphe* is resolved as sister to *Serenoa* with moderate support (Uhl *et al.* 1995, Asmussen *et al.* 2006, Baker *et al.* in review).
Common names and uses: Paurotis palm, saw cabbage palm, silver saw palm, everglades palm. Widely grown as an ornamental.
Taxonomic account: Zona (1997).
Fossil record: See comments under *Brahea*.
Notes: An attractive clustering palm with erect, rather slender, brown trunks and leaves silvery beneath, suitable for rather moist areas. Similar to *Serenoa*, but differing in an erect habit, larger thorns on the petioles, and flowers with free and overlapping sepals.

Above: ***Acoelorrhaphe wrightii***, infructescences, cultivated, Fairchild Tropical Botanic Garden, Florida. (Photo: C.E. Lewis)

Right: ***Serenoa repens***, habit, Florida. (Photo: J. Dransfield)

47. SERENOA

Thicket forming hermaphroditic fan palm with creeping, highly branched stems, occurring in pinelands, prairies and sand dunes in southeastern USA.

Serenoa Hook.f. in Benth. & Hook.f., Gen. pl. 3: 879, 926, 1228 (1883). Type: *Serenoa serrulata* (Michx.) G. Nicholson (*Chamaerops serrulata* Michaux) = **S. repens** (W. Bartram) Small (*Corypha repens* W. Bartram).
Diglossophyllum H. Wendl. ex Salomon, Palmen 155 (1887). Type: *D. serrulatum* (Michaux) H. Wendl. ex Salomon (*Chamaerops serrulata* Michx.) (= *S. repens* [W. Bartram] Small).

Commemorates American botanist, Sereno Watson (1826–1892).

Moderate, clustered, shrubby, armed, pleonanthic, hermaphroditic palm. *Stem* subterranean or prostrate and surface creeping, or rarely erect, covered with persistent leaf sheaths, axillary buds developing as either inflorescences or vegetative suckers. *Leaves* induplicate, palmate, marcescent; sheath expanding into a tattered mat of dark brown fibres; petiole flat to slightly rounded adaxially, rounded to angled abaxially, margin armed with numerous small teeth; adaxial hastula conspicuous, ± rounded, membranous, abaxial hastula semicircular, often split, membranous; blade nearly orbicular, regularly divided to below the middle into narrow stiff, shortly bifid, single-fold segments, glabrous

except for scattered caducous scales along the ribs, midribs conspicuous abaxially, transverse veinlets conspicuous, rather distant. *Inflorescences* interfoliar, erect and about equaling the leaves but often hidden by them, curved, branched to 3(–4) orders; peduncle slender, flattened, rather short; prophyll tubular, 2-keeled, with 2 triangular apical lobes; peduncular bract 1 or lacking, tightly sheathing, caducously tomentose; rachis longer than the peduncle; rachis bracts like the peduncular bract but decreasing in size distally; first-order branches with a short 2-keeled prophyll; subsequent bracts small, membranous; rachillae spreading, densely tomentose, bearing spirally arranged, small, irregularly cleft bracts subtending solitary or paired flowers. *Flowers* with tubular calyx of 3 triangular, slightly imbricate lobes; corolla tubular, split to $^2/_3$ its length into 3 lobes, valvate, inconspicuously grooved adaxially; stamens 6, filaments borne at the mouth of the corolla tube, gradually tapered, not inflexed, anthers erect in bud, elliptic, dorsifixed, somewhat versatile, latrorse; carpels 3, basally distinct, united in the attenuate stylar region to a narrow stigma, ovule anatropous. *Pollen* ellipsoidal, usually slightly asymmetric; aperture a distal sulcus; ectexine tectate, finely perforate, perforate and micro-channelled, or perforaterugulate, aperture margin slightly finer; infratectum columellate; longest axis 31–44 µm; post-meiotic tetrads tetrahedral or tetragonal, rarely rhomboidal [1/1]. *Fruit* ellipsoidal to subglobose, dark blue to black at maturity, abortive carpels basal, stigmatic scar apical or subapical; epicarp smooth, mesocarp fleshy without fibres, endocarp thin but somewhat cartilaginous. *Seed* basally attached with elongate raphe, endosperm homogeneous with a shallow lateral intrusion of seed coat; embryo lateral towards the base opposite the raphe. *Germination* remote-ligular; eophyll entire, plicate. *Cytology*: 2n = 36.

Distribution and ecology: One species in the southeastern USA. Common in pinelands, prairies, and coastal sand dunes, often forming dense swards.

Anatomy: Leaf (Tomlinson 1961), roots (Seubert 1997), floral anatomy reported by Morrow (1965) to be similar to that of *Acoelorrhaphe*.

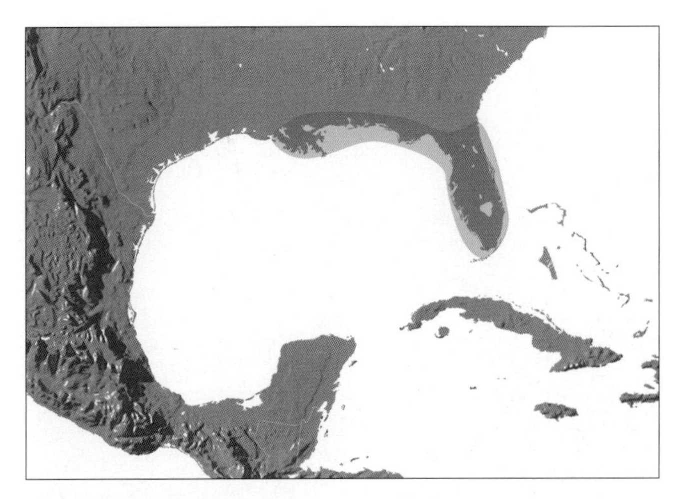

Distribution of ***Serenoa***

Relationships: For relationships, see *Acoelorrhaphe*.

Common names and uses: Saw palmetto. Regarded as a pest in the wild. The glaucous form is much prized as an ornamental. Fruits contain a bioactive ingredient, used in the treatment of benign prostate cancer and as a health supplement for men.

Taxonomic accounts: Bailey (1934) and Zona (1997).

Fossil record: An 'undoubted' palmate leaf from the Middle Cretaceous formation at Glen Cove, Long Island, New York State is described as *Serenopsis kempii* by Hollick

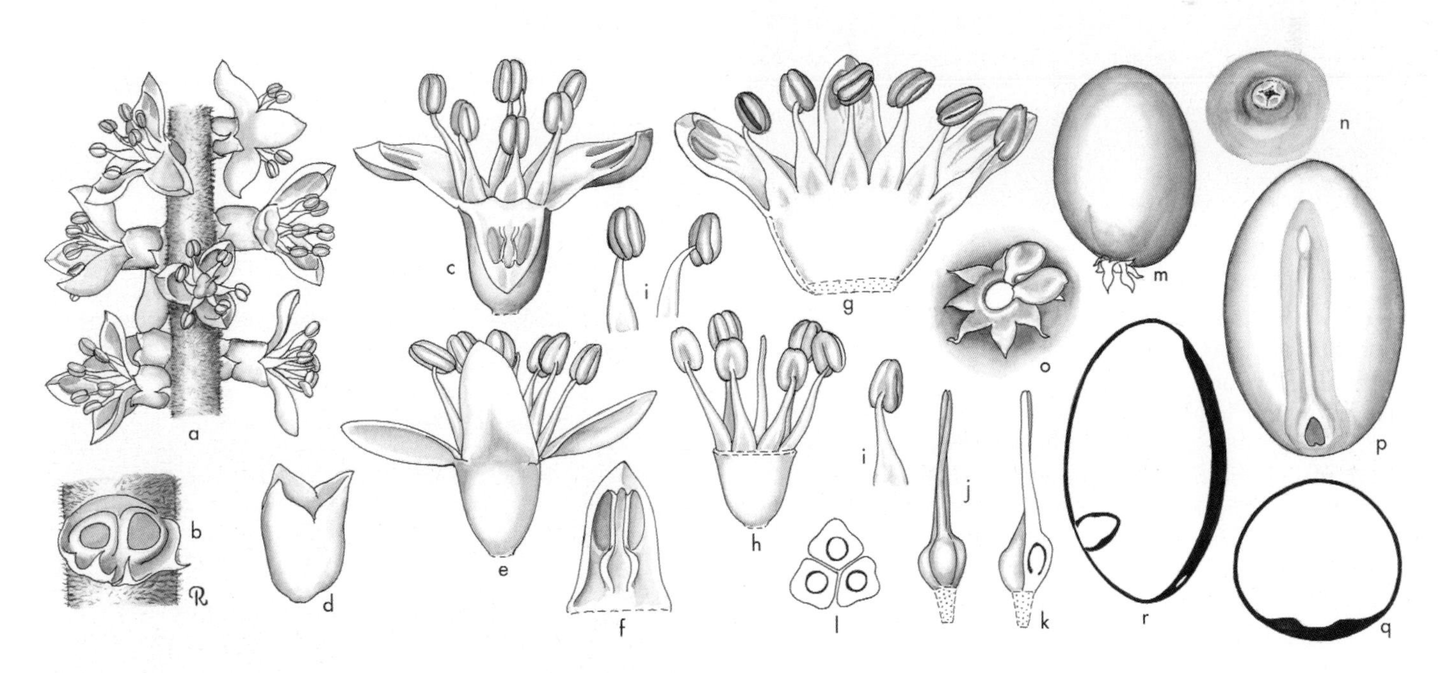

Serenoa. **a**, portion of rachilla × 3; **b**, portion of rachilla showing floral scars and bracts × 6; **c**, flower × 6; **d**, calyx × 6; **e**, flower, calyx removed × 6; **f**, free portion of petal, interior view × 6; **g**, corolla and androecium, expanded, interior view × 6; **h**, stamens and gynoecium, perianth removed × 6; **i**, stamen in 3 views × 6; **j**, gynoecium × 6; **k**, gynoecium in vertical section × 6; **l**, carpels in cross-section × 12; **m**, fruit × $1^1/_2$; **n**, apex of fruit × 12; **o**, base of fruit and abortive carpels × 6; **p**, seed × 3; **q**, seed in cross-section × 3; **r**, seed in vertical section × 3. *Serenoa repens*: all from *Read* s.n. (Drawn by Marion Ruff Sheehan)

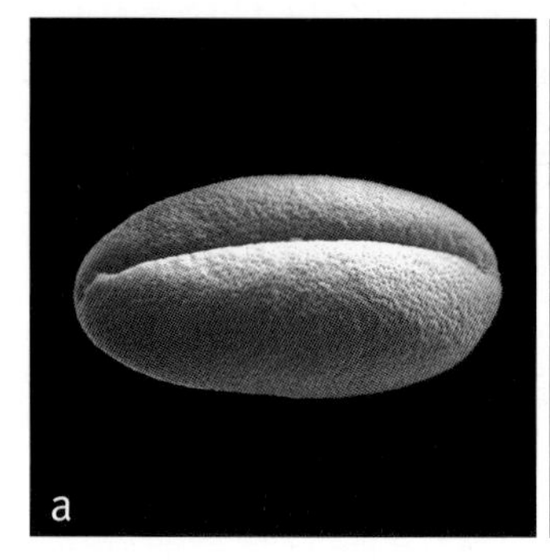

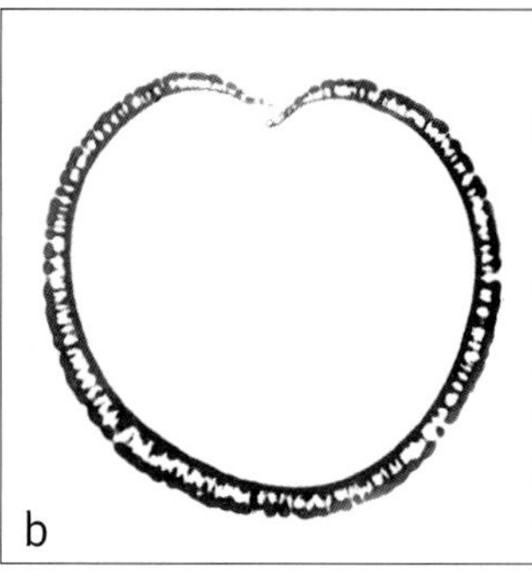

Above: ***Serenoa***
a, monosulcate pollen grain, distal face SEM × 1500;
b, ultrathin section through whole pollen grain, polar plane TEM × 2250.
Serenoa repens: **a** & **b**, *Chapman* s.n.

Top left: ***Serenoa repens***, inflorescence, cultivated, Fairchild Tropical Botanic Garden, Florida. (Photo: C.E. Lewis)

Bottom left: ***Serenoa repens***, flowers, cultivated, Fairchild Tropical Botanic Garden, Florida. (Photo: C.E. Lewis)

(1893), who considers it to be most like *Trithrinax*, *Copernicia*, *Thrinax* or *Serenoa*. However, Read and Hickey (1972) suggest that it is, "probably a cone of *Williamsonia*" (not a palm). (The long, thin, slightly sinuous leaflets of the fossil overlap near a poorly preserved central region.) Subfossil petioles and seeds ('stones') are reported by Berry (1917) from the Pleistocene of Vero (Florida). Fossil fruits are reported from southern England, Lower Eocene, *Serenoa eocenica* (Reid & Chandler 1933) and Germany (Geiseltal), Eocene, *S. carbonaria* (Mai 1976). Finely reticulate monosulcate pollen erroneously described as *Serenoa* has been reported from the Oligocene of the Isle of Wight (UK) by Pallot (1961), and from the Lower Eocene (London Clay) by Khin Sein (1961).

Notes: The unusual branching behaviour of the stem has been described in detail by Fisher and Tomlinson (1973).

48. BRAHEA

Mostly solitary hermaphroditic fan palms of Mexico and Guatemala, occurring usually on limestone in dry areas, the leaves often glaucous.

Brahea Mart. ex Endl., Gen. pl. 252 (1837). Type: **B. dulcis** (Kunth) Mart. (*Corypha dulcis* Kunth).

Erythea S. Watson, Bot. California 2: 211 (1880). Lectotype: *E. edulis* (H. Wendl.) S. Watson (*Brahea edulis* H. Wendl.) (see Cook 1915b).

Glaucothea O.F. Cook, J. Wash. Acad. Sci. 5: 237 (1915). Type: *G. armata* (S. Watson) O.F. Cook (*Brahea armata* S. Watson).

Commemorates Danish astronomer, Tycho Brahe (1546–1601).

Moderate, mostly solitary, rarely clustered, armed or unarmed, pleonanthic, hermaphroditic palms. *Stem* clothed with persistent leaf sheaths, in age becoming bare. *Leaves* induplicate, shortly costapalmate, marcescent; sheath becoming fibrous, persistent, eventually splitting basally; petiole short or long, concave, flattened, or channelled adaxially, rounded abaxially, margins unarmed or armed with sparse to dense, small or large teeth, sometimes floccose; adaxial hastula triangular to irregular, thin, membranous, at length fibrous, sometimes large, abaxial hastula a very low ridge or scarcely developed; blade nearly orbicular, regularly divided nearly to the middle or beyond into single-fold, stiff or flexible segments, deeply bifid at the apex, interfold filaments often present, surfaces glabrous, waxy or covered in caducous, floccose indumentum, midribs prominent, other veins fine, ± equal and close together giving a striate appearance, transverse veinlets inconspicuous, sometimes evident abaxially. *Inflorescences* solitary, interfoliar, nearly equalling or exceeding the leaves, erect or curving, branched to 4 orders; peduncle slender, short to medium; prophyll 2-keeled, closely sheathing, tubular, glabrous (?always), splitting irregularly abaxially; peduncular bracts 0–several, like the prophyll but single-keeled, glabrous or floccose; rachis much longer than peduncle; first-order branches distant, apparently lacking prophylls; subsequent bracts triangular, membranous, very inconspicuous; rachillae crowded, numerous, all branches and rachillae covered in a pale dense felt or deep pile of hairs. *Flowers* spirally arranged, solitary or in cincinni of 2–3, each subtended by a small bract, buds sometimes obscured by hairs until anthesis; sepals 3, distinct, imbricate, margins minutely toothed (?always); petals 3, united basally in a tube as long as the sepals, briefly imbricate, valvate apically, shallowly to deeply furrowed adaxially; stamens 6, borne at the mouth of the corolla tube, filaments connate into a 6-lobed ring, lobes triangular, abruptly narrowed at tips, anthers broadly elliptic to nearly oblong, dorsifixed, ± versatile, latrorse; carpels 3, follicular, united by the styles, ovule basal, erect, anatropous. *Pollen* ellipsoidal, slightly to extremely asymmetric; aperture a distal sulcus; ectexine tectate, finely perforate, perforate and micro-channelled, or perforate-rugulate, aperture margin slightly finer; infratectum columellate; longest axis 29–51 µm [4/10]. *Fruit* usually developing from 1 carpel, globose or ovoid, dark blue to black at maturity, abortive carpels basal, stigmatic remains apical; epicarp smooth, mesocarp fleshy, endocarp crustaceous. *Seed* basally or subbasally attached, globose or ellipsoidal, endosperm homogeneous, very shallowly to deeply penetrated by a smooth intrusion of seed coat; embryo subbasal to lateral. *Germination* remote-ligular; eophyll entire. *Cytology*: 2n = 36.

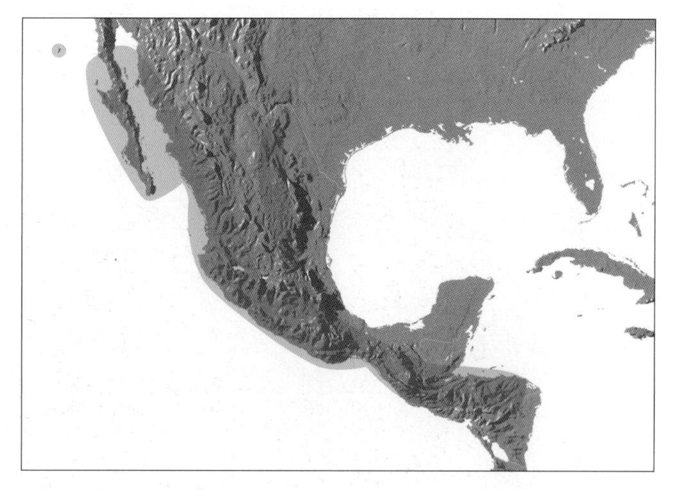

Distribution of ***Brahea***

Distribution and ecology: About 10 species in Baja California, Guadalupe Island, Mexico and Guatemala. On limestone slopes and outcrops in dry areas.

Anatomy: Leaf (Tomlinson 1961), roots (Seubert 1997), floral (Morrow 1965).

Relationships: Asmussen *et al.* (2006) found a poorly supported sister relationship between *Brahea* and subtribe Rhapidinae.

Common names and uses: Hesper palms, Guadalupe palms, rock palm, sweet brahea palm (*Brahea edulis*). The leaves are used for thatch and as a source of fibre. Fruits of some species are edible. Attractive ornamentals for drier areas.

Taxonomic accounts: Bailey (1937a, 1937b) and Henderson *et al.* (1995).

Fossil record: A study of the Middle Eocene Princeton chert of British Columbia, Canada shows palm vegetative organs to be the most common elements: five stems up to 9 cm wide with attached petiole bases and roots, plus numerous additional isolated petioles, midribs and laminae. According to Erwin and Stockey (1991) comparison with extant palms suggests that fossil stem and leaf anatomy is most similar to two coryphoid genera, *Rhapidophyllum* and *Brahea*. A hermaphroditic flower, *Palaeoraphe dominica* (Poinar 2002b) described from the Eocene Dominican amber is compared with flowers of *Brahea*, *Acoelorrhaphe* and *Colpothrinax*, although the flower is considered to resemble *Brahea* most

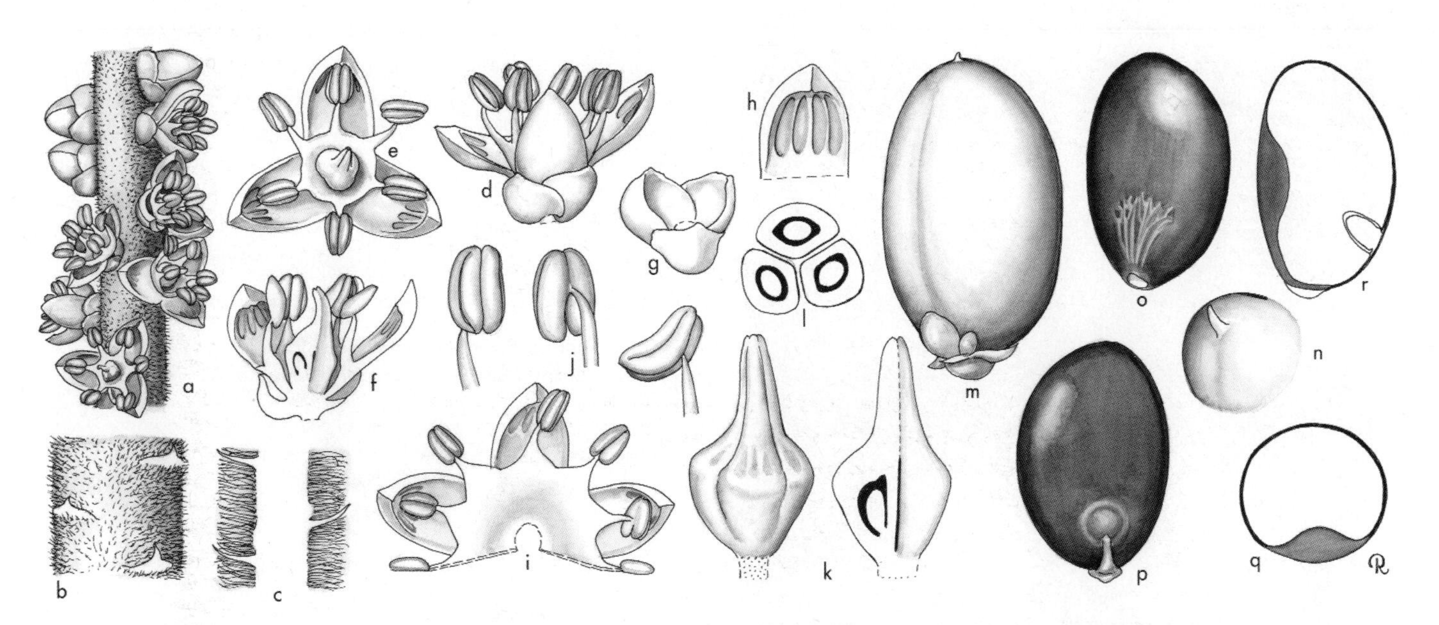

Brahea. **a**, portion of rachilla × 3; **b**, portion of rachilla showing floral scars and bracts × 6; **c**, flower × 6; **d**, flower in vertical section × 12; **e**, sepals × 12; **f**, flower, sepals removed × 6; **g**, free portion of petal, interior view × 12; **h**, corolla and androecium, expanded, interior view × 12; **i**, androecium, corolla removed × 12; **j**, stamen in 3 views × 24; **k**, gynoecium, exterior view and vertical section × 24; **l**, carpels in cross-section × 24; **m**, fruit × 1 1/2; **n**, apex of fruit × 6; **o**, base of fruit × 3; **p**, seed × 1 1/2; **q**, seed in vertical section × 1 1/2; **r**, seed in cross-section × 1 1/2. *Brahea aculeata*: all from *Read* 722. (Drawn by Marion Ruff Sheehan)

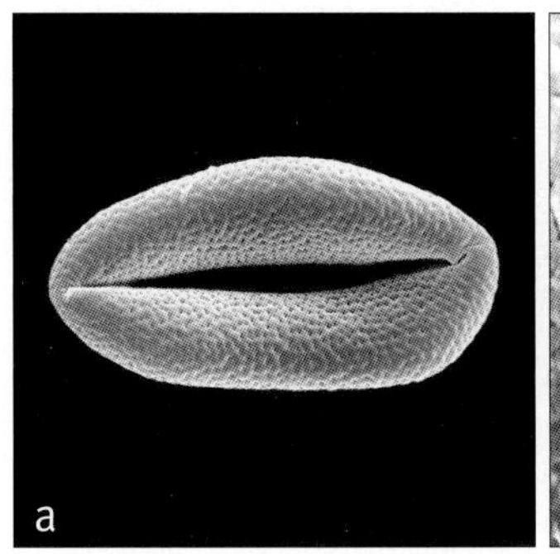

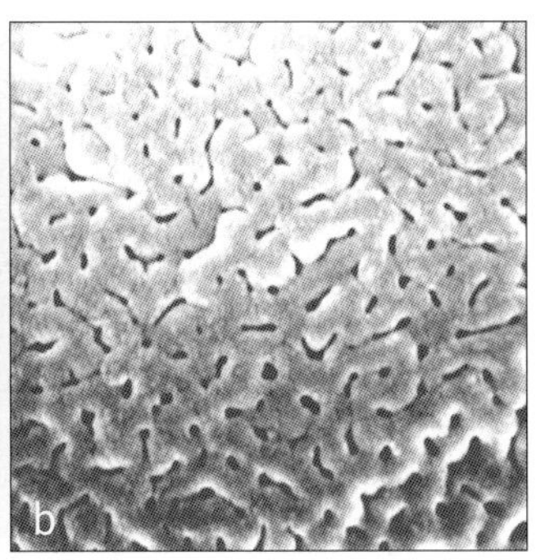

Above: ***Brahea***
a, monosulcate pollen grain, distal face SEM × 1500;
b, close-up, finely perforate-rugulate pollen surface SEM × 8000.
Brahea dulcis: **a**, *Sargent* s.n.; *B. edulis*: **b**, *Palmer* s.n.

Top left: ***Brahea brandegei***, crown, cultivated, Fairchild Tropical Botanic Garden, Florida. (Photo: C.E. Lewis)

Bottom left: ***Brahea berlandieri***, inflorescence, cultivated, Fairchild Tropical Botanic Garden, Florida. (Photo: J. Roncal)

closely, Poinar (2002b) describes differences in the style, stigma, ovary and stamens that separate the fossil flower from those of *Brahea*. The age of the amber is estimated to be somewhere between mid Eocene and mid Miocene. *Brahea*-like pollen (Graham 1976) is described from the Upper Miocene of Mexico (Paraje Solo flora, Veracruz). Small monosulcate grains, *Palmaemargosulcites fossperforatus*, from palm flower compression fossils, recovered from the Middle Eocene oil shales of Messel, Germany are compared with pollen of a number of coryphoid genera, including *Brahea* (Harley 1997).

Notes: A new taxonomic treatment is much needed.

49. COLPOTHRINAX

Moderate solitary hermaphroditic fan palms native to Cuba and Central America, closely related to *Pritchardia* but the petals either open long before anthesis or open and are not shed at anthesis.

Colpothrinax Griseb. & H. Wendl., Bot. Zeit. 37: 148 (1879). Lectotype: **C. wrightii** Griseb. & H. Wendl. ex Siebert & Voss (see Vilmorin 1895).

Combines *kolpos* — swelling, with the palm generic name *Thrinax*, in reference to the swollen trunk of *Colpothrinax wrightii*.

Moderate, solitary, unarmed, pleonanthic, hermaphroditic, tree palms. *Stem* erect, at first covered with persistent fibrous leaf sheaths, later bare, columnar (*Colpothrinax cookii*) or strongly ventricose (*C. wrightii*), marked with close leaf scars. *Leaves* induplicate, shortly costapalmate; sheath disintegrating into a coarse fibrous network or into long fine, pendulous fibres, densely tomentose; petiole long, flattened or slightly channelled adaxially, rounded abaxially, margins acute, densely scaly; adaxial hastula conspicuous, triangular or irregularly lobed, abaxial hastula absent; blade orbicular, irregularly divided sometimes beyond the middle into linear, single-fold segments, these shortly bifid at apex, thick, glabrous and waxy adaxially except along ribs where caducously scaly, abaxially densely covered with minute scales, midribs prominent, transverse veinlets very short, evident abaxially or invisible. *Inflorescences* solitary, interfoliar, several present at the same time, shorter than the leaves, branched to 4 orders; peduncle long, rounded in cross-section, enclosed in overlapping bracts, densely tomentose; prophyll short, tubular, 2-keeled laterally, splitting apically, densely scaly; peduncular bracts 4–9, tubular, with single keel, splitting apically to give a long triangular limb, densely tomentose; rachis equalling the peduncle, tomentose; rachis bracts like the peduncular, several (4–7); first-order branches with a conspicuous, somewhat inflated, brown-tomentose, 2-keeled prophyll and a similiar empty bract, subsequent bracts, membranous, triangular, very small and inconspicuous; rachillae spreading, densely hairy or glabrous, bearing spirally arranged, minute bracts each subtending a low spur bearing a solitary, sessile flower. *Flowers* with calyx cup-like, fleshy, not striate, with 3 short points; corolla considerably exceeding the calyx, fleshy, tubular at the base, divided distally into 3, ± elongate, valvate lobes, forming a deciduous cap at anthesis, adaxially grooved or petals slightly shorter than calyx, not enclosing stamens in bud and persistant; stamens 6, filaments basally connate into an epipetalous cup, adnate to and equalling or only slightly exceeding the corolla tube, free filaments broad basally, attenuate above, anthers elongate, dorsifixed near the base, connectives very narrow, light in colour, latrorse; carpels 3, follicular, ovarian parts distinct, the styles elongate, connate, stigma dot-like, ovule basal, erect, anatropous. *Pollen* ellipsoidal, usually slightly asymmetric; aperture a distal sulcus; ectexine tectate, reticulate, coarsely reticulate, or coarsely foveolate, aperture margin psilate, or scabrate and usually finely perforate; infratectum columellate; longest axis 34–66 µm [2/3]. *Fruit* globose, usually developing from 1 carpel with apical stigmatic and abortive carpel remains, perianth usually persistent; epicarp thin, smooth, mesocarp fleshy with longitudinal anastomosing fibres adjacent to the crustaceous endocarp. *Seed* subglobose, free from the

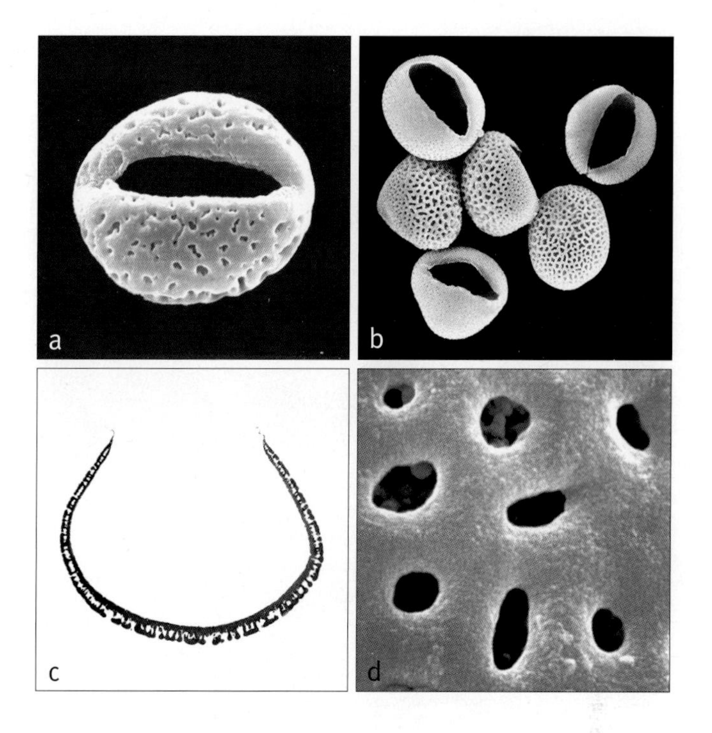

Above: ***Colpothrinax***
a, monosulcate pollen grain, distal face SEM × 1500;
b, group of pollen grains to show difference between pollen surface of proximal and distal faces SEM × 750;
c, ultrathin section through whole pollen grain, polar plane TEM × 2000;
d, close-up, foveolate pollen surface SEM × 8000.
Colpothrinax wrightii: **a**, *Wright* 3964, **d**, *Curtiss* 364; *C. aphanopetala*: **b**, *Read* 79200; *C. cookii*: **c**, *Evans* 2312.

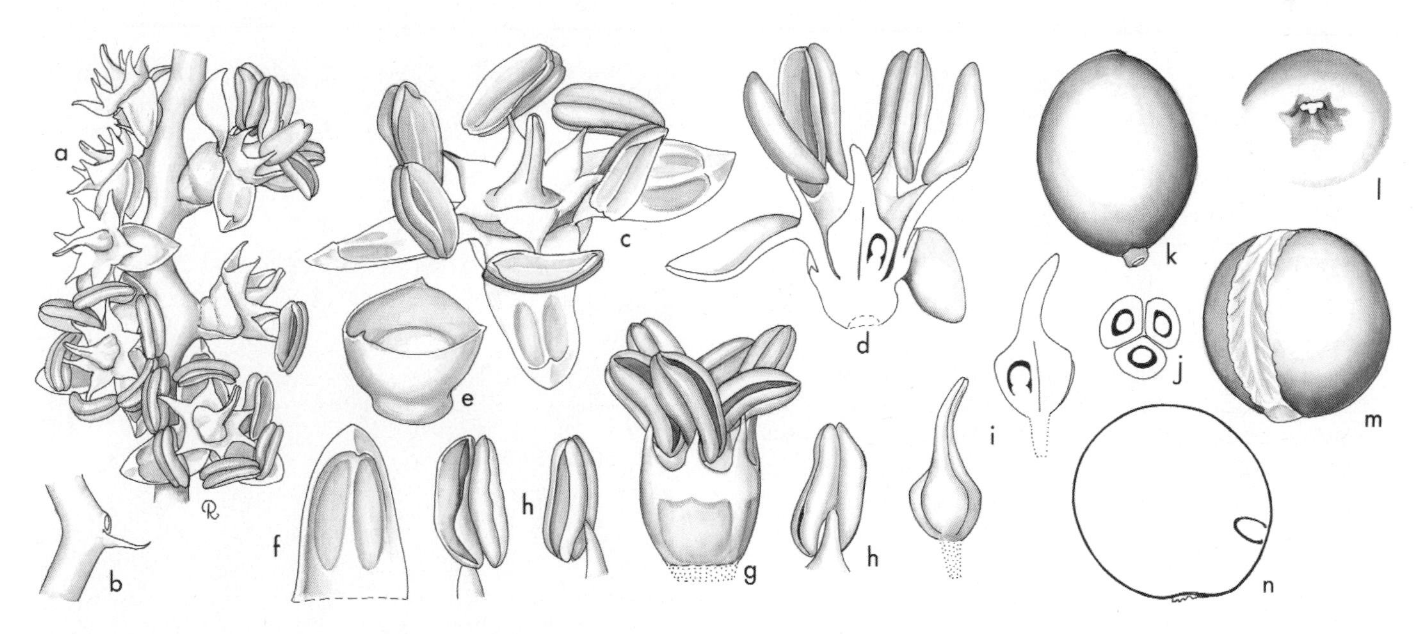

Colpothrinax. **a**, portion of rachilla × 3; **b**, portion of rachilla with floral scar and bract × 3; **c**, flower × 6; **d**, flower in vertical section × 6; **e**, calyx × 6; **f**, free portion of petal, interior view × 6; **g**, androecium, perianth removed × 6; **h**, anther in 3 views × 6; **i**, gynoecium, exterior view and vertical section × 6; **j**, carpels in cross-section × 6; **k**, fruit × $1^1/_2$; **l**, apex of fruit × 6; **m**, seed × 3; **n**, seed in vertical section × 3. *Colpothrinax wrightii*: **a–l**, *Bailey* 12504; **k–n**, *Roig* s.n. (Drawn by Marion Ruff Sheehan)

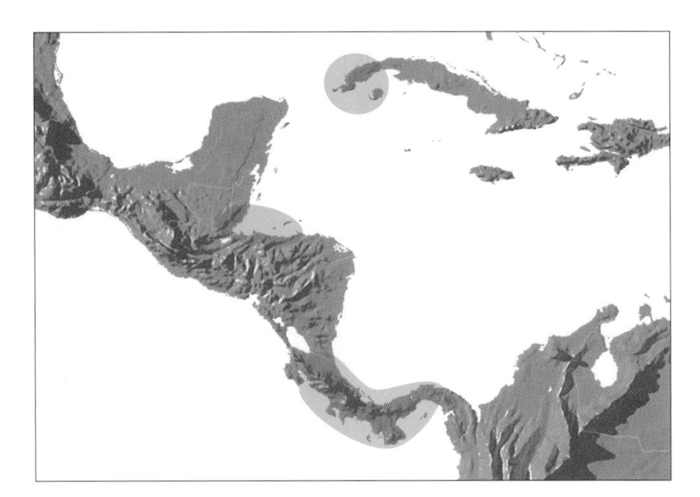

Distribution of ***Colpothrinax***

endocarp except at the small basal hilum, the raphe as long as the seed, rather broad and ± sculptured, lacking noticeable branches, endosperm homogeneous without intruded seed coat below the raphe; embryo lateral towards the base on the antirapheal side. *Germination* remote-tubular; eophyll simple. *Cytology* unknown.

Distribution and ecology: Three species, *Colpothrinax wrightii* endemic to Cuba, *C. cookii* in Belize, Guatemala and Honduras and *C. aphanopetala* in Nicaragua, Costa Rica and Panama. *Colpothrinax wrightii* occurs mostly in semi-dry savannahs and grasslands on white sand, whereas *C. cookii* and *C. aphanopetala* occur in wet premontane and lower montane rain forests up to 1,600 m above sea level.

Anatomy: Leaf (Tomlinson 1961, Read 1998), roots (Seubert 1997), floral (Morrow 1965).

Relationships: The monophyly of *Colpothrinax* has not been tested and its phylogenetic placement is unclear. *Colpothrinax* is resolved as sister to all other Trachycarpeae with low support (Baker *et al.* in review), sister to a clade of *Pritchardia* and *Copernicia* (Uhl *et al.* 1995), or sister to a clade of Rhapidinae, *Acoelorrhaphe*, *Serenoa* and *Brahea* with low support (Asmussen *et al.* 2006).

Common names and uses: Cuban belly palm, barrel palm (*Colpothrinax wrightii*). The trunks of *C. wrightii* are used for making canoes, its leaves as thatch and the fruit is eaten by pigs. All species would make handsome ornamentals.

Taxonomic account: Evans (2001).

Fossil record: See entries for *Cryosophila* and *Brahea*.

Notes: Similar to *Pritchardia* but differing in petals only rarely shed as a cap (*Colpothrinax wrightii*) and in a more shallow stamen tube. The inflorescence seems striking in the length and massiveness of the peduncle and rachis, with the branches very small in comparison.

Top right: ***Colpothrinax wrightii***, crown, cultivated, Fairchild Tropical Botanic Garden, Florida. (Photo: J. Dransfield)

Bottom right: ***Colpothrinax wrightii***, crown, cultivated, Fairchild Tropical Botanic Garden, Florida. (Photo: C.E. Lewis)

50. COPERNICIA

Moderate to massive usually solitary hermaphroditic fan palms, native to Cuba, where there is a great radiation of species, and to Hispaniola and South America; the highly branched inflorescence usually has rachillae with completely tubular bracts and the endosperm is ruminate.

Copernicia Mart. ex Endl., Gen. pl. 253 (1837). Type: *Colpothrinax cerifera* (Arruda) Mart. (*Corypha cerifera* Arruda) = **C. prunifera** (Mill.) H.E. Moore (*Palma prunifera* Mill.).

Arrudaria Macedo, Notice palm. Carnauba 5 (1867). Type: *A. cerifera* (Arruda) Macedo (illegitimate name).

Coryphomia Rojas Acosta, Bull. Acad. Intern. Géogr. Bot. 28: 158 (1918). Type: *C. tectorum* Rojas Acosta.

Commemorates the Polish astronomer, Nicolaus Copernicus (1473–1543).

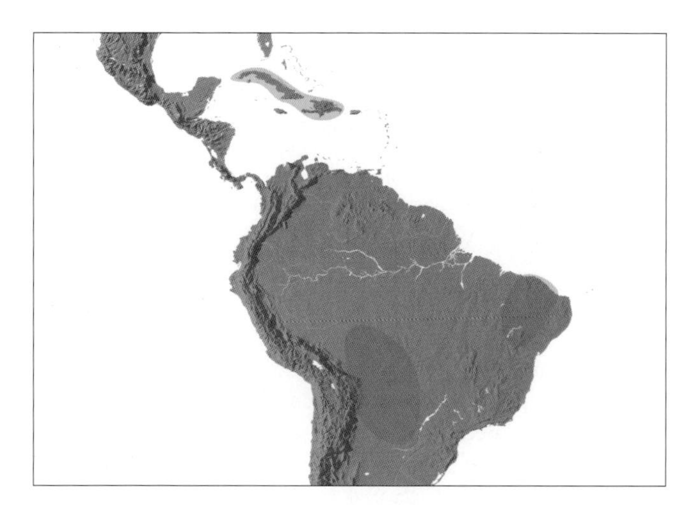

Distribution of ***Copernicia***

Moderate to tall, solitary (rarely clustered), slow-growing, armed, pleonanthic, hermaphroditic palms. *Stems* covered with persistent leaf sheaths for part or all their length, sometimes becoming bare with age, the naked portion roughened and often with close, rough, ± evident leaf scars, basally expanded or not (*Copernicia berteroana*). *Leaves* induplicate, palmate to shortly costapalmate; sheath fibrous, petiole lacking or very short to elongate, channelled or flattened adaxially, rounded abaxially, the margins armed with stout teeth; adaxial hastula short to very long, coriaceous, triangular, unarmed or spinose margined or erose, sometimes persisting after the lamina has disintegrated, abaxial hastula absent; blade wedge-shaped or orbicular, divided $^1/_4$ to $^1/_3$ to the base into single-fold pointed segments, outermost bifid at the apex, segments often spiny margined, thick, very stiff, major ribs with caducous tomentum, midribs prominent abaxially, transverse veinlets not evident. *Inflorescences* interfoliar, often exceeding the leaves, frequently densely tomentose, branched to 6 orders; peduncle elongate, narrow, elliptic in cross-section; prophyll tubular; peduncular bracts 0–1, apparently 2-winged, irregularly split apically; rachis about as long as or longer than the peduncle; rachis bracts tubular, closely sheathing, first-order branches each bearing a prophyll, subsequent bracts tubular, tightly sheathing, split apically, gradually reduced and lacking on rachillae or present and conspicuous through to the flowers, usually densely tomentose; rachillae of medium length to very short, stout or slender, often recurved, bearing spirally inserted, membranous bracts, each subtending a solitary flower or groups of 2–4 flowers, distant or very crowded, the group and each flower subtended by a membranous bracteole. *Flowers* with 3 sepals united in a thick-based, 3-lobed cup, lobes usually acute; corolla tubular below with 3 thick-tipped, valvate lobes, prominently pocketed and furrowed within; stamens 6, united by their broad filament bases into a cupule, borne at the mouth of the corolla tube, distinct filament lobes abruptly narrowed to short slender tips, these not inflexed in bud, antesepalous lobes sometimes larger than antepetalous ones, anthers usually small, ovate or oblong, dorsifixed near their bases, latrorse; carpels 3, follicular, distinct basally, styles wide basally, tapering, connate, stigma dot-like, ovule erect, basal, anatropous. *Pollen* ellipsoidal, with slight to obvious asymmetry; aperture a distal sulcus; ectexine tectate, finely perforate, perforate and micro-channelled, or perforate-rugulate, aperture margin

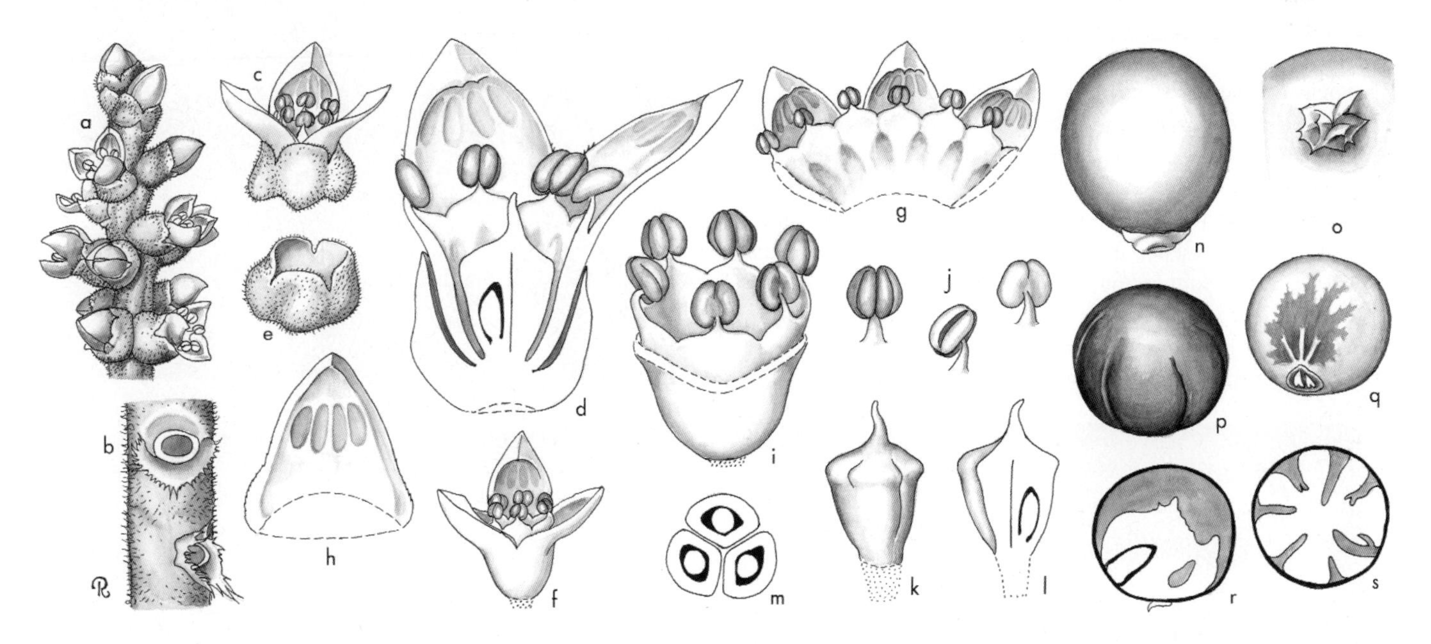

Copernicia. **a**, portion of rachilla × 3; **b**, portion of rachilla, flowers removed to show bracts × 6; **c**, flower × 6; **d**, flower in vertical section × 12; **e**, calyx × 6; **f**, flower, calyx removed × 6; **g**, corolla and androecium, interior view, expanded × 6; **h**, free portion of petal, interior view × 12; **i**, flower, petals removed to show androecium × 12; **j**, stamen in 3 views × 12; **k**, gynoecium × 12; **l**, gynoecium in vertical section × 12; **m**, carpels in cross-section × 12; **n**, fruit × $1^1/_2$; **o**, apex of fruit and abortive carpels × 3; **p**, **q**, seed in 2 views × $1^1/_2$; **r**, seed in vertical section × $1^1/_2$; **s**, seed in cross-section × $1^1/_2$. *Copernicia yarey*: all from *Read* 606. (Drawn by Marion Ruff Sheehan)

slightly finer; infratectum columellate; longest axis 24–38 µm; post-meiotic tetrads usually tetrahedral, occasionally tetragonal or, rarely, rhomboidal [5/21]. *Fruit* ovoid or spherical, usually developing from 1 carpel, carpellary remains basal, stigmatic remains apical; epicarp smooth, drying minutely roughened, mesocarp slightly fleshy with longitudinally anastomosing fibres, endocarp moderately thick, crustaceous. *Seed* ovoid or globose, basally attached, with large ovate basal hilum, raphe indistinct, narrow, branching, endosperm deeply ruminate; embryo subbasal. *Germination* remote-tubular; eophyll entire, lanceolate. *Cytology*: 2n = 36.

Distribution and ecology: Twenty-one species, three in South America, two in Hispaniola, the remainder in Cuba and several described naturally occurring hybrids. In the Caribbean, the species occur in savannahs or woodlands in the lowlands in relatively dry situations. The South American species occur in pure natural stands. *Copernicia prunifera* is found in vast natural stands in Brazil and grows in areas prone to seasonal flooding.

Anatomy: Leaf (Tomlinson 1961), roots (Seubert 1997), flower (Morrow 1965).

Relationships: The monophyly of *Copernicia* has not been tested. Uhl *et al.* (1995) and Baker *et al.* (in review) found it to be sister to *Pritchardia*. However, Asmussen *et al.* (2006) resolved the genus as sister to the Livistoninae with low support.

Common names and uses: Carnauba (*Copernicia prunifera*), petticoat palm (*C. macroglossa*), caranda palms. *Copernicia prunifera* is of great economic importance as the source of high quality carnauba wax (Johnson 1985). Other parts of all species are also used locally as leaves for thatching, stems for building, and fibres for brushes and rope. Starch from stems and fruits is edible; seedlings are used for fodder.

Taxonomic accounts: Dahlgren & Glassman (1961, 1963).

Fossil record: Monocolpate pollen from the Lower Miocene of Poland (Macko 1957) has been compared to *Copernicia* pollen, but the pollen is of too general a coryphoid type to be conclusive. See also entry for *Serenoa*.

Notes: There is considerable diversity in the genus, especially in Cuba. The very large hastulae of some species, e.g., *Copernicia macroglossa*, are most remarkable but their functional significance, if any, has yet to be explained. Another unusual feature is the presence in some species of completely tubular rachilla bracts subtending the flower clusters. Some of the largest species, such as *C. baileyana*, make most imposing ornamentals but these are notoriously slow growing.

Top: ***Copernicia berteroana***, habit, Dominican Republic. (Photo: J. Dransfield)

Above: ***Copernicia macroglossa***, crown and inflorescences, cultivated, Fairchild Tropical Botanic Garden, Florida. (Photo: J. Dransfield)

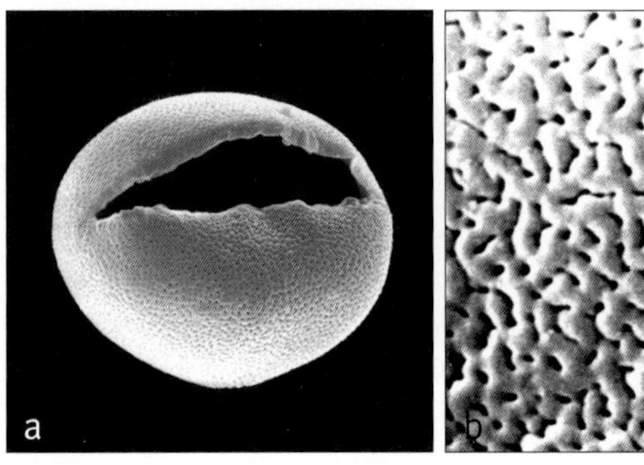

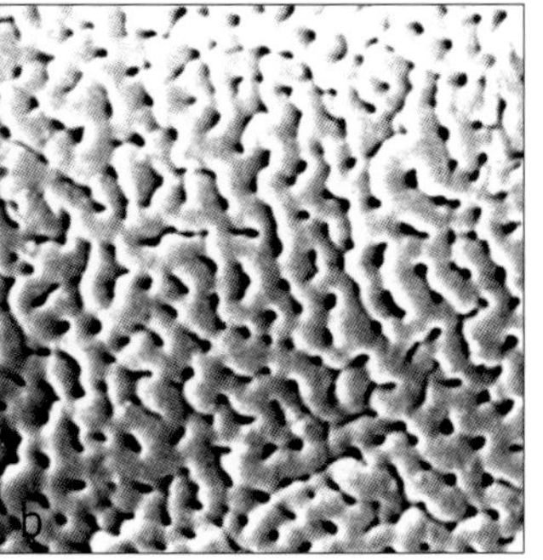

Left: ***Copernicia***

a, monosulcate pollen grain, distal face SEM × 1500;

b, close-up, finely perforate-rugulate pollen surface SEM × 8000.

Copernicia hospita: **a**, *Shafer* 2457; *C. curtissii*: **b**, *Curtiss* 435.

51. PRITCHARDIA

Moderate solitary hermaphroditic fan palms found on scattered islands through the western Pacific and with a major radiation on the Hawaiian Islands, immediately recognisable by the long peduncles and flowers in which the tip of the corolla falls off to expose the stamens.

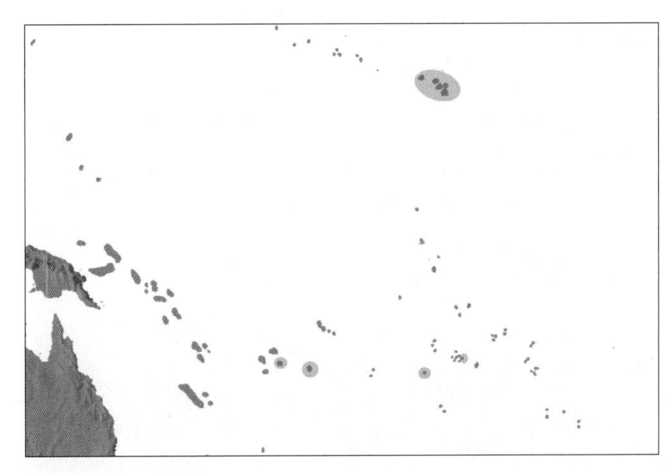

Distribution of ***Pritchardia***

Pritchardia Seem. & H. Wendl. ex H. Wendl., Bonplandia 10: 197 (1862) (conserved name). Type: **P. pacifica** Seem. & H. Wendl. ex H. Wendl..

Eupritchardia Kuntze, Revis. gen. pl. 3(3): 323 (1898). Type: *E. pacifica* (Seem. & H. Wendl. ex H. Wendl.) Kuntze (= *Pritchardia pacifica* Seem. & H. Wendl. ex H. Wendl.).

Styloma O.F. Cook, J. Wash. Acad. Sci. 5: 241 (1915). Type: *S. pacifica* (Seem. & H. Wendl. ex H. Wendl.) O.F. Cook (= *Pritchardia pacifica* Seem. & H. Wendl. ex H. Wendl.).

Honors William T. Pritchard, one-time British Consul in the Fiji Islands.

Moderate, solitary, unarmed, pleonanthic, hermaphroditic tree palms. *Stem* erect, sometimes deeply striate, ringed with close leaf scars. *Leaves* induplicate, costapalmate, marcescent in immature acaulescent individuals, deciduous in trunked individuals; sheath tomentose, soon disintegrating into a mass of fibres; petiole elongate, flattened or channelled adaxially, abaxially rounded or angular, extending into the costa without interruption, usually tomentose; adaxial hastula a ridge with a central point, abaxial hastula absent; blade divided to ca. $^1/_3$ to $^1/_2$ its radius along adaxial folds into single-fold segments, further shallowly divided along abaxial folds, interfold filaments often present, segments stiff, held in 1 plane or variously pendulous, surfaces similar or more distinctly glaucous abaxially, usually copiously tomentose along ribs, frequently with small scales on abaxial surface, transverse veinlets conspicuous or obscure. *Inflorescences* interfoliar, solitary or 2–4 together in each axil, sheathed by a common prophyll, branched to 3 orders; peduncle conspicuous, stiff, ± erect to pendulous, shorter or longer than the leaves; prophyll tubular, 2-keeled, closely sheathing, densely tomentose, sometimes disintegrating into a weft of fibres; peduncular bracts several, similar to the prophyll, tending to split along one side, irregularly tattering, sometimes inflated, adaxially glabrous, densely tomentose abaxially, rarely becoming glabrous; rachis much shorter than the peduncle; rachillae straight, curved or somewhat zigzag, tending to be crowded forming a head of flowers, glabrous or sparsely to densely hairy, bearing spirally arranged, minute bracts subtending solitary flowers. *Flowers* sessile or borne on very low tubercles, floral bracteoles apparently absent; calyx tubular, shallowly 3-lobed distally, rather thick and coriaceous; corolla considerably exceeding the calyx, coriaceous, tubular at the base, divided distally into 3 ± elongate valvate lobes, the lobes forming a cap deciduous at anthesis; stamens 6, borne near the mouth of the corolla tube, the filament bases connate to form a conspicuous tube projecting beyond the calyx, with 6 short distinct

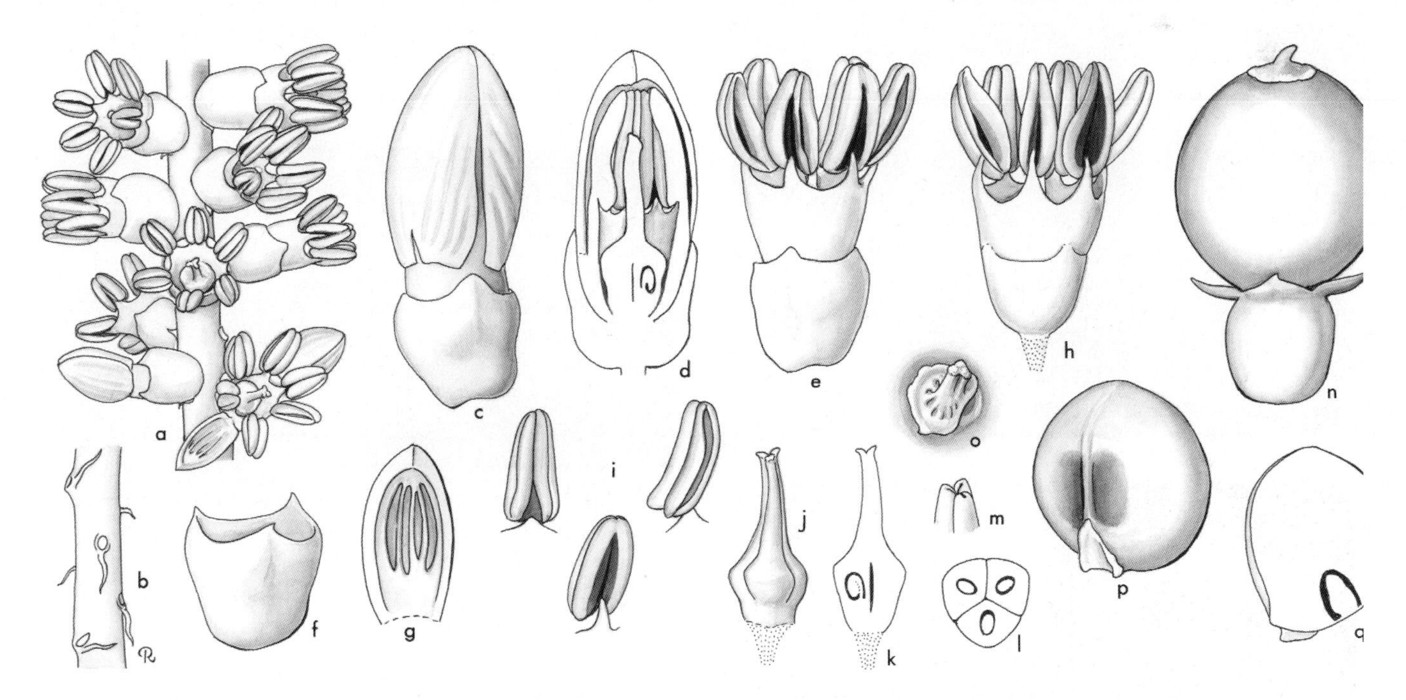

Pritchardia. **a**, portion of rachilla × 3; **b**, portion of rachilla with flowers removed to show bracteoles × 3; **c**, flower with petals already separated × 6; **d**, flower in vertical section × 6; **e**, flower with petals fallen × 6; **f**, calyx × 6; **g**, petal, interior view × 6; **h**, flower with petals fallen and sepals removed × 6; **i**, stamen in 3 views × 6; **j**, gynoecium × 6; **k**, gynoecium in vertical section × 6; **l**, carpels in cross-section × $7^1/_2$; **m**, stigmas × 12; **n**, fruit × $4^1/_2$; **o**, stigmatic remains × 6; **p**, seed × $4^1/_2$; **q**, seed in vertical section × $4^1/_2$. *Pritchardia thurstonii*: **a–m**, *Read* 662; **n–q**, *Read* 690. (Drawn by Marion Ruff Sheehan)

tips bearing oblong, ± erect, latrorse anthers; gynoecium tricarpellate, the carpels wedge-shaped, distinct in the ovarian region, connate in a common elongate style bearing a minutely 3-lobed stigma, ovule basally attached, anatropous. *Pollen* ellipsoidal, with slight to obvious asymmetry, occasionally oblate triangular; aperture a distal sulcus, less frequently a trichotomosulcus; ectexine tectate, scabrate, or perforate, aperture margin similar; infratectum columellate; longest axis 38–53 µm [7/27]. *Fruit* spherical or ovoid, developing from 1 carpel only, bearing apical stigmatic and sterile carpel remains; calyx persistent; epicarp smooth, mesocarp rather thin, fleshy, fibrous, endocarp thin, woody and rather brittle, sometimes thickened at the base. *Seed* ± spherical, basally or subbasally attached, with rounded hilum, endosperm homogeneous, the seed coat slightly thickened by the hilum but endosperm without conspicuous intrusion of the seed coat; embryo basal. *Germination* remote-tubular; eophyll entire, lanceolate, plicate. *Cytology*: 2n = 36, 36 ± 2.

Distribution and ecology: About 27 recognised species from Fiji, Tonga, Danger Islands and Hawaii. All but five species are Hawaiian endemics; many are extremely rare and endangered, or not seen in the wild for several years. Most of the Hawaiian species are found on the windward slopes of the islands in wet forested areas from sea level to over 1400 m altitude; a few species occur in dry forest on the leeward sides.

Common names and uses: Loulu palms. The large leaves are used as fans and umbrellas.

Anatomy: Leaf (Tomlinson 1961), roots (Seubert 1997), floral (Morrow 1965), fruit (Reddy & Kulkarni 1982), endocarp (Murray 1973).

Relationships: Asmussen *et al.* (2006) found strong support for the monophyly of *Pritchardia* and moderate support for its sister relationship with *Washingtonia*. There is also moderate support for a sister relationship between *Pritchardia* and *Copernicia* (Uhl *et al.* 1995, Baker *et al.* in review).

Taxonomic accounts: Beccari & Rock (1921). See also Hodel (1980, 2007).

Top: ***Pritchardia viscosa***, habit, Hawaii. (Photo: J. Dransfield)

Above left: ***Pritchardia viscosa***, crown, Hawaii. (Photo: J. Dransfield)

Above right: ***Pritchardia viscosa***, fruit, Hawaii. (Photo: J. Dransfield)

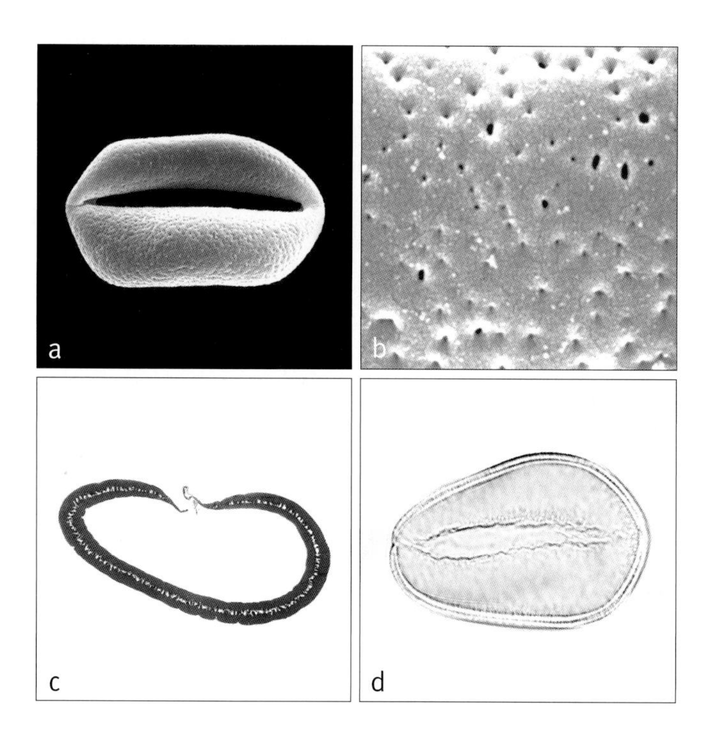

Right: ***Pritchardia***
a, monosulcate pollen grain, distal face SEM × 1000;
b, close-up, finely perforate pollen surface SEM × 8000;
c, ultrathin section through whole pollen grain, polar plane TEM × 1500;
d, monosulcate pollen grain, high-mid-focus LM × 1000.
Pritchardia minor: **a** & **b**, *Cranwell et al.* 3103; *P. pacifica*: **c**, *Christopherson* 2842; *P. martii*: **d**, *St John* 10180.

Fossil record: Leaves: *Pritchardites* (*P. wettinioides*) from the Tertiary of Italy (Bureau 1896) was placed in the synonomy of *Phoenicites* by Read and Hickey (1972). Silicified fruits that, "come very close to some species of *Pritchardia* and *Licuala* specially [sic] the latter" (Shete & Kulkarni 1985) are reported from the Indian Deccan Intertrappean, Maharashtra State (although the age span of these volcanic deposits is controversial, see Chapter 5). Numerous records of *Pritchardia* pollen and seeds are reported from the pre-human Holocene of Kaua'a Island, Hawaii (Burney *et al.* 2001). Thick-walled monosulcate pollen, with a distinctive narrow infratectum, *Palmaepollenites kutchensis*, from the Middle Eocene of Central Java (Nanggulan Formation) is compared with pollen of *Pritchardia*, and with two arecoid genera, *Basselinia* and *Burretiokentia* (Harley & Morley 1995).

Notes: Characterised by the deciduous cap of the corolla lobes and usually distinguishable by the inflorescence, which has a relatively long peduncle with flowering branches clustered at the end.

52. WASHINGTONIA

Striking solitary hermaphroditic tree fan palms native to desert oases in southwestern USA and northwestern Mexico, distinctive in the long inflorescences with bracts that split and open out, becoming almost sword-like, and the chaffy petals and sepals.

Washingtonia H. Wendl., Bot. Zeit. 37: lxi, 68, 148 (1879) (conserved name). Type: **W. filifera** (Linden ex André) H. Wendl. (*Pritchardia filifera* Linden ex André).

Neowashingtonia Sudw., U.S.D.A. Div. Forest. Bull. 14: 105 (1897) (substitute name).

Honors George Washington (1732–1799).

Robust, tall, solitary, armed, pleonanthic, hermaphroditic tree palms. *Stem* erect, usually partly or wholly covered with persistent dry leaves, ringed with close leaf scars, sometimes enlarged basally. *Leaves* induplicate, costapalmate, marcescent; sheath with a conspicuous abaxial cleft below the petiole, the margins disintegrating into a dark-brown fibrous network, the sheath densely caducous tomentose, margins becoming fibrous; petiole elongate, adaxially flattened to slightly concave, abaxially rounded, margins strongly armed with curved teeth, teeth becoming smaller and sparser distally; adaxial hastula large, membranous, triangular, irregularly margined and tattering, abaxial hastula a low ridge obscured by a mat of thick tomentum (in *Washingtonia robusta*); blade divided irregularly to ca. $^1/_3$ its length into linear single-fold segments, bifid at their apices, pendulous at maturity, filamentous at the tips, interfold filaments conspicuous, midribs prominent abaxially, transverse veinlets obscure. *Inflorescence* interfoliar, ascending, branched to 3(–4) orders, equalling or generally exceeding the leaves, curved, slender; peduncle short; prophyll tubular, closely sheathing, 2-keeled, irregularly tattered at the tip; peduncular bract 1, like the prophyll but with a single keel; rachis much longer than the peduncle; rachis bracts tubular basally, splitting longitudinally, becoming flattened and sword-like, very coriaceous; subsequent bracts minute or lacking; rachillae numerous, short, very slender, glabrous. *Flowers* solitary, elongate, spirally inserted, briefly pedicellate; calyx chaffy, tubular proximally with 3 irregularly tattered, imbricate lobes, persistent in fruit; corolla tubular for ca. $^1/_3$ its length, distinct lobes valvate, narrowly ovate, tapering to a point, reflexed at anthesis, thin, almost chaffy; stamens 6, borne at the mouth of corolla tube, filaments elongate, gradually tapering from a fleshy base, anthers elongate, medifixed, versatile, latrorse, connective narrow; gynoecium top-shaped, carpels 3, distinct basally, united through the long slender styles, ovule basal, erect, (?)anatropous. *Pollen* ellipsoidal, with slight to extreme asymmetry; aperture a distal sulcus; ectexine tectate, rugulate or reticulate, aperture margin slightly finer; infratectum columellate; longest axis 35–51 µm; post-meiotic tetrads tetrahedral, tetragonal or decussate, proportions not recorded [2/2]. *Fruit* small, broadly ellipsoidal to globose, often falling with the pedicel and unilaterally ruptured calyx tube attached, blackish, stigmatic and abortive carpel remains apical; epicarp smooth, thin, mesocarp thin, fleshy with a few flattened longitudinal fibres, endocarp thin, crustaceous, not adherent to the seed, smooth within. *Seed* ellipsoidal, somewhat compressed, hilum eccentrically basal, raphe extending $^2/_3$ the length of the shining red-brown seed coat, loosely branched laterally, seed coat intrusion very thin, endosperm homogeneous; embryo basal. *Germination* remote-ligular; eophyll entire, lanceolate. *Cytology*: 2n = 36.

Right: ***Washingtonia robusta***, habit, cultivated, Fairchild Tropical Botanic Garden, Florida. (Photo: C.E. Lewis)

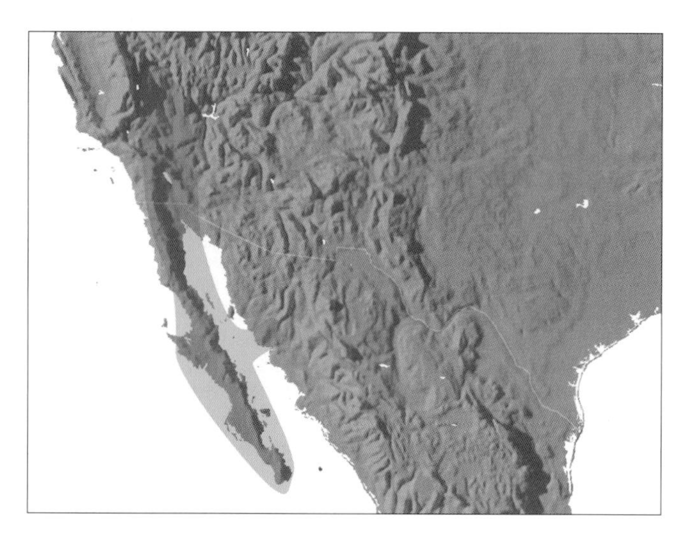

Distribution of ***Washingtonia***

Above: ***Washingtonia robusta***, infructescences, cultivated, Fairchild Tropical Botanic Garden, Florida. (Photo: C.E. Lewis)

Distribution and ecology: Two closely related species: *Washingtonia filifera* occurs in southeastern California, western Arizona, and Baja California; *W. robusta* in Baja California and Sonora, Mexico. Desert palms occurring along streams and canyons, also about springs and seepages in more open areas.
Anatomy: Leaf (Tomlinson 1961), roots (Seubert 1997), floral (Morrow 1965).
Relationships: The monophyly of *Washingtonia* has not been tested. *Washingtonia* is resolved as sister to *Pritchardia* with low support (Asmussen *et al.* 2006).
Common names and uses: Washington palms, desert fan palm (*Washingtonia filifera*), Mexican washington (*W. robusta*). Excellent ornamentals for the drier subtropics.

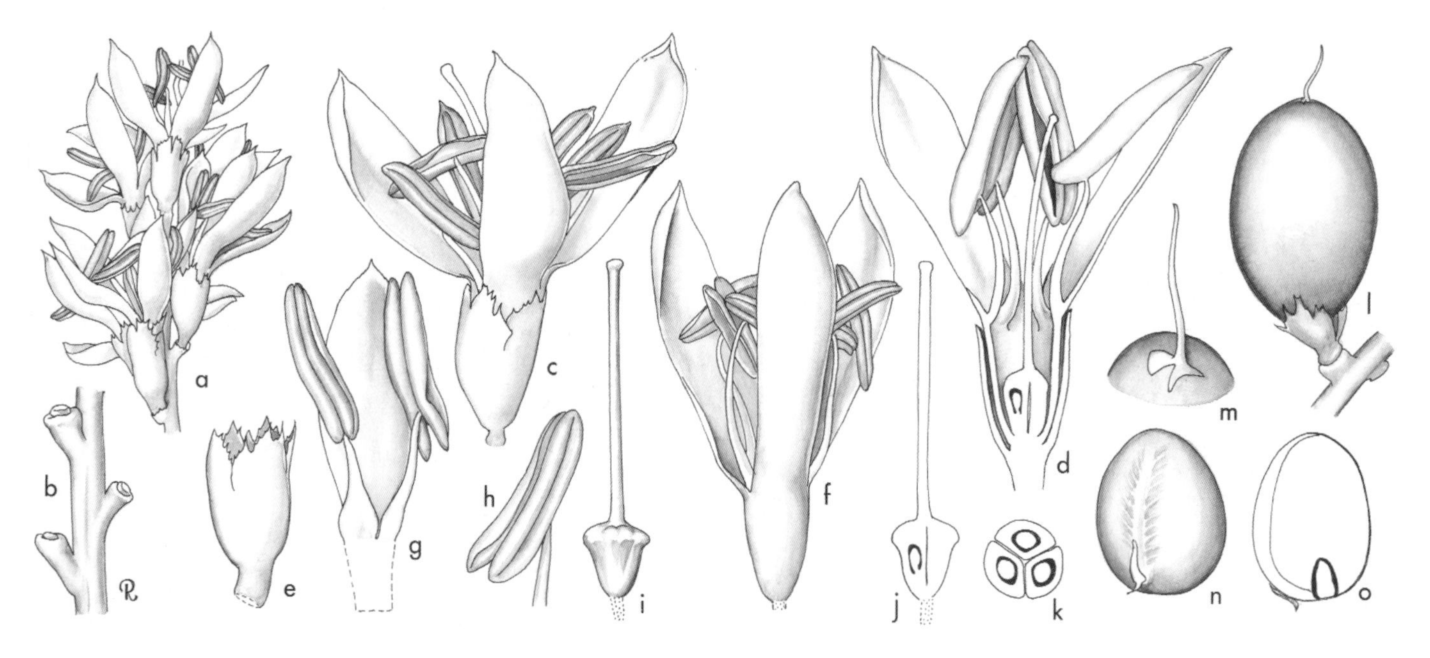

Washingtonia. **a**, portion of rachilla × 3; **b**, portion of rachilla with pedicels × 6; **c**, flower × 6; **d**, flower in vertical section × 6; **e**, calyx × 6; **f**, flower, calyx removed × 6; **g**, petal with 2 stamens, interior view × 6; **h**, anther × 6; **i**, gynoecium × 6; **j**, gynoecium in vertical section × 6; **k**, carpels in cross-section × 6; **l**, fruit × 3; **m**, apex of fruit × 6; **n**, seed × 3; **o**, seed in vertical section × 3. *Washingtonia robusta*: all from *Read* 725. (Drawn by Marion Ruff Sheehan)

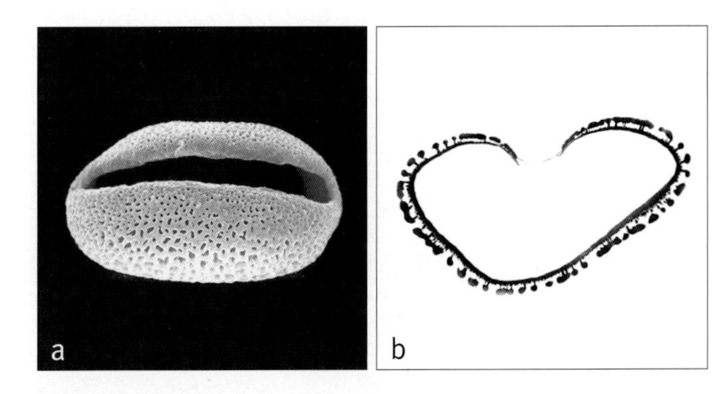

Above: ***Washingtonia***
a, monosulcate pollen grain, distal face SEM × 1500;
b, ultrathin section through whole pollen grain, polar plane TEM × 2500.
Washingtonia robusta: **a**, *Palmer* 144; *W. filifera*: **b**, *W.G. Wright* s.n.

Taxonomic accounts: Bailey (1936) and Zona (1997).
Fossil record: No generic records found.
Notes: *Washingtonia* stands apart from other members of the Livistoninae because of its unusual sword-shaped bracts and curious chaffy perianths, in particular, the large, flat, reflexed petals.

Tribe Chuniophoeniceae J. Dransf., N.W. Uhl, C. Asmussen, W.J. Baker, M.M. Harley & C. Lewis., Kew Bull. 60: 561 (2005). Type: **Chuniophoenix**.

Pleonanthic or hapaxanthic, hermaphroditic or dioecious palms; leaves induplicate, palmate or costapalmate; leaf base with or without a central split; flowers solitary or borne in simple cincinni, each subtended by a completely tubular rachilla bract; gynoecium syncarpous with three carpels united basally, the styles separate or connate throughout, if connate, then with separate stylar canals.

The inclusion of four genera in this tribe is supported by the occurrence of tubular rachilla bracts, a pedicelliform corolla base (to varying degrees) and similar gynoecium. The four genera, with a total of only five or six species, display remarkable diversity of vegetative and reproductive morphology.

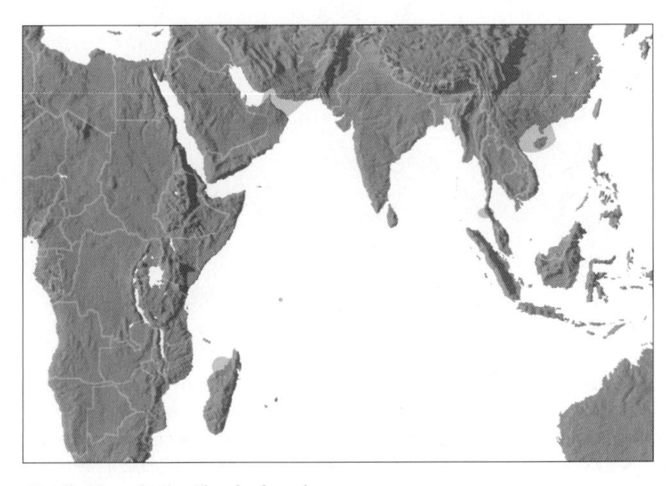

Distribution of tribe **Chuniophoeniceae**

Tribe Chuniophoeniceae is monophyletic and highly supported in all phylogenies, and often resolves on a long branch (e.g., Bayton 2005, Asmussen *et al.* 2006, Dransfield *et al.* 2008, Baker *et al.* in review). It is placed within the syncarpous clade and is sister to the remaining tribes in the group, Caryoteae, Corypheae and Borasseae (Bayton 2005, Asmussen *et al.* 2006).

Key to Genera of Chuniophoeniceae

1. Leaf blade concolorous; adaxial hastula absent; hermaphroditic; fruit smooth; endosperm homogeneous or ruminate 2
1. Leaf blade strongly discolorous; adaxial hastula present; dioecious; fruit papillate; endosperm ruminate . . . . **54. Kerriodoxa**
2. Hapaxanthic palms; inflorescence paniculate, highly branched; fruit with apical or basal stigmatic remains; endosperm homogeneous, rounded or deeply grooved . . 3
2. Pleonanthic palms; inflorescence spicate or branched to 1–2 orders; fruit with apical stigmatic remains; endosperm ruminate or homogeneous . . . . . . **53. Chuniophoenix**
3. Massive solitary unbranched palm; leaf blade with a hastula; fruit with apical stigmatic remains; seed deeply and regularly grooved . . . . **56. Tahina**
3. Shrubby palm with short erect dichotomously branching stems; leaf blade lacking a hastula; fruit with basal stigmatic remains; seed not grooved . . . . **55. Nannorrhops**

53. CHUNIOPHOENIX

Small clustering palms of southern China and northern Vietnam, distinctive in the leaf lacking a hastula and the tubular membranous rachilla bracts.

Chuniophoenix Burret, Notizbl. Bot. Gart. Berlin-Dahlem 13: 583 (1937). Type: **C. hainanensis** Burret.

Commemorates W.Y. Chun who was director of the Botanical Institute, College of Agriculture, Sun Yatsen University, Canton, China, by combining his name with phoenix — *a general name for a palm.*

Small, clustered, unarmed, pleonanthic, hermaphroditic or occasionally polygamo-dioecious palms. *Stem* slender to moderate, erect, ringed with leaf scars. *Leaves* induplicate, palmate, apparently marcescent; sheath tubular at first, later splitting opposite the petiole, in *Chuniophoenix hainanensis* with a triangular cleft at the base of the petiole, usually covered with floccose indumentum; petiole well developed, grooved, margins smooth; hastulae absent; blade irregularly divided almost to the base into single- to several-fold segments with entire or shallowly toothed apices, segments thin, with prominent ad- and abaxial ribs, and sparse floccose indumentum, blade margins decurrent on petiole, midribs prominent abaxially, transverse veinlets abundant. *Inflorescences* among the leaves, spicate or with up to 2 orders of branching; peduncle well developed, adnate at the base to the internode above the subtending node; prophyll short, persistent, tubular, 2-keeled; peduncular bracts several, persistent, tubular, tightly sheathing, distant; rachis (where present) shorter to much longer than the peduncle; rachis bracts like the peduncular; rachillae erect or spreading, bearing ± imbricate, ± spirally arranged, persistent, tubular bracts with short triangular limbs, each subtending a flower group. *Flowers* solitary or arranged in a condensed cincinnus of 1–7 flowers, each flower in turn exserted from the rachilla bract on a columnar pedicel, and bearing a 2-keeled tubular (later splitting) bracteole; calyx tubular, somewhat chaffy, shallowly 2–3-lobed, tending to split irregularly in the distal part; corolla with a long stalk-like base and 3 triangular, valvate, later reflexed, fleshy lobes; stamens 6, the antesepalous free, the antepetalous adnate to the base of the petals, filaments elongate, fleshy, the antepetalous ones much wider basally than those opposite the sepals, anthers oval to oblong, introrse; gynoecium tricarpellate, ovary somewhat stalked, elongate, with septal nectaries at the base, style 3-grooved, elongate, trifid at the tip, the stigmas somewhat divergent, ovules anatropous (very rarely 2 present in 1 carpel — ?as a monstrosity), attached to the inner carpel wall at the base; ovary or pollen aborting in polygamodioecious individuals. *Pollen* ellipsoidal, with slight to obvious asymmetry; aperture a distal sulcus; ectexine tectate, finely perforate, perforate and micro-channelled, or perforate-rugulate, aperture margin slightly finer; infratectum columellate; longest axis 33–45 µm; post-meiotic tetrads tetrahedral [2/2]. *Fruit* small, ± rounded, 1-seeded, green when immature, scarlet when ripe, with apical stigmatic remains; epicarp smooth, lustrous, or somewhat pebbled on drying, mesocarp fleshy, endocarp thin. *Seed* irregularly globose, basally attached with short hilum, grooved along raphe, and with sparse anastomosing grooves corresponding to the raphe branches, endosperm ruminate or homogeneous; embryo basal. *Germination* remote-ligular (Chavez 2003); eophyll entire. *Cytology*: 2n = 36.

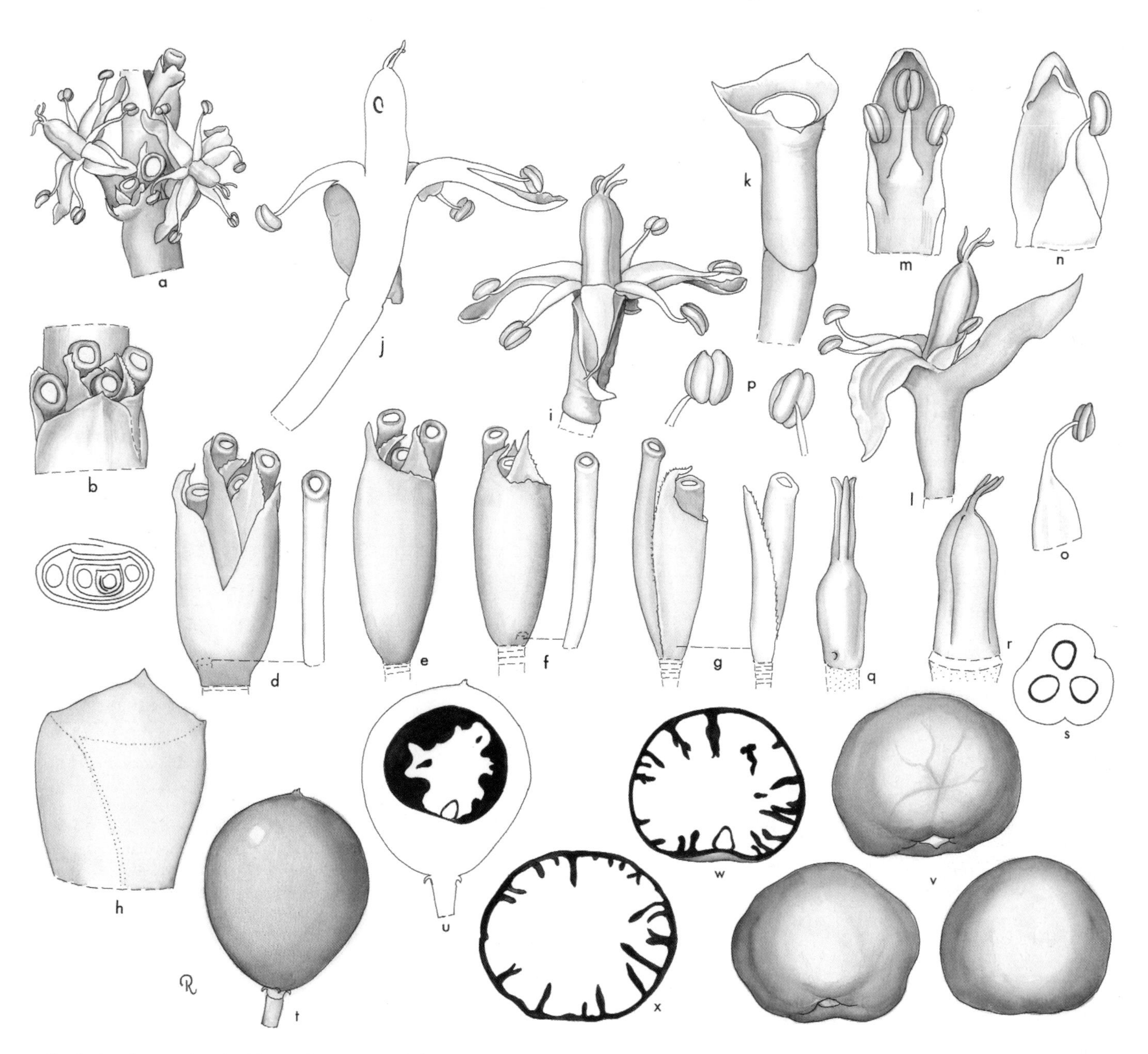

***Chuniophoenix*. a**, portion of rachilla × 3; **b**, cincinnus with floral stalks and bracts × 4 1/2; **c**, diagram of cincinnus × 4 1/2; **d–g**, dissections of cincinnus showing positions of flowers and bracteoles × 4 1/2; **h**, bract subtending cincinnus × 6; **i**, flower × 6; **j**, flower with stalk in vertical section × 6; **k**, calyx and stalk × 7 1/2; **l**, flower, calyx removed × 6; **m**, petal with epipetalous stamens × 6; **n**, petal lobe with stamen × 6; **o**, stamen × 6; **p**, anther in 2 views × 9; **q**, young gynoecium × 7 1/2; **r**, mature gynoecium × 7 1/2; **s**, ovary in cross-section × 9; **t**, fruit × 1 1/2; **u**, fruit in vertical section × 1 1/2; **v**, seed in 3 views × 2 1/4; **w**, seed in vertical section × 2 1/4; **x**, seed in cross-section × 2 1/4. *Chuniophoenix hainanensis*: **a–s**, *Wei Chao Fen* 123012 and *Whitmore* 3152; **t–x**, *Wei Chao Fen* 123194 and *Whitmore* 3152. (Drawn by Marion Ruff Sheehan)

Distribution and ecology: Three species occurring in Vietnam, south China, and Hainan Island. Undergrowth palms in forest.

Anatomy: Root (Seubert 1997).

Relationships: *Chuniophoenix* is strongly supported as monophyletic (Bayton 2005), but there is some conflict regarding its position relative to the other three genera of Chuniophoeniceae. There is high support for a sister relationship between *Chuniophoenix* and *Kerriodoxa* (Uhl *et al.* 1995, Bayton 2005), but this conflicts with other studies that place *Kerriodoxa* as sister to both *Nannorrhops* with low to moderate support (Asmusen *et al.* 2006, Baker *et al.* in review) and *Tahina* with moderate support (Dransfield *et al.* 2008). Only the study of Dransfield *et al.* contains all four genera of Chuniophoeniceae, whereas Bayton's work is based on the largest sampling of DNA regions.

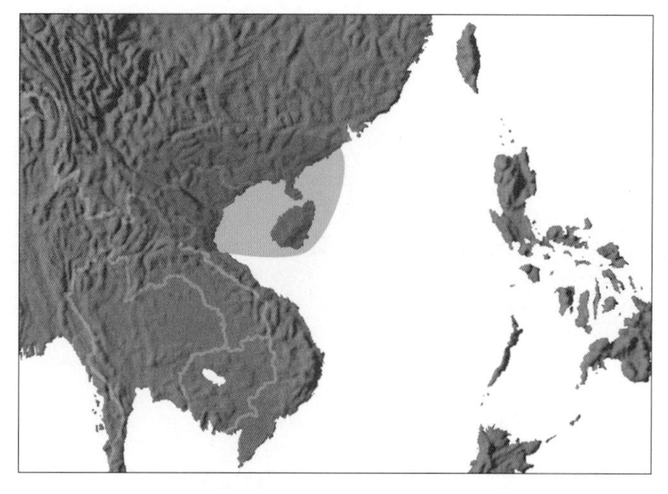
Distribution of ***Chuniophoenix***

Left: ***Chuniophoenix humilis***, habit, cultivated, China. (Photo: J. Dransfield)
Top middle: ***Chuniophoenix hainanensis***, base of leaf blade showing lack of hastula, cultivated, China. (Photo: J. Dransfield)
Top right: ***Chuniophoenix hainanensis***, crown with inflorescences, cultivated, China. (Photo: J. Dransfield)
Bottom right: ***Chuniophoenix humilis***, rachilla, cultivated, China. (Photo: J. Dransfield)

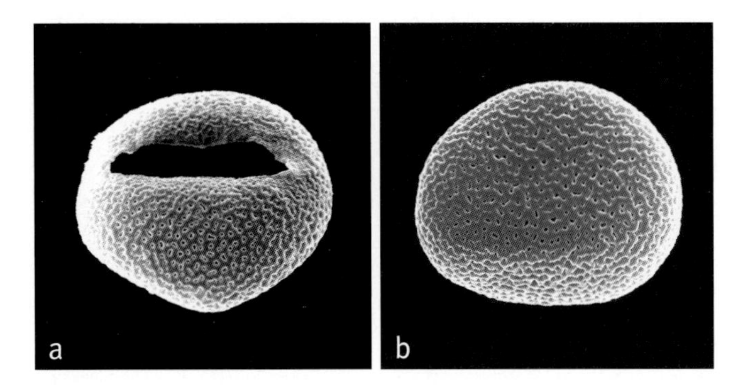

Above: ***Chuniophoenix***
a, monosucate pollen grain, distal view SEM × 1500;
b, monosucate pollen grain, proximal view SEM × 1500.
Chuniophoenix humilis: **a** & **b**, cultivated RBG Kew 1986 3018.

Common names and uses: Common names not recorded. Uses not recorded but certainly desirable ornamentals.
Taxonomic accounts: Burret (1937). See also Tang & Wu (1977).
Fossil record: No generic records found.
Notes: *Chuniophoenix* is a most remarkable genus. The two or three species are clearly congeneric but they display a range of variation in morphology unusual in such a small genus. The rachillae with their completely tubular, conspicuous bracts, have an almost calamoid appearance. The lack of a hastula is also noteworthy.

54. KERRIODOXA

Spectacular solitary short-stemmed dioecious fan palm of southern Thailand, distinctive in the large strongly discolorous leaf, the tubular rachilla bracts with much tomentum and the large fruit covered in short papillae.

Kerriodoxa J. Dransf., Principes 27: 4 (1983). Type: **K. elegans** J. Dransf.

Combines doxa — *glory, with the name of the most prolific collector of Thai plants, Arthur Francis George Kerr (1877–1942).*

Moderate, solitary, acaulescent or erect, unarmed, pleonanthic, dioecious palm. *Stem* very short, becoming erect, obscured by marcescent leaf bases, eventually becoming smooth, marked with very close leaf scars. *Leaves* induplicate, palmate, marcescent; sheath splitting opposite the petiole, not encircling the stem, not fibrous; petiole well developed, channelled adaxially, rounded abaxially, the margins hard and very sharp, surfaces bearing caducous indumentum; adaxial hastula conspicuous, abaxial hastula absent; blade regularly divided along adaxial ribs for ca. $^{1}/_{4}$–$^{1}/_{3}$ radius into single-fold segments, thin, narrow, almost herbaceous, adaxially glabrous except for caducous, scurfy indumentum along the ribs, abaxially covered with dense white indumentum, midribs evident abaxially, transverse veinlets conspicuous, interfold filaments present in expanding leaf, soon disintegrating. *Inflorescences* solitary, interfoliar, emerging from a cleft on the abaxial surface of the sheath that subtends it, staminate and pistillate dissimilar. *Staminate inflorescence* becoming curved, copiously branched to 4 orders, the whole inflorescence very condensed and congested, creamy-white at first, becoming brown with age; peduncle short; prophyll tubular, concealed within leaf sheaths; rachis longer than the peduncle, bearing up to 15 bracts, tubular near the insertion, distally with a ± expanded triangular limb, adaxially glabrous, abaxially densely tomentose; first-order branches adnate to the inflorescence axis to just below the following bract, decreasing in size distally; all axes densely tomentose, each branch above the first-order subtended by a somewhat undulate tubular bract with a triangular apiculate limb; rachillae very slender, somewhat zigzag, bearing spirally arranged, tubular bracts with undulate margins and short, triangular, apiculate limbs, each subtending a low spur bearing 2 flowers and a minute triangular bracteole. *Staminate flowers* very small, ± symmetrical, creamy-yellow at anthesis; calyx with a basal, 3-angled tube densely covered in pale tomentum, and 3 narrow, triangular, apiculate, keeled, ± glabrous lobes with somewhat undulate margins; corolla stalk-like at the base, 3-angled, lobes 3, triangular, the margins and abaxial surfaces papillose, adaxial surface somewhat wrinkled; stamens 6, borne in 2 whorls, the antesepalous filaments free, the antepetalous joined together at the base and partly adnate to the petals,

Top left: ***Kerriodoxa elegans***, habit, Thailand. (Photo: J. Dransfield)
Bottom left: ***Kerriodoxa elegans***, staminate inflorescences, cultivated, Kew, UK. (Photo: W.J. Baker)

filaments ± equal in size, elongate, gradually tapering, anthers oval, latrorse; pistillode absent. *Pollen* ellipsoidal, with slight to obvious asymmetry; aperture a distal sulcus; ectexine tectate to semitectate, reticulate with frequently interrupted angular, occasionally spinulose, muri, aperture margin similar; infratectum columellate; longest axis 22–33 µm [1/1]. *Pistillate inflorescence* erect, much more robust than the staminate, and less congested, branching to 2 orders only; peduncular bracts and rachis as in the staminate inflorescence but larger; first- and second-order branches appearing articulated, because of dense tomentum on axes and the truncate, ± glabrous bracts; rachillae somewhat zigzag, bearing low bracts with short triangular tips and glabrous margins, each subtending a short, densely tomentose spur, bearing a pair of flowers; bracteoles, if present, obscured by tomentum. *Pistillate flowers* larger than the staminate, creamy-yellow at anthesis; calyx forming a densely tomentose tube tipped with 3 short, narrow, triangular, glabrous lobes; corolla base stalk-like, densely tomentose, tipped with 3 triangular lobes, spreading at anthesis, glabrous, the margins ± translucent, denticulate or papillose; staminodes 6 with elongate filaments and flattened empty anthers; gynoecium of 3 (rarely 4) carpels, distinct at their tips, connate at the middle, stigmas short, outward curving; ovule laterally attached, anatropous. *Fruit* 1- or rarely 2-seeded, relatively large, spherical, concave depressed at base, the abortive carpels and stigmatic remains persisting at the fruit base, corolla base enlarging after fertilization; epicarp orange-yellow, covered in low pustules, mesocarp thick, soft and spongy, endocarp thin. *Seed* basally attached, endosperm shallowly ruminate; embryo subbasal. *Germination* remote-ligular; eophyll broad lanceolate, apically lobed. *Cytology* not studied.

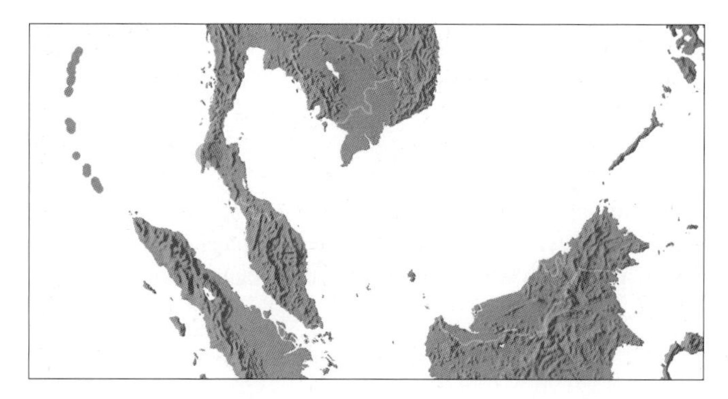

Distribution of ***Kerriodoxa***

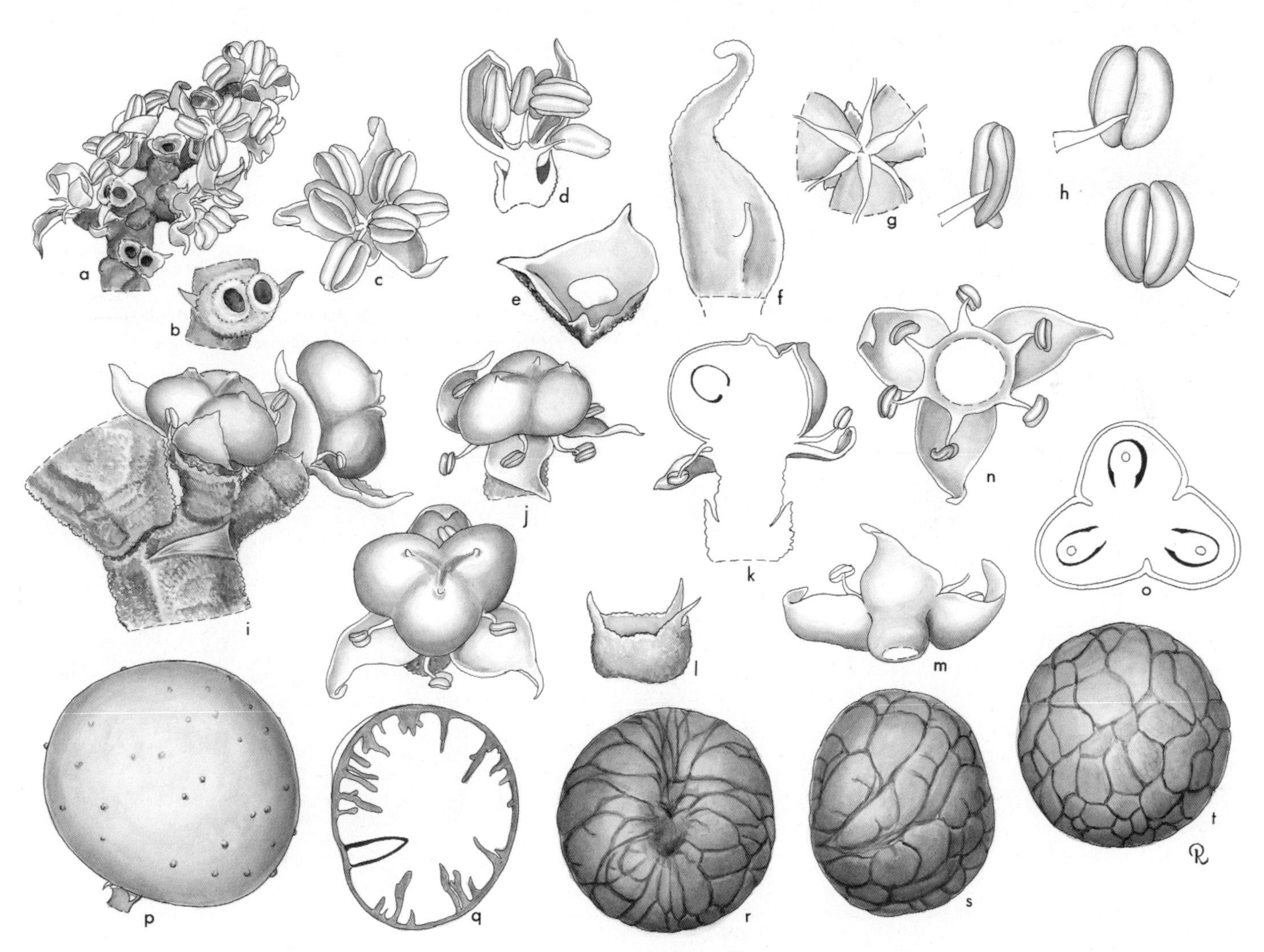

***Kerriodoxa*. a**, portion of staminate inflorescence × 4 1/2; **b**, portion of staminate inflorescence, flowers removed to show scars and bracteoles × 9; **c**, staminate flower × 9; **d**, staminate flower in vertical section × 9; **e**, staminate calyx × 15; **f**, staminate petal with adnate stamen filament, interior view × 15; **g**, stamen filaments, sepals, and parts of petals, top view × 15; **h**, stamen in 3 views × 15; **i**, portion of pistillate inflorescence × 6; **j**, pistillate flower in 2 views × 6; **k**, pistillate flower in vertical section × 6; **l**, pistillate calyx × 6; **m**, pistillate corolla × 6; **n**, pistillate corolla and staminodial ring × 6; **o**, ovary in cross-section × 7 1/2; **p**, fruit × 1; **q**, seed in vertical section × 1; **r**, **s**, **t**, seed in 3 views × 1. *Kerriodoxa elegans*: **a–o**, *Bhoonab* s.n.; **p–t**, *Dransfield* 5421. (Drawn by Marion Ruff Sheehan)

Distribution and ecology: One species known from two localities in peninsular Thailand. *Kerriodoxa elegans* grows gregariously in the undergrowth of rather dry evergreen forest on slopes of hills at altitudes of ca. 100–300 m above sea level. Little is known of its natural history.
Anatomy: Root (Seubert 1997).
Relationships: For relationships, see *Chuniophoenix*.
Common names and uses: Common names not recorded. No local uses have been recorded but the ornamental potential is great and, since its rediscovery and description, the palm has become widespread in cultivation.
Taxonomic account: Dransfield (1983).
Fossil record: No generic records found.

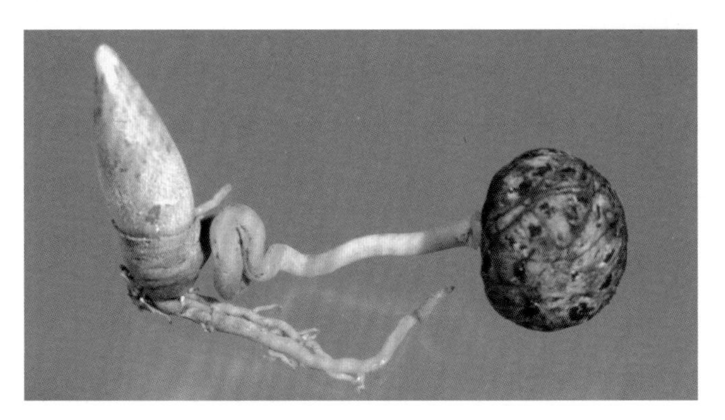

Above: ***Kerriodoxa elegans***, germinating seed, cultivated, L.H. Bailey Hortorium, USA. (Photo: H. Lyon)

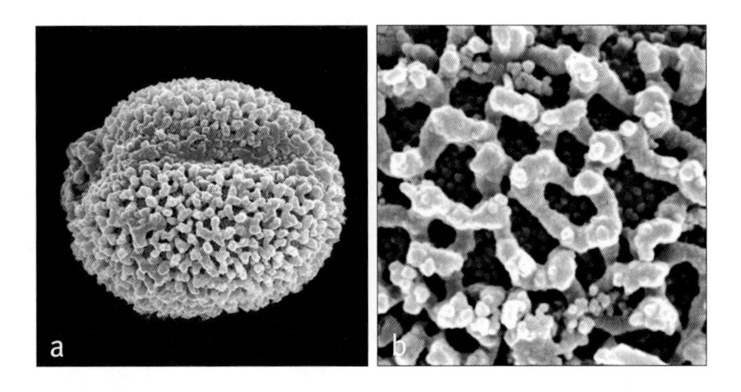

Above: ***Kerriodoxa***
a, monosulcate pollen grain, distal face SEM × 2000;
b, close-up, spinulose reticulate surface SEM × 8000.
Kerriodoxa elegans: **a** & **b**, *Dransfield* JD5423.

Notes: This is an astonishingly beautiful palm. The inflorescence in *Kerriodoxa* is most unusual, developing within the leaf base of the subtending leaf and then breaking out through an abaxial cleft. This morphology is similar to that in *Salacca* but has not been studied developmentally. The tubular rachilla bracts, the stalk-like base of the corolla and the basally fused carpels are shared with other members of the subtribe, but dioecy is not.

55. NANNORRHOPS

Shrubby hermaphroditic hapaxanthic fan palm confined to oases or seasonal water courses in Arabia, Pakistan, Iran and Afghanistan, distinctive in the leaf lacking a hastula, dichotomously branching erect stems and suprafoliar compound inflorescence.

Nannorrhops H. Wendl., Bot. Zeit. 37: 148 (1879). Type: **N. ritchiana** (Griff.) Aitch. (*Chamaerops ritchiana* Griff.).

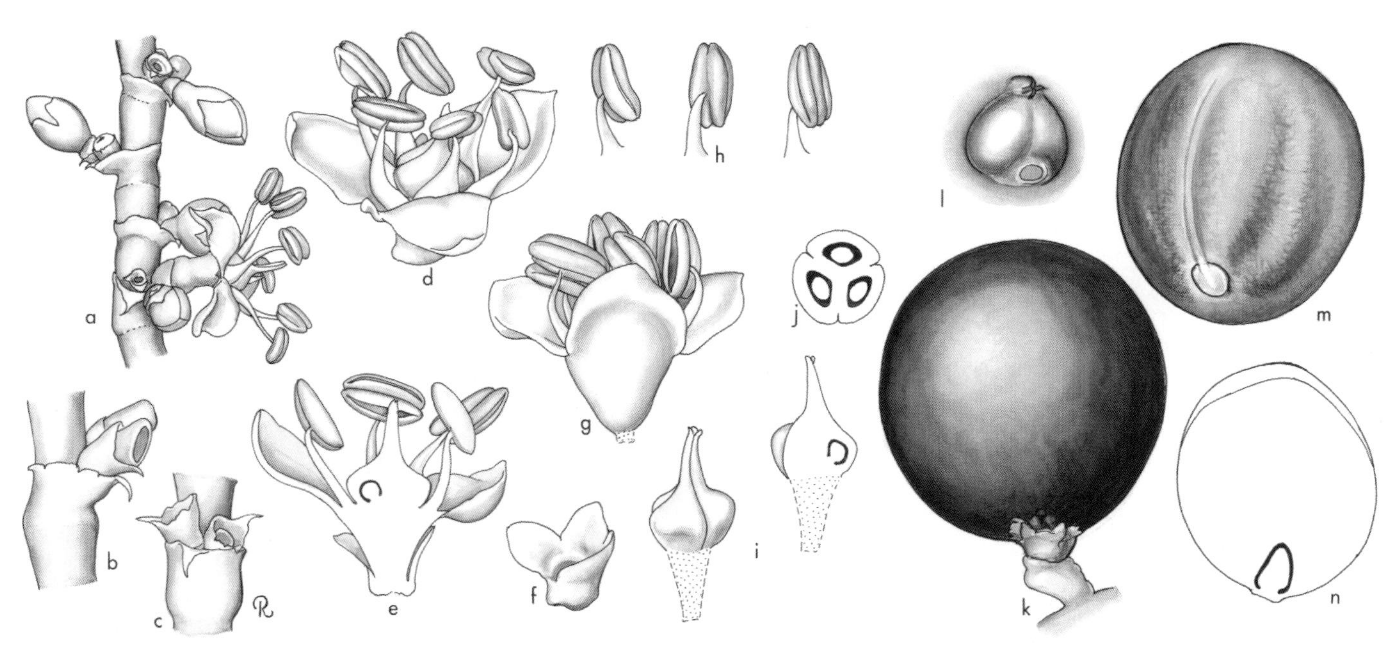

Nannorrhops. **a**, portion of rachilla × 3; **b, c**, portion of cincinnus, flowers removed to show tubular bracts × 6; **d**, flower × 6; **e**, flower in vertical section × 6; **f**, calyx × 6; **g**, flower, calyx removed × 6; **h**, stamen in 3 views × 6; **i**, gynoecium in lateral view and vertical section × 6; **j**, ovary in cross-section × 6; **k**, fruit × 3; **l**, base of fruit with stigmatic remains × 7 1/2; **m**, seed × 3; **n**, seed in vertical section × 3. *Nannorrhops ritchiana*: all from *Read* 735. (Drawn by Marion Ruff Sheehan)

Nannos — *dwarf*, rhops — *bush, in reference to the habit.*

Moderate, shrubby, clustered, unarmed, hapaxanthic, hermaphroditic palm. *Stems* branched, prostrate or erect, branching in prostrate stems axillary, in erect stems dichotomous. *Leaves* induplicate, briefly costapalmate, marcescent; sheath splitting both below and opposite the petiole, brown, woolly tomentose and margins becoming frayed; petiole elongate, shallowly channelled adaxially, rounded abaxially; hastulae absent; blade regularly divided into stiff, glaucous, single-fold segments, further divided by abaxial splits, intersegmental filaments conspicuous, midribs prominent abaxially, transverse veinlets obscure. *Inflorescences* above the leaves, compound, composed of branches equivalent to the axillary inflorescences of pleonanthic palms, each branch subtended by a leaf with reduced blade or by a tubular bract, and branched to the fourth order; prophyll tubular, 2-keeled; peduncular bracts 0 to several, similiar; bracts subtending first-order branches tubular, tips pointed, each first-order branch with a basal, tubular, 2-keeled, empty prophyll; bracts subtending second-order branches tubular; rachillae bearing conspicuous tubular bracts, variously tomentose, each subtending a flower group. *Flowers* very short pedicellate, in a condensed cincinnus of 1–3(–7) flowers, each flower bearing a minute tubular bracteole; calyx thin, tubular at the base with 3 triangular lobes; corolla with a short stalk-like base and 3 distinct lobes, imbricate in the proximal $^{2}/_{3}$, valvate in the distal $^{1}/_{3}$; stamens 6, distinct, the antesepalous with free filaments, and the antepetalous with filaments adnate at the base to the petals, filaments awl-shaped, inflexed at the tip, anthers elongate, versatile, latrorse; carpels 3, connate except at the very base, ovary distinctly 3-grooved, style single, stigma scarcely differentiated, ovule anatropous, attached ventrally and basally. *Pollen* ellipsoidal, usually slightly asymmetric; aperture a distal sulcus; ectexine tectate, reticulate or foveolate-reticulate, aperture margin psilate or scabrate; infratectum columellate; longest axis 30–39 µm [1/1]. *Fruit* subglobose to ellipsoidal, 1-seeded, stigmatic remains basal; epicarp smooth, mesocarp fleshy, endocarp thin. *Seed* globose to ovoid, with very shallow grooves corresponding to the rapheal bundles, hilum basal, endosperm homogeneous, usually with a small central hollow; embryo basal. *Germination* remote-ligular; eophyll undivided. *Cytology*: 2n = 36.

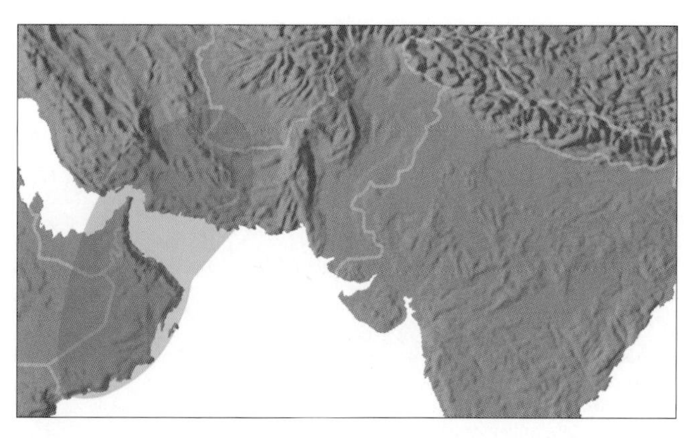

Distribution of ***Nannorrhops***

Distribution and ecology: One variable species in semidesert areas of the Middle East (Iran, Afghanistan, Pakistan and Arabia). Occurring in semi-desert areas where the water table is not too deep, but tending to avoid subtropical coastal habitats within its range; reaching to 1800 m altitude.

Anatomy: Leaf (Tomlinson 1961), root (Seubert 1997), floral (Morrow 1965, Uhl 1969a).

Relationships: For relationships, see *Chuniophoenix*.

Common names and uses: Mazari palm. The main use of *Nannorrhops ritchiana* is as a source of fibre for weaving and rope making. Though undoubtedly ornamental and one of the hardiest palms, it is only rarely grown.

Taxonomic account: Moore (1980).

Fossil record: Small monosulcate grains, *Palmaemargosulcites fossperforatus*, from palm flower compression fossils recovered from the Middle Eocene oil shales of Messel, Germany, are compared with pollen of a number of coryphoid genera, including *Nannorrhops* (Harley 1997).

Top right: ***Nannorrhops ritchiana***, habit, Oman. (Photo: J. Dransfield)

Bottom left: ***Nannorrhops ritchiana***, crown, cultivated, Montgomery Botanical Center, Florida. (Photo: J. Dransfield)

Bottom right: ***Nannorrhops ritchiana***, crown with inflorescence, cultivated, Montgomery Botanical Center, Florida. (Photo: J. Dransfield)

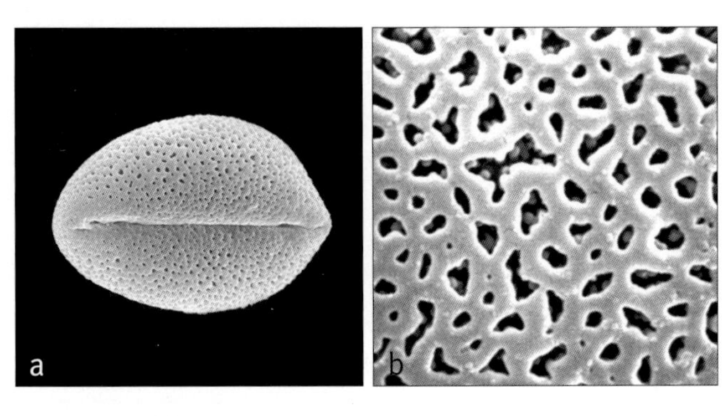

Above: ***Nannorrhops***
a, monosulcate pollen grain, distal face SEM × 1500;
b, close-up, finely reticulate surface × 8000.
Nannorrhops ritchiana: **a**, cultivated India, **b**, *Radcliffe-Smith* 5471.

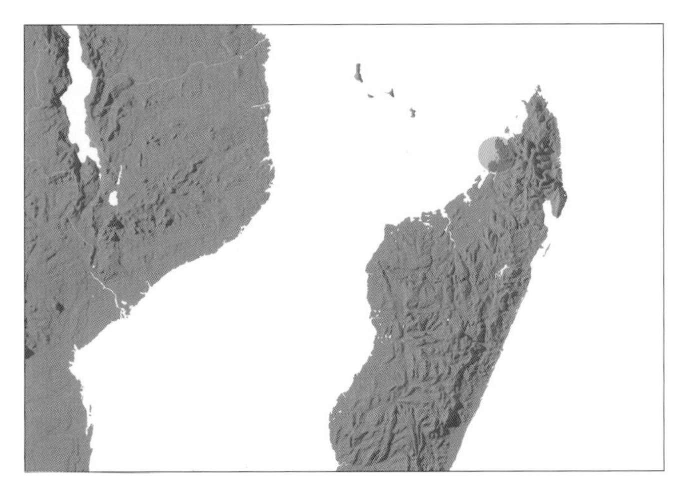

Distribution of ***Tahina***

Notes: Distinguished vegetatively by the dichotomously branching stems and large bluish-grey leaves that lack a hastula. Developmental studies of the flowers (Uhl 1969a) have shown that connation of carpels begins in the styles at a relatively late stage in ontogeny, suggesting parallels with members of the Livistoninae in which carpels are connate by styles only.

56. TAHINA

Spectacular massive solitary hapaxanthic hermaphroditic fan palm of north-western Madagascar, with huge leaves and unarmed petioles with a distinctive triangular cleft at the base of the petiole, and huge suprafoliar compound inflorescences.

Tahina J. Dransf. & Rakotoarinivo, Bot. J. Linn. Soc. 156: 81 (2008). Type: **T. spectabilis** J. Dransf. & Rakotoarinivo.

Tahina — *Malagasy for blessed, or to be protected, also the given name of the daughter of the discoverer of the palm.*

Massive, solitary, unarmed, hapaxanthic, hermaphroditic, tree palms. *Stem* erect, rather distantly ringed with leaf scars. *Leaves* induplicate, costapalmate, marcescent in immature individuals, tending to abscise under their own weight in trunked individuals; sheath with a conspicuous triangular cleft below the petiole, the margins tending to erode into broad lobes; petiole massive, long, covered with white wax, adaxially deeply channelled, abaxially rounded, margins smooth, adaxial hastula well developed, abaxial hastula a hard low rather irregular crest at the base of the lamina; blade divided to ca. $^{1}/_{2}$ its radius into multi-fold segments, these in turn more shallowly divided into single-fold segments, themselves shallowly divided along the abaxial folds, main abaxial ribs of blade very conspicuous, very crowded at the base of the blade, some much more robust than others, the less robust ribs tending to be inserted in a more adaxial position compared with the robust ribs; segments with prominent longitudinal veins and abundant irregularly arranged transverse veinlets, glabrous, the undersurface of the blade with thin white wax. *Inflorescences* above the leaves, subtended by reduced, scale like leaves, forming a massive, compound inflorescence-like structure; individual inflorescences branched to the third order, branches ending as rachillae; prophyll of inflorescences and first order branches 2-keeled, empty; bracts always obvious in all parts of the inflorescence; rachillae straight, rather rigid, bearing distichously arranged, tubular striate bracts, each enclosing a cincinnus of up to 3 flowers; first bract on cincinnus tubular, 2-keeled, 2-lobed, the wings densely lanuginose, subtending a stalked flower; second bract on the cincinnus similar, it too subtending a flower, and where present a third bract subtending a third flower, in the distal part of the rachilla second and third flowers sometimes lacking, in

Right: ***Tahina spectabilis***, in full flower, Madagascar. (Photo: J. Dransfield)

this case distal to the first and second flowers a few empty tubular bracts. *Flowers* borne on short stalks; distal to this, the flower with a stalk formed by the base of the calyx and the receptacle; calyx tubular basally, with 3 low, triangular lobes; petals ± boat-shaped, basally imbricate, the margins usually inrolled, apically somewhat cucullate, strongly reflexed at anthesis and with a glandular swelling at the base of each lobe; stamens 6, filaments terete, tapering from a fleshy base; anthers elongate, basifixed, introrse; gynoecium tricarpellate, syncarpous, triovulate, ovarypyramidal, angled and grooved, style short, slightly 3-grooved, stigma scarcely differentiated, ovule basally attached, anatropous. *Pollen* ellipsoidal, bi-symmetric or slightly asymmetric; aperture a distal sulcus; ectexine tectate, coarsely perforate-rugulate, aperture margin slightly finer; infratectum columellate; longest axis 25–55 µm; post-meiotic tetrads tetrahedral. [1/1]. *Fruit* ellipsoid, single-seeded with apical stigmatic remains; epicarp smooth, mesocarp fleshy and ± fibrous, endocarp scarcely distinguishable, membranous. *Seed* globose, with basal hilum, and deep grooves corresponding to the rapheal bundles, endosperm strongly ruminate without a central hollow; embryo subbasal. *Germination* remote-ligular; eophyll palmate. *Cytology* unknown.

Distribution and ecology: *Tahina spectabilis* is known from a single locality in north-western Madagascar where it grows at low elevation on seasonally flooded soils at the foot of a karst limestone outcrop; there are 91 individuals of varying size and a few hundred 1-leaf seedlings originating from the 2007

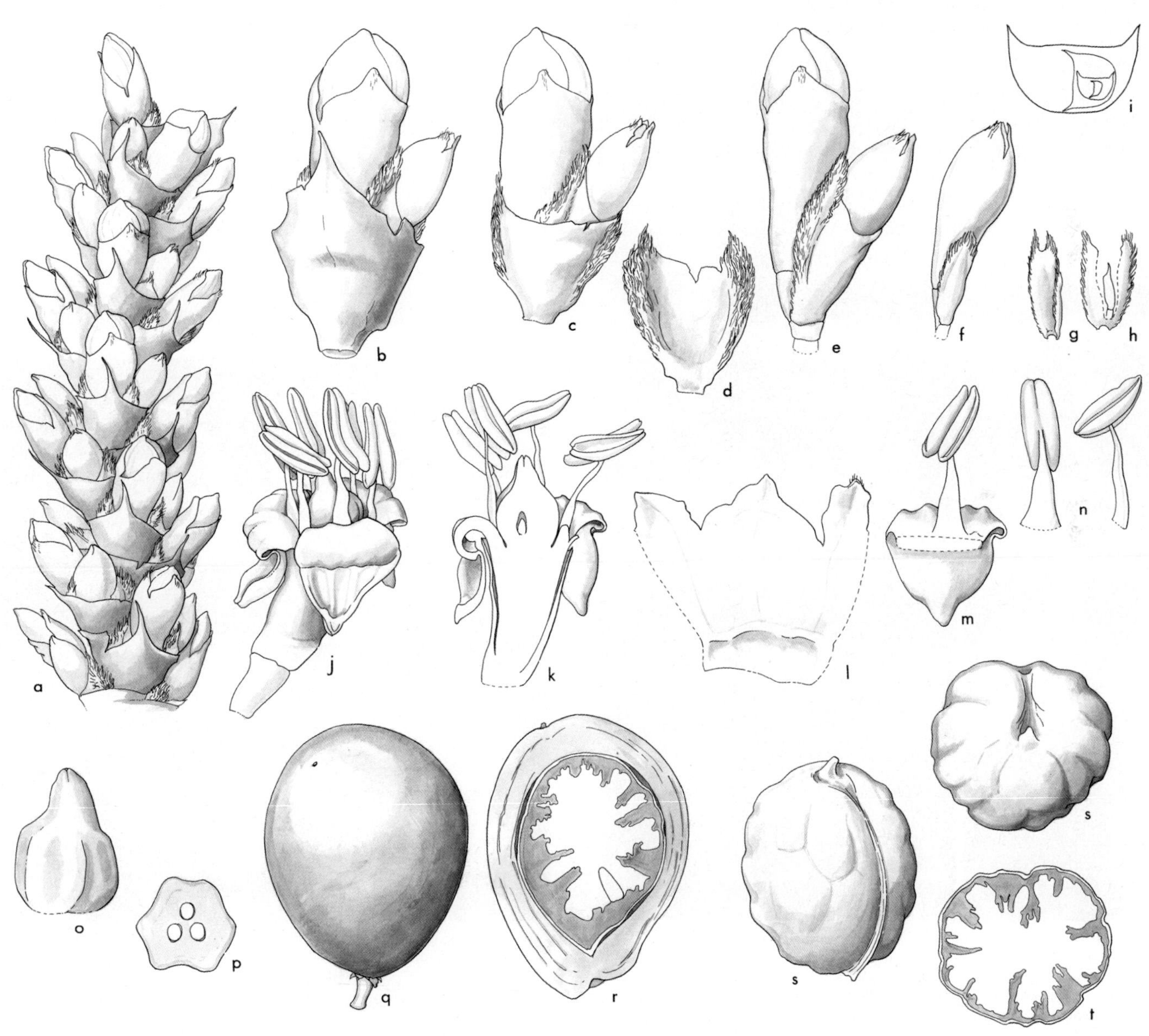

Tahina. **a**, portion of rachilla × 3; **b**, rachilla bract and flower group × 5; **c**, prophyll, abaxial view × 5; **d**, the same with rachilla bract removed × 5; **e**, flower group, prophyll removed; **f**, second flower of group × 5; **g**, third bracteole × 5; **h**, third bracteole, opened out to show abortive third flower × 5; **i**, diagram of bracteoles of flower group; **j**, flower at anthesis × 5; **k**, flower in vertical section × 5; **l**, calyx, opened out × 5; **m**, petal with stamen attached × 5; **n**, stamen, two views × 5; **o**, gynoecium × 7; **p**, gynoecium in cross-section × 7; **q**, fruit × 1 $^{1}/_{2}$; **r**, fruit in longitudinal section × 3; **s**, endocarp, two views × 3; **t**, seed in cross-section × 3. *Tahina spectabilis*: **a–o**, from *Metz*, s.n., **q–t** from *Rakotoarinivo et al.* RMJ 337. (Drawn by Lucy T. Smith)

fruiting known. When in flower, the immense compound terminal inflorescence appears to go through three successive waves of anthesis. Nectar production is so copious that the entire inflorescence appears coated with honey, attracting large numbers of bees, wasps and flies. Greater Vasa Parrots (*Coracopsis vasa*) attack the young fruit (N. & X. Metz, pers. comm.). Nothing is known of fruit dispersal.

Anatomy: Not studied.

Relationships: For relationships see *Chuniophoenix*.

Common names and uses: *Dimaka*, a name also applied to *Borassus*, *Bismarckia* and *Dypsis decipiens*. No local uses have been recorded.

Taxonomic account: Dransfield *et al.* (2008).

Fossil record: No generic records.

Notes: This astonishing palm was discovered as this book went to press. It is distinguished from all other fan-palms in Madagascar by its hapaxanthic habit. The tubular bracts on the rachillae are very reminsiscent of those of *Chuniophoenix*.

Top: ***Tahina spectabilis***, habit in full flower, Madagascar.
(Photo: J. Dransfield)

Bottom left: ***Tahina spectabilis***, rachilla in bud, Madagascar.
(Photo: J. Dransfield)

Bottom right: ***Tahina spectabilis***, flowers at anthesis, Madagascar.
(Photo: X. Metz)

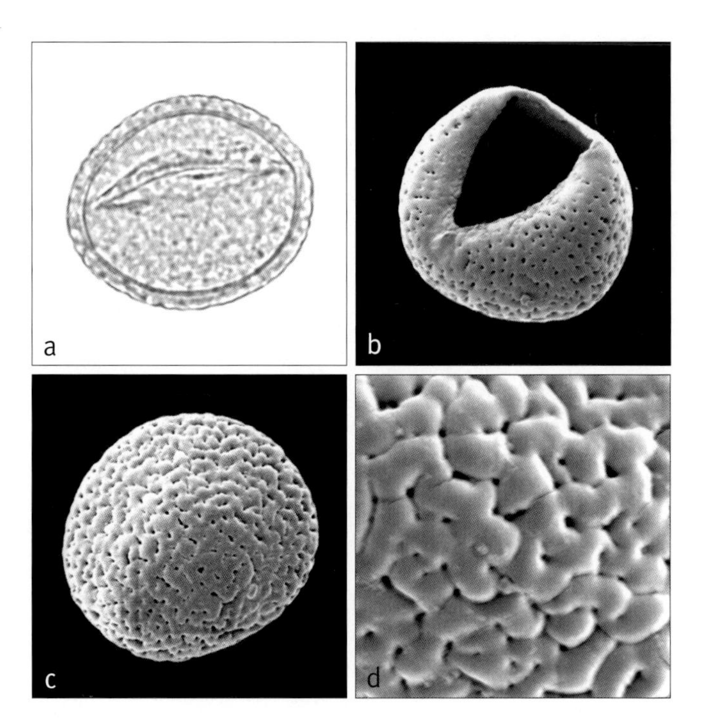

Above: ***Tahina***
a, monosulcate pollen grain, mid-focus LM × 1000;
b, monosucate pollen grain, distal view SEM × 1500;
c, monosucate pollen grain, proximal view SEM × 1500;
d, close-up, perforate rugulate surface SEM × 8000.
Tahina spectabilis: **a–d**, *Dransfield* JD7777.

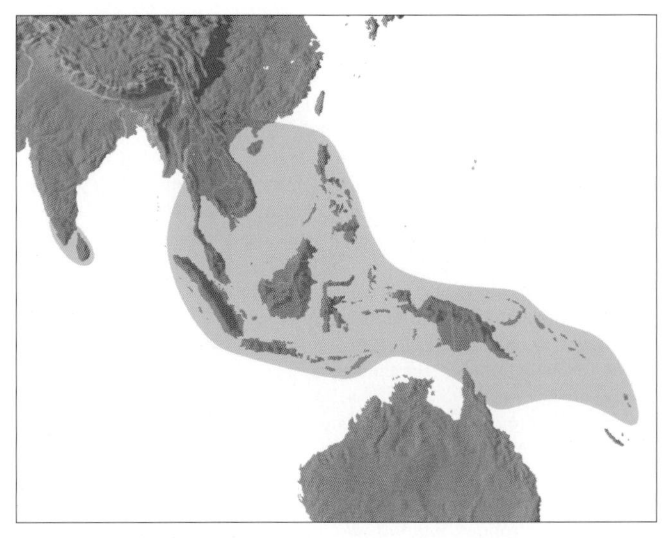

Distribution of tribe **Caryoteae**

Tribe Caryoteae Drude, Bot. Zeit. 35: 632 (1877). Type: Caryota

Slender to robust, acaulescent or stems erect; leaves pinnate or bipinnate, induplicate; leaflets praemorse; hapaxanthic (usually flowering basipetally) or pleonanthic, monoecious, (?)very rarely dioecious; inflorescences bisexual or unisexual, sometimes dimorphic, sometimes multiple, bearing a prophyll and several, usually large peduncular bracts, branched to 1(–3) orders, rarely spicate; flowers superficial, borne in triads of a central pistillate and 2 lateral staminate flowers, or derivatives of triads; pistillate flower with valvate petals; gynoecium syncarpous, trilocular, triovulate; fruit 1–3-seeded, if more than 1-seeded, not lobed.

The tribe contains three genera distributed from Mainland Asia through the Malesia to the western Pacific and Australia. The placement of Caryoteae has long been problematic. The induplicate leaf is, of course, a coryphoid feature but, except for *Phoenix*, the coryphoids have palmate leaves, whereas Caryoteae have pinnate or bipinnate leaves. Caryoteae shares striking similarities in overall inflorescence architecture with members of Iriarteeae, and flowers are borne in triads as in most Arecoideae. For these reasons, the group was included in Arecoideae in the first edition of *Genera Palmarum*. Recent phylogenetic studies, particularly those using molecular data, overwhelmingly support the inclusion of Caryoteae within Coryphoideae.

Tribe Caryoteae is monophyletic, highly supported and resolved within the syncarpous clade, on a long branch in some cases (e.g., Bayton 2005, Asmussen *et al.* 2006). It is sister to tribes Corypheae and Borasseae with strong or moderate support (Bayton 2005, Asmussen *et al.* 2006). An alternative placement as sister to Borasseae has low support (Baker *et al.* in review).

Key to Genera of Caryoteae

1. Leaf doubly pinnate; inflorescences always bearing flowers of both sexes; stamens numerous. S.E. Asia, Malesia, Australia . . . . **57. Caryota**
1. Leaf simply pinnate; inflorescences rarely bisexual, usually unisexual; stamens 3 to numerous 2
2. Basipetally hapaxanthic palms; inflorescences always unisexual; sepals of staminate flower connate in a tube; stamens (3)–6–(9–15). India, China, Indochina southwards to S. Thailand . . . . **59. Wallichia**
2. Basipetally hapaxanthic or acropetal pleonanthic palms; inflorescences sometimes bisexual; sepals of staminate flower free, imbricate; stamens (6)-numerous. China, Ryukyu, S.E. Asia, Malesia, Christmas Islands (Indian Ocean), Australia . . . . **58. Arenga**

57. CARYOTA

Solitary or clustered, monoecious hapaxanthic palms of South and Southeast Asia to the western Pacific, instantly recognisable by the doubly pinnate leaf with fishtail leaflets.

Caryota L., Sp. pl. 1189 (1753). Type: **C. urens** L.
Schunda-Pana Adans., Fam. pl. 2: 24, 602 (1763).
Thuessinkia Korth. ex Miq., Flora Ned. Ind. 3: 41 (1855) (illegitimate name). Type: *T. speciosa* Korth. ex Miq. (*Caryota furfuracea* Blume ex Mart. = *C. mitis* Lour.).

From karuotos, karyon — *a nut, nut-bearing.*

Moderate to large, solitary or clustered, hapaxanthic, monoecious palms. *Stems* with ± elongate internodes, obscured at first by persistent fibrous leaf bases and sheaths, usually becoming bare, conspicuously ringed with narrow leaf scars, striate. *Leaves* induplicately bipinnate (except in

Right: ***Caryota obtusa***, leaf, cultivated, Montgomery Botanical Center, Florida. (Photo: J. Dransfield)

juveniles where pinnate), marcescent or abscising under their own weight; sheath triangular, eroding opposite the petiole into a mass of strong black fibres, a ligule-like extension frequently present, disintegrating into strong black fibres, the sheath surface covered in a dense felt of indumentum and caducous chocolate-brown scales, sometimes in broad stripes; petiole scarcely to well developed, channelled adaxially, rounded abaxially, bearing indumentum like the sheath; secondary rachises similar in form to the primary rachis, arranged ± regularly except rarely in 1 or 2 species where the most proximal few crowded; leaflets very numerous, borne ± regularly along the secondary rachises, obliquely wedge-shaped with no distinct midrib but several major veins diverging from the swollen, sometimes stalk-like base, upper margins deeply praemorse, blade concolorous, with broad bands of caducous chocolate-brown scales abaxially, transverse veinlets obscure. *Inflorescences* bisexual, solitary, produced in a basipetal sequence, interfoliar and sometimes infrafoliar (the proximal few), usually branched to 1 order, rarely to 2 orders (*Caryota ophiopellis*) or 3 orders (*C. zebrina*) or rarely spicate (*C. monostachya*), usually pendulous; peduncle ± circular in cross-section, densely scaly; prophyll tubular at first, soon splitting, 2-keeled, relatively small, densely tomentose and/or scaly; peduncular bracts to ca. 8, conspicuous, large, enclosing the inflorescence in bud, coriaceous, tubular at first, tending to split irregularly, usually densely tomentose and/or scaly; rachis shorter or longer than the peduncle; rachillae spirally arranged, densely crowded, usually scaly, each subtended by a small, low, triangular bract; the rachilla base usually somewhat swollen, with a short to moderately long bare section above this, distal portion of rachilla bearing close or rather distant, spirally arranged, protandrous triads, each subtended by an inconspicuous rachilla bract; floral bracteoles shallow, rounded. *Staminate flowers* usually ± elongate, symmetrical; sepals 3, ± distinct, coriaceous, ± rounded, imbricate; petals 3, valvate, coriaceous, connate at the very base, considerably exceeding the sepals; stamens 6–ca. 100, the filaments short, basally sometimes connate, anthers ± linear, latrorse, the connective sometimes prolonged into a point; pistillode absent. *Pollen* ellipsoidal, ± bi-symmetric; aperture a distal sulcus; ectexine intectate, usually finely and densely clavate, less frequently spiny, spines attached to smooth upper surface of foot layer, in some species spines more numerous along aperture margin, or gemmate, occasionally with gemmae linked together to form incomplete reticulum, or coalesced into larger irregular units; longest axis ranging from 26–31 µm; post-meiotic tetrads usually tetrahedral, sometimes tetragonal or, rarely, rhomboidal [8/14]. *Pistillate flower* ± globular or elongate; sepals 3, coriaceous, rounded, imbricate, connate at the very base; petals 3, coriaceous, valvate, connate into a tube in the basal ca. $^1/_3$–$^1/_2$; staminodes 0–6; ovary rounded or somewhat 3-angled, trilocular with 1–2 locules fertile, septal glands present basally, stigma trilobed, apical, ovule hemianatropous, inserted adaxially at the base. *Fruit* globose, 1–2-seeded, with apical stigmatic remains; epicarp smooth, becoming dull, bright or dark coloured at maturity, mesocarp fleshy, filled with abundant, irritant, needle-like crystals, endocarp not differentiated. *Seeds* basally attached, irregularly spherical or hemispherical, somewhat grooved or smooth, endosperm homogeneous or ruminate; embryo lateral. *Germination* remote-tubular; eophyll bifid with rhombic, divergent, praemorse segments. *Cytology*: 2n = 34.

Below left: ***Caryota maxima***, plant in flower, China. (Photo: J. Dransfield)
Below middle: ***Caryota obtusa***, infructescence, Thailand. (Photo: P. Wilkin)
Below right: ***Caryota monostachya***, inflorescence, China. (Photo: J. Dransfield)

Distribution and ecology: About 13 species occurring from Sri Lanka, India, southern China, southwards through Southeast Asia, Malesia to northern Australia, the Solomon Islands and Vanuatu. Ranging from monsoon climates to perhumid areas, from sea level to ca. 2000 m in the mountains, in secondary forest (especially *Caryota mitis*) and in primary forest.

Anatomy: Leaf, petiole, stem, root (Tomlinson 1961), root (Seubert 1998a, 1998b), stamen development following a pattern somewhat similar to that of *Lodoicea* (Borasseae) and *Ptychosperma* (Areceae) (Uhl & Moore 1980).

Relationships: *Caryota* is resolved as a monophyletic group with moderate support (Hahn & Sytsma 1999, Asmussen *et al.* 2006) or high support (Bayton 2005). The genus is highly supported as sister to a strongly supported clade of *Wallichia* and *Arenga* (Bayton 2005, Asmussen *et al.* 2006). For interspecific relationships, see Hahn & Sytsma (1999).

Common names and uses: Fishtail palms. All species appear to be utilised in some way. The apex is edible and good. Stems provide sago, the larger species being especially favoured.

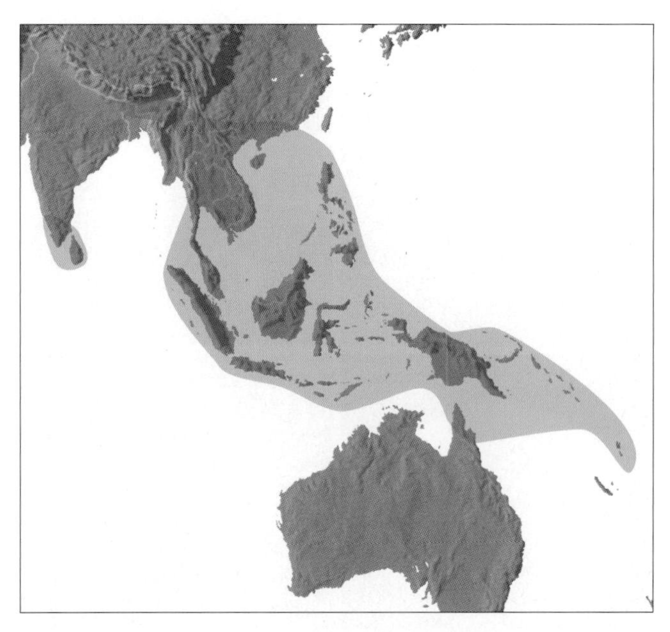

Distribution of ***Caryota***

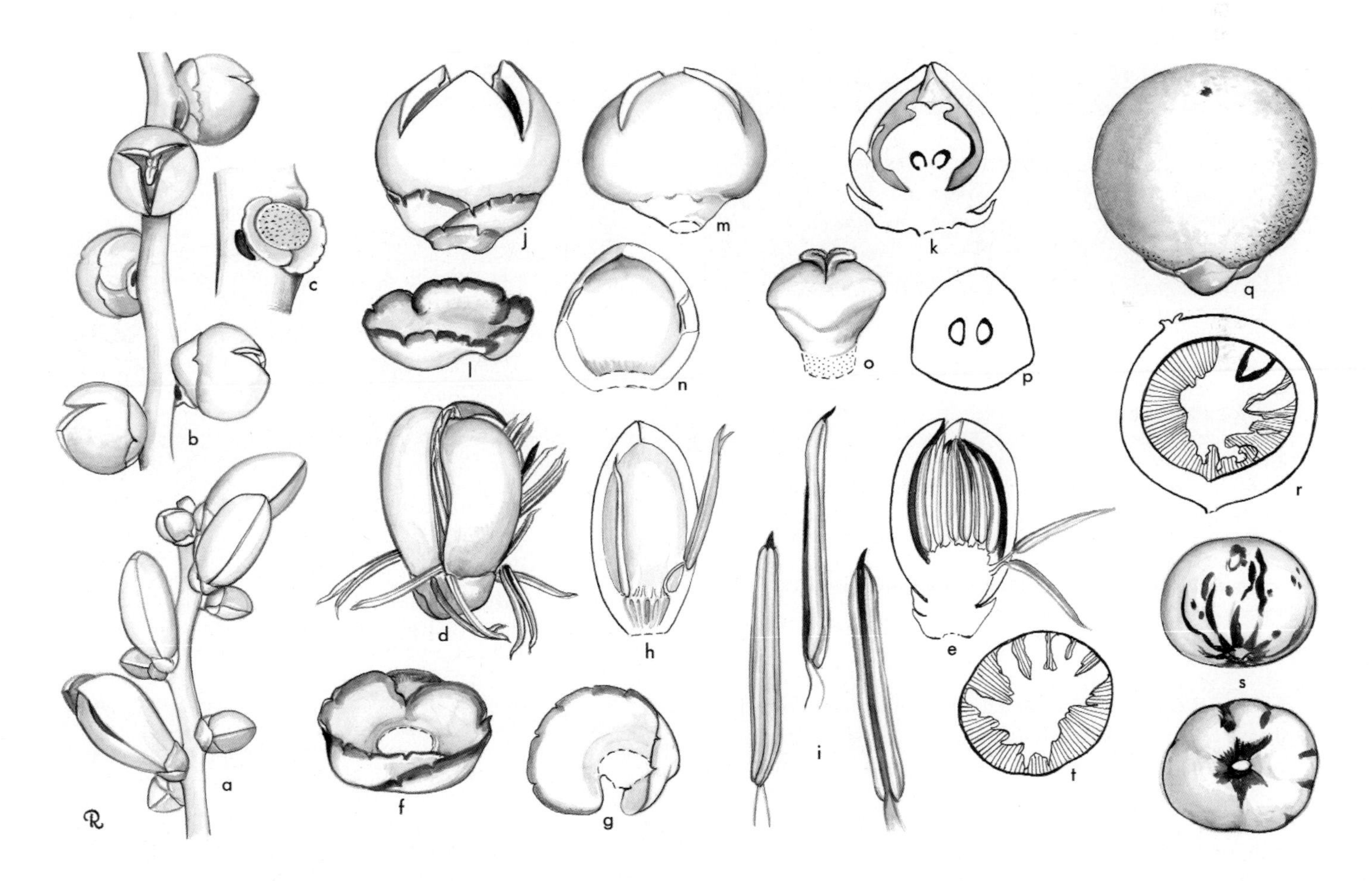

Caryota. **a,** portion of rachilla with triads × 1; **b,** portion of rachilla in pistillate flower × 1$^1/_2$; **c,** triad, flowers removed × 3; **d,** staminate flower × 1$^1/_2$; **e,** staminate flower in vertical section × 1$^1/_2$; **f,** staminate calyx × 1$^1/_2$; **g,** staminate sepal × 1$^1/_2$; **h,** staminate petal × 1$^1/_2$; **i,** stamen in 3 views × 3; **j,** pistillate flower × 3; **k,** pistillate flower in vertical section × 3; **l,** pistillate calyx × 3; **m,** pistillate flower, calyx removed × 3; **n,** pistillate corolla in section showing 1 petal and 2 staminodes × 3; **o,** gynoecium × 3; **p,** ovary in cross-section × 3; **q,** fruit × 1$^1/_2$; **r,** fruit in vertical section × 1$^1/_2$; **s,** seed in 2 views × 1$^1/_2$; **t,** seed in cross-section × 1$^1/_2$. *Caryota maxima*: all from *Read* 760. (Drawn by Marion Ruff Sheehan)

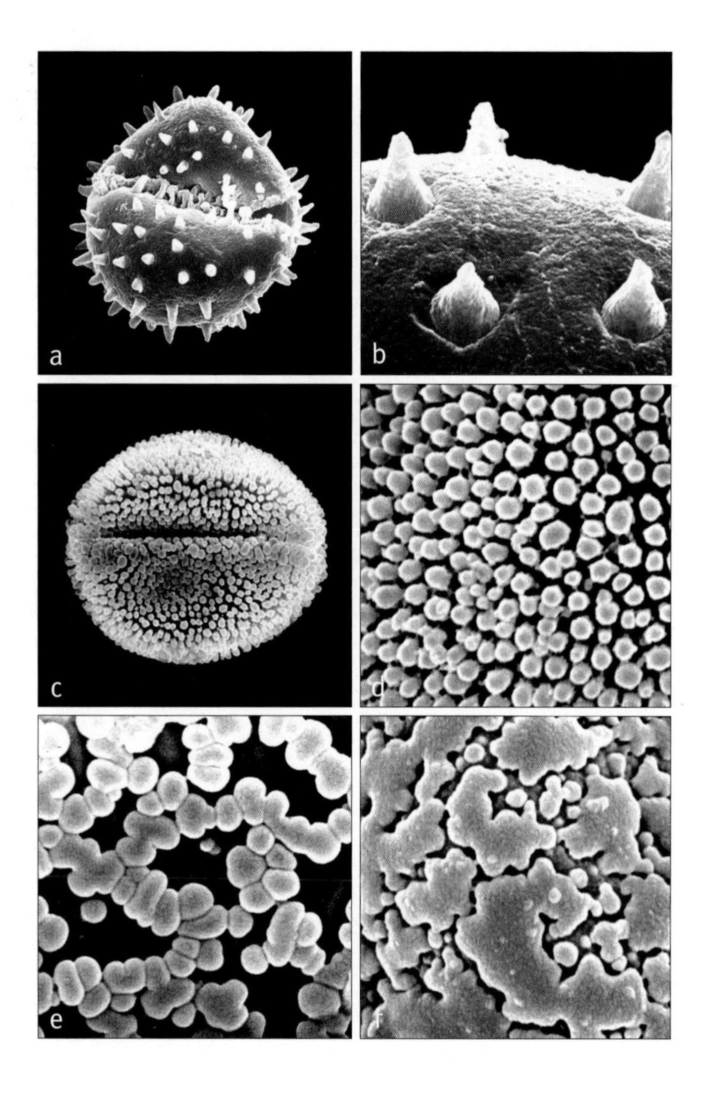

Above: ***Caryota***
a, monosulcate pollen grain, distal face SEM × 1750;
b, close-up, intectate spines SEM × 8000;
c, monosulcate pollen grain, distal face SEM × 1500;
d, close-up, finely clavate surface SEM × 8000;
e, close-up, incomplete reticulate surface, muri formed of closely linked gemmae SEM × 8000;
f, close-up, surface comprising individual gemmae, and larger 'islands' of coalesced gemmae SEM × 8000.
Caryota maxima: **a** & **b**, *Wray Jr.* 1239; *C. mitis*: **c** & **d**, *Millard* 1234; *C. ophiopellis*: **e**, *Dowe* 130; *C. zebrina*: **f**, *Van Royen & Sleumer* 6129.

Timber of *Caryota urens* is used for construction purposes. Leaf sheath fibres are extremely durable and harvested for thatch, cordage, and other purposes. The woolly indumentum on leaf sheaths, petioles, and rachis is used variously as tinder or wadding. Inflorescences, especially of *C. urens*, are tapped for palm wine or sugar. There are several other minor local uses. Many species are cultivated as ornamentals.

Taxonomic account: There is no complete recent taxonomic account of this important genus (but see Hahn 1993).

Fossil record: Seeds probably corresponding to *Caryota* from the Lower Eocene (London Clay) are described as *Caryotispermum* by Reid & Chandler (1933). A small (ca. 18 µm, long axis), finely baculate, pollen grain from the Isle of Wight Oligocene is described by Pallot (1961) as being "indistinguishable from pollen of *Caryota rumphiana*", a possibly correct identity.

Notes: The bipinnate leaf is unique in the palms. The recently described *Caryota ophiopellis* and *C. zebrina* are unusual in having inflorescences branched to more than one order and homogeneous rather than ruminate endosperm, features of *Arenga* rather than of *Caryota*. The clarification of the relationships between these two species will have to wait until a full phylogeny of the Caryoteae is completed.

58. ARENGA

Dwarf to massive, solitary or clustered monoecious or rarely dioecious pinnate-leaved palms from mainland Asia to New Guinea and Australia, distinctive in the induplicate leaflets with praemorse tips, flowers borne in triads and free sepals in the staminate flower. Flowering is usually basipetal hapaxanthic; however, a few pleonanthic species are known.

Arenga Labill. in DC., Bull. Sci. Soc. Philom. Paris 2: 162 (1800) (conserved name). Type: *A. saccharifera* Labill. = **A. pinnata** (Wurmb) Merr.

Saguerus Steck, Sagu 15 (1757). Type: *S. pinnatus* Wurmb (rejected name, see ICBN 237 [1961]).

Gomutus Correa, Ann. Mus. Nat. Hist. Nat. 9: 288 (1807). Type: *G. rumphii* Correa (= *A. pinnata* [Wurmb] Merr.).

Blancoa Blume, Rumphia 2: 128 (1843 ['1848']) (non Lindl. 1840). Type: *Caryota tremula* Blanco (= *Arenga tremula* [Blanco] Merr.).

Didymosperma H. Wendl. & Drude ex Hook.f. in Benth. & Hook.f., Gen. pl. 3: 917 (1883). Lectotype: *D. porphyrocarpum* (Blume ex Mart.) H. Wendl. & Drude ex Hook.f. (*Wallichia porphyrocarpa* Blume ex Mart.) (see Beccari & Pichi-Sermolli 1955).

Derived from aren, *the Javanese vernacular name for the sugar palm,* Arenga pinnata.

Dwarf to large, solitary or clustered, unarmed or lightly armed, pleonanthic or hapaxanthic, monoecious or very rarely apparently dioecious, acaulescent, shrubby or tree palms. *Stem* with congested or elongate internodes, usually obscured by persistent fibrous leaf bases and sheaths, more rarely becoming bare, conspicuously ringed with scars. *Leaves* flabellate and induplicately ribbed (rarely) or induplicately imparipinnate, marcescent, or rarely abscising under their own weight; sheath covered in a great variety of tomentum, scales and hairs, often extended beyond the petiole to form a ligule, eventually disintegrating into a mass of black fibres, some of which are very robust and almost spine-like; petiole usually well developed, slender to very robust, channelled or ridged (*Arenga undulatifolia*) at base adaxially, rounded abaxially, usually covered with a variety of indumentum; rachis rounded to angled adaxially, rounded to flat abaxially; leaflets single-fold (except for the terminal flabellum), regularly arranged or grouped and held in several planes, or deeply lobed and wavy, often with 1 or 2 basal auricles, the distal margins praemorse, with small sharp teeth, sometimes with a short to long, laterally compressed, basal stalk, the veins parallel to the fold, or diverging from the base, or ± pinnately arranged along the fold, adaxial surface of blade glabrescent, margins sometimes spiny, abaxial surface usually densely covered in pale indumentum with or without scattered bands of dark brown scales, midribs prominent abaxially, transverse veinlets scarcely visible. *Inflorescences* interfoliar, sometimes infrafoliar, often bursting through the leaf sheaths, produced in an acropetal sequence in pleonanthic species, in a basipetal sequence in hapaxanthic species, the distal-most inflorescences usually subtended by greatly reduced leaves, bisexual, or unisexual by sterilisation of triad components, where unisexual the pistillate tending to be distal to the staminate, the pistillate sometimes very much larger than the staminate, the staminate or bisexual inflorescences sometimes multiple, otherwise solitary, rarely spicate (*A. retroflorescens*), usually branched to 1–2 orders; peduncle very short to well developed, slender to massive, bearing a generally rather inconspicuous, basal, 2-keeled prophyll and several conspicuous, spirally arranged, peduncular bracts, soon splitting adaxially, the bract limb ± triangular, the abaxial surface usually densely covered with indumentum; rachis shorter or longer than the peduncle; rachis bracts inconspicuous, triangular; rachillae erect or pendulous, distant or crowded, very slender to extremely massive, frequently tomentose, bearing a loose to dense spiral of triads, subtended by inconspicuous low bracts. *Staminate flowers* in bisexual inflorescences opening before the pistillate; staminate flowers with sepals 3, rounded, imbricate, coriaceous, distinct, or joined very briefly at the base; corolla tubular at the very base, with 3 ovate to oblong triangular-tipped, coriaceous, valvate lobes; stamens rarely as few as 6–9, usually many more than 15, filaments short, anthers elongate, latrorse, connective sometimes prolonged into a point; pistillode absent. *Pollen* ellipsoidal, ± bi-symmetric; aperture a distal sulcus; ectexine intectate, usually spiny, spines attached to smooth upper surface of foot layer, in some species spines more numerous along aperture margin, less frequently densely clavate, apices of clavae spinulose; longest axis 27–36 µm; post-meiotic tetrads tetrahedral [13/20]. *Pistillate flowers* usually globose, sometimes massive; sepals 3, distinct, rounded, coriaceous, imbricate; petals 3, connate in the basal ca. 1/2, valvate, triangular distally; staminodes 3–0 (?sometimes more, also reported as fertile, then flower pseudohermaphroditic); ovary globose, trilocular, stigmas 2–3, low, fertile locules 2–3, septal glands present basally and opening at the ovary surface, ovules inserted adaxially at the base, hemianatropous. *Fruit* globose to ellipsoidal, often somewhat angled, 1–3 seeded with apical stigmatic remains; epicarp smooth, dull to brightly coloured, mesocarp fleshy, filled with abundant, irritant needle crystals, endocarp not differentiated. *Seeds* basally attached, smooth, endosperm homogeneous; embryo lateral. *Germination* remote-tubular; eophyll ovate to elliptic with erose margin or bifid with rhombic, divergent segments. *Cytology*: 2n = 32 (64 in one tetraploid).

Distribution and ecology: About 20 species ranging from India, South China, Ryukyus and Taiwan, through Southeast Asia, Malesia including Christmas Island (Indian Ocean) to north Australia, the greatest diversity occurring on the Sunda

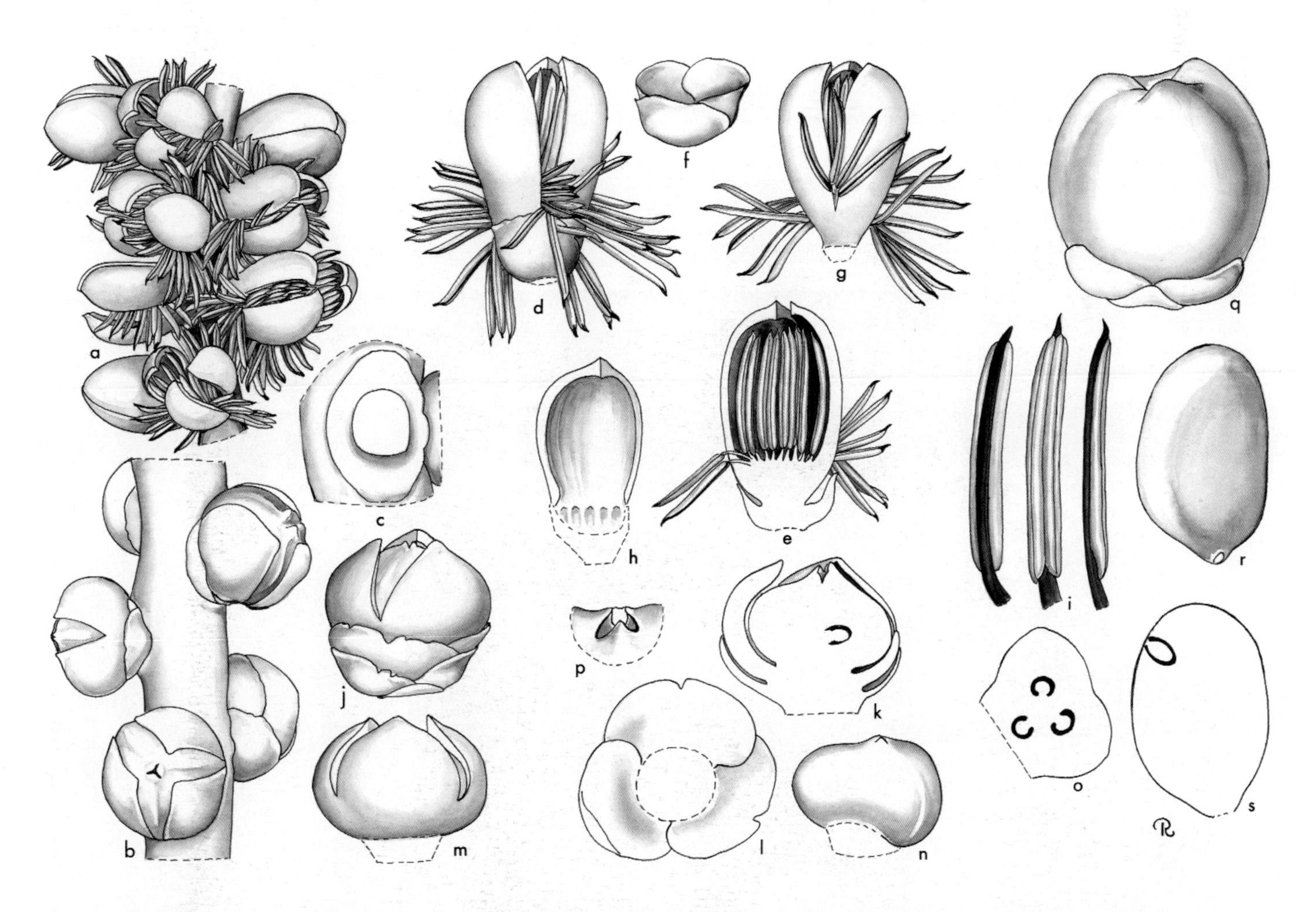

Arenga. **a**, portion of staminate rachilla × 1; **b**, portion of pistillate rachilla × 1; **c**, bracteole subtending pistillate flower × 3; **d**, staminate flower × 1 1/2; **e**, staminate flower in vertical section × 1 1/2; **f**, staminate calyx × 1 1/2; **g**, staminate flower, calyx removed × 1 1/2; **h**, staminate petal × 1 1/2; **i**, stamen in 3 views × 3; **j**, pistillate flower × 1 1/2; **k**, pistillate flower in vertical section × 1 1/2; **l**, pistillate calyx × 1 1/2; **m**, pistillate flower, calyx removed × 1 1/2; **n**, gynoecium × 1 1/2; **o**, ovary in cross-section × 1 1/2; **p**, stigmas × 9; **q**, fruit × 1; **r**, seed × 1; **s**, seed in vertical section × 1. *Arenga pinnata*: all from *Read* 618. (Drawn by Marion Ruff Sheehan)

Shelf. Most species are plants of primary forest in the lowlands and hills of the perhumid tropics; a few species are tall tree palms that grow gregariously and, with their massive leaf litter, must have a pronounced effect on forest dynamics. The smaller species, formerly included in the genus *Didymosperma*, are forest undergrowth palmlets.

Anatomy: Leaf (Tomlinson 1961), root (Seubert 1998a). Two characters distinctive in leaf anatomy: hairs with a basal cylinder of sclerotic cells surrounding 1–3 thin-walled cells and guard cells with transverse ridges on the cutinised ledges. Stegmata (Killmann & Hong 1989). Gynoecium with a basal septal nectary opening by pores on its upper surface, a vascular cylinder for each carpel evident in the gynoecial base, and raphides and tannin abundant around the locules (Uhl & Moore 1971).

Relationships: In the two studies that tested the monophyly of *Arenga*, the genus is resolved as monophyletic with strong support (Asmussen *et al.* 2006) or moderate support (Bayton 2005). Both studies find strong support for a sister relationship between *Arenga* and *Wallichia*, as do Baker *et al.* (in review).

Common names and uses: Sugar palm (*Arenga pinnata* and several other species), black fibre, *gomute*, *aren*, *enau* and *kabang* (*A. pinnata*). The more slender forest undergrowth species are scarcely used but the larger species are among the most important economic plants of Southeast Asia and Malesia. The uses of *A. pinnata* are legion; it is widely cultivated as a source of sugar, wine, fibre, thatch, sago, and many other products (Miller 1964). Other large species are often used in similar ways. Special mention may be made of *A. microcarpa* as a source of sago in some parts of the Moluccas; this seems to be a palm with considerable potential.

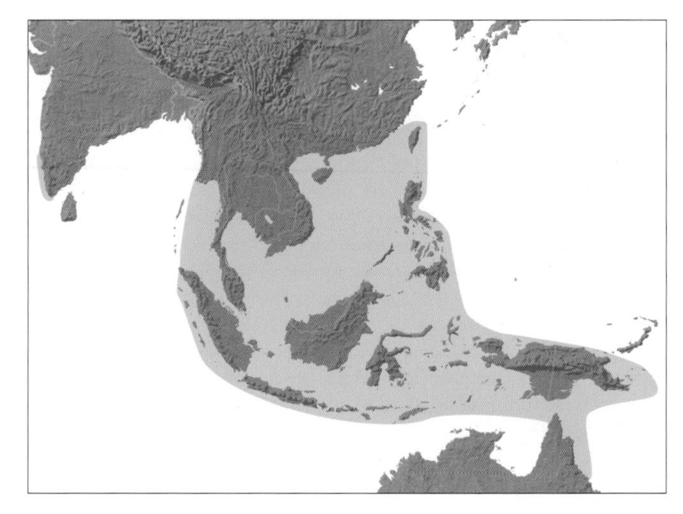

Distribution of ***Arenga***

Left: ***Arenga westerhoutii***, habit, Peninsular Malaysia. (Photo: J. Dransfield)
Middle: ***Arenga westerhoutii***, staminate inflorescence, Peninsular Malaysia. (Photo: J. Dransfield)
Right: ***Arenga caudata***, form with lobed leaf, Thailand. (Photo: J. Dransfield)

Above left: ***Arenga brevipes***, staminate flowers, Palawan. (Photo: J. Dransfield)
Above middle: ***Arenga brevipes***, pistillate flowers, Palawan. (Photo: J. Dransfield)
Above right: ***Arenga microcarpa***, fruit, Papua New Guinea. (Photo: W.J. Baker)

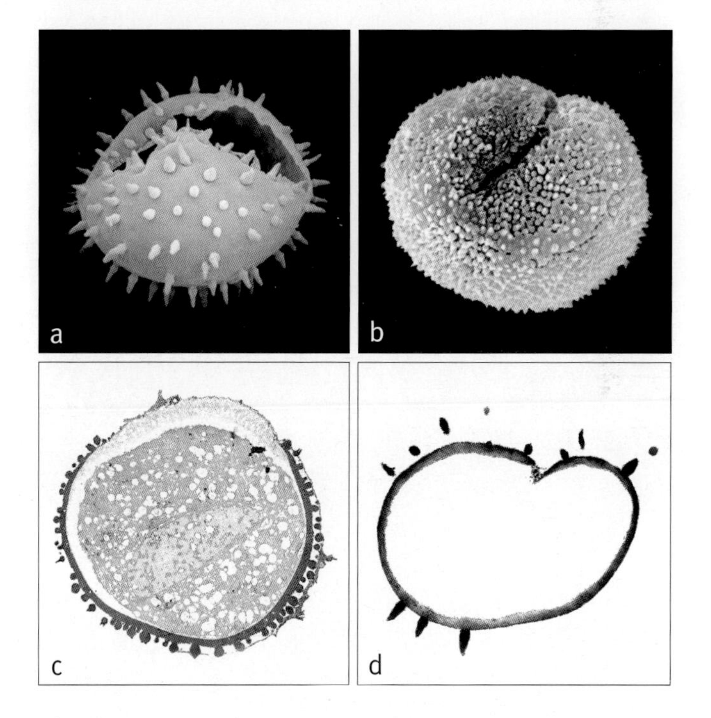

Right: ***Arenga***
a, monosulcate pollen grain, equatorial face SEM × 1500;
b, monosulcate pollen grain, short equatorial face SEM × 1750;
c, ultrathin section through whole pollen grain, internal cell fraction and intine – white band in aperture region – still present, polar plane TEM × 1750;
d, ultrathin section through whole pollen grain, polar plane TEM × 1800.
Arenga undulatifolia: **a**, *Elmer* 16237; *A. hastata*: **b** & **c**, *Dransfield* JD5801; *A. caudata*: **d**, *Kerr* 20154.

Taxonomic accounts: The genus has been monographed by J.P. Mogea; an account has yet to be published. See also Dransfield & Mogea (1984) and Mogea (2004).

Fossil record: In India a peduncle, *Palmostroboxylon arengoidum*, from the Deccan Intertrappean of Madhya Pradesh (although the age span of these volcanic deposits is controversial, see Chapter 5) is considered by Ambwani (1984) to resemble *Arenga* closely. A seed (*Iriartea collazoënsis*) recorded from the Middle Oligocene of Puerto Rico is considered to resemble closely those of *Arenga* or *Iriartea* (Hollick 1928). The earliest fossil pollen of *Arenga* is from the Lower Miocene of Borneo (Muller 1972, 1979). In India, Venkatachala & Kar (1969) recovered spinose monosulcate pollen grains, *Couperipollis kutchensis*, from the Eocene sediments of Kutch (Gujarat State). Lakhanpal (1970) later suggested that these pollen grains most probably represent *Arenga*. However, the pollen illustrated is not *Arenga*-like and the comparison was rejected by Muller (1981). The genus *Arengapollenites* was published by Kar (1985) to include, "oval, spinose and monocolpate grains where the spines are arranged on the margins alternating to close to the colpus like a crocodile jaw." This genus is known from the Lower Eocene, Naredi Formation (Kar 1985), and from the lignites of the Cambay and Kutch Basins (Kar & Bhattacharya 1992). The pollen illustrated closely resembles spiny *Arenga* pollen. *Couperipollis* was subsequently re-designated as a new genus, *Neocouperipollis* (Kar & Kumar 1986).

Notes: Within this relatively small genus, there is an astonishing range of form and flowering behaviour.

59. WALLICHIA

Dwarf to moderate, solitary or clustered monoecious pinnate-leaved palms from mainland Asia to southern Thailand, distinctive in the induplicate leaflets with praemorse tips, flowers borne in triads, very similar to *Arenga* but with united sepals in the staminate flower and always basipetal hapaxanthic.

Wallichia Roxb., Pl. Coromandel 3: 91 (1820). Type: **W. caryotoides** Roxb.

Harina Buch.-Ham., Mem. Wern. Nat. Hist. Soc. 5: 317(1826). Type: *H. caryotoides* Buch.-Ham.

Wrightea Roxb., Fl. ind. ed. 1832, 3: 621 (1832), non *Wrightia* R. Br. (1810). Type: *W. caryotoides* Roxb.

Asraoa J. Joseph, Bull. Bot. Surv. India 14: 144 (1975 ['1972']). Type: *A. triandra* J. Joseph (= *Wallichia triandra* [J. Joseph] S.K. Basu).

Commemorates Danish medical attaché and botanist, Nathanial Wallich (1786–1854) who was employed by the East India Company, eventually becoming Superintendent of the Company's garden in Calcutta.

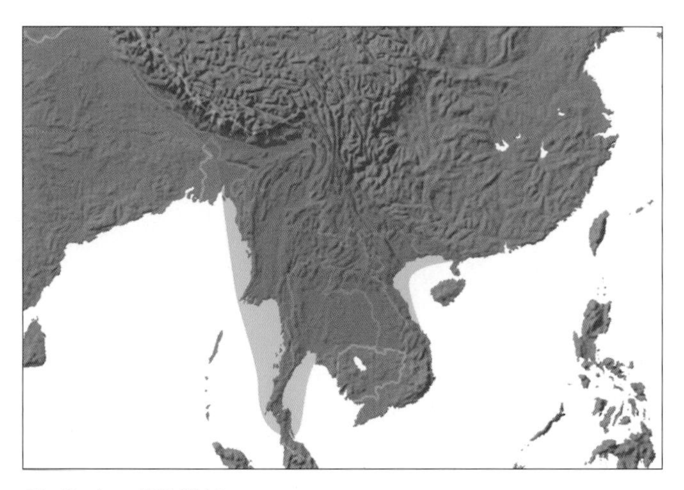

Distribution of ***Wallichia***

Dwarf to large, solitary or clustered, hapaxanthic, monoecious or ?dioecious, acaulescent, shrubby or tree palms. *Stem* with congested or elongate internodes, usually obscured by persistent fibrous leaf bases and sheaths. *Leaves* spirally or distichously arranged, induplicately imparipinnate, marcescent; sheath covered in a great variety of tomentum, scales and hairs, often extended beyond the petiole to form a ligule, eventually disintegrating into a mass of black fibres; petiole well developed, slender to robust, ± circular in cross-section or channelled adaxially, rounded abaxially, covered in a variety of scales and tomentum; rachis angled adaxially, rounded abaxially, variously tomentose; leaflets single-fold except for the terminal flabellum, regularly arranged or grouped and fanned within the groups, linear-lanceolate, irregularly rhomboid or deeply lobed, sometimes auriculate at base, the distal margins praemorse, the veins parallel to the fold or radiating from the base, or ± pinnately arranged along the fold, adaxial blade surface glabrous, abaxial surface usually densely covered in pale indumentum and scattered bands of brown scales, transverse veinlets obscure. *Inflorescences* axillary, interfoliar, solitary, bursting through leaf sheaths, produced in a basipetal sequence, branching to 1 order only, unisexual, usually dimorphic, the pistillate usually the most distal or 'terminal', with inconspicuous bracts, the staminate proximal (lateral), often hidden by very conspicuous bracts; peduncle ± circular in cross-section, usually densely covered with indumentum; prophyll small, 2-keeled, tubular only at the very base; peduncular bracts several, ± spirally arranged, much larger than the prophyll, tubular at the very base, splitting, usually densely covered in brown scales and tomentum; rachis usually longer than the peduncle; rachis bracts minute; rachillae numerous, rather slender, ± spirally arranged, usually densely covered with indumentum, bearing spirally arranged, minute bracts, subtending flowers. *Staminate flowers* paired or solitary, sometimes accompanied by the rudiments of a central pistillate flower; calyx tubular, truncate, with 3 lobes or teeth; floral receptacle elongate and stalk-like between calyx and corolla; corolla much exceeding the calyx, tubular near the base, with 3, elongate, valvate lobes distally; stamens 3–15, the filaments united basally in a short to long column, adnate partially or completely to the corolla tube, sometimes partly adnate to the lobes, anthers linear, apically obtuse or acute; pistillode absent. *Pollen* ellipsoidal, ± bi-symmetric; aperture a distal sulcus; ectexine intectate, usually spiny, spines attached to smooth upper surface of foot layer, in some species spines more numerous along aperture margins, less frequently clavate, bases of clavae often swollen; longest axis 24–27 µm [4/7]. *Pistillate inflorescence* usually erect, with fewer, more robust rachillae. *Pistillate flowers* solitary, spirally arranged, each subtended by a low bract and surrounded by 3 bracteoles; sepals 3, low, rounded, imbricate, ± distinct or joined briefly at base, petals 3 united basally to about middle, valvate distally; staminodes 0–3; gynoecium ± globose, 2–3 locular, 2–3 ovulate with a conical apical stigma, ovules inserted adaxially at the base, hemianatropous. *Fruit* ellipsoidal, small, reddish or purplish, 1–2, rarely 3-seeded, stigmatic remains apical; epicarp smooth, mesocarp fleshy, filled with irritant needle-like crystals, endocarp not differentiated. *Seeds* basally attached, ellipsoidal or hemispherical, endosperm homogeneous; embryo lateral. *Germination* remote-tubular; eophyll ovate to elliptic with erose margins. *Cytology*: 2n = 32.

Left: ***Wallichia disticha***, habit, cultivated, Fairchild Tropical Botanic Garden, Florida. (Photo: J. Dransfield)

Distribution and ecology: Nine species, from the Nepal Himalayas and upper Burma to China and southwards to peninsular Thailand. Found in humid tropical forest from near sea level to 2000 m altitude; the genus becomes rarer southwards in Thailand, suggesting that it is adapted to cooler or more seasonal climates than those found in Peninsular Malaysia. Most species are undergrowth palms, but *Wallichia disticha* is a moderate tree, recorded as growing gregariously on steep sandstone declivities in deep valleys in east Sikkim (Anderson 1869). It also occurs on limestone in Thailand.

Anatomy: Leaf (Tomlinson 1961).

Relationships: *Wallichia* is resolved as monophyletic with moderate support (Bayton 2005) and strongly supported as sister to *Arenga* (Bayton 2005, Asmussen *et al*. 2006).

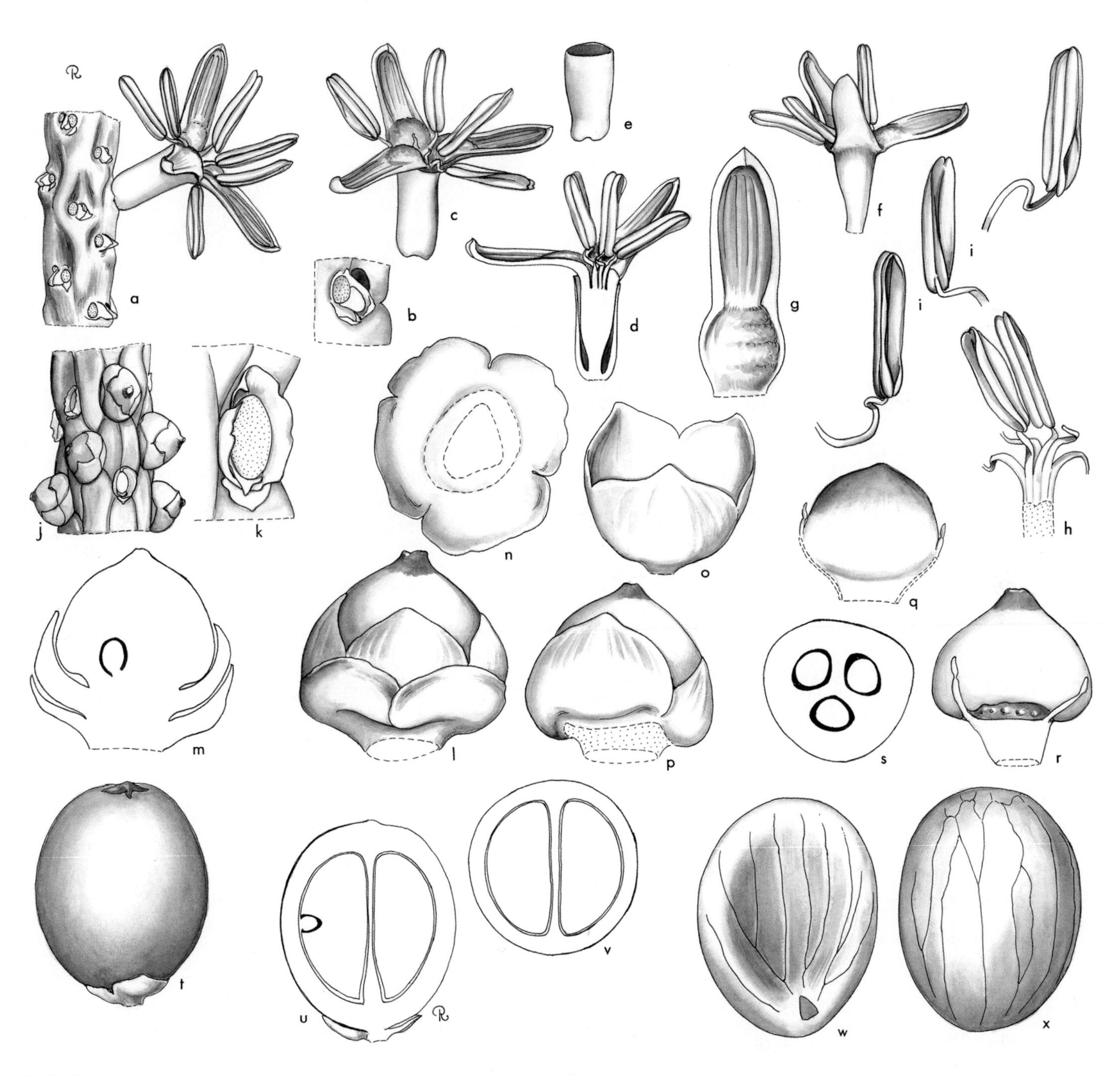

Wallichia. **a**, portion of staminate rachilla × 3; **b**, scars and bracteoles of staminate flowers × 3; **c**, staminate flower × 3; **d**, staminate flower in vertical section × 3; **e**, staminate calyx × 3; **f**, staminate flower, calyx removed × 3; **g**, staminate petal × 6; **h**, androecium × 6; **i**, stamen in 3 views × 6; **j**, portion of pistillate rachilla × $1^1/_2$; **k**, scar and bracteoles of pistillate flower × 6; **l**, pistillate flower × 6; **m**, pistillate flower in vertical section × 6; **n, o**, pistillate calyx in 2 views × 6; **p**, pistillate flower, calyx removed × 6; **q**, pistillate petal and staminodes × 6; **r**, gynoecium and staminodes × 6; **s**, ovary in cross-section × 6; **t**, fruit × 3; **u**, fruit in vertical section × 3; **v**, fruit in cross-section × 3; **w, x**, seed in 2 views × $4^1/_2$. *Wallichia oblongifolia*: **a–s**, *Moore* 9256; **t–x**, Fairchild Tropical Garden FG-58-475-B. (Drawn by Marion Ruff Sheehan)

Common names and uses: Wallich palms. Leaves of *Wallichia oblongifolia* have been used as thatch and the stem of *W. disticha* as a source of sago. All species are ornamental.
Taxonomic account: Henderson (2007).
Fossil record: No generic records found.
Notes: The morphological differences between *Arenga* and *Wallichia* are very slight.

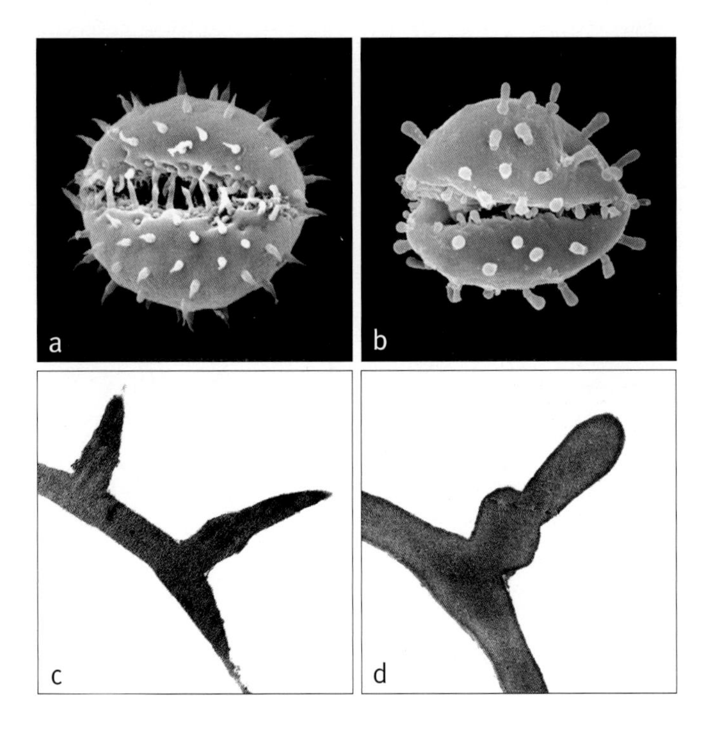

Right: ***Wallichia***
a, monosulcate pollen grain, long sharp spines with basal swelling, distal face SEM × 2000;
b, monosulcate pollen grain, long clavae with basal swelling, distal face SEM × 2000;
c, ultrathin section through spiny pollen wall TEM × 10,000;
d, ultrathin section through clavate pollen wall TEM × 10,000.
Wallichia oblongifolia: **a**, cultivated RBG Kew; *W. disticha*: **b** & **d**, no collection data; *W. caryotoides*: **c**, cultivated Calcutta.

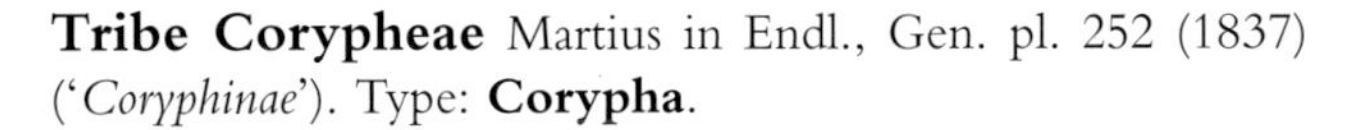

Tribe Corypheae Martius in Endl., Gen. pl. 252 (1837) ('*Coryphinae*'). Type: **Corypha**.

Hermaphroditic, hapaxanthic palms; leaves palmate, induplicate; flowering acropetal; inflorescences highly branched, the branches lacking prophylls basally; flowers superficial in cincinni; gynoecium syncarpous, trilocular, triovulate; fruit 1-seeded; embryo apical.

As circumscribed here, the tribe contains but one genus of massive, hapaxanthic palms, *Corypha*. The genus is noted for its possession of the largest compound inflorescence in the Plant Kingdom. It is distributed in mainland SE Asia and with one species occurring through Malesia to Australia.

Tribe Corypheae is monophyletic, highly supported and is often resolved on a long branch (e.g., Bayton 2005, Asmussen *et al.* 2006). It falls within the syncarpous clade as sister to tribe Borasseae with moderate to high support (Bayton 2005, Asmussen *et al.* 2006). A poorly supported placement as sister to a clade of Borasseae and Caryoteae has also been recovered (Baker *et al.* in review).

60. CORYPHA

Spectacular massive solitary hapaxanthic hermaphroditic fan palms of South and Southeast Asia, Malesia to Australia, with huge leaves that have spiny petioles with a distinctive triangular cleft at the base of the petiole, and huge suprafoliar compound inflorescences.

Corypha L., Sp. pl. 1187 (1753). Type: **C. umbraculifera** L.
Codda-Pana Adans., Fam. pl. 2: 25, 541 (1763) (type as above).
Taliera Mart., Palm. fam. 10 (1824). Type: *T. bengalensis* Spreng. (= *Corypha taliera* Roxb.).
Gembanga Blume in T. Nees, Flora 8 (2): 580, 678 (1825). Type: *G. rotundifolia* Blume (= *Corypha utan* Lam.).

Koryphe — *summit, peak, perhaps referring to the immense compound inflorescence at the stem tip.*

Massive, solitary, armed, hapaxanthic, hermaphroditic, tree palms. *Stem* erect, closely ringed with leaf scars sometimes in distinct spirals. *Leaves* induplicate, costapalmate, marcescent in immature individuals, tending to abscise under their own weight in trunked individuals; sheath sometimes with lateral lobes, later sometimes with a conspicuous triangular cleft below the petiole, the margins tending to erode into fibres; petiole massive, long, covered with caducous indumentum, adaxially deeply channelled, abaxially rounded, margins with well-defined teeth; adaxial hastula well developed, abaxial hastula rather irregular; blade regularly divided to ca. $^1/_2$ its radius into single-fold segments, these in turn shallowly divided along the abaxial folds, filaments present at upper folds in young leaves, segments with prominent longitudinal veins, abundant transverse veinlets and caducous floccose indumentum along the folds, indumentum more abundant abaxially. *Inflorescences* above the leaves, subtended by reduced, scale-like leaves, forming a massive, compound inflorescence-like structure; individual inflorescences emerging from the mouths of the bract-like leaves or through an abaxial split, branched to the third order, all branches ending as rachillae; prophyll of inflorescences 2-keeled, empty; bracts tubular, the proximal 0–several empty, other bracts inconspicuous, triangular, each subtending a first or higher order branch; rachillae bearing spirally arranged, adnate cincinni of up to 10 flowers; floral bracteoles minute. *Flowers* borne on short stalks formed by the base of the calyx and the receptacle; calyx tubular basally, with 3 low, triangular lobes; petals ± boat-shaped, basally imbricate, the margins usually inrolled, stamens 6, the 3 antesepalous free, the 3 antepetalous adnate basally to the petals, filaments tapering from a fleshy base; anthers short, somewhat sagittate basally, medifixed, latrorse; gynoecium tricarpellate, syncarpous, triovulate, ovary globose, distinctly 3-grooved, style elongate, slightly 3-grooved, stigma scarcely differentiated, ovule hemianatropous. *Pollen* ellipsoidal, usually slightly asymmetric; aperture a distal sulcus; ectexine tectate, reticulate or foveolate-reticulate, aperture margin psilate or scabrate; infratectum columellate; longest axis 28–40 µm [3/8]. *Fruit* globose, single-seeded with basal stigmatic remains; epicarp smooth, mesocarp fleshy, endocarp thin, usually remaining attached to the seed. *Seed* globose, with basal hilum, and shallow grooves corresponding to the rapheal bundles, endosperm homogeneous, with or without a central hollow; embryo apical. *Germination* remote-tubular; eophyll entire, lanceolate. *Cytology*: 2n = 36.

Distribution and ecology: Six recognised species, but probably fewer, ranging from southern India and Sri Lanka, to the Bay of Bengal, and from Indochina through Malesia to northern Australia; distribution probably much influenced by man. *Corypha* species are frequently associated with human settlements, but in the wild, they are probably a feature of open seral communities, such as alluvial plains, or submaritime storm forest; they are not found in climax tropical rain forest.

Anatomy: Leaf (Tomlinson 1961), root (Seubert 1997), floral (Uhl & Moore 1971).

Relationships: *Corypha* is a strongly supported monophyletic group (Bayton 2005, Asmussen *et al.* 2006). For relationships, see Corypheae.

Common names and uses: *Gebang* (*Corypha utan*), *talipot* (*C. umbraculifera*). *Corypha* has a wide range of uses and is intensively exploited. Leaves are used for thatch, writing material, umbrellas, buckets, etc. The stem has been used as a source of starch.

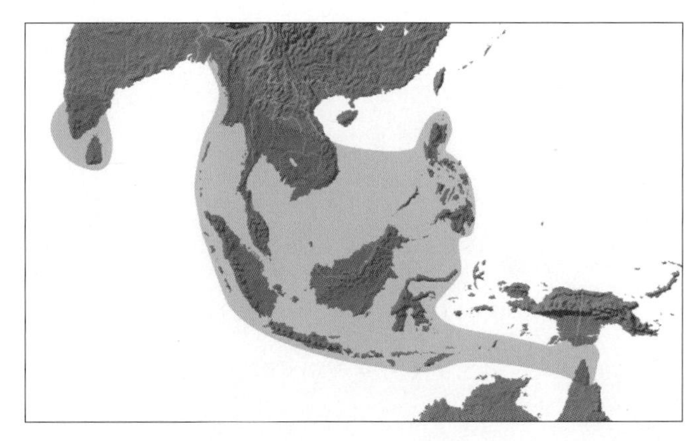

Distribution of ***Corypha***

Taxonomic accounts: Beccari (1931) and Basu (1988). A new critical revision is much needed.

Fossil record: Two species of fossil palm fruits are described from exposed Lower Palaeocene (Danian) deposits in the interior of Nûgssuaq, West Greenland: *Coryphoides poulsenii* B.E. Koch and *Coryphoicarpus globoides* B.E. Koch (Koch 1972), but their affinity needs reassessment. From the Lower

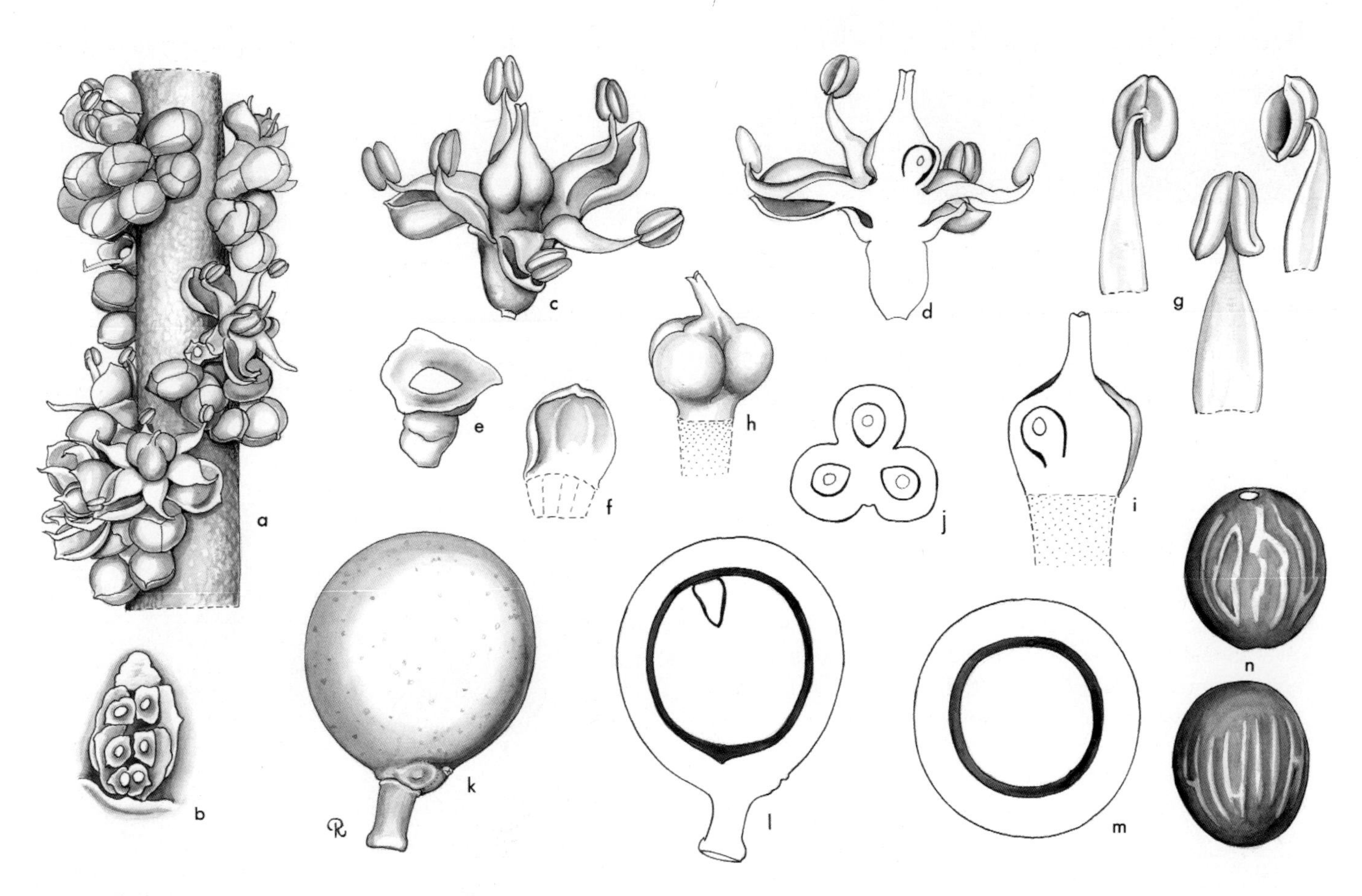

Corypha. **a**, portion of rachilla × 3; **b**, flower cluster, flowers removed to show bracteoles × 3; **c**, flower × 6; **d**, flower in vertical section × 6; **e**, calyx × 6; **f**, petal, interior view × 6; **g**, stamen in 3 views × 12; **h**, gynoecium × 12; **i**, gynoecium in vertical section × 12; **j**, ovary in cross-section × 12; **k**, fruit × 1$^1/_2$; **l**, fruit in vertical section × 1$^1/_2$; **m**, fruit in cross-section × 1$^1/_2$; **n**, seed in 2 views × 1$^1/_2$. *Corypha* sp.: **a–j**, *Read* s.n.; *C. utan*: **k–n**, *Moore & Meijer* 9223. (Drawn by Marion Ruff Sheehan)

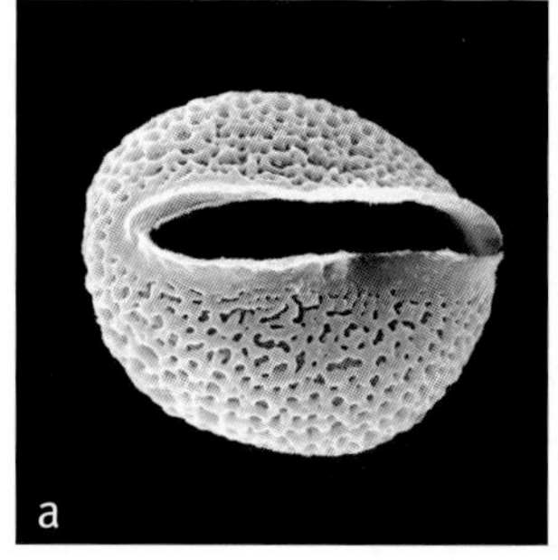

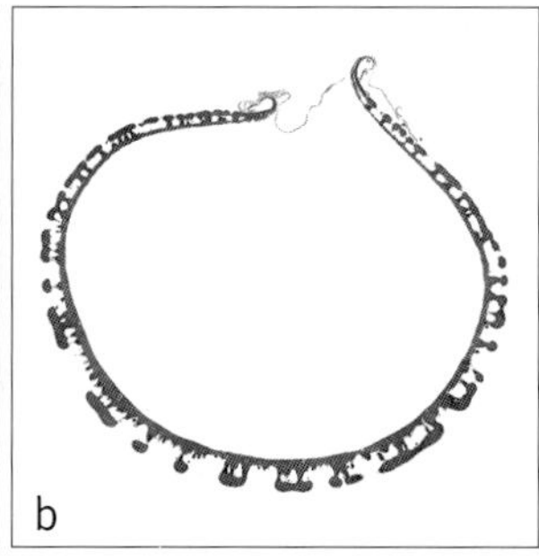

Above: ***Corypha***
a, monosulcate pollen grain, equatorial face SEM × 1500;
b, ultrathin section through whole pollen grain, polar plane × 1800.
Corypha umbraculifera: **a**, *Thwaites* 2336; *C. taliera*: **b**, *Hohenacker* 788.

Left: ***Corypha umbraculifera***, in fruit, Sri Lanka. (Photo: S.S. Hooper)

Eocene (London Clay flora) of southern England, a palm-like seed, *Palmospermum pulchrum* (Chandler 1961b, 1961c), and an internal cast of a seed, *Corypha wilkinsoni* (Chandler 1978), were said to resemble present-day Malayan '*Corypha olivaeformis*', especially *C. wilkinsoni*. However, as '*C. olivaeformis*' has never been described, the identity of the modern equivalent is not known. From the Indian Deccan Intertrappean of Madhya Pradesh (although the age span of these volcanic deposits is controversial, see Chapter 5), Ambwani & Mehrotra (1989) record a new fossil palm wood, *Palmoxylon toroides*, which they consider has affinities with *Corypha* (although comparisons of palm stem wood to generic level should always be viewed with caution). Monocolpate pollen from the Lower Miocene of Poland (Macko 1957) has been compared to *Corypha* pollen, but the pollen is of too general a coryphoid type to be conclusive.

Notes: Distinguished by toothed petiole margins, flowers in adnate cincinni, and syncarpous ovaries. Species of *Corypha* are most striking palms because of their massiveness. The compound terminal inflorescence is the largest among seed plants; the number of flowers has been estimated as 23.9 million (Fisher *et al.* 1987). See Tomlinson & Soderholm (1975) for a discussion of flowering, fruiting, and inflorescence structure.

Tribe Borasseae Mart. in Endl., Gen. pl. 250 (1837) ('*Borassinae*').

Dioecious; leaves palmate or costapalmate, induplicate; staminate and sometimes pistillate flowers borne in deep pits formed by the connation and adnation of the rachilla bracts; staminate flowers rarely solitary, usually in a cincinnus of 2 to many flowers; stamens exserted from the pits by elongation of the floral receptacle between the calyx and corolla; pistillate flowers solitary, bibracteolate; fruit 1–3-seeded with thick hard endocarp.

The eight genera of *Borasseae* are confined to the Old World where they are distributed in the lands bordering the Indian Ocean and on its islands. The tribe is divisible into two apparently natural groups. Characteristic features are the constant dioecy, staminate and sometimes pistillate flowers borne in pits in catkin-like rachillae, the staminodal ring with pointed lobes, the syncarpous gynoecium with specialised ovules and fruits with well-developed endocarp.

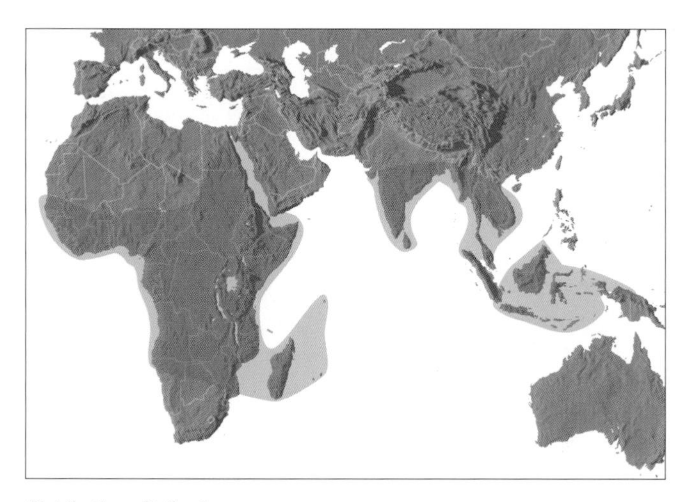

Distribution of tribe **Borasseae**

Tribe Borasseae is monophyletic, highly supported and resolves on a long branch in several phylogenies (e.g., Bayton 2005, Asmussen *et al.* 2006). It is placed within the syncarpous clade and is sister to tribe Corypheae with moderate to high support (Bayton 2005, Asmussen *et al.* 2006), although Baker *et al.* (in review) recover a weakly supported sister relationship with the Caryoteae.

Key to Subtribes of Borasseae

1. Staminate and pistillate inflorescences similar; flowers of both sexes sunken in pits; bracteoles subtending flowers bearing tufts of hairs; fruit stalked, usually 1-seeded, if 2- or 3-seeded then lobed, pyrenes not formed, stigmatic and carpellary remains basal . . . . **Hyphaeninae**
1. Staminate and pistillate inflorescences dissimilar; staminate flower clusters concealed in pits; bracteoles subtending staminate flowers not bearing tufts of hairs; pistillate flowers sessile in the axils of large leathery bracts; fruit sessile, ± symmetrical, 1–3-seeded, endocarp forming pyrenes, stigmatic remains apical . . . . **Lataniinae**

Subtribe Hyphaeninae Becc., Palmae Borasseae 1 (1924) ('*Hyphaeneae*'). Type: **Hyphaene**.

Staminate and pistillate rachillae similar; pistillate flowers pedicellate; pits of both sexes filled with hairs; fruit 1-seeded or 2–3-seeded and then 2–3-lobed; stigmatic remains basal or central between the lobes.

The four genera of this subtribe are closely related. *Bismarckia* and *Satranala* are endemic to Madagascar; *Hyphaene* is widely distributed in Africa, Madagascar and Arabia with one or possibly two species just reaching the Indian subcontinent and possibly Sri Lanka; and *Medemia* occurs in Sudan and Egypt. The genera differ largely in details of the fruits. Pyrenes are not formed; epicarp, mesocarp and endocarp develop more or less separately in each carpel, resulting in a lobed fruit if more than one carpel matures. In *Bismarckia*, the endocarp is irregularly ridged and pitted within and the endosperm is homogeneous. In *Medemia*, the endocarp is thinner and not ridged but the endosperm is ruminate. In *Hyphaene*, the fruits have unridged endocarps and homogeneous endosperm. These three genera all have an apical pore in the endocarp through which the seed germinates. In *Satranala*, the endocarp is externally flanged, internally smooth and the endosperm is ruminate. There is no pore in the endocarp; instead, the endocarp splits into two halves as the seed germinates.

Subtribe Hyphaeninae is resolved as monophyletic with high support in the most thoroughly sampled study (Bayton 2005) and with moderate support in the less densely sampled studies (Asmussen *et al.* 2006, Baker *et al.* in review). The Hyphaeninae is sister to subtribe Lataniinae with high support (Bayton 2005, Asmussen *et al.* 2006).

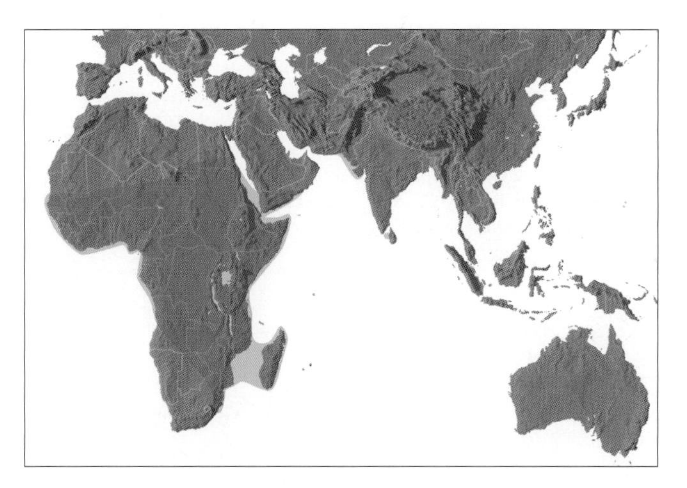

Distribution of subtribe **Hyphaeninae**

Key to Genera of Hyphaeninae

1. Seed deeply grooved. Madagascar. . . . . **61. Bismarckia**
1. Seed not grooved 2
2. Endocarp strongly flanged, lacking a circular apical pore. Madagascar . . . . **62. Satranala**
2. Endocarp not flanged and with a distinct terminal pore . . 3
3. Fruits various in shape, distally expanded, often shouldered, usually asymmetrical, rarely ovoid or spherical; seed irregular in outline, endosperm homogeneous. Africa, Madagascar, Arabia, India and ?Sri Lanka . . . . **63. Hyphaene**
3. Fruit ovoid; seed with ruminate endosperm. Egypt and Sudan . . . . **64. Medemia**

61. BISMARCKIA

Large solitary dioecious fan palm of savannahs in western Madagascar, distinctive in the ellipsoidal fruit with externally shallowly grooved endocarp, internally with numerous irregular intrusions into the homogeneous endosperm.

Bismarckia Hildebrandt & H. Wendl., Bot. Zeit. 39: 90, 93 (1881). Type: **B. nobilis** Hildebrandt & H. Wendl.

Named for Prince Otto Eduard Leopold von Bismarck-Schönhausen (1815–1898), German Chancellor responsible for the unification of the German state.

Robust, solitary, unarmed, pleonanthic, dioecious, tree palm. *Stem* erect, irregularly ringed with close leaf scars, becoming swollen basally. *Leaves* induplicate, costapalmate, marcescent in immature individuals, neatly abscising under their own weight in mature trunked individuals; sheath laterally ridged at the base, with a conspicuous triangular cleft below the petiole, with rows of scales in patches, petiole robust, adaxially channelled near the base, distally ± flattened, abaxially rounded, the surfaces greyish white, densely covered in white wax and patches of reddish, fringed caducous scales, the margins smooth; adaxial hastula often very large, distinctly lopsided, abaxial hastula absent; blade divided to ca. $^1/_4$–$^1/_3$ its length along adaxial folds into regular, stiff, single-fold segments, these further shortly bifid, interfold filaments conspicuous, surfaces obscurely striate, densely covered with wax, and along the folds with caducous scales, transverse veinlets not visible. *Inflorescence* interfoliar, solitary, shorter than the leaves, the staminate and pistillate similar; peduncle ± rounded in cross-section; prophyll short, 2-keeled, included in subtending sheath; peduncular bracts several, tubular, rather loosely sheathing, with a broad, split triangular limb, sometimes strongly keeled, and covered with caducous scales and wax; rachis longer than the peduncle; rachis bracts like the peduncular, decreasing in size distally; first-order branches crescent-shaped in cross-section, longer than the subtending bract, not bearing a prophyll, branching at the tip to produce a group of 3–7 radiating, catkin-like rachillae, the group sometimes reduced to 1; staminate rachillae usually more numerous than the pistillate, slightly sinuous, bearing a tight spiral of rounded, densely hairy, striate bracts, connate laterally and partially adnate to the axis to produce pits, densely filled with hairs, pistillate rachillae usually more massive than

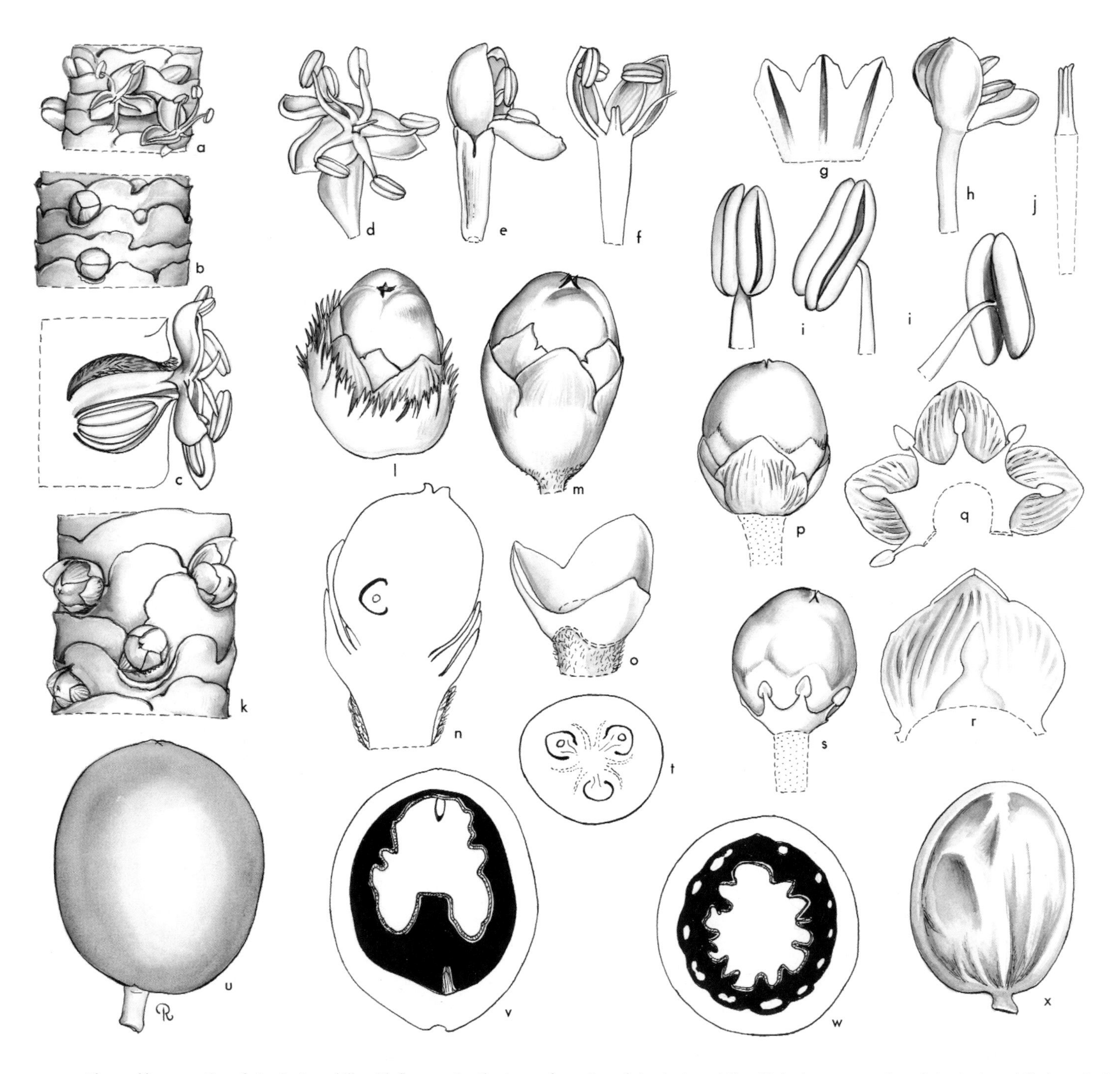

Bismarckia. **a**, portion of staminate rachilla with flowers at anthesis × 3; **b**, portion of staminate rachilla with buds × 3; **c**, portion of staminate rachilla in vertical section with flower and bud × 6; **d, e**, staminate flower in 2 views × 6; **f**, staminate flower in vertical section × 6; **g**, staminate calyx, expanded × 6; **h**, staminate flower, calyx removed × 6; **i**, stamen in 3 views × 12; **j**, pistillode × 12; **k**, portion of pistillate rachilla in flower × 3; **l**, pistillate flower and bracteoles × 6; **m**, pistillate flower × 6; **n**, pistillate flower in vertical section × 6; **o**, pistillate calyx × 6; **p**, pistillate flower, calyx removed × 6; **q**, pistillate corolla and staminodes, interior view, expanded × 6; **r**, pistillate petal, interior view × 12; **s**, gynoecium and staminodes × 6; **t**, ovary in cross-section × 6; **u**, fruit × 1; **v**, fruit in vertical section × 1; **w**, fruit in cross-section × 1; **x**, endocarp × 1. *Bismarckia nobilis*: **a–j**, *Read* 808; **k–t**, *Moore* 9597; **u–x**, *M.R. Sheehan* s.n. (Drawn by Marion Ruff Sheehan)

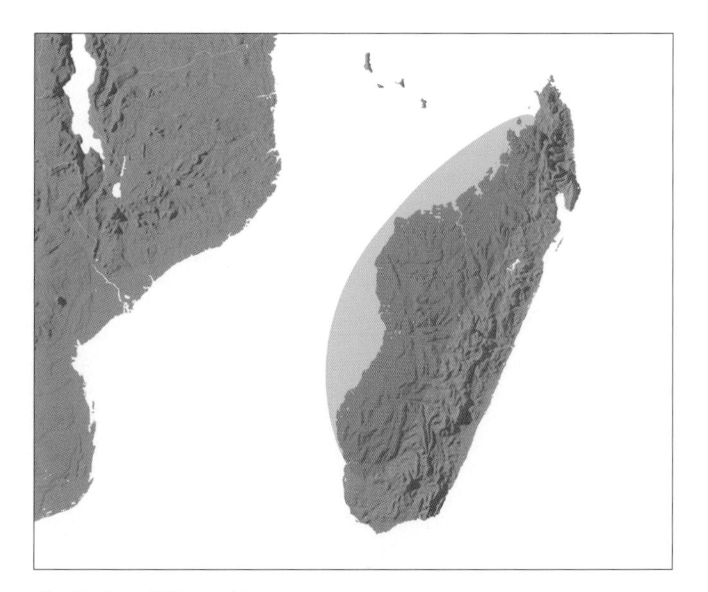

Distribution of ***Bismarckia***

the staminate, each bract subtending a solitary flower. *Staminate flowers* borne in a cincinnus of 3, embedded in the hairs, one flower emerging at a time, each bearing a membranous bracteole; calyx tubular, membranous, with 3 short, rather irregular lobes distally; corolla with a stalk-like base almost as long as the calyx lobes, bearing at its tip 3 ovate, hooded, valvate lobes; stamens 6, borne at the base of the corolla lobes, the filaments elongate, briefly united at the base, gradually tapering, anthers medifixed, versatile, latrorse; pistillode short, conical. *Pollen* ellipsoidal and ± bi-symmetric; aperture a distal sulcus; ectexine tectate, coarsely perforate-rugulate, aperture margin slightly finer; infratectum columellate; longest axis 35–46 µm; post-meiotic tetrads usually tetrahedral, rarely tetragonal or rhomboidal [1/1]. *Pistillate flowers* solitary, borne on a short hairy pedicel, the pedicel greatly elongating in fruit; sepals 3, imbricate, rounded, briefly connate basally; petals smaller than the sepals, 3, triangular, briefly connate at the base; staminodes united by their flattened triangular filaments to form a ring with 6 teeth, tipped with flattened empty, sagittate anthers; gynoecium tricarpellate, rounded, tipped with 3 low, slightly reflexed stigmas, septal nectaries present, ovule orthotropous with 2 lateral bodies, attached adaxially at the base. *Fruit* usually developing from 1 carpel, ± ellipsoidal, ovoid or rounded, the stigmatic remains and abortive carpels basal, the abortive carpels often enlarging to form 2 swellings; epicarp smooth, shiny, rich brown, somewhat speckled with lighter brown, mesocarp fibrous, ± aromatic, endocarp thick, irregularly flanged and pitted, and with a conspicuous central intrusion at the base. *Seed* basally attached, endosperm homogeneous, but grooved to match the endocarp intrusions; embryo apical. *Germination* remote-tubular; eophyll simple, strap-shaped. *Cytology*: 2n = 36.

Distribution and ecology: One species, Madagascar. *Bismarckia nobilis* occurs as a conspicuous component of savannahs in the western part of Madagascar.

Anatomy: Leaf (Tomlinson 1961), root (Seubert 1997). Somewhat resembles *Hyphaene* in leaf anatomy but hypodermal layers and fibres are less developed.

Top right: ***Bismarckia nobilis***, habit, Madagascar. (Photo: J. Dransfield)

Bottom right: ***Bismarckia nobilis***, crown, Madagascar. (Photo: J. Dransfield)

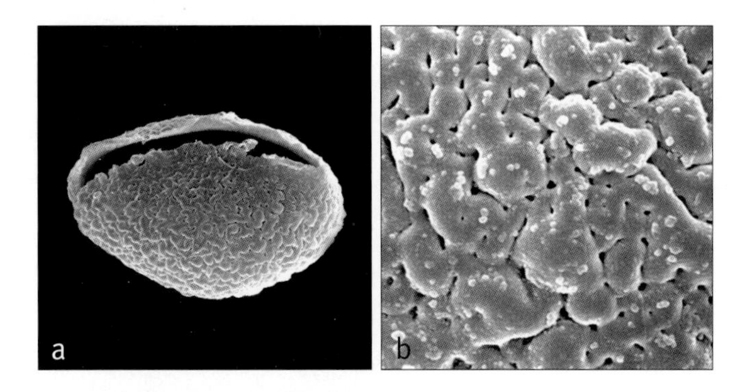

Above: ***Bismarckia***
a, monosulcate pollen grain, equatorial face SEM × 1500;
b, close-up, coarsely rugulate surface SEM × 6000.
Bismarckia nobilis: **a** & **b**, *Moore* 7576.

Relationships: *Bismarckia* is resolved as sister to *Satranala* with moderate to high support (Bayton 2005, Asmussen *et al.* 2006).

Common names and uses: Bismarck palm. The trunk is used whole or split for house construction and the leaves for thatch. The pith of the trunk produces a rather bitter sago. The stem is occasionally tapped for wine. Outside Madagascar, *Bismarckia* is a highly regarded ornamental for the drier tropics and subtropics.

Taxonomic accounts: Beccari (1924) and Dransfield & Beentje (1995b).

Fossil record: No generic records found.

Notes: Distinguished by stiff, whitish, waxy, costapalmate leaves, with large adaxial hastula and petiole with smooth margins. The internal structure of the fruit and seed distinguish it from *Medemia*, in which the genus has sometimes been included.

62. SATRANALA

Moderate solitary dioecious fan palm of rain forests in northeastern Madagascar, distinctive in the rounded fruit with externally winged and crested endocarp, internally smooth, which splits in two valves on germination, and with ruminate endosperm.

Satranala J. Dransf. & Beentje, Kew Bull. 50 (1): 87 (1995).
Type: **S. decussilvae** Beentje & J. Dransf.

Based on the Malagasy vernacular name satranala — *forest fan palm.*

Robust solitary pleonanthic tree palm. *Stem* erect, irregularly ringed with leaf scars, somewhat swollen at the base, sometimes with aerial roots above the base of the trunk. *Leaves* induplicately costapalmate, neatly abscising under their own weight in mature individuals; sheath lacking auricles, with a conspicuous triangular cleft below the petiole, abaxially with scattered scales; petiole adaxially channelled near the base, distally ± flattened, abaxially rounded, the margins sharp, bearing minute irregular teeth distally, surfaces covered in patchy hairs, scales and white wax; adaxial hastula present, abaxial hastula absent; blade divided to ca. 1/4 to 1/3 its radius along adaxial folds into induplicate segments, further divided by short splits along abaxial folds, interfold filaments caducous, lamina covered with thin white wax, transverse veinlets conspicuous, close, somewhat sinuous; lamina anatomy dorsiventral. *Staminate inflorescence* interfoliar, solitary, shorter than the leaves, branching to 2 orders; peduncle ± rounded in transverse section; prophyll short, 2-keeled, included in the subtending leaf-sheath; peduncular bracts several; rachis longer than the peduncle; rachis bracts decreasing in size distally, tubular, rather loosely sheathing, with a broad, split, triangular limb, sometimes strongly keeled, densely covered in rusty tomentum; first-order branches crescent-shaped in cross-section, longer than the subtending bract, not bearing a prophyll, branching at the tip to produce a group of 3–9 radiating, catkin-like rachillae, rarely at the inflorescence tip the group reduced to a single branch; rachillae slightly sinuous, bearing a tight spiral of rounded, densely hairy, striate bracts, connate laterally and partially adnate to the axis to produce pits, densely filled with hairs. *Staminate flowers* unknown. *Pollen* (found remaining among inflorescence bracts) ellipsoidal, ± bi-symmetric; aperture a distal sulcus; ectexine tectate, coarsely perforate-rugulate, aperture margin slightly finer; infratectum columellate; longest axis 43–50 µm [1/1]. *Pistillate inflorescence* similar to the staminate but with fewer rachillae in groups of not more than 3. *Pistillate flowers* unknown. *Fruit* developing from a single

Above: ***Satranala decussilvae***, crown with infructescence, Madagascar.
(Photo: H. Beentje)

carpel, globose, stigmatic remains basal; epicarp smooth, mesocarp fleshy and fibrous, endocarp hard, woody, externally with broad anastomosing flanges, one principal flange forming a crest along the vertical axis of the endocarp, the crest splitting during germination, allowing the cotyledonary stalk to emerge, the endocarp then splitting into two, internally the endocarp smooth, lacking a basal intrusion and lacking a germination pore opposite the embryo. *Seed* globose, basally attached; endosperm ruminate, deeply and irregularly penetrated by integumental tissue, solid, embryo apical. *Germination* remote-tubular, eophyll palmate with 2–3 segments. *Cytology* not studied.

Distribution and ecology: Eastern Madagascar in the Masoala Peninsula and the Mananara Avaratra Biosphere Reserve, growing in wet forest on shallow soils overlying ultramafic rock or quartzite, in steep-sided valleys rich in pandans and palms, at 250–300 m above sea level. All populations are small (Ravololonanahary 1999).

Anatomy: Leaf lamina dorsiventral (Rudall, pers. comm.).

Relationships: There is moderate to high support for a sister relationship between *Satranala* and *Bismarckia* (Bayton 2005, Asmussen *et al.* 2006).

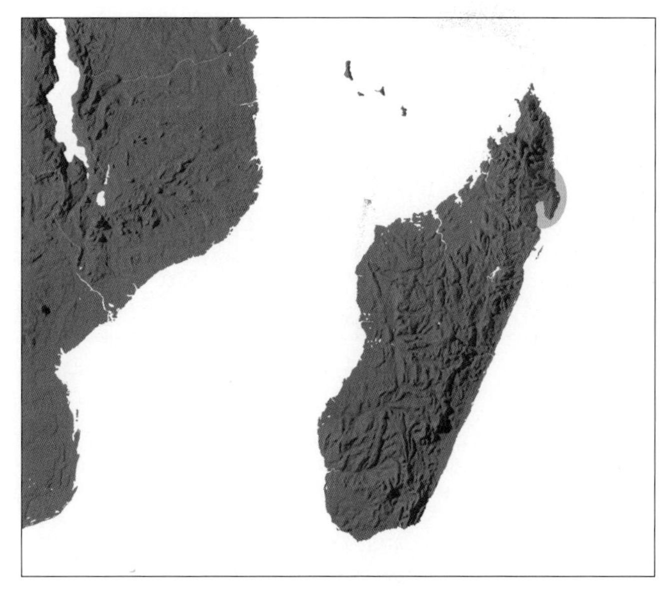

Distribution of ***Satranala***

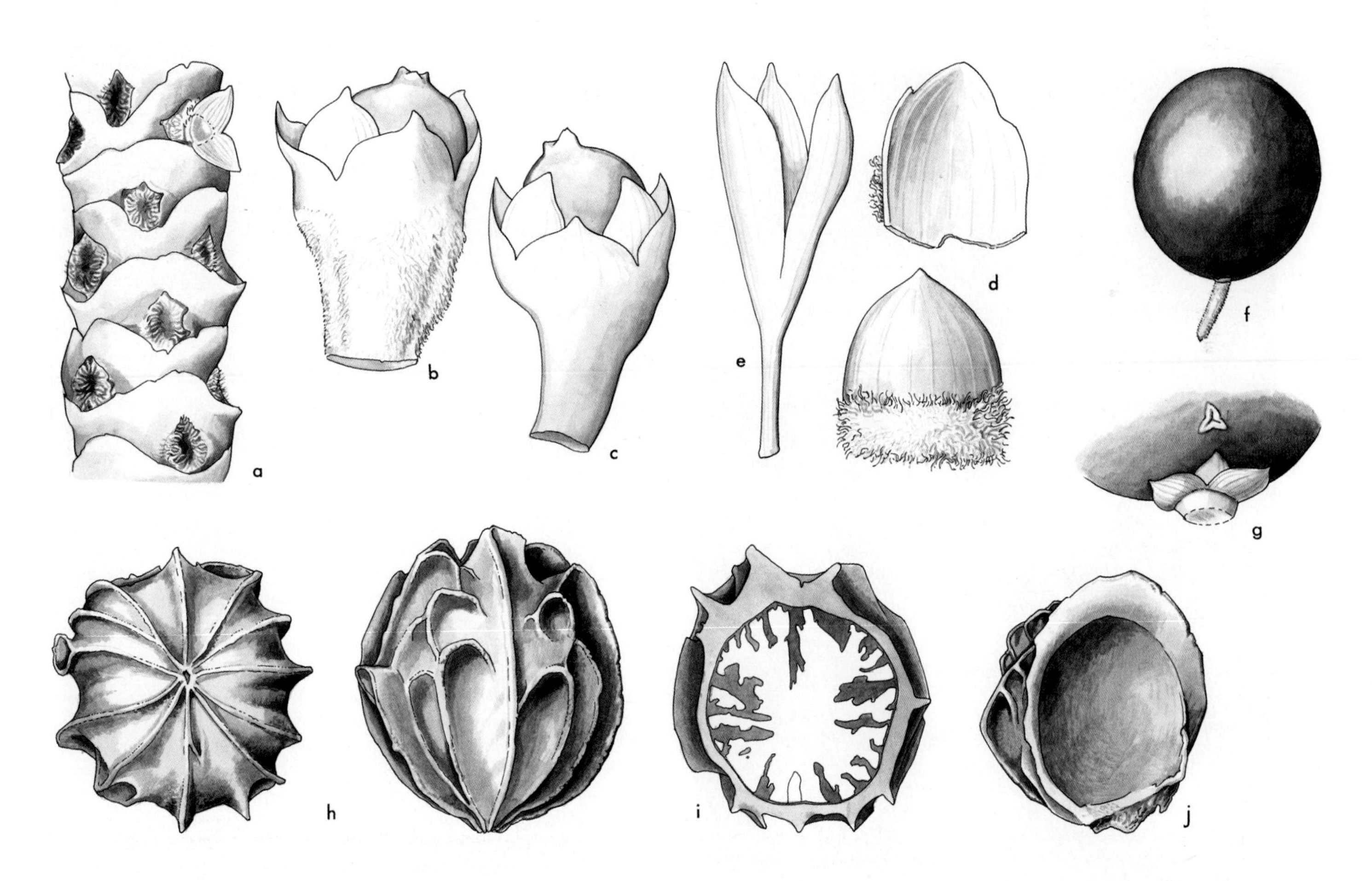

***Satranala*.** **a**, portion of old pistillate rachilla with pits, flowers removed × 2$^1/_2$; **b**, pistillate flower × 6; **c**, pistillate flower, indumentum removed × 6; **d**, petals of pistillate flower, 2 views × 10; **e**, remains of staminate, flower, calyx × 9; **f**, mature fruit × 1; **g**, detail of basal stigmatic remains × 6; **h**, endocarp, 2 views × 1; **i**, vertical section of endocarp and seed × 1; **j**, one half of old endocarp × 1. *Satranala decussilvae*: **a–d**, *Beentje et al.* 4628; **e**, *Beentje & J. Dransfield* 4628; **f**, **h–i**, *Beentje* 4474; **g**, *Beentje et al.* 4807; **j**, *Beentje* 4810. (Drawn by Lucy T. Smith)

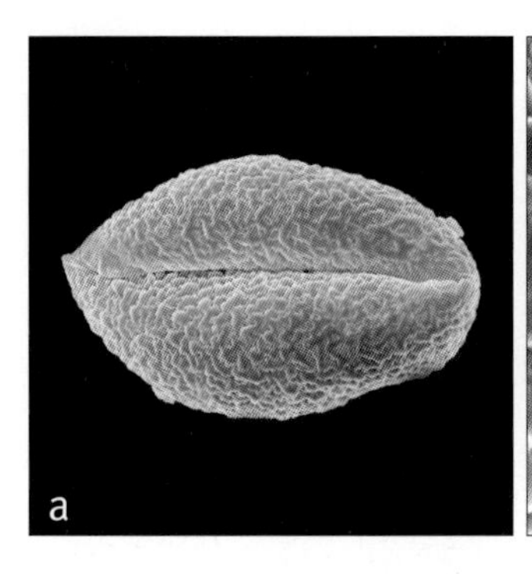
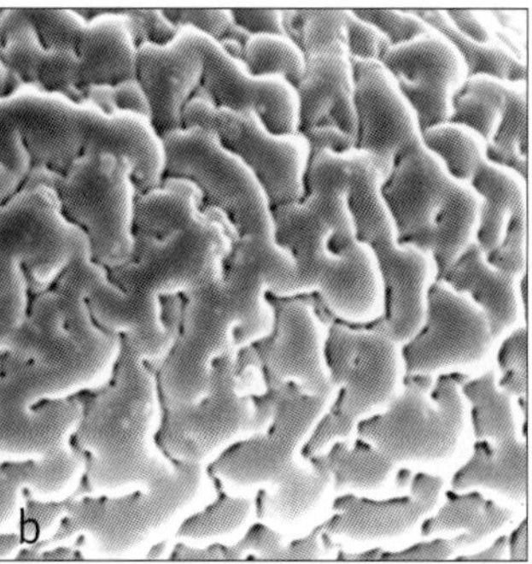

Above: ***Satranala***
a, monosulcate pollen grain, distal face SEM × 1250;
b, close-up, coarsely rugulate surface SEM × 6000.
Satranala decussilvae: **a** & **b**, *Beentje* 4628.

Below: ***Satranala decussilvae***, infructescence, Madagascar.
(Photo: J. Dransfield)

Common names and uses: *Satranabe* (Betsimisaraka). Used for thatch and the apex sometimes eaten; since its discovery, the palm has entered horticultural trade.
Taxonomic accounts: Dransfield & Beentje (1995a, 1995b).
Fossil record: No generic records found.
Notes: As the seed germinates, the cotyledonary stalk pushes its way out of the endocarp, which gapes slightly along the major crest, the endocarp thus appearing somewhat like a walnut with two valves; no other palm known to us germinates in this way.

63. HYPHAENE

Moderate solitary or clustered dioecious fan palms, usually in the drier parts of Africa and Arabia, with outliers in western India and possibly Sri Lanka, rarely in rain forest areas, often with dichotomously branched stems, distinctive in the very spiny petioles and often irregularly shaped fruit with smooth endocarp and homogeneous endosperm.

Hyphaene Gaertn., Fruct. sem. pl.1: 28 (1788). Type: **H. coriacea** Gaertn.
Cucifera Delile, Descr. Egypte, Hist. nat. 1: 53 (1809). Type: *C. thebaica* Delile (illegitimate name).
Douma Poir. in Duhamel, Traité arbr. arbust. (ed. 2) 4: 47 (1809). Type: *D. thebaica* Poir. (= *H. thebaica* [L.] Mart.).
Doma Poir. in Lam., Tabl. encycl. 4: t900 (1819), (orthographic variant of *Douma*).

Hyphaino – *entwine, in reference to the fibres in the fruit wall.*

Dwarf to large, solitary or clustered, spiny, pleonanthic, dioecious, acaulescent, creeping, shrubby or tree palms. *Stem* closely ringed with slightly raised leaf scars, usually branching several times by equal forking (dichotomy), rarely unbranched, and then sometimes the trunk ventricose; trunk surface in juveniles with a lattice of old leaf bases, later becoming bare. *Leaves* induplicate, costapalmate, marcescent, later abscising under their own weight; sheath soon becoming open, densely tomentose, later with a conspicuous triangular cleft below the petiole, margins fibrous; petiole robust, covered in caducous indumentum, adaxially channelled, abaxially rounded, the margins armed with robust, triangular, reflexed or upward pointing spines; adaxial hastula well developed, often asymmetrical, abaxial hastula absent; blade divided to about $^1/_3$ its length along the adaxial ribs into single-fold segments, these further shallowly divided along the abaxial ribs; interfold filaments often conspicuous; blade surfaces frequently glaucous with abundant wax, and also bearing minute dot-like scales and caducous indumentum, particularly along the ribs, midrib prominent, longitudinal veins close, transverse veinlets inconspicuous. *Inflorescences* interfoliar, the staminate and pistillate basically similar, though the pistillate more robust and with fewer branches; peduncle bearing a basal, 2-keeled, tubular prophyll and usually 2 empty, tubular peduncular bracts with triangular limbs, bearing abundant caducous indumentum when young; rachis longer than the peduncle; rachis bracts like the peduncular but each subtending a first-order branch; first-order branches basally bare, semicircular in cross-section, ± included in the subtending bract, terminating, in the staminate inflorescence, in a group of 1–6 or rarely more rachillae, each subtended by a low bract, in the pistillate inflorescence terminating in 1–3 rachillae; rachillae catkin-like, bearing a tight spiral of rounded, densely hairy, striate bracts, connate laterally and partially adnate to the axis to produce pits, densely filled with a pile of hairs. *Staminate flowers* borne in a cincinnus of 3 flowers, embedded in the hairs, one flower emerging at a time, each bearing a small membranous bracteole; calyx tubular at the base with 3 elongate hooded, membranous lobes; corolla with a conspicuous stalk-like base almost as large as the calyx lobes, bearing at its tip 3 ovate, hooded, valvate, striate lobes; stamens 6, borne at the base of the lobes, the filaments ± connate at their swollen bases, tapering above, anthers medifixed, versatile, latrorse to introrse; pistillode minute, 3-lobed. *Pollen* ellipsoidal, bi-symmetric; aperture a distal sulcus; ectexine tectate, finely perforate-rugulate, with psilate supratectal gemmae, aperture margin similar but with fewer gemmae; infratectum columellate; longest axis 30–44 µm; post-meiotic tetrads usually tetrahedral, sometimes tetragonal or, rarely, rhomboidal [4/10]. *Pistillate flowers* borne singly with a bracteole in each pit, on a short densely hairy pedicel, the pedicel sometimes considerably elongating after fertilisation; sepals 3, distinct, rounded, imbricate, ± membranous, striate; petals 3, similar to sepals; staminodial ring epipetalous, 6-toothed, the teeth bearing sagittate, flattened, empty anthers; gynoecium globose,

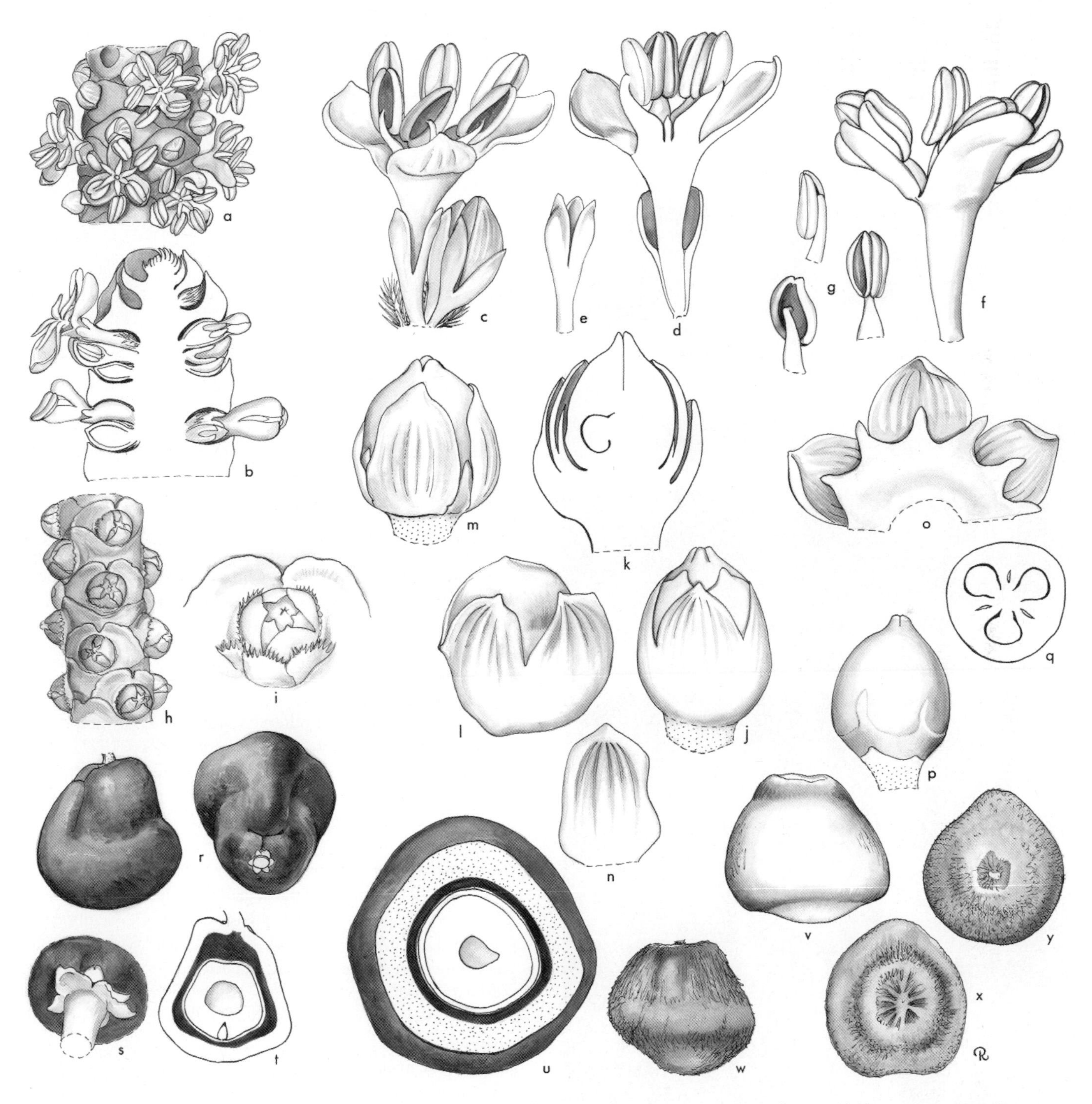

Hyphaene. **a**, portion of staminate rachilla × 1$^{1}/_{2}$; **b**, portion of staminate rachilla in vertical section × 3; **c**, staminate flower and buds × 6; **d**, staminate flower in vertical section × 6; **e**, staminate calyx × 6; **f**, staminate flower, calyx removed × 6; **g**, stamen in 3 views × 6; **h**, portion of pistillate rachilla × 1$^{1}/_{2}$; **i**, pistillate flower and bracts × 3; **j**, pistillate flower × 6; **k**, pistillate flower in vertical section × 6; **l**, pistillate calyx × 6; **m**, pistillate flower, calyx removed × 6; **n**, pistillate petal × 6; **o**, pistillate corolla and staminodes, interior view, expanded × 6; **p**, gynoecium × 6; **q**, ovary in cross-section × 6; **r**, fruit in 2 views × $^{1}/_{2}$; **s**, base of fruit × 1$^{1}/_{2}$; **t**, fruit in vertical section × $^{1}/_{2}$; **u**, fruit in cross-section × 1; **v**, endocarp × 1; **w, x, y**, seed in 3 views × 1. *Hyphaene coriacea*: all from *Read* 720. (Drawn by Marion Ruff Sheehan)

tricarpellate, triovulate, stigmas 3, short, septal nectaries present, opening by pores distally, ovules orthotropous, attached adaxially at the base of each carpel. *Fruit* borne on enlarged pedicel with persistent perianth segments, normally developing from 1 carpel, rarely 2 or 3, the fruit then 2- or 3-lobed, with basal stigmatic remains, the whole fruit very variable in shape, shouldered, distally expanded, usually asymmetrical, rarely ovoid or spherical; epicarp smooth, dull or shining, often pitted with lenticels, coloured various shades of brown, mesocarp fibrous, often aromatic, dry but sweet, endocarp well developed, hard, stony. *Seed* basally attached, endosperm homogeneous with a central hollow; embryo apical opposite a thinner area of the endocarp. *Germination* remote-tubular; eophyll simple, lanceolate, plicate. *Cytology*: 2n = 36.

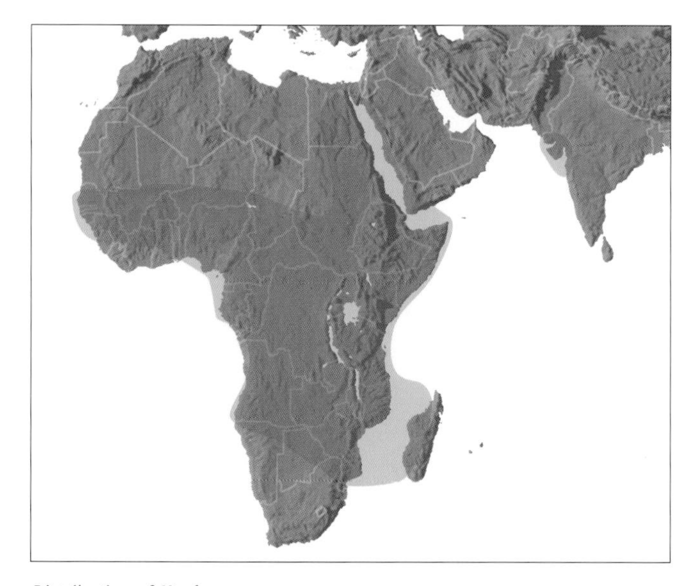

Distribution of ***Hyphaene***

Distribution and ecology: Numerous names have been published, but there are probably only about eight species distributed in the drier parts of Africa southwards to Natal, Madagascar, the Red Sea and the coasts of the Gulf of Eilat, Arabia, and western India. One species recorded for Sri Lanka may be an introduction. *Hyphaene* species tend to grow in arid or semiarid areas, in habitats where ground water is near the surface, e.g., along seasonal water-courses, coastal sand dunes and flats, and oases. In east Africa, *H. compressa* can be found inland at altitudes up to 1400 m above sea level. One species, *H. guineensis*, is found in coastal habitats in Gabon in areas with high rainfall. All species seem to be used by man; thus their distribution has been much influenced by destructive harvesting and accidental or deliberate planting. Elephants and baboons, among other wild animals, are responsible for seed dispersal. Bees have been observed visiting the flowers.

Anatomy: Leaf (Tomlinson 1961), root (Seubert 1997).

Relationships: *Hyphaene* is a highly supported monophyletic group (Bayton 2005). The genus is sister to *Medemia* with moderate to high support (Bayton 2005, Asmussen *et al.* 2006).

Common names and uses: Doum palms. The doum palms are locally very important, particularly to subsistence farmers. The leaves are used for thatch and as a source of fibre for plaiting. The apex is often semi-destructively tapped to make palm wine. Wood can be used. The fruits provide an edible mesocarp and an endosperm, edible when young, but formerly used when mature as a source of vegetable ivory. All fallen parts of the palms are used as fuel.

Taxonomic accounts: Beccari (1924) and Dransfield (1986).

Top: ***Hyphaene compressa***, habit, Kenya. (Photo: J. Dransfield)

Bottom left: ***Hyphaene coriacea***, crown, cultivated, Montgomery Botanical Center, Florida. (Photo: J. Dransfield)

Bottom right: ***Hyphaene thebaica***, infructescence, Egypt. (Photo: W.J. Baker)

Fossil record: From the Upper Oligocene of Ethiopia (Chilga) a spiny petiole, *Hyphaene kappelmanii*, and an incomplete spiny petiole, *Hyphaene* sp., have been recovered (Pan *et al.* 2006). From the Indian Deccan Intertrappean of Maharashtra State (although the age span of these volcanic deposits is controversial, see Chapter 5) a petrified palm petiole, *Palmocaulon hyphaenoides*, is described by Shete and Kulkarni (1980); it shows many similarities to petioles of *H. indica* (= *H. dichotoma*). Endocarps supposed to be from the earliest Cretaceous (Aptian) of Egypt are compared to *Hyphaene* by Vaudois-Miéja and Lejal-Nicol (1987), although one endocarp (fig. 8 in that publication) looks notably *Hyphaene*-like. An Aptian age for the formation in which these fossils were found is questionable; it is probably much younger (Late Cretaceous) and further research on this matter is needed (Schrank 1992, pers. comm.). These fossils cannot be accepted as the earliest unequivocal palm fossils. From the Indian Deccan Intertrappean, fruit (*Hyphaenocarpon indicum*) is described by Bande *et al.* (1982). Gemmate pollen from the Neogene Cauvery Basin of southern India is published as a new species of *Gemmamonocolpites*: *G. hyphaenoides* (Ramanujam *et al.* 2001); unfortunately, the pollen is not illustrated or formally described so a critical appraisal is not possible.

Notes: Distinguished by the elongate costapalmate leaf, which is often silvery, by the lack of an abaxial hastula, by petiolar spines, and by the frequent presence of dichotomous branching, long, more or less slender inflorescences, and distinctive brown fruits.

Above: ***Hyphaene***
a, monosulcate pollen grain, distal face SEM × 1500;
b, close-up, tectate perforate pollen surface with widely spaced gemmae SEM × 5000;
c, close-up, tectate perforate pollen surface with verrucae SEM × 5000;
d, ultrathin section through whole tectate gemmate pollen grain, polar plane TEM × 2000.
Hyphaene petersiana: **a**, *Gossweiler* 9778; **b**, *Trapwell* 1315; *H. thebaica*: **c**, *Wickens* 3135; *H. coriacea*: **d**, *Perrier de la Bâthie* 12829.

64. MEDEMIA

Moderate solitary dioecious fan palm of desert oases in Egypt and Sudan, distinctive in the absence of a hastula and the rounded fruit with smooth endocarp and ruminate endosperm.

Medemia Württemb. ex H. Wendl., Bot. Zeit. 39: 90, 93 (1881). Type: *M. argun* (Mart.) Württemb. ex H. Wendl. (*Hyphaene argun* Mart.).

Origin unknown.

Robust, solitary, unarmed, pleonanthic, dioecious, tree palm. *Stem* erect, ringed with close leaf scars. *Leaves* induplicately costapalmate, marcescent, or falling under their own weight; sheath soon becoming open, densely tomentose, later with a conspicuous triangular cleft below the petiole, margins fibrous; petiole well developed, flattened adaxially, rounded abaxially, the margins armed with widely spaced coarse, forward-pointing spines, mostly in the mid-section; hastulae absent; costa short, more conspicuous abaxially than adaxially; blade divided ± regularly along adaxial folds to ca. $^{2}/_{3}$ its length into single-fold segments, these further divided for a very short distance along abaxial folds, interfold filaments persistent at the adaxial sinuses, surfaces ± glaucous, with scattered dot-like scales, particularly along the ribs on the abaxial surface, longitudinal veins crowded, transverse veinlets obscure. *Inflorescences* interfoliar, becoming pendulous; prophyll and peduncular bracts, if any, not seen; rachis bracts tubular, with scattered caducous scales, and a long triangular limb; first-order branches ± pendulous, devoid of bracts except at the tip, margins sharp, the surfaces bearing scattered caducous scales, staminate inflorescence with

Right: ***Medemia argun***, habit, Egypt. (Photo: W.J. Baker)

first-order branches bearing at their tips 1–7 digitately displayed rachillae, in the pistillate bearing a single rachilla; rachillae catkin-like, bearing a tight spiral of rounded, densely hairy, imbricate bracts, connate laterally and to the axis to produce pits, filled with a dense pile of hairs. *Staminate flowers* borne in threes, each bearing a spathulate membranous bracteole included within the pit, the flowers exserted and exposed one by one from the pit; calyx stalk-like at the base with 3 narrow, spathulate, membranous, striate lobes with irregular margins; corolla with a conspicuous stalk-like base almost as long as the calyx lobes, bearing at its tip 3 oblong, membranous, striate, ± circular lobes; stamens 6, borne at the base of the corolla lobes, filaments elongate, tapering, anthers medifixed, apparently versatile, latrorse; pistillode 3-lobed, small. *Pollen* ellipsoidal, slightly asymmetric; aperture a distal sulcus; ectexine tectate, coarsely perforate-rugulate, aperture margin slightly finer; infratectum columellate; longest axis 36–49 µm [1/1]. *Pistillate flowers* solitary, borne on a short, densely hairy pedicel, lengthening after fertilisation; sepals 3, distinct, imbricate, obtuse, broad, membranous, glabrous, striate; petals 3, similar to the sepals; ?staminodes; gynoecium globose, with 3 eccentric, short, recurved stigmas, ovule probably orthotropous. *Fruit* ovoid, borne on the elongated pedicel, usually developed from only 1 carpel, rarely from 2 and then bilobed, with basal

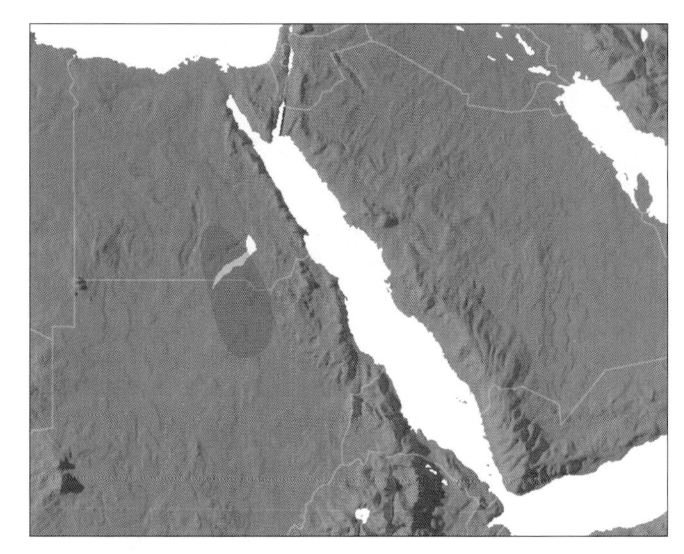

Distribution of ***Medemia***

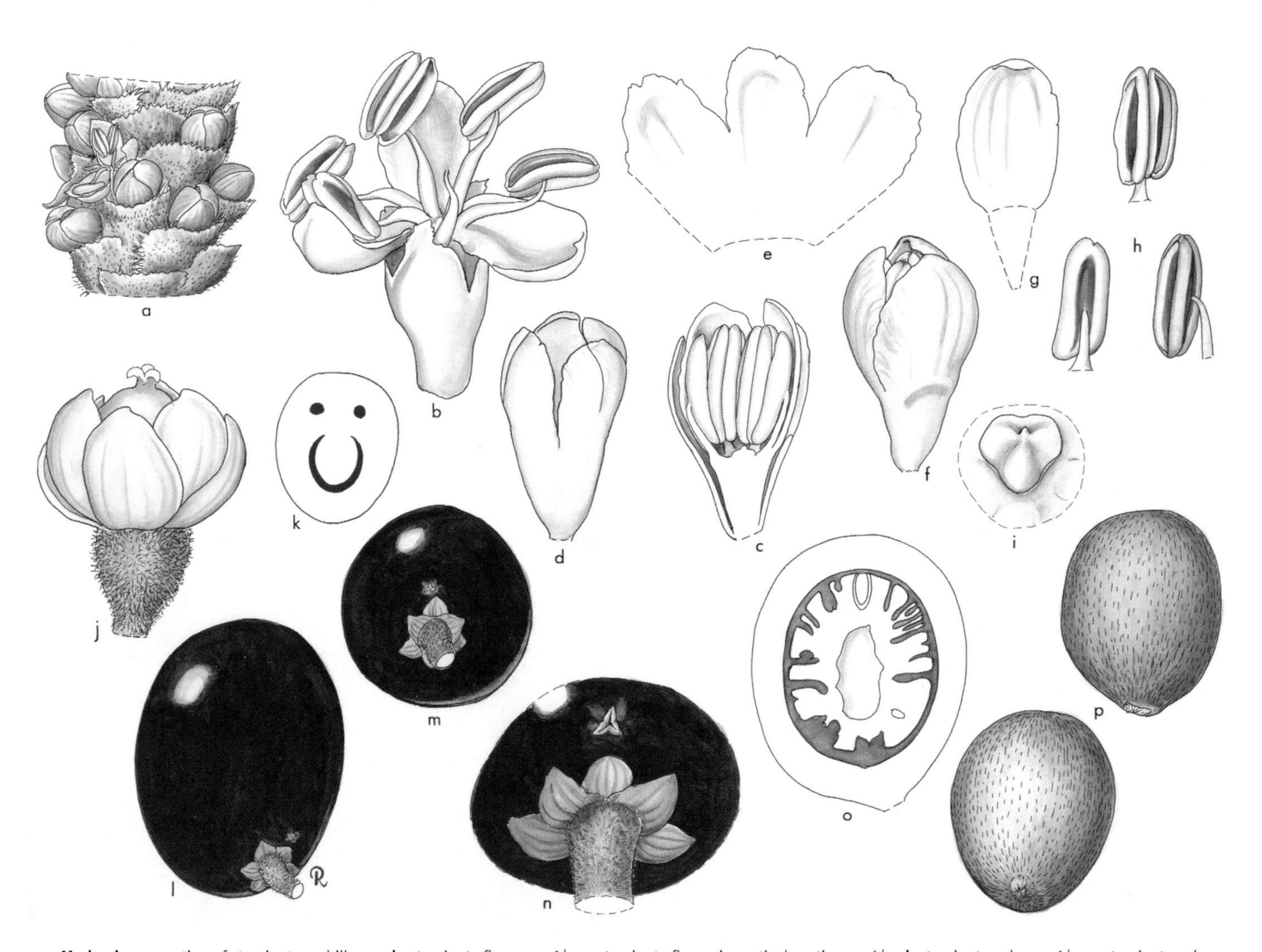

Medemia. **a**, portion of staminate rachilla × 3; **b**, staminate flower × 7 1/2; **c**, staminate flower in vertical section × 7 1/2; **d**, staminate calyx × 7 1/2; **e**, staminate calyx, expanded × 7 1/2; **f**, staminate flower, calyx removed × 7 1/2; **g**, staminate petal × 7 1/2; **h**, stamen in 3 views × 9; **i**, pistillode × 12; **j**, pistillate flower, magnification unknown; **k**, ovary in cross-section, magnification unknown; **l**, fruit × 1; **m**, base of fruit × 1; **n**, base of fruit with stigmatic remains × 2 1/4; **o**, fruit in vertical section × 1; **p**, seed in 2 views × 1. *Medemia argun*: **a–i** and **l–o**, *Talbot* s.n.; **j** & **k** redrawn from Beccari (1924); **p**, *d'Albertis* s.n. (Drawn by Marion Ruff Sheehan)

Above left: ***Medemia argun***, leaf, Egypt. (Photo: W.J. Baker)

Above middle: ***Medemia argun***, staminate inflorescences, Egypt. (Photo: W.J. Baker)

Above right: ***Medemia argun***, infructescence, Egypt. (Photo: W.J. Baker)

stigmatic remains, perianth whorls persistent; epicarp smooth, shiny, marked with scattered lenticels, mesocarp moderate, apparently ± dry at maturity, with ± radiating short fibres embedded in soft parenchyma, endocarp rather thin, crustaceous. *Seed* basally attached, ± broad, ellipsoidal, endosperm with a central hollow, conspicuously ruminate, the ruminations radial; embryo apical. *Germination* remote-tubular; eophyll entire, lanceolate. *Cytology* not studied.

Distribution and ecology: Egyptian Nubia and northeastern Sudan; two species have been described, *Medemia argun* and *M. abiadensis*, the latter differing in its smaller fruit. Boulos (1968) records variation in fruit size in Egypt and most authors have, since Beccari, accepted one species only, *M. argun*. It occurs in desert oases. Gibbons and Spanner (1996) suggest that the habitat of *M. abiadensis* appears to differ from that of *M. argun*, and that the recognition of a single species may require reassessment.

Anatomy: Leaf (Tomlinson 1961).

Relationships: *Medemia* is resolved as sister to *Hyphaene* with moderate to high support (Bayton 2005, Asmussen *et al.* 2006).

Common names and uses: Medemia or Argun palm. *Medemia* may have had the same range of uses as *Hyphaene* (Johnson 1985); over-exploitation may partly account for its rarity.

Taxonomic accounts: Beccari (1924). See also Gibbons & Spanner (1996).

Fossil record: Subfossil fruits, *Areca passalacquae*, are known from a number of Egyptian tombs (Täckholm & Drar 1950). They are named for the man who first found them (Kunth 1826, Boulos 1968, Newton 2001). However, their true identity was recognised by Unger (1859) who re-named the subfossils as *Hyphaene argun* (= *Medemia argun*).

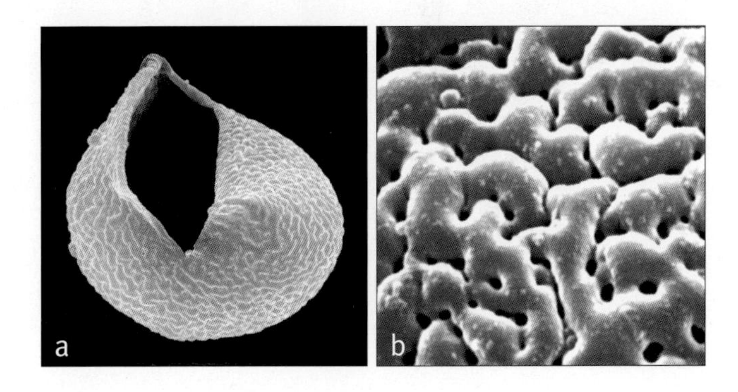

Right: ***Medemia***
a, monosulcate pollen grain, short equatorial face SEM × 1000;
b, close-up, coarsely perforate rugulate pollen surface × 5000.
Medemia argun: **a**, *Talbot* s.n., **b**, *A.F. Brown* (1903).

Notes: A substantial population of *Medemia argun* was discovered in 1995 in northern Sudan (Gibbons & Spanner 1996). The known populations in Egypt are reduced to very few individuals. It would probably thrive as an ornamental in the dry subtropics. The lack of a hastula is particularly noteworthy. This palm was well known to the ancient Egyptians. Kunth named subfossil remains as *Areca passalacquae* (Kunth 1826); although it seems almost certain this name refers to fruits of *M. argun* and is an earlier name than *Hyphaene argun*, it may be disregarded as a *nomen nudum*, notes provided by Kunth being insufficient to describe the palm.

Subtribe Lataniinae Meisner, Plant. vasc. gen. 1: 357 (1842) ('*Latanieae*'). Type: **Latania**.

Staminate and pistillate rachillae dissimilar; pistillate flowers sessile, not borne in well-defined pits; pits in staminate rachillae lacking hairs; fruit unlobed, with 1–3 pyrenes and seeds, stigmatic remains apical.

In this subtribe, *Borassus* is the most widespread genus, occurring from Africa to New Guinea. *Lodoicea* is confined to the Seychelles, *Latania* to the Mascarene Islands and *Borassodendron* to the Malay Peninsula and Borneo. Three of the four genera are remarkable in having very large pistillate flowers with thick, leathery sepals and petals. Indeed, the flowers of *Lodoicea* are the largest flowers in the family. The staminate cincinni of these genera consist of three to about 70 flowers enclosed in deep pits formed by thick leathery modified bracts; flowers are exserted one at a time by elongation of a stalk-like receptacle. The flowers of *Latania* are much smaller and the staminate flowers are borne singly in shallower pits. In the fruits of all four genera, the epicarp and mesocarp are common around all carpels, but a separate hard endocarp surrounds the locule of each carpel forming a variously fibre-covered or sculptured 'pyrene' that encloses the seed.

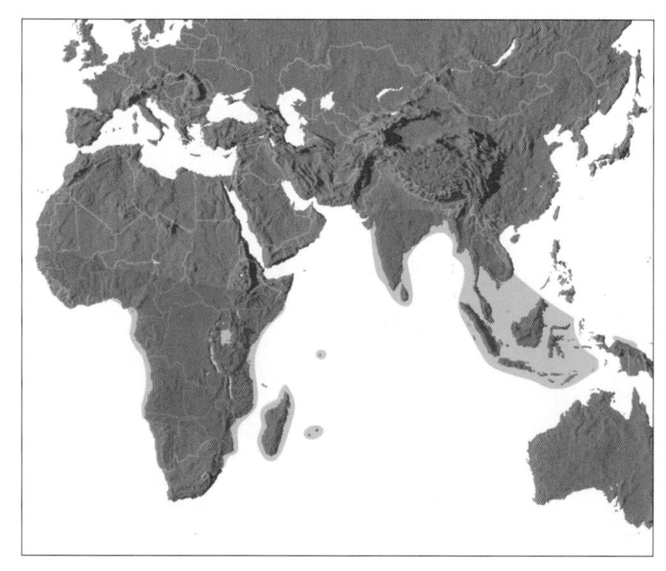

Distribution of subtribe **Lataniinae**

Subtribe Lataniinae is monophyletic and highly supported in the most densely sampled study (Bayton 2005) and moderately supported in less densely sampled studies (Uhl *et al.* 1995, Asmussen *et al.* 2006). The Lataniinae is sister to subtribe Hyphaeninae (Bayton 2005, Asmussen *et al.* 2006).

Key to Genera of Lataniinae

1. Staminate flowers solitary or in cincinni of 2–6 flowers . . 2
1. Staminate flowers numerous, in cincinni of 30–70 flowers 3
2. Bracts of staminate inflorescence imbricate at the margins; pistillode nipple-like; pyrenes densely hairy with thick, internally ridged wall; seed grooved to conform to the ridges. Malay Peninsula, Borneo . . **67. Borassodendron**
2. Bracts of the staminate inflorescence connate at the margins; pistillode columnar, prominent; pyrenes not hairy, but thin, ridged, sculptured, or apically 3-lobed; seed not grooved. Mascarene Islands . . . . . . **65. Latania**
3. Staminate rachillae several, wide; stamens 6; pyrenes not deeply lobed; seeds usually 3, wider than thick, shallowly to deeply bilobed. Africa, Madagascar, Arabia, India to New Guinea and Australia . . . . **68. Borassus**
3. Staminate rachillae only 1 or 2, very wide; stamens 15–18; pyrenes deeply bilobed, very thick; seed usually 1(–3), very large, deeply bilobed. Seychelles Islands . . . **66. Lodoicea**

65. LATANIA

Solitary moderate dioecious tree fan palms of the Mascarene Islands, distinctive in the the sculptured endocarp.

Latania Comm. ex Juss., Gen. pl. 39 (1789). Type: *L. borbonica* Lam. = **L. lontaroides** (Gaertn.) H.E. Moore (*Cleophora lontaroides* Gaertn.).

Cleophora Gaertn., Fruct. sem. pl. 2: 185 (1791). Type: *C. lontaroides* Gaertn.

Latinisation of the common name, latanier, *used in Mauritius.*

Moderate, solitary, mostly unarmed, pleonanthic, dioecious, tree palms. *Stem* erect, rough, marked with spiral, elliptic leaf scars. *Leaves* induplicate, costapalmate, marcescent in young individuals, abscising cleanly in trunked specimens; sheath narrow, inserted at an angle, asymmetrical, angled, with a flange toward the lower side, split horizontally at the base, smooth or densely tomentose; petiole robust, long, adaxially deeply channelled near the base, distally flattened, abaxially rounded, adaxial surface smooth, abaxial surface densely floccose, margin smooth or with a few shallow teeth; adaxial hastula short but conspicuous, triangular or rounded, abaxial hastula absent; blade divided to ca. $^1/_3$–$^1/_2$ its length along adaxial folds into regular, stiff, single-fold segments, these shortly bifid or not, acute to acuminate, abaxial costa and ridges of folds often densely floccose, midribs prominent abaxially, transverse veinlets not evident. *Inflorescences* interfoliar, staminate and pistillate superficially dissimilar. *Staminate inflorescence* with elongate peduncle, elliptic at base in cross-section, adaxially channelled, thin distally; prophyll short, wide, tubular basally, 2-keeled, with a sharp pointed limb about equal in length to the tubular base, abaxially densely floccose; peduncular bracts 1–2–several, loosely sheathing, resembling the prophyll but with a single keel; rachis longer than the peduncle;

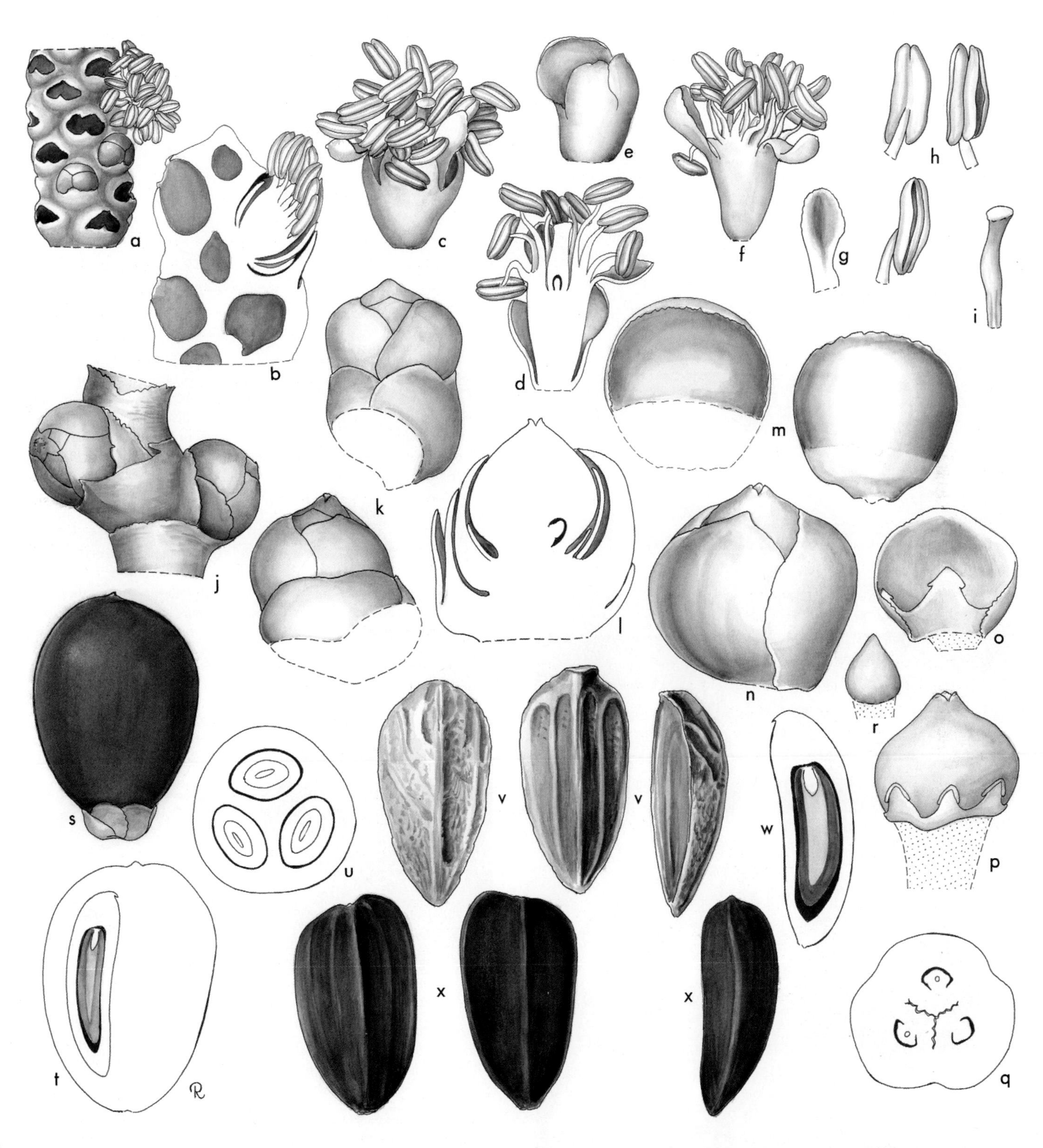

Latania. **a**, portion of staminate rachilla × $1^1/_2$; **b**, portion of staminate rachilla in vertical section × 3; **c**, staminate flower × 3; **d**, staminate flower in vertical section × 3; **e**, staminate calyx × 3; **f**, staminate flower, calyx removed × 3; **g**, staminate petal × 3; **h**, stamen in 3 views × 6; **i**, pistillode × 6; **j**, portion of pistillate rachilla × $1^1/_2$; **k**, pistillate flower with bracts in 2 views × 2; **l**, pistillate flower in vertical section × 3; **m**, pistillate sepal in 2 views × 3; **n**, pistillate flower, sepals removed × 3; **o**, pistillate petal, interior view × 3; **p**, gynoecium and staminodes × 3; **q**, ovary in cross-section × 3; **r**, ovule × 12; **s**, fruit × $^3/_4$; **t**, fruit in vertical section × $^3/_4$; **u**, fruit in cross-section × $^3/_4$; **v**, pyrene (endocarp) in 3 views × 1; **w**, pyrene (endocarp) in vertical section × 1; **x**, seed in 3 views × $1^1/_2$. *Latania verschaffeltii*: **a–i**, *Read* 617; *L. loddigesii*: **j–r**, *Read* 678; **s–x**, *Moore* s.n. (Drawn by Marion Ruff Sheehan)

rachis bracts like the peduncular; first-order branches short, not exceeding subtending bracts, flat, often wide, crescentic in cross-section, branched digitately at the tip to form several (1–14) rachillae; rachillae short or long, spike-like, terete, bearing short, crowded tubular bracts, each bract connate laterally with proximal and distal bracts to form a pit enclosing a single staminate flower. *Staminate flowers* each bearing a stiff cupular bracteole; calyx tubular basally, 3-lobed, irregularly rounded, thicker distally; corolla with a stalk-like base carrying the rest of the flower out of the pit and 3 spathulate lobes; stamens 15–30 or more, filaments short, slightly tapering, anthers basifixed, latrorse; pistillode columnar, ovarian part slightly expanded. *Pollen* ellipsoidal, usually slightly asymmetric; aperture a distal sulcus; ectexine tectate, perforate and micro-channelled, or rugulate, aperture margin similar; infratectum columellate; longest axis 34–50 µm [3/3]. *Pistillate inflorescence* with prophyll, peduncular, and first-order bracts similar to staminate, but first-order branches fewer, each bearing only 1 or 2 rachillae; rachillae wider, longer, sheathed in fewer, larger tubular bracts, lower-most and distal bracts empty, central bracts, each subtending a pistillate flower, bracts tightly surrounding base of flower but not forming pits. *Pistillate flowers* fewer, globose, much larger, widely spaced in a $^2/_5$ phyllotaxy, solitary, each bearing 2 stiff, cupular, imbricate, ± connate bracteoles; sepals 3, stiff, imbricate, rounded; petals like the sepals; staminodes 6–9, connate in a low lobed cupule, vestigial anthers sometimes present; gynoecium globose, trilocular, triovulate, style expanded, stigma undeveloped, locules uniovulate but 2 lateral bodies beside the ovule, ovule orthotropous. *Fruit* usually developing from all 3 carpels, large, oblong or obovoid, stigmatic area apical or subapical, usually 3–(1–2)-seeded, (4 carpels often present); epicarp smooth, mesocarp fleshy, endocarp comprising 3 separate pyrenes, hard, tanniniferous, pyrenes obovoid, variously ridged and sculptured, sculpturing diagnostic for species. *Seed* almond-shaped, smooth, basally attached, endosperm homogeneous; embryo apical. *Germination* remote-tubular; eophyll digitate. *Cytology*: 2n = 28.

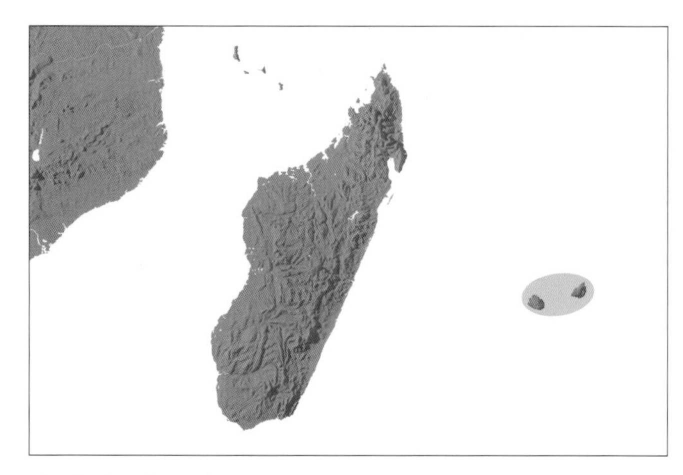

Distribution of ***Latania***

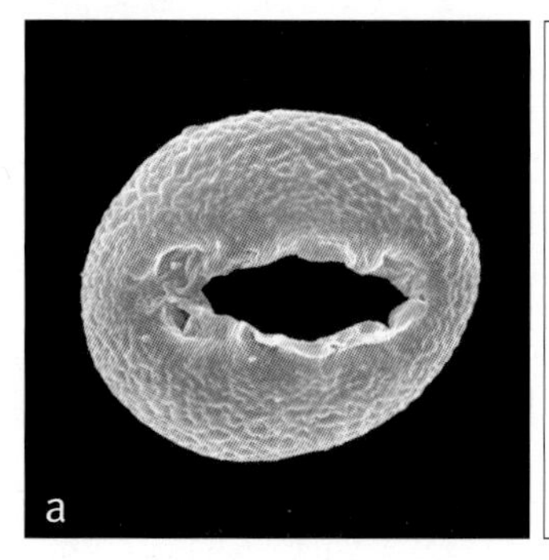

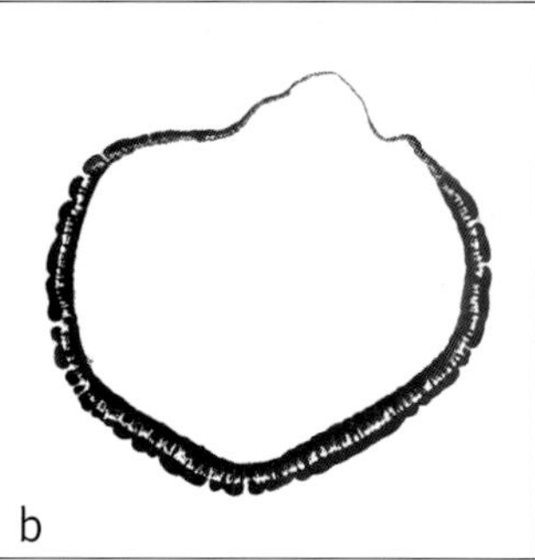

Above: ***Latania***
a, monosulcate pollen grain, distal face SEM × 1250;
b, ultrathin section through whole pollen grain TEM × 1500.
Latania lontaroides: **a** & **b**, *Vaughan* 854.

Top left: ***Latania lontaroides***, crown with staminate inflorescences, cultivated, Fairchild Tropical Botanic Garden, Florida. (Photo: C.E. Lewis)

Bottom left: ***Latania loddigesii***, crown with infructescences, cultivated, Fairchild Tropical Botanic Garden, Florida. (Photo: B. Wood)

Distribution and ecology: Three species in the Mascarene Islands. Once common on coastal cliffs, savannahs, and ravines, the species are now almost extinct in the wild, but are widely cultivated in botanic gardens where they appear to hybridise freely. Some native trees of *Latania loddigesii* are present on Round Island. *Latania lontaroides*, endemic to Reunion, is occasionally left in fields as isolated individuals whereas *L. verschaffeltii*, endemic and widespread on Rodrigues, is reduced to isolated individuals and a small population at Fond la Bonté above Baie aux Huîtres. There may be a few native *L. loddigesii* remaining on Mauritius.

Anatomy: Leaf (Tomlinson 1961), root (Seubert 1997), floral (Uhl & Moore 1971, only gynoecium studied).

Relationships: *Latania* is a strongly supported monophyletic group that is resolved as sister to a clade of *Lodoicea*, *Borassus* and *Borassodendron* with moderate support (Bayton 2005, Asmusen *et al.* 2006). For interspecific relationships, see Bayton (2005).

Common names and uses: Latan palms, *latanier*. Leaves have been used as thatch and the trunk as a source of wood; the young seeds are said to be edible. All species are handsome ornamentals.

Taxonomic account: Moore & Guého (1984).

Fossil record: No generic records found.

Notes: *Latania* is remarkable for the diversity in the sculpturing of the pyrenes; however, the adaptive significance of the pyrene form is not known. Gynoecial structure is similar to that of *Corypha* and *Nannorrhops* but carpel walls are much thicker.

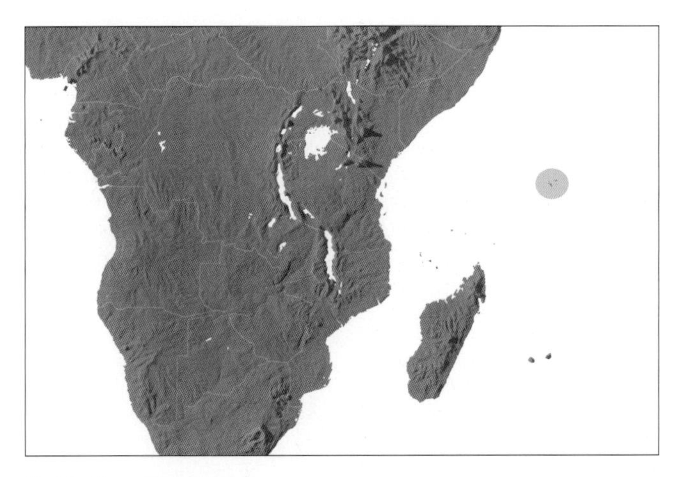

Distribution of ***Lodoicea***

66. LODOICEA

The famous double-coconut of the Seychelles Islands, a dioecious massive tree palm distinctive in the huge leaf lacking a hastula and the enormous bilobed endocarp.

Lodoicea Comm. ex DC., Bull. Sci. Soc. Philom. Paris 2(46): 171 (1800). Lectotype: *L. callypige* Comm. ex J. St.-H. = **L. maldivica** (J.F. Gmel.) Pers. (*Cocos maldivica* J.F. Gmel.).

Honoring Louis XV of France (latinised as Ludovicus).

Robust, often tall, solitary, unarmed, pleonanthic, dioecious, tree palm. *Stem* erect, slightly expanded at the base, inconspicuously ringed with leaf scars. *Leaves* induplicate, costapalmate, marcescent, later abscising under their own weight; sheath soon splitting opposite the petiole and a triangular cleft developing at the petiole base; petiole robust, deeply channelled adaxially, rounded abaxially, abaxial surface with minute black dots, irregularly tomentose, margins fibrous basally, rough to smooth distally; hastulae absent; costa long, tapering, reaching nearly to the end of the blade; blade about as long as the petiole, stiff, basally wedge-shaped, divided ca. $^1/_4$–$^1/_3$ its length into single-fold segments, these shortly bifid, free ends often drooping, adaxial surface shiny, smooth, dull abaxially with thick indumentum along abaxial ridges, midribs prominent abaxially, transverse veinlets long, conspicuous. *Inflorescences* interfoliar, massive, shorter than the leaves, pendulous, staminate and pistillate markedly dissimilar. *Staminate inflorescence* with short, narrow peduncle, often unbranched, ridged, terminating in a single rachilla or 2–3 digitately arranged rachillae; prophyll short, 2-keeled throughout its length, split ventrally but with a long closed triangular tip; peduncular bracts 1 (or more?), obscuring the peduncle, completely tubular, split ventrally above the center leaving a long solid pointed beak; rachillae massive, catkin-like, bearing several large empty, imbricate cup-like bracts basally, above these bearing spirally inserted, very tough leathery bracts, connate laterally and distally to form large pits, each containing a recurved cincinnus of 60–70 staminate flowers. *Staminate flowers* each bearing a fibrous bracteole; sepals 3, connate in an asymmetrical tube, tips distinct, imbricate, irregular; corolla with a long stalk-like base and 3 elongate lobes, unequal in width, not closed laterally around the androecium, tips of lobes thick, rounded, imbricate; stamens 17–22, borne on the surface of an elongate receptacle, filaments short, wide, variously angled, anthers elongate, tips reflexed, latrorse; pistillode columnar, trifid. *Pollen* ellipsoidal and bi-symmetric; aperture a distal sulcus; ectexine tectate, finely perforate-rugulate; aperture margin slightly finer; infratectum columellate; longest axis 61–68 µm [1/1].

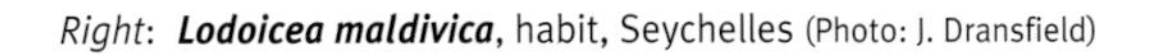

Right: ***Lodoicea maldivica***, habit, Seychelles (Photo: J. Dransfield)

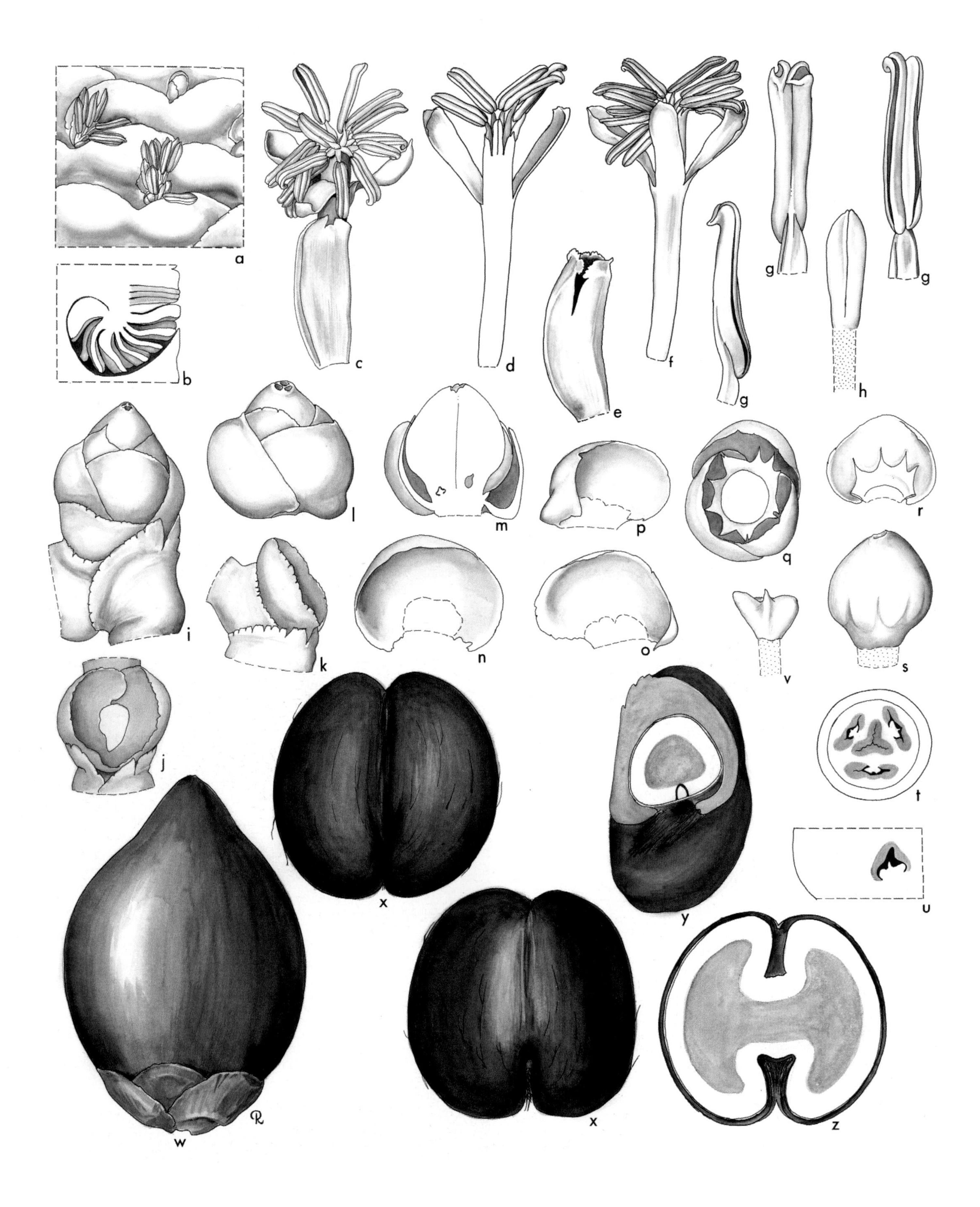
a
b
c
d
e
f
g
g
g
h
i
j
k
l
m
n
o
p
q
r
s
t
u
v
w
x
x
y
z

Pistillate inflorescence unbranched with prophyll and 2(–3 or ?several), tubular peduncular bracts, split ventrally, with long pointed tips like those of the staminate inflorescence; rachilla a direct extension of and about as long as the peduncle, short, wide, zigzag, tapered distally, bearing several empty incompletely sheathing, cupular bracts, subsequent bracts completely sheathing, large, each subtending a pistillate flower. *Pistillate flowers*, the largest flowers in the palms, each sessile, ovoid, bearing at the base 2 lateral, large, cupulate bracteoles; sepals 3, distinct, imbricate, leathery, rounded, thicker basally; petals 3, as sepals; staminodes triangular, low, briefly connate basally with several(–11) pointed tips; gynoecium ovoid, tricarpellate, trilocular with a central trilobed septal nectary, stylar regions wide, triangular, fibrous, stigmas 3, short, becoming reflexed, ovules beaked, apparently orthotropous, laterally winged and with 2 lateral bodies, perhaps vestigial ovules. *Fruit* very large, ovoid and pointed, 1(–3)-seeded; epicarp smooth, mesocarp fibrous, endocarp comprising one to three 2-lobed, thick, hard pyrenes. *Seed*, the largest known, 2-lobed, endosperm thick, relatively hard, hollow, homogeneous; embryo apical in the sinus between the 2 lobes. *Germination* remote-tubular, tube remarkably long, reported to reach ca. 4 m; eophyll shallowly lobed. *Cytology*: 2n = 34.

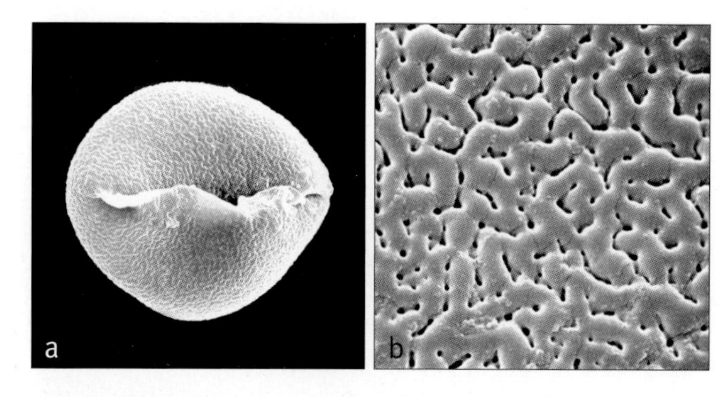

Above: ***Lodoicea***
a, monosulcate pollen grain, distal face SEM × 1000;
b, close up, perforate-rugulate pollen surface SEM × 3000.
Lodoicea maldivica: **a** & **b**, *Bailey* 433.

Below: ***Lodoicea maldivica***, crown with fruit, cultivated, Singapore Botanic Gardens, Singapore. (Photo: W.J. Baker)

Distribution and ecology: One species in the Seychelles Islands. *Lodoicea* today is restricted to hill slopes and valleys of Praslin and Curieuse but formerly may have occurred on adjacent islets. It does not reach coastal plains or main ridges. Edwards *et al.* (2002) suggest that the huge seed, the largest in the plant kingdom, may be an adaptation to establishment in shade or to sibling rivalry on an isolated island. Both hypotheses assume that *Lodoicea* evolved from a *Borassus*-like palm adapted to a drier and more savannah-like climate than that experienced by *Lodoicea* today.

Anatomy: Leaf (Tomlinson 1961), root (Seubert 1997), seed (Werker 1997).

Relationships: *Lodoicea* is resolved as sister to a clade of *Borassodendron* and *Borassus* with moderate support (Bayton 2005, Asmussen *et al.* 2006).

Common names and uses: Coco-de-Mer or double coconut. Leaves are used locally as thatch and plaiting, wood as palisades and water-troughs, seeds for dishes and vegetable ivory, and down from young leaves for stuffing pillows. The sale of nuts to tourists is an important source of revenue.

Taxonomic account: Bailey (1942).

Fossil record: No generic records found.

Notes: Recognisable by the huge, more or less diamond-shaped leaf in juvenile stages, by the lack of hastulae on both surfaces, and when present, by the huge and bizarre inflorescences, flowers, fruits and seeds. Important historically because of the many legends involving the huge seeds, and as an oddity.

Opposite: ***Lodoicea.*** **a**, portion of staminate axis at anthesis × 1 1/2; **b**, portion of staminate axis in vertical section to show arrangement of flowers and bracts × 1 1/2; **c**, staminate flower × 3; **d**, staminate flower in vertical section × 3; **e**, staminate calyx × 3; **f**, staminate flower, calyx removed × 3; **g**, stamen in 3 views × 6; **h**, pistillode × 6; **i**, portion of pistillate axis with flower × 1/2; **j**, **k**, bracts subtending pistillate flower (removed) in 2 views × 3/8; **l**, pistillate flower × 1/2; **m**, pistillate flower in vertical section × 1/2; **n**, **o**, **p**, pistillate sepals, outer, middle and inner, interior views × 1/2; **q**, pistillate petals and staminodial ring from above, pistil removed × 1/2; **r**, pistillate petal and staminodial ring portion, interior view × 1/2; **s**, gynoecium × 1/2; **t**, ovary in cross-section × 1; **u**, portion of ovary with locule and ovule in vertical section × 3; **v**, ovule × 6; **w**, fruit × 1/6; **x**, seed in two views × 1/6; **y**, seed in vertical section through groove × 1/6; seed in vertical section × 1/6. *Lodoicea maldivica*: **a–h**, *Read* 1488; **i–v**, *Moore* s.n. (Drawn by Marion Ruff Sheehan)

67. BORASSODENDRON

Large dioecious tree fan palms of tropical rain forest in South Thailand and Peninsular Malaysia and Borneo, distinctive in the leaves with razor-sharp petiole margins and large fruit with the endocarp with internal flanges that penetrate the homogeneous endosperm.

Borassodendron Becc., Webbia 4: 359 (1914). Type: **B. machadonis** (Ridl.) Becc. (*Borassus machadonis* Ridl.).

Combines the palm generic name Borassus *with* dendron – *tree*.

Robust, solitary, unarmed, pleonanthic, dioecious, tree palms. *Stem* erect, ringed with close leaf scars. *Leaves* induplicate, very briefly costapalmate, marcescent in immature individuals, neatly abscising under their own weight in mature trunked individuals; sheath becoming fibrous marginally, narrow, glabrous but tomentose abaxially along the margins, with a conspicuous triangular cleft below the petiole; petiole robust, covered with caducous indumentum, adaxially deeply channelled, abaxially rounded, the margins smooth, very hard and very sharp; adaxial hastula well developed, abaxial hastula absent; blade divided nearly to the insertion into compound segments, these further divided rather irregularly to $^1/_4$–$^2/_3$ the radius into single-fold segments, in turn shallowly divided along the abaxial folds, interfold filaments persisting, the segment surfaces similar in colour or with white indumentum beneath, blade with bands of brown caducous scales, longitudinal veins and transverse veinlets prominent. *Inflorescences* interfoliar, strongly dimorphic, ± pendulous. *Staminate inflorescence* branched to 2 orders, with a short to long peduncle; prophyll large, 2-keeled, tubular near the base, inflated distally, splitting along much of its length, densely covered in caducous indumentum, tending to disintegrate at the tip and margins into fibres; peduncular bracts 1–few, like the prophyll but with a single keel; rachis shorter or longer than the peduncle; rachis bracts like the peduncular; first-order branches distant or crowded, bare, semicircular in cross-section, with sharp edges, short or long, bearing 1–5 robust catkin-like rachillae crowded at the tip, each subtended by a small triangular bract; rachillae close or diverging, bearing a tight spiral of large scaly or hairy, imbricate bracts, connate to each other laterally and adnate to the axis to form pits, each pit containing a cincinnus of 2–6 flowers, floral bracteoles spathulate. *Staminate flowers* exserted one at a time from the pit; calyx membranous, tubular, tipped with 3 short triangular lobes, most of the calyx remaining included within the pit; corolla with a long stalk-like base carrying the rest of the flower out of the pit, petals 3, ± imbricate, elongate; stamens 6–15 with very short filaments and elongate latrorse anthers; pistillode minute or absent. *Pollen* brevi-ellipsoidal to oblate-spheroidal, bi-symmetric; aperture a distal brevi-sulcus or a single large pore; ectexine tectate, coarsely foveolate or reticulate, aperture margin psilate; infratectum columellate; longest axis 60–85 µm [2/2]. *Pistillate inflorescence* unbranched or with up to 4 branches; prophyll, peduncular bracts and rachis bracts as in the staminate inflorescence, inflorescence axis, where unbranched, terminating in a rachilla, where branched, having rachillae borne singly at the tip of bare, flattened first-order branches as in the staminate inflorescence; rachillae short to long, bearing a spiral of large hairy, frequently notched, scaly, imbricate bracts, connate to each other and adnate to the axis to form shallow pits (*Borassodendron machadonis*) or almost free but closely overlapping (*B. borneense*), each subtending a solitary pistillate flower (or abortive flower), the rachilla tip frequently bearing sterile bracts, sometimes with an apparently terminal flower, the bract margins frequently erose. *Pistillate flowers* sessile, ± superficial or partially sunken in pits, each surrounded by 2 large ovate, irregularly margined bracteoles; sepals 3, distinct, ovate, imbricate, the margins ± notched; petals 3, distinct, ovate, similar to the sepals; staminodal ring short, with 6–9 teeth bearing minute empty anthers; gynoecium globose, trilocular, triovulate, tipped with 3 fleshy approximate stigmas, ovule form unknown. *Fruit* large (1–2)–3-seeded, stigmatic remains apical; epicarp smooth, mesocarp fibrous, the interfibre parenchyma becoming sweet, fragrant, and fleshy at maturity, endocarp comprising 3 separate pyrenes with thick stony walls, walls with 8–12 shallow internal, longitudinal ridges penetrating the seed. *Seed* grooved longitudinally by the pyrene ridges, endosperm homogeneous, with a slight central hollow; embryo apical. *Germination* remote-tubular; eophyll palmate with ca. 5 segments. *Cytology* not studied.

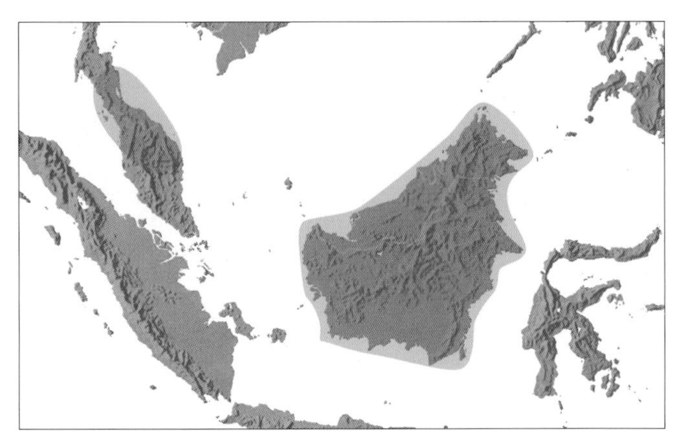

Distribution of ***Borassodendron***

Left: ***Borassodendron borneense***, habit, Brunei Darussalam (Photo: J. Dransfield)

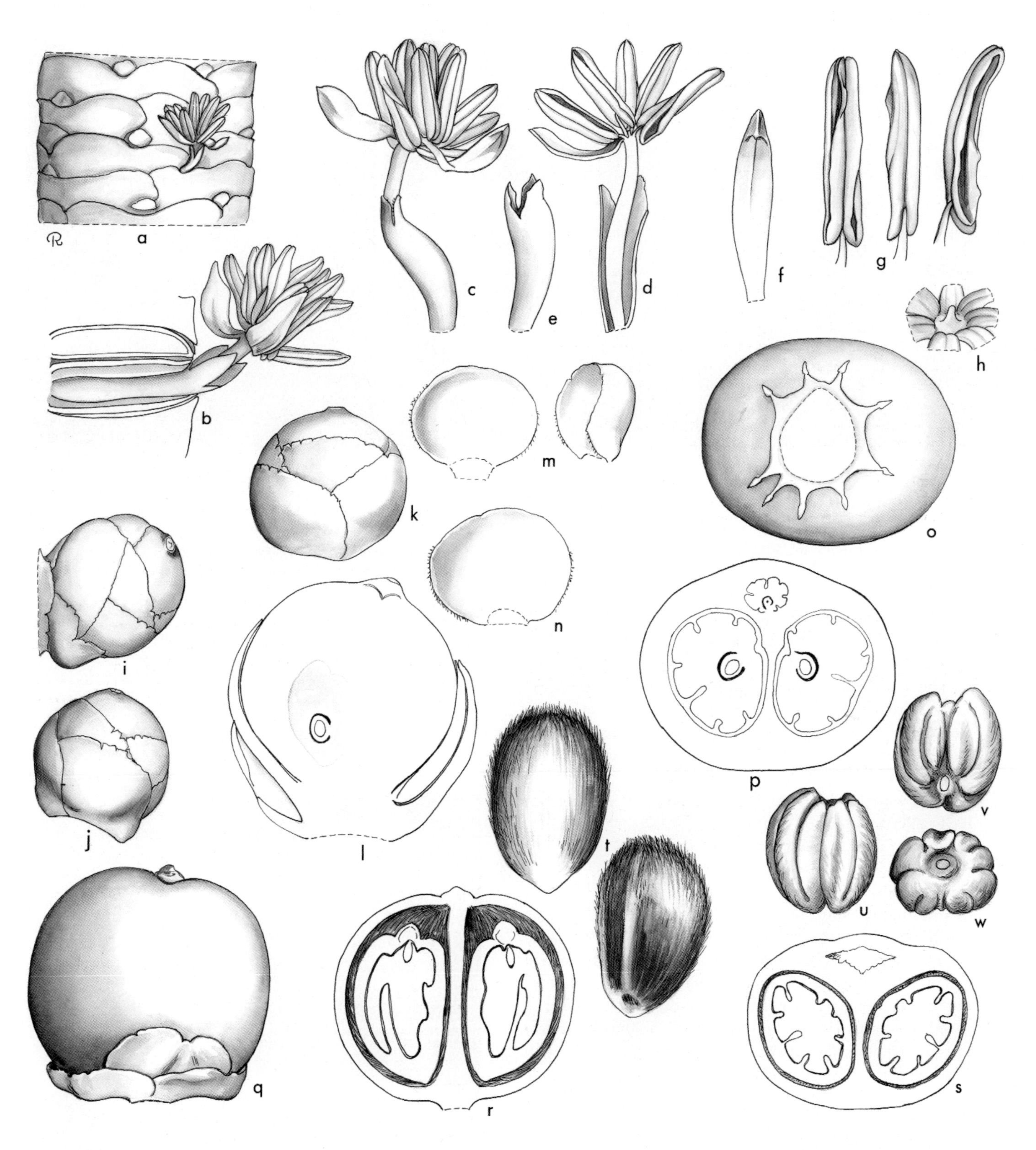

Borassodendron. **a**, portion of staminate rachilla with exserted staminate flower × 1$^1/_2$; **b**, portion of staminate rachilla in vertical section with flower × 3; **c**, staminate flower × 3; **d**, staminate flower in vertical section × 3; **e**, staminate calyx × 3; **f**, staminate petal, interior view × 3; **g**, stamen in 3 views × 6; **h**, pistillode × 12; **i**, pistillate flower and bracteoles slightly post-anthesis × 3; **j**, **k**, pistillate flower in 2 views from opposite sides × 3; **l**, pistillate flower in vertical section × 6; **m**, pistillate sepals in 2 views × 3; **n**, pistillate petal, interior view × 3; **o**, gynoecium viewed from below to show staminodes × 6; **p**, very young fruit in cross-section × 6; **q**, fruit × $^1/_2$; **r**, fruit in vertical section × $^1/_2$; **s**, fruit in cross-section × $^1/_2$; **t**, fibrous pyrene (endocarp) in 2 views × $^1/_2$; **u**, **v**, **w**, seed in 3 views × $^1/_2$. *Borassodendron machadonis:* **a–p**, *Moore* 9040; *B. borneense*: **q–w** *Moore & Meijer* 9188. (Drawn by Marion Ruff Sheehan)

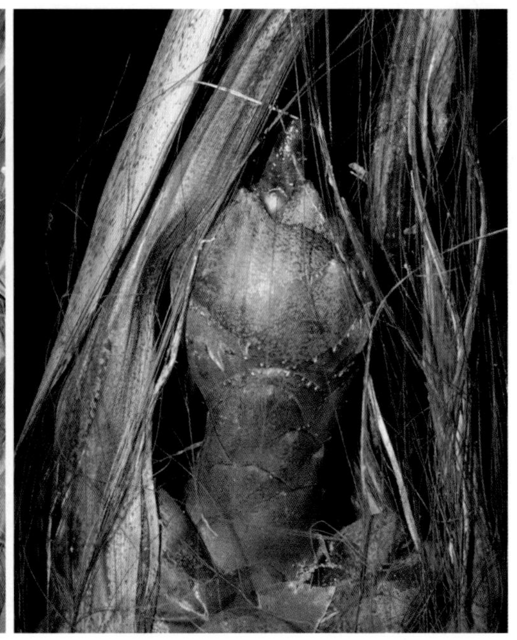

Distribution and ecology: Two species, *Borassodendron machadonis* in southern Thailand and northern Peninsular Malaysia, *B. borneense* in Borneo. *Borassodendron machadonis* is a rare palm, sometimes found in areas of deep soil on limestone hills, sometimes on ridges in hill Dipterocarp forest. In Borneo, *B. borneense* can be locally abundant on hills in the lowlands but is absent from wide areas of apparently suitable forest. The young leaves of *B. borneense* are eaten by orangutan, which can cause considerable damage; the same animals may be responsible for dispersal.

Anatomy: Leaf (Tomlinson 1961), root (Seubert 1997).

Relationships: *Borassodendron* is strongly supported as a monophyletic group (Bayton 2005). The genus is sister to *Borassus* with moderate to high support (Uhl *et al.* 1995, Bayton 2005, Asmussen *et al.* 2006).

Common names and uses: *Bindang*. The 'cabbage' of *Borassodendron borneense* is edible and is sometimes sold in Bornean village markets.

Taxonomic account: Dransfield (1972a).

Fossil record: *Jacobipollenites* Ramanujam (1966) is described from the Miocene of India (Madras). According to the revised description (Ramanujam *et al.* 1998) based on additional material, the longest axis is 40–100 μm, the pore is circular to slightly elongate (10–25 μm) and it has a 'rugged' tectate margo. The authors make convincing comparison with *Borassodendron*; the only alternative in the palms would be *Ammandra* with closely similar pollen. *Borassodendron machadonis* pollen has been reported from the Pliocene–early Quaternary of the Mahakam Delta, Kalimantan by Caratini and Tissot (1985). Maloney (2000) reviews this record and other records from the SE Asian Holocene.

Notes: Easily recognised by the large palmate but deeply split leaf blade, smooth sharp petiole, and well-developed adaxial hastula.

Above left: ***Borassodendron machadonis***, staminate inflorescence, cultivated, Fairchild Tropical Botanic Garden, Florida. (Photo: C.E. Lewis)

Above middle: ***Borassodendron machadonis***, pistillate inflorescence, cultivated, Fairchild Tropical Botanic Garden, Florida. (Photo: J. Horn)

Above right: ***Borassodendron borneense***, pistillate flower, Sarawak. (Photo: J. Dransfield)

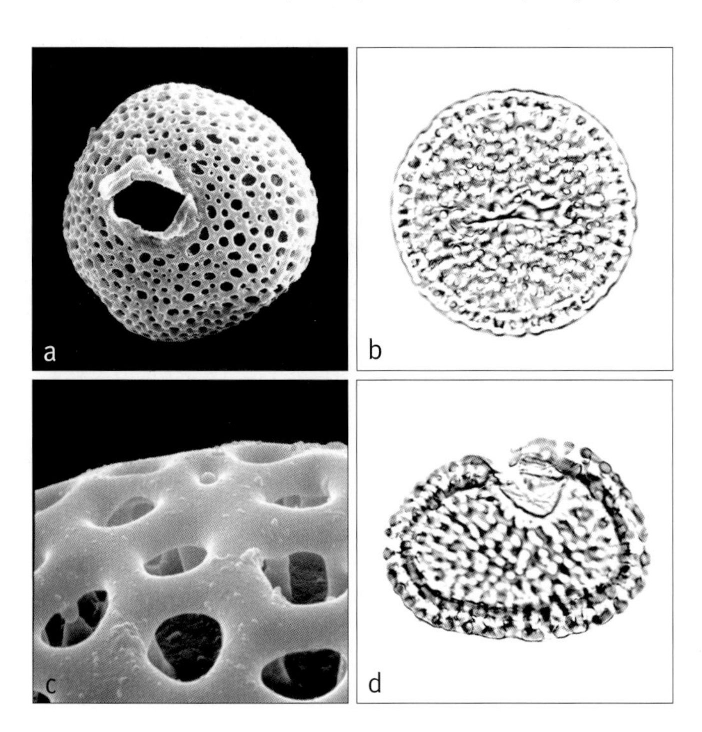

Below: ***Borassodendron***
a, monoporate/brevisulcate pollen grain, distal face SEM × 600;
b, monoporate/brevisulcate pollen grain, high-mid-focus LM × 600;
c, close-up, coarsely reticulate pollen surface SEM × 5000;
d, brevisulcate pollen grain, equatorial face, mid-focus LM × 800.
Borassodendron machadonis: **a**, *Nur* s.n.; *B. borneense*: **c** & **d**, *Dransfield* JD800.

68. BORASSUS

Large dioecious tree fan palms of Africa, Madagascar, the Indian Subcontinent and Southeast Asia, and the drier parts of Malesia; distinctive in the leaves with spiny petiole margins, the spines often very irregularly shaped, and large fruit with the endocarp usually lacking internal flanges that penetrate the homogeneous endosperm.

Borassus L., Sp. pl. 1187 (1753). Type: **B. flabellifer** L.
Lontarus Adans., Fam. pl. 2: 25, 572 (1763) (superfluous name).

Said to be derived from borassos*, an immature inflorescence of the date palm, but why Linnaeus should have used this name is not clear.*

Tall, robust, solitary, armed, pleonanthic, dioecious, tree palms. *Stem* massive, covered in a lattice of leaf bases abscising cleanly in older specimens, then rough, ringed with wide leaf scars. *Leaves* induplicate, strongly costapalmate; sheath open early in development, later with a wide triangular cleft at the base of the petiole; petiole deeply channelled adaxially, rounded abaxially, surfaces smooth to minutely rough, margins of sheath and petiole armed with coarse irregular teeth; adaxial hastula conspicuous, triangular or scalloped, abaxial hastula a low ridge (?always); blade suborbicular to flabellate, divided along adaxial folds to ca. $^{1}/_{2}$ its length into regular, stiff single-fold segments, these shortly bifid, interfold filaments present or absent, surfaces smooth, ramenta or tomentum along abaxial costa and ridges of folds, midribs prominent abaxially, transverse veinlets conspicuous, short, numerous. *Inflorescences* interfoliar, shorter than the leaves, the staminate and pistillate dissimilar. *Staminate inflorescence* branched to 2 orders; peduncle very short; prophyll 2-keeled, with long tubular base, limb short, pointed, variously split apically; (?)peduncular bracts lacking; rachis also short, rachis bracts similar to the prophyll; first-order branches long, flattened, each bearing a prophyll and branched digitately into several (1–3) rachillae; rachillae large, catkin-like, elongate, bearing spirally arranged, imbricate bracts, connate laterally and distally to form large pits, each containing a reflexed cincinnus of ca. 30 staminate flowers, exserted singly in succession from the pit mouth. *Staminate flowers* each subtended by a long membranous bracteole; sepals 3, asymmetrical, connate only basally or to $^{2}/_{3}$ their length, distinct lobes keeled, elongate, membranous, stiff; corolla with a long stalk-like base and 3 short, rounded lobes, ridged adaxially; stamens 6, filaments short, triangular, anthers medifixed, elongate, latrorse; pistillode small, conical. *Pollen* ellipsoidal, bi-symmetric; aperture a distal sulcus; ectexine tectate, reticulate or finely perforate-rugulate, rarely foveolate-reticulate, with psilate supratectal gemmae, aperture margin similar but often without supratectal gemmae; infratectum columellate; longest axis 42–85 µm [2/6]. *Pistillate inflorescence* unbranched or with a single first-order branch; peduncle short; prophyll tubular, pointed, 2-keeled, split ventrally about $^{1}/_{2}$ its length; peduncular bracts few (2 or more), if present as long as or longer than the peduncle; rachilla massive, bearing large cupular bracts, the first few empty, the subsequent each subtending a single pistillate flower, several empty bracts above the flowers. *Pistillate flowers* large, each bearing 2 lateral cup-like, rounded, leathery, bracteoles; sepals 3 distinct, imbricate, thick, rounded; petals 3, similar to sepals; staminodes triangular, connate basally in a low cupule, sterile anthers present or not; gynoecium rounded, tricarpellate, with a central, basal septal nectary, stylar region hemispherical, stigma a low knob, carpels each with a basal, orthotropous ovule, and 2 lateral bodies, perhaps vestigial ovules. *Fruit* large, rounded, sometimes wider than long, bearing 1–3 seeds, stigmatic remains apical, perianth enlarged, persistent; epicarp smooth, mesocarp thick, fibrous, often fragrant, endocarp comprising 3 hard bony pyrenes. *Seed* shallowly to deeply bilobed, pointed, basally attached, endosperm homogeneous with a central hollow; embryo apical. *Germination* remote-tubular; eophyll undivided, elliptical. *Cytology*: 2n = 36.

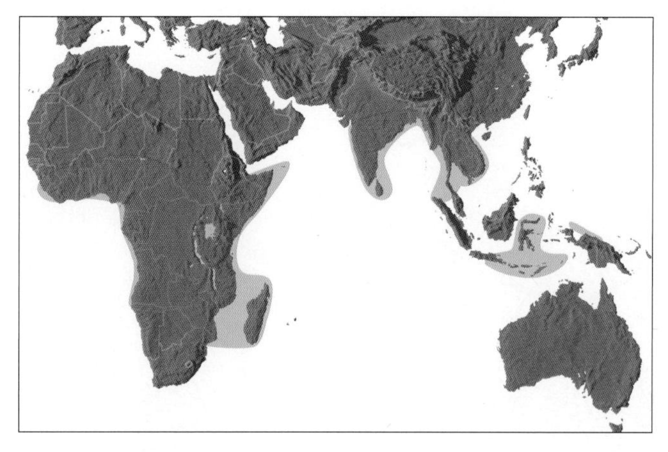

Distribution of ***Borassus***

Distribution and ecology: Six species have been recognised. They occur in Africa, Madagascar, north-eastern Arabia, through India and Southeast Asia to New Guinea and Australia. *Borassus* is one of the most widespread palm genera. *Borassus flabellifer* can occur in some mountain districts of India at elevations of 500–800 m, and is also found on banks of rivers. It is most abundant, however, on low sandy plains near sea level where exposed to sun and winds. In Africa, *B. aethiopum* occurs in open secondary forest and savannah.
Anatomy: Leaf (Tomlinson 1961), root (Seubert 1997).

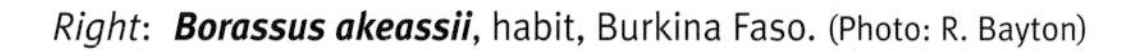

Right: ***Borassus akeassii***, habit, Burkina Faso. (Photo: R. Bayton)

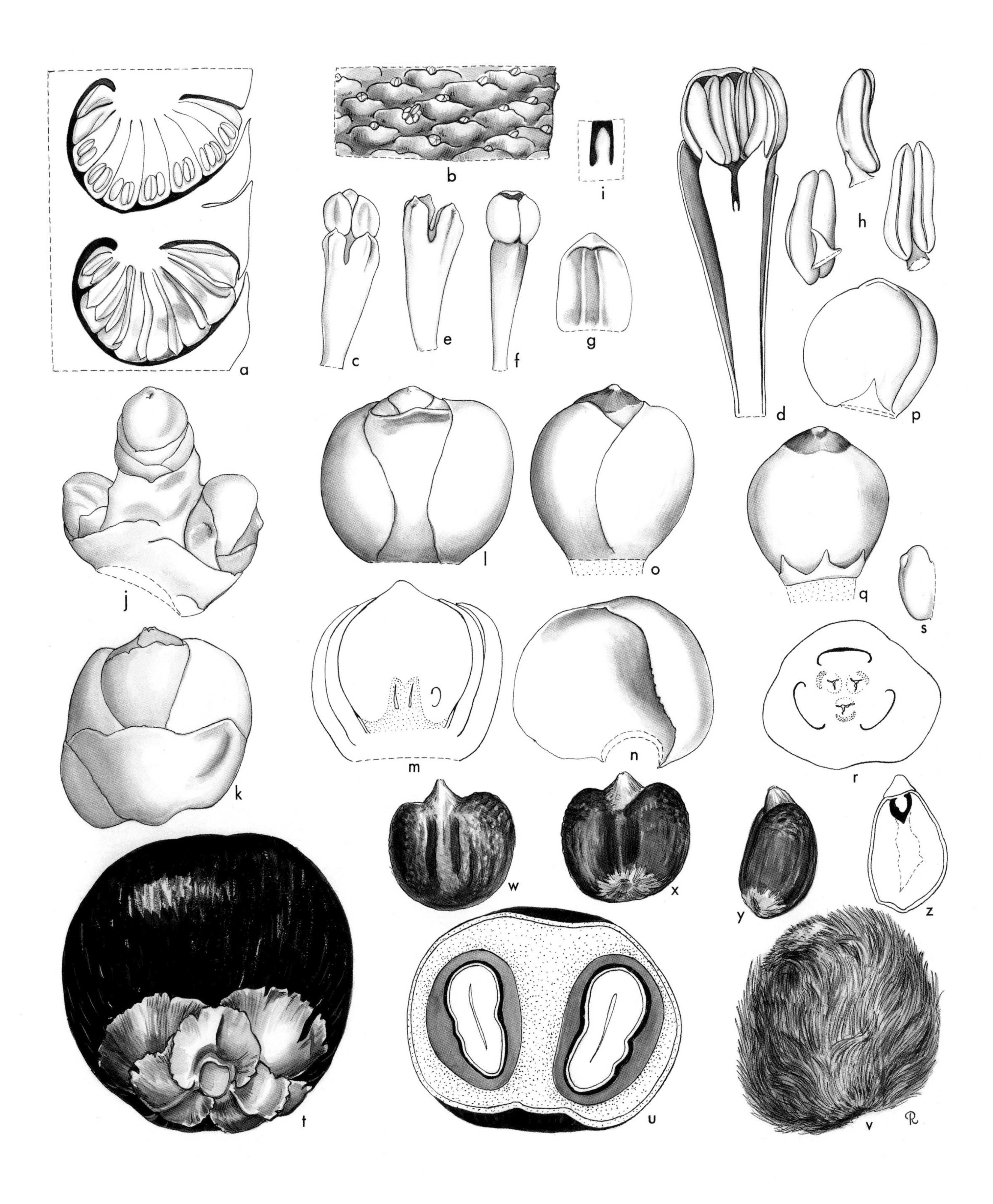
a
b
i
c
e
f
g
h
d
p
j
l
o
q
s
k
m
n
r
w
x
y
z
t
u
v

Top: ***Borassus aethiopum***, detail of staminate rachilla with several flowers at anthesis emerging from pits, cultivated, Fairchild Tropical Botanic Garden, Florida. (Photo: W.J. Baker)

Bottom left: ***Borassus aethiopum***, vertical section of staminate rachilla showing flower buds in pits, cultivated, Fairchild Tropical Botanic Garden, Florida. (Photo: W.J. Baker)

Bottom right: ***Borassus aethiopum***, pistillate inflorescence, Burkina Faso. (Photo: R. Bayton)

Relationships: *Borassus* is a strongly supported monophyletic group (Bayton 2005) that is resolved as sister to *Borassodendron* with moderate to high support (Uhl *et al*. 1995, Bayton 2005, Asmussen *et al*. 2006). For interspecific relationships, see Bayton (2005).

Common names and uses: Toddy or wine palm, *lontar*, palmyra, or *siwalan* (*Borassus flabellifer*). *Borassus flabellifer* is one of the most intensively used palms. Leaves have been used for writing; wood is valuable for building; inflorescences are tapped and the syrup, sugar, or alcohol may be a staple (Fox 1977).

Taxonomic accounts: Beccari (1924) and Bayton (2007).

Fossil record: A well-preserved leaf impression, *Amesoneuron borassoides*, from the Indian Deccan Intertrappean of Madhya Pradesh (although the age span of these volcanic deposits is controversial, see Chapter 5) is described by Bonde (1986a) and compared, particularly, with *Borassus*. An Upper Cretaceous (Senonian) seed was compared with *Borassus* (Monteillet & Lappartient 1981), but this comparison has proved to be incorrect (Uhl & Dransfield 1987; Bayton 2005). Fossil stem, *Palmoxylon aschersonii*, from Paleogene and Neogene of Algeria and Lower Miocene of Libya is compared with *Borassus aethiopum* (Louvet & Magnier 1971; Boureau 1947; Boureau *et al*. 1983), but this is considered a doubtful comparison (Bayton 2005). From the Indian Miocene, petrified stems, *Palmoxylon coronatum*, are also compared with *Borassus* (Mahabalé 1959, Sahni 1964, Roy & Ghosh 1980). Fossil palm roots from the Deccan Intertrappean of Nahwargaon, Maharashtra (Ambwani 1981) is compared with *B. flabellifer*. (Comparisons of palm stem wood or root to generic level should always be viewed with caution). A seed, Upper Senonian (Monteillet & Lapartient 1981), seems very questionable.

Notes: Can be recognised by the large stiff costapalmate leaves with both adaxial and abaxial hastulae, and by the large irregular teeth on the petiole.

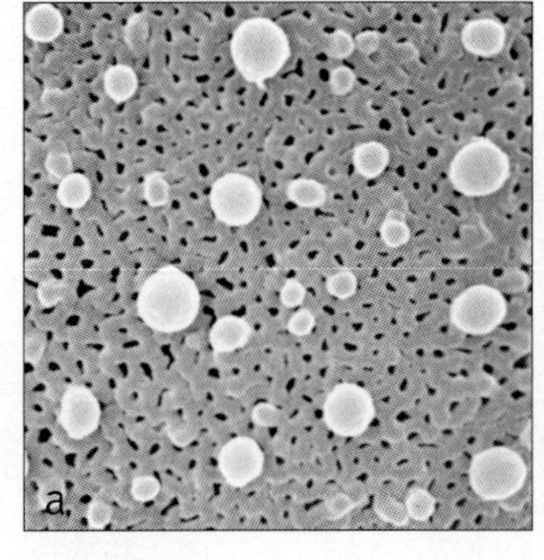

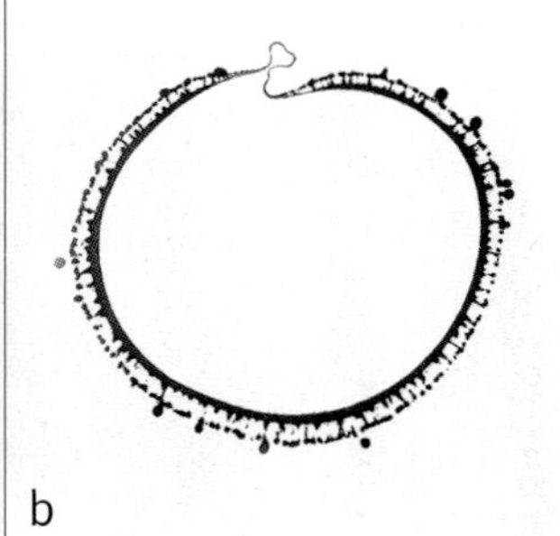

Right: ***Borassus***
a, close-up, tectate perforate pollen surface with widely spaced gemmae SEM × 3000;
b, ultrathin section through whole pollen grain TEM × 800.
Borassus aethiopum: **a** & **b**, Herb. R. Christaensen.

Opposite: ***Borassus***. **a**, portion of staminate axis in vertical section showing cincinni × 6; **b**, portion of staminate axis showing buds among overlapping bracts × 1 1/2; **c**, staminate flower × 6; **d**, staminate flower in vertical section × 12; **e**, staminate calyx × 6; **f**, staminate receptacle and corolla × 6; **g**, staminate petal, interior view × 12; **h**, stamens in 3 views × 12; **i**, pistillode × 36; **j**, portion of pistillate axis × 1/2; **k**, pistillate flower and bracts × 1; **l**, pistillate flower × 1; **m**, pistillate flower in vertical section × 1; **n**, pistillate sepal × 1; **o**, pistillate flower, sepals removed × 1; **p**, pistillate petal and staminode, interior view × 1; **q**, gynoecium and staminodes × 1; **r**, ovary in cross-section × 1 1/2; **s**, ovule × 3; **t**, fruit with perianth × 1/2; **u**, 2-seeded fruit in cross-section × 1/2; **v**, pyrene (endocarp) × 1/2; **w, x, y**, seed in 3 views × 1/2; **z**, seed in vertical section × 1/2. *Borassus aethiopum*: **a–i**, *Tomlinson* s.n.; *B. flabellifer*: **j–z**, *Read* 819. (Drawn by Marion Ruff Sheehan)

IV. CEROXYLOIDEAE

CEROXYLOIDEAE Drude, Bot. Zeitung 35: 632 (1877) ('*Ceroxylinae*'). Type: **Ceroxylon.**
Phytelephoideae Drude in Engl. & Prantl, Nat. Pflanzenfam. II, 3: 28 (1887).

Tall, moderate or acaulescent, solitary or very rarely clustered, unarmed palms; hermaphroditic or dioecious; leaves pinnate with reduplicate leaflets; crownshaft present or absent; inflorescence with a prophyll and 1–several peduncular bracts, rachis bracts much reduced; flowers sessile or on short stalks, solitary, or if clustered, then flower group monopodial, not sympodial, never in pits, when unisexual only slightly to very strongly dimorphic; gynoecium syncarpous with 3 or more locules; ovules 3 or more; endosperm homogeneous.

The subfamily includes three tribes, Cyclospatheae, Ceroxyleae and Phytelepheae. The last was previously included in its own subfamily, Phytelephantoideae. There is remarkable diversity in both vegetative and reproductive structures in the subfamily.

Subfamily Ceroxyloideae is resolved as monophyletic in numerous analyses (Asmussen & Chase 2001, Asmussen *et al.* 2006, Savolainen *et al.* 2006, Trénel *et al.* 2007, Baker *et al.* in review, in prep.), with high support in some cases. However, some studies resolve subfamily Ceroxyloideae as polyphyletic due to various alternative resolutions of *Pseudophoenix* (Uhl *et al.* 1995, Lewis & Doyle 2001, Hahn 2002b). The Ceroxyloideae is sister to subfamily Arecoideae (Asmussen & Chase 2001, Asmussen *et al.* 2006, Savolainen *et al.* 2006, Baker *et al.* in review, in prep.). This position is highly supported in the analysis of Baker *et al.* (in review) and moderately supported elsewhere (Asmussen & Chase 2001, Asmussen *et al.* 2006).

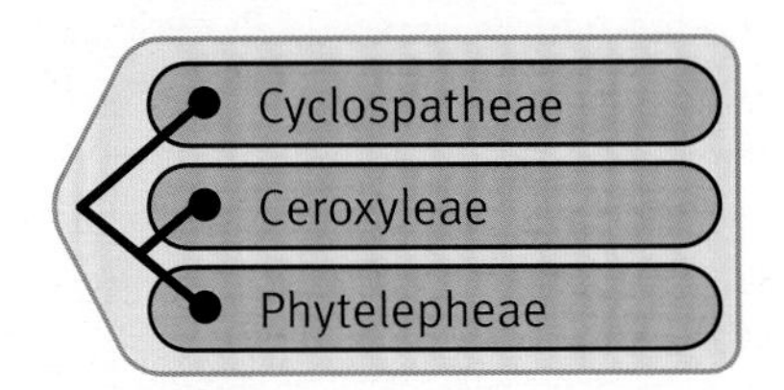

Summary tree showing relationships among tribes of the Ceroxyloideae that have been resolved with moderate to high support in multiple phylogenetic analyses. For further explanation, see discussions under tribe treatments.

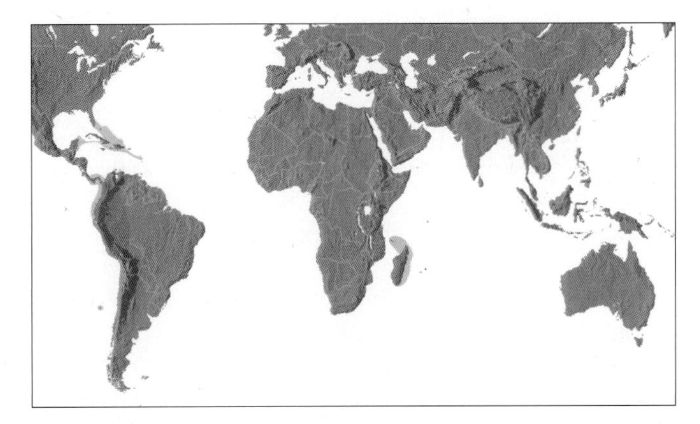

Distribution of subfamily **Ceroxyloideae**

Key to Tribes of Ceroxyloideae

1. Hermaphroditic palms; leaf sheaths remaining tubular, not fibrous, forming a crownshaft; flowers bisexual near the base of the rachillae, staminate distally; filaments apically embedded in anther connectives . . . . . . **Cyclospatheae**
1. Dioecious palms; leaf sheaths splitting, the margins usually fibrous, not forming a crownshaft. Flowers always unisexual; filaments not apically embedded in anther connectives . . 2
2. Flowers scarcely dimorphic, the staminate and pistillate superficially similar; stamens not exceeding 20; gynoecium trilocular, triovulate; fruit with smooth epicarp . . . . **Ceroxyleae**
2. Flowers strongly dimorphic; staminate flower with minute perianth lobes and usually with a very large number of stamens; pistillate flower with numerous large perianth segments; gynoecium 4–10 carpellate, 4–10 ovulate; fruit with corky-warted pericarp . . . . **Phytelepheae**

Tribe Cyclospatheae O.F. Cook, Mem. Torrey Bot. Club 12: 24 (1902). Type: *Cyclospathe* = **Pseudophoenix.**

Stems moderate, erect; leaf sheaths tubular, forming a short crownshaft; flowers hermaphroditic, spirally arranged, pseudopedicellate, borne in the axils of small bracts; filaments apically embedded in connectives.

The tribe contains the single genus *Pseudophoenix*, which is confined to the northern Caribbean islands and coastal areas of mainland North America.

Tribe Cyclospatheae is highly supported as monophyletic (Asmussen *et al.* 2006, Trénel *et al.* 2007). It resolves on a long branch as sister to a clade of tribes Ceroxyleae and Phytelepheae with moderate or high support (Asmussen *et al.* 2006, Trénel *et al.* 2007, Baker *et al.* in review, in prep.).

Opposite: ***Ravenea xerophila***, habit, Madagascar. (Photo: H. Beentje)

69. PSEUDOPHOENIX

Relatively short stemmed pinnate-leaved Caribbean tree palms, often with bottle-like trunks; crownshaft present; flowers are hermaphroditic or occasionally unisexual towards the tips of the rachillae and the ovary with three locules and three ovules. The fruit is lobed when more than one seed develops.

Pseudophoenix H. Wendl. ex Sarg., Bot. Gaz. 11: 314 (1886). Type: **P. sargentii** H. Wendl. ex Sarg

Chamaephoenix H. Wendl. ex Curtiss, Florida Farmer Fruit Grower 1(8): 57 (1887). Type: *C. sargentii* H. Wendl. ex Curtiss (illegitimate name) (= *Pseudophoenix sargentii* H. Wendl. ex Sarg.).

Sargentia H. Wendl. & Drude ex Salomon, Palmen. 160 (1887) (rejected name). Type: *S. aricocca* H. Wendl. & Drude ex Salomon (illegitimate name) (= *Pseudophoenix sargentii* H. Wendl. ex Sarg.).

Cyclospathe O.F. Cook in Northrop, Mem. Torrey Bot. Club 12: 25 (1902). Type: *C. northropii* O.F. Cook (= *Pseudophoenix sargentii* H. Wendl. ex Sarg.).

pseudo — *false*, phoenix — *the date palm, though why Wendland chose this name is not clear.*

Moderate, solitary, pleonanthic, polygamous or hermaphroditic palms. *Stem* erect, often swollen, prominently ringed with rather wide leaf scars, smooth to finely striate, grey or green, waxy. *Leaves* few, mostly ca. 10, reduplicately pinnate, deciduous by a basal abscision zone; sheath forming a short, somewhat swollen crownshaft, splitting distally opposite the petiole, waxy; petiole channelled adaxially, rounded abaxially; rachis flat to angled adaxially, rounded abaxially, ± glabrous; leaflets numerous, irregularly arranged, grouped and fanned within the groups, stiff, acute, single-fold, waxy on both surfaces, midribs evident, other veins small, transverse veinlets not evident. *Inflorescences* interfoliar, pendulous or arched, branched to the fifth order; peduncle elongate, dorsiventrally flattened; prophyll tubular, 2-keeled, flattened, leathery, persistent, opening at the apex; peduncular bracts 2, the first similar to the prophyll, the second usually collar-like; rachis longer than or about as long as the peduncle; rachillae stiffly spreading or pendulous, each subtended by a small open bract. *Flowers* borne singly, spirally arranged, each subtended by an acuminate bract, hermaphoditic proximally but the distal few staminate with a much reduced pistil, base of the flower extended in a pseudopedicel formed by fusion and elongation of the receptacle and the base of the calyx; calyx 3-lobed with rounded, apiculate tips; petals 3, valvate, thick, much longer than the calyx, basally connate in a very short tube; stamens 6, filaments thin, dilated and briefly connate in a ring basally, the apex lying in a groove in the abaxial surface of the anther to about the midpoint of the connective, then bent sharply inward, anthers large, elongate, ± pointed apically and basally, dorsifixed, latrorse; gynoecium conical, trilocular, tri-ovulate, with 3 glands at the base opposite the petals, stigmas sessile, very short, becoming recurved after fertilisation, ovules campylotropous, inserted on the adaxial side of the locule. *Pollen* ellipsoidal, usually more or less bi-symmetric; aperture a distal sulcus; ectexine tectate, foveolate or reticulate, aperture margin and proximal face similar, but lumina smaller; infratectum columellate; longest axis ranging from 43–55 µm [3/4]. *Fruit* 1–3-seeded, waxy red, globose or 2–3-lobed, stigmatic remains near the base or in a central depression in 3-seeded fruits; epicarp smooth, mesocarp fleshy, with raphides, lacking fibres, endocarp hard, brown, smooth. *Seed* not adherent to endocarp at maturity, hilum basal, raphe branches ascending and spreading in shallow grooves, endosperm homogeneous; embryo subbasal. *Germination* remote-tubular; eophyll narrow lanceolate. *Cytology*: 2n = 34.

Distribution of ***Pseudophoenix***

Distribution and ecology: Four species from Florida, the Bahama Islands, Cuba, Hispaniola, and Dominica, to Mexico and Belize. *Pseudophoenix* occurs on well-drained sand or porous limestone near the coast or inland on dry hills. The seeds are long-lived for palms, germinating after as much as two years in storage. Fruits become buoyant when dry and in *P. sargentii* may be adapted for dispersal by sea (Read 1968).

Anatomy: Leaf and floral (Read 1968), and root (Seubert 1996b). The vascular system of the carpel, which consists of three major bundles with ventral bundles in 'lateral' positions and an ovular supply associated with ventral bundles only, resembles that of ceroxyloid and chamaedoreoid palms. The ground tissue of the gynoecium, which lacks tannins but has abundant raphides, is like that of *Chamaedorea* (Read 1968, Uhl & Moore 1971).

Left: ***Pseudophoenix vinifera***, habit, Dominican Republic. (Photo: J. Dransfield)

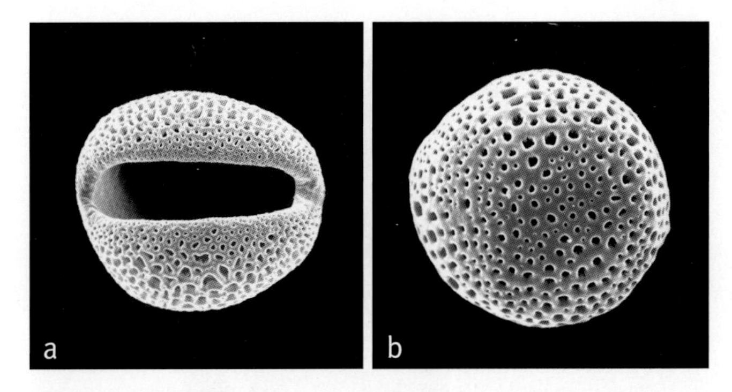

Above: ***Pseudophoenix***
a, monosulcate pollen grain, distal face SEM × 1000;
b, monosulcate pollen grain, proximal face SEM × 1000.
Pseudophoenix sargentii: **a**, *Ekman* 10802; *P. vinifera*: **b**, *Ekman* H5496.

Left: ***Pseudophoenix sargentii***, crown, cultivated, Fairchild Tropical Botanic Garden, Florida. (Photo: C.E. Lewis)

Pseudophoenix. **a**, portion of rachilla with staminate flowers × 3; **b**, staminate flower × 6; **c**, staminate calyx × 6; **d**, staminate petal, interior view × 6; **e**, androecium × 6; **f**, stamen in 2 views × 6; **g**, pistillode × 6; **h**, hermaphroditic flower × 6; **i**, hermaphroditic flower in vertical section × 3; **j**, gynoecium, exterior view and in vertical section × 6; **k**, ovary in cross-section × 6; **l**, fruit with 2 seeds × $1^1/_2$; **m**, base of fruit and stigmatic remains × 3; **n**, endocarp × $1^1/_2$; **o**, seed × 3; **p**, seed in vertical section × 3. *Pseudophoenix sargentii*: all from *Read* 759. (Drawn by Marion Ruff Sheehan)

Above left: ***Pseudophoenix vinifera***, crown with inflorescence, cultivated, Montgomery Botanical Center, Florida. (Photo: J. Dransfield)

Above right: ***Pseudophoenix vinifera***, flower, cultivated, Montgomery Botanical Center, Florida. (Photo: W.J. Baker)

Relationships: *Pseudophoenix* is monophyletic with high support and on a long branch (Asmussen *et al.* 2006, Trénel *et al.* 2007). For relationships, see tribe Cyclospatheae and for interspecies relationships, see Trénel *et al.* (2007).
Common names and uses: Cherry palm, buccaneer palm. Leaves may be used for thatch, and the fruit of some species for animal feed. In the past, juice from the trunk of *Pseudophoenix vinifera* and *P. ekmanii* was used in making a fermented drink. All species are striking ornamentals.
Taxonomic accounts: Quero (1981), Read (1968, 1969) and Zona (2002a).
Fossil record: Reticulate trichotomosulcate pollen from the Lower Eocene of Saudi Arabia is compared erroneously with *Pseudophoenix* pollen (Srivastava & Binda 1991). Reticulate monosulcate palm-like pollen, *Liliacidites tritus* Frederiksen, from the Upper Eocene (Upper Jacksonian) of Texas, is also suggested to have an affinity with *Pseudophoenix* (Frederiksen 1980, 1981). Although geographically less persuasive, this fossil is probably more like the reticulate pollen of *Ceroxylon*.
Notes: *Pseudophoenix* stands apart as the only genus of the Ceroxyloideae having hermaphroditic flowers and a well-developed crownshaft. It has unique anatomical features of the leaf and attachment of the anther, as well as an unusually elongate base or pseudopedicel on the flower. The spirally arranged single flowers on a highly branched interfoliar inflorescence, certain anatomical features, and the presence of just one single phloem strand in the central vascular bundles of the petiole ally *Pseudophoenix* with the Ceroxyleae.
Additional figures: Fig. 1.7, Glossary figs 3, 5, 14, 22.

Tribe Ceroxyleae Satake, Hikobia 3: 125 (1962). Type: Ceroxylon.

Stems often very tall, very rarely slender; dioecious; sheaths not forming crownshafts; flowers scarcely dimorphic, usually pedicellate, open from early in development, solitary, spirally or subdistichously arranged in the axils of small bracts.

The four genera of this tribe are morphologically very similar. The inflorescences have several large peduncular bracts that are frequently densely tomentose, and flowers are always borne singly along the rachillae. The flowers show very little dimorphism. The distribution of the tribe though the Southern Hemisphere is quite remarkable, with *Ravenea* in Madagascar, *Oraniopsis* in Queensland (Australia), *Juania* in Juan Fernandez and *Ceroxylon* in the Andes of South America.

Tribe Ceroxyleae is monophyletic (e.g., Uhl *et al.* 1995) and highly supported (Asmussen *et al.* 2006, Trénel *et al.* 2007, Baker *et al.* in review, in prep.). It is resolved as sister to tribe Phytelepheae with moderate to high support (Hahn 2002b, Asmussen *et al.* 2006, Trénel *et al.* 2007, Baker *et al.* in review, in prep.).

Key to Genera of Ceroxyleae

1. Stigmatic remains basal in fruit 2
1. Stigmatic remains lateral to subapical in fruit 3
2. Petals basally united; stamens 6–15 or more. Andes of South America . . . . 70. **Ceroxylon**
2. Petals free, valvate; stamens 6. Queensland, Australia . . . . 72. **Oraniopsis**
3. Pistillate flower with staminodes bearing rudimentary anthers; staminate inflorescence often multiple; prophyll incomplete. Madagascar . . . . 73. **Ravenea**
3. Pistillate flower with staminodes lacking rudimentary anthers; staminate inflorescence solitary; prophyll complete. Juan Fernandez Island . . . . 71. **Juania**

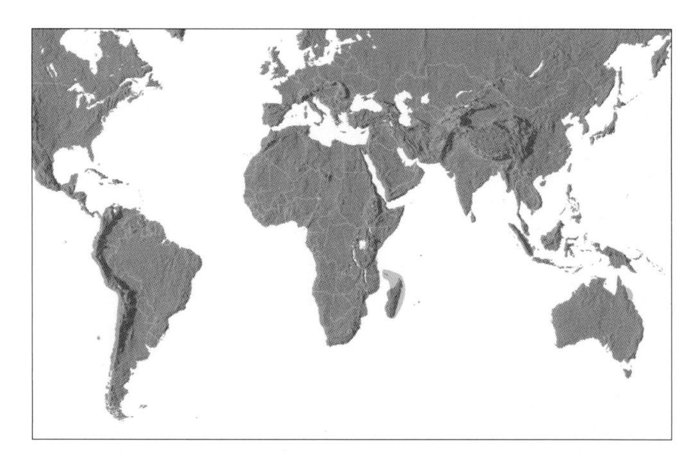

Distribution of tribe **Ceroxyleae**

70. CEROXYLON

The Andean wax palms are dioecious pinnate-leaved palms of the Andes; trunks often immensely tall, with thick wax; prophyll incomplete, petals united basally; stamens 6–15; stigmatic remains subbasal.

Ceroxylon Bonpl. ex DC., Bull. Sci. Soc. Philom. Paris 3: 239 (1804). Type: **C. alpinum** Bonpl. ex DC.

Klopstockia H. Karst., Linnaea 28: 251 (1856). Lectotype: *K. cerifera* H. Karst. (see H.E. Moore 1963c) (= *Ceroxylon ceriferum* [H. Karst.] Pittier).

Beethovenia Engel, Linnaea 33: 677 (1865). Type: *B. cerifera* Engel (= *Ceroxylon ceriferum* [H. Karst.] Pittier).

keros — *wax,* xylon — *wood, referring to the thick white wax on the trunks.*

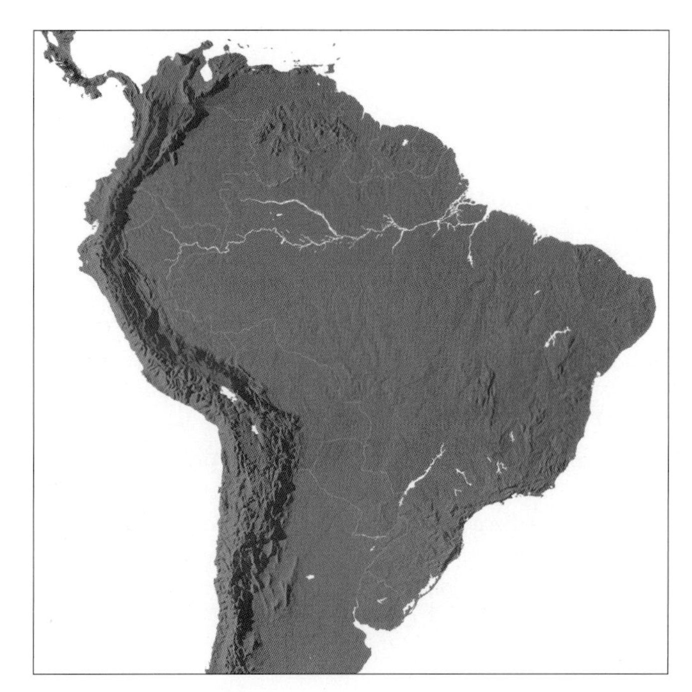

Distribution of ***Ceroxylon***

Tall to very tall, solitary, unarmed, pleonanthic, dioecious palms. *Stems* smooth, usually waxy, with prominent leaf scars. *Leaves* moderate to large, reduplicately pinnate, neatly abscising; sheath splitting opposite the petiole at maturity, usually not forming a crownshaft, leathery; petiole channeled adaxially, rounded abaxially; rachis adaxially flat to angled, glabrous, abaxially rounded, silvery-grey tomentose; leaflets acute, single-fold, evenly spaced or clustered, usually glossy adaxially, often waxy or tomentose-scaly abaxially, midrib conspicuous, larger adaxially, no other veins evident. *Inflorescences* interfoliar, solitary in the leaf axil, branched to 3–4 orders; peduncle elongate; prophyll tubular, 2-keeled, flattened, open apically, incompletely encircling the peduncle abaxially; peduncular bracts several (ca. 5–7), inserted near the base of the peduncle, the lower ones open apically, the upper 3–4 terete, beaked, completely enclosing the inflorescence in bud, splitting abaxially at anthesis, the uppermost sometimes reduced and inserted higher than the others, prophyll and peduncular bracts with indumentum or scales; rachis bracts small, open; rachillae usually flexuous or zigzag, often short, the pistillate usually shorter than the staminate, glabrous or with indumentum, bearing small, pointed bracts subtending the flowers. *Flowers* ebracteolate, open from early in development, borne singly along the rachillae, pedicellate. *Staminate flowers* with 3 sepals connate in a low, acutely or acuminately lobed cupule; petals 3, fleshy, acute or acuminate, briefly connate basally with each other and with the bases of antesepalous stamen filaments, separate above at anthesis; stamens 6–15(–17), filaments awl-shaped, not inflexed at the apex in bud, anthers basifixed, bifid basally, bifid to acute or pointed apically; pistillode minute, conic, usually minutely trifid. *Pollen* ellipsoidal or oblate triangular, asymmetric; aperture a distal sulcus

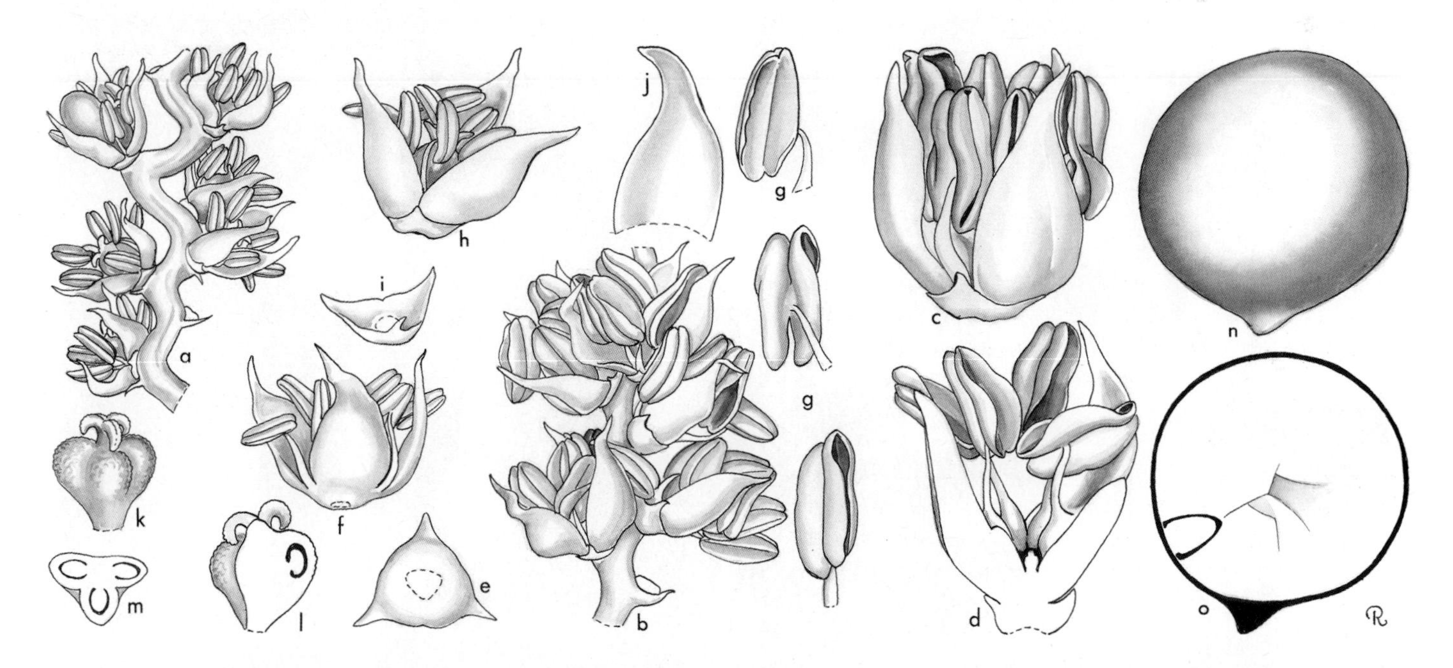

Ceroxylon. **a**, portion of pistillate rachilla × 3; **b**, portion of staminate rachilla × 3; **c**, staminate flower × 12; **d**, staminate flower in vertical section × 12; **e**, staminate calyx × 6; **f**, staminate flower, calyx removed × 6; **g**, stamen in 3 views × 6; **h**, pistillate flower × 6; **i**, pistillate calyx × 6; **j**, pistillate petal × 6; **k**, gynoecium × 6; **l**, ovary in vertical section × 6; **m**, ovary in cross-section × 6; **n**, seed × 3; **o**, seed in vertical section × 3. *Ceroxylon* sp.: all from *Cuatrecasas* 19322. (Drawn by Marion Ruff Sheehan)

or trichotomosulcus; ectexine tectate or semi-tectate, finely rugulate-reticulate, coarsely reticulate or gemmate-reticulate (muri comprise rows of gemmae), aperture margin similar or slightly finer; infratectum columellate; longest axis ranging from 32–46 µm [5/11]. *Pistillate flowers* similar to the staminate but staminodes usually smaller with halberd-shaped or sagittate abortive anthers; gynoecium ovoid, trilocular, triovulate, but 2 ovules usually aborting, stigmas 3, recurved at anthesis, ovules pendulous, hemianatropous. *Fruit* red, orange-red, or orange to purplish-black at maturity, globose, normally 1-seeded, stigmatic remains lateral near the base; epicarp smooth or minutely roughened, mesocarp fleshy with few fibres, endocarp thin, not adherent to seed. *Seed* globose, hilum basal, round, raphe branches obscure, ascending from the hilum, endosperm homogenous; embryo lateral near the base. *Germination* adjacent-ligular; eophyll elliptic or narrowly lanceolate. *Cytology*: 2n = 36.

Distribution and ecology: Eleven species occurring at high elevations in the Andes from Venezuela through Colombia, Ecuador, and Peru to Bolivia. Species of *Ceroxylon* are some of the tallest palms. They occur in premontane to low and high montane forest often among clouds for most of the time. Today trees are frequently left standing in fields where the forest has been cleared. All species are endangered (Moore 1977). *Ceroxylon parvifrons* occurs at elevations in excess of 3500 m above sea level, the palm species with the highest elevational occurrence (Borchsenius & Skov 1997).

Anatomy: Leaf (Tomlinson 1961, Roth 1990), root (Seubert 1996b), floral (Uhl 1969b) and stamen development (Uhl & Moore 1980).

Relationships: *Ceroxylon* is strongly supported as monophyletic (Trénel *et al.* 2007) and resolved as sister to *Juania* with high support (Uhl *et al.* 1995, Asmussen *et al.* 2006, Trénel *et al.* 2007, Baker *et al.* in review). For interspecies relationships, see Trénel *et al.* (2007).

Common names and uses: Andean wax palms. Stems provide wax for candles and matches; fruit are used for cattle food. Over-exploitation of young fronds for Christian religious ceremonies has seriously endangered some species. Several species have become prized but slow-growing ornamentals.

Left: ***Ceroxylon quindiuense,*** habit, Colombia. (Photo: R. Bernal)
Right: ***Ceroxylon vogelianum,*** habit, Ecuador. (Photo: F. Borchsenius)

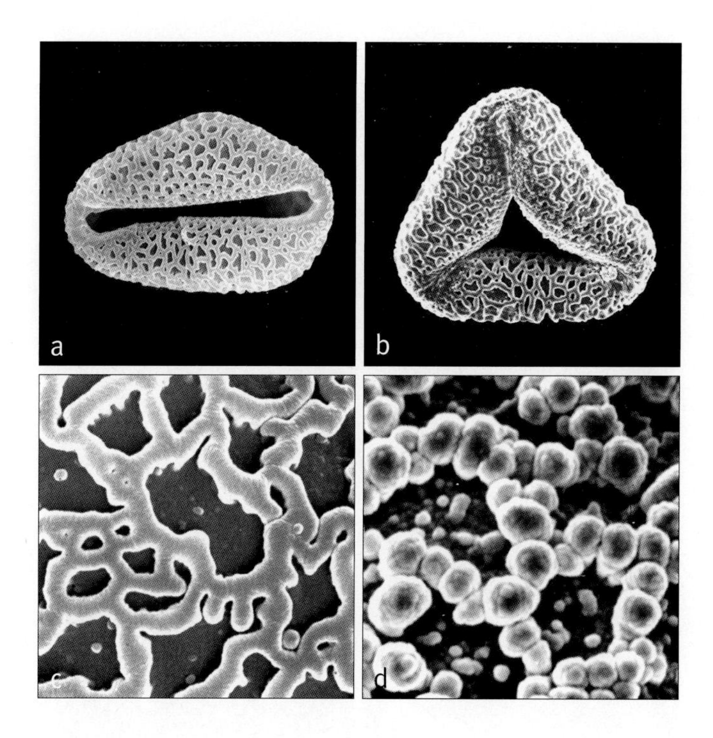

Above: ***Ceroxylon***
a, monosulcate pollen grain, distal face SEM × 1500;
b, trichotomosulcate pollen grain, distal face SEM × 1500;
c, close-up, coarsely reticulate pollen surface SEM × 6000;
d, close-up, incomplete reticulate surface, muri formed of closely linked gemmae SEM × 8000.
Ceroxylon quindiuense: **a** & **c**, *André* 2426; *C. alpinum*: **b**, *Balslev* 62512; *C.* aff. *vogelianum*: **d**, *Balslev* 62085.

Taxonomic accounts: Burret (1929), Moore & Anderson (1976), Galeano-Garces & Bernal-Gonzalez (1982) and Galeano (1995).

Fossil record: Ramanujam (1987) compares a collection of *Paravuripollis* from the Lower to Middle Miocene of Kerala with *Ceroxylon* (or *Oncosperma*); however, the fossils appear to be more or less zonasulcate and clavate, and closer to the pollen of a number of species of *Korthalsia* rather than to the monosulcate, reticulate pollen of *Ceroxylon*. See also entries for *Korthalsia* and *Pseudophoenix*.

Notes: The wax palms are a spectacular feature of Andean montane forest. Identification of species, particularly in the herbarium, is often difficult because there is rather little variation in vegetative and reproductive characters. The morphology and anatomy (Uhl 1969b) of the flowers of *Juania*, *Ravenea*, and *Ceroxylon* are very similar. The development of polyandry in *Ceroxylon* is different from that in other subfamilies (Uhl & Moore 1980).

Additional figures: Glossary fig. 20.

71. JUANIA

Dioecious pinnate-leaved palm of Juan Fernandez Island off the coast of Chile. Similar to *Ceroxylon* and *Oraniopsis* but distinct in the combination of complete prophyll, distinct petals, consistently 6 stamens and the eccentrically subapical stigmatic remains.

Juania Drude, Nachr. Königl. Ges. Wiss. Georg-Augusts-Univ. 1878 (1): 40 (1878). Type: **J. australis** (Mart.) Drude ex Hook.f. (*Ceroxylon australe* Mart.).

Named for the island of Juan Fernandez.

Solitary, moderate, unarmed, pleonanthic, dioecious palm. *Stem* stout, leaf scars oblique, internodes shorter above middle and very short at the crown, green, very smooth, with slight bloom. *Leaves* reduplicately pinnate, erect at first, then spreading; sheath fibrous, splitting opposite the petiole, not forming a crownshaft, covered with scaly tomentum when young; petiole much shorter than the rachis, channelled adaxially, rounded abaxially, with small brown scales; rachis triangular in cross-section, ridged adaxially, slightly rounded abaxially; leaflets narrow, single-fold, relatively short, bifid at tips, stiff, smooth, midribs more prominent adaxially, transverse veinlets not evident. *Inflorescences* interfoliar, solitary, usually 2 developed each year, the remainder aborting, branched to 2 orders at least proximally; peduncle elongate; prophyll short, tubular, laterally keeled, flat, open apically; peduncular bracts 3, similar to the prophyll, the second the largest and enclosing the third in bud, both the second and third larger than and inserted at some distance above the prophyll and first bract, all tubular and ± dorsiventrally

Right: ***Juania australis***, habit, Juan Fernandez. (Photo: H.E. Moore)

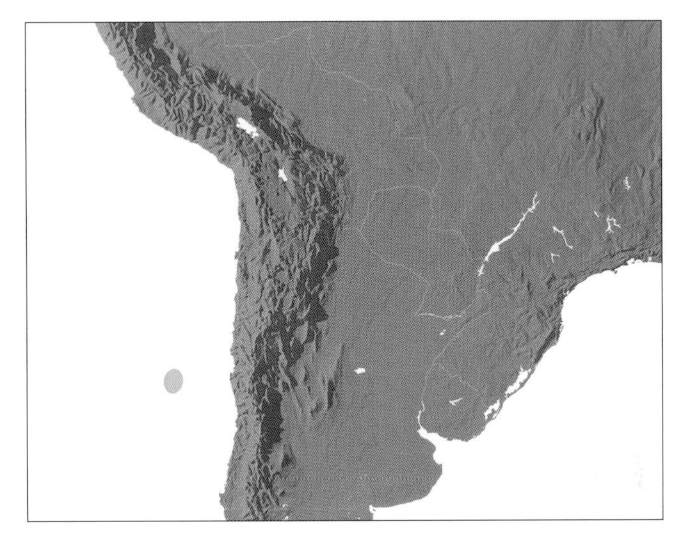

Distribution of ***Juania***

compressed in bud, splitting abaxially at anthesis, becoming pendulous, persistent in fruit and at length marcescent; rachis elongate, longer than the peduncle, bearing numerous, spirally arranged branches, those at the base once-branched into short stiff rachillae, distal branches less divided or undivided, each branch and rachilla subtended by a small, sometimes adnate bract, bracts often lacking distally. *Flowers* white, solitary, open from early in development, briefly pedicellate, the pedicel subtended by an acute bract, the individual flowers usually with a bracteole on the pedicel. *Staminate flowers* with 3 sepals, united in a 3-lobed cupule, the lobes acute, about as long as to longer than the tube; petals 3, distinct, ± asymmetrical, ovate-acute, imbricate basally, separated above and shorter than the stamens in bud and at anthesis; stamens 6, the filaments distinct, anthers basifixed, versatile, emarginate to bifid apically, sagittate but the locules not divergent basally, latrorse; pistillode minute, ovoid, with trifid apex. *Pollen* ellipsoidal, asymmetric; aperture a distal sulcus; ectexine tectate, perforate-rugulate, aperture margin broad scabrate and finely perforate; infratectum columellate; longest axis ranging from 33–40 µm [1/1]. *Pistillate flowers* with 3 sepals, united in a 3-lobed cupule, the lobes, or some of them, as long as the tube; petals 3, distinct, imbricate basally and separated above in bud and in flower; staminodes 6, awl-shaped, distinct, lacking abortive anthers; gynoecium ovoid-attenuate, trilocular with 3 ovules, only 1 normally maturing, stigmas 3, short, recurved, ovules pendulous, hemianatropous. *Fruit* globose or nearly so, orange-red at maturity with eccentrically subapical stigmatic remains; epicarp smooth, mesocarp succulent and ± orange, with a few, flat, longitudinal, unbranched, whitish fibres adjacent to the very thin, cartilaginous endocarp, this adherent to the seed. *Seed* globose with mostly simple or forked vascular strands ascending from the base, endosperm homogeneous; embryo lateral in lower $^{1}/_{3}$ or near the base. *Germination* adjacent-ligular; eophyll lanceolate, entire. *Cytology* not studied.

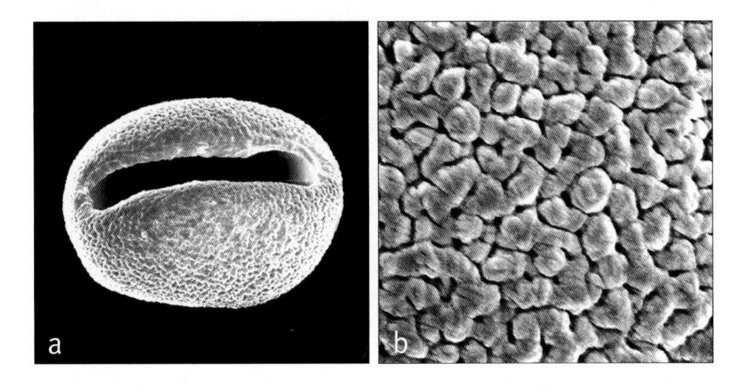

Above: ***Juania***
a, monosulcate pollen grain, distal face SEM × 1500;
b, close-up, perforate-rugulate pollen surface SEM × 8000.
Juania australis: **a** & **b**, *Skottsberg* 310.

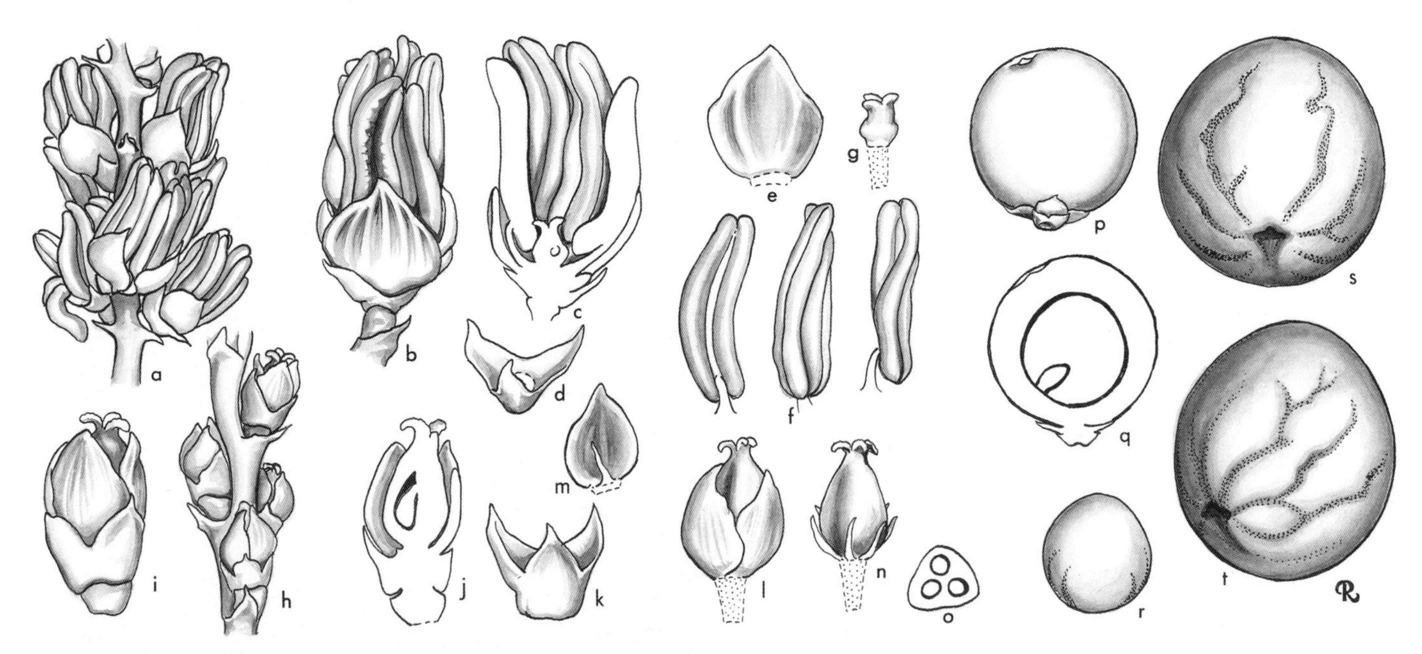

Juania. **a**, portion of staminate rachilla in bud × 3; **b**, staminate bud × 6; **c**, staminate bud in vertical section × 6; **d**, staminate calyx × 6; **e**, staminate petal, interior view × 6; **f**, stamen in 3 views × 6; **g**, pistillode × 6; **h**, portion of pistillate rachilla in bud × 3; **i**, pistillate bud × 6; **j**, pistillate bud in vertical section × 6; **k**, pistillate calyx × 6; **l**, pistillate flower, calyx removed × 6; **m**, pistillate petal and staminode, interior view × 6; **n**, gynoecium and staminodes × 6; **o**, ovary in cross-section × 6; **p**, fruit × $1^{1}/_{2}$; **q**, fruit in vertical section × $1^{1}/_{2}$; **r**, seed × $1^{1}/_{2}$; **s, t**, seed in 2 views × 3. *Juania australis*: **a–g**, *Moore et al.* 9369; **h–t**, *Moore et al.* 9368. (Drawn by Marion Ruff Sheehan)

Distribution and ecology: One species on Juan Fernandez Islands, occurring on steep slopes and ridges in lower and upper montane forest at altitudes of 200–800 m above sea level, most abundant above 500 m.

Anatomy: Leaf, stem (Tomlinson 1969), root (Seubert 1996b) and flowers (Uhl 1969b). Single phloem strands in large vascular bundles of the petiole. Patterns of floral vasculature are similar to those of *Ravenea* and *Ceroxylon* (Uhl 1969b).

Relationships: For relationships, see *Ceroxylon*.

Common names and uses: Juania palm, chonta palm. The apex is edible. In the past, the wood was used for walking sticks, cabinet work, and carvings.

Taxonomic account: Moore (1969e).

Fossil record: No generic records found.

Notes: *Juania* is threatened in its habitat by introduced goats. It has proved rather difficult to grow and remains very rare both in the wild and in cultivation.

72. ORANIOPSIS

Dioecious pinnate-leaved palm from Queensland, Australia; similar to *Ceroxylon* and *Juania* but differing in the combination of distinct petals, six stamens and basal stigmatic remains.

Oraniopsis (Becc.) J. Dransf., A.K. Irvine & N.W. Uhl, Principes 29: 57 (1985). Type: **O. appendiculata** (F.M. Bailey) J. Dransf., A.K. Irvine & N.W. Uhl (*Areca appendiculata* F.M. Bailey).

Orania Zipp. subgenus *Oraniopsis* Becc. in Becc. & Pic.Serm., Webbia 11: 172 (1955).

Orania — *the palm genus,* opsis — *similar to; originally a section of Orania with which it has, in fact, no relationship.*

Medium, solitary, unarmed, pleonanthic, dioecious palm. *Stem* erect, sometimes quite tall, becoming bare, leaf scars apparently not very conspicuous. *Leaves* numerous, reduplicately pinnate, ± upward-pointing, marcescent, several dead leaves hanging vertically for some time, forming a skirt below the crown before falling completely; sheath apparently tubular at first, soon splitting opposite the petiole, the leaf base then open; petiole short, adaxially channelled, ± glabrous, abaxially rounded, densely covered with scales and tomentum, the margins smooth and rather sharp; rachis ± stiffly held, adaxially flattened or channelled near the base, abaxially rounded, distally angled adaxially, a minute flange present at the junction between the flattened and angled areas of the rachis, both surfaces of the rachis bearing scattered scales; leaflets very numerous, single-fold, regularly arranged, ± stiff, ± linear, unevenly acute or acuminate, the basal-most few on each side short, narrow and crowded, adaxial surface ± glabrous or with scattered scales along the midrib, abaxial surface covered with dot-like scales and a dense felt of indumentum; transverse veinlets not evident. *Inflorescences* solitary, axillary, interfoliar, shorter than the leaves, staminate and pistillate superficially similar, branching to 4 orders; prophyll short, obscured by the leaf bases, incompletely tubular, 2-keeled, ± leathery, becoming fibrous and disintegrating distally, sparsely tomentose, the basal margins decurrent; peduncle elongate ± flattened and winged at the base, distally ± elliptic in cross-section, sparsely to densely tomentose; peduncular bracts 3–5, elongate, the first inserted near the prophyll, the rest ± evenly spaced along the peduncle, the distal 2–3 ± enclosing the inflorescence in bud, ± beaked, leathery, tubular at first, then splitting longitudinally and becoming flattened, sparsely to densely tomentose, eventually caducous, leaving circular or crescentic scars; rachis slightly shorter than the peduncle; rachis bracts numerous, inconspicuous, short, triangular, acute or acuminate, membranous, incomplete, each subtending a first-order branch; first-order branches with a basal bare portion, distally bearing spirally arranged second-order branches each subtended by a minute incomplete bract; rachillae crowded, ± twisted or zigzag at anthesis, the pistillate spreading but remaining rather zigzag in fruit, bearing rather

Top right: ***Oraniopsis appendiculata***, crown, Australia. (Photo: J. Dransfield)

Bottom right: ***Oraniopsis appendiculata***, staminate inflorescence, Australia. (Photo: J. Dransfield)

distant, spirally arranged or subdistichous, minute triangular bracts, each subtending a short stalk bearing a minute, membranous, incomplete, triangular bracteole and terminating in a solitary flower. *Staminate flowers* symmetrical, or somewhat misshapen from close packing, open from early in development; sepals 3, very small, triangular, membranous, connate basally and forming a cup; petals 3, distinct, fleshy, much longer than the sepals, narrow, triangular; stamens 6, almost as long as or longer than the petals, the antesepalous inserted between the petals in, apparently, the same whorl, the antepetalous epipetalous, filaments very fleshy with ± conical, swollen bases, tapering to the connective, anthers oblong, versatile, basally sagittate, latrorse; pistillode usually very much shorter than the filaments, 3-angled, apically trifid. *Pollen* ellipsoidal, slightly asymmetric; aperture a distal sulcus; ectexine tectate, foveolate, aperture margin broad scabrate and finely perforate; infratectum columellate; longest axis ranging from 33–40 µm [1/1]. *Pistillate flowers* like the staminate but with slightly broader sepals and petals; staminodes like the stamens, the empty anthers large; gynoecium tricarpellate, triovulate, conspicuously 3-lobed, stigmas apical, short, becoming recurved; ovules laterally attached, ?hemianatropous. *Fruit* developing from 1 carpel, rounded, the stigmatic and carpel remains basal; epicarp smooth, yellow at maturity; mesocarp ± fleshy, with horizontal fibres and stone cells; endocarp obsolescent. *Seed*, rounded, the integuments thick, ± woody, with a basal short spur, and few sparsely branched, impressed vascular strands; endosperm homogeneous with a narrow central hollow; embryo lateral to subbasal. *Germination* adjacent-ligular; seedling leaf bifid with entire tips. *Cytology* not studied.

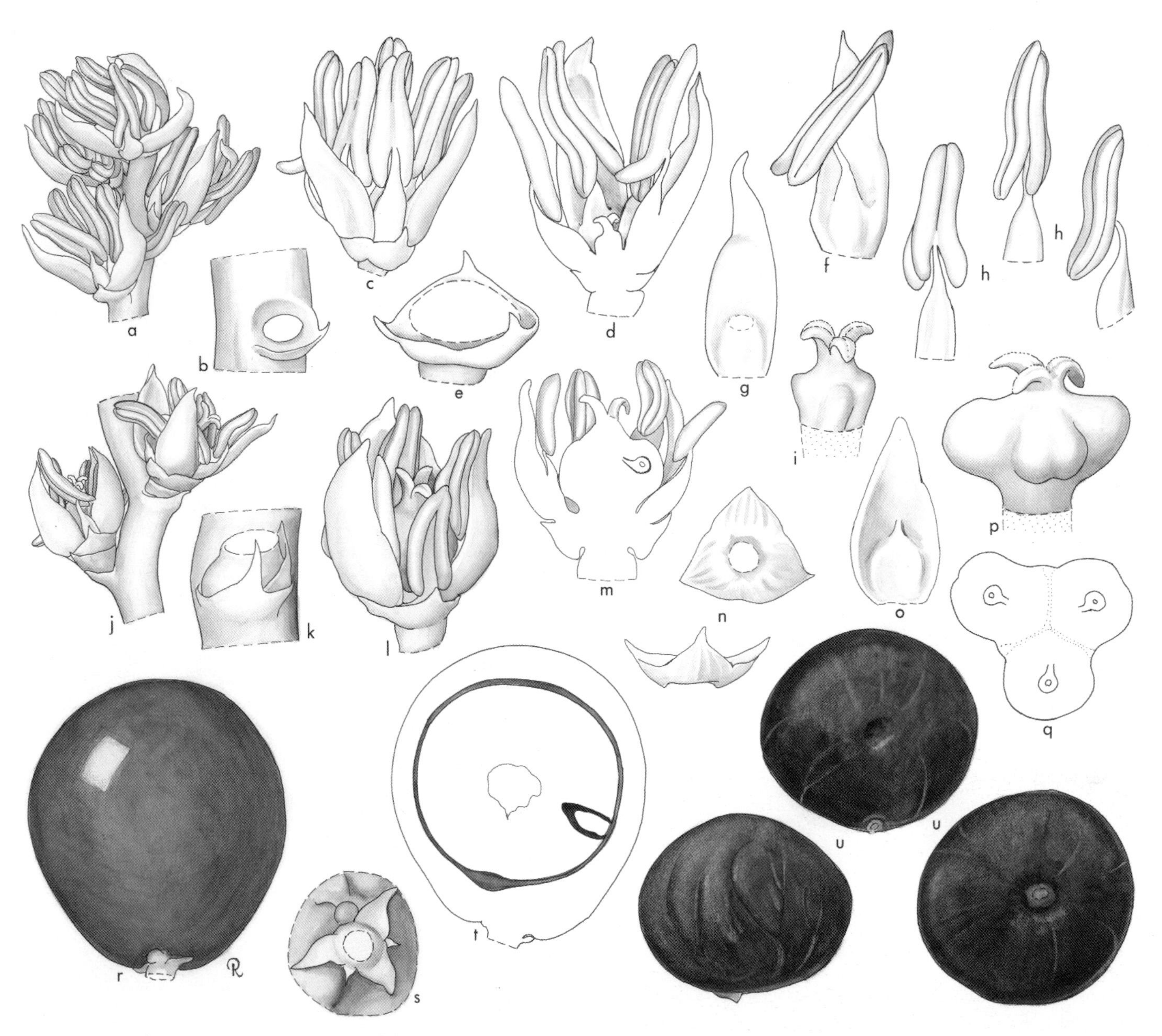

Oraniopsis. **a**, portion of staminate rachilla × 3; **b**, scar and bract of staminate flower × 6; **c**, staminate flower × 4 1/2; **d**, staminate flower in vertical section × 6; **e**, staminate calyx × 9; **f**, staminate petal and stamen × 6; **g**, staminate petal, interior view × 6; **h**, stamen in 3 views × 6; **i**, pistillode × 18; **j**, portion of pistillate rachilla × 3; **k**, portion of stalk, subtending bract, and bracteole of pistillate flower × 6; **l**, pistillate flower × 6; **m**, pistillate flower in vertical section × 6; **n**, pistillate calyx in 2 views × 6; **o**, pistillate petal, interior view × 6; **p**, gynoecium × 9; **q**, ovary in cross-section × 9; **r**, fruit × 1 1/2; **s**, base of fruit × 2 1/4; **t**, fruit in vertical section × 1 1/2; **u**, seed in 3 views × 1 1/2. *Oraniopsis appendiculata*: **a–q**, *Irvine* s.n.; **r–u**, *Moore* 9237. (Drawn by Marion Ruff Sheehan)

Distribution of ***Oraniopsis***

Distribution and ecology: Queensland, Australia; the single species *Oraniopsis appendiculata*, occurs in rain forests of mountain ranges from the upper Tully River area (15°40' S) northwards to the Big Tableland (17°50' S). The species occurs mostly above 300 m up to ca. 1500 m altitude, on soils of granitic and metamorphic origin; the palm also occurs on shallow basaltic soils with impeded drainage, but is usually absent from deep, well-drained basalt soils.
Anatomy: Root (Seubert 1996b).
Relationships: *Oraniopsis* is sister to *Ravenea* with moderate support in some studies (Asmussen *et al.* 2006, Baker *et al.* in review), but in others it is sister to a clade of *Ceroxylon* and *Juania* with moderate support (Uhl *et al.* 1995, Trénel *et al.* 2007).
Common names and uses: Not recorded. The palm has proved to be a handsome but very slow growing ornamental.
Taxonomic account: Dransfield *et al.* (1985a).
Fossil record: No generic records found.
Notes: The palm is very slow growing and seems to stay in the rosette stage for at least 20–30 years. In dense rain forest, rosettes may even be twice this age with erect leaves 3–8 m long. Unless growth rates accelerate markedly when a trunk is produced, tall-stemmed individuals must be several hundred years old.
Additional figures: Glossary fig. 19.

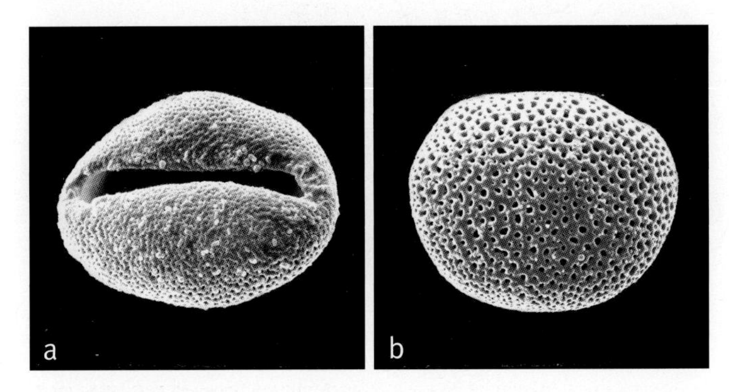

Above: ***Oraniopsis***
a, monosulcate pollen grain, distal face SEM × 1500;
b, monosulcate pollen grain, proximal surface SEM × 1500.
Oraniopsis appendiculata: **a** & **b**, *Gibbs* 6316.

73. RAVENEA

Dioecious pinnate-leaved palms of Madagascar and the Comores, distinguishable by the combination of incomplete prophyll, staminodes bearing rudimentary anthers and the fruit with subapical or apical stigmatic remains.

Ravenea Bouché ex H. Wendl. in Hook.f., Bot. Mag. 110: t6776 (1884). Type: **R. hildebrandtii** Bouché ex H. Wendl.
Ranevea L.H. Bailey, Cycl. Amer. hort. 1497 (1902). Type: as above. (Proposed as a substitute name for *Ravenea*, because of the close similarity in spelling to *Ravenia* Vell. [1825] [Rutaceae]; however, *Ravenea* Bouché is generally accepted as legitimate.)
Louvelia Jum. & H. Perrier, Compt. Rend. Hebd. Séances Acad. Sci. 155: 411 (1912). Type: *L. madagascariensis* Jum. & H. Perrier.

Named for Louis Ravené, a 19th century official in Berlin.

Solitary, slender to robust, unarmed, dioecious, pleonanthic palms. *Stem* erect, rarely very short, very rarely branching dichotomously, often tall, becoming bare, sometimes swollen at the base, conspicuously ringed with leaf scars or not. *Leaves* few to numerous, reduplicately pinnate, often upward pointing, abscising neatly or marcescent, crownshaft absent; sheaths soon disintegrating opposite the petiole into fine or coarse fibres, densely tomentose; petiole very short to moderately long, adaxially flattened or channelled, abaxially rounded, usually densely tomentose; rachis flattened adaxially, triangular abaxially, sometimes channelled laterally; leaflets numerous, single-fold, usually stiff, elongate, acute or acuminate, the proximal few sometimes very short and slender, abaxially frequently with small narrowly elliptic ramenta, rarely densely covered with white indumentum abaxially, midribs more prominent adaxially, transverse veinlets evident or not. *Inflorescences* interfoliar, 1 or several within a single prophyll in a leaf axil, usually large, sometimes highly condensed and hidden among the sheaths, the staminate and pistillate superficially similar but the staminate more slender, branched to 1–2 orders; peduncle slender to robust, very short to elongate, ± circular in cross-section distally, usually scaly or tomentose; prophyll short, 2-keeled, incomplete; peduncular bracts tubular, 3–5, 2 usually short, apically open and tattered, the 2 or more distal bracts narrow lanceolate, as long as and enclosing the inflorescence in bud, splitting

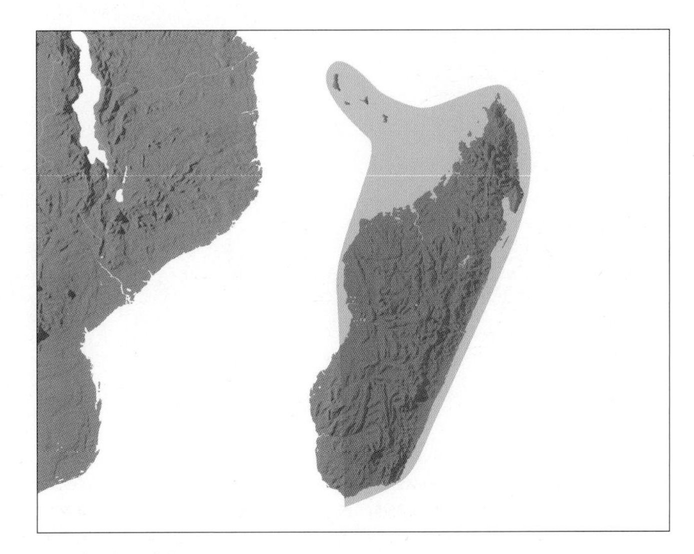

Distribution of ***Ravenea***

longitudinally, all bracts persistent, always densely scaly or tomentose abaxially, membranous, coriaceous, or almost woody; rachis usually shorter than the peduncle, bearing spirally arranged, usually numerous, first-order branches, the first-order bracts very small, acute or long acuminate, usually adnate to the subtended branch; proximal first-order branches, subdigitately branched or unbranched, the distal unbranched; rachillae very short to elongate, usually stiff, slender, sometimes eventually pendulous, the staminate shorter and more slender than the pistillate, slightly zigzag, sparsely to densely scaly, bearing spirally arranged, rather lax peg-like floral stalks, each subtended by a minute, narrow triangular rachilla bract, usually adnate to the stalk. Flowers more crowded and somewhat grouped distally, congenitally open. *Staminate flowers* ± symmetrical; sepals 3, triangular, connate in the basal $^1/_3$, adnate to the floral stalk, to the base of the stamen filaments and to the petal bases; petals 3, broadly ovate, distinct or rarely connate basally, fleshy; stamens 6, sometimes inserted in 2 series, filaments slender, short, basally expanded, shortly adnate to sepals and petals, rarely connate in an androecial ring; anthers straight or somewhat twisted, ± sagittate, latrorse. *Pollen* spheroidal, aperture a distal pore; ectexine tectate, scabrate-perforate with supratectal spines or, rarely, semi-tectate, foveolate-reticulate, pore margin similar to main tectum, or a slightly raised psilate annulus; longest axis 24–35 µm [9/17]. *Pistillate flowers* with sepals and petals similar to staminate; staminodes 6, broadly triangular with sterile anthers; gynoecium ovoid, tricarpellate, carpels connate, trilocular, triovulate, stigmas 3, fleshy, recurved, ovules pendulous, hemianatropous. *Fruit* 1–3 seeded globose to ellipsoid when 1-seeded, slightly lobed when more than 1 seed develops, yellow, orange or red, more rarely brown, purple or black, stigmatic remains subbasal, lateral, or subapical; epicarp smooth or minutely pebbled, mesocarp fleshy, endocarp thin. *Seeds*, globose, hemispherical or representing a third of a sphere, hilum basal, raphe branches indistinct, endosperm homogeneous; embryo basal. *Germination* adjacent-ligular or remote-ligular; eophyll bifid or pinnate. *Cytology*: 2n = 30, 32.

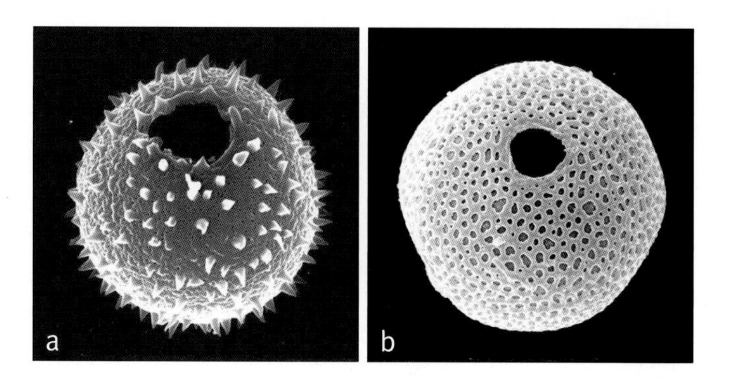

Above: ***Ravenea***
a, monoporate pollen grain, tilted equatorial face with distal pore in view SEM × 2000;
b, monoporate pollen grain, distal face SEM × 1500.
Ravenea madagascariensis: **a**, *Dransfield* JD6430; *R. albicans*: **b**, *Perrier de la Bâthie* 11939.

Distribution and ecology: Eighteen species, sixteen endemic to Madagascar, two in the Comoro Islands. Recent collecting by commercial seed collectors suggests the presence of several undescribed taxa, including a small species in the Comoros with immensely long peduncles that reach the ground. Occurring in tropical rain forest from sea level to 2000 m, including mossy montane forest, one species (*Ravenea xerophila*) in spiny forest in the driest area of Madagascar,

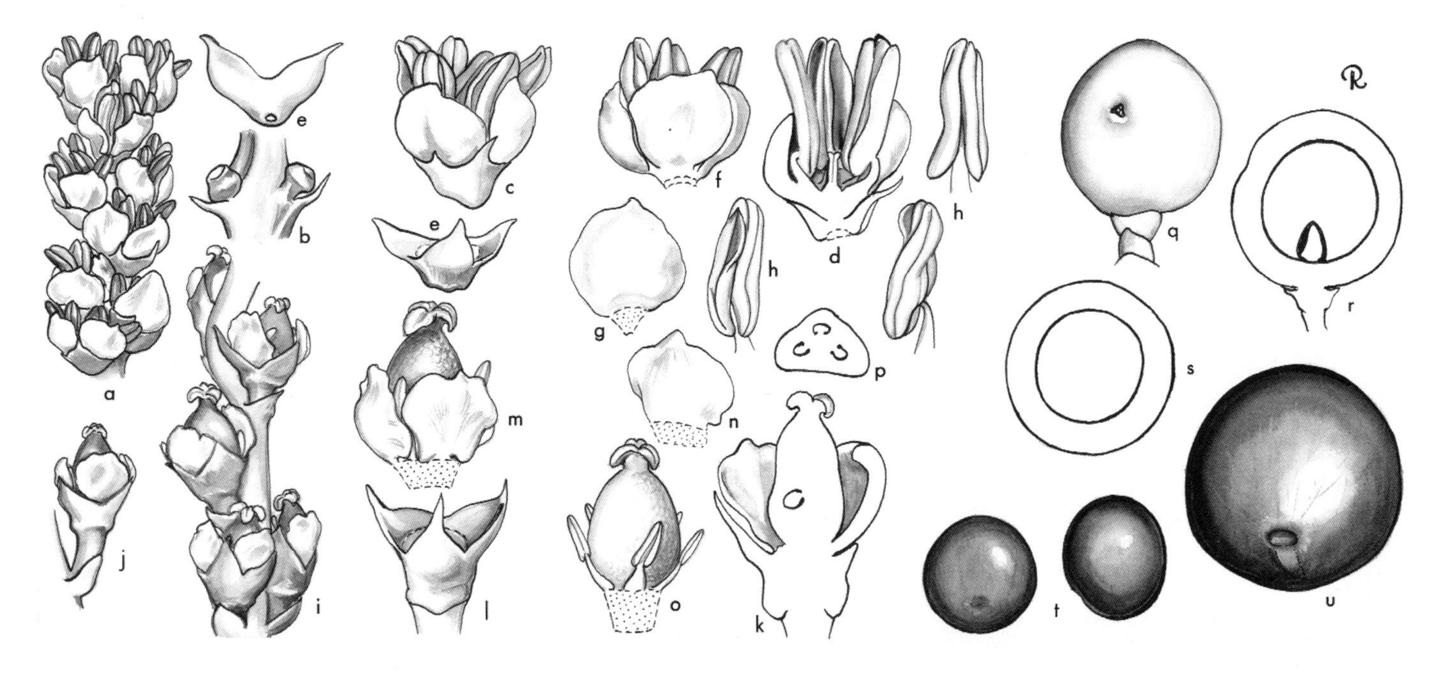

Ravenea. **a**, portion of staminate rachilla × 3; **b**, pedicels and bracteoles of staminate flowers × 6; **c**, staminate flower × 6; **d**, staminate flower in vertical section × 6; **e**, staminate calyx in 2 views × 6; **f**, staminate flower, sepals removed × 6; **g**, staminate petal × 6; **h**, stamen in 3 views × 6; **i**, portion of pistillate rachilla × 3; **j**, pistillate flower, pedicel, and bracteole × 3; **k**, pistillate flower in vertical section × 6; **l**, pistillate calyx × 6; **m**, pistillate flower, calyx removed × 6; **n**, pistillate petal × 6; **o**, gynoecium and staminodes × 6; **p**, ovary in cross-section × 6; **q**, fruit × $1^1/_2$; **r**, fruit in vertical section × $1^1/_2$; **s**, fruit in cross-section × $1^1/_2$; **t**, seed in 2 views × $1^1/_2$; **u**, seed × 3. *Ravenea madagascariensis*: **a–h**, *Moore* 9010; **i–p**, *Moore* 9019; *R. moorei*: **q–u**, *Moore* 9028. (Drawn by Marion Ruff Sheehan)

another (*R. musicalis*) growing in flowing water and starting its life as a submerged aquatic. Often gregarious, the rain-forest species sometimes grow in full light.

Anatomy: Vegetative (Tomlinson 1969), root (Seubert 1996b) and floral (Uhl 1969b).

Relationships: *Ravenea* is monophyletic with high support (Trénel *et al.* 2007). Two possible placements of *Ravenea* have been resolved: first, as sister to *Oraniopsis* with moderate support (Asmussen *et al.* 2006, Baker *et al.* in review) and, second, as sister to a clade of *Oraniopsis*, *Ceroxylon* and *Juania* (Uhl *et al.* 1995, Trénel *et al.* 2007).

Common names and uses: For Malagasy local names, see Dransfield and Beentje (1995b). The trunk of *Ravenea madagascariensis* is very hard and flexible and is used in various ways; that of *R. robustior* has abundant pith, is rich in starch, and is used as a source of sago. The 'cabbage' of most species is edible but may be bitter. Trunks of *R. musicalis* are sometimes hollowed out to make canoes. Many species are widely traded as ornamentals, the most significant being *R. rivularis* (Majesty palm).

Taxonomic accounts: Beentje (1994) and Dransfield & Beentje (1995b).

Fossil record: No generic records found.

Notes: *Ravenea* is the most diverse genus in the tribe, in numbers of species, in ecological adaptation and in vegetative characters. *Louvelia*, previously recognised as distinct, was included by Beentje (1994) in synonymy, as the characters used to separate the genus were shown to be unreliable.

Additional figures: Figs 1.7, 3.1, 8.2, Glossary fig. 18.

Top left: ***Ravenea madagascariensis,*** habit, Madagascar. (Photo: J. Dransfield)

Top right: ***Ravenea sambiranensis,*** multiple staminate inflorescences with common incomplete prophyll, Madagascar. (Photo: J. Dransfield)

Bottom: ***Ravenea sambiranensis,*** fruit, Madagascar. (Photo: J. Dransfield)

Tribe Phytelepheae Horan., Char. ess. fam. 38 (1847). Type: **Phytelephas**.

Moderate to rather large, acaulescent or erect; unarmed; pleonanthic; dioecious; leaves pinnate, reduplicate; inflorescences markedly dimorphic; prophyll rather short, usually obscured by the leaf sheath; peduncular bract 1; staminate inflorescence spike-like or racemose; pistillate inflorescence head-like; flowers multiparted, the staminate grouped in monopodial clusters, the pistillate solitary, crowded; staminate flowers with much reduced perianth and very numerous stamens; pistillate flower with large elongate fleshy sepals and petals, gynoecium of 5–10 connate carpels, styles elongate; fruit 5–10 seeded, borne in a congested head-like cluster, stigmatic remains apical; mesocarp cracking to form irregular fibrous, pyramidal warts; pyrenes smooth, hard, thin, enclosing individual seeds; endosperm very hard, homogeneous.

The tribe includes three closely related genera from South America, differing from each other in the structure of the staminate flowers. These vegetable ivory palms possess perhaps the most bizarre of all palm flowers, flowers that deviate strikingly from a trimerous monocot floral base plan. The pistillate flowers are very large with conspicuous tepals, whereas the staminate are very much smaller with scarcely evident tepals.

Tribe Phytelepheae is monophyletic (e.g., Uhl *et al.* 1995) and highly supported in all phylogenies (Hahn 2002b, Asmussen *et al.* 2006, Trénel *et al.* 2007, Baker *et al.* in review, in prep.). It is sister to tribe Ceroxyleae with moderate to high support (Hahn 2002b, Asmussen *et al.* 2006, Trénel *et al.* 2007, Baker *et al.* in review, in prep.).

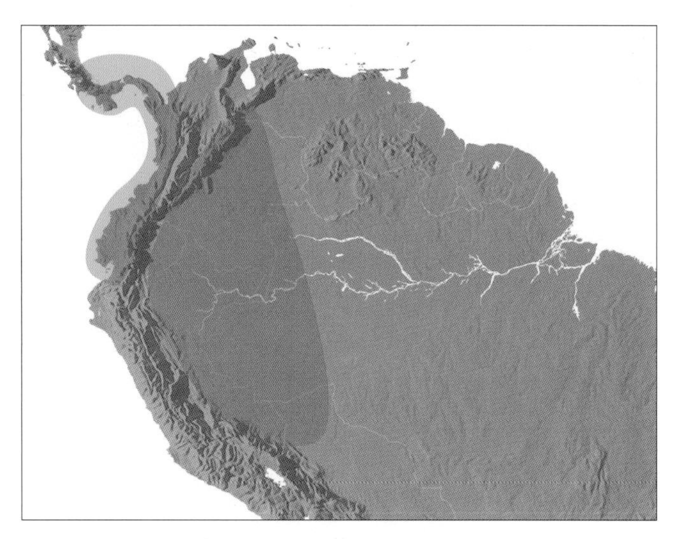

Distribution of tribe **Phytelepheae**

Key to Genera of Phytelepheae

1. Staminate flowers with minute rounded anthers borne on very short filaments on an angular polyhedral receptacle . . . . 74. **Ammandra**
1. Staminate flowers with elongate anthers borne on elongate filaments; receptacle not polyhedral 2
2. Leaf sheaths producing abundant piassava that hangs down obscuring the upper part of the trunk; staminate flowers borne in fours or fives with floral receptacles forming conspicuous funnel-shaped pseudopedicels . .75. **Aphandra**
2. Leaf sheaths fibrous but scarcely producing piassava; staminate flowers sessile or borne in groups of four on short stalks, with relatively flat or slightly rounded floral receptacles . . . . 76. **Phytelephas**

74. AMMANDRA

Remarkable dioecious stemless or short-trunked ivory palm from northern South America; distinguished by the chunky polyhedral receptacles of the staminate flowers bearing very numerous rounded anthers that appear like grains of sand.

Ammandra O.F. Cook, J. Wash. Acad. Sci. 17: 220 (1927). Type: **A. decasperma** O.F. Cook.

ammos — *sand,* aner — *man, referring to the anthers that appear like grains of sand.*

Solitary, stemless or short-trunked, unarmed, pleonanthic, dioecious palm. *Stem* extremely short, internodes short, obscured by a loose network of long, slender, straight, sheath fibres. *Leaf* pinnate; sheath soon disintegrating into a mass of long straight fibres resembling horse hair; petiole erect, long, slender, grooved adaxially at the base, becoming cylindrical distally; leaflets very regular except for the lower-most which may be irregular, stiffly horizontal, the lowest very narrow, the middle lanceolate, the terminal very short, shiny dark green, a midrib and a pair of marginal veins prominent abaxially, the submarginal veins forming a prominent ridge with a resulting outer groove along the leaflet margins, transverse veinlets not evident. *Inflorescences*, the staminate and pistillate dissimilar; staminate short, racemose, recurved at anthesis, branched to 1 order; peduncle moderate, rounded, glabrous; prophyll tubular, short, shallowly 2-keeled, rounded to a shallow point, splitting apically; complete peduncular bracts l, similar to the propyll but longer, other peduncular bracts few (5 according to Cook) large or small and shallow; rachis slightly longer than the peduncle, bearing spirally arranged, short, terete branches, each subtended by a small pointed bract; first-order branches each bearing ca. 6(–9), crowded, staminate flowers, subtending bracts small, pointed, membranous or not evident. Staminate flowers with a short terete stalk; perianth consisting of a low membranous rim or absent; floral receptacle chunky with several flat sides all bearing irregularly to somewhat spirally arranged stamens, filaments very short, appressed, or briefly elongate, anthers short, rounded or ± elongate, basifixed, latrorse; pistillode terminal, conical, whitish, irregular in position. *Pollen* brevi-ellipsoidal to spheroidal, usually ± symmetric; aperture a distal brevi sulcus or large pore; ectexine semi-tectate, foveolate or reticulate, aperture margin psilate; infratectum columellate; longest axis 70–85 µm [1/1]. *Pistillate inflorescence* head-like, unbranched; prophyll tubular, short, 2-keeled laterally, flattened, pointed, splitting along one side; peduncular bracts several, the first complete, tubular, rounded, with a short pointed tip, splitting apically on one side, the second and third bracts incomplete, short, united basally to form a tube, distal parts distinct, triangular, fourth to sixth bracts also united basally into a shorter tube with distinct, tapering tips, seventh and eight bracts united basally on one

Right: ***Ammandra decasperma***, pistillate inflorescence, Colombia.
(Photo: J. Dransfield)

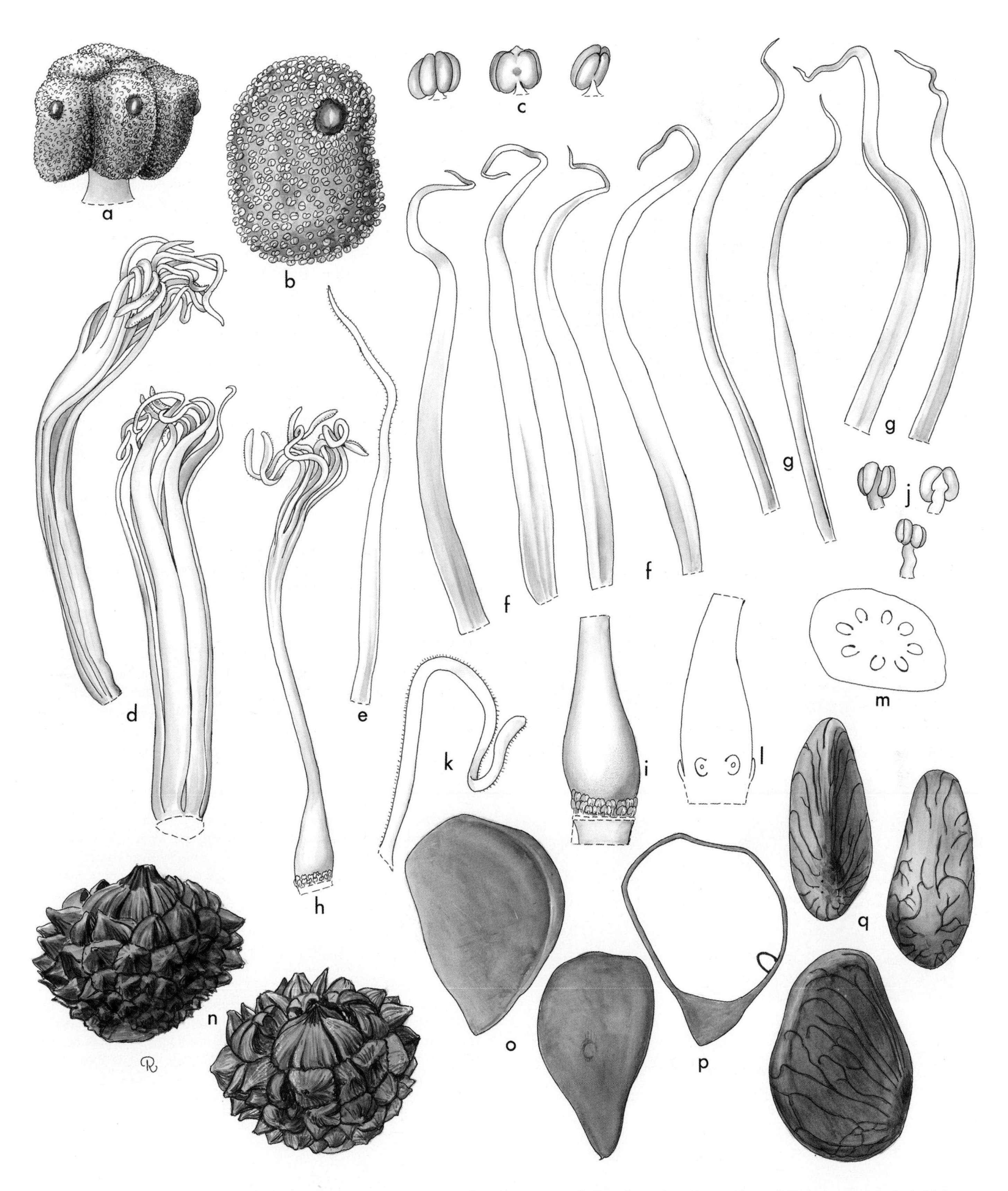

Ammandra. **a**, staminate rachilla × $1^1/_2$; **b**, staminate flower × 3; **c**, stamen in 3 views × 9; **d**, pistillate flower in 2 views × $^3/_4$; **e**, pistillate bract × $^3/_4$; **f**, pistillate sepals × $^3/_4$; **g**, pistillate petals × $^3/_4$; **h**, gynoecium and staminodes × $^3/_4$; **i**, base of gynoecium and staminodes × $1^1/_2$; **j**, staminode in 3 views × 6; **k**, stigma × $1^1/_2$; **l**, ovary in vertical section × $1^1/_2$; **m**, ovary in cross-section × 3; **n**, fruit in 2 views × $^1/_2$; **o**, endocarp in 2 views × $^3/_4$; **p**, endocarp in vertical section × $^3/_4$; **q**, seed in 3 views × $^3/_4$. *Ammandra decasperma*: all from *Moore & Parthasarathy* 9456. (Drawn by Marion Ruff Sheehan)

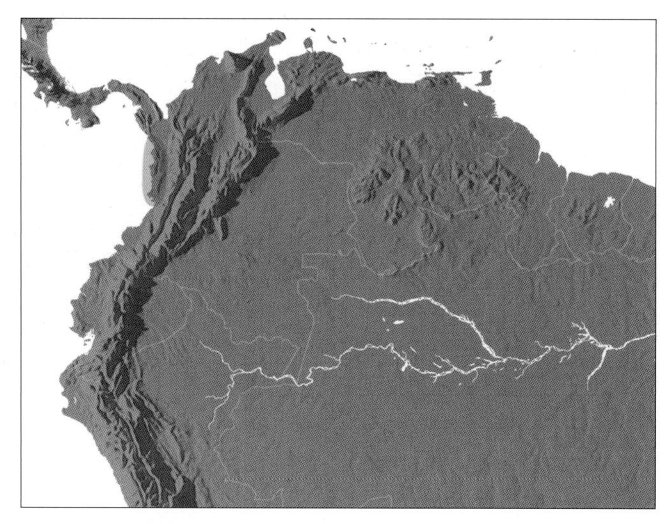

Distribution of ***Ammandra***

side, open on the other side. *Pistillate flowers* spirally arranged, closely appressed, each subtended by a bract; sepals ± 4, narrow, elongate; petals ± 4, like the sepals but longer and somewhat wider, variously wrinkled; staminodes apparently absent; gynoecium consisting of ca. 8 carpels, connate laterally, ovarian part terete, tapering into an elongate, cylindrical style and ca. 8, curly, elongate stigmas, conduplicately folded, bearing stigmatoid tissue along the margins. *Fruits* borne in large head-like clusters of 3–6, clusters smaller than those of *Phytelephas*, each fruit rounded, covered in large, pointed warts, stylar remains terminal, forming a large beak; epicarp with short, close fibres, mesocarp fibres fine, endocarp shell-like with adherent fibres enclosing each seed. *Seed* ± kidney-shaped, hilum basal, raphe fibres parallel, ascending, with short branches forming grooves in the endosperm, endosperm homogeneous, very hard; embryo lateral near the base. *Germination* remote-ligular; eophyll pinnate. *Cytology* not studied.

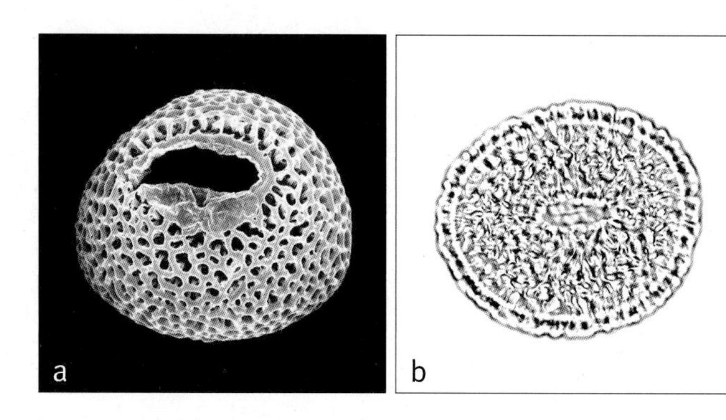

Above: ***Ammandra***
a, brevisulcate pollen grain, distal face SEM × 500;
b, brevisulcate pollen grain, high-focus LM × 500.
Ammandra decasperma: **a** & **b**, *Balslev et al.* 62070.

Distribution and ecology: One species known from the western coastal regions of Colombia and a disjunct population in eastern Colombia. An undergrowth palm in forests where rainfall is heavy and nearly continuous throughout the year. Many beetles emerged from inflorescences collected by Cook (1927).
Anatomy: Leaves (Barfod 1991) and root (Seubert 1996b).
Relationships: For relationships, see *Phytelephas*.
Common names and uses: Ivory palms, tagua, cabecita. Used for vegetable ivory and thatch.
Taxonomic account: Barfod (1991).
Fossil record: No generic records found.
Notes: For diagnostic characters see notes under *Phytelephas*.
Additional figures: Glossary fig. 20.

75. APHANDRA

Moderately tall ivory palm distinguished from other ivory palms by the abundant piassava on the leaf bases, the scaly rather than glabrous prophyll and peduncular bracts, and the staminate flowers borne in fours or fives with floral receptacles forming conspicuous funnel-shaped pseudopedicels.

Aphandra Barfod, Opera botanica 105: 44 (1991). Type: **A. natalia** (Balslev & A.J. Hend.) Barfod (*Ammandra natalia* Balslev & A.J. Hend.).

Combines the a of Ammandra *with the ph of* Phytelephas*, reflecting its close similarity to the two genera (Barfod, pers. comm.).*

Solitary, short-trunked, unarmed, pleonanthic, dioecious palm. *Stem* erect, eventually becoming bare, internodes short, obscured at first by a dense network of long, coarse, straight, sheath fibres (piassava). *Leaf* pinnate; sheath soon disintegrating into a mass of long coarse fibres resembling horse hair; petiole short, adaxially channelled, abaxially with a covering of white waxy indumentum; leaflets regularly arranged, the basal-most very narrow, the middle lanceolate, the terminal very short, midrib prominent, a pair of marginal veins less prominent, transverse veinlets inconspicuous. *Inflorescences*, the staminate and pistillate dissimilar; staminate short, racemose, recurved at anthesis, branched to 1 order; peduncle moderate, rounded, glabrous; prophyll tubular, short, shallowly 2-keeled, rounded to a shallow point, splitting apically; complete peduncular bracts 1, similar to the propyll but longer, incomplete peduncular bracts few 3–5, large or small and shallow; rachis slightly longer than the peduncle, bearing spirally arranged, short, terete branches, each subtended by a small pointed bract; first-order branches each bearing up to 4 staminate flowers, subtending bracts small, pointed, membranous or not evident. *Staminate flowers* with a conspicuous terete stalk; perianth consisting of a low membranous rim or absent; floral receptacle rounded bearing 400–650 stamens, filaments short, anthers elongate, basifixed, latrorse; pistillode rarely present, if so then minute, carpelliform. *Pollen* ellipsoidal, usually with either slight or obvious asymmetry; aperture a distal sulcus; ectexine semi-tectate, reticulate, aperture margin similar; infratectum columellate; longest axis ranges from 75–80 µm [1/1]. *Pistillate inflorescence* head-like,

Opposite: ***Aphandra***. **a**, staminate flower cluster × 2; **b**, staminate flower cluster, another view × 2; **c**, staminate flower × $2\frac{1}{3}$; **d**, staminate flower in vertical section × $2\frac{1}{3}$; **e**, stamens, in 3 views × 10; **f**, pistillate flower × $\frac{1}{2}$; **g**, bracteoles × $\frac{1}{2}$; **h**, sepals and petals × $\frac{1}{2}$; **i**, gynoecium and staminodes × $\frac{1}{2}$; **j**, ovarian part of gynoecium and staminodes × 3; **k**, vertical section of ovary × 3; **l**, cross section of ovary × 3; **m**, fruit × $\frac{1}{3}$; **n**, pyrene in 4 views × $\frac{1}{3}$; **o**, seed × $\frac{1}{3}$; **p**, seed in cross-section × $\frac{1}{3}$; **q**, seed in vertical section × $\frac{1}{3}$. *Aphandra natalia*: all from *Barford* 17. (Drawn by Marion Ruff Sheehan)

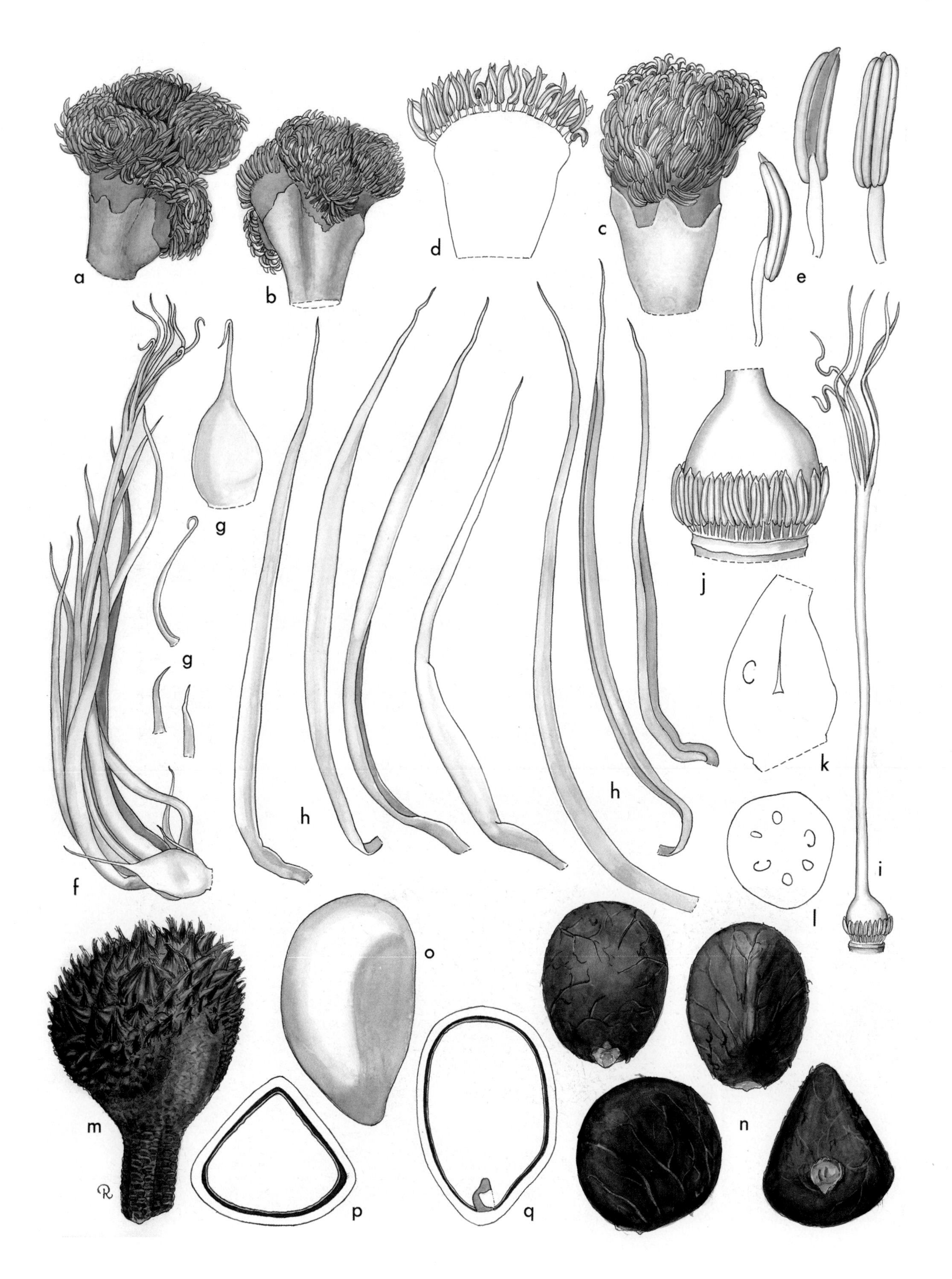
a
b
d
c
e
f
g
g
h
h
j
k
i
l
m
o
p
q
n

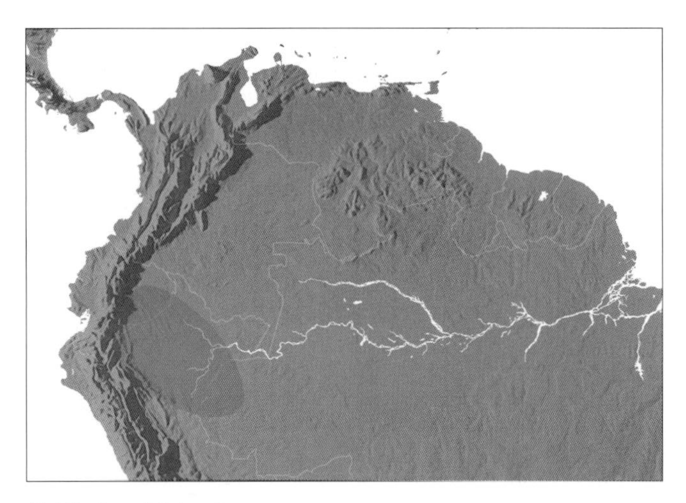

Distribution of ***Aphandra***

unbranched; prophyll tubular, short, 2-keeled laterally, flattened, pointed, splitting along one side; peduncular bracts 15–25, the first complete, tubular, rounded, with a short pointed tip, splitting apically on one side, subsequent bracts incomplete. *Pistillate flowers* spirally arranged, closely appressed, each subtended by a bract; sepals ± 4, narrow, elongate; petals ± 4, like the sepals but longer and somewhat wider, variously wrinkled; staminodes 30–50; gynoecium consisting of usually 6–8 carpels, connate laterally, ovarian part terete, tapering into an elongate, cylindrical style and 6–8, curly, elongate stigmas, conduplicately folded, bearing stigmatoid tissue along the margins. *Fruits* borne in large head-like clusters, each fruit rounded, covered in large, pointed warts, stylar remains terminal, forming a large beak; epicarp with short, close fibres, mesocarp fibres fine, endocarp shell-like with adherent fibres enclosing each seed. *Seed* ± kidney-shaped, hilum basal, raphe fibres parallel, ascending, with short branches forming grooves in the endosperm, endosperm homogeneous, very hard; embryo lateral near the base. *Germination* remote-ligular; eophyll pinnate. *Cytology*: 2n = 36.

Right: ***Aphandra***
a, monosulcate pollen grain, distal face SEM × 500;
b, close-up, coarsely reticulate pollen surface SEM × 5000.
Aphandra natalia: **a** & **b**, *Barfod* 60150.

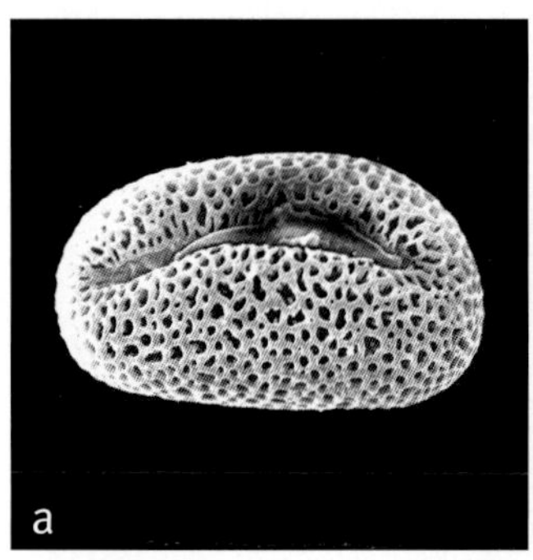

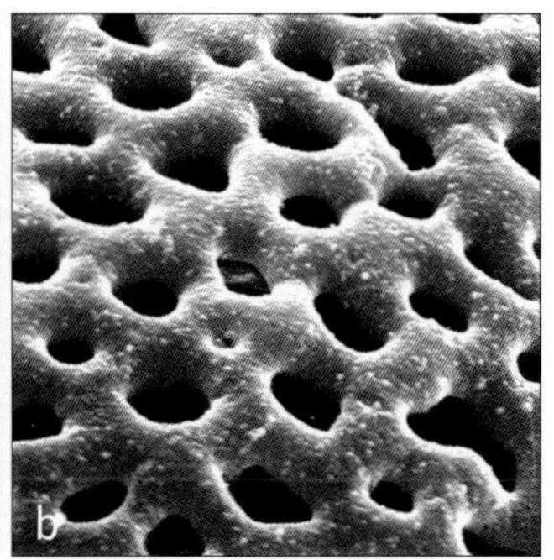

Below left: ***Aphandra natalia***, pistillate inflorescence, Ecuador (Photo: F. Borchsenius)

Below middle: ***Aphandra natalia***, crown with fruit, Ecuador (Photo: F. Borchsenius)

Below right: ***Aphandra natalia***, staminate inflorescence, Ecuador (Photo: F. Borchsenius)

Distribution: One species known from the Ecuador, Peru and Brazil, growing in lowland and premontane rain forest up to 800 m elevation, occasionally cultivated above this.
Anatomy: Leaves (Barfod 1991) and floral (Barfod & Uhl 2001).
Relationships: For relationships, see *Phytelephas*.
Common names and uses: See Barfod (1991) for common names. Piassava used for broom making and blow-pipe darts; leaves used for basket-weaving, fruit eaten and used for vegetable ivory.
Taxonomic account: Barfod (1991).
Fossil record: No generic records found.
Notes: For diagnostic characters, see notes under *Phytelephas*.
Additional figures: Fig. 1.9.

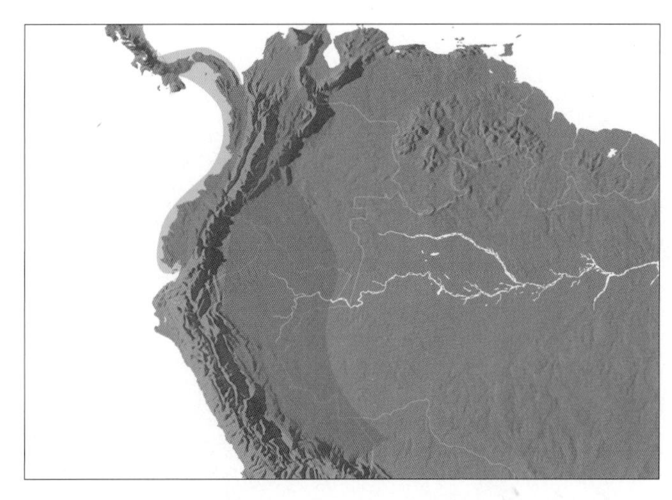

Distribution of ***Phytelephas***

76. PHYTELEPHAS

Vegetable ivory palms distinguished by the staminate flowers having relatively flat or slightly rounded floral receptacles and being sessile or borne in groups of four on short stalks.

Phytelephas Ruiz & Pavon, Syst. Veg. Fl. Peruv. Chil.: 299 (1798). Lectotype: **P. macrocarpa** Ruiz & Pavon (see O.F. Cook 1927).

Elephantusia, Willd., Sp. pl. 4(2): 890, 1156 (1806). Type: *E. macrocarpa* (Ruiz & Pavon) Willd. (*Phytelephas macrocarpa* Ruiz & Pavon).

Yarina O.F. Cook, J. Wash. Acad. Sci. 17: 223 (1927). Type: *Y. microcarpa* (Ruiz & Pav.) O.F. Cook (*Phytelephas microcarpa* Ruiz & Pavon).

Palandra O.F. Cook, J. Wash. Acad. Sci. 17: 228 (1927). Type: *P. aequatorialis* (Spruce) O.F. Cook (*Phytelephas aequatorialis* Spruce).

phyt — *plant*, elephas — *elephant, in reference to its use as a source of vegetable ivory.*

Moderate, solitary or clustered, unarmed, pleonanthic, dioecious palms. *Stem* robust or rarely rather slender, erect or procumbent, internodes short, covered with a mass of fibres and petiole bases, when bare marked by spiral, triangular, often pitted leaf scars. *Leaves* numerous or rarely few, erect, arching, evenly pinnate; marcescent; sheath tubular, sometimes with a large ligule opposite the petiole, becoming fibrous; petiole short, lacking, or rarely elongate, shallowly channelled adaxially, rounded abaxially, margins rounded or sharp; rachis triangular in section, with greyish brown scales abaxially, leaflets regularly arranged in one plane, or irregularly arranged and held in different planes to give the leaf a plumose appearance, subopposite, single-fold, pointed, often pinched in at the base, usually smaller basally and distally, glossy dark green adaxially, paler and duller beneath; tomentose abaxially along a conspicuous midrib, transverse veinlets conspicuous. *Inflorescences* interfoliar, staminate and pistillate dissimilar; staminate unbranched; peduncle short; prophyll short, tubular, 2-keeled laterally, broadly pointed, splitting apically; complete peduncular bracts 1, like the prophyll but longer, splitting abaxially, persistent above the inflorescence, subsequent peduncular bracts several (4–5), incomplete, spirally inserted below the flowers. *Staminate flowers* in groups of 4, sessile or with a conspicuous common stalk, usually lacking a subtending bract; perianth a low cupule with 3–8 points, not separable into sepals and petals at maturity (but see Uhl & Moore 1977b); stamens 36–900 or more, filaments erect, awl-shaped, anthers elongate, latrorse; pistillode lacking. *Pollen* ellipsoidal, usually with slight or, occasionally, with obvious asymmetry; aperture a distal sulcus; ectexine tectate, coarsely perforate, or perforate-rugulate, aperture margin slightly finer; infratectum columellate; longest axis 72–90 µm [3/7]. *Pistillate inflorescence* head-like; peduncle short, dorsiventrally flattened; prophyll and first peduncular bract as in the staminate, subsequent peduncular bracts numerous, larger than in the staminate, sometimes in series, elongate, pointed, ± covering the flowers. *Pistillate flowers* asymmetrical, each subtended by a pointed bract, spirally arranged, closely appressed; sepals 3 or more, triangular, ± elongate; petals 4–10, long, narrow, variously folded and wrinkled; staminodes numerous, 35 or more, like the stamens but irregular in size; gynoecium of 4–10 united carpels, ovarian part short, rounded, stigma long, narrow, cylindrical, styles as many as the carpels, long, narrow, conduplicately folded with stigmatoid tissue along the margins, ovules 1 per carpel, hemianatropous or anatropous. *Fruit* clusters, individual fruits ± rounded, 4–10-seeded, covered with large, woody, pointed warts, stylar remains terminal; epicarp woody, mesocarp fibrous, endocarp surrounding each seed bony or shell-like, bifacial

Right: ***Phytelephas aequatorialis***, habit, Ecuador. (Photo: F. Borchsenius)

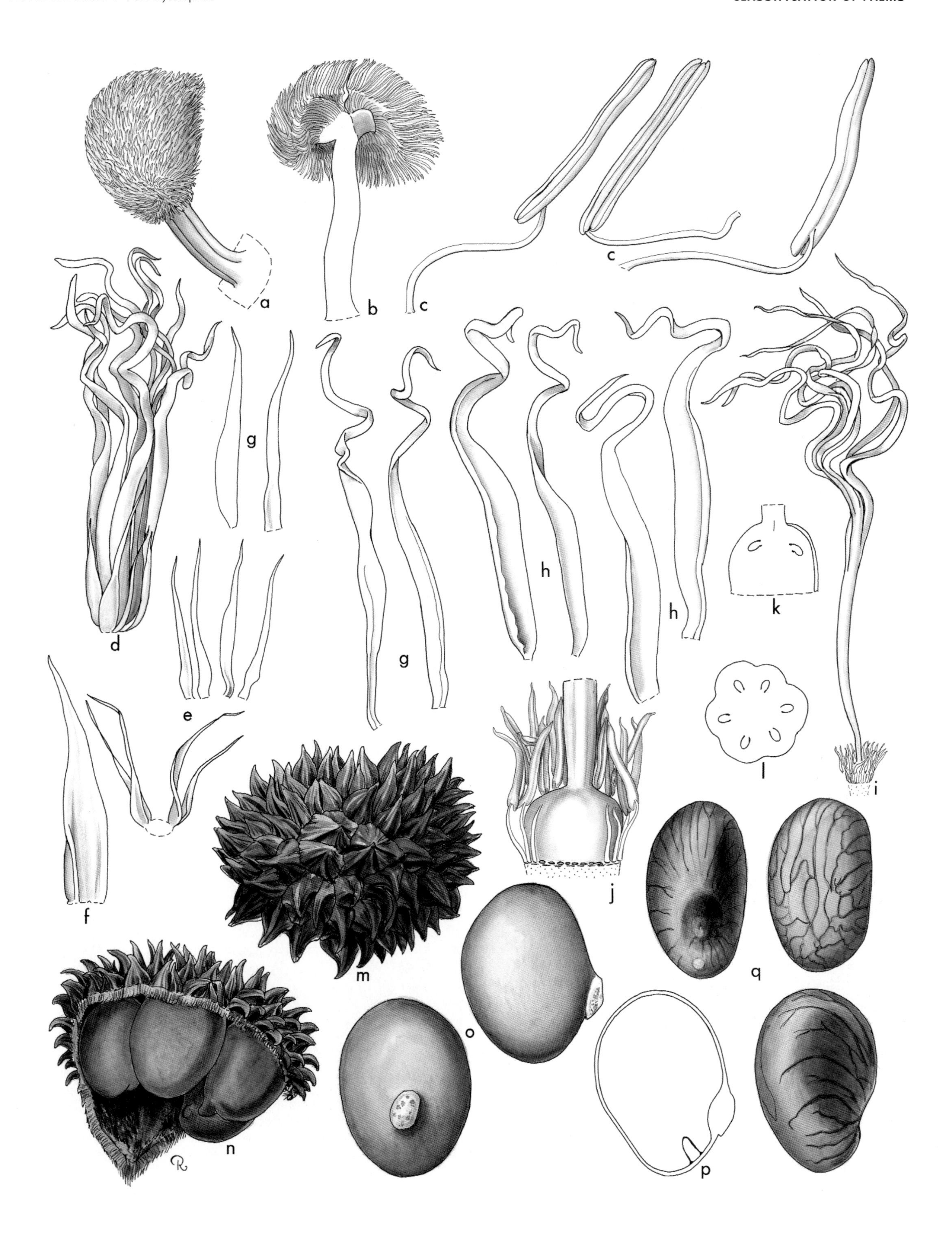
a
b
c
c
d
e
f
g
g
h
h
i
j
k
l
m
n
o
p
q

adaxially with round basal projection, rounded abaxially. *Seed* ± kidney-shaped, basally or laterally attached, hilum round, median to basal, raphe branches numerous, laterally ascending and anastomosing; endosperm homogeneous, hard (vegetable ivory), embryo basal or lateral. *Germination* remote-ligular; eophyll pinnate. *Cytology*: 2n = 36.

Distribution and ecology: Six species occurring in the Amazonian Basin in Bolivia, Ecuador and Peru, and along the northwest coast of Ecuador, Colombia and into Venezuela and Panama. Strictly confined to rain forest usually under large trees along streams and on wet hillsides.

Anatomy: Vegetative (Tomlinson 196l, Barfod 1991), root (Seubert 1996b), floral (Moore & Uhl 1973, Uhl & Moore 1977b, Uhl & Dransfield 1984) and seed (Werker 1997).

Relationships: In more densely sampled studies, *Phytelephas* is monophyletic with high or moderate support (Barfod *et al.* 1999, Trénel *et al.* 2007). Asmussen *et al.* (2006) recover a non-monophyletic *Phytelephas* with moderate support, but they only include two species. *Phytelephas* is highly supported as sister to a highly supported clade of *Aphandra* and *Ammandra* (Baker *et al.* in review, in prep., Trenél *et al.* 2007). For interspecies relationships, see Trenél *et al.* (2007).

Common names and uses: Ivory palms, tagua, yarina. Leaves are used for thatch; immature pericarp and endosperm are edible; mature endosperm is hard and used as vegetable ivory for carvings.

Taxonomic account: Barfod (1991).

Fossil record: A sandstone cast of a fruit, *Phytelephas olssonii*, is reported from the Upper Miocene-Lower Pliocene of Ecuador (Brown 1956b). Fossil stem material described from the Miocene of Antigua (Kaul 1943) is questionable (Uhl & Dransfield 1987).

Notes: The three genera of this tribe are separated largely on characters of the staminate flowers. *Phytelephas* is distinguished from *Aphandra* by sessile staminate flowers and by up to about 250 rather than more than 900 stamens, and from *Ammandra* by flat rather than club-shaped receptacles in staminate flowers.

Additional figures: Figs 1.7, 1.9.

Top: ***Phytelephas macrocarpa***, staminate inflorescence, cultivated, Fairchild Tropical Botanic Garden, Florida. (Photo: C.E. Lewis)

Bottom: ***Phytelephas macrocarpa***, infructescence, Peru. (Photo: J. Dransfield)

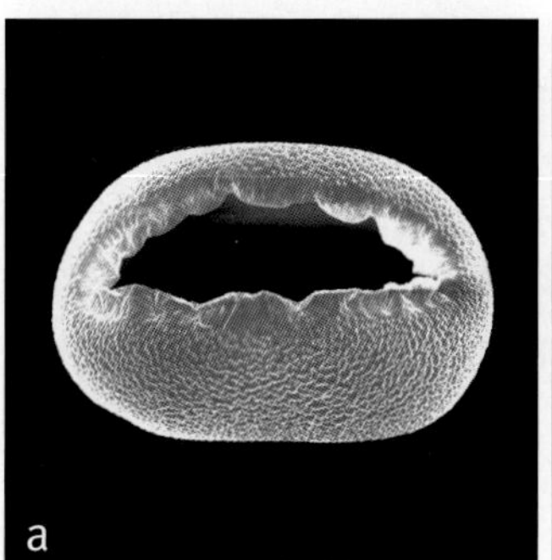

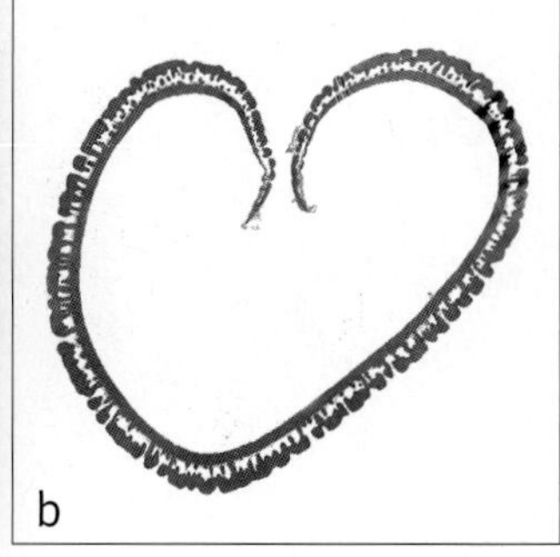

Left: ***Phytelephas***
a, monosulcate pollen grain, distal face SEM × 400;
b, ultrathin section through whole pollen grain, polar plane TEM × 600.
Phytelephas aequatorialis: **a**, *Barfod et al.* 60187, **b**, *Moore & Anderson* 10210.

Opposite: ***Phytelephas.*** **a**, staminate flower cluster with four flowers × $^3/_4$; **b**, staminate flower cluster, cut to show base of one flower on right × 1; **c**, stamen in 3 views × 6; **d**, pistillate flower × $^3/_4$; **e**, outer four bracts surrounding pistillate flower *in situ*, and dissected × 1; **f**, bracteole of pistillate flower × 1; **g**, pistillate sepals × 1; **h**, pistillate petals × 1; **i**, gynoecium and staminodes × 1; **j**, base of gynoecium, some staminodes removed × 4 $^1/_2$; **k**, ovary in vertical section × 4 $^1/_2$; **l**, ovary in cross-section × 4 $^1/_2$; **m**, fruit × $^1/_2$; **n**, fruit, some epicarp and mesocarp removed to show separate endocarps × $^1/_2$; **o**, single endocarp in 2 views × 1; **p**, endocarp in vertical section × 1; **q**, seed in 3 views × $^3/_4$. *Phytelephas aequatorialis*: all from *Moore* 10210. (Drawn by Marion Ruff Sheehan)

V. ARECOIDEAE

ARECOIDEAE

Reduplicately pinnate, pleonanthic monoecious or dioecious palms; inflorescence rachis bracts highly reduced; flowers in triads of a central pistillate and two lateral staminate flowers, or in cincinni, acervuli or (rarely by reduction) solitary; flowers with trimerous calyx and corolla; fruit lacking imbricate scales.

This is the largest and most diverse subfamily. In the first edition of *Genera Palmarum*, it was divided into six tribes. The new classification presented here recognises 13 tribes and 10 unplaced genera. The arecoids are distinguished by pinnate leaves, and most also by flowers in triads or groups reduced from triads; only members of Chamaedoreeae differ, bearing flowers in acervuli, or derivatives thereof (including solitary flowers). Rachis bracts, those that subtend the primary branches of the inflorescence, are generally very inconspicuous, a feature shared with Ceroxyloideae but rare elsewhere in the family.

Subfamily Arecoideae is resolved as monophyletic in the majority of studies (Uhl *et al.* 1995, Asmussen *et al.* 2000, 2006, Asmussen & Chase 2001, Lewis & Doyle 2002, Loo *et al.* 2006, Savolainen *et al.* 2006, Baker *et al.* in review, in prep.), with moderate to high support in several more recent analyses (Asmussen & Chase 2001, Asmussen *et al.* 2006, Baker *et al.* in review, in prep.). The subfamily is moderately to strongly supported as sister to the Ceroxyloideae (Asmussen & Chase 2001, Asmussen *et al.* 2006, Savolainen *et al.* 2006, Baker *et al.* in review, in prep.). In many phylogenies, a clade of six tribes is resolved within the Arecoideae, often with high support (Hahn 2002a, 2002b, Lewis & Doyle 2002, Norup *et al.* 2006, Baker *et al.* in review, in prep.). This group, termed here the core arecoid clade, consists of tribes Areceae, Pelagodoxeae, Euterpeae, Geonomateae, Leopoldinieae and Manicarieae.

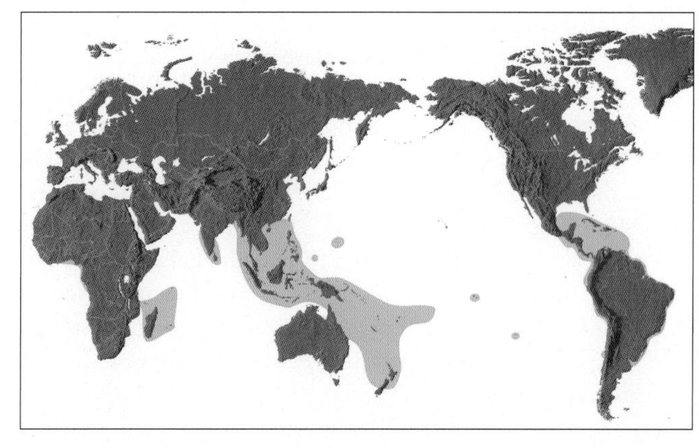

Distribution of subfamily **Arecoideae**

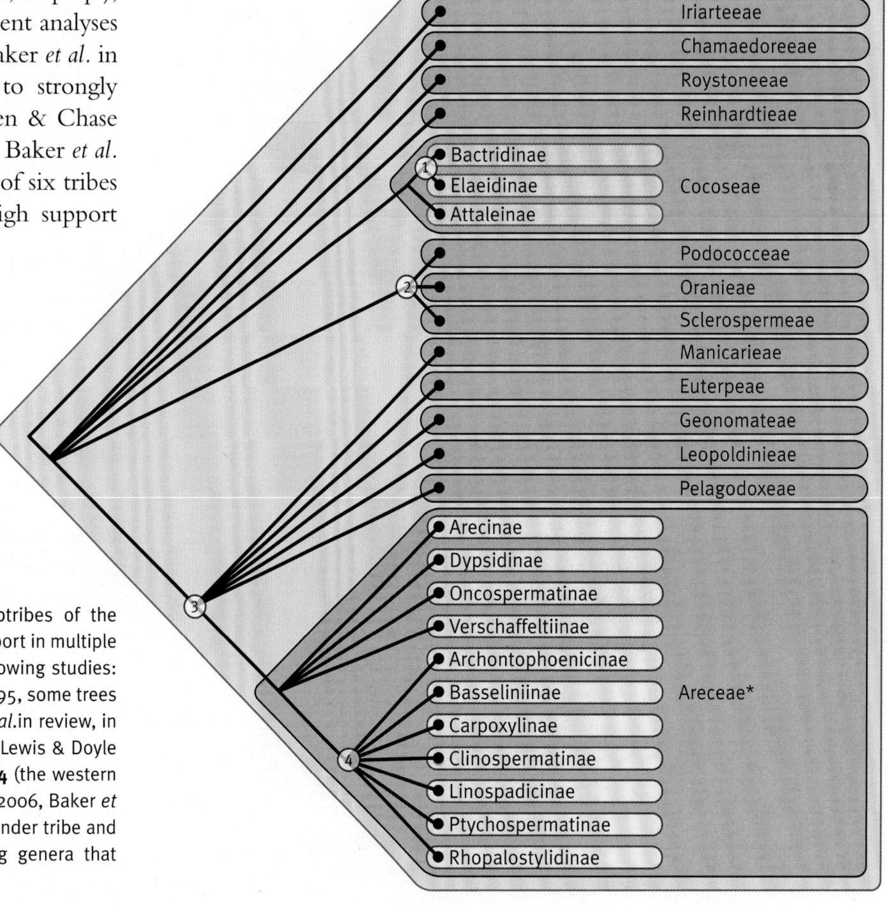

Summary tree showing relationships among tribes and subtribes of the Arecoideae that have been resolved with moderate to high support in multiple phylogenetic analyses. Labelled nodes are resolved in the following studies: **node 1** — Baker *et al.* in review, in prep.; **node 2** — Uhl *et al.* 1995, some trees of Hahn 2002a, Lewis & Doyle 2002, Loo *et al.* 2006, Baker *et al.*in review, in prep.; **node 3** (the core arecoid clade) — Hahn 2002a, 2002b, Lewis & Doyle 2002, Norup *et al.* 2006, Baker *et al.* in review, in prep.; **node 4** (the western Pacific clade) — Lewis 2002, Lewis & Doyle 2002, Norup *et al.* 2006, Baker *et al.* in review, in prep. For further explanation, see discussions under tribe and subtribe treatments. An asterisk indicates a tribe containing genera that cannot yet be placed into a subtribe.

Opposite: ***Pinanga veitchii***, habit, Brunei Darussalam. (Photo: J. Dransfield)

Key to Tribes of Arecoideae

1. Leaflets praemorse at the tips, with several principal ribs diverging from the base or, sometimes, leaflets divided longitudinally into several 1-ribbed parts; inflorescence with a prophyll and 2 or more large peduncular bracts . . 2
1. Leaflets acute or bilobed, or if praemorse or longitudinally divided, the inflorescence with (0) 1 (2) large peduncular bracts (ignore small incomplete bracts) 3
2. Inflorescence rarely spicate, usually branched to 1–2 orders; flowers sessile, not borne in pits . . . . **Iriarteeae**
2. Inflorescence spicate; flowers borne in pits and exserted at anthesis on elongated receptacles . . . . . . **Podococceae**
3. Monoecious or dioecious; flowers solitary or in acervulae, very rarely in triads; peduncular bracts usually more than 2 . . . . **Chamaedoreeae**
3. Monoecious; flowers borne in triads of a central pistillate and two lateral staminate flowers, or as paired or solitary staminate flowers by abortion of pistillate flower, or as solitary pistillate flowers by abortion of staminate flowers; peduncular bracts rarely more than 2 4
4. Gynoecium with 3 or more locules and ovules; fruit never lobed; endocarp very hard, with 3 or more pores . . . . **Cocoseae**
4. Gynoecium with 1 or 3 locules and ovules; fruit sometimes lobed; endocarp various, if very hard, lacking clearly defined pores 5
5. Flowers always sunken in pits in the rachillae; petals of both staminate and pistillate flowers connate basally in a soft tube, the lobes valvate, styles elongate, conspicuous . . . . **Geonomateae**
5. Flowers usually superficial; petals of staminate flowers usually ± free, valvate; petals of pistillate flowers imbricate with minutely to conspicuously valvate tips 6
6. Crownshaft absent; fruits corky-warted 7
6. Crownshaft present or absent; fruits not corky-warted, or if corky-warted, then crownshaft present 8
7. Prophyll and peduncular bract net-like, both enclosing the entire inflorescence at anthesis . . . . . . **Manicarieae**
7. Prophyll and peduncular bract not net-like, not enclosing the inflorescence at anthesis . . . . **Pelagodoxeae**
8. Gynoecium triovulate 9
8. Gynoecium pseudomonomerous, rarely 2 abortive carpels present 12
9. Prophyll very much smaller than the peduncular bract(s), usually obscured by the leaf sheaths, peduncular bracts large, woody, usually beaked 10
9. Prophyll and peduncular bracts similar, or peduncular bract lacking altogether; peduncular bract, when present, membranous or coriaceous but scarcely woody, peduncular bract not conspicuously beaked 11
10. Endocarp with 3 clearly defined pores . . . . . **Cocoseae**
10. Endocarp lacking pores but with a heart-shaped basal button . . . . **Oranieae**
11. Moderate palms with stems obscured with reticulate or elongate leaf sheath fibres; inflorescence with numerous spreading branches of up to the 4th order; staminate flowers rounded with 6 stamens; fruit lenticular or ovoid, stigmatic remains basal . . . . **Leopoldinieae**
11. Diminutive to moderate palms with slender stems not obscured with sheath fibres; inflorescence spicate or with few approximate branches of 1, rarely 2 orders; staminate flowers pointed with 8–40 stamens; fruit ovoid to ellipsoid, stigmatic remains apical . . . . . **Reinhardtieae**
12. Acaulescent palms of W. Africa with spicate inflorescences hidden among leaves . . . . **Sclerospermeae**
12. Palms of New World and Old World but not Africa . . 13
13. New World Palms 14
13. Old World Palms. . . . . **Areceae**
14. Petals of pistillate flower connate basally for $^{1}/_{3}$ length; staminodes connate in a cupule adnate to the petals; fruit with nearly basal stigmatic remains . . . . . **Roystoneeae**
14. Petals of pistillate flower distinct; staminodes not connate in a cupule adnate to the petals; fruit with apical stigmatic remains . . . . **Euterpeae**

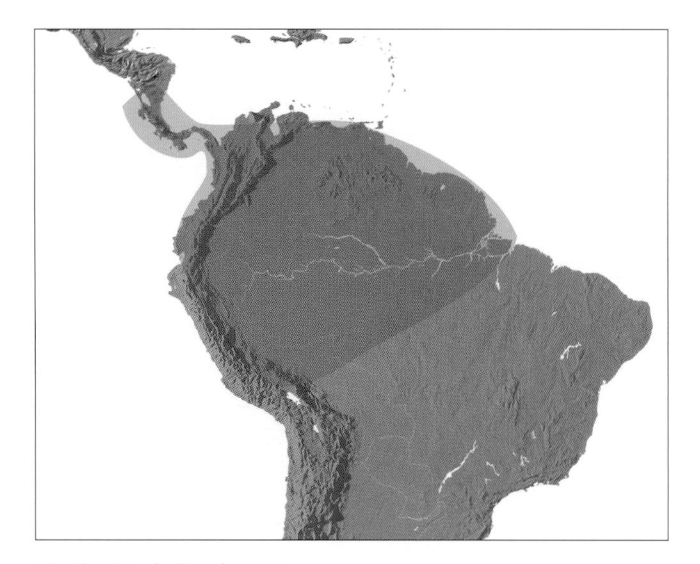

Distribution of tribe **Iriarteeae**

Tribe Iriarteeae Drude, Bot. Zeit. 35: 632 (1877). Type: **Iriartea**.

Slender to robust, erect, stilt-rooted, monoecious; leaves pinnate, with praemorse leaflets, the ribs and major veins radiating from the base, frequently longitudinally divided into many segments; sheaths forming a crownshaft; inflorescences sometimes multiple, bisexual or unisexual, spicate or branched to 1 or 2 orders only, bearing a prophyll and several usually large peduncular bracts; flowers superficial, borne in triads or triad derivatives; pistillate flowers with imbricate petals and valvate tips, or petals open from early in development; gynoecium 1–3 ovulate; fruit 1-seeded, or if more than 1-seeded, lobed.

This New World tribe includes five genera of stilt-rooted palms. In the first edition of *Genera Palmarum*, two subtribes were recognised: Iriarteinae Hook.f. and Wettiniinae Hook.f. More complete information has shown this to be unjustified. Members of the tribe are generally magnificent palms, distinctive in their stilt-roots and crownshafts, usually with relatively few leaves in the crown and praemorse leaflet. There is considerable diversity in floral and fruit morphology. These stilt-rooted palms are sometimes erroneously thought to have the ability to 'walk' along the forest floor (see Chapter 1).

Tribe Iriarteeae is resolved as monophyletic in all studies (e.g., Lewis & Doyle 2001, Hahn 2002a, 2002b, Baker *et al.* in prep.). Asmussen *et al.* (2006) and Baker *et al.* (in review) included all genera of Iriarteeae and found high bootstrap support for the tribe's monophyly. The placement of the Iriarteeae with respect to other Arecoid tribes is not clear and two alternative hypotheses have been recovered. Asmussen *et al.* (2006) and Baker *et al.* (in review) resolve the tribe as sister to the rest of the Arecoideae with moderate support, whereas Lewis and Doyle (2002), Loo *et al.* (2006) and Baker *et al.* (in prep.) resolve the Iriarteeae as sister to the core arecoid clade (Areceae, Euterpeae, Geonomateae, Leopoldinieae, Manicarieae and Pelagodoxeae) with low to moderate bootstrap support.

Key to the Genera of the Iriarteeae

1. Staminate and pistillate flowers borne on the same inflorescence in triads, or towards the tips of the rachillae, the staminate paired or solitary; inflorescences always solitary at a node 2
1. Staminate and pistillate flowers borne on separate inflorescences (pistillate flowers sometimes accompanied by abortive lateral staminate flowers), these usually several at a node, the central pistillate, the lateral staminate . . . . **81. Wettinia**
2. Stigmatic residue apical or subapical in fruit; staminate flowers with 9–100 or more stamens 3
2. Stigmatic residue near or at the base in fruit; staminate flowers with 6 stamens 4
3. Inflorescence terete and decurved in bud, the numerous (10 or more) bracts falling from the terete peduncle as the inflorescence expands, the rachillae long, slender, laxly pendulous; staminate flowers ± symmetrical, closed in bud, with imbricate sepals and 9–20 stamens; seed with lateral embryo; stilt roots mostly slender, sparsely prickly, forming a dense cone obscuring the central stem. Costa Rica to Bolivia . . . . **79. Iriartea**
3. Inflorescences somewhat dorsi-ventrally compressed and erect in bud, the 4–7 bracts splitting abaxially, subpersistent and erect on the dorsiventrally compressed peduncle, which becomes reflexed at anthesis; rachillae short, stout, rather stiffly spreading at anthesis, pendulous in fruit; staminate flowers ± asymmetrical, or at least angled by close packing, ± open in bud, sepals usually united basally at least briefly, stamens 20–100 or more; seed with apical or excentrically apical embryo; stilt roots stout, usually densely prickly forming an open supporting cone. Costa Rica to Bolivia . . . . **80. Socratea**
4. Slender undergrowth palmlets of lowland forest. Leaflets borne in one plane, undivided; inflorescence interfoliar, or becoming infrafoliar in fruit, erect, branching to 1 order only, long pedunculate, the bracts ± persistent; fruit small, ellipsoid or obovoid; seed with apical embryo. Guiana, Amazonian Peru . . . . **77. Iriartella**
4. Robust canopy palms of upland or montane forest; leaflets, at least in mid-leaf, divided into several segments displayed in different planes; inflorescences infrafoliar, erect or decurved in bud, branching to 2 orders; the bracts deciduous; fruit mostly globose; seed with basal embryo. Panama and northern South America, Andean . . . . **78. Dictyocaryum**

77. IRIARTELLA

Solitary or clustered, small pinnate-leaved palms of the forest undergrowth in the Amazon basin, distinctive in the undivided praemorse rhomboid leaflets, inflorescence branched to one order only and seed with apical embryo.

Iriartella H. Wendl., Bonplandia 8: 103, 106 (1860). Type: **I. setigera** (Mart.) H. Wendl. (*Iriartea setigera* Mart.).

Cuatrecasea Dugand, Rev. Acad. Colomb. Ci. Exact. 3(12): 392 (1940). Type: *C. vaupesana* Dugand (= *Iriartella setigera* [Mart.] H. Wendl.).

Combines palm generic name Iriartea *with the dimitutive suffix* -ella: *little* Iriartea.

Usually clustered, slender, lightly armed, pleonanthic, monoecious palms. *Stem* erect, conspicuously ringed with leaf scars, stilt roots well developed at the base, internodes densely covered in scales, hairs, and sometimes sharp black bristles, becoming smooth in age. *Leaves* few, pinnate, neatly abscising; sheaths forming a crownshaft, sparsely to densely armed with solitary or clustered black bristles and abundant scales and hairs, sometimes with a short ligule; petiole well developed, ± rounded in cross-section; armed like the sheath; rachis adaxially

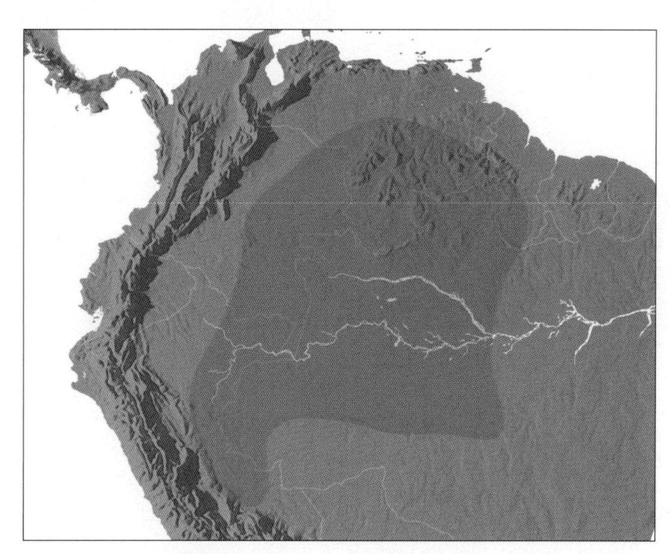

Distribution of ***Iriartella***

angled, abaxially flattened, densely hairy; leaflets regularly arranged, distant, rhombic to trapezoidal, upper margins irregularly lobed and praemorse, distal pair of leaflets truncate, joined along the rachis, ribs numerous, conspicuous abaxially, ± parallel, scaly or rough hairy abaxially, sometimes also adaxially, transverse veinlets not evident. *Inflorescences* solitary, interfoliar at first, becoming infrafoliar in fruit, branching to 1 order, protandrous; peduncle elongate, flattened, elliptic in cross-section; prophyll inserted near the base, tubular, 2-keeled, membranous, included in the leaf sheath, eventually disintegrating; peduncular bracts 3–5, exceeding the prophyll, similar but not 2-keeled, the proximal included within the leaf sheath, the distal exposed; rachis shorter than the peduncle, bearing minute triangular, incomplete, spirally arranged bracts each subtending a rachilla; rachillae slender, 3 to ca. 30, short to moderately long, bearing spirally arranged, close, slightly sunken triads throughout except at the very tip where bearing paired or solitary staminate flowers, rachilla bracts not evident. *Staminate flowers* borne in a close pair distal to the pistillate, symmetrical; sepals 3, rounded, keeled, distinct, imbricate or basally connate, both connate and distinct sepals found in different collections of *Iriartella setigera*; petals 3, distinct, oblong, valvate, about 3 times the length of the sepals; stamens 6, filaments very short, broad, fleshy, anthers oblong, basifixed, latrorse; pistillode absent. *Pollen* ellipsoidal, bi-

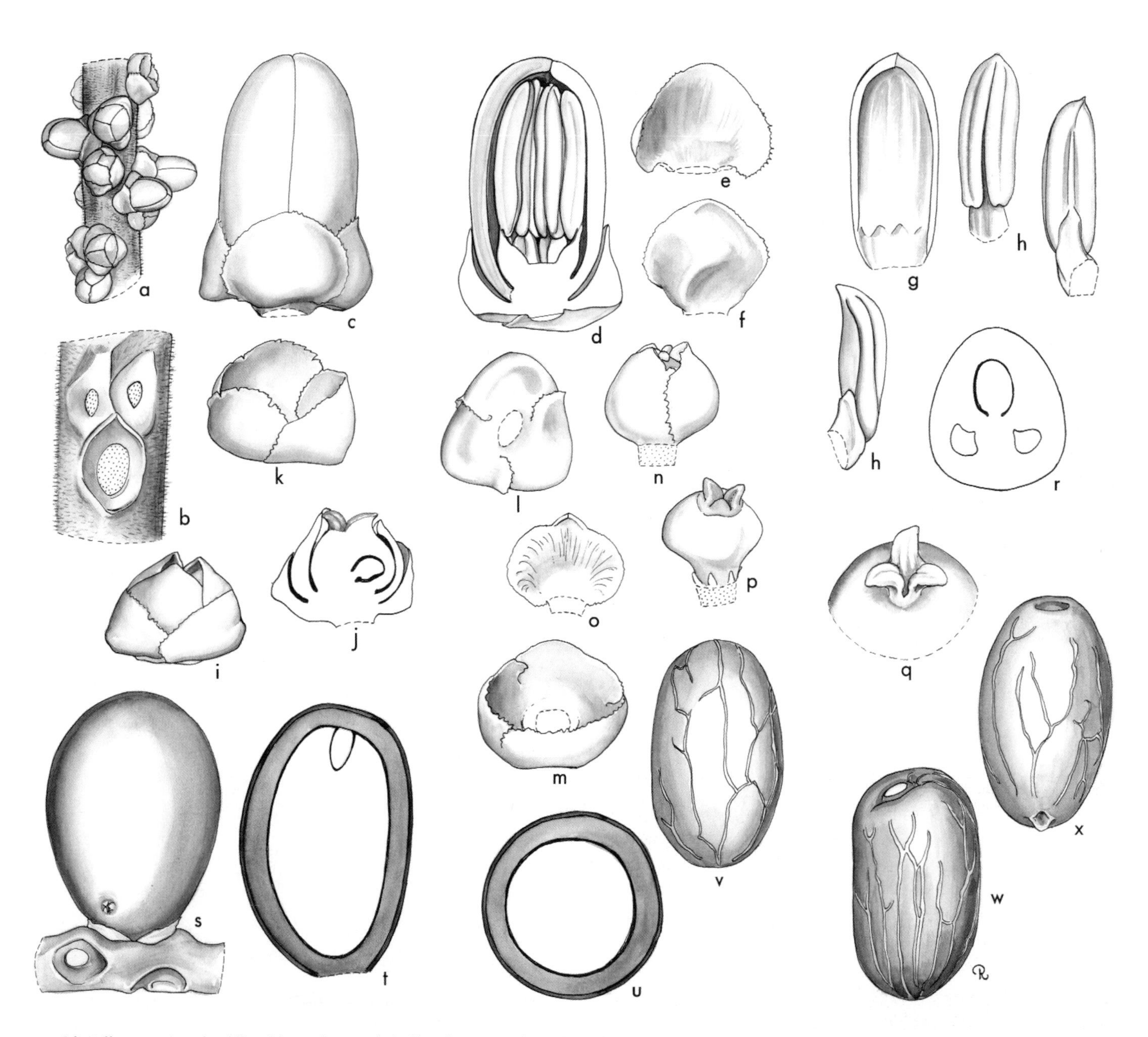

Iriartella. **a**, portion of rachilla with staminate and pistillate flowers × 3; **b**, triad with flowers removed × 6; **c**, staminate bud × 12; **d**, staminate bud in vertical section × 12; **e, f**, staminate sepal, interior (e) and exterior (f) views × 12; **g**, staminate petal, interior view × 12; **h**, stamen in 3 views × 12; **i**, pistillate flower × 12; **j**, pistillate flower in vertical section × 12; **k, l, m**, pistillate calyx in 3 views × 12; **n**, pistillate flower, calyx removed × 12; **o**, pistillate petal, interior view × 12; **p**, gynoecium and staminodes × 12; **q**, stigmas × 18; **r**, ovary in cross-section × 24; **s**, fruit × 3; **t**, fruit in vertical section × 3; **u**, fruit in cross-section × 3; **v, w, x**, seed in 3 views × 3. *Iriartella setigera*: **a–h**, *Moore et al.* 9501; **i–r**, *Schultes & Cabrera* 13729; **s–x**, *Moore et al.* 9533. (Drawn by Marion Ruff Sheehan)

symmetric; aperture comprising two parallel distal sulci, narrowly separated by an ectexinous bridge; ectexine, tectate, coarsely perforate, tectum between sulci and aperture margins similar or slightly less perforate; infratectum columellate; longest axis 22–26 µm; post-meiotic tetrads tetrahedral [2/2]. *Pistillate flowers* smaller than the staminate; sepals 3, distinct, broad, imbricate or basally connate, splitting into 3 in fruit; petals 3, imbricate basally with short triangular, valvate tips; staminodes 6, minute, tooth-like; ovary globular, trilocular, triovulate, tipped with 3 short, recurved stigmas, ovule form unknown. *Fruit* usually developing from 1 carpel, scarlet, orange, or brownish, ellipsoidal, stigmatic and carpellary remains basal; epicarp smooth, mesocarp slightly fleshy with few longitudinal fibres, endocarp thin. *Seed* ellipsoidal, attached basally, with loosely branched raphe and tannin network, endosperm homogeneous; embryo apical. *Germination* adjacent-ligular; eophyll shallowly bifid. *Cytology* unknown.

Distribution and ecology: Two species limited to the Amazonian drainage of Peru, Colombia, Brazil, Venezuela and Guyana; undergrowth palms of lowland tropical rain forests below 1000 m elevation.
Anatomy: Not studied.
Relationships: The mononophyly of *Iriartella* has not been tested. The relationships among the genera of Iriarteeae have been investigated in three studies that include all genera (Henderson 1990, Asmussen *et al.* 2006, Baker *et al.* in review) to which readers are referred. There is no agreement between the results of these studies. Only one sister relationship, between *Iriartella* and *Wettinia*, is supported by bootstrap analysis (Asmussen *et al.* 2006). New analyses are required to elucidate this group.
Common names and uses: *Palma de cerpatana* (*Iriartella setigera*). An infusion of leaf bases is used medicinally. Stems are hollowed out and used for the exterior tube of blow guns.
Taxonomic account: Henderson (1990).
Fossil record: No generic records found.
Notes: Similarities to *Podococcus* are of a superficial nature only.
Additional figures: Glossary fig. 24.

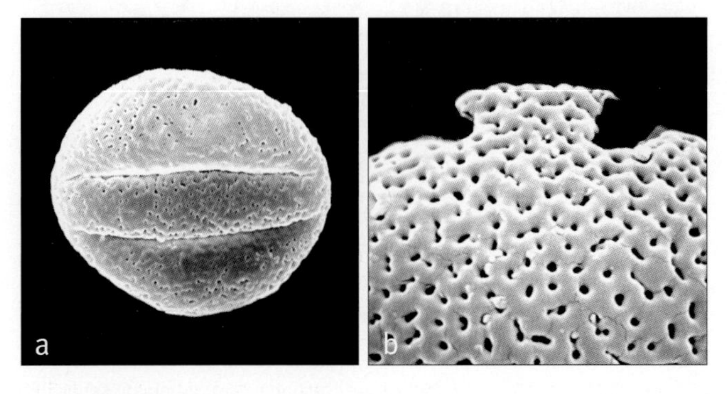

Above: ***Iriartella***
a, distal disulcate pollen grain, distal face SEM × 2000;
b, close-up, paired sulcus apices with unmodified exine between SEM × 6000.
Iriartella setigera: **a**, *Moore et al.* 9501; *I. stenocarpa*: **b**, *Moore et al.* 8367.

Above left: ***Iriartella stenocarpa***, crown, Peru. (Photo: J.-C. Pintaud)
Above right: ***Iriartella setigera*** leaves and infructescence, Peru. (Photo: J. Dransfield)

78. DICTYOCARYUM

Usually robust solitary or clustered pinnate-leaved tree palms of the Andes and foothills, the stems with stilt roots; leaflets are rhomboid, praemorse and are usually longitudinally divided to give the whole leaf an ostrich-feather appearance; fruit has basal stigmatic remains and embryo.

Dictyocaryum H. Wendl., Bonplandia 8: 106 (1860). Lectotype: *D. lamarckianum* (Mart.) H. Wendl. (*Iriartea lamarckiana* Mart.) (see H.E. Moore 1963c).
Dahlgrenia Steyerm., Fieldiana, Bot. 28: 82 (1951). Type: *D. ptariana* Steyerm. (= *Dictyocaryum fuscum* (H. Karst.) H. Wendl. [*Socratea fusca* H. Karst.]).

Dictyon — *a net*, karyon — *nut, referring to the net-like branching pattern of the raphe on the surface of the seed.*

Solitary, or very rarely clustered, moderate to robust, unarmed, pleonanthic, monoecious tree palms. *Stem* erect, slightly or rarely markedly ventricose, conspicuously ringed with leaf scars, with stilt roots bearing short somewhat sharp or cylindrical, lateral roots. *Leaves* few, pinnate, neatly abscising; sheath forming a conspicuous crownshaft, bearing scattered small scales (?always); petiole short or very short, adaxially channelled at the base, rounded or angled distally, rounded abaxially, sometimes densely tomentose; rachis angled to convex adaxially, rounded abaxially; leaflets massive with numerous ribs, longitudinally divided between the ribs to the base into narrow segments displayed in many planes giving the whole leaf a dense plumose appearance, each segment praemorse at the apex, blade strongly discolorous, abaxially green or densely covered in white indumentum and abundant unbranched hairs and/or dot-like scales, usually 1 large rib per segment, transverse veinlets not evident. *Inflorescences* solitary, infrafoliar, erect or pendulous and curved, branched to 2 orders, in bud sometimes horn-shaped, protandrous; peduncle winged or not at the base, elongate, rounded in cross-section, massive; prophyll short, 2-keeled, tubular, soon opening at the tip, eventually shed, tomentose; peduncular bracts up to 9, tubular with pointed tips, completely sheathing at first, then splitting apically to allow elongation of the peduncle, proximal few rather short, middle to distal much longer, conspicuously beaked, all shed at anthesis, prophyll and peduncular bracts coriaceous to woody; rachis ± equalling to much longer than the peduncle; rachis bracts spirally arranged, triangular, proximally conspicuous; first-order branches spreading, swollen at the base with a long bare portion, the proximal bearing about 3–4 rachillae, distal unbranched; rachillae slender, elongate, flexuous, very numerous, bearing rather distant, spirally arranged triads proximally, paired and solitary staminate flowers distally. *Staminate flowers* fleshy, sessile, ± symmetrical; sepals 3, distinct, imbricate, rounded, strongly gibbous basally; petals much longer than the sepals, 3, slightly connate at the base, ± lanceolate, valvate; stamens 6, filaments short, broad, fleshy, anthers elongate, basifixed, latrorse; pistillode short, broad, columnar, rounded or minutely trifid at the apex. *Pollen* ellipsoidal, ± bi-symmetric; aperture a distal sulcus; ectexine intectate, coarsely granular to gemmate, granulae/gemmae often coalesced into larger irregular units, aperture margin similar; longest axis 24–30 µm [3/3]. *Pistillate flowers* smaller than the staminate, sessile; sepals 3, distinct, rounded, imbricate, thick; petals ca. 3 times as long as the sepals, ± triangular, imbricate; staminodes 6, minute, strap-like or tooth-like; gynoecium tricarpellate, triovulate, rounded, tipped with 3 low stigmas, ovule probably anatropous. *Fruit* developing from 1 carpel, globose or ellipsoidal, with basal carpel and stigmatic remains; epicarp smooth, usually yellow at maturity, dark brown when dry, mesocarp thick with outer layer of sclereids and inner layer of tannin and fibres, endocarp very thin, scarcely differentiated. *Seed* spherical, basally attached, seed coat thick with a conspicuous network of raphe fibres, hilum rounded, endosperm homogeneous; embryo basal. *Germination* adjacent-ligular; eophyll bifid. *Cytology* not studied.

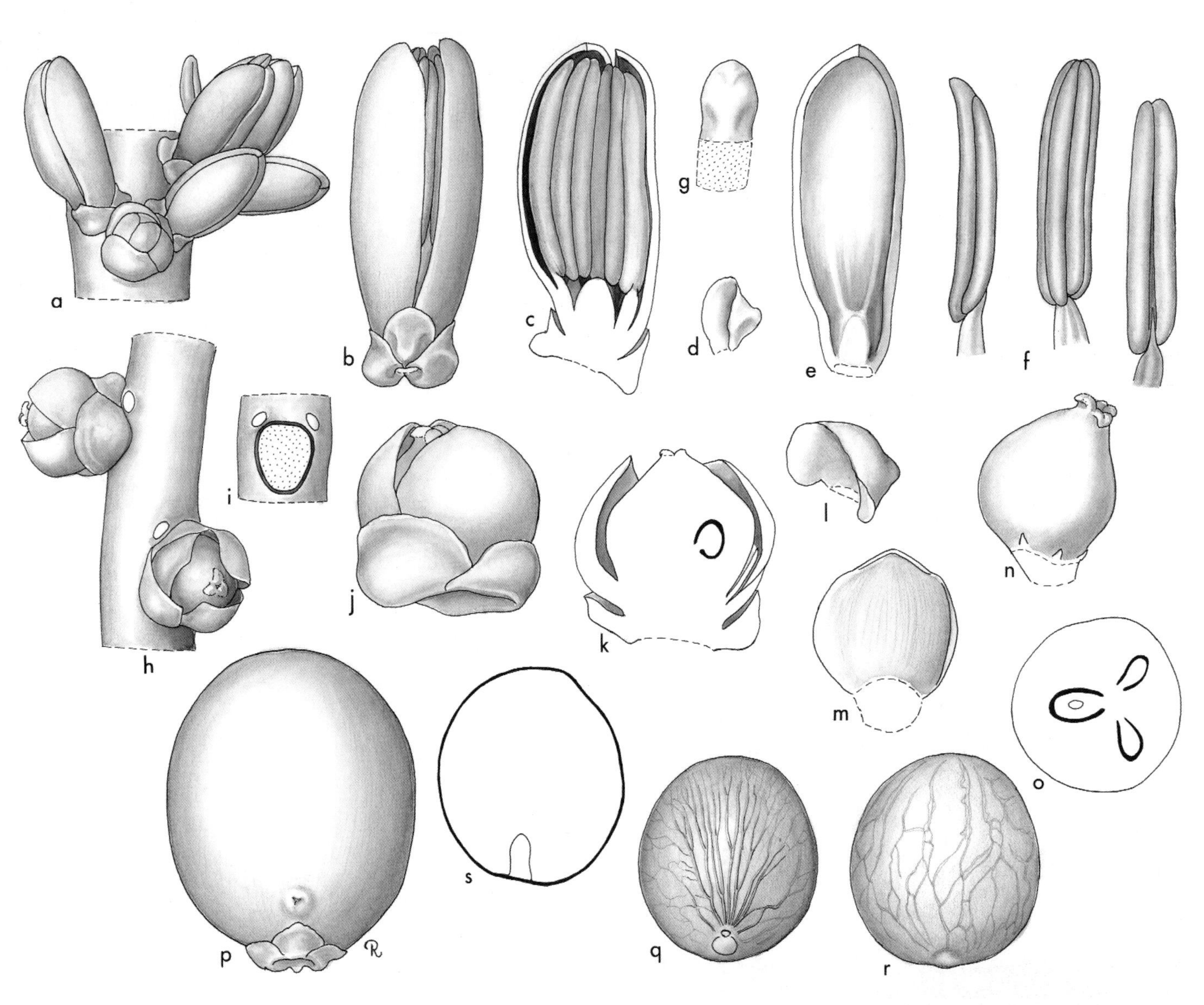

***Dictyocaryum*.** **a**, portion of rachilla with triad at staminate anthesis × 3; **b**, staminate flower × 6; **c**, staminate flower in vertical section × 6; **d**, staminate sepal × 6; **e**, staminate petal × 6; **f**, stamen in 3 views × 6; **g**, pistillode × 9; **h**, portion of rachilla at pistillate anthesis × 4 1/2; **i**, triad with scars of staminate and pistillate flowers × 4 1/2; **j**, pistillate flower × 7 1/2; **k**, pistillate flower in vertical section × 7 1/2; **l**, pistillate sepal × 7 1/2; **m**, pistillate petal, interior view × 7 1/2; **n**, gynoecium and staminodes × 7 1/2; **o**, ovary in cross-section × 9; **p**, fruit × 1 1/2; **q, r**, seed in 2 views × 1 1/2; **s**, seed in vertical section × 1 1/2. *Dictocaryum* sp.: **a–g**, *Moore et al.* 9846; *D. fuscum*: **h–s**, *Steyermark* 91502. (Drawn by Marion Ruff Sheehan)

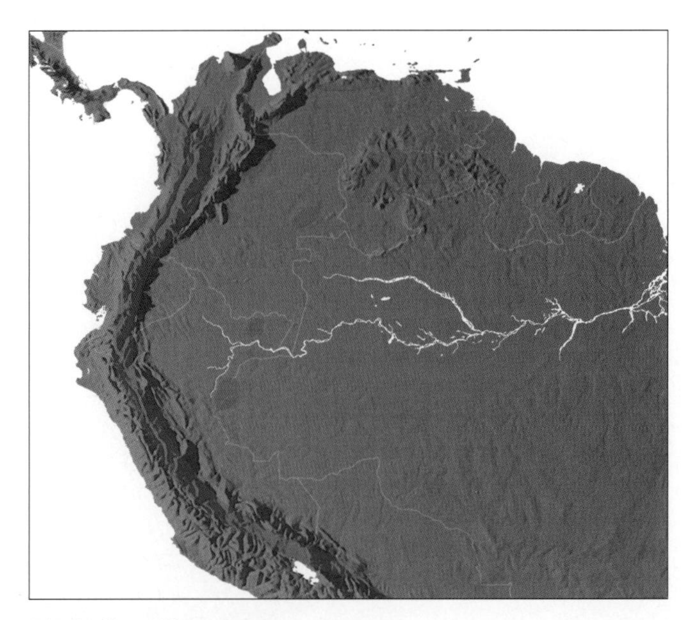

Distribution of ***Dictyocaryum***

Top: ***Dictyocaryum lamarckianun***, crown, Ecuador. (Photo: F. Borchsenius)

Above: ***Dictyocaryum ptarianum***, stilt-roots, Peru. (Photo: J. Dransfield)

Distribution and ecology: Three species described in Colombia, Ecuador, Peru, Bolivia, Brazil, Venezuela, Guyana and Panama. Usually in montane rain forest at medium elevations, on very steep slopes, often occurring in great numbers, and forming a conspicuous component of the forest canopy. *Dictyocaryum ptarianum* also rarely grows in the lowlands of the Amazon basin.

Anatomy: Leaf and seed (Roth 1990), and root (Seubert 1998a, 1998b).

Relationships: The monophyly of *Dictyocaryum* has not been tested. For relationships, see *Iriartella*.

Common names and uses: *Araque*, *palma real*. Specific uses have not been recorded.

Taxonomic account: Henderson (1990).

Fossil record: Small monosulcate grains, *Palmaemargosulcites insulatus*, from palm flower compression fossils, recovered from the Middle Eocene oil shales of Messel, Germany, are compared with pollen of *Dictyocaryum* and *Dypsis* (Harley 1997).

Notes: There are two very different inflorescence habits within the genus; the inflorescence may be more-or-less erect with wide spreading rachillae, or pendulous and curved in bud with pendulous rachillae.

Additional figures: Glossary fig. 20.

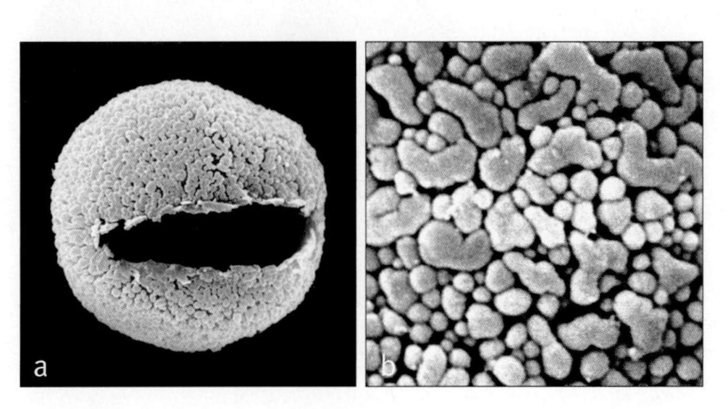

Left: ***Dictyocaryum***

a, monosulcate pollen grain, distal face SEM × 2000;
b, close-up, coarsely granular/gemmate pollen surface interspersed with larger units of coalesced granulae/gemmae SEM × 8000.
Dictyocaryum ptarianum: **a** & **b**, *Steyermark* 94146.

79. IRIARTEA

Robust solitary pinnate-leaved palm of humid rain forest in Central and South America, the stems with a dense cone of stilt roots and sometimes ventricose; leaflets are rhomboid, praemorse, longitudinally divided to give the whole leaf an ostrich-feather appearance; fruit has apical stigma remains and a lateral embryo.

Iriartea Ruiz & Pav., Fl. peruv. prodr. 149 (1794). Type: **I. deltoidea** Ruiz & Pav.

Deckeria H. Karst., Linnaea 28: 258 (1857) (non *Deckera* Schultz [1834]). Lectotype: *D. corneto* H. Karst. (= *Iriartea corneto* [H. Karst.] H. Wendl. = *I. deltoidea* Ruiz & Pav.) (see H.E. Moore 1963c).

Commemorates Bernardo de Iriarte (1735–1814), Spanish diplomat.

Solitary, robust, often very tall, unarmed, pleonanthic, monoecious tree palm. *Stem* erect, ± bellied, conspicuously ringed with leaf scars, bearing slender stilt roots forming a dense cone obscuring the stem base. *Leaves* rather few in number, pinnate, neatly abcising; sheaths forming a well-defined crownshaft; petiole rather short, adaxially channelled, abaxially rounded; rachis adaxially angled, abaxially rounded; leaflets large, asymmetrically deltoid to elliptic, the proximal margin entire for ca. $^1/_3$ its

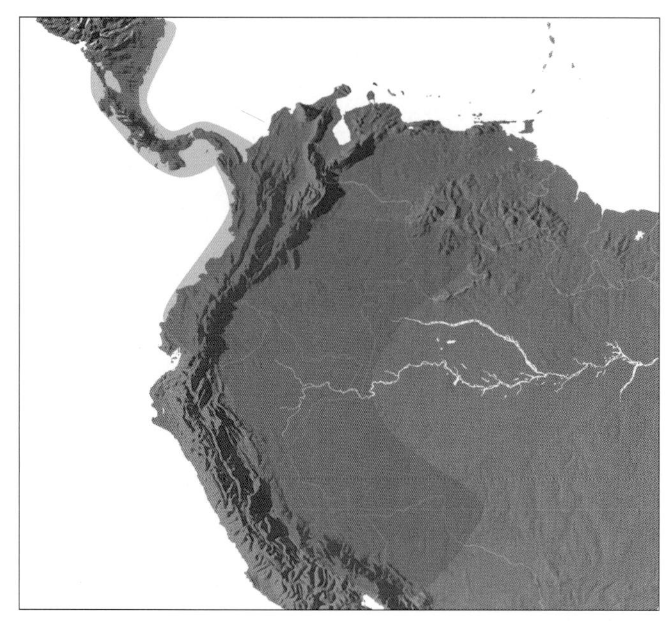

Distribution of ***Iriartea***

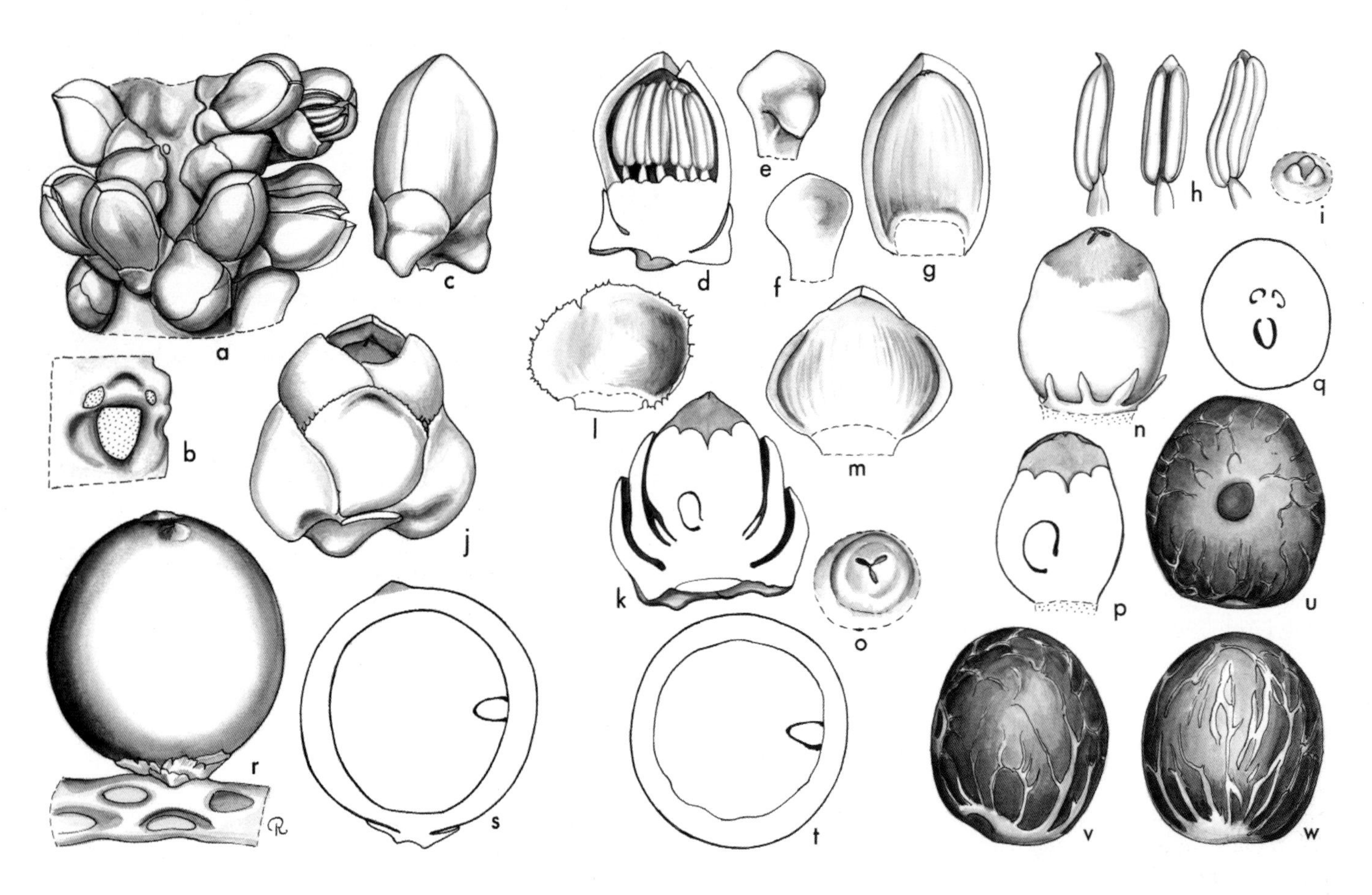

Iriartea. **a**, portion of rachilla at staminate anthesis × 3; **b**, triad, flowers removed × 6; **c**, staminate bud × 6; **d**, staminate bud in vertical section × 6; **e**, staminate sepal, exterior view from below × 6; **f**, staminate sepal, interior view × 6; **g**, staminate petal, interior view × 6; **h**, stamen in 3 views × 6; **i**, pistillode × 6; **j**, pistillate flower × 6; **k**, pistillate flower in vertical section × 6; **l**, pistillate sepal × 6; **m**, pistillate petal, interior view × 6; **n**, gynoecium and staminodes × 6; **o**, stigmas × 6; **p**, gynoecium in vertical section × 6; **q**, ovary in cross-section × 6; **r**, fruit × $1^1/_2$; **s**, fruit in vertical section × $1^1/_2$; **t**, fruit in cross-section × $1^1/_2$; **u**, **v**, **w**, seed in 3 views × $1^1/_2$. *Iriartea deltoidea*: **a–i** and **r–w**, *Moore & Parthasarathy* 9491; **j–q**, *Moore* 6524. (Drawn by Marion Ruff Sheehan)

length then praemorsely toothed, the distal margin entire for a shorter distance, then praemorsely toothed, ribs conspicuous sometimes with scaly margins, the main ribs diverging from the base to the margin, the whole leaflet usually irregularly split into linear segments displayed in different planes giving the leaf a plumose appearance, transverse veinlets not evident. *Inflorescences* solitary, infrafoliar, pendulous, strongly curved in bud, branching to 1 order distally to 2 orders proximally, protandrous; peduncle massive, ± circular in cross-section; prophyll short, tubular, 2-keeled, apically open; peduncular bracts 8–12, spirally arranged, tubular, the proximal several short, soon splitting, the distal very long, tubular, enclosing the inflorescence, all bracts variously hairy, eventually deciduous, leaving conspicuous, close annular scars; rachis equalling or slightly longer than the peduncle, bearing spirally arranged, minute, collar-like bracts; first-order branches digitately branched proximally, unbranched distally, bases of branches swollen; rachillae very long, moderately robust, bearing spirally arranged, slightly sunken, close triads throughout their length except at tips where bearing solitary or paired staminate flowers; rachilla bracts and floral bracteoles not evident. *Staminate flowers* ± symmetrical; sepals 3, distinct, gibbous, rounded, imbricate, bearing deciduous, bristle-like hairs; petals 3, 3–4 times longer than the sepals, valvate, ± boat-shaped and curved, the tips rounded to acute; stamens 9–20, filaments very short, slender, anthers elongate, acute to mucronate apically, latrorse; pistillode minute or lacking. *Pollen* ellipsoidal, ± bi-symmetric; aperture a distal sulcus; ectexine intectate, closely to densely gemmate, gemmae often coalesced into larger units, sometimes with large well-defined gemmae, surrounded by smaller gemmae, aperture margin similar; longest axis 31–35 µm [1/1]. *Pistillate flowers* smaller than the staminate; sepals 3, distinct, broadly imbricate; petals 3, distinct, broad, rounded, imbricate except at the triangular valvate tips; staminodes to 12, very small, tooth-like; gynoecium globose, trilocular, triovulate, stigmas 3, low, only 1 ovule normally maturing, basally attached, form unknown. *Fruit* mostly globose, yellow when ripe, stigmatic remains apical; epicarp smooth, mesocarp granular and fibrous, endocarp very thin. *Seed* globose, basally attached, hilum circular, raphe branches coarse anastomosing, endosperm homogeneous; embryo lateral. *Germination* adjacent-ligular; eophyll praemorse, undivided. *Cytology*: 2n = 32.

Distribution and ecology: A single species, distributed from Costa Rica and Nicaragua southwards to Colombia, Ecuador, Peru, Bolivia, Venezuela and Brazil. Frequently gregarious in lowland tropical rain forest but reaching 1200 m, often as a distinct component of the forest canopy. Pollination is by bees (Henderson 1985).

Anatomy: Leaf (Tomlinson 1961, Roth 1990), root (Seubert 1998a, 1998b, Avalos 2004), stamen development (Uhl & Moore 1980), gynoecium (Uhl & Moore 1971), seed (Roth 1990).

Relationships: For relationships, see *Iriartella*.

Common names and uses: Stilt palm, horn palm. The outer part of the trunk is extremely hard and durable and is used in the construction of dwellings and in making spears. Wallace (1853) records the use of swollen sections of the trunk as canoes (see also Johnson [1998]). Henderson *et al.* (1995) recorded the use of stems for coffins in the Choco region of Colombia. For medicinal uses, see Plotkin and Balick (1984).

Taxonomic account: Henderson (1990).

Fossil record: Pinnate leaf fragments from the Miocene of Peru are described as *Iriartites tumbezensis* (Berry 1919b), although the author comments that this is, "…a convenient form-genus for the remains of fossil palms that appear to belong to tribe Iriarteeae, but whose exact generic identity is uncertain." Furthermore, material described as *Sabalites* from the Tertiary of Venezuela (Berry 1921b) resembles a leaf of *Iriartea*, but these fossils need to be restudied. Gardner (1882) recovered an abundance of pinnate palm leaves from the Middle Eocene Bournemouth Freshwater Beds, which he considered to resemble *Iriartea* more than any other genus (Chandler 1963). A 'palm nut' from the Miocene of the Panama Canal Zone (Gatun) is described as being, "very close to the endocarp of *Iriartea*" (Berry 1921a) and a seed (*Iriartea collazoënsis*) recorded from the Middle Oligocene of Puerto Rico is considered to resemble closely those of *Arenga* and

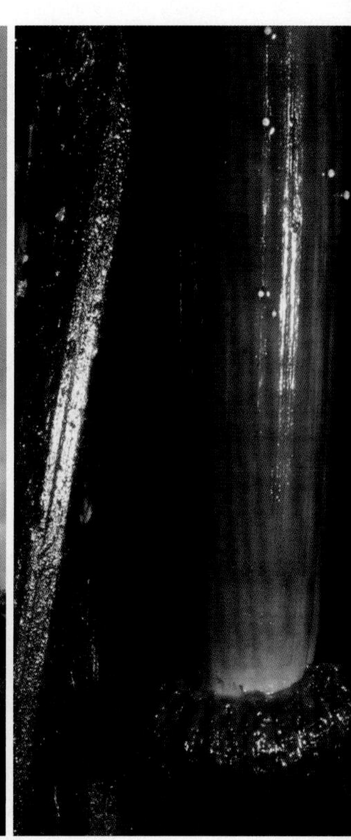

Above left: ***Iriartea deltoidea***, habit, Ecuador. (Photo: F. Borchsenius)
Above right: ***Iriartea deltoidea***, stilt-roots, Peru. (Photo: J.-C. Pintaud)

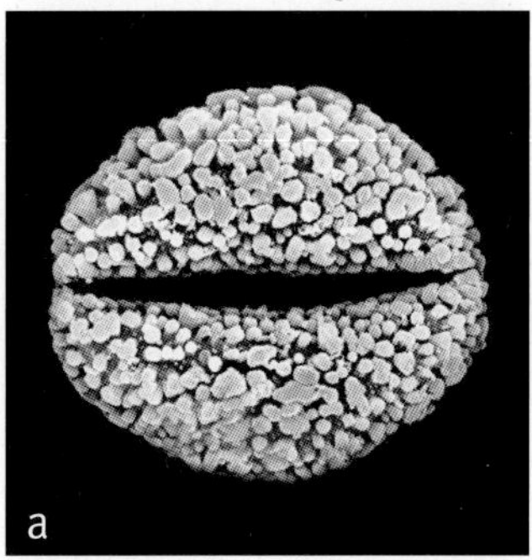

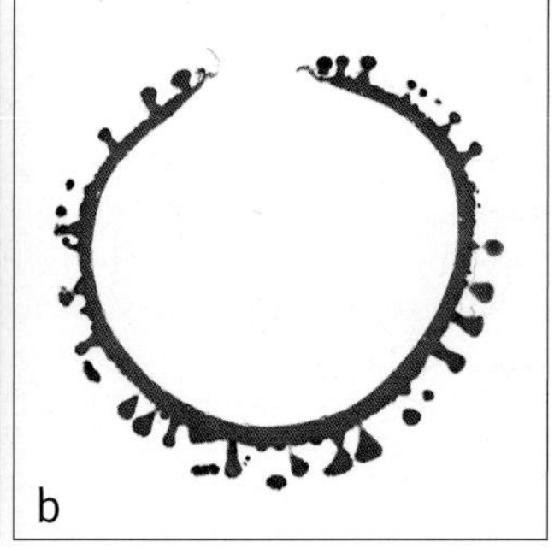

Above: ***Iriartea***
a, monosulcate pollen grain, distal view SEM × 1500;
b, ultrathin section through whole pollen grain, polar plane TEM × 1500.
Iriartea deltoidea: **a**, *Steyermark* 90777; **b**, *Moore* 9865.

Iriartea (Hollick 1928). The anatomy of a large specimen of stem wood, *Palmoxylon iriarteum*, from the West Indies (Antigua), in the collection of the Naturhistoriska Riksmuseet, Stockholm, was described in detail by Stenzel (1897) who considered it ancestral to *Iriartea*; no age is given. Comparisons of palm stem wood or root to generic level should always be viewed with caution.

Notes: *Iriartea* is distinguished from *Socratea* by the dense rather than open cone of stilt roots, the closed staminate flowers with fewer stamens, the clavate rather than spinose pollen, and the very different pollination syndrome described by Henderson (1985).

Additional figures: Glossary fig. 18.

80. SOCRATEA

Moderate solitary (very rarely clustered) pinnate-leaved tree palms of humid rain forest in Central and South America, the stems with an open cone of stilt roots; leaflets are rhomboid praemorse, sometimes longitudinally divided to give the whole leaf an ostrich-feather appearance; fruit has apical stigmatic remains and embryo.

Socratea H. Karst., Linnaea 28: 263 (1857) ('1856'). Lectotype: *S. orbigniana* (Martius) Karsten (*Iriartea orbigniana* Mart.) (see H.E. Moore 1963c) (= *S. exorrhiza*).

Metasocratea Dugand, Rev. Acad. Colomb. Ci. Exact. 8: 389 (1951). Type: *M. hecatonandra* Dugand (= *Socratea hecatonandra* [Dugand] R. Bernal).

Commemorates the great Athenian philosopher Socrates (ca. 470–399 BC).

Solitary or very rarely clustered, moderate, pleonanthic, monoecious tree palms. *Stems* erect, conspicuously ringed with leaf scars, bearing an open cone of stout, usually densely prickly, stilt roots. *Leaves* rather few, pinnate, neatly abcising; sheaths tubular forming a well-defined crownshaft; petiole short, adaxially channelled or flattened, abaxially rounded, bearing a variety of indumentum types; rachis adaxially angled, abaxially rounded; leaflets regularly arranged, asymmetrically deltoid to elliptic, proximal margin entire for much of its length, distal margin entire ca. $^1/_3$ its length, otherwise praemorse, main ribs numerous, radiating from the base, leaflet remaining entire or splitting longitudinally between the ribs into narrow segments displayed in different planes giving the leaf a plumose appearance. *Inflorescences* solitary, infrafoliar, somewhat dorsiventrally compressed and erect in bud, branching to 1 order, branches pendulous when exposed, protogynous; peduncle well developed, elliptic in cross-section, winged at base; prophyll inserted near the base, short, tubular, 2-keeled, apically open, thinly coriaceous; peduncular bracts ca. 5, tubular, tips pointed, central ones larger than proximal or distal, ± flattened, eventually deciduous after anthesis; rachis ± flattened, shorter or longer than the peduncle, bearing spirally arranged, pendulous rachillae, each subtended by a minute collar-like bract; rachillae rather robust, often somewhat flattened, elongate, bearing spirally arranged, crowded triads; rachilla bracts and bracteoles scarcely evident; staminate and pistillate flowers maturing at the same time. *Staminate flowers* open in bud, sepals 3, triangular, united basally in a low, complete or partially interrupted ring; petals 3, distinct, fleshy,

Top right: ***Socratea hecatonandra***, habit, Colombia (Photo: A.J. Henderson)

Bottom right: ***Socratea exorrhiza***, stilt roots, cultivated, Fairchild Tropical Botanic Garden, Florida (Photo: J. Dransfield)

markedly asymmetrical, lightly imbricate basally, much longer than the sepals; stamens 17–145, filaments very short, awl-shaped above expanded bases, anthers erect, basifixed, linear, acute or apiculate, latrorse; pistillode much shorter than the stamens, conical, briefly trifid. *Pollen* ellipsoidal, ± bi-symmetric; aperture a distal sulcus; ectexine intectate, upper surface of foot layer covered by fine, dense gemmae or clavae, loosely supporting short, wide-spaced, broad-based spines; longest axis 43–50 µm [2/5]. *Pistillate flowers* symmetrical, much smaller than the staminate, ± 3-angled; sepals 3, rounded, strongly imbricate, dorsally thickened; petals 3, distinct, strongly imbricate, ± rounded with a minute, triangular valvate apex; staminodes 6, minute, tooth-like; gynoecium obovoid, tricarpellate, triovulate, stigmas 3, apical, fleshy, reflexed, ovules basally attached, orthotropous, one usually larger than the others. *Fruit* separated at maturity, ellipsoidal to subglobose with eccentrically apical stigmatic remains; epicarp minutely roughened when dry, at maturity splitting into ± distinct valves at apex, exposing the rather dry white mesocarp with included reddish sclerosomes and slender fibres, endocarp thin. *Seed* ± ovoid, basally attached, hilum circular, raphe branches conspicuous, numerous, sparsely anastomosing, endosperm homogeneous; embryo eccentrically apical. *Germination* adjacent-ligular; eophyll bifid with praemorse tips. *Cytology*: 2n = 36.

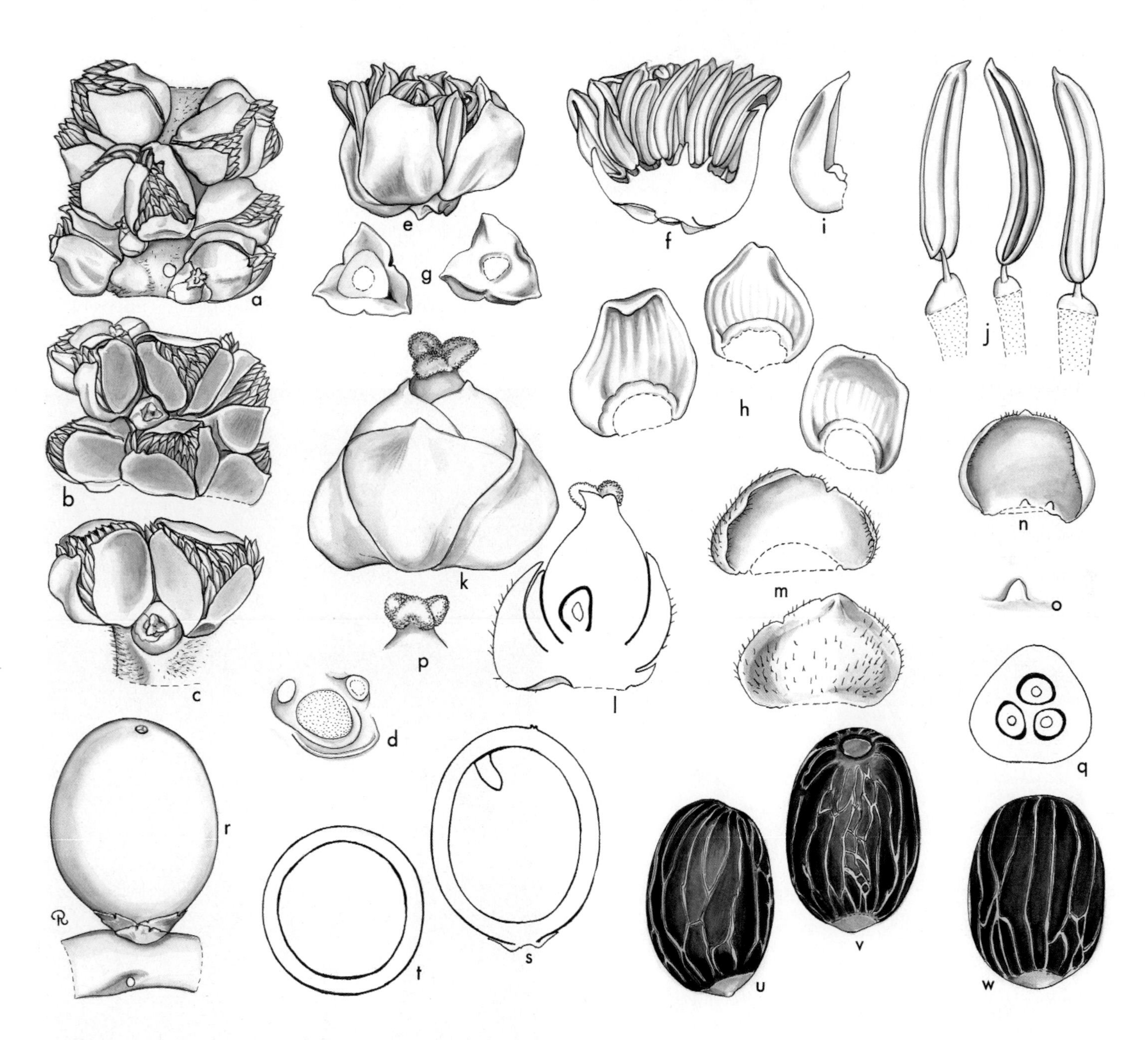

Socratea. **a**, **b**, portions of rachilla at anthesis × 1$^1/_2$; **c**, triad at anthesis × 2$^1/_4$; **d**, triad, flowers removed × 2$^1/_4$; **e**, staminate flower × 3; **f**, staminate flower in vertical section × 3; **g**, staminate calyx, interior and exterior views × 3; **h**, staminate petals, interior views × 3; **i**, staminate petal, lateral view × 3; **j**, stamen in 3 views × 6; **k**, pistillate flower × 6; **l**, pistillate flower in vertical section × 6; **m**, pistillate sepal, interior and exterior views × 6; **n**, pistillate petal, interior view with staminodes × 6; **o**, staminode × 24; **p**, stigmas × 6; **q**, ovary in cross-section × 6; **r**, fruit × 1$^1/_2$; **s**, fruit in vertical section × 1$^1/_2$; **t**, fruit in cross-section × 1$^1/_2$; **u**, **v**, **w**, seed in 3 views × 3. *Socratea exorrhiza*: **a–q**, *Moore et al.* 9550; **r–w**, *Moore et al.* 9538. (Drawn by Marion Ruff Sheehan)

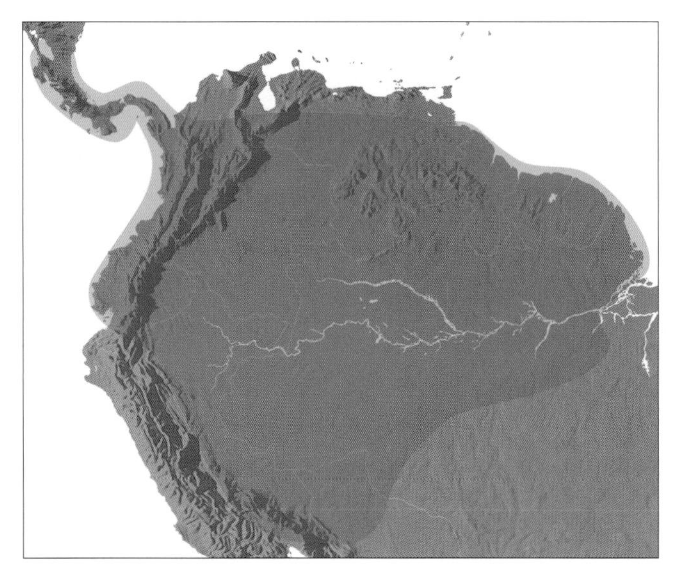

Distribution of ***Socratea***

Distribution and ecology: Five accepted species: one, *Socratea exorrhiza*, is very widely distributed from Nicaragua and Costa Rica southward to Colombia, Ecuador, Peru, Venezuela, Guyana, Surinam, Brazil and Bolivia; the other four are much more restricted in the Andes or adjacent lowland areas. Occurring in lowland and montane tropical rain forest; pollination, where known, is by beetles (Henderson 1985).

Anatomy: Leaf (Tomlinson 1961, Roth 1990), root (Tomlinson 1961, Seubert 1998a, 1998b, Avalos 2004), gynoecium (Uhl & Moore 1971), floral development (Uhl & Moore 1980), seed (Roth 1990).

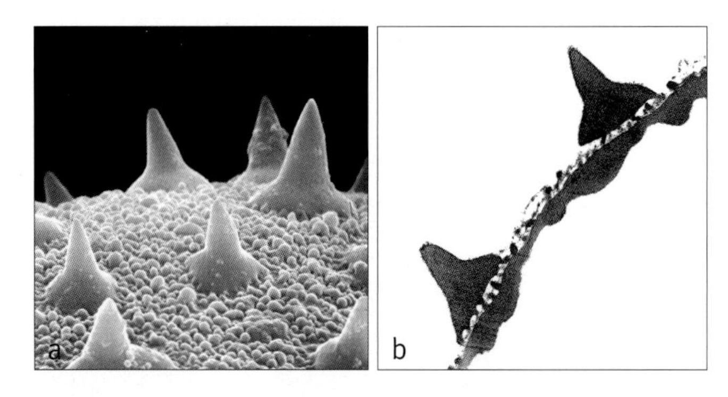

Above: ***Socratea***
a, close-up, intectate spiny pollen surface SEM × 8000;
b, ultrathin section through pollen wall TEM × 8000.
Socratea rostrata: **a**, *Balslev & Balslev* 4427; *S. exorrhiza*: **b**, *Bartlett* 16715.

Right: ***Socratea salazarii***, inflorescence, Brazil. (Photo: A.J. Henderson)

Relationships: The monophyly of *Socratea* has not been tested. For relationships, see *Iriartella*.

Common names and uses: Stilt palms. The outer layers of the trunk are extremely hard and durable and are used, split, in the construction of houses and corrals. Wallace (1853) records the use of the spiny roots as cassava graters. Older palms may be cut to make bows (Balick 1985).

Taxonomic account: Henderson (1990).

Fossil record: From Mexico (State of Chiapas), Upper Oligocene–Lower Miocene, two staminate flowers in amber are described as *Socratea brownii* (Poinar 2002a).

Notes: Wessels Boer (1965) and MacBride (1960) included *Socratea* in *Iriartea*; however, there is a whole suite of characters separating the two (see *Iriartea*). Furthermore, floral biology is significantly different in the two genera (as confirmed by Henderson 1985). The separation of *Metasocratea* was based on a misinterpretation of the position of the embryo (see Bernal 1986). There is an expansion of the floral apex into a large truncate area opposite each sepal during stamen initiation. This pattern appears characteristic of the tribe.

Pintaud and Millan (2004) describe a population of *Socratea salazarii* displaying flagelliform inflorescences that root at their tips.

81. WETTINIA

Usually moderate solitary or clustered pinnate-leaved tree palms of the Andes and foothills, the stems with stilt roots; leaflets are rhomboid praemorse, and sometimes longitudinally divided to give the whole leaf an ostrich-feather appearance; inflorescences are often multiple at each node, often unisexual, and the fruit has basal stigmatic remains and embryo.

Wettinia Poepp. in Endl., Gen. pl. 243 (1837). Type: **W. augusta** Poepp. & Endl.

Catoblastus H. Wendl., Bonplandia 8: 104, 106 (1860). Lectotype: *C. praemorsus* (Willd.) H. Wendl. (*Oreodoxa praemorsa* Willd.) (see O.F. Cook & Doyle 1913) (= *Wettinia praemorsa* (Willd.) Wess. Boer).

Wettinella O.F. Cook & Doyle, Contrib. U.S. Natl. Herb. 16: 235 (1913). Type: *W. quinaria* O.F. Cook & Doyle (= *Wettinia quinaria* [O.F. Cook & Doyle] Burret).

Acrostigma O.F. Cook & Doyle, Contrib. U.S. Natl. Herb. 16: 228 (1913). Type: *A. aequale* O.F. Cook & Doyle (= *Wettinia aequalis* [O.F. Cook & Doyle] R. Bernal).

Catostigma O.F. Cook & Doyle, Contrib. U.S. Natl. Herb. 16: 230 (1913). Type: *C. radiatum* O.F. Cook & Doyle (= *Wettinia radiata* [O.F. Cook & Doyle] R. Bernal).

Wettiniicarpus Burret, Notizbl. Bot. Gart. Berlin-Dahlem 10: 937 (1930). Type: *W. fascicularis* Burret (= *Wettinia fascicularis* [Burret] H.E. Moore & J. Dransf.).

Named for Frederick August of the House of Wettin (1750–1827), King of Saxony.

Solitary or clustered, slender, moderate or robust, unarmed, pleonanthic, monoecious tree palms. *Stem* erect, conspicuously ringed with leaf scars, bearing at the base a cone of stilt roots, covered in small sharp lateral roots. *Leaves* few in number, spirally arranged or rarely distichous, pinnate, neatly abscising or rarely marcescent; sheaths forming a well-defined crownshaft, covered with a variety of indumentum types; petiole rather short, adaxially channelled or convex, abaxially rounded; rachis adaxially angled, abaxially rounded, bearing hairs of various types; leaflets of two sorts, one undivided, elongate, asymmetrically and narrowly elliptic in outline, the proximal margin entire for ca. $^{2}/_{3}$ its length, then praemorsely toothed, the distal margin entire for ca. $^{1}/_{4}$ its length, then praemorsely toothed, conspicuously ribbed, the main ribs diverging from the base to the praemorse margin, the other leaflet type similar but with stouter ribs, and split between the ribs to the base into narrow segments displayed in several planes giving the whole leaf a plumose appearance, leaflets densely hairy abaxially, transverse veinlets not evident. *Inflorescences* unisexual, infrafoliar, 3–8(–15) at a node, maturing centrifugally, the central pistillate or staminate, the lateral staminate, or sometimes the inflorescence single by abortion of accessory buds at the node, either staminate or pistillate, spicate or branched to 1 order; peduncle prominent, shorter than or ± as long as rachis; prophyll short, tubular, 2-keeled, open at the apex; peduncular bracts 4–7, proximal 2 rather short, tubular, rounded, not flattened, open apically, distal bracts much longer, tubular, ± beaked, enclosing the inflorescence, splitting longitudinally, prophyll and all peduncular bracts very coriaceous, variously hairy or bristly, persisting long into fruiting stage; rachis where inflorescence branched, bearing small, collar-like or scarcely evident, spirally arranged bracts, the rachis and branches often coiled in bud; rachillae radiating or pendulous, bearing spirally arranged flowers. *Flowers* white or cream-coloured at anthesis, densely crowded.

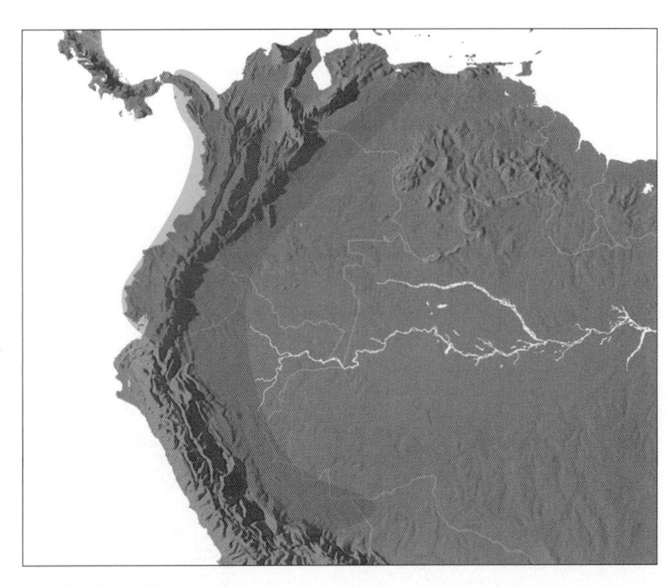

Distribution of ***Wettinia***

Staminate flowers crowded in ebracteolate pairs or solitary, open within the inflorescence bud; sepals 3(–4), briefly connate or distinct, ± narrow-triangular, small; petals much longer than the sepals, 3(–4), narrow triangular, straight or hooked at the apex, briefly valvate at the base; stamens 6–20, filaments short, slender, anthers basifixed, erect, elongate, latrorse; pistillode absent or minute and trifid. *Pollen* ellipsoidal, ± bi-symmetric; aperture a distal sulcus; ectexine intectate, upper surface of foot layer covered by fine, dense gemmae or clavae, loosely supporting short, wide-spaced, broad-based spines, aperture margin similar; longest axis 43–50 µm [5/21]. *Pistillate flowers* open in bud asymmetrical due to close packing, usually borne with 2 vestigial staminate flowers; sepals 3(–4), imbricate, or separated, or briefly connate basally, deltoid to elongate triangular; petals 3(–4), similar to but usually longer and broader than the sepals; staminodes 6, minute, tooth-like or absent; gynoecium of 1–3 minutely roughened, hairy or bristly fertile carpels and (0–)2 abortive carpels, with basal or apical, short to elongate, glabrous or hairy style, and

Right: ***Wettinia maynensis***, stilt roots, Ecuador. (Photo: H. Balslev)

3 elongate, large stigmas, persistent or deciduous in fruit, ovule laterally attached at the base, anatropous. *Fruit* developing from 1 carpel, rarely from 2, densely crowded or rather loosely arranged, 1-seeded, prismatic, irregular, ellipsoid or globose, stigmatic remains basal; epicarp minutely roughened, softly hairy, or hairy and warty, or prickly with shining straight or twisted spines, mesocarp granular, with a layer of sclereids external to a parenchymatous layer with included tannin cells and elongate fibres, endocarp very thin. *Seed* ellipsoidal or subglobose, sometimes enclosed in a gelatinous mass when fresh, basally attached with rounded hilum, raphe elongate with reticulate branches, endosperm homogeneous or ruminate; embryo basal. *Germination* adjacent-ligular; eophyll praemorse, undivided or with a brief apical split. *Cytology* not studied.

Distribution and ecology: Twenty-one species in Panama, Colombia, Peru, west Brazil and Ecuador, Greatest diversity in Colombia, west of the Andes in the Choco refugium, but also found east of the Andes. Confined to ever-wet tropical rain forest at low to medium elevations, often occurring in abundance.

Anatomy: Leaf (Tomlinson 1961, Roth 1990), root (Seubert 1998a, 1998b), gynoecium (Uhl & Moore 1971), stamen development (see *Socratea*), seed (Roth 1990, as *Catoblastus praemorsus*).

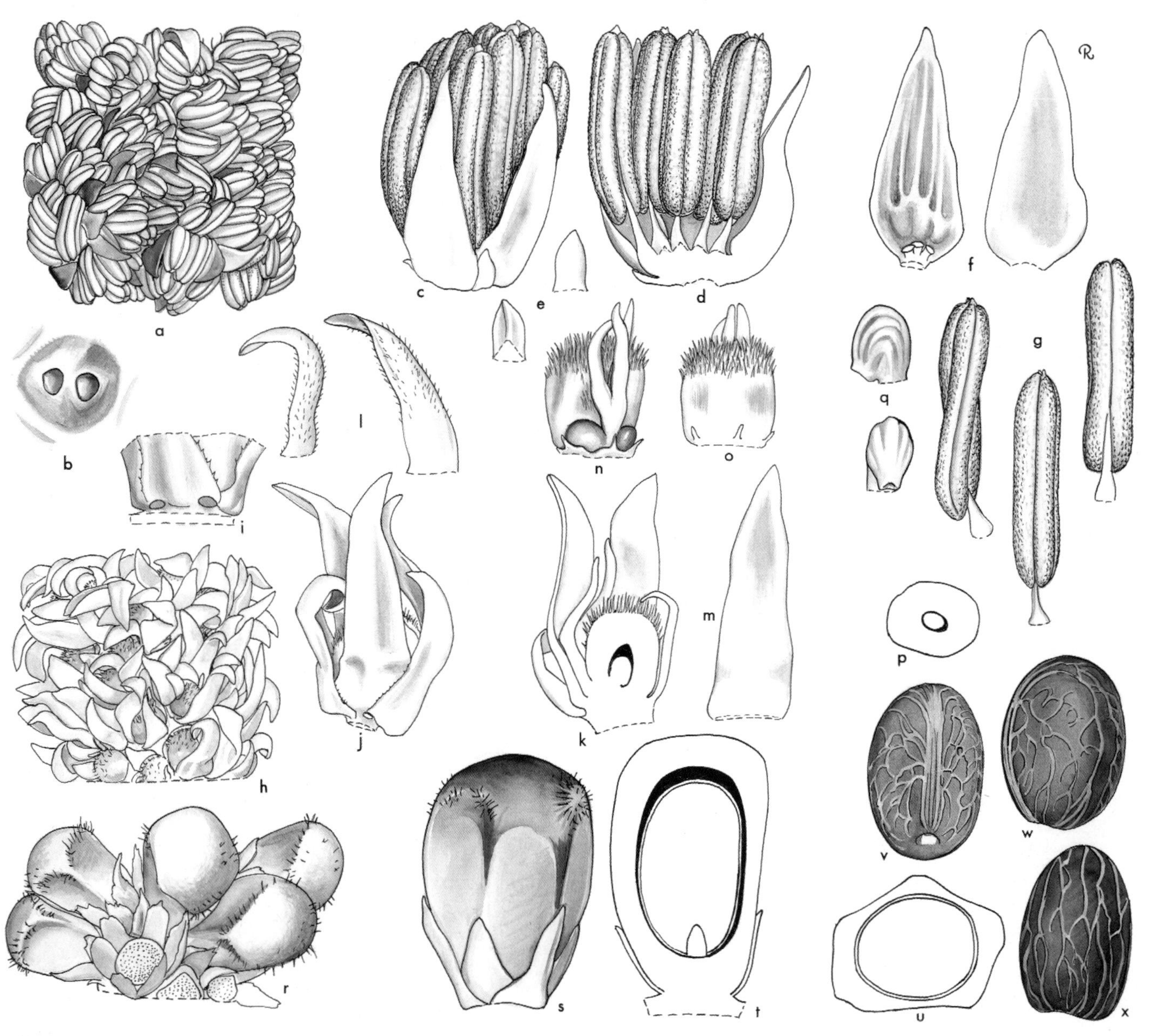

Wettinia. **a**, portion of staminate rachilla × 3; **b**, scars of staminate flowers × 6; **c**, staminate flower × 6; **d**, staminate flower in vertical section × 6; **e**, staminate sepals in exterior (upper) and interior (lower) views × 6; **f**, staminate petal in interior (left) and exterior (right) views × 6; **g**, stamen in 3 views × 6; **h**, portion of pistillate rachilla × 1 1/2; **i**, base of pistillate flower and scars of abortive staminate flowers × 3; **j**, pistillate flower × 3; **k**, pistillate flower in vertical section × 3; **l**, pistillate sepals in 2 views × 3; **m**, pistillate petal × 3; **n, o**, gynoecium in 2 views with staminodes × 3; **p**, ovary in cross-section × 3; **q**, ovule in 2 views × 12; **r**, portion of fruiting axis × 1; **s**, fruit × 1 1/2; **t**, fruit in vertical section × 1 1/2; **u**, fruit in cross-section × 1 1/2; **v, w, x**, seed in 3 views × 1 1/2. *Wettinia quinaria*: all from *Moore & Parthasarathy* 9485. (Drawn by Marion Ruff Sheehan)

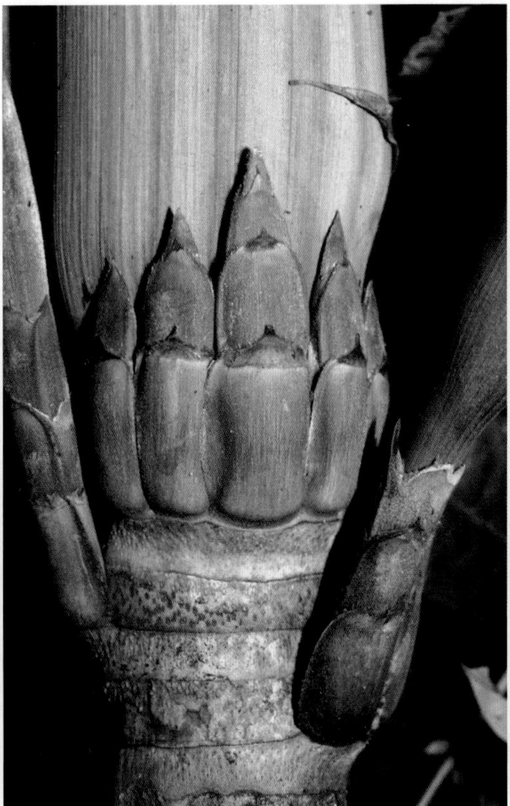

Left: ***Wettinia cladospadix***, crown, Colombia. (Photo: J. Dransfield)
Middle: ***Wettinia augusta***, multiple inflorescence buds, Colombia. (Photo: J. Dransfield)
Right: ***Wettinia drudei***, leaves, Peru. (Photo: J.-C. Pintaud)

Relationships: The monophyly of *Wettinia* has not been tested. For relationships, see *Iriartella*.
Common names and uses: Stilt palms. Leaves are used for thatching and the trunk split and used for flooring and walling.
Taxonomic accounts: Moore & Dransfield (1978), Moore (1982), Galeano-Garces & Bernal-Gonzales (1983) and Bernal (1995).
Fossil record: No generic records found.
Notes: There is striking diversity in fruit form, particularly in fruit surface; fruits can be warty, bristly or prickly.

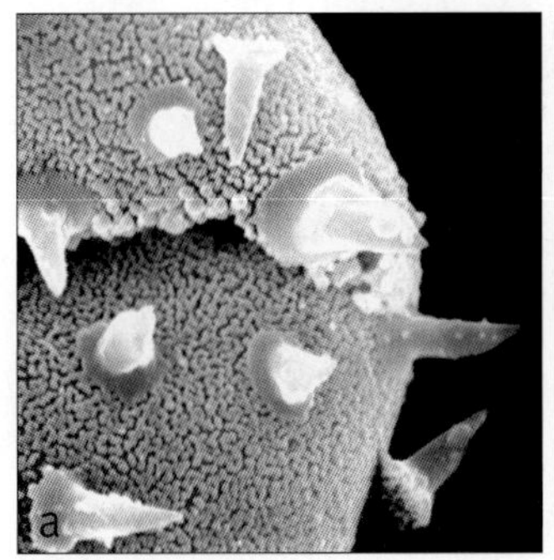

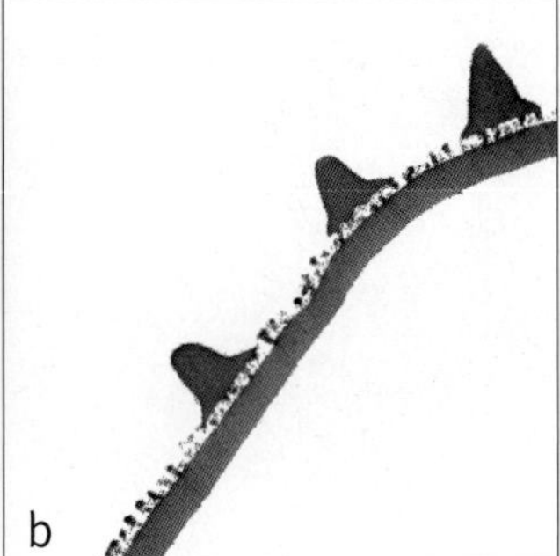

Above: ***Wettinia***
a, close-up, intectate spiny pollen surface, sulcus area SEM × 8000;
b, ultrathin section through pollen wall TEM × 8000.
Wettinia fascicularis: **a**, *Moore & Dransfield* 10231; *W.* sp.: **b**, *Steyermark et al.* 106290.

Tribe Chamaedoreeae Drude, Bot. Zeit. 35: 632 (1877).

Type: **Chamaedorea**.
Hyophorbeae Drude in Mart., Fl. bras. 3(2): 275 (1881).

Moderately robust to very slender, rarely climbing, acaulescent or stems erect; leaves pinnate or entire, leaflets almost always acute, very rarely subpraemorse; sheaths usually forming a crownshaft; monoecious or dioecious, flowers usually globose, solitary or in triads or acervuli, floral bracts scarcely evident; gynoecium tricarpellate, triovulate; fruit almost always 1-seeded, with basal stigmatic remains.

The tribe includes five genera: three ranging from Mexico south to Peru, one (*Gaussia*) in northern Central America and the Caribbean and the fifth (*Hyophorbe*) in the Mascarene Islands. There is great diversity of habit in this tribe, ranging from robust bottle palms to diminutive species that are among the smallest in the family. The small usually globose flowers are of a distinctive form and are usually arranged in an acervulus (a cincinnus of linear form) (Uhl 1978b, Uhl & Moore 1978). Although *Chamaedorea* usually has solitary flowers, flower clusters are in curved or straight rows when present.

Tribe Chamaedoreeae is monophyletic with high support in all studies (Uhl *et al.* 1995, Lewis & Doyle 2001, 2002,

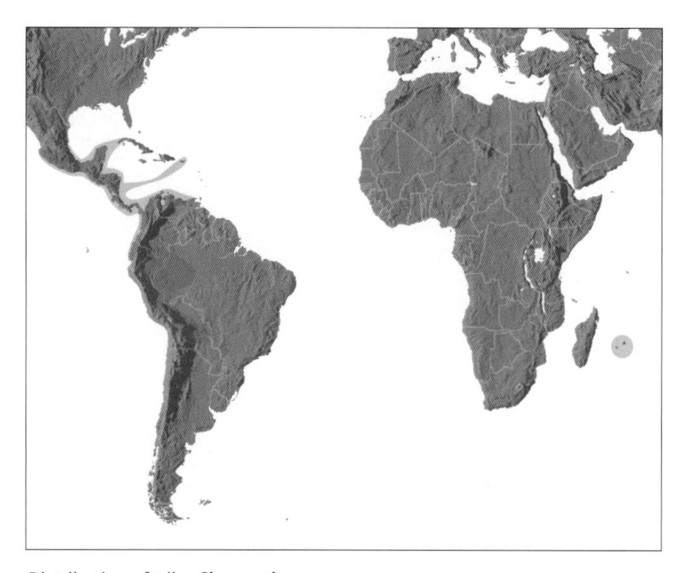

Distribution of tribe **Chamaedoreeae**

Hahn 2002a, 2002b, Asmussen *et al.* 2006, Thomas *et al.* 2006, Cuenca & Asmussen-Lange 2007, Baker *et al.* in review, in prep.). The sister-group relationships of Chamaedoreeae are not yet clear. Two studies resolve Chamaedoreeae as sister to all Arecoideae except Iriarteeae with moderate support (Hahn 2002b, Baker *et al.* in review). Other studies resolve Chamaedoreeae as sister to a clade of Iriarteeae, Areceae, Pelagodoxeae, Leopoldinieae, Euterpeae, Geonomateae and Manicarieae, but with low support (Lewis & Doyle 2002, Baker *et al.* in prep.).

Key to the Genera of the Chamaedoreeae

1. Trees of medium size 2
1. Understory palms 3
2. Crownshaft present; inflorescences infrafoliar in bud, rachillae pendulous. Mascarene Islands . . **82. Hyophorbe**
2. Crownshaft lacking; inflorescences interfoliar in bud, becoming infrafoliar, often persistent, rachillae divaricate. Cuba, Puerto Rico, Mexico, Guatemala and Belize. . . . . **86. Gaussia**
3. Dioecious understory palms 4
3. Monoecious, flowers in acervulae. Southern Mexico, Central America, and northwestern South America . . . . **84. Synechanthus**
4. Peduncular bracts several, flowers solitary or rarely in curved lines, sepals and petals free or connate, endocarp hard. Central Mexico to Brazil and Bolivia . . . . **85. Chamaedorea**
4. Peduncular bracts 1, both staminate and pistillate flowers in acervulae, sepals and petals always connate, endocarp membranous. Amazonian Peru . . . . **83. Wendlandiella**

82. HYOPHORBE

Moderate single-stemmed pinnate-leaved palms with conspicuous crownshafts and often bottle-shaped trunks, from the Mascarene Islands; inflorescence is horn-like in bud and has many tubular bracts and flowers borne in acervuli.

Hyophorbe Gaertn., De Fructibus et Seminibus Plantarum 2: 186 (1791). Type: **H. indica** Gaertn.

Sublimia Comm. ex Mart., Historia Naturalis Palmarum 3: 164 (1838). Type: *S. vilicaulis* Comm. ex Mart. (= *Hyophorbe indica* Gaertn.).

Mascarena L.H. Bailey, Gentes Herb. 6: 71 (1942). Lectotype: *M. revaughanii* L.H. Bailey (see Beccari & Pichi-Sermolli 1955) (= *Hyophorbe lagenicaulis* [L.H. Bailey] H.E. Moore).

Hys, hyos — *pig, sow,* phorbe — *food, in reference to the past use of the fruits as pig food.*

Solitary, moderate, unarmed, pleonanthic, monoecious palms. *Stem* in some species variously swollen, of ± uniform diameter in others, ringed with leaf scars, ± striate, grey. *Leaves* pinnate, neatly abscising; sheaths forming a prominent crownshaft; petiole short, robust, channelled adaxially, rounded abaxially; rachis flat adaxially, rounded abaxially; leaflets acute to acuminate, single-fold, stiff, midrib prominent with 0–2(–3) evident veins on each side, ramenta prominent on the midrib beneath. *Inflorescences* infrafoliar, solitary, branched to 3–4 orders, horn-shaped and erect in bud, becoming ± horizontal, protandrous; peduncle

Right: ***Hyophorbe verschaffeltii***, cultivated, Montgomery Botanical Center, Florida. (Photo: J. Dransfield)

stout, elongate; prophyll very short, tubular, caducous, opening at the apex; peduncular bracts 5(4–9), caducous, progressively opening apically, rachis prominent, slightly longer than the peduncle, first-order branches spirally arranged, divaricate, each with a short basal bare part; rachillae slender, elongate, pendulous at first, becoming divaricate; bracts subtending branches and rachillae not evident at anthesis. Flowers orange, yellowish, or white at anthesis, sometimes fragrant, borne in acervuli of a basal pistillate and 3–7 distal staminate, bracts subtending the acervuli and bracteoles subtending the flowers not evident at anthesis. *Staminate flowers* symmetrical or somewhat asymmetrical in bud; sepals 3, distinct, imbricate or basally connate; petals 3, valvate, briefly connate basally; stamens 6, filaments connate basally and adnate to the corolla or extending beyond the adnation in a tube, distally free, subulate, erect, anthers dorsifixed at or above the middle, bifid basally for $^1/_2$ their length or more, emarginate or briefly bifid apically, latrorse; pistillode conic-ovoid and shorter than the stamens or sometimes minute, 3-lobed. *Pollen* ellipsoidal, asymmetric, infrequently oblate triangular; aperture a distal sulcus, infrequently a trichotomosulcus; ectexine tectate, scabrate-perforate, aperture margin similar; narrow infratectum columellate; longest axis ranging from 38–58 µm; post-meiotic tetrads usually tetrahedral, occasional tetragonal or, rarely, rhomboidal [4/5]. *Pistillate flowers* symmetrical, ovoid; sepals 3, distinct and imbricate, or connate basally in a cupule, the lobes then slightly imbricate; petals 3, valvate or slightly imbricate, briefly connate basally; staminodes connate basally in a 6-lobed cupule, sometimes with minute abortive anthers; gynoecium trilocular, triovulate, with 3 recurved, minutely papillate stigmas at anthesis, ovary with septal nectary, ovules laterally attached, hemianatropous, arillate in *Hyophorbe verschaffeltii*. *Fruit* ellipsoidal to globose or obpyriform, orange to black, red, or brown, normally 1-seeded, with basal stigmatic remains, perianth persistent, thickened; epicarp smooth or drying somewhat roughened or minutely warty, mesocarp thin, fleshy, with numerous reddish tannin cells and flat fibres of various widths in more than one layer, endocarp thin. *Seed* ovoid to ellipsoidal or globose, the hilum small, basal, vasculature of few simple or little-branched strands radiating distally and laterally from the hilum, endosperm homogeneous; embryo lateral to apical. *Germination* adjacent-ligular; eophyll bifid or rarely pinnate. *Cytology*: 2n = 32.

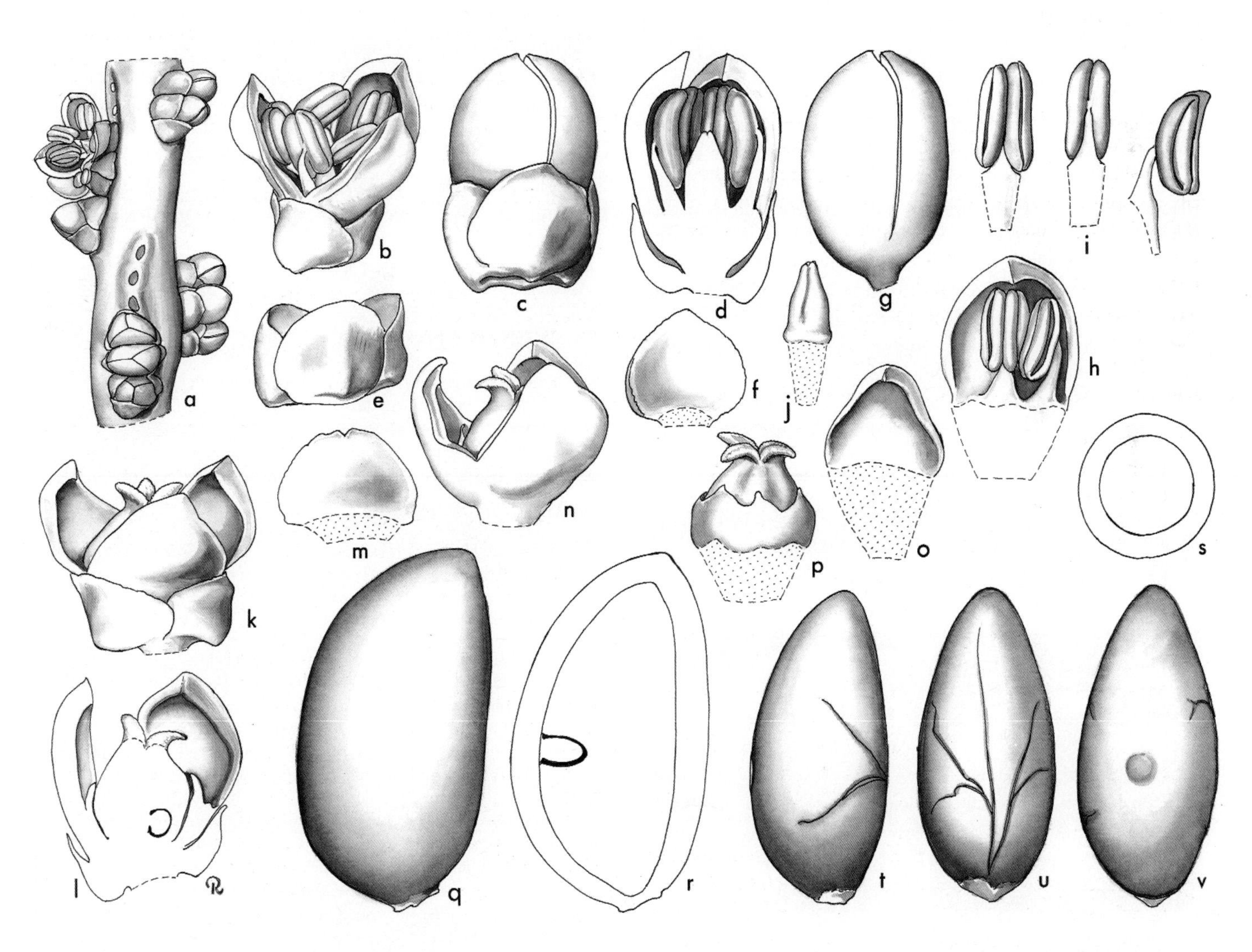

***Hyophorbe*.** **a**, portion of rachilla × $4^1/_2$; **b**, staminate flower × 12; **c**, staminate bud × 12; **d**, staminate bud in vertical section × 12; **e**, staminate calyx × 12; **f**, staminate sepal, interior view × 12; **g**, staminate corolla × 12; **h**, staminate petal with stamens attached, interior view × 12; **i**, stamen in 3 views × 12; **j**, pistillode × 12; **k**, pistillate flower × 12; **l**, pistillate flower in vertical section × 12; **m**, pistillate sepal × 12; **n**, pistillate flower, sepals removed × 12; **o**, pistillate petal, interior view × 12; **p**, staminodes and gynoecium × 12; **q**, fruit × 3; **r**, fruit in vertical section × 3; **s**, fruit in cross-section × 3; **t, u, v**, seed in 3 views × 3. *Hyophorbe verschaffeltii*: **a–p**, *Read* 1387; **q–v**, *Moore* s.n. (Drawn by Marion Ruff Sheehan)

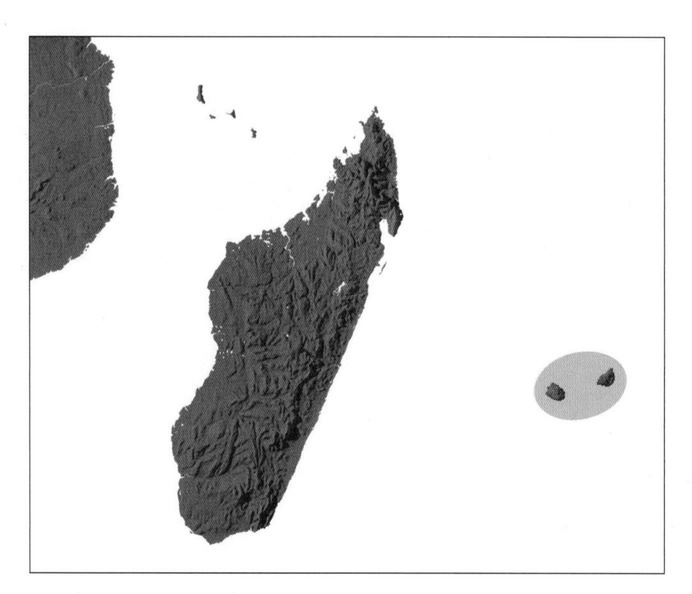

Distribution of ***Hyophorbe***

Distribution and ecology: Five species endemic to the Mascarene Islands: one each on Rodrigues and Reunion, and three on Mauritius and Round Island. These handsome palms once covered the mountains and valleys of the Mascarene Islands. All species are now nearly extinct in the wild state but formerly were apparently palms of the forest to ca. 700 m altitude or perhaps of the coastal savannah (*Hyophorbe lagenicaulis*). The last is now restricted to a few individuals on the exposed rock of Round Island; the others occur on volcanic soils or on both volcanic soils and calcarenite limestones (*H. verschaffeltii*). Only a single tree of *H. amaricaulis* now remains, growing in the botanic garden at Curepipe, Mauritius, and relatively few individuals of the other four species can be found (Moore 1978a).
Anatomy: Leaves and floral (Uhl 1978b, 1978c). Developmental studies have shown that the acervulus is an adnate cincinnus (Uhl & Moore 1978). Root (Seubert 1998a, 1998b).

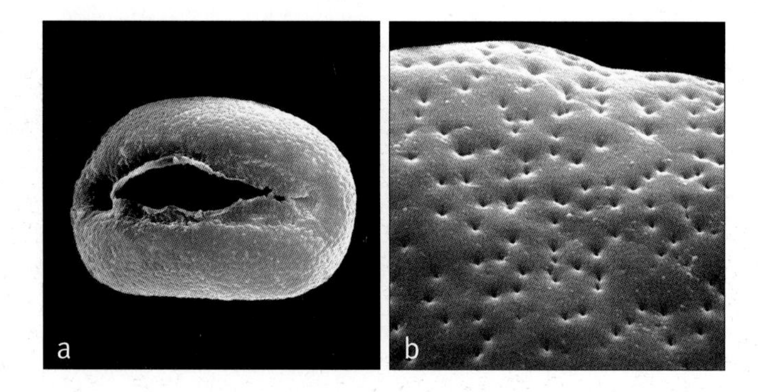

Above: ***Hyophorbe***
a, monosulcate pollen grain, distal face SEM × 1000;
b, close-up, finely perforate pollen surface SEM × 8000.
Hyophorbe lagenicaulis: **a**, *Barkly* s.n.; *H. amaricaulis*: **b**, *Lesouëf* s.n.

Relationships: *Hyophorbe* is monophyletic with high support (Cuenca & Asmussen-Lange 2007). The relationships among the five genera of tribe Chamadoreeae are still unclear despite the dense sampling in some studies. There is a discrepancy between phylogenies resulting from data from nuclear DNA and those from plastid DNA. Most studies that are based on nuclear data place *Hyophorbe* as sister to the rest of the Chamaedoreeae with moderate to high support (Thomas *et al.* 2006, Baker *et al.* in prep.), whereas *Hyophorbe* is sister to *Wendlandiella* with moderate support in studies based on plastid DNA (Asmussen *et al.* 2006, Cuenca & Asmussen-Lange 2007).
Common names and uses: Bottle palms, spindle palm (*Hyophorbe verschaffeltii*). These are important ornamentals, often commanding high prices in the horticultural trade. Some species apparently have poisonous cabbage, whereas the cabbage of others (e.g., *H. amaricaulis*) was eaten in the past despite being very bitter.
Taxonomic account: Moore (1978a).
Fossil record: No generic records found.
Notes: The combination of a prominent crownshaft, a horn-like inflorescence bud borne below the leaves, briefly connate but valvate petals, adnation and some connation of stamen filaments, connate staminodes, and large fruits distinguish *Hyophorbe* from the other genera of the Chamaedoreeae.
Additional figures: Figs 1.4, 7.3, Glossary figs 4, 15.

83. WENDLANDIELLA

Diminutive pinnate-leaved dioecious undergrowth palms from rain forest in South America; the leaves are often entire bifid; inflorescences are very small with flowers in acervuli.

Wendlandiella Dammer, Bot. Jahrb. Syst. 36 (Beibl. 80): 31 (1905). Type: **W. gracilis** Dammer.

Commemorates Hermann Wendland (1825–1903), German botanist and horticulturist who collected palms in Costa Rica and described many palms, with the diminutive iella, *perhaps distinguishing the genus from one named after H. Wendland's grandfather, J.C. Wendland, also a horticulturist.*

Dwarf, clustering, unarmed, pleonanthic, dioecious palm. *Stem* very slender, internodes long, ringed with leaf scars. *Leaves* pinnate or entire, bifid and pinnately nerved; sheaths slender, tubular, not splitting opposite the petiole; petiole slender; rachis ± triangular in cross-section; leaflets when present only 24 per side, with broad insertions, thin, smooth, lanceolate, midrib and sometimes another pair of veins larger, transverse veinlets not apparent. *Inflorescences* interfoliar, staminate and pistillate inflorescences superficially similar, branching to 1(–2) orders; peduncle elongate; prophyll tubular; peduncular bract 1, tubular, it and the prophyll included within the sheath or exserted and exceeding the peduncle; rachis shorter than the peduncle; rachillae slender, few and subdigitate to numerous, bearing low membranous bracts, floral bracteoles not evident. *Staminate flowers* in acervuli of 2–6; sepals 3, briefly connate and markedly gibbous basally, distinct distally and the lobes ± hooded at least in bud, the margins usually thin; petals 3, thin, nerveless when dry, valvate, spreading to recurved at anthesis; stamens 6, the filaments erect in bud, distinct, anthers medium with rounded ends, fleshy connectives very short; pistillodes 3, distinct or somewhat connate. *Pollen* ellipsoidal, slightly asymmetric; aperture a distal sulcus; ectexine tectate, very finely striate, aperture margin similar or slightly perforate; infratectum columellate; longest axis ranging from 26–32 µm; post-meiotic tetrads tetrahedral, occasionally tetragonal [1/1]. *Pistillate flowers* solitary or in vertical pairs; sepals 3, connate in a low, 3-lobed cupule, gibbous basally; petals 3, imbricate, gibbous, about twice as long as sepals; staminodes 3, minute; gynoecium subglobose, trilocular, triovulate, stigmas 3, reflexed, ovule pendulous, form unknown. *Fruit* ellipsoidal, with basal stigmatic remains and remnants of abortive carpels, orange-red at maturity; epicarp smooth, mesocarp thin, lacking fibres, endocarp membranous, not adherent to the seed. *Seed* with a single raphe branch ventrally curving over the top and a side branch curving around each lateral surface, endosperm homogeneous; embryo lateral slightly below the middle. *Germination* and eophyll unrecorded. *Cytology*: 2n = 28.

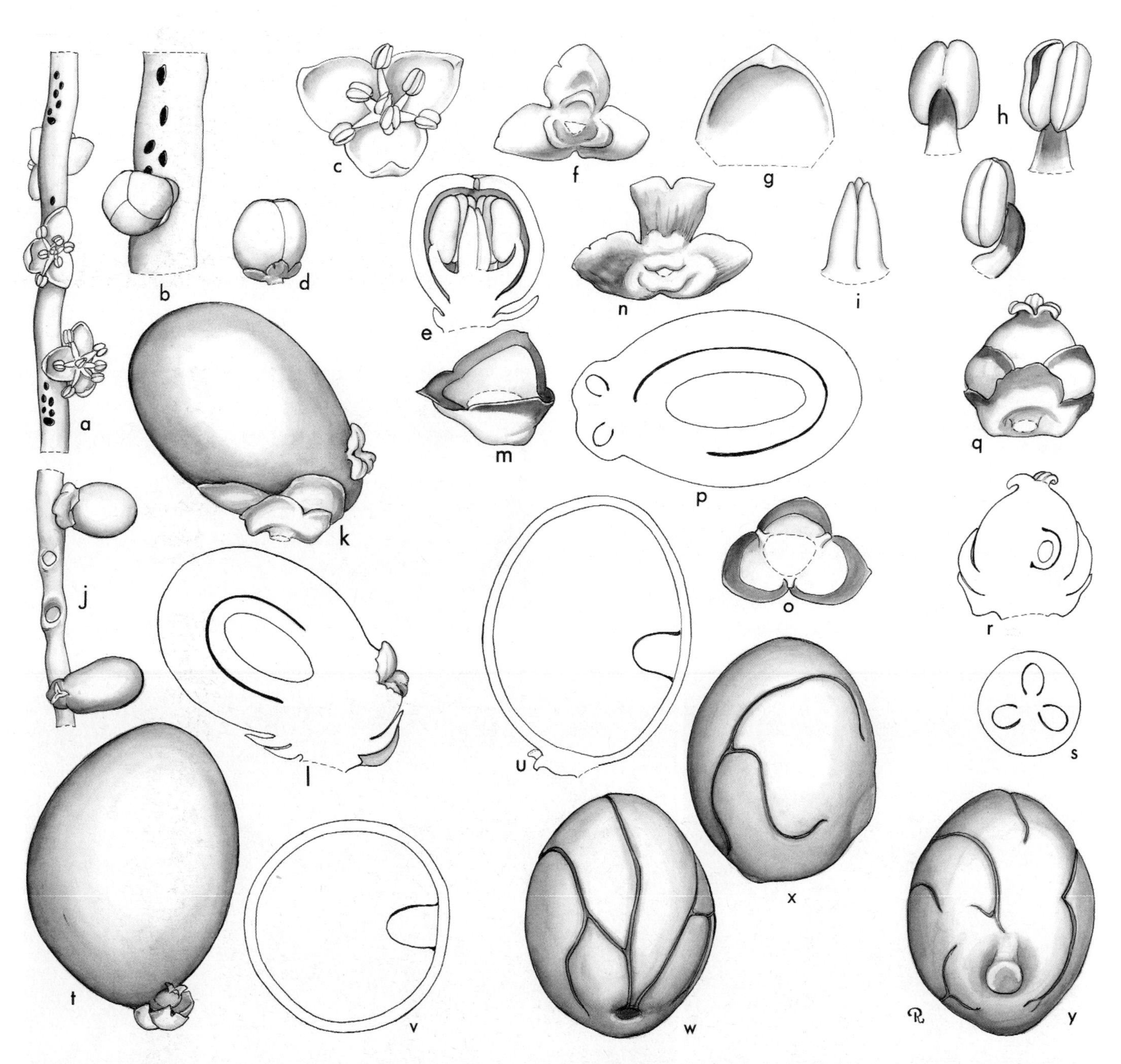

Wendlandiella. **a**, portion of staminate rachilla × 6; **b**, portion of staminate rachilla with single staminate bud and scars of others × 12; **c**, staminate flower at anthesis × 12; **d**, staminate bud × 12; **e**, staminate bud in vertical section × 24; **f**, staminate calyx, exterior view × 24; **g**, staminate petal, interior view × 24; **h**, stamen in 3 views × 24; **i**, pistillode × 24; **j**, portion of pistillate rachilla with young fruit × 3; **k**, young fruit × 12; **l**, young fruit in vertical section × 12; **m**, pistillate calyx in young fruit × 12; **n**, pistillate calyx from below × 12; **o**, pistillate corolla, interior view to show staminodes × 12; **p**, young fruit in cross-section to show abortive locules × 12; **q**, pistillate flower, reconstructed from perianth in young fruit × 12; **r**, pistillate flower in vertical section, reconstructed from young fruit × 12; **s**, ovary in cross-section, reconstructed from young fruit × 12; **t**, fruit × 6; **u**, fruit in vertical section × 6; **v**, fruit in cross-section × 6; **w**, **x**, **y**, seed in 3 views × 6. *Wendlandiella gracilis* var. *polyclada*: **a–s**, *Moore* 9511; *W. gracilis* var. *gracilis*; **t–y**, *Killip & Smith* 27775. (Drawn by Marion Ruff Sheehan)

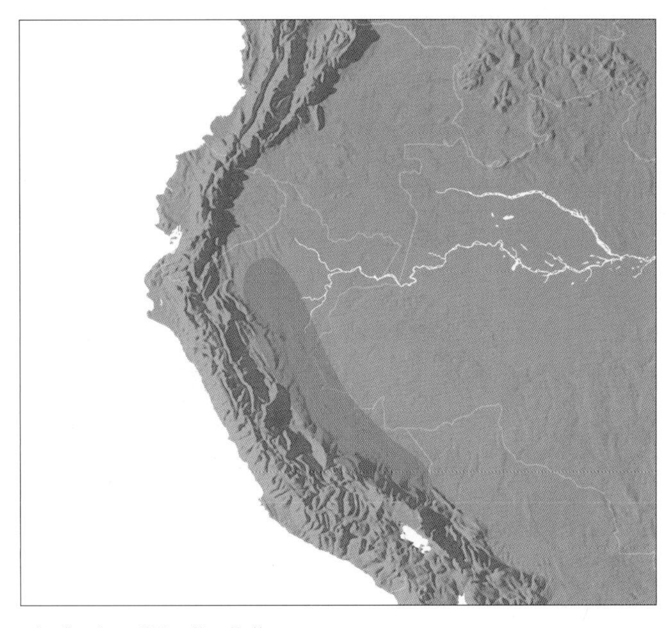

Distribution of ***Wendlandiella***

Distribution and ecology: A single species with three varieties, each confined to three separate basins of the western Amazon in Brazil, Peru and Bolivia. This is a palm of the undergrowth of lowland rain forest on *terra firme* (Henderson 1995).

Common names and uses: *Chontilla*, *ponilla* (Peru). Grown as ornamentals by a few enthusiasts.

Anatomy: Root (Seubert 1998a, 1998b).

Relationships: The placement of *Wendlandiella* is not yet clear and varies among different studies. In a well-sampled study based on nuclear DNA alone, *Wendlandiella* is found to be sister to the rest of the Chamaedoreeae, excluding *Hyophorbe*, with high support (Thomas *et al.* 2006). In the studies based on plastid DNA alone, *Wendlandiella* is found to be sister to *Hyophorbe* with moderate or low support (Asmussen *et al.* 2006, Cuenca & Asmussen-Lange 2007). Alternative placements are found in other studies based on combinations of data types: *Wendlandiella* as sister to *Chamaedorea* (Uhl *et al.* 1995) and *Wendlandiella* as sister to *Synechanthus* (Baker *et al.* in review).

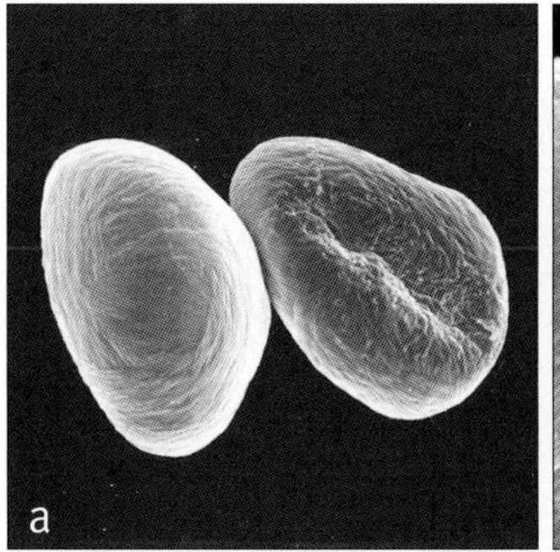

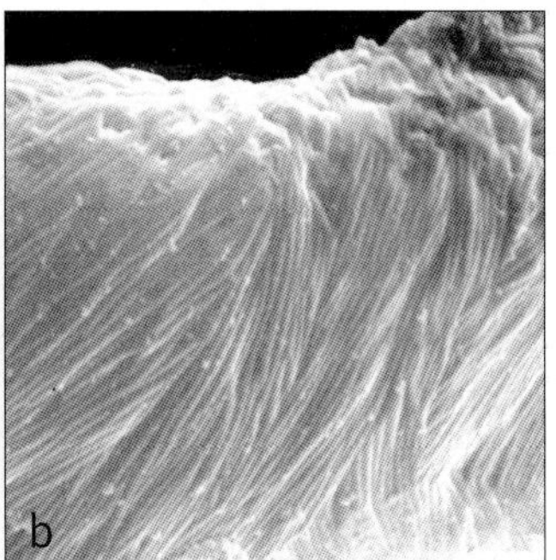

Right: **Wendlandiella**
a, two monosulcate pollen grains: proximal face left, distal face right SEM × 1500;
b, close-up of of finely striate pollen surface SEM × 15,000.
Wendlandiella gracilis: **a** & **b**, *Dransfield* s.n.

Below left: ***Wendlandiella gracilis*** var. ***polyclada***, habit, Peru. (Photo: J.-C. Pintaud)

Below right: ***Wendlandiella gracilis*** var. ***polyclada***, inflorescences, Peru. (Photo: J.-C. Pintaud)

Taxonomic account: Henderson (1995).
Fossil record: No generic records found.
Notes: *Wendlandiella* resembles *Chamaedorea* but differs in having only two bracts on the inflorescence, the staminate and sometimes pistillate flowers borne in rows (acervuli), sepals of both staminate and pistillate flowers connate in a low ring, and endocarp membranous not hard.

84. SYNECHANTHUS

Moderate solitary or clustered pinnate-leaved monecious palms from forest undergrowth in Central and northern South America; inflorescence has stiff rachillae bearing flowers in acervuli.

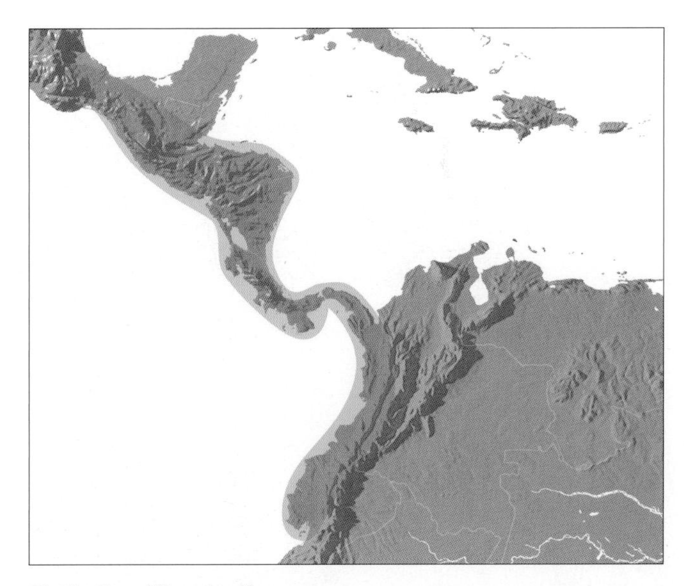

Distribution of ***Synechanthus***

Synechanthus H. Wendl., Bot. Zeit. 16: 145 (1858). Lectotype: **S. fibrosus** (H. Wendl.) H. Wendl. (see Moore 1963c).

Reineckea H. Karst., Wochenschr. Gärtnerei Pflanzenk. 1: 349 (1858). ('*Reineckia*') (non *Reineckea* Kunth, 1844) (conserved name). Type: *R. triandra* H. Karst. (= *Synechanthus warscewiczianus* H. Wendl.).

Rathea H. Karst, Wochenschr. Gärtnerei Pflanzenk. 1: 377 (1858). Type. *R. fibrosa* (H. Wendl.) H. Karst. (*Chamaedorea fibrosa* H. Wendl.) (= *Synechanthus fibrosus* [H. Wendl.] H. Wendl.).

Synechos — *continuous,* anthos — *flower, referring to the linear arrangement of the flowers (acervulus).*

Moderate, solitary or clustered, unarmed, pleonanthic, monoecious palms, sometimes flowering when still acaulescent. *Stem* slender, usually erect, rarely decumbent, smooth, yellowish or glossy deep or dark olive-green, ringed with prominent widely spaced leaf scars. *Leaves* reduplicately pinnate; sheath elongate on new leaves but soon splitting opposite the petiole and differentiated from it only by a narrow, usually fibrous, dry strip along each margin; petiole circular in cross-section; rachis angled adaxially, rounded abaxially; leaflets broadly reduplicate at insertion, acute to acuminate, slightly to markedly sigmoid or, when broad-based, the apex sickle-shaped, with 1 to several principal nerves, these elevated above, or the blade sometimes undivided except at the bifid apex. *Inflorescences* interfoliar or becoming infrafoliar, branched to 1 or 2 orders basally, erect at anthesis, curved or pendulous in fruit, solitary; peduncle long; prophyll short, tubular, sheathing, ultimately disintegrating into fibres, open apically; peduncular bracts 4–5, similar to but longer than the prophyll and inserted at increasingly greater distances, the distal-most usually exceeding the peduncle; rachis usually elongate; rachillae slender, nearly equal in length, 4-angled to markedly flattened and ± flexuous, the tips usually slender and almost spine-like. *Flowers* borne in mostly distichously arranged lines (acervuli) of a proximal pistillate and 5–13 distal, biseriate, staminate flowers, the distal flower of the acervulus opening first and subsequent flower-opening basipetal. *Staminate flowers* green in bud, golden-yellow at anthesis, depressed-triangular in bud; sepals 3, connate in a low, acutely 3-lobed cupule; petals 3, valvate, very prominently nerved in bud when dry, spreading

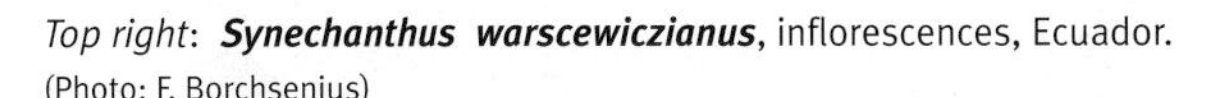

Top right: ***Synechanthus warscewiczianus***, inflorescences, Ecuador. (Photo: F. Borchsenius)

Bottom right: ***Synechanthus warscewiczianus***, infructescence, Ecuador. (Photo: F. Borchsenius)

at anthesis; stamens 6, filaments short, incurved in bud, erect at anthesis, or 3, with long filaments markedly incurved and inflexed at the apex in bud, horizontally exserted at anthesis, anthers basifixed, shallowly bifid at apex and base, latrorse; pistillode small, deltoid-ovoid, apically 3-lobed or absent. *Pollen* ellipsoidal, slightly asymmetric; aperture a distal sulcus; ectexine tectate, scabrate-perforate, aperture margin similar or very slightly finer; infratectum columellate; longest axis ranging from 25–32 µm; post-meiotic tetrads tetrahedral, rarely tetragonal or rhomboidal [2/2]. *Pistillate flowers* yellowish at anthesis; sepals 3, connate in a 3-lobed cupule; petals 3, distinct, imbricate, twice as long as the sepals or more; staminodes apparently lacking or 3, minute, or connate in a 6-lobed ring and partially adnate to the petals; gynoecium, ovoid, drying 3-angled, trilocular, triovulate, stigmas 3, short, recurved, ovules laterally attached, campylotropous but laterally elongate. *Fruit* rather large, round or elongate, yellow, becoming red at maturity, with basal stigmatic remains; epicarp smooth, mesocarp fleshy with few slender, loosely anastomosing, flat fibres against the membranous endocarp. *Seed* not adherent to endocarp, with inconspicuous basal hilum, raphe branches distinctive, large, ascending adaxially from the base, little anastomosed, curving laterally and descending abaxially; endosperm homogeneous or minutely ruminate marginally to markedly ruminate; embryo lateral above the middle to subapical. *Germination* adjacent-ligular; eophyll bifid. *Cytology*: 2n = 32.

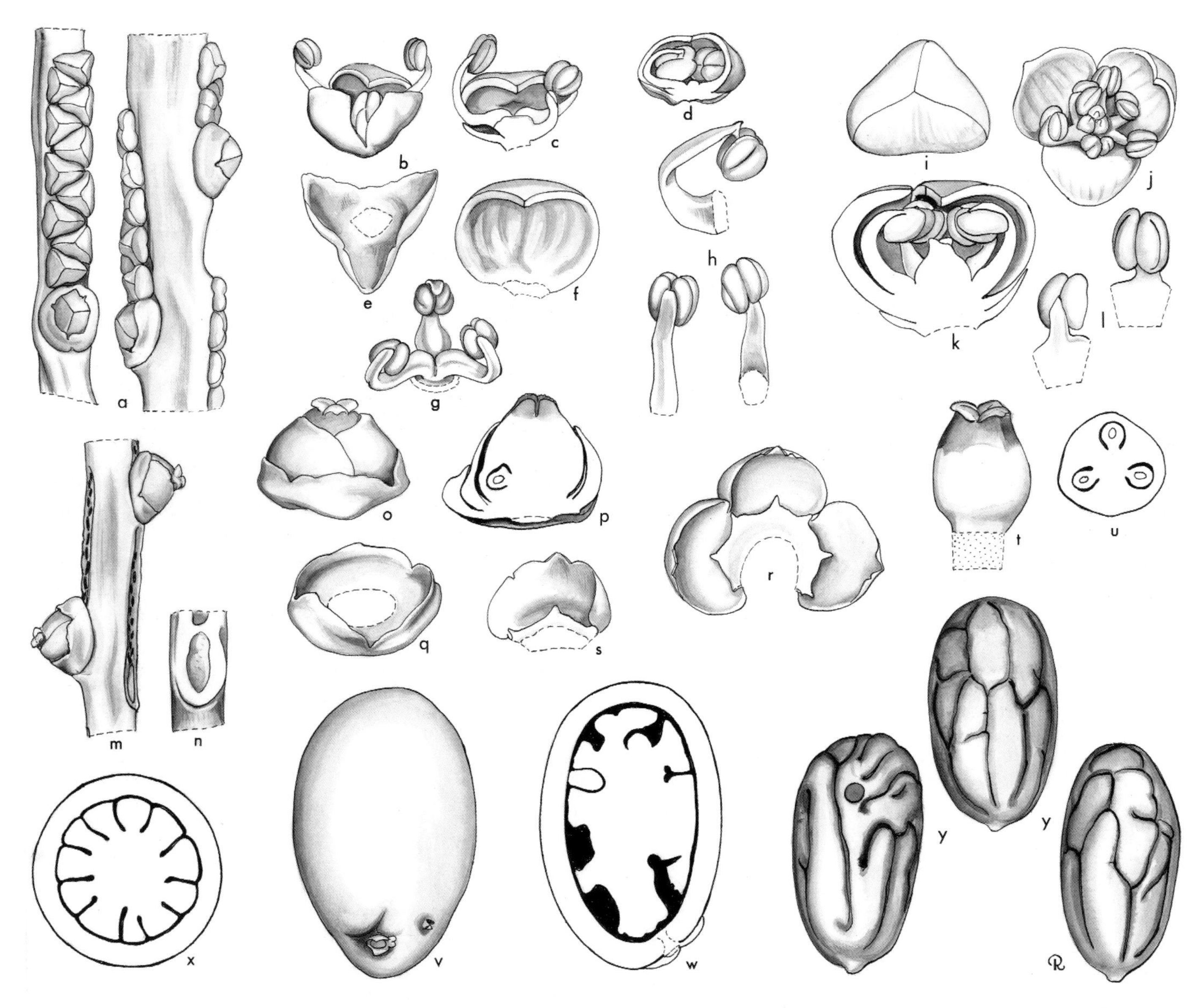

Synechanthus. **a**, portion of rachilla with acervuli in 2 views × 6; **b**, staminate flower × 12; **c**, staminate flower in vertical section × 12; **d**, staminate bud in vertical section × 12; **e**, staminate calyx, interior view × 24; **f**, staminate petal, interior view × 24; **g**, androecium × 12; **h**, stamen in 3 views × 24; **i**, staminate bud × 12; **j**, staminate flower × 12; **k**, staminate bud in vertical section × 24; **l**, stamen in 2 views × 24; **m**, portion of rachilla at pistillate anthesis × 6; **n**, portion of rachilla with scar of pistillate flower × 6; **o**, pistillate flower × 12; **p**, pistillate flower in vertical section × 12; **q**, pistillate calyx × 12; **r**, pistillate petals and staminodes, interior view, expanded × 12; **s**, pistillate petal and staminodes, interior view × 12; **t**, gynoecium × 12; **u**, ovary in cross-section × 12; **v**, fruit × $1^1/_2$; **w**, fruit in vertical section × $1^1/_2$; **x**, fruit in cross-section × $1^1/_2$; **y**, seed in 3 views × $1^1/_2$. *Synechanthus warscewiczianus*: **a–h**, *Moore & Parthasarathy* 9486; **m–u** and **w**, *Moore et al.* 9466; **v, x** and **y**, *Moore & Parthasarathy* 9409; *S. fibrosus*: **i–l**, *Hernandez X. & Sharp* x-1287. (Drawn by Marion Ruff Sheehan)

Distribution and ecology: Two species in southern Mexico, Central America, and northwestern South America. Fairly frequent in wet forests at sea level and low elevations but up to 1200 m in the mountains.
Common names and uses: *Bola*, *palmilla*, jelly bean palm. Widely cultivated as an ornamental.
Anatomy: Leaf (Tomlinson 1961), distinguished from other palms by guard cells with only one outer cutinised ledge. Root (Seubert 1998a, 1998b).
Relationships: *Synechanthus* is monophyletic with high support (Cuenca & Asmussen-Lange 2007). The sister-group relationships of *Synechanthus* have not yet been settled and all resolutions are with low support. One study based on plastid DNA places *Synechanthus* as sister to a clade of *Chamaedorea* and *Gaussia* (Asmussen *et al.* 2006), another study places *Synechanthus* as sister to *Wendlandiella* (Baker *et al.* in review), and a third analysis of the nuclear gene *rpb2* alone, places *Synechanthus* as sister to *Gaussia* (Thomas *et al.* 2005).
Taxonomic accounts: Moore (1971) and Henderson & Ferreira (2002).
Fossil record: No generic records found.
Notes: Leaf form is very variable ranging from undivided to irregularly and regularly pinnate; fruit form may be ellipsoidal or globose in the same population. Seven species were recognised at one time but these taxa are now regarded as variants of the two species listed below. The inflorescence is distinctive because of the long peduncle with erect rachillae near its end. The yellow to bright orange fruits are very attractive.
Additional figures: Figs 1.7, 1.9, Glossary fig. 18.

Above: ***Synechanthus warscewiczianus***, infructescence, Peru.
(Photo: J.-C. Pintaud)

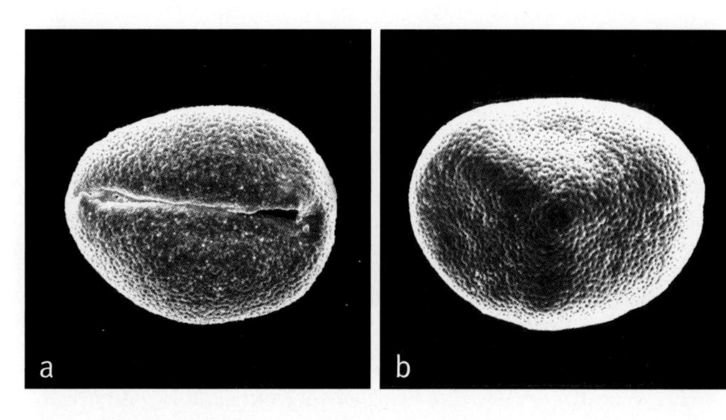

Above: ***Synechanthus***
a, monosulcate pollen grain, distal face SEM × 2000;
b, monosulcate pollen grain, proximal face SEM × 2000.
Synechanthus fibrosus: **a**, *J.D. Hooker* s.n.; **b**, *S. warscewiczianus*: *Henderson et al.* 058.

85. CHAMAEDOREA

Generally rather small, often clustering, pinnate-leaved dioecious palms from the undergrowth of rain forest from Mexico southwards to South America, very diverse and with a wide range of habits including one climbing species, inflorescence and flower form equally varied.

Chamaedorea Willd., Species Plantarum 4(2): 638, 800 (1806) (conserved name). Type: *C. gracilis* Willd. (illegitimate name) = **C. pinnatifrons** (Jacq.) Oerst. (*Borassus pinnatifrons* Jacq.).

Morenia Ruiz & Pavon, Florae Peruvianae et Chilensis Prodromus 150 (1794) (rejected name). Type: *M. fragrans* Ruiz & Pavon. (= *Chamaedorea linearis* Ruiz & Pav.) Mart.

Nunnezharia Ruiz & Pavon, Florae Peruvianae et Chilensis Prodromus 147 (1794) (rejected name). Type: *N. fragrans* Ruiz & Pavon (= *Chamaedorea fragrans* [Ruiz & Pavon] Martius).

Nunnezia Willd., Species Plantarum 4: 890, 1154 (1806) (= *Nunnezharia*).

Kunthia Humb. & Bonpl., Plantae Aequinoctiales 2: 127 (1813). Type: *K. montana* Humb. & Bonpl. (= *Chamaedorea linearis* [Ruiz & Pav.] Mart.).

Stachyophorbe (Liebm. ex Mart.) Liebm. ex Klotzsch, Allg. Gartenzeitung 20: 363 (1852). Lectotype: *S. cataractarum* (Mart.) Liebm. ex Klotzsch (= *Chamaedorea cataractarum* Mart.) (see H.E. Moore 1963c).

Chamaedorea sect. *Stachyophorbe* Liebm. ex Mart., Historia Naturalis Palmarum 3: 309 (1849).

Collinia (Liebm.) Liebm. ex Oerst., Vidensk. Meddel. Dansk Naturhist. Foren. Kjøbenhavn 1858: 5 (1859). Non Raf (1819). Lectotype: *C. elegans* (Mart.) Liebm. ex Oerst. (*Chamaedorea elegans* Mart.) (see Burret 1933a).

Chamaedorea sect. *Collinia* Liebm. in Mart., Historia Naturalis Palmarum 3: 308 (1849).

Dasystachys Oerst., Vidensk. Meddel. Dansk Naturhist. Foren. Kjøbenhavn 1858: 25 (1859). Type: *D. deckeriana* (Klotzsch) Oerst. (*Stachyophorbe deckeriana* Klotzsch) (= *Chamaedorea deckeriana* [Klotzsch] Hemsl.).

Eleutheropetalum H. Wendl. ex Oerst., Vidensk. Meddel. Dansk Naturhist. Foren. Kjøbenhavn 1858: 6 (1859). Type: *E. ernesti-augusti* (H. Wendl.) H. Wendl. ex Oerst. (*Chamaedorea ernesti-augusti* H. Wendl.).

Spathoscaphe Oerst., Vidensk. Meddel. Dansk Naturhist. Foren. Kjøbenhavn 1858: 29 (1859). Type: *S. arenbergiana* (H. Wendl.) Oerst. (*Chamaedorea arenbergiana* H. Wendl.).

Stephanostachys (Klotzsch) Klotzsch ex Oerst., Vidensk. Meddel. Dansk Naturhist. Foren. Kjøbenhavn 1858: 26 (1859). Type: *S. casperiana* (Klotzsch) Oerst. (*Chamaedorea casperiana* Klotzsch).

Chamaedorea subgenus *Stephanostachys* Klotzsch, Allg. Gartenzeitung 20: 363 (1852).

Kinetostigma Dammer, Notizbl. Königl. Bot. Gart. Berlin 4: 171 (1905). Type: *K. adscendens* Dammer (= *Chamaedorea adscendens* [Dammer] Burret).

Chamai — *on the ground,* dorea — *gift, probably in reference to the usually dwarf habit and elegant form.*

Small, sometimes moderate, erect or procumbent, rarely climbing, acaulescent or trunked, solitary or clustered, unarmed, pleonanthic, dioecious palms. *Stem* usually slender, covered wholly or partially in fibrous leaf bases or smooth, green, prominently ringed with leaf scars. *Leaves* bifid or variously pinnate, very rarely entire, reduplicate; sheath closed or becoming split, short or elongate, sometimes with a marcescent lobe opposite the petiole; petiole short to elongate, flattened adaxially, rounded abaxially, sometimes with a prominent pale green or yellow, abaxial stripe; rachis rounded, angled, or flattened adaxially, rounded abaxially; blade entire, bifid and pinnately ribbed, or regularly or irregularly pinnately divided, leaflets few or many, of 1 or several folds, narrow or broad, often oblique or sigmoid, acuminate, surfaces glabrous. *Inflorescences* among or below the leaves, solitary or several per leaf axil, unbranched or branched to 1(–2) order, sometimes forked; staminate often more branched than pistillate; peduncle short to elongate; prophyll tubular with tapering bifid tip; peduncular bracts 2–several, elongate, tubular, sheathing the peduncle, coriaceous or membranous, persistent, tips short, bifid; rachillae, long or short, slender or fleshy, sometimes ridged, lacking bracts at maturity, bearing closely appressed or rather widely spaced, spirally arranged staminate or pistillate flowers, rarely bearing curved acervuli of staminate flowers. *Flowers* sessile or partly enclosed in a cavity in the fleshy rachilla, small or minute. *Staminate flowers* symmetrical; sepals 3, entire, united basally or distinct; petals 3, distinct or variously connate, lobes valvate; stamens 6, filaments short, broad or awl-shaped; anthers dorsifixed, included, oblong or didymous; pistillode various, cylindric or expanded basally, sometimes trilobed. *Pollen* ellipsoidal, occasionally oblate triangular, bi-symmetric or slightly asymmetric; aperture a distal sulcus, occasionally a trichotomosulcus; ectexine tectate, finely rugulate, finely perforate-rugulate, finely reticulate, or reticulate, aperture margin either similar or, more frequently, broad and psilate or scabrate, in reticulate pollen, reticulum often notably finer on proximal face, less frequently proximal face psilate; infratectum columellate; longest axis 20–36 µm; post-meiotic tetrads usually tetrahedral, sometimes tetragonal or rarely rhomboidal [50/108]. *Pistillate flower* with sepals 3, as in the staminate; petals 3, usually connate, distinct lobes valvate or imbricate; staminodes present and tooth-like or absent, gynoecium ovoid, tricarpellate, syncarpous, trilocular, trilovulate, stigmas small, recurved, ovule campylotropous, laterally inserted. *Fruit* small, globose or oblong, stigmatic remains basal; epicarp smooth, mesocarp fleshy, endocarp thin. *Seed* erect, globose, or ellipsoidal, hilum small, basal, branches of raphe obscure, endosperm cartilaginous; embryo basal to subapical. *Germination* adjacent-ligular; eophyll bifid or pinnate. *Cytology*: 2n = 26, 32.

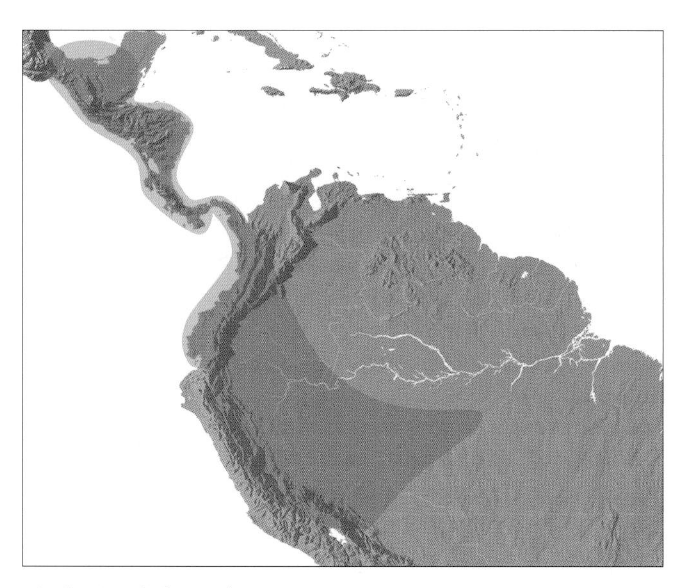

Distribution of ***Chamaedorea***

Distribution and ecology: Approximately 110 species ranging from central Mexico to Brazil and Bolivia. All species are plants of the understory. They occur in moist, wet, or mixed forest in lowlands or mountain forest. Some species occur on limestone.

Far left: ***Chamaedorea seifrizii***, habit, cultivated, Fairchild Tropical Botanic Garden, Florida (Photo: C.E. Lewis)

Left: ***Chamaedorea ernesti-augusti***, staminate flowers, Fairchild Tropical Botanic Garden, Florida. (Photo: C.E. Lewis)

Anatomy: Leaf (Tomlinson 1961, Roth 1990), root (Seubert 1998a, 1998b), seed (Roth 1990). Some features of floral anatomy, including vascularisation of the ovule by a strand from each ventral bundle and abundant raphides in styles and stigmas, are characteristic of other genera in Chamaedoreeae and Ceroxyleae (Uhl & Moore 1971).

Relationships: *Chamaedorea* is monophyletic with high support (Thomas *et al.* 2006, Cuenca & Asmussen-Lange 2007). Most studies resolve *Chamaedorea* as sister to *Gaussia* with moderate to high support (Asmussen *et al.* 2006, Thomas *et al.* 2006, Baker *et al.* in review, in prep.), although a single study resolves *Chamaedorea* as sister to *Wendlandiella* (Uhl *et al.* 1995). For species relationships, see Thomas *et al.* (2006) and Cuenca & Asmussen-Lange (2007).

Common names and uses: Parlour palm, Neanthe Bella (*Chamaedorea elegans*), bamboo palm (*C. seifrizii*). Inflorescences of a few species (e.g., *C. tepejilote*) are eaten as vegetables, and leaves of some species are used for thatch. Some are used

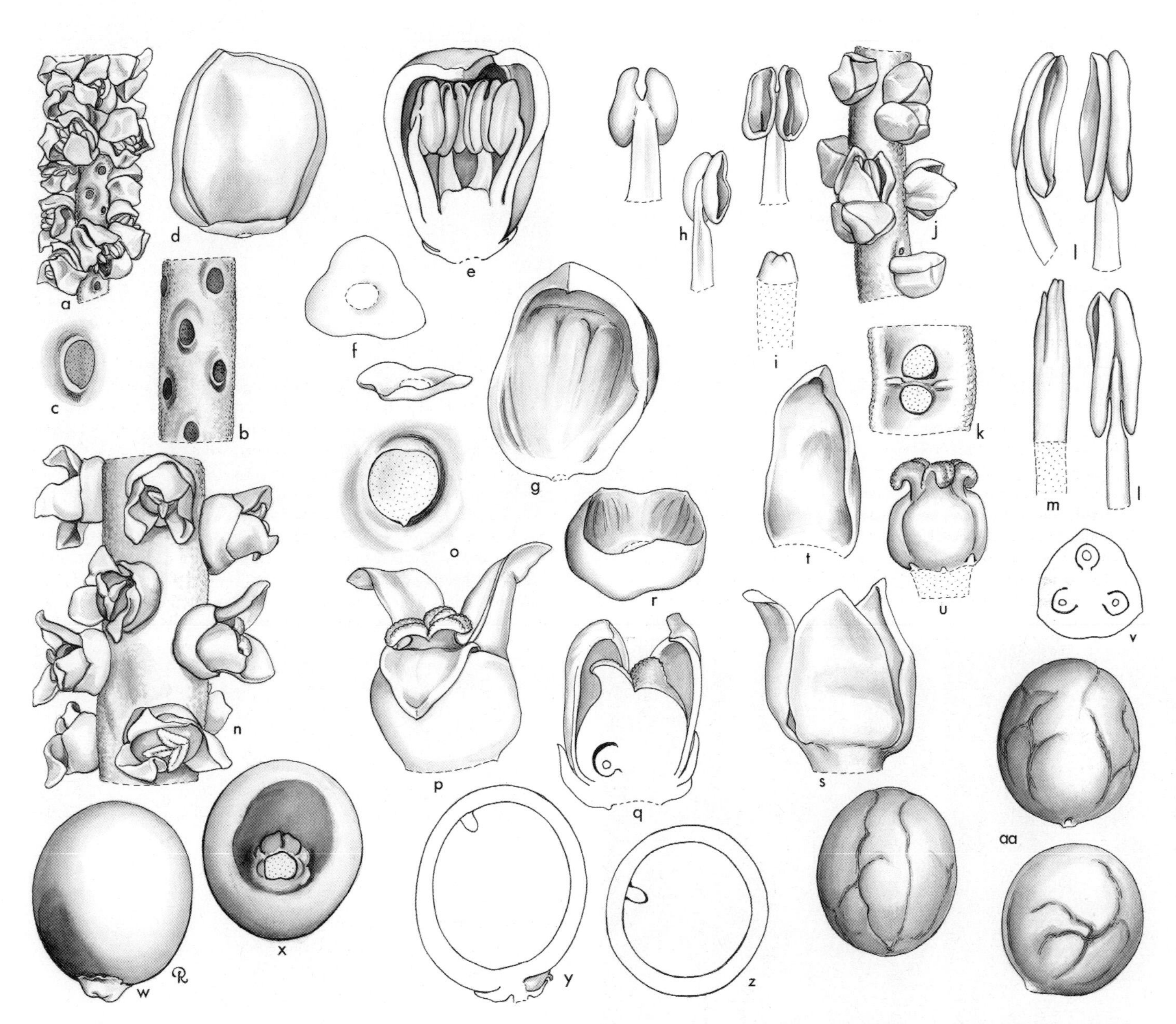

Chamaedorea. **a**, portion of staminate rachilla × 3; **b**, portion of staminate rachilla, flowers removed × 6; **c**, scar of staminate flower × 12; **d**, staminate bud × 12; **e**, staminate bud in vertical section × 12; **f**, staminate calyx, in 2 views × 12; **g**, staminate petal, interior view × 12; **h**, stamen in 3 views × 12; **i**, pistillode × 12; **j**, portion of staminate rachilla with paired flowers × 3; **k**, scars of paired staminate flowers × 12; **l**, stamen in 3 views × 12; **m**, pistillode × 12; **n**, portion of pistillate rachilla × 3; **o**, scar of pistillate flower × 6; **p**, pistillate flower × 6; **q**, pistillate flower in vertical section × 6; **r**, pistillate calyx × 6; **s**, pistillate corolla × 6; **t**, pistillate petal, interior view × 6; **u**, gynoecium and staminodes × 6; **v**, ovary in cross-section × 6; **w**, fruit × $1^1/_2$; **x**, base of fruit showing stigmatic remains and abortive carpels × $1^1/_2$; **y**, fruit in vertical section × $1^1/_2$; **z**, fruit in cross-section × $1^1/_2$; **aa**, seed in 3 views × $1^1/_2$. *Chamaedorea* sp.: **a–i** and **n–aa**, *Moore & Parthasarathy* 9488; *C. linearis*: **j–m**, *Moore et al.* 8354. (Drawn by Marion Ruff Sheehan)

medicinally (Plotkin & Balick 1984). Cut leaves of some species, harvested from the wild, are used as foliage in the cut flower trade. Commercially, several species are extremely important as pot plants, produced in vast quantities.

Taxonomic account: Hodel (1992).

Fossil record: Fossil leaves from the Eocene of southeastern United States have been described as *Chamaedorea danai* by Berry (1916b) but correspondence with the modern genus is not certain. Some *Monocolpopollenites* pollen from the Tertiary of Hungary has been compared to *Chamaedorea* pollen (Kedves & Bohoney 1966), but the pollen is too general for this suggestion to be convincing. *Chamaedorea*-like pollen (Graham 1976) is also reported from the Upper Miocene of Mexico (Paraje Solo flora, Veracruz).

Notes: Hodel (1992) published a detailed and beautifully illustrated account of the genus. In it, he recognises 96 species arranged in eight subgenera. Since then, several more taxa have been published making a grand total of 108. The subgenera are defined by characters of the arrangement of the flowers and their form. Henderson *et al.* (1995) chose to recognise 77 species, reducing some of Hodel's taxa to synonymy but without discussion.

There is great diversity in leaf, inflorescence, and flower form in the genus, which is divided into sections on the basis of floral structure. Because there are so many species, and considerable variation in leaf form within some species, identification of species is often difficult.

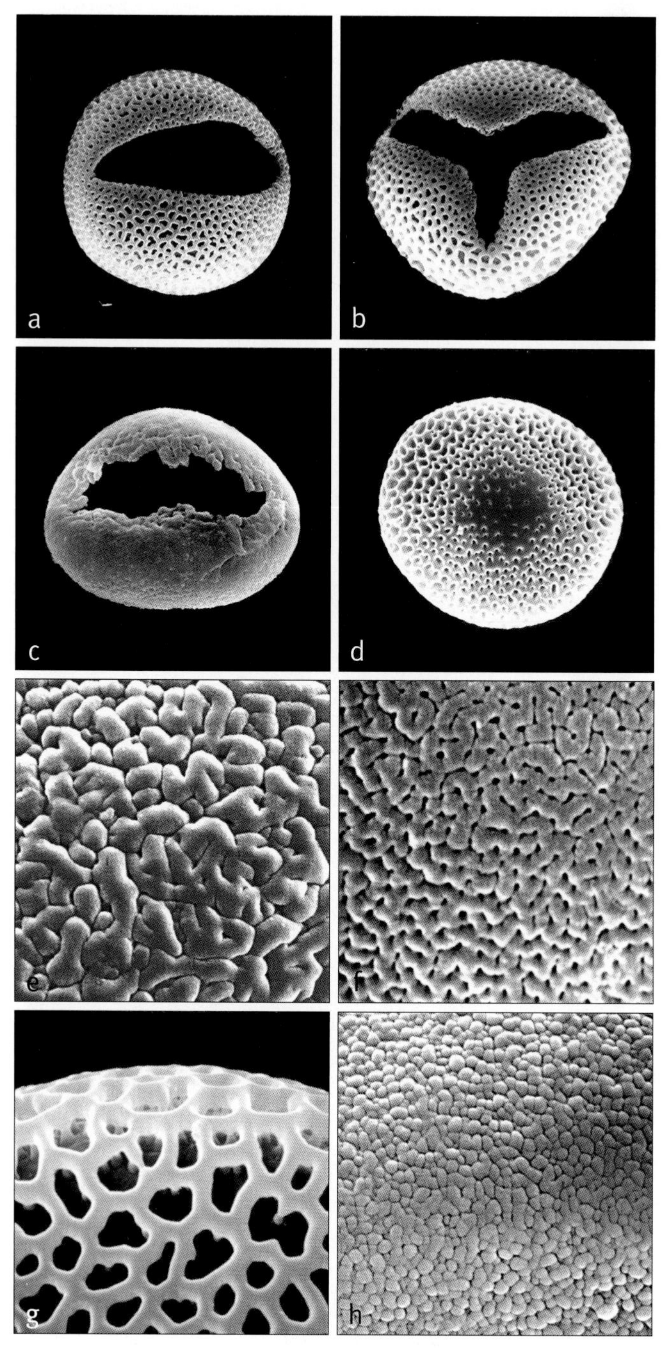

*Above: **Chamaedorea***

a, monosulcate pollen grain, distal face SEM × 2000;
b, trichotomosulcate pollen grain, distal face SEM × 2000;
c, monosulcate pollen grain, distal face SEM × 2000;
d, monosulcate pollen grain, proximal face SEM × 2000;
e, close-up, rugulate pollen surface SEM x 8000;
f, close-up, finely perforate-rugulate pollen surface SEM × 8000;
g, close-up, reticulate pollen surface SEM × 8000;
h, close-up, finely granular-rugulate surface SEM × 8000.
Chamaedorea glaucifolia: **a**, *Moore* 52030; *C.* sp.: **b**, (AH44); *C. alternans*: **c**, *Moore* 8929; *C. ibarrae*: **d**, *Skutch* 4557; *C.* sp. (AH79): **e**, *Plowman* 5540; *C. pinnatifrons*: **f**, *Purdie* s.n.; *C. oblongata*: **g**, cultivated RBG Kew H1364/68; *C. geonomiformis*: **h**, *Moore* 8255.

Below left: ***Chamaedorea benziei***, staminate inflorescence, cultivated, Fairchild Tropical Botanic Garden, Florida. (Photo: C.E. Lewis)

Below right: ***Chamaedorea seifrizii***, infructescence, cultivated, Fairchild Tropical Botanic Garden, Florida. (Photo: C.E. Lewis)

Cook (1937a, 1939b, 1943a, 1943b, 1947a, 1947b) and Cook and Doyle (1939) published the following generic names for species of *Chamaedorea* in the *National Horticultural Magazine* and in *Science*; as the names appear without Latin description and postdate 1935, they are invalid and without any botanical standing: *Anothea, Cladandra, Discoma, Docanthe, Edanthe, Ercheila, Legnea, Lobia, Lophothele, Mauranthe, Meiota, Migandra, Neanthe, Omanthe, Paranthe, Platythea*, and *Vadia*.
Additional figures: Figs 1.1, 1.7, 1.9, Ch. 2 frontispiece, 2.4, 2.6a, Glossary figs 1, 2, 8, 13, 15.

86. GAUSSIA

Solitary, moderate or tall pinnate-leaved monoecious palms from Central America and the Caribbean; stems often bear many inflorescences at the same time; flowers are borne in groups of 3–7.

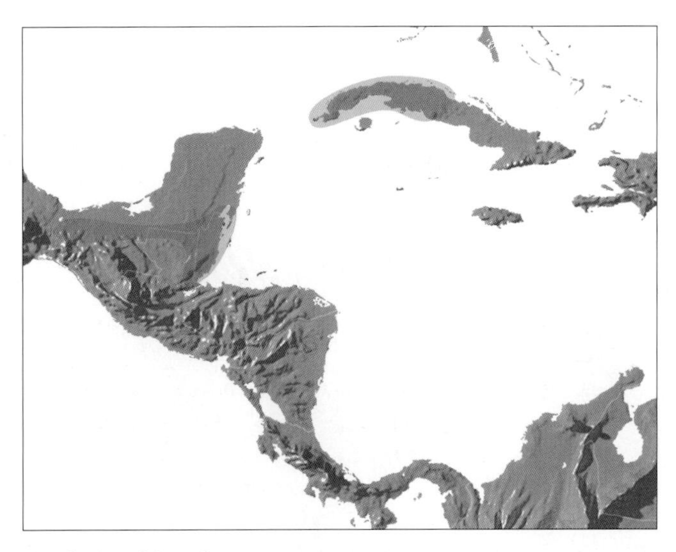

Distribution of ***Gaussia***

Gaussia H. Wendl., Nachr. Königl. Ges. Wiss. Geor g-Augusts-Univ. 327 (1865). Type: **G. princeps** H. Wendl.

Aeria O.F. Cook, Bull. Torrey Bot. Club 28: 547 (1901). Type: *A. attenuata* O.F. Cook (= *Gaussia attenuata* [O.F. Cook] Becc.)

Opsiandra O.F. Cook, J. Wash. Acad. Sci. 13: 182 (1923). Type: *O. maya* O.F. Cook (= *Gaussia maya* [O.F. Cook] H.J. Quero & Read).

Commemorates German mathematician Karl Friedrich Gauss (1777–1855).

Solitary, moderate or tall, unarmed, monoecious, pleonanthic palms. *Stem* occasionally leaning, brown or grey to whitish, straight or somewhat ventricose basally or near the middle, leaf scars prominent and broad, or inconspicuous, roots bearing small prickly lateral roots in a prominent mass at base of trunk. *Leaves* pinnate; sheath wide basally, narrowing distally into the petiole, split opposite the petiole at maturity, not forming a distinct crownshaft, upper edges reflexed, edges sometimes shredded; petiole short, subterete, only narrowly and shallowly channelled adaxially; rachis prominently ridged adaxially, ± rounded abaxially; leaflets acute, single-fold, broadly reduplicate at insertion, obscurely or prominately pulvinate above, lightly waxy, midribs prominent with 2 or more conspicuous secondary pairs of ribs, transverse veinlets not evident. *Inflorescences* interfoliar, not long-persistent, falling with or soon after the subtending leaves, branching to 2–3 orders; peduncle elongate, ascending; prophyll very short, tubular, flat, ± 2-keeled, pointed, open at apex; peduncular

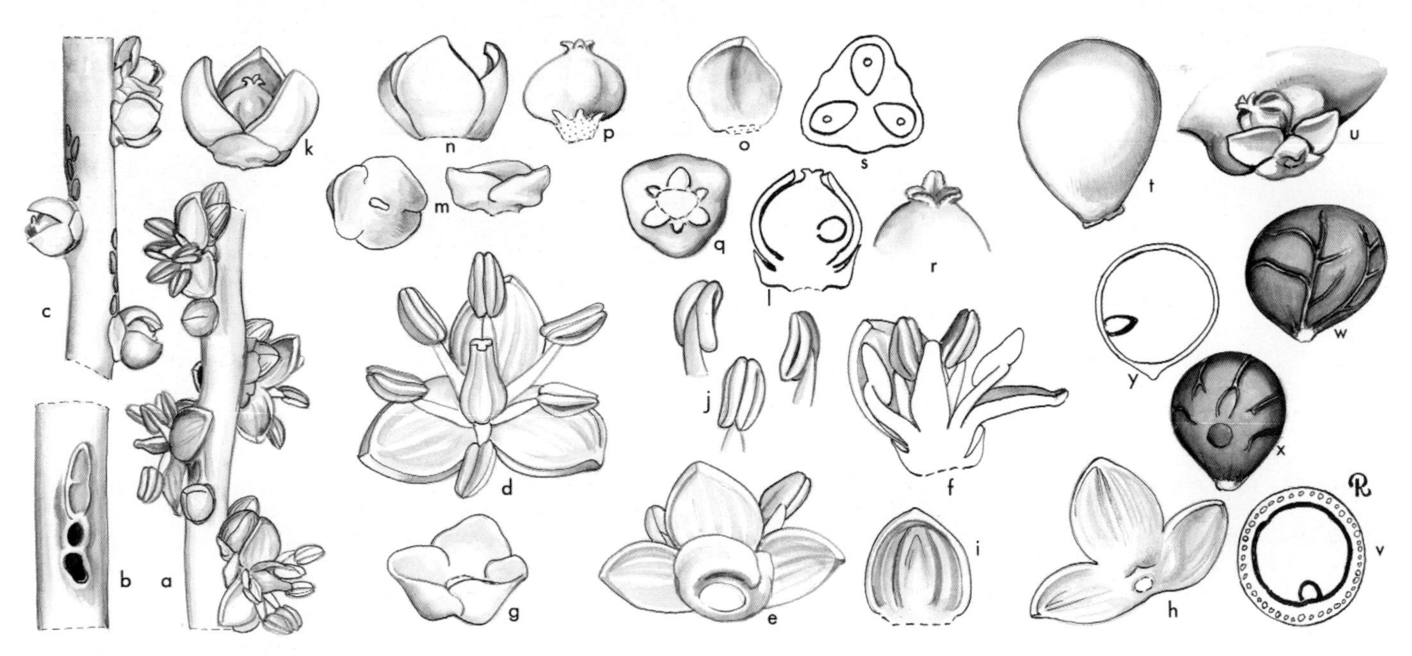

Gaussia. **a**, portion of rachilla showing acervuli with staminate flowers and pistillate buds × 3; **b**, portion of rachilla with floral scars × 3; **c**, portion of rachilla at pistillate anthesis × 3; **d**, staminate flower × 6; **e**, staminate flower from below × 6; **f**, staminate flower in vertical section × 6; **g**, staminate calyx × 6; **h**, staminate corolla, exterior view × 6; **i**, staminate petal, interior view × 6; **j**, stamen in 3 views × 6; **k**, pistillate flower × 6; **l**, pistillate flower in vertical section × 6; **m**, pistillate calyx in 2 views × 6; **n**, pistillate corolla × 6; **o**, pistillate petal × 6; **p**, gynoecium and staminodes × 6; **q**, gynoecium and staminodes from below × 6; **r**, stigmas × 12; **s**, ovary in cross-section × 6; **t**, fruit × 1 1/2; **u**, base of fruit and abortive carpels × 3; **v**, fruit in cross-section × 1 1/2; **w, x**, seed in 2 views × 1 1/2; **y**, seed in vertical section × 1. *Gaussia attenuata*: all from *Read* 757. (Drawn by Marion Ruff Sheehan)

Left: ***Gaussia attenuata***, habit, cultivated, Fairchild Tropical Botanic Garden, Florida. (Photo: J. Dransfield)

Above: ***Gaussia maya***, infructescence, cultivated, Fairchild Tropical Botanic Garden, Florida. (Photo: C.E. Lewis)

Below: ***Gaussia***
a, two monosulcate pollen grains: proximal face left, distal face right SEM × 1000;
b, close-up, finely perforate-rugulate pollen surface SEM × 8000.
Gaussia maya: **a** & **b**, cultivated, Fairchild Botanic Garden 801-58.80101.

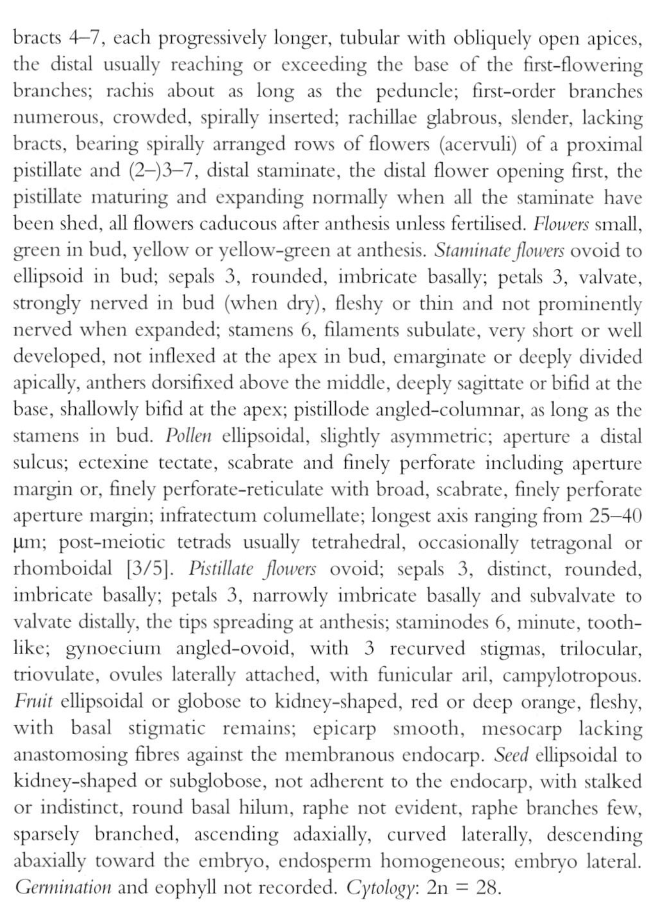

bracts 4–7, each progressively longer, tubular with obliquely open apices, the distal usually reaching or exceeding the base of the first-flowering branches; rachis about as long as the peduncle; first-order branches numerous, crowded, spirally inserted; rachillae glabrous, slender, lacking bracts, bearing spirally arranged rows of flowers (acervuli) of a proximal pistillate and (2–)3–7, distal staminate, the distal flower opening first, the pistillate maturing and expanding normally when all the staminate have been shed, all flowers caducous after anthesis unless fertilised. *Flowers* small, green in bud, yellow or yellow-green at anthesis. *Staminate flowers* ovoid to ellipsoid in bud; sepals 3, rounded, imbricate basally; petals 3, valvate, strongly nerved in bud (when dry), fleshy or thin and not prominently nerved when expanded; stamens 6, filaments subulate, very short or well developed, not inflexed at the apex in bud, emarginate or deeply divided apically, anthers dorsifixed above the middle, deeply sagittate or bifid at the base, shallowly bifid at the apex; pistillode angled-columnar, as long as the stamens in bud. *Pollen* ellipsoidal, slightly asymmetric; aperture a distal sulcus; ectexine tectate, scabrate and finely perforate including aperture margin or, finely perforate-reticulate with broad, scabrate, finely perforate aperture margin; infratectum columellate; longest axis ranging from 25–40 µm; post-meiotic tetrads usually tetrahedral, occasionally tetragonal or rhomboidal [3/5]. *Pistillate flowers* ovoid; sepals 3, distinct, rounded, imbricate basally; petals 3, narrowly imbricate basally and subvalvate to valvate distally, the tips spreading at anthesis; staminodes 6, minute, tooth-like; gynoecium angled-ovoid, with 3 recurved stigmas, trilocular, triovulate, ovules laterally attached, with funicular aril, campylotropous. *Fruit* ellipsoidal or globose to kidney-shaped, red or deep orange, fleshy, with basal stigmatic remains; epicarp smooth, mesocarp lacking anastomosing fibres against the membranous endocarp. *Seed* ellipsoidal to kidney-shaped or subglobose, not adherent to the endocarp, with stalked or indistinct, round basal hilum, raphe not evident, raphe branches few, sparsely branched, ascending adaxially, curved laterally, descending abaxially toward the embryo, endosperm homogeneous; embryo lateral. *Germination* and eophyll not recorded. *Cytology*: 2n = 28.

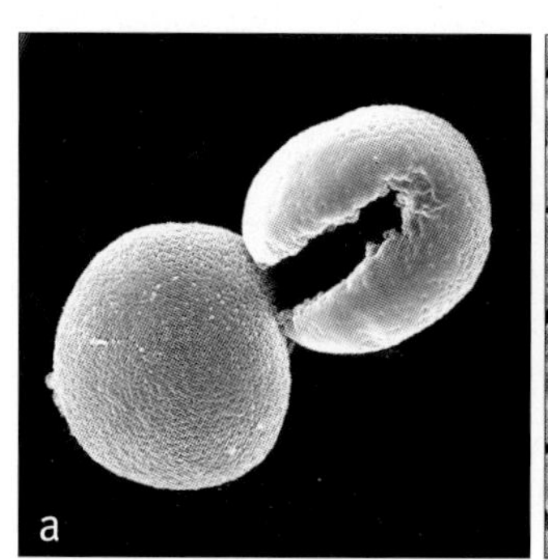

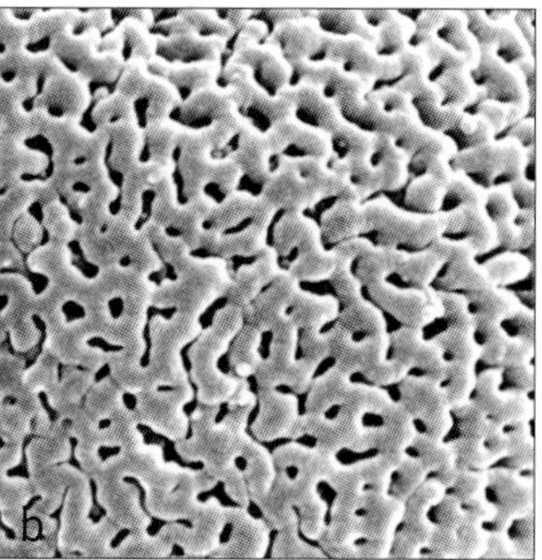

Distribution and ecology: Five species, occurring in Mexico, Guatemala, Belize, Cuba, Hispaniola and Puerto Rico. On limestone hills (mogotes), sometimes growing from crevices in very steep rocks and at low elevations on limestone escarpments or low hills, often on pyramids in the archeological region of Guatemala, less frequently on the forest floor in high forest over limestone.

Anatomy: Leaf (Tomlinson 1961), root (Seubert 1998a, 1998b).

Relationships: *Gaussia* is monophyletic with high support (Cuenca & Asmussen-Lange 2007). For relationships, see *Chamaedorea*. In addition, a phylogeny based on morphology alone places *Gaussia* as sister to the rest of tribe Chamaedoreeae (Cuenca-Navarro 2007).

Common names and uses: Llume palm (*Gaussia attenuata*), maya palm (*G. maya*). Local uses are not recorded. Species are rarely grown as ornamentals.

Taxonomic accounts: Quero & Read (1986), Beccari (1912) and Moya Lopez & Leiva (1991).
Fossil record: No generic records found.
Notes: Species of *Gaussia* superficially resemble *Pseudophoenix* spp. However, they are easily recognised because they have grouped unisexual flowers rather than solitary hermaphroditic flowers.
Additional figures: Figs, 1.7, 7.3.

Tribe Podococceae J. Dransf. & N.W. Uhl, Principes 30: 6 (1986). Type: **Podococcus**.

Slender, erect, unarmed; leaves pinnate, leaflets praemorse, ribs radiating from the base; sheaths forming a crownshaft; inflorescence bisexual, spicate, bearing a prophyll and 2–3 peduncular bracts; triads of flowers borne in deep pits, both staminate and pistillate flowers exerted by receptacular elongation between calyx and corolla; petals of pistillate flower imbricate; gynoecium trilocular, triovulate, fruit 1–3-lobed, lobes horizontal, stigmatic remains basal; seeds 1–3 (1 per lobe).

The Podococceae includes the single genus, *Podococcus*, which is confined to equatorial rain forests on the west coast of Africa. *Podococcus* bears a spicate inflorescence with triads of flowers in pits, which led Hooker to include the genus in tribe Geonomateae. However, the flowers in the triad are arranged with the pistillate flower adaxial to the staminate flowers in all genera of Geonomateae but abaxial to the staminate flowers in *Podococcus*. Pit-closing bracts are well-developed in Geonomateae but reduced in *Podococcus*. Flower and fruit structure also differ greatly.

There is moderate to high support for a clade comprising tribes Podococceae, Oranieae and Sclerospermeae (Uhl *et al.* 1995, some trees of Hahn 2002a, Lewis & Doyle 2002, Loo *et al.* 2006, Baker *et al.* in review, in prep.). This group is sister to a clade of Cocoseae, Roystoneeae and Reinhardtieae (Lewis & Doyle 2002, Baker *et al.* in prep.) or alternatively placed as sister to Areceae, Pelagodoxeae, Leopoldinieae, Euterpeae, Geonomateae and Manicarieae (Baker *et al.* in review). Within the clade, tribe Oranieae is sister to Sclerospermeae with moderate to high support (Loo *et al.* 2006, Baker *et al.* in review, in prep.). Different relationships between these three tribes have been recovered elsewhere. For example, Lewis and Doyle (2002) resolve the Sclerospermeae as sister to Podococceae with moderate support. A sister-group relationship between Podococceae and Cyclospatheae is recovered with high support in the results of Hahn (2002b), but this relationship has not been observed in any other phylogeny. Another alternative position of Podococceae, as sister to Reinhardtieae with moderate support, was recovered by Asmussen and Chase (2001) and by Hahn (2002a). Tribe Oranieae has also been resolved as sister to the Cocoseae (Hahn 2002a, 2002b) and as sister to tribes Euterpeae, Geonomateae and Areceae (Lewis & Doyle 2001).

87. PODOCOCCUS

Slender colonial monoecious palms with pinnate leaves and praemorse leaflets, sometimes acaulescent, sometimes erect, found in Equatorial West African rain forest. Inflorescence is spicate with pits and flowers borne on stalks; fruit has basal stigmatic remains.

Podococcus G. Mann & H. Wendl., Trans. Linn. Soc. London 24: 426 (1864). Type: **P. barteri** G. Mann & H. Wendl.

Podos — *foot,* kokkus — *grain, seed or berry, referring to the characteristic stalked fruit.*

Small, colonial, unarmed, acaulescent or erect, pleonanthic, monoecious palms. *Stem* erect, slender, reed-like, or subterranean, covered in reddish-brown fibrous leaf bases, eventually bare and ringed with leaf scars, axillary stolons present basally and extending horizontally and eventually vertically, developing roots, and becoming new shoots; prop roots with pneumatophores developed basally. *Leaves* few, marcescent, pinnate, the first leaves on new shoots undivided, elliptical, pinnately ribbed, margins toothed, apex very shallowly bifid; sheath tubular, becoming split opposite the petiole, margins fibrous; petiole very slender, narrowly channelled adaxially, rounded abaxially, with dot-like scales and deciduous tomentum; rachis like the petiole but longer, tapering, extending through the terminal, often minutely bifid leaflet; leaflets rhombic, single-fold, basal half of each leaflet wedge-shaped with smooth margins, the upper half triangular, margins doubly toothed, 5 large ribs divergent from the base, midrib not evident, blade glabrous

Right: ***Podococcus barteri***, habit, Cameroon. (Photo: J. Dransfield)

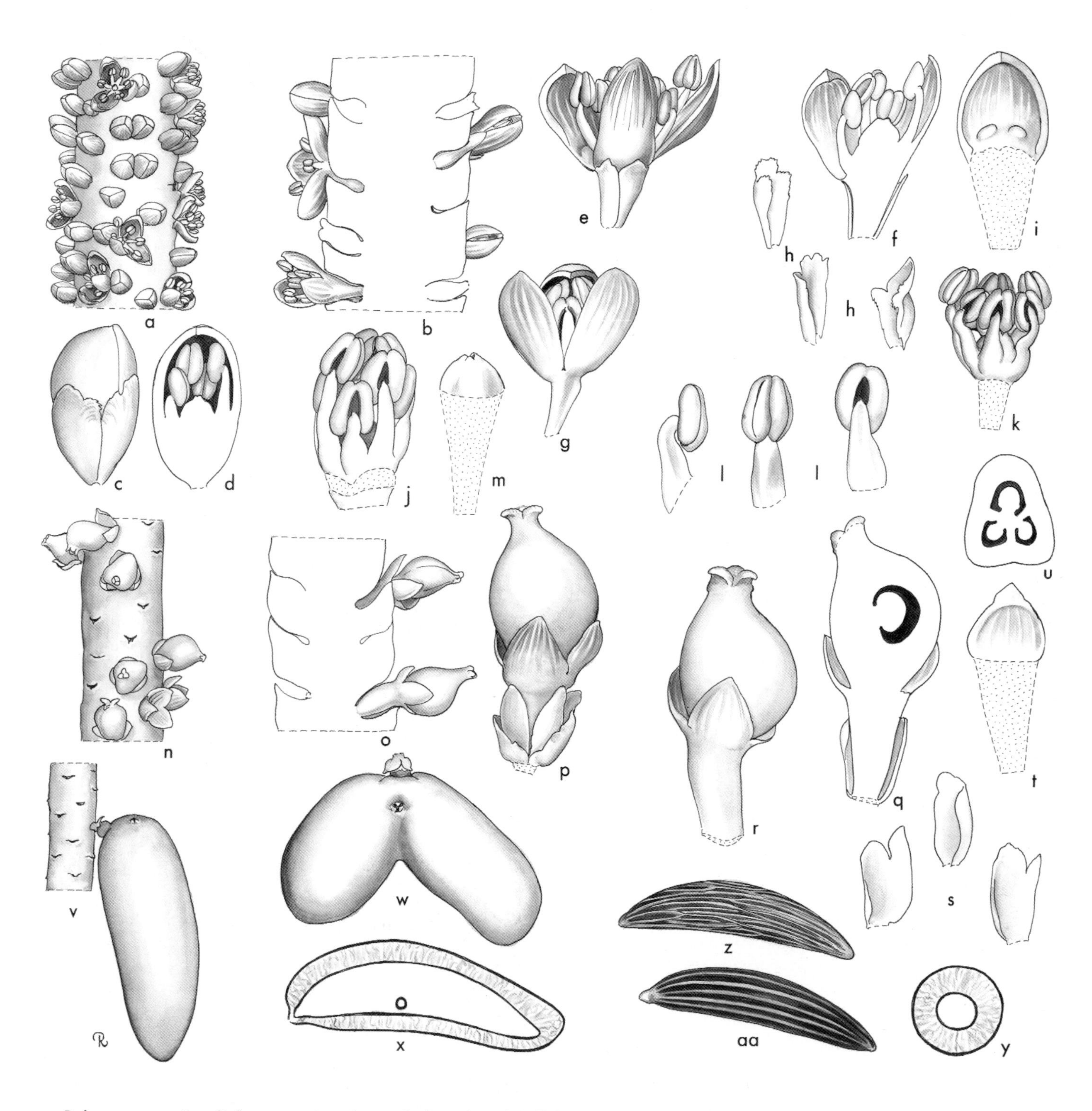

Podococcus. **a**, portion of inflorescence at staminate anthesis × 3; **b**, portion of inflorescence at staminate anthesis in vertical section × 3; **c**, staminate bud × 10; **d**, staminate bud in vertical section × 10; **e**, staminate flower at anthesis × 10; **f**, staminate flower at anthesis in vertical section × 10; **g**, staminate flower, sepals removed to show elongate base × 10; **h**, staminate sepals × 10; **i**, staminate petal, interior view × 10; **j**, androecium in bud × 10; **k**, androecium at anthesis × 10; **l**, stamen in 3 views × 10; **m**, pistillode × 10; **n**, portion of inflorescence with very young fruit × 3; **o**, portion of inflorescence with very young fruit in vertical section × 3; **p**, pistillate flower in very young fruit × 10; **q**, pistillate flower in very young fruit in vertical section × 10; **r**, pistillate flower, sepals removed to show elongate base × 10; **s**, pistillate sepals × 10; **t**, pistillate petal, interior view × 10; **u**, ovary in cross-section × 10; **v**, portion of inflorescence with 1-seeded fruit × $1^1/_2$; **w**, 2-seeded fruit × $1^1/_2$; **x**, 1-seeded fruit in vertical section × $1^1/_2$; **y**, 1-seeded fruit in cross-section × $1^1/_2$; **z**, seed with adherent fibers × $1^1/_2$; **aa**, seed with fibers removed × $1^1/_2$. *Podococcus barteri*: all from *Moore* 9900. (Drawn by Marion Ruff Sheehan)

adaxially, densely covered with slender, bulbous-based, tanniniferous hairs abaxially, transverse veinlets not evident. *Inflorescences* solitary, interfoliar, spicate, erect at first, pendulous in fruit, protandrous; peduncle very slender, with deciduous tomentum; prophyll basal, tubular, 2-keeled, disintegrating into long fibres; peduncular bracts 2–3, inserted at wide intervals above the prophyll, tubular, disintegrating as the prophyll; rachis about as long as or longer than the peduncle, bearing spirally arranged triads of flowers basally, paired or solitary staminate flowers distally, flowers enclosed in pits; rachis bracts subtending the triads evident only as raised, curved lower margins of the pits, densely covered in stellate hairs between and inside the pits; floral bracteoles membranous, irregular, mostly pointed. *Staminate flowers* ± adaxial to the pistillate; staminate sepals 3, distinct, somewhat keeled dorsally, tips irregular or rounded, imbricate basally; petals 3, valvate, about twice as long as the sepals, at anthesis adnate about $^2/_3$ their length to a solid receptacle, nearly equalling the depth of the pit, distinct lobes exserted, spreading; stamens 6, in 2 whorls, the outer opposite the sepals and shorter than the inner antepetalous whorl, filaments awl-shaped, rather short, incurved distally, anthers short, dorsifixed near the base, latrorse, connective tanniniferous; pistillode short, briefly 3-lobed. *Pollen* ellipsoidal, with slight or obvious asymmetry; aperture a distal sulcus; ectexine tectate, finely perforate, perforate and micro-channelled, or perforate-rugulate, aperture margin slightly finer; infratectum columellate; longest axis 27–32 µm [1/1]. *Pistillate flower* symmetrical; sepals 3, distinct, irregular, variously notched distally, about half as long as the petals, imbricate; petals connate and adnate basally $^1/_3$ to $^1/_2$ their length to a solid receptacle, nearly equalling the depth of the pit, apices imbricate in bud, spreading at anthesis; staminodes 6 (according to Mann & Wendland 1864), minute; gynoecium ovoid, trilocular, triovulate or 2 locules abortive, style not evident, stigmas 3, short, recurved at anthesis, ovule pendulous. *Fruit* long, ellipsoidal or with ellipsoidal lobes, often slightly curved, bright orange, 1–3-seeded, stigmatic remains central in 3-seeded fruits, lateral in 1–2-seeded fruits, the base briefly stalked, lobes developing horizontally to the floral axis; epicarp thin, smooth, rather leathery, mesocarp gelatinous with an internal layer of fibres ± adherent to the seed, endocarp crustaceous. *Seed* ellipsoidal, basally attached, hilum basal, raphe branches anastomosing from the base, endosperm homogeneous; embryo lateral near the middle, (?) ± basal relative to the plane of the seed. *Germination* adjacent-ligular; eophyll rhombic, undivided, minutely bifid, toothed above the middle. *Cytology* not known.

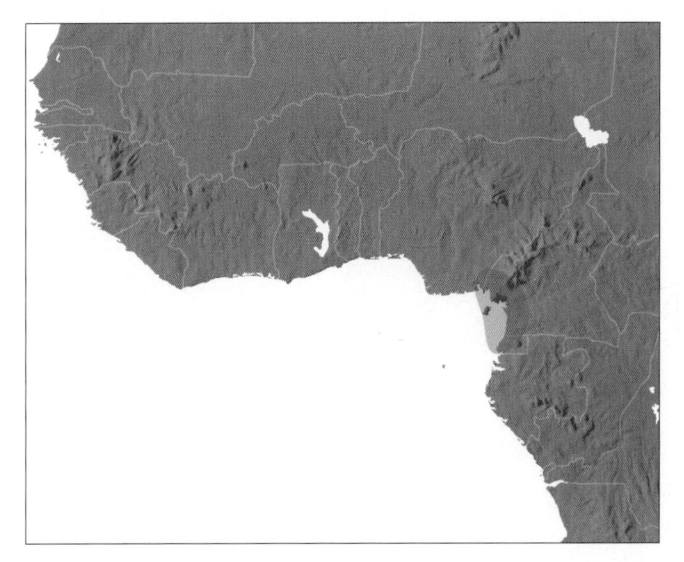

Distribution of ***Podococcus***

Distribution and ecology: Two species of western Africa from Nigeria to Gabon, ranging from the Niger Delta almost to the Congo River. Found in rain forest at low elevations. The genus apparently never occurs more than 200 km inland. Its distribution conforms roughly to the limits of the Biafran forest and closely related associations, and it is not found in post-cultivation or roadside habitats. The species mostly occur on soils of the feralitic latosol type but are not confined to one soil type (Bullock 1980a).

Below left: ***Podococcus acaulis***, crown with inflorescences, Gabon. (Photo: J. van Valkenburg)

Below right: ***Podococcus acaulis***, part of rachilla with fruits, Gabon. (Photo: J. van Valkenburg)

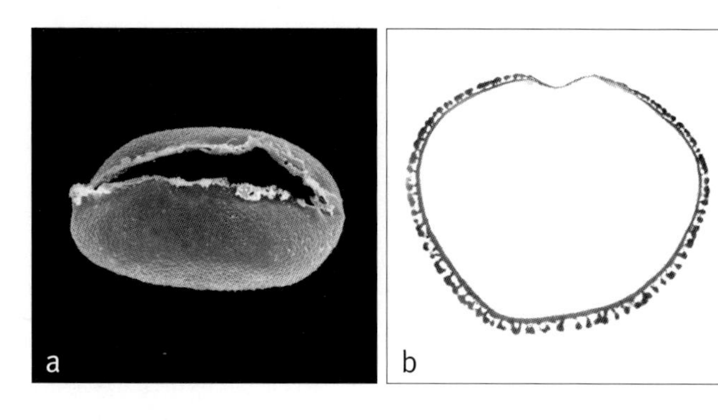

Above: **Podococcus**
a, monosulcate pollen grain, distal face SEM × 1000;
b, ultrathin section through whole pollen grain, polar plane TEM × 2000.
Podococcos barteri: **a** & **b**, *Onochie* 33423.

Anatomy: Silica bodies of stegmata spherical, central vascular bundles of petiole with two phloem strands (Parthasarathy pers. comm.); root (Seubert 1998a, 1998b); sepals of both staminate and pistillate flowers usually lacking a vascular supply or rarely second sepal in pistillate flower with 1–2 small fibrous bundles; petals with 4–5 parallel bundles in pistillate and about 9 parallel bundles in staminate flowers; stamens each with a single trace; in young gynoecium, carpels connate with a distal septal nectary and the pistillode also apparently secretory; each carpel with 1 dorsal and 2 ventral bundles and a pendulous ovule with two integuments (Uhl unpublished).

Relationships: The monophyly of *Podococcus* has not been tested. For relationships, see tribe Podococceae.

Common names and uses: Not recorded.

Taxonomic account: van Valkenburg *et al*. (2007).

Fossil record: No generic records found.

Notes: *Podococcus* is similar in leaf and gynoecial structure to some members of Iriarteeae but several characters set it apart.

Additional figures: Glossary fig. 24.

Tribe Oranieae Becc., Webbia 11: 172 (1955). Type: Orania.

Moderate to robust, unarmed; leaves pinnate, leaflets praemorse, ribs parallel; sheaths not forming a crownshaft; inflorescence bisexual; prophyll much shorter than the 1–2 peduncular bracts; triads of flowers superficial; petals of pistillate flower valvate; gynoecium trilocular, triovulate; fruit 1–3-seeded, rounded when 1-seeded, otherwise lobed, stigmatic remains basal; seeds 1–3 (1 per lobe); epicarp smooth; endocarps with a basal button; embryo lateral or subapical; endosperm homogeneous.

The tribe comprises the single genus *Orania*, with a remarkable disjunct geographic range, occurring in Madagascar and the Malesian region.

For relationships, see tribe Podococceae.

88. ORANIA

Moderate to robust single-stemmed palms with praemorse leaflets and lacking crownshafts, found in Madagascar and from south Thailand through to New Guinea, where most species are found; the peduncular bract(s) greatly exceeds the prophyll; fruits are relatively large and rounded.

Orania Zipp. in Blume, Alg. Konst-Lett.-Bode 1: 297 (1829). Type: **O. regalis** Zipp.

Arausiaca Blume, Rumphia 2: viii, t. 122 (1838–1839). Type: *A. excelsa* Blume (= *O. regalis* Zipp.).

Macrocladus Griff., Calcutta J. Nat. Hist. 5: 489 (1845). Type: *M. sylvicola* Griff. (= *O. sylvicola* [Griff.] H.E. Moore).

Sindroa Jum., Ann. Inst. Bot.-Géol. Colon. Marseille ser. 5.1 (1): 11 (1933). Type: *S. longisquama* Jum. (= *O. longisquama* [Jum.] J. Dransf. & N.W. Uhl).

Halmoorea J. Dransf. & N.W. Uhl, Principes 28(4): 164 (1984). Type: *H. trispatha* J. Dransf. & N.W. Uhl (= *O. trispatha* (J. Dransf. & N.W. Uhl) Beentje & J. Dransf.)

Commemorates F.G.L. Willem van Nassau, Prince of Orange (Oranje) and Crown Prince of the Netherlands (1792–1849).

Small to large, solitary, unarmed, pleonanthic, monoecious palms. *Stem* erect, short to tall, becoming bare, conspicuously ringed with leaf scars, and sometimes bearing corky, warty protuberances. *Leaves* distichous (*Orania disticha*, *O. ravaka*, *O. trispatha* and sometimes *O. lauterbachiana*) or spirally arranged, large, pinnate, deciduous under their own weight; crownshaft not present; sheath well developed, splitting longitudinally opposite the petiole, usually densely tomentose, distally narrowing into the petiole;

Right: ***Orania palindan***, habit, Papua. (Photo: W.J. Baker)

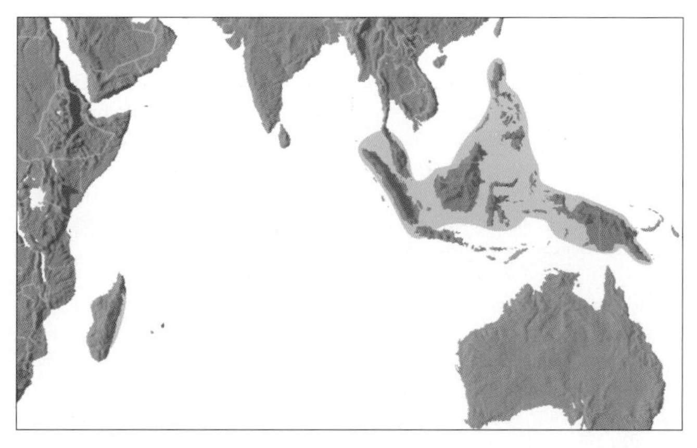

Distribution of ***Orania***

petiole usually relatively short, channelled adaxially, rounded abaxially, bearing abundant persistent or caducous tomentum; rachis much longer than the petiole; leaflets single-fold, regularly arranged and held ± in one plane, or rarely (*O. archboldiana*) grouped and held in several planes giving the leaf a plumose appearance, linear-lanceolate, often narrow, frequently somewhat plicate, apices praemorse, adaxial surface glabrous, dark green, abaxial surface covered with dense white indumentum and rarely with brown ramenta along the mainvein (Madagascar species), midrib very prominent adaxially, transverse veinlets obscure. *Inflorescences* axillary, interfoliar, solitary, often massive, branching to 1–3 orders, protandrous; prophyll short, tubular, 2-keeled, included within the subtending leaf, usually becoming frayed distally; peduncular bracts usually 1, rarely 2, borne just above the prophyll, very large and conspicuous, almost woody, tubular, completely enclosing the inflorescence in bud, before anthesis splitting along their length to expose the inflorescence, eventually deciduous, apically with a solid flattened, lanceolate beak, and bearing sparse to abundant tomentum, often grooved on drying; peduncle ± circular in cross-section, short to very long, variously tomentose; subsequent first-order bracts very inconspicuous except rarely in *O. oreophila* where well developed in one collection; rachis shorter or longer than the peduncle; first-order branches often with a basal pulvinus; further branches, where present, each subtended by an inconspicuous triangular bract; rachillae usually spreading, flexuous, (in *O. regalis* congested), glabrous or variously tomentose, bearing rather distant triads proximally and solitary or paired staminate flowers distally, more rarely with triads almost throughout, or with staminate flowers throughout; triads subdistichous or spirally arranged, ± superficial, subtended by a minute triangular rachilla bract; floral bracteoles minute or not visible. *Staminate and pistillate flowers* superficially rather similar, cream-coloured. *Staminate flowers* narrower and longer than the pistillate; calyx very short, flattened, with 3, low triangular lobes or with 3 distinct imbricate lobes; petals 3, distinct, valvate, broad to narrow-lanceolate, ± striate in dried state; stamens 3, 4, 6 or 9–32, filaments distinct or variously connate, short to moderate, rather fleshy, anthers elongate, basifixed, erect, with large connective, extrorse, or latrorse; pistillode usually lacking, minute and trilobed in *O. palindan*, sometimes present in *O. sylvicola* (minute, conical).

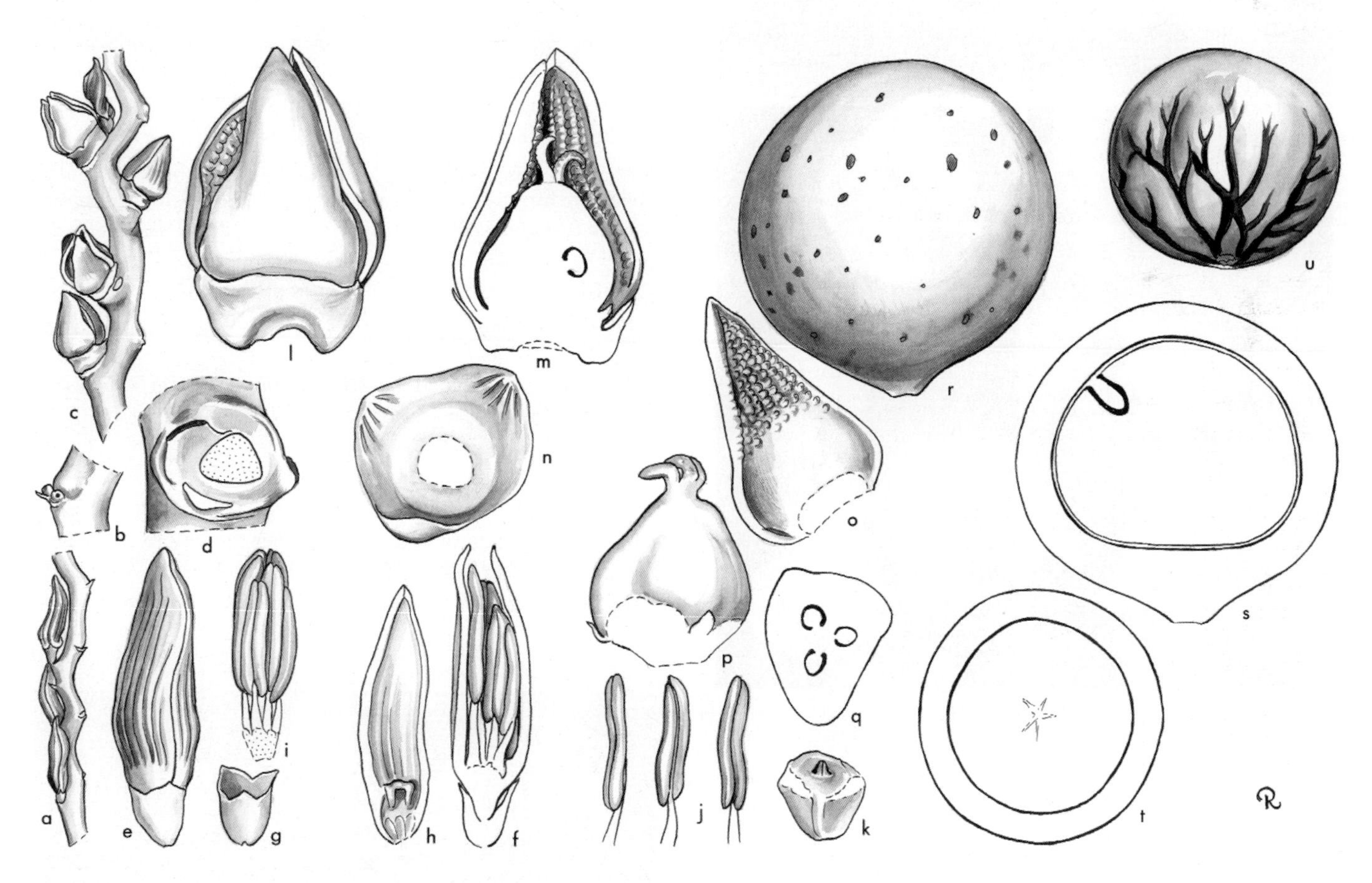

Orania. **a**, portion of rachilla with paired staminate flowers × $1\frac{1}{2}$; **b**, scars of paired staminate flowers × 3; **c**, portion of rachilla with triads in pistillate flower × $1\frac{1}{2}$; **d**, triad, flowers removed × 6; **e**, staminate bud × 6; **f**, staminate bud in vertical section × 6; **g**, staminate calyx × 6; **h**, staminate petal, interior view × 6; **i**, androecium × 6; **j**, stamen in 3 views × 6; **k**, pistillode × 12; **l**, pistillate flower × 6; **m**, pistillate flower in vertical section × 6; **n**, pistillate calyx × 6; **o**, pistillate petal, interior view × 6; **p**, gynoecium × 6; **q**, ovary in cross-section × 6; **r**, fruit × 1; **s**, fruit in vertical section × 1; **t**, fruit in cross-section × 1; **u**, seed × 1. *Orania paraguanensis*: all from *Moore & Meijer* 9222. (Drawn by Marion Ruff Sheehan)

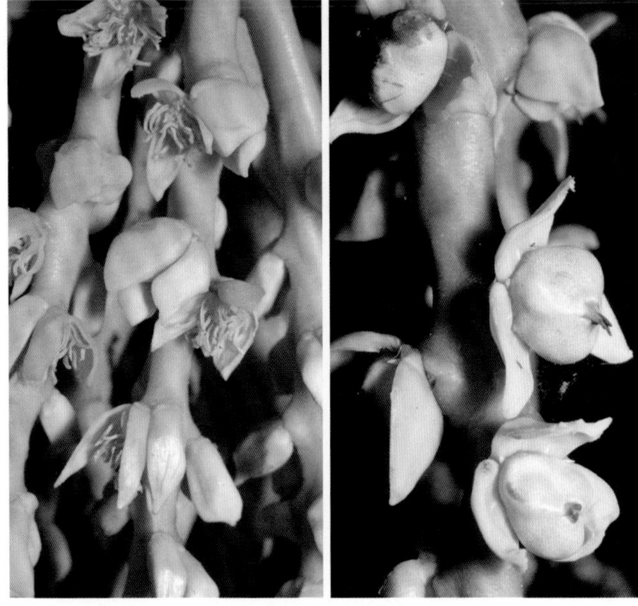

Top: ***Orania palindan***, inflorescence, Papua. (Photo: W.J. Baker)

Above left: ***Orania ravaka***, staminate flowers, Madagascar. (Photo: J. Dransfield)

Above right: ***Orania ravaka***, pistillate flowers, Madagascar. (Photo: J. Dransfield)

Pollen ellipsoidal, slight or obvious asymmetry; aperture a distal sulcus; ectexine tectate, finely perforate, perforate and micro-channelled, or perforate-rugulate, aperture margin broad, psilate-perforate; infratectum columellate; longest axis ranging from 23–40 µm (Thankaimoni 1970) [2/25]. *Pistillate flowers* ± conical or pyramidal; calyx flattened, very short, with 3 low, triangular lobes or with 3 distinct imbricate sepals; petals 3, distinct, valvate, triangular; staminodes 3–11, very short, awl-shaped, or well developed, possibly rarely producing pollen (some collections of *O. sylvicola*); gynoecium trilocular, triovulate, ± pyramidal, stigmas 3, short, recurved at anthesis, ovule form unknown. *Fruit* developing from 1, 2, or rarely 3 carpels, orange, green, or dull orange to yellowish-brown at maturity, spherical or very slightly pear- shaped, where more than 1 carpel developing, each lobe spherical, stigmatic remains subbasal; epicarp smooth, mesocarp thin or thick, fleshy, traversed by numerous short radial fibres, endocarp rather thin. Seed spherical, basally attached with a ± circular hilum, the surface of the seed somewhat grooved by a sparse network of fibres, endosperm homogeneous, sometimes with a very small central hollow; embryo subapical or lateral. *Germination* remote-tubular; eophyll bifid with praemorse apices, or rarely pinnate. *Cytology*: 2n = 32.

Distribution and ecology: About 25 species distributed in south Thailand, Malay Peninsula, Sumatra, Java, Borneo, Philippines, Sulawesi, Moluccas and New Guinea, and three species in Madagascar. The greatest diversity occurs in New Guinea, with a minor radiation in the Philippines. Most species are large tree palms of the canopy or subcanopy of humid tropical rain forest in the lowlands or hills up to ca. 1700 m; *Orania parva* and *O. oreophila* are smaller palms of the forest undergrowth. There is some evidence that *O. sylvicola* avoids the highest rainfall areas within its range of distribution. Nothing is known of pollination or dispersal.

Anatomy: Leaf (Tomlinson 1961), root (Seubert 1998a, 1998b).

Relationships: *Orania* is monophyletic with high support (Lewis & Doyle 2002, Asmussen *et al.* 2006, Baker *et al.* in review). For relationships, see tribe Oranieae.

Common names and uses: Orania palms, *ibul*, *sindro*. Outer part of the trunk is reputed to be strong and has been used to make spears. The 'cabbage' of all species seems to be poisonous and avoided by local people; *Orania sylvicola* is reputed to be very poisonous in all its parts.

Taxonomic account: Keim (in prep.).

Fossil record: No generic records found.

Notes: *Orania* is of considerable interest. Not only does it have an astonishingly disjunct distribution but the inflorescence and flowers are rather unspecialised within the Areceae.

Additional figures: Figs 1.6, Ch. 9 Frontispiece, Glossary figs 18, 20, 22.

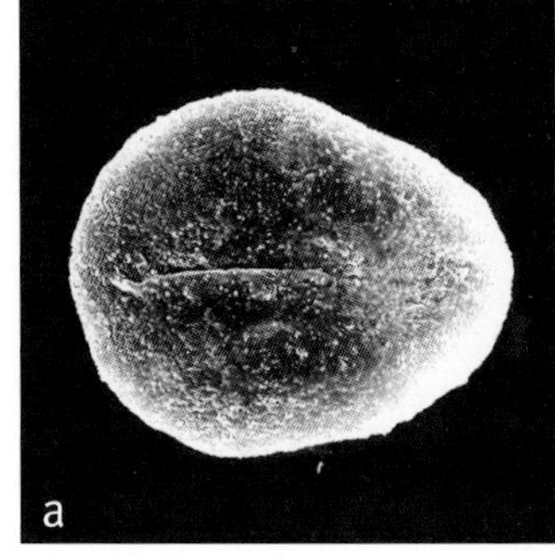

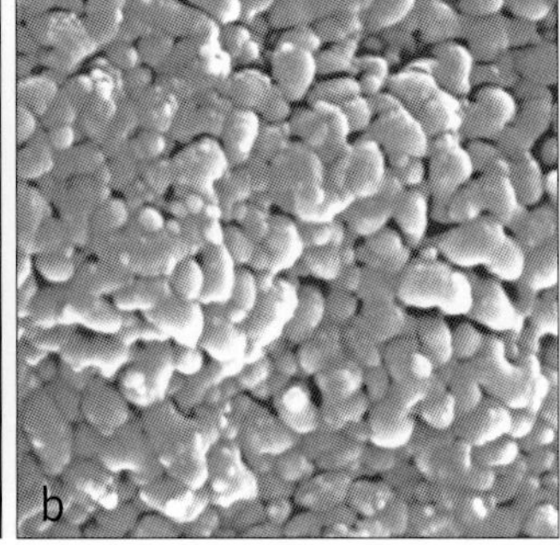

Right: ***Orania***
a, monosulcate pollen grain, distal face SEM × 1500;
b, close-up, finely rugulate pollen surface SEM × 10,000.
Orania trispatha: **a** & **b**, *Moore* 9921.

Tribe Sclerospermeae J. Dransf., N.W. Uhl, C. Asmussen, W.J. Baker, M.M. Harley & C. Lewis., Kew Bull. 60: 562 (2005). Type: **Sclerosperma**.

Moderate, unarmed, acaulescent; leaves entire or irregularly pinnate; sheaths not forming a crownshaft; margins praemorse, ribs parallel; inflorescence, spicate, bisexual, interfoliar; prophyll much shorter than the first peduncular bract; peduncular bract tending to enclose the inflorescence at anthesis, subsequent peduncular bracts small, incomplete; triads of flowers superficial; staminate flower with 60–100 stamens; gynoecium pseudomonomerous; fruit 1-seeded, stigmatic remains apical; epicarp smooth; endocarp with a basal button; embryo basal.

The unusual genus *Sclerosperma* is the only member of the tribe and is restricted to the perhumid equatorial rain forests surrounding the Gulf of Guinea in Africa.

For relationships, see tribe Podococceae.

89. SCLEROSPERMA

Acaulescent thicket-forming pinnate-leaved palms found in Equatorial West African rain forest; leaves are undivided or divided into leaflets, the blade margins praemorse, and the blade discolorous; inflorescence is short, unbranched, and hidden among leaf sheath bases.

Sclerosperma G. Mann & H. Wendl., Trans. Linn. Soc. London 24: 427 (1864). Type: **S. mannii** H. Wendl.

Skleros — *hard*, sperma — *seed, referring to the very hard endosperm.*

Short or acaulescent, clustering, unarmed, pleonanthic, monoecious palms. *Stem* if evident, creeping or erect, rather stout, closely ringed with leaf scars. *Leaves* reduplicate, bifid or divided, very large, deeply bifid in juveniles, ascending; sheath rather short, splitting opposite the petiole, margins fibrous; petiole long, slender, adaxially channelled, abaxially rounded; leaflets when present, composed of several very narrow folds, midribs prominent, marginal ribs next largest, blade adaxially dark, abaxially covered with a dense layer of amorphous white indumentum and with small scales along the veins, folds apically praemorse, margins minutely toothed, transverse veinlets not evident. *Inflorescences* interfoliar, concealed among the leaf bases and sometimes partially obscured by accumulated debris, spicate; peduncle very short, elliptic in cross-section, densely tomentose; prophyll rather short, strongly 2-keeled, becoming fibrous; peduncular bract longer than the prophyll, tubular, forming a fibrous net around the flowers, opening distally and inflorescence becoming partially exserted, 2 incomplete, pointed peduncular bracts borne laterally just below the flowers; rachis longer than the peduncle, but short, stout, bearing a few (ca. 12) triads of flowers at the base and numerous rows of staminate flowers distally, triads each subtended by a shallow pointed fibrous bract, the distal staminate flowers by small acute bracts; floral bracteoles present in triads, flat, ± rounded and partially united. *Staminate flowers* in triads ± pedicellate and asymmetrical, distal flowers sessile, symmetrical; sepals 3, distinct, imbricate basally, elongate, tapering, truncate apically or with a short central point; petals 3, distinct, valvate but tips flattened and buds truncate apically, thick; stamens 60–100, filaments very short, ± triangular, anthers elongate, basifixed, latrorse, connective prominent, apiculate; pistillode lacking. *Pollen* symmetric oblate-triangular in polar view, heteropolar; three operculate pores positioned subapically on the distal face; ectexine tectate, perforate, perforate-rugulate, rugulate or

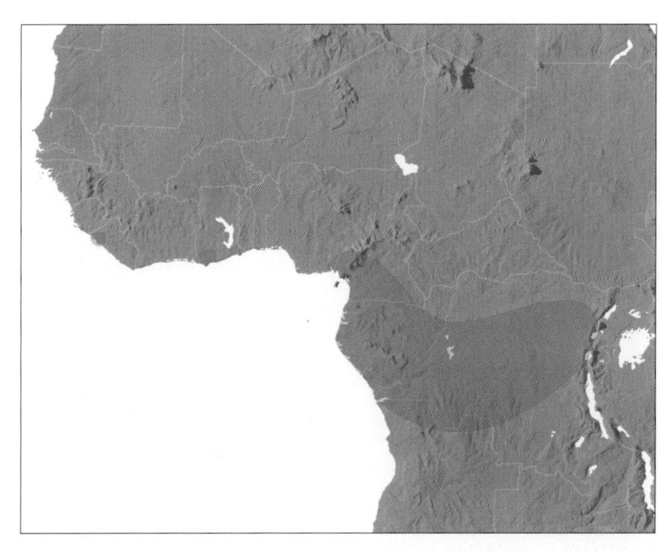

Distribution of ***Sclerosperma***

reticulate, aperture margins similar or slightly finer; infratectum columellate; longest axis 37–59 μm; post-meiotic tetrad tetrahedral [2/3]. *Pistillate flowers* larger than the staminate, broadly ovoid; sepals 3, connate in a 3-lobed, glabrous cupule or margins of 2 sepals distinct and imbricate, somewhat angled by mutual pressure; petals 3, distinct, asymmetrical, broadly imbricate with thick valvate tips; staminodes 6,

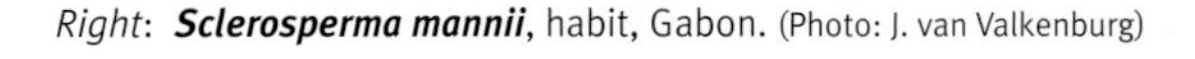

Right: ***Sclerosperma mannii***, habit, Gabon. (Photo: J. van Valkenburg)

very small, triangular or absent; gynoecium ovoid, unilocular, uniovulate, covered in thin brown scales, bearing a large, cap-like, 3-angled stigma; ovule ± pendulous, probably campylotropous. *Fruit* globose to obovoid, depressed apically around a short beak of stigmatic remains, purplish to black at maturity; epicarp thin, mesocarp thin, parenchymatous with silica(?) inclusions, endocarp bony, thick, irregularly and shallowly pitted externally, with basal pore region. *Seed* globose to obovoid, somewhat rough, hilum elongate, endosperm homogeneous; embryo basal. *Germination* remote-tubular; eophyll bifid. *Cytology* not studied.

Distribution and ecology: Three species in humid equatorial West Africa, usually occurring in low, wet, swampy areas.

Anatomy: Leaf (Tomlinson 1961).

Relationships: For relationships, see the tribe Podococceae.

Common names and uses: Common names, see van Valkenburg *et al.* (2008). Leaves are used for thatch and the seeds are eaten.

Taxonomic account: van Valkenburg *et al.* (2007, 2008).

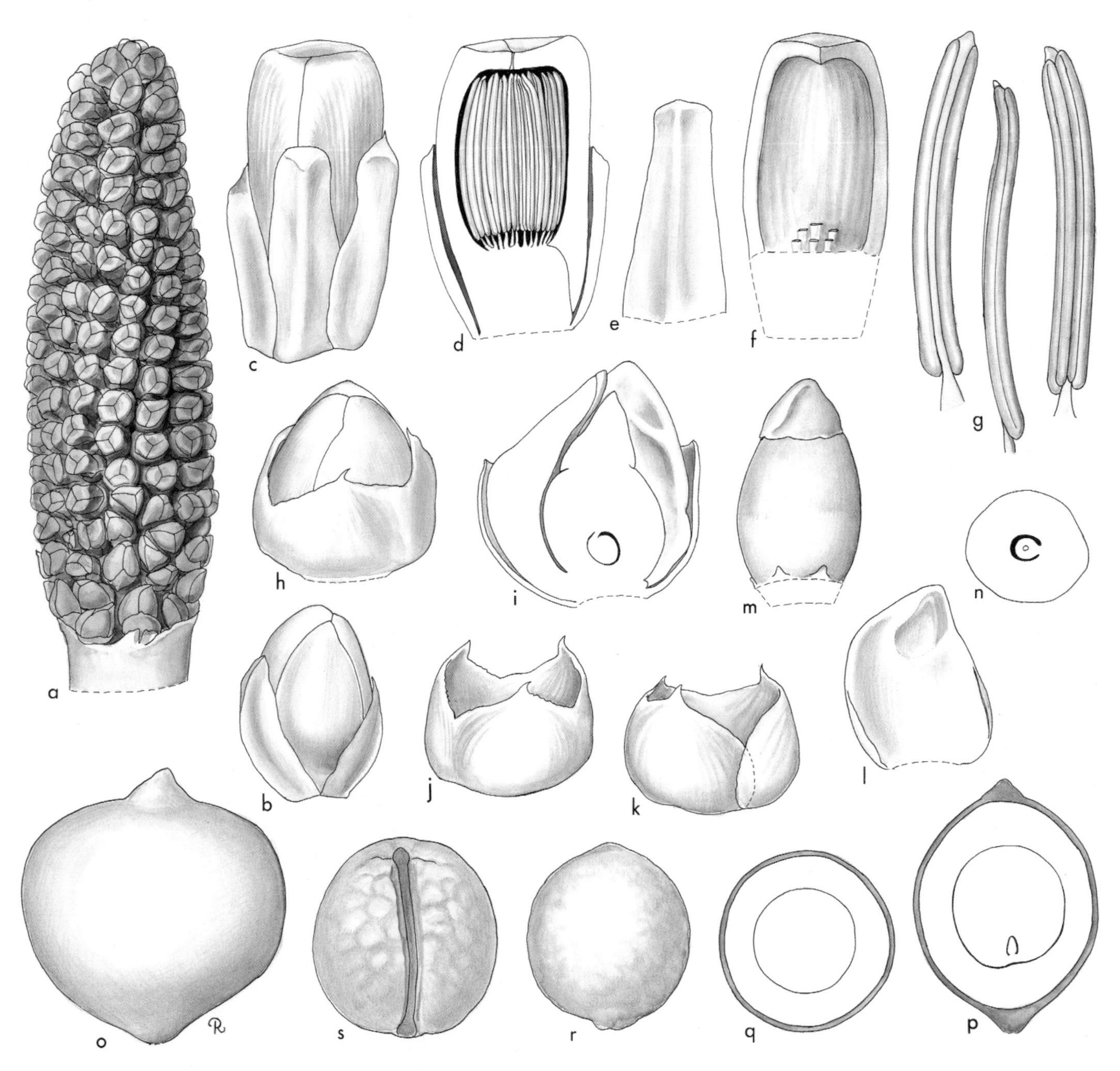

Sclerosperma. **a**, inflorescence × $^1/_2$; **b**, staminate bud from triad × 4 $^1/_2$; **c**, staminate bud from distal portion of inflorescence × 4 $^1/_2$; **d**, distal staminate bud in vertical section × 4 $^1/_2$; **e**, staminate sepal × 4 $^1/_2$; **f**, staminate petal, with stamens removed, interior view × 4 $^1/_2$; **g**, stamen in 3 views × 9; **h**, pistillate bud × 3; **i**, pistillate bud in vertical section × 4 $^1/_2$; **j, k**, pistillate calyx in 2 views × 3; **l**, pistillate petal, interior view × 3; **m**, gynoecium and staminodes × 4 $^1/_2$; **n**, ovary in cross-section × 4 $^1/_2$; **o**, fruit × 1 $^1/_2$; **p**, fruit in vertical section × 1; **q**, fruit in cross-section × 1; **r**, endocarp × 1; **s**, seed × 1 $^1/_2$. *Sclerosperma profiziana*: **a–n** from *Moore & Enti* 9883; *S. mannii*: **o–s** from *Moore* 9899. (Drawn by Marion Ruff Sheehan)

Fossil record: Problems in understanding the morphology of the triporate pollen of this genus can be traced to a drawing of a dessicated pollen grain that is described as typical (Erdtman & Singh 1957). As a result, Tertiary records of protrudent triporate-fossil pollen, *Trilatiporites*, from India have often been wrongly associated with *Sclerosperma* (Harley & Baker 2001), for example by Bande and Ambwani (1982), Misra *et al.* (1996) and Srivastava (1987–8). One of the few accurate records is that of Médus (1975), who published illustrations of triporate pollen from the Bignona borehole (Miocene) of Senegal that are comparable with pollen of *Sclerosperma*.

Notes: The erect unbranched inflorescence with congested flowers is unusual.

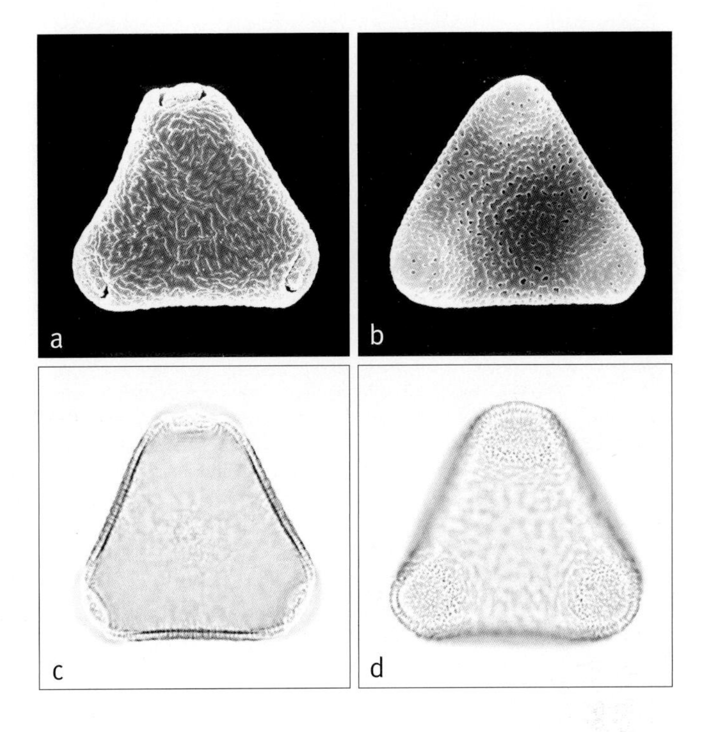

Above: ***Sclerosperma***
a, subapical, operculate triporate pollen grain, distal face SEM × 1000;
b, subapical, operculate triporate pollen grain, proximal face SEM × 1000;
c, subapical, operculate triporate pollen grain, distal face, mid-focus LM × 1000;
d, subapical, operculate triporate pollen grain, proximal face, low-focus LM × 1000.
Sclerosperma profiziana: **a, c** & **d**, *Profizi* 841; *S. mannii*: **b**, *Gillet* 2792.

Top left: ***Sclerosperma walkeri***, habit, Gabon. (Photo: T. Sunderland)

Bottom left: ***Sclerosperma mannii***, infructescence, Gabon. (Photo: J. van Valkenburg)

Tribe Roystoneeae J. Dransf., N.W. Uhl, C. Asmussen, W.J. Baker, M.M. Harley & C. Lewis., Kew Bull. 60: 562 (2005). Type: **Roystonea**.

Robust unarmed; leaves pinnate, the leaflet tips entire; sheaths forming a crownshaft; inflorescences infrafoliar, highly branched; flowers borne in triads near the base, solitary or paired staminate flowers distally; pistillate flowers with petals connate basally, valvate distally; staminodes connate in a conspicuous ring; gynoecium pseudomonomerous; fruit with basal stigmatic remains; epicarp smooth.

The tribe with its single genus, *Roystonea*, is morphologically rather isolated and is restricted to the Caribbean.

Recent phylogenetic analyses have recovered three different hypotheses for the phylogenetic placement of tribe Roystoneeae: as sister to the Reinhardtieae with high support (Lewis & Doyle 2002, Loo *et al.* 2006); as sister to a clade of Reinhardtieae and Cocoseae with low to moderate support (Baker *et al.* in review, in prep.); and as sister to the Chamaedoreeae with low support (Asmussen & Chase 2001, Hahn 2002a).

90. ROYSTONEA

The Royal Palms — spectacular majestic solitary pinnate-leaved palms from the Craibbean islands and neighbouring parts of North, Central and South America; crownshaft is very conspicuous and the inflorescence branched to at least 4 orders with rather stiff spreading rachillae.

Roystonea O.F. Cook, Science, ser. 2. 12: 479 (1900). Type: **R. regia** (Kunth) O.F. Cook (= *Oreodoxa regia* Kunth).
Oreodoxa of many authors, not Willd.

Commemorates the American engineer, General Roy Stone (1836–1905).

Tall, stout, solitary, unarmed, pleonanthic, monoecious palms. *Stem* columnar, variously tapered or swollen, tan, grey, or white, ringed by prominent or obscure leaf scars. *Leaves* pinnate; sheath tubular, large, forming a prominent crownshaft; petiole relatively short, channelled adaxially, rounded abaxially; leaflets narrow, elongate, tapering to a point, single-fold, held in one plane or variously inserted, crowded or in groups, rather thin, midrib only or midrib and other longitudinal veins raised abaxially, hairs frequent and scales prominent along the midrib, transverse veinlets evident abaxially. *Inflorescences* infrafoliar, massive, branched to 3(–4) orders; peduncle very short, stout; prophyll tubular, elongate, strongly 2-keeled laterally, truncate, leathery, green, splitting apically; peduncular bract 2 to 3 times as long as the prophyll, terete, pointed, glabrous, leathery, green, splitting longitudinally; rachis much longer than the peduncle, bearing small, pointed, spirally inserted bracts; rachillae very long, slender and pendulous or short, stout and variously divaricate, straight or undulate, white when first exposed due to copious free scales; rachilla bracts spirally arranged, small, membranous, tapered, subtending widely spaced triads of flowers proximally and paired or solitary staminate flowers distally; floral bracteoles small, thin, membranous. *Staminate flowers* nearly symmetrical, larger than the pistillate buds at anthesis; sepals 3, distinct, triangular, imbricate, very

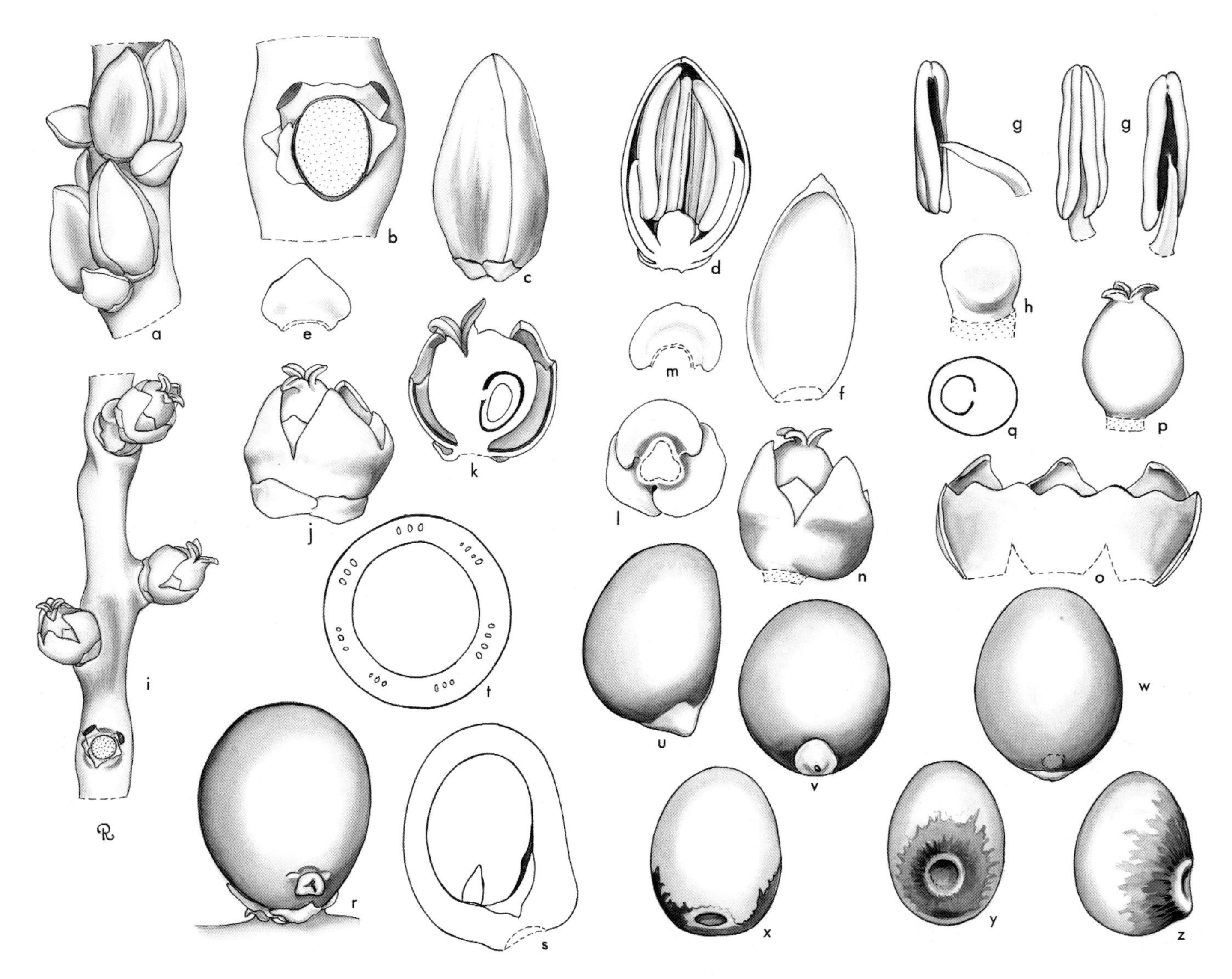

Roystonea. **a**, portion of rachilla with triads × 3; **b**, triad, flowers removed × 6; **c**, staminate bud × 6; **d**, staminate bud in vertical section × 6; **e**, staminate sepal × 6; **f**, staminate petal, interior view × 6; **g**, stamen in 3 views × 6; **h**, pistillode × 12; **i**, portion of rachilla at pistillate anthesis × 3; **j**, pistillate flower × 6; **k**, pistillate flower in vertical section × 6; **l**, pistillate calyx, interior × 6; **m**, pistillate sepal, exterior view × 6; **n**, pistillate flower, calyx removed × 6; **o**, pistillate corolla and staminodes, expanded, interior view × 6; **p**, gynoecium × 6; **q**, ovary in cross-section × 6; **r**, fruit × 3; **s**, fruit in vertical section × 3; **t**, fruit in cross-section × 3; **u**, **v**, **w**, endocarp in 3 views × 3; **x**, **y**, **z**, seed in 3 views × 3. *Roystonea* sp.: all from *Read* 602. (Drawn by Marion Ruff Sheehan)

short; petals 3, distinct, ovate, valvate, about 10 times the length of the sepals, tips thickened; stamens 6–12, filaments awl-shaped, erect in bud; anthers elongate, versatile, sagittate basally, dorsifixed near the middle, latrorse, connective tanniniferous; pistillode subglobose or trifid. *Pollen* grains ellipsoidal, occasionally oblate triangular, with slight or obvious asymmetry; aperture a distal sulcus, occasionally a trichotomosulcus; ectexine tectate, finely or coarsely perforate or perforate-rugulate, aperture margin slightly finer than main tectum; infratectum columellate; longest axis 61–66 µm [4/10]. *Pistillate flowers* nearly conical to shortly ovoid; sepals 3, distinct, very short, broadly imbricate, rounded; petals 3, ovate, connate about $^1/_2$ their length, valvate distally, more than twice as long as the sepals; staminodes 6, connate in a 6-lobed cupule adnate to the corolla basally; gynoecium subglobose, unilocular, uniovulate, style not distinct, stigmas 3, recurved, ovule laterally attached, form unknown. *Fruit* obovoid to oblong-ellipsoidal or subglobose, stigmatic remains nearly basal, perianth persistent; epicarp smooth, thin, mesocarp of pale parenchyma over a layer of thin, flat, anastomosing fibres next to the endocarp, endocarp thin, horny, fragile, somewhat operculate at the base, roughened and often ± adherent to the seed adaxially. *Seed* ellipsoidal, brown, hilum large, circular, lateral, raphe branches fine, radiating from the hilum, endosperm homogeneous; embryo nearly basal. *Germination* adjacent-ligular; eophyll entire. *Cytology*: 2n = 36, 38.

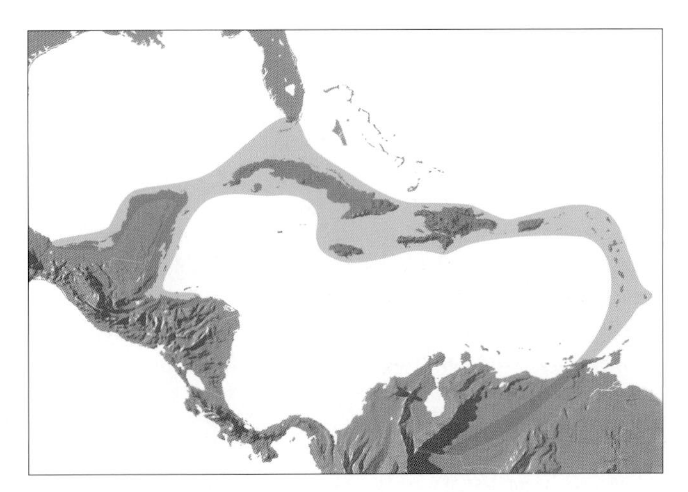

Distribution of ***Roystonea***

Distribution and ecology: Ten species found throughout the islands of the Caribbean and bordering continental areas such as Florida, Mexico, eastern Central America, and northern South America. Primarily palms of the lowlands. Some species are thought to be indicators of good soil conditions. Most of their original habitats are now cleared for agriculture.
Anatomy: Leaf (Tomlinson 1961), root (Seubert 1998a, 1998b).

Left: ***Roystonea borinquena***, crown, Dominican Republic. (Photo: J. Dransfield)

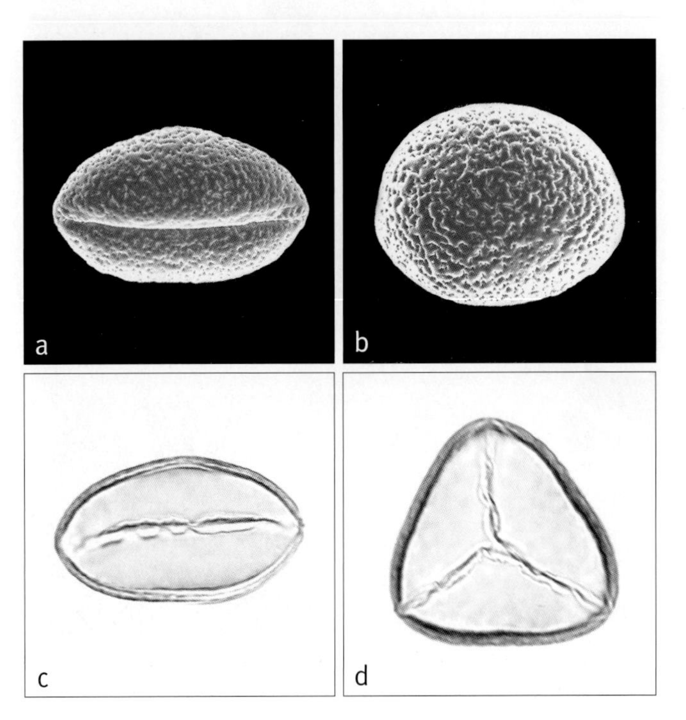

Below: ***Roystonea***
a, monosulcate pollen grain, distal face SEM × 800;
b, monosulcate pollen grain, proximal face × 800;
c, monosulcate pollen grain, mid-focus LM × 800;
d, trichotomosulcate pollen grain, mid-focus LM × 800.
Roystonea regia: **a–d**, *Curtiss* 432.

Relationships: *Roystonea* is monophyletic with high support (Baker *et al.* in review). For interspecific relationships, see Zona (1996). For sister-group relationships, see tribe Roystoneeae.
Common names and uses: Royal palms, mountain cabbage palm (*Roystonea altissima*). *Roystonea* species are among the most elegant of the large palms and are widely cultivated in both hemispheres. Fruits are high in oil content and are used as pig food. The 'cabbage' of *R. oleracea* is edible.
Taxonomic account: Zona (1996).
Fossil record: From the Indian Deccan Intertrappean, Maharashtra State (although the age span of these volcanic deposits is controversial, see Chapter 5), stem tissue from an almost complete stem trunk, *Palmoxylon kamalan* Rode, is considered to resemble that of *Roystonea regia* closely (Kulkarni & Mahabalé 1971). However, comparisons of palm stem wood or root to generic level should always be viewed with caution. Two flowers from Dominican amber, one staminate and one pistillate, estimated to be somewhere between mid-Eocene and mid-Miocene, are compared with *Roystonea* flowers (Poinar 2002b). This author notes, however, that the calyx in both flowers is too large to match the flowers with any extant species of the genus.
Notes: See subtribe.
Additional figures: Glossary figs 2, 22.

Above: ***Roystonea borinquena***, crown, Dominican Republic.
(Photo: J. Dransfield)

Tribe Reinhardtieae J. Dransf., N.W. Uhl, C. Asmussen, W.J. Baker, M.M. Harley & C. Lewis., Kew Bull. 60: 562 (2005). Type: **Reinhardtia**.

Diminutive to moderate palms; leaves pinnate or entire bifid or ± completely entire, the leaflet tips oblique, subpraemorse; sheaths not forming a crownshaft; inflorescences interfoliar, highly branched to spicate; prophyll and peduncular bract similar; flowers borne in triads near the base, solitary or paired staminate flowers distally; pistillate flowers with petals connate basally, valvate distally; staminodes connate in a conspicuous epipetalous ring; gynoecium tricarpellate, triovulate; fruit 1-seeded, with apical stigmatic remains; epicarp smooth, embryo basal.

This tribe contains the single genus *Reinhardtia*, found in the humid rain forest of Central America and northern South America, with a single species in the Dominican Republic in the Caribbean.

The phylogenetic placement of tribe Reinhardtieae is unclear. Lewis and Doyle (2002) and Loo *et al.* (2006) resolve the tribe as sister to tribe Roystoneeae with high support. Other analyses show a relationship between tribes Reinhardtieae and Cocoseae, with the two tribes resolved as sister to each other with moderate support in the studies of Hahn (2002b) and Baker *et al.* (in review). Another hypothesis, resolved by Asmussen and Chase (2001) and by Hahn (2002a), places the Reinhardtieae as sister to tribe Podococceae with moderate support.

91. REINHARDTIA

Diminutive to moderate, solitary or clustered pinnate-leaved palms with fibrous sheaths and no crownshaft, native to Central and northern South America, the leaves often with 'windows' and the fruit 1-seeded.

Reinhardtia Liebm. in Mart., Hist. nat. palm. 3: 311 (1849). Type: **R. elegans** Liebm. in Mart.
Malortiea H. Wendl., Allg. Gartenzeitung 21: 25 (1853). Type: *M. gracilis* Wendl. (= *Reinhardtia gracilis* [H. Wendl.] Drude ex Dammer).

Not explained by original author but thought to be named for a family of Danish naturalists.

Very small to moderate, solitary or clustered, unarmed, pleonanthic, monoecious palms. *Stem* erect, rarely exceeding 8 m tall, usually very much less, sometimes stilt-rooted at the base, with very short to moderately long internodes and conspicuous leaf scars. *Leaves* undivided and pinnately ribbed, with a very short or conspicuous apical notch, or pinnate, sometimes 'windowed', marcescent or abscising under their own weight; sheaths tubular but not forming a crownshaft, expanding and drying into an interwoven mass opposite the petiole, brown-scaly, produced beyond the level of the petiole into a membranous or fibrous ligule, in age the fibres often disintegrating; petiole well developed, adaxially concave or flattened, abaxially rounded or angled, or narrowed and almost winged along margins, bearing caducous brown scales; leaflets 1–several-fold, where single-fold the tips bifid, where compound, the tips appearing obliquely and sharply toothed, subpraemorse, veins conspicuous in the expanding leaf, in some species short splits ('windows') occurring next to the rachis along the abaxial folds in the otherwise unsplit compound leaflets, caducous brown scales present along the ribs in expanding leaves, transverse veinlets obscure. *Inflorescences* solitary, interfoliar, apparently protandrous, spicate or branching to 1 or 2 orders, shorter than or as long as the leaves; peduncle very slender to moderate, continuing to elongate after anthesis; prophyll tubular, membranous, 2-keeled, distally with 2 triangular lobes, usually ± included within the subtending leaf sheath (except perhaps *Reinhardtia elegans*); peduncular bract single, tubular or not, elongate, papery, at first included within the prophyll, eventually carried out by peduncular elongation and disintegrating, rarely a second peduncular bract also present; peduncle extending into a simple spike (*R. koschnyana*) or bearing a few crowded unbranched rachillae at its tip, the rachillae long, exceeding the rachis, each subtended by a narrow triangular bract, or the proximal branches once branched, all axes greenish at first, covered in caducous brown scales, after fertilization becoming orange-red to bright red; rachilla bracts spirally, subdistichously, or distichously arranged, short, triangular, each subtending a triad borne in a shallow depression, except distally subtending solitary or paired staminate flowers. *Staminate flowers* bearing a 2-keeled, irregularly lobed and split bracteole; sepals 3, distinct, imbricate, obtuse, concave, striate on drying; petals 3, to 2–3-times as long as the sepals, valvate, connate at the very base, striate when

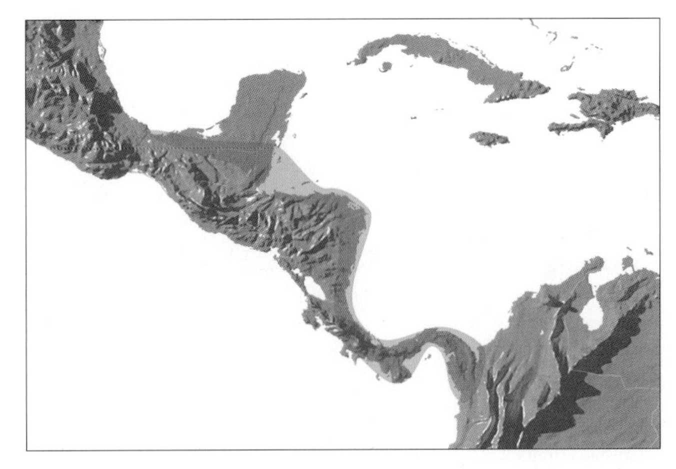

Distribution of ***Reinhardtia***

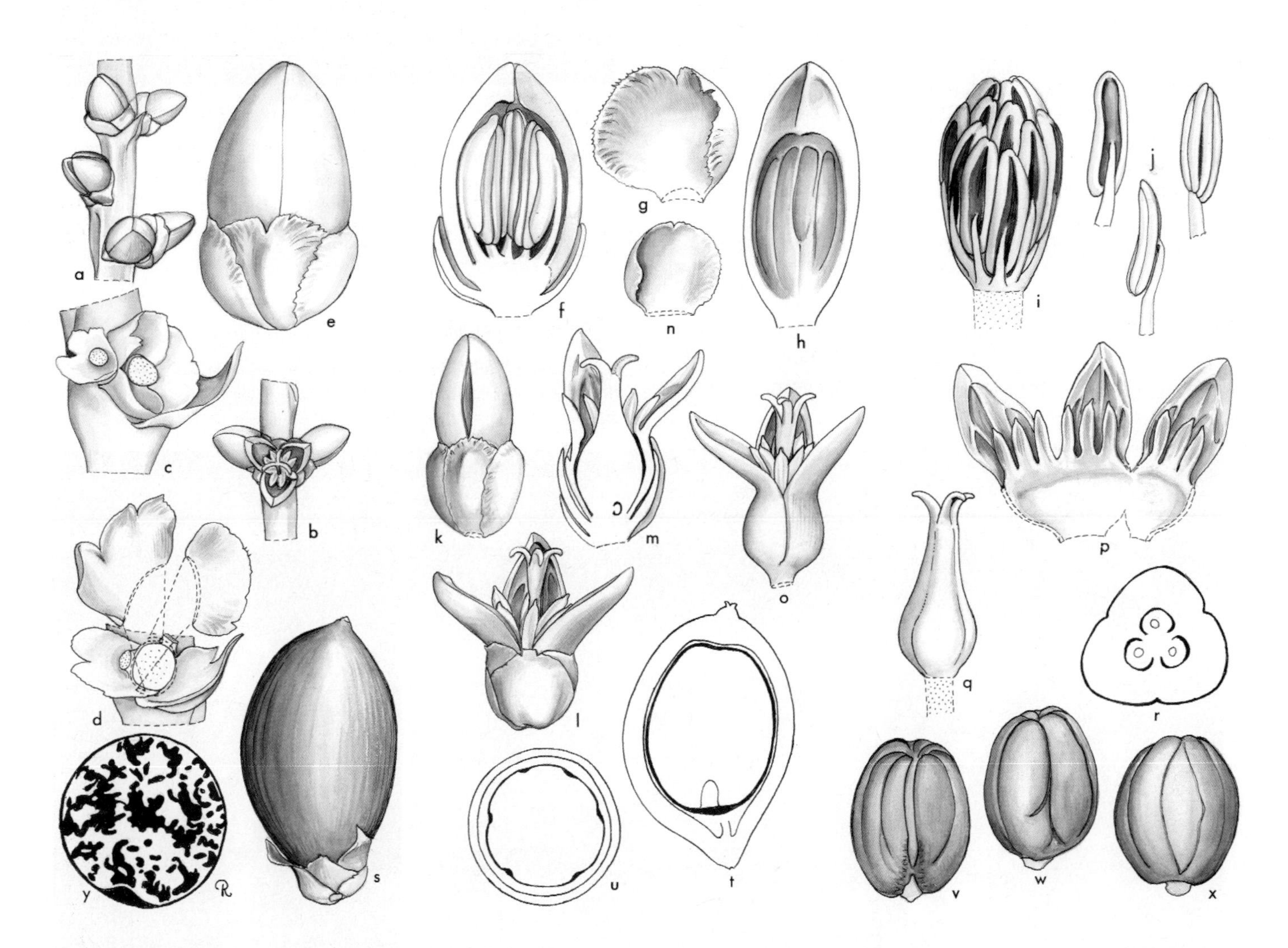

Reinhardtia. **a**, portion of rachilla with triads × 3; **b**, protogynous triad with staminate buds and expanded pistillate flower × 3; **c**, triad, flowers removed to show bracteoles × 6; **d**, triad, flowers and bracteoles around pistillate flower removed to show arrangement of flowers and the bracteole of a staminate flower × 6; **e**, staminate bud × 12; **f**, staminate bud in vertical section × 12; **g**, staminate sepal, interior view × 12; **h**, staminate petal, interior view × 12; **i**, androecium × 12; **j**, stamen in 3 views × 12; **k**, pistillate bud × 6; **l**, pistillate flower, expanded × 6; **m**, pistillate flower in vertical section × 6; **n**, pistillate sepal, interior view × 6; **o**, pistillate flower, sepals removed × 6; **p**, pistillate corolla with staminodes, expanded × 6; **q**, gynoecium × 6; **r**, ovary in cross-section × 12; **s**, fruit × 3; **t**, fruit in vertical section × 3; **u**, fruit in cross-section × 3; **v, w, x**, seed in 3 views × 3; **y**, seed in cross-section × 3. *Reinhardtia simplex*: **a–r**, BH 61-868; **s–x**, *Moore* 6529; *R. elegans*: **y**, *Hernandez & Sharp* X-1245. (Drawn by Marion Ruff Sheehan)

dry; stamens 8–40, filaments short, slender, briefly connate at the base and adnate to the base of the petals, anthers basifixed or medifixed, elongate, the apices acute or bifid, latrorse; pistillode lacking. *Pollen* ellipsoidal or oblate triangular, slight or obvious asymmetry; aperture a distal sulcus or trichotomosulcus; ectexine tectate, finely perforate, or perforate and micro-channelled, and rugulate, aperture margin similar or slightly finer; infratectum columellate; longest axis 37–53 µm [2/6]. *Pistillate flowers* bearing a prophyllar bracteole; sepals 3, distinct, subglobose, imbricate, becoming striate when dry; petals 3, exceeding the sepals, slightly imbricate and partially connate at the base, valvate distally, or valvate throughout, distally grooved on the adaxial surface, the upper ca. 1/2 of the petals spreading at anthesis; staminodes connate at the base, adnate to the petals very shortly or to 1/2 the petal length, distally each portion of the staminodal ring bearing 2–5 teeth, the teeth usually erect, projecting, conspicuous at anthesis; gynoecium ovoid or ellipsoidal, trilocular at the base, triovulate, style robust, stigmas recurved at anthesis, ovules attached slightly above the base, form unknown. *Fruit* 1-seeded, black, borne on the enlarged reddish-tinged rachillae, usually ovoid or ellipsoidal, stigmatic remains apical; epicarp smooth, mesocarp fleshy with 2 layers of flattish longitudinal fibres, endocarp thin, fragile. *Seed* ovoid or ellipsoidal, basally or laterally attached, usually furrowed by sparse vascular strands, raphe superficial or impressed, endosperm homogeneous or ruminate; embryo basal. *Germination* adjacent-ligular; eophyll simple or bifid. *Cytology* not studied.

Distribution and ecology: Six species, distributed from Mexico to Panama, one species reaching northwest Colombia. Undergrowth palms, usually found in lowland tropical rain forest. *Reinhardtia elegans* and *R. gracilis* var. *tenuissima* are found at altitudes of 1000–1500 m.

Anatomy: Tomlinson (1961), root (Seubert 1998a, 1998b).

Relationships: *Reinhardtia* is monophyletic with high support (Baker *et al.* in review). For relationships, see tribe Reinhardtieae. For species relationships, see Henderson (2002a).

Common names and uses: Window palm, reinhardtia. No local uses appear to have been recorded; all species are ornamental and have been rather widely cultivated by enthusiasts.

Taxonomic account: Moore (1957b), Henderson (2002a).

Fossil record: Two monosulcate palm pollen types are recovered from the Pliocene, Gatun Lake Formation, Panama (Graham 1991). The first of these types, asymmetrical and

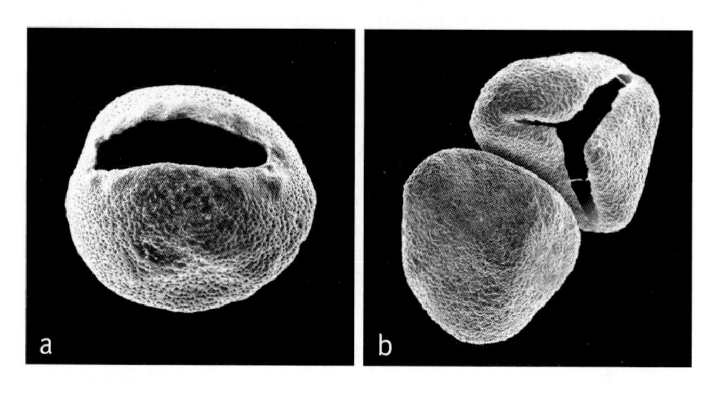

Above: ***Reinhardtia***
a, monosulcate pollen grain, distal face SEM × 1000;
b, two trichotomosulcate pollen grains: proximal bottom left, distal top right SEM × 800.
Reinhardtia koschnyana: **a**, *Henderson & Contraires* 093; *R. simplex*: **b**, *Wendland* 1857.

Below left: ***Reinhardtia gracilis***, habit, cultivated, Mexico. (Photo: T. Dyer)
Below right: ***Reinhardtia koschnyana***, inflorescence with pistillate flowers, Colombia. (Photo: J. Dransfield)

scabrate tectate, is compared with pollen of *Aiphanes*, *Manicaria*, *Reinhardtia* or, possibly, *Acrocomia*. It is a very common arecoid pollen type, and difficult to place. Of the suggested genera, *Reinhardtia* is the most probable.
Notes: Small habit, sometimes subpraemorse leaflets, and numerous stamens combine with a triovulate condition to make this a distinctive genus.
Additional figures: Glossary figs 13, 15.

Tribe Cocoseae Mart. in Endl., Gen. pl. 254 (1837). ('*Cocoinae*').

Slender to very robust, acaulescent to erect or climbing palms; crownshaft lacking; leaves pinnate or pinnately ribbed, reduplicate, leaflets acute, or obliquely lobed or praemorse; inflorescence unisexual or bisexual, spicate or branched to 1 or very rarely 2 orders, never multiple, bearing a short prophyll, usually included within the leaf sheaths and a usually very much longer, frequently very thick, woody peduncular bract; flowers superficial or occasionally in pits; petals of pistillate flowers imbricate or connate; staminodes usually connate in a conspicuous ring; gynoecium with 3 or rarely more locules and ovules; fruit sometimes very large, 1–several seeded, never lobed; endocarp with 3 or more well-defined pores.

The Cocoseae includes 18 genera arranged in three subtribes. The distribution is predominantly New World but there are a few members in the Old World. The tribe is characterised, above all else, by the presence of an endocarp with three or more clearly defined pores. In this treatment, three subtribes are recognised, rather than the five in the first edition of *Genera Palmarum*.

The monophyly of tribe Cocoseae is highly supported in most studies (Baker *et al.* 1999a, in review, Asmussen *et al.* 2000, Lewis & Doyle 2002), but some studies found only moderate to low support for the tribe's monophyly (Asmussen & Chase 2001, Hahn 2002b, Gunn 2004). Current phylogenetic hypotheses indicate three possible positions of the Cocoseae within subfamily Arecoideae. Hahn (2002b) and Baker *et al.* (in review) place the Cocoseae as sister to the Reinhardtieae with moderate support, and Lewis and Doyle (2002) and Loo *et al.* (2006) place the tribe as sister to a clade of Reinhardtieae and Roystoneeae with moderate support. Hahn (2002a) finds a sister relationship between the Cocoseae and the Oranieae with low support.

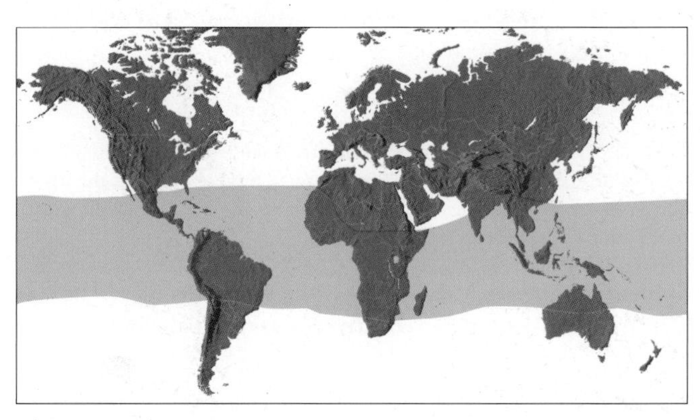

Distribution of tribe **Cocoseae**

Key to subtribes of Cocoseae

1. Prickly palms, armed in some or all parts with soft to mostly stout spines, or when rarely unarmed (*Bactris* spp.), then the perianth of pistillate flowers with connate petals; endocarp with pores at or above the middle, the pores more-or-less deeply impressed in the endocarp except rarely in *Acrocomia*, usually covered by or plugged with fibres adherent to the endocarp, lacking a clearly demarcated and visible operculum; fruit (?)always 1-seeded; endosperm always homogeneous . . . . **Bactridinae**
1 Unarmed palms except for the sometimes sharply toothed petiole margin; corolla of pistillate flowers always of broadly imbricate petals; endocarp with pores above, at, or below the middle 2
2. Pistillate flowers not or only slightly sunken in the rachillae; endocarp with pores at or below the middle; peduncular bract woody . . . . **Attaleinae**
2. Pistillate flowers deeply sunken in the rachillae; endocarp with pores at or above the middle; peduncular bract fibrous or ± woody . . . . **Elaeidinae**

Subtribe Attaleinae Drude in Engl. & Prantl, Nat. Pflanzenfam. 2, 3: 27, 78 (1887). Type: **Attalea**.

Acaulescent or erect, slender to very massive palms, lacking epidermal emergences; flowers superficial.

The delimitation of this subtribe has been broadened to include all members of the tribe Cocoseae except for the spiny cocosoids, and *Elaeis* and *Barcella*.

Subtribe Attaleinae is monophyletic with high support in studies by Gunn (2004), Baker *et al.* (in review, in prep.), Roncal *et al.* (2005) and with moderate support in work by Asmussen and Chase (2001). The polyphyly of the subtribe recovered by Asmussen *et al.* (2006), who resolve *Beccariophoenix* outside tribe Cocoseae, is likely to be an artefact. In several studies, all neotropical genera of Attaleinae (including *Cocos*) resolve as a monophyletic group (Hahn 2002b, Gunn 2004, Asmussen *et al.* 2006, Baker *et al.* in review). Subtribe Attaleinae is sister to the rest of the Cocoseae with moderate support in Baker *et al.* (in review) and with low support in Asmussen and Chase (2001). A similar result is obtained by Gunn (2004), although see discussion under Elaeidinae below.

Key to Genera of Attaleinae

1. Peduncular bract borne at the tip of the peduncle, circumscissile and usually deciduous at anthesis; stamens 15–21, the filaments connate in a torus. Madagascar . . . . **92. Beccariophoenix**
1. Peduncular bract borne near the prophyll insertion or higher, but not at the tip, not circumscissile, splitting abaxially but not deciduous at anthesis, persistent or marcescent; stamens 3-many, filaments variously free or connate 2

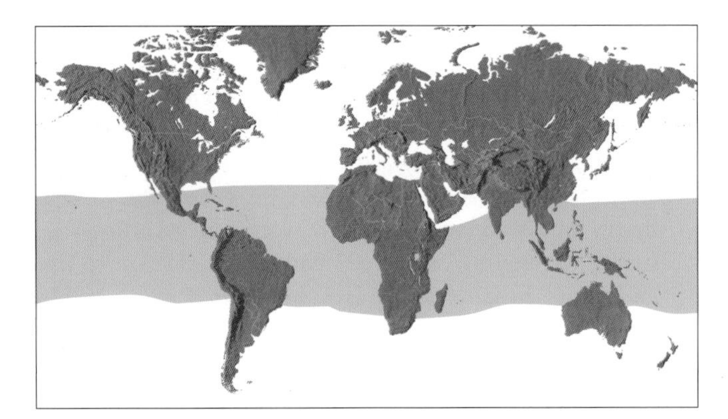

Distribution of subtribe **Attaleinae**

2. Peduncular bract nearly smooth or finely striate but not deeply grooved; pores at or below the middle, not impressed in the endocarp, each with a clearly demarcated thin operculum 3
2. Peduncular bract shallowly to deeply grooved; pores mostly near the base of and usually impressed in the endocarp, often covered by or plugged with fibres, sometimes with a clearly demarcated operculum 5
3. Staminate flowers with 6 stamens; petioles usually conspicuously toothed on the margin toward the base; fruit 1–3-seeded. Brazil to Paraguay . . . . **97. Butia**
3. Staminate flowers with more than 6 stamens; petioles not toothed along the margin 4
4. Staminate flowers stalked, the sepals connate with free-pointed tips; stamens 15–30; leaflets clustered; trunks always solitary, massive, up to 1 m in diameter or more. Coastal central Chile . . . . **99. Jubaea**
4. Staminate flowers not stalked; sepals imbricate, stamens 8–16; trunks slender and often multiple. Southeastern Africa . . . . **93. Jubaeopsis**
5. Inflorescences not normally differentiated into staminate and those bearing flowers of both sexes, instead the pistillate flowers borne at the base of, in the lower part of, or nearly throughout the rachillae or the spike, the staminate flowers lateral to the pistillate and usually pedicellate in the triads, sessile and paired or solitary in the upper part of the rachilla or spike; pistillate flowers always trilocular and triovulate 6
5. Inflorescences normally of more than one kind on the same plant, staminate and those bearing flowers of both sexes, or sometimes also pistillate; those bearing flowers of both sexes with the pistillate flowers large and few at the base of short rachillae, the pistillate flowers accompanied by lateral fertile or abortive staminate flowers and usually with some paired or solitary staminate flowers on the distal portion of the rachillae; staminate inflorescences with longer rachillae bearing staminate flowers only; pistillate flowers sometimes with more than 3 carpels, ovules, and stigmas. Mexico and Caribbean to Brazil and Bolivia . . . . **96. Attalea**
6. Triads separated on a usually branched inflorescence, not densely arranged; sepals of staminate flowers rarely as long as the petals 7
6. Triads densely arranged on the lower portion of a spike, the terminal portion with dense staminate flowers; bracteoles prominent; sepals of staminate flowers linear, acute to acuminate, more than half as long as the petals, the flowers rather markedly asymmetrical, lateral to the densely packed pistillate flowers on the lower portion, and densely crowded on the terminal portion of the spike; stamens 6–120 . . . . **95. Allagoptera**
7. Stamens 12–15 8
7. Stamens 6 9
8. Stamens ca. 15; endocarp irregularly sculptured and roughened and with 3 prominent crests at the apex, internally smooth. Ecuador and Bolivia . . . . **102. Parajubaea**
8. Stamens 12; endocarp deeply grooved externally, internally irregularly sculptured, the rounded protrusions penetrating the endosperm. Madagascar . . . . **94. Voanioala**
9. Pistillate flowers very large, globose-ovoid, the sepals and petals rounded, without valvate apices, imbricate; staminate flowers with distinct sepals and 6 stamens; fruit very large, up to 25 cm long or more at maturity, with a very thick, fibrous mesocarp and thick, bony endocarp; seed normally 1, with liquid endosperm when young and a large, hollow interior when completely mature. Pantropical . . . . **98. Cocos**
9. Pistillate flowers ovoid or conic-ovoid, the sepals acute and ± hooded, the petals usually with conspicuously valvate apices 10
10. Leaflets densely white or pale brown tomentose abaxially, very narrow, close, and regularly arranged; epicarp and mesocarp splitting regularly and longitudinally from the apex to the base into 3 sections at maturity exposing the thin endocarp, anthers versatile, medifixed, filaments inflexed at tip. Brazil . . . . **100. Lytocaryum**
10. Leaflets lacking a dense covering of tomentum abaxially (very rarely thin indument present but then leaflets grouped); epicarp and mesocarp not splitting at maturity; endocarp very thick, variously beaked, ridged, minutely pitted or invaginated; anthers only rarely versatile. South America and Lesser Antilles . . . . **101. Syagrus**

92. BECCARIOPHOENIX

Spectacular solitary unarmed pinnate-leaved palms endemic to Madagascar; distinctive in the thick peduncular bract borne at the tip of the peduncle, which splits and falls at anthesis leaving a collar-like scar, and staminate flowers with 18–21 stamens.

Beccariophoenix Jum. & H. Perrier, Ann. Fac. Sci. Marseille 23(2): 34 (1915). Type: **B. madagascariensis** Jum. & H. Perrier.

Commemorates the great Italian palm botanist, Odoardo Beccari (1843–1920) by combining his name with phoenix *— a general name for a palm.*

Robust, solitary, unarmed, pleonanthic, monoecious, tree palm. *Stem* erect, eventually becoming bare and ringed with leaf scars. *Leaves* massive, pinnate, apparently marcescent; sheath tubular at first, with a large, lateral obtuse lobe on each side, disintegrating into a mass of grey fibres; petiole absent; rachis adaxially channelled near the base, abaxially rounded, distally with 2 lateral grooves; leaflets single-fold, very numerous, ± regularly arranged, more slender and crowded at the base of the rachis than distally, ± rigid, acute, adaxially glabrous, abaxially covered with a thin layer of powdery white wax, transverse veinlets short, conspicuous. *Inflorescences* solitary, interfoliar, exserted from leaf sheaths, branching to 1(–2) orders; peduncle massive, elliptic in cross-section, densely grey-tomentose; prophyll inserted at the base of and ± equalling the peduncle in length, thick, coriaceous, persistent, disintegrating into coarse interwoven fibres, strongly 2-keeled, tubular, splitting briefly at the tip; peduncular bract inserted at the apex of the peduncle, ± the same length as the prophyll, thin to extremely thick (up to 3 cm), woody, tubular,

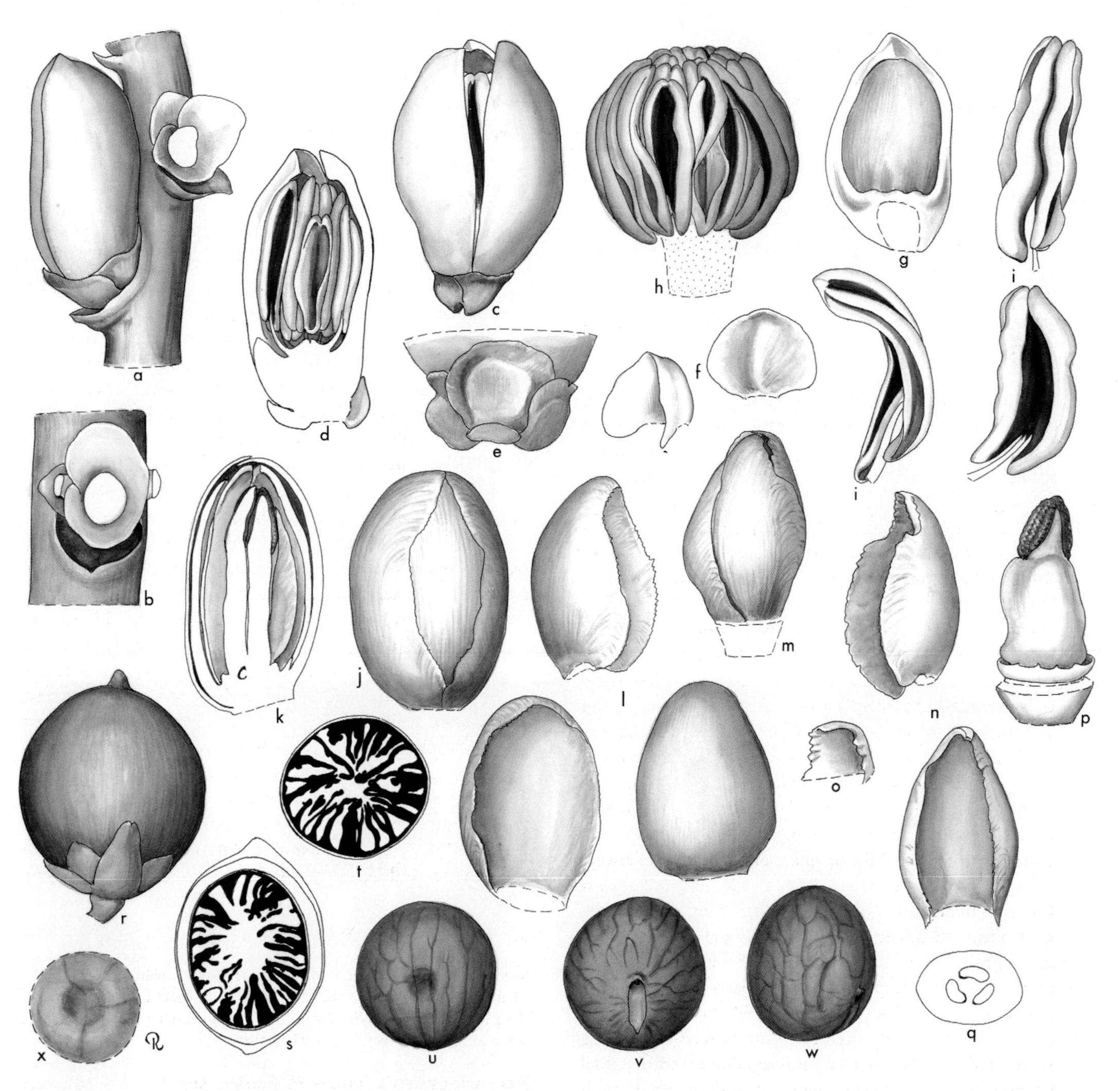

Beccariophoenix. **a**, portion of staminate rachilla × 3; **b**, triad, flowers removed to show bracteoles and scars × 3; **c**, staminate flower × 3; **d**, staminate flower in vertical section × 3; **e**, staminate calyx × 4 1/2; **f**, staminate sepal in 2 views × 4 1/2; **g**, staminate petal, interior view × 3; **h**, androecium × 4 1/2; **i**, stamen in 3 views × 4 1/2; **j**, pistillate bud × 3; **k**, pistillate bud in vertical section × 3; **l**, pistillate sepals × 3; **m**, pistillate bud, sepals removed × 3; **n**, pistillate petal in 2 views × 3; **o**, pistillate petal tip × 3; **p**, gynoecium and staminodial cupule × 3; **q**, ovary in cross-section × 6; **r**, fruit × 1 1/2; **s**, fruit in vertical section × 1 1/2; **t**, fruit in cross-section × 1 1/2; **u, v, w**, seed in 3 views × 1 1/2; **x**, indentation in seed, enlarged. *Beccariophoenix madagascariensis*: all from *Humbert* 20572. (Drawn by Marion Ruff Sheehan)

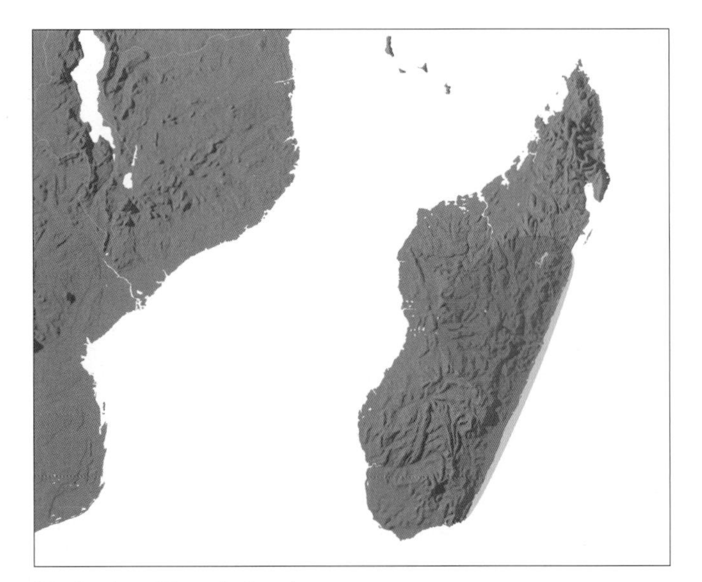

Distribution of ***Beccariophoenix***

with a solid beak, splitting or not, but circumscissile at the insertion, caducous, leaving a collar-like scar, adaxially smooth, shiny, abaxially tomentose and longitudinally shallowly grooved; rachis very short, bearing crowded, spirally arranged, short, triangular, acuminate, coriaceous bracts each subtending a first-order branch, the proximal few sometimes bearing 1–2 second-order branches; rachillae straight, rigid, rather thick, each with a large swelling at the very base, forming a spherical pulvinus, proximally with a very short bare portion, distally bearing subdistichous or strictly distichous flower groups, each subtended by a low triangular bract, the rachilla surface sparsely waxy or bare, flowers borne in triads throughout much of the rachilla length except near the tips where flowers solitary or paired and staminate, or near the base where flowers occasionally in tetrads of 2 pistillate and 2 staminate; floral bracteoles well developed, broad, rounded, striate, rather coriaceous. *Staminate flowers* relatively very large, covered with white wax (?always), subsymmetrical; sepals 3, distinct, keeled, imbricate, coriaceous, rather short; petals 3, distinct except at the very base, valvate, strongly coriaceous to woody, ± boat-shaped, much longer than the sepals, somewhat striate, adnate at the very base to the floral axis; stamens 15–21, filaments short, slender at the base, adnate to the floral axis, anthers elongate, erect, ± basifixed, sometimes irregularly sagittate; pistillode absent or minute (Beccari & Pichi-Sermolli 1955). *Pollen* ellipsoidal, usually with either slight or obvious asymmetry; aperture a distal sulcus; ectexine tectate, finely perforate, perforate and micro-channelled, or perforate-rugulate, aperture margin slightly finer; infratectum columellate; longest axis 43–52 µm [1/1]. *Pistillate flowers* only slightly larger than the staminate, covered in thin wax (?always); sepals 3, distinct, broadly imbricate, hooded, strongly coriaceous, ± striate; petals 3, distinct, imbricate, thinly coriaceous, with very brief valvate tips; staminodes connate in a brief irregularly toothed ring; gynoecium trilocular, triovulate, obpyriform, stigmas 3, tightly appressed in bud, ovule form unknown. *Fruit* relatively large, 1-seeded, ± ovoid with a short triangular beak, the perianth segments persisting as a cupule; epicarp smooth, mesocarp rather dry with abundant longitudinal fibres, the outer fine, the inner broad and flattened, easily separated from endocarp, endocarp woody, relatively thick, marked with 3 indistinct pores, 1 opposite the embryo. *Seed* broadly ovoid, attached near the base with a broad hilum, with numerous anastomosing raphe branches, endosperm deeply ruminate; embryo lateral below the equator. *Germination* adjacent-ligular; eophyll entire, lanceolate. *Cytology*: 2n = 36.

Distribution and ecology: At least two species endemic to Madagascar, occurring in coastal white sand forest at sea level, lower montane forest at 900 m and gallery forest in high altitude grassland.

Anatomy: Root (Seubert 1998a, 1998b); stamen development (Uhl 1988).

Below left: ***Beccariophoenix alfredii***, habit, Madagascar. (Photo: M. Rakotoarinivo)

Below right: ***Beccariophoenix madagascariensis***, inflorescence in bud, Madagascar. (Photo: J. Dransfield)

Relationships: The monophyly of *Beccariophoenix* has not been tested. *Beccariophoenix* is sister to the rest of the Attaleinae with moderate support (Hahn 2002b, Baker *et al.* in review).
Common names and uses: *Manarano*. The leaves are used in hat making and the cabbage is eaten. Destructive exploitation is responsible for the palm's very localised distribution.
Taxonomic account: Dransfield & Beentje (1995b).
Fossil record: No generic records found.
Notes: In cultivation, there are two sorts of seedling of *Beccariophoenix*: one has leaves with a broad terminal flabellum composed of thin, scarcely coriaceous, incompletely split leaflets, distally joined together and split proximally to produce conspicuous 'windows'; the other has very coriaceous leaflets, the apical one or two only incompletely split, and thus with insignificant 'windows'. Seed of *B. alfredii* and from the type locality of *B. madagascariensis* and populations at Vondrozo and near Tolagnaro produce the latter type of seedling; the origin of the first type appears to be from lowlands near Toamasina and this type may represent a third species. *Beccariophoenix alfredii* has been discovered recently in deep valleys in rocky grasslands west of Antsirabe (Rakotoarinivo *et al.* 2007).
Additional figures: Glossary figs 19, 20.

Top right: ***Beccariophoenix alfredii***, crown, Madagascar.
(Photo: M. Rakotoarinivo)

Bottom right: ***Beccariophoenix alfredii***, fruit, Madagascar.
(Photo: M. Rakotoarinivo)

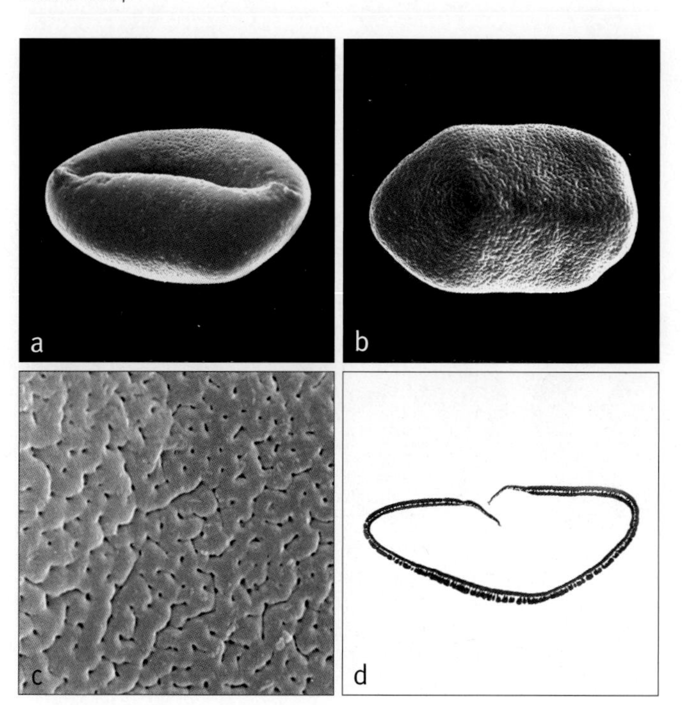

Below: ***Beccariophoenix***
a, monosulcate pollen grain, distal face SEM × 1000;
b, monosulcate pollen grain, proximal face SEM × 1000;
c, close-up, finely perforate-rugulate pollen surface SEM × 8000;
d, ultrathin section through whole pollen grain, polar plane TEM × 1500.
Beccariophoenix madagascariensis: **a**, **b** & **d**, *Humbert* 20572, **c**, *Perrier de la Bâthie* 12014.

93. JUBAEOPSIS

Clustering palm with short erect stems that frequently branch dichotomously, native to southeastern South Africa, staminate flowers with 7–16 stamens.

Jubaeopsis Becc., Webbia 4: 171 (1913). Type: **J. caffra** Becc.

Derived from the palm generic name Jubaea *and* -opsis *— similar to, in reference to the similarities in floral and fruit structure.*

Moderate, clustered, unarmed, pleonanthic, monoecious palm. *Stems* erect, branching at the base and also aerially by forking, bearing leaf sheath remains distally, eventually becoming bare, marked with close leaf scars. *Leaves* pinnate, arranged in 5 vertical rows, marcescent or neatly abscising; sheaths tubular, soon disintegrating into an interwoven mass of fibres; apparent petiole short to elongate, adaxially channelled, abaxially rounded, glabrous, the margins bearing the remains of leaf sheath fibres, or becoming smooth; rachis ± straight or curved, adaxially channelled near the base, angled distally, abaxially rounded or flattened; leaflets numerous, single-fold, close, regularly arranged, except at the very tip, stiff, held in one plane, linear, the tips mostly asymmetrically

2-lobed, except at the leaf tip where acute or sometimes hooked, thinly glaucous, adaxial surface bearing scattered, minute, dot-like scales, abaxially with scattered dot-like scales, and a few large brown ramenta along the main vein, transverse veinlets conspicuous, sinuous. *Inflorescences* solitary, interfoliar, branching to 1 order, shorter than the leaves, protandrous; peduncle elongate, round in cross-section; prophyll short, tubular, 2-keeled, enclosed within the leaf sheaths, splitting apically; peduncular bract inserted near the prophyll, tubular and entirely enclosing the inflorescence until shortly before anthesis, later splitting longitudinally along the abaxial face and expanding, becoming cowl-like, woody, smooth, abaxially somewhat striate but not grooved, apically with a short, laterally flattened beak; rachis usually shorter than the peduncle, bearing numerous, spirally arranged, rather distant, spreading rachillae, each subtended by a low triangular bract; rachillae elongate, swollen at the very base and with a short or long basal bare portion, above which bearing few to numerous spirally arranged triads, distally bearing paired or solitary staminate flowers; rachilla bracts and floral bracteoles small, inconspicuous. *Staminate flowers* rather large, asymmetrical, sessile; sepals 3, distinct, unequal, imbricate, ± triangular, keeled; petals 3, distinct, very unequal, much larger than the sepals, boat-shaped with triangular tips, valvate; stamens (7–)8–16, filaments slender, fleshy, ± cylindrical, apically inflexed, anthers linear, basally sagittate, dorsifixed, introrse; pistillode small, irregularly trifid. *Pollen* ellipsoidal, frequently elongate, usually with either slight or obvious asymmetry; aperture a distal sulcus; ectexine tectate, finely perforate, perforate and micro-channelled, or perforate-rugulate, aperture margin similar; infratectum columellate; longest axis 63–81 µm [1/1]. *Pistillate flowers* ovoid, only slightly larger than the staminate; sepals 3, distinct, broad, rounded, with pointed tips, imbricate, somewhat irregularly keeled; petals 3, distinct, ± twice as long as the sepals, broad, rounded, imbricate except at the short valvate triangular tips; staminodal ring collar-like, very briefly toothed or consisting of several irregular tooth-like lobes; gynoecium ± ovoid, trilocular, triovulate, stigmas 3, very

Distribution of ***Jubaeopsis***

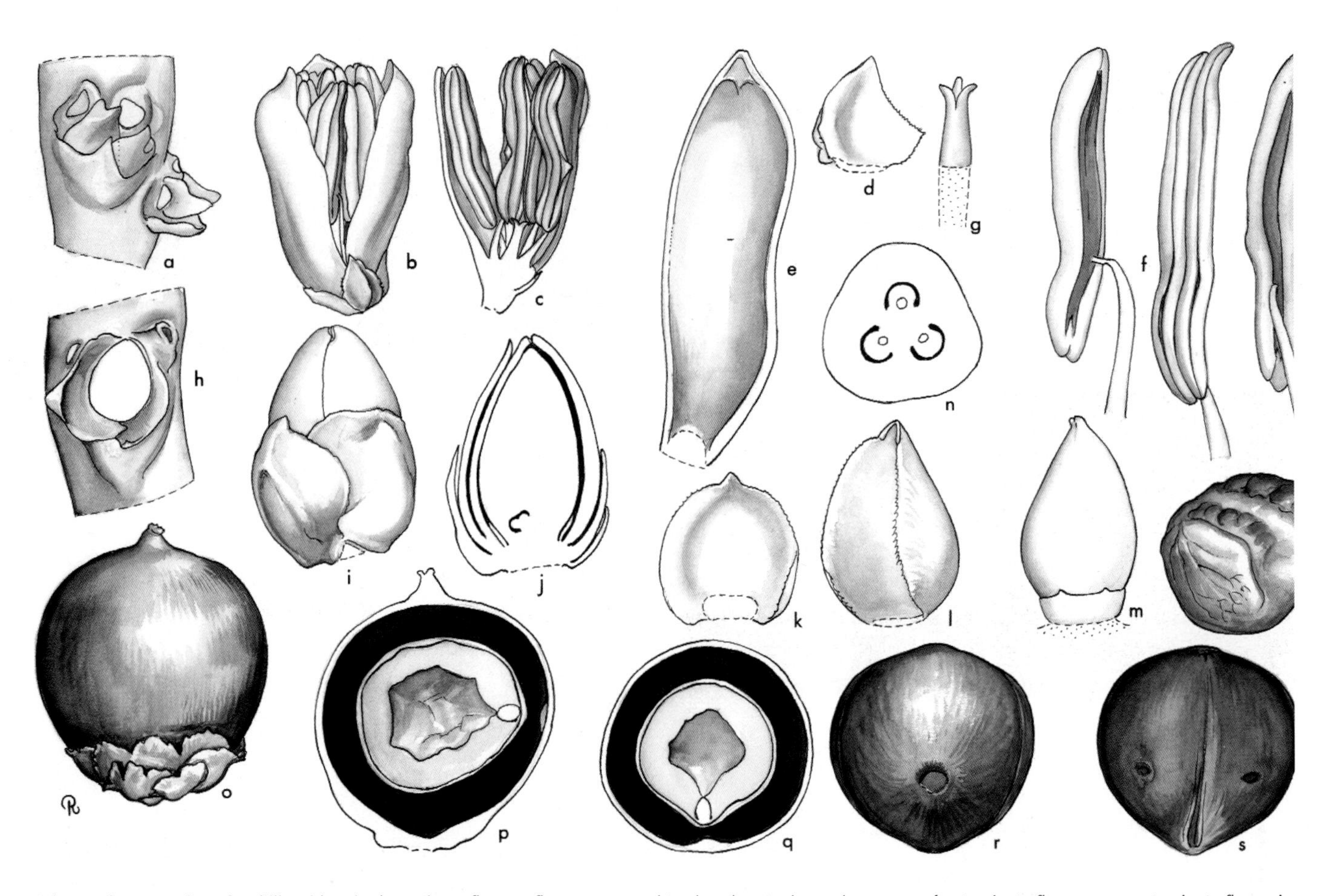

Jubaeopsis. **a**, portion of rachilla with paired staminate flowers, flowers removed to show bracteoles and scars × 3; **b**, staminate flower × 3; **c**, staminate flower in vertical section × 3; **d**, staminate sepal × 6; **e**, staminate petal × 6; **f**, stamen in 3 views × 6; **g**, pistillode × 6; **h**, triad, flowers removed to show bracteoles and scars × 3; **i**, pistillate flower × 3; **j**, pistillate flower in vertical section × 3; **k**, pistillate sepal × 3; **l**, pistillate petal × 3; **m**, gynoecium and staminodial ring × 3; **n**, ovary in cross-section × 6; **o**, fruit × 1; **p**, fruit in vertical section × 1; **q**, fruit in cross-section × 1; **r**, **s**, endocarp in 2 views × 1; **t**, seed × 1. *Jubaeopsis caffra*: **a–n**, *Prosser* s.n.; **o–t**, *Codd* s.n. (Drawn by Marion Ruff Sheehan)

short, ± triangular, ovules hemianatropous, laterally attached. *Fruit* 1-seeded, brown at maturity, globose with a short apical beak, stigmatic remains apical; epicarp smooth, mesocarp thin, fibrous, only slightly fleshy, easily separated from the endocarp at maturity, endocarp thick, bony, with 3 vertical grooves, the pores lateral just below the equator. *Seed* basally attached, somewhat irregular in shape, endosperm homogeneous with a large central cavity; embryo lateral, next to one of the endocarp pores. *Germination* remote tubular; eophyll entire, lanceolate. *Cytology*: 2n = 160–200.

Distribution and ecology: One species confined to the coastal reaches of two rivers in South Africa. *Jubaeopsis caffra* grows gregariously on the steep north, rocky banks of the rivers, near sea-level.

Anatomy: Root (Seubert 1998a, 1998b). See Robertson (1976a) for anatomy and development of fruit and seed.

Relationships: *Jubaeopsis* is sister to all other members of the Attaleinae except *Beccariophoenix* (Asmussen *et al.* 2006, Baker *et al.* in review).

Common names and uses: Pondoland palm. The endosperm may be eaten. Because of its rarity, *Jubaeopsis caffra* is sought after as a collector's item.

Top right: ***Jubaeopsis caffra***, habit, South Africa. (Photo: M.D. Ferrero)
Bottom right: ***Jubaeopsis caffra***, infructescence, South Africa. (Photo: M.D. Ferrero)

Below: ***Jubaeopsis***
a, monosulcate pollen grain, distal face SEM × 600;
b, close-up, finely perforate and microchanelled surface SEM × 600;
c, ultrathin section through whole pollen grain, polar plane TEM × 800;
d, monosulcate pollen grain, mid-focus LM × 600.
Jubaeopsis caffra: **a**, **c** & **d**, *Marloth* s.n., **b**, *Ross* s.n.

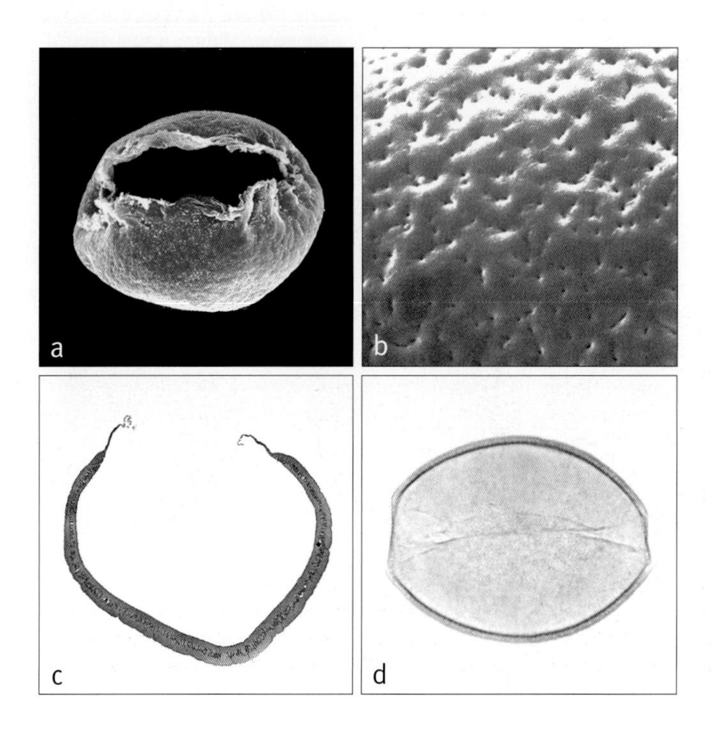

Above: ***Jubaeopsis caffra***, habit, in cultivation, Australia. (Photo: W. Donovan)

Taxonomic accounts: For a discussion of morphology see Robertson (1976b). See also Beccari (1913a) and Glassman (1987).

Fossil record: No generic records found.

Notes: *Jubaeopsis* is the only member of Attaleinae in Africa, remarkable for its dichotomous stems and high chromosome number.

94. VOANIOALA

The forest coconut — solitary pinnate-leaved tree palm from northeastern Madagascar, the staminate flowers have 12–13 stamens and the fruit has large grooved endocarps with irregular protruberances on the inside that penetrate the homogenous endosperm.

Voanioala J. Dransf., Kew Bull. 44: 192 (1989). Type: **V. gerardii** J. Dransf.

Derived from the Malagasy vernacular voanioala *meaning forest coconut.*

Solitary, unarmed, pleonanthic monoecious palms. *Stem* erect, very conspicuously stepped and ringed with oblique leaf scars. *Leaves* pinnate, cleanly abscising; leaf-sheath tubular at first, fibrous, apparently soon disintegrating to leave a massive elongate rectangular leaf base, forming an apparent petiole with sparsely fibrous margins; true petiole absent; rachis rectangular in cross-section in the mid-leaf region; leaflets numerous, regularly arranged, scarcely pendulous, coriaceous, concolorous, unevenly bilobed at the tip, adaxially glabrous apart from bands of caducous chocolate-coloured scales on areas exposed in the sword leaf, abaxially bearing scattered ramenta near the base of the midrib, transverse veinlets obscure, but surface of leaflet minutely transversely striate, thin white wax present on both surfaces. *Inflorescence* solitary, apparently protandrous, branching to 1 order; prophyll tubular, 2-keeled, fibrous, remaining hidden among the leaf bases; peduncular bract bright green and strictly tubular until shortly before anthesis, then splitting longitudinally, flattening and becoming somewhat cowl-like, abaxially deeply and longitudinally grooved, bearing scattered brown scales on the ridges between the grooves, adaxially smooth, glabrous, pale cream-coloured; peduncle ± circular in cross-section; rachis bearing spirally arranged rachillae, each subtended by a small triangular bract; rachillae numerous, most with a basal bare portion. *Staminate flowers* asymmetrical, broadly or narrowly triangular in outline; sepals 3, distinct, slightly to strongly imbricate at the base, triangular, acute to acuminate, membranous, glabrous; petals 3, distinct, unequal, valvate, glabrous, thinly coriaceous except at the thick angular tips, broadly and irregularly triangular-ovate, with acute or acuminate tips, abaxially smooth, adaxially marked with the impressions of the stamens and papillose near the thick tips; stamens 12(–13), filaments subulate, very short to moderate in length, anthers basifixed, basally sagittate, apiculate at the tips, latrorse; pistillode absent. *Pollen* bisymmetric, ellipsoidal, or less frequently oblate triangular; aperture a distal sulcus or trichotomosulcus; ectexine intectate, finely granular interspersed with larger psilate gemmae, separate or arranged in a loosely reticulate pattern, aperture margin with small gemmae arranged linearly; longest axis of ellipsoid grains 55–65 µm, trichotomosulcate grains 49–52 µm; post-meiotic tetrads tetrahedral [1/1]. *Pistillate flowers* only known as buds, much larger than the staminate, irregularly triangular; sepals 3, distinct, unequal, strongly imbricate, broadly ovate with triangular keeled tips, coriaceous, glabrous, the margins minutely toothed; petals 3, distinct, longer than the sepals, basally irregularly imbricate, conspicuously valvate at the triangular tips, abaxially with scaly indumentum towards the apex, adaxially strongly papillose towards the tip; staminodal ring high with 9 irregular triangular teeth, 0.1–0.5 mm; gynoecium syncarpous, tricarpellate, triovulate, stigmas 3, angled, papillose and scaly, ovules anatropous. Immature *fruit* green covered with dense chestnut-brown scaly indumentum. Mature fruit 1-seeded, somewhat irregularly ellipsoid, tipped with a short beak and stigmatic remains; epicarp purplish-brown, densely covered with brown scaly indumentum; mesocarp with an outer fibrous zone just below the epicarp, and an inner fleshy zone; endocarp ± ellipsoid, apically pointed, basally truncate, very heavily thickened, pale brown when fresh, becoming grey with age, deeply grooved without, with numerous embedded fibres and lacunae, and with irregular rounded intrusions, penetrating the central cavity, basally with 3 very deep intrusions, each with a pore. *Seed* irregularly ellipsoid, filling the endocarp cavity, laterally attached with a narrow irregular hilum, endosperm homogeneous but irregularly intruded by the endocarp protuberances, very hard, white, with a narrow irregular central lacuna; embryo basal, top-shaped, positioned opposite an endocarp pore. *Germination* remote-tubular; eophyll entire, lanceolate. *Cytology*: 2n = 550–606 ± 3.

Opposite: ***Voanioala***. **a**, portion of rachilla with solitary staminate flowers × 1; **b**, portion of rachilla with triads × 1; **c**, scars of paired staminate flowers and bracteole × 3; **d**, paired staminate flowers × 2; **e**, staminate bud × 4; **f**, staminate sepals × 4; **g**, staminate petal, interior view × 4; **h**, pistillode × 10; **i**, staminate bud in vertical section × 4; **j**, stamens × 8; **k**, triad of flowers removed to show bracteoles × 3; **l**, pistillate flower × 4; **m**, pistillate sepals × 4; **n**, pistillate petals × 4; **o**, pistillate flower in vertical section × 4; **p**, gynoecium and staminodial ring × 4; **q**, ovary in cross-section × 12; **r**, staminodial ring × 8; **s**, fruit × 1; **t**, fruit in vertical section × 1; **t**, fruit in cross-section × $1^1/_2$; **u**, base of endocarp showing pores × 1; **v, w**, endocarp in 2 views × 1. *Voanioala gerardii*: all from *Dransfield* 6389. (Drawn by Marion Ruff Sheehan)

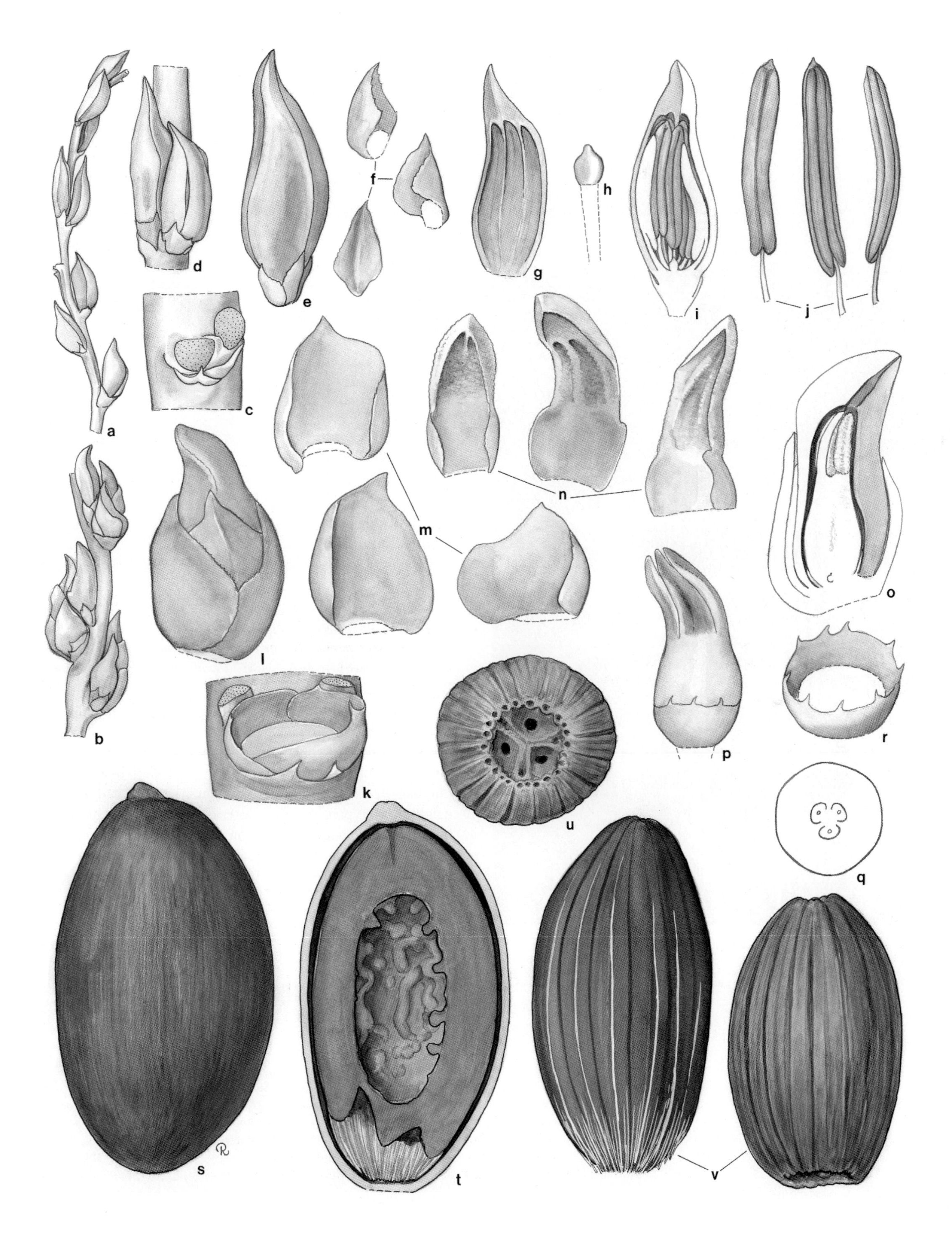
a
b
c
d
e
f
g
h
i
j
k
l
m
n
o
p
q
r
s
t
u
v

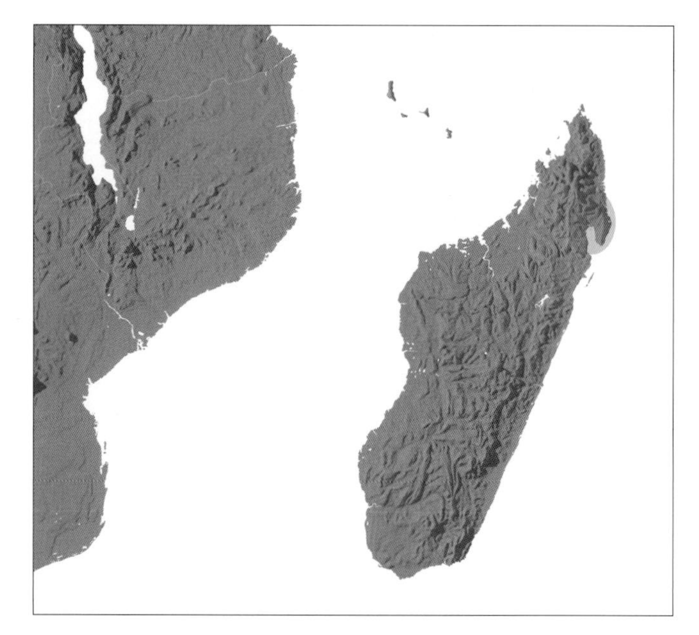

Distribution of ***Voanioala***

Distribution and ecology: A single species endemic to Madagascar, found in primary forest rich in palms and pandans in swampy valley bottoms and on gentle slopes at ca. 400 m.
Anatomy: Root (Seubert 1998a, 1998b).
Relationships: *Voanioala* is sister to a clade of Neotropical Attaleinae and *Cocos* with moderate support (Hahn 2002b, Asmussen *et al.* 2006, Baker *et al.* in review).
Common names and uses: *Voanio-ala* (forest coconut, Betsimisaraka dialect). Sometimes cut for palm hearts.
Taxonomic accounts: Dransfield (1989) and Dransfield and Beentje (1995b).
Fossil record: No generic records found.
Notes: A highly endangered palm, surviving in very low numbers in the forested interior of the Masoala Peninsula and neighbouring parts of the eastern rain forests. The large, heavily sclerified endocarps accumulate beneath the parent trees and there appears to be little or no effective dispersal. The polyploid chromosome number is the highest in the monocotyledons (550–660 ± 3).
Additional figures: Fig. 3.2.

Top: ***Voanioala gerardii***, crown and trunk, Madagascar. (Photo: J. Dransfield)

Above left: ***Voanioala gerardii***, part of peduncular bract and flower buds, Madagascar. (Photo: J. Dransfield)

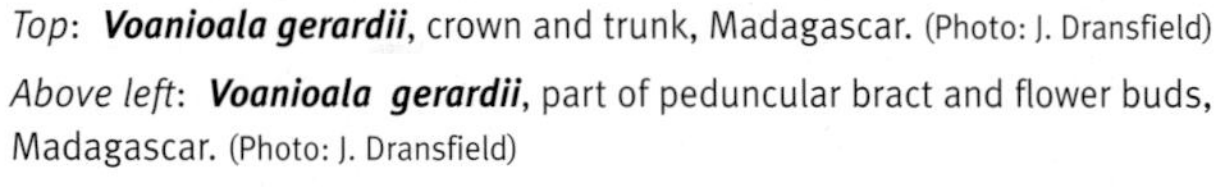

Above right: ***Voanioala gerardii***, infructescence, Madagascar. (Photo: J. Dransfield)

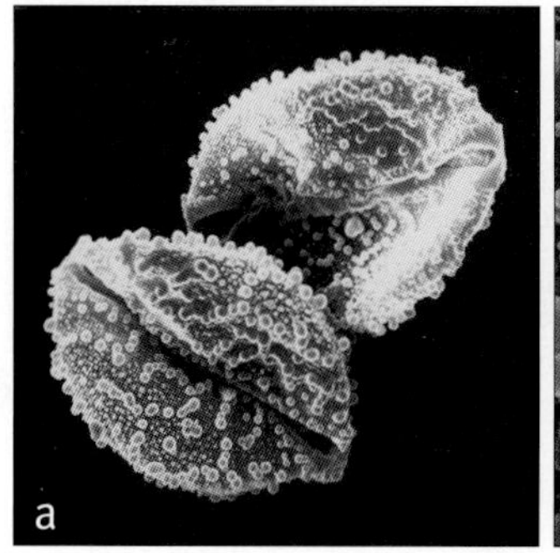

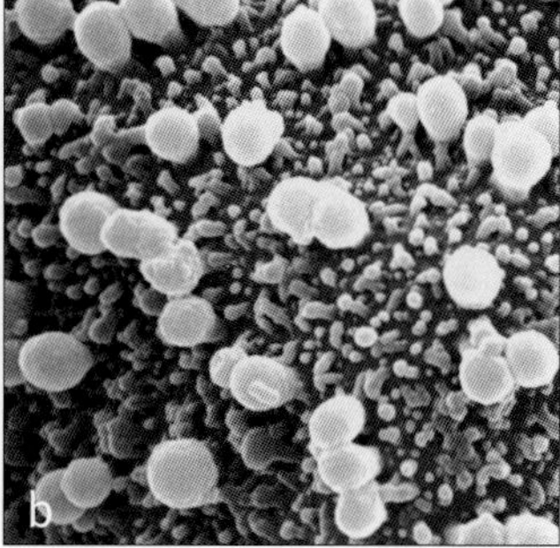

Right: ***Voanioala***
a, two monosulcate pollen grains, distal faces SEM × 750;
b, close-up, finely granular pollen surface, interspersed with larger psilate gemmae SEM × 6000. *Voanioala gerardii*: **a** & **b**, *Dransfield* 6389.

95. ALLAGOPTERA

Small acaulescent or moderate erect palms native to eastern South America, with spicate inflorescences; the staminate flowers are very densely packed and with 6 to over 100 stamens, fruit tending to be rather irregularly shaped because of close-packing.

Allagoptera Nees in M.A.P. Wied-Neuwied, Reise Bras. 2: 335 (1821). Type: *A. pumila* Nees (= **A. arenaria** [Gomes] Kuntze [*Cocos arenaria* Gomes]) (see H.E. Moore 1963c).

Diplothemium Mart., Palm. fam. 20 (1824). Lectotype: *D. maritimum* Mart. (see H. Wendl. 1854) (= *Allagoptera arenaria* [Gomes] Kuntze).

Polyandrococos Barb. Rodr., Contr. Jard. Bot. Rio de Janeiro 1: 7 (1901). Lectotype: *P. caudescens* (Mart.) Barb. Rodr. (*Diplothemium caudescens* Mart. = *Allagoptera caudescens* [Mart.] Kuntze).

Allage — *other, different,* pteron — *wing, perhaps referring to the leaflets held in different planes in* Allagoptera arenaria.

Small, moderate or tall, acaulescent or erect, solitary or clustered, armed or unarmed, pleonanthic, monoecious palms. *Stem* erect, or very short and subterranean, occasionally branching dichotomously, rough and closely ringed with leaf scars. *Leaves* pinnate, marcescent; sheath short to long, tubular but splitting adaxially when young, becoming woody or soft fibrous, slightly expanded at the base to the petiole, sometimes finely striate and covered with whitish, rusty spotted tomentum; petiole short to long, slender to robust, deeply channelled adaxially, rounded abaxially, margins smooth or roughly toothed to spiny, tomentose abaxially like the sheath; rachis arched or straight, adaxially channelled, distally flattened, abaxially rounded, or triangular in cross-section, glabrous or scaly; leaflets single-fold, inserted regularly or in groups, long, narrow, tapering, pointed, or 2-lobed and split apically, a midrib evident on both surfaces, large scales present or absent abaxially along the midrib, glabrous or glaucous throughout, or silvery beneath, transverse veinlets apparent adaxially. *Inflorescences* solitary, interfoliar, erect or pendulous, unbranched; peduncle short to very long, circular in cross-section; prophyll tubular, thin, dorsiventrally flattened, 2-keeled laterally, glabrous or scaly, becoming fibrous, opening apically; peduncular bract 1, long, basally slender, sometimes appearing stalked, inflated above, tapering to a beak, woody, ± plicate, splitting abaxially; rachis bearing close triads throughout the lower $^1/_2$ or more of its length and pairs of staminate flowers distally, staminate flowers shed early, leaving a long, bare, pointed tip on the rachis at pistillate anthesis, bracts subtending triads ovate, pointed, those subtending staminate flower pairs with longer pointed tips, bracts adnate laterally to the rachis and to the bases of adjacent bracts, forming a curved depression surrounding the flower pairs, floral bracteoles inconspicuous. *Staminate flowers* large, asymmetrical, ovoid to obovoid, angled, those of the triads sometimes borne on long, ± flat pedicels curved around the pistillate flower, distal, paired staminate flowers sessile; sepals 3, narrow, connate basally for ca. $^1/_4$ their length, widely separated, pointed, keeled, margins entire or crenate; petals 3, distinct, irregular, angled, triangular, valvate, slightly–4 times longer than the sepals, the tips thickened; stamens 6–ca. 100, filaments awl-shaped, ± united at the base, erect, sometimes flexible and variously bent and curved, anthers somewhat irregular, short to elongate, curved but not twisted, dorsifixed near the base of a prominent connective or toward the middle, sometimes versatile, latrorse or introrse; pistillode lacking or slender, conical, ca. $^1/_2$ as long as the stamens. *Pollen* ellipsoidal, may be elongate and/or pyriform, usually with either slight or obvious asymmetry; aperture a distal sulcus; ectexine tectate, finely perforate, or perforate-rugulate, aperture margin may be slightly finer; infratectum columellate; longest axis 20–50 µm; post-meiotic tetrads tetrahedral, rarely tetragonal or rhomboidal [4/5]. *Pistillate flowers* smaller or slightly larger than the staminate, globose; sepals 3, distinct, broadly imbricate, tips valvate in bud; petals 3, distinct, about as long as the sepals, broadly imbricate, tips valvate; staminodes connate in a low, shallowly lobed cupule; gynoecium ovoid or obovoid, trilocular, triovulate, stigmas narrow, recurved between the petal apices at anthesis, ovules laterally attached, anatropous. *Fruit* obovoid, angled by mutual pressure, greenish-yellow or brown, usually 1-seeded, stigmatic remains represented by an apical knob, perianth enlarged and persistent; epicarp glabrous or with woolly scales, mesocarp fibrous and fleshy, endocarp hard but relatively thin, or thick and bony, smooth, with 3 pores near the base, internally with 3, broad, shining lines. *Seed* obovoid or elongate, basally attached, hilum

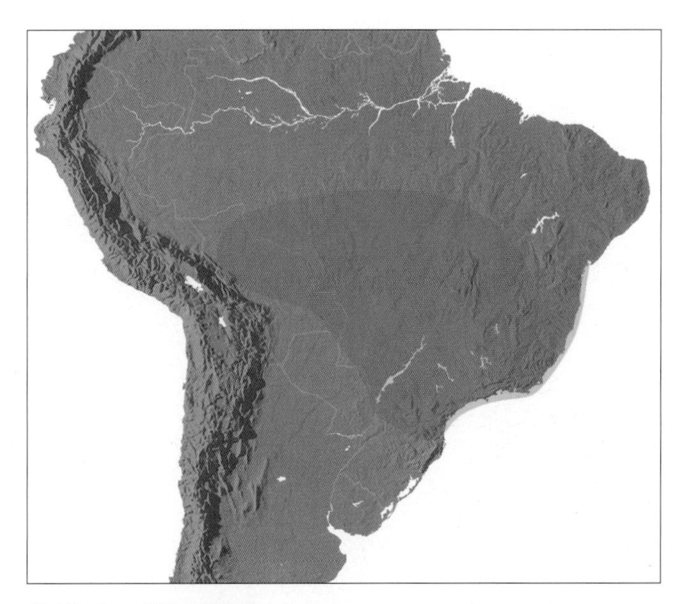

Distribution of ***Allagoptera***

Right: ***Allagoptera arenaria***, habit, cultivated, Fairchild Tropical Botanic Garden, Florida. (Photo: C.E. Lewis)

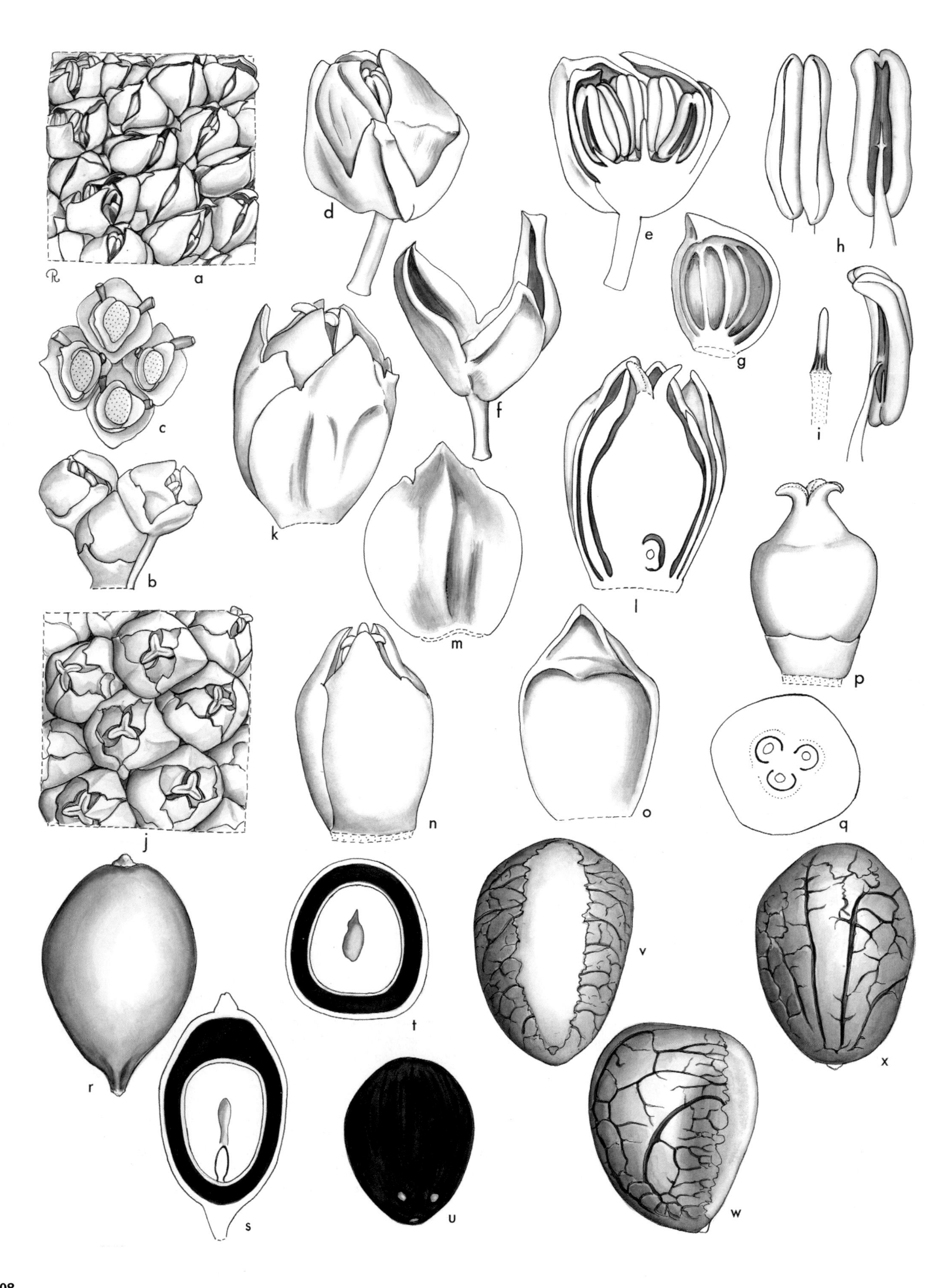
a
b
c
d
e
f
g
h
i
j
k
l
m
n
o
p
q
r
s
t
u
v
w
x

small, raphe wide with large curved and small anastomosing branches, endosperm hard, with or without a central hollow, homogeneous or shallowly ruminate; embryo basal to subbasal. *Germination* remote-tubular; eophyll entire or briefly bifid. *Cytology*: 2n = 32.

Distribution and ecology: Five species in Brazil and Paraguay. In loose sand on beaches, on dunes, in open tree and scrub woodland, among rocks on sandstone hills, in dry grassy or shrubby vegetation zones (cerrado) or in open areas in the mountains.

Anatomy: Leaf (Moraes 1996a), root (Seubert 1998a, 1998b), stamen development (Uhl 1988).

Relationships: *Allagoptera* is monophyletic with high support (Gunn 2004). The genus is resolved as sister to a clade of *Attalea*, *Lytocaryum* and *Syagrus* with low support by Baker *et al.* (in review) and sister to *Cocos* and *Attalea* with low support by Hahn (2002b).

Common names and uses: For common names, see Glassman (1972).

Taxonomic account: Moraes (1996a).

Fossil record: No generic records found.

Notes: Easily recognisable by the striking spicate inflorescence bearing closely appressed pistillate flowers basally and staminate flowers distally. The large unbranched inflorescence of *Allagoptera caudescens* at anthesis entirely covered by stamens has a striking resemblance to that of some species of *Phytelephas*, though there is no relationship.

Additional figures: Glossary fig. 12.

Above left: ***Allagoptera arenaria***, inflorescence, cultivated, Fairchild Tropical Botanic Garden, Florida. (Photo: C.E. Lewis)

Above right: ***Allagoptera caudescens***, infructescences, cultivated, Venezuela. (Photo: J. Dransfield)

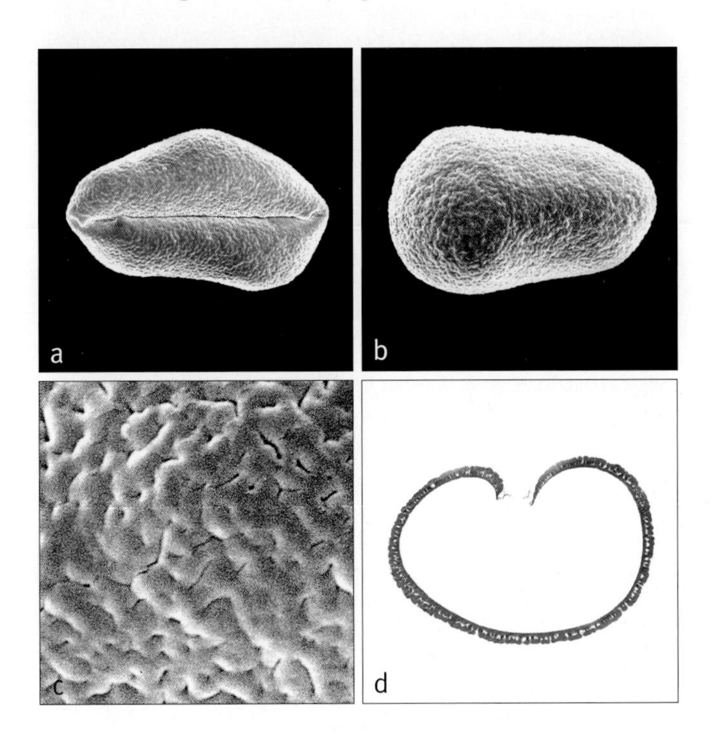

Left: ***Allagoptera***
a, monosulcate pollen grain, distal face SEM × 1000;
b, monosulcate pollen grain, proximal face SEM × 1000;
c, close-up, finely rugulate pollen surface SEM × 8000;
d, ultrathin section through whole pollen grain, polar plane TEM × 1500.
Allagoptera caudescens: **a** & **c**, cultivated RBG Kew 000-69.10235; **b** & **d**, *Glaziou* 8054.

Opposite: ***Allagoptera***. **a**, portion of spike at staminate anthesis × 3; **b**, triad × 3; **c**, triads, flowers removed to show bracteoles and staminate pedicels × 3; **d**, staminate flower × 6; **e**, staminate flower in vertical section × 6; **f**, staminate calyx and pedicel × 6; **g**, staminate petal, interior view × 6; **h**, stamen in 3 views × 12; **i**, pistillode × 12; **j**, portion of spike at pistillate anthesis × 3; **k**, pistillate flower × 6; **l**, pistillate flower in vertical section × 6; **m**, pistillate sepal, exterior view × 6; **n**, pistillate flower, sepals removed × 6; **o**, pistillate petal, interior view × 6; **p**, gynoecium and staminodial cupule × 6; **q**, ovary in cross-section × 12; **r**, fruit × 2; **s**, fruit in vertical section × 2; **t**, fruit in cross-section × 2; **u**, endocarp × 2; **v**, **w**, **x**, seed in 3 views × 4. *Allagoptera arenaria*: **a–q**, *Read* 955; **r–x**, *Wilkinson* 889. (Drawn by Marion Ruff Sheehan)

96. ATTALEA

Solitary, small to massive pinnate-leaved palms native to Central to South America and the Caribbean, with fibrous leaf sheaths, often huge leaves, and with inflorescences that are either staminate or pistillate or carry flowers of both sexes, all on the same plant; fruit is generally large with very thick endocarp, 1–3 or more seeded.

Attalea Kunth in Humb., Bonpl. & Kunth, Nov. gen. sp. 1: folio edition 248; quarto edition 309 (1816). Type: **A. amygdalina** Kunth.

Maximiliana Mart., Palm. fam. 20 (1824) (conserved name). Type: *M. martiana* H. Karst. (*M. regia* Mart. [1826], non *Maximilianea regia* Mart. [1819] = Cochlospermaceae) = *Attalea maripa* (Aubl.) Mart. (*Palma maripa* Aubl.).

Lithocarpos Ant. Targ. Tozz., Mem. Mat. Fis. Soc. Ital. Sci. Modena, Pt. Mem. Fis. 20(2): 312 (1833 [non Blume 1825–1826]). Type: *L. cocciformis* Ant. Targ. Tozz. (illegitimate name) (*Cocos lapidea* Gaertn. = *Attalea lapidea* [Gaertn.] Burret).

Orbignya Mart. ex Endl., Gen. pl. 257 (1837). Conserved name. Type: *O. phalerata* Mart. (= *Attalea speciosa* Mart.).

Scheelea H. Karst., Linnaea 28: 264 (1857) ('1856'). Lectotype: *S. regia* H. Karst. (see H.E. Moore 1963c) (= *Attalea butyracea* [Mutis ex L.f.] Wess. Boer [*Cocos butyracea* Mutis ex L.f.]).

Englerophoenix Kuntze, Revis. gen. pl. 2: 728 (1891). Type: *E. regia* (Mart.) Kuntze (illegitimate name) (*Maximiliana regia* Mart [1826, non 1819] [= *Attalea maripa* (Aubl.) Mart.]).

Pindarea Barb. Rodr., Pl. jard. Rio de Janeiro 5 (1895) ('1896'). Type: *P. concinna* Barb. Rodr. (= *Attalea concinna* [Barb.Rodr.] Burret).

Parascheelea Dugand, Caldasia 1(1): 10 (1940). Type: *P. anchistropetala* Dugand (= *Attalea luetzelburgii* (Burret) Wess. Boer).

Sarinia O.F. Cook, Natl. Hort. Mag. 21: 78, 84 (1942). Type: *S. funifera* (Mart.) O.F. Cook (*Attalea funifera* Mart.).

Ynesa O.F. Cook, Natl. Hort. Mag. 21: 71, 72, 84 (1942). Type: *Y. colenda* (*Attalea colenda*).

Commemorates Attalus III Philometor, King of Pergamum in Asia Minor, 138–133 BC, who in his later life was interested in medicinal plants.

Small to massive, solitary, acaulescent or erect, unarmed, pleonanthic, monoecious palms. *Stem* subterranean to tall, usually becoming bare, obliquely marked with leaf scars. *Leaves* massive, pinnate, marcescent; sheath thick, finely or coarsely fibrous (in *Attalea funifera* producing piassava); petiole lacking or short to elongate, adaxially channelled, abaxially rounded, variously tomentose, rachis adaxially channelled near the base, distally angled, abaxially rounded or flattened, abaxially variously tomentose; leaflets inserted on the lateral faces or in shallow grooves; leaflets numerous, linear-lanceolate, single-fold, regularly arranged or in clusters of 2–5, irregularly lobed at the tips, caducous scales abundant along the leaflet margins exposed in the sword leaf, midrib prominent, other longitudinal veins rather indistinct, transverse veinlets abundant, conspicuous. *Inflorescences* solitary, interfoliar, ± erect or becoming pendulous, entirely staminate, entirely pistillate, or with flowers of both sexes, branched to 1 order or branches short and flowers appearing ± sessile on the main axis; peduncle short to long; prophyll obscured by leaf sheaths and not known, peduncular bract tubular, entirely enclosing the inflorescence in bud with a short to long solid beak, splitting abaxially, expanding and usually becoming cowl-like, thick and woody, abaxially deeply grooved, adaxially glabrous, abaxially densely tomentose, long persistent, subsequent peduncular bracts small, incomplete, triangular, ± coriaceous; rachis shorter or longer than the peduncle, bearing spirally or unilaterally arranged rachillae, each subtended by a short triangular bract; staminate rachillae with a short to long basal bare portion, above which bearing paired or solitary flowers, spirally arranged (rarely) or in 2 rows on one side, glabrous or floccose-tomentose, bisexual rachillae of two types, either similar to the staminate but bearing a few basal pistillate flowers or bearing 1 to several triads with a short slender apical portion bearing fertile or sterile staminate flowers, in the putative pistillate rachillae lacking all trace of staminate flowers at maturity. *Staminate flowers* asymmetrical; sepals 3, distinct, triangular, very small, sometimes slightly imbricate basally; petals 3, distinct, much longer than the sepals, ovate-triangular, acute, valvate, or terete and scarcely valvate, or terete basally and distally expanded into a triangular ± valvate limb; stamens 3–75, usually much shorter, rarely much longer than the petals, filaments slender, short to long, anthers ± straight to twisted and coiled, dorsifixed or rarely medifixed, sometimes sagittate basally, introrse or latrose; pistillode minute

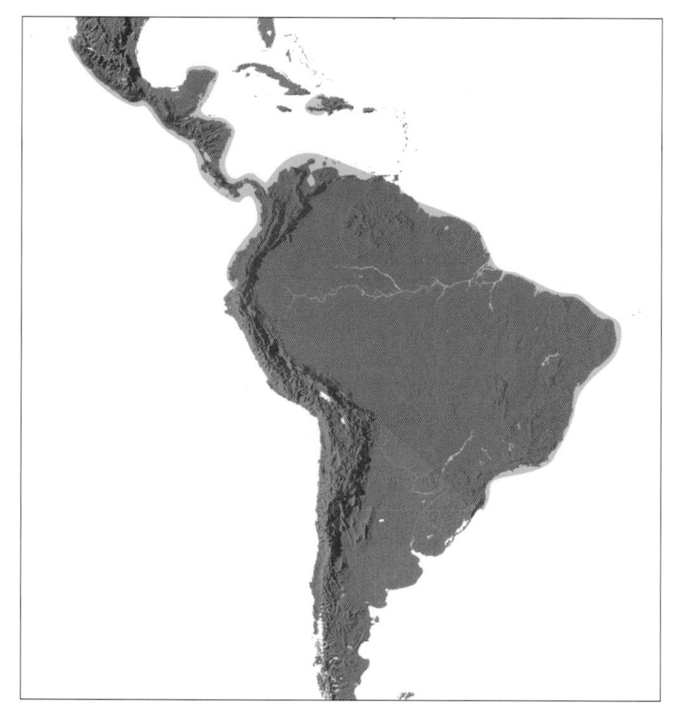

Distribution of ***Attalea***

Opposite. ***Attalea.*** **a**, portion of staminate rachilla × 1 1/2; **b**, staminate flower × 3; **c**, staminate flower in vertical section × 3; **d**, staminate sepal × 6; **e**, staminate petal × 3; **f**, staminate petal in cross-section × 6; **g**, stamen in 3 views × 6; **h**, pistillode × 6; **i**, pistillate rachilla × 1/2; **j**, pistillate flower × 1 1/2; **k**, pistillate flower in vertical section × 1 1/2; **l**, pistillate sepal × 1 1/2; **m**, pistillate petal × 1 1/2; **n**, gynoecium and staminodial cupule × 1 1/2; **o**, ovary in cross-section × 1 1/2; **p**, fruit × 1/2; **q**, fruit in cross-section × 1/2; **r, s**, endocarp in 2 views × 3/4; **t, u**, seed in 2 views × 3/4; **v–aa**, further variations in staminate flower form – **v**, staminate flower × 3; **w**, staminate flower in vertical section × 3; **x**, staminate flowers × 6; **y**, staminate flower in vertical section × 6; **z**, staminate flowers × 3; **aa**, staminate flower in vertical section × 3. *Attalea amygdalina*: **a–h**, *Moore et al.* 10196; **i–u**, *Moore et al.* 10199; *Attalea plowmanii*: **v, w**, *Moore* 9510; *Attalea maripa*: **x, y**, *Aba* 401; *Attalea cohune*: **z, aa**, *Armour* 151. (Drawn by Marion Ruff Sheehan)

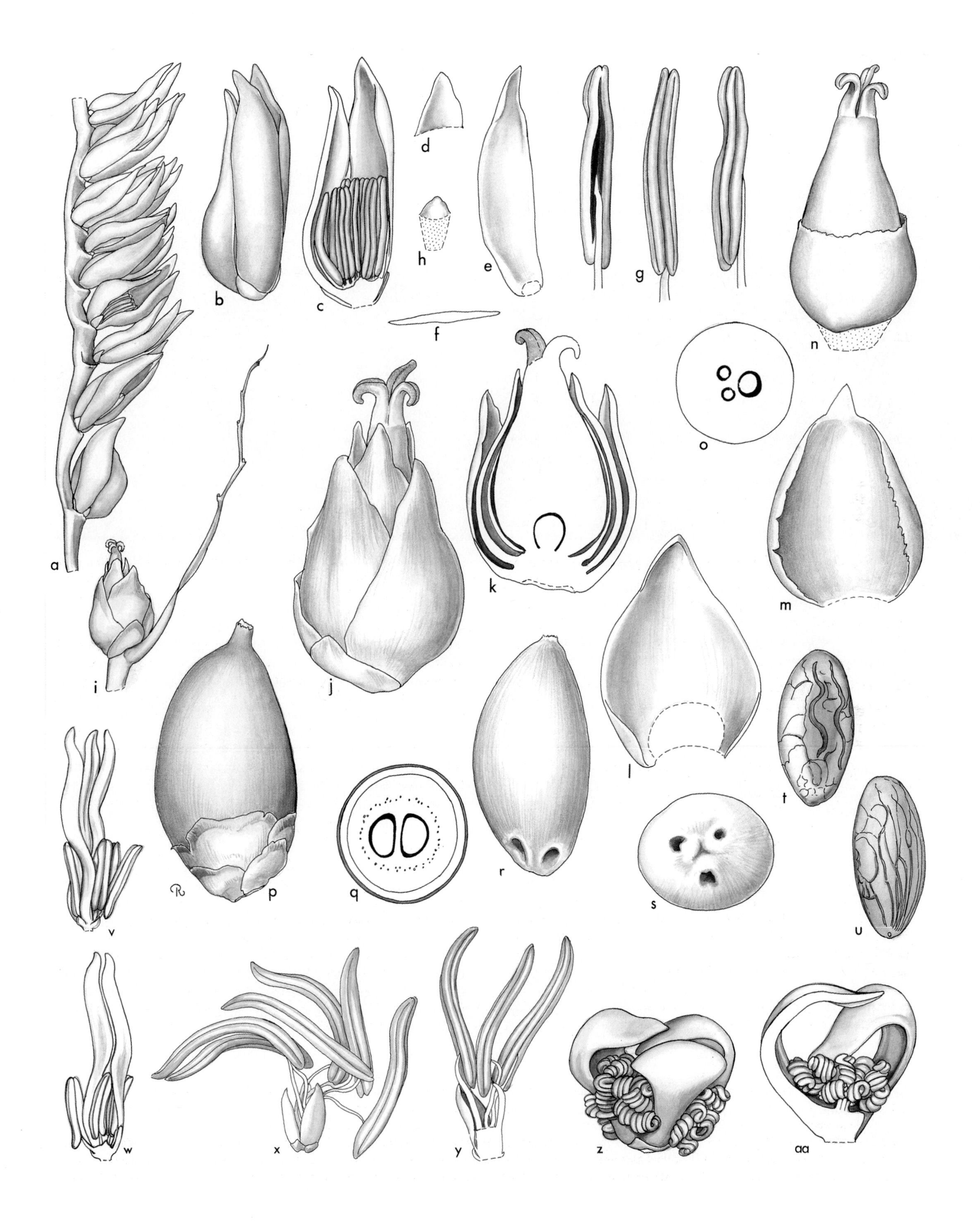
a
b
c
d
h
e
f
g
n
o
i
j
k
m
l
p
q
r
s
t
u
v
w
x
y
z
aa

Above left: ***Attalea colenda***, habit, Ecuador. (Photo: F. Borchsenius)

Above right: ***Attalea colenda***, rachillae with staminate flowers, Ecuador. (Photo: F. Borchsenius)

or absent. *Pollen* ellipsoidal, usually with either slight or obvious asymmetry, occasionally pyriform, trichotomosulcate pollen also present; aperture a distal sulcus or trichotomosulcus; ectexine tectate, finely to coarsely perforate, finely to coarsely perforate and micro-channelled, or perforate-rugulate or, unusually, tectate gemmate, aperture margin slightly finer; infratectum columellate; longest axis 32–85 µm [17/71]. *Pistillate flowers* very much larger than the staminate, generally ovoid; sepals 3, distinct, ± triangular, broadly imbricate, leathery; petals 3, distinct, rounded or ± triangular with triangular tips, glabrous or tomentose; staminodal ring large, coriaceous, tomentose; gynoecium of 3–several connate carpels, ovoid or obpyriform, style tapering, stigmatic lobes equal in number to the carpels, linear, reflexed at anthesis, ovules 1 per carpel, basal, form unknown. *Fruit* ± ovoid, sometimes asymmetrical, 1–several seeded, with a short to moderate beak and apical stigmatic remains, perianth and staminodal ring persistent and enlarging; epicarp minutely grooved, bearing scales, mesocarp usually fleshy and fibrous, endocarp very thick, stony, smooth without or closely grooved, often with included fibres, the pores subbasal, deeply impressed, ?always. *Seed* ellipsoidal or laterally somewhat flattened, basally attached with fine anastomosing raphe bundles, endosperm homogeneous, solid (?always); embryo basal. *Germination* remote-tubular; eophyll entire, lanceolate. *Cytology*: 2n = 32.

Distribution and ecology: About 69 species occurring from Mexico southwards to Bolivia and Peru, occurring in a wide range of habitats from tropical rain forest to dry 'campo rupestre' and 'cerrado'.

Anatomy: Leaf (Tomlinson 1961, Glassman 1999), root (Seubert 1998a, 1998b), gynoecium (Uhl & Moore 1971).

Relationships: *Attalea* is monophyletic with high support (Hahn 2002b, Gunn 2004). The genus is resolved as sister to a clade of *Lytocaryum* and a subclade of *Syagrus* with moderate support (Gunn 2004) or as sister to *Lytocaryum* with low support (Baker *et al.* in review).

Common names and uses: For common names see Glassman (1999). These are palms with a multiplicity of uses, the most important being as a source of oil. For medicinal uses, see Plotkin and Balick (1984).

Taxonomic accounts: Glassman (1999) and Zona (2002b).

Fossil record: A palm endocarp from the Upper Eocene of southeast North America (Florida), *Attalea gunteri*, is reported by Berry (1929). A fruit, *Attaleinites* gen. nov., is reported from the Oligocene of Hungary (Tuzson 1913). *Attalea*-like pollen (Graham 1976) is also reported from the Upper Miocene of Mexico (Paraje Solo flora, Veracruz).

Notes: Opinion is divided as to both the number of genera and species. Glassman (1999) recognises four genera, *Attalea*, *Scheelea*, *Orbignya* and *Maximiliana*. These four genera were also recognised in the first edition of *Genera Palmarum*. As the palms have become better known in the field and more herbarium material has accumulated, the characters of the staminate flowers used to differentiate the genera seem increasingly unreliable. Intermediate conditions occur (which Glassman [1999] attributes to intergeneric hybridisation) and the form of the staminate flower seems not be correlated with any other varying characters. Henderson (1995) and Henderson *et al.* (1995) included all genera in *Attalea*, arguing convincingly that the previously recognised genera are untenable. This broad generic approach is followed here. At the species level,

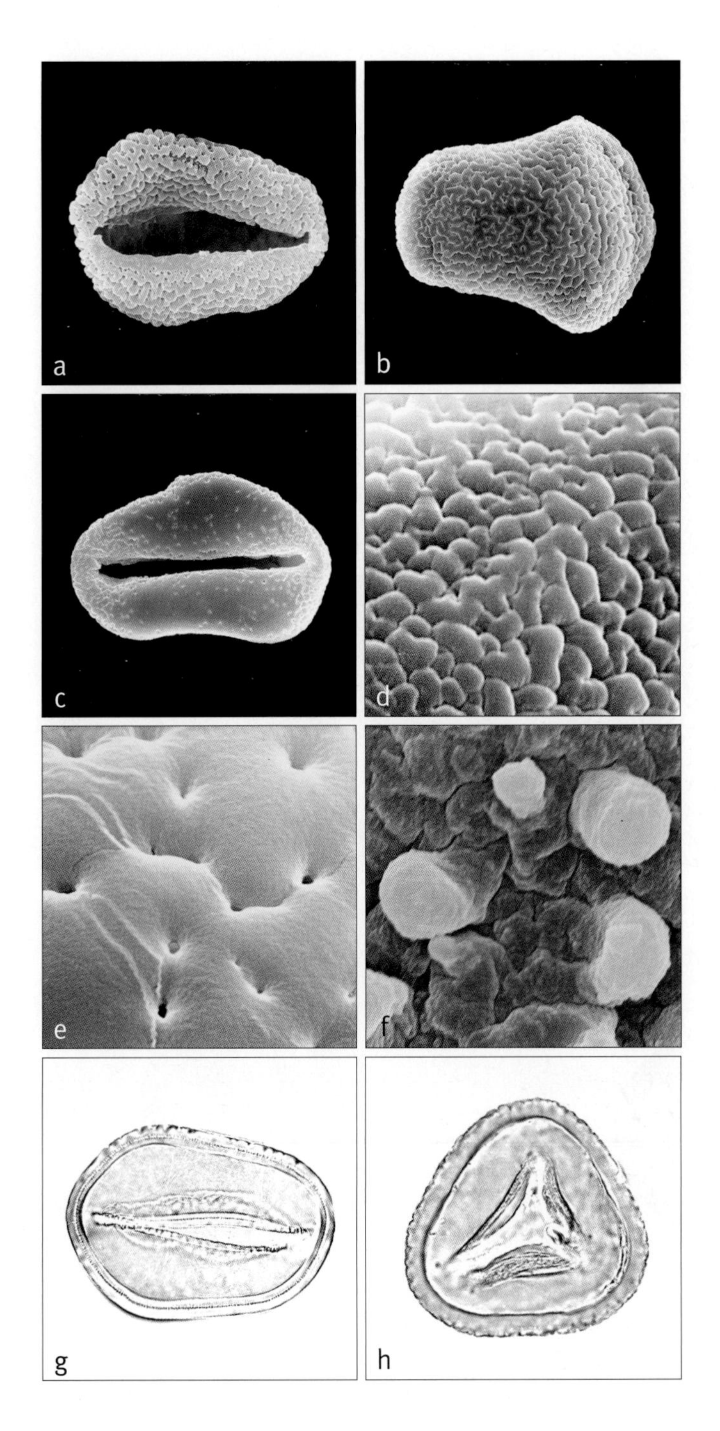

Above: ***Attalea***
a, monosulcate pollen grain, distal face SEM × 800;
b, monosulcate pollen grain, proximal face × 800;
c, monosulcate pollen grain, distal face SEM × 800;
d, close-up, rugulate pollen surface SEM × 8000;
e, close-up, pollen surface with coarse, widely spaced perforations SEM × 8000;
f, close-up, tectate gemmate pollen surface SEM × 8000;
g, monosulcate pollen grain, mid-focus LM × 800;
h, trichotomosulcate pollen grain, mid-focus LM × 800.
Attalea cohune: **a**, *Watson* s.n.; *A. cuatrecasana*: **b** & **c**, *Gentry et al.* 53665; *A. colenda*: **d**, *Skov et al.* 64828; *A. insignis*: **e** & **g**, *Krukoff* 5572; *A. crassispatha*: **f**, *Henderson & Aubry* 805; *A. luetzelburgii*: **h**, *Zarucchi* 2108.

Glassman (1999) recognises 66 species whereas Henderson *et al.* (1995) recognise 29. There is clearly scope for more detailed revisionary taxonomic work before a clear understanding of the species limits is reached. *Attalea crassispatha* from Haiti was used by O.F. Cook as the basis of his invalidly published genus *Bornoa* (Cook 1939a). He also published the invalid names *Temenia* (Cook 1939a) and *Ethnora* (Cook 1940), both for *Attalea maripa*, and *Heptantra*, for *Attalea speciosa* (Cook 1939a). Three intergeneric hybrid names have been published: *Markleya* Bondar (Arch. Jard. Bot. Rio de Janeiro 15: 50 [1957]) for a hybrid between *Orbignya phalerata* and *Maximiliana maripa*; ×*Maximbignya* Glassman (Illinois Biol. Monogr. 59: 199 [1999]) as an explicit hybrid name (*Maximbignya dahlgreniana* [Bondar] Glassman); and ×*Attabignya* Balick (A.B. Anderson & Med.-Costa, Brittonia 39: 27 [1987]) for a hybrid between *Attalea compta* and *Orbignya oleifera* (namely ×*Attabignya minarum* Balick *et al.*). With the subsuming of all genera in *Attalea*, new combinations for these hybrids in *Attalea* were published by Zona (2002b).

Blue-throated Macaws (*Ara glaucogularis*) feed on the mesocarp of *Attalea phalerata* fruit (Yamashita & de Barros 1997).
Additional figures: Glossary figs 15, 17, 19.

97. BUTIA

Small to moderate, solitary or clustered pinnate-leaved palms, native to cooler parts of south America; the petioles are usually with toothed margins, the staminate flowers have 6 stamens and the endosperm is homogeneous.

Butia (Becc.) Becc., Agric. Colon. 10: 489 (1916). Lectotype: **B. capitata** (Mart.) Becc. (*Cocos capitata* Mart.) (see H.E. Moore 1963c).
Cocos subgenus *Butia* Becc., Malpighia 1: 352 (1887).
Syagrus section *Butia* (Becc.) Glassman, Fieldiana, Bot. 32: 235 (1970) (excluding *S. vagans* and *S. schizophylla*).

Derived from a Portuguese corruption of a vernacular name, mbotiá, *said to be from* mbo *— to make, and* tiá *— those who have incurved teeth, presumably referring to the teeth on the petiole.*

Small to moderate, solitary or clustered, armed or unarmed, pleonanthic, monoecious palms. *Stem* subterranean to erect, generally not tall, obscured by remains of leaf bases, eventually becoming bare, marked with close leaf scars. *Leaves* pinnate, small to large, arching; sheaths tubular at first, disintegrating into a fibrous network, often densely tomentose; petiole short to long, channelled or flat adaxially, rounded or angled abaxially, proximally unarmed and bearing scattered fibres or armed with coarse spines decreasing in size distally until represented by short teeth, variously caducously scaly or glabrous, often glaucous; rachis usually curved, adaxially angled or flattened, rounded or flattened abaxially; leaflets single-fold, usually numerous, regularly arranged, held stiffly in the same plane, linear, acuminate, acute, obtuse or asymmetrical at the tips, frequently glaucous, usually with crowded ramenta on the abaxial surface of the main vein near the rachis, transverse veinlets obscure. *Inflorescences* solitary, interfoliar, shorter than the leaves, branching to 1 order, apparently protandrous; peduncle ± rounded in cross-section, short to long, ± glabrous or with scattered caducous scales; prophyll short, flattened, tubular, 2-keeled, usually hidden by the leaf sheaths, becoming fibrous with age, splitting at the tip, persistent; peduncular bract inserted near the prophyll, much longer, tubular, enclosing the inflorescence until shortly before anthesis, tightly sheathing proximally, beaked distally, splitting

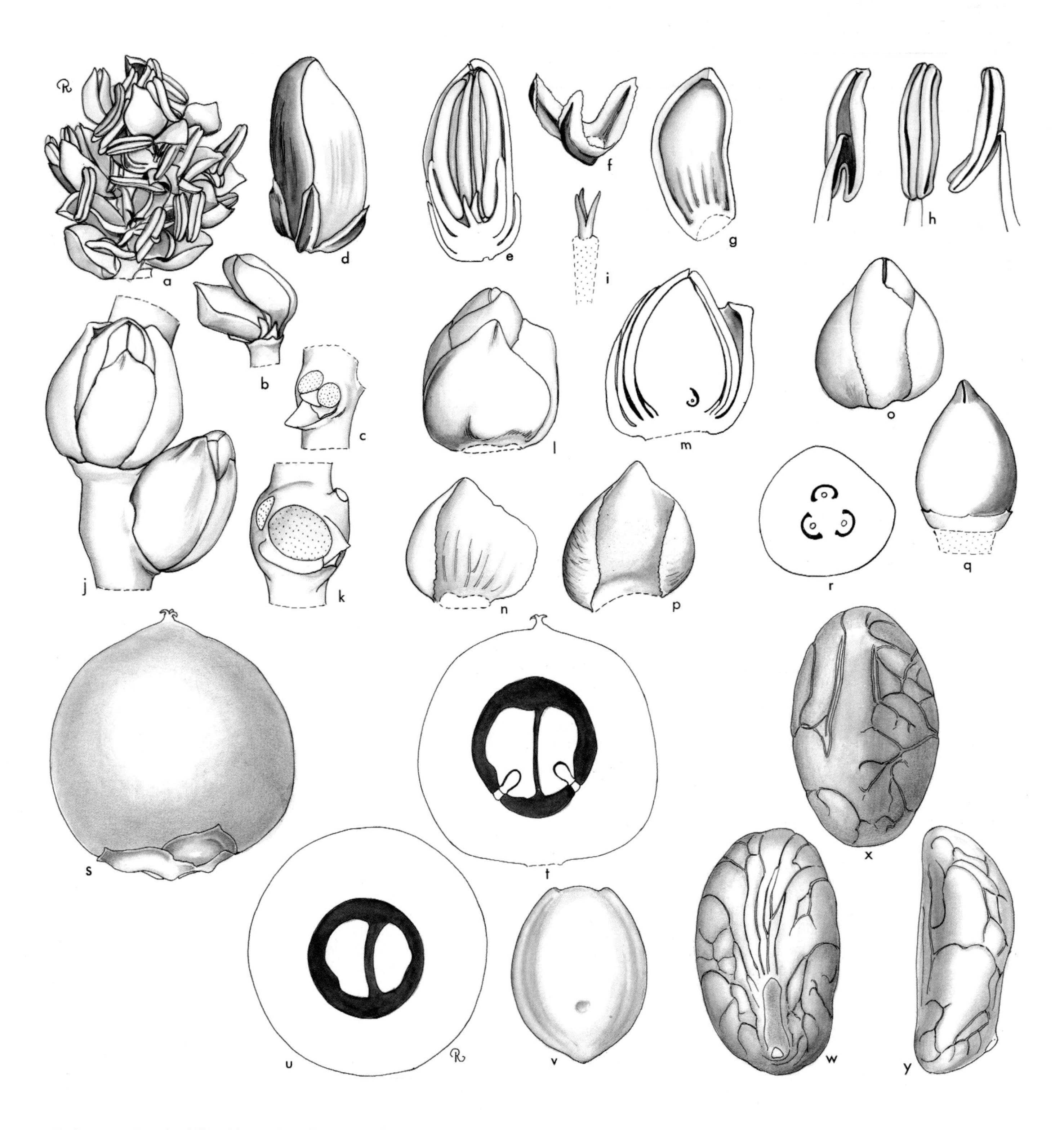

Butia. **a**, portion of rachilla with staminate flowers × 3; **b**, staminate buds × 3; **c**, scars of paired staminate flowers with bract and bracteoles × 3; **d**, staminate bud × 6; **e**, staminate bud in vertical section × 6; **f**, staminate calyx × 6; **g**, staminate petal, interior view × 6; **h**, stamen in 3 views × 6; **i**, pistillode × 6; **j**, triad with pistillate buds × 3; **k**, triad, flowers removed to show bract and bracteoles × 3; **l**, pistillate bud × 3; **m**, pistillate bud in vertical section × 3; **n**, pistillate sepal × 3; **o**, pistillate bud, 2 sepals removed × 3; **p**, pistillate petal × 3; **q**, gynoecium and staminodial ring × 3; **r**, ovary in cross-section × 6; **s**, fruit × 1 1/2; **t**, fruit in vertical section × 1 1/2; **u**, fruit in cross-section × 1 1/2; **v**, endocarp × 1 1/2; **w**, **x**, **y**, seed in 3 views × 3. *Butia capitata*: **a–i**, *Moore* s.n.; **j–r**, *Read* 906; **s–y**, *Sheehan* s.n. (Drawn by Marion Ruff Sheehan)

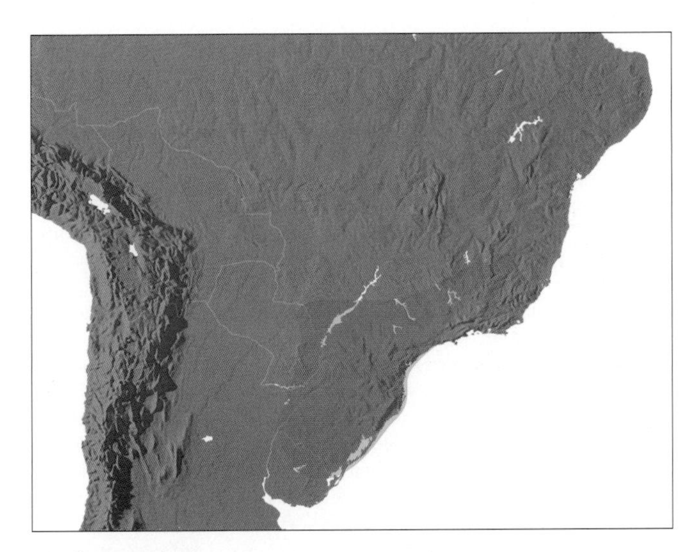

Distribution of ***Butia***

Distribution and ecology: Nine species confined to cooler, drier areas of South America, in southern Brazil, Paraguay, Uruguay and Argentina. Often gregarious in grasslands, 'campo rupestre', 'cerrado' and woodlands in the lowlands.

Anatomy: Leaf (Glassman 1979), root (Seubert 1998a, 1998b), floral (Uhl & Moore 1971).

Relationships: *Butia* is monophyletic with high support (Gunn 2004). The genus is highly supported as sister to *Jubaea* (Gunn 2004, Baker *et al.* in review).

Common names and uses: Yatay palms, jelly palms, butia palms. Mesocarp of *Butia capitata* is edible and can be made into jams; several species are widespread as slow-growing cold-tolerant ornamentals.

longitudinally along the abaxial face to expose the flowers, expanding distally and becoming cowl-like, smooth or becoming longitudinally striate with age, adaxially glabrous, abaxially glabrous, scaly, or very densely tomentose; rachis shorter or longer than the peduncle, bearing spirally arranged, relatively few to very numerous rachillae, each subtended by an inconspicuous triangular bract; rachillae rather stiff, ± zigzag, glabrous or minutely dotted, with a short to long, basal, bare portion, above which bearing few to numerous, spirally arranged triads proximally, paired or solitary staminate flowers distally, the distal-most rachillae sometimes entirely staminate, the flower groups close or relatively distant, superficial; rachilla bracts and floral bracteoles inconspicuous. *Staminate flowers* sessile or briefly pedicellate, slightly asymmetrical; sepals 3, distinct or joined at the base, narrow, triangular, membranous, ± keeled, acute; petals 3, distinct, or very briefly connate at the base, valvate, at least 3 times as long as the sepals, ± fleshy, ovate to triangular; stamens 6, filaments distinct, awl-shaped, elongate, anthers elongate, medifixed, versatile, basally sagittate, introrse; pistillode shorter than the filaments, trifid. *Pollen* ellipsoidal, frequently elongate, usually with either slight or obvious asymmetry; aperture a distal sulcus; ectexine tectate, perforate and micro-channelled, or perforate-rugulate, aperture margin may be slightly finer; infratectum columellate; longest axis 38–57 µm; post-meiotic tetrads tetrahedral, occasionally tetragonal, or rarely, rhomboidal [4/9]. *Pistillate flowers* much larger than the staminate, globose to ovoid, ± symmetrical; sepals 3, distinct, broadly imbricate, coriaceous, somewhat keeled, triangular, the tips conspicuously hooded; petals 3, ± the same length as and similar to the sepals, distinct, broadly imbricate, the tips briefly valvate; staminodal ring well developed as a free, fleshy collar surrounding the base of the ovary, irregularly minutely lobed; gynoecium ± ovoid, trilocular, triovulate, stigmas 3, conspicuous, reflexed at anthesis, ovules hemianatropous, laterally attached to the ventral angles of the locules, septal canals present, opening at the bases of the stigmas. *Fruit* 1–3-seeded, spherical, oblate, or ovoid, yellow, brown, or purplish, with a short to long beak and apical stigmatic remains; epicarp smooth, mesocarp thin to thick, pulpy or fleshy, sometimes sweet, fibrous, endocarp thick, bony, with 1–3 developed cavities, the pores lateral below the equator or subbasal. *Seed* basally attached, conforming to the shape of the endocarp cavities, endosperm homogeneous, solid; embryo opposite the endocarp pore. *Germination* and eophyll not recorded. *Cytology*: 2n = 32.

Top right: ***Butia yatay***, habit, cultivated, Montgomery Botanical Center, Florida. (Photo: J. Dransfield)

Bottom right: ***Butia yatay***, infructescence, cultivated, Montgomery Botanical Center, Florida. (Photo: J. Dransfield)

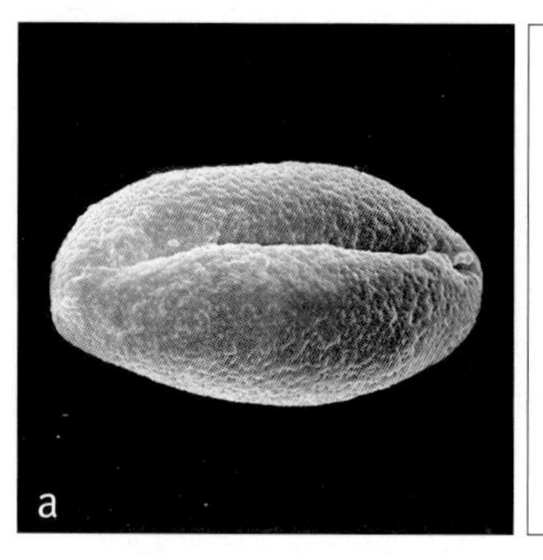

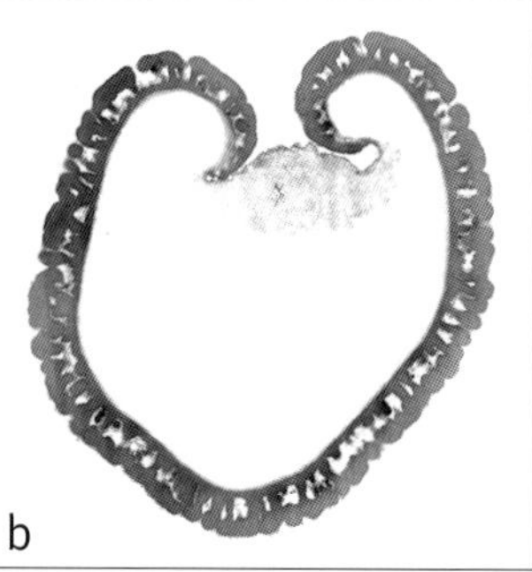

Above: ***Butia***
a, monosulcate pollen grain, distal face SEM × 1000;
b, ultrathin section through whole pollen grain, intine 'plug' still in place in sulcus region, polar plane TEM × 1750.
Butia yatay: **a**, *Schinini* 24939; *B. purpurascens*: **b**, *Schinini* 32582.

Taxonomic accounts: Glassman (1979) and Noblick (2006).
Fossil record: Two small palm endocarps included in *Palmocarpon luisii*, from the Maastrichtian of Brazil, State of Parahyba do Norte (Rio Gramame) are compared with members of the Cocoseae (Maury 1930). One is described as being about the size and form of the living *Cocos eriospatha* but more deformed. (*Cocos eriospatha* = *Butia eriospatha*; see Govaerts & Dransfield 2005.) However, the endocarps illustrated do not show three pores and could in fact belong to a quite different family of flowering plants (see also under *Syagrus*).
Notes: *Butia* differs from *Jubaea* in having only six rather than numerous stamens and in the usually conspicuously toothed petioles. *Anodorhynchus* macaws feed on the endosperm of *Butia yatay* (Yamashita & Valle 1993).
Additional figures: Fig. 1.4, Glossary fig. 20.

98. COCOS

The often slanting stems and graceful crowns of the coconut are largely responsible for palms being considered the hallmark of the tropics. Furthermore, the coconut, one of the ten most important crop trees, is the mainstay of many people.

Cocos L., Sp. pl. 1188 (1753). Type: **C. nucifera** L.
Calappa Steck, Sagu 9 (1757).
Coccus Mill., Gard. Dict. abr. ed. 4. (1754).

Latinization of the Portuguese word, coco, *originally used for a bugbear or ape, in reference to the face-like appearance of the partially dehusked endocarp.*

Moderate, solitary, unarmed, pleonanthic, monoecious palm, sometimes flowering while still without an emergent trunk. *Stem* erect, often curved or slanting, becoming bare and conspicuously ringed with leaf scars. *Leaves* numerous, pinnate, neatly abscising; sheath fibrous, forming a woven supportive network with a conspicuous, tongue-like extension opposite the petiole, eventually disintegrating and becoming open; petiole short to long, adaxially channelled, abaxially rounded, bearing caducous tomentum abaxially; rachis elongate, curved or straight, adaxially angled near the tip, abaxially rounded, with caducous tomentum abaxially; leaflets very numerous, single-fold, regularly arranged in one plane, usually rather stiff, linear, acuminate, usually bifid with slightly asymmetrical tips, adaxially glabrous, abaxially with abundant, dot-like scales and very small ramenta along the midrib, midrib prominent adaxially, transverse veinlets evident. *Inflorescences* solitary, interfoliar, axillary, branched to 1 order, protandrous; peduncle ± elliptic in cross-section, robust, elongate, bearing scattered scales; prophyll tubular, 2-keeled laterally, opening apically, becoming fibrous, tomentose, persistent, ± obscured by the leaf sheaths; peduncular bract inserted near the prophyll, very large, tubular, entirely enclosing the inflorescence until shortly before anthesis, splitting abaxially, becoming boat-shaped, beaked, thick, woody, adaxially smooth, abaxially with longitudinal, shallow grooves and caducous tomentum; rachis ± equalling the peduncle, bearing spirally arranged rachillae, each subtended by an inconspicuous triangular bract and with a swollen base; rachillae robust, ± pendulous at first, later spreading with a basal bare portion and none or a few basal triads and pairs or solitary staminate flowers distally; rachilla bracts and floral bracteoles inconspicuous. *Staminate flowers* ± asymmetrical, narrowly ovoid, moderate, sessile; sepals 3, distinct, rather unequal, imbricate, triangular, ± keeled; petals much longer than the sepals, thick, rather leathery, distinct, valvate, irregularly boat-shaped, acute; stamens 6, filaments rather short, distinct, awl-shaped, fleshy, ± erect, anthers deeply sagittate basally, shallowly so at the apex, elongate, medifixed, ± versatile, latrorse; pistillode with 3, slender, pointed lobes. *Pollen* ellipsoidal, frequently elongate and/or pyriform, usually with either slight or obvious asymmetry; aperture a distal sulcus; ectexine

Right: ***Cocos nucifera***, habit, Madagascar. (Photo: J. Dransfield)

tectate, finely perforate, perforate and micro-channelled, or perforate-rugulate, aperture margin slightly finer; infratectum columellate; longest axis 62–70 µm [1/1]. *Pistillate flowers* very large, globose in bud, becoming very broadly ovoid at anthesis; sepals 3, distinct, imbricate, ± rounded; petals similar to and somewhat longer than the sepals, lacking valvate apices, very leathery; staminodal ring low, membranous, not lobed; gynoecium trilocular at the very base, triovulate, broadly ovoid, obscurely 3-angled, extremely fibrous distally, stigmas 3, very short, borne in a slight depression, ovule anatropous, very small, laterally attached. *Fruit* very large (except in unusual forms), ellipsoidal to broadly ovoid, indistinctly 3-angled, dull green, brown, brilliant-orange, yellow, to ivory-coloured when ripe, perianth enlarging in fruit, stigmatic remains apical; epicarp smooth, mesocarp very thick and fibrous, dry, endocarp thick and woody, ± spherical to narrow ovoid, indistinctly 3-angled, with 3 longitudinal ridges, and 3, large, slightly sunken, basal pores, each with an operculum. *Seed* almost always 1 only, very large, with a narrow layer of homogeneous endosperm, and a large central cavity partially filled with fluid; embryo basal, opposite one of the endocarp pores. *Germination* adjacent-ligular; eophyll entire, broadly lanceolate. *Cytology*: 2n = 32.

Distribution and ecology: A single species widely cultivated throughout the tropics and warmer subtropics. Origin uncertain but said to be western Pacific (Harries 1978, Gruezo & Harries 1984, Buckley & Harries 1984) (but see below). *Cocos nucifera* is often regarded as a strand plant but it will flower and fruit in humid equatorial regions at altitudes up to 900 m above sea level. Its natural habitat may well have been strand vegetation.

Anatomy: Leaf, stems, root (Tomlinson 1961), phloem (Parthasarathy 1974, 1980), wood (Chen 1995), root (Seubert 1998a, 1998b), megasporogenesis (Reddy & Kulkarni 1989), fruit (Roth 1977, Reddy & Kulkarni 1985).

Relationships: *Cocos* is moderately supported as sister to *Parajubaea* (Baker *et al.* in review).

Common names and uses: Coconut, one of the most important tropical crops with a multiplicity of uses both local and commercial.

Taxonomic accounts: Glassman (1987) and Harries (1978, 1992).

Fossil record: Fruit or, more often, endocarps are the most frequently recorded fossils. The endocarps are distinguished by usually 3, sometimes more, well-defined pores. However, encocarps as large as those of *Cocos nucifera* are unknown. This suggests that the fossils may represent other genera within subtribe Attaleinae. Berry (1926b) first described *Cocos zeylandica*, a small endocarp (ca. 3.5 cm long), from Pliocene brown coal *Cocos*-bearing Beds in New Zealand (Mangonui, North Island); it was one of a number of specimens recovered. The endocarps are usually well-preserved and commonly washed up on the beach. Couper (1952), in his record of the pollen and spores of the *Cocos*-bearing Beds, notes that the rich flora of these beds had been recognised by geologists as early as 1872. From two pollen- and spore-rich samples, Couper (1952) records 4% and 8% abundance for *Rhopalostylis*/*Cocos*-type pollen (the two genera have closely similar pollen). Couper (1953) claims there can be little doubt that the commonly encountered palm pollen in the *Cocos*-bearing Beds is associated with *Cocos zeylandica*. However, although pollen of the two genera is closely similar in exine and size range (60–70 µm), whereas the pollen of other cocosoid palms with similar exine characteristics is smaller (<58 µm), it seems more reasonable to associate the fossil pollen with *Rhopalostylis*, a New Zealand palm. Further specimens of carbonised *Cocos zeylandica* were recovered from Miocene turbidites (possibly the result of a tsunami) at a different site in North Island by Ballance *et al.* (1981). From South Island, New Zealand (Otago Peninsula and

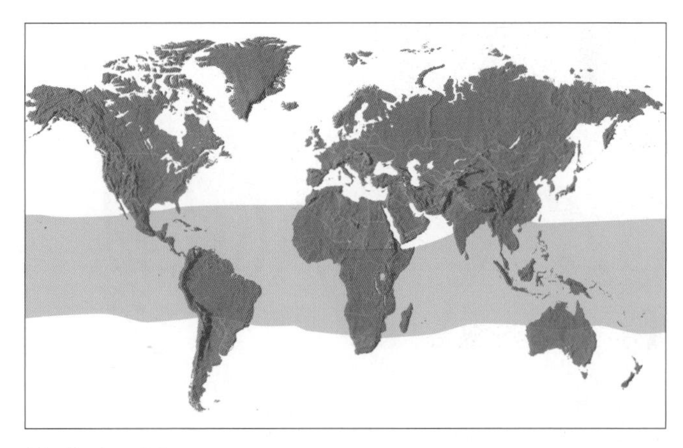

Distribution of ***Cocos***

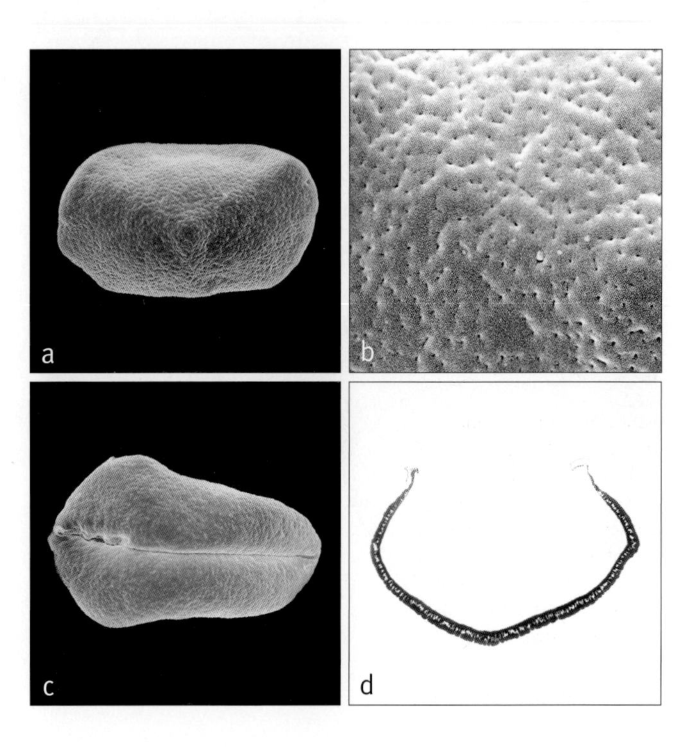

Right: ***Cocos***

a, monosulcate pollen grain, proximal face SEM × 600;
b, close-up, finely perforate pollen surface SEM × 6000;
c, monosulcate pollen grain, distal face SEM × 600;
d, ultrathin section through whole pollen grain, polar plane TEM × 1000.
Cocos nucifera: **a–c**, *S.K. Lau* 375, **d**, *Cuadra* A.1243.

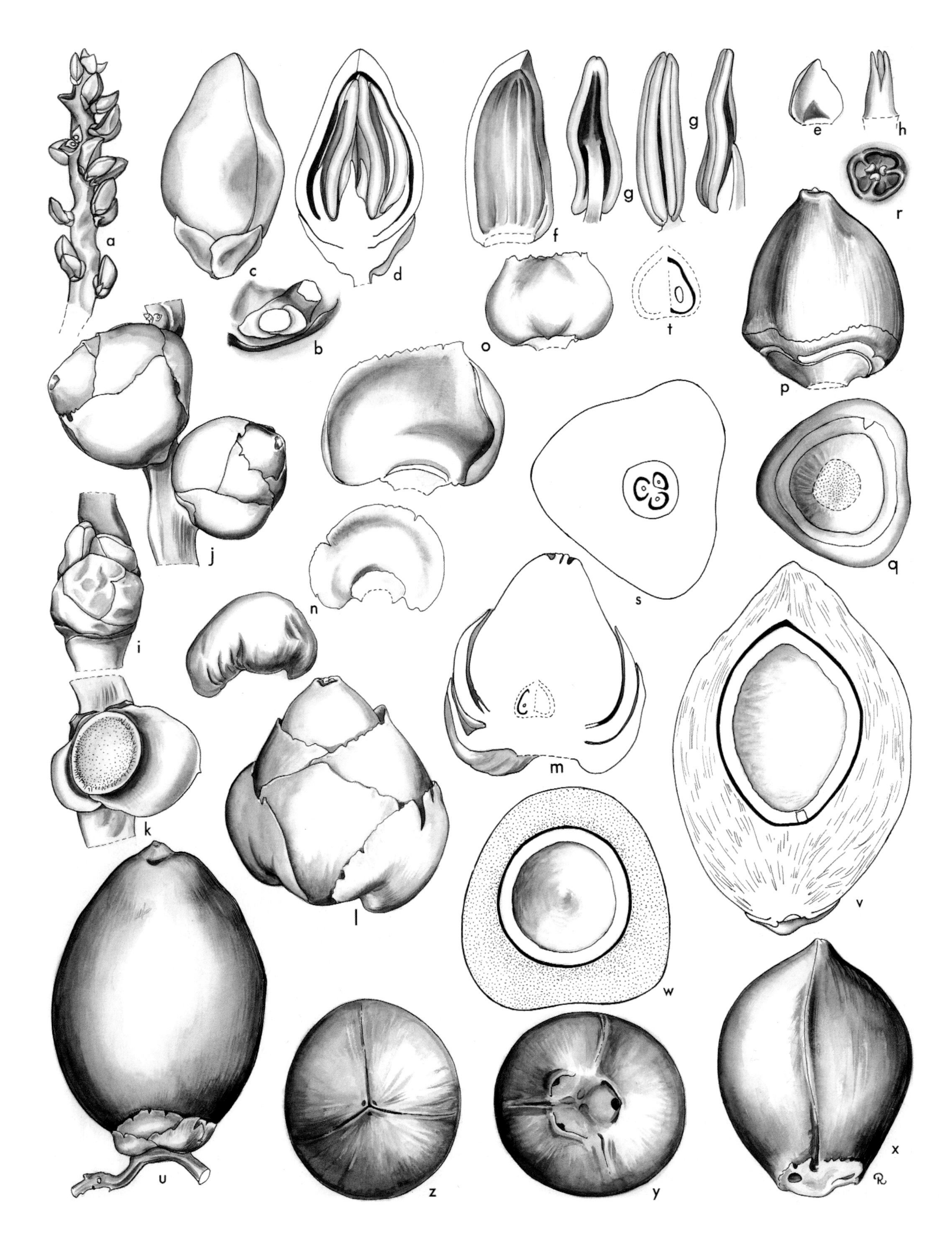
a
c
d
f
g
g
e
h
r
b
o
t
p
j
s
q
i
n
m
k
v
l
w
u
z
y
x

Canterbury), Campbell *et al.* (2000) describe fossil endocarps (5–14.5 cm long) that have distinctive cocosoid pores, with the records from South Island ranging from mid-Eocene (45 million years ago) to Lower Oligocene (ca. 35 million years ago). Other records of fruits and seeds include silicified fruit (10 cm long × up to 9.5 cm wide) from the latest Pliocene of Australia (Queensland), which Rigby (1995) considers most resemble *Cocos nucifera*. *Palmocarpon cetera*, recorded from the Middle Oligocene of Puerto Rico, is compared with *Cocos* and *Astrocaryum*, although there is insufficient detail to make a satisfactory comparison (Hollick 1928). From the upper Cenomanien of France, Sainte Menehould area, Fliche (1894) describes an endocarp that bears a strong resemblance to cocosoid endocarps (6 cm in length, a thick testa 8 mm), to which he gives the name *Cocoopsis*. He also describes *Cocoopsis ovata* and *C. zeilleri* endocarps from the Lower Cenomanian of France (Fliche 1896); these have a diameter ranging from 41–60 mm, with a thick testa (8–12 mm). From India, Eocene (Rajasthan State), *Cocos sahnii* was described by Kaul (1951); large *Cocos nucifera*-like petrified fruit, 13 × 10 × 6 cm, with a thin epicarp, wide and fibrous mesocarp, and well-developed hardened endocarp are described from the Deccan Intertrappean deposits of Amarkantak, Madhya Pradesh (Tripathi *et al.* 1999) (although the age span of these volcanic deposits is controversial, see Chapter 5). A much smaller oval fruit (5 × 3 cm) with a smooth epicarp, fibrous mesocarp and stony endocarp, *Cocos intertrappeansis*, is recorded from the Tertiary of Mohgaonkalan (Madhya Pradesh) (Patil & Uphadhye 1984). An almost-complete *Cocos*-like palm stem, *Palmoxylon sundaram*, has been described from the Deccan Intertrappean beds of India by Sahni (1946). Although stem wood is difficult to identify to generic level, this is an interesting fossil and probably worth further study. Mahabalé (1978) discusses the fossil history and origins of the coconut. Harries (1978) comments, "... if the smaller-fruited coconuts identified in the New Hebrides Islands are taken into consideration, together with the (small) coconut fossil(s) in New Zealand, it may be suggested that the centre of origin was in the region of the submerged continental fragment of the Lord Howe Rise–Norfolk Ridge complex ... isolated from Australia about 80 my ago and apparently submerged below sea level about 15 my ago ..." However, it must be emphasised that determining the generic affinity of these fossils on the basis of their endocarps alone is problematic, as most of the genera of the Attaleinae are distinguished on morphological features unlikely to be preserved in the fossil record.

Notes: *Cocos* differs from other genera in Attaleinae in having large pistillate flowers with rounded sepals and petals, in the large fruit with thick fibrous mesocarp, and in the endosperm. Gruezo and Harries (1984) and Buckley and Harries (1984) record the presence of "wild-type" coconuts in apparently natural coastal forest in the Philippines and Australia, and argue that these areas are exactly where *Cocos* might be predicted to be native. Gunn (2004), however, in her phylogenetic analysis of Cocoseae, suggests a South American origin of the lineage that eventually gave rise to the coconut (see Chapter 7).

Additional figures: Ch. 1 frontispiece, Glossary figs 18, 23.

99. JUBAEA

The Chilean Wine palm, native to Central Chile, and one of the most massive of all palms; the peduncular bract is smooth and there are 18 stamens in the staminate flowers.

Jubaea Kunth in Humb., Bonpl. & Kunth, Nova gen. et sp. 1: quarto edition 308; folio edition 247 (1816) ('1815'). Type: *J. spectabilis* Kunth = **J. chilensis** (Molina) Baill. (*Palma chilensis* Molina).

Molinaea Bertero, Mercurio Chileno 13: 606 (1829), (non Comm. ex Juss. 1789). Type: *M. micrococos* Bertero (superfluous name).

Micrococos Phil., Bot. Zeit. 17: 362 (1859). Type: *M. chilensis* (Molina) Phil. (*Palma chilensis* Molina).

Above: ***Jubaea chilensis***, habit, Chile. (Photo: F. Simonetti)

Opposite: ***Cocos.*** **a**, apical portion of rachilla with staminate buds × $^{2}/_{3}$; **b**, pair of staminate flowers, flowers removed to show scars and bracteoles × 6; **c**, staminate bud × 6; **d**, staminate bud in vertical section × 6; **e**, staminate sepal, interior view × 6; **f**, staminate petal, interior view × 6; **g**, stamen in 3 views × 6; **h**, pistillode × 6; **i**, triad with pistillate bud and staminate flowers × $1^{1}/_{2}$; **j**, lower portion of rachilla in pistillate flower × 1; **k**, triad, flowers removed to show bracteoles and scars × $1^{1}/_{2}$; **l**, pistillate flower × $1^{1}/_{2}$; **m**, pistillate flower in vertical section × $1^{1}/_{2}$; **n**, pistillate sepals, interior and exterior views × $1^{1}/_{2}$; **o**, pistillate petals, interior and exterior views × $1^{1}/_{2}$; **p**, gynoecium and staminodial ring × $1^{1}/_{2}$; **q**, basal view of gynoecium showing staminodial ring × $1^{1}/_{2}$; **r**, stigmas × 3; **s**, ovary in cross-section × $1^{1}/_{2}$; **t**, locule and ovule in vertical section × 3; **u**, fruit × $^{3}/_{8}$; **v**, fruit in vertical section × $^{3}/_{8}$; **w**, fruit in cross-section × $^{3}/_{8}$; **x, y, z**, endocarp in 3 views × $^{1}/_{2}$. *Cocos nucifera*: all from *Read* s.n. and *Moore* s.n. (Drawn by Marion Ruff Sheehan)

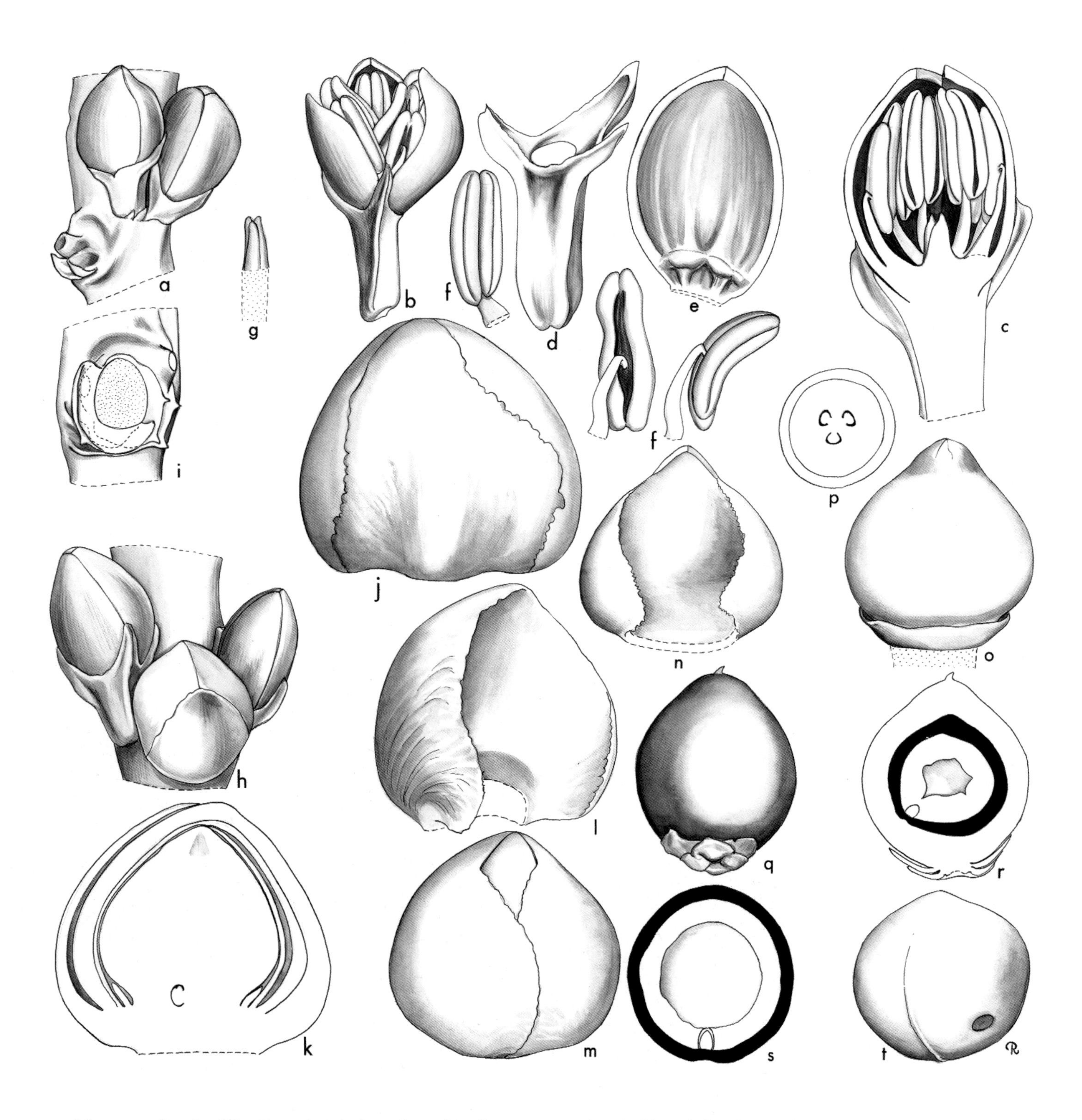

***Jubaea*.** **a**, portion of rachilla with staminate buds × 3; **b**, staminate flower × 3; **c**, staminate bud in vertical section × 6; **d**, staminate calyx and receptacle × 6; **e**, staminate petal, interior view × 6; **f**, stamen in 3 views × 6; **g**, pistillode × 6; **h**, portion of rachilla with triad × 3; **i**, triad, flowers removed to show bracteoles × 3; **j**, pistillate bud × 6; **k**, pistillate bud in vertical section × 6; **l**, pistillate sepal × 6; **m**, pistillate flower, sepals removed × 6; **n**, pistillate petal, interior view × 6; **o**, gynoecium and staminodial ring × 6; **p**, ovary in cross-section × 6; **q**, fruit × 1; **r**, fruit in vertical section × 1; **s**, endocarp and seed in cross-section × 1 1/2; **t**, endocarp × 1 1/2. *Jubaea chilensis*: **a–p**, *Moore* 9366; **q–t**, *Osborne* s.n. (Drawn by Marion Ruff Sheehan)

Honours Juba II (ca. 50 BC–24 AD), king of Numidia 29–25 BC.

Massive, solitary, unarmed, pleonanthic, monoecious palm. *Stem* erect, very stocky, eventually bare and marked with close, oblique leaf scars and vertical cracking. *Leaves* pinnate, many in the crown, neatly abscising in mature-trunked individuals; sheaths soon disintegrating into fibres and eventually becoming open; petiole short to long, sometimes hardly distinguishable from the sheath, edged with disintegrated leaf sheath fibres except near the tip where almost smooth, adaxially flattened, abaxially rounded or angled, bearing thin or thick white wax or glabrous; rachis stiff or gently curving, proximally adaxial face flattened, angled distally, abaxially rounded, bearing scattered caducous scales distally; leaflets numerous, single-fold, close but irregularly grouped, held ± all in the same plane, linear, very stiff, the tips often with a reflexed hook-like flange representing a fragment of the rein, irregularly obliquely bifid or regularly bifid, thinly glaucous, adaxially bearing caducous scales along the main vein and very few scattered scales and wax on the blade surface, abaxially with abundant caducous scales and bifid ramenta throughout the length of the main vein, transverse veinlets obscure. *Inflorescences* solitary, interfoliar, large, branching to 1 order, protandrous; peduncle elongate, ± circular in cross-section, maroon when fresh, covered in dense, caducous tomentum; prophyll short, tubular, 2-keeled, opening distally, becoming fibrous with age; peduncular bract inserted near the prophyll, much exceeding it, enclosing the entire inflorescence until shortly before anthesis, tubular, woody, with a short solid beak, at anthesis splitting down ± the entire length, expanding and becoming cowl-shaped, adaxially smooth, glabrous, creamy-yellow when fresh, abaxially not grooved, only faintly striate, densely covered in soft, brown tomentum; rachis shorter than the peduncle, bearing numerous, elongate, spreading, spirally arranged rachillae, each subtended by a short, inconspicuous triangular bract; rachillae maroon, swollen at the very base and with a short basal bare portion, above which bearing numerous, spirally arranged triads in the proximal ca. $^1/_5$–$^1/_4$ and paired or solitary staminate flowers distally, the distal-most rachillae sometimes entirely staminate; floral bracteoles very small. *Staminate flowers* slightly asymmetrical; calyx with a solid, elongate, stalk-like base and 3, narrow, triangular keeled lobes; petals 3, much longer than the calyx, distinct, valvate, ± boat-shaped with triangular tips; stamens ca. 18, filaments slender, fleshy, elongate, cylindrical, apically inflexed, anthers medifixed, versatile, ± rectangular, latrorse; pistillode small, trifid. *Pollen* ellipsoidal, frequently elongate, usually with either slight or obvious asymmetry; aperture a distal sulcus; ectexine tectate, perforate and micro-channelled, or perforate-rugulate, aperture margin similar; infratectum columellate; longest axis 46–54 µm [1/1]. *Pistillate flowers* globular, only slightly larger than the staminate; sepals 3, distinct, rounded, broadly imbricate, the outermost ± keeled; petals 3, distinct, rounded, broadly imbricate except at the short triangular valvate tips; staminodal ring low, ± shallowly lobed, forming a collar surrounding the gynoecium; gynoecium trilocular, triovulate, ± broadly ovoid, stigmas closely appressed, ovules hemianatropous, laterally attached to the ventral angle of the locules. *Fruit* usually 1-seeded, orange-yellow, ± ovoid, with a short beak and apical stigmatic remains; epicarp smooth, mesocarp thick, fleshy, sweet, endocarp smooth, thick, bony, with 3 low crests and 3 pores lateral below the equator. *Seed* basally attached, closely adhering to the endocarp, endosperm homogeneous with large central cavity; embryo opposite one of the endocarp pores. *Germination* adjacent-ligular; eophyll entire, lanceolate. *Cytology*: 2n = 32.

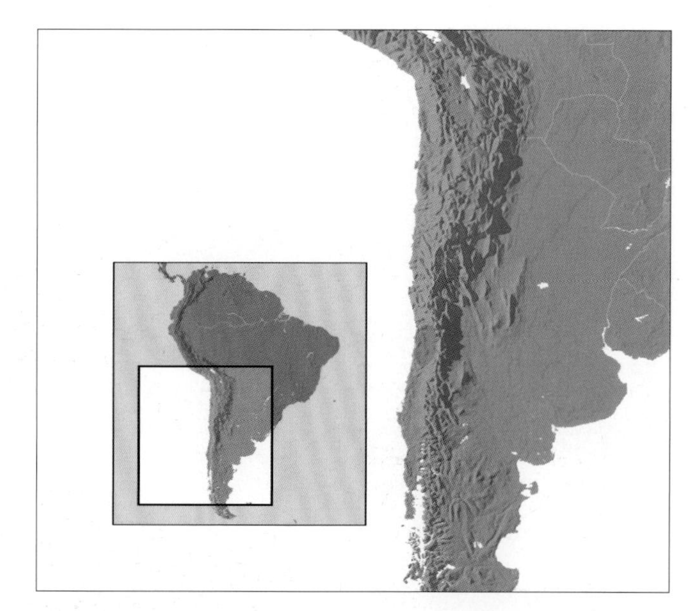

Distribution of ***Jubaea***

Distribution and ecology: One species, now much restricted and threatened in central Chile; widely cultivated in warm temperate regions. Growing on sides of ravines and ridges in dry scrubby woodland.

Anatomy: Leaf, readily identified by the anatomy of the lamina and showing some resemblances to *Butia* (Tomlinson 1961), root (Seubert 1998a, 1998b).

Relationships: For relationships, see *Butia*.

Common names and uses: Chilean wine palm. Formerly trunks of *Jubaea* were felled and tapped for wine and sugar, the yield from a single trunk being prodigious. The palm is a widespread and important ornamental in dry warm temperate regions.

Taxonomic account: Glassmann (1987).

Left: ***Jubaea chilensis***, habit, Chile. (Photo: F. Simonetti)

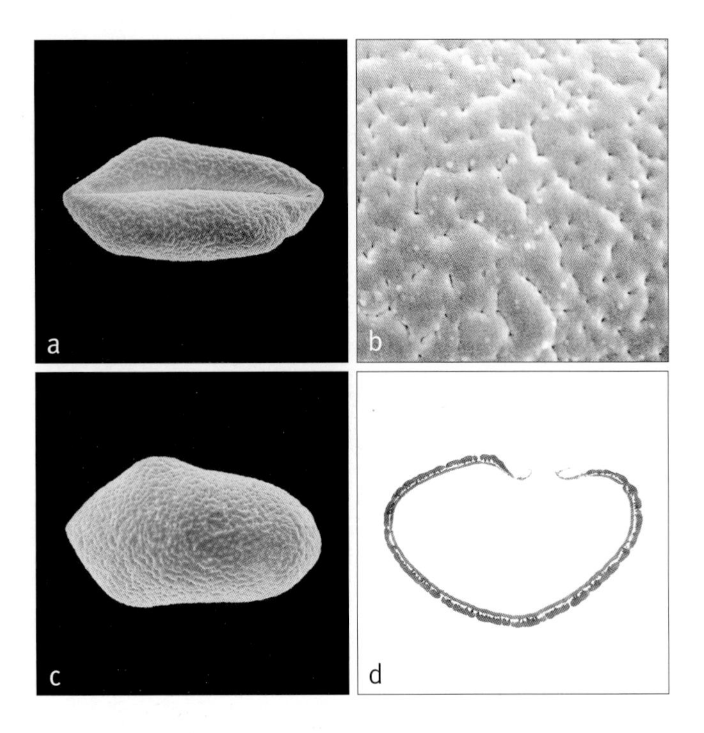

Above: ***Jubaea***
a, monosulcate pollen grain, distal face SEM × 1000;
b, close-up, perforate micro-channelled pollen surface SEM × 6000;
c, monosulcate pollen grain, distal face SEM × 1000;
d, ultrathin section through whole pollen grain, polar plane TEM × 2000.
Jubaea chilensis: **a–d**, *Philippi* s.n.

Fossil record: Subfossil endocarps, *Paschalococos disperta* (Dransfield 1991a), were described from a cave floor on Easter Island (Poike Peninsula) by Dransfield *et al.* (1984). Although represented only by endocarps, these are sufficiently well-preserved to determine them as belonging to the Attaleinae, and very closely related to *Jubaea*. However, until more remains of the extinct palm become available, it is impossible accurately to equate the palm with any known extant genus (Dransfield 1991a).
Notes: *Jubaea chilensis* is one of the most massive palms. It differs from *Butia* in the lack of spines on the petioles, in having stalked rather than sessile staminate flowers, and in having more than six stamens; and from *Jubaeopsis* in its solitary habit and stalked staminate flowers with connate sepals.
Additional figures: Fig. 7.2.

100. LYTOCARYUM

Graceful undergrowth palms from eastern Brazil with distinctive slender discolorous leaflets and fruit in which the pericarp splits longitudinally when ripe, exposing the endocarp.

Lytocaryum Toledo, Arq. Bot. Estado São Paulo ser. 2. 2(1): 6 (1944). Type: **L. hoehnei** (Burret) Toledo (*Syagrus hoehnei* Burret).
Glaziova Mart. ex Drude, in Mart., Fl. bras. 3(2): 295 (1881) (non Bureau 1868). Lectotype: *G. martiana* Glaz. ex Drude (illegitimate name) (= *Lytocaryum weddellianum* [H. Wendl.] Toledo) (see H.E. Moore 1963c).
Microcoelum Burret & Potztal, Willdenowia 1: 387 (1956). Lectotype: *M. martianum* (Glaz. ex Drude) Burret & Potztal (*Glaziova martiana* Glaz. ex Drude [illegitimate name], = *Lytocaryum weddellianum* [H. Wend.] Toledo [*Cocos weddellianum* H. Wendl., *M. weddellianum* (H. Wendl.) H.E. Moore]).

Lyton — *loosened,* caryon — *nut, referring to the way that the fruit coat splits away from the endocarp at maturity.*

Slender, solitary, unarmed, pleonanthic, monoecious palms. *Stem* erect, short, rarely exceeding 3 m, at first obscured by leaf sheath remains, later bare, closely ringed with leaf scars. *Leaves* pinnate, marcescent, graceful; leaf sheaths densely light or dark brown-hairy, with a triangular ligule-like projection opposite the petiole, later disintegrating into a fibrous network and splitting; petiole very short to elongate, adaxially flat to rounded, abaxially rounded or angled, fibrous along the margins, also with scattered thin indumentum and some coarse dark hairs; rachis neatly curved, usually bearing conspicuous, dark, coarse hairs adaxially; leaflets single-fold, numerous, slender, often extremely so, linear, close and regularly arranged, the tips asymmetrical, soft in texture, the adaxial surface dark green, abaxial surface covered with grey or pale brownish indumentum, with few to numerous ramenta along the midrib, transverse veinlets obscure. *Inflorescences* solitary, interfoliar, branching to 1 order, protandrous; peduncle short to elongate, elliptic in cross-section, sparsely to densely tomentose; prophyll tubular, flattened, 2-keeled, usually mostly concealed by the leaf sheaths, opening distally, becoming fibrous in age, light or dark brown tomentose; peduncular bract elongate, inserted just above and much longer than the prophyll, coriaceous to ± woody, entirely enclosing the inflorescence until shortly before anthesis, splitting longitudinally along the abaxial face and expanding, adaxial surface glabrous, smooth or tomentose, ± grooved, abaxial surface deeply grooved, densely light or dark brown-tomentose; rachis usually shorter than the peduncle, sparsely to densely tomentose like the peduncle,

Distribution of ***Lytocaryum***

bearing numerous, spirally arranged rachillae, each subtended by a minute triangular bract; rachillae eventually widely spreading, slender, sparsely tomentose, somewhat zigzag, with a short to long basal bare portion, above which bearing few triads proximally and paired or solitary staminate flowers distally, the distal-most rachillae shorter and sometimes entirely staminate; rachilla bracts and floral bracteoles inconspicuous. *Staminate flowers* small, ± symmetrical, sessile or borne on brief, slender pedicels; calyx with or without a solid, short to long stalk-like base, and 3, membranous, keeled, narrow triangular, acute lobes; petals 3, distinct, valvate, ovate-triangular, acute, thinly coriaceous with scattered, caducous, dot-like scales; stamens 6, filaments slender or basally thickened, very briefly epipetalous, elongate, ± inflexed, anthers slender, ± oblong, basally sagittate, apically sometimes pointed, medifixed, versatile, latrorse; pistillode conspicuous, about $^1/_2$ the height of the filaments, trifid or minute. *Pollen* ellipsoidal, frequently elongate, usually with either slight or

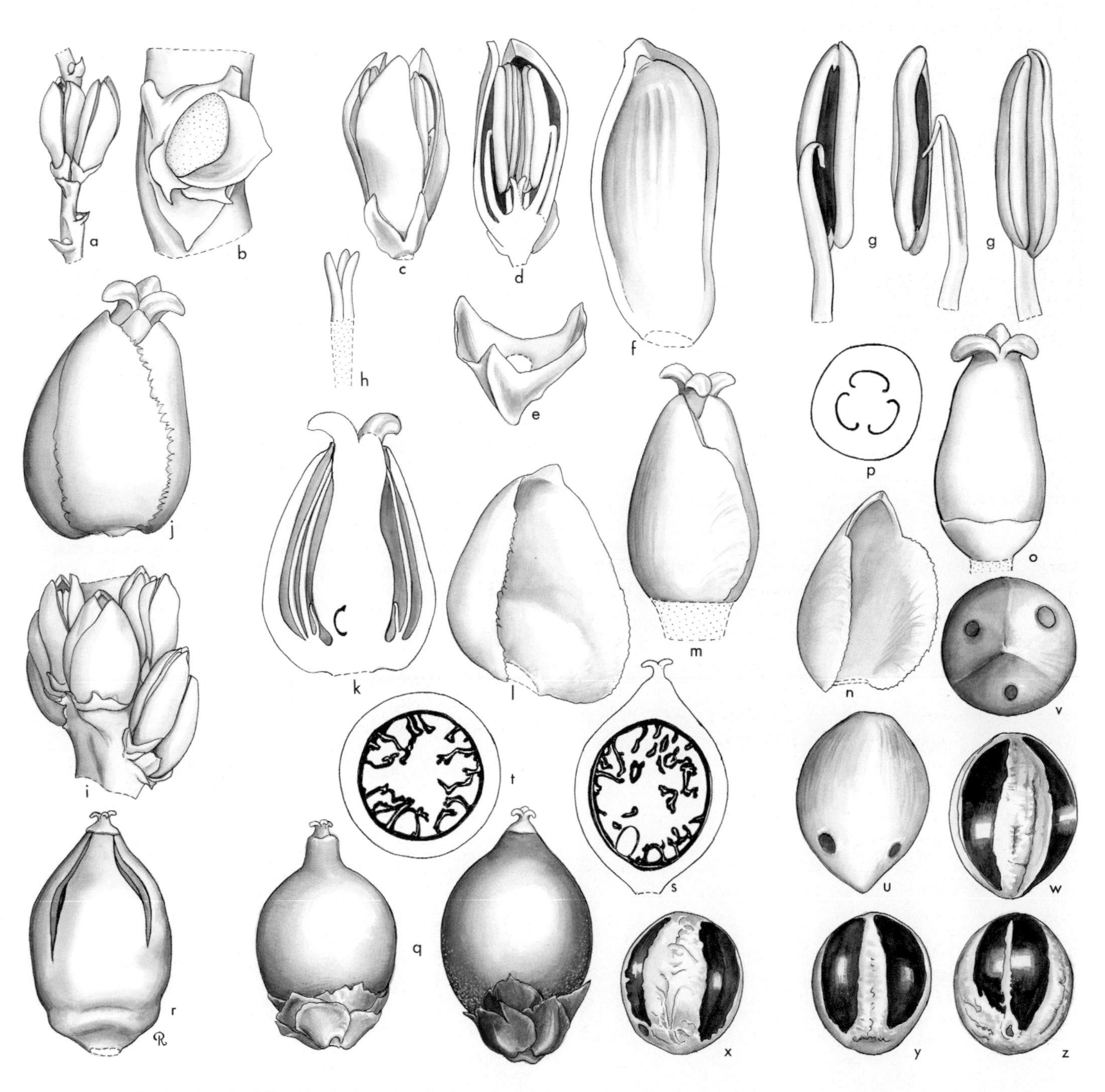

***Lytocaryum*.** **a**, distal portion of rachilla with paired staminate buds × 3; **b**, triad, flowers removed to show bracteoles × 6; **c**, staminate flower × 6; **d**, staminate flower in vertical section × 6; **e**, staminate calyx × 12; **f**, staminate petal, interior view × 12; **g**, stamen in 3 views × 12; **h**, pistillode × 12; **i**, portion of rachilla with triads × 3; **j**, pistillate flower × 6; **k**, pistillate flower in vertical section × 6; **l**, pistillate sepal × 6; **m**, pistillate flower, sepals removed × 6; **n**, pistillate petal × 6; **o**, gynoecium and staminodial ring × 6; **p**, ovary in cross-section × 18; **q**, two fruits × 1$^1/_2$; **r**, dried fruit with split pericarp × 1$^1/_2$; **s**, fruit in vertical section × 1$^1/_2$; **t**, fruit in cross-section × 1$^1/_2$; **u, v**, endocarp in 2 views × 1$^1/_2$; **w**, interior of endocar **p**, seed removed × 1; **x, y, z**, seed in 3 views × 1$^1/_2$. *Lytocaryum hoehnei*; **a–q** and **s–z**, *Eiten* s.n.; **r**, *Leandro & Toledo* 51606. (Drawn by Marion Ruff Sheehan)

obvious asymmetry; aperture a distal sulcus; ectexine tectate, finely or coarsely perforate, perforate and micro-channelled, or perforate-rugulate, aperture margin may be slightly finer; infratectum columellate; longest axis 34–54 µm [2/2]. *Pistillate flower* larger than the staminate, ± pyramidal; sepals 3, distinct, triangular, broadly imbricate, coriaceous, keeled, the tips ± hooded or not; petals ± equalling the sepals, 3, distinct, broadly imbricate at the base, abruptly narrowed at ± the midpoint to broad or narrow, tapering, valvate tips; staminodal ring thinly fleshy, irregularly 6-toothed or truncate; gynoecium ± pyramidal, trilocular, triovulate, brown-hairy, with a very short to long style and 3 stigmas appressed in bud, ovules laterally attached to the central axis, form unknown. *Fruit* globose to ovoid, 1-seeded, tinged pink or reddish, with a short beak and apical stigmatic remains; epicarp ± smooth, mesocarp thin, ± fibrous, it and the epicarp dehiscing along 3 vertical sutures to expose the endocarp, endocarp thin, rather fragile, marked with 3 vertical lines externally, internally with 3 shining broad bands, endocarp pores lateral near the base. *Seed* laterally attached with broad lateral hilum, endosperm homogeneous (*Lytocaryum weddellianum*) or deeply ruminate (*L. hoehnei*), with or without a central hollow; embryo basal opposite an endocarp pore. *Germination* adjacent-ligular; eophyll pinnate. *Cytology*: 2n = 32.

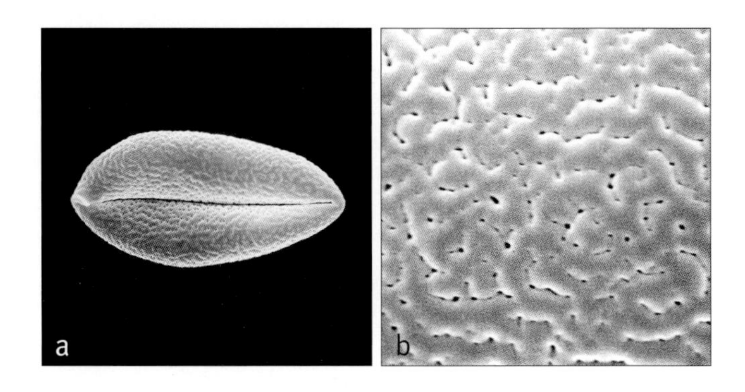

Above right: ***Lytocaryum***
a, monosulcate pollen grain, distal face SEM × 1000;
b, close-up, perforate micro-channelled pollen surface SEM × 6000.
Lytocaryum weddellianum: **a** & **b**, *Glaziou* 8057.

Below: ***Lytocaryum hoehnii***, habit, Brazil. (Photo: A.J Henderson)

Distribution and ecology: Two closely related species restricted to south-east Brazil. Found in shady forest at altitudes of 800–1800 m.

Anatomy: Root (Seubert 1998a, 1998b).

Relationships: The monophyly of *Lytocaryum* has not been tested. The genus is resolved as sister to *Syagrus romanzoffiana* with high support (Gunn 2004); however, there is low support for an alternative placement as sister to *Attalea* (Baker *et al.* in review).

Common names and uses: Common names not recorded. *Lytocaryum weddellianum* is an important pot palm, sold in large quantities in Europe.

Taxonomic account: See Glassman (1987).

Fossil record: No generic records found.

Notes: The only difference between *Microcoelum* and *Lytocaryum* is the nature of the endosperm; otherwise the two taxa are very similar. Thus, *Microcoelum* is placed in synonymy. *Lytocaryum*, in the present sense, has a combination of characters unusual in the subtribe Butiinae. *Lytocaryum* is separated from *Syagrus* by its distinctive leaves with abundant rachis tomentum, its strongly versatile anthers, fruit with epicarp and mesocarp dehiscing by vertical sutures; and its thin rather fragile endocarp.

101. SYAGRUS

Extremely variable genus native to the Caribbean (one species) and South America, where it is particularly abundant in drier areas; leaflets are concolourous and the mesocarp does not split.

Syagrus Mart., Palm. fam. 18 (1824). Type: **S. cocoides** Mart.

Langsdorffia Raddi, Mem. Mat. Fis. Soc. Ital. Sci. 18(2): 345 (1820) (illegitimate name) (non Mart. 1818, Balanophoraceae). Type: *L. pseudococos* Raddi (= *Syagrus pseudococos* [Raddi] Glassman).

Platenia H. Karst., Linnaea 28: 250 (1856). Type: *P. chiragua* H. Karst. (= *Syagrus chiragua* [H. Karst.] H. Wendl.) (see Bernal & Galeano-Garces 1989).

Cocos subgenus *Arecastrum* Drude in Mart., Fl. bras. 3 (2): 402 (1881).

Barbosa Becc., Malpighia 1: 349, 352 (1887). Type: *B. pseudococos* (Raddi) Becc. (*Langsdorffia pseudococos* Raddi) (= *Syagrus pseudococos* [Raddi] Glassman).

Rhyticocos Becc., Malpighia 1: 350, 353 (1887). Type: *R. amara* (Jacq.) Becc. (*Cocos amara* Jacq.) (= *Syagrus amara* [Jacq.] Mart.)

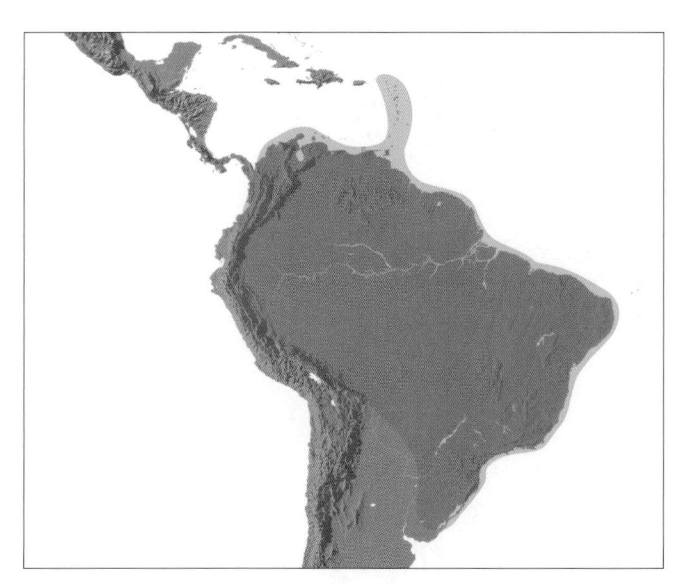

Distribution of ***Syagrus***

Arikuryroba Barb. Rodr., Pl. jard. Rio de Janeiro 1: 5 (1891). Type: *A. capanemae* Barb. Rodr. = *A. schizophylla* (Mart.) L.H. Bailey (*Cocos schizophylla* Mart.) (= *Syagrus schizophylla* [Mart.] Glassman).

Arikury Becc., Agric. Colon. 10: 445 (1916). Superfluous substitute name.

Arecastrum (Drude) Becc., Agric. Coloniale 10: 446 (1916). Type: *A. romanzoffianum* (Chamisso) Becc. (*Cocos romanzoffiana* Cham.) (= *Syagrus romanzoffiana* [Cham.] Glassman).

Chrysallidosperma H.E. Moore, Principes 7: 109 (1963). Type: *C. smithii* H.E. Moore (= *Syagrus smithii* [H.E. Moore] Glassman).

Syagrus — *the name of a kind of palm tree in Latin, apparently used by Pliny, but certainly not for members of this New World genus.*

Small to tall, solitary or clustered, rarely forking below ground (*Syagrus cearensis*), unarmed or armed, pleonanthic, monoecious palms. *Stem* very short, subterranean to erect and tall, rarely stolon-like, sometimes swollen basally, distally obscured by leaf-base remains, becoming bare, sometimes striate, and marked with inconspicuous or raised or impressed, oblique or circular conspicuous leaf scars. *Leaves* reduplicately pinnate, spirally arranged or, in *S. coronata*, ± even, scars circular, arranged in 5 ranks, marcescent or neatly abscising; sheath disintegrating into an interwoven mass of fibres, the fibres slender to robust and flattened, rarely flattened and spine-like; petiole very short to long, adaxially channelled or flattened, abaxially rounded or angled, the margins smooth or bearing short caducous fibres, and rarely also bearing coarse spine-like fibres, surfaces variously glabrous, tomentose or scaly, sometimes waxy; rachis straight or curved, short to long, variously scaly, tomentose or glabrous; leaflets single-fold, few to very numerous, regularly or irregularly arranged, held in one or several planes, linear, moderately wide to very narrow, stiff or curved, the tips acute, acuminate or obtuse, symmetrical and shallowly bifid or asymmetrical; blade adaxially glabrous or with sparse scales or hairs, sometimes waxy, abaxially usually with conspicuous ramenta along the main vein, very rarely also with dense grey indumentum, transverse veinlets often conspicuous. *Inflorescences* solitary, interfoliar, rarely spicate, usually branching to 1 order, ?protandrous, much shorter than the leaves; peduncle ± elliptic in cross-section, short to long, glabrous or variously hairy or scaly; prophyll usually mostly concealed within the leaf sheaths, tubular, flattened, 2-keeled, splitting at the tip, becoming fibrous and disintegrating with age; peduncular bract persistent, much longer than the prophyll, usually inserted just above the prophyll, tubular, enclosing the inflorescence until shortly before anthesis, very rarely shorter than the expanded inflorescence, often ± spindle-shaped before splitting longitudinally along the abaxial face for most of its length, then expanding and becoming cowl-like, beaked, usually ± woody, rarely thinly coriaceous or papery, glabrous or glaucous or pubescent, longitudinally grooved; rachis usually shorter than the peduncle or nearly equal (*S. coronata*), glabrous or hairy, bearing spirally arranged rachillae, each subtended by a short, triangular, usually coriaceous bract; rachillae few to numerous, short or elongate, slender, straight or often twisted in bud, frequently zigzag, glabrous or sparsely tomentose, bearing spirally arranged triads proximally, paired or solitary staminate flowers distally, or rarely, the distal-most rachillae bearing only staminate flowers; flower groups usually sessile, subtended by usually inconspicuous bracts; floral bracteoles minute. *Staminate flowers* usually ± asymmetrical; sepals 3, ± triangular, distinct and imbricate or briefly connate, rarely connate in a stalk-like base; petals 3, distinct, valvate, ± thinly coriaceous or fleshy, much longer than the sepals, variously lanceolate, oblong, or ovate with acute tips, glabrous, tomentose, scaly or

Right: ***Syagrus vermicularis***, habit, cultivated, Montgomery Botanical Center, Florida. (Photo: J. Dransfield)

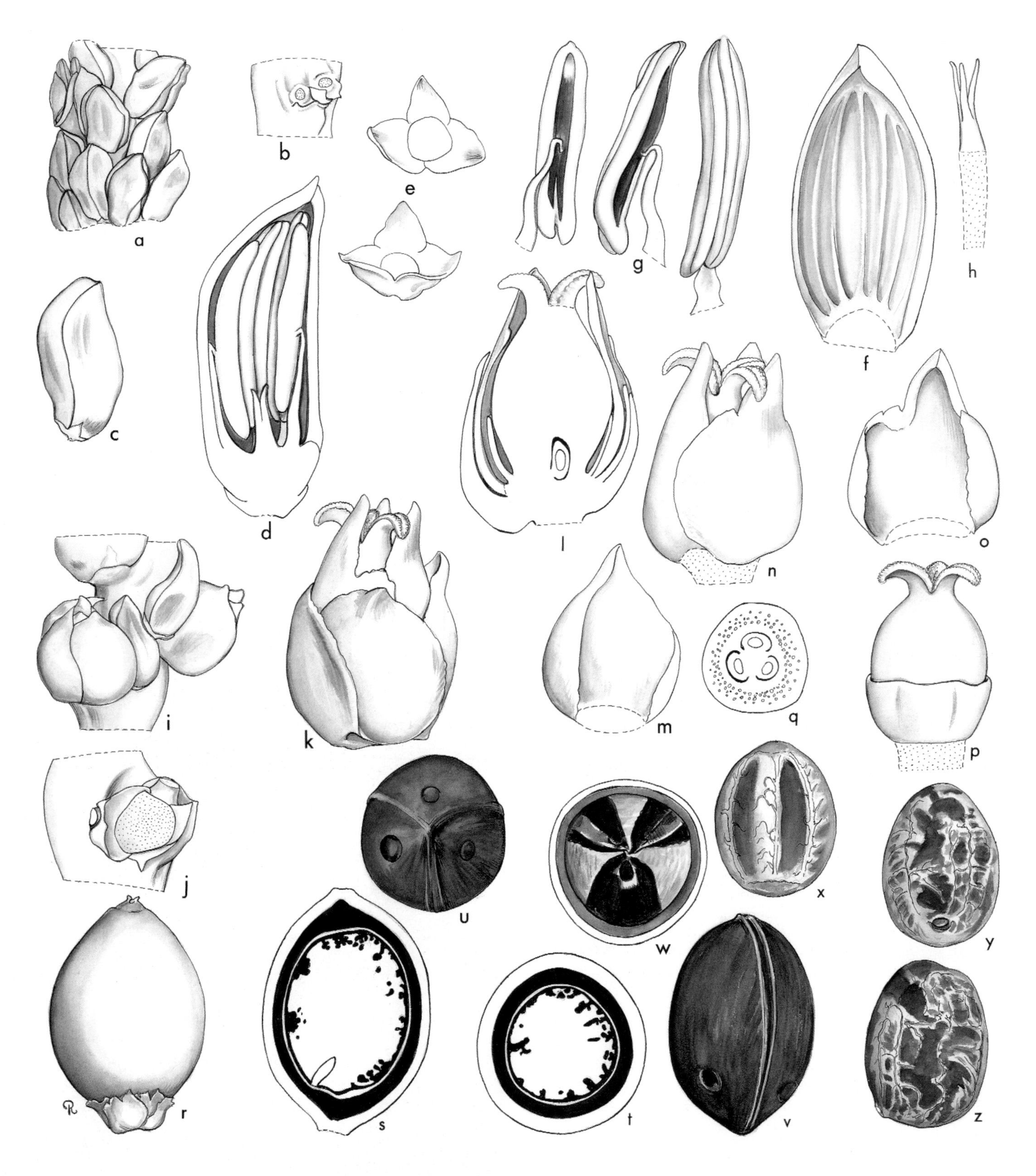

Syagrus. **a**, portion of rachilla with paired staminate flowers × 3; **b**, scars of paired staminate flowers and bracteole × 6; **c**, staminate bud × 6; **d**, staminate flower in vertical section × 12; **e**, staminate calyx in 2 views × 12; **f**, staminate petal, interior view × 12; **g**, stamen in 3 views × 12; **h**, pistillode × 12; **i**, portion of rachilla with triads × 3; **j**, triad, flowers removed to show bracts × 6; **k**, pistillate flower × 6; **l**, pistillate flower in vertical section × 6; **m**, pistillate sepal × 6; **n**, pistillate flower, sepals removed × 6; **o**, pistillate petal, interior view × 6; **p**, gynoecium and staminodial ring × 6; **q**, ovary in cross-section × 6; **r**, fruit × $1^1/_2$; **s**, fruit in vertical section × $1^1/_2$; **t**, fruit in cross-section × $1^1/_2$; **u, v, w**, endocarp in 3 views × $1^1/_2$; **x, y, z**, seed in 3 views × $1^1/_2$. *Syagrus schizophylla*: all from *Read* 824.

dotted; stamens 6, filaments distinct, or very briefly connate, relatively short, ± fleshy, anthers elongate, dorsifixed near the base or medifixed and ± versatile, introrse or latrorse; pistillode minute, trifid or absent. *Pollen* ellipsoidal, frequently elongate and/or pyriform, usually with either slight or obvious asymmetry; aperture a distal sulcus; ectexine tectate, finely perforate, perforate and micro-channelled, perforate-rugulate, rugulate-verrucate or reticulate, aperture margin similar; infratectum columellate; longest axis 36–56 µm; post-meiotic tetrads tetrahedral [11/31]. *Pistillate flowers* slightly smaller to very much larger than the staminate flowers; sepals 3, distinct, broadly imbricate, triangular to ovate, acute or obtuse, fleshy to coriaceous, sometimes tomentose or scaly; petals 3, distinct, slightly shorter to slightly longer the than the sepals, triangular or ovate, broadly imbricate at the base, with minute to moderately large and conspicuous valvate tips; staminodal ring membranous, low, ± 6-toothed, occasionally apparently absent; gynoecium columnar to conical or ovoid, trilocular, triovulate, glabrous or tomentose to scaly, the stigmas 3, reflexed, fleshy, ovules laterally attached to the central wall of the locules, ?anatropous. *Fruit* small to relatively large, 1–(rarely 2-)seeded, spherical, ovoid, or ellipsoidal, variously green, brown, yellow, or reddish, sometimes beaked, the perianth segments and staminodal ring persistent and sometimes enlarging as a cupule at the fruit base; epicarp smooth or longitudinally striate, glabrous or hairy, mesocarp fleshy or dry, with abundant longitudinal fibres, endocarp thick, woody, with 3(4) basal or subbasal pores, sometimes beaked, sometimes with 3 longitudinal ridges, rarely with 3 irregular vertical bands of minute pores, endocarp cavity irregular or more usually circular, rarely triangular in cross-section, with 3, conspicuous, vertical lines, very rarely with a curved lateral protrusion into the seed (*S. romanzoffiana*). *Seed* conforming to the shape of the endocarp cavity, subbasally attached, endosperm homogeneous or ruminate, sometimes with a central cavity; embryo basal or subbasal opposite one of the endocarp pores. *Germination* adjacent-ligular or remote tubular; eophyll entire. *Cytology*: 2n = 32.

Distribution and ecology: Thirty-one species recorded in South America from Venezuela southwards to Argentina, with the greatest number of species in Brazil; one species in the Lesser Antilles. Most species are confined to dry or semi-arid areas; these include all of the acaulescent species. A few, usually tree-like species are restricted to mesic and tropical rain forest. The acaulescent species are conspicuous components of several Brazilian arid vegetation types such as 'cerrado' and 'campo rupestre'.

Anatomy: Leaf (Tomlinson 1961, Glassman 1987, Roth 1990), root (Seubert 1998a, 1998b), fruit (Reddy & Kulkarni 1985).

Relationships: *Syagrus* is resolved as polyphyletic by Gunn (2004). For a preliminary assessment of relationships, see Gunn (2004). A detailed species-level phylogeny of Attaleinae is required to circumscribe a monophyletic *Syagrus*.

Common names and uses: Syagrus palms, licury palm (*Syagrus coronata*), Queen palm (*S. romanzoffiana*). Leaves of many species are used as thatch, those of *S. coronata* also yield wax. The mesocarp of *S. oleracea* and *S. coronata* is edible, and the endosperm of *S. cocoides* and *S. coronata* can be used as a source of palm oil. The wood is also useful. For medicinal uses, see Plotkin and Balick (1984). Many species are cultivated as ornamentals.

Taxonomic accounts: Glassman (1987) and Noblick (2004a, 2004b).

Fossil record: Two small palm endocarps (Maury 1930) that are included in *Palmocarpon luisii*, from the Maastrichtian of Brazil, State of Parahyba do Norte (Rio Gramame), are compared with members of the Cocoseae. One is described as being about the size and form of the living *Cocos datel* (sic.) "but more deformed" (*Cocos datil* Drude & Griseb. = *Syagrus romanzoffiana* [Cham.] Glassman; see Govaerts & Dransfield 2005), these are almost certainly not cocosoid in origin and probably not from palms at all.

Notes: Several genera earlier recognised as distinct because of their ruminate endosperm were included in *Syagrus* in the first edition of *Genera Palmarum*; this circumscription of the genus is followed here. The basis for including the genera is the extraordinary variability of *Syagrus* itself and the unreliability of the endosperm character for separating genera elsewhere in the family. Thus, *S. vagans* and *S. ruschiana* approach *Arikuryroba*, *S. inajai* differs from

Top right: ***Syagrus vermicularis***, inflorescences, cultivated, Montgomery Botanical Center, Florida. (Photo: J. Dransfield)

Bottom right: ***Syagrus cearensis***, fruit, cultivated, Montgomery Botanical Center, Florida. (Photo: J. Dransfield)

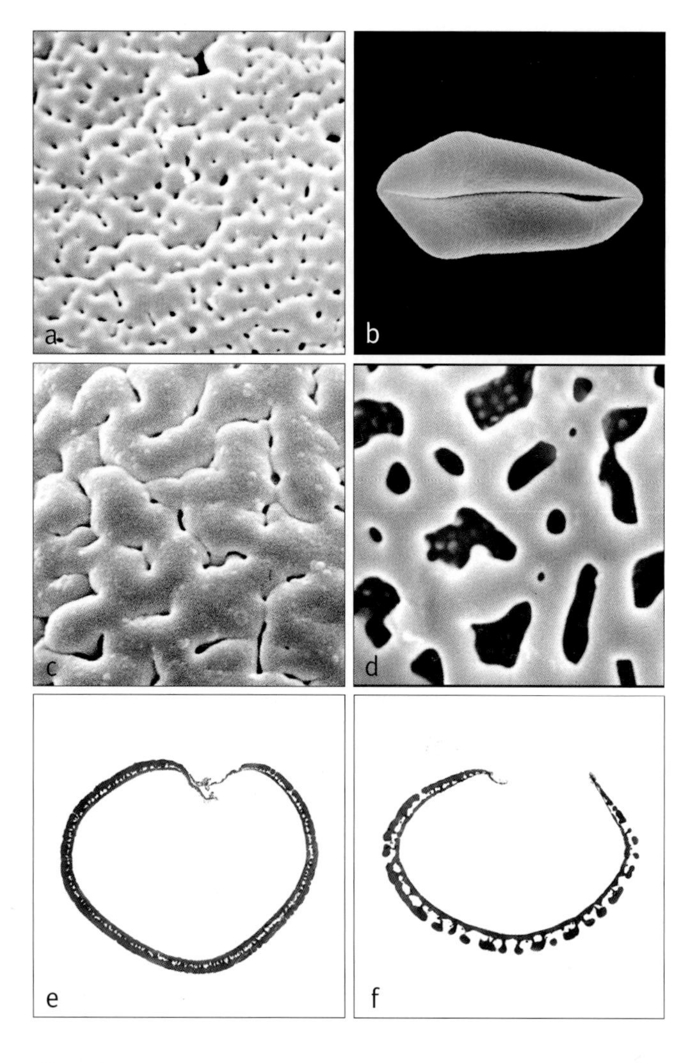

Above: ***Syagrus***
a, close-up, finely perforate-rugulate pollen surface SEM × 8000;
b, monosulcate pollen grain, distal face SEM × 1250;
c, close-up, coarsely perforate-rugulate pollen surface SEM × 8000;
d, close-up, reticulate pollen surface SEM × 8000;
e, ultrathin section through whole pollen grain, polar plane TEM × 2500;
f, ultrathin section through whole pollen grain, polar plane TEM × 2500.
Syagrus smithii: **a**, *Moore* 8516; *S. sancona*: **b**, *Eggers* 15681; *S. flexuosa*: **c**, *Gardener* 4026; *S. harleyi*: **d** and **f**, *Harley et al.* 15963; *S. pseudococos*: **e**, *Glaziou* 8048.

Chrysallidosperma smithii only in a few minor characters and in the endosperm. The connate petals of *Rhyticocos* referred to by Read (1979) have not been found on our specimens, and the palm is otherwise rather like *S. sancona* but with a large ruminate endosperm. *Barbosa pseudococos* is very similar in flower and fruit characters to *S. stratinicola*. Gunn (2004), in her preliminary phylogenetic analysis of Cocoseae, recovered trees that suggested *Syagrus* to be polyphyletic. However, she sampled just three species of *Syagrus*, and these were all at one time included in segregate genera. More work is needed.

Additional figures: Glossary figs 5, 7.

102. PARAJUBAEA

Large pinnate-leaved palms of high altitude in inter-Andean valleys in South America; staminate flowers with ca. 15 stamens and fruit with thick irregularly thickened endocarp.

Parajubaea Burret, Notizbl. Bot. Gart. Berlin-Dahlem 11: 48 (1930). Type: **P. cocoides** Burret.

Para — *beside or near,* Jubaea — *the palm generic name, suggesting similarity to Jubaea.*

Large, solitary, unarmed, pleonanthic, monoecious palms. *Stem* tall, stout or rather slender, grey, obscurely ringed with leaf scars. *Leaves* pinnate, slightly arching, sometimes twisted, marcescent; sheath not forming a crownshaft, disintegrating into a mass of rather fine to coarse brown fibres; petiole short, adaxially flat, abaxially rounded, glabrous, margins fibrous at least basally; rachis longer than the petiole, tapering, rounded adaxially, grooved laterally, rounded abaxially; leaflets numerous, regularly arranged or in groups of 2–5 in one plane, narrow, elongate, single-fold, pointed distally, tips shortly bifid to oblique, glabrous adaxially, abaxially sparsely scaly and with linear ramenta along the midribs, midrib prominent adaxially, the only large vein, transverse veinlets not evident. *Inflorescences* interfoliar, erect, becoming pendulous in fruit, branched to 1 or 2 orders; peduncle elongate, dorsiventrally flattened, glabrous; prophyll short, 2-keeled, opening apically, not exserted from the leaf sheath; peduncular bract much longer than the prophyll, narrowly tubular, tapering to a short apical beak, woody, deeply grooved, splitting abaxially and eventually marcescent, glabrous or covered with tomentum; several short, incomplete peduncular bracts present; rachis about as long as or shorter than the peduncle, bearing rather distant, spirally arranged, short, wide, centrally pointed bracts subtending rachillae; rachillae erect and closely appressed to the rachis, rather short, stout, sometimes zigzag, glabrous or with caducous tomentum, in *Parajubaea cocoides* distally with a few short branches of the 2nd order, rachillae bearing spirally arranged, very short, wide, pointed bracts subtending few triads (2–8) at the base and paired or solitary staminate flowers throughout most of the rachilla length, floral bracteoles short, rounded to pointed. *Staminate flowers* asymmetrical, ± pointed; sepals 3, distinct, imbricate basally, keeled, acute; petals 3, distinct, much larger than the sepals, angled, valvate; stamens 13–15, filaments erect, awl-shaped, ± inflexed; anthers linear, emarginate apically, sagittate basally, dorsifixed near the middle, introrse; pistillode short, briefly trifid. *Pollen* ellipsoidal, frequently elongate, usually with either slight or

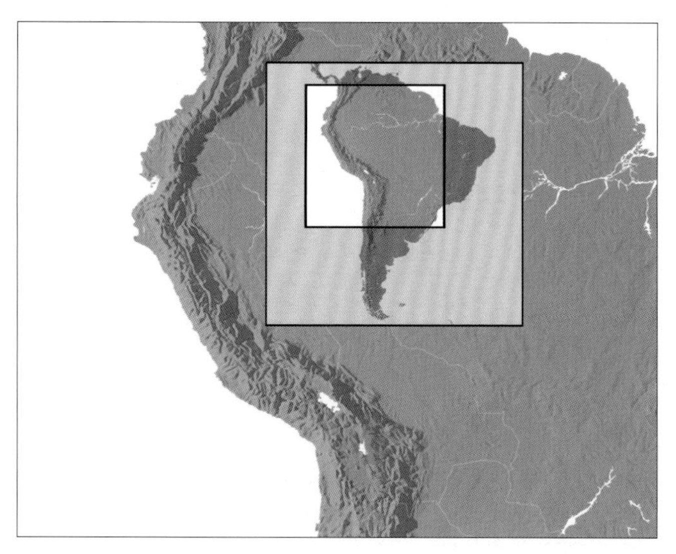

Distribution of ***Parajubaea***

obvious asymmetry; aperture a distal sulcus; ectexine tectate, finely perforate, perforate and micro-channelled or perforate-rugulate, aperture margin may be slightly finer; infratectum columellate, longest axis 41–60 µm [2/3]. *Pistillate flowers* broadly ovoid, larger than the staminate; sepals 3, distinct, imbricate, margins irregular; petals 3, distinct, imbricate with briefly pointed, valvate tips; staminodes united in a shallow cupule, unlobed or with 3 pointed tips; gynoecium broadly ovoid, trilocular, triovulate, glabrous or tomentose, stigma short, rounded, trilobed, ovules laterally attached, form unknown. *Fruit* oblong-ovoid, beaked, perianth persistent on fresh fruit; epicarp smooth, mesocarp rather thin, fibrous, endocarp thick, very hard, with shining lines internally, externally irregularly sculptured with 3 prominent ridges (*P. torallyi* and *P. sunkha*), or surface and thickness irregular, ridges not prominent (*P. cocoides*), pores 3, basal, sunken. *Seeds* 1–3, rounded, raphe elongate, lateral, endosperm homogeneous, hollow; embryo subbasal. *Germination* and eophyll not recorded. *Cytology* not known.

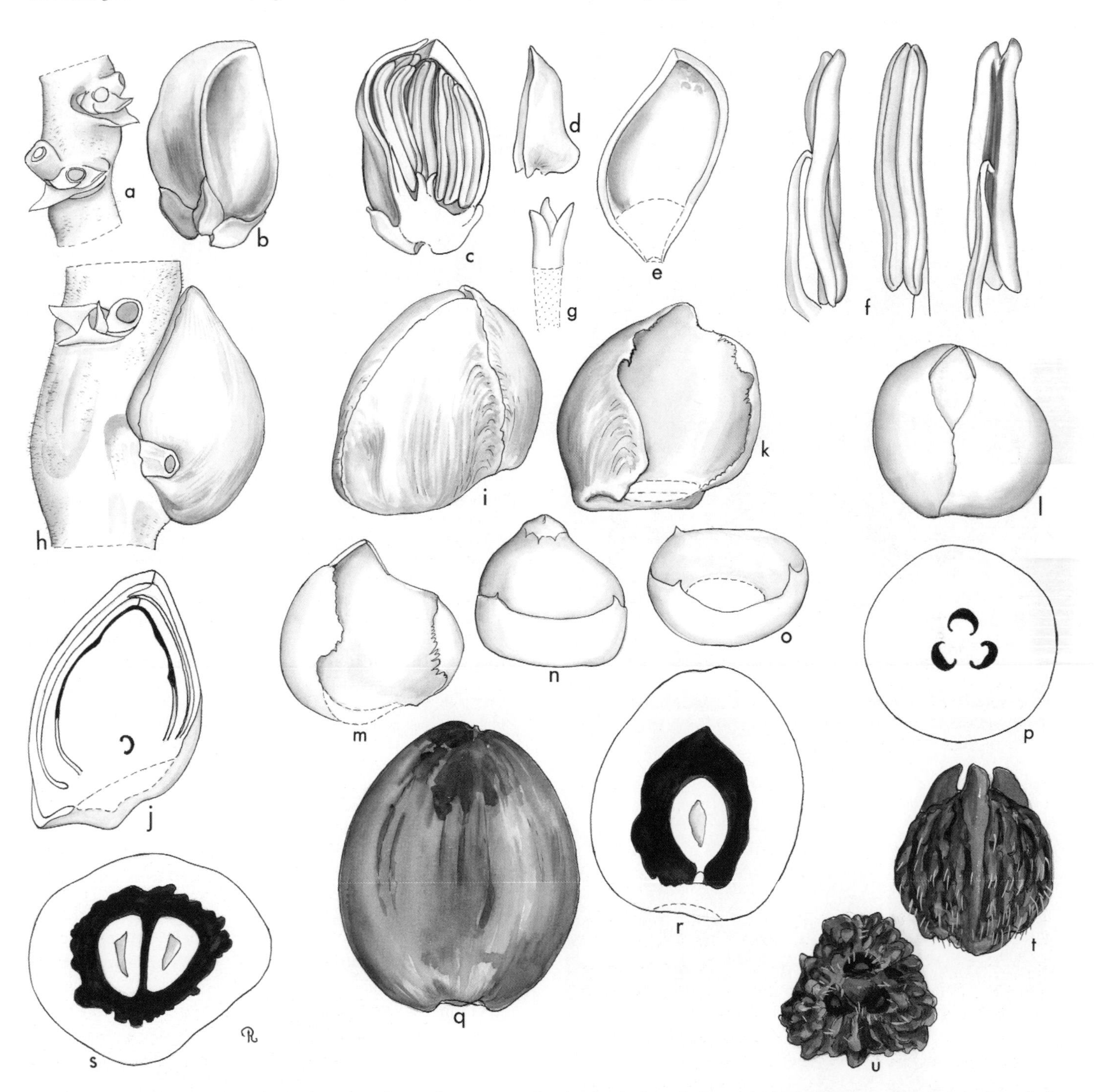

Parajubaea. **a**, portion of rachilla with paired staminate flowers, flowers removed to show bracteoles and scars × 3; **b**, staminate bud × 3; **c**, staminate bud in vertical section × 3; **d**, staminate sepal × 6; **e**, staminate petal, interior view × 3; **f**, stamen in 3 views × 6; **g**, pistillode × 6; **h**, portion of rachilla with triads, all flowers but one pistillate bud removed to show scars and bracteoles × 3; **i**, pistillate flower × 3; **j**, pistillate flower in vertical section × 3; **k**, pistillate sepal × 3; **l**, pistillate flower, sepals removed × 3; **m**, pistillate petal × 3; **n**, gynoecium and staminodial ring × 3; **o**, staminodial ring × 3; **p**, ovary in cross-section × 6; **q**, fruit × $^3/_4$; **r**, fruit in vertical section × $^3/_4$; **s**, fruit in cross-section × $^3/_4$; **t**, **u**, endocarp in 2 views × $^3/_4$. *Parajubaea torallyi*; all from *Cardenas* s.n. (Drawn by Marion Ruff Sheehan)

Distribution and ecology: Three species in Ecuador, Bolivia and Colombia. *Parajubaea torallyi* is found in humid ravines of spectacular sandstone mountains at high elevations (2400–3400 m), where it does not rain for ten months of the year, and *P. cocoides* was described from 3000 m in the eastern Andes. *Parajubaea sunkha* occurs at elevations of 1700–2200 m above sea level in semi-deciduous forest in interandean valleys in Bolivia.
Anatomy: Root (Seubert 1998a, 1998b).
Relationships: The monophyly of *Parajubaea* is untested. The genus is moderately supported as sister to *Cocos* (Baker *et al.* in review).
Common names and uses: For common names, see Moraes & Henderson (1990). The mesocarp is fleshy and sweet and is eaten; the seed contains usable oil. These palms should make handsome ornamentals in cold and dry areas.
Taxonomic accounts: Moraes & Henderson (1990) and Moraes (1996b, 2004).
Fossil record: No generic records found.
Notes: The combination of the rather congested, closely adpressed rachillae, staminate flower with 15 stamens and irregularly sculptured endocarp is not found elsewhere.

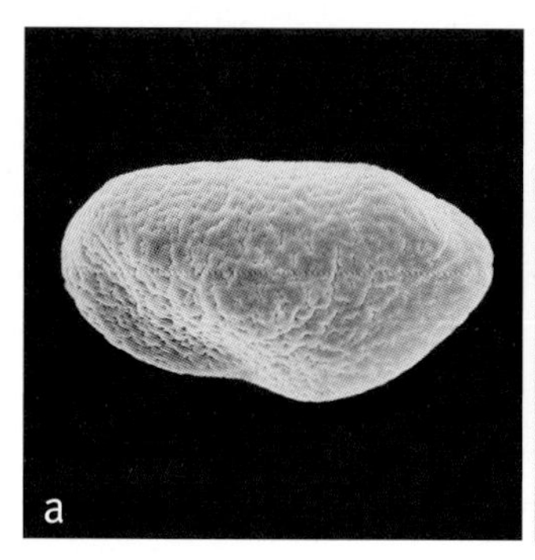

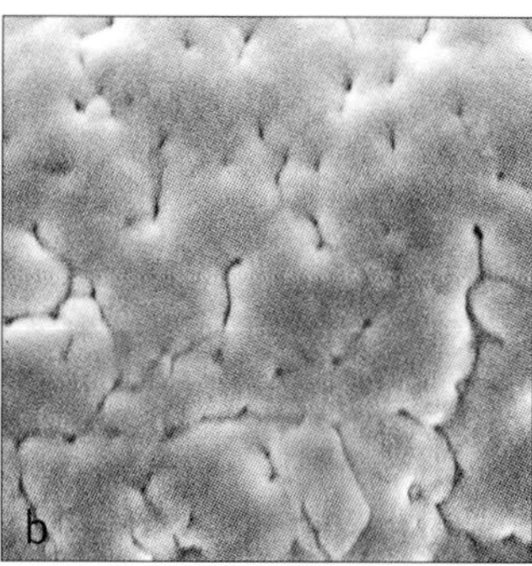

Above: ***Parajubaea***
a, monosulcate pollen grain, proximal face SEM × 1000;
b, close-up, perforate and micro-channelled-rugulate pollen surface SEM × 6000.
Parajubaea cocoides: **a** & **b**, *Bernal & Galeano* 921.

Top left: ***Parajubaea cocoides***, habit, cultivated, Ecuador. (Photo: J.-C. Pintaud)
Bottom left: ***Parajubaea cocoides***, leaf sheaths and fruit, cultivated, Ecuador. (Photo: J.-C. Pintaud)
Bottom right: ***Parajubaea cocoides***, infructescences, cultivated, Ecuador. (Photo: J.-C. PIntaud)

Subtribe Bactridinae Hook.f. in Benth. & Hook.f., Gen. pl. 3: 873, 881 (1883) ('Bactrideae'). Type: **Bactris**.

Diminutive to very large, acaulescent, erect or climbing, armed in some or all parts with soft or mostly stout spines (very rarely unarmed, but then pistillate petals connate); leaves pinnate or pinnately ribbed, leaflets acute or praemorse; peduncular bract usually thick, borne at the base of the peduncle, persistent; petals of pistillate flowers connate or imbricate; fruit 1-seeded; endocarp pores at or above middle except rarely (*Acrocomia*); endosperm always homogeneous.

This is a very diverse group consisting of five clearly demarcated genera that are restricted to the New World and particularly abundant in South America. There is great diversity of habit from massive palms with ventricose stems to diminutive undergrowth palmlets, massive acaulescent palms and climbers. The subtribe is distinguished by the presence of epidermal emergence spines on some organs of nearly all members; only a few species of *Bactris* are more-or-less unarmed, bearing a few bristles along the leaflet margins.

Bactridinae is monophyletic with low to high support (Gunn 2004, Asmussen *et al.* 2006, Baker *et al.* in review). The Bactridinae is sister to Elaeidinae with moderate to high support (Asmussen *et al.* 2006, Baker *et al.* in review, in prep.).

Key to the Genera of Bactridinae

1. Petals of pistillate flowers distinct and broadly imbricate, or if sometimes partially connate basally and with an adnate staminodal tube, then at least the margins free and imbricate; pistillate sepals distinct or connate; staminate flowers with distinct sepals and the petals distinct and valvate or adnate basally to the short floral receptacle then free and valvate; stamen filaments inflexed at the apex in bud, the anthers dorsifixed and versatile; fruit with abundant short fibres in the mesocarp, these strongly adherent to the smooth or only very shallowly pitted endocarp; staminate flowers borne lateral to the pistillate in basal triads on the rachillae, immediately above the triads in pairs, or mostly or entirely singly and subtended by membranous bractlets forming cells resembling those of a honeycomb. Cuba and Mexico to Argentina . . . . **103. Acrocomia**
1. Petals of pistillate flowers connate $^1/_3$–$^1/_2$ their length in a campanulate tube with prominent spreading or erect, valvate lobes, or more than $^1/_2$ their length in an urceolate, briefly 3-lobed, 3-toothed or even truncate tube; sepals of the pistillate flowers distinct and imbricate or connate in a shallow to deep cupule; staminate flowers and inflorescences various; fruit lacking abundant short fibres adherent to the endocarp 2
2. Leaflets praemorse; pistillate petals connate $^1/_3$–$^1/_2$ their length in a campanulate tube with prominent valvate lobes; staminodes connate and adnate to the corolla tube basally, but distinct or continued in a free, 3–6-lobed or -toothed or truncate tube, sometimes nearly equalling the stigmas above; staminate flowers with stamen filaments erect, the basifixed anthers often sagittate basally; pistillode evident. Lesser Antilles, South America from Brazil to Peru and Bolivia . . . .**105. Aiphanes**
2. Leaflets acute, or very rarely praemorse; pistillate petals connate beyond the middle or completely connate in a 3-lobed, 3-toothed, or truncate, urceolate or tubular corolla, lobes not spreading when developed; staminodes distinct or united in a short tube but not adnate to the corolla; staminate flowers with stamen filaments erect or inflexed at the apex or from nearly the middle in bud; pistillode usually lacking 3
3. Staminate flowers often associated with the pistillate in triads basally or along the upper part of the rachilla but above these triads, then paired or generally solitary and densely aggregated in a distinct terminal portion of the rachilla, each pair of flowers or each flower subtended by a prominent bracteole adnate to or coherent with adjacent bractlets to form a cupule, sometimes as high as the flowers; stamen filaments inflexed at the apex in bud; anthers dorsifixed, versatile. Mexico to Brazil . . . . **104. Astrocaryum**
3. Staminate flowers not densely aggregated in a distinct terminal portion of the rachilla but associated with the pistillate in triads or irregularly interspersed among the triads and subtended by short, distinct bracteoles 4

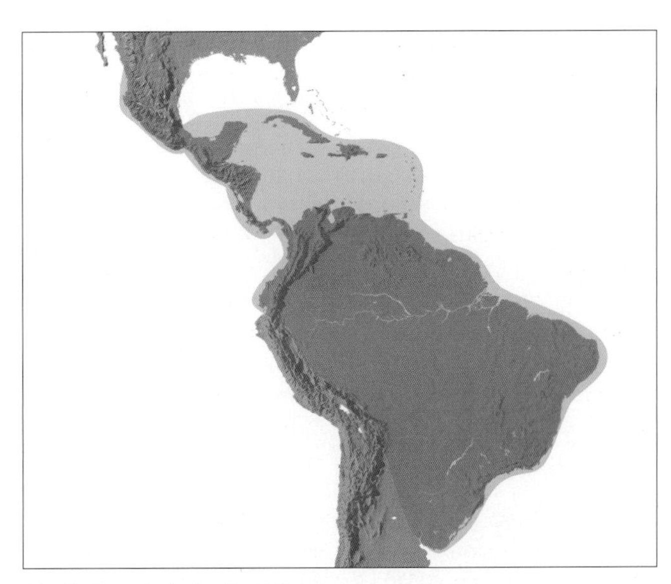

Distribution of subtribe **Bactridinae**

4. Erect plants; upper pinnae not modified reflexed climbing organs; flowers all or nearly all borne in triads or the staminate more numerous and irregularly interspersed among triads; stamen filaments inflexed at the apex or from nearly the middle in bud; anthers mostly dorsifixed, versatile; peduncular bract inserted near the prophyll at the base of the peduncle. Cuba and Mexico to Brazil . . . . . **106. Bactris**
4. Climbing plants; upper pinnae usually very distant and modified into reflexed hook-like climbing organs (acanthophylls); flowers in triads nearly throughout the rachillae; stamen filaments erect in bud, short; anthers basifixed, erect, sagittate basally; peduncular bract often inserted above the middle of the peduncle. Mexico to Brazil . . . . **107. Desmoncus**

103. ACROCOMIA

Viciously spiny solitary acaulescent or erect palms, native to South and Central America and the Caribbean.

Acrocomia Mart., Hist. nat. palm. 2: 66 (1824). Type: *A. sclerocarpa* Mart. (illegitimate name) (*Cocos aculeata* Jacq.) = **A. aculeata** (Jacq.) Lodd.

Acanthococos Barb. Rodr., Palm. hassler. 1 (1900). Type: *A. hassleri* Barb. Rodr. (= *Acrocomia hassleri* [Barb. Rodr.] W.J. Hahn).

Gastrococos Morales, Repert. Fis.-Nat. Isla Cuba 1: 57 (1865). Type: *G. armentalis* Morales = *G. crispa* (Kunth) H.E. Moore (*Cocos crispa* Kunth) (= *Acrocomia crispa* [Kunth] C.F. Baker ex Becc.).

Akros — *highest,* kome — *hair or tuft, perhaps referring to the crown of leaves at the tip of the tall stem.*

Dwarf to large, solitary, spiny, pleonanthic, monoecious palms. *Stem* very short, subterranean and geotropic, or erect, columnar or sometimes swollen and spindle-shaped, covered in persistent leaf bases and eventually becoming bare, or armed heavily with spines at first, soon becoming bare, eventually smooth, ringed with leaf scars. *Leaves* few to numerous, pinnate, marcescent or abscising neatly; sheath disintegrating into a mass of fibres, usually both spiny and finely bristly; petiole short or ± absent, adaxially channelled, abaxially rounded, usually spiny and finely bristly, often with tomentum; rachis usually curved, armed with robust spines, especially along the margins, adaxially channelled near the base, angled distally, abaxially angled or rounded; leaflets numerous, single-fold, linear, acute or shallowly bifid, sometimes plicate, subregularly arranged (rarely) or grouped, usually held in different planes giving the leaf a plumose appearance, usually coriaceous, acute or briefly bifid, adaxially usually glabrous, abaxially glabrous, glaucous or pubescent, transverse veinlets obscure. *Inflorescences* axillary, interfoliar, shorter than the leaves, arching or becoming pendulous, apparently protandrous, branching to 1 order; peduncle ± oval in cross-section, often elongate, spiny and/or tomentose, rarely unarmed; prophyll tubular, 2-keeled, closely sheathing, usually remaining ± hidden within the leaf sheaths, soon tattering and splitting irregularly, glabrous or densely shaggy hairy, sometimes also spiny; peduncular bract inserted near the prophyll, much larger than the prophyll, persistent, tubular, enclosing the rachillae in bud, strongly beaked, woody, splitting along the abaxial face, then expanded or ± cowl-like, abaxial surface densely shaggy tomentose, and often sparsely to densely spiny, spines sometimes restricted to near the beak, adaxial surface glabrous, often conspicuously pale yellow; rachis longer or shorter than the peduncle, variously spiny, tomentose, or glabrous, bearing few to numerous, spirally arranged rachillae, each subtended by a short triangular bract; rachillae short to elongate, straight or somewhat flexuous, often with a pulvinus (?nectariferous) at the base, usually cream-coloured or yellowish, with a very short to long basal, bare portion, then bearing 1–several, rather distant spirally arranged triads, distal to these a few pairs of staminate flowers, in the distal portion bearing dense spirals of solitary staminate flowers, each flower group subtended by a short-triangular bract, those subtending the staminate flowers forming shallow pits; floral bracteoles small, mostly obscured within the pits. *Staminate flowers* creamy yellow, usually rather strongly earthy smelling, symmetrical or somewhat asymmetrical, those of the triads briefly stalked, those of distal portion of rachillae sessile; sepals 3, distinct or connate, small, narrow to broadly triangular, sometimes irregularly ciliate; petals 3, distinct except at the very base, much longer than the sepals, ± boat-shaped, ± fleshy, valvate; stamens 6, filaments distinct or briefly adnate to the base of the petals, elongate, inflexed at the tip, anthers ± rectangular, dorsifixed or medifixed, latrorse; pistillode small, trifid. *Pollen* oblate triangular, occasionally oblate square, usually symmetric; aperture a distal trichotomosulcus or occasionally a tetrachotomosulcus; ectexine semi-tectate or tectate, coarsely perforate, perforations widely separated and indented, or finely reticulate, aperture margin may be slightly finer; infratectum columellate; longest axis 37–62 µm [3/3]. *Pistillate flowers* larger than the staminate, conic-ovoid; sepals 3, distinct, ± imbricate, broadly triangular or connate in a 3-lobed cupule; petals much longer than the sepals, 3, ± distinct except sometimes near the base, or connate, always with broad, imbricate, distinct margins, except for valvate tips; staminodes 6, united to form a staminodal ring, free, or briefly adnate to the petals, 6-toothed, usually bearing well-

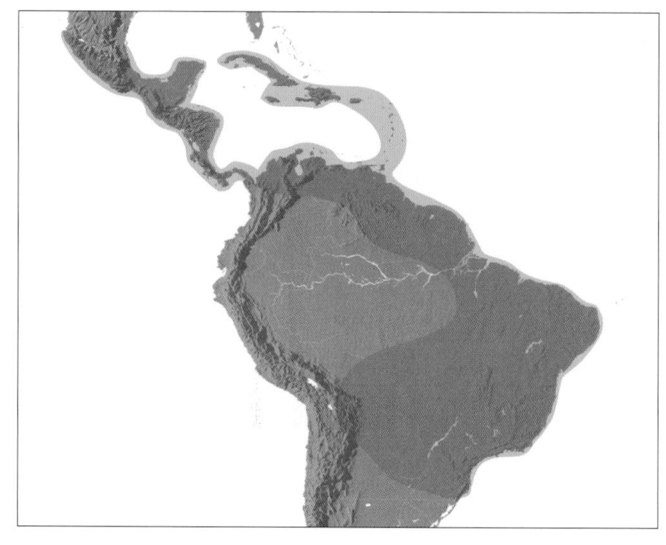

Distribution of ***Acrocomia***

Left: ***Acrocomia crispa***, stem and infructescence, cultivated, Montgomery Botanical Center, Florida. (Photo: J. Dransfield)

Opposite: ***Acrocomia***. **a**, portion of rachilla near apex with solitary staminate flowers × 3; **b**, intermediate portion of rachilla with paired staminate flowers × 6; **c**, portion of rachilla, staminate flowers removed × 6; **d**, staminate bud × 6; **e**, staminate bud in vertical section × 6; **f**, staminate sepal × 12; **g**, staminate bud, sepals removed × 6; **h**, staminate petal, interior view × 6; **i**, stamen in 3 views × 12; **j**, pistillode × 12; **k**, triad at base of rachilla × 3; **l**, triad, flowers removed × 3; **m**, pistillate bud × 3; **n**, pistillate flower in vertical section × 6; **o**, pistillate sepal, interior view × 6; **p**, pistillate petal, interior view × 6; **q**, staminodes, expanded × 6; **r**, gynoecium and staminodes × 6; **s**, stigmas × 12; **t**, ovary in cross-section × 6; **u**, fruit × 1; **v**, fruit in vertical section × 1; **w**, fruit in cross-section × 1; **x, y, z, aa**, endocarp in 4 views × 1 1/2; **bb, cc, dd, ee**, seed in 4 views × 1 1/2. *Acrocomia* sp.: **a–t**, *O.F. Cook* s.n.; *A. aculeata*: **u–ee**, *Coto Fernandez* s.n. (Drawn by Marion Ruff Sheehan)

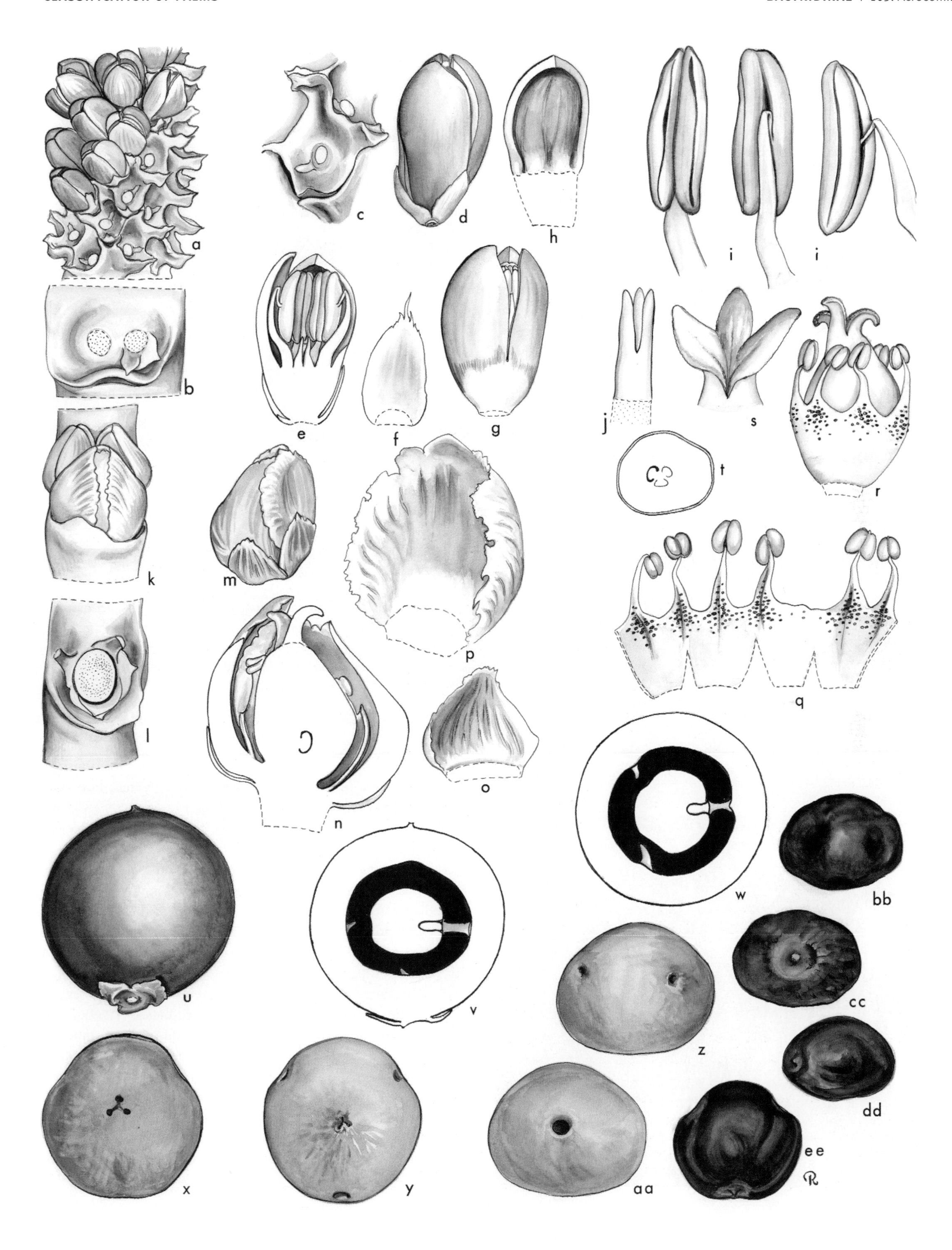
a
b
c
d
h
i
i
e
f
g
j
s
r
t
k
m
p
q
l
n
o
u
v
w
bb
cc
z
dd
x
y
aa
ee

developed but empty rounded anthers; gynoecium irregularly ovoid, trilocular, triovulate, variously scaly or tomentose, stigmas 3, fleshy, conspicuous, sometimes violet in colour, reflexed beyond the petals at maturity, ovule laterally attached, orthotropous. *Fruit* usually 1-seeded, globose, or rarely somewhat pyriform, olive-green to yellow-brown, the stigmatic remains apical; epicarp smooth, or tomentose-bristly, mesocarp fleshy, with abundant short fibres adnate to the endocarp; endocarp very thick, stony, sometimes pitted, dark brown, with 3 pores ± at the equator. *Seed* basally attached, endosperm homogeneous, sometimes with a central hollow; embryo lateral opposite one of the pores. *Germination* adjacent-ligular; eophyll not recorded. *Cytology*: 2n = 30.

Distribution and ecology: About 34 species have been described, but there are probably far fewer, distributed throughout the West Indies and from Mexico southwards to Argentina. Henderson *et al.* (1995) accept just two species, *Acrocomia aculeata* and *A. hassleri* and *Gastrococos crispa* (now *Acrocomia crispa*) however, this may be too sweeping a synonymy. *Acrocomia aculeata* usually avoids ever-wet regions and may be a conspicuous component of savannah and man-made grasslands, the seed sometimes distributed by cattle. *Acrocomia hassleri* is a palm of 'cerrado' in southern Brazil and Paraguay. *Acrocomia crispa* is endemic to Cuba, where it is confined to calcareous soils throughout the island.

Anatomy: Leaf (Tomlinson 1961), root (Seubert 1998a, 1998b), fruit (Vaughan 1960).

Relationships: *Acrocomia* is resolved as monophyletic, mostly with high support (Hahn 2002b, Gunn 2004, Asmussen *et al.* 2006, Baker *et al.* in review). Although several studies have explored the relationships between the five genera of Bactridinae (Hahn 2002b, Gunn 2004, Asmussen *et al.* 2006, Couvreur *et al.* 2007, Baker *et al.* in review), much conflict and ambiguity remains. Only one relationship, placing *Acrocomia* as sister to the remaining Bactridinae, is highly supported (Hahn 2002b, Baker *et al.* in review). Two studies provide moderate support for a sister relationship between *Astrocaryum* and *Bactris* (Gunn 2004, Baker *et al.* in review), but this is contradicted by Hahn (2002b) who resolves *Bactris* as sister to *Desmoncus*. Hahn (2002b) also places *Aiphanes* as sister to the *Bactris–Desmoncus* clade with moderate support. A number of other relationships among genera of Bactridinae have been recovered, but these are not supported by bootstrap analysis.

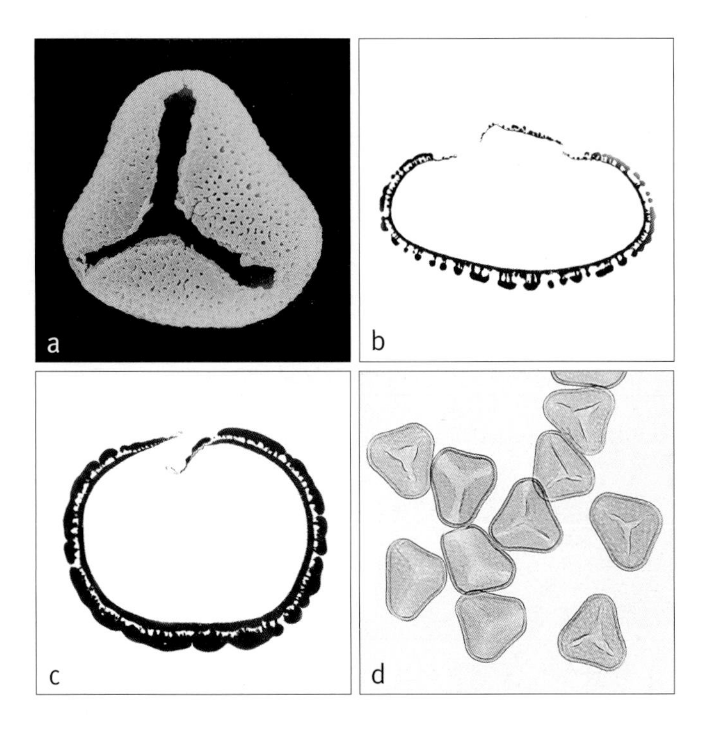

Above: ***Acrocomia***
a, trichotomosulcate pollen grain, distal face SEM × 1000;
b, ultrathin section through whole pollen grain, polar plane TEM × 1000;
c, ultrathin section through whole pollen grain, polar plane TEM × 1000;
d, group of trichotomosulcate pollen grains, mid-focus LM × 250.
Acrocomia crispa: **a** & **b**, *Elkman* 1171; *A. aculeata*: **c**, *Fiebrig* 4359; **d**, unknown collector (MH602).

Below left: ***Acrocomia aculeata***, crown and inflorescence, Colombia.
(Photo: J. Dransfield)

Common names and uses: *Macauba, gru gru* (*Acrocomia aculeata*), belly palm (*A. crispa*) *Mbocaya totai* and *mucaja*. *Acrocomia* is considered to be a genus of great potential (Lleras 1985). Starch can be extracted from the stem of *A. aculeata*; leaves are used as a source of fibre for weaving hammocks and other articles, and in Cuba, leaflets of *A. crispa* are used for making brooms and for rope fibre. Fruit of many species are used as a source of oil (from the endosperm) and animal food, and they are sometimes sold for human consumption. The cabbage is edible. For medicinal uses, see Plotkin & Balick (1984).

Taxonomic account: A massive reduction in the number of recognised species was made by Henderson *et al.* (1995); however, the genus is in need of a critical revision.

Fossil record: A fruit (*Palmocarpon acrocomioides*) recorded from the Middle Oligocene of Puerto Rico is compared with *Acrocomia crispa* (Hollick 1928). Two monosulcate palm pollen types are recovered from the Pliocene, Gatun Lake Formation, Panama (Graham 1991). The first of these types, asymmetrical

and scabrate tectate, is a very common arecoid pollen type and difficult to place. It has been compared with pollen of *Aiphanes*, *Manicaria*, *Reinhardtia* or, possibly, a "monocolpate form" of *Acrocomia*. However, pollen of *Acrocomia* is always trichotomosulcate; *Reinhardtia* is the most probable comparison.
Notes: *Acanthococos*, described to include three acaulescent species, is distinctive only in its habit and is thus included in synonymy. *Gastrococos* differs only in having connate petals in the pistillate flower; in Gunn's phylogenetic study (Gunn 2004), it resolved in a clade embedded within *Acrocomia* and for this reason is placed in synonymy. The reproductive biology of *A. aculeata* was studied by Scariot and Lleras (1991).
Additional figures: Glossary figs 7, 14, 20.

104. ASTROCARYUM

Extremely spiny pinnate-leaved palms from Central and South America, distinctive in the marked separation of pistillate flowers from the staminate part of the rachilla.

Astrocaryum G. Mey., Prim. fl. esseq. 265 (1818) (conserved name). Type: **A. aculeatum** G. Mey.

Avoira Giseke, Praelectiones in Ordines Naturales Plantarum 38, 53 (1792) (name rejected in favour of *Astrocaryum*). Lectotype: *A. vulgaris* Giseke (see O.F. Cook, J. Wash. Acad. Sci. 30: 299 [1940] [= *Astrocaryum* sp.]).

Hexopetion Burret, Notizbl. Bot. Gart. Berlin-Dahlem 12: 156 (1934). Type: *H. mexicanum* (Liebm. ex Mart.) Burret (*Astrocaryum mexicanum* Liebm. ex Mart.).

Toxophoenix Schott in Schreibers, Nachr. österr. Naturf. Bras. 2 (Anhang): 12 (1822). Type: *T. aculeatissima* Schott (= *Astrocaryum aculeatissimum* [Schott] Burret).

Astron — *star,* karyon — *nut, referring to the star-like pattern of fibres around the endocarp pores.*

Moderate to robust, solitary or clustered, sometimes acaulescent, spiny, pleonanthic, monoecious palms. *Stem* very short to tall, often slender, obscured by leaf bases, or becoming bare and conspicuously ringed with leaf scars, often armed with fierce spines pointing in several directions, sometimes losing spines with age. *Leaves* few to numerous, regularly or irregularly pinnate, neatly abscising or marcesent; sheath splitting opposite the petiole, usually fiercely armed with large and small spines, and frequently bearing abundant indumentum; petiole very short to long, adaxially channelled near the base, distally ± flattened or angled, abaxially rounded, bearing abundant spines of varying length and dense indumentum; rachis usually much longer than the petiole, adaxially ± angled, abaxially rounded, usually densely armed and tomentose like the petiole; leaflets numerous and single-fold (or rarely few and composed of many folds), regularly arranged or grouped, and usually fanned within the groups, the whole leaf then appearing plumose, sometimes ± secondarily plicate, linear, acute, usually dark green and shiny adaxially, abaxially almost always with abundant white indumentum, the leaflet margins often conspicuously armed with short spines or bristles; transverse veinlets conspicuous or obscure. *Inflorescences* solitary, interfoliar, erect at first, becoming pendulous, ?protandrous, branching to 1 order; peduncle usually elongate, ± circular in cross-section, often heavily armed with spines, sometimes with spines confined only to the area just below the bract insertion, the surface frequently densely covered in indumentum; prophyll ± membranous, tubular, 2-keeled, unarmed (?always), ± included within the leaf sheaths, soon tattering; peduncular bract much exceeding the prophyll, tubular, beaked, enclosing the rachillae in bud, splitting longitudinally along the abaxial face, arched over the rachillae, persistent or eroding, usually densely tomentose and heavily armed with spines, rarely unarmed; rachis shorter than the peduncle (often very much so) often armed as the peduncle, bearing numerous spirally arranged, crowded rachillae, each subtended by a narrow triangular bract; rachillae complex, elongate, with or without an armed or unarmed basal bare portion above which bearing a single triad or 2–5 distant triads, with or without a slender bare portion distal to the triads, distal to which the rachillae appearing cylindrical, catkin-like and bearing densely packed staminate flowers in pairs or singly, immersed in pits; rachilla bracts ± acute, forming lower lip of pits, floral bracteoles very small, sometimes partially connate with rachilla bract; after anthesis, staminate portions of the rachillae eroding away, in those species with solitary triads, the fruit then borne in a close-packed 'spike' or head, in those with several triads the fruit more loosely arranged. *Staminate flowers* small, ± symmetrical; sepals 3, very small, ± triangular, ?sometimes basally connate; petals 3, much exceeding the sepals, valvate, boat-shaped, connate basally and adnate to the receptacle; stamens (3–)6(–12, fide Wessels Boer 1965), filaments epipetalous, short, inflexed in bud, anthers ± rectangular or linear, dorsifixed, versatile, latrorse; pistillode present and trifid or absent. *Pollen* ellipsoidal, or oblate-triangular, usually with slight asymmetry; aperture a distal sulcus or trichotomosulcus; ectexine tectate, finely perforate-psilate or coarsely perforate, perforations closely or widely spaced or, perforate and micro-

Right: ***Astrocaryum chambira***, habit, Ecuador. (Photo: F. Borchsenius)

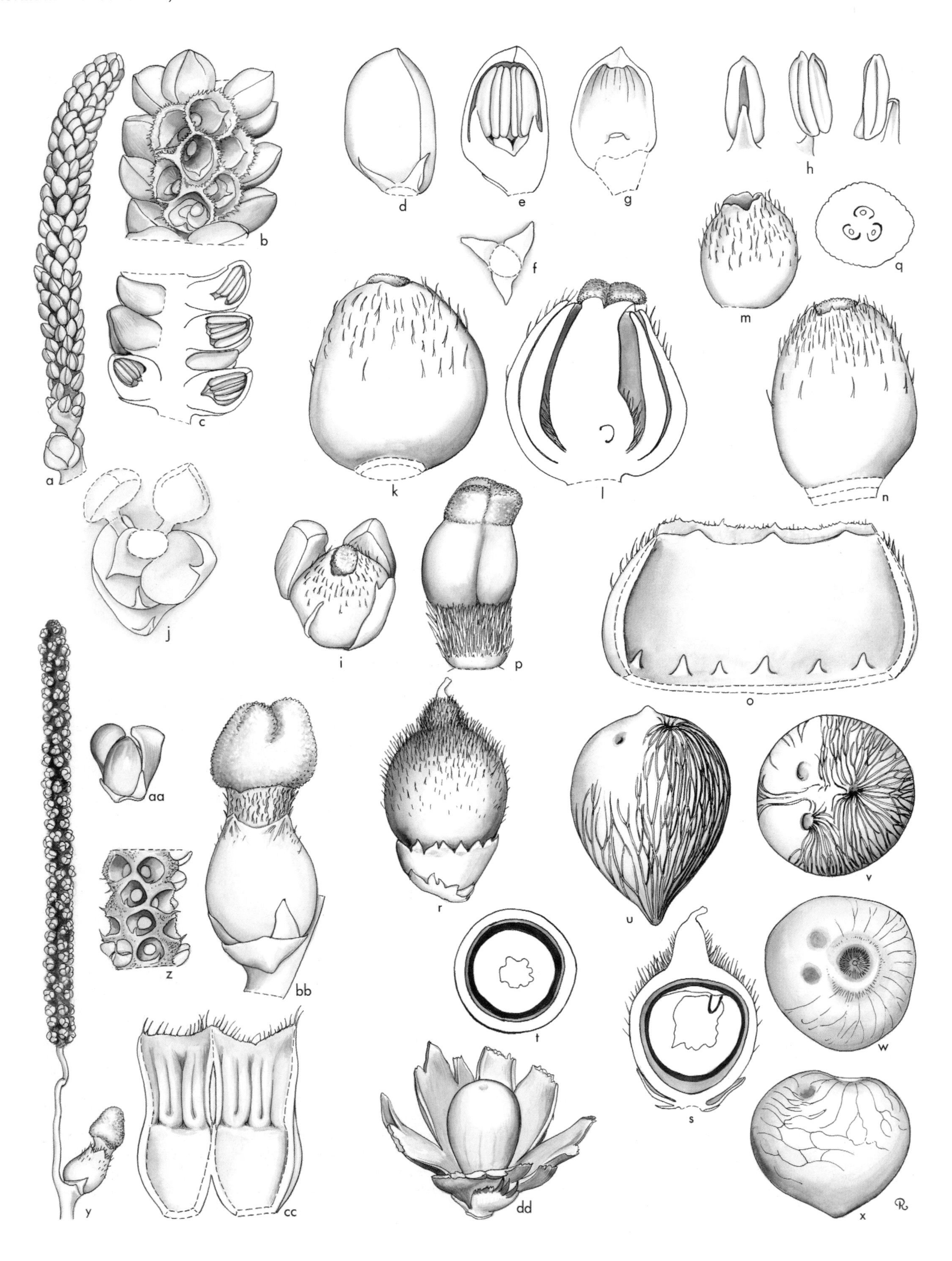
a
b
c
d
e
f
g
h
i
j
k
l
m
n
o
p
q
r
s
t
u
v
w
x
y
z
aa
bb
cc
dd

Left: ***Astrocaryum alatum***, inflorescence, cultivated, Fairchild Tropical Botanic Garden, Florida. (Photo: C.E. Lewis)
Middle: ***Astrocaryum standleyanum***, inflorescence, Peru. (Photo: J.-C. Pintaud)
Right: ***Astrocaryum mexicanum***, fruit, cultivated, Fairchild Tropical Botanic Garden, Florida. (Photo: B.J. Wood)

channelled and rugulate, aperture margin may be slightly finer; infratectum columellate; longest axis 41–78 µm [11/36]. *Pistillate flower* very much larger than the staminate; calyx urn-shaped or cup-shaped, truncate or shallowly 3-lobed, sometimes bearing numerous short spicules, usually densely tomentose; corolla not, briefly, or considerably exceeding, and similar to the calyx, or composed of 3 imbricate triangular lobes, connate basally; staminodes 6, epipetalous near the base of the corolla, connate into a low membranous ring or tooth-like; gynoecium varied in shape, trilocular, triovulate, the 3 large fleshy erect, or head-like, reflexed stigmas borne on a beak, protruding through the mouth of the corolla tube, sometimes bearing short spines and/or tomentum, ovule ?orthotropous, laterally attached. *Fruit* 1(–2)-seeded with apical stigmatic remains, beaked, spherical, top-shaped, prismatic, or ovoid, often brightly coloured, brown, yellowish or orange-red, calyx and corolla persistent, enlarged and irregularly splitting; epicarp spiny or unarmed, tomentose or glabrous, mesocarp relatively thin, fleshy or dry and starchy, and fibrous, sometimes with the epicarp irregularly splitting and spreading to expose the endocarp, endocarp thick, stony, with numerous flattened, black longitudinal fibres on the surface, conspicuously radiating from the 3 subterminal pores. *Seed* irregularly globular, basally attached, hilum circular, raphe branches anastomosing, endosperm homogeneous, usually hollow; embryo subapical, opposite one of the endocarp pores. *Germination* adjacent-ligular; eophyll bifid, usually bristly. *Cytology*: 2n = 30.

Distribution and ecology: About 36 accepted species distributed from Mexico southwards to Brazil and Bolivia; absent from the West Indies except Trinidad. Ecologically, the genus is very varied: some species are undergrowth palms of primary lowland forest; others are light-demanding and occur in secondary forest or forest margins (e.g., riverbanks). In Surinam, *Astrocaryum vulgare* is particularly frequent in white sand savannah. Most species seem to be confined to the lowlands.

Anatomy: Leaf (Tomlinson 1961, Beerling & Kelly 1996), root (Seubert 1998a, 1998b).

Relationships: Published evidence indicates that *Astrocaryum* is monophyletic with moderate support (Gunn 2004; for relationships, see *Acrocomia*). However, preliminary phylogenetic studies based on molecular data (Pintaud, pers. comm.) suggest that *Astrocaryum*, as currently delimited, may not be monophyletic. The problem could be addressed, at least in part, by removing two taxa, *Astrocaryum mexicanum* and *A. alatum*, which are sister to each other and tend to resolve elsewhere in the Bactridinae. *Astrocaryum mexicanum* was

Opposite: ***Astrocaryum***. **a**, rachilla with basal pistillate flower and staminate flowers × 1; **b**, portion of axis with staminate flowers × 3; **c**, staminate portion of rachilla in vertical section × 3; **d**, staminate bud × 6; **e**, staminate bud in vertical section × 6; **f**, staminate sepals, basal view × 6; **g**, staminate petal, interior view × 6; **h**, stamen in 3 views × 6; **i**, pistillate flower and staminate buds × 3; **j**, triad, flowers removed to show bracteoles × 3; **k**, pistillate flower × 6; **l**, pistillate flower in vertical section × 6; **m**, pistillate calyx × 3; **n**, pistillate corolla × 6; **o**, pistillate corolla with staminodes, expanded, interior view × 6; **p**, gynoecium × 6; **q**, ovary in cross-section × 6; **r**, fruit × 1; **s**, fruit in vertical section × 1; **t**, fruit in cross-section × 1; **u**, **v**, endocarp in 2 views × $1^1/_2$; **w**, **x**, seed in 2 views × 3; **y**, rachilla × 1; **z**, portion of staminate axis to show pits × 3; **aa**, staminate flower × 6; **bb**, pistillate flower × 3; **cc**, pistillate corolla with adnate staminodial cupule, interior view × 3; **dd**, fruit, dehisced × 1. *Astrocaryum mexicanum*: **a–q**, *Read* 902; **r–x**, *Moore* 6021; *A. murumuru*: **y–cc**, *Moore* 9551; *A. gynacanthum*: **dd**, *Moore* 9523. (Drawn by Marion Ruff Sheehan)

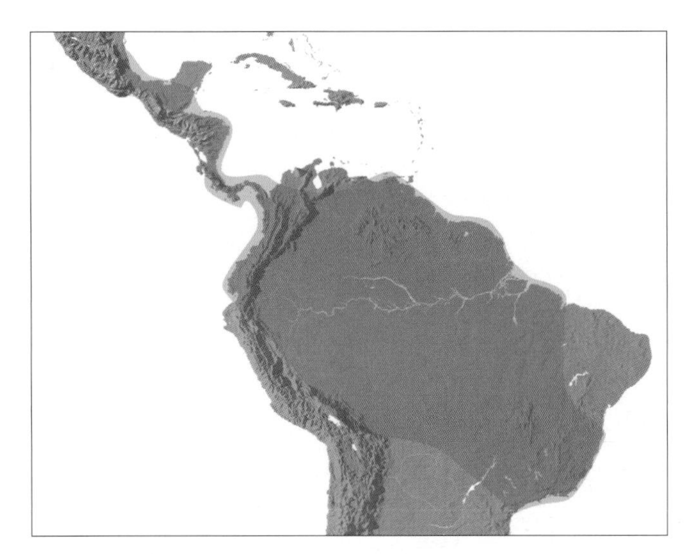

Distribution of ***Astrocaryum***

separated by Burret as the basis of a new genus, *Hexopetion*, because the staminodes are free as opposed to being cupuliform as in the rest of *Astrocaryum*. The staminodes in *A. alatum* form a cupuliform ring, however, so this character seems of no value in separating *Hexopetion* from *Astrocaryum*. The one gross morphological character shared by the two species is the fact that the staminate flowers occur directly above the single pistillate flower at the base of the rachillae, whereas in *Astrocaryum* there is a bare portion immediately distal to the pistillate flowers. There is also a single anatomical difference: the perivascular sclerified sheath in the leaf midrib is continuous in *A. alatum* and *A. mexicanum* whereas it is discontinuous in all other species of *Astrocaryum* examined. Insufficient evidence has been presented to date to warrant the recognition of *Hexopetion*. A thorough study of relationships across the Bactridinae is required before further changes in Bactridinae can be justified.

Common names and uses: *Tucuma*. For common names, see Pesce (1941, translation 1985). For local uses, see Glassman (1972) and Balick (1985). The epidermis and hypodermis of the sword leaf of *Astrocaryum vulgare* provide an important fibre used by Amerindians in manufacturing mats, hats, hammocks, fishing lines and nets. The mesocarp of some species is eaten by humans or fed to cattle. The kernel of *A. vulgare* produces a fine oil, which is excellent for eating or soap making. Fruits of *A. murumuru* and *A. aculeatum* have also been used as a source of oil, and the 'cabbage' of many species is utilised.

Taxonomic accounts: See Henderson *et al*. (1995). Numerous papers by Kahn and his associates have appeared in recent years, elucidating species and species complexes, but a modern synthesis of the whole genus is yet to be produced. See also, Kahn & Second (1999).

Fossil record: Fruits from the Middle Oligocene of Puerto Rico, *Palmocarpon cetera*, are compared with *Cocos* and *Astrocaryum*, although there is insufficient detail to make a very satisfactory comparison (Hollick 1928). From the Middle Eocene of northwestern Peru, Berry (1926a) described palm endocarps, *Astrocaryum olsoni*, with a size range of 3.75 – 5.25 cm long × 2.5 – 3.75 cm wide; they have a fibrous outer layer, and a 2–3 mm thick inner layer; their interior is filled with calcified structureless material. From the lower Cenomanian of France (Argonne), Fliche (1896) describes *Astrocaryum astrocaryopsis* and, also from the upper Cenomanien, *Astrocaryopsis* sp. from the Sainte Menhould area (Fliche 1894). *Astrocaryum*-like pollen (Graham 1976) is reported from the Upper Miocene of Mexico (Paraje Solo flora, Veracruz). Van der Hammen and Garcia de Mutis (1966) suggest that the "natural relationship" of the zonasulcate *Proxapertites* (described by the authors as having,

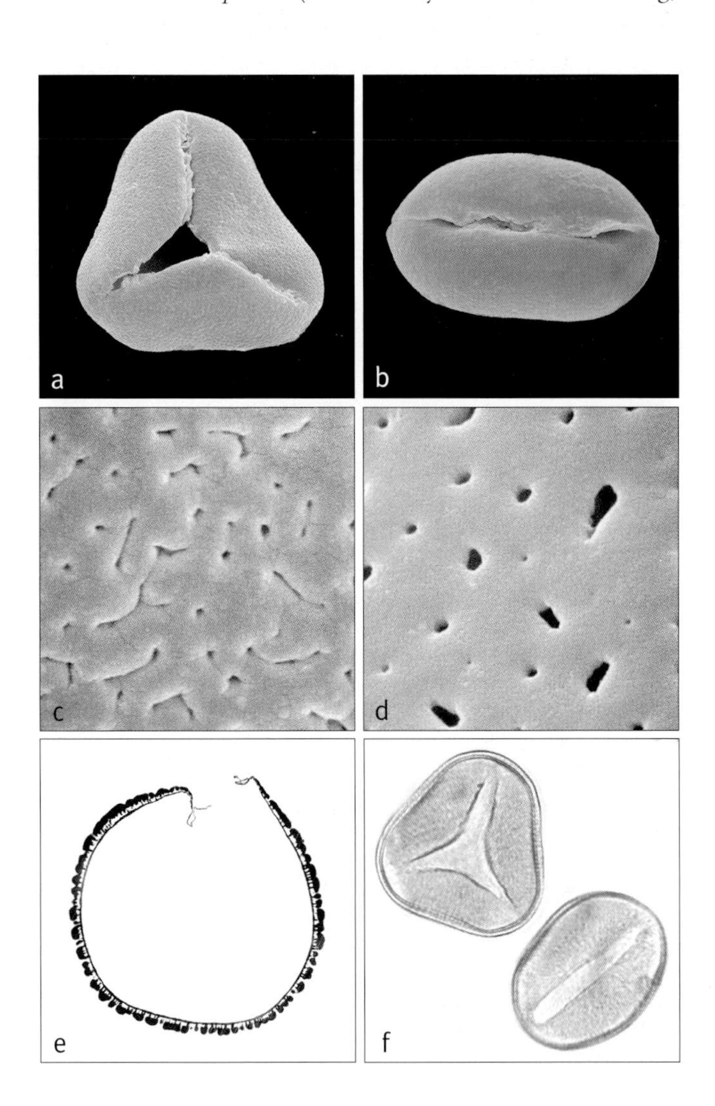

Right: ***Astrocaryum***
a, trichotomosulcate pollen grain, distal face SEM × 1000;
b, monosulcate pollen grain, distal face SEM × 1000;
c, close-up, perforate/micro-channelled pollen surface SEM × 8000; **d**, close-up, perforate pollen surface SEM × 8000;
e, ultrathin section through whole pollen grain, polar plane TEM × 1250;
f, two pollen grains: top left trichotomosulcate, bottom right monosulcate, mid-foci LM × 750.
Astrocaryum murumuru: **a, d** & **f**, *Glaziou* 17857; *A. gynacanthum*: **b**, *Krukoff* 6765; *A. mexicanum*: **c** & **e**, *Nur* s.n.

"a very large, variable, ± irregular, aperture") is *Astrocaryum acaule*, but this is unlikely because this species has mono- or trichotomosulcate pollen.

Notes: *Astrocaryum mexicanum* has been the subject of detailed long-term demographic and reproductive biological studies led by Sarukhan (e.g., Búrquez *et al.* 1987).

Additional figures: Glossary figs 20, 22.

105. AIPHANES

Very spiny solitary or clustering pinnate-leaved palms from South America and the Caribbean, instantly recognisable by the praemorse concolourous leaflets.

Aiphanes Willd., Mém. Acad. Roy. Sci. Hist. (Berlin) 1804: 32 (1807). Type: *A. aculeata* Willd. (= **A. horrida** [Jacq.] Burret).

Marara H. Karst., Linnaea 28: 389 (1857). Lectotype: *M. erinacea* H. Karst. (see H.E. Moore 1963c) (= *Aiphanes erinacea* [H. Karst.] H. Wendl.).

Curima O.F. Cook, Bull. Torrey Bot. Club 28: 561 (1901). Type: *C. colophylla* O.F. Cook (= *Aiphanes minima* [Gaertn.] Burret).

Tilmia O.F. Cook, Bull. Torrey Bot. Club 28: 565 (1901). Lectotype: *T. caryotifolia* (Kunth) O.F. Cook (*Martinezia caryotifolia* Kunth) (see H.E. Moore 1963c) (= *Aiphanes horrida* [Jacq.] Burret).

Martinezia of many authors (non *Martinezia* Ruiz & Pav. [1794] = *Prestoea*).

Derivation not explained by author, but possibly from aeiphanes – *ever-shining or ever-appearing.*

Small to moderate, solitary or clustered, sometimes stoloniferous, spiny, pleonanthic, monoecious palms. *Stem* often very short, the plant then ± acaulescent, or erect, rarely branching aerially, becoming bare, conspicuously ringed with leaf scars and usually bearing horizontal rows or rings of robust black spines. *Leaves* few to numerous, pinnate or entire bifid, spirally arranged, distichous or tristichous (*Aiphanes leiostachys*) neatly abscising; sheaths tubular at first, soon disintegrating into a weft of fibres and broad shreds, usually densely spiny and/or tomentose, distally prolonged into a tubular, tattering ligule; petiole short to long, adaxially channelled in proximal part, flattened or angled distally, abaxially rounded or angled, usually variously golden-yellow to black-spiny and sometimes also tomentose, the spines themselves often bearing caducous tomentum; rachis (or axis of entire leaf) adaxially ± angled, abaxially rounded, often variously spiny and/or tomentose or glabrous; blade where undivided, with a shallow to deep apical notch, the margins praemorse, the main ribs unarmed or spiny on abaxial and/or adaxial surfaces, leaflets, where blade divided, narrow lanceolate to broad rhomboid, regularly arranged or grouped, held in one plane or twisted into several planes, proximal margins entire, the apical praemorse, shallowly or deeply lobed, variously tomentose or bristly or glabrous, variously armed with short to long spines along veins on one or both surfaces and along margins, transverse veinlets obscure. *Inflorescences* solitary or rarely multiple, interfoliar, spicate (rarely) or branching to 1 order only, very rarely to 2, apparently protandrous; peduncle elongate, curved to pendulous, ± elliptic to circular in cross-section, unarmed or sparsely to fiercely armed with spicules and spines, glabrous or tomentose; prophyll usually lanceolate, ± beaked, flattened, 2-keeled, tubular, enclosing the inflorescence in bud, splitting longitudinally and tattering apically, but persistent, variously glabrous or tomentose, unarmed or spiny; peduncular bract inserted near the prophyll, much longer, ± terete, sometimes beaked, unarmed or variously spiny, persistent; rachis (where inflorescence branched) shorter than the peduncle, proximally often armed, distally usually unarmed, often scaly or tomentose, bearing spirally arranged, evenly spaced rachillae each subtended by a small triangular bract; rachillae slender, usually elongate, often spreading, straight or flexuous, with a short to long, basal bare

Top right: ***Aiphanes grandis***, habit, Ecuador. (Photo: F. Borchsenius)

Bottom right: ***Aiphanes eggersii***, fruit, Ecuador. (Photo: F. Borchsenius)

portion, the whole glabrous or more usually scaly or tomentose, flowers borne spirally in triads proximally, distally the rachillae bearing solitary or paired staminate flowers, rarely the rachillae bearing staminate flowers only, very rarely flowers borne in tetrads of 2 pistillate and 2 staminate (according to Read 1979), flower groups superficial or sunken in pits, the rachilla bracts forming the lower lips of the pits; floral bracteoles minute. *Staminate flowers* usually small, sessile or with a brief stalk; sepals 3, distinct, or connate and spreading in a 3-lobed ring, triangular, membranous; petals 3, distinct or minutely connate basally, valvate, triangular, rather fleshy, much longer than the calyx, ± ovate to triangular, adaxially with impressions of the stamens; stamens 6, filaments short, fleshy, wider and minutely connate basally and/or briefly epipetalous, anthers orbicular, ± rectangular, or linear, medifixed, versatile, latrorse; pistillode minute, conical or trifid. *Pollen* ellipsoidal, usually with either slight or obvious asymmetry; aperture a distal sulcus; ectexine tectate, perforate, micro-channelled and rugulate; infrequently coarsely perforate and spinose or, coarsely perforate and verrucate, aperture margin usually slightly finer; infratectum columellate; longest axis 24–34 µm [11/23]. *Pistillate flowers* larger than the staminate, sessile; sepals 3, distinct, broad, imbricate; petals 3, exceeding the sepals, fleshy, connate in the basal ca. $^1/_2$, apically with 3, triangular, valvate lobes; staminodal ring 6-toothed, adnate to the corolla tube; gynoecium ovoid,

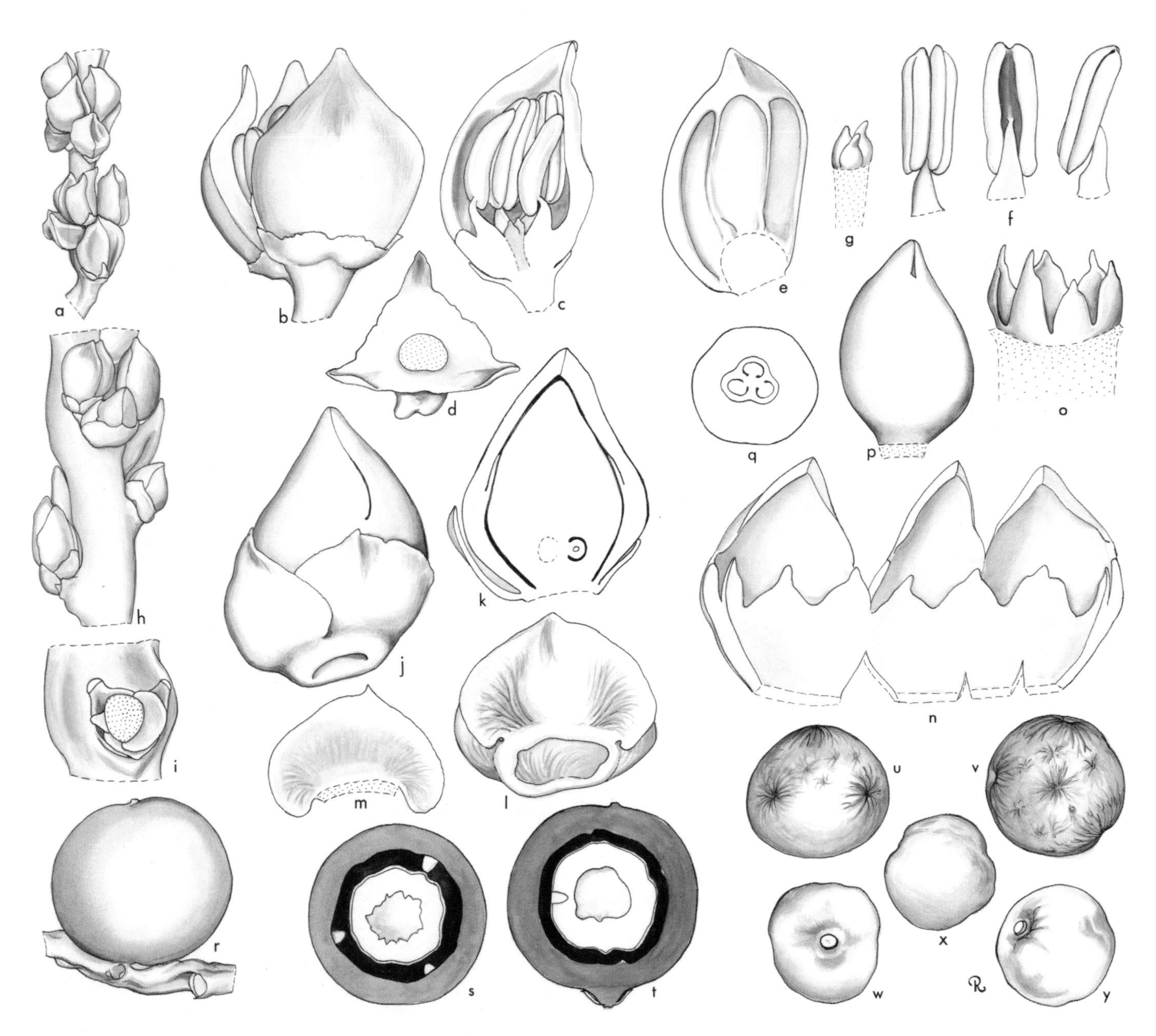

***Aiphanes*.** **a**, apical portion of rachilla with paired staminate flowers × 3; **b**, staminate flower × 12; **c**, staminate flower in vertical section × 12; **d**, staminate calyx × 12; **e**, staminate petal, interior view × 12; **f**, stamen in 3 views × 12; **g**, pistillode × 12; **h**, portion of rachilla with triads × 3; **i**, triad, flowers removed to show bracteoles × 6; **j**, pistillate flower × 12; **k**, pistillate flower in vertical section × 12; **l**, pistillate calyx, exterior view × 12; **m**, pistillate sepal, interior view × 12; **n**, pistillate corolla and staminodes, interior view, expanded × 12; **o**, staminodes × 12; **p**, gynoecium × 12; **q**, ovary in cross-section × 12; **r**, fruit × 1 $^1/_2$; **s**, fruit in cross-section × 1 $^1/_2$; **t**, fruit in vertical section × 1 $^1/_2$; **u, v**, endocarp in 2 views × 1 $^1/_2$; **w, x, y**, seed in 3 views × 1 $^1/_2$. *Aiphanes horrida*: **a–q**, *Read* 905; **r–y**, *Moore* s.n. (Drawn by Marion Ruff Sheehan)

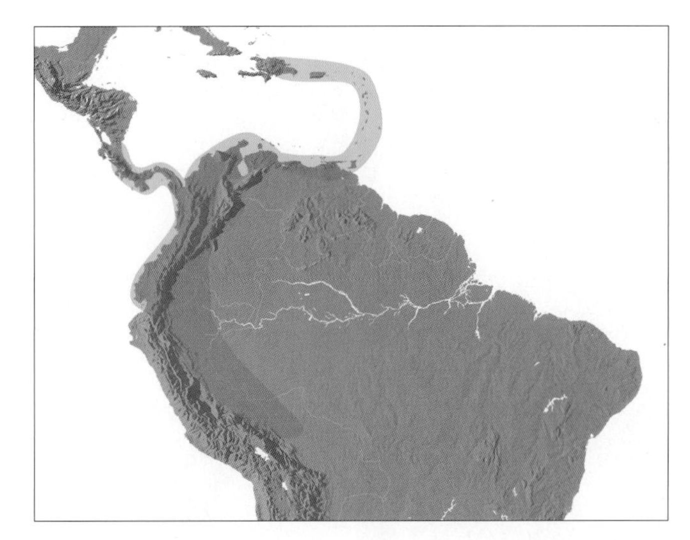

Distribution of ***Aiphanes***

trilocular, triovulate, stigmas 3, becoming reflexed at anthesis, ovules ?orthotropous, laterally attached. *Fruit* 1-seeded, ± globose, brilliant red at maturity, stigmatic remains small, apical; epicarp smooth, mesocarp thick, fleshy, fibrous, endocarp thick, very hard and woody, usually uneven, with 3, usually equatorially placed pores, surrounded by radiating fibres. *Seed* irregularly globose, basally attached, endosperm homogeneous with a central cavity; embryo lateral, opposite one of the pores. *Germination* adjacent-ligular; eophyll very shallowly to deeply bifid with praemorse tips, frequently densely spiny. *Cytology*: 2n = 30.

Distribution and ecology: Twenty-four species: a few in the West Indies, the rest in northern South America, especially diverse in Colombia. Found in a variety of habitats in the undergrowth of tropical rain forest at low elevations to montane forest.

Anatomy: Leaf (Tomlinson 1961, Borchsenius & Bernal 1996), root (Borchsenius & Bernal 1996, Seubert 1998a, 1998b), flowers (Borchsenius & Bernal 1996).

Relationships: *Aiphanes* is monophyletic with high support (Gunn 2004). For relationships, see *Acrocomia*.

Common names and uses: Ruffle palm, *coyure* (*Aiphanes acanthophylla*). Fruits of *A. horrida* are eaten; many species are very decorative.

Taxonomic account: Borchsenius & Bernal (1996).

Fossil record: Two monosulcate palm pollen types have been recovered from the Pliocene, Gatun Lake Formation, Panama (Graham 1991). The first of these types, asymmetrical and scabrate tectate, is a very common arecoid pollen type and difficult to place. It has been compared with pollen of *Manicaria*, *Reinhardtia*, *Astrocaryum* and also *Aiphanes*. However, the size range of *Aiphanes* pollen (24–34 µm, long axis) is much smaller than that of the fossil pollen (47–58 µm, long axis); *Reinhardtia* is the most probable comparison.

Top left: ***Aiphanes erinacea***, inflorescence, Ecuador. (Photo: F. Borchsenius)

Bottom left: ***Aiphanes ulei***, inflorescence, Ecuador. (Photo: F. Borchsenius)

Bottom right: ***Aiphanes chiribogensis***, staminate flowers, Ecuador. (Photo: F. Borchsenius)

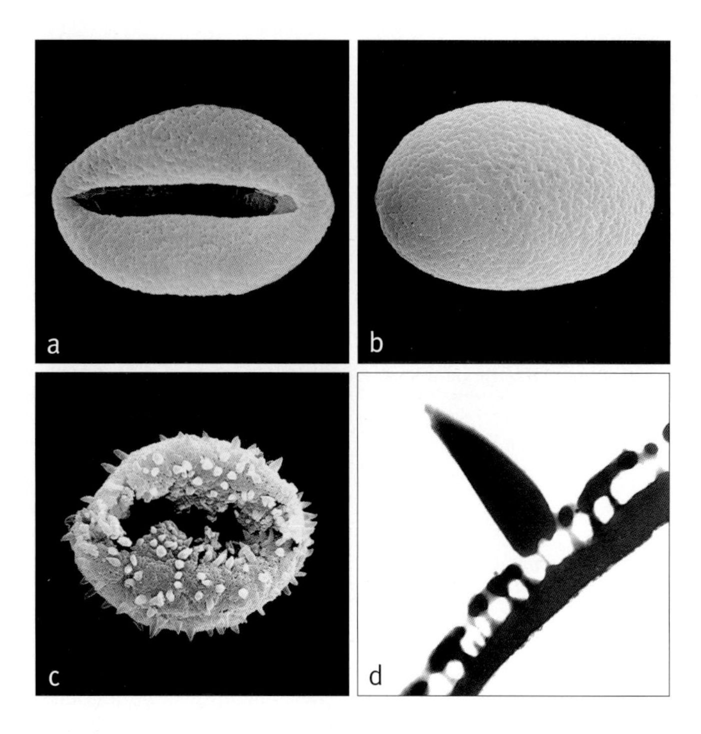

Above: ***Aiphanes***
a, monosulcate pollen grain, distal face SEM × 1500;
b, monosulcate pollen grain, proximal face SEM × 1500;
c, spiny monosulcate pollen grain, distal face SEM × 1500;
d, ultrathin section through tectate spiny pollen wall TEM × 15,000.
Aiphanes minima: **a**, *Sintenis* 2500; **b**, *Eggers* 7135; *A. hirsuta* subsp. *intermedia*: **c** & **d**, *Dransfield* JD 4854.

Notes: *Aiphanes* is unlikely to be confused with any other genus. The only other member of the Bactridinae that has praemorse leaflets is *Bactris caryotifolia*, which has a very different inflorescence with connate sepals in flowers of both sexes.
Additional figures: Glossary fig. 4.

106. BACTRIS

Extremely variable genus of spiny pinnate-leaved palms from Central and South America and the Caribbean, with almost always acute not praemorse leaflets, or entire margins. The staminate flowers are borne in triads along with the pistillate, not concentrated at the tips of the rachillae.

Bactris Jacq. ex Scop., Intro. hist. nat. 70 (1777). Lectotype: *B. minor* Jacq. (illegitimate name) (= **B. guineensis** [L.] H.E. Moore) (*Cocos guineensis* L.) (see Karsten 1857).

Guilielma Mart., Palm. fam. 21 (1824). Type: *G. speciosa* Mart. (illegitimate name) (= *G. gasipaes* [Kunth] L.H. Bailey) = *Bactris gasipaes* Kunth.

Augustinea H. Karst., Linnaea 28: 395 (1857) (non A. St.-Hil. & Naudin 1844). Type: *A. major* (Jacq.) H. Karst. (*Bactris major* Jacq.).

Pyrenoglyphis H. Karst., Linnaea 28: 607 (1857) ('1856'). Type: *P. major* (Jacq.) H. Karst. (*Bactris major* Jacq.) (substitute name for *Augustinea* H. Karst [1856], non A. St.-Hil. & Naudin 1844).

Amylocarpus Barb. Rodr., Contrib. Jard. Bot. Rio de Janeiro 3: 69 (1902) (non Currey 1859). Lectotype: *A. simplicifrons* (Mart.) Barb. Rodr. (*Bactris simplicifrons* Mart.) (see H.E. Moore 1963c).

Yuyba (Barb. Rodr.) L.H. Bailey, Gentes Herb. 7: 416 (1947). Type: *Y. simplicifrons* (Mart.) L.H. Bailey (*Bactris simplicifrons* Mart.). (Substitute name for *Amylocarpus* Barb. Rodr [1902], non Currey 1858).

Amylocarpus section *Yuyba* Barb. Rodr., Contrib. Jard. Bot. Rio de Janeiro 3: 69 (1902).

Probably derived from baktron — *stick, cane, staff, because of the slender stems of many species.*

Diminutive to large, solitary or clustered, unarmed (rarely) to very spiny, pleonanthic, monoecious palms. *Stems* subterranean and very short, to erect, very slender to moderate, with short to long internodes and, eventually, with conspicuous nodal scars, often scaly, frequently armed with short to long spines. *Leaves* pinnate or entire bifid, marcescent or neatly deciduous; sheaths usually splitting opposite the petiole, the margins smooth or becoming fibrous, unarmed to densely spiny, glabrous, scaly, hairy or bristly, a ligule-like projection sometimes also present; petiole very short to long, adaxially channelled, flat, or angled, abaxially rounded, variously unarmed to spiny; rachis usually longer than the petiole, adaxially angled except near base where channelled or not, abaxially rounded to flattened, variously armed or unarmed; blade where undivided with smooth or spiny margins, numerous ribs and an apical V-shaped notch, leaflets 1–several-fold, regularly arranged or irregularly grouped, often held in different planes within the groups, linear, lanceolate, or sigmoid, the tips very rarely praemorse (*Bactris caryotifolia*), acute or acuminate in a long drip tip, more rarely bifid or irregularly lobed, sometimes the abaxial surface covered in chalky-white indumentum, sometimes spiny along midrib on abaxial surface, the margins often bristly, blade surfaces sometimes softly hairy, midrib prominent adaxially, transverse veinlets conspicuous or obscure. *Inflorescences* interfoliar, or mostly becoming infrafoliar, solitary, spicate (rarely) or branching to 1 order, protogynous; peduncle usually relatively short, sometimes elongate, ± curved, oval in cross-section, armed or unarmed; prophyll short, tubular, 2-keeled, tightly sheathing, often concealed within the leaf sheath, usually membranous, unarmed, splitting along the abaxial face; peduncular bract inserted near the base of the peduncle, usually persistent, much longer than the prophyll, enclosing the rachillae in bud, coriaceous to woody, tightly sheathing the peduncle,

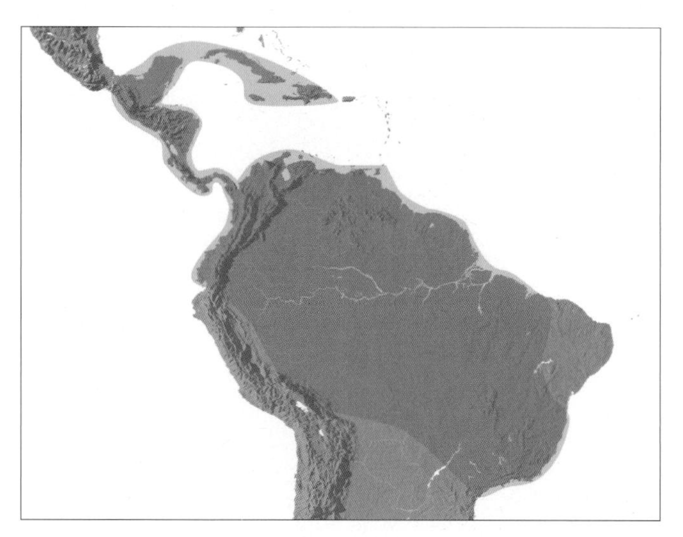

Distribution of ***Bactris***

tubular, later splitting longitudinally in distal region and often expanding and becoming boat-shaped or cowl-like, usually bearing indumentum, often bearing spines on the outer face, inner face smooth, sometimes conspicuously cream-coloured, rarely a second peduncular bract present; rachis usually shorter than the peduncle, bearing spirally arranged, rather stiff, ± glabrous, densely hairy or bristly rachillae, each subtended by an inconspicuous triangular bract; rachillae bearing spirally arranged, usually rather crowded, small triangular rachilla bracts subtending flower groups, flowers borne in triads ± throughout the rachillae, or triads scattered among paired or solitary staminate flowers ± throughout, or triads borne in proximal ca. 1/2 and solitary or paired staminate flowers distally; floral bracteoles minute. *Staminate flowers* often somewhat asymmetrical, sessile, or rarely borne on slender, unequal pedicels; calyx cupular or spreading, very short, shallowly trilobed; petals 3, fleshy, asymmetrically triangular, distally valvate, connate basally to ca. 1/2 their length and adnate basally to a fleshy floral axis; stamens (3–)6(–12), filaments slender, inflexed at the apex nearly from the middle in bud, sometimes curved, anthers usually dorsifixed, short to elongate, ± versatile, latrorse; pistillode absent. *Pollen* grains ellipsoidal, or oblate-triangular, usually with either slight or obvious asymmetry; aperture a distal sulcus or trichotomosulcus; ectexine tectate, usually, finely to coarsely rugulate, perforate and/or micro-channelled, or psilate with usually widely spaced perforations, less frequently finely perforate rugulate tectum with either supratectal spines, verrucae or gemmae, aperture margin may be slightly finer; infratectum columellate; longest axis ranges from 28–52 µm [42/73]. *Pistillate flowers* scarcely larger than the staminate; calyx annular, somewhat flattened or urn-shaped, truncate or very shallowly 3-lobed, sometimes hairy, scaly or spinulose; corolla much longer than the calyx or ± the same length, urn-shaped, truncate or very shallowly 3-lobed, variously hairy or spiny or glabrous; staminodes absent or forming a membranous ring, not adnate to the corolla; gynoecium columnar to ovoid, sometimes spiny or hairy, trilocular, triovulate, stigmas 3, very short, ovules laterally attached, orthotropous. *Fruit* usually 1-seeded, very small to large, ovoid, obpyriform, oblate, or top-shaped, yellow, red, green, brown, purple, or black; epicarp smooth, spiny, roughened or hairy, mesocarp thin to very thick, fleshy, juicy or starchy with sparse or abundant fibres, endocarp thick, bony, with 3 pores at or above the equator, sometimes with fibres radiating from the pores. *Seed* irregularly globular, basally attached, hilum circular, raphe branches sparsely anastomosing (?always) endosperm homogeneous, with or without a central hollow; embryo next to one of the endocarp pores. *Germination* adjacent-ligular; eophyll bifid or rarely pinnate, often spiny, bristly or hairy. *Cytology*: 2n = 30.

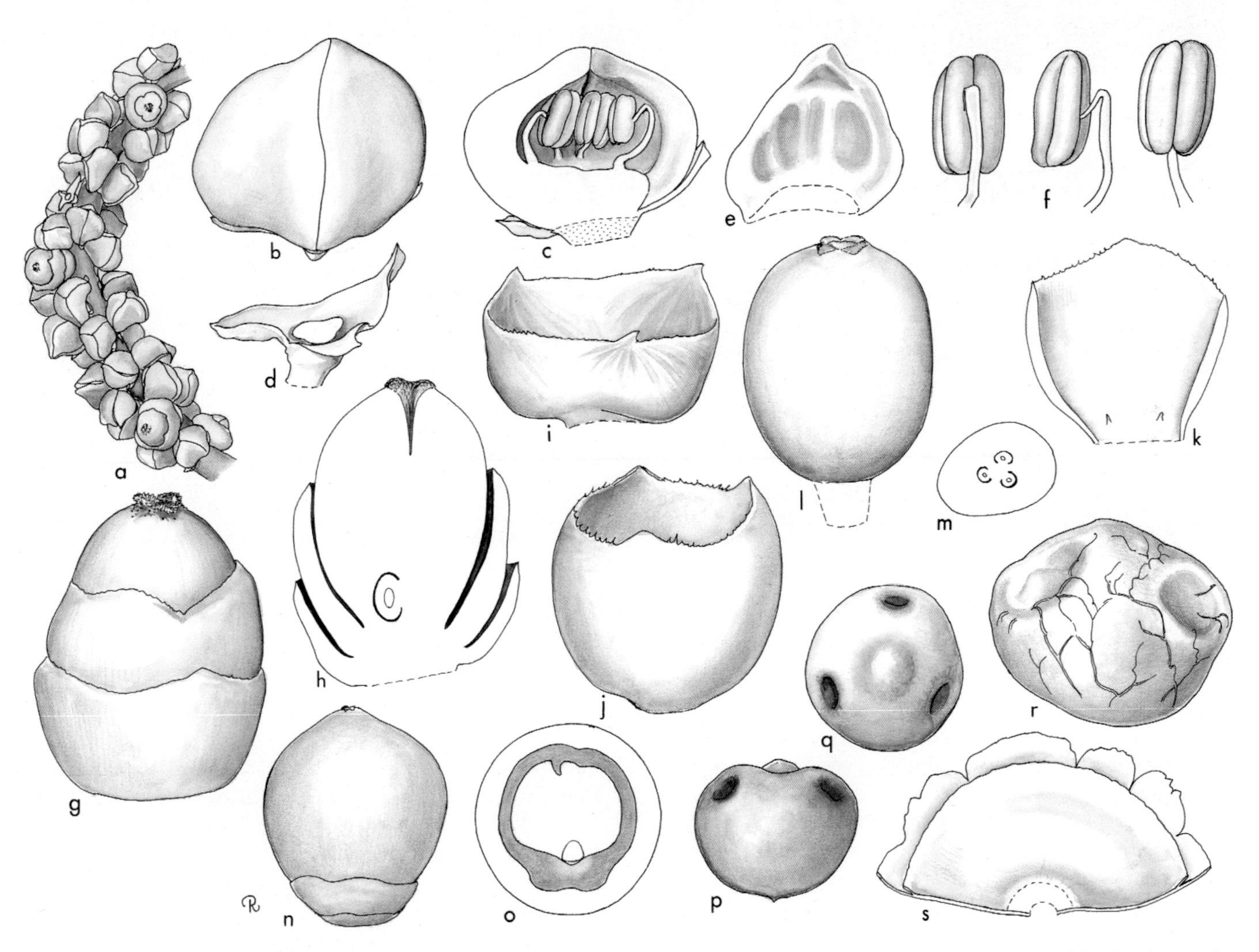

Bactris. **a,** portion of rachilla × 1 1/2; **b,** staminate bud × 9; **c,** staminate bud in vertical section × 9; **d,** staminate calyx × 9; **e,** staminate petal, interior view × 9; **f,** stamen in 3 views × 18; **g,** pistillate flower × 6; **h,** pistillate flower in vertical section × 6; **i,** pistillate calyx × 6; **j,** pistillate corolla × 6; **k,** portion of corolla of pistillate flower showing staminodes × 6; **l,** gynoecium × 6; **m,** ovary in cross-section × 6; **n,** fruit × 1 1/2; **o,** fruit in cross-section × 1 1/2; **p, q,** endocarp in 2 views × 1 1/2; **r,** seed × 3; **s,** portion of pistillate corolla, interior view, showing staminodial cupule × 2 1/4. *Bactris setulosa*: **a–r,** *Moore et al.* 9837; *B. major*: **s,** *Bailey & Bailey* s.n. (Drawn by Marion Ruff Sheehan)

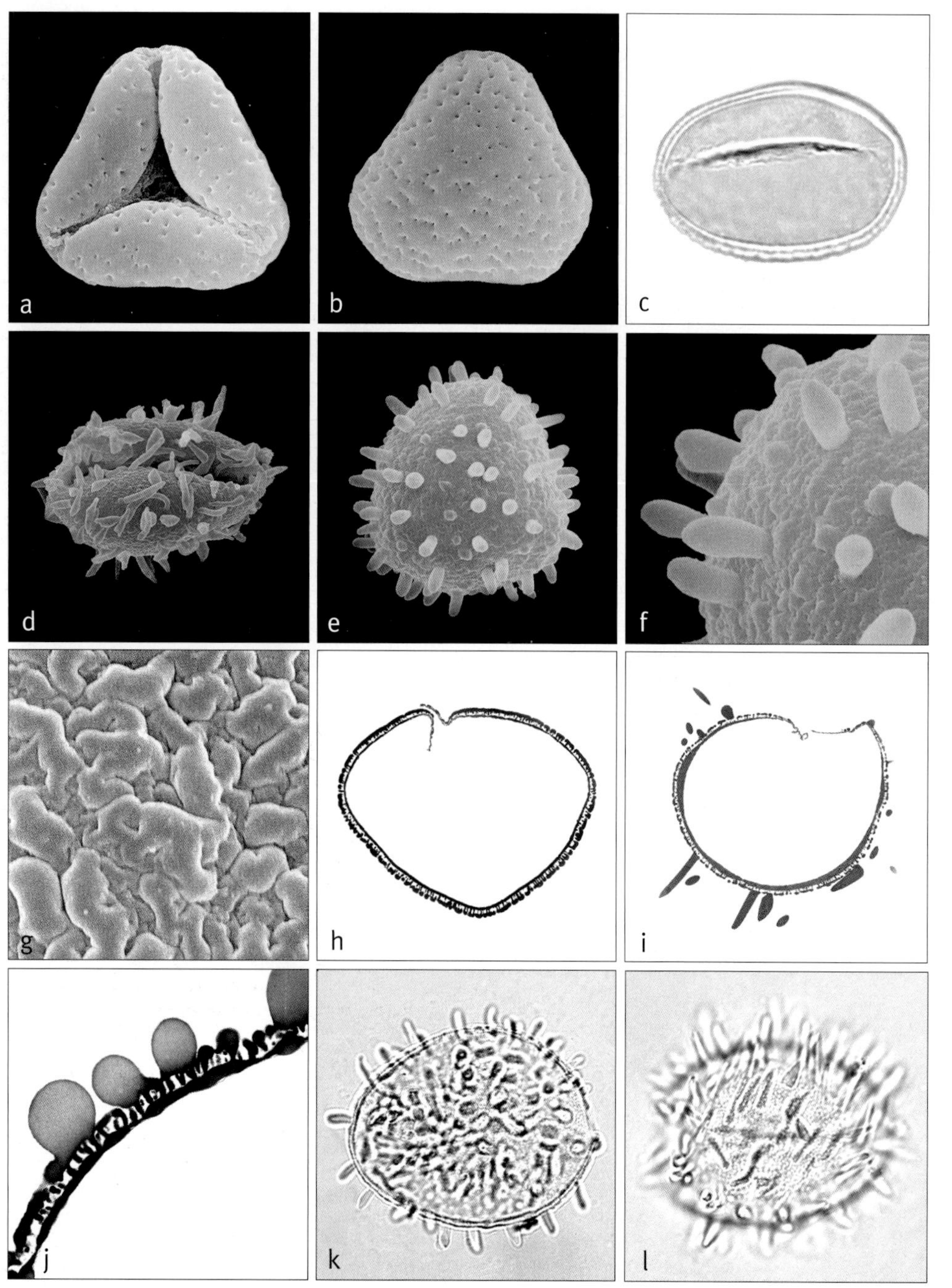

Left: ***Bactris***
a, trichotomosulcate pollen grain, distal face SEM × 1000;
b, trichotomosulcate pollen grain, proximal face SEM × 1000;
c, monosulcate pollen grain, mid-focus LM × 1000;
d, monosulcate pollen grain with long acute spines, tilted distal/equatorial face SEM × 1000;
e, trichotomosulcate pollen grain with blunt spines, proximal face SEM × 1000;
f, close-up, spiny pollen surface ('e') SEM × 2500;
g, close-up, rugulate pollen surface SEM × 6500;
h, ultrathin section through whole pollen grain, polar plane TEM × 1250;
i, ultrathin section through whole spiny pollen grain (d), polar plane TEM × 1250;
j, ultrathin section through tectate gemmate pollen wall TEM × 5000;
k, monosulcate pollen grain with blunt spines (e, f), high-focus LM × 1000;
l, monosulcate pollen grain with long acute spines (d, i), high-focus LM × 1000.
Bactris concinna: **a** & **b**, *Krukoff* 6497; *B. hirta* var. *pectinata*: **c**, *Henderson* 666; *B. longiseta*: **d, i** & **l**, *Meadland* s.n.; *B. acanthocarpoides*: **e, f** & **k**, *Balslev & Asanza* 4345; *B. caryotifolia*: **g**, *Glaziou* 8494; *B. cubensis*: **h**, *Wright* 599; *B.* sp.: **j**, *Glaziou* 8053.

Distribution and ecology: Although over 230 species have been described in the past, Henderson (2000), using a broad species concept in his recent monograph, has brought order to the genus. Seventy-seven species are currently accepted, distributed from Mexico and the West Indies south to Paraguay, with the greatest diversity in Brazil. It is not surprising that there are species of *Bactris* adapted to a very wide range of habitats in the lowlands and uplands, but the genus appears to be absent from montane forest. There are species confined to the undergrowth of tropical rain forest, others adapted to the landward fringe of mangrove, to white sand savannahs, and to freshwater swamp forest. Pollination in *Bactris*, where known, is by beetles (many studies, e.g., Essig 1971a, Beach 1984, and reviewed by Listabarth 1996).

Anatomy: Leaf morphology (Clement & Urpi 1983), anatomy (Tomlinson 1961, Roth 1990), root (Seubert 1998a, 1998b), flower (Uhl & Moore 1971, 1977a) and seed (Werker 1997).

Relationships: *Bactris* is monophyletic with high support (Couvreur *et al.* 2007). For relationships, see *Acrocomia*. For interspecies relationships, see Couvreur *et al.* (2007).

Common names and uses: Peach palm, *pejibaye*, *chonta*, *pupunha* (*Bactris gasipaes*). Economically, the most important

species is *B. gasipaes* (*Guilielma gasipaes*), which is widely cultivated and not known in the truly wild state (Mora-Urpi 1983); this species is thought to be one of the oldest of all domesticated palms and its endocarps have been found in early archaeological sites (Morcote-Rios & Bernal 2001). It produces thick mesocarp flesh that is edible and tasty after cooking, and that is sufficiently rich in nutrients and vitamins to be an important constituent of the diet of rural people. The 'cabbage' of the same species is edible and good. See also Guerrero and Clement (1982) for references on *B. gasipaes*. Other species have edible fruits and some have been used as a source of walking sticks, and for thatch and fibre.

Taxonomic account: Henderson (2000).

Fossil record: From the Middle Eocene of southeastern North America (Claiborne Group), very fragmentary pinnate fronds, some with spiny margins, are referred to *Bactrites pandanifoliolus* (Berry 1924). However, the leaves referred to *Bactrites* cannot safely be attributed to the extant genus; fossil vegetative parts, especially spines, can bear some resemblance to *Bactris* but there should be caution in equating them (Uhl & Dransfield 1987). Hollick (1928) compared fossil fruit from the Middle Oligocene of Puerto Rico, named *Bactris pseudocuesco*, with fruit of *Bactris cuesco* Engl. (= *B. corossilla*), and commented that, "the fossils can hardly be distinguished."

Notes: This genus is distinguished by staminate flowers in triads or dispersed among the triads on the rachillae and by pistillate flowers with united petals. Several attempts have been made to divide the genus into infrageneric groupings (e.g., Sanders 1991), some of which have also been elevated to generic rank. Henderson (2000) recognises none of these, preferring to use six informal groupings, which he clearly indicates as being groupings of convenience that are not necessarily natural.

Additional figures: Figs 1.1, 2.2, Glossary figs 7, 19.

Above, left to right:

Bactris gasipaes, habit, cultivated, Fairchild Tropical Botanic Garden, Florida. (Photo: C.E. Lewis);

Bactris gasipaes, fruit, cultivated, Fairchild Tropical Botanic Garden, Florida. (Photo: C.E. Lewis);

Bactris major, habit, cultivated, Fairchild Tropical Botanic Garden, Florida. (Photo: C.E. Lewis);

Bactris elegans, habit, Brazil. (Photo: A.J. Henderson)

107. DESMONCUS

Clustering spiny climbing palms of Central and South America, with reflexed acanthophylls borne on a whip at the end of the leaf.

Desmoncus Mart., Palm. fam. 20 (1824) (conserved name, but conservation superfluous). Type: **D. polyacanthos** Mart.
Atitara Kuntze, Revis. gen. pl. 2: 726 (1891). Type: *A. polyacantha* (Mart.) Kuntze (*Desmoncus polyacanthos* Mart.).

Desmos — *band*, ogkos — *hook, referring to the acanthophylls at the leaf tip*.

Slender, clustering (?always), spiny, pleonanthic, monoecious climbing palms. *Stem* covered with leaf sheaths, eventually becoming bare, with long internodes and conspicuous nodal scars, the first stem slender, not usually reaching a great height before being replaced by more robust sucker shoots (?always). *Leaves* pinnate, marcescent; sheath tubular, tightly sheathing, elongate, often tomentose and densely armed with spines in the distal exposed areas or glabrous and/or unarmed; ocrea well developed, armed or unarmed like the sheath, entire or disintegrating into a fibrous network; petiole very short to elongate, adaxially channelled, abaxially angled, usually with reflexed, bulbous-based spines; rachis elongate, usually curved, usually armed with swollen-based, reflexed spines, apically extended into a long cirrus armed with spines and pairs of small to robust, reflexed acanthophylls, acanthophylls absent on juvenile leaves, very rarely absent on adults; leaflets usually ovate, acuminate, often much narrowed at the base into a brief stalk, rather distant, ± regularly arranged or grouped, thin to coriaceous, with a conspicuous midrib and several more slender lateral veins, in *Desmoncus cirrhiferus* the main rib extended into a long flexuous tendril, margins smooth or armed with short spines, the main rib sometimes bearing spines, indumentum sometimes present in bands and along veins, transverse veinlets sometimes conspicuous. *Inflorescences* interfoliar, emerging through the leaf sheath mouths, branching to 1 order, becoming ± pendulous, apparently protandrous; peduncle elongate, slender, semicircular in cross-section; prophyll inserted some distance above the base of the peduncle, thinly coriaceous, 2-keeled, tubular, splitting longitudinally on the abaxial face and tattering, only partially exserted, persistent; peduncular bract 1, longer than and inserted far above the prophyll, thick, coriaceous to subwoody, tubular, enclosing the rachillae in bud, later splitting longitudinally, ± persistent, variously unarmed or spiny, adaxially smooth, often pale cream at anthesis, tomentose or ± glabrous abaxially; rachis shorter than the peduncle, bearing few to numerous, ± spirally arranged, flexuous, slender, short to elongate, often somewhat zig-zag rachillae, each subtended by a minute, triangular bract; rachillae very few to numerous, bearing rather distant, spiral, or subdistichous triads except in the distal ca. $^{1}/_{3}$–$^{1}/_{5}$ where bearing paired or solitary staminate flowers, each flower group subtended by an inconspicuous triangular bract; bracteoles minute. *Staminate flowers* somewhat asymmetrical; calyx cupular, short, ± membranous with 3, low or acuminate, triangular lobes; petals 3, distinct, ± fleshy, ovate-lanceolate, much exceeding the calyx, acute or acuminate; stamens 6–9, filaments irregularly adnate to the petals, the free portion very short or moderate, very slender at the tip, anthers ± rectangular,

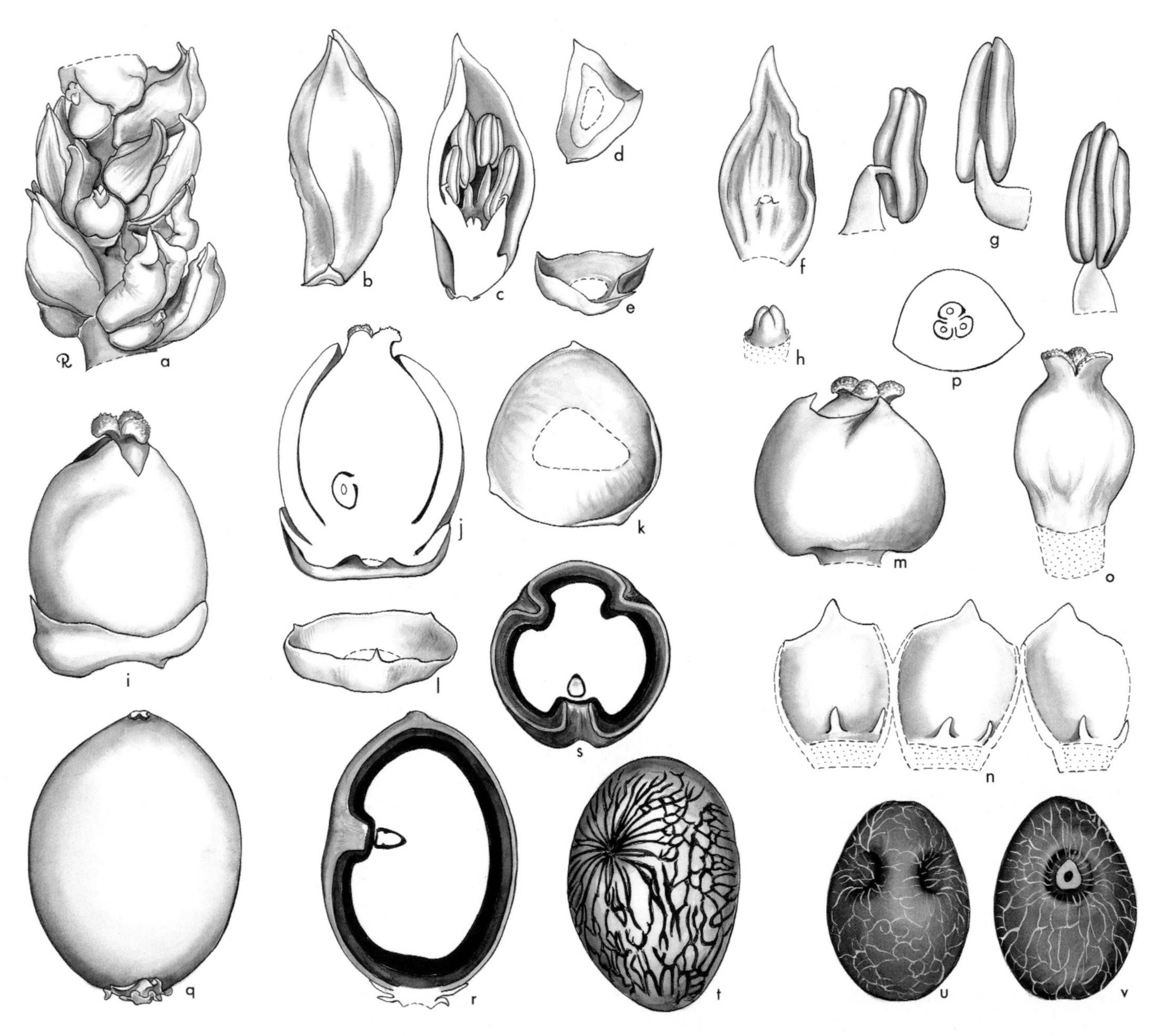

Desmoncus. **a**, portion of rachilla with triads × 3; **b**, staminate bud × 6; **c**, staminate bud in vertical section × 6; **d, e**, staminate calyx in 2 views × 12; **f**, staminate petal, interior view × 6; **g**, stamen in 3 views × 12; **h**, pistillode × 12; **i**, pistillate flower × 12; **j**, pistillate flower in vertical section × 12; **k, l**, pistillate calyx in 2 views × 12; **m**, pistillate flower, calyx removed × 12; **n**, pistillate corolla and staminodes, interior view, expanded × 12; **o**, gynoecium × 12; **p**, ovary in cross-section × 12; **q**, fruit × 3; **r**, fruit in vertical section × 3; **s**, fruit in cross-section × 3; **t**, endocarp × 3; **u, v**, seed in 2 views × 3. *Desmoncus* spp.: **a–p**, *Moore & Parthasarathy* 9471; **q–v**, *C. J. Miller* s.n. (Drawn by Marion Ruff Sheehan)

basifixed, sagittate at the base, latrorse; pistillode minute, conical, or absent. *Pollen* ellipsoidal, usually with slight asymmetry; aperture a distal sulcus; ectexine tectate, finely perforate, perforate and micro-channelled, and rugulate, aperture margin may be slightly finer; infratectum columellate; longest axis 19–41 µm [5/12]. *Pistillate flowers* ± globular or ovoid, usually smaller than or equalling the staminate; calyx cupular or tubular, sometimes ± flattened, ± membranous, very briefly trilobed; corolla much exceeding the calyx, tubular, ± membranous, shallowly trilobed or truncate, sometimes minutely ciliate along the margins; staminodes 6, minute, tooth-like, epipetalous; gynoecium ovoid or columnar, trilocular, triovulate, only slightly exceeding the corolla, stigmas 3, fleshy, reflexed, ovule laterally attached, ?orthotropous. *Fruit* 1-seeded, ± ovoid or spherical, bright red, deep purple, or black, with apical stigmatic remains; epicarp smooth, mesocarp thin, fleshy, endocarp stony with 3 pores slightly distal to the equator. *Seed* ovoid, with 3 surface hollows, basally attached, hilum basal, circular, raphe branches densely anastomosing, endosperm homogeneous; embryo lateral. *Germination* adjacent-ligular; eophyll bifid with rather broad, acute segments or pinnate (2 pairs of leaflets in *D. costaricensis*). *Cytology*: 2n = 30.

Distribution and ecology: Sixty-one species have been described but there are probably far fewer. Henderson *et al.* (1995) accept only seven species. *Desmoncus* is distributed from Mexico southwards to Brazil and Bolivia, and is absent from the West Indies except for Trinidad. Most species are palms of the lowlands, often found in open areas, swamps, on riverbanks, and more rarely in the undergrowth of tropical rain forest.

Anatomy: Leaf (Tomlinson 1961), stem (Fisher & French 1976, Tomlinson & Zimmermann 2003), and root (Seubert 1998a, 1998b).

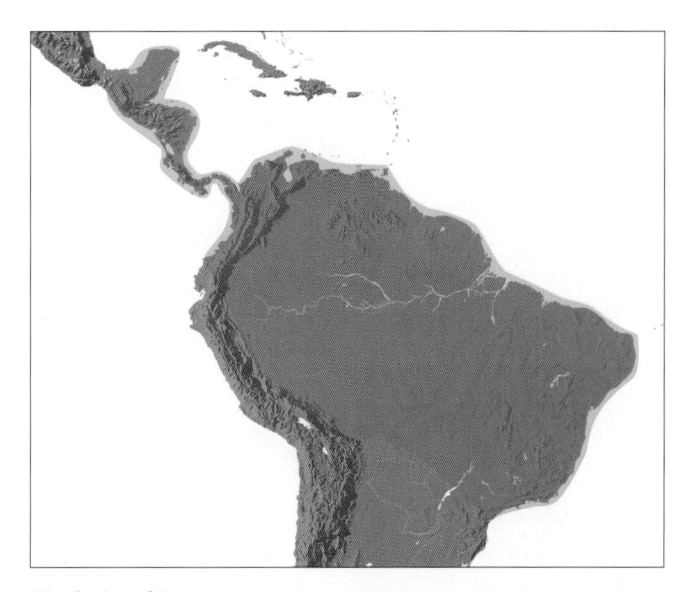

Distribution of ***Desmoncus***

Below left: ***Desmoncus polyacanthos***, inflorescences, Brazil. (Photo: A.J. Henderson)

Below middle: ***Desmoncus polyacanthos***, fruit, Brazil. (Photo: A.J. Henderson)

Below right: ***Desmoncus cirrhiferus***, fruit, Peru. (Photo: J.-C. Pintaud)

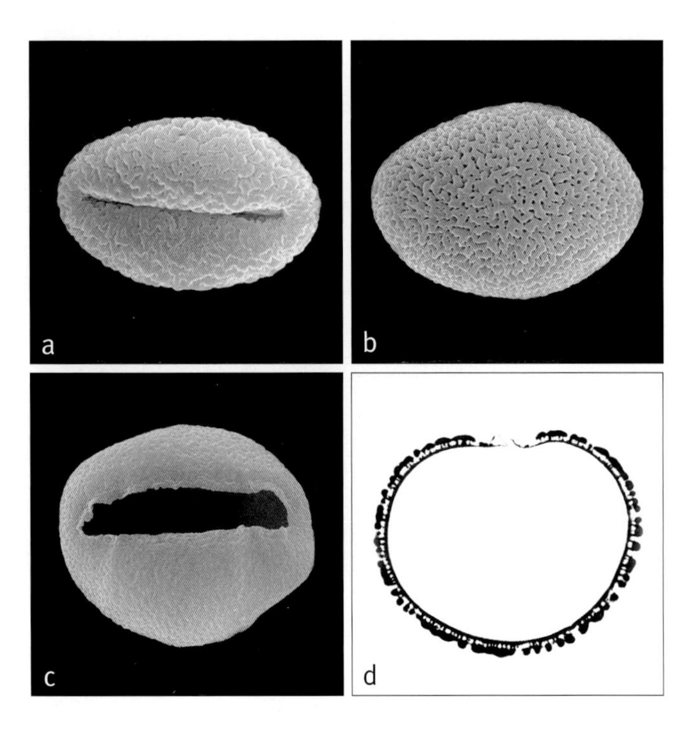

Above: ***Desmoncus***
a, monosulcate pollen grain, distal face SEM × 2000;
b, monosulcate pollen grain, proximal face SEM × 2000;
c, monosulcate pollen grain, distal face SEM × 2000;
d, ultrathin section through whole pollen grain, polar plane TEM × 2500.
Desmoncus phoenicocarpus: **a, b** & **d**, *Harley & Souza* 11230; **c**, *Harley & Souza* 10346.

Relationships: The monophyly of *Desmoncus* has not been tested. For relationships, see *Acrocomia*.
Common names and uses: For common names, see Glassman 1972. Locally, *Desmoncus* species may provide cane for cordage or rarely for inexpensive furniture; they are not, however, utilised to the same extent as the Asiatic rattans. See also Hübschmann *et al.* 2007.
Taxonomic account: Taxonomy of this genus is in disarray. A critical revision is greatly needed. In the interim, see Henderson *et al.* (1995).
Fossil record: No generic records found.
Notes: The genus represents the New World counterpart of the Asiatic and African calamoid rattans. The parallel development of grapnel acanthophylls on the cirrus of *Desmoncus* and of the calamoid genera *Eremospatha*, *Oncocalamus* and *Laccosperma* is a remarkable parallel development. Floral biology has been studied by Listabarth (1994).
Additional figures: Glossary figs 6, 14.

Subtribe Elaeidinae Hook.f. in Benth. & Hook.f., Gen. pl. 3: 873, 882 (1883). Type: *Elaeis*.

Moderate to large; petioles armed with fibre spines or not, leaves pinnate, leaflets acute; peduncular bract fibrous or long, woody, borne at the base of the peduncle; flowers in deep pits; pistillate flowers with imbricate petals; fruit usually 1-seeded; endocarp pores at or above the middle; endosperm homogeneous.

Two genera make up this subtribe: *Barcella*, which is confined to central Amazonia, and *Elaeis*, which has two species, one in Central and South America, the other in tropical Africa. The subtribe is distinguished by the presence of pits, by the stamen filaments connate in a tube and by the relatively massive stigmas.

Subtribe Elaeidinae is monophyletic with high support (Hahn 2002b, Baker *et al.* in review) and is sister to the Bactridinae (Asmussen *et al.* 2006, Baker *et al.* in review, in prep.). In the study of Gunn (2004), however, the Elaeidinae is not monophyletic, *Elaeis* being strongly supported as sister to the Bactridinae but *Barcella* resolving with low support as sister to all remaining Cocoseae. This outcome is likely to be anomalous and may not accurately reflect relationships.

Key to Genera of Elaeidinae

1. Flowers of both sexes borne on the same inflorescence, the pistillate near the bases of rachillae; inflorescence long-peduncled, laxly branched. Brazil (Paduairi River). . . . . **108. Barcella**
1. Staminate and pistillate flowers normally in separate inflorescences on the same plant; inflorescence short-peduncled, densely branched. Costa Rica to Brazil (*Elaeis oleifera*) or West Africa (*E. guineensis*) . . . . . . **109. Elaeis**

108. BARCELLA

Acaulescent unarmed pinnate-leaved palm from Amazonia, with flowers borne in pits.

Barcella (Trail) Trail ex Drude in Mart., Fl. bras. 3(2): 459 (1881). Type: **B. odora** (Trail) Trail ex Drude (*Elaeis odora* Trail).
Elaeis subgenus *Barcella* Trail, J. Bot. 15: 80 (1877).

Barca — *little boat*, -ella — *diminutive, perhaps in reference to the shape of the peduncular bract.*

Moderate, solitary, acaulescent, unarmed, pleonanthic, monoecious palm. *Stem* very short, subterranean. *Leaves* arching or erect, pinnate, marcescent; sheath abaxially with caducous tomentum; petiole short, adaxially flattened, abaxially ± angled, the margins sharp, both surfaces

Opposite: ***Barcella.*** **a**, portion of rachilla in staminate bud × 3; **b**, portion of rachilla, staminate buds removed to show pits × 6; **c**, vertical section of rachilla to show position of staminate bud in pit × 6; **d**, staminate bud × 12; **e**, staminate bud in vertical section × 12; **f**, staminate sepals in 2 views × 12; **g**, staminate bud, sepals removed × 12; **h**, staminate petal, interior view × 12; **i**, stamen in 3 views × 12; **j**, pistillode × 24; **k**, portion of rachilla with pistillate flowers and staminate pits × 3; **l**, pistillate flowers × 3; **m**, vertical section of rachilla to show position of pistillate flower in pit × 3; **n**, pistillate flower × 3; **o**, pistillate sepal × 3; **p**, pistillate petals × 3; **q**, gynoecium and staminodial cupule × 3, **r**, ovary in cross-section × 6; **s**, fruit in 2 views × 1; **t**, fruit in vertical section × 1; **u**, fibrous mesocarp × 1; **v**, endocarp in 4 views × 1; **w**, endocarp in vertical section × $1^1/_2$; **x**, seed in 3 views × 3; **y**, seed in vertical section × 3. *Barcella odora*: **a–r**, *Rodrigues & Coelho* 8394; **s–y**, *Trail* s.n. (Drawn by Marion Ruff Sheehan)

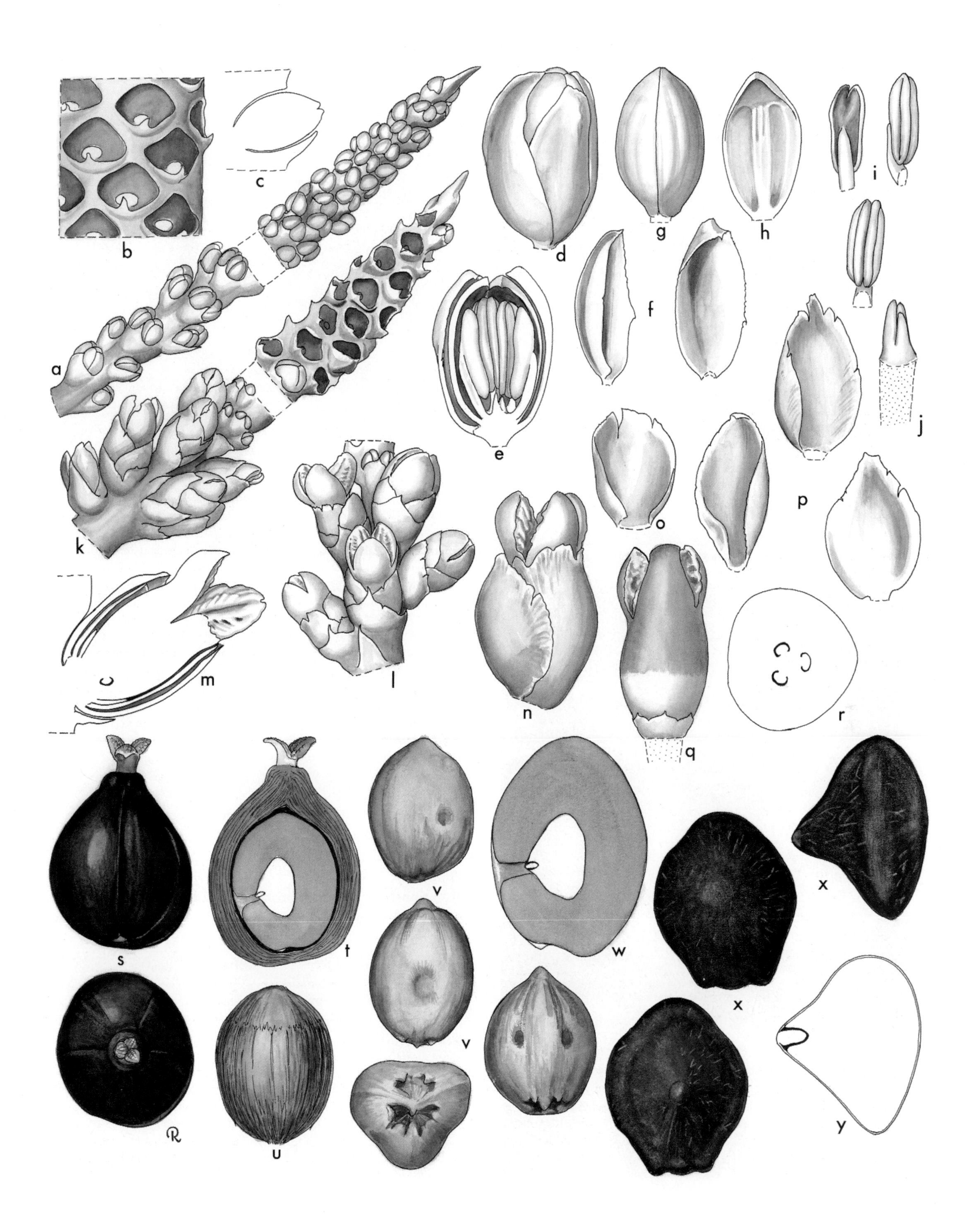
a
b
c
d
e
f
g
h
i
j
k
l
m
n
o
p
q
r
s
t
u
v
w
x
y

glabrous; rachis adaxially angled, abaxially flattened; leaflets numerous, regularly arranged, single-fold, very gradually narrowed from near the base to a long acuminate tip, ± plicate, adaxially glabrous, abaxially bearing a few small ramenta along the main vein near the base, transverse veinlets conspicuous, very short. *Inflorescences* interfoliar, solitary, branching to 1 order, apparently protandrous; peduncle elongate, curving, ± circular or semicircular in cross-section, densely tomentose; prophyll short, thinly coriaceous, 2-keeled, acute, becoming fibrous distally; peduncular bract much longer than the prophyll, ± woody, enclosing the inflorescence in bud, splitting longitudinally allowing elongation of the peduncle and rachis, beaked, striate abaxially, covered with minute scales; rachis much shorter than the peduncle, bearing up to ca. 40, crowded, rigid, spirally arranged, densely tomentose rachillae, each subtended by a triangular, acuminate bract; proximal few rachillae expanded at the base and bearing up to ca. 9 pistillate flowers, solitary or in triads, with staminate flowers distal and all other rachillae bearing dense spirals of paired staminate flowers, except at the very tips where staminate flowers solitary, the pistillate flowers ± superficial, each subtended by a triangular bract, the staminate flowers, especially distally, sunken in pits, the small triangular rachilla bracts forming the lower lips of the pits, extreme tips of rachillae bare of bracts and flowers, ± sharply pointed. *Staminate flowers* small; sepals 3, distinct, widely separated, membranous, narrow elliptic, abaxially keeled, hooded at the tips; petals 3, distinct, about the same length as the sepals, valvate, ovate, striate; stamens 6, filaments broad, fleshy, united laterally to each other to form a tube with 6, short, distinct, reflexed, abruptly narrowed tips, anthers ± rectangular, ± versatile, latrorse; pistillode much shorter than the connate filaments, columnar, trifid. *Pollen* asymmetric-ellipsoid to pyriform; aperture a distal sulcus; ectexine tectate, perforate and micro-channelled and coarsely rugulate, aperture margin finer; infratectum columellate; longest axis 25–45 µm [1/1]. *Pistillate flowers* much larger than the staminate; sepals 3, distinct, ovate, broadly imbricate, coriaceous, the margins tending to split irregularly; petals 3, distinct, ovate, broadly imbricate except at the short valvate, triangular tips, ± coriaceous except at the membranous, striate, irregular margins; staminodes forming a membranous, 6-toothed ring; gynoecium ± columnar, trilocular, triovulate, with 3, massive, fleshy, 3-angled stigmas, ovule apparently orthotropous, attached centrally. *Fruit* moderately large, 1-seeded, ovoid or basally angled by close packing, bright orange, usually with a prominent apical beak bearing the stigmatic remains; epicarp smooth, glabrous, mesocarp thick, fleshy, oily, endocarp black, woody, ± 3-angled, traversed by longitudinal fibres, with 3 lateral pores (according to Trail [1877]). *Seed* (?)basally attached, with a coarse network of fibres, endosperm homogeneous with a central cavity; embryo lateral near one of the endocarp pores. *Germination* not recorded; eophyll apparently entire, lanceolate. *Cytology* not studied.

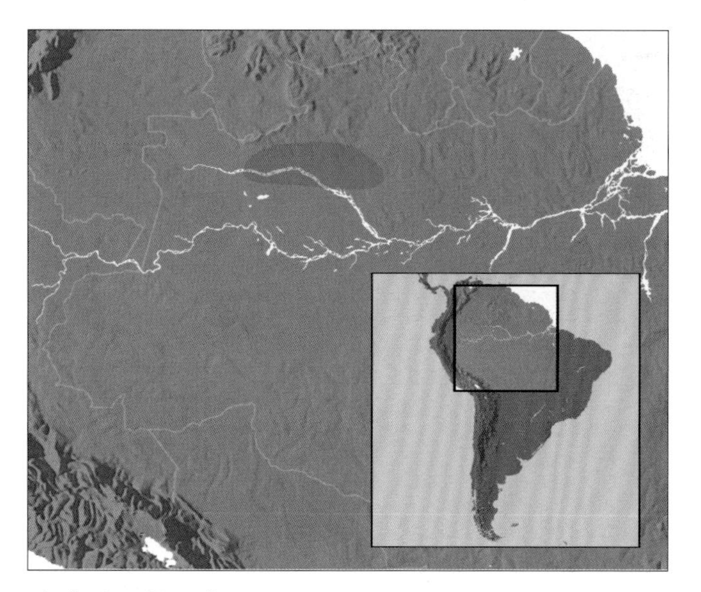

Distribution of ***Barcella***

Left: ***Barcella odora***, habit, Brazil. (Photo: A.J. Henderson)

Distribution and ecology: A single species confined to a small area of the banks of the Rio Negro and its tributaries in Brazil, occurring on river banks at low altitude.
Anatomy: Not studied.
Relationships: For relationships, see subtribe Elaeidinae.
Common names and uses: Common names not recorded. Many uses such as wood for construction and leaves for weaving, but the greatest potential of this palm lies with possible hybridization with the African or American species of *Elaeis* to introduce novel characteristics.
Taxonomic accounts: Trail (1877) and Henderson (1986b).
Fossil record: No generic records found.
Notes: *Barcella* and *Elaeis* are separated from other Cocoseae by having the rather large pistillate flowers deeply sunken in

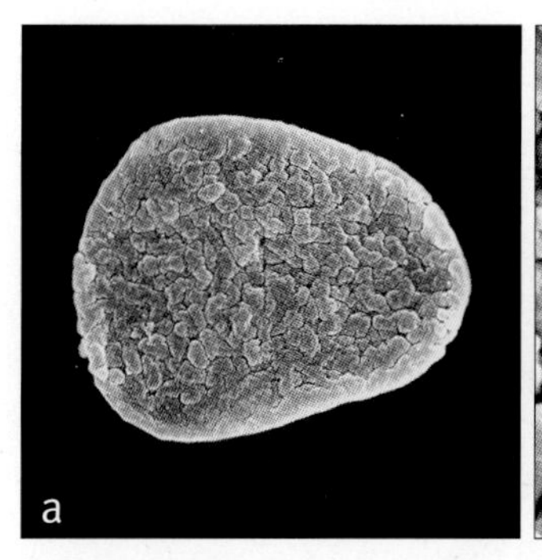

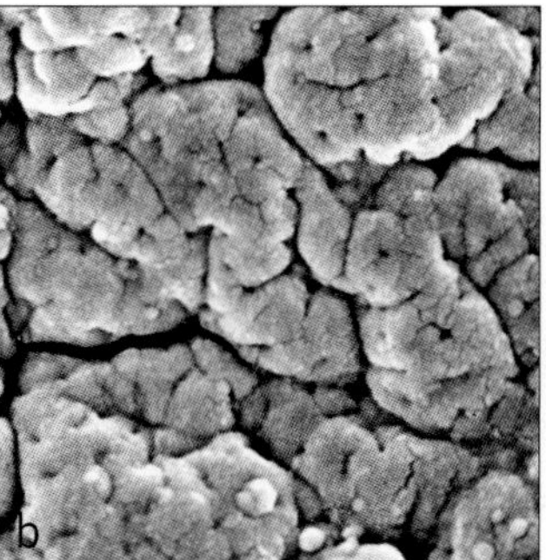

Above: ***Barcella***
a, monosulcate pollen grain, proximal face SEM × 1500;
b, close-up, coarsely rugulate pollen surface SEM × 8000.
Barcella odora: **a** & **b**, *Cordeiro et al.* 32.

Right: ***Barcella odora***, inflorescence, Brazil. (Photo: A.J. Henderson)
Far right: ***Barcella odora***, fruit, Brazil. (Photo: A.J. Henderson)

the rachillae and by endocarp pores at or above the middle. *Barcella* is distinguished from *Elaeis* by the lack of spines on the petiole and by the long peduncle, by having a woody rather than a fibrous peduncular bract, and by the consistent presence of both staminate and pistillate flowers in the inflorescence.

109. ELAEIS

Solitary pinnate-leaved palms from South and Central America and humid Tropical Africa, including the African oil palm of commerce, distinctive in fibre spines and spines formed from leaflet midribs at the base of the leaf, and highly condensed unisexual inflorescences borne among the leaf bases, both male and female borne on the same tree.

Elaeis Jacq., Select. stirp. amer. hist. 280 (1763). Type: **E. guineensis** Jacq.

Corozo Jacq. ex Giseke, Prael. ord. nat. pl. 42, 92 (1792). Lectotype: *C. oleifera* (Kunth) L.H. Bailey (*Alfonsia oleifera* Kunth) (see Bailey 1940) (= *Elaeis oleifera* [Kunth] Cortes).

Alfonsia Kunth in Humboldt, Bonpland & Kunth, Nov. gen. sp. pl. 1: folio edition 245; quarto edition 306 (1816). Type: *A. oleifera* Kunth (= *Elaeis oleifera* [Kunth] Cortes).

Elaia — *olive tree, olive, referring to the oil of the oil palm.*

Moderate to robust, solitary, short to tall, armed, pleonanthic, monoecious palms. *Stem* procumbent or erect, bearing persistent leaf bases, eventually becoming bare, the internodes short, leaf scars wide, oblique. *Leaves* many in the crown, pinnate, withering and not abscising neatly except in tall-trunked individuals; sheath tubular at first, later disintegrating into an interwoven mass of fibres, those fibres attached to the base of the petiole remaining as regularly spaced, broad, flattened spines; petiole conspicuous, adaxially channelled, abaxially angled, bearing caducous tomentum, the margins armed with regularly spaced fibre spines, distally (strictly speaking the proximal part of the rachis) with margins armed with short, triangular, bulbous-based spines representing the pulvini and midribs of the proximal few vestigial leaflets, the blades of which soon disintegrate on leaf expansion; rachis curving or straight, adaxially angled, abaxially curved or flattened; leaflets numerous, single-fold, regularly arranged or slightly grouped and held in different planes, giving the whole leaf a plumose appearance, linear, gradually tapering to acute tips, sometimes with bands of caducous scales, midribs prominent, transverse veinlets very short, inconspicuous. *Inflorescences* interfoliar, solitary, short and condensed, unisexual (except as monstrosities), usually several adjacent axils producing inflorescences of one sex followed by several producing the other sex, branching to 1 order; peduncle short, ± elliptic in cross-section; prophyll short, tubular and flattened, 2-keeled, tomentose, included within the subtending leaf sheath, thick, traversed by numerous, thick, longitudinal fibres, disintegrating distally into a mass of fibres, the larger fibres spine-like; first peduncular bract inserted some distance from the prophyll, tubular, fibrous, thinner than the prophyll, distally disintegrating into a fibrous mass, and splitting longitudinally, subsequent peduncular bracts small, not sheathing, narrow triangular, with sharp tips, striate; rachis shorter than, ± equalling, or slightly longer than the peduncle, tomentose, bearing numerous, spirally arranged, narrow triangular, membranous to coriaceous, acute bracts, each subtending a rachilla; staminate rachillae ± cylindrical, catkin-like, often somewhat angled due to close packing, tomentose, densely floriferous except at the ± spine-like tip where bare of flowers and bracts, the flowers solitary, borne in deep, spirally arranged pits, pistillate rachillae more massive than the staminate, bearing fewer flowers, the tips prolonged into a woody spine, each rachilla proximally bearing lax, ± superficial or only partially sunken, spirally arranged membranous rachilla bracts; bracts short, acute, or prolonged into a straight or flexuous spine-like tip, each subtending a solitary flower. *Staminate flowers* small, only slightly protruding from the pits at anthesis; sepals 3, distinct, unequal, ± rectangular, membranous, the edges not meeting in bud, abaxially keeled; petals 3, distinct, ± ovate, ± equalling the sepals, valvate, very thin; stamens 6, exserted at anthesis, filaments broad, fleshy, united laterally to form a tube, with 6 short, distinct, reflexed, abruptly narrowed tips, anthers ± rectangular,

Elaeis. **a**, terminal portion of staminate rachilla × 3; **b**, portion of staminate axis, flowers removed to show pits formed by bracts × 3; **c**, staminate flower × 6; **d**, staminate flower in vertical section × 6; **e**, staminate sepal × 6; **f**, staminate petal × 6; **g**, androecium × 6; **h**, androecium, expanded to show pistillode × 6; **i**, stamen in 3 views × 12; **j**, terminal portion of pistillate rachilla × 1; **k**, pistillate flower × 3; **l**, pistillate flower and portion of rachilla in vertical section × 3; **m**, pistillate bracteole × 3; **n**, pistillate sepal × 3; **o**, pistillate petal × 3; **p**, staminodial ring × 3; **q**, gynoecium × 3; **r**, gynoecium in cross-section × 6; **s**, fruit × $1^1/_2$; **t**, fruit in vertical section × $1^1/_2$; **u**, fruit in cross-section × $1^1/_2$; **v**, endocarps showing variation in shape × $1^1/_2$; **w, x, y, z**, seed in 4 views × 3. *Elaeis oleifera*: **a–r**, *Read* 1388; **s–z**, *Lindsay* s.n.
(Drawn by Marion Ruff Sheehan)

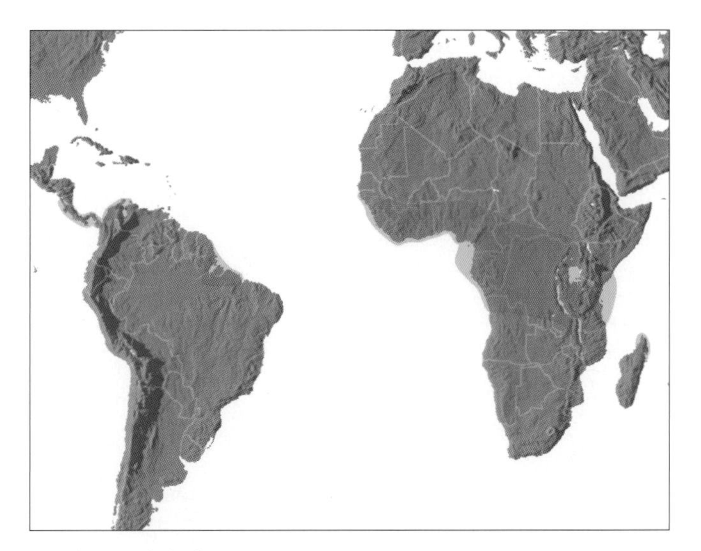

Distribution of ***Elaeis***

Anatomy: Leaf (Tomlinson 1961), phloem (Parthasarathy 1980), root (Seubert 1998a, 1998b), stegmata (Killmann & Hong 1989) and gynoecium (Uhl & Moore 1971).

Relationships: *Elaeis* is monophyletic with high support (Gunn 2004). For relationships, see subtribe Elaeidinae.

Common names and uses: African oil palm (*Elaeis guineensis*), American oil palm (*E. oleifera*). *Elaeis guineensis* is the most important commercial oil-producing plant in the tropics, and is used locally as a source of wine, thatch and building materials. Even waste endocarp has been used as road metalling. For further details, see Hartley (1988), and for references, Johnson (1983b).

Taxonomic accounts: Beccari (1914b) and Zeven (1967).

± versatile, introrse; pistillode columnar, trifid, slightly shorter than the staminal tube. *Pollen* either ellipsoidal, slight to obvious asymmetry (*Elaeis oleifera*), or oblate triangular (*E. guineensis*); aperture a distal sulcus or trichotomosulcus; ectexine perforate scabrate or perforate rugulate, aperture margin (ellipsoid pollen) similar, aperture margin (trichotomosulcate pollen) broad and psilate or psilate-perforate; infratectum columellate; longest axis ranges from 31–39 µm [2/2]. *Pistillate flowers* much larger than the staminate, borne with 2 acute or spine-tipped bracteoles; sepals 3, distinct, imbricate, rather thin; petals 3, distinct, imbricate, rather thin; staminodal ring low, 6-pointed, tanniniferous; gynoecium columnar to ovoid, trilocular, triovulate, stigmas 3, fleshy, reflexed, ± 3-angled, ovules orthotropous, attached centrally. *Fruit* 1–(rarely more)-seeded, ± ovoid but basally angled by close packing, variously orange or yellow, overlain with deep violet or black in exposed parts, apically beaked, stigmatic remains apical; epicarp smooth, mesocarp thick, fleshy, oily, fibrous, endocarp black, woody and very hard, variously ovoid, flattened or angled, with 3 apical pores. *Seed* basally attached with coarse, reticulate raphe branches, endosperm homogeneous, with or without a central cavity; embryo ± apical, opposite a pore. *Germination* adjacent-ligular; eophyll entire, lanceolate. *Cytology*: 2n = 32.

Distribution and ecology: Two species. *Elaeis guineensis* is native to the more humid areas of tropical Africa, possibly introduced in Madagascar, now widely cultivated throughout the humid tropics as the most productive perennial oil crop, and frequently naturalised. In the wild, it occurs on the margins of humid forest and along watercourses in drier areas. *Elaeis oleifera* is native to central and northern South America, and is frequent on poorly drained, sandy soils and in savannas. In Costa Rica, it is found in palm swamp and some mangrove communities (Allen 1956).

Top right: ***Elaeis guineensis***, habit, Tanzania. (Photo: J. Dransfield)

Bottom left: ***Elaeis guineensis***, leaf bases and infructescence, Tanzania. (Photo: J. Dransfield)

Bottom right: ***Elaeis guineensis***, infructescence, Tanzania. (Photo: J. Dransfield)

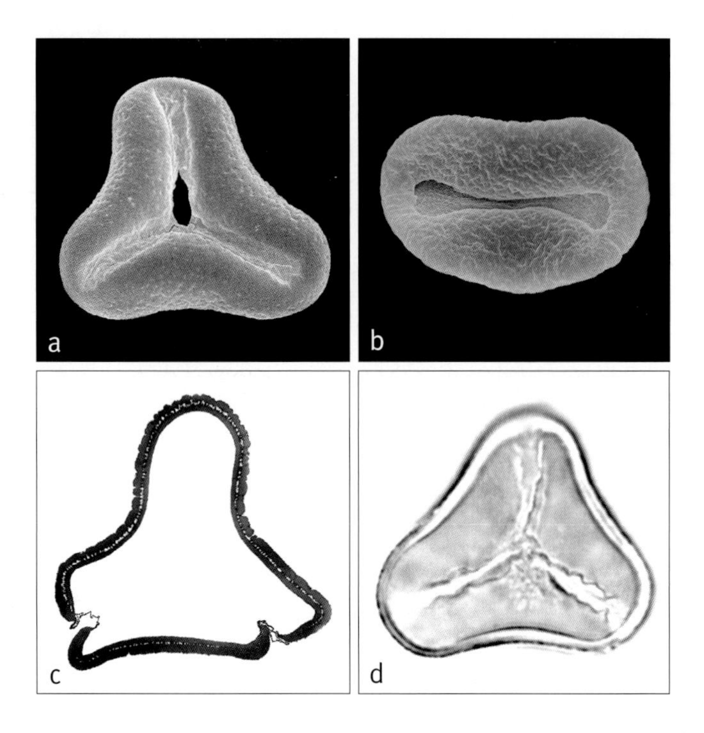

Above: ***Elaeis***
a, trichotomosulcate pollen grain, distal face SEM × 1500;
b, monosulcate pollen grain, distal face SEM × 1500;
c, ultrathin section through trichotomosulcate pollen grain, slightly oblique equatorial plane TEM × 1500;
d, trichotomosulcate pollen grain, low-focus LM × 1500.
Elaeis guineensis: **a**, *Mags* 3, **c** & **d**, *Mann* 1869; *E. oleifera*: **b**, *Cortes, Henderson & Pardini* 1508.

Fossil record: An endocarp from the Middle Oligocene of Puerto Rico, *Palmocarpon rabellii*, is cautiously compared both with the endocarp of *Elaeis guineensis* and, due to evidence of pores, with *Copernicia cerifera* (Hollick 1928). In fact, the endocarps of these two palms are very dissimilar; clearly this record should be reassessed. Pollen from the Zinguinchor borehole (Middle Eocene to Lower Miocene) of Senegal shows notable similarity to that of *E. guineensis* (Médus 1975). Ergo (1997) describes seeds of *Elaeis* from the Upper Micene of Uganda.

Notes: *Elaeis* is notable as one of only two genera of the Cocoseae present in Africa and also for its distribution on either side of the Atlantic. See further notes under *Barcella*.

Elaeis guineensis is the focus of ongoing evolutionary development research (Adam *et al.* 2005, Adam *et al.* 2006, Adam *et al.* 2007, Jouannic *et al.* 2005)

Additional figures: Glossary figs 7, 20.

Tribe Manicarieae J. Dransf., N.W. Uhl, C. Asmussen, W.J. Baker, M.M. Harley & C. Lewis., Kew Bull. 60: 562 (2005). Type: **Manicaria**.

Moderate, acaulescent to short stemmed, unarmed; leaves large, entire or irregularly pinnate, margins praemorse, ribs parallel; sheaths not forming a crownshaft; inflorescence bisexual; prophyll and peduncular bracts beaked, net-like and enclosing the inflorescence; triads of flowers somewhat sunken in the rachilla axis; petals of pistillate flower valvate; gynoecium trilocular, triovulate; fruit 1–3-seeded, rounded when 1-seeded, otherwise lobed, stigmatic remains basal; seeds 1–3 (1 per lobe); epicarp corky-warted; endocarp with a basal button; embryo basal.

The single genus of the tribe, *Manicaria*, occurs in freshwater swamps in central and northern South America.

The Manicarieae resolves in various places within the core arecoid clade. It was found to be sister to the Geonomateae with low support in Baker *et al.* (in review) and in the RFLP tree of Uhl *et al.* (1995). A relationship to the Geonomateae is also indicated by Asmussen *et al.* (2006), who resolved the Manicarieae as sister to a subclade of the Geonomateae. Lewis and Doyle (2002) and Baker *et al.* (in prep.) place the Manicarieae as sister to a clade of Euterpeae, Leopoldinieae, Pelagodoxeae, Geonomateae and Areceae. The *prk* data of Lewis and Doyle (2002) and Loo *et al.* (2006) recover a sister relationship between the Manicarieae and the Areceae with low support.

110. MANICARIA

Acaulescent palm of Central and South America with large mostly undivided leaf, held erect and instantly recognisable by the net-like prophyll and peduncular bract that cover the entire inflorescence; fruit corky-warted.

Manicaria Gaertn., Fruct. sem. pl. 2(3): 468 (1791). Type: **M. saccifera** Gaertn.

Pilophora Jacq., Fragm. Bot. 32: t. 35, 36 (1802) ('1809'). Type: *P. testicularis* Jacq. (= *Manicaria saccifera* Gaertn.).

Manicarius — *of sleeves or gloves, in reference to the peduncular bract.*

Robust, solitary or clustered, unarmed, pleonanthic, monoecious palms. *Stem* rather short, erect or leaning, sometimes dichotomously branched, conspicuously ringed with leaf scars, enlarged and with a mass of roots evident basally. *Leaves* very large, marcescent, pinnate, undivided or variously divided to or part way to the rachis, sometimes with separated leaflets; sheath splitting opposite the petiole, becoming narrow and deeply channelled distally, margins with many fibres; petiole long, deeply channelled adaxially, keeled abaxially, covered with small, rough scales abaxially; leaflets where blade divided single-fold, narrow, elongate, tips pointed, shortly bifid, midribs very prominent adaxially, intercostal ribs also prominent, hairs present or absent, scales usually present along ribs abaxially, transverse veinlets not evident. *Inflorescences* solitary, interfoliar, protandrous, branched to 1–4 orders; peduncle short, rounded in section,

Opposite: ***Manicaria***. **a**, portion of rachilla with staminate buds × 1$^1/_2$; **b**, scar of staminate flower with bract and bracteole × 3; **c**, staminate bud × 6; **d**, staminate bud in vertical section × 6; **e**, staminate calyx × 6; **f**, staminate flower, sepals removed × 6; **g**, staminate petal, interior view × 6; **h**, stamen in 3 views × 12; **i**, portion of rachilla with triads × 1$^1/_2$; **j**, triad, flowers removed to show bract and bracteoles × 3; **k**, pistillate bud × 6; **l**, pistillate bud in vertical section × 6; **m**, pistillate sepal, exterior view × 6; **n**, pistillate flower, sepals removed × 6; **o**, pistillate petal, interior view × 6; **p**, gynoecium and staminodes × 6; **q**, ovary in cross-section × 6; **r**, **s**, **t**, 1-, 2-, and 3-seeded fruits × $^1/_2$; **u**, fruit in vertical section × 1; **v**, fruit in cross-section × 1; **w**, endocarp × 1; **x**, **y**, **z**, seed in 3 views × 1. *Manicaria saccifera*: **a–q**, *Bailey* 214; **r–z**, *Moore & Parthasarathy* 9453. (Drawn by Marion Ruff Sheehan)

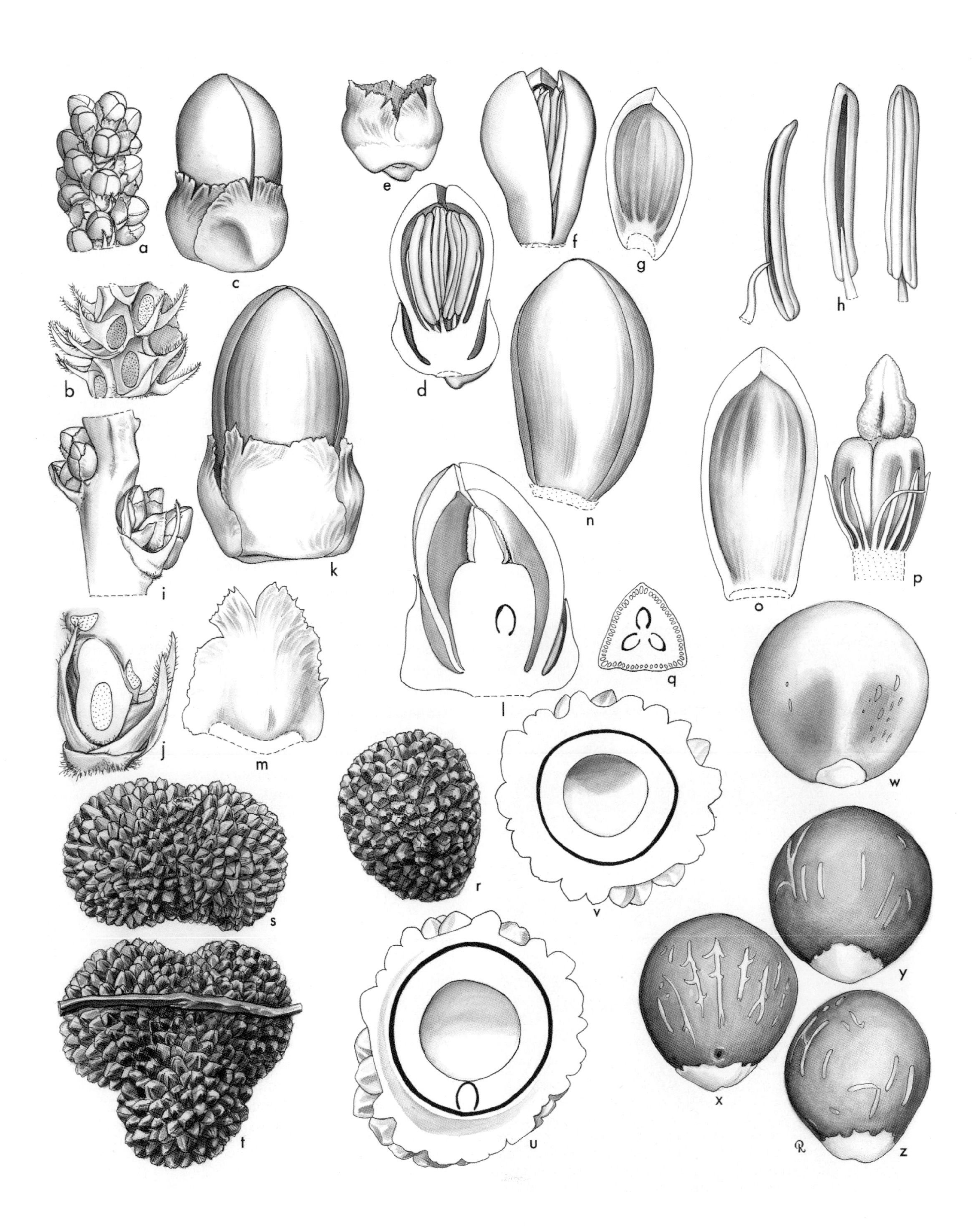
a
b
c
d
e
f
g
h
i
j
k
l
m
n
o
p
q
r
s
t
u
v
w
x
y
z

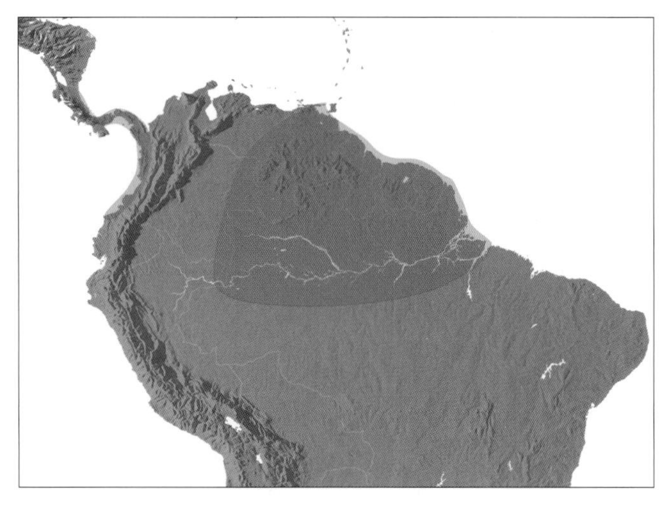

Distribution of ***Manicaria***

rather slender, covered in dense dark red tomentum; prophyll long, tubular, somewhat bulbous basally, tapering to a solid tip, completely enclosing the inflorescence, flexible, net-like, composed of thin, interwoven fibres; peduncular bract (?always present) like the prophyll but inserted near the middle of the peduncle, a few long, fibrous, incomplete peduncular bracts present above the first; rachis longer than the peduncle bearing spirally arranged, rather long, narrow, pointed bracts each subtending a rachilla; rachillae short to moderate, rather crowded, glabrous or with deciduous, dark red tomentum; rachilla bracts stiff, pointed, subtending basally a few (1–3) triads followed by closely appressed staminate flowers, each with a prominent stiff, pointed bracteole, flowers somewhat sunken, rachilla bracts and floral bracteoles persistent, surrounding rounded, shallow floral insertions giving a characteristic pattern to the rachillae after flowers are shed. *Staminate flowers* slightly asymmetrical, obovoid in bud; sepals 3, broadly rounded, united basally for nearly $^{1}/_{3}$ their length, imbricate where distinct, thick basally, margins thin and variously notched; petals 3, more than twice as long as the sepals, united with the receptacle to form a solid base, adnate to stamen filaments basally, lobes distinct, thick, valvate, grooved adaxially; stamens 30–35, filaments terete, moderate in length, variously coiled in bud, anthers elongate, dorsifixed above the base, introrse, connective tanniniferous; pistillode lacking. *Pollen* ellipsoidal or oblate triangular, with slight or obvious asymmetry; aperture a distal sulcus or trichotomosulcus; ectexine tectate, finely perforate, or perforate and micro-channelled, or perforate-rugulate, aperture margin broad, psilate-perforate; infratectum columellate; longest axis 32–40 µm [1/1]. *Pistillate flowers* shortly ovoid in bud; sepals 3, distinct, imbricate, truncate, margins variously notched; petals 3, unequal, thick, valvate; staminodes ca. 15, linear, flat, thin; gynoecium triangular in cross-section, obovoid, truncate, trilocular, triovulate, bearing 3 central, linear, connate styles ending in 3 linear stigmas, ovules laterally attached, anatropous. *Fruit* large, rounded, 1–3-lobed, 1–3-seeded, stigmatic remains subbasal; epicarp obsolescent at maturity, outer mesocarp woody, covered in wart-like projections, inner mesocarp spongy, tanniniferous, endocarp thin, smooth. *Seed* rounded, basally attached, raphe branches sunken, parallel, sparsely branched, endosperm homogeneous, hollow; embryo basal. *Germination* adjacent-ligular; eophyll bifid. *Cytology* not studied.

Distribution and ecology: A single variable species occurring from Central America, across Trinidad, the Orinoco Delta, and the Guianas to the lower Amazon River. Occurring in freshwater swamps near the coast, sometimes occurring as large dense stands.

Anatomy: Leaf (Tomlinson 1961).

Relationships: For relationships, see tribe Manicarieae.

Common names and uses: Sleeve palm, monkey cap palm, *tenudie*. Makes excellent thatch. Intensively used as food, raw material, and medicine by South American Indians (Plotkin & Balick 1984). The inflorescence bracts have been used as caps (Wilbert 1980a) and the leaves as sails (Wilbert 1980b).

Taxonomic account: Henderson (1995).

Fossil record: The leaf fossil genus *Manicarites* (*M. dantesianus*) from the Tertiary of Italy (Bureau 1896) was placed in the synonomy of *Phoenicites* by Read and Hickey (1972). A silicified fruit, *Manicaria edwardsii*, was described from Mexico (Kaul 1946). A fruiting palm inflorescence compared with *Manicaria* (Weber 1978) from the lower Maastrichtian of northeastern Mexico is possibly the earliest record of a floral structure described to date. However, Weber (1978) stated that, "The fossil inflorescence will be described elsewhere as a new genus." The publication of the new genus has not yet been traced and so its affinity with *Manicaria* cannot be assessed. Two monosulcate palm pollen types have been recovered from the Pliocene, Gatun Lake Formation, Panama (Graham 1991). The first of these types, which is asymmetrical and scabrate tectate, is compared with pollen of *Aiphanes*, *Reinhardtia* and *Manicaria*. It is a very common arecoid pollen type and difficult to place. Of the suggested genera, *Reinhardtia* is the most probable.

Left: ***Manicaria saccifera***, infructescence, Ecuador. (Photo: F. Borchsenius)

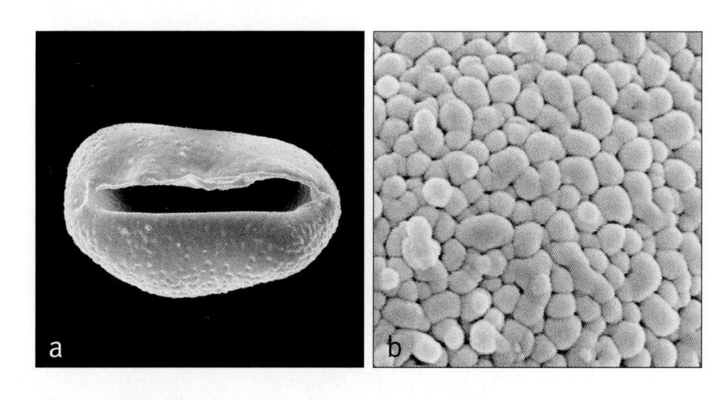

Above: ***Manicaria***
a, monosulcate pollen grain, distal face SEM × 1500;
b, close-up, coarsely granular-rugulate pollen surface SEM × 8000.
Manicaria saccifera: **a** & **b**, *King* 672.

Notes: Very striking and remarkable palm especially with regard to the huge, often more or less entire leaves, the trilocular, triovulate gynoecium and the warty fruits. The resemblance of the gynoecium to that of phytelephantoid palms is noteworthy. Stems may branch dichotomously (Fisher & Zona 2006).

Tribe Euterpeae J. Dransf., N.W. Uhl, C. Asmussen, W.J. Baker, M.M. Harley & C. Lewis., Kew Bull. 60: 562 (2005). Type: **Euterpe**.

Slender to robust, acaulescent to erect, unarmed; leaves pinnate, the leaflet tips entire, or entire-bifid; sheaths usually forming a crownshaft; inflorescences infrafoliar or more rarely interfoliar, branched to 1 order, rarely spicate; flowers superficial, borne in triads near the base, solitary or paired staminate flowers distally; pistillate flowers with petals distinct and imbricate basally, valvate distally; staminodes minute, distinct, not connate in a conspicuous ring; gynoecium pseudomonomerous; fruit with basal, lateral or apical stigmatic remains; epicarp smooth.

The tribe contains five genera, all palms of humid rain forest at low to high elevations in Central and South America and the Caribbean

Tribe Euterpeae is monophyletic with moderate support in numerous studies (Henderson 1999a, Lewis & Doyle 2001, Hahn 2002b, Asmussen *et al.* 2006, Baker *et al.* in review). Although there is wide agreement that the tribe resolves within the core arecoid clade, there are many conflicting hypotheses on its placement within the core arecoids, and most have low levels of bootstrap support. Several analyses show a relationship between the Euterpeae and the Geonomateae, with Asmussen and Chase (2001) and Cuenca and Asmussen-Lange (2007) resolving a sister relationship between the two tribes with low support. Asmussen *et al.* (2006) resolve the Euterpeae as sister to a subclade of the Geonomateae with low support. Hahn (2002b) found a moderately supported sister relationship between the Euterpeae and the Areceae. The *prk* analyses of Lewis and Doyle (2002) and Loo *et al.* (2006) resolved a sister relationship between the Euterpeae and the Pelagodoxeae with low support. The Euterpeae was found to be sister to a clade of Pelagodoxeae and Leopoldinieae with low support by Norup *et al.* (2006), and sister to the Leopoldinieae and a subclade of Areceae with low support by Hahn (2002a).

Key to Genera of Euterpeae

1. Inflorescence hippuriform (shaped like a horse's tail), adaxial surface of rachis lacking branches; rachillae all lateral and abaxial, curved and pendulous; leaflets grey abaxially. Central and South America . . . . **115. Oenocarpus**
1. Inflorescence not hippuriform; leaflets concolourous . . . 2
2. Inner stamens markedly adnate to the pistillode in staminate flowers; sepals of pistillate flowers connate basally; stigmatic remains basal in fruit. Costa Rica to Peru . . . . **111. Hyospathe**
2. Inner stamens not markedly adnate to the pistillode in staminate flowers; sepals of pistillate flowers imbricate; stigmatic remains apical or lateral 3
3. Inflorescence spicate; stigmatic remains apical. Costa Rica to Panama . . . . **114. Neonicholsonia**
3. Inflorescence usually branched to 1 order, very rarely spicate; stigmatic remains lateral or subapical 4
4. Leaf sheaths forming a conspicuous crownshaft; rachillae densely covered with hairs, not changing colour from flowering to fruiting; peduncle dorsiventrally compressed; prophyll and peduncular bract equal or subequal, papery, falling at anthesis. Lesser Antilles, Central and South America . . . . **112. Euterpe**
4. Leaf sheaths not or very rarely forming a crownshaft; rachillae not densely covered in hairs, changing colour from white at anthesis to red at fruiting; peduncle terete; prophyll shorter than the peduncular bract, coriaceous and ± persistent. West Indies, Nicaragua south to Brazil and Peru . . . . **113. Prestoea**

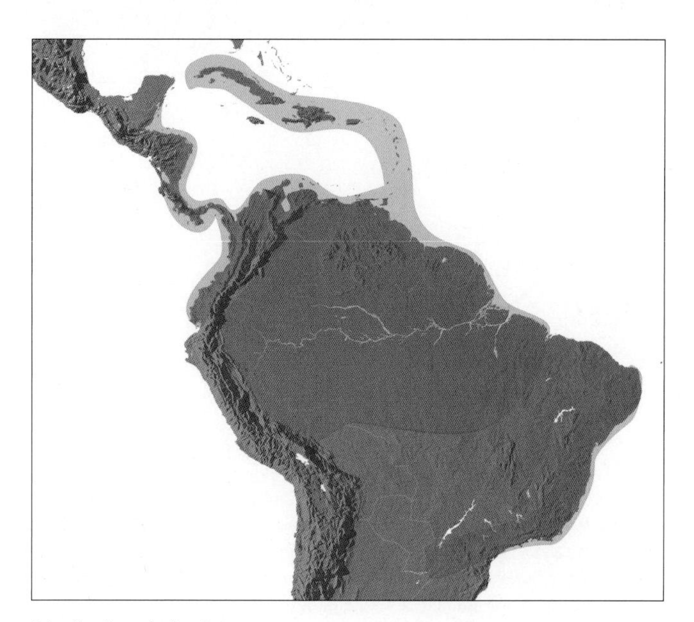

Distribution of tribe **Euterpeae**

111. HYOSPATHE

Small, solitary or clustering palms of South American rain forests, often with entire leaves or with broad pinnae: the staminate flowers have stamens adnate to the pistillode and the fruit has basal stigmatic remains.

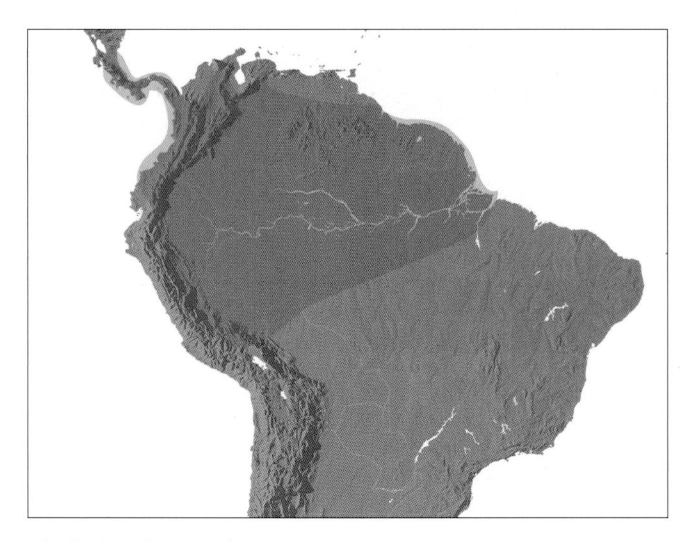

Distribution of ***Hyospathe***

Hyospathe Mart., Hist. nat. palm. 2: 1 (1823). Type: **H. elegans** Mart.

Hyo- *from* hys — *pig,* spathe — *sheath or bract, derived from the vernacular name* tajassu-ubi — *pig leaf or pork palm.*

Small or rarely moderate, solitary or clustering, graceful, unarmed, pleonanthic, monoecious palms. *Stem* slender, ringed with conspicuous, sometimes oblique, rather distant leaf scars. *Leaves* regularly pinnate, entire and bifid, or bi- or trijugate; sheath usually forming a short to long crownshaft, not splitting opposite the petiole until shed, margin irregular, chartaceous, striate, adaxially grooved, glabrous, abaxially with scattered scales; petiole moderate, slender, abaxially rounded, with hairs or deciduous scales; rachis adaxially ridged, abaxially ridged or rounded, also scaly; leaflets lanceolate to falcate, pointed, often somewhat curved apically, alternate, single-fold, several-fold or many-fold, bearing scattered deciduous scales along abaxial ribs and sometimes basally on adaxial ribs, surfaces similar or dissimilar in colour, midrib prominent, 1–2 pairs of parallel veins also conspicuous in some species, transverse veinlets conspicuous or obscure. *Inflorescence* solitary, branched to 1 order, branches stiff or pendulous; peduncle short or long, slender; prophyll 2-keeled laterally, tapering to a rather blunt point, chartaceous, splitting dorsiventrally and apically to become bifid, inserted above the base of the peduncle; peduncular bracts 1–2, terete, much longer than or about as long as the prophyll, beaked, splitting abaxially, inserted somewhat above the prophyll; rachis much shorter to rarely longer than the peduncle; rachis

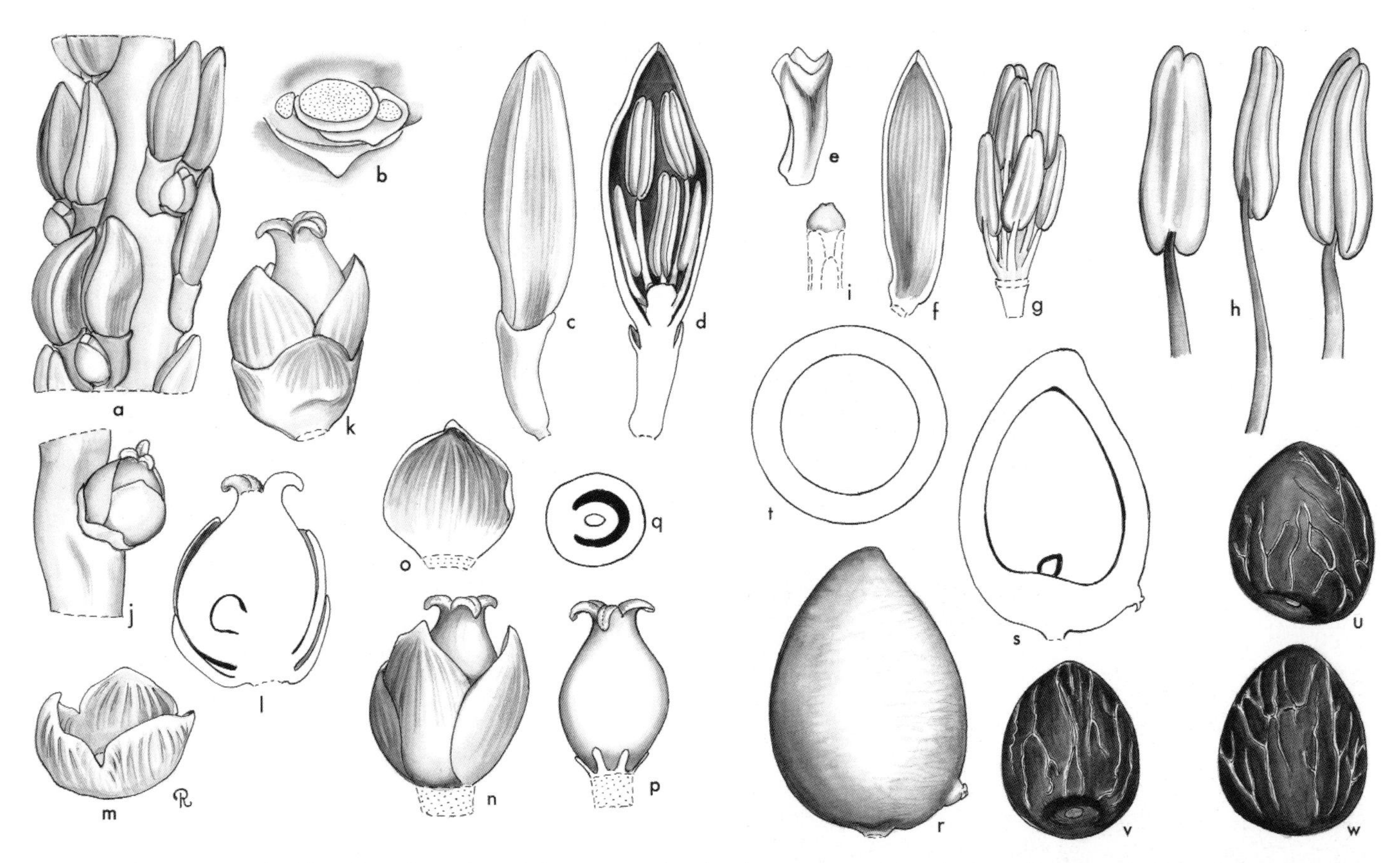

Hyospathe. **a**, portion of rachilla with triads × 3; **b**, triad, flowers removed to show bracts × 9; **c**, staminate bud × 6; **d**, staminate bud in vertical section × 6; **e**, staminate calyx × 6; **f**, staminate petal, interior view × 6; **g**, androecium × 6; **h**, stamen in 3 views × 12; **i**, pistillode × 6; **j**, portion of rachilla at pistillate anthesis × 6; **k**, pistillate flower × 12; **l**, pistillate flower in vertical section × 12; **m**, pistillate calyx × 12; **n**, pistillate flower, calyx removed × 12; **o**, pistillate petal, interior view × 12; **p**, gynoecium and staminodes × 12; **q**, ovary in cross-section × 12; **r**, fruit × 3; **s**, fruit in vertical section × 3; **t**, fruit in cross-section × 3; **u, v, w**, seed in 3 views × 3. *Hyospathe elegans*: **a–i**, *Moore et al.* 9500; **j–q**, *Guppy* G-651; **r–w**, *Moore et al.* 9508. (Drawn by Marion Ruff Sheehan)

bracts very short, obscure, subtending spirally inserted rachillae; rachillae slender, rather distant or crowded, moderate to short and stiff or long and pendulous, sometimes undulate apically, bearing spirally arranged triads of flowers nearly throughout and paired to solitary staminate flowers distally; floral bracteoles shallow, rounded. *Staminate flowers* lateral to the pistillate, sessile or stalked, narrow, elongate; sepals 3, united in a tube, adnate basally for at least $^2/_3$ their length to the receptacle to form a stalk-like base, free tips short, broadly triangular; petals 3, distinct, narrowly ovate, asymmetical, curved basally, pointed distally; stamens 6, the 3 antesepalous with shorter filaments, free or shortly joined with the pistillode, the 3 antepetalous with filaments much longer, adnate to the pistillode nearly to its apex, filaments awl-shaped, anthers moderately long, dorsifixed near the base, latrorse, united throughout; pistillode narrowly ovoid with 2 stigmatic lobes. *Pollen* ellipsoidal, with slight or obvious asymmetry; aperture a distal sulcus; ectexine tectate, finely or coarsely perforate-rugulate, aperture margin finer; infratectum columellate; longest axis 34–39 µm [1/17]. *Pistillate flower* ovoid, shorter than the staminate; sepals 3, united in a cupule for ca. $^2/_3$ their length, tips broad, pointed, striate; petals 3, distinct, ovate, moderately imbricate, striate, tips pointed; staminodes 6, small, strap-like; gynoecium ovoid, unilocular, uniovulate, narrowed to a short tubular style, stigmas 3, recurved at anthesis, ovule basal, laterally attached, form unknown. *Fruit* ovoid to cylindrical, pointed, asymmetrical, black at maturity, stigmatic remains basal; epicarp smooth, lightly mottled, mesocarp fibrous, endocarp thin, crustaceous. *Seed* narrow, ovoid, ± pointed, hilum basal, raphe branches anastomosing, endosperm homogeneous; embryo rather large, basal. *Germination* adjacent-ligular; eophyll bifid. *Cytology* not studied.

Distribution and ecology: Six species ranging from Costa Rica to Peru. In rain forest, in swamps or on dry ground at low elevations but also on the slopes of the Andes between 1000–2000 m.

Anatomy: See Skov & Balslev (1989).

Relationships: *Hyospathe* is monophyletic (Henderson 1999a). The genus is resolved as sister to the rest of the Euterpeae with moderate support (Asmussen *et al.* 2006, Baker *et al.* in review), or as sister to *Neonicholsonia*, *Oenocarpus* and *Prestoea* with low support (Henderson 1999a).

Top: ***Hyospathe elegans***, habit, Ecuador. (Photo: F. Borchsenius)

Above: ***Hyospathe elegans***, inflorescence, Ecuador. (Photo: F. Borchsenius)

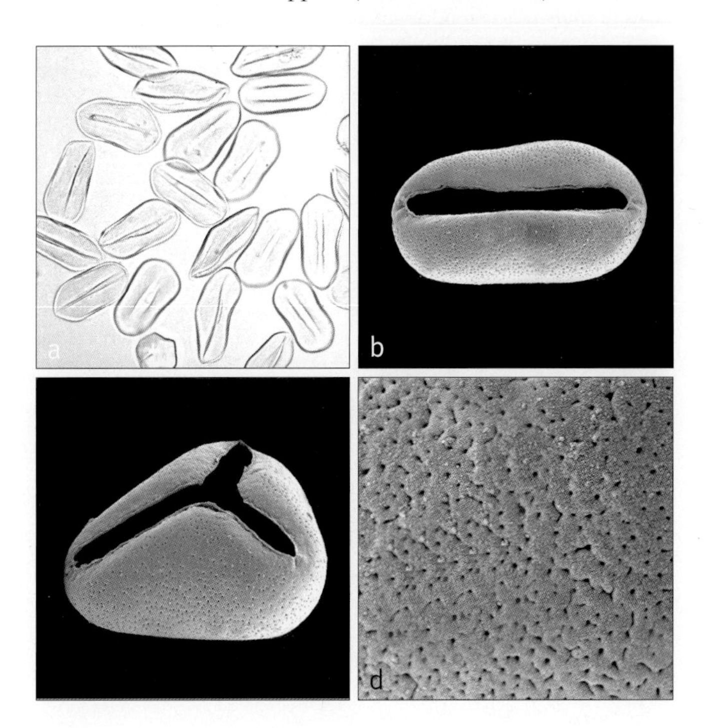

Left: ***Hyospathe***
a, group of monosulcate pollen grains, mid- and low-foci LM × 350;
b, monosulcate pollen grain, distal face SEM × 1500;
c, trichotomosulcate pollen grain, proximal face SEM × 1500;
d, close-up, perforate pollen surface SEM × 8000.
Hyospathe elegans: **a**, *Cid et al.* 83, **b–d**, *Irwin et al.* 54453.

Common names and uses: Hog palm, ubim palm (*Hyospathe elegans*). These palms would make handsome ornamentals.
Taxonomic accounts: Skov & Balslev (1989), Henderson (2004), see also Henderson (1999a).
Fossil record: No generic records found.
Notes: It seems surprising that their ornamental potential has not been exploited. The difference in length of antesepalous and antepetalous stamens and the adnation of the antepetalous filaments to the pistillode are distinctive and unusual in the family as a whole.
Additional figures: Glossary fig. 20.

112. EUTERPE

Elegant solitary or clustering pinnate-leaved palms from Central and South America and the Caribbean, with tall graceful stems, and regularly and finely pinnate leaves; the inflorescences have grey-white rachillae.

Euterpe Mart., Hist. nat. palm. 2: 28 (1823); emended 3: 165 (1837); 3: 230 (ed 2) (1845) (conserved name). Type: **E. oleracea** Mart. (conserved type).
Catis O.F. Cook, Bull. Torrey Bot. Club 28: 557 (1901). Type: *C. martiana* O.F. Cook (illegitimate name) (= *Euterpe oleracea* Martius).
Plectis O.F. Cook Bull. Torrey Bot. Club 31: 352 (1904). Type: *P. oweniana* O.F. Cook (= *Euterpe macrospadix* Oerst.).
Rooseveltia O.F. Cook, Smithsonian Misc. Collect. 98(7): 21 (1939). Type: *R. frankliniana* O.F. Cook (= *Euterpe macrospadix* Oerst.).

Named after one of the nine Muses of Greek mythology.

Moderate to large, solitary or clustered, unarmed, pleonanthic, monoecious palms. *Stem* erect, sometimes slender, obscurely to distinctly ringed with leaf scars, grey to white, base sometimes enlarged. *Leaves* few in crown, often spreading, pinnate; sheath elongate, tubular, forming a prominent crownshaft, smooth, variously glaucous, tomentose, or with scales, with or without a prominent, fibrous, adaxial ligule; petiole very short or absent, rarely elongate, slender, deeply concave adaxially, or flat with a central ridge, rounded abaxially, with scattered dark brown to blackish, branched scales or deciduous tomentum on both surfaces, usually denser adaxially; rachis slender, rounded abaxially, channelled adaxially near the base, distally angled, with dark brown to blackish scales more numerous adaxially; leaflets often ± pendulous, narrow, lanceolate, single-fold, tips long-attenuate, pointed, midrib conspicuous, 1 or 2 pairs of large veins also evident, deciduous tattered scales often prominent abaxially along midribs and larger veins, other elliptic scales present or absent abaxially and near the base adaxially, transverse veinlets not evident. *Inflorescences* axillary, infrafoliar at anthesis, erect in bud, branched to 1 order; peduncle short, often dorsiventrally compressed, covered with scales, tomentum, or hairs, minutely brown-dotted or rarely glabrous; prophyll tubular, elongate, flattened dorsiventrally, inserted obliquely near the base of the peduncle, chartaceous, with scattered sometimes black, tattered-peltate scales, or ± glabrous, margins with wide flat keels, tip usually rounded, splitting abaxially below the tip; peduncular bract about as long as or longer than the prophyll, tubular, chartaceous, with scales as on the prophyll, tip pointed, hard, a second, incomplete, rather long, pointed peduncular bract sometimes present; rachis longer than the peduncle, covered with dense white, yellow to dark red tomentum; rachillae moderate to long, often slender, becoming pendulous, usually covered with dense white, orange, or dark brownish tomentum, and bearing rather close or distant, spirally arranged bracts, the proximal somewhat elongate, pointed, the distal smaller, often rounded, each subtending a triad of flowers at the base or more distally on the rachilla a pair of staminate or a single staminate flower, the pistillate flower sunken in a pit and surrounded by 2 rounded, stiff bracteoles, one usually larger, the 2 staminate flowers of the triad in shallow indentations above the pistillate flower. *Staminate flowers* elongate, pointed in bud; sepals 3, distinct, broadly imbricate, irregular, rounded to ± pointed, margins often tattered; petals 3, distinct, unequal, asymmetrical, valvate, the tips with short solid points; stamens 6, filaments short, linear, sometimes wider basally, anthers elongate, sagittate, medifixed, latrorse; pistillode 3-lobed, columnar. *Pollen*

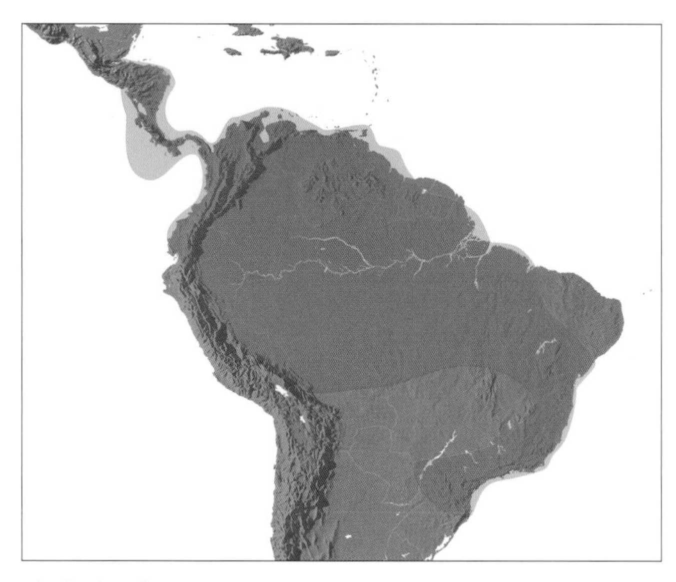

Distribution of ***Euterpe***

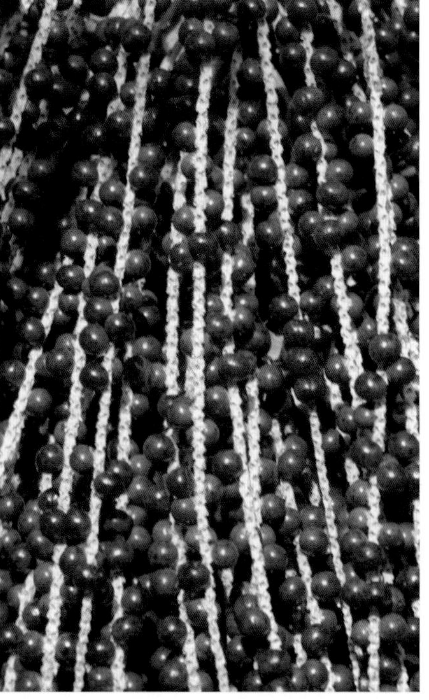

Far left: ***Euterpe precatoria***, habit, Brazil. (Photo: A.J. Henderson)

Left: ***Euterpe precatoria***, fruiting rachillae, Brazil. (Photo: A.J. Henderson)

ellipsoidal, with slight or obvious asymmetry; aperture a distal sulcus; ectexine tectate, finely perforate, or perforate and micro-channelled and rugulate, aperture margin similar or slightly finer; infratectum columellate; longest axis 35–57 µm [4/8]. *Pistillate flowers* ovoid; sepals 3, distinct, imbricate, margins often lacerate, from ca. 1/4 to 2/3 as long as the petals; petals 3, distinct, imbricate, margins irregular, tips with solid points; staminodes usually absent (present in *Euterpe luminosa*); gynoecium ovoid, unilocular, uniovulate, stigmas 3 short, fleshy, recurved, ovule probably hemianatropous, laterally attached. *Fruit* subglobose or rarely ellipsoid, small to moderate, single-seeded, stigmatic remains lateral to subapical; epicarp smooth, minutely pebbled when dry, mesocarp rather thin with radially arranged sclereid bundles and an inner layer of thin flat fibres, endocarp thin, crustaceous, tanniniferous. *Seed* globose, laterally attached, hilum elongate, ± 2-lobed, raphe branches forming a network, endosperm homogeneous or rarely ruminate; embryo subbasal. *Germination* adjacent-ligular; eophyll bifid or pinnate with narrow leaflets. *Cytology*: 2n = 36.

Distribution and ecology: Seven species from the Lesser Antilles and Central America south through Brazil to Peru and Bolivia. Occurring in lowland rain forest and montane forests and swamps, often along rivers, but sometimes at higher elevations. *Euterpe* has a wide altitudinal range occurring from swamps at very low elevation to 2500 m on mountain slopes.

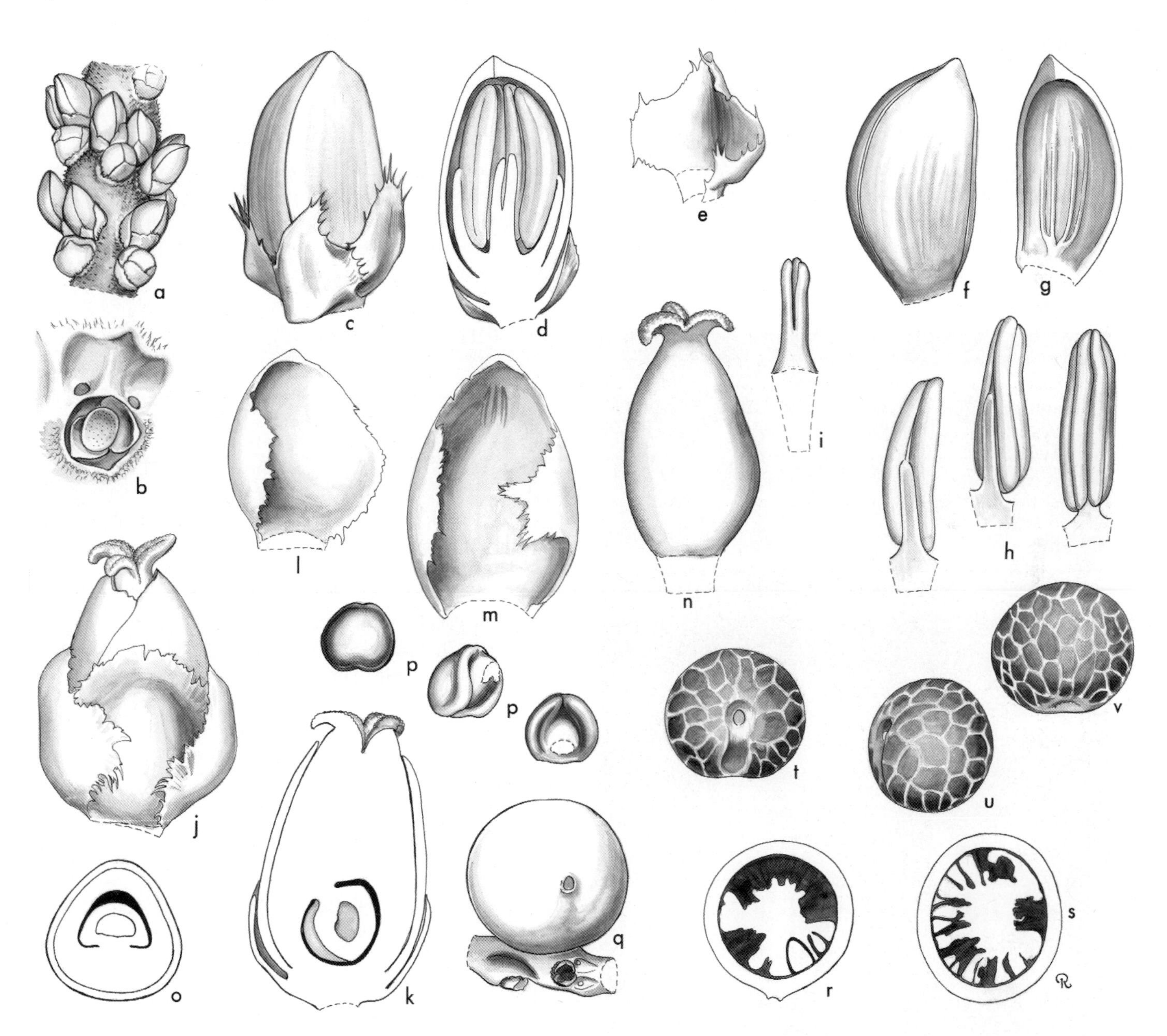

Euterpe. **a**, portion of rachilla with triads × 3; **b**, triad, flowers removed × 6; **c**, staminate bud × 12; **d**, staminate bud in vertical section × 12; **e**, staminate sepal, interior view × 12; **f**, staminate flower, sepals removed × 12; **g**, staminate petal, interior view × 12; **h**, stamen in 3 views × 12; **i**, pistillode × 12; **j**, pistillate flower × 12; **k**, pistillate flower in vertical section × 12; **l**, pistillate sepal, interior view × 12; **m**, pistillate petal, interior view × 12; **n**, gynoecium × 12; **o**, ovary in cross-section × 12; **p**, ovule in 3 views × 12; **q**, fruit × 1 1/2; **r**, fruit in vertical section × 1 1/2; **s**, fruit in cross-section × 1 1/2; **t**, **u**, **v**, seed in 3 views × 1 1/2. *Euterpe oleracea*: all from *Moore & Parthasarathy* 9404. (Drawn by Marion Ruff Sheehan)

Anatomy: Leaf (Tomlinson 1961, Roth 1990); root, leaflets, and fruits of *Euterpe oleracea* (Elias de Paula 1975); stems, leaves and roots (Henderson & Galeano 1996), root (Seubert 1998a, 1998b), and fruit (Moegenburg 2003).

Relationships: *Euterpe* is monophyletic (Henderson 1999a). The genus is resolved as sister to a clade of *Neonicholsonia*, *Oenocarpus* and *Prestoea* with moderate support (Asmussen *et al.* 2006, Baker *et al.* in review), or as sister to all Euterpeae (Henderson 1999a).

Common names and uses: Assai palms, *manaco*. The 'cabbage' is edible and much prized, being sweet and succulent and the most common source of hearts of palm in the Americas. Fruits of some species are used in preserves and beverages. Unopened inflorescences are made into pickles. The hard outer part of the stem may be used as planks, and leaves for thatch.

Taxonomic account: Henderson & Galeano (1996).

Fossil record: No generic records found.

Notes: Reproductive biology of *Euterpe precatoria* was studied by Küchmeister *et al.* (1997).

Henderson (1999a) published a phylogenetic study of Euterpeinae (= Euterpeae *sensu Genera Palmarum* 2nd edition) based on morphology and anatomy. This clarified the differences between *Euterpe* and *Prestoea*, two genera that had been much confused previously. *Euterpe* differs in its pendulous pinnae and a lack of an ocrea, whereas *Prestoea* possesses an ocrea and does not have pendulous pinnae.

Additional figures: Glossary fig. 12.

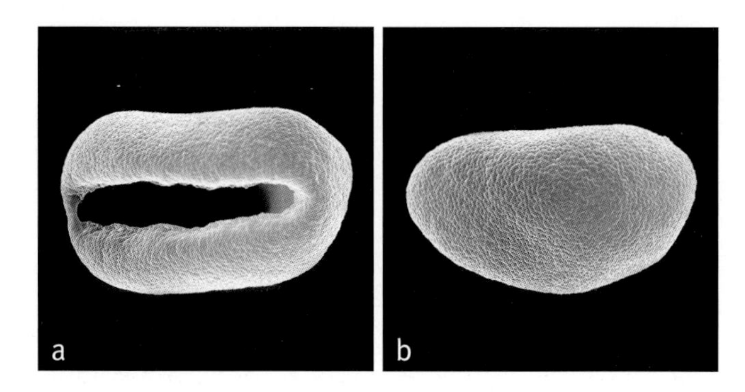

Above: ***Euterpe***
a, monosulcate pollen grain, distal face SEM × 1500;
b, monosulcate pollen grain, proximal face SEM × 1500.
Euterpe precatoria: **a** & **b**, *Britton et al.* 278.

Left: ***Prestoea acuminata***, habit, Ecuador. (Photo: F. Borchsenius)

113. PRESTOEA

Small or moderate pinnate-leaved palms from Central and South America and the Caribbean, with rather coarse leaflets, rarely the leaf undivided; rachillae are generally brightly coloured.

Prestoea Hook.f. in Benth. & Hook.f., Gen. pl. 3: 875, 899 (1883) (conserved name). Type: **P. pubigera** (Griseb. & H. Wendl.) Hook.f. (*Hyospathe pubigera* Griseb. & H. Wendl.).

Martinezia Ruiz & Pav., Fl. peruv. prodr. 148 (1794). Lectotype: *M. ensiformis* Ruiz & Pav. (see Burret 1933c) (= *Prestoea ensiformis*) (Ruiz & Pavon) H.E. Moore.

Oreodoxa Willd., Mém. Acad. Roy. Sci. Hist. (Berlin) 1804: 34 (1807). Lectotype: *O. acuminata* Willd. (see Klotzsch, Linnaea 20: 448 [1847] [rejected name]) (= *Prestoea acuminata* [Willd.] H.E. Moore).

Acrista O.F. Cook, Bull. Torrey Bot. Club 28: 555 (1901). Type: *A. monticola* O.F. Cook (= *Prestoea montana* [Graham] G. Nicholson).

Euterpe Gaertn. *sensu* Becc. (non Mart.), Pomona Coll. J. Econ. Bot. 2: 351 (1912).

Euterpe sect. *Euterpe* subsect. *Leiostachys* Burret, Bot. Jahrb. Syst. 63: 50 (1929).

Euterpe sect. *Meteuterpe* Burret, Notizbl. Bot. Gart. Berlin-Dahlem 14: 328 (1939).

Named for Henry Prestoe (1842–1923), British botanist and traveller, who collected plants in Trinidad.

Small to moderate, rarely solitary, usually clustered, unarmed, pleonanthic, monoecious tree palms. *Stem* slender or relatively stout, erect or decumbent, brown or grey, sometimes swollen basally, leaf scars prominent or obscure, adventitious roots present or absent. *Leaves* regularly or irregularly pinnate, or undivided, curving or erect; sheath tubular, splitting opposite the petiole, usually not forming a distinct crownshaft, but crownshaft sometimes evident in *Prestoea acuminata*, often scaly or tomentose, becoming glabrous; petiole usually elongate, rarely short, often slender, channelled adaxially, rounded abaxially, both surfaces densely dark tomentose or scaly, rachis channelled at the base, flat to ridged adaxially, rounded abaxially, densely tomentose (?always); leaflets long, narrow, opposite or subopposite, regularly arranged in one plane, shorter basally and distally, sometimes curved, tips pointed, edges often thickened, single-fold or several-fold distally or in partly entire leaves, midribs of folds prominent adaxially, other veins small, ± equal or 1–2 pairs slightly larger, blade adaxially ± glabrous, abaxially lightly tomentose, whitish dot-like hairs usually abundant, large to small tattered scales along the midrib and sometimes along larger veins, often waxy, transverse veinlets not evident. *Inflorescences* usually interfoliar in bud, becoming infrafoliar at anthesis or in fruit, branched to 1 order or rarely spicate; peduncle short or more usually elongate, longer than the rachis; prophyll usually persistent, markedly shorter than the peduncular bract, tubular, 2-keeled laterally, ± flat, splitting apically and dorsiventrally so as sometimes to appear bifid, inserted at the base of the peduncle, chartaceous to coriaceous, variously scaly; peduncular bract usually persistent, several times longer than the prophyll, terete, with a long hard beak, usually inserted some distance above the prophyll,

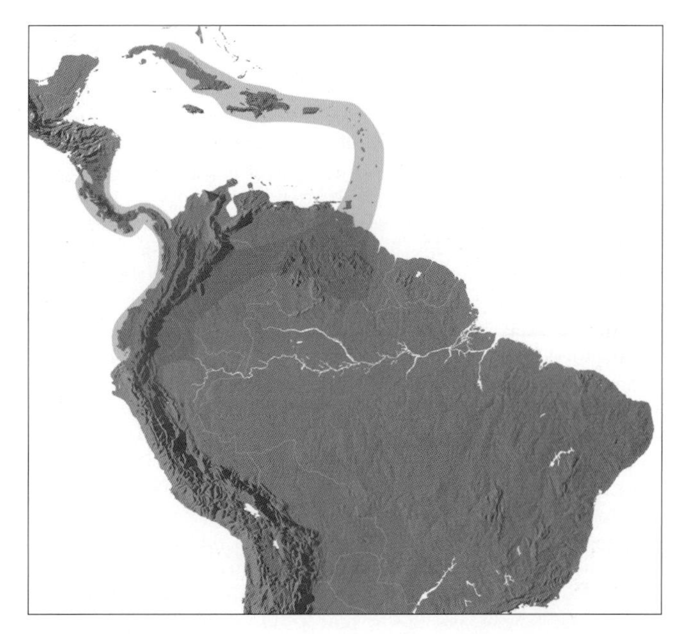

Distribution of ***Prestoea***

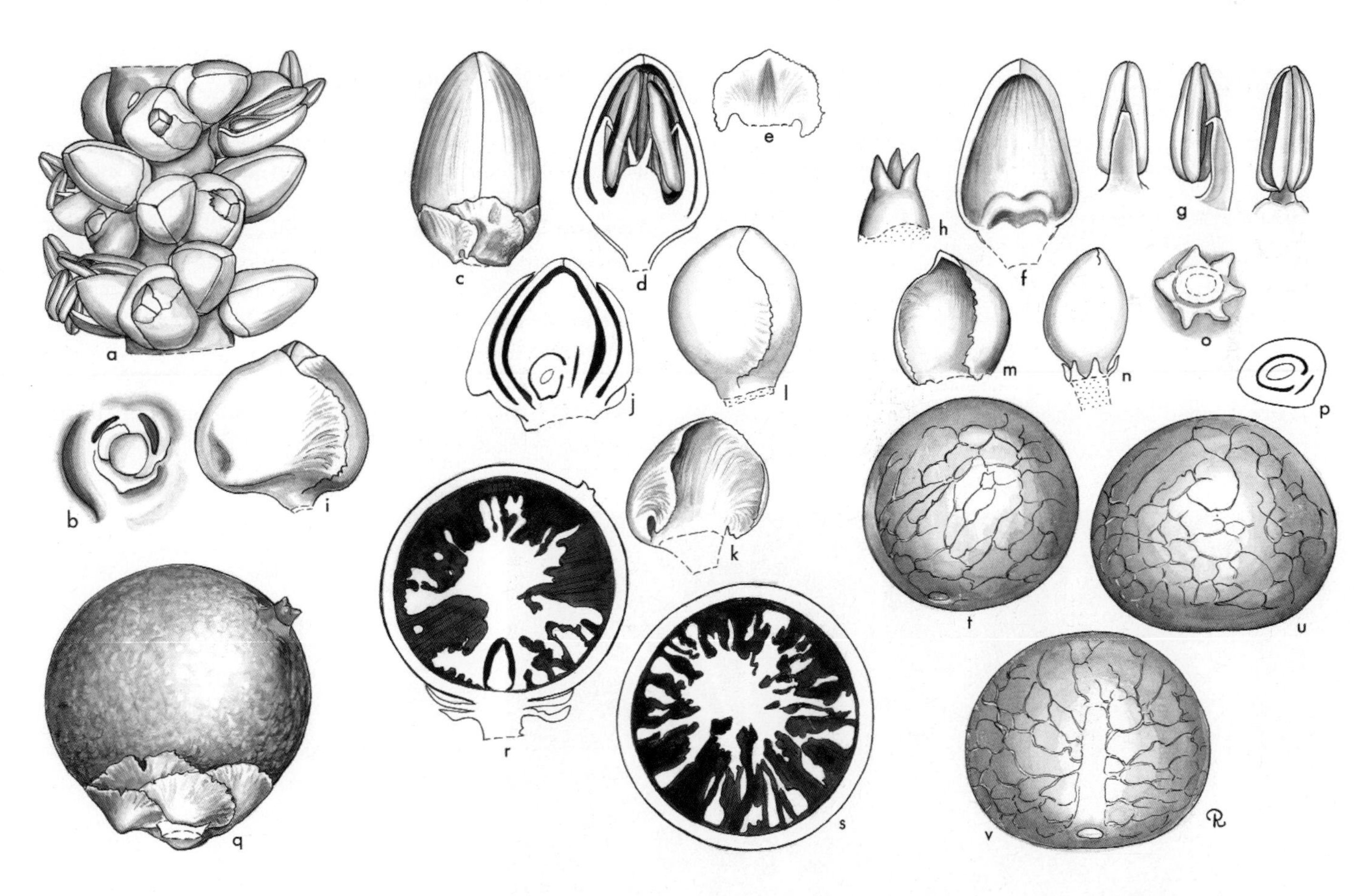

Prestoea. **a**, portion of rachilla with staminate flowers and pistillate buds × 3; **b**, triad, flowers removed × 6; **c**, staminate bud × 6; **d**, staminate bud in vertical section × 6; **e**, staminate sepal × 6; **f**, staminate petal, interior view × 6; **g**, stamen in 3 views × 6; **h**, pistillode × 6; **i**, pistillate bud × 6; **j**, pistillate bud in vertical section × 6; **k**, pistillate sepal, interior view × 6; **l**, pistillate flower, sepals removed × 6; **m**, pistillate petal × 6; **n**, gynoecium and staminodes × 6; **o**, staminodes × 6; **p**, ovary in cross-section × 6; **q**, fruit × 3; **r**, fruit in vertical section × 3; **s**, fruit in cross-section × 3; **t**, **u**, **v**, seed in 3 views × 3. *Prestoea montana*: all from Fischer & Ernst 858. (Drawn by Marion Ruff Sheehan)

chartaceous or coriaceous; rachis bearing spirally arranged, short, thin, membranous or stiff, rounded or pointed bracts, each subtending a rachilla; rachillae white at anthesis, usually becoming reddish in fruit, slender, moderate to elongate, erect at first, becoming divaricate or stiffly ascending, sometimes markedly swollen or bulbous basally, rachis and rachillae densely covered with soft pale hairs or with dark red or brown tomentum or glabrous; rachilla bracts very shallow, membranous, subtending triads of flowers basally and paired to solitary staminate flowers distally, flowers superficial or in a slight depression, triads with the staminate flowers only slightly larger than and lateral to the pistillate, rarely (*Prestoea longepetiolata* var. *cuatrecasasii*) the pistillate flower surrounded by 2 large bracteoles. *Staminate flowers* symmetrical or asymmetrical, ovoid, stalked or sessile; sepals 3, distinct and shortly imbricate basally or united in a low cupule, margins smooth or hairy with tufts of hairs at the tips, usually keeled to some extent; petals 3, ovate, distinct, valvate, tips valvate or appressed, margins smooth or hairy with tufts of hairs at the tips; stamens 6, filaments terete, briefly inflexed at the apex, anthers linear, acute or bifid apically, sagittate or free basally, dorsifixed slightly below the middle, latrorse; pistillode columnar or trifid, light or dark. *Pollen* ellipsoidal, with slight or obvious asymmetry; aperture a distal sulcus; ectexine tectate, finely perforate, or perforate-rugulate, aperture margin similar or slightly finer; infratectum columellate; longest axis 26–46 µm [7/10]. *Pistillate flowers* broadly ovoid; sepals 3, distinct, imbricate, rounded, margins smooth or hairy; petals 3, distinct, broadly imbricate, rounded, tips valvate; staminodes 6, small, tooth-like; gynoecium ovoid, asymmetrical, unilocular, uniovulate, style not evident, stigmas 3, appressed or reflexed, ovule large, basal, form unknown. *Fruit* rounded, dark purple to black at maturity (?always), perianth persistent, stigmatic remains subapical or near the middle; epicarp smooth or slightly irregular, mesocarp fleshy, with an inner layer of wide flat fibres, endocarp thin, crustaceous. *Seed* globose, laterally attached, hilum elongate, raphe branches numerous, anastomosing to form a network, endosperm ruminate or rarely (*P. longepetiolata* var. *cuatrecasasii*) homogeneous; embryo subbasal. *Germination* adjacent-ligular; eophyll bifid (e.g., *P. acuminata*) or pinnate (e.g., *P. decurrens*). *Cytology*: 2n = 36.

Distribution and ecology: Ten species distributed throughout the West Indies, from Nicaragua southward in Central America, and into Brazil, Peru and Bolivia. One taxon (*Prestoea acuminata* var. *montana*) is widespread in the Lesser Antilles. Mostly on well-drained slopes at moderate to rather high elevations.

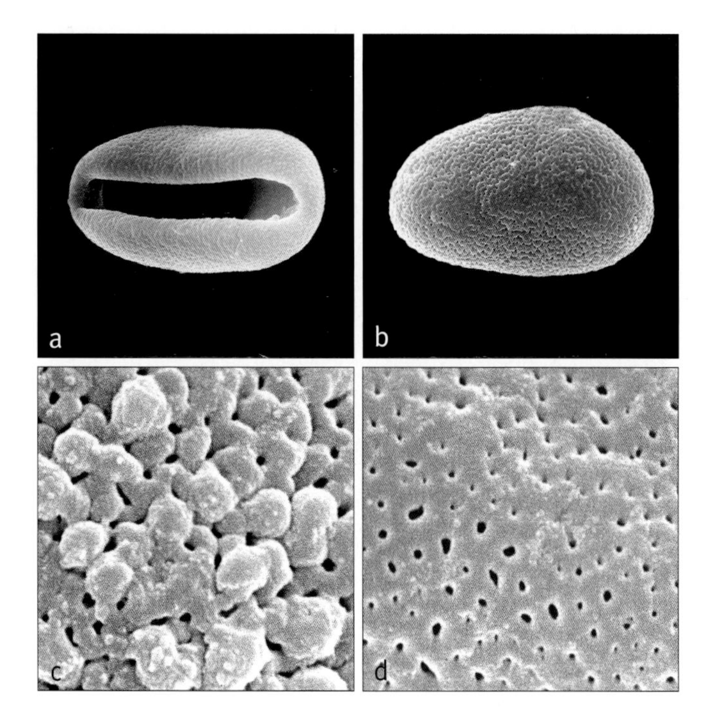

Anatomy: Stems, leaves and roots (Henderson & Galeano 1996), and root (Seubert 1998a, 1998b).

Relationships: *Prestoea* is monophyletic (Henderson 1999a). The genus is resolved as sister to *Oenocarpus* and *Neonicholsonia* with moderate support (Baker *et al.* in review), or as sister to *Oenocarpus* (Henderson 1999a).

Common names and uses: Mountain cabbage palm (*Prestoea acuminata* var. *montana*). The 'cabbage' is edible (Hodge 1965), and some species are desirable ornamentals.

Taxonomic accounts: Henderson & Galeano (1996), see also Henderson (1999a).

Fossil record: No generic records found.

Notes: For differences from *Euterpe*, see comments under *Euterpe*.

Top: ***Prestoea***
a, monosulcate pollen grain, distal face SEM × 1500;
b, monosulcate pollen grain, proximal face SEM × 1500;
c, close-up, perforate-rugulate, slightly verrucate, pollen surface SEM × 8000;
d, close-up, perforate pollen surface SEM × 8000.
Prestoea decurrens: **a**, *Moore* 6576; *P. ensiformis*: **b**, *Henderson* 501; *P. tenuiramosa*: **c**, *Henderson* 5500; *P. acuminata*: **d**, *Arguello* 399.

Far left: ***Prestoea acuminata***, inflorescence, Ecuador. (Photo: F. Borchsenius)

Left: ***Prestoea schultzeana***, inflorescence, Peru. (Photo: J.-C. Pintaud)

114. NEONICHOLSONIA

Usually acaulescent palm of forest undergrowth in Central America, remarkable for its spicate inflorescence.

Neonicholsonia Dammer, Gard. Chron. series 3. 30: 178 (1901). Lectotype: **N. watsonii** Dammer (see H.E. Moore 1951).

Bisnicholsonia Kuntze in Post & Kuntze, Lex. gen. phan., 621 and inserenda (1903) (superfluous name).

Woodsonia, L.H. Bailey, Gentes Herb. 6: 262 (1943). Type: *W. scheryi* L.H. Bailey (= *N. watsonii* Dammer).

Commemorates George Nicholson (1847–1908); neo — *new, added to distinguish from* Nicolsonia *(Fabaceae).*

Small, acaulescent or short, solitary (?always), unarmed, pleonanthic, monoecious palm. *Stem* rhizomatous, or very short, sheathed by leaf bases. *Leaves* pinnate; sheaths short, thin, fibrous, opening opposite the petiole, not forming a crownshaft; petiole moderate, slender, 4-sided; rachis 4-sided basally, distally angled adaxially, rounded abaxially, glabrous; leaflets lanceolate or linear, rather short, tapering to a point, opposite to subopposite, single-fold, distal pair sometimes united basally, thin and papery, adaxially glabrous, abaxially lightly waxy-tomentose, midrib prominent, transverse veinlets not evident or visible and short.

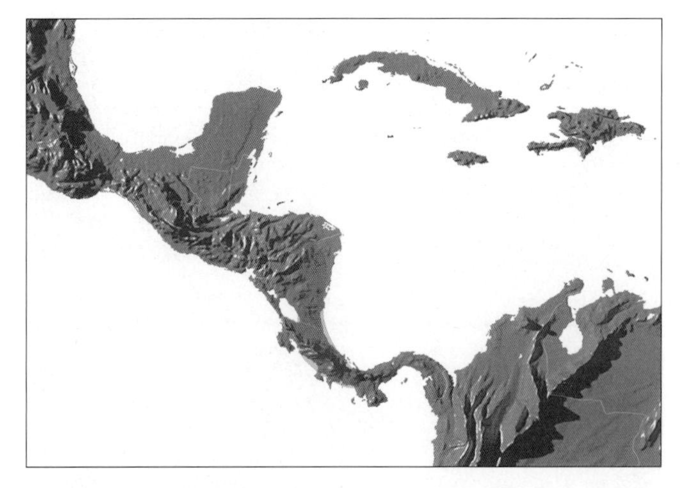

Distribution of ***Neonicholsonia***

Inflorescences solitary, interfoliar, spicate; peduncle very long, slender; prophyll short, 2-keeled laterally, splitting apically; peduncular bract much longer than the prophyll, terete, slender, beaked, splitting abaxially; flower-bearing portion also elongate but shorter than the peduncle, densely hairy becoming glabrous, bearing spirally arranged, closely crowded, shallow, pointed bracts each subtending a triad basally

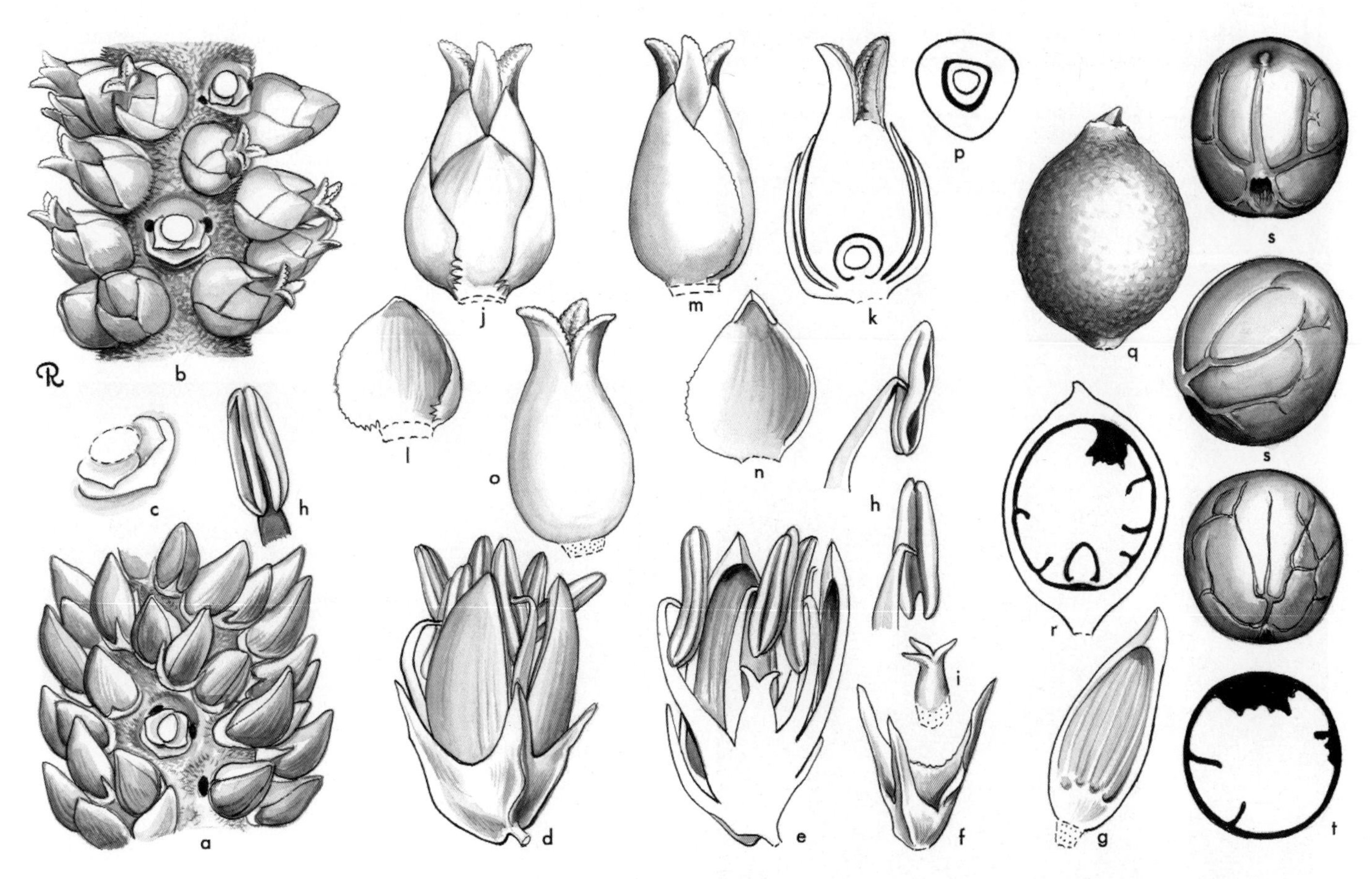

Neonicholsonia. **a**, portion of spike at staminate anthesis × 3; **b**, portion of spike at pistillate anthesis × 3; **c**, bracteoles subtending pistillate flower × 6; **d**, staminate flower × 6; **e**, staminate flower in vertical section × 6; **f**, staminate calyx × 6; **g**, staminate petal, interior view × 6; **h**, stamen in 3 views × 6; **i**, pistillode × 6; **j**, pistillate flower × 6; **k**, pistillate flower in vertical section × 6; **l**, pistillate sepal, interior view × 6; **m**, pistillate flower, sepals removed × 6; **n**, pistillate petal, interior view × 6; **o**, gynoecium × 6; **p**, ovary in cross-section × 6; **q**, fruit × 3; **r**, fruit in vertical section × 3; **s**, seed in 3 views × 3; **t**, seed in cross-section × 3. *Neonicholsonia watsonii*: all from *Huttleston* 1943. (Drawn by Marion Ruff Sheehan)

and pairs of or single staminate flowers distally; floral bracteoles conspicuous, shallow, pointed, the outer markedly larger than the inner. *Staminate flowers* lateral to the pistillate, somewhat variable, ovoid, curved in bud; sepals 3, basally connate for 1/2 their length in a short erect cupule with 3 long, narrow, pointed tips, keeled or not; petals 3, distinct, valvate, about twice as long as the sepals, striate, grooved adaxially, narrow or rather broad, tips thickened; stamens 6, distinct or briefly connate basally, filaments awl-shaped, somewhat enlarged basally, antesepalous filaments shorter, anthers elongate, medifixed, basally ± sagittate, versatile, latrorse, connective ± prolonged in a tip; pistillode short or slender and elongate, trifid apically or throughout. *Pollen* ellipsoidal, with slight or obvious asymmetry; aperture a distal sulcus; ectexine semitectate, coarsely foveolate-reticulate or reticulate, aperture margin slightly coarsely perforate or foveolate; infratectum columellate; longest axis 55–62 µm [1/1]. *Pistillate flowers* ovoid; sepals 3, distinct, partially imbricate, margins somewhat fringed, tips pointed, striate; petals like the sepals but tips thickened, valvate; staminodes lacking; gynoecium ovoid, unilocular, uniovulate, style not evident, stigma with 3 fleshy lobes, briefly reflexed, papillose adaxially, ovule basally attached, form unknown. *Fruit* ovoid, black when mature, irregular, remnants of 2 abortive carpels present, stigmatic remains apical to subapical forming a distinct beak; epicarp thin, smooth, mesocarp thin, composed of large, flat fibres, endocarp thin, crustaceous. *Seed* globose, hilum basal, raphe branches few, large, deeply sunken making the endosperm ruminate peripherally; embryo large, basal to subbasal. *Germination* adjacent ligular; eophyll bifid, the tips further divided. *Cytology*: 2n = 36.

Distribution and ecology: One variable species in Panama and Nicaragua. Found only in rain forest at low elevations from 0–250 m.

Anatomy: Stems, leaves and roots (Henderson & Galeano 1996), and root (Seubert 1998a, 1998b).

Relationships: *Neonicholsonia* is resolved as sister to *Oenocarpus* with moderate support (Baker *et al.* in review), or as sister to *Prestoea* and *Oenocarpus* (Henderson 1999a).

Common names and uses: *Coladegallo*. No uses recorded.

Taxonomic accounts: Henderson & Galeano (1996), see also Henderson (1999a).

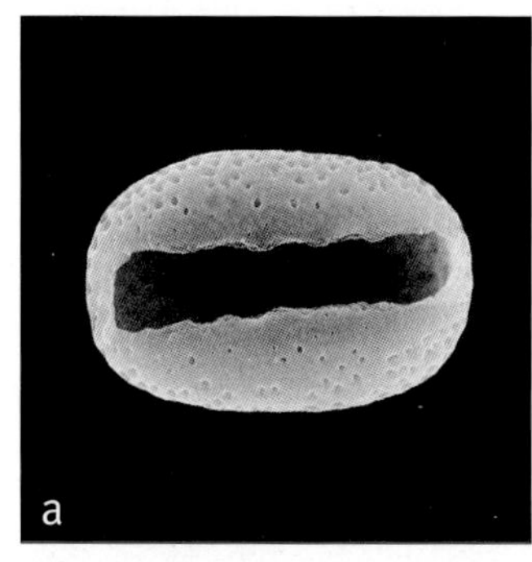

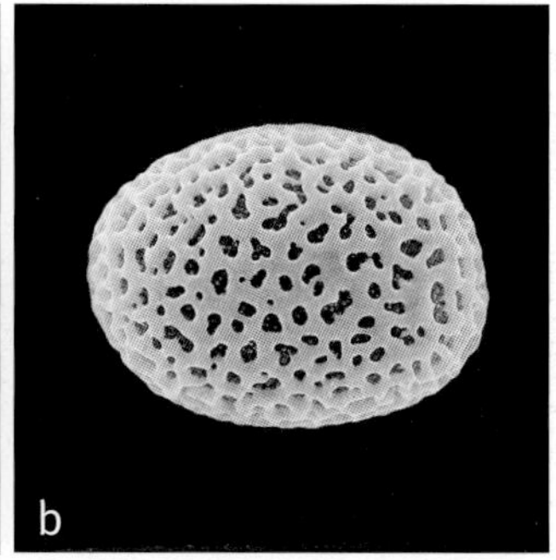

Above: ***Neonicholsonia***
a, monosulcate pollen grain, distal face SEM × 800;
b, monosulcate foveolate-reticulate pollen grain, proximal face SEM × 800.
Neonicholsonia watsonii: **a** & **b**, *Baker* 1152.

Top left: ***Neonicholsonia watsonii***, habit, Costa Rica. (Photo: A.J. Henderson)

Bottom left: ***Neonicholsonia watsonii***, spike at staminate anthesis, Costa Rica. (Photo: A.J. Henderson)

Bottom right: ***Neonicholsonia watsonii***, spike in fruit, Costa Rica. (Photo: A.J. Henderson)

Fossil record: From the Upper Oligocene–Lower Mocene of Mexico (State of Chiapas), a single flower has been found that, "has some features found in the genus *Neonicholsonia*, but cannot be placed with certainty in any extant genera" (Poinar 2002a). Two monosulcate palm pollen types are recovered from the Pliocene, Gatun Lake Formation, Panama (Graham 1991). The second of these types, bisymmetrical and finely reticulate, is compared with *Colpothrinax*, *Cryosophila* and *Neonicholsonia*. Of these, *Cryosophila* is the most probable identity.
Notes: The peripheral rumination of the seed appears to result from deeply impressed raphe branches.

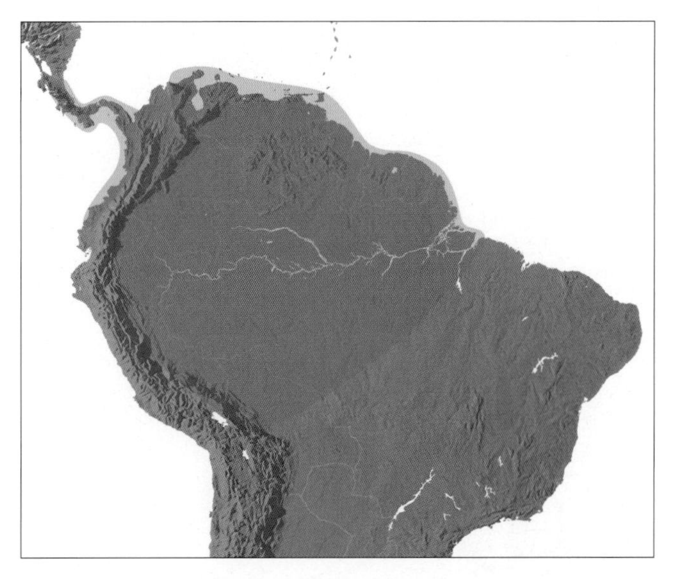
Distribution of ***Oenocarpus***

115. OENOCARPUS

Moderate to very large, solitary or clustered pinnate-leaved palms from Central and South America, with distinctive inflorescences in the form of a horse's tail.

Oenocarpus Mart., Hist. nat. palm. 2: 21 (1823). Lectotype: **O. bacaba** Mart. (see H.E. Moore 1963c).
Jessenia H. Karst., Linnaea 28: 387 (1857). Type: *J. polycarpa* H. Karst. (= *Oenocarpus bataua* Mart., *J. bataua* [Mart.] Burret).

Oinos — *wine*, karpos — *fruit, in reference to use of the fruit to make a drink.*

Moderate to massive, solitary or clustered, unarmed (except for sharp fibres of leaf sheaths), pleonanthic, monoecious palms. *Stem* erect, densely covered in fibrous leaf sheaths, when mature becoming bare except rarely (*Oenocarpus circumtextus*) fibrous network persistent, leaf scars smooth, flush with stem basally, swollen and prominent distally, a small mass of slender roots sometimes present basally. *Leaves* pinnate or entire-bifid, spirally arranged or distichous, suberect when young, becoming spreading; sheath tightly clasping but not forming a distinct crownshaft, splitting at least partially opposite the petiole, thick, leathery, lightly furrowed adaxially, glabrous or scaly abaxially, disintegrating marginally into masses of hair-like black or brown fibres and sometimes also fewer stout, sharp, knitting-needle-like fibres; petiole short, rarely elongate, channelled adaxially, rounded abaxially; leaflets regularly arranged in one plane or irregularly clustered, broadly lanceolate, acute to tapering, single-fold, blade adaxially glabrous, sparsely to densely abaxially glabrous or covered with persistent, shining, pale, straw-coloured or brownish, membranous, orbicular to transversely elliptical or sickle-shaped or needle-like medifixed scales, or with scattered, whitish, waxy, sickle-shaped hairs, midrib largest but other intermediate veins also large, transverse veinlets not evident. *Inflorescences* interfoliar in bud, becoming infrafoliar, hippuriform (shaped like a horse's tail), protandrous, branched to 1 order laterally and abaxially, adaxial branches absent; peduncle short to elongate, flattened, tomentose; prophyll short, wide, adaxially flattened, 2-keeled, splitting abaxially, margins broadly toothed; peduncular bract much longer than the prophyll, terete, beaked, scaly; rachis longer

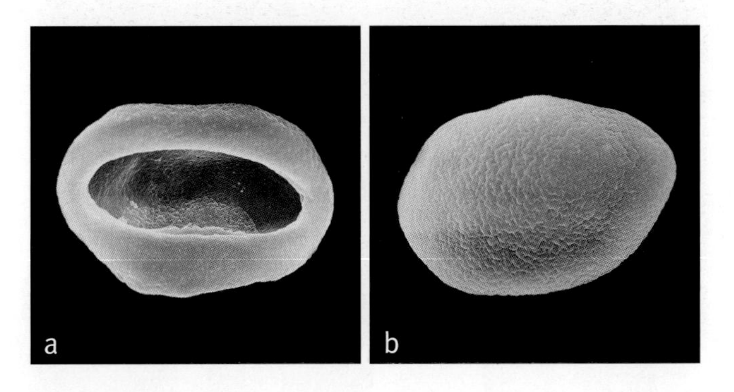

Above: ***Oenocarpus***
a, monosulcate pollen grain, distal face SEM × 1000;
b, monosulcate pollen grain, proximal face SEM × 1000.
Oenocarpus bataua: **a** & **b**, *Krukoff* 5758.

Right: ***Oenocarpus circumtextus***, habit, Colombia. (Photo: A.J. Henderson)
Far right: ***Oenocarpus mapora***, inflorescences, Brazil. (Photo: A.J. Henderson)

than the peduncle but short, tapering, bearing spirally arranged, very small, slightly sunken, pointed to scalloped, thin bracts, adaxial ones abortive and evident only in young stages, lateral and abaxial bracts subtending rachillae; rachillae ± flexuous, pendulous, short to elongate, straight to slightly undulate, slender, tapering, bearing triads of flowers basally and pairs to single staminate flowers distally, rarely completely staminate, flowers borne in shallow depressions; rachilla bracts low, rounded with a short point, slightly sunken; floral bracteoles similar to rachilla bracts. *Staminate flowers* asymmetrical, pointed in bud; sepals 3, distinct, valvate, imbricate or briefly connate basally; petals 3, distinct, ovate, somewhat asymmetrical, valvate; stamens 6 or (7–8) 9–20, filaments terete, slender, straight or variously curved and bent, distinctly inflexed at the apex, anthers elongate, basally free and sagittate, rounded or blunt apically, dorsifixed, versatile, connective not extending above locules, latrorse; pistillode bifid or trifid. *Pollen* ellipsoidal, occasionally oblate triangular, with slight or obvious asymmetry; aperture a distal sulcus, occasionally, a trichotomosulcus; ectexine tectate, finely or coarsely perforate-rugulate, aperture margin finer; infratectum columellate; longest axis 38–56 µm [6/9]. *Pistillate flowers* shorter than the staminate; sepals 3, distinct, suborbicular, imbricate, hooded; petals 3, distinct, imbricate except for valvate apices when young, otherwise like the sepals; staminodes tooth-like or lacking; gynoecium ovoid, briefly stalked, unilocular, uniovulate, style short, cylindrical, bearing 3 fleshy stigmas,

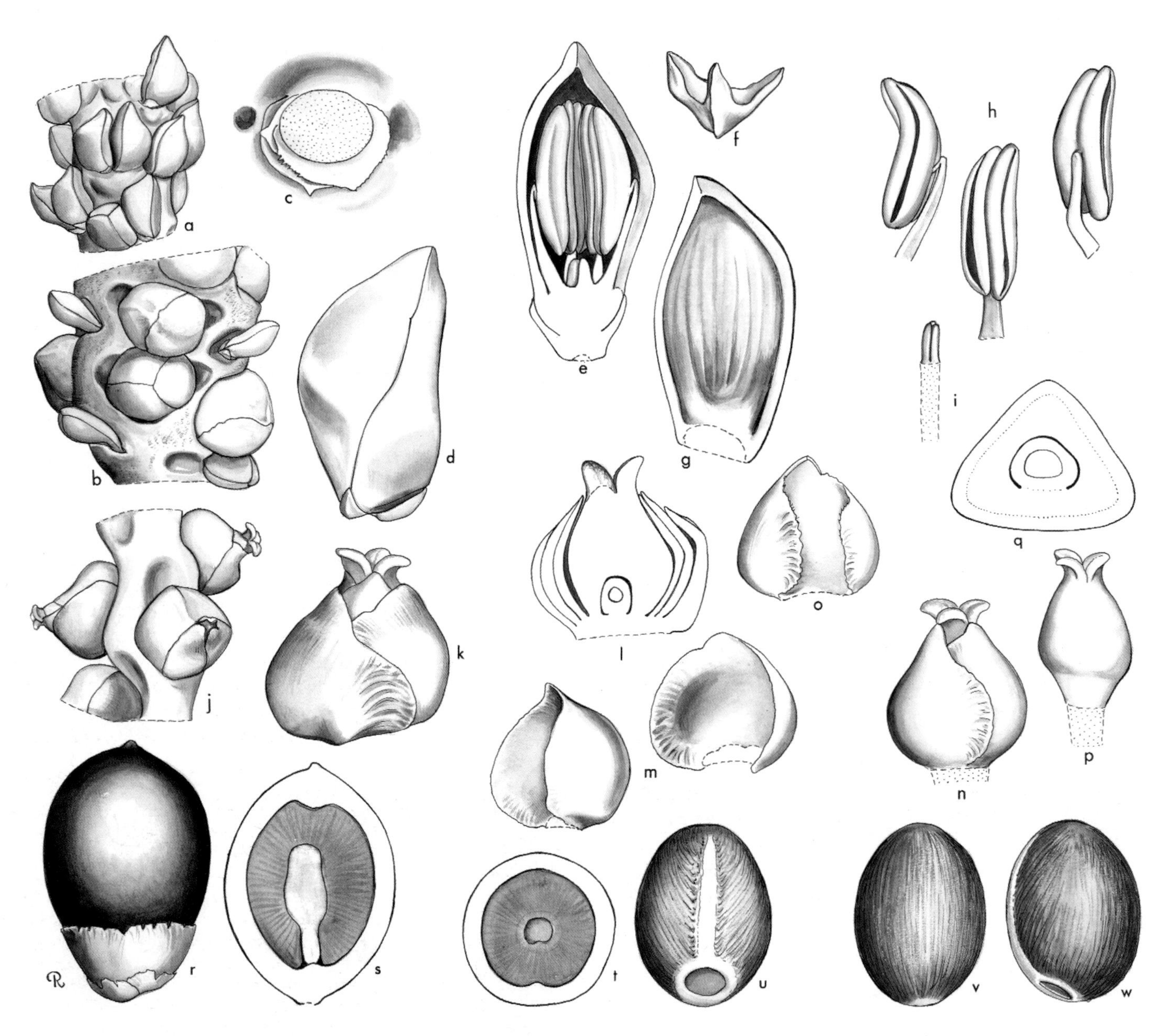

Oenocarpus. **a**, distal portion of rachilla with paired staminate flowers × 3; **b**, proximal portion of rachilla with triads × 3; **c**, triad, flowers removed to show bracteoles × 6; **d**, staminate bud × 12; **e**, staminate bud in vertical section × 12; **f**, staminate calyx × 12; **g**, staminate petal, interior view × 12; **h**, stamen in 3 views × 12; **i**, pistillode × 12; **j**, portion of rachilla at pistillate anthesis × 3; **k**, pistillate flower × 6; **l**, pistillate flower in vertical section × 6; **m**, pistillate sepals in 2 views × 6; **n**, pistillate flower, sepals removed × 6; **o**, pistillate petal × 6; **p**, gynoecium × 6; **q**, ovary in cross-section × 12; **r**, fruit × 1$^1/_2$; **s**, fruit in vertical section × 1$^1/_2$; **t**, fruit in cross-section × 1$^1/_2$; **u**, **v**, **w**, seed in 3 views × 1$^1/_2$. *Oenocarpus mapora*: **a–c**, *Moore et al.* 8412; **d–i**, *Moore et al.* 8406; **j–q**, *Moore et al.* 8415; **r–w**, *Moore & Parthasarathy* 9479. (Drawn by Marion Ruff Sheehan)

reflexed at anthesis, papillose adaxially. *Fruit* ellipsoidal to globose, dark purple when ripe, perianth persistent, stigmatic remains apical to slightly eccentric; epicarp smooth or minutely pebbled, waxy, mesocarp fleshy, oily, with internal fibres adnate to and covering the seed, endocarp apparently lacking. *Seed* ovoid-ellipsoidal to globose, hilum basal, raphe lateral, branches parallel, indistinct, endosperm homogeneous and striate, or ruminate, with central cavity; embryo basal, very large, extending through the endosperm into central cavity. *Germination* adjacent-ligular; eophyll bifid. *Cytology*: 2n = 36.

Distribution and ecology: Nine species ranging from Costa Rica and Panama to the Amazon and Orinoco Valleys in Colombia, Ecuador, Venezuela, Guyana, Surinam, French Guiana, Brazil, Peru and Bolivia. Rain forest species found on sandy soil of terra firme areas, along river margins.
Anatomy: Root (Seubert 1998a, 1998b).
Relationships: The monophyly of *Oenocarpus* has not been tested. The genus is resolved as sister to *Neonicholsonia* with moderate support (Baker *et al.* in review), or as sister to *Prestoea* (Henderson 1999a).
Common names and uses: *Bacaba*, seje palm, mille pesos palm, for local names see Balick (1985). Important for pericarp oil; the mesocarp provides a creamy drink. The 'cabbage' is edible and good, and the trunk is used for construction and spears. For more details see Balick (1980, 1985).
Taxonomic account: Balick (1980).
Fossil record: No generic records found.
Notes: *Jessenia* was separated from *Oenocarpus* by its ruminate endosperm, discolorous leaflets and stamen number greater than six; there is now good evidence for its inclusion in *Oenocarpus* (Henderson 1999a).

Above left: ***Oenocarpus bataua***, habit, Brazil. (Photo: A.J. Henderson)
Above right: ***Oenocarpus balickii***, infructescences, Brazil. (Photo: A.J. Henderson)

Tribe Geonomateae Luerss., Handb. Syst. Bot. 2: 342 (Jan 1880). Type: **Geonoma**.

Diminutive to large, acaulescent to erect, unarmed; leaves pinnate or pinnately ribbed; crownshaft absent; inflorescences spicate or branched to up to 3 orders, peduncular bracts usually 1, sometimes 2 or more, rarely absent; flowers always in triads sunken in pits; petals of the pistillate flower connate basally in a soft tube; gynoecium tri- or uniovulate, style usually slender and elongate; fruit 1-seeded, stigmatic remains basal; epicarp smooth or slightly pebbled, not warty; endosperm homogeneous.

The genera that make up this tribe are restricted to the New World humid tropics. Their habit ranges from subcanopy trees to some of the smallest of all palms. The genera show specialisations in floral structure, such as connate sepals and petals, elongate floral receptacles and styles. *Geonoma* usually has pseudomonomerous gynoecia but the other genera are trilocular and triovulate.

Tribe Geonomateae is monophyletic with high bootstrap support in most recent studies (Hahn 2002b, Lewis & Doyle 2002, Roncal *et al.* 2005, Baker *et al.* in review, in prep.) and is monophyletic with moderate support in a few earlier studies (Asmussen *et al.* 2000, Asmussen & Chase 2001). There are several conflicting, poorly supported hypotheses on the position of the tribe, although it consistently resolves within the core arecoid clade. Baker *et al.* (in review) and Uhl *et al.* (1995) resolve a sister relationship between the Geonomateae and the Manicarieae. Asmussen and Chase (2001) and Cuenca and Asmussen-Lange (2007) resolve a sister relationship between Geonomateae and Euterpeae. The tribe is sister to the Areceae in the phylogeny resolved by Lewis and Doyle (2001), and sister to a clade of Pelagodoxeae, Geonomateae, Euterpeae and Areceae in the phylogenies described by Baker *et al.* (in prep.) and by Norup *et al.* (2006).

Key to the Genera of Geonomateae

1. Moderate to large tree palms; leaves regularly pinnate . . 2
1. Acaulescent or small understory palms, stem short or if tall then slender and cane-like; leaves undivided and bifid or variously divided, rarely evenly pinnate 3
2. Leaflets broadly lanceolate, lacking a distinct midrib; inflorescence stout, rachis very short bearing a few (ca. 8) long pendulous rachillae; stamens numerous 36(–42); petals of pistillate flower connate for $^2/_3$ their length, tips valvate, glumaceous; staminodes ca. 15, awl-shaped where free . . . . **116. Welfia**
2. Leaflets narrowly lanceolate, midrib and two pairs of lateral ribs evident abaxially; inflorescence moderate, bearing many rachillae equal in length, clustered at the end of the long peduncle; stamens 6; petals of pistillate flower connate in a tube, opening by a circumscissile cap; staminodes connate forming a fleshy tube enclosing the style and stigmas . . . . **119. Calyptronoma**

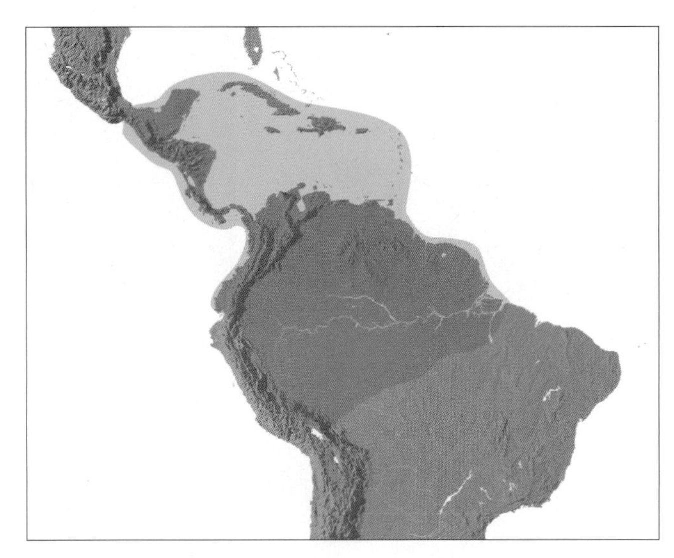

Distribution of tribe **Geonomateae**

3. Bracts covering floral pits not 'locked' into the rachilla distally but upper margins rounded, truncate, or split, lateral margins adnate to the rachilla beside the floral pits; anthers terminal on the end of the connective, inflexed in bud, thecae spread apart or parallel; ovary unilocular at anthesis . . . . **121. Geonoma**
3. Bracts covering the floral pits overlapping laterally or immersed in the rachilla; anthers sagittate or thecae terminal on a bifid connective; ovary trilocular at anthesis 4
4. Stem solitary, short to moderate; leaves ± irregularly divided into several-fold pinnae; petioles long, slender; bracts covering the floral pits overlapping laterally, densely tomentose; filaments of stamens and staminodes united in a short tube but free and awl-shaped distally . . . . **117. Pholidostachys**
4. Stem solitary, usually very short or lacking; leaves bifid or with usually unequal, several-fold pinnae; petioles short; bracts covering the floral pits not overlapping laterally, glabrous or lightly hairy, not densely tomentose 5
5. Leaves bifid; bracts covering the floral pits 'locked' in bud by a distinct rounded upper lip on the pit; theca separated on a bifid connective; staminodes free and fleshy distally . . . . **120. Asterogyne**
5. Leaves irregularly divided; floral pits without definite upper lip; anthers sagittate; staminodes united in a tube, constricted at the middle, rounded and very briefly 6-lobed distally . . . . **118. Calyptrogyne**

116. WELFIA

Moderate pinnate-leaved palm from Central and South America, distinctive in the very robust rachillae, with deep pits out of which emerge the flowers, the staminate flower with numerous stamens, and the 2-keeled fruit.

Welfia H. Wendl., Gartenflora 18: 242 (1869). Type: **W. regia** H. Wendl. ex André.

Named for the House of Welf (English Guelf *or* Guelph*, Italian* Guelpho*), dynasty of German nobles and rulers in Italy and central Europe in the Middle Ages, later including the Hanoverian Welfs, who became rulers of Great Britain.*

Moderate, solitary, unarmed, pleonanthic, monoecious palm. *Stem* erect, leaf scars conspicuous, wide, rather distant, upper trunk orange to brown. *Leaves* large, marcescent, regularly pinnate, ± erect and arching at the tip in a feather-duster crown; sheath soon splitting opposite the petiole, not forming a crownshaft, abaxially thick, somewhat ridged, densely tomentose, margins disintegrating into large fibres; petiole short, deeply channelled adaxially, rounded abaxially, densely tomentose; rachis much longer than the petiole, adaxially flattened, laterally channelled, abaxially rounded; leaflets subopposite to almost alternate, broadly lanceolate, pendulous, single-fold, glabrous and darker adaxially, pale but with a dense layer of tomentum abaxially, midrib not evident, ca. 8–several, nearly equal, rather large veins more conspicuous abaxially, transverse veinlets not evident. *Inflorescences* interfoliar and erect in bud, becoming infrafoliar and pendulous, branched to 1(–2) orders, protandrous; peduncle short, very stout, recurved; prophyll tubular, flat, wide, woody, deeply grooved abaxially, tomentose, margins with wide flat keels, often notched; complete peduncular bracts 1, like the prophyll but shorter and thinner, leathery, subsequent peduncular bracts several, short, stiff, rounded, spirally inserted above the peduncular bract; rachis short, about as long as the peduncle, bearing spirally arranged, small, rounded or pointed, ovate bracts subtending rachillae; rachillae stout, bearing 8 rows of partly sunken, stiff ovate bracts each subtending a triad of flowers borne in a pit. *Staminate flowers* sessile, borne to the outside of the pistillate flower within the pit; sepals 3, chaffy, briefly united basally to the floral receptacle, narrow, keeled, overlapping distally in bud; petals 3, elongate, connate and joined with the floral receptacle for ca. $^{1}/_{3}$ their length, free lobes ± boat-shaped, valvate, chaffy; stamens 36 (27–42) in antesepalous and antepetalous groups (see Uhl & Moore 1980), filaments short, broad, connective with a pointed tip, anthers linear-sagittate, basifixed, introrse; pistillode consisting of 3 small tubercles or lacking. *Pollen* asymmetric ellipsoidal, or pyriform; aperture a distal sulcus; ectexine tectate, finely perforate, perforate and micro-channelled, or perforate-rugulate, aperture margin slightly finer; infratectum columellate, longest axis 25–45 µm [1/1]. *Pistillate flower* rounded, pointed in bud; sepals 3, distinct, narrow, overlapping, keeled, chaffy; petals 3, connate in a tube for $^{2}/_{3}$ their length or more, distinct lobes triangular, valvate, chaffy; staminodes numerous (15–16), adnate to the corolla tube for $^{2}/_{3}$ their length, free and awl-shaped or linear-triangular above; gynoecium trilocular, triovulate, 3-angled with the adaxial side longer, style long, cylindrical, stigmas 3, recurved, ovule axile at the centre of the locule, anatropous. *Fruit* almond-shaped, slightly compressed dorsiventrally, laterally ridged, with a short apical point, dull purple, stigmatic remains and abortive carpels basal;

Opposite: ***Welfia***. **a**, portion of rachilla in bud with pits closed by bracts × $1^{1}/_{2}$; **b**, portion of rachilla in vertical section showing pits × 2; **c**, portion of rachilla with staminate flowers × $1^{1}/_{2}$; **d**, staminate flower in abaxial view × 3; **e**, staminate flower in adaxial view × 2; **f**, staminate sepal × 3; **g**, staminate flower, sepals removed × 3; **h**, staminate flower, sepals removed, in vertical section × 3; **i**, staminate petal, interior view × 3; **j**, stamen in 3 views × 6; **k**, portion of rachilla with pistillate flower × 3; **l, m**, pistillate flower in abaxial and adaxial views × 3; **n**, pistillate flower in vertical section × 3; **o**, pistillate sepal, interior view × 3; **p**, pistillate flower, sepals removed × 3; **q**, pistillate petal, interior view × 3; **r**, staminodial tube and staminodes, expanded, interior view × 3; **s**, gynoecium × 3; **t**, ovary in cross-section × 6; **u**, portion of rachilla with fruit × 1; **v**, fruit × $1^{1}/_{2}$; **w**, fruit in vertical section × $1^{1}/_{2}$; **x**, fruit in cross-section × $1^{1}/_{2}$; **y**, seed in 3 views × $1^{1}/_{2}$. *Welfia regia*: **a** & **b** and **k–y**, *Moore & Anderson* 10183; **c–j**, *Moore & Parthasarathy* 9412 and *Tomlinson* s.n. (Drawn by Marion Ruff Sheehan)

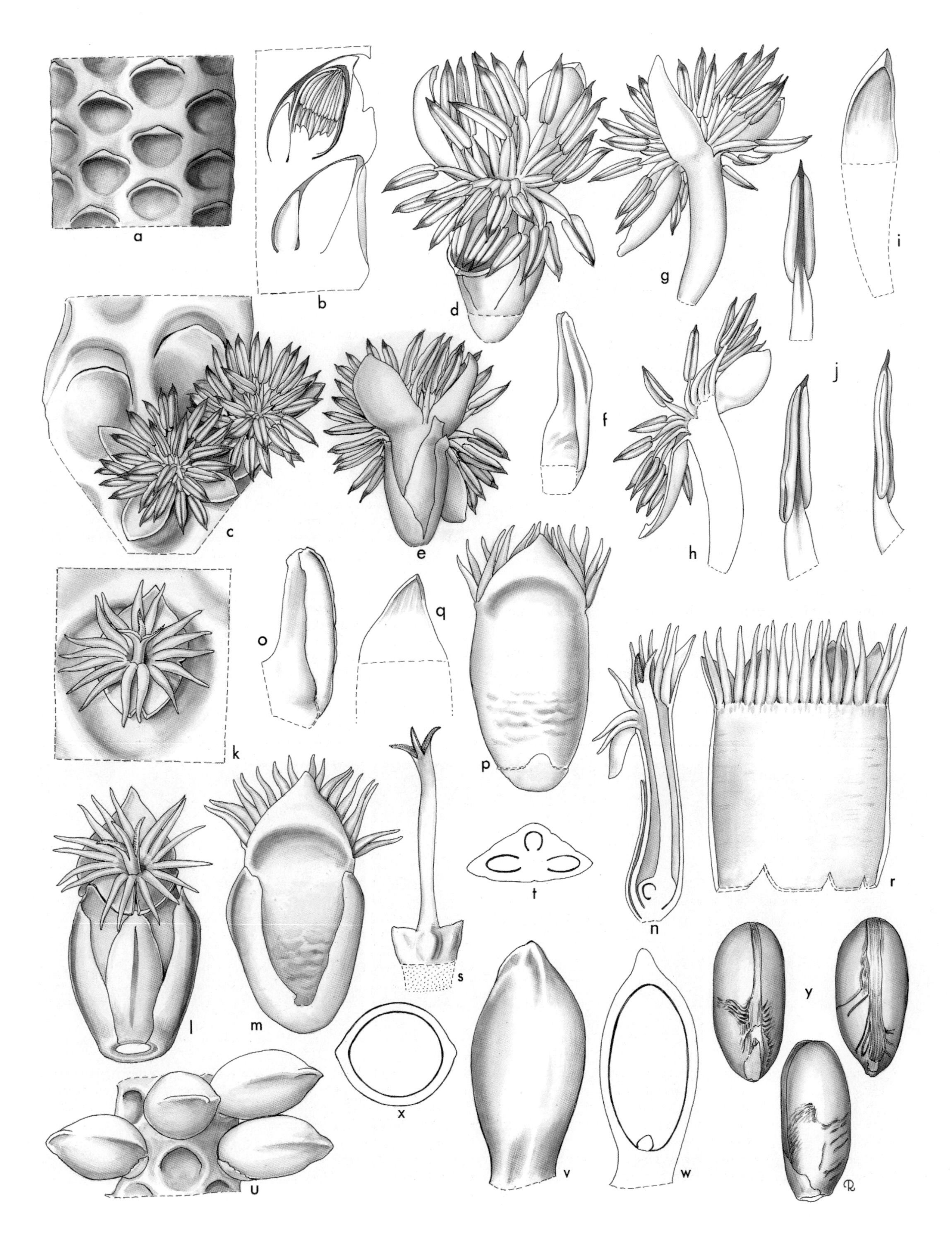
a
b
d
g
i
j
c
e
f
h
k
o
q
p
n
r
t
s
l
m
y
x
v
w
u

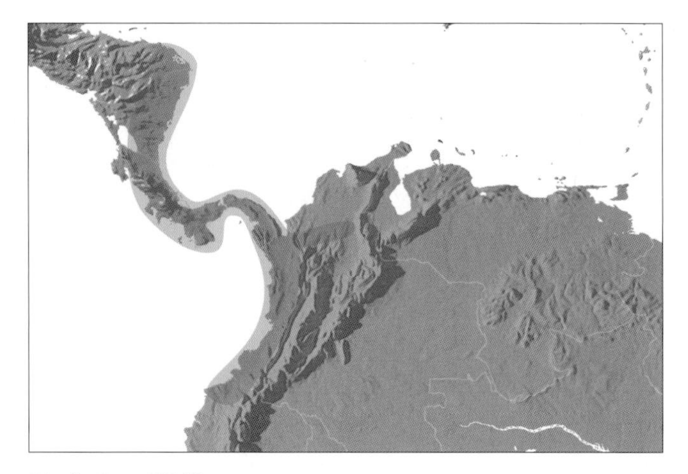

Distribution of ***Welfia***

epicarp smooth, ± shiny, mesocarp with slender parallel fibres, endocarp thin, crustaceous. *Seed* purple, ellipsoidal, rounded apically, covered in a white, sweet-tasting mucilage-like coating when ripe, hilum lateral at the base, raphe ± encircling the seed with short basal and apical branches, endosperm homogeneous; embryo eccentric at the base. *Germination* adjacent-ligular, eophyll bifid. *Cytology* unknown.

Distribution and ecology: One species, from Honduras to west and east Colombia and Ecuador. Occurring from lowlands to 2000 m above sea level in the Andes in dense rain forest.

Anatomy: Leaf (Tomlinson 1961, Wessels Boer 1968), root (Seubert 1998a, 1998b), and flower (Uhl & Moore 1980, Stauffer & Endress 2003).

Relationships: *Welfia* is resolved as sister to *Pholidostachys* with moderate support (Asmussen *et al.* 2006, Baker *et al.* in review), or as sister to a clade of *Pholidostachys*, *Calyptronoma* and *Calyptrogyne* (Asmussen 1999a).

Common names and uses: Not recorded.

Taxonomic accounts: Wessels Boer (1968) and Henderson *et al.* (1995).

Fossil record: No generic records found.

Notes: A distinctive large subcanopy tree with feather duster crown and leaves with large, several-ribbed pinnae. The numerous (about 42) stamens with filaments united in a tube basally but free and fleshy distally are diagnostic. Their development follows a pattern not recorded in any other palm genus; after sepal and petal origin, the floral apex expands in lobes opposite the sepals, and stamens arise in arcs of ca. 6 opposite each sepal and petal (Uhl & Moore 1980).

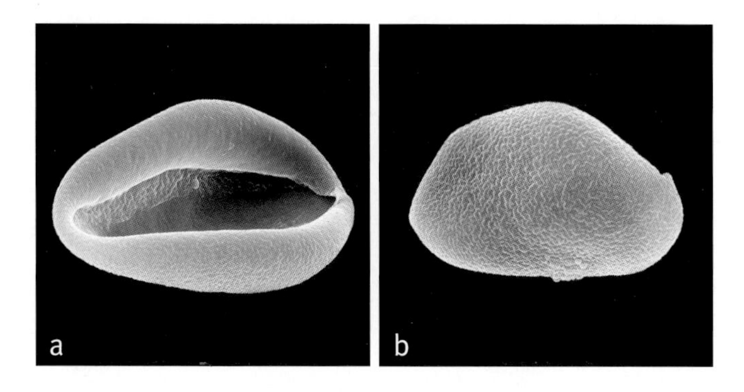

Above: ***Welfia***
a, monosulcate pollen grain, distal face SEM × 1000;
b, monosulcate pollen grain, proximal face SEM × 1000.
Welfia regia: **a** & **b**, *Henderson & Herrera* 711.

Below left: ***Welfia regia***, habit, Peru. (Photo: J.-C. Pintaud)
Below middle: ***Welfia regia***, staminate flowers, Costa Rica. (Photo: J. Dransfield)
Below right: ***Welfia regia***, part of infructescence, Peru. (Photo: J.-C. Pintaud)

117. PHOLIDOSTACHYS

Small or moderate pinnate-leaved palms from Central and South America, with distinctive long slender petioles.

Pholidostachys H. Wendl. ex Hook.f. in Benth. & Hook.f., Gen. pl. 3: 915 (1883). Lectotype: **P. pulchra** H. Wendl. ex Burret (see Burret 1930).

Calyptrogyne H. Wendl. subgenus *Pholidostachys* (H. Wendl.) Wess. Boer., Verh. Kon. Ned. Akad. Wetensch., Afd. Natuurk., Tweede Sect.: 73 (1968).

Pholidos — *scale,* stachys — *an ear of wheat, or in botanical usage, a spike, referring to the pit bracts on the rachillae.*

Small to moderate, solitary, unarmed, pleonanthic, monoecious palms. *Stems* slender, erect, closely ringed with leaf scars. *Leaves* pinnate, marcescent; sheath tubular at first, later splitting opposite the petiole, not forming a crownshaft, thin, coriaceous, margins fibrous, lightly hairy adaxially, with dense reddish-brown tomentum abaxially; petiole long, very slender, adaxially shallowly to deeply grooved, abaxially narrowly ridged, V-shaped in cross-section, ± tomentose, becoming glabrous; blade short, leaflets broadly lanceolate, tapering to a point, median leaflets much longer than the proximal or distal, all leaflets several-fold, folds narrow, midribs ca. 4–5, elevated adaxially, interfold ribs conspicuous abaxially, blade adaxially lightly tomentose near the base, abaxially sparsely covered in scales and a dense layer of wax, transverse veinlets inconspicuous. *Inflorescence* solitary, protandrous, erect, becoming pendulous and infrafoliar, spicate or digitately or paniculately branched to 1(–3) orders; peduncle short, rounded or ± flattened; prophyll short, tubular, adaxially flat and 2-keeled, keels toothed, inserted near the base of the peduncle, splitting abaxially at the tip, irregularly covered in dark reddish tomentum; peduncular bract tubular with a short solid tip, much longer (3 or more times) than the prophyll, inserted well above the prophyll, splitting adaxially near the tip, caducous or marcescent, thinly covered in dark red tomentum; other peduncular bracts several, small, spirally inserted, the lowest often shortly tubular, the others open, stiff, pointed; rachis lacking, very short, or elongate, bearing spirally inserted, short, stiff, irregular bracts each subtending a rachilla; rachillae bearing obovate bracts, alternating in 5–11 closely appressed rows, ± immersed in the axis, basally tomentose, margins thin, sometimes overlapping the margins of bracts of adjacent rows, each bract subtending a triad of flowers enclosed in a pit, floral bracteoles 3, narrow, keeled, chaffy, sepal-like. *Staminate flowers* only about $^1/_2$ exserted from the pit at anthesis; sepals 3, imbricate basally, keeled, chaffy, truncate or rounded to subacute and often toothed at the apex; petals 3, ovate with pointed tips, connate ca. $^1/_2$ their length, valvate, briefly adnate to the receptacle basally, tips chaffy; stamens 6, filaments fleshy, connate for $^2/_3$ their length in a thick tanniniferous tube, free parts angled–awl-shaped, anthers sagittate, medifixed, erect in bud, exserted and spreading at anthesis, introrse, connectives with pointed tips; pistillodes 3, minute, pointed. *Pollen* ellipsoidal, usually ± symmetric; aperture a distal sulcus; ectexine tectate or semi-tectate, finely perforate-rugulate, or finely reticulate with slightly ridged supratectal spines, aperture margin similar; infratectum columellate; longest axis 35–42 µm [2/4]. *Pistillate flowers* with only free parts of staminodes and stigmas exserted from the pit at anthesis; sepals 3, free, narrow unequal, imbricate basally, truncate or rounded to subacute, often toothed apically, keeled, chaffy; petals 3, fleshy and connate to about $^1/_3$ to $^1/_2$ their length basally (or perhaps sometimes less), with valvate, ± chaffy, free tips; staminodes 6–8, fleshy, connate in a tube $^2/_3$ their length, adnate to the corolla for a short distance basally, the free portions angled, awl-shaped, exserted and spreading at anthesis; gynoecium trilocular, triovulate, with a central elongate style terminating in 3 exserted, spreading, slender stigmas, ovules anatropous, only one maturing. *Fruit* moderate, obovoid, purple when ripe with basal remains

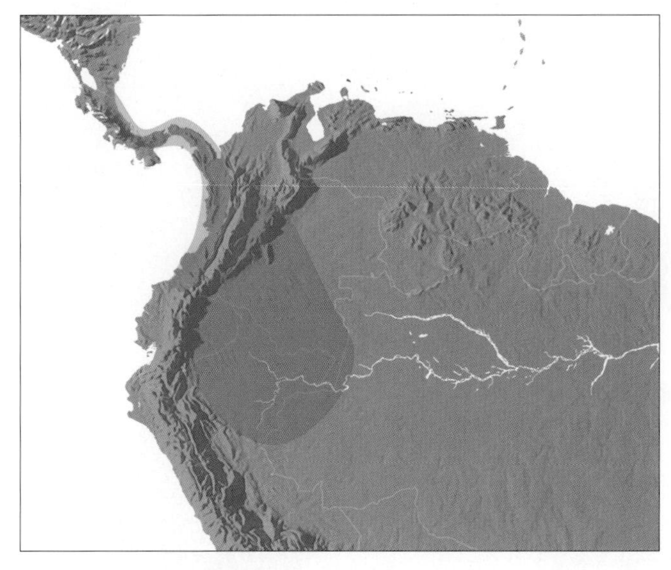

Distribution of ***Pholidostachys***

Below: ***Pholidostachys pulchra***, infructescences, Costa Rica.
(Photo: J. Dransfield)

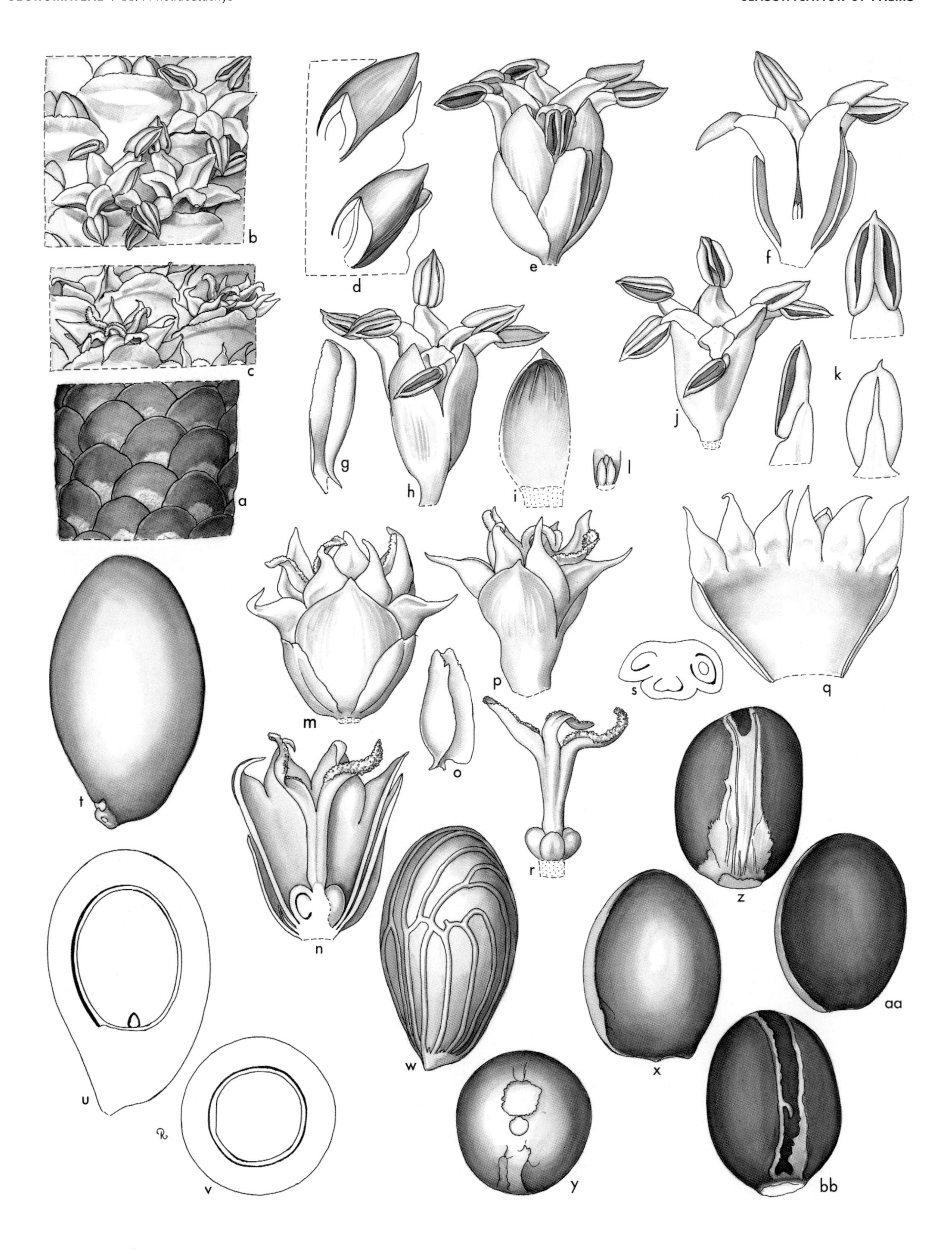
b
c
a
d
e
f
g
h
i
l
j
k
m
p
q
s
o
r
t
n
w
z
u
x
aa
v
y
bb

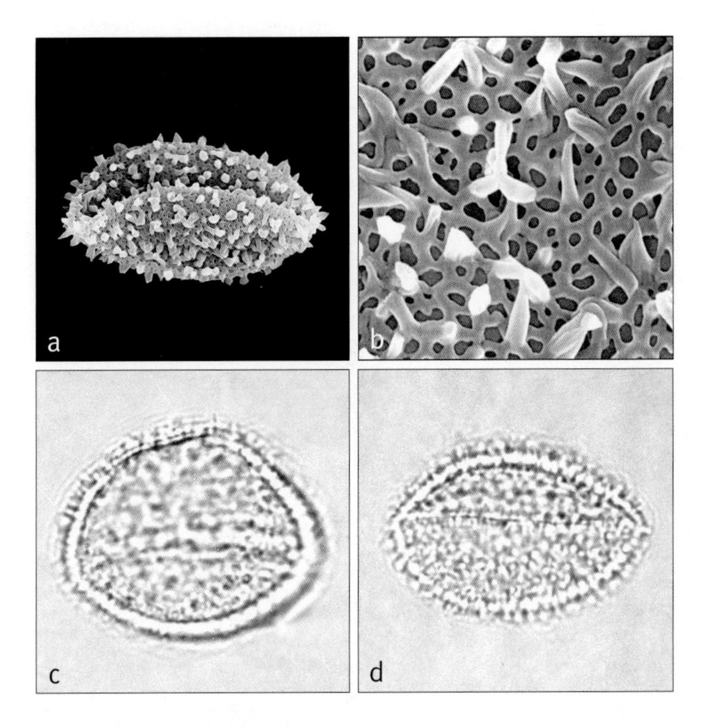

Above: ***Pholidostachys***
a, monosulcate grain, tilted distal/equatorial view SEM × 1500;
b, close-up, tectate reticulate spiny pollen surface SEM × 8000;
c, monosulcate pollen grain, mid–low-focus LM × 1500;
d, monosulcate pollen grain, high-focus LM × 1500.
Pholidostachys dactyloides: **a–d**, *Balslev et al.* 62096.

of abortive carpels and stigmas; epicarp smooth, mesocarp with outer tannin layer, fleshy granulate, with thick, curved and anastomosing included fibres, endocarp tough, whitish, thinner over the hilum, with a small operculum over the embryo. *Seed* ellipsoidal, rapheal lines arched from the rounded hilum over the apex to the base, endosperm homogeneous, sometimes with a central hollow; embryo basal. *Germination* adjacent ligular; eophyll bifid. *Cytology* not studied.

Distribution and ecology: Four species from Costa Rica to Peru. All species are found in the understory of tropical rain forest.
Anatomy: Leaf (Tomlinson 1961), leaf and stem (Wessels Boer 1968), root (Seubert 1998a, 1998b), and flower (Stauffer & Endress 2003).
Relationships: *Pholidostachys* is monophyletic with high support (Asmussen 1999b, Roncal *et al.* 2005). The genus has been resolved as sister to *Welfia* with moderate support (Asmussen *et al.* 2006, Baker *et al.* in review).
Common names and uses: Common names not recorded.
Taxonomic accounts: Wessels Boer (1968) and Henderson *et al.* (1995).
Fossil record: No generic records found.
Notes: *Pholidostachys* differs from other Geonomateae mainly in the androecium of six stamens with filaments united in a tube basally, and free and awl-shaped distally. The short to moderate stem, long petioles, and often irregularly divided leaves give it a distinctive appearance in the forest.

118. CALYPTROGYNE

Small acaulescent or short-stemmed pinnate-leaved palms of rain forest in Central and northern South America, distinctive in the inflated staminodal tube in the pistillate flower.

Calyptrogyne H. Wendl., Bot. Zeit. 17: 72 (1859). Lectotype: *C. spicigera* (Koch) H. Wendl. (*Geonoma spicigera* Koch) (see H.E. Moore 1963c) (= **Calyptrogyne ghiesbreghtiana** [Linden & H. Wendl.] H. Wendl.) (*Geonoma ghiesbreghtiana* Linden & H. Wendl.).

Kalyptra — *a lid,* gyne — *woman or female, referring to the upper part of the corolla in the pistillate flower that is pushed off like a cap.*

Small, solitary, unarmed, pleonanthic, monoecious palms. *Stem* often subterranean or short and erect, leaf scars indistinct or clearly defined, internodes sometimes rusty-brown tomentose. *Leaves* few, about 10, pinnate, usually irregular, or bifid and pinnately veined, marcescent; sheath splitting opposite the petiole, densely covered in dark caducous tomentum, margins with large fibres; petiole slender, rather short, evenly concave adaxially, angled abaxially, sparsely tomentose; leaflets irregular in width, distant, 1–several fold, tapering to pointed tips, lightly waxy-tomentose and with small scales on both surfaces, larger scales abaxially along midrib, midrib and 1–2 pairs of veins prominent abaxially, transverse veinlets usually not evident. *Inflorescence* interfoliar, slender, spicate, rarely branched to 1 order, erect peduncle very long, slender;

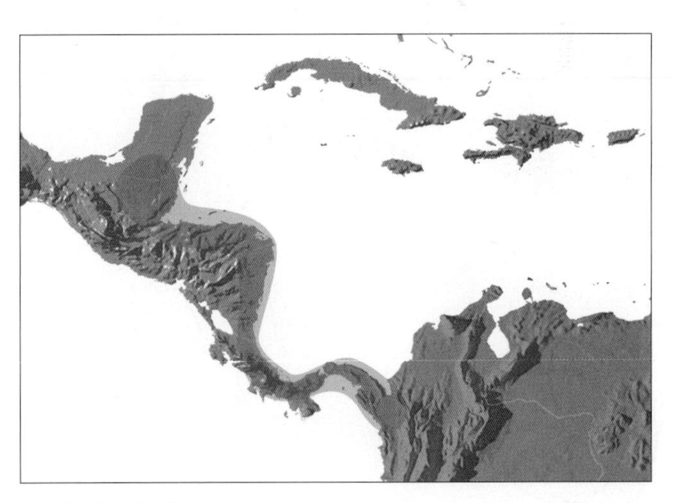

Distribution of ***Calyptrogyne***

Opposite: ***Pholidostachys*.** **a**, portion of spike in bud to show overlapping margins of bracts × 3; **b**, portion of spike at staminate anthesis × 4 1/2; **c**, portion of spike at pistillate anthesis × 3; **d**, portion of spike in vertical section to show insertion of flowers in pits × 6; **e**, staminate flower × 6; **f**, staminate flower in vertical section × 6; **g**, staminate sepal × 6; **h**, staminate flower, sepals removed × 6; **i**, staminate petal, interior view × 6; **j**, androecium, perianth removed × 6; **k**, anther and top of filaments in 3 views × 9; **l**, pistillode × 9; **m**, pistillate flower × 6; **n**, pistillate flower in vertical section × 6; **o**, pistillate sepal × 6; **p**, pistillate flower, sepals removed × 6; **q**, perianth and staminodes expanded, interior view × 6; **r**, gynoecium × 6; **s**, ovary in cross-section × 12; **t**, fruit × 2 1/4; **u**, fruit in vertical section × 2 1/4; **v**, fruit in cross-section × 2 1/4; **w**, fruit, outer layer removed to show fibers × 3; **x**, **y**, endocarp in 2 views × 3; **z**, **aa**, **bb**, seed in 3 views × 3. *Pholidostachys pulchra*: **a–s**, *Hartshorn* s.n.; **t–bb**, *Moore & Parthasarathy* 9410. (Drawn by Marion Ruff Sheehan)

prophyll tubular, rather thin, pointed, with two flat, narrow lateral keels, papery to coriaceous, striate, lightly scaly, splitting distally, inserted near the base of the peduncle; peduncular bract terete, pointed, striate, lightly scaly, membranous, splitting distally, inserted near or a short distance below the floral pits, often caducous leaving a ruffled scar; rachis very short, bearing 1–2 small empty bracts, short or long, often much larger in diameter than the peduncle, glabrous or densely brown tomentose; rachillae bearing alternate rows (7–11) of closely appressed, marginally thin, glabrous or tomentose, ovate bracts, each subtending a triad of flowers borne in a pit, floral bracteoles unequal, somewhat keeled, thin, membranous, tips pointed. *Staminate flower* slightly asymmetrical; sepals 3, free, elongate, unequal, imbricate at base or throughout in bud, narrow, tips irregular, somewhat truncate, sometimes tomentose; petals 3, asymmetrical, united in a soft tube for ca. $^1/_2$ their length, free lobes unequal, valvate, chaffy; stamens 6, filaments fleshy, united and adnate to the receptacle forming a solid stalk, free lobes thick, awl-shaped, recurved at anthesis, anthers sagittate, dorsifixed near the base, decurved with bases uppermost at anthesis, introrse; pistillode minute, deltoid. *Pollen* ellipsoidal, usually with either slight or obvious asymmetry, occasionally oblate-triangular; aperture a distal sulcus, or

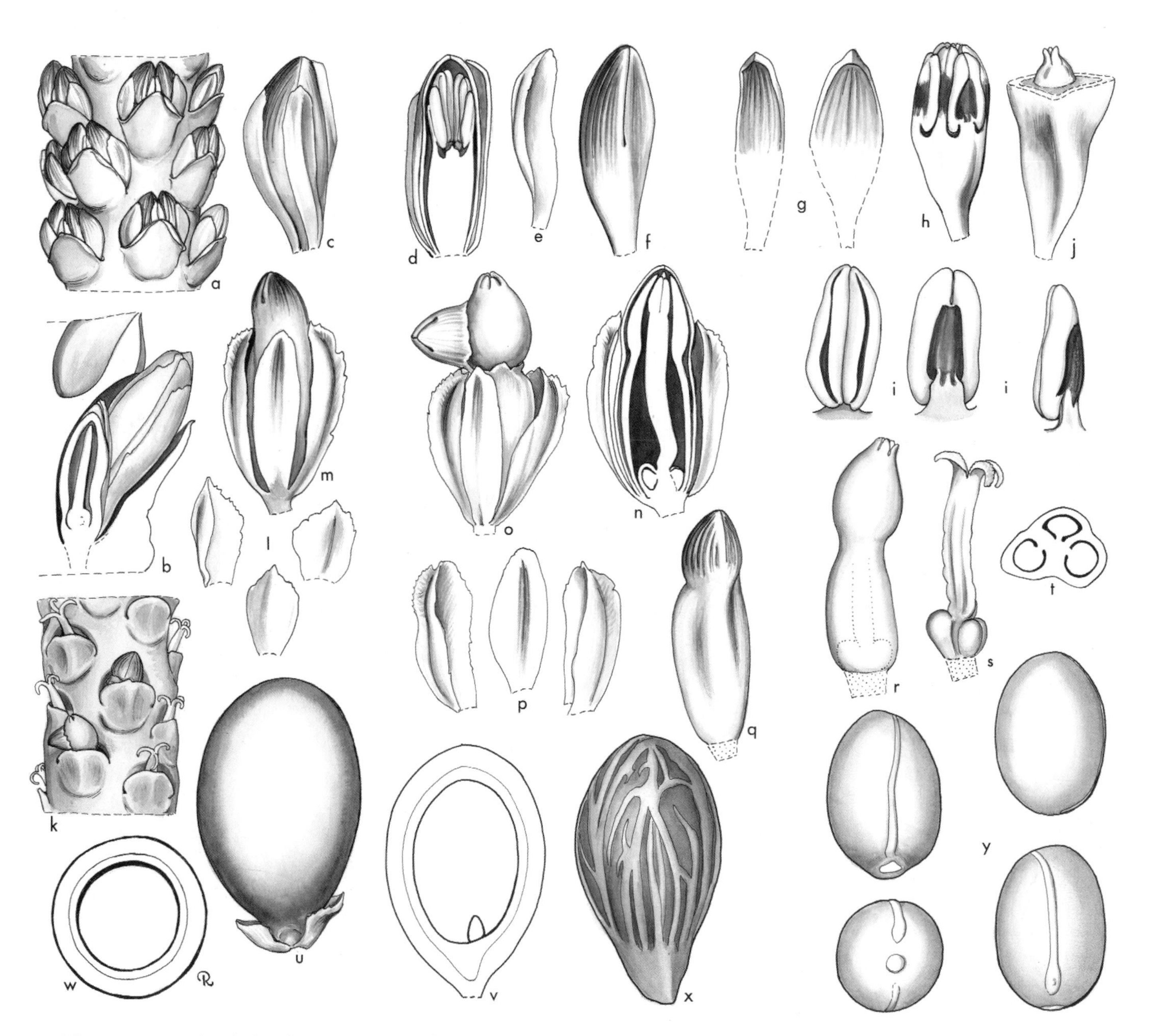

Calyptrogyne. **a**, portion of axis with staminate buds × 3; **b**, portion of axis with staminate buds in vertical section to show orientation in pit, pistillate bud in section × 6; **c**, staminate bud × 6; **d**, staminate bud in vertical section × 6; **e**, staminate sepal × 6; **f**, staminate bud, sepals removed × 6; **g**, staminate petal in 2 views × 6; **h**, androecium × 6; **i**, stamen in 3 views × 12; **j**, androecium, stamens removed to show pistillode × 12; **k**, portion of axis at pistillate anthesis with flowers in 3 stages of expansion × 3; **l**, bracteoles × 6; **m**, pistillate bud × 6; **n**, pistillate bud in vertical section × 6; **o**, pistillate flower at anthesis, calyptrate corolla apex at left, expanded staminodial tube in center × 6; **p**, pistillate sepals × 6; **q**, pistillate corolla in bud × 6; **r**, staminodial tube at anthesis × 6; **s**, gynoecium × 6; **t**, ovary in cross-section × 12; **u**, fruit × 3; **v**, fruit in vertical section × 3; **w**, fruit in cross-section × 3; **x**, endocarp × 4; **y**, seed in 4 views × 3. *Calyptrogyne ghiesbreghtiana*: **a–t**, *Moore & Parthasarathy* 9405; **u–y**, *Moore* 6785. (Drawn by Marion Ruff Sheehan)

trichotomosulcus; ectexine tectate, coarsely perforate, finely rugulate or, coarsely perforate-rugulate, aperture margin finer; infratectum columellate; longest axis 47–66 µm [6/9]. *Pistillate flower* asymmetrical, adaxial side curved to conform to the pit wall; sepals 3, free, unequal, imbricate, two lateral ones keeled, abaxial one smaller, flattened; petals united in a tube, very briefly free and valvate distally, tube striate, distal $^{1}/_{3}$ shed as a cap, lower part of tube remaining in the pit; staminodes united in a tube, constricted near the middle, very briefly 6-lobed distally, upper part shed to reveal the stigmas; gynoecium trilocular, triovulate, asymmetrically 3-lobed, style triangular in cross-section, elongate, ending in 3 linear stigmas, reflexed at anthesis, ovule anatropous, basally attached. *Fruit* obovoid, 1-seeded, purple or black when ripe, stigmatic residue and abortive carpels basal; epicarp smooth, mesocarp fleshy with inner layer of large anastomosing fibres, the largest median and completely encircling, endocarp ± transparent, tough. *Seed* ellipsoidal, basally attached, hilum short, raphe encircling, unbranched, endosperm homogeneous; embryo eccentrically basal. *Germination* adjacent ligular; eophyll bifid. *Cytology*: 2n = 28.

Distribution and ecology: Nine species from Mexico and Guatemala to Colombia. Understory palms of tropical rain forests. Abundant in swamps and along stream margins, some species at elevations below 700 m only, other species from 700–2200 m. Bats have been observed pollinating the flowers in Costa Rica (Beach 1986).

Anatomy: Leaf (Tomlinson 1961, Wessels Boer 1968), root (Seubert 1998a, 1998b), and flower (Stauffer & Endress 2003).

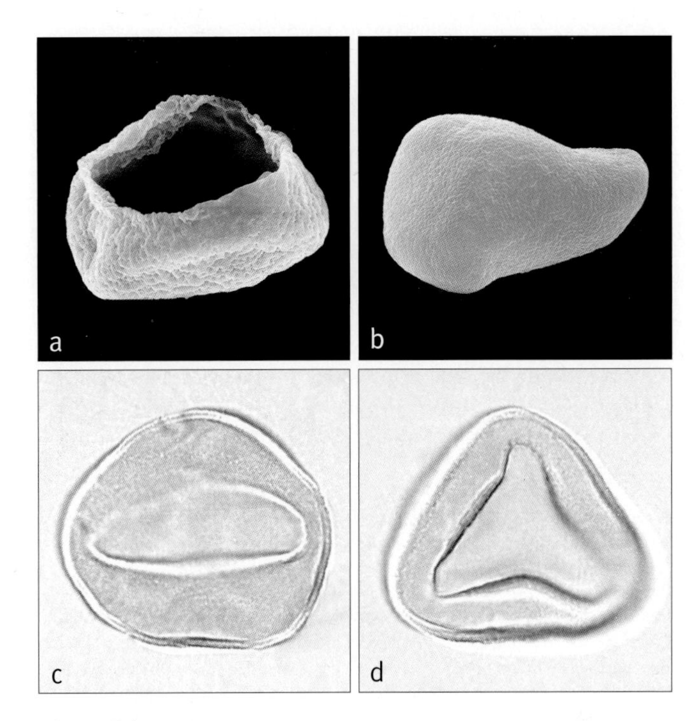

Above: ***Calyptrogyne***
a, monosulcate pollen grain, distal face SEM × 1000;
b, monosulcate pollen grain, proximal face SEM × 1000;
c, monosulcate pollen grain, mid-focus LM × 1000;
d, trichotomosulcate pollen grain, mid–low-focus LM × 1000. *Calyptrogyne ghiesbreghtiana*: **a**, *Boutin & Schlosser* 5001; *C. trichostachys*: **b–d**, *Moore et al.* 6567.

Relationships: *Calyptrogyne* is monophyletic with moderate to high support (Asmussen 1999a, Roncal *et al.* 2005). The relationships of *Calyptrogyne* are not yet resolved, despite dense species sampling and several studies of independent datasets. In the most densely sampled study to be based on nuclear genes, *Calyptrogyne* is nested within *Calyptronoma* with moderate support (Roncal *et al.* 2005). In two other studies, *Calyptrogyne* is resolved as sister to *Calyptronoma* with high support (Asmussen *et al.* 2006, Baker *et al.* in review). Finally, in a study based on one plastid DNA region, *Calyptrogyne* is resolved in a clade with *Calyptronoma* and *Pholidostachys* (Asmussen 1999b). Henderson (2005) has presented a phenetic study of the genus that may shed some light on species relationships.

Common names and uses: Common names unknown; sometimes cultivated as ornamentals.

Taxonomic accounts: Wessels Boer (1968) and De Nevers (1995).

Fossil record: No generic records found.

Notes: *Calyptrogyne* is immediately distinguished from other understory geonomoid palms by the insertion of the peduncular bract at the apex of the peduncle. Flower structure, especially the unusual staminodes, and the insertion of the peduncular bract well above the prophyll, is very similar to that of *Calyptronoma* (see comments above).

Additional figures: Fig. 1.9.

Left: ***Calyptrogyne ghiesbreghtiana***, habit, Costa Rica. (Photo: A.J. Henderson)

119. CALYPTRONOMA

Erect pinnate-leaved tree palms from the Caribbean, distinctive in the inflorescence with many rachillae radiating from the tip of the peduncle and inflated staminodal tube in the pistillate flower.

Calyptronoma Griseb., Fl. Brit. W. I. 518 (1864). Type: *C. swartzii* Griseb. (illegitimate name) = **C. occidentalis** (Sw.) H.E. Moore (*Elaeis occidentalis* Sw.).

Cocops O.F. Cook, Bull. Torrey Bot. Club 28: 568 (1901). Type: *C. rivalis* O.F. Cook (= *Calyptronoma rivalis* [O.F. Cook] L.H. Bailey).

Calyptrogyne subgenus *Calyptronoma* (Griseb.) Wess. Boer., Verh. Kon. Ned. Akad. Wetensch., Afd. Natuurk., Tweede Sect. 1968.

Kalyptra — *a lid*, nomos — *that which is in habitual use, referring to the upper part of the corolla in the pistillate flower that is pushed off like a lid or cap.*

Moderate, solitary, unarmed, pleonanthic, monoecious palms. *Stem* stout, erect, becoming bare, irregularly ringed with leaf scars. *Leaves* pinnate, erect or ± horizontal, arching, marcescent; sheath becoming open opposite the petiole, not forming a crownshaft, expanded and thicker basally, margins fibrous, both surfaces covered with waxy, caducous tomentum, thinner adaxially; petiole relatively short, deeply grooved adaxially, rounded abaxially; rachis adaxially grooved near the base, flat and laterally channelled distally; leaflets narrow, lanceolate, wider at the middle, tapering to a pointed tip, single-fold, adaxial surface lightly waxy, abaxial surface with thin waxy tomentum and scales along the midrib, midrib and 2 pairs of lateral ribs more evident abaxially, transverse veinlets not evident.

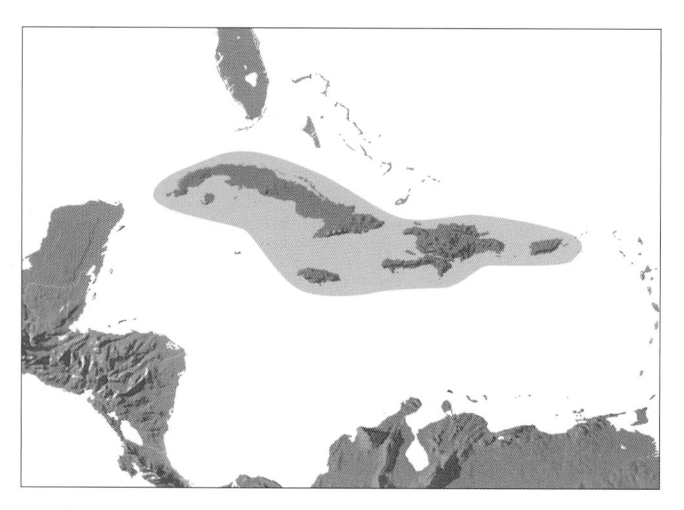

Distribution of ***Calyptronoma***

Inflorescences solitary, interfoliar, becoming infrafoliar, erect at first, pendulous later, branched to 1(–4) orders, protandrous; peduncle stout, elongate, elliptic in cross-section; prophyll tubular, dorsiventrally flattened, with 2, rather wide, lateral keels, pointed, splitting apically and ventrally, woody, lightly waxy, tomentose, inserted at the base of the peduncle; peduncular bract tubular, pointed, stiff, thinner than the prophyll, lightly waxy and tomentose, inserted about halfway along the peduncle, often shed before the prophyll leaving a ruffled scar, subsequent peduncular bracts few, short, spirally inserted; rachis ca. $^2/_3$ as long as the peduncle; rachis bracts short, wide and irregularly notched to pointed, ovate distally; lower branches with a short bare portion, then branched 1(–2) times into rachillae; rachillae moderate, rather short and crowded, bearing rows of partly sunken, short, rounded bracts, each subtending a triad of flowers borne in a pit; floral bracteoles 3, the first bracteole narrow elongate, the second short, wide, truncate, the third short, pointed. *Staminate flowers* lateral and outside of the pistillate flower in the pit; sepals 3, distinct, unequal, long, narrow, ± keeled, rounded or pointed distally, imbricate basally, chaffy; petals 3, adnate basally to the receptacle, connate for $^2/_3$ their length in a tube, tips valvate; stamens 6, filaments fleshy, connate and adnate to the receptacle for $^2/_3$ their length, the receptacle elongating at anthesis, forming a stalk-like base to the androecium, the androecium hence far-exserted distally, the filaments terete, anthers sagittate, dorsifixed near the base, introrse, versatile and horizontal at anthesis, connective tanniniferous; pistillode 3-lobed with minute tips or lacking. *Pollen* ellipsoidal, usually with either slight or obvious asymmetry; aperture a distal sulcus; ectexine tectate, perforate, perforate-rugulate or rugulate-insulate, aperture margin slightly finer or psilate; longest axis 24–49 µm. Post-meiotic tetrads tetrahedral [3/3]. *Pistillate flowers* asymmetrically triangular in bud; sepals 3, distinct, imbricate, narrow, briefly keeled, pointed, margins irregular; petals 3, connate in a tube with short valvate apices, upper $^1/_4$ shed as a circumscissile cap; staminodes completely connate, urn-shaped, inflated above the corolla tube at anthesis, white, membranous, exserted, pushing off the corolla cap, the apex briefly 6-toothed; gynoecium trilocular, triovulate, ± 3-angled, adaxial side longest, style long, cylindrical, stigmatic lobes 3, short, linear, ovule large, anatropous. *Fruit* ovoid, 1-seeded, stigmatic remains basal; epicarp smooth, granular when dry, mesocarp fleshy with stout fibres anastomosing distally, one large fibre conspicuous, endocarp thin, colourless, tough, with a circular operculum over the embryo. *Seed* globose-ellipsoidal, raphe unbranched, encircling the seed, hilum short, basal, endosperm homogeneous; embryo basal. *Germination* adjacent-ligular; eophyll bifid. *Cytology*: 2n = 28.

Left: ***Calyptronoma rivalis***, habit, cultivated, Fairchild Tropical Botanic Garden, Florida. (Photo: J. Roncal)

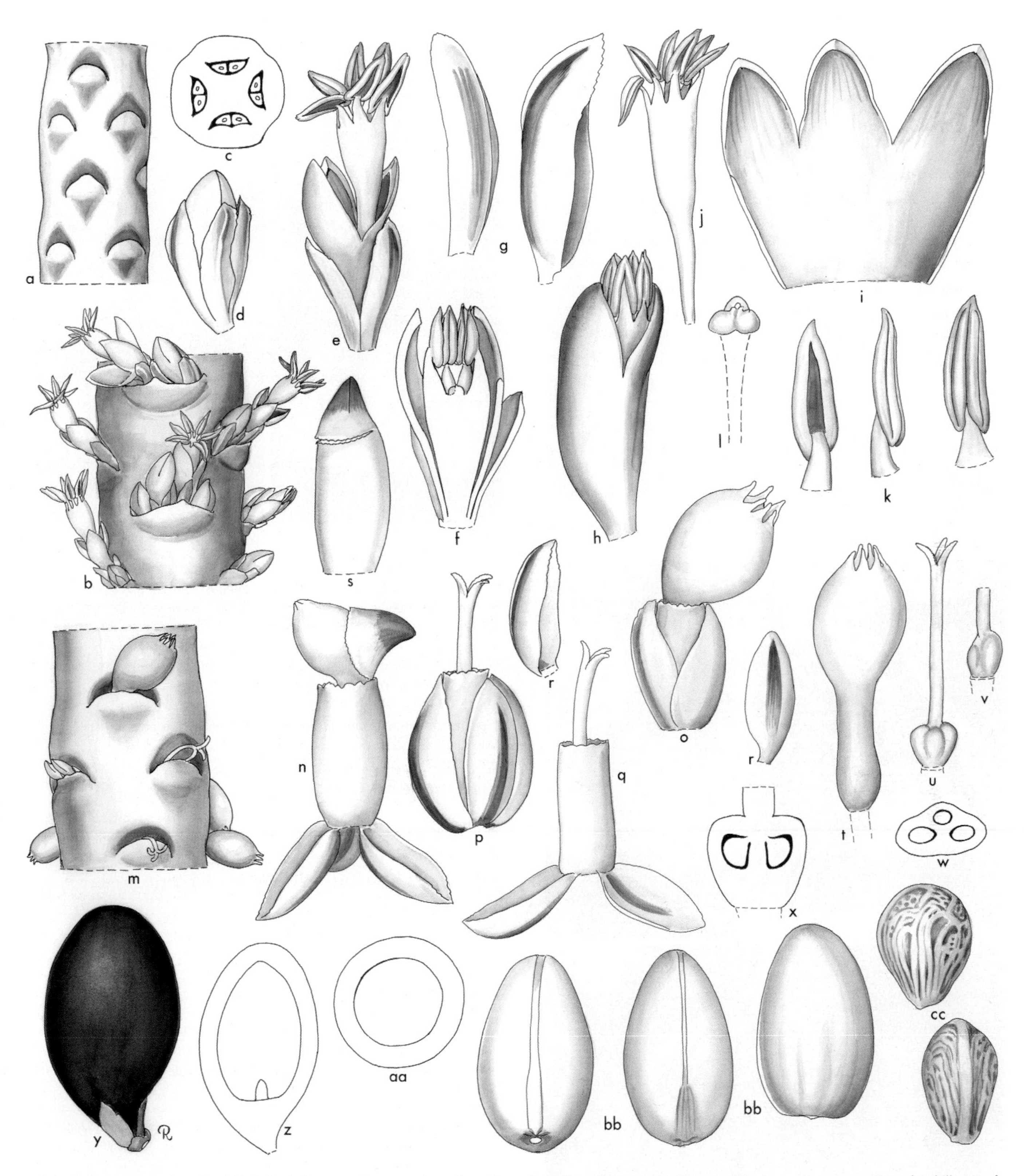

***Calyptronoma*.** **a**, portion of rachilla in bud, flowers not emergent × 3; **b**, portion of rachilla with staminate flowers and buds × 4; **c**, cross-section of rachilla × 3; **d**, staminate bud × 6; **e**, staminate flower at anthesis × 6; **f**, staminate bud in vertical section × 6; **g**, staminate sepals in 2 views × 12; **h**, staminate flower, sepals removed × 12; **i**, staminate corolla, interior view, expanded × 12; **j**, androecium × 6; **k**, anther in 3 views × 12; **l**, pistillode × 24; **m**, portion of rachilla at pistillate anthesis × 3; **n**, pistillate flower at anthesis, sepals bent downward, calyptra pushed open by distal portion of staminodial tube × 6; **o**, pistillate flower at anthesis, calyptra fallen × 6; **p, q**, pistillate flower at anthesis, calyptra and distal portion of staminodial tube fallen × 6; **r**, pistillate sepal in 2 views × 6; **s**, pistillate corolla, calyptra partially split off × 6; **t**, staminodial tube × 6; **u**, gynoecium × 6; **v**, ovary × 6; **w**, ovary in cross-section × 12; **x**, ovary in vertical section in ventral plane of 'w' × 12; **y**, fruit × 3; **z**, fruit in vertical section × 3; **aa**, fruit in cross-section × 3; **bb**, endocarp in 3 views × 3; **cc**, seed in 2 views × 3. *Calyptronoma occidentalis*: **a–l** and **y–cc**, *Read* 1610; **m–x**, *Read* 1693. (Drawn by Marion Ruff Sheehan)

Distribution and ecology: Three species in the Greater Antilles. In swamps, near the ocean and beside streams, and in wet places in the mountains.
Anatomy: Leaf (Wessels Boer 1968, Tomlinson 1961), root (Seubert 1998a, 1998b), and flower (Stauffer & Endress 2003).
Relationships: Unpublished morphological data resolve *Calyptronoma* as monophyletic (Asmussen *et al.* unpublished), whereas nuclear DNA data resolve *Calyptronoma* as paraphyletic with respect to *Calyptrogyne* with moderate support (Roncal *et al.* 2005). *Calyptronoma* is resolved in a clade with *Calyptrogyne* with moderate to high support (Asmussen 1999b, Roncal *et al.* 2005, Asmussen *et al.* 2006, unpublished, Baker *et al.* in review). Despite the dense species sampling carried out by Roncal *et al.* (2005) and Asmussen *et al.* (unpublished), more species and characters should be sampled to clarify relationships in this group.
Common names and uses: Manac palm, long thatch. The leaves resemble those of the coconut but have shorter leaflets, much used for thatch.
Taxonomic accounts: Wessels Boer (1968) and Zona (1995).
Fossil record: No generic records found.

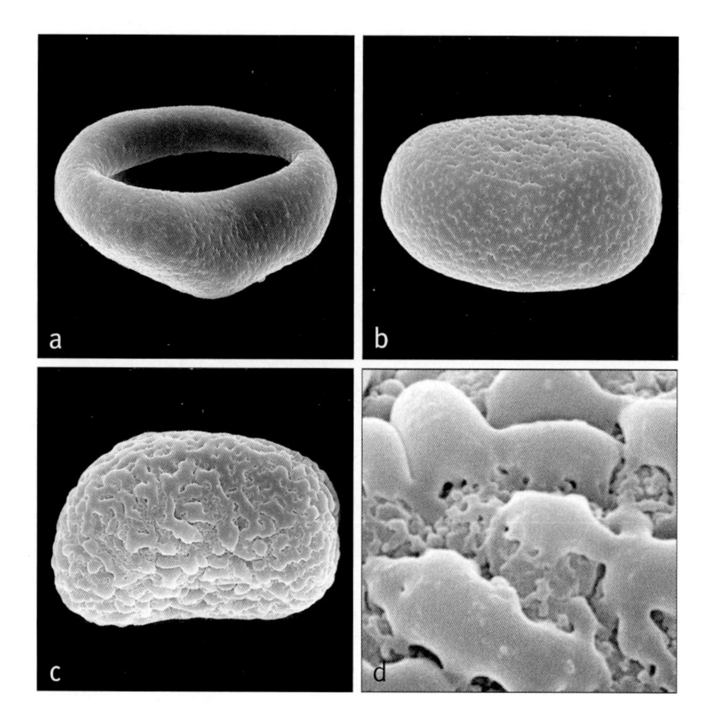

Above: ***Calyptronoma***
a, monosulcate pollen grain, tilted distal/equatorial face SEM × 1500;
b, monosulcate pollen grain, proximal face SEM × 1000;
c, monosulcate pollen grain, proximal face SEM × 1000;
d, close-up, rugulate-verrucate pollen surface SEM × 8000.
Calyptronoma plumeriana: **a**, *Wright* 3972; *C. rivalis*: **b**, *Burret* 5498;
C. occidentalis: **c** & **d**, *Harris* 9842.

Top left: ***Calyptronoma occidentalis***, inflorescence, cultivated, Fairchild Tropical Botanic Garden, Florida. (Photo: C.E. Lewis)

Bottom left: ***Calyptronoma rivalis***, part of inflorescence, cultivated, Fairchild Tropical Botanic Garden, Florida. (Photo: J. Roncal)

Notes: *Calyptronoma* is most closely related to *Calyptrogyne*. It is morphologically an unusual and striking genus because of its large size and pistillate flower with a dehiscent petal cap and staminodal body.
Additional figures: Fig. 2.8h.

120. ASTEROGYNE

Short-stemmed pinnate-leaved palms from the undergrowth of rain forest in Central and northern South America, with either spicate inflorescence or inflorescence with radiating branches, and distinctive divergent anther thecae.

Asterogyne H. Wendl. ex Hook.f. in Benth. & Hook.f., Gen. pl. 3: 914 (1883). Lectotype: **A. martiana** (H. Wendl.) H. Wendl. ex Hemsl. (*Geonoma martiana* H. Wendl.) (see Hemsley 1885).

Aristeyera H.E. Moore, J. Arnold Arbor. 47: 3 (1966). Type: *A. spicata* H.E. Moore (= *Asterogyne spicata* [H.E. Moore] Wess. Boer).

Aster — *star*, gyne — *woman or female, probably referring to the star-like corolla lobes in the pistillate flower.*

Small, solitary, unarmed, pleonanthic, monoecious palms. *Stem* short, erect, sometimes prostrate for a time, rather closely ringed with inconspicuous, narrow, oblique leaf scars. *Leaves* erect, almost always undivided, bifid, sometimes becoming split irregularly, marcescent; sheath short, tubular, eventually splitting opposite the petiole, margins with a few stiff fibres, covered with dull, reddish-brown scales; petiole short, slender, flat and glabrous adaxially; rounded, with dark brown scales, often becoming brown-dotted abaxially, blade wedge-shaped, pinnately ribbed, distinctly plicate, thin but rather tough, deep green adaxially, silvery green abaxially, glabrous adaxially, small membranous scales and trichomes along large ribs abaxially, transverse veinlets not evident, large close ribs more evident abaxially, alternating with bands of small veins on both surfaces. *Inflorescences* solitary, ± erect, becoming ± curved in fruit, interfoliar, spicate or subdigitately branched to 1 order, protandrous; peduncle long, slender, covered with dark brown, caducous tomentum; prophyll tubular, pointed, thin, papery, with 2, very narrow, lateral keels, covered with dark red, caducous tomentum; peduncular bracts 1(–2), tubular, like the prophyll but longer; rachis very short, bearing short, ovate, pointed bracts, the first 2 (or more) empty, others subtending rachillae; rachillae 1 or few (2–8), about equal in length, bearing sunken ovate bracts, each subtending

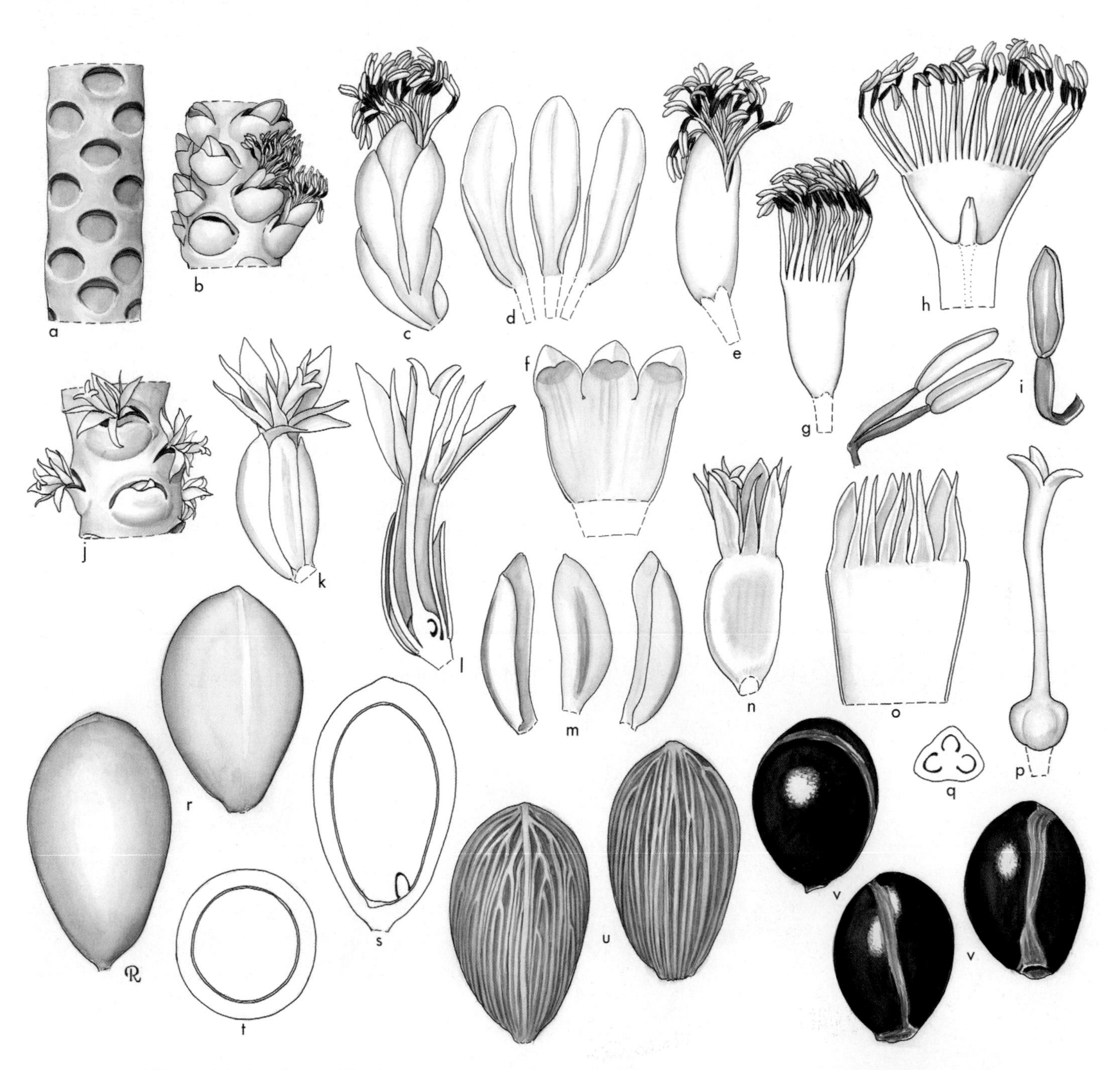

Asterogyne. **a**, rachilla in early bud × $1^1/_2$; **b**, rachilla with staminate flowers in late bud and at early anthesis × $1^1/_2$; **c**, staminate flower × 4; **d**, staminate sepals × 4; **e**, staminate flower, sepals removed × 4; **f**, staminate corolla, expanded × 4; **g**, androecium × 4; **h**, androecium, expanded to show pistillode × 4; **i**, stamens, much enlarged; **j**, rachilla with pistillate flowers at anthesis × 3; **k**, pistillate flower × 6; **l**, pistillate flower in vertical section × 6; **m**, pistillate sepals × 6; **n**, pistillate flower, sepals removed × 6; **o**, pistillate corolla and staminodes, interior view, expanded × 6; **p**, gynoecium × 8; **q**, ovary in cross-section × 8; **r**, fruit in 2 views × 3; **s**, fruit in vertical section × 3; **t**, fruit in cross-section × 3; **u**, endocarp in 2 views × 3; **v**, seed in 3 views × 3. *Asterogyne spicata*: all from *Moore* 9601. (Drawn by Marion Ruff Sheehan)

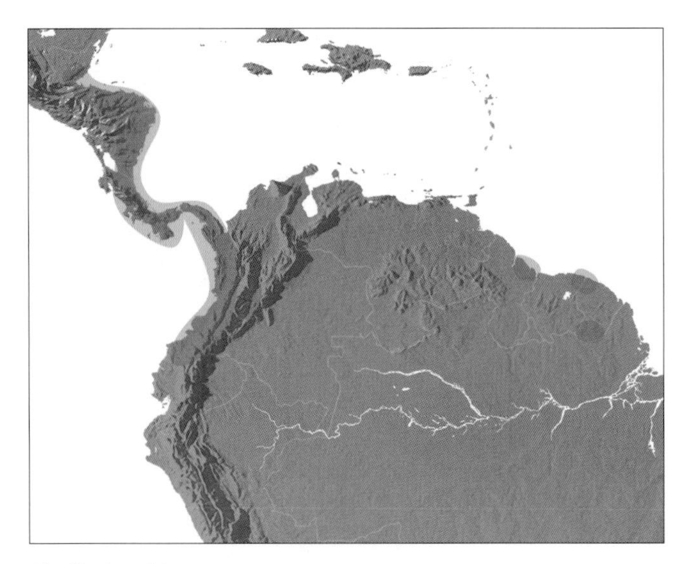

Distribution of ***Asterogyne***

a triad of flowers borne in a pit, pits with distinct upper 'lips' which lock the bract over the developing flowers, rachillae ending in long, slender, pointed tips, densely scaly; floral bracteoles 3, the outer bracteole shallow, the middle 2-keeled, the inner acute. *Staminate flowers* exserted at anthesis, borne laterally and outside the pistillate flower; sepals 3, briefly connate where adnate basally to the receptacle, free above, narrow, elongate, keeled, acute or emarginate apically; petals 3, about as long as the sepals, basally adnate to the receptacle, connate in a tube for about $^2/_3$ their length, free and valvate above; stamens 6–ca. 24, filaments connate and adnate basally to the receptacle forming a solid stalk-like base, united in a tube for an equal distance, distally free, erect, terete and tapering, connective bifid, tanniniferous, bearing separated thecae, inflexed in bud, ± erect at anthesis, introrse; pistillode shallow, irregularly 3-lobed. *Pollen* ellipsoidal, usually with obvious asymmetry, including pyriform; aperture a distal sulcus; ectexine tectate, perforate and micro-channelled, and rugulate, aperture margin slightly finer; infratectum columellate; longest axis 37–43 µm [1/5]. *Pistillate flowers* slightly asymmetrical; sepals 3, distinct, chaffy, imbricate in bud; petals 3, connate for about $^2/_3$ their length in a soft tube, valvate and chaffy distally; staminodes ca. 15, connate basally and adnate to the petal tube, free, somewhat fleshy, angled or terete distally, tips dark like the connectives; gynoecium trilocular, triovulate, 3-lobed, style elongate, grooved, ending in 3 triangular, stigmatic lobes, probably recurved at anthesis, ovule anatropous, pendulous, attached in the top of the locule, only 1 normally maturing. *Fruit* ellipsoidal-ovoid, 1-seeded, dorsiventrally compressed, slightly keeled apically, abortive carpels and stigmatic remains basal; epicarp smooth, mesocarp fleshy to dry, with an inner layer of closely appressed, longitudinal fibres, endocarp thin, crustaceous, shiny. *Seed* ellipsoidal to obovoid, slightly compressed laterally, hilum small, basal, raphe encircling the seed, somewhat impressed, unbranched to furcate or with a few parallel branches, endosperm homogeneous; embryo eccentrically basal. *Germination* adjacent-ligular; eophyll bifid. *Cytology* not studied.

Distribution and ecology: Five species in Central America and northern South America. Found only in wet forests, often at low elevations ca. 200–400 m or less, sometimes on well-drained slopes.

Anatomy: Leaf (Tomlinson 1961, 1966), root (Seubert 1998a, 1998b), and floral and inflorescence (Uhl 1966, Stauffer *et al.* 2003, Stauffer & Endress 2003).

Relationships: *Asterogyne* is monophyletic (Stauffer *et al.* 2003). For species relationships, see Stauffer *et al.* (2003). The genus has been resolved as sister to a clade of *Geonoma*, *Calyptronoma* and *Calyptrogyne* with low to moderate support (Asmussen *et al.* 2006).

Common names and uses: For local names, see Glassman (1972). *Asterogyne* is a fine ornamental and is also used for thatch. The fruit of *A. spicata* is edible.

Taxonomic account: Stauffer *et al.* (2003).

Fossil record: No generic records found.

Notes: Species of *Asterogyne* are elegant small palms distinguished by bifid leaves and by floral pits with covering bracts, which are 'locked' in bud by a distinct rounded lip. The separation of anther thecae on a bifid connective is distinctive.

Additional figures: Fig. 1.9, Glossary figs 19, 20.

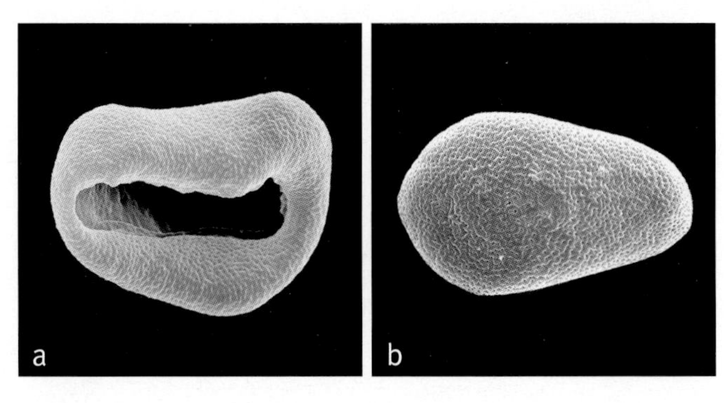

Above: ***Asterogyne***
a, monosulcate pollen grain, distal face SEM × 1500;
b, monosulcate pollen grain, proximal face SEM × 1500.
Asterogyne martiana: **a** & **b**, *Schipp* 392.

Left: ***Asterogyne martiana***, habit, cultivated, Fairchild Tropical Botanic Garden, Florida. (Photo: C.E. Lewis)

121. GEONOMA

Extremely variable genus of mostly rather small solitary or clustering palms from rain forest in Central and South America, with distinctive anthers with divergent thecae.

Geonoma Willd., Sp. pl. 4(1): 174, 593 (1805). Lectotype: **G. simplicifrons** Willd. (see H.E. Moore 1963c).

Vouay Aubl., Hist. pl. Guiane 2 (Appendix): 99 (1775). Type not designated.

Gynestum Poit., Mém Mus. Hist. Nat. 9: 387 (1822). Lectotype: *G. maximum* Poit. (see H. E. Moore 1963c) (= *Geonoma maxima* [Poit.] Kunth).

Roebelia Engel, Linnaea 33: 680 (1865). Type: *R. solitaria* Engel (= *Geonoma weberbaueri* Dammer ex Burret) (see Bernal & Galeano-Garces 1989).

Kalbreyera Burret, Bot. Jahrb. Syst. 63: 142 (1930). Type: *K. triandra* Burret (= *Geonoma triandra* [Burret] Wess. Boer).

Taenianthera Burret, Bot. Jahrb. Syst. 63: 267 (1930). Type: *T. macrostachys* (= *Geonoma macrostachys* Mart.)

Geonomos — *colonist, presumably referring to the clustering, spreading habit of many species.*

Small to moderate, solitary or clustered, unarmed, pleonanthic, monoecious palms. *Stem* very short, subterranean, erect, or creeping, slender, sometimes tall, enclosed by thin leaf sheaths, becoming bare, usually cane-like, ringed with close or distant, conspicuous or inconspicuous leaf scars. *Leaves* pinnate, regularly or irregularly divided, or entire and bifid; sheath short, splitting opposite the petiole, margins fibrous, glabrous or variously tomentose; petiole short to long, slightly grooved or flattened adaxially, rounded abaxially, glabrous or tomentose; blade bifid, or with 2 or 3 pairs of leaflets, or irregularly divided, or nearly evenly pinnate, thin and papery or somewhat leathery, usually glabrous adaxially, glabrous, tomentose or with scales abaxially, especially along the main ribs, uniseriate hairs present or absent, midribs of single folds conspicuous, transverse veinlets not evident. *Inflorescences* solitary, interfoliar or infrafoliar, spicate, forked, or branched to 3(–4) orders, protandrous where known; peduncle very short to very long, glabrous or tomentose; prophyll tubular, short to long, pointed, very briefly 2-keeled laterally, membranous or leathery, glabrous or variously tomentose; peduncular bracts (0–)1(–2), short or long, deciduous or persistent, like the prophyll; rachillae straight or folded and twisted in bud, short to moderate, bearing rounded, truncate, or distally split, ± raised bracts, laterally adnate to the branch, decussate, spiral, or whorled and in definite rows, bracts closely appressed and the rachillae larger than the peduncle in diameter, or bracts more distant and the rachillae narrow, each bract subtending a triad of flowers sunken in a pit, pits without upper lip or upper lip distinct, glabrous or hairy, pit cavity glabrous or variously hairy; floral bracteoles 3, irregular, small, membranous. *Staminate flowers* about $^1/_2$ exserted from the pit; sepals 3, distinct, chaffy, narrow, elongate, tips rounded, keeled or not; petals 3, connate for $^2/_3$ their length, tips distinct, valvate; stamens (3) 6 (rarely more), filaments united with receptacle in a stalk-like base, connate in a tube above the base, free, narrow, flat, long

Below left: ***Geonoma deversa***, habit, Venezuela. (Photo: H. Beentje)

Below middle: ***Geonoma linearis***, inflorescence at staminate anthesis, Colombia. (Photo: J. Dransfield)

Below right: ***Geonoma macrostachys***, infructescence, Peru. (Photo: J. Dransfield)

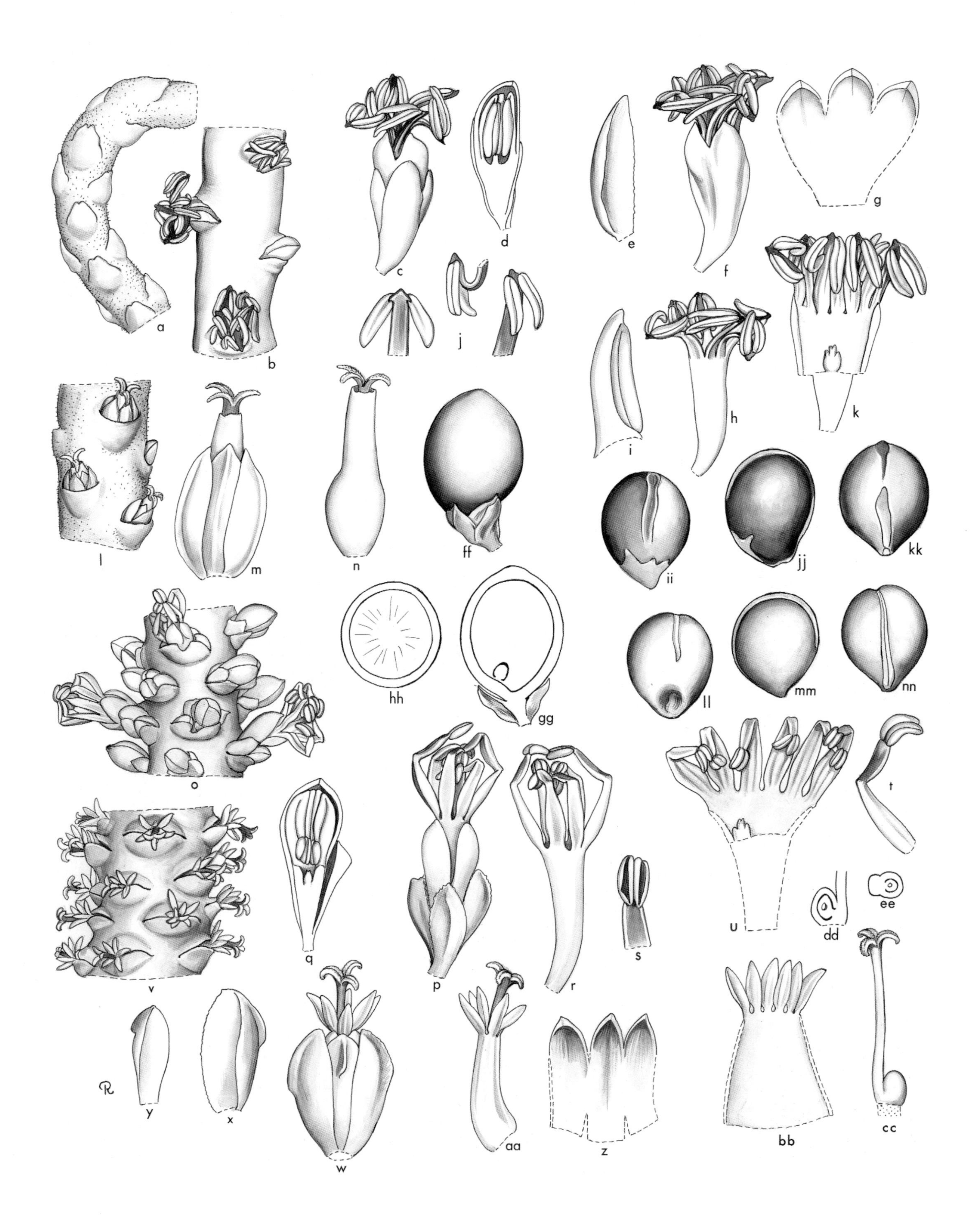
a
b
c
d
e
f
g
h
i
j
k
l
m
n
o
p
q
r
s
t
u
v
w
x
y
z
aa
bb
cc
dd
ee
ff
gg
hh
ii
jj
kk
ll
mm
nn

or short distally, inflexed near the tip in bud, anthers borne at tips of the filaments, connective divided, thecae elongate, free and divaricate, or short and united, introrse; pistillode small, round, 3-lobed. *Pollen* ellipsoidal, usually with either slight or obvious asymmetry; aperture a distal sulcus; ectexine tectate, coarsely perforate or perforate and/or micro-channelled, and rugulate, aperture margin usually slightly finer; infratectum columellate; longest axis 22–48 µm [28/59]. *Pistillate flowers* sunken in the pit with only the tips of the floral organs exserted; sepals 3, united basally and adnate to the receptacle, often keeled, free and imbricate distally; petals 3, connate in a soft tube, briefly adnate to receptacle basally, ending in 3, valvate, chaffy, spreading lobes; staminodes united in a tube, truncate, 6-toothed or 6-lobed, lobes, if present, spreading at anthesis, tubes basally adnate to the receptacle, and sometimes also the corolla tube; gynoecium tricarpellate but 2 carpels vestigial at anthesis, unilocular, uniovulate, ovule anatropous, style tubular, lateral to basal, elongate, ending in 3 linear stigmas, recurved at anthesis. *Fruit* ± globose, sometimes somewhat pointed, green, brown, or purple-black, 1-seeded, stigmatic remains basal, the rachillae often becoming brightly coloured; epicarp, thin smooth, mesocarp thin, with narrow longitudinal fibres, endocarp thin, crustaceous to membranous. *Seed* ± globose, hilum short, basal, raphe encircling the seed, endosperm homogeneous; embryo erect basal. *Germination* adjacent-ligular; eophyll bifid. *Cytology*: 2n = 28.

Distribution and ecology: Fifty-nine or more species ranging from Mexico to Brazil and Bolivia. All species are understorey rain forest palms, occurring at low to high elevations, including some of the highest elevations recorded for palms in South America. (*G. weberbaueri* has been recorded at 3150 m above sea level [Henderson *et al.* 1995].)

Anatomy: Leaf (Tomlinson 1961, Roth 1990), root (Seubert 1998a, 1998b), floral (Uhl & Moore 1971, Stauffer & Endress 2003), and leaf and fruit (Wessels Boer 1968).

Relationships: *Geonoma* is monophyletic with high support (Asmussen 1999a, Roncal *et al.* 2005). The genus is resolved as sister to a clade of *Calyptronoma* and *Calyptrogyne* with low support (Asmussen *et al.* 2006) or as sister to *Asterogyne* also with low support (Baker *et al.* in review).

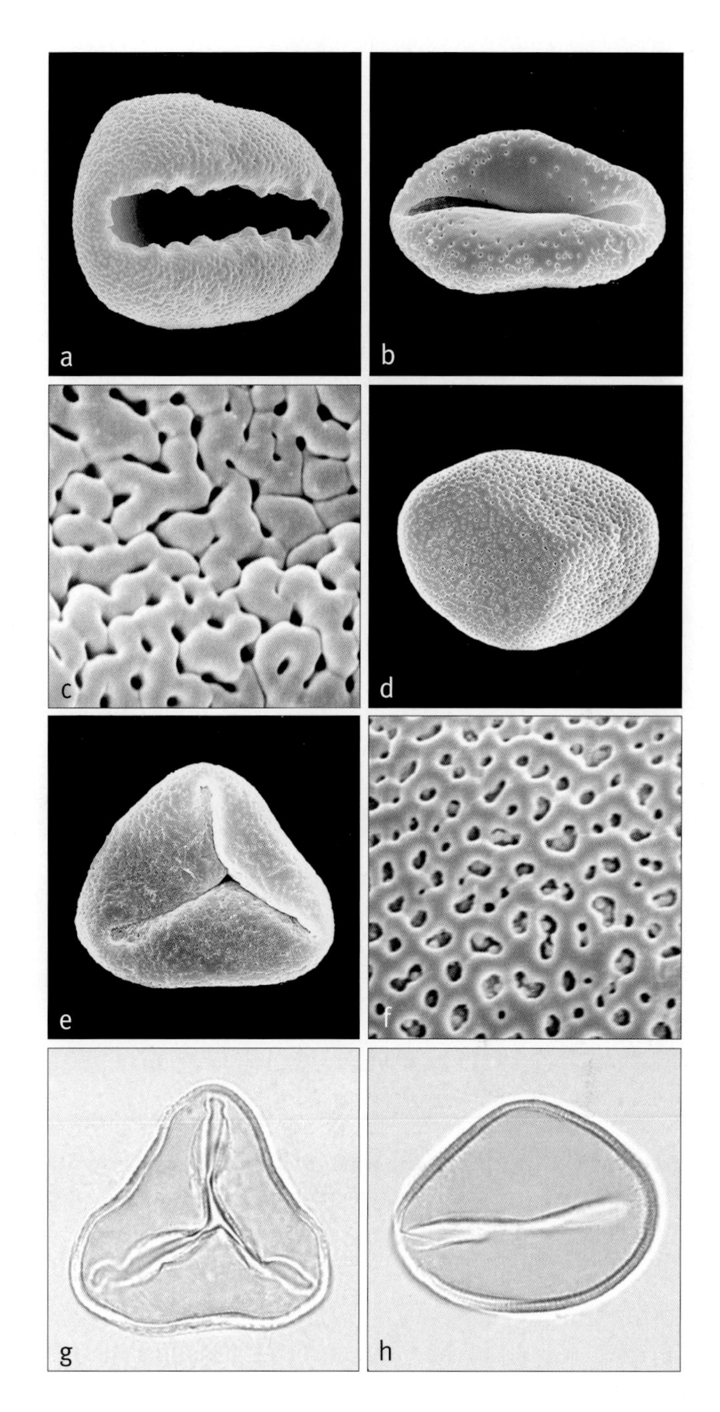

Right: ***Geonoma***
a, monosulcate pollen grain, distal face SEM × 2000;
b, monosulcate pollen grain, distal face SEM × 1500;
c, close-up, perforate-rugulate pollen surface SEM × 8000;
d, monosulcate pollen grain, proximal face SEM × 1500;
e, trichotomosulcate pollen grain, distal face SEM × 1500;
f, close-up, finely reticulate-rugulate pollen surface SEM × 8000;
g, trichotomosulcate pollen grain, mid-focus LM × 1500;
h, monosulcate pollen grain, mid-focus LM × 1500.
Geonoma macrostachys var. *acaulis*: **a**, *Ule* 5284, **b** & **c**, *Henderson et al.* 1651; *G. macrostachys*: **d** & **h**, *Prance et al.* 2801; *G. interrupta* var. *interrupta*: **e** & **g**, 1309; *G. laxiflora*: **f**, *Tye* F45.

Opposite: ***Geonoma***. **a**, portion of rachilla in bud × 3; **b**, portion of rachilla at staminate anthesis × 3; **c**, staminate flower × 6; **d**, staminate bud in vertical section × 6; **e**, staminate sepal × 6; **f**, staminate flower, sepals removed × 6; **g**, staminate corolla, expanded × 6; **h**, androecium × 6; **i**, stamen in bud × 12; **j**, stamen in 3 views × 6; **k**, androecium, expanded to show pistillode × 6; **l**, portion of rachilla at pistillate anthesis × 6; **m**, pistillate flower × 12; **n**, staminodial tube and styles × 12; **o**, portion of rachilla at staminate anthesis × 3; **p**, staminate flower × 6; **q**, staminate bud in vertical section × 6; **r**, androecium × 6; **s**, anther locules × 6; **t**, stamen in side view × 6; **u**, androecium, expanded to show pistillode × 6; **v**, portion of rachilla at pistillate anthesis × 3; **w**, pistillate flower × 6; **x**, pistillate sepal × 6; **y**, pistillate petal × 6; **z**, pistillate corolla, interior view, expanded × 6; **aa**, androecium, style, and stigmas × 6; **bb**, androecium, expanded, interior view × 6; **cc**, gynoecium × 6; **dd**, gynoecium in vertical section × 6; **ee**, gynoecium in cross-section × 6; **ff**, fruit × 3; **gg**, fruit in vertical section × 3; **hh**, fruit in cross-section × 3; **ii**, **jj**, **kk**, endocarp in 3 views × 3; **ll**, **mm**, **nn**, seed in 3 views × 3. *Geonoma interrupta*: **a–k** and **ff–nn**, *Moore & Parthasarathy* 9441, **l–n**, *Moore* 6523; *G. acaulis*: **o–ee**, *Moore et al.* 8502. (Drawn by Marion Ruff Sheehan)

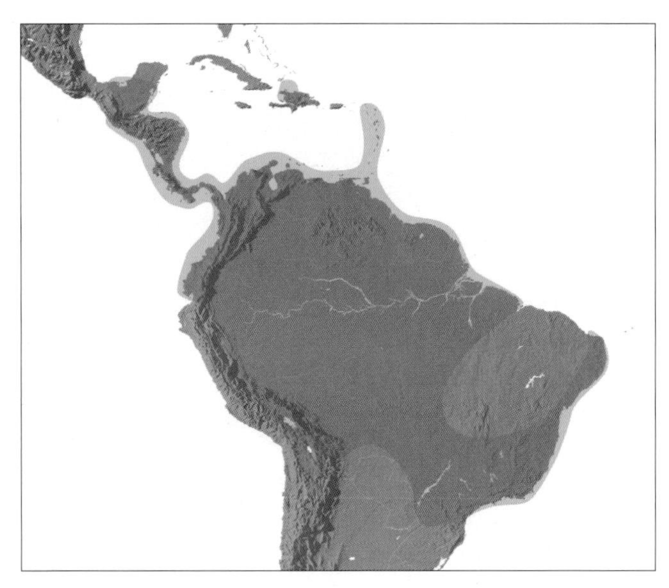

Distribution of ***Geonoma***

Common names and uses: For local names, see Glassman (1972). Many species are desirable ornamentals; some are also used for thatch. Young 'cabbage' is sometimes eaten.

Taxonomic accounts: Wessels Boer (1968); see also Henderson *et al.* (1995).

Fossil record: Pinnate leaf material (*Phoenicites/Geonomites/Hemiphoenicites*) from the Tertiary of Italy (Verona) was first described by de Visiani (1864); the genus has since been sunk into the synonomy of the redefined *Phoenicites* of Read and Hickey (1972). Other records of *Geonomites* include that of Berry (1924): *G. claibornensis* from the Middle Eocene Claiborne Flora of southeastern North America. Whether the fossil genus has any affinity with extant *Geonoma* seems doubtful.

Notes: The reduction of two locules in the gynoecium makes this, the largest genus in the tribe, immediately distinct from the other genera. Sometimes a definite 'upper lip' on the pit-closing bracts is lacking.

Additional figures: Figs 1.9, 2.6c, Glossary figs 4, 19, 20.

Tribe Leopoldinieae J. Dransf., N.W. Uhl, C. Asmussen, W.J. Baker, M.M. Harley & C. Lewis., Kew Bull. 60: 562 (2005). Type: **Leopoldinia**.

Moderate, unarmed; monoecious or rarely dioecious; leaves pinnate, the leaflet tips entire; sheaths not forming a crownshaft; inflorescences interfoliar, highly branched, sometimes unisexual; flowers superficial, in bisexual inflorescences borne in triads near the base, solitary or paired staminate flowers distally; pistillate flowers with petals distinct and imbricate basally, valvate distally; staminodes minute, distinct, not connate in a ring; gynoecium at first trilocular, triovulate, but only 1 seed developing; fruit with basal stigmatic remains; epicarp smooth.

The tribe, with its single genus *Leopoldinia*, has a narrow distribution in South America, being restricted to western Brazil, Amazonian Colombia and neighbouring southern Venezuela.

The monophyly of the Leopoldinieae has not been tested. The tribe resolves within the core arecoid clade, but its precise position is unclear. The tribe is described as sister to the Pelagodoxeae with moderate to high support by Norup *et al.* (2006) and by Baker *et al.* (in prep.). The Leopoldinieae was found to be sister to the Euterpeae with low support by Lewis and Doyle (2002) and by Loo *et al.* (2006), and sister to a clade of Euterpeae and Areceae with low support by Hahn (2002b). Tribe Leopoldinieae has been resolved as sister to the Manicarieae with low support by Asmussen and Chase (2001), as sister to a clade of Manicarieae and Geonomateae with low support by Baker *et al.* (in review), and as sister to a clade of Manicarieae and a subclade of Geonomateae with low support by Asmussen *et al.* (2006). The tribe was found to be sister to a subclade of Areceae by Hahn (2002a).

122. LEOPOLDINIA

Remarkable pinnate-leaved palms from Amazonia, the stems covered by long fibres or a broad network of fibres, the fruit lens-shaped or egg-like, with basal stigmatic remains.

Leopoldinia Mart., Hist. nat. palm. 2: 58 (1824). Lectotype: **L. pulchra** Mart. (see H.E. Moore 1963c).

Named for Maria Leopoldina Josephina Carolina of Habsburg (1797–1826), Archduchess of Austria and Empress of Brazil, whose father sponsored an expedition to Brazil during which Martius collected palms.

Moderate, solitary or clustered, unarmed, pleonanthic, monoecious (*Leopoldinia pulchra*) rarely dioecious according to Spruce [1871] palms. *Stems* erect, covered with marcescent leaf sheath fibres, eventually becoming bare, internodes short, at the base of the stem with abundant adventitious roots. *Leaves* pinnate, marcescent; sheath with a triangular ligule-like projection opposite the petiole, densely tomentose, the whole expanding and drying into an elegant interwoven mesh of broad flattened fibres, the margins remaining entire, or the whole disintegrating into extremely long black fibre bundles ('piassava') which hang down and obscure the stem; petiole well developed, adaxially flattened or convex, abaxially rounded or ± angled, bearing abundant, caducous scales; rachis longer than the petiole, adaxially angled, abaxially rounded or flattened, scaly as the petiole; leaflets single-fold, linear, acuminate or minutely bifid, numerous, regularly arranged, somewhat plicate, concolourous or discolourous, adaxially glabrous, abaxially bearing ramenta along the midrib, particularly near the base, transverse veinlets conspicuous. *Inflorescences* interfoliar, solitary, much shorter than the leaves, branching to 4 orders, the whole densely brown-tomentose, staminate inflorescences alternating with pistillate, or proximal rachillae pistillate and the distal staminate, or each rachilla pistillate at base, staminate at the tip or, rarely, plants apparently dioecious; peduncle elongate, partially obscured by subtending leaf sheaths, narrow-crescentic in cross-section; prophyll borne considerably above the base, tubular, narrowly elliptical in outline, 2-winged, ± membranous, splitting down its entire length early in development, circumscissile near the base, leaving a low membranous collar; peduncular bract 1, like the prophyll, also early caducous; rachis usually much shorter than the peduncle; first-order branches rather slender, each subtended by a very small, low, membranous, triangular bract; second, third, and fourth-order branches slender, tending to be somewhat divaricate or sinuous; rachillae rather slender, very densely tomentose, the flowers partially immersed in tomentum, where pistillate flowers borne on separate rachillae, the rachillae more robust than the staminate, pistillate flowers apparently solitary or in triads, staminate flowers usually paired or solitary. *Staminate flowers* very small, ± globular, bearing a striate chaffy bracteole; sepals 3, distinct, rounded, imbricate, ± striate; petals 3, distinct, valvate, ± triangular-ovate, marked on adaxial face by impressions of anthers; stamens 6, very small, filaments very short, connate only at the very base, rather broad, ± inflexed at the tip, anthers ± oval in outline, latrorse; pistillode barrel-shaped. *Pollen* ellipsoidal, slightly asymmetric; aperture a distal sulcus; ectexine tectate, surface very finely granular, finely perforate, or perforate and slightly rugulate, aperture margin similar; infratectum columellate; longest axis 21–26 µm [1/3]. *Pistillate flowers* larger than the staminate; sepals 3, distinct, imbricate, rounded, ± hooded, the margins ± toothed; petals 3, distinct, valvate; staminodes 6, distinct, very small, short, flat and ± truncate; gynoecium trilocular, triovulate, ± pyramidal, stigmas 3, rather obscure, sessile, ovule form unknown. *Fruit* dull red at maturity, ovoid, slightly flattened laterally, or strongly lenticular or disciform, 1-seeded, developing from 1 carpel, perianth whorls persisting, stigmatic and sterile carpellary remains basal; epicarp smooth, mesocarp composed of several complex reticulate systems of thick anastomosing fibres, embedded in fleshy parenchyma, the fibres becoming more numerous and closer towards the centre of the fruit, endocarp thin, smooth internally. *Seed* rounded or lenticular, attached opposite the stigmatic remains, with a vertical hilum running across one of the lateral faces, endosperm homogeneous; embryo subbasal. *Germination* adjacent-ligular; eophyll bifid, the segments very slender. *Cytology* not studied.

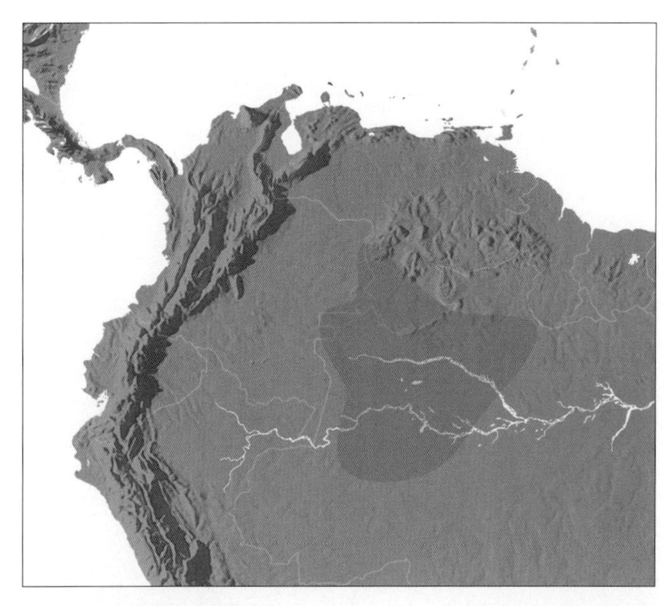

Distribution of ***Leopoldinia***

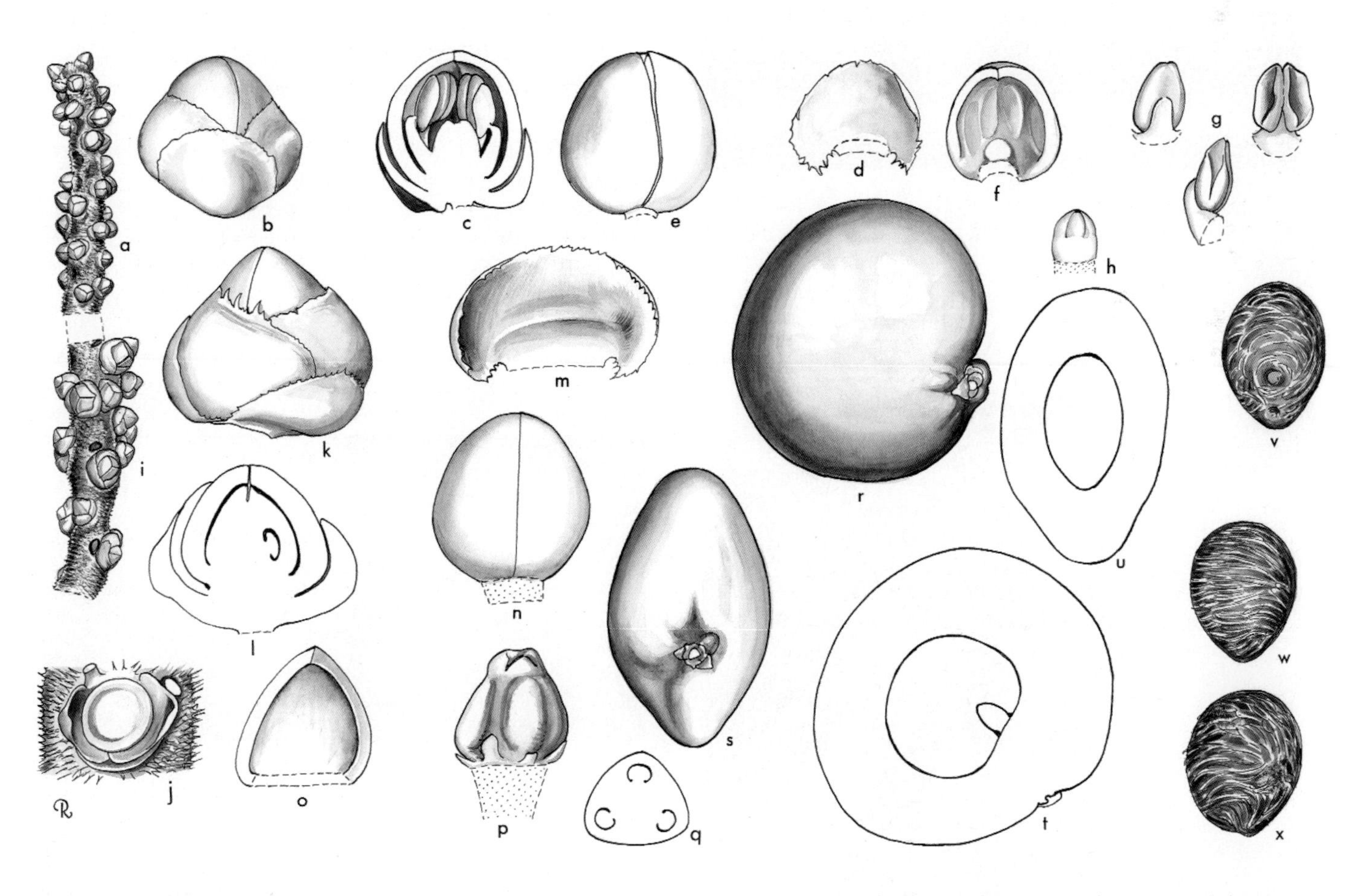

***Leopoldinia*. a**, apical portion of rachilla with paired staminate buds × 3; **b**, staminate bud × 24; **c**, staminate bud in vertical section × 24; **d**, staminate sepal, interior view × 24; **e**, staminate flower, sepals removed × 24; **f**, staminate petal, interior view × 24; **g**, stamen in 3 views × 24; **h**, pistillode × 24; **i**, lower portion of rachilla with triads × 3; **j**, triad, flowers removed × 12; **k**, pistillate bud × 24; **l**, pistillate bud in vertical section × 24; **m**, pistillate sepal, interior view × 24; **n**, pistillate flower, sepals removed × 24; **o**, pistillate petal, interior view × 24; **p**, gynoecium and staminodes × 24; **q**, ovary in cross-section × 24; **r**, **s**, fruit in 2 views × $1^1/_2$; **t**, fruit in vertical section × $1^1/_2$; **u**, fruit in cross-section × $1^1/_2$; **v**, **w**, **x**, seed in 3 views × $1^1/_2$. *Leopoldinia pulchra*: all from *Moore et al.* 9525. (Drawn by Marion Ruff Sheehan)

Distribution and ecology: Three species, confined to west Brazil, Amazonian Colombia, and southern Venezuela. All species are recorded from low lying, periodically flooded, tropical rain forest. Spruce (1871) records *Leopoldinia pulchra* and *L. major* from the banks of black-water rivers and on stony islands and *L. piassaba* from low sandy flats.

Anatomy: Leaf (Tomlinson 1961).

Relationships: The monophyly of *Leopoldinia* has not been tested. For relationships, see Leopoldinieae.

Common names and uses: Jara palms, piassava palm (*Leopoldinia piassaba*). The stems of *L. pulchra* are used as fence posts and the fruits of *L. major* burned to produce a salt substitute. However, the most useful species is undoubtedly *L. piassaba*; its leaves are used as thatch and the mesocarp crushed with water makes a creamy drink. Commercially, this species is important as a source of piassava, which is used for a variety of purposes from rope making to brooms (Putz 1979). See also Plotkin & Balick (1984) for medicinal uses.

Taxonomic accounts: Henderson (1995); see also Guanchez & Romero (1995).

Fossil record: No generic records found.

Notes: This is a remarkable genus in many respects, combining as it does a triovulate ovary with some unusual vegetative features. Most striking is the presence of two quite different leaf sheath types in the same genus; in one, the sheath is composed of a broad network of fibres, whereas in the other, there is conspicuous development of long free black fibres (piassava). The fruit shape is also variable. Kubitski (1991) postulated the dispersal of *L. major* and *L. pulchra* by fish, whereas *L. piassaba* is probably dispersed by terrestrial animals.

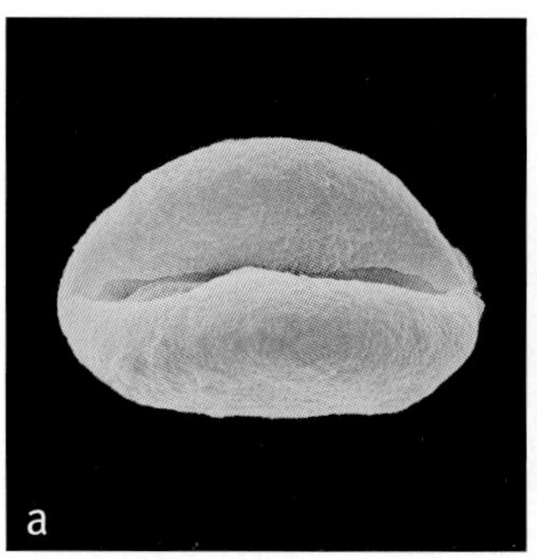

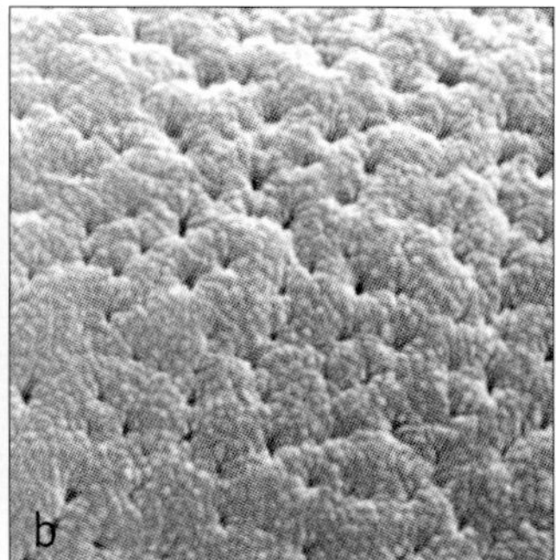

Above: ***Leopoldinia***
a, monosulcate pollen grain, distal face SEM × 2000;
b, close-up, very finely granular, perforate-rugulate pollen surface SEM × 10,000.
Leopoldinia pulchra: **a** & **b**, *Henderson et al.* 650.

Top left: ***Leopoldinia major***, habit, Brazil. (Photo: A.J. Henderson)

Bottom left: ***Leopoldinia pulchra***, crown with inflorescence, Brazil. (Photo: A.J. Henderson)

Tribe Pelagodoxeae J. Dransf., N.W. Uhl, C. Asmussen, W.J. Baker, M.M. Harley & C. Lewis., Kew Bull. 60: 563 (2005). Type: **Pelagodoxa**.

Diminutive, acaulescent or erect to moderate, solitary, monoecious, unarmed; leaves entire bifid or irregularly pinnate, the leaf margin not praemorse but lobed corresponding to the ribs; sheaths not forming a crownshaft; inflorescences interfoliar, sparsely to highly branched; flowers borne in triads near the base, solitary or paired staminate flowers distally, in shallow pits; pistillate flowers with petals distinct, imbricate basally, valvate distally; staminodes minute, distinct, not connate in a conspicuous ring; gynoecium pseudomonomerous; fruit with basal stigmatic remains; epicarp corky-warted.

This is a tribe of the western Pacific Ocean, with one member on New Guinea and the other assumed to be from the Marquesas Islands. The two genera share the same fruit character, the corky-warted epicarp. In the first edition of *Genera Palmarum*, little emphasis was given to this character as it appears scattered throughout the family, and corky-warted and smooth-fruited species can occur within the same genus (e.g., *Licuala* and *Pholidocarpus*). It was something of a surprise then to find the two genera constantly and strongly associated in phylogenies that are based on molecular evidence.

The tribe Pelagodoxeae is monophyletic with high support in all analyses (Lewis & Doyle 2002, Asmussen *et al.* 2006, Loo *et al.* 2006, Norup *et al.* 2006, Baker *et al.* in review, in prep.). Baker *et al.* (in prep.) and Norup *et al.* (2006) resolve the Pelagodoxeae as sister to the Leopoldinieae. Baker *et al.* (in review) place the tribe as sister to a clade of Geonomateae, Leopoldinieae and Manicarieae. In the *prk* analyses of Lewis and Doyle (2002) and of Loo *et al.* (2006), tribe Pelagodoxeae is placed as sister to the Euterpeae with low support.

Key to Genera of Pelagodoxeae

Moderate palm with an erect stem; inflorescence branching to 2 orders; fruit at least 30 mm in diameter at maturity, covered in dry corky warts. Marquesas . . . . **123. Pelagodoxa**

Slender acaulescent or short-stemmed palm; inflorescence branched to 1 order only; fruit not exceeding 12 mm in diameter, covered in corky warts on a fleshy mesocarp. New Guinea . . . . **124. Sommieria**

123. PELAGODOXA

Striking solitary palm of Pacific islands with a ± undivided strongly discolorous leaf and corky-warted fruit.

Pelagodoxa Becc. in Bois, Rev. Hort. series 2. 15: 302 (1917). Type: **P. henryana** Becc.

Pelagos — *the sea,* doxa — *glory, perhaps in reference to the remote oceanic island habitat.*

Moderate, solitary, unarmed, pleonanthic, monoecious palm. *Stem* erect, bare, ringed with close leaf scars. *Leaves* pinnately ribbed, undivided except for the bifid apex, but often split by wind; leaf sheaths soon splitting opposite the petiole, not forming a crownshaft, densely tomentose, with an irregular ligule at the mouth, disintegrating into fine

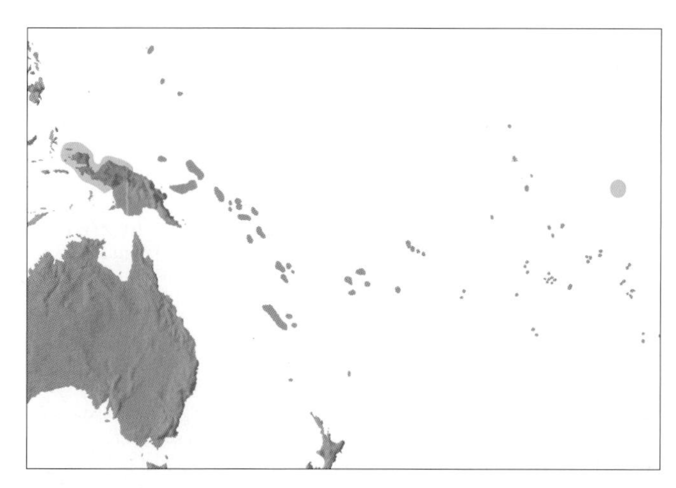

Distribution of tribe ***Pelagodoxeae***

fibres; petiole relatively short, adaxially channelled, abaxially ± rounded, densely covered in tomentum; rachis curved, densely tomentose; blade adaxially bright, shiny green, glabrous, abaxially with a thin felt of grey tomentum, distal margin shallowly lobed, the lobes corresponding to the major folds and minutely toothed, transverse veinlets obscure. *Inflorescences* solitary, interfoliar, much shorter than the leaves, branching to 2 orders, protandrous; peduncle ± oval in cross-section, exserted from the leaf sheaths; prophyll short, inserted near the base of the peduncle,

Above: ***Pelagodoxa henryana***, infructescence, cultivated, National Tropical Botanic Garden, Kauai, Hawaii. (Photo: J. Dransfield)

only partially enclosing the inflorescence in bud, splitting abaxially near the tip, beaked, strongly 2-keeled, coriaceous to sub-woody, densely tomentose, tending to disintegrate into fibres; peduncular bract longer than the prophyll, inserted just above the prophyll, tubular, beaked, enclosing the inflorescence in bud, densely tomentose; subsequent bracts rather small, incomplete, broad, triangular; rachis only slightly longer than the peduncle; all inflorescence axes tomentose; rachillae stiff, relatively thick, gradually tapering to a pointed tip, bearing spirally arranged, shallow pits, the rachilla bracts forming low triangular lips to the pits, pits in the proximal ca. $^{1}/_{4}$–$^{1}/_{3}$ of rachillae containing triads, distally containing paired or solitary staminate flowers, the rachilla bracts at the tips of the rachillae more prominent than at the base; floral bracteoles low, rounded, inconspicuous, included within the pits. *Staminate flowers* small, ± globular, ± symmetrical, only partially exserted from the pit; sepals 3, distinct, imbricate, strongly keeled, rather chaffy; petals 3, about twice as long as the sepals, united proximally and adnate to the receptacle forming a stalk-like base, distally ± triangular-ovate, striate, valvate; stamens 6, the filaments united basally to the pistillode forming a solid column above the insertion of the free portion of petals, free filaments fleshy, triangular, gradually tapering to the connective, anthers short, medifixed, basally sagittate, latrorse, the connective prolonged in a brief point; pistillode pyramidal. *Pollen* ellipsoidal slightly asymmetric; aperture a distal sulcus; ectexine tectate, perforate-rugulate, aperture margin similar or slightly finer; infratectum columellate; longest axis 22–27 µm [1/1]. *Pistillate flowers* globose, at anthesis larger than the staminate, tending to crack open the floral pits; sepals 3, distinct, broad, rounded, imbricate; petals 3, distinct, broadly imbricate except for brief, triangular valvate tips, very briefly joined basally to form a short broad stalk with the receptacle; staminodes 3–6, triangular, flattened, very small; gynoecium ± trilocular, uniovulate or rarely a second ovule present, rounded, stigmas 3, short, reflexed, ovule laterally attached, campylotropous. *Fruit* large, spherical, perianth whorls persistent, stigmatic remains basal, the fruit surface cracked into low, pyramidal, corky warts; epicarp obsolescent at maturity, mesocarp massively corky with abundant radiating fibres, endocarp thin, woody. *Seed* basally attached with rounded hilum, endosperm homogeneous with a large central hollow; embryo basal. *Germination* adjacent-ligular; eophyll bifid. *Cytology*: 2n = 32.

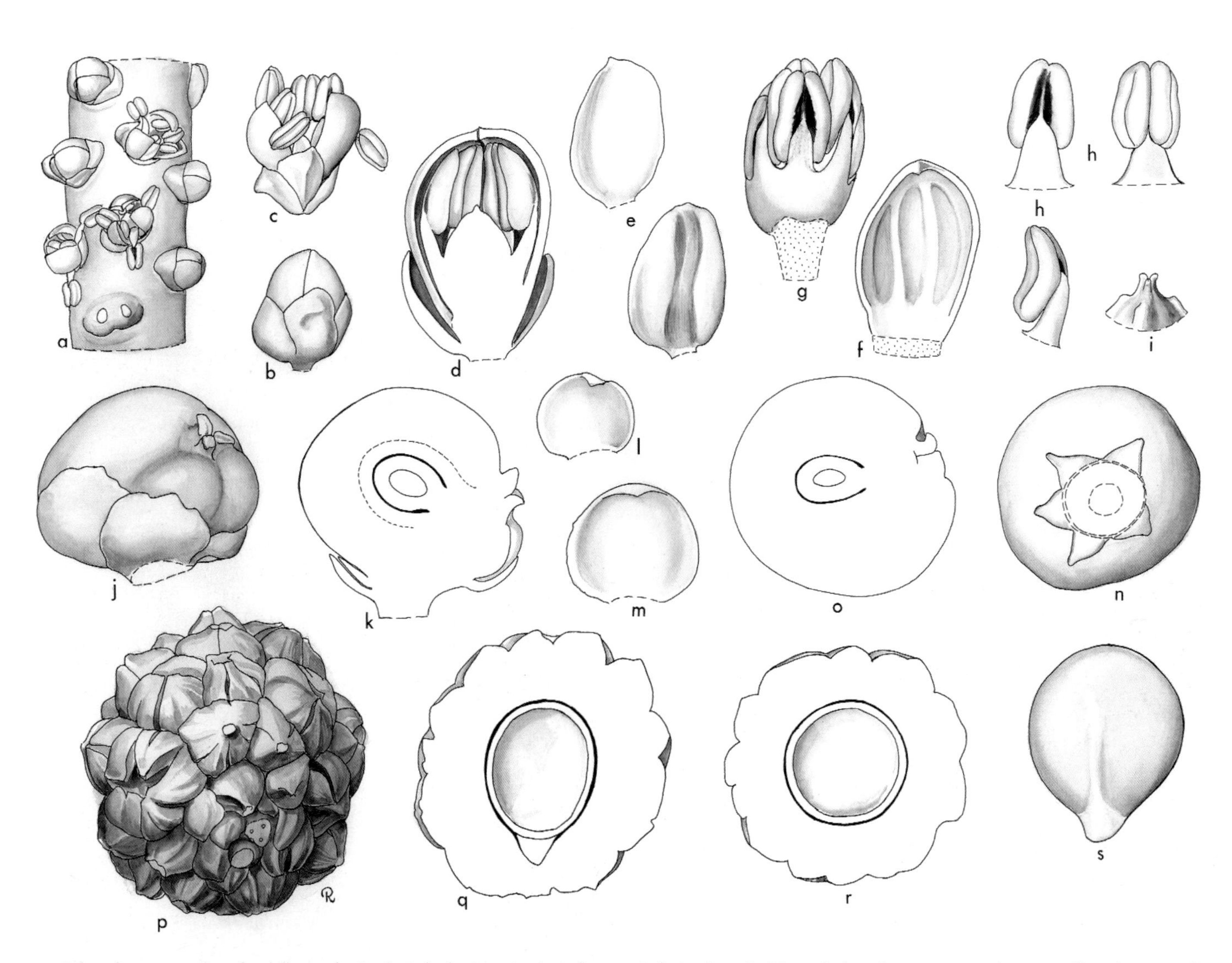

Pelagodoxa. **a**, portion of rachilla × 3; **b**, staminate bud × 6; **c**, staminate flower × 6; **d**, staminate bud in vertical section × 12; **e**, staminate sepal in 2 views × 12; **f**, staminate petal × 12; **g**, androecium × 12; **h**, stamen in 3 views × 12; **i**, pistillode × 24; **j**, pistillate flower × 6; **k**, pistillate flower in vertical section × 6; **l**, pistillate sepal × 6; **m**, pistillate petal × 6; **n**, gynoecium, bottom view to show staminodes × 6; **o**, ovary in cross-section × 6; **p**, fruit × $^{3}/_{4}$; **q**, fruit in vertical section × $^{3}/_{4}$; **r**, fruit in cross-section × $^{3}/_{4}$; **s**, seed × $^{3}/_{4}$. *Pelagodoxa henryana*: **a–o**, *Moore* 9400; **p–s** *Read* s.n. (Drawn by Marion Ruff Sheehan)

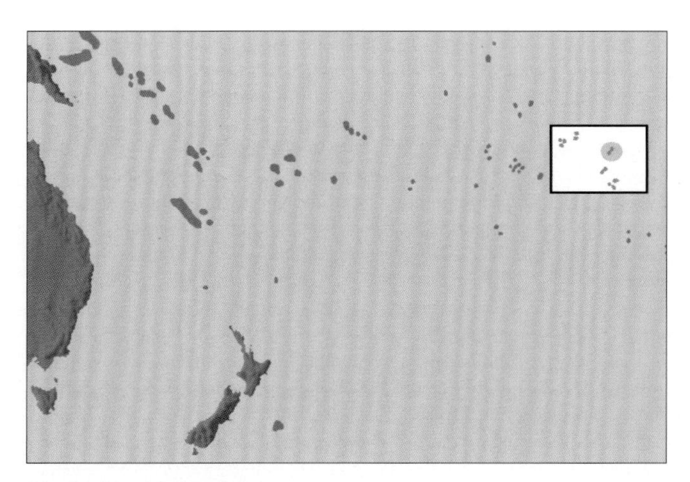

Distribution of ***Pelagodoxa***

Distribution and ecology: One species, apparently confined to the Marquesas Islands where it occurs as a few individuals in dense rain forest in a humid valley at about 135 m above sea level. In cultivation, there are two strikingly different sizes of fruit.

Anatomy: Leaf (Tomlinson 1961); floral (Stauffer *et al.* 2004), and fruit (Essig *et al.* 1999, Chapin *et al.* 2001).

Relationships: The sister-group relationship between *Pelagodoxa* and *Sommieria* is strongly supported in many analyses (Lewis & Doyle 2002, Asmussen *et al.* 2006, Loo *et al.* 2006, Norup *et al.* 2006, Baker *et al.* in review, in prep.).

Common names and uses: *Enu, vahani.* The young endosperm is said to be eaten; the palm is cultivated and much sought-after as an ornamental.

Taxonomic accounts: See Martelli (1932) and Dowe & Chapin 2006.

Fossil record: No generic records found.

Notes: This is an extraordinary genus with an unusual distribution. The huge bifid leaf, the connate petals in the staminate flower, stamens adnate to the pistillode and the corky-warted fruit represent an unusual combination of characters. *Pelagodoxa mesocarpa*, based on a single fruit, is probably conspecific with *P. henryana*. In cultivation, however, there appear to be two forms, one with a much larger fruit than the other. The larger-fruited form appears to be that known in the Marquesas. In recent years, individuals of *Pelagodoxa* have been discovered in the Solomon Islands and Vanuatu, but always associated with villages. Whether it is truly native in the Marquesas may also be debatable.

Additional figures: Fig. 1.9, Glossary fig. 22.

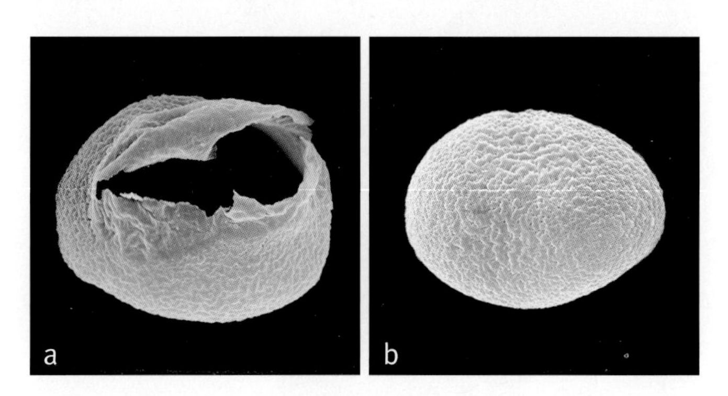

Above: ***Pelagodoxa***

a, monosulcate pollen grain, distal face, ruptured sulcus membrane still in place SEM × 2000;

b, monosulcate pollen grain, proximal face SEM × 2000.

Pelagodoxa henryana: **a**, *Moore* 9400, **b**, *Gillett* 2213.

124. SOMMIERIA

Short, sometimes acaulescent palm of forest undergrowth in western New Guinea, with entire-bifid or irregularly pinnate strongly discolourous leaves and small corky-warted fruit.

Sommieria Becc., Malesia 1: 66 (1877). Lectotype: **S. leucophylla** Becc. (see Beccari & Pichi-Sermolli 1955).

Commemorating Stephen Sommier (1848–1922), Italian natural historian and friend of Odoardo Beccari.

Small, solitary, acaulescent to short-stemmed, unarmed, pleonanthic, monoecious palms. *Stem* eventually erect, becoming bare, ringed with very close leaf scars, sometimes also bearing bunches of aerial roots. *Leaves*

Right: ***Sommieria leucophylla***, crown, Papua. (Photo: J. Dransfield)

numerous, entire, bifid, sometimes with 2 pairs of segments, pinnately ribbed, marcescent; leaf sheaths densely tomentose, eventually splitting irregularly opposite the petiole and disintegrating into an interwoven mass of fibres, the mouth (?always) prolonged into a fibrous ligule; petiole usually short, adaxially channelled, abaxially rounded, variously tomentose; rachis adaxially channelled near the base, distally angled, abaxially rounded, gradually tapering; blade divided to produce a large, bifid part and 1 pair of narrow acuminate basal segments, or simply bifid, the apical margins lobed, the lobes corresponding to the major folds, or subentire, adaxially minutely scaly (?always), abaxially glabrous or densely white-tomentose, transverse veinlets obscure. *Inflorescences* ± erect, interfoliar, solitary, ± equalling the leaves, branching to 1 order, protandrous; peduncle very long, slender, ± elliptical in cross-section; prophyll scarcely exserted from the subtending leaf sheath, 2-keeled, tubular, splitting along the abaxial face, the tip somewhat beaked, sometimes tattering into fibres at the tip; peduncular bract 1, tubular, borne at the tip of the peduncle, ± enclosing the rachillae before anthesis, membranous, splitting down one side to the base and becoming lanceolate, apparently sometimes persisting, sometimes deciduous; rachis very short; rachillae few in number (less than 12), spirally arranged, pendulous, ± stiff, slender, elongate, each subtended by a minute first-order bract, the surface of the rachilla densely dark brown-tomentose, flowers arranged in triads, sunken within pits ± throughout the entire length of the rachilla; the rachilla bracts low, minutely toothed, forming the lower lips of the pits; floral bracteoles minute. *Staminate flowers* ± symmetrical, the base somewhat stalked; sepals 3, distinct, rounded, imbricate, hooded, strongly keeled, ± striate; petals 3, ± twice as long as the sepals, distinct, ovate-triangular, valvate, scarcely opening at anthesis; stamens 6, filaments minutely connate basally, fleshy, awl-shaped, inflexed in bud (?always), the antesepalous much longer than the antepetalous, at anthesis spreading between the petals, the antepetalous included, anthers short, rectangular, medifixed, ± versatile, latrorse; pistillode ± as long as petals, columnar, ± angled. *Pollen* ellipsoidal asymmetric, occasionally

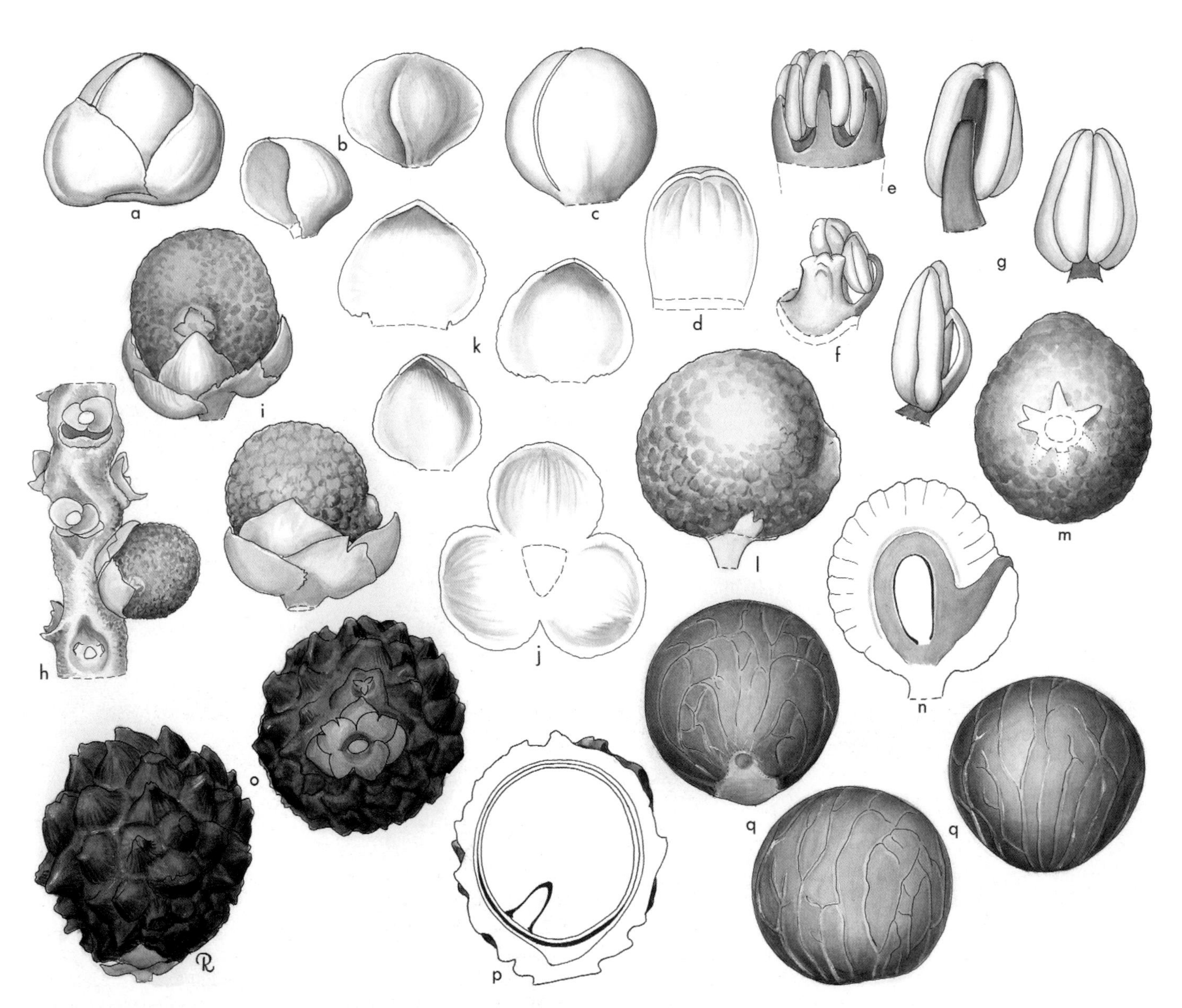

Sommieria. **a**, staminate bud × 12; **b**, staminate sepal in 2 views × 12; **c**, staminate bud, calyx removed × 12; **d**, staminate petal × 12; **e**, androecium × 12; **f**, pistillode with 2 stamens × 12; **g**, stamen in 3 views × 24; **h**, rachilla with young fruit × 6; **i**, young fruit in 2 views × 12; **j**, pistillate calyx × 12; **k**, pistillate petals × 12; **l**, gynoecium of young fruit × 12; **m**, base of gynoecium with staminodes × 12; **n**, young fruit in vertical section × 12; **o**, mature fruit in 2 views × 4; **p**, mature fruit in vertical section × 4; **q**, seed in 3 views × 5. *Sommieria leucophylla*: **a–g**, *Beccari* 607; **h–q**, source unknown. (Drawn by Marion Ruff Sheehan)

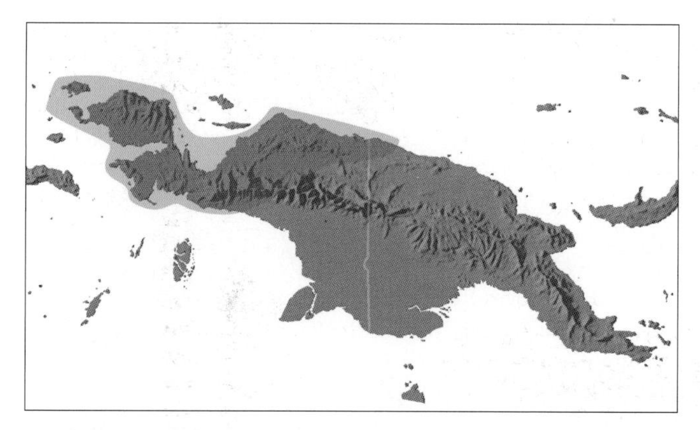

Distribution of ***Sommieria***

oblate triangular; aperture a distal sulcus, less frequently a trichotomosulcus; ectexine tectate, perforate-rugulate, aperture margin similar or slightly finer; infratectum columellate; longest axis 22–28 µm [1/1]. *Pistillate flowers* eventually larger than the staminate, ± globular; sepals 3, distinct, or briefly connate, rounded, strongly imbricate; petals ± equalling the sepals, 3, distinct or briefly connate, rounded, imbricate except for the short, triangular, valvate tips; staminodes 3–6, tooth-like; gynoecium unilocular, uniovulate, ovoid, stigmas 3, apical, reflexed, ovule form not known. *Fruit* small, spherical, perianth whorls persistent, the stigmatic remains basal; epicarp smooth, brown early in development, soon cracking, obsolescent at maturity, mesocarp cracked to form pyramidal to hexagonal, corky warts, brown-tipped, pink-sided and white-based when fresh, drying dull brown throughout, endocarp thin, bony, operculate, closely adhering to the seed. *Seed* basally attached, spherical, hilum ± circular, raphe branches sparsely anastomosing, endosperm homogeneous; embryo subbasal. *Germination* adjacent-ligular; eophyll bifid. *Cytology*: 2n = 34.

Distribution and ecology: A single species confined to New Guinea where it occurs predominantly in the western half of the island, confined to the undergrowth of humid lowland tropical rain forest.
Anatomy: Floral (Stauffer *et al.* 2004) and fruit (Essig *et al.* 1999).
Relationships: For relationships, see *Pelagodoxa*.
Common names and uses: *Mbebmega* (Hatam language), *som* (Biak), *man* (Bewani).

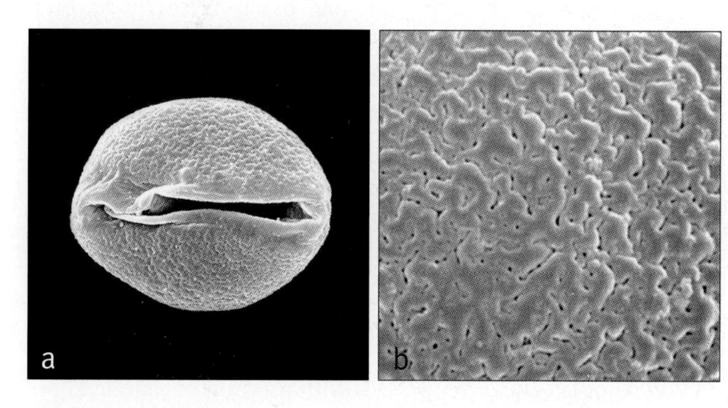

Above: ***Sommieria***
a, monosulcate pollen grain, distal face, ruptured sulcus membrane still in place SEM × 2000;
b, close-up, perforate-rugulate surface SEM × 8000.
Sommieria leucophylla: **a**, *Beccari* s.n. **b**, *Beccari* 607.

Above left: ***Sommieria leucophylla***, pistillate flowers, Papua (Photo: W.J. Baker)
Above right: ***Sommieria leucophylla***, fruit, Papua. (Photo: W.J. Baker)

Taxonomic account: Heatubun (2002).
Fossil record: No generic records found.
Notes: Heatubun (2002) showed convincingly that variation in this genus does not warrant the recognition of more than one species.
Additional figures: Fig. 3.1.

Tribe Areceae

Diminutive to robust, acaulescent, erect or very rarely climbing, unarmed or armed; leaves pinnate or entire bifid, the leaflet tips entire or praemorse; sheaths forming a crownshaft or crownshaft absent; inflorescences infrafoliar or interfoliar, spicate to highly branched; inflorescence bracts usually but not always comprising a prophyll and a single peduncular bract; flowers borne in triads near the base, solitary or paired staminate flowers distally or triads throughout, superficial or sunk in pits; pistillate flowers with petals distinct or connate basally, valvate distally; staminodes distinct, or very rarely connate in a conspicuous ring; gynoecium pseudomonomerous; fruit with basal or apical stigmatic remains; epicarp smooth.

The relationships of the ten subtribes recognised here are unclear and for that reason they are arranged in alphabetical order. There are also ten genera that clearly belong to Areceae, but their relationships remain obscure; these genera are thus listed alphabetically at the end as 'unplaced Arecoid genera'. This great diversity of palms, the Indopacific Pseudomonomerous Arecoid clade, seems to be the product of a dramatic radiation of genera across the whole Indian and western Pacific Oceans and neighbouring landmasses. In particular, there is perplexing and complex diversity on the island arcs of the western Pacific.

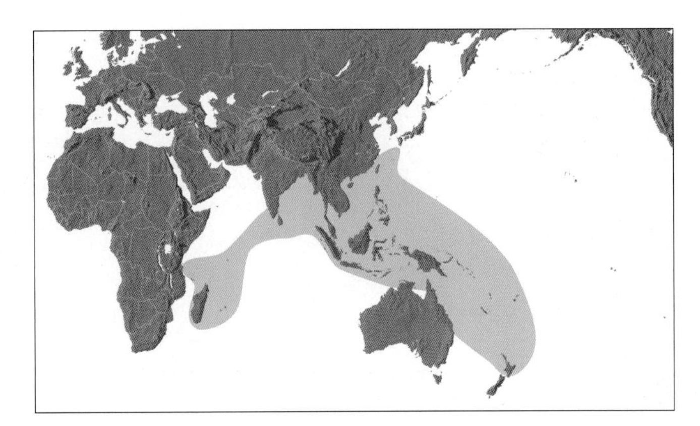

Distribution of tribe **Areceae**

Several analyses show high bootstrap support for the monophyly of the Areceae (Hahn 2002b, Lewis & Doyle 2002, Loo *et al.* 2006, Norup *et al.* 2006, Baker *et al.* in prep.). In addition, low to moderate support for the group is recovered in several other studies (Uhl *et al.* 1995, Asmussen & Chase 2001, Lewis & Doyle 2001, Asmussen *et al.* 2006, Baker *et al.* in review). There are conflicting, poorly supported hypotheses on the phylogenetic position of tribe Areceae, although there is universal agreement on its placement in the core arecoid clade. The tribe was placed as sister to the Geonomateae by Lewis and Doyle (2001), sister to the Euterpeae by Hahn (2002b), and sister to the Manicarieae by Loo *et al.* (2006) and in the *prk* analysis of Lewis and Doyle (2002). It is resolved as sister to a clade of Manicarieae and Leopoldinieae by Asmussen and Chase (2001) and sister to a clade of Pelagodoxeae, Leopoldinieae and Euterpeae by Baker *et al.* (in prep.) and by Norup *et al.* (2006). Within the Areceae, a moderately supported clade of western Pacific subtribes is resolved, comprising Rhopalostylidinae, Basseliniinae, Carpoxylinae, Linospadicinae, Clinospermatinae, Ptychospermatinae, Archontophoenicinae, *Dransfieldia* and *Heterospathe* (Lewis & Doyle 2002, Norup *et al.* 2006, Savolainen *et al.* 2006, Baker *et al.* in review).

Key to Subtribes of Areceae

1. Leaves and often stems covered with epidermal spines, these sometimes caducous 2
1. Unarmed palms 3
2. Leaves with leaflets all 1-ribbed, acute or acuminate; sheaths usually forming a conspicuous crownshaft; inflorescence usually infrafoliar; peduncular bract enclosed by the prophyll, both inserted close together, splitting and usually caducous at anthesis; peduncle short, usually ± equalling the rachis; rachillae straight or coiled and twisted in bud; staminate flowers asymmetrical, sepals acute . . . **Oncospermatinae**
2. Leaves with blades undivided, bifid, or irregularly to regularly divided into leaflets, these usually with more than 1 rib, acute, bifid or praemorse; sheaths not forming a conspicuous crownshaft; inflorescences interfoliar or becoming infrafoliar in age; peduncular bract caducous, inserted some distance from the prophyll and exceeding it; prophyll persistent; peduncle elongate, usually much longer than the rachis; rachillae straight in bud; staminate flower symmetrical or asymmetrical; sepals rounded. Seychelles . . . . **Verschaffeltiinae**
3. Peduncle bearing a single large enclosing bract (prophyll); large enclosing peduncular bracts absent . . . . **Arecinae**
3. Peduncle bearing a prophyll and one or more rarely 2 enclosing peduncular bracts 4
4. Inflorescences spicate, the spikes sometimes many in each leaf axil; flowers borne in pits or depressions; staminate flower symmetrical or nearly so; sepals rounded, imbricate; fruit with apical stigmatic remains. Moluccas to Australia and Lord Howe Island . . . . **Linospadicinae**
4. Inflorescence branched to at least one order or sometimes spicate, never multiple; flowers superficial or in pits; staminate flowers symmetrical or asymmetrical; sepals various; fruit with apical, lateral or basal stigmatic remains, if apical then flowers not in pits or inflorescences not spicate 5
5. Leaflets almost always acute; inflorescences interfoliar or infrafoliar; crownshaft present or not; stigmatic remains apical, lateral or basal. Madagascar . . . . . . **Dypsidinae**
5. Leaflets acute or praemorse; inflorescences interfoliar or infrafoliar; crownshaft present or not; stigmatic remains various. Elsewhere in Old World, not Madagascar . . . 6
6. Leaflet tips or leaf margin praemorse 7
6. Leaflet tips acute 10
7. Small undergrowth palms usually lacking a crownshaft; inflorescence with slender rachillae, flowers borne in usually rather distant pits, emerging one at a time; fruit with basal stigmatic remains. Malay Peninsula, Sumatra and Borneo . . . . unplaced genus **Iguanura**
7. Small to large palms, always with a crownshaft; rachillae lacking pits; fruit with apical stigmatic remains 8
8. Staminate flowers symmetrical, bullet-shaped; pistillode usually large and ± bottle-shaped. Philippines to Papuasia and W. Pacific . . . . **Ptychospermatinae**
8. Staminate flower asymmetrical, not bullet-shaped; pistillode absent, obscure or very short, not bottle-shaped 9
9. Inflorescence with peduncular bract distinctly shorter than the prophyll; rachillae bearing spirally arranged triads in the proximal $^1/_2 - ^3/_4$, staminate flowers only distally. Sri Lanka . . . . unplaced genus **Loxococcus**
9. Inflorescence with peduncular bract ± the same size as the prophyll, completely sheathing the inflorescence in bud; rachillae bearing opposite and decussate or whorled triads, rarely triads spiral, throughout ± their entire length. Papuasia and W. Pacific . . . . unplaced genus **Hydriastele**
10. Flowers almost always borne in pits 11
10. Flowers superficial 12
11. Pits not laterally compressed, with prominent rounded lips; staminate flowers lacking hairy pedicels; stamens 6–15; stigmatic remains apical; seeds not ridged and grooved. Malay Peninsula, New Guinea and Solomon Islands . . . . unplaced genus **Cyrtostachys**

11. Pits laterally compressed, lacking prominent rounded lips; staminate flower with long hairy pedicels; stamens 6; fruit with stigmatic remains in lower 1/4; seeds ridged and grooved. Mainland India and Nicobar Islands . . . . unplaced genus **Bentinckia**
12. Fruit with stigmatic remains basal or lateral. New Caledonia . . . . **Clinospermatinae**
12. Fruit with stigmatic remains apical, subapical or in the upper ca. 1/2 13
13. Inflorescence in bud with twisted and coiled rachillae; peduncle very short; lower branches markedly divaricate; endosperm ruminate. Nicobar Islands, Malay Peninsula, Moluccas to Solomon Islands . . . . unplaced genus **Rhopaloblaste**
13. Inflorescence in bud with straight rachillae; peduncle various, the lower branches rarely divaricate; endosperm homogeneous or ruminate 14
14. Prophyll almost always incompletely sheathing; endocarp often heavily ornamented. New Caledonia . . **Basseliniinae**
14. Prophyll completely sheathing; endocarps smooth or fibrous, not heavily ornamented (except in 1 species of *Heterospathe*) 15
15. Palms lacking crownshafts. Philippines to Papuasia and Western Pacific . . . . . . unplaced genus **Heterospathe**
15. Palms with crownshafts 16
16. Inflorescence with prophyll much shorter than peduncular bract(s), or peduncular bract exserted far beyond the prophyll tip 17
16. Inflorescence with prophyll not markedly shorter than peduncular bract(s) 18
17. Small undergrowth palm; 1 peduncular bract; staminate flower bullet-shaped; fruit with strictly apical stigmatic remains. New Guinea . . . unplaced genus **Dransfieldia**

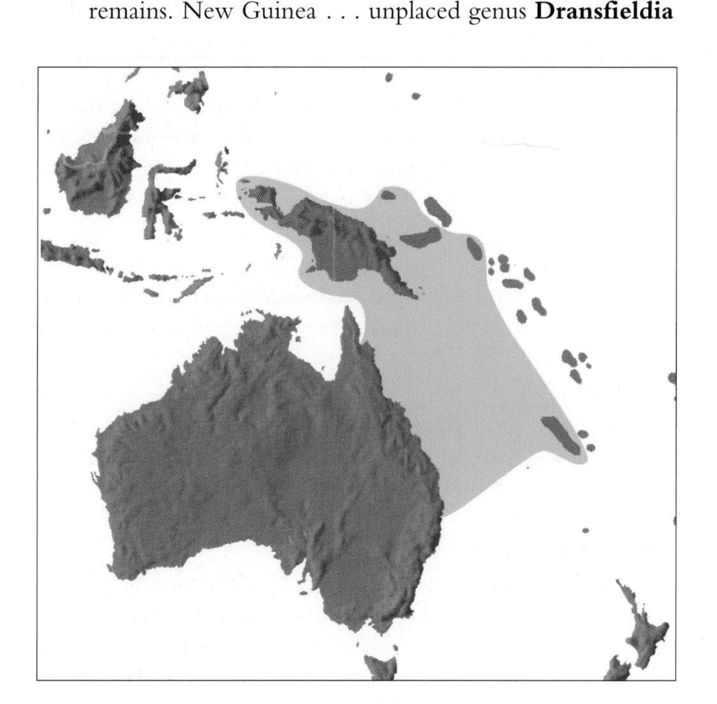

Distribution of subtribe **Archontophoenicinae**

17. Tall tree palms; peduncular bracts 1 or 2; staminate flower not bullet-shaped; fruit with apical or eccentrically apical stigmatic remains. W. Pacific . . . . **Carpoxylinae**
18. Staminate flowers with basally connate sepals or widely separated distinct, not imbricate sepals. New Zealand and S.W. Pacific . . . . **Rhopalostylidinae**
18. Staminate flowers with distinct imbricate sepals 19
19. Inflorescence lacking branches adaxially except at apex, branched to 1 order only and the lower branches ± ascending, not divaricate from the rachis at a 90° angle. Mascarenes . . . . unplaced genus **Dictyosperma**
19. Inflorescences not lacking branches adaxially, branched to 2–3 orders, lower branches spreading rather than ascending 20
20. Palms often stilt-rooted; stigmatic remains apical, subapical or lateral; stamens 6. W. Pacific . . . . unplaced genus **Clinostigma**
20. Palms not stilt-rooted; stigmatic remains strictly apical; stamens 12–55. New Guinea, New Caledonia, Australia . . . . **Archontophoenicinae**

Subtribe Archontophoenicinae J. Dransf. & N.W. Uhl, Principes 30: 8 (1986). Type **Archontophoenix.**

Moderate to robust, often tall, unarmed; leaves pinnate, leaflets entire; crownshaft always well developed; inflorescences branched to 2–3 orders; prophyll and peduncular bract similar; staminate flowers symmetrical or ± asymmetrical, filaments erect or inflexed in bud; petals of pistillate flower imbricate; gynoecium pseudomonomerous; fruit with apical stigmatic remains; endocarp lacking an operculum; embryo basal.

The genera of this subtribe are confined to New Guinea, Australia and New Caledonia.

Monophyletic groups comprising some or all of *Actinokentia*, *Archontophoenix*, *Chambeyronia* and *Kentiopsis* are resolved, sometimes with high support (Hahn 2002b, Norup *et al.* 2006, Baker *et al.* in review, in prep.). The placement of *Actinorhytis* in the subtribe is less frequently recovered (Lewis & Doyle 2002, Baker *et al.* in prep.). Although the genus is included in the Archontophoenicinae here, further phylogenetic evidence is required to substantiate this relationship. The subtribe is placed within the western Pacific clade of Areceae, but more precise relationships cannot yet be determined (Lewis & Doyle 2002, Norup *et al.* 2006, Baker *et al.* in prep.).

Key to Genera of Archontophoenicinae

1. Endosperm ruminate 2
1. Endosperm homogeneous 3
2. Staminate flower with 12–14 stamens; fruit not exceeding 2 cm in diameter; hilum basal. Australia . . . . **126. Archontophoenix**
2. Staminate flower with 24–33 or more stamens; fruit exceeding 4 cm in diameter; hilum lateral. New Guinea, Moluccas to Bougainville, cultivated elsewhere . . . . **125. Actinorhytis**

3. Pistillode lacking. New Caledonia . **128. Chambeyronia**
3. Pistillode conical, elongate or trifid 4
4. Tall moderately robust palms with robust stiff paniculate almost broom-like inflorescences, branches not divaricate; staminate flowers asymmetrical or symmetrical with pistillode conical, columnar or trifid, shorter than or equaling stamens in bud. New Caledonia . . . . **129. Kentiopsis**
4. Slender palm with small inflorescences with strongly divaricate branches; staminate flowers asymmetrical with prominent columnar pistillode, more than half as long as stamens in bud. New Caledonia . . . **127. Actinokentia**

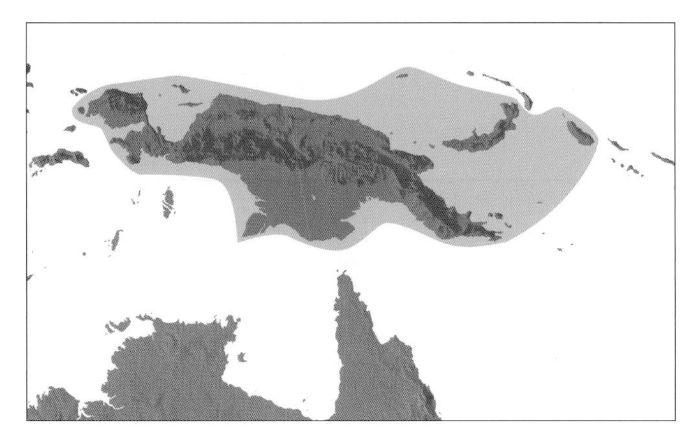

Distribution of ***Actinorhytis***

125. ACTINORHYTIS

Tall solitary tree palm of New Guinea and the Solomon Islands with slender crownshaft, strongly arching leaves and highly branched inflorescence bearing large fruit with deeply ruminate endosperm.

Actinorhytis H. Wendl. & Drude, Linnaea 39: 184 (1875). Type: **A. calapparia** (Blume) H. Wendl. & Drude ex Scheff. (*Areca calapparia* Blume).

Aktis — *ray,* rhytis — *wrinkle or fold, referring to the radiating ruminations in the endosperm.*

Tall, solitary, unarmed, pleonanthic, monoecious tree palms. *Stem* erect, bare, conspicuously marked with leaf scars, with a large mass of roots at the base. *Leaves* pinnate, arching, neatly abscising; sheaths tubular, forming a long, slender, well-defined crownshaft, bearing scattered caducous scales, the mouth with a short ligule; petiole very short in mature individuals (long in juveniles), adaxially channelled or flattened, abaxially rounded, densely caducously tomentose; rachis conspicuously down-curved toward the tip; leaflets very numerous, close, regularly arranged, single-fold, acute, acuminate or briefly bifid, the margins thickened, adaxially glabrous, abaxially with minute dot-like scales and conspicuous ramenta along the midrib, transverse veinlets obscure. *Inflorescences* infrafoliar, erect in bud, becoming horizontal or pendulous, branching to 3 orders proximally, to 1 order distally, protandrous; peduncle short, winged at the very base, grossly swollen just above the base in the centre, caducously tomentose; prophyll inserted near the base of the peduncle, tubular, beaked, 2-keeled, entirely enclosing the inflorescence in bud, sparsely scaly, splitting abaxially, deciduous; peduncular bract, inserted just above the prophyll, similar to the prophyll but scarcely 2-winged, deciduous; subsequent bracts low, triangular, inconspicuous; rachis longer than the peduncle, ± elliptic in cross-section, bearing relatively few, large, spirally arranged first-order branches, with conspicuous, bare, proximal portions; rachillae rather stiff, elongate, bearing spirally arranged triads in the proximal $^{1}/_{2}$ to $^{2}/_{3}$, and paired or solitary staminate flowers distally, or rarely, bearing only staminate flowers; rachilla bracts low, rounded, quite conspicuous, tending to form very shallow pits; floral bracteoles sepal-like. *Staminate flowers* asymmetrical in bud; sepals 3, distinct, imbricate, ± triangular-tipped, keeled; petals 3, distinct, ± ovate, valvate, ± 2–3 times as long as the sepals; stamens 24–33 or more, exserted at anthesis, filaments slender, elongate, inflexed at the tip, anthers medifixed, narrow oblong, ± versatile, latrorse; pistillode columnar, ± as long as the stamens in bud, shorter when stamens exserted. *Pollen* ellipsoidal asymmetric, occasionally elongate; aperture a distal sulcus; ectexine tectate, perforate-rugulate, aperture margin similar or slightly finer; infratectum columellate; longest axis 33–50 µm [1/1]. *Pistillate flowers* globular, at anthesis much larger than the staminate; sepals 3, distinct, imbricate, rounded; petals 3, ± twice as long as the sepals, distinct, broadly imbricate with conspicuous, triangular, valvate tips; staminodes 3, narrow triangular, flattened; gynoecium ovoid to obovoid, unilocular, uniovulate, stigmas 3, large, fleshy, recurved, ovule laterally attached near the apex of the locule, hemianatropous. *Fruit* very large, ovoid, ± beaked, green turning red at maturity, perianth whorls persistent, stigmatic remains apical; epicarp smooth, mesocarp with thin flesh and abundant anastomosing fibres adhering to the endocarp, endocarp closely adhering to the seed, thin, ± bony. *Seed* globose, with lateral, longitudinal hilum, endosperm deeply ruminate, with a central, irregular hollow; embryo basal. *Germination* adjacent-ligular; eophyll bifid. *Cytology* not studied.

Left: ***Actinorhytis calapparia*** (with *Areca catechu* on right), crown with inflorescences, cultivated, West Malaysia. (Photo: J. Dransfield)

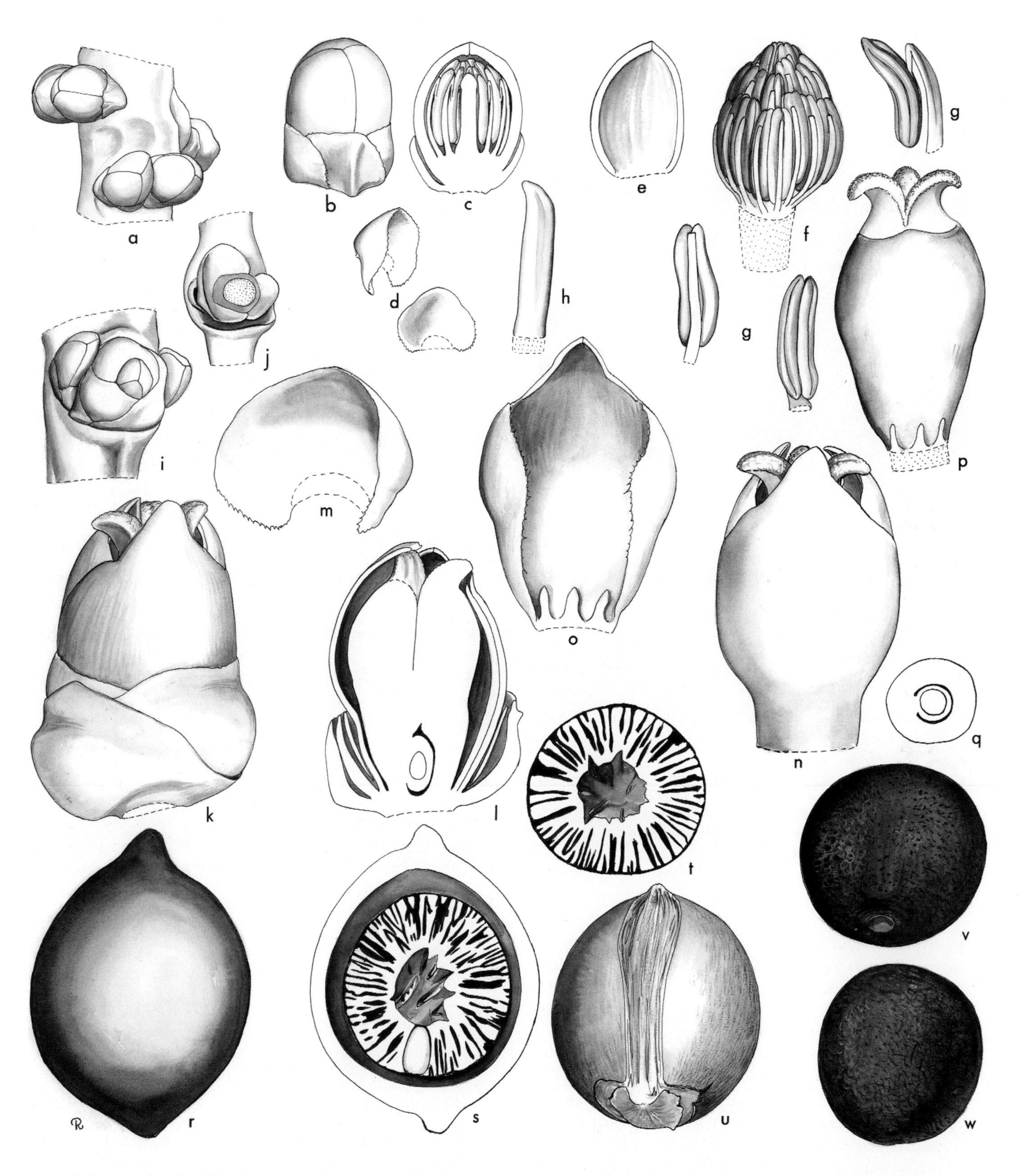

***Actinorhytis*.** **a**, portion of rachilla with paired staminate buds × 3; **b**, staminate bud × 6; **c**, staminate bud in vertical section × 6; **d**, staminate sepals in 2 views × 6; **e**, staminate petal, interior view × 6; **f**, androecium × 6; **g**, stamen in 3 views × 12; **h**, pistillode × 12; **i**, portion of rachilla with triad × 3; **j**, triad, flowers removed to show bracteoles × 3; **k**, pistillate flower × 6; **l**, pistillate flower in vertical section × 6; **m**, pistillate sepal, interior view × 6; **n**, pistillate flower, sepals removed × 6; **o**, pistillate petal with staminodes, interior view × 6; **p**, gynoecium and staminodes × 6; **q**, ovary in cross-section × 6; **r**, fruit × 1; **s**, fruit in vertical section × 1; **t**, seed in cross-section × 1; **u**, endocarp with basal operculum × 1; **v**, **w**, seed in 2 views × 1. *Actinorhytis calapparia:* **a–h** and **r–w**, *Whitmore* BSIP 4169; **i–q**, *Bailey* s.n. (Drawn by Marion Ruff Sheehan)

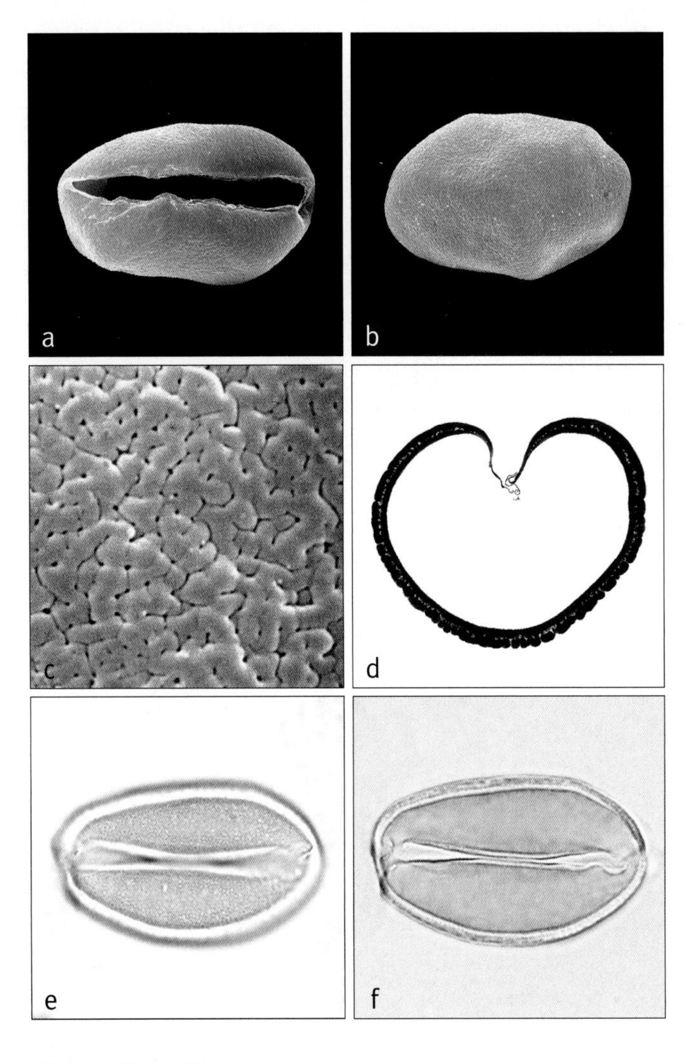

Above: ***Actinorhytis***
a, monosulcate pollen grain, distal face SEM × 1500;
b, monosulcate pollen grain, proximal face SEM × 1500;
c, close-up, perforate-rugulate pollen surface SEM × 8000;
d, ultrathin section through whole pollen grain, polar plane TEM × 2500;
e, monosulcate pollen grain, low-focus LM × 1500;
f, monosulcate pollen grain, mid-focus LM × 1500.
Actinorhytis calapparia: **a–f**, *Talbot* 1789.

Distribution and ecology: A single species, *Actinorhytis calapparia*, native to New Guinea and the Solomon Islands, now widespread as an ornamental or ceremonial plant in Southeast Asia. In the wild, it grows in lowland tropical rain forest at altitudes up to about 1000 m above sea level.

Anatomy: Leaf (Tomlinson 1961), root (Seubert 1998a, 1998b) and fruit (Essig *et al.* 1999).

Relationships: Moderate support for the placement of *Actinorhytis* as sister to all other Archontophoenicinae has been recovered in one study (Baker *et al.* in prep.), and other accounts yield compatible relationships (Lewis & Doyle 2002, Baker *et al.* in review).

Common names and uses: *Pinang penawar, pinang mawar. Actinorhytis calapparia* is widely planted in Southeast Asia and Malesia; it is very decorative, but the main reason for its cultivation by villagers is as a magic or medicinal plant. The seed may also be chewed as a betel substitute.

Taxonomic account: Two species have been described but have recently been shown to be conspecific (Wanggai, pers. comm.).

Fossil record: A fossil seed from the Upper Eocene of Hungary with a ruminate endosperm has been described as *Actinorhytis eocaenica* (Rásky 1956), but it lacks the central hollow and distinctive radiating ruminations of the modern genus. This fossil seems likely to be a palm, but its equation with the modern genus *Actinorhytis* is highly suspect. Asymmetric monosulcate pollen with a distinctive irregularly columellate infratectum, *Palmaepollenites* sp., from the Middle Eocene of Central Java (Nanggulan Formation) is compared with pollen of *Actinorhytis* and with pollen of *Cyphosperma, Cyphophoenix* and *Moratia* (= *Cyphokentia*) (Harley & Morley 1995).

Notes: This is a large palm with arching leaves, a very slender crownshaft, conical masses of roots at the base of the trunk and large, widely spreading inflorescences below the leaves. These characters and the large fruits, the largest in the subtribe, are distinctive.

126. ARCHONTOPHOENIX

Solitary pinnate-leaved tree palms from Australia, with acute leaflets and highly branched inflorescences.

Archontophoenix H. Wendl. & Drude, Linnaea 39: 182, 211 (1875). Lectotype: **A. alexandrae** (F. v. Mueller) H. Wendl. & Drude (*Ptychosperma alexandrae* F. Muell.) (see O.F. Cook [1915] and Beccari & Pichi-Sermolli [1955]).

Loroma O.F. Cook, J. Wash. Acad. Sci. 5: 117 (1915). *L. amethystina* O.F. Cook (= *Archontophoenix cunninghamiana* [H. Wendl.] H. Wendl. & Drude).

Archon — *chief, ruler,* phoenix — *date palm or palm in general, named for its regal stature and appearance.*

Distribution of ***Archontophoenix***

Moderate to tall, solitary, unarmed, pleonanthic, monoecious palms. *Stem* columnar, graceful, rather slender, slightly or strongly swollen basally, leaf scars obscure or prominent, often raised, distant or close. *Leaves* pinnate, erect or spreading, sometimes twisted about 90 degrees basally; sheaths tubular, forming a prominent crownshaft, thick, leathery, green, rusty-brown or purplish-red, often somewhat swollen basally; petiole short, grooved adaxially, rounded abaxially; rachis very long, similar to the petiole near the base, becoming flat adaxially and grooved laterally, scaly and minutely brown-dotted; leaflets lanceolate, elongate, tips irregularly pointed, single-fold, green or whitish abaxially due to very small silvery scales, ramenta large, dark-brown, often twisted or divided, medifixed or basifixed, present or lacking abaxially along the midrib, the midrib and large veins prominently or obscurely brown-dotted, midrib and several pairs of veins prominent abaxially, transverse veinlets not apparent. *Inflorescences* infrafoliar, erect in bud, becoming horizontal or drooping, with pendulous branches, branched to 3(–4) orders, protandrous; peduncle very short, stout; prophyll tubular, elongate, somewhat dorsiventrally flattened, 2-keeled laterally, briefly beaked, rather thin; peduncular bract like the prophyll but not keeled, prophyll and peduncular bract caducous; rachis moderate, tapering; rachis bracts low, ± ruffled to prominent, sharply pointed; rachillae somewhat divaricate and pendulous, bearing spirally arranged, rather thick, basally cupular, low and rounded or short pointed bracts subtending triads of flowers nearly throughout the rachillae, a few paired and solitary staminate flowers present distally; floral bracteoles low, rounded. Flowers pale lavender to purplish or cream to yellow. *Staminate flowers* asymmetrical, borne lateral to the pistillate in the triads; sepals 3, distinct, imbricate, broadly ovate, keeled, tips pointed; petals 3, distinct, ca. 5 times as long as the sepals, narrowly ovate, grooved adaxially, tips thicker, pointed; stamens ca. 12–14 (8 or 9–24 according to Hooker [1883] and Bailey [1935a], filaments short, awl-shaped, erect apically in bud, anthers elongate, linear, dorsifixed near the middle, erect in bud, later versatile, bifid basally, pointed to slightly emarginate distally, latrorse, the connective elongate, tanniniferous; pistillode more than half as long to as long as the stamens, trifid or cylindrical. *Pollen* ellipsoidal or elongate, with slight or obvious asymmetry; aperture a distal sulcus; ectexine tectate, finely perforate-rugulate, aperture margin slightly finer than main tectum; infratectum columellate; longest axis 43–65 µm [2/6]. *Pistillate flowers* symmetrical, ovoid; sepals 3, distinct, broadly imbricate, tips briefly pointed; petals 3, distinct, imbricate except for prominent valvate tips; staminodes 3 or 4, tooth-like, borne on one side of the gynoecium; gynoecium irregularly ovoid, unilocular, uniovulate, style indistinct, stigmas 3, recurved, ovule laterally attached, form unknown. *Fruit* globose to ellipsoidal, pink to red, stigmatic remains apical; epicarp smooth, mesocarp thin, soft, fleshy with flattened, conspicuously branched and interlocking, longitudinal fibres, endocarp thin, smooth, fragile, not operculate. *Seed* ellipsoidal to globose, basally attached, hilum basal, elongate, raphe branches numerous, anastomosing, endosperm ruminate; embryo basal. *Germination* adjacent-ligular; eophyll bifid. *Cytology*: 2n = 32.

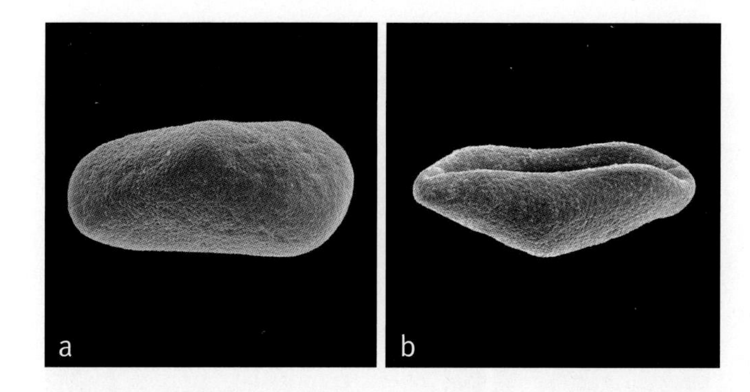

Above: ***Archontophoenix***
a, monosulcate pollen grain, proximal face SEM × 1000;
b, monosulcate pollen grain, distal face SEM × 1000.
Archontophoenix alexandrae: **a** & **b**, *Brass* 19831.

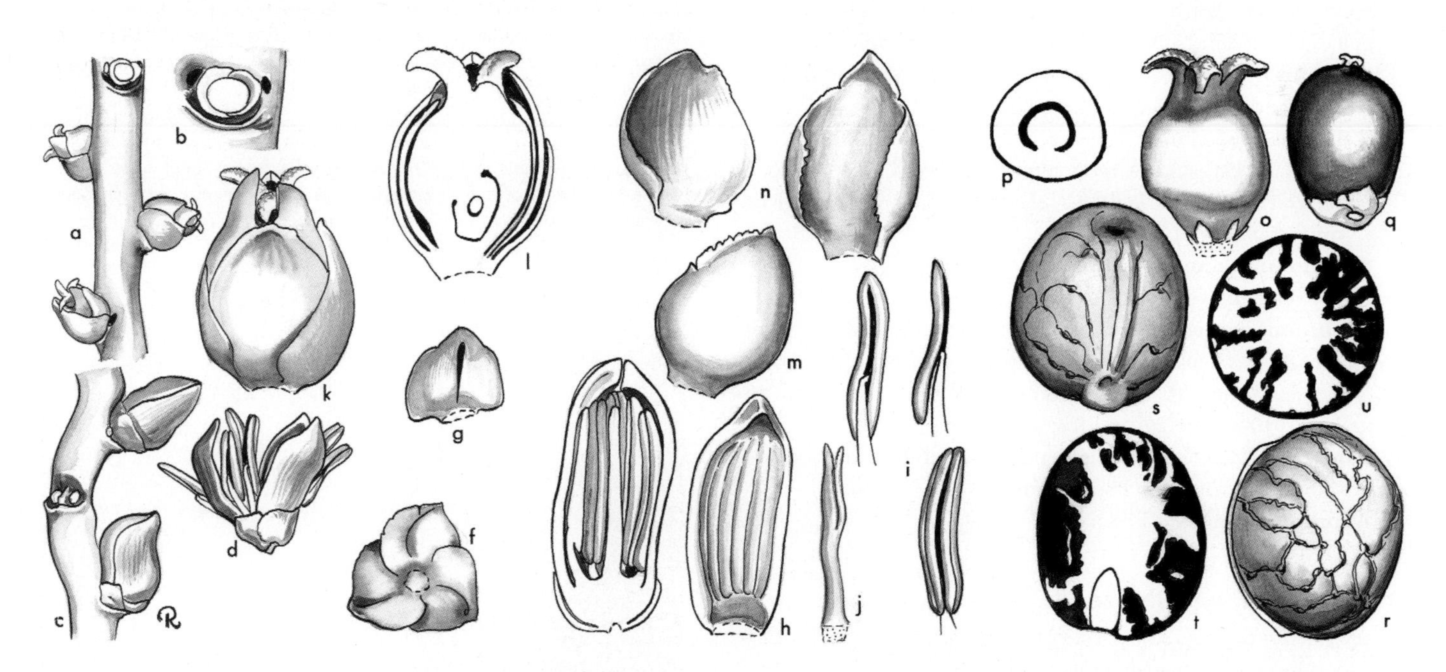

***Archontophoenix*.** **a**, portion of rachilla with triads, staminate flowers fallen × 1 1/2; **b**, triad, flowers removed × 6; **c**, portion of rachilla with paired staminate flowers × 3; **d**, staminate flower × 3; **e**, staminate flower in vertical section × 6; **f**, staminate calyx × 6; **g**, staminate sepal × 6; **h**, staminate petal, interior view × 6; **i**, stamen in 3 views × 6; **j**, pistillode × 6; **k**, pistillate flower × 6; **l**, pistillate flower in vertical section × 6; **m**, pistillate sepal, exterior view × 6; **n**, pistillate petal in 2 views × 6; **o**, gynoecium and staminodes × 6; **p**, ovary in cross-section × 6; **q**, fruit × 1 1/2; **r, s**, seed in 2 views × 3; **t**, seed in vertical section × 3; **u**, seed in cross-section × 3. *Archontophoenix alexandrae*: **a–p**, *Moore & Stephens* 9249; **q–u**, *White* 3364. (Drawn by Marion Ruff Sheehan)

Above left: ***Archontophoenix maxima***, habit, Australia. (Photo: J. Dransfield)

Above right: ***Archontophoenix myolensis***, inflorescences, cultivated, Montgomery Botanical Center, Florida. (Photo: J. Dransfield)

Distribution and ecology: Six species of eastern Australia from the southern coast of New South Wales to the northern coast of Queensland. Occurring in forest in warm-temperate to tropical regions at sea level to elevations of about 1200 m, often in wet gullies, on stream banks or edges of swamps on various soils.

Anatomy: Leaf (Tomlinson 1961), root (Seubert 1998a, 1998b), and fruit (Essig & Hernandez 2002).

Relationships: *Archontophoenix* is resolved as monophyletic (Pintaud 1999a) and is moderately supported as sister to a clade comprising *Actinokentia*, *Kentiopsis* and *Chambeyronia* (Norup *et al*. 2006).

Taxonomic account: Dowe & Hodel (1994).

Fossil record: No generic records found.

Common names and uses: Alexander palm, King palm (*Archontophoenix alexandrae*); piccabean palm, bangalow palm (*A. cunninghamiana*). Grown commercially as ornamentals in many warm-temperate and tropical regions.

Notes: Distinctive in having pendulous rachillae, an elongate pistillode and ruminate endosperm.

Additional figures: Glossary figs 12, 14.

127. ACTINOKENTIA

Small to moderate tree palms from New Caledonia with divaricate inflorescences.

Actinokentia Dammer, Bot. Jahrb. Syst. 39: 20 (1906). Lectotype: **A. divaricata** (Brongn.) Dammer (*Kentiopsis divaricata* Brongn.) (see Beccari 1920).

Combining aktis *— a ray or beam of light, with the generic name* Kentia*, named for William Kent (1779–1827), once curator of the botanic gardens at Buitenzorg, Java (now Kebun Raya Bogor).*

Solitary, small to moderate, unarmed, pleonanthic, monoecious palms. *Stem* slender, erect, prominently ringed with somewhat sunken leaf scars, sometimes with prickly roots. *Leaves* pinnate; sheaths thick, forming a crownshaft; petiole short, rounded abaxially, channelled adaxially, or elongate and terete; leaflets regularly arranged, lanceolate, acute to tapering, single-fold, adaxially waxy or glabrous, abaxially waxy tomentose with large ramenta along the midribs, midribs conspicuous, second largest ribs those along margins, transverse veinlets not evident. *Inflorescences* infrafoliar, protandrous, divaricately branched to 3 orders basally, 1–2 orders distally; peduncle short; prophyll tubular, pointed, rather thin, indistinctly 2-keeled, completely encircling the peduncle at insertion and enclosing the peduncular bract, caducous; peduncular bract like the prophyll but lacking keels; rachis longer than the peduncle bearing spirally arranged, spreading, acute bracts subtending branches and rachillae; rachilla bracts prominent, rounded, lip-like and shorter than the flowers or acute and exceeding the flowers, subtending triads basally, paired or solitary staminate flowers distally, in broadened depressions in the rachillae; bracteoles surrounding the pistillate flower sepal-like, outermost bracteole prominent, ca. $^1/_2$ as long or as long as the inner bracteoles. *Staminate flowers* symmetrical, larger at anthesis than the pistillate buds; sepals 3, distinct, broadly imbricate and rounded, scarcely longer than broad, the outer often prominently keeled or pouch-like near the apex; petals 3, distinct, valvate, boat-shaped; stamens 19–50, filaments erect or nearly so at the apex in bud, anthers erect in bud, linear, dorsifixed, slightly emarginate apically, bifid basally, latrorse, the connective elongate; pistillode as long as the stamens in bud, tapered to a slender apex from a broad base. *Pollen* ellipsoidal or asymmetric to pyriform; aperture a distal sulcus; ectexine tectate, perforate, aperture margin similar or slightly finer; infratectum columellate; longest axis 48–60 µm [1/2]. *Pistillate flowers*, buds usually well developed at staminate anthesis, symmetrical; sepals 3, distinct, broadly imbricate and rounded; petals 3, distinct, imbricate except for briefly valvate apices; staminodes 3, small, tooth-like, borne at one side of the gynoecium; gynoecium unilocular, uniovulate, stigmas 3, prominent, recurved, ovule pendulous, hemianatropous. *Fruit* ellipsoidal with apical stigmatic remains; epicarp smooth, mesocarp underlain by a shell of short, pale sclereids, elliptic in outline at surface, the sclereid shell over parenchyma with flat, anastomosing longitudinal fibres adherent to the endocarp, tannin cells lacking, or few and interspersed among the fibres, endocarp thin, fragile, not operculate. *Seed* attached by an elongate hilum, raphe branches anastomosing, endosperm homogeneous; embryo basal. *Germination* adjacent-ligular; eophyll bifid. *Cytology* not known.

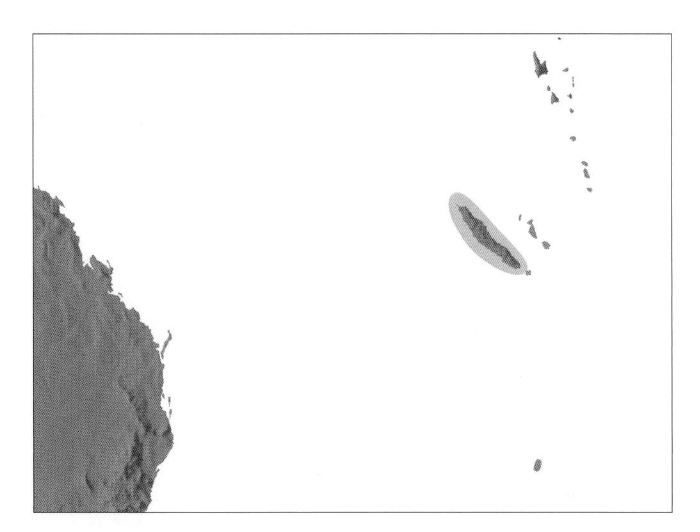

Distribution of ***Actinokentia***

Distribution and ecology: Two species in New Caledonia, occurring in wet forests on serpentine soils from 60–1000 m in the southern part of the island. Individuals are small relative to those of other genera in the Archontophoenicinae and seldom reach the forest canopy. Where *Actinokentia divaricata* and *A. huerlimannii* are sympatric on the flanks of Mont Nekando, they appear to occupy different habitats and exposures.

Anatomy: Leaf (Uhl & Martens 1980) and fruit (Essig & Hernandez 2002).

Relationships: The monophyly of *Actinokentia* is moderately to highly supported (Pintaud 1999a, Baker *et al.* in prep.). Pintaud (1999a) found weak evidence that the genus is nested within *Chambeyronia*, whereas other workers recover a moderately supported sister relationship with *Kentiopsis* (Baker *et al.* in review).

Common names and uses: Common names unknown. *Actinokentia divaricata* was reported to be introduced into cultivation in Europe more than a century ago; recent introductions have been made and both species are now cultivated as ornamentals.

Taxonomic accounts: Moore & Uhl (1984) and Hodel & Pintaud 1998.

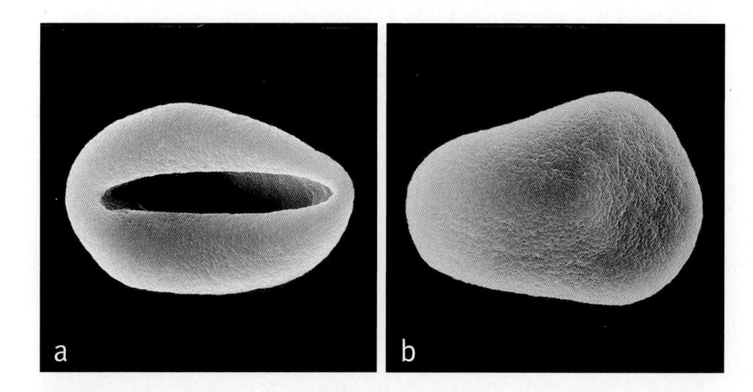

Above: ***Actinokentia***
a, monosulcate pollen grain, distal face SEM × 1000;
b, monosulcate pollen grain, proximal face × 1000.
Actinokentia divaricata: **a** & **b**, *Moore* 10419.

Fossil record: No generic records found.

Notes: Leaves of the two species are so different in morphology that *A. huerlimannii* was thought at first to belong in another genus. Foliar anatomy is strikingly similar, however, as are flowers and fruits. Small size and few leaves in the crown help to distinguish the genus.

Additional figures: Glossary fig. 19.

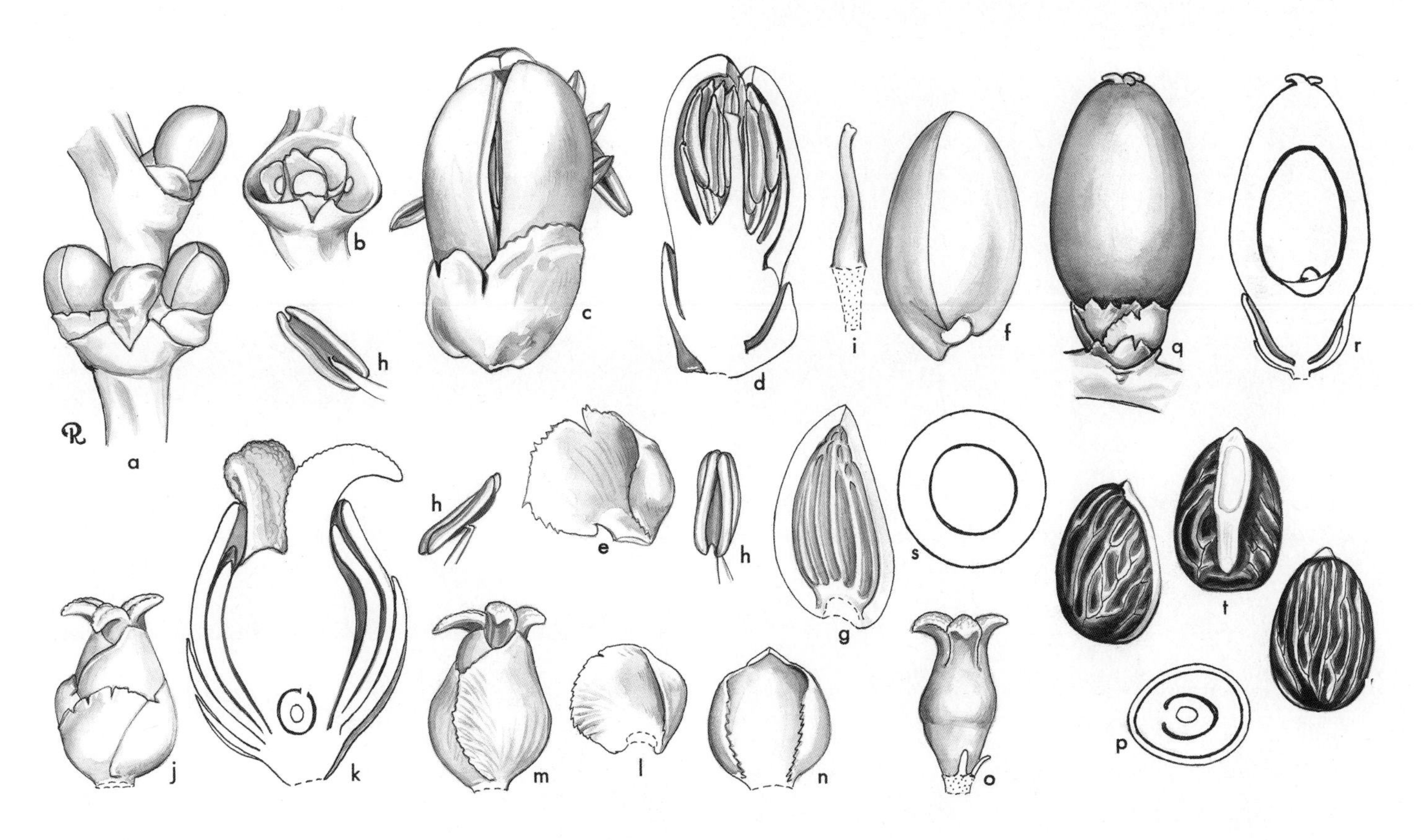

Actinokentia. **a**, portion of rachilla with triad × 3; **b**, triad, flowers removed to show bracts × 3; **c**, staminate flower × 6; **d**, staminate flower in vertical section × 6; **e**, staminate sepal × 6; **f**, staminate bud, sepals removed × 6; **g**, staminate petal, interior view × 6; **h**, stamen in 3 views × 6; **i**, pistillode × 6; **j**, pistillate flower × 3; **k**, pistillate flower in vertical section × 6; **l**, pistillate sepal × 3; **m**, pistillate flower, sepals removed × 3; **n**, pistillate petal, interior view × 3; **o**, gynoecium and staminodes × 3; **p**, ovary in cross-section × 9; **q**, fruit × $1^1/_2$; **r**, fruit in vertical section × $1^1/_2$; **s**, fruit in cross-section × $1^1/_2$; **t**, seed in 3 views × $1^1/_2$. *Actinokentia divaricata*: all from *Moore et al.* 9340. (Drawn by Marion Ruff Sheehan)

128. CHAMBEYRONIA

Moderate pinnate-leaved tree palms from New Caledonia with broad leathery-textured leaflets and relatively large frut.

Chambeyronia Vieill., Bull. Soc. Linn. Normandie, ser. 2. 6: 229 (1873). Lectotype: **C. macrocarpa** (Brongn.) Vieill. ex Becc. (*Kentiopsis macrocarpa* Brongn.) (see Beccari [1920], Beccari & Pichi-Sermolli [1955]).

Named for Charles-Marie-Léon Chambeyron (1827–1891), French naval officer and hydrographer, who mapped much of the coast of New Caledonia and assisted Vieillard in the exploration of the island.

Moderate, solitary, unarmed, pleonanthic, monoecious palms. *Stem* erect, ringed with leaf scars, enlarged at the base but roots not prominent. *Leaves* regularly pinnate, curved, spreading, often red when first exposed; sheath tubular, forming a prominent crownshaft, with or without a shallow notch opposite the petiole, glabrous adaxially, lightly or densely covered in scales abaxially; petiole channelled adaxially, rounded abaxially, with scattered small brown scales; leaflets acute to acuminate, single-fold, wide, waxy or glabrous adaxially, small brown scales scattered abaxially, midrib and marginal ribs large, transverse veinlets obscure or evident. *Inflorescences* infrafoliar, protandrous, branched to 2(–3) orders basally, to 1 order distally; peduncle very short; prophyll tubular, completely encircling the peduncle and enclosing the peduncular bract, caducous, dorsiventrally flattened, with 2 wide lateral keels, chartaceous, glabrous adaxially, lightly or densely scaly abaxially; peduncular bract lacking keels and thinner, with a more definite beak, otherwise like the prophyll; rachis longer than the peduncle bearing spirally arranged, low, or prominent pointed, bracts subtending branches or rachillae; rachillae

Top: ***Actinokentia huerlimanii***, habit, New Caledonia. (Photo: J.-C. Pintaud)
Bottom left: ***Actinokentia huerlimanii***, inflorescence, New Caledonia. (Photo: J.-C. Pintaud)
Bottom right: ***Actinokentia huerlimanii***, young fruit, New Caledonia (Photo: J.-C. Pintaud)

Below: ***Chambeyronia***
a, monosulcate pollen grain, equatorial view SEM × 1000;
b, monosulcate pollen grain, proximal face SEM × 1000;
c, close-up, finely perforate-rugulate pollen surface SEM × 8000;
d, ultrathin section through pollen wall TEM × 7500.
Chambeyronia lepidota: **a–c**, *Moore* 10462; *C. macrocarpa*: **d**, *Balansa* 2911.

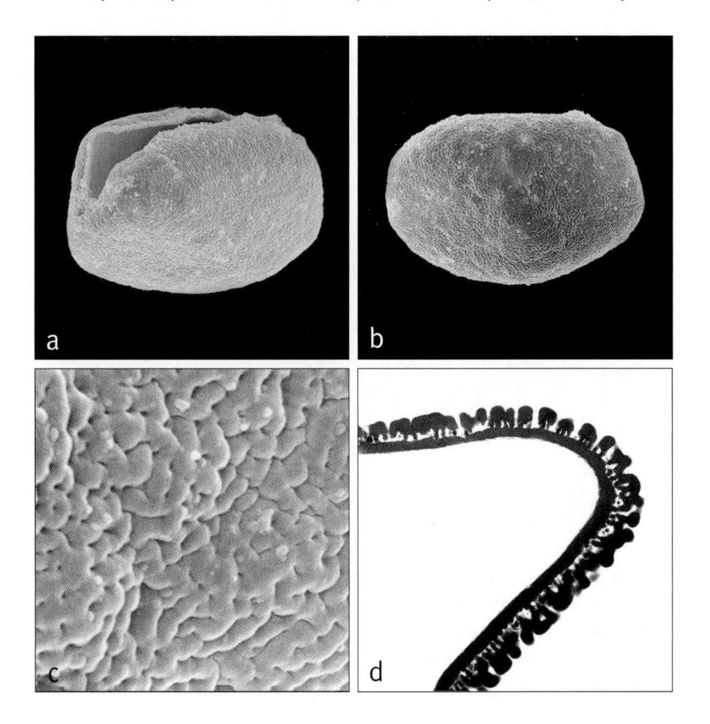

rather stout, tapering, sinuous, lightly or densely scaly; rachilla bracts prominent, spreading or ascending, spirally arranged, subtending triads of flowers basally and paired or solitary staminate flowers distally, the triads rather distant, sometimes appearing as though impressed in the axis; bracteoles surrounding the pistillate flower unequal, not sepal-like, the larger one shorter or exceeding the triad bract. *Staminate flowers* asymmetrical to subsymmetrical; sepals 3, distinct, acute to long pointed; petals 3, distinct, valvate, asymmetrical, angled, and strongly nerved to nearly symmetrical and smooth when dry; stamens 19–55, filaments awl-shaped, briefly inflexed, anthers erect in bud, linear, dorsifixed, bifid basally, emarginate apically, latrorse, the connective elongate; pistillode lacking. *Pollen* grains ellipsoidal, slight or obvious asymmetry; aperture a distal sulcus; ectexine tectate, perforate-rugulate, aperture margin similar; infratectum columellate; longest axis ranging from 45–75 µm [2/2]. *Pistillate flowers* symmetrical; sepals 3, distinct, broadly imbricate, acute; petals 3, distinct, broadly imbricate with prominently valvate apices; staminodes 3, small, tooth-like, borne at one side of the gynoecium; gynoecium with 3 spreading stigmas, unilocular, uniovulate, ovule laterally attached, hemianatropous. *Fruit* subglobose to ovoid, with apical stigmatic remains; epicarp smooth, underlain by a mesocarp of oblique, short, pale sclereids over parenchyma with dispersed tannin cells and stout, flat, longitudinal, anastomosing fibres adherent to the endocarp, endocarp thin, fragile, not operculate. *Seed* attached by an elongate lateral hilum, raphe branches numerous, anastomosing, endosperm homogeneous; embryo basal. *Germination* adjacent-ligular; eophyll bifid. *Cytology*: 2n = 32.

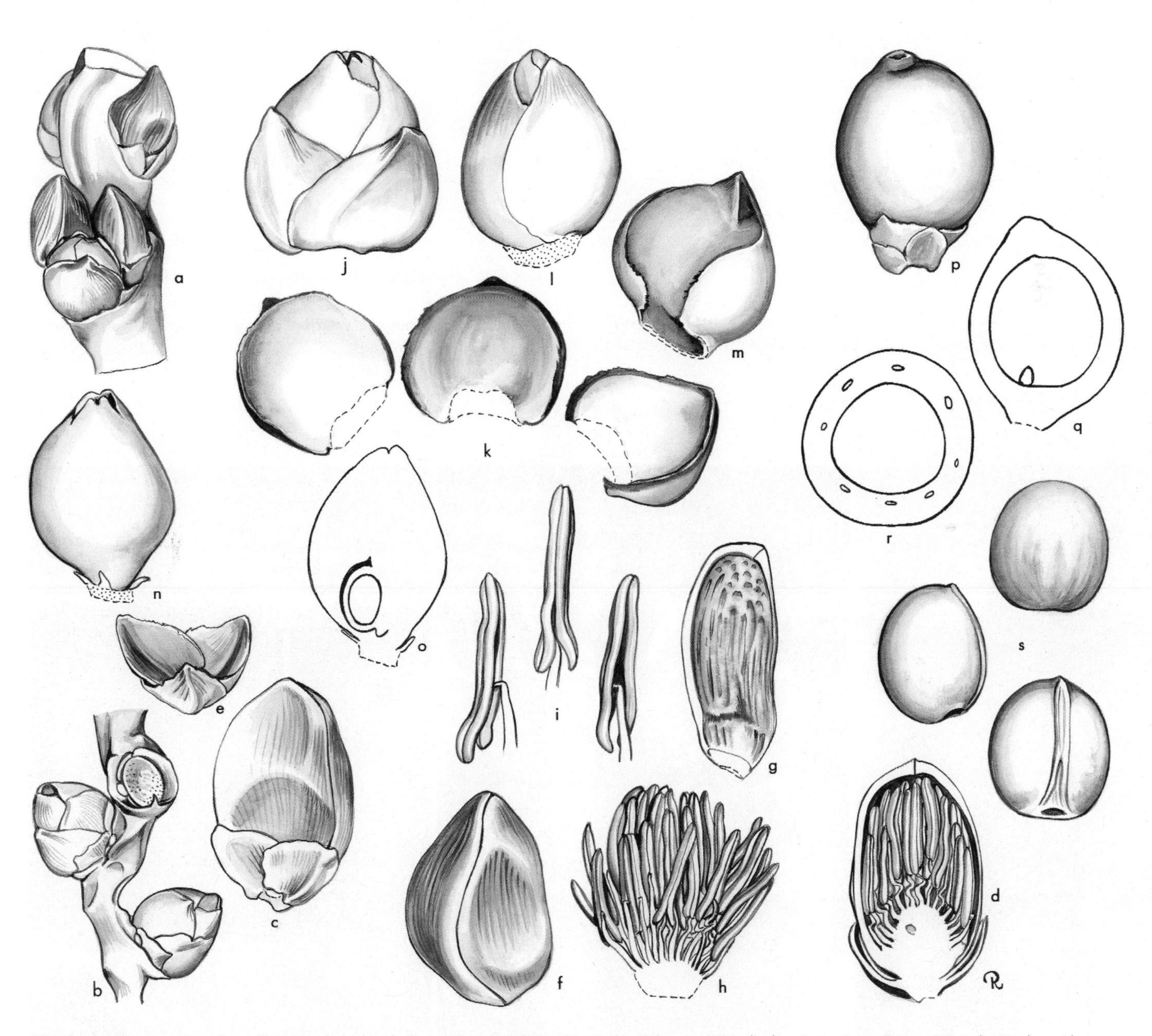

Chambeyronia. **a**, portion of rachilla with triads × 1 1/2; **b**, portion of rachilla with triads, all flowers fallen (top) and staminate flowers fallen (bottom) × 1 1/2; **c**, staminate bud × 3; **d**, staminate bud in vertical section × 3; **e**, staminate sepals × 3; **f**, staminate bud, sepals removed × 3; **g**, staminate petal, interior view × 3; **h**, androecium × 3; **i**, stamen in 3 views × 6; **j**, pistillate bud × 3; **k**, pistillate sepals × 3; **l**, pistillate bud, sepals removed × 3; **m**, pistillate petal × 3; **n**, gynoecium and staminodes × 3; **o**, gynoecium in vertical section × 3; **p**, fruit × 1 1/2; **q**, fruit in vertical section × 1 1/2; **r**, fruit in cross-section × 1 1/2; **s**, seed in 3 views × 1 1/2. *Chambeyronia macrocarpa*: all from *Moore et al.* 9335. (Drawn by Marion Ruff Sheehan)

Distribution and ecology: Two species in New Caledonia. *Chambeyronia macrocarpa* is found in wet forest or gallery forest nearly throughout New Caledonia, whereas *C. lepidota* occurs on schistose soils only in the north-eastern part of the island.
Anatomy: Leaf (Uhl & Martens 1980), root (Seubert 1998a, 1998b), and fruit (Essig & Hernandez 2002).
Relationships: Pintaud (1999a) found that *Chambeyronia* is paraphyletic with respect to *Actinokentia*, but there is low support for these relationships. Furthermore, they are not congruent with moderately supported relationships recovered elsewhere, which place the genus as sister to a clade of *Actinokentia* and *Kentiopsis* (Baker *et al.* in review).
Common names and uses: Common names unknown. *Chambeyronia macrocarpa* is probably the most widely cultivated New Caledonian palm.
Taxonomic accounts: Moore & Uhl (1984) and Hodel & Pintaud (1998).
Fossil record: No generic records found.
Notes: In leaf anatomy, *Chambeyronia macrocarpa* is distinguished from other members of the Archontophoenicinae in New Caledonia by a large fibrous strand of ca. 15 cells between each two vascular bundles. *Chambeyronia lepidota* is the only species of the Archontophoenicinae to have thick cuticles on both upper and lower leaf surfaces, and these are perhaps associated with its more exposed habitat. *Chambeyronia macrocarpa* is very variable in the wild.

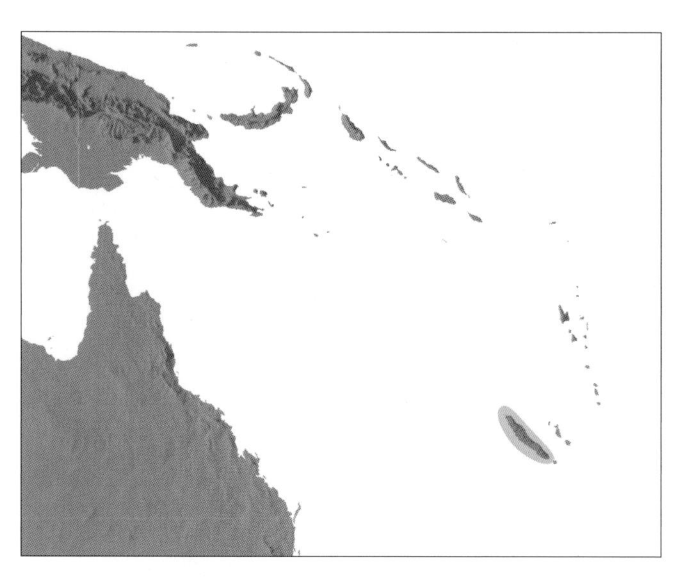

Distribution of ***Chambeyronia***

Below left: ***Chambeyronia* macrocarpa**, habit, New Caledonia. (Photo: J.-C. Pintaud)

Below middle: ***Chambeyronia macrocarpa***, crown with infructescences, New Caledonia. (Photo: J.-C. Pintaud)

Below right: ***Chambeyronia macrocarpa***, mature fruit, New Caledonia. (Photo: J.-C. Pintaud)

129. KENTIOPSIS

Tall pinnate-leaved palms from New Caledonia with broom-like inflorescences.

Kentiopsis Brongn., Compt. Rend. Hebd. Séances Acad. Sci. 77: 398 (1873). Lectotype: **K. oliviformis** (Brongn. & Gris) Brongn. (*Kentia oliviformis* Brongn. & Gris) (see Beccari 1920).

Mackeea H.E. Moore, Gentes Herb. 11: 304 (1978). Type: *M. magnifica* H.E. Moore.

Derived from the generic name Kentia, *named for William Kent (1779–1827), one-time curator of the botanic gardens at Buitenzorg, Java (now Kebun Raya Bogor) with* opsis *— resembling, from a presumed resemblance to* Kentia *Blume.*

Solitary, tall, unarmed, pleonanthic, monoecious palms. *Stem* erect, thick basally, grey, ringed with ± somewhat prominent, rather close leaf scars and with exposed roots at the base. *Leaves* pinnate, erect or spreading, neatly abscising; sheaths forming a prominent crownshaft; petiole channelled adaxially, rounded abaxially; leaflets regularly arranged, lanceolate, acute to acuminate, single-fold, adaxially glabrous and with wax, abaxially densely covered with small punctiform scales and abundant ramenta along ribs, midrib prominent, marginal ribs second in size to midrib, numerous secondary ribs conspicuous abaxially, transverse veinlets not evident. *Inflorescences* infrafoliar, branched to (2)–3–(4) orders basally, 1–2 orders distally, protandrous; peduncle very short, variously tomentose; prophyll and peduncular bract caducous, prophyll completely encircling the peduncle and enclosing the peduncular bract, briefly beaked, flat, keeled laterally, rather thin, chartaceous, both surfaces densely covered in whitish deciduous tomentum; peduncular bract like the prophyll but lacking keels; rachis elongate, longer than the peduncle, bearing spirally arranged, low, rounded, ± ruffled bracts subtending the branches and rachillae; rachillae rather slender to stout, about equal in length, straight or curved, usually glabrous, bearing spirally arranged prominent, rounded, lip-like bracts subtending flowers borne in triads of 2

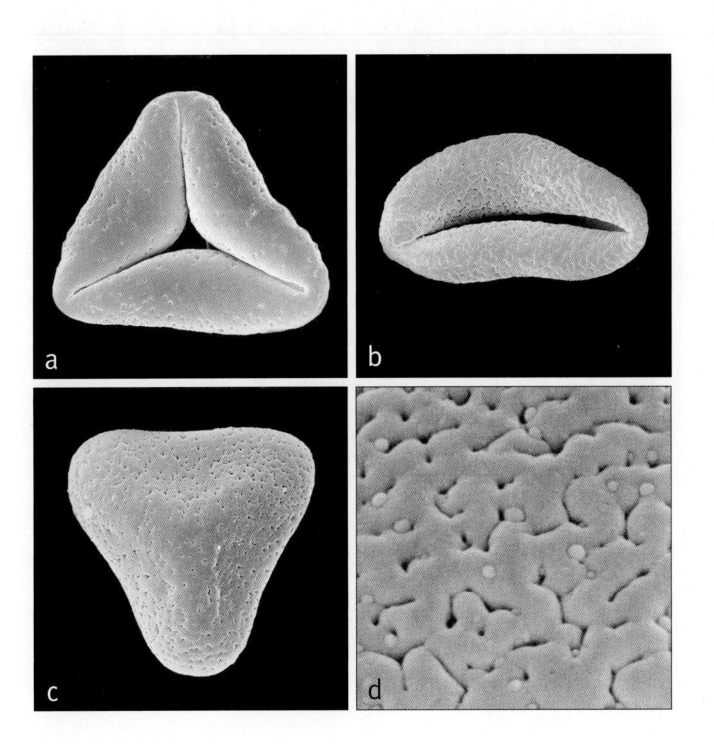

Top: ***Kentiopsis oliviformis***, crown with inflorescences, cultivated, New Caledonia. (Photo: J.-C. Pintaud)

Above: ***Kentiopsis oliviformis***, fruit, cultivated, New Caledonia. (Photo: J.-C. Pintaud)

Left: ***Kentiopsis***
a, trichotomosulcate pollen grain, distal face SEM × 1000;
b, monosulcate pollen grain, distal face SEM × 1000;
c, trichotomosulcate pollen grain, proximal face SEM × 1000;
d, close-up, finely perforate-rugulate pollen surface SEM × 8000.
Kentiopsis oliviformis: **a–d**, *Moore et al.* 9974.

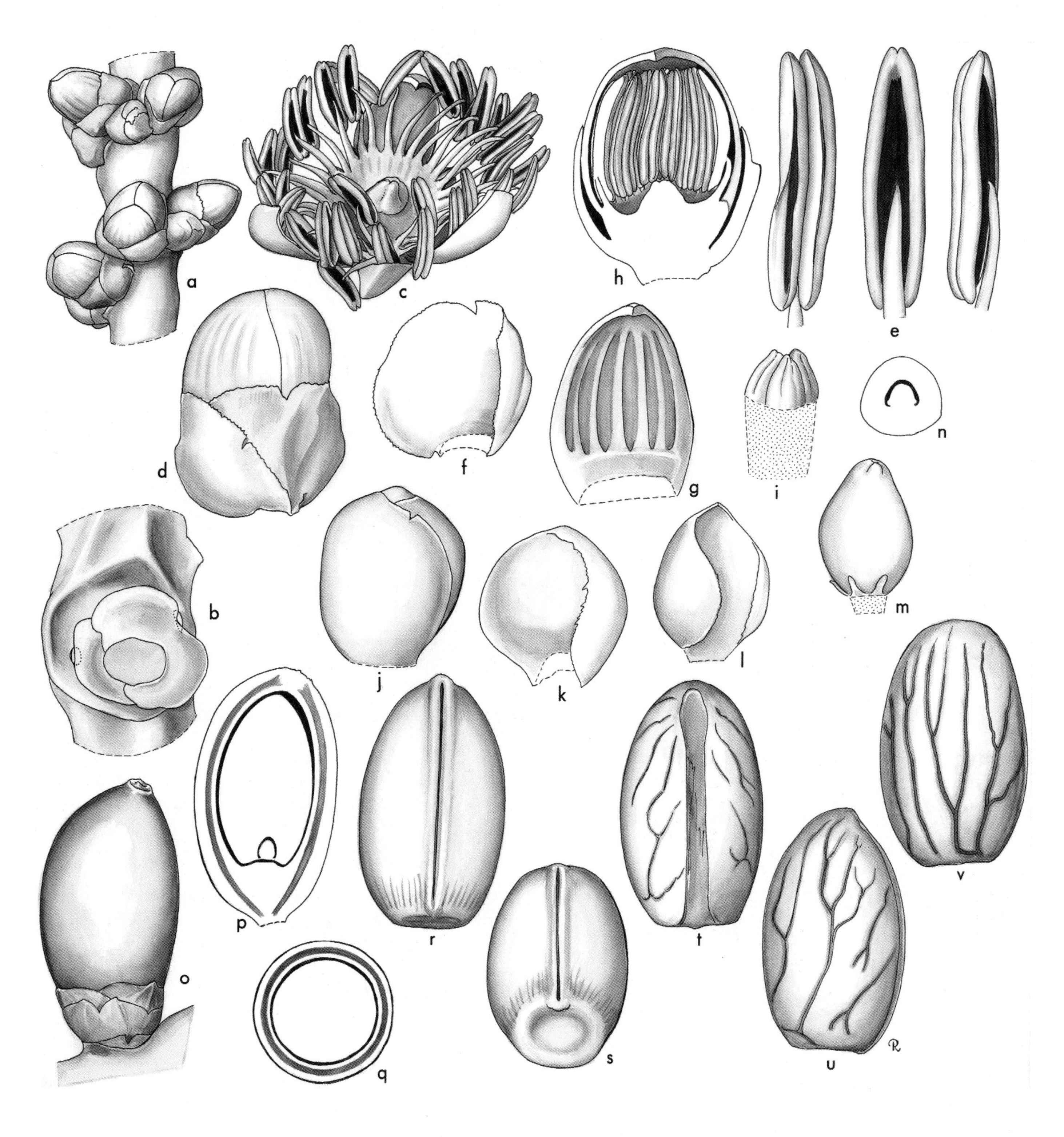

Kentiopsis. **a**, portion of rachilla with triads, flowers in bud × 2 1/4; **b**, triad, flowers removed to show bracteoles around pistillate bud × 4 1/2; **c**, staminate flowers at anthesis × 4 1/2; **d**, staminate bud × 6; **e**, staminate bud in vertical section × 6; **f**, staminate sepal × 6; **g**, staminate petal, interior view × 6; **h**, stamen in 3 views × 12; **i**, pistillode × 12; **j**, pistillate bud × 6; **k**, pistillate sepal × 6; **l**, pistillate petal × 6; **m**, gynoecium and staminodes × 6; **n**, ovary in cross-section × 6; **o**, fruit × 2 1/4; **p**, fruit in vertical section × 2 1/4; **q**, fruit in cross-section × 2 1/4; **r**, **s**, endocarp in 2 views × 3; **t**, **u**, **v**, seed in 3 views × 3. *Kentiopsis magnifica*: **a–n**, *McKee* 26471; **o–v**, *Moore & Schmid* 10054. (Drawn by Marion Ruff Sheehan)

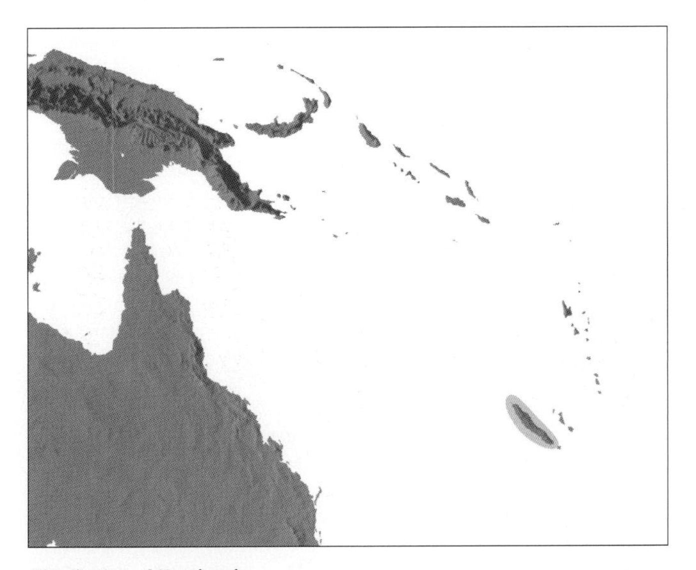

Distribution of ***Kentiopsis***

staminate and a pistillate nearly throughout the rachillae, a few paired or solitary staminate flowers present distally; bracteoles surrounding the pistillate flower low, unequal, rounded, not sepal-like. *Staminate flowers* symmetrical or somewhat asymmetrical; sepals 3, distinct, ± deltoid, ± acute, imbricate basally, scarcely higher than wide; petals 3, valvate, angled, acute; stamens 11–38, shorter than, equalling or exceeding the petals, filaments erect at the apex in bud, anthers erect in bud, linear, dorsifixed, emarginate apically, bifid basally, latrorse, the connective elongate; pistillode nearly as high as the stamens (lacking, according to Beccari), with an attenuate, sometimes briefly trifid apex. *Pollen* ellipsoidal or oblate triangular, slight or obvious asymmetry; aperture a distal sulcus or trichotomosulcus; ectexine tectate, perforate or perforate-rugulate, aperture margin similar or slightly finer; infratectum columellate; longest axis ranging from 39–52 µm [1/4]. *Pistillate flowers* symmetrical; sepals 3, distinct, broadly imbricate; petals 3, distinct, broadly imbricate with valvate apices; staminodes 3, small, tooth-like, borne at one side of the gynoecium, or 6 and connate in a ring; gynoecium pseudomonomerous, with 3 prominent, recurved stigmas, unilocular, uniovulate, ovule pendulous. *Fruit* ellipsoidal, red or purplish at maturity with apical or subapical stigmatic remains; epicarp smooth, drying minutely pebbled, mesocarp consisting of a shell of pale, short, ellipsoidal sclereids over pale parenchyma with a few included longitudinal fibres and at maturity a layer of tannin cells adjacent to flat, anastomosing fibres adherent to the endocarp, endocarp thin, fragile, not operculate. *Seed* ellipsoid or pyriform, attached by an elongate hilum, the raphe branches numerous, anastomosing, endosperm homogeneous; embryo basal. *Germination* adjacent-ligular; eophyll bifid, seedlings sometimes displaying saxophone growth, sometimes with distichous leaves. *Cytology*: 2n = 32

Distribution and ecology: Four species in New Caledonia. One species, *Kentiopsis pyriformis*, is found on ultramafic rock while the other three are found on schistose rocks. *Kentiopsis oliviformis* grows in forest transitional to semihumid forest and *K. piersoniorum* grows on exposed wet shrubby vegetation on montane ridges. All four species tend to grow gregariously, forming spectacular colonies. *Kentiopsis magnifica* is the tallest and stateliest palm in New Caledonia. For further details, see Pintaud & Hodel (1998).

Anatomy: Leaf (Uhl & Martens 1980), root (Seubert 1998a, 1998b) and fruit (Essig & Hernandez 2002).

Relationships: The monophyly of *Kentiopsis* has been resolved with low to moderate support (Baker *et al.* 1999a, Pintaud 1999a, Asmussen *et al.* 2000). The genus has been placed as sister to *Actinokentia* (Baker *et al.* in review) or alternatively as sister to a clade of *Chambeyronia* and *Actinokentia* (Pintaud 1999a).

Common names and uses: No common names recorded. Several of the species are becoming widespread in cultivation.

Taxonomic account: Pintaud & Hodel (1998a).

Fossil record: No generic records found.

Notes: *Kentiopsis* is distinct from other members of the Archontophoenicinae in its asymmetrical staminate flowers with angled petals and markedly unequal bracteoles surrounding the pistillate flower.

Additional figures: Glossary fig. 15.

Subtribe Arecinae

Diminutive to moderate, unarmed; leaves pinnate, or entire bifid, the leaflet tips and margins entire or lobed, not praemorse; sheaths almost always forming a crownshaft; inflorescences infrafoliar, spicate or branched to 1–3 orders, bearing a membranous prophyll, enclosing the inflorescence up to anthesis, peduncular bract(s) lacking; flowers borne in triads throughout the rachilla or with triads near the base, solitary or paired staminate flowers distally; fruit with apical stigmatic remains; epicarp smooth.

The subtribe includes three closely related but distinct genera, abundant in the rain forests of southeast Asia and Malesia, petering out in the western Pacific. All three genera are unusual in lacking peduncular bracts. The subtribe has remarkable diversity in pollen morphology, and displays diverse patterns of floral presentation.

This well-defined subtribe is strongly or moderately supported as monophyletic (Lewis & Doyle 2002, Asmussen *et al.* 2006, Loo *et al.* 2006, Norup *et al.* 2006, Baker *et al.* in review, in prep.). The subtribe resolves outside the western Pacific clade, but precise relationships within the Areceae remain uncertain (Lewis & Doyle 2002, Asmussen *et al.* 2006, Baker *et al.* in review, in prep.).

Key to Genera of Arecinae

1. Inflorescence protogynous, spicate or branched to 1 order only; rachillae bearing distichous or more rarely spirally arranged triads throughout their length. Southeast Asia to New Guinea . . . . **132. Pinanga**
1. Inflorescence protandrous (in a few *Areca* spp. recorded as protogynous but perhaps erroneously so), spicate or branched to 1–3 orders; rachillae bearing triads proximally and paired or solitary, spirally arranged, distichous or unilateral staminate flowers distally 2
2. Triads usually very few in number at the base of each rachilla, rarely numerous; seed basally attached with circular hilum. Southeast Asia to Solomon Islands . . . .**130. Areca**
2. Triads numerous, often occupying $^{3}/_{4}$ the rachilla length; seed laterally attached with narrow longitudinal hilum. Indochina to West Malesia . . . . **131. Nenga**

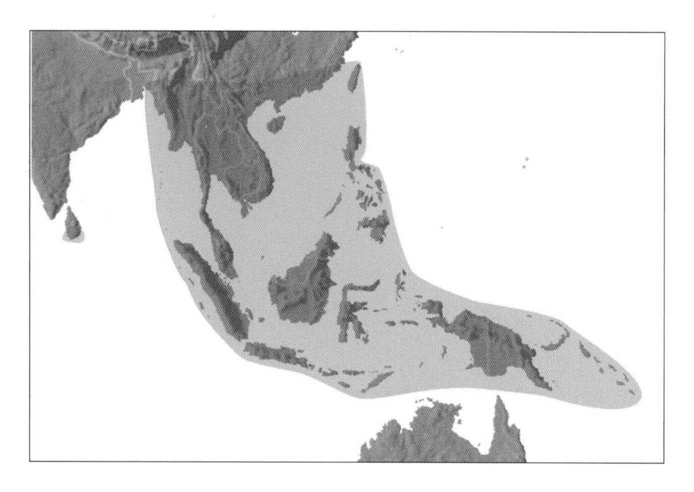

Distribution of subtribe **Arecinae**

130. ARECA

The betel-nut palm and its relatives; acaulescent, or erect, diminutive or robust palms of Southeast Asia to West Pacific, with crownshafts, with entire or lobed leaflet tips and a single large bract in the inflorescence, the pistillate flowers borne only at the rachilla bases and with basal hilum on the seed.

Areca L., Sp. pl. 1189 (1753). Type: **A. catechu** L.

Mischophloeus Scheff., Ann. Jard. Bot. Buitenzorg 1: 115, 134 (1876). Type: *M. paniculatus* (Scheff.) Scheff. (*Areca paniculata* Scheff.) (= *A. vestiaria* Giseke).

Gigliolia Becc., Malesia 1: 171 (1877). (non Barb. Rodr.). Lectotype: *G. insignis* Becc. (= *A. insignis* [Becc] J. Dransf.).

Pichisermollia H.C. Monteiro, Rodriguesia 41: 198 (1976). Type: *P. insignis* (Becc.) H.C. Monteiro (*Gigliolia insignis* Becc.) (= *A. insignis* [Becc.] J. Dransf.).

Said to be derived from a vernacular name used in Malabar, India.

Very small to moderate, solitary or clustered, acaulescent to erect, unarmed, pleonanthic, monoecious palms. *Stem* slender to moderate, occasionally stilt-rooted, internodes very short to elongate, leaf scars often conspicuous. Leaves undivided and pinnately ribbed, with or without an apical notch, or pinnate; sheaths forming a well-defined crownshaft with leaves neatly abscising, or rarely crownshaft not well developed when leaves marcescent or the sheaths partly open; petiole present or absent, adaxially channelled or rounded, abaxially rounded, glabrous or variously indumentose; leaflets regularly or irregularly arranged, 1–several fold, acute, acuminate or lobed, the lobes corresponding to the folds, the apical pair almost always lobed, held in one plane, very rarely (*Areca insignis*) with a basal auricle reflexed across the rachis, blade variously scaly or hairy, transverse veinlets obscure. *Inflorescences* erect or pendulous, mostly infrafoliar, rarely interfoliar in acaulescent species with marcescent leaves, in one species sometimes bursting through marcescent leaf sheaths (*A. jugahpunya*), branched to 3 orders basally, fewer orders distally, very rarely spicate, protandrous (or very rarely recorded as protogynous); peduncle very short to long; prophyll thin, membranous, enclosing the inflorescence in bud, quickly splitting and falling, other bracts very inconspicuous; rachis shorter or longer than the peduncle; rachillae glabrous or variously indumentose; rachilla bracts minute; triads confined to the proximal part of the main axis, or to the proximal part of each order of branching, or rarely to a subdistal part of the main axis only; rachillae otherwise bearing solitary or paired staminate flowers arranged spirally, distichously, or in 2 approximate rows on one side of the rachilla, the rachilla tips sometimes devoid of flowers. *Staminate flowers* frequently minute, sessile, or with a stalk formed from the receptacle; calyx with 3 distinct, slightly imbricate, triangular sepals, or cupular with 3 triangular lobes; corolla with 3 triangular, valvate petals, rarely briefly connate at the base, much longer than the sepals; stamens free or briefly epipetalous, 3, 6, 9 or up to 30 or more, filaments short to elongate, anthers linear or sinuous, sometimes very irregular, latrorse or rarely opening by apical pores; pistillode present and conspicuous as a trifid column as long as the stamens, or minute, or often absent. *Pollen* usually ellipsoidal, symmetric or slightly asymmetric, less frequently oblate triangular or oblate spheroidal; aperture a distal sulcus, in some species an extended sulcus, trichotomosulcus, or incomplete, presumed equatorial zonasulcus, rarely brevi or monoporate, or triporate; ectexine tectate or semi-tectate, finely to coarsely perforate, foveolate or finely reticulate, occasionally with very narrow muri, occasionally perforate-rugulate, aperture margin similar or slightly finer; infratectum columellate; longest axis 25–58 µm; post-meiotic tetrads tetrahedral, rarely tetragonal or rhomboidal [35/48]. *Pistillate flowers* sessile, usually much larger than the staminate, ± globular; sepals 3, distinct, imbricate;

Right: ***Areca minuta***, habit, Sarawak. (Photo: J. Dransfield)

petals similar to the sepals, 3, distinct, sometimes valvate at the very tip, otherwise imbricate; staminodes 3–9 or absent; gynoecium unilocular, uniovulate, globose to ovoid, stigmas 3, fleshy, triangular, ± reflexed at anthesis, ovule anatropous or campylotropous, basally attached. Rachilla distal to pistillate flowers drying after anthesis, portions bearing fruit sometimes becoming brightly coloured. *Fruit* globose, ovoid, or spindle-shaped, often brightly coloured, rarely dull brown or green, stigmatic remains apical; epicarp smooth, shiny or dull, mesocarp thin to moderately thick, fleshy or fibrous, endocarp composed of robust longitudinal fibres, usually closely appressed to the seed, becoming free at the basal end or not. *Seed* conforming to the fruit shape or slightly hollowed at the base, with basal hilum and raphe branches anastomosing, endosperm deeply ruminate; embryo basal. *Germination* adjacent-ligular; eophyll bifid or rarely entire with a minute apical cleft. *Cytology*: 2n = 32.

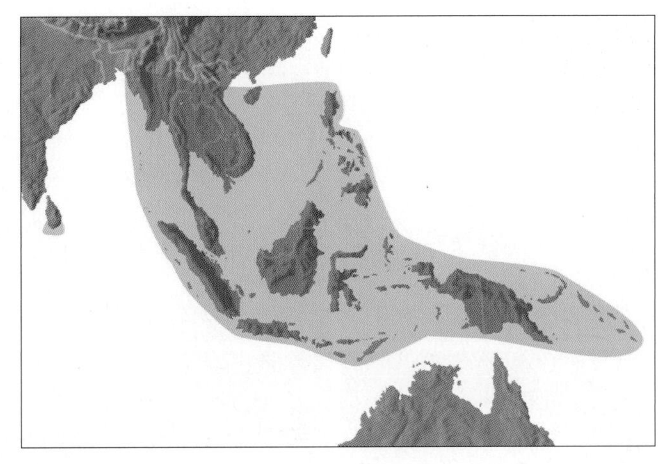

Distribution of ***Areca***

Distribution and ecology: About 47 species, distributed from India and south China through Malesia to New Guinea and the Solomon Islands. Most species are small to moderate palms of the undergrowth of tropical rain forest. *Areca catechu*, the betel nut, is very widespread as a crop plant and seems to tolerate open conditions. Some species of *Areca* have very narrow ecological limits; for example, *A. rheophytica* is confined to the banks of fast-flowing streams on ultramafic rock in Sabah, Borneo. *Areca triandra* is a polymorphic species occurring from India to Borneo. Some of the entities within this complex taxon have rather precise habitat requirements.

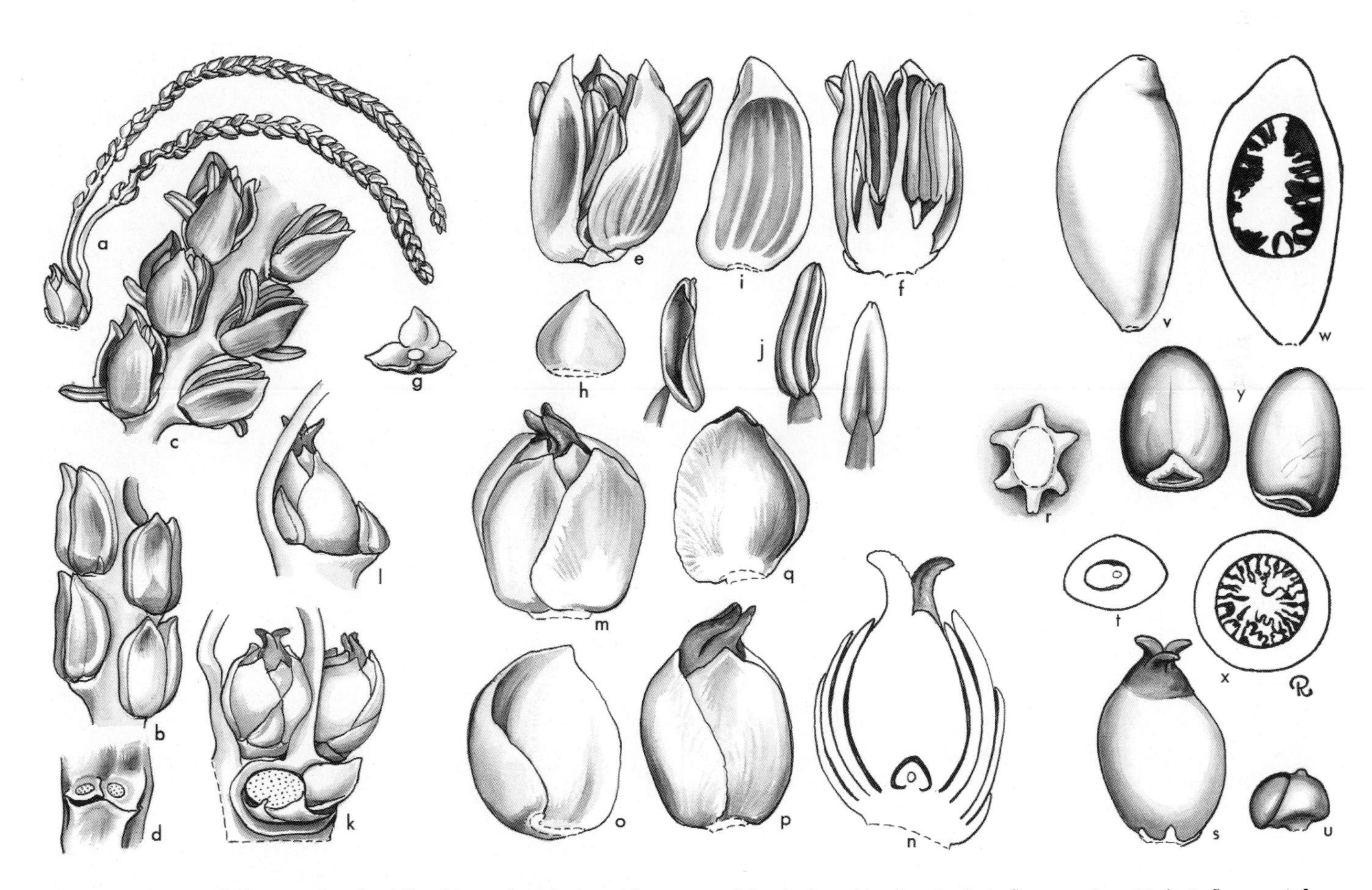

***Areca*.** **a**, rachillae × $^1/_2$; **b**, **c**, portion of rachilla with staminate buds and flowers × 3; **d**, bracteoles subtending staminate flowers × 6; **e**, staminate flower × 6; **f**, staminate flower in vertical section × 6; **g**, staminate calyx × 6; **h**, staminate sepal, exterior view × 12; **i**, staminate petal, interior view × 6; **j**, stamen in 3 views × 6; **k**, bases of rachillae with pistillate flowers × 1 $^1/_2$; **l**, pistillate flower with 2 sterile staminate flowers × 1 $^1/_2$; **m**, pistillate flower × 3; **n**, pistillate flower in vertical section × 3; **o**, pistillate sepal, interior view × 3; **p**, pistillate flower, sepals removed × 3; **q**, pistillate petal, interior view × 3; **r**, staminodes × 3; **s**, gynoecium and staminodes × 3; **t**, ovary in cross-section × 3; **u**, ovule × 6; **v**, fruit × 1; **w**, fruit in vertical section × 1; **x**, fruit in cross-section × 1; **y**, seed in 2 views × 1. *Areca macrocalyx*: all from *Moore & Whitmore* 9303. (Drawn by Marion Ruff Sheehan)

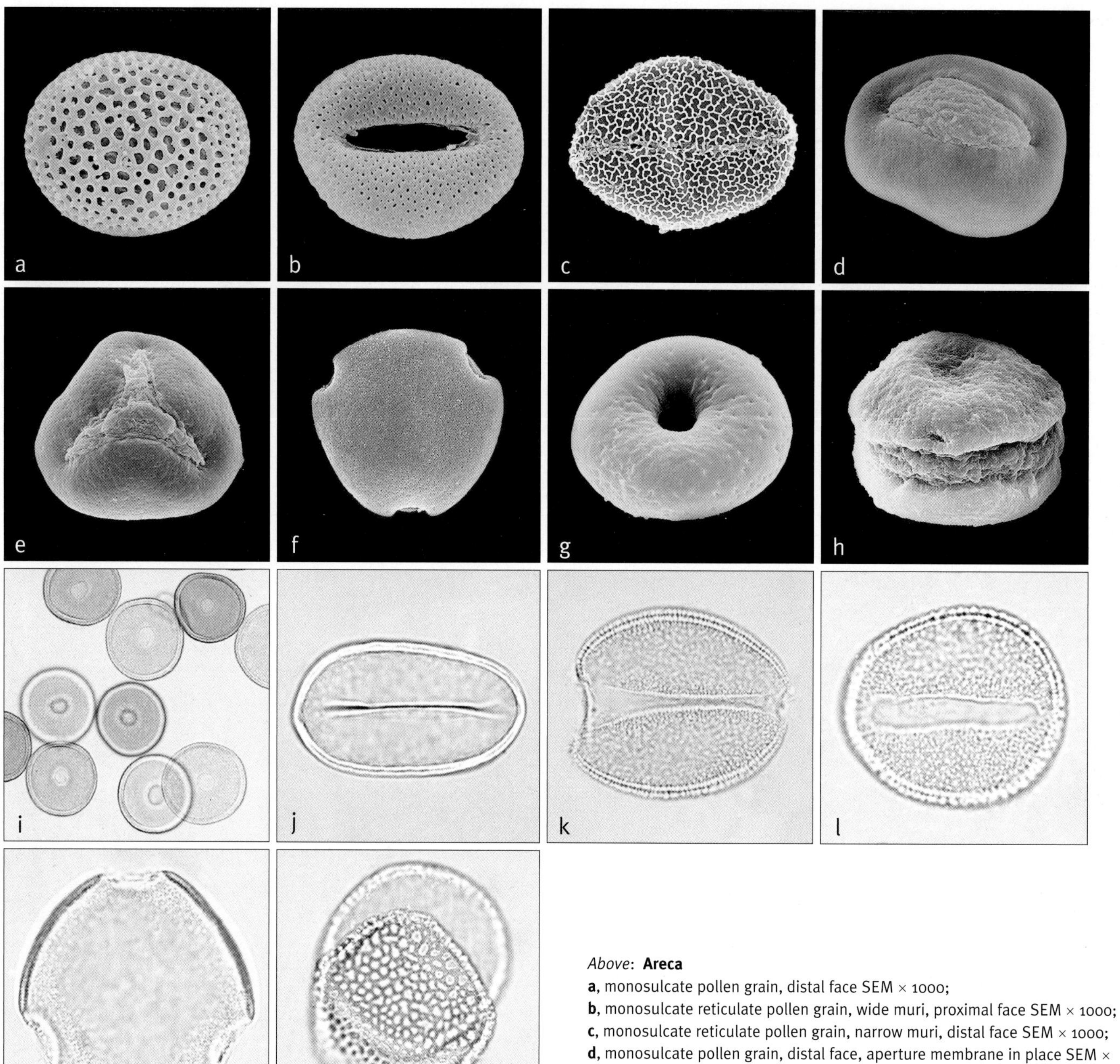

Above: **Areca**
a, monosulcate pollen grain, distal face SEM × 1000;
b, monosulcate reticulate pollen grain, wide muri, proximal face SEM × 1000;
c, monosulcate reticulate pollen grain, narrow muri, distal face SEM × 1000;
d, monosulcate pollen grain, distal face, aperture membrane in place SEM × 1500;
e, trichotomosulcate pollen grain, distal face, aperture membrane in place SEM × 1500;
f, equatorial triporate pollen grain, polar face SEM × 1500;
g, oblate monoporate pollen grain, distal polar face SEM × 1500;
h, incomplete equatorial zonasulcate pollen grain, presumed equatorial face SEM × 1000;
i, group of oblate monoporate grains, various foci LM × 750;
j, monosulcate pollen grain, mid-focus LM × 1000;
k, extended sulcate pollen grain, high–mid-focus LM × 1000;
l, monosulcate pollen grain, high–mid-focus LM × 1500;
m, equatorial triporate pollen grain, mid-focus LM × 1000;
n, incomplete equatorial zonasulcate pollen grain, high-focus LM × 1000.
Areca whitfordii: **a** & **b**, *Elmer 16240*; *A. insignis* var. *moorei*: **c**, *Dransfield* JD738; *A. minuta*: **d** & **e**, *Dransfield* JD5339; *A. klingkangensis*: **f** & **m**, *Dransfield* JD6103; *A. caliso*: **g**, *Elmer* 11898, **i**, *Fernando* EF674; *A. klingkangensis*: **h**, *George* S.42803; *A. arundinacea*: **j**, *Dransfield* JD6120; *A. jugahpunya*: **k**, *Chai* S.36065; *A. camarinensis*: **l**, *Fernando* EF509; *A. chaiana*: **n**, *Mohtar et al.* S.49276.

Anatomy: Leaf, stem, root (Tomlinson 1961), root (Seubert 1998a, 1998b), gynoecium (Uhl & Moore 1971), stegmata (Killmann & Hong 1989), and fruit (Essig & Young 1979).

Relationships: *Areca* is a strongly supported monophyletic group, though sampling in available phylogenies is limited (Loo *et al.* 2006). The genus is highly supported as sister to a clade comprising *Nenga* and *Pinanga* (Loo *et al.* 2006). The same relationships are recovered in other studies (Norup *et al.* 2006, Baker *et al.* in review, in prep.).

Common names and uses: Betel nut palm, *pinang*, *bunga*, *jambe*. *Areca catechu* is economically important and widely cultivated, sometimes on a plantation scale. The endosperm is

chewed with leaves or inflorescences of *Piper betle* L., lime and other ingredients; it contains the alkaloid arecaine, which acts as a mild narcotic. An estimated 200–400 million people use betel nut in this way, making it the fourth most widly "abused" substance after nicotine, alcohol and caffeine (Gupta & Warnakulasuriya 2002, Norton 1998). The fruit are also used as a source of tannin in dyeing, medicinally, and rarely, as toothbrushes. The apex is edible and the flowers often used as ceremonial decoration. The leaf sheath may be utilised in making containers, and other species may serve as substitutes in betel-chewing. Several species are cultivated as ornamentals.
Taxonomic accounts: Furtado (1933) and Dransfield (1984a).
Fossil record: There are a few records of reticulate monosulcate pollen from the Maastrichtian of Cameroon that have been compared with *Areca*: *Retimonocolpites pluribaculatus* nov. sp. (Salard-Cheboldaeff 1978), *Arecipites lusaticus* and *A. convexus* (Salard-Cheboldaeff 1979); *A. convexus* is closely similar to the pollen of *Areca catechu*. Muller (1981) suggested that *R. pluribaculatus* (reticulate pollen with ca. 45–50 µm long axis) is close to the reticulate monosulcate pollen of *Areca ipot* and shares a similar size range; the affinity of this fossil pollen with a species of *Areca*, a genus that includes a number of species that have reticulate pollen, certainly cannot be ruled out. As a general comment it, is noted that the name *Arecipites* is widely used for small more-or-less symmetric monosulcate palm-like dispersed fossil pollen; its use for *Areca*-like fossil pollen is exceptional.
Notes: As circumscribed here, *Areca* is a remarkably variable genus that has very distinctive sections. The variation in Borneo is particularly noteworthy.
Additional figures: Glossary figs 3, 19.

Top left: ***Areca catechu***, inflorescence, cultivated, Montgomery Botanical Center, Florida. (Photo: J. Dransfield)

Top right: ***Areca ipot***, young fruit, cultivated, Australia. (Photo: J. Dransfield)

Above: ***Areca ahmadii***, inflorescence, Sarawak. (Photo: J. Dransfield)

131. NENGA

Acaulescent or erect palms of forest undergrowth in West Malesia, Thailand and Vietnam, almost always with crownshafts, with entire or lobed leaflet tips and a single large bract in the inflorescence, the pistillate flowers borne in the basal part of the rachilla and with lateral hilum on the seed.

Nenga H. Wendl. & Drude, Linnaea 39: 182 (1875). Type: *Pinanga nenga* (Blume ex Mart.) Blume (*Areca nenga* Blume ex Mart.) (= *N. pumila* [Mart.] H. Wendl. [*Areca pumila* Mart.]).

Based on a Javanese vernacular name, nenge.

Moderate, solitary or clustered, acaulescent or erect, unarmed, pleonanthic, monoecious palms. *Stem* slender, short, rarely exceeding 5 m in height with short or elongate internodes and conspicuous leaf scars, stilt roots frequent. *Leaves* pinnate; sheaths usually forming a well-defined crownshaft with leaves neatly abscising, or leaves marcescent and crownshaft poorly developed (*Nenga gajah*); petiole usually well developed, flattened or grooved adaxially, rounded or angled abaxially; leaflets with 1–several folds, linear to sigmoid, acute or acuminate, the terminal pair obscurely lobed, lobes corresponding to the folds, the adaxial ribs often bearing ramenta on the under surface, transverse veinlets obscure. *Inflorescence* infrafoliar or interfoliar (*N. gajah*), erect or pendulous, branching to 1 order, rarely to 2 orders or unbranched, protandrous; peduncle short in species with infrafoliar inflorescences, long where inflorescences interfoliar (*N. gajah*); prophyll thin,

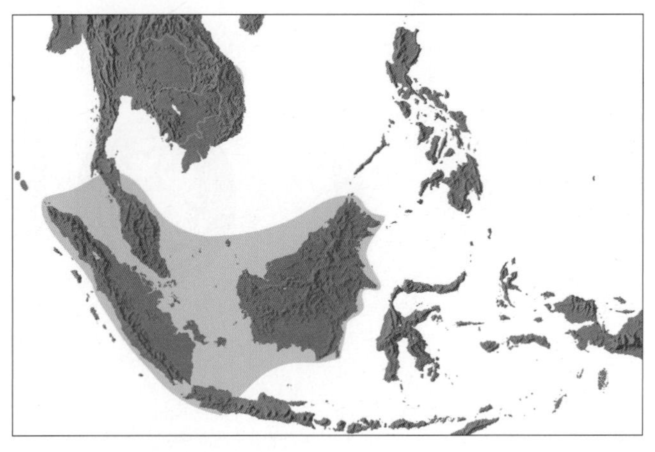

Distribution of ***Nenga***

membranous, enclosing the inflorescence in bud, splitting and falling at anthesis or thick, almost woody, persistent, eventually rotting; peduncular bracts incomplete, small, triangular; rachillae bearing spirally arranged minute bracts subtending triads proximally, solitary or paired staminate flowers distally, or triads confined to central rachilla and lateral rachillae with staminate flowers only, flowers not or only slightly sunken in the rachillae; floral bracteoles minute. *Staminate flowers* fleshy, sessile; sepals 3, connate at the very base, shorter than, almost as long as, or far exceeding the corolla; corolla with slightly stalk-like base or not, with 3 long, valvate lobes; stamens 6, borne at the base of the corolla lobes, filaments short, anthers oblong to linear, latrorse; pistillode absent. *Pollen* grains ellipsoidal to spheroidal, bi-symmetric; aperture a distal sulcus, short or same length as long axis; ectexine semi-tectate, coarsely reticulate, muri of reticulum may be spinulose, aperture margin similar; infratectum columellate; longest axis ranging from 37–72 µm [4/5]. *Pistillate flowers* sessile, globular; sepals 3, distinct, imbricate; petals 3, distinct, imbricate; staminodes absent or minute; gynoecium globose or columnar, uniloculate, uniovulate, style lacking, stigmas 3, massive, fleshy, divergent, ovule laterally attached, form unknown. Fruiting rachillae usually not differing greatly in colour from flowering ones. *Fruit* ovoid to obpyriform, dull to brightly coloured, stigmatic remains apical; epicarp smooth, dull or shiny, mesocarp thin, fleshy, sweet, endocarp composed of longitudinal fibres adhering to the seed, becoming free at both ends (*N. pumila*) or at one end only, the fibres enclosing a solid parenchymatous mass of varying size distal to the seed. *Seed* with longitudinal hilum and raphe branches anastomosing, endosperm deeply ruminate; embryo basal. *Germination* adjacent-ligular; eophyll bifid. *Cytology* not studied.

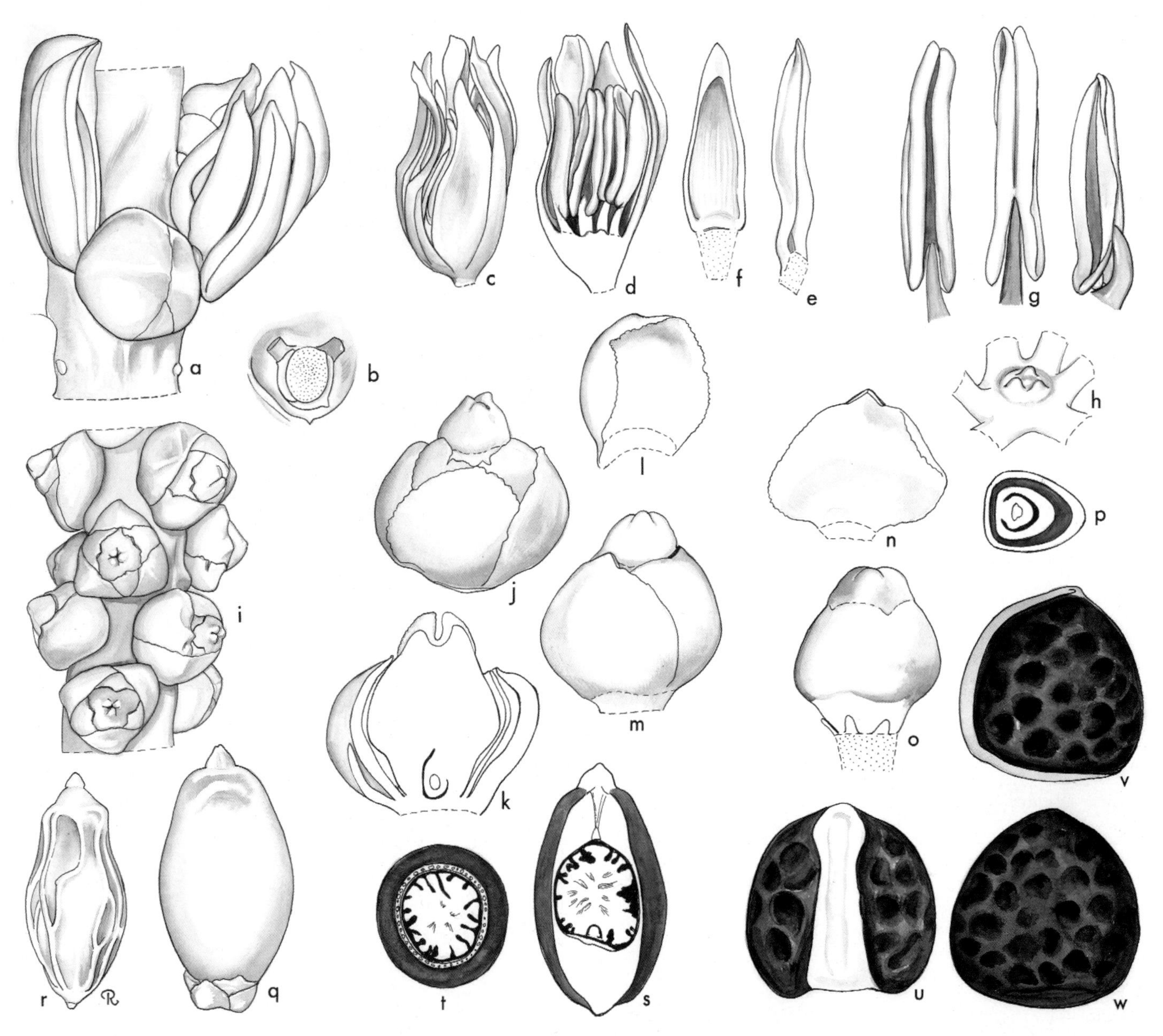

Nenga. **a**, portion of rachilla showing triad × 3; **b**, triad, flowers removed × 3; **c**, staminate flower × 3; **d**, staminate flower in vertical section × 3; **e**, staminate sepal × 3; **f**, staminate petal, interior view × 3; **g**, stamen in 3 views × 6; **h**, pistillode × 6; **i**, portion of rachilla in pistillate flower × 3; **j**, pistillate flower × 3; **k**, pistillate flower in vertical section × 3; **l**, pistillate sepal × 3; **m**, pistillate flower, sepals removed × 3; **n**, pistillate petal, interior view × 3; **o**, gynoecium and staminodes × 3; **p**, ovary in cross-section × 3; **q**, fruit × $1^1/_2$; **r**, dry fruit × $1^1/_2$; **s**, fruit in vertical section × $1^1/_2$; **t**, fruit in cross-section × $1^1/_2$; **u, v, w**, seed in 3 views × 3. *Nenga grandiflora*: **a–h,** *Moore & Pennington* 9061; *N. pumila*: **i–p,** *Moore & Pennington* 9071; **q–w,** *Moore & Pennington* 9056. (Drawn by Marion Ruff Sheehan)

Distribution and ecology: Five species ranging from Vietnam and Burma to Sumatra, the Malay Peninsula, Borneo and Java. All species are confined to primary tropical rain forest and are found from sea level to altitudes of about 1400 m. *Nenga pumila* var. *pachystachya* sometimes occurs in peat swamp forest.

Anatomy: Root (Seubert 1998a, 1998b) and fruit (Essig & Young 1979).

Relationships: *Nenga* is a strongly supported monophyletic genus (Loo *et al.* 2006). For relationships, see *Areca*.

Common names and uses: Pinang palms. Stems are sometimes used split, as laths.

Taxonomic account: Fernando (1983).

Fossil record: Monosulcate pollen with a rather loose, widely meshed reticulum characteristic of *Nenga* occurs in Borneo from the Lower Miocene upwards, according to Muller (1964, 1972, 1979); unfortunately, Muller did not publish illustrations of the fossil pollen so it is difficult to comment.

Notes: *Nenga gajah* is an extraordinary, aberrant species that nevertheless belongs in the genus; most of its peculiarities seem to be related to the interfoliar position of the inflorescence, which in turn is related to the short internodes and marcescent leaves. Such an anomalous habit is also found in a few species of *Areca* and *Pinanga*.

Additional figures: Ch. 2 frontispiece, Fig. 2.8e.

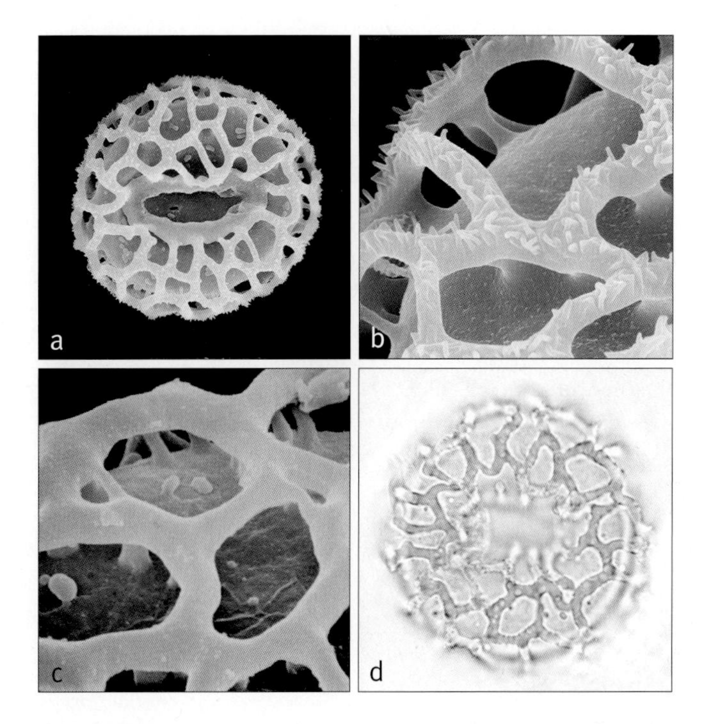

Above: ***Nenga***
a, monosulcate coarsely reticulate pollen grain, distal face SEM × 1000;
b, close-up, reticulum with spinulose muri SEM × 5750;
c, close-up, reticulum with psilate muri SEM × 6000;
d, monosulcate pollen grain, high–mid-focus LM x 1000.
Nenga gajah: **a** & **b**, *Gundersen et al.* LVG29, **d**, Gunderson *et al.* 29; *N. pumila* var. *pachystachya*: **c**, *Kian & Mosey* 33400.

Below, from left:
Nenga gajah, inflorescence, Sabah. (Photo: J. Dransfield)
Nenga macrocarpa, staminate flowers, W. Malaysia. (Photo: J. Dransfield)
Nenga grandiflora, pistillate flowers. (Photo: J. Dransfield)
Nenga pumila var. ***pumila***, infructescences, Java. (Photo: J. Dransfield)

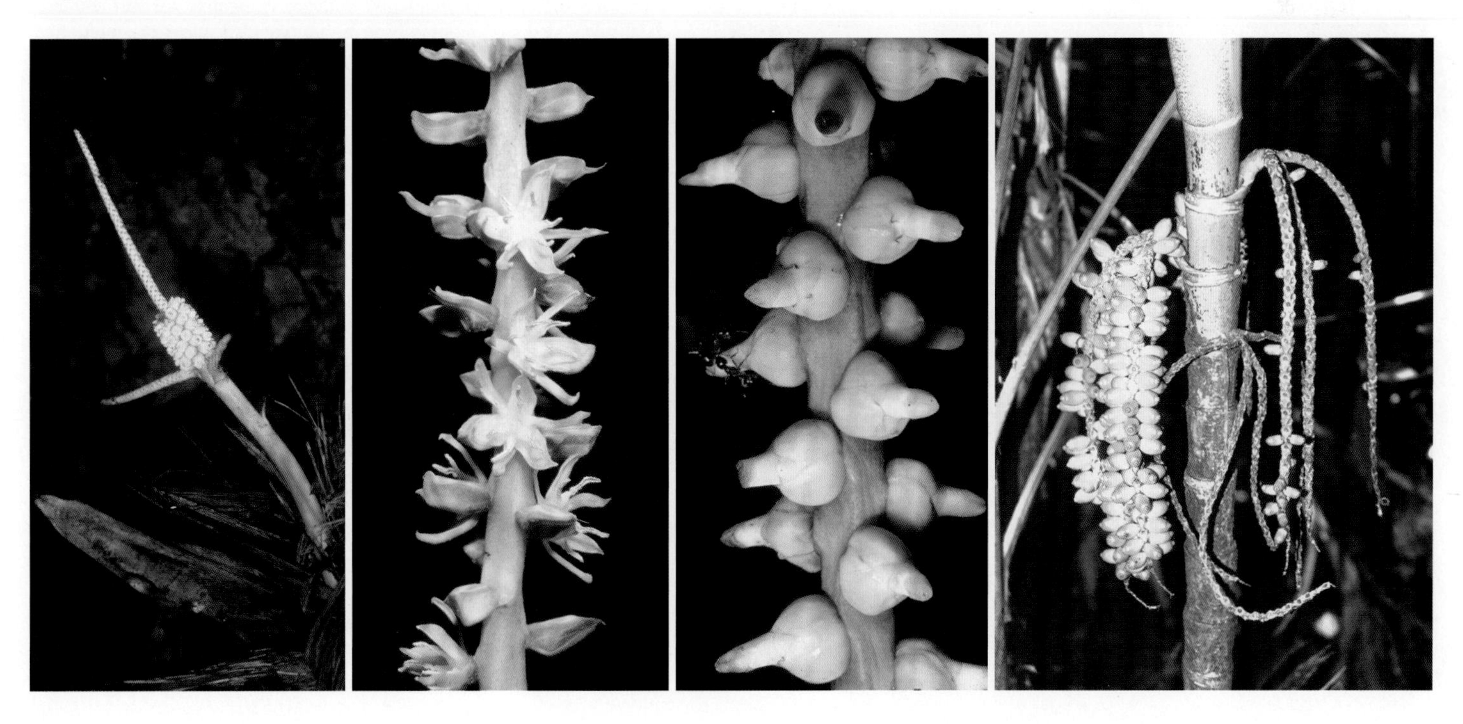

132. PINANGA

Acaulescent, or erect, diminutive or robust palms of Southeast Asia to New Guinea, with crownshafts, with entire or lobed leaflet tips and a single large bract in the inflorescence, the pistillate flowers borne throughout the rachillae, seed with basal hilum.

Pinanga Blume, Bull. Sci. Phys. Nat. Néerl. 1: 65 (1838). Lectotype: **P. coronata** (Blume ex Mart.) Blume (*Areca coronata* Blume ex Mart.) (see Beccari & Pichi-Sermolli 1955).

Cladosperma Griff., Not. pl. asiat. 3: 165 (1851). Type: *C. paradoxa* (Griff.) Becc. (*Areca paradoxa* Griff.) (= *Pinanga paradoxa* [Griff.] Scheff.).

Ophiria Becc., Ann. Jard. Bot. Buitenzorg 2: 128 (1885). Type: *O. paradoxa* (Griff.) Becc. (*Areca paradoxa* Griff.) (= *Pinanga paradoxa* [Griff.] Scheff.).

Pseudopinanga Burret, Notizbl. Bot. Gart. Berlin-Dahlem 13: 188 (1936). Type: *P. insignis* (Becc.) Burret (*Pinanga insignis* Becc.).

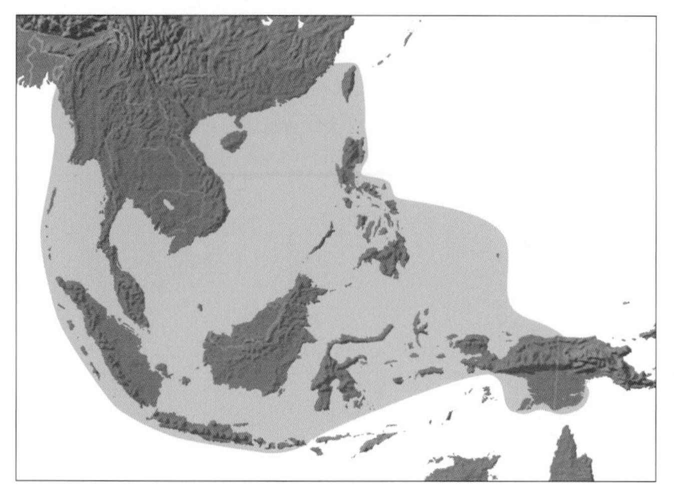

Distribution of ***Pinanga***

Latinization of the Malay vernacular name, pinang, *applied to the betel palm,* Areca catechu *and species of* Areca, Pinanga *and* Nenga *in the wild.*

Very small to robust, solitary or clustered, acaulescent or erect, unarmed, pleonanthic, monoecious palms. *Stem* very slender to moderate, with elongate or short internodes and conspicuous leaf scars, occasionally stilt-rooted. *Leaves* undivided and pinnately ribbed, with or without an apical notch, or pinnate; sheaths tubular, forming a well-defined crownshaft, with leaves neatly abscising, very rarely leaves marcescent and crownshaft not well developed; petiole present or absent, adaxially rounded or channelled, abaxially rounded, glabrous or variously indumentose; leaflets 1 to several-fold, regularly to irregularly arranged, acute, acuminate, or lobed, the lobes corresponding to the folds, the apical leaflets almost always lobed, blade occasionally mottled, sometimes paler beneath, often with a wide variety of scales and hairs, transverse veinlets usually obscure. *Inflorescence* mostly infrafoliar, rarely interfoliar in acaulescent species with marcescent leaves, very rarely bursting through marcescent leaf sheaths (*Pinanga simplicifrons*), usually rapidly becoming pendulous, occasionally erect, protogynous, unbranched or branching to 1 order only; peduncle usually short, dorsiventrally flattened, glabrous or tomentose; prophyll thin, membranous, 2-keeled, enclosing the inflorescence in bud, quickly splitting to expose the flowers except in *P. simplicifrons* and *P. cleistantha* where persistent and enclosing inflorescence up to almost mature fruiting; peduncular bracts absent; rachis usually longer than the peduncle; rachis bracts triangular, usually very inconspicuous; rachillae bearing spiral or distichous triads throughout, or triads in 4 or 6 vertical rows, or, more rarely, spiral proximally and distichous distally; triads sometimes partially sunken in the axis of the rachilla, but well-defined pits not present; floral bracteoles minute. *Staminate flowers* asymmetrical, sessile, rarely stalked at the base, very rarely the stalk of one flower much longer than the other (*P. cleistantha*); calyx cupular with 3 triangular, frequently unequal lobes; petals 3, triangular, frequently unequal, joined briefly basally, valvate in bud, much exceeding the calyx lobes, usually very fleshy; stamens rarely 6, usually 12–68, filaments short, anthers linear, latrorse; pistillode absent. *Pollen* usually ellipsoidal, occasionally oblate triangular, with at least one plane of symmetry, less frequently without symmetry; aperture either a distal sulcus, a distal trichotomosulcus, an extended sulcus or a presumed meridional zonasulcus (rare); ectexine either tectate, semitectate, or intectate; ectexine tectate or semitectate pollen finely to coarsely perforate, finely rugulate-reticulate, finely to coarsely reticulate (in some species the muri perforate or reticulate), discrete, psilate ring-like elements, dense supratectal clavae (in some species vertically striate), or finely reticulate with large smooth, broad-based supratectal spines; ectexine of intectate pollen with semi-coalesced mushroom-like pilae interspersed with dense granulae or spinulae, spines interspersed with dense granulae or small clavae, small and large gemmae interspersed or, urceolae interspersed with small dense clavae; infratectum columellate; longest axis 26–60 µm; post-meiotic tetrads tetragonal, and possibly also tetrahedral [50/128]. *Pistillate flowers* usually globose, symmetrical, much smaller than the staminate; sepals 3, membranous, striate, imbricate, distinct, or connate proximally with 3 broad, sometimes imbricate lobes distally; petals 3, distinct, imbricate, membranous; staminodes absent; gynoecium unilocular, uniovulate, globose, stigma usually convolute, sessile or on a short style, ovule basally attached, anatropous. Fruiting rachillae usually brightly coloured (reddish or orange). *Fruit* globose, or ellipsoidal to spindle-shaped, sometimes narrow spindle-shaped and curved (*P. salicifolia* and

Far left: ***Pinanga rupestris***, habit on vertical sandstone cliffs, Sarawak. (Photo: J. Dransfield)

Left: ***Pinanga veitchii***, young plant, Brunei Darussalam. (Photo: J. Dransfield)

others), bright crimson, scarlet, orange or black, very rarely dull brown or green, frequently passing through pink to crimson to black at maturity, stigmatic remains apical; epicarp usually smooth, shiny, with a silky sheen, or dull, mesocarp usually thin, fleshy, sweet, rarely greatly expanding (e.g., *P. keahii*), endocarp of longitudinal fibres, usually adhering to the seed, becoming free at the basal end only (see *Nenga*), fruit without a solid beak. *Seed* conforming to the fruit shape, but usually slightly hollowed at the base, with conspicuous basal hilum and anastomosing raphe branches, endosperm deeply ruminate or, very rarely, subruminate or homogeneous; embryo basal. *Germination* adjacent-ligular; eophyll bifid or rarely entire with a minute apical cleft. *Cytology*: 2n = 32.

Distribution and ecology: About 131 species ranging from the Himalayas and south China to New Guinea, with the greatest diversity in the wet areas of the Sunda Shelf; very poorly represented in Papuasia. Almost all species are plants of the forest undergrowth; a few massive species such as *Pinanga insignis* contribute to the lower part of the forest canopy. In altitude, the genus ranges from sea level to ca. 2800 m in the mountains. Some species may be associated with various rock types, including limestone and ultramafics, but the greatest

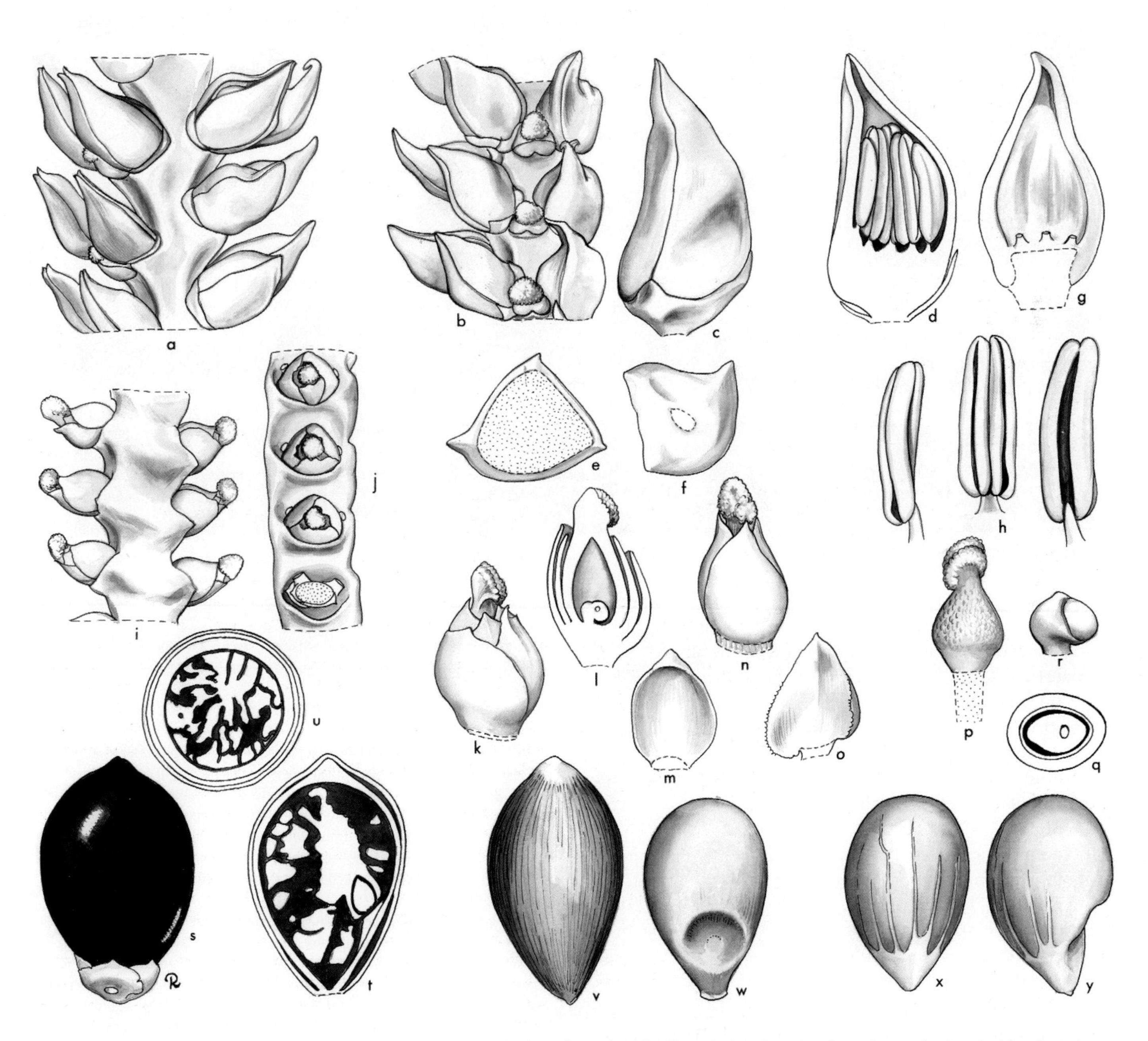

Pinanga. **a, b,** portion of rachilla with triads in 2 views × 3; **c,** staminate bud × 6; **d,** staminate bud in vertical section × 6; **e, f,** staminate calyx, interior (e) and exterior (f) views × 6; **g,** staminate petal, interior view × 6; **h,** stamen in 3 views × 12; **i, j,** portion of rachilla at pistillate anthesis in 2 views × 3; **k,** pistillate flower × 6; **l,** pistillate flower in vertical section × 6; **m,** pistillate sepal, interior view × 6; **n,** pistillate flower, calyx removed × 6; **o,** pistillate petal, interior view × 6; **p,** gynoecium × 6; **q,** ovary in cross-section × 12; **r,** ovule × 12; **s,** fruit × 3; **t,** fruit in vertical section × 3; **u,** fruit in cross-section × 3; **v,** endocarp × 3; **w, x, y,** seed in 3 views × 3. *Pinanga coronata*: **a–r,** *Read* 822 and 825; **q–y,** *Moore* s.n. (Drawn by Marion Ruff Sheehan)

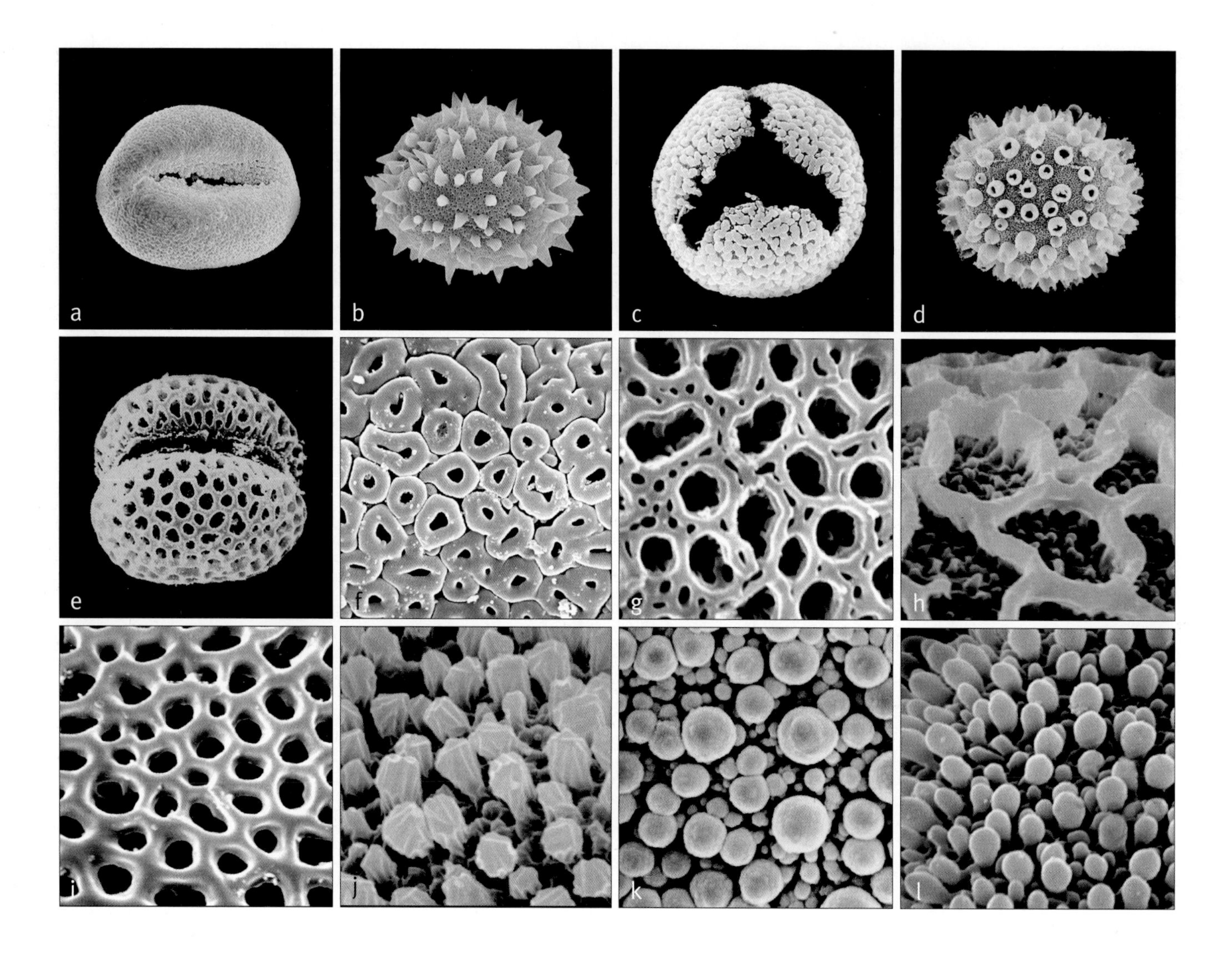

diversity appears to be in primary forest developed on sandstones on Borneo, where in some rich localities, as many as nine species may be found growing sympatrically.

Anatomy: Leaf (Tomlinson 1961), root (Seubert 1998a, 1998b), and fruit (Essig & Young 1979).

Relationships: The monophyly of *Pinanga* is strongly supported (Loo *et al.* 2006). For relationships, see *Areca*.

Common names and uses: *Pinang*, *bunga*. Stems may be used as laths, and leaves as thatch; fruit are rarely used as a betel substitute.

Taxonomic accounts: Beccari (1886) is the last monograph of this genus and many species have been described since, for example, by Furtado (1934), Dransfield (1980b, 1991c), Fernando (1994), and Hodel (1997). There is a useful regional account for Malaya (Lim 2001) and for Java and Bali (Witono *et al.* 2002). A modern monographic account is much needed.

Fossil record: No generic records found.

Notes: This is a wonderfully diverse genus as far as habit is concerned, but it seems to show rather little variation in inflorescence presentation and form. The pollen, however, is extremely varied (Ferguson *et al.* 1983). Pollination, where casually observed, appears to be by beetles.

Above: ***Pinanga***
a, monosulcate pollen grain, distal face SEM × 2000;
b, spiny monosulcate pollen grain, proximal face SEM × 1500;
c, trichotomosulcate pollen grain, distal face, gemmate-reticulate pollen surface SEM × 1500;
d, monosulcate pollen grain, proximal face, urceolate pollen surface SEM × 2000;
e, extended sulcate pollen grain, oblique distal/equatorial face SEM × 1500;
f, close-up, pollen surface comprising discrete psilate ring-like structures SEM × 3500;
g, close-up, reticulate pollen surface with perforate muri SEM × 8000;
h, close-up, reticulate pollen surface with narrow, angular muri SEM × 8000;
i, close-up, simple reticulate pollen surface with rounded lumina SEM × 8000;
j, close-up, finely reticulate pollen surface with vertically ridged, block-like, clavae SEM × 8000;
k, close-up, pollen surface comprising small and large gemmae SEM × 6000;
l, close-up, pollen surface comprising small and large clavae SEM × 8000.
Pinanga chaiana: **a**, *Wong* WKM650; *P. cleistantha*: **b**, *Dransfield* JD5179; *P. coronata*: **c**, *Read* 822; *P. aristata*: **d**, *Dransfield* JD5341; *P. maculata*: **e**, *Moore & Langlois* s.n.; *P. gracilis*: **f**, *Gamble* 7881; *P. andamanensis*: **g**, *Nair* s.n.; *P. capitata* var. *capitata*: **h**, *Clemens* 27896; *P. pectinata*: **i**, *Ridley* 7024; *P. variegata*: **j**, *Winkler* 2781; *P. disticha*: **k**, cultivated RBG Kew 085-85.01527; *Pinanga rupestris*: **l**. *Dransfield* JD5917.

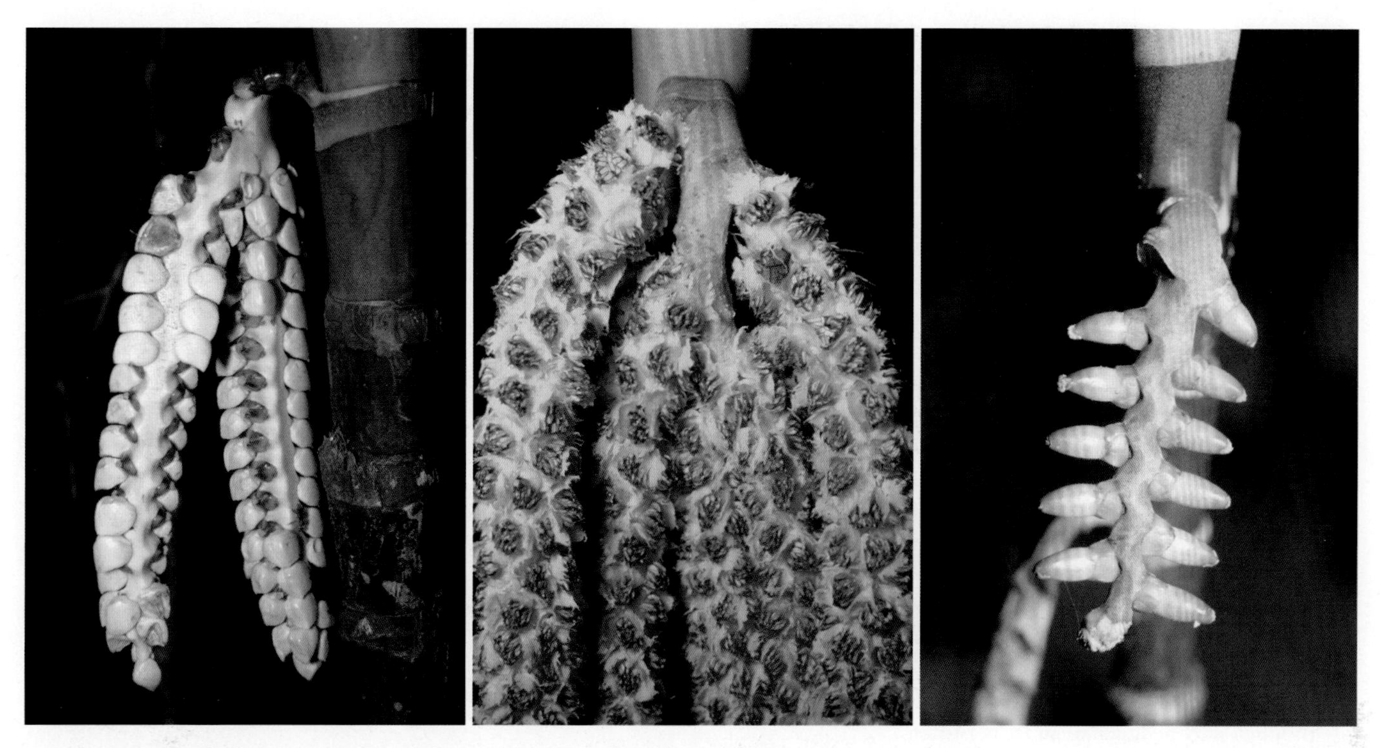

Above left: ***Pinanga mirabilis***, inflorescence at pistillate anthesis, Brunei Darussalam. (Photo: J. Dransfield)
Above middle: ***Pinanga aristata***, inflorescence at staminate anthesis, Sarawak. (Photo: J. Dransfield)
Above right: ***Pinanga yassinii***, young fruit, Brunei Darussalam. (Photo: J. Dransfield)

Additional figures: Figs. 2.5a, 2.8a, 2.8i, 2.8l, Glossary figs 5, 13, 14, 24.

Subtribe Basseliniinae J. Dransf., N.W. Uhl, C. Asmussen, W.J. Baker, M.M. Harley & C. Lewis., Kew Bull. 60: 563 (2005). Type: **Basselinia**.

Small to moderate, erect, unarmed; leaves pinnate, sometimes irregularly so, or entire bifid, the leaflet tips entire; sheaths usually forming a well-defined crownshaft, crownshaft absent in *Cyphosperma* and poorly developed in *Burretiokentia*; inflorescences infrafoliar, rarely interfoliar, highly branched; prophyll almost always not completely tubular; flowers borne in triads near the base, solitary or paired staminate flowers distally; fruit with apical to subapical or lateral stigmatic remains; endocarp frequently ornamented; epicarp smooth, very rarely roughened.

This subtribe is one of three subtribes of Areceae present in New Caledonia; it occurs elsewhere in Vanuatu, Fiji and west to the Solomon Islands and the Bismarck Archipelago. The incomplete prophyll is highly characteristic, and is present in all members except for one species of *Basselinia* and *Lepidorrhachis*, which is polymorphic, with both complete and incomplete prophylls; distinctive ornamented endocarps are often present (Pintaud & Baker 2008).

Evidence from phylogenetic analyses for the monophyly of the Basseliniinae is limited. In one family-wide study (Baker *et al.* in review), all members of the subtribe except for *Lepidorrhachis* resolve as monophyletic with low of support, although other studies recover subclades the group (Pintaud 1999b, Lewis & Doyle 2002, Norup *et al.* 2006, Baker *et al.* in prep.). Nevertheless, the subtribe is morphologically well defined, for example, by the presence of an incomplete prophyll. There are strong similarities in leaf anatomy among the New Caledonian taxa, but *Lepidorrhachis* and *Physokentia* have yet to be studied. The subtribe falls within the western Pacific clade of Areceae, but its precise relationships are uncertain (Baker *et al.* in review).

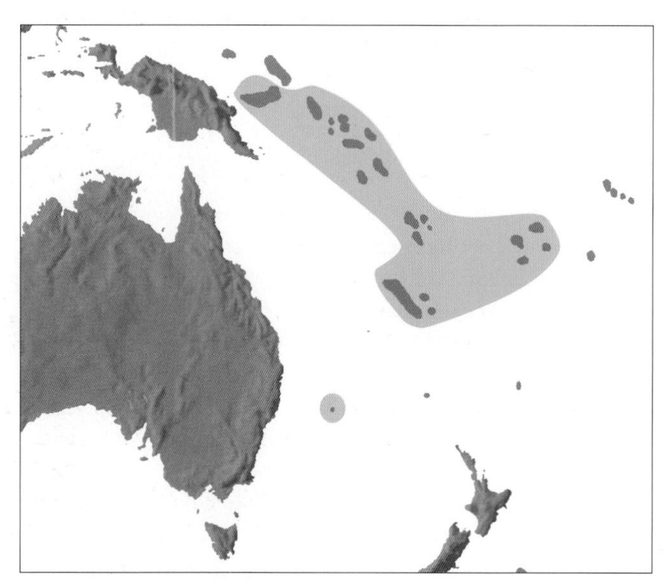

Distribution of subtribe **Basseliniinae**

Key to Genera of Basseliniinae

1. Fruit usually with apical stigmatic remains 2
1. Fruit usually with lateral stigmatic remains; endocarp smooth or pitted, not sculptured. New Caledonia . . . . **133. Basselinia**
2. Inflorescences interfoliar, apparently unisexual. Lord Howe Island . . . . **137. Lepidorrhachis**
2. Inflorescences inter- or infrafoliar, bisexual 3
3. Stilt-rooted palms 4
3. Palms lacking obvious stilt-roots 5
4. Endocarp and seed smooth. New Caledonia . . . . **135. Cyphophoenix** (in part)
4. Seed usually heavily ornamented. New Britain, Solomon Islands, Fiji and Vanuatu . . . . **138. Physokentia**
5. Seed terete ovoid, ellipsoidal . . . . **135. Cyphophoenix** (in part)
5. Seed irregular in cross-section, externally angled or intricately ridged, furrowed and sculptured 6
6. Leaf sheaths split opposite the petiole in bud, not forming a crownshaft; inflorescence interfoliar in bud, eventually infrafoliar; peduncle elongate, much exceeding the rachis; prophyll and peduncular bract persistent . . . . **136. Cyphosperma**
6. Leaf sheaths not split in bud, forming a crownshaft; inflorescence infrafoliar; peduncle shorter than the rachis; prophyll and peduncular bract deciduous 7
7. Anther locules lacking a conspicuous central tanniniferous area, pistillode much exceeding stamens in bud; fruit surface minutely pebbled. New Caledonia . . . . **135. Cyphophoenix** (in part)
7. Anther locules with a conspicuous central taniniferous area, pistillode slightly shorter than stamens in bud; fruit surface not minutely pebbled. New Caledonia . . . . **134. Burretiokentia**

133. BASSELINIA

Characteristic small to moderate pinnate-leaved palms from New Caledonia, displaying substantial variation in habit, leaf and inflorescence form, the prophyll usually incomplete and the fruit with lateral to apical stigmatic remains and ± smooth endocarp.

Basselinia Vieill., Bull. Soc. Lin. Normandie sere. 2. 6: 230 (1873). Lectotype: **B. gracilis** (Brongn. & Gris) Vieill. (*Kentia gracilis* Brongn. & Gris) (see Beccari 1920).

Microkentia H. Wendl. ex Hook.f. in Benth. & Hook.f., Gen. pl. 3: 895 (1883). Lectotype: *Microkentia gracilis* (Brongn.& Gris) Hook.f. ex Salomon (= *Basselinia gracilis* [Brongn.& Gris] Vieill. [see Pintaud & W.J. Baker 2008]).

Nephrocarpus Dammer, Bot. Jahrb. Syst. 39: 21 (1906). Type: *N. schlechteri* Dammer (= *Basselinia pancheri* [Brongn. & Gris] Vieill. [*Kentia pancheri* Brongn. & Gris]).

Alloschmidia H.E. Moore, Gentes Herbarum 11: 293 (1978). Type: *A. glabrata* (Becc.) H.E. Moore (= *Basselinia glabrata* Becc.).

Honors French fuller and poet Olivier Basselin (ca. 1400–1450).

Small to stout, solitary or clustered, unarmed, pleonanthic, monoecious palms. *Stem* erect, usually ± prominently ringed, internodes glabrous, scaly, or densely tomentose, sometimes with exposed roots at the base. *Leaves* pinnate, sometimes irregularly so, or entire and pinnately ribbed, spreading to ascending; sheaths sometimes partly open but forming a prominent crownshaft, variously scaly and tomentose; petiole short to moderate, channelled adaxially, rounded abaxially; rachis angled adaxially, abaxially rounded; leaflets soft or coriaceous when dry, ± regularly arranged, acute, single or several-fold, or the blade undivided except at the apex, bearing small dotted scales over ribs and surface abaxially (scales large and dense in *Basselinia vestita*), scales usually only on veins adaxially, midrib prominent, abaxially bearing ramenta fixed to one side, lateral and marginal veins prominent or not, transverse veinlets not evident. *Inflorescences* infrafoliar, branched to 1 or 3 orders; peduncle short or elongate; prophyll incompletely or completely encircling the peduncle, shortly 2-keeled laterally, rather thin, open abaxially; peduncular bract tubular, complete, not or somewhat exserted from the prophyll, ± beaked; rachis longer than the peduncle; rachis and rachillae glabrous to scaly or tomentose; bracts subtending the branches, rachilla, and triads low, rounded to acute, flowers sometimes obscured by hairs; rachillae

Right: ***Basselinia glabrata***, crown with inflorescences, New Caledonia. (Photo: J.-C. Pintaud)

moderate, stiff, ± spreading, bearing flowers horizontally aligned in triads in the lower $^1/_2$ –$^3/_4$ or more, and paired or solitary staminate flowers distally; bracteoles surrounding the pistillate flower equal or unequal, brown, sepal-like. *Staminate buds* symmetrical; sepals 3, distinct, imbricate, ± acute to rounded; petals 3, distinct, valvate; stamens 6, filaments connate at the very base, inflexed at the apex in bud, anthers dorsifixed, bifid at the base and apex, latrorse; pistillode nearly as high as or exceeding the stamens in bud, angled-cylindrical, narrowed to slightly expanded at the apex. *Pollen* ellipsoidal asymmetric, sometimes elongate or lozenge-shaped; aperture a distal sulcus; ectexine tectate, psilate-perforate, perforate and micro-channelled or finely perforate-rugulate, aperture margin similar or slightly finer; infratectum columellate; longest axis 29–48 µm [8/12]. *Pistillate flowers* smaller than, equaling or larger than the staminate; sepals 3, distinct, imbricate, rounded; petals 3, distinct, imbricate except for the briefly valvate apices; staminodes 3 at one side of the gynoecium, small, tooth-like; gynoecium unilocular, uniovulate, stigmas 3, prominent, recurved, ovule ± pendulous, sometimes briefly arillate, usually hemianatropous. *Fruit* globose to elongate-ellipsoidal, sometimes bilobed, red or black, with lateral to apical stigmatic remains; epicarp smooth or drying pebbled, mesocarp with a thin layer of small, irregular tannin cells external to a thin layer of short sclereids over abundant ellipsoidal tannin cells and a few flat, thin fibres, endocarp thin, vitreous, fragile, sometimes minutely reticulate, with a rounded to elongate basal operculum. *Seed* globose, kidney-shaped or ovoid-ellipsoidal, hilum and raphe short to elongate, the raphe branches anastomosing laterally, rarely scarcely anastomosing, endosperm homogeneous; embryo basal or lateral below the middle. *Germination* adjacent-ligular; eophyll bifid (where known). *Cytology* not studied.

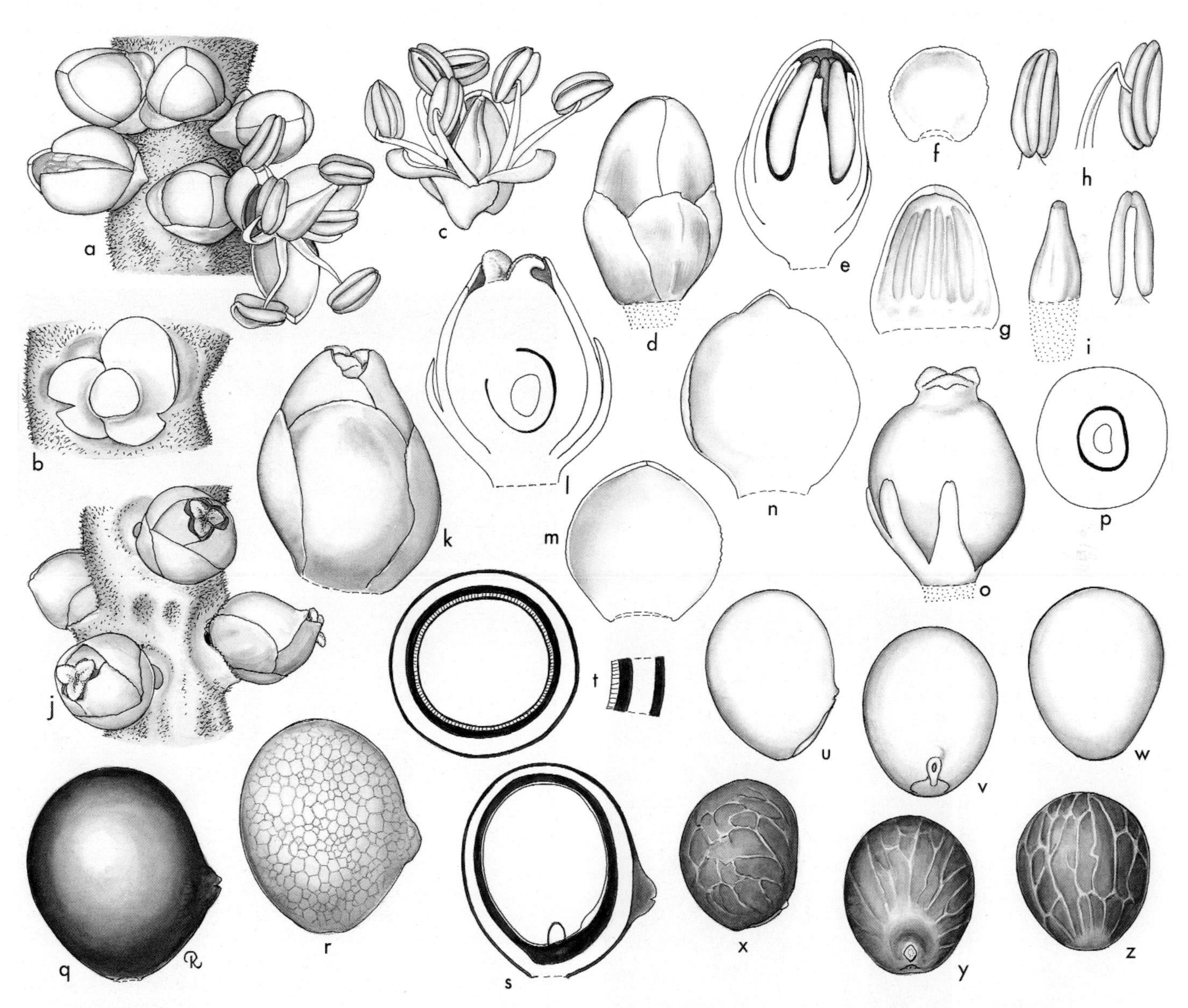

Basselinia. **a**, portion of rachilla with triads × 4 $^1/_2$; **b**, triad, flowers removed to show bracteoles × 6; **c**, staminate flower at anthesis × 4 $^1/_2$; **d**, staminate bud × 7 $^1/_2$; **e**, staminate bud in vertical section × 7 $^1/_2$; **f**, staminate sepal × 7 $^1/_2$; **g**, staminate petal, interior view × 7 $^1/_2$; **h**, stamen in 3 views × 7 $^1/_2$; **i**, pistillode × 7 $^1/_2$; **j**, portion of rachilla with pistillate flowers × 4 $^1/_2$; **k**, pistillate flower × 7 $^1/_2$; **l**, pistillate flower in vertical section × 7 $^1/_2$; **m**, pistillate sepal, interior view × 7 $^1/_2$; **n**, pistillate petal, interior view × 7 $^1/_2$; **o**, gynoecium and staminodes × 7 $^1/_2$; **p**, ovary in cross-section × 7 $^1/_2$; **q**, fruit × 3; **r**, fruit with epidermal layer removed to show sclereids × 3; **s**, fruit in vertical section × 3; **t**, fruit in cross-section × 3, and part of wall, enlarged; **u**, **v**, **w**, endocarp in 3 views × 3; **x**, **y**, **z**, seed in 3 views × 3. *Basselinia sordida*: **a–p**, *Moore et al.* 10072; *B. velutina*: **q–z**, *Moore et al.* 9964. (Drawn by Marion Ruff Sheehan)

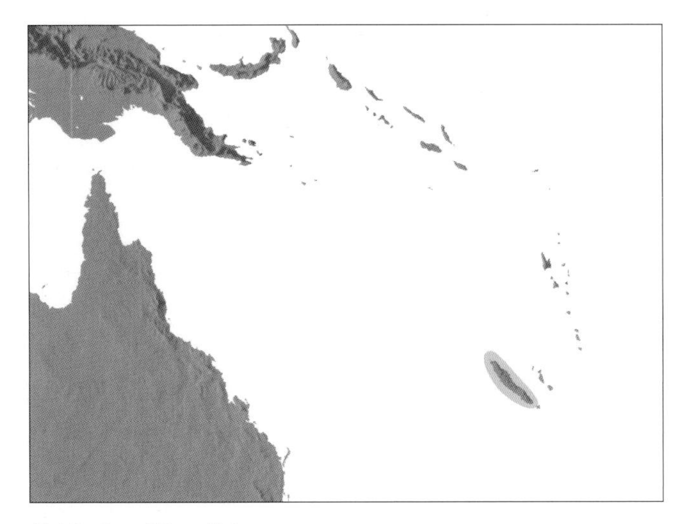

Distribution of ***Basselinia***

Distribution and ecology: Twelve species, locally or widely distributed in New Caledonia. Ten species of *Basselinia* are restricted to either serpentine or schistose soils; *B. gracilis* occurs on both soil types and is the most widely distributed palm in New Caledonia; whereas *B. glabrata* is restricted to forest on schistose rocks.

Anatomy: Leaf (Uhl & Martens 1980), root (Seubert 1998a, 1998b), and fruit (Essig *et al.* 1999).

Relationships: Several analyses provide moderate to high support for the monophyly of *Basselinia* (Asmussen *et al.* 2006, Norup *et al.* 2006, Baker *et al.* in review, in prep.). Intrageneric relationships have been explored by Pintaud (1999b). The wider relationships of the genus within the Basseliniinae are not yet clear.

Common names and uses: Common names not recorded. All species would make elegant ornamentals but apparently are difficult to grow.

Taxonomic accounts: Moore and Uhl (1984), Hodel and Pintaud (1998), Pintaud and Baker (2008).

Fossil record: Thick-walled monosulcate pollen with a distinctive narrow infratectum, *Palmaepollenites kutchensis*, from the Middle Eocene of Central Java (Nanggulan Formation) is compared with pollen of *Basselinia* and *Burretiokentia*, and with the pollen of coryphoid genus *Pritchardia* (Harley & Morley 1995).

Notes: The genus *Basselinia* is divided into two sections. The extremes are so different in general aspect that they were at one time thought to represent two or even three distinct genera. However, the species differ among themselves less than the complex as a unit does from other genera in the Basseliinae. The two sections of *Basselinia* are distinct in leaf.

Anatomy: *B. deplanchei*, *B. gracilis*, *B. pancheri* and *B. vestita* form a very coherent group that share similar epidermal, hypodermal, mesophyll, and guard cell structure, and differ only in the distribution of fibrous strands and minor

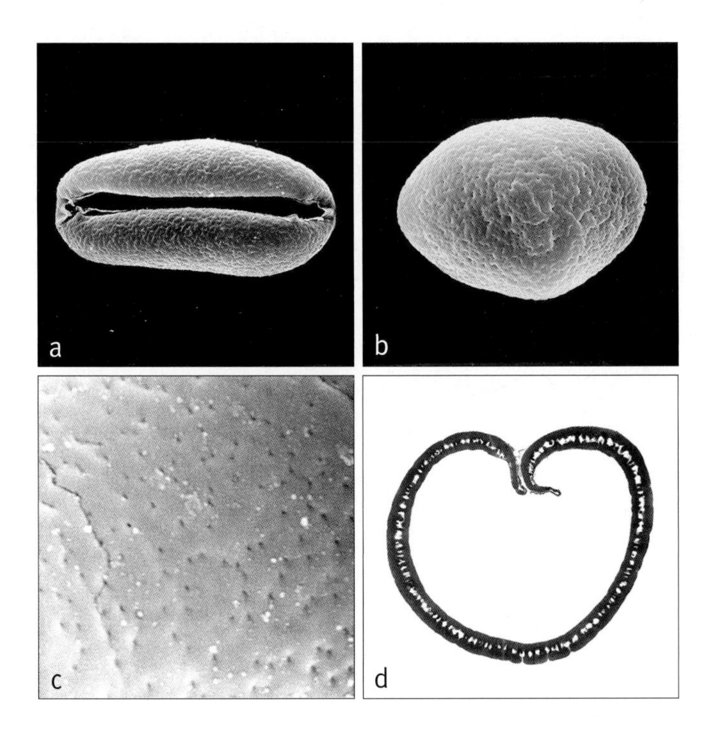

Above: ***Basselinia***
a, monosulcate pollen grain, distal face SEM × 1500;
b, monosulcate pollen grain, proximal face SEM × 1500;
c, close-up, finely perforate pollen surface SEM × 8000;
d, ultrathin section through whole pollen grain, polar plane TEM × 2750.
Basselinia humboldtiana: **a**, *Moore et al.* 10426; *B. pancheri*: **b**, *Bernardi* 10440; *B. glabrata*: **c** & **d**, *Moore et al.* 9957.

Far left: ***Basselinia deplanchei***, habit, New Caledonia. (Photo: J.-C. Pintaud)
Left: ***Basselinia tomentosa***, inflorescence, New Caledonia. (Photo: J.-C. Pintaud)

characteristics of midribs. The six species of section *Taloua*, except for *B. sordida*, are strikingly and distinctively fibrous, in contrast to the species of section *Basselinia*, which all have large amounts of tannin (Uhl & Martens 1980).

134. BURRETIOKENTIA

Moderate to large solitary pinnate-leaved palms from rain forest in New Caledonia, with infrafoliar inflorescences with incomplete prophylls, and fruit with irregularly sculptured endocarp.

Burretiokentia Pic. Serm. in Becc. & Pic. Serm., Webbia 11: 122 (1955). Type: **B. vieillardii** (Brongn. & Gris) Pic. Serm. (*Kentia vieillardii* Brongn. & Gris).

Rhynchocarpa Beccari, Palme Nuova Caledonia 37 (1920); and Webbia 5: 105 (1921) (non Schrad. ex Endl (1839). Type: *R. vieillardii* (Brongn. & Gris) Becc. (*Kentia vieillardii* Brongn. & Gris) (= *Burretiokentia vieillardii* [Brongn. & Gris] Pic. Serm.).

Honoring German palm botanist Karl Ewald Maximilian Burret (1883–1964) by combining his name with the generic name Kentia, *named for William Kent (1779–1827), one-time curator of the botanic gardens at Buitenzorg, Java (now Kebun Raya Bogor).*

Moderate to large, solitary, slender, unarmed, pleonanthic, monoecious palms. *Stem* erect with prominent nodal scars, green or brown, often slightly expanded basally, prickly adventitious roots sometimes present. *Leaves* regularly pinnate, spreading; sheaths tubular, usually forming a prominent crownshaft, more rarely split opposite the petiole, covered with scales or tomentum, obliquely lined; petiole short, shallowly channelled adaxially, rounded abaxially, margins sharp, densely scaly or minutely dotted; rachis angled adaxially, rounded abaxially, scaly or minutely dotted on both surfaces; leaflets stiff, acute, single-fold, 1(–2) lateral veins on each side prominent adaxially, the midrib and numerous veins prominent abaxially, densely scaly when young, becoming dotted in age, ramenta usually abundant along ribs, transverse veinlets not evident. *Inflorescences* infrafoliar, branched to 1(–2) orders, protandrous; peduncle short; prophyll incompletely encircling the peduncle at insertion, open abaxially, caducous; peduncular bract thin, inserted close to and not much exceeding the prophyll, completely enclosing the inflorescence in bud, also caducous; rachis angled, longer than the

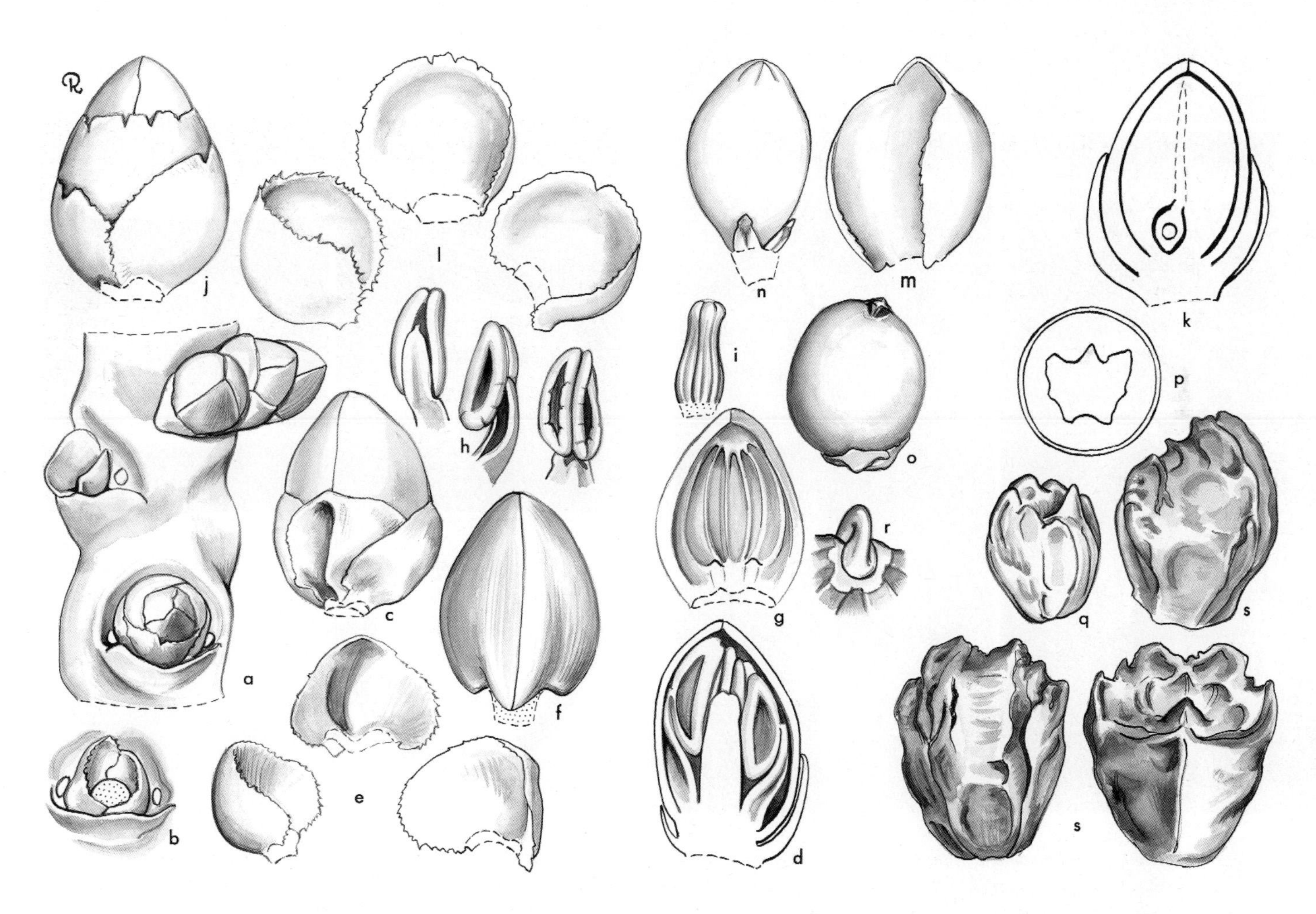

***Burretiokentia*.** **a**, portion of rachilla with complete triad, with staminate flowers fallen, and with all flowers fallen × 3; **b**, triad, flowers removed × 3; **c**, staminate bud × 6; **d**, staminate bud in vertical section × 6; **e**, staminate sepals × 6; **f**, staminate bud, sepals removed × 6; **g**, staminate petal, interior view × 6; **h**, stamen in 3 views × 6; **i**, pistillode × 6; **j**, pistillate bud × 6; **k**, pistillate bud in vertical section × 6; **l**, pistillate sepals × 6; **m**, pistillate petal × 6; **n**, gynoecium and staminodes × 6; **o**, fruit × 1 1/2; **p**, fruit in cross-section × 1 1/2; **q**, endocarp × 1 1/2; **r**, operculum of endocarp × 3; **s**, seed in 3 views × 3. *Burretiokentia vieillardii*: all from *Moore et al.* 9334. (Drawn by Marion Ruff Sheehan)

peduncle, bearing low bracts subtending glabrous to densely tomentose branches and rachillae; rachillae appearing rather stout, arching and spreading, bearing spirally arranged, low, rounded, somewhat spreading bracts subtending triads in the lower $^1/_2$–$^2/_3$ of each rachilla and paired or solitary staminate flowers distally; bracteoles surrounding the pistillate flower prominent, nearly equal, sepal-like, exceeding the subtending bract. *Staminate flowers* not at all or slightly to markedly asymmetrical, often developing well before the pistillate buds or these abortive and the inflorescence wholly staminate; sepals 3, distinct, imbricate, rounded, and the outer usually keeled dorsally; petals 3, distinct, valvate, about twice as long as the sepals, drying lined; stamens 6, filaments flattened, of uniform width or tapered distally from a broad base, inflexed at the apex in bud, anthers dorsifixed, briefly bifid at the apex, bifid nearly $^1/_2$ their length basally, the connective dark, each locule with a central sterile connective-like area marked with included raphides; pistillode ca. $^1/_2$ as long as the stamens or a little more to equalling them in bud, trilobed, sometimes deeply so, not much expanded at apex. *Pollen* ellipsoidal asymmetric; aperture a distal sulcus; ectexine tectate, psilate and sparsely perforate, aperture margin similar; infratectum columellate; longest axis 43–65 µm [2/2]. *Pistillate flowers* shorter than the staminate; sepals 3, distinct, broadly imbricate; petals 3, distinct, not much longer than the sepals, imbricate except for the briefly valvate apices; staminodes 3, tooth-like, borne at one side of the gynoecium; gynoecium unilocular, uniovulate, ovoid, stigmas 3, recurved, ovule pendulous, hemianatropous. *Fruit* globose to ellipsoidal, green when immature, becoming red at maturity, smooth when fresh, pebbled and usually irregularly shouldered and angled when dry, the stigmatic remains apical or eccentrically apical; epicarp thin, mesocarp with a dense layer of elongate, oblique sclereids over a densely tanniniferous layer and a few flat, longitudinal fibres against the endocarp, endocarp

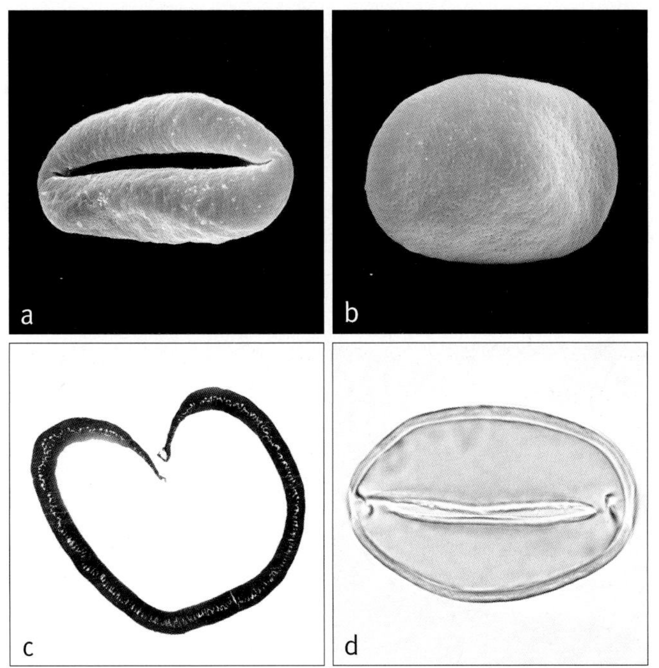

Top left: ***Burretiokentia grandiflora***, habit, New Caledonia. (Photo: J.-C. Pintaud)

Top right: ***Burretiokentia grandiflora***, rachilla at staminate anthesis, New Caledonia. (Photo: J.-C. Pintaud)

Bottom left: ***Burretiokentia vieillardii***, habit, New Caledonia. (Photo: J.-C. Pintaud)

Bottom right: ***Burretiokentia koghiensis***, rachilla at staminate anthesis, New Caledonia. (Photo: J.-C. Pintaud)

Above: ***Burretiokentia***

a, monosulcate pollen grain, distal face SEM × 1000;
b, monosulcate pollen grain, proximal face SEM × 1000;
c, ultrathin section through whole pollen grain, polar plane TEM × 1500;
d, monosulcate pollen grain, mid-focus LM × 1250.
Burretiokentia vieillardii: **a**, *Bernardi* 10392, **b**, *Bernardi* 10203, **c** & **d**, *Balansa* 1962.

Distribution of ***Burretiokentia***

irregularly sculptured, with an adaxial keel and an abaxial groove between lateral ridges, operculum basal, rounded. *Seed* irregular, sculptured like the endocarp, hilum linear, lateral, endosperm homogeneous; embryo basal. *Germination* adjacent-ligular; eophyll bifid. *Cytology* not studied.

Distribution: Five species in New Caledonia; *Burretiokentia vieillardii* is widely distributed from south-east to north-east New Caledonia in moist forest on serpentine or schists, *B. hapala* occurs in wet forests or gallery forest on calcareous and schistose soils from 50–400 m in northern New Caledonia, *B. dumasii* occurs on ultramafic rock at 600 m in west central New Caledonia, *B. grandiflora* is restricted to ultramafic rock at 200–900 m in south-east New Caledonia and *B. koghiensis* is found only on the Mt Koghi massif on ultramafic rock at 500–600 m.

Anatomy: Leaf (Uhl & Martens, 1980), root (Seubert 1998a, 1998b), and fruit (Essig *et al.* 1999).

Relationships: *Burretiokentia* is moderately supported as monophyletic (Baker *et al.* in prep.). Intrageneric relationships have been explored by Pintaud (1999b). DNA sequence data provide weak evidence that the genus is nested within *Cyphophoenix* (Baker *et al.* in prep.), while morphological evidence places it as sister to *Cyphophoenix* (Pintaud 1999b).

Common names and uses: Not recorded.

Taxonomic accounts: Moore & Uhl (1984) and Pintaud & Hodel (1998b).

Fossil record: Thick-walled monosulcate pollen with a distinctive narrow infratectum, *Palmaepollenites kutchensis*, from the Middle Eocene of Central Java (Nanggulan Formation) is compared with pollen of *Burretiokentia* and *Basselinia* and with that of the coryphoid genus *Pritchardia* (Harley & Morley 1995).

Notes: Distinguished by an incomplete prophyll and strikingly sculptured endocarp. *Burretiokentia vieillardii* is one of the most common palms in New Caledonia; the prominently ringed glossy, green or brown stems are conspicuous. In leaf anatomy, *Burretiokentia* is like *Cyphokentia* and *Basselinia* in having single adaxial and abaxial hypodermal layers. In *Burretiokentia*, adaxial hypodermal cells are very large, twice as long and four to five times as wide as the epidermal cells (Uhl & Martens 1980).

135. CYPHOPHOENIX

Moderate solitary pinnate-leaved palms from rain forest in New Caledonia, one species with spectacular stilt-roots, all with infrafoliar inflorescences with incomplete prophylls, and fruit with smooth or irregularly sculptured endocarp.

Cyphophoenix H. Wendl. ex Hook.f. in Benth. & J.D. Hook.f., Gen. pl. 3: 893 (1883). Lectotype: *Kentia elegans* Brongn. = **C. elegans** (Brongn.) H.A. Wendl. ex Salomon (see Beccari 1920).

Campecarpus H. Wendl. ex Becc., Palme Nuova Caledonia 28 (1920); and Webbia 5: 96 (1921). Type: *C. fulcitus* (Brongn.) H. Wendl. ex Becc. (*Kentia fulcita* Brongn. = *Cyphophoenix fulcita* [Brongn.] Pintaud & W.J. Baker).

Veillonia H.E. Moore, Gentes Herb. 11: 299 (1978). Type: *V. alba* H.E. Moore (= *Cyphophoenix alba* [H.E. Moore] Pintaud & W.J. Baker).

Kyphos — *bent, humped,* phoenix — *a general name for a palm, perhaps in reference to the prominent terminal stigmatic remains in the fruit.*

Moderate, solitary, unarmed, pleonanthic, monoecious palms. *Stems* erect, prominently ringed, smooth, yellowish, green or grey, sometimes enlarged at the base, sometimes with a very conspicuous cone of long stilt roots or with exposed adventitious roots. *Leaves* regularly pinnate, straight or arched, spreading or erect; sheaths forming a sometimes prominently and diagonally ribbed, somewhat inflated crownshaft, covered outside and inside with pale, grey, brown or red tomentum and persistent or deciduous scales, or white waxy; petiole short to elongate, channelled adaxially, rounded abaxially, tomentose to glabrous adaxially, with tomentum and scales abaxially; rachis flat to angled adaxially with broad margins basally, becoming nearly deltoid in section apically, rounded abaxially, minutely dotted adaxially, tomentose and scaly abaxially; leaflets single-fold, coriaceous when dry, obliquely acute to acuminate or rarely briefly bifid, margins not thickened, abaxially with medifixed ramenta with tattered and twisted margins, prominent toward the base on the midrib, sometimes present throughout and sometimes on two lateral veins, with or without small brown scales, midrib prominent and elevated adaxially, secondary

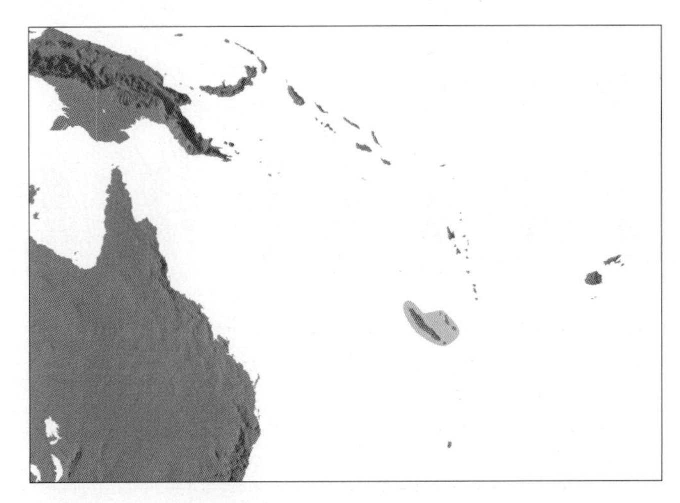

Distribution of ***Cyphophoenix***

veins scarcely prominent, transverse veinlets not evident. *Inflorescences* infrafoliar, stiffly branched to 1 or 2(–3) orders basally, fewer orders distally, protandrous; peduncle short, ± dorsiventrally compressed, sometimes white-waxy; prophyll incompletely encircling the peduncle, open abaxially, 2-keeled laterally, tomentose, sometimes splitting into 2 halves, rather thin, sometimes white-waxy, caducous; peduncular bract completely encircling the peduncle and enclosing the inflorescence in bud, slightly exceeding the prophyll, sometimes ± woody, also caducous and tomentose; rachis much exceeding the peduncle, angled, minutely striate when dry, bearing spirally arranged, low rounded or acute bracts subtending several divaricate branches, the lower ones with a prominent bare base; rachillae long, curved, often pendulous, slender or stout, bearing prominent, rounded or acute, lip-like bracts subtending triads in the lower $^1/_3$–$^2/_3$ of each rachilla, and paired or solitary staminate flowers distally; bracteoles about the pistillate flower prominent, marginally imbricate, $^1/_2$ the length of the sepals. *Staminate flowers* symmetrical, borne on very short, strap-like pedicels in triads, but ± sessile in the distal pairs; sepals 3, distinct, imbricate, broadly rounded or becoming crenulate apically, keeled dorsally and ± compressed laterally toward the base; petals 3, distinct, about twice as long as the sepals, valvate, scarcely acute at apex, ± lined when dry; stamens 6, filaments distinct, awl-shaped, the slender apex inflexed in bud, anthers versatile at anthesis, emarginate apically, bifid ca. $^1/_2$ their length basally, basifixed but appearing dorsifixed; pistillode shorter or longer than stamens, narrowly pyramidal or nearly columnar, angled and grooved, at least when dry, very briefly 3-lobed at apex. *Pollen* ellipsoidal asymmetric, less frequently oblate triangular; aperture a distal sulcus, occasionally a trichotomosulcus; ectexine tectate, psilate-perforate, perforate and micro-channelled, aperture margin similar or slightly finer; infratectum columellate; longest axis 47–60 µm [4/4]. *Pistillate flowers* larger than the staminate, symmetrical; sepals 3, distinct, broadly imbricate and rounded, nearly as high as the petals; petals 3, distinct, longer than the sepals, strongly imbricate except for the very briefly valvate or subvalvate apices; staminodes 3, distinct, awl-shaped, flat; gynoecium ovoid when fresh, unilocular, uniovulate, stigmas 3, recurved, short, the ovule pendulous, probably hemianatropous. *Fruit* ovoid, oblong-ellipsoidal or eccentrically ovoid, sometimes slightly curved at tip, yellow with black tip at maturity to dull brownish red, stigmatic remains apical or eccentrically apical; epicarp smooth, sometimes drying densely pebbled over a layer of short, pale, obliquely oriented fibres, mesocarp ± dry or fleshy, with numerous slender, elongate, red tannin cells intermixed with the fibres and beneath the fibres against the endocarp, endocarp thin, crustaceous, fragile, oblong-ellipsoidal, circular in cross-section, or with a longitudinal ridge on one side, or highly sculptured with an adaxial ridge and basal operculum framed by lateral, ± flat areas, and with 2 lateral and 2 abaxial irregular crests and a dorsal groove, operculum circular, basal. *Seed* oblong-ellipsoidal, circular in cross-section, or sculptured like the endocarp, hilum obovoid or elongate, apical, tapered toward the base, raphe branches weakly or strongly anastomosing, endosperm homogeneous; embryo basal. *Germination* adjacent-ligular; eophyll bifid. *Cytology* not studied.

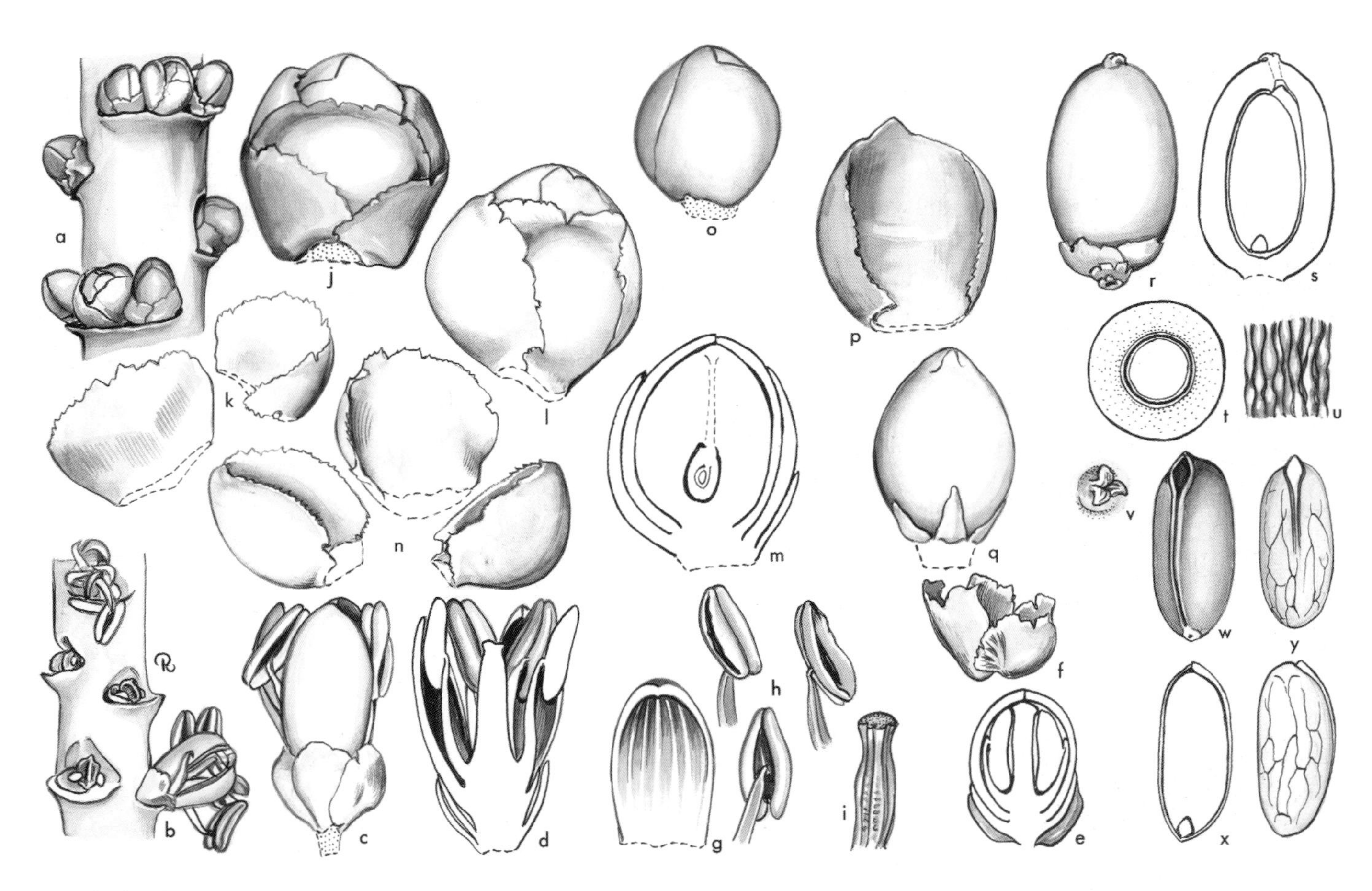

***Cyphophoenix*.** **a**, portion of rachilla with triads × 3; **b**, portion of rachilla with paired staminate flowers, most fallen × 3; **c**, staminate flower × 6; **d**, staminate flower in vertical section × 6; **e**, staminate bud in vertical section × 6; **f**, staminate sepals × 6; **g**, staminate petal, interior view × 6; **h**, stamen in 3 views × 6; **i**, pistillode × 6; **j**, pistillate bud with subtending bracteoles × 6; **k**, bracteoles subtending pistillate flower × 6; **l**, pistillate bud × 6; **m**, pistillate bud in vertical section × 6; **n**, pistillate sepals × 6; **o**, pistillate bud, sepals removed × 6; **p**, pistillate petal, interior view × 6; **q**, gynoecium and staminodes × 6; **r**, fruit × $1^1/_2$; **s**, fruit in vertical section × $1^1/_2$; **t**, fruit in cross-section × $1^1/_2$; **u**, portion of dried exocarp, much enlarged; **v**, stigmatic remains × 3; **w**, endocarp × $1^1/_2$; **x**, endocarp in vertical section × $1^1/_2$; **y**, seed in 2 views × $1^1/_2$. *Cyphophoenix elegans*: all from *Moore et al.* 9323. (Drawn by Marion Ruff Sheehan)

Distribution and ecology: Four species, three in New Caledonia and one in the adjacent Loyalty Islands. *Cyphophoenix elegans* and *C. alba* occur in forest on schistose rocks in northeastern New Caledonia. *Cyphophoenix fulcita* grows in New Caledonia in wet forest on serpentine rocks, often occurring perched on rocks on its stilt roots; *C. nucele* is restricted to the raised coral of Lifou Island.

Anatomy: Leaf (Uhl & Martens 1980), root (Seubert 1998a, 1998b), and fruit (Essig *et al.* 1999).

Relationships: The monophyly of *Cyphophoenix* is resolved in two studies, but with low support (Pintaud 1999b, Baker *et al.* in review). One study suggests that the genus is not monophyletic, but this resolution is also poorly supported (Baker *et al.* in prep.). The precise relationships of the genus within the Basseliniinae are unclear.

Common names and uses: Unknown.

Taxonomic accounts: Moore and Uhl (1984), Hodel and Pintaud (1998), Pintaud and Baker (2008).

Fossil record: Asymmetric monosulcate pollen with a distinctive irregularly columellate infratectum, *Palmaepollenites* sp., from the Middle Eocene of Central Java (Nanggulan Formation) is compared with pollen of *Cyphophoenix* as well as with that of *Cyphosperma*, *Actinorhytis* and *Moratia* (= *Cyphokentia*) (Harley & Morley 1995).

Above left: ***Cyphophoenix fulcita***, habit, New Caledonia. (Photo: J.-C. Pintaud)
Above right: ***Cyphophoenix fulcita***, stilt roots, New Caledonia. (Photo: J.-C. Pintaud)

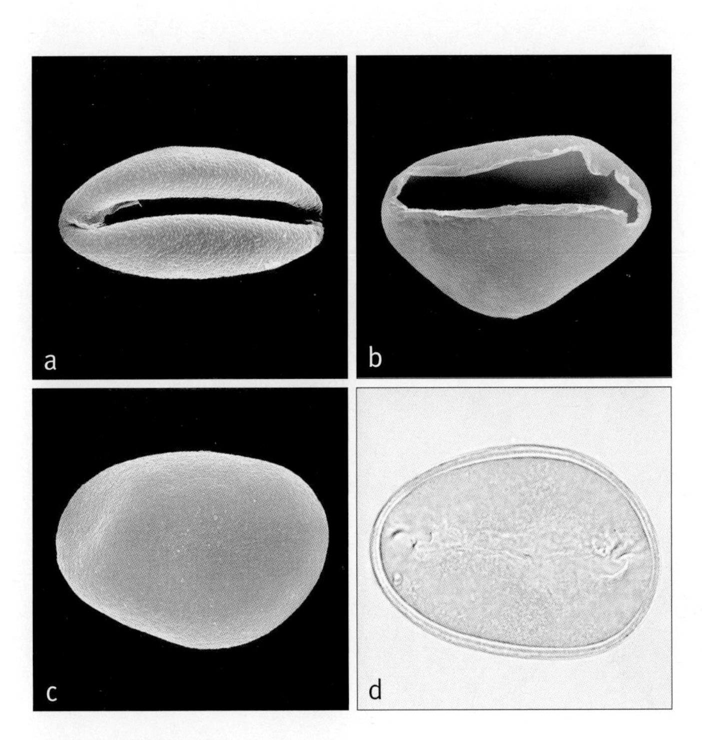

Above: ***Cyphophoenix***
a, monosulcate pollen grain, distal face SEM × 1000;
b, monosulcate pollen grain, oblique distal/equatorial face SEM × 1000;
c, monosulcate pollen grain, proximal face SEM × 1000;
d, monosulcate pollen grain, mid-focus LM × 800.
Cyphophoenix elegans: **a**, *Moore* 9323; *C. alba*: **b** & **c**, *Bernardi* 12627; *C. fulcita*: **d**, *Moore et al.* 1041.

Notes: *Cyphophoenix elegans* and *C. nucele*, are each confined to a particular edaphic situation and the two are separated by a long distance. Seed of *C. nucele* has been introduced into cultivation in the United States, where it may be expected to do especially well on the coral rocks of southern Florida. *Cyphophoenix* has the following distinctive features of leaf anatomy: vascular bundles in uneven rows, many small vascular bundles present, and lower bundle sheaths crescentic and wider than upper sheaths. It is the only genus in New Caledonia, having two-layered upper and lower hypodermal layers (Uhl & Martens 1980).

136. CYPHOSPERMA

Moderate to large solitary pinnate-leaved palms from rain forest in New Caledonia, Vanuatu and Fiji, with interfoliar inflorescences with incomplete prophylls, and fruit with irregularly sculptured endocarp.

Cyphosperma H. Wendl. ex Hook.f., in Benth. & Hook.f., Gen. pl. 3: 895 (1883). Lectotype: *Cyphokentia balansae* Brongn. = **C. balansae** (Brongn.) H. Wendl. ex Salomon (see Beccari 1920).

Taveunia Burret, Occas. Pap. Bernice Pauahi Bishop Mus. 11: 12 (1935). Type: *T. trichospadix* Burret (= *Cyphosperma trichospadix* [Burret] H.E. Moore).

Kyphos — *bent, humped*, sperma — *seed, probably referring to the irregular humps and ridges on the seed.*

Moderate, solitary, unarmed, pleonanthic, monoecious palms. *Stem* erect, ± prominently ringed with irregular leaf scars. *Leaves* regularly pinnate, or bifid or irregularly divided, ± 3-ranked, spreading or ± erect; sheaths split opposite the petiole, not forming a crownshaft, glabrous; petiole short, stout, channelled to ± flat adaxially, rounded abaxially; rachis adaxially channelled above the base, becoming nearly triangular in section with a narrow ridge toward the apex, rounded abaxially, with scales and tomentum; leaflets when present, acute, single-fold, with elevated midrib, adaxially with 2 lateral veins and many secondary veins with or without linear scales, prominently veined abaxially with deciduous tomentum, ramenta sometimes present, transverse veinlets inconspicuous. *Inflorescences* interfoliar but persisting below the leaves, arched in flower, pendulous in fruit, branched to 2 orders basally, to 1 order distally, branches and rachillae with a prominent pulvinus at the base, protandrous; peduncle somewhat dorsiventrally compressed and elliptical in cross-section, elongate, much exceeding the bracts at anthesis; prophyll tubular, short, incompletely encircling the peduncle abaxially, 2-keeled laterally, chartaceous, open apically, ± glabrous, marcescent; peduncular bract with tubular base, much longer than the prophyll, beaked, covered with deciduous tomentum, also marcescent; rachis bearing low, acute to rounded bracts subtending branches and rachillae; rachillae distant, slender, moderate, tomentose throughout or glabrous except for patches of stiff, pale brown hairs in the upper and lateral parts of pit

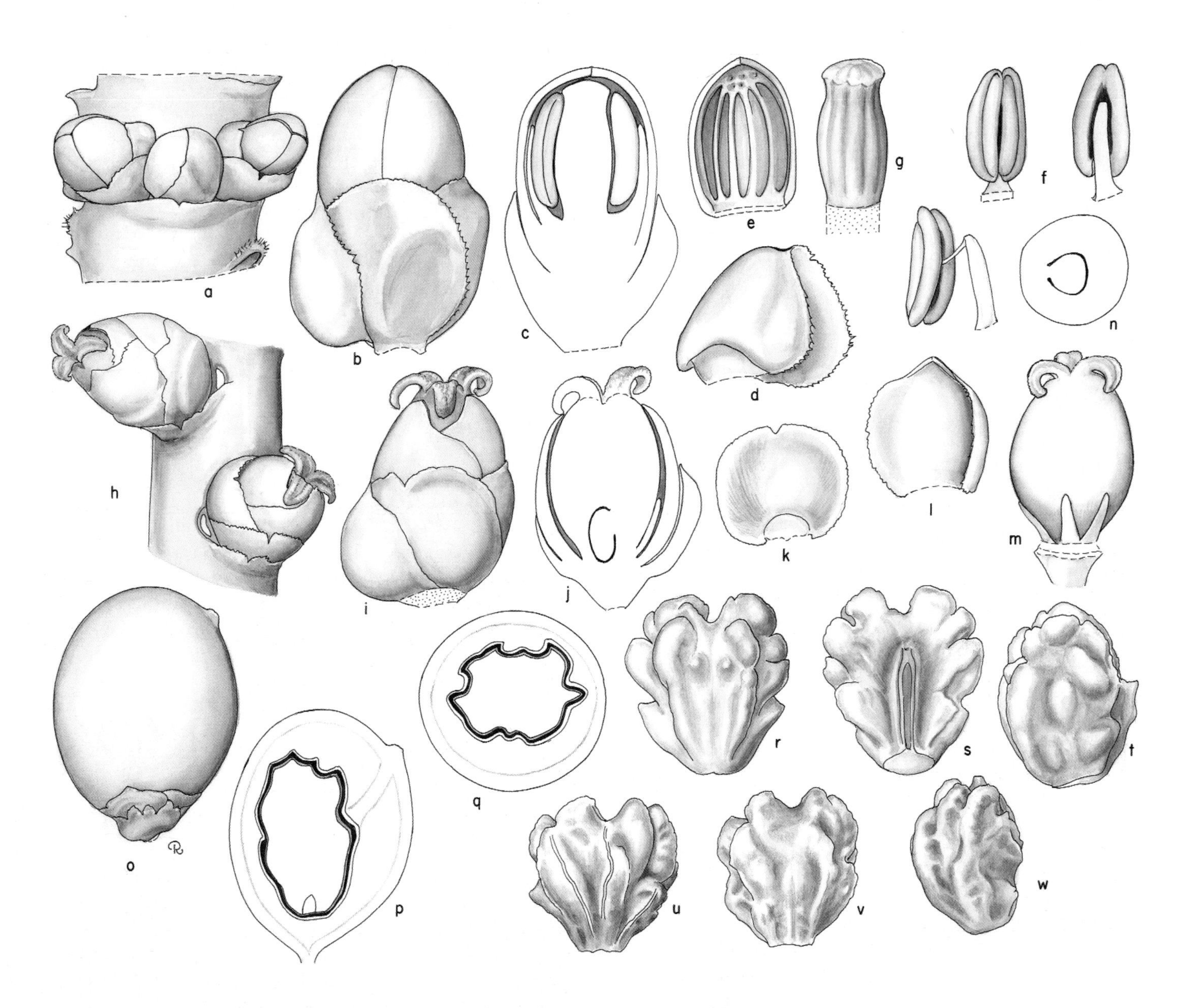

***Cyphosperma*.** **a**, portion of rachilla with triad × 6; **b**, staminate bud × 12; **c**, staminate bud in vertical section × 12; **d**, staminate sepal × 12; **e**, staminate petal, interior view × 12; **f**, stamen in 3 views × 12; **g**, pistillode × 12; **h**, portion of rachilla at pistillate anthesis, staminate flowers fallen × 6; **i**, pistillate flower × 7 $^1/_2$; **j**, pistillate flower in vertical section × 7 $^1/_2$; **k**, pistillate sepal × 7 $^1/_2$; **l**, pistillate petal × 7 $^1/_2$; **m**, gynoecium and staminodes × 7 $^1/_2$; **n**, ovary in cross section × 9; **o**, fruit × 3; **p**, fruit in vertical section × 3; **q**, fruit in cross-section × 3; **r, s, t**, endocarp in 3 views × 3; **u, v, w**, seed in 3 views × 3. *Cyphosperma balansae*: **a–g**, *Moore et al.* 10073; **h–w**, *Moore et al.* 10033. (Drawn by Marion Ruff Sheehan)

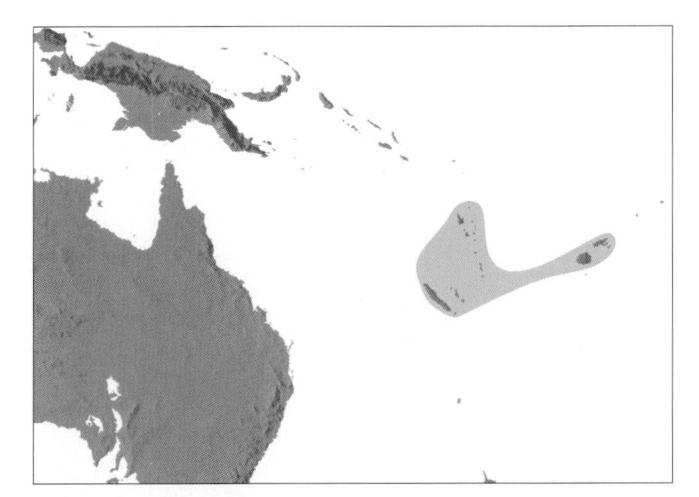

Distribution of ***Cyphosperma***

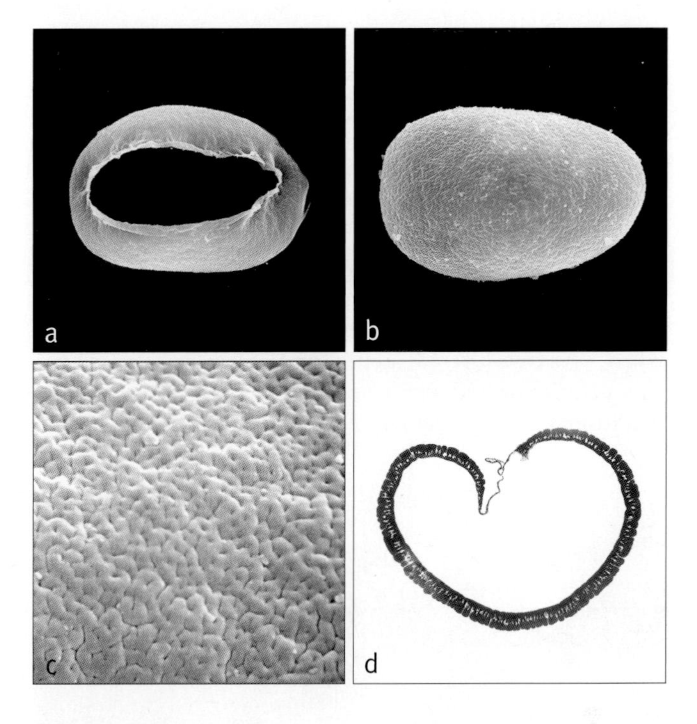

Above: ***Cyphosperma***
a, monosulcate pollen grain, distal face SEM × 1500;
b, monosulcate pollen grain, proximal face SEM × 1500;
c, close-up, finely perforate-rugulate pollen surface SEM × 8000;
d, ultrathin section through whole pollen grain, polar plane TEM × 2000.
Cyphosperma balansae: **a–c**, *Moore et al.* 10479, **d**, *Balansa* 1961.

cavities, bearing acute or rounded bracts subtending partially sunken triads nearly throughout the rachillae, with paired or solitary staminate flowers in the upper $^1/_4$ or less; bracteoles surrounding the pistillate flower nearly equal, imbricate, nearly as long as the bract subtending the triad. *Staminate flowers* symmetrical; sepals 3, distinct, rounded, imbricate, gibbous basally and centrally keeled; petals 3, distinct, valvate; stamens 6, filaments flattened, inflexed at the apex in bud, erect at anthesis, anthers oblong, dorsifixed, briefly emarginate at the base and apex; pistillode overtopping the stamens in bud, nearly columnar, apex expanded, 3-grooved. *Pollen* ellipsoidal asymmetric, occasionally lozenge-shaped; aperture a distal sulcus; ectexine tectate, perforate and micro-channelled or finely perforate-rugulate, aperture margin similar or slightly finer; infratectum columellate; longest axis 29–43 µm [1/3]. *Pistillate flowers* larger than the staminate; sepals 3, distinct, rounded, imbricate; petals 3, distinct, imbricate except for briefly valvate apices; staminodes 3, distinct, tooth-like at one side of gynoecium; gynoecium unilocular, uniovulate, ellipsoidal, stigmas 3, recurved, ovule pendulous, form not known. *Fruit* broadly ellipsoidal, with stigmatic remains lateral in upper $^1/_4$ –$^1/_3$; epicarp smooth when fresh, drying pebbled, mesocarp fleshy, with short fibre bundles nearly perpendicular to the epicarp, whitish parenchyma with dispersed, elongate, irregular tannin cells and thin, flat fibres, near but not adnate to the endocarp, endocarp thin, vitreous, fragile, irregularly sculptured, ridged, and grooved. *Seed* irregularly sculptured like the endocarp, hilum linear, endosperm homogeneous; embryo basal. *Germination* adjacent-ligular; eophyll bifid. *Cytology*: 2n = 32.

Distribution and ecology: Four species, two in Fiji, one in New Caledonia and one in Vanuatu. On schists, graywackes, and peridotites in northeastern and northwestern New Caledonia, in dense forests at elevations from 600 m or less to 900 m in Fiji and in rain forest on volcanic soils at 900–1100 m in Vanuatu.

Anatomy: Leaf (Uhl & Martens 1980), root (Seubert 1998a, 1998b), and fruit (Essig *et al.* 1999).

Relationships: The monophyly of *Cyphosperma* has not been tested. DNA sequence data reveal a moderately supported relationship with *Physokentia* (Lewis & Doyle 2002, Asmussen *et al.* 2006), whereas morphological data suggest that the genus is sister to a clade of *Basselinia*, *Burretiokentia* and *Cyphophoenix* (Pintaud 1999b).

Common names and uses: Not recorded.

Below left: ***Cyphophoenix elegans***, habit, New Caledonia. (Photo: J.-C. Pintaud)
Below right: ***Cyphosperma balansae***, crown with inflorescence, New Caledonia. (Photo: J.-C. Pintaud)

Taxonomic accounts: Beccari & Pichi-Sermolli (1955), Moore (1979), Moore & Uhl (1984), and Hodel & Pintaud (1998).
Fossil record: Asymmetric monosulcate pollen with a distinctive irregularly columellate infratectum, *Palmaepollenites* sp., from the Middle Eocene of Central Java (Nanggulan Formation) is compared with pollen of *Cyphosperma* as well as with that of *Cyphophoenix*, *Actinorhytis* and *Moratia* (= *Cyphokentia*) (Harley & Morley 1995).
Notes: In leaf anatomy, *Cyphosperma balansae* is distinguished by small fibrous strands in both upper and lower hypodermal layers. A two-layered upper and single-layered lower hypodermis is shared with *Clinosperma* (Uhl & Martens 1980).
Additional figures: Glossary fig. 18.

137. LEPIDORRHACHIS

Small to moderate pinnate-leaved palm known only from mountain forest on Lord Howe Island; leaf sheaths not forming a crownshaft, the inflorescences interfoliar, apparently unisexual, but both sexes borne on the same plant.

Lepidorrhachis (H. Wendl. & Drude) O.F. Cook, J. Heredity 18: 408 (1927) ('*Lepidorhachis*'). Type: **L. mooreana** (F. Muell.) O.F. Cook (*Kentia mooreana* F. Muell.).
Clinostigma section *Lepidorrhachis* H. Wendl. & Drude, Linnaea 39: 218 (1875). Type: *Clinostigma mooreanum* (F. Muell.) H. Wendl. & Drude.

Lepis — *scale,* rhachis — *spine or backbone, referring to the scaly leaf rachis.*

Small to moderate, solitary, unarmed, pleonanthic, monoecious palm. *Stem* thick, short, green, prominently ringed with close leaf scars. *Leaves* pinnate, ascending; sheath inflated at the base, deeply split opposite the petiole, not forming a distinct crownshaft, densely brown-scurfy toward the apex and petiole; petiole short, shallowly channelled adaxially, rounded abaxially, densely brown-scurfy; rachis adaxially channelled to ridged, abaxially rounded, densely tomentose; leaflets regularly arranged, alternate, single-fold, acute, stiff, ascending from the rachis, the midrib and 2–3 secondary nerves on each side prominent adaxially, margins thickened, bearing scales along ribs on both surfaces, transverse veinlets not evident. *Inflorescences* apparently unisexual, interfoliar in bud, becoming infrafoliar at anthesis, branched to 3 orders basally, fewer distally, densely covered with basifixed, twisted, simple to forked or tattered, brown, membranous scales; peduncle very short; prophyll usually incompletely sheathing, 2-keeled, open apically, apparently completely encircling at the base; peduncular bract tubular, beaked, exserted from the prophyll; rachis longer than the peduncle, bearing short, wide, acute bracts subtending rather distant, spirally arranged branches and rachillae; rachillae short, bearing paired or solitary staminate flowers or solitary pistillate flowers, the flowers superficial, each subtended by an acute bract. *Staminate flowers* slightly asymmetrical; sepals 3, distinct, imbricate basally, keeled, ± rounded apically; petals 3, distinct, slightly asymmetrical, about twice as long as the sepals, strongly nerved when dry; stamens 6, filaments distinct, inflexed in bud, anthers oblong in outline, latrorse; pistillode narrowly cylindrical, expanded apically, slightly longer than the stamens in bud. *Pollen* grains ellipsoidal asymmetric, occasionally lozenge-shaped; aperture a distal sulcus; ectexine tectate, perforate and micro-channelled, aperture margin similar or slightly finer; infratectum columellate; longest axis 32–44 µm [1/1]. *Pistillate flowers* symmetrical; sepals 3, distinct, broadly rounded and imbricate; petals 3, distinct, imbricate with briefly valvate apices, strongly nerved when dry; staminodes 3, on one side of the gynoecium; gynoecium unilocular, uniovulate, subglobose, stigmas recurved, ovule pendulous, probably hemianatropous. *Fruit* globose or nearly so, red at maturity, stigmatic remains lateral in the upper $^{1}/_{3}$–$^{1}/_{4}$; epicarp smooth, drying granular over included sclerosomes, mesocarp fleshy, thin, with included longitudinal fibres except along the rapheal region, endocarp thin, fragile, operculate. *Seed* globose, with elongate raphe, hilum apical, short, elliptic, vasculature prominent, branched, the branches anastomosing abaxially, endosperm homogeneous; embryo subbasal. *Germination* and eophyll unrecorded. *Cytology* not studied.

Top left: ***Lepidorrhachis mooreana***, habit, Lord Howe Island. (Photo: W.J. Baker)

Middle left: ***Lepidorrhachis mooreana***, staminate flowers, Lord Howe Island. (Photo: I. Hutton)

Bottom left: ***Lepidorrhachis mooreana***, pistillate flowers, Lord Howe Island. (Photo: I. Hutton)

Distribution and ecology: One species on Lord Howe Island, found only in low mossy forest at high elevations.

Anatomy: Gynoecium with many bundles of sclereids peripherally (Uhl unpublished); leaf anatomy not studied, and fruit (Essig *et al.* 1999).

Relationships: The sister relationships of *Lepidorrhachis* remain unresolved (Asmussen *et al.* 2006, Norup *et al.* 2006, Baker *et al.* in review, in prep.).

Common names and uses: Little mountain palm. A handsome ornamental. The fruit is favoured by invasive rats and must be protected.

Taxonomic accounts: Green (1994) and Baker & Hutton (2006).

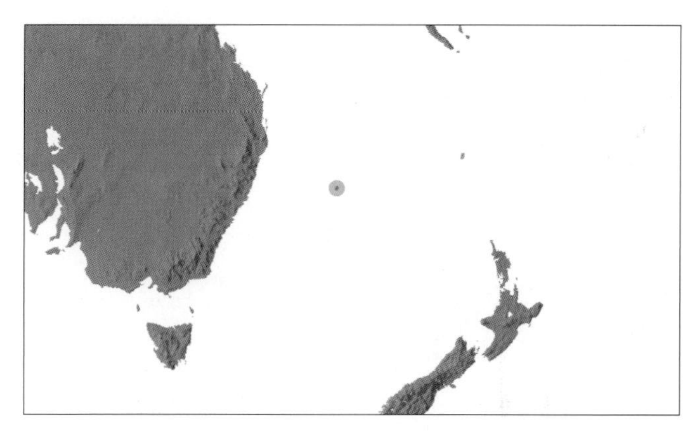

Distribution of ***Lepidorrhachis***

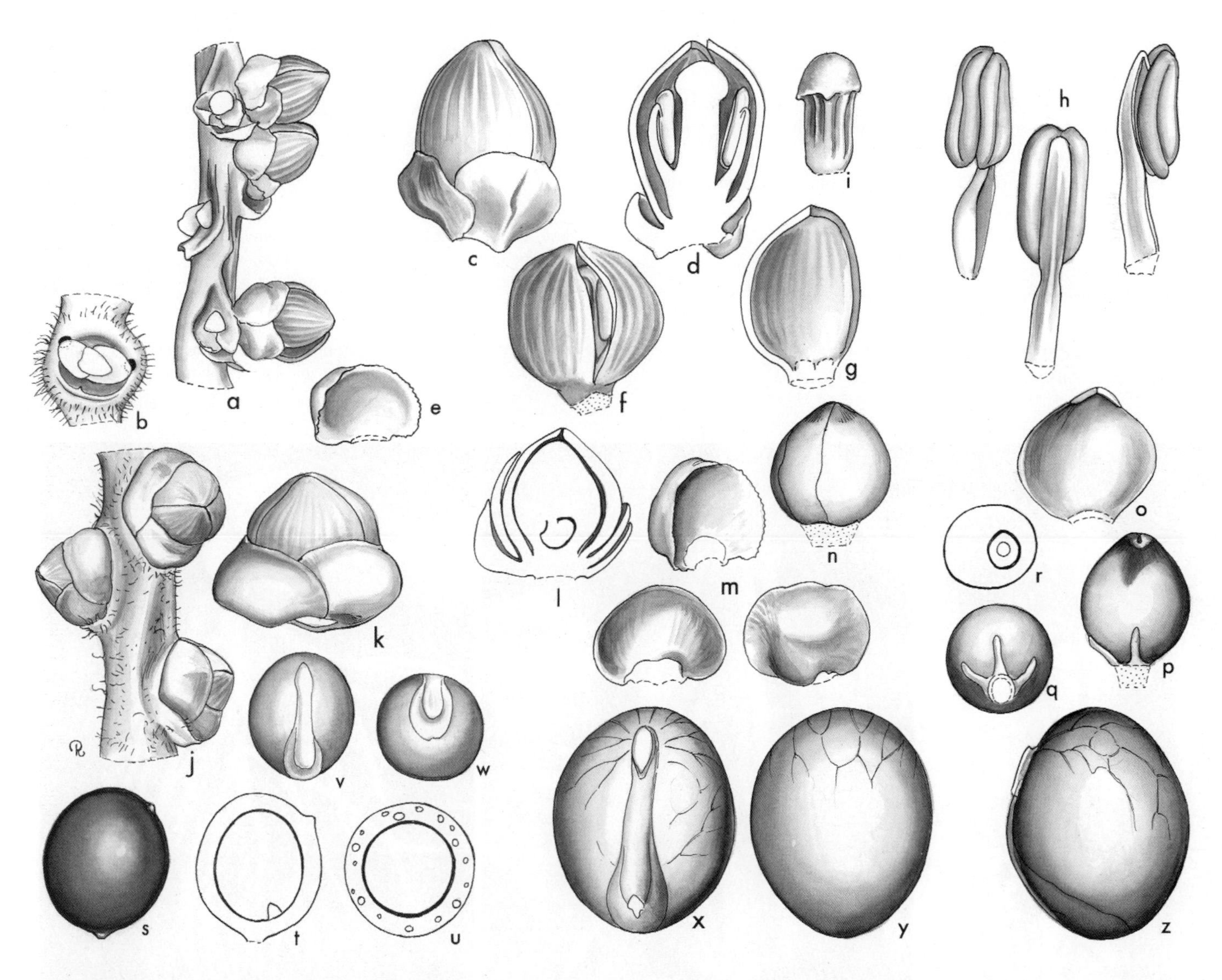

Lepidorrhachis. **a**, portion of rachilla with staminate flowers × 5; **b**, triad, flowers removed to show bracteoles and scars × 3; **c**, staminate bud × 10; **d**, staminate bud in vertical section × 10; **e**, staminate sepal × 10; **f**, staminate bud, sepals removed × 10; **g**, staminate petal, interior view × 10; **h**, stamen in 3 views × 15; **i**, pistillode × 10; **j**, rachilla with pistillate buds × 3; **k**, pistillate bud × 6; **l**, pistillate bud in vertical section × 6; **m**, pistillate sepals × 6; **n**, pistillate bud, sepals removed × 6; **o**, pistillate petal × 6; **p**, gynoecium and staminodes × 6; **q**, gynoecium and staminodes, basal view × 6; **r**, ovary in cross-section × 6; **s**, fruit × 1 $^1/_2$; **t**, fruit in vertical section × 1 $^1/_2$; **u**, fruit in cross-section × 1 $^1/_2$; **v**, **w**, endocarp with operculum in 2 views × 1 $^1/_2$; **x**, **y**, **z**, seed in 3 views × 3. *Lepidorrhachis mooreana*: all from *Moore & Schick* 9250. (Drawn by Marion Ruff Sheehan)

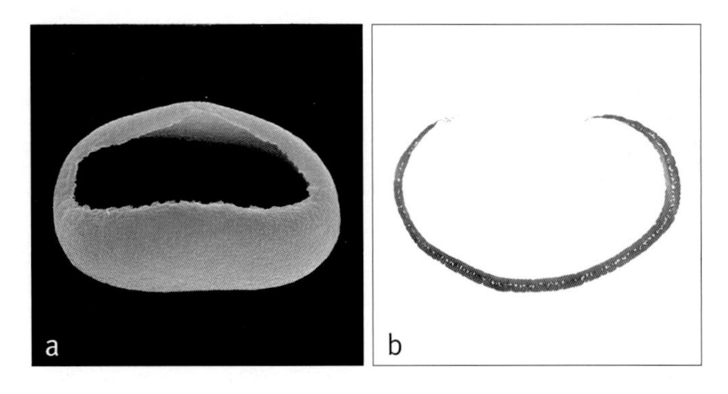

Above: ***Lepidorrhachis***
a, monosulcate pollen grain, distal face SEM × 1500;
b, ultrathin section through whole pollen grain, polar plane TEM × 2000.
Lepidorrhachis mooreana: **a** & **b**, *Mueller* 1880.

Fossil record: No generic records found.
Notes: The rather short stout stem, ascending leaflets, lack of a crownshaft, sheath with copious buff indumentum and red fruits are characteristic. The restriction of this palm to high elevations on Lord Howe Island makes it particularly vulnerable to climate change.
Additional figures: Fig. 8.4.

Below left: ***Physokentia rosea***, crown with inflorescences, Fiji. (Photo: J.L. Dowe)
Below middle: ***Physokentia rosea***, stilt roots, Fiji. (Photo: J.L. Dowe)
Below right: ***Physokentia rosea***, inflorescence, Fiji. (Photo: J.L. Dowe)

138. PHYSOKENTIA

Small to moderate pinnate-leaved palms from Fiji, Vanuatu, Solomon Islands and New Britain, distinctive in the stilt roots, the incomplete prophyll and usually highly sculptured endocarp.

Physokentia Becc. in Martelli, Atti Soc. Tosc. Sci. Nat. Pisa Mem. 44: 152 (1934). Lectotype: **P. tete** (Becc.) Becc. in Martelli (*Cyphosperma tete* Becc.) (see Burret 1935).
Goniosperma Burret, Occas. Pap. Bernice Pauahi Bishop Mus. 11(4): 10 (1935). Type: *G. vitiense* Burret (= *Physokentia thurstonii* [Becc.] Becc. [*Cyphosperma thurstonii* Becc.]).
Goniocladus Burret, Notizbl. Bot. Gart. Berlin-Dahlem 15: 86 (1940). Type: *G. petiolatus* Burret (= *Physokentia petiolata* [Burret] D. Fuller).

Combining physa — *bellows or bubble, with the generic name* Kentia, *named for William Kent (1779–1827), once curator of the botanic gardens at Buitenzorg, Java (now Kebun Raya Bogor), probably referring to the relatively large globose fruit.*

Solitary, small to moderate, unarmed, pleonanthic, monoecious palms. *Stem* erect, ringed with leaf scars, stilt roots conspicuously developed, forming a cone supporting the stem base. *Leaves* pinnate, neatly abscising; sheaths tubular, forming a conspicuous crownshaft, usually bearing a dense covering of ± caducous, floccose tomentum, the mouth lacking a ligule; petiole absent or short, ± rounded, variously clothed in scales and/or tomentum; rachis variously scaly or tomentose like the petiole; leaflets regularly arranged, single-fold, acute or acuminate (or minutely lobed in *Physokentia tete*) or blade irregularly divided into single- or several-fold, narrow to broad, ± sigmoid leaflets, the proximal acute or acuminate, the distal, including the apical pair, shallowly lobed, the lobes

corresponding to the adaxial ribs, blade adaxially glabrous or with minute, dot-like scales along the main veins, abaxially rather densely covered in minute, dot-like scales, the main ribs also bearing numerous conspicuous ramenta, transverse veinlets inconspicuous. *Inflorescences* solitary, infrafoliar, branching to 2 (rarely 3) orders proximally, to 1 order distally, apparently protandrous; peduncle short, winged at the base, narrow elliptic in cross-section; prophyll inserted near the base of the peduncle, open abaxially in bud, not completely encircling the peduncle at the insertion; peduncular bract inserted just above and exceeding the prophyll, completely encircling the peduncle, tubular, enclosing the inflorescence in bud, ± beaked, splitting abaxially, abscising with the prophyll at anthesis; rachis longer than the peduncle, but itself relatively short, bearing ca. 12–20, spirally arranged, first-order branches; rachis bracts inconspicuous; rachillae spreading, curved or ± pendulous, somewhat flexuous, ± angled, bearing spirally arranged triads of white to red flowers in the proximal ca. $^1/_3$, distally bearing solitary or paired staminate flowers, rarely rachillae bearing only staminate flowers; rachilla bracts prominent, rounded to acute, often ± reflexed; bracteoles membranous, rounded, or rarely (*P. dennisii*) each prolonged into a slender process, sometimes ciliate-margined. *Staminate flowers* briefly pedicellate, ± asymmetrical; sepals 3, distinct, imbricate, ± broad, triangular, keeled, the margins often coarsely toothed; petals 3, distinct or very briefly joined at the base, valvate; stamens 6, filaments slender, elongate, prominently inflexed at the apex in bud; anthers oblong-linear, medifixed, versatile, latrorse; pistillode conspicuous, elongate, conical or columnar, ± trifid. *Pollen* ellipsoidal asymmetric, occasionally oblate triangular; aperture a distal sulcus, less frequently a trichotomosulcus; ectexine tectate, perforate and micro-channelled or finely perforate-rugulate, aperture margin similar or slightly finer; infratectum columellate; longest axis 33–64 µm [5/8]. *Pistillate flowers* ± globular, sessile; sepals 3, distinct, imbricate; petals 3, distinct, broadly imbricate with short valvate triangular tips; staminodes 3, tooth-like; gynoecium ovoid, unilocular, uniovulate, stigmas 3, short, recurved, ovule laterally attached, hemianatropous. *Fruit* globose or subglobose, red or black at maturity, stigmatic remains eccentrically apical; perianth whorls persistent, epicarp smooth, or drying wrinkled, mesocarp fleshy with few flat fibres and numerous sclereids, easily separated from the endocarp, endocarp thin or thick, variously angled or ridged and sculptured or almost smooth and rather fragile (*P. avia*), usually with a prominent adaxial keel and a sharp to obtuse abaxial ridge, operculum rounded to 4-angled. *Seed* conforming to the endocarp shape, laterally attached with elongate, narrow hilum, raphe branches ± horizontal, loosely anastomosing, endosperm homogeneous or ruminate; embryo basal. *Germination* adjacent-ligular; eophyll bifid. *Cytology*: 2n = ca. 32.

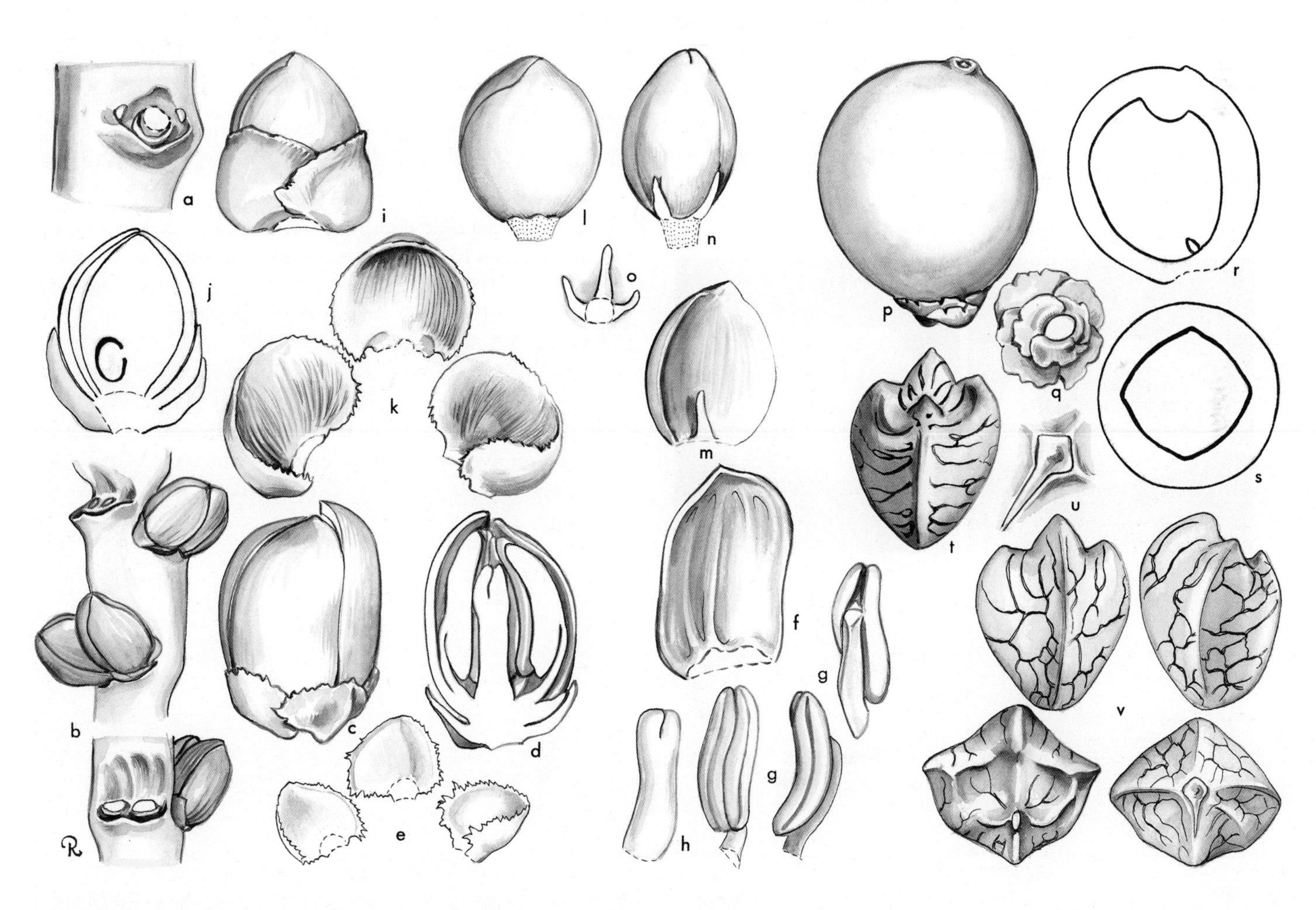

Physokentia. **a**, portion of rachilla with triad, flowers removed × 3; **b**, portion of rachilla with paired staminate flowers in 2 views × 3; **c**, staminate bud × 6; **d**, staminate bud in vertical section × 6; **e**, staminate sepals × 6; **f**, staminate petal, interior view × 6; **g**, stamen in 3 views × 6; **h**, pistillode × 6; **i**, pistillate bud × 6; **j**, pistillate bud in vertical section × 6; **k**, pistillate sepals × 6; **l**, pistillate bud, sepals removed × 6; **m**, pistillate petal and staminode, interior view × 6; **n**, gynoecium and staminodes × 6; **o**, staminodes × 6; **p**, fruit × 1$^1/_2$; **q**, fruiting perianth × 1$^1/_2$; **r**, fruit in vertical section × 1$^1/_2$; **s**, fruit in cross-section × 1$^1/_2$; **t**, endocarp × 1$^1/_2$; **u**, operculum × 3; **v**, seed in 4 views × 1$^1/_2$. *Physokentia thurstonii*: **b–h**, *Moore & Koroiveibau* 9347; **p–v**, *Moore & Koroiveibau* 9353; *P. rosea*: **a** and **i–o**, *Moore & Koroiveibau* 9363. (Drawn by Marion Ruff Sheehan)

Distribution of ***Physokentia***

Distribution and ecology: Seven species in the Fiji Islands, Vanuatu, Solomon Islands and New Britain; confined to undergrowth of rain forest at low to high elevations. *Physokentia whitmorei* is found in forest developed on ultrabasic rock, and *P. dennisii* has been recorded on soils overlying limestone.
Anatomy: Root (Seubert 1998a, 1998b) and fruit (Essig *et al.* 1999).

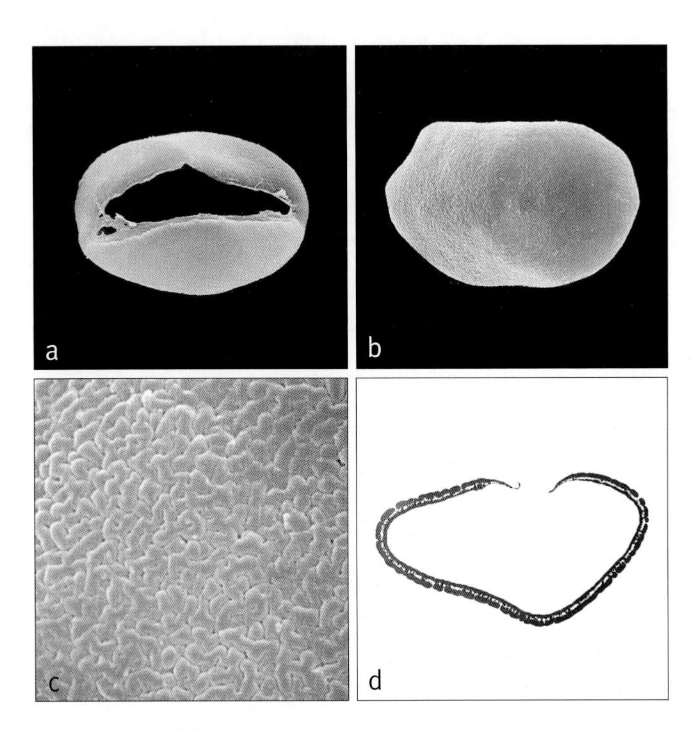

Above: ***Physokentia***
a, monosulcate pollen grain, distal face SEM × 1500;
b, monosulcate pollen grain, proximal face SEM × 1500;
c, close-up, finely perforate-rugulate pollen surface × 8000;
d, ultrathin section through whole pollen grain, polar plane TEM × 2000.
Physokentia dennisii: **a** & **b**, *Whitmore* BSIP4391; **c** & **d**, *Whitmore* 6050.

Relationships: *Physokentia* is strongly supported as monophyletic (Baker *et al.* in prep.) and has been resolved as sister to *Cyphosperma* with moderate support (Lewis & Doyle 2002, Asmussen *et al.* 2006). Weaker evidence resolves *Physokentia* within *Cyphophoenix* (Baker *et al.* in prep.).
Common names and uses: For common names, see Moore (1969d).
Taxonomic accounts: Moore (1969d); see also Fuller (1999).
Fossil record: No generic records found.
Notes: The incompletely encircling prophyll and highly sculptured endocarp resemble those of *Cyphosperma* and *Burretiokentia*. *Heterospathe longipes* also has a sculptured endocarp but the prophyll is complete.
Additional figures: Glossary fig. 24.

Subtribe Carpoxylinae J. Dransf., N.W. Uhl, C. Asmussen, W.J. Baker, M.M. Harley & C. Lewis., Kew Bull. 60: 563 (2005). Type: **Carpoxylon**.

Robust unarmed; leaves pinnate, the leaflet tips entire; sheaths forming a crownshaft; inflorescences infrafoliar, highly branched; prophyll much shorter than the peduncular bracts; enlarged peduncular bracts 1 or 2, completely enclosing the rest of the inflorescence in bud; flowers borne in triads near the base, solitary or paired staminate flowers distally; fruit with subapical stigmatic remains.

The three genera of this subtribe are geographically widely separated at opposite ends of the western Pacific Ocean, one in the Ryukyu Islands, one in Fiji and Vanuatu and the third in Vanuatu. They share similar inflorescence morphology and their association in a subtribe is supported strongly by phylogenetic studies.

Subtribe Carpoxylinae is frequently resolved as monophyletic, usually with high support (Lewis & Doyle 2002, Norup *et al.* 2006, Baker *et al.* in review, in prep.). The subtribe resolves within the western Pacific clade of Areceae, but its precise placement is uncertain (Baker *et al.* in review, in prep.).

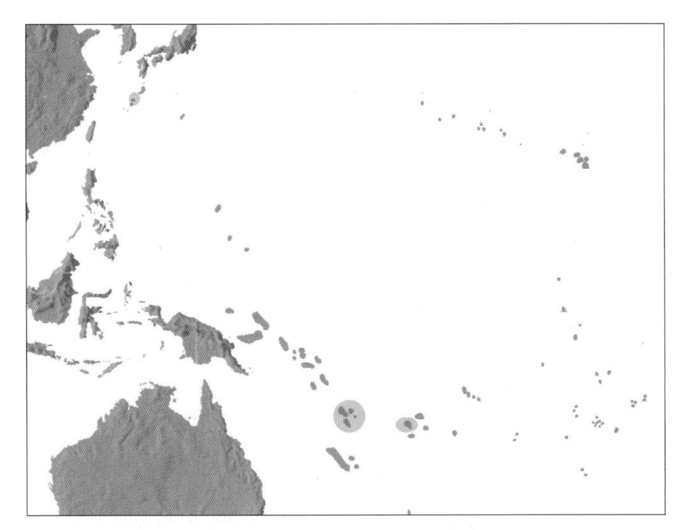

Distribution of subtribe **Carpoxylinae**

Key to Genera of Carpoxylinae

1. Inflorescence bearing a prophyll and two large completely enclosing peduncular bracts, further incompletely sheathing bracts also present; distal staminate flowers borne in horizontal pairs 2
1. Inflorescence bearing a prophyll and a single large completely enclosing peduncular bract; distal staminate flowers borne in vertical rather than horizontal pairs. Fiji and Vanuatu . . . . **141. Neoveitchia**
2. Rachillae glabrous; staminate flower strongly asymmetrical; fruit large, at least 5 cm long at maturity. Vanuatu . . . . **139. Carpoxylon**
2. Rachillae densely tomentose; staminate flower only slightly asymmetrical; fruit not exceeding 2 cm long at maturity. Ryukyu Islands . . . . **140. Satakentia**

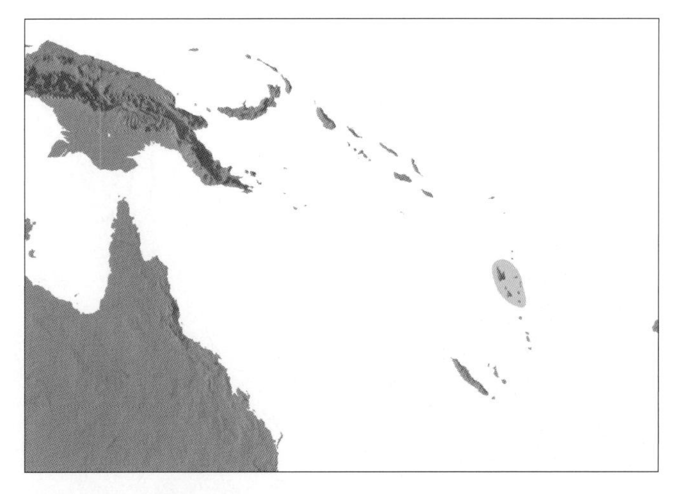

Distribution of ***Carpoxylon***

139. CARPOXYLON

Remarkable large pinnate-leaved palm recently rediscovered on Vanuatu after previously being known only from a fruit in the Natural History Museum in London; distinctive in the two large peduncular bracts and the large fruit with subapical stigmatic remains.

Carpoxylon H. Wendl. & Drude, Linnaea 39: 177 (1875).
Type: **C. macrospermum** H. Wendl. & Drude.

Karpos — *fruit*, xylon — *wood, referring to the woody endocarp.*

Moderate, solitary, unarmed, pleonanthic, monoecious palm. *Stem* erect, longitudinally fissured, swollen basally and with a boss of adventitious roots, distally prominently ringed with slightly sunken leaf scars, internodes short. *Leaves* regularly pinnate, spreading but arched towards the tips, neatly abscising; sheaths forming a crownshaft, sheaths glossy, glabrous to lightly scaly, splitting opposite the petiole; petiole short, tapering distally, ridged adaxially, rounded abaxially; rachis flexible, broadly ridged adaxially at base, narrowly ridged distally, rounded abaxially, extended beyond the apical leaflets in a flexible tip; leaflets regularly arranged, single-fold, erect, linear, tapering to an irregularly rounded, ± bifid tip, stiff, coriaceous, glabrous adaxially, midveins most prominent, marginal veins next largest, 2 other pairs of large veins conspicuous, transverse veinlets not evident. *Inflorescences* infrafoliar branched to 3 orders basally, to 1 order distally, branches stiffly spreading; peduncle short, stout, elliptical in cross-section; prophyll completely encircling peduncle at the base, tubular, 2-keeled, tapering distally, splitting abaxially, tomentose; peduncular bracts 2, longer than the prophyll, the first inserted just above the prophyll, the second an equal distance above the first, both tubular, complete, tapering to rather short pointed tips, glabrous, caducous; scars of 2–3 incomplete bracts above inner peduncular bract; rachis about twice as a long as the peduncle, rachis bracts low, each in a slit-like cavity, subtending primary branches; primary branches ca. 10, dorsiventrally flattened, with a short bare basal portion and 2 large lateral pulvini at the base, and distally bearing low bracts each in a slit-like cavity and subtending rachillae; rachillae angled, basally devoid of flowers, distally tapering and bearing spirally arranged low bracts subtending triads in the basal $^{1}/_{3}$ and distally subtending paired or solitary staminate flowers, within the triads one staminate flower often distal and one lateral to the pistillate flower, rachillae ending in a short bare portion; first bracteole large, rounded, coriaceous, the second smaller and shallower. *Staminate flowers* very asymmetrical in bud, rounded or pointed apically; sepals 3, distinct, irregular, imbricate basally, keeled, prominently ridged when dry; petals 3, distinct, valvate, tips thickened, ridged when dry; stamens 6, filaments slender, inflexed at tip, anthers ± sagittate basally, slightly bifid apically, dorsifixed just below the middle, versatile, latrorse, connective tanniniferous; pistillode columnar, as long as the stamens. *Pollen* ellipsoidal asymmetric or oblate triangular, occasionally elongate; aperture a distal sulcus or trichotomosulcus; ectexine tectate, perforate and micro-channelled or finely perforate-rugulate, aperture margin similar or slightly finer; infratectum columellate; longest axis ranging from 48–58 µm [1/1]. *Pistillate flower* in young bud rounded; sepals 3, distinct, very broadly imbricate, extremely thick basally; petals 3, very broadly imbricate, thick basally, tips thick, valvate; staminodes joined in a ring with ca. 5 broad tooth-like tips; gynoecium irregularly obovoid,

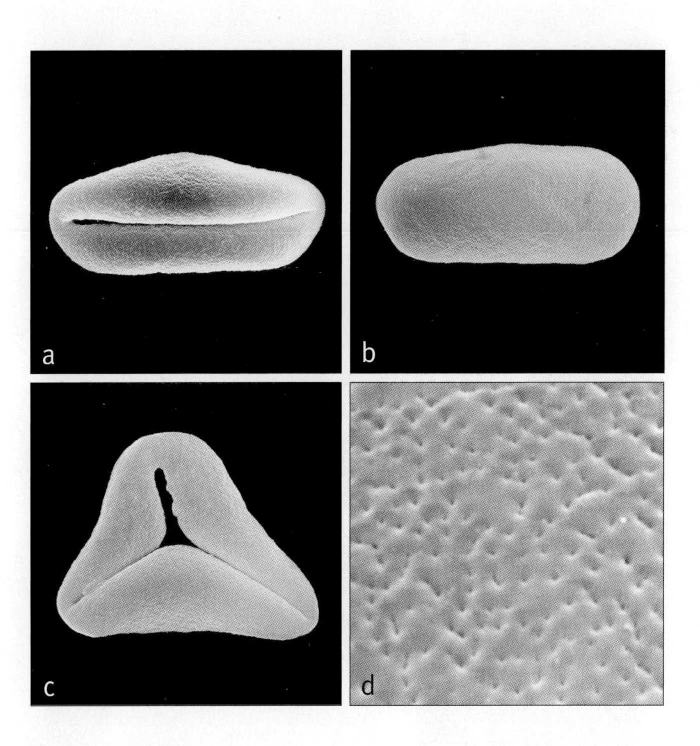

Above: ***Carpoxylon***
a, monosulcate pollen grain, distal face SEM × 1000;
b, monosulcate pollen grain, proximal face SEM × 1000;
c, trichotomosulcate pollen grain, distal face SEM × 1000;
d, close-up, finely perforate pollen surface SEM × 8000.
Carpoxylon macrospermum: **a–d**, *Dowe* 001.

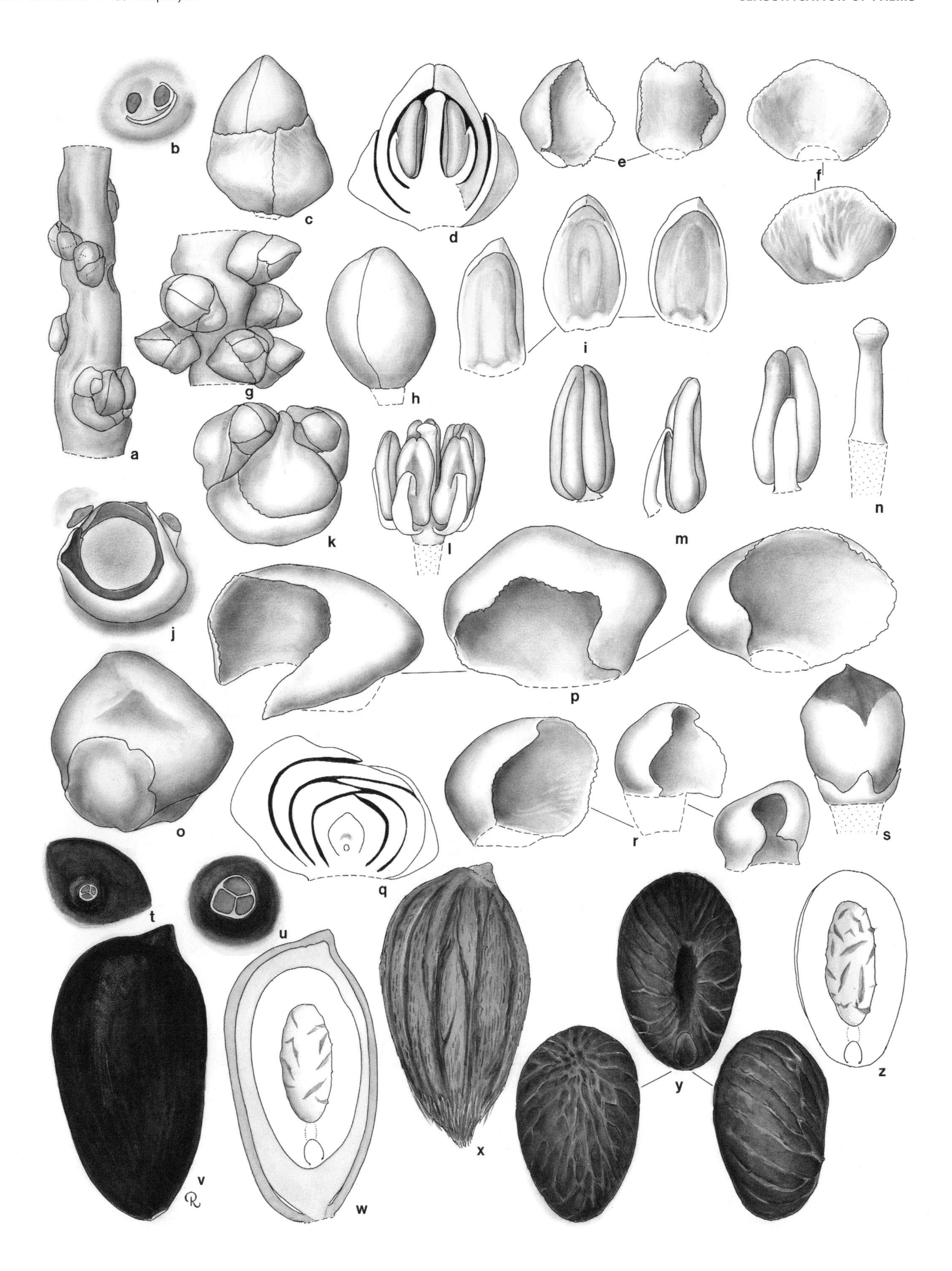
a
b
c
d
e
f
g
h
i
j
k
l
m
n
o
p
q
r
s
t
u
v
w
x
y
z

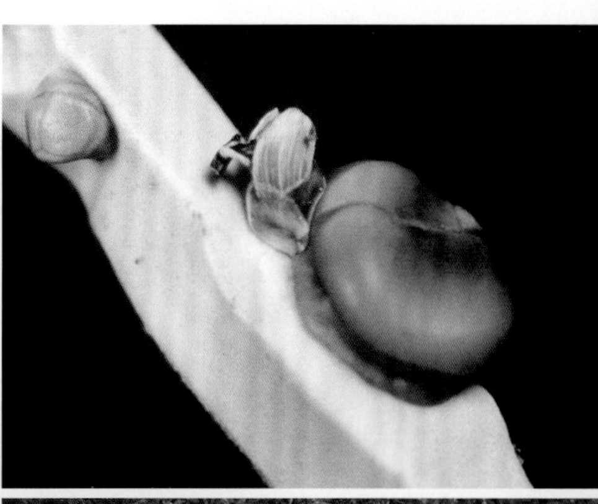

unilocular, uniovulate, stigmas 3, fleshy, ovule erect at stage studied, ?anatropous. *Fruit* large, obovoid to ellipsoid, somewhat asymmetrical, red at maturity single-seeded, with eccentrically apical, conical stigmatic remains; epicarp smooth, wrinkled basally at maturity, thin, mesocarp thick with close-packed longitudinal fibres, the innermost fibres adherent to the thick, whitish, bony, longitudinally ridged endocarp. *Seed* ± oblong-ovoid, flattened at the raphe, hilum impressed, ± subbasal, raphe branches numerous, radiating, diverging, many ascending, all anastomosing, endosperm homogeneous with a central hollow; embryo basal. *Germination* adjacent-ligular, eophyll bifid. *Cytology*: not studied.

Above left: ***Carpoxylon macrospermum***, habit, Vanuatu. (Photo: J.L. Dowe)

Above middle: ***Carpoxylon macrospermum***, crown with inflorescences, Vanuatu. (Photo: J.L. Dowe)

Top right: ***Carpoxylon macrospermum***, detail of triad, Vanuatu. (Photo: J.L. Dowe)

Bottom right: ***Carpoxylon macrospermum***, infructescence, Vanuatu. (Photo: J.L. Dowe)

Distribution and ecology: One species in Vanuatu, rediscovered in 1987, occurring in rain forest at low elevation on Aneityum, Tanna and Futuna. It can also be found cultivated in villages throughout the islands of Vanuatu.

Anatomy: Not studied.

Relationships: Several studies place *Carpoxylon* as sister to a clade of *Neoveitchia* and *Satakentia*, all relationships with high support (Norup *et al.* 2006, Baker *et al.* in review, in prep.). However, moderate support has also been obtained for a sister relationship between *Carpoxylon* and *Satakentia* (Lewis & Doyle 2002, Asmussen *et al.* 2006).

Common names and Uses: For common names, see Dowe and Cabalion (1996).

Taxonomic account: Dowe & Uhl (1989).

Fossil record: No generic records found.

Notes: The rediscovery of *Carpoxylon*, previously known only from a single fruit, was one of the most exciting palm discoveries of the late 1980s.

Opposite: ***Carpoxylon***. **a**, portion of rachilla in bud × 2 $^1/_2$; **b**, scars of staminate dyad × 6; **c**, staminate bud × 12; **d**, staminate bud in vertical section × 12; **e**, staminate sepal in 2 views × 12; **f**, staminate sepal in 2 views × 12; **g**, portion of rachilla with staminate buds × 5; **h**, staminate bud, sepals removed × 12; **i**, staminate petals, interior views × 12; **j**, triad scar × 5; **k**, triad × 5; **l**, androecium × 12; **m**, stamens, 3 views × 12; **n**, pistillode × 12; **o**, young pistillate bud × 12; **p**, pistillate sepals × 12; **q**, pistillate bud in vertical section × 12; **r**, pistillate petals × 12; **s**, gynoecium and staminodes × 12; **t**, end of stigma × 2; **u**, end of stigma enlarged × 6; **v**, fruit × 1 $^1/_3$; **w**, fruit in vertical section × 1 $^1/_3$; **x**, endocarp × 1 $^1/_3$; **y**, seed in 3 views × 1 $^1/_3$; **z**, seed in vertical section × 1 $^1/_3$. *Carpoxylon macrospermum*: all from *Dowe* 30. (Drawn by Marion Ruff Sheehan)

140. SATAKENTIA

Moderate solitary pinnate-leaved palm from the Ryukyu Islands in southern Japan, remarkable for the two large peduncular bracts and small fruit.

Satakentia H.E. Moore, Principes 13: 5 (1969). Type: **S. liukiuensis** (Hatusima) H.E. Moore (*Gulubia liukiuensis* Hatusima).

Honoring Toshihiko Satake (1910–1998), Japanese industrialist and palm hobbyist, by combining his name with the generic name Kentia, *named for William Kent (1779–1827), one-time curator of the botanic gardens at Buitenzorg, Java (now Kebun Raya Bogor).*

Moderate, solitary, unarmed, pleonanthic, monoecious palm. *Stem* erect, usually enlarged and with a mass of adventitious roots at the base, columnar above, green to brown, longitudinally striate, ringed with close leaf scars. *Leaves* pinnate, spreading; sheaths tubular, forming a prominent crownshaft and with a prominent chartaceous ligule; petiole short, adaxially channelled with a central ridge, abaxially rounded; rachis elongate, flattened adaxially, rounded abaxially, tomentose; leaflets regularly arranged, acute, single-fold, midrib evident abaxially, marginal nerves thickened, usually 2(–3) secondary ribs, and numerous tertiary veins on each side, glabrous adaxially, ramenta present abaxially near the base of the midrib, transverse veinlets not evident. *Inflorescences* infrafoliar, densely and minutely stellate-tometose, branched to 2 orders basally, to 1 order distally; peduncle short, stout; prophyll tubular, terete, 2-keeled laterally, briefly beaked, much shorter than the peduncular bracts; first peduncular bract, complete, tubular, thick, woody, terete, beaked,

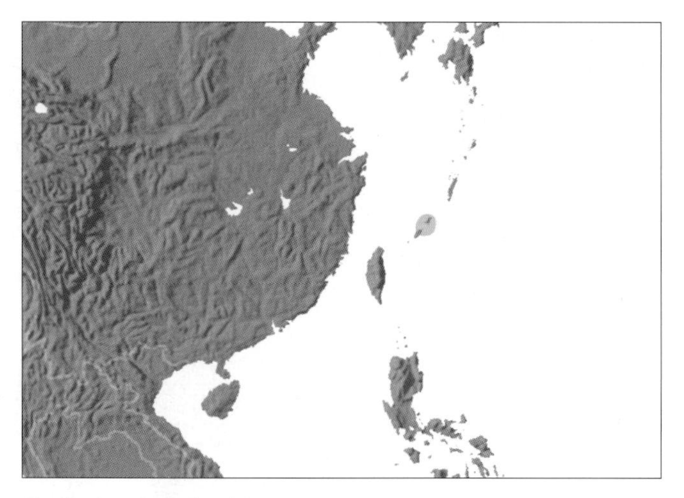

Distribution of ***Satakentia***

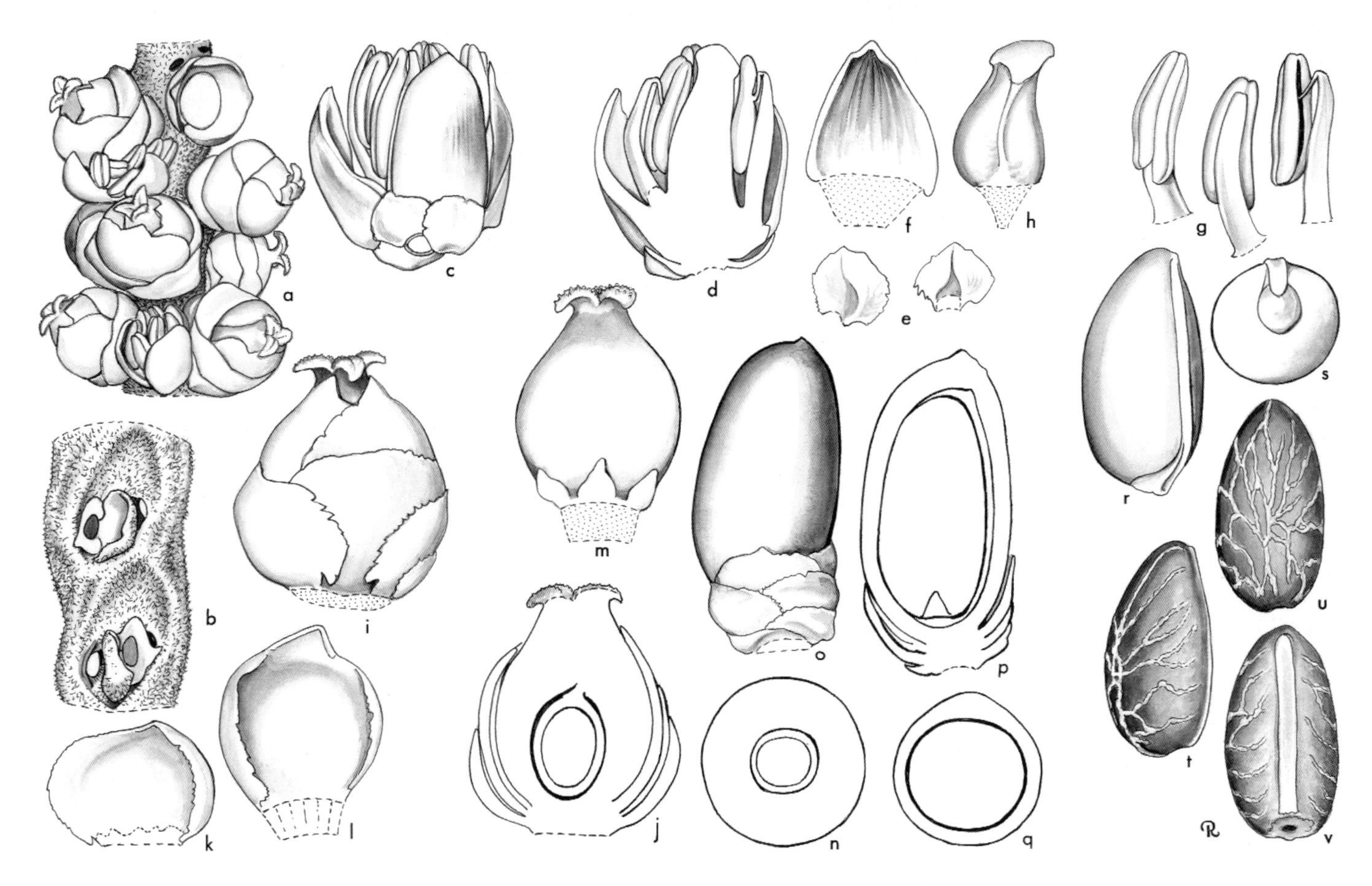

Satakentia**.** **a**, portion of rachilla with staminate and pistillate flowers × 3; **b**, triads, flowers removed to show bracteoles × 3; **c**, staminate flower × 6; **d**, staminate flower in vertical section × 6; **e**, staminate sepals, exterior and interior views × 6; **f**, staminate petal, interior view × 6; **g**, stamen in 3 views × 6; **h**, pistillode × 6; **i**, pistillate flower × 6; **j**, pistillate flower in vertical section × 6; **k**, pistillate sepal, interior view × 6; **l**, pistillate petal, interior view × 6; **m**, gynoecium and staminodes × 6; **n**, gynoecium in cross-section × 6; **o**, fruit × 3; **p**, fruit in vertical section × 3; **q**, fruit in cross-section × 3; **r**, endocarp with operculum × 3; **s**, operculum × 3; **t**, **u**, **v**, seed in 3 views × 3. *Satakentia liukiuensis*: **a** and **i–n**, *Murata* s.n.; **b–h**, *Moore et al.* 9382; **o–v**, *Yamakawa* s.n. (Drawn by Marion Ruff Sheehan)

Above left: ***Satakentia liukiuensis***, habit, cultivated, Fairchild Tropical Botanic Garden, Florida. (Photo: C.E. Lewis)

Above middle: ***Satakentia liukiuensis***, crown with inflorescence in bud, cultivated, Fairchild Tropical Botanic Garden, Florida. (Photo: J. Dransfield)

Top right: ***Satakentia liukiuensis***, inflorescence, cultivated, Fairchild Tropical Botanic Garden, Florida. (Photo: C.E. Lewis)

Bottom right: ***Satakentia liukiuensis***, rachilla at staminate anthesis, cultivated, Fairchild Tropical Botanic Garden, Florida. (Photo: C.E. Lewis)

enclosing a second almost complete and similar peduncular bract, both splitting abaxially and caducous at anthesis, a prominent but much shorter third and sometimes fourth, chartaceous incomplete peduncular bract also developed; rachis about as long as the peduncle, tapering, densely tomentose, angled, bearing spirally inserted, rather large, acute bracts subtending basal branches and smaller rounded bracts subtending distal branches; rachillae elongate, rather stout, stiff, bearing spirally arranged, low, rounded bracts subtending flowers borne in triads of 2 staminate and 1 pistillate in lower $^1/_4$ to $^1/_3$ of the rachillae, paired to solitary staminate flowers distally. *Staminate flowers* nearly symmetrical; sepals 3, distinct, imbricate, ± rounded; petals 3, distinct, valvate, more than twice as long as the sepals; stamens 6, filaments distinct, awl-shaped, inflexed at the apex in bud, anthers oblong in outline, latrorse; pistillode as long as the stamens, cylindrical, with obliquely subcapitate apex. *Pollen* ellipsoidal asymmetric; aperture a distal sulcus; ectexine tectate, perforate, aperture margin similar; infratectum columellate; longest axis 43–45 µm [1/1]. *Pistillate flowers* ovoid; sepals 3, distinct, broadly imbricate; petals 3, distinct, imbricate, with shortly valvate apices; staminodes 3, tooth-like, on one side of the gynoecium; gynoecium ovoid, unilocular, uniovulate, stigmas 3, recurved at anthesis, ovule pendulous, anatropous. *Fruit* ovoid-ellipsoidal with eccentrically apical stigmatic remains; epicarp smooth but drying longitudinally lines, mesocarp with numerous flat longitudinal fibres in thin flesh and some red-brown stone cells near the apex, endocarp thin, fragile, operculate at the base of the elongate hilar seam, not adherent to the seed. *Seed* ellipsoidal, hilum elongate, raphe branches anastomosing, endosperm homogeneous; embryo basal. *Germination* adjacent-ligular; eophyll bifid. *Cytology* not studied.

Distribution and ecology: A single species on Ishigaki Island (Yonehara) and Iriomote Island (Hoshitate, Nakam River, Sonai, and Yoeyama Group of the Ryukyus), growing on hill slopes or more rarely near the sea; often growing in dense more-or-less even-aged stands.

Anatomy: Fruit (Essig *et al.* 1999).

Relationships: For relationships, see *Carpoxylon*.

Common names and uses: *Noyashi* and *yaeyama-yashi*. Cultivated as an ornamental. The 'cabbage' is said to have been eaten during World War II.

Taxonomic accounts: Moore (1969a); see also Pintaud & Setoguchi (1999).

Fossil record: No generic records found.

Notes: Pintaud and Setoguchi (1999) were the first to recognise that the inflorescence of *Satakentia* has two peduncular bracts, a character it shares with *Carpoxylon* but not with *Neoveitchia*. However, the inflorescences and fruit of the three genera are similar.

Additional figures: Glossary fig. 25.

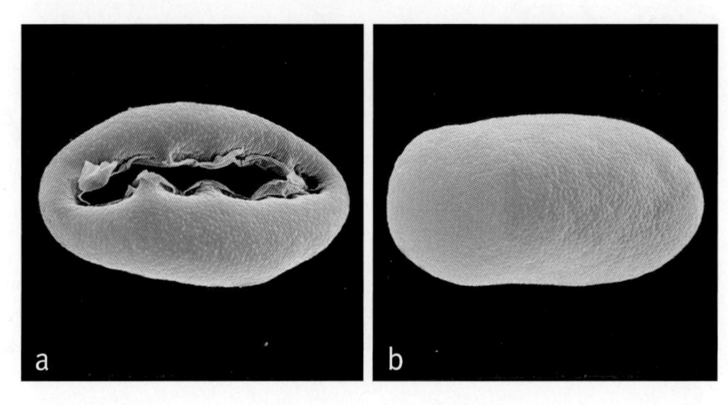

Below: ***Satakentia***

a, monosulcate pollen grain, distal face SEM × 1000;

b, monosulcate pollen grain, proximal face SEM × 1000.

Satakentia liukiuensis: **a** & **b**, *Moore 9382*.

141. NEOVEITCHIA

Moderate solitary pinnate-leaved palms from Fiji and Vanuatu, with a single large peduncular bracts and curious vertical rather than horizontal arrangement of the staminate flowers on the rachilla.

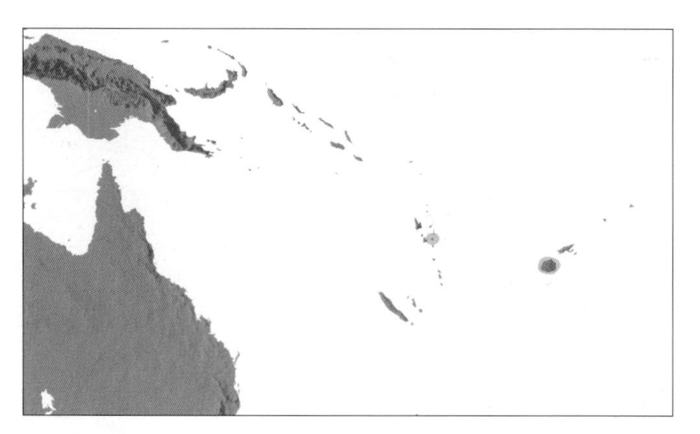

Distribution of ***Neoveitchia***

Neoveitchia Becc., Palme Nuova Caledonia 9 (1920); and Webbia 5: 77 (1921). Type: **N. storckii** (H. Wendl.) Becc. (*Veitchia storckii* H. Wendl.).

Neo — *new and* Veitchia*, the palm genus;* Neoveitchia storckii *was first described under* Veitchia *but then moved to the new genus,* Neoveitchia.

Moderate, solitary, unarmed, pleonanthic, monoecious tree palms. *Stem* erect, becoming bare, conspicuously ringed with leaf scars. *Leaves* pinnate, neatly abscising; leaf sheath ± split opposite the petiole, not forming a well-developed crownshaft, bearing scattered scales, the mouths irregularly fibrous; petiole flattened or shallowly channelled adaxially, abaxially rounded, bearing scattered caducous scales; rachis ± curved, densely minutely dotted abaxially, deciduously floccose adaxially; leaflets numerous, regularly arranged, single-fold, acute, with conspicuously thickened margins, adaxially glabrous, abaxially with few ramenta along the midrib near the base, transverse veinlets not visible. *Inflorescences* infrafoliar, branching to 3 orders, apparently protandrous; peduncle winged at the base, somewhat swollen, rapidly tapering, distally ± circular in cross-section; prophyll inserted near the base of the peduncle, tubular, flattened, strongly 2-keeled, splitting apically long before anthesis while still enclosed within the leaf sheath, allowing the peduncular bract to protrude, eventually deciduous; peduncular bract about twice as long as the prophyll, beaked, tubular, entirely enclosing the inflorescence, splitting longitudinally at anthesis, then deciduous, 1 (or 2) smaller, incomplete, ± triangular, peduncular bracts also present; rachis ± the same length as the peduncle; rachis bracts spirally arranged, small, low, triangular; first-order branches with a basal bare portion; rachillae rather

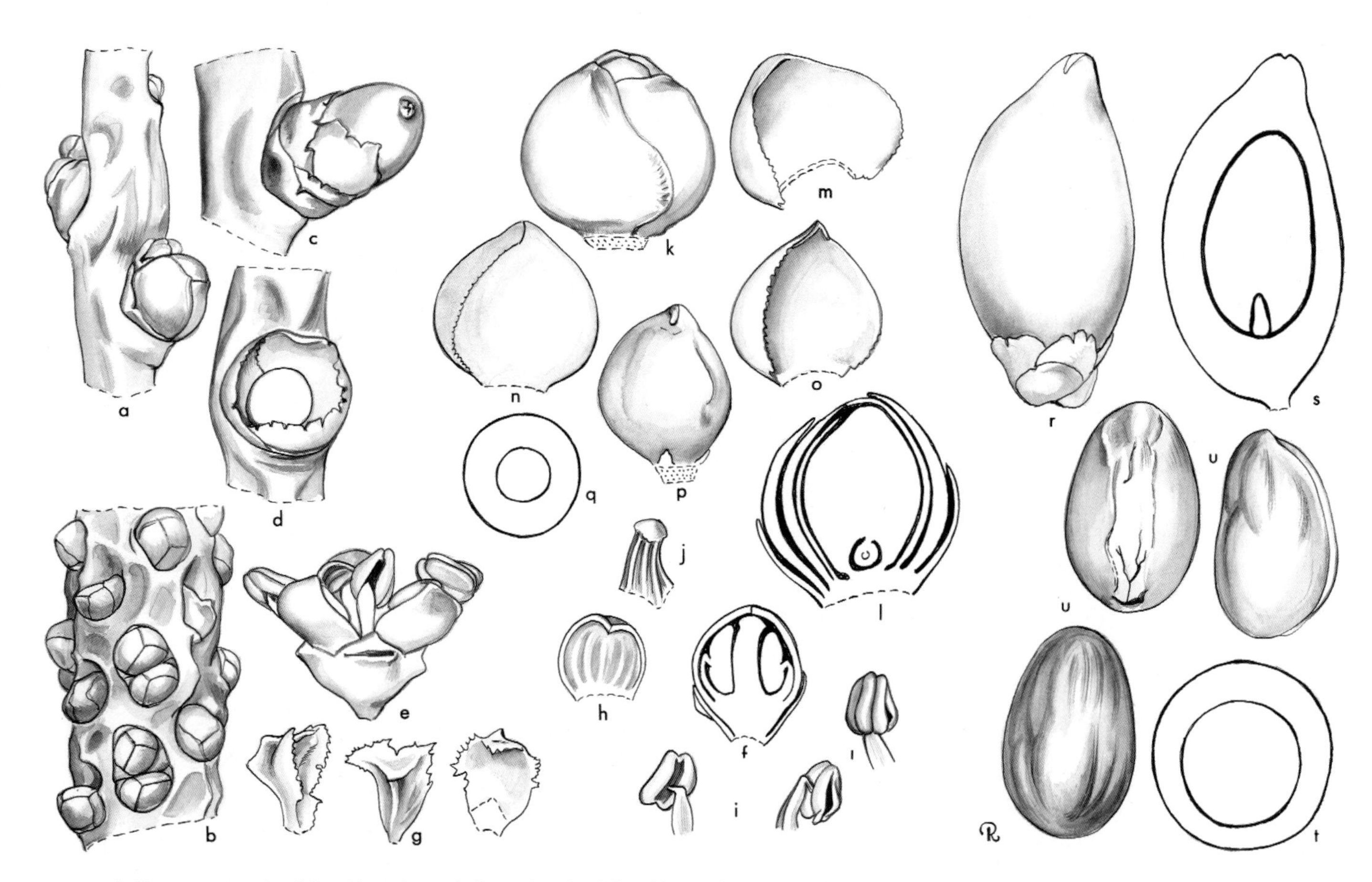

Neoveitchia. **a,** portion of rachilla with triads × 1$^1/_2$; **b,** portion of rachilla with paired, superposed, staminate buds × 3; **c,** portion of rachilla with very young fruit × 1$^1/_2$; **d,** bracteoles subtending pistillate flower × 1$^1/_2$; **e,** staminate flower, expanded × 6; **f,** staminate bud in vertical section × 6; **g,** staminate sepals × 6; **h,** staminate petal, interior view × 6; **i,** stamen in 3 views × 6; **j,** pistillode × 6; **k,** pistillate bud × 3; **l,** pistillate bud in vertical section × 3; **m,** pistillate sepal × 3; **n,** pistillate bud, sepals removed × 3; **o,** pistillate petal × 3; **p,** gynoecium and staminodes × 3; **q,** ovary in cross-section × 6; **r,** fruit × 1; **s,** fruit in vertical section × 1; **t,** fruit in cross-section × 1; **u,** seed in 3 views × 1. *Neoveitchia storckii*: all from *Moore & Koroiveibau* 9360. (Drawn by Marion Ruff Sheehan)

stiff, curved, eventually pendulous, somewhat angled, basally bearing rather few, distant, ± superficial triads, the 2 staminate flowers ± distal (rather than lateral) to the pistillate flower, distally the rachilla bearing ± 7 vertical series of floral pits, each partially enclosing a vertical (rather than lateral) pair of staminate flowers, the staminate portion of the rachilla deciduous at fruiting stage; floral bracteoles small, included within the pits. *Staminate flowers* asymmetrical, those originating from triads with flattened pedicels, those from staminate pairs ± sessile; sepals 3, strongly keeled or winged, distinct, strongly imbricate, or frequently 1 distinct, and 2 joined for $^3/_4$ their length into a 2-keeled prophyll-like structure, or all 3 joined but retaining 2 distinct imbricate margins, the margins coarsely toothed; petals 3, distinct, triangular-ovate, valvate, smooth; stamens 6, filaments distinct, rather short, narrow, fleshy, inflexed in bud, anthers short, rectangular, medifixed, ± versatile, latrorse, the connective broad, conspicuous; pistillode columnar, striate, truncate, as long as the stamens. *Pollen* ellipsoidal asymmetric, occasionally lozenge-shaped; aperture a distal sulcus; ectexine tectate, perforate and micro-channelled or finely perforate-rugulate, aperture margin similar or slightly finer; infratectum columellate; longest axis 46–58 µm [1/2]. Pistillate flowers very much larger than the staminate, ± spherical; sepals 3, distinct, broadly imbricate; petals 3, distinct, broadly imbricate except for the minute, triangular, valvate tips; staminodes 3, very small, triangular, tooth-like; gynoecium unilocular, uniovulate, ovoid, stigmas 3, very small, reflexed, ovule laterally attached, form unknown. *Fruit* ellipsoidal, rather large, becoming reddish-yellow at maturity, perianth whorls persistent, stigmatic remains apical; epicarp smooth, mesocarp fleshy with abundant close longitudinal fibres, endocarp thin, bony, closely adhering to the seed. *Seed* laterally attached, hilum elongate, running ± the length of the seed, raphe branches abundant, anastomosing, endosperm homogeneous; embryo basal. *Germination* and eophyll not recorded. *Cytology* not studied.

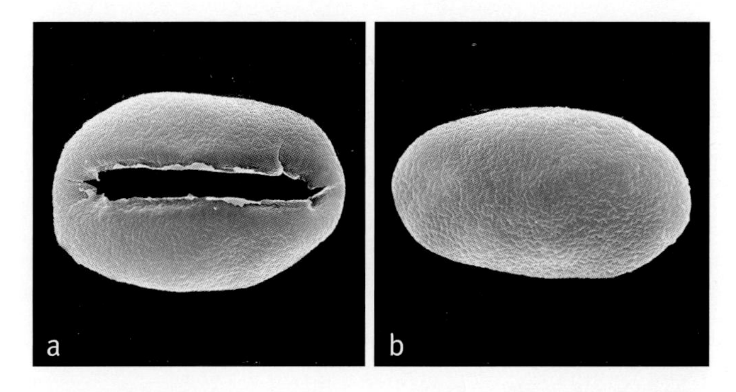

Above: ***Neoveitchia***
a, monosulcate pollen grain, distal face SEM × 1000;
b, monosulcate pollen grain, proximal face SEM × 1000.
Neoveitchia storckii: **a** & **b**, *Melville et al.* 71/948.

Distribution and ecology: Two species: *Neoveitchia storckii* is confined to a very limited area in Naitasiri Province, Viti Levu, Fiji Islands, where it is in danger of extinction, and *N. brunnea* restricted to Vanuatu. *Neoveitchia storckii* is a canopy emergent in secondary forest on alluvial plains and nearby foothills; much of its habitat has already been converted for agriculture or forestry plantation (Watling 2005). In Vanuatu, *N. brunnea* is found in rain forest in high rainfall areas, on red clay soil at about 300 m elevation (Dowe & Cabalion 1996).

Anatomy: Root (Seubert 1998a, 1998b), and fruit (Essig *et al.* 1999).

Relationships: The monophyly of *Neoveitchia* has not been tested. For relationships, see *Carpoxylon*.

Common names and uses: *Vileito*, in Fiji, where trunks have been used as posts in house construction; the young fruit is eaten.

Taxonomic accounts: Moore (1957a, 1979) and Dowe & Cabalion (1996).

Fossil record: No generic records found.

Notes: The arrangement of the staminate flowers, one above the other, is unique.

Subtribe Clinospermatinae J. Dransf., N.W. Uhl, C. Asmussen, W.J. Baker, M.M. Harley & C. Lewis., Kew Bull. 60: 563 (2005). Type: **Clinosperma**.

Moderate, erect, unarmed; leaves pinnate, the leaflet tips entire; sheaths usually forming a well-defined crownshaft; inflorescences infrafoliar, rarely interfoliar at anthesis but becoming infrafoliar, usually highly branched; prophyll completely tubular; peduncular bracts usually 2; flowers borne in triads near the base, solitary or paired staminate flowers distally; anthers usually didymous; fruit with lateral or basal stigmatic remains; endocarp not ornamented except in *Clinosperma macrocarpa* where ornamented; epicarp smooth.

This subtribe is one of three subtribes of Areceae present in New Caledonia; unlike Archontophoenicinae and Basseliniinae, the Clinospermatinae is found nowhere outside New Caldeonia.

Left: ***Neoveitchia storckii***, habit, cultivated, Fairchild Tropical Botanic Garden, Florida. (Photo: J. Dransfield)

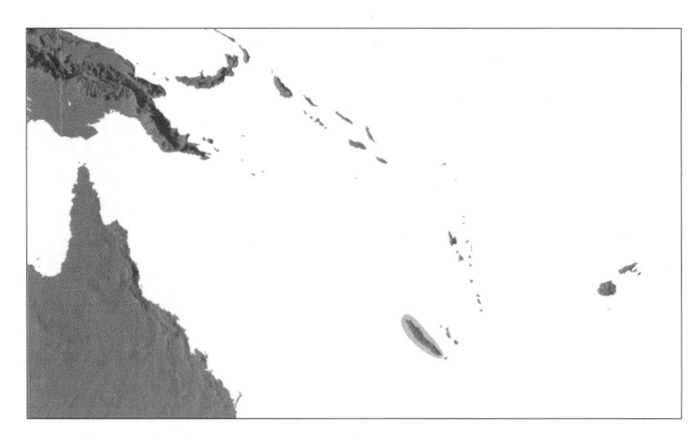

Distribution of subtribe **Clinospermatinae**

Subtribe Clinospermatinae is a robust monophyletic group that is often resolved with high support (Asmussen *et al*. 2006, Norup *et al*. 2006, Baker *et al*. in review, in prep.). Limited molecular evidence suggests a relationship with Linospadicinae, *Dransfieldia* and *Heterospathe* within the western Pacific clade of Areceae (Asmussen *et al*. 2006, Baker *et al*. in review, in prep.).

Key to Genera of Clinospermatinae

1. Stamens 6, didymous; pistillode squat, capitate. New Caledonia . . . . **143. Clinosperma**
1. Stamens 6–12, elongate; pistillode columnar. New Caledonia . . . . **142. Cyphokentia**

142. CYPHOKENTIA

Moderate solitary pinnate-leaved palms with striking white-waxy crownshafts, endemic to New Caledonia; the staminate flowers have elongate anthers.

Cyphokentia Brongn., Compt. Rend. Hebd. Séances Acad. Sci. 77: 399 (1873). Lectotype: **C. macrostachya** Brongn. (see Beccari 1920).

Dolichokentia Becc., Palme Nuova Caledonia 45 (1920); and Webbia 5: 113 (1921). Type: *D. robusta* (Brongn.) Becc. (*Cyphokentia robusta* Brongn. = *C. macrostachya* Brongn.).

Moratia H.E. Moore, Gentes Herb. 12: 18 (1980). Type: *M. cerifera* H.E. Moore (= *Cyphokentia cerifera* [H.E. Moore] Pintaud & W.J. Baker)

Combining kyphos — *hump, with the generic name* Kentia, *named for William Kent (1779–1827), one-time curator of the botanic gardens at Buitenzorg, Java (now Kebun Raya Bogor), referring to the lateral to subbasal stigmatic remains.*

Moderate, solitary (or exceptionally with 2–3 stems), unarmed, pleonanthic, monoecious palms. *Stem* erect, faintly to prominently ringed, enlarged at the base, yellow or whitish, brown-spotted. *Leaves* regularly pinnate, gracefully spreading; sheaths tubular, forming a prominent, white-waxy crownshaft; petiole short, concave adaxially, rounded abaxially, glabrous or brown scaly; rachis concave adaxially near base, becoming angled distally, rounded abaxially, often somewhat curved, glabrous or scaly as the petiole; leaflets regularly arranged, acute, slightly arched from rachis, single-fold, stiff, waxy on both surfaces but particularly so abaxially, midrib and a lateral vein on each side elevated and prominent adaxially, secondary veins numerous, large ribs clothed basally with ramenta, transverse veinlets not evident. *Inflorescences* infrafoliar, branched to 2–3 orders basally and fewer distally, branches pendulous, protandrous; peduncle very short, stout, ± recurved; prophyll wide, tubular, broadly 2-keeled laterally, completely encircling the peduncle and enclosing the peduncular bract; first peduncular bract similar, briefly beaked, both it and the prophyll caducous, a second peduncular bract usually present; rachis longer than the peduncle, bearing very low and rounded or acute to scarcely evident bracts subtending branches and rachillae; rachillae long, slender, variously compressed in bud, tapering, pendulous, distant, bearing low, rounded, lip-like bracts subtending flowers in triads nearly throughout the rachillae, and paired or solitary staminate flowers distally; bracteoles surrounding pistillate flowers sepal-like, imbricate, about as high as bracts of the triad. *Staminate buds* ± symmetrical; sepals 3, distinct, slightly imbricate basally, small, rounded, gibbous at base and keeled dorsally; petals 3, distinct, valvate, broadly ovate in outline or subacute, not markedly fibrous, grooved on inner surface; stamens 6–12, filaments incurved, awl-shaped, briefly inflexed at the apex in bud, about as long as the anthers, anthers oblong in outline, dorsifixed, briefly bifid at base and apex, latrorse; pistillode as long as the anthers in bud, somewhat 3-angled and expanded into a 3-angled, flat apex. *Pollen* ellipsoidal asymmetric, occasionally lozenge-shaped; aperture a distal sulcus; ectexine tectate,

Right: ***Cyphokentia cerifera***, habit, New Caledonia. (Photo: W.J. Baker)

perforate and micro-channelled or finely perforate-rugulate, aperture margin similar or slightly finer; infratectum columellate; longest axis 37–56 µm [2/2]. *Pistillate flowers* not seen at anthesis; sepals 3, distinct, rounded, imbricate basally in fruit; petals 3, distinct, imbricate except briefly for the valvate tips, distinctly tanniniferous, nerved in fruit; staminodes 3–5, rarely 6, tooth-like; gynoecium ovoid, unilocular, uniovulate, ovule pendulous, probably hemianatropous. *Fruit* depressed globose to ellipsoidal, dull orange to red at maturity, with stigmatic remains at or below the middle; epicarp smooth or minutely roughened, drying somewhat wrinkled but not regularly pebbled, thin, not tanniniferous, overlying a shell of very short sclereids external to a fleshy layer of mesocarp, mesocarp with longitudinal flat fibres, and a tanniniferous layer adjacent to the endocarp, endocarp vitreous, rather thick, with a round or spathulate, basal operculum and lateral beak. *Seed* subglobose to nearly ellipsoidal, hilum near the base, round, raphe branches ascending from the base, endosperm homogeneous; embryo basal. *Germination* adjacent-ligular; eophyll bifid. *Cytology* not known.

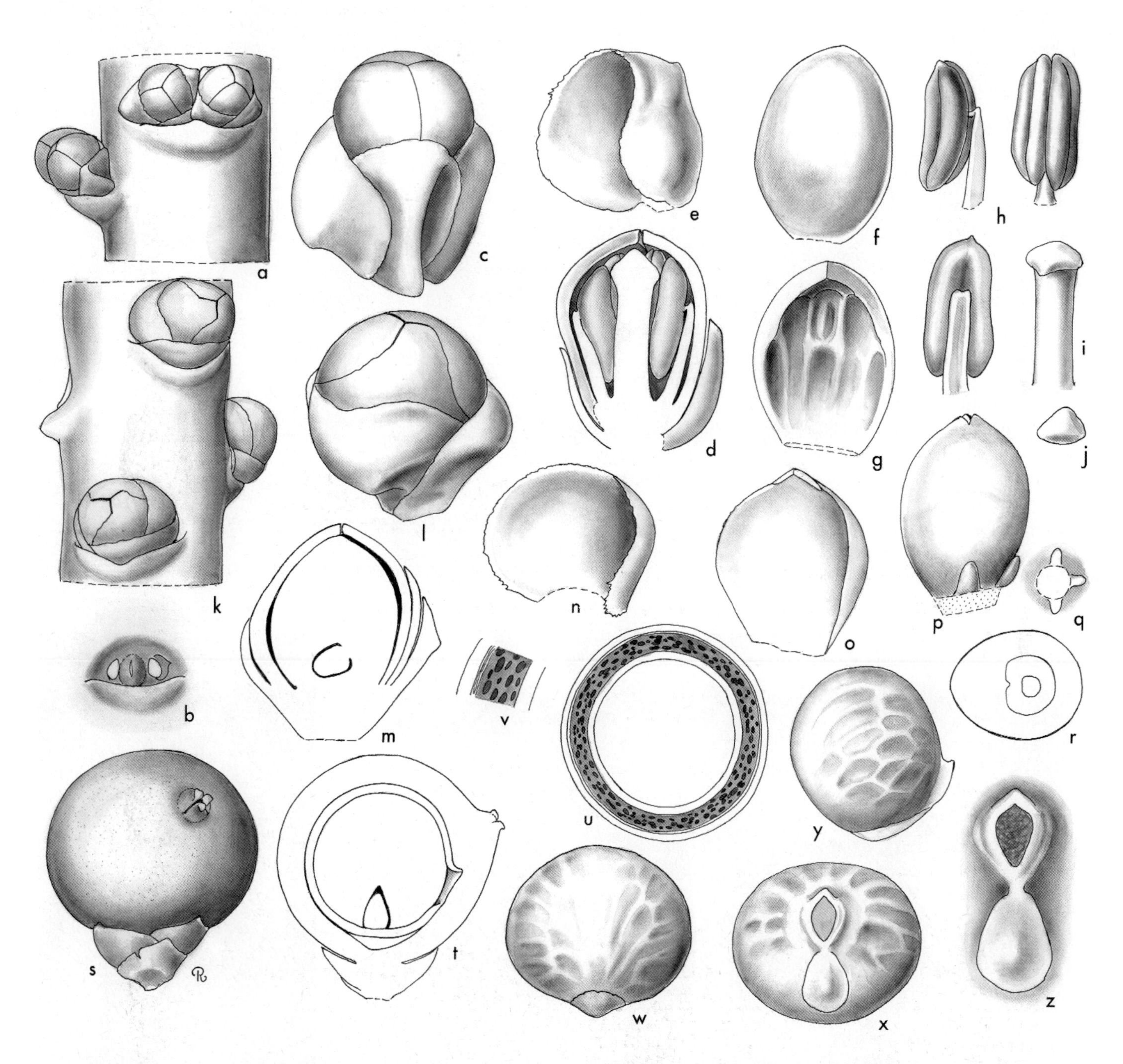

Cyphokentia. **a**, portion of rachilla in staminate bud × 4½; **b**, triad, staminate flowers removed to show bracteoles around developing pistillate bud × 4½; **c**, staminate bud × 12; **d**, staminate bud in vertical sction × 12; **e**, staminate sepal × 12; **f**, **g**, staminate petals, exterior (**f**) and intererior (**g**) views × 12; **h**, stamen in 3 views × 12; **i**, pistillode × 12; **j**, apex of pistillode × 12; **k**, portion of rachilla with pistillate buds × 4½; **l**, pistillate bud × 9; **m**, pistillate bud in vertical section × 9; **n**, pistillate sepal × 9; **o**, pistillate petal × 9; **p**, gynoecium and staminodes × 9; **r**, ovary in cross-section × 9; **s**, fruit in vertical section × 3; **t**, fruit in vertical section × 3; **u**, fruit in cross-section × 3; **v**, section of fruit wall × 6; **w**, **x**, **y**, endocarp in 3 views × 3; **z**, point of seed attachment and operculum × 4½. *Cyphokentia cerifera*: all from *Moore & Morat* 10400. (Drawn by Marion Ruff Sheehan)

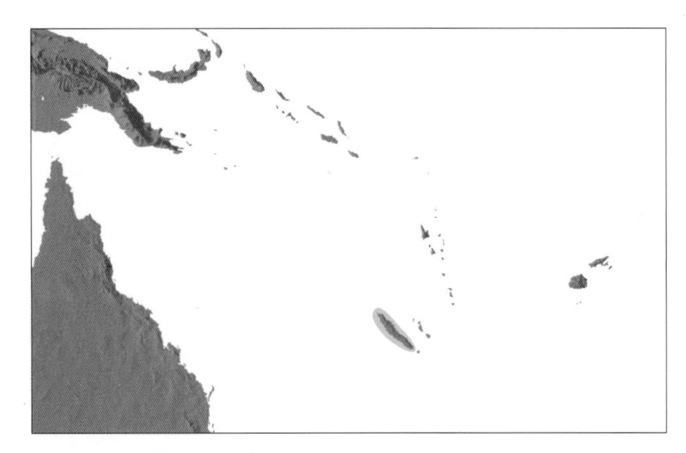

Distribution of ***Cyphokentia***

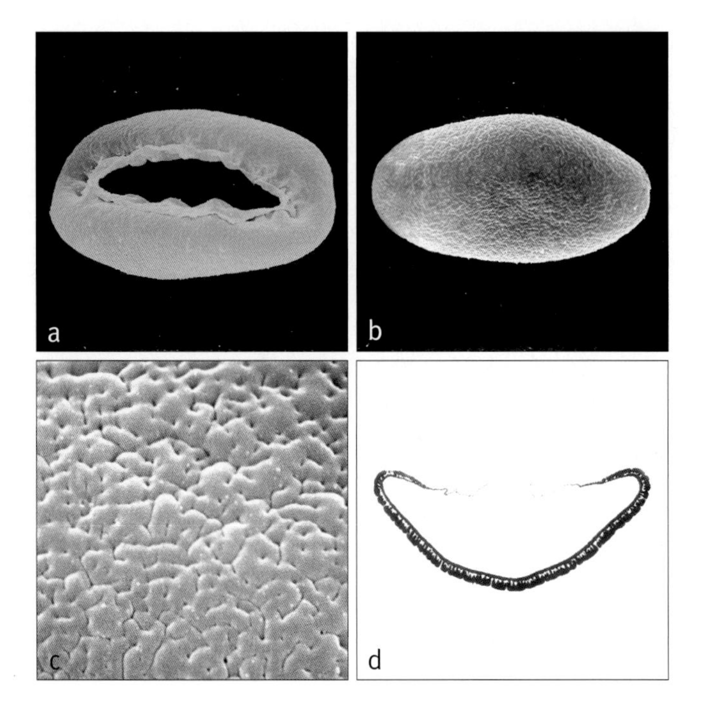

Above: ***Cyphokentia***
a, monosulcate pollen grain, distal face SEM × 1500;
b, monosulcate pollen grain, proximal face SEM × 1500;
c, close-up, finely perforate-rugulate pollen surface SEM × 8000;
d, ultrathin section through whole pollen grain, polar plane TEM × 2000.
Cyphokentia cerifera: **a**, *Moore* 10400; *C. macrostachya*: **b–d**, *Moore* 10415.

Distribution and ecology: Two species in New Caledonia. *Cyphokentia macrostachya* is a somewhat variable species that occurs over a wide range on soil derived from both serpentines and schists, while *C. cerifera* is restricted to schistose soils.
Anatomy: Leaf (Uhl & Martens 1980), root (Seubert 1998a, 1998b) and fruit (Essig *et al.* 1999).
Relationships: The monophyly of *Cyphokentia* is recovered in numerous studies with moderate to high support (Pintaud 1999b, Asmussen *et al.* 2006, Norup *et al.* 2006, Baker *et al.* in review, in prep.). The genus is resolved as sister to *Clinosperma* with high support (Pintaud 1999b, Norup *et al.* 2006, Baker *et al.* in review, in prep.).
Common names and uses: Not recorded.
Taxonomic accounts: Moore and Uhl (1984), Hodel & Pintaud (1998), Pintaud and Baker (2008).

Below left: ***Cyphokentia macrostachya***, habit, New Caledonia. (Photo: J.-C. Pintaud)

Below middle: ***Cyphokentia cerifera***, crown with inflorescence, New Caledonia. (Photo: W.J. Baker)

Below right: ***Cyphokentia macrostachya***, rachilla at staminate anthesis, New Caledonia. (Photo: J.-C. Pintaud)

Fossil record: Asymmetric monosulcate pollen with a distinctive irregularly columellate infratectum, *Palmaepollenites* sp., from the Middle Eocene of Central Java (Nanggulan Formation) is compared with pollen of *Moratia* (= *Cyphokentia*) and with that of *Actinorhytis*, *Cyphosperma* and *Cyphophoenix* (Harley & Morley 1995).
Notes: These are beautiful palms of montane forest in New Caledonia, often with striking white-waxy crownshafts.

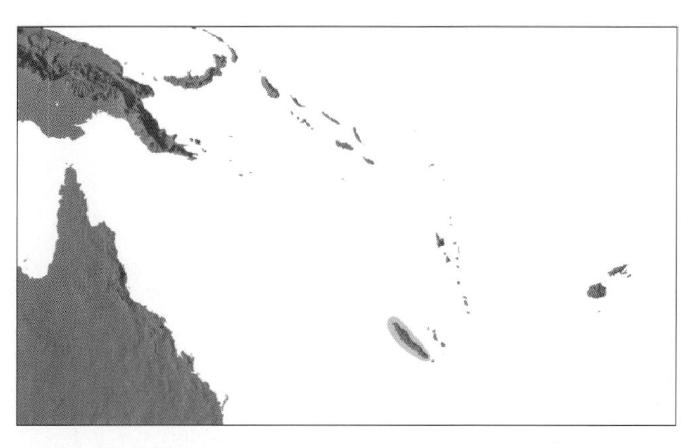

Distribution of ***Clinosperma***

143. CLINOSPERMA

Small or moderate solitary pinnate-leaved palms with or without crownshafts, endemic to New Caledonia; staminate flowers have didymous anthers.

Clinosperma Becc., Palme Nuova Caledonia 51 (1920); and Webbia 5: 119 (1921). Type: **C. bracteale** (Brongn.) Becc. (*Cyphokentia bractealis* Brongn.).

Brongniartikentia Becc., Palme Nuova Caledonia 48 (1920); and Webbia 5: 116 (1921). Type: *B. vaginata* (Brongn.) Becc. (*Cyphokentia vaginata* Brongn. = *Clinosperma vaginata* [Brongn.] Pintaud & W.J. Baker).

Lavoixia H.E. Moore, Gentes Herbarum 11: 296 (1978). Type: *L. macrocarpa* H.E. Moore (= *Clinosperma macrocarpa* [H.E. Moore] Pintaud & W.J. Baker)

Klinein — *slant or slope,* sperma — *seed, perhaps because the seed is inserted obliquely in the immature endocarp.*

Small or moderate, solitary, unarmed, pleonanthic, monoecious palm. *Stem* erect, irregularly ringed with prominent broad leaf scars and bases of old inflorescences towards the summit, often not obviously ringed below, sometimes greatly enlarged at the base, internodes elongate, brown. *Leaves* sometimes in 3 ranks, regularly pinnate, spreading; sheaths split $^{3}/_{4}$ to nearly to the base, forming a prominent crownshaft or not, glabrous adaxially, glaucous with brown, membranous, tattered scales or brown tomentum abaxially; petiole short or long, rounded in section, except directly above the sheath where shallowly concave, bearing reddish-brown tomentum or dot-like scales; rachis angled adaxially, abaxially rounded, with deciduous tomentum; leaflets regularly arranged, acute or distally briefly bifid, single-fold, midrib prominent and squared adaxially, prominent abaxially, 2 other veins elevated, all veins often dotted or with small pale scales abaxially, ramenta lacking or present along midrib abaxially, transverse veinlets not evident. *Inflorescences* interfoliar in bud and at anthesis or becoming infrafoliar at anthesis, branched to 1–3 orders basally, to 1 order distally, erect, curved or pendulous, densely scaly throughout or only in branch axils; peduncle short, or long, sometimes dorsiventrally flattened; prophyll completely encircling the peduncle at insertion, prominently 2-keeled laterally, slightly beaked, chartaceous, enclosing the peduncular bract in bud; first peduncular bract similar, inserted near to or some distance above the prophyll, briefly beaked, sometimes densely scaly, a second, large, open, pointed peduncular bract often present, prophyll and peduncular bract caducous, the latter leaving a ruff-like base, or marcescent; rachis shorter, equal to or longer than the peduncle, bearing spirally inserted sometimes ruffled bracts subtending rachillae; rachillae short to moderate, very slender, bearing low, acute bracts subtending triads in the lower $^{1}/_{3}$, more rarely in lower $^{1}/_{2}-^{5}/_{6}$, and often rounded bracts subtending paired or solitary staminate flowers distally, the flowers sometimes somewhat sunken; bracteoles surrounding the pistillate flower unequal, brown, somewhat sepal-like, the shorter about as long as the subtending bract, the longer exceeding the bract. *Staminate flowers* symmetrical, sometimes very small; sepals 3, distinct, basally imbricate, rounded, gibbous dorsally near the apex; petals 3, distinct, valvate; stamens 6, filaments very briefly connate at the base, free part flattened, not inflexed at the apex in bud, anthers nearly as long as the filaments, ± didymous from a short, darkened connective, introrse; pistillode fleshy, as long as or exceeding the stamens, about as wide as high, slightly expanded apically in a 3-lobed cap, lobes rounded. *Pollen* ellipsoidal asymmetric, occasionally oblate-triangular; aperture a distal sulcus, less frequently a trichotomosulcus; ectexine tectate, perforate, perforate and micro-channelled or finely perforate-rugulate, aperture

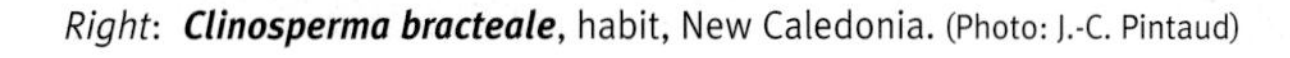

Right: ***Clinosperma bracteale***, habit, New Caledonia. (Photo: J.-C. Pintaud)

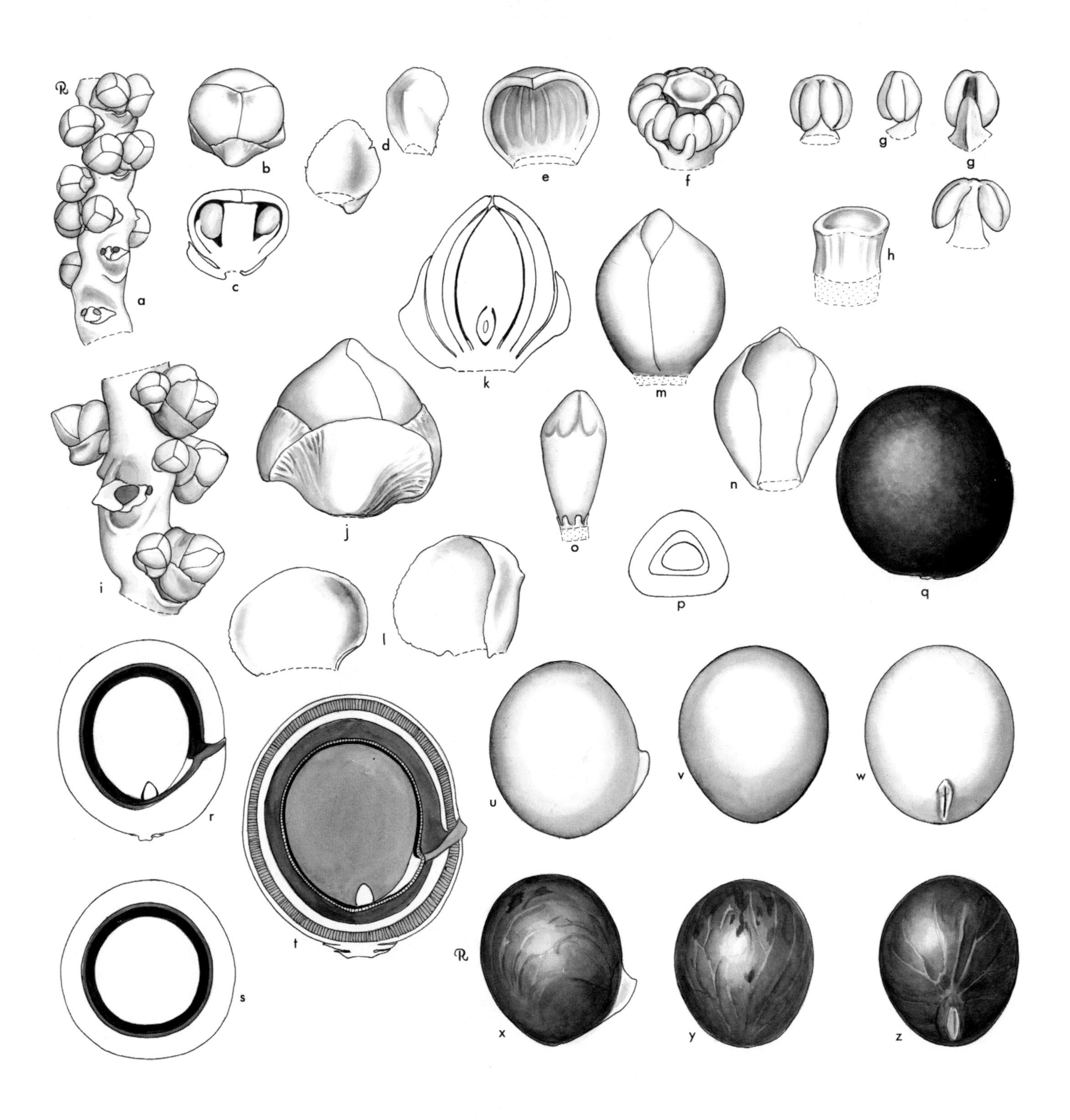

Clinosperma. **a**, portion of rachilla with paired staminate flowers × 6; **b**, staminate bud × 12; **c**, staminate bud in vertical section × 12; **d**, staminate sepals, interior and exterior views × 12; **e**, staminate petal, interior view × 12; **f**, androecium and pistillode × 18; **g**, stamen in 4 views, one with locules spread to show attachment × 18; **h**, pistillode × 18; **i**, portion of rachilla with triads × 6; **j**, pistillate bud × 12; **k**, pistillate bud in vertical section × 12; **l**, pistillate sepals in 2 views × 12; **m**, pistillate bud, sepals removed × 12; **n**, pistillate petal × 12; **o**, gynoecium and staminodes × 12; **p**, ovary in cross-section × 18; **q**, fruit × 3; **r**, fruit in vertical section × 3; **s**, fruit in cross-section × 3; **t**, fruit in vertical section, pericarp regions diagrammatic × 4 $^1/_2$; **u**, **v**, **w**, endocarp in 3 views × 3; **x**, **y**, **z**, seed in 3 views × 3. *Clinosperma bracteale*: **a–h** and **q**, *MacDaniels* s.n.; **i–p**, *Lavoix* 23; **r–z**, *Moore et al.* 9970. (Drawn by Marion Ruff Sheehan)

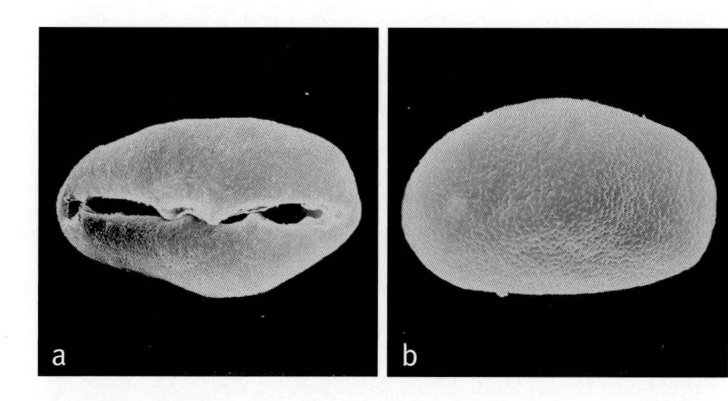

Above: ***Clinosperma***
a, monosulcate pollen grain, distal face SEM × 1500;
b, monosulcate pollen grain, proximal face SEM × 1500.
Clinosperma bracteale: **a** & **b**, *McKee* 26538.

Below: ***Clinosperma macrocarpa***, fruit, New Caledonia. (Photo: J.-C. Pintaud)

margin similar or slightly finer; infratectum columellate; longest axis 30–42 µm [3/4]. *Pistillate flowers* about twice as long as the staminate at anthesis; sepals 3, distinct, broadly imbricate, rounded, somewhat gibbous dorsally towards the apex; petals 3, distinct, imbricate except for briefly valvate apices; staminodes 3, at one side of the gynoecium; gynoecium ovoid, unilocular, uniovulate, ovule pendulous, probably hemianatropous. *Fruit* globose, eccentrically globose or ovoid ellipsoidal, sometimes very large, with stigmatic remains lateral at about the middle or basal, red, dark purplish or black; epicarp smooth, mesocarp of pale, fleshy parenchyma over a solid shell of pale sclerosomes underlain by pale parenchyma with flat, slender, ± anastomosing fibres and irregularly flattened tannin cells, endocarp fragile, vitreous, eccentrically globose, with short groove and rounded basal operculum. *Seed* globose, ellipsoidal or laterally compressed, variously indented or smooth, light brown, hilum short, raphe branches ascending and curved laterally from raphe, anastomosing somewhat abaxially, endosperm homogeneous; embryo basal. *Germination* adjacent-ligular; eophyll bifid. *Cytology* not known.

Distribution and ecology: Four species in New Caledonia found in wet forests on serpentine soils and schists.
Anatomy: Leaf (Uhl & Martens 1980) and fruit (Essig *et al.* 1999).
Relationships: The monophyly of *Clinosperma* is highly supported in several studies as is its sister relationship to *Cyphokentia* (Pintaud 1999b, Norup *et al.* 2006, Baker *et al.* in review, in prep.).
Common names and uses: Not recorded.
Taxonomic accounts: Moore and Uhl (1984), Hodel and Pintaud (1998), Pintaud and Baker (2008).
Fossil record: No generic records found.
Notes: *Clinosperma* is distinctive in its compact inflorescence with acuminate rachis bracts, triads mostly borne in the lower one-third of the rachillae, and pistillate flowers larger than staminate at staminate anthesis. Indumentum of the inflorescence is variable but characteristic minute, tattered scales are usually evident at least in axils of the branches. Tannin and fibrous strands are both lacking in leaves. Midribs of leaflets have long narrow extensions of the adaxial fibrous sheath (Uhl & Martens 1980). Floral anatomy has not been reported.
Additional figures: Glossary fig. 19.

Subtribe Dypsidinae Becc., Palme Madagascar 2 (1914). Type: Dypsis.

Lemurophoenicinae Beentje & J. Dransf., Palms Madagascar 415 (1995). Type: *Lemurophoenix*
Masoalinae Beentje & J. Dransf., Palms Madagascar 421 (1995). Type: *Masoala*.

Diminutive to very robust, acaulescent to erect or rarely climbing; leaves entire bifid to pinnate, leaflet tips entire or praemorse; sheaths often forming a crownshaft; inflorescences bisexual or rarely unisexual; stamens 3–6, rarely more than 50; stigmatic remains usually basal; epicarp smooth, very rarely corky-warted.

There is enormous morphological diversity within this subtribe. It is almost impossible to provide a diagnosis based on morphological characters. The one feature that appears to unite the group is that all genera occur in Madagascar! It appears that there has been an astonishing radiation on this island. Relationships within the group and with other members of the Areceae may be made clearer by a detailed phylogenetic analysis of *Dypsis* using molecular and morphological data.

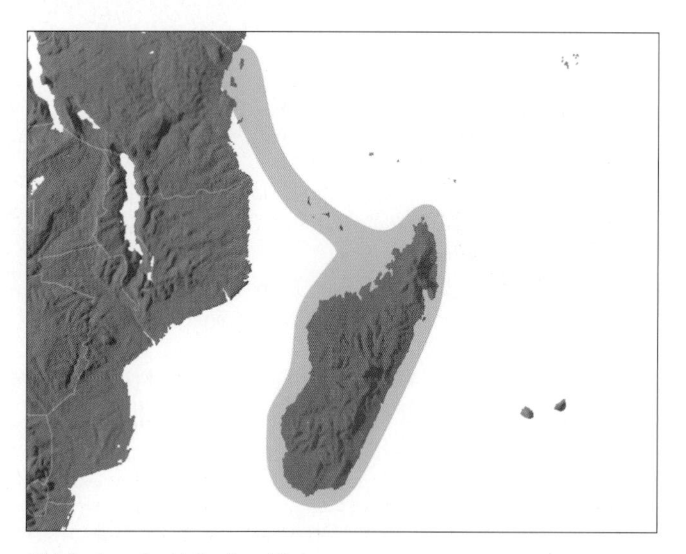

Distribution of subtribe **Dypsidinae**

The monophyly of the Dypsidinae has been recovered in the most thoroughly sampled studies, but with low support (Lewis & Doyle 2002, Baker *et al.* in prep.). Several other studies have reported alternative relationships (Lewis 2002, Norup *et al.* 2006, Baker *et al.* in review). Further work is required to test the monophyly of the group. The position of the subtribe remains uncertain within tribe Areceae (Lewis 2002, Lewis & Doyle 2002, Norup *et al.* 2006, Baker *et al.* in review, in prep.). This group needs further phylogenetic study.

Key to Genera of Dypsidinae

1. Inflorescences unisexual, both staminate and pistillate borne on the same plant. Madagascar . . **146. Marojejya**
1. Inflorescences bearing both staminate and pistillate flowers 2
2. Stamens 52–59; fruit corky-warted, endocarp with a basal heart-shaped button; endosperm very shallowly ruminate; embryo apical. Madagascar . . . . . **145. Lemurophoenix**
2. Stamens 3–6; fruit smooth; endocarp lacking a heart-shaped button; endosperm homogeneous or ruminate; embryo basal 3
3. Fruit with apical stigmatic remains. Madagascar . . . . **147. Masoala**
3. Fruit with basal stigmatic remains. Madagascar . . . . **144. Dypsis**

144. DYPSIS

A complex and highly variable genus of pinnate-leaved palms that has radiated spectacularly in Madagascar and Comores, with an outlier on Pemba off the coast of Tanzania; the genus includes towering forest giants, bottle palms, litter-trappers, stemless palms, some of the most slender of all palms, and even two climbing members; all have fruit with basal stigmatic remains.

Dypsis *Noronha ex Mart.* Hist. Nat. Palm. 3: 180 (1838). Lectotype: **D. forficifolia** Mart.

Chrysalidocarpus H. Wendl., Bot. Zeit. 36: 117 (1878). Type: *C. lutescens* H. Wendl. (= *Dypsis lutescens* [H. Wendl.] Beentje & J. Dransf.).

Phloga Noronha ex Hook.f., Benth. & Hook.f., Gen. pl. 3: 877, 909 (1883). Type: *P. nodifera* (Mart.) Noronha ex Salomon (= *Dypsis nodifera* Mart.).

Neodypsis Baill., Bull. Mens. Soc. Linn. Paris 148: 1172 (1894). Type: *N. lastelliana* Baill. (= *Dypsis lastelliana* [Baill.] Beentje & J. Dransf.).

Neophloga Baill., Bull. Mens. Soc. Linn. Paris 2: 1173 (1894). Type: *N. commersoniana* Baill. (= *Dypsis commersoniana* [Baill.] Beentje & J. Dransf.).

Haplodypsis Baill., Bull. Mens. Soc. Linn. Paris 2: 1167 (1894). Type: *H. pervillei* Baill. (= *Dypsis pervillei* [Baill.] Beentje & J. Dransf.).

Haplophloga Baill., Bull. Mens. Soc. Linn. Paris 2: 1168 (1894). Type: *H. poivreana* Baill. (= *Dypsis poivreana* [Baill.] Beentje & J. Dransf.).

Dypsidium Baill., Bull. Mens. Soc. Linn. Paris 2: 1172 (1894). Type: *D. catatianum* Baill. (see Beccari & Pichi-Sermolli 1955) (= *Dypsis catatiana* [Baill.] Beentje & J. Dransf.).

Phlogella Baill., Bull. Mens. Soc. Linn. Paris 2: 1174 (1894). Type: *P. humblotiana* Baill. (= *Dypsis humblotiana* [Baill.] Becc.).

Trichodypsis Baill., Bull. Mens. Soc. Linn. Paris 2: 1174 (1894). Type: *T. hildebrandtii* Baill. (= *Dypsis hildebrandtii* [Baill.] Becc.).

Vonitra Becc., Bot. Jahrb. Syst. 38, Beibl. 87: 18 (1906). Type: *V. fibrosa* (C.H. Wright) Becc. (*Dictyosperma fibrosum* C.H. Wright) (= *Dypsis fibrosa* [C.H. Wright] Beentje & J. Dransf.).

Adelodypsis Becc, Bot. Jahrb. Syst. 38, Beibl. 87: 18 (1906). Type: *A. gracilis* (Bory) Becc. (*Dypsis gracilis Bory*) (= *D. pinnatifrons* Mart.).

Macrophloga Becc., Palme Madagascar 47 (1914). Type: *M. decipiens* (Becc.) Becc. (= *Dypsis decipiens* Becc.).

Antongilia Jum., Ann. Inst. Bot.-Géol. Colon. Marseille sér. 4, 6 (2): 19 (1928). Type: *A. perrieri* Jum. (= *Dypsis perrieri* [Jum.] Beentje & J. Dransf.).

Derivation obscure.

Right: ***Dypsis fibrosa***, habit, Madagascar. (Photo: J. Dransfield)

Very small to very large unarmed pleonanthic monoecious palms. *Stems* solitary or clustered, very short, subterranean, creeping-rhizomatous, erect, in one species climbing, sometimes branched aerially by apparent dichotomy. *Leaves* pinnate or pinnately ribbed, neatly abscising or marcescent; sheath tubular, rarely almost open, usually forming a well-defined crownshaft, sometimes fibrous, in a few species with abundant pendulous piassava, sheath surface variously scaly and/or waxy or glabrous, auricles sometimes present; petiole absent or short to long, variously glabrous, scaly or hairy; blade entire, entire-bifid, or divided into single or multi-fold reduplicate leaflets, regularly or irregularly arranged, sometimes fanned within groups to produce a plumose appearance, leaflets usually entire, rarely praemorse, very rarely discolourous, often with abundant minute punctiform scales on both surfaces and ramenta along the main rib abaxially. *Inflorescences* mostly interfoliar, more rarely infrafoliar, spicate or branched to 1–4 orders, apparently protrandrous (?always); peduncle usually elongate, basal branches not sharply divaricate; prophyll often borne above the base of the peduncle; peduncular bract usually conspicuous, exserted and caducous; rachillae variously glabrous or scaly and hairy; rachilla bracts low, generally inconspicuous, sometimes conspicuous; flowers borne in triads of a central pistillate flower and two lateral staminate flowers, triads superficial or slightly sunken in shallow pits. *Staminate flowers* symmetrical, ± rounded to bullet-shaped, sometimes very small; sepals imbricate; petals valvate, basally briefly connate; stamens 3 (antesepalous or antepetalous) or 6 (very rarely 1, 4 or 5 as monstrosities), 3 staminodes sometimes present, these either antesepalous or antepetalous, very rarely adnate to the pistillode; pistillode present or absent. *Pollen* ellipsoidal, elongate ellipsoidal, pyriform or oblate triangular, with slight or obvious asymmetry; aperture usually a distal sulcus, occasionally a trichotomosulcus; ectexine usually tectate, occasionally semi-tectate,

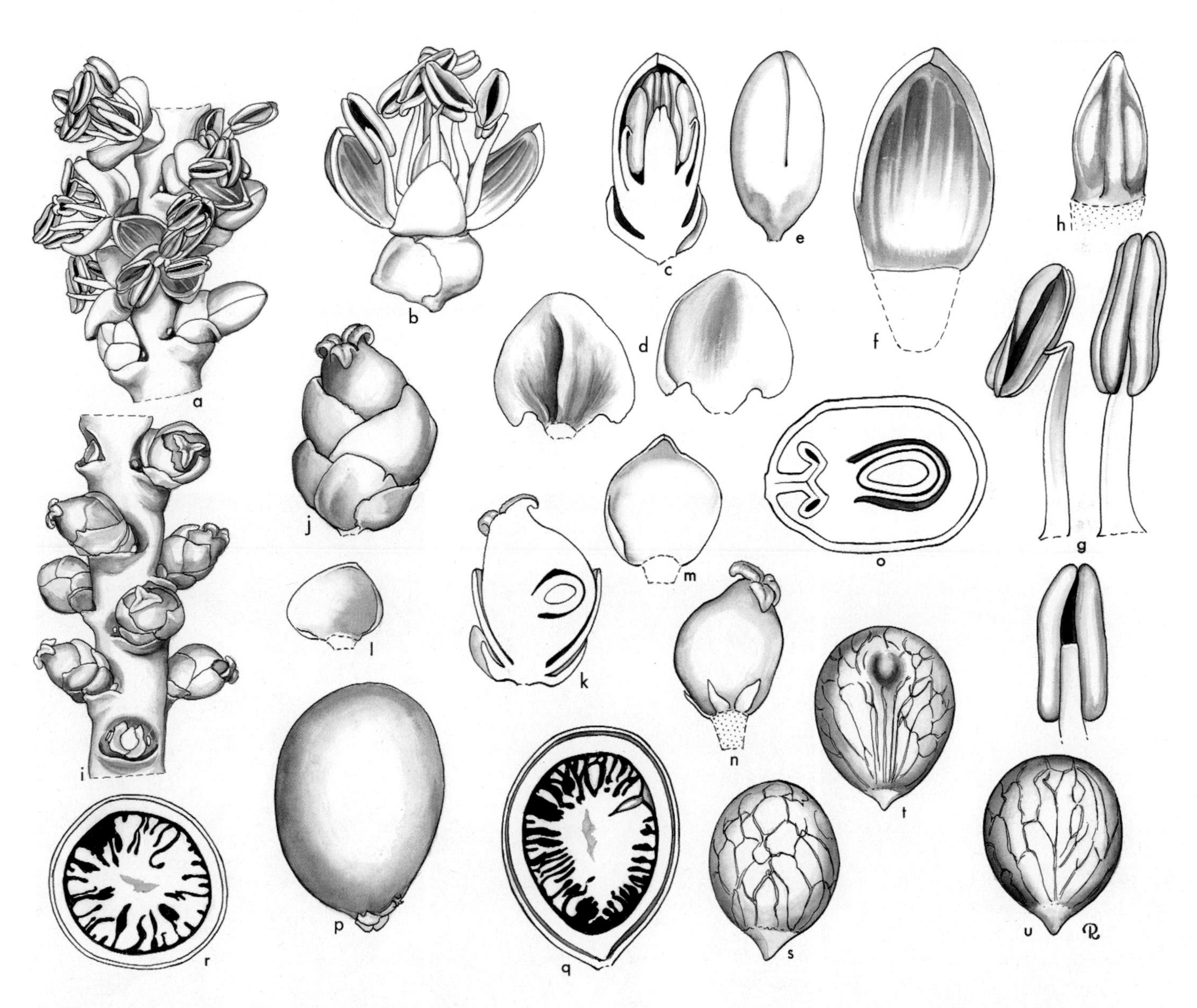

***Dypsis*.** **a**, portion of rachilla at staminate anthesis × 3; **b**, staminate flower × 6; **c**, staminate bud in vertical section × 6; **d**, staminate sepals in 2 views × 12; **e**, staminate flower, sepals removed × 6; **f**, staminate petal, interior view × 12; **g**, stamen in 3 views × 12; **h**, pistillode × 12; **i**, portion of rachilla at pistillate anthesis × 3; **j**, pistillate flower after fertilization × 6; **k**, pistillate flower in vertical section × 6; **l**, pistillate sepal, interior view × 6; **m**, pistillate petal, interior view × 6; **n**, gynoecium and staminodes × 6; **o**, ovary in cross-section × 12; **p**, fruit × 1 1/2; **q**, fruit in vertical section × 1 1/2; **r**, fruit in cross-section × 1 1/2; **s, t, u**, seed in 3 views × 1 1/2. *Dypsis decaryi*: all from *Read* 827bis. (Drawn by Marion Ruff Sheehan)

perforate, perforate and micro-channelled, perforate-rugulate, reticulate, muri of reticulum occasionally coarsely granular (rarely granular crotonoid) or spinulose, aperture margin similar or slightly finer; infratectum columellate; less frequently ectexine intectate with coarsely granular structures, sometimes coalesced into larger elements with or without spinulae, aperture margin similar; longest axis 17–65 µm; post-meiotic tetrads usually tetrahedral, rarely tetragonal or rhomboidal [30/140]. *Pistillate flowers* about the same size as the staminate; sepals rounded, broadly imbricate; petals imbricate with triangular valvate tips; staminodes usually present, minute, tooth-like, usually 3 or 6 at one side of the ovary; pistil pseudomonomerous, often strongly asymmetrical (especially in smaller species), stigmas 3, apical, sometimes eccentric, ovule form unknown. *Fruit* borne with persistent calyx and corolla, spherical, ellipsoid, fusiform or rarely curved, stigmatic remains basal, often obscured by perianth; epicarp often brightly coloured or jet black or rarely dull green or brown; mesocarp thin, fleshy or fibrous; endocarp usually thin, fibrous. *Seed* closely adhering to the endocarp, endosperm homogeneous, sometimes deeply pentrated by regular grooves, or weakly to strongly ruminate; embryo subbasal. *Germination* adjacent ligular; eophyll bifid or pinnate. *Cytology*: 2n = 32, 34.

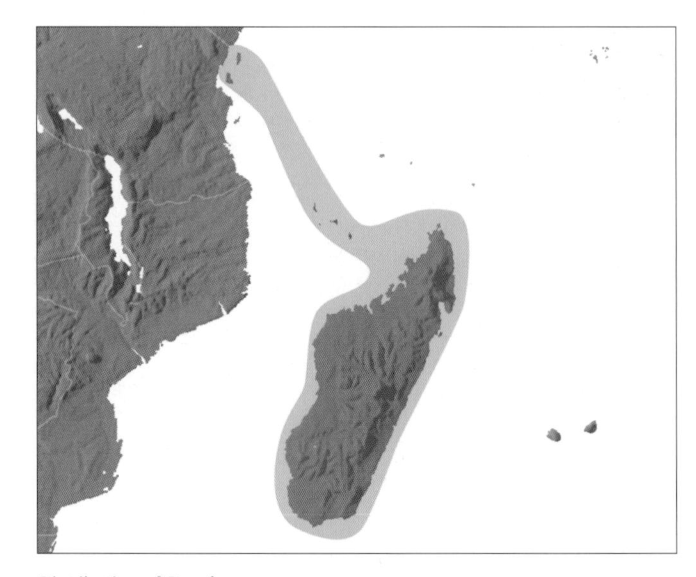

Distribution of ***Dypsis***

Distribution and ecology: Approximately 140 species confined to Madagascar, the Comores and the island of Pemba off the coast of Tanzania. At least 20 more are currently undescribed. The genus displays an extraordinary range of ecological adaptations, occurring from sea level to over 2200 m in the mountains, from rain forest to forest transitional with spiny xeromorphic scrub. Many species are palms of the forest canopy whereas others are among the smallest of all palms. *Dypsis crinita* is a rheophyte, at least as a juvenile, while *D. aquatilis* grows in relatively deep water, paralleling the remarkable *Ravenea musicalis* that grows in a nearby river system.

Below left: ***Dypsis lastelliana***, habit, Madagascar. (Photo: J. Dransfield)

Below middle: ***Dypsis prestoniana***, crown and infructescence, Madagascar. (Photo: H.Beentje)

Below right: ***Dypsis pachyramea***, habit, Madagascar. (Photo: J. Dransfield)

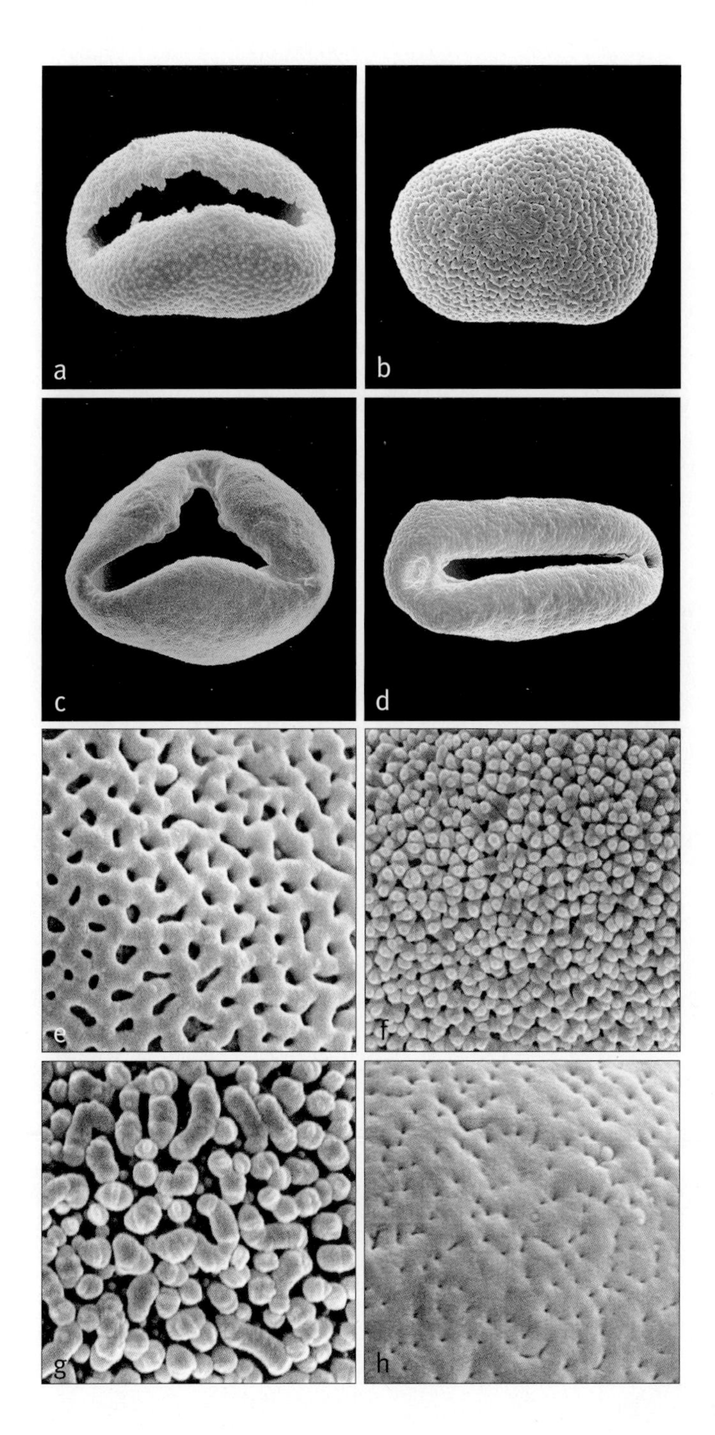

Above: ***Dypsis***

a, monosulcate pollen grain, distal face SEM × 2500;
b, monosulcate pollen grain, proximal face SEM × 2500;
c, trichotomosulcate pollen grain, distal face SEM × 2000;
d, monosulcate pollen grain, distal face SEM × 1500;
e, close-up, finely reticulate pollen surface SEM × 8000;
f, close-up, finely granular-crotonoid pollen surface SEM × 8000;
g, close-up, intectate pollen surface comprising coarse granulae or coalesced granulae SEM × 8000;
h, close-up, finely perforate pollen surface SEM × 8000.
Dypsis catatiana: **a**, *Guillaumet* 2161; *D. jumelleana*: **b**, *Leandri* 721; *D. concinna*: **c**, *Perrier de la Bâthie* 15983; *D. madagascariensis*: **d** & **h**, *Leaman & Sabohrean* 2988; *D. pusilla*: **e**, *Dransfield et al.* JD6467; *D. ambilaensis*: **f**, *Dransfield et al.* JD6444; *D. nodifera*: **g**, *Nicholl* 140.

Above: ***Dypsis decaryi***, crown, Madagascar. (Photo: J. Dransfield)

Anatomy: Leaf (Achilli 1913, Tomlinson 1961), root (Seubert 1998a, 1998b) and floral (Rudall *et al.* 2003).

Relationships: *Dypsis* has never been resolved as monophyletic in any study (Lewis & Doyle 2002, Baker *et al.* in prep.). A wide range of relationships have been recovered between *Dypsis*, *Marojejya*, *Masoala* and *Lemurophoenix* (and sometimes other Areceae), but these remain poorly supported (Lewis 2002, Lewis & Doyle 2002, Loo *et al.* 2006, Norup *et al.* 2006, Baker *et al.* in review, in prep.). Further data are required to facilitate a revision of the limits of *Dypsis*.

Common names and uses: For local names and uses see Dransfield & Beentje (1995b).

Taxonomic account: Dransfield & Beentje (1995b). **Fossil record:** From the Palaeocene-Lower Eocene, Deccan Intertrappean of India (Madhya Pradesh) (although the age span of these volcanic deposits is controversial, see Chapter 5) a petrified palm stem, *Palmoxylon ghuguensis*, is compared with *Chrysalidocarpus* (= *Dypsis*) (Ambwani & Prakash 1983). The affinity of the fossil is inconclusive. Small monosulcate grains, *Palmaemargosulcites insulatus*, from palm flower compression fossils, recovered from the Middle Eocene oil shales of Messel, Germany, are compared with pollen of *Dictyocaryum* and *Dypsis* (Harley 1997).

Notes: This is an astonishingly variable genus. Dransfield and Beentje (1995b) found the circumscription of the genera *Chrysalidocarpus*, *Neophloga*, *Phloga*, *Vonitra*, *Antongilia* and *Neodypsis* to be based on highly unreliable characters, and decided to sink them all into the single genus *Dypsis*. It may well be that after a rigorous phylogeny is produced based on both molecular and morphological characters, the circumscription of groups within *Dypsis* at the generic level may be possible and desirable.

Additional figures: Figs 1.1, 2.8b, Glossary figs 4, 20.

145. LEMUROPHOENIX

Spectacular pinnate-leaved canopy palm from northeastern Madagascar, distinctive in the staminate flower with many stamens and the large corky-warted fruit, the endocarp with a basal button, the seed shallowly ruminate with apical embryo.

Lemurophoenix J. Dransf., Kew Bull. 46: 61 (1991). Type: **L. halleuxii** J. Dransf.

Lemur — *lemur,* phoenix — *general name for a palm, in reference to its Malagasy vernacular name,* hovitra varimena, *the red lemur palm.*

Massive solitary unarmed monoecious pleonanthic tree palm. *Trunk* bare, ringed with leaf scars. *Leaf* reduplicately pinnate; sheath tubular, forming a prominent crownshaft, bearing wax and scales; petiole rather short, channelled adaxially, rounded or ridged abaxially, densely covered with caducous chocolate-brown scales; rachis adaxially somewhat channelled near the base, ridged distally, abaxially rounded or flattened, scaly as the petiole; leaflets regularly arranged, numerous, linear-lanceolate, long acuminate except near the tip where bifid; adaxial leaflet surface glabrous, abaxial bearing a few large dark brown ramenta near the base on the main vein and more numerous small ramenta on secondary veins, pale brown peltate scales abundant on all veins; transverse veinlets not visible. *Inflorescence* infrafoliar, branched to 3 orders, the whole inflorescence exposed long before anthesis, protandrous; peduncle moderate in length; prophyll splitting along one side; peduncular bract longer than the prophyll; first-order branches widely spreading, the basal few branched to the third order, the distal-most branched to the second order or unbranched; rachillae numerous, elongate, pendulous or spreading, somewhat swollen, with flowers partially embedded in shallow pits; rachilla bracts rather

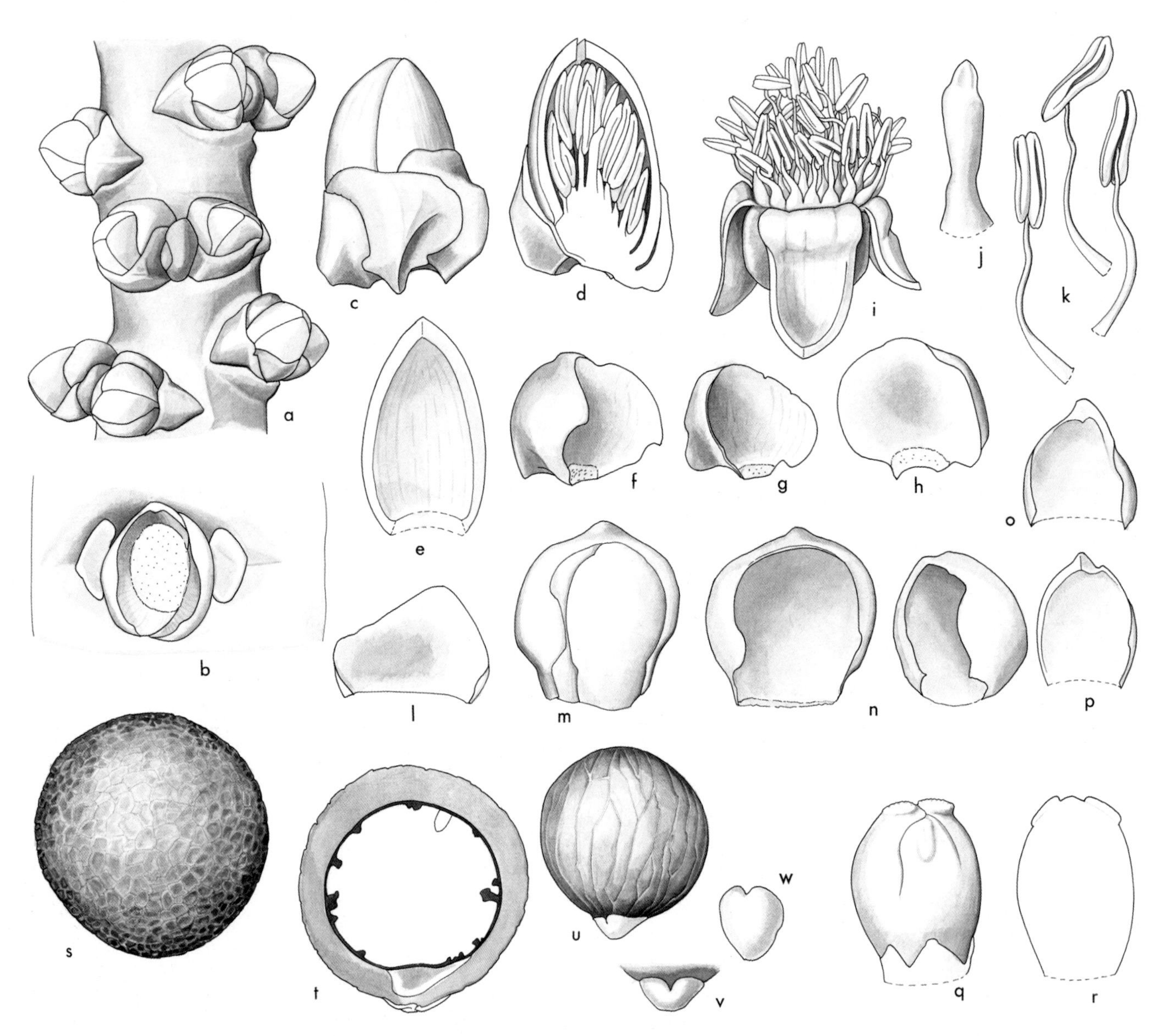

Lemurophoenix**.** **a**, portion of rachilla with triads × 3; **b**, triad bracts and scars, flowers removed × 6; **c**, staminate bud × 6; **d**, staminate bud in vertical section × 6; **e**, staminate petal × 6; **f, g, h**, staminate sepals × 6; **i**, open staminate flower × 4; **j**, pistillode × 10; **k**, stamens, 3 views × 9; **l**, bracteole × 9; **m**, pistillate bud × 9; **n**, pistillate sepals, 2 views × 9; **o, p** pistillate petals × 9; **q**, gynoecium × 14; **r**, gynoecium in vertical section × 14; **s**, fruit × 3/4; **t**, fruit in vertical section × 3/4; **u**, seed × 3/4; **v, w**, details of basal button × 1. *Lemurophoenix halleuxii*: **a–r**, *Dransfield* 6453; **s–w**, *Dransfield* 6452. (Drawn by Lucy T. Smith)

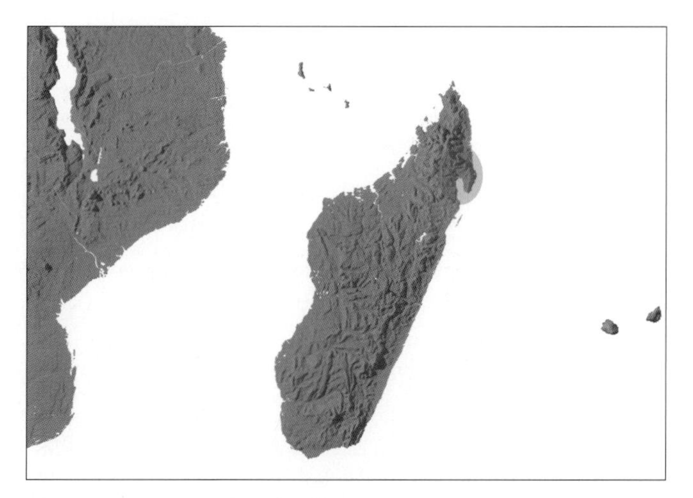

Distribution of ***Lemurophoenix***

obscure, forming the lower lip to the pits; floral bracteoles minute, included in the pits. *Flowers* borne in triads of a central pistillate and two lateral staminate for about two-thirds the rachilla length, and in pairs of staminate flowers in the distal third. *Staminate flower* in bud ± bullet-shaped; sepals 3, ± distinct, minutely connate at the base, imbricate, strongly keeled and gibbous; petals 3, ± distinct in bud, valvate, boat-shaped, adaxially grooved, glabrous, later the floral receptacle greatly enlarging carrying the petal bases above the calyx, the petals becoming reflexed by a swollen pulvinus at the petal bases; stamens 52–59, borne on the dome-shaped receptacle, filaments terete, straight or contorted in bud, rarely filaments partially connate, anthers frequently rather irregular in outline due to close-packing in the bud, basally sagittate, medifixed, latrorse; pistillode columnar, hidden among the filament bases. *Pollen* ellipsoidal, slight or obvious asymmetry; aperture a distal sulcus; ectexine tectate, finely perforate, or perforate and micro-channelled and rugulate, aperture margin slightly finer; infratectum columellate; longest axis ranging from 52–60 µm [1/1]. *Pistillate flower* known only in immature bud; sepals 3, distinct, imbricate, unequal, rounded; petals 3, distinct, basally strongly imbricate, with triangular valvate tips; staminodes 10–12, minute, tooth-like or strap-shaped, distributed evenly around the gynoecium; gynoecium pseudomonomerous, ovoid, stigmas apical, as yet scarcely developed, ovule heminanatropous, basally attached. *Fruit* large, usually borne in abundance, globose, the epicarp cracked polygonally into low corky warts, stigmatic remains basal; mesocarp rather spongy, easily separable from the endocarp; endocarp spherical, with a basal heart-shaped pale brown button; endosperm very shallowly and sparsely ruminate; embryo apical. *Germination* adjacent-ligular; eophyll bifid. *Cytology* not studied.

Distribution and ecology: A single species known from two localities in north-eastern Madagascar, occurring on hill slopes in humid rain forest at elevations of 250–450 m above sea level. Seeds exported by seed merchants in the late 1990s suggest that there may by a second species with a smaller fruit lacking corky warts.

Anatomy: Not studied.

Relationships: For relationships, see *Dypsis*. Note that the phylogeny with the greatest sampling of *Dypsis* species places *Lemurophoenix* as sister to the rest of the Dypsidinae (Baker *et al.* in prep.).

Below left: ***Lemurophoenix halleuxii***, crown, Madagascar. (Photo: D.N. Cooke)

Below middle: ***Lemurophoenix halleuxii***, infructescence, Madagascar. (Photo: J. Dransfield)

Below right: ***Lemurophoenix halleuxii***, infructescence, Madagascar. (Photo: J. Dransfield)

Local names and uses: *Hovitra vari mena.*
Taxonomic accounts: Dransfield (1991b) and Dransfield & Beentje (1995b).
Fossil record: No generic records found.
Notes: A spectacular large palm distinctive in its corky-warted fruit and multistaminate staminate flowers.

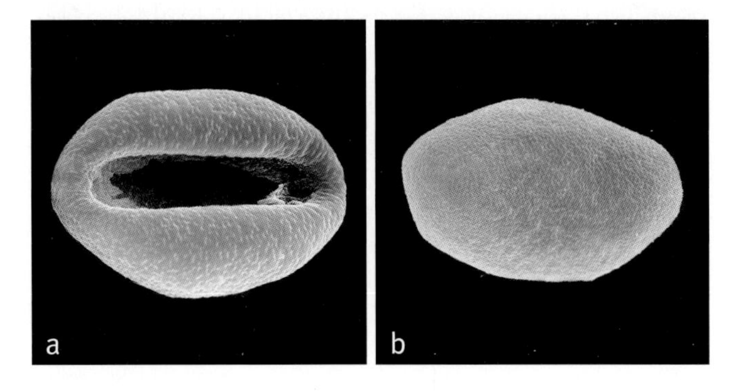

Above: ***Lemurophoenix***
a, monosulcate pollen grain, distal face SEM × 800;
b, monosulcate pollen grain, proximal face SEM × 800.
Lemurophoenix halleuxii: **a**, *Darian* s.n., **b**, *Dransfield* JD6453.

Below: ***Marojejya darianii***, habit, Madagascar. (Photo: J. Dransfield)

146. MAROJEJYA

Robust squat litter-trapping palms from rain forest in Madagascar, remarkable for their condensed unisexual inflorescences, found among the leaf sheaths, but with both sexes found on the same tree.

Marojejya Humbert, Mém. Inst. Sci. Madagascar, Sér. B, Biol. Vég. 6: 92 (1955). Type: **M. insignis** Humbert.

Named after the Marojejy Massif in northeast Madagascar, whence the palm was first collected and described.

Stout, solitary, unarmed, monoecious, pleonanthic palms. *Stem* erect, obscurely ringed with leaf scars, internodes short, sometimes with short root spines. *Leaves* numerous, massive, pinnate or ± entire, the crown often filled with fibres; sheath with or without rounded auricles; petiole absent or thick and wide at the base, gradually tapering to the rachis, adaxially channelled, abaxially rounded, densely covered with caducous, dense brown scales; rachis adaxially deeply channelled, abaxially rounded basally, becoming laterally channelled distally, densely scaly at the base, blade undivided for ca. $^{1}/_{4}$ to the entire length, where not entire, distally irregularly divided into, 1–several-nerved, obtuse leaflets with decurrent bases, the blade eventually splitting irregularly, abaxially with scattered irregular bands of caducous, chocolate-brown scales, numerous fine, longitudinal veins between the major ribs, transverse veinlets not evident. *Inflorescences* unisexual (but see below), hidden among the leaf bases beneath debris, branching to 1 order, staminate and pistillate inflorescences basically similar but staminate with more numerous slender and longer branches; peduncle large, slightly flattened, short, densely covered in dark brown scales; prophyll tubular, 2-keeled, strongly flattened, splitting longitudinally, thinly coriaceous, bearing numerous caducous, dark brown scales; basal peduncular bract similar to the prophyll but smaller, subsequent peduncular bracts numerous, crowded, spirally arranged, incomplete, acute to acuminate, stiff, ± erect, gradually diminishing in size distally; rachis shorter than the peduncle, bearing spirally arranged rachillae, each subtended by a conspicuous, acute or acuminate bract; staminate rachillae ± equal in length, stout, catkin-like, somewhat flexuous, densely covered in brown tomentum; rachilla bracts conspicuous, paired, narrow, triangular, subtending densely crowded, paired staminate flowers except near the base and at the tip where solitary staminate flowers present, flowers abortive at the very base, rarely a few pistillate flowers present at the base (as a monstrosity), distal bracts forming pits; floral bracteoles 2, well-developed, acute, ciliate margined. *Staminate flowers* rather small, somewhat asymmetrical due to close packing; sepals 3, free, unequal, narrow, ovate, keeled, chaffy, ciliate margined, tending to be widely separated; petals 3, ± boat-shaped,

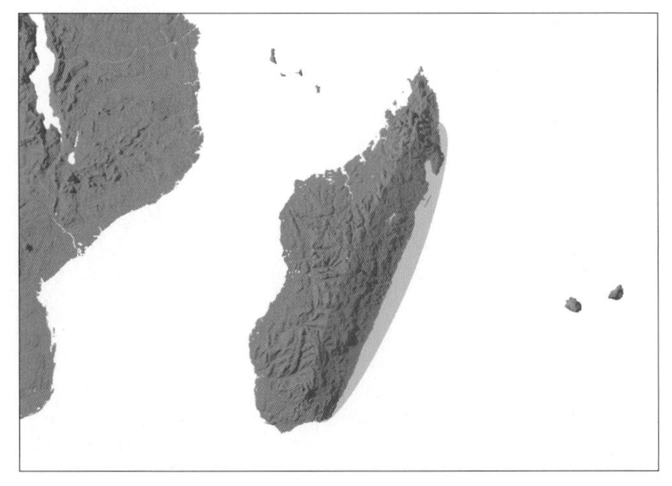

Distribution of ***Marojejya***

valvate, coriaceous, connate basally for 1/3 their length and adnate to the receptacle; stamens 6, filaments basally connate, the distinct portions flattened, tapering, elongate, inflexed at the tip, anthers medifixed, ± versatile, latrorse; pistillode small, 3 lobed. *Pollen* ellipsoidal asymmetric, occasionally lozenge-shaped, pyriform or oblate triangular; aperture a distal sulcus, infrequently a trichotomosulcus; ectexine tectate, perforate-rugulate, aperture margin similar, or slightly finer; infratectum columellate; longest axis ranging from 34–37 µm [2/2]. *Pistillate rachillae* shorter, thicker, and fewer than the staminate, densely brown tomentose, bearing crowded, spirally arranged, triangular bracts forming the lower lips of shallow pits, each pit bearing 3 membranous bracteoles, 2 very small abortive staminate flowers and a large solitary pistillate flower. *Pistillate flowers* much larger than the staminate, obpyriform, somewhat asymmetrical; sepals 3, distinct, somewhat chaffy, ovate with triangular tips, ± striate; petals 3, distinct, similar to the sepals but larger and with short, triangular, valvate tips; staminodes 6, narrow, triangular;

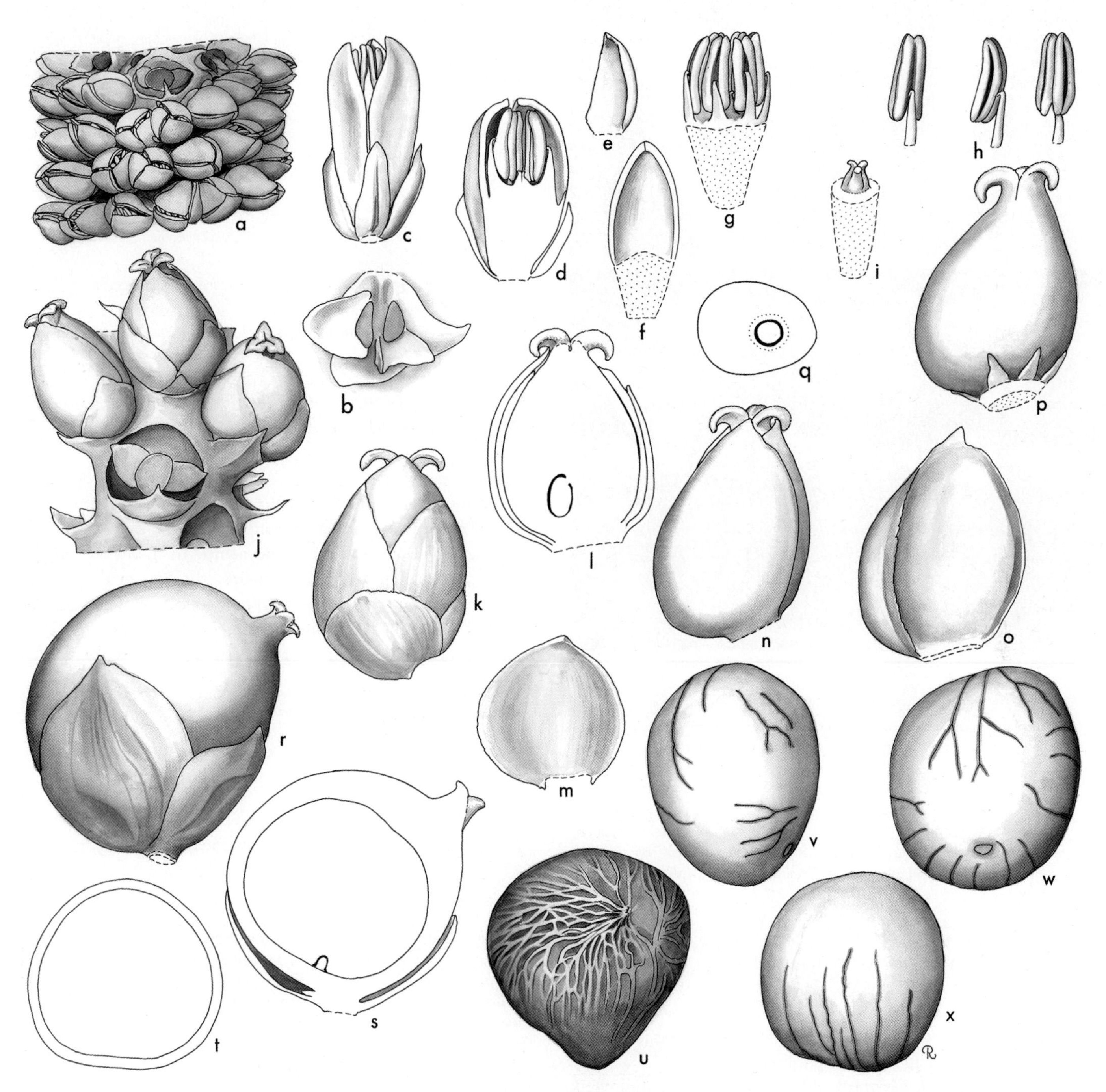

Marojejya. **a**, portion of staminate rachilla × 3; **b**, bracteoles of staminate dyad × 6; **c**, staminate flower × 6; **d**, staminate flower in vertical section × 6; **e**, staminate sepal, interior view × 6; **f**, staminate petal, interior view × 6; **g**, androecium × 6; **h**, stamen in 3 views × 6; **i**, pistillode × 6; **j**, portion of pistillate rachilla × 2 1/4; **k**, pistillate flower × 9; **l**, pistillate flower in vertical section × 9; **m**, pistillate sepal × 9; **n**, pistillate flower, sepals removed × 9; **o**, pistillate petal × 9; **p**, gynoecium and staminodes × 9; **q**, ovary in cross-section × 9; **r**, fruit × 2 1/4; **s**, fruit in vertical section × 2 1/4; **t**, fruit in cross-section × 2 1/4; **u**, endocarp and fibres × 2 1/4; **v**, **w**, **x** seed in 3 views × 2 1/4. *Marojejya insignis*: all from *Moore* 9901. (Drawn by Marion Ruff Sheehan)

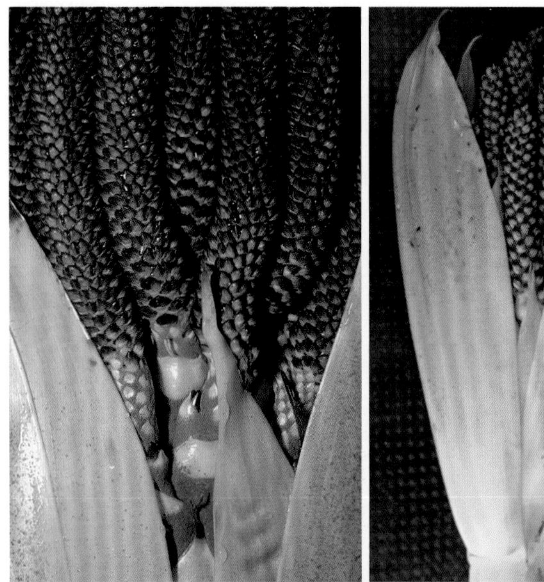

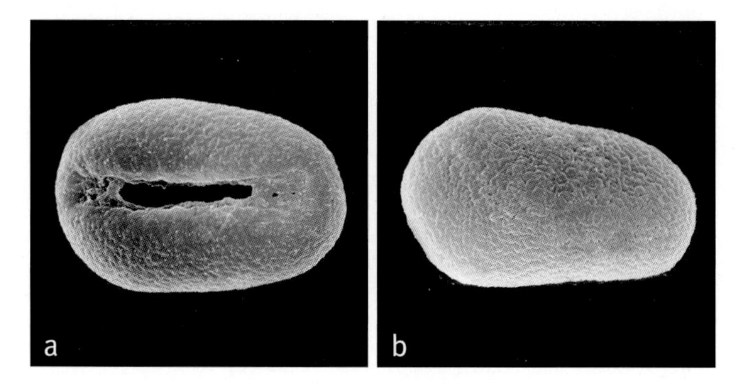

Above: ***Marojejya***
a, monosulcate pollen grain, distal face SEM × 1500;
b, monosulcate pollen grain, proximal face SEM × 1500.
Marojejya insignis: **a** & **b,** *Moore* 9901.

Top left: ***Marojejya insignis***, staminate inflorescence, Madagascar. (Photo: J. Dransfield)

Top right: ***Marojejya insignis***, pistillate inflorescence, Madagascar. (Photo: J. Dransfield)

Bottom: ***Marojejya darianii***, infructescence, Madagascar. (Photo: J. Dransfield)

gynoecium gibbous, unilocular, uniovulate, gradually tapering to 3, large, triangular, recurved stigmas, ovule large, pendulous, campylotropous. *Fruit* asymmetrically globular, perianth persistent, stigmatic remains forming a large lateral beak; epicarp smooth, mesocarp thin, granular, endocarp composed of several layers of broad, soft anastomosing fibres closely adhering to the seed. *Seed* irregularly rounded, flattened, or ± kidney-shaped, smooth or grooved and ridged, subapically attached, endosperm homogeneous; embryo basal, opposite the fruit attachment. *Germination* adjacent-ligular; eophyll bifid, with or without a petiole. *Cytology*: 2n = 32.

Distribution and ecology: Two species endemic to the rain forests of Madagascar; occurring on hill slopes and in swamps in tropical rain forest from sea leavel to about 900 m altitude.
Anatomy: Root (Seubert 1998a, 1998b), ovule with distinct tannin and fibrous and vascular bundle layer around locule, ovule unusually large (Uhl unpublished).
Relationships: *Marojejya* is highly supported as monophyletic (Asmussen *et al.* 2006, Baker *et al.* in prep.). For relationships, see *Dypsis*. Note that *Marojejya* and *Masoala* form a monophyletic group in an analysis of morphological characters (Lewis 2002), but have never been resolved as sister taxa in molecular phylogenies (Lewis & Doyle 2002, Loo *et al.* 2006, Norup *et al.* 2006, Baker *et al.* in review, in prep.).
Common names and uses: *Ravin-be* (Malagasy). Much prized ornamentals.
Taxonomic accounts: Dransfield & Beentje (1995b).
Fossil record: No generic records found.
Notes: These are litter-trapping palms. In *Marojejya darianii*, the leaf sheath auricles tightly enclose the base of the crown, preventing rain water from escaping; adventitious roots grow from the internodes into the resulting tank of water and debris.
Additional figures: Fig. 8.3, Glossary fig. 6.

147. MASOALA

Robust squat litter-trapping palms from rain forest in Madagascar, with lax erect bisexual inflorescences, and fruit with apical stigmatic remains.

Masoala Jum., Ann. Inst. Bot.-Géol. Colon. Marseille série 5. 1(1): 8 (1933). Type: **M. madagascariensis** H. Jumelle.

Named after the Masoala Peninsula in northeast Madagascar whence the palm was first collected and described.

Robust, solitary, unarmed, pleonanthic, monoecious palms. *Stem* erect, short, covered with remains of the leaf sheaths. *Leaves* large, reduplicately pinnate, erect, forming a shuttlecock-like litter-trapping crown, marcescent; sheaths tubular, sparsely tomentose, attenuate distally, the margins smooth, elongate auricles sometimes present; petiole absent or very short, adaxially deeply channelled, abaxially rounded, sparsely tomentose; rachis adaxially ridged, abaxially rounded or flattened; leaflets numerous, regularly arranged, linear, single-fold except at the very tip where sometimes 2-fold, or irregular, broad, composed of many folds, the midribs strong, the tips briefly bifid, adaxially glabrous, abaxially minutely dotted, bearing several large, brown ramenta along the midrib, transverse veinlets obscure. *Inflorescences* solitary (very rarely possibly multiple; Dransfield & Beentje 1995b), interfoliar, branching to 2 orders, protandrous or protogynous; peduncle elongate, semicircular or crescent-shaped in cross-section, sparsely tomentose; prophyll inserted some distance above the base of the peduncle, large, flattened, narrow, elliptical, beaked, strongly 2-keeled, coriaceous, sparsely tomentose, splitting longitudinally along the abaxial face, persistent; peduncular bract similar to the prophyll but not 2-keeled, caducous or persistent, incomplete peduncular bracts several, membranous, relatively large, triangular, open; rachis shorter or longer than the peduncle, bearing spirally arranged, short triangular bracts each subtending a first-order branch; proximal first-order branches with a short bare portion, distally bearing 1–2 branches; rachillae ± straight, rather thick, elongate, bearing spirally arranged, very slightly sunken triads, each subtended by a thick, coriaceous, low triangular bract through most of the rachilla length, or bearing more crowded, more sunken pairs of staminate flowers towards the rachilla tips, or distal rachillae entirely staminate; floral bracteoles conspicuous, ± triangular. *Staminate flowers* symmetrical; sepals 3, distinct, imbricate, coriaceous, triangular, strongly keeled; petals 3, distinct, triangular, about 3 times as long as the sepals, coriaceous; stamens 6, filaments slender, distinct or briefly connate at the base, anthers elongate, basifixed, latrorse; pistillode columnar, deeply grooved, ± equalling the stamens in length. *Pollen* ellipsoidal asymmetric, occasionally oblate triangular; aperture a distal sulcus, less frequently a trichotomosulcus; ectexine tectate, coarsely

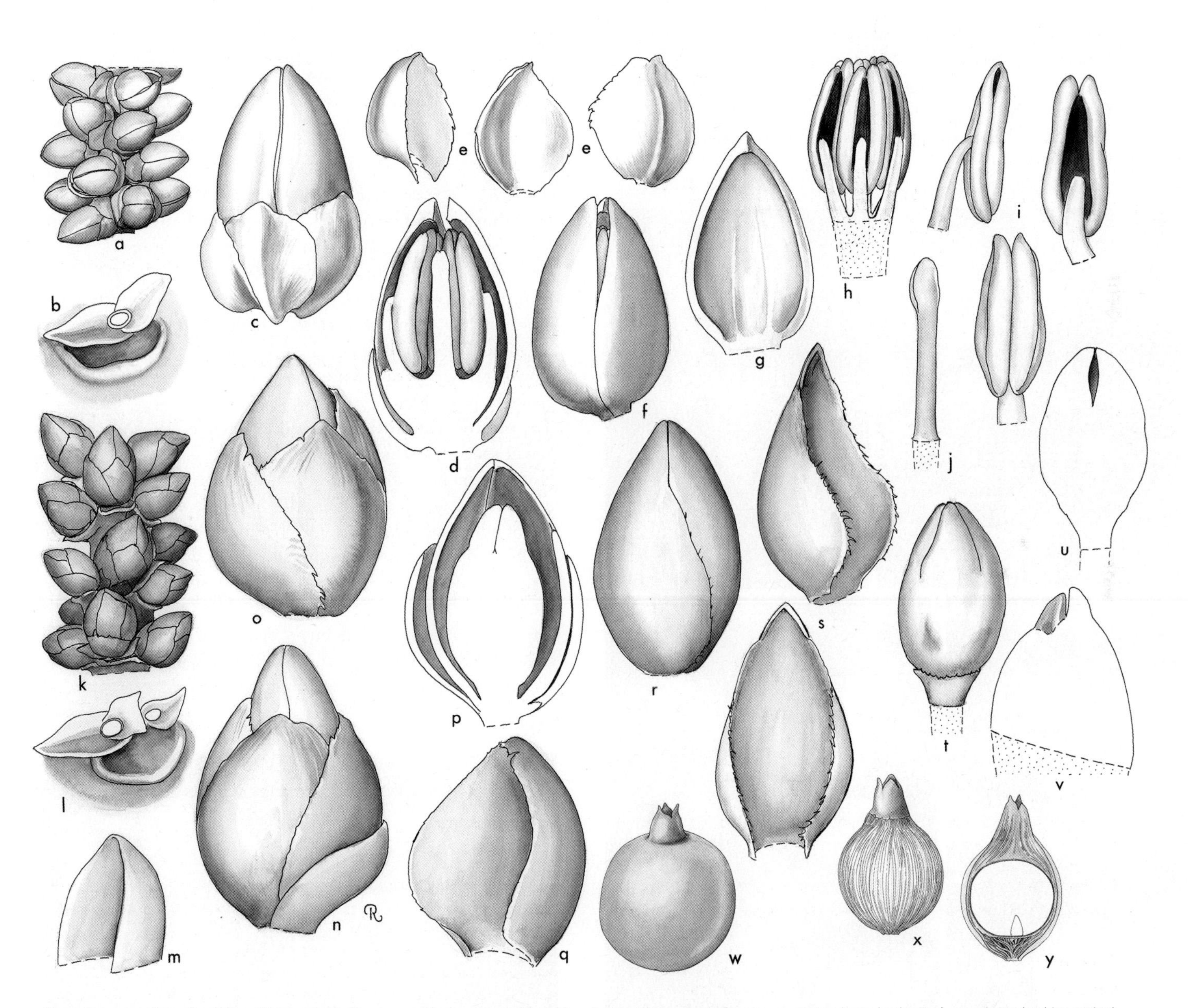

Masoala. **a**, portion of rachilla with staminate flowers × $1\,^{1}/_{2}$; **b**, scar and floral bracteoles of staminate flower × 3; **c**, staminate bud × 6; **d**, staminate bud in vertical section × 6; **e**, staminate sepals × 6; **f**, staminate flower, sepals removed × 6; **g**, staminate petal × 6; **h**, androecium × 6; **i**, stamen in 3 views × $7\,^{1}/_{2}$; **j**, pistillode × $7\,^{1}/_{2}$; **k**, portion of rachilla with triads × $1\,^{1}/_{2}$; **l**, triad, flowers removed to show scars and bracteoles, scar of left staminate flower hidden × 3; **m**, bract subtending pistillate flower × 6; **n**, pistillate bud with subtending bract × 6; **o**, pistillate bud × 6; **p**, pistillate bud in vertical section × 6; **q**, pistillate sepal × 6; **r**, pistillate bud, sepals removed × 6; **s**, pistillate petal in 2 views × 6; **t**, gynoecium × 6; **u**, gynoecium in vertical section × 6; **v**, vertical section of part of an older gynoecium; **w**, fruit × 2; **x**, endocarp × 2; **y**, longitudinal section of endocarp and seed × 2. *Masoala madagascariensis*: **a–v**, *Perrier de la Bâthie* 11938; **x–z**, *Beentje* 4632. (Drawn by Marion Ruff Sheehan)

perforate, finely perforate-rugulate or foveolate, aperture margin similar, or finer; infratectum columellate; longest axis ranging from 35–55 µm [2/2]. *Pistillate flowers* ovoid, much larger than the staminate; sepals 3, distinct, broadly imbricate, triangular, keeled, coriaceous, shiny; petals 3, distinct, imbricate with valvate tips, triangular, coriaceous, striate; staminodes 6, distinct, tooth-like; gynoecium ovoid, with a conspicuous beak, unilocular, uniovulate, stigmas 3, large, triangular, appressed in bud. *Fruit* ellipsoid, yellowish brown at maturity (*Masoala madagascariensis*), beaked and with stigmatic remains apical; epicarp smooth; mesocarp fleshy; endocarp composed of coarse longitudinal fibres. *Seed* depressed globose, basally attached; endosperm homogeneous; embryo basal. *Germination* adjacent-ligular; eophyll bifid. *Cytology*: 2n = 32.

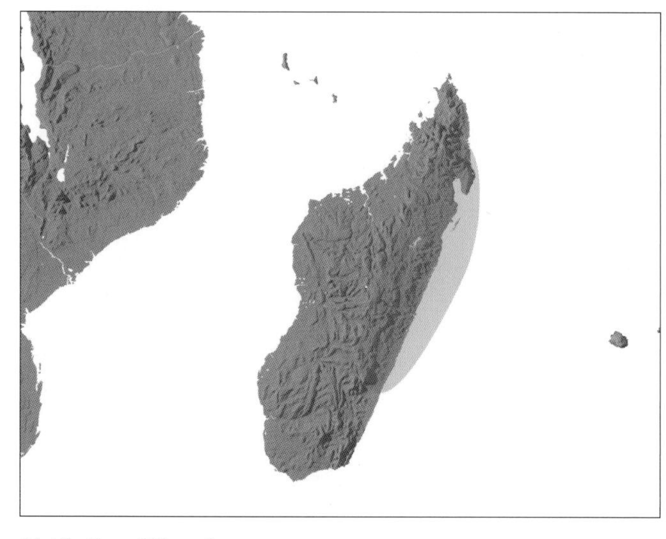
Distribution of ***Masoala***

Below left: ***Masoala madagascariensis***, habit, Madagascar. (Photo: J. Dransfield)
Below right: ***Masoala kona***, habit, Madagascar. (Photo: H. Beentje)
Bottom left: ***Masoala madagascariensis***, inflorescence, Madagascar. (Photo: J. Dransfield)
Bottom right: ***Masoala madagascariensis***, young fruit, Madagascar. (Photo: J. Dransfield)

Distribution and ecology: Two species known in eastern Madagascar, in tropical rain forest at low elevations.
Anatomy: Root (Seubert 1998a, 1998b).
Relationships: *Masoala* is moderately supported as monophyletic (Asmussen *et al.* 2006). For relationships, see *Dypsis* and *Marojejya*.
Common names and uses: *Kona*; sometimes used for thatch; growing point edible.
Taxonomic account: Dransfield & Beentje (1995b).
Fossil record: No generic records found.
Notes: Jumelle and Perrier (1945) describe a strong, truncate, undulate margined staminodal cupule to 3 mm high surrounding the ovary but we have found no trace of this; we attribute the supposed staminodal ring to possible differential shrinkage of the wall of the ovary. Like *Marojejya* spp., the two species of *Masoala* are litter-trapping palms. Superficially similar to *Marojejya*, the genus is distinguished by lax bisexual rather than congested unisexual inflorescences and by symmetrical fruit with apical stigmatic remains.

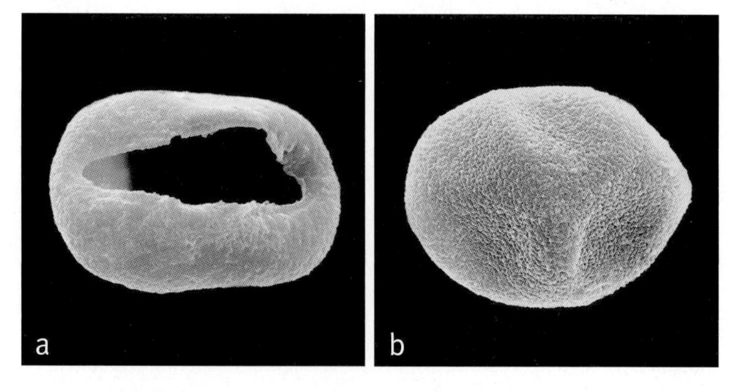

Above: ***Masoala***
a, monosulcate pollen grain, distal face SEM × 1000;
b, monosulcate pollen grain, proximal face SEM × 1000.
Masoala madagascariensis: **a** & **b**, *Perrier de la Bâthie* 11938.

Subtribe Linospadicinae Hook.f. in Benth. & Hook.f., Gen. pl. 3: 872, 876 (1883) ('*Linospadiceae*'). Type: **Linospadix**.

Small and slender to moderate, unarmed; sheaths not forming a crownshaft; leaves pinnate, sometimes irregularly so, or entire bifid; inflorescences interfoliar, sometimes multiple, always spicate; prophyll persistent, usually hidden among leaf sheaths; peduncular bract borne near base of peduncle, persistent or borne at tip of peduncle and caducous; triads borne in pits; fruit with apical stigmatic remains; embryo basal.

The subtribe contains four closely related genera occurring in New Guinea, the Moluccas, Australia and Lord Howe Island. Morphologically, the subtribe is well defined by the lack of crownshafts, the spicate inflorescences with flowers borne in deep pits and the fruit with apical stigmatic remains. However, the Linospadicinae is not resolved as monophyletic in any molecular phylogeny that includes *Calyptrocalyx*, although the relationships that place *Calyptrocalyx* outside the subtribe are not highly supported. Nevertheless, *Laccospadix*, *Linospadix* and *Howea* are usually resolved as monophyletic, often with high support (Asmussen *et al.* 2006, Norup *et al.* 2006, Baker *et al.* in review, in prep.). The position of the subtribe remains uncertain within the western Pacific clade of Areceae (Lewis 2002, Lewis & Doyle 2002, Norup *et al.* 2006, Baker *et al.* in review, in prep.).

Key to Genera of Linospadicinae

1. Peduncular bract inserted near the base of the peduncle, ± flattened, opening apically, shorter than and not enclosing the spike before anthesis, ± persistent and marcescent; filaments mostly inflexed; anthers dorsifixed; endocarp adherent to the seed. Moluccas to New Guinea . . . . **148. Calyptrocalyx**
1. Peduncular bract inserted at the tip of the peduncle, tubular, enclosing the spike in bud, marcescent or caducous, leaving a ruffled scar on the peduncle; filaments not inflexed, anthers basifixed; endocarp adherent or free from the seed 2
2. Endocarp thick, cartilaginous, not adhering to the seed; endosperm homogeneous; staminate flower with 30–70 stamens; stems solitary, stout; leaflets always 1-ribbed. Lord Howe Island . . . . **150. Howea**
2. Endocarp thin, adhering to the seed; endosperm homogeneous or ruminate; staminate flower with 6–15 stamens; stems usually clustered and slender; leaflets 1-many ribbed 3
3. Seed with ruminate endosperm, raphe extending its length, the branches reticulate; leaflets 1-ribbed. Australia . . . . **151. Laccospadix**
3. Seed with homogeneous endosperm, raphe extending 1/3 its length or less, the branches free or anatomosing; leaflets 1-several ribbed or the blade undivided. New Guinea and Australia . . . . **149. Linospadix**

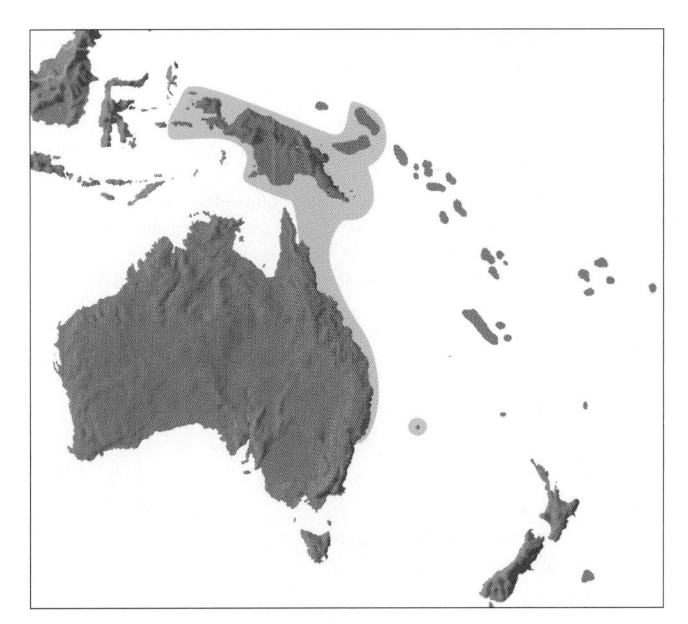

Distribution of subtribe **Linospadicinae**

148. CALYPTROCALYX

Small to moderate, solitary or clustering undergrowth palms of rain forest in the Moluccas and New Guinea, with spicate inflorescences with the peduncular bract inserted just above the prophyll at the base of the peduncle.

Calyptrocalyx Blume, Bull. Sci. Phys. Nat. Néerl. 1: 66 (1838). Type: **C. spicatus** (Lam.) Blume (*Areca spicata* Lam.).

Linospadix Becc. ex Hook.f. in Benth. & Hook.f., Gen. pl. 3: 903 (1883) (non H. Wendl. 1875). Lectotype: *L. arfakianus* Becc. (see Beccari & Pichi-Sermolli 1955) = *Calyptrocalyx arfakianus* (Becc.) Dowe & M.D. Ferrero.

Paralinospadix Burret, Notizbl. Bot. Gart. Berlin-Dahlem 12: 331 (1935). Lectotype: *P. arfakianus* (Becc.) Burret (*Linospadix arfakianus* Becc.) (see Beccari & Pichi-Sermolli 1955) = *Calyptrocalyx arfakianus* (Becc.) Dowe & M.D. Ferrero.

Calyptra — *a veil*, calyx — *outer perianth of flower, in reference to the way the outermost sepal covers the rest of the flower in bud.*

Solitary or more often clustering, small to moderate, unarmed, pleonanthic, monoecious palms. *Stem* erect, usually becoming bare, conspicuously ringed with leaf scars. *Leaves* undivided with bifid apices and pinnately ribbed, or pinnate, marcescent or neatly abscising, a crownshaft scarcely developed; sheaths soon splitting opposite the petiole, scaly or not, the margins usually fibrous; petiole absent, short or long, adaxially channelled, abaxially rounded; lamina bifid with acute or lobed tips, or leaflets 1–several ribbed, acute, acuminate or sometimes toothed, concolorous, variously scaly or glabrous. *Inflorescences* solitary or multiple, interfoliar, protandrous, spicate; peduncle winged at the base, erect or pendulous, elongate; prophyll inserted at the very base, ± included within the sheaths, common to all axes where inflorescences multiple, tubular, flattened, ± 2-winged, tending to disintegrate into fibres apically; peduncular bract inserted near the peduncle base, tubular, open apically, shorter than the spike and not enclosing it, persistent or rotting on the inflorescence; spike short to elongate bearing a dense spiral of low, rounded bracts forming the lower lips of usually deep floral pits, each enclosing a triad except at the apex where enclosing paired or solitary staminate flowers, flowers exserted one at a time; floral bracteoles small, included. *Staminate flowers* sessile or briefly pedicellate; sepals 3, distinct, imbricate, often keeled; petals 3, distinct, about twice as long as the sepals, valvate, marked within by stamen impressions; stamens 6–140, filaments elongate, linear, erect or usually inflexed at the apex in bud, distinct or, in *Calyptrocalyx doxanthus*, united for 3/4 their length to form a staminal tube, anthers dorsifixed, erect or versatile, mostly deeply sagittate basally, latrorse; pistillode sometimes lacking or slender and ± club-shaped, about as long as stamens. *Pollen* ellipsoidal, asymmetric to pyriform; aperture a distal sulcus; ectexine tectate, perforate, or perforate-rugulate, aperture margin similar or slightly finer; infratectum columellate; longest axis 36–61 µm [4/27]. *Pistillate flowers* ± globular; sepals 3, distinct, imbricate; petals 3, distinct, exceeding the sepals, broadly imbricate except at the minutely valvate tips; staminodes (2–)3–9; gynoecium unilocular, uniovulate, ellipsoidal with apical button-like or trifid stigma, ovule hemianatropous, attached laterally above the middle of the locule. *Fruit* small to large, orange, bright red, pink or purplish black at maturity, perianth whorls persistent, stigmatic remains apical or slightly eccentric; epicarp smooth, fragile, glabrous or rarely pilose or with scattered scales, mesocarp fleshy or dry, white or pink, endocarp thin, closely adhering to or separating from the seed. *Seed* subbasally to laterally attached, hilum short to elongate, raphe branches anastomosing, endosperm homogeneous or ruminate; embryo basal. *Germination* adjacent-ligular; eophyll bifid. *Cytology*: 2n = 32.

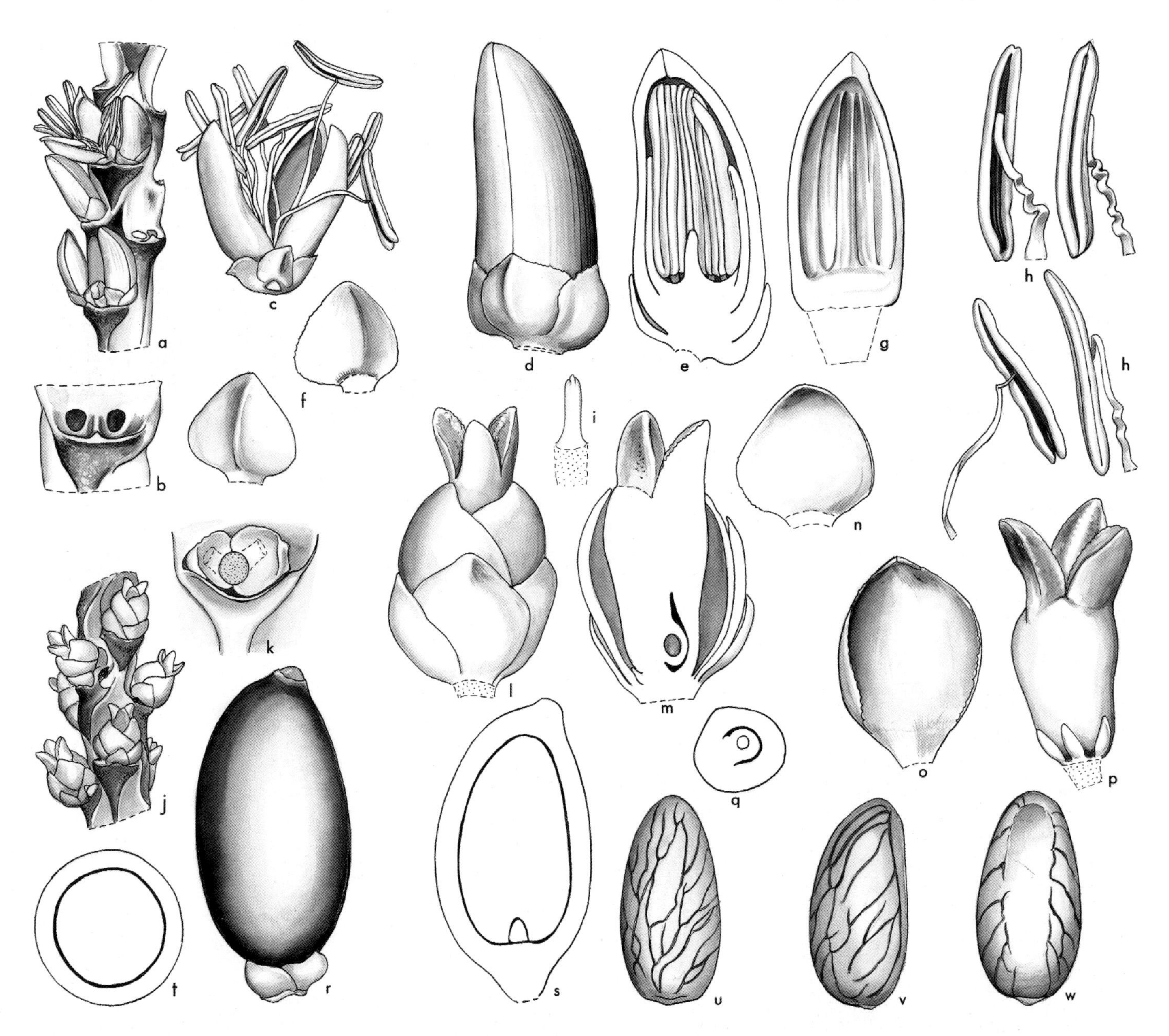

Calyptrocalyx. **a**, portion of rachilla with triads × 3; **b**, scars of paired staminate flowers × 6; **c**, staminate flower × 6; **d**, staminate bud × 12; **e**, staminate bud in vertical section × 12; **f**, staminate sepal in 2 views × 12; **g**, staminate petal, interior view × 12; **h**, stamen in 4 views × 12; **i**, pistillode × 12; **j**, portion of rachilla at pistillate anthesis × 3; **k**, triad, flowers removed to show bracteoles × 6; **l**, pistillate flower × 12; **m**, pistillate flower in vertical section × 12; **n**, pistillate sepal, interior view × 12; **o**, pistillate petal, interior view × 12; **p**, gynoecium and staminodes × 12; **q**, ovary in cross-section × 12; **r**, fruit × 3; **s**, fruit in vertical section × 3; **t**, fruit in cross-section × 3; **u, v, w**, seed in 3 views × 3. *Calyptrocalyx hollrungii*: **a–q**, *Moore et al.* 9274; **r–w**, *Moore et al.* 9265. (Drawn by Marion Ruff Sheehan)

Distribution and ecology: About 27 species, one widespread in the Moluccas, the remainder in New Guinea. Undergrowth palms of primary rain forest occurring at elevations from sea level to ca. 1000 m in mountains, usually on montane slopes with well-drained soils, more rarely along streams or sometimes gregarious in swampy or poorly drained areas.
Anatomy: Leaf (Tomlinson 1961), root (Seubert 1998a, 1998b), and fruit (Essig 2002).
Relationships: *Calyptrocalyx* is strongly supported as monophyletic (Savolainen *et al.* 2006, Baker *et al.* in prep.). See earlier discussion under subtribe Linospadicinae. The relationships between *Calyptrocalyx* and genera from outside subtribe Linospadicinae are never strongly supported. The most robust of these places *Calyptrocalyx* as sister to subtribe Archontophoenicinae with moderate support (Baker *et al.* in prep.).
Common names and uses: Not recorded.
Taxonomic account: Dowe & Ferrero (2001).
Fossil record: No generic records found.
Notes: The structure of the inflorescence and the insertion of major bracts are constant features that bind together a series of species that exhibit marked differences in habit, foliage and to some degree habitat.

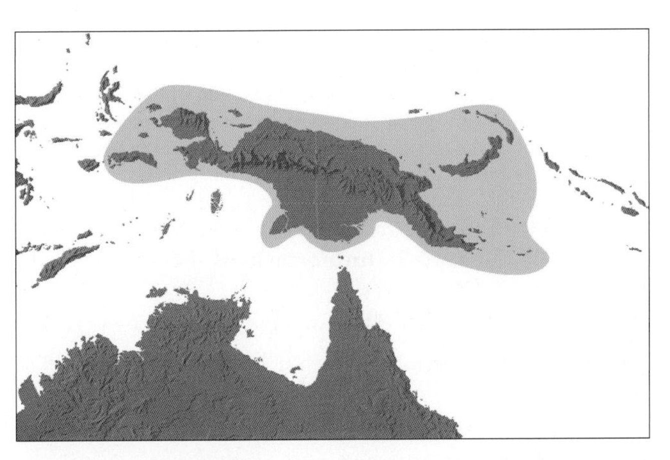

Distribution of ***Calyptrocalyx***

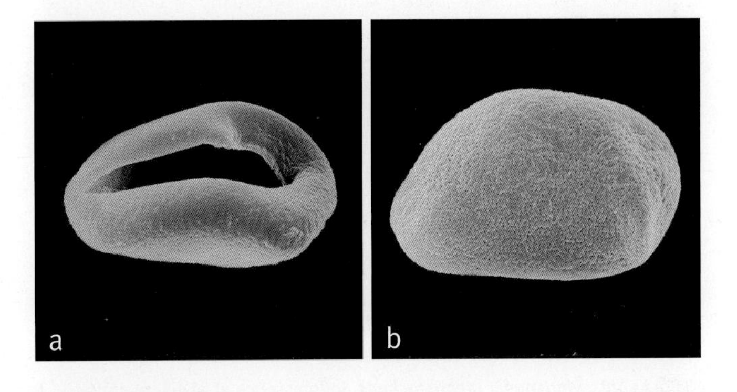

Above: ***Calyptrocalyx***
a, monosulcate pollen grain, distal face SEM × 1500;
b, monosulcate pollen grain, proximal face SEM × 1500.
Calyptrocalyx lauterbachianus: **a** & **b**, *H & Kini W.* LAE 72547.

Below left: ***Calyptrocalyx hollrungii***, habit, Papua New Guinea. (Photo: W.J. Baker)
Below middle: ***Calyptrocalyx micholitzii***, habit, Papua. (Photo: J. Dransfield)
Below right: ***Calyptrocalyx hollrungii***, infructescence, Papua New Guinea. (Photo: W.J. Baker)

149. LINOSPADIX

Small or clustering undergrowth palms of rain forest in New Guinea and eastern Australia, with spicate inflorescences with the peduncular bract inserted far above the prophyll at the base of the flower-bearing part of the inflorescence; seed with ruminate endosperm.

Linospadix H. Wendl. in H. Wendl. & Drude, Linnaea 39: 177, 198 (1875). Type: **L. monostachyos** (Mart.) H. Wendl. ('*monostachyos*') (*Areca monostachya* Mart.).

Bacularia F. Muell. ex Hook.f., Bot. Mag. 6644 (1882) superfluous name. Type: *B. monostachya* (Mart.) F. Muell. ex Hook.f. (*Areca monostachya* Mart.).

Linon — *flax or thread,* spadix — *branch or frond, but in botany, inflorescence, referring to the slender spicate inflorescence.*

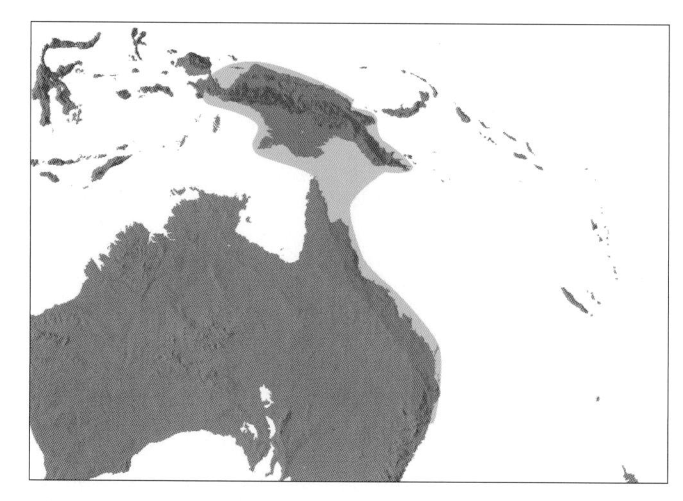
Distribution of ***Linospadix***

Small to very small, solitary or clustered, unarmed, pleonanthic, monoecious palms. *Stem* erect, slender, eventually becoming bare, conspicuously ringed with leaf scars. *Leaves* bifid to pinnate, neatly abscising or marcescent, a crownshaft not well developed; sheaths soon splitting opposite the petiole, bearing scattered scales, the margins often becoming fibrous, a tattering ligule sometimes present; petiole ± absent or very short to long, usually scaly; rachis very short to long; blade bifid with acute, acuminate or lobed tips, or divided into 1–several fold leaflets, the leaflets regular or irregular, acute, acuminate, bifid or irregularly lobed and praemorse, often bearing minute scales on both surfaces, transverse veinlets usually obscure. *Inflorescences* solitary, interfoliar, ± erect, protandrous, unbranched; peduncle winged at the base; prophyll inserted near the base of the peduncle, tubular, 2-keeled, ± included within the subtending leaf sheath, persistent, becoming tattered and fibrous; peduncular bract 1, inserted in the distal part of the peduncle or at its very tip, tubular, ± beaked, ± enclosing the spike in bud, soon splitting longitudinally, deciduous;

Below left: ***Linospadix minor***, habit, in fruit, Australia. (Photo: J. Dransfield)

Below middle: ***Linospadix monostachyos***, inflorescence at staminate anthesis, Australia. (Photo: W.J. Baker)

Below right: ***Linospadix monostachyos***, inflorescence at pistillate anthesis, Australia. (Photo: W.J. Baker)

spike short to elongate, variously scaly, bearing dense to lax, spirally arranged, broad, ± triangular bracts forming the lower lips of very shallow floral pits, each bearing a triad except at the tip where bearing solitary or paired staminate flowers, the flowers exposed, not enclosed by the pit; floral bracteoles minute. *Staminate flowers* ± sessile, ± symmetrical, both of a triad apparently developing ± at the same time; sepals 3, distinct, broadly imbricate, ± keeled; petals 3, distinct, about twice as long as the sepals, with very thick valvate tips, internally marked with stamen impressions; stamens 6–12 erect, filaments very short, anthers dorsifixed, ± linear, apically acute, basally ± sagittate, latrorse; pistillode very small, minutely 3-pointed. *Pollen* ellipsoidal, slightly asymmetric to lozenge-shaped; aperture a distal sulcus; ectexine tectate, coarsely perforate, or coarsely perforate-rugulate, aperture margin slightly finer; infratectum columellate; longest axis 24–41 µm [3/9]. *Pistillate flowers* eventually much larger than the staminate; sepals 3, distinct, broadly imbricate; petals 3, distinct, slightly exceeding the sepals, with broad imbricate bases and conspicuous, thickened, triangular, valvate tips; staminodes 3–6, irregularly lobed and tooth-like; gynoecium unilocular, uniovulate, ± ovoid, with 3 short stigmas, becoming recurved, ovule laterally attached near the base, hemianatropous (?always). *Fruit* ellipsoidal to spindle-shaped, rarely curved, bright red (?always) at maturity, perianth whorls persistent, the stigmatic remains apical; epicarp smooth, mesocarp thin, fleshy, with thin fibres next to the endocarp, endocarp very thin, closely adhering to the seed. *Seed* subbasally attached, the raphe extending ca. 1/3 the seed length, or less, the branches free or anastomosing, endosperm homogeneous; embryo basal. *Germination* adjacent-ligular; eophyll bifid. *Cytology*: 2n = 32.

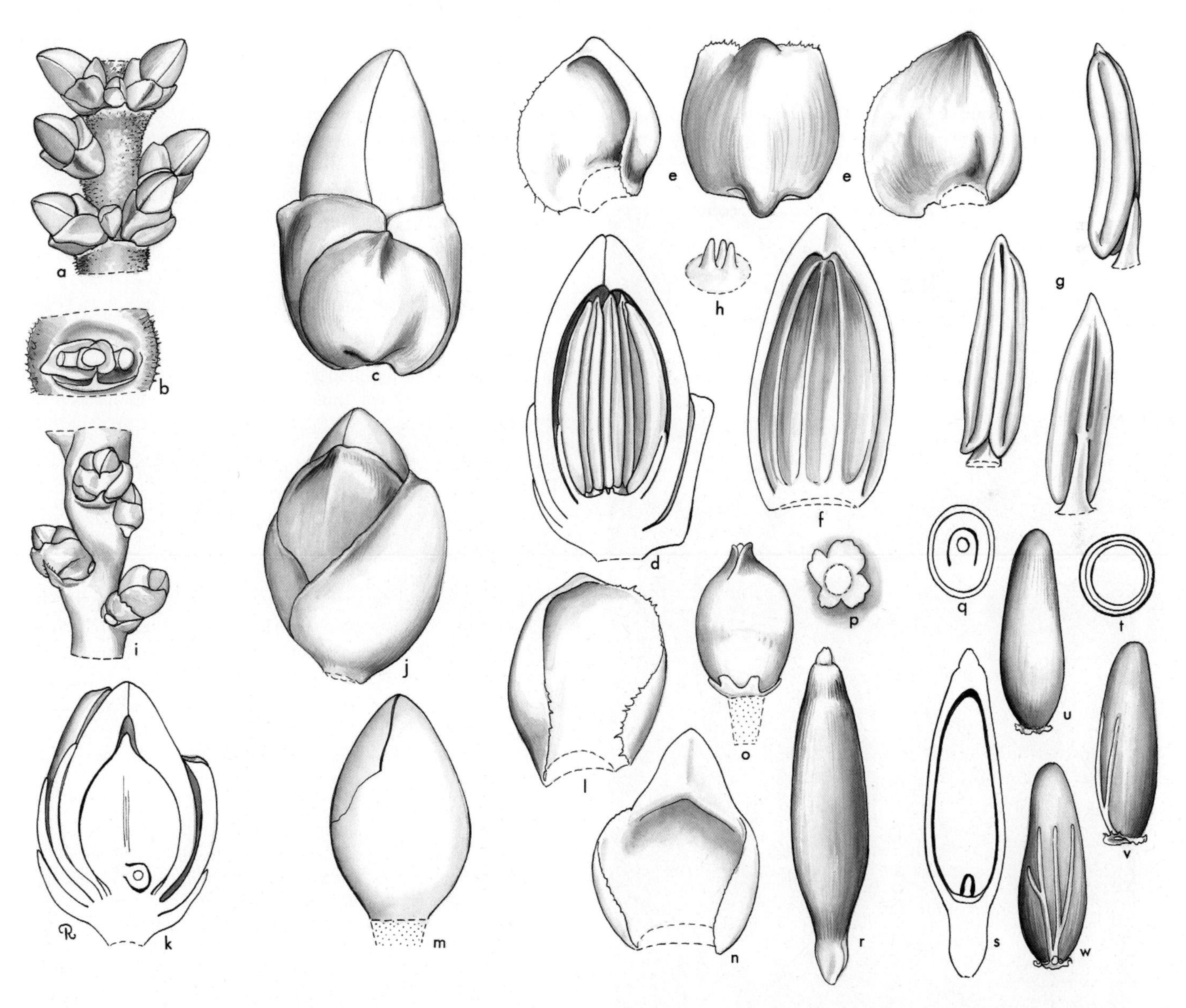

Linospadix. **a,** portion of rachilla with triads × 3; **b,** triad, flowers removed to show bracteoles × 6; **c,** staminate bud × 12; **d,** staminate bud in vertical section × 12; **e,** staminate sepals × 12; **f,** staminate petal, interior view × 12; **g,** stamen in 3 views × 12; **h,** pistillode × 25; **i,** portion of rachilla in pistillate bud × 3; **j,** pistillate bud × 12; **k,** pistillate bud in vertical section × 12; **l,** pistillate sepal, interior view × 12; **m,** pistillate bud, sepals removed × 12; **n,** pistillate petal, interior view × 12; **o,** gynoecium and staminodes × 12; **p,** staminodes from below × 12; **q,** ovary in cross-section × 12; **r,** fruit × 3; **s,** fruit in vertical section × 3; **t,** fruit in cross-section × 3; **u, v, w,** seed in 3 views × 3. *Linospadix monostachyos*: **a–q,** *Moore* 9232; *L. albertisianus*: **r–w,** *Brass* 6893. (Drawn by Marion Ruff Sheehan)

Distribution and ecology: Nine species, two species in New Guinea, the rest in Australia. Minute to small palms of the undergrowth of tropical rain forest, especially at higher elevations.
Anatomy: Root (Seubert 1998a, 1998b) and fruit (Essig 2002).
Relationships: *Linospadix* is strongly supported as monophyletic (Savolainen *et al.* 2006, Baker *et al.* in prep.) and also highly supported as sister to a moderately supported clade of *Howea* and *Laccospadix* (Savolainen *et al.* 2006, Norup *et al.* 2006, Baker *et al.* in prep.). An alternative, less robust phylogeny places *Howea* as sister to a clade of *Laccospadix* and *Linospadix* (Baker *et al.* in review).
Common names and uses: Walking stick palms. Stems of *Linospadix monostachyos* have been used as walking sticks. The 'cabbage' is edible and the mesocarp, though thin, is pleasantly acid to taste.
Taxonomic accounts: Dowe & Ferrero (2001) and Dowe & Irvine (1997).

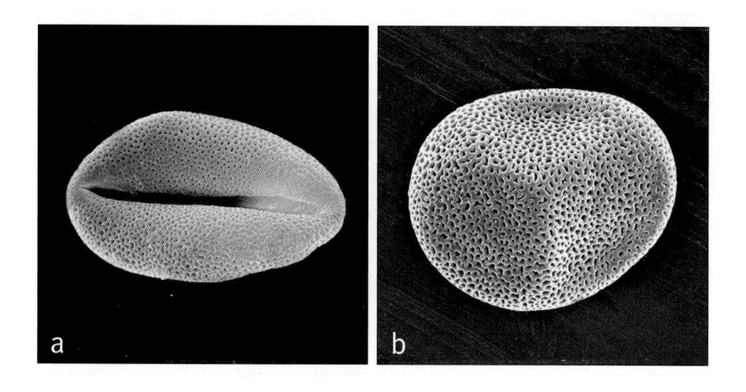

Above: ***Linospadix***
a, monosulcate pollen grain, distal face SEM × 1500;
b, monosulcate pollen grain, proximal face SEM × 1500.
Linospadix palmerianus: **a** & **b**, *Blake* 9820.

Fossil record: No generic records found.
Notes: *Linospadix* is immediately distinguished from the smaller species of *Calyptrocalyx* by the insertion of the peduncular bract towards the apex of the peduncle just below the flower-bearing portion of the spike, and by the fact that the bract is caducous.

150. HOWEA

Spectacular pinnate-leaved palms endemic to Lord Howe Island, where they occur in huge populations; distinctive in the robust pendulous spicate inflorescences and staminate flowers with large numbers of stamens.

Howea Becc., Malesia 1: 66 (1877) ('*Howeia*'). Lectotype: **H. belmoreana** (C. Moore & F. Muell.) Becc. (*Kentia belmoreana* C. Moore & F. Muell.) (see O.F. Cook [1926] and Beccari & Pichi-Sermolli [1955]).
Grisebachia Drude & H. Wendl., Nachr. Königl. Ges. Wiss. Georg-Augusts-Univ. 1875: 55 (1875) (non Klotzsch 1838). Lectotype: *G. belmoreana* (C. Moore & F. Muell.) Drude & H. Wendl. (*Kentia belmoreana* C. Moore & F. Muell.) (see O.F. Cook [1926] and Baccari & Pichi-Sermolli [1955]).
Denea O.F. Cook, J. Wash. Acad. Sci.16: 395 (1926). Type: *D. forsteriana* (F. Muell.) O.F. Cook (*Kentia forsteriana* F. Muell. = *Howea forsteriana* [F. Muell.] Becc.).

Derived from Lord Howe Island, which in turn commemorates Admiral Lord Richard Howe (1726–1799).

Moderate, solitary, unarmed, pleonanthic, monoecious palms. *Stem* erect, bare, conspicuously marked with close, horizontal or oblique leaf scars, the base sometimes expanded into a knob. *Leaves* pinnate, neatly abscising but not forming a crownshaft; sheath well developed, splitting longitudinally opposite the petiole, disintegrating into an interwoven mass of fine fibres; petiole short to moderately long, flattened or slightly channelled adaxially, abaxially ± angled, sparsely to densely scaly; rachis ± rounded to angled abaxially, adaxially angled, scaly as the petiole; leaflets numerous, single-fold, regularly arranged, curved or stiffly

Left: ***Howea belmoreana***, habit, Lord Howe Island. (Photo: W.J. Baker)

ascending, acute, acuminate or minutely bifid, adaxially with sparse scattered scales, abaxially ± glabrous or rather densely dotted with scales and bearing abundant floccose indumentum and ramenta along the midrib, transverse veinlets obscure. *Inflorescences* interfoliar, sometimes becoming infrafoliar after leaf fall, short or almost as long as the leaves, spicate, solitary or compound with up to 3–8 borne together on a common axillary boss, erect at first, later pendulous, protandrous; peduncle ± elliptic in cross-section, much shorter than or ± equalling the rachis, densely scaly; prophyll tubular, membranous; peduncular bract inserted near to or some distance from the prophyll, enclosing the inflorescence until anthesis, ± membranous, tubular, later splitting down its length, disintegrating and falling, leaving a low collar; rachis robust, scaly, densely covered with spirally arranged, ± spreading, low, rounded or triangular, rigid, coriaceous bracts, each forming a lip to a floral pit, enclosing a triad of flowers, except at the very tips where pits enclosing paired staminate flowers; floral bracteoles ± sepal-like. *Staminate flowers* partially exserted one at a time from the pit at anthesis; sepals 3, distinct, imbricate, usually keeled, ± rounded, the margins toothed; corolla with a stalk-like base ± as long as the sepals, and 3 ovate, valvate lobes; stamens 30–70 or more, filaments elongate, variously connate at the base for much of their length, the connective sometimes prolonged into a point, anthers elongate, ± latrorse; pistillode absent. *Pollen* ellipsoidal, asymmetric to pyriform; aperture a distal sulcus; ectexine tectate, finely perforate-rugulate, or granular-rugulate, especially on proximal face, aperture margin finely perforate-rugulate; infratectum columellate; longest axis 37–52 µm [2/2]. *Pistillate flowers* ± globular; sepals 3, distinct, imbricate, rounded, the margins toothed; petals 3, distinct, basally strongly imbricate, the tips briefly valvate; staminodes 3–6, forming a low, irregularly lobed, membranous ring, or irregularly separated as triangular or bifid flanges; gynoecium unilocular,

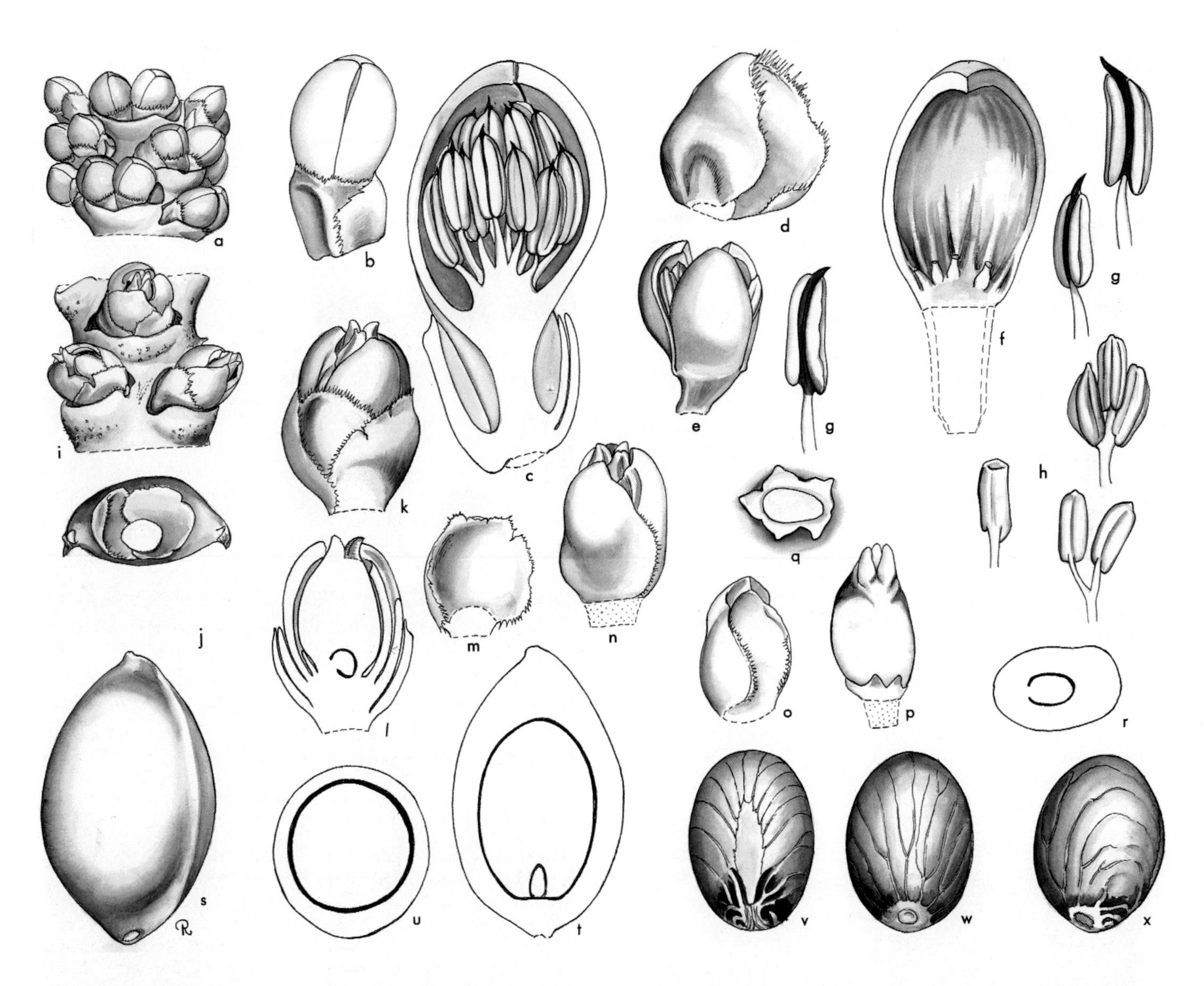

Howea. **a**, portion of spike in staminate bud × 1½; **b**, staminate bud × 3; **c**, staminate bud in vertical section × 6; **d**, staminate sepal × 6; **e**, staminate flower, sepals removed × 3; **f**, staminate petal, interior view × 6; **g**, normal stamen in 3 views × 6; **h**, unusual solitary and branched stamens with diminished connective in 3 views × 6; **i**, portion of spike at pistillate anthesis × 1½; **j**, view into pit to show bracteoles × 3; **k**, pistillate flower × 3; **l**, pistillate flower in vertical section × 3; **m**, pistillate sepal, interior view × 3; **n**, pistillate flower, sepals removed × 3; **o**, pistillate petal, interior view × 3; **p**, gynoecium and staminodes × 3; **q**, staminodes × 3; **r**, ovary in cross-section × 6; **s**, fruit × 1½; **t**, fruit in vertical section × 1½; **u**, fruit in cross-section × 1½; **v, w, x**, seed in 3 views × 1½. *Howea forsteriana*; **a–h**, *Bailey* 9053; **i–r**, Sydney Botanic Garden; **s–x**, *Moore* 9254. (Drawn by Marion Ruff Sheehan)

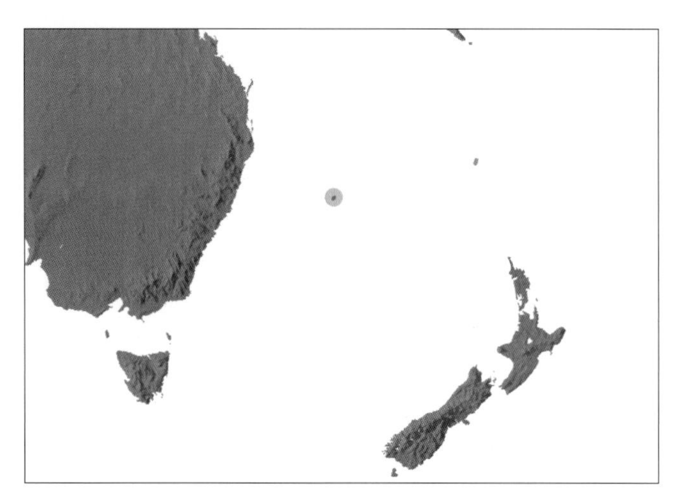

Distribution of ***Howea***

uniovulate, tipped with 3 short stigmas ± reflexed at anthesis, ovule laterally attached, campylotropous. *Fruit* ovoid, sometimes faintly ridged, 1-seeded, shiny dark green at first, turning dull yellowish-green or reddish-brown, perianth whorls persistent, stigmatic remains apical; epicarp smooth, mesocarp rather thinly fleshy with abundant longitudinal fibres, endocarp cartilaginous, not adhering to the seed. *Seed* laterally attached, raphe extending $^1/_3$ the length of the seed or less, endosperm homogeneous; embryo basal. *Germination* adjacent-ligular; eophyll bifid. *Cytology*: 2n = 32.

Distribution and ecology: Two easily distinguished species endemic to Lord Howe Island: *Howea forsteriana* is abundant on the island in lowland forest on sandy areas; *H. belmoreana* can be found as scattered individuals with *H. forsteriana*, but becomes abundant at higher elevations up to about 450 m above sea level. Wind has been shown to be the primary pollinator, one of the few proven examples in the palms (Savolainen *et al.* 2006).

Right: ***Howea forsteriana***, crown with multiple inflorescences, Lord Howe Island. (Photo: W.J. Baker)

Below: ***Howea***
a, two monosulcate pollen grains: left proximal face, right distal face SEM × 750;
b, close-up, granular-rugulate surface SEM × 8000.
Howea forsteriana: **a** & **b**, cultivated RBG Kew H 155-17.

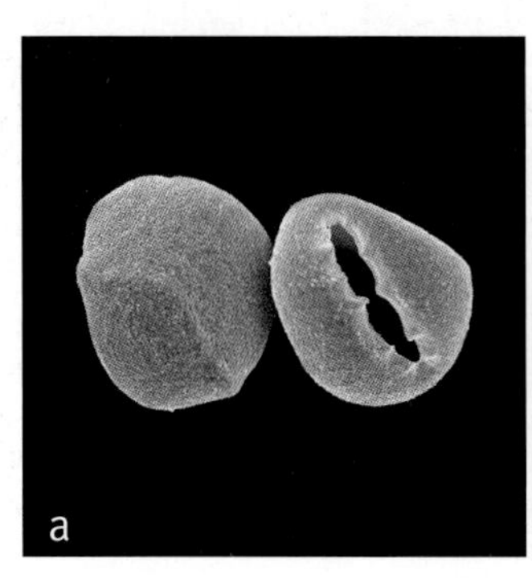

Anatomy: Leaf (Tomlinson 1961), root (Seubert 1998a, 1998b), and fruit (Essig 2002).
Relationships: *Howea* is strongly supported as monophyletic (Savolainen *et al.* 2006, Baker *et al.* in prep.). For relationships, see *Linospadix*.
Common names and uses: Kentia palms, Howea palms, sentry palms. Both species, but especially *Howea forsteriana*, are important as commercially grown ornamentals.
Taxonomic account: Bailey (1939a).
Fossil record: No generic records found.
Notes: The two species of *Howea* are the largest palms in the Linospadicinae. *Howea belmoreana* is distinguished by curved leaves with erect leaflets and by a single spike in each leaf axil, while *H. forsteriana* has rather flat leaves with drooping leaflets and several spikes in each leaf axil. Flowers and fruits mature slowly so that several inflorescences in different stages are often present on a single tree.

Recent research has demonstrated that *Howea* speciated sympatrically on Lord Howe Island and is thus a rare convincing example of this controversial mode of speciation (Savolainen *et al.* 2006).
Additional figures: Fig. 1.4, Ch. 7 frontispiece.

151. LACCOSPADIX

Short or acaulescent pinnate-leaved palms endemic to the rain forests of Queensland, Australia; distinctive in the spicate inflorescences and staminate flowers with 9–12 stamens.

Laccospadix Drude & H. Wendl., Nachr. Königl. Ges. Wiss. Georg-Augusts-Univ. 1875: 59 (1875). Type: **L. australasicus** H. Wendl. & Drude.

Lakkos — *pit*, spadix — *branch or frond, but in botany, inflorescence, referring to the spicate inflorescence with pits.*

Small to moderate, solitary or clustering, unarmed, pleonanthic, monoecious palm. *Stem* erect, becoming bare, conspicuously ringed with leaf scars. *Leaves* pinnate, marcescent; sheaths soon splitting opposite the petiole, bearing scattered scales, the margins becoming fibrous; petiole long, adaxially channelled, abaxially rounded, bearing scattered scales; rachis curved, scaly like the petiole; leaflets numerous, single-fold, conspicuously plicate, acute or acuminate, adaxially and abaxially with minute scattered scales, abaxially with conspicuous ramenta along the main ribs, transverse veinlets not visible. *Inflorescences* solitary, interfoliar, protandrous, unbranched; peduncle winged at the base, erect, elongate, bearing scattered caducous scales; prophyll inserted at the base of the peduncle, flattened, tubular, 2-winged, ± included within the leaf sheaths, tending to disintegrate into fibres apically; peduncular bract 1, inserted near the tip of the peduncle, enclosing the spike in bud, splitting longitudinally, deciduous at anthesis; rachis ± equalling the peduncle, pendulous, bearing a slightly spiralled series of prominent, rounded bracts forming the lower lips of the floral pits; floral pits enclosing triads except at the apex where enclosing paired or solitary staminate flowers only, flowers exserted one at a time; floral bracteoles 3, ± sepal-like. *Staminate flowers* borne on very short flattened pedicels, ± symmetrical; sepals 3, distinct, keeled, ± chaffy, with irregular margins; petals 3, ± twice as long as the sepals, distinct, triangular-ovate, valvate, marked within by anther impressions; stamens 9–12, erect, filaments very short, anthers ± elongate, basifixed, latrorse; pistillode minute, trifid. *Pollen* ellipsoidal, asymmetric; aperture a distal sulcus; ectexine tectate, finely perforate-rugulate, or finely granular-rugulate, especially on proximal face, aperture margin finely perforate-rugulate; infratectum columellate; longest axis 36–42 µm [1/1]. *Pistillate flowers* ± globular; sepals 3, distinct, ± chaffy, with irregular margins; petals similar to but longer than sepals, distinct, broadly imbricate except at the valvate, triangular tip; staminodes 3, small, ± triangular; gynoecium unilocular, uniovulate, ovoid-ellipsoidal, tipped with 3, short, recurved stigmas, ovule attached laterally near the base, campylotropous. *Fruit* ellipsoidal when fresh, red at maturity, perianth whorls persistent, the stigmatic scar apical; epicarp smooth, mesocarp thin, fleshy, overlying stout longitudinal fibres,

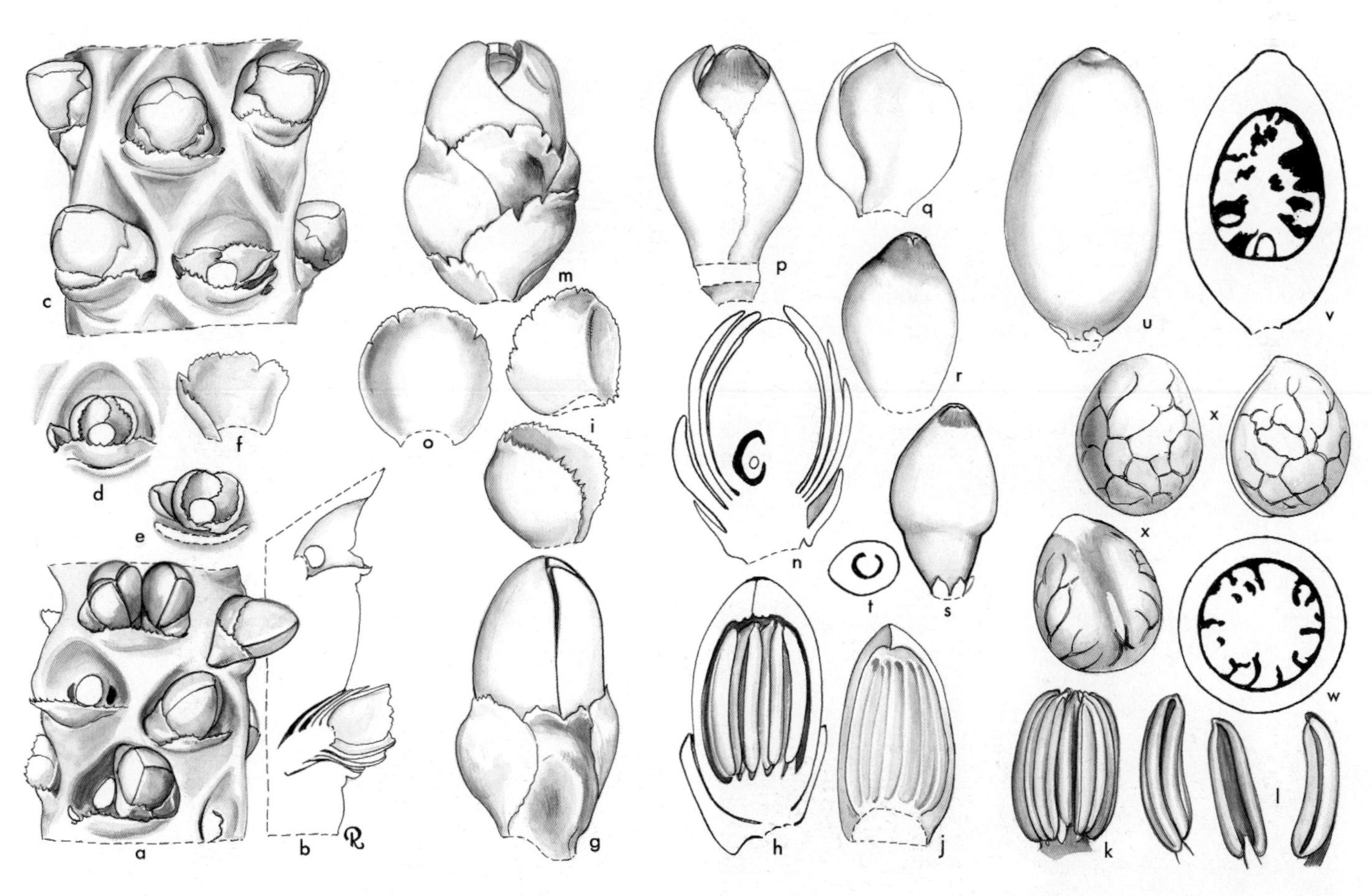

Laccospadix. **a**, portion of rachilla at staminate anthesis × 3; **b**, portion of rachilla at staminate anthesis in vertical section × 3; **c**, portion of rachilla in pistillate bud, staminate flowers fallen × 3; **d**, triad, flowers removed × 2; **e**, triad, flowers and lip removed × 3; **f**, bracteole subtending pistillate flower × 3; **g**, staminate bud × 6; **h**, staminate bud in vertical section × 6; **i**, staminate sepal in 2 views × 6; **j**, staminate petal, interior view × 6; **k**, androecium × 4; **l**, stamen in 3 views × 6; **m**, pistillate flower with bracteoles × 6; **n**, pistillate flower in vertical section × 6; **o**, pistillate sepal, interior view × 6; **p**, pistillate flower, sepals removed × 6; **q**, pistillate petal, interior view × 6; **r**, gynoecium × 6; **s**, gynoecium and staminodes × 6; **t**, ovary in cross-section × 6; **u**, fruit × 3; **v**, fruit in vertical section × 3; **w**, fruit in cross-section × 3; **x**, seed in 3 views × 3. *Laccospadix australasicus*: all from *Moore & Volk* 9240. (Drawn by Marion Ruff Sheehan)

endocarp thin, adherent to seed. *Seed* laterally attached with short oblong hilum, the raphe ± extending the length of the seed, the raphe branches anastomosing, endosperm deeply ruminate; embryo basal. *Germination* adjacent-ligular; eophyll bifid. *Cytology*: 2n = 32.

Distribution and ecology: One variable species widespread in tropical rain forest in northeastern Queensland, Australia, occurring in shaded, humid rain forest on mountain ranges and tablelands at altitudes of 800–1400 m above sea-level.
Anatomy: Root (Seubert 1998a, 1998b) and fruit (Essig 2002).
Relationships: For relationships, see *Linospadix*.
Common names and uses: Atherton palm. Sometimes cultivated as an ornamental.
Taxonomic account: There is no recent published assessment of infraspecific variation.
Fossil record: No generic records found.
Notes: A very attractive palm because of the long spikes of red fruits.

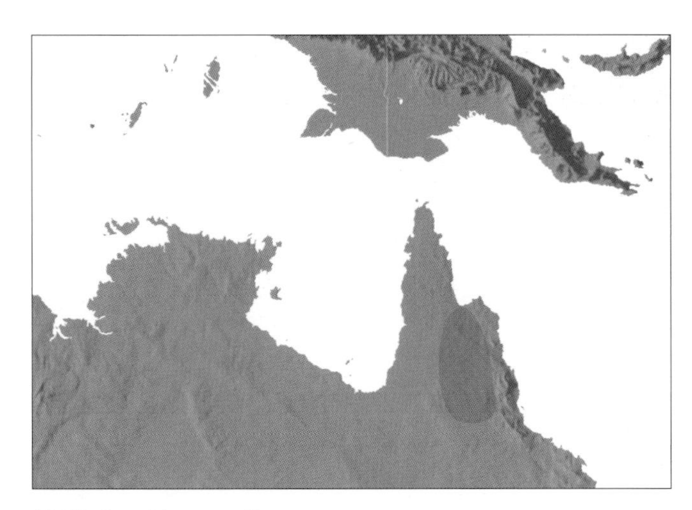

Distribution of ***Laccospadix***

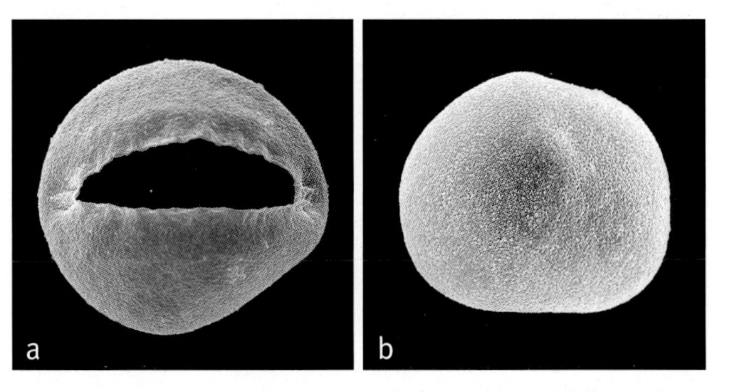

Above: ***Laccospadix***
a, monosulcate pollen grain, distal face SEM × 1500;
b, monosulcate pollen grain, proximal face SEM × 1500.
Laccospadix australasicus: **a** & **b**, *Moore & Volk* 9240.

Left: ***Laccospadix australasicus***, habit, Australia. (Photo: J. Dransfield)

Subtribe Oncospermatinae Hook.f. in Benth. & Hook.f., Gen. pl. 3: 872, 874 (1883) (‘*Oncospermae*’). Type: **Oncosperma**.

Moderate to robust, erect, armed with spines, at least when young; sheaths forming a well-defined crownshaft; leaves pinnate, leaflet tips acute; inflorescences infrafoliar, bisexual, branched to at least 2 orders; prophyll and peduncular bract similar in form and size, both usually caducous at anthesis; staminate flowers usually asymmetrical; fruit with basal, lateral or apical stigmatic remains; epicarp smooth.

Members of this subtribe are found on Indian Ocean islands and neighbouring parts of Mainland Asia and West Malesia. They are immediately distinguished by the presence of a crownshaft, infrafoliar infloresences and the presence of spines, which are epidermal emergences.

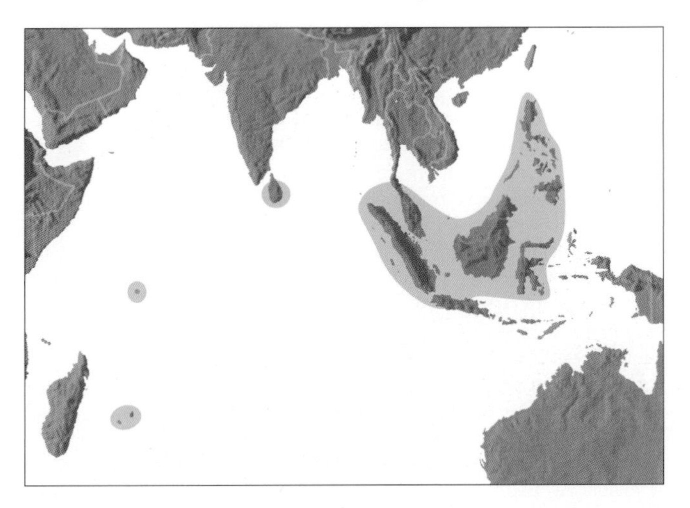

Distribution of subtribe **Oncospermatinae**

Subtribe Oncospermatinae has been resolved as monophyletic with low support (Lewis & Doyle 2002), moderate support (Asmussen *et al.* 2006, Baker *et al.* in review), and high support (Lewis 2002). The position of the subtribe remains uncertain within the Areceae, although it does not resolve within the western Pacific clade. A poorly supported topology of Lewis (2002) placed the Oncospermatinae as sister to a clade of Arecinae, Archontophoenicinae, Ptychospermatinae and unplaced genera of Areceae. Baker *et al.* (in review) resolved the Oncospermatinae as sister to a clade of Verschaffeltiinae, *Masoala* and unplaced members of Areceae with low support.

Key to Oncospermatinae

1. Fruit ovoid with apical or excentrically apical stigmatic remains; perianth ± half as high as the fruit; seed with homogeneous endosperm; triads borne in six vertical rows, each triad subtended by a saucer-like bract, the bracts obscured by the staminate flowers, but becoming conspicuous when the flowers have fallen. Mascarene Islands . . . . **155. Tectiphiala**
1. Fruit ellipsoid to globose, with lateral to basal stigmatic remains; perianth less than half as high as the fruit; endosperm homogeneous or ruminate; rachilla bracts not saucer-like 2
2. Fruit ellipsoid with basal stigmatic remains; staminate flower open early in development with ca. 9 stamens, scarcely exceeding the petals and a conspicuous, slender, elongate, trifid pistillode. Seychelles . . . . **153. Deckenia**
2. Fruit globose to ellipsoid with lateral stigmatic remains; staminate flower closed in bud, stamens 6–12; pistillode shorter than stamens and petals 3
3. Stamens exserted at anthesis; pistillode minutely trifid; seed with homogeneous endosperm. Mascarene Islands . . . . **154. Acanthophoenix**
3. Stamens included at anthesis; pistillode deeply trifid; seed with ruminate endosperm. Sri Lanka to Celebes . . . . **152. Oncosperma**

152. ONCOSPERMA

Tall clustering very spiny pinnate-leaved palms, native to Sri Lanka, Southeast Asia and West Malesia, with conspicuous crownshafts and seed with ruminate endosperm.

Oncosperma Blume, Bull. Sc. Phys. Nat. Néerl. 1: 64 (1838). Type. *O. filamentosa* Blume = **O. tigillarium** (Jack) Ridl. (*Areca tigillaria* Jack).

Keppleria Meisn., Pl. Vasc. Gen. 1: 355; 2: 266 (1842). *K. tigillaria* (Jack) Meisn. (*Areca tigillaria* Jack).

Onkos — *bulk, mass, tumour,* sperma — *seed, presumably based on the wide groove filled with spongy material at the base of the seed.*

Tall, usually clustered, spiny, pleonanthic, monoecious palms. *Stems* erect, often very tall, becoming bare, ringed with leaf scars, frequently armed with scattered or dense, robust spines, the spines tending to erode with age, in one species (*Oncosperma fasciculatum*) the stems sometimes branching aerially. *Leaves* pinnate, neatly abscising; leaf sheaths tubular, closely sheathing, forming a well-defined crownshaft, bearing abundant tomentum and spines of varying lengths, the distal margins tending to tatter irregularly; petiole usually robust, adaxially concave or flattened, abaxially rounded, usually densely armed with short to long spines and indumentum; rachis adaxially angled, rounded abaxially; marginal reins sometimes well developed and persisting as pendulous strips at the base of the rachis, leaflets very numerous, single-fold, acute or acuminate, regularly arranged, pendulous or not, or distinctly grouped and held in

Below: ***Oncosperma tigillarium***, habit, Malaysia. (Photo: C.E. Lewis)

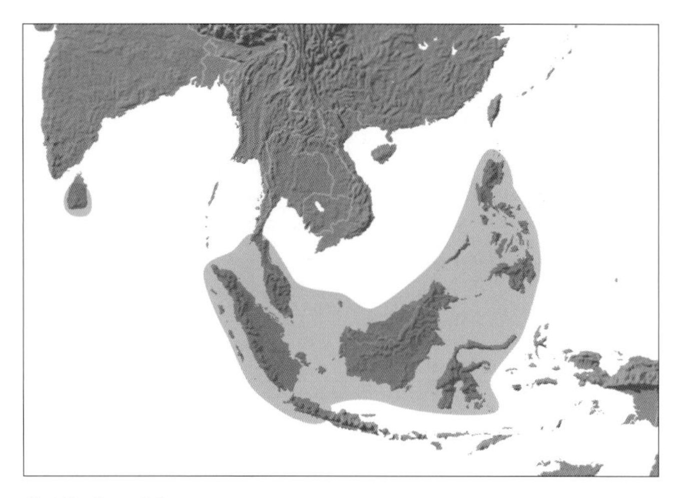

Distribution of ***Oncosperma***

different planes (*O. fasciculatum*), lacking spines, bearing bands of caducous scales adaxially, abaxially with ramenta along the main vein, and sometimes with dot-like scales. *Inflorescences* solitary, infrafoliar, branching to 2 orders near the base, to 1 order distally, protandrous, erect in bud; peduncle short, broad, winged at the base, horizontal at anthesis, sparsely to densely armed with straight spines; prophyll inserted just above the base of the peduncle, robust, tubular, flattened, 2-keeled, briefly beaked, completely enclosing the inflorescence until anthesis, coriaceous to ± woody, densely indumentose, very sparsely to very densely covered in short or long, straight or twisted spines; peduncular bract 1, inserted just above and similar to the prophyll but only slightly 2-keeled and usually more sparsely armed; rachis longer than the peduncle, bearing few to numerous, pendulous first-order branches; rachis bracts very inconspicuous; rachillae long, slender, rather robust, flexuous, pendulous, glabrous, bearing spirally arranged, low triangular bracts each subtending a triad, or in the distal portion of the rachilla, subtending solitary or paired staminate flowers. *Staminate flowers* asymmetrical, sessile or sometimes slightly stalked at the base; sepals 3, ± acute, keeled, imbricate, basally briefly connate or ± distinct; petals 3, acute or acuminate, valvate, much longer than the sepals, distinct or briefly connate at the base; stamens shorter than the petals, 6–9 (?12 recorded but not found by us), briefly epipetalous or ± free, the filaments short, thick, sometimes connate at the base in a very short ring, anthers ± medifixed, basally sagittate, latrorse; pistillode shorter or longer than the stamens, deeply trifid. *Pollen* ellipsoidal symmetric; aperture a distal sulcus; ectexine tectate, coarsely perforate-finely reticulate with supra-tectate, angular, striate, block-like clavae, aperture margin similar; infratectum columellate; longest axis 27–43 µm [4/5]. *Pistillate flowers* ± globular; sepals 3, distinct, rounded, imbricate; petals 3, distinct, rounded, imbricate at the base, the tips valvate; staminodes 6 (?always), very small, tooth-like; gynoecium unilocular, uniovulate, ± globose or ovoid, the stigmas apical, scarcely prominent, ovule form unknown. *Fruit* spherical, dark blue-black, with lateral or subapical stigmatic remains;

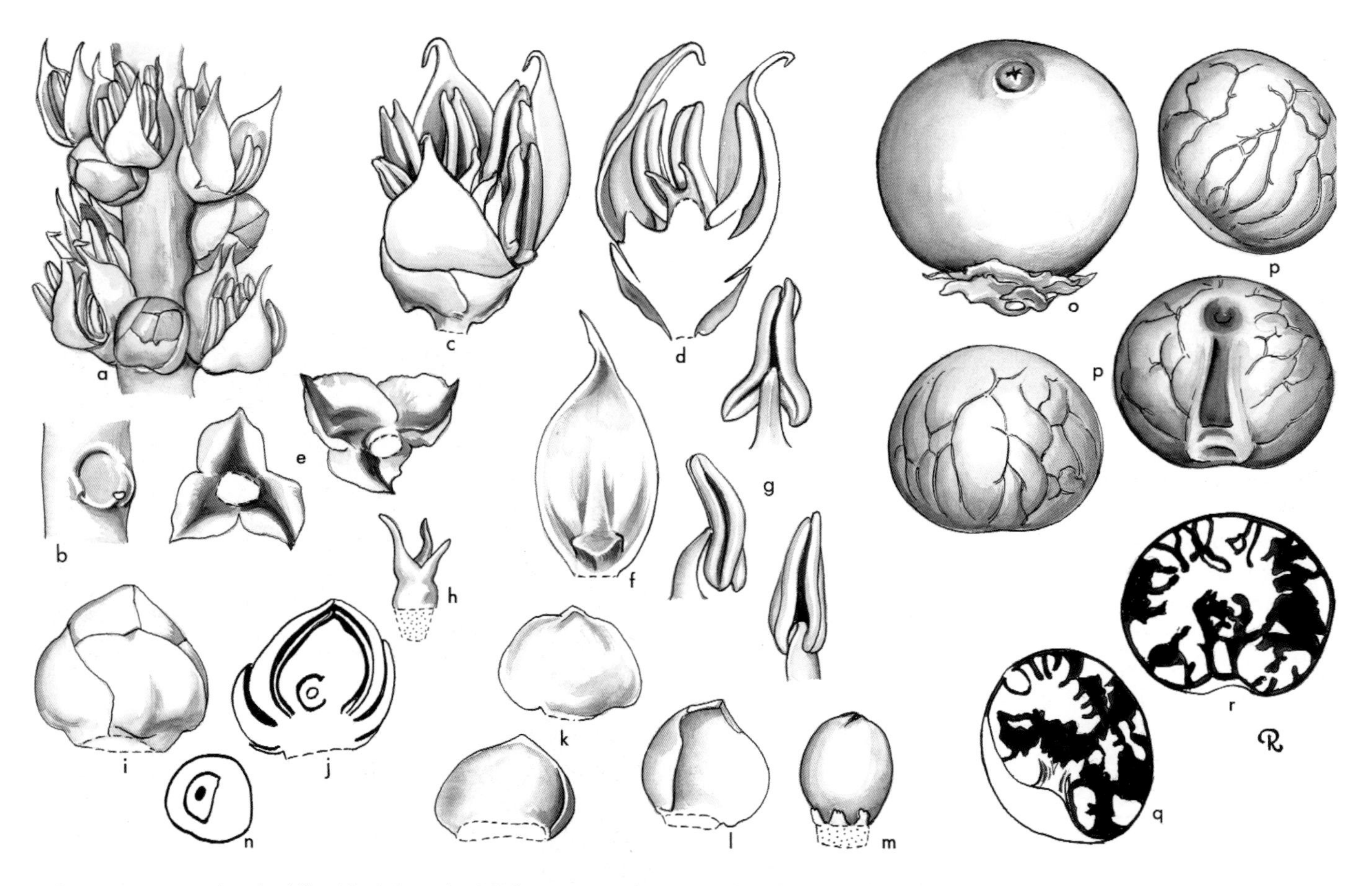

Oncosperma. **a**, portion of rachilla with triads × 3; **b**, triad, flowers removed × 3; **c**, staminate flower × 6; **d**, staminate flower in vertical section × 6; **e**, staminate calyx in 2 views × 6; **f**, staminate petal, interior view × 6; **g**, stamen in 3 views × 6; **h**, pistillode × 6; **i**, pistillate bud × 6; **j**, pistillate bud in vertical section × 6; **k**, pistillate sepal, exterior and interior views × 6; **l**, pistillate petal × 6; **m**, gynoecium and staminodes × 6; **n**, ovary in cross-section × 6; **o**, fruit × 3; **p**, seed in 3 views × 3; **q**, seed in vertical section × 3; **r**, seed in cross-section × 3. *Oncosperma tigillarium*: **a–n**, *Moore & Meijer* 9183; **o–r**, *Moore & Meijer* 9221. (Drawn by Marion Ruff Sheehan)

epicarp smooth or pebbled, mesocarp thinly fleshy, without fibres, endocarp thin, crustaceous, closely adherent to the seed, with a round basal operculum. *Seed* attached by an elongate lateral hilum, raphe branches anastomosing, partially embedded in the endosperm, endosperm deeply ruminate; embryo subbasal. *Germination* adjacent ligular; eophyll bifid. *Cytology*: 2n = 32 (± 2–4).

Distribution and ecology: Five species, one endemic to Sri Lanka, two endemic to the Philippines and two widespread in Southeast Asia and West Malesia, reaching Sulawesi, the Philippines and western Moluccas. *Oncosperma tigillarium* is characteristic of the landward fringe of mangrove forest and is also rarely found on poor sandy soils inland; *O. horridum*, *O. platyphyllum* and *O. gracilipes* are characteristic of hilly sites inland and *O. fasciculatum* occurs on steep slopes at altitudes of 300–1000 m above sea level. Species of *Oncosperma* frequently occur gregariously and are very conspicuous; their heavy spiny litter may play an important role in preventing regeneration of dicotyledonous trees below the palms. Little is known of the natural history of these common palms (see House 1984).

Anatomy: Leaf, root (Tomlinson 1961), root (Seubert 1998a, 1998b), stegmata (Killmann & Hong 1989) and fruit (Essig *et al.* 2001).

Relationships: *Oncosperma* is resolved as monophyletic with moderate to high support (Lewis 2002, Baker *et al.* in prep.). It is resolved as sister to a clade of *Deckenia*, *Acanthophoenix* and *Tectiphiala* with moderate support (Lewis 2002, Lewis & Doyle 2002). *Deckenia* is moderately to highly supported as sister to a strongly supported clade of *Acanthophoenix* and *Tectiphiala* (Lewis & Doyle 2001, 2002, Asmussen *et al.* 2006, Loo *et al.* 2006, Norup *et al.* 2006, Baker *et al.* in review, in prep.). Alternatively, *Tectiphiala* is resolved as sister to a clade of *Deckenia* and *Acanthophoenix* with low support (Lewis 2002). Note that in some analyses, *Oncosperma* is placed with other genera of tribe Areceae but with low support (e.g., Loo *et al.* 2006, Norup *et al.* 2006). For interspecific relationships within *Oncosperma*, see Lewis (2002).

Common names and uses: *Nibung* (*Oncosperma tigillarium*), *bayas* (*O. horridum*). The cabbage of most species is excellent and widely collected. The trunks of *O. tigillarium* are extremely durable and have been used (with spines removed) as telegraph poles and split as flooring. Sheaths are sometimes used as buckets. Despite their spines, all species are very elegant ornamentals.

Taxonomic account: There is no recent account.

Left: ***Oncosperma tigillarium***, crown with infructescences, cultivated, Fairchild Tropical Botanic Garden, Florida. (Photo: C.E. Lewis)

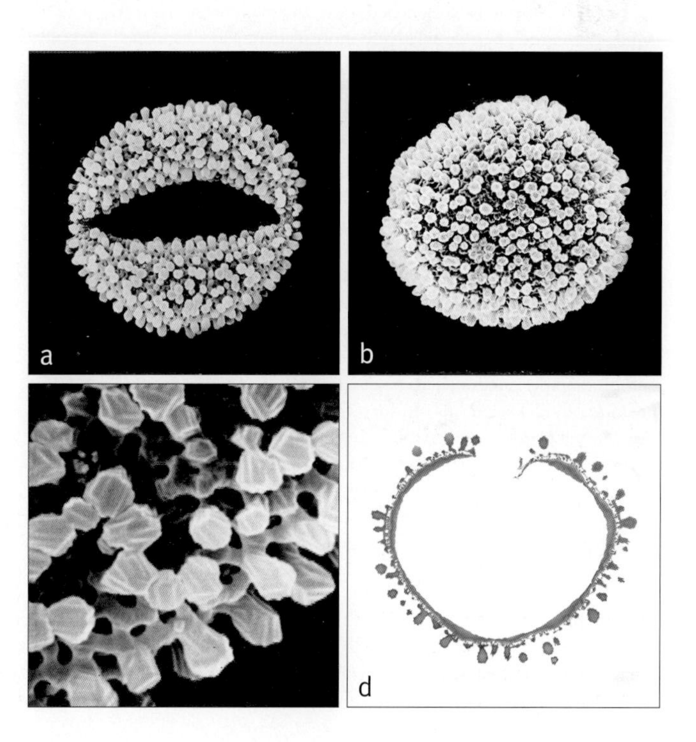

Below: ***Oncosperma***
a, monosulcate pollen grain, distal face SEM × 1500;
b, monosulcate pollen grain, proximal face SEM × 1500;
c, close-up, finely reticulate pollen surface with vertically ridged, block-like, clavae SEM × 8000;
d, ultrathin section through whole pollen grain, polar plane TEM × 1500.
Oncosperma horridum: **a**, **c** & **d**, *Beccari* 977; *O. tigillarium*: **b**, *Elmer* 12662.

Fossil record: Fruits attributed to *Oncosperma* (*O. anglicum*) are reported from the Lower Eocene of southern England (London Clay Flora) (Reid & Chandler [1933], Chandler [1964]). Clavate *Oncosperma*-like extended sulcate pollen is reported from Oligocene deposits in Borneo (Muller 1964, 1972, 1979); unfortunately, none of Muller's articles is supported by illustrations of the pollen, making it difficult to comment. From the Tertiary of India (Cannanore, northern Kerala), Srisailam and Ramanujam (1982) described extended sulcate pollen closely resembling that of *Oncosperma*. *Paravuripollis* has been compared with *Oncosperma* pollen (Ramanujam 1987), but its affinities are more likely to be with *Korthalsia*, as *Oncosperma* has extended sulcate, not zonasulcate, pollen.

Notes: Despite their spines, all species are very elegant ornamentals.

Additional figures: Nypoideae frontispiece, Glossary figs 22, 24.

153. DECKENIA

Handsome solitary spiny pinnate-leaved palms, native to the Seychelles, with conspicuous crownshafts and seed with homogeneous endosperm.

Deckenia H. Wendl. ex Seem., Gard. Chron 1870: 561 (1870). Type: **D. nobilis** H. Wendl. ex Seem.

Commemorates Baron Karl Klaus von der Decken (1833–1865), explorer, who was one of the first Europeans to climb Mt Kilimanjaro.

Robust, solitary, spiny, pleonanthic, monoecious palm. *Stem* erect, tall, spiny when juvenile, becoming distally unarmed at maturity, conspicuously ringed with leaf scars, slightly swollen at the base, with abundant adventitious roots. *Leaves* pinnate, neatly abscising; sheaths tubular, forming a well-defined crownshaft, densely spiny when young, unarmed or minutely roughened near the petiole at maturity, densely white-tomentose; petiole relatively short, flattened or channelled adaxially, abaxially rounded, bearing abundant, caducous, shaggy hairs, also spiny in juveniles; rachis robust, gradually tapering, hairy and, in juveniles, spiny like the petiole; leaflets numerous, all single-fold, regularly arranged, rather stiff, elongate, acuminate, somewhat plicate, adaxially bearing caducous tomentum when young, abaxially densely covered with minute, dot-like scales and abundant, conspicuous ramenta along the major longitudinal veins, transverse veinlets obscure. *Inflorescences* solitary, infrafoliar, erect in bud, branching to 2 orders at the base, to 1 order distally, protandrous; peduncle relatively short, winged at the base, elliptic in cross-section, glabrous; prophyll inserted just above the base of the peduncle, tubular, elliptic, very briefly beaked, 2-keeled, splitting along one face, caducous, usually very densely armed with erect, rather soft, short to long, golden-coloured spines; peduncular bract very similar to the prophyll, but less prominently 2-keeled, also caducous; rachis longer than the peduncle, it and its branches twisted in bud, becoming straight, rachis and peduncle remaining erect, the branches becoming pendulous; first-order branches numerous, rather crowded, spirally arranged; rachillae numerous, elongate, pendulous, flexuous, bearing rather crowded, spirally arranged, superficial triads throughout their length except at the extreme tips. *Staminate flowers*

Top right: ***Deckenia nobilis***, habit, Seychelles. (Photo: J. Dransfield)

Bottom right: ***Deckenia nobilis***, crown and inflorescences, Seychelles. (Photo: J. Dransfield)

± globular, open early in development; sepals 3, narrow, triangular, keeled, joined briefly at the base; petals 3, triangular, sometimes briefly connate in a somewhat stalk-like base; stamens 6–9, filaments very short and broad, adnate to the petals, anthers medifixed, tightly recurved and/or twisted, latrorse; pistillode columnar longer than the petals, trifid, very conspicuous at first, becoming less conspicuous at anthesis. *Pollen* ellipsoidal bi-symmetric, occasionally lozenge-shaped; aperture a distal sulcus; ectexine semitectate, coarsely angular striate clavae, separated or coalesced into short or long horizontal elements, aperture margin similar; infratectum columellate; longest axis 27–34 µm [1/1]. *Pistillate flowers* globular; sepals 3, distinct, rounded, imbricate; petals 3, distinct, rounded, imbricate, with minute, triangular, valvate tips; staminodes 6, minute, tooth-like; gynoecium globular or ovoid, unilocular, uniovulate, stigmas 3, reflexed, ovule laterally attached, ?hemianatropous. *Fruit* relatively small, narrowly ovoid, black at maturity, perianth whorls persistent, the stigmatic remains conspicuous as an eccentric, basally rounded pad; epicarp smooth, drying longitudinally ridged, mesocarp thin, with few, large, longitudinal fibres

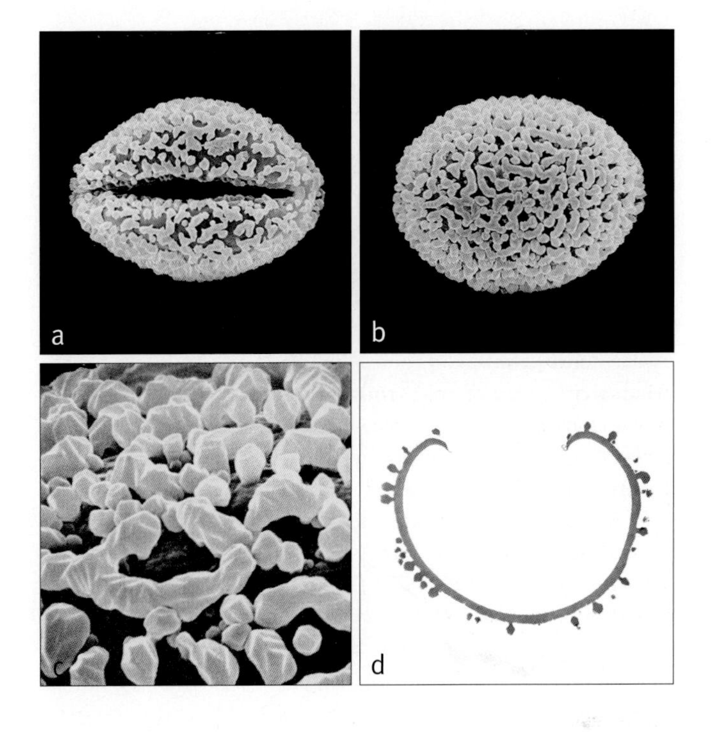

Right: ***Deckenia***
a, monosulcate pollen grain, distal face SEM × 2000;
b, monosulcate pollen grain, proximal face SEM × 2000;
c, close-up, angular striate clavae, separate or coalesced SEM × 8000;
d, ultrathin section through whole pollen grain, polar plane TEM × 2500.
Deckenia nobilis: **a–c**, *Dransfield* JD6236, **d**, *Jeffrey* 775.

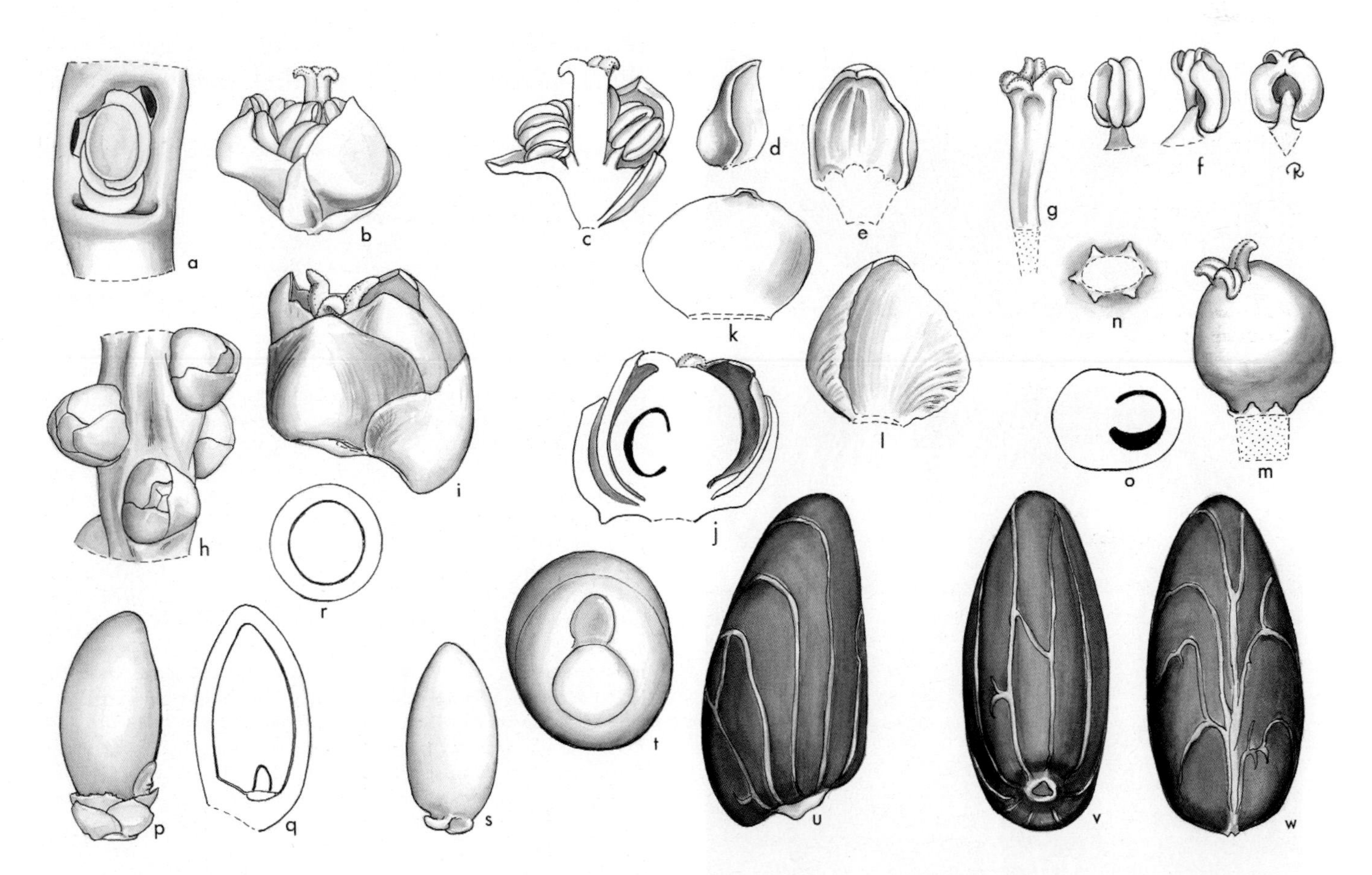

Deckenia. **a**, portion of rachilla with triad, flowers removed to show bracteoles × 6; **b**, staminate flower × 12; **c**, staminate flower in vertical section × 12; **d**, staminate sepal × 12; **e**, staminate petal, interior view × 12; **f**, stamen in 3 views × 12; **g**, pistillode × 12; **h**, portion of rachilla with pistillate buds × 6; **i**, pistillate flower × 12; **j**, pistillate flower in vertical section × 12; **k**, pistillate sepal × 12; **l**, pistillate petal, interior view × 12; **m**, gynoecium and staminodes × 12; **n**, staminodes × 12; **o**, ovary in cross-section × 12; **p**, fruit × 3; **q**, fruit in vertical section × 3; **r**, fruit in cross-section × 3; **s**, endocarp × 3; **t**, base of endocarp and operculum × 6; **u, v, w**, seed in 3 views × 6. *Deckenia nobilis*: **a–g**, *Vaughan* 806; **h–o**, *Furtado* 31123; **p–w**, *Squibb* s.n. (Drawn by Marion Ruff Sheehan)

corresponding to the ridges when dry, endocarp thin, cartilaginous, with a basal circular operculum. *Seed* very narrowly ovoid, basally attached, with a small circular hilum, and sparse raphe branches only rarely anastomosing, endosperm homogeneous; embryo basal. *Germination* adjacent ligular; eophyll bifid. *Cytology* not known.

Distribution and ecology: One species confined to the Seychelles Islands. *Deckenia nobilis* occurs at altitudes from sea level to nearly 600 m, but grows gregariously mostly on knolls and ridges at about 300 m altitude; its massive litter prevents regeneration and there is usually no undergrowth under dense stands.

Anatomy: Leaf, petiole, root (Tomlinson 1961) and fruit (Essig *et al.* 2001).

Relationships: For relationships, see *Oncosperma.*

Common names and uses: Cabbage palm, *palmiste.* The cabbage is eaten and the leaf sheaths used for making containers.

Taxonomic account: Bailey (1942).

Fossil record: No generic records found.

Notes: The staminate flowers are unusual in opening early in development and in their globular shape, also in more-or-less connate petals with adnate stamen filaments and curved or twisted anthers.

Additional figures: Fig. 2.8f.

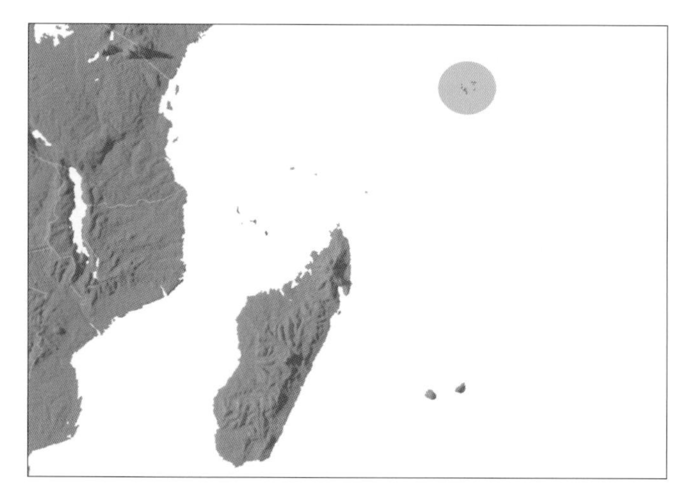

Distribution of ***Deckenia***

154. ACANTHOPHOENIX

Tall solitary very spiny pinnate-leaved palms, native to the Mascarene Islands, with conspicuous crownshafts and seed with homogeneous endosperm.

Acanthophoenix H. Wendl., Fl. Serres Jard. Eur. 16: 181 (1867). Lectotype: **A. rubra** (Bory) H. Wendl. (*Areca rubra* Bory) (see Beccari & Pichi-Sermolli 1955).

Akanthos — *spine,* phoenix — *a general name for a palm, referring to the abundant spines.*

Robust, solitary, spiny, pleonanthic, monoecious palms. *Stem* erect, usually unarmed, rarely spiny on new growth, conspicuously ringed with leaf scars, slightly swollen at the base. *Leaves* pinnate, neatly abscising; sheaths tubular, forming a well-defined crownshaft, bearing abundant tomentum and black, rather fragile spines, rarely very sparsely armed, ligule absent; petiole rather short, adaxially flattened or shallowly channelled, abaxially rounded, tomentose or glabrous, ± unarmed to densely armed with black spines, especially in juveniles; rachis glabrous or scaly, tomentose or bearing ramenta; leaflets elongate, single-fold, acute, regularly arranged, ± pendulous, with white wax abaxially, adaxially usually armed with slender, bulbous-based bristles along the midrib near the base, rarely also armed abaxially, abaxially bearing minute dot-like scales and conspicuous ramenta along the main vein near the insertion, sometimes with white wax, transverse veinlets obscure. *Inflorescences* solitary, infrafoliar, protandrous, erect in bud, branching to 2 orders near the base, to 1 order distally; peduncle short, winged at the base, tomentose to becoming glabrous, unarmed or armed with flat, dark brown, straight or sinuous spines; prophyll inserted near the base of the peduncle, tubular, elliptic, briefly beaked, 2-keeled, completely enclosing the inflorescence in bud, unarmed or rarely spiny, glabrous or tomentose, splitting along one face, deciduous; peduncular bract 1, inserted just above and similar to the prophyll but less conspicuously 2-keeled, also caducous; rachis much longer than the peduncle, rachis bracts very inconspicuous, low, triangular, glabrous to tomentose, often spiny; first-order branches spirally arranged, rather distant; rachillae ivory-white to reddish at anthesis, becoming green in fruit, elongate, pendulous, sinuous, bearing very inconspicuous to prominent, spirally arranged, triangular bracts, each subtending a triad throughout the rachilla length

Left: ***Acanthophoenix rubra***, habit, Mauritius. (Photo: C.E. Lewis)

except at the very tip where subtending solitary or paired staminate flowers; floral bracteoles low, very inconspicuous. *Staminate flowers* often briefly pedicellate, asymmetrical, white, cream-coloured to red; sepals 3, distinct or briefly connate, imbricate, short, acute, ± keeled; petals 3, basally ± briefly connate, valvate, 3–4 times as long as the sepals, ± elliptic, fleshy; stamens (4–)6–12, exserted at anthesis, filaments slender to wide, elongate, often coiled in bud but not inflexed, anthers dorsifixed, linear, basally sagittate, versatile, latrorse; pistillode ± ovoid, short, briefly to deeply trifid. *Pollen* ellipsoidal bi-symmetric; aperture a distal sulcus; ectexine tectate, psilate-foveolate, aperture margin similar or slightly finer; infratectum columellate; longest axis 35–45 µm [1/1]. *Pistillate flowers* ± ovoid; sepals 3, distinct, imbricate, broad, rounded; petals 3, distinct, imbricate except for the minute, valvate, triangular tips; staminodes 6–9, tooth-like or absent; gynoecium unilocular, uniovulate, ovoid, stigmas broad, ovule large pendulous, campylotropous. *Fruit* borne with persistent perianth whorls, ellipsoidal to subglobose, black at maturity, with lateral to subapical stigmatic remains; epicarp smooth or slightly pebbled when dry, mesocarp with tanniniferous parenchyma overlying a shell of short sclereids, external to a layer of tanniniferous parenchyma with included thin, flat, longitudinal fibres, endocarp thin, crustaceous, fragile, with circular basal operculum. *Seed* attached laterally near the base, hilum circular, raphe branches ascending and anastomosing, endosperm homogeneous; embryo basal. *Germination* adjacent-ligular; eophyll bifid or pinnate. *Cytology*: 2n = 32.

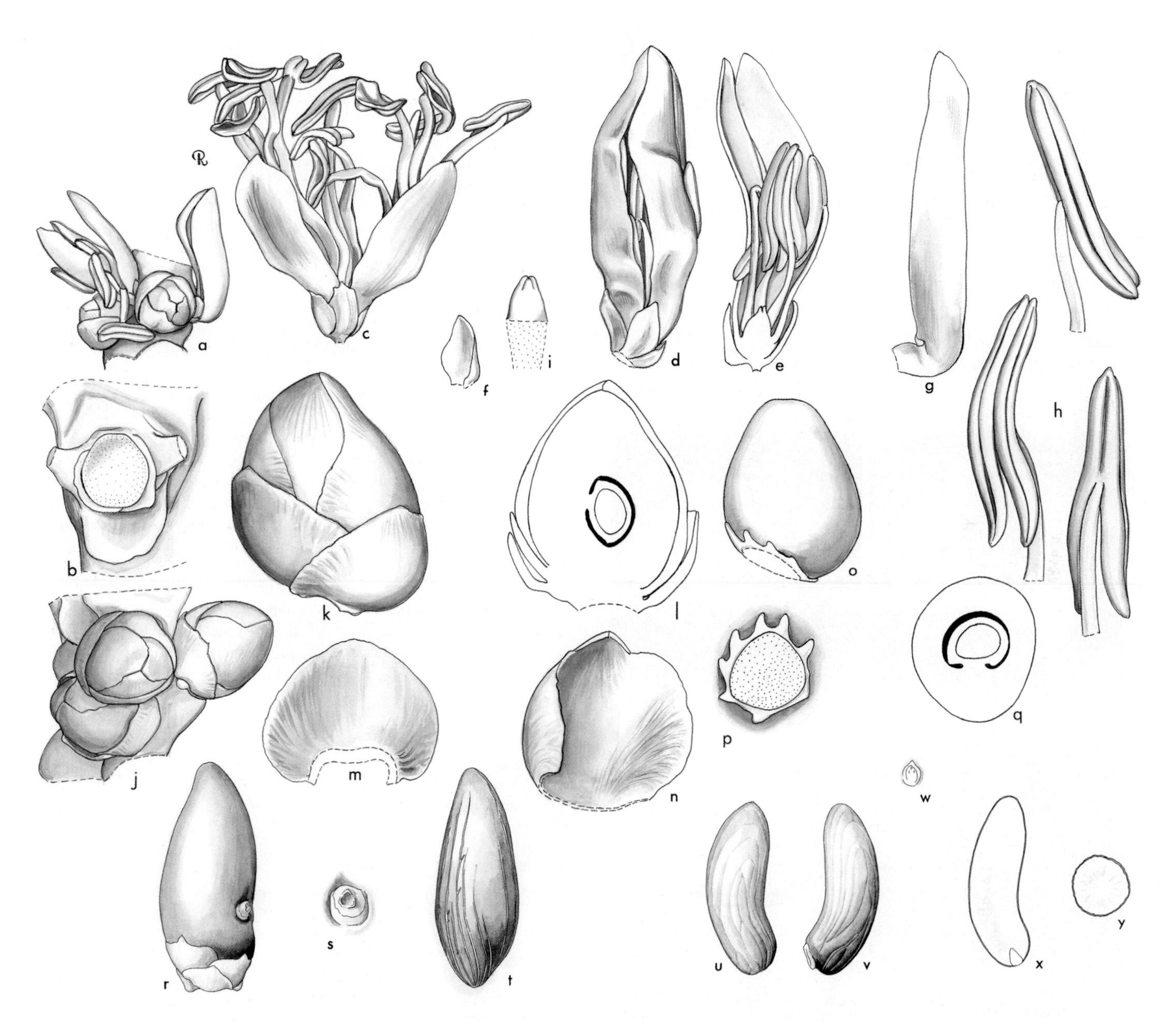

Acanthophoenix. **a**, portion of rachilla with triads × 3; **b**, triad, flowers removed to show bracteoles × 6; **c**, staminate flower × 3; **d**, staminate bud × 6; **e**, staminate bud in vertical section × 6; **f**, staminate sepal × 6; **g**, staminate petal × 6; **h**, stamen in 3 views × 12; **i**, pistillode × 12; **j**, rachilla in pistillate flower × 3; **k**, pistillate bud × 6; **l**, pistillate bud in vertical section × 6; **m**, pistillate sepal × 6; **n**, pistillate petal × 6; **o**, gynoecium and staminodes × 6; **p**, staminodes and base of gynoecium × 6; **q**, ovary in cross-section × 6; **r**, fruit × 3; **s**, detail of stigmatic remains × 7 $^{1}/_{2}$; **t**, endocarp × 3; **u**, **v**, seed × 3; **w**, detail of hilum × 3; **x**, seed in longitudinal section × 3; **y**, seed in transverse section × 3. *Acanthophoenix rubra*: **a**, **b**, and **d–i**, *Vaughan* 852; **c** and **j–q**, *Vaughan* 851a; *A. rousselii*: **r–y**, *Ludwig* s.n. (Drawn by Marion Ruff Sheehan)

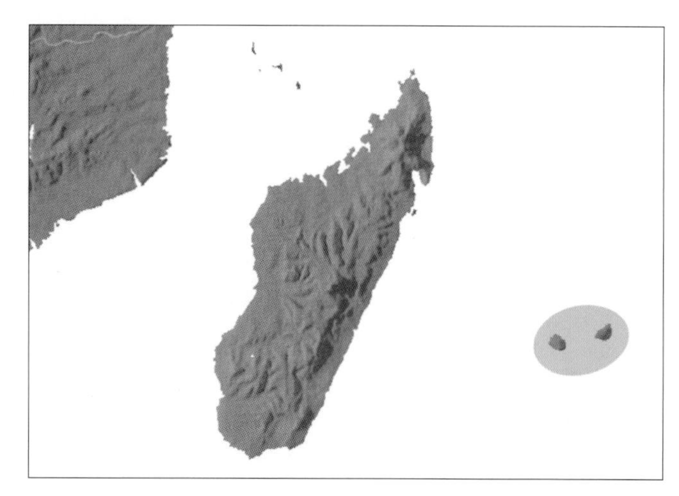

Distribution of ***Acanthophoenix***

Distribution and ecology: Three species confined to Mauritius and Réunion, once common, now exceedingly rare in the wild; elsewhere cultivated, but not very common. *Acanthophoenix rubra* grows at altitudes less than 600 m, *A. rousselii* occurs in drier forest at less than 850 m, while *A. crinita* occurs at higher elevations in montane forest. All wild populations are very small and disturbed.
Anatomy: Fruit (Essig *et al.* 2001).
Relationships: For relationships, see *Oncosperma*.

Above: ***Acanthophoenix***
a, monosulcate pollen grain, distal face SEM × 1500;
b, monosulcate pollen grain, proximal face SEM × 1500;
c, close-up, psilate-foveolate pollen surface SEM × 8000;
d, ultrathin section through whole pollen grain, polar plane TEM × 1500.
Acanthophoenix rubra: **a–d**, *Horne* s.n.

Common names and uses: *Palmiste rouge*. The cabbage is eaten, this being responsible for the near extinction of this palm.
Taxonomic accounts: Moore & Guého (1984) and Ludwig (2006).
Fossil record: No generic records found.
Notes: Ludwig's recent paper has clarified much of the confusion in the erstwhile recognised single species, *Acanthophoenix crinita*, by demonstrating the presence of good morphological characters for separating three species.
Additional figures: Glossary fig. 7.

155. TECTIPHIALA

Moderate solitary very spiny pinnate-leaved palms, native to Mauritius, with conspicuous crownshafts and distinctive cup-like pits on the rachillae.

Tectiphiala H.E. Moore, Gentes Herb. 11(4): 285 (1978). Type: **T. ferox** H.E. Moore.

Tectus — *covered*, phiala — *a broad, flat-bottomed drinking vessel, referring to the shape of the rachilla bracts and the way they are at first covered by staminate flowers*.

Moderate, solitary or clustered, spiny, pleonanthic, monoecious palm. *Stem* erect, bearing persistent leaf bases basally, distally free of leaf bases, ringed with leaf scars and abundant long spines with bulbous bases. *Leaves* pinnate, neatly abscising in mature individuals; sheaths tubular, forming a crownshaft, bearing an untidy ligule, and very densely covered in spines of varying length and abundant dark hairs; petiole rather short, adaxially with short spines, abaxially hairy; rachis bearing stiff hairs on both surfaces or on the adaxial surface alone; leaflets single-fold, very coriaceous, acute, arranged in close or distant fascicles, and fanned within the groups, adaxially glabrous, abaxially with a very dense covering of white scales, the midnerve bearing scattered ramenta, transverse veinlets obscure. *Inflorescences* solitary, infrafoliar, erect in bud, becoming ± pendulous, branching to 1 order, protandrous; peduncle covered in short spines at the base, above the insertion of the peduncular bract bearing a variety of short to very long spines; prophyll inserted just above the base of the peduncle, tubular, 2-keeled, completely enclosing the inflorescence in bud, splitting along the ventral midline and abscising, densely covered in stiff dark hairs;

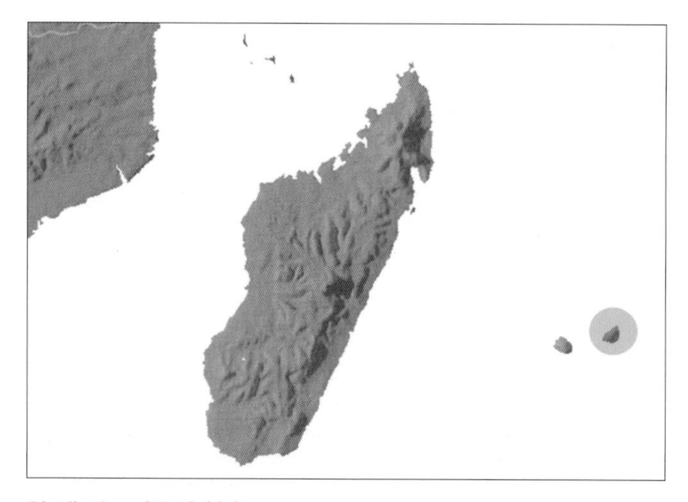

Distribution of ***Tectiphiala***

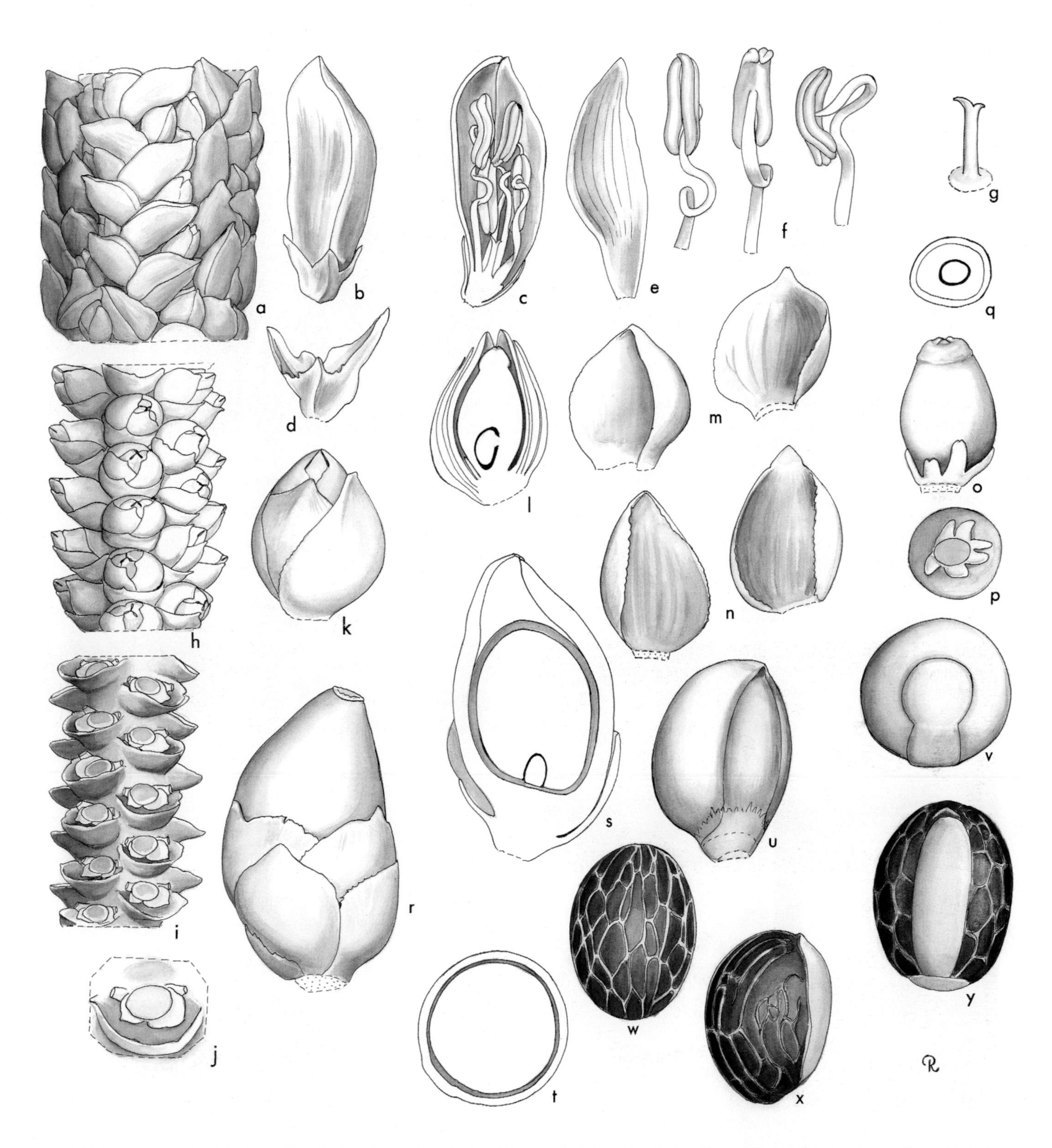

Tectiphiala. **a**, portion of rachilla in staminate bud × 2; **b**, staminate bud × 4 1/2; **c**, staminate bud in vertical section × 4 1/2; **d**, staminate calyx × 7 1/2; **e**, staminate petal, exterior view × 4 1/2; **f**, stamen in 3 views × 6; **g**, pistillode × 7 1/2; **h**, portion of rachilla in pistillate bud × 2; **i**, portion of rachilla, flowers removed to show prominent subtending bracts × 2; **j**, triad, flowers removed to show bracteoles and staminate pedicels × 4 1/2; **k**, pistillate bud × 4 1/2; **l**, pistillate bud in vertical section × 4 1/2; **m**, pistillate sepal in 2 views × 4 1/2; **n**, pistillate petal in 2 views × 4 1/2; **o**, gynoecium and staminodes × 4 1/2; **p**, staminodes at base of gynoecium × 4 1/2; **q**, ovary in cross-section × 4 1/2; **r**, fruit × 4 1/2; **s**, fruit in vertical section × 4 1/2; **t**, fruit in cross-section × 4 1/2; **u**, endocarp × 4 1/2; **v**, endocarp with operculum × 4 1/2; **w, x, y**, seed in 3 views × 4 1/2. *Tectiphiala ferox*: all from *Moore & Gueho* 9925. (Drawn by Marion Ruff Sheehan)

peduncular bract 1, inserted just above the prophyll, similarly hairy, abscising before anthesis; rachis scarcely evident; rachillae 3–5, congested at the apex of the peduncle, straight or flexuous, flattened and spiny at the base; rachilla bracts arranged in ca. 6 vertical rows throughout most of the rachilla length, prominent, approximate, projecting, saucer-like, rounded, each subtending a triad except at the very tip of the rachilla where subtending solitary or paired staminate flowers; bracteoles surrounding the pistillate flower unequal, one much larger than the other. *Staminate flowers* asymmetrical, acute, very briefly stalked, obscuring the rachilla bracts; sepals 3, often unequal, acute, briefly connate at the base; petals 3, distinct, strongly nerved when dry, angled, acute, valvate; stamens 6(–7), ± equalling the petals, filaments ± cylindrical, ± twisted and coiled, erect at the tip, anthers dorsifixed, briefly bifid at the tip, deeply bifid at the base, latrorse; pistillode usually apparent, $^1/_2$ as long as the stamens, trifid or oblique. *Pollen* ellipsoidal symmetric to asymmetric; aperture a distal sulcus; ectexine tectate, coarsely perforate and/or rugulate, aperture margin similar or slightly finer; infratectum columellate; longest axis 42–66 µm [1/1]. *Pistillate flowers* in bud ± obscured by the staminate flowers; sepals 3, distinct, broadly imbricate, ± acute; petals 3, distinct, scarcely exceeding the sepals, broadly imbricate with briefly valvate tips; staminodes 6(–7), small, tooth-like or linear; gynoecium ovoid, unilocular, uniovulate, the stigmas not prominent, ovule large, pendulous, probably hemianatropous. *Fruit* asymmetrically ovoid, dark blue-black, with apical stigmatic remains; epicarp smooth, underlain by longitudinal sclereids over a layer of tannin cells, endocarp thin with round basal operculum. *Seed* attached by an elongate elliptical hilum, raphe branches anastomosing, endosperm homogeneous; embryo basal. *Germination* and eophyll not known. *Cytology* not studied.

Above: ***Tectiphiala ferox***, habit, Mauritius. (Photo: C.E. Lewis)

Below: ***Tectiphiala***
a, monosulcate pollen grain, distal face SEM × 1000;
b, close-up, coarsely perforate-rugulate pollen surface SEM × 8000;
c, ultrathin section through whole pollen grain, polar plane TEM × 1000;
d, monosulcate pollen grain, mid-focus LM × 1000.
Tectiphiala ferox: **a–d**, *Vaughan* 12580.

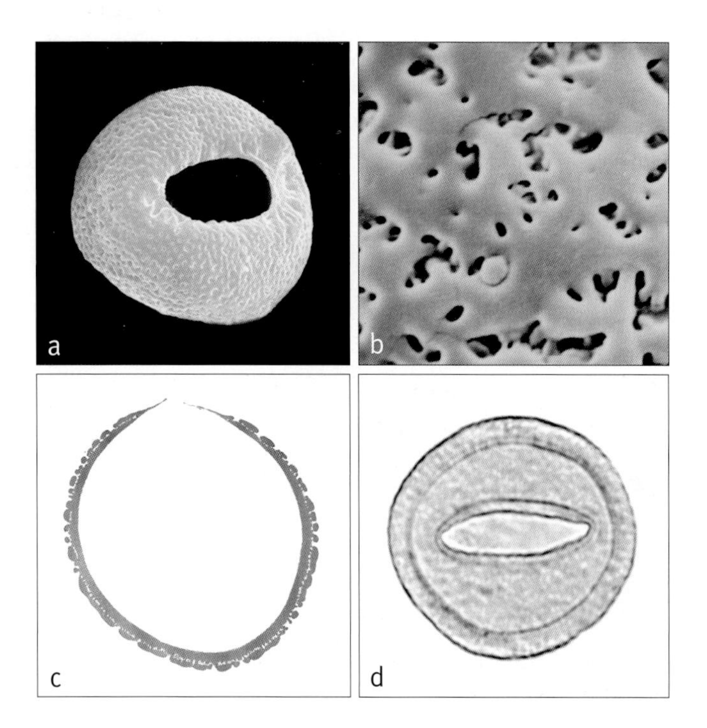

Distribution and ecology: A single species endemic to Mauritius; *Tectiphiala* grows in relict scrub in mostly wet, more-or-less acid situations at elevations of ca. 570–650 m above sea level.
Anatomy: Fruit (Essig *et al*. 2001).
Relationships: For relationships, see *Oncosperma*.
Common names and uses: Common names not recorded. The cabbage is edible.
Taxonomic account: Moore (1978b).
Fossil record: No generic records found.
Notes: *Tectiphiala* is distinguished from *Acanthophoenix* by the remarkable saucer-like bracts that subtend the triads of flowers.

Subtribe Ptychospermatinae Hook.f. in Benth. & Hook.f., Gen. pl. 3: 872, 874 (1883) ('Ptychospermeae'). Type: **Ptychosperma**.

Small, moderate or tall, unarmed; leaves pinnate, leaflet tips praemorse; crownshaft well developed; leaflets linear to broadly cuneate, sometimes longitudinally split between the main veins; inflorescences infrafoliar; prophyll and peduncular bract similar, inserted close together, or peduncular bract much larger than the prophyll, inserted significantly above and exserted through the prophyll; staminate flowers bullet-shaped, larger than the pistillate; stamens numerous; pistillode usually large, bottle-shaped or small and conical; fruit with apical stigmatic remains; epicarp smooth; endocarp often ridged, or irregularly angled; embryo basal.

This subtribe is widespread in East Malesia extending to Fiji, Samoa and Australia, with one outlier (*Adonidia*) in Palawan (Philippines) and Sabah, and species of *Ponapea* in the Caroline Islands. All genera have praemorse leaflets and distinctive generally widely branched inflorescences.

Subtribe Ptychospermatinae is monophyletic in almost all phylogenies. The Ptychospermatinae is resolved with high support by Lewis and Doyle (2002, unpublished a) and by Baker *et al.* (in review), and with moderate support by Asmussen *et al.* (2006), Norup *et al.* (2006) and Baker *et al.* (in prep.). The postition of the subtribe remains uncertain within the western Pacific clade (Lewis 2002, Lewis & Doyle 2002, Norup *et al.* 2006, Baker *et al.* in review, in prep.).

Key to Genera of Ptychospermatinae

1. Pinnae of juvenile leaves entire, cuneate, several-nerved; pinnae of mature leaves divided longitudinally into 7–17 linear segments; peduncular bract similar to and included within the prophyll, prophyll and peduncular bract caducous 2
1. Pinnae cuneate, truncate, or oblique in both juvenile and mature leaves, prophyll and peduncular bract marcescent; peduncular bract similar to or longer than the prophyll, and inserted well above and protruding from the prophyll. 3
2. Stem slightly bottled; primary leaflets regularly arranged, divided into 11–17 segments, lacking white woolly scales abaxially; outer endocarp with large, conspicuously branched, tough black fibres; endosperm homogeneous. Queensland, Australia . . . . **163. Wodyetia**
2. Stem moderate to slender, not bottled, primary leaflets clustered, divided into 7–9 segments, bearing white woolly scales abaxially; outer endocarp with sparsely branched, thin, terete, straw-coloured fibres; endosperm ruminate. Queensland, Australia . . . . **165. Normanbya**
3. Seed terete in cross-section 4
3. Seed angled or grooved in cross-section 7
4. Leaflet tips bifid; endosperm deeply ruminate. Palawan and Northern Borneo . . . . **158. Adonidia**
4. Leaflet tips not bifid; endosperm homogeneous or ruminate 5
5. Peduncular bract terete and prominently rostrate, inserted well above and becoming exserted from the prophyll . . 6
5. Peduncular bract similar to and included within the prophyll 7
6. Prophyll caducous; inflorescence axes white at anthesis. Samoa . . . . **159. Solfia**
6. Prophyll persistent; inflorescence axes green at anthesis. Halmahera, Ceram, Ambon, to New Guinea and the Solomon Ialands . . . . **164. Drymophloeus**
7. Leaflets wide in the middle or throughout, moderate to rather short, chaffy ramenta present abaxially near the base of the midribs; mesocarp with several series of more-or-less round fibrous bundles next to endocarp. Vanuatu and Fiji . . . . **161. Veitchia**
7. Leaflets narrow, long, lacking chaffy ramenta on midribs; mesocarp with a single series of large, flat fibrous bundles closely adherent to the endocarp. Northern Territory, Australia . . . . **162. Carpentaria**
6. Seed irregularly angled in cross-section, acute; fruit often angled and tapered at both ends; peduncle elongate; peduncular bract inserted well above and exserted from the prophyll. Fiji Islands, Samoa . . . . **160. Balaka**
6. Seed 3–5 sulcate in cross-section, acute and truncate or pointed on both ends; fruit ovoid; peduncle short, peduncular bract similar to and included in the prophyll 7
7. Pinnae 3-pronged at the apex, the centre prong the longest; fruit very strongly 5-angled when dry; the seed with 5 acute ridges; anther with connective tanniniferous, protruding basally from anthers or not sagittate basally, basifixed. Eastern New Guinea . . **166. Brassiophoenix**
7. Pinnae acute, oblique, truncate, or 2-pronged and concave apically; fruit rounded or irregularly ridged when dry; seed with 5 rounded or irregular ridges; anthers dorsifixed, sagittate basally 8

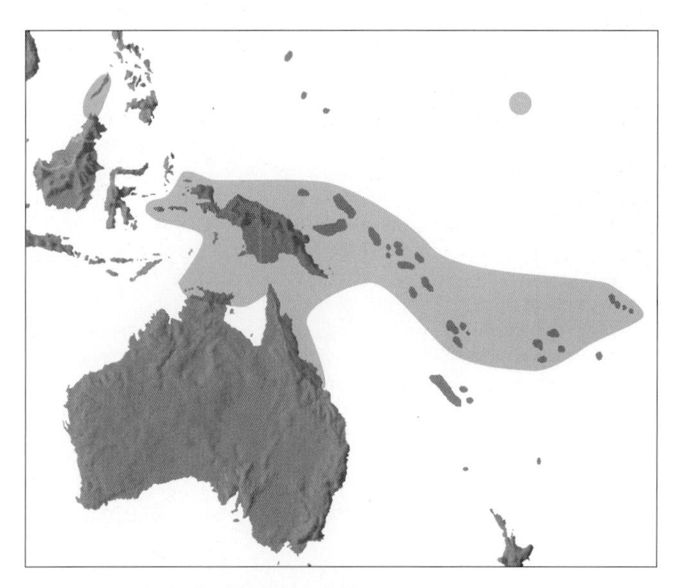

Distribution of subtribe **Ptychospermatinae**

8. Seed pointed apically, with irregular ridges; endocarp very thick, hard; pistillode always lageniform. New Guinea, Bismarck Archipelago, Solomon Islands . . . . **167. Ptychococcus**
8. Seed rounded apically and with 5 rounded ridges; endocarp thin, fibrous or if somewhat bony then not elaborated and staminate flowers with short, conic-ovoid or lageniform pistillode 9
9. Pistillode lageniform; endocarp generally thin, fibrous. Moluccas, New Guinea, northeastern Australia, D'Entrecasteaux and Louisiade Archipelagos, Solomon Islands . . . . **156. Ptychosperma**
9. Pistillode short, conic-ovoid, rarely lageniform; endocarp bony. Micronesia . . . . **157. Ponapea**

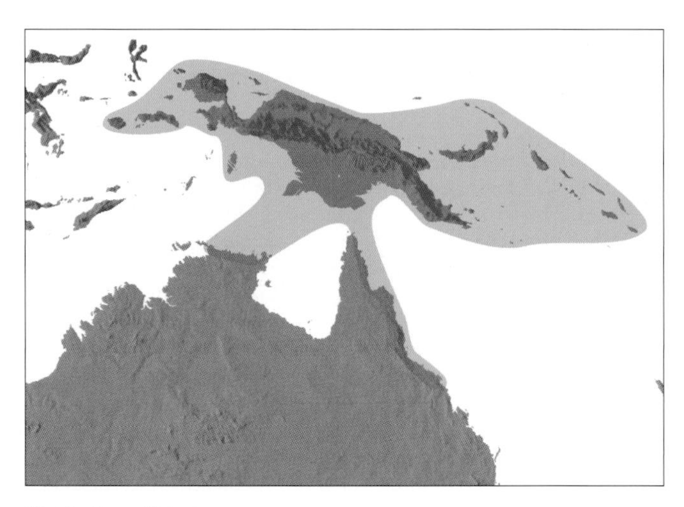

Distribution of ***Ptychosperma***

156. PTYCHOSPERMA

Variable small to moderate, solitary or clustered pinnate-leaved palms, native to the Moluccas through New Guinea to Solomon Islands and Australia, all with crownshafts and praemorse leaflets, and generally with fibrous ridged endocarp.

Ptychosperma Labill., Mém. Cl. Sci. Math. Inst. Natl. France 1808 (2): 252 (1809). Type: **P. gracile** Labill.

Seaforthia R. Br., Prodr. 267 (1810). Type: *S. elegans* R. Br. (= *Ptychosperma elegans* [R.Br.] Blume).

Drymophloeus subgenus *Actinophloeus* Becc., Malesia 1: 42 (1877).

Actinophloeus (Becc.) Becc., Ann. Jard. Bot. Buitenzorg 2: 126 (1885). Lectotype: *A. ambiguus* (Becc.) Becc. (*Drymophloeus ambiguus* Becc.) (= *Ptychosperma ambiguum* [Becc.] Becc.).

Romanowia Sander ex André, Rev. Hort. 71: 262 (1899). Type: *R. nicolai* Sander ex André (*Ptychosperma nicolai* [Sander ex André] Burret (see H.E. Moore 1963c).

Strongylocaryum Burret, Notizbl. Bot. Gart. Berlin-Dahlem 13: 95 (1936). Type: *S. macranthum* Burret (= *Ptychosperma salomonense* Burret).

Ptyx — *a fold or cleft,* sperma — *seed, referring to the grooved seed.*

Small to moderate, solitary or clustered, unarmed, pleonanthic, monoecious palms. *Stems* erect, usually slender, smooth, often grey, obscurely or conspicuously ringed with leaf scars. *Leaves* pinnate, rather short, vertical to ± horizontal, relatively few in the crown; sheath elongate, forming a prominent crownshaft, bearing tattered, peltate, or tufted scales and tomentum, the sheath apex with or without a triangular or ligulate, sometimes divided appendage opposite or to one side of the petiole; petiole short or elongate, channelled adaxially, usually rounded abaxially, tomentose or with scales; rachis longer than the petiole, adaxially ridged, abaxially rounded, variously scaly; leaflets regularly or irregularly arranged, or clustered, single-fold, wedge-shaped, linear or wider medianly, apices obliquely or concavely praemorse, or praemorse and notched, the margins extending beyond the midrib, midrib always prominent, marginal ribs thickened, usually glabrous adaxially, the large or small distal leaflets linear to broadly wedge-shaped, ramenta present or absent along abaxial ribs, transverse veinlets evident (?always). *Inflorescences* infrafoliar, branched usually to 2, 3, or 4(–6) orders, protandrous; peduncle usually short but elongate in *Ptychosperma tagulense*, angled, glabrous, or with scales and tomentum throughout; prophyll tubular, dorsiventrally flattened, keeled laterally, attached at the base of the peduncle, splitting apically, then abaxially, early caducous; peduncular bract tubular, similar to, attached close to, and enclosed by the prophyll, often with a hard short or tapering beak, splitting abaxially,

Right: ***Ptychosperma keiense***, habit, Papua. (Photo: J. Dransfield)

early caducous, an incomplete peduncular bract usually present; rachis longer than the peduncle except where peduncle elongate; rachis bracts triangular to ligulate, or short, stubby, in horizontal furrows, spirally arranged; rachillae elongate, often fleshy, bearing spirally arranged bracts similar to the rachis bracts, subtending triads basally and paired to solitary staminate flowers distally, from as few as 4 to more than 100 clusters per rachilla depending on the species; floral bracteoles short, rounded. *Staminate flowers* bullet-shaped to ovoid, lateral to the pistillate in triads; sepals 3, distinct, broadly imbricate, sometimes gibbous, margins fringed, tips shortly pointed; petals 3, distinct, ovate, rather thick, fibrous, valvate, grooved adaxially, 3–4 times as long as the sepals; stamens 9–over 100, arranged in alternating antesepalous whorls of 3 and antepetalous whorls of several (Uhl & Moore 1980), filaments short, awl-shaped, not inflexed, anthers linear-lanceolate, strongly, often unevenly sagittate basally, bifid apically, dorsifixed near the middle, versatile, latrorse, connective elongate, tanniferous; pistillode bottle-shaped with a long neck, irregularly cleft apically, or short, conic-ovoid, trifid or tripapillate apically. *Pollen* ellipsoidal asymmetric, occasionally pyriform or lozenge-shaped; aperture a distal sulcus; ectexine tectate, perforate, or perforate-rugulate, aperture margin similar or slightly finer; infratectum columellate; longest axis 32–62 µm; post-meiotic tetrads tetrahedral, sometimes tetragonal or, rarely, rhomboidal [14/28]. *Pistillate flowers* shorter than the staminate, conic-ovoid; sepals 3, distinct, broadly imbricate, sometimes gibbous, marginally fringed; petals 3, distinct, broadly imbricate, tips valvate, thick, pointed, opening slightly but not reflexed at anthesis; staminodes tooth-like, linear or united, then broad, toothed or ribbed, and scale-like; gynoecium conic-ovoid, unilocular, rarely bi- or trilocular, stigmas 3, short, reflexed at anthesis, ovule hemianatropous, 5-angled, pendulous, funicle long, bearing a short aril. *Fruit* globose to ellipsoidal, red, orange or purple-black at maturity, stigmatic remains apical, forming a beak, perianth persistent; epicarp granular due to short or long, oblique, fibrous bundles and interspersed brachysclereids, mesocarp fleshy, mucilaginous or tanniniferous, sometimes with irritant needle crystals, endocarp fibrous with vascular

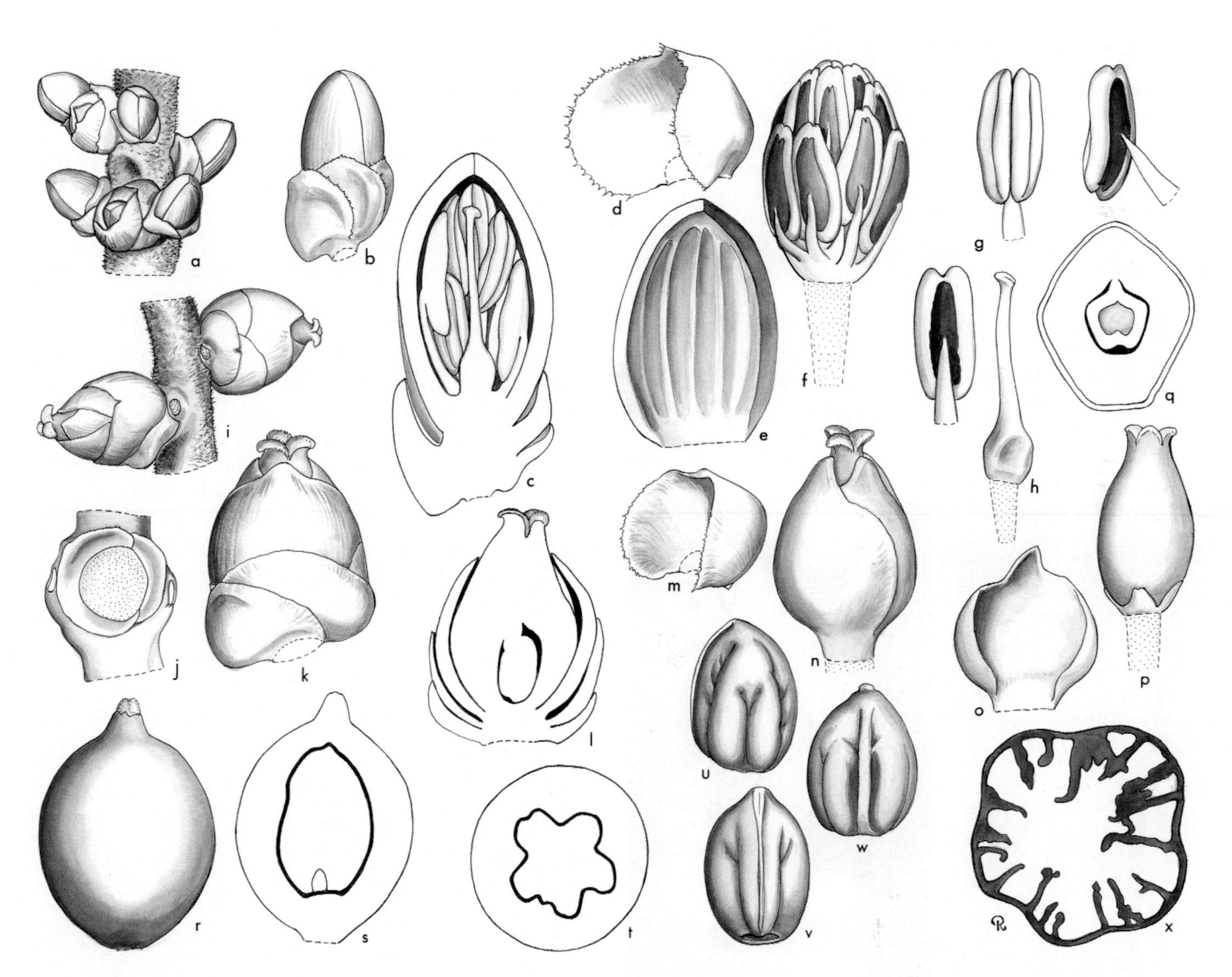

***Ptychosperma*.** **a**, portion of rachilla with triads × 3; **b**, staminate bud × 6; **c**, staminate bud in vertical section × 12; **d**, staminate sepal × 12; **e**, staminate petal, interior view × 12; **f**, androecium × 12; **g**, stamen in 3 views × 12; **h**, pistillode × 12; **i**, portion of rachilla in pistillate flower × 3; **j**, triad, flowers removed to show bracteoles × 6; **k**, pistillate flower × 6; **l**, pistillate flower in vertical section × 6; **m**, pistillate sepal × 6; **n**, pistillate flower, sepals removed × 6; **o**, pistillate petal, interior view × 6; **p**, gynoecium and staminodes × 6; **q**, ovary in cross-section × 12; **r**, fruit × 3; **s**, fruit in vertical section × 3; **t**, fruit in cross-section × 3; **u, v, w**, seed in 3 views × 3; **x**, seed in cross-section × 6. *Ptychosperma caryotoides*: **a–q**, *Moore & Womersley* 9279; *P. macarthurii*: **r–w**, *Moore* 6023; *P. elegans*: **x**, *White* 7761. (Drawn by Marion Ruff Sheehan)

bundles with large fibrous sheaths at several levels and extending into the inner mesocarp or united in a single layer (*P. salomonense*), usually adherent to the seed. *Seed* longitudinally 5-grooved or angled, rarely 3-grooved or rounded in cross-section, hilum lateral, raphe branches few, sparsely branched, endosperm homogeneous or ruminate; embryo basal. *Germination* adjacent-ligular; eophyll bifid. *Cytology*: 2n = 32.

Distribution and ecology: Twenty-nine species centred in New Guinea and the D'Entrecasteaux and Louisiade archipelagos, but extending west to east from the Moluccas to the Solomon Islands and south to north-eastern Australia. A number of other species have been described but are as yet poorly known. Each of the four subgenera and the two sections of subgenus *Actinophloeus* has a distinct range and habitat. The centre of diversity of subgenus *Ptychosperma* and the two sections of subgenus *Actinophloeus* is the mountainous south-eastern tip of New Guinea, where each of the groups has a number of endemic taxa. Some species inhabit coastal or swampy lowland forests, others the better-drained edges of these areas or foothills. *Ptychosperma vestitum* is exceptional in occurring in fresh-water swamps. Pollination in *P. macarthurii* is mostly by bees of the genus *Nomia* (Halictidae), which are attracted to flowers of both sexes by nectar (Essig 1973).

Anatomy: Leaf (Tomlinson 1961), root (Seubert 1998a, 1998b), inflorescence and flower development (Uhl 1976a, 1976b), stamen development (Uhl & Moore 1980), correlations of anatomy with pollination (Uhl & Moore 1977a), and fruit (Essig 1977).

Relationships: *Ptychosperma* is strongly supported as monophyletic with high support (Lewis *et al.* in prep., Baker *et al.* in prep.). Lewis *et al.* (in prep.) place the genus as sister to a clade of *Ponapea* and *Drymophloeus hentyi* with low support. The findings of Asmussen *et al.* (2006), Norup *et al.* (2006) and Baker *et al.* (in review), who did not include *D. hentyi* in their studies, are congruent with this relationship.

Common names and uses: Solitare palm (*Ptychosperma elegans*), Macarthur palm (*P. macarthurii*). Some species make elegant ornamentals but need plentiful moisture and protection from winds. Essig (1978) reports uses of the wood for bows, arrowheads and spears, and that the fruit of some species has been a poor substitute for betel nut.

Taxonomic account: Essig (1978).

Fossil record: No generic records found.

Notes: Large areas within the geographical range of the genus are still unexplored botanically and new species may be discovered in the future, as noted by Essig (1978). About 12 species had been introduced into cultivation through the 1950s; seven of these have been in cultivation for over 100 years and are widely distributed around the world. Growers should be aware of the possibility of hybridization, which has been common among the cultivated species (Essig 1975).

Additional figures: Glossary figs 13, 14, 15, 20.

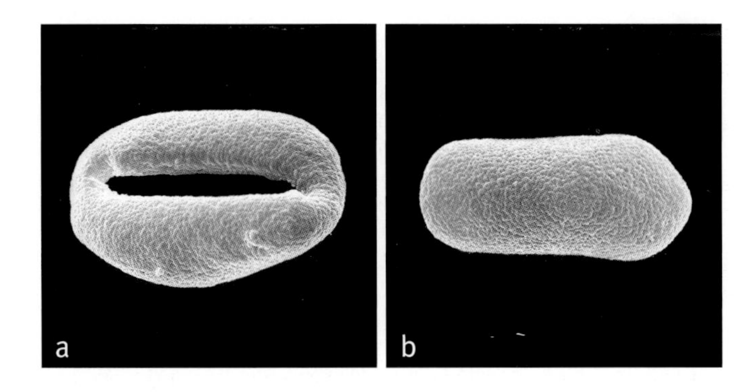

Above: ***Ptychosperma***
a, monosulcate pollen grain, distal face SEM × 1500;
b, monosulcate pollen grain, proximal face SEM × 1500.
Ptychosperma gracile: **a**, cultivated RBG Kew; *P. microcarpum*: **b**, *Darbyshire* 667.

Below left: ***Ptychosperma keiense***, crown with inflorescence, Papua. (Photo: J. Dransfield)

Below right: ***Ptychosperma macarthurii***, staminate flowers, cultivated, Malaysia. (Photo: W.J. Baker)

157. PONAPEA

Moderate, solitary pinnate-leaved palms, native to the Caroline Islands, all with crownshafts and praemorse leaflets.

Ponapea Becc., Bot. Jahrb. Syst. 59: 13 (1924). Type: **P. ledermanniana** Becc.

Named for the island of Ponape (Pohnpei).

Moderate, solitary, unarmed, pleonanthic, monoecious palm. *Stem* erect, moderate, smooth, grey, ringed with leaf scars. *Leaves* pinnate, arching; sheath forming a prominent crownshaft, abaxially densely covered with grey scales; petiole well developed, deeply channelled adaxially, rounded abaxially, covered with small brown punctiform scales; rachis channelled basally, ridged distally; leaflets broadly lanceolate, regularly arranged, tips oblique to truncate, conspicuously praemorse, adaxially glabrous, abaxially covered with numerous dark brown ramenta on major veins, transverse veinlets not apparent. *Inflorescences* infrafoliar, held horizontally, branched to 4 orders basally, to 2–1 orders distally; peduncle short, rather slender, flattened, glabrous; prophyll tubular, membranous, bearing scattered scales, rounded at the tip, splitting abaxially allowing the peduncular bract to emerge; peduncular bract borne just above the prophyll, exceeding the prophyll, longitudinally splitting, caducous, bearing scales as the prophyll; peduncular bract tubular, caducous, shortly beaked, splitting abaxially, adaxially glabrous, abaxially densely to lightly covered in stellate brown scales, scar of 1 incomplete peduncular bract present; rachis much longer than the peduncle, elongate, tapering, bearing rather widely spaced ± angled first-order branches, each subtended by a very small, ridge-like bract; rachillae white, slender, spreading, somewhat divaricate and zig-zag, glabrous, bearing subdistichous, distant triads of flowers for $^{2}/_{3}$ their length and paired to solitary staminate flowers distally; floral bracteoles large, low, rounded. *Staminate flowers* symmetrical, bullet-shaped; sepals 3, distinct, broadly imbricate, irregularly rounded, somewhat gibbous; petals 3, distinct, ovate, valvate, evenly thickened, adaxially grooved; stamens ca. 100, filaments erect in bud, short, awl-shaped, anthers oblong-elliptical, curled, deeply bifid basally and apically, dorsifixed near the base, ?latrorse, connective broad, tanniniferous; pistillode conical, much shorter than the stamens. *Pollen* ellipsoidal asymmetric; aperture a distal sulcus; ectexine tectate, finely perforate-rugulate, aperture margin similar or slightly finer; infratectum columellate; longest axis 43–51 µm [1/3]. *Pistillate flowers* ovoid; sepals 3, distinct, imbricate, rounded,

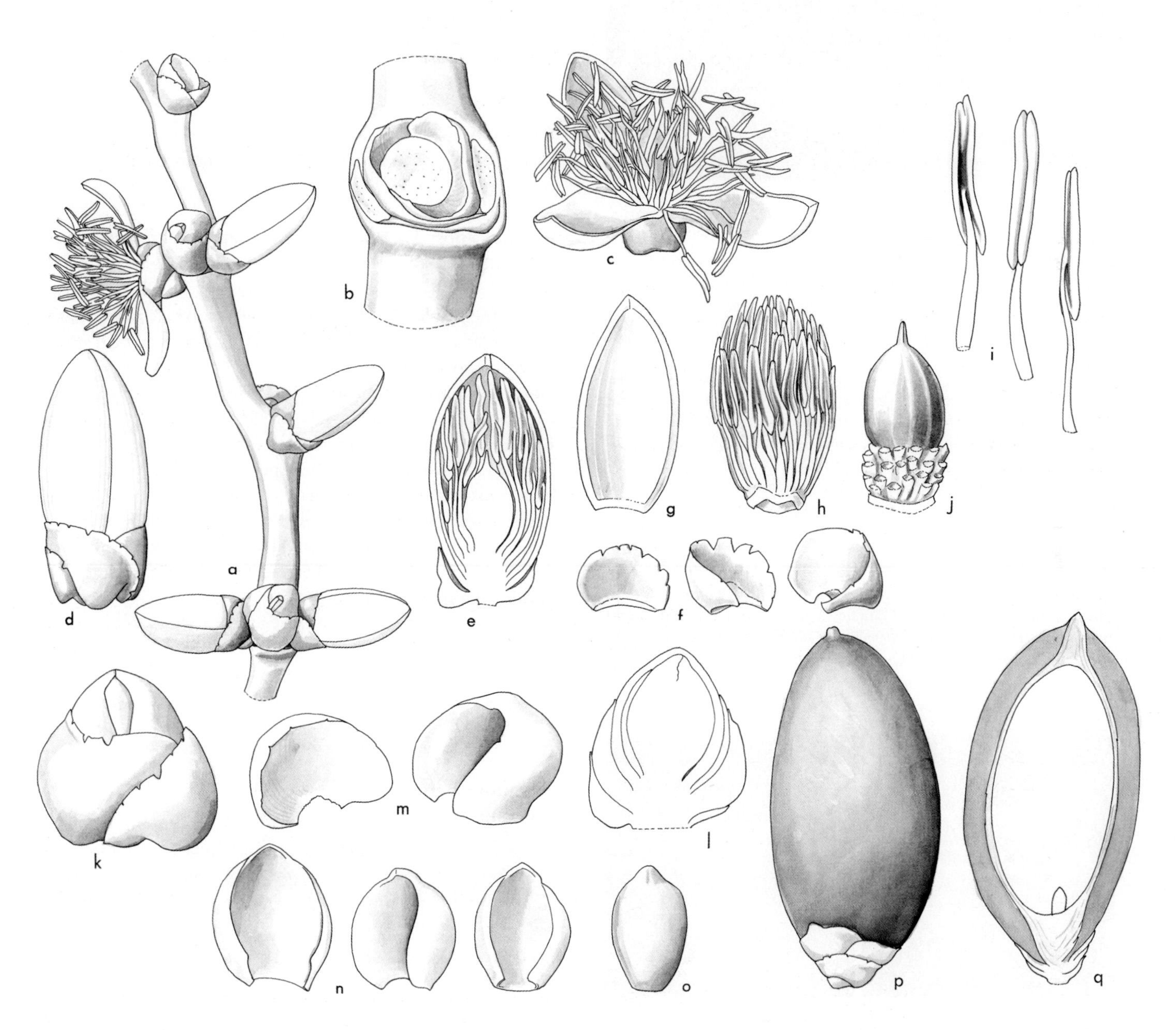

Ponapea. **a**, portion of rachilla at staminate anthesis × 2; **b**, triad scars × 6; **c**, staminate flower × 4; **d**, staminate bud × 4; **e**, staminate bud in section × 4; **f**, staminate sepals, 3 views × 4; **g**, staminate petal, interior view × 4; **h**, stamens × 4; **i**, stamens in 3 views × 8; **j**, pistillode × 6; **k**, pistillate flower bud × 5; **l**, pistillate flower bud in vertical section × 5; **m**, pistillate sepals, 2 views × 5; **n**, pistillate petals, 3 views × 5; **o**, gynoecium × 5; **p**, fruit × $1^{1}/_{2}$; **q**, fruit in vertical section × $1^{1}/_{2}$. *Ponapea ledermanniana*: all from *Zona* 878. (Drawn by Lucy T. Smith)

Above left: ***Ponapea ledermanniana***, habit, cultivated, Fairchild Tropical Botanic Garden, Florida. (Photo: C.E. Lewis)

Above right: ***Ponapea ledermanniana***, infructescence, cultivated, Fairchild Tropical Botanic Garden, Florida. (Photo: J. Roncal)

margins variously split; petals 3, distinct, broadly ovate and imbricate, tips thick, valvate, opening briefly apically to expose the stigmas at anthesis; staminodes 3, irregular, tooth-like; gynoecium asymmetrical, ovoid with a bulge on one side, unilocular, uniovulate, stigmas 3, fleshy, recurved at anthesis, ovule very large, laterally attached, form unknown. *Fruit* ovoid when fresh, drying irregularly 5-ridged, red at maturity, stigmatic remains apical, perianth persistent; epicarp smooth, becoming somewhat striate when dry, mesocarp fleshy, endocarp thick, black, conspicuously 5-ridged (*Ponapea leddermanniana*), ridged and straw-coloured (*P. hosinoi*), or terete and straw-coloured (*P. palauensis*). *Seed* attached laterally, ovoid, conforming to the shape of the endocarp, hilum elongate, raphe branches anastomosing, endosperm homogeneous; embryo basal. *Germination* adjacent-ligular; eophyll bifid with broad conspicuously praemorse lobes. *Cytology* not studied.

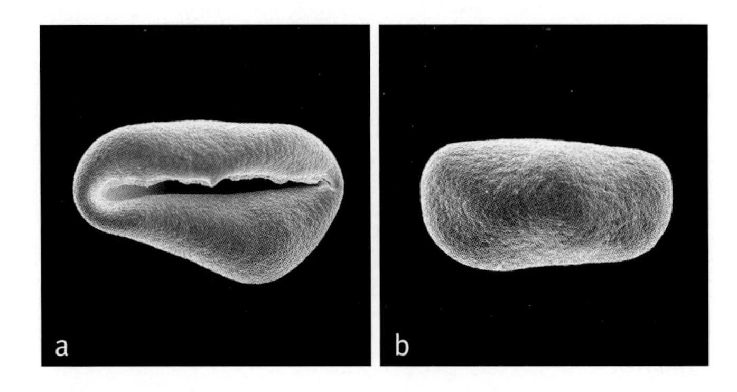

Above: ***Ponapea***
a, monosulcate pollen grain, distal face SEM × 1000;
b, monosulcate pollen grain, proximal face SEM × 1000.
Ponapea ledermanniana: **a** & **b**, *Kanehira* 1361.

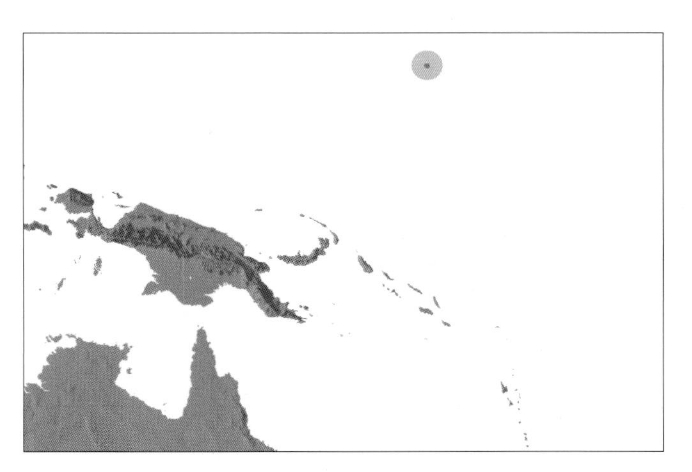

Distribution of ***Ponapea***

Distribution and ecology: Three species in the Caroline Islands (Pohnpei and Palau). Found in rain forest along banks of streams at low elevations.
Anatomy: Fruit (Essig 1977).
Relationships: *Ponapea* is a strongly supported monophyletic group that is resolved as sister to *Drymophloeus hentyi* with high support (Lewis *et al.* in prep.). Asmussen *et al.* (2006), Norup *et al.* (2006) and Baker *et al.* (in review) resolve *Ponapea* as sister to *Ptychosperma* with low to moderate support, but *D. hentyi* was not sampled in these studies.
Common names and uses: Not known.
Taxonomic account: Beccari (1885).
Fossil record: No generic records found.
Notes: The three described species have very different endocarps. Furthermore, *Ponapea ledermanniana* and *P. hosinoi* have short conical pistillodes, shorter than their stamens, whereas *P. palauensis* has a bottle-shaped pistillode, like that commonly found among other Ptychospermatoid palms.
Additional figures: Fig. 2.1b.

158. ADONIDIA

Moderate, solitary pinnate-leaved palm, native to karst limestone of Palawan and neighbouring islands in the Philippines and Malaysia, with crownshaft, praemorse leaflets and ruminate endosperm.

Adonidia Becc., Philipp. J. Sci., 14: 329 (1919). Type: **A. merrillii** (Becc.) Becc. (*Normanbya merrillii* Becc.).

Derivation not explained by author but probably from Adonis, handsome youth of Greek mythology whose blood stained the flowers of Adonis *(Ranunculaceae); perhaps Beccari called the palm* Adonidia *in reference to the bright red fruit.*

Moderate solitary, unarmed, pleonanthic, monoecious palms. *Stem* moderate, ringed with close leaf scars, becoming longitudinally striate, grey. *Leaves* pinnate, strongly curved; sheaths forming a prominent crownshaft, covered with deciduous grey tomentum and scattered dark scales, two short triangular auricles present at base of petiole; petiole short, adaxially channelled, abaxially rounded, densely covered with deciduous tomentum at least basally; rachis elongate, flat to ridged adaxially, rounded abaxially, tomentose; leaflets regularly arranged, but diverging in slightly different planes, strongly curved and rather irregularly bent near the tip,

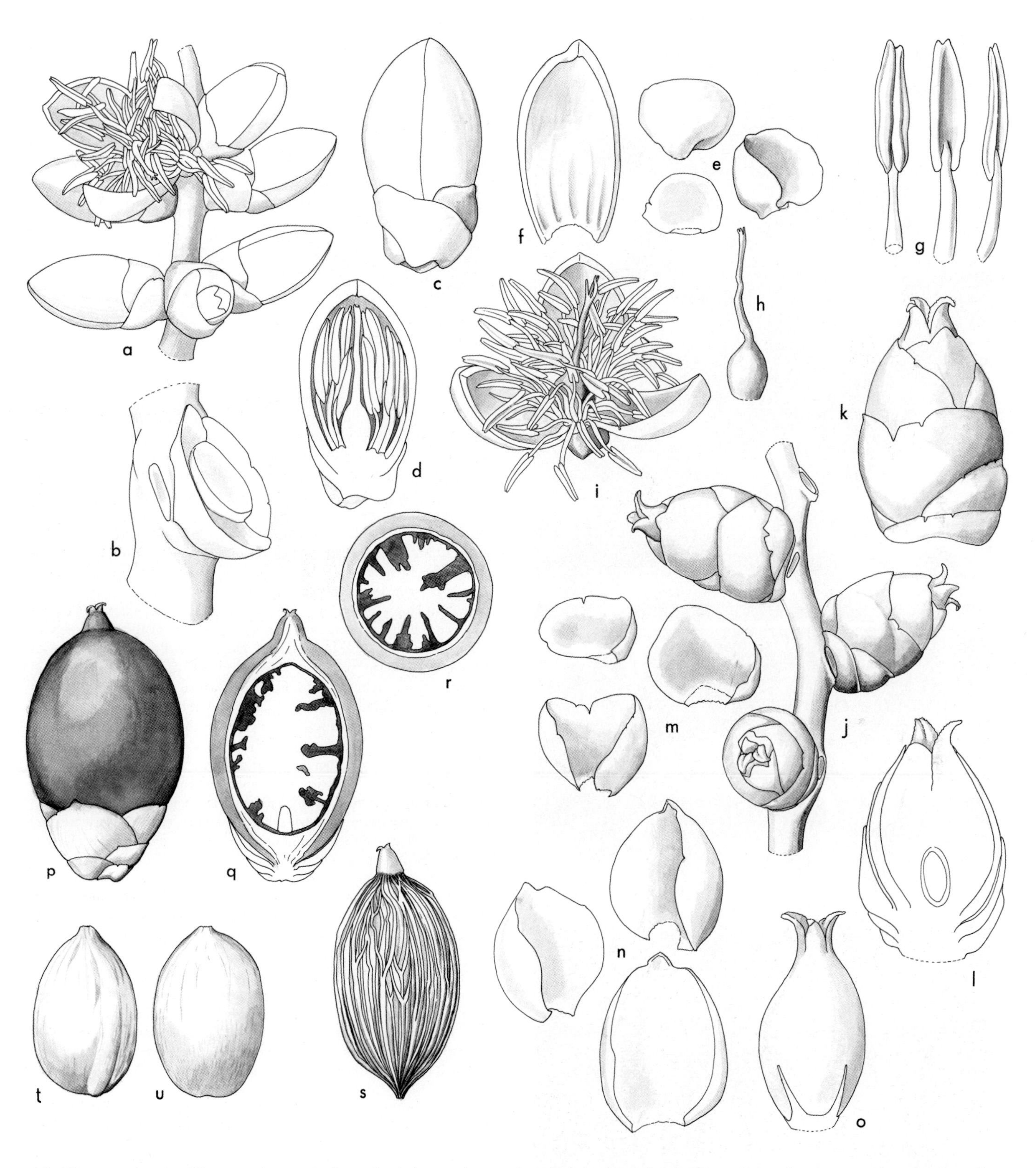

Adonidia. **a**, portion of rachilla at staminate anthesis × 3; **b**, triad scar × 6; **c**, staminate bud × 4; **d**, staminate bud in vertical section × 4; **e**, staminate sepals × 4; **f**, staminate petal, interior view × 4; **g**, stamen in 3 views × 12; **h**, pistillode × 12; **i**, open staminate flower × 3; **j**, portion of rachilla at pistillate anthesis × 2; **k**, pistillate flower × 4; **l**, pistillate flower in vertical section × 4; **m**, pistillate sepals, 3 views × 4; **n**, pistillate petals, 3 views × 4; **o**, gynoecium × 4; **p**, fruit × 1 $^1/_2$; **q**, fruit in vertical section × 1$^1/_2$; **r**, fruit in cross-section × 1$^1/_2$; **s**, endocarp × 1$^1/_2$; **t, u**, seed in 2 views × 1$^1/_2$. *Adonidia merrillii*: **a–i**, *Mackee* 4462; **j–o**, *C. Lewis* s.n.; **p–u**, *L.T. Smith* s.n. (Drawn by Lucy T. Smith)

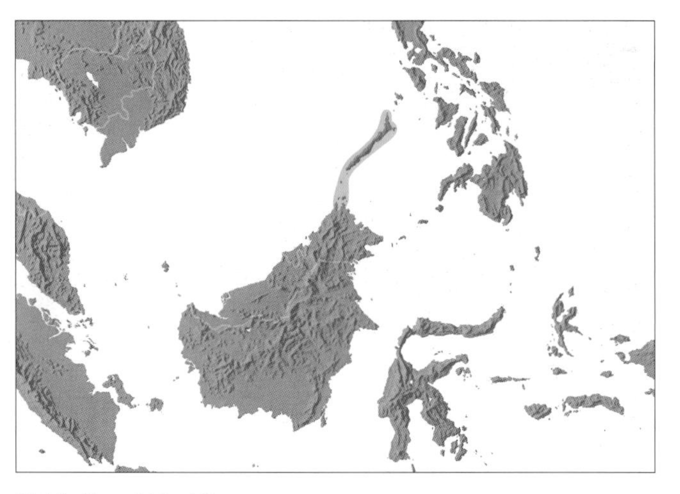

Distribution of ***Adonidia***

single-fold, tapered from middle to base and usually to the apex, apically oblique, truncate, acute, or acuminate and variously toothed, the upper margin usually longest, midrib and marginal veins prominent, midrib bearing ramenta abaxially near the base, otherwise abaxial surface covered in minute brown scales, transverse veinlets not evident. *Inflorescences* infrafoliar, branched to 3–4 orders basally, fewer orders distally, branches tomentose when newly emerged; peduncle very short, stout, dorsiventrally compressed; prophyll and peduncular bract caducous, prophyll tubular, rather thin, 2-keeled laterally; peduncular bract like the prophyll but lacking keels, briefly beaked, scar of an incomplete peduncular bract usually present; rachis longer than the peduncle, tapering, stiff, bearing very short bracts subtending branches and rachillae; rachillae rather short and slender, slightly flexuous, bearing spirally, or distally subdistichous triads throughout at least half their length, distally flower clusters of paired or solitary staminate flowers; floral bracteoles very small inconspicuous, short, rounded, margins notched. *Staminate flowers* bullet-shaped; sepals 3, distinct, imbricate, rounded, ± hooded; petals 3, slightly asymmetrical to symmetrical, valvate, more than twice as long as the sepals; stamens numerous, 45–50, filaments awl-shaped, long, slender, not inflexed apically, anthers linear, dorsifixed, the bases deeply sagittate, bifid or acute apically, latrorse; pistillode attenuate from a bulbous base, as long as the stamens, usually bifid or trifid apically. *Pollen* ellipsoidal asymmetric; aperture a distal sulcus; ectexine tectate, perforate and microchanelled, aperture margin similar; infratectum columellate; longest axis 48–58 µm [1/1]. *Pistillate flowers* ovoid; sepals 3, distinct, thick, imbricate, margins notched; petals 3, distinct, broadly imbricate, only slightly longer than the sepals, apices shortly valvate, margins notched; staminodes 6, tooth-like, variously connate; gynoecium irregularly ovoid, unilocular, uniovulate, style thick, stigmas 3, sessile, recurved at anthesis, ovule large, laterally attached, form unknown. *Fruit* ovoid, beaked, red at maturity, perianth whorls enlarged, persistent, stigmatic remains apical; epicarp thin, sometimes appearing minutely pebbled when dry, mesocarp yellowish, thin-fleshy, with 3–5 layers of slender to large fibres and ground tissue sclerified near the endocarp, endocarp thin, almost glass-like, fragile. *Seed* ovoid, truncate basally, pointed apically, attached laterally, hilum elongate, raphe branches descending from the apex and sides, simple, forked, or branched and much anastomosed, endosperm ruminate; embryo basal. *Germination* adjacent-ligular; eophyll bifid. *Cytology*: 2n = 32.

Top left: ***Adonidia merrillii***, habit, cultivated, Fairchild Tropical Botanic Garden, Florida. (Photo: C.E. Lewis)

Bottom left: ***Adonidia merrillii***, inflorescence, cultivated, Fairchild Tropical Botanic Garden, Florida. (Photo: J. Roncal)

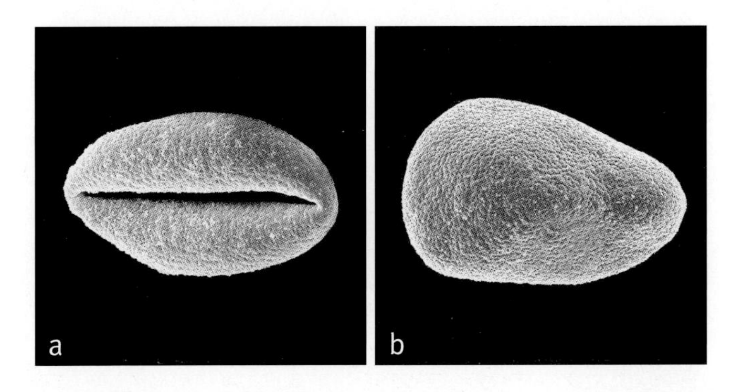

Above: ***Adonidia***
a, monosulcate pollen grain, distal face SEM × 1000;
b, monosulcate pollen grain, proximal face SEM × 1000.
Adonidia merrillii: **a** & **b**, *Bell* s.n.

Distribution and ecology: One species confined to the northern part of Palawan and its offshore islands (Philippines) and the coast of Sabah. *Adonidia merrillii* grows on karst limestone cliffs.

Anatomy: Leaf (Tomlinson 1961) and fruit (Essig 1977).

Relationships: *Adonidia* is moderately supported as sister to a clade of *Solfia*, *Balaka* and *Veitchia* (Norup *et al.* 2006, Baker *et al.* in review, Lewis *et al.* in prep.).

Common names and uses: Manila palm, Christmas palm; very widely planted throughout the tropics and subtropics.

Taxonomic account: Beccari (1919).

Fossil record: No generic records found.

Notes: This represents a remarkable outlier of the Ptychospermatinae at the edge of Sundaland.

159. SOLFIA

Moderate, solitary pinnate-leaved palm, native to Samoa, with crownshaft, praemorse leaflets and distinctive elongate peduncle.

Solfia Rech., Repert. Spec. Nov. Regni Veg. 4: 232 (1907). Type: **S. samoensis** Rech.

Commemorates Wilhelm Solf (1862–1936), one-time governor of German Samoa.

Moderate solitary, unarmed, pleonanthic, monoecious palms. *Stem* erect, slender, ringed with rather widely spaced leaf scars. *Leaves* pinnate, arching slightly, few in crown; sheath rather thick, forming a slender crownshaft, glabrous adaxially, abaxially densely covered with whitish tomentum and small grey-brown scales; petiole relatively short, slender, adaxially channelled near base, rounded distally, abaxially rounded, covered with grey scales; rachis long, slender, adaxially often sharply pointed or ridged and rounded, abaxially rounded, densely covered with grey scales or tomentum; leaflets held in one plane, regularly arranged, single-fold, lanceolate, distal pair small and narrow, leaflets distally oblique, raggedly praemorse, adaxially glabrous except for a few scales near bases of major veins, abaxially covered with small scales on veins, ramenta lacking, midvein and a pair of large veins along or close to the margins prominent, margins densely covered with caducous tomentum and scales, transverse

Top right: ***Solfia samoensis***, habit, Samoa. (Photo: D.R. Hodel)

Bottom right: ***Solfia samoensis***, inflorescences and infructescence, Samoa. (Photo: D.R. Hodel)

veinlets not evident. *Inflorescences* infrafoliar, branched to 3 orders, protandrous; peduncle relatively slender, elongate, elliptical in cross-section; prophyll apparently tardily deciduous, tubular, slender, dorsiventally flattened, with 2, narrow, lateral keels, pointed, splitting apically and for a short distance abaxially to release the peduncular bract, abaxially densely covered with multibranched or grey scales; peduncular bract marcescent, tubular, slender, much longer than the peduncle, with a long rather flat, pointed tip, densely covered with scales as the prophyll; rachis shorter to slightly longer than the peduncle, angled, tapering, bearing spirally arranged branches; rachis bracts low, rounded; rachillae slender, tapering, bearing spirally arranged triads of flowers basally, and pairs of a pistillate and one staminate, or a single staminate flower distally; floral bracteoles low, rounded or truncate. *Staminate flowers* bullet-shaped; sepals 3, distinct, imbricate, irregularly but strongly keeled, margins thin, variously notched; petals 3, distinct, ovate, valvate, evenly thickened, adaxially grooved, abaxially striate when dry, reflexed at anthesis; stamens numerous, ca. 35–37, filaments slender, awl-shaped, anthers elongate, sagittate basally, uneven, sometimes divided apically, medifixed, versatile, latrorse; pistillode absent in some flowers or ovoid attenuate as long as the

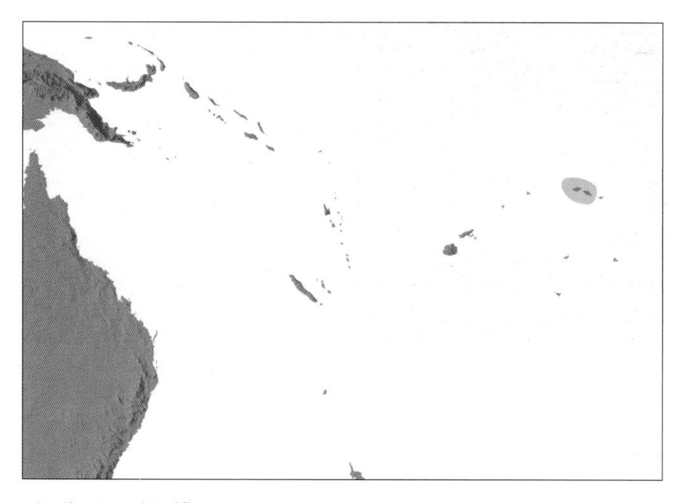

Distribution of ***Solfia***

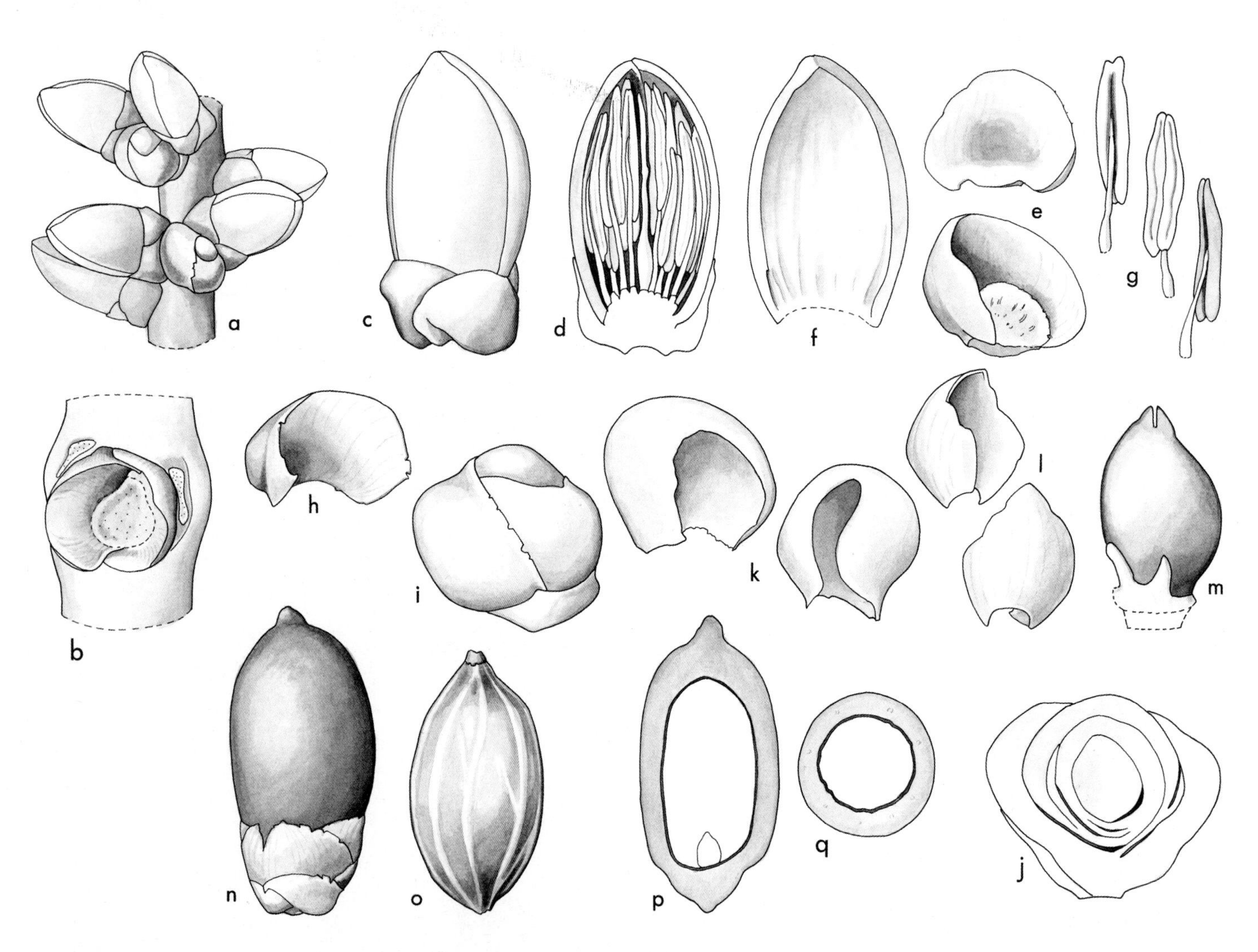

Solfia. **a**, portion of rachilla with triads × 3; **b**, triad scar × 6; **c**, staminate flower bud × 6; **d**, staminate bud in vertical section × 6; **e**, staminate sepals, 2 views × 6; **f**, staminate petal, interior view × 6; **g**, stamens, 3 views × 12; **h**, bracteole × 8; **i**, pistillate flower bud × 6; **j**, pistillate flower bud in vertical section × 10; **k**, pistillate sepals, 2 views × 8; **l**, pistillate petals, 2 views × 10; **m**, gynoecium and staminodes × 15; **n**, fruit × 2 1/2; **o**, endocarp × 2 1/2; **p**, fruit in vertical section × 2 1/2; **q**, fruit in cross-section × 2 1/2. *Solfia samoensis*: **a–m**, *Tipamaia* 02; **n–q**, *Tipamaia* 01. (Drawn by Lucy T. Smith)

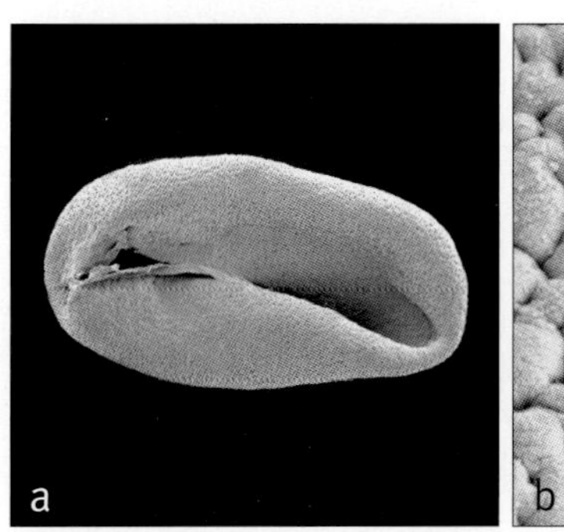

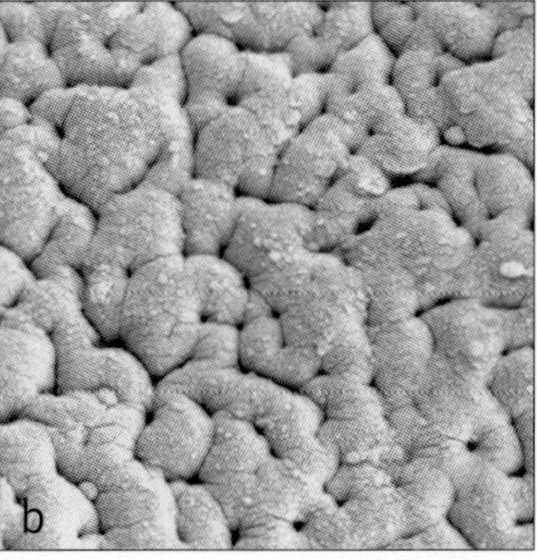

Above: ***Solfia***
a, monosulcate pollen grain, distal face SEM × 600;
b, close-up, perforate-rugulate pollen surface SEM × 8000.
Solfia samoensis: **a**, *Moore & Fasavalu* 9980, **b**, *Christophersen* 2897.

stamens, usually shortly trifid apically. *Pollen* ellipsoidal asymmetric; aperture a distal sulcus; ectexine tectate, psilate-scabrate and finely perforate, aperture margin similar; infratectum columellate; longest axis ranging from 58–73 µm [1/1]. *Pistillate flowers* broadly ovoid; sepals 3, distinct, imbricate, hooded, edges minutely toothed or variously notched; petals 3, twice as long as the sepals in late bud, distinct, ovate, imbricate, lateral margins shortly fringed, tips thick, valvate; staminodes 3, shortly joined basally, wide, truncate and bifid or uneven distally; gynoecium symmetrical, ovoid, tapering distally, unilocular, uniovulate, stigmas 3, recurved, ovule attached laterally or pendulous from the top of the locule, form unknown. *Fruit* ovoid, red at maturity, stigmatic remains eccentrically apical; epicarp thin, smooth, becoming wrinkled and somewhat pebbled when dry, mesocarp fleshy, fibrous, with stinging crystals, major fibres conspicuous adherent to the endocarp, endocarp terete, straw-coloured. *Seed* ovoid, surface smooth, hilum apical, raphe much branched, endosperm homogeneous; embryo basal. *Germination* adjacent-ligular; eophyll bifid. *Cytology*: 2n = 32.

Distribution and ecology: One species endemic to Samoa, occurring in montane forest.
Anatomy: Not studied.
Relationships: Norup *et al.* (2006), Baker *et al.* (in review) and Lewis *et al.* (in prep.) found high support for a sister relationship between *Solfia* and *Balaka.*
Common names and uses: Not recorded.
Taxonomic account: Rechinger (1907).
Fossil record: No generic records found.
Notes: The slender elongate peduncle is distinctive.

160. BALAKA

Small or moderate, solitary pinnate-leaved palms, native to Fiji and Samoa, all with crownshafts and praemorse leaflets, with long peduncles and generally with a 4–5-ridged seed.

Balaka Becc., Ann. Jard. Bot. Buitenzorg 2: 91 (1885). Lectotype: **B. perbrevis** (H. Wendl.) Becc. (*Ptychosperma perbreve* H. Wendl. [see Beccari & Pichi-Sermolli 1955] = *Balaka seemannii* [H. Wendl.] Becc.).

Derived from the Fijian vernacular name, balaka.

Small to moderate, solitary, unarmed, pleonanthic, monoecious palms. *Stem* erect, slender, ringed with rather prominent leaf scars. *Leaves* pinnate, somewhat arched and spreading; sheath tubular, forming a prominent crownshaft, covered with brown scales; petiole very short to moderate,

Below left: ***Balaka tahitensis***, habit, Samoa. (Photo: D.R. Hodel)
Below middle: ***Balaka tahitensis***, part of inflorescence and open staminate flower, Samoa. (Photo: D.R. Hodel)
Below right: ***Balaka tahitensis***, immature fruit, Samoa. (Photo: D.R. Hodel)

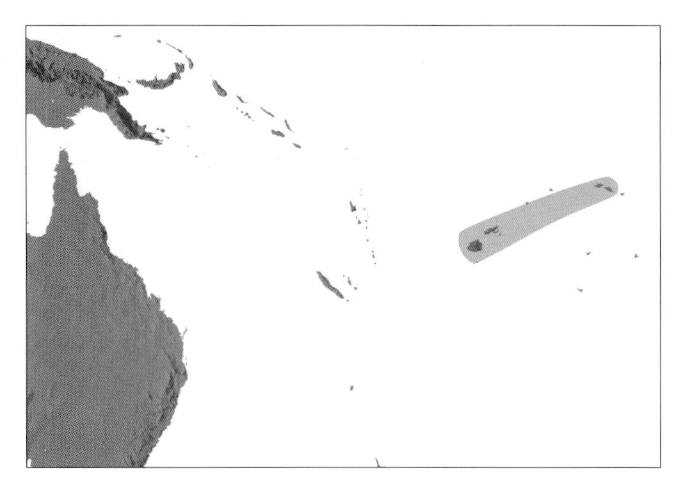

Distribution of ***Balaka***

nearly terete or channelled adaxially, rounded abaxially, covered in tattered brown scales; rachis channelled to ridged adaxially, rounded abaxially, densely covered in tattered, brown scales; leaflets elongate to sigmoid or wedge-shaped and little tapered, alternate in one plane, single-fold, oblique or truncate apically, praemorse, with brown scales along ribs abaxially and scattered on bases of ribs adaxially, several veins about equal in size to midrib and marginal veins prominent, transverse veinlets not evident. *Inflorescences* infrafoliar, branched to 2 orders basally, 1 order distally, spreading; peduncle long; prophyll rather short and slender, 2-keeled laterally, slightly beaked, opening apically; peduncular bract inserted well above and much longer than the prophyll, terete, beaked, both bracts usually caducous before anthesis and bearing brown scales or tomentum; rachis shorter than the peduncle, bearing spirally arranged, pointed bracts subtending rachillae; rachillae rather short, bearing distichously arranged, low, rounded bracts subtending triads of flowers nearly throughout, a few solitary staminate flowers distally; floral bracteoles prominent, margins somewhat jagged. *Staminate flowers* symmetrical, bullet-shaped in bud; sepals 3, distinct, glabrous or red-brown tomentose, rounded, imbricate, margins toothed; petals 3, briefly united at the base, ovate, valvate, grooved adaxially; stamens numerous (24–50), filaments erect in bud, awl-shaped, anthers elongate, dorsifixed near the base, bifid apically, briefly sagittate basally, latrorse; pistillode bottle-shaped with a long neck, often ± flexuous apically at anthesis. *Pollen* ellipsoidal asymmetric; aperture a distal sulcus; ectexine tectate, perforate, or perforate-finely rugulate, aperture margin similar or slightly finer; infratectum columellate; longest axis ranging from 33–41 µm [5/11]. *Pistillate flowers* ovoid; sepals 3, distinct, imbricate, margins toothed; petals 3, distinct, widely imbricate, toothed laterally, pointed apically; staminodes usually 6, small, ± united, tooth-like with jagged tips; gynoecium ovoid, unilocular, uniovulate, stigmas apparently short, ovule pendulous, form unknown. *Fruit* irregularly ovoid, tapered distally or at both ends, often angled, ± beaked, reddish-orange at maturity, drying

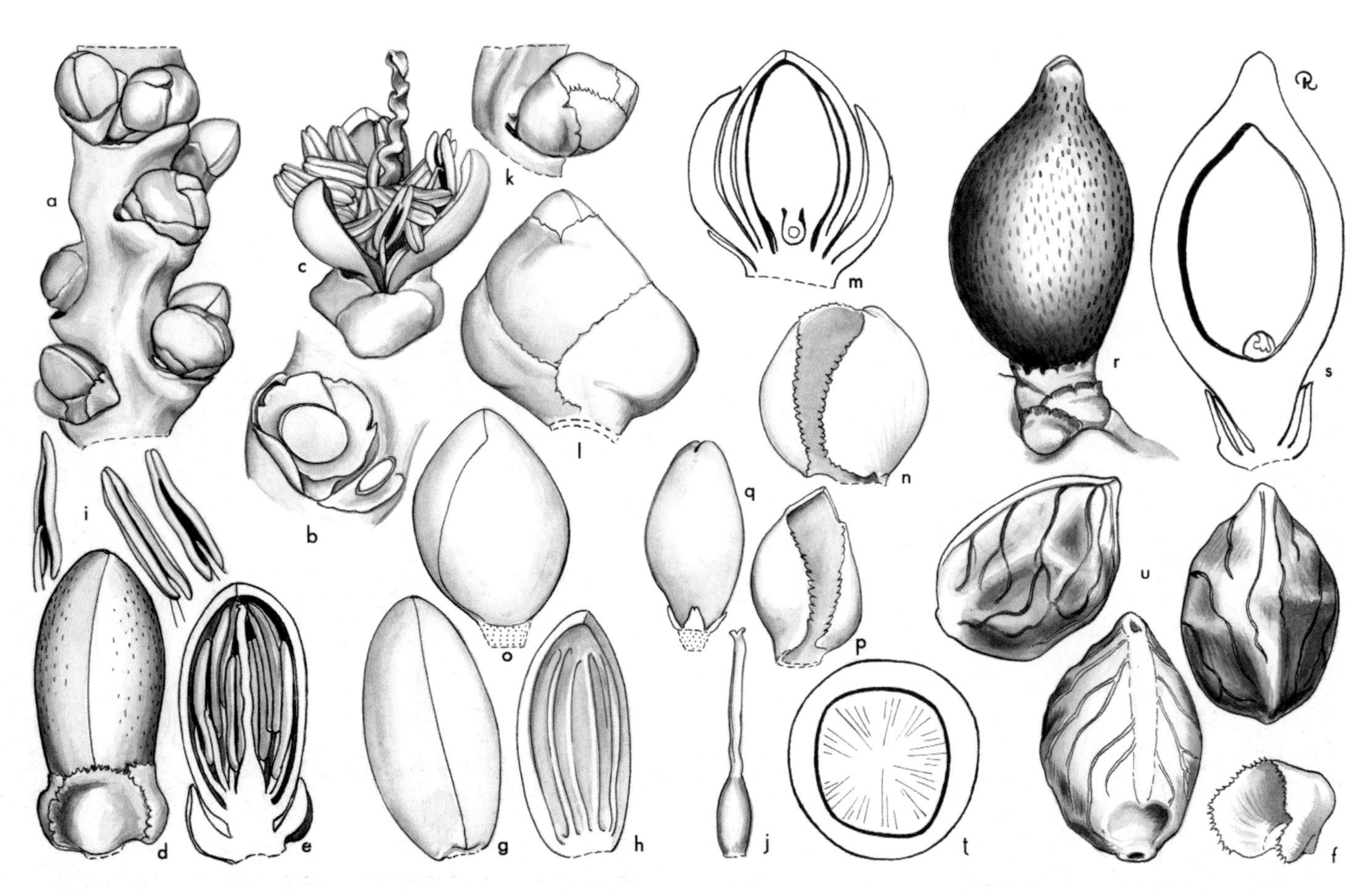

Balaka. **a**, portion of rachilla with triads × 3; **b**, triad, flowers removed × 6; **c**, staminate flower at anthesis × 6; **d**, staminate bud × 6; **e**, staminate bud in vertical section × 6; **f**, staminate sepal × 6; **g**, staminate bud, sepals removed × 6; **h**, staminate petal, interior view × 6; **i**, stamen in 3 views × 6; **j**, pistillode × 6; **k**, portion of rachilla with pistillate bud × 3; **l**, pistillate bud × 6; **m**, pistillate bud in vertical section × 6; **n**, pistillate sepal × 6; **o**, pistillate bud, sepals removed × 6; **p**, pistillate petal × 6; **q**, gynoecium and staminodes × 6; **r**, fruit × 3; **s**, fruit in vertical section × 3; **t**, fruit in cross-section × 3; **u**, seed in 3 views × 3. *Balaka longirostris*: **d–q**, *Moore & Koroiveibau* 9362; *B. microcarpa*: **a–c** and **r–u**, *Moore et al.* 9359. (Drawn by Marion Ruff Sheehan)

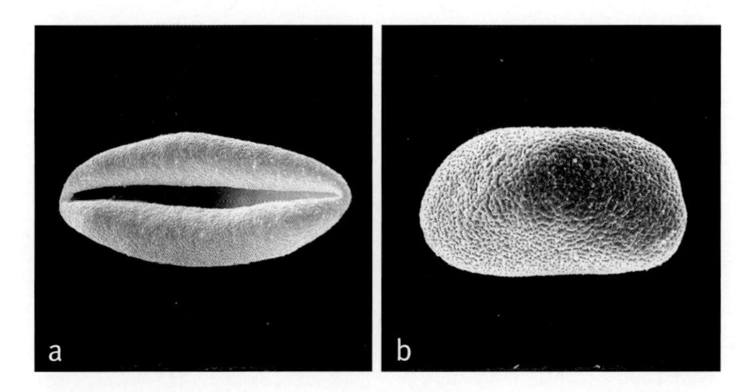

Above: ***Balaka***
a, monosulcate pollen grain, distal face SEM × 1500;
b, monosulcate pollen grain, proximal face SEM × 1500.
Balaka microcarpa: **a**, *Melville & Siwauiban* s.n.; *B. tahitensis*: **b**, *Whistler* W4173.

pebbled, stigmatic remains apical; epicarp thin, mesocarp with outer sclereid layer, inner fleshy layer, and a single series of vascular bundles with thick fibrous sheaths next to the endocarp, endocarp thin, not operculate. *Seed* elongate, pointed, ridged, 4–5-angled in cross-section, hilum elongate, lateral, raphe branches few, endosperm homogeneous, embryo basal. *Germination* and eophyll not recorded. *Cytology* not studied.

Distribution and ecology: About 11 species, six endemic to Fiji and five in Samoa. These palms occur in various types of forest up to about 1000 m altitude.
Anatomy: Root (Seubert 1998a, 1998b) and fruit (Essig 1977).
Relationships: *Balaka* is resolved as monophyletic with moderate support (Lewis *et al.* in prep.) and placed as sister to *Solfia* with high support (Norup *et al.* 2006, Baker *et al.* in review, Lewis *et al.* in prep.). One study, however, suggests that *Solfia* is nested within a paraphyletic *Balaka* (Baker *et al.* in prep.).
Common names and uses: *Balakwa* (Fiji), balaka palms. The straight stems have been used for walking sticks and spears, and the kernel is reported to be edible. Some species (e.g., *Balaka seemannii*) are widely grown as ornamentals in Fiji and elsewhere.
Taxonomic account: Moore (1979) and Whistler (1992).
Fossil record: No generic records found.
Notes: Differing from *Veitchia* in having ridged seeds and often angled and tapered fruit, and from *Ptychosperma* in having a more elongate peduncle and a peduncular bract that is exserted well above and is much longer than the prophyll.
Additional figures: Glossary fig. 19.

161. VEITCHIA

Elegant, moderate to tall, solitary pinnate-leaved palms, native to Vanuatu, Fiji and Tonga, all with crownshafts and almost always praemorse leaflets, the fruit with smooth thin endocarp and rounded seed with homogeneous endosperm.

Veitchia H. Wendl. in Seem., Flora vit. 270 (1868) (conserved name). Type: **V. joannis** H. Wendl. (conserved type).
Vitiphoenix Becc., Ann. Jard. Bot. Buitenzorg 2: 9 (1885). Type: *V. filifera* (H. Wendl.) Becc. (*Ptychosperma filiferum* H. Wendl.) (= *Veitchia filifera* [H. Wendl.] H.E. Moore).
Kajewskia Guillaumin, J. Arnold Arbor. 13: 113 (1932). Type: *K. aneityensis* Guillaumin (= *Veitchia spiralis* H. Wendl.).

Honours James Veitch (1792–1863), famous British nurseryman.

Moderate to tall, solitary, unarmed, pleonanthic, monoecious palms. *Stem* moderate, sometimes wide at the base, ringed with close leaf scars, becoming longitudinally striate, covered in grey scales or smooth, grey to brown. *Leaves* pinnate, spreading or partially erect; sheaths forming a prominent crownshaft, covered with deciduous grey to brown tomentum; petiole short, adaxially channelled, abaxially rounded, densely covered with decidous tomentum at least basally, often bearing dark brown to black, tattered scales in various, sometimes diagnostic patterns at the apex; rachis elongate, flat to ridged adaxially, rounded abaxially, tomentose; leaflets regularly or irregularly arranged, single-fold, tapered from middle to base and usually to the apex, apically oblique, truncate, acute, or acuminate and variously toothed, the upper margin usually longest, midrib and marginal veins prominent, midrib bearing ramenta abaxially, otherwise abaxial surface covered in pale scales with brown centres, shining red scales, or glabrous, transverse veinlets not evident. *Inflorescences* infrafoliar, branched to 3 or 4 orders basally, fewer orders distally, branches variously tomentose; peduncle short, stout, dorsiventrally compressed; prophyll and peduncular bract caducous, prophyll tubular, rather thin, 2-keeled laterally; peduncular bract like the prophyll but lacking keels, briefly beaked, scar of an incomplete peduncular bract usually present; rachis longer than the peduncle, tapering, stiff, bearing very short bracts subtending branches and rachillae; rachillae medium to long, slightly to markedly flexuous, bearing

Below: ***Veitchia arecina***, habit, cultivated, Montgomery Botanical Center, Florida. (Photo: J. Dransfield)

spirally, or distally subdistichous triads nearly throughout, or triads only near the base of the rachillae and distally flower clusters of paired or solitary staminate flowers; floral bracteoles very small or conspicuous, short, rounded, margins notched. *Staminate flowers* bullet-shaped; sepals 3, distinct, imbricate, rounded, ± hooded; petals 3, slightly asymmetrical to symmetrical, valvate, more than twice as long as the sepals; stamens numerous, to over 100, filaments awl-shaped, long, slender, united at the base, not inflexed apically, anthers linear, basifixed or dorsifixed and then bases deeply sagittate, acute, emarginate, or bifid apically, latrorse; pistillode attenuate from a bulbous base, as long as the stamens, usually bifid or trifid apically. *Pollen* ellipsoidal asymmetric, often elongate; aperture a distal sulcus; ectexine tectate, perforate, or perforate-finely rugulate, aperture margin similar or slightly finer; infratectum columellate; longest axis 40–73 µm; post-meiotic tetrads tetrahedral, rarely tetragonal or rhomboidal [6/8]. *Pistillate flowers* ovoid; sepals 3, distinct, thick, imbricate, margins notched; petals 3, distinct, broadly imbricate, only slightly longer than the sepals, apices shortly valvate, margins notched; staminodes 3–6, tooth-like, usually variously connate; gynoecium irregularly ovoid, unilocular, uniovulate, style thick, stigmas 3, sessile, recurved at anthesis, ovule large, laterally attached, form unknown. *Fruit* ovoid, beaked, small to moderately large, red or orange-red at maturity, perianth whorls enlarged, persistent, stigmatic remains apical; epicarp thin, sometimes appearing lined or pebbled when dry, mesocarp yellowish, thin-fleshy, with 3–5 layers of slender to large fibres and ground tissue sclerified near the endocarp, endocarp thin, chartaceous, cartilaginous or nearly glass-like, fragile, in some species

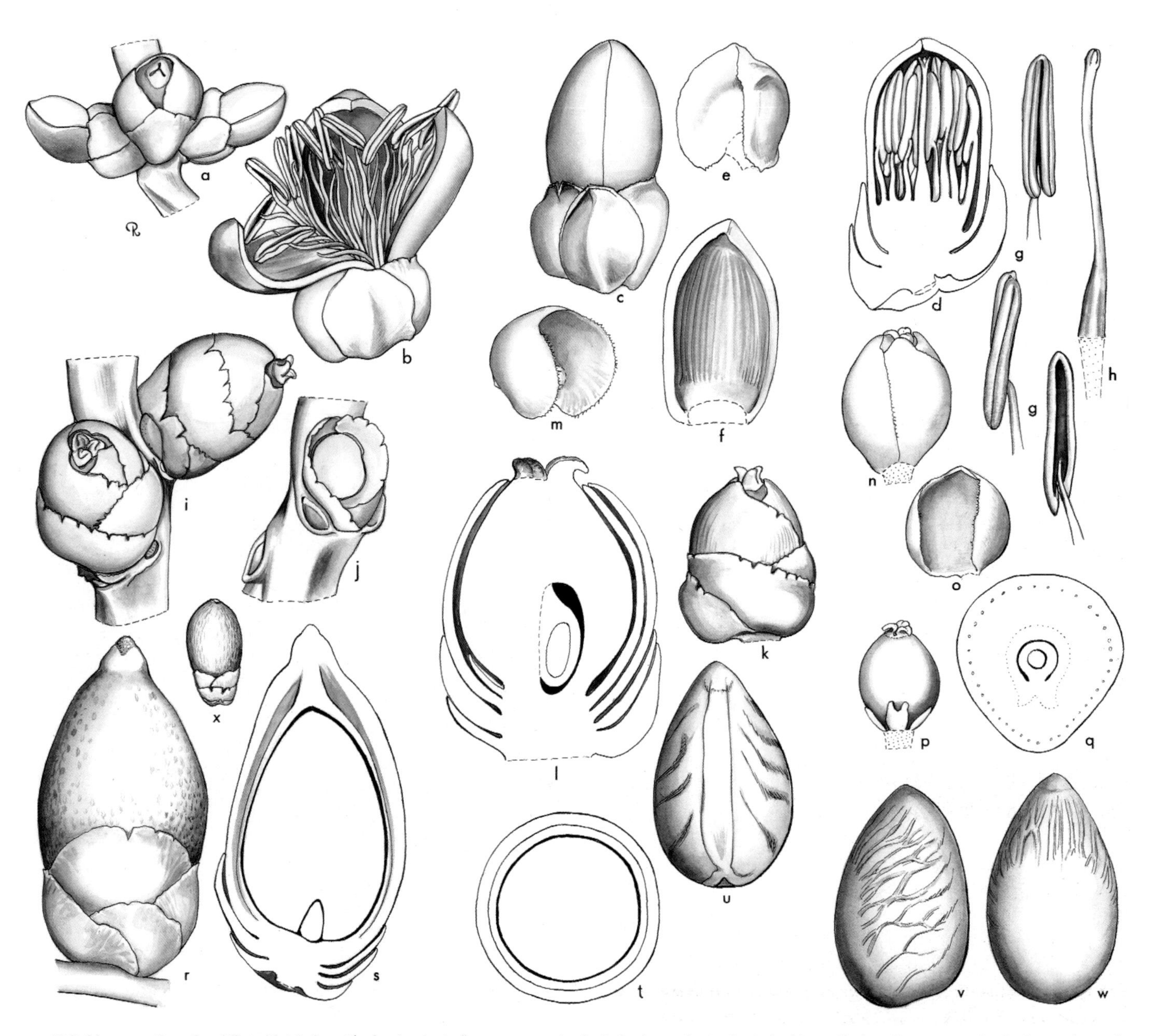

***Veitchia*.** **a**, portion of rachilla with triad × 1 ¹/₂; **b**, staminate flower × 3; **c**, staminate bud × 3; **d**, staminate bud in vertical section × 3; **e**, staminate sepal × 3; **f**, staminate petal, interior view × 3; **g**, stamen in 3 views × 6; **h**, pistillode × 6; **i**, rachilla at pistillate anthesis × 3; **j**, triad, flowers removed to show bracteoles × 3; **k**, pistillate flower × 3; **l**, pistillate flower in vertical section × 6; **m**, pistillate sepal × 3; **n**, pistillate flower, sepals removed × 3; **o**, pistillate petal × 3; **p**, gynoecium and staminodes × 3; **q**, ovary in cross-section × 6; **r**, fruit × 1; **s**, fruit in vertical section × 1; **t**, fruit in cross-section × 1; **u**, **v**, **w**, seed in 3 views × 1; **x**, fruit × 1. *Veitchia joannis*: **a–h** and **r–w**, *Moore & Koroiveibau* 9351; *V. simulans*: **i–q**, *Moore & Koroiveibau* 9352; *V. filifera*: **x**, *Moore & Koroiveibau* 9348. (Drawn by Marion Ruff Sheehan)

Distribution of ***Veitchia***

fracturing transversely. *Seed* ovoid to ellipsoidal, truncate basally, rounded to pointed apically, attached laterally, hilum elongate, raphe branches descending from the apex and sides, simple, forked, or branched and much anastomosed, endosperm homogenous; embryo basal. *Germination* adjacent-ligular; eophyll bifid. *Cytology*: 2n = 32.

Distribution and ecology: Eight species in Vanuatu, Fiji and Tonga. Found from near sea level to cloud forests at 1000 m in dense or light forest.
Anatomy: Leaf (Tomlinson 1961), root (Seubert 1998a, 1998b) and fruit (Essig 1977).
Relationships: Lewis *et al.* (in prep.) resolve *Veitchia* as monophyletic with high bootstrap support. The genus is moderately supported as sister to a clade of *Solfia* and *Balaka* (Norup *et al.* 2006, Baker *et al.* in review, in prep., Lewis *et al.* in prep.).
Common names and uses: *Niuniu*, *niusawa* (Fiji). Several species are elegant ornamentals. The leaves have been used for thatch and the trunks as wood for rafters, spears, and canoes. The heart is edible as a salad and the inflorescence and seed are also said to be edible in *Veitchia vitiensis*.
Taxonomic account: Zona & Fuller (1999).
Fossil record: No generic records found.

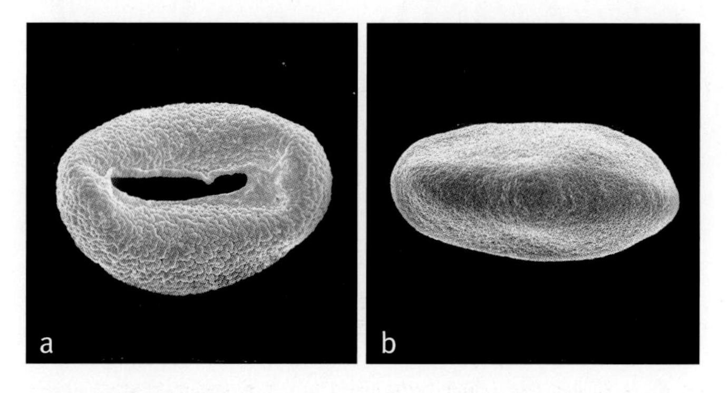

Above: ***Veitchia***
a, monosulcate pollen grain, distal face SEM × 1500;
b, monosulcate pollen grain, proximal face SEM × 1000.
Veitchia filifera: **a**, *Smith* 6635; *V. arecina*: **b**, *Raynal* RSNH16263.

Notes: The genus is distinguished from other Ptychospermatinae by the combination of similar prophyll and peduncular bracts, lanceolate leaflets, and rounded rather than angled seeds.
Additional figures: Glossary fig. 4.

162. CARPENTARIA

Elegant, moderate to tall, solitary pinnate-leaved palm, native to Northern Territory, Australia, with crownshaft and praemorse leaflets, the leaf rachis gracefully arching; the fruit has a distinctive network of black fibres next to the endocarp; the seed has homogeneous endosperm.

Carpentaria Becc., Ann. Jard. Bot. Buitenzorg 2: 128 (1885). Type **C. acuminata** (H. Wendl. & Drude) Becc. (*Kentia acuminata* H. Wendl. & Drude).

Named from the Gulf of Carpentaria in northern Australia where Carpentaria acuminata *is endemic.*

Below: ***Carpentaria acuminata***, crown and inflorescences, cultivated, Australia. (Photo: J. Dransfield)

Moderate or tall, solitary, unarmed, pleonanthic, monoecious palm. *Stem* erect, moderate, smooth, grey, ringed with leaf scars. *Leaves* pinnate, arching or drooping, tips becoming pendulous; sheath forming a prominent crownshaft; petiole very short, deeply channelled adaxially, rounded abaxially, covered with small brown scales; rachis channelled basally to ridged; leaflets abaxially shallowly convex, ± clustered basally, opposite to subopposite distally, long, narrow, lanceolate, tips oblique to truncate, praemorse, often with 2–4 longer prongs, the proximal sometimes tapering to a single or double point, adaxially lightly tomentose, abaxially covered with small brown-centred scales, midrib the only prominent nerve on both surfaces, larger abaxially, transverse veinlets not apparent. *Inflorescences* infrafoliar, horizontal, appearing rather large, branched to 3 orders basally, to 2–1 orders distally; peduncle very short, stout, flattened, bearing deciduous brown tomentum; prophyll tubular (not seen), caducous; peduncular bract tubular, caducous, shortly beaked, splitting abaxially, adaxially glabrous, abaxially densely to lightly covered in stellate brown scales, scar of 1 incomplete peduncular bract present; rachis much longer than the peduncle, elongate, tapering, bearing many (more than 20) ± angled first-order branches, each subtended by a very small, ridge-like bract; rachillae rather short, slender, spreading, bearing spirally arranged, distant triads of flowers for 2/3 their length and paired to solitary staminate flowers distally; floral bracteoles large, low, rounded. *Staminate flowers* lateral to the pistillate, symmetrical, ovoid; sepals 3, distinct, broadly imbricate, irregularly rounded, somewhat gibbous; petals 3, distinct, rather broadly ovate, valvate, evenly thickened, adaxially grooved; stamens ca. 33, filaments erect in bud, short, awl-shaped, anthers oblong-elliptical, bifid basally, emarginate apically, dorsifixed near the base, latrorse, connective broad, tanniniferous; pistillode bottle-shaped, as long as the stamens in bud. *Pollen* ellipsoidal asymmetric; aperture a distal sulcus; ectexine tectate, perforate and micro-chanelled, aperture margin slightly finer than main tectum; infratectum columellate; longest axis ranging from 39–53 µm; post-meiotic tetrads tetrahedral [1/1]. *Pistillate flowers* ovoid; sepals 3, distinct, imbricate, rounded, margins variously split; petals 3, distinct, broadly ovate and imbricate, tips thick, valvate, opening briefly apically to expose the stigmas at anthesis; staminodes 3, tooth-like, bifid apically; gynoecium asymmetrical, ovoid with a bulge on one side, unilocular, uniovulate, stigmas 3, fleshy, recurved at anthesis, ovule very large, laterally attached, form unknown. *Fruit* ovoid when fresh, red at maturity, stigmatic remains apical, perianth persistent; epicarp smooth, becoming wrinkled when dry, mesocarp fleshy over a layer of broad, flat fibres

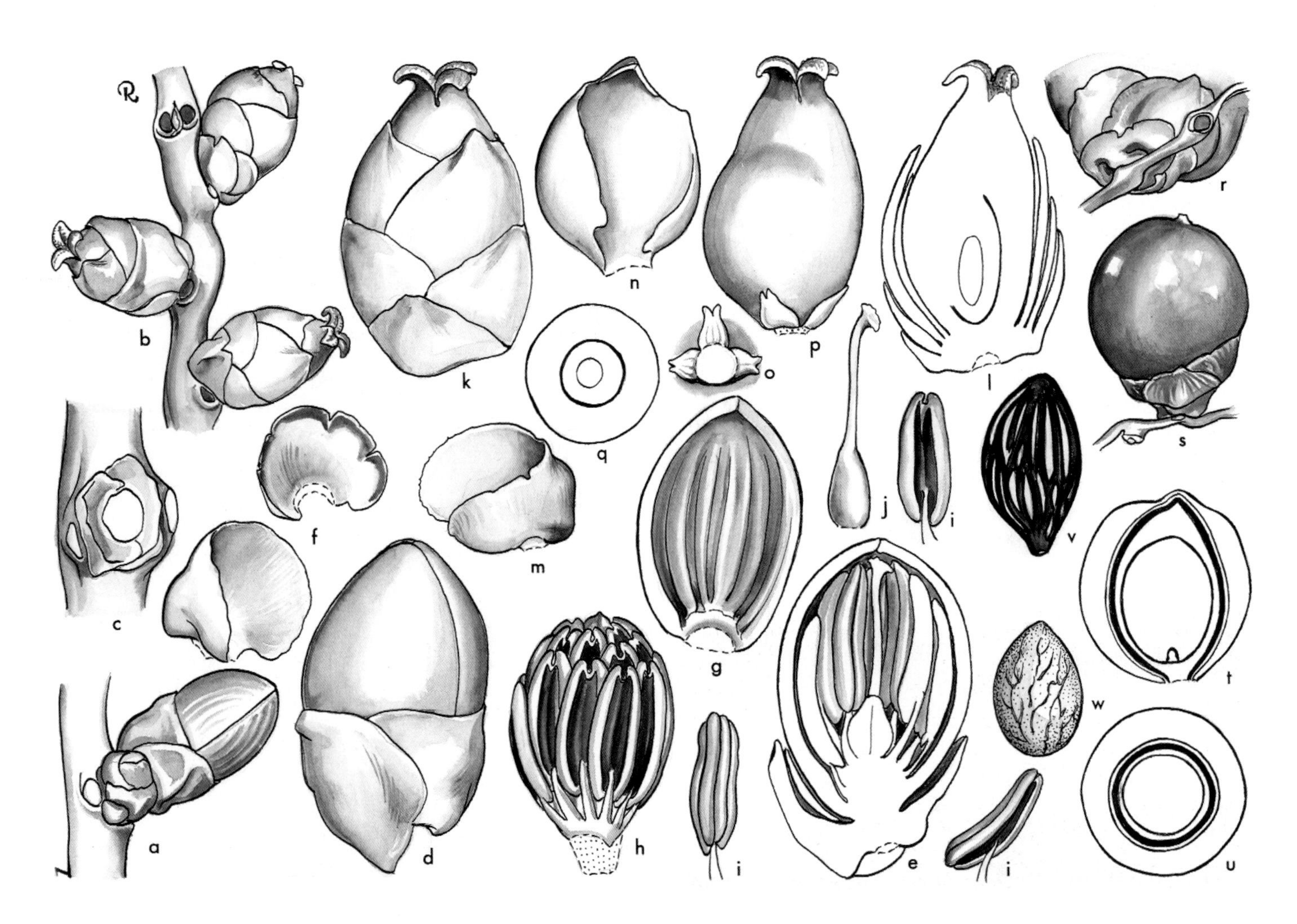

Carpentaria. **a**, portion of rachilla with triad, one staminate flower fallen × 3; **b**, portion of rachilla in pistillate flower × 3; **c**, triad, flowers removed × 6; **d**, staminate bud × 6; **e**, staminate bud in vertical section × 6; **f**, staminate sepals in 2 views × 6; **g**, staminate petal, interior view × 6; **h**, androecium × 6; **i**, stamen in 3 views × 6; **j**, pistillode × 6; **k**, pistillate flower × 6; **l**, pistillate flower in vertical section × 6; **m**, pistillate sepal × 6; **n**, pistillate petal × 6; **o**, staminodes × 6; **p**, gynoecium and staminodes × 6; **q**, ovary in cross-section × 6; **r**, perianth in fruit × 3; **s**, fruit × 1 1/2; **t**, fruit in vertical section × 1 1/2; **u**, fruit in cross-section × 1 1/2; **v**, fibers over endocarp × 1 1/2; **w**, seed × 1 1/2. *Carpentaria acuminata*: all from *Moore & Spence* 9228. (Drawn by Marion Ruff Sheehan)

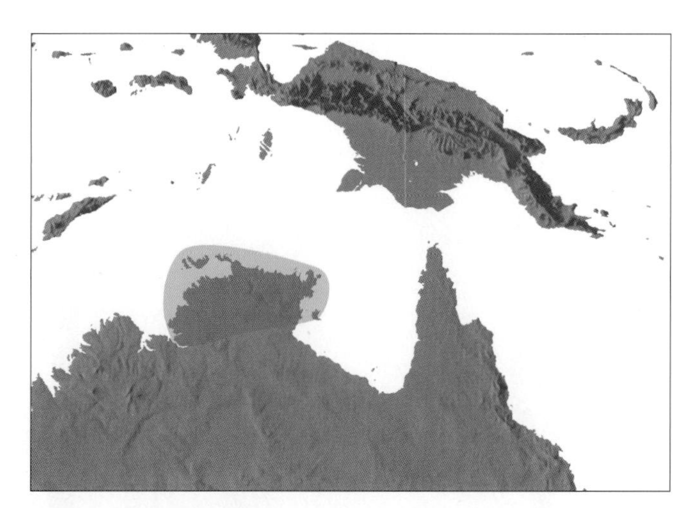

Distribution of ***Carpentaria***

anastomosing distally and closely appressed to the endocarp, endocarp thin, glass-like. *Seed* attached laterally, ovoid, ± pointed, round in cross-section, hilum elongate, raphe branches anastomosing, endosperm homogeneous; embryo basal. *Germination* adjacent-ligular; eophyll bifid. *Cytology*: 2n = 32.

Distribution and ecology: One species in Northern Territory, Australia. Found in rain forest along banks of streams at low elevations, usually near brackish water estuaries.
Anatomy: Root (Seubert 1998a, 1998b) and fruit (Essig 1977).
Relationships: A sister relationship between *Carpentaria* and *Wodyetia* is moderately to highly supported by several studies (Asmussen *et al.* 2006, Norup *et al.* 2006, Baker *et al.* in review, in prep., Lewis *et al.* in prep.)
Common names and uses: Carpentaria palm. A handsome ornamental but requires humid tropical or subtropical conditions. Widely planted in northeastern Australia and southern Florida in parks and along streets.
Taxonomic account: Beccari (1885).
Fossil record: No generic records found.

Above: ***Carpenteria***
a, monosulcate pollen grain, distal face SEM × 1500;
b, monosulcate pollen grain, proximal face SEM × 1500.
Carpenteria acuminata: **a** & **b**, *Moore & Spence* 9228.

Notes: Differs from *Veitchia*, which has a similar habit, in having thinner and narrower lanceolate leaflets, which are often four-pointed and lacking ramenta on their midribs, and in having only one series of fibres in the mesocarp.

163. WODYETIA

Spectacular, moderate solitary pinnate-leaved palm, native to northeastern Queensland, Australia, with crownshaft and praemorse leaflets, the leaflets longitudinally divided into many segments that are splayed out, giving the whole leaf a foxtail appearance; the fruit is relatively large and has a distinctive network of black fibres next to the endocarp; the seed has homogeneous endosperm.

Wodyetia A.K. Irvine, Principes 27: 161 (1983). Type: **W. bifurcata** A.K. Irvine.

Commemorates Wad-yeti, *last surviving male Aboriginal in the Bathurst Bay area of Queensland who acted as anthropological and linguistic informant to many researchers; he was also called Johnny Flinders. He died in 1978.*

Moderate, solitary, unarmed, pleonanthic, monoecious palm. *Stem* columnar, slightly bottle-shaped, closely ringed with leaf scars, light grey. *Leaves* pinnate, appearing plumose and oblong-elliptical in outline; sheath tubular forming a prominent crownshaft, elongate, green, splitting opposite the petiole, covered with a greyish-white tomentum; petiole short, stout, adaxially slightly concave to flat, abaxially rounded; rachis much longer than the petiole, adaxially becoming angled, abaxially rounded, petiole and rachis covered with greyish-white to brown, tattered-peltate scales and larger ramenta; leaflets single-fold, regularly arranged, divided into several linear segments, or deeply lobed, each segment usually with 1 (2–4) main ribs, apices coarsely praemorse sometimes obliquely so, or with 2-several small teeth, rarely pointed, tapering, terminal leaflets single or paired, adaxially glabrous, abaxially densely covered with very small whitish scales, transverse veinlets not evident. *Inflorescences* infrafoliar, horizontal in bud, becoming ± pendulous in fruit, branched to 4 orders basally, 2–1 orders distally; peduncle short, wide, ± flattened; prophyll tubular, dorsiventrally flattened, with 2 flat lateral keels, rather narrow, tapering from the base to a blunt point, caducous; peduncular bract like the prophyll but not keeled, 4–5 small, incomplete bracts present above the peduncular bract; rachis much longer than the peduncle bearing distant, stout, angled, spirally arranged branches subtended by small, pointed, or wrinkled bracts; rachillae rather short, cylindrical, bearing widely spaced triads basally and paired or solitary

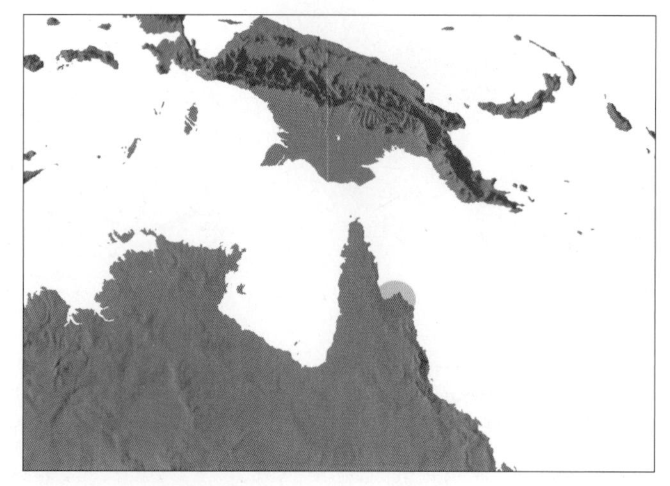

Distribution of ***Wodyetia***

staminate flowers distally; floral bracteoles small, narrow, imbricate. *Staminate flower* symmetrical, bullet-shaped in bud; sepals 3, distinct, imbricate, rounded, inflated, margins finely toothed; petals 3, distinct, broadly ovate, valvate, very hard, about twice as long as the sepals; stamens 60–71, filaments slender, terete, short, anthers elongate, narrow, slightly bifid apically, unevenly sagittate basally, dorsifixed near the base, latrorse, connective elongate, tanniniferous; pistillode bottle-shaped with a long neck, terminated by 4–5 short, linear lobes or papillae. *Pollen* ellipsoidal asymmetric; aperture a distal sulcus; ectexine tectate, perforate, or perforate-finely rugulate, aperture margin similar or slightly finer;

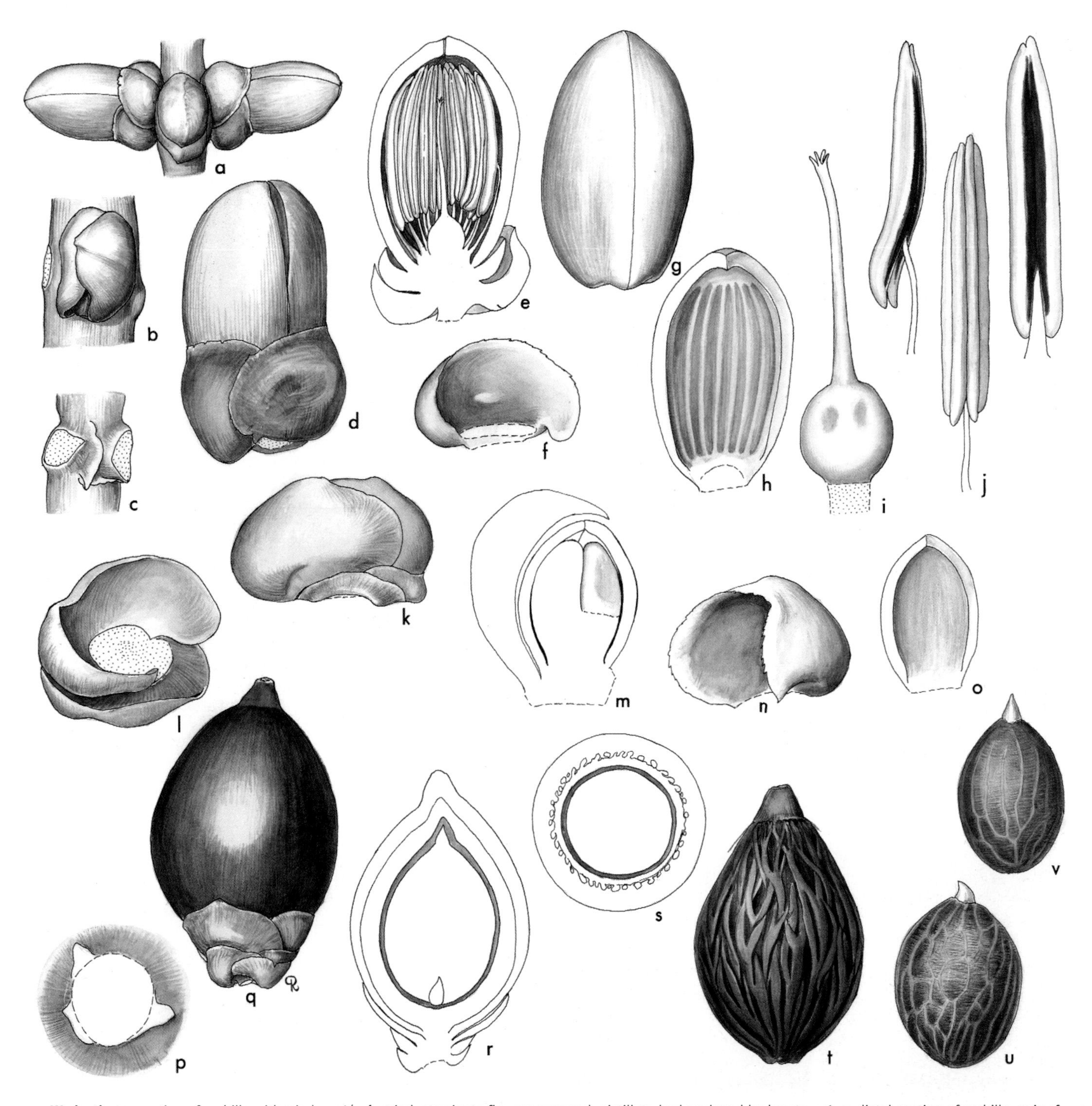

***Wodyetia*.** **a**, portion of rachilla with triad $\times 2\,^1/_4$; **b**, triad, staminate flowers removed, pistillate bud enclosed by bracts × 6; **c**, distal portion of rachilla, pair of staminate flowers removed to show scars and bracts × 6; **d**, staminate bud $\times 4\,^1/_2$; **e**, staminate bud in vertical section $\times 4\,^1/_2$; **f**, staminate sepal $\times 4\,^1/_2$; **g**, staminate bud, sepals removed $\times 4\,^1/_2$; **h**, staminate petal, interior view $\times 4\,^1/_2$; **i**, pistillode × 9; **j**, stamen in 3 views × 9; **k**, pistillate bud and 2 basal bracts $\times 4\,^1/_2$; **l**, bracts from base of pistillate flower $\times 6\,^3/_4$; **m**, pistillate flower in vertical section × 9; **n**, pistillate sepal $\times 4\,^1/_2$; **o**, pistillate petal, interior view × 9; **p**, fruit in basal view $\times 1\,^1/_2$; **q**, mature fruit × 1; **r**, fruit in vertical section × 1; **s**, fruit in cross-section × 1; **t**, outer endocarp showing large forking fibers × 1; **u**, **v**, seed in 2 views × 1. *Wodyetia bifurcata*: all from *Irvine* 2184. (Drawn by Marion Ruff Sheehan)

Above left: ***Wodyetia bifurcata***, habit, cultivated, Montgomery Botanical Center, Florida. (Photo: J. Dransfield)
Above middle: ***Wodyetia bifurcata***, crown and inflorescences, cultivated, Fairchild Tropical Botanic Garden (Photo: J. Roncal)
Above right: ***Wodyetia bifurcata***, inflorescence with some open staminate flowers, cultivated, Montgomery Botanical Center, Florida. (Photo: J. Dransfield)

infratectum columellate; longest axis ranging from 50–56 µm [1/1]. *Pistillate flowers* (buds only seen), ovoid; sepals 3, distinct, broadly imbricate, rounded, margins finely tattered; petals 3, distinct, imbricate, hooded, valvate distally; staminodes 6, very small, triangular with short filaments; gynoecium conic-ovoid, unilocular, uniovulate, stigmas 3, apices rounded. *Fruit* globose-ovoid, orange-red at maturity, stigmatic remains apical forming a conical beak; epicarp thin with very short, stout fibres below the epidermal layer, mesocarp fleshy, orange-yellow when ripe, thin with longitudinal fibres, some forked, endocarp complex with outer distinctive thick, flat, branched fibres and an inner layer of horizontal fibres. *Seed* ellipsoidal, beaked, raphe branches medium, anastomosing, slightly impressed, endosperm homogeneous; embryo basal. *Germination* adjacent-ligular; eophyll bifid. *Cytology*: 2n = 32.

Below: ***Wodyetia***
a, monosulcate pollen grain, distal face SEM × 800;
b, monosulcate pollen grain, proximal face SEM × 800.
Wodyetia bifurcata: **a** & **b**, *Irvine* 2184.

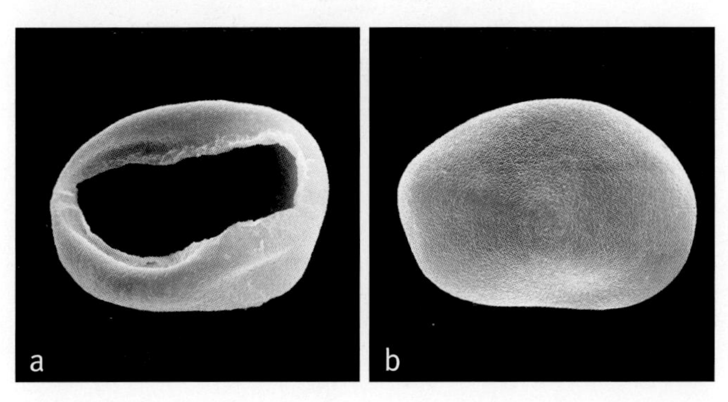

Distribution and ecology: One species in north-eastern Queensland, Australia, confined to the south-west, south and south-east sides of the Melville Range. Forming the canopy in open woodland communities of rain forest in coarse, loose granite sand, among huge granite boulders.
Anatomy: Root (Seubert 1998a, 1998b).
Relationships: For relationships, see *Carpentaria*.
Common names and uses: Foxtail palm. Since its discovery in the early 1980s, *Wodyetia* has rapidly become a highly valued and very widely dispersed ornamental.
Taxonomic account: Irvine (1983).
Fossil record: No generic records found.
Notes: The divided leaflets and large black, pericarp fibres are the distinguishing characters.
Additional figures: Glossary fig. 22.

164. DRYMOPHLOEUS

Small to moderate, solitary pinnate-leaved palms, native to Moluccas, New Guinea, the Bismarck Archipelago, Solomon Islands and Samoa, with crownshaft and praemorse leaflets, elongate peduncles and seed rounded in cross section; the endosperm can be ruminate or homogeneous.

Drymophloeus Zippel, Alg. Konst-Lett.-Bode 1829 (19): 297 (1829). Lectotype: **D. oliviformis** (Giseke) Miq. (*Areca olivaeformis* Giseke) (see Beccari & Pichi-Sermolli 1955).
Coleospadix Becc., Ann. Jard. Bot. Buitenzorg 2: 90 (1885). Lectotype: *C. litigiosa* (Beccari) Becc. (*Ptychosperma*

Top left: ***Drymophloeus oliviformis***, habit, Papua. (Photo: W.J. Baker)

Top right: ***Drymophloeus oliviformis***, crown with infructescence, Papua. (Photo: W.J. Baker)

Bottom left: ***Drymophloeus litigiosus***, rachilla with staminate flowers, Papua. (Photo: W.J. Baker)

Bottom right: ***Drymophloeus oliviformis***, infructescence, Papua. (Photo: W.J. Baker)

litigiosum Becc.) (see Beccari & Pichi-Sermolli 1955) (= *Drymophloeus litigiosus* [Becc.] H.E. Moore).

Saguaster Kuntze, Revis. gen. pl. 2: 734 (1891). Type: *S. oliviformis* (Giseke) Kuntze (= *Drymophloeus oliviformis* [Giseke] Miq.).

Rehderophoenix Burret, Notizbl. Bot. Gart. Berlin-Dahlem 13: 86 (1936). Type: *R. pachyclada* Burret (= *Drymophloeus pachycladus* [Burret] H.E. Moore).

Drymos — *wood,* phloios – *bark, but the reason for the choice of name was not explained and remains obscure.*

Small to moderate, solitary or rarely weakly clustering, unarmed, pleonanthic, monoecious palms. *Stem* erect, slender, ringed with rather widely spaced leaf scars, sometimes short stilt roots present basally. *Leaves* pinnate, arching slightly, few in crown; sheath rather thick, forming a slender crownshaft, glabrous adaxially, abaxially densely covered with whitish tomentum and small red-brown scales, becoming minutely brown-dotted; petiole absent or short, slender, adaxially deeply channelled, abaxially rounded, covered with deciduous tomentum and pale or dark scales; rachis long, slender, adaxially often sharply pointed or ridged and rounded, abaxially rounded, densely covered with brown tattered scales or tomentum; leaflets subopposite to alternate, in one plane, single-fold, broadly or narrowly wedge-shaped, or broadly lanceolate to narrowly obovate, distal pair sometimes broader and united or small and narrow, leaflets distally variously lobed or oblique, raggedly praemorse, adaxially glabrous except for a few scales near bases of major veins, abaxially densely covered with small scales on minor veins, major veins with ramenta toward the base, or lacking scales except on midrib, midvein and a pair of large veins along or close to the margins prominent, margins densely covered with caducous tomentum and scales, transverse veinlets not evident. *Inflorescences* infrafoliar, branched to 1 or basally to 2–3 orders, protandrous; peduncle relatively slender, elongate (except *Drymophloeus subdistichus*), elliptical in cross-section; prophyll deciduous, tubular, slender, dorsiventally flattened, with 2, narrow, lateral keels, pointed, splitting apically and for a short distance abaxially to release the peduncular bract, densely covered with multibranched or pale-margined brown scales; peduncular bract tubular, slender, much longer than the prophyll, with a long rather flat, pointed tip, densely covered with scales like the prophyll, a few small incomplete peduncular bracts sometimes also present; rachis shorter than the peduncle, angled, tapering, bearing rather few, distant, spirally arranged branches; rachis bracts low, rounded; rachillae short, tapering, bearing spirally arranged triads of flowers basally, and pairs of a pistillate and one staminate, or a single staminate flower distally; floral bracteoles low, rounded or truncate. *Staminate flowers* borne laterally toward the lower side of the pistillate flower in rounded indentations in the rachillae; sepals 3, distinct, imbricate, irregularly but strongly keeled,

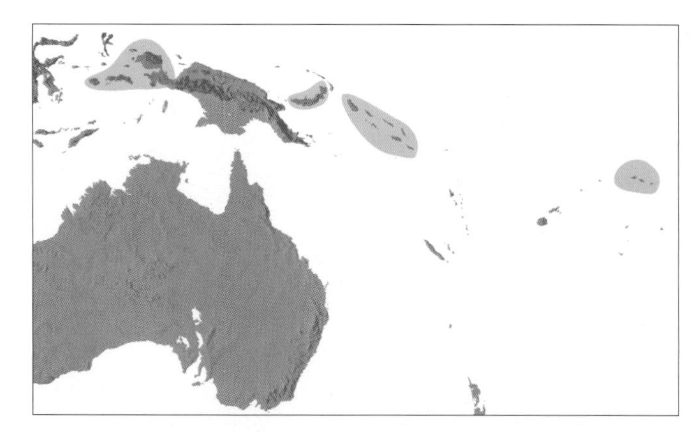

Distribution of ***Drymophloeus***

margins thin, variously notched; petals 3, distinct, ovate, valvate, evenly thickened, adaxially grooved, reflexed at anthesis; stamens numerous, 24 to more than 320, filaments moderate, awl-shaped, anthers elongate, sagittate basally, uneven, sometimes divided apically, medifixed, versatile, latrorse; pistillode with 3 short, rounded, 3-angled lobes or ovoid attenuate to $^{2}/_{3}$ as long as the stamens, usually shortly trifid apically. *Pollen* ellipsoidal or oblate triangular, asymmetric to pyriform; aperture a distal sulcus or trichotomosulcus; ectexine tectate, perforate or perforate-reticulate, aperture margin similar or slightly finer; infratectum columellate; longest axis ranging from 50–63 µm [2/7]. *Pistillate flowers* broadly ovoid; sepals 3, distinct, imbricate, hooded, edges minutely toothed or variously notched; petals 3, twice as long as the sepals in late bud, distinct, ovate, imbricate, lateral margins shortly fringed, tips thick, valvate; staminodes 3, shortly joined basally, wide, truncate and bifid or uneven distally; gynoecium symmetrical, ovoid, tapering distally, unilocular, uniovulate, stigmas 3, recurved, ovule attached laterally or pendulous from the top of the locule, form unknown. *Fruit* fusiform to ovoid, red at maturity, stigmatic remains apical; epicarp thin, smooth, becoming pebbled when dry, mesocarp fleshy, fibrous, with stinging crystals, fibres adherent to the thin, rather smooth endocarp, endocarp usually circular in cross-section, 5-lobed in *D. hentyi* and *D. subdistichus*). *Seed* ovoid, surface smooth, hilum apical, raphe much branched, branches somewhat sunken, endosperm homogeneous or ruminate; embryo basal. *Germination* adjacent-ligular; eophyll bifid, or entire, ovate, margins toothed. *Cytology*: 2n = 32.

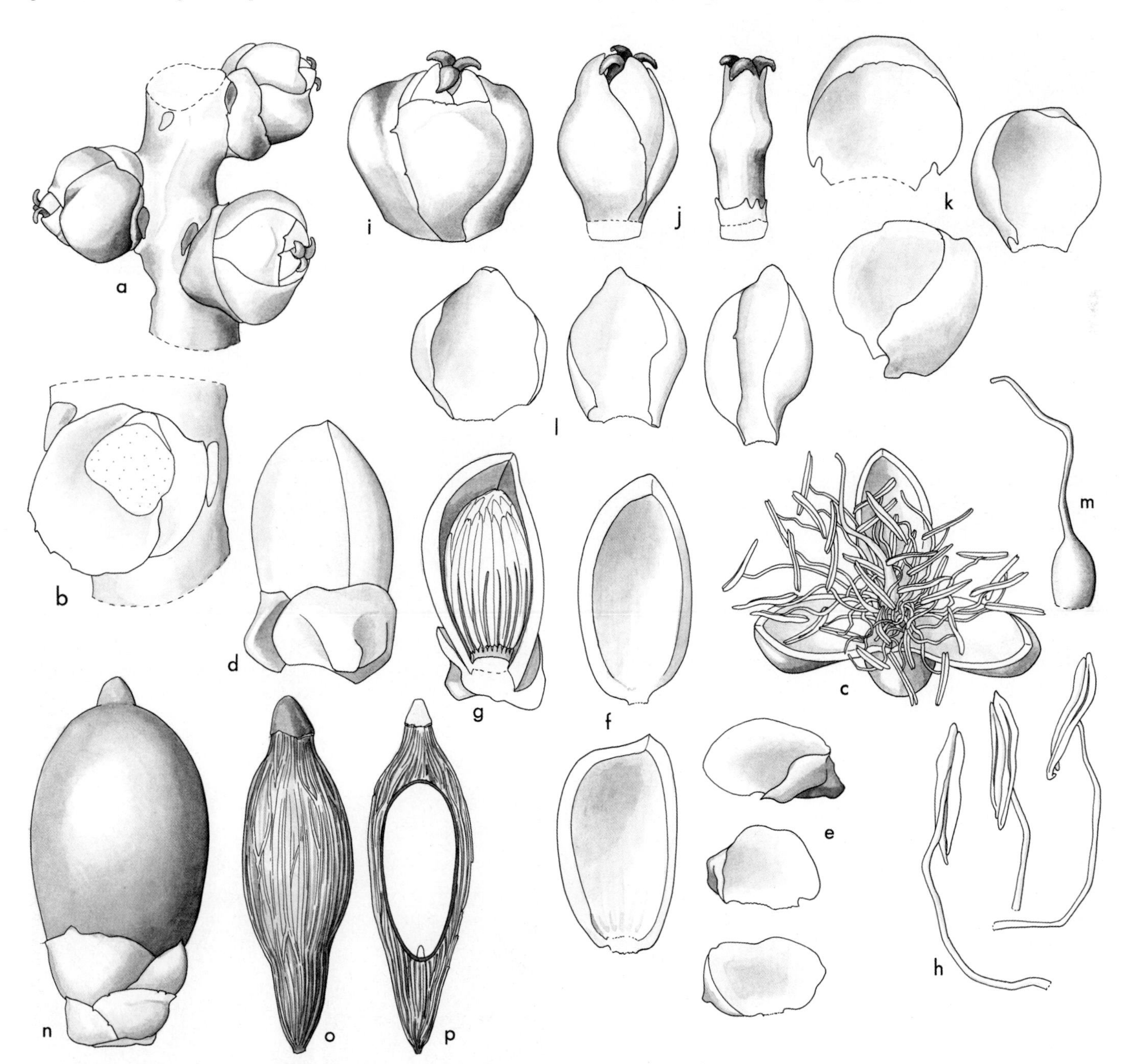

Drymophloeus. **a**, portion of rachilla at pistillate anthesis × 4; **b**, triad scar × 7 $^{1}/_{2}$; **c**, staminate flower at anthesis, × 4; **d**, staminate bud × 6; **e**, staminate sepals, 3 views × 6; **f**, staminate petal, interior view × 6; **g**, staminate bud, 1 sepal and 1 petal removed × 6; **h**, stamens, 3 views × 8; **i**, pistillate flower × 6; **j**, pistillate flower, sepals removed × 6; **k**, pistillate sepals, 3 views × 6; **l**, pistillate petals, 3 views × 6; **m**, gynoecium × 6; **n**, fruit × 4; **o**, endocarp × 2 $^{1}/_{2}$; **p**, endocarp and seed in vertical section × 2 $^{1}/_{2}$. *Drymophloeus oliviformis*: **a, b** and **i–m**, *Zona* 764, **c**, *Zona* 684, **d–h**, *Mogea* 6237, **n–p**, *Zona & J. Dransfield* 684. (Drawn by Lucy T. Smith)

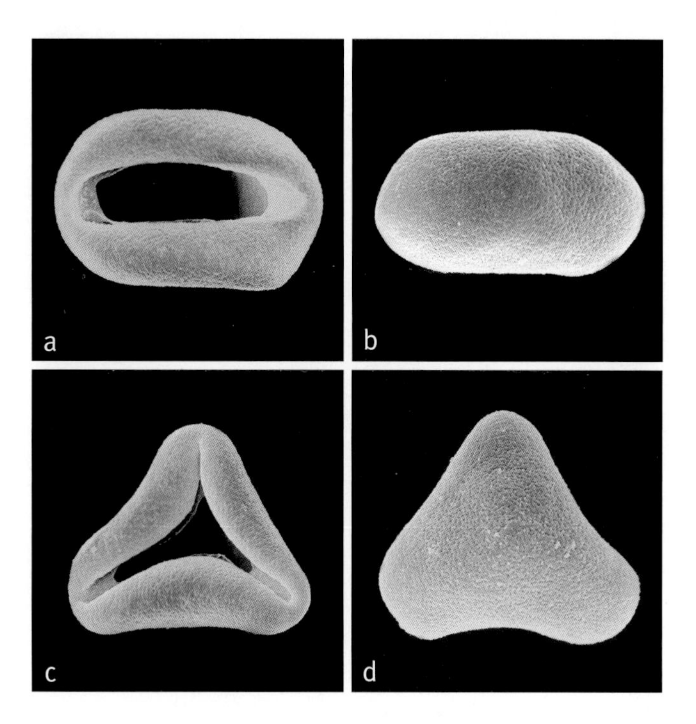

Above: ***Drymophloeus***
a, monosulcate pollen grain, distal face SEM × 800;
b, monosulcate pollen grain, proximal face SEM × 800;
c, trichotomosulcate pollen grain, distal face SEM × 800;
d, trichotomosulcate pollen grain, proximal face SEM × 800.
Drymophloeus lepidotus: **a–d**, *Dennis* 53.

Distribution and ecology: Eight species extending from the Moluccas and New Guinea, through the Bismarck Archipelago and Solomon Islands to Samoa. Understory palms in rain forest in areas of high rainfall, variously occurring from sea level to 1200 m.
Anatomy: Leaf (Tomlinson 1961), root (Seubert 1998a, 1998b) and fruit (Essig 1977).
Relationships: Lewis *et al.* (in prep.) resolved *Drymophloeus* as polyphyletic. *Drymophloeus litigiosus* and *D. oliviformis* are resolved in a clade with *Normanbya*, *Carpentaria* and *Wodyetia* with low support. *Drymophloeus hentyi* is resolved as sister to *Ponapea* with high support. Norup *et al.* (2006) resolve a clade with low support comprising *D. litigiosus* and *Normanbya*.
Common names and uses: Common names not recorded. The black wood of the trunk is used for making spears, arrowheads, and other items. Some species are grown as ornamentals in moist tropical areas.
Taxonomic account: Zona (1999b).
Fossil record: No generic records found.
Notes: Zona (1999a) demonstrated good evidence for removing *Drymophloeus samoensis* and returning it to the genus *Solfia* and some evidence for dividing the genus into two groups: undergrowth species (*Drymophloeus*) and emergent species (*Rehderophoenix*). However, *Drymophloeus* in the broader sense is retained here. Zona (1999a) suggested that *Drymophloeus* is closely related to *Brassiophoenix*, *Balaka* and *Solfia*, with *Balaka* as the sister genus.

165. NORMANBYA

Spectacular, moderate solitary pinnate-leaved palm, native to northeastern Queensland, Australia, with crownshaft and praemorse leaflets, the leaflets longitudinally divided into many segments that are splayed out, giving the whole leaf a foxtail appearance; the fruit is relatively large and has thin pale fibres next to the endocarp; the seed has ruminate endosperm.

Normanbya F. Muell. ex Becc., Ann. Jard. Bot. Buitenzorg 2: 91, 170, 171 (1885). Lectotype: *N. muelleri* Becc. (illegitimate name) (*Cocos normanbyi* W. Hill = **Normanbya normanbyi** [W. Hill] L.H. Bailey) (see H.E. Moore 1963c).

Named in honour of Sir George Augustus Constantine Phipps, Second Marquis of Normanby (1819–1890).

Moderate to tall, solitary, unarmed, pleonanthic, monoecious palm. *Stem* erect, moderate, ringed with distinct leaf scars, vertically striate, grey, bulbous basally. *Leaves* pinnate, plumose, loosely arching; sheath forming a prominent crownshaft, pale, ashy grey, brownish near the top; petiole short or nearly lacking, channelled adaxially, rounded abaxially, densely covered with whitish tomentum and scattered brown, tattered scales; rachis long, arching, ± rounded adaxially and abaxially, densely covered in tattered brown scales; leaflets single-fold, irregularly arranged, divided nearly to the base into 7–9 linear segments, with or without midribs and

Below: ***Normanbya normanbyi***, habit, Australia. (Photo: J.L. Dowe)

with 1–3 large veins, apices of segments praemorse, only outermost 2 of each group with thickened margins, blade appearing dark green adaxially, bluish-white abaxially, adaxial surface glabrous, abaxial densely covered with uniseriate, medifixed scales, transverse veinlets not evident. *Inflorescences* infrafoliar, divaricate, somewhat pendulous in fruit, branched to 2 (or more) orders; peduncle short; prophyll tubular, rather narrow, 2-keeled laterally; peduncular bract like the prophyll, both deciduous; rachis bracts low, ridge-like, subtending spirally arranged, stout, angled branches and terete rachillae; rachilla bracts low, ridge-like, subtending triads basally, paired and solitary staminate flowers distally; floral bracteoles low, rounded. *Staminate flowers* symmetrical, bullet-shaped in bud, borne lateral to the pistillate on short, laterally flattened stalks; sepals 3, distinct, imbricate, rounded, upper margins ± truncate, minutely toothed; petals 3, distinct, valvate, ovate, evenly thickened; stamens 24–40, filaments short, awl-shaped, anthers elongate, shortly bifid apically, dorsifixed almost at the base, ± introrse, connective elongate, tanniniferous; pistillode flask-shaped with long narrow neck, slightly longer than the stamens, apically expanded. *Pollen* ellipsoidal asymmetric, occasionally pyriform; aperture a distal sulcus; ectexine tectate, perforate, aperture margin similar; infratectum columellate; longest axis 64–83 µm [1/1]. *Pistillate flowers* ovoid; sepals 3, distinct, broadly imbricate, rounded with short pointed tips, margins slightly fringed; petals 3, like the sepals but longer and with short valvate tips; staminodes 3, broadly tooth-like; gynoecium narrowly ovoid, unilocular, uniovulate, narrowing shortly above the ovarian region to 3, large reflexed stigmas, ovule pendulous, form unknown. *Fruit* ovoid to obpyriform, pointed distally, dull salmon-pink to purplish-brown at maturity, stigmatic remains apical forming a short beak; epicarp somewhat fleshy, drying wrinkled, mesocarp rather thin, with longitudinal, branched, straw-coloured fibres adherent to the smooth endocarp. *Seed* laterally attached with a long unbranched raphe, hilum lateral, endosperm ruminate; embryo basal. *Germination* adjacent-ligular; eophyll bifid. *Cytology* not studied.

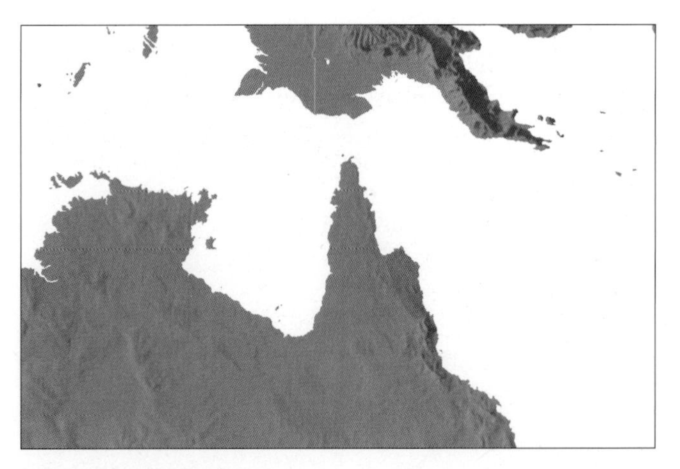
Distribution of ***Normanbya***

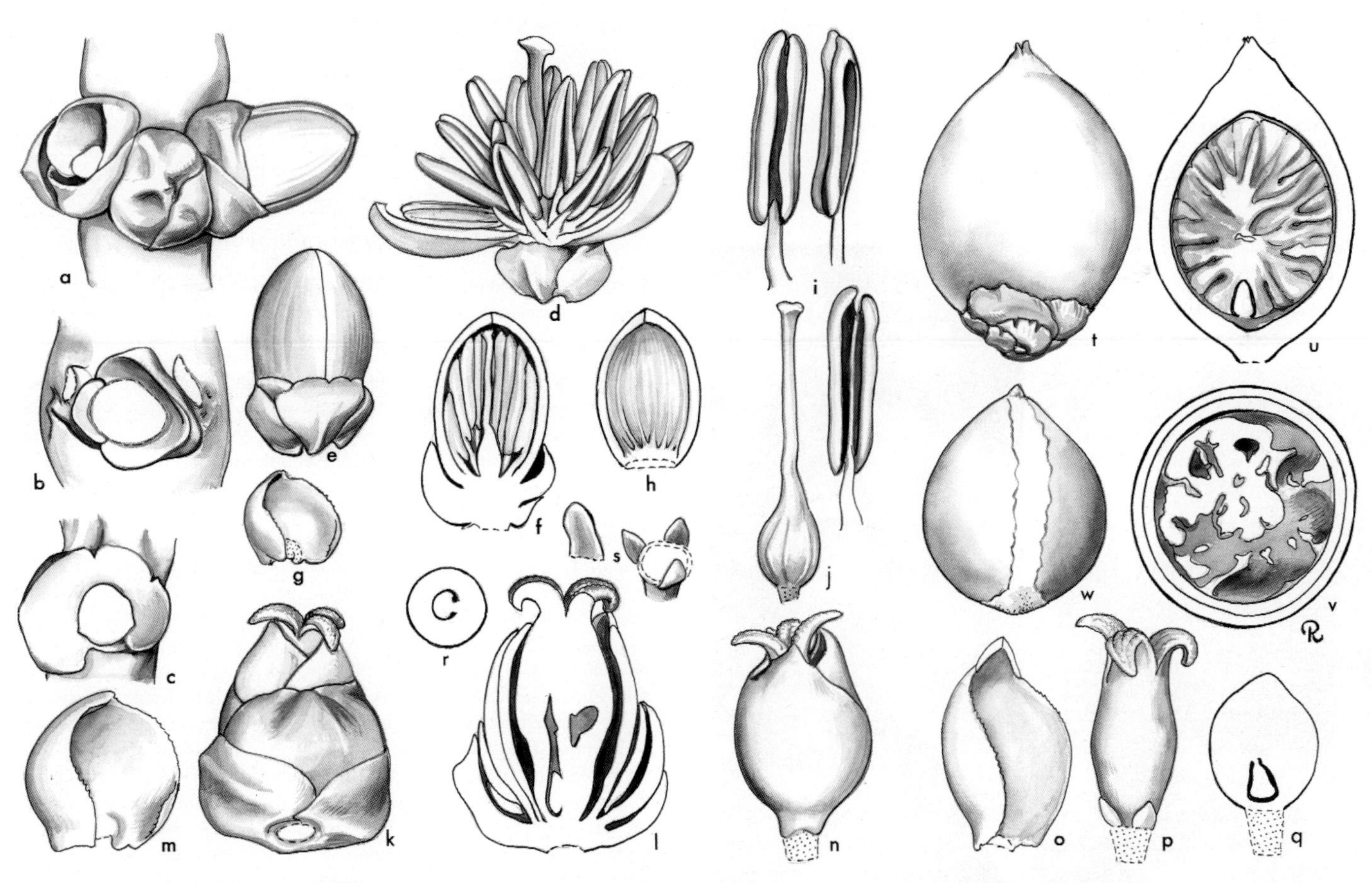

***Normanbya*.** **a**, portion of rachilla showing triad × 3; **b**, triad, flowers removed to show bracts × 3; **c**, scar of pistillate flower with enlarged bracteole × 3; **d**, staminate flower × 3; **e**, staminate bud × 3; **f**, staminate bud in vertical section × 3; **g**, staminate sepal × 3; **h**, staminate petal, interior view × 3; **i**, stamen in 3 views × 6; **j**, pistillode × 6; **k**, pistillate flower × 3; **l**, pistillate flower in vertical section × 3; **m**, pistillate sepal × 3; **n**, pistillate flower, sepals removed × 3; **o**, pistillate petal × 3; **p**, gynoecium and staminodes × 3; **q**, gynoecium in vertical section × 3; **r**, ovary in cross-section × 3; **s**, staminodes × 6; **t**, fruit × 1; **u**, fruit in vertical section × 1; **v**, fruit in cross-section × 1; **w**, seed × 1. *Normanbya normanbyi*: **a**, **b** and **e–j**, *Moore et al.* 9246; **c**, **d** and **k–r**, *Nur* s.n.; **s–w**, *Moore* 7015. (Drawn by Marion Ruff Sheehan)

Distribution and ecology: One species in the rain forest of northern Queensland, Australia. *Normanbya* grows in moist complex and simple mesophyll vine forest with an annual rainfall of 3000 mm, close to rivers and streams, often in swampy areas and usually in gravelly alluvial soils, in areas where dry periods are not more than 40 days.

Anatomy: Leaf (Tomlinson 1961), root (Seubert 1998a, 1998b) and fruit (Essig 1977).

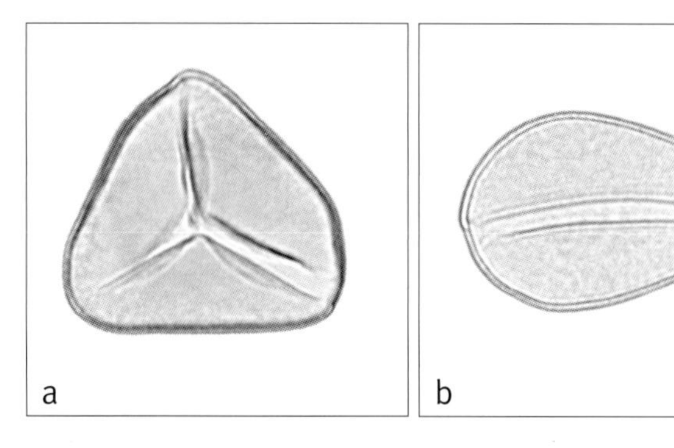

Above: ***Normanbya***
a, trichotomosulcate pollen grain mid–low-focus LM × 500;
b, monosulcate pollen grain, mid-focus LM × 500.
Normanbya normanbyi: **a** & **b**, *Fitzalan & Mueller* (1880).

Below left: ***Normanbya normanbyi***, crown, Australia. (Photo: J.L. Dowe)

Below right: ***Normanbya normanbyi***, fruit and abaxial leaf surface, Australia. (Photo: J.L. Dowe)

Relationships: The relationships of *Normanbya* to the other genera of Ptychospermatinae are unclear, but Lewis *et al.* (in prep.) resolve the genus in a clade with *Drymophloeus*, *Carpentaria* and *Wodyetia*. Norup *et al.* (2006) resolve *Normanbya* as sister to *D. litigiosus* with low support

Common names and uses: Black palm. The hard, dark wood was used by aborigines for making spears. It is a handsome ornamental requiring a warm moist, somewhat protected location.

Taxonomic account: Bailey (1935b).

Fossil record: No generic records found.

Notes: *Normanbya* can be immediately recognised by the irregularly arranged, longitudinally divided leaflets with 7–9 segments. *Wodyetia* has a similar leaf but with regularly arranged leaflets divided into more (11–17) segments. In *Normanbya*, the leaflets also bear white woolly scales abaxially, the outer endocarp has a few thin fibres rather than many large fibres and the endosperm is ruminate rather than homogeneous. Structurally, the leaves become plumose by the longitudinal splitting of a single leaflet. Such division of leaflets is unusual in palms, occurring elsewhere only in certain genera of the Iriarteeae.

166. BRASSIOPHOENIX

Small to moderate, solitary pinnate-leaved palm, endemic to New Guinea, with crownshaft and praemorse leaflets, the leaflets conspicuously 3-pronged; the fruit has a deeply grooved endocarp; the seed has homogeneous endosperm.

Brassiophoenix Burret, Notizbl. Bot. Gart. Berlin-Dahlem 12: 345 (1935). Type: **B. drymophloeoides** Burret.

Commemorates botanist, Leonard J. Brass (1900–1971), prolific collector in New Guinea, by combining his name with phoenix *— a general name for a palm.*

Small to moderate, solitary, unarmed, pleonanthic, monoecious palms. *Stem* slender, erect, ringed with leaf scars, light grey to brown. *Leaves* pinnate, spreading to erect; sheath forming a crownshaft, sometimes bearing a triangular appendage opposite the petiole, glaucous or densely white tomentose and minutely dotted; petiole very short, slender, deeply channelled adaxially, rounded abaxially, densely white tomentose, minutely brown-dotted, and bearing dark curly ramenta; rachis slender, adaxially ridged in the centre, rounded abaxially, densely covered in dark or red-brown tattered scales; leaflets single-fold, wedge-shaped, apically trilobed, the centre lobe much prolonged, all lobes toothed, basally bearing dark chaffy scales as the rachis, glaucous abaxially, midrib and marginal ribs large, transverse veinlets not evident. *Inflorescences* infrafoliar, branched to 2(–3) orders basally, densely dark or white-woolly tomentose throughout; peduncle short, dorsiventrally flattened; prophyll tubular, short, 2-keeled laterally, splitting apically; peduncular bract twice as long as the prophyll, tubular, exserted at maturity, 1 or 2 additional incomplete, ribbon-like or triangular to elongate peduncular bracts present; rachis longer than the peduncle bearing very short, spirally arranged bracts subtending branches and rachillae; rachillae rather short, thick in the middle, bearing very short, spirally arranged bracts subtending triads of flowers nearly throughout the rachillae, triads distant, staminate flowers projecting laterally, much wider than the rachillae; floral bracteoles 3, the first small, pointed, the second 2 large, rounded. *Staminate flowers* bullet-shaped in bud; sepals 3, distinct, short, rounded, imbricate, gibbous basally, edges toothed; petals 3, distinct, ovate, valvate, thick, tapering to a blunt point; stamens numerous (ca. 100–230), inserted on a conical

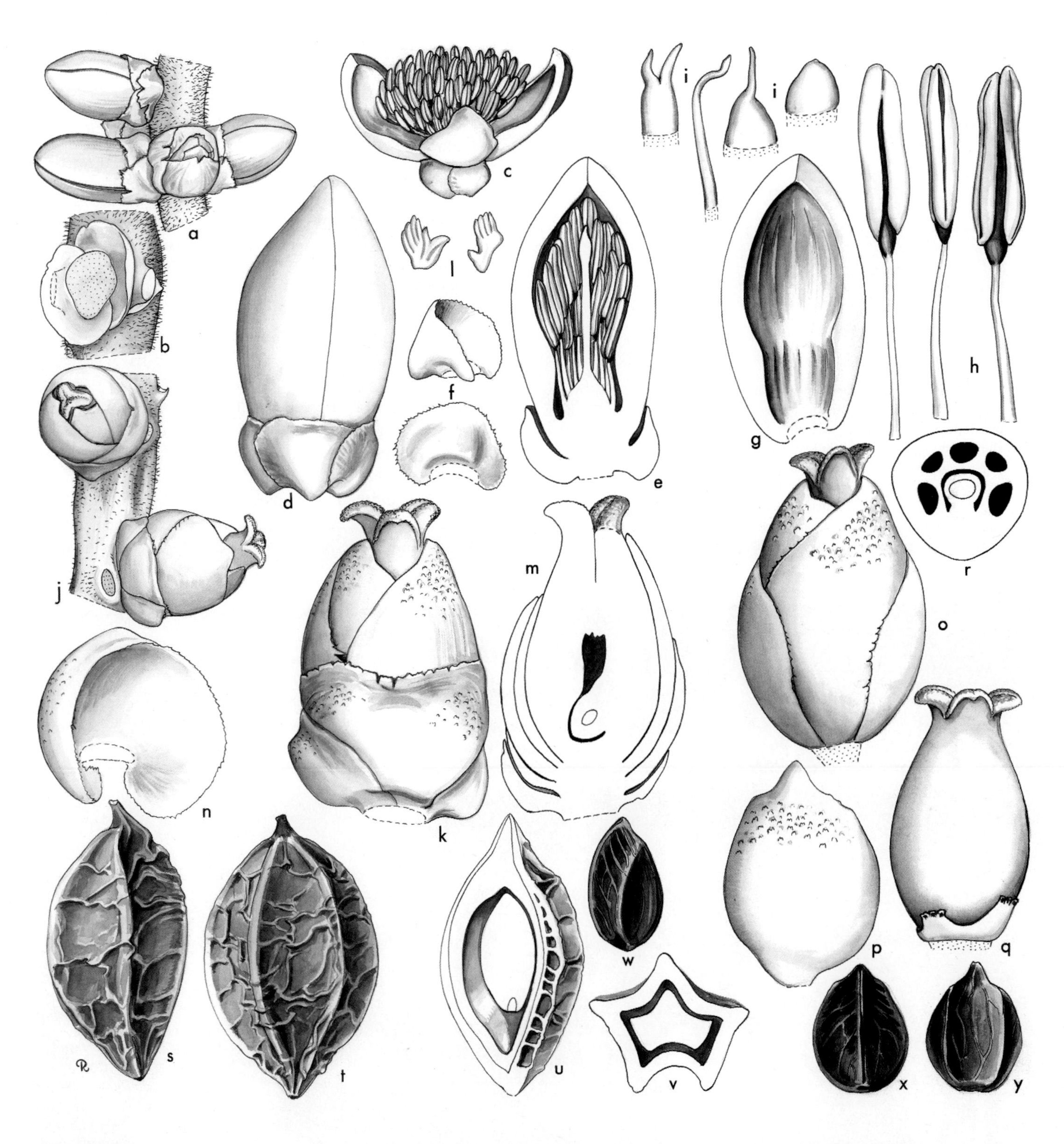

Brassiophoenix. **a**, portion of rachilla with triads × 3; **b**, triad, flowers removed to show bracteoles × 3; **c**, staminate flower × 3; **d**, staminate bud × 6; **e**, staminate bud in vertical section × 6; **f**, staminate sepals in 2 views × 6; **g**, staminate petal, interior view × 6; **h**, stamen in 3 views × 18; **i**, pistillodes of various forms × 6; **j**, portion of rachilla in pistillate flower × 3; **k**, pistillate flower × 6; **l**, scales from pistillate perianth, much enlarged; **m**, pistillate flower in vertical section × 6; **n**, pistillate sepal × 6; **o**, pistillate flower, sepals removed × 6; **p**, pistillate petal, exterior view × 6; **q**, gynoecium and staminodes × 6; **r**, ovary in cross-section × 6; **s**, **t**, fruit in 2 views × 1$^1/_2$; **u**, fruit in vertical section × 1$^1/_2$; **v**, fruit in cross-section × 1$^1/_2$; **w**, **x**, **y**, seed in 3 views × 1$^1/_2$. *Brassiophoenix schumannii*: **a**, *Read* 1476; *Brassiophoenix* sp.: **b–r**, *Read* 809; *B. drymophloeoides*: **s–y**, *Darbyshire* 964. (Drawn by Marion Ruff Sheehan)

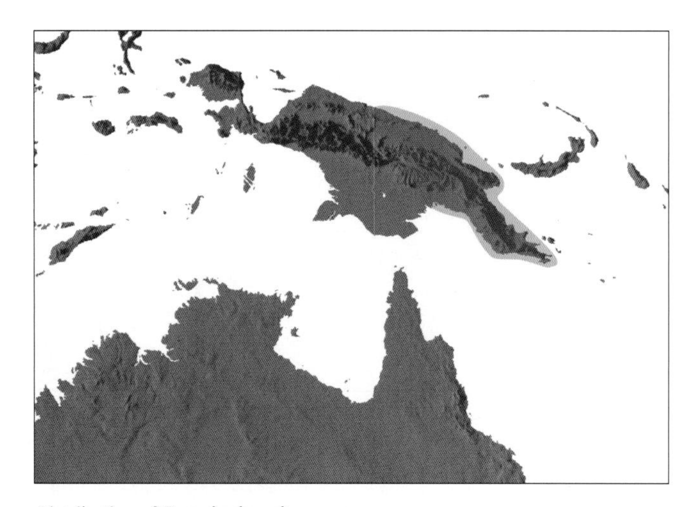

Distribution of ***Brassiophoenix***

receptacle, filaments long, slender, anthers elongate, basifixed, latrorse, connective tanniniferous, prolonged basally between the anthers; pistillode small, conical, sometimes with a short terete neck. *Pollen* ellipsoidal asymmetric; aperture a distal sulcus; ectexine tectate, perforate, or perforate-rugulate, aperture margin similar or slightly finer; infratectum columellate; longest axis ranging from 31–56 µm [2/2]. *Pistillate flowers* ovoid; sepals 3, distinct, thickened dorsally, broadly imbricate, bearing large scales, margins split irregularly; petals 3, distinct, about twice as long as the sepals at anthesis, otherwise like the sepals, tips thick, shortly valvate; staminodes joined basally into a shallow ring bearing short 3-lobed projections; gynoecium ovoid, unilocular, uniovulate, style not differentiated, stigma 3-lobed, fleshy, papillose, reflexed at anthesis, ovule laterally attached, pendulous, form unknown. *Fruit* ellipsoidal, tapered at both ends, wrinkled and ridged when dry, pale yellow-orange or red at maturity, stigmatic remains apical; epicarp with short, single, oblique fibrous bundles and interspersed brachysclereids, mesocarp fleshy, endocarp hard, thick, ridged. *Seed* laterally attached, irregular with 5 ridges, pointed distally, hilum elongate, raphe branches curved, somewhat anastomosing, endosperm homogeneous; embryo basal. *Germination* adjacent-ligular; eophyll bifid, tips toothed. *Cytology*: 2n = 32.

Distribution and ecology: Two species in New Guinea. Both species are inhabitants of mixed lowland rain forest.

Anatomy: Root (Seubert 1998a, 1998b) and fruit (Essig 1977).

Relationships: *Brassiophoenix* is a moderately supported monophyletic group (Lewis *et al.* in prep.) that is resolved, again with moderate support, as sister to *Ptychococcus* (Asmussen *et al.* 2006, Norup *et al.* 2006, Baker *et al.* in review, in prep., Lewis *et al.* in prep.).

Common names and uses: Common names unknown. Handsome ornamentals.

Taxonomic account: Zona & Essig (1999).

Fossil record: No generic records found.

Notes: Distinguished from other genera of Ptychospermatinae by the three-pronged leaflets and by the basifixed anthers with the connective basally enlarged. The endocarp is hard,

Far left: ***Brassiophoenix schumannii***, habit, Papua New Guinea. (Photo: A. Kjaer)

Left: ***Brassiophoenix drymophloeoides***, habit, Papua New Guinea. (Photo: W.J. Baker)

Below: ***Brassiophoenix schumannii***, staminate flowers and buds, Papua New Guinea. (Photo: W.J. Baker)

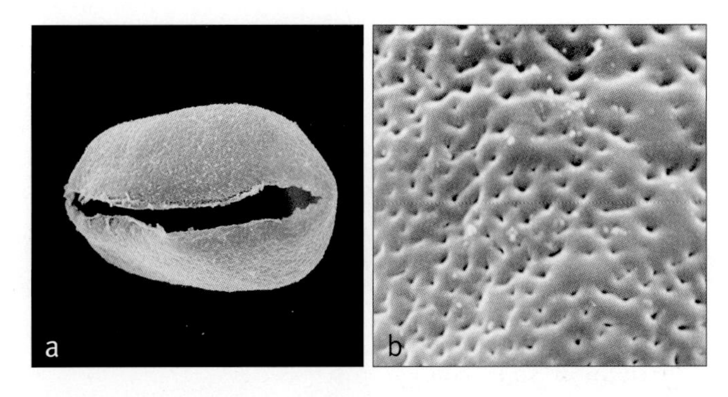

Above: ***Brassiophoenix***
a, monosulcate pollen grain, distal face SEM × 800;
b, close-up, perforate pollen surface SEM × 8000.
Brassiophoenix drymophloeoides: **a** & **b**, *Womersley* 19272.

as in *Ptychococcus*, but lacks unsheathed vascular bundles and is thus distinct anatomically (Essig 1977). In leaf form, *Brassiophoenix* seems closest to *Drymophloeus*, but it resembles *Ptychosperma* in inflorescence structure and *Ptychococcus* in fruit, and therefore represents a different combination of characters within the subtribe.

Additional figures: Glossary figs 14, 19.

167. PTYCHOCOCCUS

Moderate solitary pinnate-leaved palm, native to New Guinea, with crownshaft and praemorse leaflets; the fruit is relatively large and has an elaborated grooved and ridged black endocarp; the seed has homogeneous or ruminate endosperm.

Ptychococcus Becc., Ann. Jard. Bot. Buitenzorg 2: 100 (1885). Lectotype: **P. paradoxus** (Scheff.) Becc. (*Drymophloeus paradoxus* Scheff.) (see Burret 1928).

Ptyx — *groove,* coccus —*seed, referring to the grooved seed.*

Moderate, solitary, unarmed, pleonanthic, monoecious palms. *Stem* slender to moderate, ringed with leaf scars, grey or brownish. *Leaves* pinnate, spreading to ascending, blade ± horizontal basally but twisting toward an arched apex; sheath forming a long crownshaft, margins oblique lacking a ligule, covered in dense, partially deciduous scales or tomentum; petiole very short, channelled adaxially, rounded abaxially, densely tomentose; rachis adaxially flat basally, with a central ridge at midlength to ridged distally, abaxially rounded, densely tomentose; leaflets lanceolate, single-fold, apically very oblique, truncate, or somewhat pointed, and toothed, red-brown or pale scales on both surfaces, denser abaxially and along veins, midrib prominent, marginal ribs also large and sometimes (?always) densely tomentose, transverse veinlets not evident. *Inflorescences* solitary, infrafoliar, stiff, usually several clustered below the crownshaft, horizontal in flower, drooping in fruit, branched to 3 orders basally, fewer orders distally, all branches densely scaly, becoming glabrous; peduncle short, stout, dorsiventrally compressed; prophyll tubular, 2-keeled laterally, slightly beaked, bearing deciduous tomentum; peduncular bract like the prophyll but lacking keels, second partial peduncular bract present in young stages;

Top right: ***Ptychococcus paradoxus***, habit, Papua New Guinea.
(Photo: W.J. Baker)

Bottom right: ***Ptychococcus paradoxus***, habit, Papua New Guinea.
(Photo: W.J. Baker)

Above left: ***Ptychococcus paradoxus***, fruit in cross-section, Papua New Guinea. (Photo: W.J. Baker)

Above right: ***Ptychococcus paradoxus***, fruit in vertical section, Papua New Guinea. (Photo: W.J. Baker)

Distribution of ***Ptychococcus***

rachis longer than the peduncle bearing very short, wide, rounded bracts subtending distant branches and rachillae; rachillae rather short, sometimes zigzag, bearing very short rounded bracts subtending distant triads of large flowers, flowers borne nearly throughout the rachillae, projecting on either side of the rachillae; floral bracteoles irregular, rounded, those surrounding the pistillate flower of medium height; scar of pistillate flower very large compared to those of staminate flowers. *Staminate flowers* slightly asymmetrical; sepals 3, distinct, imbricate, keeled dorsally toward a gibbous base, margins split and bearing hairs; petals 3, distinct, valvate, ovate, glabrous or densely covered in small membranous scales; stamens numerous (up to 100), filaments short, awl-shaped, anthers elongate, notched apically, deeply bifid basally, dorsifixed near the base, ± introrse, connective tanniniferous; pistillode bottle-shaped with neck as long as the stamens, apically with a few short pointed tips. *Pollen* ellipsoidal asymmetric; aperture a distal sulcus; ectexine tectate, perforate, aperture margin similar; infratectum columellate; longest axis 46–66 µm [2/3]. *Pistillate flowers* ovoid, smaller than the staminate at anthesis; sepals 3, distinct, imbricate, rounded, sometimes covered with small hairs; petals 3, distinct, broadly imbricate with short, thick valvate tips, sometimes densely scaly; staminodes 3, ± united in a low semicupule; gynoecium ovoid, unilocular, uniovulate, style not differentiated, stigma 3-lobed, ovule pendulous, 5-angled, form unknown. *Fruit* ovoid, prominently wrinkled and angled when dry, orange to red at maturity, perianth persistent as a large cupule, stigmatic scar slightly eccentric; epicarp with short oblique fibrous bundles and interspersed brachysclereids, mesocarp fleshy with tanniniferous cells and vascular bundles lacking fibrous sheaths, endocarp with large keel, deeply grooved between 3 lateral and 2 ventral ridges, wall hard, thick. *Seed* 5-lobed, like the endocarp in shape, hilum round, apical,

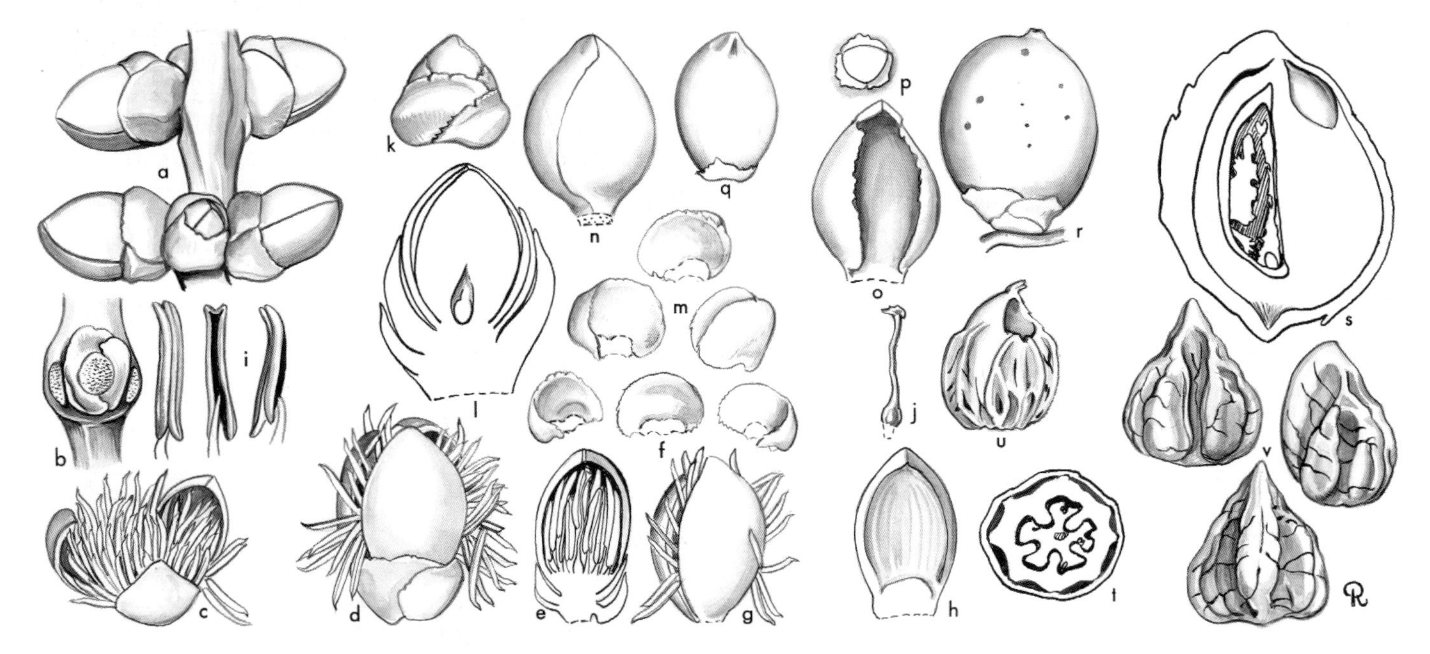

Ptychococcus. **a**, portion of rachilla with triads × 1$^1/_2$; **b**, triad, flowers removed × 1$^1/_2$; **c**, staminate flower from above × 1$^1/_2$; **d**, staminate flower, lateral view × 1$^1/_2$; **e**, staminate bud in vertical section × 1$^1/_2$; **f**, staminate sepals × 1$^1/_2$; **g**, staminate flower, sepals removed × 1$^1/_2$; **h**, staminate petal, interior view × 1$^1/_2$; **i**, stamen in 3 views × 3; **j**, pistillode × 1$^1/_2$; **k**, pistillate bud × 1$^1/_2$; **l**, pistillate bud in vertical section × 3; **m**, pistillate sepals × 1$^1/_2$; **n**, pistillate bud, sepals removed × 3; **o**, pistillate petal × 3; **p**, staminodes × 3; **q**, gynoecium and staminodes × 3; **r**, fruit × $^1/_2$; **s**, fruit in vertical section × 1; **t**, fruit in cross-section × $^1/_2$; **u**, endocarp × $^1/_2$; **v**, seed in 3 views × 1. *Ptychococcus lepidotus*: all from *Moore et al.* 9259. (Drawn by Marion Ruff Sheehan)

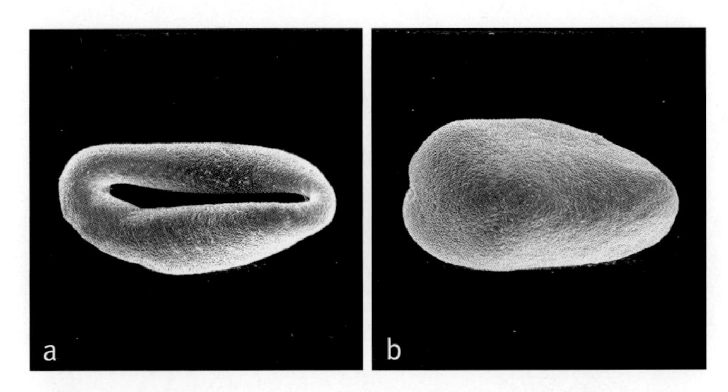

Above: ***Ptychococcus***
a, monosulcate pollen grain, distal face SEM × 1000;
b, monosulcate pollen grain, proximal face SEM × 1000.
Ptychococcus paradoxus: *Beccari* 423.

raphe branches thin, endosperm homogeneous with shallow marginal ruminations and a deep intrusion in the rapheal lobes or deeply ruminate; embryo basal. *Germination* adjacent-ligular; eophyll bifid, tips toothed. *Cytology*: 2n = 32.

Distribution and ecology: Two species, in New Guinea Bismarck Archipelago and Bougainville in rain forest along rivers in the lowlands and on mountain ridges.
Anatomy: Leaf (Tomlinson 1961) and fruit (Essig 1977).
Relationships: The monophyly of *Ptychococcus* has never been tested in a phylogenetic analysis. For relationships, see *Brassiophoenix*.
Common names and uses: Common names not recorded. The wood is extremely tough and used for bows, arrows, and building. The seed is said to be edible.
Taxonomic account: Zona (2003).
Fossil record: No generic records found.
Notes: The thick hard endocarp is highly distinctive, being pointed at both ends with sharp or irregular, not rounded, ridges. The variability in endosperm structure, from completely homogeneous to deeply ruminate in the same species, *Ptychococcus paradoxus*, is remarkable (Zona 2003).
Additional figures: Glossary fig. 24.

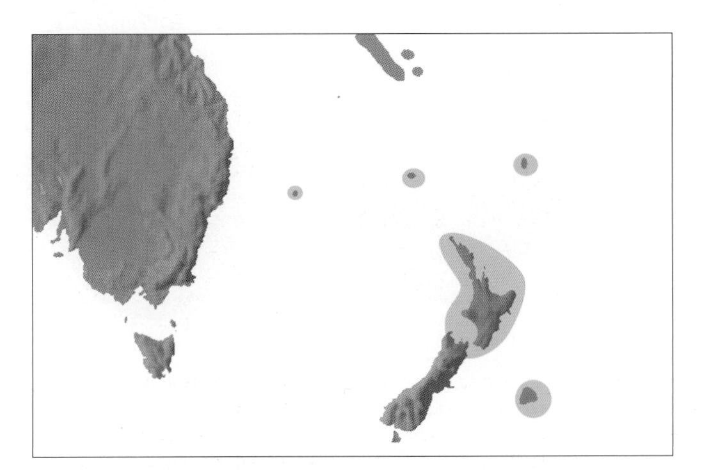

Distribution of subtribe **Rhopalostylidinae**

Subtribe Rhopalostylidinae J. Dransf, N.W. Uhl, C. Asmussen, W.J. Baker, M. Harley & C. Lewis, Kew Bull. 60: 563 (2005). Type: **Rhopalostylis**.

Moderate, unarmed, pleonanthic monoecious palms with pinnate leaves, the leaflets with entire tips, leaf sheaths forming a well-defined short stout crownshaft; inflorescences infrafoliar, peduncle very short, the prophyll and peducular bract enclosing the inflorescence in bud; staminate flowers asymmetrical with narrow not imbricate sepals, stamens 6–10, pistillode narrow columnar exceeding the stamens; fruit with subapical stigmatic remains.

The subtribe comprises two genera previously included in Archontophoenicinae: *Rhopalostylis* in New Zealand, Norfolk, Chatham and Raoul Islands in the south Pacific, and *Hedyscepe* in Lord Howe Island.

The subtribe is monophyletic in several phylogenetic studies, although it resolves with varying degrees of support (Pintaud 1999a, Asmussen *et al.* 2006, Baker *et al.* in review). *Hedyscepe* and *Rhopalostylis* are placed within the western Pacific clade of Areceae (Norup *et al.* 2006, Baker *et al.* in review, in prep.). Some phylogenies resolve a moderately supported relationship between *Hedyscepe* and *Basselinia*, thus rendering the subtribe Rhopalostylidinae non-monophyletic (Norup *et al.* 2006, Baker *et al.* in prep.), but this may be a reflection of molecular evolutionary phenomena rather than phylogenetic relationships.

Key to Genera of Rhopalostylidinae

1. Triads borne throughout ± the entire length of the rachilla; staminate flowers with 6 stamens and pistillode rounded at base. New Zealand, Chatham, Norfolk and Raoul Islands . . . . **168. Rhopalostylis**
1. Triads borne in the basal part of the rachillae only, paired or single staminate flowers distally; staminate flowers with 9–10 stamens and pistillode cylindrical throughout. Lord Howe Island . . . . **169. Hedyscepe**

168. RHOPALOSTYLIS

Elegant feather-duster pinnate-leaved palms native to New Zealand, Norfolk Island, Raoul Island and the Chatham Islands, with rather stiff leaves and bulbous crownshafts.

Rhopalostylis H. Wendl. & Drude, Linnaea 39: 180, 234 (1875). Type: **R. baueri** (Hook.f. ex Lem.) H. Wendl. & Drude (*Areca baueri* Hook.f ex Lem.).
Eora O.F. Cook, J. Heredity 18: 409 (1927). Type: *E. sapida* (H. Wendl. & Drude) O.F. Cook (= *R. sapida* [Sol. ex G. Forst.] H. Wendl. & Drude).

Rhopalon — *club*, stylis — *column or pillar, but botanically the style, referring to the shape of the pistillode in the staminate flower.*

Moderate to tall, solitary, unarmed, pleonanthic, monoecious palms. *Stem* erect, smooth, green to grey, closely ringed with prominent leaf scars, sometimes enlarged and bearing exposed roots at the base. *Leaves* pinnate, ascending and erect or somewhat arched, often twisted; sheath forming a prominent, rather short and somewhat bulbous crownshaft, enlarged basally or ± straight, sheath with definite diagonal nerves, and small brown scales; petiole very short, channelled adaxially, rounded abaxially; rachis like the petiole proximally, becoming channelled laterally and ridged adaxially, densely covered with small to medium brown scales; leaflets subopposite, forward pointing, stiff, lanceolate, tapering, often curved distally, single-fold, adaxially glabrous, abaxially and marginally somewhat tomentose, large scales present along the midrib, midrib evident adaxially, midrib and 1 pair of veins near the margins raised abaxially, transverse veinlets not evident. *Inflorescence* infrafoliar, spreading, ± pendulous in fruit, branched to 3 orders; peduncle very short, dorsiventrally flattened, stout; prophyll tubular, elongate, rather short and broad, with 2 flat, lateral keels, broadly pointed distally, thin, papery, deciduous; peduncular bract like the prophyll but not keeled, deciduous with the prophyll; rachis longer than the peduncle, stout, tapering, bearing spirally arranged, sometimes prominent bracts, bracts basally wide, tapering to a point, subtending rachillae; rachillae divaricate, short, stout, with or without a short bare portion, bearing triads of flowers basally and paired to solitary staminate flowers on a short distal portion; rachilla bracts

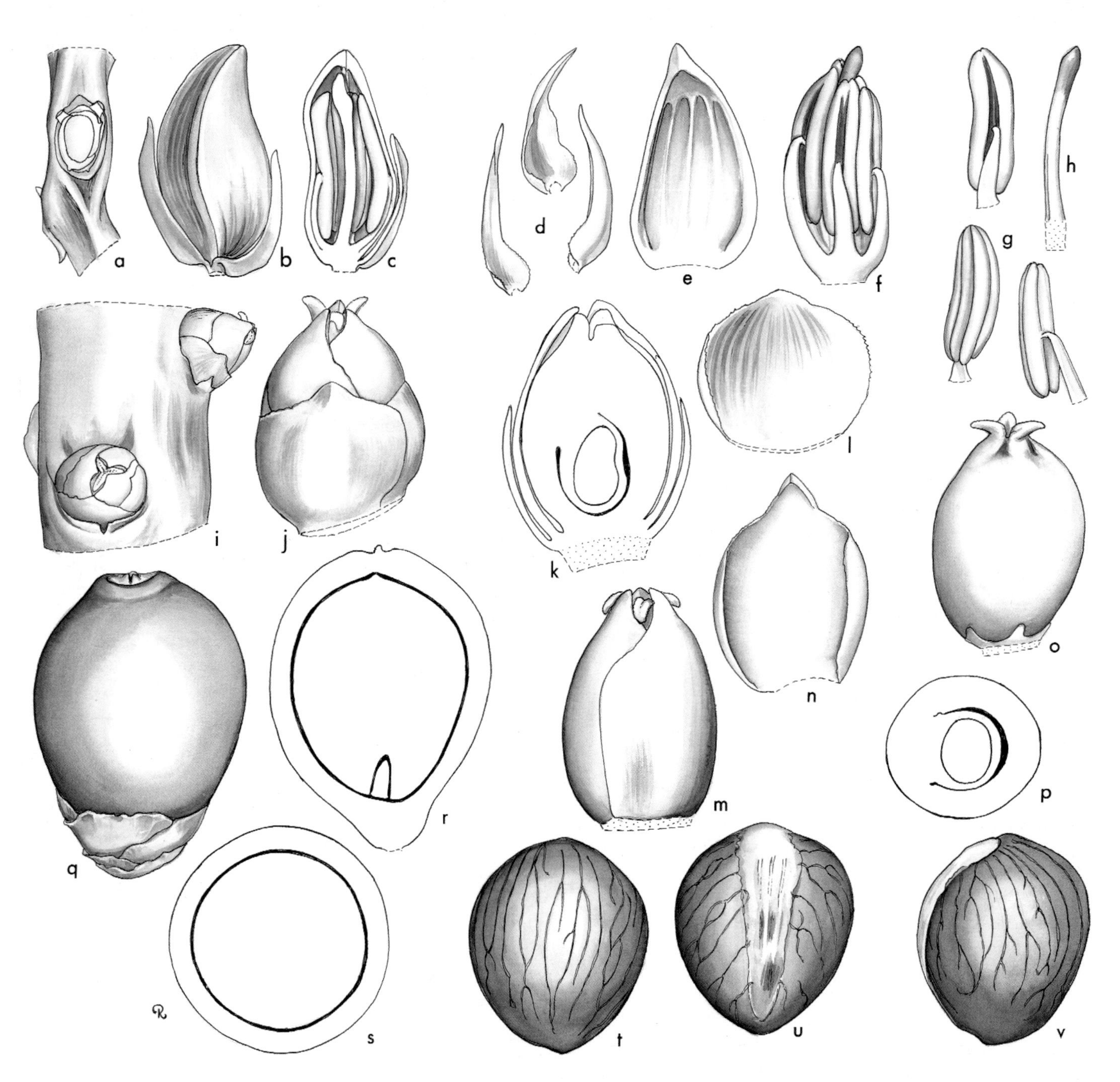

***Rhopalostylis*.** **a**, portion of rachilla with triad, flowers removed × 3; **b**, staminate bud × 6; **c**, staminate bud in vertical section × 6; **d**, staminate sepals × 6; **e**, staminate petal, interior view × 6; **f**, androecium and pistillode × 6; **g**, stamen in 3 views × 6; **h**, pistillode × 6; **i**, portion of rachilla at pistillate anthesis × 3; **j**, pistillate flower × 6; **k**, pistillate flower in vertical section × 6; **l**, pistillate sepal, interior view × 6; **m**, pistillate flower, sepals removed × 6; **n**, pistillate petal, interior view × 6; **o**, gynoecium and staminodes × 6; **p**, ovary in cross-section × 6; **q**, fruit × 3; **r**, fruit in vertical section × 3; **s**, fruit in cross-section × 3; **t, u, v**, seed in 3 views × 3. *Rhopalostylis baueri*: **a–h**, *Leyden* s.n.; **i–v**, *Osbourne* s.n. (Drawn by Marion Ruff Sheehan)

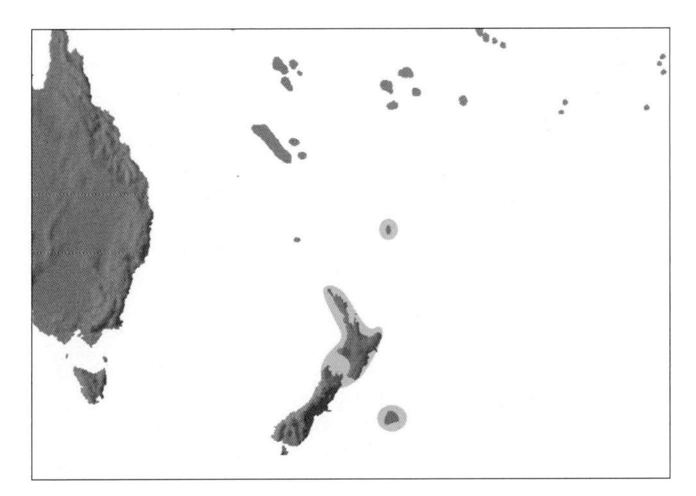

Distribution of ***Rhopalostylis***

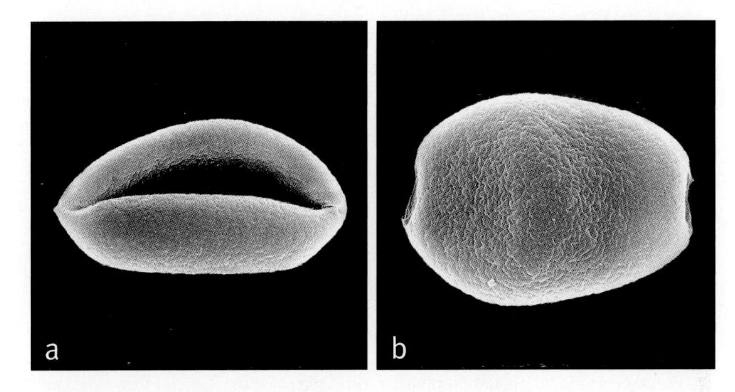

Above: ***Rhopalostylis***
a, monosulcate pollen grain, distal face SEM × 800;
b, monosulcate pollen grain, proximal face SEM × 800.
Rhopalostylis baueri: **a** & **b**, cultivated RBG Kew 1973-12601.

Below left: ***Rhopalostylis sapida***, habit, New Zealand. (Photo: S. Andrews)
Below right: ***Rhopalostylis baueri***, rachillae, cultivated, California. (Photo: J. Dransfield)
Bottom: ***Rhopalostylis baueri***, ripe fruit, cultivated, California. (Photo: J. Dransfield)

pointed, thick, striate, semicircular, tightly enclosing the lower half of the pistillate flower; floral bracteoles short, pointed. *Staminate flowers* asymmetrical, borne laterally and somewhat distally to the pistillate; sepals 3, distinct, briefly imbricate basally, narrow, tapering, somewhat keeled; petals 3, distinct, only slightly or markedly longer than the sepals, curved, adaxially grooved, tips thick, pointed; stamens 6, filaments markedly inflexed in bud, terete, joined briefly basally, anthers erect in bud, linear, dorsifixed near the middle, emarginate apically, bifid basally, latrorse, connective elongate, tanniniferous; pistillode slender, cylindrical, about as long as the stamens. *Pollen* ellipsoidal or pyriform, less frequently oblate triangular, slight or obvious asymmetry; aperture a distal sulcus, less frequently a trichotomosulcus; ectexine tectate, perforate, aperture margin similar; infratectum columellate; longest axis ranging from 55–73 µm [2/3]. *Pistillate flowers* symmetrical, ovoid; sepals 3, distinct, imbricate, truncate with a central point; petals 3, distinct, broadly imbricate basally with prominent valvate apices; staminodes lacking or tooth-like; gynoecium ellipsoidal, unilocular, uniovulate, style not distinct, stigmas 3, recurved, conspicuous, ovule very large, laterally attached, form unknown. *Fruit* globose or ellipsoidal, red when ripe, stigmatic remains apical, perianth persistent, spreading; epicarp smooth, mesocarp with a thin layer of sclereids and many flat, longitudinal fibres adherent to the endocarp, endocarp thin, fragile, not operculate, white or ± tan coloured. *Seed* ellipsoidal or globose, laterally attached by an elongate, rather wide, tapering hilum, raphe branches few or numerous, moderate to fine, anastomosing, endosperm homogeneous; embryo basal, often exactly so. *Germination* adjacent-ligular; eophyll bifid. *Cytology*: 2n = 32.

Distribution and ecology: Variously considered as three species or two species, one with two varieties; distributed in New Zealand, Chatham Islands, Norfolk Island and Kermadec Islands. Chiefly in dense lowland forests, usually not far from the sea in a mild, warm-temperate climate with abundant moisture.

Anatomy: Root (Seubert 1998a, 1998b), leaf (Tomlinson 1961) and fruit (Essig & Hernandez 2002).

Relationships: The monophyly of *Rhopalostylis* has not been tested. For relationships, see Rhopalostylidinae.

Common names and uses: Nikau palm (*Rhopalostylis sapida*), norfolk palm (*R. baueri*). Used extensively by the Maoris for thatch and basket weaving. The terminal bud is edible. The palms are striking ornamentals in suitably moist climates.

Taxonomic account: Bailey (1935b).

Fossil record: In New Zealand (North Island), pollen of *Rhopalostylis* (*R. sapida*) occurs from the Lower Miocene to Recent where it is described as rare but easily recognisable (Couper 1953). *Cocos* has similar pollen (see entry for *Cocos*).
Notes: Very similar and closely related to *Hedyscepe*; see notes under *Hedyscepe*.
Additional figures: Fig. 7.2.

169. HEDYSCEPE

Moderate, solitary pinnate-leaved palm endemic to Lord Howe Island, with a conspicuous white crownshaft.

Hedyscepe H. Wendl. & Drude, Linnaea 39: 178, 203 (1875). Type: **H. canterburyana** (C. Moore & F. Muell.) H. Wendl. & Drude (*Kentia canterburyana* C. Moore & F. Muell.).

Hedys — *pleasant,* skepe — *shade, in allusion to its local name, umbrella palm.*

Solitary, moderate, unarmed, pleonanthic, monoecious palm. *Stem* rather stout, conspicuously ringed with narrow, raised, whitish leaf scars. *Leaves* pinnate, stiff, arching; sheath forming a prominent crownshaft, thick, striate, densely covered with white wax and scattered small brown scales; petiole very short, stout, adaxially grooved, abaxially rounded; leaflets subopposite, rather short, lanceolate, pointed, proximal and distal leaflets distinctly narrower and shorter, stiff, ± erect, single-fold, adaxially glabrous, abaxially tomentose along the margins and scaly along the ribs on both surfaces, midrib and 3 pairs of veins raised adaxially, transverse veinlets not evident. *Inflorescences* solitary, infrafoliar, ?protandrous, branched to 1(–3) orders; peduncle short, ± flat, stout, horizontal; prophyll tubular, dorsiventally flattened, pointed but not tapering, with 2 flat lateral keels, chartaceous, deciduous; peduncular bract like the prophyll but not keeled, also caducous; ca. 3 wide ridge-like bracts above the peduncular bract; rachis rather short but longer than the peduncle, bearing low, pointed, spirally arranged bracts subtending rachillae; rachillae divaricate, stout, bearing triads of flowers nearly throughout, distally with a few pairs and single staminate flowers; floral bracteoles, low, pointed. *Staminate flowers* asymmetrical; sepals 3, distinct, slightly imbricate and bulbous or connate basally, narrow, elongate, widely separated, tapering distally, keeled; petals 3, distinct, variously angled, valvate; stamens 9–10 (–12 according to J.D. Hooker [1883]) in 2 series, antesepalous stamens solitary, antepetalous stamens paired, filaments awl-shaped, markedly inflexed at the apex in bud, anthers erect in bud, large, variously curved, linear, emarginate apically, bifid basally, latrorse, connective elongate, tanniniferous; pistillode with a rounded ovarian part and long terete style about as high as the stamens. *Pollen* grains ellipsoidal, slight asymmetry; aperture a distal sulcus; ectexine tectate, perforate, aperture margin similar; infratectum columellate; longest axis ranging from 55–60 µm [1/1]. *Pistillate flowers* symmetrical, ovoid; sepals 3, distinct, imbricate, rounded; petals 3, distinct, imbricate with

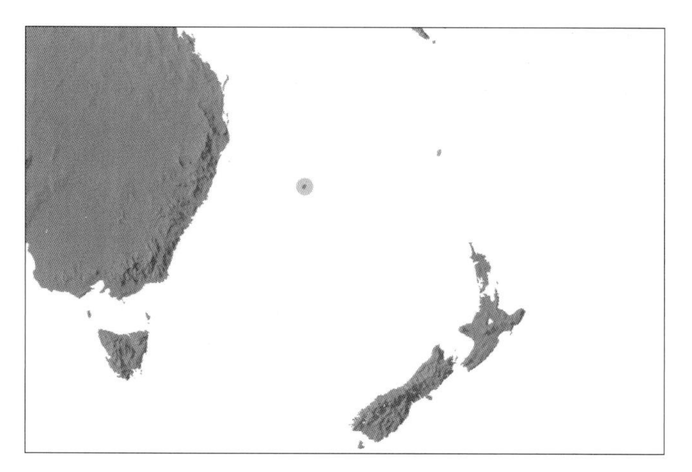

Distribution of ***Hedyscepe***

Below left: ***Hedyscepe canterburyana***, young fruit and a Lord Howe Pied Currawong (*Strepera graculina crissalis*), Lord Howe Island. (Photo: I. Hutton)
Below right: ***Hedyscepe canterburyana***, parts of rachillae showing staminate flowers in bud and at anthesis and pistillate buds, Lord Howe Island. (Photo: I. Hutton)

briefly valvate apices; staminodes 3, small, tooth-like, (?)borne on one side of the gynoecium; gynoecium unilocular, uniovulate, ovoid with 3 recurved stigmas, ovule laterally attached, form unknown. *Fruit* broadly ellipsoidal, deep dull red when ripe, stigmatic remains apical; epicarp smooth, mesocarp with a prominent layer of longitudinal, slender fibres over parenchyma and dispersed tannin cells and flat, long fibres adherent to the endocarp, endocarp crustaceous, thin, fragile, not operculate. *Seed* laterally attached by a broad elongate hilum, raphe branches numerous, anastomosing, endosperm homogeneous; embryo basal. *Germination* adjacent-ligular; eophyll bifid. *Cytology* not studied.

Distribution and ecology: One species endemic to Lord Howe Island. Found on high cliffs above the sea, most common on the more exposed ridges at 600–750 m elevation.

Anatomy: Root (Seubert 1998a, 1998b) and fruit (Essig & Hernandez 2002).

Relationships: For relationships, see Rhopalostylidinae.

Common names and uses: Big mountain palm, canterbury umbrella palm. Occasionally cultivated.

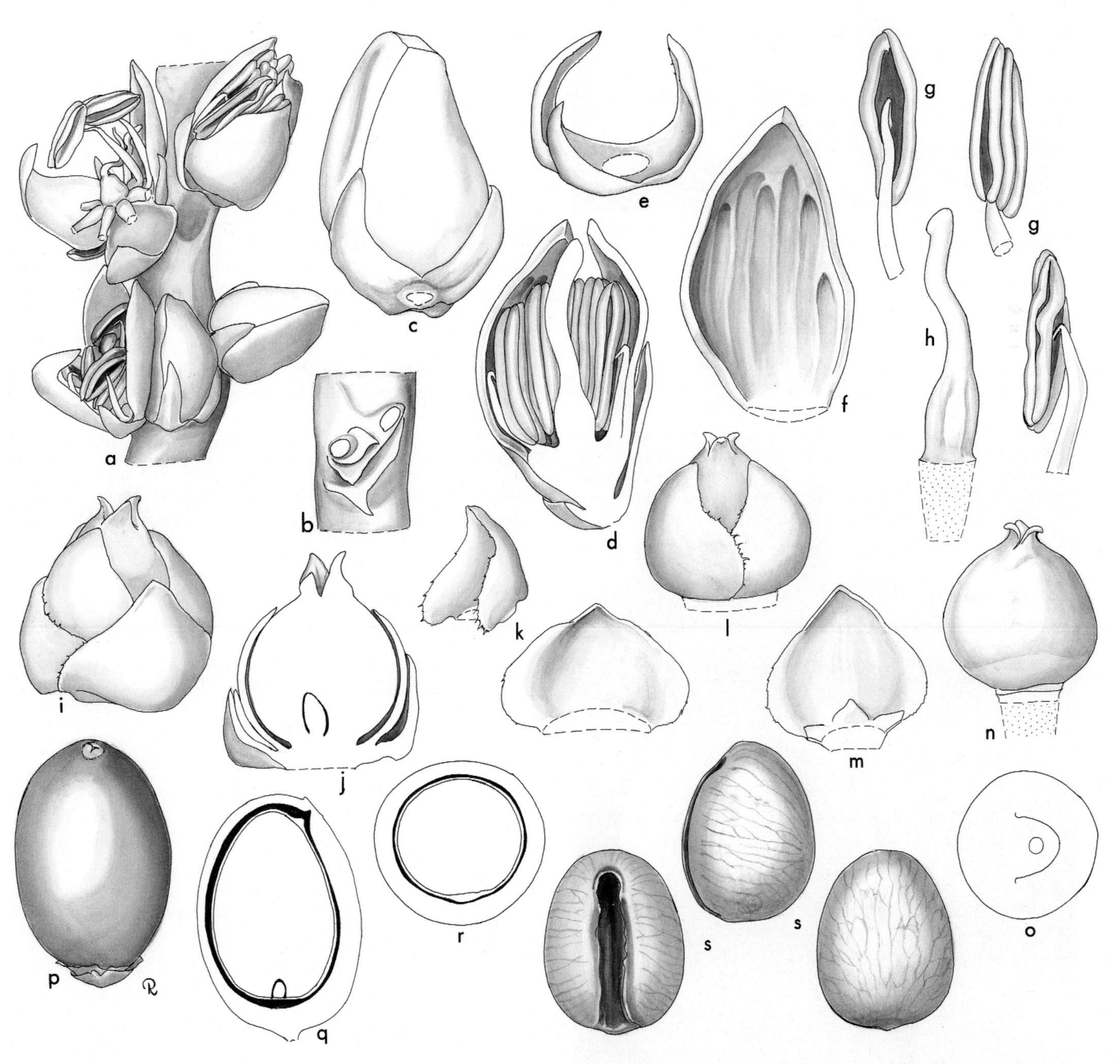

***Hedyscepe*.** **a**, portion of rachilla with paired staminate flowers × 3; **b**, portion of rachilla, paired staminate flowers removed to show scars and bracteoles × 6; **c**, staminate bud × 6; **d**, staminate bud in vertical section × 6; **e**, staminate calyx × 6; **f**, staminate petal, interior view × 6; **g**, stamen in 3 views × 6; **h**, pistillode × $7^1/_2$; **i**, pistillate flower × $4^1/_2$; **j**, pistillate flower in vertical section × $4^1/_2$; **k**, pistillate sepal in 2 views × $4^1/_2$; **l**, pistillate flower, sepals removed × $4^1/_2$; **m**, pistillate petal with staminodes, interior view × $4^1/_2$; **n**, gynoecium × $4^1/_2$; **o**, ovary in cross-section × $4^1/_2$; **p**, fruit × 1; **q**, fruit in vertical section × 1; **r**, fruit in cross-section × 1; **s**, seed in 3 views × 1. *Hedyscepe canterburyana*: **a–h**, *Schick* s.n.; **i–o**, *Darian* s.n.; **p–s**, *Moore & Schick* 9252. (Drawn by Marion Ruff Sheehan)

Taxonomic account: Green (1994).
Fossil record: No generic records found.
Notes: Closely related to *Rhopalostylis*. Differs in having triads nearly throughout the rachillae, more than six stamens and a pistillode with a rounded ovarian base. Also differs in being protandrous, *Rhopalostylis* being protogynous.

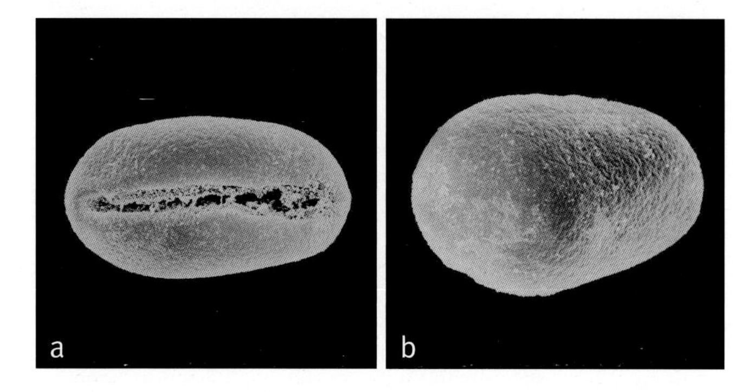

Above: ***Hedyscepe***
a, monosulcate pollen grain, distal face SEM × 800;
b, monosulcate pollen grain, proximal face SEM × 800.
Hedyscepe canterburyana: **a** & **b**, *Wendland & Drude* C.4357.

Below: ***Hedyscepe canterburyana***, habit, Lord Howe Island. (Photo: I. Hutton)

Subtribe Verschaffeltiinae J. Dransf, N.W. Uhl, C. Asmussen, W.J. Baker, M. Harley & C. Lewis, Kew Bull. 60: 564 (2005). Type: **Verschaffeltia**.

Small to moderate, erect, armed with spines, at least when young; sheaths not usually forming a well-defined crownshaft (crownshaft present in *Roscheria*); leaves pinnate, sometimes irregularly so, or entire bifid, leaflets tips and leaf margin praemorse or entire; inflorescences interfoliar, bisexual, branched to at least 2 orders; prophyll much shorter than the peduncular bract, the prophyll usually persisting, the peduncular bract inserted some distance from prophyll; staminate flowers usually symmetrical; fruit with basal or lateral stigmatic remains; epicarp smooth.

This subtribe is formed of four monotypic genera from the Seychelles, previously included in Oncospermatinae. The heterogeneity of Oncospermatinae (*sensu* Uhl & Dransfield 1987) has been indicated in the past; recent studies using molecular data strongly support the division of this subtribe into two, not necessarily related, groups. Fruit anatomy provides confirmatory data.

Subtribe Verschaffeltiinae is resolved as a monophyletic group in several studies with moderate to high support (Lewis 2002, Lewis & Doyle 2002, Baker *et al.* in review). However, Norup *et al.* (2006) resolved the subtribe as polyphyletic, with *Roscheria* and *Verschaffeltia* positioned with low support as sister to some members of the Oncospermatinae and *Nephrosperma*, and with *Phoenicophorium* unresolved within the Areceae. The group is placed outside the western Pacific clade, but its precise position is unclear. Baker *et al.* (in review) resolve the subtribe as sister to *Iguanura*, an unplaced genus of tribe Areceae. Lewis (2002) places the Verschaffeltiinae as sister to *Masoala* and *Marojejya*.

Key to Genera of Verschaffeltiinae

1. Leaflets or blade praemorse; staminate flowers with petals about twice as long as the sepals; stamens 6; pistillode large, truncate, 3-angled and lobed, about as high as the petals 2
1. Leaflets or blade lobes acute or acuminate; staminate flower with petals about 4 times as long as the sepals; stamens 18 or more; pistillode not truncate, angled or lobed 3
2. Leaf blade almost appearing undivided, the leaflets scarcely separated; fruit large, 2–2.5 cm diameter, the endocarp conspicuously ridged and crested; seed ± ridged, raphe branches anastomosing. Seychelles . . **173. Verschaffeltia**
2. Leaflets clearly separated; fruit small, ca. 5 mm diameter or less; endocarp not ridged; seed not ridged, the raphe branches few, ascending. Seychelles . . . **172. Roscheria**
3. Leaf blade usually undivided though lobed marginally, the lobes bifid; inflorescence branching to 2 orders; staminate flower asymmetrical; stamens ca. 18, the filaments tapered, anthers ± versatile; pistillode small, slender; fruit ovoid. Seychelles . . . . **171. Phoenicophorium**

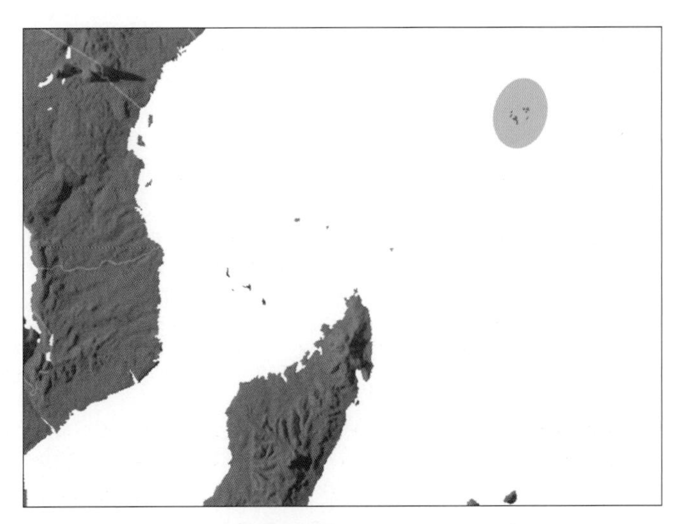

Distribution of subtribe **Verschaffeltiinae**

3. Leaf pinnate, with rather distant, mostly 2–3-ribbed acute or acuminate leaflets; inflorescence branching to 1 order; staminate flower symmetrical; stamens 40–50, the filaments expanded distally, anthers not versatile; pistillode ovoid, minutely trifid; fruit subglobose. Seychelles . . . . **170. Nephrosperma**

170. NEPHROSPERMA

Moderate solitary spiny pinnate-leaved palm endemic to Seychelles; the leaf sheaths do not form a crownshaft and the inflorescences have long peduncles; the seed is kidney-shaped.

Nephrosperma Balf.f. in Baker, Fl. Mauritius 386 (1877). Type: **N. van-houtteanum** (H. Wendl. ex Van Houtte) Balf.f. (*Oncosperma van-houtteanum* H. Wendl. ex Van Houtte).

Nephros — *kidney,* sperma — *seed, in reference to the shape of the seed.*

Moderate, solitary, spiny when young, unarmed or only very sparsely armed when mature, pleonanthic, monoecious palm. *Stem* erect, becoming bare, conspicuously ringed with leaf scars, unarmed. *Leaves* pinnate, neatly abscising; sheaths tubular, becoming open, not forming a well-defined crownshaft, densely tomentose, bearing abundant, black spines in juveniles, ± unarmed or very sparsely armed in mature individuals, sheath margin irregularly ligule-like, tattering; petiole well developed, adaxially channelled, abaxially rounded, bearing white indumentum and scattered scales, and few bristles near the base; rachis curved; leaflets rather regularly arranged, neatly curved, distant, composed usually of 2–3 folds, acute or acuminate, adaxially glabrous, abaxially with numerous, minute, dot-like scales and abundant ramenta along the adaxial ribs, transverse veinlets obscure; expanding leaf flushed red. *Inflorescences* solitary, interfoliar, branching to 1 order only, protandrous; peduncle very long (± $^1/_2$ the length of the leaves or more), erect at first, becoming curved, winged at the base, ± oval in cross-section, unarmed, scaly; prophyll inserted near the base of the peduncle, persistent, coriaceous, tubular, 2-keeled, the wings tending to be irregularly split or toothed, splitting apically for a short distance, covered with rather dense scales and scattered white wax, armed with short weak bristles and spines or rarely unarmed; peduncular bract, inserted a short distance from the prophyll, elongate, with a conspicuous long beak, tubular at first, then splitting along its length, deciduous, scaly and spiny as the prophyll; rachis shorter than the peduncle, scaly, bearing rather lax, spirally arranged rachillae; rachis bracts minute; rachillae scaly, with a pronounced swelling and bare section at the base, very long, spreading, slender, bearing distant, spirally arranged, superficial triads throughout, except at the tip where bearing solitary or paired staminate flowers; floral bracteoles minute. *Staminate flowers* symmetrical; sepals 3, distinct, imbricate, rounded, irregularly splitting, keeled; petals about 3–4 times as long as the sepals, 3, distinct, valvate, ± boat-shaped; stamens ca. 40–50, filaments elongate, anthers very small, rounded, with a broad connective, not versatile, latrorse; pistillode ovoid, conspicuous, minutely but clearly trifid at its tip. *Pollen* ellipsoidal asymmetric, occasionally elongate or pyriform; aperture a distal sulcus; ectexine tectate, finely perforate-rugulate, aperture margin similar; infratectum columellate; longest axis 35–43 µm [1/1]. *Pistillate flowers* globular; sepals 3, distinct, imbricate, low, ± rounded, thick, tending to split irregularly; petals 3, distinct, imbricate, ± rounded, with short triangular, valvate tips; staminodes 6, small, tooth-like; gynoecium ± obpyriform, unilocular, uniovulate, with minute apical stigmas, ovule laterally attached, form unknown. *Fruit* relatively small, spherical to somewhat kidney-shaped, red, perianth whorls persistent, stigmatic remains lateral; epicarp shiny, smooth, mesocarp thinly fleshy with a layer of slender fibres next to the endocarp, endocarp very thin, cartilaginous, with a thin, ± rounded operculum. *Seed* ± globose, somewhat kidney-shaped, attached laterally near the base, with an oblong hilum, raphe branches distant, slightly embedded in the endosperm, endosperm deeply ruminate; embryo basal. *Germination* adjacent-ligular; eophyll bifid. *Cytology*: 2n = 32.

Right: ***Nephrosperma van-houtteanum***, habit, Seychelles. (Photo: J. Dransfield)

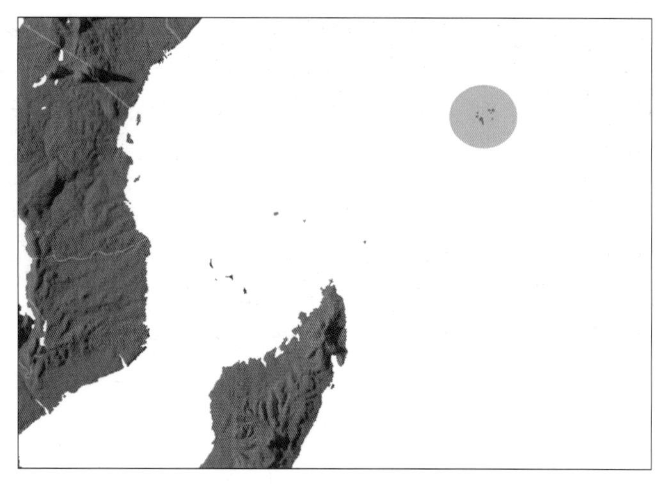

Distribution of ***Nephrosperma***

Distribution and ecology: One species confined to the Seychelles Islands; elsewhere widely cultivated. A lowland species, not occurring above 500 m altitude, growing on rocky slopes; also found in some secondary forest types.

Anatomy: Leaf (Tomlinson 1961), root (Seubert 1998a, 1998b) and fruit (Essig *et al.* 2001).

Relationships: *Nephrosperma* is resolved as sister to *Phoenicophorium* with high support (Lewis & Doyle 2002, Loo *et al.* 2006, Norup *et al.* 2006, Baker *et al.* in review, in prep.).

Common names and uses: *Latanier millepattes.* Not recorded apart from its use as an ornamental.

Taxonomic account: Bailey (1942).

Fossil record: No generic records found.

Notes: The staminate flowers are distinctive because of the large number of stamens (40–50) and the large pistillode.

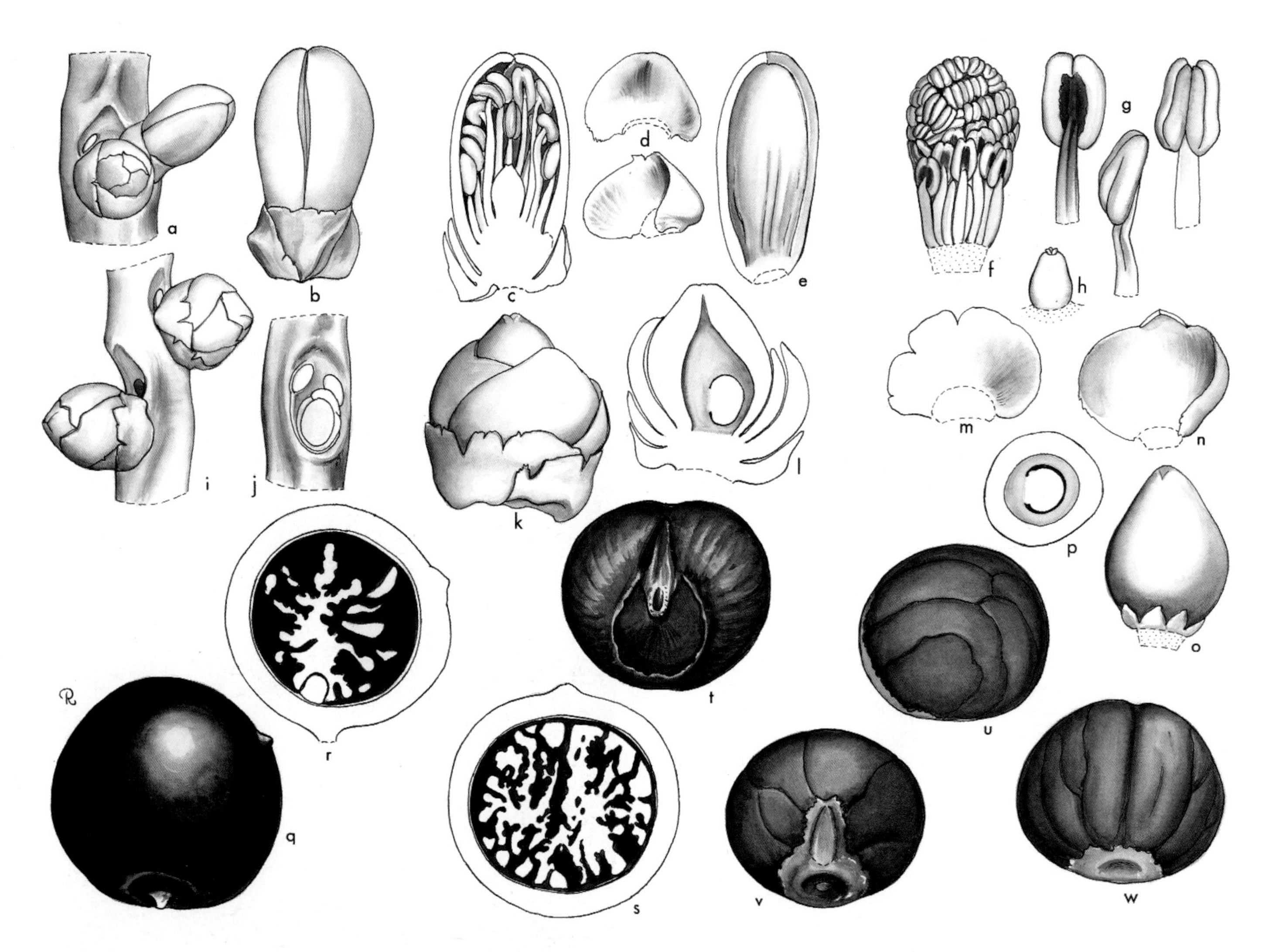

Nephrosperma. **a**, portion of axis with triad, one staminate flower fallen × 3; **b**, staminate bud × 6; **c**, staminate bud in vertical section × 6; **d**, staminate sepal in 2 views × 6; **e**, staminate petal, interior view × 6; **f**, androecium × 6; **g**, stamen in 3 views × 12; **h**, pistillode × 6; **i**, portion of rachilla at pistillate anthesis × 3; **j**, triad, flowers removed × 3; **k**, pistillate flower × 6; **l**, pistillate flower in vertical section × 6; **m**, pistillate sepal, interior view × 6; **n**, pistillate petal, interior view × 6; **o**, gynoecium and staminodes × 6; **p**, ovary in cross-section × 6; **q**, fruit × 3; **r**, fruit in vertical section × 3; **s**, fruit in cross-section × 3; **t**, endocarp and operculum × 3; **u**, **v**, **w**, seed in 3 views × 3. *Nephrosperma van-houtteanum*: **a–h** and **q–w**, *Read* 1557; **i–p**, *Moore* 9033. (Drawn by Marion Ruff Sheehan)

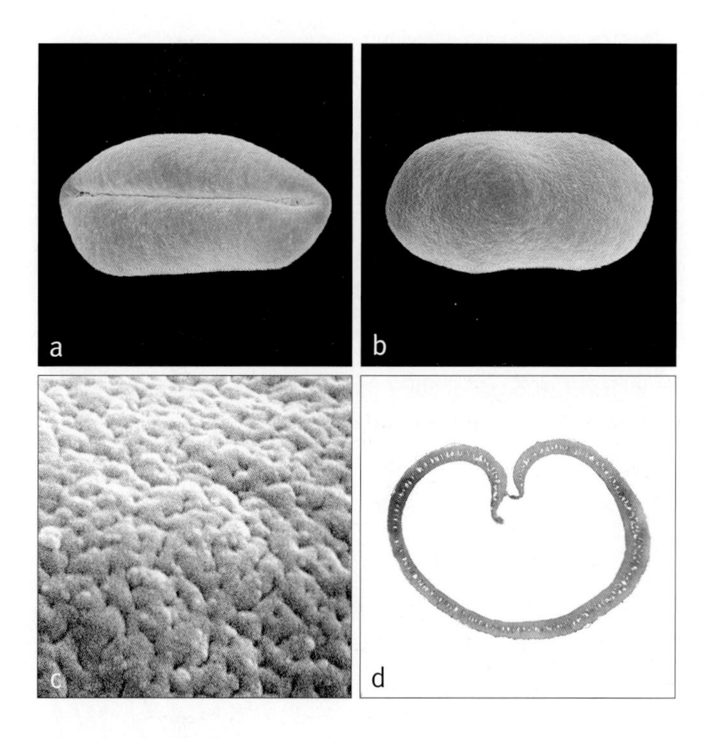

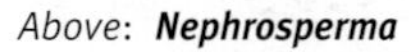

Above: ***Nephrosperma***
a, monosulcate pollen grain, distal face SEM × 1500;
b, monosulcate pollen grain, proximal face SEM × 1500;
c, close-up, finely perforate-rugulate pollen surface SEM × 8000;
d, ultrathin section through whole pollen grain, polar plane TEM × 3000.
Nephrosperma van-houtteanum: **a–d**, *Jeffrey* 1218.

Above right: ***Nephrosperma van-houtteanum***, infructescence, Seychelles.
(Photo: J. Dransfield)

171. PHOENICOPHORIUM

Moderate solitary spiny palm endemic to Seychelles, lacking a crownshaft and with ± undivided leaves; inflorescences have long peduncles.

Phoenicophorium H. Wendl., Ill. Hort. 12 (Misc.): 5 (1865). Type: *P. sechellarum* H. Wendl. (illegitimate name) (= **P. borsigianum** [K. Koch] Stuntz) (*Astrocaryum borsigianum* K. Koch).

Stevensonia Duncan ex Balf.f. in Baker, Fl. Mauritius 388 (1877). Type: *S. grandifolia* Duncan ex Balf.f. (= *Phoenicophorium borsigianum* [K. Koch] Stuntz).

Phoenix — *a general name for a palm,* phorios — *stolen, in reference to the fact that the original plant in Kew that was to have been presented to Wendland, was stolen by another German gardener.*

Moderate, solitary, spiny when young, sparsely armed or ± unarmed at maturity, pleonanthic, monoecious palm. *Stem* erect, becoming bare, conspicuously ringed with leaf scars, the juvenile bearing abundant black spines, usually ± unarmed at maturity. *Leaves* large, bifid, lobed and pinnately ribbed but not pinnatifid, neatly abscising; sheaths becoming open, not forming a well-defined crownshaft, covered in abundant tomentum, easily detached spicules, and scattered, large black spines when young, becoming markedly less spiny as maturity is reached, sheath margin irregularly ligule-like, tattering; petiole well developed, adaxially channelled, abaxially rounded, sparsely scaly, in juveniles bearing abundant, large, black spines abaxially; rachis bearing black spines in the proximal region in juveniles; blade bifid, with very conspicuous, pinnate ribs, splitting along abaxial ribs to $^1/_8$ to $^1/_3$ the rib length into briefly bifid lobes, sometimes split further by wind, blade bright green, often reddish-tinged, adaxially glabrous, abaxially with abundant dot-like scales and large ramenta along the adaxial folds near the rachis, transverse veinlets obscure. *Inflorescences* solitary, interfoliar, branching to 1–2 orders proximally, to 1 order distally, protandrous; peduncle elongate, winged at the base, oval in cross-section, unarmed, glabrous; prophyll inserted some distance from the base of the peduncle, persistent, very coriaceous, tubular, 2-keeled, the keels tending to be irregularly split or toothed, splitting apically for a short distance, scaly or sparsely tomentose, unarmed or armed with sparse to abundant, short, black spines; peduncular bract 1, inserted some distance above the prophyll but included within it, ± woody, conspicuously beaked, deciduous, unarmed, minutely scaly, tubular at first then splitting along ± its entire length; rachis much shorter than the peduncle; rachis bracts minute, triangular, inconspicuous; first-order branches spirally arranged, crowded, ± pendulous, their bases somewhat swollen; rachillae glabrous, slender, elongate, flexuous, with a very

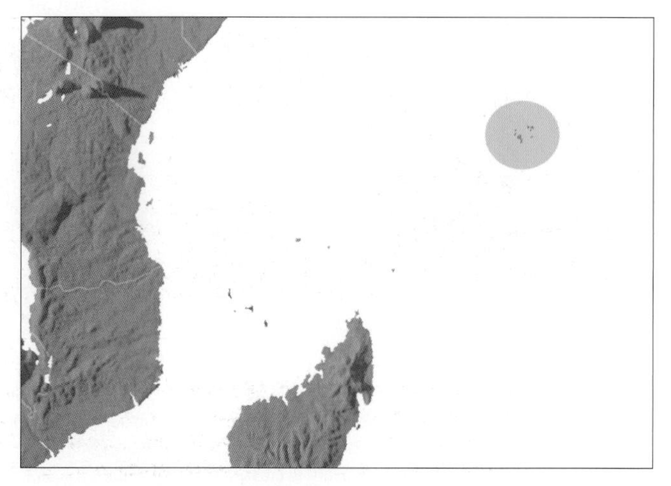

Distribution of ***Phoenicophorium***

short, bare area at the base, above this bearing spirally arranged, superficial triads, except near the tip where bearing solitary or paired staminate flowers; bracteoles minute, rounded. *Staminate flowers* slightly asymmetrical; sepals 3, distinct, imbricate, low, ± rounded, keeled; petals about 4–5 times as long as the sepals, 3, distinct, valvate; stamens 15–18, filaments short in bud, at anthesis becoming elongate, slender, anthers elongate, medifixed, basally sagittate, latrorse; pistillode absent. *Pollen* ellipsoidal bi-symmetric; aperture a distal sulcus; ectexine tectate, coarsely perforate with broad-based supratectal spines, often basally branched, aperture margin similar; infratectum columellate; longest axis 23–30 µm [1/1]. *Pistillate flowers* ± globular, about the same size as the staminate; sepals 3, distinct, imbricate, rounded, ± keeled; petals 3, distinct, rounded, imbricate, with short, triangular, valvate tips; staminodes 6, tooth-like; gynoecium asymmetrically ovoid, unilocular, uniovulate, stigmas 3, apical, very low, ovule laterally attached, ?campylotropous. *Fruit* 1-seeded, relatively small, ovoid to ellipsoidal, red, with persistent perianth whorls, and subbasal stigmatic remains; epicarp shiny, mesocarp thinly fleshy with a thick layer of tannin cells external to the endocarp, endocarp thin, cartilaginous, with basal round operculum. *Seed* ovoid, basally attached, with rounded hilum and sparse, anastomosing raphe branches, endosperm deeply ruminate; embryo basal. *Germination* adjacent-ligular; eophyll lanceolate, entire. *Cytology*: 2n = 32.

Distribution and ecology: One species, widespread in the Seychelles Islands.

Phoenicophorium borsigianum occurs abundantly from sea level to altitudes of about 300 m, above which it becomes much rarer. It occasionally occurs in pure stands, and seems to be relatively tolerant of disturbance. It is frequently planted as an ornamental.

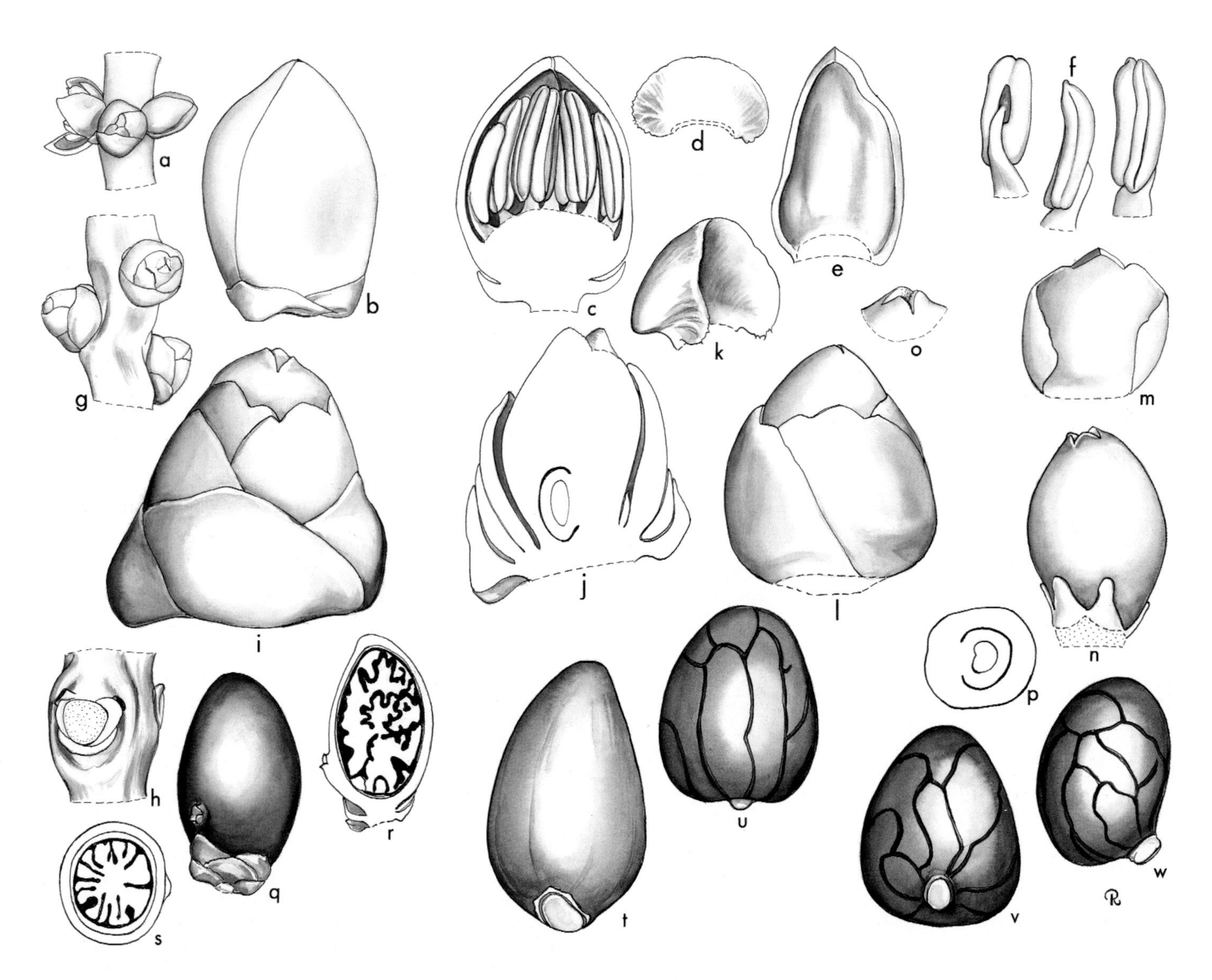

Phoenicophorium. **a**, portion of rachilla with triad × 3; **b**, staminate bud × 12; **c**, staminate bud in vertical section × 12; **d**, staminate sepal × 12; **e**, staminate petal, interior view × 12; **f**, stamen in 3 views × 12; **g**, portion of rachilla at pistillate anthesis × 3; **h**, triad, flowers removed to show scars and bracteoles × 6; **i**, pistillate flower × 12; **j**, pistillate flower in vertical section × 12; **k**, pistillate sepal × 12; **l**, pistillate flower, sepals removed × 12; **m**, pistillate petal × 12; **n**, gynoecium and staminodes × 12; **o**, stigmas × 12; **p**, ovary in cross-section × 12; **q**, fruit × 3; **r**, fruit in vertical section × 3; **s**, fruit in cross-section × 3; **t**, endocarp and operculum × 6; **u**, **v**, **w**, seed in 3 views × 6. *Phoenicophorium borsigianum*: all from *Read* 1550. (Drawn by Marion Ruff Sheehan)

Top: ***Phoenicophorium borsigianum***, habit, Seychelles. (Photo: J. Dransfield)

Above: ***Phoenicophorium borsigianum***, young plant, Seychelles. (Photo: J. Dransfield)

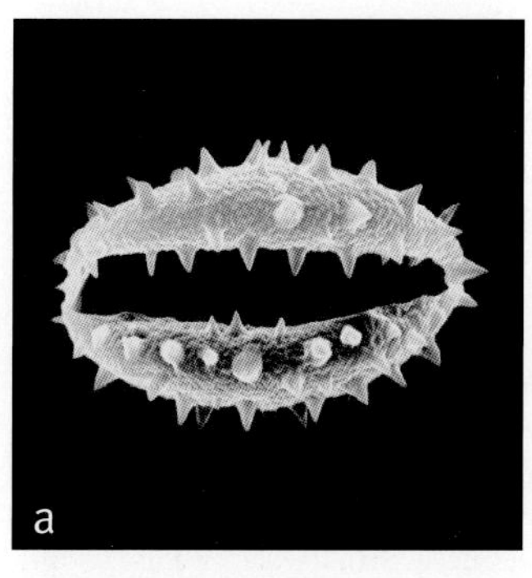

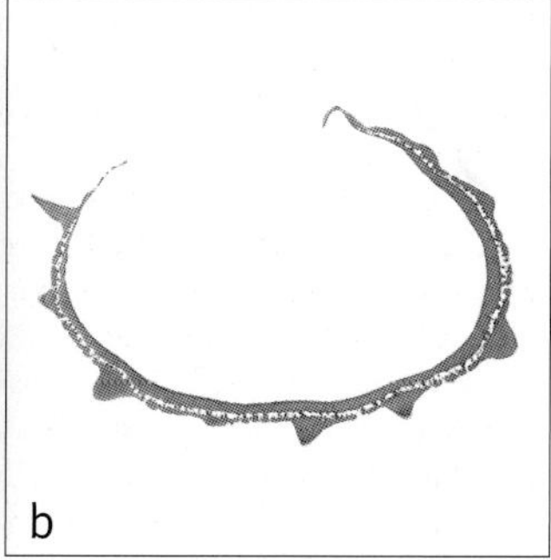

Above: ***Phoenicophorium***
a, spiny monosulcate pollen grain, distal face SEM × 2000;
b, ultrathin section through whole pollen grain, polar plane TEM × 3000.
Phoenicophorium borsigianum: **a** & **b**, *Jeffrey* 565.

Anatomy: Leaf (Tomlinson 1961 as *Stevensonia*), root (Seubert 1998a, 1998b) and fruit (Essig *et al*. 2001).
Relationships: For relationships, see *Nephrosperma*.
Common names and uses: *Latanier feuille*. A much-prized ornamental, used in the Seychelles for thatch.
Taxonomic account: Bailey (1942).
Fossil record: No generic records found.
Notes: The leaf is very distinctive being undivided but lobed marginally, the lobes acuminate or acute not praemorse as in *Verschaffeltia* and *Roscheria*.

172. ROSCHERIA

Small or moderate solitary spiny palm endemic to Seychelles, with a conspicuous crownshaft and with leaves divided into few broad segments with praemorse tips; inflorescences have long peduncles and very small flowers.

Roscheria H. Wendl. ex Balf.f. in Baker, Fl. Mauritius. 386 (1877). Type: **R. melanochaetes** (H. Wendl.) H. Wendl. ex Balf.f. (*Verschaffeltia melanochaetes* H. Wendl.).

Commemorates Dr Albrecht Roscher (1836–1860), German explorer who followed close on Burton, Speke and Grant in exploring East Africa.

Small or moderate, solitary, spiny when young, sparsely armed at maturity, pleonanthic, monoecious palm. *Stem* erect, becoming bare, conspicuously ringed with leaf scars, the juvenile bearing rings of black spines, at maturity unarmed or with very few, scattered, weak spines, the stem base sometimes with aerial roots. *Leaves* irregularly pinnate, neatly abscising; sheaths tubular, forming a well-defined crownshaft, covered in scattered brown scales and armed with scattered, short, black spines, spines much more densely near the base of the petiole; petiole adaxially deeply channelled, abaxially rounded, densely armed along the margins near the base with short, easily detached, black spines; rachis unarmed, bearing scattered scales; blade entire, bifid in juveniles, at maturity irregularly divided into 1–several-fold leaflets, those with single ribs acuminate, those with several ribs truncate, praemorse, adaxially glabrous, abaxially bearing abundant minute, dot-like scales and conspicuous ramenta along the major ribs, transverse veinlets obscure. *Inflorescences* solitary, interfoliar at first, becoming infrafoliar after leaf fall, copiously branching to 3, very rarely to 4 orders proximally, to fewer orders distally; peduncle elongate, crescentic in cross-section, winged at the base; prophyll inserted some distance from the base, membranous to coriaceous, tubular, persistent, flattened, 2-keeled, briefly split at the tip, unarmed, bearing scattered small scales; peduncular bract inserted some distance above the prophyll, usually much exceeding it, very sparsely scaly, tubular at first, later splitting along much of its length and expanding, eventually falling; rachis usually shorter than the peduncle; rachis bracts all minute, triangular, very inconspicuous; branches of all orders with a short to long portion free of branches or flowers, all axes scaly; first-order branches subdistichous, the proximal longer than the distal; rachillae slender, rather short, flexuous, bearing spirally arranged, minute bracts subtending triads in the middle portion and paired or solitary staminate flowers distally, the flowers not at all sunken. *Staminate flowers* very small, globular in bud; sepals 3, ± distinct, imbricate, broad, triangular, ± keeled; petals 3, distinct, valvate, more than twice as long as the sepals, broad, triangular; stamens 6, filaments connate basally to form a low ring, the distinct portions short, broad, triangular, anthers very small, rounded, medifixed, ± versatile, latrorse; pistillode relatively large, truncate, 3-angled. *Pollen* ellipsoidal asymmetric; aperture a distal sulcus; ectexine tectate, rugulate-reticulate with finely ridged muri, aperture margin similar; infratectum columellate; longest axis 28–33 µm

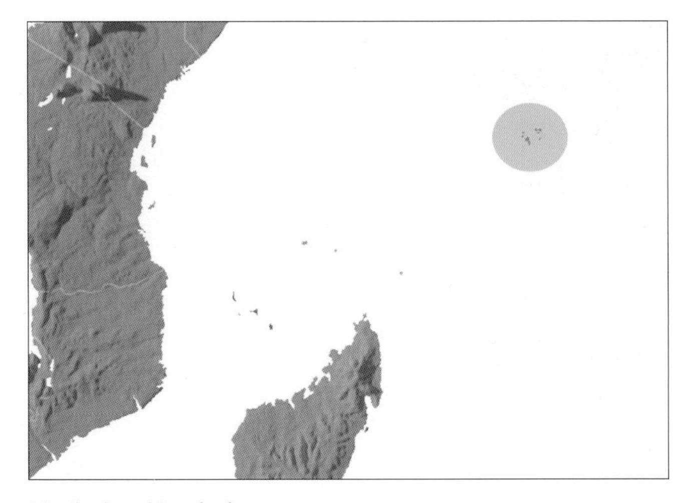

Distribution of ***Roscheria***

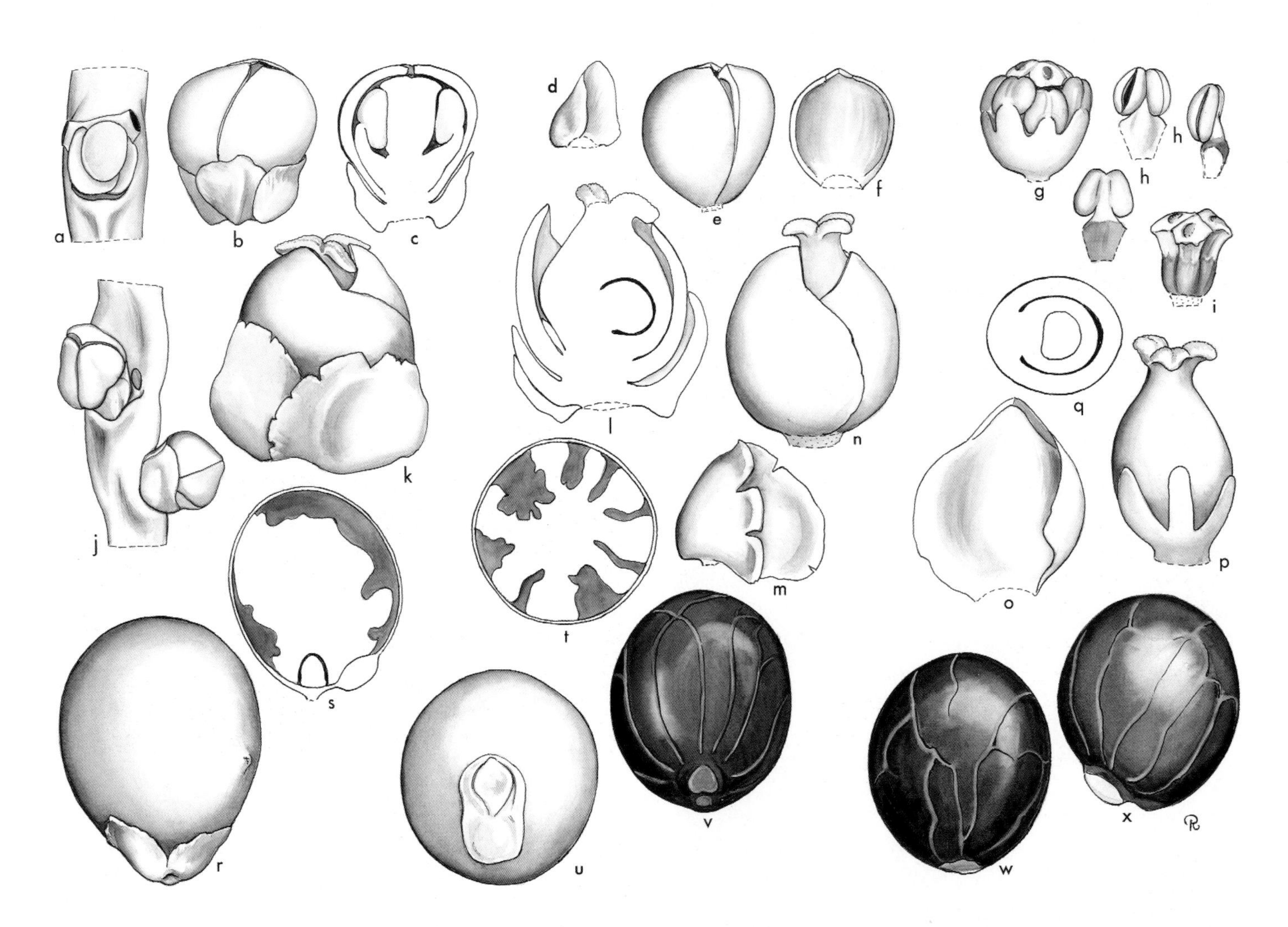

***Roscheria*.** **a**, portion of rachilla with triad, flowers removed to show bracteoles × 12; **b**, staminate bud × 18; **c**, staminate bud in vertical section × 18; **d**, staminate sepal × 18; **e**, staminate flower, sepals removed × 18; **f**, staminate petal × 18; **g**, androecium × 18; **h**, stamen in 3 views × 18; **i**, pistillode × 18; **j**, portion of rachilla with pistillate buds × 12; **k**, pistillate flower × 18; **l**, pistillate flower in vertical section × 18; **m**, pistillate sepal × 18; **n**, pistillate flower, sepals removed × 18; **o**, pistillate petal, interior view × 18; **p**, gynoecium and staminodes × 18; **q**, ovary in cross-section × 18; **r**, fruit × 6; **s**, fruit in vertical section × 6; **t**, fruit in cross-section × 6; **u**, endocarp and operculum × 6; **v**, **w**, **x**, seed in 3 views × 6. *Roscheria melanochaetes*: all from *Squibb* s.n. (Drawn by Marion Ruff Sheehan)

Top left: ***Roscheria melanochaetes***, habit, Seychelles. (Photo: J. Dransfield)

Top right: ***Roscheria melanochaetes***, rachillae with flower buds, Seychelles. (Photo: J. Dransfield)

Above: ***Roscheria melanochaetes***, young fruit, Seychelles. (Photo: J. Dransfield)

[1/1]. *Pistillate flowers* larger than the staminate; sepals 3, distinct, thick, imbricate, broad, rounded, the margins irregularly splitting; petals 3, distinct, broad, triangular, the tips briefly valvate, otherwise imbricate; staminodes 6, tooth-like, ± united basally; gynoecium asymmetrical, ovoid, unilocular, uniovulate, stigma apical, 3-lobed, ovule laterally attached, form unknown. *Fruit* small, globular or ellipsoidal, red at maturity, perianth whorls persistent, stigmatic remains subbasal; epicarp smooth, mesocarp very thin, with a layer of anastomosing fibres and abundant raphides, endocarp relatively thick, bony, smooth, with basal operculum. *Seed* basally attached with rounded hilum, raphe branches sparse, anastomosing, endosperm deeply ruminate; embryo basal. *Germination* and eophyll not recorded. *Cytology*: 2n = 32.

Distribution and ecology: Monotypic, confined to Mahé and Silhouette Islands in the Seychelles. An undergrowth palm of mountain rain forest, very rarely found below 500 m altitude; at ca. 750 m it can form pure stands on very steep slopes in the undergrowth of *Northia* forest.

Anatomy: Root (Seubert 1998a, 1998b) and fruit (Essig *et al.* 2001).

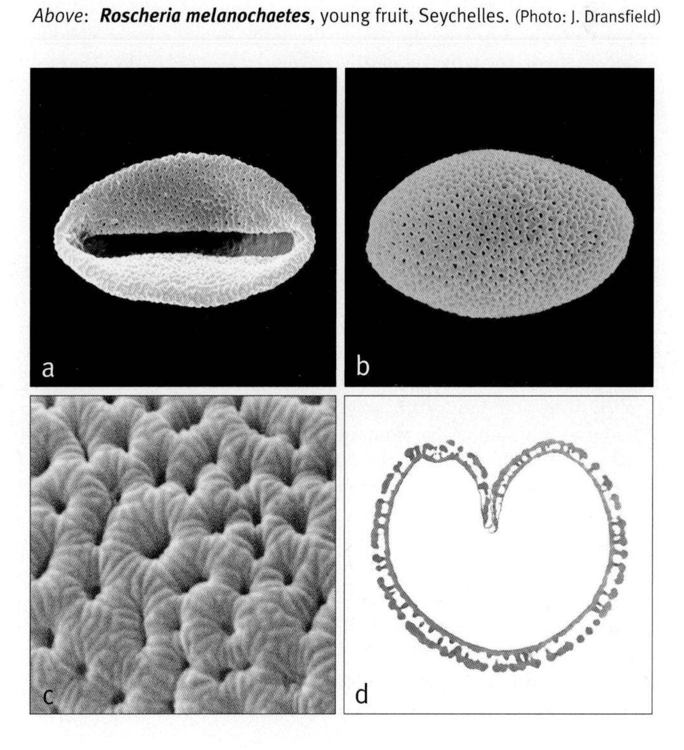

Right: ***Roscheria***

a, monosulcate pollen grain, distal face SEM × 2000;
b, monosulcate pollen grain, proximal face SEM × 2000;
c, close-up, tectate, rugulate-reticulate surface with finely ridged muri SEM × 10,000;
d, ultrathin section through whole pollen grain, polar plane TEM × 3000.
Roscheria melanochaetes: **a–d**, *Jeffrey* 728.

Relationships: *Roscheria* is resolved as sister to *Verschaffeltia* with high support (Lewis & Doyle 2002, Lewis 2002, Loo *et al*. 2006, Norup *et al*. 2006, Baker *et al*. in review, in prep.).
Common names and uses: *Latanier hauban*. Uses not recorded, apart from being a prized ornamental.
Taxonomic account: Bailey (1942).
Fossil record: Reticulate trichotomosulcate pollen from the Lower Eocene of Saudi Arabia is compared erroneously with *Roscheria melanochaetes* pollen (Srivastava & Binda 1991).
Notes: *Roscheria* is the only member of the Verschaffeltiinae to possess a crownshaft. Nevertheless, the inflorescences are interfoliar, at least at anthesis.
Additional figures: Fig. 2.8k.

173. VERSCHAFFELTIA

Spectacular moderate solitary spiny stilt-rooted palm endemic to Seychelles, lacking a crownshaft and with ± undivided leaves with praemorse margins; infloresncnces have long peduncles.

Verschaffeltia H. Wendl., Ill. Hort. 12 (misc.): 5 (1865). Type: **V. splendida** H. Wendl.

Commemorates Ambroise Colette Alexandre Verschaffelt (1825–1886), Belgian nurseryman and founder of the horticultural journal L'Illustration Horticole.

Moderate, solitary, spiny when young, becoming less spiny at maturity, pleonanthic, monoecious palm. *Stem* erect, becoming bare, conspicuously ringed with leaf scars, the juvenile very densely armed with long black spines, the adult more sparsely armed with rings of reflexed spines, the base of the trunk supported on a cone of robust stilt roots. *Leaves* large, neatly abscising, bifid, pinnately ribbed and lobed, irregularly split into approximate several-fold leaflets; sheath becoming open, not forming a crownshaft, very densely black-spiny in juveniles, unarmed in adults, the margins irregularly ligule-like; petiole short, glabrous, heavily armed as is the rachis in juveniles, unarmed in adults, adaxially deeply grooved, abaxially rounded; blade unsplit in juveniles, irregularly split in adults, the margins deeply lobed to up to 1/2 the blade depth into reduplicate segments, the segment tips irregularly praemorse, blade adaxially glabrous, abaxially with sparse, minute, dot-like scales and conspicuous, large, tattered ramenta along adaxial ribs, transverse veinlets obscure. *Inflorescences* solitary, interfoliar, branching to 2 (rarely 3) orders proximally, 1 order distally, protandrous; peduncle elongate, ± rounded in cross-section, densely tomentose like other inflorescence axes, winged at the base; prophyll inserted some distance from the base of the peduncle, very large, ± persistent, coriaceous, tubular, 2-keeled, the keels tending to become irregularly split or toothed, apically splitting to almost 1/2 the length, unarmed, bearing scattered scales; peduncular bract 1, inserted some distance above the prophyll, deciduous, similar to the prophyll but thinner and not 2-keeled; rachis shorter than the peduncle; rachis bracts minute, triangular, inconspicuous; first-order branches numerous, with a swollen base and short bare section; rachillae numerous, spreading, flexuous or rigid, somewhat sinuous, bearing spirally arranged superficial triads except at the very tip where bearing solitary or paired staminate flowers. *Staminate flowers* small, globular symmetrical; sepals 3, distinct, imbricate, rounded; petals 3, distinct, ± broadly triangular, valvate, ± twice as long as sepals; stamens 6, filaments distinct, ± fleshy, rather short,

Top right: ***Verschaffeltia splendida***, crown with infructescence, cultivated, Fairchild Tropical Botanic Garden, Florida. (Photo: C.E. Lewis)

Bottom right: ***Verschaffeltia splendida***, young plant, cultivated, Fairchild Tropical Botanic Garden, Florida. (Photo: C.E. Lewis)

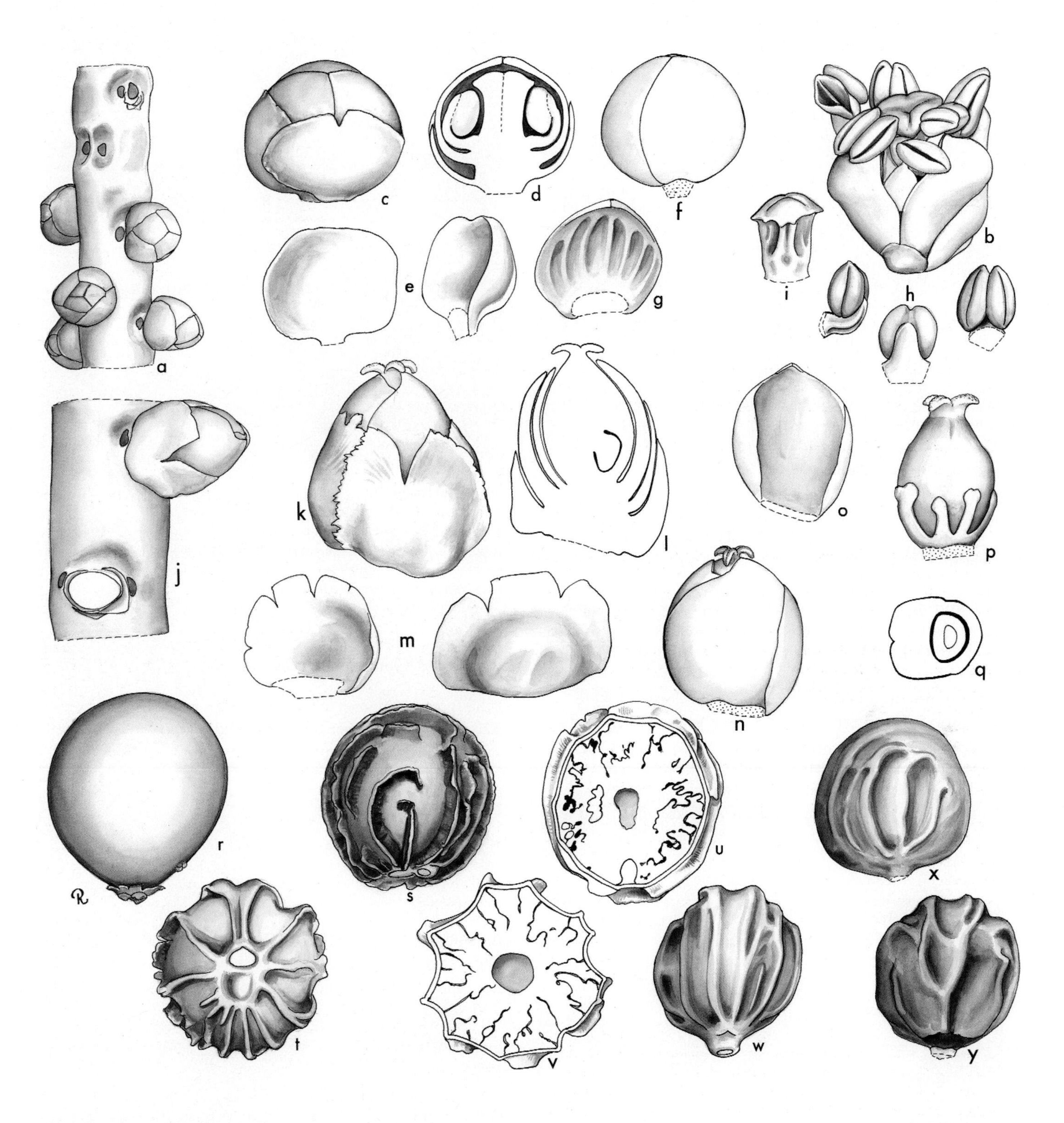

***Verschaffeltia*.** **a**, distal portion of rachilla with staminate buds and scars × 4 $^1/_2$; **b**, staminate flower at anthesis × 12; **c**, staminate bud × 12; **d**, staminate bud in vertical section × 12; **e**, staminate sepals in 2 views × 12; **f**, staminate bud, sepals removed × 12; **g**, staminate petal, interior view × 12; **h**, stamen in 3 views × 12; **i**, pistillode × 12; **j**, basal portion of rachilla with triads, staminate flowers removed (top), all flowers removed (bottom), to show scars and bracteoles × 7 $^1/_2$; **k**, pistillate flower × 12; **l**, pistillate flower in vertical section × 12; **m**, pistillate sepal in 2 views × 12; **n**, pistillate flower, sepals removed × 12; **o**, pistillate petal × 12; **p**, gynoecium and staminodes × 12; **q**, ovary in cross-section × 12; **r**, fruit × 1 $^1/_2$; **s**, endocarp × 1 $^1/_2$; **t**, base of endocarp and operculum × 1 $^1/_2$; **u**, endocarp in vertical section × 1 $^1/_2$; **v**, endocarp in cross-section × 1 $^1/_2$; **w, x, y**, seed in 3 views × 1 $^1/_2$. *Verschaffeltia splendida*: **a–q**, *Moore* 10098; **r**, *Squibb* 4963; **s–y**, *Moore* s.n. (Drawn by Marion Ruff Sheehan)

anthers rounded, medifixed, ± versatile, latrorse; pistillode large, truncate, 3-angled, trifid, about as long as the petals. *Pollen* ellipsoidal asymmetric; aperture a distal sulcus; ectexine tectate, perforate-rugulate, aperture margin similar or slightly finer; infratectum columellate; longest axis 41–47 µm [1/1]. *Pistillate flowers* larger than the staminate, ± globular; sepals 3, distinct, imbricate, ± broad, rounded, irregularly splitting; petals 3, distinct, broad, rounded, imbricate, with very short, triangular, valvate tips; staminodes 6, with flattened, ribbon-like filaments and wide, flattened tips; gynoecium ovoid, unilocular, uniovulate, stigmas 3, short, reflexed, ovule laterally attached, ?hemianatropous. *Fruit* 1-seeded, moderate, spherical, brownish green, perianth whorls persistent, stigmatic remains basal; epicarp smooth, roughened when dry, mesocarp with a crustose layer just below the epicarp, and relatively thin flesh beneath, fibres sparse attached to endocarp, endocarp thin, cartilaginous, bearing conspicuous, irregular flanges and ridges, and a rounded basal operculum. *Seed* conforming to the endocarp shape, conspicuously ridged when immature, slightly ridged at maturity, basally attached with rounded hilum, raphe branches sparse, anastomosing, endosperm deeply ruminate, with a small central hollow; embryo basal. *Germination* adjacent-ligular; eophyll bifid, spiny. *Cytology*: 2n = 36.

Distribution and ecology: A single species confined to the islands of Mahé, Silhouette and Praslin in the Seychelles; widespread in cultivation. *Verschaffeltia splendida* grows in relic forest on slopes between 300 and 600 m above sea level; more rarely, it occurs in river valleys below 300 m altitude. It seems to be confined to steep hillsides and precipitous ravines, where it usually occurs as solitary individuals, rarely in colonies.
Anatomy: Leaf, root (Tomlinson 1961), root (Seubert 1998a, 1998b) and fruit (Essig *et al.* 2001).
Relationships: There is strong support for a sister relationship between *Verschaffeltia* and *Roscheria* (Lewis & Doyle 2002, Lewis 2002, Loo *et al.* 2006, Norup *et al.* 2006, Baker *et al.* in review, in prep.).
Common names and uses: *Latanier latte.* Trunks have been used in house construction. The genus is also widespread in cultivation as an ornamental.
Taxonomic account: Bailey (1942).
Fossil record: No generic records found.
Notes: *Verschaffeltia* is distinguished by a nearly undivided leaf with praemorse margins and by large fruits that have a ridged endocarp.

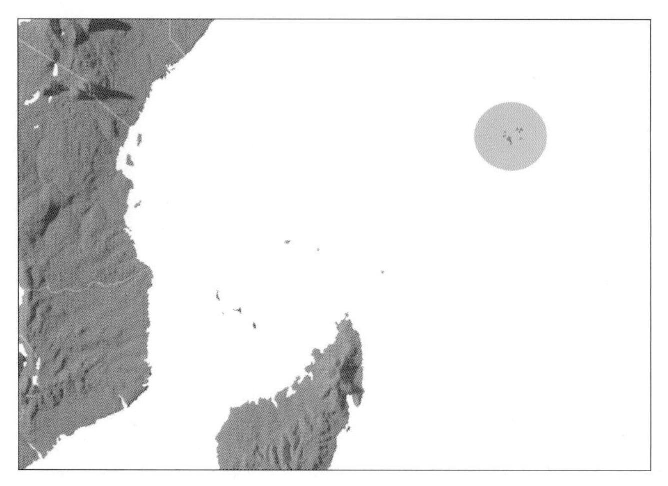

Distribution of ***Verschaffeltia***

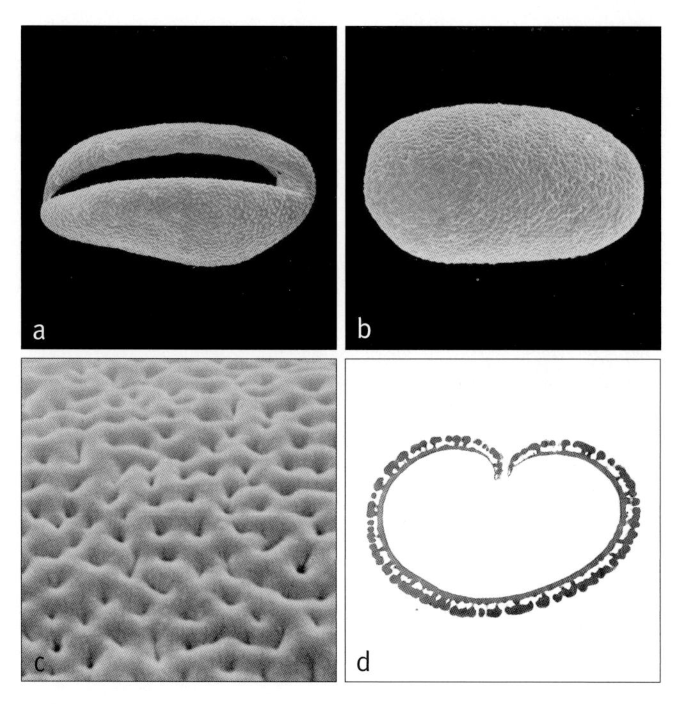

Above: ***Verschaffeltia***
a, monosulcate pollen grain, distal face SEM × 1000;
b, monosulcate pollen grain, proximal face SEM × 1000;
c, close-up, finely perforate-rugulate surface SEM × 8000;
d, ultrathin section through whole pollen grain, polar plane TEM × 1500.
Verschaffeltia splendida: **a–d**, *Horne* s.n.

Unplaced Genera in Areceae

It has proved impossible with currently available evidence to place the following genera in any of the subtribes recognised above. They are listed alphabetically.

174. BENTINCKIA

Moderate solitary pinnate-leaved palms endemic to the Western Ghats in India and the Nicobar Islands, distinctive in conspicuous crownshafts and highly branched inflorescences with rachillae bearing flowers in laterally compressed pits.

Bentinckia Berry in Roxb., Fl. ind. ed. 1832, 3: 621 (1832). Type: **B. condapanna** Berry.
Keppleria Mart. ex Endl., Gen. pl. 251 (1837). Type not designated.

Commemorates Lord William Henry Cavendish Bentinck (1774–1839), Governor-General of India 1828–1835.

Moderate, solitary, unarmed, pleonanthic, monoecious palms. *Stem* erect, moderately robust, brown, leaf scars clearly defined, close. *Leaves* pinnate, somewhat arching to spreading, becoming pendulous, neatly abscising; sheaths thick, striate, ± glabrous, tubular, forming a conspicuous crownshaft; petiole very short, stout, adaxially channelled, abaxially rounded; rachis elongate, angled adaxially, rounded abaxially; leaflets single-fold (basal leaflets sometimes united), lanceolate, acute or acuminate, tips bifid, both surfaces with small brown scales, long, pale ramenta near the base adaxially, along ribs abaxially, midrib raised adaxially, transverse veinlets not evident. *Inflorescences* infrafoliar, branched to 3 orders basally, fewer distally, branches somewhat pendulous at anthesis; peduncle very short, dorsiventrally flattened; prophyll inserted close to base of peduncle, tubular, splitting abaxially, caducous, chartaceous, tomentose, rather wide, 2-keeled laterally, tapering slightly to a blunt tip; peduncular bract inserted close to prophyll, similar but beaked and lacking keels, also caducous; rachis longer than the peduncle, bearing rather distant, spirally arranged, short, sometimes pointed bracts subtending branches and rachillae, glabrous except for a dense tuft of short hairs in bract axils; rachillae rather stiff, moderate, tapering, bearing spirally arranged, low, rounded bracts subtending triads of flowers nearly throughout, a few paired or solitary staminate flowers distally; flowers borne in vertical, laterally compressed pits, inner surfaces of pits densely hairy; floral bracteoles about equal, shallow, rounded. *Staminate flowers* slightly asymmetrical, borne on hairy pedicels; sepals 3, distinct, scarcely imbricate, narrow, ± acute, membranous; petals 3, asymmetrical and angled, ± strongly nerved; stamens 6, those opposite the sepals usually shorter than those opposite the petals, filaments awl-shaped, inflexed at the apex in bud, inflexed portion very slender, anthers elliptic to oblong or nearly quadrate, basifixed, latrorse, the connective very short; pistillode, in bud as long as the stamens, ovoid with expanded capitate tip when fresh. *Pollen* ellipsoidal asymmetric, occasionally lozenge-shaped or pyriform; aperture a distal sulcus; ectexine tectate, perforate and micro-channelled or finely perforate-rugulate, aperture margin similar or slightly finer; infratectum columellate; longest axis 23–44 µm [2/2]. *Pistillate flowers* ± symmetrical; sepals 3, distinct, imbricate, ± rounded apically, nearly as long as the petals in bud, glumaceous; petals 3, broadly imbricate with very briefly valvate tips; staminodes 3–6, awl-shaped or narrowly deltoid; gynoecium ellipsoid, asymmetrical, unilocular but vestigial locules evident, uniovulate, stigmas 3, recurved, papillose, ovule probably hemianatropous, pendulous. *Fruit*

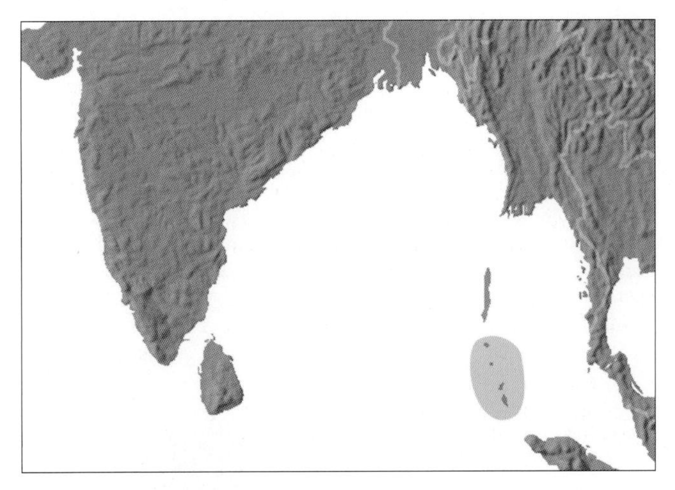

Distribution of ***Bentinckia***

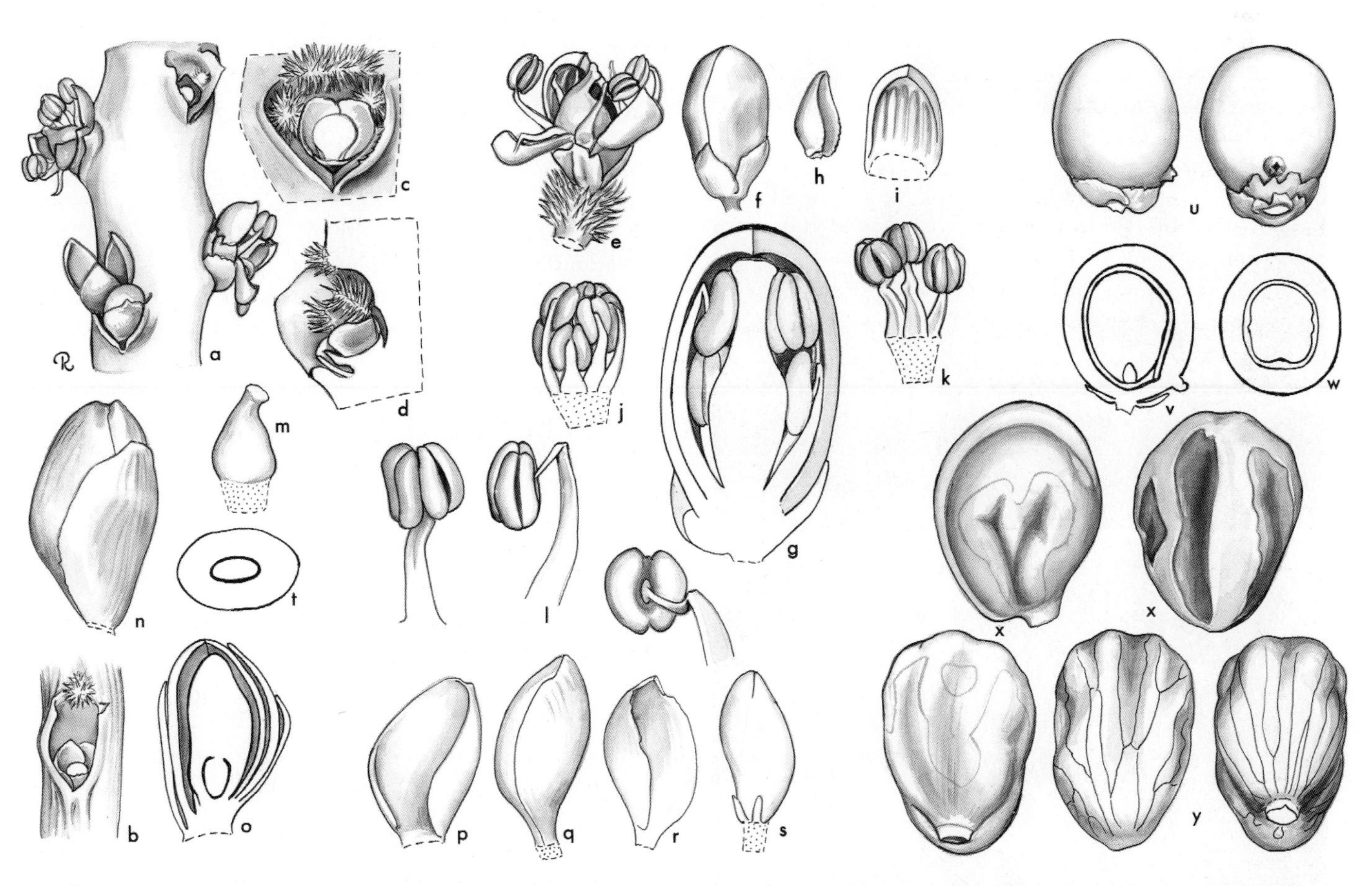

Bentinckia. **a**, portion of rachilla with staminate flowers in pits × 3; **b**, pit, flowers removed to show bracts × 3; **c**, pit, expanded × 3; **d**, pit in vertical section × 3; **e**, staminate flower and pedicel × 6; **f**, staminate bud × 6; **g**, staminate bud in vertical section × 12; **h**, staminate sepal × 6; **i**, staminate petal, interior view × 6; **j**, androecium × 6; **k**, portion of androecium, interior view × 6; **l**, stamen in 3 views × 12; **m**, pistillode × 6; **n**, pistillate bud × 6; **o**, pistillate bud in vertical section × 6; **p**, pistillate sepal × 6; **q**, pistillate flower, sepals removed × 6; **r**, pistillate petal × 6; **s**, gynoecium and staminodes × 6; **t**, ovary in cross-section × 12; **u**, fruit × 1 1/2; **v**, fruit in vertical section × 1 1/2; **w**, fruit in cross-section × 1 1/2; **x**, endocarp in 3 views × 3; **y**, seed in 2 views × 3. *Bentinckia nicobarica*: **a–m** and **u–y**, *Moore* 6097; **n–t**, *Bailey* 582. (Drawn by Marion Ruff Sheehan)

globose-obovoid, black or purplish at maturity, stigmatic remains near the base in lower $^1/_4$; epicarp smooth but drying dimpled, mesocarp fleshy with sclerosomes, the principal fibres 4, 1 short from base to stigmatic remains, 1 looped over the endocarp and 2 laterally branched and anastomosing toward the apex, endocarp operculate, rather thick, less fragile than in most genera, grooved abaxially from operculum to apex and laterally, attached directly to the operculum. *Seed* shining brown, conspicuously grooved abaxially and laterally, raphe branches ascending adaxially, arched over the seed and laterally, anastomosing adaxially, hilum rounded; endosperm homogeneous; embryo basal. *Germination* adjacent ligular; eophyll bifid. *Cytology*: 2n = 32.

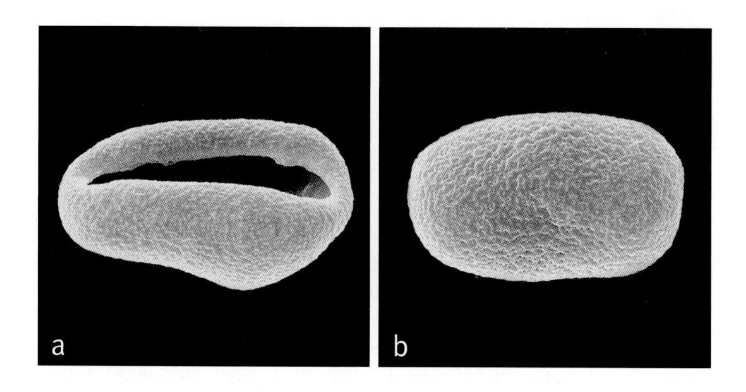

Above: ***Bentinckia***
a, monosulcate pollen grain, distal face SEM × 1500;
b, monosulcate pollen grain, proximal face SEM × 1500.
Bentinckia condapanna: **a** & **b**, *Wright* 8601.

Top left: ***Bentinckia nicobarica***, young plant, cultivated, Fairchild Tropical Botanic Garden, Florida. (Photo: C.E. Lewis)

Bottom left: ***Bentinckia nicobarica***, inflorescences, cultivated, Fairchild Tropical Botanic Garden, Florida. (Photo: C.E. Lewis)

Distribution and ecology: Two species, one in Tranvancore, India, and one on the Nicobar Islands. *Bentinckia condapanna* grows at 1000–1400 m at the edge of high peaks in the Travancore hills, where it is threatened by forest clearance and the grazing of young foliage by elephants. *Bentinckia nicobarica* apparently grows at somewhat lower elevations along with *Areca catechu*, *Pinanga manii* and *Rhopaloblaste augusta*.
Anatomy: Leaf (Tomlinson 1961), root (Seubert 1998a, 1998b) and fruit (Essig *et al*. 1999).
Relationships: The monophyly of *Bentinckia* has not been tested. In one study, the genus resolves as sister to *Clinostigma* with moderate support (Baker *et al*. in review).
Common names and uses: Bentinckia palm. The inflorescence of *Bentinckia condapanna* has been used in religious ceremonies.
Taxonomic account: No recent treatment exists. See Gamble (1935).
Fossil record: No generic records found.
Notes: The rachillae of *Bentinckia* have unusual pits out of which the staminate flowers are projected on short hairy pedicels. The endocarp is irregularly ridged and grooved.

175. CLINOSTIGMA

Elegant tall pinnate-leaved palms of the western Pacific Islands with conspicuous crownshafts and inflorescences with short bulbous peduncles.

Clinostigma H. Wendl., Bonplandia 10: 196 (1862). Type: **C. samoense** H. Wendl.

Exorrhiza Becc., Ann. Jard. Bot. Buitenzorg 2: 128 (1885). Type: *E. wendlandiana* Becc.

Bentinckiopsis Becc., Palme Nuova Caledonia 45 (1920); and Webbia 5: 113 (1921). Lectotype: *B. carolinensis* (Becc.) Becc. (*Cyphokentia carolinensis* Becc.) (see Beccari & Pichi-Sermolli 1955). (= *Clinostigma carolinense* [Becc.] H.E. Moore & Fosberg).

Clinostigmopsis Becc. in Martelli, Atti Soc. Tosc. Sci. Nat. Pisa Mem. 44: 161 (1934). Lectotype: *C. thurstonii* (Becc.) Becc., (= *Clinostigma thurstonii* Becc.) (see Beccari & Pichi-Sermolli 1955).

Klinein — *to bend,* stigma — *mark or, in botany, the point of the gynoecium that receives pollen, perhaps referring to the eccentrically apical stigmatic remains in the fruit.*

Tall, robust, solitary, unarmed, pleonanthic, monoecious palms. *Stem* erect, often longitudinally fissured, densely and conspicuously ringed with leaf scars, new internodes glaucous, sometimes with prominent prickly stilt roots. *Leaves* pinnate; sheaths tubular, forming a prominent crownshaft, ± glaucous; petiole mostly short, concave adaxially, rounded abaxially, glaucous; rachis flat adaxially, rounded abaxially, glaucous when young; leaflets single-fold, regularly arranged, ± arched to pendulous from the rachis, with a prominent elevated midrib and 1–2 secondary ribs on each side above, all the veins rather densely covered abaxially with minute, pale-margined, brown-centred, membranous scales and often with large ramenta on the midrib below, margins nearly parallel, tapering gradually to an acuminate apex, this often frayed and bifid in age. *Inflorescences* infrafoliar, branched to 3 orders basally, to 2–1 orders distally, or to 1 order only; peduncle short, wide, flat, glaucous, frequently becoming swollen; prophyll tubular, thin, 2-keeled, completely encircling the peduncle, enclosing the thin, beaked peduncular bract; both caducous; rachis longer than the peduncle, tapering, glaucous when young, bearing spirally arranged, conspicuous, pointed bracts each subtending a first-order branch; first-order branches with a short bare portion basally, bearing very small pointed, spirally arranged bracts subtending rachillae; rachillae long, very slender, bearing rather close, spirally arranged acute bracts subtending triads of flowers nearly throughout or only in the lower $^1/_3$ to $^1/_2$, staminate flowers paired or solitary toward the apex, the axis somewhat impressed above the triad; floral bracteoles unequal in size, the third largest, often exceeding the triad bract. *Staminate flowers* markedly asymmetrical, one

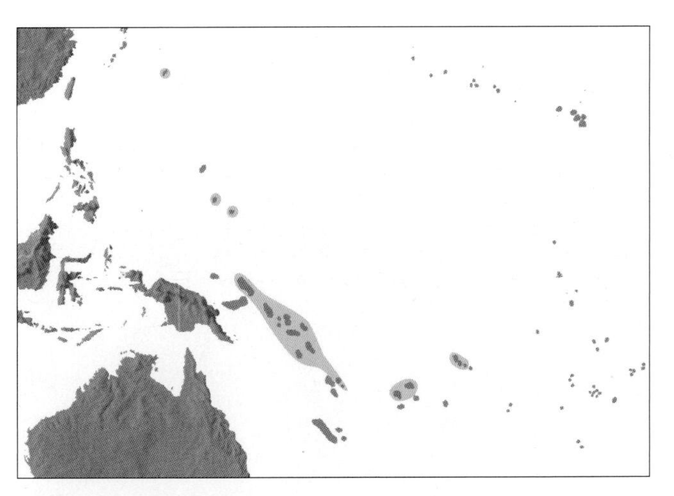

Distribution of ***Clinostigma***

ebracteolate, the other subtended by a low spreading bracteole; sepals 3, distinct, basally imbricate, acute, laterally compressed and dorsally keeled toward the base; petals 3, distinct, valvate, asymmetrical, acute, prominently lined centrally but the margins ± veinless and pale when dry; stamens 6, filaments distinct, awl-shaped, equal, or those opposite the sepals inserted lower than those opposite the petals, the filaments inflexed at the apex in bud, the anthers versatile at anthesis, acute to emarginate or deeply bifid at the apex, equally or unequally bifid at the base; pistillode short, broadly conical, trifid. *Pollen* grains ellipsoidal asymmetric, occasionally oblate triangular; aperture a distal sulcus, less frequently a trichotomosulcus; ectexine tectate, perforate, perforate and micro-channelled or finely perforate-rugulate, aperture margin similar or slightly finer; infratectum columellate; longest axis 38–60 µm [6/13]. *Pistillate flowers* symmetrical,

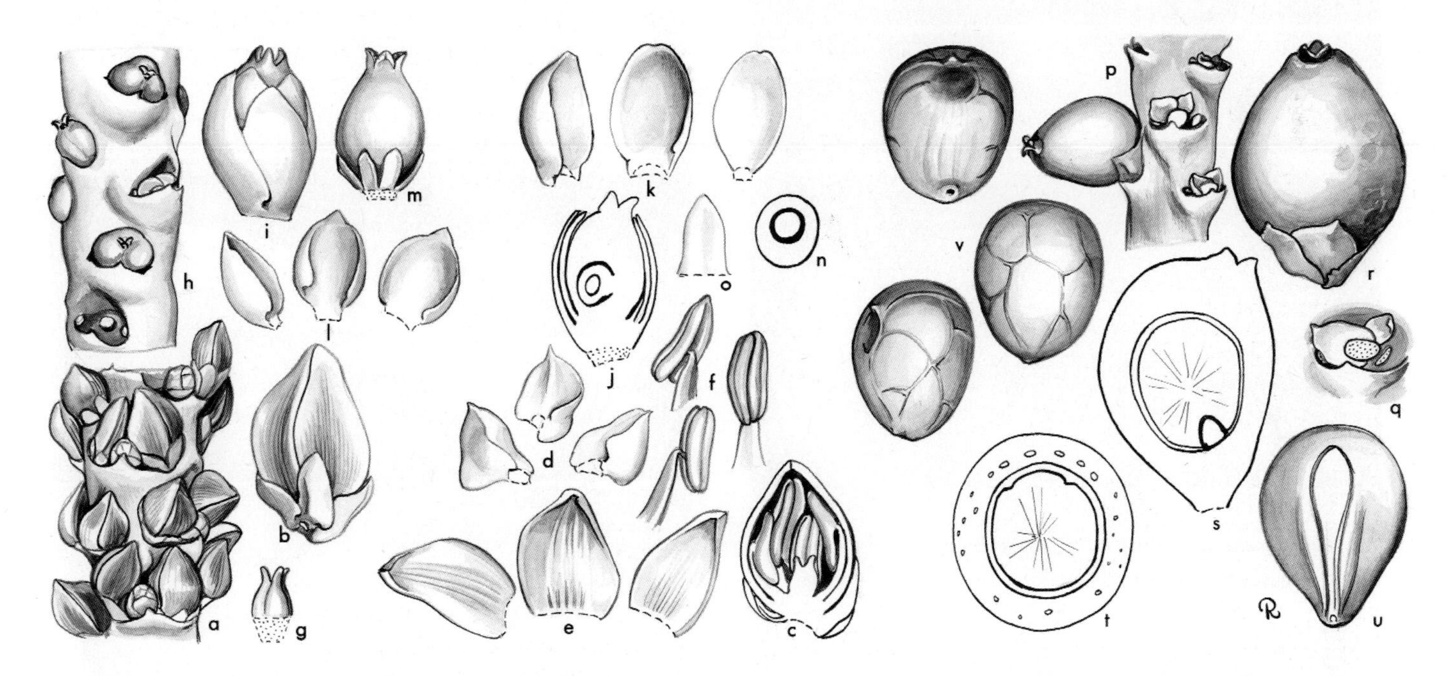

Clinostigma. **a**, portion of rachilla with triads × 3; **b**, staminate bud × 6; **c**, staminate bud in vertical section × 6; **d**, staminate sepals × 6; **e**, staminate petals × 6; **f**, stamen in 3 views × 6; **g**, pistillode × 6; **h**, portion of rachilla in young fruit × 3; **i**, pistillate flower × 6; **j**, pistillate flower in vertical section × 6; **k**, pistillate sepals × 6; **l**, pistillate petals × 6; **m**, gynoecium and staminodes × 6; **n**, ovary in cross-section × 6; **o**, staminode × 12; **p**, portion of rachilla in fruit × 3; **q**, triad, fruit removed to show floral scars and bracteoles × 6; **r**, fruit × 6; **s**, fruit in vertical section × 6; **t**, fruit in cross-section × 6; **u**, endocarp × 6; **v**, seed in 3 views × 6. *Clinostigma exorrhizum*: all from *Moore & Koroiveibau* 9355. (Drawn by Marion Ruff Sheehan)

Top: ***Clinostigma warburgii***, habit, Western Samoa. (Photo: D.R. Hodel)

Above: ***Clinostigma savaiiense***, infructescence, Samoa. (Photo: D.R. Hodel)

ovoid; sepals 3, distinct, ± boat-shaped, broadly imbricate; petals 3, distinct, slightly longer than the sepals, broadly imbricate with very briefly valvate apices; staminodes (5–)6, membranous, tooth-like; gynoecium ovoid, unilocular, uniovulate, with 3, short, recurved stigmas, the ovule laterally or apically attached, form unknown. *Fruit* ovoid to ellipsoidal and terete to laterally compressed, red when mature, the stigmatic remains eccentrically apical to lateral or rarely basal; epicarp smooth but often drying granulose-wrinkled over fibres, mesocarp with prominent, sometimes greatly thickened (*Clinostigma ponapense*), longitudinal fibres and a thin layer of red sclerosomes (*C. exorrhizum*) over a thin, crustaceous, fragile endocarp, this neither angled nor sculptured except for a thickened apical cap opposite the hilum, tapered downward to a narrow operculum opposite the embryo. *Seed* obovoid to ellipsoidal, sometimes somewhat compressed laterally, hilum rounded to elongate, basal or extending along one side of the seed, raphe branches few, unbranched or loosely anastomosing; endosperm homogeneous; embryo basal. *Germination* adjacent-ligular; eophyll bifid or entire. *Cytology*: 2n = 32.

Distribution and ecology: About 11 species from the Bonin and Caroline Islands to Samoa, Fiji Islands, Vanuatu (Banks Group), the Solomon Islands and New Ireland. The species are found in montane rain forest, usually in dense forests on crests and ridges; *Clinostigma harlandii* is not known below 1000 m and is almost perpetually in clouds or mist.

Anatomy: Fruit (Essig *et al.* 1999).

Relationships: *Clinostigma* is a strongly supported monophyletic genus (Baker *et al.* in prep.). *Clinostigma* resolves as sister to *Cyrtostachys* with moderate support in several studies (Lewis & Doyle 2002, Norup *et al.* 2006, Baker *et al.* in prep.), but also as sister to *Bentinckia* with moderate support elsewhere (Baker *et al.* in review).

Common names and uses: *Niuniu* (Fiji). Split stems are used for house building. All species are very elegant and would make fine ornamentals.

Taxonomic accounts: Moore (1969b, 1979) and Dransfield (1982a).

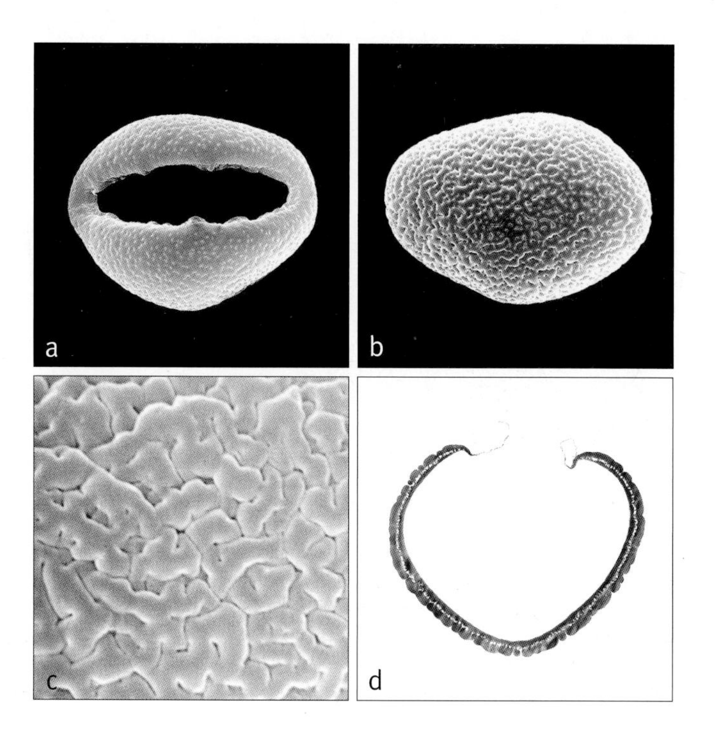

Right: ***Clinostigma***
a, monosulcate pollen grain, distal face SEM × 1500;
b, monosulcate pollen grain, proximal face SEM × 1500;
c, close-up, perforate-rugulate pollen surface × 8000;
d, ultrathin section through whole pollen grain, polar plane TEM × 2000.
Clinostigma exorrhizum: **a**, **b** & **d**, *Seemann* 660; *C. samoense*: **c**, *Moore et al.* 10538.

Fossil record: No generic records found.

Notes: *Clinostigma* occurs in a great arc through the western Pacific. The staminate flowers are asymmetrical and the base of the peduncle is usually grossly swollen.

176. CYRTOSTACHYS

Solitary or clustered, moderate to very robust pinnate-leaved palms of West Malesia and New Guinea and the Solomon Islands, with conspicuous crownshafts, inflorescences with short peduncles and flowers generally borne in pits; fruits have apical stigmatic remains.

Cyrtostachys Blume, Bull. Sc. Phys. Nat. Néerl. 1: 66 (1838). Type: **C. renda** Blume.

Cyrtos — *curved*, stachys — *ear of grain or spike, perhaps referring to the curved rachillae*.

Solitary or clustered, moderate to tall, unarmed, pleonanthic, monoecious palms. *Stems* erect, bare, conspicuously ringed with leaf scars, often bearing a mass of adventitious roots at the base, where clustering, the clump rather close, or more rarely diffusely spreading by stolons. *Leaves* pinnate, neatly abscising; sheaths tubular, forming a well-defined crownshaft, brilliantly orange-red coloured in 1 species (*Cyrtostachys renda*), glabrous or scaly; petiole short to long, adaxially channelled or flattened, abaxially rounded or angled, glabrous or scaly; rachis like the petiole but angled adaxially; leaflets always single-fold, acute or acuminate, regularly arranged, often stiff, sometimes ascending, sometimes slightly paler beneath, ± glabrous adaxially, abaxially often with ramenta along the midvein and sometimes minutely dotted between the veins, transverse veinlets conspicuous or obscure. *Inflorescence* apparently protandrous, infrafoliar, highly branched to 3 orders, rather diffuse and spreading; peduncle usually very short, ± oval in cross-section; prophyll enclosing the inflorescence until leaf fall, borne just above the winged base of the peduncle, tubular, 2-keeled, ± lanceolate, with winged margins, splitting, soon caducous; peduncular bract borne just above the prophyll, completely enclosing the inflorescence, splitting longitudinally like the prophyll, caducous; subsequent bracts very inconspicuous, incomplete, low, triangular; rachis longer than the peduncle; first-order branches robust, spreading, with a short bare portion at the base, then branching to produce diverging rachillae or second-order branches; second-order branches, when not rachillae, also with short bare portion and then branching to produce rachillae; rachillae elongate, cylindrical, rather robust, glabrous, papillose, minutely roughened or indumentose, often brightly coloured, expanding long before anthesis; rachilla bracts low, triangular, spirally arranged, rather crowded, each partially enclosing a shallow pit bearing a triad of flowers, triads borne throughout the length of the rachillae; floral bracteoles membranous, very small and inconspicuous. *Staminate flowers* with 3, distinct, imbricate, broad, strongly keeled sepals with minutely toothed margins (?always); petals

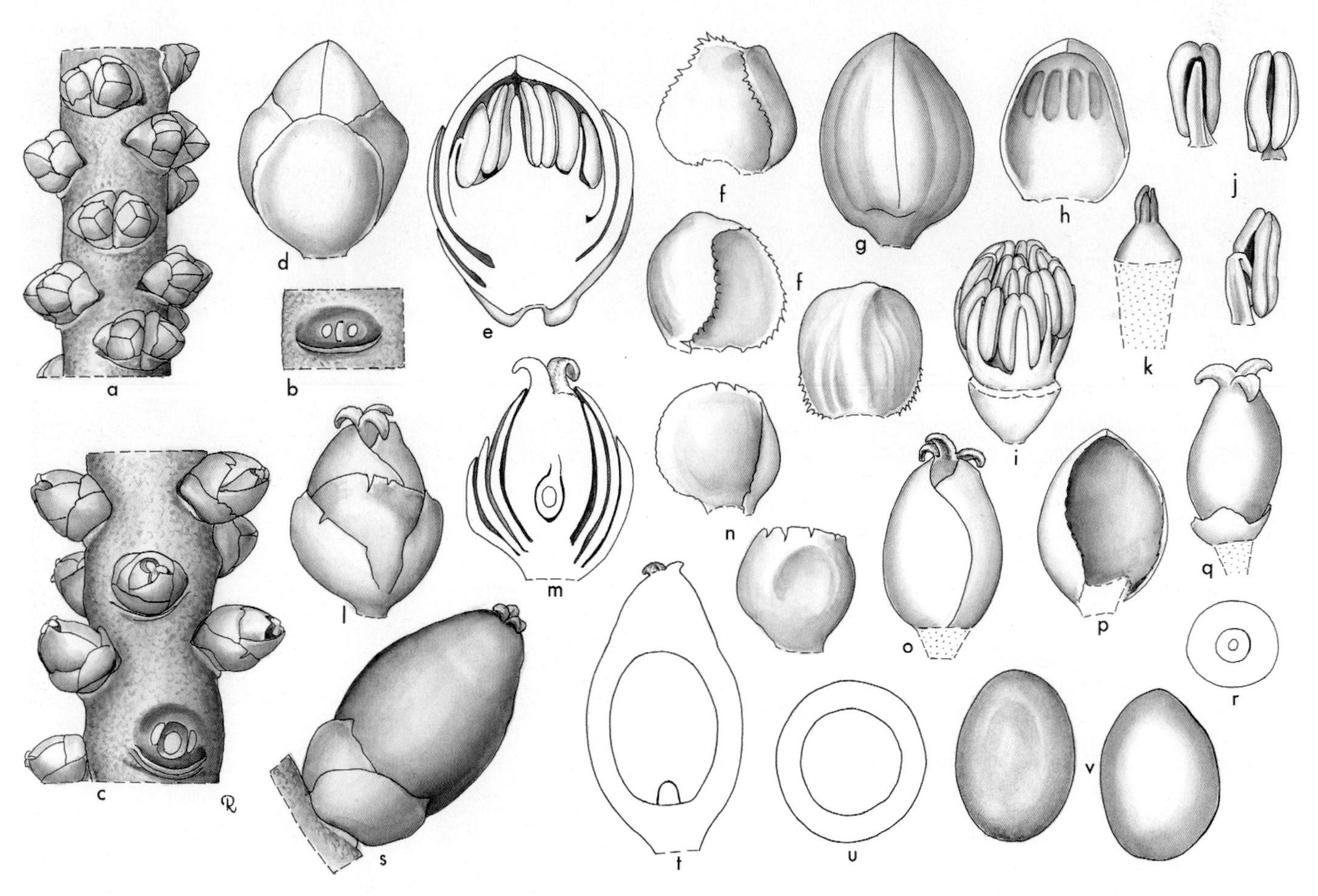

Cyrtostachys. **a**, portion of rachilla with paired staminate flowers in pit × 3; **b**, detail of pit, flowers removed × 3; **c**, portion of rachilla in pistillate flower × 3; **d**, staminate bud × 9; **e**, staminate bud in vertical section × 12; **f**, staminate sepals × 9; **g**, staminate bud, sepals removed × 9; **h**, staminate petal, interior view × 9; **i**, androecium × 9; **j**, stamen in 3 views × 12; **k**, pistillode × 12; **l**, pistillate flower × 6; **m**, pistillate flower in vertical section × 6; **n**, pistillate sepals in 2 views × 6; **o**, pistillate flower, sepals removed × 6; **p**, pistillate petal × 6; **q**, gynoecium and staminodes × 6; **r**, ovary in cross-section × 6; **s**, fruit × 4; **t**, fruit in vertical section × 4; **u**, fruit in cross-section × 4; **v**, seed in 2 views × 4. *Cyrtostachys renda*: all from *Moore & Meijer* 9157. (Drawn by Marion Ruff Sheehan)

Top left: ***Cyrtostachys loriae***, habit, Papua. (Photo: W.J. Baker)
Top right: ***Cyrtostachys elegans***, inflorescence, Papua. (Photo: W.J. Baker)
Bottom left: ***Cyrtostachys loriae***, part of rachilla with pistillate flower buds, Papua New Guinea. (Photo: W.J. Baker)
Bottom right: ***Cyrtostachys loriae***, part of rachilla with ripe fruit, Papua New Guinea. (Photo: W.J. Baker)

about twice as long as sepals, united at the very base to ca. $^{1}/_{3}$ their length, distally with 3 triangular, valvate tips; stamens 9–15, the filaments awl-shaped, connate basally, apically inflexed in bud, anthers apically and basally slightly bilobed, dorsifixed, latrorse; pistillode almost as long as filaments, narrow, elongate, trifid. *Pollen* ellipsoidal, less frequently, oblate triangular, symmetric or slightly asymmetric; aperture a distal sulcus or trichotomosulcus; ectexine tectate, perforate rugulate, in some species with verrucate or gemmate supratectal processes, aperture margin similar; infratectum columellate; longest axis 27–56 µm; post-meiotic tetrads tetrahedral [2/11]. *Pistillate flowers* about the same size as or slightly larger than the staminate; sepals 3, distinct, rounded, imbricate, the margins minutely toothed (?always); petals 3, slightly larger than the sepals, distinct, imbricate proximally, asymmetrical, rounded with short triangular valvate tips; staminodal ring membranous, very low, bearing short truncate or irregularly triangular teeth; gynoecium unilocular, ellipsoidal with 3 short recurved stigmas, ovule pendulous from the apex of the locule, form unknown. *Fruit* 1-seeded, broad to narrow-ellipsoidal, usually black, the perianth whorls persistent, stigmatic remains apical; epicarp smooth, contrasting with the rachilla, mesocarp thin, oily, with abundant longitudinal fibre bundles, endocarp thin, closely adhering to the seed. *Seed* globose or ellipsoidal, apically attached, the hilum orbicular, endosperm homogeneous; embryo basal. *Germination* adjacent-ligular; eophyll bifid with narrow lobes. *Cytology*: 2n = 32.

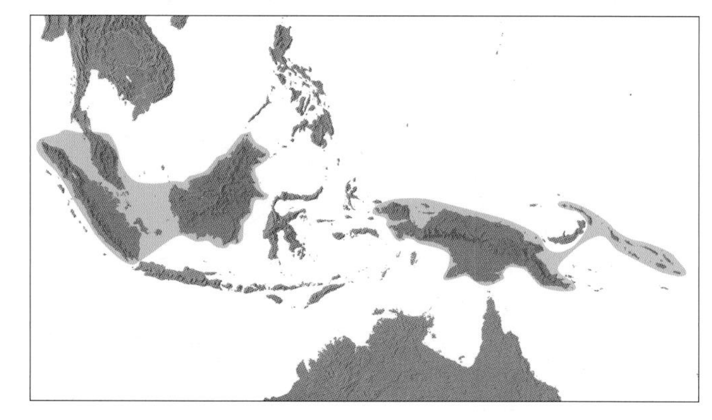

Distribution of ***Cyrtostachys***

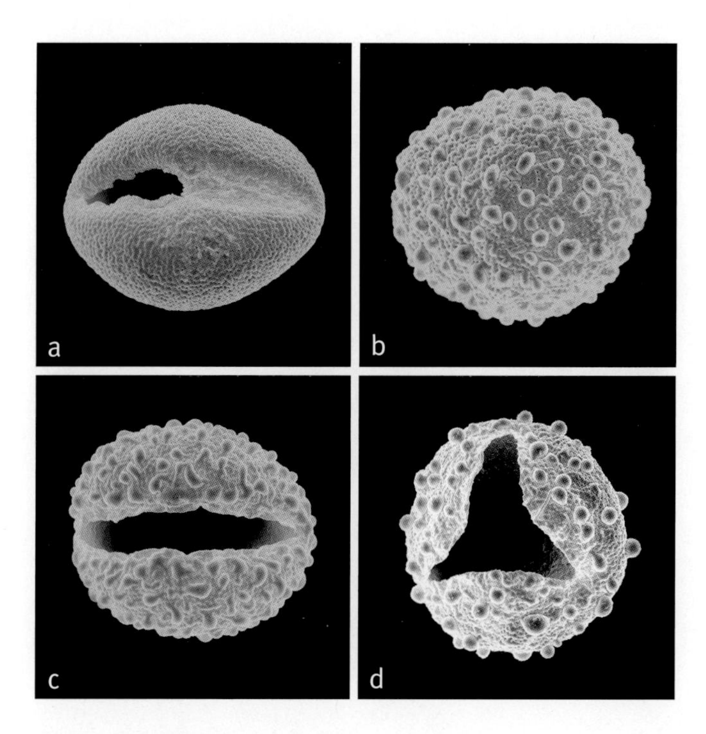

Above: ***Cyrtostachys***
a, monosulcate pollen grain, distal face SEM × 2000;
b, monosulcate pollen grain, proximal face SEM × 2000;
c, monosulcate pollen grain, distal face SEM × 2000;
d, trichotomosulcate pollen grain, distal face SEM × 2000.
Cyrtostachys renda: **a**, *Keith* 2491; *C. loriae*: **b** & **c**, *Darbyshire* 867; **d**, *Whitmore* 3945.

Distribution and ecology: Seven species: *Cyrtostachys renda* in the Malay Peninsula, Sumatra and Borneo, and very widely cultivated; all other species in New Guinea and Melanesia. *Cyrtostachys renda* is exclusive to peat swamp forest, usually near the coast, where it can be a conspicuous component of the vegetation; other species may be found in lowland rain forest and at altitudes of up to about 500 m.
Anatomy: Leaf (Tomlinson 1961) and root (Seubert 1998a, 1998b).
Relationships: *Cyrtostachys* is strongly supported as monophyletic (Heatubun *et al.* in press) *Cyrtostachys* resolves as sister to *Clinostigma* with moderate support in a number of studies (Lewis & Doyle 2002, Norup *et al.* 2006, Baker *et al.* in prep.).
Common names and uses: Sealing wax palm, *pinang rajah* (*Cyrtostachys renda*); this species is a commercially important ornamental. Locally, its trunk is also used as a source of laths for supporting *Nypa* leaf thatch. The larger New Guinea species may supply timber.
Taxonomic account: Heatubun *et al.* in press.
Fossil record: Pollen referable to *Cyrtostachys* has been recorded from Upper Miocene deposits in Borneo (Muller 1972); furthermore, Morley (2000) found *Cyrtostachys* pollen from a Middle Miocene coal from Berakas, Brunei. Unfortunately, these fossil records have no supporting illustration of the pollen.

Notes: The relationships of *Cyrtostachys* are not obvious. Leaves of *Cyrtostachys* can be distinguished from those of other arecoid palms by the sinuous epidermal cell walls and fibrous hypodermis (Tomlinson 1961).
Additional figures: Fig. 2.6b.

177. DICTYOSPERMA

Beautiful moderate solitary pinnate-leaved palm native to Mascarene Islands, with conspicuous crownshaft, and inflorescence shaped like a horse's tail.

Dictyosperma H. Wendl. & Drude, Linnaea 39: 181 (1875). Type: **D. album** (Bory) H. Wendl. & Drude ex Scheff. (*Areca alba* Bory).
Dicrosperma H. Wendl. & Drude ex Watson, Gard. Chron. series 2, 24: 362 (1885). (Type as above.)
Linoma O.F. Cook, J. Wash. Acad. Sci. 7: 123 (1917). Type: *L. alba* (Bory) O.F. Cook (*Areca alba* Bory = *Dictyosperma album* [Bory] H. Wendl. & Drude).

Diktyon — *net,* sperma — *seed, referring to the net-like raphe branches on the surface of the seed.*

Stout, solitary, unarmed, pleonanthic, monoecious palm. *Stem* erect, sometimes enlarged at the base, brown or grey, often vertically fissured. *Leaves* pinnate, spreading with arched, ± pendulous leaflets; sheath tubular, forming a prominent crownshaft, thinly to densely white, grey or brown

Above: ***Dictyosperma album***, habit, cultivated, Fairchild Tropical Botanic Garden, Florida. (Photo: C.E. Lewis)

tomentose; petiole very short, nearly flat to slightly rounded adaxially, abaxially rounded, tomentose marginally and adaxially, the scales with blackish centres and pale, tattered-twisted margins, deciduous, leaving dot-like scars; rachis adaxially angled with scales as on the petiole, abaxially rounded; leaflets lanceolate, acute, single-fold, subopposite, regularly arranged, abaxially with pale to dark brown, twisted, basifixed or medifixed membranous ramenta on the midrib, midrib conspicuous adaxially, transverse veinlets not evident. *Inflorescences* infrafoliar, protandrous, as many as 6 present below the crownshaft, erect in bud, branched to 1 order, branches stiff, ascending, ± unilateral; peduncle very short, dorsiventrally compressed, grey-brown tomentose; prophyll tubular, bluntly pointed, rather wide, shortly 2-keeled laterally, glabrous or with deciduous grey tomentum, splitting adaxially, caducous; peduncular bract like the prophyll, lacking keels, beaked, rather thin, fibrous; rachis longer than the peduncle, stout, tapering, ± densely covered with twisted hairs or glabrous, bearing acute or low rounded bracts subtending rachillae on 3 sides, lacking rachillae adaxially, at least toward the base; rachillae moderate, tapering, stiff, ascending, becoming

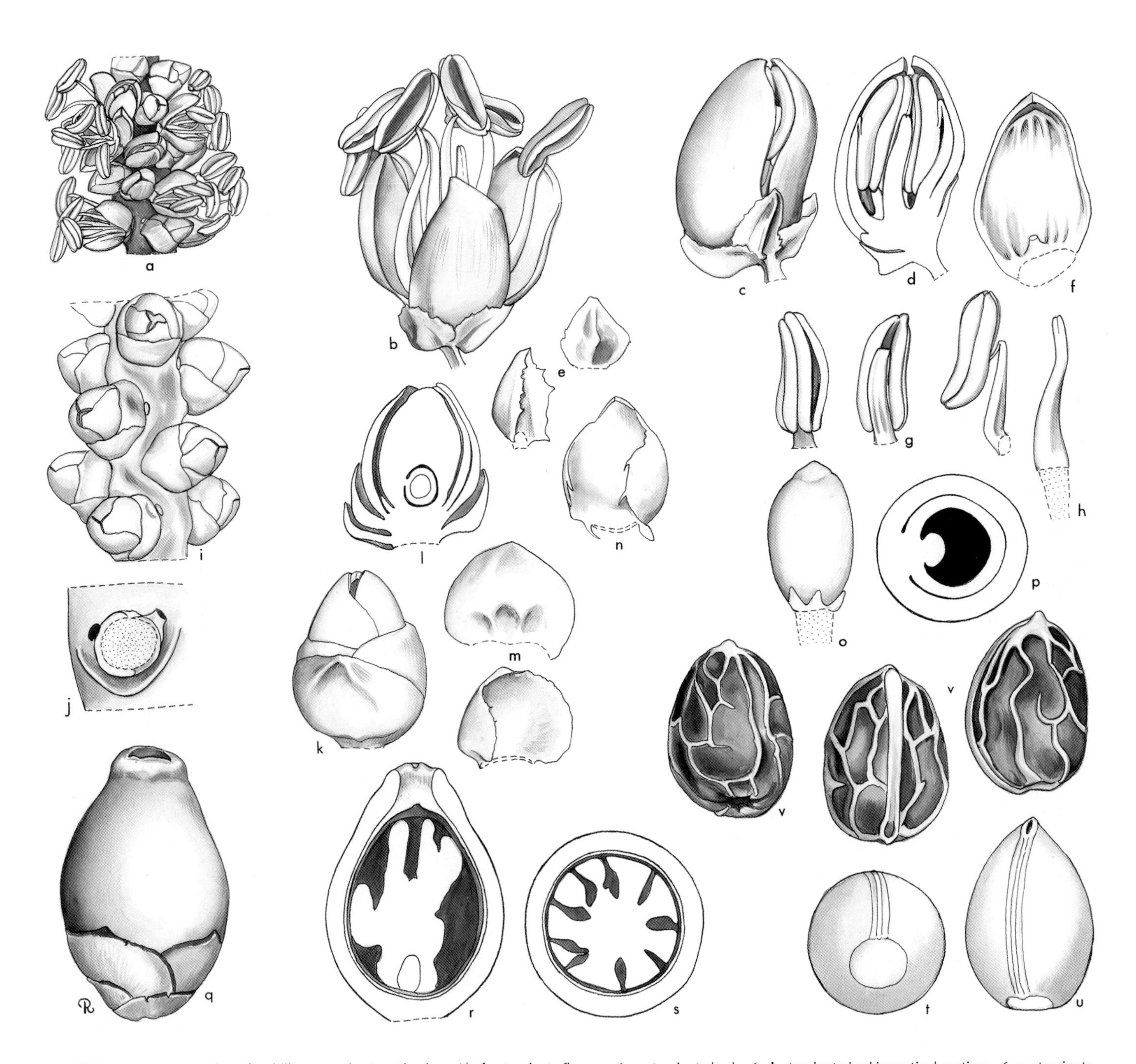

***Dictyosperma*.** **a**, portion of rachilla at staminate anthesis × $1^1/_2$; **b**, staminate flower × 6; **c**, staminate bud × 6; **d**, staminate bud in vertical section × 6; **e**, staminate sepal in 2 views × 6; **f**, staminate petal, interior view × 6; **g**, stamen in 3 views × 6; **h**, pistillode × 6; **i**, portion of rachilla at pistillate anthesis × 3; **j**, triad, flowers removed × 6; **k**, pistillate flower × 6; **l**, pistillate flower in vertical section × 6; **m**, pistillate sepals in 2 views × 6; **n**, pistillate petal, interior view × 6; **o**, gynoecium and staminodes × 6; **p**, ovary in cross-section × 12; **q**, fruit × 3; **r**, fruit in vertical section × 3; **s**, fruit in cross-section × 3; **t**, **u**, endocarp in 2 views showing operculum × 3; **v**, **w**, **x**, seed in 3 views × 3. *Dictyosperma album*: all from *Read* 831. (Drawn by Marion Ruff Sheehan)

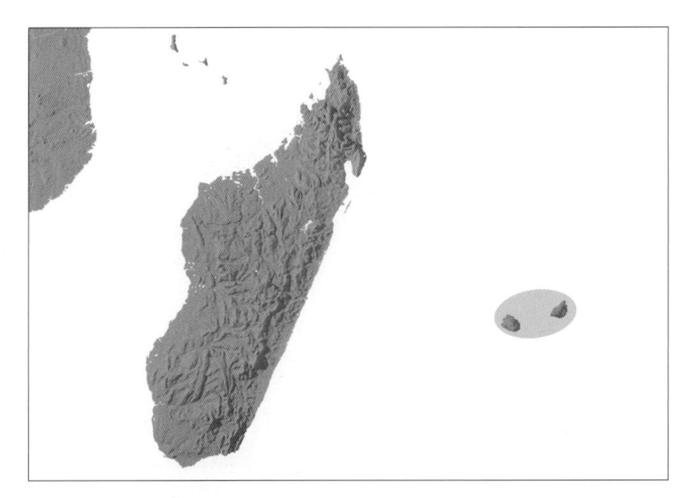

Distribution of ***Dictyosperma***

recurved, glabrous or hairy only at the base, bearing triads in the lower $^1/_2$–$^4/_5$, and paired staminate flowers distally; floral bracteoles low, flattened, not sepal-like. *Staminate flowers* sessile or briefly pedicellate, yellow to maroon, asymmetrical; sepals 3, distinct, slightly imbricate at the base only, acute, keeled and ± gibbous towards the base; petals 3, distinct, valvate, ovate, acute, with an outer layer of thick fibres and tannin cells; stamens 6, filaments stout, inflexed at the apex in bud, anthers dorsifixed near the middle, becoming versatile, linear-lanceolate, briefly bifid at the apex, more deeply bifid basally, the locules in bud separated by a very narrow sterile portion, latrorse; pistillode nearly as long as the stamens, tapered to a slender tip from a broad base. *Pollen* ellipsoidal symmetric or slightly asymmetric; aperture a distal sulcus; ectexine tectate, psilate and sparsely perforate (in some collections with small sparsely distributed club-shaped spinulae) aperture margin similar; infratectum columellate; longest axis 40–60 µm [1/1]. *Pistillate flowers* ovoid, smaller than the staminate; sepals 3, distinct, imbricate, broadly rounded; petals 3, distinct, broadly imbricate, very briefly valvate at the apex; staminodes 3, small, triangular, at one side of the gynoecium; gynoecium ovoid, unilocular, uniovulate, stigmas scarcely differentiated, not exserted or recurved, ovule large,

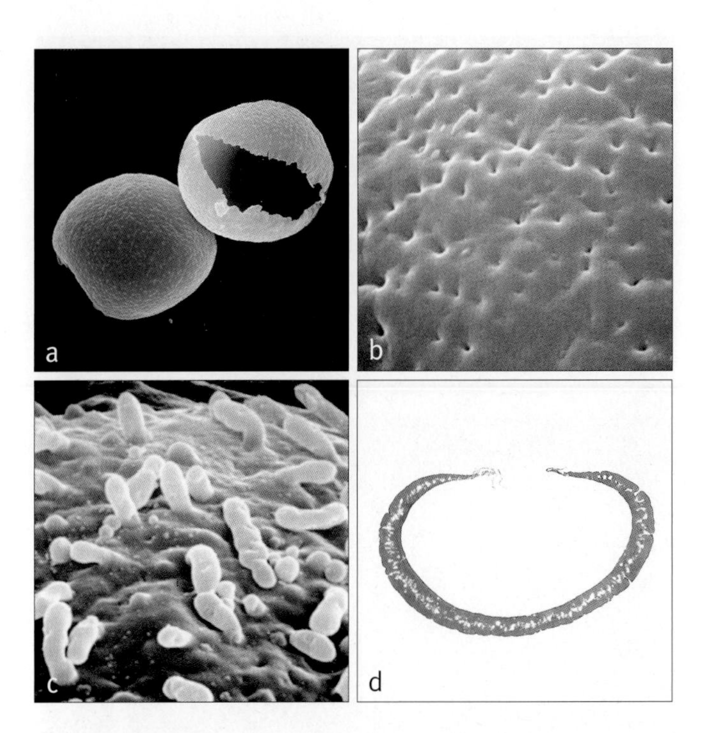

Above: ***Dictyosperma***
a, two monosulcate pollen grains: proximal face, bottom left; distal face, top right SEM × 600;
b, close-up, perforate pollen surface SEM x 8000;
c, close-up, small sparsely distributed club-shaped spinulae SEM × 8000;
d, ultrathin section through whole pollen grain, polar plane TEM × 1000.
Dictyosperma album: **a** & **b**, *Horne* 1882, **c**, *Bernardi* 14741, **d**, *Horne* 25.

Top right: ***Dictyosperma album***, inflorescences, cultivated, Montgomery Botanical Center, Florida. (Photo: J. Dransfield)

Bottom right: ***Dictyosperma album***, rachilla with open staminate flowers and buds, cultivated, Montgomery Botanical Center, Florida. (Photo: J. Dransfield)

prominently vascularised, attached laterally in upper part of locule, hemianatropous. *Fruit* ovoid or ovoid-ellipsoidal, black or purplish at maturity, stigmatic remains apical; epicarp smooth when fresh, wrinkled but not pebbled when dry, mesocarp with an external and an internal layer of elongate, vertically oriented parenchyma cells with some flat, thin, longitudinal fibres between the layers, cells of the inner layer longer and overlying a layer of contiguous thickened fibres and a thin layer of ± isodiametric tannin cells adherent to the endocarp, endocarp horny, with round basal operculum and scar of seed attachment extending the length of the adaxial side to the acute apex. *Seed* ovoid-ellipsoidal, acute, brown, with elongate hilum and only slightly anastomosing raphe branches descending from the apex, endosperm deeply ruminate; embryo basal. *Germination* adjacent-ligular; eophyll bifid. *Cytology*: 2n = 32.

Distribution and ecology: One species in the Mascarene Islands where it grows from sea level to 600 m or more, nearly extinct in the wild state, cultivated on Mauritius, Reunion and elsewhere.
Anatomy: Leaf (Tomlinson 1961) and fruit (Essig *et al.* 1999).
Relationships: *Dictyosperma* is moderately supported as sister to *Rhopaloblaste* (Norup *et al.* 2006, Baker *et al.* in review, in prep.).
Common names and uses: Princess palm, hurricane palm. Frequently grown as an ornamental.
Taxonomic accounts: Moore and Guého (1980, 1984).
Fossil record: No generic records found.
Notes: This is a very isolated genus in both distribution and relationships.
Additional figures: Glossary fig. 13.

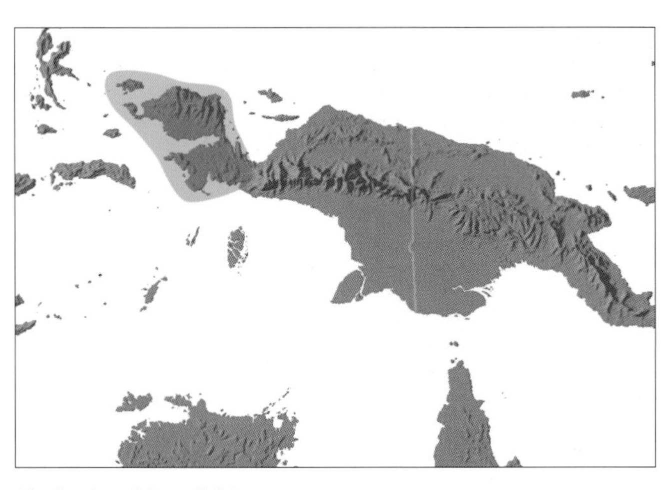

Distribution of ***Dransfieldia***

178. DRANSFIELDIA

Small to moderate, solitary or clustering pinnate-leaved palm from western New Guinea, with crownshaft, acute leaflets and inflorescence with the peduncular bract deciduous but the prophyll persistent.

Dransfieldia W.J. Baker & Zona, Syst. Bot. 31: 61 (2006). Type: **D. micrantha** (Becc.) W.J. Baker & Zona (*Ptychosperma micranthum* Becc.)

Honours John Dransfield, co-author of Genera Palmarum.

Small to moderate, clustering or solitary, unarmed, pleonanthic, monoecious, understory tree palm. *Stem* erect, slender, ringed with prominent leaf scars. *Leaves* pinnate, few in the crown; sheath strictly tubular, forming a well-defined crownshaft, glabrous adaxially, abaxially with sparse to dense indumentum of brown-black irregular scales of various sizes and brown to white, matted, fibrous scales; petiole present, slender, adaxially channelled, abaxially rounded, covered with indumentum as on the leaf sheath; rachis long, slender, adaxially forming a ridge, abaxially rounded, with indumentum as on the petiole; leaflets subopposite to alternate, arranged regularly, in one plane, single-fold, spreading, appearing corrugated due to the presence of conspicuous raised ridges on the adaxial surface of major veins, linear to narrowly elliptic, attenuating to a narrowly acute apex, sometimes with a few, widely separated, shallow indentations to one side of the apex, distal leaflets with apex acute and usually notched, apical leaflet pair not united at the base, with brown medifixed ramenta scattered on the abaxial surface of major veins and more numerous near the leaflet bases, with scales as on the rachis on both surfaces of leaflet base, minute white scales sparsely distributed throughout both surfaces visible in fresh material, transverse veinlets not evident. *Inflorescence* infrafoliar, branched to 2 (rarely 3) orders, protandrous, somewhat deflexed at anthesis, divaricate, with sparse to densely matted, brown, stellate scales throughout inflorescence axes, bracts and bracteoles; peduncle relatively slender, elongate, elliptical in cross-section; prophyll

Rght: ***Dransfieldia micrantha***, habit, Papua. (Photo: W.J. Baker)

persistent, though sometimes tattering or caducous, splitting apically or subapically by emerging inflorescence; peduncular bracts few, first peduncular bract similar to prophyll, but lacking keels, tubular, attached midway up peduncle, exserted from prophyll and enclosing inflorescence prior to expansion, splitting abaxially and distally on inflorescence expansion, typically caducous, though sometimes persistent and tattering, other peduncular bracts inconspicuous, triangular, incomplete; rachis shorter than peduncle, angled, tapering; rachis bracts low, rounded; primary branches several, spirally arranged; rachillae fleshy, tapering, usually bearing spirally arranged triads of flowers throughout, rarely pistillate flowers absent from triads throughout inflorescence; rachilla bracts inconspicuous; floral bracteoles low, rounded or truncate. *Staminate flowers*

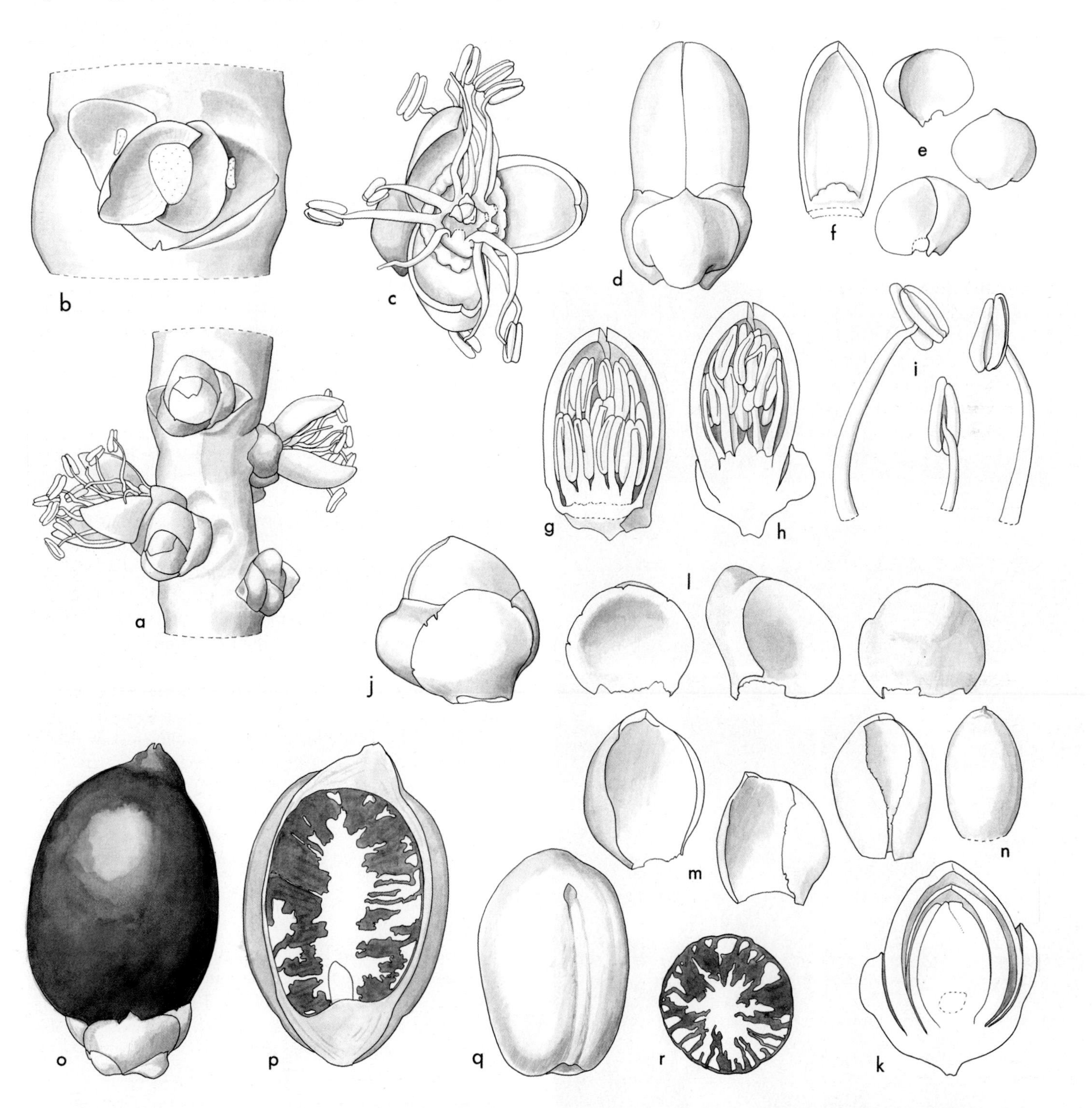

***Dransfieldia*.** **a**, portion of rachilla at staminate anthesis × 4; **b**, triad scar × 9; **c**, staminate flower at anthesis, some stamens removed to show pistillode × 8; **d**, staminate bud × 9; **e**, staminate sepals, 3 views × 9; **f**, staminate petal, interior view × 9; **g**, staminate bud, sepals and 1 petal removed × 9; **h**, staminate bud in vertical section × 9; **i**, stamens in 3 views × 12; **j**, pistillate flower bud × 9; **k**, pistillate flower in vertical section × 9; **l**, pistillate sepals, 3 views × 9; **m**, pistillate petals, 3 views × 9; **n**, gynoecium × 9; **o**, fruit × 4; **p**, fruit in vertical section × 4; **q**, seed × 4; **r**, seed in cross section × 4. *Dransfieldia micrantha*: **a–b**, **d–i**, **o–r**, *Baker* 1067; **c**, **j–n**, *Baker* 1066. (Drawn by Lucy T. Smith)

borne laterally toward the upper side of the pistillate flower in rounded indentations in the rachillae, symmetrical, bullet-shaped in bud, glabrous or with scattered scales as inflorescence; sepals 3, distinct, strongly imbricate, orbicular, spathulate, coriaceous, thickened abaxially, thinning towards margin, margins minutely ciliate; corolla united basally, corolla lobes 3, valvate, ovate, indurated; stamens numerous, up to 19, filaments awl-shaped, outer whorl irregularly inflexed in bud and basally adnate to the petals, inner whorl erect in bud, anthers ellipsoidal, dorsifixed, versatile, connective dark, dehiscence latrorse; pistillode trilobed or papilla-like. *Pollen* ellipsoidal slightly asymmetric, occasionally pyriform; aperture a distal sulcus; ectexine tectate, finely perforate-rugulate, aperture margin similar; infratectum columellate; longest axis ranging from 30–40 µm [1/1]. *Pistillate flowers* symmetrical, subglobose, glabrous or with scattered scales as inflorescence; sepals 3, distinct, strongly imbricate, closely resembling staminate sepals; petals 3, strongly imbricate, resembling sepals, but thinner and with acute apex; staminodes 3–4, shortly joined basally, truncate; gynoecium ovoid, symmetrical, pseudomonomerous, unilocular, uniovulate, stigmas 3, ovule located near base of gynoecium, laterally attached, ?campylotropous. *Fruit* ellipsoidal, stigmatic remains apical, perianth persistent and clasping; epicarp thin, smooth, mesocarp fibrous, endocarp circular in cross-section, closely adpressed to seed, comprising two layers of closely adhering fibres. *Seed* ellipsoidal with flattened base, surface smooth, hilum basal, raphe lateral, endosperm deeply ruminate, embryo basal. *Germination* adjacent ligular, eophyll bifid. *Cytology* not studied.

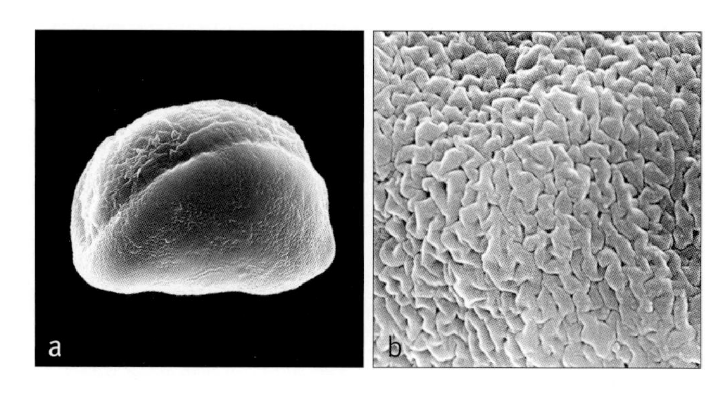

Above: ***Dransfieldia***
a, monosulcate pollen grain, equatorial face SEM × 1500;
b, close-up, finely perforate-rugulate pollen surface × 8000.
Dransfieldia micrantha: **a**, *Baker* 1067, **b**, *Baker* 1066.

Distribution and ecology: Restricted to far western Papua province in Indonesian New Guinea. Known from Waigeo Island in the Raja Ampat Archipelago, the Kepala Burung (Sorong and Bintuni Bay), the lower slopes of the Wondiwoi Mountains and the vicinity of Etna Bay. Grows in lowland forests and forest on slopes and ridge tops, 10–180 m elevation. Palm growers have reported that the single species of this genus occurs in Papua New Guinea (Migliaccio 2001). We have seen no confirmation of this, and suspect that the origin of the seed source has been misinterpreted.

Anatomy: Not studied.

Relationships: The relationships of *Dransfieldia* remain uncertain. *Dransfieldia* has been resolved as sister to *Linospadix* and *Laccospadix* of the Linospadicinae, forming a clade that in turn is sister to *Heterospathe*, but these relationship have low bootstrap support (Asmussen *et al.* 2006).

Common names and uses: *Ititohoho* (Jamur), *Kapis* (Biak-Raja Ampat) and *Tama'e* (Wondama). Stems used for harpoons. Leaves used for thatch. Unspecified parts used for sewing thatch. The species is grown as an ornamental in the USA and Australia, but is not yet widely available in the horticultural trade.

Taxonomic account: Baker *et al.* (2006).

Fossil record: No generic records found.

Notes: The colourful new leaves and inflorescences of this palm, along with its petite habit, make it highly desirable among palm collectors (Migliaccio 2001).

Top: ***Dransfieldia micrantha***, crown with inflorescences, Papua. (Photo: W.J. Baker)

Bottom left: ***Dransfieldia micrantha***, staminate buds and open flowers, Papua. (Photo: W.J. Baker)

Bottom right: ***Dransfieldia micrantha***, pistillate flowers, Papua. (Photo: W.J. Baker)

179. HETEROSPATHE

Very variable small to moderate, solitary or clustered pinnate-leaved palms of the Philippines, the Moluccas, New Guniea and western Pacific Islands; lacking a conspicuous crownshaft, and fruit with lateral to apical stigmatic remains.

Heterospathe Scheff., Ann. Jard. Bot. Buitenzorg 1: 141 (1876). Type: **H. elata** Scheff.

Ptychandra Scheff., Ann. Jard. Bot. Buitenzorg 1: 140 (1876). Type: *P. glauca* Scheff. (= *Heterospathe glauca* [Scheff.] H.E. Moore).

Barkerwebbia Becc., Webbia 1: 281 (1905). Type: *B. elegans* Becc. (= *Heterospathe elegans* [Becc.] Becc.).

Alsmithia H.E. Moore, Principes 26: 122 (1982). Type: *A. longipes* H.E. Moore (= *Heterospathe longipes* [H.E. Moore] Norup).

Heteros — *different,* spathe — *sheath, referring to the different sizes and shapes of the prophyll and peduncular bract.*

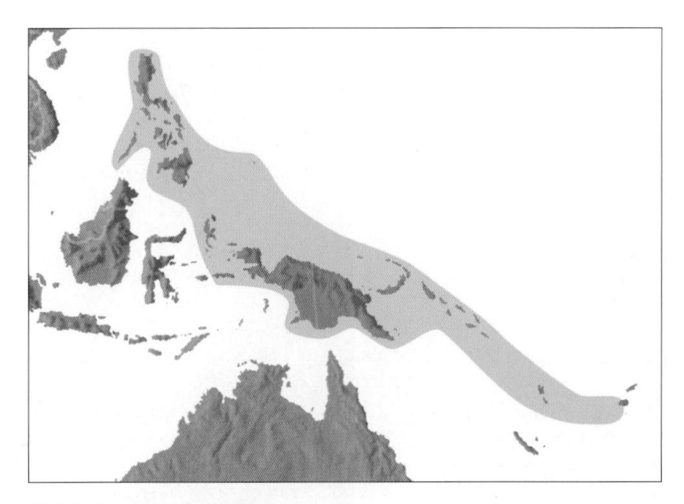

Distribution of ***Heterospathe***

Dwarf to moderate, solitary or sometimes clustered, unarmed, pleonanthic monoecious palms. *Stem* creeping or erect, sometimes basally expanded, grey-green to brown, leaf scars prominent. *Leaves* pinnate, rarely entire bifid, erect, becoming spreading, often reddish when young; sheath splitting abaxially and not forming a well-defined crownshaft, margins fibrous, acute, glaucous or not; petiole short to elongate, usually deeply channelled adaxially, rounded abaxially, variously indumentose; rachis straight or curved, basally channelled adaxially, distally ridged, rounded abaxially, variously indumentose; leaflets, when present, single-fold, acute to acuminate, prominent, midrib elevated, marginal ribs often thickened, veins adaxially ± waxy or glabrous, abaxially tomentose or brown-dotted, with or without basifixed ramenta on midrib. *Inflorescences* interfoliar or infrafoliar at anthesis, branched to 1–4 orders basally, fewer distally, often with reddish-brown, deciduous tomentum; peduncle prominent, elongate, elliptic in cross-section; prophyll persistent, attached near the base and completely encircling the peduncle, tubular, 2-keeled laterally, more-or-less dorsiventrally flattened, splitting abaxially, and apically; peduncular bract 1 or rarely 2 (*Heterospathe trispatha*), attached below or sometimes above the middle of the peduncle, terete, beaked, enclosing the inflorescence in bud, greatly exceeding the prophyll, splitting abaxially and caducous or marcescent as the inflorescence matures; rachis short to elongate, bearing spirally arranged, short, pointed bracts subtending a few simple rachillae, or several branches with basal bare portions; rachillae slender, bearing sessile or slightly depressed, spirally arranged triads subtended by spreading lip-like bracts throughout the rachillae, or with paired or solitary staminate flowers toward the apex of the rachillae;

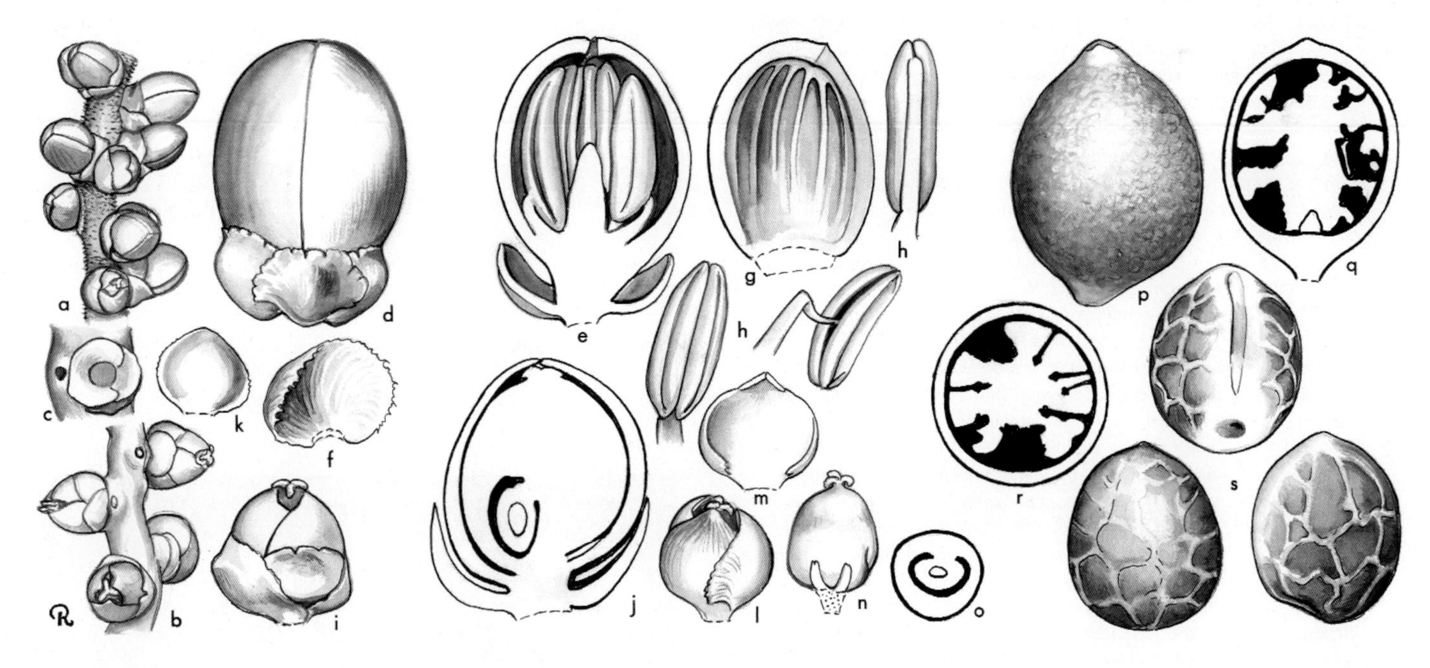

Heterospathe. **a**, portion of rachilla with triads × 3; **b**, portion of rachilla, staminate flowers fallen × 3; **c**, triad, flowers removed × 6; **d**, staminate bud × 12; **e**, staminate bud in vertical section × 12; **f**, staminate sepal, interior view × 12; **g**, staminate petal, interior view × 12; **h**, stamen in 3 views × 12; **i**, pistillate flower × 6; **j**, pistillate flower in vertical section × 12; **k**, pistillate sepal × 6; **l**, pistillate flower, sepals removed × 6; **m**, pistillate petal, interior view × 6; **n**, gynoecium and staminodes × 6; **o**, ovary in cross-section × 6; **p**, fruit × 3; **q**, fruit in vertical section × 3; **r**, fruit in cross-section × 3; **s**, seed in 3 views × 3. *Heterospathe humilis*: all from *Moore et al.* 926. (Drawn by Marion Ruff Sheehan)

bracteoles of the staminate flowers small, bracteoles surrounding the pistillate flower 2, spreading to cupular and imbricate. *Staminate flowers* symmetrical or slightly to markedly asymmetrical; sepals 3, distinct, broadly imbricate and rounded, ± keeled dorsally and gibbous basally; petals 3, distinct, valvate, usually about twice as long as the sepals, prominently lined when dry, ± acute, one usually somewhat larger than the others; stamens 6–36 or more, distinct, the filaments awl-shaped and strongly inflexed at the apex, anthers oblong in outline, dorsifixed and versatile at anthesis, latrorse; pistillode either small and conical, or columnar, prominent, nearly as long as the stamens, sometimes with an expanded apex. *Pollen* ellipsoidal asymmetric, occasionally oblate triangular; aperture a distal sulcus, infrequently a trichotomosulcus; ectexine tectate, perforate, perforate and micro-channelled or finely perforate-rugulate, aperture margin similar or slightly finer; infratectum columellate; longest axis ranging from 26–54 µm [12/32]. *Pistillate flowers* symmetrical, ± same size as the staminate; sepals 3, distinct, broadly imbricate, rounded; petals 3, distinct, broadly imbricate with briefly valvate apices; staminodes 3, tooth-like; gynoecium unilocular, uniovulate, short, soft, expanded upward into a thick stylar region below 3 recurved, short stigmas, the ovule lateral at top of locule, pendulous, hemianatropous. *Fruit* globose to ellipsoidal, small to large, orange to red when mature, stigmatic remains apical, eccentrically apical or subapical to lateral; epicarp smooth but drying granular or with irregular lines over short sclerosomes in the thinly to thickly fleshy mesocarp, with flattened anastomosing fibres, endocarp thin, operculate, smooth, shining within, or with thickened adnate fibres, irregularly sculptured, ridged and grooved, beaked at the apex, with a mass of slender fibres within a framework of thickened fibres at the base. *Seed* not adherent to endocarp, globose to ellipsoidal or with angled and with 3 rounded ridges laterally and abaxially, attached apically and laterally by the elongate hilum extending nearly the length of the seed, raphe branches simple to anastomosing, endosperm ruminate or rarely homogeneous (*H. longipes* and *H. uniformis*); embryo basal. *Germination* adjacent-ligular; eophyll bifid where known. *Cytology*: 2n = 32.

Distribution and ecology: About 40 species from the Philippines and Micronesia to eastern Indonesia and to the Solomon Islands, Fiji and Vanuatu, including 16 species in New Guinea. Inhabitants of lowland and montane rain forest. Many species are undergrowth palms; a few contribute to the forest canopy.

Above left: ***Heterospathe* sp.**, habit, Papua New Guinea. (Photo: W.J. Baker)

Above middle: ***Heterospathe delicatula***, inflorescence, cultivated, Lyon Arboretum, Hawaii. (Photo: W.J. Baker)

Above right: ***Heterospathe humilis***, crown with inflorescences, Papua New Guinea. (Photo: W.J. Baker)

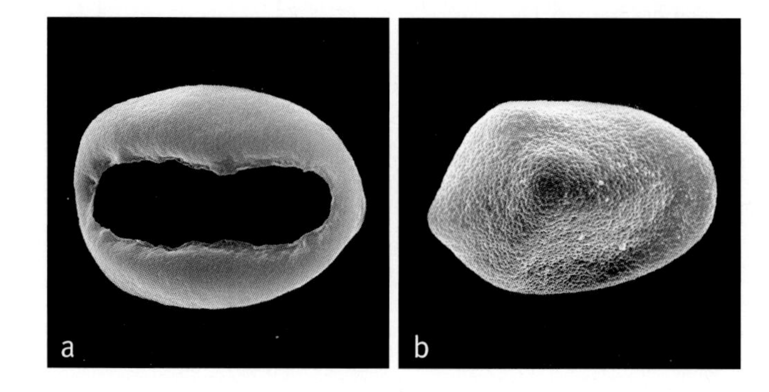

Below: ***Heterospathe***
a, monosulcate pollen grain, distal face SEM × 1500;
b, monosulcate pollen grain, proximal face SEM × 1500.
Heterospathe elata: **a**, *Rodin* 737; *H. intermedia*: **b**, *Fernando* EF534.

Anatomy: Leaf (*Heterospathe elata*; Tomlinson 1961), root (Seubert 1998a, 1998b) and fruit (Essig *et al.* 1999).

Relationships: *Heterospathe* is moderately to strongly supported as monophyletic (Norup 2005, Asmussen *et al.* 2006, Norup *et al.* 2006, Baker *et al.* in review, in prep.). The genus is moderately supported as sister to *Linospadix*, which was the sole representative of the Linospadicinae in a study by Lewis and Doyle (2002). Other data place *Heterospathe* as sister to a poorly supported clade of *Dransfieldia*, *Linospadix* and *Laccospadix* (Asmussen *et al.* 2006).

Common names and uses: Sagisi palm. Fruit of *Heterospathe elata* is chewed as a betel substitute in the Philippines; the cabbage is said to be edible and the split petioles and leaflets are used in weaving.

Taxonomic accounts: Moore (1969c) and Fernando (1990). No satisfactory key to species exists.

Fossil record: No generic records found.

Notes: A widespread genus of the western fringe of the Pacific; although it occurs in Palawan (*Heterospathe dransfieldii*), the genus has not yet been found in Borneo.

Additional figures: Glossary fig. 3.

180. HYDRIASTELE

Very variable small to very robust, solitary or clustered pinnate-leaved palms from Sulawesi eastwards to Fiji and Australia, with conspicuous crownshafts and often conspicuous praemorse leaflets; the inflorescences bear triads (and hence fruit) throughout the length of the rachillae.

Hydriastele H. Wendl. & Drude, Linnaea 39: 180, 208 (1875). Type: **H. wendlandiana** (F. Muell.) H. Wendl. & Drude).

Adelonenga Hook.f. in Benth. & Hook.f., Gen. pl. 3: 885 (1883). Lectotype: *A. variabilis* (Becc.) Becc. (*Nenga variabilis* Becc.) (= *H. variabilis* [Becc.] Burret) (see Beccari & Pichi-Sermolli 1955).

Gronophyllum Scheff., Ann. Jard. Bot. Buitenzorg 1: 135 (1876). Type: *G. microcarpa* Scheff. (= *H. microcarpa* [Scheff.] W.J. Baker & Loo).

Gulubia Becc., Ann. Jard. Bot. Buitenzorg 2: 128, 131, 134 (1885). Lectotype: *G. moluccana* (Becc.) Becc. (*Kentia moluccana* Becc.) (see Beccari & Pichi-Sermolli 1955) (= *H. moluccana* [Becc.] W.J. Baker & Loo).

Gulubiopsis Becc., Bot. Jahrb. Syst. 59: 11 (1924). Type: *G. palauensis* Becc. (= *Hydriastele palauensis* [Becc.] W.J. Baker & Loo).

Leptophoenix Becc., Ann. Jard. Bot. Buitenzorg 2: 82 (1885). Lectotype: *L. pinangoides* (Becc.) Becc. (*Nenga pinangoides* Becc.) (see Beccari & Pichi-Sermolli 1955) (= *Hydriastele pinangoides* [Becc.] W.J. Baker & Loo).

Nengella Becc., Malesia 1: 32 (1877). Lectotype: *N. montana* Becc. (see Burret 1936b) (= *Hydriastele montana* [Becc.] W.J. Baker & Loo).

Paragulubia Burret, Notizbl. Bot. Gart. Berlin-Dahlem 13: 84 (1936). Type: *P. macrospadix* Burret (= *Hydriastele macrospadix* [Burret] W.J. Baker & Loo).

Siphokentia Burret, Notizbl. Bot. Gart. Berlin-Dahlem 10: 198 (1927). Type: *S. beguinii* Burret (*Hydriastele beguinii* [Burret] W.J. Baker & Loo)

Kentia Blume, Bull. Sci. Phys. Nat. Néerl. 1: 64 (1838) (non *Kentia* Adans. 1763). Type: *K. procera* Blume (= *Hydriastele procera* [Blume] W.J. Baker & Loo).

Hydrias — *water nymph*, stele — *column or pillar, perhaps referring to the erect slender stems of those species growing near water.*

Small, moderate or tall, solitary or clustered, unarmed, pleonanthic, monoecious palms. *Stems* erect, slender to robust, bare, conspicuously ringed with leaf scars. *Leaves* entire-bifid or pinnate, neatly abscising; sheaths elongate, forming a well-defined crownshaft, usually densely scaly or tomentose, and/or waxy, a ligule-like prolongation sometimes present opposite or at the base of the petiole; petiole short to long, adaxially channelled, abaxially rounded, usually conspicuously scaly; rachis adaxially channelled or angled near the base, distally angled, abaxially rounded, usually scaly as the petiole; leaflets regularly arranged, or grouped, pendulous or horizontal or ascending, straight or curved, single-fold or several-fold, the terminal pair usually broad, several-fold, the rest parallel sided or somewhat wedge-shaped, apically acute, bifid or conspicuously praemorse, adaxial and abaxial surfaces bearing scattered minute scales, abaxially sometimes with scattered ramenta along the main veins, sometimes also with bands of deciduous chaffy

Right: ***Hydriastele longispatha***, habit, Papua. (Photo: W.J. Baker)

scales along major ribs, transverse veinlets conspicuous or obscure. *Inflorescences* infrafoliar, branching to 1–3 orders or rarely spicate, usually horsetail-like, protandrous or protogynous; peduncle short, winged at the base, sometimes becoming swollen; prophyll compressed, entirely enclosing the inflorescence in bud, 2-keeled, with a conspicuous apical beak, thin, papery when dry, glabrous or scaly, soon drying on exposure, splitting longitudinally on the abaxial face and abscising together with the peduncular bract; peduncular bract 1 rarely 2, similar to and entirely enclosed by the prophyll, tubular, enclosing the inflorescence in bud; subsequent bracts inconspicuous; rachis (where present) longer or shorter than the peduncle, bearing inconspicuous rachis bracts subtending few to many crowded, ± spirally arranged first-order branches, the proximal bearing a few branches or all unbranched; rachillae elongate, usually ± straight or curved, of ± equal length, tending to curve downwards, bearing throughout their length spirally arranged or opposite and decussate pairs of triads of cream-coloured or pinkish-tinged flowers, except at the very tip where bearing solitary or paired staminate flowers; rachilla bracts very inconspicuous, low, ± rounded. *Staminate flowers* fleshy, asymmetrical; calyx sessile or with a short stalk-like base, sepals 3, short, triangular, ± distinct or joined into a cup for ca. $^1/_2$ their length; petals 3, fleshy, distinct, except at the very base, valvate except in *Hydriastele palauensis* where margins not meeting in bud, 4–5 times as long as the calyx, narrow, triangular, 1 usually larger than the other 2; stamens 6–24, epipetalous, filaments very short, fleshy, variously epipetalous and connate, anthers elongate, erect, basifixed, latrorse, connective sometimes prolonged into a short point; pistillode absent. *Pollen* ellipsoidal, bi-symmetric; aperture a distal sulcus, brevi-, ± same length as long axis or, frequently, extended; ectexine semi-tectate and coarsely (rarely finely) reticulate, muri of reticulum sometimes perforate, aperture margins similar; or pollen ellipsoidal or oblate-triangular, asymmetric; aperture a distal sulcus or trichotomosulcus; ectexine tectate, coarsely perforate, foveolate, coarsely perforate-rugulate or rarely scabrate verrucate, aperture margin similar; longest axis ranging from 33–70 µm; post-meiotic tetrads tetragonal or tetrahedral [22/47]. *Pistillate flowers* globose or ± conical in bud, smaller than the staminate; sepals 3, distinct, rounded or triangular, broadly imbricate or connate in a ring with 3 low triangular lobes; petals 3, distinct or connate, not more than to at least twice as long as the sepals, rounded or triangular, basally broadly imbricate or connate in a ring,

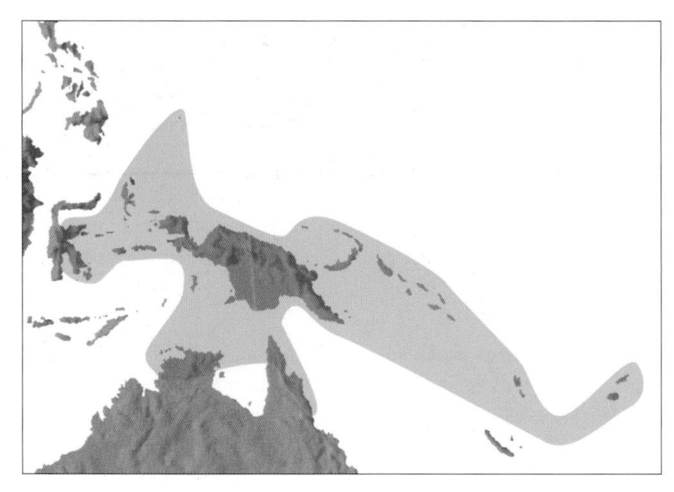

Distribution of ***Hydriastele***

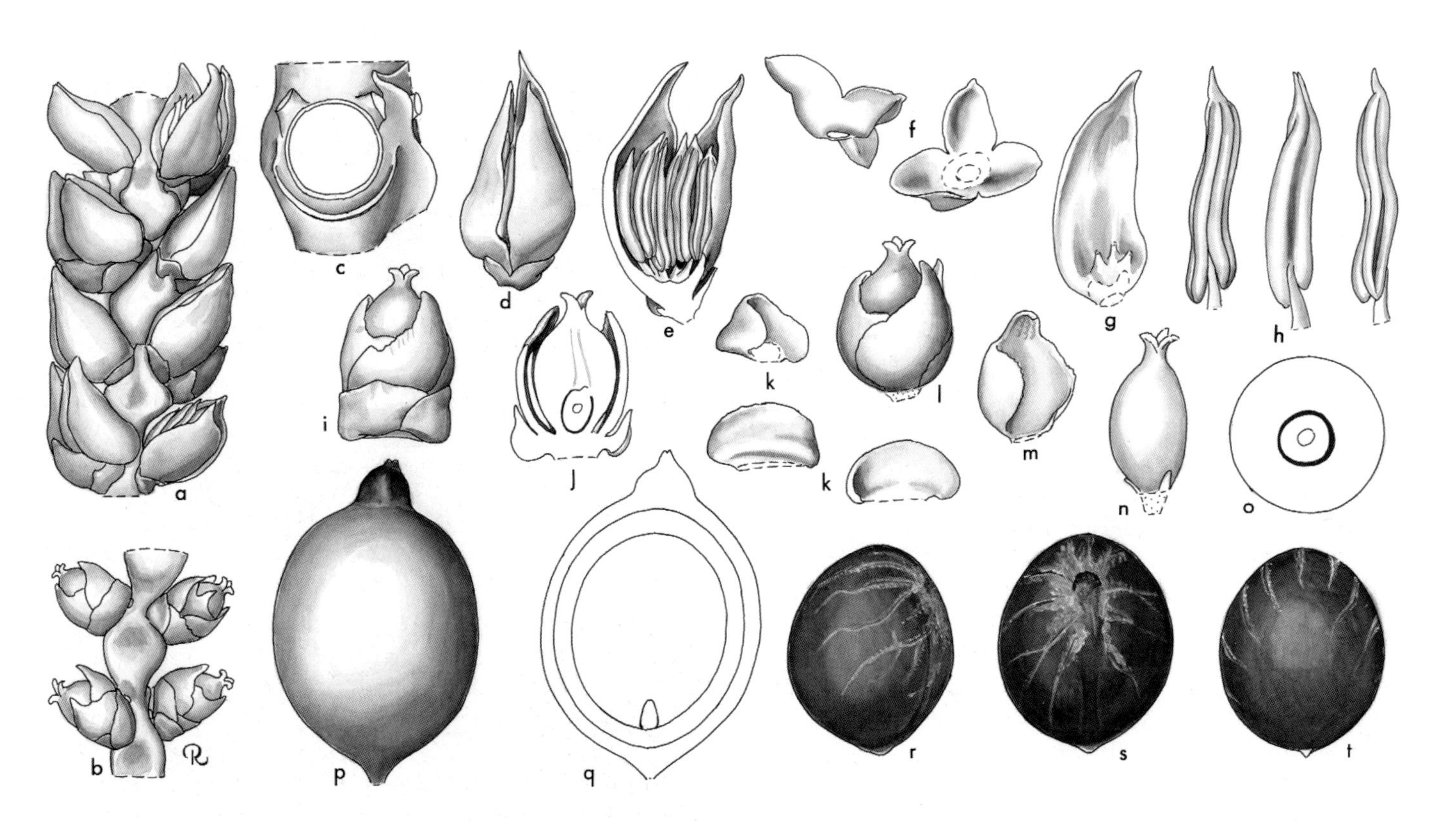

Hydriastele. **a**, portion of rachilla in staminate flower × 1$^1/_2$; **b**, portion of rachilla in pistillate flower × 1$^1/_2$; **c**, triad, flowers removed × 6; **d**, staminate flower × 3; **e**, staminate flower in vertical section × 3; **f**, staminate calyx in 2 views × 6; **g**, staminate petal, interior view × 3; **h**, stamen in 3 views × 6; **i**, pistillate flower × 3; **j**, pistillate flower in vertical section × 3; **k**, pistillate sepals × 3; **l**, pistillate flower, sepals removed × 3; **m**, pistillate petal, interior view × 3; **n**, gynoecium and staminodes × 3; **o**, ovary in cross-section × 6; **p**, fruit × 3; **q**, fruit in vertical section × 3; **r**, **s**, **t**, seed in 3 views × 3. *Hydriastele chaunostachys*: all from *Moore et al.* 9258 and *Hoogland & Craven* 10934. (Drawn by Marion Ruff Sheehan)

apically rounded except for very small triangular valvate tips or with conspicuous triangular valvate tips, closely appressed in bud, the tips persisting or eroding into fibres in fruit; staminodes 3(–6), tooth-like, minute; gynoecium ± globose or ovoid, unilocular, uniovulate, stigmas 3, low, sessile or fleshy, reflexed, ovule laterally attached near apex of locule, hemianatropus (?always). *Fruit* globose to narrowly ellipsoidal, straight or curved, bright red to purplish-black, sometimes drying ridged, sometimes briefly beaked, stigmatic remains apical, perianth whorls persistent, the petal tips sometimes reflexed or appressed to the fruit; epicarp smooth or slightly pebbled, mesocarp thin, with abundant tannin cells, and longitudinal fibre bundles, endocarp thin, crustose or obsolescent. *Seed* ovoid or globose, laterally or basally attached with elongate or rounded hilum, raphe branches sparse, anastomosing, endosperm homogeneous or shallowly to deeply ruminate; embryo basal. *Germination* adjacent-ligular; eophyll bifid with entire or minutely to strongly praemorse tips. *Cytology*: 2n = 32.

Distribution and ecology: About 47 species in Sulawesi, Moluccas, New Guinea, Bismarck Archipelago, northern Australia, Fiji, Vanuatu and Palau. Lowland to upland tropical rain forest. One species, *Hydriastele rheophytica*, occurs as a rheophyte in western New Guinea, and has very slender leaflets (Dowe & Ferrero 2000). Several species are recorded from limestone and others from ultramafic rock. Pollination has been studied by Essig (1973) who showed that curculionid beetles are probably the pollinators in *H. microspadix*.

Anatomy: Fruit (Essig 1982).

Relationships: *Hydriastele* is moderately to highly supported as a monophyletic genus following the recent inclusion of three genera, *Gronophyllum*, *Gulubia* and *Siphokentia*, in synonymy (Baker & Loo 2004, Asmussen *et al.* 2006, Loo *et al.* 2006, Norup *et al.* 2006, Baker *et al.* in prep.). The relationships of *Hydriastele* remain unclear, but it is worth noting that, in the most densely sampled studies, the genus does not resolve within the western Pacific clade of Areceae, despite its distribution (Norup *et al.* 2006, Baker *et al.* in prep.). Lewis and Doyle (2002) resolve *Hydriastele* as sister to the western Pacific clade.

Common names and uses: *Pinang salea* (*Hydriastele microcarpa*). Essig (1982) records the use of trunks for floorboards and side panels of houses in New Guinea. Stems have been split and used as spears. Several species are cultivated as ornamentals.

Taxonomic accounts: Baker and Loo (2004). See also Essig (1982), Baker *et al.* (2000d), Burret (1936a, 1936b) and Essig & Young (1985).

Fossil record: Fossil leaf material referred to *Kentites* (= *Hydriastele*) from the Tertiary of Italy (Bureau 1896) was placed in the synonomy of *Phoenicites* by Read and Hickey (1972). Any link with extant *Hydriastele* is almost certainly spurious, particularly as the generic name *Kentia* has been so misapplied in the past.

Notes: Uhl and Dransfield (1987) indicated that generic delimitation among the Papuasian members of Arecinae was problematic. In the first edition of *Genera Palmarum*, four Papuasian genera were assigned to Arecinae, together with *Areca*, *Nenga*, *Pinanga* and *Loxococcus*. Uhl and Dransfield

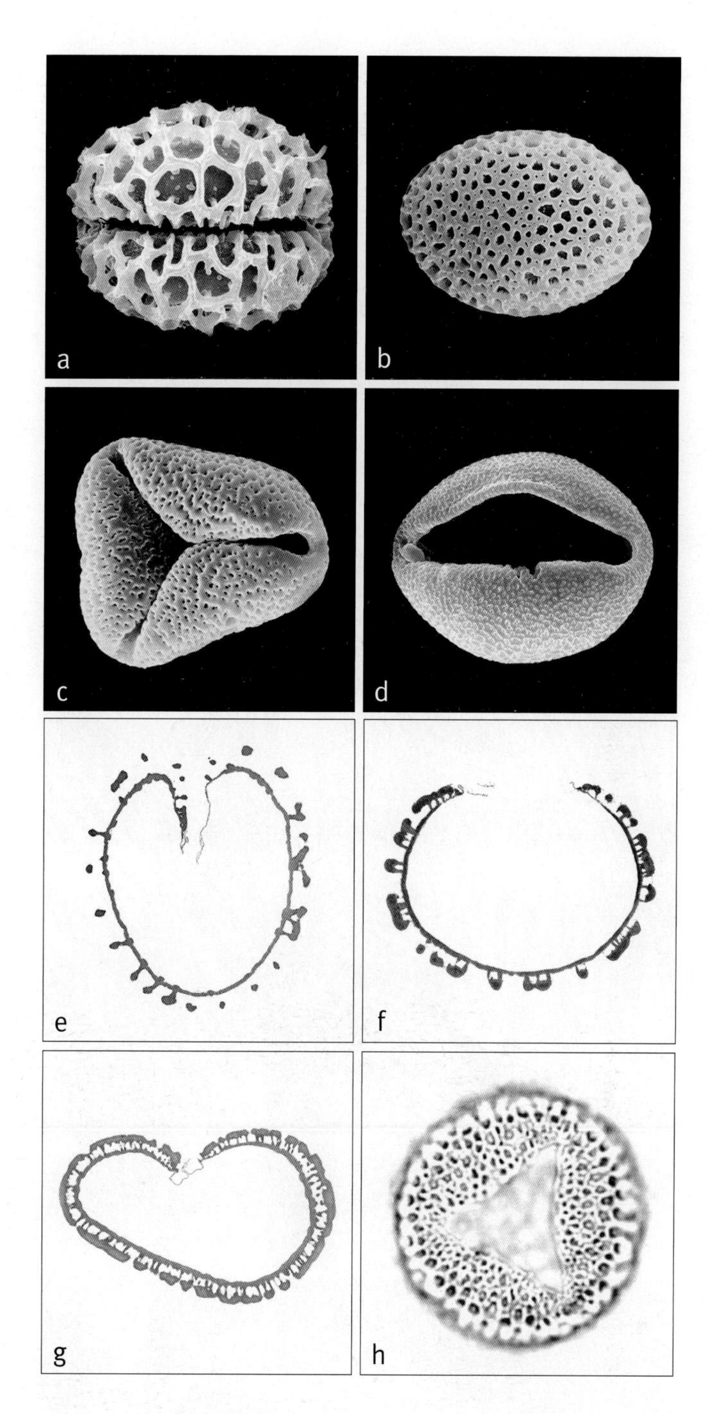

Above: ***Hydriastele***
a, extended sulcate pollen grain, distal face SEM × 2000;
b, monosulcate pollen grain, proximal face SEM × 2000;
c, trichotomosulcate pollen grain, distal face SEM × 1500;
d, monosulcate pollen grain, distal face SEM × 1500;
e, ultrathin section through whole pollen grain, polar plane TEM × 1000;
f, ultrathin section through whole pollen grain, polar plane TEM × 1000;
g, ultrathin section through whole pollen grain, polar plane TEM × 2000;
h, trichotomosulcate pollen grain, high-focus LM × 1000.
Hydriastele pinangoides: **a**, *Boden Kloss* 3100; *H. microspadix*: **b**, cultivated RBG Kew 079-64.07901; *H. ramsayi*: **c** & **g**, *Specht* 1113; *H. costata*: **d**, *Volkar* 1873; *H. microcarpa*: **e**, *Robinson* 1610; *H.* sp.: **f** & **h**, *Millar* NGF 23072.

placed emphasis on aspects of floral morphology that seemed to be correlated with protandry and protogyny. Species in which the pistillate flowers have large triangular petals closely adpressed in bud (*Gronophyllum* and *Siphokentia*) seemed to be protandrous, whereas those with inconspicuous rounded petals with minute triangular tips (*Hydriastele* and *Gulubia*) seemed to be protogynous. This apparently neat correlation was, however, based on very few observations of the sequence of flowering and was in the most part inferred. Furthermore, this delimitation resulted in genera of sometimes disparate habit. Recent phylogenetic studies (Loo *et al.* 2006) strongly support the monophyly of the Papuasian clade. Within the clade, however, only *Hydriastele* is monophyletic; *Gronophyllum* and *Gulubia* are polyphyletic with members of both genera resolving in a number of separate, highly supported groups with members of other genera within the clade. The position of *Siphokentia* is ambiguous but this genus, along with *Hydriastele*, is deeply nested within the clade. It has not been possible to find morphological characters that differentiate most of the groupings resolved in the molecular studies, and thus it is not possible to alter the generic delimitation to reflect the molecular phylogeny. For these reasons, we have followed Baker and Loo's proposal (2004) to accept a single genus, for which the earliest name is *Hydriastele*.

Additional figures: Glossary fig. 15.

Top: ***Hydriastele costata***, crown, cultivated, Lae Botanic Gardens, Papua New Guinea. (Photo: W.J. Baker)

Above: ***Hydriastele rhopalocarpa***, inflorescence, Papua. (Photo: W.J. Baker)

181. IGUANURA

Small solitary or clustered pinnate-leaved palms of the forest undergrowth of Malay Peninsula, southern Thailand, Sumatra and Borneo; often lacking crowshafts, always with praemorse leaflets or leaf margins, and the flowers borne in pits; stigmatic remains in the fruit are basal.

Iguanura Blume, Bull. Sci. Phys. Nat. Néerl. 1: 66 (1838). Type: **I. leucocarpa** Blume.

Slackia Griffith, Calcutta J. Nat. Hist. 5: 468 (1845). Type: *S. geonomiformis* (Mart.) Griff. (= *Iguanura geonomiformis* Mart.).

Iguana — *a lizard,* ura — tail, *referring to the rachilla bracts that give a scaly appearance to the inflorescence axes of some species.*

Small, solitary or clustered, acaulescent or erect, unarmed, pleonanthic, monoecious palms. *Stems* rarely exceeding about 4 m in height, with very short to elongate internodes, leaf scars inconspicuous, stilt roots sometimes present. *Leaves* undivided and pinnately ribbed, with or without an apical notch, or regularly to irregularly pinnate, marcescent or neatly abscising; sheaths usually splitting opposite the petiole and persistent, not forming a crownshaft, or rarely neatly abscising and a crownshaft developed, variously tomentose, a tattering ligule sometimes present; petiole present or absent, variously indumentose as the rachis; expanding blades frequently reddish-tinged, leaflets single- or several-fold, the tips or margins in entire blades irregularly praemorse, the main ribs frequently bearing scales or hairs, scattered or in bands, the blade with conspicuous longitudinal veins, transverse veinlets inconspicuous. *Inflorescences* usually interfoliar or infrafoliar (in species with a crownshaft), solitary, protandrous, spicate or branching to 1 or 2 orders, emerging long before anthesis, erect at first, becoming curved or pendulous; peduncle very short to very long; prophyll 2-keeled, tubular, short, usually enclosed within the leaf sheath, frequently marcescent, often indumentose; peduncular bract attached a short distance above the prophyll, usually much exceeding it, otherwise similar, marcescent, rarely neatly abscising, subsequent peduncular bracts very inconspicuous; rachis usually much shorter than the peduncle; rachillae rarely exceeding about 20 in number, usually fewer, very slender to moderately robust, glabrous or densely tomentose, bearing distant or dense, spirally arranged triads, superficial or, more usually, sunken in pits, each subtended by a low rachilla bract, forming the lower lip of the pit, an upper lip sometimes also differentiated, the pits frequently bearing sparse or abundant hairs, distal-most bracts subtending paired or solitary staminate flowers; floral bracteoles very small, included within the pits; flowers opening singly, in pit-bearing species the flowers exserted one at a time. *Staminate flowers* sessile, ± globular in bud, symmetrical, abscising after anthesis usually leaving the calyx behind; sepals 3, distinct, imbricate, membranous, ± striate, often keeled, often ciliate-margined, sometimes tomentose, scarcely exserted from the pit; petals 3, valvate, twice as long as the sepals, somewhat hooded, very briefly joined at the base; stamens 6, filaments slender,

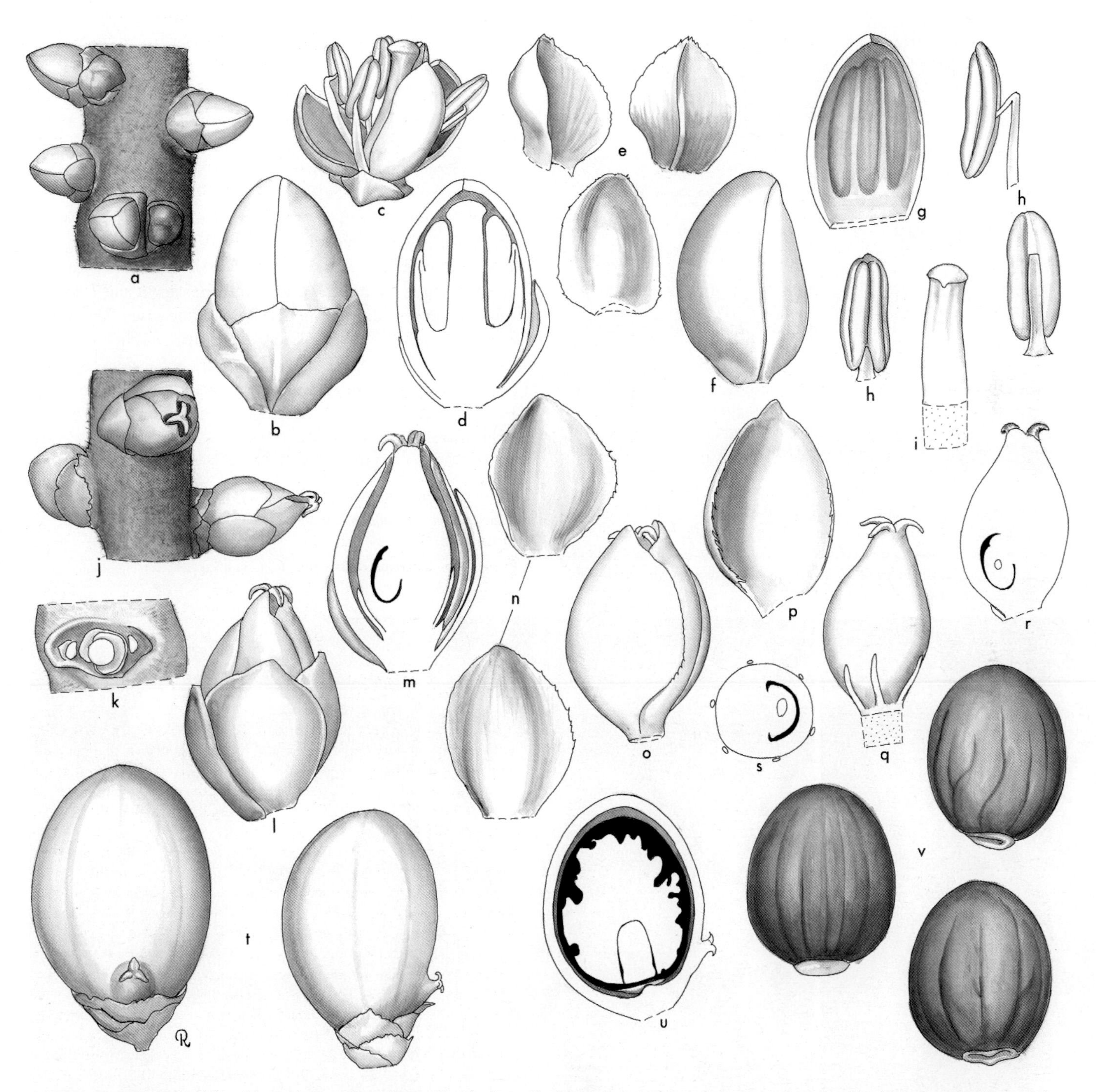

Iguanura. **a,** portion of rachilla with staminate and smaller pistillate buds × 4 ½; **b,** staminate bud × 12; **c,** staminate flower × 7 ½; **d,** staminate bud in vertical section × 12; **e,** staminate sepals × 12; **f,** staminate bud, sepals removed × 12; **g,** staminate petal, interior view × 12; **h,** stamen in 3 views × 12; **i,** pistillode × 12; **j,** portion of rachilla with pistillate flowers × 4 ½; **k,** triad, flowers removed to show bracts and scars × 7 ½; **l,** pistillate flower × 7 ½; **m,** pistillate flower in vertical section × 7 ½; **n,** pistillate sepal in 2 views × 7 ½; **o,** pistillate flower, sepals removed × 7 ½; **p,** pistillate petal × 7 ½; **q,** gynoecium and staminodes × 7 ½; **r,** gynoecium in vertical section × 7 ½; **s,** ovary in cross-section × 7 ½; **t,** fruit in 2 views × 4 ½; **u,** fruit in vertical section × 4 ½; **v,** seed in 3 views × 4 ½. *Iguanura wallichiana*: **a–i,** *Moore* 9063; **j–s,** *Moore* 9072; **t–v,** *Kunstler* 7999. (Drawn by Marion Ruff Sheehan)

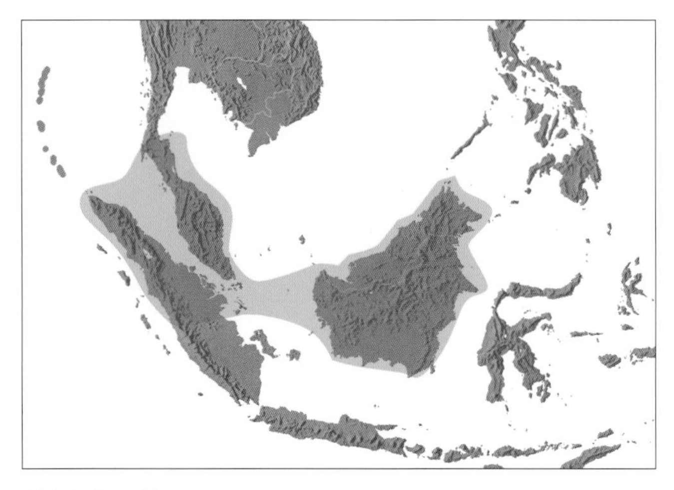

Distribution of ***Iguanura***

elongate, fleshy, inflexed in bud, anthers ± oblong, the margins sometimes undulate, ± versatile, latrorse; pistillode conspicuous, columnar, as long as the petals, fleshy. *Pollen* ellipsoidal asymmetric, occasionally pyriform or lozenge-shaped, less frequently oblate triangular; aperture a distal sulcus, infrequently a trichotomosulcus; ectexine tectate, perforate and micro-channelled or finely perforate-rugulate, aperture margin similar or slightly finer; infratectum columellate; longest axis 27–45 µm; post-meiotic tetrads tetrahedral [5/18]. *Pistillate flower* ± globular, slightly larger than the staminate; sepals 3, distinct, broadly imbricate, rounded, ± striate; petals 3, distinct, exceeding the sepals, broadly imbricate, with minute, triangular, valvate tips, ± striate; staminodes 6, slender, flattened; gynoecium unilocular, uniovulate, slightly asymmetrical, ± ovoid, stigmas 3, large, fleshy, reflexed, ovule laterally attached, hemianatropous. *Fruit* ovoid, ellipsoidal, bilobed, or narrowly spindle-shaped and straight or curved, or even flat and 5-pointed, smooth when fresh, smooth or ridged when dry, green, white, brownish or pink turning brilliant red at maturity, the stigmatic remains basal, perianth whorls persistent; epicarp smooth, shiny, mesocarp fleshy, endocarp well developed, woody, smooth or variously ridged, with a basal rounded operculum. *Seed* conforming to the shape of the endocarp, attached at one side near the base, the hilum ± circular, raphe branches spirally ascending, anastomosing, endosperm homogeneous or ruminate; embryo basal. *Germination* adjacent-ligular; eophyll ± entire, with praemorse margins, with or without a small apical notch. *Cytology*: 2n = 32.

Distribution and ecology: About 32 species in Sumatra, the Malay Peninsula (including south Thailand) and Borneo, some of them very local and poorly known. All species are plants of the undergrowth of primary tropical rain forest and often occur gregariously; they may be found at altitudes from sea level to about 1200 m in the mountains. In Malaya, *Iguanura wallichiana* is extraordinarily widespread, yet in Sumatra, it is local, for no obvious reason. In Borneo, many taxa are known in only very restricted areas. *Iguanura melinauensis* and *I. elegans* are found in rich alluvial soil developed at the base of limestone hills, but are not confined to this habitat, neither do they appear to be true calcicoles. Ants, flies, bees and wasps are frequent visitors to the staminate flowers of *I. wallichiana* (Kiew 1972), but of these, ants seem to be the most consistent visitors of pistillate flowers.

Below left: ***Iguanura melinauensis***, habit, Sarawak. (Photo: J. Dransfield)

Below middle: ***Iguanura minor***, form with undivided leaves, Sarawak. (Photo: J. Dransfield)

Below right: ***Iguanura palmuncula***, fruit, Sarawak. (Photo: J. Dransfield)

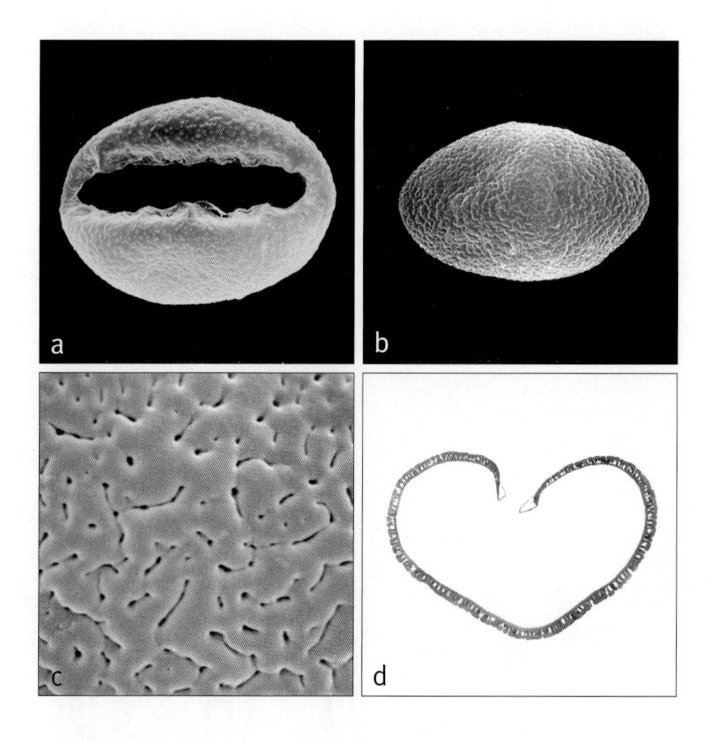

Above: ***Iguanura***
a, monosulcate pollen grain, distal face SEM × 1500;
b, monosulcate pollen grain, proximal face SEM × 1500;
c, close-up, finely perforate-rugulate pollen surface SEM × 8000;
d, ultrathin section through whole pollen grain, polar plane TEM × 2500.
Iguanura palmuncula: **a**, *Dransfield* JD6065, **d**, *Clemens* 22090; *I. bicornis*: **b**, *Kunstler* 6375; *I. wallichiana*: **c**, *Furtado* 33070.

Anatomy: Root (Seubert 1998a, 1998b) and fruit (Essig *et al.* 1999).
Relationships: The monophyly of *Iguanura* has not been tested. The genus resolves in numerous alternative positions, all with low support (Lewis & Doyle 2001, Loo *et al.* 2006, Norup *et al.* 2006, Baker *et al.* in review, in prep.).
Common names and uses: *Pinang*. Species of *Iguanura* are seldom used except in a casual way, e.g., the leaves may be used for thatching overnight shelters. Roots and fruit of *I. wallichiana* have been attributed with contraceptive properties (Burkill 1966). All species are very decorative, but appear to be difficult to cultivate.
Taxonomic accounts: Kiew (1976, 1979); see also Lim (1998).
Fossil record: No generic records found.
Notes: The small size and usual lack of a crownshaft help to distinguish this genus. The most remarkable feature of the genus is undoubtedly the range of fruit shape. Although the genus has been recently monographed; many species, including the type, are insufficently known and represented in herbaria by few specimens.
Additional figures: Glossary fig. 19.

182. LOXOCOCCUS

Moderate solitary pinnate-leaved palm from rain forest in Sri Lanka, with conspicuous crownshaft, praemorse leaflets and asymmetrical staminate flowers with 12 stamens; the seed is deeply ruminate.

Loxococcus H. Wendl. & Drude, Linnaea 39: 185 (1875). Type: **L. rupicola** (Thwaites) H. Wendl. & Drude ex Hook.f. (*Ptychosperma rupicolum* Thwaites).

Loxos — *slanting*, kokkos — *seed, referring to the oblique development of fruit and seed.*

Moderate, solitary, unarmed, pleonanthic, monoecious palm. *Stem* erect, rather slender, conspicuously ringed with leaf scars, sometimes slightly swollen at the base. *Leaves* stiff, pinnate, neatly abscising; sheaths forming a well-defined crownshaft, bearing sparse scales; petiole usually very short, adaxially channelled, abaxially rounded, sparsely scaly; rachis adaxially becoming angled distally, abaxially rounded; leaflets mostly single-fold except for the terminal pair and, rarely, the basal pair, regularly arranged, generally rather stiff and coriaceous, linear, close, the apices praemorse, truncate or oblique, glabrous adaxially, abaxially paler with very thin, white, caducous indumentum and conspicuous ramenta along the midrib near the base, midrib prominent, transverse veinlets not visible. *Inflorescences* infrafoliar, rather short, stiff, spreading, branching to 2 orders basally, to 1 order distally, apparently protandrous; peduncle short, the base bulbous and with 2 clasping wings; prophyll borne just above the base of the peduncle, narrow, ovate, beaked, laterally 2-keeled, entirely enclosing the inflorescence until leaf fall, then

Right: ***Loxococcus rupicola***, habit, Sri Lanka. (Photo: A.J. Henderson)

splitting longitudinally, bearing abundant, scattered scales; peduncular bract borne just above the prophyll, much shorter than the prophyll, lanceolate, acuminate, apparently not completely sheathing the inflorescence, incompletely encircling the peduncle, scaly as the prophyll; 1 or 2 small, triangular, acuminate, open, peduncular bracts sometimes present; rachis stiffly projecting upward, much longer than the peduncle, bearing spirally arranged small, triangular, acuminate bracts subtending branches and rachillae; rachillae rather flexuous, short, stout, deep crimson at anthesis, glabrous, bearing spirally arranged, very small, low, triangular bracts subtending flower groups, flowers borne in triads for $^{1}/_{2}$–$^{3}/_{4}$ the rachilla length, with solitary or paired staminate flowers distally; floral bracteoles minute. *Staminate flowers* somewhat asymmetrical, ± fleshy; sepals 3, distinct, imbricate, broadly triangular, keeled, the margins minutely toothed; petals 3, elongate, unequal, briefly connate basally, valvate, tips triangular; stamens 12, filaments short, slender, distinct, anthers ± sagittate at the base, elongate, latrorse; pistillode ± conical, domed, or 3-angled with 3 short slender appendages (?vestigial stamens). *Pollen* ellipsoidal asymmetric; aperture a distal sulcus; ectexine tectate, psilate-perforate, aperture margin similar; infratectum columellate; longest axis ranging from 40–44 µm [1/1]. *Pistillate flowers* ± globular; sepals 3, distinct, imbricate, short, broad, keeled; petals 3, distinct, imbricate, the tips minutely valvate at anthesis, about twice as long as sepals; staminodal ring low, membranous, with ca. 9, irregular, triangular lobes; gynoecium unilocular, uniovulate, spherical, stigmas 3,

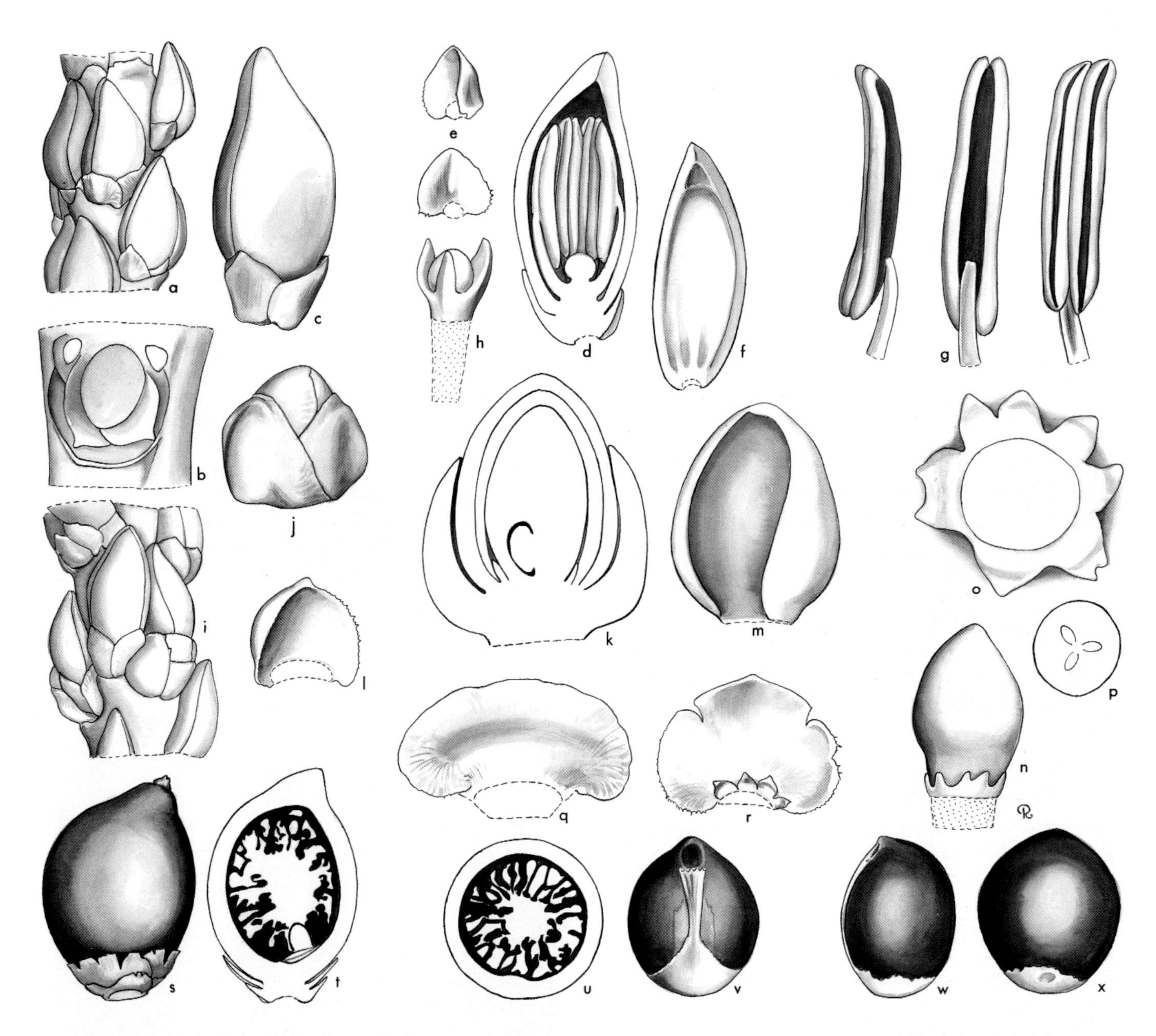

Loxococcus. **a**, distal portion of rachilla × 3; **b**, triad, flowers removed to show bracteoles × 6; **c**, staminate bud × 6; **d**, staminate bud in vertical section × 6; **e**, staminate sepal in 2 views × 6; **f**, staminate petal, interior view × 6; **g**, stamen in 3 views × 12; **h**, pistillode × 12; **i**, triads from inflorescence in bud × 3; **j**, pistillate bud × 6; **k**, pistillate bud in vertical section × 12; **l**, pistillate sepal, interior view × 6; **m**, pistillate petal × 12; **n**, gynoecium and staminodes × 12; **o**, staminodes × 24; **p**, ovary in cross-section × 12; **q**, sepal in fruit, interior view × 6; **r**, petal and staminodes in fruit, interior view × 6; **s**, fruit × 1$^{1}/_{2}$; **t**, fruit in vertical section × 1$^{1}/_{2}$; **u**, fruit in cross-section × 1$^{1}/_{2}$; **v**, **w**, **x**, seed in 3 views × 1. *Loxococcus rupicola*: all from *Moore et al.* 9031. (Drawn by Marion Ruff Sheehan)

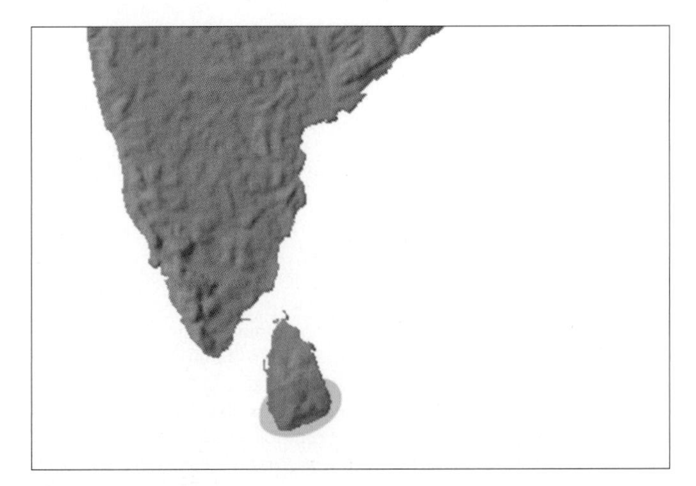

Distribution of ***Loxococcus***

reflexed apically, ovule laterally attached, form unknown. *Fruit* reddish-brown at maturity, ± spherical with a short, broad, slightly eccentric beak tipped with the stigmatic remains, perianth whorls persistent; epicarp smooth, mesocarp thin with numerous longitudinal, pale fibres, becoming free basally after disintegration of epicarp, endocarp thin, not adhering to the seed. *Seed* globose, basally and laterally attached with a ± circular, basal, slender, lateral hilum running ± the length of the seed, endosperm deeply ruminate; embryo basal. *Germination* and eophyll not recorded. *Cytology*: 2n = 32.

Distribution and ecology: One species in Sri Lanka. On cliffs, rocks and steep slopes in humid rain forest at altitudes of 300–1600 m.

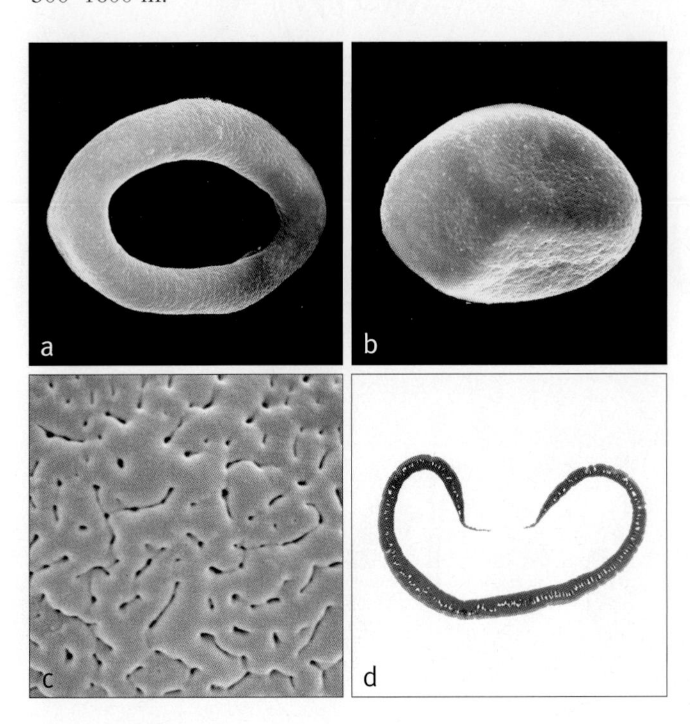

Above: ***Loxococcus***
a, monosulcate pollen grain, distal face SEM × 1000;
b, monosulcate pollen grain, proximal face SEM × 1000;
c, close-up, finely perforate pollen surface SEM × 8000;
d, ultrathin section through whole pollen grain, polar plane TEM × 1250.
Loxococcus rupicola: **a–d**, *Thwaites* CP 2732.

Above: ***Loxococcus rupicola***, crown, Sri Lanka. (Photo: M.D. Ferrero)

Anatomy: Fruit (Essig &Young 1979).
Relationships: It is striking that this Indian Ocean palm resolves as sister to the western Pacific clade with moderate support (Loo *et al.* 2006, Baker *et al.* in prep.). *Loxococcus* has also been poorly supported as sister to *Hydriastele* (Baker *et al.* in review).
Common names and uses: *Dotalu*. Seed is rarely used as a substitute for betel.
Taxonomic account: Beccari and Pichi-Sermolli (1955).
Fossil record: No generic records found.
Notes: *Loxococcus* is a rather poorly known genus. Though handsome, it seems to be in only a few collections.

183. RHOPALOBLASTE

Mostly tall (one species diminutive) pinnate-leaved tree palms from Malay Peninsula and Nicobar Islands, and the Moluccas, New Guniea and the Solomon islands, with distinctive rachillae that are coiled like intestines before emergence and with dark reddish-brown hairs along midrib.

Rhopaloblaste Scheff., Ann. Jard. Bot. Buitenzorg 1: 137 (1876). Type: *R. hexandra* Scheff. = **R. ceramica** (Miq.) Burret (*Bentinckia ceramica* Miq.).

Ptychoraphis Beccari, Ann. Jard. Bot. Buitenzorg 2: 90 (1885). Type: *P. singaporensis* (Becc.) Becc. (*Ptychosperma singaporense* Becc.) (= *Rhopaloblaste singaporensis* [Becc.] H.E. Moore).

Rhopalon — *club*, blaste — *bud, referring to the large club-shaped embryo*.

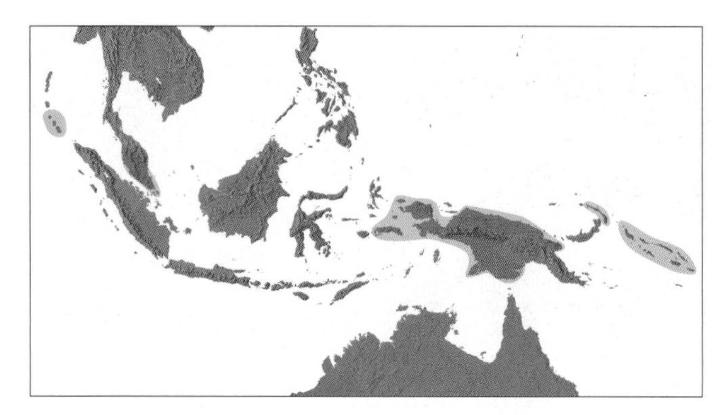

Distribution of ***Rhopaloblaste***

Small to large, solitary or clustered, unarmed, pleonanthic, monoecious palms. *Stem* short or tall, often enlarged at the base but uniform and moderate to slender above, clearly ringed with leaf scars, brown or grey. *Leaves* pinnate; sheaths tubular, forming a crownshaft, crownshaft sometimes obscured by dead leaves, ± densely covered with deciduous tomentum; petiole short to elongate, channelled adaxially, rounded abaxially; rachis rounded abaxially, angled above towards the apex, the sheath, lower surface of petiole and rachis usually densely covered with peltate scales with tattered interlocking margins, the upper surface of petiole and rachis usually with basifixed, twisted, entire or tattered, membranous scales persisting about the bases of the leaflets and where protected elsewhere; leaflets spreading or pendulous, with a pulvinus at the base, linear, acutely to acuminately and obliquely bifid apically, the midrib and one or more secondary ribs on each side prominent abaxially, minutely brown-dotted and at least the midrib with prominent dull brown, basifixed or medifixed, twisted, membranous scales basally or throughout, transverse veinlets not evident. *Inflorescences* borne below the leaves, branched to 3 orders basally, fewer orders distally, peduncle short; the prophyll 2-keeled, tubular, enclosing the similar peduncular bract, both usually ± tomentose at least when young, and caducous; rachis short to prominent but as long as or longer than the peduncle; basal branches usually abruptly divaricate, spreading at an angle of about 90° to the rachis (in *Rhopaloblaste singaporensis* often at an acute angle); bracts subtending the branches often prominent, pointed, rachilla bracts prominent or not, subtending triads nearly throughout; bracteoles surrounding the pistillate flower subequal or unequal, prominent and sepal-like. *Staminate flowers* symmetrical or nearly so at anthesis but in bud the outer sepal prominent and largely enfolding the remainder of the perianth; sepals 3, distinct, broadly imbricate at

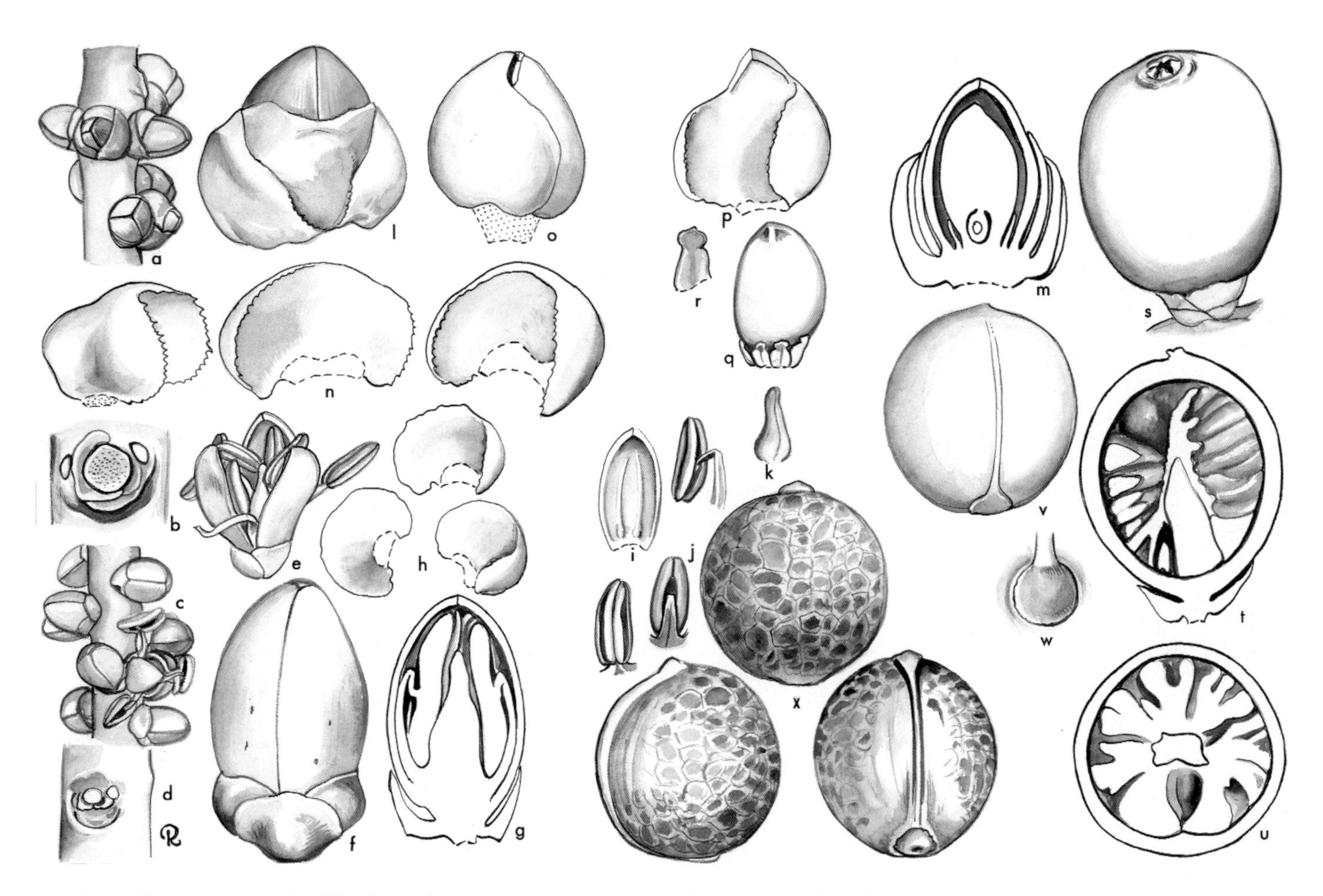

Rhopaloblaste. **a**, portion of rachilla with triads × 1$^{1}/_{2}$; **b**, triad, flowers removed × 3; **c**, portion of rachilla with paired and solitary staminate flowers × 1$^{1}/_{2}$; **d**, scars and bracteoles of paired staminate flowers × 3; **e**, staminate flower at anthesis × 3; **f**, staminate bud × 6; **g**, staminate bud in vertical section × 6; **h**, staminate sepals × 6; **i**, staminate petal, interior view × 3; **j**, stamen in 3 views × 3; **k**, pistillode × 3; **l**, pistillate bud × 6; **m**, pistillate bud in vertical section × 6; **n**, pistillate sepals × 6; **o**, pistillate bud, sepals removed × 6; **p**, pistillate petal × 6; **q**, gynoecium and staminodes × 6; **r**, staminode × 12; **s**, fruit × 1$^{1}/_{2}$; **t**, fruit in vertical section × 1$^{1}/_{2}$; **u**, fruit in cross-section × 1$^{1}/_{2}$; **v**, endocarp × 1$^{1}/_{2}$; **w**, operculum × 3; **x**, seed in 3 views × 1$^{1}/_{2}$. *Rhopaloblaste elegans*: all from *Moore & Whitmore* 9310. (Drawn by Marion Ruff Sheehan)

anthesis, rounded, ± gibbous and keeled dorsally; petals 3, distinct, valvate; stamens 6–9, the filaments very briefly connate basally or ± distinct, strap-shaped, narrowed and prominently inflexed at the apex in bud, the anthers narrowly elliptic in outline, medifixed, emarginate apically and basally, the connective prominent the entire length of the anther, latrorse; pistillode conical to columnar and ± angled, the apex briefly 3-lobed and sometimes somewhat expanded. *Pollen* ellipsoidal asymmetric; aperture a distal sulcus; ectexine tectate, finely to coarsely perforate and micro-channelled, aperture margin similar or slightly finer; infratectum columellate; longest axis 30–51 µm [4/6]. *Pistillate flowers* broader than high in bud and with the outer sepal usually enfolding the remainder of the perianth as in the staminate; sepals 3, distinct, broadly imbricate, rounded; petals 3, distinct, broadly imbricate basally, the short valvate apices erect and scarcely exceeding the sepals at anthesis but the petals in fruit generally nearly twice as long as the sepals; staminodes mostly 6, obtuse, ± deltoid, membranous, often united in pairs or irregularly united or united in a membranous, lobed ring; gynoecium irregularly ovoid, unilocular, uniovulate, stigmas erect to recurved between the valvate apices of the petals at anthesis, the ovule (in *R. ceramica*) hemianatropous, broader laterally, attached adaxially (in the ventral angle) and pendulous from the top of the locule. *Fruit* ovoid or ellipsoidal to subglobose, orange-yellow to red at maturity, with apical stigmatic remains; epicarp smooth, mesocarp lacking fibre sclereids or tannin cells, with flattened longitudinal fibres in one or usually more

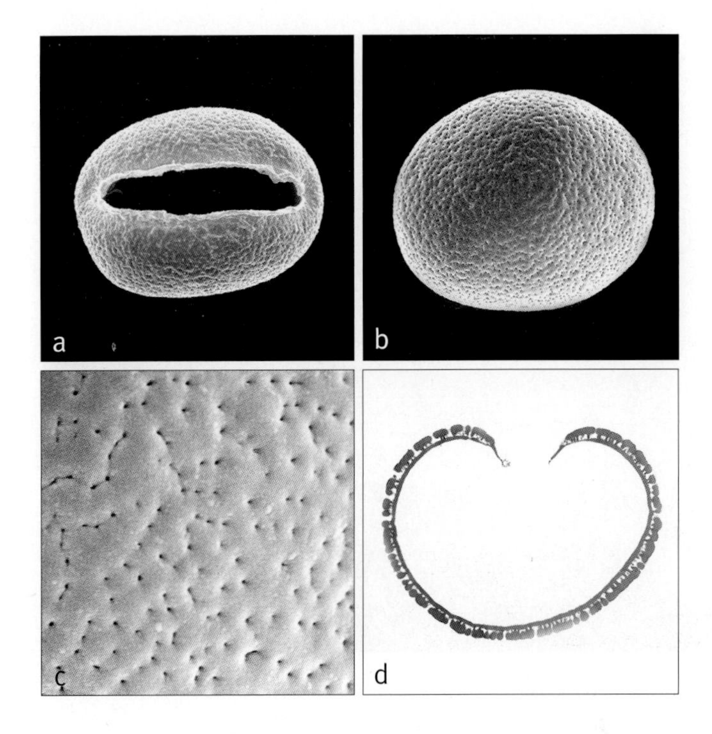

Above: ***Rhopaloblaste***
a, monosulcate pollen grain, distal face SEM × 1500;
b, monosulcate pollen grain, proximal face SEM × 1500;
c, close-up, finely perforate pollen surface SEM × 8000;
d, ultrathin section through whole pollen grain, polar plane TEM × 2000.
Rhopaloblaste augusta: **a–c**, *Furtado* 30946; *R. singaporensis*: **d**, *Lobb* 96.

Left: ***Rhopaloblaste ledermanniana***, habit, Papua. (Photo: W.J. Baker)

Below: ***Rhopaloblaste ledermanniana***, inflorescence, Papua. (Photo: S. Barrow)

Above: ***Rhopaloblaste ledermanniana***, inflorescence bud opened to show coiled and twisted branches, Papua. (Photo: W.J. Baker)

than one layer against the yellowish, fragile endocarp, this impressed over the hilum and with a round basal operculum. *Seed* brown, with a lightly to deeply impressed hilum along the adaxial side, raphe branches anastomosing, endosperm deeply ruminate; embryo basal, large. *Germination* adjacent-ligular, eophyll finely pinnate. *Cytology*: 2n = 32.

Distribution and ecology: Six species in the Nicobar Islands, Malay Peninsula and Singapore, the Moluccas, New Guinea and Solomon Islands. All species inhabit rain forests at relatively low elevations.

Anatomy: Leaf (Tomlinson 1961), root (Seubert 1998a, 1998b) and fruit (Essig *et al.* 1999).

Relationships: *Rhopaloblaste* is a robustly supported monophyletic genus (Norup *et al.* 2006). The genus has been resolved as sister to *Dictyosperma* with moderate support (Norup *et al.* 2006, Baker *et al.* in review, in prep.).

Common names and uses: *Kerinting* (*Rhopaloblaste singaporensis*). Widely grown as ornamentals.

Taxonomic account: Banka & Baker (2004).

Fossil record: No generic records found.

Notes: Few collections have been made from the wild and more are desirable. The distribution pattern is unusual.

Additional figures: Fig. 2.8g.

GENERIC NAMES OF DUBIOUS APPLICATION

The following generic names have been validly published but their application is in doubt because the original descriptions are too meagre and type specimens, if they still exist, have not yet been located.

Micronoma H. Wendl ex Hook.f. in Benth. & Hook.f., Gen. pl. 3: 882. 1883. Type not designated.

Paripon Voigt, Syll. Ratisb. 2: 51 (1828). Type: **P. palmiri** Voigt

ADDITIONAL GENERIC NAMES OF NO FORMAL STANDING

The following names, listed by Burret and Potztal (1956) as genera, were never validly published at the rank of genus, or are clearly printer's errors when published.

Anaclasmus Griff. (published as a section of *Areca* including species now referable to *Nenga* and *Pinanga*).

Anoosperma Kunze (typographical error = *Oncosperma*).

Araucasia Benth. & Hook.f. (typographical error for *Arausiaca* = *Orania*).

Blumea Zipp. ex Miq. (= *Arenga*).

Calycodon H. Wendl. (= *Hyospathe*).

Diploriphis Drude (published as a section of *Mauritia* including species now referable to *Mauritiella*).

Drypsis Duch. (typographical error = *Dypsis*).

Jatitara Marcgr. (= *Desmoncus*).

Oniscophora H. Wendl. ex Burret (? = *Dypsis*).

Opsicocos H. Wendl. (? = *Actinorhytis*).

GEOGRAPHICAL LISTING OF GENERA

Arabia
Hyphaene
Livistona
Nannorrhops
Phoenix

Argentina
Acrocomia
Allagoptera
Butia
Copernicia
Euterpe
Syagrus
Trithrinax

Australia (see also Lord Howe Island)
Archontophoenix
Arenga
Calamus
Carpentaria
Caryota
Cocos
Corypha
Hydriastele
Laccospadix
Licuala
Linospadix
Livistona
Normanbya
Nypa
Oraniopsis
Ptychosperma
Wodyetia

Bolivia
Acrocomia
Aiphanes
Astrocaryum
Attalea
Bactris
Ceroxylon
Chamaedorea
Chelyocarpus
Copernicia
Desmoncus
Dictyocaryum
Euterpe
Geonoma
Hyospathe
Iriartea
Iriartella
Mauritia
Mauritiella
Oenocarpus
Parajubaea
Phytelephas
Prestoea
Socratea
Syagrus
Trithrinax
Wendlandiella
Wettinia

Borneo
Adonidia
Areca
Arenga
Borassodendron
Calamus
Caryota
Ceratolobus
Corypha
Cyrtostachys
Daemonorops
Eleiodoxa
Eugeissona
Iguanura
Johannesteijsmannia
Korthalsia
Licuala
Livistona
Nenga
Nypa
Oncosperma
Orania
Pholidocarpus
Pinanga
Plectocomia
Plectocomiopsis
Pogonotium
Retispatha
Salacca

Brazil
Acrocomia
Aiphanes
Aphandra
Attalea
Allagoptera
Astrocaryum
Bactris
Barcella
Butia
Chamaedorea
Chelyocarpus
Copernicia
Desmoncus
Dictyocaryum
Elaeis
Euterpe
Geonoma
Hyospathe
Iriartea
Iriartella
Itaya
Leopoldinia
Lepidocaryum
Lytocaryum
Manicaria
Mauritia
Mauritiella
Oenocarpus
Pholidostachys
Phytelephas
Prestoea
Raphia
Socratea
Syagrus
Trithrinax
Wendlandiella
Wettinia

Central America
Acoelorrhaphe
Acrocomia
Aiphanes
Asterogyne
Astrocaryum
Attalea
Bactris
Brahea
Calyptrogyne
Chamaedorea
Colpothrinax
Cryosophila
Desmoncus
Elaeis
Euterpe
Gaussia
Geonoma
Hyospathe
Iriartea
Manicaria
Neonicholsonia
Oenocarpus
Pholidostachys
Phytelephas
Prestoea
Pseudophoenix
Raphia
Reinhardtia
Roystonea
Sabal
Schippia
Socratea
Synechanthus
Welfia

Chile (see also Juan Fernandez Islands)
Jubaea

China
Arenga
Borassus
Calamus
Caryota
Chuniophoenix
Corypha
Daemonorops
Guihaia
Licuala
Livistona
Nypa
Phoenix
Pinanga
Plectocomia
Rhapis
Salacca
Trachycarpus
Wallichia

Colombia
Acrocomia
Aiphanes
Ammandra
Asterogyne
Astrocaryum
Attalea
Bactris
Calyptrogyne
Ceroxylon
Chamaedorea
Chelyocarpus
Copernicia
Cryosophila
Desmoncus
Dictyocaryum
Elaeis
Euterpe
Geonoma
Iriartea
Iriartella
Itaya
Hyospathe
Leopoldinia
Lepidocaryum

Manicaria
Mauritia
Mauritiella
Oenocarpus
Pholidostachys
Phytelephas
Prestoea
Raphia
Reinhardtia
Roystonea
Sabal
Socratea
Syagrus
Synechanthus
Welfia
Wettinia

Ecuador

Aiphanes
Aphandra
Astrocaryum
Attalea
Bactris
Ceroxylon
Chamaedorea
Chelyocarpus
Desmoncus
Dictyocaryum
Elaeis
Euterpe
Geonoma
Hyospathe
Iriartea
Manicaria
Mauritia
Mauritiella
Oenocarpus
Parajubaea
Pholidostachys
Phytelephas
Prestoea
Socratea
Syagrus
Synechanthus
Welfia
Wettinia

Europe, N. Africa, Egypt, Asia Minor

Chamaerops
Hyphaene
Medemia
Phoenix

Fiji and Samoa

Balaka
Calamus
Clinostigma
Cyphosperma
Drymophloeus
Heterospathe
Metroxylon
Neoveitchia
Physokentia
Pritchardia
Solfia
Veitchia

Greater Antilles (Cuba, Hispaniola, Jamaica, Puerto Rico)

Acoelorrhaphe
Acrocomia
Attalea
Bactris
Calyptronoma
Coccothrinax
Colpothrinax
Copernicia
Gaussia
Geonoma
Hemithrinax
Leucothrinax
Prestoea
Pseudophoenix
Roystonea
Sabal
Thrinax
Zombia

Guianas

Acrocomia
Astrocaryum
Attalea
Bactris
Desmoncus
Dictyocaryum
Elaeis
Euterpe
Geonoma
Hyospathe
Iriartella
Lepidocaryum
Manicaria
Mauritia
Mauritiella
Oenocarpus
Prestoea
Roystonea
Socratea
Syagrus

Hawaii

Pritchardia

India, including Andamans and Nicobars

Areca
Arenga
Bentinckia
Borassus
Calamus
Caryota
Corypha
Daemonorops
Hyphaene
Korthalsia
Licuala
Livistona
Nannorrhops
Nypa
Phoenix
Pinanga
Plectocomia
Rhopaloblaste
Trachycarpus
Wallichia

Indochina

Areca
Arenga
Borassus
Calamus
Caryota
Corypha
Chuniophoenix
Daemonorops
Guihaia
Korthalsia
Licuala
Livistona
Myrialepis
Nenga
Nypa
Oncosperma
Phoenix
Pinanga
Plectocomia
Plectocomiopsis
Rhapis
Salacca
Trachycarpus
Wallichia

Iran, Afghanistan, Pakistan

Hyphaene
Nannorrhops
Phoenix

Java

Areca
Arenga
Borassus
Calamus
Caryota
Ceratolobus
Corypha
Daemonorops
Korthalsia
Licuala
Livistona
Nenga
Nypa
Oncosperma
Orania
Pinanga
Plectocomia
Salacca

Juan Fernandez Islands

Juania

Lesser Antilles

Acrocomia
Aiphanes
Coccothrinax
Desmoncus
Euterpe
Geonoma
Leucothrinax
Prestoea
Pseudophoenix
Roystonea
Sabal
Syagrus

Lesser Sunda Islands

Borassus
Calamus
Caryota
Corypha
Daemonorops
Licuala
Nypa

Lord Howe Island

Hedyscepe
Howea
Lepidorrhachis

Madagascar and Comores Islands

Beccariophoenix
Bismarckia
Borassus
Dypsis
Elaeis
Hyphaene
Lemurophoenix
Marojejya
Masoala
Orania
Phoenix
Raphia
Ravenea
Satranala
Tahina
Voanioala

Malay Peninsula

Areca
Arenga
Borassodendron
Calamus
Caryota
Ceratolobus
Corypha
Cyrtostachys
Daemonorops
Eleiodoxa
Eugeissona

Iguanura
Johannesteijsmannia
Korthalsia
Licuala
Livistona
Maxburretia
Myrialepis
Nenga
Nypa
Oncosperma
Orania
Phoenix
Pholidocarpus
Pinanga
Plectocomia
Plectocomiopsis
Pogonotium
Rhopaloblaste
Salacca

Marquesas

Pelagodoxa

Mascarenes

Acanthophoenix
Dictyosperma
Hyophorbe
Latania
Tectiphiala

Mexico

Acoelorrhaphe
Acrocomia
Astrocaryum
Attalea
Bactris
Brahea
Calyptrogyne
Chamaedorea
Coccothrinax
Cryosophila
Desmoncus
Gaussia
Geonoma
Pseudophoenix
Reinhardtia
Roystonea
Sabal
Synechanthus
Thrinax
Washingtonia

Micronesia

Clinostigma
Heterospathe
Hydriastele
Livistona
Metroxylon
Nypa
Pinanga
Ponapea

Moluccas

Areca
Arenga
Borassus
Calamus
Calyptrocalyx
Caryota
Corypha
Daemonorops
Drymophloeus
Heterospathe
Hydriastele
Licuala
Livistona
Metroxylon
Nypa
Oncosperma
Orania
Pholidocarpus
Pigafetta
Pinanga
Ptychosperma
Rhopaloblaste

Myanmar

Areca
Arenga
Borassus
Calamus
Caryota
Corypha
Daemonorops
Korthalsia
Licuala
Livistona
Myrialepis
Nypa
Oncosperma
Phoenix
Pinanga
Plectocomia
Salacca
Trachycarpus
Wallichia

New Caledonia

Actinokentia
Basselinia
Burretiokentia
Chambeyronia
Clinosperma
Cyphokentia
Cyphophoenix
Cyphosperma
Kentiopsis
Pritchardiopsis

New Guinea and the Bismarck Archipelago

Actinorhytis
Areca
Arenga
Borassus
Brassiophoenix
Calamus
Calyptrocalyx
Caryota
Clinostigma
Corypha
Cyrtostachys
Daemonorops
Dransfieldia
Drymophloeus
Heterospathe
Hydriastele
Korthalsia
Licuala
Linospadix
Livistona
Metroxylon
Nypa
Orania
Physokentia
Pigafetta
Pinanga
Ptychococcus
Ptychosperma
Rhopaloblaste
Sommieria

New Hebrides

Calamus
Carpoxylon
Caryota
Clinostigma
Cyphosperma
Heterospathe
Hydriastele
Licuala
Metroxylon
Neoveitchia
Physokentia
Veitchia

New Zealand

Rhopalostylis

Paraguay

Acrocomia
Allagoptera
Attalea
Bactris
Butia
Copernicia
Geonoma
Syagrus
Trithrinax

Peru

Aiphanes
Aphandra
Astrocaryum
Attalea
Bactris
Ceroxylon
Chamaedorea
Chelyocarpus
Desmoncus
Dictyocaryum
Euterpe
Geonoma
Hyospathe
Iriartella
Iriartea
Itaya
Lepidocaryum
Mauritia
Mauritiella
Oenocarpus
Pholidostachys
Phytelephas
Prestoea
Socratea
Syagrus
Wendlandiella
Wettinia

Philippines

Adonidia
Areca
Arenga
Livistona
Calamus
Caryota
Cocos
Corypha
Daemonorops
Heterospathe
Korthalsia
Licuala
Nypa
Oncosperma
Orania
Phoenix
Pinanga
Plectocomia
Salacca

Ryukyu Islands

Arenga
Livistona
Nypa
Phoenix
Satakentia

Seychelles

Deckenia
Lodoicea
Nephrosperma
Phoenicophorium
Roscheria
Verschaffeltia

Solomon Islands

Actinorhytis
Areca

Calamus
Caryota
Clinostigma
Cyrtostachys
Drymophloeus
Heterospathe
Hydriastele
Licuala
Livistona
Metroxylon
Nypa
Physokentia
Ptychosperma
Rhopaloblaste

Southern Africa

Borassus
Hyphaene
Jubaeopsis
Phoenix
Raphia

Sri Lanka (Ceylon)

Areca
Borassus
Calamus
Caryota
Corypha
Loxococcus
Oncosperma
Phoenix

Sulawesi

Areca
Arenga
Borassus
Calamus
Caryota
Corypha
Daemonorops
Hydriastele
Korthalsia
Licuala
Livistona
Nypa
Oncosperma
Orania
Pholidocarpus
Pigafetta
Pinanga

Sumatra

Areca
Arenga
Calamus
Caryota
Ceratolobus
Corypha
Cyrtostachys
Daemonorops
Eleiodoxa
Iguanura
Johannesteijsmannia
Korthalsia
Licuala
Livistona
Myrialepis
Nenga
Nypa
Oncosperma
Orania
Phoenix
Pholidocarpus
Pinanga
Plectocomia
Plectocomiopsis
Rhapis
Salacca

Thailand

Areca
Arenga
Borassodendron
Borassus
Calamus
Caryota
Ceratolobus
Corypha
Cyrtostachys
Daemonorops
Eleiodoxa
Eugeissona
Iguanura
Johannesteijsmannia
Kerriodoxa
Korthalsia
Licuala
Livistona
Maxburretia
Myrialepis
Nenga
Nypa
Oncosperma
Orania
Phoenix
Pinanga
Plectocomia
Plectocomiopsis
Rhapis
Salacca
Trachycarpus
Wallichia

Tropical East Africa

Borassus
Elaeis
Eremospatha
Hyphaene
Laccosperma
Livistona
Medemia
Phoenix
Raphia

Tropical West Africa

Borassus
Elaeis
Eremospatha
Hyphaene
Laccosperma
Oncocalamus
Phoenix
Podococcus
Raphia
Sclerosperma

Uruguay

Butia
Syagrus
Trithrinax

USA (continental)

Acoelorrhaphe
Coccothrinax
Leucothrinax
Pseudophoenix
Rhapidophyllum
Roystonea
Sabal
Serenoa
Thrinax
Washingtonia

Venezuela, Trinidad and Tobago

Acrocomia
Aiphanes
Asterogyne
Astrocaryum
Attalea
Bactris
Ceroxylon
Chamaedorea
Coccothrinax
Copernicia
Desmoncus
Dictyocaryum
Euterpe
Geonoma
Hyospathe
Iriartea
Iriartella
Leopoldinia
Lepidocaryum
Manicaria
Mauritia
Mauritiella
Oenocarpus
Prestoea
Roystonea
Sabal
Socratea
Syagrus
Wettinia

GLOSSARY

Abaxial — the side of an organ that faces away from the axis which bears it, e.g., the under surface of a leaf, the lower surface of the petiole, or the outer surface of a tubular bract (Glossary figs 1, 10g, 19).

Abscission — separation, often by a distinct layer.

Acanthophyll — a spine, often large, derived from a leaflet (Glossary fig. 6).

Acaulescent — lacking a visible stem (Glossary fig. 2e).

Acervulus — a group of flowers borne in a line (Fig. 1.8, Glossary fig. 18).

Acetolysis/acetolyse — a process that is used to clear pollen grains of their internal cellular fraction, the intine, and any external lipidic coatings. The acetolysis mixture comprises nine parts acetic anhydride and one part sulphuric acid.

Acropetal — toward the apex (Fig. 1.1).

Acuminate — tapering to a point (Glossary fig. 15a,c,f)

Acute — distinctly and sharply pointed but not drawn out (Glossary fig. 15b).

Adaxial — the side of an organ towards the axis that bears it; e.g., the upper side of a leaf, the upper surface of a petiole, or the inner surface of a tubular bract (Glossary figs 1, 19).

Adjacent-ligular — in germination, the new shoot developing next to the seed and enclosed by a ligule (Glossary fig. 25).

Adnate — attached to, usually of one kind of organ to another kind, as bract to axis (Glossary fig. 16).

AFLP — amplified fragment length polymorphism, a type of DNA data used to study relationships among populations or species.

Anastomose — to unite, usually forming a network (Glossary fig. 24).

Anatropous — an ovule bent parallel to its stalk so that the micropyle is adjacent to the hilum (Glossary fig. 21).

Androecium — collective term for the stamens (Fig. 1.9, Glossary fig. 17).

Anther — the part of a stamen that contains the pollen, usually two-chambered (Glossary figs 17, 19).

Anthesis — time of flower opening, when the flower is ready for fertilisation, i.e., shedding of pollen or receptivity of stigma.

Aperture — see 'Germinal aperture'.

Apiculate — bearing a short, sharp but not stiff point (Glossary fig. 19).

Apocarpous — with free carpels (Fig. 1.9, Glossary fig. 20).

Aril — an outgrowth of the stalk of the ovule, sometimes resembling a third integument (Glossary fig. 20).

Armed — bearing some form of spines.

Asymmetric/asymmetrical — without symmetry.

Auricle — an ear-like lobe (Glossary fig. 6).

Axillary — borne in an axil, which is the angle between the stem and the leaf or another organ that arises from the stem, such as a bract.

Basifixed — attached by the base (Glossary fig. 19).

Basipetal — developing from the apex toward the base (Figs 1.1, 1.4).

Bicarinate — having two keels (Glossary fig. 16).

Bifid — divided in two, usually equal, parts (Fig. 1.3, Glossary figs 13d, 15d,e,i).

Bijugate — used for a pinnate leaf with two pairs of leaflets (Glossary fig. 13e).

Bipinnate — doubly pinnate (Fig. 1.3, Glossary fig. 8c).

Bi-symmetric/bi-symmetrical — with two planes of symmetry.

Bootstrap support — an estimate of confidence in an individual clade within a phylogenetic tree. High bootstrap support indicates a high level of confidence that the clade will be resolved as new data are analysed. Low bootstrap support indicates a possibility that the clade may not be resolved by new data.

Bract — modified leaf associated with the inflorescence (Glossary fig. 18).

Bracteole — a small bract borne on a flower stalk, often present even when the flower is essentially sessile (Glossary fig. 18).

Bremer support — a measurement of the reliability of an individual clade with respect to alternative hypotheses. Bremer support is negatively affected by conflict in a dataset, decreasing when alternative hypotheses are nearly as good as the best hypothesis.

Brevisulcate — of a pollen grain having a very short sulcus. (Fig. a, p.157)

Caducous — dropping off early or gradually.

Caespitose — clustered (Fig. 1.1).

Callose envelope — callose is a structural polysaccharide found in higher plants. It has a blocking function — in the early stages of pollen development, a thick layer of callose is deposited around each pollen mother cell (meiocyte) separating them from each other (Fig. 2.4).

Calyx — the outermost or lower-most whorl of floral organs, the sepals (Glossary figs 17, 18, 19, 20).

Campylotropous — an ovule curved with the micropyle close to the hilum and the embryo sac also curved (Glossary fig. 21).

Capitate — head-like; in pollen, of columellae with expanded apices.

Carpel — the single unit of the gynoecium (Fig. 1.9, Glossary fig. 20).

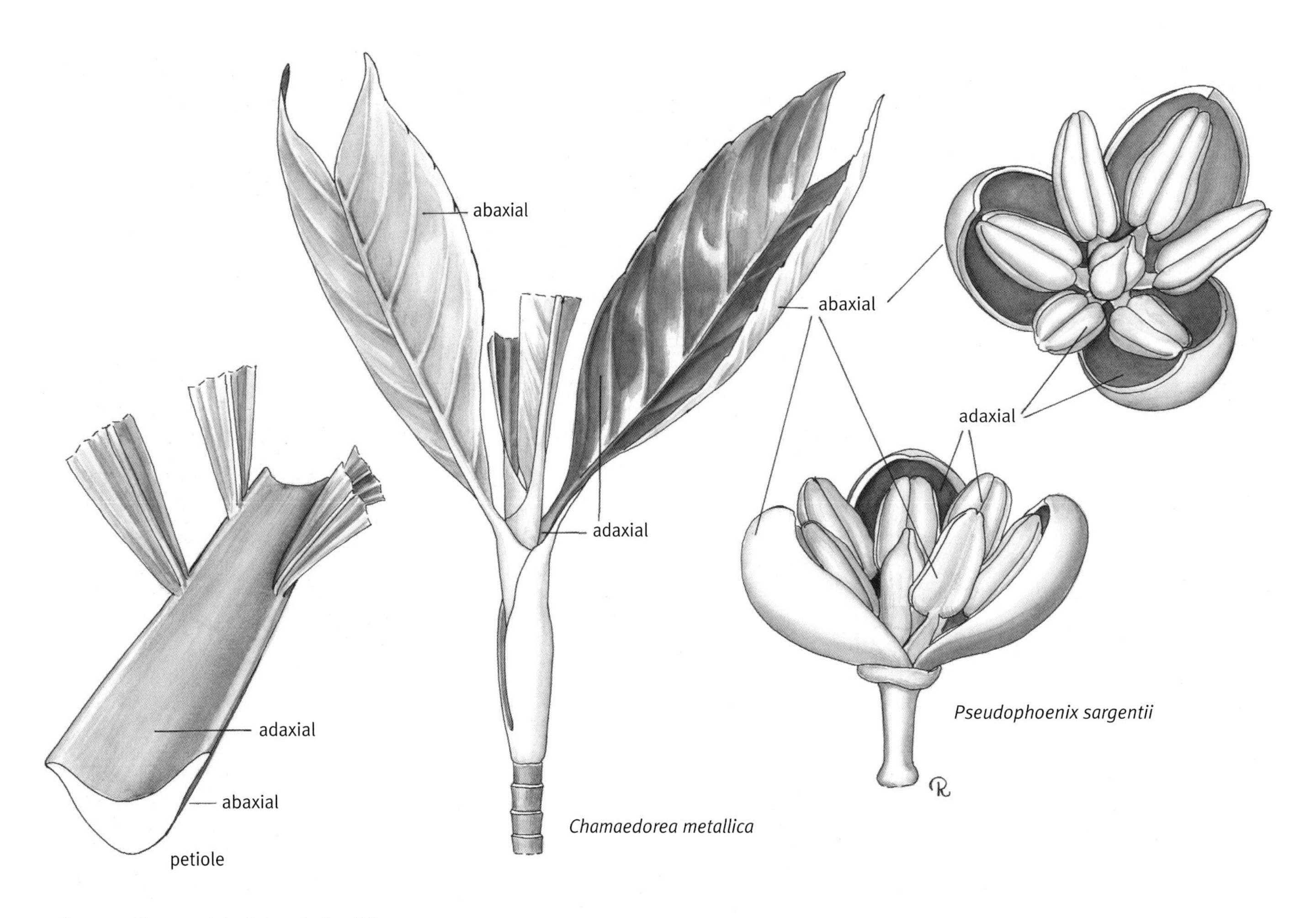

Glossary fig. 1 — Adaxial and abaxial (Drawn by Marion Ruff Sheehan)

Catkin-like — used to describe a cylindrical rachilla resulting when flowers are densely crowded (Glossary fig. 16).

Chartaceous — paper-like, thin and stiff.

Ciliate — bearing a fringe of hairs (Glossary fig. 18).

Cincinnus — a flower cluster wherein each successive flower arises in the axil of a bracteole borne on the stalk of the previous flower (Glossary fig. 18).

Cirrate — bearing a cirrus (Glossary fig. 6).

Cirrus — a climbing organ, structurally a whip-like extension of the leaf rachis, armed with reflexed spines and/or acanthophylls, cf. flagellum (Glossary fig. 6).

Clade — a monophyletic group. A branch on a phylogenetic tree.

Clava/clavae/clavate — a club-shaped element of the ectexine that is higher than 1 µm, with a diameter smaller than height and thicker at the apex than the base (Figs a, b, p.263; a–d, p.569)

Clustered — with several stems (Fig. 1.1, Glossary fig. 2b).

Columella/columellae — a rod-like element of the ectexine supporting the tectum (Fig. c, p.145).

Columellar layer — synonym of infratectum.

Concolorous — when two surfaces (usually leaves) are of the same colour.

Connate — united, used when organs of the same kind are joined, as sepals connate; see 'tubular calyx' (Fig. 1.9, Glossary fig. 20).

Connective — the part of a stamen that connects the anthers, usually distinct from the filament (Glossary fig. 19).

Consensus tree — a phylogenetic tree that summarises more than one hypothesis of relationships. A strict consensus tree only includes clades that are found in all hypotheses.

Coriaceous — leathery.

Corolla — the second whorl of flower organs, the petals, inside or above the calyx (Fig. 1.9, Glossary figs 19, 20).

Cortex — the ground tissue between the vascular cylinder and the epidermis.

Costa — a rib, when single, the midrib or vein.

Costapalmate — shaped like the palm of a hand and having a short midrib or costa (Fig. 1.3, Glossary figs 8d, 10g,h).

Crenulate — shallowly scalloped or bearing rounded teeth (Glossary fig. 20).

Crotonoid — a characteristic type of ectexine ornamentation comprising rings of five or six (sometimes more) raised, often triangular, elements arranged around a circular area, usually formed by capitate columellae.

Crown — the cluster of leaves borne at the tip of the stem.

Glossary fig. 2 — Habits

a, solitary (*Roystonea regia*);
b, clustered (*Chamaedorea* sp.);
c, **d**, dichotomously branched (*Hyphaene coriacea*);
e, acaulescent (*Sabal etonia*);
f, decumbent (*Serenoa repens*);
g, climbing (*Korthalsia echinometra*).

(All photos C.E. Lewis)

Glossary fig. 3 — Stem bases and roots

a, cylindrical (*Sabal domingensis*);
b, basally swollen (*Borassus flabellifer*);
c, ventricose (*Pseudophoenix vinifera*);
d, roots exposed (*Heterospathe elata*);
e, stilt roots (*Areca novohibernica*);
f, root boss (*Thrinax radiata*).

(All photos C.E. Lewis)

Crownshaft — a conspicuous cylinder formed by tubular leaf sheaths at the top of the stem (Fig. 1.4, Glossary fig. 5a,b).
Cupule — a small cup-like structure (Glossary fig. 20).
Cuticle — a layer of wax covering the epidermis.
Cytokinesis — the process of cytoplasmic division.
Deciduous — shed periodically, falling.
Decumbent — reclining, but with the apex or tip ascending (Glossary fig. 2f).
Decussate tetrad — a multi-planar tetrad of pollen grains or spores arranged in two pairs lying across one another, more or less at right angles.
Deltoid — shaped like an equilateral triangle.
Dentate — toothed (Glossary fig. 20).
Determinate — bearing a terminal organ, as in an inflorescence when a branch ends in a flower bud.
Dichotomous (stem) — equally forking (Fig. 1.1, Glossary fig. 2c,d).
Didymous — of anthers where the connective is almost absent (Glossary fig. 19).
Digitate — like fingers.
Dimorphic — with two different forms (Glossary fig. 17).
Dioecious — when staminate and pistillate flowers are borne on different plants.
Diporate — of a pollen grain having two pori (germinal apertures) (Figs a, b, p.172).
Discolorous — when two surfaces (usually leaves) are of a different colour. (c.f. concolorous)
Distal — farthest from the place of attachment; in pollen, applied to features of the mature pollen grain that face outward during the tetrad stage.
Distal sulcate — of a pollen grain with paired sulci lying parallel to the long axis of the pollen grain on the distal face (Figs a, p.249; a, b, p.259)
Distant — widely separated.
Distichous — arranged in two ranks (Glossary figs 14, 16).
Disulcate — of a pollen grain having two sulci (germinal apertures). In palms, usually positioned on the short equatorial faces of the pollen grain but see also 'distal disulcate'.
Divaricate — spread widely (Glossary fig. 19).
Dorsifixed — attached by the abaxial (usually outer) side (Glossary fig. 19).
Dyad — a pair (Glossary fig. 18).
Ectexine — the outer layer of the pollen exine, in TEM, it has higher electron density than endexine (Fig. 2.7).
Elliptic — oblong with regularly rounded ends (Glossary fig. 19).
Emarginate — with a notch cut out, often at the apex (Glossary fig. 15i).
Embryo — the rudimentary plant present in a seed (Fig. 1.2, Glossary figs 23, 24).
Endexine — the inner layer of the pollen exine, it has lower electron density than ectexine.
Endocarp — the innermost layer of the fruit wall (Glossary figs 23, 24).
Endosperm — in palms, the nutritive body of a seed (Glossary figs 23, 24).
Entire (leaf) — undivided (Fig. 1.3, Glossary fig. 10b,f).
Eophyll — in a seedling, the first leaf having a blade (Glossary fig. 24).
Epicarp — the outermost layer of the fruit wall (Glossary figs 22, 23).
Epipetalous — united with, often appearing to be borne on, the petals (Glossary fig. 19).
Equator/equatorial — the dividing line between the distal and proximal faces of a pollen grain or spore.
Exine — the outer usually sporopollenin layer of a pollen grain, which has an important protective function during pollen transfer from anther to stigma.
Exposed columellae — columellae not overlain by a tectum, usually in the form of spines or clavae in intectate pollen (Figs a–d, p.306).
Extended sulcate — of a pollen grain having a single sulcus, the apices of which extend beyond the equator towards the proximal face (Figs a, d, p.264).
Extrorse — of anthers, opening abaxially, away from the centre of the flower (Glossary fig. 19).
Fenestrate — with holes or openings, sometimes like windows.
Fibrous — composed of or including fibres.
Filament — of a stamen, the stalk which bears the anther (Glossary figs 17, 19).
Filamentous — thread-like.
Flabellate — fan-shaped or wedge-shaped (Glossary fig. 15h).
Flagellum — a whip-like climbing organ derived from an inflorescence, bearing reflexed spines, cf. 'cirrus' (Glossary fig. 6).
Floccose — bearing soft, uneven hairs or scales (Glossary fig. 7a,b).
Foot layer — the inner layer of the ectexine (Fig. 2.7).
Foveola/foveolae/foveolate — large rounded holes or depressions (lumina) in or through the tectum, which are too large to be described as perforations (>1 μm in diameter), separated by broad areas of tectum similar in width, or wider than, the foveolae (Fig. c, p.264).
Funiculus — the stalk of an ovule (Glossary fig. 21).
Gamo — as a prefix, united or fused (see sepals and petals).
Gemma/gemmae/gemmate — sexine elements that are constricted at the base, higher than 1 μm, and approximately the same width as height (Figs a, b, p.406; j, p.444)
Germinal aperture — the functional opening, covered by an aperture membrane *in vivo*, through which the pollen tube emerges (in palms a sulcus or pore).
Gibbous — more convex on one side than another, as a not quite full moon.
Glabrous — smooth, lacking hairs, scales or other indument.
Glaucous — covered with a bluish gray or greenish bloom.
Glomerule — a knob-like cluster of flowers (see *Livistona*).

Glossary fig. 4 — Stems

a, reed-like (*Geonoma* sp.);
b, ringed (*Veitchia joannis*);
c, ringed (*Dypsis lutescens*);
d, obliquely ringed (*Borassus aethiopum*);
e, large flat scars (*Phoenix canariensis*);
f, inflorescence scars (*Hyophorbe verschaffeltii*);
g, striate (*Sabal domingensis*);
h, rough-ringed with eroding leaf bases (*Sabal mexicana*);
i, corky (*Schippia concolor*);
j, spiny (*Aiphanes aculeata*);
k, root spines (*Cryosophila stauracantha*);
l, root spines (*Mauritiella armata*).

(All photos C.E. Lewis)

Glossary fig. 5 — Leaf sheaths and bases

a, crownshaft (*Pinanga elmeri*);
b, crownshaft with vertical abscission zone (*Pseudophoenix vinifera*);
c, persistent fibrous leaf bases (*Syagrus botryophora*);
d, leaf sheaths net-like (*Coccothrinax salvatoris*);
e, whorls of leaf sheath spines (*Zombia antillarum*);
f, sheaths disintegrating into rigid fibres (*Arenga undulatifolia*);
g, leaf bases spiny (*Syagrus schizophylla*);
h, leaf bases unarmed (*Copernicia alba*);
i, bulbous leaf bases (*Livistona benthamii*);
j, leaf bases spirally arranged (*Corypha utan*);
k, leaf bases split (*Latania loddigesii*);
l, marcescent leaves (*Copernicia macroglossa*).

(All photos C.E. Lewis)

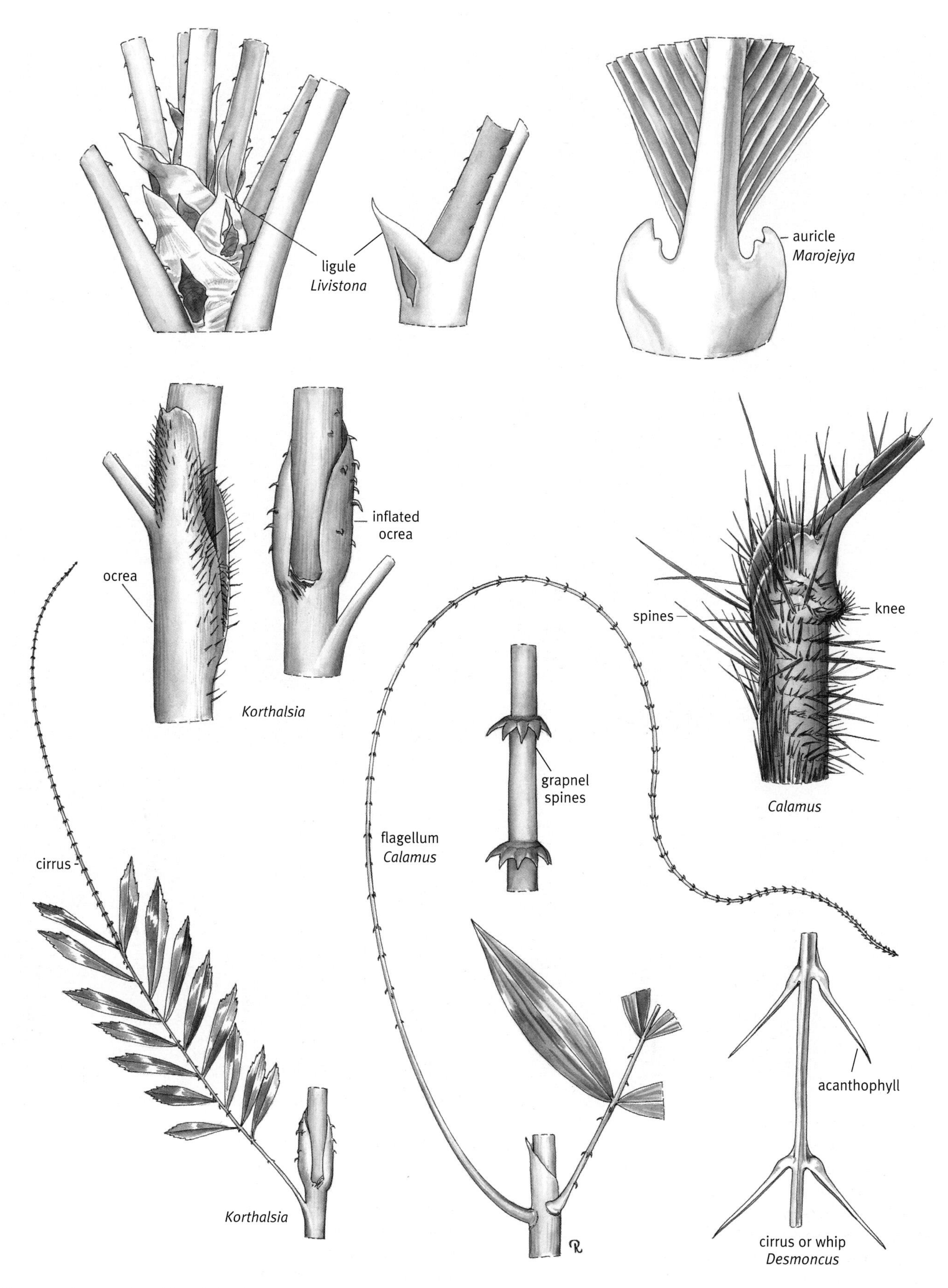

Glossary fig. 6 — Special leaf structures (Drawn by Marion Ruff Sheehan)

Glossary fig. 7 — Indumentum and spines

a, floccose (*Pritchardia hillebrandii*);
b, floccose (*Bismarckia nobilis*);
c, ramenta (*Nypa fruticans*);
d, spiny leaflets and rachis (*Acrocomia crispa*);
e, sheath spines (*Acanthophoenix rubra*);
f, sheath and petiole spines (*Daemonorops curranii*);
g, petiole spines (*Elaeis guineensis*);
h, petiole spines (*Phoenix rupicola*);
i, petiole spines (*Syagrus schizophylla*);
j, petiole spines (*Livistona robinsoniana*);
k, petiole spines (*Serenoa repens*);
l, petiole spines (*Chamaerops humilis*);
m, petiole spines (*Licuala spinosa*);
n, petiole spines (*Phoenix paludosa*);
o, petiole spines (*Salacca zalacca*);
p, petiole spines (*Bactris gasipaes*).

(All photos C.E. Lewis)

Glossary fig. 8 — Leaf form

a, pinnate (*Chamaedorea tepejilote*);
b, palmate (*Coccothrinax jamaicensis*);
c, bipinnate (*Caryota mitis*);
d, costapalmate (*Sabal maritima*).

(All photos C.E. Lewis)

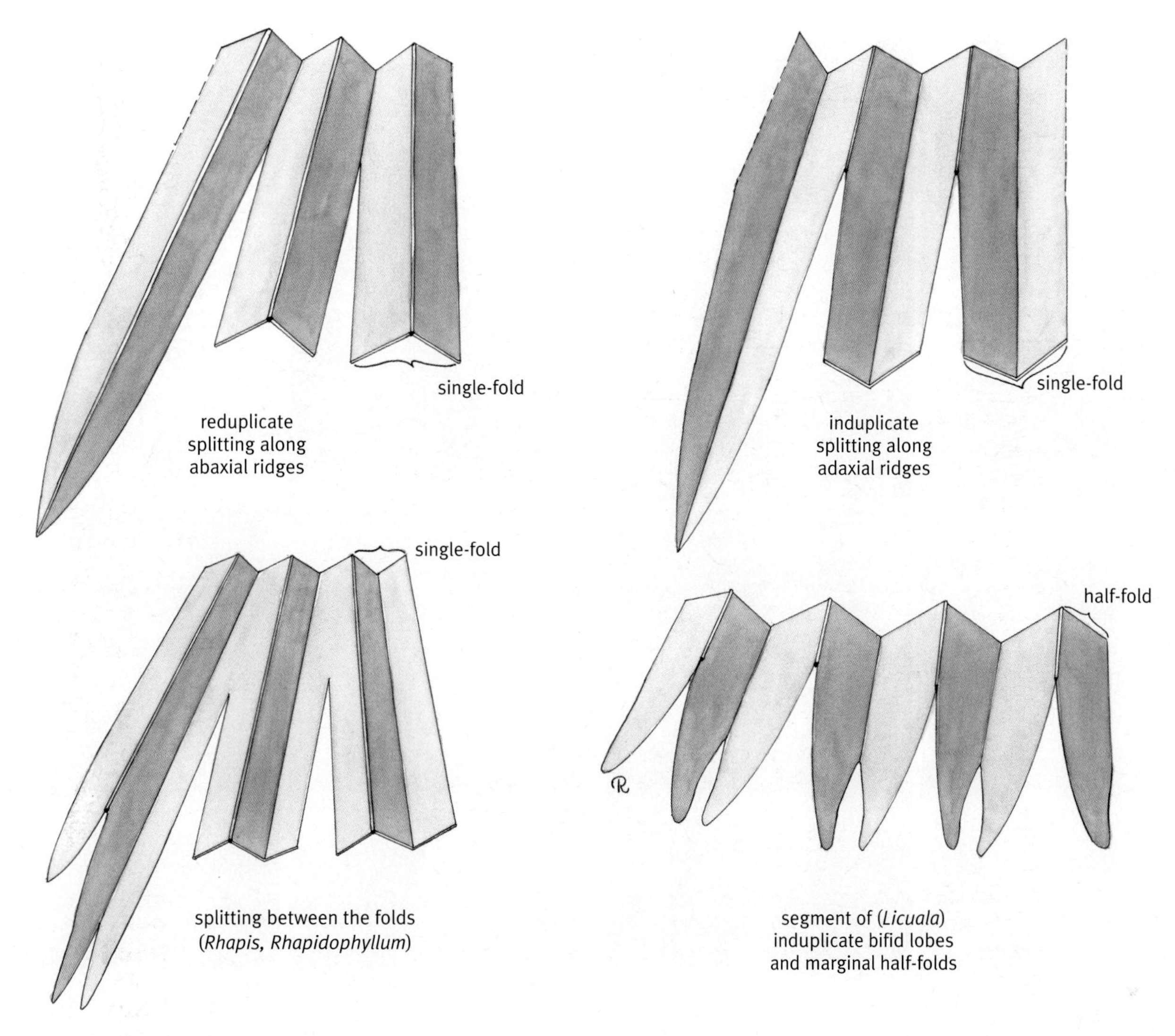

Glossary fig. 9 — Leaf division (Drawn by Marion Ruff Sheehan)

Granulae/granular — general word for small rounded elements.

Grapnel — a small anchor or hook with three or more flukes (Glossary fig. 6).

Gynoecium — the ovule-bearing organ of the flower, composed of one to several carpels and divisible into an ovary, a style, and one or several stigmas (Fig. 1.9, Glossary figs 17, 20).

Hapaxanthic — describing shoots that flower then die, cf. 'pleonanthic' (Figs 1.1, 1.2, 1.4)

Haploid — a nucleus or individual containing only one representative of each chromosome of the chromosome complement.

Hastula — a flap of tissue borne at the insertion of the blade on the petiole on the upper, lower, or both leaf surfaces (Glossary fig. 11).

Hemianatropous — an ovule turned so that the micropyle is at right angles to its stalk (funiculus) (Glossary fig. 21).

Hermaphrodite — flowers having both stamens and gynoecium (Fig. 1.9, Glossary fig. 17).

Heteropolar — pollen grains or spores where the distal and proximal faces differ; for example, in the monosulcate pollen grains of palms.

Hilum — the scar left on the seed where it was attached (Glossary fig. 24).

Homogeneous — uniform, the same throughout (Glossary fig. 24).

Imbricate — overlapping, cf. 'valvate' (Glossary figs 19, 20, 22).

Imparipinnate — unevenly pinnate, bearing a terminal leaflet.

Impression mark — a mark on the proximal face of the pollen grain retained from the post-meiotic stage. This mark can be linear from tetragonal or decussate tetrads or Y-shaped from tetrahedral tetrads (Figs 2.5c; 2.6c).

Inaperturate — without a germinal aperture (Fig. b, p.181).

Incomplete equatorial — of a pollen grain having a very elongated (extended) sulcus, extending almost completely around the equator so that the apices all but meet (Fig. n, p.510).

Indeterminate — not bearing a terminal flower or other organ.

Glossary fig. 10 — Palmate leaves

a, deeply split, segments several-fold (*Licuala spinosa*);
b, entire, many fold (*Licuala grandis*);
c, deep central split (*Cryosophila kalbreyeri*);
d, segments single-fold (*Schippia concolor*);
e, intersegmental fibres (*Medemia argun*);
f, leaf entire (*Johannesteijsmannia altifrons*);
g, costapalmate abaxial view (*Corypha umbraculifera*);
h, briefly costapalmate, adaxial view (*Mauritia flexuosa*).

(All photos C.E. Lewis)

Indument — any covering as hairs or scales.

Induplicate — V-shaped in cross-section, trough-shaped (Fig. 1.3, Glossary figs 9, 12j).

Inflexed — bent or curved inward toward the centre (Glossary fig. 19).

Inflorescence — the branch that bears the flowers, including all its bracts and branches (Figs 1.4, 1.6, Glossary fig. 16).

Infrafoliar — borne below the leaves (Fig. 1.4).

Infratectum — a general term for the ectexine layer beneath the tectum, and above the foot layer (Fig. 2.7).

Infructescence — as inflorescence but in fruit.

Intectate — without a tectum but with sculptural elements (Figs a–d, p.306).

Integument — a covering or envelope enclosing the ovule (Glossary fig. 21).

Interfoliar — borne among the leaves (Fig. 1.4).

Internal cellular fraction — the internal part of the pollen grain including the generative and vegetative cells and surrounding cytoplasm and organelles (Fig. 2.7).

Internode — the space or part of a stem between the attachments of two leaves.

Interrupted reticulum/reticulate — a reticulum where the muri are not continuous, as in a net, but appear 'broken up' or fragmented (Fig. b, p.292).

Intersegmental fibres — fibres resulting from disintegration of the ribs between the segments of palmate leaves.

Intine — the inner layer, of the pollen grain wall ('sporoderm'), which has an important protective function during pollen germination and pollen tube growth (Fig. 2.7).

Introrse — of anthers, opening toward the centre of the flower (Glossary fig. 19).

Glossary fig. 11 — Hastulas

a, b, adaxial, abaxial (*Cryosophila williamsii*);
c, d, adaxial, abaxial (*Acoelorrhaphe wrightii*);
e, f, adaxial, abaxial (*Guihaia argyrata*);
g, h, adaxial, abaxial (*Kerriodoxa elegans*);
i, j, adaxial, abaxial (*Livistona australis*);
k, l, adaxial, abaxial (*Zombia antillarum*);
m, n, adaxial, abaxial (*Sabal etonia*);
o, p, adaxial, abaxial (*Latania verschaffeltii*).

(All photos C.E. Lewis)

Glossary fig. 12 — Pinnate leaves

a, b, regularly arranged, 2-ranked (*Archontophoenix myolensis*) (abaxial view);
c, d, several-ranked (*Phoenix reclinata*);
e, f, pendulous, leaflets subopposite (*Euterpe oleracea*);
g, h, grouped, fanned within the group (*Allagoptera arenaria*);
i, reduplicate;
j, induplicate.

(All photos C.E. Lewis)

Glossary fig. 13 — Some pinnate leaf forms

a, leaflets subopposite (*Chamaedorea radicalis*);
b, distichous paripinnate (*Pinanga dicksonii*);
c, irregularly pinnate (*Ptychosperma microcarpum*);
d, bifid (*Chamaedorea brachypoda*);
e, bijugate (*Reinhardtia latisecta*);
f, leaflets joined by marginal rein (*Dictyosperma album*).

(All photos C.E. Lewis)

Jacknife support — an estimate of confidence in an individual clade within a phylogenetic tree. See **Bootstrap support** for interpretation of support values.

Keeled — bearing a ridge, like the keel of a boat (Glossary fig. 20).

Knee — a swelling on the leaf sheath at the base of the petiole, present in most rattans (Glossary fig. 6).

Lamellae/lamellar — a general term for a thin layer — in palm pollen, these layers refer to a lamellated foot layer as in Calamoideae, they are unusual elsewhere in the family (Figs b, p.159; b, p.166).

Lamina — blade.

Lanceolate — narrow, tapering at both ends, the basal end often broader (Glossary fig. 14c,d).

Latrorse — of anthers, opening lateral to the filament (Glossary fig. 19).

Leaflets — as used in this book, divisions of pinnate leaves (Glossary fig. 14).

Glossary fig. 14 — Leaflet shapes

a, linear (*Pseudophoenix vinifera*);
b, linear, spiny (*Acrocomia crispa*);
c, lanceolate (*Archontophoenix tuckeri*);
d, lanceolate, with tail-like projection (*Desmoncus cirrhiferus*);
e, undulate, sinuous (*Arenga undulatifolia*);
f, truncate tip (*Ptychosperma lauterbachii*);
g, multi-fold, sigmoid (*Pinanga coronata*);
h, flabellate, praemorse (*Caryota rumphiana*);
i, rhomboid, praemorse (*Arenga caudata*);
j, concavely praemorse (*Ptychosperma cuneata*);
k, tri-lobed, praemorse (*Brassiophoenix drymophloeoides*).

(All photos C.E. Lewis)

Glossary fig. 15 — Leaf tips

a, linear acuminate (*Phoenix reclinata*);
b, acute (*Hyophorbe lagenicaulis*);
c, acuminate notched (*Brahea* sp.);
d, deeply bifid (*Nannorrhops ritchiana*);
e, deeply bifid with persistent fibre (*Sabal palmetto*);
f, acuminate with intersegmental fibres (*Washingtonia filifera*);
g, oblique (*Kentiopsis oliviformis*);
h, oblique, notched (*Attalea crassispatha*);
i, bifid emarginated acute (*Chamaedorea metallica*);
j, praemorse truncate (*Reinhardtia* sp.);
k, rhomboid, toothed (*Wallichia oblongifolia*);
l, concavely praemorse (*Ptychosperma macarthurii*);
m, unevenly praemorse (*Hydriastele pleurocarpa*).

(All photos C.E. Lewis)

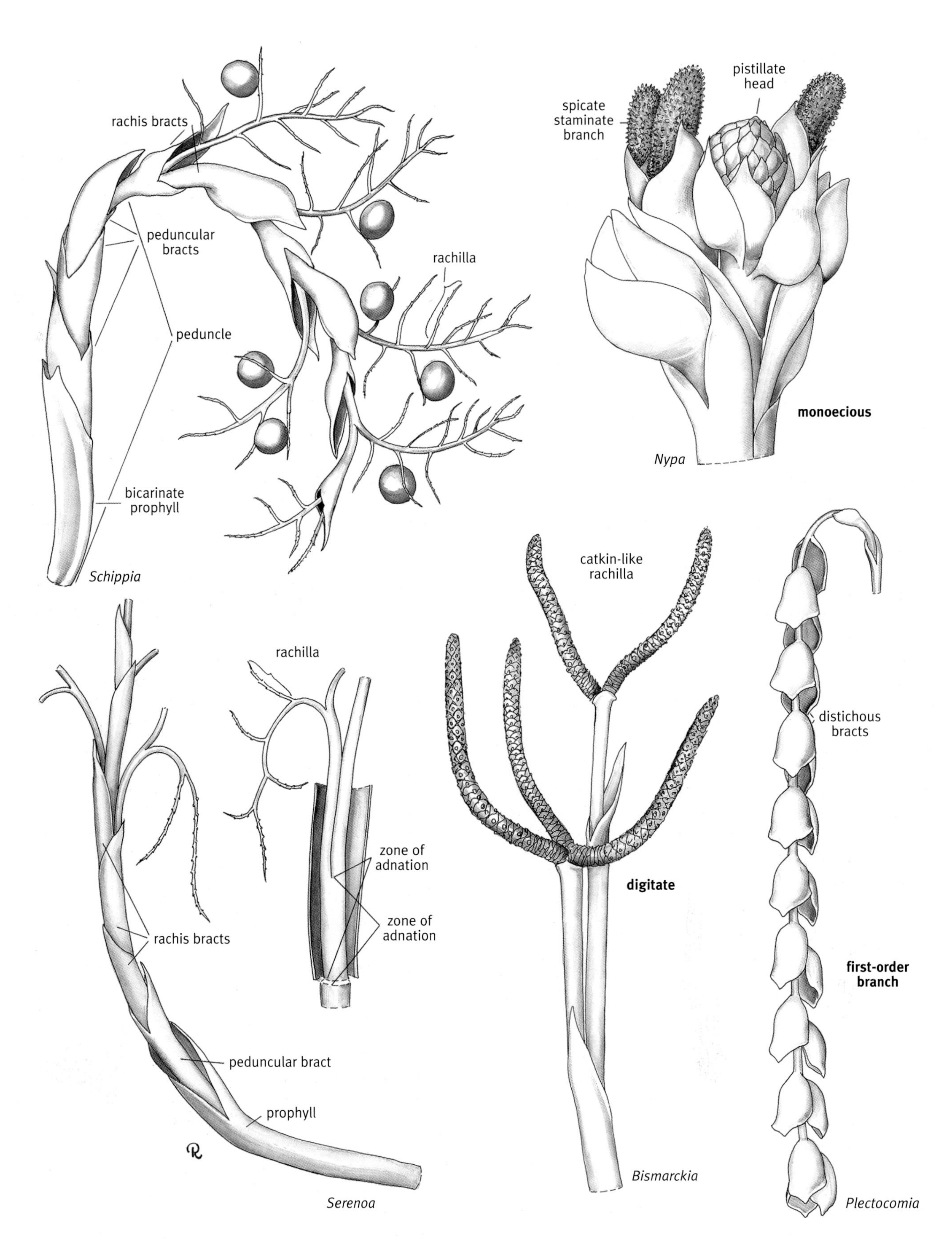

Glossary fig. 16 — Inflorescences (Drawn by Marion Ruff Sheehan)

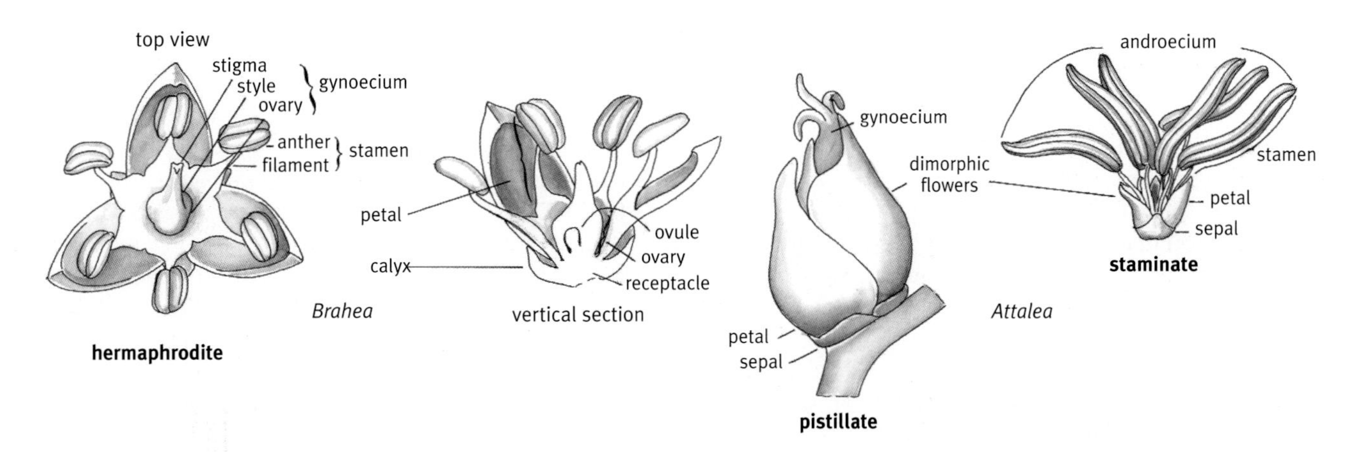

Glossary fig. 17 — Parts of flowers (Drawn by Marion Ruff Sheehan)

Ligule — a distal projection of the leaf sheath (Glossary figs 6, 25).

Linear — several times longer than wide, usually narrow (Glossary figs 14a,b, 15a).

Linear impression mark — a mark on the proximal face of the pollen grain retained from the post-meiotic stage of tetragonal or decussate tetrad (Fig. 2.5c).

Lipid-based coatings — 'pollenkitt'/'pollen coat' — sticky material produced by the tapetum that is considered to have a range of probable functions (Hesse 2000).

Locule — the cavity in which the ovule is borne (Glossary fig. 20).

Long axis — the widest dimension of the pollen grain, particularly used for heteropolar pollen.

Lumina/lumen — the space(s) enclosed/surrounded by the muri in a reticulate ectexine (Fig. c, p.513).

Marcescent — withering before being shed (Glossary fig. 1).

Maximum likelihood — a criterion for choosing among different hypotheses of relationships, favouring the hypothesis that best fits a defined model of evolution.

Medifixed — attached at the middle.

Meridional/meridionally — describing longitudinal features on a pollen grain or spore oriented perpendicular to the equator.

Meristem — the growing region of a plant, a special area of undifferentiated cells wherein new cells and organs are developed.

Mesocarp — the middle layer of the fruit wall (Glossary fig. 23).

Micro-channels — small channels in the tectum that penetrate to the infratectum (Fig. c, p.641).

Micro-fossulae — channel-like indentations in the tectum that do not penetrate to the infratectum.

Micropyle — an aperture through the integuments of the ovule (Glossary fig. 21).

Microsatellites — regions of repetitive DNA that are highly variable and useful for genetic analyses of populations or species.

Microsporogenesis — in pollen, the development from the sporogenous tissue of the anther to maturity as (usually) individual pollen grains.

Mitochondrial DNA — DNA within mitochondria, which are organelles found inside the cells of plants and many other kinds of organisms.

Moderate — medium or intermediate in size.

Monocarpic — fruiting once, then dying completely.

Monoecious — describing a plant bearing both staminate and pistillate flowers (Glossary fig. 16).

Monomerous — formed of a single member.

Monophyletic group — a taxonomic group that includes all descendants of a common ancestor. A branch on a phylogenetic tree. Monophyletic groups are the basis of modern, phylogenetic classifications.

Monopodial — with a single main axis.

Monoporate — of a pollen grain, having a single, presumed distal, pore (germinal aperture) (Figs a, b, p.344).

Monosulcate — of a pollen grain, having a single sulcus (germinal aperture) (Fig. 2.1a).

Morphology/morphological — the study of form, particularly of external structures — the introduction of the word is attributed to the poet and botanist Johann Wolfgang von Goethe.

Muri/murus — a continuous ridge that separates the lumina of a reticulate ectexine (Fig. c, p.513).

Node — the area of a stem where a leaf is attached.

Nuclear DNA — the majority of DNA found in plant cells, located within the nucleus.

Nucleotide substitution — a mutation that replaces one nucleotide in a segment of DNA with another. When present in more than one taxon, these mutations may provide evidence of phylogenetic relationship.

Oblate — ± spherical but flattened at the poles; of pollen, a rather flattened spheroidal or trianguloid pollen grain (Fig. g, p.510).

Obovoid — egg-shaped, broader distally (Glossary fig. 24).

Obpyriform — pear-shaped but attached at the broad end.

Ocrea — an extension of the leaf sheath beyond the petiole insertion (Glossary fig. 6).

Ontogeny — the development of an individual through its various stages.

Operculum — a lid or cover (Glossary fig. 24); in pollen, a distinctly delimited ectexinous lid-like structure that covers all or part of an aperture (Fig. a, p.341).

Orbicule/orbiculae (or Ubisch body) — a distinctive granule, usually orbicular, of sporopollenin produced by the tapetum. These may be deposited on the outer exine of maturing pollen grains.

Order (and its extensions, first, etc.) — a sequence, as of branching: a first-order branch branches to produce a second-order branch, etc.

Orthotropous — an erect ovule with the micropyle distal and hilum basal (Glossary fig. 21).

Ovary — the ovule-bearing part of the gynoecium (Fig. 1.9, Glossary fig. 17).

Ovoid — egg-shaped.

Ovule — the young or developing seed (Glossary figs 17, 20).

Palman — the undivided central part of a palmate leaf.

Palmate — shaped like the palm of the hand, all ribs or leaflets arising from a central area (Fig. 1.3, Glossary figs 8b, 10).

Papillose — bearing nipple-like projections.

Paraphyletic — a taxonomic group that includes some, but not all descendants of a common ancestor.

Paripinnate — evenly pinnate, lacking a terminal leaflet (Glossary fig. 13b).

Parsimony — a criterion for choosing among different hypotheses of relationships. Parsimony favours the hypothesis that offers the simplest explanation for observed data.

PCR — polymerase chain reaction, a process that allows specific DNA regions to be targeted and copied for many types of genetic analysis, including sequencing.

Pedicel — a floral stalk (Glossary figs 18, 19).

Peduncle — the lower unbranched part of an inflorescence (Figs 1.5, 1.6, Glossary fig. 16).

Peduncular bracts — empty bracts on the main inflorescence axis between the prophyll and the first rachis bract (Figs 1.4, 1.5, 1.6, Glossary fig. 16).

Peltate — round, shield-like, attached in the centre.

Perforate/perforation — in pollen morphology, applied to holes through the tectum of less than 1 µm in diameter (Fig. c, p.645).

Perianth — the sepals and the petals together (Fig. 1.9).

Petal — one unit of the inner floral envelope or corolla (Fig. 1.9, Glossary figs 1, 17, 18, 19, 20).

Petiole — the stalk of a leaf (Glossary figs 1, 7).

Phylogeny — a hypothesis of relationships that can be expressed in the form of a branching diagram or tree (phylogenetic tree).

Pinna(e) — a leaflet (leaflets) of a pinnate leaf.

Pinnate — feather-like, lateral ribs or leaflets arising from a central axis (Fig. 1.3, Glossary fig. 8a).

Pinnule — the leaflet of a bipinnate leaf.

Pistillate — bearing a pistil (gynoecium), the ovule-bearing organ of the flower (Fig. 1.9, Glossary figs 16, 17, 18).

Pistillode — a sterile gynoecium (Fig. 1.9, Glossary fig. 19).

Pit — a cavity formed by united bracts, enclosing flowers (Glossary fig. 18).

Pith — the parenchymatous or often spongy centre of a stem.

Plastid DNA — DNA within plastids, a class of organelles found in plant cells.

Pleonanthic — describing shoots flowering continuously, not dying after flowering, cf. 'hapaxanthic' (Figs 1.1, 1.2, 1.4).

Plicate — pleated.

Plumose — softly feathered.

Pneumatophores — Above-ground, 'breathing' roots that allow gas exchange in habitats with inundated or waterlogged soil.

Podsol — a zoned soil with a layer of humus overlying an acidic, highly leached layer, and a basal layer with iron deposition.

Polar/polarity — the condition of having distinct poles. Polarity of pollen grains is determined by their orientation at tetrad stage or by inference from the distribution of apertures and other features.

Pollen mother cell (PMC) — synonym of microsporocyte or meiocyte: a cell that divides by meiosis to produce haploid pollen or spores (Fig. 2.4).

Pollen viability — the potential of a pollen grain to germinate and fertilise ovules.

Polygamo-dioecious — bearing hermaphrodite and either male or female flowers.

Polyphyletic — a taxonomic group that does not have a single recent common ancestor. Two or more unrelated branches on a phylogenetic tree.

Pontoperculate/pontoperculum — a type of operculum that is not completely isolated from the remainder of the ectexine, but linked to it between the ends of the apertures as in *Chamaerops* and *Iriartella* (Figs a, p.249; a & b, p.359).

Porate — of a pollen grain, having one or more germinal pores.

Poricidal — opening by pores (Glossary fig. 19).

Praemorse — jaggedly toothed, as if bitten (Glossary figs 14f,h,i,j,k, 15j,l,m).

Prophyll — the first bract borne on the inflorescence, usually 2-keeled (Figs 1.5, 1.6, Glossary fig. 16).

Prostrate — lying flat (Fig. 1.1).

Protandrous — of flowers or inflorescences, pollen shed before the stigma is receptive.

Protogynous — of flowers or inflorescences, the stigma receptive before the pollen is shed.

Proximal — nearest to the attachment, basal; in pollen, applied to features of the mature pollen grain that face inward at tetrad stage.

Pseudomonomerous — appearing to be of one member but actually of several, as a gynoecium with one fertile carpel and one locule but parts of two other carpels present (Fig. 1.9, Glossary fig. 20).

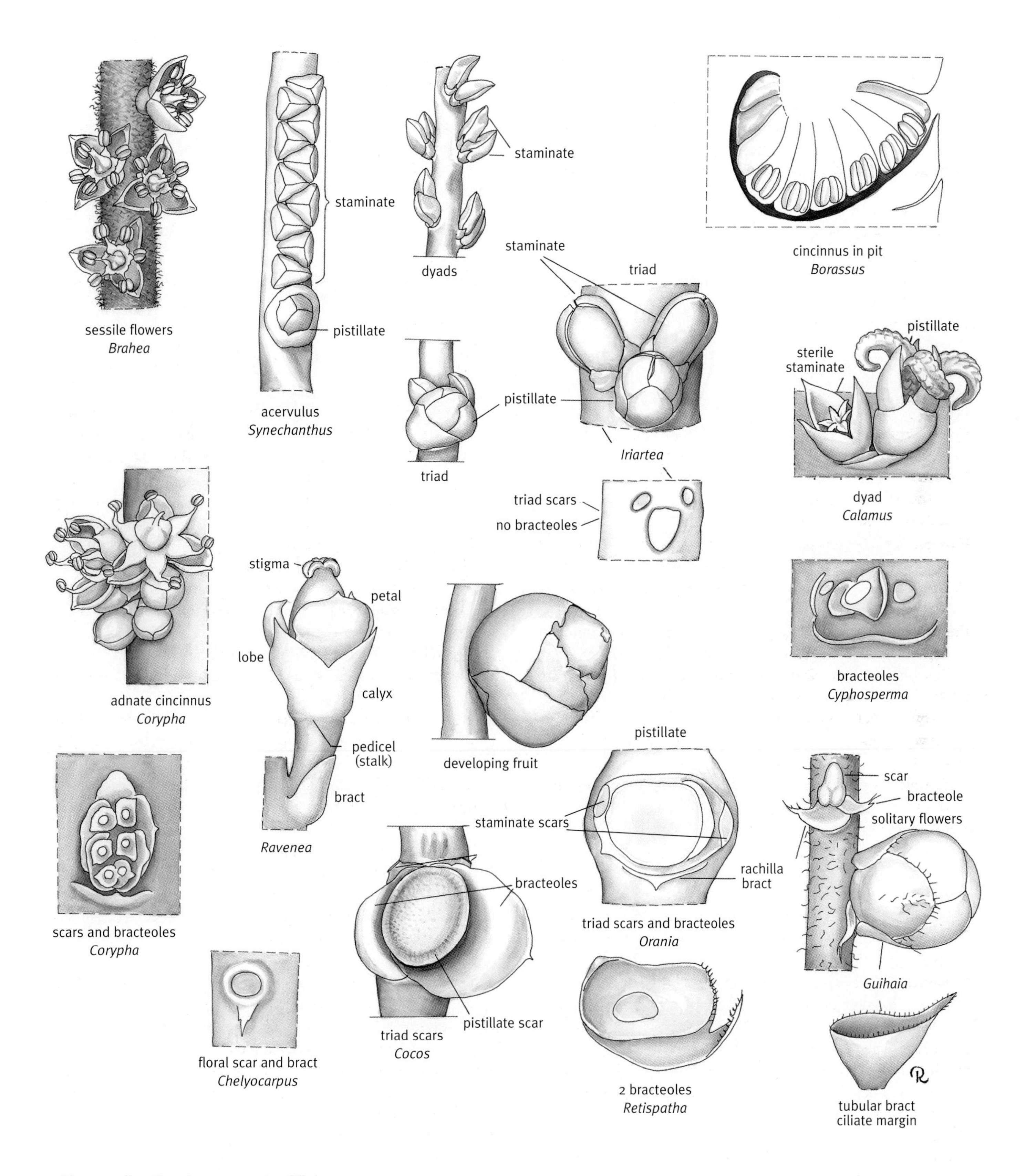

Glossary fig. 18 — Arrangements of flowers (Drawn by Marion Ruff Sheehan)

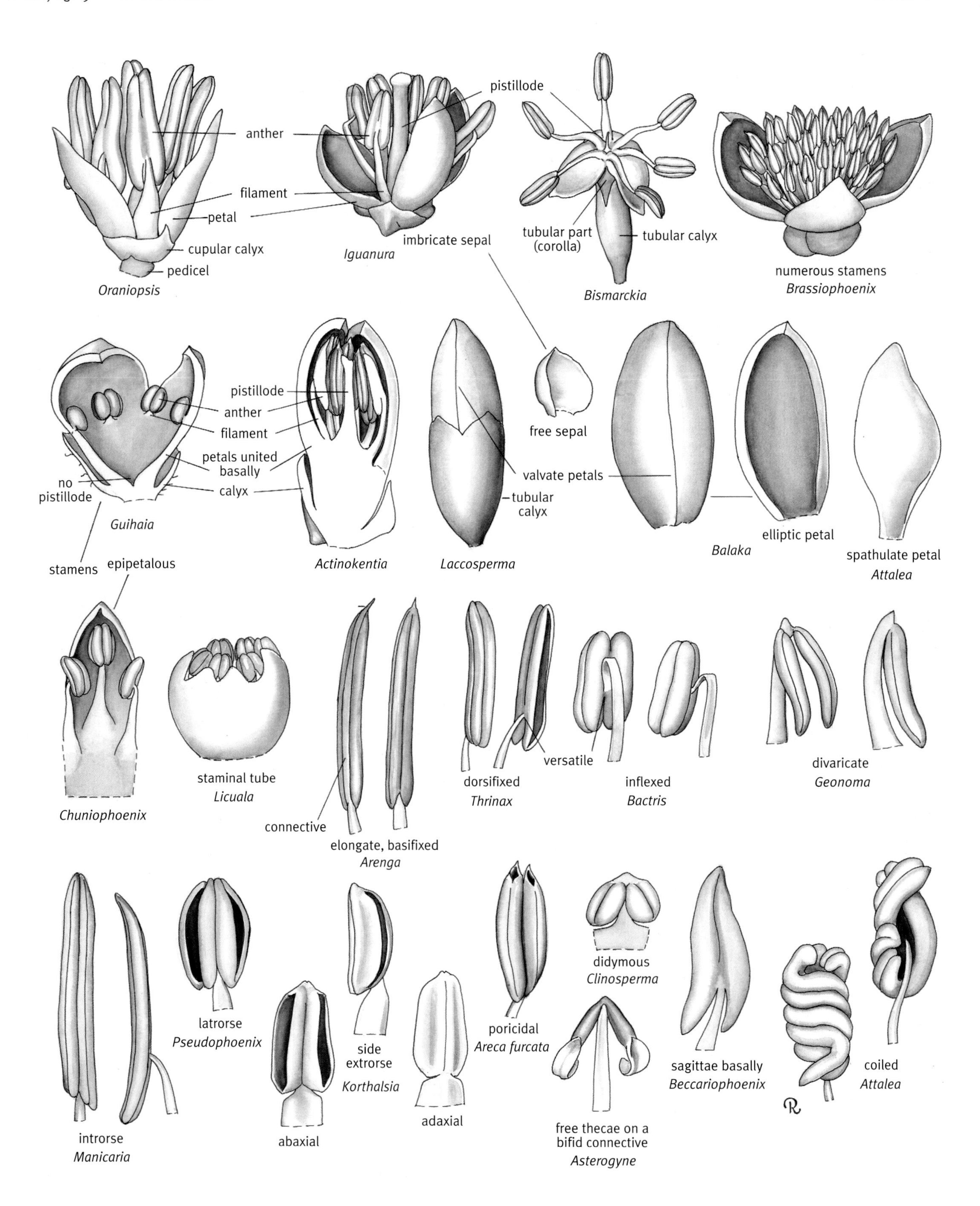

Glossary fig. 19 — Staminate flowers (Drawn by Marion Ruff Sheehan)

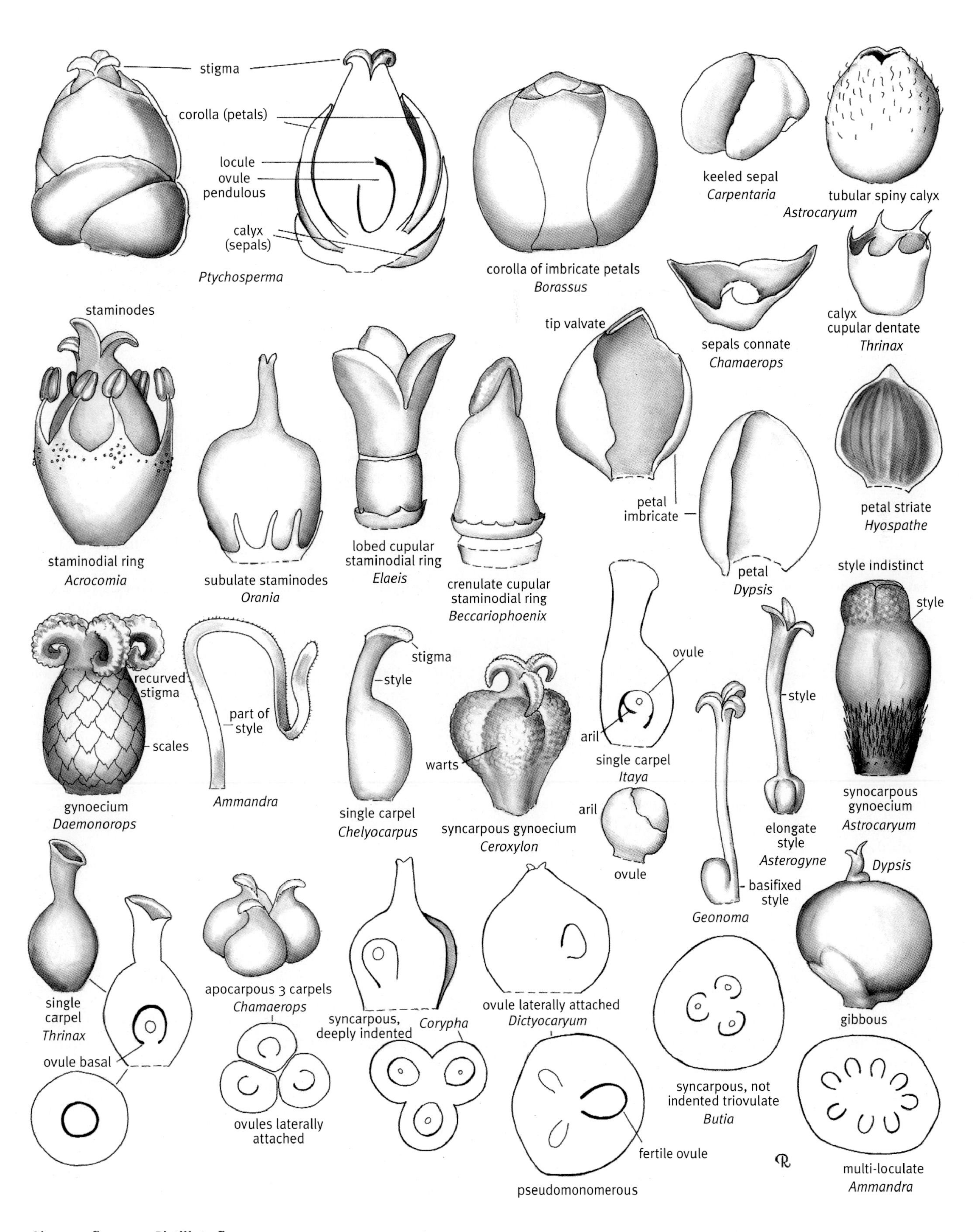

Glossary fig. 20 — Pistillate flowers (Drawn by Marion Ruff Sheehan)

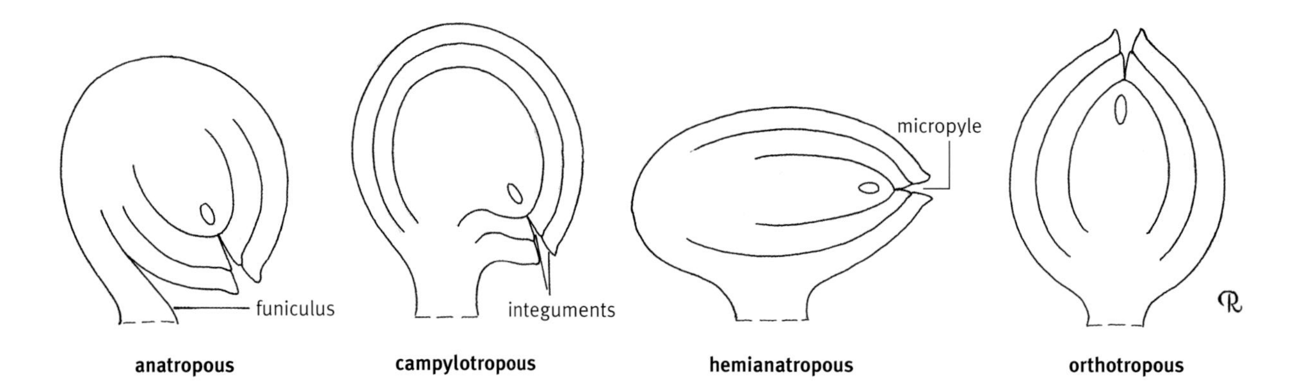

Glossary fig. 21 — Types of ovule (Drawn by Marion Ruff Sheehan)

Psilate — smooth.

Punctiform — dot-like.

Pyrene — a seed-like body formed by a hard, often sculptured layer of endocarp that surrounds the seed (see Borasseae and accompanying generic drawings).

Pyriform — pear-shaped (Glossary fig. 22f).

Rachilla — the branch that bears the flowers (Figs 1.5, 1.6, Glossary fig. 16)

Rachis — the axis of a leaf beyond the petiole; or the axis of an inflorescence beyond the first branches, i.e., beyond the peduncle (Figs 1.5, 1.6, Glossary figs 7d,j, 16).

Rachis bracts — bracts subtending first-order branches of the inflorescence (Figs 1.5, 1.6, Glossary fig. 16).

Radicle — the first root formed by the embryo (Glossary fig. 25).

Ramenta — rather thin scales with ragged edges, often large and irregular in shape (Glossary fig. 7c).

Rank — a row, usually a vertical one.

Raphe — a ridge or depression on the seed, usually the source of fibrovascular branches (Glossary fig. 24).

Raphides — bundles of needle-shaped crystals of calcium oxalate.

Receptacle — the central axis of a flower to which the organs (sepals, petals, stamens, carpels) are attached (Glossary fig. 17).

Reduplicate — of leaflets Λ-shaped in cross-section (Fig. 1.3, Glossary figs 9, 12i).

Rein — in palms, a narrow marginal strip that is shed when the compound leaf unfolds (Glossary fig. 13f).

Remote-ligular — in germination, the young plant connected to the seed by a long tubular cotyledonary petiole, bearing a ligule (Glossary fig. 25).

Remote-tubular — in germination, the young plant connected to the seed by a long tubular cotyledonary petiole, lacking a ligule (Glossary fig. 25).

Resolution — an arrangement of taxa in a phylogenetic tree, e.g., "Tribe Caryoteae is resolved within subfamily Coryphoideae." Also used to indicate the level of ambiguity in a phylogenetic hypothesis, e.g., "Relationships within tribe Areceae are poorly resolved."

Reticulum/reticulate — large rounded or, more frequently, angular holes (lumina) through the tectum. Unlike foveolae, the lumina are separated by areas of tectum (muri) that are narrower than the lumina; the overall appearance is net-like (Figs a–d, p.513).

RFLP — restriction fragment length polymorphism, a type of DNA data used to study phylogenetic relationships at all taxonomic levels.

Rheophyte — a plant adapted to growing in or on the banks of fast-flowing rivers; leaflets of rheophytic palms are usually very narrow, often linear.

Rhizome — underground stem (Fig. 1.1).

Ribosomal DNA — regions of repetitive DNA that encode structural RNA molecules involved in protein synthesis.

Ring murate — superficially similar to reticulate but each lumen in the tectum (sometimes two or three small lumina together) is surrounded by a discrete ring or loop of tectum pressed against neighbouring rings (Fig. f, p.516).

Root boss — a rounded basal protuberance from which lateral roots emerge (Glossary fig. 3f).

Root spines — spines developed from short roots (Glossary fig. 4k,l).

Rugose — wrinkled.

Rugulae/rugulate — in pollen, elongate irregular crumpled/wrinkled elements of the tectum, sometimes with perforations or microfossulae/microchannels between (Figs e, p.380; g, p.444; b, p.451).

Ruminate — darkly streaked due to infolding of the seed coats (Glossary figs 23, 24).

Sagittate — enlarged at the base into two acute, straight lobes like the barbed head of an arrow (Glossary fig. 19).

Sarcotesta — a fleshy layer developed (in palms) from the outer seed coat (Glossary fig. 23).

Scabrate — (of pollen) roughened or with elements smaller than 1 µm.

Scale leaf — a reduced leaf (Glossary fig. 25).

Segment — the division of a palmate or costapalmate leaf (Glossary fig. 10).

Glossary fig. 22 — Fruits

a, beaked, apical stigmatic remains (*Wodyetia bifurcata*);
b, basal stigmatic remains, vestigial carpels (*Orania palindan*);
c, subapical stigmatic remains, pebbled epicarp (*Oncosperma tigillarium*);
d, imbricate scales (*Raphia farinifera*);
e, subbasal stigmatic remains (*Roystonea regia*);
f, pyriform (*Hyphaene coriacea*);
g, spiny (*Astrocaryum alatum*);
h, three-seeded (*Pseudophoenix sargentii*);
i, corky-warted (*Pelagodoxa henryana*).

(All photos C.E. Lewis)

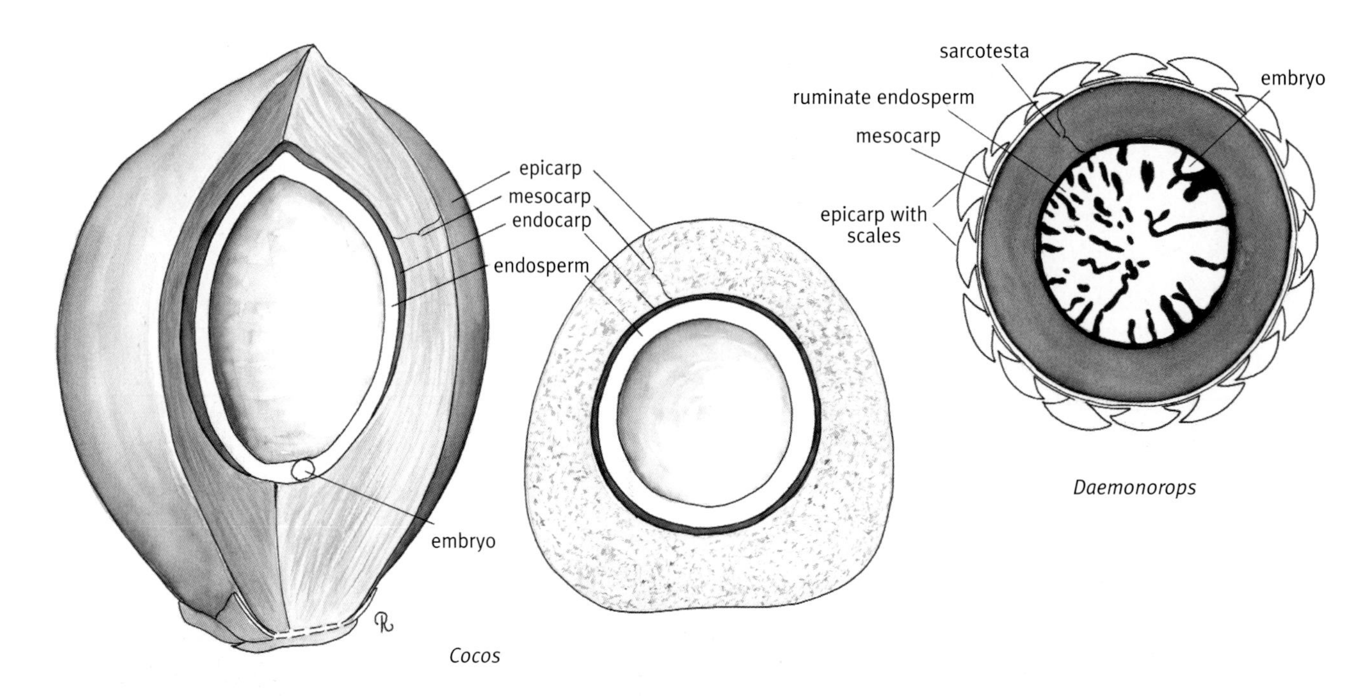

Glossary fig. 23 — Fruit structure (Drawn by Marion Ruff Sheehan)

Semitectate — a partial tectum in which tectal interruptions to the infratectum are wider than the surrounding tectum (e.g., reticulate).

Sepal — a single part of the outermost whorl of floral organs the calyx (Fig. 1.9, Glossary figs 17, 19, 20).

Seral — a temporary or developing vegetation type.

Sessile — lacking a stalk (Glossary fig. 18).

Sheath — the lowest part, or the base of the leaf that is always tubular at first but often splits during or after maturity (Glossary figs 5, 7e,f)

Short axis — the shortest dimension of the pollen grain, particularly used for heteropolar pollen.

Sigmoid — S-shaped (Glossary fig. 15g).

Simultaneous cytokinesis — the dividing cell plates advance centripetally, and all cell plates are formed only after the second meiotic division.

Sinuous — wavy (Glossary fig. 15e).

Sister group — the closest relative in a phylogenetic tree.

Solitary — single stemmed, not clustering (Fig. 1.1, Glossary fig. 2a).

Spadix — in palms, the whole inflorescence. Not used in this book because of ambiguities with other families.

Spathe — a large sheathing bract usually either the prophyll or peduncular bract. Not used in this book because of its ambiguity within palms.

Spatulate — shaped like a small spatula, oblong with an extended basal part (Glossary fig. 19).

Spicate — spike-like, the inflorescence unbranched, the flowers apparently borne directly on the axis (Glossary fig. 16).

Spicule — a very slender brittle, needle-like spine.

Spines/spinulose — used in pollen morphology for long, usually tapering elements; in palm pollen, length is defined as >2 μm (Figs b, p.213; c, d, p.442; a, b, p.615).

Spinulae/spinulose — small pointed elements extending from the tectum or foot layer (in intectate pollen); in palm pollen, height is defined as <2 μm (Fig. b, p.513).

Spinules — very small spines; in pollen, small pointed elements extending from the foot layer or tectum; in palm pollen, height is defined as <2 μm.

Sporopollenin — the material comprising the tough outer pollen wall of most flowering plants. It comprises carbon, hydrogen and oxygen in an approximate ratio of 4:6:1. Recent results confirm the presence of fatty, aromatic and minimal carboxylic acid components. Although the components are consistent, the ratio of the components is apparently not consistent through all plant groups. Sporopollenin is probably "a randomly cross-linked biomacromolecule without a repetitive large-scale structure" and, furthermore, this is "a characteristic which would inherently make this material resistant to enzymic attack, as well as to many laboratory procedures designed to reduce/return it to its principle components". This being so would account for the extraordinary preservational qualities of pollen exine.

Spur — a short, often curved and tapered projection (see *Kerriodoxa*).

Stamen — the male organ of a flower; a stalk or filament bearing an anther containing pollen (Fig. 1.9, Glossary fig. 17).

Staminate — bearing stamens, the pollen bearing organs of a flower (Fig. 1.1, Glossary figs. 16, 17, 18).

Staminode — a vestigial stamen often greatly modified in form (Fig. 1.9, Glossary fig. 20).

Stem — the main axis.

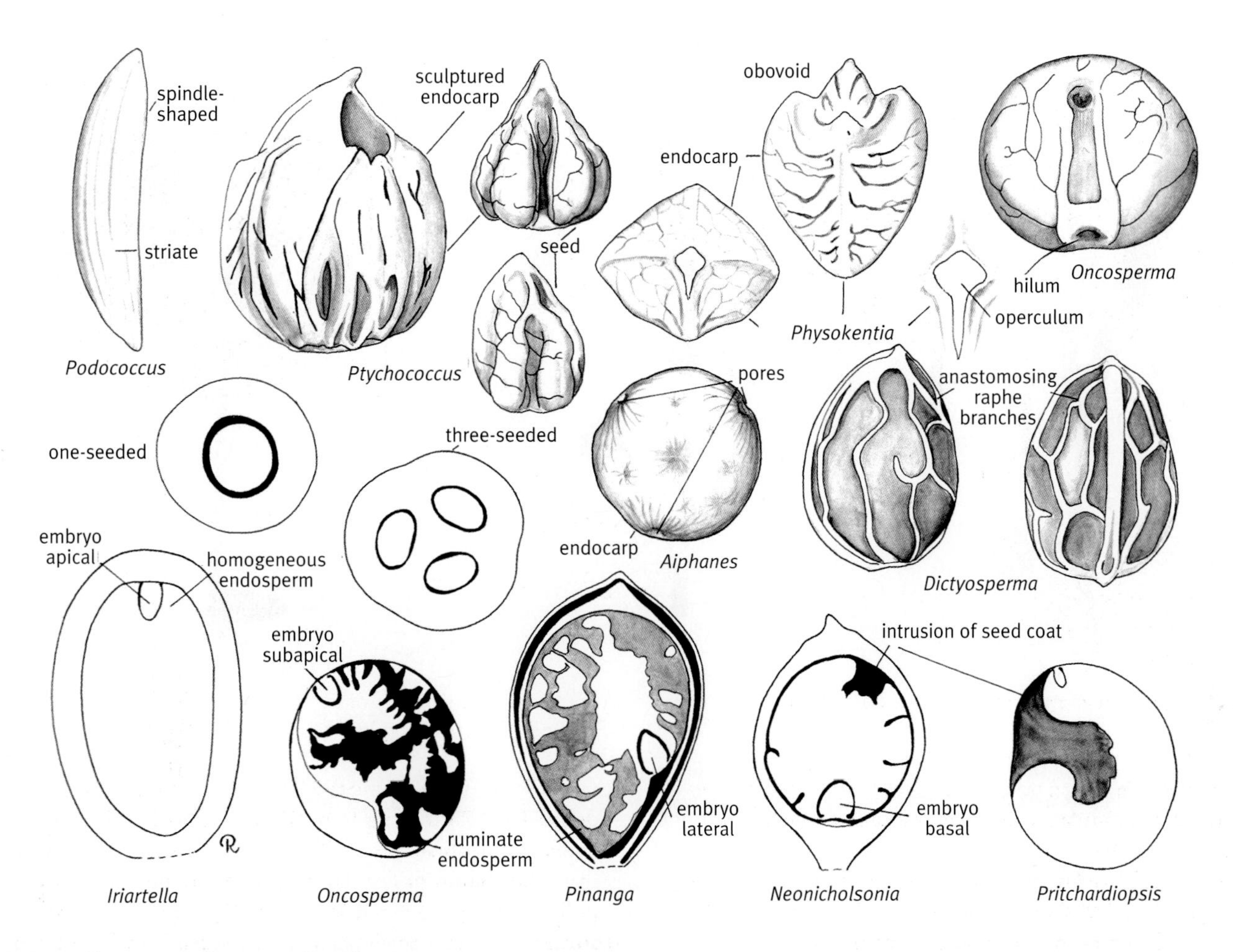

Glossary fig. 24 — Endocarps and seeds (Drawn by Marion Ruff Sheehan)

Stigma — the pollen receptor on the gynoecium, usually distal (Fig. 1.9, Glossary figs 17, 20, 22).

Stilt roots — oblique, lateral roots, often large as in *Socratea*, also called prop roots (Glossary fig. 3e).

Striate — lined (Glossary figs. 4g, 20, 24); in pollen, narrow, elongate, closely set and generally parallel elements of the tectum (Fig. b, p.174).

Style — the often attenuate part of a carpel or gynoecium between the ovary and the stigma (Fig. 1.9, Glossary figs 17, 20).

Sub — as a prefix, meaning nearly or almost, e.g., subopposite equals nearly opposite.

Subequatorial — (of pollen) below the equatorial line, towards the proximal face.

Subulate — awl-shaped (Glossary fig. 20).

Successive cytokinesis — the dividing cell plates extend centrifugally, and the cell plates are formed at the first and second meiotic division, giving a distinct dyad stage.

Sulcate/sulcus/sulci — grooved or furrowed; in pollen, an elongate or slit-like aperture with a distal position, and generally associated monocotyledons. In the calamoid palms, however, there are two sulci positioned on the short equatorial axes of the pollen grain

Sulcus/sulci — an elongate aperture usually situated either at the distal long axis face of the pollen grain or, less commonly, at the equatorial short axes of the pollen grain.

Supratectate — above the tectum: for elements on top of, but attached to, the tectum (Figs b, d, p.317; c, d, p.442).

Symmetric — with at least one plane of symmetry.

Sympodial — a stem made of superimposed branches, lacking a single main axis.

Syncarpous — with united carpels (Fig. 1.9, Glossary fig. 20).

Tectate — with a tectum (Fig. 2.7).

Tectum — the uppermost layer of the ectexine, usually subtended by an infratectum (Fig. 2.7).

Terete — circular in cross-section, usually cylindrical.

Testa — the outer coat of the seed.

Tetrachotomosulcate/tetrachotomosulcus — of pollen having a four-armed sulcus.

Tetrad — a group of four united pollen grains or spores, either as a developmental stage or as a dispersal unit (Fig. 2.4).

Tetragonal tetrad — a uniplanar tetrad in which all four members are in contact at the centre of the tetrad so that, in correct orientation, the adjacent walls form a cross (Figs 2.5a,b).

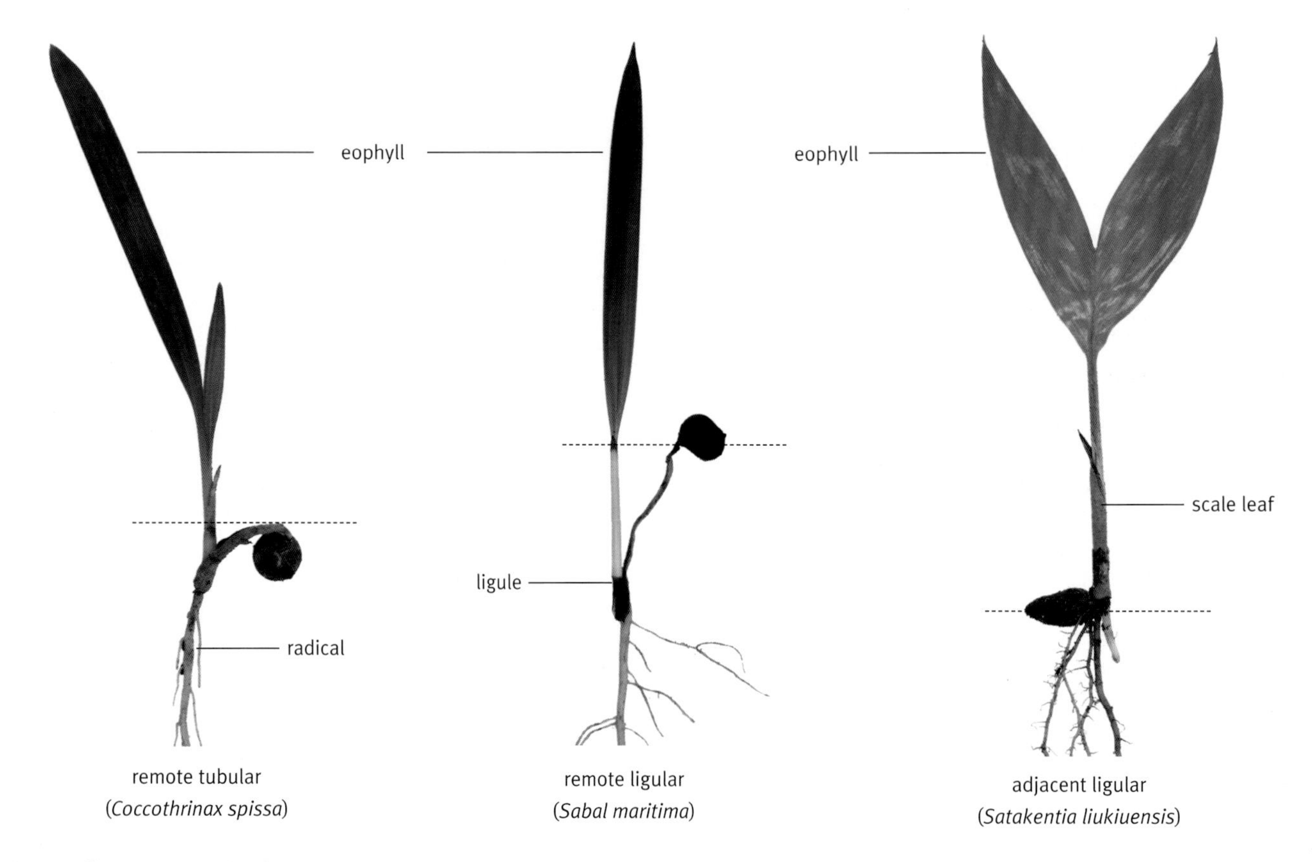

Glossary fig. 25 — Germination (All photos C.E. Lewis)

Tetrahedral tetrad — a multiplanar tetrad in which each of the four members is in contact with the other three members, so that the centres of the grains define a tetrahedron.

Thecae — the locules of an anther (Glossary fig. 19).

Tomentose — densely covered with short hairs, scales, wool, or down.

Topology — the arrangement of branches on a phylogenetic tree.

Total evidence — a phylogenetic analysis of all available data.

Triad — a special group of two lateral staminate and a central pistillate flower, structurally a short cincinnus (Glossary fig. 18).

Trichotomosulcate/trichotomosulcus — of pollen having a three-armed sulcus (Fig. 2.2).

Trijugate — bearing three pairs of leaflets.

Trilocular — having three chambers, each usually bearing an ovule or seed (Glossary fig. 20).

Triovulate — gynoecium with three ovules, one in the locule of each carpel (Glossary fig. 20).

Triporate — of pollen having three pores (Figs a–d, p.391).

Truncate — as though cut off nearly straight across (Glossary figs 14f, 15j).

Tubercles — short, stout, persistent floral stalks, appearing as small humps, in coryphoid palms.

Ultrastructure/ultrastructural — structural details of cells above the limit of resolution of the light microscope, and only revealed by electron microscopy.

Undulate — waved (Glossary fig. 14e).

Unilocular — with a single cavity (Glossary fig. 20).

Uniovulate — with a single ovule (Glossary fig. 20).

Urceolae/urceolate — urn-shaped elements situated on the foot layer (Fig. d, p.516).

Valvate — meeting exactly without overlapping, cf. 'imbricate' (Glossary figs 19, 20).

Ventricose — swollen (Glossary fig. 3c).

Verrucae/verrucate — a wart-like element of ectexine, more than 1 µm wide, broader than it is high and not constricted at the base (Figs j, p.195; d, p.480).

Versatile — of anthers, attached near the centre and movable on the filament (Glossary fig. 19).

Whip — a climbing organ in some Calamoideae, general term for cirrus and flagellum (Glossary fig. 6).

Y-shaped impression mark — a mark on the proximal face of the pollen grain retained from the post-meiotic stage of tetrahedral tetrads (Fig. 2.6c).

Zonasulcus/zonasulcate — a ring-like sulcus. Strictly, 'zonasulcus' equals meridional orientation and 'zonasulculus' equals equatorial orientation. However, the two words are easily confused by the non-specialist and, in the context of this book 'meridional zonasulcus/zonasulcate' or 'equatorial zonasulcus/zonasulcate' are used (Figs b, p.168; a, d, p.213).

LITERATURE CITED

ACHILLI, J. 1913. Contribution a l'étude anatomique des Dypsidées de Madagascar. Ann. Mus. Colon. Marseille, sér. 3, I: 99–147.

ADAM, H., S. JOUANNIC, J. ESCOUTE, Y. DUVAL, J.-L. VERDEIL AND J.W. TREGEAR. 2005. Reproductive developmental complexity in the African oil palm (*Elaeis guineensis*, Arecaceae). Amer. J. Bot. 92: 1836–1852.

ADAM, H., S. JOUANNIC, F. MORCILLO, F. RICHAUD, Y. DUVAL AND J.W. TREGEAR. 2006. MADS box genes in oil palm (*Elaeis guineensis*): patterns in the evolution of the SQUAMOSA, DEFICIENS, GLOBOSA, AGAMOUS, and SEPALLATA subfamilies. J. Molec. Evol. 62: 15–31.

ADAM, H., S. JOUANNIC, F. MORCILLO, J.-L. VERDEIL, Y. DUVAL AND J.W. TREGEAR. 2007. Determination of flower structure in *Elaeis guineensis*: do palms use the same homeotic genes as other species? Ann. Bot. 100: 1–12.

AINSLIE, J.R. 1926. The physiography of Sothern Nigeria and its effect on the forest flora of the country. Oxford Forest. Mem. 5: 1–36.

AKHMETIEV, M.A. 1989. New findings of palms from the Darrydagh Oligocene deposits (Nakhichevan ASSR). Byull. Moskovsk. Obshch. Isp. Prir., Otd. Geol. 64: 51–67.

ALLEN, P.H. 1956. The rain forests of the Golfo Dulce. Univ. of Florida Press, Gainesville.

AL-MAYAH, A.-R.A. 1986. Cytological studies of the date palm *Phoenix dactylifera* L. Acta Bot. Hung. 32: 177–181.

AL-RAWI, A. 1945. Blütenmorphologische und zytologische Untersuchungen an Palmen der Unterfamilie der Ceroxyloideae. Arbeiten Inst. Allg. Bot. Univ. Zürich. Series 3, No. 6.

AMBWANI, K. 1981. Borassoid fossil palm root from the Deccan Intertrappean Beds of Nawargaon in Wardha District, Maharashtra. Geophytology 11: 13–15.

AMBWANI, K. 1983. *Palmoxylon shahpuraensis* sp. nov., a fossil palm resembling *Licuala* from the Deccan Intertrappean beds of Mandla District, Madhya Pradesh. Palaeobotanist 31: 52–59.

AMBWANI, K. 1984. *Palmostroboxylon arengoidum* sp. nov.: a fossil palm peduncle resembling *Arenga* from the Deccan Intertrappean Beds of Shahpura, Madhya Pradesh. Palaeobotanist 32: 134–139.

AMBWANI, K. AND D. DUTTA 2005. Seed-like structure in dinosaurian coprolite of Lameta Formation (Upper Cretaceous) at Pisdura, Maharashtra, India. Curr. Sci. 88: 352–354.

AMBWANI, K. AND M. KUMAR. 1993. Pollen morphology of the coryphoid genus *Licuala* (Palmae). Grana, 32: 164–168.

AMBWANI, K. AND R.C. MEHROTRA. 1989. A new fossil palm wood from the Deccan Intertrappean beds of Shahpura, Mandla District, Madhya Pradesh. Geophytology 19: 70–75.

AMBWANI, K. AND U. PRAKASH. 1983. *Palmoxylon ghuguensis* sp. nov. resembling *Chrysalidocarpus* from the Deccan Intertrappean beds of Mandla District in Madhya Pradesh. Palaeobotanist 31: 76–81.

ANCIBOR, E. 1995. Palmeras fosiles del Cretacico Tardio de la Patagonia Argentina (Bajo de Santa Rosa, Rio Negro). Ameghiniana 32: 287–299.

ANDERSON, T. 1869. An enumeration of the palms of Sikkim. J. Linn. Soc., Bot. 11: 4–14.

ANDREÁNSZKY, G. 1949. Reste einer neuen Tertiären Palme aus Ungarn. Hung. Acta Biol. 1: 31–36.

ANON. 1981. Palm research. Principes 25:182.

ANSTETT, M.-C. 1998. An experimental study of the interaction between the dwarf palm (*Chamaerops humilis*) and its floral visitor *Derelomus chamaeropsis* throughout the life cycle of the weevil. Acta Oecol. 20: 551–558.

ARNOLD, C.A. 1952. Tertiary plants from North America. 1. A *Nipa* fruit from the Eocene of Texas. Palaeobotanist 1: 73–74. [Not seen cited according to Gee (1990).]

ARNONE, A., G. NASINI AND O. VAJNA DE PAVA. 1997. Constituents of Dragon's blood. 5. Dracoflavans B1, B2, C1, C2, D1, and D2, new A-type deoxyproanthocyanidins. J. Nat. Prod. (Lloydia) 60: 971–975.

ARUNIKA, H.L., A.N. GUNAWARDENA AND N.G. DENGLER. 2006. Alternative modes of leaf dissection in monocotyledons. Bot. J. Linn. Soc. 150: 25–44.

ASAMI, A., Y. HIRAI AND J. SHOJI. 1991. Studies on the constituents of palmae plants. VI: steroid saponins and flavonoids of leaves of *Phoenix canariensis* hort. ex Chabaud, *P. humilis* Royle var. *hanceana* Becc., *P. dactylifera* L., and *Licuala spinosa* Wurmb. Chem. Pharm. Bull. 39: 2053–2056.

ASMUSSEN, C.B. 1999a. Toward a chloroplast DNA phylogeny of the tribe Geonomeae (Palmae). Pages 121–129 *in* Henderson, A. and F. Borchsenius (*eds*). Evolution, variation and classification of palms. The New York Botanical Garden Press, Bronx, NY.

ASMUSSEN, C.B. 1999b. Relationships of the tribe Geonomeae (Arecaceae) based on plastid *rps*16 DNA sequences. Acta Bot. Venez. 22: 65–76.

ASMUSSEN, C.B., W.J. BAKER AND J. DRANSFIELD. 2000. Phylogeny of the palm family (Arecaceae) based on *rps*16 intron and *trn*L-*trn*F plastid DNA sequences. Pages 525–537 *in* Wilson, K.L. and D.A. Morrison (*eds*). Monocots: systematics and evolution. CSIRO, Melbourne.

ASMUSSEN, C.B. AND M.W. CHASE. 2001. Coding and noncoding plastid DNA in palm systematics. Amer. J. Bot. 88: 1103–1117.

ASMUSSEN, C. B., J. DRANSFIELD, V. DEICKMANN, A.S. BARFOD, J.-C. PINTAUD AND W.J. BAKER. 2006. A new subfamily classification of the palm family (Arecaceae): evidence from plastid DNA phylogeny. Bot. J. Linn. Soc. 151: 15–38.

ASPINALL, G.O., E.L. HIRST, E.G.V. PERCIVAL AND I.R. WILLIAMSON. 1953. The mannans of ivory nut (*Phytelephas macrocarpa*). Part I. The methylation of mannan A and mannan B. J. Chem. Soc. 1953: 3184–3188.

ASPINALL, G.O., R.B. RASHBROOK AND G. KESSLER. 1958. The mannans of ivory nut (*Phythelephas macrocarpa*). Part II. The partial acid hydrolysis of mannans A and B. J. Chem. Soc. 1958: 215–221.

AZUMA H., M. TOYOTA, Y. ASAKAWA, T. TAKASO AND H. TOBE. 2002. Floral scent chemistry of mangrove plants. J. Pl. Res. 115: 47–53.

AVALOS, G. 2004. Production of a second set of stilt roots in arborescent palms: a solution to the puzzle. Palms 48: 83–85.

Babajide Salami, M. 1985. Upper Senonian and Lower Tertiary pollen grains from the southern Nigeria sedimentary basin. Rev. Esp. Micropaleontol. 17: 5–26.

Bachman, S., W.J. Baker, N. Brummitt, J. Dransfield and J. Moat. 2004. Elevational gradients, area and tropical island diversity: an example from the palms of New Guinea. Ecography 27: 299–310.

Bacon, C.D., F.A. Feltus, A.H. Paterson, and C.D. Bailey. 2008. Novel nuclear intron-spanning primers for Arecaceae evolutionary biology. Mol. Ecol. Notes. 8: 211–214.

Bacon, P.R. 2001. Germination of *Nypa fruticans* in Trinidad. Palms 45: 57–61.

Bailey, L.H. 1934. American palmettos. Gentes Herb. 3: 274–339.

Bailey, L.H. 1935a. The king palms of Australia — *Archontophoenix*. Gentes Herb. 3: 389–409.

Bailey, L.H. 1935b. Certain ptychospermate palms of horticulturists. Gentes Herb. 3: 410–437.

Bailey, L.H. 1936. *Washingtonia*. Gentes Herb. 4: 53–82.

Bailey, L.H. 1937a. *Erythea* — the Hesper Palms. Gentes Herb. 4: 85–118.

Bailey, L.H. 1937b. Notes on *Brahea*. Gentes Herb. 4: 119–125.

Bailey, L.H. 1939a. *Howea* in cultivation — the Sentry Palms. Gentes Herb. 4: 188–198.

Bailey, L.H. 1939b. New Haitian genus of palms. Gentes Herb. 4: 239–246.

Bailey, L.H. 1940. The generic name *Corozo*. Gentes Herb. 4: 357–360.

Bailey, L.H. 1942. Palms of the Seychelles Islands. Gentes Herb. 6: 3–48.

Bailey, L.H. and H.E. Moore. 1949. Palms uncertain and new. Gentes Herb. 8: 93–205.

Baker, W.J. 1997. Rattans and rheophytes — palms of Mubi River. Principes 41: 148–157.

Baker, W.J., C.B. Asmussen, S.C. Barrow, J. Dransfield and T.A. Hedderson. 1999a. A phylogenetic study of the palm family (Palmae) based on chloroplast DNA sequences from the *trn*L-*trn*F region. Pl. Syst. Evol. 219: 111–126.

Baker, W.J., M.J.E. Coode, J. Dransfield, S. Dransfield, M.M. Harley, P. Hoffmann and R.J. Johns. 1998. Patterns of distribution of Malesian vascular plants. Pages 243–258 *in* Hall, R. and J.D. Holloway (*eds*). Biogeography and geological evolution of SE Asia. Backhuys, Leiden.

Baker, W.J., J. Dransfield, M.M. Harley and A. Bruneau. 1999b. Morphology and cladistic analysis of subfamily Calamoideae (Palmae). Pages 307–324 *in* Henderson, A. and F. Borchsenius (*eds*). Evolution, variation and classification of palms. The New York Botanical Garden Press, Bronx, NY.

Baker, W.J. and J. Dransfield. 2000. Towards a biogeographic explanation of the calamoid palms. Pages 545–553 *in* Wilson, K.L. and D.A. Morrison (*eds*). Monocots: Systematics and Evolution. CSIRO, Melbourne.

Baker, W.J. and J. Dransfield. 2008. *Calospatha* subsumed in *Calamus* (Arecaceae: Calamoideae). Kew Bull. 63: 161–162.

Baker, W.J., J. Dransfield and T.A. Hedderson. 2000a. Phylogeny, character evolution and a new classification of the Calamoid palms. Syst. Bot. 25: 297–322.

Baker, W.J., T.A. Hedderson and J. Dransfield. 2000b. Molecular phylogenetics of subfamily Calamoideae (Palmae) based on nrDNA ITS and cpDNA *rps*16 intron sequence data. Molec. Phylogen. Evol. 14: 195–217.

Baker, W.J., T.A. Hedderson and J. Dransfield. 2000c. Molecular phylogenetics of *Calamus* (Palmae) and related rattan genera based on 5S nrDNA spacer sequence data. Molec. Phylogen. Evol. 14: 218–231.

Baker, W.J. and I. Hutton. 2006. *Lepidorrhachis*. Palms 50: 33–38.

Baker, W.J. and A.H.B. Loo. 2004. A synopsis of the genus *Hydriastele* (Arecaceae). Kew Bull. 59: 61–68.

Baker, W.J., R.A. Maturbongs, J. Wanggai and G.G. Hambali. 2000d. *Siphokentia*. Palms 44: 175–181.

Baker, W.J., S. Zona, C.D. Heatubun, C.E. Lewis, R.A. Maturbongs and M.V. Norup. 2006. *Dransfieldia* (Arecaceae) — a new palm genus from western New Guinea. Syst. Bot. 31: 60–68.

Balick, M.J. 1980. The biology and economics of the *Oenocarpus–Jessenia* (Palmae) complex. PhD thesis. Harvard University, Cambridge, MA.

Balick, M.J. 1981. *Mauritiella* (Palmae) reconsidered. Brittonia 33: 459–460.

Balick, M.J. 1985. The indigenous palm flora of "Las Gaviotas," Colombia, including observations on local names and use. Bot. Mus. Leafl. 30: 1–34.

Balick, M.J. and D. Johnson. 1994. The conservation status of *Schippia concolor* in Belize. Principes 38: 124–128.

Ballance, P.F., M.R. Gregory and G.W. Gibson. 1981. Coconuts in Miocene turbidites in New Zealand: possible evidence for tsunami origin of some turbidity currents. Geology 9: 592–595.

Balslev, H., J. Luteyn, B. Øllgaard and L.B. Holm-Nielsen. 1987. Composition and structure of adjacent unflooded and flood-plain forest in Amazonian Ecuador. Opera Bot. 92: 37–57.

Bandaranayake, W.M. 2002. Bioactivities, bioactive compounds and chemical constituents of mangrove plants. Wetlands Ecol. Managem. 10: 421–452.

Bande, M.B. and K. Ambwani. 1982. *Sclerosperma*-type pollen grains from the Neyveli lignite of India. Palaeobotanist 30: 63–67.

Bande, M.B., U. Prakash and K. Ambwani. 1982. A fossil palm fruit *Hyphaeneocarpon indicum* gen. et sp. nov. from the Deccan Intertrappean beds, India. Palaeobotanist 30: 303–309.

Banka, R. and W.J. Baker. 2004. A monograph of the genus *Rhopaloblaste* (Arecaceae). Kew Bull. 59: 47–60.

Barathan G., T.E. Goliber, C. Moore, S. Kessler, T. Pham and N.R. Sinha. 2002. Homologies in leaf form inferred from *KNOX1* gene expression during development. Science 296: 1858–1860.

Barbosa Rodrigues, J. 1903. Sertum palmarum Brasiliensium. 2 vols. Veuve Monnom, Brussels.

Barfod, A.S. 1988. Pollen morphology of *Ammandra*, *Palandra* and *Phytelephas* (Arecaceae). Grana, 27: 239–242.

Barfod, A.S. 1991. A monographic study of the subfamily Phytelephantoideae (Arecaceae). Opera Bot. 105:1–73.

Barfod, A.S., T. Burholt and F. Borchsenius. 2003. Contrasting pollination modes in three species of *Licuala* (Arecaceae: Coryphoideae). Telopea 10: 207–223.

Barfod, A.S., F. Ervik and R. Bernal. 1999. Recent evidence on the evolution of phytelephantoid palms (Palmae). Pages 266–277 *in* Henderson, A. and F. Borchsenius (*eds*). Evolution, variation and classification of palms. The New York Botanical Garden Press, Bronx, New York.

Barford, A.S. and N.W. Uhl. 2001. Floral development in *Aphandra* (Arecaceae). Amer. J. Bot. 88:185–195.

BARROW, S. 1994. In search of *Phoenix roebelenii*, the Xishuangbanna palm. Principes 38: 177–181.

BARROW, S. 1996. A monograph of *Phoenix* L. (Palmae: Coryphoideae). PhD Thesis, University of Reading.

BARROW, S.C. 1998. A revision of *Phoenix* L. (Palmae: Coryphoideae). Kew Bull. 53: 513–575.

BARROW, S.C. 1999. Systematic studies in *Phoenix* L. (Palmae: Coryphoideae). Pages 215–223 *in* Henderson, A. and F. Borchsenius (*eds*). Evolution, variation and classification of palms. The New York Botanical Garden Press, Bronx, NY.

BASU, S.K. 1988. *Corypha* palms in India. J. Econ. Taxon. Bot. 11: 477–486.

BAVAPPA, K.V.A. AND V.S. RAMAN. 1965. Cytological studies in *Areca catechu* Linn. and *Areca triandra*. Roxb. J. Indian Bot. Soc. 44: 495–505.

BAYTON, R.P. 2005. *Borassus* L. and the borassoid palms: systematics and evolution. Ph.D. Thesis, The University of Reading.

BAYTON, R.P. 2007. A revision of *Borassus* L. (Arecaceae). Kew Bull. 62: 561–586.

BEACH, J.H. 1984. The reproductive biology of the Peach or "Pejibaye" palm (*Bactris gasipaes*) and a wild congener (*B. porschiana*) in the Atlantic lowlands of Costa Rica. Principes 28: 107–119.

BEACH, J.H. 1986. Pollination biology of spadices and spicate inflorescences in Cyclanthaceae and Palmae. Amer. J. Bot. 73: 615–616.

BEAL, J.M. 1937. Cytological studies in the genus *Phoenix*. Bot. Gaz. 99: 400.

BECCARI, O. 1885. Reliquiae Schefferianae. Illustrazione di alcune palme viventi nel Giardino Botanico di Buitenzorg. Ann. Jard. Bot. Buitenzorg 2: 77–171.

BECCARI, O. 1886. Nuovi studi sulle palme asiatiche. Malesia 3: 58–149.

BECCARI, O. 1908. Asiatic palms — Lepidocaryeae. Part I. The species of *Calamus*. Ann. Roy. Bot. Gard. Calcutta 11: 1–518.

BECCARI, O. 1910. Studio monografica del genere *Raphia*. Webbia 3: 37–130.

BECCARI, O. 1911. Asiatic palms — Lepidocaryeae. Part I. The species of *Daemonorops*. Ann. Roy. Bot. Gard. Calcutta 12(1): 1–237.

BECCARI, O. 1912. The palms indigenous to Cuba I. Pomona Coll. J. Econ. Bot. 2: 253–276.

BECCARI, O. 1913a. Contributi alla conoscenza delle palme. Webbia 4: 143–240.

BECCARI, O. 1913b. Asiatic palms — Lepidocaryeae. Supplement to Part I. The species of *Calamus*. Ann. Roy. Bot. Gard. Calcutta. 11(App): 1–142.

BECCARI, O. 1914a. Palme del Madagascar, Instituto Micrografico Italiano, Firenze.

BECCARI, O. 1914b. Contributo alla conoscenza della palma a olio (*Elaeis guineensis*). Agric. Colon. 8: 5–37, 108–118, 201–212, 255–270.

BECCARI, O. 1918. Asiatic palms — Lepidocaryeae. Part III. Ann. Roy. Bot. Gard. Calcutta 12(2): 1–231.

BECCARI, O. 1919. The palms of the Philippine Islands. Philipp. J. Sc. 14: 295–362.

BECCARI, O. 1920. Le Palme della Nuova Caledonia. M. Ricci, Firenze. Reprinted from Webbia 5: 71–146.

BECCARI, O. 1924. Palme della tribù Borasseae (*ed*. U. Martelli). G. Passeri, Firenxe.

BECCARI, O. 1931. Asiatic palms — Corypheae (*ed*. U. Martelli). Ann. Roy. Bot. Gard. Calcutta 13: 1–356.

BECCARI, O. AND J.F. ROCK 1921. A monographic study of the genus *Pritchardia*. Mem. Bernice Pauahi Bishop Mus. 8: 1–77.

BECCARI, O. AND R.G. PICHI-SERMOLLI. 1955. Subfamiliae Arecoidearum palmae gerontogeae tribuum et generum conspectus. Webbia 11: 1–187.

BEENTJE, H.J. 1993. A new aquatic palm from Madagascar. Principes 37: 197–202.

BEENTJE, H.J. 1994. A monograph of *Ravenea* (Palmae: Ceroxyloideae). Kew Bull. 49: 623–671.

BEERLING, D.J. AND C.K. KELLY. 1996. Evolutionary comparative analyses of the relationships between leaf structure and function. New Phytol. 134: 35–51.

BEHNKE, H.-D. 2000. Forms and sizes of sieve-element plastids and evolution of the monocotyledons. Pages 163–188 *in* K.L. Wilson and D.A Morrison (*eds*). Monocots: systematics and evolution. CSIRO, Melbourne.

BEILSCHMIED, C.T. 1833. J. Lindley's characters distinctive oder Hauptkennzeichen der natürlichen Pflanzenfamilien. Flora 16 (Beibl. 2): 49–112.

BENNETT, M.D. AND I.J. LEITCH. 2005. Angiosperm DNA C-values database (release 6.0, October 2005). http://www.rbgkew.org.uk/cval/homepage.html.

BERNAL, R.G. 1986. The genus *Metasocratea* (Palmae). Kew Bull. 41: 151–152.

BERNAL, R. 1995. Nuevas especies y combinaciones en la subtribu Wettiniinae (Palmae). Caldasia 17(82–85): 367–378.

BERNAL, R. 1998. The growth form of *Phytelephas seemannii* — a potentially immortal solitary palm. Principes 42: 15–23.

BERNAL, R., G. RAMÍREZ AND R.I. MORALES. 2001. Notes on the genus *Ammandra* (Palmae). Palms 45: 123–126.

BERNAL, R.G. AND G. GALEANO-GARCES 1989. The identity of *Roebelia* and *Platenia* (Palmae). Kew Bull. 44: 321–328.

BERRY, E.W. 1905. A palm from the mid-Cretaceous. Torreya, 5: 30–33.

BERRY, E.W. 1911. Contributions to the Mesozoic flora of the Atlantic coastal plain — VII. Bull. Torrey Bot. Club, 38: 399–424.

BERRY, E.W. 1914a. Fruits of a date palm in the Tertiary deposits of eastern Texas. Amer. J. Sci. 37: 403–406.

BERRY, E.W. 1914b. The Upper Cretaceous and Eocene floras of South Carolina and Georgia. United States Geological Survey, Professional paper 84. 200 pp.

BERRY, E.W. 1914c. A Nipa-palm in the North American Eocene. Amer. J. Science 37: 57–60.

BERRY, E.W. 1916a. A petrified palm from the Cretaceous of New Jersey. Amer. J. Sci. 41: 193–197.

BERRY, E.W. 1916b. Lower Eocene floras of southeastern North America. U.S. Geol. Surv. Profess. Paper 91: 1–481.

BERRY, E.W. 1917. The fossil plants from Vero, Florida. Florida State Geol. Surv., 9th Ann. Report, pp. 19–53.

BERRY, E. W. 1919a. Upper Cretaceous floras of the eastern Gulf region in Tennessee Mississippi, Alabama, and Georgia. Prof. Pap. U.S. Geol. Surv., 112, 177 pp., 23 pls.

BERRY, E.W. 1919b. Miocene fossil plants from Peru. Proc. U.S. Natl. Mus. 55: 279–294.

BERRY, E.W. 1921a. A palm nut from the Miocene of the Canal Zone. Proc. U.S. Natl. Mus. 59: 21–22.

BERRY, E.W. 1921b. Tertiary fossil plants from Venezuela. Proc. U.S. Natl. Mus. 59: 553–579.

BERRY, E.W. 1924. The middle and upper Eocene floras of southeastern North America. U.S. Geol. Surv. Profess. Paper 92. 206 pp.

BERRY, E.W. 1926a. A fossil palm fruit from the Middle Eocene of Northwestern Peru. Proc. U.S. Natl. Mus. 70, art. 3: 1–4.

BERRY, E.W. 1926b. *Cocos* and *Phymatocaryon* in the Pliocene of New Zealand. Amer. J. Sci. 212: 181–184.

BERRY, E.W. 1929. A palm nut of *Attalea* from the Upper Eocene of Florida. J. Wash. Acad. Sci. 19: 252–255.

BERRY, E.W. 1937. *Gyrocarpus* and other fossil plants from the Cumarebo field in Venezuela. J. Wash. Acad. Sci. 27: 501–506.

BHARATHAN, G., G. LAMBERT AND D.W. GALBRAITH. 1994. Nuclear DNA content of monocotyledons and related taxa. Amer. J. Bot. 81: 381–386.

BININDA-EMONDS, O.R.P., J.L. GITTLEMAN AND M.A. STEEL. 2002. The (Super)tree of life: procedures, problems, and prospects. Annual Rev. Ecol. Syst. 33: 265–289.

BIONDI, E. AND R. FILIGHEDDU. 1990. A palm fossil closely related to *Chamaerops humilis* L. from the Lower Miocene of Sardinia. Giorn. Bot. Ital. 124: 711–724.

BJORHOLM, S., J.C. SVENNING, F. SKOV AND H. BALSLEV. 2005. Environmental and spatial controls of palm (Arecaceae) species richness across the Americas. Global Ecol. Biogeogr. 14: 423–429.

BJORHOLM, S., J.C. SVENNING, W.J. BAKER, F. SKOV AND H. BALSLEV. 2006. Historical legacies in the geographical diversity patterns of New World palm (Arecaceae) subfamilies. Bot. J. Linn. Soc. 151: 113–125.

BLACKMORE, S., C.A. MCCONOCHIE AND R.B. KNOX. 1987. Phylogenetic analysis of the male ontogenetic program in aquatic and terrestial monocotyledons. Cladistics 3: 333–347.

BODLEY, J.H. AND F.C. BENSON. 1980. Stilt-root walking by an iriarteoid palm in the Peruvian Amazon. Biotropica 12: 67–71.

BØGH, A. 1996a. Abundance and growth of rattans in Khao Chong National Park, Thailand. Forest Ecol. Managem. 84: 71–80.

BØGH, A. 1996b. The reproductive phenology and pollination biology of four *Calamus* (Arecaceae) species in Thailand. Palms 40: 5–15.

BOLTENHAGEN, E. 1967. Spores et pollen du Crétacé Supérieur du Gabon. Pollen & Spores 9: 335–355.

BONDE, S.D. 1986a. *Amesoneuron borassoides* sp. nov. — a borassoid palm leaf from the Deccan Intertrappean bed at Mohgaonkalan, India. Biovigyanan 12: 89–91.

BONDE, S.D. 1986b. *Sabalophyllum livistonoides* gen. et sp. nov.: a petrified palm leaf segment from Deccan Intertrappean bed at Nawargaon District Wardha, Maharashtra, India. Biovigyanan 12: 113–118.

BONDE, S.D., M.S. KUMBHOJKAR AND R.T. AHER. 2000. *Phoenicicaulon mahabalei* gen. et sp. nov., a sheathing leaf base of *Phoenix* from the Deccan Intertrappean beds of India. Geophytology 29: 11–16.

BONNET, E. 1904. Sur un *Nipadites* de l'Eocéne d'Egypte. Bull. Mus. Hist. Nat. (Paris) 10: 499–502.

BORCHSENIUS, F. AND R. BERNAL. 1996. *Aiphanes* (Palmae). Fl. Neotrop. Monogr. No 70: 1–94.

BORCHSENIUS, F. AND F. SKOV. 1997. Ecological amplitudes of Ecuadorian palms. Principes 41: 179–183.

BORHIDI, A. AND O. MUÑIZ 1985. Adiciones al catálogo de las palmas de Cuba. Acta Bot. Hung. 31: 225–230.

BOSCH, E. 1947. Blütenmorphologische und zytologische Untersuchungen an Palmen. Ber. Schweiz. Bot. Ges. 57: 37–100.

BOUILLENNE, R. 1930. Un voyage botanique dans le bas-Amazone. Arch, Inst. Bot. Univ. Liège 8: 1–185.

BOULOS, L. 1968. The discovery of *Medemia* palm in the Nubian Desert of Egypt. Bot. Not. 121: 117–120.

BOUREAU, E. 1947. Sur la presence du *Palmoxylon aschersoni* Schenk dans les couches tertiaires de la vallee du Chelif (Algerie). Bull. Mus. Nat. Hist. Nat. Sér. 2, 19: 230–236.

BOUREAU, E., M. SALARD-CHEBOLDAEFF, M, J.C. KOENIGUER AND P. LOUVET. 1983. Evolution des Flores et de la vegetation tertiares en Afrique, au nord de l'Equateur. Proc. 10th AETFAT Congress, Pretoria, 1982. Bothalia 14: 355–367.

BOWDEN, W.M. 1945. A list of chromosome numbers in higher plants. II. Menispermaceae to Verbenaceae. Amer. J. Bot. 32: 191–202.

BOWERBANK, J.S. 1840. A History of the fossil fruits and seeds of the London Clay, pp. 1–44. John Van Vorst, London.

BOYD, L. 1932, Monocotyledonous seedlings: morphological studies in the post-seminal development of the embryo. Trans. & Proc. Bot. Soc. Edinburgh 31: 1–224.

BREMER, K. 2000. Early Cretaceous lineages of monocot flowering plants. Proc. Natl. Acad. Sci. U.S.A. 97: 4707–4711.

BRIDSON, G.D.R. (*ed*.). 1991. B-P-H/S. Botanico-Periodicum-Huntianum/Supplementum. Hunt Institute for Botanical Documentation, Pittsburgh, PA.

BROSCHAT, T.K. AND A. MEEROW. 2000. Ornamental palm horticulture. University Press of Florida, Florida, USA.

BROWN, K.E. 1976. Ecological studies of the Cabbage Palm, *Sabal palmetto*. Principes 20: 3–10.

BROWN, R.W. 1956a. Palm-like plants from the Delores Formation (Triassic) in southwestern Colorado. U.S. Geol. Surv. Prof. Paper, 274: 205–209.

BROWN, R.W. 1956b. Ivory-nut palm from the late Tertiary of Ecuador. Science 123: 1131–1132.

BROWN, R.W. 1962. Paleocene flora of the Rocky Mountains and Great Plains. US Geol. Surv. Prof. Paper 375. 199 pp.

BROWN, W.H. AND E.D. MERRILL. 1919. Philippine palms and palm products. Bull. Bur. Forest Philipp. Islands No. 18. Manila.

BRUMMITT, R.K. AND C.E. POWELL (*eds*). 1992. Authors of plant names. A list of authors of scientific names of plants, with recommended standard forms of their names, including abbreviations. Royal Botanic Gardens, Kew.

BÜCHLER, W. 1990. Eine fossile Flora aus dem oberen Oligozän von Ebnat-Kappel. Bot. Helv. 100: 133–166.

BUCKLEY, R. AND H.C. HARRIES. 1984. Self-sown wild-type coconuts from Australia. Biotropica 16: 148–151.

BUDANTSEV, L.YU. 1979. The find of a *Sabal* palm in the Eocenic beds in northwestern Kamchatka Peninsula. Bot. Zhurn. S.S.S.R. 64: 1777–1783, + 5 pls. [in Russian]

BULLOCK, S. 1980a. Demography of an undergrowth palm in littoral Cameroon. Biotropica 12: 247–255.

BULLOCK, S. 1980b. Dispersal of a desert palm by opportunistic frugivores. Principes 24: 29–32.

BUREAU, E. 1896. Sur quelques palmiers fossiles d'Italie. Bull. Mus. Hist. Nat. (Paris) 2: 280–285.

BURKILL, I.H. 1966. A dictionary of the economic products of the Malay Peninsula. 2nd Ed. Ministry of Agriculture and Cooperatives, Kuala Lumpur, Malaysia.

BURNETT, G.T. 1835. Outlines of botany, including a general history of the vegetable kingdom, in which plants are arranged according to the system of natural affinities. H. Renshaw, London.

Burney, D.A., H.F. James, L.P. Burney, S.L. Olson, W. Kikuchi, W.L. Wagner, M. Burney, D. McClosky, D. Kikuchi, F.V. Grady, R. Gage II and R. Nishek. 2001. Fossil evidence for a diverse biota from Kaua'i and its transformation since human arrival. Ecol. Monogr. 71: 615-641.

Búrquez, A., J. Sarukhán and A. Pedroza. 1987. Floral biology of a rain forest palm, *Astrocaryum mexicanum* Liebm. Bot. J. Linn. Soc. 94: 407–419.

Burret, M. 1928. Beitrage zur Kenntnis der Palmen von Malesia, Papua und der Südsee. Repert. Spec. Nov. Regni Veg. 24: 253–296.

Burret, M. 1929. Die gattung *Ceroxylon* Humb. et Bonpl. Notizbl. Bot. Gart. Berlin-Dahlem 10: 841–854.

Burret, M. 1930. Geonomae americanae. Bot. Jahrb. Syst. 63: 123–270.

Burret, M. 1933a. *Chamaedorea* Willd. und verwandte Palmengattungen. Notizbl. Bot. Gart. Berlin-Dahlem. 11: 724–768.

Burret, M. 1933b. *Schippia*, eine neue Palmengattung aus Brit. Honduras. Notizbl. Bot. Gart. Berlin-Dahlem. 11: 867–869.

Burret, M. 1933c. *Bactris* und verwandte Palmengattungen. Repert. Spec. Nov. Regni Veg. 34: 167–184.

Burret, M. 1934. Die Palmengattung *Astrocaryum* G.F.W. Meyer. Repert. Spec. Nov. Regni Veg. 35: 114–158.

Burret, M. 1935. New palms from Fiji. Occas. Pap. Bernice Pauahi Bishop Mus. 11: 1–14.

Burret, M. 1936a. Die Palmengattung *Gronophyllum* Scheff. Notizbl. Bot. Gart. Berlin-Dahlem 13: 200–205.

Burret, M. 1936b. Die Palmengattungen *Nengella* Becc. und *Leptophoenix* Becc.. Notizbl. Bot. Gart. Berlin-Dahlem 13: 312–317.

Burret, M. 1937. Palmae chinenses. Notizbl. Bot. Gart. Berlin-Dahlem 13: 582–606.

Burret, M. 1941. Die Palmengattung *Tessmaniodoxa* gen. nov. Notizbl. Bot. Gart. Berlin-Dahlem. 15: 336–338.

Burret, M. 1942. Neue Palmen aus der Gruppe der Lepidocaryoideae. Notizbl. Bot. Gart. Berlin-Dahlem 15: 728–755.

Burret and Potztal. 1956. Systematische Übersicht über die Palmen (Fortsetzung). Willdenowia 1: 350–385.

Bůžek, C. 1977. Date-palm seeds from the Lower Miocene of Central Europe. Vestn. Ústředn. Ústavu Geol. 52: 159–168.

Campbell, J., R. Fordyce, A. Grebneff and P. Maxwell. 2000. Fossil coconuts from mid-Cenozoic shallow marine sediments in southern New Zealand. Otago University, Geology Dept., Dunedin, New Zealand. [Poster abstract]

Caratini, C. and C. Tissot. 1985. Le Sondage Misedor, etude palynologique. Etudes de Géographie Tropical, Bordeaux: Centre Nationale de Recherche Scientifique.

Carillo, L., R. Orellana and L. Varela. 2002. Mycorrhizal associations in three species of palms of the Yucatan Peninsula, Mexico. Palms 46: 39–46.

Carlquist, S. and E.L. Schneider. 2002. The tracheid-vessel element transition in angiosperms involves multiple independent features: cladistic consequences. Amer. J. Bot. 89: 185–195.

Carr, S.G.M. and D.J. Carr. 1961. The functional significance of syncarpy. Phytomorphology 11: 249–256.

Castilho, A., A. Vershinin and J.S. Heslop-Harrison. 2000. Repetitive DNA and the chromosomes in the genome of oil palm (*Elaeis guineensis*). Ann. Bot. (Oxford) 85: 837–844.

Chandler, M.E.J. 1954. Some Upper Cretaceous and Eocene fruits from Egypt. Bull. Brit. Mus. (Nat. Hist.) Geol. 2: 147–187.

Chandler, M.E.J. 1957. The Oligocene Flora of the Bovey Tracey Lake Basin, Devonshire. Bull. Brit. Mus. (Nat. Hist.) Geol. 3: 71–123; + pls 11–17.

Chandler, M.E.J. 1961a. Flora of the Lower Headon Beds of Hampshire and the Isle of Wight. Bull. Brit. Mus. (Nat. Hist.) Geology 5: 91–158.

Chandler, M.E.J. 1961b. The Lower Tertiary Floras of Southern England. Volume I. Palaeocene Floras, London Clay Flora. Brit. Mus. (Nat. Hist.), London.

Chandler, M.E.J. 1961c. The Lower Tertiary Floras of Southern England. Volume I. Palaeocene Floras, London Clay Flora. Supplement — Atlas of plates. Brit. Mus. (Nat. Hist.), London.

Chandler, M.E.J. 1962. The Lower Tertiary Floras of Southern England. Volume II. Flora of the Pipe Clay Series of Dorset (Lower Bagshot). Brit. Mus. (Nat. Hist.), London.

Chandler, M.E.J. 1963. The Lower Tertiary Floras of Southern England. Volume III. Flora of the Bournemouth Beds, the Boscombe, and the Highcliff Sands. Brit. Mus. (Nat. Hist.), London.

Chandler, M.E.J. 1964. The Lower Tertiary Floras of Southern England. Volume IV. A summary and survey of findings in the light of recent botanical observations. Brit. Mus. (Nat. Hist.), London.

Chandler, M.E.J. 1978. Supplement to The Lower Tertiary Floras of Southern England. Part 5. Tertiary Research Special Paper #4. The Tertiary Research Group, London.

Chapin, M.H., F.B. Essig and J.-C. Pintaud. 2001. The morphology and histology of the fruits of *Pelagodoxa* (Arecaceae): taxonomic and biogeographical implications. Syst. Bot. 26: 779–785.

Chase, M.W. 2004. Monocot relationships: an overview. Amer. J. Bot. 91: 1645–1655.

Chase, M.W., D.E. Soltis, P.S. Soltis, P.J. Rudall, M.F. Fay, W.J. Hahn, S. Sullivan, J. Joseph, M. Molvray, P.J. Kores, T.J. Givnish, K.J. Sytsma and J.C. Pires. 2000. Higher-level systematics of the monocotyledons: an assessment of current knowledge and a new classification. Pages 3–16 *in* Wilson, K.L. and D.A. Morrison (*eds*). Monocots: systematics and evolution. CSIRO, Melbourne.

Chase, M.W., D.E. Soltis, R.G. Olmstead, D. Morgan, D.H. Les, B.D. Mishler, M.R. Duvall, R.A. Price, H.G. Hills, Y.L. Qiu, K.A. Kron, J.H. Rettig, E. Conti, J.D. Palmer, J.R. Manhart, K.J. Sytsma, H.J. Michaels, W.J. Kress, K.G. Karol, W.D. Clark, M. Hedren, B.S. Gaut, R.K. Jansen, K.J. Kim, C.F. Wimpee, J.F. Smith, G.R. Furnier, S.H. Strauss, Q.Y. Xiang, G.M. Plunkett, P.S. Soltis, S.M. Swensen, S.E. Williams, P.A. Gadek, C.J. Quinn, L.E. Eguiarte, E. Golenberg, G.H. Learn, S.W. Graham, S.C.H. Barrett, S. Dayanandan and V.A. Albert. 1993. Phylogenetics of seed plants — an analysis of nucleotide sequences from the plastid gene *rbc*L. Ann. Missouri Bot. Gard. 80: 528–580.

Chase, M.W., M.R. Duvall, H.G. Hills, J.G. Conran, A.V. Cox, L.E. Eguiarte, J. Hartwell, M.F. Fay, L.R. Caddick, K.M. Cameron and S. Hoot. 1995a. Molecular phylogenetics of Lilianae. Pages 109–137 *in* Rudall, P.J., P.J. Cribb, D.F. Cutler and C.J. Humphries (*eds*). Monocotyledons: systematics and evolution. Royal Botanic Gardens, Kew.

Chase, M.W., D.W. Stevenson, P. Wilkin and P.J. Rudall. 1995b. Monocot systematics: a combined analysis. Pages 685–730 *in* P.J. Rudall, P.J. Cribb, D.F. Cutler and C.J. Humphries (*eds*). Monocotyledons: systematics and evolution. Royal Botanic Gardens, Kew.

CHASE, M.W., M.F. FAY, D.S. DEVEY, O. MAURIN, N. RONSTED, T.J. DAVIES, Y. PILLON, G. PETERSEN, O. SEBERG, M.N. TAMURA, C.B. ASMUSSEN, K. HILU, T. BORSCH, J.I. DAVIS, D.W. STEVENSON, J.C. PIRES, T.J. GIVNISH, K.J. SYTSMA, M.A. MCPHERSON, S.W. GRAHAM AND H.S. RAI. 2006. Multigene analyses of monocot relationships: a summary. Aliso 22: 63–75.

CHAVEZ, F.M. 2003. Morphological, anatomical and phylogenetic study of palm germination and seedlings. Ph.D. Thesis, The City University of New York.

CHAZDON, R. 1991. Plant size and form in the understory palm genus *Geonoma*: are species variations on a theme? Amer. J. Bot. 78: 680–694.

CHEADLE, V.I. 1943. Vessel specialization in the late metaxylem of the various organs in the Monocotyledoneae. Amer. J. Bot 30: 484–490.

CHEADLE, V.I. 1944. Specialization of vessels within the xylem of each organ in the Monocotyledoneae. Amer. J. Bot. 31: 81–92.

CHEADLE, V.I. 1948. Observations on the phloem in the Monocotyledoneae. II. Additional data on the occurrence and phylogenetic specialization in structure of the sieve tubes in the metaphloem. Amer. J. Bot. 35: 129–131.

CHEADLE, V.I. AND N.B. WHITFORD. 1941. Observations on the phloem in the Monocotyledoneae. I. The occurrence and phylogenetic specialization in structure of the sieve tubes in the metaphloem. Amer. J. Bot. 28: 623–627.

CHEN 1995 mentioned in Anatomy of Cocos

CHURCHILL, D.M. 1973. The ecological significance of tropical mangroves in the Early Tertiary floras of southern Australia. Special publication, Geological Society of Australia 4: 79–86. Canberra. [Not seen, cited according to Gee (1990).]

CLEMENT, C.R. AND J.M. URPI. 1983. Leaf morphology of the Pejibaye palm (*Bactris gasipaes* H.B.K.). Rev. Biol. Trop. 31: 103–112.

COBB, A.R. 2006. Water relations of rattans. PhD thesis, Harvard University, Cambridge, Mass.

COCUCCI, A.E. 1964. Sobre la embriologia de *Butia paraguayensis* (Barb. Rodr.) Bailey (Palmae). Trab. Mus. Bot. Córdoba 3: 548–551.

COE, M.J., D.L. DILCHER, J.O. FARLOW, D.M. JARZEN AND D.A. RUSSELL. 1987. Dinosaurs and land plants. Pages 225–258 *in* E.M. Friis, W.G. Chaloner and P.R. Crane (*eds*). The origins of Angiosperms and their biological consequences, Cambridge University Press, Cambridge.

COLLINSON, M.E. 1983. Fossil plants of the London Clay. The Palaeontological Association, London.

COLLINSON, M.E., M.C. BOULTER, AND P.L. HOLMES. 1993. Magnoliophyta ('angiospermae'). Pages 809–841 *in* M.J. Benton (*ed*.). The Fossil Record 2. Chapman-Hall, London.

COMPTON, J.A., A. CULHAM AND S.L. JURY. 1998. Reclassification of *Actaea* to include *Cimicifuga* and *Souliea* (Ranunculaceae): phylogeny inferred from morphology, nrDNA ITS, and cpDNA *trn*L-F sequence variation. Taxon 47: 593–634.

CONWENTZ, H.W. 1886. Die Flora des Bernsteins. Die angiospermen des Bernsteins. *In* Goeppert, H.R. and A. Menge. Die Flora des Bernsteins und ihre Beziehungen zur Flora der Tortiärformation und der Gegenwart. Danzig, Germany.

COOK, O.F. 1915a. A new genus of palms allied to *Archontophoenix*. J. Washington Acad. Sci. 5: 116–122.

COOK, O.F. 1915b. *Glaucothea*, a new genus of palms from Lower California. J. Washington Acad. Sci. 5: 236–241.

COOK, O.F. 1926. A new genus of palms based on *Kentia forsteriana*. J. Washington Acad. Sci. 16: 392–397.

COOK, O.F. 1927. New genera and species of ivory palms from Colombia, Ecuador and Peru. J. Washington Acad. Sci. 17: 218–230.

COOK, O.F. 1937a. New household palm, *Neanthe bella*. Science 86: 120–122.

COOK, O.F. 1937b. Hurricane palms in Florida, including a new genus *Simpsonia*. Science 85: 332–333.

COOK, O.F. 1939a. *Bornoa*, an endemic palm of Haiti. Natl. Hort. Mag. 18: 254–280.

COOK, O.F. 1939b. A second household palm, *Omanthe costaricana*. Science 90: 298–299.

COOK, O.F. 1940. Aublet the botanist, a pioneer against slavery, with a memorial genus of palms. J. Washington Acad. Sci. 30: 294–299.

COOK, O.F. 1943a. Household palms and related genera. Part I. Natl. Hort. Mag. 22: 83–102.

COOK, O.F. 1943b. Household palms and related genera. Part II. Natl. Hort. Mag. 22: 134–152.

COOK, O.F. 1947a. Cascade palms of southern Mexico. Natl. Hort. Mag. 26: 10–34.

COOK, O.F. 1947b. Climbing and creeping palms in Mexico and Guatemala, related to household palms. Natl. Hort. Mag. 26: 215–231.

COOK, O.F. AND C.B. DOYLE. 1939. The edible pacaya palm of Alta Verapaz. Natl. Hort. Mag. 18: 161–179.

COOKSON, I.C. AND A. EISENACK. 1967. Some early Tertiary plankton and pollen grains from a deposit near Strahan, western Tasmania. Proc. Roy. Soc. Victoria 80: 131–140. Melbourne. [Not seen cited according to Gee (1990).]

CORNER, E.J.H. 1966. The natural history of palms. Weidenfeld and Nicolson, London.

CORNET, B. 1986. The leaf venation and reproductive structures of a Late Triassic angiosperm, *Sanmiguelia lewisii*. Evol. Theory, 7: 231–309.

CORNET, B. 1989a. The reproductive morphology and biology of *Sanmiguelia lewisii*, and its bearing on angiosperm evolution in the Late Triassic. Evol. Trends Pl. 3: 25–51.

CORNET, B. 1989b. Late Triassic angiosperm-like pollen from the Richmond Rift Basin of Virginia, U.S.A. Palaeontographica Abt. B, 213: 37–87.

CORNETT, J.W. 1985. Germination of *Washingtonia filifera* seeds eaten by coyotes. Principes 29: 19.

COUPER, R.A. 1952. The spore and pollen flora of the *Cocos*-bearing beds, Mangonui, North Auckland. Trans. Roy. Soc. New Zealand, Bot. 79: 340–348.

COUPER, R.A. 1953. Upper Mesozoic and Cainozoic spores and pollen grains from New Zealand. Palaeontol. Bull. New Zealand Geol. Surv. 22. 77 pp., 3 tab., 3 text figs, 9 pls. Wellington.

COUVREUR, T.L.P., W.J. HAHN, J.-J. DE GRANVILLE, J.-L. PHAM, B. LUDEÑA AND J.-C. PINTAUD. 2007. Phylogenetic relationships of the cultivated Neotropical palm *Bactris gasipaes* (Arecaceae) with its wild relatives inferred from non-coding chloroplastic sequences and nuclear microsatellite polymorphism. Syst. Bot. 32: 519–530.

CRABTREE, D.R. 1987. Angiosperms of the northern Rocky Mountains: Albian to Campanian (Cretaceous) megafossil floras. Ann. Missouri Bot. Gard. 74: 707–747.

CRIÉ, L. 1892. Recherches sur les Palmiers silicifiés des terrains Crétacés de l'Anjou. Bull. Soc. Études Sci. Angers 21: 97–103.

CRONQUIST, A. 1981. An integrated system of classification of flowering plants. Columbia University Press, New York.

CUENCA-NAVARRO, A. 2007. Systematics, biogeography and evolution of morphological characters of the palm tribe Chamaedoreeae. Ph.D. Thesis, Faculty of Life Sciences, University of Copenhagen.

CUENCA, A. AND C.B. ASMUSSEN-LANGE. 2007. Phylogeny of the palm tribe Chamaedoreeae (Arecaceae) based on plastid DNA sequences. Syst. Bot. 32: 250–263.

CUENCA, A., C.B. ASMUSSEN-LANGE AND F. BORCHSENIUS. 2008. A dated phylogeny of the palm tribe Chamaedoreeae supports Eocene dispersal between Africa, North and South America. Molec. Phylogen. Evol. 46: 760–775.

CUNNINGHAM, S.A. 1995. Ecological constraints on fruit initiation by *Calyptrocalyx ghiesbreghtiana* (Arecaceae): floral herbivory, pollen availability, and visitation by pollinating bats. Amer. J. Bot. 82: 1527–1536.

CZECZOTT, H.O. AND K. JUCHNIEWICZ.1975. Monocotyledonae: Palmae. Flora Kopalna Turowa kolo Bogatyni, II. Prace Muz. Ziemi 24: 57–64; + 4 pls.

CZECZOTT, H.O. AND K. JUCHNIEWICZ.1980. Palmae II. Flora Kopalna Turowa kolo Bogatyni, II. Prace Muz. Ziemi 33: 26–29; + 5 pls.

D'ARCY, W.G. 1996. Anthers and stamens and what they do. Pages 1–24 *in* D'Arcy, W.G. and R.C. Keating (*eds*). The anther: form, function and phylogeny. Cambridge University Press.

D'ARCY, W.G., R.C. KEATING AND S.L. BUCHMANN. 1996. The calcium oxalate package or so-called resorption tissue in some angiosperm anthers. Pages 159–191 *in* D'Arcy, W.G. and R.C. Keating (*eds*). The anther: form, function and phylogeny. Cambridge University Press.

DAGHLIAN, C.P. 1978. Coryphoid palms from the Lower and Middle Eocene of southeastern North America. Palaeontographica, Abt. B 166: 44–82.

DAHLGREN, B.E. AND S.F. GLASSMAN. 1961. A revision of the genus *Copernicia* 1. South American species. Gentes Herb. 9: 3–40.

DAHLGREN, B.E. AND S.F. GLASSMAN. 1963. A revision of the genus *Copernicia* 2. West Indian species. Gentes Herb. 9: 43–232.

DAHLGREN, R.M.T. AND H.T. CLIFFORD. 1982. The monocotyledons: a comparative study. Academic Press, London.

DAHLGREN, R.M.T., H.T. CLIFFORD AND P.F. YEO. 1985. The families of the monocotyledons. Springer-Verlag, Berlin.

DARLINGTON, C.D. AND E.K. JANAKI AMMAL. 1945. Chromosome atlas of cultivated plants. George Allen & Unwin Ltd., London.

DARLINGTON, C.D. AND A.P. WYLIE. 1955. Chromosome atlas of flowering plants. George Allen & Unwin Ltd., London.

DAUMANN, E. 1970. Das Blütennektarium der Monocotyledonen unter besonderer Berücksichtigung seiner systematischen und Phylogenetischen Bedeutung. Feddes Repert. 80: 463–590.

DAVIS, J.I., D.W. STEVENSON, G. PETERSEN, O. SEBERG, L.M. CAMPBELL, J.V. FREUDENSTEIN, D.H. GOLDMAN, C.R. HARDY, F.A. MICHELANGELI, M.P. SIMMONS, C.D. SPECHT, F. VERGARA-SILVA AND M. GANDOLFO. 2004. A phylogeny of the monocots, as inferred from *rbc*L and *atp*A sequence variation, and a comparison of methods for calculating jackknife and bootstrap values. Syst. Bot. 29: 467–510.

DAVIS, J.I., G. PETERSEN, O. SEBERG, D.W. STEVENSON, C.R. HARDY, M.P. SIMMONS, F.A. MICHELANGELI, D.H. GOLDMAN, L.M. CAMPBELL, C.D. SPECHT AND J.I. COHEN. 2006. Are mitochondrial genes useful for the analysis of monocot relationships? Taxon 55: 857–870.

DE CARVALHO, M.D.F. AND G. BANDEL. 1986. Citogénetica do babaçu (*Orbignya* spp.). Ci. & Cult. 38: 930.

DE GRANVILLE, J.J. 1974. Aperçu sur la structure des pneumatophores de deux espèces des sols hydromorphes en Guyane. *Mauritia flexuosa* L. et *Euterpe oleracea* Mart. (Palmae). Généralisation au système respiratoire racinaire d'autres palmiers. Cah. O.R.S.T.O.M., sér. Biol. 23: 3–22.

DE MADEIROS-COSTA, J. AND S. PANIZZA. 1983. Palms of the cerrado vegetation formation of São Paulo State, Brazil. Principes 27: 118–125.

DE MAGNANO, S.M. 1973. *Trithrinax campestris* (Palmae): inflorescentia y flor con especial referncia al gineceo. Kurtziana 7: 137–152.

DE MASON, D.A., K.W. STOLTE AND B. TISSERAT. 1982. Floral development in *Phoenix dactylifera*. Canad. J. Bot. 60: 1437–1446.

DE NEVERS, G. 1995. Notes on Panama palms. Proc. Calif. Acad. Sci. 48: 329–342.

DE QUEIROZ, A. 2005. The resurrection of oceanic dispersal in historical biogeography. Trends Ecol. Evol. 20: 68–73.

DE QUEIROZ, A. AND J. GATESY. 2007. The supermatrix approach to systematics. Trends Ecol. Evol. 22: 34–41.

DE VISIANI R. 1864. Palmae pinnatae tertiariae agri ceneti. Mem. Reale Inst. Veneto Sci. 11: 435–460.

DELPINO, F. 1870. Ulteriori osservazzioni e considerazioni sulla dicogamia nel regno vegetale. Atti Soc. Ital. Sci. Nat. 13: 167–205.

DENGLER, N.G., R.E. DENGLER AND D.R. KAPLAN. 1982. The mechanism of plication inception in palm leaves: histogenetic observations on the pinnate leaf of *Chrysalidocarpus lutescens*. Canad. J. Bot. 60: 82–95.

DESMOND, A.J. 1975. The hot-blooded dinosaurs — a revolution in palaeontology. Blond & Briggs Ltd, London.

DI FULVIO, T.E. 1966. Cromosomas gameticos de *Trithrinax campestris* (Palmae). Kurtziana 3: 233–234.

DIGGLE, P.K. 2003. Architectural effects on floral form and function: a review. Pages 63–80 *in* Stuessy, T.F., V. Mayer and E. Horandi (*eds*). Deep morphology: toward a renaissance of morphology in plant systematics. A.R.G. Gantner Verlag, Ruggell, Liechtenstein.

DOLIANITI, E. 1955. Fructos de *Nipa* no Paleoceno de Pernambuco, Brasil. Bol. Minist. Agric. (Rio de Janeiro) 158: 1–36 & pls. 102. [Not seen cited according to Tralau (1964).]

DORF, E. 1942. Upper Cretaceous floras of the Rocky Mountain region. Contributions to paleontology, Carneg. Inst, publ. no. 508. 168 pp.

DOWE, J.L. AND P. CABALION. 1996. A taxonomic account of Arecaceae in Vanuatu, with descriptions of three new species. Austral. Syst. Bot. 9: 1–60.

DOWE, J.L. AND M.H. CHAPIN. 2006. Beccari's "Grande Nouveauté": the discovery, taxonomic history and typification of *Pelagodoxa henryana*. Palms 50: 185–192.

DOWE, J.L AND M.D. FERRERO. 2000. A new species of rheophytic palm from New Guinea. Palms 44: 194–197.

DOWE, J.L. AND M.D. FERRERO. 2001. Revision of *Calyptrocalyx* and the New Guinea species of *Linospadix* (Linospadicinae: Arecoideae: Arecaceae). Blumea 46: 207–251.

DOWE, J.L. AND D.R. HODEL. 1994. A revision of *Archontophoenix* H. Wendl. & Drude (Arecaceae). Austrobaileya 4: 227–244.

DOWE, J.L. AND A.K. IRVINE. 1997. A revision of *Linospadix* in Australia, with the description of a new species. Principes 41(4): 192–197, 211–217.

Dowe, J.L. and N.W. Uhl. 1989. *Carpoxylon macrospermum*. Principes 33: 68–73.

Dransfield, J. 1970. Studies in the Malayan palms *Eugeissona* and *Johannesteijsmannia*. PhD. Thesis. University of Cambridge.

Dransfield, J. 1972a. The genus *Borassodendron* in Malesia. Reinwardtia 8: 351–363.

Dransfield, J. 1972b. The genus *Johannesteijsmannia* H.E. Moore Jr. Gard Bull. Singapore 26: 63–83.

Dransfield, J. 1974. Notes on *Caryota no* Becc. and other Malesian *Caryota* spp. Principes 18: 87–93.

Dransfield, J. 1976b. A note on the habitat of *Pigafetta filaris* in North Celebes. Principes 20: 48.

Dransfield, J. 1977. *Calamus caesius* and *Calamus trachycoleus* compared. Gard. Bull. Singapore 30: 75–78.

Dransfield, J. 1978a. The genus *Maxburretia* (Palmae). Gentes Herb. 11(4): 187–199.

Dransfield, J. 1978b. The growth forms of rain forest palms. Pages 247–268 *in* Tomlinson, P.B. and M.H. Zimmermann (*eds*). Tropical trees as living systems. Cambridge University Press. New York.

Dransfield, J. 1979a. A manual of the rattans of the Malay Peninsula. Mal. For. Records 29. 270 pp. Forest Dept. Malaysia.

Dransfield, J. 1979b. A monograph of the genus *Ceratolobus* (Palmae). Kew Bull. 34: 1–33.

Dransfield, J. 1979c. *Retispatha*, a new genus of Palmae (Lepidocaryoideae) from Borneo. Kew Bull. 34: 529–536.

Dransfield, J. 1980a. *Pogonotium* (Palmae: Lepidocaryoideae) a new genus related to *Daemonorops*. Kew Bull. 34: 761–768.

Dransfield, J. 1980b, Systematic notes on *Pinanga* (Palmae) in Borneo. Kew Bull. 34: 769–788.

Dransfield, J. 1981a. A synopsis of *Korthalsia* (Palmae Lepidocaryoideae). Kew Bull. 36: 163–194.

Dransfield, J. 1981b. Palms and Wallace's line. Pages 43–56 *in* T. C. Whitmore (*ed*.). Wallace's line and plate tectonics. Clarendon Press, Oxford.

Dransfield, J. 1981c. A house in Thailand. Principes 25: 36–38.

Dransfield, J. 1982a. *Clinostigma* in New Ireland. Principes 26: 73–76.

Dransfield, J. 1982b. A reassessment of the genera *Plectocomiopsis*, *Myrialepis* and *Bejaudia*. Kew Bull. 37: 237–254.

Dransfield, J. 1982c. *Pogonotium moorei*, a new species from Sarawak. Principes 26: 174–177.

Dransfield, J. 1983. *Kerriodoxa*, a new coryphoid palm genus from Thailand. Principes 27: 3–11.

Dransfield, J. 1984a. The rattans of Sabah. Forest Department, Sabah.

Dransfield, J. 1984b. The genus *Areca* (Palmae: Arecoideae) in Borneo. Kew Bull. 39: 1–22.

Dransfield, J. 1986. Palmae. Flora of Tropical East Africa. A.A. Balkema, Rotterdam.

Dransfield, J. 1987. Bicentric distributions in Malesia as exemplified by palms. Pages 60–72 *in* Whitmore, T.C. (*ed*.). Biogeographical evolution of the Malay Archipelago. Clarendon Press, Oxford.

Dransfield, J. 1988a. *Beccariophoenix madagascariensis*. Principes 32: 59–68.

Dransfield, J. 1988b. The palms of Africa and their relationships. Pages 95–103 *in* Goldblatt, P. and P.P. Lowry (*eds*). Proceedings of the Eleventh Plenary Meeting of the Association for the Study of the Flora of Tropical Africa. Missouri Botanical Garden, St. Louis.

Dransfield, J. 1989. *Voanioala* (Arecoideae: Cocoeae: Butiinae), a new palm genus from Madagascar. Kew Bull. 44: 191–198.

Dransfield, J. 1991a. 123. *Paschalococos disperta* J. Dransfield, gen. et sp. nov. Pages 64–65 *in* Zizka, G. Flowering plants of Easter Island. Hortus Francofurtensis, Wissenschaftliche Berichte, Palmengarten, Frankfurt.

Dransfield, J. 1991b. *Lemurophoenix* (*Palmae*: *Arecoideae*), a new genus from Madagascar. Kew Bull. 46: 61–68.

Dransfield, J. 1991c. Notes on *Pinanga* (Palmae) in Sarawak. Kew Bull. 46: 691–698.

Dransfield, J. 1992a. Observations on rheophytic palms in Borneo. Bull. Inst. Fr. Études Andines 21: 415–432.

Dransfield, J. 1992b. Morphological considerations: the structure of rattans. Pages 11–26 *in* Wan Razali, J. Dransfield and N. Manokaran (*eds*). A guide to rattan planting. Forest Research Institute, Malaysia.

Dransfield, J. 1992c. The rattans of Sarawak. Royal Botanic Gardens Kew and Sarawak Forest Department. 213 pp.

Dransfield, J. 1998. *Pigafetta*. Principes 42: 34–40.

Dransfield, J. 1999. Species and species concepts in Old World palms. Pages 5–20 *in* Henderson, A. and F. Borchsenius (*eds*). Evolution, variation and classification of palms. The New York Botanical Garden Press, Bronx, NY.

Dransfield, J., W.J. Baker, C.H. Heatubun and J. Witono. 2000. The palms of Mount Jaya. Palms 44: 202–208.

Dransfield, J. and W.J. Baker. 2003. An account of the Papuasian species of *Calamus* (Arecaceae) with paired fruit. Kew Bull. 58: 371–387.

Dransfield, J. and H.J. Beentje. 1995a. *Satranala* (Coryphoideae: Borasseae: Hyphaeninae), a new palm genus from Madagascar. Kew Bull. 50: 85–92.

Dransfield, J. and H.J. Beentje. 1995b. The palms of Madagascar. Royal Botanic Gardens, Kew and the International Palm Society.

Dransfield, J., I.K. Ferguson and N.W. Uhl. 1990. The coryphoid palms — patterns of variation and evolution. Ann. Missouri Bot. Gard. 77: 802–815.

Dransfield, J., J.R. Flenley, S.M. King, D.D. Harkness and S. Rapu. 1984. A recently extinct palm from Easter Island. Nature 312 (No 5996): 750–752.

Dransfield, J., A.K. Irvine and N.W. Uhl. 1985a. *Oraniopsis appendiculata*, a previously misunderstood Queensland Palm. Principes 29: 56–63.

Dransfield, J., S.K. Lee and F.N. Wei. 1985b. *Guihaia*, a new coryphoid genus from China and Vietnam. Principes 29: 3–12.

Dransfield, J. and J.P. Mogea. 1984. The flowering behaviour of *Arenga* (Palmae: Caryotoideae). Bot. J. Linn.Soc. 88: 1–10.

Dransfield, J. and H.E. Moore. 1982. The Martian Correlation — two editions of Martius' Historia Naturalis Palmarum compared. Kew Bull. 37: 91–116.

Dransfield, J., M. Rakotoarinivo, W.J. Baker, R.P. Bayton, J.B. Fisher, J.W. Horn, B. Leroy and X. Metz. 2008. A new Coryphoid palm genus from Madagascar. Bot. J. Linn. Soc. 156: 79–91.

Dransfield, J. and N.W. Uhl. 1983a. The transfer of *Livistona kingiana* (Palmae) to *Pholidocarpus*. Kew Bull. 38: 197–198.

Dransfield, J. and N.W. Uhl. 1983b. *Wissmannia* (Palmae) reduced to *Livistona*. Kew Bull. 38: 199–200.

Dransfield, J. and N.W. Uhl. 1986. An outline of a classification of palms. Principes 30: 3–11.

Dransfield, J. and N.W. Uhl. 1998. Palmae. *In* Kubitski, K. (*ed*.). The families and genera of vascular plants. IV. Flowering plants. Monocotyledons. Springer, Berlin.

DRANSFIELD, J., N.W. UHL, C.B. ASMUSSEN, W.J., BAKER, M.M. HARLEY AND C.E. LEWIS. 2005. A new phylogenetic classification of the palm family, Arecaceae. Kew Bull. 60: 559–569.

DRANSFIELD, J., S.K. LEE AND F.N. WEI. 1985b. *Guihaia*, a new coryphoid genus from China and Vietnam. Principes 29: 3–12.

DRESCHER, J. AND A. DUFAY. 2002. Importation of mature palms: a threat to native and exotic palms in the Mediterranean countries. Palms 46: 179–184.

DRUDE, O. 1882. Palmae. Pages 253–583 *in* von Martius, K.F.P., Flora Brasiliensis 3(2).

DRUDE, O. 1887. Palmae. Pages 1–93 *in* Engler, A. and K. Prantl. Die Natürlichen Pflanzenfamilien 2(3).

DUFAY, M., M. HOSSAERT-MCKEY AND M.-C. ANSTETT. 2003. When leaves act like flowers: how dwarf palms attract their pollinators. Ecol. Letters 6: 28–34.

DUKE, N.C. 1991. *Nypa* in the mangroves of Central America: introduced or relict? Principes 35: 127–132.

EAMES, A.J. 1953. Neglected morphology of the palm leaf. Phytomorphology 3: 172–189.

EAMES, A.J. 1961. Morphology of the angiosperms. McGraw Hill, New York.

EDIGER, V.S., Z. BATI AND C. ALISAN. 1990. Paleopalynology and paleoecology of *Calamus*-like disulcate pollen grains. Rev. Palaeobot. Palynol. 62: 97–105.

EDWARDS, P.J., J. KOLLMANN AND K. FLEISCHMANN. 2002. Life history evolution in *Lodoicea maldivica* (Arecaceae). Nordic J. Bot. 22: 227–237.

EICHORN, A. 1953. Étude caryologique des palmiers. I. Rev. Cytol. Biol. Vég. 14: 13–29.

EICHORN, A. 1957. Nouvelles contribution á l'étude caryologique des palmiers. Rev. Cytol. Biol. Vég. 18: 139–151.

EITEN, G. 1972. The cerrado vegetation of Brazil. Bot. Rev. (Lancaster) 38: 201–341.

ELIAS DE PAULA, J. 1975. Anatomia de *Euterpe oleracea* Mart. Acta Amazon. 5: 265–278.

ENDRESS, P.K. 1994. Diversity and evolutionary biology of tropical flowers. Cambridge University Press.

ENDRESS, P.K. 1996. Diversity and evolutionary trends in anthers. Pages 92–110 *in* D'Arcy, W.G. and R.C. Keating (*eds*). The anther: form, function and phylogeny. Cambridge University Press.

ENDRESS, P.K. 2003. What should a "complete" morphological phylogenetic analysis entail? Pages 131–164 *in* Stuessy, T.F., V. Mayer and E. Horandi (*eds*). Deep morphology: toward a renaissance of morphology in plant systematics. A.R.G. Gantner Verlag, Ruggell, Liechtenstein.

ENDRESS, P.K. AND A. IGERSHEIM. 2000. Gynoecium structure and evolution in basal angiosperms. Int. J. Pl. Sci. 161: 8211–8223.

ENDT, D. 1998. The Chatham Islands: home of the most southern naturally occurring palm in the world, *Rhopalostylis* "Chatham". Principes 42: 145–147.

ERDTMAN, G. 1934. Über die Verwendung von essigsäurean-hybrid bei Pollenuntersuchungen. Svensk Bot. Tidskr. 28: 354–358.

ERDTMAN, G. 1960. The acetolysis method, a revised description. Svensk Bot. Tidskr. 54:561–564.

ERDTMAN, G. AND G. SINGH. 1957. On the pollen morphology in *Sclerosperma mannii*. Bull. Jard. Bot. État Bruxelles 27: 217–220.

ERGO, A.B. 1997. Nouvelle evidence d'origine africaine de l'*Elaeis guineensis* Jacq. Par la decouverte de graines fossils en Uganda. Ann. Gembloux 102: 191–201.

ERVIK, F., L. TOLLSTEN AND J. KNUDSEN. 1999. Floral scent chemistry and pollination ecology in phytelephantoid palms (Arecaceae). Pl. Syst. Evol. 217: 279–297.

ERWIN, D.M. AND R.A. STOCKEY. 1991. Silicified monocotyledons from the Middle Eocene Princeton chert (Allenby Formation) of British Columbia, Canada. Rev. Palaeobot. Palynol. 70: 147–162.

ESSIG, F.B. 1970. New chromosome counts in *Chamaedorea* (Palmae). Principes 14: 136–137.

ESSIG, F.B. 1971a. Observations on pollination in *Bactris*. Principes 15: 20–24, 35.

ESSIG, F.B. 1971b. Palm briefs. New chromosome counts. Principes 15: 126.

ESSIG, F.B. 1973. Pollination in some New Guinea palms. Principes 17: 75–83.

ESSIG, F.B. 1975. A systematic study of the genus *Ptychosperma* (Palmae). PhD Thesis, Cornell University.

ESSIG, F.B. 1977. A systematic histological study of palm fruits. I. The *Ptychosperma* alliance. Syst. Bot. 2: 151–168.

ESSIG, F.B. 1978. A revision of the genus *Ptychosperma*. Allertonia 1: 415–478.

ESSIG, F.B. 1982. A synopsis of the genus *Gulubia*. Principes 26: 159–173.

ESSIG, F.B. 1999. Trends of specialization in the palm pericarp. Pages 73–77 *in* Henderson, A. and F. Borchsenius (*eds*). Evolution, variation and classification of palms. The New York Botanical Garden Press, Bronx, NY.

ESSIG, F.B. 2002. A systematic histological study of palm fruits. VI. Subtribe Linospadicinae (Arecaceae). Brittonia: 54: 196–201.

ESSIG, F.B., L. BUSSARD AND N. HERNANDEZ. 2001. A systematic histological study of palm fruits. IV. Subtribe Oncospermatinae (Arecaceae). Brittonia 53: 466–471.

ESSIG, F.B. AND N. HERNANDEZ. 2002. A systematic histological study of palm fruits: 5. Subtribe *Archontophoenicinae* (*Arecaceae*). Brittonia 54: 65–71.

ESSIG, F.B. AND L. LITTEN. 2004. A systematic histological study of palm fruits. VII. The Cyrtostachidinae (Arecaceae). Brittonia 56: 375–379.

ESSIG, F.B., T. MANKA AND L. BUSSARD. 1999. A systematic histological study of palm fruits. III. Subtribe Iguanurinae (Arecaceae). Brittonia 51: 307–325.

ESSIG, F.B. AND B.E. YOUNG. 1985. A reconsideration of *Gronophyllum* and *Nengella* (Arecoideae). Principes 29: 129–137.

ESSIG, F.B. AND B.E. YOUNG. 1979. A systematic histological study of palm fruits. II. The *Areca* alliance. Syst. Bot. 4: 16–28.

EVANS, R.J. 1995. Systematics of *Cryosophila* (Palmae). Syst. Bot. Monogr. 46: 1–70.

EVANS, R.J. 2001. Monograph of *Colpothrinax*. Palms 45: 177–195.

EVANS, T. AND K. SENGDALA. 2001. The Indochinese rattan *Calamus acanthophyllus* — a fire-loving palm. Palms 45: 25–28.

EVANS, T.D., K. SENGDALA. B. THAMMAVONG, O.V. VIENGKHAM AND J. DRANSFIELD. 2002. A synopsis of the rattans (Arecaceae: Calamoideae) of Laos and neighbouring parts of Indochina. Kew Bulletin 57: 1–84.

EYRE-WALKER, A. AND B.S. GAUT. 1997. Correlated rates of synonymous site evolution across plant genomes. Molec. Biol. Evol. 14: 455–460.

FELSENSTEIN, J. 2004. Inferring phylogenies. Sinauer, Sunderland, Massachusetts.

FERGUSON, I.K. 1987. Observations on the variation in pollen morphology of Palmae and its significance. Canad. J. Biol. 64: 3079–3090.

FERGUSON, I.K., J. DRANSFIELD AND I. FLAWN. 1988. A review of the pollen morphology and systematics of the genera *Ravenea* and *Louvelia* (Ceroxyleae: Ceroxyloideae: Palmae). J. Palynol., 23–4: 65–72.

FERGUSON, I.K., J. DRANSFIELD, F.C. PAGE AND G. THANIKAIMONI. 1983. Notes on the pollen morphology of *Pinanga* with special reference to *P. aristata* and *P. pilosa* (Palmae: Arecoideae). Grana, 22: 65–72.

FERGUSON, I.K. AND M.M. HARLEY. 1993. The significance of new and recent work on pollen morphology of the Palmae. Kew Bull. 48: 205–243.

FERGUSON, I.K., A.J. HAVARD AND J. DRANSFIELD. 1987. The pollen morphology of the tribe Borasseae (Palmae: Coryphoideae). Kew Bull. 42: 405–422.

FERNANDO, E.S. 1983. A revision of the genus *Nenga*. Principes 27: 55–70.

FERNANDO, E.S. 1990. The genus *Heterospathe* (Palmae: Arecoideae) in the Philippines. Kew Bull. 45: 219–234.

FERNANDO, E.S. 1994. New species of *Pinanga* (Palmae: Arecoideae) from Luzon Island, Philippines. Kew Bull. 49: 775–784.

FERRAZ-VICENTINI, K.R. AND M.L. SALGADO-LABOURIAU. 1996. Palynological analysis of a palm swamp in Central Brazil. J. S. Amer. Earth Sci. 9: 207–219.

FERREIRA, E. 1999. The phylogeny of Pupunha (*Bactris gasipaes* Kunth, Palmae) and allied species. Pages 225–236 *in* Henderson, A. and F. Borchsenius (*eds*). Evolution, variation and classification of palms. The New York Botanical Garden Press, Bronx, NY.

FERRY, M. AND S. GÓMEZ. 2002. The Red Palm Weevil in the Mediterranean area. Palms 46: 172–178.

FISHER, J. 1978. A quantitative description of shoot development in three rattan palms. Malaysian Forester 41:280–293.

FISHER, J., J.N. BURCH AND L.R. NOBLICK. 1996. Stem structure of the Cuban belly palm (*Gastrococos crispa*). Principes 40: 15–128.

FISHER J.B. AND J. DRANSFIELD. 1977. Comparative morphology and development of inflorescence adnation in rattan palms. Bot. J. Linn. Soc. 75: 119–140.

FISHER J.B. AND J. DRANSFIELD. 1979. Development of axillary and leaf-opposed buds in the rattan palm *Daemonorops*. Ann. Bot. (Oxford) 44(1): 57–66.

FISHER, J. AND J. FRENCH. 1976. The occurrence of intercalary and uninterrupted meristems in the internodes of tropical monocotyledons. Amer. J. Bot. 63: 510–525.

FISHER, J. AND J. FRENCH. 1978. Internodal meristems of monocotyledons: further studies and a general taxonomic study. Ann. Bot. (Oxford) 42: 41–50.

FISHER, J. AND K.J. MAIDMAN. 1999. Branching and architecture in palms: value for systematics. Pages 35–46 *in* Henderson, A. and F. Borchsenius (*eds*). Evolution, variation and classification of palms. The New York Botanical Garden Press, Bronx, NY.

FISHER, J. AND J.P. MOGEA. 1980. Intrapetiolar inflorescence buds in *Salacca* (Palmae): development and significance. Bot. J. Linn Soc. 81: 47–59.

FISHER, J., AND H.E. MOORE. 1977. Multiple inflorescences in palms: their development and significance. Bot. Jahrb. Syst. 98: 573–611.

FISHER, J.B., R.W. SANDERS AND N. EDMONSON. 1987. The flowering and fruiting of *Corypha umbraculifera* in Miami, Florida. Principes 31: 68–77.

FISHER, J.B. AND P.B. TOMLINSON. 1973. Branch and inflorescence production in saw palmetto (*Serenoa repens*). Principes 17: 10–19.

FISHER, J.B. AND S. ZONA. 2006. Unusual branching in *Manicaria*. Palms 50: 99–102.

FLICHE, P. 1894. Sur les fruits de palmiers trouvés dans le cénomanien, environs de Sainte-Menehould. Compt. Rend. Hebd. Séances Acad. Sci. 118: 889–890.

FLICHE, P. 1896. Étude sur la flore fossile de l'Argonne Albien-Cenomanien. Bull. Soc.Sci. Nancy 14: 114–306.

FONG, F.W. 1987. An unconventional alcohol fuel crop. Principes 31: 64–67.

FONG, F.W. 1989. The apung palm; traditional techniques of sugar-tapping and alcohol extraction in Sarawak. Principes 33: 21–26.

FOX, J.J. 1977. Harvest of the Palm; Ecological Change in Eastern Indonesia. Harvard Univ. Press, Cambridge, Massachusetts.

FREDERIKSEN, N.O. 1980. Sporomorphs from the Jackson Group (Upper Eocene) and adjacent strata of Mississippi and western Alabama. U.S. Geol. Surv. Reston, VA Prof. Paper 1084: 1–75.

FREDERIKSEN, N.O. 1981. Middle Eocene to Early Oligocene plant communities of the Gulf Coast. Pages 493–549 *in* J. Gray, A.J. Boucot and W.B. Berry (*eds*). Communities of the Past. Hutchinson Ross Publ. Co., Stronsburg PA.

FREDERIKSEN, N.O. 1994. Middle and late Paleocene angiosperm pollen from Pakistan. Palynology 18: 91–137.

FREDERIKSEN, N.O., V.D. WIGGINS, I.K. FERGUSON, J. DRANSFIELD and C.M. AGER. 1985. Distribution, paleoecology, and botanical affinity of the Eocene pollen genus *Diporoconia* n. gen. Palynology 9: 37–60.

FRIIS, E.M., K.R. PEDERSEN AND P.R. CRANE. 2004. Araceae from the Early Cretaceous of Portugal: Evidence on the emergence of monocotyledons. Proc. Natl. Acad. Sci. U.S.A. 101: 16565–16570.

FRITEL, P.H. 1921. Sur la découverte, au Sénégal, de deux fruits fossils appartenant aux genres *Kigelia* DC et *Nipadites* Bowerb. Compt. Rend. Hebd. Séances Acad. Sci. 173: 245–246.

FROHLICH, J. AND K.D. HYDE. 1999. Biodiversity of palm fungi in the tropics: are global fungal diversity estimates realistic? Biodiver. & Conservation 8: 977–1004.

FROHLICH, J., K.D. HYDE AND O. PETRINI. 2000. Endophytic fungi associated with palms. Mycol. Res. 104: 1202–1212.

FULLER, D. 1999. The lost palm of Fiji — a resolution of *Goniocladus* and a preliminary cladistic analysis of *Physokentia* (Palmae). Pages 203–213 *in* Henderson, A. and F. Borchsenius (*eds*). Evolution, variation and classification of palms. The New York Botanical Garden Press, Bronx, NY.

FURLEY, P.A. 1975. The significance of the cohune palm *Orbignya cohune* (Mart.) Dahlgren on the nature and in the development of the soil profile. Biotropica 7: 32–36.

FURNESS, C.A. AND P.J. RUDALL. 2003. Apertures with lids: distribution and significance of operculate pollen in monocotyledons. Int. J. Pl. Sci. 164: 835–854.

FURTADO, C.X. 1933. The limits of *Areca* L. and its sections. Repert. Spec. Nov. Regni Veg. 33: 217–239.

FURTADO, C.X. 1934. Palmae Malesicae. Repert. Spec. Nov. Regni Veg. 35: 18–25.

FURTADO, C.X. 1940. Palmae Malesicae VIII. — The genus *Licuala* in the Malay Peninsula. Gard. Bull Straits Settlem. 11: 31–73.

FURTADO, C.X. 1949. Palmae Malesicae X. The Malayan species of *Salacca*. Gard. Bull. Singapore 12: 378–403.

FURTADO, C.X. 1953. The species of *Daemonorops* in Malaya. Gard. Bull. Singapore 14: 49–147.

GALEANO, G. 1995. Novedades en el genero *Ceroxylon* (Palmae). Caldasia 17(82–85): 395–408.

GALEANO, G. AND F. SKOV. 1989. *Geonoma linearis* — a rheoophytic palm from Colombia and Ecuador. Principes 33: 108–112.

GALEANO-GARCES, G. AND R. BERNAL-GONZALEZ .1982. Two new species of *Ceroxylon* from Colombia. Caldasia 13(65): 693–699.

GALEANO-GARCES, G. AND R. BERNAL-GONZALES. 1983. Novedades de las palmas de Colombia. Caldasia 13(65): 693–699.

GALLETTI, M., C.I. DONATTI, A.S. PIRES, P.R. GUIMARÃES AND P. JORDANO. 2006. Seed survival and dispersal of an endemic Atlantic forest palm: the combined effects of defaunation and forest fragmentation. Bot. J. Linn. Soc. 151: 141–149.

GAMBLE, J.S. 1935. Flora of the Presidency of Madras. Botanical Survey of India, Calcutta 3: 1085–1086.

GANDOLFO, M.A., K.C. NIXON AND W.L. CREPET. 2000. Monocotyledons: A review of their early Cretaceous record. Pages 44–51 *in* K.L. Wilson and D.A Morrison (*eds*) Monocots: systematics and evolution. CSIRO, Melbourne.

GARDNER, J.S. 1882. Description and correlation of the Bournemouth Beds. Part II. Lower or Freshwater Series. Quart. J. Geol. Soc. London 38: 1–15.

GASSNER, G.G. 1941. Über bau der männlichen Blüten under Pollentwicklung einiger Palmen der Unterfamilie Ceroxylinae. Beih. Bot. Centralbl., Abt. 1 61: 237–276.

GAUT, B.S., B.R. MORTON, B.C. MCCAIG AND M.T. GLEGG. 1996. Substitution rate comparisons between grasses and palms: synonymous rate differences at the nuclear gene *Adh* parallel rate differences at the plastid gene *rbc*L. Proc. Natl. Acad. Sci. U.S.A. 93: 10274–10279.

GAUT, B.S., S.V. MUSE, W.D. CLARK AND M.T. CLEGG. 1992. Relative rates of nucleotide substitution at the *rbcL* locus of monocotyledonous plants. J. Molec. Evol. 35: 292–393.

GE, C.-J. AND Y.-K. LI. 1989. Observations on the chromosome numbers of medicinal plants from Shandong Province (II). Chin. Tradit. Herbal Drugs 20: 34–35.

GEE, C.T. 1993. On the fossil occurrence of the mangrove palm *Nypa*. Pages 315–319 *in* Knobloch, E. and Z. Kvacek (*eds*). Paleofloristics and paleoclimatic changes in the Cretaceous and Tertiary. Geologicky ustav Dionyza Stura, Prague.

GEE, C.T. 2001. The mangrove palm Nypa in the geologic past of the New World. Wetlands Ecol. Managem. 9: 181–194.

GIBBONS, M. 2001. *Trithrinax*: trials and tribulations. Palms 45: 74–79.

GIBBONS, M. AND T.W. SPANNER. 1996. *Medemia argun* lives! Principes 40: 65–74.

GIBBONS, M. AND T.W. SPANNER. 1997. *Trachycarpus oreophilus:* the Thailand *Trachycarpus*. Principes 41: 201–207.

GIBBONS, M. AND T.W. SPANNER. 1998. *Trachycarpus latisectus:* the Windamere Palm. Principes 42: 24–29.

GIBBONS, M., T.W. SPANNER AND S.Y. CHEN. 1995. *Trachycarpus princeps*, the stone gate palm, an exciting new species from China. Principes 39: 65–74.

GIBBONS, M, T.W. SPANNER, V.D. NGUYEN AND P. ANH. 2003. *Trachycarpus geminisectus,* the eight peaks fan palm, a new species from Vietnam. Palms 47: 143–148.

GISEKE, P.D. 1792. Palmae. Pages 21–122 *in* Linnaeus, C. Praelectiones in Ordinales Naturales. B.G. Hoffmann, Hamburg.

GLASSMAN, S. 1972. A revision of B.E. Dahlgren's index of American palms. Phanerogamarum Monographiae, Tomus VI. Cramer, Germany.

GLASSMAN, S.F. 1979. Re-evaluation of the genus *Butia* with a description of a new species. Principes 23: 65–79.

GLASSMAN, S. 1987. Revisions of the palm genus *Syagrus* Mart. and other selected genera in the *Cocos* alliance. Urbana, IU., Chicago, IU., University of Illinois Press. ix, 230p. (Illinois biological monographs, 56).

GLASSMAN, S.F. 1999. A taxonomic treatment of the palm subtribe *Attaleinae* (tribe *Cocoeae*). Urbana: University of Illinois Press. 414 pp. (Illinois Biol. Monogr. 59).

GLASSMAN, S.F., J.B. HARBORNE, J. ROBERTSON AND P.J. HOLLOWAY. 1981. Chemotaxonomic studies of selected cocosoid palms. Principes 25: 54–62.

GONZÁLEZ-GUZMÁN, E. 1967. A palynological study on the Upper los Cuervos and Mirador Formations (Lower and Middle Eocene; Tibú Area, Colombia). 65 pp. Brill, Leiden.

GONZÁLEZ-CERVANTES, E., A. MARTÍNEZ MENA, H.J. QUERO AND J. MÁRQUENEZ-GUZMÁN. 1997. Embryology of *Chamaedorea elegans* (Arecaceae): microsporangium, microsporogenesis, and microgametogenesis. Principes 41: 131–137.

GOTTSBERGER, G. 1978. Seed dispersal by fish in the inundated regions of Humaita, Amazonia. Biotropica 10: 170–183.

GOVAERTS, R. AND J. DRANSFIELD. 2005. World checklist of palms. Royal Botanic Gardens, Kew.

GRAHAM, A. 1976. Studies in neotropical botany II. The Miocene communities of Veracruz, Mexico. Ann. Missouri Bot. Gard. 63: 787–842.

GRAHAM, A. 1991. Studies in neotropical American paleobotany. VIII. The Pliocene communities of Panama — introduction and ferns, gymnosperms, angiosperms (monocots). Ann. Missouri Bot. Gard. 78: 190–200.

GRAHAM, S.W., J.M. ZGURSKI, M.A. MCPHERSON, D.M. CHERNIAWSKY, J.M. SAARELA, E.F.C. HORNE, S.Y. SMITH, W.A. WONG, H.E. O'BRIEN, V.C. BIRON, J.C. PIRES, R.G. OLMSTEAD, M.W. CHASE AND H.S. RAI. 2006. Robust inference of monocot deep phylogeny using an expanded multigene plastid data set. Aliso 22: 3–21.

GREEN, P.S. 1994. Arecaceae. Flora of Australia 49: 407–412.

GREENWOOD, D.R. AND J.G. CONRAN. 2000. The Australian Cretaceous and Tertiary monocot fossil record. Pages 52–59 *in* Wilson, K.L. and D.A. Morrison (*eds*). Monocots systematics and evolution. CSIRO Publishing, Melbourne.

GREGOR, H.J. 1980. Zum vorkommen fossiler Palmenreste im Jungtertiär Europas unter besonderer Berücksichtigung der Ablagerungender Oberen Süsswasser-Molasse Süd-Deutschlands. Ber. Bayer. Bot. Ges. 51: 135–144.

GREGOR, H.J. 1982. Palmae *in* Die Jungtertiären Floren Süddeutschlands. Ferdinand Enke Verlag, Stuttgart.

GREGOR H.J. AND H. HAGN 1982. Fossil fructifications from the Cretaceous-Palaeocene boundary of Egypt (Danian Bir Abu Mungar). Tertiary Res. 4: 121–147.

GRIFFITH, W. 1844. The palms of British India. Calcutta J. Nat. Hist. 5: 1–103, 311–355.

GRIFFITH, W. 1845. The palms of British India. Calcutta J. Nat. Hist. 5: 445–491.

GRUAS-CAVAGNETTO, C. 1976. Étude palynologique du Paléogène du sud de l'Angleterre. Cahier de Micropaléontologie, CNRS Paris. 49 pp. 10 pls.

GRUEZO, W.M. AND H.C. HARRIES. 1984. Self-sown, wild-type coconuts in the Philippines. Biotropica 16: 140–147.

GUANCHEZ, F.J. AND G.A. ROMERO. 1995. The flowers and unusual inflorescences of *Leopoldinia*. Principes 39: 152–158.

GUERRA, M.D.S. 1986. Citogénetica de Angiospermas coletadas en Pernambuco, I. Revista Brasil. Genét. 9: 21–40.

GUERRERO, A.M.A. DE AND C.R. CLEMENT 1982. Bibliografia sobre Pejibaye (Bactris gasipaes). Centro Agronómico Tropical de Investigación y Enseñanza unidad de Recursos Genétios CATIE/GT2. Turrialba, Costa Rica.

GULERIA, J.S., M.B. BANDE AND N. AWASTHI. 1996. Fossil records and antiquity of some common plants in India. Rheedea 6: 13–27.

GUNN, B.F. 2004. The phylogeny of the Cocoeae (Arecaceae) with emphasis on *Cocos nucifera*. Ann. Missouri Bot. Gard. 91: 505–522.

GUPTA, P.C. AND S. WARNAKULASURIYA. 2002. Global epidemiology of areca nut usage. Addict. Biol. 7: 77–83.

HAFFER, J. 1969. Speciation in Amazonian forest birds. Science 165: 131–137.

HAHN, W. 1993. Biosystematics and evolution of the genus *Caryota* (Palmae: Arecoideae). PhD Thesis, University of Wisconsin, Madison, Wisconsin.

HAHN, W.J. 1999. Molecular systematic studies of the Palmae. Pages 47–60 *in* Henderson, A. and F. Borchsenius (*eds*). Evolution, variation and classification of palms. The New York Botanical Garden Press, Bronx, NY.

HAHN, W.J. 2002a. A molecular phylogenetic study of the Palmae (Arecaceae) based on *atp*B, *rbc*L, and 18S nrDNA sequences. Syst. Biol. 51: 92–112.

HAHN, W.J. 2002b. A phylogenetic analysis of the Arecoid Line of palms based on plastid DNA sequence data. Molec. Phylogen. Evol. 23: 189–204.

HAHN, W.J. AND K.J. SYTSMA. 1999. Molecular systematics and biogeography of the Southeast Asian genus *Caryota* (Palmae). Syst. Bot. 24: 558–580.

HALL, R. 1998. The plate tectonics of Cenozoic SE Asia and the distribution of land and sea. Pages 99–131 *in* Hall, R. and J.D. Holloway (*eds*). Biogeography and Geological Evolution of SE Asia. Backhuys, Leiden.

HALLÉ, F. 1977. The longest leaf in palms? Principes 21: 18.

HALLÉ, F. 2004. Architectures de plantes. JPC Edition.

HALLÉ, F., R.A.A. OLDEMAN AND P.B. TOMLINSON. 1978. Tropical trees and forests: an architectural analysis. Springer Verlag, Berlin.

HARBORNE, J.B. AND C.A. WILLIAMS. 1991. Distribution and evolution of flavonoids in the Palmae and related monocot families. Bot. Jahrb. Syst. 113: 237–254.

HARLEY, M.M. 1989. Pollen morphology of *Voanioala gerardii* (Palmae: Arecoideae: Cocoeae: Butiinae) from Madagascar. Kew Bull. 44: 199–205.

HARLEY, M.M. 1990. Occurrence of simple, tectate, monosulcate or trichotomosulcate pollen grains within the Palmae. Rev. Palaeobot. Palynol. 64: 137–147.

HARLEY, M.M. 1996. Palm pollen and the fossil record. PhD thesis, University of East London in collaboration with RBG, Kew.

HARLEY, M.M. 1997. Ultrastructure of pollen from some Eocene palm flowers (Messel, Germany). Meded. Rijks Geol. Dienst, 58: 193–209.

HARLEY, M.M. 1998. Pollen morphology of the Indian palm genus, *Bentinckia* (Iguanurinae: Areceae). Pages 343–351 *in* Dutta, N.M., S. Gupta-Bhattacharya, S. Mandal and K. Bhattacharya (*eds*). Current concepts in pollen, spore and biopollution research: Prof. Sunirmal Chanda 60th birthday anniversary felicitation volume. Research Periodicals and Book Publishing House, Texas, USA.

HARLEY, M.M. 1999a. Tetrad variation: its influence on pollen form and systematics in the Palmae. Pages 289–304 *in* Kurmann, M.H. and A. Hemsley (*eds*). The evolution of plant architecture. Royal Botanic Gardens, Kew.

HARLEY, M.M. 1999b. Palm pollen: overview and examples of taxonomic value at species level. Pages 95–120 *in* Henderson, A. and F. Borchsenius (*eds*). Evolution, variation and classification of palms. The New York Botanical Garden Press, Bronx, NY.

HARLEY, M.M. 2006. A summary of fossil records for Arecaceae. Bot. J. Linn. Soc. 151: 39–67.

HARLEY, M.M. AND W.J. BAKER. 2001. Pollen aperture morphology in Arecaceae: application within phylogenetic analyses, and a summary of the fossil record of palm-like pollen. Grana 40: 45–77.

HARLEY, M.M. AND J. DRANSFIELD. 2003. Triporate pollen in the Arecaceae. Grana 41: 3–19.

HARLEY, M.M., M.H. KURMANN AND I.K. FERGUSON. 1991. Systematic implications of comparative morphology in selected fossil and extant pollen from the Palmae and the Sapotaceae. Pages 225–238 *in* Blackmore, S. and S. Barnes (*eds*). Pollen and spores: patterns of diversification. Clarendon Press, Oxford.

HARLEY, M.M. AND R.J. MORLEY. 1995. Ultrastructural studies of some fossil and extant palm pollen, with notes on the geological history of subtribes Iguanurinae and Calaminae. Rev. Palaeobot. Palynol. 85: 153–182.

HARRIES, H.C. 1978. Evolution, dissemination and classification of *Cocos nucifera* L. Bot. Rev. (Lancaster) 44: 265–319.

HARRIES H.C. 1992. Biogeography of the coconut *Cocos nucifera* L. Principes 36: 155–162.

HARTLEY, C.W.S. 1977. The oil palm. 2nd Ed. Longman, London.

HARTLEY, C.W.S. 1988. The oil palm. 3rd Ed. Longman, London.

HASELDONCKX, P. 1972. The presence of *Nypa* palms in Europe: a solved problem. Geol. & Mijnb. 51: 645–650.

HASTINGS, L.H. 2003. A revision of *Rhapis*, the Lady Palms. Palms 47: 62–78.

HASTON, E., J.E. RICHARDSON, P.F. STEVENS, M.W. CHASE AND D.J. HARRIS. 2007. A linear sequence of Angiosperm phylogeny Group II families. Taxon 56: 7–12.

HEATUBUN, C.D. 2002. A monograph of *Sommieria* (*Arecaceae*). Kew Bull. 57: 599–611.

HEATUBUN, C.D., J. DRANSFIELD, J.P. MOGEA, W.J. BAKER, M.M. HARLEY AND S.S. TJITROSOEDIRDJO. A monograph of *Cyrtostachys* Blume (Arecaceae). Kew Bull. (in press).

HEMSLEY, U.B. 1885. Godman and Sakvin. Biologia Centrali-Americana. Botany 4: 276.

HENDERSON, A. 1984. Observations on pollination of *Cryosophila albida* (Palmae). Principes 28: 120–126.

HENDERSON, A. 1985. Pollination of *Socratea exorrhiza* and *Iriartea ventricosa*. Principes 29: 64–71.

HENDERSON, A. 1986a. A review of pollination studies in the Palmae. Bot. Rev. (Lancaster) 52: 221–259.

HENDERSON, A. 1986b. *Barcella odora*. Principes 30: 74–76.

HENDERSON, A. 1990. Arecaceae. Part I. Introduction and the Iriarteinae. Fl. Neotrop. Monogr. 53: 1–100.

HENDERSON, A. 1995. The palms of the Amazon. Oxford University Press, New York.

HENDERSON, A. 1999a. A phylogenetic analysis of the Euterpeinae (Palmae, Arecoideae, Areceae) based on morphology and anatomy. Brittonia 51: 106–113.

HENDERSON, A. 1999b. Species, species concepts, and palm taxonomy in the New World. Pages 21–28 *in* Henderson, A. and F. Borchsenius (*eds*). Evolution, variation and classification of palms. The New York Botanical Garden Press, Bronx, New York.

HENDERSON, A. 2000. *Bactris* (Palmae). Fl. Neotrop. Monogr. 79: 1–181.

HENDERSON, A. 2002a. Phenetic and phylogenetic analysis of *Reinhardtia* (Palmae). Amer. J. Bot. 89: 1491–1502.

HENDERSON, A. 2002b. Evolution and ecology of palms. The New York Botanical Garden Press. Bronx, New York.

HENDERSON, A. 2004. A multivariate analysis of *Hyospathe* (Palmae). Amer. J. Bot. 91(6): 953–965.

HENDERSON, A. 2005. A multivariate study of *Calyptrogyne* (Palmae). Syst. Bot. 30: 60–83.

HENDERSON, A.J. 2007. A revision of *Wallichia* (Palmae). Taiwania 52: 1–11.

HENDERSON A. AND F. BORCHSENIUS (*eds*). 1999. Evolution, variation, and classification of palms. The New York Botanical Garden Press. New York.

HENDERSON, A. AND E. FERREIRA. 2002. A morphometric study of *Synechanthus* (Palmae). Syst. Bot. 27: 693–702.

HENDERSON, A. AND G. GALEANO. 1996. *Euterpe*, *Prestoea* and *Neonicholsonia* (Palmae). Fl. Neotrop. Monogr. 72: 1–90.

HENDERSON, A., G. GALEANO AND R. BERNAL. 1995. A field guide to the palms of the Americas. Princeton University Press, Princeton, New Jersey.

HENDERSON, F.M. AND D.W. STEVENSON. 2006. A phylogenetic study of Arecaceae based on seedling morphological and anatomical data. Aliso 22: 251–264.

HENDERSON, S.A., N. BILLOTTE AND J.-C. PINTAUD. 2006. Genetic isolation of Cape Verde Island *Phoenix atlantica* (Arecaceae) revealed by microsatellite markers. Conservation Genet. 7: 213–223.

HERENDEEN, P.S. AND P.R. CRANE. 1995. The fossil history of the monocotyledons. Pages 1–21 *in* Rudall, P.J., P.J. Cribb, D.F. Cutler and C.J. Humphries (*eds*). Monocotyledons: systematics and evolution. Royal Botanic Gardens, Kew.

HERMANN, A. AND J. KVAČEK. 2002. Campanian Grünbach Flora of Lower Austria: preliminary floristics and palaeoclimatology. Ann. Naturhist. Wien, A. 103: 1-121.

HERMINA, M., E. KLITZSCH AND F.K. LIST (*eds*). 1989. Stratigraphic lexicon and explanatory notes to the geological map of Egypt 1:500 000. Conoco Inc., Cairo. 264 pp.

HESSE, M., M. WEBER AND H.-M. HALBRITTER. 2000. A comparative study of the polyplicate pollen types in Arales, Laurales, Zingiberales and Gnetales. Pages 227–239 *in* Harley, M.M., C.M. Morton and S. Blackmore (*eds*). Pollen and spores: morphology and biology, Royal Botanic Gardens, Kew.

HEYWOOD, V.H. 1978. Flowering plants of the world. Batsford, London.

HILGEMAN, R.H. 1954. The differentiation, development and anatomy of the axillary bud, inflorescence and offshoot in the date palm. Date Grower's Inst. Rep. 31: 6–10.

HIRAI Y., S. SANADA, Y. IDA AND J. SHOJI. 1984a. Studies on the constituents of Palmae plants. I: The constituents of *Trachycarpus fortunei* (Hook.) H. Wendl. Chem. Pharm. Bull. 32: 295–301.

HIRAI Y., S. SANADA, Y. IDA AND J. SHOJI. 1984b. Studies on the constituents of Palmae plants. II: The constituents of *Rhapis exelsa* Henry and *R. humilis* Bl. Chem. Pharm. Bull. 32: 4003–4011.

HIRAI Y., S. SANADA, Y. IDA AND J. SHOJI. 1986. Studies on the constituents of Palmae plants. III: The constituents of *Chamaerops humilis* L. and *Trachycarpus wagnerianus* Becc. Chem. Pharm. Bull. 34: 82–87.

HNATIUK, R.J. 1977. Population structure of *Livistona eastonii* Gardn., Mitchell Plateau, Western Australia. Austral. J. Ecol. 2: 461–466.

HODEL, D.R. 1980. Notes on *Pritchardia* in Hawaii. Principes 24: 65–81.

HODEL, D.R. 1992. *Chamaedorea* palms: the species and their cultivation. International Palm Society, Kansas.

HODEL, D.R. 1997. New species of palms from Thailand, part 2. Palm J. 136: 7–20.

HODEL, D.R. 2007. A review of the genus *Pritchardia*. Palms 51. Supplement to Vol. 4. 1–53.

HODEL, D.R. AND J.-C. PINTAUD. 1998. The palms of New Caledonia. Kampon Tansacha, Nong Nooch Tropical Garden, Pattaya, Thailand.

HODGE, W.H. 1965. Palm cabbage. Principes 9: 124–131.

HOEKEN-KLINKENBERG, P.M.J. 1964. A palynological investigation of some Upper-Cretaceous sediments in Nigeria. Pollen & Spores 6: 209–231.

HOFMANN, C.-C. AND R. ZETTER. 2001. Palynological investigation of the Krappfeld area, Palaeocene/Eocene, Carintha (Austria). Palaeontographica B 259: 47–64.

HOLDSWORTH, D.K., R.A. JONES AND R. SELF. 1998. Volatile alkaloids from *Areca catechu*. Phytochemistry 48: 581–582.

HOLLICK, A. 1893. A new fossil palm from the Cretaceous formation at Glen Cove, Long Island. Bull. Torrey Bot. Club 10: 168–169.

HOLLICK, A. 1928. Palaeobotany of Porto Rico. Pages 177–193 *in* Scientific Survey of Porto Rico and the Virgin Islands 7.

HOLLICK, A. 1936. The Tertiary floras of Alaska. U.S. Geological Survey Paper 182: 1–185. [Not seen, cited according to Read & Hickey (1972).]

HOLTTUM, R. 1955. Growth habits of monocotyledons — variations on a theme. Phytomorphology 5: 399–413.

HONG, Y.J., F.A. TOMAS-BARBERAN, A.A. KADER AND A.E. MITCHELL. 2006. The flavonoid glycosides and procyanidin composition of Deglet Noor dates (*Phoenix dactylifera*). J. Agric. Food Chem. 54: 2405–2411.

HOOKER, J.D. 1883. Palmae. Pages 870–948 *in* G. Bentham and J.D. Hooker, Genera Plantarum 3. L. Reeve and Co., London.

HOPPE, L.E. 2005. The pollination biology and biogeography of the mangrove palm *Nypa fruticans* Wurmb (Arecaceae). Master's thesis, University of Aarhus.

HOROWITZ, A. 1992. Palynology of Arid Lands. Elsevier Science, Amsterdam.

HOUSE, A.P.N. 1984. The ecology of *Oncosperma horridum* on Siberut Isalnd, Indonesia. Principes 28: 83–89.

HOWARD, F.W, D. MOORE, R.M. GIBLIN-DAVIS AND R.G. ABAD. 2001. Insects on palms. CABI Publishing, Wallingford U.K.

HSU, C.-C. 1967. Preliminary chromosome studies on the vascular plants of Taiwan (I). Taiwania 13: 117–129.

HSU, C.-C. 1972. Preliminary chromosome studies on the vascular plants of Taiwan (V). Cytotaxonomy on some monocotyledons. Taiwania 17: 48–65.

HUANG, S.-F., Z.-Y. CHEN, S.-J. CHEN, Q.-Y. QI AND X.-H. SHI. 1986a. Plant chromosome counts (2). Subtrop. Forest. Sci. Technol. 3: 41–47.

HUANG, S.-F., Z.-Y. CHEN, S.-J. CHEN, X.-X. HUANG, Q.-Y. QI AND X.-H. SHI. 1986b. Plant chromosome counts (3). Subtrop. Forest. Sci. Technol. 4: 50–56.

HUANG, S.-F., Z.-F. ZHAO, Z.-Y. CHEN, S.-J. CHEN AND X.-X. HUANG. 1985. Preliminary report on chromosome numbers of plants. Asian Forest. Sci. Technol. 1: 1–15.

HUANG, S.-F., Z.-F. ZHAO, Z.-Y. CHEN, S.-J. CHEN AND X.-X. HUANG. 1989. Chromosome counts of one hundred species and infraspecifc taxa. Acta Bot. Austro Sin. 5: 161–176.

HUARD, J. 1967. Restes epineux de palmier lepidocaryoide du Neogene des Landes. Naturalia Monspel., Sér. Bot. 18: 319–346.

HÜBSCHMANN, L.K., C. GRANDEZ AND H. BALSLEV. 2007. Uses of *Vara Casha* – a neotropical liana palm, *Desmoncus polyacanthos* – in Iquitos, Peru. Palms 51: 167–176.

HUGHES, N.F. 1994. The Enigma of Angiosperm origins. Cambridge Paleobiology Series I. Cambridge University Press. 303 pp.

HUYNH, K.-L. 1991. New data on the taxonomic position of *Pandanus eydouxia* (Pandanaceae), a species of the Mascarene Islands. Bot. Helv. 101: 29–37.

IDAKA, K., Y. HIRAI AND J. SHOJI. 1988. Studies on the constituents of Palmae plants. IV: The constituents of the leaves of *Sabal causiarum* Becc. Chem. Pharm. Bull. 36: 1783–1790.

IDAKA, K., Y. HIRAI AND J. SHOJI. 1991. Studies on the constituents of Palmae plants. V, Steroid saponins and flavonoids of leaves of *Phoenix rupicola* T. Anderson, *P. loureirii* Kunth, *P. reclinata* N.J. Jacquin, and *Arecastrum romanzoffianum* Beccari. Chem. Pharm. Bull. 36: 1455–1461.

IGERSHEIM, A, M. BUZGO AND P.K. ENDRESS. 2001. Gynoecium diversity and systematics in basal monocots. Bot. J. Linn. Soc. 136: 1–65.

IMCHANITZKAJA, N. 1985. Arecaceae seu Palmae. Leningrad Branch Soviet Science Press. NAUKA.

IRVINE, A.K. 1983. *Wodyetia*, a new arecoid genus from Australia. Principes 27: 158–167.

ISNARD, S. 2006. Biomechanics and development of rattans: what is special about *Plectocomia himalayana* Griff. (Calamoideae, Plectocomiinae)? Bot. J. Linn. Soc. 151: 83–91.

ISNARD, S., T. SPECK and N.P. ROWE. 2005. Biomechanics and development of the climbing habit in two species of the South American palm genus *Desmoncus* (Arecaceae). Amer. J. Bot. 92: 1444–1456.

JABLONSZKY, J. 1914. A tarnóci mediterrán koru flora (Die mediterrane Flora von Tarnóc). Magyar Királyi állami Földtani intézet évkönyve, Budapest 22: 228–273.

JAHNICHEN, H. 1990. The oldest record of *Nypa burtini*. International Organisation of Palaeobot. Newsletter 42: 12.

JAIZME-VEGA M.C. AND M.A. DÍAZ-PÉREZ. 1999. Effect of *Glomus intraradices* on *Phoenix roebelenii* during the nursery stage. Acta Hort. 486: 199–202.

JAN DU CHÊNE, R.E., M.S. ONYIKE AND M.A. SOWUNMI. 1978. Some new Eocene pollen on the Ogwashi-Asaba Formation, South-Eastern Nigeria, Rev. Micropaléontol. 10: 285–322.

JANOS, D.P. 1977. Vesicular-arbuscular mycorrhizae affect the growth of *Bactris gasipaes*. Principes 21: 12–21.

JANSSEN, T. AND K. BREMER. 2004. The age of major monocot groups inferred from 800+ *rbcL* sequences. Bot. J. Linn. Soc. 146: 385–398.

JANZEN, D.H. 1971. The fate of *Scheelea rostrata* fruits beneath the parent tree: predispersal attack by bruchids. Principes 15: 89–101.

JARAMILLO, C. AND D.L. DILCHER. 2001. Middle Paleogene palynology of Central Colombia, South America: a study of pollen and spores from tropical latitudes. Palaeontographica Abt. B 258: 87–213.

JARZEN, D.M. 1978. Some Maestrichtian palynomorphs and their phytogeographical and palaeoecological implications. Palynology 2: 29–38.

JOHNSON, D.V. 1983a. A bibliography of graduate theses on the date and other *Phoenix* spp. Date Palm J. 2: 257–267.

JOHNSON, D.V. 1983b. A bibliography of major works on the oil palm. Oil Palm News 27: 4–6.

JOHNSON, D.V. 1984. Additional graduate theses on the date palm and other *Phoenix* spp. Date Palm J. 3: 437.

JOHNSON, D.V. 1985. Present and potential economic uses of palms in arid and semi-arid areas. *In* Plants for arid lands. Royal Botanic Gardens, Kew.

JOHNSON, D.V. (*ed*.). 1996. Palms: their status and sustained utilization. IUCN, Gland, Switzerland and Cambridge, UK.

JOHNSON, D.V. 1998. The making of a dugout canoe from the trunk of the palm *Iriartea deltoidea*. Principes 42: 201–205, 208.

JOHNSON, M.A.T. 1979. The chromosomes of *Ceratolobus* (Palmae). Kew Bull. 34: 35–36.

JOHNSON, M.A.T. 1985. New chromosome counts in the Palmae. Kew Bull. 40: 109–114.

JOHNSON, M.A.T. 1989. An unusually high chromosome number in *Voanioala gerardii* (Palmae: Arecoideae: Cocoeae: Butiinae) from Madagascar. Kew Bull. 44: 207–210.

JOHNSON, M.A.T., A.Y. KENTON, M.D. BENNETT AND P.E. BRANDHAM. 1989. *Voanioala gerardii* has the highest known chromosome number in the monocotyledons. Genome 32: 328–333.

JONES, L., D. BARFIELD, J. BARRETT, A. FLOOK, K. POLLOCK AND P. ROBINSON. 1982. Cytology of oil palm cultures and regenerant plants. Pages 727–728 *in* A. Fujiwara (*ed*.). Proceedings of the 5th International Congress of Plant Tissue and Cell Culture. Japanese Association for Plant Tissue Culture, Tokyo.

JOUANNIC, S., X. ARGOUT, F. LECHAUVE, C. FIZAMES, A. BORGEL, F. MORCILLO, F. ABERLENC-BERTOSSI, Y. DUVAL AND J. TREGEAR. 2005. Analysis of expressed sequence tags from oil palm (*Elaeis guineensis*). FEBS Lett. 579: 2709–2714.

JUMELLE, H. AND H. PERRIER DE LA BÂTHIE. 1945. Palmiers *in* H. Humbert (*ed*.). Flore de Madagascar et des Comores. 30e famille. Imprimerie officiella, Atananarivo.

KAEMPFER, E. 1712. Amoenitatum exoticarum. Lemgo, Lippe.

KAHN, F. AND K. MEJIA. 1988. A new species of *Chelyocarpus* (Palmae, Coryphoideae) from Peruvian Amazonia. Principes 32: 69–72.

KAHN, F. AND K. MEJIA. 1991. The palm communities of two terra firme forests in Peruvian Amazonia. Principes 35: 22–26.

KAHN, F. AND G. SECOND. 1999. The genus *Astrocaryum* (*Palmae*) in Amazonia: classical taxonomy and DNA analysis (AFLP). Mem. New York Bot. Gard. 83: 179–184.

KAPLAN, D.R., N.G. DENGLER AND R.E. DENGLER. 1982a. The mechanism of plication inception in palm leaves; problem and developmental morphology. Canad. J. Bot. 60: 2939–2975.

KAPLAN, D.R., N.G. DENGLER AND R.E. DENGLER. 1982b. The mechanism of plication inception in palm leaves: histogenetic observations on the palmate leaf of *Rhapis excelsa*. Canad. J. Bot. 60: 2999–3106.

KAR, R.K. 1985. The fossil floras of Kachchh — IV. Tertiary palynostratigraphy. Palaeobotanist, 34: 1–279.

KAR, R.K. AND M. BHATTACHARYA. 1992. Palynology of Rajpardi lignite, Cambay Basin and Gujra Dam and Akri lignite, Kutch Basin. Palaeobotanist 39: 250–263.

KAR, R.K. AND M. KUMAR. 1986. *Neocouperipollis* - a new name for *Couperipollis* Venkatachala & Kar. Palaeobotanist, 35: 171–174.

KAUL, K.N. 1943. A palm stem from the Miocene of Antigua, W.I. *Phytelephas seewardii* sp. nov. Proc. Linn. Soc. London 155: 3–4.

KAUL, K.N. 1946. A silicified palm fruit from Mexico, *Manicaria edwardsii* sp. nov. J. Indian Bot. Soc. (M.O.P. Iyengar Comm. vol.) 327–330. [Not seen.]

KAUL, K.N. 1951. A palm fruit from Kapurdi (Jodhpur, Rajasthan Desert), *Cocos sahnii* sp. nov. Curr. Sci. 20: 138.

KAUL, K.N. 1960. The anatomy of the stem of palms and the problem of the artificial genus *Palmoxylon* Schenk — I. Bull. Natl. Bot. Gard. 51. 52 pp.

KEDVES, M. 1980. Morphological investigation of recent Palmae pollen grains. Acta Bot. Acad. Sci. Hung. 26: 339–373.

KEDVES, M. AND E. BOHONY. 1966. Observations sur quelques pollens de Palmiers provenant des couches Tertiares de Hongrie. Pollen & Spores 8: 141–147.

KEIM, A.P. (in prep.). A monograph of *Orania* Zipp. (Arecaceae: Arecoideae).

KELLOG, E.A. 2000. A model of inflorescence development. Pages 84–88 *in* Wilson, K.L. and D.A. Morrison (*eds*). Monocots: systematics and evolution. CSIRO, Melbourne.

KHAN, A.M. 1976. Palynology of Tertiary sediments from Papua New Guinea. I. New form genera and species from Upper Tertiary sediments. Austral. J. Bot. 24: 753–781.

KHIN SEIN, M. 1961. Fossil Spores of the London Clay. 401 pp. Unpublished PhD thesis. University of London.

KIEW, R. 1972. The natural history of *Iguanura geonomiformis* Martius, a Malayan undergrowth palmlet. Principes 16: 3–10.

KIEW, R. 1976. The genus *Iguanura* Bl. (Palmae) Gard. Bull. Singapore 28: 191–226.

KIEW, R. 1979. New species and records of *Iguanura* (Palmae) from Sarawak and Thailand. Kew Bull. 34: 143–145.

KIMNACH, M. 1977. The species of *Trachycarpus*. Principes 21: 155–160.

KLITZSCH, E., F.K. LIST AND G. PÖHLMANN (*eds*). 1987. Geological map of Egypt 1: 500 000. Sheet NF 36 NW El-Saad El-Ali. Conoco Coral, Cairo.

KLOTZ, L.H. 1978a. Form of the perforation plates in the wide vessels of metaxylem in palms. J. Arnold Arbor. 59: 105–128.

KLOTZ, L.H. 1978b. The number of wide vessels in petiolar vascular bundles of palms: an anatomical feature of systematic significance. Principes 22: 64–69.

KLOTZ, L.H. 1978c. Observations on diameters of vessels in palms. Principes 22: 99–106.

KNOBLOCH, E., M. KONZALOVÁ AND Z. KVACEK. 1996. Die obereozäne Flora der Staré Sedlo-Schichtenfolge in Böhmen (Mitteleuropa). Ceský Geol. Ustav. Praha. 260 pp.

KNUDSEN, J. 1999a. Floral scent chemistry in geonomoid palms (Palmae: Geonomeae) and its importance in maintaining reproductive isolation. Mem. New York Bot. Gard. 83: 141–157.

KNUDSEN, J. 1999b. Floral scent differentiation among coflowering, sympatric species of Geonoma (Arecaceae). Pl. Spec. Biol. 14: 137–142.

KNUDSEN, J.T., L. TOLLSTEN AND F. ERVIK. 2001. Flower scent and pollination in selected neotropical palms. Pl. Biol. (Stuttgart) 3: 642–653.

KOCH, B.E. 1972. Coryphoid palm fruits and seeds from the Danian of Nûgssuaq, West Greenland. Medd. Grønland Kommiss. Vidensk. Unders. Grønland 193: 1–38.

KONZALOVÁ, M. 1971. Arecales (Palmae) in the North Bohemian Tertiary. Sbornik Geol. Paleont. Prada P. 13: 143–158.

KOWALSKA, M.T., R.W. SANDERS AND C.E. NAUMAN. 1991. Phenolic constituents of *Coccothrinax* (Palmae). Principes 35: 142–146.

KRABBE, N. 2000. Overview of conservation priorities for parrots in the Andean region with special consideration for Yellow-eared Parrot. Int. Zoo Yearbook 37: 283–288.

KRÄUSEL, R. 1923. *Nipadites borneensis* n. sp. eine fossile palmenfrucht aus Borneo. Senckenbergiana 5: 3–4.

KRÄUSEL, R. 1939. Ergebnisse der Forschungsreisen Prof. E. Stromers in der Wüsten Ägyptens. IV. Die fossilen Floren Ägyptens. 3. Die fossilen Pflanzen Ägyptens. Abhandlungen der Bayerischen Akademie der Wissenschaften. Mathematisch-naturwissenschaftliche Abteilung (NS) 47. 140 pp. + 23 pls.

KREFT, H., J.H. SOMMER AND W. BARTHLOTT. 2006. The significance of geographic range size for spatial diversity patterns in Neotropical palms. Ecography 29: 21–30.

KRYSHTOFOVICH, A.N. 1918. Occurrence of the palm, *Sabal nipponica*, n. sp. in the Tertiary rocks of Hokkaido and Kyushu. J. Geol. Soc. Tokyo 25: 59–66.

KRYSHTOFOVICH, A.N. 1927. Palm remains (*Nipadites burtinii* Brongniart) from the Eocene at Viosnessensk in the Odessa Region. Izv. Geol. Kom. 45: 6.

KUBITSKI, K. 1991. Dispersal and distribution in *Leopoldinia* (Palmae). Nordic J. Bot. 11: 429–432.

KÜCHMEISTER, H., G. GOTTSBERGER AND I. SILBERBAUER-GOTTSBERGER. 1993. Pollination biology of *Orbignya spectabilis*, a "monoecious" Amazonian palm. Pages 67–75 *in* Barthlott, W., C.M. Naumann, K. Schmidt-Loske and K.-L. Schuchmann (*eds*). Animal-Plant Interactions in Tropical Environments. Bonn, Zoologisches Forschungsinstitut und Museum Alexander Koenig.

KÜCHMEISTER, H., I. SILBERBAUER-GOTTSBERGER AND G. GOTTSBERGER. 1997. Flowering, pollination, nectar standing crop, and nectarines of *Euterpe precatoria* (Arecaceae), an Amazonian rain forest palm. Pl. Syst. Evol. 206: 71–97.

KULKARNI, A. AND T.S. MAHABALÉ. 1971. *Palmoxylon kamalam* Rode from Kondhali, District Nagpur, M.S., and its resemblance with other palms. Palaeobotanist 20: 170–178 +2 pls.

KULKARNI, A.R. AND N.R. PHADTARE. 1980. Leaf epidermis of *Nypa* from lignitic beds of Ratnagiri District, Maharashtra. Geophytology 10, 125–128.

KULKARNI, A.R. AND N.R. PHADTARE. 1981. Pollen of *Nypa* from lignitic beds of Ratnagiri District, Maharashtra. Phytomorphology 31: 48–51.

KUNTH, C.S. 1826. Recherche sur les plantes trouvées dans les tombeaux egyptiens par M. Passalacqua. Ann. Sci. Nat. (Paris) 8: 418–423.

KVAČEK, Z. 1998. Bílina: a window on Early Miocene marshland environments. Rev. Palaeobot. Palynol. 101: 111–123.

KVAČEK, J. AND A.B. HERMANN. 2004. Monocotyledons from the Early Campanian (Cretaceous) of Grünbach, Lower Austria. Rev. Palaeobot. Palynol. 128: 323–353.

LACROIX, A. 1896. Sur la découverte d'un gisement d'empreintes végétales dans les cendres volcaniques anciennes de l'île de Phira (Santorin). Compt. Rend. Hebd. Séances Acad. Sci. Serie D. 123: 656–659.

LAKHANPAL, R.N. 1970. Tertiary floras of India and their bearing on the historical geology of the region. Taxon 19: 675–694.

LAKHANPAL, R.N., J.S. GULERIA AND N. AWASTHI. 1984. The fossil floras of Kachchh. III. Tertiary megafossils. Palaeobotanist 33: 228–319.

LAMOTTE, R.S. 1952. Catalogue of the Cenozoic plants of North America through 1950. Mem. Geol. Soc. Amer. 51: 1–381.

LEE, Y.F. 1995. Pollination in the rattans *Calamus subinermis* and *C. caesius* (Palmae: Calamoideae). Sandakania 6: 15–39.

LEON, H. 1939. Contribucion al studio de las palmas de Cuba. III. Género *Coccothrinax*. Mém. Soc. Cub. Hist. Nat. "Felipe Poey" 13: 107–156.

LEON, H. 1941. Contribucion al studio de las palmas de Cuba V. Novedades en el genero *Hemithrinax*. Mem. Soc. Cub. Hist. Nat. "Felipe Poey" 15: 379–383.

LEON, H. 1946. Palmaceas. *In* Flora de Cuba I. Cultural, S.A., La Habana.

LEOPOLD, A.C. 2000. Many modes of movement. Science 288: 2131–2132.

LEPESME, P. 1947. Les insectes des palmiers. Paul Lechevalier Editeur, Paris.

LEWIS, C.E. 2002. A phylogenetic analysis of the palm subtribe Oncospermatinae (Arecaceae) based on morphological characters. Brittonia 54: 78–91.

LEWIS, C.E. AND J.J. DOYLE. 2001. Phylogenetic utility of the nuclear gene malate synthase in the palm family (Arecaceeae). Molec. Phylogen. Evol. 19: 409–420.

LEWIS, C.E. AND J.J. DOYLE. 2002. A phylogenetic analysis of tribe Areceae (Arecaceae) using two low-copy nuclear genes. Pl. Syst. Evol. 236: 1–17.

LEWIS, C.E. AND S. ZONA. 2000. A survey of cyanogenesis in palms (Arecaceae). Biochem. Syst. Ecol. 28: 219–228.

LEWIS, C.E. AND S. ZONA. 2008. *Leucothrinax morrisii*, a new name for a familiar Carribean palm. Palms 52: 84–88.

LI, H.L. AND J.J. WILLAMAN. 1968. Distribution of alkaloids in Angiosperm phylogeny. Econ. Bot. 22: 239–252.

LIM, C.K. 1998. Unravelling *Iguanura* Bl. (*Palmae*) in Peninsular Malaysia. Gard. Bull. Singapore 48: 1–64.

LIM, C.K. 2001. Unravelling *Pinanga* Blume (*Palmae*) in Peninsular Malaysia. Folia Malaysiana 2: 219–276.

LINDLEY, J. 1830. An introduction to the natural system of botany: or, A systematic view of the organisation, natural affinities, and geographical distribution, of the whole vegetable kingdom: together with the uses of the most important species in medicine, the arts, and rural or domestic economy. Longman, Rees, Orme, Brown and Green, London.

LINNAEUS, C. 1753. Species plantarum. 2 vols. Stockholm.

LISTABARTH, C. 1994. Pollination and pollinator breeding in *Desmoncus*. Principes 38: 13–23.

LISTABARTH, C. 1996. Pollination of *Bactris* by *Phyllotrox* and *Epurea*. Implications of the palm breeding beetles on pollination at the community level. Biotropica 28: 69–81.

LITKE, R. 1966. Kutikularanalytische Untersuchungen im Niederlausitzer Unterflöz. Paläontol. Abh., Abt. B. Paläobot. 2: 327–413 + 39 pls.

LLERAS, E. 1985. *Acrocomia*, um gênero com grande potencial. Noticário 1: 3–5. EMBRAPA/CENARGEN, Brasilia.

LOO, A.H.B., J. DRANSFIELD, M.W. CHASE AND W.J. BAKER. 2006. Low copy nuclear DNA, phylogeny and the evolution of dichogamy in the betel nut palms and their relatives (Arecinae, Arecaceae). Molec. Phylogen. Evol. 39: 598–618.

LOOK, S.L. 2007. Population genetics and phylogeny of the Malesian palm genus *Johannesteijsmannia* H.E. Moore. PhD Thesis, National University of Singapore.

LORENTE, M.A. 1986. Palynology and palinofacies of the Upper Tertiary of Venezuela. Dissertaciones Botanicae Bd 99. J. Cramer, Berlin, Stuttgart.

LOUVET, P. AND P. MAGNIER. 1971. Confirmation de la derive du continent Africain au Tertiare par la paléobotanique. 96e Congrès national des sociétés savants, Toulouse, 1971, sciences 5: 177–189.

LUDWIG, N. 2006. *Acanthophoenix* in Réunion, Mascarene Islands. Palms 50: 82–98.

MACBRIDE, J.F. 1960. Flora of Peru. Publ. Field Mus. Nat. Hist., Bot. Ser. 13: 321–418.

MACKO, S. 1957. Lower Miocene pollen flora from the valley of Klodnica near Gliwece (Upper Silesia). Prace Wroclawsk. Towarz. Nauk. 88: 1–314.

MADULID, D.A. 1981. A monograph of *Plectocomia* (Palmae: Lepidocaryoideae). Kalikasan 10: 1–94.

MAHABALÉ, T.S. 1958. Resolution of the artificial palm genus, *Palmoxylon*: a new approach. Palaeobotanist 7: 76–83.

MAHABALÉ, T.S. 1978. The origin of the coconut. Palaeobotanist 25: 238–248.

MAHABALÉ, T.S. AND M.S. CHENNAVEERAIAH. 1958. Studies on *Hyphaene indica* Becc. I. Morphology. Phytomorphology 7: 184–194.

MAI, D.H. 1964. Die Mastixioideen–Floren im Tertiär der Oberlausitz. Paläeontol. Abh., Palaeobot. (Berlin) Abt. B, 2: 1–192. [Not seen cited according to Gregor (1980) and Collinson (1983)].

MAI, D.H. 1976. Fossile Früchte und Samen des Geiseltales aus den Mitteleozän. Abh. Zentr. Geol. Inst., Paläontol. Abh. 26: 93–149.

MAI, D.H. AND H. WALTHER. 1978. Die Floren der Haselbacher Serie im Weißelster-Becken (Bezirk Leipzig, DDR). Abh. Staatl. Mus. Min. Geol. (Dresden) 28: 1–200.

MALONEY, B.K. 2000. *Borassodendron* (Palmae) in the Southeast Asian pollen record. Blumea 45: 427–432.

MARKLEY, K.S. 1955. Caranday — a source of palm wax. Econ. Bot. 9: 39–52.

MARTELLI, U. 1932. *Pelagodoxa henryana* Becc., palma della Isole Marquesas. Nuovo Giorn. Bot. Ital. n.s. 39: 243–250.

MARTIUS, C.F.P. VON. 1823–50. Historia Naturalis Palmarum. 3 vols. Munich.

MARTIUS, C.F.P. VON. 1824. Palmarum familia ejusque genera denuo illustrata/indicit C.F.P. de Martius. M. Lindauer. Munich.

MAUNDER, M., B. LYTE, J. DRANSFIELD AND W.J. BAKER. 2001. The conservation value of botanic garden palm collections. Biol. Conservation 98: 259–271.

MAURY, C.J. 1930. O Cretaceo da Parahyba do Norte. As floras do Cretaceo Superior da America do Sul. Monogr. Serv. Geol. Mus. Brasil 8: 1–305.

MCCLATCHEY, W.C. 1999. Phylogenetic analysis of morphological characters of *Metroxylon* section *Coelococcus* (Palmae) and resulting implications for studies of other Calamoideae genera. Pages 285–306 *in* Henderson, A. and F. Borchsenius (*eds*). Evolution, variation and classification of palms. The New York Botanical Garden Press, Bronx, NY.

MCKAMEY, L. 1983. Secret of the Orient: dwarf *Rhapis excelsa*. Grunwald Printing Co., Corpus Christi, Texas.

MÉDUS, J. 1975. Palynologie de sédiments Tertiaires du Sénégal méridional. Pollen & Spores, 17: 545–601.

MEEKIJJAROENROJ, A. AND M.-C. ANSTETT. 2003. A weevil pollinating the Canary Islands date palm: between parasitism and mutualism. Naturwissenschaften 90: 452–455.

MEEROW, A.W. 2005. Betrock's cold hardy palms. Betrock Information Systems, Hollywood, Florida.

Mehrotra, R.C., R.P. Tiwari and B.I. Mazumder. 2003. *Nypa* megafossils from the Tertiary sediments of northeast India. Geobios 36: 83–92.

Mendis, N.M., I.K. Ferguson and J. Dransfield. 1987. The pollen morphology of the subtribe Oncospermatinae (Palmae: Arecoideae: Arecaeae). Kew Bull. 42: 47–63.

Menendez, C.A. 1969. Die fossilen Floren Südamerikas. Monogr. Biol. 19: 519–561.

Migliaccio, C. 2001. The genus *Heterospathe* in cultivation. Palms 45: 15–21.

Mildenhall, D.C. and D.T. Pocknall. 1989. Miocene-Pleistocene spores and pollen from Central Otago, South Island, New Zealand. Palaeontol. Bull. New Zealand Geol. Surv. 59: 1–128.

Miller, R.H. 1964. The versatile sugar palm. Principes 8: 115–147.

Misra, B.K., A. Singh and C.G.K. Ramanujam. 1996. Trilatiporate pollen from Indian Palaeogene and Neogene sequences: evolution, migration and continental drift. Rev. Palaeobot. Palynol. 91: 331–352.

Moegenburg, S. 2003. The functions of hooked fibers on *Euterpe* endocarps. Palms 47: 16–20.

Mogea, J.P. 1978. Pollination in *Salacca edulis*. Principes 22: 56–63.

Mogea, J.P. 1980. The flabellate species of *Salacca* (Palmae). Reinwardtia 9: 461–479.

Mogea, J.P. 2004. Four new species of *Arenga* (Palmae) from Indonesia. Reinwardtia 12: 181–189.

Monteillet, J. and J.R. Lappartient. 1981. Fruits et graines du Crétace superieur des carriers de Paki (Sénégal). Rev. Palaeobot. Palynol. 34: 331–344.

Montufar, R. and J. C. Pintaud. 2006. Variation in species composition, abundance and microhabitat preferences among western Amazonian terra firme palm communities. Bot. J. Linn. Soc. 151: 127–140.

Moore, H.E. 1951. Critical studies. 7. *Neonicholsonia* vs. *Woodsonia*. Gentes Herb. 8: 239–243.

Moore, H.E. 1957a. Synopses of various genera of Arecoideae. 22. *Neoveitchia*. Gentes Herb. 8: 537–540.

Moore, H.E. 1957b. Synopses of various genera of Arecoideae. 23. *Reinhardtia*. Gentes Herb. 8: 541–576.

Moore, H.E. 1960. A new subfamily of palms — the Caryotoideae. Principes 4: 102–117.

Moore, H.E. 1963a. Two new palms from Peru. Principes 7: 107–115.

Moore, H.E. 1963b. The typification and species of *Palma* Miller (1754). Gentes Herb. 9: 235–244.

Moore, H.E. 1963c. The types and lectotypes of some palm genera. Gentes Herb. 9: 245–274.

Moore, H.E. 1969a. *Satakentia* — a new genus of Palmae-Arecoideae. Principes 13: 3–12.

Moore, H.E. 1969b. New palms from the Pacific II. Principes 13: 67–76.

Moore, H.E. 1969c. New palms from the Pacific III. Principes 13: 99–108.

Moore, H.E. 1969d. A synopsis of the genus *Physokentia* (Palmae-Arecoideae). Principes 13: 120–136.

Moore, H.E. 1969e. The genus *Juania* (Palmae: Arecoideae). Gentes Herb. 10: 385–393.

Moore, H.E. 1971. The genus *Synechanthus* (Palmae). Principes 15: 10–19.

Moore, H.E. 1972. *Chelyocarpus* and its allies *Cryosophila* and *Itaya* (Palmae). Principes 16: 67–88.

Moore, H.E. 1973a. The major groups of palms and their distribution. Gentes Herb. 11: 27–140.

Moore, H.E. 1973b. Palms in the tropical forest ecosystems of Africa and South America. Pages 63–88 *in* Meggers, B.J., E.S. Ayensu and W.D. Duckworth (*eds*). Tropical forest ecosystems in Africa and South America: a comparative review. Smithsonian Institute Press, Washington.

Moore, H.E. 1977. Endangerment at the specific and generic levels in palms. Pages 267–282 *in* Prance, G.T. and T.S. Elias (*eds*). Extinction is forever: the status of threatened and endangered plants in the Americas. New York Botanical Garden, Bronx.

Moore, H.E. 1978a. The genus *Hyophorbe* (Palmae). Gentes Herb. 11: 212–245.

Moore, H.E. 1978b *Tectiphiala*, a new genus of Palmae from Mauritius. Gentes Herb. 11: 284–290.

Moore, H.E. 1979. Arecaeae. Pages 392–438 *in* Smith, A.C. Flora Vitiensis Nova 1. Pacific Tropical Garden, Lawaii, Kauai, Hawaii.

Moore, H.E. 1980. Palmae. Pages 1–6 *in* Rechinger, K.H. Flora Iranica 146. Akademische Druck, Austria.

Moore, H.E. 1982. A new species of *Wettinia* (Palmae) from Ecuador. Principes 26: 42–43.

Moore, H.E. and A.B. Anderson. 1976. *Ceroxylon alpinum* and *Ceroxylon quindiuense* (Palmae). Gentes Herb. 11: 168–185.

Moore, H.E. and F.R. Fosberg. 1956. The palms of Micronesia and the Bonin Islands. Gentes Herb. 8: 423–478.

Moore, H.E. and L.J. Guého. 1980. *Acanthophoenix* and *Dictyosperma* (Palmae) in the Mascarene Islands. Gentes Herb. 12: 1–16.

Moore, H.E. and L.J. Guého. 1984. Flore des Mascareignes. Palmiers 189: 1–34.

Moore, H.E. and J. Dransfield. 1978. A new species of *Wettinia* and notes on the genus. Notes Roy. Bot. Gard, Edinburgh 36: 259–267.

Moore, H.E. and J. Dransfield. 1979. The typification of Linnean palms. Taxon 28: 59–70.

Moore, H.E. and W. Meijer. 1965. A new species of *Arenga* from Borneo. Principes 9: 100–103.

Moore, H.E. and N.W. Uhl. 1973. The monocotyledons: their evolution and comparative biology. VI. Palms and the origin and evolution of monocotyledons. Quart. Rev. Biol. 48: 414–436.

Moore, H.E. and N.W. Uhl. 1982. Major trends of evolution in palms. Bot. Rev. (Lancaster) 48: 1–69.

Moore, H.E. and N.W. Uhl. 1984. The indigenous palms of New Caledonia. Allertonia 3: 313–402.

Moraes, M. 1996a. *Allagoptera* (Palmae). Fl. Neotrop. Monogr 73: 1–34. New York Botanical Garden, New York.

Moraes, M. 1996b. Novelties of the genera *Parajubaea* and *Syagrus* (Palmae) from Interandean valleys of Bolivia. Novon 6: 85–92.

Moraes, M. 2004. Flora de Palmeras de Bolivia. Herbario Nacional de Bolivia, La Paz, Bolivia.

Moraes, M. and A. Henderson. 1990. The genus *Parajubaea* (Palmae). Brittonia 42: 92–99.

Mora-Urpi, J. 1983. El Pejibaye (*Bactris gasipaes* H.B.K.): origen, biologia floral y manejo agronónomico. Palmeras poco utilizadas de America Tropical: Informe de la Reunión de Consulta Organizada por FAO y Catie: 118–160. Turrialba, Costa Rica.

Morcote-Rios, G. and R. Bernal. 2001. Remains of palms (Palmae) at archaeological sites in the New World: a review. Bot. Rev. (Lancaster) 67: 309–350.

MORLEY, R.J. 1998. Palynological evidence for Tertiary plant dispersals in the SE Asian region in relation to plate tectonics and dispersal. Pages 211–234 *in* R. Hall and J. D. Holloway (*eds*). Biogeography and geological evolution of SE Asia. Backhuys, Leiden.

MORLEY, R.J. 2000. Origin and evolution of tropical rain forests. Wiley, Chichester.

MORLEY, R.J. 2003. Interplate dispersal paths for megathermal angiosperms. Perspect. Pl. Ecol. Evol. Syst. 6: 5–20.

MORROW, L.O. 1965. Floral morphology and anatomy of certain Coryphoideae (Palmae). Ph.D. thesis, Cornell University.

MORTON, B.R., B.S. GAUT AND M.T. CLEGG. 1996. Evolution of alcohol dehydrogenase genes in the palm and grass families. Proc. Natl. Acad. Sci. U.S.A. 93: 11735–11739.

MOYA LOPEZ, C.E. AND A. LEIVA. 1991. *Gaussia spirituana* Moya et Leiva, sp. nov.: una nueva palma de Cuba Central. Revista Jard. Bot. Nac. Univ. Habana 12: 15–20.

MULLER, J. 1964. Palynological contributions to the history of Tertiary vegetation in N.W. Borneo. Xth Int. Bot. Congress, Edinburgh. Abstracts: 271.

MULLER, J. 1968. Palynology of the Pedawan and Plateau sandstone formations (Cretaceous — Eocene) in Sarawak, Malaysia. Micropaleontology, 14: 1–37.

MULLER, J. 1972. Palynological evidence for change in geomorphology, climate and vegetation in the Mio-Pliocene of Malesia. *In* Trans. II Aberdeen-Hull Symp. Malesian Ecol. Univ. Hull. Dept. Geograph. Miscell. Series no. 13: 6–16.

MULLER, J. 1979. Reflections on fossil palm pollen. *In* Proc. IV Palynol. Conf. Lucknow I: 568–579.

MULLER, J. 1981. Fossil pollen records of extant angiosperms. Bot. Rev. (Lancaster) 47: 1–142.

MUÑIZ, O. AND A. BORHIDI. 1982. Catálogo de las palmas de Cuba. Acta Bot. Hung. 28: 309–345.

MURÍN, A. AND I.I. CHAUDHRI. 1970. *In* Löve, A. (*ed*.). IOPB chromosome number reports XXVI. Taxon 19: 264–269.

MURRAY, S. G. 1971. The developmental anatomy of certain palm fruits. Ph.D. Thesis, Cornell University.

MURRAY, S.G. 1973. The formation of endocarp in palm fruits. Principes 17: 91–102.

NAIR, M.K. AND M.J. RATNAMBAL. 1978. Cytology of *Areca macrocalyx* Becc. Curr. Sci. 47: 172–173.

NEWTON, C. 2001. Le palmier Argoun: *Medemia argun* (Mart.) Württemb ex Wendl. Encyclopédie religieuse de l'Univers vegetal Croyances phytoreligieuses de l'Egypte ancienne (ERUV) II OrMonsp XI: 141–153.

NICHOLS, D.J., H. TATE AMES AND A. TRAVERSE. 1973. On *Arecipites* Wodehouse, *Monocolpopollenites* Thomson & Pflug, and the species "*Monocolpopollenites tranquillus*". Taxon 22: 241–256.

NINAN, C.A. AND T.G. RAVEENDRANATH. 1975. A study of the karyotype of *spicata* and *typica* varieties of coconut and its bearing on the breeding behaviour of *spicata* palms. New Botanist, Int. Quart. J. Pl. Sci. Res. 2: 81–87.

NOBLICK, L.R. 2004a. *Syagrus cearensis*, a twin-stemmed new palm from Brazil. Palms 48: 70–76.

NOBLICK, L.R. 2004b. *Syagrus vermicularis*, a fascinating new palm from northern Brazil. Palms 48: 109–116.

NOBLICK, L.R. 2006. The grassy *Butia*: two new species and a new combination. Palms 50: 167–178.

NOÉ, A.C. 1936. Fossil palms. *In* Dahlgren, B.E. (*ed*.). Index of American palms. Publ. Field Mus. Nat. Hist. Bot. Ser. 14: 441–456.

NORTON, S.A. 1998. Betel: consumption and consequences. J. Amer. Acad. Dermatol. 38: 81–88.

NORUP, M.V. 2005. *Alsmithia* subsumed in *Heterospathe* (Arecaceae, Arecoideae). Novon 15: 455–457.

NORUP, M.V., J. DRANSFIELD, M.W. CHASE, A.S. BARFOD, E.S. FERNANDO AND W.J. BAKER. 2006. Homoplasious character combinations and generic delimitations: a case study from the Indo-Pacific arecoid palms (Arecaceae: Areceae). Amer. J. Bot. 93: 1065–1080.

NUR SUPARDI MD NOR, J. DRANSFIELD AND B. PICKERSGILL. 1998. Preliminary observations on the species diversity of palms in Pasoh Forest Reserve, Negri Sembilan. Pages 105–104 *in* Lee, S.S. *et al.* (*eds*). Conservation, management and development of forest resources. Proceedings of Malaysia-UK Programme Workshop, Forest Research Institute Malaysia.

OHTSUKI T., M. SATO, T. KOYANO, T. KOWITHAYAKORN, N. KAWAHARA, Y. GODA AND M. ISHIBASHI. 2006. Steroidal saponins from *Calamus insignis*, and their cell growth and cell cycle inhibitory activities. Bioorg. Med. Chem. 14: 659–665.

OKOLO, E.C. 1988. Chromosome counts on Nigerian species of the genus *Raphia*. Principes 32: 156–159.

OLLIVIER-PIERRE, M.-F., C. GRUAS-CAVAGNETTO, E. ROCHE AND M. SCHULER. 1987. Elements de flore de type tropical et variations climatiques au Paleogene dans quelques basins d'Europe nord-occidentale. Mém. Trav. Inst. Montpellier École Prat. Hautes Études 17: 173–205.

OLSZEWSKA, M.J. AND R. OSIECKA. 1982. The relationship between 2C DNA content, life-cycle type, systematic position, and the level of DNA endoreplication in nuclei of parenchyma cells during growth and differentiation of roots in some monocotyledonous species. Biochem. Physiol. Pflanzen 177: 319–336.

ONO, M. AND Y. MASUDA. 1981. Chromosome numbers of some endemic species of the Bonin Islands II. Ogasawara Res. 4: 1–24.

OOI, P.A.C. 1982. The oil palm pollinating weevil. Nat. Malaysiana 8: 12–15.

OTEDOH, M. 1977. The African origin of *Raphia taedigera* — Palmae. Nigerian Field 42: 11–16.

OTEDOH, M. 1982. A revision of the genus *Raphia* Beauv. (Palmae). J. Nigerian Inst. Oil Palm Res. 6: 145–189.

ÔYAMA, T. AND H. MATSUO. 1964. Notes on palmaean leaf from Ôarai flora (Upper Cretaceous), Oarai Machi, Ibaraki Prefecture, Japan. Trans. & Proc. Palaeontol. Soc. Jap. n.s. 55: 241–246.

PADMANABAN, D. 1998. Concepts in the developmental morphology of the palm leaf — a review. Phytomorphology 48: 1–33.

PALLOT, J.M. 1961. Plant microfossils from the Isle of Wight. Unpublished PhD thesis. 216 pp. University of London.

PALOMINO, G. AND H.J. QUERO. 1992. Karyotype analysis of three species of *Sabal* L. (Palmae: Coryphoideae). Cytologia 557: 485–489.

PAN, A.D., B.F. JACOBS, J. DRANSFIELD AND W.J. BAKER. 2006. The fossil history of palms in Africa and new records from the Late Oligocene (~28 — 27 Myr) of northwestern Ethiopia. Bot. J. Linn. Soc. 151: 69–81.

PARTHASARATHY, M.V. 1968. Observations on metaphloem in the vegetative parts of palms. Amer. J. Bot. 55: 1140–1168.

PARTHASARATHY, M.V. 1974. Ultrastructure of phloem in palms. III. Mature phloem. Protoplasma 79: 265–315.

PARTHASARATHY, M.V. AND L.H. KLOTZ. 1976. Palm "wood." I. Anatomical aspects. Wood Sci. Technol. 10: 215–229.

Parthasarathy, M.V. and P.B. Tomlinson. 1967. Anatomical features of metaphloem in stems of *Sabal, Cocos* and two other palms. Amer. J. Bot. 54: 1143–1151.

Patil, G.V. and E.V. Uphadhye. 1984. *Cocos*-like fruit from Mohgaonkalan and its significance towards the stratigraphy of Mohgaonkalan Intertrappean beds. Pages 541–554 *in* Sharma, A.K., G.C. Mitra and M. Banerjee (*eds*). Proc. Symp. Evolutionary Botany and Biostratigraphy. Univ. Calcutta 1979. Today & Tomorrow's Printers & Publ., New Delhi.

Peking Institute of Botany and Nanjing Institute of Geology and Palaeontology. 1978. Chinese fossils of all groups. Fossil plants of China. Fascicle III Cenozoic plants of China. Academia Sinica, pp. 159–161. Science Press, Peking.

Pennington, R.T. and C.W. Dick. 2004. The role of immigrants in the assembly of the South American rainforest tree flora. Phil. Trans., Ser. B 359: 1611–1622.

Periasamy, K. 1962. Morphological and ontogenetic studies in palms I. Development of the plicate condition in the palm leaf. Phytomorphology 12: 54–64.

Periasamy, K. 1965. Morphological and ontogenetic studies in palms. II. Growth pattern of the leaves of *Cocos nucifera* and *Borassus flabellifer* after the initiation of plications. Austral. J. Bot. 13: 225–234.

Periasamy, K. 1966a. Morphological and ontogenetic studies in palms. III. Growth pattern of the leaves of *Caryota* and *Phoenix* after initiation of plications. Phytomorphology 16: 474–490.

Periasamy, K. 1966b. Morphological and ontogenetic studies in palms. IV. Ontogeny of the vascular patterns in four genera. Austral. J. Bot. 14: 277–291.

Periasamy, K. 1967. Morphological and ontogenetic studies in palms. Early ontogeny and vascular architecture of the leaf of *Rhapis flabelliformis*. Austral. J. Bot. 15: 151–159.

Pesce, C. 1941 (Translation 1985). Oil palms and other oil seeds of the Amazon. Translated by D.V. Johnson. Reference Publ. Inc., Algonac, Michigan.

Peters, H.A., A. Pauw, M.R. Silman and J.W. Terborgh. 2004. Falling palm fronds structure Amazonian rainforest sapling communities. Proc. R. Soc. London, Ser. B. Biol. Sci. (Suppl.) 271: S367–S369.

Petersen, G., O. Seberg, J.I. Davis, D.H. Goldman, D.W. Stevenson, L.M. Campbell, F.A. Michelangeli, C.D. Specht, M.W. Chase, M.F. Fay, J.C. Pires, J.V. Freudenstein, C.R. Hardy and M.P. Simmons. 2006. Mitochondrial data in monocot phylogenetics. Aliso 22: 52–62.

Pflanzl, G. 1956. Das Alter der Braunkohlen des Meissners der Flöze 2 und 3 des Hirschberges und eines benachbarten Kohlenlagers bei Laudenbach. Notizbl. Hess. Land. Bodenforsch. (Wiesbaden) 84: 232–244.

Pfleger, F.L. and R.G. Linderman (*eds*). 1994. Mycorrhizae and plant health. Amer. Phytopath. Soc. Press, St. Paul, Minn.

Phadtare, N.R. and A.R. Kulkarni. 1980. Palynological investigations of Ratnagiri lignite, Maharashtra. Geophytology 10: 158–179.

Phadtare, N.R. and A.R. Kulkarni. 1984. Affinity of the genus *Quilonipollenites* with the Malaysian palm *Eugeissona* Griffith. Pollen & Spores, 26: 217–226.

Pinheiro, C.U.B. 2001. Germination strategies of palms: the case of *Schippia concolor* Burret in Belize. Brittonia 53: 519–527.

Pintaud, J.-C. 1999a. A cladistic analysis of the Archontophoenicinae (Palmae, Areceae) based on morphological and anatomical characters. Pages 279–284 *in* Henderson, A. and F. Borchsenius (*eds*). Evolution, variation and classification of palms. The New York Botanical Garden Press, Bronx, NY.

Pintaud, J.-C. 1999b. Phylogénie, biogéographie et écologie des palmiers de Nouvelle-Calédonie. Ph.D. thesis, Université Paul Sabatier, Toulouse.

Pintaud, J.-C. and W.J. Baker. 2008. A revision of the palm genera (Arecaceae) of New Caledonia. Kew Bull. 63: 61–73.

Pintaud, J.-C. and B. Millan. 2004. Vegetative transformation of inflorescences in *Socratea salazarii*. Palms 48: 86–89.

Pintaud, J.-C. and D.R. Hodel. 1998a. A revision of *Kentiopsis,* a genus endemic to New Caledonia. Principes 42: 32–33, 41–53.

Pintaud, J.-C. and D.R. Hodel. 1998b. Three new species of *Burretiokentia*. Principes 42: 152–155, 160–166.

Pintaud, J.-C, T. Jaffré and J.-M. Veillon. 1999. Conservation status of New Caldeonia palms. Pacific Conservation Biol. 5: 9–15.

Pintaud, J.-C. and H. Setoguchi. 1999c. *Satakentia* revisited. Palms 43(4): 194–199.

Plotkin, M.J. and M.J. Balick. 1984. Medicinal uses of South American palms. J. Ethnopharmacology 10: 157–179.

Pohl, F. 1941. Über Rhapiden pollen und seine blütenëkologische Bedeutung. Oesterr. Bot. Z. 90: 81–96.

Poinar, G. Jr. 2002a. Fossil flowers in Dominican and Mexican amber. Bot. J. Linn. Soc. 138: 57–61.

Poinar, G. Jr. 2002b. Fossil flowers in Dominican and Baltic amber. Bot. J. Linn. Soc. 139: 361–367.

Pole, M. 1994. The New Zealand Flora — entirely long-distance dispersal? J. Biogeogr. 21: 625–635.

Pole, M.S. and M.K. McPhail. 1996. Eocene *Nypa* from Regatta Point, Tasmania. Rev. Palaeobot. Palynol. 92: 55–67.

Potonié, R. 1960. Sporologie der eozänen Kohle von Kalewa in Burma. Senckenberg. Leth. 41: 451–481.

Potter, F. 1976. Investigations of angiosperms from the Eocene of southeastern North America: pollen assemblages from Miller Pit, Henry County, Tennessee. Palaeontographica Abt. B Paläophytol. 157: 44–96.

Potztal, E. 1964. Reihe Principes. Pages 579–588 *in* Melchior, H. (*ed*.). A. Engler's Syllabus der Pflanzenfamilien. Ed 12, 2. Gebrüder Borntraeger, Berlin.

Prakash, U. 1974. Palaeogene angiospermous woods. Pages 306–320 *in* Surange, K.R., R.N. Lakhanpal and D.C. Bharadwaj (*eds*). Aspects and appraisal of Indian palaeobotany. Birbal Sahni Inst. Palaeobotany, Lucknow.

Prakash, U. and K. Ambwani. 1980. A petrified *Livistona*-like palm stem, *Palmoxylon livistonoides* sp. nov. from the Deccan Intertrappean beds of India. Palaeobotanist 26: 297–306.

Prance, G.T. 1982. A review of the phytogeographic evidences for pleistocene climate changes in the Neotropics. Ann. Missouri Bot. Gard. 69: 594–624.

Prasad, V., C.A.E. Strömberg, H. Alimohammadian and A. Sahni. 2005. Dinosaur coprolites and the early evolution of grasses and grazers. Science 18: 1177–1180.

Pritchard, H.W., C.B. Wood, S. Hodges, and H.J. Vautier. 2004. 100-seed test for desiccation tolerance and germination: a case study on eight tropical palm species. Seed Sci. Technol. 32: 393–403.

Prychid, C.J., P.J. Rudall and M. Gregory. 2004. Systematics and biology of silica bodies in monocotyledons. Bot. Rev. (Lancaster) 60: 377–440.

PUNT, W. AND J.G. WESSELS-BOER. 1966a. A palynological study in cocoid palms. Acta Bot. Neerl. 15: 255–265.

PUNT, W. AND J.G. WESSELS-BOER. 1966b. A palynological study in geonomoid palms. Acta Bot. Neerl. 15: 266–275.

PUTZ, F.E. 1979. Biology and human use of *Leopoldinia piassaba.* Principes 23: 149–156.

PUTZ, F.E. 1990. Growth habits and trellis requirements of climbing palms (*Calamus* spp.) in North-eastern Queensland. Austral. J. Bot. 38: 603–608.

QUERO, H.J. 1980. *Coccothrinax readii*, a new species from the peninsula of Yucatan, Mexico. Principes 24: 118–124.

QUERO, H.J. 1981. *Pseudophoenix sargentii* in the Yucatan Peninsula, Mexico. Principes 25: 63–72.

QUERO, H.J. 1991. *Sabal gretheriae*, a new species of palm from the Yucatan Peninsula, Mexico. Principes 35: 219–224.

QUERO, H.J. AND R.W. READ. 1986. A revision of the palm genus *Gaussia.* Syst. Bot. 11: 145–154.

RÁKOSI, L. 1976. A Magyarországi Eocén mangrove palinológiai adatai. M. Áll. Föld. Int. Évi Jel., pp. 357–372 + summary in French, pp. 372–374.

RAKOTOARINIVO, M., T. RANARIVELO AND J. DRANSFIELD. 2007. A new species of *Beccariophoenix* from the High Plateau of Madagascar. Palms 51: 63–75.

RAMANUJAM, C.G.K. 1953. *Palmoxylon arcotense* sp. nov., a fossil palm resembling the living genus *Livistona* from South India. Palaeobotanist 2: 89–91. [Not seen]

RAMANUJAM, C.G.K. 1966. Palynology of the Miocene lignite from South Arcot District, Madras, India. Pollen & Spores, 8: 149–203.

RAMANUJAM, C.G.K. 1987. Palynology of the Neogene Warkalli Beds of Kerala State in South India. J. Palaeontol. Soc. India 32: 26–46.

RAMANUJAM, C.G.K. 2004. Palms through the ages in southern India — a reconnaissance. Palaeobotanist 53: 1–4.

RAMANUJAM, C.G.K., H. RAMAKRISHNA AND C. MALLESHAM. 1986. Palynoassemblage of the subsurface Miocene sediments of the East Coast of Southern India — its floristic and environmental significance. Proc. Spl. Ind. Geo. Con., Poona, pp. 113–117.

RAMANUJAM, C.G.K. AND K.P. RAO. 1977. A palynological approach to the study of Warkalli deposits of Kerala in South India. Geophytology 7: 160–164.

RAMANUJAM, C.G.K., G.M. RAO AND P.R. REDDY. 1991a. Palynological studies of subsurface sediments at Mynagapally, Quilon District, Kerala State. Biovigyanam 17:1–11.

RAMANUJAM, C.G.K., AND P.R. REDDY. 1984. Palynoflora of Neyveli lignite — floristic and palaeoenvironmental analysis. J. Palynol. 20:58–74.

RAMANUJAM, C.G.K., P.R. REDDY AND H. RAMAKRISHNA. 1997. Dicolpate palm pollen from the Neogene deposits of Godavari – Krishna Basin, A.P. J. Palynol. 33: 129–136.

RAMANUJAM, C.G.K., P.R. REDDY AND H. RAMAKRISHNA. 1998. Botanical affinities of *Jacobipollenites* (Ramanujam) Singh & Misra. Geophytology 27: 111–113.

RAMANUJAM, C.G.K., P.R. REDDY AND H. RAMAKRISHNA. 2001. Pollen types of Arecaceae from the Tertiary deposits of Southern India — a critical appraisal. J. Swamy Bot. Cl. 18: 51–63.

RAMANUJAM, C.G.K., P.R. REDDY AND G.M. RAO. 1991b. Palynoassemblages of the subsurface Tertiary strata at Pattanakad, Alleppey District, Kerala State. J. Palaeontol. Soc. India 36: 51–58.

RAMANUJAM, C.G.K., P.R. REDDY AND G.M. RAO. 1992. Palynology of Tertiary subcrops of Kalaikode Borewell in Kerala State. Indian J. Earth Sci. 19: 18–27.

RANGASAMY, S.R.S. AND P. DEVASAHAYAM. 1972. Cytology and sex determination in palmyrah palm (*Borassus flabellifer* Linn.). Cellule 69: 129–135.

RAO, A.R. AND V. ACHUTHAN. 1973. A review of fossil palm remains for India. Palaeobotanist, 20: 190–202.

RAO, C.V. 1959a. Contributions to the embryology of Palmae: I. Sabaleae. Proc. Natl. Inst. Sci. India, B 25: 143–168.

RAO, C.V. 1959b. II. Contributions to the embryology of Palmae: Ceroxylineae. Proc. Natl. Inst. Sci. India, B 38: 46–75.

RAO, K.P. AND C.G.K. RAMANUJAM. 1975. A palynological approach to the study of Quilon Beds of Kerala State in South India. Curr. Sci., 44: 730–732.

RAO, K.P. AND C.G.K. RAMANUJAM. 1978. Palynology of the Neogene Quilon Beds of Kerala State in South India I. Spores of pteridophytes and pollen of monocotyledons. Palaeobotanist, 25: 397–427.

RÁSKY, K. 1949. *Nipadites burtini* Brong. Termése Duddarról. Földt. Közl. 79: 1–4.

RÁSKY, K. 1956. Fosszilis növények a Martinovics-hegyi (Budapest) felsöecénból. Földt. Közl. 86: 295–298.

RAUWERDINK, J.B. 1985. An essay on *Metroxylon*, the sago palm. Principes 30: 165–180.

RAVOLOLONANAHARY, H. 1999. The conservation status of *Satranala decussilvae* in the Ianobe Valley, Masoala National Park, Madagascar. Palms 43: 145–148.

READ, R.W. 1963. Palm chromosomes. Principes 7: 85–88.

READ, R.W. 1964. Palm chromosome studies facilitated by pollen culture on a colchicine-lactose medium. Stain Technol. 39: 99–106.

READ, R.W. 1965a. Chromosome numbers in the Coryphoideae. Cytologia 30: 385–391.

READ, R.W. 1965b. Palm chromosomes by airmail. Principes 9: 4–10.

READ, R.W. 1966. New chromosome counts in the Palmae. Principes 10: 55–61.

READ, R.W. 1967. More chromosomes counts by mail. Principes 11: 77.

READ, R.W. 1968. A study of *Pseudophoenix* (Palmae). Gentes Herb. 10: 169–213.

READ, R.W. 1969. Some notes on *Pseudophoenix* and a key to the species. Principes 13: 13–22.

READ, R.W. 1975. The genus *Thrinax* (Palmae: Coryphoideae). Smithsonian Contr. Bot. 19: 1–98.

READ, R.W. 1979. Palmae. Pages 320–368 *in* R.A. HOWARD (*ed.*) Flora of the Lesser Antilles 3. Arnold Arboretum, Harvard Univ.

READ, R.W. 1980. Notes on Palmae, I. Phytologia 46: 285–287.

READ, R.W. 1998. Utilization of indigenous palms in the Caribbean (in relation to their abundance). Advances Econ. Bot. 6: 137–143.

READ, R.W. AND L.J. HICKEY. 1972. A revised classification of fossil palm and palm-like leaves. Taxon 21: 129–137.

RECHINGER, K. 1907. Plantae novae pacificae. Repert. Spec. Nov. Reg. Veg. 4: 232.

REDDY, G.N. AND A.R. KULKARNI. 1982. Developmental fruit anatomy of some coryphoid palms. Geophytology 12: 233–244.

REDDY, G.N. AND A.R. KULKARNI. 1985. Contribution to the anatomy of palm fruits — Cocosoid palms. Proc. Indian Acad. Sci., Pl. Sci. 95: 153–165.

REDDY, G.N. AND A.R. KULKARNI. 1989. Megasporogenesis in *Cocos nucifera* L. — a reinvestigation. Curr. Sci. 58: 156–157.

REID, E.M. AND M.E.J. CHANDLER. 1926. The Bembridge Flora. Volume 1. British Museum (Natural History) London.

REID, E.M. AND M.E.J. CHANDLER. 1933. The flora of the London Clay. Brit. Mus. (Nat. Hist.) Publication 8. 561 pp. + 35 pls.

RENDLE, A.B. 1894. Revision of the genus *Nipadites* Bowerbank. J. Linn. Soc., Bot. 30: 143–154.

RENNER, S. 2004. Plant dispersal across the tropical Atlantic by wind and sea currents. Int. J. Pl. Sci. 165: S23–S33.

RICH, P.M. 1987a. Developmental anatomy of the stem of *Welfia Georgia*, *Iriartea gigantea*, and other arborescent palms: implications of mechanical support. Amer. J. Bot. 74: 792–802.

RICH, P.M. 1987b. Mechanical structure of the stem of arborescent palms. Bot. Gaz. 148: 42–50.

RICH, P.M., M. HOLBROOK AND N. LUTTINGER. 1995. Leaf development and crown geometry of two iriarteoid palms. Amer. J. Bot. 82: 328–336.

RICHARDS, P.W. 1952. The tropical rainforest. Cambridge University Press, Cambridge.

RICKSON, F.R. AND M.M. RICKSON. 1986. Nutrient acquisition facilitated by litter collection and ant colonies on two Malaysian palms. Biotropica 18: 337–343.

RIGBY, J.F. 1995. A fossil *Cocos nucifera* L. fruit from the latest Pliocene of Queensland Australia. Pages 379–381 *in* Pant, D.D. (ed.). Birbal Sahni Centenary Volume. Allahabad University.

ROBERTSON, B.L. 1976a. Embryology of *Jubaeopsis caffra* Becc.: 1. Microsporangium, microsporogenesis and microgametogenesis. J. S. African Bot. 42: 97–108.

ROBERTSON, B.L. 1976b. Embryology of *Jubaeopsis caffra* Becc.: 2. Megasporangium, megasporogenesis and megagametogenesis. J.S. African Bot. 42: 173–184.

ROCHE, E. 1982. Etude palynologique (pollen et spores) de l'Eocene de Belgique. Service Geologique de Belgique — Professional paper N° 193. Ministere des Affaires Economiques.

ROMO, V., G. PALOMINO AND H. QUERO. 1988. *In* Löve, A. (*ed.*). IOPB chromosome number reports XCVIII. Taxon 37: 194–196.

RONCAL, J., J. FRANCISCO-ORTEGA, C.B. ASMUSSEN AND C.E. LEWIS. 2005. Molecular phylogenetics of tribe Geonomeae (Arecaceae) using nuclear DNA sequences of phosphoribulokinase and RNA polymerase II. Syst. Bot. 30: 275–283.

RONCAL, J., S. ZONA AND C.E. LEWIS. 2008. Molecular phylogenetic studies of Caribbean palms (Arecaceae) and their relationships to biogeography and conservation. Bot. Rev. 74: 78–102.

RÖSER, M. 1993. Variation and evolution of karyotype characters in palm subfamily Coryphoideae s.l. Bot. Acta 106: 170–182.

RÖSER, M. 1994. Pathways of karyological differentiation in palms (Arecaceae). Pl. Syst. Evol. 189: 83–122.

RÖSER, M. 1995. Trends in karyo-evolution of palms. Pages 249–265 *in* Brandham, P.E. and M.D. Bennett (*eds*). Kew Chromosome Conference IV. Royal Botanic Gardens, Kew.

RÖSER, M. 1999. Chromosome structures and karyotype rearrangement in palms (Palmae). Pages 61–71 *in* Henderson, A. and F. Borchsenius (*eds*). Evolution, variation and classification of palms. The New York Botanical Garden Press, Bronx, NY.

RÖSER, M. 2000. DNA amounts and qualitative properties of nuclear genomes in palms (Arecaceae). Pages 538–544 *in* Wilson, K.L. and D.A. Morrison (*eds*). Monocots: systematics and evolution. CSIRO, Melbourne.

RÖSER, M., M.A.T. JOHNSON AND L. HANSON. 1997. Nuclear DNA amounts in palms (Arecaceae). Bot. Acta 110: 79–89.

ROTH, I. 1990. Leaf structure of a Venezuelan cloud forest in relation to the microclimate. Handbuch der Pflanzenanatomie XIV, 1. Gebrüder Borntraeger: Berlin & Stuttgart. 244 pp.

ROTINSULU, W.C. 2001. The use of *Pigafetta elata* for making furniture in Indonesia. Palms 45: 39–41.

ROWE, N., S. ISNARD and T. SPECK. 2004. Diversity of mechanical architectures in climbing plants: an evolutionary perspective. J. Pl. Growth Regulat. 23: 108–128.

ROY, S.K. AND P.K. GHOSH. 1980. On the occurrence of *Palmoxylon coronatum* in West Bengal, India. Ameghiniana 17: 130–134.

RUDALL, P.J., K. ABRANSON, J. DRANSFIELD AND W. BAKER. 2003. Floral anatomy in *Dypsis* (Arecaceae–Areceae): a case of complex synorganization and stamen reduction. Bot. J. Linn. Soc. 143: 115–133.

RUDALL, P.J. AND M. BURZGO. 2002. Evolutionary history of the monocot leaf. Pages 431–458 *in* Cronk, Q.C.B., R.M. Bateman and J.A. Hawkins (*eds*). Developmental genetics and plant evolution. Taylor & Francis, London.

RULL, V. 1998. Biogeographical and evolutionary considerations of *Mauritia* (Arecaceae), based on palynological evidence. Rev. Palaeobot. Palynol. 100: 109–122.

RUMPHIUS, G.E. 1741–1755. Herbarium Amboinense. 6 vols. J. Burmann, Meinard Uytwerf, Amsterdam.

RUSSELL, T.A. 1965. The *Raphia* palms of West Africa. Kew Bull. 19: 173–196.

SAAKOV, S.G. 1954. Palms and their culture in the USSR. Acad. Sci. USSR, Moscow and Leningrad.

SAHNI, B. 1946. A silicified *Cocos*-like palm stem, *Palmoxylon* (*Cocos*) *sundaram*, from the Deccan Intertrappean beds. J. Indian Bot. Soc., M.O.P. Iyengar Comm. vol., pp. 361–374.

SAHNI, B. 1964. Revisions of Indian fossil plants. III. Monocotyledons. Monograph 1, Birbal Sahni Institute of Palaeobotany, Lucknow pp. 1–89.

SALARD-CHEBOLDAEFF, M. 1978. Sur la palynoflore Maestrichtienne et Tertiare du bassin sédimentaire littoral du Cameroun. Pollen & Spores, 10: 215–260.

SALARD-CHEBOLDAEFF, M. 1979. Palynologie Maestrichtienne et Tertiaire du Cameroun. Etude qualitative et repartition verticale ds principales espèces. Rev. Palaeobot. Palynol. 28: 365–387.

SALARD-CHEBOLDAEFF, M. 1981. Palynologie Maestrichtienne et Tertiaire du Cameroun. Rev. Palaeobot. Palynol. 32: 401–439.

SALUJHA, S.K., G.S. KINDRA AND K. REHMAN. 1973a. Palynology of the South Shillong Front, part I: The Palaeogene of Garo Hills. Proc. Sem. Paleopalynol. Ind. Stratigr. 1971, Calcutta, pp. 265–291.

SALUJHA, S.K., K. REHMAN AND G.S. KINDRA. 1973b. Distinction between the Bhuban and Bokabil sediments on the southern edge of Shillong Plateau based on palynofossil assemblages. Bull. O.N.G.C. 10: 109–117.

SALZMAN, V.T. AND W.S. JUDD. 1995. A revision of the Greater Antillean species of *Bactris* (Bactridinae: Arecaceae). Brittonia 47: 345–371.

SANDERS, R. 1991. Cladistics of *Bactris* (Palmae): survey of characters and refution of Burret's classification. Selbyana 12: 105–133.

SANDERSON, M.J., J.L. THORNE, N. WIKSTRÖM AND K. BREMER. 2004. Molecular evidence on plant divergence times. Amer. J. Bot. 91: 1656–1665.

Sannier, J., C.B. Asmussen-Lange, M.M. Harley, and S. Nadot. 2007. Evolution of microsporogenesis in palms (Arecaceae). Int. J. Pl. Sci. 168: 877–888.

Sannier, J., S. Nadot, A. Forchioni, M.M. Harley, and B. Albert. 2006. Variations in the microsporogenesis of monosulcate palm pollen. Bot. J. Linn. Soc. 151: 93–102.

Saporta, G. Le Marquis De. 1865. Études sur la végétation du Sud-Est de la France a l'époque Tertiare. Armissan près Narbonne (Aude). Ann. Sci. Nat., Bot. Sér. 5, 4: 5–264.

Saporta, G. le Marquis de. 1878. Essai descriptiv sur plantes fossiles des arkoses de Brives pres de Puy-en-Velay. Ann. Soc. Agric. Puy 33: 1–72.

Saporta, G. Le Marquis De. 1879. Le Monde des Plantes avant l'Apparition de l'Homme. G. Masson, Paris.

Saporta, G. le Marquis de. 1889. Des palmiers fossils. Rev. Gén. Bot. 1: 229–243; pls 11–12.

Sarkar, A.K. 1986. Karyomorphological studies of *Calamus* L. (Palmae) to ascertain their taxonomic affinities. Proc. Indian. Sci. Congr. 73 (3, VI): 157.

Sarkar, A.K. 1987. Cytological investigation of *Arenga* Labill. and *Areca* L. of Palmae to ascertain their taxonomic affinities. Proc. Indian. Sci. Congr. 74 (3, VI): 199–200.

Sarkar, A.K. and N. Datta. 1985. Cytology of *Calamus* L. as an aid to their taxonomy. Cell Chromosome Res. 8: 69–73.

Sarkar, A.K., N. Datta, R. Mallick and U. Chatterjee. 1976. *In* Löve, A. (*ed*.). IOPB chromosome number reports LIV. Taxon 25: 631–649.

Sarkar, A.K., R. Mallick, N. Datta and U. Chatterjee. 1977. *In* Löve, A. (*ed*.). IOPB chromosome number reports LVII. Taxon 26: 443–452.

Sarkar, A.K., N. Datta and U. Chatterjee. 1978a. *In* Löve, A. (*ed*.). IOPB chromosome number reports LXII. Taxon 27: 519–535.

Sarkar, A.K., N. Datta, R. Mallick and U. Chatterjee. 1978b. *In* Löve, A. (*ed*.). IOPB chromosome number reports LIX. Taxon 27: 53–61.

Sarkar, S.K. 1957. Sex chromosomes in palms. Genét. Ibér. 9: 133–142.

Sarkar, S.K. 1958. Chromosome studies in palms. Proc. Indian. Sci. Congr. 45 (3, VI): 296–297.

Sarkar, S.K. 1970. Palmales. Res. Bull. Cytogen. Lab. Dept. Bot. Univ. Calcutta 2: 22–23.

Sarma, P.S., P.R. Reddy and K. Srisailam. 1984. Palynomorphs referable to monocotyledons from the Neyveli lignite deposit of Tamilnadu. Indian J. Bot. 7: 201–209.

Sarukhán, J. 1978. Studies on the demography of tropical trees. Pages 163–184 *in* Tomlinson, P.B. and M.H. Zimmermann (*eds*). Tropical trees as living systems. Cambridge University Press, New York.

Satake, T. 1962. A new system of the classification of Palmae. Hikobia 3: 112–133.

Satô, D. 1946. Karyotype alteration and phylogeny, VI. Karyotype analysis in Palmae. Cytologia 14: 174–186.

Savage, A.J.P. and P.S. Ashton. 1983. The population structure of the double coconut and some other Seychelles palms. Biotropica 15: 15–25.

Savolainen, V., M.-C. Anstett, C. Lexer, I. Hutton, J.J. Clarkson, M.V. Norup, M.P. Powell, D. Springate, N. Salamin and W.J. Baker. 2006. Sympatric speciation in palms on an oceanic island. Nature 441: 210–213.

Saw, L.G. 1997. A revision of *Licuala* (Palmae) in the Malay Peninsula. Sandakania 10: 1–95.

Sawant, N.A. 1975. Cytological observations in *Wallichia disticha* T. Anders. Proc. Indian. Sci. Congr. 62 (3, VI): 121–122.

Saxena, R.K. 1992. Neyveli lignites and associated sediments — their palynology, palaeoecology, correlation and age. Palaeobotanist 40: 345–353.

Saxena, R.K. and N.K. Misra. 1990. Palynological investigation of the Ratnagiri beds of Sindhu Durg District, Maharashtra. Palaeobotanist 38: 263–276.

Scariot, A.O. and E. Lleras. 1991. Reproductive biology of the palm *Acrocomia aculeata* in Central Brazil. Bioitropica 23: 12–22.

Schaarschmidt, F. and V. Wilde. 1986. Palmenblüten und -blätter aus dem Eozän von Messel. Courier Forschungsinst. Senckenberg, 86: 177–202.

Scheffer, R.H.C.C. 1873. Sur quelques palmiers du groupe des Arecinées. Natuurk. Tijdschr. Ned.-Indie 32: 149–193.

Scheffer, R.H.C.C. 1876. Sur quelques palmiers du groupe des Arecinées. Ann. Jard. Bot. Buitenzorg 1: 103–164.

Schenk, A. 1869. Beiträge zur Flora der Vorwelt III. Der fossilen Pflanzen der Wernsdorfer Schichten in den Nordkarpathen. Palaeontographica 19: 1–34.

Schmid, R. 1970. Notes on the reproductive biology of *Asterogyne martiana* (Palmae). II. Pollination by Syrphid flies. Principes 14: 39–49.

Schmid, R. 1985. Functional interpretations of the morphology and anatomy of septal nectaries. Acta Bot. Neerl. 34: 125–128.

Schmidt, M.E. 1994. The distribution and characteristics of Louisiana petrified palmwood. Principes 38: 142–145.

Schmitt, U., G. Weiner and W. Liese. 1995. The fine structure of the stegmata in *Calamus axillaris* during maturation. I.A.W.A. J. 16: 61–68.

Schrank, E. 1992. Nonmarine Cretaceous correlations in Egypt and northern Sudan: palynological and palaeobotanical evidence. Cretaceous Res. 13: 351–368.

Schrank, E. 1994. Palynology of the Yesomma Formation in Northern Somalia: a study of pollen spores and associated phytoplankton from the late Cretaceous Palmae Province. Palaeontographica, Abt. B 231: 63–112.

Schuler, M. and J. Doubinger. 1970. Observations palynologiques dans le Bassin d'Amaga (Colombie). Pollen & Spores 12: 429–450.

Schuler, M., C. Cavelier, C. Dupuis, E. Steurbaut, N. Vandenberghe, J. Riveline, E. Roche and M.-J. Soncini. 1992. The Paleogene of the Paris and Belgian Basins. Standard-stages and regional stratotypes. Cah. Micropaléontol. 7: 29–92.

Schweingruber, F.H. 1990. Anatomie europaischer Hölzer. Verlag Paul Haupt: Bern & Stuttgart. 800 pp.

Scotland, R.W., R.G. Olmstead and J.R. Bennett. 2003. Phylogeny reconstruction: the role of morphology. Syst. Biol. 52: 539–548.

Scott, R.A., P.L. Williams, L.C. Craig, E.S. Barghoorn, L.J. Hickey and H.D. MacGinitie. 1972. "Pre-Cretaceous" angiosperms from Utah: evidence for Tertiary age of the palm woods and roots. Amer. J. Bot. 59: 886–896.

Seubert, E. 1996a. Root anatomy of palms, II. Calamoideae. Feddes Repert. 107: 43–59.

Seubert, E. 1996b. Root anatomy of palms, III. Ceroxyloideae, Nypoideae, Phytelepheae. Feddes Repert. 107: 597–619.

Seubert, E. 1997. Root anatomy of palms, I. Coryphoideae. Flora 192: 81–103.

Seubert, E. 1998a. Root anatomy of palms, IV. Arecoideae, part I. General remarks and descriptions of the roots. Feddes Repert. 109: 89–127.

Seubert, E. 1998b. Root anatomy of palms, IV. Arecoideae, part II. Systematic implications. Feddes Repert. 109: 231–247.

Shapcott, A. 1998a. The patterns of genetic diversity in *Carpentaria acuminata* (Arecaceae), and rainforest history in northern Australia. Mol. Ecol. 7: 833–847.

Shapcott, A. 1998b. The genetics of *Ptychosperma bleeseri*, a rare palm from the Northern Territory, Australia. Biol. Conservation 85: 203–209.

Shapcott, A. 1999. Comparison of the population genetics and densities of five *Pinanga* palm species at Kuala Belalong, Brunei. Mol. Ecol. 8: 1641–1654.

Shapcott, A. 2000. Conservation and genetics in the fragmented monsoon rainforest in the Northern Territory, Australia: a case study of three frugivore-dispersed species. Austral. J. Bot. 48: 397–407.

Shapcott, A., M. Rakotoarinivo, R.J. Smith, G. Lysaková, M. Fay and J. Dransfield. 2007. Can we bring Madagascar's critically endangered palms back from the brink? Genetics, ecology and conservation of the critically endangered palm *Beccariophoenix madagascariensis*. Bot. J. Linn. Soc. 154: 589–608.

Sharma, A.K. and S.K. Sarkar. 1956. Cytology of different species of palms and its bearing on the solution of the problems of phylogeny and speciation. Genetica 28: 361–488.

Shete, R.H. and A.R. Kulkarni. 1980. *Palmocaulon hyphaeneoides* sp. nov. from the Deccan Intertrappean beds of Wardha District, Maharashtra, India. Palaeontographica Abt. B. 172: 117–124.

Shete, R.H. and A.R. Kulkarni. 1985. *Palmocarpon coryphoidium* sp. nov. A coryphoid palm fruit from Deccan Intertrappean beds of Wardha District, Maharashtra. J. Indian Bot. Soc. 64: 45–50.

Shibata, K. 1962. Estudios citologicos de plantas Colobianas silvestres y cultivadas. J. Agric. Sci. (Tokyo) 8: 49–62.

Shinde, N.W. and A.R. Kulkarni. 1989. Fruits of *Nyssa* and *Eugeissona* from lignitic exposures of Ratnagiri District, Maharashtra. Proceedings of Special Indian Geophytology Conference, Poona, pp. 165–169. [Not seen cited according to Guleria (1992)].

Shuey, A.G. and R.P. Wunderlin. 1977. The needle palm: *Rhapidophyllum hystrix*. Principes 21: 47–59.

Silberbauer-Gottsberger, I. 1990. Pollination and evolution in palms. Phyton (Horn) 30: 213–233.

Simmonds, N.W. 1954. Chromosome behaviour in some tropical plants. Heredity 8: 139–146.

Singh, H.P. and M.R. Rao. 1990. Tertiary palynology of Kerala Basin — an overview. Palaeobotanist 38: 256–262.

Singh, R.S. 1999. Diversity of *Nypa* in the Indian subcontinent: Late Cretaceous to recent. Palaeobotanist 48: 147–154.

Skov, F. and F. Borchsenius. 1997. Predicting plant species distribution patterns using simple climatic parameters: a case study of Ecuadorian palms. Ecography 20: 347–355.

Skov, F. and H. Balslev. 1989. A revision of *Hyospathe* (Arecaceae). Nordic J. Bot. 9: 189–202.

Small, I.K. 1923. The needle palm *Rhapidophyllum hystrix*. J. New York Bot. Gard. 24: 105–114.

Smets, E.F., L.-P. Ronse Decraene, P. Caris and P.J. Rudall. 2000. Floral nectaries in monocotyledons: Distribution and evolution. Pages 230–240 *in* Wilson, K.L. and D.A. Morrison (*eds*). Monocots: systematics and evolution. CSIRO, Melbourne.

Smith, A.G., D.G. Smith and B.M. Funnell. 1994. Atlas of Mesozoic and Cenozoic coastlines. Cambridge University Press, Cambridge.

Söderberg, E. 1919. Über die Pollentwicklung bei *Chamaedorea corallinae*. Svensk Bot. Tidskr. 13: 204–205.

Sole de Porta, N. 1961. Contribución al studio palinológico del terciario de Colombia. Bol. Geol. Univ. Industrial Santander, Bucaramanga. 7: 55–81.

Soliman, A.S. and A.-R.A. Al-Mayah. 1978. Chromosome studies in the date palm *Phoenix dactylifera* L. Microscop. Acta 80: 145–148.

Soltis, D.E., P.S. Soltis, P.K. Endress and M.W. Chase. 2005. Phylogeny and evolution of angiosperms. Sinauer, Sunderland, Massachusetts.

Song, Z.C., M. Li, W. Wang, C. Zhao, Z. Zhu, Y. Zheng, Y. Zhang, D. Wang, S. Zhou and Y. Zhao. 1999. Fossil spores and pollen of China: vol. 1: the late Cretaceous and Tertiary spores and pollen. Science Press, China.

Sowunmi, M.A. 1968. Pollen morphology in the Palmae, with special reference to trends in aperture development. Rev. Palaeobot. Palynol. 7: 45–53.

Sowunmi, M.A. 1972. Pollen morphology of the Palmae and its bearing on taxonomy. Rev. Palaeobot. Palynol. 13: 1–80.

Spanner, T.W., H.J. Noltie and M. Gibbons. 1997. A new species of *Trachycarpus* (*Palmae*) from W. Bengal, India. Edinburgh J. Bot. 54: 257–259.

Spruce, R. 1871. Palmae Amazonicae. J. Linn. Soc. Bot. 11: 65–183.

Srisailam, K., and C.G.K. Ramanujam. 1982. Fossil palm pollen grains from the Warkalli sediments of Cannanore area. Rec. Geol. Surv. India 114: 123–131.

Srivastava, S.K. 1987–8. *Ctenolophon* and *Sclerosperma* paleogeography and Senonian plate position. J. Palynol. 23–4: 239–253.

Srivastava, S.K. and P.L. Binda. 1991. Depositional history of the Early Eocene Shumaysi Formation, Saudi Arabia. Palynology 15: 47–61.

Stafleu, F. and R. Cowan (with E. Mennega). 1976–1998. Taxonomic literature — a selective guide to botanical publications and collections with dates, commentaries and types. Second edition. IDC and IAPT.

Stauffer, F.W. and P.K. Endress. 2003. Comparative morphology of female flowers and systematics in Geonomeae (Arecaceae). Pl. Syst. Evol. 242: 171–203.

Stauffer, F.W., C.B. Asmussen, A. Henderson and P.K. Endress. 2003. A revision of *Asterogyne* (Arecaceae, Arecoideae, Geonomeae). Brittonia 55: 326–356.

Stauffer, F.W., W.J. Baker, J. Dransfield and P.K. Endress. 2004. Comparative floral structure and systematics of *Pelagodoxa* and *Sommieria* (Arecaceae). Bot. J. Linn. Soc. 146: 27–39.

Stenzel, G. 1897. *Palmoxylon iriarteum* n. sp., ein fossiles Palmenholz aus Antigua. Bih. Kongl. Svenska Vetensk.-Akad. Handl. 22 III No.11. 18 pp.

Stenzel, G. 1904. Fossile Palmenhölzer. *In* Uhlig, V. and G. von Arthaber (*eds*). Beiträge zur Paläontologie und Geologie Österreich-Ungarns und des Orients 15. K.U.K. Hof-und Universitäts-Buchhändler, Wien and Leipzig.

Stephen, J. 1974. Cytological investigations on the endosperm of *Borassus flabellifer* Linn. Cytologia. 39: 195–207.

Stephyrtza, A.G. 1972. Conifers, box and palm of Miocene flora of Bursuk in Moldavia. Bot. Zhurn. S.S.S.R. 57: 458–468.

STEWART, W.N. AND G.W. ROTHWELL. 1993. Palaeobotany and the Evolution of Plants. Second Edition. Cambridge University Press, Cambridge.

STOVER, L.E. AND P.R. EVANS. 1973. Upper Cretaceous-Eocene spore-pollen zonation, offshore Gippsland Basin, Australia. Spec. Publs Geol. Soc. Austral. 44: 55–72.

STOVER, L.E. AND A.D. PARTRIDGE. 1973. Tertiary and Late Cretaceous spores and pollen from the Gippsland Basin, Southeastern Australia. Proc. Roy. Soc. Victoria 85–86: 237–285; pls 13–27.

STÜHRK, C. 2006. Molekularsystematische Studien in der Subtribus Thrinacinae, mit besonderer Berücksichtigung der Gattung *Trachycarpus* H. Wendl. (Arecaceae). Diplomarbeit im Studienfach Biologie (Masters Thesis), University of Hamburg.

SUESSENGUTH, K. 1921. Beiträge zur frage der systematischen ausschlusses der monokotyl. Beih. Bot. Centrbl., Abt. 2 38: 1–70.

SUNDERLAND, T.C.H. 2001. The taxonomy, ecology and utilisation of African rattans (Palmae: Calamoideae). PhD thesis, University College, London.

SUNDERLAND, T.C.H. 2002. Hapaxanthy and pleonanthy in African rattans (Palmae: Calamoideae). J. Bamboo Rattan 1: 131–139.

SUNDERLAND, T.C.H. 2007. Field Guide to the rattans of Africa. Kew Publishing, Royal Botanic Gardens, Kew.

SUZUKI, M. 1989. A new record of *Palmoxylon* fossil wood from the Lower Miocene of Kanazawa. J. Phytogeog. Taxon. 37: 127–128.

SVENNING, J.C. 1999. Microhabitat specialization in a species-rich palm community in Amazonian Ecuador. J. Ecol. 87: 55–65.

SVENNING, J.C. 2000. Small canopy gaps influence plant distributions in the rain forest understory. Biotropica 32: 252–261.

SVENNING, J.C. 2001a. On the role of microenvironmental heterogeneity in the ecology and diversification of neotropical rain-forest palms (Arecaceae). Bot. Rev. (Lancaster) 67: 1–53.

SVENNING, J.C. 2001b. Environmental heterogeneity, recruitment limitation and the mesoscale distribution of palms in a tropical montane rain forest (Maquipucuna, Ecuador). J. Trop. Ecol. 17: 97–113.

SYED, R.A. 1979. Studies on oil palm (*Elaeis guineensis*) pollination by insects. Bull. Entomol. Res. 69: 213–224.

SZAFER, W. 1961. Mioceńka flora ze starych Gliwic na Ślaşku. Prace Inst. Geol. (Warsawa) 33: 1–205.

TÄCKHOLM, V. AND M. DRAR. 1950. Flora of Egypt. II. Angiospermae, part Monocotyledones: Cyperaceae-Juncaceae. Fouad I University Press.

TAKAHASHI, K. 1982. Miospores from the Eocene Nanggulan Formation in the Yogyakarta region, Central Java. Trans. Proc. Japan Soc. N.S. 126: 303–353.

TAKHTAJAN, A. 1958. A taxonomic study of the Tertiary fan palms of the USSR. Bot. Zhurn. 43: 1661–1674.

TAN, C.F and W.C. WOON. 1992. Economic of cultivation of small-diameter rattan. Pages 177–203 *in* Wan Razali, J. Dransfield and N. Manokaran (*eds*). A guide to rattan planting. Forest Research Institute Malaysia.

TANG, C.Z. AND T.L. WU. 1977. A new species of *Chuniophoenix* (Palmae) from Hainan. Acta Phytotax. Sin. 15: 111–112.

TAYLOR, W.T. AND L.J. HICKEY. 1996. Evidence of and implications of an herbaceous origin for angiosperms. Pages 232–266 *in* Taylor, W.T. and L.J. Hickey (*eds*). Flowering plant origin, evolution and phylogeny. Chapman & Hall, New York.

TEODORIDIS, V. 2003. Early Miocene carpological material from the Czech part of the Zittau Basin. Acta Palaeobot. 43: 9–49.

THANIKAIMONI, G. 1966. Contribution á l'étude palynologique des palmiers. Trav. Sect. Sci. Techn. Inst. Franc. Pondichèry, 5: 1–92.

THANIKAIMONI, G. 1970. Les Palmiers: palynologie et systematique. Trav. Sect. Sci. Techn. Inst. Franç. Pondichèry 11: 1–286.

THEILE-PFEIFFER, H. 1988. Die Mikroflora aus dem Mitteleozänen Ölschiefer von Messel bei Darmstadt. Palaeontographica (Stuttgart) B 211: 1–86.

THOMAS, M.M., N.C. GARWOOD, W.J. BAKER, S.A. HENDERSON, S.J. RUSSEL, D.R. HODEL AND R.M. BATEMAN. 2006. Molecular phylogeny of the palm genus *Chamaedorea*, based on the low-copy nuclear genes *PRK* and RPB2. Molec. Phylogen. Evol. 38: 398–415.

THORNE, R.F. 1983. Proposed new alignments in the angiosperms. Nordic J. Bot. 3: 85–117.

TIDWELL, W.D., S.R. RUSHFORTH, J.L. REVEAL AND H. BEHUNIN. 1970. *Palmoxylon simperi* and *Palmoxylon pristina*: two pre-Cretaceous angiosperms from Utah. Science 168 (3933): 835–840.

TIDWELL, W.D., A.D. SIMPER AND G.F. THAYN. 1977. Additional information concerning the controversial Triassic plant: *Sanmiguelia*. Palaeontographica Abt B, 163: 143–151.

TILLICH, H.J. 2000. Ancestral and derived character states in seedlings of monocotyledons. Pages 221–229 *in* Wilson, K.L. and D.A Morrison (*eds*). Monocots: systematics and evolution. CSIRO, Melbourne.

TIMELL, T.A. 1957. Vegetable ivory as a source of a mannan polysaccharide. Canad. J. Chem. 35: 333–338.

TOMLINSON, P.B. 1960. Seedling leaves in palms and their morphological significance. J. Arnold Arbor. 41: 414–428.

TOMLINSON, P.B. 1961. Palmae. *In* Metcalfe, C.R. (*ed*.). Anatomy of the monocotyledons II. Clarendon Press, Oxford.

TOMLINSON, P.B. 1964. The vascular skeleton of the coconut leaf. Phytomorphology 14: 218–230.

TOMLINSON, P.B. 1966. Notes on the vegetative anatomy of *Aristeyera spicata* (Palmae). J. Arnold Arbor. 47: 23–29.

TOMLINSON, P.B. 1969. The anatomy of vegetative organs of *Juania australis* (Palmae). Gentes Herb. 10: 412–424.

TOMLINSON, P.B. 1971. Flowering in *Metroxylon* (the sago palm). Principes 15: 49–62.

TOMLINSON, P.B. 1979. Systematics and ecology of the Palmae. Ann. Rev. Ecol. Syst. 10: 85–107.

TOMLINSON, P.B. 1990. The structural biology of palms. Clarendon Press, Oxford.

TOMLINSON, P.B. 1995. Non-homology of vascular organisation in monocotyledons and dicotyledons. Pages 589–622 *in* Rudall, P.J., P.J.Cribb, D.F. Cutler, and C.J. Humphries (*eds*). Monocotyledons: systematics and evolution. Royal Botanical Gardens, Kew.

TOMLINSON, P.B. 2006a. The uniqueness of palms. Bot. J. Linn. Soc. 151: 5–14.

TOMLINSON, P.B. 2006b. Stem anatomy of climbing palms in relation to long-distance water transport. Aliso 22: 265–277.

TOMLINSON, P.B., J.B. FISHER, R.E. SPANGLER AND R.A. RICHER. 2001. Stem vascular architecture in the rattan *Calamus* (Arecoideae-Calamoideae-Calaminae). Amer. J. Bot. 88: 797–809.

TOMLINSON, P.B. AND H.E. MOORE. 1968. Inflorescence in *Nannorrhops ritchiana* (Palmae). J. Arnold Arbor. 49: 16–34.

TOMLINSON, P.B. AND P.K. SODERHOLM. 1975. The flowering and fruiting of *Corypha elata* in South Florida. Principes 19: 83–99.

TOMLINSON, P.B. AND R.E. SPANGLER. 2002. Developmental features of the discontinuous stem vascular system in the rattan palm *Calamus* (Arecaceae-Calaminae) Amer. J. Bot. 89: 1128–1141.

TOMLINSON, P.B. AND M.H. ZIMMERMANN. 2003. Stem vascular architecture in the American climbing palm *Desmoncus* (Arecaceae-Arecoideae-Bactridinae). Bot. J. Linn. Soc. 142: 243–254.

TRAIL, J.W.H. 1877. Descriptions of new species of palms collected in the valley of the Amazon in north Brazil in 1874. J. Bot. 15: 75–81.

TRALAU, H. 1964. The genus *Nypa* van Wurmb. Kungl. Svenska Vetenskapsakad. Handl. Ser. 4 10: 5–29.

TRÉNEL, P., M.H.G. GUSTAFSSON, W.J. BAKER, C.B. ASMUSSEN-LANGE, J. DRANSFIELD AND F. BORCHSENIUS. 2007. Mid-Tertiary dispersal, not Gondwanan vicariance explains distribution patterns in the wax palm subfamily (Ceroxyloideae: Arecaceae). Molec. Phylogen. Evol. 45: 272-288.

TRIPATHI, R.P., S.N. MISHRA AND B.D. SHARMA. 1999. *Cocos nucifera* like petrified fruit from the Tertiary of Amarkantak, M.P., India. Palaeobotanist 48: 251–255.

TRIPP, E.A. AND K.G. DEXTER. 2006. *Sabal minor* (Arecaceae): a new northern record of palms in eastern North America. Castanea 71: 172–177.

TRIVEDI, B.S. AND C.L. VERMA. 1979–80 (1981). *Sabalocaulon intertrappeum* gen. et sp. nov. from the Deccan Intertrappean beds of Madhya Pradesh, India. Palaeobotanist 28–29: 329–337.

TRUSWELL, E.M. AND J.A. OWEN. 1988. Eocene pollen from Bungonia, New South Wales. Mem. Ass. Australas. Palaeontol. 5: 259–284.

TSCHAPKA, M. 2003. Pollination of the understorey palm, *Calyptrocalyx ghiesbreghtiana* by hovering and perching bats. Biol. J. Linn. Soc 80: 281–288.

TSCHUDY, R.H. 1973. Stratigraphic distribution of significant Eocene palynomorphs of the Mississippi Embayment. Geol. Surv. Prof. Paper, 743-B: 1–24 + 4 pls.

TUCKER, R. 1991. Hapaxanthy: dying in order to succeed. Mooreana 1: 15–24.

TULEY, P. 1995. The palms of Africa. Trendrine Press, Zennor, Cornwall.

TUZSON, J. 1913. Adatok Magyarország fossilis flórájához. Additamenta add florum fossilem Hungariae III. Magyar Kir. Allamí Foldt. Intez. Evk. 21: 210–233.

UHL, N.W. 1966. Morphology and anatomy of the inflorescence axis and flowers of a new palm, *Aristeyera spicata*. J. Arnold Arbor. 47: 9–22.

UHL, N.W. 1969a. Anatomy and ontogeny of the cincinni and flowers in *Nannorrhops ritchiana* (Palmae). J. Arnold Arbor. 50: 411–431.

UHL, N.W. 1969b. Floral anatomy of *Juania*, *Ravenea*, and *Ceroxylon* (Palmae-Arecoideae). Gentes Herb. 10: 394–411.

UHL, N.W. 1972a. Inflorescence and flower structure in *Nypa fruticans* (Palmae). Amer. J. Bot. 59: 729–743.

UHL, N.W. 1972b. Floral anatomy of *Chelyocarpus*, *Cryosophila*, and *Itaya* (Palmae). Principes 16: 89–100.

UHL, N.W. 1972c. Leaf anatomy in the *Chelyocarpus* alliance. Principes 16: 101–110.

UHL, N.W. 1976a. Developmental studies in *Ptychosperma* (Palmae). I. The inflorescence and the flower cluster. Amer. J. Bot. 63: 82–96.

UHL, N.W. 1976b. Developmental studies in *Ptychosperma* (Palmae). II. The staminate and pistillate flowers. Amer. J. Bot. 63: 97–109.

UHL, N.W. 1978a. Floral anatomy of *Maxburretia* (Palmae). Gentes Herb. 11: 200–211.

UHL, N.W. 1978b. Floral anatomy of the five species of *Hyophorbe* (Palmae). Gentes Herb. 11: 245–267.

UHL, N.W. 1978c. Leaf anatomy in the species of *Hyophorbe* (Palmae). Gentes Herb. 11: 268–283.

UHL, N.W. 1988. Floral organogenesis in palms. Pages 25–44 *in* Leins, P., S.C. Tucker and P.K. Endress (*eds*). Aspects of floral development. Berlin.

UHL, N.W. AND J. DRANSFIELD. 1984. Development of the inflorescence, androecium and gynoecium with reference to palms. Pages 397–449 *in* White, R.A. and W.C. Dickison (*eds*). Contemporary problems in plant anatomy. Academic Press, New York.

UHL, N.W. AND J. DRANSFIELD. 1987. Genera Palmarum, a classification of palms based on the work of Harold E. Moore Jr. L.H. Bailey Hortorium and the International Palm Society, Lawrence, Kansas.

UHL, N.W., J. DRANSFIELD, J.I. DAVIS, M.A. LUCKOW, K.S. HANSEN AND J.J. DOYLE. 1995. Phylogenetic relationships among palms: cladistic analyses of morphological and chloroplast DNA restriction site variation. Pages 623–661 *in* Rudall, P.J., P.J. Cribb, D.F. Cutler and C.J. Humphries (*eds*). Monocotyledons: systematics and evolution. Royal Botanic Gardens, Kew.

UHL, N.W. AND J. MARTENS. 1980. Systematic leaf anatomy of New Caledonian palms. Bot. Soc. Amer. Misc. Ser. Publ. 158: 120.

UHL, N.W., L.O. MORROW AND H.E. MOORE, JR. 1969. Anatomy of the palm *Rhapis excelsa*, VII. Flowers. J. Arnold Arbor. 50: 138–152.

UHL, N.W. AND H.E. MOORE. 1971. The palm gynoecium. Amer. J. Bot. 58: 945–992.

UHL, N.W. AND H.E. MOORE. 1973. The protection of pollen and ovules in palms. Principes 17: 111–149.

UHL, N.W. AND H.E. MOORE. 1975. Observations on the structure of the palm androecium. Abstracts. XII International Botanical Congress, Leningrad, p. 238.

UHL, N.W. AND H.E. MOORE. 1977a. Correlations of inflorescence, flower structure, and floral anatomy with pollination in some palms. Biotropica 9: 170–190.

UHL, N.W. AND H. E. MOORE. 1977b. Centrifugal stamen initiation in phytelephantoid palms. Amer. J. Bot. 64: 1152–1161.

UHL, N.W. AND H.E. MOORE. 1978. The structure of the acervulus, the flower cluster of chamaedoreoid palms. Amer. J. Bot. 65: 197–204.

UHL, N.W. AND H.E. MOORE. 1980. Androecial development in six polyandrous genera representing five major groups of palms. Ann. Bot. (Oxford) ser. 2, 445: 57–75.

UHL, N.W., L.O. MORROW AND H.E. MOORE. 1969. Anatomy of the palm *Rhapis excelsa*. VII. Flowers. J. Arn. Arb. 50: 138–152.

UNGER, F. 1859. Botanishe Streifzüge auf dem Gebiete der Culturgeschichte. IV. Die Pflanzen des alten Ägypten. Sitzungsber. Kaisersl. Akad. Wiss. Wien, Math.-Naturwiss. Cl. 38: 69–140.

URQUHART, G. R. 1997. Paleoecological evidence of *Raphia* in the pre-Columbian neotropics. J. Trop. Ecol. 13: 783–792.

VAN DER BURGH, J. 1984. Some palms in the Miocene of the Lower Rhenish Plain. Rev. Palaeobot. Palynol. 40: 359–374.

VAN DER HAMMEN, T. 1954. El desarollo de la Flora Colombiana en los periodos geologicos. I. Maestrichtiano hasta Terciaro mas inferior. Bol. Geol. (Bogotà), 2: 49–106.

van der Hammen, T. 1956. Descripción de algunos géneros y especies de polen y esporas fósiles. Bol. Geol. (Bogotà) 4: 111–117.

van der Hammen, T. and C. Garcia de Mutis. 1966. The Paleocene pollen flora of Colombia. Leidse Geologische Mededlingen 35: 105–116.

van Heel, W.A. 1977. On the morphology of ovules in *Salacca* (Palmae). Blumea 23: 371–375.

van Heel, W.A. 1988. On the development of some gynoecia with septal nectaries. Blumea 33: 477–504.

van Rheede tot Drakenstein, H. 1678–1693. Hortus Indicus Malabaricus. 12 vols. J. v. Someran and J. v. Dyck, Amsterdam.

van Steenis, C.G.G.J. 1981. Rheophytes of the World. An account of flood-resistant flowering plants and ferns and the theory of autonomous evolution. Sijthoff and Noordhoff, Alphen aan den Rijn.

Van Valen, L. 1975. Life, death and energy of a tree. Biotropica 7: 259–269.

van Valkenburg, J., T.C.H. Sunderland, L.N. Banak and Y. Issembé. 2007. *Sclerosperma* and *Podococcus* in Gabon. Palms 51: 77–83.

van Valkenburg, J., T.C.H. Sunderland and T. Couvreur. 2008. A revision of the genus *Sclerosperma* (Arecaceae). Kew Bull. 63: 75–86.

Vandermeer, J.H., J. Stout and S. Risch. 1979. Seed dispersal of a common Costa Rican rainforest palm (*Welfia georgii*). Trop. Ecol. 20: 17–26.

Varma, Y.N.R., C.G.K. Ramanujam and R.S. Patil. 1986. Palynoflora of Tertiary sediments of Tonakkal area, Kerala. J. Palynol. 22: 39–53.

Vaudois-Miéja, N. and A. Lejal-Nicol. 1987. Paléocarpologies africaine: apparition dès l'Aptien en Égypte d'un Palmier (*Hyphaeneocarpon aegyptiacum* n. sp.). Compt. Rendu Acad. Sci. Paris, 304 (Ser. 2) 6: 233–238.

Vaughan, J.G. 1960. Structure of *Acrocomia* fruit. Nature 188 (4744): 81.

Venkatachala, B.S. and R.K. Kar. 1969. Palynology of the Tertiary sediments of Kutch. 1. Spores and pollen from Bore-hole No. 14. Palaeobotanist, 17: 157–178.

Venkatasubban, K.R. 1945. Cytological studies in Palmae. Pt. I. Chromosome numbers in a few species of palms of British India and Ceylon. Proc. Indian Acad. Sci., B 22: 193–207.

Verma, C.L. 1974. Occurrence of fossil *Nypa* root from the Deccan Intetrappean beds of M.P. India. Curr. Sci. 43: 289–290.

Vormisto, J. 2002. Palms as rainforest resources: how evenly are they distributed in Peruvian Amazonia? Biodiver. & Conservation 11: 1025–1045.

Vormisto, J., O.L. Phillips, K. Ruokolainen, H. Tuomisto and R. Vasquez. 2000. A comparison of fine-scale distribution patterns of four plant groups in an Amazonian rainforest. Ecography 23: 349–359.

Vormisto, J., H. Tuomisto and J. Oksanen. 2004a. Palm distribution patterns in Amazonian rainforest: what is the role of topographic variation? J. Veg. Sci. 15: 485–494.

Vormisto, J., J.C. Svenning, P. Hall and H. Balslev. 2004b. Diversity and dominance in palm (Arecaceae) communities in terra firme forests in the western Amazon basin. J. Ecol. 92: 577–588.

Walker, J.W. and A.G. Walker. 1984. Ultrastructure of lower Cretaceous angiosperm pollen and the origin and early evolution of flowering plants. Ann. Missouri Bot. Gard. 71: 464–521.

Walker, J.W. and A.G. Walker. 1986. Ultrastructure of Early Cretaceous angiosperm pollen and its evolutionary implications. Pages 203–217 *in* Blackmore, S. and I.K. Ferguson (*eds*). Pollen and spores: form and function. Academic Press, London and New York.

Wallace, A.R. 1853. The palm trees of the Amazon and their uses. John van Voorst, London.

Walther, G R. 2003. Are there indigenous palms in Switzerland? Bot. Helv. 113: 159–180.

Waterhouse, J.T. and C.J. Quinn. 1978. Growth patterns in the stem of the palm *Archontophoinis cunninghamiana*. Bot. J. Linn. Soc. 77: 73–93.

Watling, D. 2005. Palms of the Fiji Islands. Environmental Consultants, Fiji.

Weber, R. 1978. Some aspects of the Upper Cretaceous angiosperm flora of Coahuila, México. Courier Forschungsinst. Senckenberg 30: 38–46.

Weberling, F. 1998. Morphology of flowers and inflorescences. Cambridge University Press, Cambridge.

Weiner, G. and W. Liese. 1993. Morphologcal characteristics of the epidermis of rattan palms. J. Trop. Forest Sc. 6: 197–201.

Wendland, H. 1854. Palmae. Pages 1–42 in Index Palmarum, Cyclanthearum, Pandanearum, Cycadearum, quae in hortis europaeis coluntur synonymis gravioribus interpositis. Hannover.

Werker, E. 1997. Seed anatomy. *In* Handbuch der Pflanzenanatomie, vol x.3. Gebruder Borntraeger, Berlin and Stuttgart. 424 pp.

Wessels Boer, J.G. 1965. The indigenous palms of Suriname. E.J. Brill, Leiden.

Wessels Boer, J.G. 1968. The geonomoid palms. Verh. Kon. Akad. Wetensch., Afd. Natuurk., Sect. 2 58: 1–202.

Whistler, W.A. 1992. The palms of Samoa. Mooreana 2: 24–29.

Whitmore, T.C. 1984. Tropical rain forests of the Far East (Second Edition). Oxford Science Publications, Clarendon Press, Oxford.

Whitmore, T.C. 1998. An introduction to tropical rain forests. Oxford University Press, Oxford.

Whitten, A.J. 1980. *Arenga* fruit as a food for gibbons. Principes 24: 143–146.

Wikstrom, N., V. Savolainen and M.W. Chase. 2001. Evolution of the angiosperms: calibrating the family tree. Proc. Roy. Soc. London., Ser. B, Biol. Sci. 268: 2211–2220.

Wilbert, J. 1980a. The Temiche cap. Principes 24: 105–109.

Wilbert, J. 1980b. The palm-leaf sail of the Warao Indians. Principes 24: 162–169.

Williams, C.A., J.B. Harborne and H.T. Clifford. 1973. Negatively charged flavones and tricin as chemosystematic markers in the Palmae. Phytochemistry 12: 2417–2430.

Williams, C.A., J.B. Harborne and S.F. Glassman. 1983. Flavonoids as taxonomic markers in some cocosoid palms. Pl. Syst. Evol. 142: 157–169.

Wilson, M.A., B. Gaut and M.T. Clegg. 1990. Chloroplast DNA evolves slowly in the palm family (Arecaceae). Molec. Biol. Evol. 7: 303–314.

Wing, S.L. and B.H. Tiffney. 1987. Interactions of angiosperms and herbivorous tetrapods through time. Pages 203–224 *in* Friis, E.M., W.G. Chaloner and P.R. Crane (*eds*). The origins of Angiosperms and their biological consequences, Cambridge University Press, Cambridge.

Witono, J.R., J.P. Mogea and S. Somadikarta. 2002. *Pinanga* in Java and Bali. Palms 46: 193–201.

YAMASHITA, C. AND M. DE P. VALLE. 1993. On the linkage between *Anodorhynchus* macaws and palm nuts, and the extinction of the Glaucous Macaw. Bull. B.O.C. 113: 53–60.

YAMASHITA, C. AND Y.M. DE BARROS. 1997. The Blue Throated Macaw *Ara glaucogularis*; characterization of its distinctive habitats in savannahs of the Beni, Bolivia. Ararajuba 5: 141–150.

YEATEN, R.I. 1979. Intraspecific competition in a population of the stilt palm *Socratea durissima* (Oerst.) Wendl. on Barro Colorado Island, Panama. Biotropica 11: 155–158.

ZETTER, R. AND C.-CH. HOFMANN. 2001. New aspects of the palynoflora of the lowermost Eocene in Austria (Krappfeld area). Pages 473–507 *in* Piller, E.W. and M.W. Rasser (*eds*). Paleogene of Austria. Schriftenreihe de Erdwissenschaftlichen Kommission, Austria Academy of Science.

ZEVEN, A.C. 1967. The semi-wild oil palm and its industry in Africa. Agricultural Research Repports 689. Centre for Agricultural Publications and Documentation, Wageningen, The Netherlands.

ZIMMERMANN, M.H. AND P.B. TOMLINSON. 1974. Vascular patterns in palm stems: variations of the Rhapis principle. J. Arnold Arbor. 55: 402–424.

ZONA, S. 1990. A monograph of *Sabal* (Arecaceae: Coryphoideae). Aliso 12: 583–666.

ZONA, S. 1995. A revision of *Calyptronoma* (Arecaceae). Principes 39: 140–151.

ZONA, S. 1996. *Roystonea* (Arecaceae, Arecoideae). Fl. Neotrop. Monogr 71. The New York Botanical Garden Press, Bronx, NY.

ZONA, S. 1997. The genera of Palmae (Arecaceae) in the southeastern United States. Harvard Pap. Bot. 2: 71–107.

ZONA, S. 1999a. New perspectives on generic limits and relationships in the Ptychospermatinae (Palmae: Arecoideae). Pages 255–263 *in* Henderson, A. and F. Borchsenius (*eds*). Evolution, variation and classification of palms. The New York Botanical Garden Press, Bronx, NY.

ZONA, S. 1999b. Revision of *Drymophloeus* (Arecaceae: Arecoideae). Blumea 44: 1–24.

ZONA, S. 2001. Starchy pollen grains in Commelinoid Monocots. Ann. Bot. (Oxford) 87: 109–116.

ZONA, S. 2002a. A revision of *Pseudophoenix*. Palms 46: 19–38.

ZONA, S. 2002b. Name changes in *Attalea*. Palms 46: 132–133.

ZONA, S. 2003. Endosperm condition and the paradox of *Ptychococcus paradoxus*. Telopea 10: 179–185.

ZONA, S. 2004. Raphides in palm embryos and their systematic distribution. Ann. Bot. (Oxford) 93: 415–421.

ZONA, S. AND F.B. ESSIG. 1999. How many species of *Brassiophoenix*? Palms 43: 45–48.

ZONA, S. AND D. FULLER. 1999. A revision of *Veitchia* (Arecaceae–Arecoideae). Harvard Pap. Bot. 4: 543–560.

ZONA, S. AND A. HENDERSON. 1989. A review of animal-mediated seed dispersal of palms. Selbyana 11: 6–21.

ZONNEVELD, B.J.M., I.J. LEITCH and M.D. BENNETT. 2005. First nuclear DNA amounts in more than 300 angiosperms. Ann. Bot. (Oxford) 96: 229–244.

INDEX TO SCIENTIFIC NAMES

Names in *italics* are synonyms; **bold** numbers refer to the generic descriptions.

INDEX TO COMMON NAMES

SUBJECT INDEX